ICU intensive care unit
ID intradermal
IF intrinsic factor
IFN interferon
Ig immunoglobin
IM intramuscular; infectious mononucleosis
IN internist
IOP intraocular pressure
IPPA inspection, palpation, percussion, auscultation
IPSP inhibitory postsynaptic potential
IUD intrauterine device
IV intravenous
IVC inferior vena cava
IVF in vitro fertilization
IVP intravenous pyelogram
IVT intravenous transfusion
JGA juxtaglomerular apparatus
KS Kaposi's sarcoma
KUB kidneys, ureters, bladder
LBB left breast biopsy
LDL low-density lipoprotein
LFT liver function test
LG laryngectomy
LH luteinizing hormone
LLQ left lower quadrant
LMP last menstrual period
LOC loss of consciousness
LP lumbar puncture
LPN licensed practical nurse
LRI lower respiratory infection
LUQ left upper quadrant
LVAD left ventricular assist device
MAb monoclonal antibody
mEq/l milliequivalents per liter
MG myasthenia gravis
MI myocardial infarction
MLT medical laboratory technologist
mm^3 cubic millimeter
mm Hg millimeters of mercury
MOA medical office assistant
MRI magnetic resonance imaging
MS multiple sclerosis
MSH melanocyte-stimulating hormone
MSOF multisystem organ failure
MVP mitral valve prolapse
MVR minute volume of respiration
NBM nothing by mouth
ND natural death
NE norepinephrine
NGU nongonococcal urethritis
NLMC nocturnal leg muscle cramping
NMG neuromuscular junction
NPN nonprotein nitrogen
NREM nonrapid eye movement
NSAID nonsteroidal antiinflammatory drug
NSU nonspecific urethritis
NTG nitroglycerin
NTP normal temperature and pressure
OB/GYN obstetrician-gynecologist; obstetrics-gynecology
OC oral contraceptive
OD overdose; right eye
OHS open heart surgery
OI opportunistic infection

OR operating room
ORT operating room technician
OT oxytocin
OTC over-the-counter
OV office visit
P pressure
PABA para-aminobenzoic acid
PCP *Pneumocystis carinii* pneumonia
PCV packed cell volume
PD Parkinson's disease
PE pulmonary embolism; physical examination
PED pediatrics; pediatrician
Peff effective filtration pressure
PEG pneumoencephalogram
PEMF pulsating electromagnetic fi...
PET positron emission tomogra...
PG prostaglandin
pH hydrogen-ion concentr...
PID pelvic inflammatory...
PKU phenylketonuria
PMH past medical hist...
PMN polymorphonu...
PMP plasma membra...
PMS premenstrual syndr...
PNS peripheral nervous system
PRL prolactin
PROG progesterone
PT prothrombin time; physical therapist
PTCA percutaneous transluminal coronary angioplasty
PTD permanent and total disability
PTH parathyroid hormone
PTT partial thromboplastin time
PTX pneumothorax
PUBS percutaneous umbilical blood sampling
PUL percutaneous ultrasonic lithotripsy
Px prognosis; pneumothorax
PX physical examination
R roentgen (unit of x radiation)
RA rheumatoid arthritis
RAD radiation absorbed dose
RAS reticular activating system
RBB right breast biopsy
RBC red blood cell; red blood count
RBOW rupture of bag of waters
RDA recommended daily allowance
RDS respiratory distress syndrome
REM rapid eye movement
Rh *Rhesus*
RHC respirations have ceased
RIA radioimmunoassay
RK radial keratotomy
RLQ right lower quadrant
RLX relaxin
RM radical mastectomy
RN registered nurse
RNA ribonucleic acid
ROS review of symptoms
RR respiratory rate
RRR regular rate and rhythm (heart)
RS Reye's syndrome
RT radiotherapy; radiologic technologist

RUQ right upper quadrant
Rx prescription
SA sinoatrial (sinuatrial)
SBP systolic blood pressure
SC subcutaneous
SCA sickle-cell anemia
SCD sudden cardiac death
SCID severe combined immunodeficiency syndrome
SDS same-day surgery
SF synovial fluid
SG skin graft; specific gravity
SH social history
SIDS sudden infant death syndrome
sigmoidoscopy; sigmoidoscope
SIW self-inflicted wound
SLE systemic lupus erythematosus
SMD senile macular degeneration
SNS somatic nervous system
SOB shortness of breath
SPF sun protection factor
S/S (Sx) signs and symptoms
STD sexually transmitted disease
SubQ or SQ subcutaneous
SV stroke volume
SVC superior vena cava
T temperature
TAH total artificial heart
TB tuberculosis
TBI total body irradiation
TIA transient ischemic attack
TLI total lymphoid irradiation
Tm transport maximum
TM transcendental meditation
TMJ temporomandibular joint
TND transient neurologic deficit
TOP termination of pregnancy
t-PA tissue plasminogen activator
TPE therapeutic plasma exchange
TPN total parenteral nutrition
TPR temperature, pulse, and respiration
TSH thyroid-stimulating hormone
TSS toxic shock syndrome
Tx treatment
UA urinalysis
UDO undetermined origin
UG urogenital
URI upper respiratory infection
US ultrasound; ultrasonography
UTI urinary tract infection
UV ultraviolet
VD venereal disease
VDRL venereal disease research laboratory test (blood test for syphilis)
VF ventricular fibrillation
VPC ventricular premature contraction
VS vital signs
VT ventricular tachycardia
VV varicose veins; vulva and vagina
WBC white blood cell; white blood count
WNL within normal limits
X-match cross-match
XRT x-ray therapy

The following is a complete list of all HarperCollins Life Sciences texts available for faculty review and student purchase:

ALLIED HEALTH

Bastian
ILLUSTRATED REVIEW OF ANATOMY AND PHYSIOLOGY
  Series (1993)
  BASIC CONCEPTS OF CHEMISTRY, THE CELL, AND TISSUES
  THE MUSCULAR AND SKELETAL SYSTEMS
  THE NERVOUS SYSTEM
  THE ENDOCRINE SYSTEM
  THE CARDIOVASCULAR SYSTEM
  THE LYMPHATIC AND IMMUNE SYSTEMS
  THE RESPIRATORY SYSTEM
  THE DIGESTIVE SYSTEM
  THE URINARY SYSTEM
  THE REPRODUCTIVE SYSTEM

Kreier/Mortensen
PRINCIPLES OF INFECTION, RESISTANCE, AND IMMUNITY
  (1990)

Telford/Bridgman
INTRODUCTION TO FUNCTIONAL HISTOLOGY (1989)

Tortora
INTRODUCTION TO THE HUMAN BODY, Second Edition (1991)

Tortora
PRINCIPLES OF HUMAN ANATOMY, Sixth Edition (1992)

Tortora/Grabowski
PRINCIPLES OF ANATOMY AND PHYSIOLOGY, Seventh Edition
  (1993)

Volk
BASIC MICROBIOLOGY, Seventh Edition (1992)

LIFE SCIENCES

Beck/Liem/Simpson
LIFE: AN INTRODUCTION TO BIOLOGY, Third Edition (1991)

Harris
CONCEPTS IN ZOOLOGY (1992)

Herrmann
CELL BIOLOGY (1989)

Hill/Wyse
ANIMAL PHYSIOLOGY, Second Edition (1988)

Jenkins
HUMAN GENETICS, Second Edition (1989)

Kaufman
PLANTS: THEIR BIOLOGY AND IMPORTANCE (1990)

Kleinsmith/Kish
PRINCIPLES OF CELL BIOLOGY (1987)

Mix/Farber/King
BIOLOGY: THE NETWORK OF LIFE (1992)

Nickerson
GENETICS: A GUIDE TO BASIC CONCEPTS AND PROBLEM
  SOLVING (1989)

Nybakken
MARINE BIOLOGY: AN ECOLOGICAL APPROACH, Third Edition
  (1993)

Penchenik
A SHORT GUIDE TO WRITING ABOUT BIOLOGY, Second Edition
  (1993)

Rischer/Easton
FOCUS ON HUMAN BIOLOGY (1992)

Russell
GENETICS, Third Edition (1992)

Shostak
EMBRYOLOGY: AN INTRODUCTION TO DEVELOPMENTAL
  BIOLOGY (1991)

Wallace
BIOLOGY: THE WORLD OF LIFE, Sixth Edition (1992)

Wallace/Sanders/Ferl
BIOLOGY: THE SCIENCE OF LIFE, Third Edition (1991)

Webber/Thurman
MARINE BIOLOGY, Second Edition (1991)

ECOLOGY AND ENVIRONMENTAL STUDIES

Kaufman/Franz
BIOSPHERE 2000: PROTECTING OUR GLOBAL ENVIRONMENT
  (1993)

Krebs
ECOLOGY: THE EXPERIMENTAL ANALYSIS OF DISTRIBUTION
  AND ABUNDANCE, Third Edition (1984)

Krebs
ECOLOGICAL METHODOLOGY (1988)

Krebs
THE MESSAGE OF ECOLOGY (1987)

Pianka
EVOLUTIONARY ECOLOGY, Fourth Edition (1988)

Smith
ECOLOGY AND FIELD BIOLOGY, Fourth Edition (1991)

Smith
ELEMENTS OF ECOLOGY, Third Edition (1992)

Yodzis
INTRODUCTION TO THEORETICAL ECOLOGY (1989)

LABORATORY MANUALS

Beishir
MICROBIOLOGY IN PRACTICE: A SELF-INSTRUCTIONAL
  LABORATORY COURSE, Fifth Edition (1991)

Donnelly
LABORATORY MANUAL FOR HUMAN ANATOMY: WITH CAT
  DISSECTIONS, Second Edition (1993)

Donnelly/Wistreich
LABORATORY MANUAL FOR ANATOMY AND PHYSIOLOGY:
  WITH CAT DISSECTIONS, Fourth Edition (1993)

Donnelly/Wistreich
LABORATORY MANUAL FOR ANATOMY AND PHYSIOLOGY:
  WITH FETAL PIG DISSECTIONS (1993)

Eroschenko
LABORATORY MANUAL FOR HUMAN ANATOMY WITH
  CADAVERS (1990)

Tietjen
THE HARPERCOLLINS BIOLOGY LABORATORY MANUAL (1991)

Tietjen
LABORATORY MANUAL TO ACCOMPANY BIOLOGY: THE
  NETWORK OF LIFE (1992)

Tietjen/Harris
HARPERCOLLINS ZOOLOGY LABORATORY MANUAL (1992)

COLORING BOOKS

Diamond/Scheibel/Elson
THE HUMAN BRAIN COLORING BOOK (1985)

Elson
THE ZOOLOGY COLORING BOOK (1982)

Griffin
THE BIOLOGY COLORING BOOK (1987)

Kapit/Macey/Meisami
THE PHYSIOLOGY COLORING BOOK (1987)

Kapit/Elson
THE ANATOMY COLORING BOOK, Second Edition (1993)

Niesen
THE MARINE BIOLOGY COLORING BOOK (1982)

Young
THE BOTANY COLORING BOOK (1982)

PRINCIPLES OF ANATOMY AND PHYSIOLOGY

About the Authors

Gerard J. Tortora

Jerry Tortora is a professor of biology and teaches human anatomy and physiology and microbiology at Bergen Community College in Paramus, New Jersey. He has just completed 30 years as a teacher, the past 24 at Bergen CC. He received his B.S. in biology from Fairleigh Dickinson University in 1962 and his M.A. in biology from Montclair State College in 1965. He has also taken graduate courses in education and science at Columbia University and Rutgers University. He belongs to numerous biology organizations, such as the Human Anatomy and Physiology Society (HAPS), the American Association for the Advancement of Science (AAAS), the American Association of Microbiology (ASM), and the Metropolitan Association of College and University Biologists (MACUB). Jerry is the author of a number of best-selling anatomy and physiology, anatomy, and microbiology textbooks and several laboratory manuals.

Sandra Reynolds Grabowski

Sandy Grabowski is an instructor in the Department of Biological Sciences at Purdue University in West Lafayette, Indiana, where she has been a faculty member since 1977. During that time, she has taught human anatomy and physiology to students in a wide range of academic programs. In 1992 she was selected by the students as one of the top 10 teachers in the School of Science at Purdue. Sandy received her B.S. in biology and Ph.D. in neurophysiology from Purdue. She is an active member of the Human Anatomy and Physiology Society (HAPS), served as editor of *HAPS News* from 1990 to 1992, and is HAPS President-Elect for 1992–1993. In addition she belongs to the American Association for the Advancement of Science (AAAS), the Association for Women in Science (AWIS), the National Science Teachers Association (NSTA), the Society for College Science Teachers (SCST), and the Indiana Association of Biology Teachers (IABT).

Evan J. Colella

Personal Touch Photo Studio

Seventh Edition

Principles of Anatomy and Physiology

Gerard J. Tortora

Bergen Community College

Sandra Reynolds Grabowski

Purdue University

 HarperCollins*CollegePublishers*

To Lynne Marie, Gerard Joseph Jr., and Kenneth Stephen Tortora, who make it all worthwhile.

G. J. T.

To Helen McDaniel Reynolds and Glenn Edward Reynolds, my parents, who encouraged me to discover the true joy of learning.

S. R. G.

Acquisitions Editor: Bonnie Roesch

Developmental Editor: Meryl R. G. Muskin

Project Editor: Thomas R. Farrell

Art Director: Teresa J. Delgado

Text Design: Baseline Graphics

Cover Design: Jaye Zimet

Cover Photo: © 1989 Howard Sochurek, The Stock Market

Photo Researcher: Mira Schachne

Production Administrator: Jeffrey Taub

Compositor: Arcata Graphics/Kingsport

Printer and Binder: RR Donnelley & Sons Company

Cover Printer: The Lehigh Press, Inc.

Principles of Anatomy and Physiology, *Seventh Edition*

Library of Congress Cataloging-in-Publication Data

Tortora, Gerard J.
    Principles of anatomy and physiology / Gerard J. Tortora, Sandra
Reynolds Grabowski. — 7th ed.
        p.   cm.
    Includes index.
    ISBN 0–06–046702–9 (student edition)—ISBN 0–06–501713–7 (student edition)
    1. Human physiology.   2. Human anatomy.   I. Grabowski. Sandra
Reynolds.   II. Title
QP34.5.T67 1992   92–18562
612—dc20   CIP

94 95 9 8 7 6 5 4

CONTENTS IN BRIEF

CONTENTS IN DETAIL

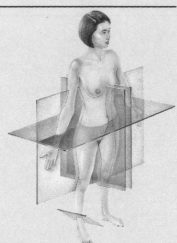

Chapter 4   The Tissue Level of Organization   96

Chapter 5   The Integumentary System   126

UNIT 2

PRINCIPLES OF SUPPORT AND MOVEMENT   144

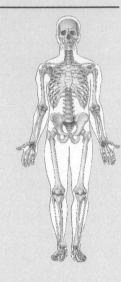

Chapter 6   Bone Tissue   146

Chapter 7   The Skeletal System: The Axial Skeleton   165

Chapter 8   The Skeletal System: The Appendicular Skeleton   198

UNIT 3

CONTROL SYSTEMS OF THE HUMAN BODY   344

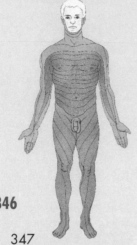

Chapter 12   Nervous Tissue   346

Chapter 13   The Spinal Cord and Spinal Nerves   375

Chapter 14   The Brain and Cranial Nerves   404

UNIT 4

MAINTENANCE OF THE HUMAN BODY   564

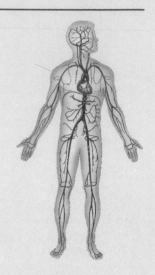

Chapter 19 The Cardiovascular System: The Blood   566

UNIT 5

CONTINUITY   920

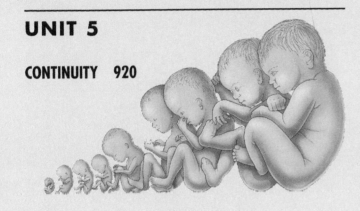

PREFACE

With the seventh edition of *Principles of Anatomy and Physiology,* a classic text is born anew. We have, of course, retained the organization, features, and accuracy of previous editions upon which anatomy and physiology instructors and students have depended for almost twenty years. The text has been completely revised, however, from the first page to the last, and includes hundreds of new and revised figures, up-to-date physiology, helpful new learning devices, and interesting clinical applications.

As a newly created author team for this edition of the text, we had clearly defined goals and challenges for the revision. Jerry Tortora's experience as a respected author and teacher ensured that we maintained the clarity of explanations, consistency of terminology, accuracy of presentation, and completeness of coverage that were the hallmarks of past editions. Sandy Grabowski's expertise as a physiologist enabled us to achieve an even better balance between anatomy and physiology and to incorporate recent advances in physiology. Together, we streamlined difficult discussions and pruned unnecessary details, producing a book that aims once again to set new standards for anatomy and physiology texts.

THEMES AND KEY FEATURES

Principles of Anatomy and Physiology is designed for an introductory course in anatomy and physiology and assumes no prior knowledge of the human body by the student. It is geared to students preparing for careers in health-related professions, such as nursing, occupational therapy, physical therapy, medical technology, medicine, and dentistry. Because of its scope, the text is also useful for students in the biological sciences, science technology, science education, and physical education programs.

As in the past, the new edition provides a solid foundation for understanding the structure and function of the human body. In addition, a wealth of clinical material deals with predicting the problems that arise when normal anatomy and physiology fail. These **dual themes** of **homeostasis and homeostatic imbalances** are established in Chapter 1 and expanded throughout the text. Several **new negative feedback figures** emphasize the common features of body control systems. Students will see how dynamic counterbalancing forces act to maintain normal anatomy and physiology.

Clinical applications within the text have been carefully chosen to pique the reader's interest and illustrate the importance of understanding normal mechanisms. **Discussions of homeostatic imbalances** at the end of most chapters provide additional explanations of important medical problems. These include answers to many of the questions that students commonly ask about disorders and diseases.

Highly effective pedagogical features, first introduced in previous editions and since used successfully by over a million students, have been maintained and enhanced. These learning devices include complete chapter outlines, student objectives, phonetic pronunciations, word roots, summary exhibits, study outlines, review questions, and definitions of medical terminology. New to this edition are the questions with figures. Each of these outstanding features is further explained in our overview, *For Students: Making the Most of This Text,* and highlighted in the visual guide to the text's features following the preface.

ORGANIZATION

The book follows the same unit and topic sequence as its six earlier editions. It is divided into five principal areas of concentration. *Unit 1, Organization of the Human Body,* provides an understanding of the structural and functional levels of the body, from molecules to organ systems. *Unit 2, Principles of Support and Movement,* analyzes the anatomy and physiology of the skeletal system, articulations, and the muscular system. *Unit 3, Control Systems of the Human Body,* emphasizes the importance of neural communication in the immediate maintenance of homeostasis, the role of sensory receptors in providing information about the internal and external environment, and the significance of hormones in maintaining long-range homeostasis. The autonomic nervous system now immediately precedes the endocrine system so that these two major controllers of homeostasis may be covered sequentially. *Unit 4, Maintenance of the Human Body,* explains how body systems function to maintain homeostasis on a day-to-day basis through the processes of circulation, respiration, digestion, cellular metabolism, urinary functions, and buffer systems.

Unit 5, Continuity, covers the anatomy and physiology of the reproductive systems, development, and the basic concepts of genetics and inheritance.

NEW TO THIS EDITION

There are five global changes in the seventh edition: **more physiology, a completely revised art program, questions with the figures, a higher ratio of figures to text material, and improved readability.** There are many other improvements throughout the text, including updated terminology, new and current clinical information, and an expanded number of highly praised summary exhibits, many of which are now illustrated to improve clarity. To list every change in every chapter here would be exhausting. A detailed list of changes in each chapter is included in the Instructor's Resource Manual and is also available from your HarperCollins representative to assist you in preparing your syllabus.

Balance Between Anatomy and Physiology

Each instructor approaches the teaching of anatomy and physiology from a different perspective and background. The advantage of two authors with very different backgrounds working together became obvious early in the revision process. The result is a text that presents a body of knowledge filtered and refined both by an anatomist (Jerry) and a physiologist (Sandy). Thus the balance between anatomy and physiology is better than ever. In addition, we have emphasized correlations between normal physiology and pathophysiology, normal anatomy and pathology, and homeostasis and homeostatic imbalances.

As noted, throughout the text physiological coverage has been increased and brought up to date. In Chapter 1, unique negative feedback diagrams are introduced (see Figs. 1.3 and 1.4 on pages 11 and 12). This type of figure is used throughout the book to reinforce and clarify the verbal explanations. For specific examples of the enhanced physiological coverage that is integrated throughout the entire text, take a look at the discussion of titin, the third most prevalent protein in muscle (Chapter 10, page 241) and the discussion of current views on oxygen consumption after exercise (Chapter 10, page 252); coverage of membrane potential and action potential (Chapter 12, pages 356 and 358); the discussion of cardiac output (Chapter 20, page 608); the thoroughly revised descriptions of immune responses (Chapter 22, pages 705–709), including current coverage of AIDS (Chapter 22, page 712); and the new overview of metabolic pathways (Chapter 25, pages 841–843).

The Illustration Program

Previous editions of this text have always been praised for the clarity of their artwork. For this edition, we realized that we would be adding many new pieces because of the increased physiology coverage. As a result, there are over 100 conceptually new pieces of art. We also believed that it was time to take a close look at each piece of art in the text to make sure that it was well conceptualized and also consistent in style and presentation throughout. To help plan for the revision of figures, HarperCollins conducted a number of focus groups with the specific goal of addressing the needs of the visual learner. The publisher commissioned premier medical illustrators to render anatomical and histological art. The result is a superb art program that clearly conveys body structures in a realistic, three-dimensional manner. Most anatomical figures now include small diagrams that provide orientation for viewing the figure. Over 80% of the art is new or redrawn for this edition.

To emphasize structural and functional relations, colors are used in a consistent and meaningful manner throughout the text. For example, sensory structures, sensory neurons, and sensory regions of the brain are shades of blue ■, whereas motor structures are light red ■. Membrane phospholipids are gray ■ and orange ■, the cytosol is sand ■, and extracellular fluid is blue ■. Negative and positive feedback loops also use color cues to aid the students in understanding and recognizing the concept. Stimulus and response are both orange ■ since they both alter the controlled condition, the controlled condition is green ■, the receptor is blue ■, the control center is purple ■, and the effector is red ■. Such color cues provide additional help for students who are trying to learn complex anatomical and physiological concepts.

Questions with Figures

A key new feature, and one unique to our anatomy and physiology textbook, is **Questions with Figures.** These questions are designed to help students interpret the figures and correlate them with the text material. They serve as a "self-check" as the reader progresses through a discussion, and the answers are given at the end of each chapter. Often the questions are easily answered by careful examination of the figure. Sometimes the question requires the reader to synthesize verbal and visual information. Occasionally, the question stimulates prediction of conclusions that will be reached a little later in the chapter. Both outstanding and average students who class-tested manuscript chapters appreciated the questions. A typical comment was, "The questions with figures are good because they encourage me to *think* about what I am looking at and reading to make sure I understand the material."

Verbal and Visual Material

Since many students are "visual learners," prime goals of the seventh edition revision were to add new figures, refine existing figures, and prune unnecessary words, thereby creating a better ratio of art to words. As mentioned above, there are over 100 completely new figures in this edition, while there are 20,000 fewer words. The text has been carefully designed to ensure that the text and art are presented together, enhancing the integration of these separate but interdependent information sources.

Readability

Previous editions of the text have earned praise for their clear descriptions. By any measure, however, this new edition is easier than ever to read and understand. Sentences are more concise, and the active voice is used more often. Pronunciations are given for more terms, including all skeletal muscles (Chapter 11), nerves (Chapter 13), and blood vessels (Chapter 21). Students will appreciate how easy it is to use the system of phonetic pronunciations pioneered in the third edition of *Principles of Anatomy and Physiology*. Readability is also enhanced by reorganization of some sections, addition of new subheads, and presentation of related points in a numbered list format. Both students and instructors reviewed the manuscript and pointed out difficult passages. The result is better, more polished writing than ever before.

H.E.A.R.T.

Principles of Anatomy and Physiology, Seventh Edition, is much more than a textbook. It is the hub of an extensive, integrated system of ancillary materials called **H.E.A.R.T.** (Human Anatomy and Physiology Education And Resource Tools). **H.E.A.R.T.** comprises printed materials, visual aids, videos, a laserdisc, and software. Each item has been carefully produced to enhance the learning environment. The variety of ancillaries ensures that all teaching and learning styles—whether auditory, visual, or verbal—are supported.

For complete information on any of the following items, please call either your local HarperCollins representative or the HarperCollins customer communication line **(1-800-8-HEART-1),** or write:

Marketing Manager—Allied Health Sciences
HarperCollins College Publishers
10 East 53rd Street
New York, NY 10022–5299

● **Instructor's Resource Manual** Written by Judith Lanum Mohan of Case Western Reserve University, each chapter contains lecture outlines, lecture enhancements, teaching tips, suggested resources, and supplementary teaching materials such as reading lists and medical tests that can be reproduced for student use.

● **Concept Maps** Prepared by Janice Smith of Tarrant County Junior College, these concept maps correlate with the learning objectives of each chapter. Each map is prepared so that it can easily be reproduced as an overhead for classroom use or duplicated for individual student study.

● **Testbank** Prepared by Lanier Byrd and Fred Richards of St. Phillip's College, this all-new test bank includes approximately 4000 questions in a variety of forms (multiple choice, essay, short answer, true–false, and matching). It is available in printed form as well as on *Test Master* for use with IBM-compatible or Macintosh computers.

● *Essays on Wellness* Written by Barbara A. Brehm of Smith College, this brief, full-color collection of essays relates the concept of wellness to topics in the textbook. Our goal has always been to provide students with a strong foundation that will enable them to understand the benefits of wellness and health. This optional supplement reflects our philosophy that acquiring the knowledge for staying well and cultivating behaviors that prevent illness are just as important as understanding disease. Each essay correlates with a chapter of the text and offers another perspective to help students see the relevance of their study to everyday topics of interest such as vegetarian diets, weight control, and stress management.

● **Transparency Acetates** A set of 400 full-color acetates is available to adopters of the text. Illustrations from the text, clearly labeled for classroom use, highlight those topics and concepts that benefit most from visual reinforcement.

● **Slides** Prepared by Victor P. Eroschenko of the University of Idaho, this new set of 157 histology slides was created specifically to accompany this text. They reflect all tissue types and are correlated to all the major body systems.

● **Learning Guide** Written by Kathleen Schmidt Prezbindowski of the College of Mount St. Joseph with Jerry Tortora, this highly successful student guide includes numerous and varied exercises, labeling and coloring diagrams, and mastery tests. Each chapter begins with a *framework* that helps students visualize the relationships among key concepts and terms. *Wordbytes, checkpoints,* and *clinical challenges* all offer opportunities for students to enhance their understanding of the textual material.

● **Laboratory Manuals** Three laboratory manuals are available, depending on the dissection animal you have chosen. Patricia J. Donnelly and George A. Wistreich of East Los Angeles College have revised their successful manual with cat dissections. For the first time, they have now also produced a manual with fetal pig dissection. Victor Eroschenko of the University of Idaho has written a lab manual to use with prosected cadavers or models. This manual includes diagrams suitable for coloring as well as study. Instructor's manuals are available for use with the laboratory manuals.

● **Coloring Books** Many students benefit from the

interactive and enjoyable activity of studying with one or more of the very popular HarperCollins coloring books, including *The Anatomy Coloring Book,* by Kapit and Elson; *The Physiology Coloring Book,* by Kapit, Masey, and Meisami; and the *The Human Brain Coloring Book,* by Diamond, Scheibel, and Elson.

● **Software**   Several software packages are available that are fully coordinated with the new edition of this text. They support both anatomical and physiological study and can be operated on a variety of computers, including IBM compatibles and Macintosh. Demonstration disks are available for review. Please contact your HarperCollins representative for details on the various packages available for instructor or student use and to receive a demo for review.

● **Visual Media**   A selection of videos and a laserdisc are available to adopters. As with the software, these items support both anatomical and physiological study, and your local representative will be happy to share details about their content with you.

ACKNOWLEDGMENTS

This book reflects the dedication and efforts of many people who have labored tirelessly to ensure that it is a work of the highest quality possible in college publishing. The guiding spirit throughout the creation of the seventh edition has been Bonnie Roesch, Acquisitions Editor. We are greatly indebted to her for the vision she had for the continued success of a classic text and for her consistent encouragement and support. Meryl R. G. Muskin, Developmental Editor, shepherded the manuscript through each revision cycle and recruited a truly outstanding group of reviewers. The beautiful line drawings by medical illustrators Leonard Dank, Sharon Ellis, Lauren Keswick, Lynn O'Kelley, Hilda Muinos, Nadine Sokol, Kevin Somerville, and Beth Willert make this edition truly a work of art. The computer graphics expertise of Jared Schneidman Design, Muinos Medical Illustrations, and Jean Jackson have contributed greatly to the visual impact of many physiological figures. Mira Schachne's work researching photographs for the text has once again been of the highest quality. Our thanks to Teresa Delgado, Director of Design, and Claudia Durrell, Art Coordinator, for their care in overseeing the all-important visual elements of this text. Thanks also go to Thomas Farrell, Project Editor, for keeping track of the thousands of details involved in the production of the text.

Dr. Michael C. Kennedy of Hahnemann University reviewed the entire illustration program for accuracy of drawing, labeling, and description—an especially valuable contribution.

Sandy appreciates the contributions of her colleagues at Purdue University. J. Alfred Chiscon, an exceptional teacher, inspired her to become a biology major and continues to be an important mentor and friend. William L. Pak provided guidance and encouragement in research and advice on the resulting early writing endeavors. Charles F.

Babbs sparked an interest both in teaching and in writing about human anatomy and physiology. During this revision, Anna W. Berkovitz, C. David Bridges, Paul V. Malven, Joann J. Otto, and Edward H. Simon provided greatly appreciated, expert opinions on selected topics in their areas of specialty.

We are most grateful to the focus group participants and reviewers listed below. Their insights have substantially enhanced this revision. We sincerely thank them for their outstanding contributions to the seventh edition of *Principles of Anatomy and Physiology*.

Focus Group Participants

Robert Anthony, Triton College
Charles Bennett, Kentucky State University
Jerry Button, Portland Community College
Wayne Carley, Lamar University
Gurment Dhaliwal, Malcolm X College
Victor Eroschenko, University of Idaho
Sharon Fowler, Duchess Community College
John Galligan, Westchester Community College
Freda Glaser, University of Maryland–Baltimore County
Jeff Hochbaum, Middlesex Community College
Beth Howard, Rutgers University
Gary Johnson, Madison Area Technical College
Michael Kennedy, Hahnemann University School of
   Medicine
Kathleen Lee, University of Vermont
Jim Love, College of Du Page
Linda Nichols, Santa Fe Community College
Antonios Pappantoniou, Lehman College
Philip Poli, Brookdale Community College
Sister Bernadette Agnes Proudfoot, Caldwell College
Henry Ruschin, Humber College
James Siniscalchi, Jersey City State College
Caryl Tickner, Stark Technical College
Steve Trautwein, Southeast Missouri State University
Donna Van Wynsberghe, University of Wisconsin–
   Milwaukee
Jane Wallace, Chattanooga State Technical Community
   College
Richard Welton, Southern Oregon State College
David Williams, Purdue University

Seventh Edition Reviewers

Attilio Alacchi, Cuyahoga Community College
Anne Betts, Royal College of Nursing (U.K.)
Jerry Button, Portland Community College
Kenneth Bynum, University of North Carolina at Chapel
   Hill
Lanier Byrd, St. Philip's College
Robert Callahan, William Paterson College
Wayne Carley, Lamar University
Anthony Chee, Houston Community College
Alice Chelich, Purdue University Calumet
Redding I. Corbett, Midlands Technical College
Sammy Culpepper, Meridian Community College
Emmet Dennis, Rutgers University
Danielle DesRoches, William Paterson College

Keith Dupre, University of Nevada Las Vegas
John Dustman, Indiana University Northwest
Victor Eroschenko, University of Idaho
Sue Fyfe, Curtin University of Technology, Australia
Carol Gerding, Cayahoga Community College
Dinah Gould, King's College (U.K.)
Amy Haddad, R. N., Creighton University
Clare Hays, Metropolitan State College
Jean Helgeson, Collin County Community College
James Horwitz, Palm Beach Community College
Ken F. Howells, Oxford Polytechnic (U.K.)
Gary Johnson, Madison Area Technical College
Michael Kennedy, Hahnemann University
William Kleinelp Jr., Middlesex County College
Pamela Langley, New Hampshire Technical Institute
Kathleen Lee, University of Vermont
Audric Lewis, Houston Community College, SE
Jerri Lindsey, Tarrant County Junior College, NE
Bonnie Lustigman, Montclair State College
Christine Martin, Stark Technical College
Douglas Merrill, Rochester Institute of Technology
William Miller III, Grambling State University
Richard Mortensen, Ohio State University
Dorothy Moses, University of Akron
W. Brian O'Connor, University of Massachusetts, Amherst
Brent Palmer, Ohio University
Izak Paul, Mount Royal College (Canada)
Joseph Adam Pearson, Scottsdale Community College
Robert Pollack, Nassau Community College
Diane Recker, R. N., Bellin Hospital
Tom Reeves, Midlands Technical College
Virginia Rivers, Truckee Meadows Community College
Gerald Robinson, University of South Florida
Henry Ruschin, Humber College (Canada)
Frank P. Sivik, Broward Community College
Mark Shoop, Macon College
David Smith, San Antonio College
Roberta O'Dell Smith, University of New Orleans
Francis Sullivan, Front Range Community College
Eric Sun, Macon College
Henry Tamar, Indiana State University
Don Terpening, Ulster County Community College
Caryl Tickner, Stark Technical College

Steve Trautwein, Southeast Missouri State University
Jean Tribulski, R. N., Rutgers University
Frank Veselovsky, South Puget Sound Community College
John M. Wakeman, Louisiana Tech University
Jane Wallace, Chattanooga State Technical Community
  College
Richard Welton, Southern Oregon State College
Horst Welzel, Nassau Community College
Ben Wise, Keene State College
Donna Van Wynsberghe, University of Wisconsin–
  Milwaukee

Multimedia Consultants

Merlyn Anderberg, Spokane Falls Community College
Robert Benno, William Paterson College
Jerry Button, Portland Community College
Jane Wallace, Chattanooga State Technical Community
  College

Finally, we would like to express deep appreciation to the thousands of instructors and 1.4 million students who have used this text in previous editions. Over the past two decades, you have been *"the heart of Tortora"* because of the feedback—comments, criticisms, requests, and reviews—you have provided. We invite all readers and users of this seventh edition of our text to continue the tradition of sending your suggestions to us, so that we can include them in our plans for subsequent editions.

GERARD J. TORTORA
Natural Sciences and Mathematics S229
Bergen Community College
400 Paramus Road
Paramus, NJ 07652

SANDRA REYNOLDS GRABOWSKI
Department of Biological Sciences
1392 Lilly Hall of Life Sciences
Purdue University
West Lafayette, IN 47907–1392

FOR STUDENTS: MAKING THE MOST OF THIS TEXT

A variety of special learning aids are designed into this book to help you build a strong foundation in anatomy and physiology. Each of these has been developed and refined based on feedback from students using previous editions of this text. With the help of their comments and suggestions, we have created a pedagogical plan for each chapter that will help you understand the content with greater ease and more enjoyment. They're described here so you can familiarize yourself with them before you begin.

CHAPTER OUTLINES AND STUDENT OBJECTIVES

Each chapter begins with **Chapter Contents at a Glance** and a listing of **Student Objectives**. Before you read the chapter, please review these carefully. Each objective states a skill or a piece of knowledge that you should acquire. To meet these objectives, you will have to perform several activities. Obviously, you must read the chapter carefully. If there are sections that you do not understand after one reading, you should reread those sections before continuing. As you read, pay particular attention to the figures and exhibits; they have been carefully coordinated with the textual narrative.

QUESTIONS WITH FIGURES

A new feature of the seventh edition is **Questions with Figures.** If you try to answer these questions as you go along, they will serve as self-checks to help you evaluate how well you're understanding the material. Often it will be possible to answer a question by examining the figure. Such questions reinforce a visual message by putting it into words. In other cases, a question will encourage you to integrate the knowledge you've gained by reading the associated text with the information presented in the figure. Some questions may prompt you to think critically about the topic at hand or predict a consequence in advance of its description in the text. If you find you can't immediately come up with an answer, first reexamine the verbal description of the figure in the text. Then, read the next section of the text to see if it offers a clue. We've provided answers at the end of each chapter so you can confirm that you're on the right track.

SUMMARY EXHIBITS

These exhibits summarize several preceding pages of text material. Here, in a nutshell, are the highlights of most chapters in a format that is perfect for reviewing and organizing the concepts.

CROSS-REFERENCING

Many cross-references have been added to help you relate concepts to previously learned material. In addition, some cross-references direct you ahead for a more complete discussion of topics under consideration.

END-OF-CHAPTER ACTIVITIES

At the end of each chapter are two and sometimes three other learning guides that you may find useful. The first, a **Study Outline,** is a concise summary of important topics discussed in the chapter. This section is designed to consolidate the essential points covered in the chapter so that you may recall and relate them to one another. For convenience, page numbers are listed next to key topics so you can easily refer to specific passages in the text for clarification or amplification. **Review Questions** are next. These are a series of questions designed specifically to help you master the objectives. Again, page numbers help you locate the answers. After you have answered the review questions, you should return to the beginning of the chapter and reread the objectives to determine whether you have achieved the goals. A third aid, **Medical Terminology,** appears in some chapters. This is a listing of terms designed to build your medical vocabulary. Every medical term is accompanied by a phonetic pronunciation.

APPENDIXES

There are several useful appendixes: **Measurements** summarizes U.S., metric, and apothecary units of length, mass, volume, and time; **Normal Values for Selected Blood and Urine Tests** contains a listing of reference values for the principal constituents of these fluids; and a **Periodic Table** is included for easy reference. In addition, the endpapers of the text include **Lists of Symbols and Medical Abbreviations, Eponyms Used in the Text,** and **Selected Terms Used in Writing Prescriptions.**

GLOSSARIES

Two glossaries appear at the end of the book. The first presents combining forms, word roots, prefixes, and suffixes. The second is a comprehensive glossary of terms.

PRONUNCIATIONS

As a further aid, we have included pronunciations for many terms that may be new to you. These appear in parentheses immediately following new words, and most are repeated in the Glossary of Terms at the back of the book. Look at these words carefully and say them out loud several times. Learning to pronounce a new word will help you remember it and make it a useful part of your medical vocabulary. Take a few minutes now to read the following pronunciation key, so it will be familiar as you encounter new words. The key is repeated at the beginning of the Glossary of Combining Forms, Word Roots, Prefixes, and Suffixes.

Pronunciation Key

1.   The most strongly accented syllable appears in capital letters, for example, bilateral (bī-LAT-er-al) and diagnosis (dī-ag-NŌ-sis).

2.   If there is a secondary accent, it is noted by a prime mark ('), for example, constitution (kon′-sti-TOO-shun) and physiology (fiz′-ē-OL-ō-jē). Any additional secondary accents are also noted by a single prime mark, for example, decarboxylation (dē′-kar-bok′-si-LĀ-shun).

3.   Vowels marked with a line above the letter are pronounced with the long sound as in the following common words:

ā as in *māke*
ē as in *bē*
ī as in *īvy*
ō as in *pōle*

4.   Vowels not so marked are pronounced with the short sound as in the following words:

e as in *bet*
i as in *sip*
o as in *not*
u as in *bud*

5.   Other phonetic symbols are used to indicate the following sounds:

a as in *above*
oo as in *sue*
yoo as in *cute*
oy as in *oil*

We wish you a rewarding experience in your study of anatomy and physiology. It is our sincere hope that you find this text useful now and throughout your journey toward your career goal. Please feel free to write to us with questions, comments, or suggestions as many of the students who preceded you have done. *To your success!*

GERARD J. TORTORA
Natural Sciences and Mathematics S229
Bergen Community College
400 Paramus Road
Paramus, NJ 07652

SANDRA REYNOLDS GRABOWSKI
Department of Biological Sciences
1392 Lilly Hall of Life Sciences
Purdue University
West Lafayette, IN 47907–1392

A Visual Guide To

PRINCIPLES OF ANATOMY AND PHYSIOLOGY

Seventh Edition

Beginning Each Chapter

Each chapter begins with a pair of proven pedagogical engine starters: **Chapter Contents at a Glance** and **Student Objectives.** By examining each of these listings carefully before diving into a chapter, students can get a quick overview of the material they're about to encounter and an idea of how the chapter is structured. It's a good way of "warming up" before "working out." Later, students can refer back to these lists to make sure they've mastered each objective.

Chapter 20
THE CARDIOVASCULAR SYSTEM: THE HEART

Chapter Contents at a Glance

LOCATION AND SIZE OF THE HEART
PERICARDIUM
HEART WALL
CHAMBERS OF THE HEART
BLOOD FLOW THROUGH THE HEART
VALVES OF THE HEART
  Atrioventricular Valves
  Semilunar Valves
HEART BLOOD SUPPLY
  Coronary Arteries
  Coronary Veins
  Reperfusion Damage
CONDUCTION SYSTEM AND
PACEMAKER
  Autorhythmic Cells: The Conduction
  System
  Timing of Atrial and Ventricular
  Excitation
PHYSIOLOGY OF CARDIAC MUSCLE
CONTRACTION

ELECTROCARDIOGRAM
CARDIAC CYCLE
  Phases of the Cardiac Cycle
  Timing of Systole and Diastole
  Heart Sounds
CARDIAC OUTPUT
  Regulation of Stroke Volume
    Preload: Effect of Stretching •
    Contractility • Afterload
  Regulation of Heart Rate
    Autonomic Control of Heart Rate •
    Chemical Regulation of Heart Rate
    • Other Factors
  Help for Failing Hearts
RISK FACTORS IN HEART DISEASE
PLASMA LIPIDS AND HEART DISEASE
  Lipoproteins in Blood
  Blood Cholesterol
EXERCISE AND THE HEART

DEVELOPMENTAL ANATOMY OF THE
HEART
DISORDERS: HOMEOSTATIC
IMBALANCES
  Coronary Artery Disease •
  Congenital Defects • Arrhythmias
CLINICAL APPLICATIONS
  Cardiopulmonary Resuscitation •
  Pericarditis and Cardiac Tamponade
  • Rheumatic Fever • Angina Pectoris
  and Myocardial Infarction • Heart
  Murmurs • Congestive Heart Failure
MEDICAL TERMINOLOGY
STUDY OUTLINE
REVIEW QUESTIONS
ANSWERS TO QUESTIONS WITH
  FIGURES

Student Objectives

1. Describe the location of the heart and the structure and functions of the wall, chambers, great vessels, and valves of the heart.
2. Describe the blood supply of the heart.
3. Explain the structural and functional features of the conduction system of the heart.
4. Describe the physiology of cardiac muscle contraction.
5. Explain the meaning of an electrocardiogram (ECG) and its diagnostic importance.

6. Describe the phases, timing, and sounds associated with a cardiac cycle.
7. Define cardiac output (CO) and describe the factors that affect it.
8. Explain how heart rate is regulated.
9. List and explain the risk factors involved in heart disease.
10. Explain the relationship between plasma lipids and heart disease.
11. Describe the developmental anatomy of the heart.

12. Explain the benefits of regular exercise on the heart.
13. Define the following disorders: coronary artery disease (CAD), congenital defects, and arrhythmias.
14. Define medical terminology associated with the heart.

Homeostasis

Homeostasis serves as one of the text's major unifying themes. It's introduced immediately in Chapter 1, then integrated and amplified throughout the entire book. **New negative feedback figures** help make potentially confusing concepts much easier to understand. They're used whenever applicable to help students grasp the dynamic counterbalancing act that systems must perform to maintain normal anatomy and physiology.

FIGURE 1.3 Components of a feedback system (loop).

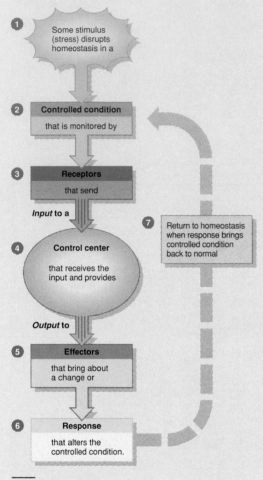

Question: What are two ways that negative and positive feedback systems differ?

FIGURE 18.7 Negative feedback system for regulation of T$_3$ (triiodothyronine) concentration in the blood.

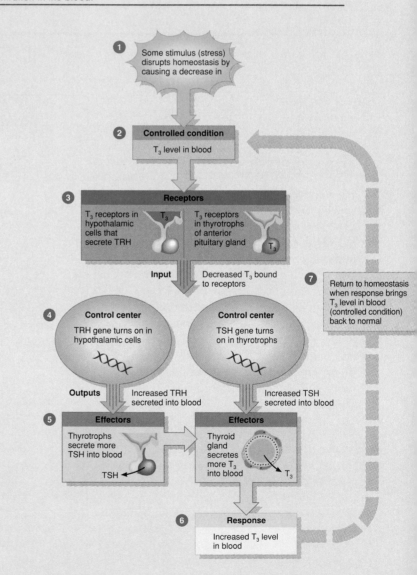

Question: How can you tell that this is a *negative* feedback system?

Homeostatic Imbalances

To further illustrate the relevance of normal mechanisms, disorders are explored through in-text **Clinical Applications** and chapter-ending **Disorders: Homeostatic Imbalances** sections. These discussions also answer many of the questions commonly asked by students about medical disorders and disease.

Retinal is the light-absorbing portion of all visual photopigments. In the human retina, there are four different opsins, one for each cone photopigment and one for rhodopsin. Small variations in the amino acid sequences of the different opsins permit the rods and cones to absorb different colors of incoming light. Rhodopsin absorbs blue to green light most effectively whereas the three cone photopigments most effectively absorb blue, green, or yellow to red light.

A very important characteristic of retinal is that it exists in two forms. In darkness, retinal has a bent shape, called *cis*-retinal, which snugly fits against the opsin portion of the photopigment (Fig. 16.12). When it absorbs light, *cis*-retinal straightens out to a shape called *trans*-retinal. This *cis* to *trans* conversion is called **isomerization** and it is the first step in transduction. Forming a visual image begins with isomerization of particular photopigments in certain rods and cones. After retinal isomerizes, several unstable substances form and disappear. In about a minute, *trans*-retinal completely separates from opsin. The final products look colorless, so the whole process is called **bleaching** of photopigment.

In darkness, an enzyme called **retinal isomerase** can reconvert *trans*- to *cis*-retinal, which then binds to opsin and reforms a functional photopigment. Resynthesis of a photopigment is called **regeneration.** The pigment epithelium, adjacent to the photoreceptors, stores a large quantity of vitamin A and contributes to the regeneration process in rods. The extent of rhodopsin regeneration decreases drastically if the retina detaches from the pigment epithelium. Cone photopigments regenerate much more quickly than does rhodopsin and are less dependent on the pigment epithelium. After complete bleaching, it takes 5 minutes to regenerate half of the rhodopsin but only 1½ minutes to regenerate half of the cone photopigments. Full regeneration of bleached rhodopsin takes 30 to 40 minutes.

FIGURE 16.12 Bleaching and regeneration of photopigment.

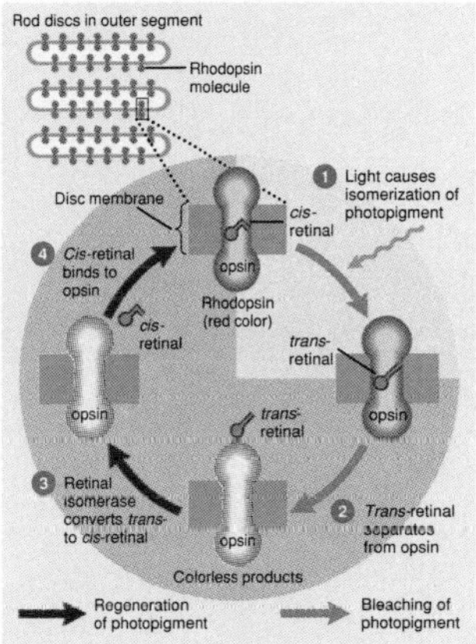

Question: What is the conversion of *cis*- to *trans*-retinal called?

CLINICAL APPLICATION

COLOR BLINDNESS

Most forms of **color blindness** (inability to distinguish certain colors) result from absence or deficiency of one of the three cone photopigments. The most common type is **red–green color blindness,** in which a photopigment sensitive to red or green light is missing. As a result, the person has no way to distinguish between red and green. Color blindness is an inherited condition that affects males far more often than females. The inheritance of the condition is discussed on page ■■■ and illustrated in Fig. 29.20.

Light and Dark Adaptation

When you emerge from dark surroundings, such as a tunnel, into the sunshine, your visual system adjusts in seconds to the brighter environment by decreasing its sensitivity. This is called **light adaptation.** On the other hand, when you enter a darkened room, such as a theater, your visual sensitivity increases slowly over many minutes. This is called **dark adaptation.** Bleaching and regeneration of the photopigments account for much but not all of the sensitivity change during light and dark adaptation.

As the light level increases, more and more photopigment is bleached so more light is needed to stimulate the remaining unbleached photopigment. At the same time that light is bleaching some photopigment molecules, however, others are being regenerated. Regeneration of rhodopsin occurs slowly enough that in daylight it is bleached as fast as it is regenerated. For this reason, rods contribute little to daylight vision. Regeneration of cone photopigments, on the other hand, is fast enough that some of the *cis* form is always present, even in very bright light.

If the light level decreases abruptly, sensitivity increases rapidly at first and then more slowly. In complete darkness, full regeneration of the cone photopigments occurs during the first 8 minutes of dark adaptation. During this time interval, a threshold light flash (barely perceptible) is seen

DISORDERS: HOMEOSTATIC IMBALANCES

CEREBROVASCULAR ACCIDENT

The most common brain disorder is a **cerebrovascular accident (CVA)**, also called a **stroke**. CVAs are classified into two principal types: (1) *ischemic*, the most common type, due to a decreased blood supply, and (2) *hemorrhagic*, due to rupture of a blood vessel in the brain. Common causes of CVAs are intracerebral hemorrhage (from a blood vessel in the pia mater or brain), emboli (blood clots), and atherosclerosis (formation of cholesterol-containing plaques that block blood flow) of the cerebral arteries. A CVA is characterized by abrupt onset of persisting neurological symptoms that arise from destruction of brain tissue.

Among the risk factors implicated in CVAs are high blood pressure, high blood cholesterol, heart disease, narrowed carotid arteries, transient ischemic attacks, diabetes, smoking, obesity, and excessive alcohol intake.

TRANSIENT ISCHEMIC ATTACK

A **transient ischemic attack (TIA)** is an episode of temporary cerebral dysfunction caused by impaired blood flow to the brain. Symptoms include dizziness, weakness, numbness, or paralysis in a limb or in half of the body; drooping of one side of the face; headache; slurred speech or difficulty understanding speech; or a partial loss of vision or double vision. Sometimes nausea or vomiting also occur. The onset of symptoms is sudden and reaches maximum intensity almost immediately. A TIA usually persists for a few minutes and only rarely lasts as long as 24 hours; it leaves no persistent neurologic deficits. The causes of the impaired blood flow that lead to TIAs are blood clots, atherosclerosis, and certain blood disorders.

It is estimated that about one-third of patients who experience a TIA will have a CVA within five years. Therapy for TIAs includes agents such as aspirin that block the aggregation of blood cells involved in clotting (platelets); anticoagulants; cerebral artery bypass grafting; and carotid endarterectomy (removal of the cholesterol-containing plaques and inner lining of an artery).

ALZHEIMER'S DISEASE

Alzheimer's (ALTZ-hī-merz) **disease (AD)** is a disabling neurological disorder that afflicts about 11% of the population over age 65. Its causes are unknown, its effects are irreversible and devastating, and it has no cure. In the U.S., AD afflicts four million people and claims over 100,000 lives a year, making it the fourth leading cause of death among the elderly, following heart disease, cancer, and stroke.

Victims of AD initially have trouble remembering recent events. Next they become more confused and forgetful, often repeating questions or getting lost while traveling to previously familiar places. Disorientation grows, memories of past events disappear, and there may be episodes of paranoia, hallucination, or violent changes in mood. As their minds continue to deteriorate, they lose their ability to read, write, talk, eat, or walk. Finally, the disease culminates in dementia, the loss of reason and ability to care for oneself. A person with AD usually dies of some complication that afflicts bedridden patients, such as pneumonia.

Diagnosis of AD is difficult because there is no definitive medical test and the mental deficits are slight in the early stages of the disease. Only when the disease is well-advanced and all other possible causes of the dementia are ruled out can a probable diagnosis of AD be made. Examination of the brain at autopsy confirms the diagnosis. Brains
tims show three distinct structural abnormalitie
loss of neurons in specific regions, (2) plaques
proteins deposited outside neurons, and (3) tan
filaments within neurons. Some AD brains also
vated levels of aluminum, an element that serv
function in the human body.

Neuronal loss is severe in regions such as the hippocampus and cerebral cortex that are important for memory and learning. Many of the lost neurons are thought to use acetylcholine as a neurotransmitter because the level of an enzyme needed to synthesize ACh (choline acetyltransferase) may be greatly reduced in damaged regions. This finding prompted research on treating AD patients with precursors of ACh, which were not helpful. Some success is being seen using drugs that inhibit acetylcholinesterase (AChE), the enzyme that inactivates ACh. AChE blocking drugs such as tacrine (Cognex) and slow-release physostigmine (Synapton) are being tested on AD patients.

Besides neuronal destruction, there is an accumulation of **amyloid plaques**, which consist of degenerating axons and axon terminals and fibrils of an abnormal protein called **beta amyloid**. This 40 to 42 amino acid fragment is part of a membrane glycoprotein called amyloid precursor protein (APP) that is coded by a gene on chromosome 21. People with Down syndrome, which is caused by an extra chromosome 21 (Chapter 29), are mentally retarded and their brains also exhibit beta amyloid-containing plaques. This protein is found only rarely in elderly people with normal mental abilities. Several pieces of evidence suggest that beta amyloid is neurotoxic. It causes degeneration of neurons, especially neurons responsible for learning and memory. Also, it has been discovered that substance P can slow or prevent beta amyloid-induced neuron death. This finding raises hopes for stopping the progression of AD.

The third feature of AD is the presence of **neurofibrillary tangles**, which are abnormal bundles of protein filaments inside cells in affected brain regions. These tangles contain a protein called A68 that also appears in the brains of older Down syndrome individuals. A68 appears to be an altered version of a protein called tau that normally associates with microtubules of the cytoskeleton. It also may be involved in neuronal death.

The causes of AD are not known. In a few cases, there is a genetic mutation in the gene that codes for amyloid precursor protein. Among identical twins, however, one may have AD while the other does not. Other proposed causes of AD are a slow-acting virus, environmental toxins such as aluminum, genetic defects in mitochondria, and decreased blood flow to the brain that results in inadequate oxygen and glucose.

BRAIN TUMORS

A **brain tumor** refers to any benign or malignant growth within the cranium. Tumors may arise from neuroglial cells in the cerebrum, brain stem, and cerebellum or from supporting or neighboring structures such as cranial nerve coverings, meninges, and the pituitary gland. Pressure in the brain from a growing tumor or edema (tissue swelling) associated with the tumor produces the characteristic signs and symptoms. Among these are headache, altered consciousness, vomiting, seizures, visual problems, cranial nerve abnormalities, hormonal syndromes, personality changes, dementia, and sensory or motor deficits. Treatment of brain tumors involves surgery, radiation therapy, and chemotherapy.

Outstanding Art Program

This edition's art program, as usual, sets a new standard for the discipline. It is fresh, conceptually consistent, and crystal clear. Over 80% of the art is either completely new or redrawn by the industry's premier medical illustrators. Body structures are depicted in a realistic three-dimensional manner.

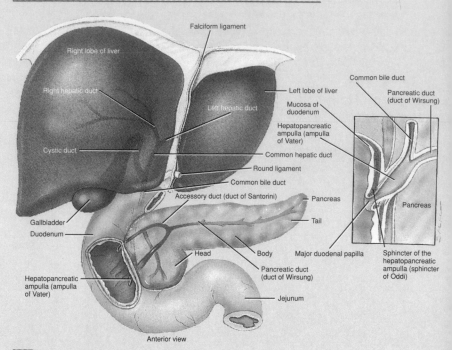

FIGURE 24.17 Relation of pancreas to liver, gallbladder, and duodenum. The inset shows details of the common bile duct and pancreatic duct forming the hepatopancreatic ampulla (of Vater) and emptying into the duodenum.

Question: What type of fluid would you find in the pancreatic duct? The common bile duct? The hepatopancreatic ampulla?

FIGURE 12.11 Changes in voltage-gated channels during action potential depolarization and repolarization. (a) Resting state: voltage-gated Na+ channels are in resting state and voltage-gated K+ channels are closed. (b) Depolarization to threshold (about –55 mV) opens Na+ channel activation gates. The inward flow of Na+ further depolarizes the membrane until its polarity is reversed (the inside becomes more positive than the outside). (c) More slowly, the depolarization also opens voltage-gated K+ channels, which permit outflow of K+. At the same time the Na+ channel inactivation gates close. Repolarization begins. (d) Outflow of K+ restores the resting membrane potential. Not illustrated are the leakage channels and the Na+/K+ pumps that maintain a low concentration of Na+ inside the cell. This figure exaggerates the relative numbers of Na+ that enter and K+ that exit during a single action potential.

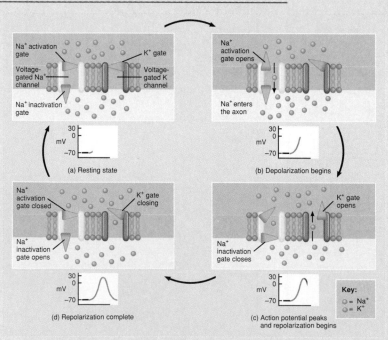

Question: Recalling that there are leakage channels for both K+ and Na+, could repolarization occur if the voltage-gated K+ channels did not exist?

Figure 11.15 (continued)

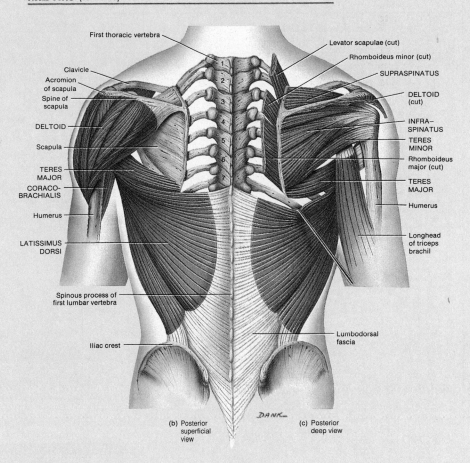

First thoracic vertebra

Levator scapulae (cut)

Rhomboideus minor (cut)

Clavicle

SUPRASPINATUS

Acromion
of scapula

Spine of
scapula

DELTOID
(cut)

DELTOID

INFRA-
SPINATUS

TERES
MINOR

Scapula

Rhomboideus
major (cut)

TERES
MAJOR

TERES
MAJOR

CORACO-
BRACHIALIS

Humerus

Humerus

Longhead
of triceps
brachil

LATISSIMUS
DORSI

Spinous process of
first lumbar vertebra

Lumbodorsal
fascia

Iliac crest

DANK

(b) Posterior
superficial
view

(c) Posterior
deep view

Question: Which muscles flex the arm?

Figure 10.17 Histology of cardiac muscle tissue.

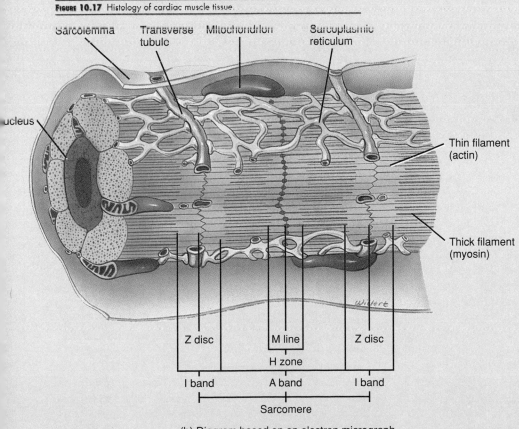

Sarcolemma

Transverse
tubule

Mitochondrion

Sarcoplasmic
reticulum

Nucleus

Thin filament
(actin)

Thick filament
(myosin)

Widert

Z disc

M line

Z disc

H zone

I band

A band

I band

Sarcomere

(b) Diagram based on an electron micrograph

FIGURE 7.4 Skull.

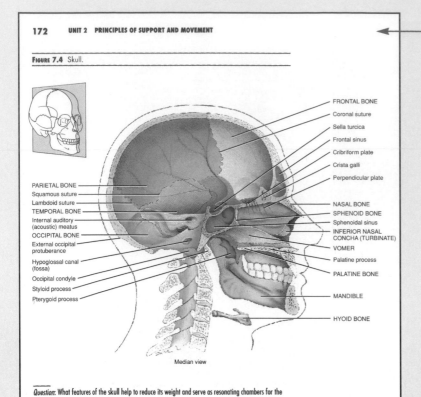

FRONTAL BONE
Coronal suture
Sella turcica
Frontal sinus
Cribriform plate
Crista galli
Perpendicular plate

NASAL BONE
SPHENOID BONE
Sphenoidal sinus
INFERIOR NASAL
CONCHA (TURBINATE)
VOMER
Palatine process
PALATINE BONE

MANDIBLE

HYOID BONE

PARIETAL BONE
Squamous suture
Lambdoid suture
TEMPORAL BONE
Internal auditory
(acoustic) meatus
OCCIPITAL BONE
External occipital
protuberance
Hypoglossal canal
(fossa)
Occipital condyle
Styloid process
Pterygoid process

Median view

Question: What features of the skull help to reduce its weight and serve as resonating chambers for the voice?

Questions with Figures

Each illustration in the text is now accompanied by one or more questions designed to serve as an on-going self-check system. Students can readily evaluate their level of comprehension at any time, so troublesome material can be reread before moving on to new concepts. The questions are closely correlated to the text material, and many of them require students to synthesize verbal and visual information, think critically, or draw conclusions. Answers are given at the end of each chapter.

Exhibits

These handy capsule reviews appear periodically throughout the text, summarizing the material of the preceding pages. They're especially helpful for review, mental organization of concepts, and continual self-checking.

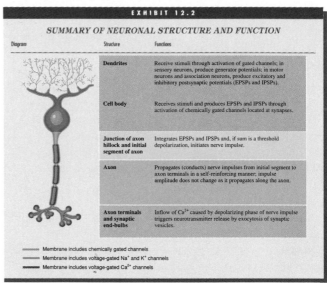

EXHIBIT 12.2

SUMMARY OF NEURONAL STRUCTURE AND FUNCTION

Diagram	Structure	Functions
	Dendrites	Receive stimuli through activation of gated channels; in sensory neurons, produce generator potentials; in motor neurons and association neurons, produce excitatory and inhibitory postsynaptic potentials (EPSPs and IPSPs).
	Cell body	Receives stimuli and produces EPSPs and IPSPs through activation of chemically gated channels located at synapses.
	Junction of axon hillock and initial segment of axon	Integrates EPSPs and IPSPs and, if sum is a threshold depolarization, initiates nerve impulse.
	Axon	Propagates (conducts) nerve impulses from initial segment to axon terminals in a self-reinforcing manner; impulse amplitude does not change as it propagates along the axon.
	Axon terminals and synaptic end-bulbs	Inflow of Ca^{2+} caused by depolarizing phase of nerve impulse triggers neurotransmitter release by exocytosis of synaptic vesicles.

Membrane includes chemically gated channels
Membrane includes voltage-gated Na^+ and K^+ channels
Membrane includes voltage-gated Ca^{2+} channels

rons and by some CNS neurons. ACh is an excitatory neurotransmitter at the neuromuscular junction, where it acts directly to open chemically gated cation channels. It is also known to be an inhibitory neurotransmitter at other synapses, where its effects on ion channels appear to occur indirectly via receptors that link to a G-protein. One example is parasympathetic fibers of the vagus nerve (cranial nerve X) that innervate the heart; ACh slows heart rate by these inhibitory synapses.

Several amino acids are CNS neurotransmitters. **Glutamate** (glutamic acid) and **aspartate** (aspartic acid) have powerful excitatory effects. Two others, **gamma aminobutyric** (GAM-ma am-i-nō-byoo-TIR-ik) **acid (GABA)** and **glycine** are important inhibitory neurotransmitters. Both cause IPSPs by opening chemically gated Cl⁻ channels. Although GABA is an amino acid, it is not incorporated into proteins in your body. It is found only in the brain, where it is the most common inhibitory neurotransmitter.

Perhaps a third of all brain synapses use GABA. Glycine is more prevalent in the spinal cord than in the brain.

Catecholamine neurotransmitters (norepinephrine, epinephrine, and dopamine) are synthesized from the amino acid tyrosine. They are excitatory at some synapses and inhibitory at others.

Neurotransmitters in the CNS are discussed in more detail on page ■■■. Certain disorders such as Parkinson's disease, Alzheimer's disease, depression, anxiety, and schizophrenia involve problems relating to neurotransmitters.

CLINICAL APPLICATION

STRYCHNINE POISONING

The importance of inhibitory neurons can be appreciated by observing what happens when normal inhibition is blocked.

Medical Terminology

These helpful and interesting vocabulary-builders are accompanied by phonetic pronunciations.

Study Outlines

These concise, easy-to-follow summaries are designed to consolidate all the essential points covered in a chapter. They appear at the end of each chapter, with page numbers listed next to key topics to facilitate review.

MEDICAL TERMINOLOGY

Arthralgia (ar-THRAL-jē-a; *arth* = joint; *algia* = pain) Pain in a joint.

Arthrosis (ar-THRŌ-sis) Refers to an articulation; also a disease of a joint.

Bursectomy (bur-SEK-tō-mē; *ectomy* = removal of) Removal of a bursa.

Chondritis (kon-DRĪ-tis; *chondro* = cartilage) Inflammation of cartilage.

Synovitis (sin'-ō-VĪ-tis; *synov* = joint) Inflammation of a synovial membrane in a joint.

Study Outline

Classification of Joints (p. 217)

1. An articulation (joint) is a point of contact between two or more bones.
2. Structural classification is based on the presence of a synovial (joint) cavity and type of connecting tissue. Structurally, joints are classified as fibrous, cartilaginous, or synovial.
3. Functional classification of joints is based on the degree of movement permitted. Joints may be synarthroses (immovable), amphiarthroses (partially movable), or diarthroses (freely movable).

Synarthrosis (Immovable Joint) (p. 217)

1. Types include suture, gomphosis, and synchondrosis.
2. A suture is a fibrous joint composed of a thin layer of dense fibrous connective tissue that unites skull bones. An example is the coronal suture.
3. A gomphosis is a fibrous joint in which a cone-shaped peg fits into a socket. An example is the root of a tooth in its socket.
4. A synchondrosis is a cartilaginous joint in which the connecting material is hyaline cartilage. An example is the epiphyseal plate.

Amphiarthrosis (Slightly Movable Joint) (p. 218)

1. An amphiarthrosis may be a syndesmosis or a symphysis.
2. A syndesmosis is a fibrous joint in which there is more fibrous connective tissue than in a suture. An example is the distal articulation of the tibia and fibula.
3. A symphysis is a cartilaginous joint in which the connecting material is a disc of fibrocartilage. An example is the pubic symphysis.

Diarthrosis (Freely Movable Joint) (p. 218)

Structure (p. 218)

1. A diarthrosis contains a space between bones called the synovial (joint) cavity. All diarthroses are synovial joints.
2. Other characteristics of a diarthrosis are the presence of articular cartilage and an articular capsule (fibrous capsule and synovial membrane).
3. Many diarthroses also contain accessory ligaments (extracapsular and intracapsular), articular discs (menisci), and bursae.

Contact and Movement (p. 220)

1. Among the factors that keep the articular surfaces of diarthroses in contact are the shape of the articulating bones,

strength and tension of ligaments, arrangement and tension of muscles, and apposition of soft parts.
2. Joint flexibility may also be affected by hormones, such as relaxin, which relaxes the pubic symphysis during childbirth.

Types of Diarthroses (p. 220)

1. Gliding joints: articulating surfaces are flat; bone glides back-and-forth and side-to-side; found between carpals and tarsals.
2. Hinge joint: convex surface of one bone fits into concave surface of another; movement is flexion–extension (monaxial); example is elbow.
3. Pivot joint: a round or pointed surface of one bone fits into a ring formed by another bone and a ligament; movement is rotation (monaxial); example is when atlas rotates around axis.
4. Ellipsoidal: an oval-shaped condyle of one bone fits into an elliptical cavity of another; movements are flexion–extension and abduction–adduction (biaxial) and circumduction; example is between carpals and radius.
5. Saddle: articular surface of one bone is shaped like a saddle and the other is shaped like a rider sitting in the saddle; movements are flexion–extension and abduction–adduction (biaxial) and circumduction; example is between trapezium and metacarpal of thumb.
6. Ball-and-socket: ball-shaped surface of one bone fits into cup-like depression of another; movements are flexion–extension, abduction–adduction, and rotation (triaxial) and circumduction; example is the shoulder.

Special Movements at Diarthroses (p. 224)

1. In addition to flexion–extension, hyperextension, abduction–adduction, rotation, and circumduction, special movements occur at particular joints.
2. Examples are elevation–depression, protraction–retraction, inversion–eversion, dorsiflexion–plantar flexion, and supination–pronation.

Knee (Tibiofemoral) Joint (p. 226)

1. The knee (tibiofemoral) joint is formed by the patella and femur and by the tibia and femur.
2. Other joints are described in Exhibit 9.3.
3. Arthroscopy is a procedure for examining and repairing the interior of a joint.
4. Arthroplasty refers to surgical replacement of joints.

Review Questions

1. Define an articulation. What factors determine the degree of movement at joints? (p. 216)
2. Classify joints on the basis of structure and function. (p. 217)
3. Set up an outline for the structural classification of joints, list and define the various subtypes, and give an example of each. (p. 217)
4. Describe the structure of a diarthrosis (freely movable joint). (p. 218)
5. What is an accessory ligament? Define the two principal types. (p. 218)
6. What is an articular disc? Why are they important? (p. 220)
7. Explain how the articulating bones in a synovial joint are held together. (p. 220)

8. List the six subtypes of diarthroses, define each, describe the movements at each, and give an example of each.
9. Describe the various special movements that occur at diarthroses. (p. 224)
10. Have another person assume the anatomical position and note each of the movements at joints discussed in this chapter. Reverse roles and see if you can execute the same movements.
11. Be sure that you can name the bones that form the joint, identify the joint by type, and list the anatomical components of the joint for each of the following: shoulder (humeroscapular), hip (coxal), and knee (tibiofemoral) joints. (p. 231)

Answers to Questions with Figures

9.1 Synovial joint.
9.2 Hinge and pivot joints.
9.3 They permit movement in a single plane.
9.4 Movement of a bone around its longitudinal axis.

9.5 When you adduct your arm or leg, you bring it closer to the midline of the body, thus "adding" it to the trunk.
9.6 They occur only at specific joints.
9.7 Articular discs (menisci).

Review Questions

Situated at the end of each chapter, these special questions are specifically designed to help students master the learning objectives stated at the outset of the chapter. As with the Study Outlines, convenient page references link the questions to specific text passages, making study and review more efficient.

Principles of Anatomy and Physiology

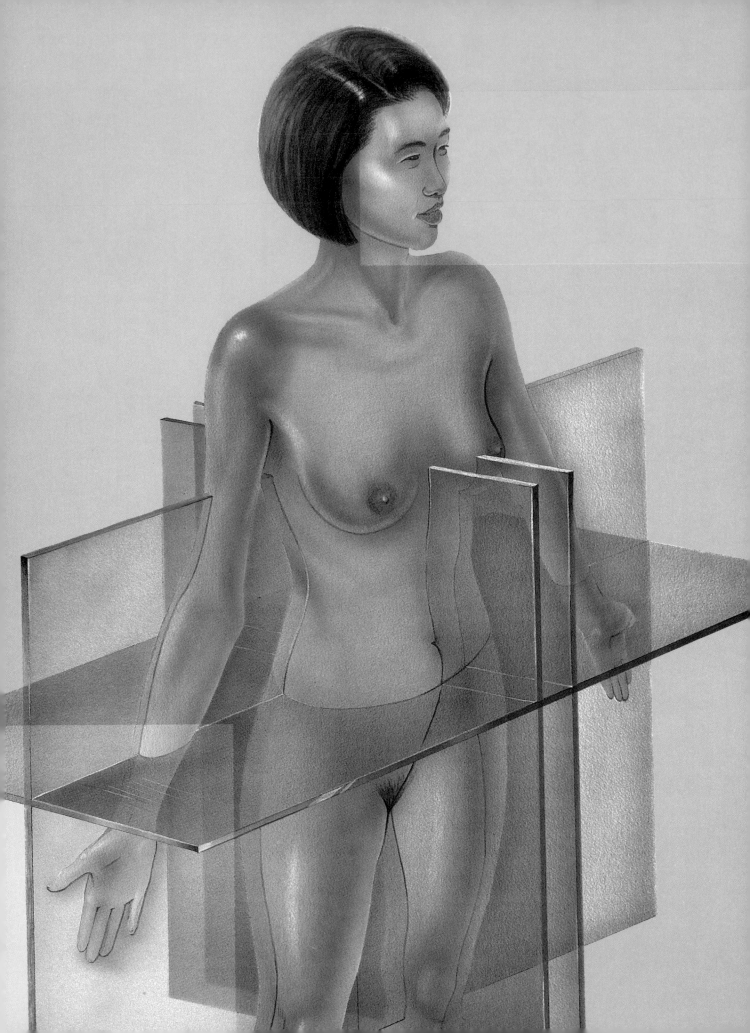

Unit 1
ORGANIZATION OF THE HUMAN BODY

This unit is designed to show you how your body is organized at different levels. After you study the various regions and parts of your body, you will discover the importance of the chemicals that make it up. You will then find out how your cells, tissues, and organs form the systems that keep you alive and healthy.

Chapter 1

AN INTRODUCTION TO THE HUMAN BODY

Chapter Contents at a Glance

Student Objectives

1. Define anatomy and physiology and their subdivisions.

2. Define each of the following levels of structural organization that make up the human body: chemical, cellular, tissue, organ, system, and organismic.

3. Identify the principal systems of the human body, list representative organs of each system, and describe the function of each system.

4. List and define several important life processes of humans.

5. Define homeostasis and explain the effects of stress on homeostasis.

6. Define a feedback system and explain its role in homeostasis.

7. Contrast the operation and results of a negative feedback system and a positive feedback system.

8. Explain the relationship between homeostasis and disease.

9. Define the anatomical position and compare common and anatomical terms used to describe various regions of the human body.

10. Define several directional terms and anatomical planes used in association with the human body.

11. List by name and location the principal body cavities and the organs contained within them.

12. Describe the principle and importance of selected medical imaging techniques in the diagnosis of disease.

You are beginning a study of the human body to learn how it is organized and how it functions. Once you have a basic understanding of how the body is organized, how its different parts normally work, and how various conditions affect the operation of its different parts to maintain life and health, you will be able to understand what happens when your body is injured, diseased, or placed under extreme stress.

In this chapter, you will be introduced to the various systems that compose the human body. You will also learn how these systems generally cooperate with each other to maintain the health of the body as a whole. In later chapters, when you have studied body systems in some detail, it will be pointed out how body systems interact to keep you healthy.

ANATOMY AND PHYSIOLOGY DEFINED

Two branches of science that will help you understand your body's parts and functions are anatomy and physiology. **Anatomy** (a-NAT-ō-mē; *anatome* = to dissect) refers to the study of **structure** and the relationships among structures.

Whereas anatomy and its branches deal with structures of the body, **physiology** (fiz′-ē-OL-ō-jē) deals with **functions** of the body parts, that is, how they work. Since function cannot be completely separated from structure, you will learn about the human body by studying its anatomy and physiology together. You will see how each structure of the body is designed to carry out a particular function and how the structure of a part often reveals the functions it can perform. For example, the hairs lining the nose filter air that we inhale. The fused bones of the skull protect the brain but permit no movement, whereas the spaces (joints) between long bones in the arms and legs permit various types of movement. The external part of the ear is shaped in such a way as to collect sound waves, which results in hearing. The air sacs of the lungs are so thin that oxygen and carbon dioxide readily pass between the sacs and the blood.

Several subdivisions of anatomy and physiology are described in Exhibit 1.1.

LEVELS OF STRUCTURAL ORGANIZATION

The human body consists of several levels of structural organization that are associated with one another in various ways (Fig. 1.1). The lowest level of organization, the **chemical level,** includes all atoms and molecules essential for maintaining life. Certain atoms, such as carbon (C), hydrogen (H), oxygen (O), nitrogen (N), calcium (Ca), potassium (K), and sodium (Na), are essential for maintaining life. Atoms combine to form molecules. Some examples of molecules are proteins, carbohydrates, fats, and vitamins.

Molecules, in turn, combine to form structures at the next higher level of organization: the **cellular level. Cells** are the basic structural and functional units of an organism. Among the many kinds of cells in your body are muscle cells, nerve cells, and blood cells. Figure 1.1 shows several isolated cells from the lining of the stomach. Each has a different structure, and each performs a different function. As you will see on page 69, cells contain specialized structures called **organelles,** such as the nucleus, mitochondria, and lysosomes, that perform specific functions.

The next higher level of structural organization is the **tissue level. Tissues** are groups of similar cells (and the substance surrounding them) that usually arise from common ancestor cells and work together to perform a particular function. The four basic types of tissue in your body are *epithelial tissue, muscle tissue, connective tissue,* and *nervous tissue.* When the cells shown in Figure 1.1 come together, they form an epithelial tissue that lines the stomach. Each type of cell in the tissue has a specific function. *Parietal cells* produce the hydrochloric acid (HCl) found in the stomach's gastric juice. *Mucous cells* produce mucus, a slippery, thick secretion that protects the stomach lining from being damaged by its own acidic juice. *Zymogenic cells* produce an enzyme that starts the digestion of proteins.

Throughout the body, different kinds of tissues are joined to form the next higher level of organization: the **organ level. Organs** are structures that are composed of two or more different tissues, have specific functions, and usually have recognizable shapes. Examples of organs are the heart, liver, lungs, brain, and stomach. Figure 1.1 shows several tissues that make up the stomach. (1) The outer covering is the *serosa,* a layer of epithelial tissue and connective tissue that protects the stomach and reduces friction when the stomach moves and rubs against other body organs. (2) The *muscle tissue layers* contract to churn and mix food and push it on to the next digestive organ (the small intestine). (3) The *epithelial tissue layer* lining the stomach, as noted earlier, produces acid, mucus, and an enzyme.

The next level of structural organization in the body is the **system level. A system** consists of several related organs that have a common function. For example, the organs of the digestive system, which break down and absorb food, include the mouth, pharynx (throat), esophagus, stomach, small intestine, and large intestine. Other organs—the salivary glands (which produce saliva), liver, and pancreas—are also part of the digestive system because they secrete materials needed for digestion. Sometimes an organ is part of more than one system. The pancreas, for example, is part of both the digestive system and the hormone-producing endocrine system.

The highest level is the **organismic level.** All the parts of the body functioning with one another comprise the total **organism**—one living individual.

In the chapters that follow, you will examine the

EXHIBIT 1.1

SELECTED SUBDIVISIONS OF ANATOMY AND PHYSIOLOGY

Subdivisions of Anatomy	Description
Surface anatomy	Study of the form (morphology) and markings of the surface of the body.
Gross (macroscopic) anatomy	Study of structures that can be examined without using a microscope.
Systemic (systematic) anatomy	Study of specific systems of the body such as the nervous or respiratory system.
Regional anatomy	Study of a specific region of the body such as the head or chest.
Radiographic (rā′-dē-ō-GRAF-ik; *radio* = ray; *graph* = to write) **anatomy**	Study of the structure of the body that includes the use of x-rays.
Developmental anatomy	Study of development from the fertilized egg to adult form.
Embryology (em′-brē-OL-ō-jē; *embryon* = embryo; *logos* = study of)	Study of development from the fertilized egg through the eighth week in utero.
Histology (hiss′-TOL-ō-jē; *histio* = tissue)	Microscopic study of the structure of tissues.
Cytology (sī-TOL-ō-jē; *cyto* = cells)	Chemical and microscopic study of the structure of cells.
Pathological (path′-ō-LOJ-i-kal; *patho* = disease) **anatomy**	Study of structural changes (from gross to microscopic) associated with disease.

Subdivisions of Physiology	Description
Cell physiology	Study of the functions of cells.
Systems physiology	Study of the operation of organ systems.
Pathophysiology (PATH-ō-fiz-ē-ol′-ō-jē; *patho* = disease)	Study of functional changes associated with disease and aging.
Exercise physiology	Study of changes in cell and organ functions during muscular activity.
Neurophysiology (NOO-rō-fiz-ē-ol′-ō-jē; *neuro* = nerve or nervous system)	Study of functional characteristics of nerve cells.
Endocrinology (en′-dō-kri-NOL-ō-jē; *endo* = within; *crin* = to secrete)	Study of hormones (chemical regulators in the blood) and how they control body functions.
Cardiovascular (kar-dē-ō-VAS-kyoo-lar; *cardio* = heart; *vascular* = blood vessels) **physiology**	Study of functions of the heart and blood vessels.
Immunology (im′-yoo-NOL-ō-jē; *immunis* = free)	Study of body defense mechanisms.
Respiratory (re-SPĪ-ra-tō′-rē; *respiratio* = to breathe) **physiology**	Study of functions of the air passageways and lungs.
Renal (RĒ-nal) **physiology**	Study of the functions of the kidneys.
Reproductive (rē-prō-DUK-tiv) **physiology**	Study of the functions of reproductive structures.

---

anatomy and physiology of the major body systems. Exhibit 1.2 describes the components and functions of these systems in the order they are discussed in the book.

LIFE PROCESSES

All living forms carry on certain processes that distinguish them from nonliving things. Following are some of the important life processes of humans:

1.   Metabolism is the sum of all the chemical processes that occur in the body. One phase of metabolism, called **catabolism,** provides the energy needed to sustain life by breaking down large, complex molecules. The other phase, called **anabolism,** uses the energy from catabolism to build the body's structural and functional components. An example of anabolism is the synthesis of the proteins that make up muscle.

2.   Responsiveness is the ability to detect and respond to changes in the external or internal environment. Different cells detect different sorts of changes and respond in characteristic ways. Neurons (nerve cells) respond by generating electrical signals, known as action potentials (nerve impulses), and sometimes carrying them over long distances, such as between your big toe and brain. Muscle fibers (cells) respond by contracting, that is, becoming shorter, to move body parts. Endocrine cells

FIGURE 1.1 Levels of structural organization in the human body.

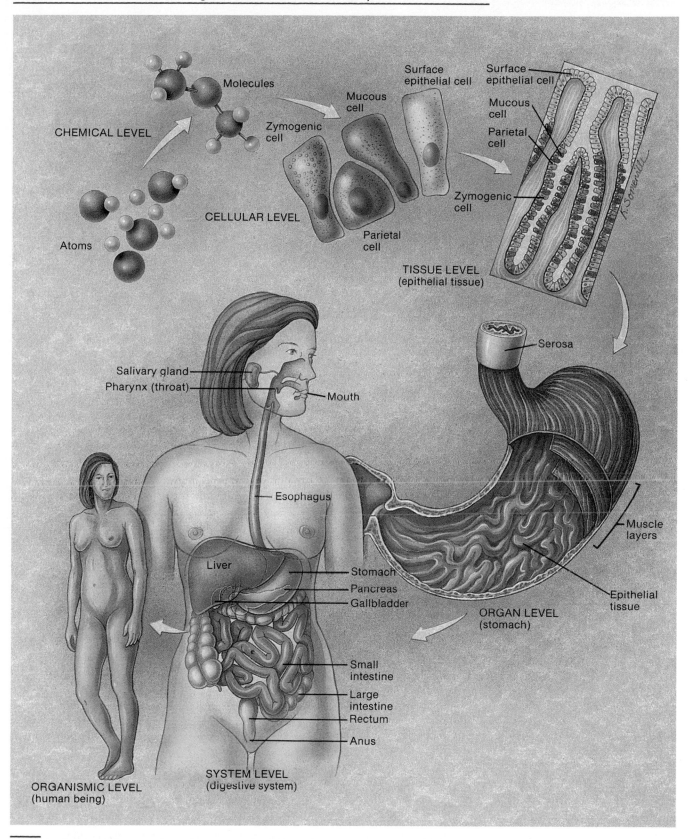

Question: Which level of structural organization is composed of two or more tissues and usually has a recognizable shape?

EXHIBIT 1.2

MAJOR SYSTEMS OF THE HUMAN BODY

1. Integumentary system
Components: The skin and structures derived from it, such as hair, nails, and sweat and oil glands.
Function: Helps regulate body temperature, protects the body, eliminates some wastes, helps produce vitamin D, and receives certain stimuli such as temperature, pressure, and pain.
Reference: See Fig. 5.1.

2. Skeletal system
Components: All the bones of the body, their associated cartilages, and the joints of the body.
Function: Supports and protects the body, assists in body movements, houses cells that give rise to blood cells, and stores minerals.
Reference: See Fig. 7.1.

3. Muscular system
Components: Specifically refers to skeletal muscle tissue. Other muscle tissues are smooth and cardiac.
Function: Participates in bringing about movement, maintains posture, and produces heat.
Reference: See Fig. 11.3.

4. Nervous system
Components: Brain, spinal cord, nerves, and special sense organs, such as the eye and ear.
Function: Regulates body activities through action potentials (nerve impulses) by detecting changes in the internal and external environment, interpreting the changes, and responding to the changes by inducing muscular contractions or glandular secretions.
Reference: See Figs. 13.2 and 14.1.

5. Endocrine system
Components: All hormone-producing glands and cells such as the pituitary gland, thyroid gland, and pancreas.
Function: Regulates body activities through hormones, chemicals transported in the blood to various target organs of the body.
Reference: See Fig. 18.1.

6. Cardiovascular system
Components: Blood, heart, and blood vessels.
Function: Distributes oxygen and nutrients to cells, carries carbon dioxide and wastes away from cells, helps maintain the acid–base balance of the body, protects against disease, prevents hemorrhage by forming blood clots, and helps regulate body temperature.
Reference: See Figs. 20.1, 21.20, and 21.25.

7. Lymphatic and immune systems
Components: Lymph, lymphatic vessels, and structures or organs containing lymphatic tissue (large numbers of white blood cells called lymphocytes), such as the spleen, thymus gland, lymph nodes, and tonsils.
Function: Returns proteins and plasma to the cardiovascular system, transports fats from the gastrointestinal tract to the cardiovascular system, filters body fluid, site of maturation and proliferation of certain white blood cells, and helps protect against disease through the production of proteins called antibodies, as well as other responses.
Reference: See Fig. 22.1.

8. Respiratory system
Components: Lungs and a series of associated passageways leading into and out of them.
Function: Supplies oxygen, eliminates carbon dioxide, helps regulate the acid–base balance of the body, and produces vocal sounds (phonation).
Reference: See Fig. 23.1.

9. Digestive system
Components: A long tube called the gastrointestinal tract and associated organs such as the salivary glands, liver, gallbladder, and pancreas.
Function: Performs the physical and chemical breakdown and absorption of food for use by cells and eliminates solid and other wastes.
Reference: See Fig. 24.1.

10. Urinary system
Components: Organs such as the kidneys, ureters, urinary bladder, and urethra that together produce, store, and eliminate urine.
Function: Regulates the volume and chemical composition of blood, eliminates wastes, regulates fluid and electrolyte balance, helps maintain the acid–base and calcium balance of the body, and helps regulate red blood cell production.
Reference: See Fig. 26.1.

11. Reproductive system
Components: Organs (testes and ovaries) that produce reproductive cells or gametes (sperm and ova) and other organs such as the uterine (Fallopian) tubes and uterus in females and the epididymis, ductus (vas) deferens, and penis in males that transport and store reproductive cells.
Function: Reproduces the organism.
Reference: See Figs. 28.1 and 28.10.

---

in the pancreas respond to elevated blood glucose (sugar) level by secreting the hormone insulin, which lowers blood sugar level back to normal.

3.   Movement includes motion of the whole body, individual organs, single cells, or even organelles inside cells. For example, the coordinated contraction of several leg muscles moves your whole body from one place to another when you walk or run. After you eat a meal that contains fats, your gallbladder contracts and releases bile to help in the digestion of fats. When a body tissue is damaged or infected, certain white blood cells move from the blood into the tissue to help clean up and repair the area. And inside individual cells, various cell parts move from one position to another.

4.   Growth refers to an increase in size and complexity. It is due to an increase in the number or size of cells or both. Sometimes, a tissue increases in size because the substance between cells increases in amount. In growing bone, for example, mineral deposits accumulate around the bone cells.

5.   Differentiation is the change that a cell undergoes from an unspecialized one to a specialized one. Specialized cells have structural and functional characteristics that differ from their undifferentiated ancestor cells. Through differentiation, a fertilized egg normally develops into an embryo, and then into a fetus, infant, child, and finally an adult.

6.   Reproduction refers to either the formation of new cells for growth, repair, or replacement, or the production of a new individual. Through reproduction, life continues from one generation to the next.

HOMEOSTASIS: MAINTAINING PHYSIOLOGICAL LIMITS

As we have seen, the human body is composed of various systems and organs, each of which consists of millions of cells. These cells need relatively stable conditions to function effectively and contribute to the survival of the body as a whole. The maintenance of stable conditions for its cells is an essential function of every many-celled organism. Physiologists call such relative stability homeostasis, and it is one of the major themes of this textbook. Another theme is pathology—disease conditions that occur when homeostasis is disrupted.

Homeostasis (hō'-mē-o-STĀ-sis; *homeo* = same; *stasis* = standing still) is a condition in which the body's internal environment remains within certain physiological limits. For the body's cells to survive, the composition of the surrounding fluids must be precisely maintained at all times. Fluid outside body cells is called **extracellular** (*extra* = outside) **fluid (ECF)** and is found in two principal places. The ECF filling the narrow spaces between cells of tissues is called **interstitial** (in'-ter-STISH-al; inter = between) **fluid, intercellular fluid,** or **tissue fluid.** The ECF in blood vessels is termed **plasma** (Fig. 1.2). Fluid within cells is called **intracellular** (*intra* = within, inside) **fluid (ICF).** Plasma circulates from arteries to arterioles to microscopic vessels called blood capillaries. Certain components of plasma leave the blood through capillaries and the fluid circulates in the spaces between body cells. Here it is called interstitial fluid. Most of this fluid is returned to capillaries as plasma and passes into venules and veins. Some of the interstitial fluid passes into microscopic lymphatic vessels called lymph capillaries. Here the fluid is called lymph. Eventually, the lymph is returned to blood. Since interstitial fluid surrounds all body cells, it is often called the body's **internal environment.** Among the substances dissolved in the water of ECF and ICF are gases, nutrients, and electrically charged chemical particles called *ions,* such as sodium ions (Na^+) and chloride ions (Cl^-), all needed to maintain life.

An organism is said to be in homeostasis when its internal environment (1) contains the optimum concentration of gases, nutrients, ions, and water; (2) has an optimal temperature; and (3) has an optimal volume for the health of the

---

FIGURE 1.2 Internal environment of the body. Extracellular fluid (ECF) is found in blood vessels as plasma and between cells as interstitial fluid.

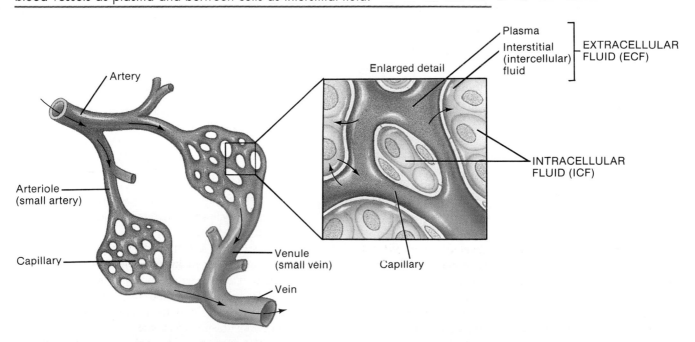

Artery

Arteriole (small artery)

Capillary

Venule (small vein)

Vein

Enlarged detail

Plasma

Interstitial (intercellular) fluid

EXTRACELLULAR FLUID (ECF)

INTRACELLULAR FLUID (ICF)

Capillary

Question: What substances are present in ECF and ICF?

cells. When homeostasis is disturbed, illness may result. If the body fluids are not eventually brought back into homeostasis, death may occur.

Stress and Homeostasis

Homeostasis in all organisms is continually disturbed by **stress,** any stimulus that tends to create an imbalance in the internal environment. The stress may come from the external environment in the form of stimuli such as heat, cold, loud noises, or lack of oxygen. Or the stress may originate within the body in the form of stimuli such as low blood sugar level, increased acidity of the extracellular fluid, pain, or unpleasant thoughts. Most stresses are mild and routine, and the responses of body cells quickly restore balance in the internal environment. Poisoning, overexposure to temperature extremes, and severe infection are examples of extreme stresses, situations in which homeostasis may fail.

Fortunately, the body has many regulating (homeostatic) devices that oppose the forces of stress and bring the internal environment back into balance. Some people live in deserts where the daytime temperatures easily reach 49°C (120°F). Others work outside all day in subzero weather. Yet everyone's internal body temperature remains near 37°C (98.6°F). Mountain climbers exercise strenuously at high altitudes, where the oxygen content of the air is low. But once they adjust to the new altitude, they usually do not suffer from oxygen shortage. The extremes in temperature and in oxygen content of the air are external stresses, and the exercise performed is an internal stress, yet the body compensates and remains in homeostasis.

Walter B. Cannon (1871–1945), an American physiologist who coined the term *homeostasis*, noted that heat produced by the muscles during strenuous exercise would curdle and inactivate the body's proteins if the body did not dissipate heat quickly. Besides heat, muscles also produce lactic acid during exercise. If the body did not have a homeostatic mechanism for reducing the amount of acid, the extracellular fluid would become too acidic and destroy the cells. Every body structure, from the cellular to the systemic level, contributes in some way to keeping the internal environment within normal limits.

Regulation of Homeostasis by Nervous and Endocrine Systems

The homeostatic responses of the body are regulated by the nervous system and the endocrine system working together or independently. The nervous system regulates homeostasis by detecting deviations from the balanced state and then sending messages in the form of nerve impulses to the proper organs to counteract the stress. For instance, when muscle fibers (cells) are active, they take a great deal of oxygen from the blood. They also give off a lot of carbon dioxide, which enters the blood. Certain nerve cells detect these chemical changes in the blood and send impulses to the brain. In response, the brain sends impulses to the heart to cause it to pump blood more quickly and forcefully to the lungs so the blood can give up carbon dioxide and take on oxygen more rapidly. Simultaneously, the brain sends nerve impulses to the muscles that control breathing to contract more often. As a result, more carbon dioxide is exhaled and more oxygen is inhaled.

The endocrine system—a group of glands that secrete chemical regulators, called **hormones,** into the blood—also regulates homeostasis. Whereas nerve impulses cause rapid changes, hormones usually work more slowly. Both means of regulation work toward the same end—achieving homeostasis.

Feedback Systems (Loops)

A **feedback system (loop)** is a cycle of events in which information about the status of a condition is continually monitored and fed back (reported) to a central control region (Fig. 1.3). A feedback system consists of three basic components — control center, receptor, and effector.

1. The **control center** determines the point at which some aspect of the body, called a **controlled condition,** should be maintained. In the body, there are hundreds of controlled conditions. A few examples are heart rate, blood pressure, acidity of the blood, blood sugar level, body temperature, and breathing rate. The control center receives information about the status of a controlled condition from a receptor and then determines an appropriate course of action.

2. The **receptor** monitors changes in the controlled condition and then sends the information, called the *input*, to the control center. Any stress that changes a controlled condition is called a **stimulus.** For example, a stimulus such as exercise raises the body temperature (the controlled condition), and thermal (heat) receptors send input to the control center, which in this case is in the brain.

3. The **effector** receives information, called the *output*, from the control center and produces a **response** (effect). So while you are exercising, your brain (control center) signals for increased secretions by your sweat glands (effectors). As sweat evaporates from the skin, body temperature drops back to normal.

The response that occurs is continually monitored by the receptor and fed back to the control center. If the response reverses the original stimulus, as in the example just described, the system is a **negative feedback system (loop).** If the response enhances the original stimulus, the system is a **positive feedback system (loop).**

Negative feedback systems tend to maintain conditions that require frequent monitoring and adjustment within physiological limits, for example, body temperature or blood sugar level. Positive feedback systems, on the other hand, are important for conditions that do not occur often

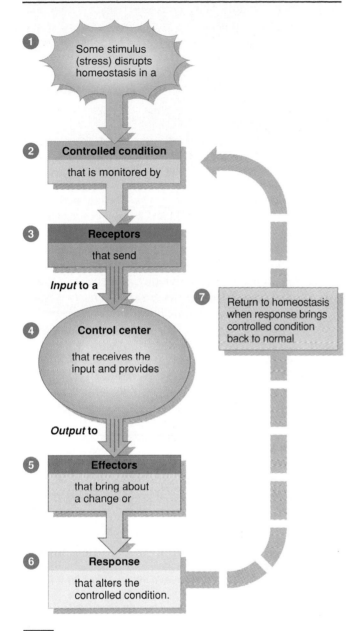

FIGURE 1.3 Components of a feedback system (loop).

1. Some stimulus (stress) disrupts homeostasis in a

2. **Controlled condition**
   that is monitored by

3. **Receptors**
   that send

Input to a

4. **Control center**
   that receives the input and provides

Output to

5. **Effectors**
   that bring about a change or

6. **Response**
   that alters the controlled condition.

7. Return to homeostasis when response brings controlled condition back to normal

Question: What are two ways that negative and positive feedback systems differ?

Having considered the components and operation of feedback systems in general, we can now look at the relationship of feedback systems to homeostasis in the body. As examples, we will describe the homeostasis of blood pressure, a negative feedback system, and labor contractions, a positive feedback system.

Homeostasis of Blood Pressure: Negative Feedback

Blood pressure (BP) is the force exerted by blood as it presses against the walls of the blood vessels, especially the arteries. When the heart beats faster or harder, BP increases; when blood volume increases, BP also rises.

If some stimulus (stress), either internal or external, causes blood pressure (controlled condition) to rise, the following sequence of events occurs (Fig. 1.4). The higher pressure is detected by pressure-sensitive nerve cells (the receptors) in the walls of certain arteries. They send nerve impulses (input) to the brain (control center), which interprets the impulses and responds by sending nerve impulses (output) to the heart (effector). Heart rate decreases and blood pressure drops (response). This returns blood pressure (controlled condition) to normal, and homeostasis is restored.

A second set of effectors also contributes to maintaining normal blood pressure. Small arteries, called *arterioles,* have muscular walls that can constrict (narrow) or dilate (widen) upon receiving appropriate signals from the brain. When a stimulus causes blood pressure to increase, pressure-sensitive nerve cells (receptors) in certain arteries send nerve impulses (input) to the brain (control center). The brain interprets the messages and responds by sending fewer nerve impulses (output) to the arterioles. This causes the arterioles (effectors) to dilate (response). Thus the blood flowing through the wider arterioles meets less resistance, blood pressure drops back to normal, and homeostasis is restored.

Homeostasis of Labor Contractions: Positive Feedback

The hormone oxytocin is produced in a portion of the brain called the hypothalamus. One of its functions is to enhance muscular contraction (controlled condition) of the pregnant uterus (see Fig. 18.11). When labor begins, the uterus is stretched (stimulus) and pressure-sensitive nerve cells in the uterine wall (receptors) send nerve impulses (input) to the hypothalamus (control center) in the brain. The hypothalamus responds by causing the release of oxytocin (output). Oxytocin enters the blood, is carried to the uterus, and stimulates the uterus (effector) to contract more forcefully (response). As the baby's head moves down into the birth canal, further stretching of the uterus occurs, causing the release of more oxytocin, and producing even more forceful uterine contractions. Thus this is a positive feedback cycle.

and do not require continual fine-tuning. Since positive feedback systems tend to intensify a controlled condition, they usually are shut off by some mechanism outside the system if they are part of a normal physiological response.

Given the characteristics of both negative and positive feedback systems, it is not surprising that most feedback systems in the body are negative. Positive feedback systems can be destructive and result in various disorders, yet some are normal and beneficial. Examples are blood clotting, which helps stop loss of blood from a cut, and labor contractions during birth of a baby.

The cycle is broken by the birth of the baby, which decreases uterine stretching and inhibits the release of oxytocin.

FIGURE 1.4 Homeostasis of blood pressure by a negative feedback system. Note that the response is fed back into the system, and the system continues to lower blood pressure until there is a return to homeostasis.

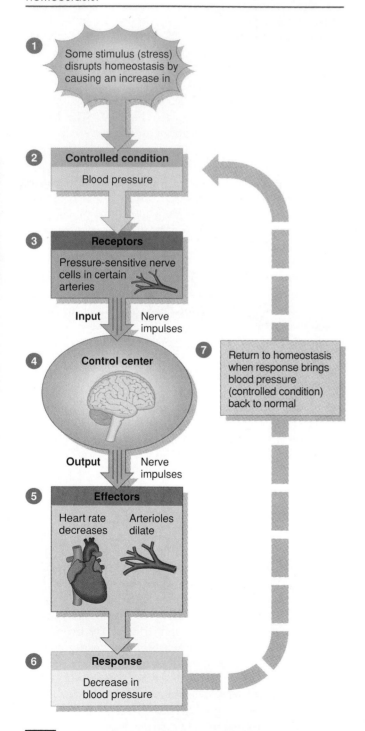

Question: What are two examples of positive feedback systems?

Disease: Homeostatic Imbalance

As long as the various body processes remain within normal physiological limits, body cells function efficiently and homeostasis (health) is maintained. When one or more components of the body lose their ability to contribute to homeostasis, however, body processes do not function efficiently. If the homeostatic imbalance is moderate, disease may result; if it is severe, death may result.

Disease is any change from a state of health in which part or all of the body is not carrying on its normal functions. A **local disease** is one that affects one part or a limited area of the body. A **systemic disease** affects either the entire body or several parts. Each disease alters body structures and functions in particular ways. A patient may experience certain **symptoms.** Symptoms are *subjective* changes in body functions that are not apparent to an observer, for example, headache or nausea. *Objective* changes that a clinician can observe and measure are called **signs.** Signs can be either anatomical or physiological changes: swelling, fever, rash, paralysis, and so forth.

The science that deals with why, when, and where diseases occur and how they are transmitted in a human community is known as **epidemiology** (ep′-i-dē′-mē-OL-ō-jē; *epi* = on or among; *demos* = people; *logos* = study of). The science that deals with the effects and uses of drugs in the treatment of disease is called **pharmacology** (far′-ma-KOL-ō-jē; *pharmakon* = drug or poison).

CLINICAL APPLICATION

DIAGNOSIS OF DISEASE

Diagnosis (dī′-ag′-NŌ-sis; *dia* = through; *gnosis* = knowledge) is the art of distinguishing one disease from another or determining the nature of a disease. It is an early step in evaluating a disease, usually after a medical history is taken and a physical examination is given. A **medical history** consists of information that is collected about past events that might be related to a patient's illness (chief complaint, history of present illness, past medical problems, family medical problems, social history, and review of symptoms). A **physical examination** is a methodical evaluation that includes inspection (looking at or into a patient with various instruments); palpation (touching to feel irregularities); auscultation (listening); percussion (striking gently); measuring vital signs (temperature, pulse, respiratory rate, and blood pressure); and sometimes laboratory tests.

ANATOMICAL POSITION

In anatomy, descriptions of any region or part of the human body assume that the body is in a specific position called the **anatomical position.** In the anatomical position, the subject stands erect (upright position) facing the observer, with feet flat on the floor, arms placed at the sides, and the palms of the hands turned forward (Fig. 1.5). Once the body is in the anatomical position, it is easier to visualize and

FIGURE 1.5 Anatomical position. The common names and anatomical terms, in parentheses, are indicated for many of the regions of the body. For example, the chest is the thoracic region.

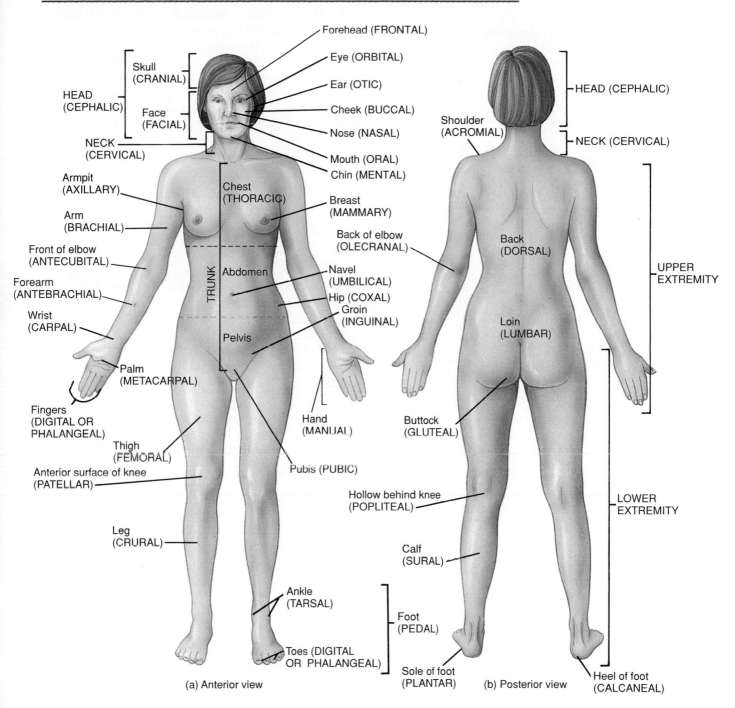

HEAD (CEPHALIC)
Skull (CRANIAL)
Face (FACIAL)
NECK (CERVICAL)

Forehead (FRONTAL)
Eye (ORBITAL)
Ear (OTIC)
Cheek (BUCCAL)
Nose (NASAL)
Mouth (ORAL)
Chin (MENTAL)

Armpit (AXILLARY)
Arm (BRACHIAL)
Front of elbow (ANTECUBITAL)
Forearm (ANTEBRACHIAL)
Wrist (CARPAL)

Chest (THORACIC)
TRUNK
Abdomen
Pelvis

Breast (MAMMARY)
Back of elbow (OLECRANAL)
Navel (UMBILICAL)
Hip (COXAL)
Groin (INGUINAL)

Palm (METACARPAL)
Fingers (DIGITAL OR PHALANGEAL)
Thigh (FEMORAL)
Anterior surface of knee (PATELLAR)
Leg (CRURAL)

Hand (MANUAL)
Pubis (PUBIC)

Ankle (TARSAL)
Toes (DIGITAL OR PHALANGEAL)

(a) Anterior view

Shoulder (ACROMIAL)

HEAD (CEPHALIC)
NECK (CERVICAL)

Back (DORSAL)
Loin (LUMBAR)

UPPER EXTREMITY

Buttock (GLUTEAL)

Hollow behind knee (POPLITEAL)
Calf (SURAL)

LOWER EXTREMITY

Foot (PEDAL)
Sole of foot (PLANTAR)
Heel of foot (CALCANEAL)

(b) Posterior view

Question: What are four characteristics of the anatomical position?

understand how it is organized into various regions. Having one standard anatomical position allows directional terms to be clear; any body part or region can be described relative to any other part.

REGIONAL NAMES

The common names and anatomical terms of the principal body regions are given in Fig. 1.5. These terms designate particular areas of the body, such as the arm (brachial region) or the nose (nasal region).

DIRECTIONAL TERMS

To locate various body structures in relation to one another, anatomists use certain **directional terms.** Such terms are precise and avoid the use of unnecessary words. There are many pairs of directional terms with opposite meanings, for example, superior and inferior. Important directional terms are defined in Exhibit 1.3, and the parts of the body referred to in the examples are labeled in Fig. 1.6. If you study the exhibit and the figure together, the directional relations among various body parts will be clear.

FIGURE 1.6 Anatomical terms. Study Exhibit 1.3 with this figure to understand the directional terms: *superior, inferior, anterior, posterior, medial, lateral, intermediate, ipsilateral, contralateral, proximal,* and *distal.*

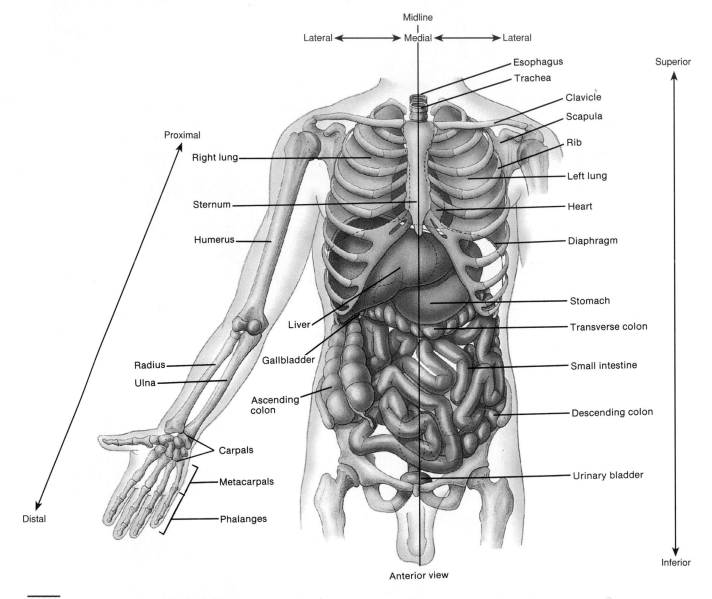

Anterior view

Question: Is the radius proximal to the humerus? Is the esophagus anterior to the trachea? Are the ribs superficial to the liver? Is the urinary bladder medial to the ascending colon? Is the sternum lateral to the descending colon?

EXHIBIT 1.3

DIRECTIONAL TERMS[a]

Term	Definition	Example
Superior (soo′-PEER-ē-or) **(cephalic or cranial)**	Toward the head or the upper part of a structure.	The heart is superior to the liver.
Inferior (in′-FEER-ē-or) **(caudal)**	Away from the head or toward the lower part of a structure.	The stomach is inferior to the lungs.
Anterior (an-TEER-ē-or) **(ventral)**	Nearer to or at the front of the body. In the **prone position,** the body lies anterior side down. In the **supine position,** the body lies anterior side up.	The sternum (breastbone) is anterior to the heart.
Posterior (pos-TEER-ē-or) **(dorsal)**	Nearer to or at the back of the body.	The esophagus is posterior to the trachea.
Medial (MĒ-dē-al) **(mesial)**	Nearer to the midline of the body or a structure. The **midline** is an imaginary vertical line that divides the body into equal left and right sides.	The ulna is on the medial side of the forearm.
Lateral (LAT-er-al)	Farther from the midline of the body or a structure.	The lungs are lateral to the heart.
Intermediate (in′-ter-MĒ-dē-at)	Between two structures.	The ring finger is intermediate between the little and middle fingers.
Ipsilateral (ip-si-LAT-er-al)	On the same side of the body.	The gallbladder and ascending colon of the large intestine are ipsilateral.
Contralateral (CON-tra-lat-er-al)	On the opposite side of the body.	The ascending and descending colons of the large intestine are contralateral.
Proximal (PROK-si-mal)	Nearer to the attachment of an extremity (limb) to the trunk or a structure; nearer to the point of origin.	The humerus is proximal to the radius.
Distal (DIS-tal)	Farther from the attachment of an extremity (limb) to the trunk or a structure; farther from the point of origin.	The phalanges are distal to the carpals (wrist bones).
Superficial (soo′-per-FISH-al)	Toward or on the surface of the body.	The muscles of the thoracic wall are superficial to the organs in the thoracic cavity. (See Fig. 1.10.)
Deep (DĒP)	Away from the surface of the body.	The ribs are deep to the skin of the chest. (See Fig. 1.10.)
Parietal (pa-RĪ-e-tal)	Pertaining to or forming the outer wall of a body cavity.	The parietal pleura forms the outer layer of the pleural sacs that surround the lungs. (See Fig. 23.8.)
Visceral (VIS-er-al)	Pertaining to the covering of an organ (viscus) within the ventral body cavity.	The visceral pleura forms the inner layer of the pleural sacs and covers the external surface of the lungs. (See Fig. 23.8.)

[a] Study this exhibit with Figs. 1.5 and 1.6 to visualize the examples given.

PLANES AND SECTIONS

You will also study parts of the body relative to **planes** (imaginary flat surfaces) that pass through it (Fig. 1.7). A **sagittal** (SAJ-i-tal) **plane** is a vertical plane that divides the body or an organ into right and left sides. More specifically, if such a plane passes through the midline of the body or organ and divides it into *equal* right and left sides, it is called a **midsagittal (median) plane.** If the sagittal plane does not pass through the midline but instead divides the body or an organ into *unequal* right and left sides, it is called a **parasagittal** (*para* = near) **plane.** A **frontal (coronal;** kō-RŌ-nal) **plane** divides the body or organ into anterior (front) and posterior (back) portions. A **transverse** (cross-sectional or horizontal) **plane** divides the body or organ into superior (top) and inferior (bottom) portions. These four planes are all at right angles to one another. An **oblique plane,** on the other hand, passes through the body or organ at an angle between the transverse plane and either the midsagittal, parasagittal, or frontal plane. When you study a body region, you will often view it in section, meaning that you look at only one flat surface of the three-dimensional structure. It is important to know the plane of the section so that you can understand the anatomical relationship of one part to another. Figure 1.8 indicates how three different sections—a *transverse (cross) section,* a *frontal section,* and a *midsagittal section*—give different views of the brain.

BODY CAVITIES

Body cavities are confined spaces within the body that contain internal organs. The cavities help to protect, separate, and support the organs. The various body cavities may be separated from each other by structures such as muscles, bones, or ligaments. Figure 1.9 shows the two principal ones, the dorsal and ventral body cavities. The **dorsal body cavity** is located near the dorsal (back) surface of the body. It is further subdivided into a **cranial cavity,** which is formed by the cranial (skull) bones and contains the brain, and a **vertebral (spinal) canal,** which is formed by the vertebrae of the backbone and contains the spinal cord and the beginnings (roots) of spinal nerves.

The other principal body cavity is the **ventral body cavity.** This cavity is located on the ventral (front) aspect of the body. A thin, slippery tissue called a **serous membrane** lines the wall of the ventral body cavity and covers the organs within it. The organs inside are called **viscera** (VIS-er-a). The ventral body cavity also has two principal subdivisions—an upper portion, called the **thoracic** (thor-AS-ik) **cavity** (or chest cavity), and a lower portion, called the **abdominopelvic** (ab-dom′-i-nō-PEL-vik) **cavity.** The struc-

FIGURE 1.7 Planes of the human body.

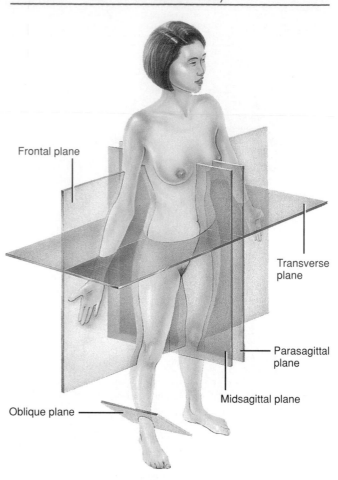

Frontal plane

Transverse plane

Parasagittal plane

Midsagittal plane

Oblique plane

Right anterolateral view

Question: Which plane divides the brain into anterior and posterior portions?

ture that divides the ventral body cavity into the thoracic and abdominopelvic cavities is the diaphragm (DĪ-a-fram; *diaphragma* = partition or wall), an important muscle for breathing.

The thoracic cavity has three compartments: two pleural cavities and one pericardial cavity. Each of the two **pleural cavities** surrounds a lung (Fig. 1.10). Each pleural cavity is a small, fluid-filled space between the part of the serous membrane that covers the lung and the part that lines the wall of the thoracic cavity. The serous membrane associated with the lungs is called the **pleura.** Between the lungs is the **pericardial** (per′-i-KAR-dē-al; *peri* = around; *cardi* = heart) **cavity.** It is a fluid-filled space between the part of the serous membrane that covers the heart and the part that lines the thoracic cavity. The serous membrane associated with the heart is called the **pericardium.**

The **mediastinum** (mē′-dē-as-TĪ-num; *media* = middle;

FIGURE 1.8 Planes and sections through different parts of the brain. The planes are shown in the diagrams on the left and the resulting sections are shown in the photographs on the right.

Superior (top)

Transverse
(cross-sectional
or horizontal)
plane

Anterior
(front)

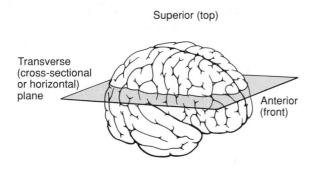
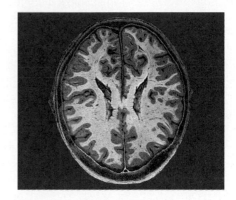

(a) Transverse (cross) section

Frontal plane

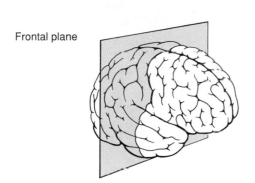
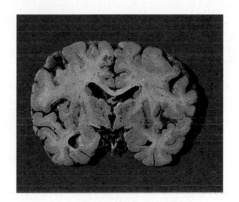

(b) Frontal section

Midsagittal
plane

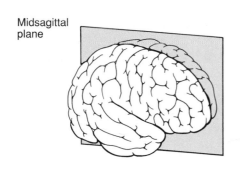
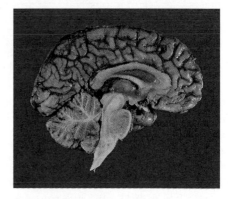

(c) Midsagittal section

Question: Which plane divides the brain into equal right and left sides?

stare = stand in) is a region between the lungs, extending from the sternum (breastbone) to the vertebral column or backbone (Fig. 1.10). The mediastinum includes all the contents of the thoracic cavity except the lungs themselves. Among the structures in the mediastinum are the heart, esophagus, trachea, thymus gland, and many large blood and lymphatic vessels.

The **abdominopelvic cavity,** as the name suggests, is divided into two portions, although no wall separates them (see Fig. 1.9). The serous membrane that lines the abdom-inopelvic cavity and covers the organs within it is called the **peritoneum.** The upper portion of the abdominopelvic cavity, the **abdominal** (*abdere* means to hide, because it hides the viscera) **cavity,** contains the stomach, spleen, liver, gallbladder, pancreas, small intestine, and most of the large intestine. The lower portion, the **pelvic cavity,** contains the urinary bladder, portions of the large intestine, and the internal organs of reproduction. The pelvic cavity is the region between two imaginary planes, shown by dashed lines in Fig. 1.9a.

FIGURE 1.9 Body cavities.

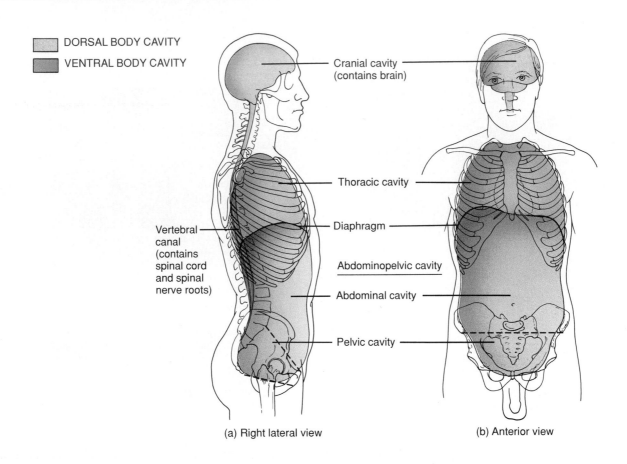

DORSAL BODY CAVITY

VENTRAL BODY CAVITY

Cranial cavity (contains brain)

Thoracic cavity

Diaphragm

Vertebral canal (contains spinal cord and spinal nerve roots)

Abdominopelvic cavity

Abdominal cavity

Pelvic cavity

(a) Right lateral view

(b) Anterior view

Question: In which cavities (T = thoracic, A = abdominal, or P = pelvic) are the following organs located: urinary bladder, stomach, heart, pancreas, small intestine, lungs, internal female reproductive organs, thymus gland, spleen, rectum, liver?

FIGURE 1.10 Thoracic cavity. The two pleural cavities surround the right and left lungs and the pericardial cavity surrounds the heart. The mediastinum is found between the lungs and extends from the sternum to the backbone.

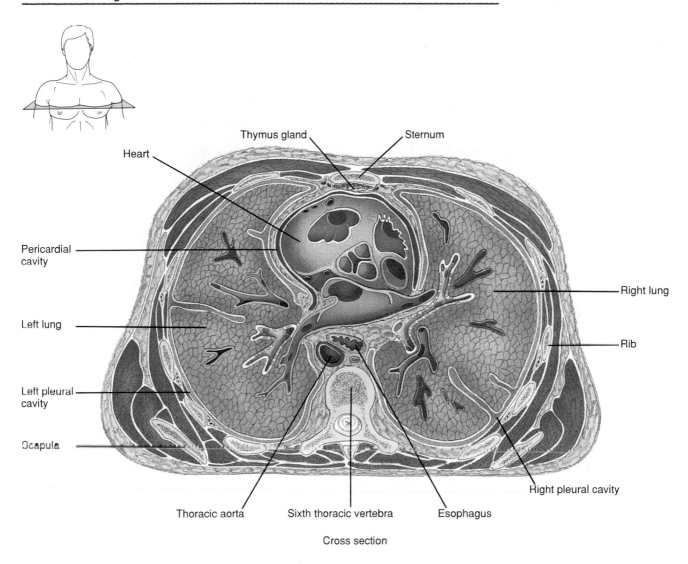

Cross section

Question: Which of the following structures is contained in the mediastinum (circle your answers): thymus gland, right lung, heart, esophagus, thoracic aorta, rib, left pleural cavity?

ABDOMINOPELVIC REGIONS AND QUADRANTS

To describe the location of organs easily, the abdominopelvic cavity is divided into nine **abdominopelvic regions** (Fig. 1.11a). Note which organs and parts of organs are in the different regions by carefully examining Fig. 1.11b,c. Although some parts of the body in the illustrations may be unfamiliar to you at this point, they will be discussed in detail in later chapters.

The abdominopelvic cavity is divided more simply into **quadrants** (*quad* = four). These are shown in Fig. 1.12. In this method, often used by clinical staff, a horizontal line and a vertical line are passed through the umbilicus (navel). Whereas the nine region division is more widely used for anatomical studies, the quadrant division is more commonly used for locating the site of an abdominopelvic pain, tumor, or other abnormality.

CLINICAL APPLICATION

AUTOPSY

To determine the cause of death accurately, it is necessary to perform an **autopsy** (AW-top-sē; *auto* = self; *opsis* = to see with one's own eyes). An autopsy can uncover the exis-

Figure 1.11 Abdominopelvic cavity. (a) The nine regions. The **subcostal** (top horizontal) **line** is drawn just inferior to the bottom of the rib cage, across the lower portion of the stomach; the **transtubercular** (bottom horizontal) **line** is drawn just inferior to the tops of the hipbones. The **left** and **right midclavicular** (two vertical) **lines** are drawn through the midpoints of the clavicles (collar bones), just medial to the nipples. The four lines divide the abdominopelvic cavity into a larger middle section and smaller left and right sections. (b) The greater omentum has been removed. (c) Many anterior organs have been removed, exposing the posterior structures. The internal reproductive organs in the pelvic cavity are shown in Figs. 28.1 and 28.10.

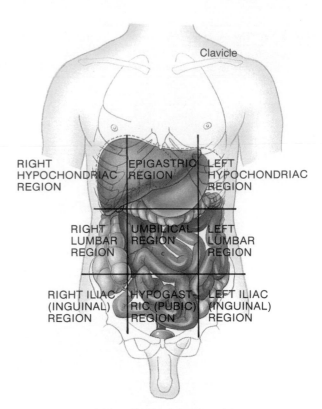

(a) Location of abdominopelvic regions, anterior view

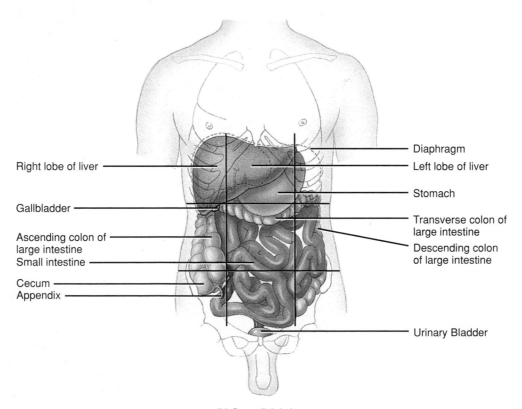

(b) Superficial view

Figure 1.11 (*continued*)

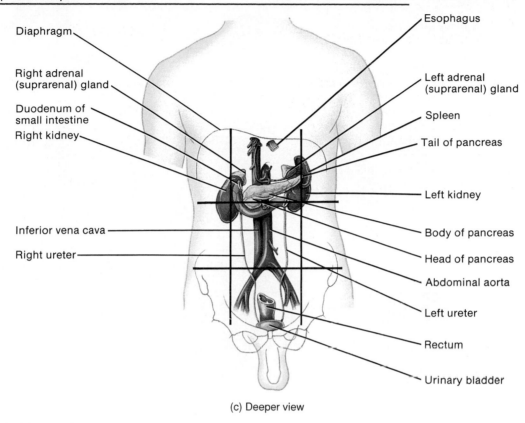

Diaphragm

Right adrenal
(suprarenal) gland

Duodenum of
small intestine

Right kidney

Inferior vena cava

Right ureter

Esophagus

Left adrenal
(suprarenal) gland

Spleen

Tail of pancreas

Left kidney

Body of pancreas

Head of pancreas

Abdominal aorta

Left ureter

Rectum

Urinary bladder

(c) Deeper view

Question: In which abdominopelvic region is each of the following found: most of the liver, transverse colon, urinary bladder, spleen?

Figure 1.12 Quadrants of the abdominopelvic cavity. The two lines intersect at right angles at the umbilicus.

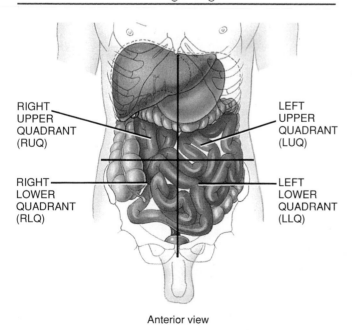

RIGHT
UPPER
QUADRANT
(RUQ)

LEFT
UPPER
QUADRANT
(LUQ)

RIGHT
LOWER
QUADRANT
(RLQ)

LEFT
LOWER
QUADRANT
(LLQ)

Anterior view

Question: In which quadrant would the pain from appendicitis be felt?

tence of diseases not detected during life. It also may support the accuracy of diagnostic tests, establish the beneficial and adverse effects of drugs, discover the impact of environmental influences on the body, and educate health-care students. Moreover, an autopsy can reveal conditions that may affect offspring or siblings (such as congenital heart defects). An autopsy may be needed in a criminal investigation, but it may also help resolve disputes between beneficiaries and insurance companies.

MEDICAL IMAGING

Various kinds of **medical imaging** techniques allow physicians to peer inside our bodies. The images provide clues to both abnormal anatomy and deviations from normal physiology. They are increasingly helpful for precise diagnosis of a wide range of disorders. The grandfather of all medical imaging techniques is conventional radiography, in use since the late 1940s. The newer techniques not only contribute to diagnosis of disease, but they also are advancing our understanding of normal physiology. Exhibit 1.4 describes some commonly used medical imaging techniques.

EXHIBIT 1.4

SUMMARY OF MEDICAL IMAGING PROCEDURES

Procedure	Description and example
Conventional radiography	A single barrage of x-rays passes through the body and produces a two-dimensional image of the interior of the body called a **radiograph** (RĀ-dē-ō-graf′). *Comment:* Overlap of structures can make diagnosis difficult, and subtle differences in tissue density cannot always be discerned.

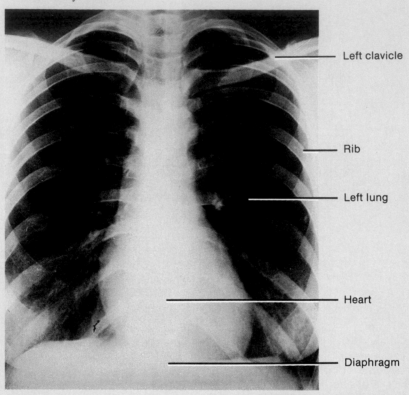

— Left clavicle

— Rib

— Left lung

— Heart

— Diaphragm

Radiograph of chest

| **Computed tomography (CT) scanning** [formerly called computerized axial tomography (CAT) scanning] | An x-ray beam moves in an arc around the body, producing a cross-sectional image called a **CT scan** on a video monitor hooked to a computer. By stacking images one on top of the other, the computer can construct three-dimensional images that can be rotated, for example, to plan plastic surgery.
Comment: Quick, painless CT scans have replaced exploratory surgery in many cases. Detailed images reveal tumors, aneurysms (bulges in blood vessels), kidney stones, gallstones, infections, tissue damage, and deformities. |

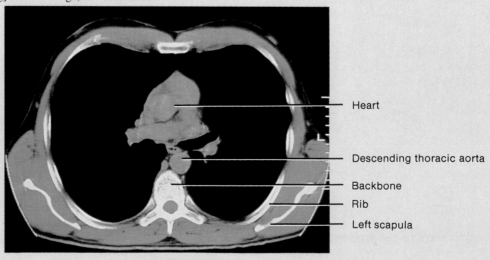

Heart

Descending thoracic aorta

Backbone

Rib

Left scapula

CT scan of chest

Procedure	Description and example
Dynamic spatial reconstruction (DSR)	A highly sophisticated x-ray machine produces moving, three-dimensional, life-size images from any view. *Comment*: Image can be rotated, tipped, ''sliced open,'' enlarged, replayed, and viewed in slow motion or at high speeds; good procedure for heart, lung, and blood vessel imaging, measuring movements and volumes, and assessing tissue damage.

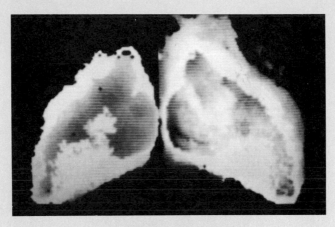

DSR Image of ''opened'' heart

Digital subtraction angiography (DSA)	A computer compares an x-ray of a region of the body before and after a contrast dye has been injected into a blood vessel. Tissues around the blood vessel in the first image can be subtracted (erased) from the second image, leaving an unobstructed view of the blood vessel. *Comment*: DSA is used primarily to study blood vessels in the brain and heart.

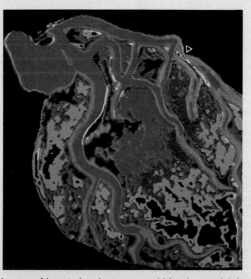

DSA image of heart showing narrowed blood vessel (triangle)

Exhibit continues

SUMMARY OF MEDICAL IMAGING PROCEDURES

Procedure	Description and example
Positron emission tomography (PET)	A radioactive substance that emits positively charged particles called positrons is injected into the body. The collision of positrons with negatively charged electrons in body tissues releases gamma rays, which are similar to x-rays. A computer receives signals from gamma cameras positioned around the patient and constructs an image called a **PET scan** that can be displayed in color on a video monitor. The PET scan shows where the radioactive substance is being used in the body. *Comment*: PET scans provide information on function as well as structure and are very useful in detecting metabolic changes in healthy or diseased organs such as the brain and heart.

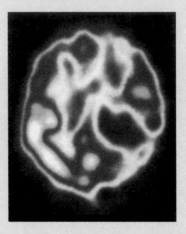

PET image showing blood flow through brain

Procedure	Description and example
Magnetic resonance imaging (MRI) [formerly called nuclear magnetic resonance (NMR) imaging]	Measures the response of protons (small positive particles within atoms, such as hydrogen) to a pulse of radio waves while they are being magnetized. The colored image that results is a two- or three-dimensional blueprint of cellular chemistry. *Comment*: Due to the magnetism needed, MRI can't be used in persons with artificial pacemakers or metal joints. It is used to detect tumors and artery-clogging fatty plaques, assess mental disorders, reveal brain changes, measure blood flow, and study metabolism. Although MRI is noninvasive and employs radiation thought to be harmless (radio waves), it isn't recommended for pregnant women.

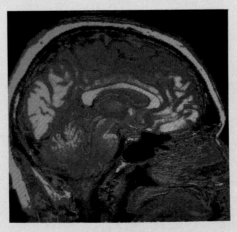

MRI of brain

Procedure	Description and example
Ultrasound (US)	High-frequency sound waves produced by a hand-held device reflect from body tissues and are detected by the same instrument, which transmits the signals to a video monitor. The image, called a **sonogram** (sō-nō-gram), may be still or moving. *Comment*: US is used to image the fetus during amniocentesis (see Fig. 29.11a) and to study abdominal and pelvic organs, blood flow, the heart (echocardiography), and the developing fetus (fetal ultrasound). The procedure uses sound waves, which show no obvious harmful effects, is noninvasive, painless, and uses no dyes. Obesity and scars interfere with US.

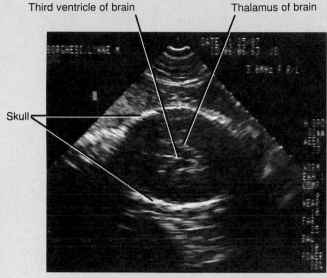

Sonogram of fetal head at 33 weeks. These structures can be seen in an adult brain in Figure 14.8a on page 416.

Study Outline

Anatomy and Physiology Defined (p. 5)

1. Anatomy is the study of structure and the relationship among structures.
2. Subdivisions of anatomy include surface anatomy (form and markings of surface features), gross (macroscopic) anatomy, systemic (systematic) anatomy (systems), regional anatomy (regions), radiographic anatomy (x-rays), developmental anatomy (development from fertilization to adulthood), embryology (development from fertilized egg through eighth week in utero), histology (tissues), cytology (cells), and pathological anatomy (disease).
3. Physiology is the study of how body structures function.
4. Subdivisions of physiology include cell physiology, systems physiology, pathophysiology (disease), exercise physiology, neurophysiology (nerve cells), endocrinology (hormones), cardiovascular physiology (heart and blood vessels), immunology (defense mechanisms), respiratory physiology (air passageways and lungs), renal physiology (kidneys), and reproductive physiology (reproductive structures).

Levels of Structural Organization (p. 5)

1. The human body consists of several levels of structural organization; among these are the chemical, cellular, tissue, organ, system, and organismic levels.
2. Cells are the basic structural and functional units of an organism.

3. Tissues consist of groups of similarly specialized cells and the substance surrounding them that usually arise from a common ancestor and perform certain special functions.
4. Organs are structures of definite form that are composed of two or more different tissues and have specific functions.
5. Systems consist of associations of organs that have a common function.
6. The human organism is a collection of structurally and functionally integrated systems.
7. The systems of the human body are the integumentary, skeletal, muscular, nervous, endocrine, cardiovascular, lymphatic and immune, respiratory, digestive, urinary, and reproductive (see Exhibit 1.2).

Life Processes (p. 6)

1. All living forms have certain characteristics that distinguish them from nonliving things.
2. Among the life processes in humans are metabolism, responsiveness, movement, growth, differentiation, and reproduction.

Homeostasis: Maintaining Physiological Limits (p. 9)

1. Homeostasis is a condition in which the internal environment (interstitial fluid) of the body remains within physiological

limits in terms of chemical composition, temperature, and volume.
2. All body systems attempt to maintain homeostasis.

Stress and Homeostasis (p. 10)

1. Stress is any external or internal stimulus that creates an imbalance in the internal environment.
2. If a stress acts on the body, homeostatic mechanisms attempt to counteract the effects of the stress and bring the condition back to normal.

Regulation of Homeostasis by Nervous and Endocrine Systems (p. 10)

1. Homeostasis is regulated by the nervous and endocrine systems acting together or separately.
2. The nervous system detects body changes and sends nerve impulses to counteract the stress.
3. The endocrine system regulates homeostasis by secreting hormones.

Feedback Systems (Loops) (p. 10)

1. A feedback system consists of (1) a control center that determines the point at which a controlled condition should be maintained, (2) receptors that monitor changes in the controlled condition and send the information (input) to the control center, and (3) effectors that receive information (output) from the control center and produce a response (effect).
2. If a response reverses the original stimulus, the system is a negative feedback system.
3. If a response enhances the original stimulus, the system is a positive feedback system.

Homeostasis of Blood Pressure: Negative Feedback (p. 11)

1. If a stimulus (stress) causes blood pressure (controlled condition) to rise, pressure-sensitive nerve cells (receptors) in arteries send impulses (input) to the brain (control center). The brain sends impulses (output) to the heart (effector), causing the heart rate to decrease (response) and the return of blood pressure to normal (restoration of homeostasis).
2. In a second negative feedback system, pressure-sensitive nerve cells (receptors) send nerve impulses (input) to the brain (control center). Impulses from the brain (output) are sent to arteries (effectors), causing them to dilate (response). Blood meets less resistance and blood pressure decreases back to normal (restoration of homeostasis).

Homeostasis of Labor Contractions: Positive Feedback (p. 11)

1. When labor begins, the uterus is stretched (stimulus) and stretch-sensitive nerve cells in the cervix of the uterus (receptors) send nerve impulses (input) to the hypothalamus (control center).
2. The hypothalamus causes the release of oxytocin (output), which stimulates the uterus (effector) to contract more forcefully (response).
3. Movement of the baby's head down the birth canal causes further stretching, the release of more oxytocin, and even more forceful contractions.
4. The cycle is broken with the birth of the baby.

Disease: Homeostatic Imbalance (p. 12)

1. Disruptions of homeostasis can lead to disease and death.
2. Disease is any change from a state of health, characterized by symptoms and signs.
3. The diagnosis of disease involves a medical history and physical examination.

Anatomical Position (p. 12)

1. Descriptions of any region of the body assume the body is in the anatomical position.
2. When in the anatomical position, the subject stands erect facing the observer, the upper extremities are placed at the sides, the palms of the hands are turned forward, and the feet are flat on the floor.

Regional Names (p. 14)

1. Regional names are terms given to specific regions of the body for reference.
2. Examples of regional names include cranial (skull), thoracic (chest), brachial (arm), patellar (knee), cephalic (head), and gluteal (buttock).

Directional Terms (p. 14)

1. Directional terms indicate the relationship of one part of the body to another.
2. Commonly used directional terms are summarized in Exhibit 1.3.

Planes and Sections (p. 16)

1. Planes are imaginary flat surfaces that are used to divide the body or organs into definite areas. A midsagittal (median) plane divides the body or organs into equal right and left sides; a parasagittal plane divides the body or organs into unequal right and left sides; a frontal plane divides the body or organs into anterior and posterior portions; a transverse plane divides the body or organs into superior and inferior portions; and an oblique plane passes through the body or organs at an angle between a transverse plane and either a midsagittal, parasagittal, or frontal plane.
2. Sections are flat surfaces resulting from cuts through body structures. They are named according to the plane on which the cut is made and include transverse, frontal, and midsagittal sections.

Body Cavities (p. 16)

1. Spaces in the body that contain internal organs are called body cavities.
2. The dorsal and ventral cavities are the two principal body cavities. The dorsal cavity contains the brain and spinal cord.
3. The dorsal cavity is subdivided into the cranial cavity, which contains the brain, and the vertebral (spinal) canal, which contains the spinal cord and beginnings of spinal nerves.
4. The ventral body cavity is subdivided by the diaphragm into an upper thoracic cavity and a lower abdominopelvic cavity.
5. The thoracic cavity contains two pleural cavities and a mediastinum, which includes the pericardial cavity.
6. The mediastinum is a broad, median partition—actually, a mass of tissues—between the lungs that extends from the sternum to the vertebral column; it contains all contents of the thoracic cavity except the lungs.
7. The abdominopelvic cavity is divided into a superior abdominal and an inferior pelvic cavity.
8. Viscera of the abdominal cavity include the stomach, spleen, pancreas, liver, gallbladder, small intestine, and most of the large intestine.
9. Viscera of the pelvic cavity include the urinary bladder, portions of the large intestine, and internal female and male reproductive structures.

Abdominopelvic Regions and Quadrants (p. 19)

1. To describe the location of organs easily, the abdominopelvic cavity may be divided into nine regions by drawing four imaginary lines (left midclavicular, right midclavicular, subcostal, and transtubercular).
2. The names of the nine abdominopelvic regions are epigastric, right hypochondriac, left hypochondriac, umbilical, right lumbar, left lumbar, hypogastric (pubic), right iliac (inguinal), and left iliac (inguinal).
3. To locate the site of an abdominopelvic abnormality in clinical studies, the abdominopelvic cavity may be divided into quadrants by passing imaginary horizontal and vertical lines through the umbilicus.
4. The names of the abdominopelvic quadrants are right upper quadrant (RUQ), left upper quadrant (LUQ), right lower quadrant (RLQ), and left lower quadrant (LLQ).

Medical Imaging (p. 21)

1. Medical imaging techniques allow physicians to peer inside the body to diagnose abnormal anatomy and deviations from normal physiology.
2. Exhibit 1.4 summarizes several imaging techniques.

Review Questions

1. Define anatomy. List and define the various subdivisions of anatomy. (p. 5)
2. Define physiology. List and define the various subdivisions of physiology. (p. 5)
3. Give several examples of how structure and function are related. (p. 5)
4. Define each of the following terms: cell, tissue, organ, system, and organism. (p. 5)
5. Using Exhibit 1.2 as a guide, outline the functions of each system of the body, then list several organs that compose each system. (p. 8)
6. List and define the life processes of humans. (p. 6)
7. Define homeostasis. What is interstitial fluid? Why is it called the internal environment of the body? (p. 9)
8. Under what conditions is the internal environment said to be in homeostasis? (p. 9)
9. What is a stress? Give several examples. How is stress related to homeostasis? (p. 10)
10. What systems of the body control homeostasis? Explain. (p. 10)
11. Define a feedback system. List and describe the components of a feedback system (p. 10)
12. Distinguish between a negative and a positive feedback system. Give an example of each. (p. 10)
13. Define disease and distinguish between a symptom and a sign. (p. 12)
14. What is meant by diagnosis? On what basis is a diagnosis made? (p. 12)
15. Define the anatomical position. Why is the anatomical position used? (p. 12)
16. Review Fig. 1.5. See if you can locate each region on your own body and name each by its common and anatomical term. (p. 13)
17. What is a directional term? Why are these terms important? Use each of the directional terms listed in Exhibit 1.3 in a complete sentence. (p. 14)
18. Define the various planes that may be passed through the body. Explain how each plane divides the body. Describe the meaning of transverse, frontal, and midsagittal sections. (p. 16)
19. Define a body cavity. List the body cavities discussed and tell which major organs are located in each. What landmarks separate the various body cavities from one another? What is the mediastinum? (p. 16)
20. Describe how the abdominopelvic area is subdivided into nine regions. Name and locate each region and list the organs, or parts of organs, in each. (p. 19)
21. Describe how the abdominopelvic cavity is divided into quadrants and name each quadrant. (p. 19)
22. Why is an autopsy performed? (p. 19)
23. Explain the principle and uses for the following medical imaging techniques: conventional radiography, computed tomography (CT) scanning, dynamic spatial reconstruction (DSR), digital subtraction angiography (DSA), positron emission tomography (PET), magnetic resonance imaging (MRI), and ultrasound (US). (p. 21)

Answers to Questions with Figures

1.1 Organ level.
1.2 Water, gases, nutrients, and ions.
1.3 (1) In negative feedback systems, the response reverses the original stimulus; in positive feedback systems, the response exhances the original stimulus. (2) Negative feedback systems tend to regulate conditions that require frequent monitoring and adjustment within physiological limits; positive feedback systems are involved with conditions that do not occur often and do not require continual fine-tuning.
1.4 Blood clotting and labor contractions.
1.5 The subject stands erect facing the observer, the feet are flat on the floor, the arms are at the sides, and the palms are turned forward.
1.6 No, No, Yes, Yes, No.
1.7 Frontal plane.
1.8 Midsagittal plane.
1.9 P, A, T, A, A, T, P, T, A, P, A.
1.10 Thymus gland, heart, esophagus, thoracic aorta.
1.11 Epigastric, umbilical, hypogastric, left hypochondriac.
1.12 Right lower quadrant (RLQ).

Chapter 2

THE CHEMICAL LEVEL OF ORGANIZATION

Chapter Contents at a Glance

Student Objectives

1. Distinguish between matter and energy.
2. Describe the structure of an atom.
3. Identify by name and symbol the principal chemical elements of the human body.
4. Explain how ionic, covalent, and hydrogen bonds form.
5. Define a chemical reaction and explain the basic differences between synthesis, decomposition, exchange, and reversible chemical reactions.
6. List and compare the properties of water and inorganic acids, bases, and salts.
7. Define pH and explain the role of a buffer system as a homeostatic mechanism that maintains the pH of a body fluid.
8. Compare the structure and functions of carbohydrates, lipids, proteins, deoxyribonucleic acid (DNA), ribonucleic acid (RNA), adenosine triphosphate (ATP), and cyclic AMP.
9. Describe the characteristics and importance of enzymes.

The common substances we eat and drink—water, sugar, table salt, cooking oil—play vital roles in keeping us alive. In this chapter, you will learn something about how these substances function in your body. Since chemicals compose your body and all body activities are chemical in nature, it is important to become familiar with the language and fundamental ideas of chemistry.

THE LANGUAGE OF CHEMISTRY

The structure and function of your body result from thousands of interactions that take place at the chemical level of organization. You will see that all life processes involve chemical reactions. For example, calcium participates in muscle contraction for movement and sodium and potassium are necessary for nerve impulses related to responsiveness. To understand the nature of the matter that composes your body and the changes it undergoes in health and in disease, you will need to know how it is organized and how different elements of matter interact with one another.

MATTER AND ENERGY

All living and nonliving things consist of **matter,** which is anything that occupies space and has mass. Matter exists in one of three states: solid, liquid, or gas. Although the terms mass and weight often are used interchangeably, there is a distinction. *Mass* is the amount of matter that a substance contains. *Weight* is the force of gravity acting on a mass. Although your mass is the same, your weight depends on where you are—on a beach in Florida, high in the Rocky Mountains, or orbiting the earth in a space station. Objects weigh less when they are farther from the earth's core because the pull of gravity is weaker. In outer space, weight is close to zero but mass remains the same as it was on Earth at sea level.

Energy is the capacity to do work, that is, to put mass into motion. Mass and energy can be neither created nor destroyed, but one can be converted into the other. The two principal kinds of energy are **potential energy** (inactive or stored energy) and **kinetic energy** (energy of motion). The energy stored in a battery, a tightly coiled spring, or a high water tower is potential energy. Any object in motion, from a molecule, to a tennis ball, to a river, has kinetic energy. Energy, whether potential or kinetic, exists in several different forms.

Chemical energy is the energy absorbed or released in the breaking apart or forming of chemicals. The building processes of the body—the construction of bones, the growth of hair and nails, the replacement of injured cells—all require energy. When foods are broken down, however, energy is available in a form that can be used for building processes.

Radiant energy, such as heat and light, is energy that travels in waves. The waves may be spaced closer together (short wavelength) or farther apart (long wavelength) depending on the type of radiant energy. From longest to shortest wavelength, forms of radiant energy are radio waves, microwaves (used in microwave ovens), infrared waves (heat), visible light waves, ultraviolet waves (cause sunburn), and x-rays and gamma rays (used for medical imaging).

Electrical energy results from the flow of electrons or other charged particles such as ions. Action potentials (impulses) in nerve and muscle cells are examples of electrical energy. The various forms of energy can be transformed from one form into another.

Chemical Elements

All forms of matter are made up of a limited number of building blocks called chemical elements, substances that cannot be split into simpler substances by ordinary chemical reactions. Scientists now recognize 109 different elements, of which 92 occur in nature. (Physicists can artificially create the rest in experiments using powerful machines called accelerators, but the existence of these elements is brief.)

Elements are given letter abbreviations, usually derived from the first or first and second letters of the English or Latin name for the element. Such letter abbreviations are called **chemical symbols.** Examples of chemical symbols are H (hydrogen), C (carbon), O (oxygen), N (nitrogen), Ca (calcium), Na (*natrium* = sodium), K (*kalium* = potassium), Fe (*ferrum* = iron), and P (phosphorus). Twenty-six of the 92 naturally occurring elements are present in your body. Oxygen, carbon, hydrogen, and nitrogen make up about 96% of the body's mass. Other elements, such as calcium, phosphorus, potassium, sulfur, sodium, chlorine, magnesium, iodine, and iron, make up about 3.9% of the body's mass. Thirteen other chemical elements, called **trace elements** because they are present in minute concentrations, compose about 0.1% of total body mass. Exhibit 2.1 lists the major and trace elements.

Structure of Atoms

Each element is made up of **atoms,** the smallest units of matter that enter into chemical reactions. An **element** is simply a quantity of matter composed of atoms, all of the same type. A bit of the element carbon, such as pure coal dust or a diamond, contains only carbon atoms. A tank of oxygen contains only oxygen atoms. The smallest atoms (hydrogen) are less than 0.00000001 cm (1/250,000,000 in.) in diameter, and the largest atoms are only five times larger. If 50 million of the largest atoms were placed end to end, they would span approximately 2.5 cm (1 in.).

An atom consists of three major types of subatomic particles: electrons, neutrons, and protons (Fig. 2.1).

EXHIBIT 2.1

CHEMICAL ELEMENTS PRESENT IN THE BODY

Chemical Element (Symbol)	Percentage of Total Body Mass	Comment
Oxygen (O)	65.0	Constituent of water and organic molecules (carbon-containing, made by a living system); needed for cellular respiration, which produces adenosine triphosphate (ATP), an energy-rich chemical in cells.
Carbon (C)	18.5	Found in every organic molecule.
Hydrogen (H)	9.5	Constituent of water, all foods, and most organic molecules; contributes to acidity when it is positively charged (H^+).
Nitrogen (N)	3.2	Component of all proteins and nucleic acids. The nucleic acids are deoxyribonucleic acid (DNA) and ribonucleic acid (RNA).
Calcium (Ca)	1.5	Contributes to hardness of bone and teeth; needed for many body processes, for example, blood clotting, movement of structures inside cells, release of neurotransmitters, and contraction of muscle.
Phosphorus (P)	1.0	Component of many proteins, nucleic acids, and adenosine triphosphate (ATP), an energy-rich chemical in cells; required for normal bone and tooth structure.
Potassium (K)	0.4	Most abundant cation (positively charged particle) inside cells; important in conduction of nerve impulses and muscle contraction.
Sulfur (S)	0.3	Component of many proteins, especially the contractile proteins of muscle.
Sodium (Na)	0.2	Most plentiful cation outside cells; essential in blood to maintain water balance; needed for conduction of nerve impulses and muscle contraction.
Chlorine (Cl)	0.2	Most plentiful anion (negatively charged particle) outside cells; essential in blood and interstitial fluid to maintain water balance.
Magnesium (Mg)	0.1	Needed for many enzymes to function properly.
Iodine (I)	0.1	Vital to production of hormones by the thyroid gland.
Iron (Fe)	0.1	Essential component of hemoglobin (oxygen-carrying protein in blood) and some enzymes needed for ATP production.
Aluminum (Al) Boron (B) Chromium (Cr) Cobalt (Co) Copper (Cu) Fluorine (F) Manganese (Mn) Molybdenum (Mo) Selenium (Se) Silicon (Si) Tin (Sn) Vanadium (V) Zinc (Zn)		These elements are called trace elements because they are present in minute concentrations.

Compose about 96% of total body mass.

Compose about 3.9% of total body mass.

Compose about 0.1% of total body mass.

FIGURE 2.1 Structure of an atom. In this simplified diagram of a carbon atom, note the central location of the nucleus. The nucleus contains six neutrons and six protons, although all are not visible in this view. The six electrons move about the nucleus in regions called electron shells, shown here as circles.

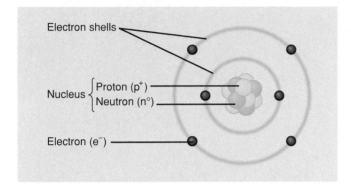

Question: What is the atomic number of this atom?

Negatively charged **electrons (e⁻)** orbit the dense central core or **nucleus.** Within the nucleus are positively charged particles called **protons (p⁺)** and uncharged (neutral) particles called **neutrons (n⁰).** The electron orbits drawn in diagrams like Fig. 2.1 are very simplified, showing just the average (or most probable) distance between the nucleus and a particular group of orbiting electrons. Actually, the fast-moving electrons follow no particular path around the nucleus. You can picture the electrons as forming a cloud of negative charge around the nucleus.

The number of electrons in an atom of an element always equals the number of protons. Since each electron carries one negative charge, the negatively charged electrons and the positively charged protons balance each other. Thus each atom is electrically neutral; its total charge is zero.

The standard unit for measuring the mass of atoms and their subatomic particles is a **dalton** or an **atomic mass unit (amu).** A neutron has a mass of 1.008 daltons, and a proton has a mass of 1.007 daltons. The mass of an electron is 0.0005 daltons, about 1/2000 the mass of a neutron or proton. Adding up the masses of all the protons, neutrons, and electrons in an atom gives the **atomic mass (atomic weight).**

The *number of protons* in the nucleus makes the atoms of one element different from those of another. Figure 2.2 shows that the hydrogen atom contains one proton; the helium atom contains two; the carbon atom has six; and so on. Each different kind of atom has a different number of protons in its nucleus. The number of protons in an atom is called the atom's **atomic number.** Therefore each kind of atom, or element, has a different atomic number. For example, oxygen has an atomic number of 8, because its nucleus has 8 protons, while sodium has an atomic number of 11 because its nucleus has 11 protons.

The **mass number** of an atom is the total number of protons and neutrons. For sodium, with 12 neutrons, the mass number is 23 (11 protons plus 12 neutrons = 23). Since electrons have so little mass, the atomic mass of an atom may be nearly equal to its mass number.

Atoms of one element, however, although chemically alike, may have different mass numbers because they have different numbers of neutrons. The atomic mass assigned to an element is thus an average, reflecting contributions from heavier and lighter atoms. The atomic mass of sodium, for example, is 22.99. Although most sodium atoms have 12 neutrons (mass number = 23), a few have only 11 neutrons (mass number 22).

Different atoms of an element that have the same number of protons but different numbers of neutrons are called **isotopes.** All isotopes of an element have the same *chemical properties* because they have the same number of electrons. (As you will see shortly, the chemical properties of an atom are a function of its electrons.) In a sample of oxygen, for example, most atoms have 8 neutrons, but a few have 9 or 10, even though all have 8 protons and 8 electrons. The isotopes of oxygen are designated as ¹⁶O, ¹⁷O, and ¹⁸O (or O-16, O-17, and O-18). The numbers indicate the mass number (total number of protons and neutrons) in each isotope.

Certain isotopes called **radioactive isotopes (radioisotopes)** are unstable: their nuclear structure decays or changes to a more stable configuration. In decaying, they emit radiation (alpha or beta particles or gamma rays) that can be monitored with radiation detectors. Such instruments can not only detect the emissions from a radioactive isotope but with the aid of computers can also form an image of its distribution within the body. The decay of a particular radioactive isotope may be slow or fast, taking thousands of years or a fraction of a second. Each one has a characteristic **half-life,** that is, the time required for the radioactive isotope to emit half of the original amount of radiation.

CLINICAL APPLICATION

MEDICAL IMAGING USING RADIOACTIVE ISOTOPES

Cells don't distinguish between radioactive isotopes and stable isotopes because all isotopes have the same number of electrons and thus the same chemical behavior. Radioactive isotopes can be used to study both the structure and function of particular tissues. For example, radioactive iodine (I-131) can be used to visualize the size, position, and activity of the thyroid gland (thyroid scan). Thallium-201 is used to image the heart (thallium imaging) and to assess adequacy of blood flow to the heart muscle. Various other radioactive isotopes are also used to study bones (bone scan), the brain (PET scan), the lungs (lung scan), and the kidneys (kidney scan).

FIGURE 2.2 Atomic structures of several atoms that have important roles in living systems.

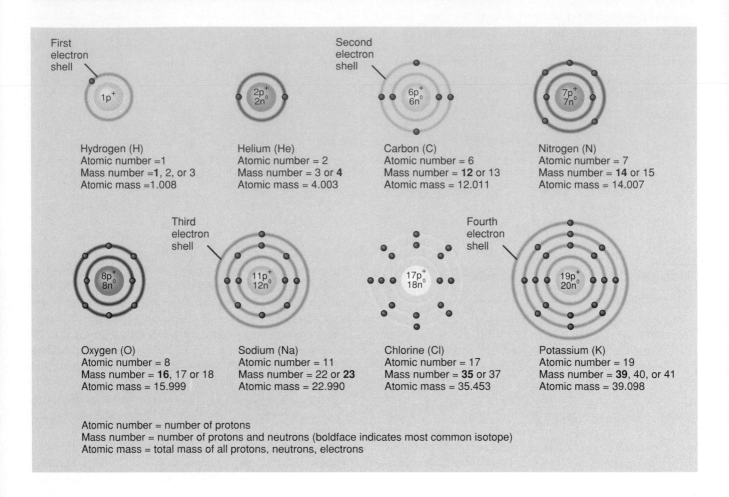

First electron shell

Hydrogen (H)
Atomic number = 1
Mass number = **1**, 2, or 3
Atomic mass = 1.008

Helium (He)
Atomic number = 2
Mass number = 3 or **4**
Atomic mass = 4.003

Second electron shell

Carbon (C)
Atomic number = 6
Mass number = **12** or 13
Atomic mass = 12.011

Nitrogen (N)
Atomic number = 7
Mass number = **14** or 15
Atomic mass = 14.007

Third electron shell

Oxygen (O)
Atomic number = 8
Mass number = **16**, 17 or 18
Atomic mass = 15.999

Sodium (Na)
Atomic number = 11
Mass number = 22 or **23**
Atomic mass = 22.990

Chlorine (Cl)
Atomic number = 17
Mass number = **35** or 37
Atomic mass = 35.453

Fourth electron shell

Potassium (K)
Atomic number = 19
Mass number = **39**, 40, or 41
Atomic mass = 39.098

Atomic number = number of protons
Mass number = number of protons and neutrons (boldface indicates most common isotope)
Atomic mass = total mass of all protons, neutrons, electrons

Question: Which of these elements is inert?

Electrons and Chemical Reactions

When atoms combine with or break apart from other atoms, a **chemical reaction** occurs. In the process, new products with different chemical properties are formed. Chemical reactions are the foundation of all life processes, and electron interactions are the basis of all chemical reactions.

In their motion around the nucleus, electrons tend to spend most of the time in specific atomic regions. Figure 2.2 depicts these most likely regions as circles lying at varying distances from the nucleus. Each circle represents one **electron shell,** which can hold a certain maximum number of electrons. For instance, the electron shell nearest the nucleus never holds more than two electrons, no matter what the element. This is the first shell. The second shell holds a maximum of eight electrons, while the third can hold up to 18 electrons. Higher shells (there are as many as

seven) can contain many more electrons. Iodine, the largest element present in the human body, holds 18 electrons in the fourth shell and seven in the fifth.

To achieve stability, atoms tend to either empty their outermost shell or fill it to the maximum. In so doing, they may *give up, accept, or share* electrons with other atoms—whichever is easiest. The **valence** (combining capacity) is the number of extra or deficient electrons in the **valence shell,** the outermost shell. Take a look at the chlorine atom in Fig. 2.2. Its outermost shell, which happens to be the third shell, has seven electrons. Since the third shell would be stable with eight electrons, chlorine is one electron short. Sodium, by contrast, has only one electron in its valence shell. This again happens to be the third level. It is much easier for sodium to get rid of the one electron than to fill the third shell by taking on seven more electrons. Atoms of a few elements, like helium and neon, have completely filled outer shells and do not tend to gain or lose electrons. Such elements are called **inert elements** and do not usually participate in chemical reactions.

Atoms with incompletely filled outer shells, such as sodium and chlorine, tend to combine with each other in chemical reactions. During a reaction the atoms can exchange or share valence electrons and thereby fill or empty their outer shells. When two or more atoms combine in a chemical reaction, the resulting combination is called a **molecule** (MOL-e-kyool). A molecule may contain two atoms of the same kind, as in the hydrogen molecule: H_2. The subscript 2 shows there are two hydrogen atoms in the molecule. Two or more different kinds of atoms may also react to form a molecule, as in the hydrogen chloride molecule: HCl. Here one atom of hydrogen is attached to one atom of chlorine.

A **compound** is a substance that can be broken down into two or more different elements by chemical means. The molecules of a compound always contain atoms of two or more different elements. Hydrogen chloride, which is dissolved by water to form hydrochloric acid in the digestive juices of the stomach, is a compound. A molecule of hydrogen is not.

CLINICAL APPLICATION

LASERS

The **laser** (acronym for **l**ight **a**mplification by **s**timulated **e**mission of **r**adiation) is based on the principle that certain atoms, molecules, or ions (charged particles) can be excited by absorption of thermal, electrical, or light energy. After such energy absorption, the atoms, molecules, or ions give off a beam of synchronized light waves. This **laser beam** is a narrow, intense, and monochromatic (single color) light beam that can be used for a variety of purposes, for example, to stop bleeding, make incisions, and remove tissue.

Chemical Bonds

The atoms of a molecule are held together by forces of attraction called **chemical bonds,** a form of potential energy. In chemical reactions that occur in the body, breaking bonds usually requires energy, and forming bonds usually releases energy. Here we will consider ionic bonds, covalent bonds, and hydrogen bonds.

Ions and Ionic Bonds

Atoms are electrically neutral because the number of positively charged protons equals the number of negatively charged electrons. When an atom gains or loses electrons, however, this balance is upset. If the atom gains electrons, it acquires a negative charge; if it loses electrons, it is left with a positive charge. Such a negatively or positively charged particle is called an **ion**. Ions in solution are called **electrolytes** (e-LEK-trō-līts) because the ionic solution is capable of conducting an electric current. (The chemistry and importance of electrolytes are discussed in detail in Chapter 27.)

Consider the sodium ion (Fig. 2.3a). An atom of sodium has 11 protons and 11 electrons, with 1 electron in its

FIGURE 2.3 Ions and ionic bond formation. (a) Sodium (Na) tends to give up the single electron in its valence shell; it is an electron donor. (b) Chlorine (Cl) tends to pick up one electron to completely fill its valence shell; it is an electron acceptor. (c) Ionic bonds hold oppositely charged ions together. Na$^+$ joins Cl$^-$ to form sodium chloride, ordinary table salt.

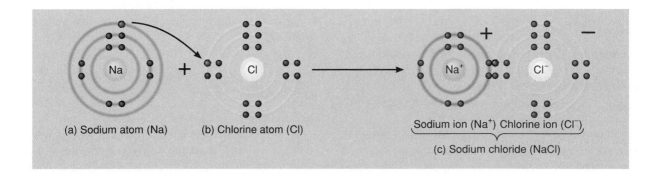

(a) Sodium atom (Na)   (b) Chlorine atom (Cl)   Sodium ion (Na$^+$) Chlorine ion (Cl$^-$)

(c) Sodium chloride (NaCl)

Question: Will potassium (K) be more likely to be an electron donor or an electron acceptor? (Look back to Fig. 2.2 for the atomic structure of K.)

valence (outermost) shell. When sodium gives up its valence electron, it is left with 11 protons and only 10 electrons. It is an **electron donor** because it gives up an electron. The atom now has a positive charge of one (+1). This positively charged sodium atom is called a sodium ion (written Na+). Note that an ion is symbolized by writing the chemical abbreviation followed by the number of positive (+) or negative (−) charges the ion acquires.

Another example is the formation of the chloride ion (Fig. 2.3b). Chlorine has a total of 17 electrons, 7 of them in its valence shell. Since this shell can hold up to eight electrons, chlorine tends to pick up an electron that another atom has lost. Thus chlorine is an **electron acceptor.** By accepting an electron, chlorine acquires a total of 18 electrons. However, it still has only 17 positively charged protons in its nucleus. The chloride ion therefore has a negative charge of one (−1) and is written as Cl−.

The positively charged sodium ion (Na+) and the negatively charged chloride ion (Cl−) attract each other: unlike charges attract. The attraction, called an **ionic bond,** holds the two ions together, and a molecule of sodium chloride (NaCl), or table salt, is formed (Fig. 2.3c). Thus ionic bonds form when oppositely charged ions associate.

Generally, if the outer electron shell is less than half-filled, an atom loses electrons and forms a positively charged ion called a **cation** (KAT-ī-on). Examples of cations are the potassium ion (K+), sodium ion (Na+), and calcium ion (Ca^{2+}). By contrast, if the outer shell is more than half-filled, an atom tends to accept electrons and form a negatively charged ion called an **anion** (AN-ī-on). Examples of anions include the iodide ion (I−), chloride ion (Cl−), and sulfide ion (S^{2-}).

Hydrogen is an example of an atom whose outer electron shell is exactly half-filled. The first shell can hold two electrons, but in hydrogen atoms it contains only one. Hydrogen may lose its electron and become a positive ion (H+). This is precisely what happens when hydrogen combines with chlorine to form hydrogen chloride (HCl). However, hydrogen is equally capable of forming another kind of bond called a covalent bond.

Covalent Bonds

The second type of chemical bond is the **covalent bond**. This kind of bond is much more common in the body than an ionic bond. When a covalent bond forms, neither of the combining atoms loses or gains electrons. Instead, the two atoms *share* one, two, or three pairs of electrons.

One way a hydrogen atom can fill its outer shell is to combine with another hydrogen atom to form the molecule H$_2$ (Fig. 2.4a). In H$_2$ molecules, the two atoms share a pair of electrons. Each hydrogen atom has its own electron plus a shared electron from the other atom. The two electrons orbit the nuclei (single protons) of both atoms, but they are most often in the region between the two nuclei. The attrac-

tion between the negatively charged electrons in the middle and the positively charged protons in the nuclei holds the hydrogen molecule together.

When two atoms share one pair of electrons, as in the H$_2$ molecule, a **single covalent bond** is formed. A single covalent bond is written as a single line between the atoms (H—H). When two atoms share two pairs of electrons, the result is a double covalent bond, written as two parallel lines (O=O in Fig. 2.4b). A **triple covalent bond**, written as three parallel lines, occurs when three pairs of electrons are shared (N≡N in Fig. 2.4c).

The same principles that apply to covalent bonding between atoms of the same element also apply to atoms of different elements. Methane (CH$_4$), also known as marsh gas, is an example of covalent bonding between atoms of different elements (Fig. 2.4d). The outer shell of the carbon atom can hold eight electrons but has only four of its own. Each hydrogen atom can hold two electrons but has only one of its own. In the methane molecule the carbon atom shares four pairs of electrons. One pair is shared with each hydrogen atom. Each of the four carbon electrons orbits both the carbon nucleus and a hydrogen nucleus. Each hydrogen electron likewise orbits both its own nucleus and the carbon nucleus.

In some covalent bonds, atoms share the electrons equally; that is, one atom does not attract the shared electrons more strongly than the other atom. This is called a **nonpolar covalent bond**. The bonds between two identical atoms always are nonpolar covalent bonds (Fig. 2.4a–c). Another example of a nonpolar covalent bond is the single covalent bond between carbon and hydrogen (Fig. 2.4d).

In other covalent bonds, the sharing of electrons between atoms is unequal; that is, one atom attracts the shared electrons more strongly than the other. This type of covalent bond is a **polar covalent bond**. The most important example in living systems is the bond between oxygen and hydrogen in a molecule of water (see Fig. 2.5a). When polar covalent bonds form, the resulting molecule has a partial negative charge (written as δ−) near the atom that attracts electrons more strongly and a partial positive charge (written as δ+) near the other atoms. As we will discuss in more detail later in the chapter, polar covalent bonds allow water to dissolve many molecules that are important to life.

Hydrogen Bonds

In a **hydrogen bond** two other atoms (usually oxygen or nitrogen) associate with a hydrogen atom. A hydrogen atom that is covalently bonded to one oxygen or nitrogen atom is also attracted to the partial negative charge of another covalently bonded nitrogen or oxygen atom. Because hydrogen bonds are weak, only about 5% as strong as covalent bonds, they do not bind atoms into molecules. However, they do serve as bridges between molecules, for example, between

Figure 2.4 Covalent bond formation. When two atoms share a pair of valence electrons, they form a covalent bond. To the right are simpler ways to represent these molecules. In structural formulas, each covalent bond is written as a straight line between the symbols for two atoms. In molecular formulas, the number of atoms in each molecule is noted by subscripts.

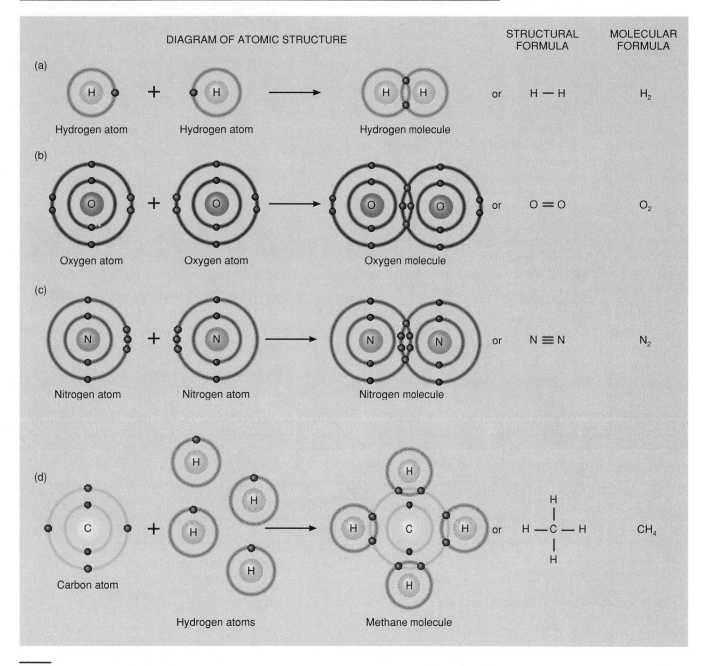

Question: What is a covalent bond?

water molecules (see Fig. 2.5b), or between various parts of the same molecule (see Fig. 2.16). Even though single hydrogen bonds are weak, large molecules may contain hundreds of these bonds. Thus they confer considerable strength and stability. In addition, hydrogen bonds determine the three-dimensional shape of large molecules, such as proteins and nucleic acids. Usually, an exact molecular shape is needed for a particular molecule to carry out a precise function.

Chemical Reactions

Chemical reactions are nothing more than the making or breaking of bonds between atoms. Through hundreds of reactions, body structures are built and body functions are carried out. The term **metabolism** refers to all chemical reactions occurring in an organism.

Atoms, ions, and molecules all are continuously moving and colliding with one another; they have kinetic energy. A sufficiently forceful collision disrupts electron interactions and breaks an existing chemical bond or forms a new one. Several factors determine whether a collision will occur and cause a chemical reaction.

1. **Concentration.** The more particles present in a confined region, the greater the chance that they will collide.
2. **Speed.** Up to a point, the higher the particles' speed, the greater the chance that their collision will result in a reaction. A higher temperature increases the particles' speeds. At the same temperature, smaller particles have higher speeds than larger ones.
3. **Energy.** Each chemical reaction requires a specific level of energy. The collision energy required for a chemical reaction is its **activation energy.** This is the amount of energy needed to disrupt the stable electronic configuration of a specific molecule so that the electrons can be rearranged.
4. **Proper orientation.** Even if colliding particles have the minimum energy needed to react, no reaction will take place unless the particles are properly oriented toward each other.

After a chemical reaction, the total number of atoms is the same, but because they are rearranged, there are new molecules with new properties. In this section, we will look at the main types of chemical reactions common to all living cells. Once you have learned them, you will be able to understand the chemical reactions discussed later.

Synthesis Reactions: Anabolism

When two or more atoms, ions, or molecules combine to form new and larger molecules, the process is called a **synthesis reaction.** The word *synthesis* means "to put together," and synthesis reactions involve the *forming of*

new bonds. Synthesis reactions can be expressed in the following way:

$$A + B \xrightarrow{\text{Combine to form}} AB$$

Atom, ion, or molecule A · · · Atom, ion, or molecule B · · · New molecule AB

The combining substances, A and B, are called the **reactants;** the substance formed by the combination is the **product.** The arrow indicates the direction in which the reaction proceeds. An example of a synthesis reaction is

$$N_2 + 3H_2 \longrightarrow 2NH_3$$

One nitrogen molecule · · · Three hydrogen molecules · · · Two ammonia molecules

All the synthesis reactions that occur in your body are collectively called anabolic reactions, or simply **anabolism** (a-NAB-ō-lizm). Recall that anabolism is one of the phases of metabolism; catabolism is the other. Combining amino acids to form proteins is an example of anabolism. The importance of anabolism is considered in detail on page 823.

Decomposition Reactions: Catabolism

The reverse of a synthesis reaction is a **decomposition reaction**. The word *decompose* means to break down into smaller parts. In a decomposition reaction, *bonds are broken.* Large molecules are broken down into smaller molecules, ions, or atoms. A decomposition reaction occurs in this way:

$$AB \xrightarrow{\text{Breaks down into}} A + B$$

Molecule AB · · · Atom, ion, or molecule A · · · Atom, ion, or molecule B

For example, under the proper conditions, methane can decompose into carbon and hydrogen molecules:

$$CH_4 \longrightarrow C + 2H_2$$

One methane molecule · · · One carbon atom · · · Two hydrogen molecules

All the decomposition reactions that occur in your body are collectively called catabolic reactions, or simply **catabolism** (ka-TAB-ō-lizm). The digestion of food molecules is an example. When chemical bonds are broken, the energy stored within them often is released for use by cells. The importance of catabolism is also considered in detail on page 823.

Exchange Reactions

All chemical reactions are based on synthesis or decomposition processes. In other words, chemical reactions are simply the making or breaking of ionic or covalent bonds. Many reactions, such as **exchange reactions,** are partly synthesis and partly decomposition. One type of exchange reaction works like this:

$$AB + CD \longrightarrow AD + BC$$

The bonds between A and B and between C and D break in a decomposition process. New bonds then form between A and D and between B and C in a synthesis process. As you will see in chapter 27, buffer reactions, in which the body attempts to maintain its normal acid–base balance, are exchange reactions.

Reversible Reactions

When chemical reactions are reversible, the product can revert to the original reactants. A **reversible reaction** is indicated by two arrows:

$$A + B \underset{\text{Breaks down into}}{\overset{\text{Combines to form}}{\rightleftharpoons}} AB$$

Some reactions are reversible only under special conditions:

$$A + B \underset{\text{Water}}{\overset{\text{Heat}}{\rightleftharpoons}} AB$$

The breakdown and reformation of ATP is a reversible reaction.

Whatever is written above or below the arrows indicates the condition needed for the reaction to occur. In this case, A and B react to produce AB only when heat is applied, and AB breaks down into A and B only when water is added.

CHEMICAL COMPOUNDS AND LIFE PROCESSES

Most of the chemicals in your body exist in the form of compounds. Biologists and chemists divide these compounds into two principal classes: inorganic compounds and organic compounds. **Inorganic compounds** usually lack carbon. In living systems, the important inorganic compounds usually are small molecules. They include water, oxygen, carbon dioxide, and many salts, acids, and bases. **Organic compounds,** on the other hand, always contain carbon and hydrogen. Carbon is a unique element in metabolism because it has four electrons in its outer (valence) shell. It can combine with a variety of atoms, including other carbon atoms, to form rings and straight or branched chains. Carbon chains are the backbone for many substances of living cells. Organic compounds are held together mostly or entirely by covalent bonds. Important categories of organic compounds in the body include carbohydrates, lipids, proteins, nucleic acids, and adenosine triphosphate (ATP).

INORGANIC COMPOUNDS

Water

One of the most important, as well as the most abundant, inorganic substances in all living systems is **water.** It is by far the most abundant substance in your body. With a few exceptions, such as tooth enamel and bone, water makes up most of the volume of cells and body fluids. For example, about 60% of red blood cells, 75% of muscle tissue, and 92% of blood plasma (liquid portion of blood) is water. The following functions of water explain why it is such a vital compound in living systems.

1. **Water is an excellent solvent and suspending medium.** Alchemists in medieval times tried to find a universal solvent, a substance that would dissolve all other materials. They found nothing that worked as well as water. Although it is the most versatile solvent known, it is not a "universal solvent." If it were, no container could hold it; the container would dissolve.

A **solvent** is a liquid or gas in which some other material (solid, liquid, or gas), called a **solute,** has been dissolved. The combination of solvent plus solute is called a **solution.** One common example of a solution is salt water. A solute, such as salt, does not settle out and accumulate on the bottom of the container. In a **suspension,** by contrast, the suspended material may mix with the liquid or suspending medium for some time, but it will eventually settle out. An example of a suspension is cornstarch and water. If the two materials are shaken together, a milky mixture forms. After the mixture sits for awhile, however, the water clears at the top, and the cornstarch settles to the bottom.

The versatility of water as a solvent is due to its polar covalent bonds, in which electrons are not shared equally between atoms. In a molecule of water, there are both positive and negative areas (Fig. 2.5a). When the two hydrogen atoms bond covalently to an oxygen atom, the shared electrons spend more time around oxygen than hydrogen. Since electrons have a negative charge, the unequal sharing causes the oxygen atom to have a partial negative charge (δ^-) and each hydrogen atom to have a partial positive charge (δ^+).

To understand the dissolving capability of water, consider what happens when a crystal of an ionic compound such as sodium chloride (NaCl) is placed in water. The sodium and chloride ions at the surface of the salt crystal are exposed to the polar water molecules. The negatively charged oxygen portion of water molecules is attracted to the sodium ion (Na^+) of the salt. And the positively charged hydrogen portions of water molecules are attracted to the chloride ions (Cl^-) of the salt (Fig. 2.5c). Soon, water molecules surround some Na^+ and Cl^- and separate them from each other. In this way, water molecules take apart—or dissolve—the salt.

Figure 2.5 Polar water molecules dissolve ionic compounds and other polar substances. Because the oxygen nucleus attracts the shared electrons more strongly (a), the oxygen end of a water molecule has a partial negative charge (written δ^-), and the hydrogen ends have a partial positive charge (written δ^+). When a crystal of sodium chloride is dropped into a glass of water (c), the slightly negative oxygen end of water molecules is attracted to the positive sodium ions (Na^+); the slightly positive hydrogen portions of the water molecules are attracted to the negative chloride ions (Cl^-).

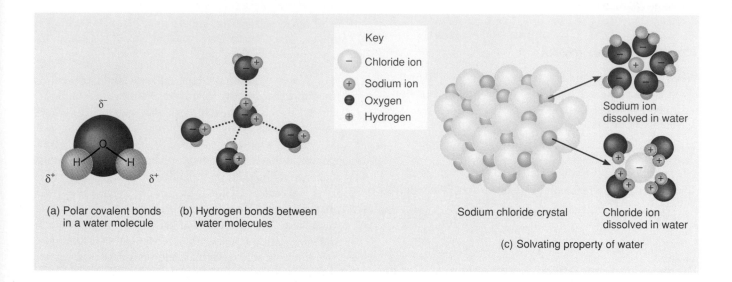

Key
- – Chloride ion
- + Sodium ion
- – Oxygen
- + Hydrogen

(a) Polar covalent bonds in a water molecule

(b) Hydrogen bonds between water molecules

Sodium chloride crystal

Sodium ion dissolved in water

Chloride ion dissolved in water

(c) Solvating property of water

Question: Table sugar (sucrose) easily dissolves in water but is not an ionic compound like sodium chloride. Is it likely that all of the covalent bonds between atoms in table sugar are nonpolar bonds?

The solvating and suspending properties of water are essential to health and survival. Since water can dissolve or suspend so many different substances, it is an ideal medium for metabolic reactions. Because they are together in a common fluid, reactants and other necessary materials, such as ATP and enzymes, readily collide to form products. Water also dissolves waste products and flushes them out of the body in the urine.

2.   Water participates in chemical reactions. During digestion, for example, water can be added to large nutrient molecules to break them down into smaller molecules (see Fig. 2.8a). This kind of breakdown (called hydrolysis and described shortly) is necessary if the body is to use the energy in nutrients. On the other hand, when two smaller molecules join to form one larger one in a synthesis reaction, a water molecule often is removed. Such reactions occur in the production of proteins and other large molecules (see Fig. 2.13).

3.   Water absorbs and releases heat very slowly. In comparison to other substances, water requires a large amount of heat to increase its temperature and a great loss of heat to decrease its temperature. It has a high heat capacity. Thus the presence of a large amount of water lessens the effects of environmental temperature changes and thereby helps to maintain homeostasis of body temperature.

4.   Water requires a large amount of heat to change from a liquid to a gas. When water (perspiration) evaporates from the skin, it takes with it large quantities of heat and provides an excellent cooling mechanism. Its heat of vaporization is high.

5.   Water serves as a lubricant. It is a major part of mucus and other lubricating fluids in which protein is dissolved to make it more viscous (thicker). Lubrication is especially necessary in the chest and abdomen, where internal organs touch and slide over each other. It is also needed at joints, where bones, ligaments, and tendons rub against one another. In the gastrointestinal tract, water in mucus moistens foods to ensure their smooth passage.

Inorganic Acids, Bases, and Salts

When molecules of inorganic acids, bases, or salts dissolve in water, they undergo **ionization** (ī'-on-i-ZĀ-shun) or **dissociation** (dis'-sō-sē-Ā-shun); that is, they separate into ions.

An **acid** (Fig. 2.6a) may be defined as a substance that dissociates into one or more **hydrogen ions (H^+)** and one or more *anions* (negative ions). Since H^+ is a single proton with a charge of +1, an acid may also be defined as a proton donor. A **base**, by contrast (Fig. 2.6b), dissociates into one or more **hydroxide ions (OH^-)** and one or more *cations* (positive ions). A base may also be viewed as a proton acceptor. Hydroxide ions have a strong attraction for protons.

FIGURE 2.6 Ionization of inorganic acids, bases, and salts.

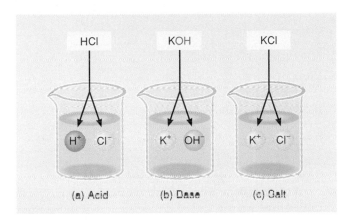

(a) Acid (b) Base (c) Salt

Question: The compound $CaCO_3$ (calcium carbonate) dissociates into a calcium ion (Ca^{2+}) and a carbonate ion (CO_3^{2-}). Is it an acid, a base, or a salt? What about H_2SO_4, which dissociates into two H^+ and one SO_4^{2-}?

A **salt**, when dissolved in water, dissociates into cations and anions, neither of which is H^+ or OH^- (Fig. 2.6c). Acids and bases react with one another to form salts. For example, the combination of hydrochloric acid (HCl), an acid, and potassium hydroxide (KOH), a base, produces potassium chloride (KCl), a salt, and water (H_2O). This exchange reaction can be written

$$\text{HCl} + \text{KOH} \longrightarrow \text{KCl} + \text{H}_2\text{O}$$
$$\text{Acid} \qquad \text{Base} \qquad \text{Salt} \qquad \text{Water}$$

Many salts are present in the body. Some are in cells, in the intracellular fluid, whereas others are in extracellular fluids such as lymph, blood, and the interstitial fluid of tissues. The ions of salts provide many essential chemical elements.

Acid–Base Balance: The Concept of pH

Body fluids must constantly contain balanced quantities of acids and bases. In solutions such as those found inside or outside body cells, acids dissociate into hydrogen ions (H^+) and anions. Bases, on the other hand, dissociate into hydroxide ions (OH^-) and cations. The more hydrogen ions that exist in a solution, the more acidic the solution; conversely, the more hydroxide ions, the more basic (alkaline) the solution.

Biochemical reactions—those that occur in living systems—are very sensitive to even small changes in acidity or alkalinity. Any departure from the narrow limits of normal H^+ and OH^- concentrations may greatly modify cell functions and disrupt homeostasis. For this reason, the acids and bases that are constantly formed in the body must be kept in balance.

A solution's acidity or alkalinity is expressed on the **pH scale,** which runs from 0 to 14 (Fig. 2.7). This scale is

---

FIGURE 2.7 The pH scale. At pH 7, a neutral solution, the concentrations of H^+ and OH^- are equal (10^{-7} moles per liter). The lower the numerical value of the pH, the more acidic the solution is because the H^+ concentration becomes progressively greater. The higher the pH, the more basic the solution.

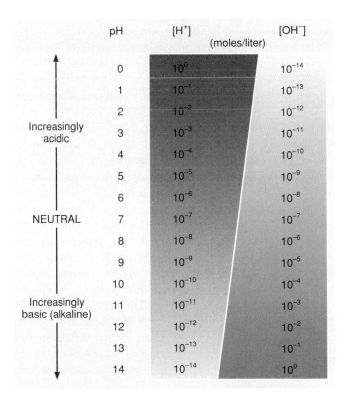

Question: What is the concentration of H^+ and OH^- at pH 6? Which pH is more acidic, 6.82 or 6.91? Which pH is closer to neutral, 8.41 or 5.59?

based on the concentration of H^+ in a solution expressed in chemical units called **moles per liter (moles/liter).** A pH of 7 means that a solution contains one ten-millionth (0.0000001) of a mole* of hydrogen ions per liter. The number 0.0000001 is written as 10^{-7} in scientific notation (exponential form), which indicates that the number is 1 with the decimal point moved 7 places to the left. To convert this value to pH, the negative exponent (−7) is changed to a positive number (7). A solution with a H^+ concentration of 0.0001 (10^{-4}) moles per liter has a pH of 4; a solution with a H^+ concentration of 0.000000001 (10^{-9}) moles per liter has a pH of 9; and so on.

The midpoint in the scale is 7, where the concentrations of H^+ and OH^- are equal. A substance with a pH of 7, such as distilled (pure) water, is neutral. A solution that has more H^+ than OH^- is an **acidic solution** and has a pH below 7. A solution that has more OH^- than H^+ is a **basic (alkaline) solution** and has a pH above 7. A change of one whole number on the pH scale represents a 10-fold change from the previous concentration. A pH of 1 denotes 10 times more H^+ than a pH of 2. A pH of 3 indicates 10 times fewer H^+ than a pH of 2 and 100 times fewer H^+ than a pH of 1.

Maintaining pH: Buffer Systems

Although the pH of body fluids may differ, the normal limits for the various fluids are generally quite specific and narrow. Exhibit 2.2 shows the pH values for certain body fluids compared with common substances. Homeostatic mechanisms maintain the pH of blood between 7.35 and 7.45, slightly more basic than pure water. Saliva is slightly acidic, and semen is slightly basic. Since the kidneys help remove excess acid from the body, urine can be quite acidic. Even though strong acids and bases are continually taken into and formed by the body, the pH of fluids inside and outside cells remains almost constant. One important reason is the presence of **buffer systems**.

The function of a buffer system is to convert strong acids or bases (relatively unstable, ionize easily) into weak acids or bases (relatively stable, do not ionize easily). This ensures that the strong acids or bases do not alter pH very much. Strong acids (or bases) dissociate (ionize) easily and contribute many H^+ (or OH^-) to a solution. They therefore change the pH drastically, which may cause considerable damage. Weak acids (or bases) do not dissociate as much. They contribute fewer H^+ (or OH^-) and have less effect on the pH. The chemicals that replace strong acids or bases with weak ones are called **buffers.** Most buffers in the human body consist of a weak acid and the salt of that acid.

* A mole of any substance is the weight, in grams, of the combined atomic weights of the atoms that make up a molecule of the substance. Example: A mole of NaCl weighs 58.5 grams (23 for the sodium atoms + 35.5 for the chlorine atoms).

EXHIBIT 2.2

pH VALUES OF SELECTED SUBSTANCES

Substance	pH Value
Gastric juice (digestive juice of the stomach)[a]	1.2–3.0
Lemon juice	2.2–2.4
Grapefruit juice, vinegar, wine	3.0
Carbonated soft drink	3.0–3.5
Vaginal fluid[a]	3.5–4.5
Pineapple juice, orange juice	3.5
Tomato juice	4.2
Coffee	5.0
Urine[a]	4.6–8.0
Saliva[a]	6.35–6.85
Milk	6.6–6.9
Distilled (pure) water	7.0
Blood[a]	7.35–7.45
Semen (fluid containing sperm)[a]	7.20–7.60
Cerebrospinal fluid (fluid associated with nervous system)[a]	7.4
Pancreatic juice (digestive juice of the pancreas)[a]	7.1–8.2
Eggs	7.6–8.0
Bile (liver secretion that aids fat digestion)[a]	7.6–8.6
Milk of magnesia	10.0–11.0
Lye	14.0

[a] Substances in the human body.

ORGANIC COMPOUNDS

As noted earlier, organic compounds always contain carbon, which can form four covalent bonds, and hydrogen, which can form one covalent bond. Other elements most often found in organic compounds are oxygen (two bonds) and nitrogen (three bonds). Sulfur (two bonds) and phosphorus (five bonds) appear less often. Additional elements are present but only in a few organic compounds.

Carbon has several properties that make it particularly useful to living organisms. For one thing, it can react with one to thousands of other carbon atoms to form large molecules of many different shapes. This means that the body can build many compounds out of carbon, hydrogen, and oxygen. Each compound can be expressly suited for a particular structure or function. The large size of most carbon-containing molecules and the fact that some do not dissolve easily in water make them useful materials for building body structures.

Carbon compounds are mostly or entirely held together by covalent bonds and tend to decompose easily. This means that organic compounds are also a good source of energy. Ionic compounds are not good energy sources because they form new ionic bonds as soon as the old ones are broken.

Small organic molecules can combine into very large molecules called **macromolecules**. Macromolecules are usually **polymers**, that is, large molecules formed by covalent bonding of many repeating small molecules called **subunits** or **monomers**. When two monomers join together, the reaction usually involves the elimination of a molecule of water. This type of reaction is called **dehydration** (*dehydration* = loss of water) **synthesis** because a molecule of water is released. Macromolecules such as carbohydrates, lipids, proteins, and nucleic acids are assembled in the cell by dehydration synthesis. It is also possible for macromolecules to be broken down into monomers by the addition of water. This reaction is called **hydrolysis** (hī-DROL-i-sis), which means to split apart by using water.

Carbohydrates

Carbohydrates include sugars, starches, glycogen, and cellulose. They are a large and diverse group of organic compounds and have several functions, although they represent only 2 to 3% of your total body weight. Plants store carbohydrate as starch and use the carbohydrate cellulose to build the cell wall. Cellulose is the most plentiful organic substance on earth. Although humans eat cellulose, they cannot digest it. However, it does create bulk, which helps to move waste materials along the gastrointestinal tract. In animals the principal function of carbohydrates is to provide a readily available source of energy to drive metabolic reactions. Only a few carbohydrates form structural units in animals. One example is deoxyribose, a type of sugar. It is a building block of deoxyribonucleic acid (DNA), the molecule that carries hereditary information (see Fig. 2.16a).

Some carbohydrates are converted to other substances, which are used to build structures and to generate ATP. For example, the sugar in a donut you eat today could be part of one of your bones or any other tissue by tomorrow. Other carbohydrates function as food reserves. The prime storage carbohydrate in animals is glycogen.

Carbon, hydrogen, and oxygen are the elements found in carbohydrates. The ratio of hydrogen to oxygen atoms is usually 2 : 1, the same as in water. Although there are exceptions, the general rule for carbohydrates is one carbon atom (C) for each water molecule (H_2O), which is why they are called carbohydrates (= watered carbon). Carbohydrates are divided into three major groups based on size: monosaccharides, disaccharides, and polysaccharides. Monosaccharides and disaccharides also are known as **simple sugars.**

1. **Monosaccharides.** Monosaccharides (mon′-ō-SAK-a-rīds) contain from three to seven carbon atoms. Those with three carbons are called trioses. The number of carbon atoms in the molecule is indicated by the prefix, *tri*. There are also tetroses (four-carbon sugars), pentoses (five-carbon sugars), hexoses (six-carbon sugars), and heptoses (seven-carbon sugars). Glucose, a hexose, is the main energy-supplying molecule of the body.

2. **Disaccharides.** Two monosaccharides combine by dehydration synthesis to form one **disaccharide** (dī-SAK-a-rīd) molecule and a molecule of water. For example, molecules of the monosaccharides glucose and fructose combine to form a molecule of the disaccharide sucrose (table sugar) as follows:

$$\underset{\substack{\text{Glucose}\\ \text{(monosaccharide)}}}{C_6H_{12}O_6} + \underset{\substack{\text{Fructose}\\ \text{(monosaccharide)}}}{C_6H_{12}O_6} \longrightarrow \underset{\substack{\text{Sucrose}\\ \text{(disaccharide)}}}{C_{12}H_{22}O_{11}} + \underset{\text{Water}}{H_2O}$$

You may be puzzled to see that glucose and fructose have the same chemical formula. Actually, they are different monosaccharides, since the relative positions of the oxygen and carbon atoms are different. Look at Fig. 2.8a to appreciate this difference. Also, the formula for sucrose is $C_{12}H_{22}O_{11}$, not $C_{12}H_{24}O_{12}$, because a molecule of water is split out as two monosaccharides are joined. In every dehydration synthesis reaction, a molecule of water is lost.

Disaccharides can also be separated into smaller, simpler molecules by adding water. This reverse chemical reaction is called hydrolysis. A molecule of sucrose, for example, may be hydrolyzed into its components of glucose and fructose by the addition of water. Fig. 2.8a also shows this reaction.

Another important disaccharide is lactose, or milk sugar. It consists of the monosaccharides glucose and galactase.

3. **Polysaccharides.** The third major group of carbohydrates is the **polysaccharide** (pol′-ē-SAK-a-rīd; *poly* = many) family. These large carbohydrates contain tens or hundreds of monosaccharides joined through dehydration synthesis reactions. The principal polysaccharide in the human body is glycogen, which is composed of glucose units linked to each other and is stored in the liver and skeletal muscles. Like disaccharides,

FIGURE 2.8 The monosaccharides glucose and fructose and the disaccharide sucrose. (a) In dehydration synthesis (read from left to right), the two smaller molecules, glucose and fructose, are joined to form a larger molecule of sucrose. Note the loss of a water molecule. In hydrolysis (read from right to left), the larger sucrose molecule is broken down into the two smaller molecules, glucose and fructose. Here, a molecule of water is added to sucrose for the reaction to occur. (b) Ways to express chemical structures.

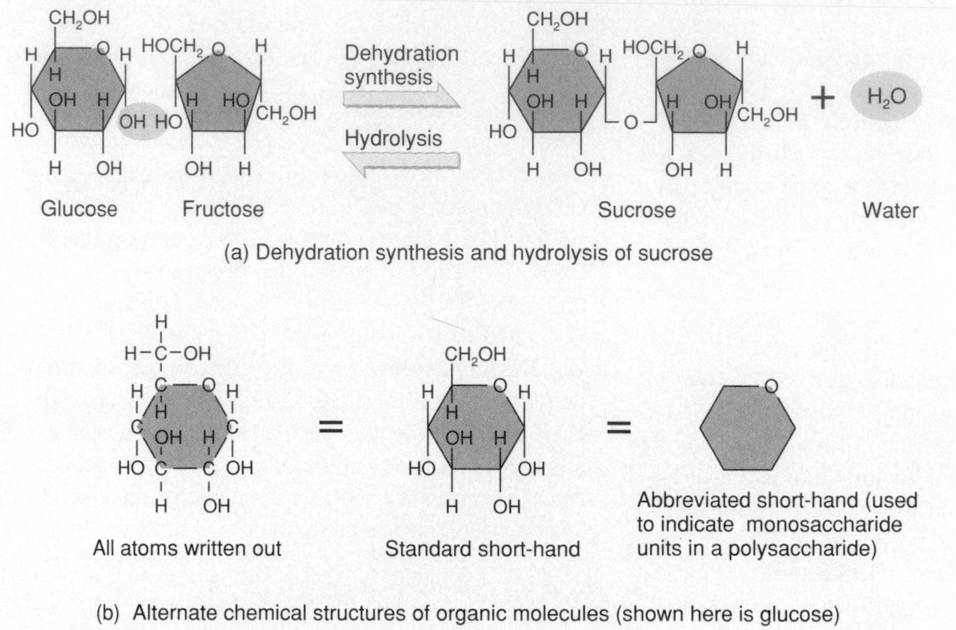

(a) Dehydration synthesis and hydrolysis of sucrose

All atoms written out · Standard short-hand · Abbreviated short-hand (used to indicate monosaccharide units in a polysaccharide)

(b) Alternate chemical structures of organic molecules (shown here is glucose)

Question: Is dehydration synthesis anabolic or catabolic? How many carbons can you count in fructose? In sucrose?

polysaccharides can be broken down into monosaccharides through hydrolysis reactions. For example, when blood glucose level falls, liver cells have the ability to break down glycogen into glucose and release it into the blood. In this way, glucose is made available to body cells where it can be broken down to provide energy. Unlike simple sugars such as fructose and sucrose, however, polysaccharides usually are not sweet and are not soluble in water.

Lipids

A second group of important organic compounds is **lipids** (*lipos* = fat). Lipids comprise about 18 to 25% of body weight in lean adults. Like carbohydrates, lipids contain carbon, hydrogen, and oxygen. Unlike carbohydrates, they do not have a 2 : 1 ratio of hydrogen to oxygen. The amount of oxygen in lipids is usually less than that in car-

bohydrates, so there are fewer polar covalent bonds. As a result, most lipids are insoluble in polar solvents such as water; they are **hydrophobic** (*hydro* = water; *phobic* = fearing). Nonpolar solvents such as chloroform and ether, however, readily dissolve lipids.

Lipids are diverse and include triglycerides (neutral fats), phospholipids (lipids that contain phosphorus), steroids (cholesterol, cortisol, vitamin D, sex hormones), carotenes (the yellow-orange pigments in carrots), vitamins A, E, and K, and eicosanoids. Since they are hydrophobic, lipids cannot travel freely in the watery blood. For efficient transport in blood, lipids combine with proteins to form water-soluble **lipoproteins.** Exhibit 2.3 lists the various types of lipids and highlights their roles in the human body.

1. Triglycerides. The most plentiful lipids in your body and in your diet are the **triglycerides** (trī-GLE-cer-īdes) or **neutral fats**. At room temperature, triglycerides may be either solids (fats) or liquids (oils), and they are the body's most highly concentrated source of chemical

EXHIBIT 2.3

TYPES OF LIPIDS IN THE BODY

Lipid	Functions
TRIGLYCERIDES (NEUTRAL FATS)	Protection, insulation, source of energy, major energy storage molecules in the body.
PHOSPHOLIPIDS	Major lipid component of cell membranes; found in high concentrations in nerves and brain tissue.
STEROIDS	
Cholesterol	Constituent of all animal cell membranes; precursor of bile salts, vitamin D, and steroid hormones.
Bile salts	Substances that emulsify or suspend fats before their digestion and absorption; needed for absorption of fat-soluble vitamins (A, D, E, and K).
Vitamin D	Produced in skin on exposure to ultraviolet radiation; aids in regulation of calcium concentration in the body; necessary for bone growth, development, and repair.
Sex hormones	Estrogens and progesterone (produced in large quantities by females) and testosterone (produced in large quantity by males) stimulate reproductive functions and sexual characteristics.
LIPOPROTEINS	Several types of lipid and protein particles in the blood that help transport lipids to the liver and to adipose (fat) tissue, carry cholesterol to tissues, and remove excess cholesterol from the blood.
EICOSANOIDS	Membrane-associated lipids made by most cells in the body that have diverse effects on blood clotting, inflammation, immunity, stomach acid secretion, airway diameter, lipid breakdown, and smooth muscle contraction in uterus and gastrointestinal tract.
OTHER LIPID SUBSTANCES	
Carotenes	Pigment in egg yolk, carrots, and tomatoes; vitamin A is formed from carotenes; retinal, formed from vitamin A, is part of all photopigments (visual pigments) in the retina of the eye.
Vitamin E	May promote wound healing, prevent scarring, and contribute to the normal structure and function of the nervous system; functions as an antioxidant, by preventing the oxidation of certain molecules in the body.
Vitamin K	Required in the synthesis of certain proteins needed for blood clotting.

energy. Triglycerides provide more than twice as much energy per gram as either carbohydrates or proteins. Our capacity to store triglycerides in fat (adipose) cells is unlimited, for all practical purposes. Excess carbohydrates, proteins, or fats in the diet all have the same fate; they are converted to triglycerides and stored in adipose (fat) tissue.

A triglyceride consists of two types of building blocks: glycerol and fatty acids. A three-carbon **glycerol** molecule forms the backbone of a triglyceride (Fig. 2.9). Three **fatty acids** are attached, by dehydration synthesis reactions, one to each carbon of the glycerol backbone. A reverse reaction, hydrolysis, breaks down a single molecule of a triglyceride into three fatty acids and glycerol.

In later chapters, we will refer to saturated, monounsaturated, and polyunsaturated fats. **Saturated** fats are triglycerides that contain only single covalent bonds between fatty acid carbon atoms. Each carbon bonds to the maximum number of hydrogen atoms; thus each fatty acid is *saturated with hydrogen atoms* (for example, palmitic acid and stearic acid in Fig. 2.9c). Triglycerides with many saturated fatty acids tend to be solid at room temperature and occur mostly in animal tissues. They also occur in a few plant products, such as cocoa butter, palm oil, and coconut oil. **Monounsaturated** fats contain fatty acids with *one* double covalent bond between two carbon atoms and thus are not completely saturated with hydrogen atoms (for example, oleic acid in Figure 2.9c). Olive oil and peanut oil are rich in triglycerides with monounsaturated fatty acids. **Polyunsaturated** fats contain *more than one* double covalent bond between fatty acid carbons. An example is linoleic acid. Corn oil, safflower oil, sunflower oil, cottonseed oil, sesame oil, and soybean oil contain a high percentage of polyunsaturated fatty acids.

FIGURE 2.9 Structure and reactions of glycerol and fatty acids. Each time a glycerol (a) and a fatty acid (b) are joined in dehydration synthesis, a molecule of water is lost. (c) Triglycerides (neutral fats) consist of one molecule of glycerol joined to three molecules of fatty acids, which vary in length and the number and location of double bonds between carbon atoms (C=C). Shown here is a molecule of a triglyceride that contains two saturated fatty acids and a monounsaturated fatty acid. The kink (bend) in the oleic acid occurs at the double bond.

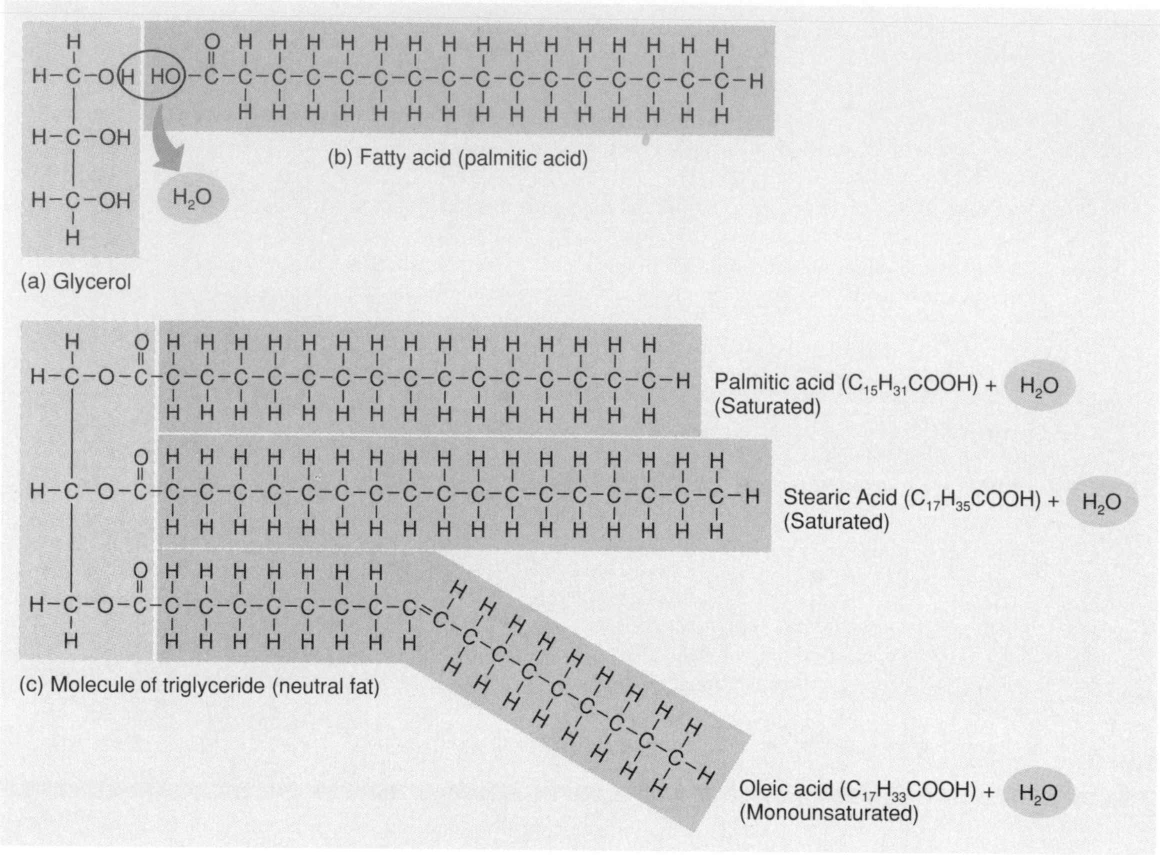

(b) Fatty acid (palmitic acid)

(a) Glycerol

Palmitic acid ($C_{15}H_{31}COOH$) + H_2O (Saturated)

Stearic Acid ($C_{17}H_{35}COOH$) + H_2O (Saturated)

(c) Molecule of triglyceride (neutral fat)

Oleic acid ($C_{17}H_{33}COOH$) + H_2O (Monounsaturated)

Question: During a dehydration synthesis reaction, a molecule of water is removed. Does the oxygen in the water come from the glycerol or from a fatty acid?

CLINICAL APPLICATION

SATURATED FATS AND ATHEROSCLEROSIS

Atherosclerosis is a disorder in which fatty plaques (deposits) form in the walls of arteries. This narrows the passage and severely limits blood flow in advanced cases. Although many factors contribute, people who eat a diet high in saturated fats run a greater risk of developing atherosclerosis than do people who eat a diet lower in saturated fats. Also, a diet high in saturated fats causes blood cholesterol (another lipid, described shortly) to rise, and blood cholesterol contributes to fatty plaque formation. To reduce dietary intake of saturated fats and cholesterol, eat less red meat (beef, pork, and lamb) and fat-laden dairy products (whole milk, butter, and cheese).

2. Phospholipids. Like triglycerides, phospholipids have a glycerol backbone. Two fatty acid chains are attached to the first two carbons. In the third position, a phosphate group (PO_4^{3-}) links a small charged group that usually contains nitrogen (N) to the backbone (Fig. 2.10). This portion of the molecule (the "head") is polar and can form hydrogen bonds with water molecules. The two fatty acids (the "tails"), on the other hand, are nonpolar and can interact only with other lipids. Molecules that have both polar and nonpolar portions are said to be **amphipathic** (am-fi-PATH-ic). Phospholipids, which have this unusual property, line up tails-to-tails in a double row to make up much of the membrane that surrounds each cell (Fig. 2.10c).

FIGURE 2.10 Phospholipid. (a) In the synthesis of phospholipids, two fatty acids attach to the first two carbons of the glycerol backbone. A phosphate group links a small charged group to the third carbon in glycerol. The result is an amphipathic molecule, one that has both polar and nonpolar portions. In (b), the circle represents the polar head region and the two wavy lines represent the two nonpolar tails. Double bonds in the fatty acid hydrocarbon chain often form a kink in the tail.

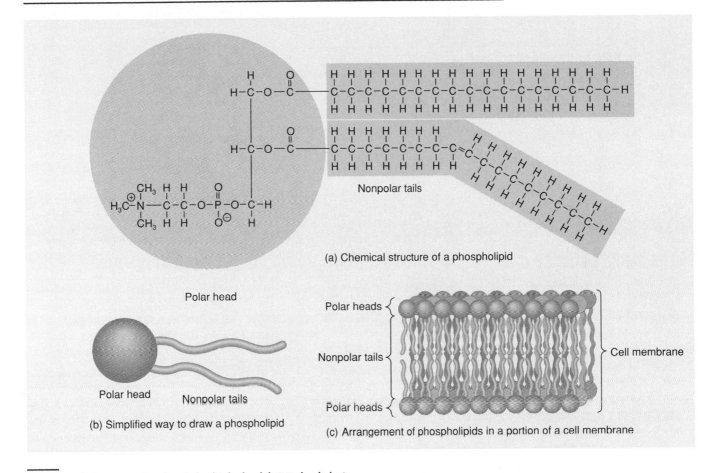

(a) Chemical structure of a phospholipid

(b) Simplified way to draw a phospholipid

(c) Arrangement of phospholipids in a portion of a cell membrane

Question: Which portion of a phospholipid is hydrophilic? Hydrophobic?

3. Steroids. Steroids such as cholesterol, sex hormones, cortisol, bile salts, and vitamin D have four rings of carbon atoms (Fig. 2.11). Their structure differs considerably from that of the triglycerides, but they also are nonpolar, fat-soluble molecules. Although cholesterol has been much maligned in the past several years, because of its role in promoting atherosclerosis, it also serves important functions—as a component of animal cell membranes and as starting material for synthesis of other steroids.

4. Eicosanoids (ī-KŌ-sa-noids). These lipids are derived from a 20-carbon fatty acid called arachidonic acid. The two principal subclasses of eicosanoids are called *prostaglandins* (pros′-ta-GLAN-dins) and *leukotrienes* (loo′-kō-TRĪ-ēns). Prostaglandins are potent substances that are involved in a wide variety of body functions. They modify responses to hormones, contribute to the inflammatory response (see page 695), prevent stomach ulcers, dilate (open) airways to the lungs, regulate body temperature, and influence formation of blood clots, to name just a few effects. Leukotrienes participate in allergic and inflammatory responses.

Proteins

Proteins, a third principal group of organic compounds, are much more complex in structure and have a larger range of functions than carbohydrates or lipids. A normal, lean adult body is about 12 to 18% protein. Some proteins have a structural role: they are cellular building materials. Other proteins have physiological roles. For example, proteins in

FIGURE 2.11 Steroids. All steroids contain four rings of carbon atoms, here labeled A, B, C, and D.

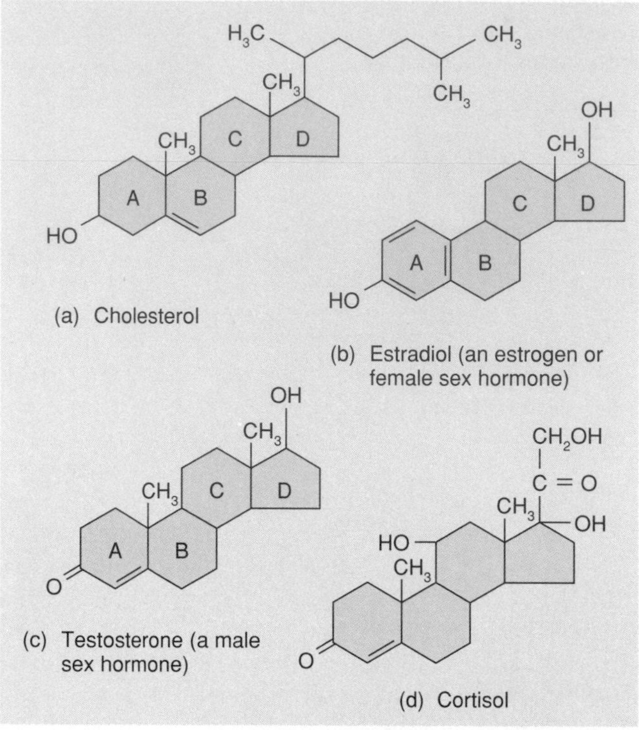

(a) Cholesterol

(b) Estradiol (an estrogen or female sex hormone)

(c) Testosterone (a male sex hormone)

(d) Cortisol

Question: How is the structure of estradiol different from that of testosterone?

the form of enzymes speed up most essential biochemical reactions. Other proteins provide the machinery for muscle contraction. Antibodies are proteins that defend against invading microbes. And some hormones that regulate homeostasis are also proteins. Exhibit 2.4 shows a classification of proteins based on their functions.

Amino Acids and Polypeptides

Chemically, proteins always contain carbon, hydrogen, oxygen, and nitrogen. Many proteins also contain sulfur. Just as monosaccharides are the building blocks of polysaccharides, **amino acids** are the building blocks of proteins. Each of the 20 different amino acids has three important groups attached to a central carbon atom (Fig. 2.12a): (1) an amino group ($-NH_2$), (2) a carboxyl (acid) group ($-COOH$), and (3) a side chain (R group). At the normal pH of body fluids, both the amino group and the carboxyl group are ionized (Fig. 2.12b). The distinctive side chain gives each amino acid its individual identity (Fig. 2.12c).

Synthesis of a protein takes place in stepwise fashion: one amino acid is joined to a second, a third is then added to the first two, and so on. The covalent bond between each pair of amino acids is called a **peptide bond.** It always

EXHIBIT 2.4

CLASSIFICATION OF PROTEINS BY FUNCTION

Type of Protein	Functions
Structural	Form the structural framework of various parts of the body. Examples: collagen in bone and other connective tissues (see page 149) and keratin in the skin, hair, and fingernails (see page 130).
Regulatory	Function as hormones, regulate various physiological processes, control growth and development, and mediate responses of the nervous system. Examples: the hormone insulin, which regulates blood sugar level (see page 545), and the neurotransmitter substance P, which allows the sensation of pain (see page 431).
Contractile	Allow shortening of muscle tissue, which produces movement. Examples: myosin and actin (see page 243).
Immunological	Aid responses that protect the body against foreign substances and invading microbes. Examples: antibodies and interleukins (see pages 702 and 705).
Transport	Carry vital substances throughout the body. Example: hemoglobin, which transports oxygen and carbon dioxide in the blood (see page 746).
Catalytic	Act as enzymes and function in regulating biochemical reactions. Examples: salivary amylase, lipase, and lactase (see page 803).

forms between the carboxyl group ($-COOH$) of one amino acid and the amino group ($-NH_2$) of another. During formation of a peptide bond, a molecule of water is removed (Fig. 2.13). Thus this is a dehydration synthesis reaction. When two amino acids combine, a **dipeptide** results. Adding another amino acid to a dipeptide produces a **tripeptide.** Further additions of amino acids result in the formation of a chainlike **polypeptide** that may contain from 10 to more than 2000 amino acids. A protein may have only one polypeptide chain or several. For example, there are

Figure 2.12 Amino acids. (a) In keeping with their name, amino acids have an amino group and a carboxyl (acid) group plus a side chain (R) that is different in each amino acid. (b) At pH close to 7, the amino group and the carboxyl group both are ionized. (c) Glycine is the simplest amino acid; the R side chain is merely an H atom. Cysteine is one of two amino acids that contain sulfur (S). The R side chain in tyrosine contains a six-carbon ring. Lysine has a second amino group at the end of the side chain.

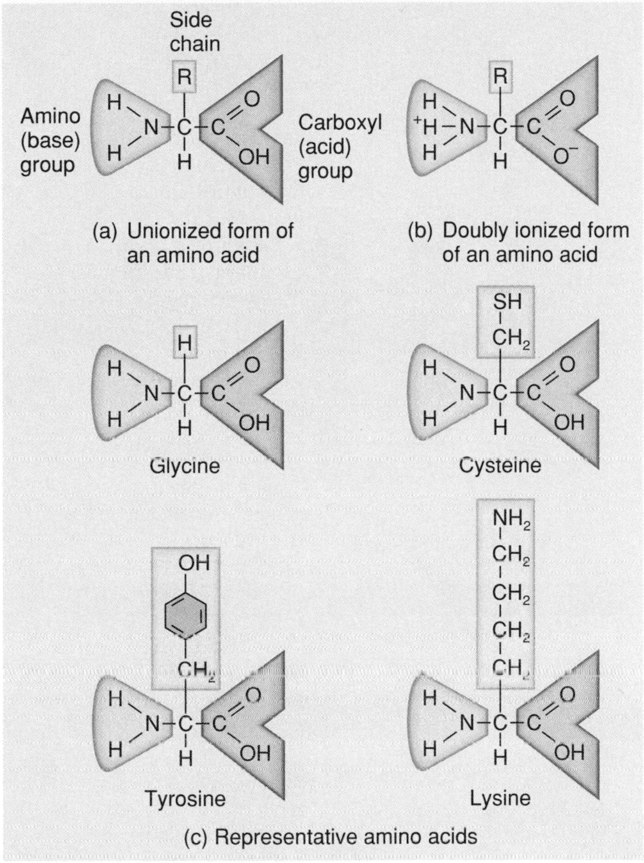

four polypeptide chains in hemoglobin—the molecule that transports oxygen in the blood (see Fig. 19.3b).

A great variety of proteins is possible because each variation in the number or sequence of amino acids can produce a different protein. The situation is similar to using an alphabet of 20 letters to form words. Each letter could be compared to a different amino acid, and each word would be a different protein.

Levels of Structural Organization

Proteins exhibit four levels of structural organization. The **primary structure** is the unique sequence of amino acids making up a polypeptide strand (Fig. 2.14a). It is genetically determined. A change in the primary structure can have serious consequences. In sickle cell anemia, for example, a nonpolar amino acid (valine) replaces a polar amino acid (glutamate) at just two locations in each hemoglobin molecule. This substitution diminishes hemoglobin's water solubility. As a result, the altered hemoglobin tends to form crystals inside red blood cells, producing deformed, sickle-shaped cells that cannot squeeze properly through narrow blood vessels.

The **secondary structure** of a protein is the repeated twisting or folding of neighboring amino acids in the polypeptide chain (Fig. 2.14b). Common secondary structures are clockwise spirals, called alpha (α) helixes, and pleated sheets. The secondary structure of a protein is due to hydrogen bonds found at regular intervals along the polypeptide backbone. The **tertiary** (TUR-shē-er′-ē) **structure** refers to the three-dimensional shape of a polypeptide chain. The tertiary folding pattern may allow amino acids at opposite ends of the chain to be close neighbors (Fig. 2.14c). The side chains (R groups) of the amino acids may bond to each other by hydrogen bonds, ionic bonds, covalent bonds between sulfur atoms called disulfide bridges, and hydrophobic interactions. The tertiary structure is

(c) Representative amino acids

Question: In an amino acid, what is the minimum number of carbon atoms? Nitrogen atoms?

Figure 2.13 Formation of a peptide bond. The bond that forms between two amino acids during dehydration synthesis is called a peptide bond. In this example, glycine is joined to alanine (forming a dipeptide), and the peptide bond forms at the point where a molecule of water is removed.

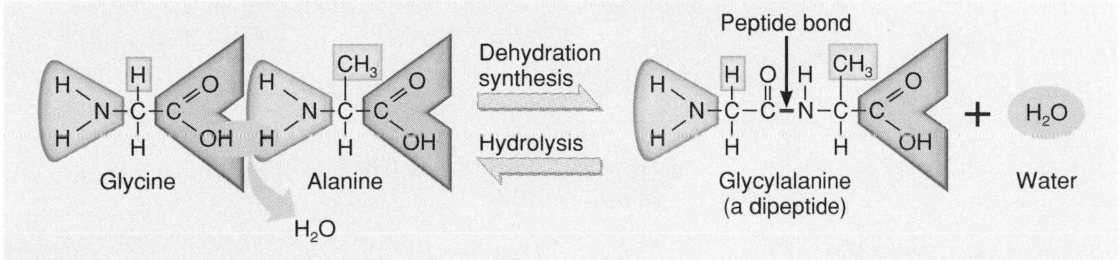

Question: What type of reaction is used in protein catabolism?

FIGURE 2.14 Levels of structural organization in proteins. (a) The primary structure is the sequence of amino acids in the polypeptide. (b) Common secondary structures include helixes, produced by hydrogen bonding between amino acids four positions apart, and pleated sheets, produced by hydrogen bonding between polypeptide chains. (c) The tertiary structure is the overall folding pattern that produces a distinctive, three-dimensional shape. (d) Quaternary structure in a protein is the arrangement of two or more polypeptide chains relative to how they bond to one another.

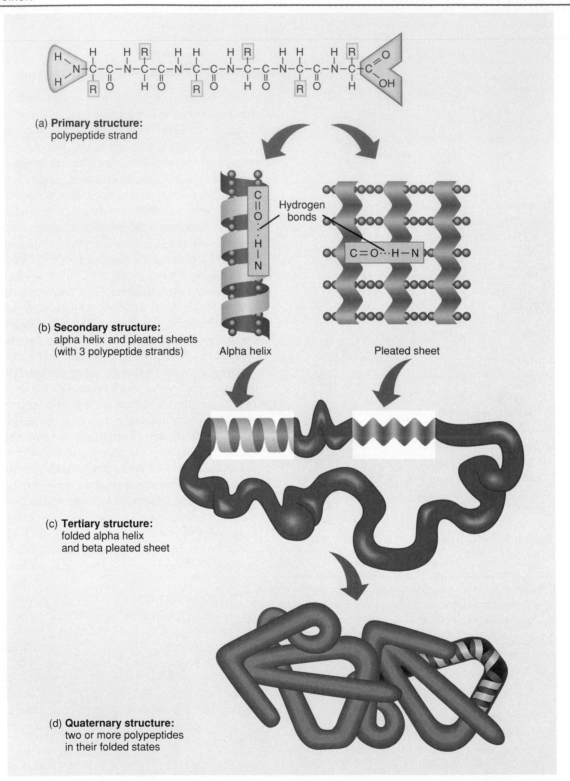

(a) **Primary structure:**
   polypeptide strand

(b) **Secondary structure:**
   alpha helix and pleated sheets
   (with 3 polypeptide strands)

Alpha helix   Pleated sheet

Hydrogen bonds

(c) **Tertiary structure:**
   folded alpha helix
   and beta pleated sheet

(d) **Quaternary structure:**
   two or more polypeptides
   in their folded states

Question: What types of bonds hold the side chains (R groups) of amino acids to one another to form the tertiary structure?

unique for each protein, and it determines how a particular protein will function.

When a protein contains more than one polypeptide chain, it also has a quaternary structure. The **quaternary** (KWA-ter-ner´-ē) **structure** describes the arrangement of the individual polypeptide chains and how they bond to each other. The bonds that hold a quaternary structure together are basically the same as those that maintain tertiary structure (Fig. 2.l4d).

Proteins vary tremendously in structure. Different proteins have different architectures and different three-dimensional shapes. This variation in structure and shape is directly related to their diverse functions. When a cell makes a protein, the polypeptide chain folds spontaneously to assume a certain shape. One reason for folding of the polypeptide is that some of its parts are attracted to water (hydrophilic), and other parts are repelled by it (hydrophobic). In practically every case, the function of a protein depends on its ability to recognize and bind to some other molecule, as a key fits in a lock. For example, an enzyme binds specifically with its substrate, that is, the molecule on which the enzyme acts (see Fig. 2.15). A hormone binds to a specific protein on a cell whose function it will alter. An antibody protein binds to a foreign substance (antigen) that has invaded the body. The unique shape of a protein permits it to interact with other specific molecules to carry out specific functions.

If a protein encounters a hostile environment in which temperature, pH, or electrolyte concentration is altered, it may unravel and lose its characteristic shape (secondary, tertiary, and quaternary structure). This process is called **denaturation.** Denatured proteins are no longer functional. A common example of denaturation is seen in frying an egg. In a raw egg the egg white protein (albumin) is soluble and appears as a clear, viscous fluid. When heat is applied to the egg, however, the protein changes shape, becomes insoluble, and looks white.

Enzymes

As we have seen, chemical reactions occur when chemical bonds are made or broken as atoms, ions, or molecules collide with one another. Normal body temperature and pressure are too low for chemical reactions to occur at a rate rapid enough to maintain life. Although raising the temperature, pressure, and the number of reacting particles can increase the frequency of collisions and also increase the rate of chemical reactions, such changes could denature proteins and damage or kill cells.

Enzymes are the living cell's solution to this problem. They speed up chemical reactions by increasing the frequency of collisions, lowering the activation energy, and properly orienting the colliding molecules. And they do this without increasing the temperature or pressure—in other words, without disrupting or killing the cell. Substances

that can speed up chemical reactions by increasing the frequency of collisions or by lowering the activation energy, *without themselves being altered,* are called **catalysts.** In living cells, **enzymes** function as catalysts. Most enzymes consist of a protein portion, called the **apoenzyme**, and a nonprotein portion called a **cofactor.** The cofactor may be a metal ion, such as iron, magnesium, zinc, or calcium, or an organic molecule called a **coenzyme.** Coenzymes often are vitamin derivatives. Together, the apoenzyme and cofactor form a **holoenzyme,** or whole enzyme. Although enzymes catalyze selected reactions, they do so with great efficiency and with many built-in controls.

1. Specificity. Enzymes are highly specific catalysts. Each particular enzyme affects only specific **substrates**—molecules on which the enzyme acts. A portion of the enzyme, called the **active site,** is thought to "fit" the substrate like a key fits in a lock (see Fig. 2.15). There is some evidence, however, that not all active sites are rigid receptors for enzymes. In some cases, the active site changes its shape to fit snugly around the substrate once the substrate enters the active site. This is known as an **induced fit.** Of the more than 1000 known enzymes, each has a characteristic three-dimensional shape with a specific surface configuration, which allows it to recognize and bond to certain substrates. When an enzyme is denatured, the active site loses its unique shape and can no longer fit together with its substrate.

Not only does an enzyme select a particular substrate, it also catalyzes a specific reaction. From the large number of diverse molecules in a cell, an enzyme must "find" the correct substrate and then take it apart or merge it with another substrate to form one or more specific products.

2. Efficiency. Under optimal conditions, enzymes can catalyze reactions at rates that are from 100 million (10^8) to 10 billion (10^{10}) times more rapid than those of comparable reactions occurring without enzymes. The **turnover number** (number of substrate molecules converted to product per enzyme molecule in 1 second) is generally between 1 and 10,000 and can be as high as 600,000.

3. Control. Enzymes are subject to a variety of cellular controls. Their rate of synthesis and their concentration at any given time are under the control of a cell's genes. Substances within the cell may either enhance or inhibit activity of a given enzyme. Many enzymes occur in both active and inactive forms in cells. The rate at which the inactive form becomes active or vice versa is determined by the environment inside the cell. Some enzymes require cofactors or coenzymes for effective function.

Exactly how enzymes lower the activation energy of a reaction is not completely understood. However, an enzyme is thought to work as shown in Fig. 2.15. (1) The surface of

FIGURE 2.15 How an enzyme works.

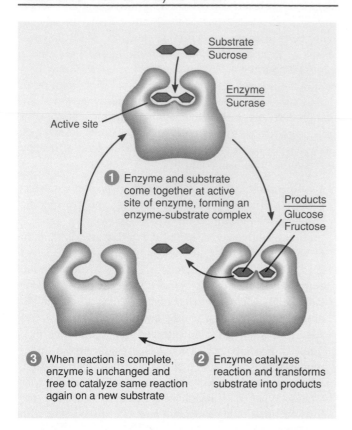

1 Enzyme and substrate come together at active site of enzyme, forming an enzyme-substrate complex

2 Enzyme catalyzes reaction and transforms substrate into products

3 When reaction is complete, enzyme is unchanged and free to catalyze same reaction again on a new substrate

Question: What type of reaction is illustrated here? One product of this reaction is not illustrated. What is it?

the substrate makes contact with the active site on the surface of the enzyme molecule, forming a temporary intermediate compound called the **enzyme–substrate complex.** (2) The substrate molecule is transformed by rearrangement of existing atoms, breakdown of the substrate molecule, or combination of several substrate molecules. The transformed substrate molecules are called the **products** of the reaction. (3) After the reaction is completed and the products of the reaction move away from the enzyme, the unchanged enzyme is free to attach to another substrate molecule.

The names of enzymes usually end in the suffix **-ase.** All enzymes can be grouped according to the types of chemical reactions they catalyze. For example, *oxidases* add oxygen, *kinases* add phosphate, *dehydrogenases* remove hydrogen, *hydrolases* add water, *isomerases* rearrange atoms within a molecule, *transferases* transfer groups of atoms, *proteases* break down proteins, and *lipases* break down lipids.

GALACTOSEMIA

Galactosemia (ga-lak-tō-SĒ-mē-a) is an inherited disorder in which galactose cannot be converted to glucose because the needed enzyme is missing. An infant with the disorder will fail to thrive within a week after birth due to anorexia (loss of appetite), vomiting, and diarrhea unless galactose and lactose are removed from the diet. The disaccharide lactose (milk sugar) is broken down into galactose and glucose. If treatment is delayed, the baby remains physically small and becomes mentally retarded. Treatment consists of eliminating galactose and lactose from the diet of the patient.

Nucleic Acids: Deoxyribonucleic Acid and Ribonucleic Acid

Nucleic (noo-KLĒ-ic) **acids,** compounds named because they were first discovered in the nuclei of cells, are huge organic molecules that contain carbon, hydrogen, oxygen, nitrogen, and phosphorus. Nucleic acids are of two varieties. **Deoxyribonucleic** (dē-ok′-sē-rī-bō-noo-KLĒ-ik) **acid (DNA)** forms the genetic code inside each cell. Each **gene** is a segment of a DNA molecule. Our genes determine which traits we inherit, and by controlling protein synthesis, they regulate most of the activities that take place in our cells throughout a lifetime. When a cell divides, its hereditary information passes on to the next generation of cells. **Ribonucleic acid (RNA)**, the second type of nucleic acid, relays instructions from the genes to guide each cell's assembly of amino acids into proteins.

The basic units of nucleic acids are nucleotides. A molecule of DNA is a chain composed of repeating nucleotide units. Each nucleotide consists of three building blocks (Fig. 2.16a):

1.   A base (nitrogenous base). Four nitrogen-containing structures containing atoms of C, H, O, and N called **bases** or **nitrogenous bases** are present in DNA and RNA. In DNA they are adenine (A), thymine (T), cytosine (C), and guanine (G). The nucleotides are named according to the base that is present. Thus a nucleotide containing thymine is called a *thymine nucleotide.* One containing adenine is called an *adenine nucleotide,* and so on.

2.   A pentose sugar. A five-carbon sugar called **deoxyribose** attaches to each base in DNA.

3.   A phosphate group. Alternating phosphate groups (PO_4^{3-}) and pentoses form the backbone of a DNA strand, while bases protrude from the backbone chain (Fig. 2.16b).

In 1953 F. H. C. Crick of Great Britain and J. D. Watson, a young American scientist, published a brief paper describing how these three components might be arranged in DNA. Their insights into data gathered by others led them to construct a model so elegant and simple that the scientific world immediately knew it was correct. In the Watson–Crick **double helix model**, DNA resembles a twisted rope ladder (Fig. 2.16b). Two strands of alternating phosphate groups and deoxyriboses form the uprights

FIGURE 2.16 DNA molecule. (a) A nucleotide consists of a base, a pentose sugar, and a phosphate group. (b) The paired bases project toward the center of the double helix. The structure is stabilized by hydrogen bonds between each base pair. There are two hydrogen bonds between adenine and thymine and three between cytosine and guanine.

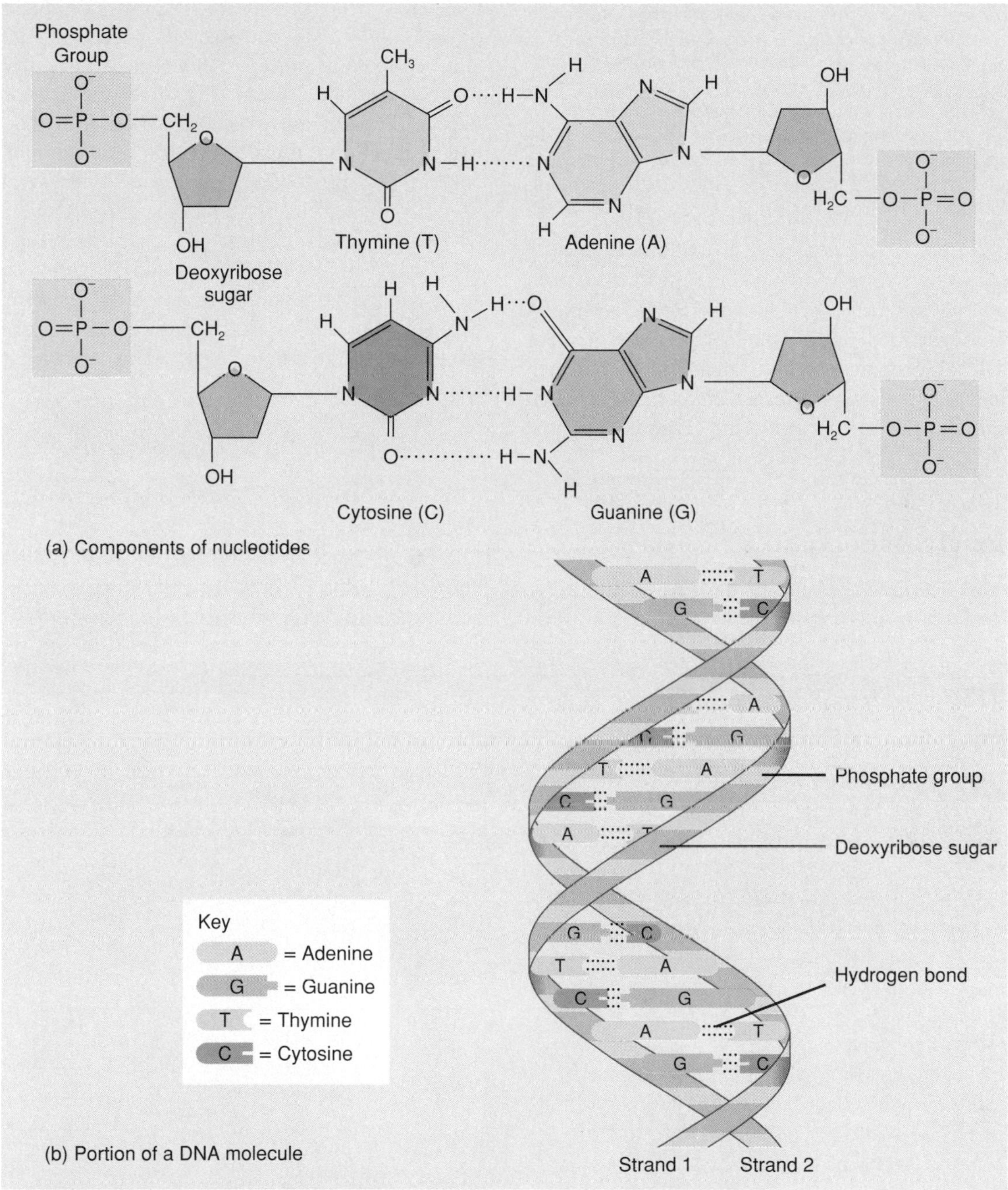

Phosphate Group

Deoxyribose sugar

Thymine (T) Adenine (A)

Cytosine (C) Guanine (G)

(a) Components of nucleotides

Key

A = Adenine
G = Guanine
T = Thymine
C = Cytosine

Phosphate group

Deoxyribose sugar

Hydrogen bond

Strand 1 Strand 2

(b) Portion of a DNA molecule

Question: Which bases always pair with one another?

of the ladder. Paired bases, held together by hydrogen bonds, form the rungs. Adenine always pairs with thymine, and cytosine always pairs with guanine. If you know the sequence of bases in one strand of DNA, you can predict the sequence on the complementary strand. Each time DNA is copied, as an organism adds cells, the two strands unwind. Each strand serves as the template or mold on which to construct a new second strand (see Fig. 3.23). Most DNA assumes the shape of a double helix twisted to the right.

DNA FINGERPRINTING

A technique called **DNA fingerprinting** is now used in research and a variety of legal situations to help identify individuals. In each person, certain DNA segments contain base sequences that are repeated several times. Both the number of repeat copies in one region and the number of regions subject to repeat are different from one person to another. Except for identical twins, the chance that two people would have the same DNA fingerprint is about 1 in 10 billion, about twice the population of the world. DNA fingerprinting can be done with minute quantities of DNA, for example, from a single strand of hair, a drop of semen, or a spot of blood. DNA fingerprinting is also used to determine paternity.

RNA, the second variety of nucleic acid, differs from DNA in several respects. In humans, RNA is single-stranded, the sugar in the RNA nucleotide is the pentose **ribose,** and RNA contains the base uracil (U) rather than thymine. Cells contain three different kinds of RNA; each has a specific role to perform in carrying out the instructions coded in DNA (see page 78).

Adenosine Triphosphate

A molecule that is indispensable to the life of the cell is **adenosine** (a-DEN-ō-sēn) **triphosphate (ATP).** This sub-stance is found universally in living systems and has the essential function of providing energy for various cellular activities. Among the cellular activities are muscular contractions, movement of chromosomes during cell division, movement of cytoplasm within cells, transporting substances across cell membranes, and putting together larger molecules from smaller ones during synthetic reactions. Structurally, ATP consists of three phosphate groups attached to an adenosine unit composed of adenine and the five-carbon sugar ribose (Fig. 2.17). ATP is the "energy currency" of living systems. When a reaction requires energy, ATP can transfer just the right amount because it contains two, high-energy phosphate bonds, symbolized ~.

When the terminal phosphate group (PO_4^{3-}), symbolized by Ⓟ in the following discussion, is hydrolyzed by addition of a water molecule, the reaction liberates energy. This energy is used by the cell to power its activities. Removal of the terminal phosphate group leaves a molecule called **adenosine diphosphate (ADP).** This reaction may be represented as follows:

$$\text{ATP} \longrightarrow \text{ADP} + Ⓟ + E$$

Adenosine   Adenosine   Phosphate   Energy
triphosphate   diphosphate group

The energy supplied by the catabolism of ATP into ADP is constantly being used by the cell. Since the supply of ATP at any given time is limited, a mechanism exists to replen-

FIGURE 2.17 Structure of ATP and ADP. The two high-energy bonds are indicated by "squiggles" (~).

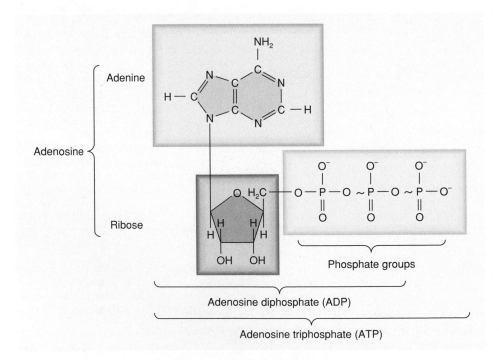

Question: What are some cellular activities that depend on energy supplied by ATP?

ish it: a phosphate group is added to ADP to manufacture more ATP. The reaction may be represented as follows:

$$\text{ADP} \; + \; \textcircled{P} \; + \; \text{E} \longrightarrow \text{ATP}$$

| Adenosine | Phosphate | Energy | Adenosine |
| diphosphate | group | | triphosphate |

Of course, energy is required to make ATP. The energy needed to attach a phosphate group to ADP is supplied mainly by the breakdown of glucose in a process called cellular respiration. Cellular respiration has two phases:

1. **Anaerobic.** In the absence of oxygen, glucose is partially broken down by a process called glycolysis into pyruvic acid. Each glucose that is converted into a pyruvic acid molecule yields two molecules of ATP.
2. **Aerobic.** In the presence of oxygen, glucose is completely broken down into carbon dioxide and water. These reactions generate heat and a large number of ATP molecules—36 to 38 ATP from each glucose.

The details of cellular respiration will be discussed in Chapters 10 and 25.

Study Outline

Matter and Energy (p. 29)

1. Matter is anything that occupies space and has mass.
2. Energy is the capacity to do work and is of two principal kinds: potential (stored) and kinetic (energy of motion).
3. Forms of energy include chemical, radiant, and electrical.

Chemical Elements (p. 29)

1. All forms of matter are composed of chemical elements.
2. Oxygen, carbon, hydrogen, and nitrogen make up 96% of body weight. These elements plus calcium and phosphorus make up about 98.5% of total body weight.

Structure of Atoms (p. 29)

1. Units of matter of all chemical elements are called atoms.
2. Atoms consist of a nucleus, which contains protons and neutrons, and electrons that move about the nucleus in different energy shells.
3. The atomic mass (atomic weight) of an atom is the sum of the masses of its protons, neutrons, and electrons.
4. Atomic number is the number of protons in an atom.
5. Mass number is the total number of protons and neutrons.
6. Different atoms of an element that have the same number of protons but different numbers of neutrons are called isotopes.

Electrons and Chemical Reactions (p. 32)

1. When atoms combine or break apart from other atoms, a chemical reaction occurs.
2. Atoms always attempt to empty their outermost shell or fill it to the maximum to achieve stability.
3. When two or more atoms combine in a chemical reaction, a molecule is formed.
4. A compound is a substance that can be broken down into two or more different elements by chemical means.

Chemical Bonds (p. 33)

1. The atoms of a molecule are held together by forces of attraction called chemical bonds.
2. In an ionic bond, valence electrons are transferred from one atom to another. The transfer forms ions, whose unlike charges attract each other and form ionic bonds.
3. In a covalent bond, there is a sharing of pairs of valence electrons. Covalent bonds may be nonpolar or polar.
4. In a hydrogen bond, two other atoms (usually oxygen or nitrogen) associate with a hydrogen atom.

Chemical Reactions (p. 36)

1. Several factors determine if a chemical reaction will occur: concentration of particles, speed of particles, activation energy, and proper orientation of the particles.
2. Synthesis reactions involve the combination of reactants to produce a new molecule. The reactions are anabolic: bonds are formed.
3. In decomposition reactions, a substance breaks down into other substances. The reactions are catabolic: bonds are broken.
4. Exchange reactions involve the replacement of one atom or atoms by another atom or atoms.
5. In reversible reactions, end products can revert to the original combining molecules.

Chemical Compounds and Life Processes (p. 37)

1. Inorganic substances usually lack carbon and are small molecules.
2. Organic substances always contain carbon and hydrogen. Most organic substances contain covalent bonds.

Inorganic Compounds (p. 37)

1. Water is the most abundant substance in the body. It is an excellent solvent, lubricating and suspending medium, participates in chemical reactions, and absorbs and releases heat slowly.
2. Inorganic acids, bases, and salts dissociate into ions in water. An acid ionizes into hydrogen ions (H^+); a base ionizes into hydroxide ions (OH^-). A salt ionizes into neither H^+ nor OH^- ions. Cations are positively charged ions; anions are negatively charged ions.
3. The pH of body fluids must remain fairly constant for the body to maintain homeostasis. On the pH scale, 7 represents neutrality. Values below 7 indicate acid solutions, and values above 7 indicate basic solutions.
4. The pH values of different parts of the body are maintained by buffer systems, which usually consist of a weak acid and a weak base. Buffer systems eliminate excess H^+ and excess OH^- ions in order to maintain pH homeostasis.

Organic Compounds (p. 40)

1. Carbohydrates are sugars or starches that provide most of the energy needed for life. They may be monosaccharides, disaccharides, or polysaccharides. Carbohydrates, and other

organic molecules, are joined together to form larger molecules with the loss of water by a process called dehydration synthesis. In the reverse process, called hydrolysis, large molecules are broken down into smaller ones upon the addition of water.

2. Lipids are a diverse group of compounds that include triglycerides (neutral fats), phospholipids, steroids, carotenes, vitamins A, D, E, and K, and eicosanoids. Triglycerides protect, insulate, provide energy, and are stored. Phospholipids are important membrane components. Eicosanoids (prostaglandins and leukotrienes) modify hormone responses, contribute to inflammation, dilate airways, and regulate body temperature.

3. Proteins are constructed from amino acids. They give structure to the body, regulate processes, provide protection, help muscles to contract, transport substances, and serve as enzymes. Structural levels of organization among proteins include primary, secondary, tertiary, and quaternary. Enzymes are catalysts that are highly specific in terms of substrates with which they react, efficient in terms of the number of substrate molecules with which they react, and subject to a variety of cellular controls.

4. Deoxyribonucleic acid (DNA) and ribonucleic acid (RNA) are nucleic acids consisting of nitrogenous bases, sugar, and phosphate groups. DNA is a double helix and is the primary chemical in genes. RNA differs in structure and chemical composition from DNA and is mainly concerned with protein synthesis reactions.

5. The principal energy-storing molecule in the body is adenosine triphosphate (ATP). When its energy is liberated, it is decomposed to adenosine diphosphate (ADP) and Ⓟ. ATP is manufactured from ADP and Ⓟ using the energy supplied by various decomposition reactions, particularly of glucose.

Review Questions

1. What is the relationship of matter and energy to the body? Distinguish the various forms of energy. (p. 29)
2. Define a chemical element. List the chemical symbols for 10 different chemical elements. Which chemical elements make up the bulk of the human organism? (p. 29)
3. What is an atom? Diagram the positions of the nucleus, protons, neutrons, and electrons in an atom of oxygen and in an atom of nitrogen. (p. 29)
4. What is an atomic number? Compare it to an atomic mass and mass number. (p. 31)
5. What is an isotope? A radioisotope? Describe several medical uses of radioisotopes. (p. 31)
6. What is an electron shell? How are electron shells related to chemical reactions? (p. 32)
7. How are chemical bonds formed? Distinguish between an ionic bond and a covalent bond. Give at least one example of each. (p. 33)
8. Can you determine how a molecule of $MgCl_2$ is ionically bonded? Magnesium has two electrons in its outer energy level. Construct a diagram to verify your answer. (p. 33)
9. Refer to Fig. 2.4b, c. See if you can determine why there is a double covalent bond between atoms in an oxygen molecule (O_2) and a triple covalent bond between atoms in a nitrogen molecule (N_2). (p. 35)
10. Define a hydrogen bond. Why are hydrogen bonds important? (p. 34)
11. What factors determine if a collision will occur to cause a chemical reaction? (p. 36)
12. What are the four principal kinds of chemical reactions? How are anabolism and catabolism related to synthesis and decomposition reactions, respectively? (p. 36)
13. Identify what kind of reaction each of the following represents:
   a. $H_2 + Cl_2 \rightarrow 2HCl$
   b. $3NaOH + H_3PO_4 \rightarrow Na_3PO_4 + 3H_2O$
   c. $CaCO_3 + CO_2 + H_2O \rightarrow Ca(HCO_3)_2$
   d. $HNO_3 \rightarrow H^+ + NO_3^-$
   e. $NH_3 + H_2O \rightleftharpoons NH_4^+ + OH^-$
14. How do inorganic compounds differ from organic compounds? List and define the principal inorganic and organic compounds that are important to the human body. (p. 37)
15. What are the essential functions of water in the body? Explain the solvating property of water. (p. 37)
16. Define an inorganic acid, a base, and a salt. How does the body acquire some of these substances? List some functions of the chemical elements furnished as ions of salts. (p. 39)
17. What is pH? Why is it important to maintain a relatively constant pH? What is the pH scale? (p. 39)
18. List the normal pH values of some common fluids, biological solutions, and foods. Refer to Exhibit 2.2 and select the two substances whose pH values are closest to neutrality. Is the pH of milk or of cerebrospinal fluid closer to 7? Is the pH of bile or of urine further from neutrality? (p. 40)
19. What are the components of a buffer system? What is the function of a buffer system? How is buffering an example of homeostasis? (p. 40)
20. Define a carbohydrate. Why are carbohydrates essential to the body? How are carbohydrates classified? (p. 41)
21. Compare dehydration synthesis and hydrolysis. Why are they significant? (p. 41)
22. How do lipids differ from carbohydrates? Explain the importance of the following lipids to the body: triglycerides, phospholipids, steroids, lipoproteins, and eicosanoids. (p. 42)
23. Distinguish among saturated, monounsaturated, and polyunsaturated fats. How do saturated fats relate to atherosclerosis? (p. 43)
24. Define a protein. What is a peptide bond? Discuss the classification of proteins on the basis of function and levels of structural organization. (p. 45)
25. What is an enzyme? What are some characteristics of enzymes? (p. 49)
26. What is a nucleic acid? How do deoxyribonucleic acid (DNA) and ribonucleic acid (RNA) differ with regard to chemical composition, structure, and function? (p. 50)
27. What is adenosine triphosphate (ATP)? What is the essential function of ATP in the human body? How is this function accomplished? (p. 52)
28. What is a laser? Describe several clinical applications for lasers. (p. 33)
29. What is galactosemia? (p. 50)
30. What is DNA fingerprinting? What is its value? (p. 52)

Answers to Questions with Figures

2.1 Six.

2.2 Helium (He), because it has a completely filled valence (outermost) shell.

2.3 K is an electron donor; when it ionizes, it becomes a cation, K^+.

2.4 A bond in which atoms share one, two, or three pairs of electrons.

2.5 Since it easily dissolves in a polar solvent (water), table sugar must have several polar covalent bonds.

2.6 $CaCO_3$ is a salt, and H_2SO_4 is an acid.

2.7 $[H^+] = 10^{-6}$ moles/liter; $[OH^-] = 10^{-8}$ moles/liter. A pH of 6.82 is more acidic than a pH of 6.91. Both pH = 8.41 and pH = 5.59 are 1.41 pH units from neutral (pH = 7).

2.8 Anabolic; there are 6 carbons in fructose, 12 in sucrose.

2.9 The fatty acid.

2.10 The polar head region is hydrophilic, and the nonpolar tail region is hydrophobic.

2.11 The only difference is the number of double bonds and the type of side group attached to ring A.

2.12 Two carbons, one nitrogen (see structure of glycine in Fig. 2.12c).

2.13 Hydrolysis.

2.14 Hydrogen bonds, ionic bonds, covalent bonds (disulfide bridges), and hydrophobic interactions.

2.15 Hydrolysis. Water.

2.16 Thymine always pairs with adenine and cytosine always pairs with guanine.

2.17 Muscular contractions, movement of chromosomes, transportation of substances across cell membranes, and synthesis reactions.

Chapter 3

THE CELLULAR LEVEL OF ORGANIZATION

Chapter Contents at a Glance

Student Objectives

1. Define a cell and list its principal parts.
2. Explain the structure and functions of the plasma membrane.
3. Describe the various passive and active processes by which materials move across plasma membranes.
4. Describe the structure and functions of the following cellular structures: cytosol, nucleus, ribosomes, endoplasmic reticulum, Golgi complex, lysosomes, peroxisomes, mitochondria, cytoskeleton, flagella, cilia, and centrosome.
5. Define a cell inclusion and give several examples.
6. Define a gene and explain the sequence of events involved in protein synthesis.
7. Discuss the stages, events, and significance of somatic and reproductive cell division.
8. Describe cancer (CA) as a homeostatic imbalance of cells.
9. Explain the relationship of aging to cells.
10. Define medical terminology associated with cells.

I t is at the cellular level of organization that activities essential to life occur and disease processes originate. A **cell** is the basic, living, structural, and functional unit of the body. **Cytology** (sī-TOL-ō-jē; *cyt* = cell; *logos* = study of) is the branch of science concerned with the study of cells. This chapter concentrates on the structure, functions, and reproduction of cells.

GENERALIZED ANIMAL CELL

The **generalized animal cell** illustrated in Fig. 3.1 is a composite of many different cells in the body. While most cells have many of the features shown in this diagram, few cells have all the illustrated features. For ease of study, we can divide a cell into four principal parts:

1. Plasma (cell) membrane. Outer, limiting membrane separating the cell's internal components from the extracellular materials and external environment. Extracellular materials, which are substances external to the cell surface, will be examined together with tissues in Chapter 4.

2. Cytosol. The term **cytoplasm** refers to all cellular contents located between the plasma membrane and the nucleus. The thick semifluid portion of the cytoplasm is the **cytosol** (SĪ-tō-sol), which is intracellular fluid. The cytosol contains many soluble proteins and enzymes, nutrients, ions, and other small molecules, which all participate in various phases of metabolism. Organelles and inclusions are suspended in the cytosol.

3. Organelles. Highly organized structures with characteristic shapes that are highly specialized for specific cellular activities.

4. Inclusions. Temporary structures that contain secretions and storage products of the cell.

PLASMA (CELL) MEMBRANE

The thin barrier that separates the internal components of a cell from the extracellular materials and external environment is the **plasma membrane**, **cell membrane**, or **plasmalemma** (Fig. 3.2). The plasma membrane is the gatekeeper that regulates passage of substances into and out of the cell.

The **fluid mosaic model** of membrane structure describes the molecular arrangement of the plasma membrane and other membranes in living organisms. A mosaic is a pattern of many small pieces fitted together. According to this model, the membrane is a mosaic of proteins floating like icebergs in a sea of lipids.

Membrane Chemistry and Anatomy

Plasma membranes of typical animal cells are about a 50 : 50 mix by weight of proteins and lipids that are held together by noncovalent interactions. Since proteins are larger and more massive than lipids, however, there are about 50 lipid molecules for each protein molecule.

Membrane Lipids

About 75% of the lipids are **phospholipids**, lipids that contain phosphorus. Present in smaller amounts are cholesterol (a steroid) and glycolipids, which are lipids with one or more sugar groups attached. The phospholipids line up in two parallel rows, forming a **phospholipid (lipid) bilayer** (Fig. 3.2). This arrangement occurs because the phospholipids are amphipathic; that is, they have both polar and nonpolar regions (see Fig. 2.12). The polar part is the phosphate-containing "head," which is hydrophilic (mixes with water). The nonpolar parts are the two fatty acid "tails," which are hydrophobic (do not mix with water). The molecules orient in the bilayer so that the heads face outward on either side, toward the watery cytosol and extracellular fluid (fluid outside cells). The tails face each other in the membrane's interior.

Glycolipids, about 5% of membrane lipids, are also amphipathic. They appear only in the layer that faces the extracellular fluid. Although membrane glycolipids are the target of certain bacterial toxins (poisons), their normal functions are still being unraveled. They are important for adhesion among cells and tissues, may mediate cell-to-cell recognition and communication, and contribute to regulation of cellular growth and development.

The remaining 20% of membrane lipids are **cholesterol** molecules, which are located among the phospholipids in both sides of the bilayer in animal cells. The stiff steroid rings of cholesterol strengthen the membrane but decrease its flexibility. Plant cell membranes lack cholesterol.

The phospholipid bilayer is dynamic because the lipids can move sideways and exchange places in their own layer. Flip-flop of molecules between layers, however, rarely occurs. The bilayer is also self-healing; if a needle is pushed through it and pulled out, the puncture site seals.

Membrane Proteins

Membrane proteins are of two types: integral and peripheral (Fig. 3.2). **Integral proteins** extend across the phospholipid bilayer among the fatty acid tails. Most (maybe all) integral proteins are **glycoproteins,** combinations of sugar and protein. The sugar portion of a glycoprotein faces the extracellular fluid. **Peripheral proteins** do not extend across the phospholipid bilayer. They are loosely attached to the inner and outer surfaces of the membrane and are easily separated from it.

To a large extent, membrane proteins and glycoproteins determine what functions a cell can perform. In general, the types of lipids present vary only slightly from one membrane to another. On the other hand, the variety of proteins that can be included in a given membrane is enormous, and

FIGURE 3.1 Generalized animal cell based on electron microscopic studies.

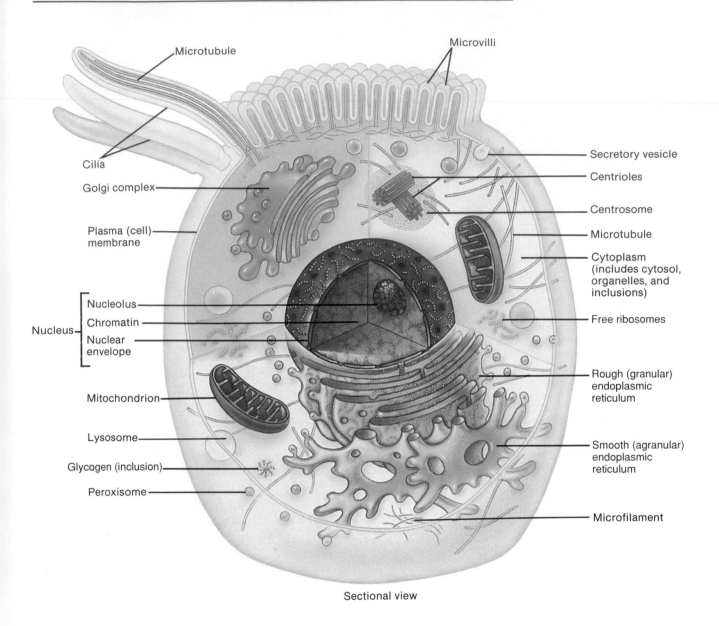

Sectional view

Question: List the four principal parts of a cell.

---

the membrane proteins have many roles (Fig. 3.3). Some integral proteins (or glycoproteins) form tiny **channels (pores)** through which certain substances flow into or out of the cell. Others act as **transporters** (carriers) to move a substance from one side of the membrane to the other (described shortly).

Integral proteins also serve as recognition sites called **receptors.** These are molecules that can identify and attach to a specific molecule, such as a hormone, a neurotransmitter, or a nutrient, that is important for some cellular function. A molecule that specifically binds to a receptor by forces other than covalent bonds

is called a **ligand** (LĪ-gand) of that receptor.

Some integral and peripheral proteins are **enzymes.** Other peripheral proteins facing the cytosol serve as **cytoskeleton anchors,** forming an attachment between the plasma membrane and filaments of the cytoskeleton. Membrane glycoproteins and glycolipids often are **cell identity markers.** They may enable a cell to recognize other cells of its own kind during tissue formation or to recognize and respond to potentially dangerous foreign cells. The ABO blood type markers are one example of cell identity markers. When you receive a transfusion, the blood must be of the same type as your own.

FIGURE 3.2 Plasma membrane. Diagram of the fluid mosaic arrangement of molecules in the plasma membrane.

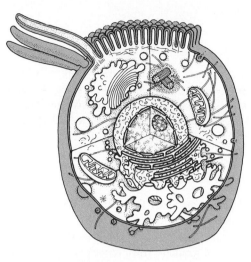

Generalized cell showing location of plasma membrane

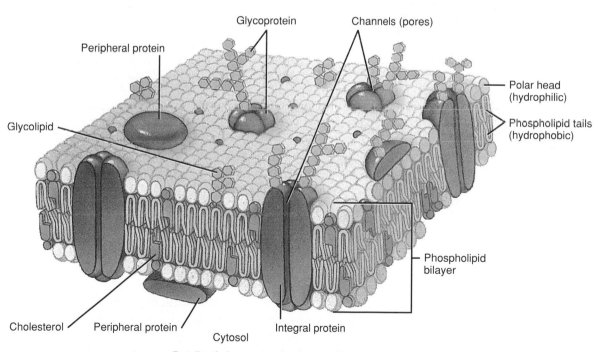

Details of plasma membrane structure

Question: What types of lipids are present in plasma membranes of animal cells?

FIGURE 3.3 Functions of membrane proteins.

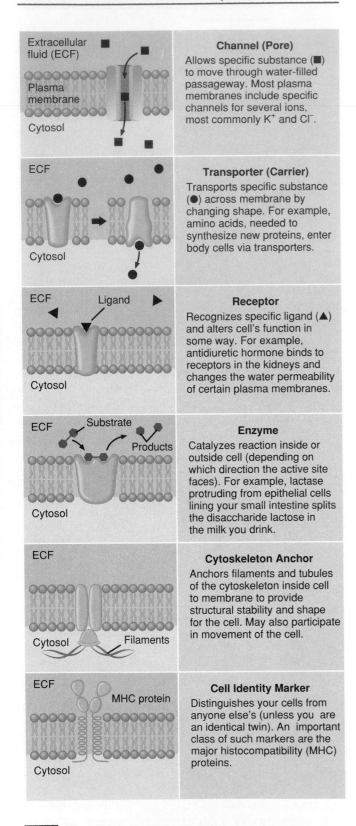

Extracellular fluid (ECF) / Plasma membrane / Cytosol	**Channel (Pore)** Allows specific substance (■) to move through water-filled passageway. Most plasma membranes include specific channels for several ions, most commonly K^+ and Cl^-.
ECF / Cytosol	**Transporter (Carrier)** Transports specific substance (●) across membrane by changing shape. For example, amino acids, needed to synthesize new proteins, enter body cells via transporters.
ECF / Ligand / Cytosol	**Receptor** Recognizes specific ligand (▲) and alters cell's function in some way. For example, antidiuretic hormone binds to receptors in the kidneys and changes the water permeability of certain plasma membranes.
ECF / Substrate / Products / Cytosol	**Enzyme** Catalyzes reaction inside or outside cell (depending on which direction the active site faces). For example, lactase protruding from epithelial cells lining your small intestine splits the disaccharide lactose in the milk you drink.
ECF / Cytosol / Filaments	**Cytoskeleton Anchor** Anchors filaments and tubules of the cytoskeleton inside cell to membrane to provide structural stability and shape for the cell. May also participate in movement of the cell.
ECF / MHC protein / Cytosol	**Cell Identity Marker** Distinguishes your cells from anyone else's (unless you are an identical twin). An important class of such markers are the major histocompatibility (MHC) proteins.

Question: When a cell is stimulated by the hormone insulin, the insulin first binds to a protein in the plasma membrane. Which of the illustrations best represents this membrane protein?

Membrane Physiology

Knowing the structure of the plasma membrane, we can now describe its several important physiological properties.

Communication

The plasma membrane functions in cellular communication. This includes interactions with other body cells, foreign cells, and ligands such as hormones, neurotransmitters, enzymes, nutrients, and antibodies in the extracellular fluid.

Electrochemical Gradient

The plasma membrane encloses the cellular contents and separates them from the extracellular fluid or, sometimes, from the external environment. The membrane maintains an electrical and chemical gradient (difference), simply called an **electrochemical gradient**, between the inside and outside of the cell. The chemical portion of the electrochemical gradient arises because the membrane maintains very different chemical compositions in the cytosol and the extracellular fluid. Although chemicals other than ions are present in different concentrations inside and outside living cells, for now we note only the differences in ion concentrations (Fig. 3.4). In extracellular fluid, the main cation (positively charged ion) is Na^+ and the main anion (negatively charged ion) is Cl^-. In cytosol, the main cation is K^+, while the two dominant anions are organic phosphates (PO_4^{3-} groups attached to organic molecules such as ATP) and negatively charged amino acids in proteins.

The electrical gradient arises because the inside surface of the membrane is more negatively charged than the outside surface in most cells. As a result, there is a voltage called the **membrane potential** across the membrane. *Voltage* is a form of potential energy that occurs when positive and negative charges are separated; an example of such charge separation is found in an ordinary flashlight or car battery. In living cells, a small separation of ions (charged particles) by the membrane gives rise to the membrane potential. This separation of charges occurs just near the plasma membrane. Overall, both the cytosol and extracellular fluid are electrically neutral; if you added up all the positive and negative charges in each fluid, they would balance. The voltage across the plasma membrane of cells throughout your body usually is between −20 and −200 millivolts (mV). (A mV is equal to one-thousandth of a volt.) The negative sign in front of the number means that the inside is negative relative to the outside. Under certain conditions, some cells exhibit a positive membrane potential. The electrochemical gradient and the resulting membrane potential are important for the proper functioning of most cells. How the membrane potential arises is discussed on page 356.

FIGURE 3.4 Electrochemical gradient. The inner surface of the plasma membrane of most cells is negative relative to the outer surface.

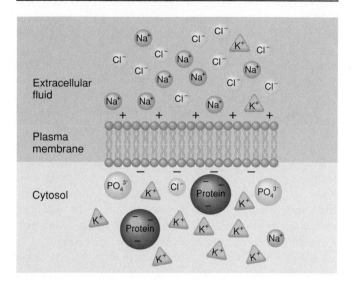

Question: Which ions congregate closest to the membrane on the cytosolic (inner) side?

Selective Permeability

The plasma membrane regulates entry and exit of materials. It permits passage of certain substances and restricts the passage of others. This property of membranes is called **selective permeability.**

A membrane is said to be *permeable* to a substance if it allows the free passage of that substance. Although plasma membranes are not completely permeable to any substance, they do permit some substances to pass more readily than others. Water, for example, usually passes more easily than most other substances. The permeability of a plasma membrane to different substances depends on several factors that relate to the structure of the membrane.

1. **Lipid solubility.** Substances that dissolve in lipids (nonpolar, hydrophobic molecules) pass easily through the membrane. This is because the phospholipid bilayer is a major part of the plasma membrane.
2. **Size.** Most large molecules cannot pass through the plasma membrane. A few very small, uncharged polar molecules can pass through the phospholipid bilayer.
3. **Charge.** The phospholipid bilayer portion of the plasma membrane is impermeable to all charged molecules and ions. Some charged substances do pass through the membrane, however, by moving through a channel (pore) or by being carried from one side to the other by a transport protein. The negative membrane potential of most cells aids the inflow of cations and hinders the inflow of anions.

4. **Presence of channels and transporters.** Cell membranes are permeable to a variety of polar and charged substances, including ions, that cannot cross the phospholipid bilayer. Proteins in the membrane help several substances to cross the membrane, much as a ferry boat aids people and cars to cross a lake. These integral membrane proteins increase membrane permeability in two ways. Some proteins form water-filled channels (pores) through the membrane. Other proteins act as transporters, picking up a substance on one side of the membrane and shuttling it through to the other side before releasing it. Most channel proteins and transport proteins are very selective, allowing only a specific solute to cross the membrane.

MOVEMENT OF MATERIALS ACROSS PLASMA MEMBRANES

The mechanisms that enable substances to move across cell membranes are essential to the life of the cell. Certain substances, for example, must move into the cell to support needed biochemical reactions, while waste materials or harmful substances must be moved out. Mechanisms that move substances across a membrane without using cellular energy (released by splitting ATP) are **passive processes.** In **active processes,** the cell uses energy from the splitting of ATP to move the substance across the membrane.

Passive Processes

The passive processes that we will discuss are diffusion, osmosis, filtration, and facilitated diffusion. These transport processes depend on pressure or concentration differences or the process of diffusion.

Simple Diffusion

Since all substances have kinetic energy (energy of motion), they are continually moving about, colliding with one another, and moving off in various directions. The random mixing of ions and molecules in a solution due to their kinetic energy is called **diffusion** (*diffus* = spreading) and occurs as follows.

If a particular ion or molecule is present in high concentration in one area and in low concentration in another area, the difference in concentration between the two areas forms a **concentration gradient.** When two such areas are connected, more particles diffuse from the region of high concentration to the region of low concentration than diffuse in the opposite direction. This difference in diffusion between two regions having different concentrations is called **net diffusion.** Substances undergoing net diffusion (from a high to a low concentration) are said to move *down* or *with* their concentration gradient. After a period of time, they become evenly distributed. This point of even distribution is called **equilibrium.** At equilibrium, although random molecular motion continues, there is no further *net* diffusion.

Since diffusion depends on the kinetic energy of the particles, diffusion occurs more rapidly when temperature increases. Also, a larger concentration gradient (difference) produces faster diffusion. And smaller molecules diffuse more rapidly than larger ones.

As an example of diffusion, think what happens if you place a crystal of dye in a water-filled container (Fig. 3.5). Just next to the dye, the color is intense because the dye concentration is greatest there. At increasing distances, the color is lighter and lighter because the dye concentration is less. Some time later, at equilibrium, the solution of water and dye has a uniform color. Both the dye molecules and the water molecules have diffused down their concentration gradients until they are evenly dispersed among one another.

In the examples just noted, no membrane was involved. Substances may also diffuse through a membrane, if the membrane is permeable to them. Water and lipid-soluble molecules, such as oxygen, carbon dioxide, nitrogen, steroids, and fat-soluble vitamins (A, E, D, and K), urea, glycerol, small alcohols, and ammonia diffuse through the phospholipid bilayer of the plasma membrane, into and out of cells (Fig. 3.6a). Surprisingly, water also diffuses easily through the phospholipid bilayer, as has been shown in many experiments using artificially made, pure phospholipid membranes. Diffusion is important in the movement of oxygen and carbon dioxide between blood and body cells and between blood and air within the lungs during breathing. It also governs the absorption of some nutrients and excretion of wastes by body cells, activities that contribute to homeostasis.

Small substances that are not lipid-soluble also may diffuse into or out of cells through small water-filled channels (pores) formed by some of the integral membrane proteins (Fig. 3.6b). Important examples include sodium ions (Na^+), potassium ions (K^+), calcium ions (Ca^{2+}), chloride ions (Cl^-), and bicarbonate ions (HCO_3^-). Water itself also diffuses into and out of cells through protein-lined channels. Diffusion of substances through channels generally is much slower than diffusion through the phospholipid bilayer.

Osmosis

Another passive process is **osmosis.** It is the net movement of a solvent, which is water in living systems, through a

FIGURE 3.5 Principle of diffusion. Crystals of dye in a beaker of water dissolve and then diffuse down their concentration gradient from a region of high dye concentration to a region of low dye concentration. At the same time water molecules are diffusing down their concentration gradient. At equilibrium, net diffusion stops but simple diffusion continues.

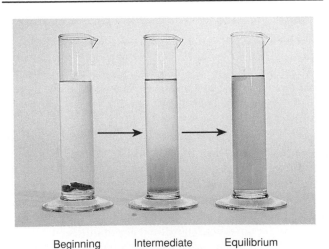

Beginning   Intermediate   Equilibrium

Question: How would a fever affect body processes that involve diffusion?

FIGURE 3.6 Diffusion through the plasma membrane.

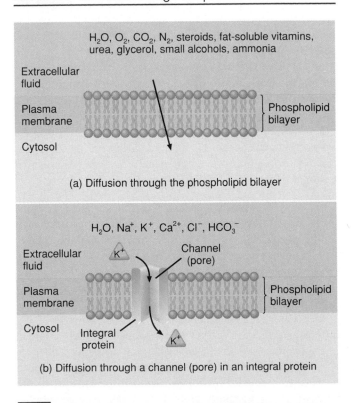

(a) Diffusion through the phospholipid bilayer

(b) Diffusion through a channel (pore) in an integral protein

Question: Which term better describes most molecules that can diffuse through the lipid bilayer: hydrophobic or hydrophilic?

selectively permeable membrane. Water moves by osmosis across a membrane from an area of higher water concentration to an area of lower water concentration. A simple apparatus may be used to demonstrate osmosis (Fig. 3.7). A sac made of cellophane, a permeable membrane that permits water but not sugar molecules to pass, is filled with a 20% sugar (sucrose) solution. The upper portion of the cellophane sac is wrapped tightly about a stopper through which a glass tube is fitted. The sac is then placed into a beaker containing distilled (pure) water.

The water concentration on the two sides of the cellophane membrane is different. There is a lower water concentration inside the sac because the addition of sugar molecules has decreased the water concentration. As a result, osmosis occurs. Water moves across the membrane down its concentration gradient, from the beaker into the cellophane sac. You can see that osmosis is occurring by observing the rising fluid level in the glass tube.

● **Osmotic Pressure** The movement of water through a selectively permeable membrane generates a pressure called osmotic pressure. **Osmotic pressure** is the pressure required to prevent the net movement of water from one solution to another when the solutions are separated by a membrane permeable only to water. In other words, osmotic pressure is the pressure needed to stop the flow of water across the membrane (see Fig. 3.7b).

What happens to the concentration and volume of the solution inside the cellophane sac? There is no movement of sugar from the sac into the beaker since the cellophane is impermeable to molecules of sugar. They are too large to pass through the pores of the membrane. As water moves into the cellophane sac, the sugar solution becomes more dilute, and the increased volume and pressure force the solution up the glass tubing. At equilibrium, the pressure of the water that has accumulated in the cellophane sac and the glass tube will be equal to the osmotic pressure. Then, the fluid level in the tube will not rise further because water molecules will be leaving and entering the cellophane sac at the same rate.

Osmotic pressure is an important force in the movement of water between various compartments of the body. Normally, the osmotic pressure of the cytosol inside cells is the same as the osmotic pressure of the extracellular fluid outside. Because the osmotic pressure on both sides of the membrane is equal, the cell volume remains relatively constant.

● **Tonicity** Osmosis may also be understood by considering the effects of different water concentrations on cell shape. To maintain the normal shape of a cell, for example, a red blood cell (RBC), the cell must be bathed in an **isotonic** (*iso* = same) **solution** (Fig. 3.8a). This is a solution in which the total concentrations of water molecules (solvent) and impermeable solute particles are the same on both sides. The concentrations of water and solute in the fluid outside the RBC must be the same as the concentration of the fluid inside the cell. Under ordinary circumstances, a 0.9% NaCl (salt) solution, called a **normal saline solution,** is isotonic for RBCs. Whereas the membrane of a RBC permits the water to move back and forth, it

FIGURE 3.7 Principle of osmosis. In (a) the cellophane sac (a selectively permeable membrane) contains a 20% sugar (sucrose) solution and is immersed in a beaker of distilled (pure) water. The arrows indicate that water molecules can pass freely into the sac. Sugar molecules, however, cannot pass out of the sac. As water moves into the sac by osmosis, the sugar solution becomes more dilute, and its volume increases. At equilibrium (b), the sugar solution has moved part way up the glass tubing. At this point, the number of water molecules entering and leaving the cellophane sac is equal. The osmotic pressure is the force exerted by the fluid rising in the tubing.

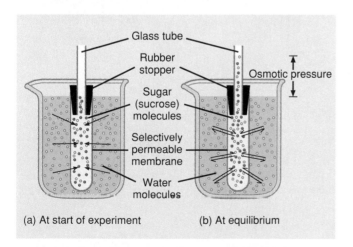

(a) At start of experiment   (b) At equilibrium

Question: Will the fluid level in the tube rise until the sugar concentration is the same in the sac and in the beaker?

FIGURE 3.8 Tonicity and red blood cells.

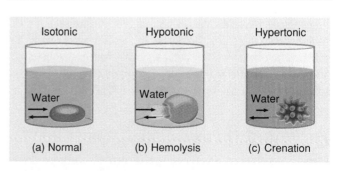

Question: When red blood cells are drawn from the body for laboratory testing, into which type of solution would they be placed to maintain their normal shape? Why?

is nearly impermeable to NaCl, the solute. When RBCs are bathed in 0.9% NaCl, water molecules enter and exit the cells at the same rate, allowing the RBCs to maintain their normal shape and volume.

A different situation results if RBCs are placed in a solution that has a *lower* concentration of solutes and therefore a higher concentration of water. This is called a **hypotonic** (*hypo* = less than, under) **solution** (Fig. 3.8b). In this condition, water molecules enter the cells faster than they leave, causing the RBCs to swell and eventually burst. The rupture of red blood cells in this manner is called **hemolysis** (hē-MOL-i-sis). Distilled water is strongly hypotonic.

A **hypertonic** (*hyper* = greater than, above) **solution** has a *higher* concentration of solutes and a lower concentration of water than the RBCs (Fig. 3.8c). One example of a hypertonic solution is a 2% NaCl solution. In such a solution, water molecules move out of the cells faster than they enter. This causes the cells to shrink. The shrinkage of RBCs in this manner is called **crenation** (kri-NĀ-shun). Red blood cells and other body cells may be damaged or destroyed if exposed to solutions that deviate significantly from isotonicity. It is for this reason that most intravenous (IV) solutions given to patients are isotonic to their red blood cells.

Filtration

Another passive process for moving materials in and out of cells is **filtration.** In filtration, water (the solvent) and some dissolved substances (solutes) move across a membrane due to gravity or hydrostatic (water) pressure. Such movement is always from an area of higher pressure to an area of lower pressure and continues as long as a pressure gradient (difference) exists. In the body, the driving pressure usually is blood pressure, generated by the pumping action of the heart. Most small- to medium-sized molecules such as nutrients, gases, ions, hormones, and vitamins can be forced through cell membranes, but large proteins cannot.

One site of filtration is in the kidneys. Blood pressure forces water and small molecules such as creatinine (a waste product) through plasma membranes of cells lining microscopic blood vessels (capillaries). The filtered liquid then enters the kidneys. Protein molecules remain in the blood since they are too large to be forced through the plasma membranes. Molecules of many harmful substances and waste products, however, are small enough to be filtered. They then can be eliminated in the urine.

Facilitated Diffusion

Certain substances are too large to diffuse through membrane channels (pores) and too lipid-insoluble to diffuse through the phospholipid bilayer. Such substances, however, can cross the plasma membrane by **facilitated diffusion**. In this process, the substance moves down the concentration gradient from a region of higher concentration to a region of lower concentration with the help of integral proteins in the membrane that serve as **transporters (carriers)**.

The most important substance that enters many body cells by facilitated diffusion is glucose (Fig. 3.9). After glucose attaches to the transporter on the outside of the membrane, the transporter changes shape. Glucose passes through the membrane and is released inside the cell. By itself, glucose cannot penetrate the membrane. Soon after glucose enters a cell, a kinase (enzyme that attaches a phosphate group) converts glucose into a substance called glucose 6-phosphate, using ATP as the source of phosphate. Since glucose 6-phosphate is different from glucose itself, this reaction keeps the intracellular concentration of glucose very low. Thus the concentration gradient always favors facilitated diffusion of glucose into, not out of, cells.

Facilitated diffusion can be considerably faster than simple diffusion and depends on (1) the difference in concentration of the substance on the two sides of the membrane, (2) the number of transporters available to transport the substance, and (3) how quickly the transporter and substance combine. Facilitated diffusion of glucose is greatly accelerated by insulin, a hormone produced by the pancreas. One of the effects of insulin is to lower the blood glucose level by speeding the facilitated diffusion of glucose from the blood into body cells. This is an example of how the plasma membrane's selective permeability can change to achieve homeostasis.

FIGURE 3.9 Facilitated diffusion of glucose.

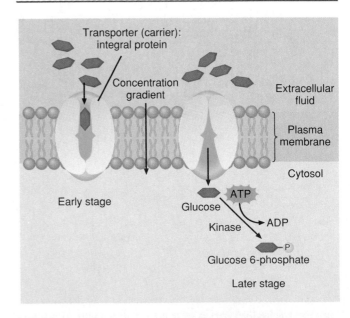

Question: Is energy from ATP required in this process? Explain.

Active Processes

Some substances that need to enter or leave body cells cannot move across plasma membranes passively. Either they are too big, have the wrong charge, or must move against the concentration gradient. Such substances can cross the membrane by **active processes.** These are the energy-consuming processes of active transport and bulk transport. In active transport, particular integral membrane proteins act as ATP-driven pumps to push certain ions and some smaller molecules across the membrane. Two general types of bulk transport processes—endocytosis and exocytosis—provide ways for moving more massive materials into and out of cells. Large particles, such as whole bacteria and red blood cells, and large molecules, like polysaccharides and proteins, may enter cells by endocytosis and leave by exocytosis. Active transport and bulk transport both require energy derived from splitting ATP.

Active Transport

The list of actively transported substances includes several ions such as Na^+, K^+, H^+, Ca^{2+}, I^-, Cl^-; amino acids; and monosaccharides. There are two types of **active transport**. In **primary active transport,** energy derived from splitting ATP *directly* moves a substance across the membrane. The cell uses the energy to induce changes in shape in transport (pump) proteins in the plasma membrane. In **secondary active transport**, the energy stored in ion gradients (differences) drives substances across the membrane. Since the ion gradients themselves are established by primary active transport pumps, secondary active transport *indirectly* uses energy obtained from splitting ATP. A typical body cell expends about 40% of the ATP it generates to run primary active transport pumps. Drugs that turn off ATP production, for example, the poison cyanide, are lethal because they shut down active transport in cells throughout the body.

● **Primary Active Transport: The Sodium Pump** The most prevalent primary active transport pump is the sodium pump. It maintains a low concentration of sodium ions (Na^+) in the cytosol by pumping them out against their concentration gradient. It also moves potassium ions (K^+) into cells against their concentration gradient. The sodium pump must work continually because K^+ and Na^+ slowly leak across the plasma membrane through leakage channels (pores).

Because of the ions it moves, this primary active transport pump is called the **Na^+/K^+ pump** or, more simply, the **sodium pump.** All cells have hundreds of sodium pumps in each square micrometer (μm^2) of membrane surface. Since the pump is a protein that acts as an enzyme to split the needed ATP, it is also called **Na^+/K^+ ATP-ase.** Figure 3.10 illustrates our current understanding of the sodium pump, which is actually a group of similar pumps with slightly different properties. Some operate as diagrammed in Fig. 3.10,

expelling three Na^+ each time they import two K^+. Others exchange one Na^+ for each K^+.

● **Secondary Active Transport: Symports and Antiports** The sodium pump maintains a large concentration difference of Na^+ across the plasma membrane. These ions have stored energy just as water behind a dam has stored energy. Accordingly, if Na^+ can leak back in, some of the stored energy can be used to transport other substances *against* their concentration gradients. Normally, the plasma membrane is fairly impermeable to Na^+ because there are few leakage channels (pores) that admit Na^+. Secondary active transport pumps harness the energy in the Na^+ gradient by providing easy routes for Na^+ to leak into cells. This is like allowing water in a reservoir to flow over a dam and turn a turbine to produce electrical power. The pump works by simultaneously binding to Na^+ and another substance. The larger the Na^+ gradient, the faster the secondary active transport.

Sometimes, two substances (usually Na^+ and another substance) move in the same direction across a plasma membrane. This process is called **symport (cotransport)**. For example, glucose, fructose, and amino acids enter cells lining the gastrointestinal tract and the kidney tubules via symports that use Na^+ (see Fig. 26.14). These pumps provide for absorption of dietary monosaccharides and amino acids. They also return to the blood nutrients that have been filtered by the kidneys so they are not lost in the urine. Two substances (usually Na^+ and another substance) may also move in opposite directions across the plasma membrane. This process is called **antiport (countertransport)**. See Fig. 26.17a. In most cells, Na^+/Ca^{2+} antiports keep Ca^{2+} concentration low in the cytosol. Likewise, Na^+/H^+ antiports help regulate the cytosol's pH (H^+ concentration) by using the Na^+ gradient to expel H^+.

CLINICAL APPLICATION

DIGITALIS

Because it strengthens the heartbeat, digitalis (Lanoxin) often is given to patients with heart failure, a condition of weakened pumping action by the heart. Digitalis exerts its effect by slowing the sodium pump, which allows more Na^+ to accumulate inside heart muscle cells. The result is a smaller Na^+ concentration difference across the membrane. The decreased Na^+ gradient, in turn, slows the Na^+/Ca^{2+} antiport. As a result, more Ca^{2+} remains inside heart muscle cells, and this increases the force of heart muscle contraction. The balance between the concentrations of Na^+ and Ca^{2+} in the cytosol and extracellular fluid is crucial to the normal functioning of nerve and muscle cells.

Bulk Transport

We will now examine how *larger* substances move across plasma membranes by several types of bulk transport: phagocytosis, pinocytosis, receptor-mediated endocytosis,

FIGURE 3.10 The sodium pump. Sodium ions (Na⁺) are expelled and potassium ions (K⁺) are imported. The pump will not work unless Na⁺ and ATP are present in the cytosol and K⁺ is present in the extracellular fluid. Note that the pump protein probably does not rotate or "flip-flop" across the membrane. Rather it pumps Na⁺ out and K⁺ in by changing its shape.

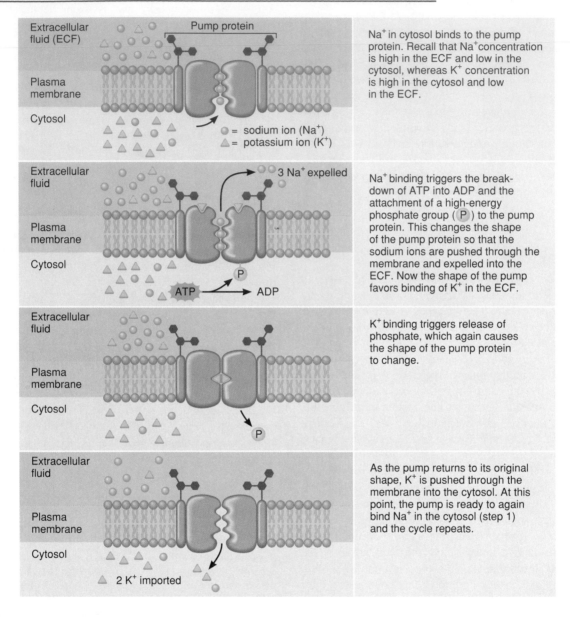

Na⁺ in cytosol binds to the pump protein. Recall that Na⁺ concentration is high in the ECF and low in the cytosol, whereas K⁺ concentration is high in the cytosol and low in the ECF.

Na⁺ binding triggers the break-down of ATP into ADP and the attachment of a high-energy phosphate group (P) to the pump protein. This changes the shape of the pump protein so that the sodium ions are pushed through the membrane and expelled into the ECF. Now the shape of the pump favors binding of K⁺ in the ECF.

K⁺ binding triggers release of phosphate, which again causes the shape of the pump protein to change.

As the pump returns to its original shape, K⁺ is pushed through the membrane into the cytosol. At this point, the pump is ready to again bind Na⁺ in the cytosol (step 1) and the cycle repeats.

Question: What is the role of ATP in the operation of this pump?

and exocytosis. Phagocytosis, pinocytosis, and receptor-mediated endocytosis are different types of **endocytosis** (*endo* = into; *cyt* = cell; *osis* = process); that is, they are processes that bring substances *into* cells. In endocytosis, a segment of the plasma membrane surrounds the substance to be taken in, encloses it, and brings it into the cell. **Exocytosis** (*exo* = out of) is a reverse process; it discharges substances *from* cells.

Endocytosis and exocytosis are important because mole-cules and particles of material that would normally be restricted from crossing the plasma membrane because of their large size can be brought in or removed from the cell.

● **Phagocytosis** In **phagocytosis** (fag′-ō-sī-TŌ-sis) or "cell eating," projections of the plasma membrane and cyto-plasm, called **pseudopods** (SOO-dō-pods), surround large solid particles outside the cell and then engulf them (Fig. 3.11a). Once the particle is surrounded, the pseudopods fuse, forming a membrane sac around the particle called a

FIGURE 3.11 Phagocytosis.

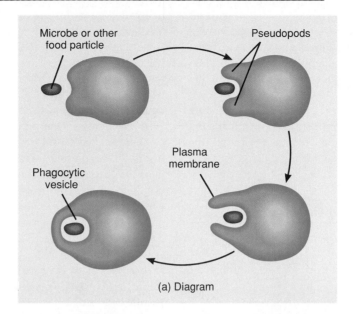

Microbe or other food particle

Pseudopods

Plasma membrane

Phagocytic vesicle

(a) Diagram

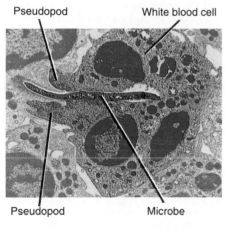

Pseudopod

White blood cell

Pseudopod

Microbe

White blood cell engulfs microbe

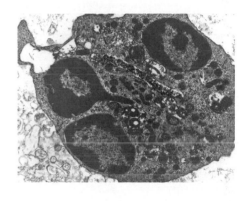

White blood cell destroys microbe

(b) Color-enhanced photomicrographs

Question: Is phagocytosis an example of endocytosis or exocytosis?

phagocytic vesicle (**phagosome**), which enters the cytoplasm. The solid material inside the vesicle is digested by enzymes provided by lysosomes (described shortly). Phagocytic white blood cells and cells in other tissues engulf and destroy bacteria and other foreign substances (Fig. 3.11b), which constitutes a vital defense mechanism to protect us from disease (see page 694).

● **Pinocytosis** In **pinocytosis** (pi′-no-si-TŌ-sis) or "cell drinking," the engulfed material is a tiny droplet of extracellular fluid rather than a solid. Moreover, no pseudopods are formed. Instead, the membrane folds inward, forming a **pinocytic vesicle** that allows the liquid to flow inward and then surrounds the liquid. The pinocytic

vesicle next detaches from the rest of the intact membrane. While only certain types of cells are capable of phagocytosis, most cells carry on pinocytosis.

● **Receptor-Mediated Endocytosis** Although similar to pinocytosis, **receptor-mediated endocytosis** is a highly selective process in which cells can take up specific molecules or particles. The extracellular fluid that bathes cells contains a large number of chemicals, most of which are present in low concentration. Some of these substances are ligands that bind to specific protein receptors in the plasma membrane. Ligands serve many functions. For example, cholesterol, iron, and vitamins are needed for chemical reactions that sustain life. Other ligands are either hormones that deliver messages to cells so cells can respond in specific ways or waste products that certain cells have the ability to break down. Although receptor-mediated endocytosis exists to import *needed* materials, some viruses can enter body cells in this way. For example, the virus that causes AIDS enters cells by attaching to a glycoprotein receptor called CD4 that is present on certain types of white blood cells.

Receptor-mediated endocytosis occurs as follows (Fig. 3.12). On the extracellular side of the plasma membrane the binding of a ligand to its specific receptor (1) causes the membrane to fold inward, forming an **endocytic vesicle** (2).

FIGURE 3.12 Receptor-mediated endocytosis.

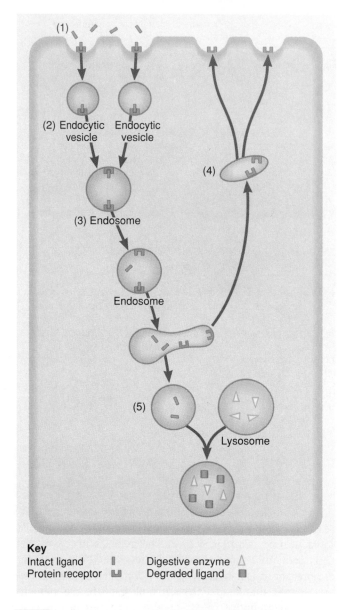

Key

| Intact ligand | ▮ | Digestive enzyme | △ |
| Protein receptor | ⊔ | Degraded ligand | ▪ |

Question: How does receptor-mediated endocytosis differ from pinocytosis?

After a vesicle moves inward from the plasma membrane, it merges with other endocytic vesicles to form a larger structure called an **endosome.** (3) Within the endosome, receptors separate from their ligand. The portion of the endosome that contains the receptors recycles them back to the plasma membrane by exocytosis. (4) The other portion of the endosome, containing the ingested material, merges with a lysosome and the ingested materials are broken down by digestive enzymes (5).

● **Exocytosis** During **exocytosis** membrane-enclosed structures called **secretory vesicles** that form inside the cell fuse with the plasma membrane and release their contents into the extracellular fluid (see Fig. 3.17). Exocytosis occurs in all cells but is especially important in nerve cells, which release their neurotransmitter substances by this process. It is also important in secretory cells, for example, cells that secrete digestive enzymes or protein hormones, such as insulin.

The various passive and active processes by which substances move across plasma membranes are summarized in Exhibit 3.1.

CLINICAL APPLICATION

LIPOSOMES AND DRUG THERAPY

Despite the impact of antibiotics and other drugs in the treatment of disease, they generally have disadvantages in that they are cleared from the blood rapidly and may be toxic to healthy cells as well as microorganisms and diseased cells. In an attempt to overcome these problems, scientists are experimenting with phospholipid sacs called **liposomes** (LIP-ō-sōms). When mixed with a drug, the phospholipids enclose it and form a sac. Liposome-containing drugs can be injected, sprayed in aerosols, or applied to the skin. It is believed that liposomes release their drugs slowly to maintain a more even blood level and reduce toxicity. The liposomes are taken into cells by diffusion or endocytosis.

By advancing the concept a step further, scientists are experimenting with incorporating anticancer drugs into low-density lipoprotein (LDL) particles, which enter body cells by receptor-mediated endocytosis. LDL is a natural particle so it may escape being attacked by the body's immune system that might otherwise destroy the LDL particles before they could deliver their drugs.

CYTOSOL

The substance inside the cell's plasma membrane and external to the nucleus is called **cytoplasm** (SĪ-tō-plazm′). The semifluid portion of the cytoplasm, in which organelles and inclusions are suspended and solutes are dissolved, is called the **cytosol,** or **intracellular fluid** (see Fig. 3.l). Physically, cytosol is a viscous, transparent, gel-like fluid containing suspended particles and a series of minute tubules and filaments that form a cytoskeleton (described shortly).

Chemically, cytosol is 75 to 90% water plus solid components. Proteins, carbohydrates, lipids, and inorganic substances comprise most of the solid components. Inorganic

SUMMARY OF MEMBRANE TRANSPORT PROCESSES

Process	Description
PASSIVE PROCESSES	Substances move down a concentration gradient from an area of higher to lower concentration or pressure; cell does not expend energy.
Simple diffusion	Movement of molecules or ions due to their kinetic energy.
Osmosis	Movement of solvent (water) molecules across a selectively permeable membrane from an area of higher to lower concentration of water until an equilibrium is reached. The solute concentration produces osmotic pressure.
Filtration	Movement of solvents (such as water) and solutes (such as glucose) across a selectively permeable membrane as a result of gravity or hydrostatic (water) pressure from an area of higher to lower pressure.
Facilitated diffusion	Diffusion of molecules across a selectively permeable membrane aided by membrane proteins that serve as transporters.
ACTIVE PROCESSES	Cell expends energy by splitting ATP to move substance, sometimes against a concentration gradient.
Primary active transport	Movement of ions or molecules across a selectively permeable membrane, often from a region of lower to higher concentration by pump proteins that split ATP.
Secondary active transport	Simultaneous movement of two substances across the membrane, one of which is Na^+. Uses energy supplied by the Na^+ concentration gradient, which is maintained by primary active transport pumps.[a] Symports (cotransporters) move Na^+ and the second substance in the same direction across the membrane while antiports (countertransporters) move Na^+ and the second substance in opposite directions.
Phagocytosis	"Cell eating"; movement of solid particles through the plasma membrane; pseudopods extend around the substance, enclose it, and bring it into the cell, forming a phagocytic vesicle.
Pinocytosis	"Cell drinking", movement of extracellular fluid droplets into cell by infolding of plasma membrane, forming pinocytic vesicle.
Receptor- mediated endocytosis	Mechanism for selected substances (ligands) to move into cells; involves binding of ligand to receptor at extracellular surface of plasma membrane; membrane then folds inward, forming endocytic vesicle.
Exocytosis	Export of substances from the cell in which vesicles fuse with the plasma membrane and release their contents into the extracellular fluid.

[a] Symports and antiports using ions other than Na^+ exist but are less common.

substances and smaller organic substances, such as simple sugars and amino acids, are soluble in water and are present as solutes. Larger organic compounds, for example, proteins and the polysaccharide glycogen, are found as **colloids**—particles that remain suspended in the surrounding medium. The colloids bear electrical charges that repel each other and thus remain suspended and separated.

Functionally, cytosol is the medium in which many metabolic reactions occur. The cytosol receives raw materials from the external environment by way of extracellular fluid and obtains usable energy from them by decomposition reactions.

ORGANELLES

Despite many chemical activities occurring at the same time in the cell, there is little interference of one reaction with another. This is because the cell has many different compartments provided by its **organelles** ("little organs"). These are specialized structures that have characteristic appearances and specific roles in growth, maintenance, repair, and control. The numbers and types of organelles vary in different kinds of cells, depending on their functions.

Nucleus

The **nucleus** (NOO-klē-us) is usually a spherical or oval organelle and is the largest structure in the cell (Fig. 3.13). It contains the hereditary units of the cell, called genes, which control cellular structure and direct many cellular activities. Most body cells contain a nucleus, although some, such as mature red blood cells, do not. Skeletal muscle fibers and a few other cells contain several nuclei.

A double membrane called the **nuclear envelope (membrane)** separates the nucleus from the cytoplasm. Both the inner and outer nuclear membranes are phospholipid bilayers similar to the plasma membrane. Water-filled **nuclear pores** in the envelope allow most ions and water-soluble molecules to shuttle between the nucleus and the cytoplasm. Nuclear pores are about ten times larger in diameter than channels in the plasma membrane and thus permit passage of large molecules such as RNA and various proteins.

One or more spherical bodies called **nucleoli** (noo-KLĒ-ō-lī′; singular, nucleolus) are present inside the nucleus. These are aggregations of protein, DNA, and RNA that are not bounded by a membrane. Nucleoli disperse and disappear during cell division and reorganize once new cells are formed. Nucleoli are the sites of assembly of ribosomes (described shortly), which contain a type of RNA called ribosomal RNA. Ribosomal RNA plays a key role in protein synthesis.

In a nondividing cell, DNA and associated proteins are

FIGURE 3.13 Nucleus.

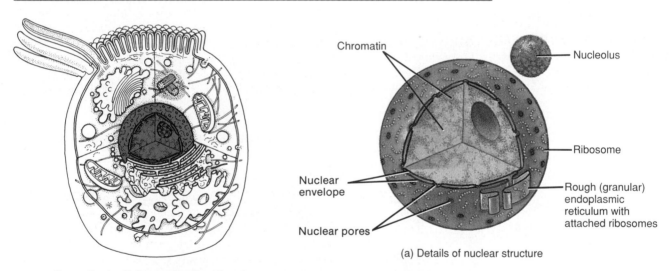

Generalized cell showing location of nucleus

(a) Details of nuclear structure

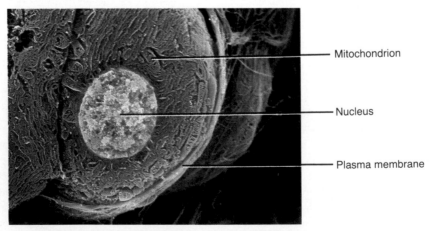

(b) Scanning electron micrograph of a nucleus, 7800x

Question: Why is the nucleus called the control center of the cell?

loosely packed and form a network called **chromatin.** During cell division, DNA and certain proteins condense and coil into rod-shaped bodies called **chromosomes.** Chromosomes contain a very large amount of DNA relative to their size. Each chromosome is thought to be a single DNA molecule, highly folded and coiled, combined with a variety of protein molecules. Through an electron microscope, chromatin appears like "beads-on-a-string." Each "bead" is a **nucleosome**, consisting of double-stranded DNA wrapped twice around a core of eight proteins called **histones** (Fig. 3.14). The "string" between "beads" is **linker DNA**, which holds adjacent nucleosomes together. Another histone promotes folding of nucleosomes into a larger diameter structure called a **chromatin fiber.** Chromatin fibers, in turn, fold into large **loops.** In cells that are not dividing, this is the extent of DNA packing. Before cell division, the DNA

duplicates and the chromatin strands subsequently shorten and coil even more to form **chromatids.** During certain stages of a cell in division, a pair of chromatids makes a chromosome.

Ribosomes

Ribosomes (RĪ-bō-sōms) are tiny spheres that contain **ribosomal RNA (rRNA)** and several ribosomal proteins. The rRNA is synthesized by DNA in the nucleolus. Ribosomes were so named because of their high content of *ribo*nucleic acid. This RNA was subsequently named rRNA to distinguish it from the other types of RNA in the cell. Structurally, a ribosome consists of two subunits, one about half the size of the other (see Fig. 3.15b). Functionally, ribosomes are the sites of protein synthesis, which is discussed later in the chapter.

Some ribosomes, called **free ribosomes,** float in the cytosol; they have no attachments to other organelles (see Fig. 3.1). The free ribosomes occur singly or in clusters, and they are concerned primarily with synthesizing proteins for use inside the cell. Other ribosomes attach to a cellular structure called the endoplasmic reticulum (Fig. 3.15a). These ribosomes are involved in the synthesis of proteins destined for insertion in the plasma membrane or for export from the cell.

Endoplasmic Reticulum

The **endoplasmic reticulum** (en'-dō-PLAS-mik rē-TIK-yoo-lum) or **ER** is a system of membrane-enclosed channels of varying shapes called **cisterns** (SIS-terns) or **cisternae** (Fig. 3.15a). The ER is continuous with the nuclear envelope. Based on its association with ribosomes, the ER is divided into two types. **Rough (granular) ER** is studded with ribosomes; **smooth (agranular) ER** has no ribosomes. The ER provides a surface area for chemical reactions and various products are transported from one portion of the cell to another via the ER.

Ribosomes associated with rough ER synthesize proteins. The rough ER also serves as a temporary storage area for newly synthesized molecules and may add sugar groups to certain proteins, thus forming a glycoprotein. Together, the rough ER and the Golgi complex (another organelle, described next) synthesize and package molecules that will be secreted from the cell.

Smooth ER is the site of fatty acid, phospholipid, and steroid synthesis. Also, in certain cells, enzymes within the smooth ER can inactivate or detoxify a variety of chemicals, including alcohol, pesticides, and carcinogens. In muscle cells, calcium ions released from the sarcoplasmic reticulum, which is analogous to the smooth ER, trigger contraction.

FIGURE 3.14 Packing of DNA into chromosomes.

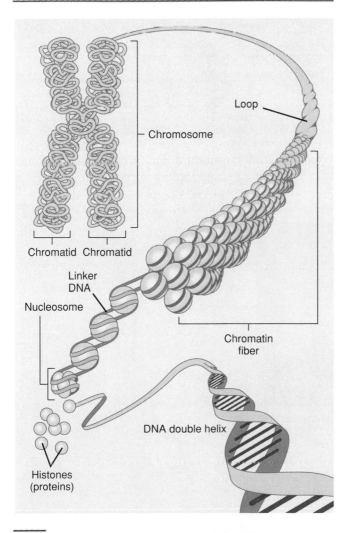

Question: What are the components of a nucleosome?

FIGURE 3.15 Endoplasmic reticulum and ribosomes.

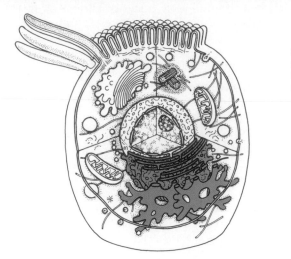

Generalized cell showing location of
endoplasmic reticulum and ribosomes

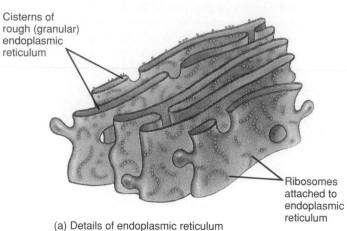

Cisterns of
rough (granular)
endoplasmic
reticulum

Ribosomes
attached to
endoplasmic
reticulum

(a) Details of endoplasmic reticulum

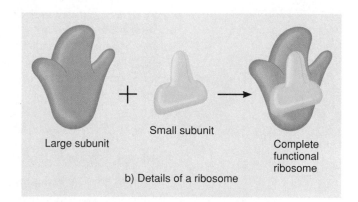

Large subunit

Small subunit

Complete
functional
ribosome

b) Details of a ribosome

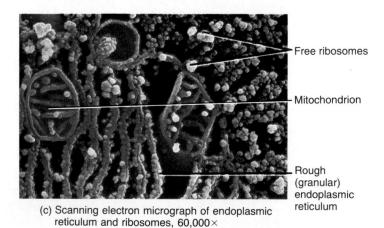

Free ribosomes

Mitochondrion

Rough
(granular)
endoplasmic
reticulum

(c) Scanning electron micrograph of endoplasmic
reticulum and ribosomes, 60,000×

Question: What is the difference between smooth (agranular) and rough (granular) ER?

Golgi Complex

One or more organelles called the **Golgi** (GOL-jē) **complex (apparatus)** are near the nucleus. In cells with high secretory activity, the Golgi complex is extensive. Usually, it consists of four to six flattened sacs called **cisterns (cisternae)**, stacked upon each other like a pile of plates with expanded bulges at their ends (Fig. 3.16). Associated with the cisterns are small **Golgi vesicles**, which cluster along the expanded ends of the cisterns.

The Golgi complex processes, sorts, packages, and delivers proteins and lipids to the plasma membrane, lysosomes, and secretory vesicles. All proteins destined for export from the cell follow a similar route: ribosomes (site of protein synthesis) → rough ER → transport vesicles → Golgi complex → secretory vesicles → release to exterior of the cell by exocytosis. Proteins and lipids destined for inclusion in the plasma membrane or for use inside lysosomes also pass through the Golgi complex.

A trip through a Golgi complex normally occurs as follows (Fig. 3.17). At the rough ER, proteins (and glycoproteins) become surrounded by a piece of the ER membrane, which forms a **transport vesicle**. The transport vesicle buds off the ER, moves to the Golgi complex, and fuses with the side of the Golgi stack closest to the ER [the *cis* (entry) cistern]. As a result of this fusion, the proteins enter the Golgi

FIGURE 3.16 Golgi complex.

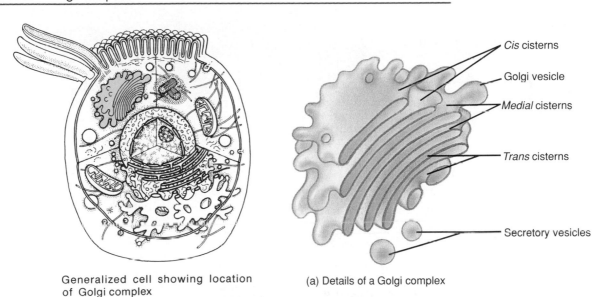

Generalized cell showing location of Golgi complex

(a) Details of a Golgi complex

Cis cisterns
Golgi vesicle
Medial cisterns
Trans cisterns
Secretory vesicles

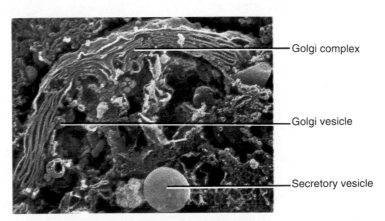

Golgi complex
Golgi vesicle
Secretory vesicle

(b) Scanning electron micrograph of Golgi complex, 20,000x

Question: What is the main function of the Golgi complex?

FIGURE 3.17 Packaging of synthesized proteins into secretory vesicles and lysosomes by the Golgi complex.

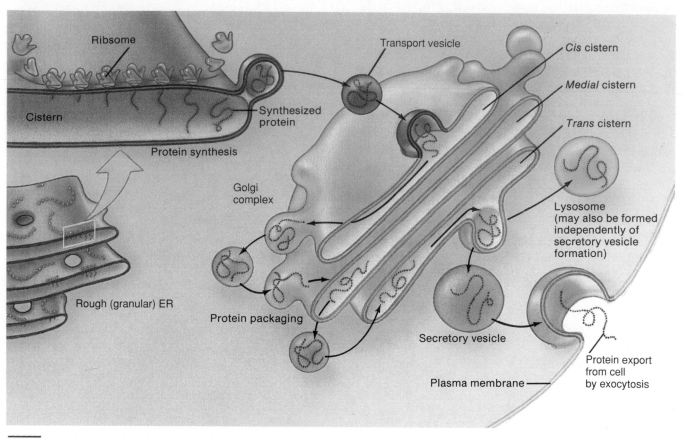

Question: What is the advantage of packaging digestive enzymes into vesicles?

complex. Once inside, they move through *medial* (middle) cisterns toward the *trans* (exit) cistern, probably by way of vesicles that bud from the ends of each cistern in the stack. As the proteins pass through the Golgi cisterns, they are modified in various ways depending on their function and destination and packaged into vesicles.

Some vesicles become secretory vesicles, which undergo exocytosis to discharge their contents into the extracellular fluid. Certain cells of the pancreas secrete digestive enzymes in this way. The vesicle membrane prevents the enzymes within from digesting the contents of the pancreatic cell itself. Other vesicles that leave the Golgi complex are loaded with digestive enzymes intended for use within the cell. They become organelles called lysosomes (described next).

CLINICAL APPLICATION

CYSTIC FIBROSIS

Sometimes a disorder results from faulty instructions for routing of a molecule. Such is apparently the case in cystic fibrosis, a deadly inherited disease (see page 760). In some cases, the product of the mutated cystic fibrosis gene fails to reach the plasma membrane, where it should be inserted to help pump chloride ions (Cl^-) out of certain cells. Evidently, the pump protein becomes stuck in the endoplasmic reticulum or Golgi complex and never reaches its correct destination. The result is an imbalance in the transport of fluid and ions across the plasma membrane and excessive build up of mucus outside certain types of cells. The mucus clogs the lining of the airways to the lungs, causing breathing difficulty.

Lysosomes

Lysosomes (LĪ-sō-sōms; *lysis* = dissolution; *soma* = body) are membrane-enclosed vesicles (see Fig. 3.17) that form in the Golgi complex. Inside are as many as 40 kinds of powerful digestive (hydrolytic) enzymes capable of breaking down a wide variety of molecules. Some disorders are caused by faulty lysosomes. For example, Tay–Sachs disease is an inherited absence of a single lysosomal enzyme. This enzyme normally breaks down a membrane glycolipid (called ganglioside GM_2) that is especially prevalent in nerve cells. As GM$_2$ accumulates, the nerve cells function less efficiently. The child becomes blind, demented, and uncoordinated and dies, usually before the age of 5.

Lysosomal enzymes work best at an acidic pH. The lysosomal membrane includes hydrogen ion (H$^+$) active transport pumps that drive H$^+$ into the lysosomes. As a result, while the cytosol has a pH close to 7, the interior of a lysosome has a pH of 5.

Lysosomal enzymes digest bacteria and other substances that enter the cell in phagocytic vesicles (phagosomes) during phagocytosis, pinocytic vesicles during pinocytosis, or endosomes during receptor-mediated endocytosis. Eventually, the products of digestion are small enough to pass out of the lysosome into the cytosol. There they are recycled to synthesize various molecules needed by the cell.

Lysosomes also use their enzymes to recycle the cell's own structures. A lysosome can engulf another organelle, digest it, and return the digested components to the cytosol for reuse. In this way, old organelles are continually replaced. The process by which worn-out organelles are digested is called **autophagy** (aw-TOF-a-jē; *auto* = self; *phagio* = to eat). A human liver cell recycles about half its contents every week.

Lysosomes also function in extracellular digestion. Lysosomal enzymes released at sites of injury help digest cellular debris, which prepares the injured area for effective repair.

CLINICAL APPLICATION

AUTOLYSIS

Normally, the lysosomal membrane is impermeable to its own enzymes. This prevents digestion of the cellular contents. However, sometimes **autolysis**, or self-destruction, by lysosomes occurs. For example, during human embryological development, the fingers and toes are webbed. The normal development of individual fingers and toes requires the selective removal of the cells of the web between the digits. This involves autolysis, in which lysosomal enzymes digest the tissue between the digits. It is actually a programmed destruction of cells.

Peroxisomes

Another group of organelles similar in structure to lysosomes, but smaller, are called **peroxisomes** (pe-ROKS-i-sōms). See Fig. 3.l. They are so named because they usually contain one or more enzymes that use molecular oxygen to oxidize (remove hydrogen atoms from) various organic substances. Such reactions produce hydrogen peroxide (H$_2$O$_2$). One of the enzymes in peroxisomes, called *catalase*, uses the H$_2$O$_2$ generated by other enzymes to oxidize a variety of other substances, including phenol, formic acid, formaldehyde, and alcohol, toxic substances that may enter the bloodstream. This type of oxidation is especially important in liver and kidney cells whose peroxisomes detoxify the potentially harmful substances.

Mitochondria

The next organelles to be considered are called **mitochondria** (mī-tō-KON-drē-a). Because of their function in generating ATP, an energy-rich molecule, they are called the "powerhouses" of the cell. A mitochondrion consists of two membranes, each of which is similar in structure to the plasma membrane (Fig. 3.18). The outer mitochondrial membrane is smooth, but the inner membrane is arranged in a series of folds called **cristae** (KRIS-tē). The central cavity of a mitochondrion, enclosed by the inner membrane and cristae, is called the **matrix.**

The elaborate folds of the cristae provide an enormous surface area for a group of chemical reactions known as **cellular respiration** (see page 826). Enzymes that catalyze these reactions are located on the cristae. Cellular respiration occurs only if oxygen (O$_2$) is present. It results in the catabolism of nutrient molecules, such as glucose, to generate ATP. Active cells, such as muscle, liver, and kidney tubule cells, have a large number of mitochondria and use ATP at a high rate.

Mitochondria self-replicate; that is, they divide to increase in number. The replication process is controlled by DNA that is part of the mitochondrial structure. Self-replication usually occurs in response to increased cellular need for ATP and at the time of cell division.

The Cytoskeleton

Cellular shape and the capability to carry out a variety of coordinated cellular movements depend on a complex internal network of filamentous proteins in the cytoplasm called the **cytoskeleton**. The cytoskeleton is responsible for movement of whole cells, such as phagocytes, and for movement of organelles and chemicals within the cell. Three main types of protein filaments—microfilaments, microtubules, and intermediate filaments—comprise the cytoskeleton (see Fig. 3.l).

Microfilaments are rodlike structures of varying length, formed from the protein actin. In muscle tissue, actin filaments (thin filaments) and myosin filaments (thick filaments) slide past one another to produce contraction (shortening) of muscle fibers (see page 246). In nonmuscle cells, actin filaments provide support and shape. They also assist in cell movement (for example, in phagocytes and cells of developing embryos) and movements within cells (secretion, phagocytosis, pinocytosis).

Microtubules are larger than microfilaments. They are relatively straight, slender, cylindrical structures that consist of a protein called **tubulin.** Together with microfilaments, microtubules help support and shape cells. Microtubules also function like a conveyor belt to move various substances and organelles through the cytosol, and they assist in the movement of pseudopods that are characteristic of phagocytes. As you will see shortly, microtubules form the structure of flagella and cilia (cellular appendages

FIGURE 3.18 Mitochondria.

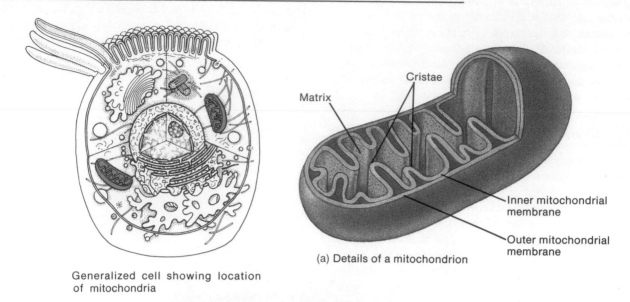

Generalized cell showing location
of mitochondria

(a) Details of a mitochondrion

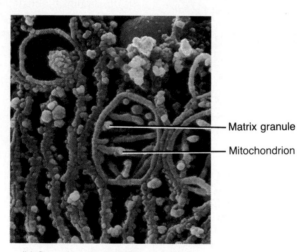

(b) Scanning electron micrograph of a
mitochondrion, 30,000x

Question: How do the cristae of a mitochondrion contribute to its energy-producing function?

involved in motility), centrioles, and the mitotic spindle (a structure that is involved in cell division).

Intermediate filaments, so-named because their size is in between that of microfilaments and microtubules, are exceptionally strong and tough. Several proteins form intermediate filaments. They provide structural reinforcement inside cells, hold organelles (such as the nucleus) in place, and associate closely with microtubules to give shape to the cell.

Flagella and Cilia

Some body cells have projections for moving the entire cell or for moving substances along the surface of the cell. These projections contain cytoplasm and are bounded by the plasma membrane. If the projections are few and long in proportion to the size of the cell (occurring singly or in pairs), they are called **flagella** (fla-JEL-a). The only example of a flagellum in the human body is the tail of a sperm cell, used for locomotion (see Fig. 28.5). If the projections are numerous and short, resembling many hairs, they are called **cilia** (SIL-ē-a). Ciliated cells of the respiratory tract remove mucus that has trapped foreign particles from the airways (see Fig. 23.5). Their movement is paralyzed by the nicotine in cigarette smoke. Both flagella and cilia contain nine pairs of microtubules that form a ring around two single microtubules in the center.

Centrosome and Centrioles

Near the nucleus is a dense area of cytoplasmic material called a **centrosome**. Within the centrosome is a pair of cylindrical structures called **centrioles** (Fig. 3.19). Each centriole is composed of nine clusters of three microtubules (triplets) arranged in a circular pattern. Centrioles lack the two central single microtubules found in flagella and cilia. The long axis of one centriole is at a right angle to the long axis of the other. Centrosomes serve as centers for organizing microtubules in nondividing cells and the mitotic spindle during cell division (described shortly). Like mitochondria, centrioles contain DNA that controls their self-replication.

CELL INCLUSIONS

Cell inclusions are a large and diverse group of chemical substances produced by cells. Although some have recognizable shapes, they are not bounded by a membrane. These products are principally organic molecules and may appear or disappear at various times in the life of a cell. Examples include melanin, glycogen, and triglycerides (neutral fats). **Melanin** is a pigment stored in certain cells of the skin, hair, and eyes. It protects the body by screening out harmful ultraviolet rays from the sun. **Glycogen** is a polysaccharide that is stored in liver, skeletal muscle, and the inner lining of the uterus and vagina. When the body requires quick energy, liver cells can break down the glycogen into glucose and release it into the blood. **Triglycerides,** which are stored in adipocytes (fat cells), may be broken down to synthesize ATP.

The major parts of a cell and their functions are summarized in Exhibit 3.2.

GENE ACTION

Although cells synthesize many chemicals to maintain homeostasis, much of the cellular machinery promotes protein production. Cells are basically protein factories that constantly synthesize large numbers of diverse proteins. The proteins, in turn, determine the physical and chemical characteristics of cells and therefore of organisms. Some proteins are structural, helping form plasma membranes, microfilaments, microtubules, centrioles, flagella, cilia, mitochondria, and other parts of cells. Other proteins serve as hormones, antibodies, and contractile elements in muscle tissue. Still other proteins are enzymes that regulate the rates of the myriad chemical reactions that occur in cells.

FIGURE 3.19 Centrosome and centrioles.

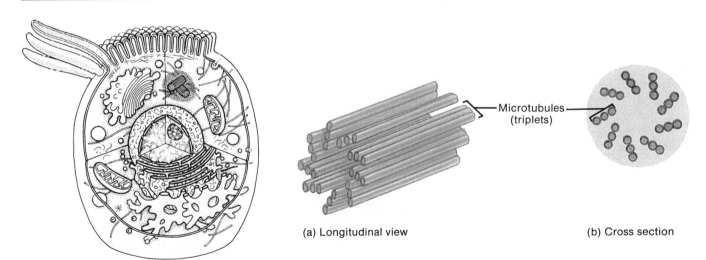

(a) Longitudinal view

(b) Cross section

Microtubules (triplets)

Generalized cell showing location of centrosome and centrioles

Question: If you observed that a cell did not have a centrosome, what could you predict about its capacity for cell division?

EXHIBIT 3.2

CELL PARTS AND THEIR FUNCTIONS

Part	Functions
PLASMA MEMBRANE	Protein-studded phospholipid bilayer that surrounds the cell; protects cellular contents; makes contact with other cells; provides receptors for hormones, enzymes, and antibodies; mediates the entrance and exit of materials.
CYTOSOL	Viscous, transparent, gel-like intracellular fluid containing water, ions, and many types of organic molecules such as enzymes; medium in which many of the cell's chemical reactions occur.
ORGANELLES	Organized compartments within the cytoplasm that have specific structures and functions.
Nucleus	Surrounded by the nuclear envelope; contains genes and controls cellular activities.
Ribosomes	Sites of protein synthesis; may be either free in the cytosol or attached to endoplasmic reticulum.
Endoplasmic reticulum (ER)	Membrane-bounded network of channels that provides surface area for many types of chemical reactions; ribosomes attached to rough ER synthesize proteins that will be secreted; smooth ER (no attached ribosomes) synthesizes lipids and detoxifies certain molecules.
Golgi complex	Stack of disc-shaped membrane-enclosed cisterns; packages proteins and lipids into secretory vesicles for export or insertion into the plasma membrane; forms lysosomes.
Lysosomes	Contain digestive enzymes that break down molecules and microbes.
Peroxisomes	Contain enzymes that use molecular oxygen to oxidize various organic substances.
Mitochondria	Sites of production of most ATP during cellular respiration.
Cytoskeleton	Includes three types of protein filaments (microfilaments, microtubules, and intermediate filaments) that give the cell shape and allow coordinated movements of organelles; in some cells the cytoskeleton is responsible for movement of the cell itself.
Flagella and cilia	Allow movement of entire cell (flagella) or movement of substances along surface of cell (cilia).
Centrosome	Helps organize microtubules in nondividing cell and mitotic spindle during cell division.
INCLUSIONS	Chemical substances produced by cells that are not surrounded by a membrane, for example, melanin, glycogen, and triglycerides (neutral fats).

The instructions for making proteins are found in DNA (deoxyribonucleic acid). Cells make proteins by translating the genetic information encoded in DNA into specific proteins. In this process, the information in a region of DNA is *transcribed* (copied) to produce a specific molecule of RNA (ribonucleic acid). Then, the information contained in RNA is *translated* into a corresponding specific sequence of amino acids in a newly produced protein molecule. Let us take a look at how DNA directs protein synthesis by considering the two principal steps in protein synthesis: transcription and translation.

Transcription

Transcription is the process by which genetic information encoded in DNA is copied onto a strand of RNA called **messenger RNA (mRNA)**. By using a specific portion of the cell's DNA (gene) as a template, the genetic information stored in the sequence of bases (nitrogenous bases) in DNA nucleotides is rewritten so that the same information appears in a complementary sequence of bases in mRNA nucleotides (Fig. 3.20).

Transcription of DNA is catalyzed by the enzyme **RNA polymerase**. During transcription, bases pair in the same manner as they do in the DNA double helix (see Fig. 2.20). The base cytosine (C) in the DNA template dictates the base guanine (G) in the new mRNA strand; a G in the DNA template dictates a C in the mRNA strand; and a thymine (T) in the DNA template dictates an adenine (A) in the mRNA. Since RNA contains uracil (U) instead of thymine, an A in the DNA template dictates a U in the mRNA. As an example, if the template portion of DNA has the base sequence ATGCAT, the newly transcribed mRNA strand would have the complementary base sequence UACGUA.

Only one of the two DNA strands serves as a template

FIGURE 3.20 Transcription. In this process, the genetic information in DNA is copied to mRNA. When RNA synthesis is complete, mRNA leaves the nucleus and enters the cytoplasm, where translation occurs. DNA also directs the synthesis of rRNA and tRNA.

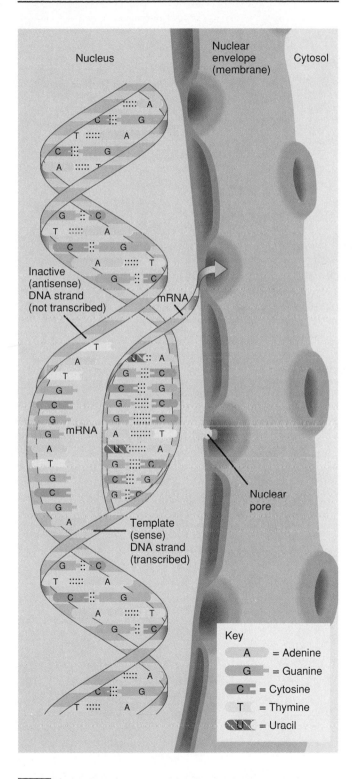

Nucleus

Nuclear envelope (membrane)

Cytosol

Inactive (antisense) DNA strand (not transcribed)

mRNA

mRNA

Template (sense) DNA strand (transcribed)

Nuclear pore

Key

A = Adenine

G = Guanine

C = Cytosine

T = Thymine

U = Uracil

Question: If the DNA template had the base sequence AGCT, what would be the mRNA base sequence?

for RNA synthesis. This strand is referred to as the **sense strand.** The other strand, the one not transcribed, is the complement of the sense strand and is called the **antisense strand.**

Within DNA are regions, called **introns**, that *do not* code for synthesis of part of a protein. Introns are located between regions called **exons**, regions that *do* code for parts of proteins. The initial RNA transcript includes both introns and exons. The RNA regions corresponding to DNA introns, however, are deleted (cut out) and the exons are rejoined before mRNA leaves the nucleus and enters the cytoplasm to proceed to the next step of protein synthesis. This process, called **mRNA splicing**, deletes at least three-fourths of the original RNA. The spliced mRNA, called **processed mRNA,** moves to ribosomes. A single gene may be able to code for more than one protein if the initial RNA transcript can be spliced in different ways to make different mRNAs. During cell differentiation, for example, splicing patterns change, and different proteins are produced.

Besides serving as the template for the synthesis of mRNA, DNA also synthesizes two other kinds of RNA. One is **ribosomal RNA (rRNA),** which together with ribosomal proteins makes up ribosomes. The other is **transfer RNA (tRNA).** Each tRNA can bind specifically to one of the 20 different types of amino acids. Once synthesized, mRNA, rRNA, and tRNA leave the nucleus of the cell. In the cytoplasm, they participate in the next principal step in protein synthesis—translation.

Translation

Just as DNA provides the template for mRNA to be made, so mRNA provides a template for protein synthesis. **Translation** is the process whereby information in the nucleotide sequence of mRNA specifies the amino acid sequence of a protein. In the mRNA molecule, each set of three consecutive nucleotides is called a **codon** and specifies one amino acid. The key events involved in translation are as follows (Fig. 3.21).

1. In the cytoplasm, the small ribosomal subunit binds one end of the newly synthesized mRNA molecule (Fig. 3.21a) and finds the **start codon**, a sequence where translation will begin. The large ribosomal subunit then joins in.

2. In the cytosol, transfer RNAs bind to specific amino acids and bring them to the ribosome. One end of the tRNA is coupled to a specific amino acid, and another part of each tRNA has a triplet of nucleotides called an **anticodon** (Fig. 3.21b).

3. By base pairing, the tRNA anticodon recognizes and attaches to a complementary codon on mRNA (Fig. 3.21c). For example, if the mRNA codon is AUG, then a tRNA having the anticodon UAC would attach. In the process, the tRNA brings along the specific amino acid.

FIGURE 3.21 Translation. Note that during protein synthesis the ribosomal subunits join together. When protein synthesis ceases, they separate.

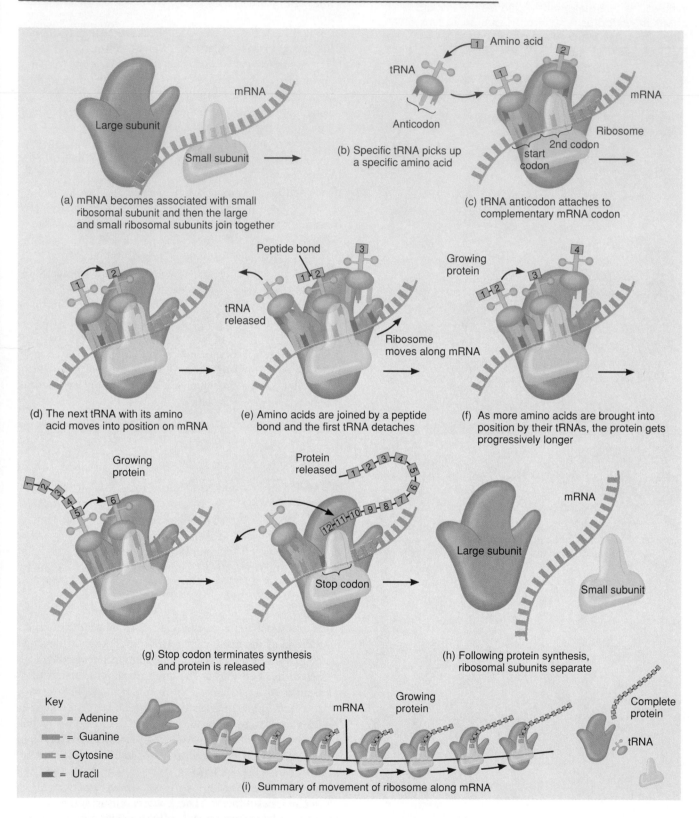

(a) mRNA becomes associated with small ribosomal subunit and then the large and small ribosomal subunits join together

(b) Specific tRNA picks up a specific amino acid

(c) tRNA anticodon attaches to complementary mRNA codon

(d) The next tRNA with its amino acid moves into position on mRNA

(e) Amino acids are joined by a peptide bond and the first tRNA detaches

(f) As more amino acids are brought into position by their tRNAs, the protein gets progressively longer

(g) Stop codon terminates synthesis and protein is released

(h) Following protein synthesis, ribosomal subunits separate

Key
= Adenine
= Guanine
= Cytosine
= Uracil

(i) Summary of movement of ribosome along mRNA

Question: Briefly, what is translation?

4. Once the first tRNA has attached to mRNA, the ribosome moves exactly three nucleotides along the mRNA, and the next tRNA with its amino acid moves into position (Fig. 3.21d).

5. When an amino acid attaches to its tRNA, ATP is used to form a high-energy bond between the amino acid and the tRNA. Such an activated amino acid can form a peptide bond with a second amino acid, if the second amino acid is brought into correct position by its own tRNA (Fig. 3.21e). The larger ribosomal subunit contains the enzymes needed for peptide bond formation. With formation of the peptide bond, tRNA detaches from the first amino acid.

6. As the proper amino acids are brought into line, one by one, peptide bonds form between them, and the protein progressively lengthens (Fig. 3.21f). Each time the ribosome moves one codon along the mRNA, an "empty" tRNA is ejected. The released tRNA can now pick up another molecule of the same amino acid.

7. When the specified protein is complete, synthesis is terminated by a special **stop codon.** The assembled protein is released from the ribosome, and the ribosome splits into its component subunits (Fig. 3.21g,h).

As each ribosome moves along the mRNA strand, it "reads" the information coded in mRNA and synthesizes a protein according to that information. Thus the ribosome synthesizes the protein by translating the nucleotide codon sequence into an amino acid sequence. Protein synthesis progresses at a rate of about 15 amino acids per second. As the ribosome moves along the mRNA and before it completes synthesis of the whole protein, another ribosome may attach behind it and begin translation of the same mRNA strand. In this way, several ribosomes may be attached to the same mRNA strand. This assembly is called a **polyribosome.** Several ribosomes moving in tandem along the same mRNA molecule permit the translation of a single mRNA strand into several identical proteins almost simultaneously.

From the description of protein synthesis just presented, we can appropriately define a **gene** as a group of nucleotides on a DNA molecule that serves as the master mold for manufacturing a specific protein (or small family of similar proteins, if RNA splicing varies). The processed mRNA contains from 300 to 3000 nucleotides. Since each set of three nucleotides codes for one amino acid, most proteins contain between 100 and 1000 amino acids. No two people have the same DNA molecules, unless they are identical twins. The nucleotide sequence is the key to heredity.

Remember that the base sequence of the gene determines the sequence of bases in mRNA, which in turn determines the order of amino acids in a given protein. Thus each gene is responsible for making a particular protein as follows:

DNA $\xrightarrow{\text{Transcription}}$ RNA $\xrightarrow{\text{Translation}}$ Protein

CLINICAL APPLICATION

GENETIC ENGINEERING

Different kinds of cells make different proteins following instructions encoded in the DNA of their genes. Since 1973, scientists have been able to alter those instructions in bacterial cells by adding genes from other organisms to the bacterial genes. This causes the bacteria to produce proteins they normally do not synthesize. Bacteria so altered are called **recombinants,** and their DNA, a combination of DNA from different sources, is called **recombinant DNA.** When recombinant DNA is properly introduced into a bacterium, the bacterium will synthesize the proteins specified by the new gene it has acquired. The new technology that has arisen from manipulating the genetic material is called **genetic engineering.** Yeast cells are also being used in recombinant DNA research.

The practical applications of recombinant DNA technology are enormous. Strains of recombinant bacteria are now producing many important therapeutic substances. These include *human growth hormone* (hGH), required for growth during childhood and important in metabolism of adults; *insulin,* a hormone that helps regulate blood sugar level and is used by diabetics; *interferon* (IFN), an antiviral (and possibly anticancer) substance; *factor VIII,* a blood clotting factor missing in people with hemophilia A; *erythropoietin,* a hormone that stimulates formation of red blood cells; *monoclonal antibodies* to diagnose and treat cancer and assist in AIDS research; and many other substances. Scientists are also using recombinant DNA techniques to develop vaccines against several viruses, including those that cause AIDS, hepatitis B, herpes, and influenza.

NORMAL CELL DIVISION

Most of the cell activities mentioned thus far maintain the life of the cell on a day-to-day basis. However, somatic (body) cells become damaged, diseased, or worn out and then die. New cells must be produced by cell division as replacements and for growth. In a 24-hour period, the average adult loses billions of cells from different parts of the body. Cells that have a short life span, such as cells of the outer layer of skin, are completely replaced every few days. In addition, germ cells, those that are specialized to form egg cells and sperm, must be produced by cell division.

Cell division is the process by which cells reproduce themselves. It consists of a nuclear division and a cytoplasmic division. Because nuclear division can be of two types, two kinds of cell division are recognized.

In the first kind of division, **somatic cell division,** a single starting cell called a **parent cell** divides to produce two identical cells called **daughter cells.** This process consists of a nuclear division called **mitosis** and a cytoplasmic division called **cytokinesis.** The process ensures that each daughter cell has the same *number* and *kind* of chromosomes as the original parent cell. This kind of cell division replaces dead or injured cells and adds new ones for body growth.

The second type of cell division, **reproductive cell division,** is the mechanism by which sperm and egg cells are produced. These are the cells needed to form a new organism. The process consists of a nuclear division called **meiosis** plus **cytokinesis.** We will first discuss somatic cell division.

Somatic Cell Division

Human cells, except for eggs and sperm, contain 23 pairs of chromosomes. As you will see later, two chromosomes that belong to a pair, one contributed by the mother and one contributed by the father, are called *homologous chromosomes.* They have a similar gene order. When a cell reproduces, it must replicate (produce duplicates of) all of its chromosomes so that its hereditary traits may be passed on to succeeding generations of cells.

Interphase

The **cell cycle** is a series of activities through which a cell passes from the time it is formed until it reproduces. It is the growth and division of a single cell into daughter cells. The cell cycle consists of two major activities: interphase and cell division (Fig. 3.22). When a cell is between divisions, it is said to be in **interphase.** It is during this stage that the replication of DNA, centrosomes, and centrioles occurs and the RNA and protein needed to produce structures required for doubling all cellular components are manufactured.

Interphase consists of three distinct phases: S, G_1, and G_2 (Fig. 3.22). The period of interphase during which chromosomes are replicated is called the **S-** (for synthesis) **phase.** The S-phase is preceded by a **G_1-** (for gap or growth) **phase,** during which cells are engaged in growth, metabolism, and the production of substances required for division. After chromosomal replication in the S-phase, there is another growth phase called the **G_2-phase.** Since the G-phases are periods when there are no events related to chromosomal replication, they are thought of as gaps or interruptions in DNA synthesis. Cells that are destined to never divide again are permanently arrested in the G_1-phase. Most nerve cells are in this state. Once a cell enters the S-phase, it is committed to go through the divisional cycle.

When DNA replicates, its helical structure partially uncoils (Fig. 3.23) and the two strands separate at the points where hydrogen bonds connect base pairs. Each exposed base then picks up a complementary base (with its associated sugar and phosphate group). This uncoiling and complementary base pairing continues until each of the two original DNA strands is matched and joined with one newly formed DNA strand. The original DNA molecule has become two DNA molecules.

A microscopic view of a cell during interphase shows a clearly defined nuclear envelope, nucleoli, and chromatin (Fig. 3.24a). The absence of visible chromosomes is another physical characteristic of interphase. Once a cell completes its replication of DNA, centrosomes, and centrioles and its production of RNA and proteins during interphase, mitosis begins.

Cell Division: Mitosis

The events that take place during mitosis and cytokinesis are plainly visible under a microscope because of condensation of chromatin into chromosomes. The process called **mitosis** is the distribution of the two sets of chromosomes into two separate and equal nuclei. It results in the *exact* duplication of genetic information. For convenience, biologists divide the process into four stages: prophase, metaphase, anaphase, and telophase. Mitosis is a continuous process, however, one stage merging imperceptibly into the next.

FIGURE 3.22 The cell cycle. Relative lengths of time devoted to each phase are indicated by the size of the compartments.

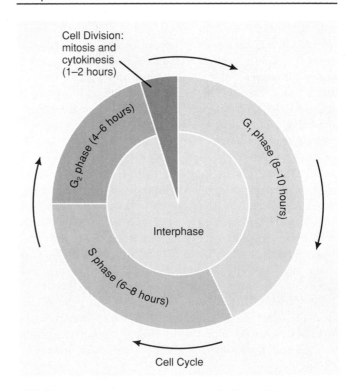

Cell Division: mitosis and cytokinesis (1–2 hours)

G_2 phase (4–6 hours)

G_1 phase (8–10 hours)

Interphase

S phase (6–8 hours)

Cell Cycle

Question: During which phase do the chromosomes replicate?

FIGURE 3.23 Replication of DNA. The two strands of the double helix separate by breaking the hydrogen bonds between nucleotides. New nucleotides attach at the proper sites, and a new strand of DNA is synthesized alongside each of the original strands. After replication, the two DNA molecules, each consisting of a new and an old strand, return to their helical structure.

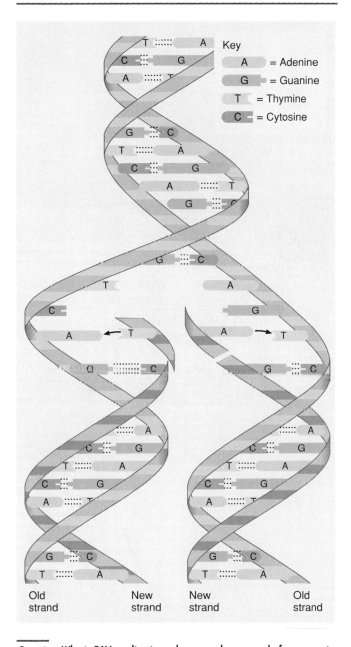

Key

A	= Adenine
G	= Guanine
T	= Thymine
C	= Cytosine

Old strand   New strand   New strand   Old strand

Question: Why is DNA replication a key step that occurs before somatic cell division?

● **Prophase** The first stage of mitosis is called **prophase** (*pro* = before) (Fig. 3.24b). During early prophase, the chromatin condenses and shortens into chro-

mosomes. This condensation is necessary because long strands might entangle, and this would interfere with their movement. Since DNA replication took place during interphase, each prophase chromosome contains a pair of identical double-stranded DNA molecules called **chromatids.** Each chromatid pair is held together by a small spherical body called a **centromere** that is required for the proper segregation of chromosomes. Attached to the outside of each centromere is a protein complex known as the **kinetochore** (ki-NET-ō-kor), whose function will be described shortly.

Later in prophase, the nucleoli disappear, during which time synthesis of RNA is temporarily halted, and the nuclear envelope breaks down and is absorbed in the cytosol. In addition, a centrosome and its centrioles each move to opposite poles (ends) of the cell. As they do so, the centrosomes start to form the **mitotic spindle**, a football-shaped assembly of microtubules that are responsible for the movement of chromosomes. The lengthening of microtubules between centrosomes pushes the centrosomes to the poles of the cell so that the spindle extends from pole to pole. As the mitotic spindle continues to develop, three types of microtubules form: (1) **nonkinetochore microtubules** grow from centrosomes, extend inward, but do not bind to kinetochores; (2) **kinetochore microtubules** grow from centrosomes, extend inward, and attach to kinetochores; and (3) **aster microtubules** grow out of centrosomes but radiate outward from the mitotic spindles. The spindle is an attachment site for chromosomes, and it also distributes chromosomes to opposite poles of the cell.

● **Metaphase** During **metaphase** (*meta* = after), the second stage of mitosis, the centromeres of the chromatid pairs line up at the exact center of the mitotic spindle. This midpoint region is called the **metaphase plate** or **equatorial plane region** (Fig. 3.24c).

● **Anaphase** The third stage of mitosis, **anaphase** (*ana-* upward), is characterized by the splitting and separation of centromeres and the movement of the two sister chromatids of each pair toward opposite poles of the cell (Fig. 3.24d). Once separated, the sister chromatids are referred to as daughter chromosomes. The movement of chromosomes is due to shortening of kinetochore microtubules and elongation of the nonkinetochore microtubules, processes that increase the distance between separated chromosomes. As the chromosomes move during anaphase, they appear V-shaped as the centromeres lead the way and seem to drag the trailing parts of the chromosomes toward opposite poles of the cell.

● **Telophase** The final stage of mitosis, **telophase** (*telo* = far or end), begins as soon as chromosomal movement stops. Telophase is essentially the opposite of prophase. During telophase, the identical sets of chromosomes at opposite poles of the cell uncoil and revert to their threadlike chromatin form. Kinetochore microtubules disappear, and nonkinetochore microtubules elongate even more. A new nuclear envelope re-forms around each chro-

FIGURE 3.24 Cell division: mitosis and cytokinesis. Start looking at the sequence at (a) and read clockwise until you complete the process.

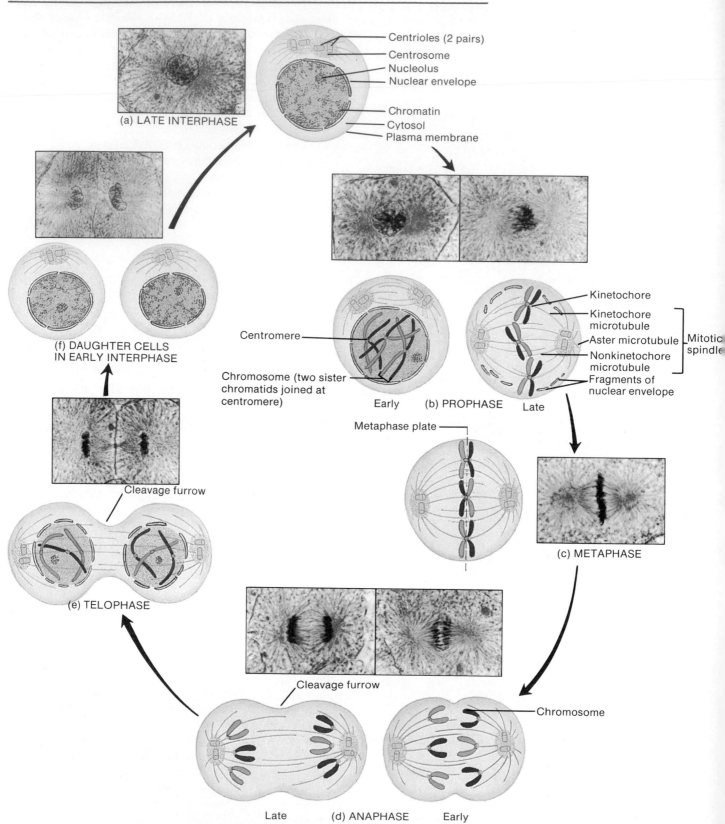

(a) LATE INTERPHASE

Centrioles (2 pairs)
Centrosome
Nucleolus
Nuclear envelope
Chromatin
Cytosol
Plasma membrane

(f) DAUGHTER CELLS IN EARLY INTERPHASE

Centromere

Chromosome (two sister chromatids joined at centromere)

Early   (b) PROPHASE   Late

Kinetochore
Kinetochore microtubule
Aster microtubule
Nonkinetochore microtubule
Fragments of nuclear envelope

Mitotic spindle

Cleavage furrow

(e) TELOPHASE

Metaphase plate

(c) METAPHASE

Cleavage furrow

Chromosome

Late   (d) ANAPHASE   Early

Question: When does cytokinesis begin and end?

matin mass; new nucleoli reappear in the daughter nuclei; and eventually the mitotic spindle breaks up.

Cell Division: Cytokinesis

Division of a parent cell's cytoplasm and organelles is called **cytokinesis** (sī-tō-ki-NĒ-sis, *kinesis* = motion). It begins in late anaphase or early telophase with formation of a **cleavage furrow,** a slight indentation of the plasma membrane that extends around the center of the cell (Fig. 3.24 d-f). The furrow gradually deepens until opposite surfaces of the cell make contact and the cell is split in two. When cytokinesis is complete, interphase begins. The result of cytokinesis is two separated daughter cells, each with separate portions of cytoplasm and organelles and its own set of identical chromosomes.

If we consider the cell cycle in its entirety, the sequence of events is G_1-phase $\rightarrow$ S-phase $\rightarrow$ G_2-phase $\rightarrow$ mitosis $\rightarrow$ cytokinesis (see Fig. 3.22). A summary of the events that occur during interphase and cell division is presented in Exhibit 3.3.

Length of Cell Cycle

The time required for mitosis varies with the kind of cell, its location, and other factors such as temperature. Further-more, the different stages of mitosis are not equal in duration. Mammalian cells studied in laboratory cultures often have the following time intervals during a cell cycle (see Fig. 3.22). The G_1-phase is highly variable, ranging from almost nonexistent in rapidly dividing cells to days, weeks, or years. However, it typically takes 8 to 10 hours. The S-phase takes about 6 to 8 hours, G_2-phase about 4 to 6 hours, and mitosis and cytokinesis about 1 to 2 hours. Within the mitosis and cytokinesis time interval, prophase lasts longest and anaphase is shortest. As you can see, mitosis and cytokinesis represent only a small part of the life cycle of a cell. Together, the various phases of the cell cycle require about 18 to 24 hours in many cultured mammalian cells.

Reproductive Cell Division

In sexual reproduction, each new organism is produced by the union and fusion of two different sex cells, one produced by each parent. The sex cells, called **gametes,** are the ovum (egg cell), produced in the female gonads (ovaries), and the sperm, produced in the male gonads (testes). The union and fusion of gametes is called **fertilization,** and the cell thus produced is known as a **zygote.** The zygote contains one set of chromosomes (DNA) from each parent and, through repeated mitotic divisions, develops into a new organism.

EXHIBIT 3.3

SUMMARY OF EVENTS ASSOCIATED WITH THE CELL CYCLE

Phase or Stage	Activity
INTERPHASE	Cell is between divisions; chromosomes are not visible.
G_1-phase	Cell engages in growth, metabolism, and production of substances required for division; no chromosomal replication.
S-phase	Chromosomal replication occurs.
G_2-phase	Same as for G_1 period.
CELL DIVISION	Single parent cell produces two chromosomally identical daughter cells.
Prophase	Chromatin shortens and coils into visible chromosomes (chromatids), nucleoli and nuclear envelope disappear, a centrosome and its centrioles each move to opposite poles of cell, and centrosomes form mitotic spindle.
Metaphase	Centromeres of chromatid pairs line up on metaphase plate of cell.
Anaphase	Centromeres divide and identical sets of chromosomes move to opposite poles of cell.
Telophase	Nuclear envelope reappears and encloses chromosomes, chromosomes resume chromatin form, nucleoli reappear, and mitotic spindle disappears.
Cytokinesis	Cleavage furrow forms around center of cell, progresses inward, and separates cytoplasm into two separate and usually equal portions.

Gametes

Gametes differ from all other body cells (somatic cells) with respect to the number of chromosomes in their nuclei. Somatic cells, such as brain cells, stomach cells, kidney cells, and others, contain 23 pairs of chromosomes, or a total of 46 chromosomes. One member of each pair is inherited from each parent. A gamete, on the other hand, has only 23 chromosomes, just one member of each pair. The two chromosomes that make up a pair (one from the mother and one from the father) are called **homologous** (hō-MOL-ō-gus) **chromosomes,** or **homologs.** They contain similar genes arranged in the same or almost the same order. When examined under the light microscope (see Fig. 29.18), homologous chromosomes look very similar. The exception to this general rule is a pair of chromosomes called the **sex chromosomes,** designated X and Y. The X chromosome is much larger than the Y chromosome and has more genes, but the Y chromosome contains the gene, called SRY, that is mainly responsible for the male pattern of development. When this gene is absent, a female develops (see page 995). The other 22 pairs of chromosomes are called **autosomes.**

The symbol n is used to designate the number of different chromosomes. In humans, $n = 23$. Since somatic cells contain two sets of chromosomes, they are $2n$ and are called **diploid** (DIP-loyd; di = two) **cells.** Gametes, with a single set of chromosomes, are $1n$ and are called **haploid** (HAP-loyd; ha = one) **cells.** If gametes had the same number of chromosomes as somatic cells, the zygote formed from their fusion would have double the normal number ($4n$). With each succeeding generation, the number of chromosomes would double. Chromosome number does not double with each generation, however, because of a special nuclear division called **meiosis,** or reduction division. Meiosis occurs only in the development of gametes, and it results in the production of cells that contain only 23 chromosomes. As you will see later, meiosis also accounts for genetic variation in daughter cells. Mitosis, by contrast, ensures exact genetic copies.

Meiosis

The formation of haploid sperm cells in the testes of the male occurs in several phases as part of the process called **spermatogenesis.** The formation of haploid ova (eggs) in the ovaries of the female occurs as part of **oogenesis** and also involves several phases. Both spermatogenesis and oogenesis are discussed on pages 925 and 937, respectively. At this point, we will examine only the essentials of meiosis, which is a common feature of both.

Meiosis occurs in two successive nuclear divisions referred to as **reduction division (meiosis I)** and **equatorial division (meiosis II).** During the interphase that precedes reduction division of meiosis, the chromosomes replicate. This replication is similar to that in the interphase preceding the mitosis of somatic cell division. Once chromosomal replication is complete, reduction division begins. It consists of four phases: prophase I, metaphase I, anaphase I, and telophase I (Fig. 3.25).

Prophase I is an extended phase in which the chromosomes shorten and thicken, the nuclear envelope and nucleoli disappear, and the mitotic spindle appears. Unlike the prophase of mitosis, however, a unique event occurs in prophase I of meiosis. The chromosomes become arranged in homologous pairs. This pairing is called **synapsis.** The four chromatids of each homologous pair are termed a **tetrad.** Another unique event of meiosis occurs within a tetrad. Portions of one chromatid may be exchanged with portions of another, a process called **crossing-over** (Fig. 3.26). This process, among others, permits an exchange of genes among homologous chromatids so that the resulting daughter cells are unlike each other genetically and unlike the parent cell that produced them. This phenomenon accounts for part of the great genetic variation among humans and other organisms that form gametes by meiosis.

In metaphase I, the homologous pairs of chromosomes line up along the metaphase plate of the cell, with the homologues side by side. Recall that there is no pairing of homologous chromosomes during the metaphase of mitosis. The centromeres of each chromatid pair form kinetochore microtubules that attach the centromeres to opposite poles of the cell. During anaphase I the members of each homologous pair separate, with one member of each pair moving to an opposite pole of the cell. The centromeres do not split and the paired chromatids, held by a centromere, remain together. (Recall that during the anaphase of mitosis, the centromeres split and the sister chromatids separate.) Telophase I and cytokinesis are similar to telophase and cytokinesis of mitosis. The net effect of reduction division is that each resulting daughter cell contains the haploid number of chromosomes; each cell contains only one member of each pair of the original homologous chromosomes in the starting parent cell.

The interphase between reduction division and equatorial division is either brief or lacking altogether. It does differ from the interphase preceding reduction division in that there is no replication of DNA between the reduction and equatorial divisions.

The equatorial division of meiosis consists of four phases: prophase II, metaphase II, anaphase II, and telophase II. These phases are similar to those that occur during mitosis since the centromeres split and the sister chromatids separate and move toward opposite poles of the cell.

In reviewing the overall process, note that reduction division starts with a parent cell with the diploid number and ends up with two daughter cells, each with the haploid number. During equatorial division, each haploid cell formed during reduction division divides, and the net result is four haploid cells that are all genetically different. As you will see later, all four haploid cells develop into sperm cells

FIGURE 3.25 Meiosis. See text for details.

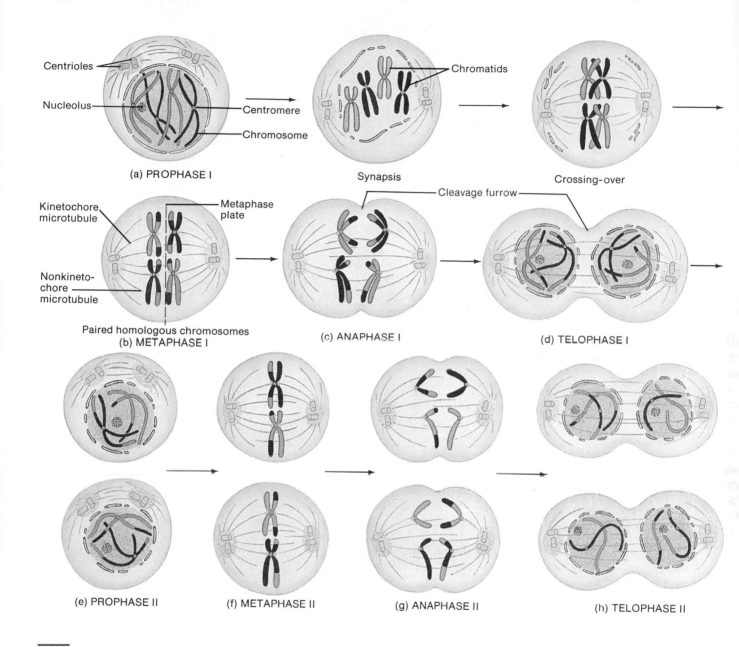

(a) PROPHASE I — Centrioles, Nucleolus, Centromere, Chromosome

Synapsis — Chromatids

Crossing-over

(b) METAPHASE I — Kinetochore microtubule, Nonkinetochore microtubule, Metaphase plate, Paired homologous chromosomes

(c) ANAPHASE I — Cleavage furrow

(d) TELOPHASE I

(e) PROPHASE II

(f) METAPHASE II

(g) ANAPHASE II

(h) TELOPHASE II

Question: What is the net result of reproductive cell division?

in the testes of the male. Only one of the haploid cells has the potential to develop into an ovum in the female. The others become structures called *polar bodies* that do not function as gametes.

Although reduction division reduces the number of chromosomes, equatorial division is necessary to separate the chromatids within each chromosome, to produce normal offspring. Failure to do so will result in genetic disorders such as Down syndrome (see page 996).

A very simplified comparison of mitosis and meiosis is illustrated in Fig. 3.27.

Control of Cell Division

How a cell knows when to divide and when to stop division is one of the fundamental mysteries in biology. Lately, however, pieces to the puzzle are beginning to fall into

FIGURE 3.26 Crossing-over within a tetrad.

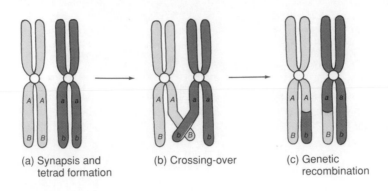

(a) Synapsis and tetrad formation

(b) Crossing-over

(c) Genetic recombination

Question: How does crossing-over affect the genetic content of daughter cells?

FIGURE 3.27 Comparison between (a) mitosis and (b) meiosis.

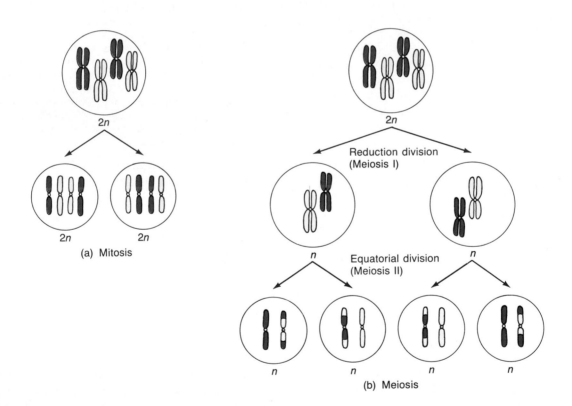

2n

2n

2n   2n

(a) Mitosis

Reduction division (Meiosis I)

n   n

Equatorial division (Meiosis II)

n   n   n   n

(b) Meiosis

Question: Where in the body does meiosis occur?

place. Knowing the factors that regulate the cell cycle has implications in two important areas. First, it might be possible to find new ways to speed up cell division to accelerate wound healing or even stimulate nerve cells to divide. Second, it might be possible to uncover mechanisms for slow-ing or stopping cell division to prevent the growth of cancer cells.

One of the key players in the regulation of cell division (both mitosis and meiosis) is a substance called **maturation promoting factor (MPF)**. When MPF becomes activated, it induces cell division. One component of MPF is a group of proteins (enzymes) called **cdc2 proteins** because they participate in the cell division cycle (cdc). These enzymes transfer a phosphate group from ATP to proteins to help regulate cell activities. Another component of MPF is a protein called **cyclin**, so named because its level rises and falls during the cell cycle. Cyclin builds up in the cell during interphase and activates cdc2 proteins and thus MPF. The result is that the cell undergoes mitosis (or meiosis). By the end of mitosis, cyclin levels are low. Accordingly, cdc2 proteins are not activated, MPF is not activated, and mitosis (or meiosis) stops. Then, the cell enters another interphase.

ABNORMAL CELL DIVISION: CANCER

When cells in some area of the body divide without control, the excess of tissue that develops is called a **tumor, growth,** or **neoplasm.** The study of tumors is called **oncology** (on-KOL-ō-jē; *onco* = swelling or mass; *logos* = study of), and a physician who specializes in this field is called an **oncologist.** Tumors may be cancerous and sometimes fatal, or they may be quite harmless. A cancerous growth is called a **malignant tumor,** or **malignancy.** A noncancerous growth is called a **benign tumor.** Benign tumors do not spread to other parts of the body, but they may be removed if they interfere with a normal body function or are disfiguring.

Growth and Spread of Malignant Tumors

Cells of malignant tumors duplicate continuously and very often quickly and without control. Such an increase in the number of cells due to an increase in the frequency of cell division is called **hyperplasia** (hī-per-PLĀ-zē-a). One property of a malignant tumor is its ability to undergo **metastasis** (me-TAS-ta-sis), the spread of cancerous cells to other parts of the body. Initially, malignant cells invade surrounding tissues. As the cancer grows, it expands and begins to compete with normal tissues for space and nutrients. Eventually, the normal tissue decreases in size (atrophies) and dies. When normal cells of the body divide and migrate (for example, skin cells that multiply to heal a superficial cut), their further migration is inhibited by contact on all sides with other skin cells. This is called **contact inhibition.** Malignant cells do not conform to the rules of contact inhibition; they have the ability to invade healthy body tissues with very few restrictions.

Following the nearby invasion, some of the malignant cells may detach from the initial (primary) tumor and invade a body cavity (abdominal or thoracic) or enter the blood or lymph. This latter condition can lead to widespread metastasis. In the next step in metastasis, those malignant cells that survive in the blood or lymph invade adjacent body tissues and establish secondary tumors.

In the final stage of metastasis, the secondary tumors become vascularized; that is, new networks of blood vessels develop to provide nutrients for their further growth. Any new tissue, whether it results from repairing a wound, normal growth, or tumors, requires a blood supply. Proteins that serve as chemical triggers for blood vessel growth are called **tumor angiogenesis factors (TAFs).** In all stages of metastasis, the malignant cells resist the antitumor defenses of the body. The pain associated with cancer develops when the growth puts pressure on nerves or blocks a passageway so that secretions build up pressure.

Causes of Cancer

What triggers a perfectly normal cell to lose control and become abnormal? Scientists believe there are several reasons. First, there are environmental agents: substances in the air we breathe, the water we drink, and the food we eat. A chemical or other environmental agent that produces cancer is called a **carcinogen.** The World Health Organization estimates that carcinogens may be associated with 60 to 90% of all human cancers. Examples of carcinogens are hydrocarbons found in cigarette tar, radon gas from the earth, and ultraviolet (UV) radiation in sunlight. Adopting a cancer-prevention life-style and avoiding environmental carcinogens can help decrease your risk of getting cancer.

Viruses are a second cause of cancer. These agents are tiny packages of nucleic acids, either DNA or RNA, that are capable of infecting cells and converting them to virus-producers. Although the link between viruses and cancer is strongly established for a variety of animal cancers, the relation in human cancers is less predictable. For a few cancers, there is good evidence for a viral role (Exhibit 3.4)

EXHIBIT 3.4

VIRUSES IMPLICATED IN HUMAN CANCERS

Virus	Type of Cancer
Human T-cell leukemia-lymphoma virus–1 (HTLV–1)	Leukemia, a malignant disease of blood-forming tissues, and lymphoma, a cancer of lymphoid tissue.
Human immunodeficiency virus (HIV)	Kaposi's sarcoma, cancer of the inner lining of some blood vessels often seen in those who have acquired immune deficiency syndrome (AIDS), which is caused by HIV.
Epstein–Barr virus (EBV)	Burkitt's lymphoma, a cancer of white blood cells, nasopharyngeal carcinoma (common in Chinese males), and Hodgkin's disease, a lymphatic system cancer; EBV causes infectious mononucleosis.
Hepatitis B virus (HBV)	Liver cancer.
Type 2 herpes simplex virus	Implicated in cancer of the cervix of the uterus; causes genital herpes.
Papilloma virus	Associated with cancer of the cervix, vagina, penis, and colon; causes warts (benign growths).

although the number of people infected with these viruses is much larger than the number who develop cancer.

Intensive research efforts are now directed toward studying **oncogenes** (ON-kō-jenz). These genes have the ability to transform a normal cell into a cancerous cell when they are *inappropriately activated*. Oncogenes develop from normal genes that regulate growth and development, called **proto-oncogenes.** These genes may undergo some change that either causes them to produce an abnormal product or disrupts their control so that they are expressed inappropriately, making their products in excessive amounts or at the wrong time. It is believed that some oncogenes cause excessive production of growth factors, chemicals that stimulate cell growth. Other oncogenes may cause changes in a surface receptor, causing it to send signals as though it were being activated by a growth factor. As a result, the growth pattern of the cell becomes abnormal.

Every cell contains proto-oncogenes. Apparently, proto-oncogenes carry out normal cellular functions until a malignant change occurs. It appears that some proto-oncogenes are activated to oncogenes by various types of mutations in which the DNA of the proto-oncogene is altered. A **mutation** is a permanent structural change in a gene. Carcinogens may induce such mutations. Other proto-oncogenes are activated by a rearrangement of the chromosomes so that segments of DNA are exchanged. Rearrangement is sufficient to activate proto-oncogenes by placing them near genes that enhance their activity. Burkitt's lymphoma, malignant tumors of the colon and rectum, and one type of lung cancer are linked to oncogenes. Viruses are believed to cause cancer by inserting their own oncogenes or proto-oncogenes into the host cell's DNA.

Researchers have also found that some cancers are not caused by oncogenes but rather by genes called **anti-oncogenes** or **tumor-suppressing genes.** These genes may produce proteins that normally oppose the action of an oncogene or inhibit cell division. Mutation of an anti-oncogene called p53 on chromosome 17 is the most common genetic change in a wide variety of cancerous cells. The normal p53 protein appears to prevent cell growth and division.

Carcinogenesis: A Multistep Process

Carcinogenesis (kar′-si-nō-JEN-e-sis), the process by which carcinomas develop, is a multistep process in which as many as ten distinct mutations may have to accumulate in a cell before it becomes cancerous. The progression of gene changes leading to cancer is best understood for colon (and rectal) cancer. Such cancers, as well as lung and breast cancer, take years or decades to develop. In colon cancer, the tumor begins as an area of increased cell proliferation that results from one genetic alteration. This growth then progresses to abnormal, but noncancerous, growths called adenomas. As two or three additional mutations occur, the adenomas develop through successive stages of increasing severity. Finally, with mutation of p53, a carcinoma develops. The fact that so many mutations are needed for cancer development indicates that cells normally control their growth with many sets of checks and balances.

The identification of mutations underlying cancers, such as those that affect the colon, lungs, and breasts may aid scientists in preventing, diagnosing, and treating these malignancies. For example, if it is known that a particular cancer requires several mutations in a multistep process and if people could be identified after only one or two mutations had taken place, they could be counseled to avoid carcinogens that would produce future mutations—mutations that could drive cells into malignancy. Also, by understanding the steps involved in the progression of cancer, scientists may be able to design better treatments.

Treatment of Cancer

Treating cancer is difficult because it is not a single disease and because all the cells in a single population (tumor) do not behave in the same way. Although most cancers are thought to derive from a single abnormal cell, by the time a tumor reaches a clinically detectable size, the cancer may contain a diverse population of cells. For example, some cells metastasize and others do not. Some cells divide and others do not. Some are sensitive to drugs and some are resistant. Because of differences in drug resistance, a single chemotherapeutic drug may destroy susceptible cells but permit resistant cells to proliferate.

Certain tumor cells may be simultaneously resistant to completely unrelated drugs, a phenomenon called **multi-drug resistance.** This broad-based resistance to chemotherapeutic drugs is due to a plasma membrane protein called P-glycoprotein (PGP) that accumulates in resistant tumor cells. PGP actively transports drugs out, and the cells are not killed. Scientists think that if they can find ways to inhibit the action of PGP, anticancer drugs can work more effectively to kill tumor cells.

Another stumbling block encountered by blood-borne anticancer drugs is the physical barrier that solid tumors, such as those that arise in the breasts, lungs, colon, and other organs, develop. Deep within such tumors are high-pressure areas that collapse blood vessels in the tumor. This makes it difficult, if not impossible, for blood-borne anticancer agents to penetrate the tumor. Besides chemotherapy, radiation therapy, surgery, hyperthermia (abnormally high temperature), and immunotherapy (bolstering the body's own defenses) may be used alone or in combination.

CELLS AND AGING

Aging is a normal process accompanied by a progressive alteration of the body's homeostatic adaptive responses. It is a general response that produces observable changes in structure and function and increased vulnerability to environmental stress and disease. The specialized branch of medicine that deals with the medical problems and care of elderly persons is called **geriatrics** (jer′-ē-AT-riks; *geras* = old age; *iatrike* = surgery, medicine).

The obvious characteristics of aging are well known: graying and loss of hair, loss of teeth, wrinkling of skin, decreased muscle mass, and increased fat deposits. The physiological signs of aging are gradual deterioration in function and capacity to respond to environmental stresses. Metabolism slows, as does the ability to maintain a constant internal environment (homeostasis) in response to changes in temperature, diet, and oxygen supply. These signs of aging are related to a net decrease in the number of cells in the body and to the dysfunctioning of the cells that remain.

The extracellular components of tissues (see page 108) also change with age. Collagen fibers, responsible for the strength in tendons, increase in number and change in quality with aging. These changes in the collagen of arterial walls are as much responsible for their loss of extensibility as are the deposits associated with atherosclerosis, the deposition of fatty materials in arterial walls. Elastin, another extracellular component, is responsible for the elasticity of blood vessels and skin. It thickens, fragments, and acquires a greater affinity for calcium with age—changes that may also be associated with the development of atherosclerosis.

Glucose, the most abundant sugar in the body, may play a role in the aging process. According to one hypothesis, glucose is added, haphazardly, to proteins inside and outside cells, forming irreversible cross-links between adjacent protein molecules. As a person ages, more cross-links are formed, and this probably contributes to the stiffening and loss of elasticity that occurs in aging tissues.

Although many millions of new cells normally are produced each minute, several kinds of cells in the body—heart cells, skeletal muscle fibers, nerve cells—cannot be replaced. Experiments have shown that many other cell types have only a limited capability to divide. Cells grown outside the body divide only a certain number of times and then stop. The number of divisions correlates with the donor's age and with the normal life span of the different species from which the cells are obtained. These observations provide strong evidence for the hypothesis that cessation of mitosis is a normal, genetically programmed event. According to this view, an "aging" gene is part of the genetic blueprint at birth, and it turns on at a preprogrammed time, slowing down or halting processes vital to life.

Another theory of aging is the **free radical theory.** Free radicals are electrically charged molecules that have an unpaired electron. Such molecules are unstable and highly reactive and can easily damage proteins. Some effects are wrinkled skin, stiff joints, and hardened arteries. Free radicals may also damage DNA. Among the factors that produce free radicals are air pollution, radiation, and certain foods we eat. Other substances in the diet such as vitamin E, vitamin C, beta-carotene, and selenium are antioxidants and inhibit free radical formation. The free radical theory of aging is bolstered by two recent discoveries. Strains of fruit flies bred for longevity produce larger-than-normal amounts of an enzyme called superoxide dismutase, which functions to neutralize free radicals. Also, injection of genes that lead to production of superoxide dismutase into fruit fly embryos prolongs their average lifetime.

Whereas some theories of aging explain the process at the cellular level, others concentrate on regulatory mechanisms operating within the entire organism. For example, the immune system, which manufactures antibodies against foreign invaders, may start to attack the body's own cells. This autoimmune response might be caused by changes in the surfaces of cells, causing antibodies to attach and mark the cell for destruction. As surface changes in cells increase, the autoimmune response intensifies, producing the well-known signs of aging.

The effects of aging on the various body systems are discussed in their respective chapters.

MEDICAL TERMINOLOGY

NOTE TO THE STUDENT

Each chapter in this text that discusses a major system of the body is followed by a glossary of **medical terminology.** Both normal and pathological conditions of the system are included in these glossaries. You should familiarize yourself with the terms, since they will play an essential role in your medical vocabulary.

Anaplasia (an'-a-PLĀ-zē-a) Reversion to a more primitive or undifferentiated form.

Atrophy (AT-rō-fē; *a* = without; *tropho* = nourish) A decrease in the size of cells with subsequent decrease in the size of the affected tissue or organ; wasting away.

Biopsy (BĪ-op-sē; *bio* = life; *opsis* = vision) The removal and microscopic examination of tissue from the living body for diagnosis.

Dysplasia (dis-PLĀ-zē-a; *dys* = abnormal; *plas* = to grow) Alteration in the size, shape, and organization of cells due to chronic irritation or inflammation; may progress to neoplasia (tumor formation, usually malignant) or revert to normal if the stress is removed.

Hyperplasia (hī-per-PLĀ-zē-a; *hyper* = over) Increase in the number of cells due to an increase in the frequency of cell division.

Hypertrophy (hī-PER-trō-fē) Increase in the size of cells without cell division.

Metaplasia (met'-a-PLĀ-zē-a; *meta* = change) The transformation of one type of cell into another.

Metastasis (me-TAS-ta-sis; *stasis* = standing still) The transfer of disease from one part of the body to another that is not directly connected with it.

Necrosis (ne-KRŌ-sis; *necros* = death; *osis* = condition) Death of a group of cells.

Neoplasm (NĒ-ō-plazm; *neo* = new) Any abnormal formation or growth, usually a malignant tumor.

Progeny (PROJ-e-nē; *progignere* = to bring forth) Offspring or descendants.

Study Outline

Generalized Animal Cell (p. 57)

1. A cell is the basic, living, structural and functional unit of the body.
2. A generalized cell is a composite that represents various cells of the body.
3. Cytology is the science concerned with the study of cells.
4. The principal parts of a cell are the plasma (cell) membrane, cytosol, organelles, and inclusions.

Plasma (Cell) Membrane (p.57)

Membrane Chemistry and Anatomy (p. 57)

1. The plasma (cell) membrane surrounds the cell and separates it from other cells and the external environment.
2. It is composed primarily of phospholipids and proteins. The proteins are integral and peripheral.
3. According to the fluid mosaic model, the membrane is a mosaic of proteins floating like icebergs in a sea of lipids.

Membrane Physiology (p .60)

1. The plasma membrane functions in cellular communication, establishment of an electrochemical gradient, and selective permeability.

2. The membrane's selectively permeable nature restricts the passage of certain substances. Substances can pass through the membrane depending on their lipid solubility, size, electrical charges, and the presence of channels and transporters.

Movement of Materials Across Plasma Membranes (p. 61)

1. Passive processes depend on the concentration of substances and their kinetic energy.
2. Simple diffusion is the net movement of molecules or ions from an area of higher concentration to an area of lower concentration until an equilibrium is reached.
3. Osmosis is the movement of water through a selectively permeable membrane from an area of higher water concentration to an area of lower water concentration.
4. In an isotonic solution, red blood cells maintain their normal shape; in a hypotonic solution, they undergo hemolysis; in a hypertonic solution, they undergo crenation.
5. Filtration is the movement of water and dissolved substances across a membrane due to gravity or hydrostatic pressure.
6. In facilitated diffusion, certain molecules, such as glucose, move through the membrane with the help of a transporter.
7. Active processes depend on the use of ATP by the cell. The two principal types are active transport and bulk transport.

8. Active transport is the movement of a substance across a cell membrane from lower to higher concentration using energy derived from ATP either directly or indirectly.
9. The most prevalent primary active transport pump is the sodium pump.
10. Secondary active transport mechanisms may operate as symports or antiports.
11. Phagocytosis is the ingestion of solid particles. It is an important process used by white blood cells to destroy bacteria that enter the body.
12. Pinocytosis is the ingestion of fluid. In this process, the fluid becomes surrounded by a pinocytic vesicle.
13. Receptor-mediated endocytosis is the selective uptake of large molecules and particles (ligands) by cells.

Cytosol (p. 68)

1. Cytoplasm is the substance inside the cell between the plasma membrane and nucleus. The semifluid portion of cytoplasm that contains inclusions and dissolved solutes is called the cytosol.
2. Cytosol is composed mostly of water plus proteins, carbohydrates, lipids, and inorganic substances. The chemicals in cytosol are either in solution or in a colloid (suspended) form.
3. Functionally, cytosol is the medium in which chemical reactions occur.

Organelles (p. 69)

1. Organelles are specialized structures suspended in the cytosol that have characteristic appearances and functions.
2. They play specific roles in cellular growth, maintenance, repair, and control.

Nucleus (p. 70)

1. Usually the largest organelle, the nucleus, controls cellular activities and contains the genetic information.
2. Most body cells have a single nucleus; some (red blood cells) have none, whereas others (skeletal muscle fibers) have several.
3. The parts of the nucleus include the double nuclear envelope, nucleoli, and chromosomes.
4. Chromosomes consist of subunits called nucleosomes that are composed of DNA (genetic material) and histones. The various levels of DNA packing are represented by nucleosomes, chromatin fibers, loops, chromatids, and chromosomes.

Ribosomes (p. 71)

1. Ribosomes are tiny spheres consisting of ribosomal RNA and ribosomal proteins.
2. They occur free (singly or in clusters) or together with endoplasmic reticulum.
3. Functionally, ribosomes are the sites of protein synthesis.

Endoplasmic Reticulum (ER) (p. 71)

1. The ER is a network of membrane-enclosed channels continuous with the nuclear membrane.
2. Rough or granular ER has ribosomes attached to it. Smooth or agranular ER does not contain ribosomes.
3. The ER transports substances, stores newly synthesized molecules, synthesizes and packages molecules, detoxifies chemicals, and releases calcium ions involved in muscle contraction.

Golgi Complex (p. 73)

1. The Golgi complex consists of four to eight stacked, flattened membranous sacs (cisterns) referred to as *cis, medial,* and *trans.*

2. The principal function of the Golgi complex is to process, sort, and deliver proteins and lipids to the plasma membrane, lysosomes, and secretory vesicles.

Lysosomes (p. 74)

1. Lysosomes are spherical structures that contain digestive enzymes. They are formed by the Golgi complex.
2. They are found in large numbers in white blood cells, which carry on phagocytosis.
3. Lysosomes function in intracellular digestion, digestion of worn-out organelles (autophagy), digestion of cellular contents (autolysis) during embryological development, and extracellular digestion.

Peroxisomes (p. 75)

1. Peroxisomes are similar to lysosomes but smaller.
2. They contain enzymes (for example, catalase) that use molecular oxygen to oxidize various organic substances.

Mitochondria (p. 75)

1. Mitochondria consist of a smooth outer membrane and a folded inner membrane surrounding the interior matrix. The inner folds are called cristae.
2. The mitochondria are called "powerhouses" of the cell because ATP is produced in them.

The Cytoskeleton (p. 75)

1. Together, microfilaments, microtubules, and intermediate filaments form the cytoskeleton.
2. The cytoskeleton provides organization for chemical reactions and assists in transporting chemicals and organelles through the cytosol.

Flagella and Cilia (p. 76)

1. These cellular projections have the same basic structure and are used in movement.
2. If projections are few (typically occurring singly or in pairs) and long, they are called flagella. If they are numerous and hairlike, they are called cilia.
3. The flagellum on a sperm cell moves the entire cell. The cilia on cells of the respiratory tract move foreign matter trapped in mucus along the cell surfaces toward the throat for elimination.

Centrosome and Centrioles (p. 77)

1. The dense area of cytoplasm containing the centrioles is called a centrosome.
2. Centrioles are paired cylinders arranged at right angles to one another.
3. Centrosomes serve as centers for organizing microtubules in interphase cells and the mitotic spindle during cell division.

Cell Inclusions (p. 77)

1. Cell inclusions are chemical substances produced by cells. They are usually organic and may have recognizable shapes.
2. Examples are melanin, glycogen, and triglycerides (neutral fats).

Gene Action (p. 77)

1. Most of the cellular machinery is concerned with synthesizing proteins.
2. Cells make proteins by translating the genetic information encoded in DNA into specific proteins. This involves transcription and translation.

3. In transcription, genetic information encoded in DNA is copied by a strand of messenger RNA (mRNA).
4. DNA also synthesizes ribosomal RNA (rRNA) and transfer RNA (tRNA).
5. The process of using the information in the base sequence of mRNA to dictate the amino acid sequence of a protein is known as translation.
6. mRNA associates with ribosomes, which consist of rRNA and protein.
7. Specific amino acids attach to molecules of tRNA. Another portion of the tRNA has a triplet of nitrogenous bases called an anticodon; a codon is a segment of three bases of mRNA.
8. tRNA delivers a specific amino acid to the codon; the ribosome moves along an mRNA strand as amino acids are joined to form a growing polypeptide.

Normal Cell Division (p. 81)

1. Cell division is the process by which cells reproduce themselves. It consists of nuclear division (mitosis or meiosis) and cytoplasmic division (cytokinesis).
2. Cell division that results in an increase in body cells is called somatic cell division and involves a nuclear division called mitosis plus cytokinesis.
3. Cell division that results in the production of sperm and eggs is called reproductive cell division and consists of a nuclear division called meiosis plus cytokinesis.

Somatic Cell Division (p. 82)

1. Prior to mitosis and cytokinesis, the DNA molecules, or chromosomes, replicate themselves so the same chromosomal complement can be passed on to future generations of cells.
2. A cell between divisions carrying on every life process except division is said to be in interphase (metabolic phase). It consists of three phases: G_1, S, and G_2.
3. Mitosis is the distribution of two sets of chromosomes into separate and equal nuclei following their replication.
4. It consists of prophase, metaphase, anaphase, and telophase.
5. Cytokinesis usually begins in late anaphase and ends in telophase.
6. A cleavage furrow forms at the cell's metaphase plate and progresses inward, cutting through the cell to form two separate portions of cytoplasm.

Reproductive Cell Division (p. 85)

1. Gametes contain the haploid (n) chromosome number, and uninucleated somatic cells contain the diploid ($2n$) chromosome number.
2. Meiosis is the process that produces haploid gametes. It consists of two successive nuclear divisions called reduction division (meiosis I) and equatorial division (meiosis II).
3. During reduction division, homologous chromosomes undergo synapsis and crossing-over; the net result is two haploid daughter cells.
4. During equatorial division, the two haploid daughter cells undergo mitosis, and the net result is four haploid cells.

Control of Cell Division (p. 87)

1. Maturation promoting factor (MPF) induces cell division.
2. MPF consists of cdc2 proteins and cyclin.
3. Cyclin builds up during interphase and activates cdc2 proteins and MPF, which induce mitosis (or meiosis).

Abnormal Cell Division: Cancer (p. 89)

1. Cancerous tumors are referred to as malignant; noncancerous tumors are called benign; the study of tumors is called oncology.
2. Malignant cells undergo hyperplasia.
3. The spread of cancer from its initial site is called metastasis.
4. Carcinogens are chemicals or environmental agents that can produce cancer. Viruses are also implicated.
5. Oncogenes are genes that can transform normal cells into cancerous cells; their normal counterparts are called proto-oncogenes.
6. Carcinogenesis is a multistep process involving mutation of oncogenes and anti-oncogenes.
7. Treating cancer is difficult because all the cells in a tumor do not behave in the same way.

Cells and Aging (p. 91)

1. Aging is a normal process accompanied by progressive alteration of the body's homeostatic adaptive responses.
2. Many theories of aging have been proposed, including genetically programmed cessation of cell division, excessive immune responses, and buildup of free radicals, but none successfully answers all the experimental objections.

Review Questions

1. Define a cell. What are the four principal portions of a cell? What is meant by a generalized cell? (p. 57)
2. Discuss the chemistry of the plasma membrane. (p. 57)
3. Describe the various functions of the plasma membrane. What determines selective permeability? (p. 60)
4. What are the major differences between passive processes and active processes in moving substances across plasma membranes? (p. 61)
5. Define and give an example of each of the following: simple diffusion, osmosis, filtration, facilitated diffusion, primary and secondary active transport, phagocytosis, pinocytosis, and receptor-mediated endocytosis. (p. 61)
6. Compare the effect of isotonic, hypertonic, and hypotonic solutions on red blood cells. What is osmotic pressure? (p. 63)
7. Discuss the chemical composition and physical nature of the cytosol. What is its function? (p. 68)
8. What is an organelle? (p. 69) By means of a labeled diagram, indicate the parts of a generalized animal cell.
9. Describe the structure and functions of the nucleus of a cell. Describe how DNA is packed into chromosomes. (p. 70)
10. Discuss the distribution and function of ribosomes. (p. 71)
11. Distinguish between rough and smooth endoplasmic reticulum (ER). What are the functions of ER? (p. 71)
12. Describe the structure and functions of the Golgi complex. (p. 73)
13. List and describe the functions of lysosomes. (p. 74)
14. What is the importance of peroxisomes? (p. 75)
15. Why are mitochondria referred to as "powerhouses" of the cell? (p. 75)
16. Contrast the structure and functions of microfilaments, microtubules, and intermediate filaments. (p. 75)
17. What are the structural and functional differences between flagella and cilia? (p. 76)

18. Describe the structure and function of the centrosome. (p. 77)
19. Define a cell inclusion. Give examples. (p. 77)
20. Summarize the steps involving gene action in protein synthesis. (p. 78)
21. Distinguish between the two types of cell division. Why is each important? (p. 81)
22. Define interphase. How does DNA replicate itself? (p. 82)
23. Describe the principal events of each stage of mitosis. (p. 82)
24. Distinguish between haploid (*n*) and diploid (*2n*) cells. (p. 86)
25. Define meiosis and contrast the principal events of reduction division and equatorial division. (p. 86)

26. How is cell division controlled? (p. 87)
27. What is a tumor? Distinguish between malignant and benign tumors. (p. 89)
28. Define metastasis. What factors contribute to metastasis? (p. 89)
29. Discuss several possible causes of cancer (CA). Distinguish oncogenes, proto-oncogenes, and anti-oncogenes. (p. 89)
30. Describe how carcinogenesis is a multistep process. (p. 90)
31. How is cancer treated? What are some of the problems with respect to treating cancer? (p. 91)
32. What is aging? List some of the characteristics of aging. (p. 91)

Answers to Questions with Figures

3.1 Plasma membrane, cytosol, organelles, inclusions.
3.2 Phospholipids, cholesterol, and glycolipids.
3.3 Receptor.
3.4 The major intracellular anions: proteins and organic phosphates.
3.5 Since a fever represents an increase in body temperature, all diffusion processes would be increased.
3.6 Hydrophobic.
3.7 No, that can never happen, since the beaker always contains pure water. Sugar cannot cross the cellophane membrane.
3.8 An isotonic solution should be used so net osmosis does not occur.
3.9 Facilitated diffusion is a passive process, so no energy is expended to transport glucose across the membrane. Once inside the cell, however, ATP is required to attach the phosphate group to glucose to form glucose 6-phosphate.
3.10 ATP adds phosphate to the pump protein, which changes its three dimensional shape.
3.11 Endocytosis, because it is a process that brings material into the cell.
3.12 Pinocytosis is a less specific process than receptor-mediated endocytosis.
3.13 It controls cellular activity by regulating which proteins will be synthesized.

3.14 Double-stranded DNA wrapped twice around a core of eight histones (proteins).
3.15 Rough ER has attached ribosomes whereas smooth ER does not. Rough ER synthesizes proteins that will be exported from the cell, while smooth ER is associated with lipid synthesis and other metabolic reactions.
3.16 It packages proteins into vesicles.
3.17 The membrane of the vesicle holds the digestive enzymes inside so they cannot inappropriately break down molecules within the cytosol.
3.18 They increase surface area for chemical reactions.
3.19 It probably will not be able to undergo cell division.
3.20 UCGA.
3.21 Translation is synthesizing protein from mRNA.
3.22 S-phase of interphase.
3.23 Because each of the new daughter cells must have a complete genome.
3.24 Usually starts in anaphase and ends in telophase.
3.25 The gametes produced contain the haploid number of chromosomes.
3.26 The daughter cells are unlike each other and unlike the parent cell that produced them.
3.27 In testes of males and ovaries of females.

Chapter 4
THE TISSUE LEVEL OF ORGANIZATION

Chapter Contents at a Glance

Student Objectives

1. Define a tissue and classify the tissues of the body into four major types.
2. Describe the structure and functions of the three principal types of cell junctions.
3. Describe the general features of epithelial tissue.
4. List the structure, location, and function for the following types of epithelium: simple squamous, simple cuboidal, simple columnar (nonciliated and ciliated), stratified squamous, stratified cuboidal, stratified columnar, transitional, and pseudostratified columnar.
5. Define a gland and distinguish between exocrine and endocrine glands.
6. Describe the general features of connective tissue.
7. Discuss the cells, ground substance, and fibers that compose connective tissue.
8. List the structure, function, and location of mesenchyme; mucous connective tissue; areolar connective tissue; adipose tissue; reticular connective tissue; dense regular and irregular connective tissue; elastic connective tissue; cartilage; bone; and blood.
9. Define an epithelial membrane and list the location and function of mucous, serous, cutaneous, and synovial membranes.
10. Contrast the three types of muscle tissue with regard to structure, location, and modes of control.
11. Describe the structural features and functions of nervous tissue.
12. Describe tissue repair in restoring homeostasis.

Cells are highly organized, but they do not function as isolated units in your body. They work together in groups called tissues. A **tissue** is a group of similar cells that usually have a common embryonic origin and function together to carry out specialized activities. The structure and properties of a specific tissue are influenced by such things as the nature of the extracellular material that surrounds the tissue cells and connections between the cells that compose the tissue. The science that deals with the study of tissues is **histology** (hiss′-TOL-ō-jē; *histio* = tissue; *logos* = study of).

CLINICAL APPLICATION

MEDICAL STUDIES OF TISSUES

A **pathologist** (pa-THOL-ō-gist; *pathos* = disease) is a physician who specializes in laboratory studies of cells and tissues to help other physicians reach accurate diagnoses. One of the principal functions of a pathologist is to examine tissues for any changes that might indicate disease. Pathologists also conduct autopsies (see page 19) and supervise laboratory personnel who examine blood and other body fluids.

A **biopsy** (BĪ-op-sē) is the removal of a sample of living tissue for microscopic examination. It is used to help diagnose many disorders, especially cancer, and to discover the cause of unexplained infections and inflammations. Once the tissue sample is removed, it may be preserved, stained to highlight special properties, and cut into thin sections for microscopic observation.

TYPES OF TISSUES AND THEIR ORIGINS

Body tissues can be classified into four principal types according to their function and structure:

1. **Epithelial tissue,** which covers body surfaces; lines hollow organs, body cavities, and ducts; and forms glands.
2. **Connective tissue,** which protects and supports the body and its organs; binds organs together; stores energy reserves as fat; and provides immunity.
3. **Muscle tissue,** which is responsible for movement and generation of force.
4. **Nervous tissue,** which initiates and transmits action potentials (nerve impulses) that help coordinate body activities.

About 8 days after a sperm fertilizes an egg, the mass of cells that results from several cell divisions embeds in the lining of the uterus and begins to form three primary **germ layers: ectoderm, endoderm,** and **mesoderm.** These are the embryonic tissues from which all tissues and organs of the body develop. Epithelial tissues develop from all three germ layers. Connective tissues and muscle tissues all derive from mesoderm. Ectoderm gives rise to nervous tissue. (See Exhibit 29.1 on page 978 for a more detailed list of structures derived from the primary germ layers.)

Epithelial tissue and connective tissue, except for bone and blood, are discussed in detail in this chapter. The general features of bone tissue and blood will be introduced here, but their detailed discussion occurs later in Chapters 6 and 19, respectively. Similarly, the structure and function of muscle tissue and nervous tissue are examined in detail in Chapters 10 and 12, respectively.

Surrounding all body cells is **extracellular fluid (ECF),** which provides a medium for dissolving and mixing solutes, transporting substances, and carrying out chemical reactions. The two major subdivisions of ECF are **interstitial (intercellular) fluid,** the fluid that fills the microscopic spaces (interstitial spaces) between cells in tissues, and **plasma,** the liquid portion of blood, found in blood vessels (see Fig. 1.10).

Normally, most cells within a tissue remain in place, anchored to other cells, basement membranes (described shortly), and connective tissues. A few cells, such as phagocytes, routinely move through the body, searching for invaders. Before birth, certain cells migrate extensively as part of the growth and development process.

CELL JUNCTIONS

Most epithelial cells, some muscle cells, and some nerve cells are tightly joined to form a close functional unit. The points of contact between adjacent plasma membranes are called **cell junctions.** Three types of cell junctions serve distinct functions: (1) **tight junctions** form fluid-tight seals between cells like the seal on a Ziploc sandwich bag; (2) **anchoring junctions** fasten cells to one another or to the extracellular material; and (3) **communicating junctions** permit electrical or chemical signals to pass from cell to cell.

Tight junctions are common among epithelial cells that line the stomach, intestines, and urinary bladder (Fig. 4.1a). They prevent fluid in a cavity from leaking into the body by passing between cells.

Anchoring junctions are common in tissues subjected to friction and stretching. Examples include the outer layer of the skin, the muscle tissue of the heart, the neck of the uterus (which is greatly stretched during childbirth), and the epithelial lining of the gastrointestinal tract. Anchoring junctions include adherens junctions, desmosomes, and hemidesmosomes.

Adherens junctions connect to microfilaments of the cytoskeleton (Fig. 4.1b) and link cells to one another or anchor cells to extracellular materials. **Desmosomes** form firm attachments between cells somewhat like spot welds (Fig. 4.1c). On the cytoplasmic surface of a desmosome, intermediate filaments of the cytoskeleton attach to a dense plaque of proteins. **Hemidesmosomes** (*hemi* = half) look like half a desmosome (Fig. 4.1d). They anchor the basal (bottom) epithelial cell plasma membranes to extracellular materials at the junction between epithelial and connective tissues.

FIGURE 4.1 Cell junctions.

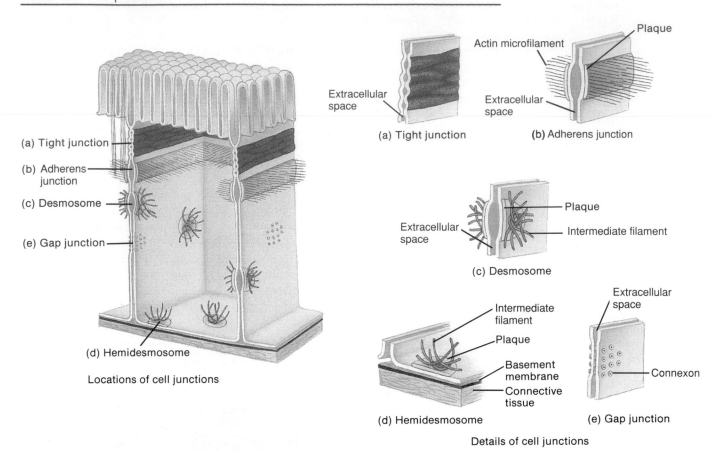

(a) Tight junction
(b) Adherens junction
(c) Desmosome
(e) Gap junction
(d) Hemidesmosome

Locations of cell junctions

Extracellular space
Actin microfilament
Plaque
Extracellular space

(a) Tight junction   (b) Adherens junction

Extracellular space
Plaque
Intermediate filament

(c) Desmosome

Intermediate filament
Plaque
Basement membrane
Connective tissue

Extracellular space
Connexon

(d) Hemidesmosome   (e) Gap junction

Details of cell junctions

Question: Which type of junction functions in communication between adjacent cells?

Communicating junctions allow the rapid spread of action potentials from one cell to the next in some parts of the nervous system and in muscle of the heart and gastrointestinal tract. In a developing embryo, chemical and electrical signals that regulate growth and differentiation may travel via communicating junctions. The microscopic structure that functions as a communicating junction is the **gap junction** (Fig. 4.1e). At a gap junction, adjacent plasma membranes approach each other, not fusing but leaving a gap of 2 to 3 nm. Spanning the gap are proteins called **connexons** that form minute fluid-filled tunnels. Through the connexons, ions and small molecules, such as glucose and amino acids, can pass directly from the cytosol of one cell into the cytosol of the next without crossing the plasma membrane. Note that cancer cells do not have gap junctions and therefore cannot communicate with each other. As a result, cell division is not coordinated and occurs in an uncontrolled manner.

EPITHELIAL TISSUE

Epithelial (ep-i-THĒ-lē-al) **tissue,** or **epithelium** (plural is **epithelia**), may be divided into two types: (1) **covering and lining epithelium** and (2) **glandular epithelium.** Covering and lining epithelium forms the outer layer of the skin and the outer layer of some internal organs. It forms the inner lining of blood vessels, ducts, body cavities, and the interiors of the respiratory, digestive, urinary, and reproductive systems. Glandular epithelium constitutes the secreting portion of glands, such as sweat glands and the thyroid gland. Epithelial tissue also combines with nervous tissue to make up special sense organs for smell, hearing, vision, and touch.

General Features of Epithelial Tissue

Following are the general features of epithelial tissue:

1. Epithelium consists largely or entirely of closely packed cells with little extracellular material between adjacent cells.

2. Epithelial cells are arranged in continuous sheets, in either single or multiple layers.

3. Epithelial cells have an **apical** (free) **surface,** which is exposed to a body cavity, lining of an internal organ, or the exterior of the body, and a **basal surface,** which is attached to the basement membrane (Fig. 4.2). The basement membrane is discussed shortly.

4. Cell junctions are plentiful, providing secure attachments among the cells.

5. Epithelia are **avascular** (*a* = without; *vascular* = blood vessels). The vessels that supply nutrients and remove wastes are located in the adjacent connective tissue. The exchange of materials between epithelium and connective tissue is by diffusion.

6. Epithelia adhere firmly to nearby connective tissue, which holds the epithelium in position and prevents it from being torn. The attachment between the epithelium and the connective tissue is a thin extracellular layer called the **basement membrane** (Fig. 4.2). It consists of two layers. The **basal lamina** contains collagen, laminin, and proteoglycans secreted by the epithelium. Cells in the connective tissue secrete the second layer, the **reticular lamina,** which contains reticular fibers, fibronectin, and glycoproteins. The basement membrane provides physical support for epithelium, provides for cell attachment, serves as a filter in the kidneys, and guides cell migration during development and tissue repair.

7. Epithelia have a nerve supply.

8. Since epithelium is subject to a certain amount of wear and tear and injury, it has a high capacity for renewal (high mitotic rate).

9. Epithelia are diverse in origin. They are derived from all three primary germ layers (ectoderm, mesoderm, and endoderm).

10. Functions of epithelia include protection, filtration, lubrication, secretion, digestion, absorption, transportation, excretion, sensory reception, and reproduction.

Covering and Lining Epithelium

Arrangement of Layers

The arrangement of covering and lining epithelium reflects its location and function. If the function of the epithelium is absorption or filtration and it is in an area that has minimal wear and tear, the cells of the tissue form a single layer. Such an arrangement is called **simple epithelium.** If the epithelium is in an area with a high degree of wear and tear, then the cells are stacked in several layers. This type of tissue is called **stratified epithelium.** A third, less common arrangement of epithelium is **pseudostratified epithelium.** Like simple epithelium, pseudostratified epithelium has only one layer of cells. However, some of the cells do not reach the surface, which gives the tissue a multilayered, or stratified, appearance.

Cell Shapes

Besides classifying covering and lining epithelium according to the number of its layers, we may also categorize it by cell shape. The cells may be flat, cubelike, columnar, or a combination of shapes. **Squamous** (SKWĀ-mus) cells are flattened and scalelike. **Cuboidal** cells are usually cube-shaped in cross section. **Columnar** cells are tall and cylindrical or somewhat rectangular. **Transitional** cells readily change shape and are found when there is a great degree of distention (stretching) in the body. Transitional cells in the basal (bottom) layer of an epithelial tissue may range in shape from cuboidal to columnar. In the intermediate layer, they may be cuboidal or polyhedral (having many sides). In the superficial layer, they may range from cuboidal to squamous, depending on how much they are stretched during certain body functions.

Classification

Considering layers and cell shapes in combination, we may classify covering and lining epithelium as follows:

I. Simple
  A. Squamous
  B. Cuboidal
  C. Columnar

II. Stratified
  A. Squamous*
  B. Cuboidal*
  C. Columnar*
  D. Transitional

III. Pseudostratified columnar

Each of the epithelial tissues described in the following sections is illustrated in Exhibit 4.1. Along with the illustrations are the descriptions, locations, and functions of the tissues.

FIGURE 4.2 Surfaces of epithelial cells and the structure and location of the basement membrane.

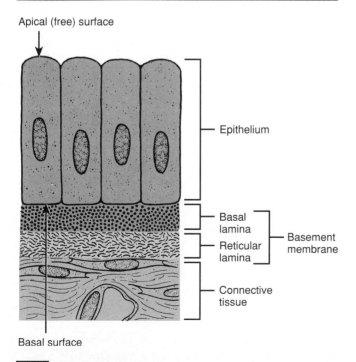

Apical (free) surface

— Epithelium

— Basal lamina

— Reticular lamina

— Basement membrane

— Connective tissue

Basal surface

Question: What are the functions of the basement membrane?

* This classification is based on the shape of the apical surface cells.

EXHIBIT 4.1

EPITHELIAL TISSUES

COVERING AND LINING EPITHELIUM

Simple Squamous Epithelium

Description: Single layer of flat, scalelike cells.
Location: Lines air sacs of lungs, glomerular (Bowman's) capsule of kidneys, and inner surface of the tympanic membrane (eardrum) of ear.
Function: Filtration, diffusion, osmosis and secretion in serous membranes.

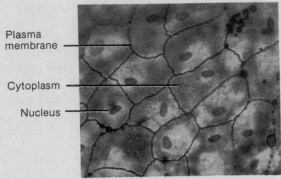

Plasma membrane

Cytoplasm

Nucleus

Surface view of mesothelial lining of peritoneal cavity (243 x)

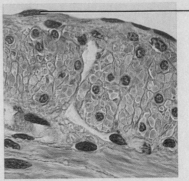

Nucleus of simple squamous cell

Sectional view of intestinal serosa (245 ×)

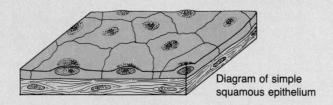

Diagram of simple squamous epithelium

Simple Cuboidal Epithelium

Description: Single layer of cube-shaped cells.
Location: Covers surface of ovary, lines anterior surface of capsule of the lens of eye, forms the pigmented epithelium at the back of the eye, and lines kidney tubules and smaller ducts of many glands.
Function: Secretion and absorption.

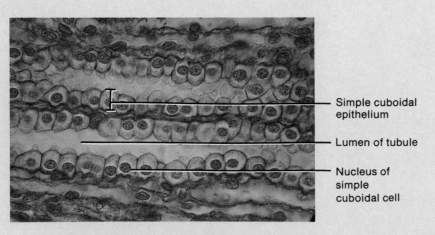

Simple cuboidal epithelium

Lumen of tubule

Nucleus of simple cuboidal cell

Sectional view of kidney tubules, (575 ×)

Diagram of simple cuboidal epithelium

Nonciliated Simple Columnar Epithelium

Description: Single layer of nonciliated rectangular cells; contains goblet cells and microvilli in some locations.
Location: Lines the gastrointestinal tract from the stomach to the anus, ducts of many glands, and gallbladder.
Function: Secretion and absorption.

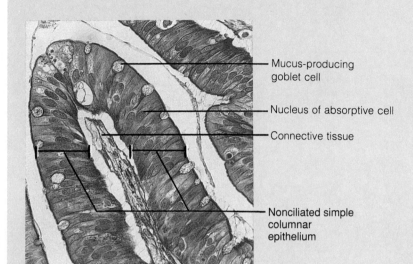

Mucus-producing goblet cell

Nucleus of absorptive cell

Connective tissue

Nonciliated simple columnar epithelium

Sectional view of epithelium from the lining of small intestine (140x)

Diagram of nonciliated simple columnar epithelium

Ciliated Simple Columnar Epithelium

Description: Single layer of ciliated columnar cells; contains goblet cells in some locations.
Location: Lines a few portions of upper respiratory tract, uterine (Fallopian) tubes, uterus, some paranasal sinuses, and central canal of spinal cord.
Function: Moves fluids or particles along a passageway by ciliary action.

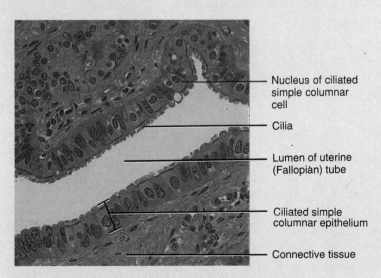

Nucleus of ciliated simple columnar cell

Cilia

Lumen of uterine (Fallopian) tube

Ciliated simple columnar epithelium

Connective tissue

Sectional view of uterine (Fallopian) tube (100 ×)

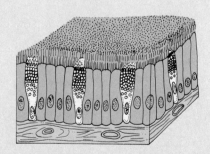

Diagram of ciliated simple columnar epithelium

Exhibit continues

EXHIBIT 4.1 (continued)

EPITHELIAL TISSUES

Stratified Squamous Epithelium

Description: Several layers of cells; cuboidal to columnar shape in deep layers; squamous cells in superficial layers; basal cells replace surface cells as they are lost.
Location: Keratinized variety forms outer layer of skin; nonkeratinized variety lines wet surfaces such as lining of the mouth, esophagus, part of epiglottis, and vagina and covers the tongue.
Function: Protection.

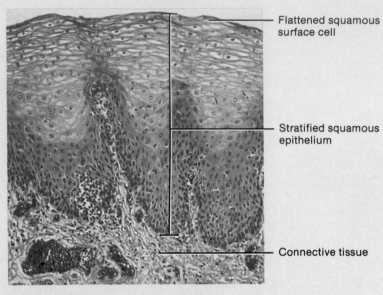

— Flattened squamous surface cell

— Stratified squamous epithelium

— Connective tissue

Sectional view of vagina (200 x)

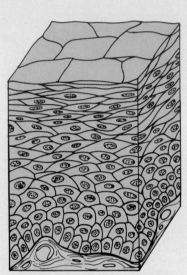

Diagram of stratified squamous epithelium

Stratified Cuboidal Epithelium

Description: Two or more layers of cells in which the surface cells are cube-shaped.
Location: Ducts of adult sweat glands and part of male urethra.
Function: Protection.

Stratified cuboidal epithelium

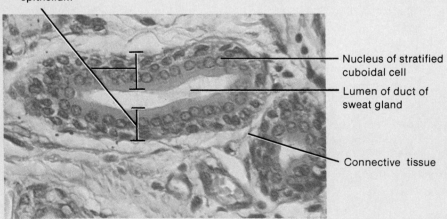

Nucleus of stratified cuboidal cell

Lumen of duct of sweat gland

Connective tissue

Sectional view of the duct of a sweat gland (450x)

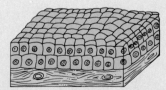

Diagram of stratified cuboidal epithelium

Stratified Columnar Epithelium

Description: Several layers of polyhedral cells; columnar cells only in superficial layer.
Location: Lines part of male urethra, large excretory ducts of some glands, small areas in anal mucous membrane and portion of conjunctiva of eye.
Function: Protection and secretion.

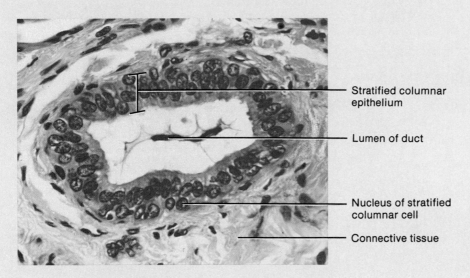

Stratified columnar epithelium

Lumen of duct

Nucleus of stratified columnar cell

Connective tissue

Sectional view of the duct of the submandibular salivary gland (495 ×)

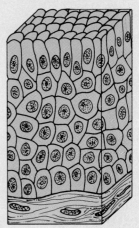

Diagram of stratified columnar epithelium

Transitional Epithelium

Description: Appearance is variable (transitional), ranging from stratified squamous to stratified cuboidal, depending on the degree of distention (stretching)
Location: Lines urinary bladder and portions of ureters and urethra.
Function: Permits distention.

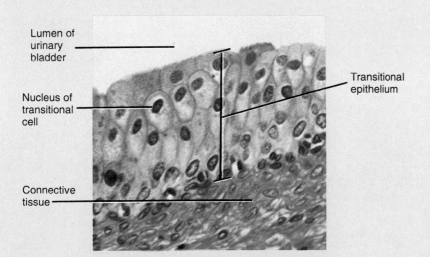

Lumen of urinary bladder

Nucleus of transitional cell

Transitional epithelium

Connective tissue

Sectional view of urinary bladder in relaxed state (240x)

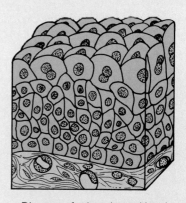

Diagram of relaxed transitional epithelium

Exhibit continues

EPITHELIAL TISSUES

Pseudostratified Columnar Epithelium

Description: Not a true stratified tissue; nuclei of cells at different levels; all cells attached to basement membrane, but not all reach surface.

Location: Lines larger ducts of many large glands, epididymis, part of male urethra, and auditory (Eustachian) tubes; ciliated variety with goblet cells lines most of the upper respiratory tract.

Function: Secretion and movement of mucus by ciliary action.

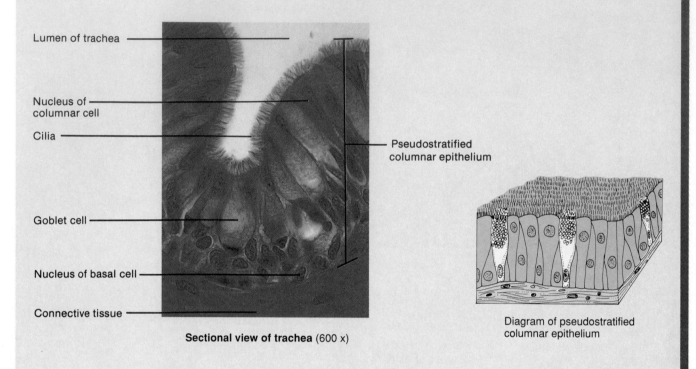

Lumen of trachea

Nucleus of columnar cell

Cilia

Goblet cell

Nucleus of basal cell

Connective tissue

Pseudostratified columnar epithelium

Sectional view of trachea (600 x)

Diagram of pseudostratified columnar epithelium

GLANDULAR EPITHELIUM

Exocrine Glands

Description: Secretory products released into ducts.

Location: Sweat, oil, ear wax, and mammary glands of the skin; digestive glands such as salivary glands that secrete into mouth cavity and pancreas that secretes into the small intestine.

Function: Produce mucus, perspiration, oil, ear wax, milk, or digestive enzymes.

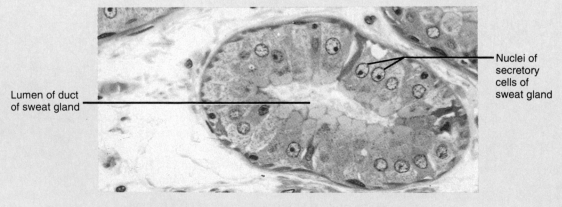

Lumen of duct of sweat gland

Nuclei of secretory cells of sweat gland

Sectional view of the secretory portion of a sweat gland (1532×)

Endocrine Glands

Description: Secretory products (hormones) diffuse into blood after passing through extracellular fluid.
Location: Examples include pituitary gland at base of brain, thyroid and parathyroid glands near larynx (voice box), adrenal glands above kidneys, ovaries in pelvic cavity, testes in scrotum, and thymus gland in thoracic cavity.
Function: Produce hormones that regulate various body activities.

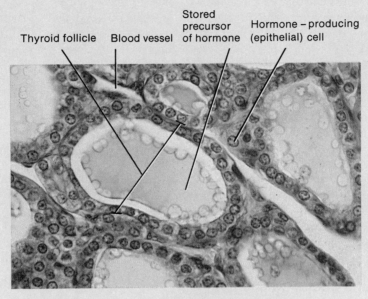

Thyroid follicle   Blood vessel   Stored precursor of hormone   Hormone – producing (epithelial) cell

Sectional view of thyroid gland (500 ×)

Simple Epithelium

● **Simple Squamous Epithelium**   This consists of a single layer of flat, scalelike cells. Its surface resembles a tiled floor. The nucleus of each cell is oval or spherical. Since simple squamous epithelium has only one layer of cells, it is highly adapted for diffusion, osmosis, and filtration. Simple squamous epithelium is found in body parts that are subject to little wear and tear.

Simple squamous epithelium that lines the heart, blood vessels, and lymphatic vessels and forms the walls of capillaries is known as **endothelium.** Simple squamous epithelium that forms the epithelial layer of serous membranes is called **mesothelium.** Endothelium and mesothelium both derive from the embryonic mesoderm.

● **Simple Cuboidal Epithelium**   The cuboidal nature of the cells is obvious only when the tissue is sectioned at right angles. The nuclei are usually round. Simple cuboidal epithelium performs the functions of secretion and absorption. **Secretion** is the production and release by cells of a fluid that may contain a variety of substances such as mucus, perspiration, or enzymes. **Absorption** is the intake of fluids or other substances by cells.

● **Simple Columnar Epithelium**   When sectioned at right angles to the surface, the cells appear somewhat rectangular with oval nuclei. Simple columnar epithelium exists in two forms: **nonciliated simple columnar epithelium** and **ciliated simple columnar epithelium.** The nonciliated type contains microvilli and goblet cells. **Microvilli** (see Fig. 24.23d) are microscopic fingerlike projections that serve to increase the surface area of the plasma membrane and thereby allow larger amounts of digested nutrients and fluids to be absorbed into the body. **Goblet cells** secrete mucus that accumulates in the upper half of the cell, causing that area to bulge out. The whole cell then resembles a goblet or wine glass. The secreted mucus serves as a lubricant and protects.

Another modification of columnar epithelium is found in cells with hairlike processes called **cilia** (*cillio* = to move). In a few parts of the upper respiratory tract, ciliated columnar cells are interspersed with goblet cells. Mucus secreted by the goblet cells forms a film over the respiratory surface that traps foreign particles that are inhaled. The cilia wave in unison and move the mucus, with any foreign particles, toward the throat, where it can be swallowed or spit out.

Stratified Epithelium

In contrast to simple epithelium, stratified epithelium has at least two layers of cells. Thus it is more durable and can protect underlying tissues from the external environment and from wear and tear. Some cells of stratified epithelia also produce secretions. The name of the specific kind of stratified epithelium depends on the shape of the *surface* cells.

● **Stratified Squamous Epithelium** In the superficial layers of this type of epithelium, the cells are flat, whereas in the deep layers, cells vary in shape from cuboidal to columnar. The basal (bottom) cells continually replicate by cell division. As new cells grow, the cells of the basal layer continually shift upward and outward and push the surface cells outward. As they move farther from the deep layer and their blood supply (in the underlying connective tissue), they become dehydrated, shrunken, and harder. At the surface, the cells lose their cell junctions and are rubbed off. Old cells are sloughed off and replaced as new cells continually emerge. The Pap smear, a screening test for precancer and cancer of the uterus, cervix, and vagina (see page 943), consists of examining cells that are sloughed off.

Stratified squamous epithelium exists in two forms: keratinized and nonkeratinized. In **keratinized stratified squamous epithelium,** a tough layer of keratin is deposited in the surface cells. **Keratin** is a protein that is waterproof and resistant to friction and helps repel bacteria. **Nonkeratinized stratified squamous epithelium** does not contain keratin.

● **Stratified Cuboidal Epithelium** This fairly rare type of epithelium sometimes consists of more than two layers of cells. Its function is mainly protective.

● **Stratified Columnar Epithelium** Like stratified cuboidal epithelium, this type of tissue also is uncommon in the body. Usually the basal layer or layers consist of shortened, irregularly polyhedral cells. Only the superficial cells are columnar in form. It functions in protection and secretion.

● **Transitional Epithelium** This kind of epithelium is variable in appearance, depending on whether it is relaxed or distended (stretched). In the relaxed state, it looks similar to stratified cuboidal epithelium except that the upper cells tend to be large and rounded. This allows the tissue to be distended without the outer cells breaking apart from one another. When stretched, they are drawn out into squamous-shaped cells, giving the appearance of stratified squamous epithelium. Because of this capability, transitional epithelium lines hollow structures that are subjected to expansion from within, such as the urinary bladder. Its function is to help prevent rupture of these organs.

Pseudostratified Columnar Epithelium

The third category of covering and lining epithelium is called pseudostratified columnar epithelium. The nuclei of the cells are at varying depths. Even though all the cells are attached to the basement membrane in a single layer, some cells do not reach the free surface. These features give the false impression of a multilayered tissue, the reason for the name *pseudo*stratified epithelium. The cells that reach the surface either secrete mucus or bear cilia that sweep away mucus and trapped foreign particles for eventual elimination from the body.

Glandular Epithelium

The function of glandular epithelium is secretion, accomplished by glandular cells that often lie in clusters deep to the covering and lining epithelium. A **gland** may consist of one cell or a group of highly specialized epithelial cells that secrete substances into ducts, onto a surface, or into the blood. The production of such substances always requires active work by the cells and results in an expenditure of energy. All glands of the body are classified as exocrine or endocrine.

Exocrine glands secrete their products into ducts (tubes) that empty at the surface of covering and lining epithelium or directly onto a free surface. The product of an exocrine gland may be released at the skin surface or into the lumen of a hollow organ. The secretions of exocrine glands include mucus, perspiration, oil, wax, and digestive enzymes. Examples of exocrine glands are sweat glands, which eliminate perspiration to cool the skin, and salivary glands, which secrete mucus and a digestive enzyme.

Endocrine glands are ductless; their secretory products enter the extracellular fluid and diffuse into the blood. The secretions of endocrine glands are always hormones, chemicals that regulate various physiological activities. The pituitary, thyroid, and adrenal glands are examples of endocrine glands. Endocrine glands will be described in detail in Chapter 18.

Structural Classification of Exocrine Glands

Exocrine glands are classified into two structural types: unicellular and multicellular. **Unicellular glands** are single-celled. A good example of a unicellular gland is a goblet cell (see Exhibit 4.1, nonciliated, simple columnar epithelium). Although goblet cells do not contain ducts, they are often classified as unicellular mucus-secreting exocrine glands. Goblet cells are found in the epithelial lining of the digestive, respiratory, urinary, and reproductive systems. They produce mucus to lubricate the free surfaces of these tissues.

Multicellular glands are many-celled glands and occur in several different forms.

Functional Classification of Exocrine Glands

The functional classification of exocrine glands is based on whether a secretion is a product of a cell or consists of entire or partial glandular cells themselves. **Holocrine** (HŌ-lō-krin) **glands** accumulate a secretory product in their cytosol. The cell then dies and is discharged with its contents as the glandular secretion (Fig. 4.3a). The discharged cell is replaced by a new cell. One example of a holocrine gland is a sebaceous (oil) gland of the skin. **Merocrine** (MER-ō-krin) **glands** simply form the secretory product and discharge it from the cell by exocytosis (Fig. 4.3b). Most exocrine glands of the body are merocrine. Examples

of merocrine glands are the salivary glands and pancreas. **Apocrine** (AP-ō-krin) **glands** accumulate their secretory product at the apical (free) surface of the secreting cell. That portion of the cell pinches off from the rest of the cell to form the secretion (Fig. 4.3c). The remaining part of the cell repairs itself and repeats the process. The mammary glands are apocrine glands.

CONNECTIVE TISSUE

The most abundant and most widely distributed tissue in the body is **connective tissue.** It binds together, supports, and strengthens other body tissues, protects and insulates internal organs, and compartmentalizes structures such as skeletal muscles. Blood, a fluid connective tissue, is the major transport system within the body, while adipose (fat) tissue is the major site of stored energy reserves.

General Features of Connective Tissue

Following are the general features of connective tissue.

1. Connective tissue consists of three basic elements: cells, ground substance, and fibers. Together, the ground substance and fibers, which are outside the cells, form the **matrix.** Unlike epithelial cells, connective tissue cells rarely touch one another; they are separated by a considerable amount of matrix.

FIGURE 4.3 Functional classification of multicellular exocrine glands.

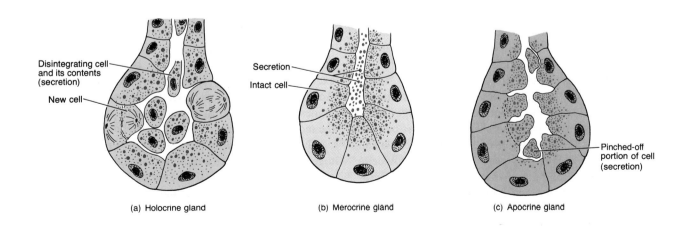

Disintegrating cell and its contents (secretion)

New cell

(a) Holocrine gland

Secretion

Intact cell

(b) Merocrine gland

Pinched-off portion of cell (secretion)

(c) Apocrine gland

Question: Which of the following—mammary gland, oil gland, salivary gland—is a holocrine gland? Merocrine gland? Apocrine gland?

consistency. This feature is of clinical importance because the reduced viscosity hastens the absorption and diffusion of injected drugs and fluids through the tissue and thus can lessen tension and pain. White blood cells, sperm, and some bacteria produce hyaluronidase. Combined with adipose tissue, areolar connective tissue forms the **subcutaneous** (sub′-kyoo-TĀ-nē-us; *sub* = under; *cut* = skin) **layer**—the layer of tissue that attaches the skin to underlying tissues and organs.

CLINICAL APPLICATION

MARFAN SYNDROME

Marfan (MAR-fan) **syndrome** is an inherited disorder that results in abnormalities of connective tissue, most likely a defect in collagen or elastin, especially in the skeleton, eyes, and cardiovascular system. Marfan victims tend to be tall, with disproportionately long arms. They have an unusually long lower half of the body, very long fingers and toes, and other skeletal defects. Based on his physical appearance, Abraham Lincoln is believed to have had Marfan syndrome. Among the serious cardiovascular problems are a weakened area in the aorta (main artery that emerges from the heart), which can suddenly burst, and weakened heart valves. Flo Hyman, an outstanding American female volleyball player, died of problems associated with Marfan syndrome in January 1986.

● **Adipose Tissue** Adipose tissue is a loose connective tissue in which the cells, called **adipocytes,** are specialized for storage of triglycerides (neutral fats). Adipocytes are derived from fibroblasts. Because of the accumulation of a single large triglyceride droplet, the cytoplasm and nucleus are pushed to the edge of the cell. Adipose tissue is found wherever areolar connective tissue is located. Adipose tissue is a good insulator and can therefore reduce heat loss through the skin. It is a major energy reserve and generally supports and protects various organs.

Most of the fat in adults is **white fat,** the type just described. There is another type called **brown fat.** The adipocytes of brown fat are smaller, and the fat is stored in several droplets rather than one large one as in white fat. The brown color is due to a very rich blood supply and numerous mitochondria, which contain colored cytochrome pigments which participate in cellular respiration. Although brown fat is widespread in the fetus and infant, in adults only small amounts are present. It may remain in the subcutaneous tissue between the shoulder blades, in the mediastinum, and near the kidneys. Brown fat generates considerable heat (thermogenesis) and probably helps to maintain proper body temperature in the newborn. The heat generated by the many mitochondria is carried away by the extensive blood supply to help maintain body temperature.

● **Reticular Connective Tissue** Reticular connective tissue consists of fine interlacing reticular fibers and reticular cells. Reticular connective tissue helps to bind together the cells of smooth muscle.

Dense Connective Tissue

Dense connective tissue contains more numerous and thicker fibers but considerably fewer cells than loose connective tissue. The types are dense regular connective tissue, dense irregular connective tissue, and elastic connective tissue.

● **Dense Regular Connective Tissue** In this tissue, bundles of collagen fibers have an orderly, parallel arrangement that confers great strength. The tissue structure withstands pulling in one direction. Fibroblasts, which produce the fibers and ground substance, appear in rows between the fibers. The tissue is silvery white, tough, yet somewhat pliable.

● **Dense Irregular Connective Tissue** This tissue contains collagen fibers that are interwoven without regular orientation and is found in parts of the body where tensions are exerted in various directions. The tissue usually occurs in sheets. Heart valves and the perichondrium, a membrane around cartilage, although classified as dense irregular connective tissue, have a fairly orderly arrangement of collagen fibers.

● **Elastic Connective Tissue** Elastic connective tissue has a predominance of freely branching elastic fibers. These fibers give the unstained tissue a yellowish color. Fibroblasts are present in the spaces between the fibers. Elastic connective tissue can be stretched and will snap back into shape (elasticity). Elastic connective tissue provides stretch and strength, allowing structures to perform their functions efficiently.

Cartilage

Cartilage is capable of enduring considerably more stress than the connective tissues just discussed. Cartilage consists of a dense network of collagen fibers and elastic fibers firmly embedded in chondroitin sulfate, a jellylike component of the ground substance. Whereas the strength of cartilage is due to its collagen fibers, its resilience (ability to assume its original shape after deformation) is due to chondroitin sulfate.

The cells of mature cartilage, called **chondrocytes** (KON-drō-sīts), occur singly or in groups within spaces called **lacunae** (la-KOO-nē) in the matrix. The surface of cartilage is surrounded by a membrane of dense irregular connective tissue called the **perichondrium** (per′-i-KON-drē-um; *peri* = around; *chondro* = cartilage). Unlike other connective tissues, cartilage has no blood vessels or nerves, except for those in the perichondrium. Three kinds of cartilage are recognized: hyaline cartilage, fibrocartilage, and elastic cartilage (see Exhibit 4.2).

EXHIBIT 4.2

CONNECTIVE TISSUES

EMBRYONIC CONNECTIVE TISSUE

MESENCHYME

Description: Consists of irregularly shaped mesenchymal cells embedded in a semifluid ground substance that contains delicate reticular fibers.
Location: Under skin and along developing bones of embryo; some mesenchymal cells found in adult connective tissue, especially along blood vessels.
Function: Forms all other kinds of connective tissue.

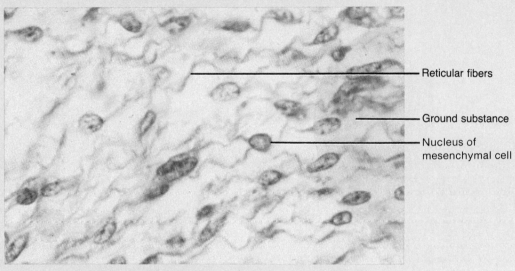

Reticular fibers

Ground substance

Nucleus of mesenchymal cell

Sectional view of mesenchyme from a developing fetus (1008×)

MUCOUS CONNECTIVE TISSUE

Description: Consists of star-shaped cells embedded in a viscous, jellylike ground substance that contains fine collagen fibers.
Location: Umbilical cord of fetus.
Function: Support

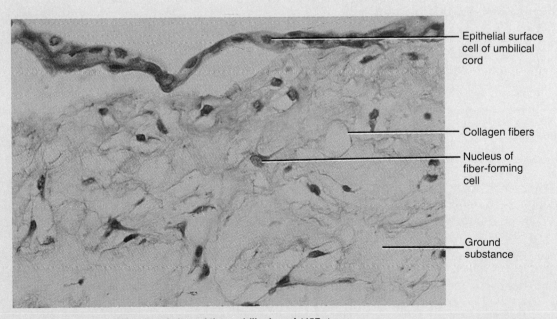

Epithelial surface cell of umbilical cord

Collagen fibers

Nucleus of fiber-forming cell

Ground substance

Sectional view of the umbilical cord (457×)

Exhibit continues

EXHIBIT 4.2 **(continued)**

CONNECTIVE TISSUES

MATURE CONNECTIVE TISSUES

LOOSE CONNECTIVE TISSUE

Areolar Connective Tissue

Description: Consists of fibers (collagen, elastic, and reticular) and several kinds of cells (fibroblasts, macrophages, plasma cells, adipocytes, and mast cells) embedded in a semifluid ground substance.

Location: Subcutaneous layer of skin, papillary (superficial) region of dermis of skin, mucous membranes, blood vessels, nerves, and around body organs.

Function: Strength, elasticity, and support.

Collagen fibers

Elastic fibers

Sectional view of subcutaneous tissue (224x)

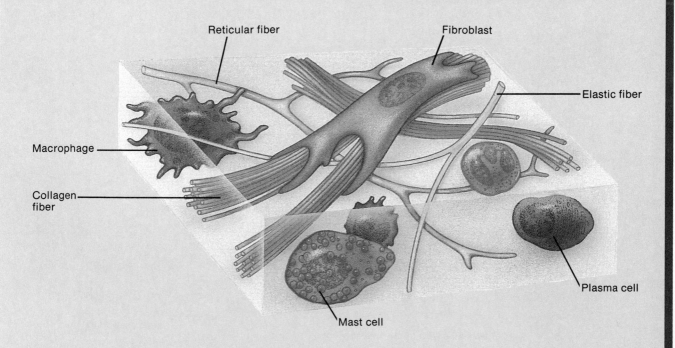

Reticular fiber

Fibroblast

Elastic fiber

Macrophage

Collagen fiber

Mast cell

Plasma cell

Adipose Tissue

Description: Consists of adipocytes, cells with peripheral nuclei, that are specialized for storage of triglycerides (neutral fats).
Location: Subcutaneous layer of skin, around heart and kidneys, yellow marrow of long bones, and padding around joints and behind eyeball in eye socket.
Function: Reduces heat loss through skin, serves as an energy reserve, supports, and protects; in newborns, brown fat generates considerable heat (thermogenesis) that probably helps to maintain proper body temperature.

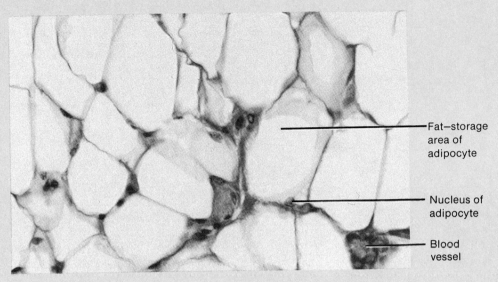

Fat–storage area of adipocyte

Nucleus of adipocyte

Blood vessel

Sectional view of adipocytes of white fat of the pancreas (1720×)

Reticular Connective Tissue

Description: Consists of a network of interlacing reticular fibers and reticular cells.
Location: Stroma (framework) of liver, spleen, lymph nodes; portion of bone marrow that gives rise to blood cells, reticular lamina of the basement membrane, and around blood vessels and muscle.
Function: Forms stroma of organs; binds together smooth muscle tissue cells.

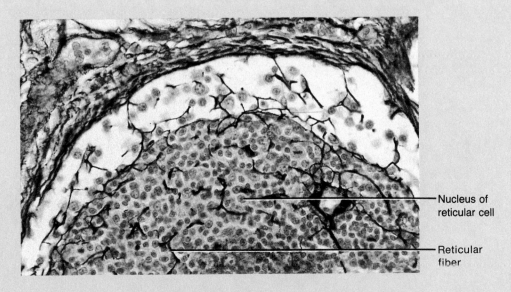

Nucleus of reticular cell

Reticular fiber

Sectional view of lymph node (496x)

Exhibit continues

EXHIBIT 4.2 (continued)

CONNECTIVE TISSUES

DENSE CONNECTIVE TISSUE

Dense Regular Connective Tissue

Description: Matrix looks shiny white; consists of predominantly collagen fibers arranged in bundles; fibroblasts present in rows between bundles.
Location: Forms tendons (attach muscle to bone), ligaments (attach bone to bone), and aponeuroses (sheetlike tendons that attach muscle to muscle or muscle to bone).
Function: Provides strong attachment between various structures.

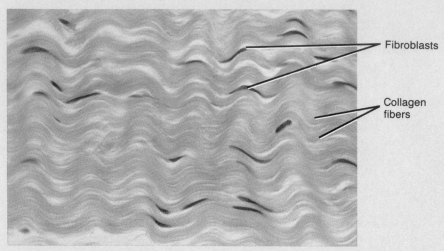

Fibroblasts

Collagen fibers

Sectional view of a tendon (250x)

Dense Irregular Connective Tissue

Description: Consists of predominantly collagen fibers, randomly arranged, and a few fibroblasts.
Location: Fasciae, reticular (deeper) region of dermis of skin, periosteum of bone, joint capsules, membrane capsules around various organs (kidneys, liver, testes, lymph nodes), and heart valves.
Function: Provides strength.

Collagen fibers

Blood vessels

Sectional view of dermis of skin (275x)

Elastic Connective Tissue

Description: Consists of predominantly freely branching elastic fibers; fibroblasts present in spaces between fibers.
Location: Lung tissue, walls of elastic arteries, trachea, bronchial tubes, true vocal cords, suspensory ligament of penis, and ligamenta flava of vertebrae (ligaments between vertebrae).
Function: Allows stretching of various organs.

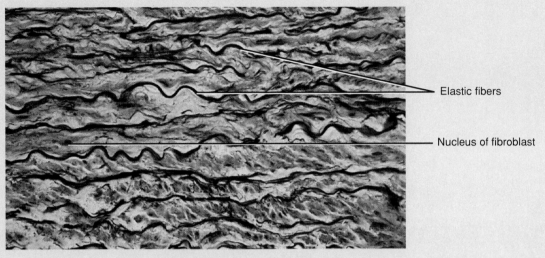

Elastic fibers

Nucleus of fibroblast

Sectional view of aorta (largest artery in the body) (335x)

CARTILAGE

Hyaline Cartilage

Description: Also called gristle; consists of a bluish white and glossy ground substance with fine collagen fibers; contains numerous chondrocytes; most abundant type of cartilage.
Location: Ends of long bones, anterior ends of ribs, nose, parts of larynx, trachea, bronchi, bronchial tubes, and embryonic skeleton.
Function: Provides movement at joints, flexibility, and support.

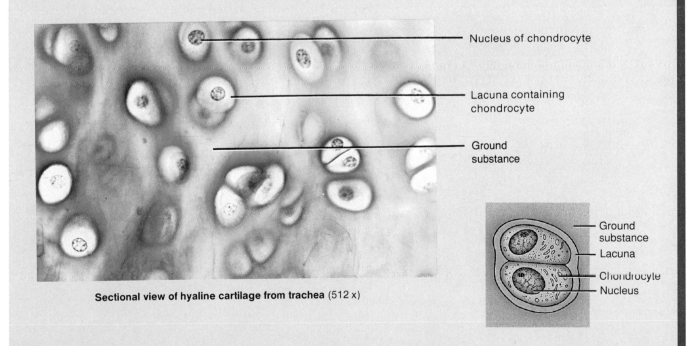

Nucleus of chondrocyte

Lacuna containing chondrocyte

Ground substance

Ground substance
Lacuna
Chondrocyte
Nucleus

Sectional view of hyaline cartilage from trachea (512 x)

Exhibit continues

EXHIBIT 4.2 (continued)

CONNECTIVE TISSUES

Fibrocartilage

Description: Consists of chondrocytes scattered among bundles of collagen fibers.
Location: Pubic symphysis (point where hipbones join anteriorly), intervertebral discs (discs between vertebrae), and menisci of knee.
Function: Support and fusion.

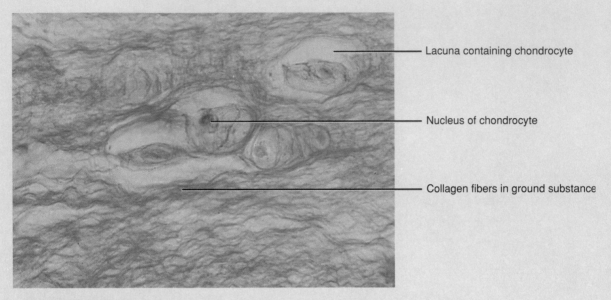

— Lacuna containing chondrocyte

— Nucleus of chondrocyte

— Collagen fibers in ground substance

Sectional view of fibrocartilage from patellar tendon insertion (742x)

Elastic Cartilage

Description: Consists of chondrocytes located in a threadlike network of elastic fibers.
Location: Epiglottis of larynx, external ear (auricle), and auditory (Eustachian) tubes.
Function: Gives support and maintains shape.

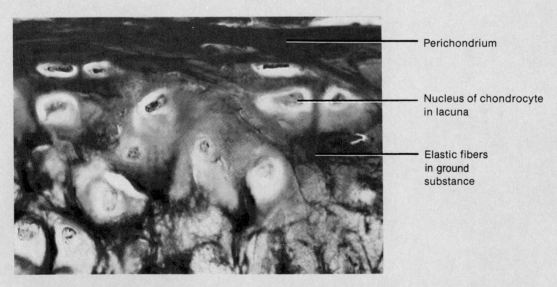

— Perichondrium

— Nucleus of chondrocyte in lacuna

— Elastic fibers in ground substance

Sectional view of elastic cartilage from auricle of ear (742x)

BONE (OSSEOUS) TISSUE

Description: Compact bone consists of osteons (Haversian systems) that contain lamellae, lacunae, osteocytes, canaliculi, and central (Haversian) canals. See also Fig. 6.3a. Spongy bone consists of thin plates called trabeculae; spaces between trabeculae are filled with red marrow. See Fig. 6.3b,c.
Location: Both compact and spongy bone comprise the various parts of bones of the body.
Function: Support, protection, storage, houses blood-forming tissue, and serves as levers that act together with muscle tissue to provide movement.

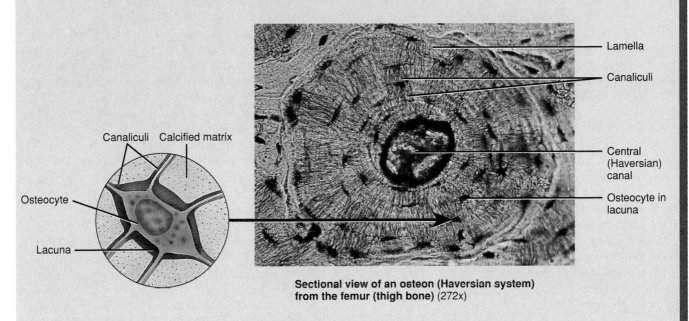

Canaliculi   Calcified matrix

Osteocyte

Lacuna

Lamella

Canaliculi

Central (Haversian) canal

Osteocyte in lacuna

Sectional view of an osteon (Haversian system) from the femur (thigh bone) (272x)

BLOOD (VASCULAR TISSUE)

Description: Consists of plasma and formed elements. The formed elements are erythrocytes (red blood cells), leukocytes (white blood cells), and thrombocytes (platelets). See also Fig. 19.2.
Location: Within blood vessels (arteries, arterioles, capillaries, venules, and veins).
Function: Erythrocytes transport oxygen and carbon dioxide; leukocytes carry on phagocytosis and are involved in allergic reactions and immunity; thrombocytes are essential for the clotting of blood.

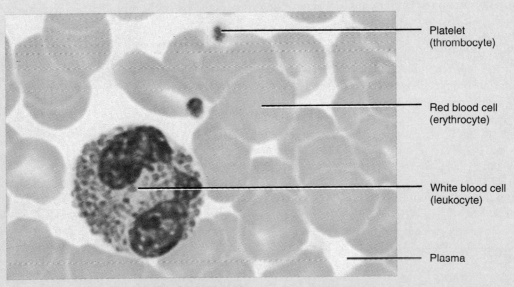

Platelet (thrombocyte)

Red blood cell (erythrocyte)

White blood cell (leukocyte)

Plasma

Blood smear (2350x)

● **Hyaline Cartilage** This cartilage, also called **gristle,** contains a resilient gel as its ground substance and appears in the body as a bluish-white, shiny substance. The fine collagen fibers, although present, are not visible with ordinary staining techniques, and the prominent chondrocytes are found in lacunae. Hyaline cartilage is the most abundant kind of cartilage in the body. Hyaline cartilage affords flexibility and support and, at joints, reduces friction and absorbs shock.

● **Fibrocartilage** Chondrocytes are scattered among clearly visible bundles of collagen fibers within the matrix of this type of cartilage. This tissue combines strength and rigidity.

● **Elastic Cartilage** In this tissue, chondrocytes are located in a threadlike network of elastic fibers within the matrix. Elastic cartilage provides strength and elasticity and maintains the shape of certain organs.

● **Growth and Repair of Cartilage** Cartilage growth is slow; metabolically, it is an inactive tissue. When injured or inflamed, cartilage repair proceeds slowly, in large part because cartilage is avascular. Substances needed for repair and blood cells that participate in tissue repair must diffuse or migrate from the perichondrium into the cartilage. Bone, which has a rich blood supply, heals much more rapidly than cartilage.

The growth of cartilage follows two basic patterns. In **interstitial (endogenous) growth,** the cartilage increases rapidly in size through the division of existing chondrocytes and continuous deposition of increasing amounts of matrix by the chondrocytes. The formation of new chondrocytes and their production of new matrix cause the cartilage to expand from within—thus the term *interstitial growth.* This growth pattern occurs while the cartilage is young and pliable—during childhood and adolescence.

In **appositional (exogenous) growth,** activity of cells in the inner chondrogenic layer of the perichondrium leads to growth. The deeper cells of the perichondrium, the fibroblasts, divide. Some differentiate into chondroblasts (immature cells) and then into chondrocytes. As differentiation occurs, the chondroblasts surround themselves with matrix and become chondrocytes. As a result, matrix accumulates on the surface of the cartilage, increasing its size. A new layer of cartilage forms beneath the perichondrium on the surface of the cartilage, causing it to grow in width. Appositional growth starts later than interstitial growth and continues throughout life.

As part of the aging process, cartilage tends to calcify; that is, minerals deposit in it. Also, cartilage may ossify, that is, change into bone (see page 152). As a result, chondrocytes die because they are poorly nourished.

Bone (Osseous) Tissue

Together, cartilage, joints, and **bone (osseous; OS-ē-us; tissue)** comprise the skeletal system. In addition to bone tissue, other components of bone that are also connective tissues include the periosteum (covering around a bone), red and yellow bone marrow, and the endosteum (lining of a space in a bone that stores yellow marrow).

Bone tissue is classified as either compact (dense) or spongy (cancellous), depending on how the matrix and cells are organized. The basic unit of compact bone is called an **osteon (Haversian system).** Each osteon is composed of **lamellae,** concentric rings of matrix that consist of mineral salts (mostly tricalcium phosphate and calcium carbonate) that give bone its hardness, and collagen fibers that give bone its strength; **lacunae,** small spaces between lamellae that contain mature bone cells called **osteocytes; canaliculi,** minute canals that project from lacunae and provide numerous routes so that nutrients can reach osteocytes and wastes can be removed from them; and a **central (Haversian) canal** that contains blood vessels and nerves. Spongy bone has no osteons; instead, it consists of plates of bone called **trabeculae** which contain lamellae, osteocytes, lacunae, and canaliculi. Spaces between lamellae are filled with red marrow. The details of the histology of bone are presented on page 147.

The skeletal system has several functions. It supports soft tissues, protects delicate structures, and works with skeletal muscles to generate movement. Bone stores calcium and phosphorus, houses red marrow, which produces several kinds of blood cells, and houses yellow marrow, which contains lipids as an energy source.

Blood (Vascular Tissue)

Blood (vascular tissue) is a connective tissue with a liquid matrix called plasma. Suspended in the plasma are formed elements—cells and cell fragments. **Plasma** is a straw-colored liquid that consists mostly of water with a wide variety of dissolved substances (nutrients, wastes, enzymes, hormones, respiratory gases, and ions). The formed elements are red blood cells (erythrocytes), white blood cells (leukocytes), and platelets (thrombocytes). See Fig. 19.2. **Erythrocytes** function in transporting oxygen to body cells and removing carbon dioxide from them. **Leukocytes** are involved in phagocytosis, immunity, and allergic reactions. **Platelets** function in blood clotting.

The details of blood are considered in Chapter 19.

MEMBRANES

The combination of an epithelial layer and an underlying connective tissue layer constitutes an **epithelial membrane.** The principal epithelial membranes of the body are mucous membranes, serous membranes, and the cutaneous membrane, or skin. The skin is an organ of the integumentary system and is discussed in the next chapter. Another kind of membrane, a **synovial membrane,** contains connective tissue rather than epithelium.

Mucous Membranes

A **mucous membrane,** or **mucosa,** lines a body cavity that opens directly to the exterior. Mucous membranes line the entire digestive, respiratory, and reproductive systems (see Fig. 24.2) and much of the urinary system. They consist of a lining layer of epithelium and an underlying layer of connective tissue.

The epithelial layer of a mucous membrane is an important aspect of the body's defense mechanisms. It is a barrier that microbes and other pathogens have difficulty penetrating. Usually, tight junctions connect the cells, so materials cannot leak in between. Certain cells of the epithelial layer of a mucous membrane secrete mucus, which prevents the cavities from drying out. It also traps particles in the respiratory passageways and lubricates food as it moves through the gastrointestinal tract. In addition, the epithelial layer secretes some of the enzymes needed for digestion and is the site of food absorption in the gastrointestinal tract. As you will see later, the epithelium of a mucous membrane varies greatly in different parts of the body. For example, the epithelium of the mucous membrane of the small intestine is simple columnar, while that of the large airways to the lungs is pseudostratified columnar.

The connective tissue layer of a mucous membrane is called the **lamina propria** (LAM-i-na; PRŌ-prē-a). The lamina propria is so named because it belongs to the mucous membrane (*proprius* = one's own). It binds the epithelium to the underlying structures and allows some flexibility of the membrane. It also holds the blood vessels in place and protects underlying muscles from abrasion or puncture. Oxygen and nutrients diffuse from the lamina propria to the epithelium covering it, while carbon dioxide and wastes diffuse in the opposite direction.

Serous Membranes

A **serous membrane,** or **serosa,** lines a body cavity that does not open directly to the exterior, and it covers the organs that lie within the cavity. Serous membranes consist of thin layers of areolar connective tissue covered by a layer of mesothelium, and they are composed of two portions. The part attached to the cavity wall is called the **parietal** (pa-RĪ-e-tal) **portion.** The part that covers and attaches to the organs inside these cavities is the **visceral portion.** The serous membrane lining the thoracic cavity and covering the lungs is called the **pleura** (see Fig. 1.7). The serous membrane lining the heart cavity and covering the heart is the **pericardium** (*cardio* = heart). The serous membrane lining the abdominal cavity and covering the abdominal organs and some pelvic organs is called the **peritoneum.**

The epithelial layer of a serous membrane secretes a lubricating fluid, called **serous fluid,** that allows the organs to glide easily against one another or against the walls of the cavities. The connective tissue layer of the serous membrane consists of a thin layer of areolar connective tissue.

Synovial Membranes

Synovial membranes line the cavities of the freely movable joints (see Fig. 9.1a). Like serous membranes, they line structures that do not open to the exterior. Unlike mucous, serous, and cutaneous membranes, they do not contain epithelium and are therefore not epithelial membranes. They are composed of areolar connective tissue with elastic fibers and varying amounts of fat. Synovial membranes secrete **synovial fluid,** which lubricates the cartilage at the ends of bones during their movements and nourishes the cartilage covering the bones at joints. These are articular synovial membranes. Other synovial membranes line cushioning sacs, called bursae, and tendon sheaths in our hands and feet that ease the movement of muscle tendons.

MUSCLE TISSUE

Muscle tissue consists of fibers (cells) that are beautifully constructed to generate force for contraction. As a result of this characteristic, muscle tissue provides motion, maintains posture, and generates heat. Based on location and certain structural and functional characteristics, muscle tissue is classified into three types: skeletal, cardiac, and smooth (Exhibit 4.3).

Skeletal muscle tissue is named for its location— attached to bones. It is also **striated;** that is, the fibers (cells) contain alternating light and dark bands (striations) that are perpendicular to the long axes of the fibers. The striations are visible under a microscope. Skeletal muscle is also **voluntary** because it can be made to contract or relax by conscious control. A single skeletal muscle fiber is roughly cylindrical. Within a muscle, the individual fibers are parallel to each other.

Cardiac muscle tissue forms the bulk of the wall of the heart. Like skeletal muscle, it is striated. However, unlike skeletal muscle tissue, it is **involuntary;** its contraction is usually not under conscious control. Cardiac muscle fibers are branched cylinders. The fibers usually have only one nucleus that is centrally located; sometimes there are two nuclei. Cardiac muscle fibers attach to each other by transverse thickenings of the sarcolemma called **intercalated discs,** which contain both gap junctions and desmosomes. Intercalated discs are unique to cardiac muscle. The desmosomes strengthen the tissue, and the gap junctions provide a route for quick conduction of muscle action potentials (impulses) throughout the heart.

Smooth muscle tissue is located in the walls of hollow internal structures such as blood vessels, airways to the lungs, the stomach, intestines, gallbladder, and urinary bladder. Its contraction helps break down food, move food and fluids through the body, and eliminate wastes. Smooth muscle fibers are usually **involuntary,** and they are **nonstriated (smooth).** A smooth muscle fiber is spindle-shaped (thickest in the middle with each end tapering to a point) and contains a single, centrally located nucleus.

EXHIBIT 4.3

MUSCLE TISSUE

Skeletal Muscle Tissue *Description:* Cylindrical, striated fibers with many peripheral nuclei; voluntary control.
Location: Usually attached to bones.
Function: Motion, posture, heat production.

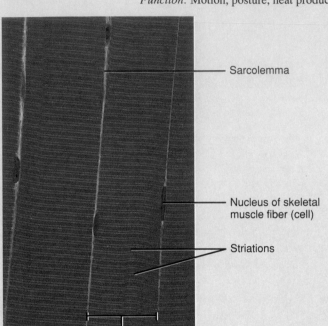

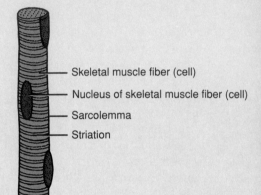

Sarcolemma

Nucleus of skeletal muscle fiber (cell)

Striations

Skeletal muscle fiber (cell)

Skeletal muscle fiber (cell)
Nucleus of skeletal muscle fiber (cell)
Sarcolemma
Striation

Section of skeletal muscle (600x)

Cardiac Muscle Tissue *Description:* Branched cylindrical, striated fibers with one or two centrally located nuclei; contains intercalated discs; mainly involuntary control.
Location: Heart wall.
Function: Pumps blood to all parts of the body.

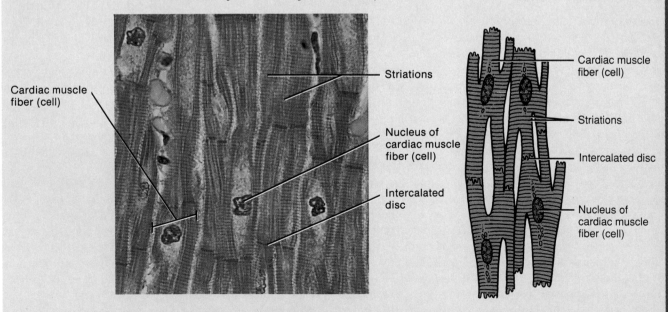

Cardiac muscle fiber (cell)

Striations

Nucleus of cardiac muscle fiber (cell)

Intercalated disc

Cardiac muscle fiber (cell)

Striations

Intercalated disc

Nucleus of cardiac muscle fiber (cell)

Section of cardiac muscle (700 x)

Smooth Muscle Tissue *Description:* Spindle-shaped, nonstriated fibers with one centrally located nucleus; usually involuntary control.
Location: Walls of hollow internal structures such as blood vessels, airways to the lungs, stomach, intestines, gallbladder, and urinary bladder.
Function: Motion (constriction of blood vessels and airways, propulsion of foods through gastrointestinal tract; contraction of urinary bladder and gallbladder).

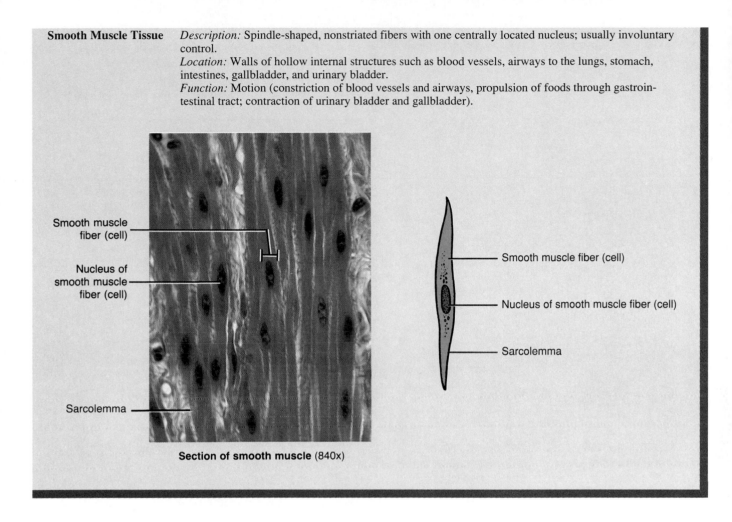

Smooth muscle fiber (cell)

Nucleus of smooth muscle fiber (cell)

Sarcolemma

Smooth muscle fiber (cell)

Nucleus of smooth muscle fiber (cell)

Sarcolemma

Section of smooth muscle (840x)

Many individual fibers in some smooth muscle tissue, for example, in the wall of the intestines, are connected by gap junctions and thus contract as a single unit. In other locations, for example, the iris of the eye, smooth muscle fibers contract individually, like skeletal muscle fibers, because gap junctions are absent.

Chapter 10 provides a more detailed discussion of muscle tissue.

NERVOUS TISSUE

Despite the awesome complexity of the nervous system, it consists of only two principal kinds of cells: neurons and neuroglia. **Neurons,** or nerve cells, are highly specialized cells. They are sensitive to various stimuli, they convert stimuli into nerve impulses (nerve action potentials), and they conduct these impulses to other neurons, muscle fibers, or glands. Neurons are the structural and functional units of the nervous system. Most consist of three basic portions: a cell body and two kinds of processes called dendrites and axons (Exhibit 4.4). The **cell body** contains the nucleus and other organelles. **Dendrites** are highly

branched, tapered processes of the cell body. **Axons** are single, long processes of the cell body that usually conduct nerve impulses away from the cell body.

Neuroglia do not generate or conduct nerve impulses. They do, however, have many important functions (see Exhibit 12.1 on page 349). They are of clinical interest because they are often the sites of tumors of the nervous system.

The detailed structure and function of neurons and neuroglia are considered in Chapter 12.

TISSUE REPAIR: RESTORING HOMEOSTASIS

During embryonic development, muscle tissue and nervous tissue become highly differentiated (specialized) and lose their capacity for mitosis. Epithelial tissues, which in some locations endure considerable wear and tear and even injury, and connective tissues generally have a continuous capacity for renewal. In some cases, immature, undifferentiated cells called **stem cells** divide to replace lost or dam-

EXHIBIT 4.4

NERVOUS TISSUE

Description: Neurons (nerve cells) consist of a cell body and processes extending from the cell body called dendrites or axons. Neuroglia do not generate or conduct nerve impulses, but have other important functions (see Exhibit 12.1 on page 349).
Location: Nervous system.
Function: Exhibits sensitivity to various types of stimuli, converts stimuli into nerve impulses, and conducts nerve impulses to other neurons, muscle fibers, or glands.

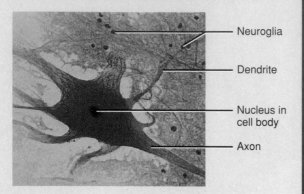

— Neuroglia

— Dendrite

— Nucleus in cell body

— Axon

Motor (efferent) neuron from the spinal cord (100x)

The cardinal factor in tissue repair is the capacity of parenchymal tissue to regenerate. This capacity, in turn, depends on the ability of the parenchymal cells to replicate quickly.

CLINICAL APPLICATION

ADHESIONS

Scar tissue can cause **adhesions,** abnormal joining of tissues. They are common in the abdomen and may occur around a site of previous inflammation such as an inflamed appendix, or they can follow surgery. Although adhesions do not always cause problems, they can affect tissue flexibility, cause obstruction (such as in the intestine), and make a later operation more difficult. Surgery may be required to remove adhesions.

Repair Process

If injury to a tissue, for example, the skin, is slight, drainage and reabsorption of pus (accumulation of leukocytes and fluid resulting from inflammation), followed by parenchymal regeneration, may lead to complete repair. When an area of skin loss is great, fluid moves out of the capillaries and the area becomes dry. Fibrin, an insoluble protein in a blood clot, seals the open tissue by hardening into a **scab.**

When tissue and cell damage are extensive and severe, as in large, open wounds, both the connective tissue stroma and the parenchymal cells are active in repair. This repair involves the rapid cell division of many fibroblasts and the manufacture of new collagen fibers to provide strength. Blood capillaries also sprout new buds to serve the healing tissue. All these processes create an actively growing connective tissue called **granulation tissue.** This new tissue forms across a wound or surgical incision to provide a framework (stroma). The framework supports the epithelial cells that migrate into the open area and fill it. The newly formed granulation tissue also secretes a fluid that kills bacteria.

Conditions Affecting Repair

Three factors affect tissue repair: nutrition, blood circulation, and age. Nutrition is vital since the healing process places a great demand on the body's store of nutrients. Adequate protein in the diet is important because most of the cell structure is made from proteins. Vitamins also play a direct role in wound healing. Among the vitamins involved and their roles in wound healing are the following:

1. Vitamin A is essential in the replacement of epithelial tissues, especially in the respiratory tract.
2. The B vitamins—thiamine, nicotinic acid, ribo-

aged cells. For example, stem cells reside in protected positions in the epithelium of the skin and gastrointestinal tract to replenish cells sloughed from the surface. And stem cells in bone marrow continually provide new red and white blood cells. In other cases, mature, differentiated cells can duplicate themselves by mitosis, for example, hepatocytes in the liver and endothelial cells in blood vessels.

Tissue repair is the process by which tissues replace dead or damaged cells. New cells originate by cell division from the **stroma,** the supporting connective tissue, or from the **parenchyma** (pa-REN-ki-ma), cells that form the organ's functioning part. The restoration of an injured organ or tissue to normal structure and function depends entirely on which type of cell—parenchymal or stromal—is active in the repair.

If parenchymal elements accomplish the repair, tissue **regeneration** is possible. A perfect or near-perfect reconstruction of the injured tissue may occur. However, if fibroblasts of the stroma are active in the repair, the replacement tissue will be new connective tissue called scar tissue. In this process, fibroblasts synthesize collagen and other matrix materials that aggregate to form scar tissue. The process of scar tissue formation is known as **fibrosis.** Since scar tissue is not specialized to perform the functions of the parenchymal tissue, the function of the tissue is impaired.

flavin—are coenzymes needed by many enzyme systems in cells. These vitamins may relieve pain and are necessary for division of the cells that accomplish repair.

3. Vitamin C directly affects the normal production and maintenance of matrix materials, especially collagen. Vitamin C also strengthens and promotes the formation of new blood vessels. With vitamin C deficiency, even superficial wounds fail to heal, and the walls of the blood vessels become fragile and are easily ruptured.

4. Vitamin D is necessary for the proper absorption of calcium from the intestine. Calcium gives bones their hardness and is necessary for the healing of fractures.

5. Vitamin E is believed to promote healing of injured tissues and may prevent scarring.

6. Vitamin K is needed for production of proteins that participate in blood clotting and thus prevent the injured person from bleeding to death.

In tissue repair, proper blood circulation is essential to transport oxygen, nutrients, antibodies, and many defensive cells to the injured site. The blood also plays an important role in the removal of tissue fluid, bacteria, foreign bodies, and debris. These elements would otherwise interfere with healing.

Generally, tissues heal faster and leave less obvious scars in the young than in the aged. The younger body is generally in a better nutritional state, its tissues have a better blood supply, and its cells have a faster metabolic rate. Thus cells can synthesize needed materials and divide more quickly.

Study Outline

Types of Tissues and Their Origins (p. 97)

1. A tissue is a group of similar cells that usually have a similar embryological origin and are specialized for a particular function.
2. Depending on their function and structure, the various tissues of the body are classified into four principal types: epithelial, connective, muscular, and nervous.
3. Extracellular fluid is external to body cells. Examples are interstitial fluid and plasma.

Cell Junctions (p. 97)

1. Cell junctions are points of contact between adjacent plasma membranes.
2. Types are tight junctions (form fluid-tight seals between cells), anchoring junctions (fasten cells to one another or the matrix), and communicating junctions (permit electrical or chemical signals to pass between cells).

Epithelial Tissue (p. 98)

1. The subtypes of epithelium include covering and lining epithelium and glandular epithelium.
2. Some general characteristics of epithelium are: consists mostly of cells with little extracellular material, arranged in sheets, attached to a basement membrane, avascular, has a nerve supply, high capacity for renewal, and derived from all three primary germ layers.

Covering and Lining Epithelium (p. 99)

1. Layers are arranged as simple (one layer), stratified (several layers), and pseudostratified (one layer that appears as several); cell shapes include squamous (flat), cuboidal (cubelike), columnar (rectangular), and transitional (variable).
2. Simple squamous epithelium consists of a single layer of flat, scalelike cells. It is adapted for diffusion and filtration and is found in lungs and kidneys. Endothelium lines the heart and blood vessels. Mesothelium lines the thoracic and abdominopelvic cavities and covers the organs within them.
3. Simple cuboidal epithelium consists of a single layer of cube-shaped cells. It is adapted for secretion and absorption. It is found covering ovaries, in kidneys and eyes, and lining some glandular ducts.
4. Nonciliated simple columnar epithelium consists of a single layer of nonciliated rectangular cells. It lines most of the gastrointestinal tract. Specialized cells containing microvilli perform absorption. Goblet cells secrete mucus. Ciliated simple columnar epithelium consists of a single layer of ciliated rectangular cells. It is found in a few portions of the upper respiratory tract where it moves foreign particles trapped in mucus out of the body.
5. Stratified squamous epithelium consists of several layers of cells in which the top layer is flat. It is protective. Nonkeratinized variety lines the mouth; keratinized variety forms outer layer of skin.
6. Stratified cuboidal epithelium consists of several layers of cells in which the top layer is cube-shaped. It is found in adult sweat glands and a portion of male urethra.
7. Stratified columnar epithelium consists of several layers of cells in which the top layer is rectangular. It protects and secretes. It is found in a portion of male urethra and large excretory ducts.
8. Transitional epithelium consists of several layers of cells whose appearance is variable. It lines the urinary bladder and is capable of stretching.
9. Pseudostratified columnar epithelium has only one layer but gives the appearance of many. It lines larger excretory ducts, part of male urethra, auditory (Eustachian) tubes, and most upper respiratory structures, where it protects and secretes.

Glandular Epithelium (p. 106)

1. A gland is a single cell or a mass of epithelial cells adapted for secretion.
2. Exocrine glands (sweat, oil, and digestive glands) secrete into ducts or directly onto a free surface.
3. Structural classification includes unicellular and multicellular glands.
4. Functional classification includes holocrine, merocrine, and apocrine glands.
5. Endocrine glands secrete hormones into the blood.

Connective Tissue (p. 107)

1. Connective tissue is the most abundant body tissue.
2. Some general characteristics of connective tissue: consists of cells, ground substance, and fibers; abundant matrix with relatively few cells; does not occur on free surfaces; has a nerve supply (except for cartilage); and highly vascular (except for cartilage and tendons).

Connective Tissue Cells (p. 108)

1. Cells in connective tissue are derived from mesenchyme.
2. Types include fibroblasts (secrete matrix), macrophages (phagocytes), plasma cells (secrete antibodies), mast cells (produce histamine), adipocytes, and white blood cells.

Connective Tissue Matrix (p. 108)

1. The ground substance and fibers comprise the matrix.
2. Substances found in the ground substance include hyaluronic acid, chondroitin sulfate, dermatan sulfate, and keratan sulfate.
3. The ground substance supports, binds, provides a medium for the exchange of materials, and is active in influencing cell functions.
4. The fibers provide strength and support and are of three types.
5. Collagen fibers (composed of collagen) are found in bone, tendons, and ligaments; elastic fibers (composed of elastin) are found in skin, blood vessels, and lungs; and reticular fibers (composed of collagen and glycoprotein) are found around fat cells, nerve fibers, and skeletal and smooth muscle cells.

Embryonic Connective Tissue (p. 109)

1. Mesenchyme forms all other connective tissues.
2. Mucous connective tissue is found in the umbilical cord of the fetus, where it gives support.

Mature Connective Tissue (p. 109)

1. Mature connective tissue is connective tissue that differentiates from mesenchyme and exists in the newborn and does not change after birth. It is subdivided into several kinds: connective tissue proper, cartilage, bone tissue, and blood. Subtypes include loose connective tissue, dense connective tissue, cartilage, bone, and blood.
2. Loose connective tissue includes areolar connective tissue, adipose tissue, and reticular connective tissue.
3. Areolar connective tissue consists of the three types of fibers, several cells, and a semifluid ground substance. It is found in the subcutaneous layer and mucous membranes and around blood vessels, nerves, and body organs.
4. Adipose tissue consists of adipocytes that store triglycerides. It is found in subcutaneous layer, around organs, and in the yellow marrow of long bones.
5. Reticular connective tissue consists of reticular fibers and reticular cells and is found in the liver, spleen, and lymph nodes.
6. Dense connective tissue includes dense regular connective tissue, dense irregular connective tissue, and elastic connective tissue.
7. Dense regular connective tissue consists of bundles of collagen fibers and fibroblasts. It forms tendons, ligaments, and aponeuroses.
8. Dense irregular connective tissue consists of randomly arranged collagen fibers and a few fibroblasts. It is found in fasciae, dermis of skin, and membrane capsules.
9. Elastic connective tissue consists of elastic fibers and fibroblasts. It is found in the lungs, wall of arteries, and bronchial tubes.
10. Cartilage has a jellylike matrix (chondroitin sulfate) containing collagen and elastic fibers and chondrocytes.
11. Hyaline cartilage is found in the embryonic skeleton, at the ends of bones, in the nose, and in respiratory structures. It is flexible, allows movement, and provides support.
12. Fibrocartilage is found in the pubic symphysis, intervertebral discs, and menisci.
13. Elastic cartilage maintains the shape of organs such as the epiglottis of the larynx, auditory (Eustachian) tubes, and external ear.
14. The growth of cartilage is accomplished by interstitial growth (from within) and appositional growth (from without).
15. Bone (osseous tissue) consists of mineral salts and collagen fibers that contribute to the hardness of bone and cells called osteocytes. It supports, protects, helps provide movement, stores minerals, and houses blood-forming tissue.
16. Blood (vascular tissue) consists of plasma and formed elements (erythrocytes, leukocytes, and thrombocytes). Functionally, its cells transport, carry on phagocytosis, participate in allergic reactions, provide immunity, and bring about blood clotting.

Membranes (p. 118)

1. An epithelial membrane is an epithelial layer overlying a connective tissue layer. Examples are mucous, serous, and cutaneous membranes.
2. Mucous membranes line cavities that open to the exterior, such as the gastrointestinal tract.
3. Serous membranes (pleura, pericardium, peritoneum) line closed cavities and cover the organs in the cavities. These membranes consist of parietal and visceral portions.
4. The cutaneous membrane is the skin.
5. Synovial membranes line joint cavities, bursae, and tendon sheaths and do not contain epithelium.

Muscle Tissue (p. 119)

1. Muscle tissue is modified for contraction and thus provides motion, maintenance of posture, and heat production.
2. Skeletal muscle tissue is attached to bones, is striated, and is voluntary.
3. Cardiac muscle tissue forms most of the heart wall, is striated, and is usually involuntary.
4. Smooth muscle tissue is found in the walls of hollow internal structures (blood vessels and viscera), is nonstriated, and is usually involuntary.

Nervous Tissue (p. 121)

1. The nervous system is composed of neurons (nerve cells) and neuroglia (protective and supporting cells).
2. Most neurons consist of a cell body and two types of processes called dendrites and axons.
3. Neurons are sensitive to stimuli, convert stimuli into nerve impulses, and conduct nerve impulses.

Tissue Repair: Restoring Homeostasis (p. 121)

1. Tissue repair is the replacement of damaged or destroyed cells by healthy ones.
2. It begins during the active phase of inflammation and is not completed until after harmful substances in the inflamed area have been neutralized or removed.

Repair Process (p. 122)

1. If the injury is superficial, tissue repair involves pus removal (if pus is present), scab formation, and parenchymal regeneration.
2. If damage is extensive, granulation tissue is involved.

Conditions Affecting Repair (p. 122)

1. Nutrition is important to tissue repair. Various vitamins (A, some B, D, C, E, and K) and a protein-rich diet are needed.
2. Adequate circulation of blood is needed.
3. The tissues of young people repair rapidly and efficiently; the process slows down with aging.

Review Questions

1. Define a tissue. What are the four basic kinds of human tissue? (p. 97)
2. What is extracellular fluid (ECF)? Why is it important? (p. 97)
3. Describe the three basic types of cell junctions and the functions of each. (p. 97)
4. Distinguish covering and lining epithelium from glandular epithelium. What characteristics are common to all epithelial tissue? epithelia? Describe the structure of the basement membrane. (p. 98)
5. Describe the various layering arrangements and cell shapes of epithelium. (p. 99)
6. How is epithelium classified? List the various types. (p. 99)
7. For each of the following kinds of epithelium, briefly describe the microscopic appearance, location in the body, and functions: simple squamous, simple cuboidal, simple columnar (nonciliated and ciliated), stratified squamous (keratinized and nonkeratinized), stratified cuboidal, stratified columnar, transitional, and pseudostratified columnar. (pp. 100–105)
8. Define the following terms: endothelium, mesothelium, secretion, absorption, goblet cell, and keratin. (p. 105)
9. What is a gland? Distinguish between endocrine and exocrine glands. (p. 106)
10. Describe the classification of exocrine glands according to structure and function and give at least one example of each. (p. 107)
11. Enumerate the ways in which connective tissue differs from epithelium. (p. 107)
12. Describe the cells, ground substance, and fibers that comprise connective tissue. (p. 107)
13. How are connective tissues classified? List the various types. (p. 109)
14. How are embryonic connective tissue and mature connective tissue distinguished? (p. 109)
15. Describe the following connective tissues with regard to microscopic appearance, location in the body, and function: areolar connective tissue, adipose tissue, reticular connective tissue, dense regular connective tissue, dense irregular connective tissue, elastic connective tissue, hyaline cartilage, fibrocartilage, elastic cartilage, bone (osseous tissue), and blood (vascular tissue). (p. 109)
16. Define the following terms: matrix, ground substance, hyaluronic acid, chondroitin sulfate, dermatan sulfate, keratan sulfate, collagen fiber, elastic fiber, reticular fiber, fibroblast, macrophage, plasma cell, mast cell, adipocyte, chondrocyte, lacuna, osteocyte, lamella, and canaliculus. (p. 108)
17. Distinguish between the interstitial and appositional growth of cartilage. (p. 118)
18. Define the following kinds of membranes: mucous, serous, cutaneous, and synovial. Where is each located in the body? What are their functions? (p. 119)
19. How is muscle tissue classified? What are its functions? (p. 119)
20. Distinguish between neurons and neuroglia. Describe the structure and function of neurons. (p. 121)
21. Following are some descriptions of various tissues of the body. For each description, name the tissue described.
    a. An epithelium that permits distention (stretching).
    b. A single layer of flat cells concerned with filtration and absorption.
    c. Forms all other kinds of connective tissue.
    d. Specialized for fat storage.
    e. An epithelium with waterproofing qualities.
    f. Forms the framework of many organs.
    g. Produces perspiration, wax, oil, or digestive enzymes.
    h. Cartilage that shapes the external ear.
    i. Contains goblet cells and lines the intestine.
    j. Most widely distributed connective tissue.
    k. Forms tendons, ligaments, and aponeuroses.
    l. Specialized for the secretion of hormones.
    m. Provides support in the umbilical cord.
    n. Lines kidney tubules and is specialized for absorption and secretion.
    o. Permits extensibility of lung tissue.
    p. Stores red marrow, protects, supports.
    q. Nonstriated, usually involuntary muscle tissue.
    r. Consists of granulocytes and agranulocytes.
    s. Composed of a cell body, dendrites, and axon.
22. What is meant by tissue repair? Distinguish between stromal and parenchymal repair. What is the importance of granulation tissue? (p. 121)
23. What conditions affect tissue repair? (p. 122)
24. Define the following: biopsy (p. 197), Marfan syndrome (p. 110), and adhesions (p. 122).

Answers to Questions with Figures

4.1 Gap junctions allow the spread of electrical and chemical signals from cell to cell.
4.2 Provides physical support for epithelium, provides for cell attachment, serves as a filter in the kidneys, and guides cell migration during development and tissue repair.
4.3 Mammary gland is an apocrine gland, oil gland is a holocrine gland, and salivary gland is a merocrine gland.

Chapter 5

THE INTEGUMENTARY SYSTEM

Chapter Contents at a Glance

Student Objectives

1. Describe the anatomy and physiology of the skin.

2. Explain the basis for skin color.

3. Compare the anatomy, distribution, and physiology of hair, sebaceous (oil), sudoriferous (sweat), and ceruminous glands.

4. Outline the steps involved in epidermal wound healing and deep wound healing.

5. Explain the role of the skin in helping to maintain the homeostasis of normal body temperature.

6. Describe the effects of aging on the integumentary system.

7. Describe the development of the epidermis, its derivatives, and the dermis.

8. Describe the causes and effects for the following skin disorders: burns, sunburn, skin cancer, acne, and pressure sores.

9. Define medical terminology associated with the integumentary system.

group of tissues that performs a specific function is an **organ.** The next higher level of organization is a **system**—a group of organs working together toward common goals. The organs that make up the **integumentary** (in′-teg-yoo-MEN-tar-ē) **system** of the body are the skin and its derivatives, such as hair, nails, glands, and nerve endings.

The developmental anatomy of the integumentary system is considered at the end of the chapter.

Of all the body's organs, none is more easily inspected or more exposed to infection, disease, and injury than the skin. Because of its visibility, skin reflects our emotions and some aspects of normal physiology, as evidenced by frowning, blushing, and sweating. Changes in skin color may indicate homeostatic imbalances in the body; for example, a bluish skin color is one sign of heart failure. Abnormal skin eruptions or rashes such as chickenpox, cold sores, or measles may reveal systemic infections or diseases of internal organs. Other disorders may involve just the skin itself, such as warts, age spots, or pimples. The skin's location makes it vulnerable to damage from trauma, sunlight, microbes, and pollutants in the environment.

Many interrelated factors affect both the appearance and health of the skin, including nutrition, hygiene, circulation, age, immunity, genetic traits, psychological state, and drugs. So important is the skin to one's image that people spend much time and money to restore skin to a more normal or youthful appearance.

SKIN

The **skin** is an organ because it consists of different tissues that are joined to perform specific activities. It is one of the largest organs of the body in surface area and weight. In adults, the skin covers an area of about 2 square meters (22 square feet), and weighs 4.5 to 5 kg (10 to 11 lb). It ranges in thickness from 0.5 to 4.0 mm, depending on location. The skin is not just a simple, thin coat that keeps the body together and provides protection. It performs several essential functions. **Dermatology** (der′-ma-TOL-ō-jē; *dermato* = skin; *logos* = study of) is the medical specialty that deals with diagnosing and treating skin disorders.

Anatomy

Structurally, the skin consists of two principal parts (Fig. 5.1). The outer, thinner portion, which is composed of *epithelium,* is called the **epidermis.** The epidermis is attached to the inner, thicker, *connective tissue* part called the **dermis.** Beneath the dermis is a **subcutaneous (subQ) layer.** This layer, also called the **superficial fascia** or **hypodermis,** consists of areolar and adipose tissues. Fibers from the dermis extend down into the subcutaneous layer and anchor the skin to it. The subcutaneous layer, in turn, attaches to underlying tissues and organs.

Physiology

Skin serves several functions, which are introduced in this chapter and explained in more detail in later chapters.

1. Regulation of body temperature. In response to high environmental temperature or strenuous exercise, the evaporation of sweat from the skin surface helps lower an elevated body temperature to normal (see Fig. 5.7). In response to low environmental temperature, production of sweat is decreased, which helps conserve heat. Changes in the flow of blood to the skin also help regulate body temperature (see page 137).

2. Protection. The skin covers the body and provides a physical barrier that protects underlying tissues from physical abrasion, bacterial invasion, dehydration, and ultraviolet (UV) radiation. Hair and nails also have protective functions, as described shortly.

3. Sensation. The skin contains abundant nerve endings and receptors that detect stimuli related to temperature, touch, pressure, and pain (see page 446).

4. Excretion. Besides removing heat and some water from the body, sweat also is the vehicle for excretion of a small amount of salts and several organic compounds.

5. Immunity. Certain cells of the epidermis are important components of the immune system, which fends off foreign invaders (see page 701).

6. Blood reservoir. The dermis of the skin houses extensive networks of blood vessels that carry 8 to 10% of the total blood flow in a resting adult. In moderate exercise, skin blood flow may increase, which helps dissipate heat from the body. During hard exercise, however, skin blood vessels constrict (narrow) somewhat, and more blood is able to circulate to contracting muscles.

7. Synthesis of vitamin D. **Vitamin D** is a group of closely related compounds. Synthesis of vitamin D begins with activation of a precursor molecule in the skin by ultraviolet (UV) rays in sunlight. Enzymes in the liver and kidneys then modify the molecule, finally producing calcitriol, the *most active form* of vitamin D. Calcitriol contributes to the homeostasis of body fluids by aiding absorption of calcium in foods. According to the synthesis sequence just described, vitamin D is a hormone, since it is produced in one location in the body, transported by the blood, and then exerts its effect in another location. In this respect, the skin may be considered an endocrine organ.

CLINICAL APPLICATION

AVOIDING VITAMIN D DEFICIENCY

During most of the year, an hour per week in the sunlight with the hands, arms, and face exposed meets the body's needs for activation of the vitamin D precursor. Additional sun exposure merely increases one's risk of skin cancer. Also, all milk sold in the U.S. is fortified with a D vitamin, so most

FIGURE 5.1 Structure of the skin and underlying subcutaneous tissue. The stratum lucidum shown is not present on hairy skin but is included so that you can see the relationship of all five epidermal strata.

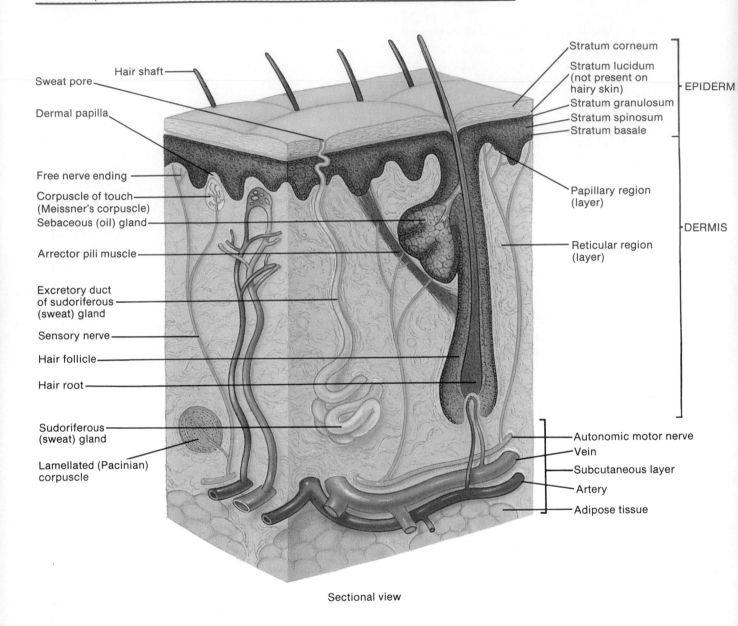

Sectional view

Question: What stratum of the epidermis is continually shed? Capable of cell division?

people take in an adequate amount. Vitamin D deficiency can arise in those who completely cover their skin when outdoors (for example, Muslim women), in those who rarely venture outside, or in individuals who drink little or no milk. Elderly people living in nursing homes or confined to a hospital bed have double the risk for developing vitamin D deficiency. They may stay indoors for long periods, and they may not be able to digest milk. The enzyme lactase, which is needed to split the milk sugar lactose, often declines as we age, producing milk intolerance.

Epidermis

The **epidermis** is composed of stratified squamous epithelium and contains four principal types of cells (Fig. 5.2). About 90% of the epidermal cells are **keratinocytes** (ker-a-TIN-ō-sīts; *kerato* = horny). They produce the protein keratin that helps waterproof and protect the skin and underlying tissues. Anchoring junctions called desmosomes (see Fig. 4.1) weld keratinocytes to one another.

Figure 5.2 Cell types and layers in the epidermis of thick skin. Thin skin is similar, but the stratum lucidum is absent and the stratum corneum is thinner.

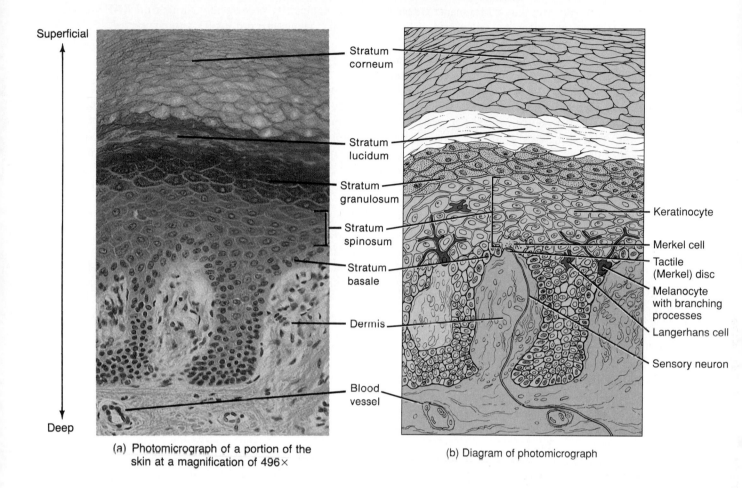

Superficial

Stratum corneum

Stratum lucidum

Stratum granulosum

Stratum spinosum

Stratum basale

Dermis

Blood vessel

Deep

Keratinocyte

Merkel cell

Tactile (Merkel) disc

Melanocyte with branching processes

Langerhans cell

Sensory neuron

(a) Photomicrograph of a portion of the skin at a magnification of 496×

(b) Diagram of photomicrograph

Question: What hormone stimulates renewal of epidermal cells?

Melanocytes (MEL-a-nō-sīts), which produce the pigment melanin, comprise about 8% of the epidermal cells. Their long, slender projections extend between and transfer granules of melanin to keratinocytes. **Melanin** (*melan* = black) is a brown-black pigment that contributes to skin color and absorbs ultraviolet (UV) light. Once inside keratinocytes, the melanin granules cluster to form a protective veil over the nucleus, on the side toward the skin surface. In this way they shield the genetic material from damaging UV light.

The third type of cell in the epidermis is known as a **Langerhans** (LANG-er-hans) **cell**. These cells arise from bone marrow and migrate to the epidermis. They interact with white blood cells called helper T cells in immune responses and are easily damaged by UV radiation.

A fourth type of cell found in the epidermis is called a **Merkel cell**. These cells are located in the deepest layer (stratum basale) of the epidermis of hairless skin, where

they are attached to keratinocytes by desmosomes. Merkel cells make contact with the flattened portion of the ending of a sensory neuron (nerve cell), called a **tactile (Merkel) disc**, and are thought to function in the sensation of touch.

Four or five distinct layers of cells form the epidermis. In most regions of the body the epidermis is about 0.1 mm thick and has four layers. Where exposure to friction is greatest, such as in the palms and soles, the epidermis is thicker (1 to 2 mm) and has five layers (Fig. 5.2). Constant exposure of thin or thick skin to friction or pressure stimulates formation of a **callus,** an abnormal thickening of the epidermis.

The names of the five layers (strata), from the deepest to the most superficial, are:

1. Stratum basale. This single layer of cuboidal to columnar cells contains stem cells, which are capable of continued cell division, and melanocytes. The stem cells multiply, producing keratinocytes, which push up toward the surface and become part of the more superficial layers. The nuclei of the keratinocytes degenerate, and the cells die. Eventually, the cell remnants are shed from the top layer of the epidermis. Other stem cells in the stratum basale migrate into the dermis and give rise to sweat and oil glands and hair follicles. The stratum basale is sometimes referred to as the **stratum germinativum** (jer′-mi-na-TĒ-vum) to indicate its role in germinating new cells. The stratum basale also contains tactile (Merkel) discs that are sensitive to touch.

2. Stratum spinosum. This layer of the epidermis contains 8 to 10 rows (sheets) of polyhedral (many-sided) cells that fit closely together. The cells here appear to be covered with prickly spines (*spinosum* = prickly) because the cells shrink apart when the tissue is prepared for microscopic examination. At each spinelike projection, filaments of the cytoskeleton insert into desmosomes, which tightly join the cells to one another. Long projections of the melanocytes extend among the keratinocytes, which take in melanin by phagocytosis of these melanocyte processes.

3. Stratum granulosum. The third layer of the epidermis consists of three to five rows of flattened cells that develop darkly staining granules of a substance called **keratohyalin** (ker′-a-tō-HĪ-a-lin). This compound is the precursor of **keratin,** a protein found in the outer layer of the epidermis. Keratin forms a barrier that protects deeper layers from injury and microbial invasion and makes the skin waterproof. The nuclei of the cells in the stratum granulosum are in various stages of degeneration. As their nuclei break down, the cells can no longer carry on vital metabolic reactions, and they die.

4. Stratum lucidum. Normally, only the thick skin of the palms and soles has this layer. It consists of three to five rows of clear, flat, dead cells that contain droplets of an intermediate substance that is formed from keratohyalin and is eventually transformed to keratin.

5. Stratum corneum. This layer consists of 25 to 30 rows of flat, dead cells completely filled with keratin. These cells are continuously shed and replaced by cells from deeper strata. The stratum corneum serves as an effective barrier against light and heat waves, bacteria, and many chemicals.

In the process of **keratinization,** cells newly formed in the basal layers undergo a developmental process as they are pushed to the surface. As the cells relocate, they accumulate keratin. At the same time the cytoplasm, nucleus, and other organelles disappear, and the cells die. Eventually, the keratinized cells slough off and are replaced by underlying cells that, in turn, become keratinized. The whole process by which a cell forms in the basal layer, rises to the surface, becomes keratinized, and sloughs off takes two to four weeks.

Epidermal growth factor (EGF) is a protein hormone that stimulates growth of epithelial and epidermal cells during tissue development, repair, and renewal. Certain oncogenes, genes that can turn normal cells into cancerous ones, cause tumors by permanently turning on EGF stimulation of cells, which then proliferate without control.

Dermis

The second principal part of the skin, the **dermis,** is composed of connective tissue containing collagen and elastic fibers (see Fig. 5.1). The few cells in the dermis include fibroblasts, macrophages, and adipocytes. The dermis is very thick in the palms and soles and very thin in the eyelids, penis, and scrotum. It also tends to be thicker on the dorsal than the ventral aspects of the body and thicker on the lateral than the medial aspects of the extremities. Blood vessels, nerves, glands, and hair follicles are embedded in the dermis.

The outer portion of the dermis, about one-fifth of the thickness of the total layer, is named the **papillary region (layer).** It consists of areolar connective tissue containing fine elastic fibers. Its surface area is greatly increased by small, fingerlike projections called **dermal papillae** (pa-PIL-ē; *papilla* = nipple). These nipple-shaped structures indent the epidermis, and many contain loops of capillaries. Some dermal papillae also contain tactile receptors called **corpuscles of touch (Meissner's corpuscles),** nerve endings that are sensitive to touch. Dermal papillae cause ridges in the overlying epidermis. It is these ridges that leave fingerprints on objects that are handled.

The deeper portion of the dermis is called the **reticular** (*rete* = net) **region (layer).** It consists of dense, irregular connective tissue containing interlacing bundles of collagen and coarse elastic fibers. Within the reticular region, bundles of collagen fibers interlace in a netlike manner. Spaces between the fibers are occupied by a small quantity of adipose tissue, hair follicles, nerves, oil glands, and the ducts of sweat glands. Varying thicknesses of the reticular region contribute to differences in the thickness of skin.

The combination of collagen and elastic fibers in the reticular region provides the skin with strength, extensibility, and elasticity. (**Extensibility** is the ability to stretch; **elasticity** is the ability to return to original shape after stretching.) The ability of the skin to stretch can readily be seen in pregnancy, obesity, and edema. Small tears that occur in the dermis during extreme stretching are initially red and remain visible afterward as silvery white streaks called **striae** (STRĪ-ē) or stretch marks.

The reticular region is attached to underlying organs, such as bone and muscle, by the subcutaneous layer, also called the **hypodermis** or **superficial fascia.** The subcutaneous layer also contains nerve endings called **lamellated** or **Pacinian** (pa-SIN-ē-an) **corpuscles** that are sensitive to pressure (see Fig. 15.1). Nerve endings sensitive to cold are found in and just below the dermis, while those sensitive to heat are located in the middle and outer dermis.

LINES OF CLEAVAGE AND SURGERY

In certain regions of the body, collagen fibers tend to orient more in one direction than another. **Lines of cleavage (tension lines)** in the skin indicate the predominant direction of underlying collagen fibers. The lines are especially evident on the palmar surfaces of the fingers, where they are arranged parallel to the long axis of the digit. Lines of cleavage are of particular interest to a surgeon because an incision running parallel to the collagen fibers will heal with only a fine scar. An incision made across the rows of fibers disrupts the collagen, and the wound tends to gape open and heal in a broad, thick scar.

Skin Color

Three pigments—melanin, carotene, and hemoglobin—give skin a wide variety of colors. Melanin is located mostly in the epidermis; carotene is mostly in the dermis; and hemoglobin is in red blood cells within capillaries in the dermis. The amount of **melanin** varies the skin color from pale yellow to black. Melanocytes are most plentiful in the mucous membranes, penis, nipples of the breasts and the area just around the nipples (areola), face, and extremities. The *number* of melanocytes is about the same in all races. Differences in skin color are due mainly to the *amount of pigment* the melanocytes produce and disperse to keratinocytes. In some people, melanin-filled cells tend to cluster in patches called **freckles.** As one grows older, **liver (age) spots** may develop. These are flat skin patches that look like freckles and range in color from light brown to black. Like freckles, liver spots are clusters of melanocytes, and neither one tends to become cancerous.

Melanocytes synthesize melanin from the amino acid *tyrosine* in the presence of an enzyme called *tyrosinase.* Synthesis occurs in an organelle called a **melanosome.** Exposure to ultraviolet radiation increases the enzymatic activity within melanosomes and leads to increased melanin production. Both the amount and the darkness of melanin increase, which tans the skin and further protects the body against UV radiation. Thus melanin serves a protective function.

An inherited inability of an individual of any race to produce melanin results in **albinism** (AL-bin-izm; *albus* = white). Most albinos (al-BĪ-nōs) have melanocytes but the cells are unable to synthesize tyrosinase. Melanin is thus absent in hair, eyes, and skin. In another condition, called **vitiligo** (vit-i-LĪ-gō), the partial or complete loss of melanocytes from patches of skin produces irregular white spots.

Carotene (KAR-o-tēn), a yellow-orange pigment, gives carrots and egg yolks their color. It is the precursor of vitamin A, which is used to synthesize pigments needed for vision. In people of Asian ancestry, carotene is also found in the stratum corneum and fatty areas of the dermis and subcutaneous layer. Together, carotene and melanin account for the yellowish hue of their skin.

In Caucasians, the epidermis is translucent because little melanin is present. The epidermis itself has no blood vessels, a characteristic of all epithelia. The skin of Caucasians appears pink to red, depending on the amount and quality of the blood moving through capillaries in the dermis. The red color is due to **hemoglobin,** the pigment that carries oxygen in blood.

SKIN COLOR CLUES

The color of skin and mucous membranes can provide clues for diagnosing certain problems. When blood is not picking up an adequate amount of oxygen in the lungs, such as in a baby who has stopped breathing, mucous membranes, nail beds, and light-colored skin appears bluish or **cyanotic** (sī-an-OT-ic; *cyan* = blue). This is because hemoglobin that is depleted of oxygen looks deep, purplish blue. **Jaundice** (JON-dis; *jaune* = yellow), a yellowed appearance of the whites of the eyes and of light-colored skin, usually indicates liver disease. **Erythema** (er-e-THĒ-ma; *erythros* = red), redness of the skin, is caused by engorgement of capillaries in the dermis with blood. In people who have light-colored skin, exercise and embarrassment may cause noticeable erythema, especially of the face. Exercise and embarrassment produce the same, though often unobservable, reaction in dark-colored skin. Erythema also occurs with skin injury, infection, inflammation, or allergic reactions.

Epidermal Ridges

The outer surface of the skin of the palms and fingers and soles and toes is marked by a series of ridges and grooves. They appear either as straight lines or as a pattern of loops and whorls, as on the tips of the digits. **Epidermal ridges** develop during the third and fourth fetal months as the epidermis conforms to the contours of the underlying dermal papillae (see Fig. 5.1). The function of the ridges is to increase the grip of the hand or foot by increasing friction and acting like tiny suction cups. Since the ducts of sweat glands open on the tops of the epidermal ridges as sweat pores, the sweat helps form fingerprints (or footprints) when a smooth object is touched. The ridge pattern, which is genetically determined, is unique for each individual. Normally, it does not change throughout life, except to enlarge, and thus can serve as the basis for identification through fingerprints or footprints.

SKIN GRAFTS

Sometimes, the germinal portion of epidermis is destroyed, and new skin cannot regenerate. Such wounds require **skin grafts**. The most successful type of skin graft involves the transplantation of a segment of skin from a donor site to a recipient site of the same individual (autograft) or an identical twin (isograft).

If skin loss is so extensive that conventional grafting is impossible, a self-donation procedure called **autologous** (aw-TOL-o-gus) **skin transplantation** may be employed. In this procedure, used most often on severely burned patients, small amounts of an individual's epidermis are removed and grown in the laboratory to produce thin sheets of skin. Then, they are transplanted back to the patient where they adhere to burn wounds and generate a permanent skin to cover burned areas. The process normally takes three to four weeks. Since hair, sweat glands, and oil glands develop from the epidermis (described shortly), they can assist in replacing the epidermis removed for an autologous skin transplant. The obvious advantage to this procedure is that a lot of new epithelium can be grown from a very small skin sample. During the waiting period, wound dressing or skin from another person (homograft), such as a cadaver, or animal (heterograft) may be used to help protect the patient from potentially fatal fluid loss and infection. Although homografts and heterografts are temporary because they are ultimately rejected by the immune system, they prevent loss of fluid, electrolytes, and protein from burn sites while also decreasing pain and increasing the patient's mobility.

One type of synthetic skin that is used as part of autologous skin transplantation consists of a protective plastic layer that serves as the temporary epidermis and an underlying layer composed of collagen fibers and cartilage derived from animals that serves as the temporary dermis. After the synthetic skin is applied to the burned area, fibroblasts produce collagen fibers to replace those in the synthetic skin, which decompose. After a few months, when the reconstruction of the dermis is complete, sheets of the patient's newly grown epidermis are transplanted over the wound area.

EPIDERMAL DERIVATIVES

Organs that develop from the embryonic epidermis—hair, glands, nails—have a host of important functions. Hair and nails protect the body. The sweat glands help regulate body temperature. The enamel of teeth is also an epidermal derivative and is discussed on page 774.

Hair

Hairs, or **pili** (PI-lē), are growths of the epidermis variously distributed over the body. Their primary function is protection. Although the protection is limited, hair on the head guards the scalp from injury and the sun's rays. It also decreases heat loss. Eyebrows and eyelashes protect the eyes from foreign particles. Hair in the nostrils protects against inhaling insects and foreign particles. Hair serves a similar protective function in the external ear canal. Touch receptors associated with hair follicles (hair root plexuses) are activated whenever a hair is even slightly moved. Normal hair loss in an adult scalp is about 70 to 100 hairs per day. Both the rate of growth and the replacement cycle may be altered by illness, diet, high fever, surgery, blood loss, or

severe emotional stress. Rapid weight-loss diets that severely restrict calories or protein increase hair loss. An increase in the rate of shedding can also occur for three to four months after childbirth and with certain drugs and radiation therapy for cancer.

Anatomy

A hair is composed of columns of dead, keratinized cells welded together. The **shaft** is the superficial portion of the hair, which projects from the surface of the skin (see Fig. 5.1). The **root** is the portion of the hair below the surface that penetrates into the dermis and sometimes into the subcutaneous layer (see Fig. 5.1). The shaft and root both consist of three concentric layers (Fig. 5.3a). The inner **medulla** is composed of two or three rows of polyhedral cells containing pigment granules and air spaces. The middle **cortex** forms the major part of the shaft and consists of elongated cells that contain pigment granules in dark hair but mostly air in white hair. The **cuticle of the hair,** the outermost layer, consists of a single layer of thin, flat, scalelike cells that are the most heavily keratinized. Cuticle cells are arranged like shingles on the side of a house, with their free edges pointing toward the top of the hair (Fig. 5.3b).

Surrounding the root of the hair is the **hair follicle,** which is made up of an external root sheath and an internal root sheath. The **external root sheath** is a downward continuation of the epidermis. Near the surface, it contains all the epidermal layers. At the bottom of the hair follicle, the external root sheath contains only the stratum basale. The **internal root sheath** forms a cellular tubular sheath between the external root sheath and the hair.

At the base of each hair follicle is an enlarged, layered structure, the **bulb.** This structure houses a nipple-shaped indentation, the **papilla of the hair,** which contains areolar connective tissue. The papilla of the hair contains many blood vessels and provides nourishment for the growing hair. The bulb also contains a ring of cells called the **matrix,** which is the germinal layer. The cells of the matrix derive from the stratum basale. They are responsible for the growth of existing hairs and produce new hairs by cell division when older hairs are shed. This replacement occurs within the same follicle. Matrix cells also give rise to the cells of the internal root sheath.

Sebaceous (oil) glands and a bundle of smooth muscle cells are also associated with hairs. Details of the sebaceous glands will be discussed shortly. The smooth muscle is called **arrector pili** (plural is **arrectores pilorum);** it extends from the dermis of the skin to the side of the hair follicle. In its normal position, hair emerges at an angle to the surface of the skin. The arrector pili muscles contract under stresses of fright, cold, and emotions and pull the hairs into a vertical position. This response makes a furry animal look larger, and thus it may appear more threatening

FIGURE 5.3 Hair. The relationship of a hair to the epidermis, sebaceous (oil) glands, and arrector pili muscle should also be noted in Fig. 5.1.

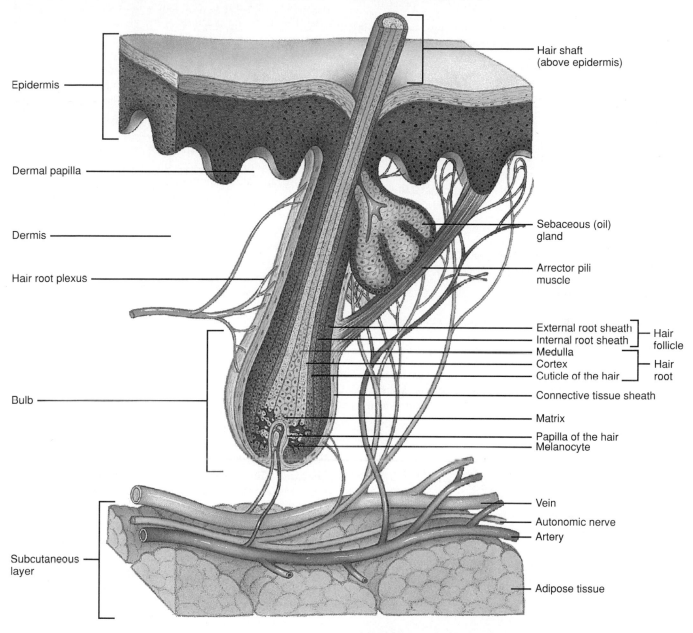

Epidermis

Dermal papilla

Dermis

Hair root plexus

Bulb

Subcutaneous layer

Hair shaft (above epidermis)

Sebaceous (oil) gland

Arrector pili muscle

External root sheath ⎤
Internal root sheath ⎦ Hair follicle
Medulla ⎤
Cortex ⎥ Hair root
Cuticle of the hair ⎦

Connective tissue sheath

Matrix

Papilla of the hair

Melanocyte

Vein

Autonomic nerve

Artery

Adipose tissue

(a) Longitudinal section of a hair root

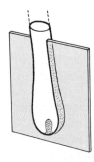

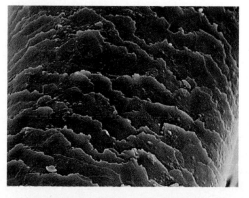

(b) Scanning electron micrograph of the surface of a hair shaft showing the shingle-like cuticular scales at a magnification of 2150x

Question: Why does it hurt when you pluck a hair out but not when you have a haircut?

to an aggressor. In people, with little body hair, it serves no apparent purpose, merely causing "goose bumps" or "gooseflesh" because the skin around the shaft forms slight elevations.

Around each hair follicle are nerve endings, called **hair root plexuses,** that are sensitive to touch. They respond if a hair shaft is moved.

Color

The color of hair is due primarily to melanin. It is synthesized by melanocytes scattered in the matrix of the bulb and passes into cells of the cortex and medulla (Fig. 5.3a). Dark-colored hair contains mostly true melanin. Blond and red hair contain variants of melanin in which there is iron and more sulfur. Graying of hair is the loss of pigment believed to be the result of a progressive decline in tyrosinase, an enzyme necessary for synthesis of melanin. White hair results from accumulation of air bubbles in the medullary shaft.

CLINICAL APPLICATION

HAIR AND HORMONES

At puberty, when their testes begin secreting significant quantities of male sex hormones (androgens), males develop the typical male pattern of hair growth, including a beard and a hairy chest. In females, both the ovaries and the adrenal glands produce small quantities of androgens. Occasionally, a tumor of one of these glands oversecretes androgens and causes **hirsutism** (hur-SOO-tiz-um). This is a condition of excessive hairiness of the upper lip, chin, chest, inner thighs, and abdomen in females or prepubertal males.

Surprisingly, androgens also must be present for the most common form of baldness, **male-pattern baldness,** to occur. In genetically predisposed males, androgens somehow inhibit hair growth. In 1979 a report noted that a new blood pressure medicine had the side effect of causing hair growth. The result was an immediate stampede to test its effectiveness in treating baldness. The drug is minoxidil (Rogaine), a potent vasodilator, that is, a drug that widens blood vessels and increases circulation. When applied topically, it does stimulate some hair regrowth in some persons with thinning hair due to male-pattern baldness. For many, however, the hair growth is meager, and it does not help individuals who already are bald.

Glands

Several kinds of glands are associated with the skin: sebaceous (oil) glands, sudoriferous (sweat) glands, ceruminous glands, and mammary glands. The anatomy and physiology of the mammary glands, which are modified sudoriferous glands, will be discussed on page 946 as components of the female reproductive system.

Sebaceous (Oil) Glands

Sebaceous (se-BĀ-shus) **glands,** or **oil glands,** with few exceptions, are connected to hair follicles (see Figs. 5.1 and 5.3a). The secreting portions of the glands lie in the dermis and open into the necks of hair follicles or directly onto a skin surface (lips, glans penis, labia minora, and tarsal glands of the eyelids). Absent in the palms and soles, sebaceous glands vary in size and shape in other regions of the body. For example, they are small in most areas of the trunk and extremities, but large in the skin of the breasts, face, neck, and upper chest.

Sebaceous glands secrete an oily substance called **sebum** (SĒ-bum), a mixture of fats, cholesterol, proteins, and inorganic salts. Sebum helps keep hair from drying and becoming brittle, prevents excessive evaporation of water from the skin, keeps the skin soft and pliable, and inhibits the growth of certain bacteria. When sebaceous glands of the face become enlarged because of accumulated sebum, **blackheads** develop. Since sebum is nutritive to certain bacteria, **pimples** or **boils** often result. The color of blackheads is due to melanin and oxidized oil, not dirt.

Sudoriferous (Sweat) Glands

Three to four million **sudoriferous** (soo'-dor-IF-er-us; *sudor* = sweat; *ferre* = to bear) or **sweat glands** empty their secretions onto the skin surface (see Fig. 5.1). They are divided into two principal types, eccrine and apocrine, based on their structure, location, and type of secretion.

Eccrine sweat glands are much more common than apocrine sweat glands. They are distributed throughout the skin except for the margins of the lips, nail beds of the fingers and toes, glans penis, glans clitoris, labia minora, and eardrums. Eccrine sweat glands are most numerous in the skin of the palms and the soles; their density can be as high as 450 per square centimeter (3000 per square inch) in the palms. The secretory portion of eccrine sweat glands is located in the subcutaneous layer, and the excretory duct projects upward through the dermis and epidermis. It ends as a pore at the surface of the epidermis (see Fig. 5.1). Eccrine sweat glands function throughout life and produce a secretion that is more watery than that of apocrine sweat glands.

Apocrine sweat glands are found mainly in the skin of the axilla (armpit), pubic region, and areolae (pigmented areas) of the breasts. The secretory portion of apocrine sweat glands is located in the dermis or subcutaneous layer, and the excretory duct opens into hair follicles. Apocrine sweat glands begin to function at puberty and produce a more viscous secretion than eccrine sweat glands. They are stimulated during emotional stresses and sexual excitement.

Sweat (perspiration) is the fluid produced by sweat glands, mainly eccrine sweat glands because they are so much more numerous. It is a mixture of water, salts (mostly NaCl), urea, uric acid, amino acids, ammonia, sugar, lactic acid, and ascorbic acid. Its principal function is to help regulate body temperature by providing a cooling mechanism (described shortly). As sweat evaporates, large quantities of heat energy leave the body surface. It also plays a small role in eliminating wastes.

Ceruminous Glands

In the ear, modified sweat glands called **ceruminous** (se-ROO-mi-nus; *cera* = wax) **glands** produce a waxy secretion. The secretory portions of the ceruminous glands lie in the subcutaneous layer, deep to sebaceous glands. Their excretory ducts open either directly onto the surface of the external auditory meatus (ear canal) or into ducts of sebaceous glands. The combined secretion of the ceruminous and sebaceous glands is called **cerumen.** Cerumen, together with hairs in the external auditory meatus, provides a sticky barrier that prevents the entrance of foreign bodies.

Nails

Nails are plates of tightly packed, hard, keratinized cells of the epidermis. The cells form a clear, solid covering over the dorsal surfaces of the terminal portions of the fingers and toes. Each nail (Fig. 5.4) consists of a nail body, a free edge, and a nail root. The **nail body** is the portion of the nail that is visible, the **free edge** is the part that may extend past the distal end of the digit, and the **nail root** is the portion that is buried in a fold of skin. Most of the nail body is pink because of blood flowing through underlying capillaries. The whitish semilunar area of the proximal end of the body is called the **lunula** (LOO-nyoo-la; *lunula* = little moon). It appears whitish because the vascular tissue underneath does not show through owing to the thickened stratum basale in the area. The **eponychium** (ep'-ō-NIK-ē-um) or **cuticle** is a narrow band of epidermis that extends from

the margin of the nail wall (lateral border), adhering to it. It occupies the proximal border of the nail and consists of stratum corneum.

The epithelium under the nail root is known as the **nail matrix.** Its function is to bring about the growth of nails. Growth occurs by the transformation of superficial cells of the matrix into nail cells. In the process, the outer, harder layer is pushed forward over the stratum germinativum. The average growth in the length of fingernails is about 1 mm (0.04 in.) per week. The growth rate is somewhat slower in toenails. The longer the digit, the faster the nail grows. Adding supplements such as gelatin to an otherwise healthy diet has no effect on making nails grow faster or stronger.

Functionally, nails help us grasp and manipulate small objects in various ways and provide protection against trauma to the ends of the digits.

THE SKIN AND HOMEOSTASIS

The skin affords two excellent examples of how a part of the body maintains homeostasis. One involves wound healing. When the skin is damaged by various stimuli (stresses), certain mechanisms go into operation to restore it to its normal or near-normal structure and functions. The other example is concerned with body temperature. The skin is one of the major organs that restores a normal body temperature should some stimulus raise or lower it.

Skin Wound Healing: Homeostasis of Skin Structure

Here we discuss the basic mechanisms by which skin wounds are repaired. First we consider wounds that affect primarily the epidermis. Then we describe wounds that extend into the dermis and subcutaneous layer.

FIGURE 5.4 Structure of nails. Shown is a fingernail.

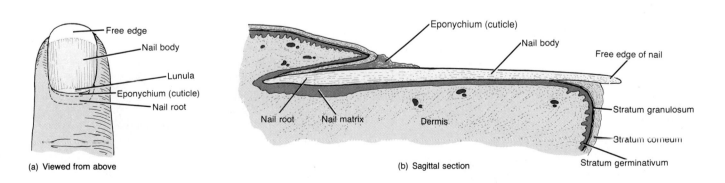

(a) Viewed from above

(b) Sagittal section

Question: Why are nails so hard?

Epidermal Wound Healing

The exposed location of the skin (and mucous membranes) makes it vulnerable to trauma as a result of physical and chemical stimuli (stresses). One common type of epidermal wound is an **abrasion** (scraped away area) such as might be experienced in the form of a skinned knee or elbow. Another type is a first-degree or second-degree burn (see page 139). In an epidermal wound, the central portion of the wound may extend to the dermis, while the edges of the wound usually involve only slight damage to superficial epidermal cells.

In response to injury, basal epidermal cells in the area of the wound break their contacts with the basement membrane. These cells then enlarge and migrate across the wound (Fig. 5.5). The cells appear to migrate as a sheet until advancing cells from opposite sides of the wound meet. When epidermal cells encounter each other, their continued migration is stopped by **contact inhibition.** According to this phenomenon, when one epidermal cell encounters another, its direction of movement changes until it encounters another like cell, and so on. Continued migration of the epidermal cell stops when it is finally in contact on all sides with other epidermal cells. Contact inhibition appears to occur only among like cells; in other words, contact inhibition does not occur between epidermal cells and other types of cells. Malignant cells do not conform to the rules of contact inhibition (page 89), and so they have the ability to invade body tissues with few restrictions.

Simultaneous with the migration of some basal epidermal cells, stationary basal stem cells divide to replace the migrated ones. Migration continues until the wound is resurfaced. Following this, the migrated cells themselves divide to form new strata, thus thickening the new epidermis. The events involved in epidermal wound healing occur within 24 to 48 hours after wounding.

Epidermal growth factor (EGF) is a protein hormone found naturally in wounds that stimulates the growth of epidermal cells and fibroblasts.

Deep Wound Healing

When an injury extends to tissues deep to the epidermis, the repair process is more complex than epidermal healing, and scar formation results.

The first step in deep wound healing involves inflammation, a vascular and cellular response that serves to dispose of microbes, foreign material, and dying tissue in preparation for repair. During the **inflammatory phase,** a blood clot forms in the wound and loosely unites the wound edges. Vasodilation and increased permeability of blood vessels enhance delivery of white blood cells called neutrophils and monocytes (macrophages) that phagocytize microbes, and mesenchymal cells, which develop into fibroblasts (Fig. 5.6a).

In the next phase, the **migratory phase,** the clot becomes a scab and epithelial cells migrate beneath the scab to bridge the wound. Fibroblasts migrate along fibrin threads and begin synthesizing scar tissue (collagen fibers and glycoproteins), and damaged blood vessels begin to regrow. During this phase, tissue filling the wound is called **granulation tissue.**

The **proliferative phase** is characterized by extensive growth of epithelial cells beneath the scab, deposition of collagen fibers in random patterns by fibroblasts, and continued growth of blood vessels.

In the final phase, the **maturation phase,** the scab sloughs off once the epidermis is restored to normal thickness. Collagen fibers become more organized, fibroblasts decrease in number, and blood vessels are restored to normal (Fig. 5.6b).

FIGURE 5.5 Epidermal wound healing.

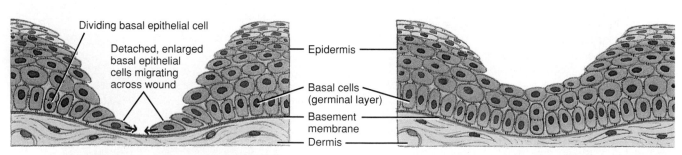

(a) Division of basal cells and migration across wound

(b) Resurfacing of wound

Dividing basal epithelial cell

Detached, enlarged basal epithelial cells migrating across wound

Epidermis

Basal cells (germinal layer)

Basement membrane

Dermis

Question: Would you expect an epidermal wound to bleed? Why?

FIGURE 5.6 Deep wound healing.

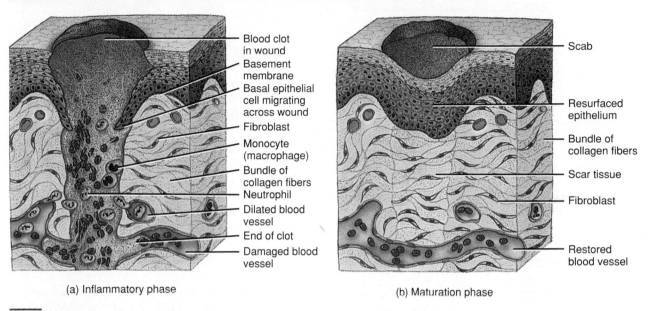

Blood clot in wound
Basement membrane
Basal epithelial cell migrating across wound
Fibroblast
Monocyte (macrophage)
Bundle of collagen fibers
Neutrophil
Dilated blood vessel
End of clot
Damaged blood vessel

Scab
Resurfaced epithelium
Bundle of collagen fibers
Scar tissue
Fibroblast
Restored blood vessel

(a) Inflammatory phase

(b) Maturation phase

Question: Why is the arrival of phagocytic white blood cells (neutrophils and monocytes) at the scene of a wound helpful?

As noted in Chapter 4, the process of scar tissue formation is called **fibrosis.** Sometimes, so much scar tissue is formed that a raised scar results, that is, one that is elevated above the normal epidermal surface. Such a scar may remain within the boundaries of the original wound (**hypertrophic scar),** or it may extend beyond the boundaries of the original wound into normal surrounding tissues (**keloid scar).** Scar tissue differs from normal skin in that its collagen fibers are more densely arranged. Also, it has fewer blood vessels and may not contain hair, skin glands, or sensory neurons.

Thermoregulation: Homeostasis of Body Temperature

The skin plays a major role in thermoregulation, that is, the homeostasis of body temperature. As warm-blooded animals, we are able to maintain our body temperature at a remarkably constant 37°C (98.6°F) even though the environmental temperature varies greatly. Negative feedback systems (see Fig. 1.3) ensure that body temperature (a controlled condition) fluctuates very little.

Suppose you are in a place where the temperature is 38°C (101°F). Heat (the stimulus) continually flows from the environment to your body, raising body temperature. To counteract this change in a controlled condition, a sequence of events is set into operation (Fig. 5.7). Temperature-sensitive receptors (nerve endings) in the skin called thermoreceptors detect the stimulus and send nerve impulses (input) to your brain (control center). A temperature control region

in the brain (called the hypothalamus, see Fig. 14.1) then sends nerve impulses (output) to the sweat glands (effectors), which produce perspiration more rapidly. As the sweat evaporates from the surface of your skin, heat is lost, and your body temperature drops to normal (return to homeostasis). As noted earlier, when environmental temperature is low, sweat glands produce less perspiration.

Your brain also sends output to blood vessels (a second set of effectors), dilating those in the dermis so skin blood flow increases. As more warm blood flows through capillaries close to the body surface, more heat can be lost to the environment by radiation, which lowers body temperature. Radiation is the transfer of heat without physical contact between a warmer and a cooler object. When you stand in the sun or hold your hand over a hot stove, you gain heat by radiation. In these two ways—evaporation and radiation—heat is lost from the body, and body temperature falls to the normal value to restore homeostasis. In response to low environmental temperature, blood vessels in the dermis constrict, blood flow decreases, and less heat is lost by radiation.

Note that temperature regulation by the skin is a *negative* feedback system because the response (cooling) is opposite to the stimulus (heating) that started the cycle. Also, the thermoreceptors continually monitor body temperature and feed back information to keep the brain informed. The brain, in turn, continues to send impulses to the sweat glands and blood vessels until the temperature returns to 37°C (98.6°F).

Regulating the rate of sweating and changing dermal blood flow are only two mechanisms by which we can

FIGURE 5.7 Thermoregulation: homeostasis of body temperature by the skin.

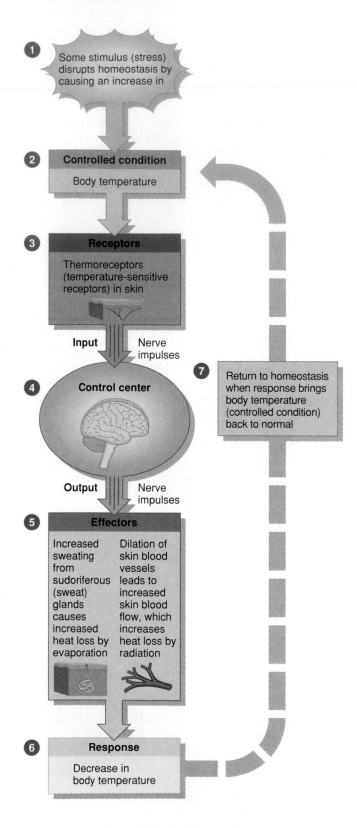

1 Some stimulus (stress) disrupts homeostasis by causing an increase in

2 **Controlled condition**

Body temperature

3 **Receptors**

Thermoreceptors (temperature-sensitive receptors) in skin

Input   Nerve impulses

4 **Control center**

7 Return to homeostasis when response brings body temperature (controlled condition) back to normal

Output   Nerve impulses

5 **Effectors**

| Increased sweating from sudoriferous (sweat) glands causes increased heat loss by evaporation | Dilation of skin blood vessels leads to increased skin blood flow, which increases heat loss by radiation |

6 **Response**

Decrease in body temperature

Question: Why is this a negative feedback cycle?

adjust body temperature. These and other mechanisms for thermoregulation are discussed in more detail in Chapter 25.

AGING AND THE INTEGUMENTARY SYSTEM

Although skin is constantly aging, pronounced effects do not occur until a person reaches the late forties. Around that time, collagen fibers decrease in number, stiffen, break apart, and form into a shapeless, matted tangle. Elastic fibers lose some of their elasticity, thicken into clumps, and fray. As a result, the skin forms crevices and furrows known as wrinkles. Fibroblasts, which produce both collagen and elastic fibers, decrease in number, and macrophages become less efficient phagocytes. With increased age, the hair and nails grow more slowly. Langerhans cells dwindle in number, thus decreasing the immune responsiveness of older skin. Decreased size of sebaceous (oil) glands leads to dry and broken skin that is more susceptible to infection. Production of sweat diminishes, which probably contributes to the increased incidence of heat stroke in the elderly. There is a decrease in the number of functioning melanocytes, resulting in gray hair and atypical skin pigmentation. An increase in the size of some melanocytes produces pigmented blotching (liver spots). Blood vessels in the dermis become thicker walled and less permeable, and subcutaneous fat is lost. In general, aged skin is thinner than young skin, especially the dermis, and migration of cells from the basal layer to the epidermal surface slows considerably. Aged skin also heals poorly and becomes more susceptible to pathological conditions such as skin cancer, itching, pressure sores, and shingles.

CLINICAL APPLICATION

PHOTODAMAGE

Nearly everyone has experienced the effect of acute overexposure to sunlight—a sunburn (see page 141). Even if sunburn does not occur, the ultraviolet (UV) rays in sunlight cause photodamage of DNA in epidermal cells and extracellular matrix materials, such as collagen and elastic fibers, in the dermis. Over the long term, chronic UV exposure accelerates the aging of skin (photoaging) and is an important factor in development of nearly all skin cancers (see page 141). Protecting your skin from UV rays is a wise approach.

A recent treatment for wrinkles and liver spots in photodamaged skin is the drug **tretinoin (Retin-A)**. It is a derivative of vitamin A that has been used to treat acne since the l960s. Among the reported benefits of using tretinoin are increased thickness of the epidermis, smoother stratum corneum, diminished number and size of melanocytes, increased production of collagen and elastin, dilation of blood vessels in the dermis, and regression of precancerous lesions. Dermatologists caution that tretinoin does not improve advanced changes associated with aging, and it may increase the risk of cancers induced by UV light exposure.

DEVELOPMENTAL ANATOMY OF THE INTEGUMENTARY SYSTEM

Throughout this book, the developmental anatomy of the body systems will be discussed usually at the end of each appropriate chapter. The principal features of embryonic development are not treated in detail until Chapter 29. To help you understand the development of organ systems, we will explain and review a few terms at this point.

As part of the early development of a fertilized egg, a portion of the developing embryo differentiates into three layers of tissue called **primary germ layers.** They are the embryonic tissues from which all tissues and organs of the body will eventually develop (see Exhibit 29.1 on page 978).

The *epidermis* is derived from the **ectoderm,** the outermost primary germ layer. At the beginning of the second month, the ectoderm consists of simple cuboidal epithelium. These cells become flattened and are known as the **periderm.** By the fourth month, all layers of the epidermis are formed and each layer assumes its characteristic structure.

Nails develop during the third month. Initially, they consist of a thick layer of epithelium called the **primary nail field.** The nail itself is keratinized epithelium and grows forward from its base. It is not until the ninth month that the nails reach the distal tips of the digits.

Hair follicles develop between the third and fourth months as downgrowths or invaginations of the stratum basale of the epidermis into the dermis. The downgrowths soon differentiate into the bulb, papilla of the hair, beginnings of the epithelial portions of sebaceous (oil) glands, and other structures associated with hair follicles. By the fifth or sixth month, the follicles produce delicate fetal hair called **lanugo** (la-NOO-gō), first on the head and then on other parts of the body. The lanugo is usually shed before birth.

The epithelial (secretory) portions of *sebaceous glands* develop from the sides of the hair follicles and remain connected to the follicles.

The epithelial portions of *sudoriferous (sweat) glands* are also derived from downgrowths of the stratum basale of the epidermis into the dermis. They appear during the fourth month on the palms and soles and a little later in other regions. The connective tissue and blood vessels associated with the glands develop from **mesoderm,** the middle primary germ layer.

The *dermis* is derived from wandering **mesodermal cells.** The mesenchyme becomes arranged in a zone beneath the ectoderm and there undergoes changes into the connective tissues that form the dermis.

DISORDERS: HOMEOSTATIC IMBALANCES

BURNS

Tissue damage from excessive heat, electricity, radioactivity, or corrosive chemicals that destroy (denature) proteins in the exposed cells is called a **burn.** Burns disrupt homeostasis because they destroy the protection afforded by the skin. They permit microbial invasion and infection, loss of fluid, and loss of thermoregulation. The injury to tissues directly or indirectly in contact with the damaging agent, such as the skin or the linings of the respiratory and gastrointestinal tracts, is the local effect of a burn. Generally, however, the systemic effects of a burn are a greater threat to life. They may include (1) a large loss of water, plasma, and plasma proteins, which causes shock; (2) bacterial infection; (3) reduced circulation of blood; (4) decreased production of urine; and (5) diminished immune responses.

A **first-degree burn** involves only the surface epidermis. It is characterized by mild pain and erythema (redness) but no blisters. Skin functions remain intact. The pain and damage caused by a first-degree burn may be lessened by immediately flushing it with cold water. Generally, a first-degree burn will heal in about two to three days and may be accompanied by flaking or peeling. A typical sunburn is an example of a first-degree burn.

A **second-degree burn** involves the entire epidermis and possibly parts of the dermis. Some skin functions are lost. In a superficial second-degree burn, the deeper layers of the epidermis are injured, and there is redness, blister formation, edema, and pain. (Blister formation is due to separation of the epidermis from the dermis with tissue fluid accumulation between.) Such an injury usually heals within 7 to 10 days with only mild scarring. In a deep second-degree burn, there is destruction of both the epidermis and the upper levels of the dermis. Epidermal derivatives, such as hair follicles, sebaceous glands, and sweat glands, are usually not injured. If there is no infection, deep second-degree burns heal without grafting in about three to four weeks. Scarring may result. First- and second-degree burns are collectively referred to as **partial-thickness burns.**

A **third-degree burn** or **full-thickness burn** destroys the epidermis, dermis, and the epidermal derivatives. Skin functions are lost. Such burns vary in appearance from marble-white to mahogany colored to charred, dry wounds. There is marked edema, and the burned region is numb because sensory nerve endings have been destroyed. Regeneration is slow, and much granulation tissue forms before being covered by epithelium.

The seriousness of a burn is determined by its depth, extent, and area involved, as well as the person's age and general health. When the burn area exceeds 70%, more than half the victims die. A quick means for estimating the extent of a burn is the **rule of nines** (Fig. 5.8a).

1. If the anterior and posterior surfaces of the head and neck are affected, the burn covers 9% of the body surface.
2. The anterior and posterior surfaces of each shoulder, arm, forearm, and hand constitute 9% of the body surface.
3. The anterior and posterior surfaces of the trunk, including the buttocks, constitute 36%.

Continues

DISORDERS: HOMEOSTATIC IMBALANCES (Continued)

FIGURE 5.8 Methods for determining the extent of a burn. Whereas the rule of nine provides a quick means for determining the extent of a burn, the Lund–Browder method is more accurate. In (b), the numbers give the relative proportions of various body regions.

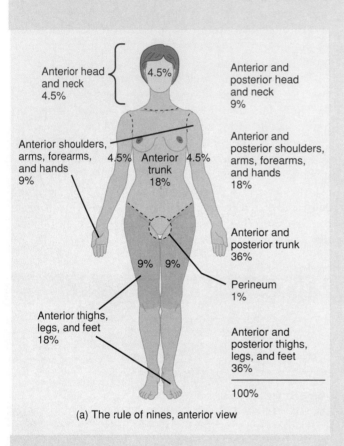

Anterior head and neck 4.5%

4.5%

Anterior and posterior head and neck 9%

Anterior shoulders, arms, forearms, and hands 9%

4.5%   Anterior trunk 18%   4.5%

Anterior and posterior shoulders, arms, forearms, and hands 18%

9%   9%

Anterior and posterior trunk 36%

Perineum 1%

Anterior thighs, legs, and feet 18%

Anterior and posterior thighs, legs, and feet 36%

100%

(a) The rule of nines, anterior view

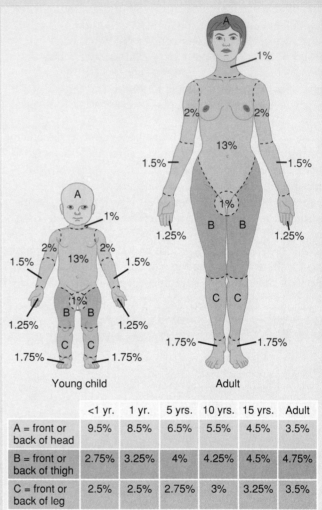

Young child

Adult

	<1 yr.	1 yr.	5 yrs.	10 yrs.	15 yrs.	Adult
A = front or back of head	9.5%	8.5%	6.5%	5.5%	4.5%	3.5%
B = front or back of thigh	2.75%	3.25%	4%	4.25%	4.5%	4.75%
C = front or back of leg	2.5%	2.5%	2.75%	3%	3.25%	3.5%

(b) The Lund-Browder method, anterior view

Question: Why would burns to the head of a child be more life threatening than similar burns in an adult?

4. The anterior and posterior surfaces of each foot, leg, and thigh as far up as the buttocks total 18%.
5. The perineum (per-i-NĒ-um) represents 1%. The perineum includes the anal and urogenital regions.

A more accurate way to estimate the amount of surface area affected by a burn is the **Lund–Browder method.** This method estimates the extent by measuring the areas affected against the percentage of total surface area for body parts shown in Fig. 5.8b. For example, if the anterior of the head and neck and the whole right hand of an adult are affected, the burn covers 7% (3.5% for anterior head + 1% for anterior neck + 2.5% for anterior and posterior aspects of hand) of the body surface. Because the proportions of the body change with growth, the percentages vary for different ages. Thus the extent of burn damage can be estimated fairly accurately for any age group.

SUNBURN

Sunburn is injury to the skin as a result of acute overexposure to the UV rays in sunlight. The damage to skin cells caused by sunburn is due to inhibition of DNA and RNA synthesis, which leads to cell dysfunction or death. There can also be damage to blood vessels and other structures in the dermis. Continual overexposure results in a leathery skin texture, wrinkles, skin folds, sagging skin, warty growths, freckling, a yellow discoloration due to abnormal elastic tissue, premature aging of the skin, and skin cancer.

SKIN CANCER

Excessive sun exposure can result in **skin cancer,** and everyone, regardless of skin pigmentation, is a potential victim of skin cancer if exposure to sunlight is sufficiently intense and prolonged. Natural skin pigment can never give complete protection. A **solar keratosis** (ker'-a-TŌ-sis; *kera* = horn) is a precancerous skin lesion induced by sunlight. These lesions commonly form on skin areas exposed to the sun and may be round or irregularly shaped, with a rough, scaly surface.

The three most common forms of skin cancer, all caused in part by chronic sun exposure, are the following: (1) **Basal cell carcinomas** account for over 75% of all skin cancers. The tumors arise from the epidermis and rarely metastasize (spread to other tissues). (2) **Squamous cell carcinomas** also arise from the epidermis, and they have a variable tendency to metastasize. Most arise from preexisting lesions on sun-exposed skin. (3) **Malignant melanomas** (*melano* = dark-colored; *oma* = tumor) arise from melanocytes and are the most prevalent life-threatening cancer in young women. Malignant melanomas metastasize rapidly and can kill a person within months of diagnosis. Often, but not always, there is a 10- to 20-year delay between photodamage and cancer detection.

It is a good practice to examine one's skin periodically for moles that may enlarge or develop highly irregular borders, uneven surfaces, or a mixture of colors. These signs of changing appearance or bleeding may indicate a developing melanoma. Fortunately, most skin cancers involve basal and squamous cells and can be treated by surgical excision.

Among the risk factors for skin cancer are:

1. **Skin type.** Persons with light-colored skin and red or blond hair who never tan but always burn are at high risk.
2. **Sun exposure.** People who live in areas with many days of sunlight per year (near the equator) and at high-altitudes have a high incidence of skin cancer. Likewise, people engaged in outdoor occupations and those who have suffered three or more severe sunburns have a higher risk.
3. **Family history.** Skin cancer rates are higher in some families than in others.

4. **Age.** Older people are more prone to skin cancer owing to longer total exposure to sunlight.
5. **Immunologic status.** Persons who are immunosuppressed have a higher incidence of skin cancer.

In recent years, **tanning salons** have become very popular, and this has become a concern to physicians. Many salons claim to use "safe" wavelengths of UV light, that is, longer (A) portions of the UV spectrum (UVA). According to most medical authorities, however, UVA is just as harmful as shorter UV wavelengths (UVB). Among the harmful effects are increased risk for skin cancer, premature aging of the skin, suppression of the immune system (changes in and decreased numbers of Langerhans cells), and damage to the lens or retina of the eye. Another problem that can develop is chemical photosensitivity, that is, adverse skin reactions to certain chemicals, including some drugs, that are applied topically or taken internally at the same time a person is exposed to UV radiation.

ACNE

Acne is an inflammation of sebaceous (oil) glands that usually begins at puberty when the sebaceous glands grow in size and increase production of sebum. Although testosterone, a male sex hormone, appears to be the most potent circulating hormone for sebaceous gland stimulation, adrenal and ovarian hormones stimulate sebaceous secretions in females. Acne occurs predominantly in sebaceous follicles that have been colonized by bacteria, which may thrive in the lipid-rich sebum. When this occurs, a cyst or sac of connective tissue cells can destroy and displace epidermal cells, resulting in permanent scarring, a condition called **cystic acne.**

Popular opinion to the contrary, most research has failed to establish a clear link between diet and acne. At one time it was a commonly held belief that foods such as chocolate, cola drinks, nuts, and dairy products could make acne worse. Some acne patients (and their physicians) claim to have isolated foods that do affect acne severity, so it remains possible that diet, especially one high in fat and refined carbohydrates, may be a contributing factor in some cases.

PRESSURE SORES

Pressure sores, also known as **decubitus** (dē-KYOO-bi-tus) **ulcers,** are caused by a constant deficiency of blood to tissues overlying a bony projection that has been subjected to prolonged pressure against an object such as a bed, cast, or splint. The deficiency results in tissue ulceration. Small breaks in the epidermis become infected, and the sensitive subcutaneous and deeper tissues are damaged. Eventually, the tissue is destroyed. Pressure sores are seen most often in patients who are bedridden for long periods of time.

MEDICAL TERMINOLOGY

Athlete's (ATH-lēts) **foot** A superficial fungus infection of the skin of the foot.

Chickenpox Highly contagious disease that begins in the respiratory system and is caused by the varicella-zoster virus and characterized by vesicular eruptions on the skin that fill with pus, rupture, and form a scab before healing. Also called **varicella** (var'-i-SEL-a). Shingles is caused by reactivation of latent chickenpox viruses.

Cold sore (KŌLD sor) A lesion, usually in oral mucous membrane, caused by type 1 herpes simplex virus (HSV), transmitted by oral or respiratory routes. Triggering factors include UV radiation, hormonal changes, and emotional stress. Also called a **fever blister.**

Contusion (kon-TOO-shun; *contundere* = to bruise) Condition in which tissue below the skin is damaged, but the skin is not broken.

Medical Terminology continues

MEDICAL TERMINOLOGY (Continued)

Corn (KORN) A painful conical thickening of the skin found principally over toe joints and between the toes. It may be hard or soft, depending on the location. Hard corns are usually found over toe joints, and soft corns are usually found between the fourth and fifth toes.

Cyst (SIST; *cyst* = sac containing fluid) A sac with a distinct connective tissue wall, containing a fluid or other material.

German measles Highly contagious disease that begins in the respiratory system and is caused by the rubella virus and characterized by a rash of small red spots on the skin. Also called **rubella** (roo-BEL-a).

Hemangioma (hē-man′-jē-Ō-ma; *hemo* = blood; *angio* = blood vessel; *oma* = tumor) Localized tumor of the skin and subcutaneous layer that results from an abnormal increase in blood vessels; one type is a **portwine stain**, a flat, pink, red, or purple lesion present at birth, usually at the nape of the neck.

Hives (HĪVZ) Condition of the skin marked by reddened elevated patches that are often itchy. Most commonly caused by infections, physical trauma, medications, emotional stress, food additives, and certain foods. Also called **urticaria** (yoor-ti-KAR-ē-a).

Impetigo (im′-pe-TĪ-go) Superficial skin infection caused by staphylococci or streptococci; most common in children.

Intradermal (in′-tra-DER-mal; *intra* = within) Within the skin. Also called **intracutaneous**.

Laceration (las′-er-Ā-shun; *lacerare* = to tear) Wound or irregular tear of the skin.

Measles Highly contagious disease caused by the measles virus that begins in the respiratory system and is characterized by a papular rash on the skin. Also called **rubeola** (roo-bē-Ō-la).

Nevus (NE-vus) A round, pigmented, flat or raised skin area that may be present at birth or develop later. Varying in color from yellow-brown to black. Also called a **mole** or **birthmark**.

Pruritus (proo-RĪ-tus; *pruire* = to itch) Itching, one of the most common dermatological disorders. It may be caused by skin disorders (infections), systemic disorders (cancer, kidney failure), or psychogenic factors (emotional stress).

Topical (TOP-i-kal) Pertaining to a definite area; local. Also in reference to a medication, applied to the surface rather than ingested or injected.

Wart (WORT) Mass produced by uncontrolled growth of epithelial skin cells; caused by a virus (papilloma virus). Most warts are noncancerous.

Study Outline

Skin (p. 127)

1. The skin and the organs derived from it (hair, glands, and nails) constitute the integumentary system.
2. The skin is one of the larger organs of the body. The principal parts of the skin are the outer epidermis and inner dermis. The dermis overlies the subcutaneous layer.
3. Following are the functions of the skin: regulating body temperature, protection, sensation, excretion, immunity, blood reservoir, and synthesis of vitamin D.
4. The epidermis consists of keratinocytes, melanocytes, Langerhans cells, and Merkel cells.
5. The epidermal layers, from deepest to most superficial, are the strata basale, spinosum, granulosum, lucidum, and corneum. The stratum basale undergoes continuous cell division and produces all other layers.
6. The dermis consists of a papillary region and a reticular region. The papillary region is areolar connective tissue containing fine elastic fibers, dermal papillae, and corpuscles of touch (Meissner's corpuscles). The reticular region is irregular connective tissue containing collagen and elastic fibers, adipose tissue, hair follicles, nerves, sebaceous (oil) glands, and ducts of sudoriferous (sweat) glands.
7. Lines of cleavage indicate the direction of collagen fiber bundles in the dermis and are considered during surgery.
8. The color of skin is due to melanin, carotene, and hemoglobin in blood in capillaries in the dermis.
9. Epidermal ridges increase friction for better grasping ability and provide the basis for fingerprints and footprints.

Epidermal Derivatives (p. 132)

1. Epidermal derivatives are structures developed from the embryonic epidermis.

2. Among the epidermal derivatives are hair, skin glands (sebaceous, sudoriferous, and ceruminous), and nails.

Hair (p. 132)

1. Hairs are epidermal growths that function in protection.
2. Hair consists of a shaft above the surface, a root that penetrates the dermis and subcutaneous layer, and a hair follicle.
3. Associated with hairs are sebaceous (oil) glands, arrectores pilorum muscles, and hair root plexuses.
4. New hairs develop from cell division of the matrix in the bulb; hair replacement and growth occur in a cyclic pattern. "Male-pattern" baldness is caused by androgens and heredity.

Glands (p. 134)

1. Sebaceous (oil) glands are usually connected to hair follicles; they are absent in the palms and soles. Sebaceous glands produce sebum, which moistens hairs and waterproofs the skin. Enlarged sebaceous glands may produce blackheads, pimples, and boils.
2. Sudoriferous (sweat) glands are divided into apocrine and eccrine. Apocrine sweat glands are limited in distribution to the skin of the axilla, pubis, and areolae; their ducts open into hair follicles. Eccrine sweat glands have an extensive distribution; their ducts terminate at pores at the surface of the epidermis. Sudoriferous glands produce perspiration, which carries small amounts of wastes to the surface and assists in maintaining body temperature.
3. Ceruminous glands are modified sudoriferous glands that secrete cerumen. They are found in the external auditory meatus.

Nails (p. 135)

1. Nails are hard, keratinized epidermal cells over the dorsal surfaces of the terminal portions of the fingers and toes.
2. The principal parts of a nail are the body, free edge, root, lunula, eponychium, and matrix. Cell division of the matrix cells produces new nails.

The Skin and Homeostasis (p. 135)

Skin Wound Healing: Homeostasis of Skin Structure (p. 135)

1. In an epidermal wound, the central portion of the wound usually extends deep down to the dermis, whereas the wound edges usually involve only superficial damage to the epidermal cells.
2. Epidermal wounds are repaired by enlargement and migration of basal cells, contact inhibition, and division of migrating and stationary basal cells.
3. During the inflammatory phase, a blood clot unites the wound edges, epithelial cells migrate across the wound, vasodilation and increased permeability of blood vessels deliver phagocytes, and fibroblasts form.
4. During the migratory phase, epithelial cells beneath the scab bridge the wound, fibroblasts begin to synthesize scar tissue, and damaged blood vessels begin to regrow.
5. During the proliferative phase, the events of the migratory phase intensify, and the open wound tissue is called granulation tissue.
6. During the maturation phase, the scab sloughs off, the epidermis is restored to normal thickness, collagen fibers become more organized, fibroblasts begin to disappear, and blood vessels are restored to normal.

Thermoregulation: Homeostasis of Body Temperature (p. 137)

1. One of the functions of the skin is the maintenance of a normal body temperature of 37°C (98.6°F).
2. If environmental temperature is high, skin receptors sense the stimulus (heat) and generate impulses (input) that are transmitted to the brain (control center). The brain then sends impulses (output) to sweat glands and blood vessels (effectors) to produce perspiration and vasodilation. As the perspiration evaporates, and more warm blood flows to the skin, the skin is cooled and body temperature returns to normal.
3. The skin-cooling response is a negative feedback mechanism.

Aging and the Integumentary System (p. 138)

1. Most effects of aging occur when an individual reaches the late forties.
2. Among the effects of aging are wrinkling, loss of subcutaneous fat, atrophy of sebaceous glands, and decrease in the number of melanocytes and Langerhans cells.

Developmental Anatomy of the Integumentary System (p. 139)

1. The epidermis, hair, nails, and skin glands are epidermal derivatives.
2. The dermis is derived from wandering mesodermal cells.

Review Questions

1. What is the integumentary system? (p. 127)
2. List the seven principal functions of the skin. (p. 127)
3. Compare the structure of epidermis and dermis. What is the subcutaneous layer? (p. 128)
4. Describe the various cells that comprise the epidermis. List and describe the epidermal layers from the deepest outward. What is the importance of each layer? (p. 128)
5. Compare the structure of the papillary and reticular regions of the dermis. (p. 130)
6. Explain the factors that produce skin color. What is an albino? (p. 131)
7. How are epidermal ridges formed? Why are they important? (p. 131)
8. List the receptors in the epidermis, dermis, and subcutaneous (SC) layer and indicate the location and role of each. (p. 131)
9. Describe the structure of a hair. How are hairs moistened? What produces "goose bumps" or "gooseflesh"? (p. 132)
10. Contrast the locations and functions of sebaceous (oil) glands, sudoriferous (sweat) glands, and ceruminous glands. (p. 134)
11. Distinguish between apocrine and eccrine sweat glands. (p. 134)
12. From what layer of the skin do nails form? Describe the principal parts of a nail. (p. 135)
13. Outline the steps involved in epidermal wound healing and deep wound healing. (p. 135)
14. Explain with a labeled diagram how the skin helps maintain normal body temperature. (p. 137)
15. Describe the effects of aging on the integumentary system. (p. 138)
16. Describe the origin of the epidermis, its derivatives, and the dermis. (p. 139)
17. Define the following: lines of cleavage (p. 131), freckle (p. 131), liver (age) spot (p. 131), cyanosis, jaundice, erythema (p. 131), skin graft (p. 131), male-pattern baldness (p. 134), and photodamage (p. 138).

Answers to Questions with Figures

5.1 Stratum corneum; stratum basale.
5.2 Epidermal growth factor (EGF).
5.3 Plucking a hair stimulates hair root plexuses in the dermis, some of which are sensitive to pain. Since the cells of a hair shaft are already dead, and since there are no nerves in the hair shaft, cutting hair is not painful.
5.4 They are composed of tightly packed, hard, keratinized epidermal cells.
5.5 Epidermal wounds do not bleed because there are no blood vessels in the epidermis.
5.6 In addition to phagocytizing microbes, they can help clean up cellular debris that results from the wound.
5.7 The result of the effectors (lowering body temperature) is opposite to the initial stimulus (rising body temperature).
5.8 The head represents a larger percentage of the total body surface in children.

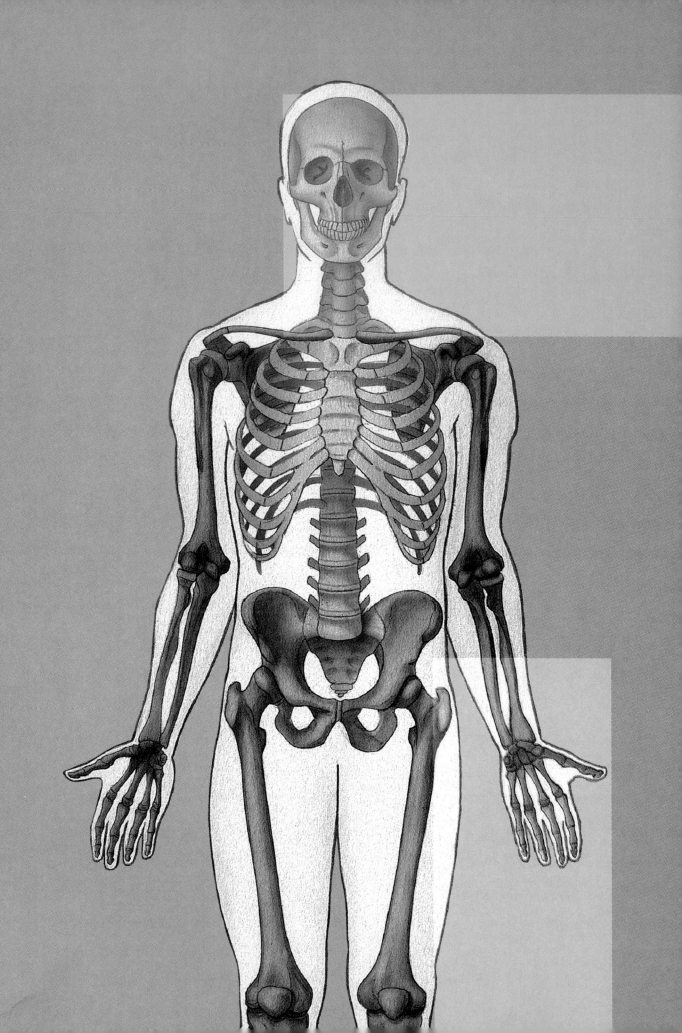

Unit 2
PRINCIPLES OF SUPPORT AND MOVEMENT

This unit considers two primary themes: support and movement. You will study the various ways in which the body is supported and the different movements it can perform. Both support and movement are made possible by the cooperative effort of bones, joints, and muscles.

Chapter 6
Bone Tissue

Chapter Contents at a Glance

Student Objectives

1. Discuss the functions of bone.
2. Identify the parts of a long bone.
3. Describe the histological features of compact and spongy bone tissue.
4. Contrast the steps involved in intramembranous and endochondral ossification.
5. Describe the processes involved in bone remodeling.
6. Define a fracture, describe several common kinds of fractures, and describe the sequence of events involved in fracture repair.
7. Describe the role of bone in calcium homeostasis.
8. Explain the effects of exercise and aging on the skeletal system.
9. Describe the development of the skeletal system.
10. Contrast the causes and clinical symptoms associated with osteoporosis and Paget's disease.
11. Define medical terminology associated with bone tissue.

Bone (osseous) tissue forms most of the skeleton, the framework that supports and protects our organs and allows us to move. Remarkably strong but lightweight, bone is a dynamic, ever-changing tissue. Throughout life, it is continually being broken down and reformed. **Osteology** (os-tē-OL-ō-jē; *osteon* = bone; *logos* = study of) is the study of bone structure and treatment of bone disorders.

The individual bones that form the skeleton and the joints that connect bones in a movable fashion are the topics of Chapters 7, 8, and 9. Here we focus on bone tissue itself. The developmental anatomy of bone is considered at the end of the chapter.

PHYSIOLOGY: FUNCTIONS OF BONE

Bone tissue and the skeletal system perform several basic functions.

1.   **Support.** Bone provides a framework for the body by supporting soft tissues and providing points of attachment for many skeletal muscles.
2.   **Protection.** Bones protect many internal organs from injury. For example, cranial bones protect the brain, vertebrae surround the spinal cord, the rib cage encloses the heart and lungs, and the hipbones guard internal reproductive organs.
3.   **Movement.** Skeletal muscles attach to bones. When muscles contract, they pull on bones and together they produce movement. Movements of muscles, bones, and joints are discussed in detail in later chapters.
4.   **Mineral homeostasis.** Bone tissue stores several minerals, especially calcium and phosphorus, which are important in muscle contraction and nerve activity, among other functions. On demand, bone releases minerals into the blood to maintain critical mineral balances and to distribute them to other parts of the body.
5.   **Site of blood cell production.** Within certain parts of bones, a connective tissue called red marrow produces blood cells, a process called **hemopoiesis** (hēm-ō-poy-Ē-sis). **Red marrow,** which is one type of bone marrow, consists of blood cells in immature stages, adipose cells, and macrophages. Red marrow produces red blood cells, white blood cells, and platelets. Hemopoiesis is discussed in detail on page 569.
6.   **Storage of energy.** Lipids stored in cells of a second type of bone marrow called yellow marrow are an important chemical energy reserve. **Yellow marrow** consists primarily of adipose cells and a few scattered blood cells .

ANATOMY: STRUCTURE OF BONE

Structurally, the skeletal system consists of cartilage, bone (osseous) tissue, bone marrow, and the periosteum, the membrane around bones. We described the microscopic structure of cartilage on page 110. Here the focus is the anatomy and histology of bone tissue.

The structure of bone may be analyzed by first considering the parts of a long bone such as the humerus, the arm bone (Fig. 6.1). A long bone is one that has greater length than width. A typical long bone consists of the following parts:

1.   **Diaphysis** (dī-AF-i-sis; *dia* = through; *physis* = growth). The shaft or long, main portion of the bone.
2.   **Epiphyses** (e-PIF-i-sēz; *epi* = above; *physis* = growth). The extremities or ends of the bone (singular is **epiphysis**).
3.   **Metaphysis** (me-TAF-i-sis). The region in a mature bone where the diaphysis joins the epiphysis. In a growing bone, it is the region that includes the epiphyseal plate where cartilage is replaced by bone (described later in the chapter). The main blood supply to the diaphysis, metaphysis, and bone marrow is through a series of **nutrient arteries.** Each enters the diaphysis of a bone through a hole called a **nutrient foramen.** The **epiphyseal arteries** supply the epiphyses.
4.   **Articular cartilage.** A thin layer of hyaline cartilage covering the epiphysis where the bone forms a joint with another bone. The cartilage reduces friction and absorbs shock at freely movable joints.
5.   **Periosteum** (per′-ē-OS-tē-um). The periosteum (*peri* = around; *osteo* = bone) is a membrane around the surface of the bone not covered by articular cartilage. It consists of two layers (see Fig. 6.3a). The outer **fibrous layer** is composed of dense, irregular connective tissue containing blood vessels, lymphatic vessels, and nerves that pass into the bone. The inner **osteogenic** (os′-tē-ō-JEN-ik) **layer** contains elastic fibers, blood vessels, and various types of bone cells. The periosteum is essential for bone growth in diameter, repair, and nutrition. It also serves as a point of attachment for ligaments and tendons.
6.   **Medullary** (MED-yoo-lar′-ē) or **marrow cavity.** This is the space within the diaphysis that contains the fatty yellow marrow in adults.
7.   **Endosteum** (end-OS-tē-um). Lining the medullary cavity is the endosteum, a membrane that contains osteoprogenitor cells (described shortly).

HISTOLOGY OF BONE TISSUE

Like other connective tissues, **bone,** or **osseous** (OS-ē-us), **tissue** contains an abundant matrix surrounding widely separated cells. The matrix is about 25% water, 25% protein fibers, and 50% mineral salts. There are four types of cells in bone tissue: osteoprogenitor (osteogenic) cells, osteoblasts, osteocytes, and osteoclasts (Fig. 6.2).

1.   **Osteoprogenitor** (os′-tē-ō-prō-JEN-i-tor; *osteo* = bone; *pro* = precursor; *gen* = to produce) **cells** are unspecialized cells derived from mesenchyme, the tissue from

FIGURE 6.1 Parts of a long bone.

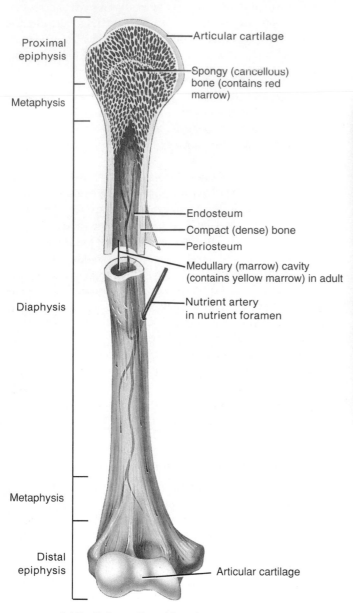

- Proximal epiphysis
- Metaphysis
- Diaphysis
- Metaphysis
- Distal epiphysis

- Articular cartilage
- Spongy (cancellous) bone (contains red marrow)
- Endosteum
- Compact (dense) bone
- Periosteum
- Medullary (marrow) cavity (contains yellow marrow) in adult
- Nutrient artery in nutrient foramen
- Articular cartilage

(a) Partially sectioned long bone

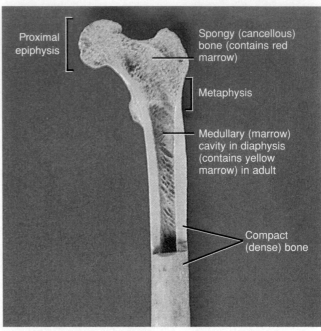

- Proximal epiphysis
- Spongy (cancellous) bone (contains red marrow)
- Metaphysis
- Medullary (marrow) cavity in diaphysis (contains yellow marrow) in adult
- Compact (dense) bone

(b) Photograph of a portion of a partially sectioned femur

Question: What are the basic functions of bone tissue?

which all connective tissues are derived. They can undergo mitosis and develop into osteoblasts. Osteoprogenitor cells are found in the inner portion of the periosteum, in the endosteum, and in canals (perforating and central) in bone that contain blood vessels.

2. Osteoblasts (OS-tē-ō-blasts′; *blast* = germ or bud) are the cells that form bone, but they have lost the ability to divide by mitosis. They secrete collagen and other organic components needed to build bone tissue.

3. Osteocytes (OS-tē-ō-sīts′; *cyte* = cell) are mature bone cells that are derived from osteoblasts; they are the

principal cells of bone tissue. Like osteoblasts, osteocytes have no mitotic potential. Osteoblasts are found on the surfaces of bone, but as they surround themselves with matrix materials they become osteocytes. Osteocytes no longer secrete matrix materials. Whereas osteoblasts initially form bone tissue, osteocytes maintain daily cellular activities of bone tissue, such as the exchange of nutrients and wastes with the blood.

4. Osteoclasts (OS-tē-ō-clasts′; *clast* = to break) are believed to develop from circulating monocytes (one type of white blood cell). They settle on the surfaces of

FIGURE 6.2 Types of cells in bone tissue.

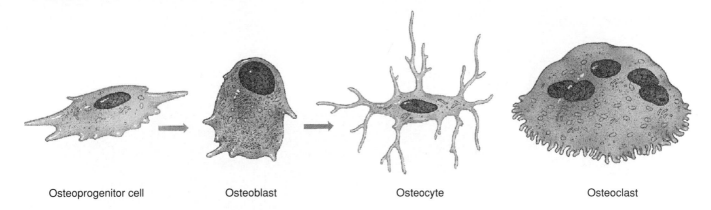

Osteoprogenitor cell   Osteoblast   Osteocyte   Osteoclast

Question: Which cell is concerned with bone formation? Bone resorption?

bone and function in bone resorption (destruction of matrix), which is important in the development, growth, maintenance, and repair of bone.

In terms of number and distribution, there are relatively more cells in immature bone than in mature (adult) bone. Immature bone refers to bone deposited initially in the skeleton of a developing fetus or in adult bone that is being broken down and reformed. Also, the cells tend to be more randomly arranged than in mature bone.

Unlike other connective tissues, the matrix of bone contains abundant mineral salts, primarily a crystallized form of tricalcium phosphate $[Ca_3(PO_4)_2 \cdot (OH)_2]$ called **hydroxyapatite** and some calcium carbonate ($CaCO_3$). In addition, there are small amounts of magnesium hydroxide, fluoride, and sulfate. As these salts are deposited in the framework formed by the collagen fibers of the matrix, crystallization occurs and the tissue hardens. This process is called **calcification** or **mineralization**. In general, mature bone contains more matrix than immature bone.

Although the *hardness* of bone depends on the crystallized inorganic mineral salts, without the organic collagen fibers it would be very brittle. Like reinforcing metal rods in concrete, the collagen fibers and other organic molecules provide bone with great pliability and *tensile strength*, which is resistance to being stretched or torn apart. Collagen fibers make bone less brittle than other calcium-based products, such as egg shells and oyster shells. If the inorganic minerals are removed, for example, by soaking a bone in weak acid such as vinegar, the result is a rubbery, flexible structure.

At one time, it was thought that calcification simply occurred when enough mineral salts were present to form crystals. Now, however, it is known that the process occurs only in the presence of collagen. Mineral salts accumulate in microscopic spaces between collagen fibers. There the salts crystallize and become hardened. Then, after the

spaces are filled, mineral salts deposit around the collagen fibers, where the salts again crystallize and harden. The combination of crystallized salts and collagen is responsible for the hardness that is characteristic of bone.

Bone is not completely solid but has many small spaces between its hard components. The spaces provide channels for blood vessels that supply bone cells with nutrients. The spaces also make bones lighter. Depending on the size and distribution of the spaces, the regions of a bone may be categorized as compact or spongy.

Compact Bone Tissue

Compact (dense) bone tissue contains few spaces. It forms the external layer of all bones of the body and the bulk of the diaphyses of long bones. Compact bone tissue provides protection and support and helps the long bones resist the stress of weight placed on them.

Compare the differences between spongy and compact bone tissues by looking at the highly magnified section in Fig. 6.3a. (See Fig. 6.1 also.) One main difference is that adult compact bone has a concentric-ring structure, whereas spongy bone appears as an irregular latticework. Blood vessels, lymphatic vessels, and nerves from the periosteum penetrate the compact bone through **perforating (Volkmann's) canals.** The blood vessels and nerves of these canals connect with blood vessels and nerves of the medullary cavity and periosteum and those of the **central (Haversian) canals.** The central canals run longitudinally through the bone. Around the canals are **concentric lamellae** (la-MEL-ē— rings of hard, calcified matrix. Between the lamellae are small spaces called **lacunae** (la-KOO-nē; *lacuna* = little lake), which contain osteocytes. **Osteocytes,** as noted earlier, are mature bone cells that no longer secrete matrix materials.

Radiating in all directions from the lacunae are minute canals called **canaliculi** (kan'-a-LIK-yoo-lī), which are filled with extracellular fluid. Inside the canaliculi are

FIGURE 6.3 Histology of bone.

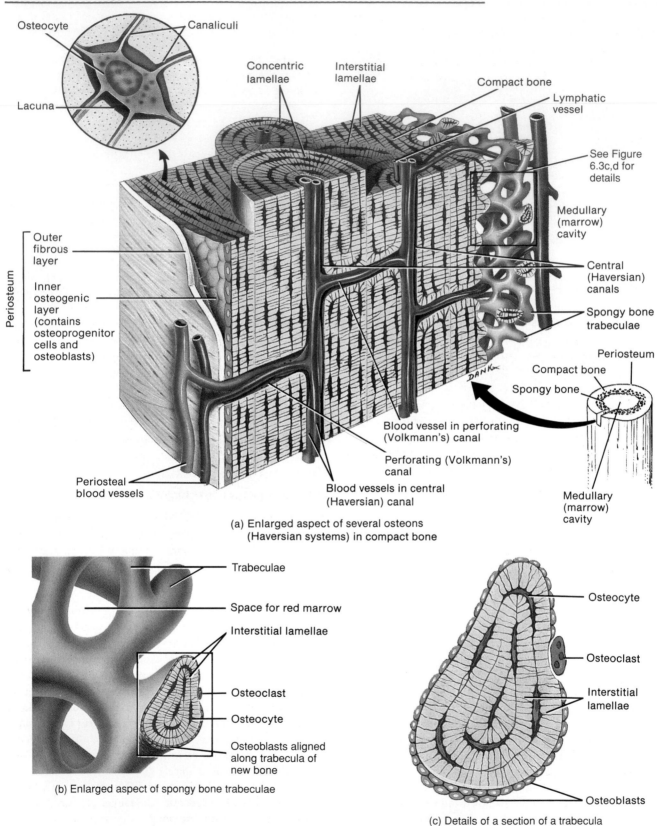

(a) Enlarged aspect of several osteons
(Haversian systems) in compact bone

(b) Enlarged aspect of spongy bone trabeculae

(c) Details of a section of a trabecula

Question: As people age, some central (Haversian) canals may become blocked. What effect would this have on osteocytes?

slender fingerlike processes of osteocytes (Fig. 6.3a). The canaliculi connect lacunae with one another and, eventually, with the central canals. Thus there is an intricate, miniature canal system throughout the bone. This branching network of canaliculi provides many routes for nutrients and oxygen to reach the osteocytes and wastes to diffuse away. Osteocytes from neighboring lacunae form gap junctions with each other, facilitating easy movement of materials from cell to cell.

Each central canal, with its surrounding lamellae, lacunae, osteocytes, and canaliculi, forms an **osteon (Haversian system)**. Osteons are characteristic of adult compact bone. The areas between osteons contain **interstitial lamellae**. These also have lacunae with osteocytes and canaliculi, but their lamellae are usually not connected to the osteons. Interstitial lamellae are fragments of older osteons that have been partially destroyed during bone replacement.

Spongy Bone Tissue

In contrast to compact bone, spongy bone does not contain true osteons (Fig. 6.3b, c). See Fig. 6.1 also. It consists of lamellae arranged in an irregular latticework of thin plates of bone called **trabeculae** (tra-BEK-yoo-lē). The spaces between the trabeculae of some bones are filled with red marrow, which produces blood cells. Within the trabeculae are osteocytes that lie in lacunae. Radiating from the lacunae are canaliculi. Blood vessels from the periosteum penetrate through to the spongy bone. Osteocytes in the trabeculae receive nourishment directly from the blood circulating through the marrow cavities. Osteons are not necessary in spongy bone because osteocytes are not deeply buried (as in compact bone) and have access to nutrients directly from the blood.

Spongy bone makes up most of the bone tissue of short, flat, and irregularly shaped bones and most of the epiphyses of long bones. Spongy bone tissue in the hipbones, ribs, breastbone (sternum), backbones (vertebrae), skull, and ends of some long bones is the only site of red marrow storage and thus hemopoiesis in adults.

Most people think of all bone as a very hard, rigid material. Yet the bones of an infant are quite soft and become rigid only after growth stops during late adolescence. Even then, bone is constantly broken down and rebuilt. It is a dynamic, living tissue. Let us now see how bones are formed and how they grow.

PHYSIOLOGY OF BONE FORMATION: OSSIFICATION

The process by which bone forms is called **ossification** (os´-i-fi-KĀ-shun). The "skeleton" of a human embryo is composed of fibrous connective tissue membranes formed by embryonic connective tissue (mesenchyme) and hyaline cartilage that are loosely shaped like bones. They provide the supporting structures for ossification. Ossification begins around the sixth or seventh week of embryonic life and continues throughout adulthood. Bone formation follows one of two patterns.

1. **Intramembranous** (in´-tra-MEM-bra-nus; *intra* = within; *membranous* = membrane) **ossification** refers to the formation of bone directly on or within the fibrous connective tissue membranes. As you will see on page 169, the fontanels ("soft spots") of an infant's skull, which are composed of fibrous connective tissue membranes, are also eventually replaced by bone through intramembranous ossification.

2. **Endochondral** (en´-dō-KON-dral; *endo* = within; *chondro* = cartilage) **ossification** refers to the formation of bone in hyaline cartilage.

These two kinds of ossification do *not* lead to differences in the structure of mature bones. They are simply different methods of bone formation. Both mechanisms involve the replacement of a preexisting connective tissue with bone.

The first stage in the development of bone is the migration of embryonic mesenchymal cells into the area where bone formation is about to begin. These cells increase in number and size and become osteoprogenitor cells. In some skeletal structures where capillaries are lacking, they become chondroblasts; in others where capillaries are present, they become osteoblasts. The **chondroblasts** are responsible for cartilage formation. Osteoblasts form bone tissue by intramembranous or endochondral ossification.

Intramembranous Ossification

The flat bones of the skull, lower jawbone (mandible), and collarbones (clavicles) develop directly on or within fibrous connective tissue membranes formed by mesenchymal cells, that is, by **intramembranous ossification**. The essentials of the process are as follows:

1. At the site where the bone will develop, mesenchymal cells become vascularized, cluster, and differentiate first into osteoprogenitor cells and then into osteoblasts. The site of such a cluster is called a **center of ossification**. Osteoblasts secrete the organic matrix of bone but cease to form matrix once they are completely surrounded by it. The cells, now called osteocytes, lie in lacunae with canaliculi that radiate out in all directions. Later, calcium and other mineral salts are deposited and the tissue hardens or calcifies. Thus calcification is just one aspect of ossification.

2. As the bone matrix forms, it develops into **trabeculae**. As trabeculae develop in various ossification centers, they fuse with one another to create the open latticework appearance of spongy bone. The spaces between trabeculae fill with vascularized connective tissue, which differentiates into red bone marrow.

3. On the outside of the bone, vascularized mesenchyme develops into the periosteum. Eventually, some of the surface layers of the spongy bone are replaced by compact bone. Much of this newly formed bone will be remodeled (destroyed and reformed) so the bone may reach its final adult size and shape.

Endochondral Ossification

The replacement of cartilage by bone is called **endochondral (intracartilaginous) ossification.** Most bones of the body are formed in this way, but this type of ossification is best observed in a long bone. It proceeds as follows (Fig. 6.4):

1. **Development of the cartilage model.** At the site where the bone is going to form, mesenchymal cells crowd together in the shape of the future bone. The mesenchymal cells differentiate into chondroblasts that produce cartilage matrix so the model consists of hyaline cartilage (Fig. 6.4a). In addition, a membrane called the **perichondrium** (per-i-KON-drē-um) develops around the cartilage model.

2. **Growth of the cartilage model.** The cartilage model grows in length by continual cell division of chondrocytes accompanied by further secretion of cartilage matrix by the daughter cells (Fig. 6.4b). This pattern of growth that results in an increase in length is called **interstitial** (in′-ter-STISH-al) **growth,** that is, growth from within. Growth of the cartilage in thickness is mainly due to the addition of more matrix to its periphery by new chondroblasts that develop from the perichondrium. This growth pattern of cartilage in which matrix is deposited on its surface is called **appositional** (a-pō-ZISH-a-nal) **growth.**

As the cartilage model continues to grow, chondrocytes in its midregion hypertrophy (increase in size), probably because they accumulate glycogen for energy and produce enzymes to catalyze further chemical reactions. The cells burst, changing the pH of the matrix, which triggers calcification. Once the cartilage becomes calcified, nutrients required by the cartilage cells no longer diffuse quickly enough through the matrix, and the cartilage cells die. The lacunae of the cells that have been killed are now empty, and the thin partitions between them break down (Fig. 6.4b).

In the meantime, a nutrient artery penetrates the perichondrium and then the bone through a hole in the bone (nutrient foramen). This occurs in the midregion of the model, stimulating osteoprogenitor cells in the perichondrium to differentiate into osteoblasts. The cells lay down a thin shell of compact bone under the perichondrium called the **periosteal bone collar** (Fig. 6.4b). Once the perichondrium starts to form bone, it is known as the **periosteum** (Fig. 6.4c).

3. **Development of the primary ossification center.** Near the middle of the model, capillaries of the periosteum grow into the disintegrating calcified cartilage. These vessels, and the associated osteoblasts, osteoclasts, and red marrow cells, are known as the **periosteal bud.** On growing into the cartilage model, the capillaries produce a **primary ossification center,** a region where bone tissue will replace most of the cartilage (Fig. 6.4c). In the center, osteoblasts begin to deposit bone matrix over the remnants of calcified cartilage, forming spongy bone trabeculae. As the center enlarges toward the ends of the bone, osteoclasts break down the newly formed spongy bone trabeculae, leaving a cavity, the medullary (marrow) cavity, in the center of the model. The cavity then fills with red marrow.

4. **Development of the diaphysis and epiphysis.** The diaphysis (shaft), which was once a solid mass of hyaline cartilage, is replaced by compact bone, the central part of which contains a red marrow-filled medullary cavity. When blood vessels (epiphyseal arteries) enter the epiphyses, **secondary ossification centers** develop, usually around the time of birth (Fig. 6.4d). In these centers, bone formation is similar to that in the primary ossification centers with a few exceptions: (1) spongy bone is retained in the interior of the epiphyses; (2) no medullary cavities are formed in the epiphyses; and (3) hyaline cartilage remains covering the epiphyses as the articular cartilage and between the diaphysis and epiphysis as the epiphyseal plate, which is responsible for the lengthwise growth of long bones (Fig. 6.4e).

PHYSIOLOGY OF BONE GROWTH

To understand how a bone grows in length, you will need to know some of the details of the structure of the epiphyseal plate.

The **epiphyseal** (ep′-i-FIZ-ē-al) **plate** consists of four zones (Fig. 6.5). The **zone of resting cartilage** is near the epiphysis and consists of small, scattered chondrocytes. The cells do not function in bone growth (thus the term "resting"); they anchor the epiphyseal plate to the bone of the epiphysis.

The **zone of proliferating cartilage** consists of slightly larger chondrocytes arranged like stacks of coins. Chondrocytes divide to replace those that die at the diaphyseal surface of the epiphyseal plate.

The **zone of hypertrophic** (hī-per-TRŌF-ik) or **maturing cartilage** consists of even larger chondrocytes that are also arranged in columns. The lengthwise expansion of the epiphyseal plate is the result of cell divisions in the zone of proliferating cartilage and maturation of the cells in the zone of hypertrophic cartilage.

The **zone of calcified cartilage** is only a few cells thick and consists mostly of dead cells because the matrix around them has calcified. The calcified matrix is taken up by osteoclasts, and the area is invaded by osteoblasts and capillaries from the bone in the diaphysis. These cells lay down bone on the calcified cartilage that persists. As a result, the

FIGURE 6.4 Endochondral ossification of the shinbone (tibia).

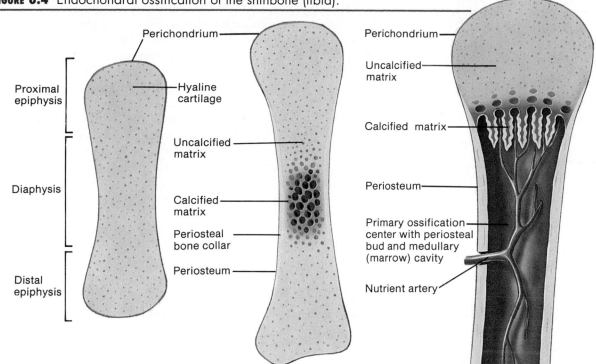

(a) Mesenchymal cells differentiate into chondroblasts which form the hyaline cartilage model

(b) Cartilage model grows by interstitial and appositional growth, chondrocytes in midregion calcify the matrix, vacated lacunae form small cavities, osteoblasts in perichondrium produce periosteal bone collar

(c) With development of periosteal bud, primary ossification center and medullary cavity form

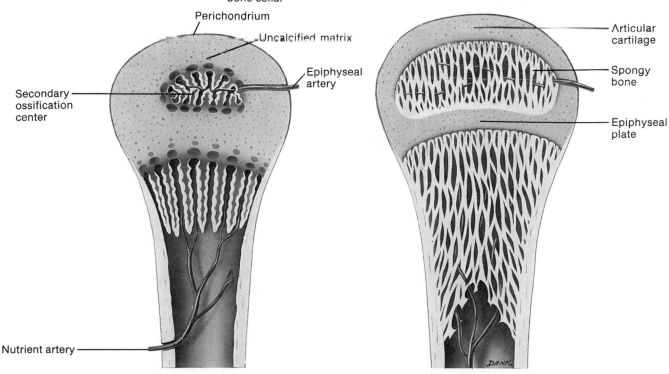

(d) Development of secondary ossification center in epiphysis. A secondary ossification center also develops in the distal epiphysis of a long bone

(e) Remnants of hyaline cartilage as articular cartilage and epiphyseal plate

Question: X-ray films of an 18-year-old basketball star do not yet show any evidence of epiphyseal lines. Is he likely to grow taller?

FIGURE 6.5 Epiphyseal plate.

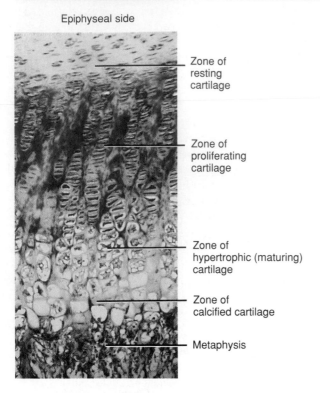

Epiphyseal side

Zone of
resting
cartilage

Zone of
proliferating
cartilage

Zone of
hypertrophic (maturing)
cartilage

Zone of
calcified cartilage

Metaphysis

Diaphyseal side

Photomicrograph of epiphyseal
plate (360x)

Question: What accounts for the lengthwise expansion of the epiphyseal plate?

diaphyseal border of the epiphyseal plate is firmly cemented to the bone of the diaphysis.

The region between the diaphysis and epiphysis of a bone where the calcified matrix is replaced by bone is called the **metaphysis** (me-TAF-i-sis). See Fig. 6.1. The activity of the epiphyseal plate is the only mechanism by which the diaphysis can increase in length. Unlike cartilage, which can grow by both interstitial and appositional growth, bone can grow in diameter only by appositional growth.

The epiphyseal plate allows the diaphysis of the bone to increase in length until early adulthood. It also shapes the articular surfaces. The rate of growth is controlled by hormones such as human growth hormone (hGH) produced by the pituitary gland and sex hormones produced by the ovaries and testes. As a child grows, cartilage cells are produced by mitosis on the epiphyseal side of the plate. They are then destroyed, and the cartilage is replaced by bone on the diaphyseal side of the plate. In this way, the thickness of the epiphyseal plate remains almost constant, but the bone on the diaphyseal side increases in length. If a bone fracture also damages the epiphyseal plate, the fractured bone will

be shorter than its normal counterpart once adult stature is reached. This is because damage to cartilage, which is avascular, accelerates closure of the epiphyseal plate, and growth in length of the bone is inhibited. A bone fracture that does not involve the epiphyseal plate usually heals quite well due to the rich blood supply of bone.

Eventually, the epiphyseal cartilage cells stop dividing, and bone replaces the cartilage. The newly formed bony structure is called the **epiphyseal line,** a remnant of the once active epiphyseal plate. With the appearance of the epiphyseal line, bone stops growing in length. The clavicle is the last bone to stop growing. Ossification of most bones is usually completed by age 25. In general, lengthwise growth in bones in females is completed before that in males.

Growth in diameter occurs along with growth in length. In this process, the bone lining the medullary cavity is destroyed by osteoclasts so that the cavity increases in diameter. At the same time, osteoblasts from the periosteum add new bone tissue to the outer surface (appositional growth). Initially, diaphyseal and epiphyseal ossification produce only spongy bone. Later, the outer region of spongy bone is reorganized into compact bone.

BONE HOMEOSTASIS

Bone, like skin, forms before birth but continually renews itself thereafter. **Remodeling** is the ongoing replacement of old bone tissue by new bone tissue. Bone is never metabolically at rest; it constantly remodels and redistributes its matrix along lines of mechanical stress. Compact bone is formed from spongy bone. However, even after bones have reached their adult shapes and sizes, old bone is continually destroyed, and new bone tissue is formed in its place. Remodeling also removes worn and injured bone, replacing it with new tissue. It allows bone to serve as the body's reservoir for calcium (discussed shortly). Many other tissues need calcium to perform their functions. Several hormones continually regulate exchanges of calcium between blood and bones.

Remodeling

Remodeling takes place at different rates in various body regions. The distal portion of the thighbone (femur) is replaced about every four months. By contrast, bone in certain areas of the shaft will not be completely replaced during the individual's life. Osteoclasts are responsible for bone resorption (destruction of matrix). A delicate homeostasis exists between the actions of the osteoclasts in removing minerals and collagen and of the bone-making osteoblasts in depositing minerals and collagen. Should too much new tissue be formed, the bones become abnormally thick and heavy. If too much mineral is deposited in the bone, the surplus may form thick bumps, or spurs, on the bone that interfere with movement at joints. A loss of too much calcium or tissue weakens the bones, and they break, as occurs in osteoporosis, or become too flexible, as in osteomalacia. Abnormal acceleration of the remodeling process results in a condition called Paget's disease (see page 162).

In the process of resorption, osteoclasts put forth projections that secrete protein-digesting lysosomal enzymes and several acids (lactic, carbonic, and citric). The enzymes are thought to digest collagen and other organic substances, while the acids apparently dissolve the bone minerals. Under a microscope, osteoclasts have a phagocytic appearance, and it is presumed that they also phagocytize whole fragments of collagen and bone minerals.

Minerals Needed for Bone Remodeling

Normal bone growth in the young and bone replacement in the adult depend on the presence of several minerals. Sufficient amounts of calcium and phosphorus, components of hydroxyapatite, the primary salt that makes bone matrix hard, must be included in the diet. Magnesium deficiency inhibits the activity of osteoblasts. Recent studies also suggest that boron may be a factor in bone growth by inhibiting calcium loss and increasing levels of estrogens. Manganese may also be important in bone growth. It has been shown that manganese deficiency significantly inhibits laying down of new bone tissue.

Vitamins Needed for Bone Remodeling

Several vitamins, including vitamins D, C, A, and B_{12}, play a role in bone remodeling. The most active form of vitamin D is calcitriol. Acting as a hormone, calcitriol promotes the removal of calcium from bone. On the other hand, calcitriol greatly increases intestinal absorption of dietary calcium and retards its loss in the urine, which makes more calcium available for deposit in bone matrix. Vitamin C helps to maintain the matrix of bone and other connective tissues. Vitamin C deficiency leads to decreased collagen production, which retards bone growth and delays healing of fractures. Vitamin A helps to control the activity, distribution, and coordination of osteoblasts and osteoclasts during development. Its deficiency results in a decreased rate of growth in the skeleton. Vitamin B_{12} may play a role in osteoblast activity.

Hormonal Regulation of Bone Growth and Remodeling

A variety of hormones contribute to normal bone tissue activity. (See Chapter 18 for more details.) Human growth hormone (hGH), secreted by the pituitary gland, is responsible for the general growth of all tissues in the body, including bones. Too much or too little of this hormone

during childhood makes the adult abnormally tall or short. The sex hormones (such as estrogens and testosterone) aid osteoblastic activity and thus promote the growth of new bone. The sex hormones, however, act as a double-edged sword. They aid in the growth of new bone, but they also bring about the degeneration of all the cartilage cells in epiphyseal plates. Because of the sex hormones, the typical adolescent experiences a spurt of growth during puberty, when sex hormone levels start to increase. The individual then quickly completes the growth process as the epiphyseal cartilage disappears. One of the negative effects of taking anabolic steroids (testosterone-like hormones) is premature closing of the epiphyseal plate, which then results in short stature. Also, premature puberty, which might be triggered by a testosterone-secreting tumor, often prevents attainment of an average adult height due to the simultaneous premature degeneration of the epiphyseal plates. Insulin produced by the pancreas and the thyroid hormones, triiodothyronine (T_3) and tetraiodothyronine (T_4), produced by the thyroid gland are also important for normal bone growth and maturity. As you will see shortly, other hormones related to bone formation and breakdown are parathyroid hormone secreted by the parathyroid glands and calcitonin secreted by the thyroid gland.

Fracture and Repair of Bone

A **fracture** is any break in a bone. Although fractures may be classified in several different ways, the scheme shown in Exhibit 6.1 is useful. See also Fig. 6.6. Usually, the fractured ends of a bone can be reduced (restored to their normal positions) by manipulation without surgery. This procedure of setting a fracture is called **closed reduction.** In other cases, the fracture must be exposed by surgery before the break is rejoined. This procedure is known as **open reduction.**

Although bone has a generous blood supply, healing sometimes takes months. Sufficient calcium and phosphorus to strengthen and harden new bone is deposited only gradually. Bone cells generally also grow and reproduce slowly. Moreover, in a fractured bone the blood supply is decreased, which helps to explain the difficulty in the healing of an infected bone. Recall that cartilage injuries heal even more slowly, however, because capillaries are present only in the perichondrium of cartilage. The following steps occur in the repair of a bone fracture (Fig. 6.7):

1. As a result of the fracture, blood vessels crossing the fracture line are broken. These vessels are found in the periosteum, osteons (Haversian systems), and medullary cavity. As blood pours from the torn ends of the vessels, it forms a clot in and about the site of the fracture. This clot, called a **fracture hematoma** (hē'-ma-TŌ-ma), usually occurs 6 to 8 hours after the injury (Fig. 6.7a). Since the circulation of blood ceases when the fracture hematoma forms, bone cells and periosteal cells

EXHIBIT 6.1

SUMMARY OF SELECTED FRACTURES

Type of Fracture	Definition
Partial	The break across the bone is incomplete.
Complete	The break across the bone is complete, so that the bone is broken into two or more pieces.
Closed (simple)	The bone does not break through the skin.
Open (compound)	The broken ends of the bone protrude through the skin (see Fig. 6.6f).
Comminuted (KOM-i-nyoo´-ted)	The bone has splintered at the site of impact, and smaller fragments of bone lie between the two main fragments (see Fig. 6.6a).
Greenstick	A partial fracture in which one side of the bone is broken and the other side bends; occurs only in children (see Fig. 6.6e).
Spiral	The bone usually is twisted apart.
Transverse	A fracture at right angles to the long axis of the bone.
Impacted	One fragment is firmly driven into the other (see Fig. 6.6c).
Displaced	The anatomical alignment of the bone fragments is not preserved.
Nondisplaced	The anatomical alignment of the bone fragments is preserved.
Stress	Microscopic fractures resulting from inability to withstand repeated stress. They are usually the result of repeated impact activities, such as running, basketball, jumping, or aerobic dancing. About 25% of all stress fractures involve the shinbone (tibia).
Pathologic	Weakening of a bone caused by disease processes such as neoplasia, osteomyelitis, osteoporosis, or osteomalacia.
Pott's	A fracture of the distal end of the lateral leg bone (fibula), with serious injury of the distal tibial articulation (see Fig. 6.6d).
Colles' (KOL-ēz)	A fracture of the distal end of the lateral forearm bone (radius) in which the distal fragment is displaced posteriorly (see Fig. 6.6b).

at the fracture site die. The hematoma serves as a focus for the cellular invasion that follows. Swelling and inflammation occur after formation of the fracture hematoma, and there is a considerable amount of cell death and debris. Blood capillaries grow into the blood clot and phagocytes (neutrophils and macrophages) plus osteoclasts begin to remove the traumatized tissue in and around the fracture hematoma. This may take up to several weeks.

2. The infiltration of blood capillaries into the fracture hematoma helps organize it into granulation tissue (see page 122), now called a **procallus.** Next, fibroblasts from the periosteum and osteoprogenitor cells from the periosteum, endosteum, and marrow invade the procal-

lus. The fibroblasts produce collagen fibers, which help connect the broken ends of the bones. Osteoprogenitor cells develop into chondroblasts in areas farther away from healthy bone tissue where the environment is avascular. Here chondroblasts begin to produce fibrocartilage and the procallus is transformed into a **fibrocartilaginous (soft) callus** (Fig. 6.7b). A **callus** is actually a mass of repair tissue that bridges the broken ends of the bones. The stage of the fibrocartilaginous callus lasts about 3 weeks.

3. In areas closer to healthy bone tissue where the environment is more vascular, osteoprogenitor cells develop into osteoblasts, which begin to produce spongy bone trabeculae. The trabeculae join living and dead portions

FIGURE 6.6 Types of fractures.

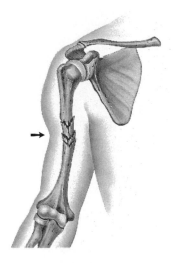

(a) Comminuted fracture

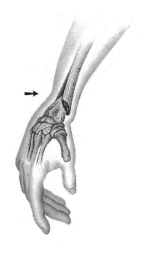

(b) Colles' fracture

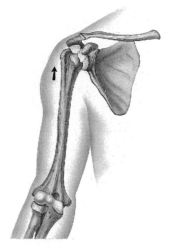

(c) Impacted fracture

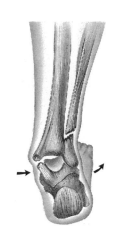

(d) Pott's fracture

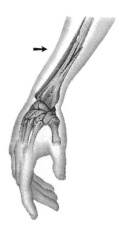

(e) Greenstick fracture

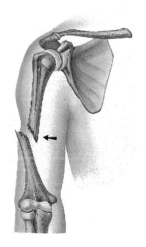

(f) Open fracture

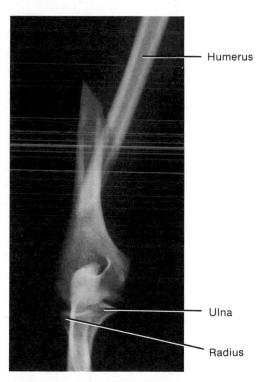

Humerus

Ulna

Radius

(g) X-ray of compound fracture of the humerus

Question: What is a callus?

FIGURE 6.7 Steps involved in fracture repair.

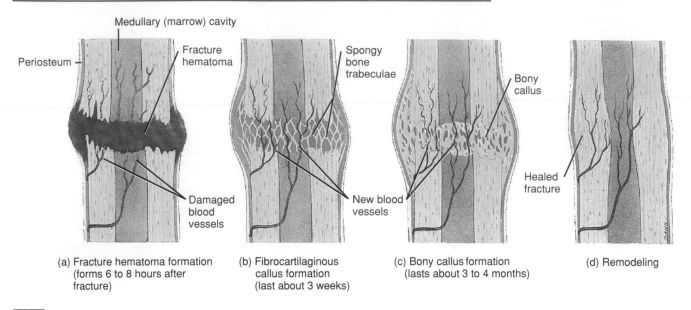

(a) Fracture hematoma formation
(forms 6 to 8 hours after
fracture)

(b) Fibrocartilaginous
callus formation
(last about 3 weeks)

(c) Bony callus formation
(lasts about 3 to 4 months)

(d) Remodeling

Question: Why does it sometimes take months for a fracture to heal?

of the original bone fragments. In time, the fibrocartilage is converted to spongy bone and the callus is then referred to as a **bony (hard) callus** (Fig. 6.7c). The stage of the bony callus lasts about 3 to 4 months.

4.   The final phase of fracture repair is the **remodeling** of the callus (Fig. 6.7d). Dead portions of the original fragments are gradually resorbed by osteoclasts. Compact bone replaces spongy bone around the periphery of the fracture. Sometimes, the healing is so complete that the fracture line is undetectable, even by x-ray. However, a thickened area on the surface of the bone usually remains as evidence of a fracture site.

Bone's Role in Calcium Homeostasis

Bone is the major reservoir of calcium in the body, containing 500 to 1000 times more than all other tissues combined. The blood level of calcium ions (Ca^{2+}) is very closely regulated; even small changes in Ca^{2+} concentration are deadly. The heart may stop (cardiac arrest) if the concentration goes too high or breathing may cease (respiratory arrest) if the level falls too low. Most functions of nerve cells depend on just the right level of Ca^{2+}. Also, many enzymes require Ca^{2+} as a cofactor, and blood needs Ca^{2+} to clot. The role of bone in calcium homeostasis is to "buffer" blood calcium level, releasing Ca^{2+} to the blood when the level decreases and taking Ca^{2+} back when the level rises. Hormones regulate these exchanges.

An important hormone regulating Ca^{2+} exchange between bone and blood is **parathyroid hormone (PTH)**, secreted by the parathyroid glands (see Fig. 18.18). PTH secretion is linked to several negative feedback systems that adjust blood Ca^{2+} concentration (controlled condition). Look at the negative feedback cycle in Fig. 6.8. If some stimulus causes blood Ca^{2+} level to fall, parathyroid gland cells (receptors) detect this change. The control center is the gene for PTH within the nucleus of a parathyroid gland cell. One input signal to the control center is increased level of a molecule known as cyclic AMP (adenosine monophosphate) in the cytosol. Cyclic AMP accelerates reactions that "turn on" the PTH gene, PTH synthesis speeds up, and more PTH (output) is released into the blood. PTH increases the number and activity of osteoclasts (effectors), which step up the pace of bone resorption. The resulting release of Ca^{2+} (and phosphate) from bone into blood (response) returns the blood Ca^{2+} level to normal.

PTH also affects the kidneys. It promotes (1) recovery of Ca^{2+} so it is not lost in the urine, (2) elimination of phosphate in the urine, and (3) formation of calcitriol. These kidney effects of PTH all augment the bone effect—they elevate blood Ca^{2+} concentration.

Another hormone also contributes to the homeostasis of blood Ca^{2+}. **Calcitonin (CT)** is secreted by certain thyroid gland cells (parafollicular cells) when blood Ca^{2+} rises above normal. It inhibits osteoclastic activity, speeds Ca^{2+} uptake by bone from blood, and accelerates Ca^{2+} deposit into bones. The net result is that calcitonin promotes bone formation and decreases blood Ca^{2+} level.

A summary of hormones related to the homeostasis of bone tissue is presented in Exhibit 6.2.

FIGURE 6.8 Negative feedback system for the regulation of blood calcium (Ca^{2+}) concentration.

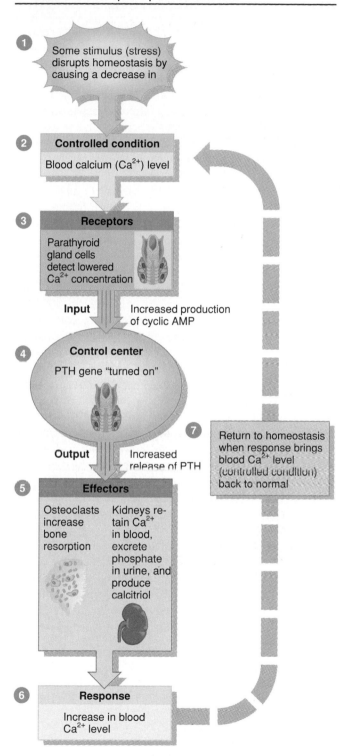

1 Some stimulus (stress) disrupts homeostasis by causing a decrease in

2 **Controlled condition**
Blood calcium (Ca^{2+}) level

3 **Receptors**
Parathyroid gland cells detect lowered Ca^{2+} concentration

Input | Increased production of cyclic AMP

4 **Control center**
PTH gene "turned on"

Output | Increased release of PTH

5 **Effectors**
Osteoclasts increase bone resorption | Kidneys retain Ca^{2+} in blood, excrete phosphate in urine, and produce calcitriol

6 **Response**
Increase in blood Ca^{2+} level

7 Return to homeostasis when response brings blood Ca^{2+} level (controlled condition) back to normal

Question: What body functions depend on proper levels of Ca^{2+}?

EXHIBIT 6.2

SUMMARY OF HORMONES THAT AFFECT THE HOMEOSTASIS OF BONE TISSUE

Hormone	Functions
Human growth hormone (hGH)	General growth of all body tissues, including bone.
Sex hormones (several estrogens and testosterone)	Increase bone-building activity of osteoblasts.
Insulin and thyroid hormones (triiodothyronine and tetraiodothyronine)	Promote normal bone growth and maturity.
Parathyroid hormone (PTH)	Increases the number and activity of osteoclasts, promotes recovery of Ca^{2+} from urine, and promotes formation of calcitriol.
Calcitonin (CT)	Inhibits activity of osteoclasts, speeds up Ca^{2+} absorption from blood, and accelerates Ca^{2+} deposit by bones.

CLINICAL APPLICATION

HELPING BONE HEAL

In the past when a fracture failed to unite, the patient could either wait, hoping that nature and time would solve the problem, or choose surgery. Now there is another alternative, called **pulsating electromagnetic fields (PEMFs),** that involves electrotherapy to stimulate bone repair.

The fracture is exposed to a weak electrical current generated from coils that are fastened around the cast. Osteoblasts adjacent to the fracture site become more active metabolically in response to the stimulation. The increased activity enhances calcification, vascularization, and endochondral ossification, accelerating fracture repair. It was noted earlier that parathyroid hormone (PTH) increases osteoclastic activity and thus stimulates bone destruction. One hypothesis suggests that electricity also heals fractures by keeping PTH from acting on osteoclasts, thus increasing bone formation and repair.

EXERCISE AND BONE

Within limits, bone has the ability to alter its strength in response to mechanical stress. When placed under mechanical stress, bone tissue becomes stronger, through increased deposition of mineral salts and production of collagen fibers. Another effect of stress is to increase the production of calcitonin, which inhibits bone resorption. Without mechanical stress, bone does not remodel normally since resorption outstrips bone formation. Removal of mechanical stress weakens bone through demineralization (loss of bone minerals) and collagen reduction. The main mechanical stresses on bone are those that result from the pull of skeletal muscles and the pull of gravity. If a person is bedridden or has a fractured bone in a cast, the strength of the unstressed bones diminishes. Astronauts subjected to the weightlessness of space also lose bone mass. In both cases, the bone loss can be dramatic, as much as 1% per week. Bones of athletes, which are repetitively and highly stressed, become notably thicker than those of nonathletes. Weight-bearing activities, such as walking or moderate weight lifting, help build and retain bone mass.

AGING AND BONE TISSUE

There are two principal effects of aging on bone tissue. The first is the loss of calcium and other minerals from bone matrix (demineralization). This loss usually begins after age 30 in females, accelerates greatly around age 40 to 45 as levels of estrogens decrease, and continues until as much as 30% of the calcium in bones is lost by age 70. In males, calcium loss typically does not begin until after age 60. The loss of calcium from bones is one of the problems in a condition called osteoporosis (see page 162).

The second principal effect of aging on the skeletal system is a decrease in the rate of protein synthesis. This results in decreased ability to produce the organic portion of bone matrix, mainly collagen, which normally gives bone its tensile strength. As a consequence, inorganic minerals gradually constitute a greater proportion of the bone matrix. The loss of tensile strength causes the bones to become very brittle and susceptible to fracture. In some elderly people protein synthesis slows, in part, due to diminished production of human growth hormone.

DEVELOPMENTAL ANATOMY OF BONE AND THE SKELETAL SYSTEM

As noted earlier, both intramembranous and endochondral ossification begin when **mesenchymal (mesodermal) cells** migrate into the area where bone formation will occur. In some skeletal structures, mesenchymal cells develop into **chondroblasts** that form *cartilage.* In other skeletal structures, mesenchymal cells develop into **osteoblasts** that form *bone tissue* by intramembranous or endochondral ossification (discussed earlier in the chapter).

Discussion of the development of the skeletal system provides us with an excellent opportunity to trace the development of the extremities. The *extremities* make their appearance about the fifth week as small elevations at the sides of the trunk called **limb buds** (Fig. 6.9a). They consist of masses of general **mesoderm** covered by **ectoderm.** At this point, a mesenchymal skeleton exists in the limbs; some of the masses of mesoderm surrounding the developing bones will become the skeletal muscles of the extremities.

By the sixth week, the limb buds develop a constriction around the middle portion. The constriction produces distal segments of the upper buds called **hand plates** and distal segments of the lower buds called **foot plates.** These plates represent the beginnings of the *hands* and *feet,* respectively. At this stage of limb development, a cartilaginous skeleton is present. By the seventh week (Fig. 6.9b), the *arm, forearm,* and *hand* are evident in the upper limb bud, and the *thigh, leg,* and *foot* appear in the lower limb bud. Endochondral ossification has begun. By the eighth week (Fig. 6.9c), as the *shoulder, elbow,* and *wrist* areas become apparent, the upper limb bud is appropriately called the *upper extremity* and the lower limb bud is referred to as the *lower extremity.*

The **notochord** is a flexible rod of tissue that lies in a position where the future vertebral column will develop (see Fig. 10.17b). As the vertebrae develop, the notochord becomes surrounded by the developing vertebral bodies. The notochord eventually disappears except for remnants that persist as the *nucleus pulposus* of the intervertebral discs (see Fig. 7.19).

FIGURE 6.9 External features of a developing human embryo at various stages of development. Many of the labeled structures are discussed in later chapters; they are indicated here to help you orient to the figure.

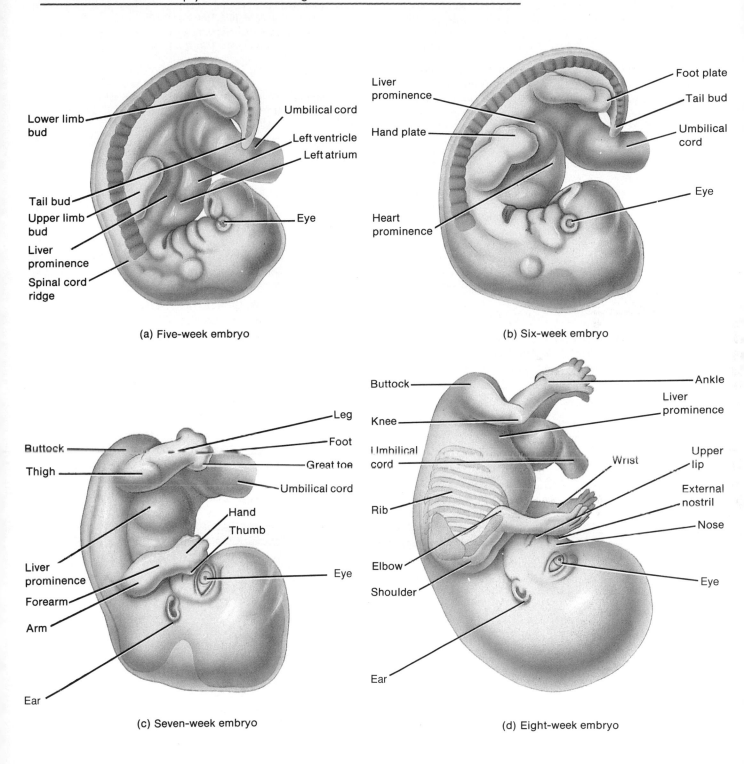

(a) Five-week embryo

(b) Six-week embryo

(c) Seven-week embryo

(d) Eight-week embryo

Question: The skeletal system develops from which basic embryonic tissue?

DISORDERS: HOMEOSTATIC IMBALANCES

OSTEOPOROSIS

Twenty million people in the U.S. suffer from **osteoporosis** (os′-tē-ō-pō-RŌ-sis), literally a condition of porous bones. It is characterized by decreased bone mass and increased susceptibility to fractures. The basic problem is that bone resorption outpaces bone formation. The disorder primarily affects middle-aged and elderly people—women more than men and whites more than blacks. Between puberty and mid-life, sex hormones (several estrogens in women and testosterone in men) and other hormones maintain bone tissue by stimulating the osteoblasts to form new bone. After menopause, women produce much smaller amounts of estrogen, and both men and women produce smaller amounts of the sex hormones as they age. As a result, the osteoblasts become less active, and there is a decrease in bone mass.

Although most common in women over age 50, osteoporosis can also occur in female runners and ballet dancers, male marathoners whose caloric intake is inadequate, teenagers on junk-food diets, young women suffering from eating disorders, people allergic to dairy products, nursing mothers, and in those exposed to prolonged treatment with cortisone or high levels of thyroid hormones. Often, the first symptom of osteoporosis is a pathological fracture. Bone mass becomes so depleted that the skeleton can no longer withstand the mechanical stresses of everyday living. For example, a hip fracture might result from sitting down too quickly. Osteoporosis causes more than 250,000 hip fractures a year, and complications from osteoporosis are the 12th leading cause of death in the U.S. It is responsible for shrinkage of the backbone (vertebrae) and height loss, hunched backs, bone fractures, and considerable pain. Osteoporosis afflicts the entire skeletal system.

Risk factors for developing osteoporosis, besides race and gender, are (1) body build (short females are at greater risk since they have less total bone mass), (2) weight (thin females and those who overdo exercise are at greater risk since adipose tissue is a source of estrone, an estrogen that retards bone loss), (3) smoking (smoking decreases blood estrogen levels), (4) calcium deficiency and malabsorption, (5) vitamin D deficiency, (6) exercise (sedentary people are more likely to develop bone loss), (7) certain drugs (alcohol, some diuretics, cortisone, and tetracycline promote bone loss), (8) premature menopause (which may be due to excessive exercise), and (9) a family history of osteoporosis (daughters of women with osteoporosis have reduced bone mass).

Adequate diet and exercise are the mainstays for preventing osteoporosis. In postmenopausal women, estrogen replacement therapy (ERT; low doses of both estrogen and progesterone, another sex hormone), calcium supplements, and weight-bearing exercise help prevent or retard the development of osteoporosis. The most important aspect of treatment is prevention. Adequate calcium intake and exercise in her early years may be more beneficial to a woman than ERT and calcium supplements when she is older.

PAGET'S DISEASE

Paget's disease is characterized by a greatly accelerated remodeling process in which osteoclastic resorption is massive and new bone formation by osteoblasts is extensive. As a result, there is an irregular thickening and softening of the bones and greatly increased vascularity, especially in bones of the skull, pelvis, and extremities.

MEDICAL TERMINOLOGY

Osteoarthritis (os′-tē-ō-ar-THRĪ-tis; *arthro* = joint) The degeneration of cartilage, allowing the bony ends to touch and, from the friction of bone against bone, worsening the condition; usually associated with the elderly.

Osteogenic (os′-tē-Ō-JEN-ik) **sarcoma** (sar-KŌ-ma; *sarcoma* = connective tissue tumor) Bone cancer that primarily affects osteoblasts and occurs most often in the bones of teenagers during their growth spurt; most common sites are the metaphyses of the thighbone (femur), shinbone (tibia), and arm bone (humerus); metastases occur most often in lungs; treatment consists of multidrug chemotherapy and removal of the malignant growth or amputation of the limb.

Osteomyelitis (os′-tē-ō-mī-el-Ī-tis; *myelos* = marrow) Inflammation of a bone, especially the marrow, caused by a pathogenic organism, especially *Staphylococcus aureus*.

Osteopenia (os′-tē-ō-PĒ-nē-a; *osteo* = bone; *penia* = poverty) Reduced bone mass due to a decrease in the rate of bone synthesis to a level insufficient to compensate for normal bone breakdown; any decrease in bone mass below normal. An example is osteoporosis.

Osteosarcoma (os′-tē-ō-sar-KŌ-ma) A malignant tumor composed of bone tissue.

Study Outline

Physiology: Functions of Bone (p. 147)

1. Bone (osseous) tissue and the skeletal system function in support, protection, movement, mineral homeostasis, blood cell production, and storage of energy.

Anatomy: Structure of Bone (p. 147)

1. Structurally, the skeletal system contains cartilage, bone, bone marrow, and the periosteum.
2. Parts of a typical long bone are the diaphysis (shaft), epiphyses (ends), metaphysis, articular cartilage, periosteum, medullary (marrow) cavity, and endosteum.

Histology of Bone Tissue (p. 147)

1. Bone tissue consists of widely separated cells surrounded by large amounts of matrix.
2. The four principal types of cells are osteoprogenitor cells, osteoblasts, osteocytes, and osteoclasts.
3. The matrix of bone contains abundant mineral salts (mostly hydroxyapatite) and collagen fibers.

Compact Bone Tissue (p. 149)

1. Compact (dense) bone consists of osteons (Haversian systems) with little space between them.
2. Compact bone lies over spongy bone and composes most of the bone tissue of the diaphysis. Functionally, compact bone protects, supports, and resists stress.

Spongy Bone Tissue (p. 151)

1. Spongy (cancellous) bone does not contain osteons. It consists of trabeculae surrounding many red marrow-filled spaces.
2. It forms most of the structure of short, flat, and irregular bones, and the epiphyses of long bones.
3. Functionally, spongy bone stores red marrow and provides some support.

Physiology of Bone Formation: Ossification (p. 151)

1. Bone forms by a process called ossification (osteogenesis), which begins when mesenchymal cells become transformed into osteoprogenitor cells. These undergo cell division and give rise to cells that differentiate into osteoblasts and osteoclasts.
2. The process begins during the sixth or seventh week of embryonic life and continues throughout adulthood. The two types of ossification, intramembranous and endochondral, involve the replacement of a preexisting connective tissue with bone.
3. Intramembranous ossification occurs within fibrous connective tissue membranes of the embryo and the adult.
4. Endochondral ossification occurs within a cartilage model. The primary ossification center of a long bone is in the diaphysis. Cartilage degenerates, leaving cavities that merge to form the medullary cavity. Osteoblasts lay down bone. Next, ossification occurs in the epiphyses, where bone replaces cartilage, except for the epiphyseal plate.

Physiology of Bone Growth (p. 152)

1. The anatomical zones of the epiphyseal plate are the zones of resting cartilage, proliferating cartilage, hypertrophic cartilage, and calcified cartilage.
2. Because of the activity of the epiphyseal plate, the diaphysis of a bone increases in length.
3. Bone grows in diameter as a result of the addition of new bone tissue by periosteal osteoblasts around the outer surface of the bone (appositional growth).

Bone Homeostasis (p. 154)

Remodeling (p. 154)

1. Remodeling is the replacement of old bone tissue by new bone tissue.
2. Old bone is constantly destroyed by osteoclasts, whereas new bone is constructed by osteoblasts.
3. Remodeling requires minerals (calcium, phosphorus, magnesium, and manganese), vitamins (D, C, A, and B_{12}), and hormones (human growth hormone, sex hormones, insulin, thyroid hormones, parathyroid hormone, and calcitonin).

Fracture and Repair of Bone (p. 155)

1. A fracture is any break in a bone.
2. The types of fractures include partial, complete, closed (simple), open (compound), comminuted, greenstick, spiral, transverse, impacted, Pott's, Colles', displaced, nondisplaced, and stress.
3. Fracture repair involves formation of a fracture hematoma, fibrocartilaginous callus, bone callus, and remodeling.

Bone's Role in Calcium Homeostasis (p. 158)

1. Bone is the major reservoir for calcium (Ca^{2+}) in the body.
2. Parathyroid hormone (PTH) secreted by the parathyroid gland increases blood Ca^{2+} level.
3. Calcitonin (CT) secreted by the thyroid gland decreases blood Ca^{2+} level.

Exercise and Bone (p. 159)

1. Bone can alter its strength in response to mechanical stress by increasing deposition of mineral salts and production of collagen fibers.
2. Removal of mechanical stress weakens bone through demineralization and collagen reduction.

Aging and Bone Tissue (p. 160)

1. The principal effect of aging is a loss of calcium from bones, which may result in osteoporosis.
2. Another effect is a decreased production of matrix (mostly collagen), which makes bones more susceptible to fracture.

Developmental Anatomy of Bone and the Skeletal System (p. 160)

1. Bone forms from mesoderm by intramembranous or endochondral ossification.
2. Extremities develop from limb buds, which consist of mesoderm and ectoderm.

Review Questions

1. List and describe the six principal functions of bone tissue. (p. 147)
2. Diagram the parts of a long bone and list the functions of each part. (p. 147)
3. Describe the four types of cells in bone tissue. (p. 147)
4. What is the composition of the matrix of bone tissue? (p. 149)
5. Distinguish between spongy and compact bone in terms of microscopic appearance, location, and function. (p. 149)
6. Diagram the microscopic appearance of compact bone and indicate the functions of the various components. (p. 149)
7. What is meant by ossification? Describe the initial events of ossification. (p. 151)
8. Outline the major events involved in intramembranous and endochondral ossification and explain the main differences. (p. 151)
9. Describe the histology of the various zones of the epiphyseal plate. How does the plate grow? What is the significance of the epiphyseal line? (p. 152)
10. Define remodeling and describe the factors that contribute to the process. (p. 154)
11. What is a fracture? Distinguish several principal kinds. Outline the three basic steps involved in fracture repair. (p. 155)
12. Describe the role of bone tissue in calcium homeostasis. (p. 158)
13. Explain the effects of exercise and aging on the skeletal system. (p. 159)
14. Describe the development of the skeletal system. (p. 160)
15. What are the principal symptoms and causes of osteoporosis? (p. 160)

Answers to Questions with Figures

6.1 Support, protection, movement, mineral homeostasis, blood cell production, storage of energy.
6.2 Osteoblast; osteoclast.
6.3 Since the central canals are the main blood supply to the osteocytes of an osteon, their blockage would lead to death of the osteocytes.
6.4 Yes. The epiphyseal lines are indications of growth zones that have ceased to function. The absence of epiphyseal lines indicates that bone lengthening is still occurring.
6.5 Cell divisions in the zone of proliferating cartilage and maturation of the cells in the zone of hypertrophic cartilage.
6.6 A mass of repair tissue that bridges the broken ends of bones.
6.7 Calcium and phosphorus deposition is a slow process, bone cells generally grow and reproduce slowly, and decreased blood supply slows healing of an infected bone.
6.8 Heartbeat, respiration, nerve cell functioning, enzyme functioning, and blood clotting, to name just a few.
6.9 Mesoderm.

Chapter 7
THE SKELETAL SYSTEM: THE AXIAL SKELETON

Chapter Contents at a Glance

Student Objectives

1. Classify the principal types of bones, on the basis of shape and location.

2. Describe the various markings on the surfaces of bones.

3. Identify the bones of the skull and the major markings associated with each.

4. Identify the principal sutures, fontanels, paranasal sinuses, and foramina of the skull.

5. Identify the bones of the vertebral column and their principal markings.

6. Identify the bones of the thorax and their principal markings.

7. Contrast herniated (slipped) disc, abnormal curves, and spina bifida as disorders associated with the skeletal system.

Without the skeletal system, you would be unable to perform movements such as walking or grasping. The slightest jar to your head or chest could damage the brain or heart. It would even be impossible to chew food. The framework of bones and cartilage that protects organs and allows movement is called the **skeletal** (*skeletos* = dried up) **system.** Since the skeletal system forms the framework of the body, a familiarity with the names, shapes, and positions of individual bones will help you understand some of the other organ systems. For example, the radial artery, the site where pulse is usually taken, is named for its proximity to the radius, the lateral bone of the forearm.

Movements such as throwing a ball, typing, and walking require the coordinated use of bones and muscles. To understand how muscles produce different movements, you need to learn where on bones the muscles attach and the types of joints acted on by the contracting muscles. The respiratory system is highly dependent on bone structure. Bones in the nasal cavity form a series of passageways that direct inhaled air toward the lungs while cleaning, moistening, and warming it. Furthermore, the bones of the thorax are specially shaped and positioned so the chest can expand during inhalation. Many bones also serve as anatomical and surgical landmarks. For example, parts of certain bones serve to locate structures within the skull and to outline the lungs and heart, and abdominal and pelvic viscera. Blood vessels and nerves often run parallel to bones. These structures can be located more easily if the bone is identified first.

The specialized branch of medicine that deals with the preservation and restoration of the skeletal system, joints, and associated structures is called **orthopedics** (or′-thō-PĒ-diks; *ortho* = correct or straighten; *pais* = child).

TYPES OF BONES

Almost all the bones of the body may be classified into four principal types on the basis of shape: long, short, flat, and irregular. **Long bones** (see Fig. 6.1) have greater length than width and consist of a shaft and a variable number of extremities (ends). They are slightly curved for strength. A curved bone absorbs the stress of the body weight at several different points so the stress is evenly distributed. If such bones were straight, the weight of the body would be unevenly distributed and the bone would easily fracture. Long bones consist mostly of *compact bone*, which is dense and has few spaces, but also contain considerable amounts of *spongy bone,* which has larger spaces (described shortly). Long bones include those in the thigh (femur), leg (tibia and fibula), toes (phalanges), arm (humerus), forearm (ulnar and radius), and fingers (phalanges).

Short bones are somewhat cube-shaped and nearly equal in length and width. They are spongy bone except at the surface, where there is a thin layer of compact bone.

Examples of short bones are the wrist (carpal) and ankle (tarsal) bones (see Figs. 8.6 and 8.14).

Flat bones are generally thin and composed of two nearly parallel plates of compact bone enclosing a layer of spongy bone. Flat bones afford considerable protection and provide extensive areas for muscle attachment. Flat bones include the cranial bones (which protect the brain), the breastbone (sternum) and ribs (which protect organs in the thorax and are shown in Fig. 7.21), and the shoulder blades (scapular).

Irregular bones have complex shapes and cannot be grouped into any of the three categories just described. They also vary in the amount of spongy and compact bone present. Such bones include the backbones (vertebrae), shown in Fig. 7.16, and certain facial bones.

There are two additional types of bones that are not included in this classification by shape, but instead are classified by location. **Sutural** (SOO-chur-al) or **Wormian bones** are small bones located between the joints (sutures) of certain cranial bones (see Fig. 7.5). Their number varies greatly from person to person. **Sesamoid bones** are small bones wrapped in tendons where considerable pressure develops, for instance, in the wrist. These, like sutural bones, are also variable in number. Two sesamoid bones, the kneecaps (patellae), are present in all individuals (see Fig. 8.12).

SURFACE MARKINGS

The surfaces of bones have various structural features adapted to specific functions. These features are called **surface markings.** Long bones that bear a lot of weight have large, rounded ends that can form sturdy joints and provide adequate surface area for the attachment of ligaments and muscles. Other bones have depressions that receive the rounded ends. Rough areas serve as points of attachment for muscles, tendons, and ligaments. Grooves on the surfaces of bones provide for the passage of blood vessels. Openings occur where blood vessels and nerves pass through the bone. Exhibit 7.1 describes the different markings and their functions.

DIVISIONS OF THE SKELETAL SYSTEM

The adult human skeleton consists of 206 named bones grouped in two principal divisions: the **axial skeleton** and the **appendicular skeleton.** Refer to Fig. 7.1 to see how the two divisions join to form the complete skeleton. The bones of the axial skeleton are shown in blue. The longitudinal **axis,** or center, of the human body is a straight line that runs through the body's center of gravity. This imaginary line extends through the head and down to the space between the feet. The axial division consists of the bones that lie

EXHIBIT 7.1

SURFACE MARKINGS

Marking	Description	Example
DEPRESSIONS AND OPENINGS		
Fissure (FISH-ur)	A narrow, cleftlike opening between adjacent parts of bones through which blood vessels or nerves pass.	Superior orbital fissure of the sphenoid bone (Fig. 7.13).
Fontanel (fon′-ta-NEL = little fountain)	Dense connective tissue-filled space between skull bones at birth.	Anterior fontanel between frontal and parietal bones (Fig. 7.6).
Foramen (fō-RĀ-men; *foramen* = hole)	An opening through which blood vessels, nerves, or ligaments pass.	Infraorbital foramen of the maxilla (Fig. 7.2).
Fossa (*fossa* = basinlike depression)	A depression in or on a bone.	Mandibular fossa of the temporal bone (Fig. 7.7).
Groove or sulcus (*sulcus* = ditchlike groove)	A furrow or depression that accommodates a soft structure such as a blood vessel, nerve, or tendon.	Intertubercular sulcus of the humerus (Fig. 8.4).
Meatus (mē-Ā-tus; *meatus* = canal) or **canal**	A tubelike passageway running within a bone.	External auditory (acoustic) meatus of the temporal bone (Fig. 7.3).
Paranasal sinus (*sin* = cavity)	An air-filled cavity within a bone connected to the nasal cavity.	Frontal sinus of the frontal bone (Fig. 7.11).
PROCESSES	Any prominent projection.	Mastoid process of the temporal bone (Fig. 7.3).
PROCESSES THAT FORM JOINTS		
Condyle (KON-dīl; *condylus* = knuckle-like process)	A large, rounded articular prominence.	Medial condyle of the femur (Fig. 8.10).
Facet	A smooth, flat surface.	Articular facet of a vertebra for the tubercle of rib (Fig. 7.18).
Head	A rounded articular projection supported on the constricted portion (neck) of a bone.	Head of the femur (Fig. 8.10).
PROCESSES TO WHICH TENDONS, LIGAMENTS, AND OTHER CONNECTIVE TISSUES ATTACH		
Crest	A prominent border or ridge.	Iliac crest of the coxal (hip) bone (Fig. 8.7).
Epicondyle (*epi* = above)	A prominence above a condyle.	Medial epicondyle of the femur (Fig. 8.10).
Line (linea)	A less prominent ridge than a crest.	Linea aspera of the femur (Fig. 8.10).
Spinous process (spine)	A sharp, slender process.	Spinous process of a vertebra (Fig. 7.16).
Trochanter (trō-KAN-ter)	A large projection found only on the femur.	Greater trochanter of the femur (Fig. 8.10).
Tubercle (TOO-ber-kul; *tube* = knob)	A small, rounded process.	Greater tubercle of the humerus (Fig. 8.4).
Tuberosity	A large, rounded, usually roughened process.	Ischial tuberosity of the hipbone (Fig. 8.7).

FIGURE 7.1 Divisions of the skeletal system. The axial skeleton is indicated in blue.

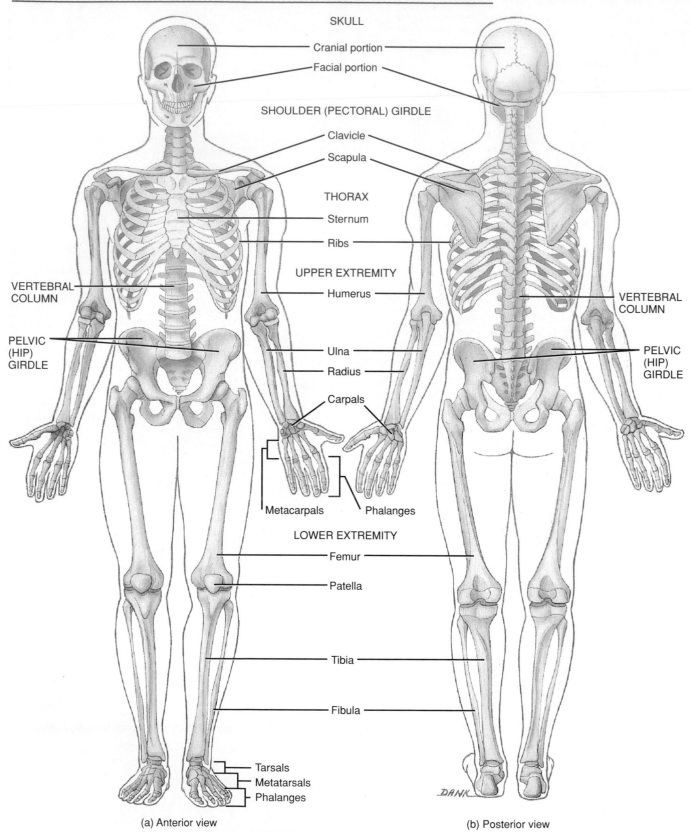

SKULL
Cranial portion
Facial portion

SHOULDER (PECTORAL) GIRDLE
Clavicle
Scapula

THORAX
Sternum
Ribs

UPPER EXTREMITY
Humerus
Ulna
Radius
Carpals
Metacarpals
Phalanges

LOWER EXTREMITY
Femur
Patella
Tibia
Fibula
Tarsals
Metatarsals
Phalanges

VERTEBRAL COLUMN

PELVIC (HIP) GIRDLE

VERTEBRAL COLUMN

PELVIC (HIP) GIRDLE

DANK

(a) Anterior view

(b) Posterior view

Question: Is each bone part of the axial or the appendicular skeleton? Skull, clavicle, vertebral column, shoulder girdle, humerus, pelvic girdle, and femur.

around the axis: skull bones, auditory ossicles, hyoid bone, ribs, breastbone, and bones of the backbone.

The appendicular division contains the bones of the **upper** and **lower extremities (limbs),** plus the bones called **girdles** that connect the extremities to the axial skeleton. Exhibit 7.2 presents the standard grouping of the 80 bones of the axial division and the 126 bones of the appendicular division.

We will study bones by examining the various regions of the body. First, we will look at the skull and see how its bones relate to each other. We will then move on to the vertebral column (backbone) and the chest. In Chapter 8 we will examine the bones of the shoulder girdle and upper extremities, then the pelvic girdle and lower extremities. This regional approach will allow you to see how the many bones of the body relate to one another.

SKULL

The **skull,** which contains 22 bones, rests on the superior end of the vertebral column. It includes two sets of bones: cranial bones and facial bones. The **cranial bones** form the cranial cavity and enclose and protect the brain. The 8 cranial bones are the frontal bone, parietal bones (2), temporal bones (2), occipital bone, sphenoid bone, and ethmoid bone. There are 14 **facial bones:** nasal bones (2), maxillae or maxillas (2), zygomatic bones (2), mandible, lacrimal bones (2), palatine bones (2), inferior nasal conchae (2), and vomer. Be sure you can locate all the skull bones in the various views of the skull (Figs. 7.2 through 7.5).

Sutures

A **suture** (SOO-chur; *sutura* = seam) is an immovable joint found only between skull bones. Very little connective tissue is found in a suture. Sutures begin to ossify between ages 20 and 30 and by age 60 almost all sutures are ossified. Four prominent sutures (Figs. 7.3, 7.5, and 7.6) are:

1. **Coronal suture** between the frontal bone and the two parietal bones.
2. **Sagittal suture** between the two parietal bones.
3. **Lambdoid** (LAM-doyd) **suture** between the parietal bones and the occipital bone.
4. **Squamous** (SKWĀ-mos) **sutures** between the parietal bones and the temporal bones.

There may be sutural bones in the sagittal or lambdoid suture.

Fontanels

The skeleton of a newly formed embryo consists of cartilage or fibrous connective tissue membranes structures shaped like bones. Gradually, bone replaces the cartilage or

EXHIBIT 7.2

DIVISIONS OF THE ADULT SKELETAL SYSTEM

Regions of the Skeleton	Number of Bones
AXIAL SKELETON	
SKULL	
Cranium	8
Face	14
HYOID	1
AUDITORY OSSICLES[a]	6
VERTEBRAL COLUMN	26
THORAX	
Sternum	1
Ribs	24
	Subtotal = 80
APPENDICULAR SKELETON	
PECTORAL (SHOULDER) GIRDLES	
Clavicle	2
Scapula	2
UPPER EXTREMITIES	
Humerus	2
Ulna	2
Radius	2
Carpals	16
Metacarpals	10
Phalanges	28
PELVIC (HIP) GIRDLE	
Hip, pelvic, or coxal bone	2
LOWER EXTREMITIES	
Femur	2
Fibula	2
Tibia	2
Patella	2
Tarsals	14
Metatarsals	10
Phalanges	28
	Subtotal = 126
	Total = 206

[a] Although the auditory ossicles are not considered part of the axial or appendicular skeleton, but rather as a separate group of bones, they are placed with the axial skeleton for convenience. The middle portion of each ear contains three auditory ossicles held together by a series of ligaments. The auditory ossicles are exceedingly small bones named for their shapes. Their names are the malleus, incus, and stapes, commonly called the hammer, anvil, and stirrup, respectively. They vibrate in response to sound waves that strike the eardrum and have a key role in the mechanism of hearing. This is described in detail in Chapter 16.

FIGURE 7.2 Skull.

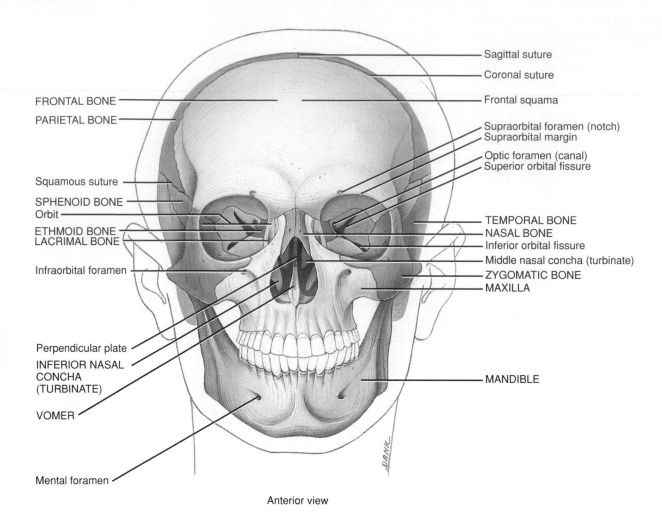

FRONTAL BONE
PARIETAL BONE

Squamous suture
SPHENOID BONE
Orbit
ETHMOID BONE
LACRIMAL BONE

Infraorbital foramen

Perpendicular plate
INFERIOR NASAL
CONCHA
(TURBINATE)

VOMER

Mental foramen

Sagittal suture
Coronal suture
Frontal squama

Supraorbital foramen (notch)
Supraorbital margin
Optic foramen (canal)
Superior orbital fissure

TEMPORAL BONE
NASAL BONE
Inferior orbital fissure
Middle nasal concha (turbinate)
ZYGOMATIC BONE
MAXILLA

MANDIBLE

Anterior view

Question: Which of the bones shown here are cranial bones?

FIGURE 7.3 Skull. Although the hyoid bone is not part of the skull, it is included in the illustration for reference.

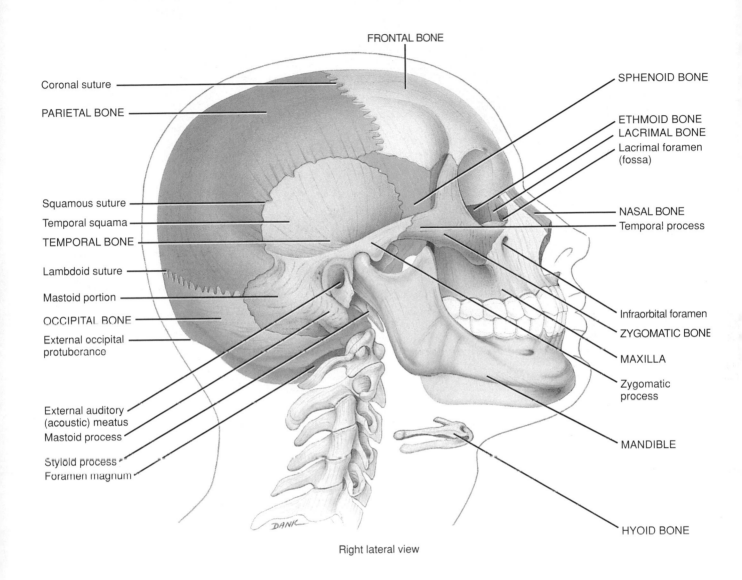

Right lateral view

Question: What are the major bones on either side of the squamous suture, the lambdoid suture, and coronal suture?

FIGURE 7.4 Skull.

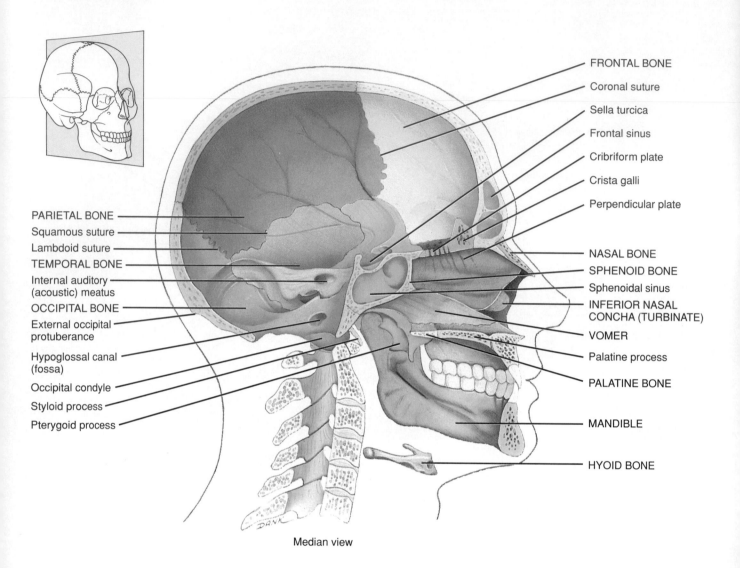

FRONTAL BONE
Coronal suture
Sella turcica
Frontal sinus
Cribriform plate
Crista galli
Perpendicular plate

NASAL BONE
SPHENOID BONE
Sphenoidal sinus
INFERIOR NASAL
CONCHA (TURBINATE)
VOMER
Palatine process
PALATINE BONE

MANDIBLE

HYOID BONE

PARIETAL BONE
Squamous suture
Lambdoid suture
TEMPORAL BONE
Internal auditory
(acoustic) meatus
OCCIPITAL BONE
External occipital
protuberance
Hypoglossal canal
(fossa)
Occipital condyle
Styloid process
Pterygoid process

Median view

Question: What features of the skull help to reduce its weight and serve as resonating chambers for the voice?

FIGURE 7.5 Skull. The sutures are exaggerated for emphasis.

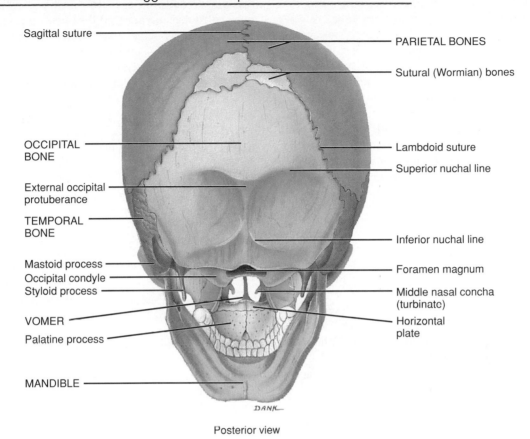

Posterior view

Question: Which sutures may contain sutural bones?

FIGURE 7.6 Fontanels of the skull at birth.

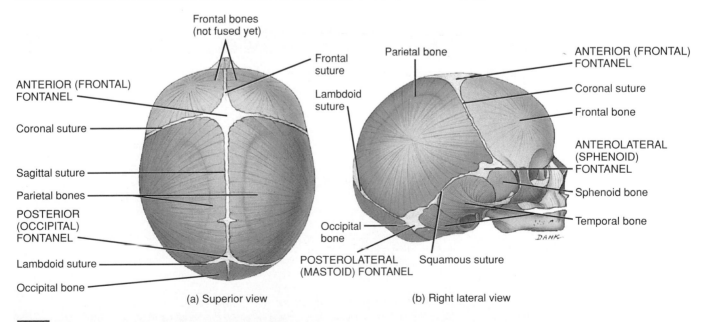

(a) Superior view   (b) Right lateral view

Question: Which fontanel is bordered by four different skull bones?

fibrous connective tissue membranes in a process called ossification. At birth, membrane-filled spaces called **fontanels** (fon'-ta-NELZ; = little fountains) are found between cranial bones (Fig. 7.6). These "soft spots" are areas of fibrous connective tissue membranes that will eventually be replaced with bone by intramembranous ossification. Functionally, the fontanels (1) enable the fetal skull to modify its size and shape as it passes through the birth canal and (2) permit rapid growth of the brain during infancy. In addition, fontanels help a physician to gauge the degree of brain development by their state of closure and serve as landmarks (anterior fontanel) for withdrawal of blood from the superior sagittal sinus (a large vein around the brain). Although an infant may have many fontanels at birth, the form and location of six are fairly constant (Exhibit 7.3).

Cranial Bones

Frontal Bone

The **frontal bone** forms the forehead (the anterior part of the cranium), the roofs of the **orbits** (eye sockets), and most of the anterior part of the cranial floor. Soon after birth, the left and right sides of the frontal bone are united by the **frontal suture** (see Fig. 7.6a), which usually disappears by age 6. If it persists throughout life, it is referred to as the **metopic suture.**

If you examine Fig. 7.2, you will note the **frontal squama** (SKWĀ-ma; *squam* = scale), or **vertical plate.** This scalelike plate, which corresponds to the forehead, gradually slopes down from the coronal suture, then turns abruptly downward. Above the orbits, the frontal bone thickens, forming the **supraorbital margin.** From this margin the frontal bone extends posteriorly to form the roof of the orbit and part of the floor of the cranial cavity. Within the supraorbital margin, slightly medial to its midpoint, is a hole called the **supraorbital foramen (notch).** As the various foramina associated with cranial bones are discussed, refer to Exhibit 7.4 on page 184 to note which structures pass through them. The **frontal sinuses** lie deep to the frontal squama. These mucus-lined cavities act as sound chambers that give the voice resonance.

CLINICAL APPLICATION

BLACK EYE

Just above the supraorbital margin is a sharp ridge. A blow to the ridge often lacerates the skin over it, resulting in bleeding. Bruising of the skin over the ridge causes tissue fluid and blood to accumulate in the surrounding connective tissue and gravitate into the upper eyelid. The resulting swelling and discoloration are called a **black eye.**

Parietal Bones

The two **parietal** (pa-RĪ-e-tal; *paries* = wall) **bones** form the greater portion of the sides and roof of the cranial cavity (see Figs. 7.3 and 7.4). The internal surfaces of the parietal bones contain many protrusions and depressions that accommodate the blood vessels supplying the outer meninx (covering) of the brain called the dura mater.

Temporal Bones

The two **temporal** (*tempora* = temples) **bones** form the inferior sides of the cranium and part of the cranial floor.

EXHIBIT 7.3

FONTANELS

Fontanel	Location	Description
Anterior (frontal)	Between the angles of the two parietal bones and the two segments of the frontal bone.	Roughly diamond-shaped, the largest of the six fontanels; usually closes 18 to 24 months after birth.
Posterior (occipital)	Situated between the two parietal bones and the occipital bone.	Diamond-shaped, considerably smaller than the anterior fontanel; generally closes about 2 months after birth.
Anterolateral (sphenoid)	One on each side of the skull at the junction of the frontal, parietal, temporal, and sphenoid bones.	Small and irregular in shape; normally close about 3 months after birth.
Posterolateral (mastoid)	One lies on each side of the skull at the junction of the parietal, occipital, and temporal bones.	Irregularly shaped; begin to close 1 or 2 months after birth, but closure is not generally complete until 12 months.

In the lateral view of the skull (see Fig. 7.3), note the **temporal squama.** This is a thin, flat portion of the temporal bone that forms the anterior and superior part of the temple. Projecting from the inferior portion of the temporal squama is the **zygomatic process,** which articulates with the temporal process of the zygomatic bone. The zygomatic process of the temporal bone and the temporal process of the zygomatic bone constitute the **zygomatic arch** (see Fig. 7.7).

At the floor of the cranial cavity (see Fig. 7.8) is the **petrous portion** of the temporal bone. This portion is triangular and located at the base of the skull between the sphenoid and occipital bones. The petrous portion houses the internal and middle ear, structures involved in hearing and

equilibrium (balance). It also contains the **carotid foramen (canal)** (see Fig. 7.7). Posterior to the carotid foramen and anterior to the occipital bone is the **jugular foramen.**

Between the squamous and petrous portions of the temporal bone is a socket called the **mandibular (glenoid) fossa.** Anterior to the mandibular fossa is a rounded eminence, the **articular tubercle.** The mandibular fossa and articular tubercle articulate with the condylar process of the mandible (lower jawbone) to form the temporomandibular joint (TMJ). The mandibular fossa and articular tubercle are seen best in Fig. 7.7.

In the lateral view of the skull (see Fig. 7.3), you will see the **mastoid portion** of the temporal bone. It is located posterior and inferior to the **external auditory (acoustic)**

FIGURE 7.7 Skull.

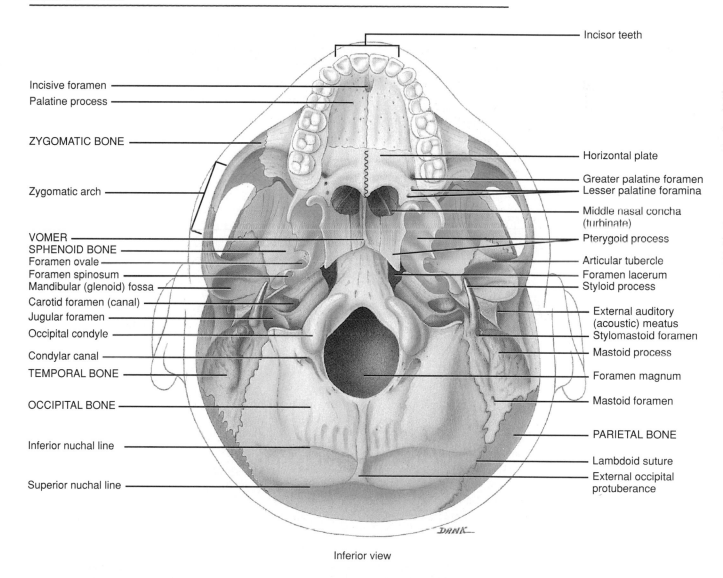

Incisor teeth

Incisive foramen
Palatine process

ZYGOMATIC BONE

Zygomatic arch

VOMER
SPHENOID BONE
Foramen ovale
Foramen spinosum
Mandibular (glenoid) fossa
Carotid foramen (canal)
Jugular foramen
Occipital condyle
Condylar canal
TEMPORAL BONE

OCCIPITAL BONE

Inferior nuchal line

Superior nuchal line

Horizontal plate
Greater palatine foramen
Lesser palatine foramina
Middle nasal concha (turbinate)
Pterygoid process
Articular tubercle
Foramen lacerum
Styloid process
External auditory (acoustic) meatus
Stylomastoid foramen
Mastoid process
Foramen magnum
Mastoid foramen
PARIETAL BONE
Lambdoid suture
External occipital protuberance

DANK

Inferior view

Question: Which bone contains the carotid foramen, stylomastoid foramen, and mastoid foramen? Which of these foramina is the largest opening?

meatus, or ear canal. In the adult, this portion of the bone contains several **mastoid air "cells."** These are air spaces, separated from the brain only by thin bony partitions. If **mastoiditis** (inflammation of these bony cells) occurs, the infection may spread to the brain or its outer covering.

The **mastoid process** is a rounded projection of the temporal bone posterior to the external auditory meatus. It serves as a point of attachment for several neck muscles. Near the posterior border of the mastoid process is the **mastoid foramen** (see Fig. 7.7). The external auditory (acoustic) meatus is the canal in the temporal bone that leads to the middle ear. The **internal auditory (acoustic) meatus** is superior to the jugular foramen (see Fig. 7.8a). The **styloid process** (see Fig. 7.3) projects downward from the undersurface of the temporal bone and serves as a point of attachment for muscles and ligaments of the tongue and neck.

Occipital Bone

The **occipital** (ok-SIP-i-tal) **bone** forms the posterior part and most of the base of the cranium (Figs. 7.3 and 7.7).

The **foramen magnum** is a large hole in the inferior part of the bone. The medulla oblongata (the part of the brain that is continuous with the spinal cord) and the vertebral and spinal arteries pass through this foramen.

The **occipital condyles** are oval processes with convex surfaces, one on either side of the foramen magnum, that articulate (form a joint) with depressions on the first cervical vertebra. Superior to each condyle is the **hypoglossal canal (fossa)** (see Fig. 7.8).

The **external occipital protuberance** is a prominent projection on the posterior surface of the bone just superior to the foramen magnum. You can feel this structure as a definite bump on the back of your head, just above your neck. The protuberance is also visible in Fig. 7.4. A large fibrous and elastic ligament, the **ligamentum nuchae**, extends from the external occipital protuberance to the seventh cervical vertebra. Extending laterally from the protuberance are two curved lines, the **superior nuchal lines,** and below these two **inferior nuchal lines,** which are areas of muscle attachment (see Fig. 7.5).

Sphenoid Bone

The **sphenoid** (SFĒ-noyd; *speno* = wedge-shaped) **bone** lies at the middle part of the base of the skull (Figs. 7.4 and 7.8). This bone is called the keystone of the cranial floor because it articulates with all the other cranial bones, holding them together. Viewing the floor of the cranium from above, note the sphenoid articulations: anteriorly with the frontal bone, laterally with the temporal bones, and posteriorly with the occipital bone. It lies posterior and slightly superior to the nasal cavity and forms part of the floor, sidewalls, and rear wall of the orbit (see Fig. 7.13). The shape of the sphenoid resembles a bat with outstretched wings

(Fig. 7.8b). The **body** of the sphenoid is the cubelike central portion between the ethmoid and occipital bones. It contains the **sphenoidal sinuses,** which drain into the nasal cavity (see Fig. 7.11). On the superior surface of the body of the sphenoid is a depression called the **sella turcica** (SEL-a TUR-si-ka; = Turkish saddle), which cradles the pituitary gland.

The **greater wings** of the sphenoid project laterally from the body, forming the anterolateral floor of the cranium. The greater wings also form part of the lateral wall of the skull just anterior to the temporal bone. The **lesser wings** are anterior and superior to the greater wings. They form part of the floor of the cranium and the posterior part of the orbit.

Between the body and lesser wing, you can locate the **optic foramen (canal).** Lateral to the body between the greater and lesser wings is a somewhat triangular slit called the **superior orbital fissure.** This fissure may also be seen in the anterior view of the orbit in Fig. 7.13.

You can see the **pterygoid** (TER-i-goyd) **processes** (Fig. 7.8b) by looking on the inferior part of the sphenoid bone. These structures project inferiorly from the points where the body and greater wings unite. Some of the muscles that move the mandible attach to the pterygoid processes.

Ethmoid Bone

The **ethmoid bone** is a light, spongelike bone located in the anterior part of the cranial floor between the orbits. It is anterior to the sphenoid and posterior to the nasal bones (Fig. 7.9). The ethmoid bone forms (1) part of the anterior portion of the cranial floor, (2) the medial wall of the orbits, (3) the superior portions of the nasal septum, a partition that divides the nasal cavity into right and left sides, and (4) most of the sidewalls of the nasal roof. The ethmoid is a major supporting structure of the nasal cavity.

Its **lateral masses (labyrinths)** compose most of the wall between the nasal cavity and the orbits. They contain several air spaces, or "cells," ranging in number from 3 to 18. It is from these "cells" that the bone derives its name (*ethmos* = sieve). The ethmoidal cells together form the **ethmoidal sinuses.** The sinuses are shown in Fig. 7.11. The **perpendicular plate** forms the superior portion of the nasal septum (see Fig. 7.10). The **cribriform (horizontal) plate** lies in the anterior floor of the cranium and forms the roof of the nasal cavity. The cribriform plate contains the **olfactory foramina.** Projecting upward from the cribriform plate is a triangular process called the **crista galli** (= cock's comb). This structure serves as a point of attachment for the membranes (meninges) that cover the brain.

The lateral masses contain two thin, scroll-shaped projections on either side of the nasal septum. These are called the **superior nasal concha** (KONG-ka; *concha* = shell) or **turbinate** and the **middle nasal concha (turbinate).** The conchae cause turbulence in inhaled air, which results in

FIGURE 7.8 Sphenoid bone.

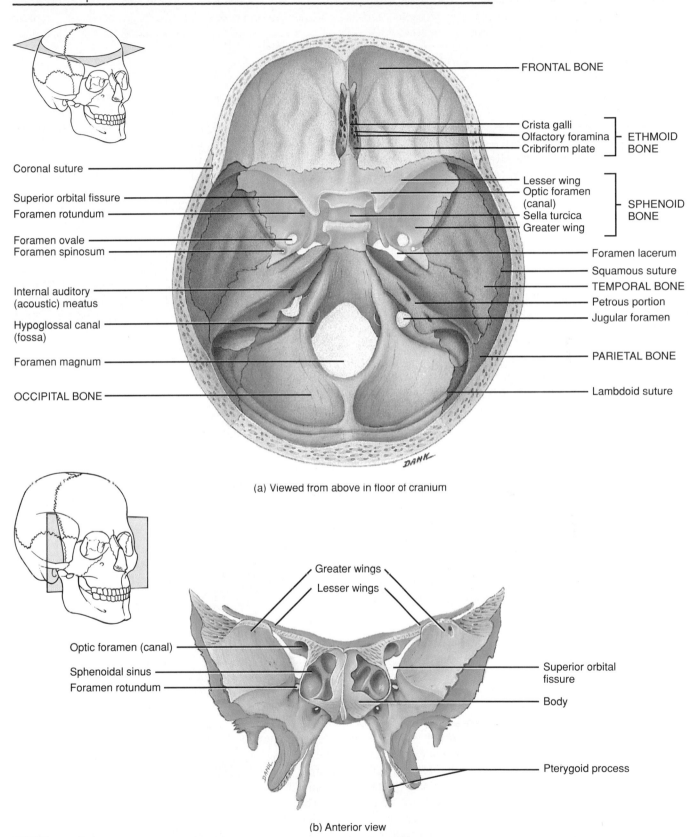

FRONTAL BONE

Crista galli
Olfactory foramina — ETHMOID
Cribriform plate   BONE

Coronal suture

Lesser wing
Optic foramen
Superior orbital fissure   (canal)   SPHENOID
Foramen rotundum   Sella turcica   BONE
Greater wing

Foramen ovale
Foramen spinosum

Foramen lacerum

Squamous suture

Internal auditory
(acoustic) meatus   TEMPORAL BONE

Petrous portion

Jugular foramen

Hypoglossal canal
(fossa)

Foramen magnum

PARIETAL BONE

OCCIPITAL BONE

Lambdoid suture

(a) Viewed from above in floor of cranium

Greater wings
Lesser wings

Optic foramen (canal)

Sphenoidal sinus
Foramen rotundum

Superior orbital
fissure

Body

Pterygoid process

(b) Anterior view

Question: Starting at the crista galli of the ethmoid bone and going in a clockwise direction, name the bones that join the sphenoid bone.

FIGURE 7.9 Ethmoid bone.

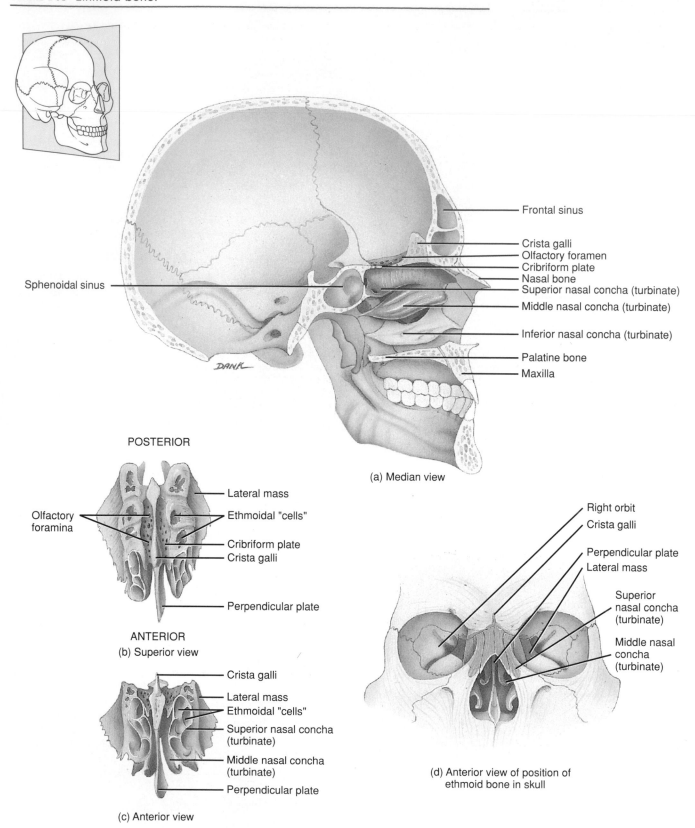

Frontal sinus

Crista galli
Olfactory foramen
Cribriform plate
Nasal bone
Superior nasal concha (turbinate)

Middle nasal concha (turbinate)

Inferior nasal concha (turbinate)

Palatine bone
Maxilla

Sphenoidal sinus

(a) Median view

POSTERIOR

Lateral mass

Olfactory
foramina

Ethmoidal "cells"

Cribriform plate
Crista galli

Perpendicular plate

ANTERIOR
(b) Superior view

Crista galli

Lateral mass
Ethmoidal "cells"
Superior nasal concha
(turbinate)
Middle nasal concha
(turbinate)
Perpendicular plate

(c) Anterior view

Right orbit
Crista galli

Perpendicular plate
Lateral mass

Superior
nasal concha
(turbinate)

Middle nasal
concha
(turbinate)

(d) Anterior view of position of
ethmoid bone in skull

Question: What part of the ethmoid bone forms the superior part of the nasal septum?

many inhaled particles striking and becoming trapped in the mucus found on the lining of the nasal passageways. This turbulence thus cleanses inhaled air before it passes into the rest of the respiratory tract.

Facial Bones

The shape of the face changes dramatically during the first two postnatal years. The brain and cranial bones expand, the teeth form and erupt, and the paranasal sinuses increase in size. Growth of the face ceases at about 16 years of age.

Nasal Bones

The paired **nasal bones** meet at the middle and superior part of the face (see Figs. 7.2 and 7.10) and form part of the bridge of the nose. The major portion of the nose consists of cartilage.

Maxillae

The paired **maxillae** (mak-SIL-ē; *macerae* = to chew) unite to form the upper jawbone (Fig. 7.10) and articulate with every bone of the face except the mandible, or lower jaw-bone. They form part of the floors of the orbits, part of the lateral walls and floor of the nasal cavity, and most of the hard palate. The hard palate is a bony partition formed by the maxillae and palatine bones that forms the roof of the mouth.

Each maxilla contains a **maxillary sinus** that empties into the nasal cavity (see Fig. 7.11). The **alveolar** (al-VĒ-ō-lar; *alveolus* = hollow) **process** is an arch that contains the **alveoli** (sockets) for the maxillary (upper) teeth. The **palatine process** is a horizontal projection of the maxilla that forms the anterior three-quarters of the hard palate. The maxillary bones unite, and the fusion is normally completed before birth.

A fissure associated with the maxilla and sphenoid bone is the **inferior orbital fissure.** It separates the greater wing of the sphenoid and the maxilla (see Fig. 7.13).

CLINICAL APPLICATION

CLEFT PALATE AND CLEFT LIP

Usually the palatine processes of the maxillary bones unite during weeks 10 to 12 of embryonic development. Failure to do so can result in a condition called **cleft palate**. The condition may also involve incomplete fusion of the horizontal

FIGURE 7.10 Maxillae.

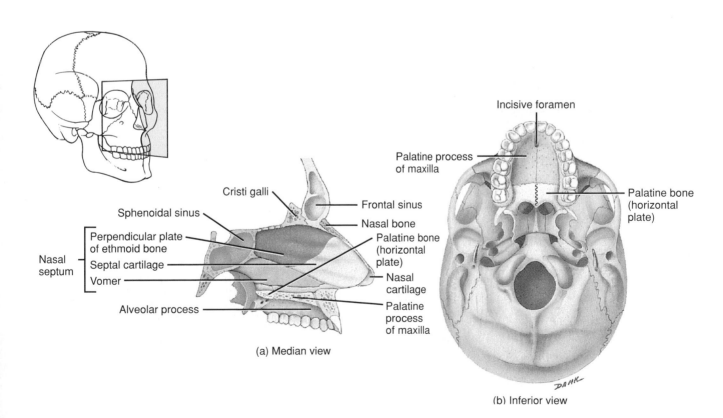

(a) Median view

Cristi galli
Sphenoidal sinus
Perpendicular plate of ethmoid bone
Septal cartilage
Vomer
Nasal septum
Alveolar process
Frontal sinus
Nasal bone
Palatine bone (horizontal plate)
Nasal cartilage
Palatine process of maxilla

(b) Inferior view

Incisive foramen
Palatine process of maxilla
Palatine bone (horizontal plate)

Question: **What bones form the hard palate?**

plates of the palatine bones (see Fig. 7.7). Another form of this condition, called **cleft lip,** involves a split in the upper lip. Cleft lip and cleft palate often occur together. Depending on the extent and position of the cleft, speech and swallowing may be affected. Facial and oral surgeons recommend closure of cleft lip during the first year of life, and surgical results are excellent. Repair of cleft palate is done between the first and second year of life, ideally before the child begins to talk. Orthodontic therapy may be needed to align the teeth. Here again, results are usually excellent.

Paranasal Sinuses

Although they are not cranial or facial bones, this is an appropriate point to discuss the **paranasal** (*para* = beside) **sinuses.** These are paired cavities in certain cranial and facial bones near the nasal cavity (Fig. 7.11). The paranasal sinuses are lined with mucous membranes that are continuous with the lining of the nasal cavity. Skull bones containing paranasal sinuses are the frontal, sphenoid, ethmoid,

and maxillae. (The paranasal sinuses were described in the discussion of each of these bones.) Besides producing mucus, the paranasal sinuses lighten the skull bones and serve as resonating chambers for sound as we speak or sing.

CLINICAL APPLICATION

SINUSITIS

Secretions produced by the mucous membranes of the paranasal sinuses drain into the nasal cavity. An inflammation of the membranes due to an allergic reaction or infection is called **sinusitis.** If the membranes swell enough to block drainage into the nasal cavity, fluid pressure builds up in the paranasal sinuses, and a sinus headache results.

Zygomatic Bones

The two **zygomatic bones (malars),** commonly called cheekbones, form the prominences of the cheeks and part of

FIGURE 7.11 Paranasal sinuses.

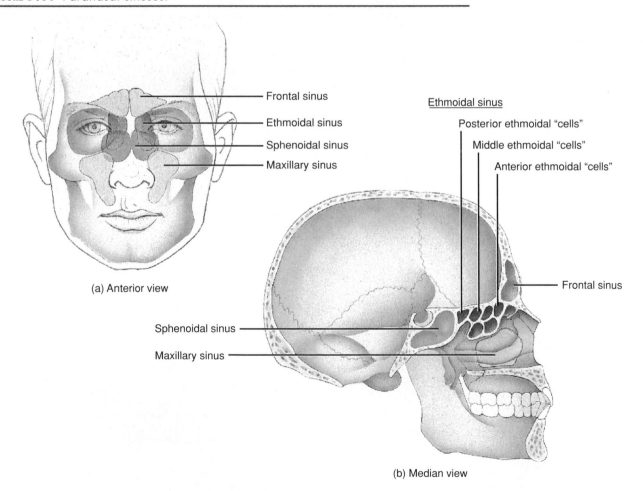

(a) Anterior view

Frontal sinus
Ethmoidal sinus
Sphenoidal sinus
Maxillary sinus

Ethmoidal sinus
Posterior ethmoidal "cells"
Middle ethmoidal "cells"
Anterior ethmoidal "cells"

Frontal sinus

Sphenoidal sinus
Maxillary sinus

(b) Median view

Question: What are the three main functions of the paranasal sinuses?

the lateral wall and floor of each orbit (see Fig. 7.13). They articulate with the frontal, maxilla, sphenoid, and temporal bones.

The **temporal process** of the zygomatic bone projects posteriorly and articulates with the zygomatic process of the temporal bone. These two processes form the **zygomatic arch** (see Fig. 7.7).

Mandible

The **mandible** (*mandere* = to chew), or lower jawbone, is the largest, strongest facial bone (Fig. 7.12). It is the only movable skull bone (other than the auditory ossicles).

In the lateral view, you can see that the mandible consists of a curved, horizontal portion, the **body**, and two perpendicular portions, the **rami** (singular is **ramus**). The **angle** of the mandible is the area where each ramus meets the body. Each ramus has a posterior **condylar** (KON-di-lar) **process** that articulates with the mandibular fossa and articular tubercle of the temporal bone to form the temporomandibular joint (TMJ). It also has an anterior **coronoid** (KOR-ō-noyd) **process** to which the temporalis muscle attaches. The depression between the coronoid and condylar processes is called the **mandibular notch**. The **alveolar process** is an arch containing the **alveoli** (sockets) for the mandibular (lower) teeth.

The **mental** (*mentum* = chin) **foramen** is approximately below the second premolar tooth. It is near this foramen that dentists sometimes reach the mental nerve when injecting anesthetics. Another foramen associated with the mandible is the **mandibular foramen** on the medial surface of the ramus, another site often used by dentists to inject anesthetics. The mandibular foramen is the beginning of the **mandibular canal,** which runs forward in the ramus deep to the roots of the teeth.

TEMPOROMANDIBULAR JOINT SYNDROME

One problem associated with the temporomandibular joint (TMJ) is **TMJ syndrome.** It is characterized by dull pain around the ear, tenderness of the jaw muscles, a clicking or popping noise when opening or closing the mouth, limited or abnormal opening of the mouth, headache, tooth sensitivity, and abnormal wearing of the teeth. TMJ syndrome can be caused by improperly aligned teeth, grinding or clenching the teeth, trauma to the jaw, or arthritis. Treatment may involve application of moist heat or ice, a soft diet, taking aspirin, muscle retraining, adjusting or reshaping the teeth, orthodontic treatment, or surgery.

Lacrimal Bones

The paired **lacrimal** (LAK-ri-mal; *lacrima* = tear) **bones** are thin and roughly resemble a fingernail in size and shape. They are the smallest bones of the face. These bones are posterior and lateral to the nasal bones and they form a part of the medial wall of each orbit. The lacrimal bones each contain a **lacrimal foramen (fossa)** through which a nasolacrimal (tear) duct passes (see Fig. 7.3). They can be seen in the anterior and lateral views of the skull in Figs. 7.2 and 7.3.

FIGURE 7.12 Mandible.

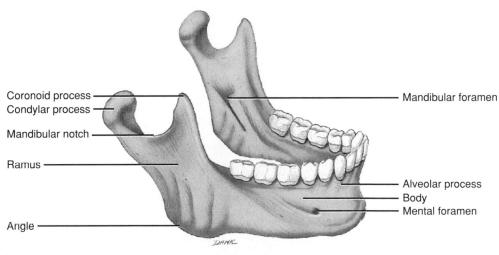

Coronoid process
Condylar process
Mandibular notch
Ramus
Angle

Mandibular foramen

Alveolar process
Body
Mental foramen

Right lateral view

Question: What is the distinctive feature of the mandible among all the skull bones?

Palatine Bones

The two **palatine** (PAL-a-tīn) **bones** are L-shaped. They form the posterior portion of the hard palate, part of the floor and lateral wall of the nasal cavity, and a small portion of the floors of the orbits. The posterior portion of the hard palate, which separates the nasal cavity from the oral cavity, is formed by the **horizontal plates** of the palatine bones. These can be seen in Fig. 7.7.

Inferior Nasal Conchae

Refer to the views of the skull in Figs. 7.2 and 7.9a. The two **inferior nasal conchae** (KONG-kē) or **turbinates** are scroll-like bones that form a part of the lateral wall of the nasal cavity and project into the nasal cavity inferior to the superior and middle nasal conchae of the ethmoid bone. They serve the same function as the superior and middle nasal conchae of the ethmoid bone; that is, they promote turbulent circulation and filtration of air before it passes into the lungs. The inferior nasal conchae are separate bones and not part of the ethmoid.

Vomer

The **vomer** (= plowshare) is a roughly triangular bone that forms the inferior and posterior part of the nasal septum. It is clearly seen in the anterior view of the skull in Fig. 7.2 and the inferior view in Fig. 7.7.

The inferior border of the vomer articulates with the septal cartilage of the nasal septum that divides the external nose into right and left sides. Its superior border articulates with the perpendicular plate of the ethmoid bone. The structures that form the **nasal septum,** or partition, are the perpendicular plate of the ethmoid, septal cartilage, vomer, and parts of the palatine bones and maxillae (Fig. 7.10a).

CLINICAL APPLICATION

DEVIATED NASAL SEPTUM

A **deviated nasal septum (DNS)** is one that is deflected laterally from the midline of the nose. The deviation usually occurs at the junction of bone with the septal cartilage. A DNS may occur as a result of a developmental abnormality or trauma. If the deviation is severe, it may entirely block the nasal passageway. Even a partial blockage may lead to infection. If inflammation occurs, it may cause nasal congestion, blockage of the paranasal sinus openings, chronic sinusitis, headache, and nosebleeds.

Orbits

Each **orbit** (eye socket) is a pyramid-shaped space that contains the eyeball and associated structures. It is formed by seven bones of the skull (Fig. 7.13) and has four walls that

FIGURE 7.13 Details of orbit.

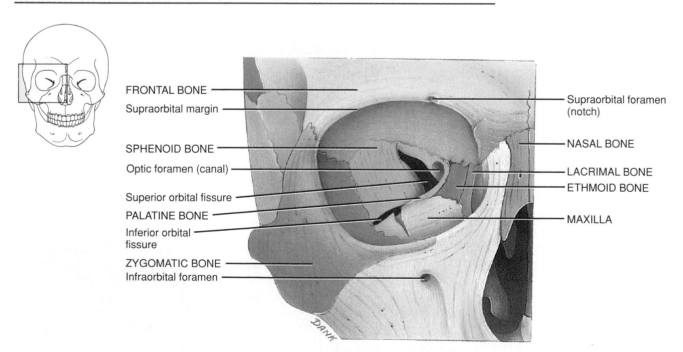

FRONTAL BONE

Supraorbital margin

SPHENOID BONE

Optic foramen (canal)

Superior orbital fissure

PALATINE BONE

Inferior orbital fissure

ZYGOMATIC BONE

Infraorbital foramen

Supraorbital foramen (notch)

NASAL BONE

LACRIMAL BONE

ETHMOID BONE

MAXILLA

Anterior view of right orbit

Question: Name the bones that form the orbit.

converge posteriorly to form an apex (back end). The roof of the orbit consists of parts of the frontal and sphenoid bones. Portions of the zygomatic and sphenoid bones form the lateral wall. The floor of the orbit is formed by parts of the maxilla, zygomatic, and palatine bones. Portions of the maxilla, lacrimal, ethmoid, and sphenoid bones form the medial wall.

The structures that pass through the various openings of the orbit are indicated in Exhibit 7.4.

Foramina

Many **foramina** (singular is **foramen**) associated with the skull were mentioned along with the descriptions of the cranial and facial bones they penetrate. As preparation for studying other systems of the body, especially the nervous and cardiovascular systems, these foramina, plus some additional ones, and the structures passing through them are listed in Exhibit 7.4. For your convenience and for future reference, the foramina are listed alphabetically.

HYOID BONE

The single **hyoid bone** (*hyoedes* = U-shaped) is a unique component of the axial skeleton because it does not articulate with any other bone (see Fig. 7.3). Rather, it is suspended from the styloid processes of the temporal bones by ligaments and muscles. The hyoid is located in the neck between the mandible and larynx. It supports the tongue and provides attachment for some of its muscles and for muscles of the neck and pharynx. The hyoid consists of a horizontal **body** and paired projections called the **lesser horns (cornua)** and the **greater horns** or **cornua** (Fig. 7.14). Muscles and ligaments attach to these paired projections.

The hyoid bone is often fractured during strangulation. As a result, it is carefully examined in an autopsy when strangulation is suspected.

VERTEBRAL COLUMN

Divisions

The **vertebral column (spine),** together with the sternum and ribs, forms the skeleton of the **trunk** of the body. The vertebral column makes up about two-fifths of the total height of the body and is composed of a series of bones called **vertebrae.** The length of the column is about 71 cm (28 in.) in an average adult male and about 61 cm (24 in.) in an average adult female. In effect, the vertebral column is a strong, flexible rod that bends anteriorly, posteriorly, and laterally and rotates. It encloses and protects the spinal cord, supports the head, and serves as a point of attachment for the ribs and the muscles of the back. Between vertebrae

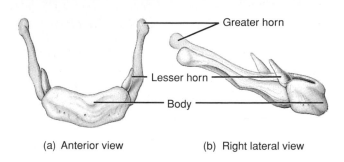

FIGURE 7.14 Hyoid bone.

Greater horn

Lesser horn

Body

(a) Anterior view   (b) Right lateral view

Question: How is the hyoid bone distinguished from all other bones of the axial skeleton?

are openings called **intervertebral foramina.** The nerves that connect the spinal cord to various parts of the body pass through these openings.

The adult vertebral column typically contains 26 vertebrae (Fig. 7.15a,b). These are distributed as follows: 7 **cervical vertebrae** (*cervix* = neck) in the neck region; 12 **thoracic vertebrae** (*thorax* = chest) posterior to the thoracic cavity; 5 **lumbar vertebrae** (*lumbus* = loin) supporting the lower back; 5 **sacral vertebrae** fused into one bone called the **sacrum** (SĀ-krum); and usually 4 **coccygeal** (kok-SIJ-ē-al) **vertebrae** fused into one or two bones called the **coccyx** (KOK-six). Before fusion of the sacral and coccygeal vertebrae, the total number of vertebrae is 33. Whereas the cervical, thoracic, and lumbar vertebrae are movable, the sacrum and coccyx are immovable.

Between adjacent vertebrae from the first vertebra (atlas) to the sacrum are **intervertebral discs.** Each disc has an outer fibrous ring consisting of fibrocartilage called the **annulus fibrosus** and an inner soft, pulpy, highly elastic structure called the **nucleus pulposus** (Fig. 7.15d; see Fig. 7.23). The discs form strong joints, permit various movements of the vertebral column, and absorb vertical shock. Under compression, they flatten, broaden, and bulge from their intervertebral spaces.

Normal Curves

When viewed from the side, the vertebral column shows four **normal curves** (Fig. 7.15b). The cervical and lumbar curves are anteriorly convex (bulging out) while the thoracic and sacral curves are anteriorly concave (cupping in). The curves of the vertebral column, like the curves in a long bone, are important because they increase its strength. The curves also help maintain balance in the upright position, absorb shocks from walking, and help protect the column from fracture.

In the fetus, there is only a single anteriorly concave

EXHIBIT 7.4

SUMMARY OF FORAMINA OF THE SKULL

Foramen	Location	Structures Passing Through
Carotid (Fig. 7.7)	Petrous portion of temporal bone.	Internal carotid artery.
Greater palatine (Fig. 7.7)	Posterior angle of hard palate.	Greater palatine nerve and greater palatine vessels.
Hypoglossal (Fig. 7.8)	Superior to base of occipital condyles.	Cranial nerve XII (hypoglossal) and branch of ascending pharyngeal artery.
Incisive (Fig. 7.10b)	Posterior to incisor teeth.	Branches of greater palatine vessels and nasopalatine nerve.
Inferior orbital (Fig. 7.13)	Between greater wing of sphenoid bone and maxilla.	Maxillary branch of cranial nerve V (trigeminal), zygomatic nerve, and infraorbital vessels.
Infraorbital (Fig. 7.2)	Inferior to orbit in maxilla.	Infraorbital nerve and artery.
Jugular (Fig. 7.7)	Posterior to carotid canal between petrous portion of temporal bone and occipital bone.	Internal jugular vein, cranial nerves IX (glossopharyngeal), X (vagus), and XI (accessory).
Lacerum (Fig. 7.8a)	Bounded anteriorly by sphenoid bone, posteriorly by petrous portion of temporal bone, and medially by the sphenoid bone and occipital bone.	Branch of ascending pharyngeal artery.
Lacrimal (Fig. 7.3)	Lacrimal bone.	Lacrimal (tear) duct.
Lesser palatine (Fig. 7.7)	Posterior to greater palatine foramen.	Lesser palatine nerves and artery.
Magnum (Fig. 7.7)	Occipital bone.	Medulla oblongata and its membranes (meninges), cranial nerve XI (accessory), and vertebral and spinal arteries.
Mandibular (Fig. 7.12)	Medial surface of ramus of mandible.	Inferior alveolar nerve and vessels.
Mastoid (Fig. 7.7)	Posterior border of mastoid process of temporal bone.	Emissary vein to transverse sinus and branch of occipital artery to dura mater.
Mental (Fig. 7.12)	Inferior to second premolar tooth in mandible.	Mental nerve and vessels.
Olfactory (Fig. 7.8a)	Cribriform plate of ethmoid.	Cranial nerve I (olfactory).
Optic (Fig. 7.8a)	Between upper and lower portions of small wing of sphenoid bone.	Cranial nerve II (optic) and ophthalmic artery.
Ovale (Fig. 7.8a)	Greater wing of sphenoid bone.	Mandibular branch of cranial nerve V (trigeminal).
Rotundum (Fig. 7.8a)	Junction of anterior and medial parts of sphenoid bone.	Maxillary branch of cranial nerve V (trigeminal).
Spinosum (Fig. 7.8a)	Posterior angle of sphenoid bone.	Middle meningeal vessels.
Stylomastoid (Fig. 7.7)	Between styloid and mastoid processes of temporal bone.	Cranial nerve VII (facial) and stylomastoid artery.
Superior orbital (Fig. 7.13)	Between greater and lesser wings of sphenoid bone.	Cranial nerves III (oculomotor), IV (trochlear), ophthalmic branch of V (trigeminal) and VI (abducens).
Supraorbital (Fig. 7.2)	Supraorbital margin of orbit.	Supraorbital nerve and artery.
Zygomaticofacial (Fig. 7.2)	Zygomatic bone.	Zygomaticofacial nerve and vessels.

FIGURE 7.15 Vertebral column. In (d), the relative size of the disc has been enlarged for emphasis. A "window" has been cut in the annulus fibrosus so that the nucleus pulposus can be seen.

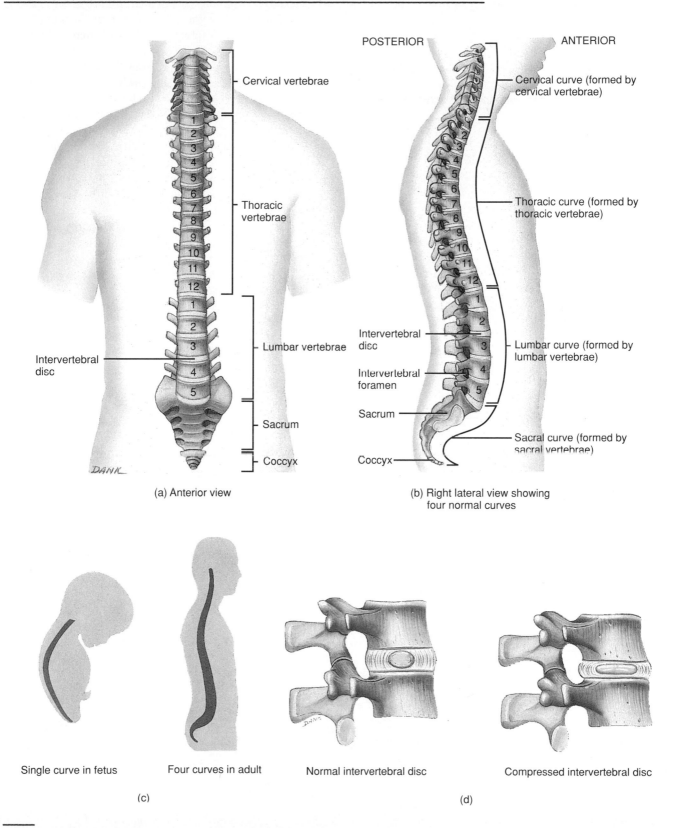

(a) Anterior view

(b) Right lateral view showing four normal curves

Single curve in fetus

Four curves in adult

(c)

Normal intervertebral disc

Compressed intervertebral disc

(d)

Question: Which curves are concave (relative to the anterior side of the body)?

curve (Fig. 7.15c). At approximately the third month after birth, when an infant begins to hold its head erect, the **cervical curve** develops. Later, when the child sits up, stands, and walks, the **lumbar curve** develops.

Typical Vertebra

Vertebrae in different regions of the spinal column vary in size, shape, and detail, but they are similar enough that we can discuss the parts and functions of a typical vertebra (Fig. 7.16).

1.   The **body (centrum)** is the thick, disc-shaped anterior portion that is the weight-bearing part of a vertebra. Its superior and inferior surfaces are roughened for the attachment of intervertebral discs. The anterior and lateral surfaces contain nutrient foramina for blood vessels.

2.   The **vertebral (neural) arch** extends posteriorly from the body of the vertebra. With the body of the vertebra, it surrounds the spinal cord. It is formed by two short, thick processes, the **pedicles** (PED-i-kuls), which project posteriorly from the body to unite with the laminae. The **laminae** (LAM-i-nē) are the flat parts that join to form the posterior portion of the vertebral arch. The space that lies between the vertebral arch and body contains the spinal cord. This space is known as the **vertebral foramen.** The vertebral foramina of all vertebrae together form the **vertebral (spinal) canal.** The pedicles exhibit superior and inferior notches. When they are stacked on top of one another, there is an opening between vertebrae on each side of the column. Each opening, called an **intervertebral foramen,** permits the passage of a single spinal nerve.

3.   Seven **processes** arise from the vertebral arch. At the point where a lamina and pedicle join, a **transverse process** extends laterally on each side. A single **spinous process (spine)** projects posteriorly and inferiorly from the junction of the laminae. These three processes serve as points of attachment for muscles. The remaining four processes form joints with other vertebrae. The two **superior articular processes** of a vertebra articulate with the vertebra immediately superior to them. The two **inferior articular processes** of a vertebra articulate with the vertebra inferior to them. The articulating surfaces of the articular processes are referred to as **facets.**

Cervical Region

When viewed from above, it can be seen that the bodies of **cervical vertebrae** are smaller than those of thoracic vertebrae (Fig. 7.17). The vertebral arches, however, are larger.

FIGURE 7.16 Typical vertebra. In (b), only one spinal nerve has been included and it has been extended beyond the intervertebral foramen for clarity.

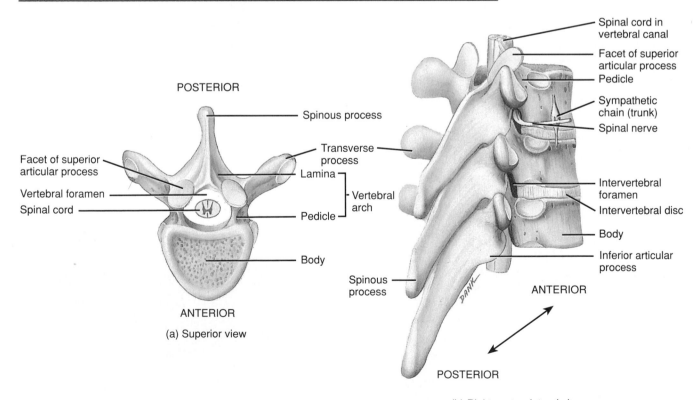

(a) Superior view

(b) Right posterolateral view

Question: What are the functions of the vertebral foramen and the intervertebral foramina?

FIGURE 7.17 Cervical vertebrae.

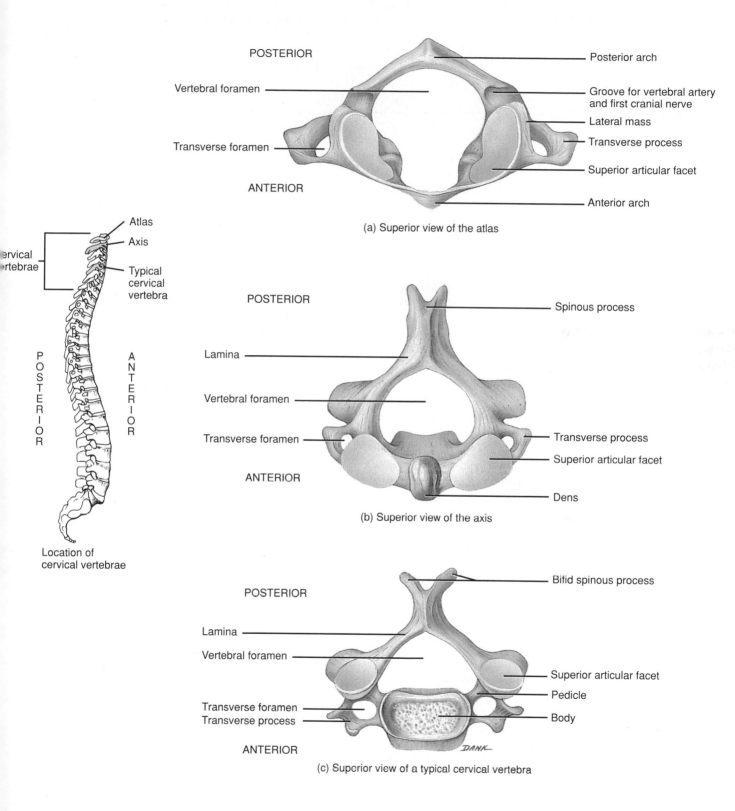

POSTERIOR

Vertebral foramen

Transverse foramen

ANTERIOR

Posterior arch

Groove for vertebral artery and first cranial nerve

Lateral mass

Transverse process

Superior articular facet

Anterior arch

(a) Superior view of the atlas

Atlas

Axis

Typical cervical vertebra

Cervical vertebrae

POSTERIOR

ANTERIOR

Location of cervical vertebrae

POSTERIOR

Lamina

Vertebral foramen

Transverse foramen

ANTERIOR

Spinous process

Transverse process

Superior articular facet

Dens

(b) Superior view of the axis

POSTERIOR

Lamina

Vertebral foramen

Transverse foramen
Transverse process

ANTERIOR

Bifid spinous process

Superior articular facet

Pedicle

Body

(c) Superior view of a typical cervical vertebra

DANK

Figure continues

FIGURE 7.17 (continued)

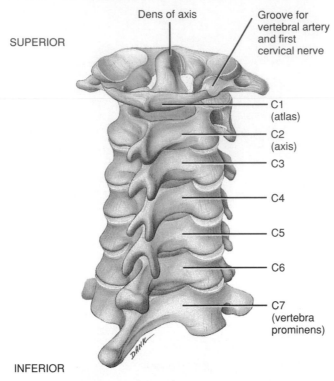

(d) Posterior view of articulated cervical vertebrae

Question: Which bones permit movement of the head to signify no?

The spinous processes of the second through sixth cervical vertebrae are often *bifid,* that is, with a cleft. All cervical vertebrae have three foramina: the vertebral foramen and two transverse foramina. The vertebral foramina of cervical vertebrae are the largest in the spinal column because they house the cervical enlargement of the spinal cord. Each cervical transverse process contains a **transverse foramen** through which the vertebral artery and its accompanying vein and nerve fibers pass.

The first two cervical vertebrae differ considerably from the others. The first cervical vertebra (C1), the **atlas,** is named for its support of the head. The atlas is a ring of bone with **anterior** and **posterior arches** and large **lateral masses.** It lacks a body and a spinous process. The superior surfaces of the lateral masses, called **superior articular facets,** are concave and articulate with the occipital condyles of the occipital bone. This articulation permits the movement seen when moving the head to signify yes. The inferior surfaces of the lateral masses, the **inferior articular facets,** articulate with the second cervical vertebra. The transverse processes and transverse foramina of the atlas are quite large.

The second cervical vertebra (C2), the **axis,** does have a body. A peglike process called the **dens** (*dens* = tooth) or **odontoid process** projects up through the anterior portion of the vertebral foramen of the atlas. The dens makes a

pivot on which the atlas and head rotate as in moving the head to signify no. This arrangement permits side-to-side rotation of the head. In various instances of trauma, the dens of the axis may be driven into the medulla oblongata of the brain. When whiplash injuries result in death, this type of injury is the usual cause.

The third through sixth cervical vertebrae (C3–C6) correspond to the structural pattern of the typical cervical vertebra previously described.

The seventh cervical vertebra (C7), called the **vertebra prominens,** is somewhat different. It is marked by a large, nonbifid spinous process that may be seen and felt at the base of the neck.

Thoracic Region

Thoracic vertebrae (T1–T12) are considerably larger and stronger than cervical vertebrae (Fig. 7.18). In addition, the spinous process on each vertebra is long, laterally flattened, and directed inferiorly. Thoracic vertebrae also have longer and heavier transverse processes than cervical vertebrae.

Except for the eleventh and twelfth thoracic vertebrae, the transverse processes have **facets** for articulating with the tubercles of the ribs. The bodies of thoracic vertebrae also have whole facets or **demifacets** (half-facets) for

FIGURE 7.18 Thoracic vertebrae.

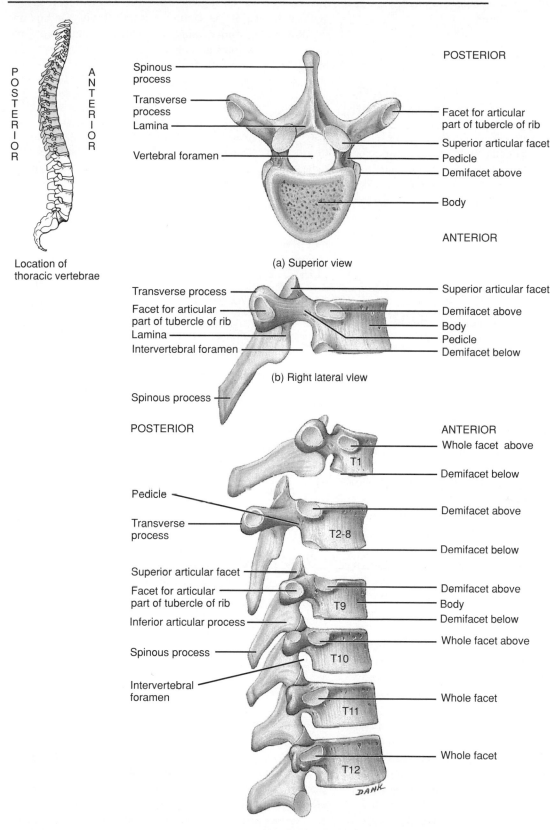

Location of thoracic vertebrae

(a) Superior view

(b) Right lateral view

(c) Right lateral view of articulated thoracic vertebrae

Question: What parts of a thoracic vertebra articulate with a rib?

articulation with the heads of the ribs. Movements of the thoracic region are limited by thin intervertebral discs and the attachment of ribs to the sternum.

Lumbar Region

The **lumbar vertebrae** (L1–L5) are the largest and strongest in the column because the body weight supported by the vertebrae increases toward the lower end of the backbone (Fig. 7.19). Their various projections are short and thick. The superior articular processes are directed medially instead of superiorly. The inferior articular processes are directed laterally instead of inferiorly. The spinous processes are quadrilateral in shape, thick, and broad and project nearly straight posteriorly. The spinous processes are well-adapted for the attachment of the large back muscles.

Sacrum

The **sacrum** (= sacred or holy bone) is a triangular bone formed by the union of five sacral vertebrae. These are indicated in Fig. 7.20 as S1–S5. Fusion begins between 16 and 18 years of age and is usually completed by the mid-twenties. The sacrum serves as a strong foundation for the pelvic girdle. It is positioned at the posterior portion of the pelvic cavity between the two hipbones.

The concave anterior side of the sacrum faces the pelvic cavity. It is smooth and contains four **transverse lines (ridges)** that mark the joining of the sacral vertebral bodies. At the ends of these lines are four pairs of **anterior (pelvic) sacral foramina.** The lateral portion of the superior surface contains a smooth surface called the **sacral ala** (wing), which is formed by the fused transverse processes of the first sacral vertebra (S1).

The convex, posterior surface of the sacrum contains a **median sacral crest,** the fused spinous processes of the upper sacral vertebrae; a **lateral sacral crest,** the transverse processes of the sacral vertebrae; and four pairs of **posterior (dorsal) sacral foramina.** These foramina communicate with the anterior sacral foramina through which nerves and blood vessels pass. The **sacral canal** is a continuation of the vertebral canal. The laminae of the fifth sacral vertebra, and sometimes the fourth, fail to meet. This leaves an inferior entrance to the vertebral canal called the **sacral hiatus** (hī-Ā-tus). On either side of the sacral hiatus are the **sacral cornua,** the inferior articular processes of the fifth sacral vertebra. They are connected by ligaments to the coccyx.

The superior border of the sacrum exhibits an anteriorly projecting border, the **sacral promontory** (PROM-on-tō′-rē). It is an obstetrical landmark for measurements of the pelvis. On both lateral surfaces, the sacrum has a large **auricular surface** for articulating with the ilium of each hipbone. Posterior to the auricular surface is a roughened surface, the **sacral tuberosity,** that contains depressions for the attachment of ligaments. The sacral tuberosity is another surface of the sacrum that unites with the hipbone to form the sacroiliac joint. The **superior articular processes** of the sacrum articulate with the fifth lumbar vertebra.

Coccyx

The **coccyx** is also triangular in shape and is formed by the fusion of the usually four coccygeal vertebrae. These are indicated in Fig. 7.20 as Co1–Co4. Fusion generally occurs between 20 and 30 years of age. On the lateral surfaces of the coccyx are a series of **transverse processes,** the first pair being the largest. The coccyx articulates superiorly with the sacrum.

CLINICAL APPLICATION

EPIDURAL ANESTHESIA

Anesthetic agents that act on the sacral and coccygeal nerves are sometimes injected through the sacral hiatus, a procedure called **epidural anesthesia** that is used most often in obstetrics. Since the sacral hiatus is between the sacral cornua, the cornua are important bony landmarks for locating the hiatus. Anesthetic agents may also be injected through the posterior (dorsal) sacral foramina.

THORAX

The term **thorax** refers to the entire chest. The skeletal portion of the thorax is a bony cage formed by the sternum, costal cartilages, ribs, and the bodies of the thoracic vertebrae (Fig. 7.21).

The thoracic cage is roughly cone-shaped, the narrow portion being superior and the broad portion inferior. It is flattened from front to back. The thoracic cage encloses and protects the organs in the thoracic cavity and upper abdominal cavity. It also provides support for the bones of the shoulder girdle and upper extremities.

Sternum

The **sternum,** or breastbone, is a flat, narrow bone measuring about 15 cm (6 in.) in length. It is located in the median line of the anterior thoracic wall and consists of three basic portions (Fig. 7.21): the **manubrium** (ma-NOO-brē-um), the superior portion; the **body,** the middle, largest portion; and the **xiphoid** (ZĪ-foyd) **process,** the inferior, smallest portion. The junction of the manubrium and body forms the **sternal angle.** The manubrium has a depression on its superior surface called the **suprasternal (jugular) notch.** On each side of the suprasternal notch are **clavicular notches** that articulate with the medial ends of the clavicles. The manubrium also articulates with the first and second ribs.

FIGURE 7.19 Lumbar vertebrae.

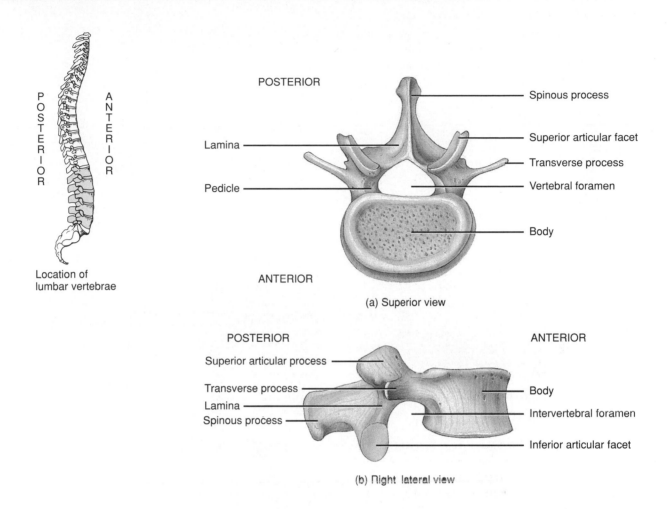

POSTERIOR   ANTERIOR

Location of
lumbar vertebrae

POSTERIOR

Lamina

Pedicle

ANTERIOR

Spinous process

Superior articular facet

Transverse process

Vertebral foramen

Body

(a) Superior view

POSTERIOR   ANTERIOR

Superior articular process

Transverse process

Lamina

Spinous process

Body

Intervertebral foramen

Inferior articular facet

(b) Right lateral view

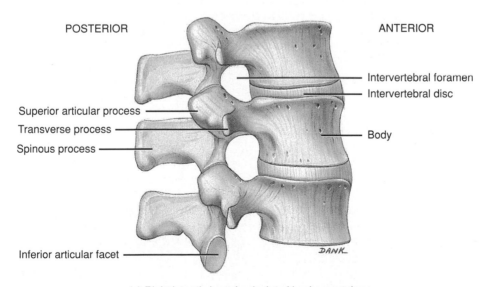

POSTERIOR   ANTERIOR

Intervertebral foramen

Intervertebral disc

Superior articular process

Transverse process

Spinous process

Body

Inferior articular facet

DANK

(c) Right lateral view of articulated lumbar vertebrae

Question: Why are the lumbar vertebrae the largest and strongest in the vertebral column?

FIGURE 7.20 Sacrum and coccyx.

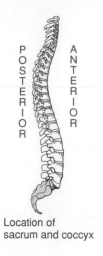

Location of
sacrum and coccyx

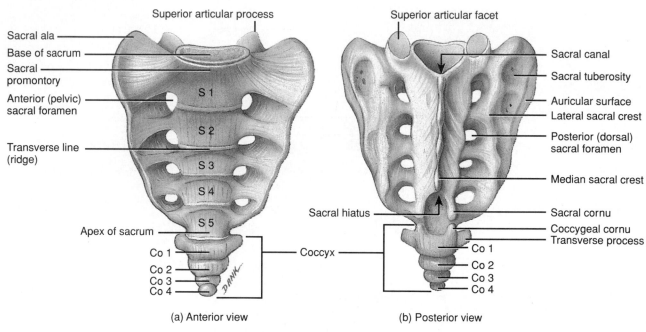

(a) Anterior view

(b) Posterior view

Question: How many foramina pierce the sacrum and what is their function?

FIGURE 7.21 Skeleton of the thorax.

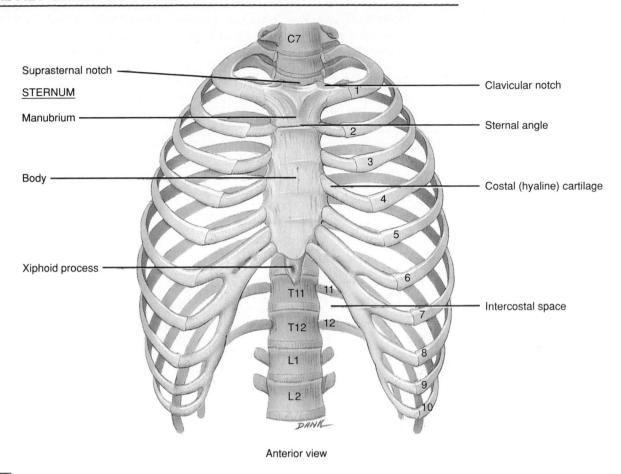

Anterior view

Question: Which ribs are true ribs, false ribs, and floating ribs?

The body of the sternum articulates directly or indirectly with the second through tenth ribs. The xiphoid process has no ribs attached to it but provides attachment for some abdominal muscles. The xiphoid process consists of hyaline cartilage during infancy and childhood and does not ossify completely until about age 40. If the hands of a rescuer are incorrectly positioned during cardiopulmonary resuscitation (CPR), there is danger of fracturing the xiphoid process, and driving it into internal organs.

CLINICAL APPLICATION

STERNAL PUNCTURE

Since the sternum houses red bone marrow throughout life and because it is readily accessible and has thin compact bone, it is a common site for withdrawal of marrow for biopsy. Under a local anesthetic, a wide-bore needle is introduced into the marrow cavity of the sternum for aspiration of a sample of red bone marrow. This procedure is called a **sternal puncture.**

Ribs

Twelve pairs of **ribs** make up the sides of the thoracic cavity (Fig. 7.21). The ribs increase in length from the first through seventh, then decrease in length to the twelfth rib. Each articulates posteriorly with its corresponding thoracic vertebra.

The first through seventh pairs of ribs have a direct anterior attachment to the sternum by a strip of hyaline cartilage called **costal cartilage** (*costa* = rib). These ribs are called **true (vertebrosternal) ribs.** The remaining five pairs of ribs are referred to as **false ribs** because their costal cartilages either attach indirectly to the sternum or do not attach to the sternum at all. The cartilages of the eighth, ninth, and tenth pairs of ribs attach to each other and then to the cartilages of the seventh pair of ribs. These false ribs are called **vertebrochondral ribs.** The eleventh and twelfth pairs of ribs are false ribs designated as **floating (vertebral) ribs** because their anterior ends do not attach to the sternum at all. They attach only posteriorly to the thoracic vertebrae.

We will examine the parts of a typical (third through ninth) rib (Fig. 7.22). The **head** is a projection at the posterior end of the rib. It consists of one or two **facets** that articulate with facets on the bodies of adjacent thoracic vertebrae. The **neck** is a constricted portion just lateral to the head. A knoblike structure on the posterior surface where the neck joins the body is called a **tubercle** (TOO-ber-kul). It consists of a **nonarticular part** that affords attachment to the ligament of the tubercle and an **articular part** that articulates with the facet of a transverse process of the inferior of the two vertebrae to which the head of the rib is connected. The **body (shaft)** is the main part of the rib. A short distance beyond the tubercle, there is an abrupt change in the curvature of the shaft. This point is called the **costal**

angle. The inner surface of the rib has a **costal groove** that protects blood vessels and a small nerve.

The posterior portion of the rib is connected to a thoracic vertebra by its head and articular part of a tubercle. The facet of the head fits into a facet on the body of a vertebra, and the articular part of the tubercle articulates with the facet of the transverse process of the vertebra.

Spaces between ribs, called **intercostal spaces,** are occupied by intercostal muscles, blood vessels, and nerves (see Fig. 7.21). Surgical access to the lungs or other structures in the thoracic cavity is commonly undertaken through an intercostal space. Special rib retractors are used to create a wide separation between ribs. The costal cartilages are sufficiently elastic to permit considerable bending.

FIGURE 7.22 Typical rib.

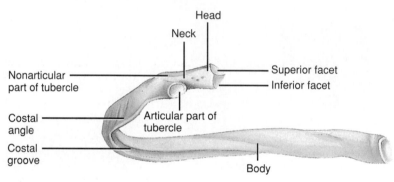

(a) Left rib viewed from behind

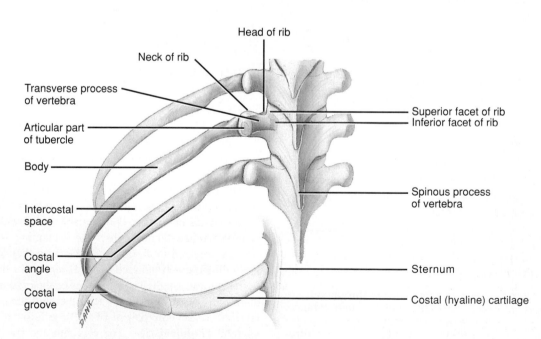

(b) Left rib articulated with a thoracic vertebra and the sternum

---

Question: How does a rib articulate with a thoracic vertebra?

DISORDERS: HOMEOSTATIC IMBALANCES

HERNIATED (SLIPPED) DISC

In their function as shock absorbers, intervertebral discs are constantly being compressed. If the anterior and posterior ligaments of the discs become injured or weakened, the pressure developed in the nucleus pulposus may be great enough to rupture the surrounding fibrocartilage (annulus fibrosus). If this occurs, the nucleus pulposus may herniate (protrude) posteriorly or into one of the adjacent vertebral bodies. This condition is called a **herniated (slipped) disc.**

Most often the nucleus pulposus slips posteriorly toward the spinal cord and spinal nerves (Fig. 7.23). This movement exerts pressure on the spinal nerves, causing considerable, sometimes very acute, pain. If the roots of the sciatic nerve, which passes from the spinal cord to the foot, are pressured, the pain radiates down the back of the thigh, through the calf, and occasionally into the foot. If pressure is exerted on the spinal cord itself, some of its neurons may be destroyed.

ABNORMAL CURVES

Various conditions may exaggerate the normal curves of the vertebral column or the column may acquire a lateral bend, resulting in **abnormal curves** of the spine.

Scoliosis (skō′-lē-Ō-sis; *scolio* = bent) is a lateral bending of the vertebral column, usually in the thoracic region. This is the most common of the abnormal curves. It may result from congenitally (present at birth) malformed vertebrae, chronic sciatica, paralysis of muscles on one side of the backbone, poor posture, or one leg being shorter than the other.

Kyphosis (kī-FŌ-sis; *kypho* = hunchback) is an exaggeration of the thoracic curve of the vertebral column. In tuberculosis of the spine, vertebral bodies may partially collapse, causing an acute angular bending of the vertebral column. In the elderly, degeneration of the intervertebral discs leads to kyphosis. Kyphosis may also be caused by rickets and poor posture. It ia also common in females with advanced osteoporosis. The term *round-shouldered* is an expression for mild kyphosis.

Lordosis (lor-DŌ-sis; *lordo* = swayback) is an exaggeration of the lumbar curve of the vertebral column. It may result from increased weight of the abdomen as in pregnancy or extreme obesity, poor posture, rickets, and tuberculosis of the spine.

SPINA BIFIDA

Spina bifida (SPĪ-na BIF-i-da) is a congenital defect of the vertebral column in which laminae fail to unite at the midline. In serious cases, protrusion of the membranes (meninges) around the spinal cord or the spinal cord itself produces perilous problems, such as partial or complete paralysis, partial or complete loss of urinary bladder control, and the absence of reflexes. Spina bifida may be diagnosed prenatally by a test of the mother's blood, sonography, or amniocentesis (withdrawal of amniotic fluid for analysis).

FIGURE 7.23 Herniated (slipped) disc.

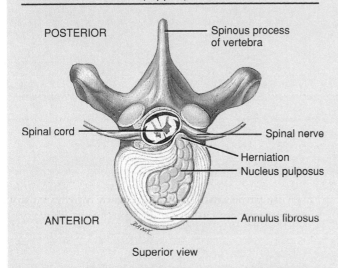

POSTERIOR

Spinous process of vertebra

Spinal cord

Spinal nerve

Herniation

Nucleus pulposus

ANTERIOR

Annulus fibrosus

Superior view

Study Outline

Types of Bones (p. 166)

1. On the basis of shape, bones are classified as long, short, flat, or irregular.
2. Sutural (Wormian) bones are found between the sutures of certain cranial bones. Sesamoid bones develop in tendons or ligaments.

Surface Markings (p. 166)

1. Surface markings are structural features visible on the surfaces of bones.
2. Each marking is structured for a specific function—joint formation, muscle attachment, or passage of nerves and blood vessels.

3. Terms that describe markings include fissure, foramen, meatus, fossa, process, condyle, head, facet, tuberosity, crest, and spine.

Divisions of the Skeletal System (p. 166)

1. The axial skeleton consists of bones arranged along the longitudinal axis. The parts of the axial skeleton are the skull, hyoid bone, vertebral column, sternum, and ribs.
2. The appendicular skeleton consists of the bones of the girdles and the upper and lower extremities. The parts of the appendicular skeleton are the pectoral (shoulder) girdles, bones of the upper extremities, pelvic (hip) girdle, and bones of the lower extremities.

Skull (p. 169)

1. The skull consists of the cranium and the face. It is composed of 22 bones.
2. Sutures are immovable joints between bones of the skull. Examples are coronal, sagittal, lambdoid, and squamous sutures.
3. Fontanels are dense, connective tissue membrane-filled spaces between the cranial bones of fetuses and infants. The major fontanels are the anterior, posterior, anterolaterals, and posterolaterals.
4. The 8 cranial bones include the frontal, parietal (2), temporal (2), occipital, sphenoid, and ethmoid.
5. The 14 facial bones are the nasal (2), maxillae (2), zygomatic (2), mandible, lacrimal (2), palatine (2), inferior nasal conchae (2), and vomer.
6. Paranasal sinuses are cavities in bones of the skull that communicate with the nasal cavity. They are lined by mucous membranes. The cranial bones containing the paranasal sinuses are the frontal, sphenoid, ethmoid, and maxillae.
7. The orbits (eye sockets) are formed by seven bones of the skull.
8. The foramina of the skull bones provide passages for nerves and blood vessels.

Hyoid Bone (p. 183)

1. The hyoid bone is a U-shaped bone that does not articulate with any other bone.

2. It supports the tongue and provides attachment for some of its muscles as well as some neck muscles and muscles of the pharynx.

Vertebral Column (p. 183)

1. The vertebral column, sternum, and ribs constitute the skeleton of the trunk.
2. The bones of the adult vertebral column are the cervical vertebrae (7), thoracic vertebrae (12), lumbar vertebrae (5), sacrum (5, fused), the coccyx (4, fused).
3. The vertebral column contains normal primary curves (thoracic and sacral) and normal secondary curves (cervical and lumbar). These curves give strength, support, and balance.
4. The vertebrae are similar in structure, each consisting of a body, vertebral arch, and seven processes. Vertebrae in the different regions of the column vary in size, shape, and detail.

Thorax (p. 190)

1. The thoracic skeleton consists of the sternum, ribs and costal cartilages, and thoracic vertebrae.
2. The thoracic cage protects vital organs in the chest area and upper abdomen.

Review Questions

1. Describe the importance of the skeletal system to the body. (p. 166)
2. What are the four principal types of bones? Give an example of each. Distinguish between a sutural (Wormian) and a sesamoid bone. (p. 166)
3. What are surface markings? Describe and give an example of each. (p. 166)
4. Distinguish between the axial and appendicular skeletons. What subdivisions and bones are contained in each? (p. 166)
5. What are the bones that compose the skull? The cranium? The face? (p. 169)
6. Define a suture. What are the four prominent sutures of the skull? Where are they located? (p. 169)
7. What is a fontanel? Describe the location of the six fairly constant fontanels. (p. 169)
8. What is a paranasal sinus? What cranial bones contain paranasal sinuses? (p. 180)
9. Describe the components of each orbit. (p. 182)
10. What bones form the skeleton of the trunk? Distinguish between the number of nonfused vertebrae found in the adult vertebral column and that of a child. (p. 183)
11. What are the normal curves in the vertebral column? How are primary and secondary curves differentiated? What are the functions of the curves? (p. 183)
12. What are the principal distinguishing characteristics of the bones of the various regions of the vertebral column? (p. 186)
13. What bones form the skeleton of the thorax? What are the functions of the thoracic skeleton? (p. 190)
14. How are ribs classified on the basis of their attachment to the sternum? (p. 193)
15. Define the following: black eye (p. 174), cleft palate and cleft lip (p. 179), sinusitis (p. 180), temporomandibular joint (TMJ) syndrome (p. 181), deviated nasal septum (DNS) (p. 182), epidural anesthesia (p. 190), and sternal puncture (p. 193).

Answers to Questions with Figures

7.1 Axial skeleton: skull, vertebral column. Appendicular skeleton: clavicle, shoulder girdle, humerus, pelvic girdle, femur.
7.2 Frontal, parietal, sphenoid, ethmoid, and temporal.
7.3 Squamous suture separates the parietal and temporal bones. Lambdoid suture separates the parietal and occipital bones. Coronal suture separates frontal and parietal bones.
7.4 The paranasal sinuses, cavities in certain skull bones (frontal, sphenoid, ethmoid, and maxillae), lighten the weight of the skull and resonate sounds produced by the vocal cords.
7.5 Sagittal and lambdoid.
7.6 Anterolateral (sphenoid).
7.7 The temporal bone; the carotid foramen.
7.8 Crista galli, frontal, parietal, temporal, occipital, temporal, parietal, frontal, crista galli.
7.9 Perpendicular plate.

7.10 Maxillae and palatine bones.

7.11 They produce mucus, lighten skull bones, and serve as resonating chambers for vocalization.

7.12 The mandible is the only movable skull bone, other than the auditory ossicles.

7.13 Frontal, sphenoid, zygomatic, maxilla, lacrimal, ethmoid, and palatine.

7.14 It does not articulate with any other bone.

7.15 The thoracic and sacral curves are concave.

7.16 The vertebral foramen encloses the spinal cord while the intervertebral foramina provide spaces for spinal nerves to exit the vertebral column.

7.17 Atlas and axis.

7.18 Facets and demifacets.

7.19 The body weight supported by vertebrae increases toward the lower end of the backbone.

7.20 There are four pairs of foramina, for a total of eight. Each anterior sacral foramen joins a posterior sacral foramen at the intervertebral foramen. Nerves and blood vessels pass through these tunnels in the bone.

7.21 True ribs (pairs 1–7), false ribs (pairs 8–12), and floating ribs (pairs 11 and 12).

7.22 The facet of the head fits into a facet on the body of a vertebra and the articular part of the tubercle articulates with the facet of the transverse process of a vertebra.

Chapter 8

THE SKELETAL SYSTEM: THE APPENDICULAR SKELETON

Chapter Contents at a Glance

Student Objectives

1. Identify the bones of the pectoral (shoulder) girdle and their major markings.

2. Identify the upper extremity, its component bones, and their markings.

3. Identify the components of the pelvic (hip) girdle and their principal markings.

4. Identify the lower extremity, its component bones, and their markings.

5. Define the structural features and importance of the arches of the foot.

6. Compare the principal structural differences between female and male skeletons, especially those that pertain to the pelvis.

his chapter discusses the bones of the appendicular skeleton, that is, the bones of the pectoral (shoulder) and pelvic (hip) girdles and upper and lower extremities. The differences between female and male skeletons are also described.

PECTORAL (SHOULDER) GIRDLE

The **pectoral** (PEK-tō-ral) or **shoulder girdles** attach the bones of the upper extremities to the axial skeleton (Fig. 8.1). Each of the two pectoral girdles consists of two bones:

FIGURE 8.1 Right pectoral (shoulder) girdle and upper extremity.

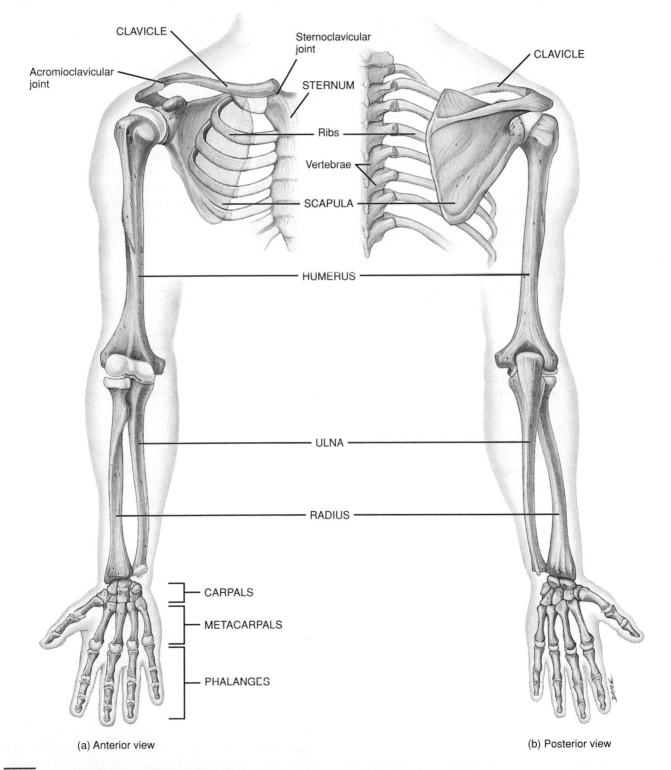

(a) Anterior view

(b) Posterior view

Question: What is the function of the pectoral girdle?

a clavicle and a scapula. The clavicle is the anterior component and articulates with the sternum at the sternoclavicular joint. The posterior component, the scapula, which is held in position by complex muscle attachments, articulates with the clavicle and humerus. The pectoral girdles have no articulation with the vertebral column. Thus, the shoulder joints are not very stable, but they are freely movable and allow movement in many directions.

Clavicle

Each **clavicle** (KLAV-i-kul), or **collarbone,** is a long, slender S-shaped bone with two curves, one convex and one concave (Fig. 8.2). The medial one-third of the clavicle is convex anteriorly, whereas the lateral one-third is concave anteriorly. Since the junction of two curves is the weakest point of a structure, this is the most frequent site of clavicular fractures. The clavicles lie horizontally in the superior and anterior part of the thorax superior to the first rib.

The medial end of the clavicle, the **sternal extremity,** is rounded and articulates with the sternum to form the **sternoclavicular joint.** The broad, flat, lateral end, the **acromial** (a-KRŌ-mē-al) **extremity,** articulates with the acromion of the scapula. This joint is called the **acromioclavicular joint.** (Refer to Fig. 8.1 for a view of these articulations.) The **conoid tubercle** on the inferior surface of the lateral end of the bone serves as a point of attachment for a ligament. The **costal tuberosity** on the inferior surface of the medial end also serves as a point of attachment for a ligament.

CLINICAL APPLICATION

FRACTURED CLAVICLE

Because of its position, the clavicle transmits forces from the upper extremity to the trunk. If such forces are excessive, as in falling on one's outstretched arm, a **fractured clavicle** may result. It is the most frequently broken bone in the body.

Scapula

Each **scapula** (SCAP-yoo-la), or **shoulder blade,** is a large, triangular, flat bone situated in the posterior part of the thorax between the levels of the second and seventh ribs (Fig. 8.3). The medial borders of the scapulae (plural) lie about 5 cm (2 inches) from the vertebral column.

A sharp ridge, the **spine,** runs diagonally across the posterior surface of the flattened, triangular **body** of the scapula. The end of the spine projects as a flattened, expanded process called the **acromion** (a-KRŌ-mē-on), easily felt as the high point of the shoulder. This process articulates with the clavicle. Inferior to the acromion is a depression called the **glenoid cavity (fossa).** This cavity articulates with the head of the humerus to form the shoulder joint.

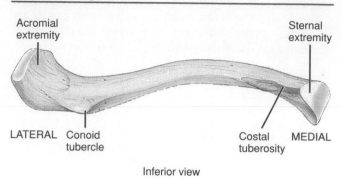

FIGURE 8.2 Right clavicle.

Acromial extremity — Sternal extremity

LATERAL   Conoid tubercle   Costal tuberosity   MEDIAL

Inferior view

Question: Which part of the clavicle is its weakest point?

The thin edge of the body near the vertebral column is the **medial (vertebral) border.** The thick edge closer to the arm is the **lateral (axillary) border.** The medial and lateral borders join at the **inferior angle.** The superior edge of the scapular body, called the **superior border,** joins the vertebral border at the **superior angle.** The **scapular notch** is a prominent indentation along the superior border through which the suprascapular nerve passes.

At the lateral end of the superior border is a projection of the anterior surface called the **coracoid** (KOR-a-koyd) **process** to which muscles attach. Above and below the spine are two fossae: the **supraspinous** (soo′-pra-SPĪ-nus) **fossa** and the **infraspinous fossa,** respectively. Both serve as surfaces of attachment for shoulder muscles. On the ventral (costal) surface is a slightly hollowed-out area called the **subscapular fossa,** also a surface of attachment for shoulder muscles.

UPPER EXTREMITY

The **upper extremities** consist of 60 bones. Figure 8.1 shows the skeleton of the right upper extremity. Each upper extremity includes a humerus in the arm, ulna and radius in the forearm, and carpals (wrist bones), metacarpals (palm bones), and phalanges (fingers) in the hand.

Humerus

The **humerus** (HYOO-mer-us), or arm bone, is the longest and largest bone of the upper extremity (Fig. 8.4). It articulates proximally with the scapula and distally at the elbow with both the ulna and radius.

The proximal end of the humerus features a **head** that articulates with the glenoid cavity of the scapula. It also has an **anatomical neck,** the former site of the epiphyseal plate, which is an oblique groove just distal to the head. The **greater tubercle** is a lateral projection distal to the neck. It is the most laterally palpable bony landmark of the shoulder region. The **lesser tubercle** is an anterior projection.

FIGURE 8.3 Right scapula.

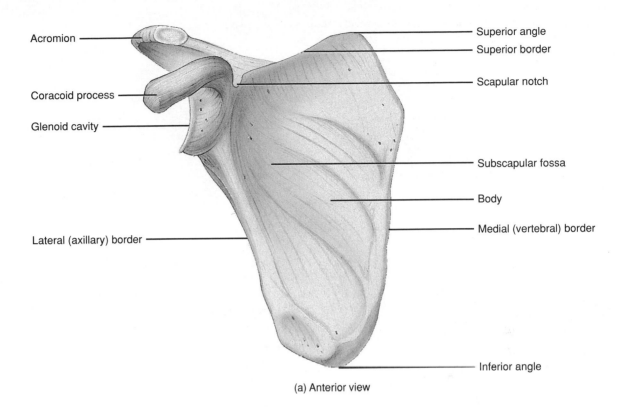

Acromion

Coracoid process

Glenoid cavity

Lateral (axillary) border

Superior angle

Superior border

Scapular notch

Subscapular fossa

Body

Medial (vertebral) border

Inferior angle

(a) Anterior view

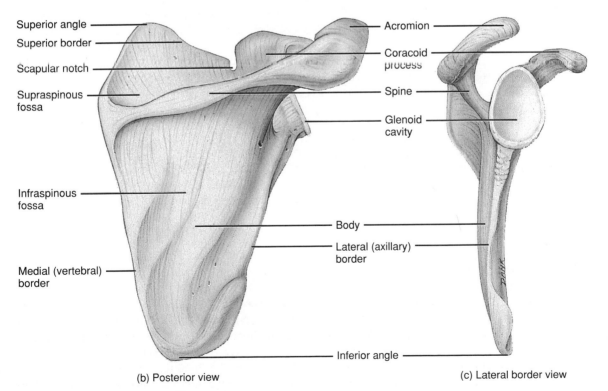

Superior angle

Superior border

Scapular notch

Supraspinous fossa

Infraspinous fossa

Medial (vertebral) border

Acromion

Coracoid process

Spine

Glenoid cavity

Body

Lateral (axillary) border

Inferior angle

(b) Posterior view

(c) Lateral border view

Question: Which part of the scapula forms the high point of the shoulder?

FIGURE 8.4 Right humerus in relation to the scapula, ulna, and radius.

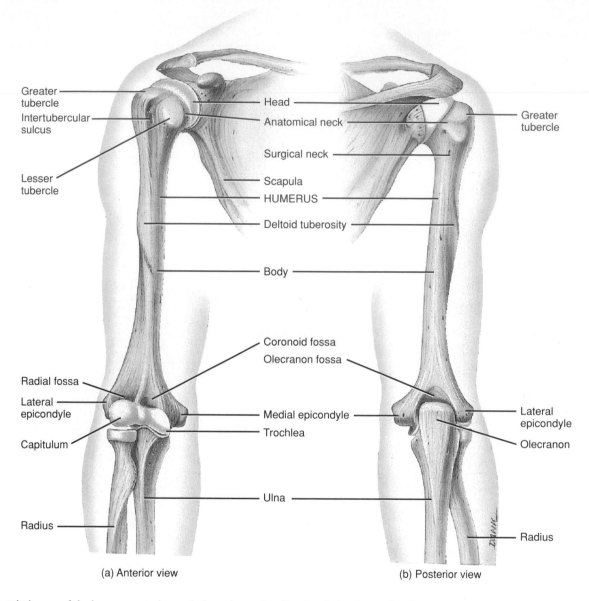

Greater
tubercle

Intertubercular
sulcus

Lesser
tubercle

Radial fossa

Lateral
epicondyle

Capitulum

Radius

Head

Anatomical neck

Surgical neck

Scapula

HUMERUS

Deltoid tuberosity

Body

Coronoid fossa

Olecranon fossa

Medial epicondyle

Trochlea

Ulna

Greater
tubercle

Lateral
epicondyle

Olecranon

Radius

(a) Anterior view

(b) Posterior view

Question: Which parts of the humerus articulate with the radius at the elbow? With the ulna at the elbow?

Between these tubercles runs an **intertubercular sulcus (bicipital groove).** The **surgical neck** is a constricted portion just distal to the tubercles and is so named because fractures often occur here.

The **body (shaft)** of the humerus is cylindrical at its proximal end. It gradually becomes triangular and is flattened and broad at its distal end. Laterally, at the middle portion of the shaft, there is a roughened, V-shaped area called the **deltoid tuberosity.** This area serves as a point of attachment for the deltoid muscle.

The following parts are found at the distal end of the humerus. The **capitulum** (ka-PIT-yoo-lum) is a rounded knob that articulates with the head of the radius. The **radial**

fossa is an anterior depression that receives the head of the radius when the forearm is flexed (bent). The **trochlea** (TRŌK-lē-a), located medial to the capitulum, is a pulley-like surface that articulates with the ulna. The **coronoid fossa** is an anterior depression that receives part of the ulna when the forearm is flexed. The **olecranon** (ō-LEK-ra-non) **fossa** is a posterior depression that receives the olecranon of the ulna when the forearm is extended (straightened). The **medial epicondyle** and **lateral epicondyle** are rough projections on either side of the distal end to which most muscles of the forearm are attached. The ulnar nerve lies on the posterior surface of the medial epicondyle and may easily be rolled between the finger and the medial epicondyle.

Ulna and Radius

The **ulna** is located on the medial aspect (little finger side) of the forearm and is longer than the radius (Fig. 8.5). At the proximal end of the ulna is the **olecranon (olecranon process),** which forms the prominence of the elbow. The **coronoid process** is an anterior projection that, together with the olecranon, receives the trochlea of the humerus. Just below the coronoid process is the **ulna tuberosity**. The **trochlear (semilunar) notch** is a large curved area between the olecranon and the coronoid process. The trochlea of the humerus fits into this notch. The **radial notch** is a depression located laterally and inferiorly to the trochlear notch. It receives the head of the radius. The distal

FIGURE 8.5 Right ulna and radius in relation to the humerus and carpals.

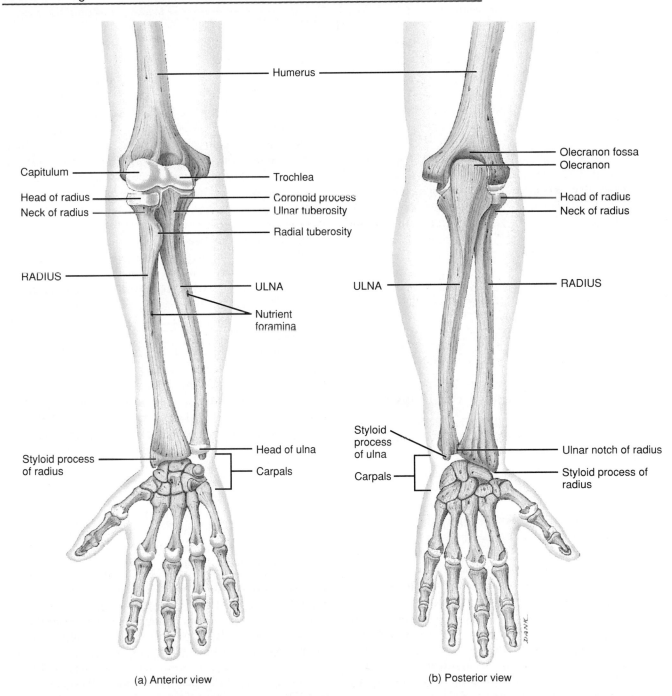

(a) Anterior view

(b) Posterior view

Figure continues

FIGURE 8.5 (continued)

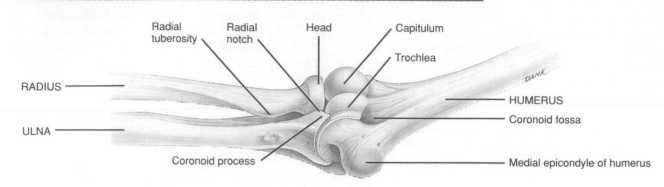

(c) Medial view in relation to humerus

Question: Which part of the ulna forms the elbow?

end of the ulna consists of a **head** that is separated from the wrist by a fibrocartilage disc. A **styloid process** is on the posterior side of the distal end.

The **radius** is located on the lateral aspect (thumb side) of the forearm. The proximal end of the radius has a disc-shaped **head** that articulates with the capitulum of the humerus and radial notch of the ulna. It also has a raised, roughened area on the medial side called the **radial tuberosity.** This is a point of attachment for the biceps brachii muscle. Below the head is the constricted **neck.** The shaft of the radius widens distally to form a concave inferior surface that articulates with two bones of the wrist called the lunate and scaphoid bones. Also at the distal end is a **styloid process** on the lateral side and a medial, concave **ulnar notch** for articulation with the distal end of the ulna.

Carpals, Metacarpals, and Phalanges

The skeleton of the hand has three regions: (1) proximal carpus, (2) intermediate metacarpus, and (3) distal phalanges.

The **carpus** (wrist) consists of eight small bones, the **carpals,** joined to one another by ligaments (Fig. 8.6). The carpals are arranged in two transverse rows, with four bones in each row, and they are named for their shapes. In the anatomical position, the carpals in the proximal row, from the lateral to medial position, are the **scaphoid** (resembles a boat), **lunate** (resembles a crescent moon in its anteroposterior aspect), **triquetrum** (has three articular surfaces), and **pisiform** (pea-shaped). In about 70% of carpal fractures, only the scaphoid is broken. The carpals in the distal row, from the lateral to medial position, are the **trapezium** (four-sided), **trapezoid** (also four-sided), **capitate** (its rounded projection, the head, articulates with the lunate), and **hamate** (= hooked; named for a large hook-shaped projection on its anterior surface).

Together, the concavity formed by the pisiform and hamate (on the ulnar side) and the scaphoid and trapezium

(on the radial side) plus the flexor retinaculum (deep fascia) constitute a space called the **carpal tunnel.** Through it pass the long flexor tendons of the digits and thumb and the median nerve. Narrowing of the carpal tunnel gives rise to a condition called carpal tunnel syndrome (see page 394).

The five bones of the **metacarpus** (*meta* = after), called **metacarpals,** constitute the palm of the hand. Each metacarpal bone consists of a proximal **base,** an intermediate **shaft,** and a distal **head.** The metacarpal bones are numbered I to V, starting with the one proximal to the thumb. The bases articulate with the distal row of carpal bones. The heads articulate with the proximal phalanges of the fingers. The heads of the metacarpals are commonly called the "knuckles" and are readily visible when the fist is clenched.

The **phalanges** (fa-LAN-jēz), or bones of the fingers, number 14 in each hand. A single bone of the finger (or toe) is referred to as a **phalanx** (FĀ-lanks). Each phalanx consists of a proximal **base,** an intermediate **shaft,** and a distal **head.** There are two phalanges in the thumb (**pollex**) and three phalanges in each of the other fingers. In order from the thumb, they are commonly referred to as the index finger, middle finger, ring finger, and little finger. The first row of phalanges, the **proximal row,** articulates with the metacarpal bones and second row of phalanges. The second row of phalanges, the **middle row,** articulates with the proximal row and the third row. The third row of phalanges, the **distal row,** articulates with the middle row. The thumb has no middle phalanx.

PELVIC (HIP) GIRDLE

The **pelvic (hip) girdle** consists of the two **hipbones** or **coxal** (KOK-sal; *coxa* = hip) **bones** (see Fig. 8.15). The pelvic girdle provides a strong and stable support for the lower extremities on which the weight of the body is carried. The hipbones are united to each other anteriorly at a joint called the **pubic symphysis** (PYOO-bik SIM-fi-sis). They unite posteriorly to the sacrum.

FIGURE 8.6 Right wrist and hand in relation to the ulna and radius.

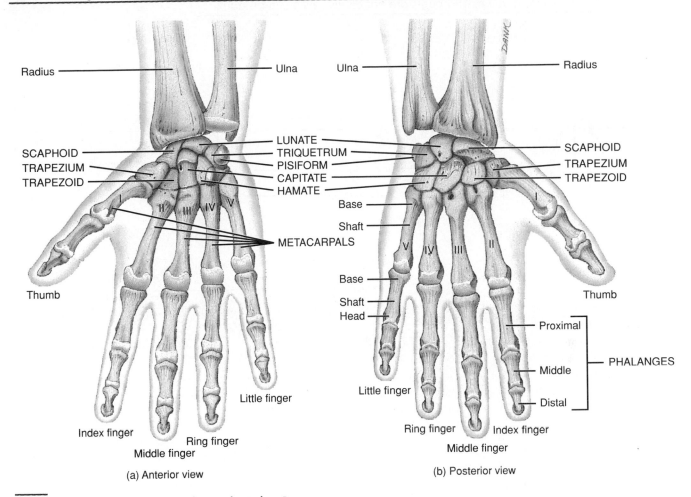

(a) Anterior view

(b) Posterior view

Question: Which is the most frequently fractured wrist bone?

Each of the two hipbones of a newborn consists of three components: a superior **ilium,** an inferior and anterior **pubis,** and an inferior and posterior **ischium** (is'-KĒ-um) (Fig. 8.7c). Eventually, the three separate bones fuse into one. The area of fusion is a deep, lateral fossa called the acetabulum, which serves as the socket for the head of the femur. Although the adult hipbones are both single bones, it is common to discuss the bones as if they still consisted of three portions.

The ilium is the largest of the three subdivisions of the hipbone. Its superior border, the **iliac crest,** ends anteriorly in the **anterior superior iliac spine.** Below it is the **anterior inferior iliac spine.** Posteriorly, the iliac crest ends in the **posterior superior iliac spine.** Below it is the **posterior inferior iliac spine.** The spines serve as points of attachment for muscles. Below the posterior inferior iliac spine is the **greater sciatic** (sī-AT-ik) **notch.** The medial surface of the ilium contains the **iliac fossa.** It is a concavity where the iliacus muscle attaches. Posterior to this fossa is the **auricular surface,** which articulates with the sacrum to form the **sacroiliac joint** (see Fig. 8.15).

The ischium is the inferior, posterior portion of the hipbone. It contains a prominent **ischial spine,** a **lesser sciatic notch** below the spine, and an **ischial tuberosity.** This prominent tuberosity may hurt someone's thigh when you sit on their lap. The rest of the ischium, the **ramus,** joins with the pubis, and together they surround the **obturator** (OB-too-rā'-ter) **foramen,** the largest foramen in the skeleton.

The pubis is the anterior and inferior part of the hipbone. It consists of a **superior ramus,** an **inferior ramus,** and a **body** between the rami that contributes to the formation of the pubic symphysis. The anterior border of the body is known as the **pubic crest** and at its lateral end is a projection, the **pubic tubercle.**

As noted earlier, the pubic symphysis is the joint between the two hipbones (see Fig. 8.15). It consists of a pad of fibrocartilage. The **acetabulum** (as'-e-TAB-yoo-lum) is the fossa formed by the ilium, ischium, and pubis. It is the socket for the head of the femur.

Together with the sacrum and coccyx, the two hipbones of the pelvic girdle form the basin-like structure called the **pelvis** (Fig. 8.8). The superior and inferior portions of the

FIGURE 8.7 Right hipbone. In the three divisions of the hipbone shown in (c), the lines of fusion of the ilium, ischium, and pubis are not always visible in an adult.

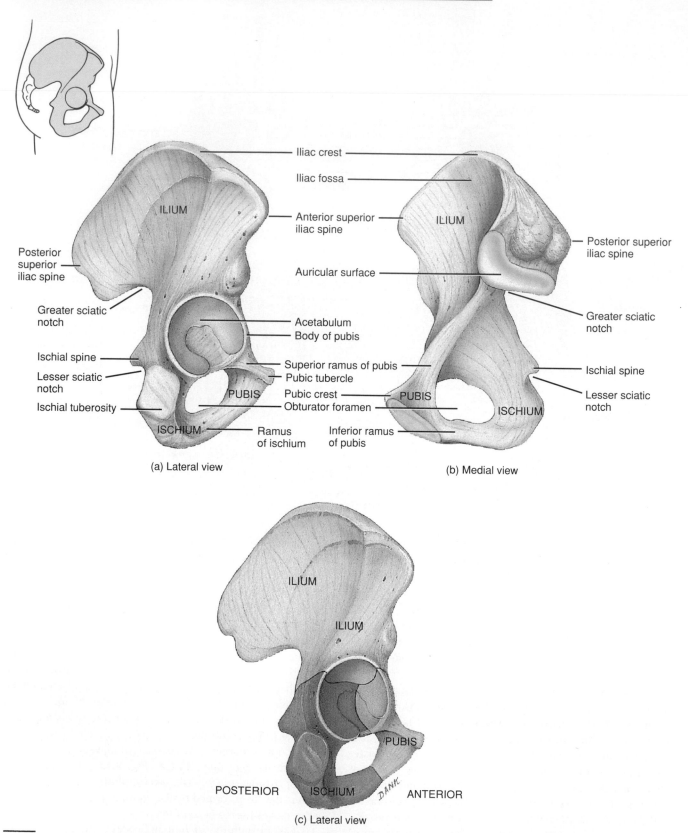

(a) Lateral view

(b) Medial view

(c) Lateral view

Question: Which part of the hipbone articulates with the femur? With the sacrum?

FIGURE 8.8 Female pelvis.

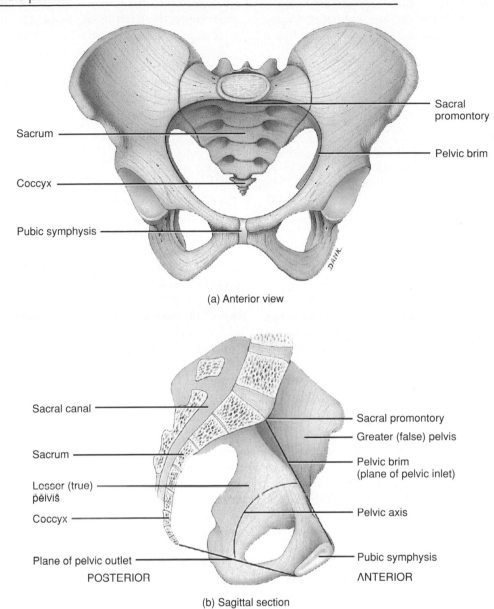

(a) Anterior view

Sacrum

Coccyx

Pubic symphysis

Sacral promontory

Pelvic brim

Sacral canal

Sacrum

Lesser (true) pelvis

Coccyx

Plane of pelvic outlet

POSTERIOR

Sacral promontory

Greater (false) pelvis

Pelvic brim (plane of pelvic inlet)

Pelvic axis

Pubic symphysis

ANTERIOR

(b) Sagittal section

Question: Which part of the pelvis surrounds the pelvic viscera in the pelvic cavity?

pelvis are separated from each other by a plane that connects the sacral promontory on the posterior side, pubic symphysis on the anterior side, and several points on the lateral side. The circumference of this oblique plane is called the **pelvic brim**.

The portion of the pelvis above the pelvic brim is called the **greater (false) pelvis.** It is bordered by the lumbar vertebrae (posteriorly), the upper portions of the hipbones (laterally), and the abdominal wall (anteriorly). The greater pelvis is actually part of the abdomen and does not contain any pelvic organs, except for the urinary bladder when it is

full and the uterus during pregnancy. The greater pelvis is not an obstacle to childbirth.

The portion of the pelvis below the pelvic brim is called the **lesser (true) pelvis.** It is bounded by the sacrum and coccyx (posteriorly), inferior portions of the ilium and ischium (laterally) and pubic bones (anteriorly). The lesser pelvis surrounds the pelvic cavity (see Fig. 1.9). The superior opening of the lesser pelvis, actually the pelvic brim, is called the **pelvic inlet;** the inferior opening is called the **pelvic outlet** (Fig. 8.8b). Whereas the pelvic inlet in females is wide and oval-shaped, in males it is narrower and heart-shaped. The **pelvic axis** is an imaginary curved line passing through the lesser pelvis at right angles to the center of the planes of the pelvic inlet and outlet. During childbirth it is the course taken by the baby's head as it descends through the pelvis.

CLINICAL APPLICATION

PELVIMETRY

Pelvimetry is the measurement of the size of the inlet and outlet of the birth canal. Measurement of the pelvic cavity in pregnant females is important because the fetus must pass through the narrower opening of the lesser pelvis at birth.

LOWER EXTREMITY

The **lower extremities** are composed of 60 bones (Fig. 8.9). Each lower extremity includes the femur in the thigh, patella (kneecap), fibula and tibia in the leg, and tarsals (ankle bones), metatarsals, and phalanges (toes) in the foot.

Femur

The **femur,** or thighbone, is the longest, heaviest, and strongest bone in the body (Fig. 8.10). Its proximal end articulates with the hipbone. Its distal end articulates with the tibia. The **body (shaft)** of the femur angles medially as it approaches the femur of the opposite thigh. As a result, the knee joints are brought nearer at the midline. The degree of convergence is greater in females because the female pelvis is broader.

The proximal end of the femur consists of a rounded **head** that articulates with the acetabulum of the hipbone. The **neck** of the femur is a constricted region distal to the head. A fairly common fracture in the elderly occurs at the neck of the femur, frequently as a result of osteoporosis. Apparently, the neck becomes so weak that it fails to support the weight of body. The **greater trochanter** (trō-KAN-ter) and **lesser trochanter** are projections that serve as points of attachment for some of the thigh and buttock muscles. The greater trochanter is the prominence felt and seen anterior to the hollow on the side of the hip. The lesser trochanter is inferior and medial to the greater trochanter. Between the trochanters on the anterior surface is a narrow **intertrochanteric line.** Between the trochanters on the posterior surface is an **intertrochanteric crest.**

The **body (shaft)** of the femur contains a rough vertical ridge on its posterior surface called the **linea aspera.** This ridge serves for the attachment of several thigh muscles.

The distal end of the femur is expanded and includes the **medial condyle** and **lateral condyle.** These articulate with the tibia. Superior to the condyles are the **medial epi-**

condyle and **lateral epicondyle.** A depressed area between the condyles on the posterior surface is called the **intercondylar** (in'-ter-KON-di-lar) **fossa.** The **patellar surface** is located between the condyles on the anterior surface.

Pathologic changes in the angle of the neck of the femur result in abnormal posture of the lower limbs. A decreased angle produces "knock-knee" condition (**genu valgum**). An abnormally large angle produces "bowleg" condition (**genu varum**). Either condition places an abnormal strain on the knee joints.

Patella

The **patella,** or kneecap, is a small, triangular bone located anterior to the knee joint (Fig. 8.11). It is a sesamoid bone that develops in the tendon of the quadriceps femoris muscle. The broad superior end of the patella is called the **base.** The pointed inferior end is the **apex.** The posterior surface contains two **articular facets,** one for the medial condyle and the other for the lateral condyle of the femur. The function of the patella is to increase the leverage of the tendon of the quadriceps femoris muscle (see Fig. 11.20) and to maintain the position of the tendon when the knee is bent (flexed).

Tibia and Fibula

The **tibia,** or shinbone, is the larger, medial bone of the leg (Fig. 8.12). It bears the weight of the leg. The tibia articulates at its proximal end with the femur and fibula, and at its distal end with the fibula of the leg and talus bone of the ankle.

The proximal end of the tibia is expanded into a **lateral condyle** and a **medial condyle.** These articulate with the condyles of the femur. The inferior surface of the lateral condyle articulates with the head of the fibula. The slightly concave condyles are separated by an upward projection called the **intercondylar eminence.** The **tibial tuberosity** on the anterior surface is a point of attachment for the patellar ligament.

The medial surface of the distal end of the tibia forms the **medial malleolus** (mal-LĒ-ō-lus). This structure articulates with the talus bone of the ankle and forms the prominence that can be felt on the medial surface of your ankle. The **fibular notch** articulates with the distal end of the fibula.

The **fibula** is parallel and lateral to the tibia. It is considerably smaller than the tibia. The **head** of the fibula, the proximal end, articulates with the inferior surface of the lateral condyle of the tibia below the level of the knee joint. The distal end has a projection called the **lateral malleolus** that articulates with the talus bone of the ankle. This forms the prominence on the lateral surface of the ankle. As noted, the fibula also articulates with the tibia at the fibular notch. A fracture of the lower end of the fibula with injury

FIGURE 8.9 Right pelvic (hip) girdle and lower extremity.

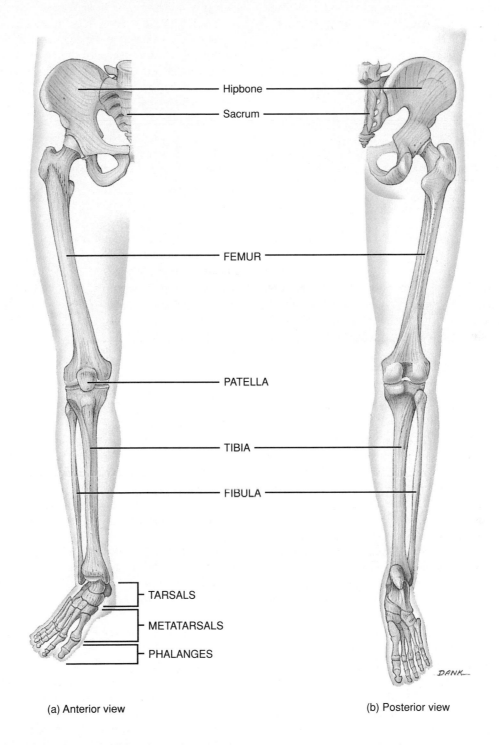

Hipbone

Sacrum

FEMUR

PATELLA

TIBIA

FIBULA

TARSALS

METATARSALS

PHALANGES

(a) Anterior view

(b) Posterior view

Question: What is the function of the pelvic girdle?

FIGURE 8.10 Right femur in relation to the hipbone, patella, tibia, and fibula.

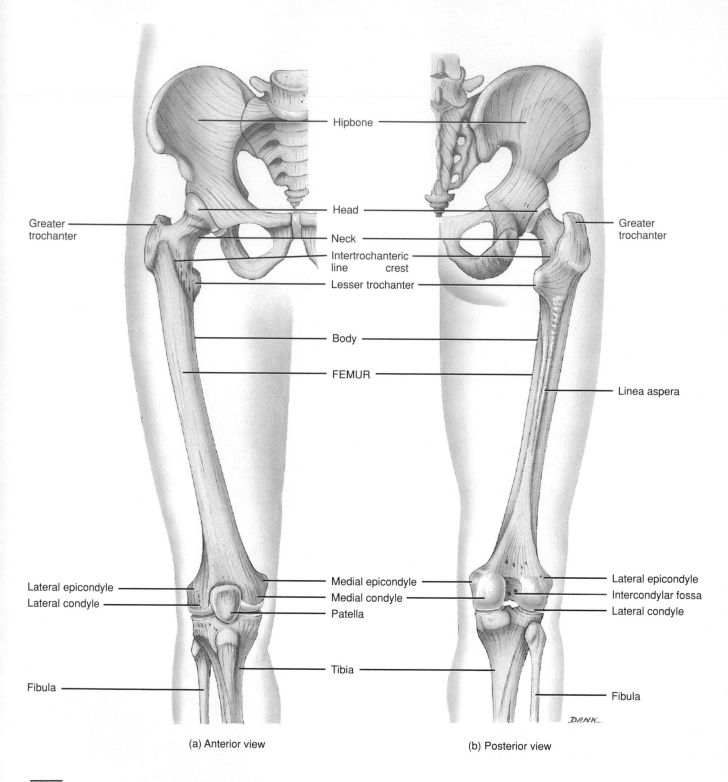

Hipbone

Head

Neck

Intertrochanteric
line   crest

Lesser trochanter

Body

FEMUR

Greater
trochanter

Greater
trochanter

Linea aspera

Lateral epicondyle

Lateral condyle

Medial epicondyle

Medial condyle

Patella

Lateral epicondyle

Intercondylar fossa

Lateral condyle

Tibia

Fibula

Fibula

DANK

(a) Anterior view

(b) Posterior view

Question: Changes in what part of the femur are related to "knock-knee" and "bowleg"?

FIGURE 8.11 Right patella.

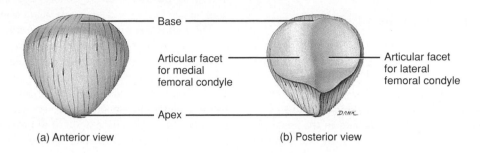

Base

Articular facet
for medial
femoral condyle

Articular facet
for lateral
femoral condyle

Apex

(a) Anterior view

(b) Posterior view

Question: Since the patella develops in the tendon of the quadriceps femoris muscle, it is classified as which type of bone?

FIGURE 8.12 Right tibia and fibula in relation to the femur, patella, and talus.

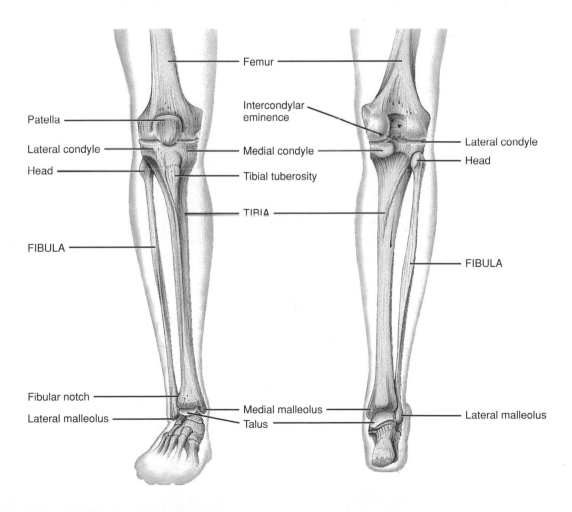

Femur

Intercondylar
eminence

Patella

Lateral condyle

Medial condyle

Lateral condyle

Head

Head

Tibial tuberosity

TIBIA

FIBULA

FIBULA

Fibular notch

Lateral malleolus

Medial malleolus

Talus

Lateral malleolus

Question: Which leg bone bears most of the weight of the leg?

to the tibial articulation is called a **Pott's fracture** (see Fig. 6.6d).

Tarsals, Metatarsals, and Phalanges

The skeleton of the foot has three regions: (1) the proximal tarsus, (2) the intermediate metatarsus, and (3) the distal phalanges.

The **tarsus** is a collective term for the seven **tarsal bones** of the ankle (Fig. 8.13). They include the **talus** (TĀ-lus) and **calcaneus** (kal-KĀ-nē-us), located in the posterior part of the foot. The anterior tarsal bones are the **cuboid, navicular,** and three **cuneiform** (*cuneiform* = wedge-shaped) **bones** called the **first (medial), second (intermediate),** and **third (lateral) cuneiforms.** The talus, the uppermost tarsal bone, is the only bone of the foot that articulates with the fibula and tibia, and it has no muscle attachments. It is surrounded on one side by the medial malleolus of the tibia and on the other side by the lateral malleolus of the fibula. During walking, the talus initially

bears the entire weight of the body. About half the weight is then transmitted to the calcaneus. The remainder is transmitted to the other tarsal bones. The calcaneus, or heel bone, is the largest and strongest tarsal bone.

The **metatarsus** consists of five **metatarsal bones** numbered I to V from the medial to lateral position. Like the metacarpals of the palm of the hand, each metatarsal consists of a proximal **base,** an intermediate **shaft,** and a distal **head.** The metatarsals articulate proximally with the first, second, and third cuneiform bones and with the cuboid. Distally, they articulate with the proximal row of phalanges. The first metatarsal is thicker than the others because it bears more weight.

The **phalanges** of the foot resemble those of the hand both in number and arrangement. Each also consists of a proximal **base,** an intermediate **shaft,** and a distal **head.** The great or big toe (**hallux**), has two large, heavy phalanges called proximal and distal phalanges. The other four toes each have three phalanges—proximal, middle, and distal.

FIGURE 8.13 Right foot.

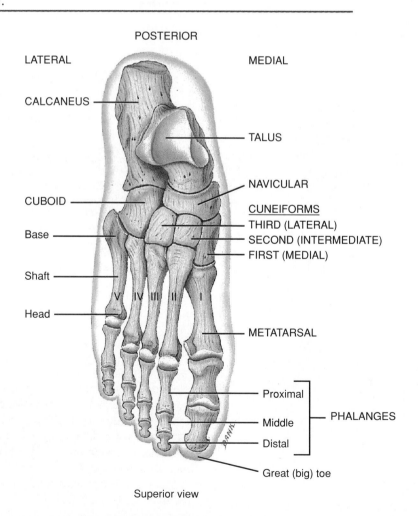

Superior view

Question: Which is the only tarsal bone to articulate with the tibia and fibula?

FIGURE 8.14 Arches of the right foot.

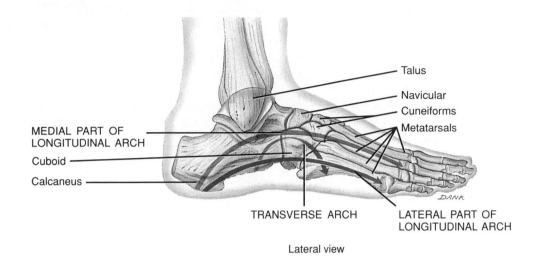

Lateral view

Question: Weakening of the ligaments and tendons of which arch results in flatfoot?

Arches of the Foot

The bones of the foot are arranged in two **arches** (Fig. 8.14). These arches enable the foot to support the weight of the body, provide an ideal distribution of body weight over the hard and soft tissues of the foot, and provide leverage while walking. The arches are not rigid. They yield as weight is applied and spring back when the weight is lifted, thus helping to absorb shocks. Usually, the arches are fully developed by age 12 or 13.

The **longitudinal arch** has two parts. Both consist of tarsal and metatarsal bones arranged to form an arch from the anterior to the posterior part of the foot. The **medial** (inner) **part** of the longitudinal arch originates at the calcaneus. It rises to the talus and descends through the navicular, the three cuneiforms, and the heads of the three medial metatarsals. The **lateral** (outer) **part** of the longitudinal arch also begins at the calcaneus. It rises at the cuboid and descends to the heads of the two lateral metatarsals.

The **transverse arch** is formed by the navicular, three cuneiforms, and the bases of the five metatarsals.

CLINICAL APPLICATION

FLATFOOT, CLAWFOOT, AND BUNIONS

The bones composing the arches are held in position by ligaments and tendons. If these ligaments and tendons are weakened, the height of the medial longitudinal arch may decrease or "fall." The result is **flatfoot.** Causes include excessive weight, postural abnormalities, weakened supporting tissues, and genetic predisposition. A custom-designed arch support (orthotic) often is prescribed to treat flatfoot.

Clawfoot is a condition in which the medial longitudinal arch is abnormally elevated. It is frequently caused by muscle imbalance, such as may result from poliomyelitis.

A **bunion (hallux valgus;** *valgus* = bent outward) is a deformity of the great toe. Although the condition may be inherited, it is typically caused by wearing tightly fitting shoes and is characterized by lateral deviation of the proximal phalanx of the great toe and medial displacement of metatarsal I. Arthritis of the first metatarsophalangeal joint may also be a predisposing factor. The condition produces inflammation of bursae (fluid-filled sacs at the joint), bone spurs, and calluses.

FEMALE AND MALE SKELETONS

The bones of the male are generally larger and heavier than those of the female. The articular ends are thicker in relation to the shafts. In addition, since certain muscles of the male are larger than those of the female, the points of attachment—tuberosities, lines, ridges—are larger in the male skeleton.

Many significant structural differences between female and male skeletons occur in the pelvis. Most are structural adaptations for pregnancy and childbirth. Typical differences are illustrated in Fig. 8.15 and listed in Exhibit 8.1.

FIGURE 8.15 Comparison of female and male pelvises.

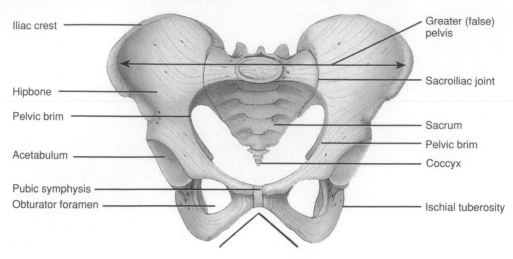

Iliac crest

Greater (false) pelvis

Sacroiliac joint

Hipbone

Pelvic brim

Sacrum

Acetabulum

Pelvic brim

Coccyx

Pubic symphysis

Obturator foramen

Ischial tuberosity

Pubic arch (greater than 90°)

(a) Anterior view of female pelvis

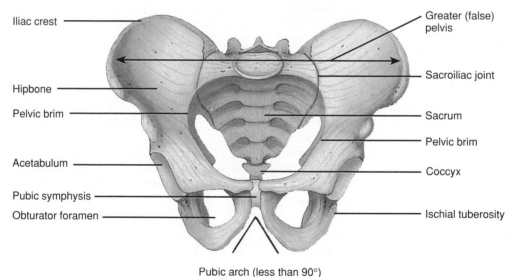

Iliac crest

Greater (false) pelvis

Sacroiliac joint

Hipbone

Pelvic brim

Sacrum

Acetabulum

Pelvic brim

Coccyx

Pubic symphysis

Obturator foramen

Ischial tuberosity

Pubic arch (less than 90°)

(b) Anterior view of male pelvis

Question: Most structural differences between the female and male pelvis are related to what functions?

EXHIBIT 8.1		
COMPARISON OF TYPICAL FEMALE AND MALE PELVIS		
Point of Comparison	Female	Male
Greater (false) pelvis (see Fig. 8.8b)	Shallow.	Deep.
Pelvic brim (pelvic inlet)	Larger and more oval.	Heart-shaped.
Pubic arch (arch formed by pubic rami)	Greater than a 90° angle.	Less than a 90° angle.

Point of Comparison	Female	Male
Pubic symphysis	Relatively shallow.	Relatively deep.
Ilium	Less vertical.	More vertical.
Iliac fossa (see Fig. 8.7b)	Shallow.	Deep.
Iliac crest	Less curved.	More curved.
Acetabulum	Small.	Large.
Obturator foramen	Oval.	Round.

Study Outline

Pectoral (Shoulder) Girdle (p. 199)

1. Each pectoral (shoulder) girdle consists of a clavicle and scapula.
2. Each attaches an upper extremity to the trunk.

Upper Extremity (p. 200)

1. There are 60 bones in the upper extremities.
2. The bones of each upper extremity include the humerus, ulna, radius, carpals, metacarpals, and phalanges.

Pelvic (Hip) Girdle (p. 204)

1. The pelvic (hip) girdle consists of two hipbones.
2. It attaches the lower extremities to the trunk at the sacrum.
3. Each hipbone consists of three fused components—ilium, pubis, and ischium.

Lower Extremity (p. 208)

1. There are 60 bones in the lower extremities.
2. The bones of each lower extremity include the femur, tibia, patella, fibula, tarsals, metatarsals, and phalanges.
3. The bones of the foot are arranged in two arches, the longitudinal arch and the transverse arch, to provide support and leverage.

Female and Male Skeletons (p. 213)

1. Male bones are generally larger and heavier than female bones and have more prominent markings for muscle attachment.
2. The female pelvis is adapted for pregnancy and childbirth. Differences in pelvic structure are listed in Exhibit 8.1.

Review Questions

1. What is the pectoral (shoulder) girdle? Why is it important? (p. 199)
2. What are the bones of the upper extremity? (p. 200)
3. What is the pelvic (hip) girdle? Why is it important? (p. 204)
4. What are the bones of the lower extremity? What is a Pott's fracture? (pp. 208, 212)
5. What is pelvimetry? What is the clinical importance of pelvimetry? (p. 208)
6. In what ways do the upper extremity and lower extremity differ structurally? (p. 208)
7. Describe the structure of the longitudinal and transverse arches of the foot. What is the function of an arch? (p. 213)
8. What are the principal structural differences between typical female and male skeletons? Use Exhibit 8.1 as a guide in formulating your response. (p. 214)

Answers to Questions with Figures

8.1 It attaches the upper extremity to the axial skeleton.
8.2 The junction of the two curves.
8.3 Acromion.
8.4 Radius: capitulum and radial fossa; ulna: trochlea, coronoid fossa, and olecranon fossa.
8.5 Olecranon.
8.6 Scaphoid.
8.7 Femur: acetabulum; sacrum: auricular surface.

8.8 Lesser (true) pelvis.
8.9 It attaches the lower extremity to the axial skeleton.
8.10 The angle of the neck.
8.11 Sesamoid.
8.12 Tibia.
8.13 Talus.
8.14 Medial longitudinal arch.
8.15 Pregnancy and childbirth.

Chapter 9

ARTICULATIONS

Chapter Contents at a Glance

Student Objectives

1. Define an articulation (joint) and identify the factors that determine the types and degree (range) of movement at a joint.

2. Classify joints on the basis of structure and function.

3. Contrast the structure, kind of movement, and location of immovable, slightly movable, and freely movable joints.

4. Describe the structure, types, and movements of freely movable joints.

5. Describe selected articulations of the body with respect to the bones that enter into their formation, structural classification, and anatomical components.

6. Describe the causes and symptoms of common joint disorders,

including rheumatism, rheumatoid arthritis (RA), osteoarthritis (OA), gouty arthritis, Lyme disease, bursitis, ankylosing spondylitis, dislocation, sprain, and strain.

7. Define medical terminology associated with articulations.

Bones are too rigid to bend without being damaged. Fortunately, flexible connective tissues form joints that hold bones together while still permitting some degree of movement, in most cases. Since most body movements occur at joints. You can appreciate the importance of joints if you imagine how a cast over the knee joint makes walking difficult or how a splint on a finger limits the ability to manipulate small objects. A few joints do not permit any movement at all but do provide a great degree of protection.

An **articulation (joint)** is a point of contact between bones, between cartilage and bones, or between teeth and bones. The scientific study of joints is called **arthrology** (ar-THROL-ō-jē; *arthro* = joint; *logos* = study of).

A joint's structure determines how it functions. Generally, the closer the bones fit at the point of contact, the stronger the joint. Tightly fitted joints, however, restrict movement. The looser the fit, the greater the movement. Unfortunately, loosely fitted joints are prone to dislocation (displacement). Several other factors also affect joint movement: (1) the precise manner in which the articulating bones fit together; (2) the flexibility (tension or tautness) of the tissues that bind the bones together; and (3) the position of ligaments, muscles, and tendons (discussed in detail later).

CLASSIFICATION OF JOINTS

Based on anatomic characteristics, joints can be categorized into structural classes, or based on the type of movement they permit, joints can be categorized into functional classes.

Structural Classification

The structural classification of joints is based on the presence or absence of a space between the articulating bones called a joint cavity (described shortly) and the type of connective tissue that binds the bones together. Structurally, a joint is classified as (1) **fibrous,** if there is no joint cavity and the bones are held together by fibrous connective tissue; (2) **cartilaginous,** if there is no joint cavity and the bones are held together by cartilage; or (3) **synovial,** if there is a joint cavity and the bones forming the joint are united by a surrounding articular capsule and frequently by accessory ligaments (described in detail later). We will discuss the joints of the body based on their functional classification, referring to their structural classification as well.

Functional Classification

The functional classification of joints takes into account the degree of movement they permit. Functionally, a joint is classified as follows:

1. A **synarthrosis** (sin′-ar-THRŌ-sis; *syn* = together; *arthros* = joint) is an immovable joint; **synarthroses** is plural.

2. An **amphiarthrosis** (am′-fē-ar-THRŌ-sis; *amphi* = on both sides) is a slightly movable joint; **amphiarthroses** is plural.

3. A **diarthrosis** (dī-ar-THRŌ-sis; *diarthros* = movable joint) is a freely movable joint; **diarthroses** is plural.

SYNARTHROSIS (IMMOVABLE JOINT)

A synarthrosis, or immovable joint, may be one of three types: suture, gomphosis, or synchondrosis.

Suture

A **suture** (SOO-cher; *sutura* = seam) is a fibrous joint composed of a thin layer of dense fibrous connective tissue and unites the bones of the skull. An example is the coronal suture between the frontal and parietal bones (see Fig. 7.3). The irregular, interlocking edges of sutures give them added strength and decrease their chance of fractures. Some sutures, although present during childhood, are replaced by bone in the adult. Such a suture is called a **synostosis** (sin′-os-TŌ-sis; *syn* = together; *osteon* = bone), or bony joint—a joint in which there is a complete fusion of bone across the suture line. An example is the frontal suture between the left and right sides of the frontal bone that begins to fuse during infancy (see Fig. 7.6a).

Gomphosis

A **gomphosis** (gom-FŌ-sis; *gomphosis* = to bolt together) is a type of fibrous joint in which a cone-shaped peg fits into a socket. The substance between the two is the periodontal ligament. The only examples are the articulations of the roots of the teeth with the alveoli (sockets) of the maxillae and mandible (see Fig. 24.6).

Synchondrosis

A **synchondrosis** (sin′-kon-DRŌ-sis; *syn* = together; *chondra* = cartilage) is a cartilaginous joint in which the connecting material is hyaline cartilage. The most common type of synchondrosis is the epiphyseal plate (see Fig. 6.5). Such a joint connects the epiphysis and diaphysis of a growing bone. Since the hyaline cartilage is eventually replaced by bone when growth ceases, the joint is temporary. It is replaced by a synostosis. Another example of a synchondrosis is the joint between the first rib and the sternum. The cartilage in this joint undergoes ossification during adult life.

AMPHIARTHROSIS (SLIGHTLY MOVABLE JOINT)

An amphiarthrosis, or slightly movable joint, may be of two types: syndesmosis or symphysis.

Syndesmosis

A **syndesmosis** (sin'-dez-MŌ-sis; *syndesmo* = band or ligand) is a fibrous joint in which there is considerably more fibrous connective tissue than in a suture. The fit between the bones thus is not quite as tight. The fibrous connective tissue forms an interosseous membrane or ligament that permits some flexibility and movement. An example of a syndesmosis is the distal articulation of the tibia and fibula (see Fig. 8.12).

Symphysis

A **symphysis** (SIM-fi-sis; *symphysis* = growing together) is a cartilaginous joint in which the connecting material is a broad, flat disc of fibrocartilage. This type of joint is found between bodies of vertebrae (see Fig. 7.15). A portion of the intervertebral disc is fibrocartilaginous material. The pubic symphysis between the anterior surfaces of the hipbones is another example (see Fig. 8.15).

DIARTHROSIS (FREELY MOVABLE JOINT)

A diarthrosis, or freely movable joint, has a variety of shapes and permits several different types of movements. First, we discuss the general structure of a diarthrosis and then consider the different types.

Structure of a Diarthrosis

A distinguishing anatomical feature of a diarthrosis is the space, called a **synovial** (si-NŌ-vē-al) **(joint) cavity** (Fig. 9.1), that separates the articulating bones. Thus diarthroses are also called **synovial joints.** Another characteristic of such joints is the presence of **articular cartilage.** Articular cartilage (which is the hyaline type) covers the surfaces of the articulating bones but does not bind the bones together.

A sleevelike **articular capsule** surrounds a diarthrosis, encloses the synovial cavity, and unites the articulating bones. The articular capsule is composed of two layers. The outer layer, the **fibrous capsule,** usually consists of dense, irregular connective tissue. It attaches to the periosteum of the articulating bones at a variable distance from the edge of the articular cartilage. The flexibility of the fibrous capsule permits movement at a joint, whereas its great tensile strength resists dislocation. The fibers of some fibrous cap-

sules are arranged in parallel bundles and are therefore highly adapted to resist recurrent strain. Such bundles of fibers are called **ligaments** (*ligare* = to bind) and are given special names. The strength of the ligaments is one of the principal factors in holding bone to bone. Diarthroses are freely movable joints because of the synovial cavity and the arrangement of the articular capsule and ligaments.

The inner layer of the articular capsule is formed by a **synovial membrane.** The synovial membrane is composed of areolar connective tissue with elastic fibers and a variable amount of adipose tissue. It secretes **synovial fluid (SF),** which fills the synovial cavity, lubricates the joint, and provides nourishment for the articular cartilage. Synovial fluid also contains phagocytic cells that remove microbes and debris resulting from wear and tear in the joint. Synovial fluid consists of hyaluronic acid and an interstitial fluid formed from blood plasma. It is similar in appearance and consistency to uncooked egg white. When there is no joint movement, the fluid is quite viscous, but as movement increases, the fluid becomes less viscous. The amount of synovial fluid varies according to the size of the joint. For example, a large joint, such as the knee, may contain 3 to 4 ml (about ⅛ oz) of fluid. It forms a thin, viscous film over the surfaces within the articular capsule. Synovial fluid reduces friction in the joint. It also supplies nutrients to and removes metabolic wastes from the chondrocytes of the articular cartilage. (Recall that cartilage is avascular.)

An interesting feature of some synovial joints is their ability to produce a **cracking sound** when pulled apart. A possible mechanism is as follows. When a synovial joint is first pulled on, the pressure of the synovial fluid decreases since the fluid suddenly occupies a larger volume. When a certain pressure is reached, the fluid evaporates and a gas bubble appears in the fluid. Then, the opposing articular surfaces abruptly separate, until limited by the articular capsule. Once the surfaces are separated, the pressure within the joint exceeds that in the bubble and the bubble collapses, producing the cracking noise. The collapse of the large bubble creates a series of smaller bubbles, which gradually go back into solution. Until the small bubbles disappear and the gas is completely dissolved, usually 20 to 30 minutes, the joint cannot be cracked again. Synovial joints whose surfaces are more congruent, such as those between phalanges and metacarpals, crack more easily than joints whose surfaces are less congruent.

Many diarthroses also contain **accessory ligaments,** called extracapsular ligaments and intracapsular ligaments. **Extracapsular ligaments** lie outside the articular capsule. An example is the fibular collateral ligament of the knee joint (see Fig. 9.7a). **Intracapsular ligaments** occur within the articular capsule but are excluded from the synovial cavity by folds of the synovial membrane. Examples are the cruciate ligaments of the knee joint (see Fig. 9.7e).

Inside some synovial joints are pads of fibrocartilage that lie between the articular surfaces of the bones and are attached by their margins to the fibrous capsule. These pads

FIGURE 9.1 Diarthrosis. Generalized structure.

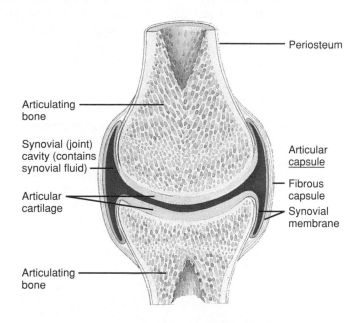

Periosteum

Articulating bone

Synovial (joint) cavity (contains synovial fluid)

Articular capsule

Articular cartilage

Fibrous capsule

Synovial membrane

Articulating bone

(a) Frontal section of a generalized diarthrotic (synovial) joint

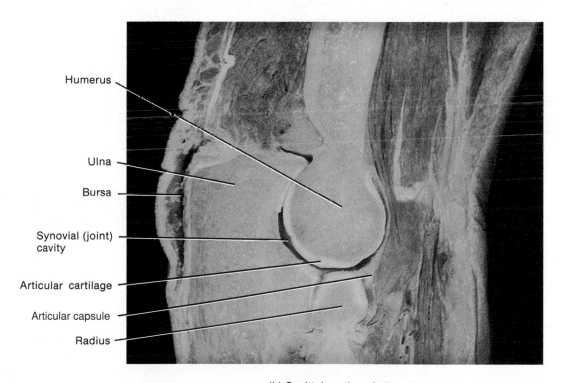

Humerus

Ulna

Bursa

Synovial (joint) cavity

Articular cartilage

Articular capsule

Radius

(b) Sagittal section of elbow joint

Question: Structurally, this type of joint is classified as which type of joint?

are called **articular discs (menisci).** See Fig. 9.7e. The discs usually subdivide the synovial cavity into two separate spaces. Articular discs allow two bones of different shapes to fit tightly; they modify the shape of the joint surfaces of the articulating bones. Articular discs also help to maintain the stability of the joint and direct the flow of synovial fluid to areas of greatest friction.

CLINICAL APPLICATION

TORN CARTILAGE

A tearing of articular discs in the knee, commonly called **torn cartilage,** occurs often among athletes. Such damaged cartilage requires surgical removal (meniscectomy), or it will begin to wear and cause arthritis. At one time, knee joint surgery for torn cartilage required cutting through layers of healthy tissue and removing much, if not all, of the cartilage. This procedure is usually painful and expensive and does not always provide full recovery. These problems have largely been overcome by arthroscopy (described on page 233).

The various motions of the body create friction between moving parts. Saclike structures called **bursae** (*bursa* = pouch or purse) are strategically situated to alleviate friction in some spots. Bursae resemble joint capsules in that their walls consist of connective tissue lined by a synovial membrane. They are also filled with a fluid similar to synovial fluid. Bursae are located between the skin and bone in places where skin rubs over bone. They are also found between tendons and bones, muscles and bones, and ligaments and bones and within articular capsules. Such fluid-filled sacs cushion the movement of one part of the body over another. Inflammation of a bursa is called **bursitis** (see p. 234).

Contact and Movement at a Diarthrosis

Several factors contribute to keeping the articular surfaces of diarthroses in contact. How the surfaces contact one another, in turn, determines the type and extent of motion that is possible.

1. First is the **structure or shape of the articulating bones,** which determines how they fit together. An interlocking shape is very obvious at the hip joint, where the head of the femur articulates with the acetabulum of the hipbone. This type of fit allows rotational movement. Other shapes permit a diversity of motions, described shortly.

2. A second factor is the **strength and tension (tautness) of the joint ligaments.** The different components of a fibrous capsule are tense only when the joint is in certain positions. Tense ligaments not only restrict the range of movement but also direct the movement of the articulating bones with respect to each other. In the knee joint, for example, the major ligaments are lax when the knee is bent but tense when the knee is straightened. Also, when the knee is straightened, the surfaces of the articulating bones are in fullest contact with each other.

3. A third factor that holds joints together but also restricts movement is the **arrangement and tension of the muscles** around the joint. Muscle tension reinforces the restraint placed on a joint by ligaments. A good example of the effect of muscle tension on a joint is seen at the hip joint. When the thigh is raised with the knee straight, the movement is restricted by the tension of the hamstring muscles on the posterior surface of the thigh. But if the knee is bent, the tension on the hamstring muscles is lessened, and the thigh can be raised further.

4. In a few joints, the **apposition** (coming together) **of soft parts** may limit mobility. For example, if you bend your arm at the elbow, it can move no further as the anterior surface of the forearm presses against the arm.

5. Joint flexibility may also be affected by **hormones.** For example, relaxin, a hormone produced by the placenta and ovaries, relaxes the pubic symphysis and ligaments between the sacrum and hipbone and sacrum and coccyx toward the end of pregnancy. This allows expansion of the birth canal (pelvic outlet), which eases delivery of the baby.

Types of Diarthroses

Though all synovial joints have a generally similar structure, the shape of the articulating surfaces varies. Accordingly, diarthroses are divided into six subtypes: gliding, hinge, pivot, ellipsoidal, saddle, and ball-and-socket joints (Fig. 9.2).

Gliding Joint

The articulating surfaces of bones in a **gliding** or **arthrodial** (ar-THRŌ-dē-al) **joint** are usually flat. A gliding movement is the simplest kind that can occur at a joint. Only side-to-side and back-and-forth movements are permitted (Fig. 9.2a). Twisting and rotation are inhibited at gliding joints, generally because ligaments or adjacent bones restrict the range of movement. Some joints that glide are those between the carpals and between the tarsals. The heads and tubercles of ribs glide on the bodies and transverse processes of vertebrae. Also, the clavicle glides on the sternum and the scapula.

Hinge Joint

In a **hinge** or **ginglymus** (JIN-gli-mus) **joint,** the convex surface of one bone fits into the concave surface of another bone. Hinge joints include the knee, elbow, ankle, and

FIGURE 9.2 Subtypes of diarthroses. For each subtype shown, there is a drawing of the actual joint and a simplified diagram.

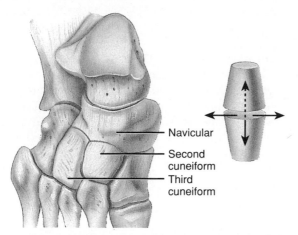

(a) Gliding joint between the navicular and second and third cuneiforms of the tarsus in the foot

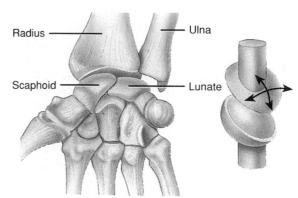

(d) Ellipsoidal joint between radius and scaphoid and lunate bones of the carpus (wrist)

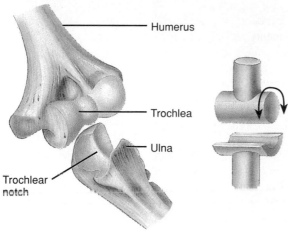

(b) Hinge joint between trochlea of humerus and trochlear notch of ulna at the elbow

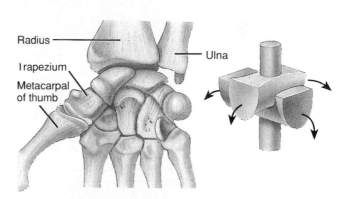

(e) Saddle joint between trapezium of carpus (wrist) and metacarpal of thumb

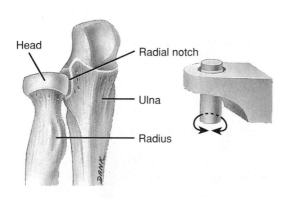

(c) Pivot joint between head of radius and radial notch of ulna

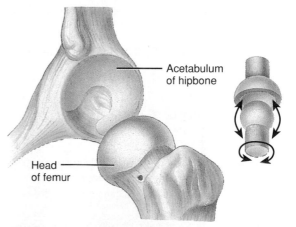

(f) Ball-and-socket joint between head of the femur and acetabulum of the hipbone

Question: Which joints shown are monaxial?

interphalangeal joints and the joint between the occipital bone and atlas in the neck. Movement is primarily in a single plane, and the joint is therefore known as **monaxial** or **uniaxial** (Fig. 9.2b). The motion is similar to that of a hinged door. Movement is usually flexion and extension. **Flexion** decreases the angle between articulating bones. For example, it occurs when you bend your knee or elbow (Fig. 9.3c). **Extension** increases the angle between articulating bones, often to restore a body part to its anatomical position after it has been flexed (Fig. 9.3). Some hinge joints are capable of **hyperextension** (Fig. 9.3a,b,d), continuation of extension beyond the anatomical position, such as when the

head bends backward. Flexion, extension, and hyperextension are called *angular movements* since the angle between bones changes.

Pivot Joint

In a **pivot** or **trochoid** (TRŌ-koyd) **joint,** a rounded or pointed surface of one bone articulates within a ring formed partly by another bone and partly by a ligament (Fig. 9.2c). The primary movement permitted is **rotation**, where a bone moves in a single plane around its longitudinal axis. The

---

FIGURE 9.3 Movements at hinge joints: flexion, extension, and hyperextension.

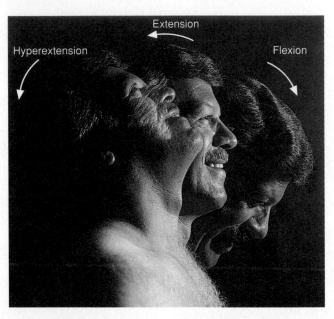

(a) Atlanto-occipital and cervical intervertebral joints

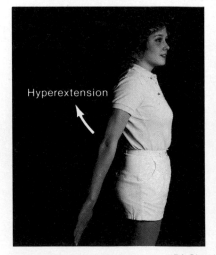
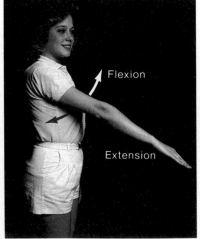

(b) Shoulder joint

FIGURE 9.3 (continued)

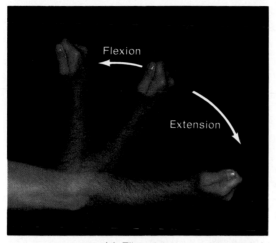

(c) Elbow joint

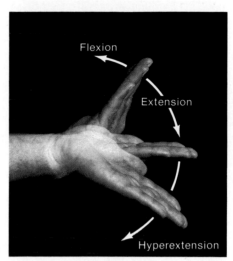

(d) Wrist joint

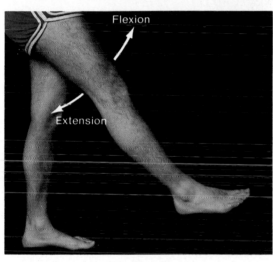

(e) Hip joint

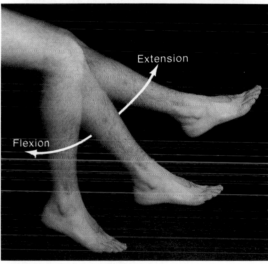

(f) Knee joint

Question: Why are hinge joints classified as monaxial?

joint is therefore monaxial. During rotation, no other motion is permitted. The atlas rotates around the dens of the axis when you turn your head from side to side to indicate "no" (Fig. 9.4). Another pivot joint is found between the proximal ends of the ulna and radius (see Fig. 9.2c) and it allows us to turn the palms forward (or upward) and backward (or downward). In **medial rotation,** the anterior surface of a bone or extremity moves toward the midline. In **lateral rotation,** the anterior surface moves away from the midline.

Ellipsoidal Joint

In an **ellipsoidal** or **condyloid** (KON-di-loyd) **joint,** an oval-shaped condyle of one bone fits into an elliptical cavity of another bone (Fig. 9.2d). The joint at the wrist between the radius and carpals is ellipsoidal. The movement permitted by such a joint is in two planes, side-to-side and back-and-forth, as when you flex and extend and abduct and adduct your wrist. Such a joint is **biaxial.** At an ellipsoidal joint, it is possible to combine the movements of

FIGURE 9.4 Movement at a pivot joint: rotation.

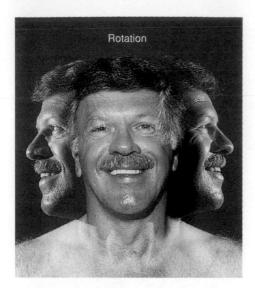

Rotation

Rotation of the head at the atlantoaxial joint

Question: What is rotation?

flexion–extension and abduction–adduction in succession to produce a movement called **circumduction**. In circumduction, the distal end of a part of the body moves in a circle. For example, if you hold your arm straight out, you can circumduct your hand at the wrist joint by turning your hand in a circle (see Fig. 9.5e). Since circumduction is a combined movement, it is not considered to be a separate plane of movement.

Abduction usually refers to movement of a bone away from the midline of the body (Fig. 9.5a,b). An example is moving the arms upward from the sides of the trunk and away from the body. With the fingers and toes, however, the midline of the body is not used as the line of reference. Abduction of the fingers (not the thumb) is a movement away from an imaginary line drawn through the middle finger; in other words, it is spreading the fingers (Fig. 9.5c). Abduction of the thumb moves the thumb away from the plane of the palm at a right angle to the palm. Abduction of the toes is relative to an imaginary line drawn through the second toe.

Adduction usually refers to movement toward the midline of the body (Fig. 9.5a,b). An example of adduction is returning the arm to the side after abduction. As in abduction, adduction of the fingers is relative to the middle finger (Fig. 9.5c), and adduction of the toes is relative to the second toe. In adduction of the thumb, the thumb moves toward the plane of the palm at a right angle to the palm. Abduction and adduction are also angular movements.

Saddle Joint

In a **saddle** or **sellaris** (sel-A-ris) **joint,** the articular surface of one bone is saddle-shaped, and the articular surface of the other bone is shaped like a rider sitting in the saddle. The joint between the trapezium of the carpus and metacarpal of the thumb is an example of a saddle joint. Movements at a saddle joint are side-to-side and back-and-forth (Fig. 9.2e). The saddle joint is a modified ellipsoidal joint in which the movement is somewhat freer. Such joints are biaxial and also permit circumduction. In circumduction, the thumb is moved in a circle.

Ball-and-Socket Joint

A **ball-and-socket** or **spheroid** (SFĒ-royd) **joint** consists of a ball-like surface of one bone fitted into a cuplike depression of another bone. The only examples of ball-and-socket joints are the shoulder joint and hip joint (Fig. 9.2f). Such joints are **triaxial** because they permit three types of movement: flexion–extension, abduction–adduction, and rotation. The shoulder joint also permits circumduction, as when winding up to pitch a ball (Fig. 9.5d).

A summary of joints based on functional classification is presented in Exhibit 9.1.

Special Movements at Diarthroses

In addition to gliding movements, flexion, extension, hyperextension, rotation, abduction, adduction, and circumduction, several other movements also occur at synovial joints. These are called **special movements** and occur only at particular joints.

Elevation is an upward movement of a part of the body, and **depression** is a downward movement of a part of the body (Fig. 9.6a,b). You elevate your mandible when you close your mouth and depress your mandible when you open your mouth. You also can elevate and depress your shoulders. **Protraction** is the movement of the mandible or shoulder girdle forward on a plane parallel to the ground. Thrusting the jaw outward is protraction of the mandible (Fig. 9.6c). Bringing your arms forward until the elbows touch requires protraction of the shoulder girdle. **Retraction** is the movement of a protracted part of the body backward on a plane parallel to the ground. Pulling the lower jaw back in line with the upper jaw is retraction of the mandible (Fig. 9.6d).

Six special movements relate to the foot and hand. **Inversion** is the movement of the sole of the foot inward (medially) so that the soles face toward each other (Fig. 9.6e). **Eversion** is the movement of the sole outward (laterally) so that the soles face away from each other (Fig. 9.6f). Whereas **dorsiflexion** involves bending of the foot in the direction of the dorsum (upper surface), **plantar flexion** involves bending the foot in the direction of the plantar

FIGURE 9.5 Abduction, adduction, and circumduction.

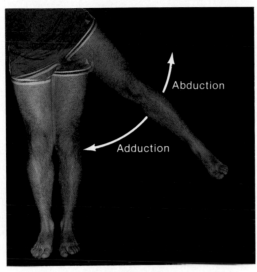

(a) Hip joint

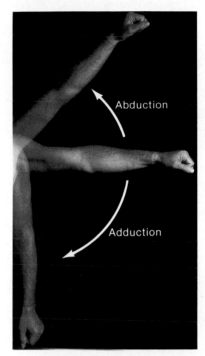

(b) Shoulder joint

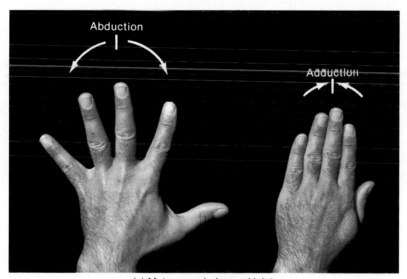

(c) Metacarpophalangeal joints

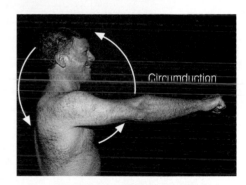

(d) Shoulder joint

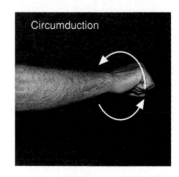

(e) Wrist joint

Question: You can remember adduction as "*add*ing your limb to your trunk." Why is this an accurate learning device?

EXHIBIT 9.1

SUMMARY OF JOINTS BASED ON FUNCTIONAL CLASSIFICATION

Type	Description	Examples
SYNARTHROSIS (IMMOVABLE)		
Suture	Found only between bones of the skull; articulating bones separated by a thin layer of fibrous tissue.	Coronal suture between frontal and parietal bones.
Gomphosis	Cone-shaped peg fits into a socket; articulating bones separated by periodontal ligament.	Roots of teeth in alveoli (sockets).
Synchondrosis	Connecting material is hyaline cartilage.	Temporary joint between the diaphysis and epiphysis of a long bone and permanent joint between true ribs and sternum.
AMPHIARTHROSIS (SLIGHTLY MOVABLE)		
Syndesmosis	Articulating bones united by dense fibrous tissue.	Distal ends of tibia and fibula.
Symphysis	Connecting material is a broad, flat disc of fibrocartilage.	Intervertebral joints and pubic symphysis.
DIARTHROSIS (FREELY MOVABLE)		
Gliding	Articulating surfaces usually flat.	Intercarpal and intertarsal joints.
Hinge	Spool-like surface fits into a concave surface. Monaxial (flexion–extension).	Elbow, ankle, and interphalangeal joints.
Pivot	Rounded, pointed, or concave surface fits into a ring formed partly by bone and partly by a ligament. Monaxial (rotation).	Joint between atlas and axis and joint at proximal ends of radius and ulna.
Ellipsoidal	Oval-shaped condyle fits into an elliptical cavity. Biaxial (flexion–extension; abduction–adduction). Circumduction also occurs.	Joint between radius and carpals.
Saddle	Articular surface of one bone is saddle-shaped, and the articular surface of the other bone is shaped like a rider sitting in the saddle. Biaxial (flexion–extension; abduction–adduction). Circumduction also occurs.	Joint between trapezium of carpus and metacarpal of thumb.
Ball-and-socket	Ball-like surface fits into a cuplike depression. Triaxial (flexion–extension; abduction–adduction; rotation). Circumduction also occurs.	Shoulder and hip joints.

surface (sole) (Fig. 9.6g). **Supination** is a movement of the forearm in which the palm of the hand is turned anteriorly or superiorly; **pronation** is a movement of the forearm in which the palm is turned posteriorly or inferiorly (Fig. 9.6h).

A summary of movements that occur at synovial joints is presented in Exhibit 9.2.

Knee (Tibiofemoral) Joint

The structure of the knee joint illustrates the complexity of a diarthrosis. The **knee (tibiofemoral) joint** is the largest joint of the body. It actually consists of three joints: (1) an

intermediate patellofemoral joint between the patella and the patellar surface of the femur; (2) a lateral tibiofemoral joint between the lateral condyle of the femur, lateral meniscus, and lateral condyle of the tibia; and (3) a medial tibiofemoral joint between the medial condyle of the femur, medial meniscus, and medial condyle of the tibia. The patellofemoral joint is a gliding joint, and the lateral and medial tibiofemoral joints are modified hinge joints.

The anatomical components of the knee joint are as follows (Fig. 9.7):

1. **Articular capsule.** No complete, independent capsule unites the bones. The ligamentous sheath surround-

FIGURE 9.6 Special movements.

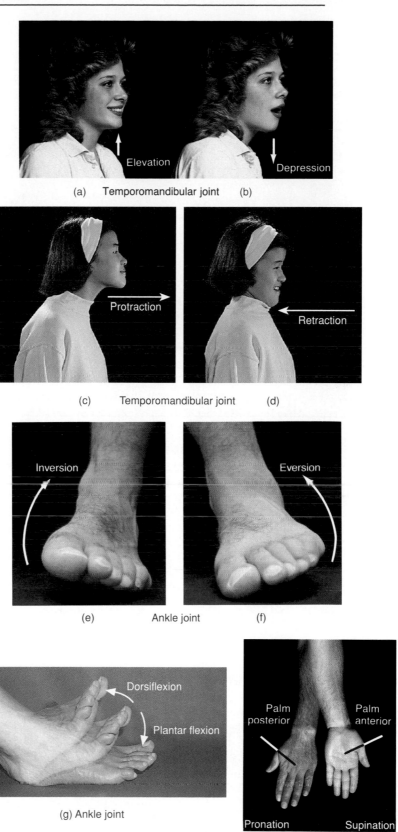

(a)   Temporomandibular joint   (b)

(c)   Temporomandibular joint   (d)

(e)   Ankle joint   (f)

(g) Ankle joint

(h) Proximal radioulnar joint

Question: Why are these movements called special movements?

EXHIBIT 9.2

SUMMARY OF MOVEMENTS AT SYNOVIAL JOINTS

Movement	Definition
GLIDING	One surface moves back and forth and from side to side over another surface without angular or rotary motion.
ANGULAR	There is an increase or decrease at the angle between bones.
Flexion	Involves a decrease in the angle between the surfaces of articulating bones.
Extension	Involves an increase in the angle between the surfaces of articulating bones.
Hyperextension	Continuation of extension beyond the anatomical position.
Abduction	Movement of a bone away from the midline.
Adduction	Movement of a bone toward the midline.
Circumduction	A combination of flexion–extension and abduction–adduction in succession in which a part of the body moves in a circle.
ROTATION	Movement of a bone around its longitudinal axis; may be medial or lateral.
SPECIAL	Occur at specific joints.
Inversion	Movement of the sole of the foot inward so that the soles face toward each other.
Eversion	Movement of the sole of the foot outward so that the soles face away from each other.
Dorsiflexion	Bending the foot in the direction of the dorsum (upper surface).
Plantar flexion	Bending the foot in the direction of the plantar surface (sole).
Protraction	Movement of the mandible or shoulder girdle forward on a plane parallel to the ground.
Retraction	Movement of a protracted part backward on a plane parallel to the ground.
Supination	Movement of the forearm in which the palm is turned anteriorly or superiorly.
Pronation	Movement of the forearm in which the palm is turned posteriorly or inferiorly.
Elevation	Movement of a part of the body upward.
Depression	Movement of a part of the body downward.

ing the joint consists mostly of muscle tendons or expansions of them. There are, however, some capsular fibers connecting the articulating bones.

2. Medial and lateral patellar retinacula. Fused tendons of insertion of the quadriceps femoris muscle and the fascia lata (deep fascia of the thigh) that strengthen the anterior surface of the joint.

3. Patellar ligament. Central portion of the common tendon of insertion of the quadriceps femoris muscle that extends from the patella to the tibial tuberosity. This ligament also strengthens the anterior surface of the joint. An **infrapatellar fat pad** separates the posterior surface of the ligament from the synovial membrane of the joint.

4. Oblique popliteal ligament. Broad, flat ligament that extends from the intercondylar fossa of the femur to the head of the tibia. The tendon of the semimembranosus muscle is superficial to the ligament and passes from the medial condyle of the tibia to the lateral condyle of the femur. The ligament and tendon afford strength for the posterior surface of the joint.

5. Arcuate popliteal ligament. Extends from the lateral condyle of the femur to the styloid process of the head of the fibula. It strengthens the lower lateral part of the posterior surface of the joint.

6. Tibial (medial) collateral ligament. Broad, flat ligament on the medial surface of the joint that extends from the medial condyle of the femur to the medial condyle of the tibia. The ligament is crossed by tendons of the sartorius, gracilis, and semitendinosus muscles, all of which strengthen the medial aspect of the joint.

7. Fibular (lateral) collateral ligament. Strong, rounded ligament on the lateral surface of the joint that extends from the lateral condyle of the femur to the lateral side of the head of the fibula. The ligament is covered by the tendon of the biceps femoris muscle. The tendon of the popliteal muscle is deep to ligament.

FIGURE 9.7 Knee (tibiofemoral) joint.

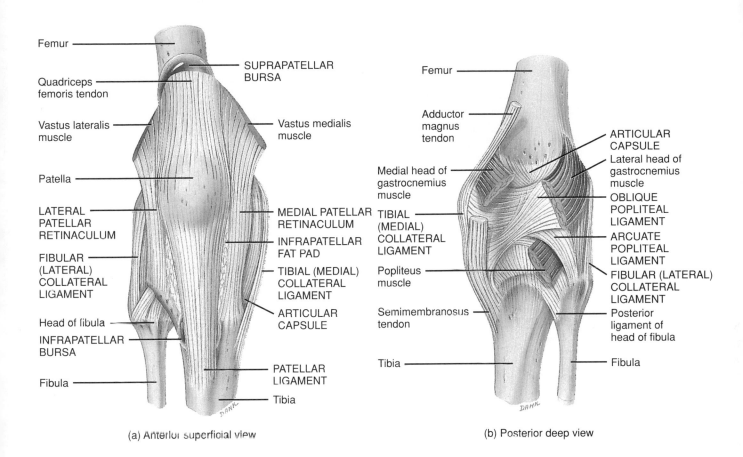

(a) Anterior superficial view

(b) Posterior deep view

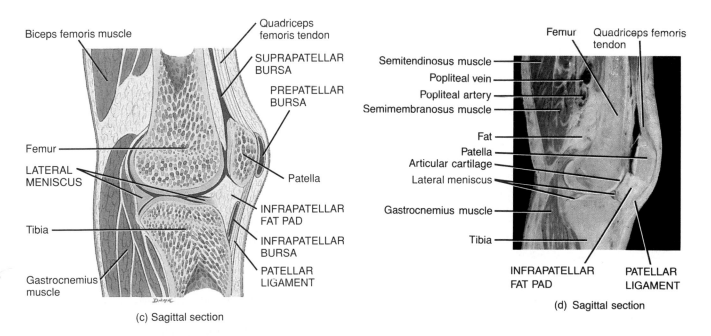

(c) Sagittal section

(d) Sagittal section

Figure continues

FIGURE 9.7 (continued)

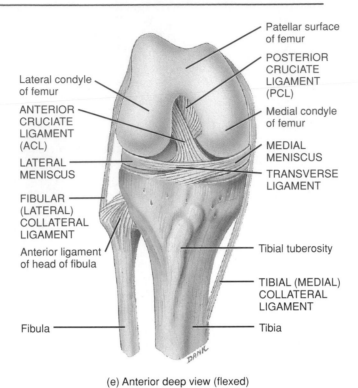

(e) Anterior deep view (flexed)

Question: What structures are involved in the knee injury called "torn cartilage"?

8. Intracapsular ligaments. Ligaments within the capsule that connect the tibia and femur.

a. Anterior cruciate ligament (ACL). Extends posteriorly and laterally from the area anterior to the intercondylar eminence of the tibia to the posterior part of the medial surface of the lateral condyle of the femur. This ligament is stretched or torn in about 70% of all serious knee injuries. Both O. J. Simpson and Gale Sayers had their careers in professional football ended by torn anterior cruciate ligaments.

b. Posterior cruciate ligament (PCL). Extends anteriorly and medially from a depression on the posterior intercondylar area of the tibia and lateral meniscus to the anterior part of the medial surface of the medial condyle of the femur.

9. Articular discs. Fibrocartilage discs between the tibial and femoral condyles. They help to compensate for the incongruence of the articulating bones.

a. Medial meniscus. Semicircular piece of fibrocartilage (C-shaped). Its anterior end is attached to the area anterior to the intercondylar eminence of the tibia, in front of the anterior cruciate ligament. Its posterior end is attached to the area posterior to the intercondylar eminence of the tibia between the attachments of the posterior cruciate ligament and lateral meniscus.

b. Lateral meniscus. Nearly circular piece of fibrocartilage (approaches an incomplete O in shape). Its anterior end is attached anterior to the intercondylar eminence of the tibia and lateral and posterior to the anterior cruciate ligament. Its posterior end is attached posterior to the intercondylar eminence of the tibia and anterior to the posterior end of the medial meniscus. The medial and lateral menisci are connected to each other by the **transverse ligament** and to the margins of the head of the tibia by the **coronary ligaments.**

10. The principal **bursae** of the knee include the following:

a. Anterior bursae. (1) Between the patella and skin (**prepatellar bursa**), (2) between the upper part of tibia and the patellar ligament (**infrapatellar bursa**), (3) between the lower part of the tibial tuberosity and the skin, and (4) between the lower part of the femur and the deep surface of quadriceps femoris muscle (**suprapatellar bursa**).

b. Medial bursae. (1) Between the medial head of the gastrocnemius muscle and the articular capsule; (2) superficial to the tibial collateral ligament between the ligament and tendons of the sartorius, gracilis, and semitendinosus muscles; (3) deep to the tibial collateral ligament between the ligament and the tendon of the semimembranosus muscle; (4) between the tendon of the semimembranosus muscle and the head of the tibia; and (5) between the tendons of the semimembranosus and semitendinosus muscles.

  c.   Lateral bursae. (1) Between the lateral head of the gastrocnemius muscle and articular capsule, (2) between the tendon of the biceps femoris muscle and fibular collateral ligament, (3) between the tendon of the popliteal muscle and fibular collateral ligament, and (4) between the lateral condyle of the femur and the popliteal muscle.

CLINICAL APPLICATION

KNEE INJURIES

The knee joint is the most vulnerable joint to damage because of the stresses to which it is subjected and because there is no interlocking of the articulating bones; reinforcement is strictly by ligaments and tendons. The most common type of **knee injury** in football is rupture of the tibial (medial) collateral ligament, often associated with tearing of the anterior cruciate ligament and medial meniscus (torn cartilage). Usually, a blow to the lateral side of the knee causes the injury. When a knee is examined for such an injury, the three Cs are kept in mind: collateral ligament, cruciate ligament, and cartilage.

A **swollen knee** may occur immediately or hours after an injury. Immediate swelling is due to escape of blood from rupture of the anterior cruciate ligament, torn menisci, fractures, or collateral ligament sprains. Delayed swelling is due to excessive production of synovial fluid as a result of conditions that irritate the synovial membrane.

A **dislocated knee** refers to the displacement of the tibia relative to the femur. Accordingly, such dislocations are classified as anterior, posterior, medial, lateral, or rotatory. The most common type is anterior dislocation, resulting from hyperextension of the knee. A frequent consequence of a dislocated knee is damage to the popliteal artery.

In 1987 a group of orthopedic surgeons at the Hospital of the University of Pennsylvania performed the **first transplant of an entire human knee.** The surgery was necessitated by a potentially malignant tumor on the knee of a 32-year-old female patient. After removing the tumor, the surgeons completely removed the patient's knee joint. The donor knee joint consisted of the lower portion of a femur that was connected to the recipient's femur by a rod; the upper portion of the tibia and head of the fibula that were connected by a metal plate; the patella; medial and lateral menisci; intracapsular and extracapsular ligaments; and certain tendons. The recipient's own muscles, nerves, and blood vessels were used.

Exhibit 9.3 lists selected joints of the body, relating their articular components, classification, and movements.

EXHIBIT 9.3

REPRESENTATIVE JOINTS ACCORDING TO ARTICULAR COMPONENTS, CLASSIFICATION, AND MOVEMENTS

TEMPOROMANDIBULAR JOINT (TMJ)

Definition	Joint formed by condylar process of the mandible and mandibular fossa and articular tubercle of temporal bone. The TMJ is the only movable joint between skull bones; all other skull joints are sutures and therefore immovable.
Type of joint	Synovial; combined hinge (ginglymus) and gliding (arthrodial) type.
Movements	Only the mandible moves since the maxilla is firmly anchored to other bones by sutures. Accordingly, the mandible may function in depression (jaw opening), elevation (jaw closing), protraction, retraction, lateral displacement, and slight rotation.

ATLANTO-OCCIPITAL JOINTS

Definition	Joints formed by the superior articular facets of the atlas and the occipital condyles of the occipital bone.
Type of joint	Synovial; ellipsoidal (condyloid) type.
Movements	Flexion (forward bending of head), extension (backward bending of head), and slight lateral tilting of head to either side.

INTERVERTEBRAL JOINTS

Definition	Joints formed between (1) vertebral bodies and between (2) vertebral arches.
Type of joint	Joints between vertebral bodies—cartilaginous (fibrocartilage), symphysis type. Joints between vertebral arches—synovial, gliding (arthrodial) type.
Movements	Flexion (bending the backbone forward), extension (bending the backbone backward), lateral flexion (bending the backbone to either side), and rotation.

Exhibit continues

EXHIBIT 9.3 (continued)

LUMBOSACRAL JOINT

Definition	Joint formed by the body of the fifth lumbar vertebra and the superior articular facet of the first sacral vertebra of the sacrum.
Type of joint	Joint between the fifth lumbar vertebra and the first sacral vertebra—cartilaginous joint, symphysis type. Joint between the articular processes—synovial joint, gliding (arthrodial) type.
Movements	Similar to those of intervertebral joints.

SHOULDER (HUMEROSCAPULAR OR GLENOHUMERAL) JOINT

Definition	Joint formed by the head of the humerus and the glenoid cavity of the scapula.
Type of joint	Synovial joint, ball-and-socket (spheroid) type.
Movements	Flexion (humerus drawn forward), extension (humerus drawn backward), abduction (humerus drawn away from midline), adduction (humerus drawn toward midline), medial rotation, lateral rotation, and circumduction.

ELBOW JOINT

Definition	Joint formed by the trochlea of the humerus, the trochlear notch of the ulna, and the head of the radius.
Type of joint	Synovial joint, hinge (ginglymus) type.
Movements	Flexion and extension of forearm.

WRIST (RADIOCARPAL) JOINT

Definition	Joint formed by the distal end of the radius, the distal surface of the articular disc separating the carpal and distal radioulnar joint, and the scaphoid, lunate, and triquetrum carpal bones.
Type of joint	Synovial joint, ellipsoidal (condyloid) type.
Movements	Flexion, extension, abduction, adduction, and circumduction.

HIP (COXAL) JOINT

Definition	Joint formed by the head of the femur and the acetabulum of the hipbone.
Type of joint	Synovial, ball-and-socket (spheroid) type.
Movements	Flexion, extension, abduction, adduction, circumduction, and rotation.

KNEE (TIBIOFEMORAL) JOINT

Definition	The largest joint of the body, actually consisting of three joints: (1) an intermediate patellofemoral joint between the patella and the patellar surface of the femur; (2) a lateral tibiofemoral joint between the lateral condyle of the femur, lateral meniscus, and lateral condyle of the tibia; and (3) a medial tibiofemoral joint between the medial condyle of the femur, medial meniscus, and medial condyle of the tibia.
Type of joint	Patellofemoral joint—partly synovial, gliding (arthrodial) type. Lateral and medial tibiofemoral joints—synovial, hinge (ginglymus) type.
Movements	Flexion, extension, slight medial rotation, and lateral rotation in flexed position.

ANKLE (TALOCRURAL) JOINT

Definition	Joints between (1) the distal end of the tibia and its medial malleolus and the talus and (2) the lateral malleolus of the fibula and the talus.
Type of joint	Both joints—synovial, hinge (ginglymus) type.
Movements	Dorsiflexion and plantar flexion.

Examining and Repairing Diarthroses

Arthroscopy

Arthroscopy (ar-THROS-kō-pē; *arthro* = joint; *skopein* = to view) is a procedure that involves examination of the interior of a joint, usually the knee, using an arthroscope, a lighted instrument the diameter of a pencil. It is used to determine the nature and extent of damage following knee injury; to remove torn cartilage and repair cruciate ligaments in the knee; to perform surgery on other joints and obtain tissue samples for analysis; and to monitor the progression of disease and the effects of therapy.

After injecting a local or general anesthetic, the orthopedic surgeon inserts the arthroscope into the knee joint through an incision as small as 5 mm (about ¼ in.). Additional small incisions may be made to insert a tube through which a saline (salt) solution is injected into the joint or to insert an instrument that shaves off and reshapes the damaged cartilage and then suctions out the cartilage shavings along with the salt solution.

Some orthopedic surgeons attach a video camera to the arthroscope so that the image from inside the knee can be enlarged and projected onto a screen; the patient often receives a videotape of the operation. Since arthroscopy requires only small incisions, recovery is more rapid than with conventional surgery, although a cast or splint may be worn for several days, depending on the extent of the procedure.

Arthroplasty

Arthroplasty (AR-thrō-plas′-tē; *arthro* = joint; *plasty* = plastic repair of) refers to surgical replacement of joints. Each year in the U.S. over 150,000 people undergo **total hip replacement (hip arthroplasty)**, a procedure involving both the acetabulum and head of the femur. Thousands of partial hip replacements, involving only the femur, are also performed annually.

In total hip replacement, prefabricated prostheses (artificial devices) replace the damaged portions of the acetabulum and head of the femur. The acetabular component consists of polyethylene, while the femoral component is made of cobalt-chrome. These materials are designed to withstand a high degree of stress. Once the appropriate components are selected, they are attached to the healthy portion of bone with acrylic cement and screws.

Total hip replacement may benefit individuals with osteoarthritis, rheumatoid arthritis, hip fractures, dislocations, metabolic bone diseases (osteoporosis and osteomalacia), and congenital and developmental deformities of the hip.

DISORDERS: HOMEOSTATIC IMBALANCES

RHEUMATISM

Rheumatism refers to any painful state of the supporting structures of the body—bones, ligaments, joints, tendons, or muscles. Arthritis is a form of rheumatism in which the joints have become inflamed.

ARTHRITIS

The term **arthritis** refers to many different diseases, most of which are characterized by inflammation of one or more joints. Inflammation, pain, and stiffness may also be present in adjacent parts of the body, such as the muscles near the joint.

Rheumatoid Arthritis (RA)

Rheumatoid (ROO-ma-toyd) **arthritis (RA)** is an autoimmune disease in which the immune system of the body attacks its own tissues, in this case its own cartilage and joint linings. It is characterized by inflammation of the joint, swelling, pain, and loss of function. Usually, this form occurs bilaterally: if your left wrist is affected, your right wrist is also likely to be affected, although usually not to the same degree.

The primary symptom of rheumatoid arthritis is inflammation of the synovial membrane (Fig. 9.8). If untreated, the membrane thickens and synovial fluid accumulates. The resulting pressure causes pain and tenderness. The membrane then produces an abnormal granulation tissue, called *pannus*, that adheres to the surface of the articular cartilage and sometimes erodes the cartilage completely. When the cartilage is destroyed, fibrous tissue joins the

FIGURE 9.8 Rheumatoid arthritis. X-ray of the right hand and wrist showing changes characteristic of rheumatoid arthritis. The arrows indicate fusion of the bones resulting in obliteration of the joint cavities.

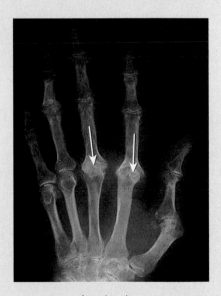

Anterior view

Continues

exposed bone ends. The tissue ossifies and fuses the joint so that it is immovable—the ultimate crippling effect of rheumatoid arthritis. The growth of the granulation tissue causes the distortion of the fingers that is so typical of hands that have been affected by this disease.

Osteoarthritis (OA)

Osteoarthritis (os′-tē-ō-ar-THRĪ-tis) is a degenerative joint disease that apparently results from a combination of aging, irritation of the joints, and wear and abrasion. It is commonly known as "wear-and-tear" arthritis.

Osteoarthritis is a noninflammatory, progressive disorder of movable joints, particularly weight-bearing joints. It is characterized by the deterioration of articular cartilage and by formation of new bone in the subchondral areas and at the margins of the joint. The cartilage slowly degenerates, and as the bone ends become exposed, small bumps, or **spurs,** of new osseous tissue are deposited on them. These spurs decrease the space of the joint cavity and restrict joint movement. Unlike rheumatoid arthritis, osteoarthritis usually affects only the articular cartilage. The synovial membrane is rarely destroyed, and other tissues are unaffected. A major distinction between osteoarthritis and rheumatoid arthritis is that the former strikes the larger joints (knees, hips) first, whereas the latter first strikes smaller joints.

Gouty Arthritis

Uric acid (a substance that gives urine its name) is a waste product produced during the metabolism of nucleic acid (DNA and RNA) subunits. A person who suffers from **gout** either produces excessive amounts of uric acid or is not able to excrete as much as normal. The result is a buildup of uric acid in the blood. This excess acid then reacts with sodium to form a salt called sodium urate. Crystals of this salt accumulate in soft tissues. Typical sites are the kidneys and the cartilage of the ears and joints.

In **gouty** (GOW-tē) **arthritis,** sodium urate crystals are deposited in the soft tissues of the joints. The crystals irritate the cartilage, causing inflammation, swelling, and acute pain. Eventually, the crystals destroy all the joint tissues. If the disorder is not treated, the ends of the articulating bones fuse, and the joint becomes immovable.

Gouty arthritis occurs primarily in middle-aged and older males. Studies indicate that an abnormal gene sometimes causes the disease. As a result of this gene, the body manufactures unusually large amounts of uric acid. Diet and environmental factors such as stress and climate are also suspected causes of gouty arthritis.

LYME DISEASE

In 1975 a cluster of disease cases that were first diagnosed as rheumatoid arthritis were reported near the city of Lyme, Connecticut. Today this disease is known as **Lyme disease,** and it has spread into most states across the country. It is caused by a bacterium (*Borrelia burgdorferi*) transmitted to humans by ticks (mainly deer ticks) that are so small their bites often go unnoticed.

Within a few weeks of the tick bite, a rash may appear at the site. The rash often resembles a bull's eye, but there are many variations, and some people never develop a rash. Other symptoms include joint stiffness, fever and chills, headache, stiff neck, nausea, and low back pain. In the next stage, some patients develop cardiac abnormalities, indicated by weakness, lightheadedness, chest pain, and an irregular heart beat. Other patients develop neurologic problems, most often Bell's palsy (sudden paralysis of one side of the face). In the third stage, arthritis is the principal complication. It usually involves the larger joints such as the knee, ankle, hip, elbow, or wrist. Lyme disease responds well to antibiotics, especially if they are given in the early stages, but some symptoms may linger for months.

BURSITIS

An acute or chronic inflammation of a bursa is called **bursitis.** The condition may be caused by trauma, by an acute or chronic infection (including syphilis and tuberculosis), or by rheumatoid arthritis. Repeated excessive friction often results in a bursitis with local inflammation and the accumulation of fluid. Bunions are often associated with a friction bursitis over the head of the first metatarsal bone. Symptoms include pain, swelling, tenderness, and limited motion.

ANKYLOSING SPONDYLITIS

Ankylosing spondylitis (ang′-ki-LŌ-sing spon′-di-LĪ-tis; *ankyle* = stiff; *spondyl* = vertebra) is an inflammatory disease that affects joints between vertebrae and between the sacrum and hipbone (sacroiliac joint). The cause is unknown. The disease is more common in males and has its onset between the ages of 20 and 40 years. It is characterized by pain and stiffness in the hips and lower back that progresses upward along the backbone. Inflammation can lead to ankylosis (severe or complete loss of movement at a joint) and kyphosis (hunchback). Treatment consists of anti-inflammatory drugs, heat, massage, and supervised exercise.

DISLOCATION

A **dislocation,** or **luxation** (luks-Ā-shun), is the displacement of a bone from a joint with tearing of ligaments, tendons, and articular capsules. A partial or incomplete dislocation is called a **subluxation.** The most common dislocations are those involving the fingers or shoulders. Dislocations of the mandible, elbow, knee, or hip are less common. Symptoms include loss of motion, temporary paralysis of the involved joint, pain, swelling, and sometimes shock. A dislocation is usually caused by a blow or fall, although unusual physical effort may lead to this condition.

SPRAIN AND STRAIN

A **sprain** is the forcible wrenching or twisting of a joint with partial rupture or other injury to its attachments without dislocation. It occurs when the attachments are stressed beyond their normal capacity. There may be damage to the associated blood vessels, muscles, tendons, ligaments, or nerves. Severe sprains may be so painful that the joint cannot be moved. There is considerable swelling, and pain may occur owing to underlying hemorrhage from ruptured blood vessels. The ankle joint is most often sprained; the low back area is another frequent location for sprains. A sprain is more serious than a **strain,** which is the overstretching of a muscle.

MEDICAL TERMINOLOGY

Arthralgia (ar-THRAL-jē-a; *arth* = joint; *algia* = pain) Pain in a joint.

Arthrosis (ar-THRŌ-sis) Refers to an articulation; also a disease of a joint.

Bursectomy (bur-SEK-tō-mē; *ectomy* = removal of) Removal of a bursa.

Chondritis (kon-DRĪ-tis; *chondro* = cartilage) Inflammation of cartilage.

Synovitis (sin′-ō-VĪ-tis; *synov* = joint) Inflammation of a synovial membrane in a joint.

Study Outline

Classification of Joints (p. 217)

1. An articulation (joint) is a point of contact between two or more bones.
2. Structural classification is based on the presence of a synovial (joint) cavity and type of connecting tissue. Structurally, joints are classified as fibrous, cartilaginous, or synovial.
3. Functional classification of joints is based on the degree of movement permitted. Joints may be synarthroses (immovable), amphiarthroses (partially movable), or diarthroses (freely movable).

Synarthrosis (Immovable Joint) (p. 217)

1. Types include suture, gomphosis, and synchondrosis.
2. A suture is a fibrous joint composed of a thin layer of dense fibrous connective tissue that unites skull bones. An example is the coronal suture.
3. A gomphosis is a fibrous joint in which a cone-shaped peg fits into a socket. An example is the root of a tooth in its socket.
4. A synchondrosis is a cartilaginous joint in which the connecting material is hyaline cartilage. An example is the epiphyseal plate.

Amphiarthrosis (Slightly Movable Joint) (p. 218)

1. An amphiarthrosis may be a syndesmosis or a symphysis.
2. A syndesmosis is a fibrous joint in which there is more fibrous connective tissue than in a suture. An example is the distal articulation of the tibia and fibula.
3. A symphysis is a cartilaginous joint in which the connecting material is a disc of fibrocartilage. An example is the pubic symphysis.

Diarthrosis (Freely Movable Joint) (p. 218)

Structure (p. 218)

1. A diarthrosis contains a space between bones called the synovial (joint) cavity. All diarthroses are synovial joints.
2. Other characteristics of a diarthrosis are the presence of articular cartilage and an articular capsule (fibrous capsule and synovial membrane).
3. Many diarthroses also contain accessory ligaments (extracapsular and intracapsular), articular discs (menisci), and bursae.

Contact and Movement (p. 220)

1. Among the factors that keep the articular surfaces of diarthroses in contact are the shape of the articulating bones, strength and tension of ligaments, arrangement and tension of muscles, and apposition of soft parts.
2. Joint flexibility may also be affected by hormones, such as relaxin, which relaxes the pubic symphysis during childbirth.

Types of Diarthroses (p. 220)

1. Gliding joints: articulating surfaces are flat; bone glides back-and-forth and side-to-side; found between carpals and tarsals.
2. Hinge joint: convex surface of one bone fits into concave surface of another; movement is flexion–extension (monaxial); example is elbow.
3. Pivot joint: a round or pointed surface of one bone fits into a ring formed by another bone and a ligament; movement is rotation (monaxial); example is when atlas rotates around axis.
4. Ellipsoidal: an oval-shaped condyle of one bone fits into an elliptical cavity of another; movements are flexion–extension and abduction–adduction (biaxial) and circumduction; example is between carpals and radius.
5. Saddle: articular surface of one bone is shaped like a saddle and the other is shaped like a rider sitting in the saddle; movements are flexion–extension and abduction–adduction (biaxial) and circumduction; example is between trapezium and metacarpal of thumb.
6. Ball-and-socket: ball-shaped surface of one bone fits into cup-like depression of another; movements are flexion–extension, abduction–adduction, and rotation (triaxial) and circumduction; example is the shoulder.

Special Movements at Diarthroses (p. 224)

1. In addition to flexion-extension, hyperextension, abduction–adduction, rotation, and circumduction, special movements occur at particular joints.
2. Examples are elevation–depression, protraction–retraction, inversion–eversion, dorsiflexion–plantar flexion, and supination–pronation.

Knee (Tibiofemoral) Joint (p. 226)

1. The knee (tibiofemoral) joint is formed by the patella and femur and by the tibia and femur.
2. Other joints are described in Exhibit 9.3.
3. Arthroscopy is a procedure for examining and repairing the interior of a joint.
4. Arthroplasty refers to surgical replacement of joints.

Review Questions

1. Define an articulation. What factors determine the degree of movement at joints? (p. 216)
2. Classify joints on the basis of structure and function. (p. 217)
3. Set up an outline for the structural classification of joints, list and define the various subtypes, and give an example of each. (p. 217)
4. Describe the structure of a diarthrosis (freely movable joint). (p. 218)
5. What is an accessory ligament? Define the two principal types. (p. 218)
6. What is an articular disc? Why are they important? (p. 220)
7. Explain how the articulating bones in a synovial joint are held together. (p. 220)
8. List the six subtypes of diarthroses, define each, describe the movements at each, and give an example of each. (p. 220)
9. Describe the various special movements that occur at diarthroses. (p. 224)
10. Have another person assume the anatomical position and execute each of the movements at joints discussed in the text. Reverse roles and see if you can execute the same movements.
11. Be sure that you can name the bones that form the joint, identify the joint by type, and list the anatomical components of the joint for each of the following: shoulder (humeroscapular), hip (coxal), and knee (tibiofemoral) joints. (p. 231)

Answers to Questions with Figures

9.1 Synovial joint.
9.2 Hinge and pivot joints.
9.3 They permit movement in a single plane.
9.4 Movement of a bone around its longitudinal axis.

9.5 When you adduct your arm or leg, you bring it closer to the midline of the body, thus "adding" it to the trunk.
9.6 They occur only at specific joints.
9.7 Articular discs (menisci).

Chapter 10

Muscle Tissue

Chapter Contents at a Glance

Student Objectives

1. List the characteristics and functions of muscle tissue.

2. Compare the location, microscopic appearance, nervous control, functions, and regenerative capacities of the three kinds of muscle tissue.

3. Describe the structure and importance of a neuromuscular junction and a motor unit.

4. Describe the principal events associated with the sliding filament mechanism of muscle contraction.

5. Identify the sources of energy for muscular contraction.

6. Explain the roles played by muscle in homeostasis of body temperature.

7. Explain how muscle tension can be varied.

8. Describe the different types of skeletal muscle fibers and compare them to cardiac and smooth muscle fibers.

9. Explain the effects of aging on muscle tissue.

10. Describe the development of the muscular system.

11. Define such common muscular disorders as fibromyalgia, muscular dystrophies, myasthenia gravis (MG), spasm, tremor, fasciculation, fibrillation, and tic.

12. Define medical terminology associated with the muscular system.

lthough bones and joints provide leverage and form the framework of the body, they cannot move the body by themselves. Motion results from alternating contraction (shortening) and relaxation of muscles, which constitute 40 to 50% of total body weight. Your muscular strength reflects the prime function of muscle—changing chemical energy (in the form of ATP) into mechanical energy to generate force, perform work, and produce movement. The scientific study of muscles is known as **myology** (mī-OL-ō-jē; *myo* = muscle; *logos* = study of).

TYPES OF MUSCLE TISSUE

The three kinds of muscle tissue—skeletal, cardiac, and smooth—differ from one another in their microscopic anatomy, location, and control by the nervous and endocrine systems.

Skeletal muscle tissue is so-named because it is attached primarily to bones, and it moves parts of the skeleton. (Some skeletal muscles are also attached to skin, other muscles, or deep fascia.) Skeletal muscle tissue is called **striated** muscle because alternating light and dark bands (striations) are visible when the tissue is examined under a microscope (see Exhibit 4.3 on page 120). It is a **voluntary** muscle tissue because it can be made to contract and relax by conscious control.

Cardiac muscle tissue forms most of the heart. It is also **striated** muscle but is **involuntary;** that is, its contraction is usually not under conscious control. Cardiac muscle includes a pacemaker system that causes the heart to beat; this built-in rhythm is called **autorhythmicity.** Adjustments in heart rate are possible because various hormones and neurotransmitters influence the pacemaker.

Smooth muscle tissue is located in the walls of hollow internal structures, such as blood vessels, the stomach, and the intestines, as well as most other abdominal organs. It is also found in the skin attached to hair follicles. Under the microscope, this tissue looks **nonstriated** or **smooth.** It is usually **involuntary** muscle tissue, often has autorhythmicity, and is also influenced by certain hormones and neurotransmitters. (Regulation of cardiac and smooth muscle by hormones and neurotransmitters is discussed in Chapter 17.)

Thus all muscle tissues are classified in the following way: (1) skeletal, striated, voluntary muscle tissue; (2) cardiac, striated, involuntary muscle tissue; or (3) smooth, nonstriated, involuntary muscle tissue (see Exhibit 10.3).

FUNCTIONS OF MUSCLE TISSUE

Through sustained contraction or alternating contraction and relaxation, muscle tissue has three key functions: producing motion, providing stabilization, and generating heat.

1. **Motion.** Motion is obvious in movements such as walking and running, and in localized movements, such as grasping a pencil or nodding the head. These movements rely on the integrated functioning of bones, joints, and skeletal muscles. Less noticeable kinds of motion are produced by cardiac and smooth muscles. Examples are the beating of the heart to move blood throughout the body, churning of food in the stomach, and contraction of the urinary bladder to expel urine.

2. **Stabilizing body positions and regulating organ volume.** Besides producing movements, skeletal muscle contractions maintain the body in stable positions, such as standing or sitting. Postural muscles display sustained contractions when a person is awake; for example, partially contracted neck muscles hold the head upright. In a similar manner, sustained contractions of smooth muscles may prevent outflow of the contents of a hollow organ. Temporary storage of food in the stomach or urine in the urinary bladder is possible because smooth muscles close off the exit route.

3. **Thermogenesis (generation of heat).** As skeletal muscle contracts to perform work, a by-product is heat. Much of the heat released by muscle is used to maintain normal body temperature. Muscle contractions are thought to generate as much as 85% of all body heat. This is why active cheering helps warm you up during a cold weather football game.

CHARACTERISTICS OF MUSCLE TISSUE

Muscle tissue has four principal characteristics that enable it to carry out its functions and thus contribute to homeostasis.

1. **Excitability (irritability),** a property of both muscle and nerve cells (neurons), is the ability to respond to certain stimuli by producing electrical signals called action potentials (impulses). For muscle, the stimuli that trigger action potentials are chemicals—neurotransmitters, released by neurons, or hormones distributed by the blood.

2. **Contractility** is the ability of muscle tissue to shorten and thicken (contract), thus generating force to do work. Muscle contracts in response to one or more muscle action potentials.

3. **Extensibility** means that muscle can be extended (stretched) without damaging the tissue. Most skeletal muscles are arranged in opposing pairs. While one is contracting, the other not only is relaxed but usually is being stretched.

4. **Elasticity** means that muscle tissue tends to return to its original shape after contraction or extension.

Skeletal muscle is the focus of much of this chapter. Cardiac muscle and smooth muscle are described briefly here, but in more detail later (with discussions of the heart in Chapter 20, the various organs containing smooth muscle, and the autonomic nervous system in Chapter 17). The developmental anatomy of skeletal muscle is considered at the end of the chapter.

ANATOMY AND INNERVATION OF SKELETAL MUSCLE TISSUE

To understand how skeletal muscle contraction produces movement, one needs some knowledge of its nerve and blood supply, connective tissue components, and microscopic anatomy.

Nerve and Blood Supply

Skeletal muscles are well supplied with nerves and blood vessels (see Fig. 11.21). Those neurons that stimulate muscle to contract are called **motor neurons**. When muscle contracts, it uses a good deal of ATP and therefore needs large amounts of nutrients and oxygen for ATP production. Moreover, the waste products of the reactions that produce ATP must be eliminated. Thus prolonged muscle action depends on a rich blood supply to deliver nutrients and oxygen and remove wastes and heat. Microscopic blood vessels called capillaries are plentiful in muscle tissue; each muscle fiber (cell) is in close contact with one or more capillaries.

Connective Tissue Components

Connective tissue surrounds and protects muscle tissue. The term **fascia** (FASH-ē-a; *fascia* = bandage) refers to a sheet or broad band of fibrous connective tissue beneath the skin or around muscles and other organs of the body. **Superficial fascia (subcutaneous layer)** is immediately deep to the skin (see Fig. 11.21). It is composed of adipose tissue and areolar connective tissue and has four important functions: (1) It stores water and fat. Much of the fat of an overweight person is in the superficial fascia. (2) It forms a layer of insulation that protects the body from heat loss. (3) It provides mechanical protection against traumatic blows. (4) It provides a pathway for nerves and blood vessels to enter and exit muscles.

Deep fascia is dense, irregular connective tissue that lines the body wall and extremities and holds muscles together, separating them into functional groups. Deep fascia allows free movement of muscles, carries nerves and blood and lymphatic vessels, and fills spaces between muscles.

Three layers of dense, irregular connective tissue extend from the deep fascia to further protect and strengthen skeletal muscle (Fig. 10.1). The outermost layer, encircling the whole muscle, is the **epimysium** (ep-i-MĪZ-ē-um; *epi* = upon). **Perimysium** (per-i-MĪZ-ē-um; *peri* = around) then surrounds bundles (fasciculi or fascicles) of 10 to 100 or more individual muscle fibers. Penetrating the interior of each fascicle and separating muscle fibers from one another is **endomysium** (en′-dō-MĪZ-ē-um; *endo* = within).

Epimysium, perimysium, and endomysium are all continuous with and contribute collagen fibers to the connective tissue that attaches the muscle to other structures, such as bone or other muscle. All three may extend beyond the muscle fibers as a **tendon** (*tendere* = to stretch out)—a cord of dense connective tissue that attaches a muscle to the periosteum of a bone. When the connective tissue elements

FIGURE 10.1 Relationships of connective tissue to skeletal muscle showing the relative positions of the epimysium, perimysium, and endomysium.

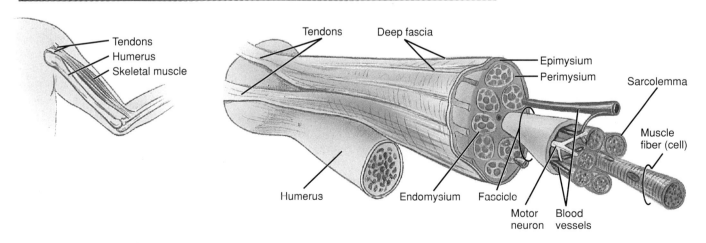

Question: Starting with the connective tissue that surrounds an individual muscle fiber (cell), list the connective tissue layers in order.

extend as a broad, flat layer, the tendon is called an **aponeurosis** (*apo* = from; *neuron* = a tendon). This structure also attaches to the coverings of a bone, another muscle, or the skin. An example of an aponeurosis is the galea aponeurotica on top of the skull (see Fig. 11.4). Certain tendons, especially those of the wrist and ankle, are enclosed by tubes of fibrous connective tissue called **tendon (synovial) sheaths** (see Fig. 11.17). They are similar in structure to bursae and contain a film of synovial fluid. Tendon sheaths permit tendons to slide back and forth more easily.

CLINICAL APPLICATION

TENOSYNOVITIS

Tenosynovitis (ten'-ō-sin-ō-VĪ-tis) is an inflammation of the tendons, tendon sheaths, and synovial membranes surrounding certain joints. The tendons most often affected are at the wrists, shoulders, elbows (called tennis elbow), finger joints (called trigger finger), ankles, and feet. The affected sheaths may or may not become visibly swollen because of fluid accumulation. Tenderness and pain are frequently associated with movement of the body part. The condition often follows trauma, strain, or excessive exercise.

The Motor Unit

A motor neuron delivers the stimulus that ultimately causes a muscle fiber to contract. A motor neuron plus all the muscle fibers it stimulates is called a **motor unit** (Fig. 10.2).

A single motor neuron makes contact with an average of 150 muscle fibers. This means that activation of one neuron causes the simultaneous contraction of about 150 muscle fibers. All the muscle fibers of a motor unit contract and relax together. Muscles that control precise movements, such as voice production by the larynx, have as few as two to three muscle fibers per motor unit. Muscles of the body that are responsible for powerful gross movements, such as the biceps brachii in the arm and gastrocnemius in the leg, may have as many as 2000 muscle fibers per motor unit. Stimulation of a motor neuron produces a contraction in all the muscle fibers of a particular motor unit. Accordingly, the total strength of a contraction is varied in part by adjusting the number of motor units that are activated.

The Neuromuscular Junction

Excitable cells (neurons and muscle fibers) make contact and communicate at specialized regions called **synapses**. At most synapses a small gap, called the **synaptic cleft,** separates the two excitable cells. Since the cells do not physically touch, the action potential from one cell cannot "jump the gap" to excite the next cell. Rather, the first cell communicates with the second by releasing a chemical called a **neurotransmitter**. The particular type of synapse formed between a motor neuron and a skeletal muscle fiber is called the **neuromuscular junction (NMJ)** or **myoneural junction.**

Close to its target skeletal muscle fibers, the axon of a motor neuron branches into clusters of bulb-shaped **axon terminals** (Fig. 10.3a,b). The region of the *muscle fiber membrane* that is adjacent to the axon terminals has special features and is called the **motor end plate.** The term neuromuscular junction includes both the axon terminals of the motor neuron and the motor end plate of the muscle fiber.

The distal end of an axon terminal contains many membrane-enclosed sacs called **synaptic vesicles** (Fig. 10.3c). Inside each synaptic vesicle are thousands of neurotransmitter molecules. Although many different neurotransmitters exist, the one present in motor neuron synaptic vesicles and released at the NMJ is **acetylcholine,** abbreviated **ACh.**

When a nerve impulse (action potential) reaches the axon terminal, it triggers exocytosis of synaptic vesicles. In this process, the synaptic vesicles fuse with the plasma membrane and liberate ACh, which diffuses into the synaptic cleft between the motor neuron and the motor end plate. On the muscle side of the synaptic cleft, the motor end plate contains **acetylcholine receptors** (Fig. 10.3d). These are integral membrane proteins that recognize and bind specifically to ACh. At a typical NMJ there are 30 to 40 million ACh receptors. The binding of ACh to its receptor opens a channel that passes small cations, most importantly Na$^+$. The resulting change in the resting membrane potential (see page 356) triggers a muscle action potential that travels along the muscle cell membrane (sarcolemma) and initiates the events leading to muscle contraction. (Chapter 12 presents details of how action potentials are generated.)

In most skeletal muscle fibers, there is only one neuromuscular junction for each muscle fiber, located near the fiber's midpoint. The muscle action potential spreads from the center of the fiber toward both ends. This arrangement permits nearly simultaneous contraction of all parts of the fiber.

FIGURE 10.2 Motor unit. Shown are two motor neurons, one in red and one in green, each supplying the muscle fibers of its motor unit.

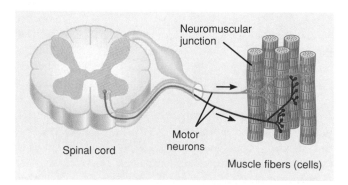

Neuromuscular junction

Spinal cord

Motor neurons

Muscle fibers (cells)

Question: How will the size of motor units influence strength of contraction?

ELECTROMYOGRAM

The recording of electrical activity in resting and contracting muscles is called **electromyography** (e-lek′-trō-mī-OG-ra-fē; *electro* = electricity; *myo* = muscle; *graph* = to write). An **electromyogram (EMG)** is a record of skeletal muscle electrical activity. It is similar to the more familiar electrocardiogram (ECG), which is a record of the heart's electrical signals (see Fig. 20.8). To record EMGs, a flat metal plate is placed on the skin over the muscle to be tested. Then, a thin sterile needle attached by wires to a recording machine is inserted through the skin into the muscle. The electrical activity of the muscle is recorded at rest and during contraction and displayed as electrical waves on an oscilloscope at the same time that the activity is reproduced as sounds over a speaker.

EMGs may be used to determine the cause of muscular weakness or paralysis, to evaluate involuntary muscle twitching, to determine why abnormal levels of muscle enzymes (such as creatine phosphokinase) appear in blood, and to serve as a component of biofeedback studies.

Nerve conduction studies are sometimes done together with EMG testing. A nerve is stimulated electrically through the skin, while a recording device detects the muscle's response. Such studies can help determine if nerve damage is the cause of muscle weakness.

Microscopic Anatomy of Muscle

Microscopic examination of a typical skeletal muscle reveals hundreds or thousands of very long, cylindrical cells called **muscle fibers** (Fig. 10.4b). The muscle fibers lie parallel to one another and range from 10 to 100 μm in diameter. While a typical length is 100 μm, some are up to 30 cm (12 in.) long. The **sarcolemma** (*sarco* = flesh; *lemma* = sheath) is a muscle fiber's plasma membrane, and it surrounds the muscle fiber's cytoplasm or **sarcoplasm** (Fig. 10.4c). Because skeletal muscle fibers arise from the fusion of many smaller cells during embryonic development, each fiber has many nuclei to direct synthesis of new proteins. The nuclei are at the periphery of the cell, next to the sarcolemma, conveniently out of the way of the contractile elements (described shortly). The mitochondria lie in rows throughout the muscle fiber, strategically close to the muscle proteins that use ATP to carry on contraction processes.

At high magnification the sarcoplasm appears stuffed with little threads. These small structures are the **myofibrils** (Fig. 10.4d). Although myofibrils extend lengthwise within the muscle fiber, their prominent alternating light and dark bands make the whole muscle cell look striped or striated. The bands are called **cross-striations**.

Myofibrils

Myofibrils are the contractile elements of skeletal muscle. They are 1 to 2 μm in diameter and contain three types of even smaller structures called **filaments (myofilaments)**.

The diameter of the **thin filaments** is about 8 nm, while that of the **thick filaments** is about 16 nm (Fig. 10.4e). The thick and thin filaments overlap one another, to a greater or lesser extent, depending on whether the muscle is contracting or relaxing. The pattern of their overlap causes the cross-striations seen both in single myofibrils and in whole muscle fibers. A third type of filament, the elastic filament, will be described shortly.

The filaments inside a myofibril do not extend the entire length of a muscle fiber. They are arranged in compartments called **sarcomeres** (Figs. 10.5 and 10.6a). Narrow plate-shaped regions of dense material called **Z discs (lines)** separate one sarcomere from the next. Within a sarcomere, the darker area, called the **A (anisotropic) band**, extends from one end to the other of the thick filaments and includes portions of the thin filaments where they overlap the thick filaments. A lighter, less dense area called the **I (isotropic) band** contains the rest of the thin filaments but no thick filaments. The Z disc passes through the center of each I band. The alternating dark A bands and light I bands give the muscle fiber its striated appearance. A narrow **H zone** in the center of each A band contains thick but not thin filaments. Dividing the H zone is the **M line**, formed by protein molecules that connect adjacent thick filaments.

The two *contractile* proteins in muscle are myosin and actin. About 200 molecules of the protein **myosin** form a single thick filament (Fig. 10.6b). Each myosin molecule is shaped like two golf clubs twisted together. The tails (golf club handles) point toward the M line in the center of the sarcomere. The projecting myosin heads, called **cross bridges**, extend out toward the thin filaments. Tails of neighboring myosin molecules lie parallel to one another, forming the shaft of the thick filament. The heads project from all around the shaft in a spiraling fashion.

Thin filaments extend from anchoring points within the Z discs. Their main component is **actin**. Also present in the thin filament are smaller amounts of two *regulatory* proteins, **tropomyosin** and **troponin** (Fig. 10.6c). Individual actin molecules appear to be shaped like kidney beans. They join to form an actin filament that is twisted into a helix. (For many years it was thought that two helical strands intertwined to form an actin filament. Recent evidence suggests there is just a single chain of bean-shaped actin molecules.) On each actin molecule is a **myosin-binding site**, a site where a cross bridge can attach. In relaxed muscle, tropomyosin covers the myosin-binding sites on actin and thus blocks attachment of myosin cross bridges to actin.

A more recently recognized component of the sarcomere is the **elastic filament** (Fig. 10.6a). It is composed of the protein titin (connectin), the third most plentiful protein in skeletal muscle (after actin and myosin). Titin anchors thick filaments to the Z discs and thereby helps stabilize the position of the thick filaments. It may also play a role in recovery of the resting sarcomere length when a muscle is stretched or during relaxation. The protein was named titin

Figure 10.3 Neuromuscular junction (NMJ).

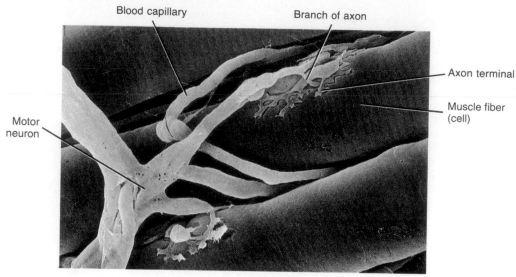

Blood capillary

Branch of axon

Axon terminal

Muscle fiber (cell)

Motor neuron

(a) Scanning electron micrograph (1650x)

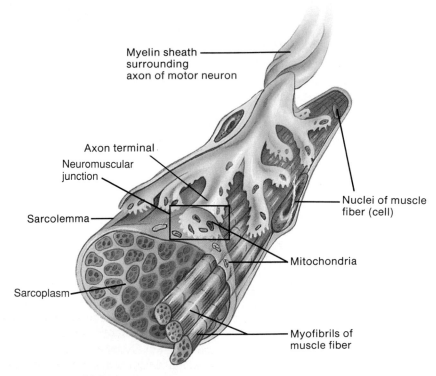

Myelin sheath surrounding axon of motor neuron

Axon terminal

Neuromuscular junction

Sarcolemma

Nuclei of muscle fiber (cell)

Mitochondria

Sarcoplasm

Myofibrils of muscle fiber

(b) Diagram based on photomicrograph

FIGURE 10.3 (continued)

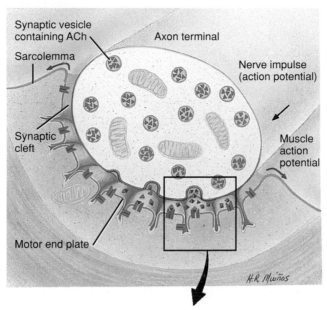

(c) Enlarged view of the neuromuscular junction

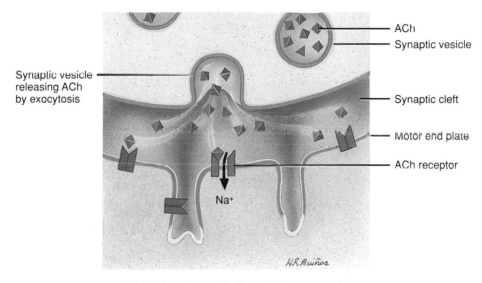

(d) Binding of acetylcholine to ACh receptors in motor end plate.

Question: What is a motor end plate?

because it has a huge (titanic) molecular weight (or connectin because of its connecting function).

Exhibit 10.1 reviews the types of filaments in skeletal muscle fibers.

Sarcoplasmic Reticulum and Transverse Tubules

A fluid-filled system of cisterns called the **sarcoplasmic reticulum (SR)** encircles each myofibril (see Fig. 10.4d).

This elaborate tubular system is similar to smooth endoplasmic reticulum in nonmuscle cells. In a relaxed muscle fiber, the sarcoplasmic reticulum stores Ca^{2+}. Release of Ca^{2+} from the sarcoplasmic reticulum into the sarcoplasm around the thick and thin filaments triggers muscle contraction. The calcium ions pass out through special pores in the sarcoplasmic reticulum called **Ca^{2+} release channels.**

The **transverse tubules (T tubules)** are tunnel-like infoldings of the sarcolemma. They penetrate the muscle fiber at right angles to the sarcoplasmic reticulum and the

FIGURE 10.4 Organization of a skeletal muscle from gross to molecular levels.

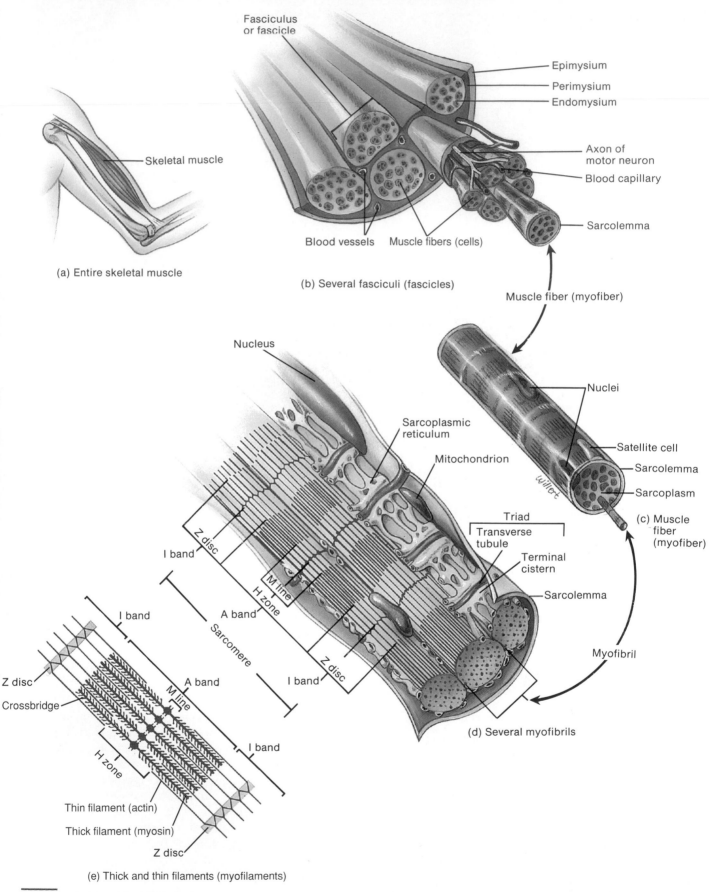

(a) Entire skeletal muscle

(b) Several fasciculi (fascicles)

(c) Muscle fiber (myofiber)

(d) Several myofibrils

(e) Thick and thin filaments (myofilaments)

Question: Which bands look dark? Light?

FIGURE 10.5 Transmission electron micrograph of mammalian skeletal muscle. The Z discs separate one sarcomere from another. Note how the overlap of thick and thin filaments produces an alternating pattern of light and dark bands.

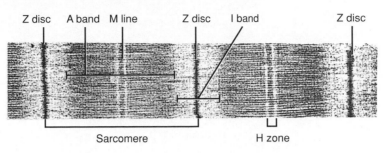

Electron micrograph of two sarcomeres (20,000x)

Question: Which filaments connect into the Z disc?

FIGURE 10.6 Detailed structure of muscle filaments. (a) The relation of thick (myosin), thin (actin), and elastic (titin) filaments in a sarcomere. Note that actin filaments are anchored directly at the Z discs while myosin filaments are connected to the Z discs by titin (also known as connectin). (b) About 200 myosin molecules comprise a thick filament. The myosin tails all point toward the center of the sarcomere. (c) Thin filaments contain actin, troponin, and tropomyosin.

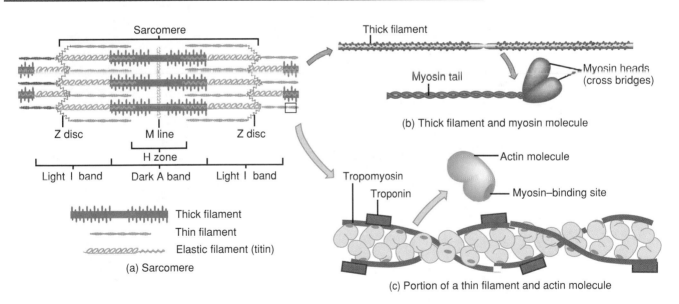

(a) Sarcomere

(b) Thick filament and myosin molecule

(c) Portion of a thin filament and actin molecule

Question: Which proteins are present in the A band and the I band?

EXHIBIT 10.1

TYPES OF FILAMENTS IN SKELETAL MUSCLE FIBERS

Filament	Diameter	Protein Composition (percent of total protein)	Functions
Thick filament	16 nm	Myosin (44%)	Myosin heads (cross bridges) move thin filaments toward center of sarcomere during contraction.
Thin filament	8 nm	Actin (22%), troponin (5%), tropomyosin (5%)	Contains cross bridge binding sites; slides along thick filament during contraction.
Elastic filament	Less than 1 nm	Titin (connectin) (9%)	Anchors thick filaments to Z discs and stabilizes them during contraction and relaxation.

myofilaments. T tubules are open to the outside of the fiber and are filled with extracellular fluid. On both sides of a transverse tubule are dilated end sacs of the sarcoplasmic reticulum called **terminal cisterns.** The term **triad** refers to a transverse tubule and the terminal cisterns on either side of it.

CONTRACTION OF MUSCLE

In the mid-1950s Jean Hanson and Hugh Huxley had a revolutionary insight into the mechanism of muscle contraction. Previously, scientists had imagined that muscle contraction must be a folding process, somewhat like closing an accordion. Hanson and Huxley proposed, however, that skeletal muscle shortens during contraction because the thick and thin filaments *slide* past one another. Their model is known as the **sliding filament mechanism** of muscle contraction.

Sliding Filament Mechanism

During muscle contraction, myosin cross bridges pull on the thin filaments, causing them to slide inward toward the H zone (Fig. 10.7). As the cross bridges pull on (apply force to) the thin filaments, the thin filaments meet at the center of the sarcomere. The myosin cross bridges may even pull the thin filaments of each sarcomere so far inward that their ends overlap (Fig. 10.7c). As the thin filaments slide inward, the Z discs come toward each other, and the sarcomere shortens, but the lengths of the thick and thin filaments do not change. The sliding of the filaments and

shortening of the sarcomeres cause shortening of the whole muscle fiber and ultimately the entire muscle.

FIGURE 10.7 Sliding filament mechanism of muscle contraction. For simplicity, the elastic filaments are not illustrated.

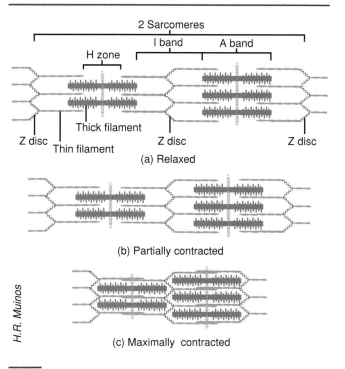

(a) Relaxed

(b) Partially contracted

(c) Maximally contracted

Question: What happens to the I band and H zone as muscle contracts? Do the lengths of the thick and thin filaments change?

Role of Calcium and Regulator Proteins

You may be wondering what starts and stops filament sliding. An increase in Ca^{2+} concentration in the sarcoplasm starts filament sliding, while a decrease turns off the sliding process.

When a muscle fiber is relaxed (not contracting), the concentration of Ca^{2+} in the sarcoplasm is low (Fig. 10.8a). This is because the sarcoplasmic reticulum (SR) membrane contains **Ca^{2+} active transport pumps** that remove Ca^{2+} from the sarcoplasm. Ca^{2+} is stored or sequestered inside the SR. As a muscle action potential travels along the sarcolemma and into the transverse tubule system, however, Ca^{2+} release channels open in the SR membrane (Fig. 10.8b). The result is a flood of Ca^{2+} from within the sarcoplasmic reticulum into the sarcoplasm around the thick and thin filaments. The calcium ions released from the sarcoplasmic reticulum combine with troponin, causing it to change shape. This shape change slides the troponin–tropomyosin complex away from the myosin-binding sites on actin (Fig. 10.8b).

The Power Stroke and the Role of ATP

As we have seen, muscle contraction requires Ca^{2+}. It also requires energy, in the form of ATP. ATP attaches to ATP-binding sites on the myosin cross bridges (heads). A portion of each myosin head acts as an ATPase, an enzyme that splits the ATP into ADP + $\textcircled{P}$ through a hydrolysis reaction. This reaction transfers energy from ATP to the myosin head, even before contraction begins. The myosin cross bridges are thus in an activated (energized) state. Such activated myosin heads spontaneously bind to the myosin-bind-

FIGURE 10.8 Regulation of contraction by troponin and tropomyosin when Ca^{2+} level changes. (a) The level of Ca^{2+} in the sarcoplasm is low during relaxation because it is pumped into the sarcoplasmic reticulum (SR) by Ca^{2+} active transport pumps. (b) A muscle action potential traveling along a transverse tubule opens calcium release channels in the SR and Ca^{2+} flows into the sarcoplasm. Note contraction is occurring since the thin filaments are closer in the center of the sarcomere.

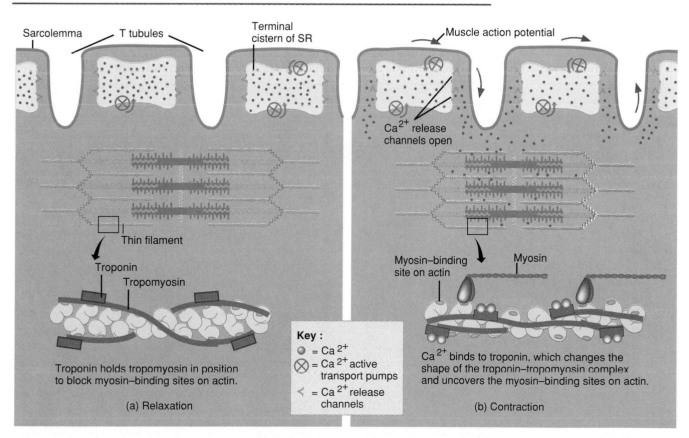

Sarcolemma   T tubules   Terminal cistern of SR   Muscle action potential

Ca^{2+} release channels open

Thin filament

Troponin

Tropomyosin

Myosin–binding site on actin   Myosin

Key :
○ = Ca^{2+}
⊗ = Ca^{2+} active transport pumps
< = Ca^{2+} release channels

Troponin holds tropomyosin in position to block myosin–binding sites on actin.

Ca^{2+} binds to troponin, which changes the shape of the troponin–tropomyosin complex and uncovers the myosin–binding sites on actin.

(a) Relaxation

(b) Contraction

Question: Why are troponin and tropomyosin called regulator proteins?

ing sites on actin when the Ca^{2+} level rises and tropomyosin slides away from its blocking position (Fig. 10.9, step 1). The shape change that occurs when myosin binds to actin produces the **power stroke** of contraction (step 2). During the power stroke, the myosin cross bridges swivel toward the center of the sarcomere, like the oars of a boat. This action draws the thin filaments past the thick filaments toward the H zone. As the myosin heads swivel, they release ADP.

Once the power stroke is complete, ATP again combines with the ATP-binding sites on the myosin cross bridges (step 3). As ATP binds, the myosin head detaches from actin. Again, ATP is split, imparting its energy to the myosin head, which returns to its original upright position (step 4). It is then ready to combine with another myosin-binding site further along the thin filament. The cycle

repeats over and over. The myosin cross bridges keep moving back and forth like the cogs of a ratchet with each power stroke, moving the thin filaments toward the H zone. At any one instant, about half of the myosin cross bridges are bound to actin and are swiveling. The other half are detached and preparing to swivel again. Contraction is analogous to running on a nonmotorized treadmill. One foot (myosin head) strikes the belt (thin filament) and pushes it backward (toward the H zone). Then the other foot comes down and imparts a second push. The belt soon moves smoothly while the runner (thick filament) remains stationary. And like the legs of a runner, the myosin heads need a constant supply of energy to keep going!

The power stroke repeats as long as ATP is available and the Ca^{2+} level near the thin filament is high. This continual movement applies the force that draws the Z discs

FIGURE 10.9 Role of ATP and the power stroke of muscle contraction. Sarcomeres shorten through repeated cycles in which the myosin heads attach to actin, swivel toward the center of the sarcomere, and detach.

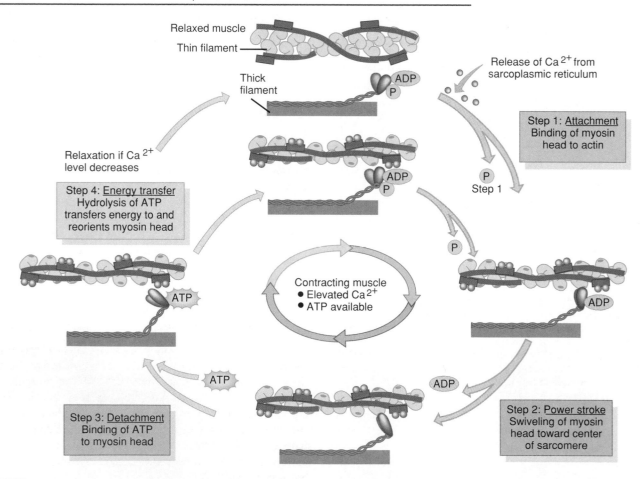

Question: What would happen if ATP were suddenly not available after the sarcomere had started to shorten?

toward each other, and the sarcomere shortens. The myofibrils thus contract, and the whole muscle fiber shortens. During a maximal muscle contraction, the distance between Z discs can decrease to half the resting length. But the power stroke does not always result in shortening of the muscle fibers and the whole muscle. Contraction without shortening is called an **isometric contraction.** An example is trying to lift a very heavy object. The cross bridges generate force, but the filaments do not slide past one another.

Relaxation

Two changes permit a muscle fiber to relax after it has contracted. First, acetylcholine is rapidly broken down by an enzyme called **acetylcholinesterase (AChE).** AChE is present in the synaptic cleft. When action potentials cease in the motor neuron, there is no new release of ACh, and AChE rapidly breaks down the ACh already present in the synaptic cleft. This stops the generation of muscle action potentials, and the Ca^{2+} release channels in the sarcoplasmic reticulum close.

Second, Ca^{2+} active transport pumps rapidly remove Ca^{2+} from the sarcoplasm into the sarcoplasmic reticulum. These pumps work so vigorously that they can keep the concentration of Ca^{2+} 10,000 times lower in the sarcoplasm of a relaxed muscle fiber than inside the SR. Also, molecules of a calcium-binding protein, appropriately called **calsequestrin,** bind to calcium ions inside the SR. This reaction takes Ca^{2+} out of solution and allows even more Ca^{2+} to be sequestered within the SR. As Ca^{2+} level drops in the sarcoplasm, the tropomyosin–troponin complex slides back over the myosin-binding sites on actin. This prevents further cross bridge binding to actin, and the thin filaments slip back to their relaxed positions. Figure 10.10 summarizes the events associated with contraction and relaxation of a muscle fiber.

CLINICAL APPLICATION

RIGOR MORTIS

After death, chemical deterioration in muscle fibers permits Ca^{2+} to leak out of the sarcoplasmic reticulum. The Ca^{2+} binds to troponin and triggers sliding of the filaments. ATP synthesis has ceased, however, so the myosin cross bridges cannot detach from actin. The resulting condition, in which muscles are in a state of rigidity (cannot contract or stretch), is called **rigor mortis** (rigidity of death). Rigor mortis lasts about 24 hours but disappears as tissues begin to disintegrate.

Muscle Tone

Sustained, small contractions give a firmness to a relaxed skeletal muscle that is known as **muscle tone.** At any instant, a few muscle fibers are contracted while most are relaxed. This small amount of contraction firms up a muscle without producing movement and is essential for maintaining posture. Asynchronous firing of motor units allows muscle tone to be sustained continuously. When the muscles in the back of the neck are in tonic contraction, for example, they keep the head upright and prevent it from slumping forward onto the chest, but they do not apply enough force to pull the head back into hyperextension.

CLINICAL APPLICATION

HYPOTONIA AND HYPERTONIA

Hypotonia refers to decreased or lost muscle tone. Such muscles are said to be **flaccid** (FLAK-sid or FLAS-sid). Flaccid muscles are loose and appear flattened rather than rounded; the affected limbs are hyperextended. Certain disorders of the nervous system may result in **flaccid paralysis,** which is characterized by loss of muscle tone, loss or reduction of tendon reflexes, and atrophy (wasting away) and degeneration of muscles.

Hypertonia refers to increased muscle tone and is expressed in two ways: spasticity or rigidity. **Spasticity** is characterized by increased muscle tone (stiffness) associated with an increase in tendon reflexes and pathological reflexes (such as the Babinski sign, which is described on page 389). Certain disorders of the nervous system may result in **spastic paralysis,** partial paralysis in which the muscles exhibit spasticity. **Rigidity** refers to increased muscle tone, although reflexes are not affected.

Muscle Metabolism

Contraction of muscle requires energy, but surprisingly little ATP is present inside muscle fibers; there is just enough to power contraction for a few seconds. Unlike most cells of the body, skeletal muscle fibers often function in a "stop and go" manner; they switch between virtual inaction and great activity. If strenuous exercise is to continue for more than a few seconds, additional ATP must be produced. On demand, the muscle fiber's metabolic machinery goes into high gear to step up ATP production.

Phosphagen System

Muscle fibers have a unique molecule called **creatine phosphate (phosphocreatine)** that can transfer its high-energy phosphate group to ADP (adenosine diphosphate), thus forming ATP (adenosine triphosphate) and creatine (Fig. 10.11a). Creatine phosphate is about three to five times more plentiful than ATP. Together, creatine phosphate and ATP constitute the **phosphagen system** and provide enough energy for muscles to contract maximally for about 15 seconds. This energy system is used for maximal short bursts of energy, for example, to run a 100-meter dash.

FIGURE 10.10 Summary of the events of contraction and relaxation.

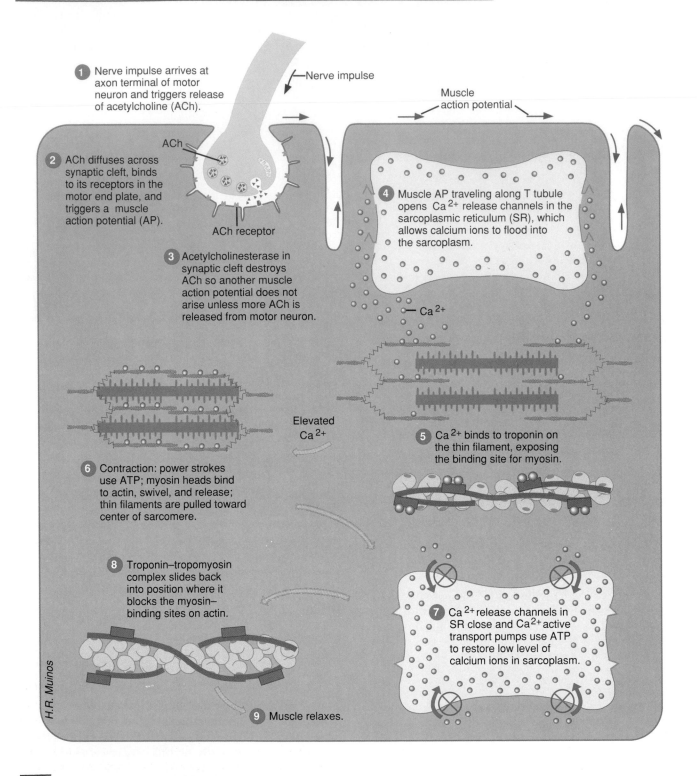

1 Nerve impulse arrives at axon terminal of motor neuron and triggers release of acetylcholine (ACh).

Nerve impulse

Muscle action potential

ACh

2 ACh diffuses across synaptic cleft, binds to its receptors in the motor end plate, and triggers a muscle action potential (AP).

ACh receptor

3 Acetylcholinesterase in synaptic cleft destroys ACh so another muscle action potential does not arise unless more ACh is released from motor neuron.

4 Muscle AP traveling along T tubule opens Ca^{2+} release channels in the sarcoplasmic reticulum (SR), which allows calcium ions to flood into the sarcoplasm.

Ca^{2+}

Elevated Ca^{2+}

5 Ca^{2+} binds to troponin on the thin filament, exposing the binding site for myosin.

6 Contraction: power strokes use ATP; myosin heads bind to actin, swivel, and release; thin filaments are pulled toward center of sarcomere.

8 Troponin–tropomyosin complex slides back into position where it blocks the myosin–binding sites on actin.

7 Ca^{2+} release channels in SR close and Ca^{2+} active transport pumps use ATP to restore low level of calcium ions in sarcoplasm.

9 Muscle relaxes.

H.R. Muinos

Question: What are three functions of ATP in skeletal muscle contraction?

FIGURE 10.11 Muscle metabolism during contraction. (a) Creatine phosphate, formed while the muscle is relaxed, transfers a high-energy phosphate group to ADP, forming ATP. (b) Breakdown of muscle glycogen into glucose and metabolism of glucose via glycolysis produce both ATP and lactic acid. Since no oxygen is needed, this is an anaerobic pathway. (c) Within mitochondria, pyruvic acid produced by glycolysis, amino acids, and fatty acids liberated from adipose cells are used to produce ATP.

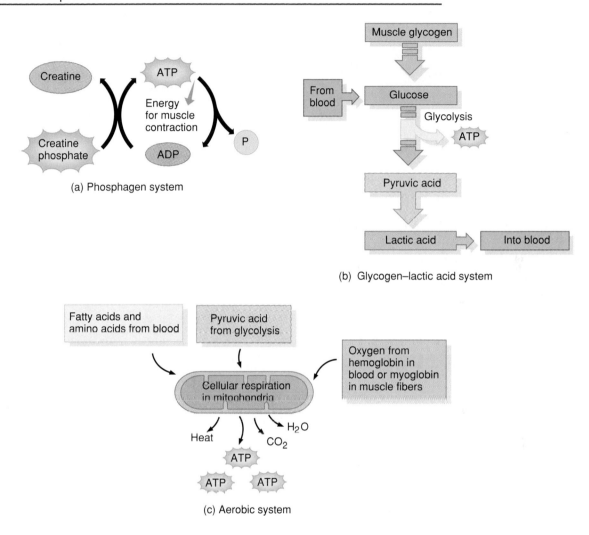

Question: Where in the muscle fiber are the events shown here occurring?

Glycogen–Lactic Acid System

When muscle activity continues and the supply of creatine phosphate is depleted, glucose is catabolized to generate ATP. Glucose easily passes into contracting muscle fibers from the blood (via facilitated diffusion) and also is produced by breakdown of glycogen within muscle fibers (Fig. 10.11b). A series of ten reactions known as **glycolysis** quickly splits each glucose molecule into two molecules of pyruvic acid and forms two molecules of ATP (see Fig. 25.2). Since glycolysis does not require oxygen, it is said to be an **anaerobic process**.

Ordinarily, the pyruvic acid formed by glycolysis enters mitochondria. There, its oxidation produces a large amount of ATP from ADP. During some activities, however, there is not enough oxygen to completely break down pyruvic acid. When this happens, most of the pyruvic acid is converted to lactic acid. About 80% of the lactic acid produced in this way diffuses from the skeletal muscles into the blood. Heart muscle fibers, kidney cells, and liver cells can use lactic acid to produce ATP. Also, liver cells can convert some of the lactic acid back to glucose. This conversion has two benefits—providing new glucose molecules and reducing acidity. However, some lactic acid accumulates in blood and in muscle tissue. The glycogen–lactic acid system just described can provide enough energy for about 30 to 40 seconds of maximal muscle activity, for example, to run a 300-meter race. Eventually, muscle glycogen must also be restored. This is accomplished mainly through eating carbohydrate-rich foods and may take several days, depending on the intensity of exercise.

Aerobic System

Muscular activity that lasts longer than half a minute depends increasingly on **aerobic processes**, that is, reactions requiring oxygen. If sufficient oxygen is present, enzymes in the mitochondria can completely oxidize pyruvic acid to carbon dioxide, water, ATP, and heat (Fig. 10.11c). This process, called **cellular respiration** or **biological oxidation**, is described in more detail in Chapter 25. While slower than glycolysis, it yields much more energy, about 36 molecules of ATP from each glucose molecule.

Muscle tissue has two sources of oxygen: (1) oxygen that diffuses into muscle fibers from the blood and (2) oxygen that is released by **myoglobin** inside muscle fibers. Both myoglobin and hemoglobin (in red blood cells) are red-colored, oxygen-binding proteins; they bind oxygen when it is plentiful and release it in times of scarcity.

The aerobic system will provide enough ATP for prolonged activity so long as sufficient oxygen and nutrients are available. Besides glucose, these nutrients include fatty acids (from the breakdown of triglycerides in adipose cells) and amino acids (from the breakdown of protein). In activities that last more than 10 minutes, the aerobic system provides more than 90% of the needed ATP. During a long-term event, such as a marathon race, close to 100% of the ATP is produced aerobically.

The maximum rate of oxygen consumption during the aerobic catabolism of pyruvic acid is called **maximal oxygen uptake.** It is influenced by gender (higher in males), age (highest at about age 20), and size (increases with body size). Highly trained athletes can have maximal oxygen uptakes that are twice those of untrained people, owing to a combination of both heredity and training. As a result, they are capable of greater muscular feats than untrained people.

Oxygen Consumption After Exercise

During muscular exercise, blood vessels in muscles dilate, blood flow increases, and oxygen delivery increases. Up to a point, the available oxygen is sufficient to meet the energy needs of the contracting muscles. When muscular exertion is very great, however, oxygen cannot be supplied to muscle fibers fast enough and cellular respiration cannot produce enough ATP.

After exercise has stopped, heavy breathing continues for a period of time, and oxygen consumption is above the resting level. Depending on the intensity of the exercise, the recovery period may be just a few minutes or several hours. In 1922 A. V. Hill coined the term **oxygen debt** for the added oxygen that is taken into the body after exercise, over and above the resting oxygen consumption. He proposed that this extra oxygen was all used to "pay back" or restore metabolic conditions to the resting level. In this view, the extra oxygen is used to (1) convert lactic acid back into pyruvic acid, (2) reestablish the glycogen stores, (3) resynthesize creatine phosphate and ATP, and (4) replace the oxygen removed from myoglobin.

The metabolic changes that occurred *during exercise*, however, account for only some of the extra oxygen used *after exercise*. Postexercise oxygen use also is boosted by ongoing changes. First, body temperature is elevated after strenuous exercise. This increases the pace of chemical reactions throughout the body. Faster reactions use ATP more rapidly, and more oxygen is needed to produce ATP. Second, the heart and muscles used in breathing are still working harder than they were at rest and thus consume more ATP. Third, tissue repair processes are occurring at an increased pace. For these reasons, a better term than oxygen debt for the elevated oxygen use after exercise is **recovery oxygen consumption.**

Muscle Fatigue

If a skeletal muscle or group of skeletal muscles is overstimulated, the strength of contraction becomes progressively weaker until the muscle no longer responds. The inability of a muscle to maintain its strength of contraction or tension is called **muscle fatigue.** It occurs when a muscle cannot produce enough ATP to meet its needs. Several factors appear to contribute to muscle fatigue, including insufficient oxygen, depletion of glycogen, buildup of lactic acid, failure of action potentials in the motor neuron to release enough acetylcholine, and unexplained fatigue mechanisms in the central nervous system.

CLINICAL APPLICATION

EXERCISE-INDUCED MUSCLE DAMAGE

Electron micrographs of muscle tissue taken from athletes before and after a marathon race or other types of severe exercise reveal considerable damage, including a torn sarcolemma in some muscle fibers, damaged myofibrils, and disrupted Z discs. Microscopic muscle damage after exercise also is indicated by increases in blood levels of myoglobin and certain enzymes (lactic acid dehydrogenase and creatine phosphokinase, for example) that normally are confined inside muscle fibers. From 12 to 48 hours after a bout of strenuous exercise, skeletal muscles often become sore. Such delayed onset muscle soreness (DOMS) is accompanied by stiffness, tenderness, and swelling. Although the causes of DOMS are not completely understood, microscopic muscle damage appears to be a major contributing factor.

Homeostasis of Body Temperature

Both smooth and skeletal muscles play important roles in maintaining the body's thermal homeostasis. The main contribution of smooth muscle is in regulation of blood vessel diameter. When smooth muscle in the walls of skin arterioles relaxes, the arterioles dilate, and more blood flows to the skin. This permits greater transfer of heat from the warm blood through the skin to the environment. On the other hand, when heat conservation is needed, smooth muscle in the skin blood vessels contracts. As a result, the vessels constrict, less blood flows through the skin, and less heat is lost.

During contraction of skeletal muscles, only a small amount of the energy stored in body chemicals is used for mechanical work (movement). As much as 85% is released as heat (thermogenesis), some of which is utilized to help maintain a normal body temperature. Excess heat is eliminated through the skin and lungs. If body temperature (controlled condition) decreases, one result is **shivering,** which is involuntary thermogenesis. This increase in skeletal muscle tone can raise heat production by several hundred percent. Shivering is initiated by a part of the brain called the hypothalamus (control center). It acts via a negative feedback system to produce enough heat to raise body temperature back to normal (Fig. 10.12). Other mechanisms for regulation of body temperature are discussed on page 854.

ADJUSTING MUSCLE TENSION

A single action potential in a motor neuron elicits a single contraction in all the muscle fibers of its motor unit. The contraction is said to be **all-or-none** because individual muscle fibers will contract to their fullest extent. In other words, muscle fibers do not partially contract. The force of their contraction can vary only slightly, depending on local chemical conditions (for example, nutrient and oxygen availability) and whether or not the motor unit has just previously contracted.

You know, however, that a muscle as a whole can have graded contractions (varying sizes) to perform different tasks. For example, your arm muscles do not contract to the same extent when you lift a textbook compared to a piece of paper. But they do contract in a smooth, graded fashion to lift either object. The amount of tension (force) that a skeletal muscle can develop depends on the frequency of stimulation of muscle fibers by motor neurons, the length of muscle fibers just before they contract, the number of muscle fibers contracting (number of motor units recruited and size of individual motor units), and structural components of the muscle itself.

Twitch

A **twitch contraction** is a brief contraction of all the muscle fibers in a motor unit in response to a single action

FIGURE 10.12 Negative feedback regulation of body temperature by the shivering mechanism.

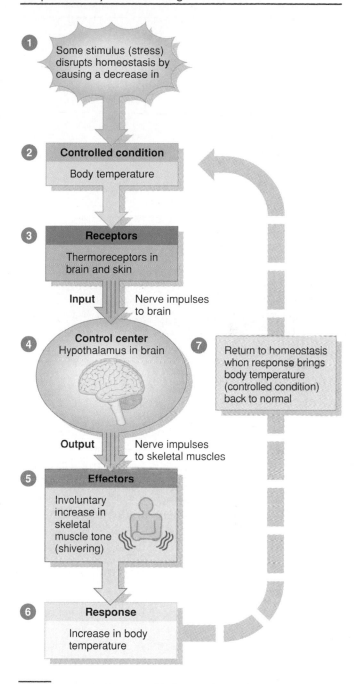

① Some stimulus (stress) disrupts homeostasis by causing a decrease in

② **Controlled condition**
Body temperature

③ **Receptors**
Thermoreceptors in brain and skin

Input Nerve impulses to brain

④ **Control center**
Hypothalamus in brain

⑦ Return to homeostasis when response brings body temperature (controlled condition) back to normal

Output Nerve impulses to skeletal muscles

⑤ **Effectors**
Involuntary increase in skeletal muscle tone (shivering)

⑥ **Response**
Increase in body temperature

Question: Can you draw a similar diagram for regulation of body temperature through sweating?

potential in its motor neuron. In the laboratory, a twitch also can be produced by direct electrical stimulation of a motor neuron or its muscle fibers. Much of our knowledge of muscle contraction has come from experiments performed on isolated, excised muscle.

Figure 10.13 is a graph of a twitch contraction. The

FIGURE 10.13 Myogram of a twitch contraction. The arrow at 0 indicates the point at which the stimulus is applied.

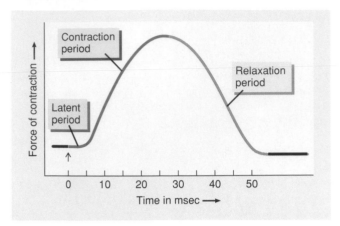

Question: What is occurring during the latent period?

record of a muscle contraction is called a **myogram.** Note that a brief period exists between application of the stimulus and the beginning of contraction; this is the **latent period.** During this time, Ca^{2+} is being released from the sarcoplasmic reticulum, the filaments start to exert force (usually taking some slack out of the system), and finally shortening starts. The latent period lasts about 2 milliseconds (msec; 1 msec = $\frac{1}{1000}$ sec). The second phase, the **contraction period,** lasts from 10 to 100 msec and is indicated by the upward tracing. The third phase, the **relaxation period,** also lasts about 10 to 100 msec and is indicated by the downward tracing. It is caused by the active transport of Ca^{2+} back into the sarcoplasmic reticulum, which results in relaxation. The duration of these periods varies with the muscle involved. Some fibers, such as those that move the eyes, are fast-twitch fibers (described shortly) and have contraction times as brief as 10 msec and an equally brief relaxation time. Others, such as those that move the legs, are slow-twitch fibers, with contraction and relaxation times of 100 msec or so.

If additional stimuli are applied to the muscle after the initial stimulus, other responses may be noted. For example, if two stimuli are applied one immediately after the other, the muscle will respond to the first stimulus but not to the second. When a muscle fiber receives enough stimulation to contract, it temporarily loses its excitability and cannot respond again until its responsiveness is regained. This period of lost excitability is called the **refractory period** and is a characteristic of both nerve and muscle cells (see page 361). The duration of the refractory period varies with the muscle involved. Skeletal muscle has a short refractory period of about 5 msec (0.005 sec). Cardiac muscle has a long refractory period of about 300 msec (0.30 sec).

Frequency of Stimulation

When two stimuli are applied and the second is delayed until the refractory period is over, the skeletal muscle will respond to both stimuli. In fact, if the second stimulus is applied after the refractory period, but before the muscle fiber has finished relaxing, the second contraction will be stronger than the first. This phenomenon, in which stimuli arrive at different times and cause larger contractions, is called **wave (temporal) summation** (Fig. 10.14a).

FIGURE 10.14 Myograms showing effect of frequency of stimulation. (a) The second stimulus (long arrow) is applied before the muscle has finished relaxing. The second contraction is stronger than the first because the forces sum. (The broken line indicates the force expected from a single twitch.) (b) In incomplete tetanus the curve looks jagged due to partial relaxation of the muscle in between stimuli. (c) In complete tetanus, the contraction force is sustained until after the last stimulus.

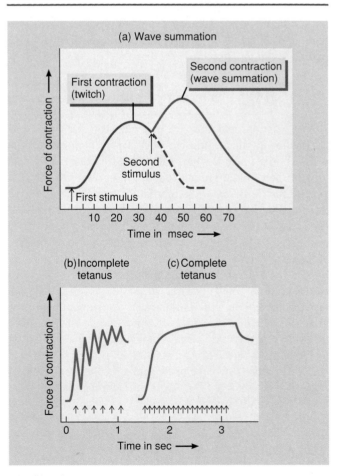

Question: In (a), would the maximum force be larger or smaller if the second stimulus were applied at 25 msec rather than at 35 msec as indicated in the figure?

Tetanus

If a muscle is stimulated at a rate of 20 to 30 times per second, it can only partly relax between stimuli. The result is a sustained contraction called **incomplete (unfused) tetanus** (Fig. 10.14b). Stimulation at an increased rate (80 to 100 stimuli per second) results in **complete (fused) tetanus,** a sustained contraction that lacks even partial relaxation between stimuli (Fig. 10.14c).

Both kinds of tetanus result from the addition of Ca^{2+} released from the sarcoplasmic reticulum by the second, and subsequent, stimuli to the Ca^{2+} still in the sarcoplasm from the first stimulus. Relaxation is either partial or does not occur at all. Most of your voluntary muscular contractions involve short-term tetanic contractions and are thus smooth, sustained contractions.

Staircase Effect (Treppe)

When a muscle has been relaxed for some time and then is stimulated to contract by several identical stimuli that are too far apart for wave summation to occur, each of the first few contractions is a little stronger than the last. This phenomenon is known as the **staircase effect** or **treppe** (TREP-eh; = staircase). After the first few contractions, the muscle reaches its peak of performance and can undergo its strongest contractions. The explanation for the staircase effect may be the same as for tetanus—a progressive buildup of Ca^{2+} in the sarcoplasm. Successive stimuli cause calcium ions to flow out of the sarcoplasmic reticulum faster than the active transport pumps take them back in. Up to a certain point, as Ca^{2+} builds up and binds to troponin, more power strokes can occur, and filament sliding intensifies. Also, other internal conditions in the muscle, such as temperature, pH, and viscosity, have changed. A rise in temperature, for example, could provoke stronger contractions. One advantage of "warming up" for athletes may be to take advantage of the staircase effect to improve performance.

Length of Muscle Fibers

As you already know, a skeletal muscle fiber contracts when myosin cross bridges of thick filaments connect with actin on the thin filaments. A muscle fiber develops its greatest tension when there is optimal overlap between thick and thin filaments (Fig. 10.15). At the optimum sarcomere length, the number of myosin cross bridges making contact with thin filaments brings about a maximal force of contraction.

If the sarcomeres of a muscle fiber are stretched to a longer length, fewer myosin cross bridges can make contact with thin filaments, and the force of contraction decreases. If a skeletal muscle fiber is stretched to 175% of its optimal length, no myosin cross bridges can bind to thin filaments

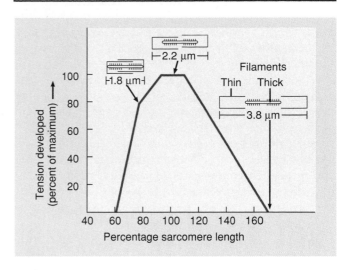

FIGURE 10.15 Length–tension relationship in skeletal muscle fibers. Maximum tension is produced at a sarcomere length of 2.2 µm (= 100%).

Question: Why is maximum tension possible at a sarcomere length of 2.2 µm?

and no contraction occurs. At sarcomere lengths less than the optimum, the force of contraction also decreases. This is because thick filaments crumple as they are compressed by the Z discs, resulting in fewer myosin cross bridge contacts with thin filaments. In the intact body, resting muscle fiber length is rarely less than 70% or more than 130% of optimum. This is because skeletal muscles are firmly anchored to bones and other inelastic tissues.

Number of Muscle Fibers Contracting

The process of increasing the number of active motor units is called **recruitment (multiple motor unit summation).** The various motor neurons to a whole muscle fire asynchronously. While some motor units are active (contraction), others are inactive (relaxed). This pattern of firing of motor neurons prevents fatigue while maintaining contraction by allowing a brief rest for the inactive units. The alternating motor units relieve one another so that the contraction can be sustained for long periods.

Recruitment also is one factor responsible for producing smooth movements rather than a series of jerky movements. As indicated before, the number of muscle fibers innervated by one motor neuron varies greatly. Precise movements require tiny changes in muscle contraction. Therefore, in such muscles, the motor units are small. In this way, when a motor unit is recruited or turned off, slight but controlled changes occur in muscle contraction. On the other hand, large motor units are employed where maintaining a constant position or posture is important and precision is not.

Role of Structural Components of Muscle Fibers

As thin filaments start to slide past thick filaments, they pull on the Z discs, which in turn pull on the rest of the muscle fibers and their connective tissue wrappings. Some of these structural components are elastic: they stretch slightly before they start to relay the force or tension being generated by the sliding filaments. The elastic components include the elastic filaments, connective tissue around the muscle fibers (epimysium, perimysium, and endomysium), and tendons that attach muscle to bone.

The tension generated by contractile elements (thin and thick filaments) is called **active tension**. The tension generated by elastic elements is called **passive tension** and is not related to muscular contraction. It depends on the degree of muscle stretch; within limits, the more a muscle is stretched, the greater its passive tension.

When a skeletal muscle contracts, it first pulls on its connective tissue coverings and tendons. As a result, the coverings and tendons stretch, they become taut, and the tension passed through the tendons pulls on the bones to which they are attached. The result is movement of a part of the body.

The role of elastic elements can be made clearer by imagining that you are trying to move a heavy object along the ground by pulling on a spring attached to the object. When you pull on the spring, it stretches, but the object stays put until the tension in the stretched spring equals the weight of the object. When the tension exceeds the weight of the object, the object moves. If the weight to be moved is light, only a slight tug on the spring is required to move it, and it will move quickly. On the other hand, if the weight is heavy, a greater tug is required because the spring must stretch farther, and the weight moves slowly.

The stretch of elastic elements is also related to wave summation and tetanus (see Fig. 10.14). During wave summation, elastic elements are not given much time to spring back between contractions, and thus they remain taut. While in this state, the elastic elements do not require very much stretching before the beginning of the next muscular contraction. The combination of the tautness of the elastic elements and partially contracted state of filaments enables the force of another contraction to be added immediately to the one before.

Isotonic and Isometric Contractions

Isotonic (*iso* = equal; *tonos* = tension) **contractions** occur when you move a constant load through the range of motions possible at a joint. During such a contraction, the tension remains almost constant. There are two types of isotonic contractions. In a **concentric contraction**, the muscle shortens and pulls on another structure, such as a bone, to produce movement and to reduce the angle at a joint (Fig. 10.16a). Picking up a book involves concentric contractions

---

FIGURE 10.16 Comparison between isotonic (concentric and eccentric) and isometric contractions.

(a) Concentric contraction   (b) Eccentric contraction   (c) Isometric contraction

Question: What kind of contraction is typical in your neck muscles while you are reading your class assignment?

of the biceps brachii muscle in the arm. When the overall length of a muscle increases during a contraction, it is called an **eccentric contraction** (Fig. 10.16b). As you lower the book to place it back on the table, the previously shortened biceps gradually lengthens while it continues to contract. For reasons that are not well understood, repeated eccentric contractions produce more muscle damage and more delayed onset muscular soreness than concentric contractions.

An **isometric contraction** occurs when the muscle does not or cannot shorten, but the *tension* on the muscle increases greatly (Fig. 10.16c). An example would be holding a book in a steady position. The book pulls the arm downward, stretching the shoulder and arm muscles. The isometric contraction of the shoulder and arm muscles counteracts the stretch. The two forces—contraction and stretching—applied in opposite directions create the tension. Although isometric contractions do not result in body movement, energy is still expended. Most activities include both isotonic and isometric contractions.

Regular aerobic workouts that feature many isotonic contractions, for example, jogging or aerobic dancing, increase the blood supply of skeletal muscles and thus improve blood flow. Activities that include many isometric contractions, for example, weight lifting, rely more on anaerobic production of ATP. Such activities also stimulate synthesis of muscle proteins, and the result, over a period of time, is an increase in muscle size (muscle hypertrophy). Aerobic training builds endurance for prolonged activities while anaerobic training builds muscle strength for short-term feats. **Interval training** is a workout regimen that incorporates both types of training, for example, alternating sprints (anaerobic activity) with jogging (aerobic activity).

CLINICAL APPLICATION

MUSCULAR ATROPHY AND HYPERTROPHY

Muscular atrophy (A-tro-fē) is a wasting away of muscles. Individual muscle fibers decrease in size owing to a progressive loss of myofibrils. Muscles atrophy if they are not used. This is termed **disuse atrophy.** Bedridden individuals and people with casts experience atrophy because the flow of nerve impulses to inactive muscle is greatly reduced. If the nerve supply to a muscle is cut, the muscle undergoes **denervation atrophy.** In about 6 months to 2 years, the muscle will be one-quarter its original size and the muscle fibers will be replaced by fibrous tissue. The transition to fibrous tissue, when complete, cannot be reversed.

Muscular hypertrophy (hī-PER-trō-fē), the opposite of atrophy, refers to an increase in the diameter of muscle fibers owing to the production of more myofibrils, mitochondria, sarcoplasmic reticulum, and so forth. It results from very forceful, repetitive muscular activity, such as strength training. Hypertrophied muscles are capable of more forceful contractions.

TYPES OF SKELETAL MUSCLE FIBERS

Skeletal muscle fibers are not all alike in structure or function. For example, they vary in color depending on their content of myoglobin, the oxygen-storing protein. Skeletal muscle fibers that have a high myoglobin content are termed **red muscle fibers.** Those that have a low content of myoglobin, on the other hand, are called **white muscle fibers.** Red muscle fibers also have more mitochondria and more blood capillaries than white muscle fibers.

As mentioned earlier, skeletal muscle fibers contract and relax with different velocities. Whether a fiber is slow-twitch or fast-twitch depends on how rapidly it splits ATP. Fast-twitch fibers split ATP more quickly. In addition, skeletal muscle fibers vary in the metabolic reactions they use to generate ATP and in how quickly they fatigue. Based on these structural and functional characteristics, skeletal muscle fibers are classified into three types:

1. **Slow oxidative (type I) fibers.** These fibers, also called **slow-twitch** or **fatigue-resistant fibers,** contain large amounts of myoglobin, many mitochondria, and many blood capillaries. They thus look red and have a high capacity to generate ATP by the aerobic system, which is why they are called *oxidative* fibers. They split ATP at a slow rate and, as a result, have a slow contraction velocity. These fibers are very resistant to fatigue. Such fibers are found in large numbers in the postural muscles, for example, in neck muscles that hold the head upright.

2. **Fast oxidative (type II A) fibers.** These fibers, also called **fast-twitch A** or **fatigue-resistant fibers,** contain large amounts of myoglobin, many mitochondria, and many blood capillaries. They are red and have a high capacity for generating ATP by oxidative processes. Such fibers also split ATP at a very rapid rate, and as a result, contraction velocity is fast. Fast oxidative fibers are resistant to fatigue but not quite as much as slow oxidative fibers. Sprinters tend to have a large proportion of fast oxidative fibers in their leg muscles.

3. **Fast glycolytic (type II B) fibers.** These fibers, also called **fast-twitch B** or **fatigable fibers,** have a low myoglobin content, relatively few mitochondria, and relatively few blood capillaries. They do, however, contain large amounts of glycogen. Fast glycolytic fibers are white and geared to generate ATP by anaerobic processes (glycolysis) that are not able to supply skeletal muscle fibers continuously with sufficient ATP. Accordingly, these fibers fatigue easily. They are the largest diameter fibers, and they split ATP at a fast rate so that contraction is strong and rapid. Muscles of the arms contain many of these fibers.

Most skeletal muscles of the body are a mixture of all three types of skeletal muscle fibers, but their proportion varies depending on the usual action of the muscle. For example, postural muscles of the neck, back, and legs have a high proportion of slow oxidative fibers. Muscles of the shoulders and arms are not constantly active but are used intermittently, usually for short periods of time, to produce large amounts of tension such as in lifting and throwing. These muscles have a high proportion of fast glycolytic fibers. Leg muscles not only support the body but are also used for walking and running. Such muscles have large numbers of both slow and fast oxidative fibers.

Even though most skeletal muscles are a mixture of all three types of skeletal muscle fibers, the skeletal muscle fibers of any one motor unit are all the same. But, the different motor units in a muscle may be used in various ways, depending on need. For example, if only a weak contraction is needed to perform a task, only slow oxidative motor units are activated. If a stronger contraction is needed, the motor units of fast glycolytic fibers are recruited. And if a maximal contraction is required, motor units of fast oxidative fibers are also called into action. Activation of various motor units is determined in the brain and spinal cord.

The total number of skeletal muscle fibers usually does not change, but the characteristics of those present can be altered to some extent. Various types of exercises can induce changes in the fibers in a skeletal muscle. Endurance-type exercises, such as running or swimming, cause a gradual transformation of some fast glycolytic (type II B) fibers into fast oxidative (type II A) fibers. The transformed muscle fibers show slight increases in diameter, mitochondria, blood capillaries, and strength. Endurance exercises result in cardiovascular and respiratory changes that cause skeletal muscles to receive better supplies of oxygen and nutrients but do not increase muscle mass. On the other hand, exercises that require great strength for short periods of time, such as weight lifting, produce an increase in the size and strength of fast glycolytic fibers. The increase in size is due to increased synthesis of thin and thick filaments. The overall result is muscle enlargement.

The exact way that muscle grows in size with training is not clear. Some studies show increases in size of individual muscle fibers while others claim increases in overall number of muscle fibers. However, the number of muscle fibers is not thought to increase significantly after birth. During childhood, the increase in the *size* of muscle fibers appears to be at least partially under the control of human growth hormone, which is produced by the anterior pituitary gland. In males, a further increase in the size of muscle fibers is due to the hormone testosterone, produced by the testes.

A summary of the factors that influence muscle tension is presented in Exhibit 10.2.

CLINICAL APPLICATION

ANABOLIC STEROIDS

Steroids are lipids derived from cholesterol that serve many useful functions in the body (discussed in later chapters). In

---

EXHIBIT 10.2

FACTORS INFLUENCING MUSCLE TENSION

Individual muscle fibers	Frequency of stimulation by motor neurons (wave summation and tetanus)
	Length of fibers just before contraction (length–tension relation)
	Diameter of fiber (larger diameter fast glycolytic fibers contract more forcefully than smaller diameter slow oxidative and fast oxidative fibers)
	Extent of fatigue (availability of nutrients and oxygen versus buildup of lactic acid)
Whole muscle	Number of muscle fibers contracting at one time
	Number of fibers per motor unit (motor unit size)
	Number of motor units recruited (multiple motor unit summation)
	Active tension generated by contractile elements
	Passive tension generated by elastic components

---

recent years, the illegal use of **anabolic steroids** by athletes has received widespread attention. These steroids, similar to the male sex hormone testosterone, are taken to increase muscle size and therefore to increase strength and endurance during athletic events.

With an optimal training program, however, there is little evidence that anabolic steroids confer significant additional strength or endurance when taken in *moderate* doses. The *large* doses needed to see an effect produce damaging and devastating side effects. These include liver cancer, kidney damage, increased risk of heart disease, stunted growth in young people, increased irritability and aggressive behavior, and wide mood swings. Females may become sterile, develop facial hair, experience deepening of the voice and atrophy of the breasts and uterus, and suffer menstrual irregularities. Males may suffer atrophy of the testes, diminished hormone secretion and sperm production by the testes, and baldness.

CARDIAC MUSCLE TISSUE

The principal tissue in the heart wall is **cardiac muscle tissue.** Although it is striated like skeletal muscle, it is involuntary like smooth muscle. Also, certain cardiac muscle fibers (and some smooth muscle fibers) display autorhythmicity, establishing an inherent rhythm for alternating contraction and relaxation.

Anatomy

The fibers of cardiac muscle tissue are roughly quadrangular, about 14 μm in diameter, and usually have only a single centrally located nucleus (see Exhibit 4.3 on page 120). The sarcolemma of cardiac muscle fibers is similar to that of skeletal muscle, but the sarcoplasm is more abundant and the mitochondria are larger and more numerous. Cardiac muscle fibers have the same arrangement of actin and myosin and the same bands, zones, and Z discs as skeletal muscle fibers (Fig. 10.17). The transverse tubules of cardiac muscle are larger and the sarcoplasmic reticulum is more scanty than in skeletal muscle.

Cardiac muscle fibers branch and interconnect with each other, but form two separate networks. The muscular walls and partition of the upper chambers of the heart (atria) compose one network. The muscular walls and partition of the lower chambers of the heart (ventricles) compose the other network. Each fiber in a network is connected to its neighbors by irregular transverse thickenings of the sarcolemma called **intercalated** (in-TER-ka-lāt-ed) **discs.** The discs contain desmosomes, which hold the fibers together, and gap junctions, which allow muscle action potentials to spread from one muscle fiber to another (see Fig. 4.1). As a consequence, when a single fiber of either network is stimulated, all the other fibers in the network become stimulated as well. Thus each network contracts as a functional unit. When the fibers of the atria contract as a unit, blood moves into the ventricles. Then, when the ventricular fibers contract as a unit, blood is pumped out of the heart into arteries.

Physiology

Under normal resting conditions, cardiac muscle tissue contracts and relaxes rapidly, continuously, and rhythmically about 75 times a minute without stopping. This is a major physiological difference between cardiac and skeletal muscle tissue. Accordingly, cardiac muscle tissue requires a constant supply of oxygen.

In cardiac muscle fibers, the mitochondria are larger and more numerous than in skeletal muscle fibers. This microscopic observation provides a clue that cardiac muscle depends greatly on the aerobic system to generate ATP. Cardiac muscle generates little ATP anaerobically by the glycogen–lactic acid system. Moreover, cardiac muscle fibers can use lactic acid produced by skeletal muscle fibers to make ATP, a benefit during exercise.

Another difference is the source of stimulation. Skeletal muscle tissue contracts only when stimulated by acetylcholine released by an action potential in a motor neuron. In contrast, cardiac muscle tissue can contract without extrinsic (outside) nerve or hormonal stimulation. Its source of stimulation is a conducting network of specialized cardiac muscle fibers within the heart. Nerve stimulation merely causes the conducting fibers to increase or decrease their rate of discharge. Some types of smooth muscle fibers and nerve cells in the brain and spinal cord also possess such autorhythmicity, which we will discuss in detail in Chapter 20.

Another difference between cardiac and skeletal muscle tissue is that cardiac muscle tissue remains contracted 10 to 15 times longer than skeletal muscle tissue. This is because there is a prolonged delivery of Ca^{2+} into the sarcoplasm. In cardiac muscle fibers, Ca^{2+} enters the sarcoplasm both from the sarcoplasmic reticulum (as in skeletal muscle fibers) and from extracellular fluid.

Cardiac muscle tissue also has a long refractory period, lasting several tenths of a second, that allows time for the heart to relax between beats. The long refractory period permits heart rate to increase significantly but prevents the heart from undergoing tetanus. If heart muscle could undergo tetanus, blood flow would cease.

SMOOTH MUSCLE TISSUE

Anatomy

Like cardiac muscle tissue, **smooth muscle tissue** is usually involuntary. When examined under the light microscope, no myofibrils are apparent. Smooth muscle fibers are considerably smaller than skeletal muscle fibers. A single smooth muscle fiber is 30 to 200 μm long, thickest in the middle (3 to 8 μm), and tapered at each end. Within each fiber is a single, oval, centrally located nucleus (Fig. 10.18; also see Exhibit 4.3 on page 121). The sarcoplasm of smooth muscle fibers contains both **thick filaments** and **thin filaments,** but they are not arranged in orderly sarcomeres as in striated muscle. In smooth muscle fibers, there are 10 to 15 thin filaments for each thick filament; in skeletal muscle fibers, the ratio is 2:1. Smooth muscle fibers also contain **intermediate filaments.** Since the various filaments have no regular pattern of organization and since there are no A or I bands, smooth muscle fibers have no cross-striations. This is the reason for the name *smooth*.

In smooth muscle fibers, intermediate filaments attach to structures called **dense bodies,** which are similar to Z discs in striated muscle fibers. Some dense bodies are dispersed throughout the sarcoplasm; others are attached to the sarcolemma. Bundles of intermediate filaments stretch from one dense body to another (Fig. 10.18c). During contraction, the sliding filament mechanism involving thick and thin filaments generates tension that is transmitted to intermediate filaments. These, in turn, pull on the dense bodies attached to the sarcolemma, causing a lengthwise shortening of the muscle fiber. Note that shortening of the muscle fiber produces a bubblelike expansion of the sarcolemma. Evidence suggests that a smooth muscle fiber contracts like a corkscrew turns; the fiber twists in a helix as it shortens and rotates in the opposite direction as it lengthens.

FIGURE 10.17 Histology of cardiac muscle tissue.

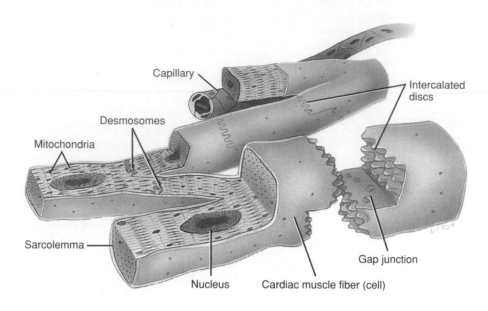

(a) Cardiac muscle fibers

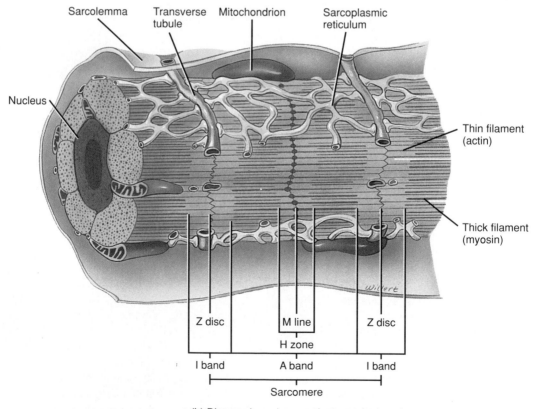

(b) Diagram based on an electron micrograph

Question: What is the function of desmosomes and gap junctions?

FIGURE 10.18 Histology of smooth muscle tissue.

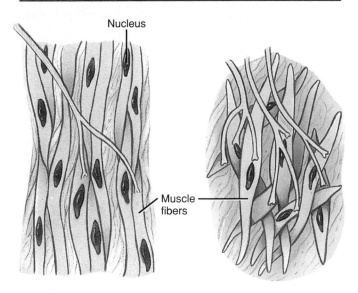

(a) Visceral (single-unit) smooth
muscle tissue

(b) Multiunit smooth
muscle tissue

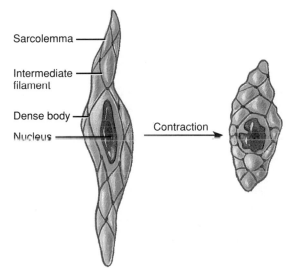

(c) Details of a smooth muscle fiber

Question: Which type of smooth muscle is more like cardiac muscle, with respect to both its structure and function?

Two kinds of smooth muscle tissue, visceral and multiunit, are recognized (Fig. 10.18a,b). The more common type is **visceral (single-unit) smooth muscle tissue.** It is found in wraparound sheets that form part of the walls of small arteries and veins and hollow viscera such as the stomach, intestines, uterus, and urinary bladder. The fibers in visceral muscle tissue form large networks because they contain gap junctions. These allow muscle action potentials to spread throughout the network. When a neurotransmitter, hormone, or autorhythmic signal stimulates one fiber, the

muscle action potential spreads to neighboring fibers, which then contract as a single unit.

The second kind of smooth muscle tissue is **multiunit smooth muscle tissue.** It consists of individual fibers, each with its own motor neuron terminals and with few gap junctions between neighboring fibers. Whereas stimulation of a single visceral muscle fiber causes contraction of many adjacent fibers, stimulation of a single multiunit fiber causes contraction of only that fiber. Multiunit smooth muscle tissue is found in the walls of large arteries, in large airways to the lungs (bronchioles), in the arrector pili muscles that attach to hair follicles, and in the radial and circular muscles of the iris that adjust pupil diameter.

Physiology

Although the principles of contraction are similar in smooth and striated muscle tissue, smooth muscle tissue exhibits several important physiological differences. In comparison with contraction in a striated muscle fiber, contraction in a smooth muscle fiber starts more slowly and lasts much longer. Moreover, smooth muscle can both shorten and stretch to a greater extent than can striated muscle.

Role of Calmodulin and Myosin Light Chain Kinase

An increase in the concentration of Ca^{2+} in smooth muscle sarcoplasm initiates contraction, just as in striated muscle. Sarcoplasmic reticulum (the reservoir for Ca^{2+} in striated muscle) is scanty in smooth muscle. Calcium ions flow into smooth muscle sarcoplasm from both the extracellular fluid and sarcoplasmic reticulum. Since there are no transverse tubules in smooth muscle fibers, it takes longer for Ca^{2+} to reach the filaments in the center of the fiber and trigger the contractile process. This accounts, in part, for the slow onset and prolonged contraction of smooth muscle.

In smooth muscle, the regulator protein that binds Ca^{2+} in the cytosol is **calmodulin**. (Recall that troponin has this role in striated muscle fibers.) After binding to Ca^{2+}, calmodulin activates an enzyme called **myosin light chain kinase**. This enzyme uses ATP to phosphorylate (add phosphate to) a portion of the myosin head. Once phosphate is attached, the myosin head can bind to actin, and contraction can occur. The kinase works rather slowly, thus contributing to the slowness of smooth muscle contraction. Other mechanisms also are known to cause contraction or relaxation of smooth muscle.

Smooth Muscle Tone

Not only does Ca^{2+} enter smooth muscle fibers slowly, it also moves slowly out of the muscle fiber when excitation declines. This delays relaxation, and the prolonged stay of

Ca^{2+} in the fibers provides for smooth muscle tone, a state of continued partial contraction. Smooth muscle tissue can sustain long-term tone, which is important in the gastrointestinal tract where the walls maintain a steady pressure on the contents of the tract and in the walls of blood vessels called arterioles that maintain a steady pressure on blood.

Regulation of Smooth Muscle Contraction

Most smooth muscle fibers contract or relax in response to action potentials from the autonomic (involuntary) nervous system. Thus smooth muscle normally is not under voluntary control. In addition, many smooth muscle fibers contract or relax in response to hormones or local factors such as changes in pH, oxygen and carbon dioxide levels, temperature, and ion concentrations. For example, the hormone epinephrine, released by the adrenal medulla, causes relaxation of smooth muscle in the airways and some blood vessel walls (those that have so-called β_2 receptors; see Exhibit 17.3 on page 512).

Unlike striated muscle fibers, smooth muscle fibers can stretch considerably and still maintain their contractile function. When smooth muscle fibers are stretched, they initially contract, developing increased tension. Within a minute or so, the tension decreases. This phenomenon is termed the **stress-relaxation response.** It allows smooth muscle to undergo great changes in length while still retaining the ability to contract effectively. Thus the smooth muscle in the walls of blood vessels and hollow organs such as the stomach, intestines, and urinary bladder can stretch, but the pressure on the contents within increases very little. After the organ empties, on the other hand, the smooth muscle in the wall rebounds, and the wall is firm, not flabby.

Exhibit 10.3 summarizes the principal characteristics of the three types of muscle tissue.

REGENERATION OF MUSCLE TISSUE

Skeletal muscle fibers have little potential to divide. After the first year of life, growth of skeletal muscle is due to enlargement of existing cells (hypertrophy), rather than an increase in the number of fibers (hyperplasia). Skeletal muscle fibers, however, can be replaced on an individual basis by new fibers derived from **satellite cells,** dormant stem cells found in association with skeletal muscle fibers (see Fig. 10.4c). During rapid postnatal growth, satellite cells lengthen existing skeletal muscle fibers by fusing with them. They also persist as a lifelong source of cells that can fuse with each other to form new skeletal muscle fibers. However, the number of new skeletal muscle fibers formed by this mechanism is not sufficient to compensate for any significant skeletal muscle damage. In cases of such dam-

age, skeletal muscle tissue undergoes **fibrosis,** replacement of muscle fibers by fibrous scar tissue. For this reason, skeletal muscle tissue has only limited powers of regeneration.

Cardiac muscle fibers, like those of skeletal muscle tissue, do not appear to divide in the body. (They have been experimentally induced to undergo division under certain laboratory conditions.) The fibers can undergo hypertrophy, however. Healing of damaged cardiac muscle tissue is by scar formation (fibrosis) because cardiac muscle tissue does not regenerate.

Smooth muscle tissue, like skeletal and cardiac muscle tissue, can undergo hypertrophy. In addition, certain smooth muscle fibers, such as those in the uterus, retain their capacity for division and thus can grow by hyperplasia. Also, new smooth muscle fibers can arise from cells called **pericytes,** stem cells found in association with the endothelium of blood capillaries and small veins. It is also known that smooth muscle fibers can proliferate in certain pathological conditions such as occur in the development of atherosclerosis (see page 616). Compared with the other two types of muscle tissue, smooth muscle tissue has a considerably higher power of regeneration. It is still limited when compared with other tissues, such as epithelium.

AGING AND MUSCLE TISSUE

Beginning at about 30 years of age, there is a progressive loss of skeletal muscle mass that is replaced largely by fat. In part, this decline is due to increasing inactivity. Accompanying the loss of muscle mass is a decrease in maximal strength and a slowing of muscle reflexes. In some muscles, there may be a selective loss of muscle fibers of a given type. With aging, the relative number of slow oxidative fibers (type I) appears to increase. This could be due either to atrophy of the other fiber types or their conversion into slow oxidative fibers. Whether this is an effect of aging itself or merely reflects the more limited physical activity of older people is still an unresolved question.

DEVELOPMENTAL ANATOMY OF THE MUSCULAR SYSTEM

In this brief discussion of the development of the human muscular system, we concentrate mostly on skeletal muscles. Except for the muscles of the iris of the eyes and the arrector pili muscles attached to hairs, all muscles of the body are derived from **mesoderm.** As the mesoderm develops, a portion of it becomes arranged in dense columns on either side of the developing nervous system. These columns of mesoderm undergo segmentation into a series of blocks of cells called **somites** (Fig. 10.19a). The first pair

EXHIBIT 10.3

SUMMARY OF THE PRINCIPAL FEATURES OF MUSCLE TISSUE

Characteristic	Skeletal Muscle	Cardiac Muscle	Smooth Muscle
Microscopic appearance	Striations, many nuclei, unbranched fibers.	Striations, single nucleus, branched fibers with intercalated discs.	No striations, single nucleus, spindle-shaped fibers.

Location	Attached primarily to bones.	Heart.	Walls of hollow viscera, blood vessels, iris, arrector pili.
Nervous control	Voluntary.	Involuntary.	Involuntary.
Sarcomeres	Yes.	Yes.	No.
Transverse tubules	Yes.	Yes.	No.
Gap junctions between fibers	No.	Yes.	Yes, in visceral smooth muscle. No, in multiunit smooth muscle.
Fiber diameter	Very large (10 to 100 μm).	Large (14 μm).	Small (3 to 8 μm).
Fiber length	100 μm to 30 cm.	50 to 100 μm.	30 to 200 μm.
Source of calcium	Sarcoplasmic reticulum.	Sarcoplasmic reticulum, extracellular fluid.	Sarcoplasmic reticulum, extracellular fluid.
Regulator proteins	Troponin and tropomyosin.	Troponin and tropomyosin.	Calmodulin and myosin light chain kinase.
Speed of contraction	Fast.	Moderate.	Slow.
Capacity for regeneration	Limited.	None, in the body.	Considerable compared with other muscle tissues but limited compared with tissues such as epithelium.

FIGURE 10.19 Development of the muscular system.

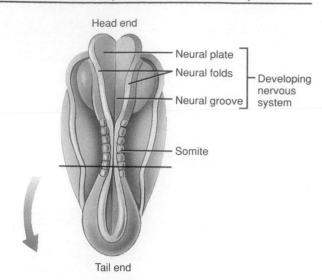

(a) Dorsal aspect of an embryo showing somites

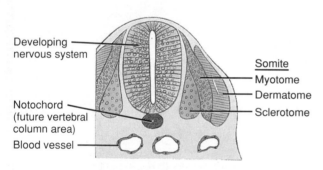

(b) Cross section of a somite

of somites appears on the 20th day of embryologic development. Eventually, 44 pairs of somites are formed by the 30th day.

With the exception of the skeletal muscles of the head and extremities, *skeletal muscles* develop from the **mesoderm of somites.** Since there are very few somites in the head region of the embryo, most of the skeletal muscles there develop from the **general mesoderm** in the head region. The skeletal muscles of the limbs develop from masses of general mesoderm around developing bones in embryonic limb buds (origins of future extremities: see Fig. 6.9a).

The cells of a somite are differentiated into three regions: (1) **myotome,** which forms most of the skeletal muscles; (2) **dermatome,** which forms the connective tissues, including the dermis; and (3) **sclerotome,** which gives rise to the vertebrae (Fig. 10.19b). *Fasciae, ligaments,* and *aponeuroses* may form as a result of degeneration of all or parts of myotomes.

Cardiac muscle develops from **mesodermal cells** that migrate to and envelop the developing heart while it is still in the form of primitive heart tubes (see Fig. 20.13).

Smooth muscle develops from **mesodermal cells** that migrate to and envelop the developing gastrointestinal tract and viscera.

DISORDERS: HOMEOSTATIC IMBALANCES

FIBROMYALGIA

Fibromyalgia (*algia* = painful condition) refers to a group of common nonarticular rheumatic disorders that are 15 times more common in women (usually appearing between the ages of 25 and 50) than in men. A striking sign is pain that results from gentle pressure at specific "tender spots." Even without pressure, there is pain, tenderness, and stiffness of muscles, tendons, and surrounding soft tissues. The disorder specifically affects the fibrous connective tissue components of muscles, tendons, and ligaments. Fibromyalgia may be caused or aggravated by physical or mental stress, trauma, exposure to dampness or cold, poor sleep, or a rheumatic condition. Often, a gentle aerobic fitness program is beneficial.

Frequent sites at which fibromyalgia occurs include the lumbar region **(lumbago),** neck, chest, and thighs **(charleyhorse).** Charleyhorse is a slang term that refers to a painful, localized cramp that sometimes is associated with pregnancy

or may occur in otherwise healthy persons. In some cases, local muscle spasms are noted. The condition usually is relieved with heat, massage, and rest and completely disappears.

MUSCULAR DYSTROPHIES

Muscular dystrophies (*dystrophy* = degeneration) are inherited muscle-destroying diseases that are characterized by degeneration of individual muscle fibers, which leads to a progressive atrophy of the skeletal muscle. Usually, the voluntary skeletal muscles are weakened equally on both sides of the body, whereas the internal muscles, such as the diaphragm, are not affected.

The most common form of muscular dystrophy is called **Duchenne** (doo-SHĀN) **muscular dystrophy.** It strikes boys almost exclusively (1 in 3500), usually appearing between the ages of 3 and 5. Using the techniques of genetic engineering, scientists have discovered that a protein named

dystrophin is present in the sarcolemma of normal muscle fibers but absent in persons with Duchenne muscular dystrophy. According to one hypothesis, lack of dystrophin may result in leakage of calcium ions into the sarcoplasm. This, in turn, may activate an enzyme (phospholipase A) that causes muscle fibers to degenerate.

MYASTHENIA GRAVIS

Myasthenia (mī-as-THĒ-nē-a) **gravis** is a weakness of skeletal muscles. It is caused by an abnormality at the neuromuscular junction that partially blocks contraction. Recall that motor neurons stimulate skeletal muscle fibers to contract by releasing acetylcholine (ACh). Myasthenia gravis is an autoimmune disorder caused by antibodies directed against ACh receptors. The antibodies bind to the receptors and hinder the attachment of ACh at the motor end plate (see Fig. 10.3). As the disease progresses, more ACh receptors are affected. The muscle becomes increasingly weaker and may eventually cease to function.

Myasthenia gravis, like most autoimmune diseases, is more common in females, occurring most often between the ages of 20 and 50. Its incidence is 2 to 5 new cases per 100,000. The muscles of the face and neck are most often affected. Initial symptoms include a weakness of the eye muscles, which may produce double vision, and difficulty in swallowing. Later, the individual has difficulty chewing and talking. Eventually, the muscles of the limbs may become involved. Death may result from paralysis of the respiratory muscles, but often the disorder does not progress to this stage.

Anticholinesterase drugs such as neostigmine have been the primary treatment. They act as inhibitors of acetylcholinesterase, the enzyme that breaks down ACh. Thus they raise the level of ACh that is available to bind with still functional receptors. More recently, steroid drugs, such as prednisone, have been used with success to reduce antibody levels. Another treatment involves plasmapheresis, a procedure that removes the antibodies from the blood. Sometimes, surgical removal of the thymus gland (thymectomy) is helpful.

ABNORMAL CONTRACTIONS OF SKELETAL MUSCLE

One kind of abnormal muscular contraction is a **spasm,** a sudden involuntary contraction of a single muscle in a large group of muscles. (Cerebral palsy is characterized by generalized spastic contractions.) A painful spasmodic contraction is known as a **cramp. Tremor** is a rhythmic, involuntary, purposeless contraction of opposing muscle groups. A **fasciculation** is an involuntary, brief twitch of a muscle visible under the skin. It occurs irregularly and is not associated with movement of the affected muscle. Fasciculations may be seen in multiple sclerosis (see p. 439) or amyotrophic lateral sclerosis, also called Lou Gehrig's disease. A **fibrillation** is similar to a fasciculation except that it is not visible under the skin. It is recorded by electromyography. A **tic** is a spasmodic twitching made involuntarily by muscles that are ordinarily under voluntary control. Twitching of the eyelid and facial muscles are examples. In general, tics are of psychological origin.

Gas gangrene (GANG-rēn; *gangraena* = an eating sore) Death of a soft tissue, such as muscle, that results from interruption of its blood supply. One type is caused by various species of *Clostridium,* bacteria that live anaerobically in the soil.

Myoma (mī-Ō-ma; *oma* = tumor) A tumor consisting of muscle tissue.

Myomalacia (mī'-ō-ma-LĀ-shē-a; *malaco* = soft) Softening of a muscle.

Myopathy (mī-OP-a-thē; *pathos* = disease) Any abnormal condition or disease of muscle tissue.

Myositis (mī'-ō-SĪ-tis; *itis* = inflammation of) Inflammation of muscle fibers (cells).

Myotonia (mī-ō-TŌ-nē-a; *tonia* = tension) Increased muscular excitability and contractility with decreased power of relaxation; tonic spasm of the muscle.

Paralysis (pa-RAL-a-sis; *para* = beyond; *lyein* = to loosen) Loss or impairment of motor (muscular) function resulting from a lesion of nervous or muscular origin.

Volkmann's contracture (FOLK-manz kon-TRAK-tur; *contra* = against) Permanent shortening of a muscle due to replacement of destroyed muscle fibers with fibrous tissue that lacks ability to stretch. Destruction of muscle fibers may occur from interference with circulation caused by a tight bandage, a piece of elastic, or a cast.

16. What is muscle fatigue and how is it caused? (p. 250)
17. Describe how muscular contractions contribute to the homeostasis of body temperature. (p. 250)
18. What is a myogram? Describe the latent period, contraction period, and relaxation period of a twitch muscle contraction. Construct a diagram to illustrate your answer. (p. 254)
19. Define the refractory period. How does it differ between skeletal and cardiac muscle? What is wave summation? (p. 254)
20. Distinguish between incomplete tetanus, complete tetanus, and the staircase effect. (p. 255)

21. What is recruitment? Why is it important? (p. 255)
22. Compare active and passive tension. (p. 256)
23. How do isotonic and isometric contractions differ? (p. 256)
24. Contrast the structure and functions of slow oxidative, fast oxidative, and fast glycolytic skeletal muscle fibers. (p. 257)
25. Compare skeletal, cardiac, and smooth muscle with regard to differences in structure, physiology, and capacity for division and regeneration. (p. 263)
26. Describe the effects of aging on muscle tissue. (p. 262)
27. Describe how the muscular system develops. (p. 262)

Answers to Questions with Figures

10.1 Endomysium, perimysium, epimysium, deep fascia, and superficial fascia.
10.2 Motor units having a large number of muscle fibers will be capable of more forceful contractions than those having a small number of fibers. Thus both recruitment and size of motor units influence strength of contraction.
10.3 The region of the muscle plasma membrane (sarcolemma), including acetylcholine receptors, that lies directly under the axon terminals of a motor neuron.
10.4 Dark = A bands; light = I bands.
10.5 Thin (actin) and elastic (titin) filaments.
10.6 A band: myosin, actin, troponin, and tropomyosin. I band: titin, actin, troponin, and tropomyosin.
10.7 They disappear. No.
10.8 Their shape changes and thus regulates whether or not the thick and thin filaments are sliding.
10.9 The myosin heads would not be able to detach from actin.
10.10 (1) Its splitting (hydrolysis) by an ATPase activates the myosin head so it can bind to actin and swivel; (2) its binding to myosin detaches the cross bridge from actin after the power stroke; and (3) it powers the pumps that transport Ca^{2+} from the sarcoplasm back into the sarcoplasmic reticulum.

10.11 Cytosol: glycolysis, exchange of phosphate between creatine phosphate and ADP, glycogen breakdown (glycogenolysis). Mitochondria: oxidation of pyruvic acid and fatty acids (cellular respiration).
10.12 See Fig. 5.7.
10.13 Release of Ca^{2+} from SR and binding to troponin; myosin heads binding to actin and starting to swivel.
10.14 A little larger since the force has already decreased a bit by 35 msec.
10.15 This is the sarcomere length that gives maximum overlap of the thick and thin filaments.
10.16 Isometric—holding your head upright in one position without movement.
10.17 Desmosomes hold cells tightly together while gap junctions allow ions and small molecules to flow from the cytosol of one cell to the cytosol of another without crossing the plasma membrane (sarcolemma).
10.18 Visceral smooth muscle and cardiac muscle are similar in that both include gap junctions, which allow action potentials to spread from one cell to its neighbors.

Chapter 11
THE MUSCULAR SYSTEM

Chapter Contents at a Glance

Student Objectives

1. Describe the relationship between bones and skeletal muscles in producing body movements.

2. Define a lever and fulcrum and compare the three classes of levers on the basis of placement of the fulcrum, effort, and resistance.

3. Identify the various arrangements of muscle fibers in a skeletal muscle and relate the

arrangements to strength of contraction and range of motion.

4. Discuss most body movements as activities of groups of muscles by explaining the roles of the prime mover, antagonist, synergist, and fixator.

5. Define the criteria employed in naming skeletal muscles.

6. Identify the principal skeletal muscles in different regions of the body by name, origin, insertion, action, and innervation.

7. Discuss the administration of drugs by intramuscular (IM) injections.

8. Describe several injuries related to running.

The term **muscle tissue** refers to all the contractile tissues of the body: skeletal, cardiac, and smooth muscle. The **muscular system**, however, refers to the *skeletal* muscle system: the skeletal muscle tissue and connective tissues that make up individual muscle organs, such as the biceps brachii muscle. Cardiac muscle tissue is located in the heart and is therefore considered part of the cardiovascular system. Smooth muscle tissue of the intestines is part of the digestive system, whereas smooth muscle tissue of the urinary bladder is part of the urinary system, and so on. In this chapter, we discuss only the muscular system. We will see how skeletal muscles produce movement, and we will describe the principal skeletal muscles.

HOW SKELETAL MUSCLES PRODUCE MOVEMENT

Origin and Insertion

Skeletal muscles produce movements by exerting force on tendons, which in turn pull on bones or other structures, such as skin. Most muscles cross at least one joint and are attached to the articulating bones that form the joint (Fig. 11.1). When such a muscle contracts, it draws one articulating bone toward the other. The two articulating bones usually do not move equally in response to the contraction. One is held nearly in its original position because other

FIGURE 11.1 Relationship of skeletal muscles to bones. (a) Skeletal muscles produce movements by pulling on bones. (b) Bones serve as levers, and joints act as fulcrums for the levers. Here the lever–fulcrum principle is illustrated by the movement of the forearm. Note where the resistance and effort are applied in this example.

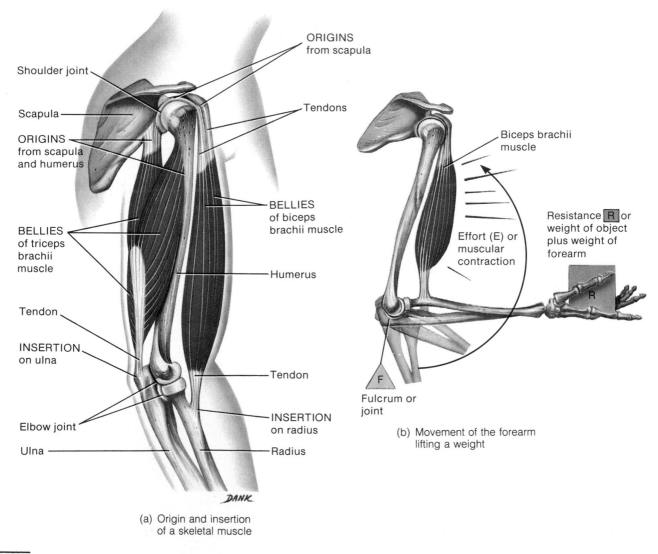

Shoulder joint

Scapula

ORIGINS from scapula and humerus

BELLIES of triceps brachii muscle

Tendon

INSERTION on ulna

Elbow joint

Ulna

ORIGINS from scapula

Tendons

BELLIES of biceps brachii muscle

Humerus

Tendon

INSERTION on radius

Radius

DANK

(a) Origin and insertion of a skeletal muscle

Biceps brachii muscle

Effort (E) or muscular contraction

Resistance R or weight of object plus weight of forearm

R

Fulcrum or joint

F

(b) Movement of the forearm lifting a weight

Question: What is the attachment of a muscle to a stationary bone called?

muscles contract to pull it in the opposite direction or because its structure makes it less movable. Ordinarily, the attachment of a muscle tendon to the stationary bone is called the **origin**. The attachment of the other muscle tendon to the movable bone is the **insertion**. A good analogy is a spring on a door. In this example, the part of the spring attached to the door represents the insertion; the part attached to the frame is the origin. The fleshy portion of the muscle between the tendons of the origin and insertion is called the **belly (gaster)**. The origin is usually proximal and the insertion distal, especially in the extremities. In addition, muscles that move a body part often do not cover the moving part. Figure 11.1a shows that although one of the functions of the biceps brachii muscle is to move the forearm, the belly of the muscle lies over the humerus, not the forearm.

Lever Systems and Leverage

In producing a body movement, bones act as levers and joints function as fulcrums of these levers. A **lever** may be defined as a rigid rod that moves about on some fixed point called a **fulcrum**. A fulcrum may be symbolized as ◭. A lever is acted on at two different points by two different forces: the **resistance** R and the **effort** (E). Whereas resistance is the force that opposes movement, effort is the force exerted to achieve an action. The resistance may be the weight of a part of the body that is to be moved. The effort is the muscular contraction. Motion is produced when the effort is applied to the bone at the insertion and the effort exceeds the resistance (load). Consider the biceps brachii flexing the forearm at the elbow as a weight is lifted (Fig. 11.1b). When the forearm is raised, the elbow is the fulcrum. The weight of the forearm plus the weight in the hand is the resistance. The shortening due to the force of contraction of the biceps brachii pulling the forearm up is the effort.

Levers are categorized into three types according to the positions of the fulcrum, the effort, and the resistance.

1. **First-class levers** have the fulcrum between the effort and resistance (Fig. 11.2a). This is symbolized

FIGURE 11.2 Classes of levers. Each is defined on the basis of the placement of the fulcrum (F), effort (E), and resistance (R). A third-class lever is also shown in Fig. 11.1b.

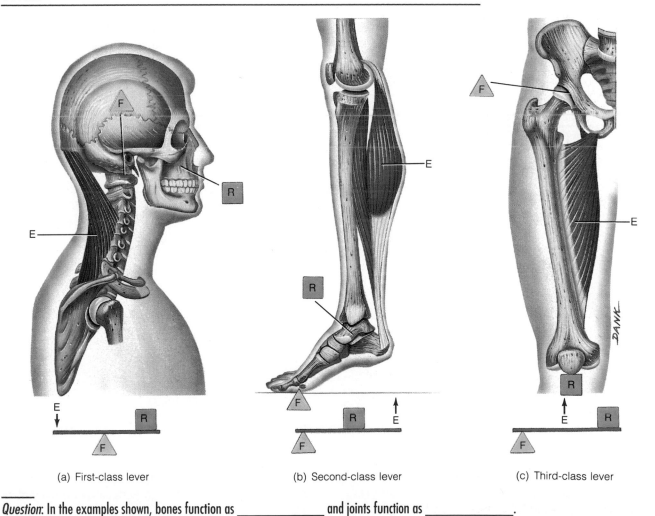

(a) First-class lever   (b) Second-class lever   (c) Third-class lever

Question: In the examples shown, bones function as _____ and joints function as _____.

EFR. An example of a first-class lever is a seesaw. There are not many first-class levers in the body. One example is the head resting on the vertebral column. When the head is raised, the facial portion of the skull is the resistance. The joints between the atlas and occipital bone (atlanto-occipital joints) form the fulcrum. The contraction of the muscles of the back is the effort.

2. **Second-class levers** have the fulcrum at one end, the effort at the opposite end, and the resistance between them (Fig. 11.2b). This is symbolized FRE. They operate like a wheelbarrow. Most authorities agree that there are very few examples of second-class levers in the body. One example is raising the body on the toes. The body is the resistance, the ball of the foot is the fulcrum, and the contraction of the calf muscles to pull the heel upward is the effort.

3. **Third-class levers** consist of the fulcrum at one end, the resistance at the opposite end, and the effort between them (Fig. 11.2c). This is symbolized FER. They are the most common levers in the body. One example is adduction of the thigh, in which the weight of the thigh is the resistance, the hip joint is the fulcrum, and contraction of the adductor muscles is the effort. Another example is flexing the forearm at the elbow. As we have seen, the weight of the forearm is the resistance, the contraction of the biceps brachii is the effort, and the elbow joint is the fulcrum (see Fig. 11.1b).

Leverage, the mechanical advantage gained by a lever, is largely responsible for a muscle's strength and range of motion (ROM), that is, the maximum ability to move the bones of a joint through an arc. Consider strength first. Suppose we have two muscles of the same strength crossing and acting on the same joint. Assume also that one is attached farther from the joint and one is nearer. The muscle attached farther away will produce the more powerful movement. Thus strength of movement depends on the placement of muscle attachments.

In considering ROM, again assume that we have two muscles of the same strength crossing and acting on the same joint and that one is attached farther from the joint than the other. The muscle inserting closer to the joint will produce the greater ROM and speed of movement. Thus ROM also depends on the placement of muscle attachments. Since strength increases with distance from the joint and ROM decreases, maximal strength and maximal range are incompatible; strength and range vary inversely.

Arrangement of Fasciculi

Recall from Chapter 10 that skeletal muscle fibers (cells) are arranged within the muscle in bundles called fasciculi (fascicles). The muscle fibers are arranged in a parallel fashion within each bundle, but the arrangement of the fasciculi with respect to the tendons may take several characteristic patterns (Exhibit 11.1).

Fascicular arrangement is correlated with the power of a muscle and range of motion. When a muscle fiber contracts, it shortens to a length just slightly greater than half of its resting length. Thus the longer the fibers in a muscle, the greater the range of motion it can produce. By contrast, the strength of a muscle depends on the total number of fibers it contains, since a short fiber can contract as forcefully as a long one. Because a given muscle can contain either a small number of long fibers or a large number of short fibers, fascicular arrangement represents a compromise between power and range of motion. Pennate muscles, for example, have a large number of fasciculi distributed over their tendons, giving them greater power, but a smaller range of motion. Parallel muscles, on the other hand, have comparatively few fasciculi that extend the length of the muscle. Thus they have a greater range of motion but less power.

Group Actions

Most movements require several skeletal muscles acting in groups rather than individually. Also, most skeletal muscles are arranged in opposing (antagonistic) pairs at joints, that is, flexors–extensors, abductors–adductors, and so on. Consider flexing the forearm at the elbow, for example. A muscle that causes a desired action is called the **prime mover (agonist)**. In this instance, the biceps brachii is the prime mover (see Fig. 11.1a). Simultaneously with the contraction of the biceps brachii, another muscle, called the **antagonist**, is relaxing. In this movement, the triceps brachii serves as the antagonist (see Fig. 11.1a). The antagonist has an effect opposite to that of the prime mover; that is, the antagonist relaxes and yields to the movement of the prime mover. You should not assume, however, that the biceps brachii is always the prime mover and the triceps brachii is always the antagonist. For example, when extending the forearm at the elbow, the triceps brachii serves at the prime mover and the biceps brachii functions as the antagonist; their roles are reversed. Note that if the prime mover and antagonist contracted simultaneously with equal force, there would be no movement.

In addition to prime movers and antagonists, most movements also involve muscles called **synergists**, which serve to steady a movement, thus preventing unwanted movements and helping the prime mover function more efficiently. For example, flex your hand at the wrist and then make a fist. Note how difficult this is to do. Now, extend your hand at the wrist and then make a fist. Note how much easier it is to clench your fist. In this case, the extensor muscles of the wrist act as synergists in cooperation with the flexor muscles of the fingers acting as prime movers. The extensor muscles of the fingers serve as antagonists (see Fig. 11.17c).

Some synergistic muscles in a group also act as **fixators**, which stabilize the origin of the prime mover so that the prime mover can act more efficiently. For example, the scapula is a freely movable bone in the pectoral (shoulder)

EXHIBIT 11.1

ARRANGEMENT OF FASCICULI

Arrangement	Description	Example
PARALLEL	Fasciculi are parallel with longitudinal axis of muscle and terminate at either end in flat tendons.	Stylohyoid muscle (see Fig. 11.7). 
FUSIFORM	Fasciculi are nearly parallel with longitudinal axis of muscle and terminate at either end in flat tendons, but muscle tapers toward tendons where the diameter is less than that of the belly.	Digastric muscle (see Fig. 11.8). 
PENNATE	Fasciculi are short in relation to muscle length and the tendon extends nearly the entire length of the muscle.	
Unipennate	Fasciculi are arranged on only one side of tendon.	Extensor digitorum longus muscle (see Fig. 11.22c). 
Bipennate	Fasciculi are arranged on both sides of a centrally positioned tendon.	Rectus femoris muscle (see Fig. 11.20a). 
Multipennate	Fasciculi attach obliquely from many directions to several tendons.	Deltoid muscle (see Fig. 11.10a). 

Exhibit continues

EXHIBIT 11.1 (continued)

ARRANGEMENT OF FASCICULI

Arrangement	Description	Example
CIRCULAR	Fasciculi are arranged in a circular pattern and enclose an orifice (opening).	Orbicularis oris muscle (see Fig. 11.4).

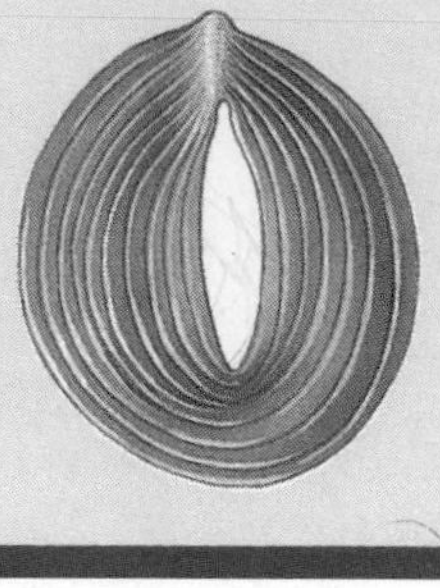

---

girdle that serves as a firm origin for several muscles that move the arm. However, for the scapula to do this, it must be held steady. This is accomplished by fixator muscles that hold the scapula firmly against the back of the chest. In abduction of the arm, the deltoid muscle serves as the prime mover, whereas fixators (pectoralis minor, rhomboideus major, rhomboideus minor, trapezius, subclavius, and serratus anterior muscles) hold the scapula firmly (see Fig. 11.14). These fixators stabilize the scapula that serves as the attachment site for the origin of the deltoid muscle, whereas the insertion of the muscle pulls on the humerus to abduct the arm. Under different conditions and depending on the movement and which point is fixed, many muscles act, at various times, as prime movers, antagonists, synergists, or fixators.

NAMING SKELETAL MUSCLES

The names of most of the nearly 700 skeletal muscles are based on several types of characteristics. Learning the terms used to indicate specific characteristics will help you remember the names of muscles (Exhibit 11.2).

PRINCIPAL SKELETAL MUSCLES

Exhibits 11.3 through 11.23 list the principal muscles of the body with their origins, insertions, actions, and innervations. (By no means have all the muscles of the body been

---

EXHIBIT 11.2

NAMING SKELETAL MUSCLES

Characteristic	Description	Example
Direction of muscle fibers	Direction of muscle fibers relative to the midline of the body. **Rectus** means the fibers run parallel to the midline. **Transverse** means the fibers run perpendicular to the midline. **Oblique** means the fibers run diagonally to the midline.	Rectus abdominis (see Fig. 11.10b). Transversus abdominis (see Fig. 11.10b). External oblique (see Fig. 11.10a).
Location	Structure near which a muscle is found. A muscle near the frontal bone. A muscle near the front of the tibia.	Frontal (see Fig. 11.4). Tibialis anterior (see Fig. 11.22d).

Characteristic	Description	Example
Size	Relative size of the muscle. **Maximus** means largest. **Minimus** means smallest. **Longus** means longest. **Brevis** means short.	 Gluteus maximus (see Fig. 11.20c). Gluteus minimus (see Fig. 11.20c). Adductor longus (see Fig. 11.20a). Peroneus brevis (see Fig. 11.22c,d).
Number of origins	Number of tendons of origin. **Biceps** means two origins. **Triceps** means three origins. **Quadriceps** mean four origins.	 Biceps brachii (see Fig. 11.16a). Triceps brachii (see Fig. 11.16b). Quadriceps femoris (see Fig. 11.20a).
Shape	Relative shape of the muscle. **Deltoid** means triangular. **Trapezius** means trapezoid. **Serratus** means saw-toothed. **Rhomboideus** means rhomboid or diamond-shaped.	 Deltoid (see Fig. 11.10a). Trapezius (see Fig. 11.14c). Serratus anterior (see Fig. 11.14b). Rhomboideus major (see Fig. 11.14c).
Origin and insertion	Sites where muscle originates and inserts.	Sternocleidomastoid originates on sternum and clavicle and inserts on mastoid process of temporal bone (see Fig. 11.14c). Stylohyoid originates on styloid process of temporal bone and inserts on hyoid bone (see Fig. 11.8).
Action	Principal action of the muscle. **Flexor** (FLEK-sor): decreases the angle at a joint. **Extensor** (eks-TEN-sor): increases the angle at a joint. **Abductor** (ab-DUK-tor): moves a bone away from the midline. **Adductor** (ad-DUK-tor): moves a bone closer to the midline. **Levator** (le-VĀ-tor): produces an upward movement. **Depressor** (de-PRES-or): produces a downward movement. **Supinator** (soo′-pi-NĀ-tor): turns the palm upward or anteriorly. **Pronator** (prō-NĀ-tor): turns the palm downward or posteriorly. **Sphincter** (SFINGK-ter): decreases the size of an opening. **Tensor** (TEN-sor): makes a body part more rigid. **Rotator** (RŌ-tāt-or): moves a bone around its longitudinal axis.	 Flexor carpi radialis (see Fig. 11.17a). Extensor carpi ulnaris (see Fig. 11.17c). Abductor pollicis brevis (see Fig. 11.17c). Adductor longus (see Fig. 11.20a). Levator scapulae (see Fig. 11.14d). Depressor labii inferioris (see Fig. 11.4b). Supinator (see Fig. 11.17b). Pronator teres (see Fig. 11.17a). External anal sphincter (see Fig. 11.12). Tensor fasciae latae (see Fig. 11.20a). Obturator externus (see Fig. 11.20b).

included.) An **overview** section in each exhibit provides a general orientation to the muscles under consideration. Refer to Chapters 7 and 8 to review bone markings, since they serve as points of origin and insertion for muscles. Students often have difficulty in pronouncing names of skeletal muscles and understanding how they get their names. To make this task easier, we have provided you with phonetic pronunciations and derivations that indicate how the muscles get their names. If you have mastered the naming of the muscles, their actions will have more meaning.

The muscles are divided into groups according to the part of the body on which they act. Figure 11.3 shows general anterior and posterior views of the muscular system. Do not try to memorize all these muscles yet. As you study groups of muscles in the following exhibits, refer to Fig. 11.3 to see how each group is related to all others.

The figures that accompany the exhibits contain superficial and deep, anterior and posterior, or medial and lateral views to show each muscle's position as clearly as possible. An attempt has been made to show the relationship of the muscles under consideration to other muscles in the area you are studying.

As a further aid to your learning efforts, the list on page 278 is provided so you can see the order in which the skeletal muscles will be studied according to region.

FIGURE 11.3 Principal superficial skeletal muscles.

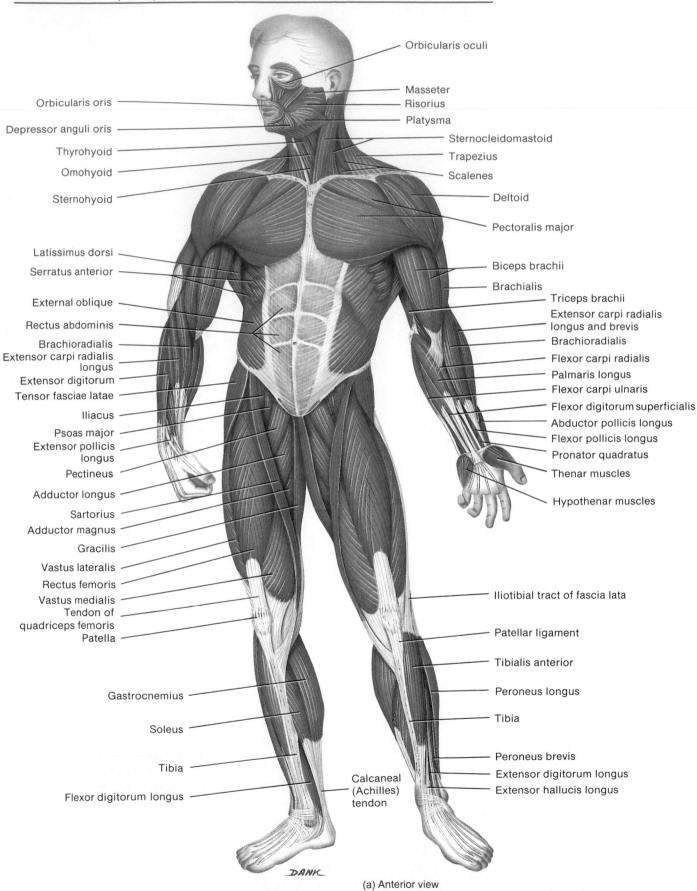

Orbicularis oculi

Masseter

Risorius

Platysma

Sternocleidomastoid

Trapezius

Scalenes

Deltoid

Pectoralis major

Biceps brachii

Brachialis

Triceps brachii

Extensor carpi radialis longus and brevis

Brachioradialis

Flexor carpi radialis

Palmaris longus

Flexor carpi ulnaris

Flexor digitorum superficialis

Abductor pollicis longus

Flexor pollicis longus

Pronator quadratus

Thenar muscles

Hypothenar muscles

Iliotibial tract of fascia lata

Patellar ligament

Tibialis anterior

Peroneus longus

Tibia

Peroneus brevis

Extensor digitorum longus

Extensor hallucis longus

Orbicularis oris

Depressor anguli oris

Thyrohyoid

Omohyoid

Sternohyoid

Latissimus dorsi

Serratus anterior

External oblique

Rectus abdominis

Brachioradialis

Extensor carpi radialis longus

Extensor digitorum

Tensor fasciae latae

Iliacus

Psoas major

Extensor pollicis longus

Pectineus

Adductor longus

Sartorius

Adductor magnus

Gracilis

Vastus lateralis

Rectus femoris

Vastus medialis

Tendon of quadriceps femoris

Patella

Gastrocnemius

Soleus

Tibia

Flexor digitorum longus

Calcaneal (Achilles) tendon

DANK

(a) Anterior view

FIGURE 11.3 (continued)

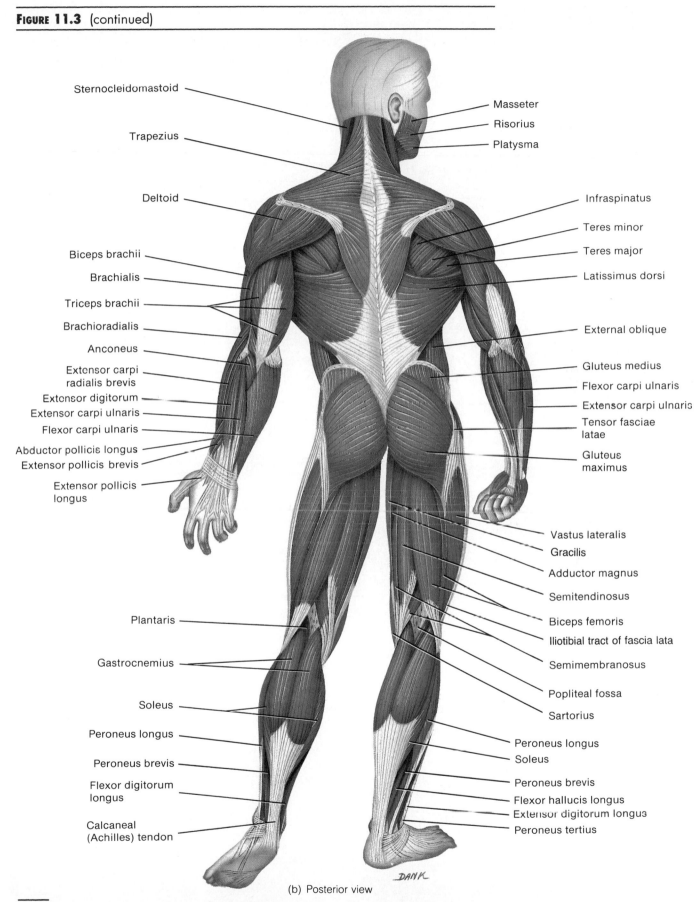

Sternocleidomastoid

Trapezius

Deltoid

Biceps brachii

Brachialis

Triceps brachii

Brachioradialis

Anconeus

Extensor carpi radialis brevis

Extensor digitorum

Extensor carpi ulnaris

Flexor carpi ulnaris

Abductor pollicis longus

Extensor pollicis brevis

Extensor pollicis longus

Plantaris

Gastrocnemius

Soleus

Peroneus longus

Peroneus brevis

Flexor digitorum longus

Calcaneal (Achilles) tendon

Masseter

Risorius

Platysma

Infraspinatus

Teres minor

Teres major

Latissimus dorsi

External oblique

Gluteus medius

Flexor carpi ulnaris

Extensor carpi ulnaris

Tensor fasciae latae

Gluteus maximus

Vastus lateralis

Gracilis

Adductor magnus

Semitendinosus

Biceps femoris

Iliotibial tract of fascia lata

Semimembranosus

Popliteal fossa

Sartorius

Peroneus longus

Soleus

Peroneus brevis

Flexor hallucis longus

Extensor digitorum longus

Peroneus tertius

DANK

(b) Posterior view

Question: Select a muscle named for the following characteristics: direction of fibers, shape, action, size, origin and insertion, location, and number of heads.

EXHIBIT 11.3

MUSCLES OF FACIAL EXPRESSION (Fig. 11.4)

Overview: The muscles in this group provide humans with the ability to express a wide variety of emotions, including surprise, fear, and happiness. The muscles themselves lie within the layers of superficial fascia. They usually arise from the fascia or bones of the skull and insert into the skin. Thus the muscles of facial expression move the skin rather than a joint when they contract.

Muscle	Origin	Insertion	Action	Innervation
Epicranius (ep-i-KRĀ-nē-us; *epi* = over; *crani* = skull)	This muscle is divisible into two portions: the frontalis over the frontal bone and the occipitalis over the occipital bone. The two muscles are united by a strong aponeurosis (sheetlike tendon), the **galea aponeurotica,** which covers the superior and lateral surfaces of the skull.			
Frontalis (fron-TA-lis; *front* = forehead)	Galea aponeurotica.	Skin superior to supraorbital margin.	Draws scalp forward, raises eyebrows, and wrinkles skin of forehead horizontally.	Facial (VII) nerve.
Occipitalis (ok-si′-pi-TA-lis; *occipito* = base of skull)	Occipital bone and mastoid process of temporal bone.	Galea aponeurotica.	Draws scalp backward.	Facial (VII) nerve.
Orbicularis oris (or-bi′-kyoo-LAR-is OR-is; *orb* = circular; *or* = mouth)	Muscle fibers surrounding opening of mouth.	Skin at corner of mouth.	Closes lips, compresses lips against teeth, protrudes lips, and shapes lips during speech.	Facial (VII) nerve.
Zygomaticus (zī-gō-MA-ti-kus) **major** (*zygomatic* = cheek bone; *major* = greater)	Zygomatic bone.	Skin at angle of mouth and orbicularis oris.	Draws angle of mouth upward and outward as in smiling or laughing.	Facial (VII) nerve.

EXHIBIT 11.3 (continued)

MUSCLES OF FACIAL EXPRESSION (Fig. 11.4)

Muscle	Origin	Insertion	Action	Innervation
Levator labii superioris (le-VĀ-tor LA-bē-ī soo-per′-ē-OR-is; *levator* = raises or elevates; *labii*=lip; *superioris*=upper)	Superior to infraorbital foramen of maxilla.	Skin at angle of mouth and orbicularis oris.	Elevates (raises) upper lip.	Facial (VII) nerve.
Depressor labii inferioris (de-PRE-ser LA-bē-ī infer′-ē-OR-is; *depressor* = depresses or lowers; *inferioris* = lower)	Mandible.	Skin of lower lip.	Depresses (lowers) lower lip.	Facial (VII) nerve.
Buccinator (BUK-si-nā′-tor; *bucc* = cheek)	Alveolar processes of maxilla and mandible and pterygomandibular raphe (fibrous band extending from the pterygoid process to the mandible).	Orbicularis oris.	Major cheek muscle; compresses cheek as in blowing air out of mouth and causes cheeks to cave in, producing the action of sucking.	Facial (VII) nerve.
Mentalis (men-TA-lis; *mentum* = chin)	Mandible.	Skin of chin.	Elevates and protrudes lower lip and pulls skin of chin up as in pouting.	Facial (VII) nerve.
Platysma (pla-TIZ-ma; *platy* = flat, broad)	Fascia over deltoid and pectoralis major muscles.	Mandible, muscles around angle of mouth, and skin of lower face.	Draws outer part of lower lip downward and backward as in pouting; depresses mandible.	Facial (VII) nerve.
Risorius (ri-ZOR-ē-us; *risor* = laughter)	Fascia over parotid (salivary) gland.	Skin at angle of mouth.	Draws angle of mouth laterally as in tenseness.	Facial (VII) nerve.
Orbicularis oculi (or-bi′-kyoo-LAR-is Ō-kyoo-lī; *ocul* = eye)	Medial wall of orbit.	Circular path around orbit.	Closes eye.	Facial (VII) nerve.
Corrugator supercilii (KOR-a-gā′-tor soo-per-SI-lē-ī; *corrugo* = wrinkle; *supercilium* = eyebrow)	Medial end of superciliary arch of frontal bone.	Skin of eyebrow.	Draws eyebrow downward as in frowning.	Facial (VII) nerve.
Levator palpebrae superioris (le-VĀ-tor PAL-pe-brē soo-per′-ē-OR-is; *palpebrae* = eyelids) (see also Fig. 11.6a)	Roof of orbit (lesser wing of sphenoid bone).	Skin of upper eyelid.	Elevates upper eyelid.	Oculomotor (III) nerve.

FIGURE 11.4 Muscles of facial expression.

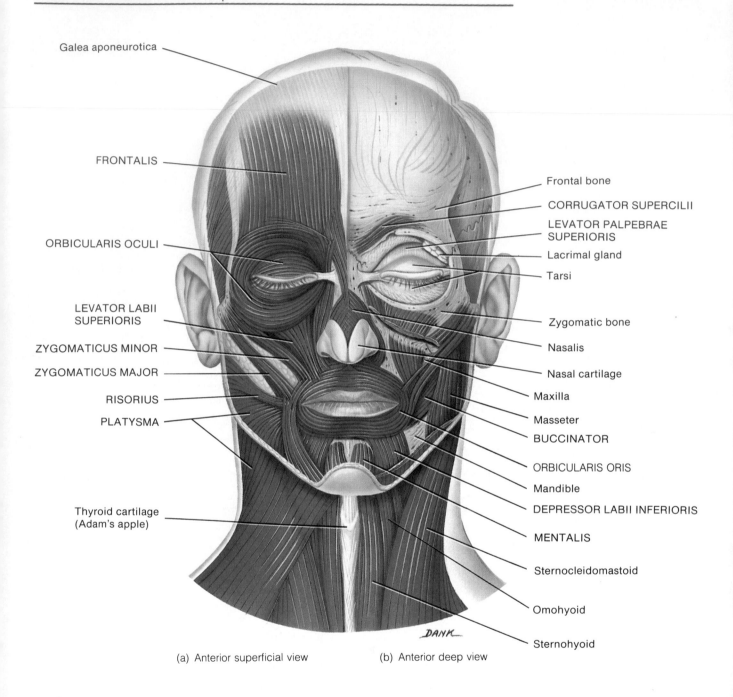

Galea aponeurotica

FRONTALIS

ORBICULARIS OCULI

LEVATOR LABII
SUPERIORIS

ZYGOMATICUS MINOR

ZYGOMATICUS MAJOR

RISORIUS

PLATYSMA

Thyroid cartilage
(Adam's apple)

Frontal bone

CORRUGATOR SUPERCILII

LEVATOR PALPEBRAE
SUPERIORIS

Lacrimal gland

Tarsi

Zygomatic bone

Nasalis

Nasal cartilage

Maxilla

Masseter

BUCCINATOR

ORBICULARIS ORIS

Mandible

DEPRESSOR LABII INFERIORIS

MENTALIS

Sternocleidomastoid

Omohyoid

Sternohyoid

DANK

(a) Anterior superficial view   (b) Anterior deep view

FIGURE 11.4 (continued)

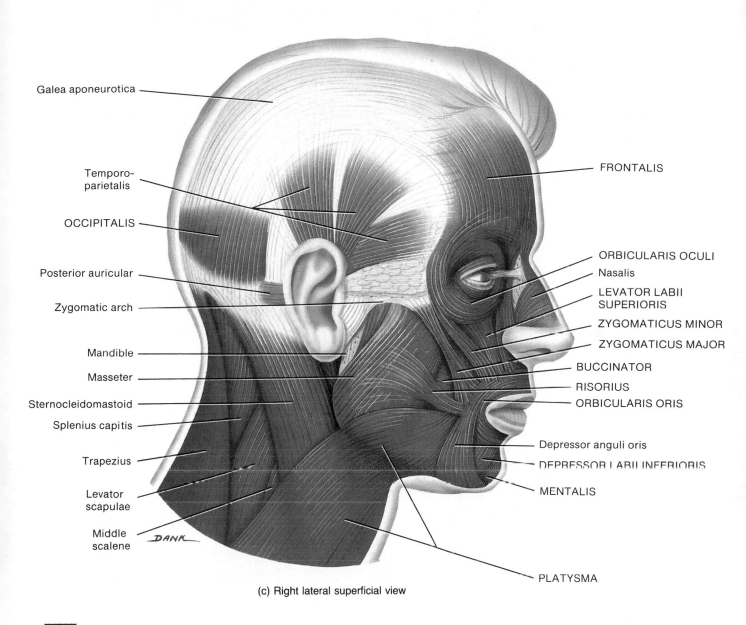

Galea aponeurotica

Temporo-
parietalis

OCCIPITALIS

Posterior auricular

Zygomatic arch

Mandible

Masseter

Sternocleidomastoid

Splenius capitis

Trapezius

Levator
scapulae

Middle
scalene

FRONTALIS

ORBICULARIS OCULI

Nasalis

LEVATOR LABII
SUPERIORIS

ZYGOMATICUS MINOR

ZYGOMATICUS MAJOR

BUCCINATOR

RISORIUS

ORBICULARIS ORIS

Depressor anguli oris

DEPRESSOR LABII INFERIORIS

MENTALIS

PLATYSMA

DANK

(c) Right lateral superficial view

Question: What major muscle causes frowning, smiling, pouting, and squinting?

EXHIBIT 11.4

MUSCLES THAT MOVE THE LOWER JAW (Fig. 11.5)

Overview: Muscles that move the lower jaw are also known as muscles of mastication because they are involved in biting and chewing. These muscles also assist in speech.

Muscle	Origin	Insertion	Action	Innervation
Masseter (MA-se-ter; *maseter* = chewer)	Maxilla and zygomatic arch.	Angle and ramus of mandible.	Elevates mandible as in closing mouth, assists in side to side movement of mandible, and protracts (protrudes) mandible.	Mandibular division of trigeminal (V) nerve.
Temporalis (tem´-por-A-lis; *tempora* = temples)	Parietal bone.	Coronoid process of mandible.	Elevates and retracts mandible and assists in side to side movement of mandible.	Mandibular division of trigeminal (V) nerve.

FIGURE 11.5 Muscles that move the lower jaw.

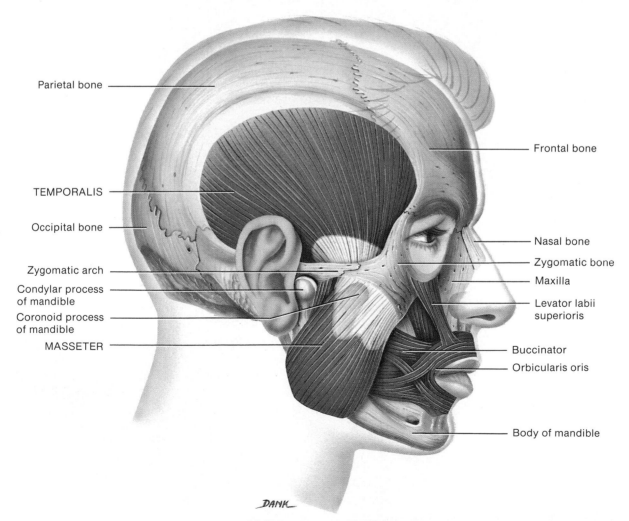

(a) Right lateral superficial view

Muscle	Origin	Insertion	Action	Innervation
Medial Pterygoid (TER-i-goid; *medial* = closer to midline; *pterygoid* = like a wing; pterygoid process of sphenoid bone)	Medial surface of lateral portion of pterygoid process of sphenoid; maxilla.	Angle and ramus of mandible.	Elevates and protracts mandible and moves mandible from side to side.	Mandibular division of trigeminal (V) nerve.
Lateral Pterygoid (TER-I-Goid; *lateral* = farther from midline)	Greater wing and lateral surface of lateral portion of pterygoid process of sphenoid.	Condyle of mandible; temporomandibular articulation.	Protracts mandible, opens mouth, and moves mandible from side to side.	Mandibular division of trigeminal (V) nerve.

FIGURE 11.5 (continued)

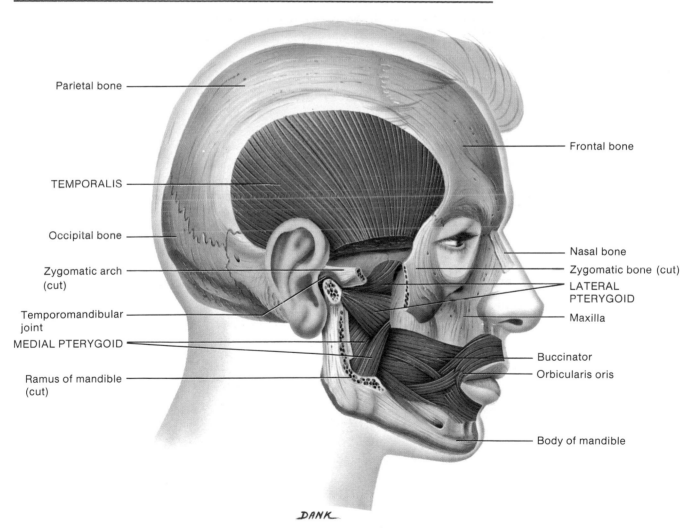

(b) Right lateral deep view

Question: Which muscles protract (protrude) the mandible?

EXHIBIT 11.5

MUSCLES THAT MOVE THE EYEBALLS—EXTRINSIC MUSCLES (Fig. 11.6)

Overview: Muscles associated with the eyeball are of two principal types: extrinsic and intrinsic. **Extrinsic muscles** originate outside the eyeball and insert on its outer surface (sclera). **Intrinsic muscles** originate and insert entirely within the eyeball. Movements of the eyeballs are controlled by three pairs of extrinsic muscles. Two pairs of rectus muscles move the eyeball in the direction indicated by their respective names—superior, inferior, lateral, and medial. One pair of muscles, the oblique muscles—superior and inferior—rotate the eyeball on its axis. The extrinsic muscles of the eyeballs are among the fastest contracting and most precisely controlled skeletal muscles of the body.

Muscle	Origin	Insertion	Action	Innervation
Superior rectus (REK-tus; *superior* = above; *rectus* = straight, in this case, muscle fibers running parallel to long axis of eyeball)	Tendinous ring attached to bony orbit around optic foramen.	Superior and central part of eyeball.	Rolls eyeball upward.	Oculomotor (III) nerve.
Inferior rectus (REK-tus; *inferior* = below)	Same as above.	Inferior and central part of eyeball.	Rolls eyeball downward.	Oculomotor (III) nerve.
Lateral rectus (REK-tus)	Same as above.	Lateral side of eyeball.	Rolls eyeball laterally.	Abducens (VI) nerve.
Medial rectus (REK-tus)	Same as above.	Medial side of eyeball.	Rolls eyeball medially.	Oculomotor (III) nerve.
Superior oblique (ō-BLĒK; *oblique* = slanting, in this case, muscle fibers running diagonally to long axis of eyeball)	Same as above.	Eyeball between superior and lateral recti. The muscle inserts in a round tendon that moves through a ring of fibrocartilaginous tissue called the trochlea (*trochlea* = pulley).	Rotates eyeball on its axis; directs cornea downward and laterally.	Trochlear (IV) nerve.
Inferior oblique (ō-BLĒK)	Maxilla (front of orbital cavity).	Eyeball between inferior and lateral recti.	Rotates eyeball on its axis; directs cornea upward and laterally.	Oculomotor (III) nerve.

FIGURE 11.6 Extrinsic muscles of the eyeball.

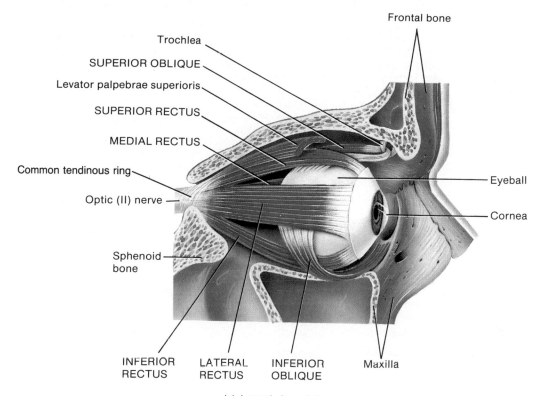

(a) Lateral view of right eyeball

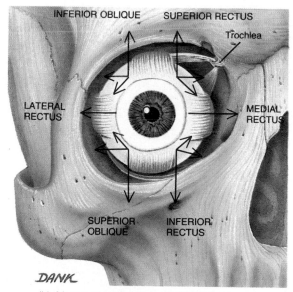

(b) Movements of right eyeball in response to
contraction of its extrinsic muscles

Question: Which muscles roll the eyeballs laterally?

EXHIBIT 11.6

MUSCLES THAT MOVE THE TONGUE—EXTRINSIC MUSCLES (Fig. 11.7)

Overview: The tongue is divided into lateral halves by a median fibrous septum. The septum extends throughout the length of the tongue and is attached inferiorly to the hyoid bone. Like the muscles of the eyeball, muscles of the tongue are of two principal types—extrinsic and intrinsic. **Extrinsic muscles** originate outside the tongue and insert into it. **Intrinsic muscles** originate and insert within the tongue. The extrinsic and intrinsic muscles of the tongue are arranged in both lateral halves of the tongue.

Muscle	Origin	Insertion	Action	Innervation
Genioglossus (jē′-nē-ō-GLOS-us; *geneion* = chin; *glossus* = tongue)	Mandible.	Undersurface of tongue and hyoid bone.	Depresses tongue and thrusts it forward (protraction).	Hypoglossal (XII) nerve.
Styloglossus (stī′-lō-GLOS-us; *stylo* = stake or pole; styloid process of temporal bone)	Styloid process of temporal bone.	Side and undersurface of tongue.	Elevates tongue and draws it backward (retraction).	Hypoglossal (XII) nerve.
Palatoglossus (pal′-a-tō-GLOS-us; *palato* = palate)	Anterior surface of soft palate.	Side of tongue.	Elevates posterior portion of tongue and draws soft palate down on tongue.	Pharyngeal plexus.
Hyoglossus (hī′-ō-GLOS-us)	Body of hyoid bone.	Side of tongue.	Depresses tongue and draws down its sides.	Hypoglossal (XII) nerve.

FIGURE 11.7 Muscles that move the tongue.

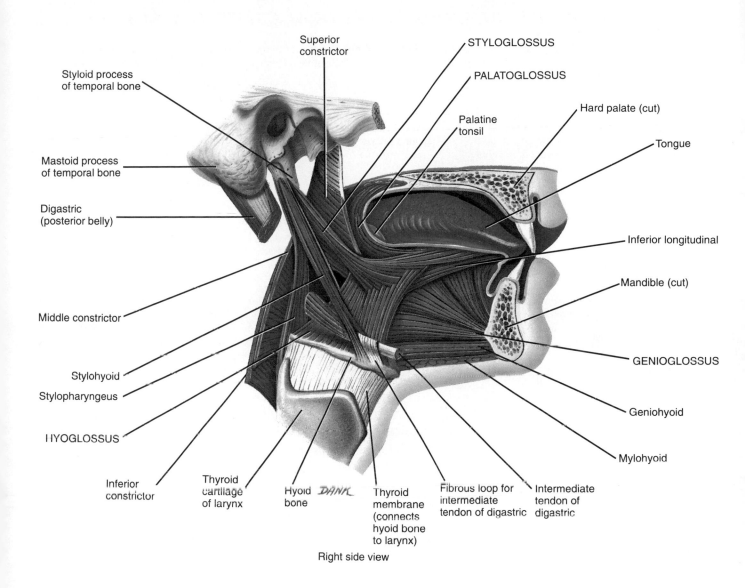

Right side view

Question: Which muscles depress the tongue?

EXHIBIT 11.7

MUSCLES OF THE FLOOR OF THE ORAL CAVITY (Fig. 11.8)

Overview: As a group, these muscles are referred to as **suprahyoid muscles**. They lie superior to the hyoid bone, and all insert into it. The digastric muscle consists of an anterior belly and a posterior belly united by an intermediate tendon that is held in position by a fibrous loop (see also Fig. 11.7).

Muscle	Origin	Insertion	Action	Innervation
Digastric (dī′-GAS-trik; *di* = two; *gaster* = belly)	Anterior belly from inner side of lower border of mandible; posterior belly from mastoid process of temporal bone.	Body of hyoid bone via an intermediate tendon.	Elevates hyoid bone and depresses mandible as in opening the mouth.	Anterior belly from mandibular division of trigeminal (V) nerve; posterior belly from facial (VII) nerve.
Stylohyoid (stī′-lō-HĪ-oid; *stylo* = stake or pole, styloid process of temporal bone; *hyoedes* = U-shaped, pertaining to hyoid bone)	Styloid process of temporal bone.	Body of hyoid bone.	Elevates hyoid bone and draws it posteriorly.	Facial (VII) nerve.
Mylohyoid (mī′-lō-HĪ-oid)	Inner surface of mandible.	Body of hyoid bone.	Elevates hyoid bone and floor of mouth and depresses mandible.	Mandibular division of trigeminal (V) nerve.
Geniohyoid (je′-nē-ō-HĪ-oid; *geneion* = chin)	Inner surface of mandible.	Body of hyoid bone.	Elevates hyoid bone, draws hyoid bone and tongue anteriorly, and depresses mandible.	Cervical nerve C1.

FIGURE 11.8 Muscles of the floor of the oral cavity and front of the neck.

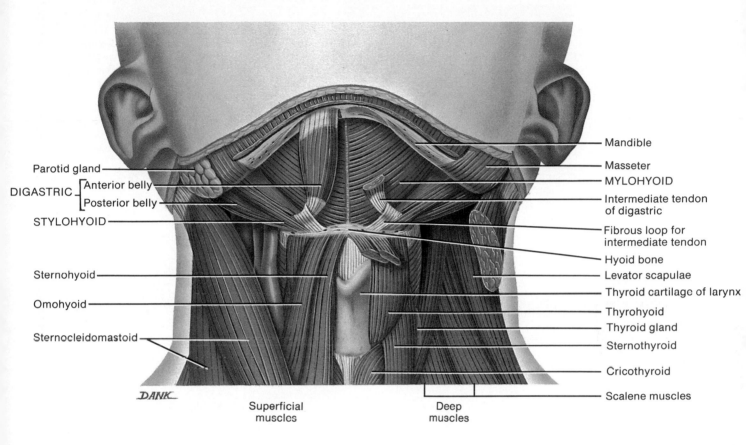

Parotid gland

DIGASTRIC
- Anterior belly
- Posterior belly

STYLOHYOID

Sternohyoid

Omohyoid

Sternocleidomastoid

DANK

Superficial
muscles

Mandible

Masseter

MYLOHYOID

Intermediate tendon
of digastric

Fibrous loop for
intermediate tendon

Hyoid bone

Levator scapulae

Thyroid cartilage of larynx

Thyrohyoid

Thyroid gland

Sternothyroid

Cricothyroid

Scalene muscles

Deep
muscles

Question: Which muscle does not depress (lower) the mandible?

EXHIBIT 11.8

MUSCLES OF THE LARYNX (Fig. 11.9)

Overview: The muscles of the larynx, like those of the eyeballs and tongue, are grouped into extrinsic and intrinsic. The extrinsic muscles of the larynx listed here are together referred to as **infrahyoid (strap) muscles.** They lie inferior to the hyoid bone. The omohyoid muscle, like the digastric muscle, is composed of two bellies and an intermediate tendon. In this case, however, the two bellies are referred to as superior and inferior, rather than anterior and posterior. Extrinsic muscles of the larynx not included here are the **stylopharyngeus** (stī-lō-fa-RIN-jē-us), **inferior constrictor** (kon-STRIK-tor), and **middle constrictor,** which are illustrated in Fig. 11.7, plus the **palatopharyngeus** (pal′-a-tō-fa-RIN-jē-us), which is not illustrated.

Muscle	Origin	Insertion	Action	Innervation
EXTRINSIC				
Omohyoid (ō-mō-HĪ-oid; *omo* = relationship to the shoulder; *hyoedes* = U-shaped; pertaining to hyoid bone)	Superior border of scapula and superior transverse ligament.	Body of hyoid bone.	Depresses hyoid bone.	Branches of ansa cervicalis (C1–C3).
Sternohyoid (ster′-nō-HĪ-oid; *sterno* = sternum)	Medial end of clavicle and manubrium of sternum.	Body of hyoid bone.	Depresses hyoid bone.	Branches of ansa cervicalis (C1–C3).
Sternothyroid (ster′-nō-THĪ-roid; *thyro* = thyroid gland)	Manubrium of sternum.	Thyroid cartilage of larynx.	Depresses thyroid cartilage.	Branches of ansa cervicalis (C1–C3).
Thyrohyoid (thī′-rō-HĪ-oid)	Thyroid cartilage of larynx.	Greater horn of hyoid bone.	Elevates thyroid cartilage and depresses hyoid bone.	Branches of ansa cervicalis (C1–C2) and descending hypoglossal (XII) nerve.
INTRINSIC				
Cricothyroid (kri-kō-THĪ-roid; *crico* = cricoid cartilage of larynx)	Anterior and lateral portion of cricoid cartilage of larynx.	Anterior border of thyroid cartilage of larynx and posterior part of inferior border of thyroid cartilage.	Elongates and places tension on vocal folds.	External laryngeal branch of vagus (X) nerve.
Posterior cricoarytenoid (kri′-kō-ar′-i-TĒ-noid; *arytaina* = shaped like a jug)	Posterior surface of cricoid cartilage.	Posterior surface of arytenoid cartilage of larynx.	Opens rima glottidis (space between vocal folds).	Recurrent laryngeal branch of vagus (X) nerve.
Lateral cricoarytenoid (kri′-kō-ar′-i-TĒ-noid)	Superior border of cricoid cartilage.	Anterior surface of arytenoid cartilage.	Closes rima glottidis (space between vocal folds).	Recurrent laryngeal branch of vagus (X) nerve.
Arytenoid (ar′-i-TĒ-noid)	Posterior surface and lateral border of one arytenoid cartilage.	Corresponding parts of opposite arytenoid cartilage.	Closes rima glottidis (space between vocal folds).	Recurrent laryngeal branch of vagus (X) nerve.
Thyroarytenoid (thī′-rō-ar′-i-TĒ-noid)	Inferior portion of thyroid cartilage and middle of cricothyroid ligament.	Base and anterior surface of arytenoid cartilage.	Shortens and relaxes vocal folds.	Recurrent laryngeal branch of vagus (X) nerve.

FIGURE 11.9 Muscles of the larynx.

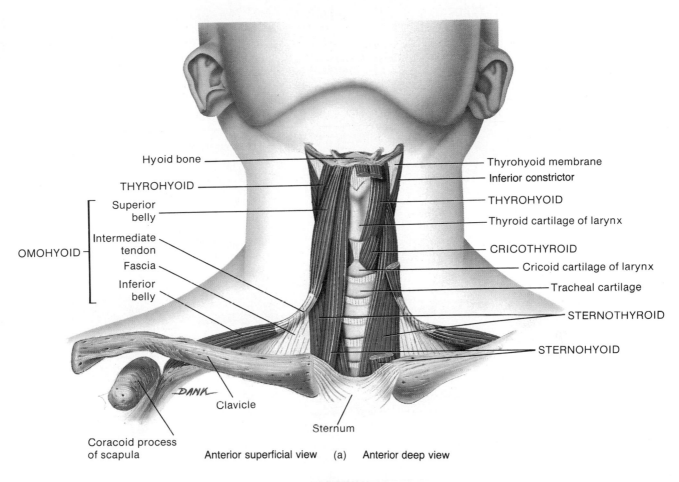

Hyoid bone

THYROHYOID

OMOHYOID
  Superior belly
  Intermediate tendon
  Fascia
  Inferior belly

Coracoid process of scapula

Clavicle

Sternum

Thyrohyoid membrane

Inferior constrictor

THYROHYOID

Thyroid cartilage of larynx

CRICOTHYROID

Cricoid cartilage of larynx

Tracheal cartilage

STERNOTHYROID

STERNOHYOID

Anterior superficial view   (a)   Anterior deep view

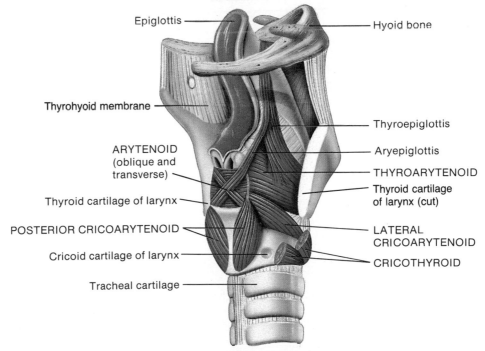

Epiglottis

Hyoid bone

Thyrohyoid membrane

ARYTENOID (oblique and transverse)

Thyroid cartilage of larynx

POSTERIOR CRICOARYTENOID

Cricoid cartilage of larynx

Tracheal cartilage

Thyroepiglottis

Aryepiglottis

THYROARYTENOID

Thyroid cartilage of larynx (cut)

LATERAL CRICOARYTENOID

CRICOTHYROID

(b) Right posterolateral view

Question: Which muscle tenses the vocal folds? Relaxes the vocal folds?

EXHIBIT 11.9

MUSCLES THAT MOVE THE HEAD

Overview: The cervical region is divided by the sternocleidomastoid muscle into two principal triangles—anterior and posterior. The **anterior triangle** is bordered superiorly by the mandible, inferiorly by the sternum, medially by the cervical midline, and laterally by the anterior border of the sternocleidomastoid muscle. The **posterior triangle** is bordered inferiorly by the clavicle, anteriorly by the posterior border of the sternocleidomastoid muscle, and posteriorly by the anterior border of the trapezius muscle. Subsidiary triangles exist within the two principal triangles.

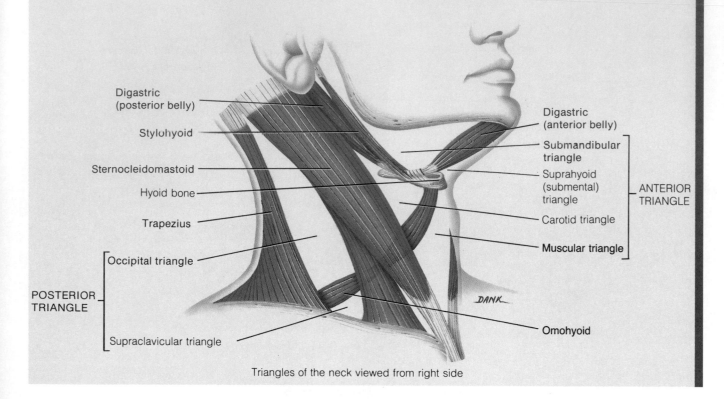

Triangles of the neck viewed from right side

Muscle	Origin	Insertion	Action	Innervation
Sternocleidomastoid (ster′-nō-klī′-dō-MAS-toid; *sternum* = breastbone; *cleido* = clavicle; *mastoid* = mastoid process of temporal bone) (see Fig. 11.14)	Sternum and clavicle.	Mastoid process of temporal bone.	Contraction of both muscles flexes the cervical part of the vertebral column and draws the head forward; contraction of one muscle rotates face toward side opposite contracting muscle.	Accessory (XI) nerve; cervical nerves C2–C3.
Semispinalis capitis (se′-mē-spi-NA-lis KAP-i-tis; *semi* = half; *spine* = spinous process; *caput* = head) (see Fig. 11.19)	Articular processes of fourth, fifth, and sixth cervical vertebrae and transverse processes of seventh cervical and first six or seven thoracic vertebrae.	Occipital bone between superior and inferior nuchal lines.	Both muscles extend head; contraction of one muscle rotates face toward side opposite contracting muscle.	Dorsal rami of cervical nerves.
Splenius capitis (SPLĒ-nē-us KAP-i-tis; *splenion* = bandage) (see Fig. 11.19)	Ligamentum nuchae and spinous processes of seventh cervical vertebra and first three or four thoracic vertebrae.	Occipital bone and mastoid process of temporal bone.	Both muscles extend head; contraction of one laterally flexes and rotates it to same side as contracting muscle.	Dorsal rami of middle and lower cervical nerves.
Longissimus capitis (lon-JIS-i-mus KAP-i-tis; *longissimus* = longest) (see Fig. 11.19)	Transverse processes of upper four thoracic vertebrae and articular processes of last four cervical vertebrae.	Mastoid process of temporal bone.	Extends head and rotates face toward same side as contracting muscle.	Dorsal rami of middle and lower cervical nerves.

EXHIBIT 11.10

MUSCLES THAT ACT ON THE ABDOMINAL WALL (Fig. 11.10)

Overview: The anterolateral abdominal wall is composed of skin, fascia, and four pairs of flat, sheetlike muscles: rectus abdominis, external oblique, internal oblique, and transversus abdominis. The anterior surfaces of the rectus abdominis muscles are interrupted by three transverse fibrous bands of tissue called **tendinous intersections**, believed to be remnants of septa that separated myotomes during embryological development. The aponeuroses of the external oblique, internal oblique, and transversus abdominis muscles form the **rectus sheath**, which encloses the rectus abdominis muscle, and meet at the midline to form the **linea alba** (white line), a tough, fibrous band that extends from the xiphoid process of the sternum to the pubic symphysis. The inferior free border of the external oblique aponeurosis, plus some collagen fibers, forms the **inguinal ligament**, which runs from the anterior superior iliac spine to the pubic tubercle (see Fig. 11.20a). The ligament demarcates the thigh and body wall.

Just superior to the medial end of the inguinal ligament is a triangular slit in the aponeurosis referred to as the **superficial inguinal ring**, the outer opening of the **inguinal canal** (see Fig. 28.8). The canal contains the spermatic cord and ilioinguinal nerve in males and round ligament of the uterus and ilioinguinal nerve in females.

The posterior abdominal wall is formed by the lumbar vertebrae, parts of the ilia of the hipbones, psoas major muscle (described in Exhibit 11.20), quadratus lumborum muscle, and iliacus muscle (also described in Exhibit 11.20). Whereas the anterolateral abdominal wall is contractile and distensible, the posterior abdominal wall is bulky and stable by comparison.

Muscle	Origin	Insertion	Action	Innervation
Rectus abdominis (REK-tus ab-DOM-in-us; *rectus* = fibers parallel to midline; *abdomino* = abdomen)	Pubic crest and pubic symphysis.	Cartilage of fifth to seventh ribs and xiphoid process.	Compresses abdomen to aid in defecation, urination, forced expiration, and childbirth and flexes vertebral column.	Branches of thoracic nerves T7–T12.
External oblique (ō-BLĒK; *external* = closer to surface; *oblique* = fibers diagonal to midline).	Lower eight ribs.	Iliac crest and linea alba (midline aponeurosis).	Contraction of both compresses abdomen; contraction of one side alone bends vertebral column laterally; laterally rotates vertebral column.	Branches of thoracic nerves T7–T12 and iliohypogastric nerve.
Internal oblique (ō-BLĒK; *internal* = farther from surface)	Iliac crest, inguinal ligament, and thoracolumbar fascia.	Cartilage of last three or four ribs and linea alba.	Compresses abdomen; contraction of one side alone bends vertebral column laterally; laterally rotates vertebral column.	Branches of thoracic nerves T8–T12, iliohypogastric, and ilioinguinal nerves.
Transversus abdominis (tranz-VER-sus ab-DOM-in-us; *transverse* = fibers perpendicular to midline)	Iliac crest, inguinal ligament, lumbar fascia, and cartilages of last six ribs.	Xiphoid process, linea alba, and pubis.	Compresses abdomen.	Branches of thoracic nerves T8–T12, iliohypogastric, and ilioinguinal nerves.
Quadratus lumborum (kwod-RĀ-tus lum-BOR-um; *quad* = four; *lumbo* = lumbar region) (see Fig. 11.11)	Iliac crest and iliolumbar ligament.	Lower border of twelfth rib and transverse processes of first four lumbar vertebrae.	During forced expiration, it pulls downward on the twelfth rib; during deep inspiration, it fixes the twelfth rib to prevent its elevation; contraction of one side bends vertebral column laterally.	Branches of thoracic nerve T12 and lumbar nerves L1–L3 or L1–L4.

FIGURE 11.10 Muscles of the male anterior abdominal wall.

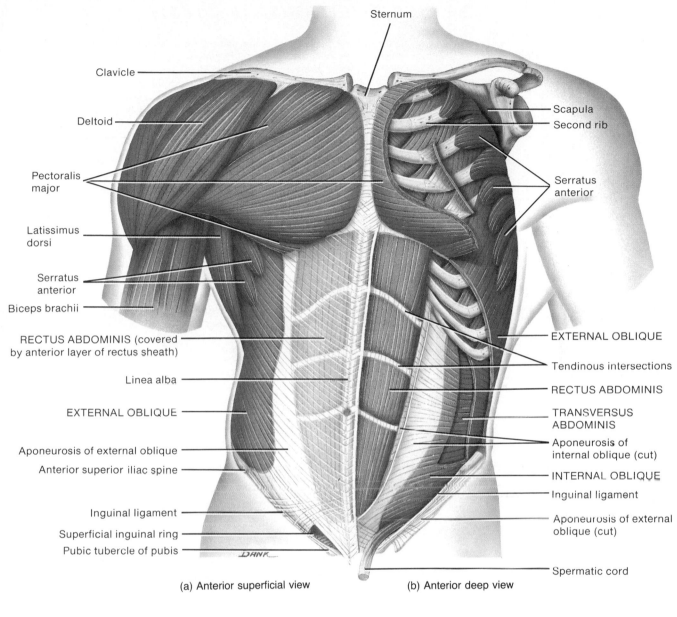

(a) Anterior superficial view   (b) Anterior deep view

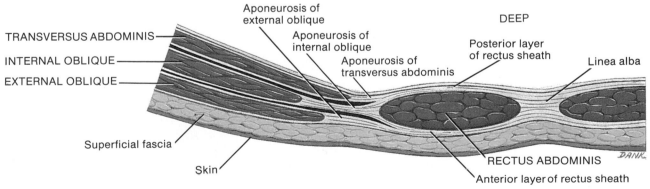

(c) Cross section of anterior abdominal wall above the umbilicus (navel)

Question: Which abdominal muscle aids in urination?

EXHIBIT 11.11

MUSCLES USED IN BREATHING (Fig. 11.11)

Overview: The muscles described here are attached to the ribs and by their contraction and relaxation alter the size of the thoracic cavity during normal breathing. In forced breathing, other muscles are involved as well. Essentially, inspiration occurs when the thoracic cavity increases in size. Expiration occurs when the thoracic cavity decreases in size.

The diaphragm, one of the muscles used in breathing, is dome-shaped and has three major openings through which various structures pass between the thorax and abdomen. These structures include the aorta along with the thoracic duct and azygos vein, which pass through the **aortic hiatus**, the esophagus with accompanying vagus (X) nerves, which pass through the **esophageal hiatus**, and the inferior vena cava, which passes through the **foramen for the vena cava**. As you will see on page 816, in a condition called a hiatus hernia, the stomach protrudes upward through the esophageal hiatus.

Muscle	Origin	Insertion	Action	Innervation
Diaphragm (DĪ-a-fram; *dia* = across; *phragma* = wall)	Xiphoid process, costal cartilages of last six ribs, and lumbar vertebrae.	Central tendon (strong aponeurosis that serves as the tendon of insertion for all muscular fibers of the diaphragm).	Forms floor of thoracic cavity; pulls central tendon downward during inspiration and as dome of diaphragm flattens increases vertical length of thorax.	Phrenic nerve.
External intercostals (in'-ter-KOS-tals; *external* = closer to surface; *inter* = between; *costa* = rib)	Inferior border of rib above.	Superior border of rib below.	May elevate ribs during inspiration and thus increase lateral and anteroposterior dimensions of thorax.	Intercostal nerves.
Internal intercostals (in'ter-KOS-tals; *internal* = farther from surface)	Superior border of rib below.	Inferior border of rib above.	May draw adjacent ribs together during forced expiration and thus decrease lateral and anteroposterior dimensions of thorax.	Intercostal nerves.

FIGURE 11.11 Muscles used in breathing as seen in a male.

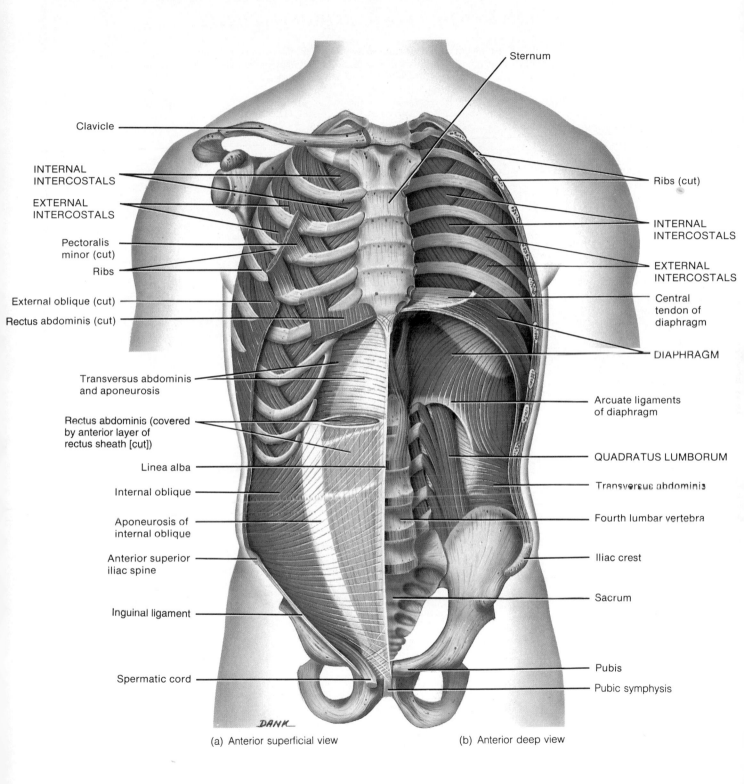

Sternum

Clavicle

INTERNAL
INTERCOSTALS

EXTERNAL
INTERCOSTALS

Pectoralis
minor (cut)

Ribs

External oblique (cut)

Rectus abdominis (cut)

Transversus abdominis
and aponeurosis

Rectus abdominis (covered
by anterior layer of
rectus sheath [cut])

Linea alba

Internal oblique

Aponeurosis of
internal oblique

Anterior superior
iliac spine

Inguinal ligament

Spermatic cord

Ribs (cut)

INTERNAL
INTERCOSTALS

EXTERNAL
INTERCOSTALS

Central
tendon of
diaphragm

DIAPHRAGM

Arcuate ligaments
of diaphragm

QUADRATUS LUMBORUM

Transversus abdominis

Fourth lumbar vertebra

Iliac crest

Sacrum

Pubis

Pubic symphysis

DANK

(a) Anterior superficial view

(b) Anterior deep view

Question: Which muscle is innervated by the phrenic nerve?

EXHIBIT 11.12

MUSCLES OF THE PELVIC FLOOR (Fig. 11.12)

Overview: The muscles of the pelvic floor, together with the fasciae covering their external and internal surfaces, are referred to as the **pelvic diaphragm**. The diaphragm is funnel-shaped and forms the floor of the abdominopelvic cavity, where it supports the pelvic viscera. It is pierced by the anal canal and urethra in both sexes and also by the vagina in the female.

Muscle	Origin	Insertion	Action	Innervation
Levator ani (le-VĀ-tor Ā-nē; *levator* = raises; *ani* = anus)	This muscle is divisible into two parts, the pubococcygeus muscle and the iliococcygeus muscle.			
Pubococcygeus (pu′-bo-kok-SIJ-ē-us; *pubo* = pubis; *coccygeus* = coccyx)	Pubis.	Coccyx, urethra, anal canal, central tendon of perineum, and anococcygeal raphe (narrow fibrous band that extends from anus to coccyx).	Supports and slightly raises pelvic floor, resists increased intra-abdominal pressure, and draws anus toward pubis and constricts it.	Sacral nerves S3–S4 or S4 and perineal branch of pudendal nerve.
Iliococcygeus (il′-ē-o-kok-SIJ-ē-us; *ilio* = ilium)	Ischial spine.	Coccyx.	Supports and slightly raises pelvic floor, resists increased intra-abdominal pressure, and draws anus toward pubis and constricts it.	Sacral nerves S3–S4 or S4 and perineal branch of pudendal nerve.
Coccygeus (kok-SIJ-ē-us)	Ischial spine.	Lower sacrum and upper coccyx.	Supports and slightly raises pelvic floor, resists intra-abdominal pressure, and pulls coccyx forward following defecation or parturition (childbirth).	Sacral nerve S3 or S4.

FIGURE 11.12 Muscles of the pelvic floor seen in the female perineum.

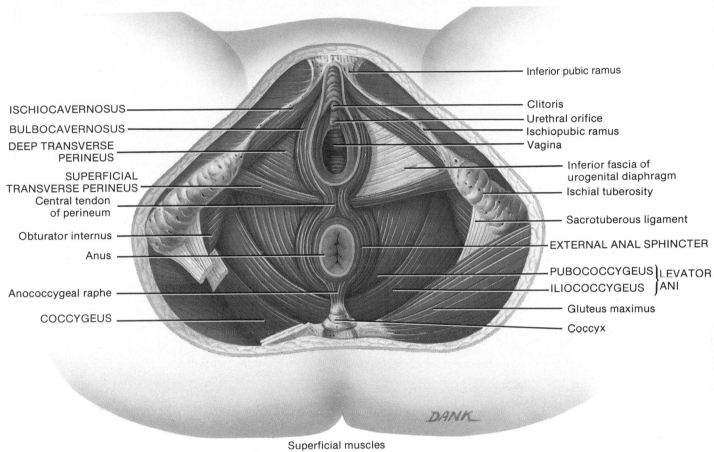

ISCHIOCAVERNOSUS

BULBOCAVERNOSUS

DEEP TRANSVERSE
PERINEUS

SUPERFICIAL
TRANSVERSE PERINEUS

Central tendon
of perineum

Obturator internus

Anus

Anococcygeal raphe

COCCYGEUS

Inferior pubic ramus

Clitoris

Urethral orifice

Ischiopubic ramus

Vagina

Inferior fascia of
urogenital diaphragm

Ischial tuberosity

Sacrotuberous ligament

EXTERNAL ANAL SPHINCTER

PUBOCOCCYGEUS } LEVATOR
ILIOCOCCYGEUS } ANI

Gluteus maximus

Coccyx

DANK

Superficial muscles

Question: Which muscle assists in parturition (childbirth)?

EXHIBIT 11.13

MUSCLES OF THE PERINEUM (Fig. 11.13; see also Fig. 11.12)

Overview: The **perineum** is the entire outlet of the pelvis. It is a diamond-shaped area at the lower end of the trunk between the thighs and buttocks. It is bordered anteriorly by the pubic symphysis, laterally by the ischial tuberosities, and posteriorly by the coccyx. A transverse line drawn between the ischial tuberosities divides the perineum into an anterior **urogenital triangle** that contains the external genitals and a posterior **anal triangle** that contains the anus (see Fig. 28.19). The deep transverse perineus, the urethral sphincter, and a fibrous membrane constitute the **urogenital diaphragm**. It surrounds the urogenital ducts and helps to strengthen the pelvic floor.

Muscle	Origin	Insertion	Action	Innervation
Superficial transverse perineus (per-i-NĒ-us; *superficial* = closer to surface; *transverse* = across; *perineus* = perineum)	Ischial tuberosity.	Central tendon of perineum (fibromuscular tissue in the midline between anus and vagina in female and anus and bulb of penis in male).	Helps to stabilize the central tendon of the perineum.	Perineal branch of pudendal nerve.
Bulbocavernosus (bul´-bō-ka´-ver-NŌ-sus; *bulbus* = bulb; *caverna* = hollow place)	Central tendon of perineum.	Inferior fascia of urogenital diaphragm, corpus spongiosum of penis, and deep fascia on dorsum of penis in male; pubic arch and root and dorsum of clitoris in female.	Helps expel last drops of urine during micturition, helps propel semen along urethra, and may assist in erection of the penis in male; decreases vaginal orifice and assists in erection of clitoris in female.	Perineal branch of pudendal nerve.
Ischiocavernosus (is´-kē-o-ka´-ver-NŌ-sus; *ischion* = hip)	Ischial tuberosity and ischial and pubic rami.	Corpus cavernosum of penis in male and clitoris in female.	May maintain erection of penis in male and clitoris in female.	Perineal branch of pudendal nerve.
Deep transverse perineus (per-i-NĒ-us; *deep* = farther from surface)	Ischial rami.	Central tendon of perineum.	Helps expel last drops of urine and semen in male and urine in female.	Perineal branch of pudendal nerve.
Urethral sphincter (yoo-RĒ-thral SFINGK-ter; *sphincter* = circular muscle that decreases size of an opening; *urethrae* = urethra)	Ischial and pubic rami.	Median raphe in male and vaginal wall in female.	Helps expel last drops of urine and semen in male and urine in female.	Perineal branch of pudendal nerve.
External anal sphincer (Ā-nal)	Anococcygeal raphe.	Central tendon of perineum.	Keeps anal canal and orifice closed.	Sacral nerve S4 and inferior rectal branch of pudendal nerve.

FIGURE 11.13 Muscles of the male perineum.

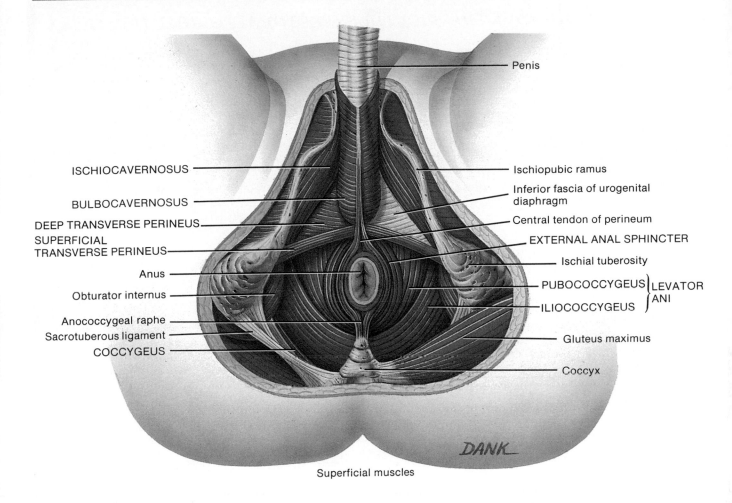

Superficial muscles

Question: Which muscles are related to erection of the penis and clitoris?

EXHIBIT 11.14

MUSCLES THAT MOVE THE SHOULDER (PECTORAL) GIRDLE (Fig. 11.14)

Overview: Muscles that move the shoulder (pectoral) girdle originate on the axial skeleton and insert on the clavicle or scapula. The muscles can be distinguished into **anterior** and **posterior** groups. The principal action of the muscles is to stabilize the scapula so that it can function as a stable point of origin for most of the muscles that move the humerus (arm).

Muscle	Origin	Insertion	Action	Innervation
ANTERIOR				
Subclavius (sub-KLĀ-vē-us; *sub* = under; *clavius* = clavicle)	First rib.	Clavicle.	Depresses clavicle.	Nerve to subclavius.
Pectoralis (pek′-tor-A-lis) **minor** (*pectus* = breast, chest, thorax; *minor* = lesser)	Third through fifth ribs.	Coracoid process of scapula.	Depresses and moves scapula anteriorly and elevates third through fifth ribs during forced inspiration when scapula is fixed.	Medial pectoral nerve.
Serratus (ser-Ā-tis) **anterior** (*serratus* = sawtoothed; *anterior* = front)	Upper eight or nine ribs.	Vertebral border and inferior angle of scapula.	Rotates scapula upward and laterally and elevates ribs when scapula is fixed.	Long thoracic nerve.
POSTERIOR				
Trapezius (tra-PĒ-zē-us; *trapezoides* = trapezoid-shaped)	Superior nuchal line of occipital bone, ligamentum nuchae, and spines of seventh cervical and all thoracic vertebrae.	Clavicle and acromion and spine of scapula.	Elevates clavicle, adducts scapula, rotates scapula upward, elevates or depresses scapula, and extends head.	Accessory (XI) nerve and cervical nerves C3–C4.
Levator scapulae (le-VĀ-tor SKA-pyoo-lē; *levator* raises; *scapulae* = scapula)	Upper four or five cervical vertebrae.	Superior vertebral border of scapula.	Elevates scapula and slightly rotates it downward.	Dorsal scapular nerve and cervical nerves C3–C5.
Rhomboideus (rom-BOID-ē-us) **major** (*rhomboides* = rhomboid or diamond-shaped)	Spines of second to fifth thoracic vertebrae.	Vertebral border of scapula below spine.	Adducts scapula and slightly rotates it downward.	Dorsal scapular nerve.
Rhomboideus (rom-BOID-ē-us) **minor**	Spines of seventh cervical and first thoracic vertebrae.	Vertebral border of scapula above spine.	Adducts scapula and slightly rotates it downward.	Dorsal scapular nerve.

Figure 11.14 Muscles that move the shoulder (pectoral) girdle.

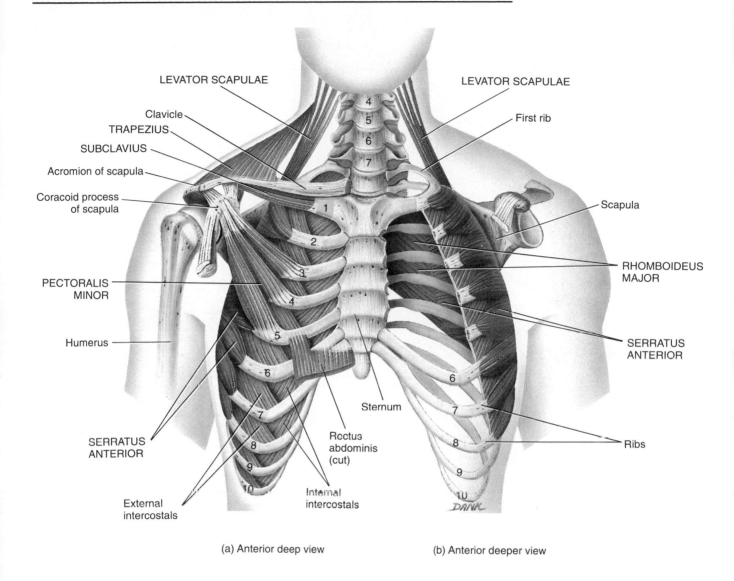

(a) Anterior deep view   (b) Anterior deeper view

Figure continues

FIGURE 11.14 (continued)

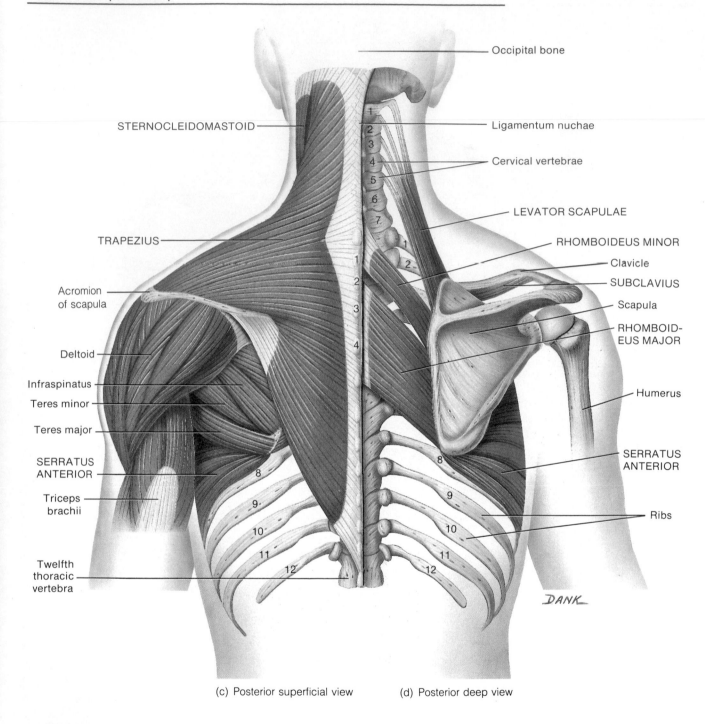

(c) Posterior superficial view   (d) Posterior deep view

Question: Which muscles originate on the ribs?

EXHIBIT 11.15

MUSCLES THAT MOVE THE ARM (HUMERUS) (Fig. 11.15)

Overview: Of the nine muscles that cross the shoulder joint, only two of them (pectoralis major and latissimus dorsi) do not originate on the scapula. These two muscles are thus designated as **axial muscles**, since they originate on the axial skeleton. The remaining seven muscles, the **scapular muscles**, arise from the scapula. The strength and stability of the shoulder joint are not provided by the shape of the articulating bones or its ligaments. Instead, four deep muscles of the shoulder and their tendons—subscapularis, supraspinatus, infraspinatus, and teres minor—strengthen and stabilize the shoulder joint. The muscles and their tendons are so arranged as to form a nearly complete circle around the joint. This arrangement is referred to as the **rotator (musculotendinous) cuff** and is a common site of injury in baseball pitchers, especially tearing of the supraspinatus muscle tendon. This tendon is especially predisposed to wear-and-tear changes because of its location between the head of the humerus and acromion of the scapula, which compresses the tendon during shoulder movements.

After you have studied the muscles in this exhibit, arrange them according to the following actions: flexion, extension, abduction, adduction, medial rotation, and lateral rotation. (The same muscle can be used more than once.)

Muscle	Origin	Insertion	Action	Innervation
AXIAL				
Pectoralis (pek´-tor-A-lis) **major** (see also Fig. 11.10a)	Clavicle, sternum, cartilages of second to sixth ribs.	Greater tubercle and intertubercular sulcus of humerus.	Flexes, adducts, and rotates arm medially.	Medial and lateral pectoral nerve.
Latissimus dorsi (la-TIS-i-mus DOR-sī; *latissimus* = widest; *dorsum* = back)	Spines of lower six thoracic vertebrae, lumbar vertebrae, crests of sacrum and ilium, lower four ribs.	Intertubercular sulcus of humerus.	Extends, adducts, and rotates arm medially; draws arm downward and backward.	Thoracodorsal nerve.
SCAPULAR				
Deltoid (DEL-toyd; *delta* = triangular)	Acromial extremity of clavicle and acromion and spine of scapula.	Deltoid tuberosity of humerus.	Abducts, flexes, extends, and medially and laterally rotates arm.	Axillary nerve.
Subscapularis (sub-scap´-yoo-LA-ris; *sub* = below; *scapularis* = scapula)	Subscapular fossa of scapula.	Lesser tubercle of humerus.	Rotates arm medially.	Upper and lower subscapular nerves.
Supraspinatus (soo´-pra-spi-NĀ-tus; *supra* = above; *spinatus* = spine of scapula)	Supraspinous fossa of scapula.	Greater tubercle of humerus.	Assists deltoid muscle in abducting arm.	Suprascapular nerve.
Infraspinatus (in´-fra-spi-NĀ-tus; *infra* = below)	Infraspinous fossa of scapula.	Greater tubercle of humerus.	Rotates arm laterally; adducts arm.	Suprascapular nerve.
Teres (TE-rēz) **major** (*teres* = long and round)	Inferior angle of scapula.	Intertubercular sulcus of humerus.	Extends arm; assists in adduction and medial rotation of arm.	Lower subscapular nerve.
Teres (TE-rēz) **minor**	Inferior lateral border of scapula.	Greater tubercle of humerus.	Rotates arm laterally; extends and adducts arm.	Axillary nerve.
Coracobrachialis (kor´-a-kō-BRĀ-kē-a´-lis; *coraco* = coracoid process)	Coracoid process of scapula.	Middle of medial surface of shaft of humerus.	Flexes and adducts arm.	Musculocutaneous nerve.

FIGURE 11.15 Muscles that move the arm (humerus).

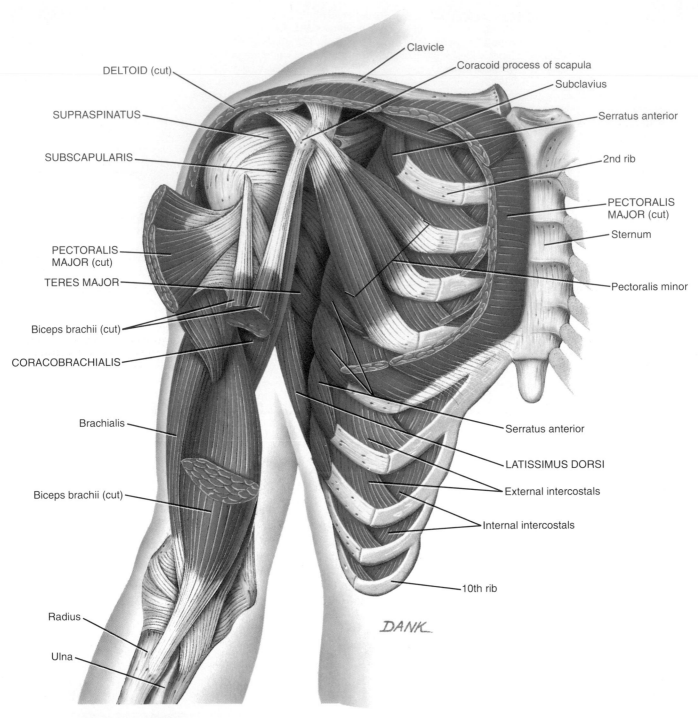

(a) Anterior deep view

FIGURE 11.15 (continued)

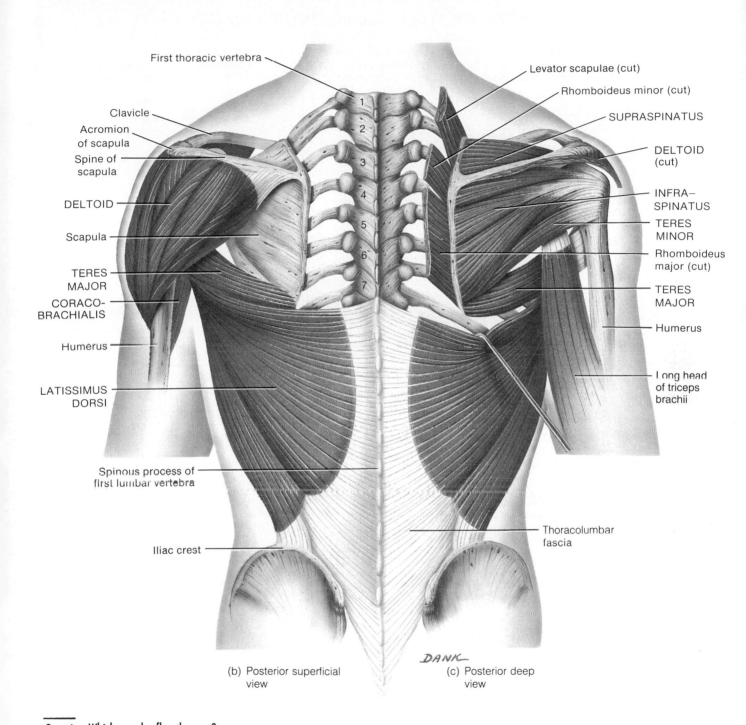

First thoracic vertebra

Levator scapulae (cut)

Rhomboideus minor (cut)

SUPRASPINATUS

Clavicle

Acromion
of scapula

DELTOID
(cut)

Spine of
scapula

INFRA−
SPINATUS

DELTOID

TERES
MINOR

Scapula

Rhomboideus
major (cut)

TERES
MAJOR

TERES
MAJOR

CORACO-
BRACHIALIS

Humerus

Humerus

Long head
of triceps
brachii

LATISSIMUS
DORSI

Spinous process of
first lumbar vertebra

Thoracolumbar
fascia

Iliac crest

DANK

(b) Posterior superficial
view

(c) Posterior deep
view

Question: Which muscles flex the arm?

EXHIBIT 11.16

MUSCLES THAT MOVE THE FOREARM (RADIUS AND ULNA) (Fig. 11.16)

Overview: Most of the muscles that move the forearm (radius and ulna) are divided into **flexors** and **extensors**. Recall that the elbow joint is a hinge joint, capable only of flexion and extension under normal conditions. Whereas the biceps brachii, brachialis, and brachioradialis are flexors of the elbow joint, the triceps brachii and anconeus are extensors. Other muscles permit pronation and supination of the forearm.

Muscle	Origin	Insertion	Action	Innervation
FLEXORS				
Biceps brachii (BĪ-ceps BRĀ-kē-ī; *biceps* = two heads of origin; *brachion* = arm)	Long head originates from tubercle above glenoid cavity; short head originates from coracoid process of scapula.	Radial tuberosity and bicipital aponeurosis (broad aponeurosis from tendon of insertion of biceps brachii muscle that descends medially across brachial artery and fuses with deep fascia over forearm flexor muscles).	Flexes and supinates forearm; flexes arm.	Musculocutaneous nerve.
Brachialis (brā'-kē-A-lis)	Distal, anterior surface of humerus.	Ulnar tuberosity and coronoid process of ulna.	Flexes forearm.	Musculocutaneous and radial nerves.
Brachioradialis (brā'-kē-ō-rā'-dē-A-lis; *radialis* = radius) (see also Fig. 11.17)	Medial and lateral borders of distal end of humerus.	Superior to styloid process of radius.	Flexes forearm; semisupinates and semipronates forearm.	Radial nerve.
EXTENSORS				
Triceps brachii (TRĪ-ceps BRĀ-kē-ī; *triceps* = three heads of origin)	Long head originates from a projection below glenoid cavity (infraglenoid tubercle) of scapula; lateral head originates from lateral and posterior surface of humerus superior to radial groove; medial head originates from entire posterior surface of humerus inferior to a groove for the radial nerve.	Olecranon of ulna.	Extends forearm; extends arm.	Radial nerve.
Anconeus (an-KŌ-nē-us; *anconeal* = pertaining to elbow) (see Fig. 11.17)	Lateral epicondyle of humerus.	Olecranon and superior portion of shaft of ulna.	Extends forearm.	Radial nerve.
PRONATORS				
Pronator teres (PRŌ-na'-ter TE-rēz; *pronation* = turning palm downward or posteriorly) (see Fig. 11.17)	Medial epicondyle of humerus and coronoid process of ulna.	Midlateral surface of radius.	Pronates forearm and hand and flexes forearm.	Median nerve.
Pronator quadratus (PRŌ-nā'-ter kwod-RĀ-tus; *quadratus* = squared, four-sided) (see Fig. 11.18a,b)	Distal portion of shaft of ulna.	Distal portion of shaft of radius.	Pronates forearm and hand.	Median nerve.

Muscle	Origin	Insertion	Action	Innervation
SUPINATOR **Supinator** (SOO-pi-nā-tor; *supination* = turning palm upward or anteriorly) (see Fig. 11.18b)	Lateral epicondyle of humerus and ridge near radial notch of ulna (supinator crest).	Lateral surface of proximal one-third of radius.	Supinates forearm and hand.	Deep radial nerve.

FIGURE 11.16 Muscles that move the forearm (radius and ulna).

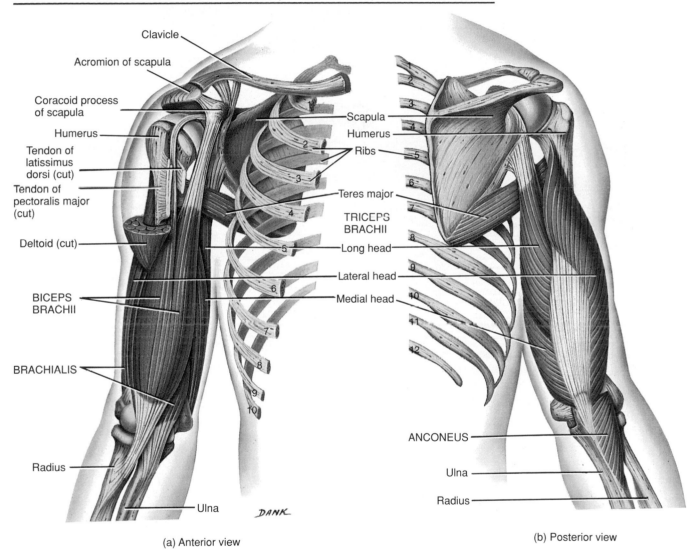

(a) Anterior view

(b) Posterior view

Figure continues

FIGURE 11.16 (continued)

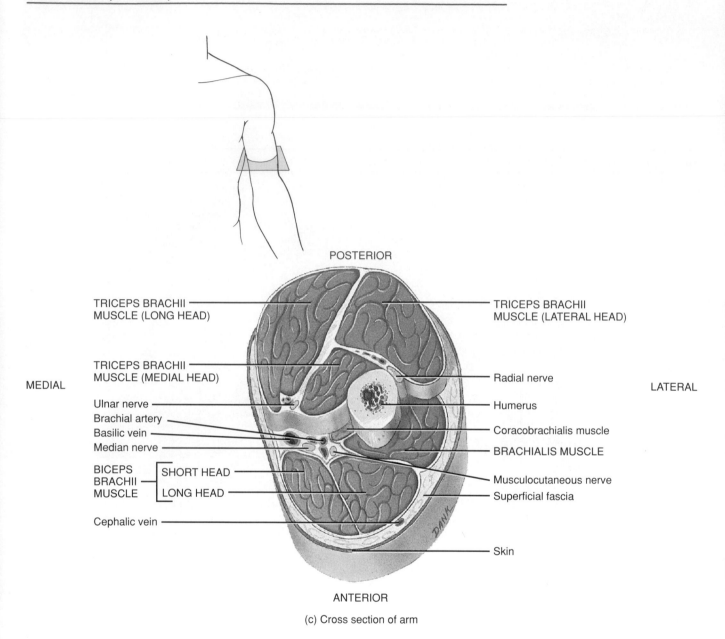

(c) Cross section of arm

Question: Which muscles extend the forearm?

EXHIBIT 11.17

MUSCLES THAT MOVE THE WRIST, HAND, AND FINGERS (Fig. 11.17)

Overview: Muscles that move the wrist, hand, and fingers are many and varied. However, as you will see, their names for the most part give some indication of their origin, insertion, or action. On the basis of location and function, the muscles are divided into two groups—anterior and posterior. The **anterior muscles** function as flexors. They originate on the humerus and typically insert on the carpals, metacarpals, and phalanges. The bellies of these muscles form the bulk of the proximal forearm. The **posterior muscles** function as extensors. These muscles arise on the humerus and insert on the metacarpals and phalanges. Each of the two principal groups is also divided into superficial and deep muscles.

The tendons of the muscles of the forearm that attach to the wrist or continue into the hand, along with blood vessels and nerves, are held close to bones by strong fascial structures. The tendons are also surrounded by tendon sheaths. At the wrist, the deep fascia is thickened into fibrous bands called retinacula (*retinere* = retain). The **flexor retinaculum (transverse carpal ligament)** is located over the palmar surface of the carpal bones. Through it pass the long flexor tendons of the digits and wrist and the median nerve. The **extensor retinaculum (dorsal carpal ligament)** is located over the dorsal surface of the carpal bones. Through it pass the extensor tendons of the wrist and digits.

After you have studied the muscles in this exhibit, arrange them according to the following actions: flexion, extension, abduction, adduction, supination, and pronation. (The same muscles can be used more than once.)

Muscle	Origin	Insertion	Action	Innervation
ANTERIOR GROUP (FLEXORS)				
SUPERFICIAL				
Flexor carpi radialis (FLEK-sor KAR-pē rā′ dē-A-lis; *flexor* = decreases angle at joint; *carpus* = wrist; *radialis* = radius)	Medial epicondyle of humerus.	Second and third metacarpals.	Flexes and abducts wrist.	Median nerve.
Palmaris longus (pal-MA-ris LON-gus; *palma* = palm; *longus* = long)	Medial epicondyle of humerus.	Flexor retinaculum and palmar aponeurosis (deep fascia in center of palm).	Flexes wrist.	Median nerve.
Flexor carpi ulnaris (FLEK-sor KAR-pe ul-NAR-is; *ulnaris* = ulna)	Medial epicondyle of humerus and upper posterior border of ulna.	Pisiform, hamate, and fifth metacarpal.	Flexes and adducts wrist	Ulnar nerve.
Flexor digitorum superficialis (FLEK-sor di′-ji-TOR-um soo′-per-fish′-ē-A-lis; *digit* = finger or toe; *superficialis* = closer to surface)	Medial epicondyle of humerus, coronoid process of ulna, and a ridge along lateral margin of anterior surface (anterior oblique line) of radius.	Middle phalanges.	Flexes middle phalanges of each finger.	Median nerve.
DEEP				
Flexor digitorum profundus (FLEK-sor di′ji-TOR-um pro-FUN-dus; *profundus* = deep)	Anterior medial surface of body of ulna.	Bases of distal phalanges.	Flexes distal phalanges of each finger.	Median and ulnar nerves.
Flexor pollicis longus (FLEK-sor POL-li-kis LON-gus; *pollex* = thumb)	Anterior surface of radius and interosseous membrane (sheet of fibrous tissue that holds shafts of ulna and radius together).	Base of distal phalanx of thumb.	Flexes thumb.	Median nerve.

Exhibit continues

EXHIBIT 11.17 (continued)

MUSCLES THAT MOVE THE WRIST, HAND, AND FINGERS (Fig. 11.17)

Muscle	Origin	Insertion	Action	Innervation
POSTERIOR GROUP (EXTENSORS)				
SUPERFICIAL				
Extensor carpi radialis longus (eks-TEN-sor KAR-pē rā′-dē-A-lis LON-gus; *extensor* = increases angle at joint)	Lateral epicondyle of humerus.	Second metacarpal.	Extends and abducts wrist.	Radial nerve.
Extensor carpi radialis brevis (eks-TEN-sor KAR-pē rā′-dē-A-lis BREV-is; *brevis* = short)	Lateral epicondyle of humerus.	Third metacarpal.	Extends and abducts wrist.	Radial nerve.
Extensor digitorum (eks-TEN-sor dī′-ji-TOR-um)	Lateral epicondyle of humerus.	Second through fifth distal and middle phalanges.	Extends phalanges.	Radial nerve.
Extensor digiti minimi (eks-TEN-sor DIJ-i-tē MIN-i-mē; *minimi* = little finger)	Tendon of extensor digitorum.	Tendon of extensor digitorum on fifth phalanx.	Extends little finger.	Deep radial nerve.
Extensor carpi ulnaris (eks-TEN-sor KAR-pē ul-NAR-is)	Lateral epicondyle of humerus and posterior border of ulna.	Fifth metacarpal.	Extends and adducts wrist.	Deep radial nerve.
DEEP				
Abductor pollicis longus (ab-DUK-tor POL-li-kis LON-gus; *abductor* = moves part away from midline)	Posterior surface of middle of radius and ulna and interosseous membrane.	First metacarpal.	Extends thumb and abducts wrist.	Deep radial nerve.
Extensor pollicis brevis (eks-TEN-sor POL-li-kis BREV-is)	Posterior surface of middle of radius and interosseous membrane.	Base of proximal phalanx of thumb.	Extends thumb and abducts wrist.	Deep radial nerve.
Extensor pollicis longus (eks-TEN-sor POL-li-kis LON-gus)	Posterior surface of middle of ulna and interosseous membrane.	Base of distal phalanx of thumb.	Extends thumb and abducts wrist.	Deep radial nerve.
Extensor indicis (eks-TEN-sor IN-di-kis; *indicis* = index)	Posterior surface of ulna.	Tendon of extensor digitorum of index finger.	Extends index finger.	Deep radial nerve.

FIGURE 11.17 Muscles that move the wrist, hand, and fingers.

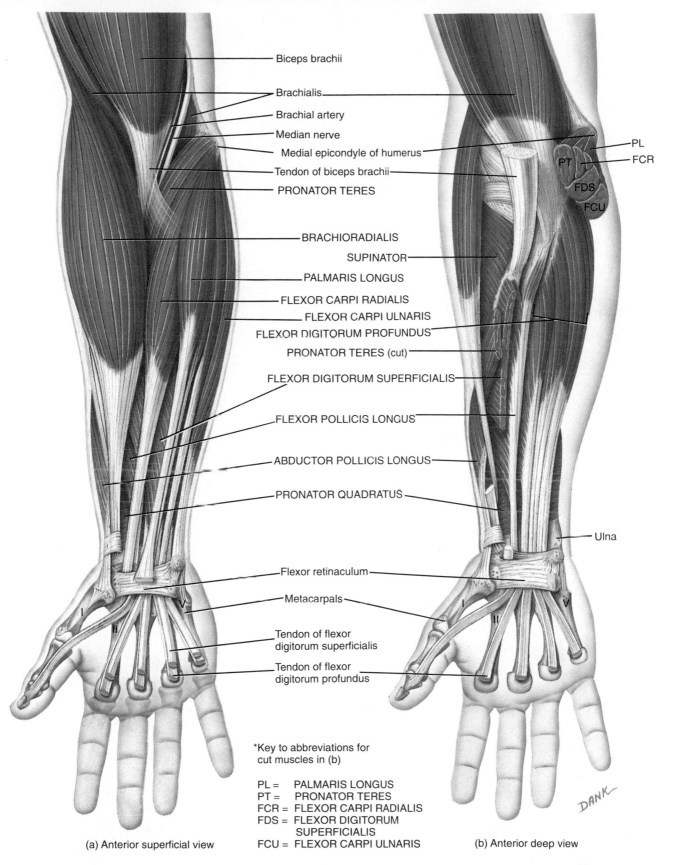

Biceps brachii

Brachialis

Brachial artery

Median nerve

Medial epicondyle of humerus

Tendon of biceps brachii

PRONATOR TERES

BRACHIORADIALIS

SUPINATOR

PALMARIS LONGUS

FLEXOR CARPI RADIALIS

FLEXOR CARPI ULNARIS

FLEXOR DIGITORUM PROFUNDUS

PRONATOR TERES (cut)

FLEXOR DIGITORUM SUPERFICIALIS

FLEXOR POLLICIS LONGUS

ABDUCTOR POLLICIS LONGUS

PRONATOR QUADRATUS

Flexor retinaculum

Metacarpals

Tendon of flexor
digitorum superficialis

Tendon of flexor
digitorum profundus

PL

PT

FCR

FDS

FCU

Ulna

*Key to abbreviations for
cut muscles in (b)

PL = PALMARIS LONGUS
PT = PRONATOR TERES
FCR = FLEXOR CARPI RADIALIS
FDS = FLEXOR DIGITORUM
        SUPERFICIALIS
FCU = FLEXOR CARPI ULNARIS

(a) Anterior superficial view

(b) Anterior deep view

DANK

Figure continues

FIGURE 11.17 (continued)

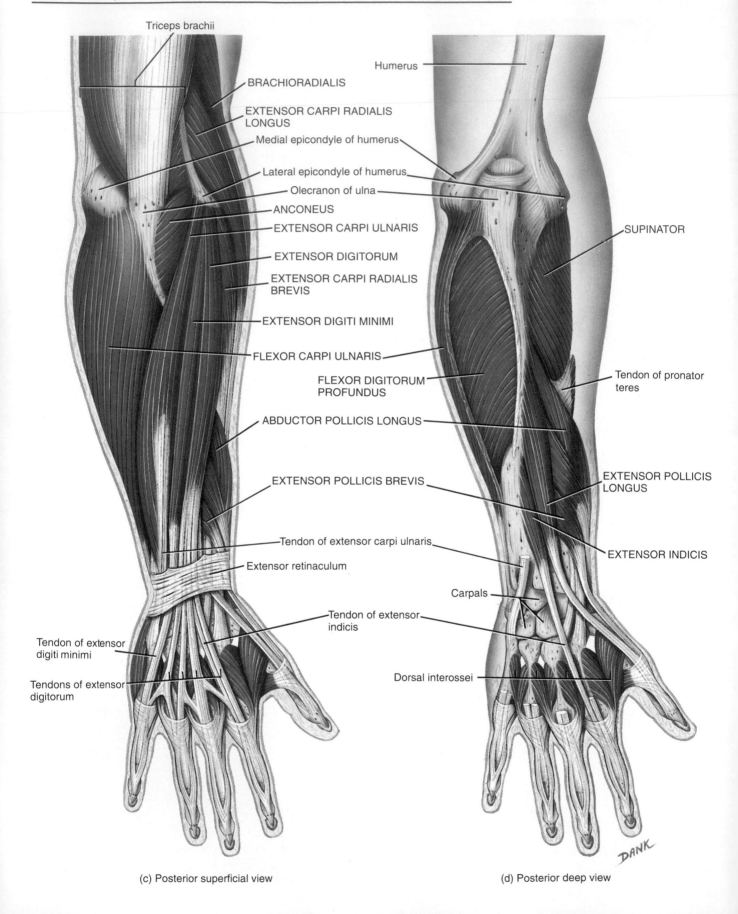

(c) Posterior superficial view   (d) Posterior deep view

FIGURE 11.17 (continued)

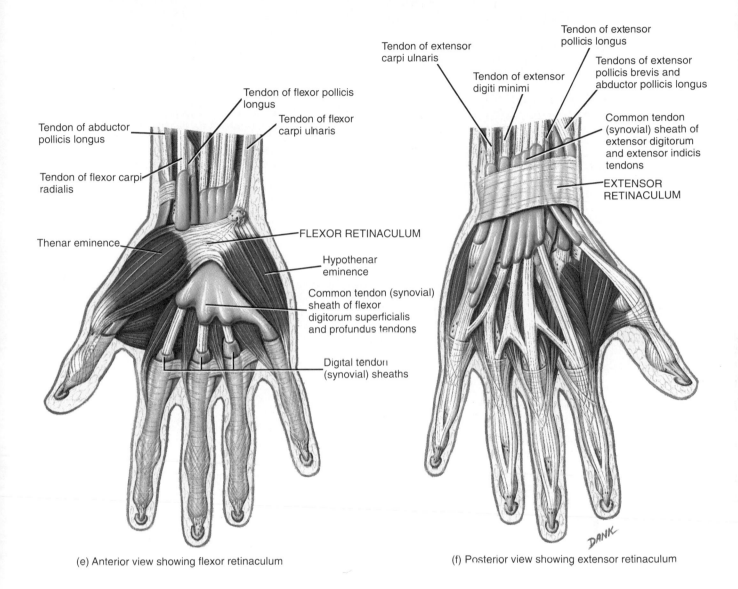

(e) Anterior view showing flexor retinaculum

(f) Posterior view showing extensor retinaculum

Question: Do anterior muscles flex or extend the wrist?

EXHIBIT 11.18

INTRINSIC MUSCLES OF THE HAND (Fig. 11.18)

Overview: Several of the muscles discussed in Exhibit 11.17 help to move the digits in various ways. In addition, there are muscles in the palmar surface of the hand called **intrinsic muscles** that also help to move the digits. Such muscles are so named because their origins and insertions are both within the hands. These muscles assist in the intricate and precise movements that are characteristic of the human hand.

The intrinsic muscles of the hand are divided into three principal groups—thenar, hypothenar, and intermediate. The four **thenar muscles** act on the thumb and form the **thenar eminence**. The four **hypothenar muscles** act on the little finger and form the **hypothenar eminence**. The 11 **intermediate (midpalmar) muscles** act on all the digits, except the thumb.

The functional importance of the hand is readily apparent when one considers that certain hand injuries can result in permanent disability. Most of the dexterity of the hand depends on the movements of the thumb. The general activities of the hand are free motion, power grip (forcible movement of the fingers and thumb against the palm, as in squeezing), precision handling (a change in position of a handled object that requires exact control of finger and thumb positions, as in winding a watch or threading a needle), and pinch (compression between the thumb and index finger or between the thumb and first two fingers).

Movement of the thumb is very important in the precise activities of the hand. The five principal movements of the thumb, illustrated below, are flexion (movement of the thumb at right angles to the fingers), extension (movement in the opposite direction), abduction (movement of the thumb anteriorly and laterally), adduction (movement of the thumb toward the index finger), and opposition (movement of the thumb across the palm so that the tip of the thumb meets the tips of the fingers). Opposition is the single most distinctive digital movement that gives humans their characteristic tool-making and tool-using capacity.

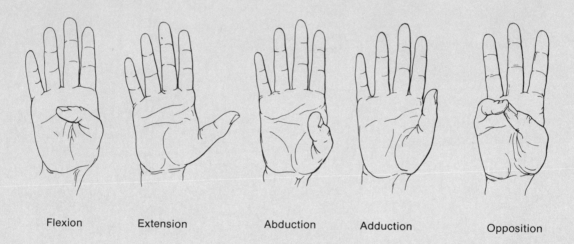

| Flexion | Extension | Abduction | Adduction | Opposition |

Muscle	Origin	Insertion	Action	Innervation
THENAR (THĒ-nar)				
Abductor pollicis brevis (ab-DUK-tor POL-li-kis BREV-is; *abductor* = moves part away from middle; *pollex* = thumb; *brevis* = short)	Flexor retinaculum, scaphoid, and trapezium.	Proximal phalanx of thumb.	Abducts thumb.	Median nerve.
Opponens pollicis (o-PŌ-nenz POL-li-kis; *opponens* = opposes)	Flexor retinaculum and trapezium.	Metacarpal of thumb.	Draws thumb across palm to meet little finger (opposition).	Median nerve.
Flexor pollicis brevis (FLEK-sor POL-li-kis BREV-is; *flexor* = decreases angle at joint)	Flexor retinaculum, trapezium, and first metacarpal.	Proximal phalanx of thumb.	Flexes and adducts thumb.	Median and ulnar nerve.
Adductor pollicis (ab-DUK-tor POL-li-kis; *adductor* = moves part toward midline)	Capitate and second and third metacarpals.	Proximal phalanx of thumb.	Adducts thumb.	Ulnar nerve.

Muscle	Origin	Insertion	Action	Innervation
HYPOTHENAR (HĪ-pō-thē′-nar)				
Palmaris brevis (pal-MA-ris BREV-is; *palma* = palm) (not illustrated)	Flexor retinaculum and palmar aponeurosis.	Skin on ulnar border of palm of hand.	Draws skin toward middle of palm as in clenching fist.	Ulnar nerve.
Abductor digiti minimi (ab-DUK-tor DIJ-i-tē MIN-i-mē; *digit* = finger or toe; *minimi* = little finger)	Pisiform and tendon of flexor carpi ulnaris.	Proximal phalanx of little finger.	Abducts little finger.	Ulnar nerve.
Flexor digiti minimi brevis (FLEK-sor DIJ-i-tē MIN-i-mē BREV-is)	Flexor retinaculum and hamate.	Proximal phalanx of little finger.	Flexes little finger.	Ulnar nerve.
Opponens digiti minimi (o-PŌ-nenz DIJ-i-tē MIN-i-mē)	Flexor retinaculum and hamate.	Metacarpal of little finger.	Draws little finger across palm to meet thumb.	Ulnar nerve.

Exhibit continues

FIGURE 11.18 Palmar view of the intrinsic muscles of the right hand.

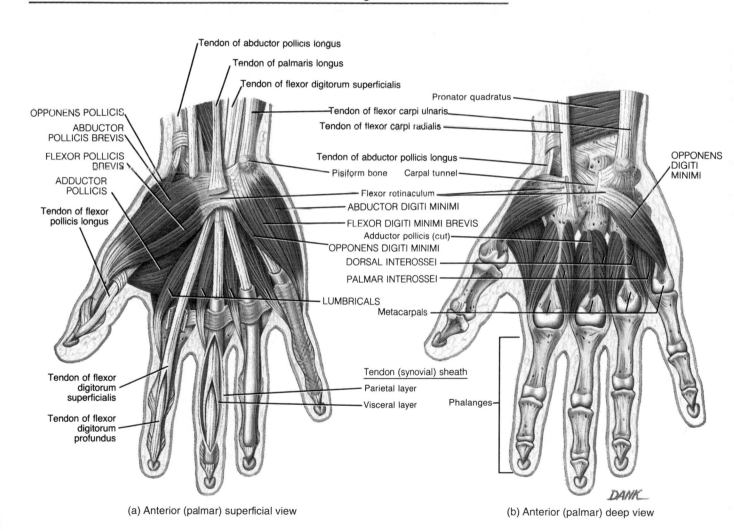

(a) Anterior (palmar) superficial view

(b) Anterior (palmar) deep view

Question: Which nerve innervates all hypothenar muscles?

EXHIBIT 11.18 (continued)

INTRINSIC MUSCLES OF THE HAND (Fig. 11.18)

Muscle	Origin	Insertion	Action	Innervation
INTERMEDIATE (MIDPALMAR)				
Lumbricals (LUM-bri-kals; four muscles)	Tendons of flexor digitorum profundus.	Tendons of extensor digitorum.	Extend interphalangeal joints and flex metacarpophalangeal joints.	Median and ulnar nerve.
Dorsal interossei (DOR-sal in′-ter-OS-ē-ī; four muscles; *dorsal* = back surface; *inter* = between; *ossei* = bones)	Adjacent sides of metacarpals.	Proximal phalanx of second, third, and fourth fingers.	Abduct fingers from middle finger; flex metacarpophalangeal joints; and extend interphalangeal joints.	Ulnar nerve.
Palmar interossei (PAL-mar in′-ter-OS-ē-ī; three muscles)	Medial side of second metacarpal and lateral sides of fourth and fifth metacarpals.	Proximal phalanx of same finger.	Adduct fingers toward middle finger; flex metacarpophalangeal joints; and extend interphalangeal joints.	Ulnar nerve.

EXHIBIT 11.19

MUSCLES THAT MOVE THE VERTEBRAL COLUMN (Fig. 11.19)

Overview: The muscles that move the vertebral column are quite complex because they have multiple origins and insertions and there is considerable overlap among them. One way to group the muscles is on the basis of the general direction of the muscle bundles and their approximate lengths. For example, the **splenius muscles** arise from the midline and run laterally and superiorly to their insertions. The **erector spinae (sacrospinalis) muscle** arises from either the midline or more laterally but usually runs almost longitudinally, with neither a marked outward nor inward direction as it is traced superiorly. The **transversospinalis muscles** arise laterally but run toward the midline as they are traced superiorly. Deep to these three muscle groups are small **segmental muscles** that run between spinous processes or transverse processes of vertebrae. Since the scalene muscles also assist in moving the vertebral column, they are included in this exhibit. Note in Exhibit 11.10 that the rectus abdominis, external oblique, internal oblique, and quadratus lumborum muscles also play a role in moving the vertebral column.

Muscle	Origin	Insertion	Action	Innervation
SPLENIUS (SPLĒ-nē-us)				
Splenius capitis (SPLĒ-nē-us KAP-i-tis; *splenium* = bandage; *caput* = head)	Ligamentum nuchae and spinous processes of seventh cervical vertebra and first three or four thoracic vertebrae.	Occipital bone and mastoid process of temporal bone.	Acting together, they extend the head and neck; acting singly, each laterally flexes and rotates head to same side.	Dorsal rami of middle cervical nerves.
Splenius cervicis (SPLĒ-nē-us SER-vi-kis; *cervix* = neck)	Spinous processes of third through sixth thoracic vertebrae.	Transverse processes of first two or four cervical vertebrae.	Acting together, they extend the head and neck; acting singly, each laterally flexes and rotates head to same side.	Dorsal rami of lower cervical nerves.
ERECTOR SPINAE (e-REK-tor SPI-nē) (SACROSPINALIS)	This is the largest muscular mass of the back and consists of three groupings—iliocostalis, longissimus, and spinalis. These groups, in turn, consist of a series of overlapping muscles. The iliocostalis group is laterally placed, the longissimus group is intermediate in placement, and the spinalis group is medially placed.			

Muscle	Origin	Insertion	Action	Innervation
ILIOCOSTALIS (LATERAL) GROUP				
Iliocostalis lumborum (il′-ē-ō-kos-TAL-is lum-BOR-um; *ilium* = flank; *costa* = rib)	Iliac crest.	Lower six ribs.	Extends lumbar region of vertebral column.	Dorsal rami of lumbar nerves.
Iliocostalis thoracis (il′-ē-ō-kos-TAL-is thō-RA-kis; *thorax* = chest)	Lower six ribs.	Upper six ribs.	Maintains erect position of spine.	Dorsal rami of thoracic (intercostal) nerves.
Iliocostalis cervicis (il′-ē-ō-kos-TAL-is SER-vi-kis)	First six ribs.	Transverse processes of fourth to sixth cervical vertebrae.	Extends cervical region of vertebral column.	Dorsal rami of cervical nerves.
LONGISSIMUS (INTERMEDIATE) GROUP				
Longissimus thoracis (lon-JIS-i-mus thō-RA-kis; *longissimus* = longest)	Transverse processes of lumbar vertebrae.	Transverse processes of all thoracic and upper lumbar vertebrae and ninth and tenth ribs.	Extends thoracic region of vertebral column.	Dorsal rami of spinal nerves.
Longissimus cervicis (lon-JIS-i-mus SER-vi-kis)	Transverse processes of fourth and fifth thoracic vertebrae.	Transverse processes of second to sixth cervical vertebrae.	Extends cervical region of vertebral column.	Dorsal rami of spinal nerves.
Longissimus capitis (lon-JIS-i-mus KAP-i-tis)	Transverse processes of upper four thoracic vertebrae and articular processes of last four cervical vertebrae.	Mastoid process of temporal bone.	Extends head and rotates it to same side.	Dorsal rami of middle and lower cervical nerves.
SPINALIS (MEDIAL) GROUP				
Spinalis thoracis (spi-NA-lis thō-RA-kis; *spinalis* = vertebral column)	Spinous processes of upper lumbar and lower thoracic vertebrae.	Spinous processes of upper thoracic vertebrae.	Extends vertebral column.	Dorsal rami of spinal nerves.
Spinalis cervicis (spi-NA-lis SER-vi-kis)	Ligamentum nuchae and spinous process of seventh cervical vertebra.	Spinous process of axis.	Extends vertebral column.	Dorsal rami of spinal nerves.
Spinalis capitis (spi-NA-lis KAP-i-tis)	Arises with semispinalis capitis.	Inserts with semispinalis capitis.	Extends vertebral column.	Dorsal rami of spinal nerves.
TRANSVERSOSPINALIS (trans-ver′-sō-spi-NA-lis)				
Semispinalis thoracis (sem′-ē-spi-NA-lis thō-RA-kis; *semi* = partially or one-half)	Transverse processes of sixth to tenth thoracic vertebrae.	Spinous processes of first four thoracic and last two cervical vertebrae.	Extends vertebral column and rotates it to opposite side.	Dorsal rami of thoracic and cervical spinal nerves.
Semispinalis cervicis (sem′-ē-spi-NA-lis SER-vi-kis)	Transverse processes of first five or six thoracic vertebrae.	Spinous processes of first to fifth cervical vertebrae.	Extends vertebral column and rotates it to opposite side.	Dorsal rami of thoracic and cervical spinal nerves.
Semispinalis capitis (sem′-ē-spi-NA-lis KAP-i-tis)	Transverse processes of first six or seven thoracic vertebrae and seventh cervical vertebra and articular processes of fourth, fifth, and sixth cervical vertebrae.	Occipital bone.	Extends vertebral column and rotates it to opposite side.	Dorsal rami of cervical nerves.

Exhibit continues

MUSCLES THAT MOVE THE VERTEBRAL COLUMN (Fig. 11.19)

Muscle	Origin	Insertion	Action	Innervation
Multifidus (mul-TIF-i-dus; *multi* = many; *findere* = to split)	Sacrum, ilium, transverse processes of lumbar, thoracic, and lower four cervical vertebrae.	Spinous process of a higher vertebra.	Extends vertebral column and rotates it to opposite side.	Dorsal rami of spinal nerves.
Rotatores (rō′-ta-TŌ-rēz; *rotate* = turn on an axis)	Transverse processes of all vertebrae.	Spinous process of vertebra above the one of origin.	Extends vertebral column and rotates it to opposite side.	Dorsal rami of spinal nerves.
SEGMENTAL (seg-MEN-tal)				
Interspinales (in-ter-SPĪ-nāl-ēz; *inter* = between)	Superior surface of all spinous processes.	Inferior surface of spinous process of vertebra above the one of origin.	Extends vertebral column.	Dorsal rami of spinal nerves.
Intertransversarii (in′-ter-trans-vers-AR-ē-ī; *inter* = between)	Transverse processes of all vertebrae.	Transverse process of vertebra above the one of origin.	Laterally flexes vertebral column.	Dorsal and ventral rami of spinal nerves.
SCALENE (SKĀ-lēn)				
Anterior scalene (SKĀ-lēn; *anterior* = front; *skalenos* = uneven)	Transverse processes of third through sixth cervical vertebrae.	First rib.	Flexes and rotates neck and assists in inspiration.	Ventral rami of fifth and sixth cervical nerves.
Middle scalene (SKĀ-lēn)	Transverse processes of last six cervical vertebrae.	First rib.	Flexes and rotates neck and assists in inspiration.	Ventral rami of third through eighth cervical nerves.
Posterior scalene (SKĀ-lēn)	Transverse processes of fourth through sixth cervical vertebrae.	Second rib.	Flexes and rotates neck and assists in inspiration.	Ventral rami of last three cervical nerves.

FIGURE 11.19 Muscles that move the vertebral column.

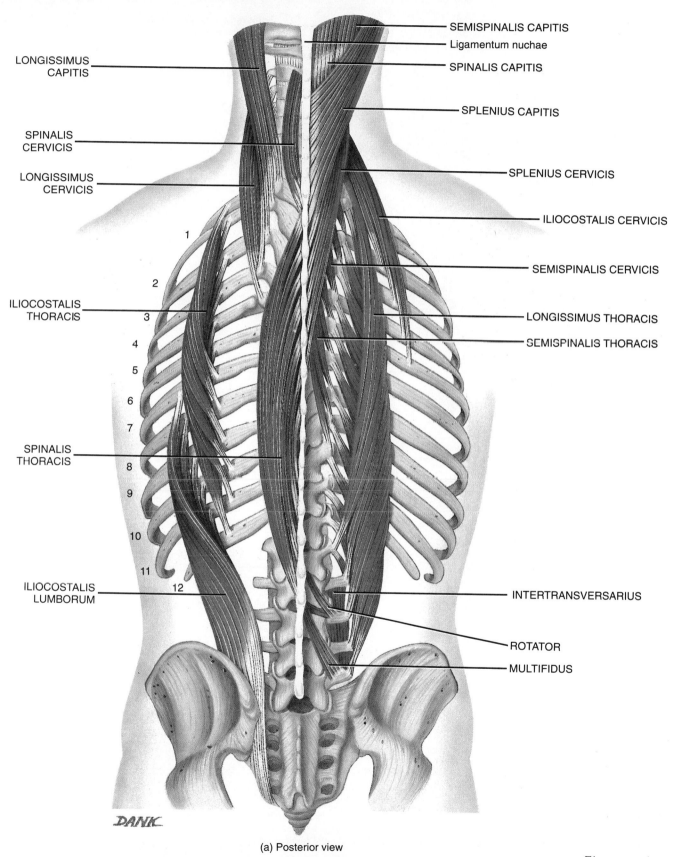

LONGISSIMUS CAPITIS

SPINALIS CERVICIS

LONGISSIMUS CERVICIS

ILIOCOSTALIS THORACIS

SPINALIS THORACIS

ILIOCOSTALIS LUMBORUM

SEMISPINALIS CAPITIS

Ligamentum nuchae

SPINALIS CAPITIS

SPLENIUS CAPITIS

SPLENIUS CERVICIS

ILIOCOSTALIS CERVICIS

SEMISPINALIS CERVICIS

LONGISSIMUS THORACIS

SEMISPINALIS THORACIS

INTERTRANSVERSARIUS

ROTATOR

MULTIFIDUS

DANK

(a) Posterior view

Figure continues

FIGURE 11.19 (continued)

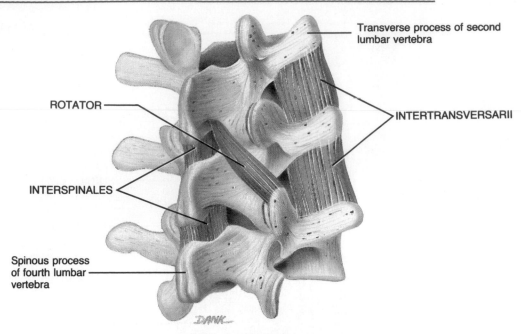

Transverse process of second lumbar vertebra

ROTATOR

INTERSPINALES

INTERTRANSVERSARII

Spinous process of fourth lumbar vertebra

(b) Posterolateral view

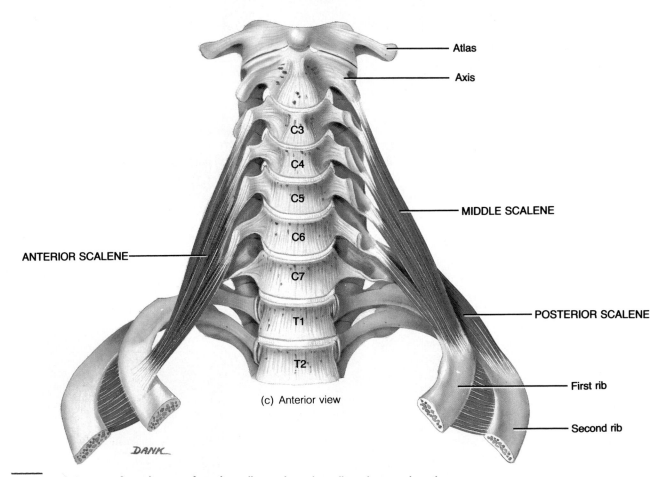

Atlas

Axis

C3

C4

C5

C6

C7

T1

T2

ANTERIOR SCALENE

MIDDLE SCALENE

POSTERIOR SCALENE

First rib

Second rib

(c) Anterior view

Question: Which group of muscles arises from the midline and runs laterally and superiorly to their insertion?

EXHIBIT 11.20

MUSCLES THAT MOVE THE THIGH (FEMUR) (Fig. 11.20)

Overview: As you will see, muscles of the lower extremities are larger and more powerful than those of the upper extremities since lower extremity muscles function in stability, locomotion, and maintenance of posture. Upper extremity muscles are characterized by versatility of movement. In addition, muscles of the lower extremities, often cross two joints and act equally on both.

The majority of muscles that act on the thigh (femur) originate on the pelvic (hip) girdle and insert on the femur. The anterior muscles are the psoas major and iliacus, together referred to as the iliopsoas muscle. The remaining muscles (except for the pectineus, adductors, and tensor fasciae latae) are posterior muscles. Technically, the pectineus and adductors are components of the medial compartment of the thigh, but they are included in this exhibit because they act on the thigh. The tensor fasciae latae muscle is laterally placed. The **fascia lata** is a deep fascia of the thigh that encircles the entire thigh. It is well developed laterally, where together with the tendons of the gluteus maximus and tensor fasciae latae muscles it forms a structure called the **iliotibial tract**. The tract inserts into the lateral condyle of the tibia.

After you have studied the muscles in this exhibit, arrange them according to the following actions: flexion, extension, abduction, adduction, medial rotation, and lateral rotation. (The same muscles can be used more than once.)

Muscle	Origin	Insertion	Action	Innervation
Psoas (SŌ-as) **major** (*psoa* = muscle of loin)	Transverse processes and bodies of lumbar vertebrae.	Lesser trochanter of femur.	Flexes and rotates thigh laterally; flexes vertebral column.	Lumbar nerves L2–L3.
Iliacus (il'-ē-AK-us; *iliac* = ilium)	Iliac fossa.	Tendon of psoas major.	Flexes and rotates thigh laterally and flexes vertebral column.	Femoral nerve.
Gluteus maximus (GLOO-tē-us MAK-si-mus; *glutos* = buttock; *maximus* = largest; strongest single muscle in body)	Iliac crest, sacrum, coccyx, and aponeurosis of sacrospinalis.	Iliotibial tract of fascia lata and lateral part of linea aspera under greater trochanter (gluteal tuberosity) of femur.	Extends and rotates thigh laterally.	Inferior gluteal nerve.
Gluteus medius (GLOO-tē-us MĒ-dē-us; *media* = middle)	Ilium.	Greater trochanter of femur.	Abducts and rotates thigh medially.	Superior gluteal nerve.
Gluteus minimus (GLOO-tē-us MIN-i-mus; *minimus* = smallest)	Ilium.	Greater trochanter of femur.	Abducts and rotates thigh medially.	Superior gluteal nerve.
Tensor fasciae latae (TEN-sor FA-shē-ē LĀ-tē; *tensor* = makes tense; *fascia* = band; *latus* = wide)	Iliac crest.	Tibia by way of the iliotibial tract.	Flexes and abducts thigh.	Superior gluteal nerve.
Piriformis (pir-i-FOR-mis; *pirum* = pear; *forma* = shape)	Anterior sacrum.	Superior border of greater trochanter of femur.	Rotates thigh laterally and abducts it.	Sacral nerves S2 or S1–S2.
Obturator internus (OB-too-rā'-tor in-TER-nus; *obturator* = obturator foramen; *internus* = inside)	Inner surface of obturator foramen, pubis, and ischium.	Greater trochanter of femur.	Rotates thigh laterally and abducts it.	Nerve to obturator internus.
Obturator externus (OB-too-rā'-tor ex-TER-nus; *externus* = outside)	Outer surface of obturator membrane.	Deep depression below greater trochanter (trochanteric fossa) of femur.	Rotates thigh laterally.	Obturator nerve.

Exhibit continues

EXHIBIT 11.20 (continued)

MUSCLES THAT MOVE THE THIGH (FEMUR) (Fig. 11.20)

Muscle	Origin	Insertion	Action	Innervation
Superior gemellus (jem-EL-lus; *superior* = above; *gemellus* = twins)	Ischial spine.	Greater trochanter of femur.	Rotates thigh laterally and abducts it.	Nerve to obturator internus.
Inferior gemellus (jem-EL-lus; *inferior* = below)	Ischial tuberosity.	Greater trochanter of femur.	Rotates thigh laterally and abducts it.	Nerve to quadratus femoris.
Quadratus femoris (kwod-RĀ-tus FEM-or-is; *quad* = four; *femoris* = femur)	Ischial tuberosity.	Elevation above mid-portion of intertro-chanteric crest (quadrate tubercle) on posterior femur.	Laterally rotates and adducts thigh.	Nerve to quadratus femoris.
Adductor longus (LONG-us; *adductor* = moves part closer to midline; *longus* = long)	Pubic crest and pubic symphysis.	Linea aspera of femur.	Adducts, medially rotates, and flexes thigh.	Obturator nerve.
Adductor brevis (BREV-is; *brevis* = short)	Inferior ramus of pubis.	Upper half of linea aspera of femur.	Adducts, medially rotates, and flexes thigh.	Obturator nerve.
Adductor magnus (MAG-nus; *magnus* = large)	Inferior ramus of pubis and ischium to ischial tuberosity.	Linea aspera of femur.	Adducts and medially rotates thigh (anterior part flexes; posterior part extends).	Obturator and sciatic nerves.
Pectineus (pek-TIN-ē-us; *pecten* = comb-shaped)	Superior ramus of pubis.	Pectineal line of femur, between lesser trochanter and linea aspera.	Flexes and adducts thigh.	Femoral nerve.

Figure 11.20 Muscles that move the thigh (femur).

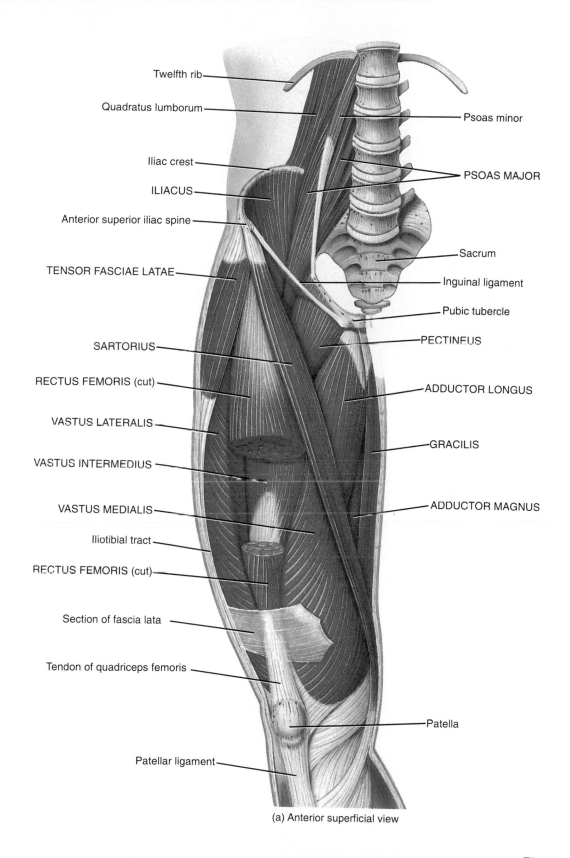

(a) Anterior superficial view

Figure continues

FIGURE 11.20 (continued)

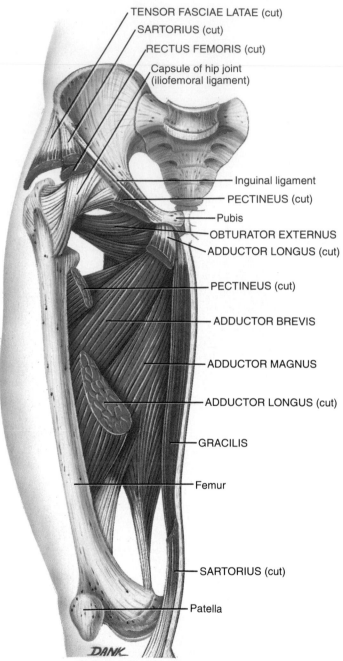

TENSOR FASCIAE LATAE (cut)
SARTORIUS (cut)
RECTUS FEMORIS (cut)
Capsule of hip joint
(iliofemoral ligament)

Inguinal ligament
PECTINEUS (cut)
Pubis
OBTURATOR EXTERNUS
ADDUCTOR LONGUS (cut)

PECTINEUS (cut)

ADDUCTOR BREVIS

ADDUCTOR MAGNUS

ADDUCTOR LONGUS (cut)

GRACILIS

Femur

SARTORIUS (cut)

Patella

DANK

(b) Anterior deep view (femur rotated laterally)

Figure 11.20 (continued)

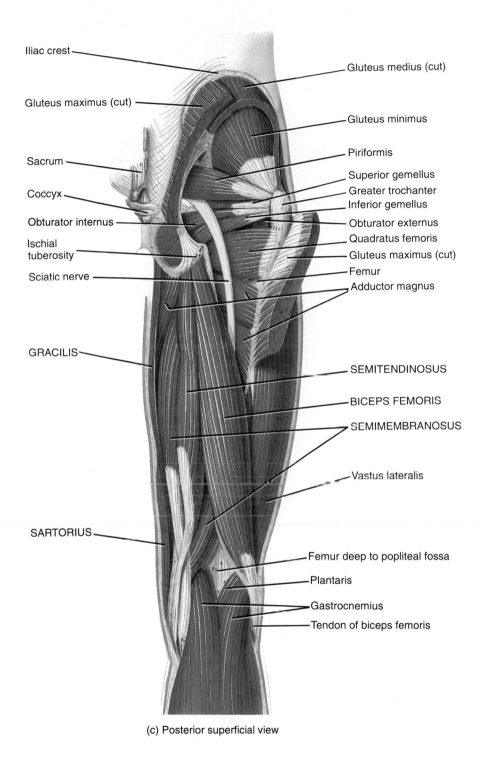

(c) Posterior superficial view

Figure continues

FIGURE 11.20 (continued)

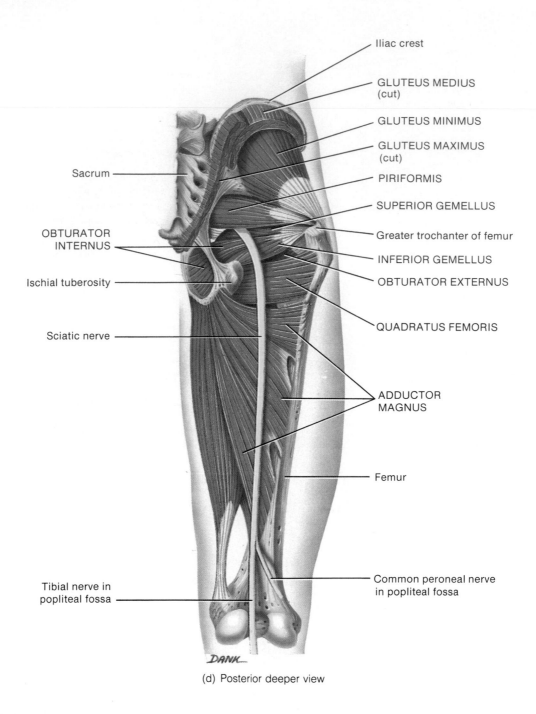

Iliac crest

GLUTEUS MEDIUS (cut)

GLUTEUS MINIMUS

GLUTEUS MAXIMUS (cut)

PIRIFORMIS

SUPERIOR GEMELLUS

Greater trochanter of femur

INFERIOR GEMELLUS

OBTURATOR EXTERNUS

QUADRATUS FEMORIS

ADDUCTOR MAGNUS

Femur

Common peroneal nerve in popliteal fossa

Sacrum

OBTURATOR INTERNUS

Ischial tuberosity

Sciatic nerve

Tibial nerve in popliteal fossa

DANK

(d) Posterior deeper view

Question: Most muscles that act on the thigh originate where? Insert where?

EXHIBIT 11.21

MUSCLES THAT ACT ON THE LEG (TIBIA AND FIBULA) (Figs. 11.20 and 11.21)

Overview: The muscles that act on the leg (tibia and fibula) originate in the hip and thigh and are separated into compartments by deep fascia. The **medial (adductor) compartment** is so named because its muscles adduct the thigh. It is innervated by the obturator nerve. As noted earlier, the adductor magnus, adductor longus, adductor brevis, and pectineus muscles, components of the medial compartment, are included in Exhibit 11.20 because they act on the femur. The gracilis, the other muscle in the medial compartment, not only adducts the thigh but also flexes the leg. For this reason, it is included in this exhibit.

The **anterior (extensor) compartment** is so designated because its muscles act to extend the leg, and some also flex the thigh. It is composed of the quadriceps femoris and sartorius muscles and is innervated by the femoral nerve. The quadriceps femoris muscle is a composite muscle that includes four distinct parts, usually described as four separate muscles (rectus femoris, vastus lateralis, vastus intermedius, and vastus medialis). The common tendon for the four muscles is known as the **patellar ligament** and attaches to the tibial tuberosity. The rectus femoris and sartorius muscles are also flexors of the thigh.

The **posterior (flexor) compartment** is so named because its muscles flex the leg (but also extend the thigh). It is innervated by branches of the sciatic nerve. Included are the hamstrings (biceps femoris, semitendinosus, and semimembranosus). The hamstrings are so named because their tendons are long and stringlike in the popliteal area. The **popliteal fossa** is a diamond-shaped space on the posterior aspect of the knee bordered laterally by the tendons of the biceps femoris and medially by the semitendinosus and semimembranosus muscles.

Muscle	Origin	Insertion	Action	Innervation
MEDIAL (ADDUCTOR) COMPARTMENT				
Adductor magnus (MAG-nus)	See Exhibit 11.20			
Adductor longus (LONG-us)	See Exhibit 11.20			
Adductor brevis (BREV-is)	See Exhibit 11.20			
Pectineus (pek-TIN-ē-us)	See Exhibit 11.20			
Gracilis (gra-SIL-is; *gracilis* = slender)	Pubic symphysis and pubic arch.	Medial surface of body of tibia.	Adducts thigh and flexes leg.	Obturator nerve.
ANTERIOR (EXTENSOR) COMPARTMENT				
QUADRICEPS FEMORIS (KWOD-ri-ceps FEM-or-is; *quadriceps* = four heads of origin; *femoris* = femur)				
Rectus femoris (REK-tus FEM-or-is; *rectus* = fibers parallel to midline)	Anterior inferior iliac spine.	Upper border of patella.	All four heads extend leg; rectus portion alone also flexes thigh.	Femoral nerve.
Vastus lateralis (VAS-tus lat′-er-A-lis; *vastus* = large; *lateralis* = lateral)	Greater trochanter and linea aspera of femur.	Tibial tuberosity through patellar ligament (tendon of quadriceps).	All four heads extend leg; rectus portion alone also flexes thigh.	Femoral nerve.
Vastus medialis (VAS-tus mē′-dē-A-lis; *medialis* = medial)	Linea aspera of femur.	Tibial tuberosity through patellar ligament (tendon of quadriceps).	All four heads extend leg; rectus portion alone also flexes thigh.	Femoral nerve.
Vastus intermedius (VAS-tus in′-ter-MĒ-de-us; *intermedius* = middle)	Anterior and lateral surfaces of body of femur.	Tibial tuberosity through patellar ligament (tendon of quadriceps).	All four heads extend leg; rectus portion alone also flexes thigh.	Femoral nerve.

Exhibit continues

EXHIBIT 11.21

MUSCLES THAT ACT ON THE LEG (TIBIA AND FIBULA)
(Figs. 11.20 and 11.21)

Muscle	Origin	Insertion	Action	Innervation
Sartorius (sar-TOR-ē-us; *sartor* = tailor; contracts when you sit in the cross-legged position of a tailor; longest muscle in body)	Anterior superior iliac spine.	Medial surface of body of tibia.	Flexes leg; flexes thigh and rotates it laterally, thus crossing leg.	Femoral nerve.

POSTERIOR (FLEXOR) COMPARTMENT

HAMSTRINGS	A collective designation for three separate muscles.			
Biceps femoris (BĪ-ceps FEM-or-is; *biceps* = two heads of origin)	Long head arises from ischial tuberosity; short head arises from linea aspera of femur.	Head of fibula and lateral condyle of tibia.	Flexes leg and extends thigh.	Tibial and common peroneal nerves from sciatic nerve.
Semitendinosus (sem´-ē-TEN-di-nō-sus; *semi* = half; *tendo* = tendon)	Ischial tuberosity.	Proximal part of medial surface of body of tibia.	Flexes leg and extends thigh.	Tibial nerve from sciatic nerve.
Semimembranosus (sem´-ē-MEM-bra-nō-sus; *membran* = membrane)	Ischial tuberosity.	Medial condyle of tibia.	Flexes leg and extends thigh.	Tibial nerve from sciatic nerve.

FIGURE 11.21 Muscles that act on the leg.

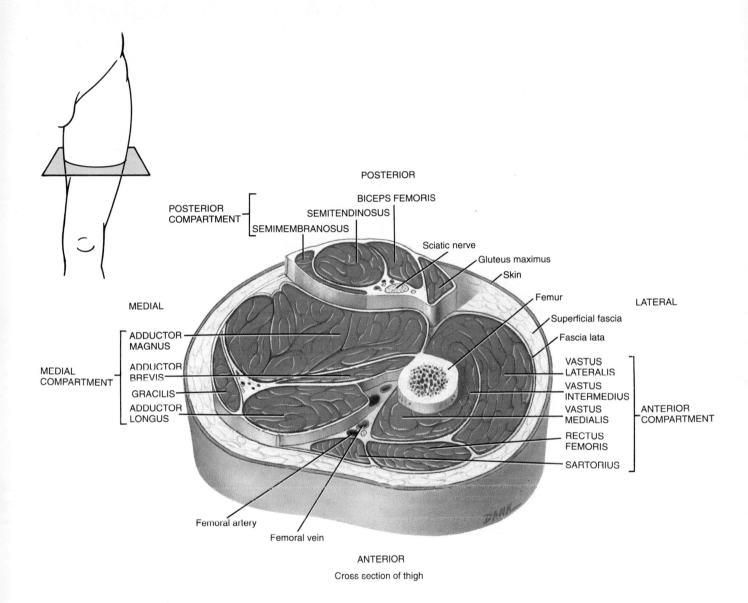

Cross section of thigh

_____
Question: Which muscles compose the quadriceps femoris muscle?

EXHIBIT 11.22

MUSCLES THAT MOVE THE FOOT AND TOES (Fig. 11.22)

Overview: The musculature of the leg, like that of the thigh, is divided into three compartments by deep fasica. In addition, all the muscles in a given compartment are innervated by the same nerve. The **anterior compartment** consists of muscles that dorsiflex the foot and are innervated by the deep peroneal nerve. In a situation analogous to the wrist, the tendons of the muscles of the anterior compartment are held firmly to the ankle by thickenings of deep fascia called the **superior extensor retinaculum (transverse ligament of the ankle)** and **inferior extensor retinaculum (cruciate ligament of the ankle)**.

The **lateral (peroneal) compartment** contains two muscles that plantar flex and evert the foot. They are supplied by the superficial peroneal nerve.

The **posterior compartment** consists of muscles that are divisible into superficial and deep groups. All are innervated by the tibial nerve. All three superficial muscles share a common tendon of insertion, the calcaneal (Achilles) tendon that inserts into the calcaneus bone of the ankle. The superficial muscles are plantar flexors of the foot. Of the four deep muscles, three plantar flex the foot.

Muscle	Origin	Insertion	Action	Innervation
ANTERIOR COMPARTMENT				
Tibialis (tib'-ē-A-lis) **anterior** (*tibialis* = tibia; *anterior* = front)	Lateral condyle and body of tibia and interosseous membrane (sheet of fibrous tissue that holds shafts of tibia and fibula together).	First metatarsal and first (medial) cuneiform.	Dorsiflexes and inverts foot.	Deep peroneal nerve.
Extensor hallucis longus (HAL-a-kis LON-gus; *extensor* = increases angle at joint; *hallucis* = hallux or great toe; *longus* = long)	Anterior surface of fibula and interosseous membrane.	Distal phalanx of great toe.	Dorsiflexes and inverts foot and extends great toe.	Deep peroneal nerve.
Extensor digitorum longus (di'-ji-TOR-um LON-gus)	Lateral condyle of tibia, anterior surface of fibula, and interosseous membrane.	Middle and distal phalanges of four outer toes.	Dorsiflexes and everts foot and extends toes.	Deep peroneal nerve.
Peroneus tertius (per'-ō-NĒ-us TER-shus; *perone* = fibula; *tertius* = third)	Distal third of fibula and interosseous membrane.	Fifth metatarsal.	Dorsiflexes and everts foot.	Deep peroneal nerve.
LATERAL (PERONEAL) COMPARTMENT				
Peroneus longus (per'-ō-NĒ-us LON-gus)	Head and body of fibula and lateral condyle of tibia.	First metatarsal and first cuneiform.	Plantar flexes and everts foot.	Superficial peroneal nerve.
Peroneus brevis (per'-ō-NĒ-us BREV-is; *brevis* = short)	Body of fibula.	Fifth metatarsal.	Plantar flexes and everts foot.	Superficial peroneal nerve.

Muscle	Origin	Insertion	Action	Innervation
POSTERIOR COMPARTMENT				
SUPERFICIAL				
Gastrocnemius (gas'-trok-NĒ-mē-us; *gaster* = belly; *kneme* = leg)	Lateral and medial condyles of femur and capsule of knee.	Calcaneus by way of calcaneal (Achilles) tendon.[a]	Plantar flexes foot and flexes leg.	Tibial nerve.
Soleus (SŌ-lē-us; *soleus* = sole of foot)	Head of fibula and medial border of tibia.	Calcaneus by way of calcaneal (Achilles) tendon.	Plantar flexes foot.	Tibial nerve.
Plantaris (plan-TA-ris; *plantar* = sole of foot)	Femur above lateral condyle.	Calcaneus by way of calcaneal (Achilles) tendon.	Plantar flexes foot.	Tibial nerve.
DEEP				
Popliteus (pop-LIT-ē-us; *poples* = posterior surface of knee)	Lateral condyle of femur.	Proximal tibia.	Flexes and medially rotates leg.	Tibial nerve.
Flexor hallucis longus (HAL-a-kis LON-gus; *flexor* = decreases angle at joint)	Lower two-thirds of fibula.	Distal phalanx of great toe.	Plantar flexes and inverts foot and flexes great toe.	Tibial nerve.
Flexor digitorum longus (di'-ji-TOR-um LON-gus; *digitorum* = finger or toe)	Posterior surface of tibia.	Distal phalanges of four outer toes.	Plantar flexes and inverts foot and flexes toes.	Tibial nerve.
Tibialis posterior (tib'-ē Λ lis) (*posterior* = back)	Tibia, fibula, and interosseous membrane.	Second, third, and fourth metatarsals; navicular; all three cuneiforms; and cuboid.	Plantar flexes and inverts foot.	Tibial nerve.

[a]The calcaneal (Achilles) tendon, the **strongest tendon of the body**, is able to withstand a 1000-pound force without tearing. Despite this, however, the calcaneal tendon ruptures more frequently than any other tendon because of the tremendous pressures placed on it during competitive sports.

FIGURE 11.22 Muscles that move the foot and toes.

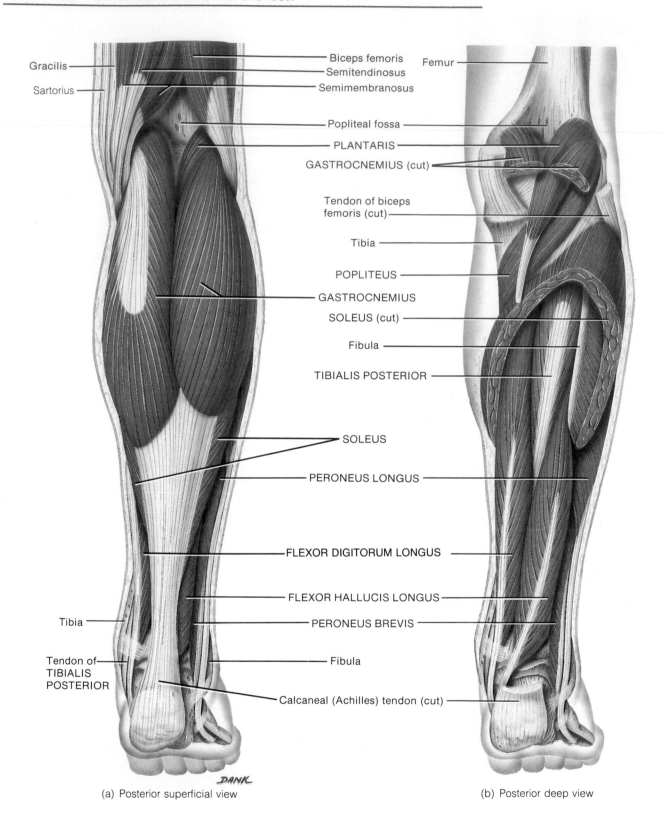

Gracilis

Sartorius

Biceps femoris
Semitendinosus
Semimembranosus

Femur

Popliteal fossa

PLANTARIS

GASTROCNEMIUS (cut)

Tendon of biceps femoris (cut)

Tibia

POPLITEUS

GASTROCNEMIUS

SOLEUS (cut)

Fibula

TIBIALIS POSTERIOR

SOLEUS

PERONEUS LONGUS

FLEXOR DIGITORUM LONGUS

FLEXOR HALLUCIS LONGUS

Tibia

PERONEUS BREVIS

Tendon of
TIBIALIS
POSTERIOR

Fibula

Calcaneal (Achilles) tendon (cut)

DANK

(a) Posterior superficial view

(b) Posterior deep view

FIGURE 11.22 (continued)

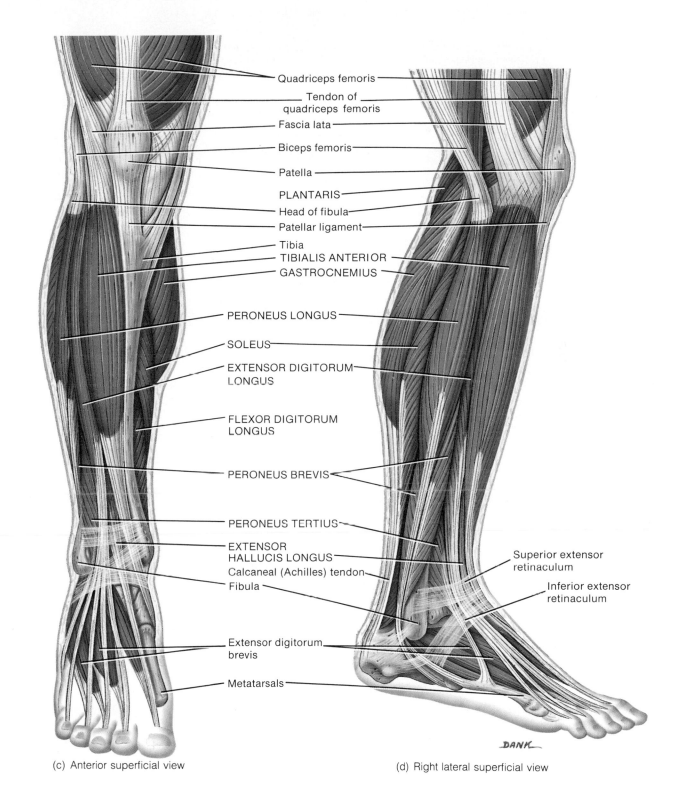

Quadriceps femoris
Tendon of quadriceps femoris
Fascia lata
Biceps femoris
Patella
PLANTARIS
Head of fibula
Patellar ligament
Tibia
TIBIALIS ANTERIOR
GASTROCNEMIUS
PERONEUS LONGUS
SOLEUS
EXTENSOR DIGITORUM LONGUS
FLEXOR DIGITORUM LONGUS
PERONEUS BREVIS
PERONEUS TERTIUS
EXTENSOR HALLUCIS LONGUS
Calcaneal (Achilles) tendon
Fibula
Extensor digitorum brevis
Metatarsals

Superior extensor retinaculum
Inferior extensor retinaculum

DANK

(c) Anterior superficial view

(d) Right lateral superficial view

Question: The muscles in which compartment dorsiflex the foot?

EXHIBIT 11.23

INTRINSIC MUSCLES OF THE FOOT (Fig. 11.23)

Overview: The intrinsic muscles of the foot are, for the most part, comparable to those in the hand. Whereas the muscles of the hand are specialized for precise and intricate movements, those of the foot are limited to support and locomotion. The deep fascia at the foot forms the **plantar aponeurosis (fascia)** that extends from the calcaneus to the phalanges. The aponeurosis supports the longitudinal arch of the foot and encloses the flexor tendons of the foot.

The intrinsic musculature of the foot is divided into two groups—dorsal and plantar. There is only one **dorsal muscle**. The **plantar muscles** are arranged in four layers, the most superficial layer being referred to as the first layer.

Muscle	Origin	Insertion	Action	Innervation
DORSAL				
Extensor digitorum brevis (di′-ji-TOR-um BREV-is; *extensor* = increases angle at joint; *digit* = finger or toe; *brevis* = short) (see Fig. 11.22c,d)	Calcaneus.	Tendon of extensor digitorum longus and proximal phalanx of great toe.	Extends first through fourth toes.	Deep peroneal nerve.
PLANTAR				
FIRST (SUPERFICIAL) LAYER				
Abductor hallucis (HAL-a-kis; *abductor* = moves part away from midline; *hallucis* = hallux or great toe)	Calcaneus and plantar aponeurosis.	Proximal phalanx of great toe	Abducts great toe and flexes metatarsophalangeal joint.	Medial plantar nerve.
Flexor digitorum brevis (di′-ji-TOR-um BREV-is; *flexor* = decreases angle at joint)	Calcaneus and plantar aponeurosis.	Middle phalanx of second through fifth toes.	Flexes second through fifth toes.	Medial plantar nerve.
Abductor digiti minimi (DIJ-i-tē MIN-i-mē; *minimi* = small toe)	Calcaneus and plantar aponeurosis.	Proximal phalanx of small toe.	Abducts and flexes small toe.	Lateral plantar nerve.
SECOND LAYER				
Quadratus plantae (quod-RĀ-tus PLAN-tē; *quad* = four; *planta* = sole of foot)	Calcaneus.	Tendon of flexor digitorum longus.	Flexes second through fifth toes.	Lateral plantar nerve.
Lumbricals (LUM-bri-kals)	Tendons of flexor digitorum longus.	Tendons of extensor digitorum longus.	Extend interphalangeal joints and flex metatarsophalangeal joints of second through fifth toes.	Medial and lateral plantar nerves.
THIRD LAYER				
Flexor hallucis brevis (HAL-a-kis BREV-is)	Cuboid and third (lateral) cuneiform.	Proximal phalanx of great toe.	Flexes great toe.	Medial plantar nerve.
Adductor hallucis (HAL-a-kis)	Second through fourth metatarsals and ligaments of metatarsophalangeal joints.	Proximal phalanx of great toe.	Adducts and flexes great toe.	Lateral plantar nerve.
Flexor digiti minimi brevis (DIJ-i-tē MIN-i-mē BREV-is)	Fifth metatarsal.	Proximal phalanx of small toe.	Flexes small toe.	Lateral plantar nerve.

Muscle	Origin	Insertion	Action	Innervation
FOURTH (DEEP) LAYER				
Dorsal interossei (in′-ter-OS-ē-ī)	Adjacent side of metatarsals.	Proximal phalanges, both sides of second toe, lateral side of third and fourth toes.	Abduct toes and flex proximal phalanges.	Lateral plantar nerve.
Plantar interossei (in′-ter-OS-ē-ī)	Third, fourth, and fifth metatarsals.	Proximal phalanges of same toes.	Adduct third, fourth, and fifth toes and flex proximal phalanges.	Lateral plantar nerve.

FIGURE 11.23 Plantar views of intrinsic muscles of the foot.

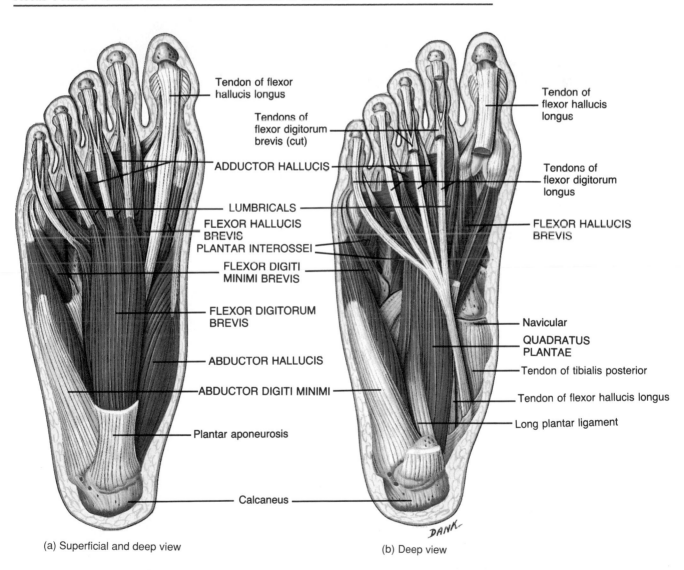

(a) Superficial and deep view

(b) Deep view

Figure continues

FIGURE 11.23 (continued)

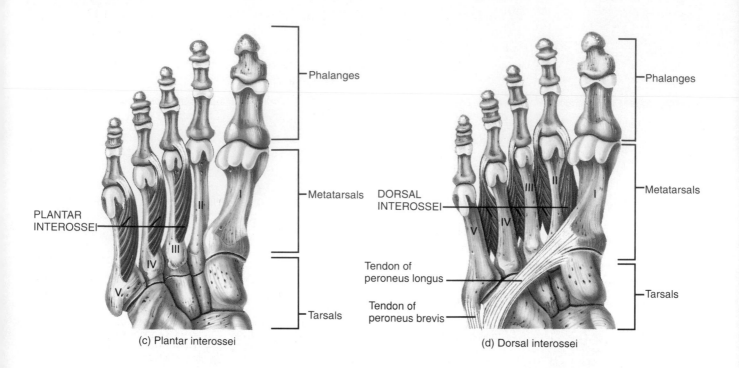

(c) Plantar interossei

(d) Dorsal interossei

Question: Which muscles of the foot flex the great toe?

INTRAMUSCULAR (IM) INJECTIONS

An **intramuscular (IM) injection** penetrates the skin and subcutaneous tissue to enter the muscle itself. Intramuscular injections are preferred when prompt absorption is desired, when larger doses than can be given subcutaneously are indicated, or when the drug is too irritating to give subcutaneously. The common sites for intramuscular injections include the buttock, lateral side of the thigh, and the deltoid region of the arm. Muscles in these areas, especially the gluteal muscles in the buttock, are fairly thick and absorption is promoted by the extensive blood supply to such large muscles. To avoid injury, intramuscular injections should be given deep within the muscle and away from major nerves and blood vessels.

For many intramuscular injections, the preferred site is the **gluteus medius muscle** of the buttock (Fig. 11.24a). The buttock is divided into quadrants, and the upper outer quadrant is used as the injection site. The iliac crest serves as a landmark for this quadrant. The spot for injection in an adult is usually about 5 to 7½ cm (2 to 3 in.) below the iliac crest. The upper outer quadrant is chosen because the muscle in this area is quite thick and has few nerves. Injection in this area thus reduces the chance of injury to the sciatic nerve, which could cause paralysis of the lower extremity. The probability of injecting the drug into a blood vessel is also remote in this area. After the needle is inserted into the gluteus medius muscle, the plunger is pulled up for a few seconds. If the syringe fills with blood, the needle is in a blood vessel, and a different injection site on the opposite buttock is chosen.

Injections may also be given in the lateral side of the thigh in the midportion of the **vastus lateralis muscle** (Fig. 11.24b). This site is determined by using the knee and greater trochanter of the femur as landmarks. The midportion of the muscle is located by measuring a handbreadth above the knee and a handbreadth below the greater trochanter.

A **deltoid** injection is given in the midportion of the muscle about two to three fingerbreadths below the acromion of the scapula and lateral to the axilla (Fig. 11.24c). Deltoid IM injections in an adult are usually for immunization (tetanus toxoid booster, pneumococcal vaccine), whereas subcutaneous injections in the deltoid region include typhoid immunization and inactivated poliovirus vaccine.

RUNNING INJURIES

It is estimated that nearly 70% of those who jog or run will sustain some type of running-related injury. Even though most such injuries are minor, such as sprains and strains, some are quite serious; moreover, untreated or inappropriately treated minor injuries may become chronic. Among runners, the knee is the most common site of injury, accounting for about 40% of injuries. Other common sites of injury include the calcaneal (Achilles) tendon, medial aspect of the tibia, hip area, groin area, foot and ankle, and back.

Running injuries are frequently related to faulty training techniques. This may involve improper or lack of warm-up, running too much, or running too soon. Or it might involve running on hard and/or uneven surfaces. Poorly constructed or worn-out running shoes can also contribute to injury. Any biomechanical problems aggravated by running can also cause injuries.

CLINICAL APPLICATION

TREATING SPORTS INJURIES

Most sports injuries should be treated initially with **RICE** therapy, which stands for Rest, Ice, Compression, and Elevation. Immediately apply ice and rest and elevate the injured part. Then apply an elastic bandage, if possible, to compress the injured tissue. Continue using RICE for 2 to 3 days, and resist the temptation to apply heat, which may worsen the swelling. Follow-up treatment may include alternating moist heat and ice massage to enhance blood flow in the injured area. Sometimes, nonsteroidal anti-inflammatory drugs (NSAIDs) or local injections of corticosteroids are needed. During the recovery period, it is important to keep active with an alternate fitness program. And careful exercise is needed to rehabilitate the injured area itself.

Hip, Buttock, and Back Injuries

When back pain occurs in runners, it is often due to a preexisting degenerative condition that is aggravated by an increase in mileage or hill running. Such pain is usually due to an injury in the buttocks, pelvis, or lumbar spine. For example, the pain may be caused by strain of the distal attachments of the abductors (especially the gluteus medius) or the proximal attachment of the abductors to the iliac crest, both of which contribute to a **"pulled groin"** or strain or partial tear of the proximal hamstrings (**hamstring strain** or **"pulled hamstrings"**). Hamstring strains are common sports injuries in individuals who run very hard. Sometimes the violent muscular exertion required to perform a feat tears off part of the tendinous origins of the hamstrings, especially the biceps femoris, from the ischial tuberosity. This is usually accompanied by a contusion (bruising) and tearing of some of the muscle fibers and rupture of blood vessels, producing a hematoma (collection of blood) and pain. Adequate training with good balance between the quadriceps femoris and hamstrings and stretching exercises before running or competing are important in preventing this injury.

Knee Injuries

Patellofemoral stress syndrome ("runner's knee") is the single most common problem in runners. During normal

FIGURE 11.24 Intramuscular injections. Shown are the three common sites for intramuscular injections.

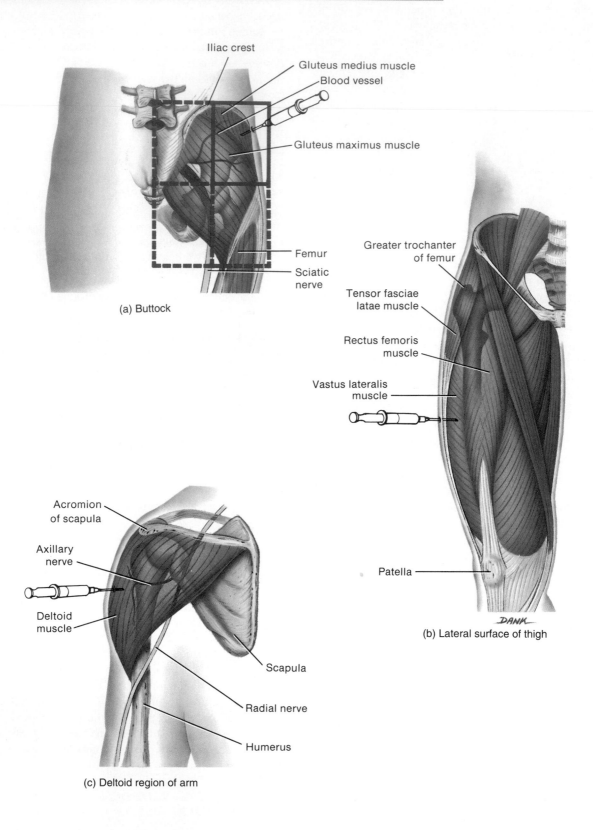

Iliac crest

Gluteus medius muscle

Blood vessel

Gluteus maximus muscle

Femur

Sciatic nerve

(a) Buttock

Greater trochanter of femur

Tensor fasciae latae muscle

Rectus femoris muscle

Vastus lateralis muscle

Patella

DANK

(b) Lateral surface of thigh

Acromion of scapula

Axillary nerve

Deltoid muscle

Scapula

Radial nerve

Humerus

(c) Deltoid region of arm

Question: When are intramuscular injections given?

flexion and extension of the knee, the patella tracks (glides) up and down in the groove between the femoral condyles. In patellofemoral stress syndrome, normal tracking does not occur; instead, the patella tracks laterally, and the increased pressure of abnormal tracking causes the associated pain. The pain is usually described as an aching or tenderness around or under the patella. The pain typically occurs after a person has been sitting for a while, especially after exercise. A common cause of runner's knee is constantly walking, running, or jogging on the same side of the road. Since roads are high in the middle and slope down on the sides, the slope stresses the knee that is closer to the center of the road.

Besides RICE, treatment of patellofemoral syndrome includes cessation of running; avoidance of kneeling, stair climbing, and prolonged sitting; hamstring muscle stretching exercises; short-arc quadriceps strengthening exercises; and use of an orthotic device for the foot, knee wraps, and knee braces.

Leg and Foot Injuries

Shinsplint Syndrome (Shinsplints)

Shinsplint syndrome, or simply **shinsplints**, refers to pain or soreness along the tibia, specifically the medial, distal two-thirds. It may be caused by tendinitis of the tibialis posterior muscle or toe flexors, inflammation of the periosteum (periostitis) around the tibia, or stress fractures of the tibia. The tendinitis usually occurs when poorly conditioned runners run on hard or banked surfaces with poorly supportive running shoes. The condition may also occur as a result of vigorous activity of the legs following a period of rela

tive inactivity. The muscles in the anterior compartment (mainly the tibialis anterior) can be strengthened to balance the stronger posterior compartment muscles. Patients who do not respond to RICE may be given local injections of corticosteroids or may have to undergo minor surgery to release pressure in the soft tissues around the bone.

Plantar Fasciitis (Painful Heel Syndrome)

Plantar fasciitis (painful heel syndrome) is an inflammatory reaction due to chronic irritation of the plantar aponeurosis (fascia) at its origin on the calcaneus (heel bone) (see Fig. 11.23a). The condition is the most common cause of heel pain in runners and arises in response to the repeated impact of walking or running. Treatment consists of RICE, heel pads, exercises that gently stretch calf muscles, NSAIDs, and, in some cases, corticosteroid injections.

Stress Fractures

Stress fractures are partial fractures that result from inability to withstand repeated stress owing to a change in training, harder surfaces, longer distances, greater speed, or an existing pathology. Such fractures can occur in the bodies of lumbar vertebrae, sacroiliac joint, pubic symphysis, iliac crest, femoral neck and body, fibula, lateral malleolus, and metatarsals. About 25% of all stress fractures involve the fibula, specifically the distal third. With all stress fractures, running must stop temporarily and immobilization may be needed.

Study Outline

How Skeletal Muscles Produce Movement (p. 270)

1. Skeletal muscles produce movement by pulling on bones.
2. The attachment to the stationary bone is the origin. The attachment to the movable bone is the insertion.
3. Bones serve as levers and joints serve as fulcrums. The lever is acted on by two different forces: resistance and effort.
4. Levers are categorized into three types—first-class, second-class, and third-class—according to the position of the fulcrum, effort, and resistance on the lever.
5. Fascicular arrangements include parallel, convergent, pennate, and circular. Fascicular arrangement is correlated with the power of a muscle and the range of motion.
6. The prime mover produces the desired action. The antagonist produces an opposite action. The synergist assists the prime mover by reducing unnecessary movement. The fixator stabilizes the origin of the prime mover so that it can act more efficiently.

Naming Skeletal Muscles (p. 274)

1. Skeletal muscles are named on the basis of distinctive criteria: direction of fibers, location, size, number of origins (or heads), shape, origin and insertion, and action.

Principal Skeletal Muscles (p. 278)

1. The principal skeletal muscles of the body are grouped according to region in Exhibits 11.3 through 11.23.

Intramuscular (IM) Injections (p. 339)

1. Advantages of intramuscular injections are prompt absorption, use of larger doses than can be given subcutaneously, and minimal irritation.
2. Common sites for intramuscular injections are the buttock, lateral side of the thigh, and deltoid region of the arm.

Running Injuries (p. 339)

1. Most running injuries involve the knee. Other commonly injured sites are the calcaneal (Achilles) tendon, medial aspect of tibia, hip, groin, foot, ankle, and back.

2. Running injuries are initially treated by rest, ice, compression, and elevation (RICE). Nonsteroidal anti-inflammatory drugs (NSAIDs), moist heat, and an alternate fitness program may also be included.

Review Questions

1. What is meant by the muscular system? (p. 270)
2. Using the terms origin, insertion, and belly in your discussion, describe how skeletal muscles produce body movements by pulling on bones. (p. 270)
3. What is a lever? Fulcrum? Apply these terms to the body and indicate the nature of the forces that act on levers. Describe the three classes of levers and provide one example for each in the body. (p. 271)
4. Describe the various arrangements of fasciculi. How is fascicular arrangement correlated with the strength of a muscle and its range of motion? (p. 272)
5. Define the role of the prime mover (agonist), antagonist, synergist, and fixator in producing body movements. (p. 272)
6. Select at random several muscles presented in Exhibits 11.3 through 11.23 and see if you can determine the criterion or criteria employed for naming each. In addition, refer to the prefixes, suffixes, roots, and definitions in each exhibit as a guide. Select as many muscles as you wish, as long as you feel you understand the concept involved.
7. What muscles would you use to do the following: (a) frown, (b) pout, (c) show surprise, (d) show your upper teeth, (e) pucker your lips, (f) squint, (g) blow up a balloon, (h) smile? (p. 279)
8. What are the principal muscles that move the mandible? Give the function of each. (p.282)
9. What would happen if you lost tone in the masseter and temporalis muscles? (p. 282)
10. What extrinsic muscles move the eyeball? In which direction does each muscle move the eyeball? (p. 284)
11. Describe the action of each of the muscles acting on the tongue. (p. 286)
12. What tongue, facial, and mandibular muscles would you use when chewing food? (p. 286)
13. Describe the muscles involved, and their actions, in moving the hyoid bone. (p. 289)
14. Describe the actions of the extrinsic and intrinsic muscles of the larynx. (p. 290)
15. What muscles are responsible for moving the head? And how do they move the head? (p. 293)
16. What muscles would you use to signify "yes" and "no" by moving your head? (p. 293)
17. Describe the composition of the anterolateral and posterior abdominal wall. (p. 294)
18. What muscles accomplish compression of the abdominal wall? (p. 294)
19. What are the principal muscles involved in breathing? What are their actions? (p. 296)
20. Describe the actions of the muscles of the pelvic floor. What is the pelvic diaphragm? (p. 298)
21. Describe the actions of the muscles of the perineum. What is the urogenital diaphragm? (p. 300)
22. In what directions is the pectoral (shoulder) girdle drawn? What muscles accomplish these movements? (p. 302)

23. What muscles are used to raise your shoulders, lower your shoulders, join your hands behind your back, and join your hands in front of your chest? (p. 302)
24. What movements are possible at the shoulder joint? What muscles accomplish these movements? (p. 305)
25. Distinguish axial and scapular muscles involved in moving the arm. (p. 305)
26. What muscles move the arm? In which directions do these movements occur? (p. 305)
27. Organize the muscles that move the forearm into flexors and extensors. What muscles move the forearm and what actions are used when striking a match? (p. 308)
28. Discuss the various movements possible at the wrist, hand, and fingers. What muscles accomplish these movements? (p. 311)
29. How many muscles and actions of the wrist, hand, and fingers that are used when writing can you list? (p. 311)
30. What is the flexor retinaculum? Extensor retinaculum? (p. 311)
31. What muscles form the thenar eminence? What are their functions? (p. 316)
32. What muscles form the hypothenar eminence? What are their functions? (p. 316)
33. Discuss the various muscles and movements of the vertebral column. How are the muscles grouped? (p. 318)
34. What muscles accomplish movements of the thigh? What actions are produced by these muscles? What is the iliotibial tract? (p. 323)
35. Organize the muscles that act on the leg into medial, anterior, and posterior compartments. What is the popliteal fossa? (p. 323)
36. What muscles act at the knee joint? What kinds of movements do these muscles perform? (p. 323)
37. Determine the muscles and their actions listed in Exhibit 11.23 that you would use to climb a ladder to a diving board, dive into the water, swim the length of a pool, and then sit at pool side.
38. Name the muscles that plantar flex, evert, pronate, and dorsiflex the foot. What is the superior extensor retinaculum? Inferior extensor retinaculum? (p. 332)
39. How are the intrinsic muscles of the foot organized? (p. 336)
40. In which directions are the toes moved? What muscles bring about these movements? (p. 336)
41. What are the advantages of intramuscular (IM) injections? Describe how you would locate the sites for an intramuscular injection in the buttock, lateral side of the thigh, and deltoid region of the arm. (p. 339)
42. List the common sites for running injuries. How are most running injuries treated? (p. 339)
43. Define the following: "pulled hamstrings" (p. 339), "pulled groin" (p. 339), shinsplint syndrome (p. 341), plantar fasciitis (p. 341), and stress fracture (p. 341).

Answers to Questions with Figures

11.1 Origin.

11.2 Levers; fulcrum.

11.3 Possible response: direction of fibers—external oblique; shape—deltoid; action—extensor digitorum; size—gluteus maximus; origin and insertion—sternocleidomastoid; location—tibialis anterior; number of heads—biceps brachii.

11.4 Frowning—corrugator supercilii and frontalis; smiling—zygomaticus major; pouting—mentalis and platysma; squinting—orbicularis oculi.

11.5 Masseter and lateral pterygoid.

11.6 Lateral rectus, superior oblique, and inferior oblique.

11.7 Genioglossus and hyoglossus.

11.8 Stylohyoid.

11.9 Tenses—cricothyroid; relaxes—thyroarytenoid.

11.10 Rectus abdominis.

11.11 Diaphragm.

11.12 Coccygeus.

11.13 Bulbocavernosus and ischiocavernosus.

11.14 Subclavius, pectoralis minor, and serratus anterior.

11.15 Pectoralis major and deltoid.

11.16 Triceps brachii and anconeus.

11.17 Flex.

11.18 Ulnar.

11.19 Splenius.

11.20 Origin—pelvic (hip) girdle; insertion—femur.

11.21 Rectus femoris, vastus lateralis, vastus medialis, and vastus intermedius.

11.22 Anterior.

11.23 Flexor hallucis brevis and adductor hallucis.

11.24 When prompt absorption is desired, when larger doses are required, and when the drug is too irritating to give subcutaneously.

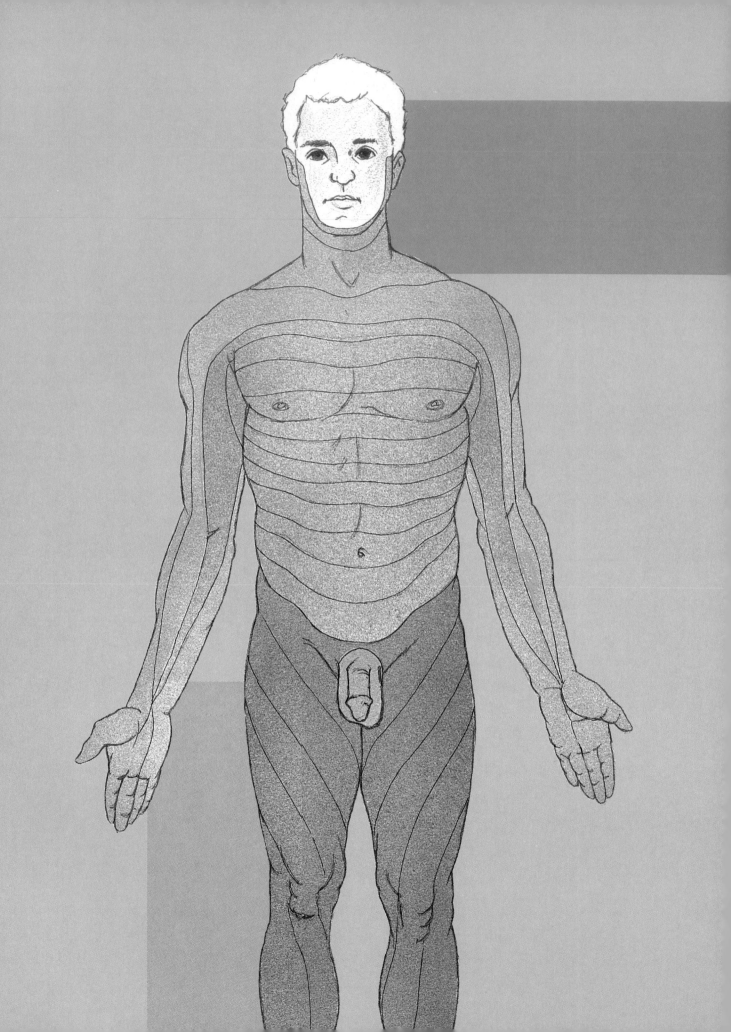

Unit 3
CONTROL SYSTEMS OF THE HUMAN BODY

This unit will show you the significance of the nerve impulse in making rapid adjustments for maintaining homeostasis. You will learn how the nervous system detects changes in the environment, selects a course of action, and responds to the changes. We will also investigate the role of hormones in maintaining long-term homeostasis.

Chapter 12
NERVOUS TISSUE

Chapter Contents at a Glance

Student Objectives

1. Identify the three basic functions of the nervous system in maintaining homeostasis.
2. Classify the organs of the nervous system into central and peripheral divisions.
3. Contrast the histological characteristics and functions of neuroglia and neurons.
4. Describe the functions of neuroglia.
5. Describe the structure and functions of neurons.
6. Define gray and white matter and give examples of each.
7. Describe the cellular properties that permit communication among neurons and muscle fibers.
8. Describe the factors that contribute to generation of a resting membrane potential.
9. Compare the basic types of ion channels and explain how they relate to action potentials and graded potentials.
10. List the sequence of events involved in generation of a nerve impulse.
11. Explain the events of synaptic transmission.
12. Distinguish between spatial and temporal summation.
13. Give examples of excitatory and inhibitory neurotransmitters and describe how they may act.
14. List four ways that synaptic transmission may be enhanced or blocked.
15. Describe the various types of neuronal circuits in the nervous system.
16. List the necessary conditions for regeneration of nervous tissue.
17. Describe events of damage and repair of peripheral neurons.
18. Describe the symptoms and causes of epilepsy.

I

n Chapter 1 we introduced the concept of homeostasis. Through homeostasis, the various processes that occur in the body are maintained within normal physiological limits. It was also pointed out that two body systems, the nervous and endocrine systems, share the responsibility for maintaining homeostasis. Their objective is the same—to keep controlled conditions within limits that maintain health. Their means of achieving that objective differ. Whereas the nervous system employs nerve impulses (action potentials) and responds rapidly to stimuli to adjust body processes, the endocrine system employs hormones and responds more slowly, though no less effectively. In this chapter, we describe the organization and operation of nervous tissue.

Your **nervous system** has three basic functions: sensory, integrative, and motor. First, it *senses* certain changes (stimuli), both within your body (the internal environment), such as stretching of your stomach or an increase in blood acidity, and outside your body (the external environment), such as a raindrop landing on your arm or the aroma of a rose; this is its sensory function. Second, it *analyzes* the sensory information, *stores* some aspects, and *makes decisions* regarding appropriate behaviors; this is its integrative function. Third, it may *respond* to stimuli by initiating muscular contractions or glandular secretions; this is its motor function.

The branch of medical science that deals with the normal functioning and disorders of the nervous system is called **neurology** (noo-ROL-ō-jē; *neuro* = nerve or nervous system; *logos* = study of).

The developmental anatomy of the nervous system is considered in Chapter 14.

NERVOUS SYSTEM DIVISIONS

The two principal divisions of the nervous system are the **central nervous system (CNS)** and the **peripheral** (pe-RIF-er-al) **nervous system (PNS)** (Fig. 12.1).

The CNS consists of the brain and spinal cord. Within the CNS, various sorts of incoming sensory information are integrated and correlated, thoughts and emotions are generated, and memories are formed and stored. Most nerve impulses that stimulate muscles to contract and glands to secrete originate in the CNS.

The CNS is connected to sensory receptors, muscles, and glands in peripheral parts of the body by the PNS. The PNS consists of **cranial nerves** that arise from the brain and **spinal nerves** that emerge from the spinal cord. Portions of these nerves carry nerve impulses into the CNS while other portions carry impulses out of the CNS.

The input component of the PNS consists of nerve cells, called **sensory** or **afferent** (AF-er-ent; *ad* = toward; *ferre* = to carry) **neurons.** They conduct nerve impulses from sensory receptors in various parts of the body to the CNS and end within the CNS. The output component consists of nerve cells, called **motor** or **efferent** (EF-er-ent; *ex* = away from; *ferre* = to carry) **neurons**. They originate within the CNS and conduct nerve impulses from the CNS to muscles and glands.

Based primarily on the part of the body that responds, the PNS may be subdivided into a **somatic** (*soma* = body) **nervous system (SNS)** and an **autonomic** (*auto* = self; *nomos* = law) **nervous system (ANS).** See Fig. 12.1b. The SNS consists of sensory neurons that convey information from cutaneous and special sense receptors primarily in the head, body wall, and extremities to the CNS and motor neurons from the CNS that conduct impulses to skeletal muscles only. Because these motor responses can be consciously controlled, this portion of the SNS is *voluntary*.

The ANS consists of sensory neurons that convey information from receptors primarily in the viscera to the CNS and motor neurons from the CNS that conduct impulses to smooth muscle, cardiac muscle, and glands. Since its motor responses are not normally under conscious control, the ANS is *involuntary*.

The motor portion of the ANS consists of two branches, the **sympathetic division** and the **parasympathetic division.** With few exceptions, the viscera receive instructions

FIGURE 12.1 Organization of the nervous system.

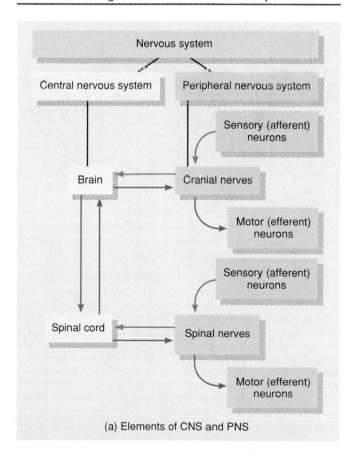

(a) Elements of CNS and PNS

Figure continues

FIGURE 12.1 (continued)

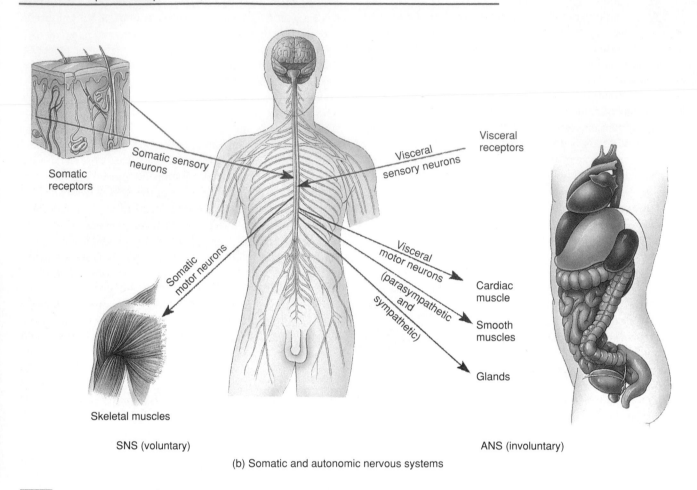

(b) Somatic and autonomic nervous systems

Question: Which neurons are said to be voluntary? Why?

from both. Usually, the two divisions have opposing actions. Processes promoted by sympathetic neurons often involve expenditure of energy while those promoted by parasympathetic neurons restore and conserve body energy. For example, sympathetic neurons speed the heartbeat while parasympathetic neurons slow it down (see page 513).

FUNCTIONAL ANATOMY OF NERVOUS TISSUE

Despite the complexity of the nervous system, it consists of only two principal kinds of cells: neuroglia and neurons. Neuroglia support, nurture, and protect the neurons, maintaining homeostasis of the fluid that bathes neurons. Neurons are responsible for most special functions attributed to the nervous system: sensing, thinking, remembering, controlling muscle activity, and regulating glandular secretions. When injured, mature neurons have little capability for repair because they do not normally undergo mitosis.

Neuroglia

About half the space in the CNS is filled by cells called **neuroglia** (noo-ROG-lē-a; *neuro* = nerve; *glia* = glue) or **glia.** They are generally smaller than neurons and outnumber them by 5 to 50 times. Unlike neurons, neuroglia can multiply and divide in the mature nervous system. In cases of traumatic injury, neuroglia multiply to fill in the spaces formerly occupied by neurons. Brain tumors commonly involve neuroglia. Such tumors, called **gliomas,** tend to be highly malignant and rapidly enlarging.

Types of Neuroglia

Four types of neuroglia—astrocytes, oligodendrocytes, microglia, and ependymal cells—are found in the CNS. **Astrocytes** (AS-trō-sīts; *astro* = star; *cyte* = cell) are star-shaped cells with many processes. They participate in the metabolism of neurotransmitters, maintain the proper balance of K$^+$ for generation of nerve impulses, participate in brain development, help form the blood–brain barrier, which regulates entry of substances into the brain (see page 411), and provide a link between neurons and blood vessels.

Oligodendrocytes (OL-i-gō-den′-drō-sīts; *oligo* = few; *dendro* = tree) have few processes and are smaller than astrocytes. They form a supporting network by twining around neurons and produce a lipid and protein wrapping called a myelin sheath (described shortly).

Microglia (mī-KROG-lē-a; *micro* = small; *glia* = glue) are small, phagocytic neuroglia derived from monocytes. They protect the central nervous system from disease by engulfing invading microbes and clearing away debris from dead cells.

Ependymal (ep-EN-di-mal; *ependyma* = upper garment) **cells** range in shape from squamous to columnar and many are ciliated. They line the brain ventricles (spaces that form and circulate cerebrospinal fluid) and the central canal of the spinal cord.

Two types of neuroglia or supporting cells are found in the PNS—**neurolemmocytes** (noo′-rō-LE-mō-sīts) or **Schwann cells**, which produce myelin sheaths around PNS neurons, and **satellite** (SAT-i-līt) **cells**, which support neurons in ganglia (clusters of neuron cell bodies) of the PNS. Exhibit 12.1 summarizes the types of neuroglia.

Myelination

The axons of most mammalian neurons are surrounded by a multilayered lipid and protein covering produced by neuroglia that is called the **myelin sheath**. The sheath electrically insulates the axon of a neuron and increases the speed of nerve impulse conduction. Axons with such a covering are said to be **myelinated,** whereas those without it are **unmyelinated.** Electron micrographs reveal that even unmyelinated axons are surrounded by a thin coat of neuroglial plasma membrane (Fig. 12.2).

Two types of neuroglia produce myelin sheaths: neurolemmocytes and oligodendrocytes. In the PNS, neurolemmocytes (Schwann cells) form myelin sheaths around axons during fetal development and the first year of life. Each neurolemmocyte wraps about 1 millimeter (1 mm = 0.04 in.) of an axon's length by spiraling many times around the axon (Fig. 12.3). Up to 500 neurolemmocytes participate in forming a myelin sheath around the longest axons in your

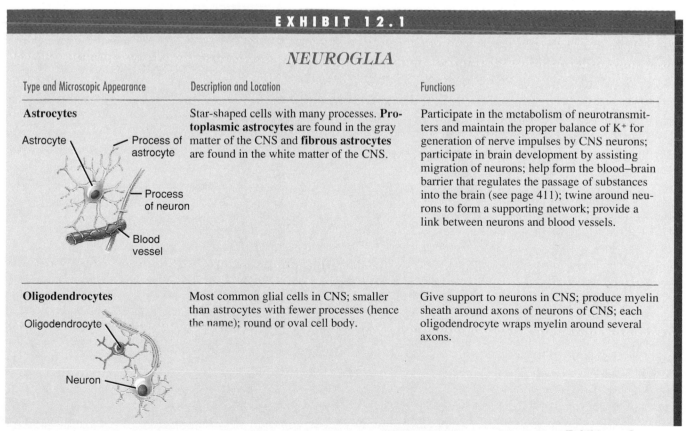

EXHIBIT 12.1

NEUROGLIA

Type and Microscopic Appearance	Description and Location	Functions
Astrocytes	Star-shaped cells with many processes. **Protoplasmic astrocytes** are found in the gray matter of the CNS and **fibrous astrocytes** are found in the white matter of the CNS.	Participate in the metabolism of neurotransmitters and maintain the proper balance of K$^+$ for generation of nerve impulses by CNS neurons; participate in brain development by assisting migration of neurons; help form the blood–brain barrier that regulates the passage of substances into the brain (see page 411); twine around neurons to form a supporting network; provide a link between neurons and blood vessels.
Oligodendrocytes	Most common glial cells in CNS; smaller than astrocytes with fewer processes (hence the name); round or oval cell body.	Give support to neurons in CNS; produce myelin sheath around axons of neurons of CNS; each oligodendrocyte wraps myelin around several axons.

Exhibit continues

EXHIBIT 12.1 (continued)

NEUROGLIA

Type and Microscopic Appearance	Description and Location	Functions
Microglia	Small cells with few processes; derived from monocytes; normally stationary but may migrate to site of injury.	Engulf and destroy microbes and cellular debris in the CNS; may migrate to area of injured nervous tissue; function as macrophages in the CNS.

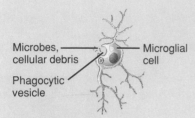

Microbes, cellular debris — Microglial cell
Phagocytic vesicle

| **Ependymal cells** | Epithelial cells arranged in a single layer and ranging in shape from squamous to columnar; many are ciliated. | Form a continuous epithelial lining for the ventricles of the brain (spaces that form and circulate cerebrospinal fluid) and the central canal of the spinal cord; probably assist in the circulation of cerebrospinal fluid (CSF) in these areas. |

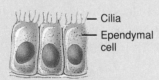

Cilia
Ependymal cell

| **Neurolemmocytes (Schwann cells)** | Flattened cells arranged around axons in PNS. | Produce myelin sheath around axons of PNS neurons. |

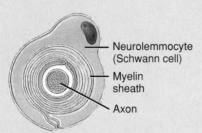

Neurolemmocyte (Schwann cell)
Myelin sheath
Axon

| **Satellite cells** | Flattened cells arranged around the cell bodies of neurons in ganglia (collections of neuronal cell bodies in the PNS). | Support neurons in ganglia of PNS. |

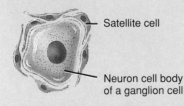

Satellite cell
Neuron cell body of a ganglion cell

FIGURE 12.2 Unmyelinated axons.

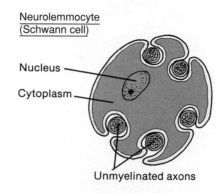

Neurolemmocyte
(Schwann cell)

Nucleus

Cytoplasm

Unmyelinated axons

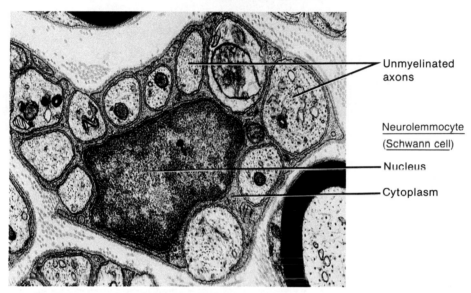

Unmyelinated
axons

Neurolemmocyte
(Schwann cell)

Nucleus

Cytoplasm

Electron micrograph of a cross section of several unmyelinated
axons

Question: Which CNS cells serve a function similar to neurolemmocytes?

FIGURE 12.3 Myelinated axons.

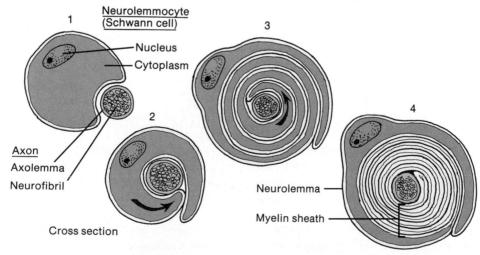

Neurolemmocyte
(Schwann cell)

Nucleus
Cytoplasm

Axon
Axolemma
Neurofibril

Cross section

Neurolemma

Myelin sheath

(a) Stages in the formation of a myelin sheath

Figure continues

FIGURE 12.3 (continued)

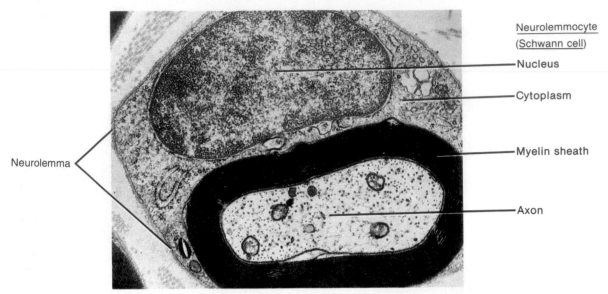

(b) Electron micrograph of a cross section of a myelinated axon

Question: What is the functional advantage of myelination?

body. Eventually, multiple layers of glial cell membrane surround the axon, with the neurolemmocyte cytoplasm and nucleus forming the outermost layer. The inner portion, consisting of up to 100 layers of neurolemmocyte membrane, is the myelin sheath. The outer nucleated cytoplasmic layer of the neurolemmocyte, which encloses the myelin sheath, is called the **neurolemma (sheath of Schwann)**. A neurolemma is found only around axons in the PNS. When an axon is injured, the neurolemma aids regeneration by forming a tube that guides and stimulates regrowth of the axon (see Fig. 12.19). At intervals along an axon, the myelin sheath has gaps called **neurofibral nodes (nodes of Ranvier**; RON-vē-ā (see Fig. 12.4).

In the CNS **oligodendrocytes** myelinate many axons in somewhat the same manner that neurolemmocytes myelinate PNS axons (see Exhibit 12.1). They merely deposit a myelin sheath, however, without forming a neurolemma. Neurofibral nodes are present, but they are fewer in number. In contrast to a neurolemmocyte, which myelinates a portion of only one PNS axon, a single oligodendrocyte myelinates portions of about 15 CNS axons. Axons in the CNS display little regrowth after injury. This is thought to be due, in part, to the absence of a neurolemma and in part to an inhibitory influence exerted by CNS neuroglia.

The amount of myelin increases from birth to maturity, and its presence greatly increases the speed of nerve impulse conduction. Since myelination is still in progress during infancy, an infant's responses to stimuli are not as rapid or coordinated as those of an older child or an adult. Certain diseases such as multiple sclerosis (see page 439) and Tay–Sachs disease (see page 74) involve destruction of myelin sheaths.

Neurons

Some neurons are tiny and relay signals over a short distance (less than 1 mm) within the CNS. Others are the longest cells in your body. Motor neurons that command muscles to wiggle your toes, for example, extend from the lumbar region of your spinal cord (just above waist level) to your foot. Some sensory neurons are even longer. Those that allow you to feel the position of your wiggling toes stretch all the way from your foot to the lower portion of your brain. Nerve impulses travel these great distances at speeds ranging from 0.5 to 130 meters per second (1 to 280 mi/hr).

The functional contact between two neurons or between a neuron and an effector (muscle or gland) cell is called a **synapse**. The synapse between a motor neuron and a muscle fiber is called a **neuromuscular junction** (see Fig. 10.3) whereas the synapse between a neuron and glandular cells is called a **neuroglandular junction**.

Several proteins called **neurotropins** regulate the normal growth and development of neurons. They include nerve growth factor (NGF), brain-derived neurotropic factor, and neurotropins-3, -4, and -5. Scientists are just beginning to unravel some of the effects of these crucial molecules. The one that is best understood at present is NGF. It is needed for the normal growth, survival, and development of sympathetic and sensory neurons. Whereas NGF stimulates sensory neurons only during a short period of embryonic development, it stimulates sympathetic neurons into adulthood. NGF also plays a role in the brain, where it increases synthesis of acetylcholine (as′-ē-til-KŌ-lēn), stimulates neuronal growth, and may help maintain neuronal function. Undoubtedly, other trophic factors contribute to growth and development of the nervous system.

Parts of a Neuron

Most neurons have three parts: (1) cell body, (2) dendrites, and (3) axon (Fig. 12.4). The **cell body (soma** or **perikaryon)** contains a nucleus surrounded by cytoplasm that includes typical organelles such as lysosomes, mitochondria, and a Golgi complex. Many neurons also contain cytoplasmic inclusions such as **lipofuscin** pigment that occurs as clumps of yellowish brown granules. Lipofuscin is probably an end-product of lysosomal activity, which collects as one ages but does not seem to harm the neuron.

Also part of the cytoplasm are structures characteristic of neurons: chromatophilic substance and neurofibrils. The **chromatophilic substance (Nissl bodies)** is an orderly arrangement of rough endoplasmic reticulum, the site of protein synthesis. Newly synthesized proteins replace those being recycled and are used for growth of neurons and regeneration of damaged peripheral nerve axons. **Neurofibrils,** composed of intermediate filaments, form the cytoskeleton, which provides support and shape for the cell.

Neurons have two kinds of processes: dendrites and axons. **Dendrites** (*dendro* = tree) are usually short, tapering, and highly branched. Often, the dendrites form a tree-shaped array of processes that emerge from the cell body. Usually, dendrites are not myelinated. Their cytoplasm contains chromatophilic substance, mitochondria, and other organelles.

The second type of process, the **axon,** is a long, thin, cylindrical projection that may be myelinated. It joins the cell body at a cone-shaped elevation called the **axon hillock.** The first portion of the axon is called the **initial segment.** Except in sensory neurons, nerve impulses arise at the junction of the axon hillock and initial segment, which is called the **trigger zone,** and then conduct along the axon toward another neuron, muscle fiber, or gland cell. An axon contains mitochondria and neurofibrils but no rough endoplasmic reticulum; thus it does not carry on protein synthesis. Its cytoplasm, called **axoplasm,** is surrounded by a plasma membrane known as the **axolemma** (*lemma* = sheath or husk). Along the length of an axon, side branches called **axon collaterals** may branch off, typically at a right angle to the axon. The axon and its collaterals end by dividing into many fine processes called **axon terminals.**

The tips of the axon terminals swell into bulb-shaped structures called **synaptic end-bulbs.** They contain many minute membrane-enclosed sacs called **synaptic vesicles** that store a chemical substance known as a **neurotransmitter.** (Different neurons have different neurotransmitters.) Neurons influence the activity of other neurons, muscle fibers, or gland cells by releasing neurotransmitter from synaptic vesicles at synapses.

Neurons display great diversity in size and shape. For example, their cell bodies range in diameter from 5 μm (smaller than a red blood cell) up to 135 μm (barely large enough to see with the naked eye). The pattern of dendritic branching is varied and distinctive for neurons in different parts of the nervous system. A few small neurons lack an axon and many others have very short axons. The longest neurons, however, have axons that extend for a meter (3.2 ft) or more.

Nerve fiber is a general term for any neuronal process (dendrite or axon). Most often, it refers to an axon and its sheaths. A **nerve** is a bundle of many nerve fibers that course along the same path in the PNS, such as the ulnar nerve in the arm or the sciatic nerve in the thigh. Most nerves include bundles of both sensory and motor fibers and are surrounded by connective tissue coats. Nerve cell bodies in the PNS generally are clustered together to form **ganglia** (GANG-lē-a; *ganglion* = knot). A **tract** is a bundle of nerve fibers, without connective tissue elements, in the CNS. Tracts may interconnect different regions of the brain or extend long distances up or down the spinal cord and connect with specific regions of the brain.

Axonal Transport

The cell body of a neuron is the site of most synthetic reactions plus recycling of worn-out molecules into new components. However, some substances are needed in the axon or at the axon terminals. Two types of transport systems carry materials from the cell body to the axon terminals and back. The slower one, which moves materials about 1 to 5 mm per day, is called **slow axonal transport (axoplasmic flow).** It conveys axoplasm in one direction only—from the cell body toward the axon terminals. It supplies new axoplasm for developing or regenerating axons and renews axoplasm in growing and mature axons.

The faster system, moving materials a distance of 200 to 400 mm per day, is called **fast axonal transport.** It uses molecular "motors" to transport materials in both directions—away from and toward the cell body—along the surfaces of microtubules. Fast axonal transport moves various organelles and materials that form the membranes of the axolemma, synaptic end-bulbs, and synaptic vesicles. Some materials returned to the cell body are degraded or recycled and others influence its growth. Still others, such as certain viruses, may even be harmful to the cell body.

FIGURE 12.4 Structure of a typical neuron. Arrows indicate the direction of information flow. The break indicates that the axon actually is longer than shown.

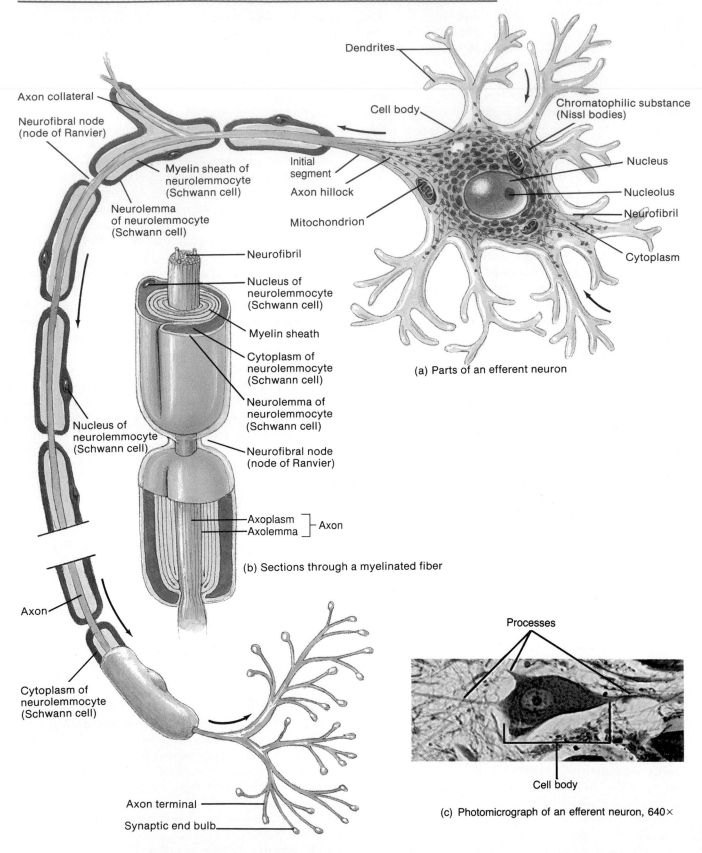

Dendrites

Cell body

Chromatophilic substance (Nissl bodies)

Axon collateral

Neurofibral node (node of Ranvier)

Myelin sheath of neurolemmocyte (Schwann cell)

Neurolemma of neurolemmocyte (Schwann cell)

Initial segment

Axon hillock

Mitochondrion

Nucleus

Nucleolus

Neurofibril

Cytoplasm

Neurofibril

Nucleus of neurolemmocyte (Schwann cell)

Myelin sheath

Cytoplasm of neurolemmocyte (Schwann cell)

Neurolemma of neurolemmocyte (Schwann cell)

Nucleus of neurolemmocyte (Schwann cell)

Neurofibral node (node of Ranvier)

Axoplasm ⎤ Axon
Axolemma ⎦

(a) Parts of an efferent neuron

(b) Sections through a myelinated fiber

Axon

Cytoplasm of neurolemmocyte (Schwann cell)

Processes

Cell body

Axon terminal

Synaptic end bulb

(c) Photomicrograph of an efferent neuron, 640×

Question: What roles do the dendrites, cell body, and axon play in communication of signals?

AXONAL TRANSPORT IN DISEASE AND THERAPY

Fast axonal transport is the route by which herpes viruses and rabies viruses make their way from axon terminals near skin cuts to nerve cell bodies, where they can multiply and cause their damage. The toxin produced by tetanus bacteria reaches the CNS by the same route. The time delay between the release of the toxin and the first appearance of symptoms is, in part, due to the time required for movement of the toxin to the cell body. This is why a tetanus-prone injury of the head or neck, or a bite in this region by a rabid dog, bat, or other animal, is a more serious matter than a similar injury in an extremity. The closer to the brain, the shorter the transit time, and the faster must be the treatment to prevent the disease.

Viruses that are able to hop aboard the fast axonal transport system may become a tool for treating genetic disorders of the CNS. Through the techniques of genetic engineering, scientists can insert corrective genes into a viral messenger, and they can also disable potentially harmful viral genes. Studies in process will reveal if good copies of defective genes can be shuttled into a person's CNS by viruses without adverse side effects.

Classification of Neurons

Both structural and functional features are used to classify the different neurons in the body. *Structural classification* is based on the number of processes extending from the cell body (Fig. 12.5). **Multipolar neurons** usually have several dendrites and one axon (see also Fig. 12.4). Most neurons

in the brain and spinal cord are of this type. **Bipolar neurons** have one dendrite and one axon and are found in the retina of the eye, inner ear, and olfactory area of the brain. **Unipolar (pseudounipolar) neurons** have just one process extending from the cell body and are always sensory neurons. They originate in the embryo as bipolar neurons, but during development the axon and dendrite fuse into a single process. The single process divides into two branches a short distance from the cell body. Dendrites at the tip of the peripheral segment monitor certain environmental changes, sometimes aided by structures known as receptors. Nerve impulses arise at the first neurofibral node (the trigger zone for sensory neurons) and conduct along the peripheral segment, past the cell body, along the central segment, and into the CNS.

The direction in which they transmit nerve impulses is the basis for *functional classification* of neurons. **Afferent neurons** transmit sensory nerve impulses from receptors in the skin, sense organs, muscles, joints, and viscera into the brain and spinal cord. **Efferent neurons** convey motor nerve impulses from the brain and spinal cord to effectors, which may be either muscles or glands. **Association neurons** or **interneurons** carry nerve impulses from one neuron to another. Most neurons in the body, perhaps 90%, are association neurons.

Spinal and cranial nerves contain fibers in seven functional categories, four afferent and three efferent.

1.   General somatic (*soma* = body) **afferent neurons** convey impulses for pain, temperature, touch, vibration,

FIGURE 12.5 Structural classification of neurons based on number of processes extending from the cell body.

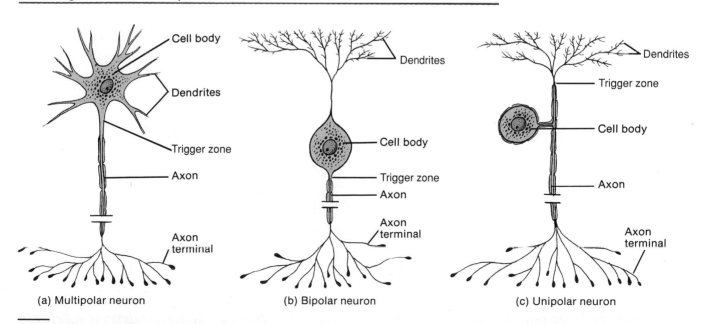

(a) Multipolar neuron   (b) Bipolar neuron   (c) Unipolar neuron

Question: **What process occurs at a trigger zone?**

and pressure from the skin and for position from joints and muscles via spinal nerves and some cranial nerves.

2.   Special somatic afferent neurons relay impulses for vision, hearing, and balance via cranial nerves.

3.   General visceral afferent neurons convey visceral information such as distention of organs and chemical conditions within the body into the CNS via both cranial and spinal nerves.

4.   Special visceral afferent neurons convey impulses for taste and smell (olfaction) via cranial nerves.

5.   General somatic efferent neurons conduct impulses to most skeletal muscles via spinal nerves and some cranial nerves.

6.   General visceral efferent neurons conduct impulses from the CNS to smooth muscle, cardiac muscle, and glands via cranial and spinal nerves.

7.   Special visceral efferent neurons conduct impulses from the CNS to skeletal muscles that control facial expression and position of the jaw, neck, larynx, and pharynx via cranial nerves.

Just two of the thousands of types of association neurons are shown in Fig. 12.6. Neurons often are named for the histologist who first described them, such as **Purkinje** (pur-KIN-jē) **cells** in the cerebellum or **Renshaw cells** in the

---

FIGURE 12.6 Two types of association neurons (interneurons). Arrows indicate direction in which information flows.

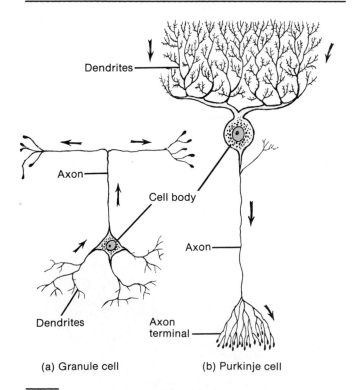

(a) Granule cell   (b) Purkinje cell

Question: Are the dendrites of association neurons receiving input or providing output?

spinal cord. Or, they are named for some aspect of their shape or appearance, as in the case of **pyramidal** (pi-RAM-i-dal) **cells**, found in the cerebral cortex (the outer layer of the cerebrum), which have a cell body shaped like a pyramid.

Gray and White Matter

In a freshly dissected section of the brain or spinal cord, some regions look white and glistening whereas others appear gray (see Fig. 12.7). **White matter** refers to aggregations of myelinated processes from many neurons. The whitish color of myelin gives white matter its name. The **gray matter** of the nervous system contains either nerve cell bodies, dendrites, and axon terminals or bundles of unmyelinated axons and neuroglia. They look grayish, rather than white, because there is no myelin in these areas.

In the spinal cord, the white matter surrounds an inner core of gray matter shaped like a butterfly or the letter H (Fig. 12.7). In the brain, a thin outer shell of gray matter covers the surface of the largest portion of the brain, the two cerebral hemispheres (Fig. 12.7). There also are many nuclei of gray matter lying deep within the brain. A **nucleus** is a mass of nerve cell bodies and dendrites inside the CNS. Since they consist mainly of nerve cell bodies, nuclei are masses of gray matter. Most nerves in the PNS and all tracts in the CNS are white matter. The arrangement of gray and white matter in the spinal cord and brain is described more extensively in Chapters 13 and 14.

NEUROPHYSIOLOGY

Communication by neurons depends on two basic properties of their plasma membranes. (1) There is an electrical voltage, called the **resting membrane potential** (see page 357) across the membrane. (2) Their plasma membranes contain a variety of **ion channels** (pores) that may be closed or open. When they are open, specific ions in the intracellular fluid (cytosol) or extracellular fluid can flow across the membrane. Part of the protein that forms such a channel may act as a gate or door, opening and closing on demand. Depending on the types of channels that are present, a portion of a neuron may be able to produce either a graded potential or an action potential (impulse). These responses are described shortly.

Resting Membrane Potential

The resting membrane potential (RMP) occurs because there is a small buildup of negative charges just inside the membrane and an equal buildup of positive charges on the outside (Fig. 12.8). Such a separation of positive and negative electric charges is a form of potential energy, which is measured in volts or millivolts (1 mV = $\frac{1}{1000}$ V). The greater

FIGURE 12.7 Distribution of gray and white matter in the spinal cord and brain.

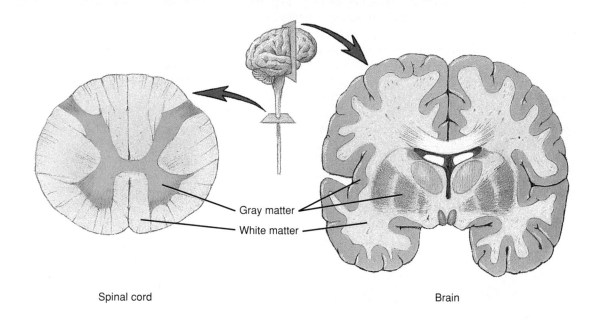

Spinal cord

Brain

Question: What is responsible for the white appearance of white matter?

the difference in charge across the membrane, the larger the membrane potential (voltage). Note in Fig. 12.8 that the buildup of charge is just in fluid that is very close to the membrane. Elsewhere in the intracellular fluid or extracellular fluid, there are equal numbers of positive and negative charges.

In neurons, the RMP ranges from −40 to −90 mV. A typical value is −70 mV. *The minus sign indicates that the inside is negative relative to the outside.* A cell that exhibits a membrane potential is said to be **polarized**. Most body cells are polarized, with the membrane voltage varying from +5 mV to −100 mV in different types of cells.

The RMP is like voltage stored in a battery. If you connect the positive and negative terminals of a battery with a piece of wire, electrons will flow along the wire. Flow of electric charge is called **current**. In a living cell, ions, rather than electrons, carry most of the current. Because the phospholipid bilayer of the plasma membrane is a good insulator, the main paths for current flow across the membrane are through ion channels. Thus when ion channels open or close in the plasma membrane, current flows and this changes the membrane potential.

Two main factors contribute to the resting membrane potential.

1. Distribution of ions across the plasma membrane. Recall from Chapter 3 (see Fig. 3.4) that the major anions and cations are different outside and inside cells. Extracellular fluid (outside cells) is rich in Na⁺ (sodium ions) and Cl⁻ (chloride ions). In intracellular fluid (inside cells), on the other hand, the main cation is K⁺ (potassium ions), and the two dominant anions are organic phosphates and amino acids in proteins.

2. Relative permeability of the plasma membrane to Na⁺ and K⁺. The permeability of the plasma membrane to K⁺ is 50 to 100 times greater than its permeability to Na⁺ in a resting neuron or muscle fiber.

First consider what would happen if the membrane were permeable only to K⁺. These positive ions would tend to leak out of the cell into the extracellular fluid along the

FIGURE 12.8 Resting membrane potential (RMP).

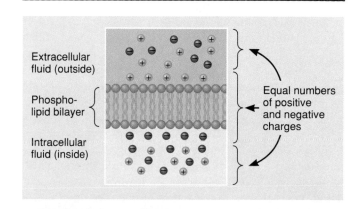

Extracellular fluid (outside)

Phospho-lipid bilayer

Intracellular fluid (inside)

Equal numbers of positive and negative charges

Question: What is a typical value for the RMP of a neuron?

concentration gradient. But, as more and more exited, the interior of the membrane would become increasingly negative. The resulting electrical gradient (inside negative) would then start to pull K^+ back into the cell. Eventually, just as many K^+ would be entering due to the electrical gradient as would be exiting due to the concentration (chemical) gradient. The membrane potential that just balances the K^+ concentration gradient is –90 mV and is called the **potassium equilibrium potential**. The RMP (–70 mV) is close to but not equal to the potassium equilibrium potential, which means that the membrane must be slightly permeable to other ions.

Actually, the membrane is moderately permeable to K^+ and Cl^- and very slightly permeable to Na^+. One way to balance the electrical effect of K^+ outflow might be Na^+ inflow, in effect exchanging one positive particle for another. But inward leakage of Na^+ is far too slow to keep pace with outward leakage of K^+. Also, although positively charged ions leaving the cell should promote the tandem exit of anions, most anions in the cell are not free to leave. They are attached either to large proteins or to other organic molecules, such as phosphates in ATP. Finally, inward leakage of Cl^- along its concentration gradient cannot balance the electrical effect of K^+ outflow. It will not happen because it would only make the inside more negative.

In summary, the result of the low membrane permeability to Na^+ and to anions inside the cell is that fluid just next to the inner surface of the plasma membrane becomes more and more negatively charged as K^+ leaves.

Note that both the electrical and concentration gradients promote Na^+ inflow: the negative interior attracts these positive ions and the concentration of Na^+ is higher outside. Even though membrane permeability to Na^+ is very low, a slow leak would eventually destroy the electrochemical gradient because there is no gradient to push Na^+ back out. The small inward Na^+ leak is taken care of by Na^+/K^+ active transport pumps (see Fig. 3.10). They maintain the resting membrane potential by pumping out Na^+ as fast as it leaks in. At the same time, the pumps bring in K^+. However, the K^+ merely redistributes according to the electrochemical gradient, as previously described. Thus the critical job of Na^+/K^+ pumps is to export Na^+.

Some Na^+/K^+ pumps expel three Na^+ for each two K^+ imported. Such pumps are said to be *electrogenic*, which means they contribute to the negativity of the resting membrane potential. The total effect of such pumps, however, is very small, no more than –3 mV of the total –70 mV resting membrane potential in a typical neuron.

Ion Channels

Ion channels are of two basic types: leakage and gated. **Leakage (nongated) channels** are always open, whereas **gated channels** open and close in response to some sort of stimulus. The plasma membrane of a neuron or muscle fiber has many more K^+ leakage channels than it has Na^+ leakage channels. This is why membrane permeability to K^+ is much higher. Four categories of stimuli operate gated ion channels: (1) voltage, (2) chemicals, (3) mechanical pressure, and (4) light.

The first type of gated ion channel opens in response to a direct change in the membrane potential (voltage) and is called a **voltage-gated (voltage-regulated) ion channel** (Fig. 12.9a). The presence of voltage-gated ion channels in nerve and muscle plasma membranes gives these cells the property of *excitability (irritability)* that is, the ability to respond to certain stimuli by producing impulses. The trigger zone for a particular neuron is the place on the membrane where voltage-gated channels are clustered most densely.

A **chemically gated ion channel** opens and closes in response to a specific chemical stimulus. A wide variety of chemical ligands, such as neurotransmitters, hormones, and ions such as H^+ and Ca^{2+}, regulate chemically gated ion channels. The neurotransmitter acetylcholine, for example, opens cation channels that pass Na^+, K^+, and Ca^{2+} (Fig. 12.9b).

Chemically gated ion channels operate in two basic ways. The chemical may *directly* change the membrane permeability to one or more ions, as in the example of acetylcholine. Or it may act *indirectly* via a type of membrane protein called a G-protein and a second messenger system that uses molecules in the cytosol. Hormones often work by second messenger systems (see Fig. 18.4), and so do some neurotransmitters.

Mechanically gated ion channels open or close in response to mechanical vibration or pressure, such as sound waves or the pressure of a touch. Finally, **light-gated ion channels** close in response to light. These two types of channels are found in sensory receptors that detect mechanical distortions or light.

The presence of chemically, mechanically, or light-gated ion channels in a membrane permits the appropriate stimulus to cause a **graded potential.** These electrical responses vary in size, being larger or smaller depending on how many gated ion channels have opened and how long each one is open.

Action Potential (Impulse)

During an action potential (Fig. 12.10), two types of voltage-gated ion channels open and then close, first channels for Na^+ and then channels for K^+. Rapid opening of voltage-gated Na^+ channels brings about **depolarization,** the loss and then reversal of membrane polarization. The slower opening of voltage-gated K^+ channels and closing of previously open Na^+ channels leads to **repolarization,** the recovery of the resting membrane potential. Together, the depolarization phase and repolarization phase comprise an action potential, which lasts about 1 msec ($1/1000$ sec) in a typical neuron.

FIGURE 12.9 Gated ion channels. (a) A change in the membrane potential opens certain voltage-gated Na⁺ channels during an action potential. (b) Chemical stimuli, for example, the neurotransmitter acetylcholine, open chemically gated ion channels.

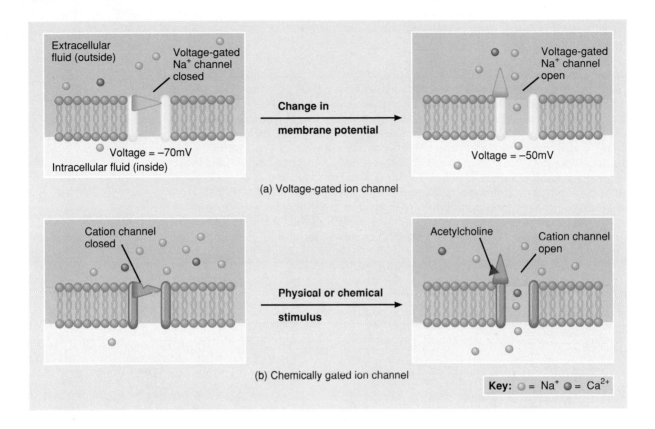

(a) Voltage-gated ion channel

(b) Chemically gated ion channel

Question: Which type of gated ion channel is activated by a touch on your arm?

FIGURE 12.10 Action potential (impulse) in a neuron.

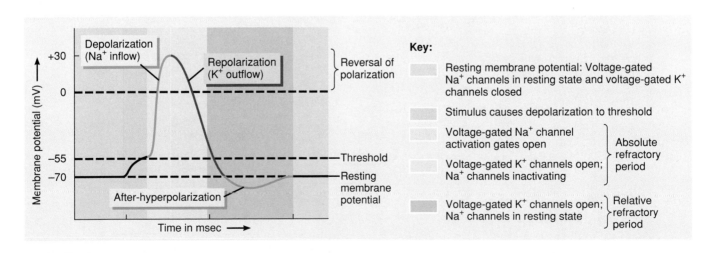

Question: Which channels are open during depolarization? During repolarization?

Depolarization

If a graded potential causes the membrane to depolarize to a critical level, called **threshold** (about –55 mV), the voltage-gated Na+ channels open. Na+ rushes inward, driven by both the electrical and the concentration (chemical) gradients. The membrane potential changes from –70 mV toward

0, and then to +30 mV. Throughout depolarization, Na+ continues to diffuse inward until the resting membrane potential *reverses:* the inside becomes 30 mV more positive than the outside.

Each voltage-gated Na+ channel has two separate gates, an *activation gate* and an *inactivation gate* (Fig. 12.11a). In a resting membrane, the inactivation gate is open, but the

FIGURE 12.11 Changes in voltage-gated channels during action potential depolarization and repolarization. (a) Resting state: voltage-gated Na+ channels are in resting state and voltage-gated K+ channels are closed. (b) Depolarization to threshold (about –55 mV) opens Na+ channel activation gates. The inward flow of Na+ further depolarizes the membrane until its polarity is reversed (the inside becomes more positive than the outside). (c) More slowly, the depolarization also opens voltage-gated K+ channels, which permit outflow of K+. At the same time the Na+ channel inactivation gates close. Repolarization begins. (d) Outflow of K+ restores the resting membrane potential. Not illustrated are the leakage channels and the Na+/K+ pumps that maintain a low concentration of Na+ inside the cell. This figure exaggerates the relative numbers of Na+ that enter and K+ that exit during a single action potential.

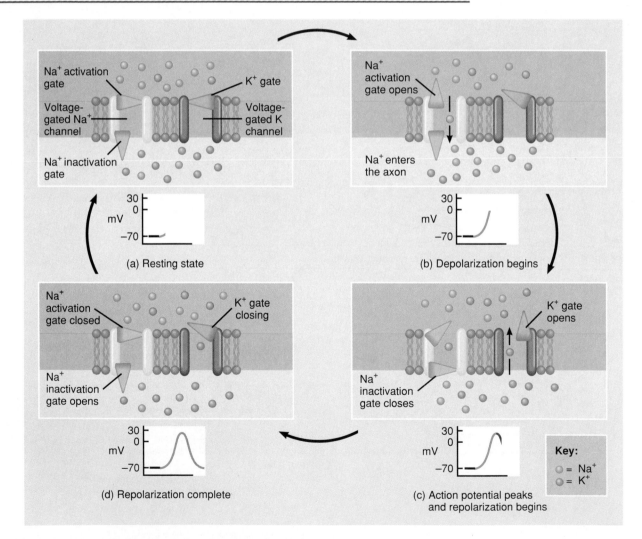

Question: Recalling that there are leakage channels for both K+ and Na+, could repolarization occur if the voltage-gated K+ channels did not exist?

activation gate is closed. As a result, Na⁺ cannot diffuse into the cell through these channels. This is the *resting* state of a voltage-gated Na⁺ channel. At threshold, many voltage-gated Na⁺ channels suddenly change from the resting to the *activated* state. In this state, both the activation and inactivation gates in the channel are open, and Na⁺ moves inward (Fig. 12.11b). As more channels open, more Na⁺ moves inward, the membrane depolarizes further, and so on. The continued inflow of Na⁺ magnifies the initial stimulus; it occurs as part of a positive feedback system. Different neurons may have different thresholds for generation of an action potential, but the threshold in any one neuron usually is constant.

The same depolarization that opens activation gates also closes inactivation gates (Fig. 12.11c). This is called the *inactivated* state of the channel. But the inactivation gate closes a few ten-thousandths of a second after the activation gate opens. Thus a voltage-gated Na⁺ channel is open for a few ten-thousandths of a second. While the channel is open, about 20,000 Na⁺ flow through the membrane. Since this number represents only one of every million Na⁺ in the fluid just outside the membrane at this site, the change in the Na⁺ concentration is minute. Since only a few Na⁺ enter during a single action potential, the Na⁺/K⁺ pumps easily bail them out and maintain the low concentration of Na⁺ inside the cell.

Repolarization

A threshold depolarization not only opens voltage-gated Na⁺ channels but also opens voltage-gated K⁺ channels (Fig. 12.11c,d). K⁺ channels open more slowly, however, so their opening occurs at about the same time that the voltage-gated Na⁺ channels are closing. Na⁺ channel inactivation slows Na⁺ inflow and K⁺ channel opening accelerates K⁺ outflow. The membrane potential thus changes from +30 mV to 0 to –70 mV. Repolarization restores the resting membrane potential and allows inactivated Na⁺ channels to revert to their resting state.

While the voltage-gated K⁺ channels are open, outflow of K⁺ may be large enough that an **after-hyperpolarization** occurs (see Fig. 12.10). **Hyperpolarization** is polarization more negative than the resting level. Because the membrane now is even more permeable to K⁺ than in the resting state, the membrane potential drifts toward the potassium equilibrium potential (about –90 mV). As the voltage-gated K⁺ channels close, however, the membrane potential returns to the resting level. In contrast to voltage-gated Na⁺ channels, most voltage-gated K⁺ channels do not exhibit an inactivated state. They flip back and forth between closed (resting) and open (activated) states.

Refractory Period

The period of time during which an excitable cell cannot generate another action potential is called the **refractory**

period (see Fig. 12.10). The **absolute refractory period** refers to the time period during which a second action potential cannot be initiated, even with a very strong stimulus. It coincides with Na⁺ channel activation and inactivation. Inactivated Na⁺ channels cannot reopen. They first must return to the resting state.

Large diameter axons have an absolute refractory period of about 0.4 msec (1/2500 sec). Thus a second nerve impulse can arise 0.4 msec after the first—up to 2500 impulses per second. Small diameter axons, on the other hand, have absolute refractory periods as long as 4 msec (1/250 sec). Thus they can transmit only 250 impulses per second. Under normal body conditions, the frequency of impulses conducted over nerve fibers ranges between 10 and 1000 impulses per second.

The **relative refractory period** is the period of time during which a second action potential can be initiated, but only by a suprathreshold (larger than threshold) stimulus. It coincides with the period when the voltage-gated K⁺ channels are still open after inactivated Na⁺ channels have returned to their resting state.

Propagation (Conduction) of Action Potentials

Nerve impulses communicate information from one part of the body to another. To do this, they must travel from where they arise, at a trigger zone, to axon terminals. The special mode of impulse travel is called **propagation (conduction)** and depends on positive feedback (Fig. 12.12). As Na⁺ flows in, depolarization increases, and the depolarization opens voltage-gated Na⁺ channels in adjacent patches of membrane. Thus the nerve impulse self-propagates along the membrane. The situation is like toppling a long row of dominoes by pushing on the first one in the line. Also, since the membrane is refractory behind the leading edge of an action potential, an impulse normally moves only in one direction from where it arises at the trigger zone.

CLINICAL APPLICATION

LOCAL ANESTHETICS

Drugs such as procaine (Novocaine) or lidocaine are used to block pain and other sensations, for example, in the skin during suturing of a gash or in the mouth during dental work. They prevent opening of voltage-gated Na⁺ channels so nerve impulses cannot pass the obstructed region. Smaller diameter axons, such as pain fibers, are more sensitive than larger diameter fibers to low doses of these anesthetic drugs.

The All-or-None Principle

A single neuron, or a single muscle fiber, generates an action potential according to the **all-or-none principle:** if depolarization reaches threshold (about –55 mV), voltage-

FIGURE 12.12 Positive feedback during depolarization of an action potential.

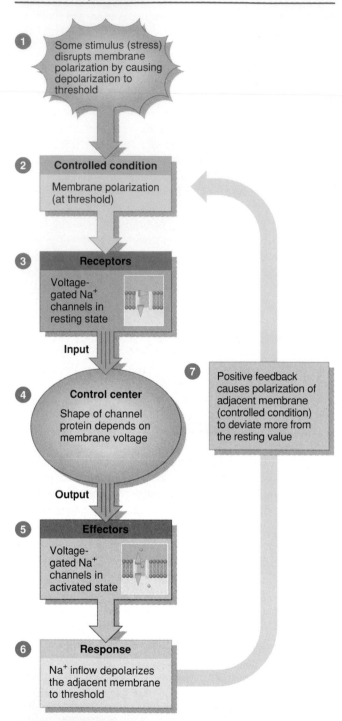

① Some stimulus (stress) disrupts membrane polarization by causing depolarization to threshold

② **Controlled condition**

Membrane polarization (at threshold)

③ **Receptors**

Voltage-gated Na⁺ channels in resting state

Input

④ **Control center**

Shape of channel protein depends on membrane voltage

Output

⑤ **Effectors**

Voltage-gated Na⁺ channels in activated state

⑥ **Response**

Na⁺ inflow depolarizes the adjacent membrane to threshold

⑦ Positive feedback causes polarization of adjacent membrane (controlled condition) to deviate more from the resting value

Question: Many positive feedback systems are destructive because they have no turn-off mechanism. What interrupts and turns off depolarization in a neuron?

gated channels open and an action potential (impulse) arises. Each time an action potential arises, it has a constant and maximum strength, unless conditions such as toxic materials or fatigue alter the membrane properties. An analogy helps in understanding the all-or-none principle and propagation of an impulse. Suppose a long trail of gunpowder is spilled along the ground. It could be ignited at one end (trigger zone) to send a blazing signal down the entire length of the trail (axon). It would not matter how tiny or great the triggering spark or flame; the blazing signal moving along the trail of gunpowder would be just the same, a maximal one (or none at all if it never started).

Saltatory Conduction

The type of impulse conduction considered thus far is typical of muscle fibers or unmyelinated nerve axons. The step-by-step depolarization of each adjacent area of the plasma membrane of the axon is called **continuous conduction** (Fig. 12.13a). In myelinated nerve axons, conduction is somewhat different. The myelin sheath acts as an electrical insulator to block ionic currents across the membrane. At intervals, however, are neurofibral nodes (nodes of Ranvier) that interrupt the myelin sheath. The nodes have a very high density of voltage-gated Na⁺ channels. Here is where membrane depolarization can occur and current carried by Na⁺ and K⁺ can flow into and out of the axon.

When a nerve impulse propagates along a myelinated fiber, ionic current flows through the extracellular fluid surrounding the myelin sheath and through the axoplasm from one node to the next (Fig. 12.13b). Thus the impulse appears to jump from node to node as each nodal area depolarizes to threshold and so conducts the impulse as it arises anew at each node. This type of impulse conduction, characteristic of myelinated fibers, is called **saltatory conduction** (*saltare* = leaping).

Since the impulse jumps long intervals as current flows from one node to the next in saltatory conduction, the impulse travels much faster than it would by continuous conduction in an unmyelinated fiber of equal diameter. This is especially important in situations where quick responses are necessary. Saltatory conduction also is more energy efficient. Since only small regions of the membrane depolarize, there is minimal inflow of Na⁺ each time a nerve impulse passes by. The result is less ATP used by Na⁺/K⁺ pumps.

Speed of Impulse Propagation

The propagation speed of a nerve impulse is not related to stimulus strength. The diameter of the fiber and the presence or absence of a myelin sheath are the most important factors that determine the speed of impulse propagation. Also, nerve fibers conduct impulses at higher speeds when warmed and lower speeds when cooled. Localized cooling

FIGURE 12.13 Propagation (conduction) of a nerve impulse after it arises at the trigger zone (initial segment). Dotted lines indicate ionic current flow. (a) In continuous conduction along an unmyelinated axon, ionic currents flow through each adjacent portion of the membrane. (b) In saltatory conduction along a myelinated axon, the nerve impulse at the first node generates ionic currents in the axoplasm and extracellular fluid that open voltage-gated Na+ channels at the second node. At the second node, the ion flows trigger a nerve impulse. Then the nerve impulse from the second node generates an ionic current that opens voltage-gated Na+ channels at the third node, and so on. Each node repolarizes after it depolarizes. The insets show the path of the current flow.

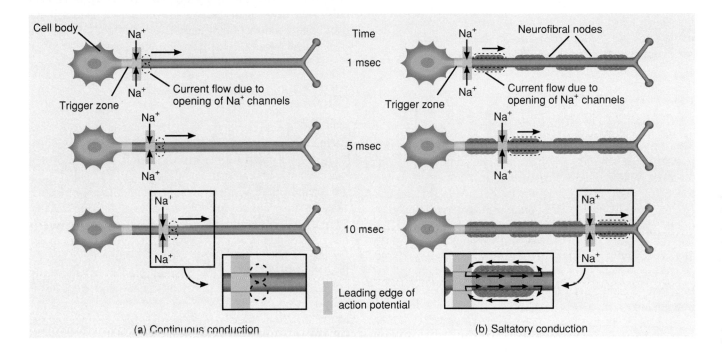

(a) Continuous conduction

(b) Saltatory conduction

Question: Where are most of the voltage-gated channels in a myelinated axon?

of a nerve can block impulse conduction. Pain resulting from injured tissue can be reduced by the application of ice because the nerve fibers carrying the pain sensations are partially blocked.

Large diameter fibers conduct impulses faster than smaller ones. The largest diameter fibers (about 5 to 20 μm diameter) are called **A fibers** and are all myelinated. The A fibers have a brief absolute refractory period and conduct impulses at speeds from 12 to 130 m/sec (27 to 280 mi/hr). The axons of large sensory nerves that relay impulses associated with touch, pressure, position of joints, heat, and cold plus nerve fibers that convey impulses to skeletal muscles are A fibers. Sensory A fibers generally connect the brain and spinal cord with receptors that may detect danger in the outside environment. Motor A fibers innervate the muscles that can do something about the situation. A fibers exist where quick reaction may mean survival.

Smaller fibers, called B and C fibers, conduct impulses more slowly and are generally found where instantaneous response is not a life-and-death matter. **B fibers** have a diameter 3 μm or less and a somewhat longer absolute refractory period than A fibers. They also are myelinated and exhibit saltatory conduction at speeds up to 15 m/sec (32 mi/hr). B fibers are found in nerves that transmit impulses from the viscera to the brain and spinal cord. They also constitute all the axons of the general visceral efferent neurons that extend from the brain and spinal cord to relay stations called autonomic ganglia.

C fibers have the smallest diameters (0.5 to 1.5 μm) and the longest absolute refractory periods. Nerve impulse conduction along a C fiber ranges from 0.5 to 2 m/sec (1.1 to 4.3 mi/hr). These unmyelinated fibers conduct some impulses for pain, touch, pressure, heat, and cold from the skin and pain impulses from the viscera. Visceral efferent fibers that extend from autonomic ganglia to stimulate the heart, smooth muscle, and glands are C fibers. Motor functions of B and C fibers include constricting and dilating the pupils, increasing and decreasing the heart rate, and contracting and relaxing the urinary bladder—functions of the autonomic nervous system.

Coding of Stimulus Intensity

If all action potentials are the same size, then how can your sensory and motor systems respond in a varying manner? Why does a light touch feel different from firmer pressure, and how can you control muscle action to be light and delicate or strong and vigorous? The main factor is the frequency of impulses, that is, how often they are generated at the trigger zone. Thus a light touch generates widely spaced nerve impulses. Firmer pressure, on the other hand, elicits nerve impulses passing down the axon at high frequency.

Comparison of Nerve and Muscle Action Potentials

The initiation and conduction of nerve and muscle action potentials are similar, although there are some notable differences. Whereas the typical resting membrane potential of a neuron is –70 mV, it is closer to –90 mV in skeletal and cardiac muscle fibers. The duration of a nerve impulse is 0.5 to 2 msec, but a muscle action potential is considerably longer—about 1.0 to 5.0 msec for skeletal muscle fibers and 10 to 300 msec for cardiac and smooth muscle fibers. Finally, the velocity of conduction of a nerve impulse can be about 18 times faster than conduction of a muscle action potential.

Transmission at Synapses

In Chapter 10 we described events occurring at the neuromuscular junction. The focus here is synaptic communication among the billions of neurons in the nervous system. Synapses are essential for homeostasis because they allow information to be integrated and filtered. Certain signals are transmitted while others are blocked. Some diseases of the brain and many psychiatric disorders result from a disruption of synaptic communication. Synapses also are the sites of action for many drugs that affect the brain, both therapeutic and addictive substances. At a synapse, the neuron sending the signal is called the **presynaptic neuron**, and the neuron receiving the message is called the **postsynaptic neuron.** Most synapses are **axodendritic** (from axon to dendrite), **axosomatic** (from axon to soma), or **axoaxonic** (from axon to axon).

There are two types of synapses, electrical and chemical, which have both structural and functional differences.

Electrical Synapses

At an **electrical synapse,** ionic current spreads directly from one cell to another through **gap junctions** (see Fig.

4.1). Each gap junction contains a hundred or so tubular, protein structures called **connexons** that form tunnels to connect the cytosol of the two cells. This provides a path for ionic current flow. Gap junctions are common in visceral (single unit) smooth muscle, cardiac muscle, and a developing embryo. They also occur in the CNS.

Electrical synapses have two obvious advantages: (1) they allow faster communication than do chemical synapses, since impulses conduct across gap junctions, and (2) they can synchronize the activity of a group of neurons or muscle fibers. The utility of synchronized action potentials in the heart or in visceral smooth muscle is to achieve coordinated contraction of these fibers. In the CNS, the role of electrical synapses is still an unsolved puzzle.

Chemical Synapses

Although the presynaptic and postsynaptic neurons of a **chemical synapse** are close, their membranes do not touch. They are separated by the **synaptic cleft,** a 20- to 50-nm space filled with extracellular fluid. Impulses cannot jump the synaptic cleft, so there must be an alternate way for the signal to cross this space. What happens is this (Fig. 12.14): the presynaptic neuron releases a neurotransmitter that diffuses across the synaptic cleft and acts on receptors in the plasma membrane of the postsynaptic neuron to produce a **postsynaptic potential** (a type of graded potential). In essence, the presynaptic electrical signal (nerve impulse) is converted into a chemical signal (liberated neurotransmitter). The postsynaptic neuron receives the chemical signal and, in turn, generates an electrical signal (postsynaptic potential). The time required for these processes at a chemical synapse—the **synaptic delay**—is about 0.5 msec. This is why chemical synapses relay messages a little more slowly than electrical synapses.

When a nerve impulse arrives at a synaptic end-bulb of a presynaptic neuron, the depolarizing phase opens voltage-gated Ca^{2+} channels in addition to the voltage-gated Na^+ channels normally opened. Because it is more concentrated in the extracellular fluid, Ca^{2+} flows inward; an increase of Ca^{2+} inside the cell triggers exocytosis of synaptic vesicles. As vesicle membranes merge with the plasma membrane, neurotransmitter molecules inside the vesicles enter the synaptic cleft. Each synaptic vesicle may contain several thousand molecules of neurotransmitter. The neurotransmitter then diffuses across the synaptic cleft to the postsynaptic membrane.

At a chemical synapse, there can be only *one-way information transfer*—from a presynaptic neuron to a postsynaptic neuron, muscle fiber, or gland cell. This is because only synaptic end-bulbs of presynaptic neurons can release neurotransmitter and only the postsynaptic membrane has the correct receptor proteins to recognize and bind that neurotransmitter. As a result, graded potentials and action potentials must move forward over their pathways. They cannot

FIGURE 12.14 Elements of a chemical synapse.

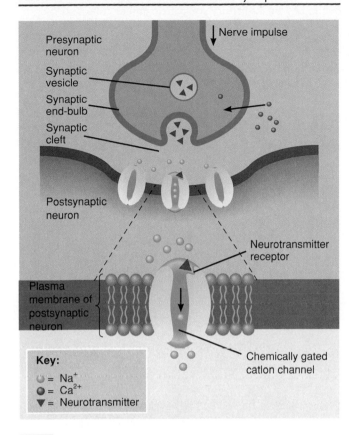

Presynaptic
neuron

Synaptic
vesicle

Synaptic
end-bulb

Synaptic
cleft

Nerve impulse

Postsynaptic
neuron

Neurotransmitter
receptor

Plasma
membrane of
postsynaptic
neuron

Chemically gated
cation channel

Key:
○ = Na$^+$
● = Ca^{2+}
▼ = Neurotransmitter

Question: Why may electrical synapses work in two directions but chemical synapses transmit information only in one direction?

back up into another presynaptic neuron, a situation that would seriously disrupt homeostasis.

Excitatory and Inhibitory Postsynaptic Potentials

If the neurotransmitter causes *depolarization* of the postsynaptic membrane, it is excitatory because it brings the membrane closer to threshold (Fig. 12.15a). A depolarizing postsynaptic potential is called an **excitatory postsynaptic potential (EPSP)**. Often, EPSPs result from opening of chemically gated cation channels. These channels allow the three most plentiful cations (Na$^+$, K$^+$, and Ca^{2+}) to pass through, but Na$^+$ inflow is greater than either Ca^{2+} inflow or K$^+$ outflow because both the electrical gradient and a large concentration gradient promote its inward movement. Although a single EPSP normally does not initiate a nerve impulse, the postsynaptic neuron does become more excitable. It is already partially depolarized and thus more likely to reach threshold when the next EPSP occurs. This effect is called **summation**.

On the other hand, if the neurotransmitter causes *hyperpolarization* of the postsynaptic membrane it is inhibitory. It increases the membrane potential by making the inside more negative, and generation of a nerve impulse is more difficult than usual. This is because the membrane potential is even farther from threshold than it was in its resting state. A hyperpolarizing postsynaptic potential is inhibitory and is termed an **inhibitory postsynaptic potential (IPSP)** (Fig. 12.15b). IPSPs often result from opening of chemically gated Cl$^-$ or K$^+$ channels. When Cl$^-$ channels open, Cl$^-$ tends to diffuse inward at a greater pace, but the negativity inside the neuron retards its inflow somewhat. When K$^+$ channels open, the membrane permeability to K$^+$ increases, more K$^+$ diffuses outward, and the inside becomes even more negative.

Removal of Neurotransmitter

Removal of the neurotransmitter from the synaptic cleft is essential for normal synaptic function. If a neurotransmitter

FIGURE 12.15 Postsynaptic potentials (PSPs).

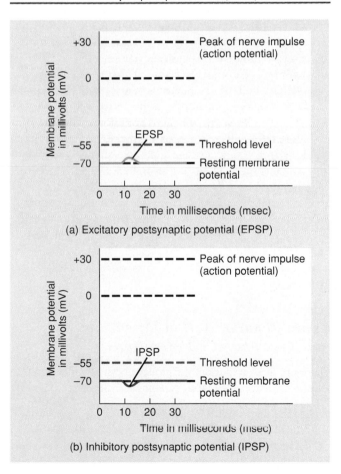

(a) Excitatory postsynaptic potential (EPSP)

(b) Inhibitory postsynaptic potential (IPSP)

Question: Suppose a neurotransmitter binds to a receptor and *closes* K$^+$ channels. Will this produce an EPSP or an IPSP?

could linger in the synaptic cleft, it would influence the postsynaptic neuron, muscle fiber, or gland cell indefinitely. Neurotransmitter is removed in three basic ways:

1. **Diffusion.** Some portion of all neurotransmitters diffuse out of the synaptic cleft.
2. **Enzymatic degradation.** A neurotransmitter may be inactivated through enzymatic degradation. For example, the enzyme acetylcholinesterase breaks down acetylcholine in the synaptic cleft.
3. **Uptake into cells.** Often, neurotransmitters are actively transported back into the neuron that released them (reuptake) or are transported into neighboring neuroglia. Norepinephrine, for example, is rapidly taken up and recycled by the same neurons that release it. The membrane proteins that accomplish such uptake are called **neurotransmitter transporters.** One reason that cocaine produces intense pleasurable feelings (euphoria) is that it blocks transporters for uptake of the neurotransmitter dopamine. This allows dopamine to linger longer in synaptic clefts, producing excessive stimulation of certain brain regions.

Presynaptic Facilitation and Inhibition

Certain synapses can modify the quantity of neurotransmitter released at other synapses. **Presynaptic facilitation** increases the amount of neurotransmitter released whereas **presynaptic inhibition** decreases it. In these situations, a synaptic end-bulb of one neuron synapses with the synaptic end-bulb of another neuron, making an axoaxonal synapse (Fig. 12.16). If neuron 1 releases excitatory neurotransmitter before an impulse arrives at a synaptic end-bulb of neuron 2, more synaptic vesicles undergo exocytosis from neuron 2. Because more neurotransmitter has been released, neuron 3 is then stimulated more strongly. This is presynaptic facilitation. If neuron 1 releases inhibitory neurotransmitter, neuron 2 releases less neurotransmitter at its synapse with neuron 3. This is presynaptic inhibition. Presynaptic facilitation and inhibition may last for several minutes or hours. They are thought to be important in learning and memory.

Comparison of Action Potentials and Postsynaptic Potentials

There are three important differences between action potentials (APs) and postsynaptic potentials (PSPs).

1. **Propagation.** APs propagate whereas PSPs do not. PSPs arise and then fade away within a millimeter or so of the synapse. They permit communication over short distances; action potentials carry messages over long distances.
2. **Amplitude.** APs are all-or-none (same size) whereas PSPs are graded in amplitude, that is, they have

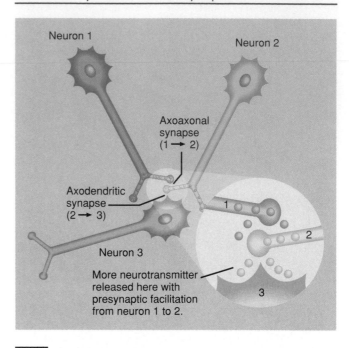

FIGURE 12.16 Presynaptic facilitation and inhibition. An excitatory or inhibitory neurotransmitter released by neuron 1 can facilitate or inhibit release of neurotransmitter by neuron 2 at its synapse with neuron 3.

Neuron 1

Neuron 2

Axoaxonal synapse (1 → 2)

Axodendritic synapse (2 → 3)

Neuron 3

More neurotransmitter released here with presynaptic facilitation from neuron 1 to 2.

Question: Suppose that neuron 2 carries pain information. Will more or less pain be felt if neuron 1 provides presynaptic inhibition?

variable sizes. The amplitude of a PSP varies according to how much neurotransmitter is released.

3. **Refractory period.** APs exhibit a refractory period whereas PSPs do not.

Spatial and Temporal Summation of PSPs

An EPSP lasts a few milliseconds, and a typical neuron in the CNS receives input from 1000 to 10,000 synapses. Integration of these inputs is known as **summation** and occurs at the trigger zone. The greater the summation, if it is a depolarization, the greater the probability a nerve impulse will be initiated.

When summation results from buildup of neurotransmitter released by *several* presynaptic end-bulbs, it is called **spatial summation** (Fig. 12.17a). When summation results from buildup of neurotransmitter released by a *single* presynaptic end-bulb firing two or more times in rapid succession, it is called **temporal summation** (Fig. 12.17b). Since a typical EPSP lasts about 15 msec, the second firing must occur soon after the first one if temporal summation is to occur.

A single postsynaptic neuron receives input from many presynaptic neurons. Some presynaptic end-bulbs produce

FIGURE 12.17 Summation. (a) When two presynaptic axons (a and b) separately cause EPSPs (arrows) in a postsynaptic neuron (c), threshold level is not reached in the postsynaptic neuron. When axons a and b, plus many others, act simultaneously on the postsynaptic cell, their EPSPs sum to reach the threshold level and trigger a nerve action potential (spatial summation). (b) Stimuli applied to the same axon (arrows) sufficiently close together in time cause overlapping EPSPs that sum (temporal summation). When depolarization reaches the threshold level, a nerve action potential is triggered.

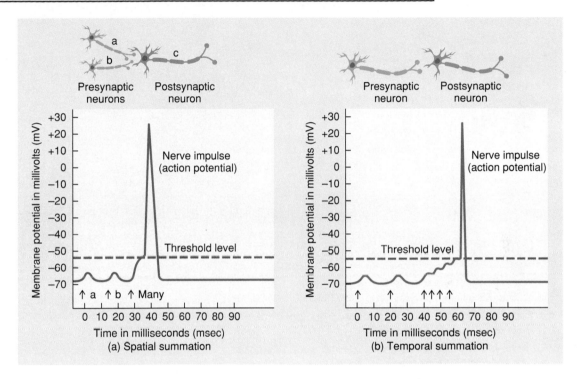

Question: What would be the result if, in addition to the four EPSPs indicated by arrows, an IPSP occurred at time 55 msec in (b)?

excitation and some produce inhibition. The sum of all the effects, excitatory and inhibitory, determines the effect on the postsynaptic neuron. It may respond in the following ways:

1. **EPSP.** If the excitatory effect is greater than the inhibitory effect, but less than the threshold level of stimulation, the result is a small EPSP. Subsequent stimuli can more easily generate a nerve impulse through summation because the neuron is partially depolarized.

2. **Impulse(s).** If the excitatory effect is greater than the inhibitory effect and reaches or surpasses the threshold level of stimulation, the result is a threshold or suprathreshold EPSP. This spreads to the initial segment of the axon and triggers one or more nerve impulses. Impulses continue to be generated as long as the EPSP is above the threshold level.

3. **IPSP.** If the inhibitory effect is greater than the excitatory effect, the membrane hyperpolarizes (IPSP).

The result is inhibition of the postsynaptic neuron and thus an inability to generate a nerve impulse.

Exhibit 12.2 summarizes the structural and functional elements of a neuron.

Neurotransmitters

Excitatory and inhibitory neurotransmitters are present in both the PNS and CNS. Sometimes the same neurotransmitter is excitatory in one location but inhibitory in another. Which response occurs depends on the nature of the receptor that binds the neurotransmitter. For many years, neurons were thought to liberate just one type of neurotransmitter at all their synaptic end-bulbs. Now we know that two or even three neurotransmitters are present in many neurons.

The best-studied neurotransmitter is **acetylcholine (ACh).** It is a neurotransmitter released by many PNS neu-

EXHIBIT 12.2

SUMMARY OF NEURONAL STRUCTURE AND FUNCTION

Diagram	Structure	Functions
	Dendrites	Receive stimuli through activation of gated channels; in sensory neurons, produce generator potentials; in motor neurons and association neurons, produce excitatory and inhibitory postsynaptic potentials (EPSPs and IPSPs).
	Cell body	Receives stimuli and produces EPSPs and IPSPs through activation of chemically gated channels located at synapses.
	Junction of axon hillock and initial segment of axon	Integrates EPSPs and IPSPs and, if sum is a threshold depolarization, initiates nerve impulse.
	Axon	Propagates (conducts) nerve impulses from initial segment to axon terminals in a self-reinforcing manner; impulse amplitude does not change as it propagates along the axon.
	Axon terminals and synaptic end-bulbs	Inflow of Ca^{2+} caused by depolarizing phase of nerve impulse triggers neurotransmitter release by exocytosis of synaptic vesicles.

— Membrane includes chemically gated channels

— Membrane includes voltage-gated Na^+ and K^+ channels

— Membrane includes voltage-gated Ca^{2+} channels

rons and by some CNS neurons. ACh is an excitatory neurotransmitter at the neuromuscular junction, where it acts directly to open chemically gated cation channels. It is also known to be an inhibitory neurotransmitter at other synapses, where its effects on ion channels appear to occur indirectly via receptors that link to a G-protein. One example is parasympathetic fibers of the vagus nerve (cranial nerve X) that innervate the heart; ACh slows heart rate by these inhibitory synapses.

Several amino acids are CNS neurotransmitters. **Glutamate** (glutamic acid) and **aspartate** (aspartic acid) have powerful excitatory effects. Two others, **gamma aminobutyric** (GAM-ma am-i-nō-byoo-TIR-ik) **acid (GABA)** and **glycine** are important inhibitory neurotransmitters. Both cause IPSPs by opening chemically gated Cl^- channels. Although GABA is an amino acid, it is not incorporated into proteins in your body. It is found only in the brain, where it is the most common inhibitory neurotransmitter.

Perhaps a third of all brain synapses use GABA. Glycine is more prevalent in the spinal cord than in the brain.

Catecholamine neurotransmitters (norepinephrine, epinephrine, and dopamine) are synthesized from the amino acid tyrosine. They are excitatory at some synapses and inhibitory at others.

Neurotransmitters in the CNS are discussed in more detail on page 429. Certain disorders such as Parkinson's disease, Alzheimer's disease, depression, anxiety, and schizophrenia involve problems relating to neurotransmitters.

CLINICAL APPLICATION

STRYCHNINE POISONING

The importance of inhibitory neurons can be appreciated by observing what happens when normal inhibition is blocked.

Normally, inhibitory neurons in the spinal cord called *Renshaw cells* release glycine at inhibitory synapses with motor neurons that cause contraction of skeletal muscle fibers. This inhibitory input prevents excessive muscular contraction. Strychnine binds to and blocks glycine receptors and causes massive tetanic contractions. All skeletal muscles, including the diaphragm, contract fully and remain contracted. Because the diaphragm cannot relax, the person cannot breathe. The normal delicate balance between excitation and inhibition in the CNS is disturbed and motor neurons are firing without restraint.

Alteration of Impulse Conduction and Synaptic Transmission

A neuron's chemical and physical environment influences both impulse conduction and synaptic transmission. *Alkalosis,* an increase in pH above 7.45, results in increased excitability of neurons that can cause lightheadedness, numbness around the mouth, tingling in the fingertips, nervousness, muscle spasms, and convulsions. *Acidosis,* a decrease in pH below 7.35, results in a progressive depression of neuronal activity that can produce apathy, weakness, and coma. If a nerve is subjected to excessive or prolonged *pressure,* as when crossing one's legs, nerve impulse conduction is blocked. That part of the body may "go to sleep."

Hypnotics, tranquilizers, and *anesthetics* depress impulse conduction and synaptic transmission by increasing the threshold for excitation of neurons, whereas *caffeine, benzedrine,* and *nicotine* reduce the threshold for excitation of neurons and result in facilitation.

There are several ways to modify chemical synaptic transmission: (1) stimulate or inhibit *neurotransmitter synthesis,* (2) block or enhance *neurotransmitter release,* (3) stimulate or inhibit the *transmitter removal,* and (4) block or activate the *receptor site.* An agent that enhances synaptic transmission or mimics the effect of a natural neurotransmitter is an **agonist,** whereas one that blocks the action of a neurotransmitter is an **antagonist.**

Botulinum toxin is made by *Clostridium botulinum* bacteria, which may proliferate in improperly canned food. The toxin inhibits the release of acetylcholine, thus weakening muscle contractions. Even a small amount is very poisonous. Yet patients with certain overly strong or uncontrollable muscle contractions may be helped by injections of this toxin. It is used to treat people who have strabismus (crossed eyes) or blepharospasm (uncontrollable winking) and may hold promise as a therapy for stuttering.

Neostigmine and *physostigmine* are anticholinesterase agents that inactivate acetylcholinesterase for several hours. The result is very slow removal of acetylcholine from its receptors. As noted on page 265, *myasthenia gravis* results from antibodies that block acetylcholine receptors at neuromuscular junctions and weaken skeletal muscle contraction. Both neostigmine and physostigmine can be used to treat myasthenia gravis. Another anticholinesterase agent is *diisopropyl fluorophosphate,* a very powerful nerve gas that is also the active ingredient of many insecticides. It acts for up to several weeks, making it a particularly lethal poison. It may cause nausea, diarrhea, sweating, bronchial constriction, excess respiratory mucus, and generalized weakness.

Substances that are agonists or antagonists of neurotransmitters have great potential for benefit or harm. The plant derivative *curare* (used by South American Indians on poisoned arrows and blow-gun darts) blocks acetylcholine receptors and thus can cause muscular paralysis. Curare-like drugs, however, often are used in surgery to relax muscles. Neostigmine is an antidote for curare. It can be used to terminate the effects of curare after surgery or in cases of curare poisoning.

Neuronal Circuits

The CNS contains billions of neurons organized into complicated patterns called **neuronal pools.** Each pool differs from all others and has its own role in regulating homeostasis. A neuronal pool may contain thousands or even millions of neurons.

Neuronal pools in the CNS are arranged in patterns called **circuits** over which the nerve impulses are conducted. In **simple series circuits** a presynaptic neuron stimulates only a single neuron in a pool. The single neuron then stimulates another, and so on. Most circuits, however, are more complex.

A single presynaptic neuron may synapse with several postsynaptic neurons. Such an arrangement, called **divergence,** permits one presynaptic neuron to influence several postsynaptic neurons or several muscle fibers or gland cells at the same time. In a **diverging circuit,** the nerve impulse from a single presynaptic neuron causes the stimulation of increasing numbers of cells along the circuit (Fig. 12.18a). An example is a single motor neuron in the brain stimulating many other motor neurons in the spinal cord. Thus a single motor neuron activated in the brain can signal to a pool of motor neurons in the spinal cord. Sensory signals also feed into diverging circuits and are often relayed to several regions of the brain.

In another arrangement, called **convergence,** several presynaptic neurons synapse with a single postsynaptic neuron. This arrangement permits more effective stimulation or inhibition of the postsynaptic neuron. In one type of **converging circuit** (Fig. 12.18b), the postsynaptic neuron receives nerve impulses from several different sources. For example, a single motor neuron that synapses with skeletal muscle fibers at neuromuscular junctions receives input from several pathways that originate in different brain regions.

Some circuits in your body are constructed so that once the presynaptic cell is stimulated, it will cause the postsynaptic cell to transmit a series of nerve impulses. One such circuit is called a **reverberating (oscillatory) circuit** (Fig.

FIGURE 12.18 Examples of neuronal circuits.

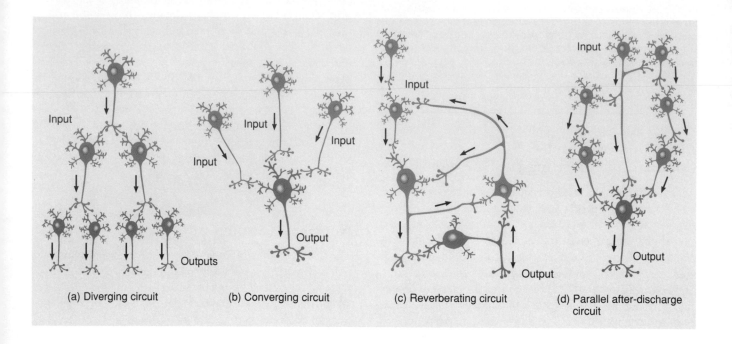

(a) Diverging circuit (b) Converging circuit (c) Reverberating circuit (d) Parallel after-discharge circuit

Question: A motor neuron in the spinal cord typically receives input from neurons that originate in several different regions of the brain. Is this an example of convergence or divergence?

12.18c). In this pattern, the incoming impulse stimulates the first neuron, which stimulates the second, which stimulates the third, and so on. Branches from later neurons synapse with earlier ones, however, sending the impulse back through the circuit again and again. The output signal may last from a few seconds to many hours, depending on the number of synapses and arrangement of neurons in the circuit. Inhibitory neurons may turn off a reverberating circuit after a period of time. Among the body responses thought to be the result of output signals from reverberating circuits are breathing, coordinated muscular activities, waking up, sleeping (when reverberation stops), and short-term memory. One form of epilepsy (grand mal) is probably caused by abnormal reverberating circuits.

A fourth type of circuit is the **parallel after-discharge circuit** (Fig. 12.18d). In this circuit, a single presynaptic cell stimulates a group of neurons, each of which synapses with a common postsynaptic cell. A differing number of synapses between the first and last neurons imposes varying synaptic delays so that the last neuron exhibits multiple EPSPs or IPSPs. If the input is excitatory, the postsynaptic neuron then can send out a stream of impulses in quick succession. It is thought that parallel after-discharge circuits may be employed for precise activities such as mathematical calculations.

REGENERATION OF NERVOUS TISSUE

Unlike the cells of epithelial tissue, mammalian neurons have very limited powers of **regeneration**, that is, the capability to replicate or repair themselves. Around 6 months of age, virtually all developing neurons lose their ability to undergo mitosis. Thus when a neuron is damaged or destroyed, it cannot be replaced by daughter cells from other neurons. A neuron destroyed is permanently lost, and only some types of damage may be repaired.

In the PNS, damage to myelinated axons and dendrites may be repaired if the cell body remains intact and if the neurolemmocytes (Schwann cells) that perform the myelination remain active. Substantial regrowth may occur even in a completely severed nerve if it is surgically reattached. The neurolemmocytes form a tube that appears to aid regeneration (see Fig. 12.19d).

In the CNS there is little or no repair of damage to neurons. Axons in the CNS are myelinated by oligodendrocytes that do not form neurolemmas. An added complication in the CNS is that the neuroglia appear to provide an environment that stops axon regrowth. Perhaps this is the same mechanism that stops axonal growth during develop-

FIGURE 12.19 Damage and repair of a peripheral neuron.

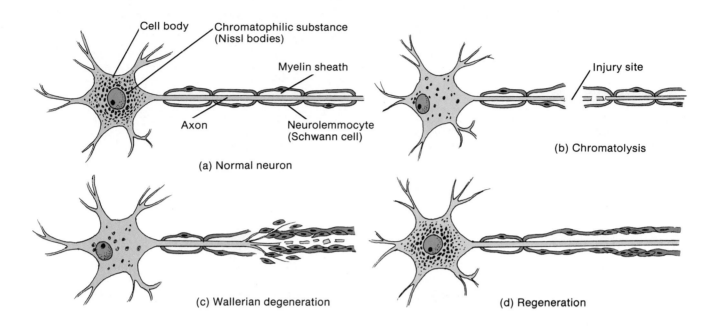

(a) Normal neuron

Cell body
Chromatophilic substance (Nissl bodies)
Myelin sheath
Axon
Neurolemmocyte (Schwann cell)

(b) Chromatolysis

Injury site

(c) Wallerian degeneration

(d) Regeneration

Question: What is the role of the neurolemma in regeneration?

ment once a target region has been reached. Also, after axonal damage, nearby astrocytes proliferate rapidly, forming a type of scar tissue. The scar tissue is a physical barrier to regeneration. Thus injury of the brain or spinal cord usually is permanent.

CLINICAL APPLICATION

ENCOURAGING NEURONAL REGROWTH

Certain types of neuronal tumor cells can grow and replicate in tissue culture. Also, in the brains of some songbirds, new neurons appear and disappear every year. The nearly complete lack of regenerative success in the mammalian CNS seems to result from two factors: (1) inhibitory influences from neuroglia and (2) absence of growth cues that were present during development.

Some developmental cues are electrical in nature. A large research effort is directed at finding a way to promote regrowth of damaged neurons using electrical stimulation. Other developmental cues are chemical in nature, involving both substances that trigger mitosis (mitogenic factors) and those that regulate growth (neurotropins). In 1992 Canadian researchers published their unexpected finding that cells taken from the brains of *adult* mammals could be encouraged to proliferate into both neurons and astrocytes. They cultured small pieces of mouse brain tissue in dishes and added **epidermal growth factor (EGF)**. Previously, EGF was known to trigger mitosis in a variety of nonneuronal cells and promote wound healing and tissue regeneration (see pages 130, 136). These experiments challenge the dogma that stem (progenitor) cells are not present in adult brains. A tantalizing possibility is that scientists may be able to find ways of stimulating these dormant stem cells to replace neurons lost through damage or disease. Also, such tissue-cultured neurons might be useful for transplantation purposes.

DISORDERS: HOMEOSTATIC IMBALANCES

DAMAGE AND REPAIR OF PERIPHERAL NEURONS

As we have seen, axons and dendrites that are associated with a neurolemma may undergo repair if the cell body is intact, the neurolemmocytes (Schwann cells) are functional, and scar tissue formation does not occur too rapidly. Most nerves in the PNS consist of processes that are covered with a neurolemma. A person who injures neurons in a nerve in the upper extremity, for example, has a good chance of regaining nerve function.

Continues

DISORDERS: HOMEOSTATIC IMBALANCES (continued)

When there is damage to an axon, usually there are changes that occur in the cell body of the affected neuron and in the portion of the axon distal to the site of injury. Changes also may occur in the portion of the axon proximal to the site of injury.

Chromatolysis

About 24 to 48 hours after injury to a process of a central or peripheral neuron, the chromatophilic substance (Nissl bodies), which normally is arranged in an orderly fashion in an uninjured cell body, breaks down into finely granular masses (Fig. 12.19b). This alteration is called **chromatolysis** (krō'-ma-TOL-i-sis; *chromo* = color; *lysis* = dissolution). It begins between the axon hillock and nucleus but spreads throughout the cell body. As a result of chromatolysis, the cell body swells, reaching a maximum between 10 and 20 days after injury.

Degeneration

The part of the process distal to the damaged region becomes slightly swollen and then breaks up into fragments by the third to fifth day. The myelin sheath also deteriorates (Fig. 12.19c). Degeneration of the distal portion of the neuronal process and myelin sheath is called **Wallerian degeneration.** Following degeneration, macrophages phagocytize the remains.

The changes in the proximal portion of the fiber, called **retrograde degeneration,** are similar to those that occur during Wallerian degeneration. The main difference in retrograde degeneration is that the changes only extend to the first neurofibral node (node of Ranvier).

Regeneration

Following chromatolysis, there are signs of recovery in the cell body. Synthesis of RNA and protein accelerates, which favors rebuilding or **regeneration** of the axon. Recovery often takes several months.

Even though the neuronal process and myelin sheath degenerate, the neurolemma remains. The neurolemmocytes on either side of the injured site multiply by mitosis, grow toward each other, and attempt to form a tube across the injured area (Fig. 12.19d). The tube guides growth of new processes from the proximal area across the injured area into the distal area previously occupied by the original nerve fiber. The growth of new axons will not occur if the gap at the site of injury is too large or if the gap becomes filled with collagen fibers.

During the first few days following damage, buds of regenerating axons begin to invade the tube formed by the neurolemmocytes (Fig. 12.19c). Axons from the proximal area grow at the rate of about 1.5 mm (0.06 in.) per day across the area of damage, find their way into the distal neurolemmal tubes, and grow toward the distally located receptors and effectors. Thus sensory and motor connections are reestablished. In time, the neurolemmocytes form a new myelin sheath. However, function is never completely restored after a nerve is severed.

EPILEPSY

The second most common neurological disorder after stroke (rupture or blockage of a brain blood vessel) is **epilepsy,** which afflicts about 1% of the population. It is characterized by short, recurrent, periodic attacks of motor, sensory, or psychological malfunction. The attacks, called **epileptic seizures,** are initiated by abnormal, synchronous electrical discharges from millions of neurons in the brain, perhaps resulting from abnormal reverberating circuits. The discharges stimulate many of the neurons to send nerve impulses over their conduction pathways. As a result, a person undergoing an attack may contract skeletal muscles involuntarily. Lights, noise, or smells may be sensed when the eyes, ears, and nose have not been stimulated. The electrical discharges may also inhibit certain brain centers. For instance, the waking center in the brain (the reticular activating system, or RAS, see page 463) may be depressed so that the person loses consciousness. When RAS activity resumes, the person regains consciousness.

Epilepsy has many causes, including brain damage at birth, the most common cause; metabolic disturbances (hypoglycemia, hypocalcemia, uremia, hypoxia); infections (encephalitis or meningitis); toxins (alcohol, tranquilizers, hallucinogens); vascular disturbances (hemorrhage, hypotension); head injuries; and tumors and abscesses of the brain. Most epileptic seizures, however, are idiopathic; that is, they have no demonstrable cause. Epilepsy almost never affects intelligence.

Epileptic seizures can be eliminated or alleviated by drugs that depress neuronal excitability. One such drug is valproic acid, which increases the quantity of the inhibitory neurotransmitter gamma aminobutyric acid (GABA).

Study Outline

Introduction (p. 347)

1. The nervous system helps regulate homeostasis and integrate all body activities by sensing changes (sensory), interpreting them (integrative), and reacting to them (motor).

Nervous System Divisions (p. 347)

1. The central nervous system (CNS) consists of the brain and spinal cord.

2. The peripheral nervous system (PNS) consists of cranial and spinal nerves. It has sensory (afferent) and motor (efferent) components.

3. The PNS also is subdivided into somatic (voluntary) and autonomic (involuntary) nervous systems.

4. The somatic nervous system (SNS) consists of neurons that conduct impulses from cutaneous and special sense receptors to the CNS and motor neurons from the CNS to skeletal muscle tissue.

5. The autonomic nervous system (ANS) contains sensory neurons from visceral organs and motor neurons that convey impulses from the CNS to smooth muscle tissue, cardiac muscle tissue, and glands.

Functional Anatomy of Nervous Tissue (p. 347)

Neuroglia (p. 347)

1. Neuroglia are specialized tissue cells that support neurons, attach neurons to blood vessels, produce the myelin sheath around axons, and carry out phagocytosis.
2. Neuroglia include astrocytes, oligodendrocytes, microglia, ependyma, neurolemmocytes (Schwann cells), and satellite cells.
3. Two types of neuroglia produce myelin sheaths: oligodendrocytes myelinate axons in the CNS and neurolemmocytes myelinate axons in the PNS.

Neurons (p. 352)

1. Most neurons, or nerve cells, consist of a cell body (soma), many dendrites, and usually a single axon. The axon conducts nerve impulses from the neuron to the dendrites or cell body of another neuron or to an effector organ of the body (muscle or gland).
2. On the basis of structure, neurons are multipolar, bipolar, and unipolar.
3. On the basis of function, sensory (afferent) neurons conduct impulses from receptors to the CNS; association (connecting) neurons conduct impulses to other neurons, including motor neurons; and motor (efferent) neurons conduct impulses to effectors.

Gray and White Matter (p. 356)

1. White matter is aggregations of myelinated processes whereas gray matter contains nerve cell bodies, dendrites, and axon terminals or bundles of unmyelinated axons and neuroglia.
2. In the spinal cord, gray matter forms an H-shaped inner core, surrounded by white matter. In the brain, a thin outer shell of gray matter covers the cerebral hemispheres.

Neurophysiology (p. 356)

Resting Membrane Potential (p. 356)

1. The membrane of a nonconducting neuron is positive outside and negative inside owing to the distribution of different ions across the membrane and the relative permeability of the membrane toward Na^+ and K^+.
2. A typical value for the resting membrane potential (RMP) is −70 mV, and the membrane is said to be polarized.
3. The sodium–potassium pumps compensate for slow leakage of Na^+ into the cell by pumping it back out.

Ion Channels (p. 358)

1. The two basic types of ion channels are leakage (nongated) and gated.
2. There are four types of gated channels: voltage-gated, chemically gated, mechanically gated, and light-gated.

Action Potential (Impulse) (p. 358)

1. During an action potential (impulse), voltage-gated Na^+ and K^+ channels open in sequence. This results first in depolarization, loss and then reversal of membrane polarization (from −70 to 0 to +30 mV), and then in repolarization, recovery of the RMP (from +30 to −70 mV).
2. During the refractory period (RP) another impulse cannot be generated at all (absolute RP) or can be triggered only by a suprathreshold stimulus (relative RP).

3. An action potential (impulse) conducts or propagates (travels) from point to point along the membrane.
4. According to the all-or-none principle, if a stimulus is strong enough to generate an action potential, the impulse travels at a constant and maximum strength for the existing conditions. A stronger stimulus will not cause a larger impulse.
5. Nerve impulse conduction in which the impulse jumps from neurofibral node to node is called saltatory conduction.
6. Fibers with larger diameters conduct impulses faster than those with smaller diameters; myelinated fibers conduct impulses faster than unmyelinated fibers.
7. The intensity of a stimulus is coded in the frequency of action potentials.

Transmission at Synapses (p. 364)

1. A synapse is the functional junction between one neuron and another or between a neuron and an effector such as a muscle or gland.
2. Two types of synapses are electrical and chemical.
3. At a chemical synapse, there is only one-way information transfer from a presynaptic neuron to a postsynaptic neuron.
4. An excitatory neurotransmitter is one that can depolarize or make less negative the postsynaptic neuron's membrane, bringing the membrane potential closer to threshold. An inhibitory neurotransmitter hyperpolarizes the membrane of the postsynaptic neuron.
5. Neurotransmitter is removed from the synaptic cleft in three ways: diffusion, enzymatic degradation, and uptake into cells (neurons and neuroglia).
6. Presynaptic facilitation increases the amount of neurotransmitter released by a presynaptic neuron whereas presynaptic inhibition decreases it.
7. If several presynaptic end-bulbs release their neurotransmitter at about the same time, the combined effect may generate a nerve impulse, due to summation. Summation may be spatial or temporal.
8. The postsynaptic neuron is an integrator. It receives signals, integrates them, and then responds accordingly.

Neurotransmitters (p. 367)

1. Both excitatory and inhibitory neurotransmitters are present in both the CNS and PNS. The same neurotransmitter may be excitatory in some locations and inhibitory in others.
2. Important neurotransmitters include acetylcholine, glutamate, aspartate, gamma aminobutyric acid (GABA), glycine, norepinephrine, epinephrine, and dopamine.

Alteration of Impulse Conduction and Synaptic Transmission (p. 369)

1. A neuron's chemical and physical environment influences both impulse conduction and synaptic transmission.
2. Chemical synaptic transmission may be stimulated or blocked by affecting neurotransmitter synthesis, release, or removal or by affecting the receptor site.

Neuronal Circuits (p. 369)

1. Neurons in the central nervous system are organized into patterns called neuronal pools. Each pool differs from all others and has its own role in regulating homeostasis.
2. Neuronal pools are organized into circuits. These include simple series, diverging, converging, reverberating (oscillatory), and parallel after-discharge circuits.

Regeneration of Nervous Tissue (p. 370)

1. Around 6 months of age, the neuronal cell body loses its mitotic apparatus and is no longer able to divide.
2. Nerve fibers that have a neurolemma are capable of regeneration.

Review Questions

1. Describe the three basic functions of the nervous system that are necessary to maintain homeostasis. (p. 347)
2. Distinguish between the central and peripheral nervous systems and describe the functions of each subdivision. (p. 347)
3. Relate the terms *voluntary* and *involuntary* to the nervous system. (p. 347)
4. What are neuroglia? List the principal types and their functions. Why are they important clinically? (p. 347)
5. What is a myelin sheath? How is it formed? (p. 349)
6. Define a neuron. Diagram and label a neuron. Next to each part, list its function. (p. 352)
7. Distinguish between slow and fast axonal transport. Why is axonal transport important clinically? (p. 353)
8. Define the neurolemma. Why is it important? (p. 352)
9. Discuss the structural classification of neurons. Give an example of each. (p. 355)
10. What are the functional differences between a typical afferent and efferent neuron? Define association neuron. (p. 355)
11. Describe the seven functional types of afferent and efferent neurons. (p. 355)
12. Distinguish between gray and white matter. Where is each found? (p. 356)
13. Describe the factors that give rise to the resting membrane potential. (p. 357)
14. Describe the two basic types of ion channels. (p. 358)
15. What types of stimuli regulate gated channels? (p. 358)
16. Outline the principal steps in the generation and conduction of a nerve impulse. (p. 360)
17. Define the following: resting membrane potential (p. 356), depolarization (p. 358), repolarization (p. 358), polarized membrane (p. 357), nerve impulse (nerve action potential) (p. 358), depolarized membrane (p. 361), repolarized membrane (p. 361), and refractory period. (p. 361)
18. What is the all-or-none principle? (p. 361)
19. What is saltatory conduction? Why is it important? (p. 362)
20. What factors determine the speed of propogation of nerve impulses? (p. 362)
21. How is the intensity of a stimulus coded in the nervous system? (p. 364)
22. Describe the events of chemical synaptic transmission. (p. 364)
23. Distinguish between excitatory and inhibitory transmission. (p. 365)
24. How is neurotransmitter removed from the synaptic cleft? (p. 365)
25. Describe presynaptic facilitation and presynaptic inhibition. (p. 366)
26. Compare action potentials and postsynaptic potentials. (p. 366)
27. Describe spatial and temporal summation. (p. 366)
28. What is a neurotransmitter? List several probable neurotransmitters, indicate their locations in the nervous system and whether or not they may lead to excitation or inhibition. (p. 367)
29. In what ways can impulse conduction or synaptic transmission be altered? (p. 369)
30. What is a neuronal circuit? Distinguish among simple series, diverging, converging, reverberating (oscillatory), and parallel after-discharge circuits. (p. 369)
31. What factors determine neuron regeneration? (p. 370)

Answers to Questions with Figures

12.1 General somatic efferent (motor) neurons because they can be consciously controlled.
12.2 Oligodendrocytes.
12.3 It increases the speed of nerve impulse conduction.
12.4 Dendrites receive (motor or association neurons) or generate (sensory neurons) inputs; cell body also receives input signals; axon conducts action potentials and transmits the message to another neuron or effector cell by releasing neurotransmitter at its synaptic end-bulbs.
12.5 Site where nerve impulses arise.
12.6 Receiving input.
12.7 Myelin.
12.8 −70 mV.
12.9 Mechanically gated.
12.10 Voltage-gated Na^+ channels are open during depolarization and voltage-gated K^+ channels are open during repolarization.
12.11 Yes, because the leakage channels would still allow K^+ to exit more rapidly than Na^+ could enter the axon. In fact, mammalian myelinated axons have few voltage-gated K^+ channels.
12.12 The Na^+ channel inactivation gates close automatically, thus limiting the duration of depolarization.
12.13 At the nodes.
12.14 In some electrical synapses (gap junctions) ions flow equally well in either direction, so either neuron may be the presynaptic one. At a chemical synapse, one neuron releases the neurotransmitter and the other neuron has receptors that bind this chemical. Thus the information can flow in only one direction.
12.15 Closing of K^+ channels will slow the outflow of K^+ so the inside will become more positive, thus produce a depolarizing EPSP.
12.16 Less.
12.17 Probably threshold depolarization would not be reached and an impulse would not be generated.
12.18 Convergence.
12.19 Provides a tube to guide regrowth of axon.

Chapter 13
THE SPINAL CORD AND SPINAL NERVES

Chapter Contents at a Glance

Student Objectives

1. Describe the protection, gross anatomical features, and cross sectional structure of the spinal cord.

2. Describe the functions of the principal sensory and motor tracts of the spinal cord.

3. Describe the components of a reflex arc and its relationship to homeostasis.

4. List and describe several clinically important reflexes.

5. Describe the composition and coverings of a spinal nerve.

6. Define a plexus and describe the composition and distribution of nerves of the cervical, brachial, lumbar, and sacral plexuses.

7. Describe spinal cord injury and list the immediate and long-range effects.

8. Explain the causes and symptoms of neuritis, sciatica, shingles, and poliomyelitis.

ogether, the spinal cord and spinal nerves contain neuronal circuits that mediate some of your quickest reactions to environmental changes. If you pick up something hot, for example, the grasping muscles may relax and you may drop it even before the sensation of extreme heat or pain reaches your conscious perception. This is an example of a spinal cord reflex, a quick, automatic response to certain kinds of stimuli that needs only neurons (nerve cells) in the spinal nerves and spinal cord. Besides processing reflexes, the spinal cord also is the site for integration (summing) of nerve impulses that arise locally or arrive from the periphery and brain. Moreover, the spinal cord is the highway traveled by sensory nerve impulses headed for the brain and by motor nerve impulses destined for spinal nerves. Keep in mind that the spinal cord is continuous with the brain and that together they constitute the central nervous system (CNS).

SPINAL CORD ANATOMY

Protection and Coverings

Two types of connective tissue coverings, the bony vertebrae and tough meninges, plus a cushion of cerebrospinal fluid (produced by the brain) surround and protect the delicate nervous tissue of the spinal cord and brain.

Vertebral Column

The spinal cord is located within the vertebral (spinal) canal of the vertebral column. The vertebral foramina of all the vertebrae, stacked one on top of the other, form the canal. The surrounding ring of vertebral bone provides a sturdy shelter for the enclosed spinal cord (see Fig. 13.1b). The vertebral ligaments, meninges, and cerebrospinal fluid (see page 405) provide additional protection.

Meninges

The **meninges** (me-NIN-jēz; singular, **meninx**) are connective tissue coverings that encircle the spinal cord and brain. They are called, respectively, the **spinal meninges** (Fig. 13.1) and the **cranial meninges** (see Fig. 14.4a). The outermost of the three spinal meninges is the **dura mater** (DYOO-ra MĀ-ter; *dura* = tough, *mater* = mother), composed of dense, irregular connective tissue. It forms a sac from the level of the foramen magnum of the occipital bone, where it is continuous with the inner layer of the dura mater of the brain, to the second sacral vertebra where it is close-ended. The spinal cord is also protected by a cushion of fat and connective tissue located in the **epidural space,** a space between the dura mater and the wall of the vertebral canal.

The middle meninx is an avascular covering called the **arachnoid** (a-RAK-noyd; *arachne* = spider) because of its spider's web arrangement of delicate collagen fibers and some elastic fibers. It lies inside the dura mater and is also continuous with the arachnoid of the brain. Between the dura mater and the arachnoid is a thin **subdural space,** which contains interstitial fluid.

The innermost meninx is the **pia mater** (PĪ-a MĀ-ter; *pia* = delicate), a thin transparent connective tissue layer that adheres to the surface of the spinal cord and brain. It consists of interlacing bundles of collagen fibers and some fine elastic fibers and contains many blood vessels that supply nutrients and oxygen to the spinal cord. Between the arachnoid and the pia mater is the **subarachnoid space,** which contains cerebrospinal fluid. Inflammation of the meninges is known as **meningitis.**

All three spinal meninges cover the spinal nerves up to the point of exit from the spinal column through the intervertebral foramina. Triangular-shaped membranous extensions of the pia mater suspend the spinal cord in the middle of its dural sheath. These extensions, called **denticulate** (den-TIK-yoo-lāt) **ligaments,** are thickenings of the pia mater. They project laterally and fuse with the dura mater along the length of the cord between the ventral and dorsal nerve roots of spinal nerves on either side. They protect the spinal cord against shock and sudden displacement.

External Anatomy of the Spinal Cord

The **spinal cord** is roughly cylindrical but flattened slightly in its anterior–posterior dimension. In the adult, it extends from the medulla, the most inferior part of the brain, to the upper border of the second lumbar vertebra (Fig. 13.2). In a newborn infant, it extends to the third or fourth lumbar vertebra. During early childhood, both the spinal cord and the vertebral column grow longer as part of overall body growth. Around age 4 or 5, however, elongation of the spinal cord stops. Since the vertebral column continues to elongate, the spinal cord does not extend the entire length of the vertebral column. The length of the adult spinal cord ranges from 42 to 45 cm (16 to 18 in.). Its diameter is about 2 cm (¾ in.) in the midthoracic region, somewhat larger in the lower cervical and midlumbar regions, and smallest at the inferior tip.

When the spinal cord is viewed externally, two conspicuous enlargements can be seen. The superior enlargement, the **cervical enlargement,** extends from the fourth cervical to the first thoracic vertebra. Nerves to and from the upper extremities arise from the cervical enlargement. The inferior enlargement, called the **lumbar enlargement,** extends from the ninth to the twelfth thoracic vertebra. Nerves to and from the lower extremities arise from the lumbar enlargement.

Below the lumbar enlargement, the spinal cord tapers to a conical portion known as the **conus medullaris** (KŌ-nus med-yoo-LAR-is), which ends at the level of the interverte-

FIGURE 13.1 Spinal meninges.

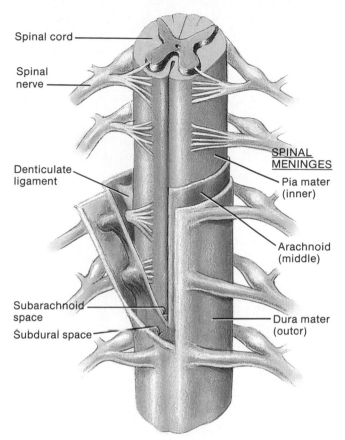

Spinal cord

Spinal nerve

Denticulate ligament

Subarachnoid space

Subdural space

SPINAL MENINGES

Pia mater (inner)

Arachnoid (middle)

Dura mater (outor)

(a) Sections through spinal cord

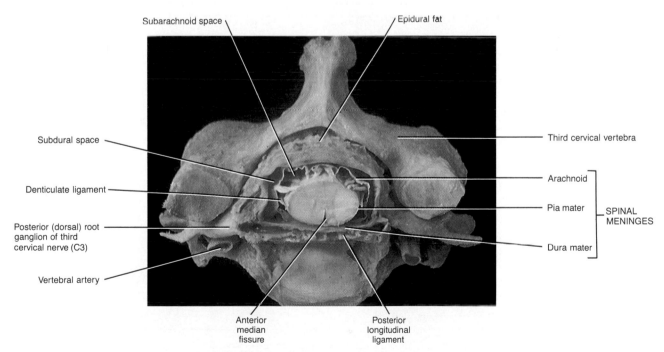

Subarachnoid space

Epidural fat

Subdural space

Denticulate ligament

Posterior (dorsal) root ganglion of third cervical nerve (C3)

Vertebral artery

Third cervical vertebra

Arachnoid

Pia mater

Dura mater

SPINAL MENINGES

Anterior median fissure

Posterior longitudinal ligament

(b) Photograph of a cross section of the spinal cord

Question: What are the superior and inferior limits of the spinal dura mater?

FIGURE 13.2 Spinal cord and spinal nerves.

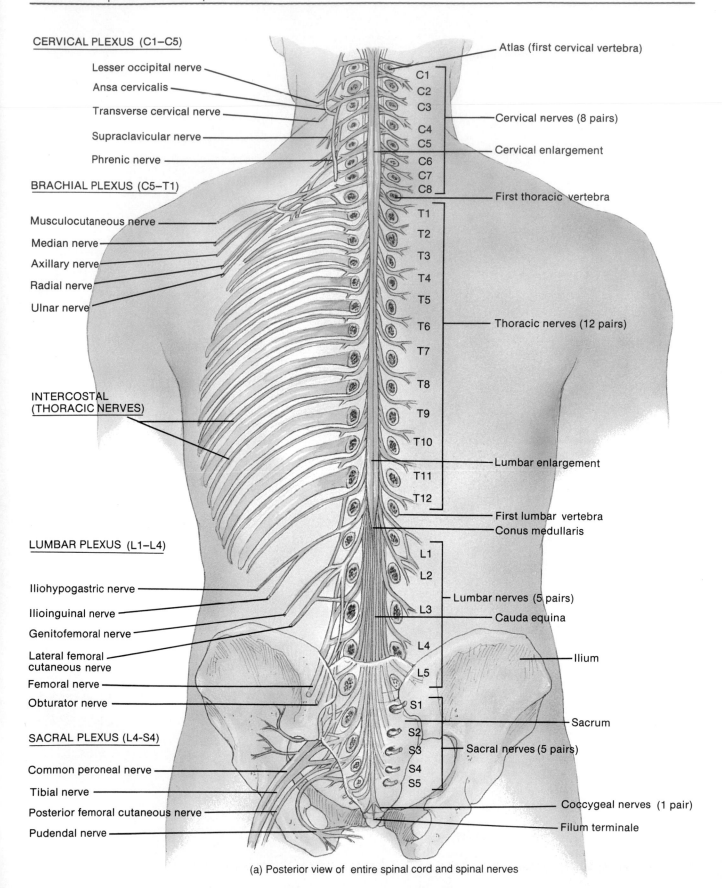

CERVICAL PLEXUS (C1–C5)

Lesser occipital nerve
Ansa cervicalis
Transverse cervical nerve
Supraclavicular nerve
Phrenic nerve

BRACHIAL PLEXUS (C5–T1)

Musculocutaneous nerve
Median nerve
Axillary nerve
Radial nerve
Ulnar nerve

INTERCOSTAL
(THORACIC NERVES)

LUMBAR PLEXUS (L1–L4)

Iliohypogastric nerve
Ilioinguinal nerve
Genitofemoral nerve
Lateral femoral
cutaneous nerve
Femoral nerve
Obturator nerve

SACRAL PLEXUS (L4–S4)

Common peroneal nerve
Tibial nerve
Posterior femoral cutaneous nerve
Pudendal nerve

Atlas (first cervical vertebra)

C1
C2
C3
C4
C5
C6
C7
C8

Cervical nerves (8 pairs)
Cervical enlargement

First thoracic vertebra

T1
T2
T3
T4
T5
T6
T7
T8
T9
T10
T11
T12

Thoracic nerves (12 pairs)

Lumbar enlargement

First lumbar vertebra
Conus medullaris

L1
L2
L3
L4
L5

Lumbar nerves (5 pairs)
Cauda equina

Ilium

S1
S2
S3
S4
S5

Sacrum

Sacral nerves (5 pairs)

Coccygeal nerves (1 pair)
Filum terminale

(a) Posterior view of entire spinal cord and spinal nerves

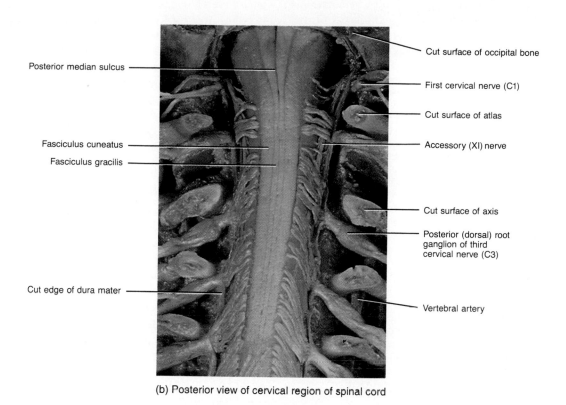

Posterior median sulcus

Fasciculus cuneatus

Fasciculus gracilis

Cut edge of dura mater

Cut surface of occipital bone

First cervical nerve (C1)

Cut surface of atlas

Accessory (XI) nerve

Cut surface of axis

Posterior (dorsal) root ganglion of third cervical nerve (C3)

Vertebral artery

(b) Posterior view of cervical region of spinal cord

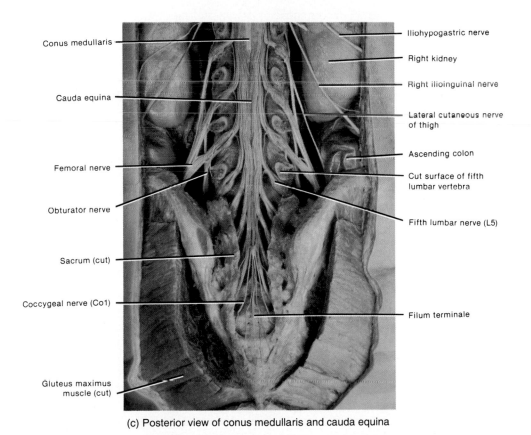

Conus medullaris

Cauda equina

Femoral nerve

Obturator nerve

Sacrum (cut)

Coccygeal nerve (Co1)

Gluteus maximus muscle (cut)

Iliohypogastric nerve

Right kidney

Right ilioinguinal nerve

Lateral cutaneous nerve of thigh

Ascending colon

Cut surface of fifth lumbar vertebra

Fifth lumbar nerve (L5)

Filum terminale

(c) Posterior view of conus medullaris and cauda equina

Question: What portion of the spinal cord serves as the point of origin for nerves that innervate the upper extremity?

bral disc between the first and second lumbar vertebrae in an adult. Arising from the conus medullaris is the **filum terminale** (FĪ-lum ter-mi-NAL-ē), an extension of the pia mater that extends inferiorly and attaches the spinal cord to the coccyx.

CLINICAL APPLICATION

SPINAL TAP (LUMBAR PUNCTURE)

A **spinal tap (lumbar puncture)** is used to withdraw cerebrospinal fluid (CSF) for diagnostic purposes; introduce antibiotics, contrast media for myelography, and anesthetics; administer chemotherapy; measure CSF pressure; and evaluate the effects of treatment.

In an adult, a spinal tap is normally performed between the third and fourth or fourth and fifth lumbar vertebrae. This region is below the lowest portion of the spinal cord and provides relatively safe access. (A line drawn across the highest points of the iliac crests passes through the spinous process of the fourth lumbar vertebra.) A local anesthetic is given, and a long needle is inserted into the subarachnoid space. Once the needle is in place, a sample of CSF, usually 5 to 10 ml, is carefully withdrawn for analysis. Leakage of fluid from the site of the spinal tap can lead to severe headache, and for this reason, the patient is asked to remain lying down for 8 to 24 hours after the procedure to minimize the leakage of CSF from the puncture site.

Some nerves that arise from the lower part of the cord do not leave the vertebral column at the same level as they exit from the spinal cord. The roots (points of attachment to the spinal cord) of these nerves angle inferiorly in the vertebral canal from the end of the cord like wisps of hair. Appropriately, the roots of these nerves are named the **cauda equina** (KAW-da ē-KWĪ-na), meaning "horse's tail."

Although the spinal cord is not actually segmented, it appears that way due to the presence of 31 pairs of spinal nerves that emerge from it. Each pair of spinal nerves is said to arise from a **spinal segment**. Two grooves divide the cord into right and left sides (see Fig. 13.3a). The **anterior median fissure** is a deep, wide groove on the anterior (ventral) side, and the **posterior median sulcus** is a shallower, narrow groove on the posterior (dorsal) surface.

Internal Anatomy of the Spinal Cord

The gray matter of the spinal cord is shaped like the letter H or a butterfly and is surrounded by white matter (Fig. 13.3). The gray matter consists primarily of cell bodies of neurons and neuroglia and unmyelinated axons and dendrites of association and motor neurons. The white matter consists of bundles of myelinated axons of motor and sensory neurons. The **gray commissure** (KOM-mi-shur) forms the cross-bar of the H. In the center of the gray commissure is a small space called the **central canal.** This canal extends the entire length of the spinal cord. At its superior end, the central canal is continuous with the fourth ventricle (a space that

contains cerebrospinal fluid) in the medulla oblongata of the brain. Anterior to the gray commissure is the **anterior (ventral) white commissure,** which connects the white matter of the right and left sides of the spinal cord.

The gray matter on each side of the spinal cord is subdivided into regions called **horns.** Those closer to the front of the cord are called **anterior (ventral) gray horns** while those closer to the back are **posterior (dorsal) gray horns.** Between the anterior and posterior gray horns are the **lateral gray horns.** Lateral gray horns are present only in the thoracic, upper lumbar, and sacral segments of the cord.

The gray matter of the cord also contains several nuclei that serve as neural processing centers for nerve impulses and origins for certain nerves. Nuclei are clusters of mainly neuron cell bodies in the spinal cord and brain.

The white matter, like the gray matter, also is organized into regions. The anterior and posterior gray horns divide the white matter on each side into three broad areas called **columns (funiculi): anterior (ventral) white columns, posterior (dorsal) white columns,** and **lateral white columns.** Each column, in turn, contains distinct bundles of nerve fibers having a common origin or destination and carrying similar information. These bundles are called **tracts (fasciculi).** Tracts may extend long distances up or down the spinal cord. **Ascending (sensory) tracts** consist of axons that conduct nerve impulses upward to the brain. Tracts consisting of axons that carry nerve impulses down the cord are called **descending (motor) tracts.** Sensory and motor tracts of the spinal cord are continuous with sensory and motor tracts in the brain.

SPINAL CORD PHYSIOLOGY

The tracts in *white matter* of the spinal cord are highways for nerve impulse conduction. Along these highways sensory impulses flow from the periphery to the brain and motor impulses flow from the brain to the periphery. The *gray matter* of the spinal cord receives and integrates incoming and outgoing information. Both functions of the spinal cord are essential to maintaining homeostasis.

Sensory and Motor Tracts

Often, the name of a tract indicates its position in the white matter, where it begins and ends, and the direction of nerve impulse conduction. For example, the anterior spinothalamic tract is located in the *anterior* white column, it begins in the *spinal cord,* and it ends in the *thalamus* (a region of the brain). Since it conveys nerve impulses from the cord upward to the brain, it is a sensory (ascending) tract. Figure 13.4 shows the principal sensory and motor tracts in the spinal cord. These tracts are described in detail in Exhibit 15.1 on page 459.

Sensory information from receptors travels up the spinal cord to the brain along two main routes on each side of the

FIGURE 13.3 Spinal cord. (a) The organization of gray and white matter in the spinal cord as seen in cross section. Note the microscopic components of the posterior root ganglion, posterior root of the spinal nerve, anterior root of the spinal nerve, and the spinal nerve. In this and other illustrations of cross sections of the spinal cord, dendrites are not shown in relation to cell bodies of motor or association neurons for purposes of simplicity.

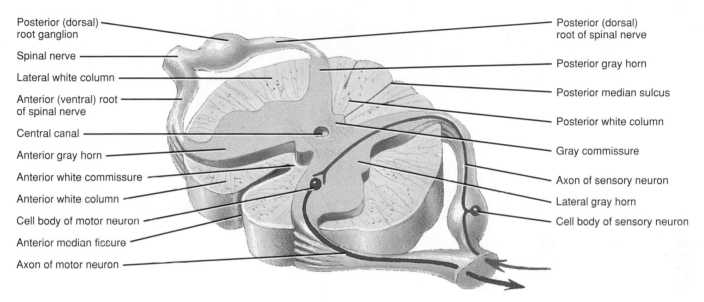

Posterior (dorsal) root ganglion

Spinal nerve

Lateral white column

Anterior (ventral) root of spinal nerve

Central canal

Anterior gray horn

Anterior white commissure

Anterior white column

Cell body of motor neuron

Anterior median fissure

Axon of motor neuron

Posterior (dorsal) root of spinal nerve

Posterior gray horn

Posterior median sulcus

Posterior white column

Gray commissure

Axon of sensory neuron

Lateral gray horn

Cell body of sensory neuron

(a) Sections through the thoracic spinal cord

POSTERIOR

Posterior white column

Lateral white column

Anterior white column

Posterior median sulcus

Posterior gray horn

Central canal

Gray commissure

Anterior gray horn

Anterior median fissure

ANTERIOR

(b) Photograph of a cross section of the thoracic region (20x)

Question: What is the difference between a horn and a column?

FIGURE 13.4 Selected sensory and motor tracts of the spinal cord. Both types of tracts actually are present on both sides of the spinal cord.

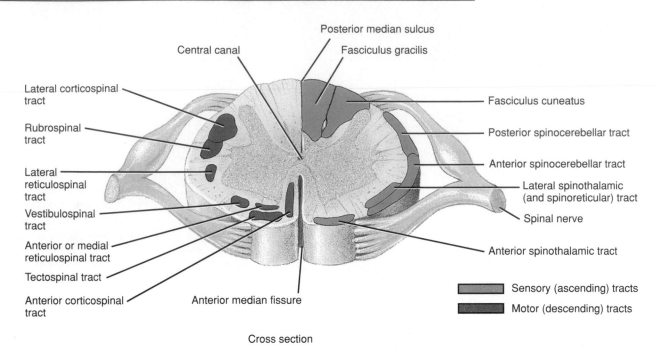

Cross section

Question: Based on its name, what is the origin, destination, and position in the cord of the anterior corticospinal tract? Is this a sensory or a motor tract?

cord: the spinothalamic tracts and the posterior column tract. The **spinothalamic tracts** convey impulses for sensing pain, temperature, crude (poorly localized) touch, and deep pressure. The **posterior column tracts** (the fasciculus gracilis and fasciculus cuneatus) carry nerve impulses for sensing: (1) proprioception, awareness of the movements of muscles, tendons, and joints; (2) discriminative touch, the ability to feel exactly what part of the body is touched; (3) two-point discrimination, the ability to distinguish that two different points on the skin are touched even though they are close together; (4) pressure; and (5) vibrations.

The sensory systems keep the CNS informed of changes in the external and internal environments. Responses to this information are brought about by motor systems, which enable us to move about and change our relationship to the world around us. As sensory information is conveyed to the CNS, it becomes part of a large pool of sensory input. Each piece of incoming information is integrated with all the other information arriving from activated sensory neurons.

With the help of association neurons, the integration process occurs not just once but in many regions of the CNS. It occurs within the spinal cord, the brain stem (which is continuous with the spinal cord), the cerebellum, and the cerebrum of the brain. As a result, motor responses to make a muscle contract or a gland secrete can be initiated at any of these levels. Most regulation of involuntary activities of

smooth muscle, cardiac muscle, and glands by the autonomic nervous system originates in the brain stem and a nearby brain region called the hypothalamus. Details are given in Chapter 17.

The cerebral cortex (outer region of the cerebrum) plays a major role in controlling precise, voluntary muscular movements whereas other brain regions provide important integration for regulation of automatic movements, such as arm swinging during walking. Motor output to skeletal muscles travels down the spinal cord in two types of descending tracts, the pyramidal tracts and the extrapyramidal tracts. **Pyramidal tracts** (the lateral corticospinal, anterior corticospinal, and corticobulbar tracts) convey nerve impulses destined to cause precise, voluntary movements of skeletal muscles. **Extrapyramidal tracts** (rubrospinal, tectospinal, and vestibulospinal tracts) convey nerve impulses that program automatic movements, help coordinate body movements with visual stimuli, maintain skeletal muscle tone and posture, and play a major role in equilibrium by regulating muscle tone in response to movements of the head.

Reflexes

The second principal function of the spinal cord related to homeostasis is to serve as an integrating center for **spinal** reflexes. **Reflexes** are fast, predictable, automatic responses to changes in the environment that help maintain homeostasis. (The integrating center for a few reflexes is the brain stem. Such reflexes involve cranial nerves and are called **cranial reflexes.**) You are most aware of **somatic reflexes,** which involve contraction of skeletal muscles. Equally important, however, are the **visceral (autonomic) reflexes,** which generally are not consciously perceived. They involve responses of smooth muscle, cardiac muscle, and glands. As you will see in Chapter 17, body functions such as heart rate, respiration, digestion, urination, and defecation are controlled by the autonomic nervous system through visceral reflexes.

Spinal nerves are the paths of communication between the spinal cord and most of the body. Figure 13.3a shows the two separate points of attachment called **roots** that connect each spinal nerve to a segment of the cord. The **posterior** or **dorsal (sensory) root** contains sensory nerve fibers and conducts nerve impulses from the periphery into the spinal cord. Each posterior root also has a swelling, the **posterior** or **dorsal (sensory) root ganglion,** which contains the cell bodies of the sensory neurons from the periphery.

The other point of attachment of a spinal nerve to the cord is the **anterior** or **ventral (motor) root.** It contains motor neuron axons and conducts impulses from the spinal cord to the periphery. The cell bodies of motor neurons are located in the gray matter of the cord. If a motor neuron supplies a skeletal muscle, its cell body is located in the anterior (ventral) gray horn. If, however, a motor neuron supplies smooth muscle, cardiac muscle, or a gland through

the autonomic nervous system, its cell body is located in the lateral gray horn.

Reflex Arc and Homeostasis

The route followed by a series of nerve impulses from their origin in the dendrites or cell body of a neuron in one part of the body to their arrival elsewhere in the body is called a **pathway.** Pathways are specific neuronal circuits and thus include at least one synapse. The simplest kind of pathway is known as a **reflex arc.** A reflex arc includes five functional components, as follows (Fig. 13.5).

1. **Receptor.** The distal end of a sensory neuron (dendrite) or an associated sensory structure serves as a receptor. It responds to a specific stimulus, a change in the internal or external environment, by producing a generator (or receptor) potential (see page 445). If a generator potential reaches threshold depolarization, it will trigger one or more nerve impulses.

2. **Sensory neuron.** The nerve impulses are conducted from the receptor to the axon terminals of the sensory neuron, located in the gray matter of the spinal cord or brain stem.

3. **Integrating center.** This is a region within the CNS. The integrating center for the simplest type of reflex is a single synapse, between a sensory neuron and a motor neuron. A reflex pathway having only one synapse is termed a **monosynaptic reflex arc.** More often, however, the integrating center consists of one or more association neurons, which may relay the impulse

---

FIGURE 13.5 General components of a reflex arc. The arrows show the direction of nerve impulse conduction.

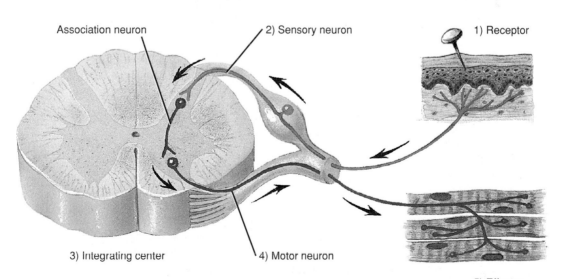

Association neuron   2) Sensory neuron   1) Receptor

3) Integrating center   4) Motor neuron

5) Effector

Question: What initiates an impulse in a sensory neuron? Which division of the nervous system includes all reflex integrating centers?

to other association neurons as well as to a motor neuron. A **polysynaptic reflex arc** is one that involves more than two types of neurons and more than one synapse.

4. **Motor neuron.** Impulses triggered by the integrating center propagate along a motor neuron to the part of the body that will respond.

5. **Effector.** The part of the body that responds to the motor nerve impulse, such as a muscle or gland, is the effector. Its action is called a reflex. If the effector is skeletal muscle, the reflex is a somatic reflex. If the effector is smooth muscle, cardiac muscle, or a gland, the reflex is a visceral reflex.

Reflexes permit the body to make exceedingly rapid adjustments to homeostatic imbalances. Examine Fig. 13.6 to note how reflexes operate in a general way to maintain homeostasis. As our discussion progresses, specific examples will be given.

Let us now examine some important somatic spinal reflexes: the stretch reflex, tendon reflex, flexor (withdrawal) reflex, and crossed extensor reflex.

Physiology of the Stretch Reflex

The **stretch reflex** is a monosynaptic reflex arc. Only two types of neurons (one sensory and one motor) are involved, and there is only one synapse in the pathway (Fig. 13.7). This reflex results in the contraction of a muscle (effector) when it is stretched suddenly. Slight stretching of a muscle stimulates receptors in the muscle called **muscle spindles** (see Fig. 15.3a). The spindles monitor changes in the length of the muscle.

In response to a stretch, a muscle spindle produces one or more nerve impulses that are propagated along a somatic sensory neuron through the posterior root of the spinal nerve into the spinal cord. The sensory neuron makes an excitatory synapse with a motor neuron in the anterior gray horn. If the excitation is strong enough, an impulse arises in the motor neuron and is conducted along its axon, which projects from the spinal cord into the anterior root. The axon terminals of the motor neuron form neuromuscular junctions with typical skeletal muscle fibers of the same muscle that contains the activated muscle spindle. Once the nerve impulse reaches the stretched muscle, a muscle action potential is generated, and the muscle contracts. Thus muscle stretch is followed by contraction, which shortens the muscle that had been stretched.

In addition to the motor neurons that innervate the typical muscle fibers, there are smaller diameter motor neurons that innervate smaller, specialized muscle fibers within the muscle spindles themselves. The brain can regulate muscle spindle sensitivity through pathways to these smaller motor neurons. This ensures proper muscle spindle signaling over a wide range of muscle lengths during voluntary and reflex contractions. Also, by adjusting how vigorously a muscle spindle responds to stretching, the brain can set an overall

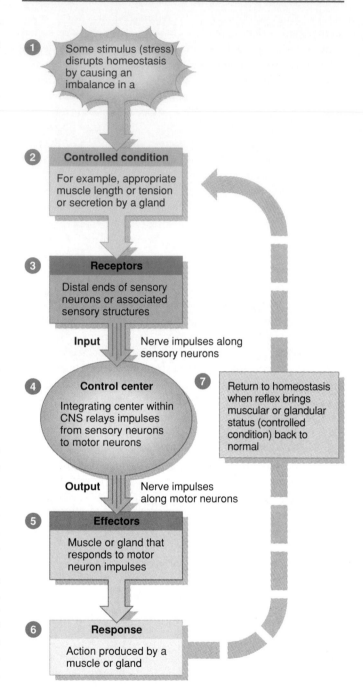

FIGURE 13.6 Relationship of reflexes to homeostasis.

1 Some stimulus (stress) disrupts homeostasis by causing an imbalance in a

2 **Controlled condition**
For example, appropriate muscle length or tension or secretion by a gland

3 **Receptors**
Distal ends of sensory neurons or associated sensory structures

Input Nerve impulses along sensory neurons

4 **Control center**
Integrating center within CNS relays impulses from sensory neurons to motor neurons

7 Return to homeostasis when reflex brings muscular or glandular status (controlled condition) back to normal

Output Nerve impulses along motor neurons

5 **Effectors**
Muscle or gland that responds to motor neuron impulses

6 **Response**
Action produced by a muscle or gland

Question: Is this a positive or negative feedback cycle?

level of **muscle tone**, the small degree of contraction present while the muscle is at rest.

In the reflex arc just described, the sensory nerve impulse enters the spinal cord on the same side that the motor nerve impulse leaves it. This arrangement is called an **ipsilateral** (ip′-si-LAT-er-al) **reflex arc.** All monosynaptic reflex arcs are ipsilateral.

FIGURE 13.7 Stretch reflex. The monosynaptic stretch reflex pathway does not contain an association neuron. The polysynaptic reflex arc to antagonistic muscles also is illustrated. Note that a monosynaptic reflex has only one synapse and two different neurons—sensory neuron and motor neuron. The synapse is between the sensory neuron from the receptor and motor neuron to the effector. A polysynaptic pathway includes at least two synapses and at least one association neuron.

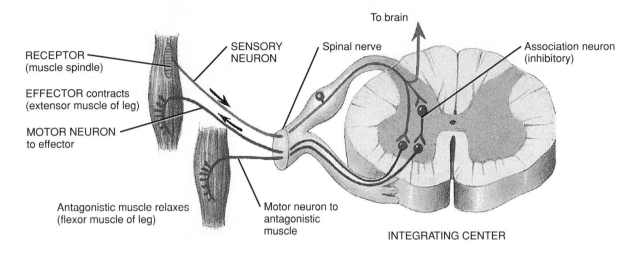

Question: Why is this an ipsilateral reflex?

In addition to maintaining proper muscle tone, the stretch reflex adjusts muscle performance during exercise. Since the stimulus for the reflex is stretching of muscle, it also helps prevent injury by opposing overstretching of muscles. Moreover, it is the basis for several tests used in neurological examinations (see page 388). One such reflex is the **patellar reflex,** in which tapping the patellar tendon causes extension of the knee joint. Tapping the patellar tendon stretches muscle fibers in the quadriceps femoris muscle, thus stimulating receptors (muscle spindles) in that muscle. Sensory nerve impulses from the receptors conduct into the spinal cord. The returning motor nerve impulses generate muscle action potentials that cause contraction of the quadriceps femoris muscle and extension of the leg at the knee.

Although the stretch reflex pathway itself is monosynaptic (just two neurons and one synapse), a polysynaptic reflex arc to the antagonistic muscle involves three neurons and two synapses. An axon collateral (branch) from the muscle spindle sensory neuron also synapses with an inhibitory association neuron in the integrating center. In turn, the association neuron synapses with and inhibits a motor neuron that normally excites the antagonistic muscles (Fig. 13.7). Thus when the stretched muscle contracts during a stretch reflex, antagonistic muscles that oppose the contraction relax.

This type of neural circuit, which provides for simultaneous contraction of one muscle and relaxation of its antagonists, is termed **reciprocal innervation.** It avoids conflict between prime movers and antagonists and is vital in coordinating body movements. You may recall from Chapter 11 that skeletal muscles act in groups rather than alone. Each muscle in the group (agonist, antagonist, synergist, fixator) has a specific role in bringing about a particular movement.

Axon collaterals of the muscle spindle sensory neuron also relay nerve impulses to the brain over specific ascending pathways. In this way, your brain is provided with input about the state of stretch or contraction of skeletal muscles that enables it to coordinate muscular movements and posture. The nerve impulses that pass to the brain also allow you to consciously perceive that the reflex has occurred.

Physiology of the Tendon Reflex

Whereas the stretch reflex operates as a feedback mechanism to control muscle *length* by causing muscle contraction, the **tendon reflex** operates as a feedback mechanism to control muscle *tension* by causing muscle relaxation. It protects tendons and their associated muscles from excessive tension. The tendon reflex, like the stretch reflex, is ipsilateral. The receptors for this reflex are called **tendon (Golgi tendon) organs** (see Fig. 15.3b). Tendon organs lie within a tendon near its junction with a muscle. Whereas muscle spindles are sensitive to changes in muscle length, tendon organs detect and respond to changes in muscle tension caused by passive stretch or muscular contraction.

When an increase in tension is applied to a tendon, the tendon organ is stimulated (depolarized to threshold). Nerve impulses are generated and these propagate into the spinal cord along a sensory neuron (Fig. 13.8). Within the spinal cord, the sensory neuron synapses with an inhibitory association neuron, which then synapses with and inhibits (hyperpolarizes) a motor neuron that innervates the muscle associated with the tendon organ. Thus as tension on the tendon organ increases, the frequency of inhibitory impulses increases, and the inhibition of the motor neurons to the muscle developing excess tension causes relaxation of the muscle. In this way, the tendon reflex protects the tendon and muscle from damage due to excessive tension.

The sensory neuron from the tendon organ also synapses with a stimulatory association neuron in the spinal cord. The stimulatory association neuron, in turn, synapses with motor neurons controlling antagonistic muscles. Thus while the tendon reflex brings about relaxation of the muscle attached to the tendon organ, it also brings about contraction of the antagonists. This is another example of reciprocal innervation. The sensory neuron also relays nerve impulses to the brain by way of sensory tracts, thus informing the brain about the state of muscle tension throughout the body.

Physiology of the Flexor (Withdrawal) Reflex and Crossed Extensor Reflexes

Another example of a reflex based on a polysynaptic reflex arc is the **flexor (withdrawal) reflex** (Fig. 13.9). Suppose you step on a tack. As a result of the painful stimulus, you immediately withdraw your foot. What has happened? A sensory neuron conducts nerve impulses from the stimulated pain receptor to the spinal cord. A second volley of impulses arises in an association neuron, which generates impulses in a motor neuron. A motor neuron stimulates the flexor muscles of your foot, and you withdraw it. Again, the reflex is protective. Contraction of flexor muscles moves a limb to avoid pain. Also, the extensor muscles (antagonists) relax. The relaxation of the extensor muscles occurs as inhibitory association neurons act to inhibit (hyperpolarize) the motor neurons that innervate the extensor muscles.

The flexor reflex, like the stretch reflex, is ipsilateral. The incoming and outgoing impulses are on the same side of the spinal cord. The flexor reflex also illustrates another feature of polysynaptic reflex arcs. When you withdraw your entire lower or upper limb from a painful stimulus, more than one muscle group is involved. Several motor neurons must simultaneously convey impulses to several upper or lower limb muscles. This happens because nerve impulses from one sensory neuron ascend and descend, activating association neurons in different segments of the spinal cord. So this type of reflex is called an **intersegmental reflex arc.** Through intersegmental reflex arcs, a single sensory neuron can activate several motor neurons and thereby cause stimulation of more than one effector. In the monosynaptic stretch reflex, the returning motor impulse affects only muscles receiving nerve impulses from one spinal cord segment.

Something else may happen when you step on a tack. You may start to lose your balance as your body weight shifts to the other foot. Automatically, however, reflexes

---

FIGURE 13.8 Tendon reflex. This reflex arc is polysynaptic since there is more than one synapse and more than three different neurons involved in the pathway. The sensory neuron synapses with two association neurons. The inhibitory association neuron causes relaxation of the effector and the stimulatory association neuron causes contraction of the antagonistic muscle.

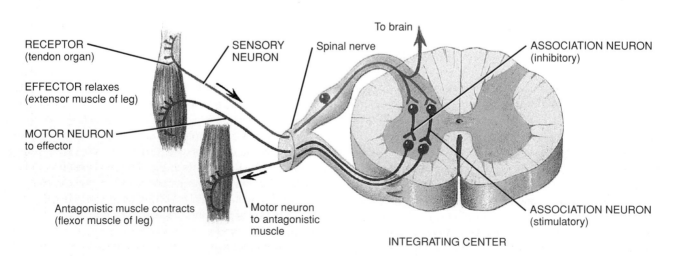

RECEPTOR (tendon organ)

SENSORY NEURON

Spinal nerve

To brain

ASSOCIATION NEURON (inhibitory)

EFFECTOR relaxes (extensor muscle of leg)

MOTOR NEURON to effector

Antagonistic muscle contracts (flexor muscle of leg)

Motor neuron to antagonistic muscle

ASSOCIATION NEURON (stimulatory)

INTEGRATING CENTER

Question: What is reciprocal innervation?

FIGURE 13.9 Flexor (withdrawal) reflex. This reflex arc is polysynaptic and ipsilateral.

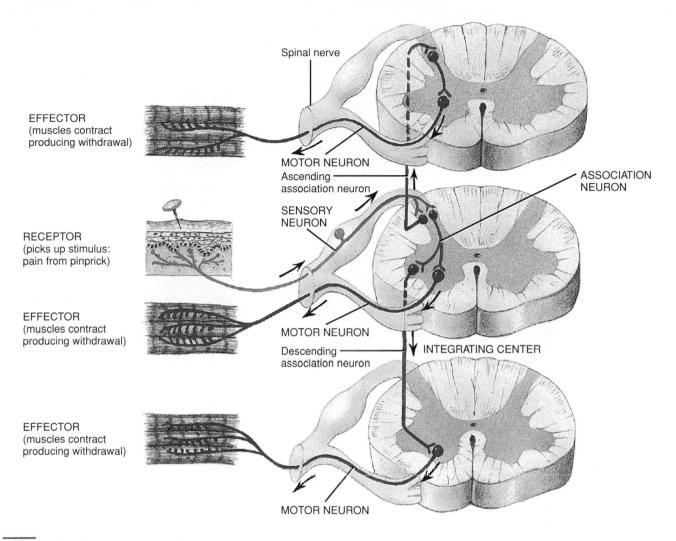

EFFECTOR
(muscles contract
producing withdrawal)

MOTOR NEURON
Ascending
association neuron

SENSORY
NEURON

RECEPTOR
(picks up stimulus:
pain from pinprick)

EFFECTOR
(muscles contract
producing withdrawal)

MOTOR NEURON
Descending
association neuron

EFFECTOR
(muscles contract
producing withdrawal)

MOTOR NEURON

Spinal nerve

ASSOCIATION
NEURON

INTEGRATING CENTER

Question: Why is this an intersegmental reflex arc?

occur to regain balance so you do not fall. The pain impulses from stepping on the tack not only initiate the flexor reflex that causes you to withdraw the extremity but also initiate a balance-maintaining **crossed extensor reflex,** as follows (Fig. 13.10). Incoming pain impulses cross to the opposite side of the spinal cord through association neurons at that level and several levels above and below the point of entry into the spinal cord. From these levels, your unstimulated limb receives motor impulses that cause extension at the knee, hip, and ankle. This allows you to place weight on the foot of the limb that must now support the entire body. Unlike the flexor reflex, which is an ipsilateral reflex, the crossed extensor reflex involves a **contralateral** (kon′-tra-LAT-er-al) **reflex arc**: the sensory impulses enter one side of the spinal cord and motor impulses exit on the opposite side. Thus a crossed extensor reflex causes synchronized

extension of the joints in one limb and flexion of the joints in the opposite limb.

Reciprocal innervation also occurs in both the flexor reflex and the crossed extensor reflex. In the flexor reflex, when the flexor muscles of your painfully stimulated lower extremity are contracting, the extensor muscles of the same extremity are being relaxed to some degree. If both sets of muscles contracted at the same time, you would not be able to flex your extremity because the two sets of muscles would pull in opposite directions on the bones. But because of reciprocal innervation, one set of muscles contracts while the other relaxes. In the crossed extensor reflex, while you are contracting the flexor muscles of the extremity that has been stimulated by the tack, the stimulated extensor muscles of your other lower extremity are producing extension to help maintain balance.

FIGURE 13.10 Crossed extensor reflex. The flexor reflex is shown on the left of the diagram so that you can correlate it with the crossed extensor reflex on the right.

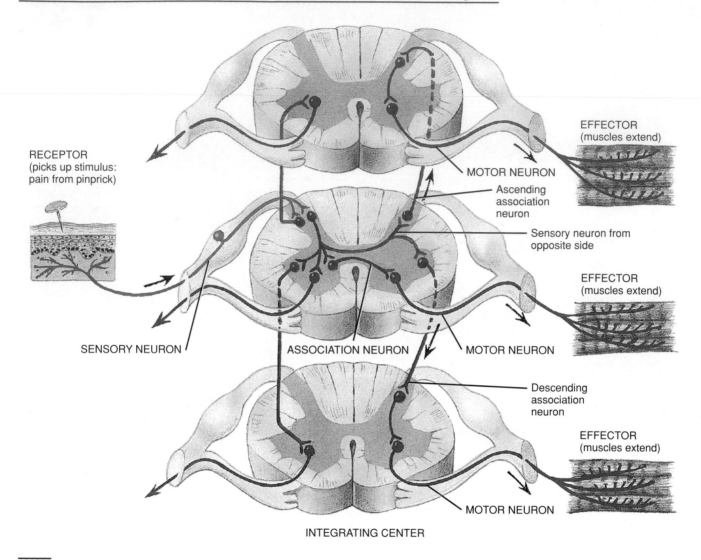

RECEPTOR
(picks up stimulus:
pain from pinprick)

EFFECTOR
(muscles extend)

MOTOR NEURON

Ascending
association
neuron

Sensory neuron from
opposite side

EFFECTOR
(muscles extend)

SENSORY NEURON   ASSOCIATION NEURON   MOTOR NEURON

Descending
association
neuron

EFFECTOR
(muscles extend)

MOTOR NEURON

INTEGRATING CENTER

Question: Why is the crossed extensor reflex classified as a contralateral reflex arc?

Reflexes and Diagnosis

Reflexes are often used for diagnosing disorders of the nervous system and locating injured tissue. If a reflex is absent, or abnormal, the damage may be somewhere along a particular conduction pathway. Visceral reflexes, however, are usually not practical tools for diagnosis. It is difficult to stimulate visceral receptors, since they are deep in the body. In contrast, many somatic reflexes can be tested simply by tapping or stroking the body surface.

CLINICAL APPLICATION

REFLEXES AND NEUROLOGICAL IMPAIRMENT

Several reflexes have clinical significance, and are used to assess certain conditions:

1. **Patellar reflex (knee jerk).** This stretch reflex involves extension of the knee joint by contraction of the quadriceps femoris muscle in response to tapping the patellar tendon (see Fig. 13.7). The reflex is blocked by damaged sensory or motor nerves to the muscle or to integrating centers in the second, third, or fourth lumbar segments of the spinal cord. Often, it is absent in people with chronic diabetes mellitus or neurosyphilis, which cause degeneration of nerves. It is exaggerated in disease or injury involving certain motor tracts descending from the higher centers of the brain to the spinal cord. This reflex may also be exaggerated by applying a second stimulus (for example, a sudden loud noise) while tapping the patellar tendon.
2. **Achilles reflex (ankle jerk).** This stretch reflex involves extension (plantar flexion) of the foot by contraction of the gastrocnemius and soleus muscles in response to tapping the calcaneal (Achilles) tendon. Absence of the Achilles reflex indicates damage to the

nerves supplying the posterior leg muscles or to the nerve cells in the lumbosacral region of the spinal cord. This reflex is also absent in people with chronic diabetes, neurosyphilis, alcoholism, and subarachnoid hemorrhages. An exaggerated Achilles reflex indicates cervical cord compression or a lesion of the motor tracts of the first or second sacral segments of the cord.

3. Babinski sign. This reflex results from gentle stroking of the outer margin of the sole of the foot. The great toe extends, with or without fanning of the other toes. This phenomenon occurs in normal children under 1½ years of age and is due to incomplete myelination of fibers in the corticospinal tract. A positive Babinski sign after age 1½ is abnormal and indicates an interruption of the corticospinal tract as the result of a lesion of the tract, usually in the upper portion. The normal response after 1½ years of age is the **plantar flexion reflex,** or negative Babinski—a curling under of all the toes, accompanied by a slight turning in and flexion of the anterior part of the foot.

4. Abdominal reflex. This reflex involves contraction of the muscles that compress the abdominal wall in response to stroking the side of the abdomen. The response is an abdominal muscle contraction that causes a lateral deviation of the umbilicus to the side opposite the stimulus. Absence of this reflex is associated with lesions of the corticospinal tracts. It may also be absent because of lesions of the peripheral nerves, lesions of integrating centers in the thoracic part of the cord, and multiple sclerosis.

SPINAL NERVES

Spinal nerves connect the CNS to receptors, muscles, and glands and are part of the peripheral nervous system (PNS). The 31 pairs of spinal nerves are named and numbered according to the region and level of the spinal cord from which they emerge (see Fig. 13.2a). The first cervical pair emerges between the atlas (first cervical vertebra) and the occipital bone. All other spinal nerves emerge from the vertebral column through the intervertebral foramina between adjoining vertebrae. There are 8 pairs of cervical nerves, 12 pairs of thoracic, 5 pairs of lumbar, 5 pairs of sacral, and 1 pair of coccygeal nerves.

Not all spinal cord segments are in line with their corresponding vertebrae. Recall that the spinal cord ends near the level of the upper border of the second lumbar vertebra. Thus the roots of the lower lumbar, sacral, and coccygeal nerves descend at an angle to reach their respective foramina before emerging from the vertebral column. This arrangement constitutes the cauda equina (horse's tail). See Fig. 13.2a.

Composition and Coverings

A typical **spinal nerve** has two separate points of attachment to the cord: a posterior root and an anterior root (see Fig. 13.3a). The posterior and anterior roots unite to form a spinal nerve at the intervertebral foramen. Since the posterior root contains sensory fibers and the anterior root contains motor fibers, a spinal nerve is a **mixed nerve,** at least

at its origin. The posterior (dorsal) root contains a ganglion in which cell bodies of sensory neurons are located.

In Fig. 13.11, you can see the coverings that surround spinal nerves (and cranial nerves). The individual fibers, whether myelinated or unmyelinated, are wrapped in a connective tissue called the **endoneurium** (en'-dō-NYOO-rē-um). Groups of fibers with their endoneurium are arranged in bundles called **fascicles,** and each bundle is wrapped in connective tissue called the **perineurium** (per'-i-NYOO-rē-um). The outermost covering around the entire nerve is the **epineurium** (ep'-i-NYOO-rē-um). The dura mater of the spinal meninges fuses with the epineurium as the nerve exits through the intervertebral foramen. Note the many blood vessels, which nourish nerves, among the connective tissue coverings (Fig. 13.11b). You may recall from Chapter 10 that the connective tissue coverings of skeletal muscles—endomysium, perimysium, and epimysium—are similar in organization to those of nerves. This organization is common to excitable tissues.

Distribution of Spinal Nerves

Branches

A short distance after passing through its intervertebral foramen, a spinal nerve divides into several branches (Fig. 13.12). These branches are known as **rami** (RĀ-mī; singular, **ramus**). The **dorsal ramus** (RĀ-mus) serves the deep muscles and skin of the dorsal surface of the trunk. The **ventral ramus** serves the muscles and structures of the upper and lower extremities and the lateral and ventral trunk. In addition to dorsal and ventral rami, spinal nerves also give off a **meningeal branch.** This branch reenters the spinal canal through the intervertebral foramen and supplies the vertebrae, vertebral ligaments, blood vessels of the spinal cord, and meninges. Other branches of a spinal nerve are the **rami communicantes** (kō-myoo-ni-KAN-tēz), components of the autonomic nervous system whose structure and function are discussed in Chapter 17.

Plexuses

The ventral rami of spinal nerves, except for thoracic nerves T2–T12, do not go directly to the body structures they supply. Instead, they form networks on both left and right sides of the body by joining with varying numbers of fibers from ventral rami of adjacent nerves. Such a network is called a **plexus** (*plexus* = braid). The principal plexuses are the cervical plexus, brachial plexus, lumbar plexus, and sacral plexus. Look back at Fig. 13.2a to see their relationships to one another. Emerging from the plexuses are nerves bearing names that are often descriptive of the general regions they serve or the course they take. Each of the nerves, in turn, may have several branches named for the specific structures they innervate.

The principal plexuses are summarized in Exhibits 13.1 through 13.4.

FIGURE 13.11 Coverings of a spinal nerve.

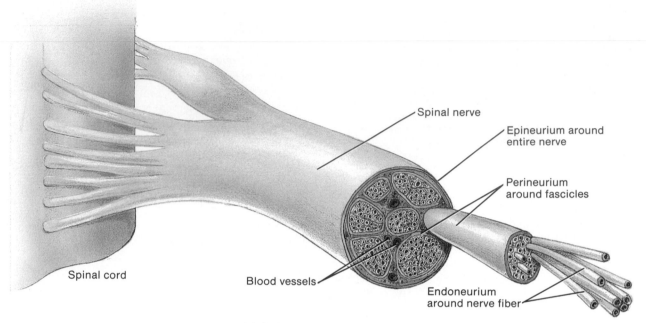

Spinal nerve

Epineurium around entire nerve

Perineurium around fascicles

Spinal cord

Blood vessels

Endoneurium around nerve fiber

(a) Cross section

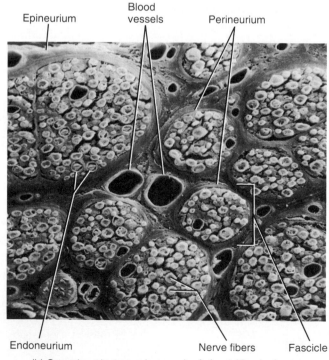

Epineurium

Blood vessels

Perineurium

Endoneurium

Nerve fibers

Fascicle

(b) Scanning electron micrograph of about 10 spinal nerves in cross section, 900 ×

Question: Why are all spinal nerves classified as mixed nerves?

FIGURE 13.12 Branches of a typical spinal nerve.

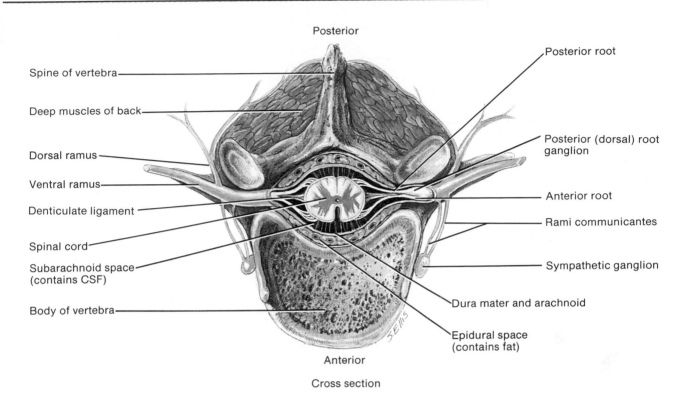

Posterior

Spine of vertebra

Deep muscles of back

Dorsal ramus

Ventral ramus

Denticulate ligament

Spinal cord

Subarachnoid space
(contains CSF)

Body of vertebra

Posterior root

Posterior (dorsal) root
ganglion

Anterior root

Rami communicantes

Sympathetic ganglion

Dura mater and arachnoid

Epidural space
(contains fat)

Anterior

Cross section

Question: Which spinal branch serves the upper and lower extremities?

Intercostal (Thoracic) Nerves

The ventral rami of spinal nerves T2–T12 do not enter into the formation of plexuses and are known as **intercostal (thoracic) nerves.** These nerves directly innervate the structures they supply in the intercostal spaces (see Fig. 13.2a). After leaving its intervertebral foramen, the ventral ramus of nerve T2 supplies the intercostal muscles of the second intercostal space and the skin of the axilla and posteromedial aspect of the arm. Nerves T3–T6 pass in the costal grooves of the ribs and then to the intercostal muscles and skin of the anterior and lateral chest wall. Nerves T7–T12 supply the intercostal muscles and the abdominal muscles and overlying skin. The dorsal rami of the intercostal nerves supply the deep back muscles and skin of the dorsal aspect of the thorax.

EXHIBIT 13.1

CERVICAL PLEXUS

Overview: The **cervical** (SER-vi-kul) **plexus** is formed by the ventral rami of the first four cervical nerves (C1–C4) with contributions from C5. There is one on each side of the neck alongside the first four cervical vertebrae. The roots of the plexus indicated in the diagram are the ventral rami.

   The cervical plexus supplies the skin and muscles of the head, neck, and upper part of the shoulders. Branches of the cervical plexus also connect with cranial nerves XI (accessory) and XII (hypoglossal). The phrenic nerves arise from the cervical plexuses and supply motor fibers to the diaphragm.

Nerve	Origin	Distribution
SUPERFICIAL OR SENSORY BRANCHES		
Lesser occipital (ok′-SIP-i-tal)	C2.	Skin of scalp behind and above ear.

Exhibit continues

EXHIBIT 13.1 **(continued)**

CERVICAL PLEXUS

Nerve	Origin	Distribution
Great auricular (aw-RIK-yoo-lar)	C2–C3.	Skin in front, below, and over ear and over parotid glands.
Transverse cervical (SER-vi-kul)	C2–C3.	Skin over anterior aspect of neck.
Supraclavicular (soo´-pra-kla-VIK-yoo-lar)	C3–C4.	Skin over upper portion of chest and shoulder.

DEEP OR LARGELY MOTOR BRANCHES

Ansa cervicalis (AN-sa ser-vi-KAL-is)		This nerve is divided into a superior root and an inferior root.
Superior root	C1.	Infrahyoid and geniohyoid muscles of neck.
Inferior root	C2–C3.	Infrahyoid muscles of neck.
Phrenic (FREN-ik)	C3–C5.	Diaphragm between thorax and abdomen.
Segmental (seg-MEN-tal) **branches**	C1–C5.	Prevertebral (deep) muscles of neck, levator scapulae, and middle scalene muscles.

Damage of Cervical Spinal Cord: Damage to the spinal cord above the origin of the phrenic nerves (C3, C4, and C5) causes respiratory arrest. Breathing stops since the phrenic nerves no longer send nerve impulses to the diaphragm.

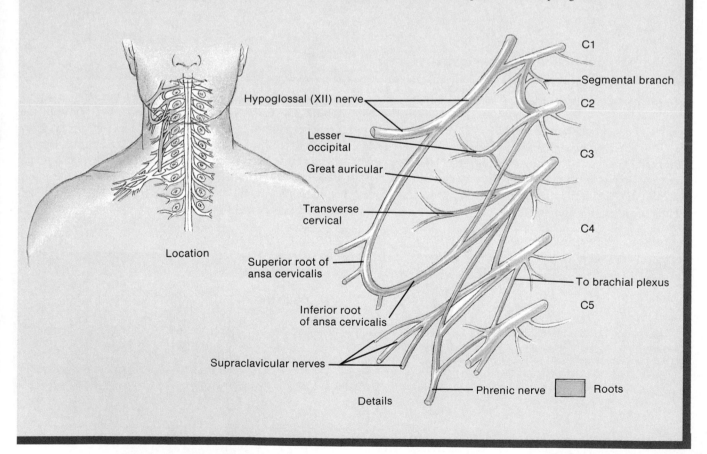

EXHIBIT 13.2

BRACHIAL PLEXUS

Overview: The ventral rami of spinal nerves C5–C8 and T1 form the **brachial** (BRĀ-kē-al) **plexus**. The brachial plexus extends downward and laterally on either side of the last four cervical and first thoracic vertebrae. It passes over the first rib behind the clavicle and then enters the axilla.

The brachial plexus provides the entire nerve supply of the shoulder and upper limb. Five important nerves arise from the brachial plexus. (1) The AXILLARY NERVE supplies the deltoid and teres minor muscles. (2) The MUSCULOCUTANEOUS NERVE supplies the flexors of the arm and forearm. (3) The RADIAL NERVE supplies the muscles on the posterior aspect of the arm and forearm. (4) The MEDIAN NERVE supplies most of the muscles of the anterior forearm and some of the muscles of the palm. (5) The ULNAR NERVE supplies the anteromedial muscles of the forearm and most of the muscles of the palm.

Nerve	Origin	Distribution
Dorsal scapular (SKAP-yoo-lar)	C5.	Levator scapulae, rhomboideus major, and rhomboideus minor muscles.
Long thoracic (thō-RAS-ik)	C5–C7.	Serratus anterior muscle.
Nerve to subclavius (sub-KLĀ-vē-us)	C5–C6.	Subclavius muscle.
Suprascapular (soo'-pra-SKAP-yoo-lar)	C5–C6.	Supraspinatus and infraspinatus muscles.
Musculocutaneous (mus'-kyoo-lō-kyoo-TĀN-ē-us)	C5–C7.	Coracobrachialis, biceps brachii, and brachialis muscles.
Median (lateral head)	C5–C7.	See distribution for **Median (medial head)** in this exhibit.
Lateral pectoral (PEK-tō-ral)	C5–C7.	Pectoralis major muscle.
Upper subscapular (sub-SKAP-yoo-lar)	C5–C6.	Subscapularis muscle.
Thoracodorsal (thō-RA-kō-dor-sal)	C6–C8.	Latissimus dorsi muscle.
Lower subscapular (sub-SKAP-yoo-lar)	C5–C6.	Subscapularis and teres major muscles.
Axillary (AK-si-lar-ē) **or circumflex** (SER-kum-fleks)	C5–C6.	Deltoid and teres minor muscles; skin over deltoid and upper posterior aspect of arm.
Radial (RĀ-dē-al)	C5–C8 and T1.	Extensor muscles of arm and forearm; skin of posterior arm and forearm, lateral two-thirds of dorsum of hand, and fingers over proximal and middle phalanges.
Medial pectoral (PEK-tō-ral)	C8–T1.	Pectoralis major and pectoralis minor muscles.
Medial brachial (BRĀ-kē-al) **cutaneous** (kyoo'-TĀ-nē-us)	C8–T1.	Skin of medial and posterior aspects of lower third of arm.
Medial antebrachial cutaneous (an'-tē-BRĀ-kē-al kyoo'-TĀ-nē-us)	C8–T1.	Skin of medial and posterior aspects of forearm.
Median (medial head)	C5–C8 and T1.	Medial and lateral heads of median nerve form median nerve. Distributed to flexors of forearm, except flexor carpi ulnaris; skin of lateral two-thirds of palm of hand and fingers.
Ulnar (UL-nar)	C8–T1.	Flexor carpi ulnaris and flexor digitorum profundus muscles; skin of medial side of hand, little finger, and medial half of ring finger.

Exhibit continues

EXHIBIT 13.2 (continued)

BRACHIAL PLEXUS

Injuries to the Brachial Plexus: Prolonged use of a crutch that presses into the axilla may result in injury to a portion of the brachial plexus. The usual **crutch palsy** involves the posterior cord of the brachial plexus or, more often, just the radial nerve, which, in general, supplies extensors.

   Radial nerve damage is indicated by wrist drop, inability to extend the hand at the wrist. Care must be taken not to injure the radial and axillary nerves when intramuscular injections are given into the deltoid. The radial nerve may also be injured when a cast is applied too tightly around the midhumerus. **Median nerve damage** is indicated by numbness, tingling, and pain in the palm and fingers; weak thumb movements; and inability to pronate the forearm and difficulty in flexing the wrist properly. **Carpal tunnel syndrome** results from compression of the median nerve inside the carpal tunnel. This narrowed passageway is formed anteriorly by the flexor retinaculum (transverse carpal ligament) and posteriorly by the carpal bones (see Fig. 11.18). It may be caused by any condition that aggravates compression of the contents of the carpal tunnel, such as trauma, edema, and repetitive flexion of the wrist as a result of activities such as playing video games, keyboarding at a computer terminal or typing, driving a car, cutting hair, and playing a piano. **Ulnar nerve damage** is indicated by an inability to adduct or abduct the four fingers (not the thumb), weakness in flexing and adducting the wrist, and loss of sensation over the little finger.

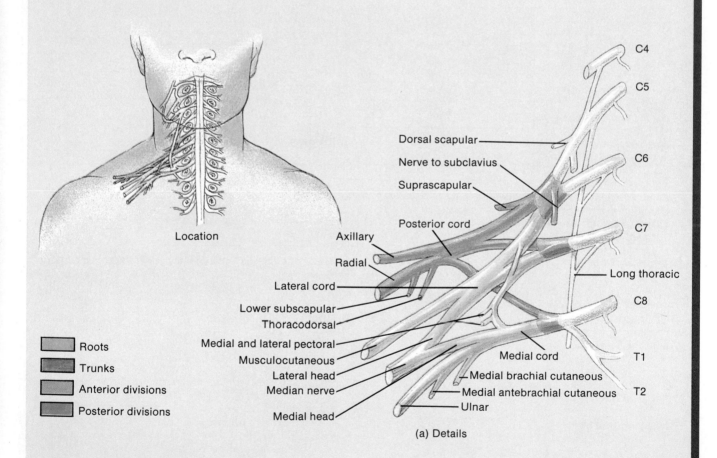

Location

C4
C5
C6
C7
C8
T1
T2

Dorsal scapular
Nerve to subclavius
Suprascapular
Posterior cord
Axillary
Radial
Lateral cord
Lower subscapular
Thoracodorsal
Medial and lateral pectoral
Musculocutaneous
Lateral head
Median nerve
Medial head
Long thoracic
Medial cord
Medial brachial cutaneous
Medial antebrachial cutaneous
Ulnar

Roots
Trunks
Anterior divisions
Posterior divisions

(a) Details

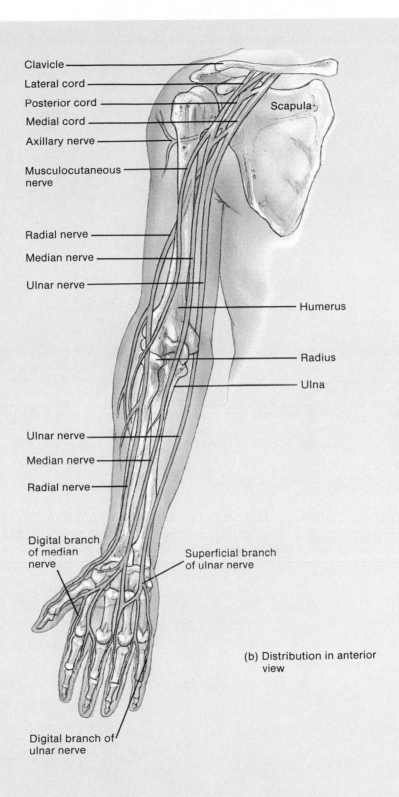

Clavicle

Lateral cord

Posterior cord

Scapula

Medial cord

Axillary nerve

Musculocutaneous nerve

Radial nerve

Median nerve

Ulnar nerve

Humerus

Radius

Ulna

Ulnar nerve

Median nerve

Radial nerve

Digital branch of median nerve

Superficial branch of ulnar nerve

(b) Distribution in anterior view

Digital branch of ulnar nerve

EXHIBIT 13.3

LUMBAR PLEXUS

Overview: Ventral rami of spinal nerves L1–L4 form the **lumbar** (LUM-bar) **plexus**. It differs from the brachial plexus in that there is no intricate intermingling of fibers. On either side of the first four lumbar vertebrae, the lumbar plexus passes obliquely outward behind the psoas major muscle and anterior to the quadratus lumborum muscle. It then gives rise to its peripheral nerves.

The lumbar plexus supplies the anterolateral abdominal wall, external genitals, and part of the lower extremity.

Nerve	Origin	Distribution
Iliohypogastric (il′-ē-ō-hī-pō-GAS-trik)	L1.	Muscles of anterolateral abdominal wall; skin of lower abdomen and buttock.
Ilioinguinal (il′-ē-ō-IN-gwi-nal)	L1.	Muscles of anterolateral abdominal wall; skin of upper medial aspect of thigh, root of penis and scrotum in male, and labia majora and mons pubis in female.
Genitofemoral (jen′-i-tō-FEM-or-al)	L1–L2.	Cremaster muscle; skin over middle anterior surface of thigh, scrotum in male, and labia majora in female.
Lateral femoral cutaneous (FEM-or-al kyoo′-TĀ-nē-us)	L2–L3.	Skin over lateral, anterior, and posterior aspects of thigh.
Femoral (FEM-or-al)	L2–L4.	Flexor muscles of thigh; skin on front and over medial aspect of thigh and medial side of leg and foot.
Obturator (OB-too-rā-tor)	L2–L4.	Adductor muscles of leg; skin over medial aspect of thigh.

Femoral Nerve Injury: The largest nerve arising from the lumbar plexus is the femoral nerve. Injury to the femoral nerve is indicated by an inability to extend the leg and by loss of sensation in the skin over the anteromedial aspect of the thigh.

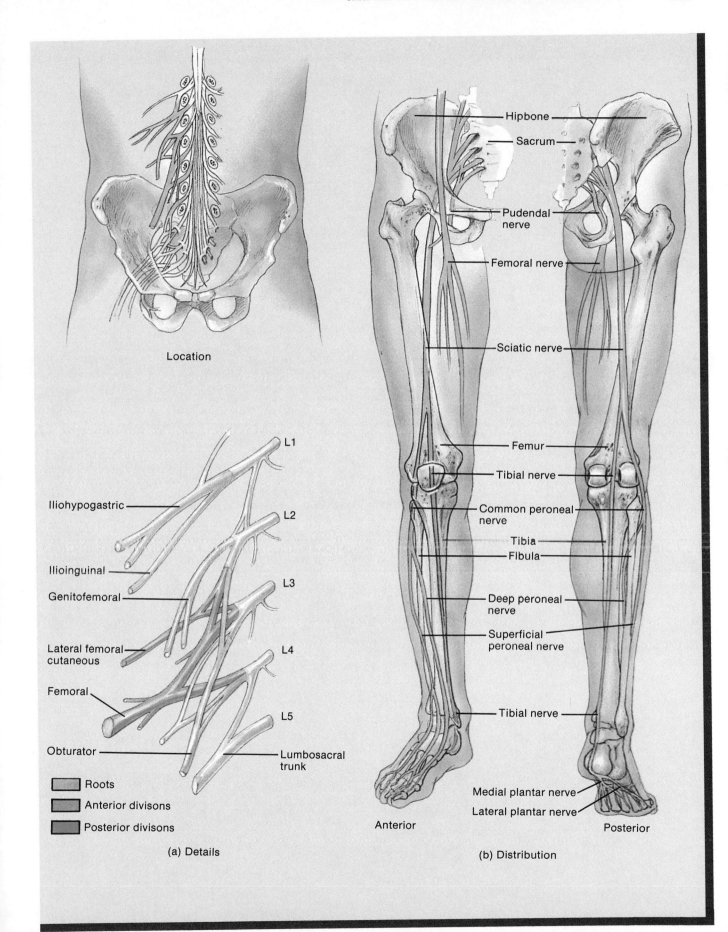

Location

Iliohypogastric

Ilioinguinal

Genitofemoral

Lateral femoral cutaneous

Femoral

Obturator

L1

L2

L3

L4

L5

Lumbosacral trunk

Roots

Anterior divisons

Posterior divisons

(a) Details

Hipbone

Sacrum

Pudendal nerve

Femoral nerve

Sciatic nerve

Femur

Tibial nerve

Common peroneal nerve

Tibia

Fibula

Deep peroneal nerve

Superficial peroneal nerve

Tibial nerve

Medial plantar nerve

Lateral plantar nerve

Anterior

Posterior

(b) Distribution

EXHIBIT 13.4

SACRAL PLEXUS

Overview: The ventral rami of spinal nerves L4–L5 and S1–S4 form the **sacral** (SĀ-kral) **plexus**. It is situated largely in front of the sacrum. The sacral plexus supplies the buttocks, perineum, and lower extremities. The largest nerve in the body—the sciatic nerve—arises from the sacral plexus.

Nerve	Origin	Distribution
Superior gluteal (GLOO-tē-al)	L4–L5 and S1.	Gluteus minimus and gluteus medius muscles and tensor fasciae latae.
Inferior gluteal (GLOO-tē-al)	L5–S2.	Gluteus maximus muscle.
Nerve to piriformis (pir-i-FORM-is)	S1–S2.	Piriformis muscle.
Nerve to quadratus femoris (quod-RĀ-tus FEM-or-is)	L4–L5 and S1.	Quadratus femoris and inferior gemellus muscles.
Nerve to obturator internus (OB-too-rā′-tor in-TER-nus)	L5–S2.	Obturator internus and superior gemellus muscles.
Perforating cutaneous (PER-fō-rā-ting kyoo′-TĀ-nē-us)	S2–S3.	Skin over lower medial aspect of buttock.
Posterior femoral cutaneous (FEM-or-al-kyoo′-TĀ-nē-us)	S1–S3.	Skin over anal region, lower lateral aspect of buttock, upper posterior aspect of thigh, upper part of calf, scrotum in male, and labia majora in female.
Sciatic (sī-AT-ik)	L4–S3.	Actually two nerves: tibial and common peroneal, bound together by common sheath of connective tissue. It splits into its two divisions, usually at knee. (See below for distributions.) As sciatic nerve descends through thigh, it sends branches to hamstring muscles and adductor magnus.
Tibial (TIB-ē-al)	L4–S3.	Gastrocnemius, plantaris, soleus, popliteus, tibialis posterior, flexor digitorum longus, and flexor hallucis longus muscles. Branches of tibial nerve in foot are medial plantar nerve and lateral plantar nerve.
Medial plantar (PLAN-ter)		Abductor hallucis, flexor digitorum brevis, and flexor hallucis brevis muscles; skin over medial two-thirds of plantar surface of foot.
Lateral plantar (PLAN-ter)		Remaining muscles of foot not supplied by medial plantar nerve; skin over lateral third of plantar surface of foot.
Common peroneal (per′-ō-NĒ-al)	L4–S2.	Divides into a superficial peroneal and a deep peroneal branch.
Superficial peroneal (per′-ō-NĒ-al)		Peroneus longus and peroneus brevis muscles; skin over distal third of anterior aspect of leg and dorsum of foot.
Deep peroneal (per′-ō-NĒ-al)		Tibialis anterior, extensor hallucis longus, peroneus tertius, and extensor digitorum longus and brevis muscles; skin on adjacent sides of great and second toes.
Pudendal (pyoo-DEN-dal)	S2–S4.	Muscles of perineum; skin of penis and scrotum in male and clitoris, labia majora, labia minora, and lower vagina in female.

Sciatic Nerve Injury: Injury to the sciatic nerve (common peroneal portion) and its branches results in pain that may extend from the buttock down the back of the leg. Other symptoms may include foot drop, an inability to dorsiflex the foot, and loss of sensation over the leg and foot. This nerve may be injured because of a herniated (slipped) disc, dislocated hip, osteoarthritis of the lumbosacral spine, pressure from the uterus during pregnancy, or an improperly administered gluteal intramuscular injection.

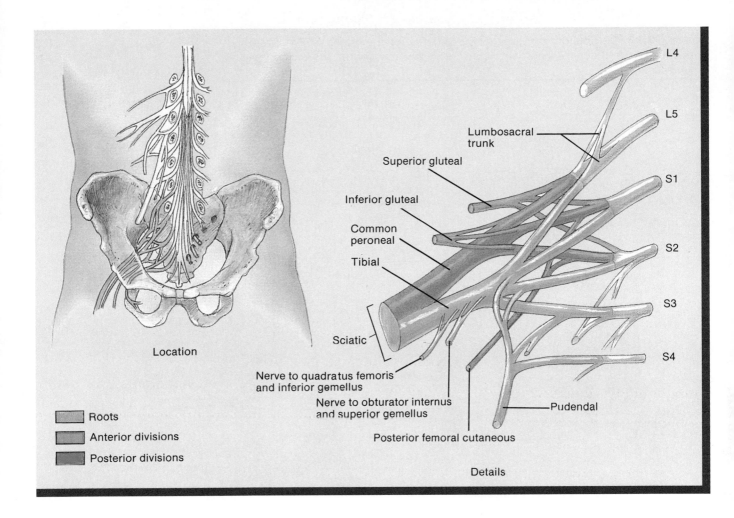

Location

Roots
Anterior divisions
Posterior divisions

L4
L5
S1
S2
S3
S4

Lumbosacral trunk
Superior gluteal
Inferior gluteal
Common peroneal
Tibial
Sciatic
Nerve to quadratus femoris and inferior gemellus
Nerve to obturator internus and superior gemellus
Posterior femoral cutaneous
Pudendal

Details

FIGURE 13.13 Distribution of spinal nerves to dermatomes. An illustration can only approximate and cannot clearly indicate the degree of overlap of the cutaneous nerve innervation.

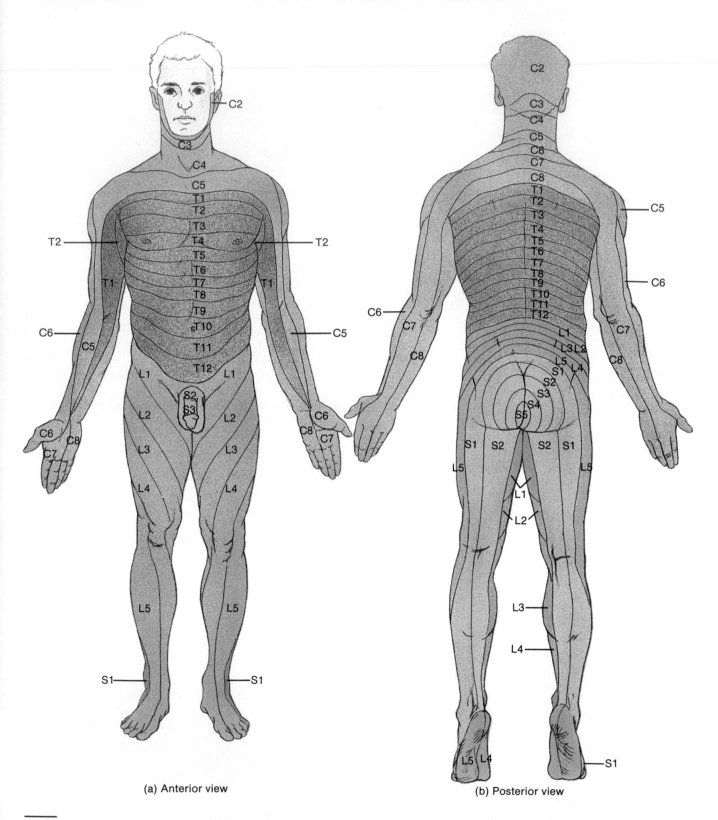

(a) Anterior view

(b) Posterior view

Question: Which is the only spinal nerve that does not supply branches to the skin?

Dermatomes

The skin over the entire body is supplied by spinal nerves that carry somatic sensory nerve impulses into the spinal cord. Each spinal nerve serves a specific, constant segment of the skin. All spinal nerves except C1 supply branches to the skin. The area of the skin that provides sensory input to the dorsal roots of one pair of spinal nerves or to one spinal cord segment is a **dermatome** (Fig. 13.13). The nerve supply in adjacent dermatomes overlaps somewhat; in some cases the overlap is considerable. Consequently, there may be little loss of sensation if only the single nerve supply to a dermatome is damaged. Most of the skin of the face and scalp is served by a single cranial nerve, the trigeminal (V) nerve.

Since physicians know which spinal cord segments supply each dermatome, it is possible to determine which segment of the spinal cord or spinal nerve is malfunctioning. If a dermatome is stimulated, and the sensation is not perceived, it can be assumed that the nerves supplying the dermatome are involved. Also, to produce a region of complete anesthesia, at least three adjacent spinal nerves must be cut or blocked by an anesthetic drug.

DISORDERS: HOMEOSTATIC IMBALANCES

SPINAL CORD INJURY

The spinal cord may be damaged by compression from a variety of causes. These include a tumor within or adjacent to the spinal cord, herniated intervertebral discs, blood clots, penetrating wounds caused by projectile fragments, or other traumatic events such as automobile accidents. Depending on the location and extent of the injury, paralysis may occur. **Paralysis** is the total loss of voluntary motor function that results from damage to nervous or muscle tissue. Paralysis may be classified as follows: **monoplegia** (*mono* = one; *plege* = stroke), paralysis of one extremity only; **diplegia** (*di* = two), paralysis of both upper extremities or both lower extremities; **paraplegia** (*para* = beyond), paralysis of both lower extremities; **hemiplegia** (*hemi* = half), paralysis of the upper extremity, trunk, and lower extremity on one side of the body; and **quadriplegia** (*quad* = four), paralysis of the two upper and two lower extremities.

Complete transection of the spinal cord means that the cord is severed from one side to the other, thus cutting all ascending and descending tracts. It results in a loss of all sensations and voluntary movement *below* the level of the transection. **Hemisection** is a partial transection of the cord on either the right or left side. Below the level of the hemisection there is a loss of proprioception and tactile sensations on the same side as the injury if the posterior column is cut, paralysis on the same side if the lateral corticospinal tract is cut, and loss of pain and temperature sensations on the opposite side if the spinothalamic tract is cut.

Following transection, there is an initial period of **spinal shock** that lasts from a few days to several weeks. During this period, all reflex activity disappears, a condition called **areflexia** (a'-rē-FLEK-sē-a). In time, however, there is a return of reflex activity. The first reflex to return is a stretch reflex (for example, the knee jerk). Its reappearance may take several days. Next, the flexion reflexes return, over a period of up to several months. Then the crossed extensor reflexes return. In some cases, males may not be able to attain an erection or ejaculation. Moreover, urinary bladder and bowel functions are no longer under voluntary control.

Recent studies have shown that patients with spinal cord injury have an improved outcome if they are given an anti-inflammatory drug called methylprednisolone, a corticosteroid, within 8 hours of the injury.

NEURITIS

Inflammation of one or several nerves is termed **neuritis.** It may result from irritation to the nerve produced by direct blows, bone fractures, contusions, or penetrating injuries. Additional causes include vitamin deficiency (usually thiamine) or poisons such as carbon monoxide, carbon tetrachloride, heavy metals, and some drugs.

SCIATICA

Sciatica (sī-AT-i-ka) is a type of neuritis characterized by severe pain along the path of the sciatic nerve or its branches. Inflammation or injury of the nerve causes pain that passes from the back or thigh down its length into the leg, foot, and toes. Probably the most common cause of sciatica is a herniated (slipped) intervertebral disc.

SHINGLES

Shingles is an acute infection of the peripheral nervous system. It is caused by herpes zoster (HER-pēz ZOS-ter). This is the virus that causes chickenpox and after a person recovers, the virus retreats to a posterior (dorsal) root ganglion. If the virus is reactivated, the immune system usually prevents it from spreading. However, from time to time, the rekindled virus overcomes a weakened immune system, leaves the ganglion, and travels down sensory neurons by fast axonal transport (see page 353). The result is pain, discoloration of the skin, and a characteristic line of skin blisters. The line of blisters marks the distribution (dermatome) of a particular cutaneous sensory nerve.

POLIOMYELITIS

Poliomyelitis (infantile paralysis), or simply **polio,** is most common during childhood and is caused by a virus called poliovirus. The onset of the disease is marked by fever, severe headache, a stiff neck and back, deep muscle pain and weakness, and loss of certain somatic reflexes. The virus can be ingested in drinking water contaminated with feces containing the virus. In its most serious form, the virus produces paralysis by destroying motor neuron cell bodies, specifically those in the anterior horns of the spinal cord and in the nuclei of the cranial nerves. Injury to the spinal cord gray matter is the basis for the name of this disease (*polio* = gray matter; *myel* = spinal cord). Polio can cause death from respiratory or heart failure if the virus invades the brain cells of the vital medullary centers. The incidence of polio in the U.S. has decreased markedly since the availability of polio vaccines.

Study Outline

Spinal Cord Anatomy (p. 376)

Protection and Coverings (p. 376)

1. The spinal cord is protected by the vertebral column, meninges, cerebrospinal fluid, and vertebral ligaments.
2. The meninges are three coverings that run continuously around the spinal cord and brain: dura mater, arachnoid, and pia mater.

External Anatomy (p. 376)

1. The spinal cord begins as a continuation of the medulla oblongata and ends at about the second lumbar vertebra in an adult.
2. It contains cervical and lumbar enlargements that serve as points of origin for nerves to the extremities.
3. The tapered inferior portion of the spinal cord is the conus medullaris, from which arise the filum terminale and cauda equina.
4. The spinal cord is partially divided into right and left sides by the anterior median fissure and posterior median sulcus.
5. The gray matter in the spinal cord is divided into horns and the white matter into columns.
6. In the center of the spinal cord is the central canal, which runs the length of the spinal cord.

Internal Anatomy (p. 380)

1. Parts of the spinal cord observed in cross section are the gray commissure; central canal; anterior, posterior, and lateral gray horns; anterior, posterior, and lateral white columns; and ascending and descending tracts.
2. The spinal cord conveys sensory and motor information by way of the ascending and descending tracts, respectively.

Spinal Cord Physiology (p. 380)

Sensory and Motor Tracts (p. 380)

1. A major function of the spinal cord is to convey nerve impulses from the periphery to the brain (sensory tracts) and to conduct motor impulses from the brain to the periphery (motor tracts).
2. Sensory information travels along two main routes: posterior column tract and spinothalamic tracts.
3. Motor information travels along two main routes: pyramidal tracts and extrapyramidal tracts.

Reflexes (p. 382)

1. Another function of the spinal cord is to serve as an integrating center for spinal reflexes. This occurs in the gray matter.
2. A reflex is a fast, predictable, automatic response to changes in the environment that helps to maintain homeostasis.
3. Reflexes may be spinal or cranial and somatic or visceral.

Reflex Arc and Homeostasis (p. 383)

1. A reflex arc is the simplest type of pathway.
2. Its components are receptor, sensory neuron, integrating center, motor neuron, and effector.
3. Somatic spinal reflexes include the stretch reflex, tendon reflex, flexor (withdrawal) reflex, and crossed extensor reflex; all exhibit reciprocal innervation.

4. A two-neuron or monosynaptic reflex arc contains one sensory and one motor neuron. A stretch reflex, such as the patellar reflex, is an example.
5. The stretch reflex is ipsilateral and is important in maintaining muscle tone and muscle coordination during exercise.
6. A polysynaptic reflex arc contains a sensory, association, and motor neuron. The tendon reflex, flexor (withdrawal) reflex, and crossed extensor reflexes are examples.
7. The tendon reflex is ipsilateral and prevents damage to muscles and tendons as a result of stretching; the flexor reflex is ipsilateral and is a withdrawal reflex; the crossed extensor reflex is contralateral.
8. Among clinically important somatic reflexes are the patellar reflex, the Achilles reflex, the Babinski sign, and the abdominal reflex.

Spinal Nerves (p. 389)

1. The 31 pairs of spinal nerves are named and numbered according to the region and level of the spinal cord from which they emerge.
2. There are 8 pairs of cervical, 12 pairs of thoracic, 5 pairs of lumbar, 5 pairs of sacral, and 1 pair of coccygeal nerves.

Composition and Coverings (p. 389)

1. Spinal nerves typically are attached to the spinal cord by a posterior root and an anterior root. All spinal nerves include both sensory and motor fibers (mixed).
2. Spinal nerves are covered by endoneurium, perineurium, and epineurium.

Distribution of Spinal Nerves (p. 389)

1. Branches of a spinal nerve include the dorsal ramus, ventral ramus, meningeal branch, and rami communicantes.
2. The ventral rami of spinal nerves, except for T2–T12, form networks of nerves called plexuses.
3. Emerging from the plexuses are nerves bearing names that are often descriptive of the general regions they supply or the course they take.
4. The cervical plexus supplies the skin and muscles of the head, neck, upper part of the shoulders, and diaphragm, and connects with some cranial nerves; and supplies the diaphragm.
5. The brachial plexus constitutes the nerve supply for the upper extremities and several neck and shoulder muscles.
6. The lumbar plexus supplies the anterolateral abdominal wall, external genitals, and part of the lower extremities.
7. The sacral plexus supplies the buttocks, perineum, and part of the lower extremities.
8. Ventral rami of nerves T2–T11 do not form plexuses and are called intercostal (thoracic) nerves. They are distributed directly to the structures they supply in intercostal spaces.

Dermatomes (p. 401)

1. All spinal nerves except C1 innervate specific, constant segments of the skin. The skin segments are called dermatomes.
2. Knowledge of dermatomes helps a physician to determine which segment of the spinal cord or a spinal nerve is malfunctioning.

Review Questions

1. Describe the bony covering of the spinal cord. (p. 376)
2. Explain the location and composition of the spinal meninges. Describe the location of the epidural, subdural, and subarachnoid spaces. Define meningitis. (p. 376)
3. Describe the location of the spinal cord. What are the cervical and lumbar enlargements? (p. 376)
4. Define conus medullaris, filum terminale, and cauda equina. What is a spinal segment? How is the spinal cord partially divided into a right and left side? (p. 380)
5. Based on your knowledge of the structure of the spinal cord in cross section, define the following: gray commissure, central canal, anterior gray horn, lateral gray horn, posterior gray horn, anterior white column, lateral white column, posterior white column, ascending tract, and descending tract. (p. 380)
6. Describe the function of the spinal cord as a conduction pathway. (p. 380)
7. Describe how the spinal cord serves as an integrating center for reflexes. (p. 383)
8. What is a reflex arc? List and define the components of a reflex arc. (p. 383)
9. Define a reflex. How are reflexes related to the maintenance of homeostasis? (p. 384)
10. Describe the mechanism and function of a stretch reflex, tendon reflex, flexor (withdrawal) reflex, and crossed extensor reflex. (p. 384)
11. Define the following terms related to reflex arcs: monosynaptic, ipsilateral, polysynaptic, intersegmental, contralateral, and reciprocal innervation. (p. 384)
12. Why are reflexes important in diagnosis? Indicate the clinical importance of the following reflexes: patellar, Achilles, Babinski sign, and abdominal. (p. 388)
13. Define a spinal nerve. How are spinal nerves named and numbered? Why are all spinal nerves classified as mixed nerves? (p. 389)
14. Describe how a spinal nerve is attached to the spinal cord. (p. 389)
15. Explain how a spinal nerve is enveloped by its connective tissue coverings. (p. 389)
16. Describe the branches and innervations of a typical spinal nerve. (p. 389)
17. What is a plexus? Describe the principal plexuses and the regions they supply. (p. 389)
18. What are intercostal (thoracic) nerves? (p. 391)
19. Define a dermatome. Why is a knowledge of dermatomes important? (p. 401)
20. Distinguish the following types of paralysis: monoplegia, diplegia, paraplegia, hemiplegia, and quadriplegia. (p. 401)
21. Define complete transection and hemisection. What are the consequences of each? What is spinal shock? (p. 401)
22. Describe nerve injuries to the brachial plexus, femoral nerve, and sciatic nerve. (pp. 394, 396, 398)
23. Why is a spinal tap (lumbar puncture) performed? Describe the procedure. (p. 380)

Answers to Questions with Figures

13.1 The foramen magnum of the occipital bone; the second sacral vertebra.
13.2 Cervical enlargement.
13.3 A horn is an area of gray matter and a column is a region of white matter.
13.4 The anterior corticospinal tract originates in the cortex of the cerebrum, ends in the spinal cord, and is located on the anterior side of the cord. It contains descending fibers and is a motor tract.
13.5 A receptor produces a generator potential, which triggers a nerve impulse if the generator potential reaches threshold. Reflex integrating centers are in the CNS.
13.6 Negative.
13.7 In an ipsilateral reflex, the sensory and motor neurons are on the same side of the spinal cord.
13.8 A neuronal circuit that provides for simultaneous contraction of one muscle and relaxation of its antagonist.
13.9 Because impulses go out over motor neurons located in several spinal nerves, each arising from a different segment of the spinal cord.
13.10 In a contralateral reflex arc, the motor impulses leave the spinal cord on the opposite side that the sensory impulses entered.
13.11 The posterior root contains sensory fibers and the anterior root contains motor fibers. The two roots unite to form the spinal nerve.
13.12 Ventral ramus.
13.13 C1.

Chapter 14

THE BRAIN AND CRANIAL NERVES

Chapter Contents at a Glance

Student Objectives

1. Identify the principal parts of the brain and describe how the brain is protected.
2. Explain the formation and circulation of cerebrospinal fluid (CSF).
3. Describe the blood supply to the brain and the blood–brain barrier (BBB).
4. Compare the structure and functions of the brain stem, diencephalon, cerebrum, and cerebellum.
5. Discuss the various neurotransmit-

ters found in the brain and the different types of neuropeptides and their functions.
6. Define a cranial nerve and identify the 12 pairs of cranial nerves by name, number, type, location, and function.
7. Describe the effects of aging on the nervous system.
8. Describe the development of the nervous system.
9. List the clinical symptoms of these disorders of the nervous system:

cerebrovascular accident (CVA), transient ischemic attack (TIA), Alzheimer's disease, brain tumors, cerebral palsy (CP), Parkinson's disease (PD), multiple sclerosis (MS), dyslexia, headache, and Reye's syndrome (RS).
10. Define medical terminology associated with the central nervous system.

To understand the anatomy and physiology of the human nervous system, in particular the human brain, is an enormous challenge. The brain is the center for registering sensations, correlating them with one another and with stored information, making decisions, and taking action. It is also the center for intellect, emotions, behavior, and memory. But the brain encompasses yet a larger domain: it directs our behavior toward others. With ideas that excite, or artistry that dazzles, or rhetoric that mesmerizes, one person may influence and shape the lives of many others. We are still at the frontier of being able to describe many brain functions.

In this chapter we will consider the principal parts of the brain, how the brain is protected and nourished, and how it is related to the spinal cord and to the 12 pairs of cranial nerves. The developmental anatomy of the brain is explained in detail at the end of the chapter.

BRAIN

Information processing by the central nervous system (CNS) takes place at several different levels. As you learned in Chapter 13, localized reflex responses to sensory input occur in the spinal cord. Lower regions of the brain, such as the brain stem, basal ganglia, and cerebellum, control most subconscious activities of the body. Examples are the regulation of blood pressure, respiratory rate, and muscle tone plus coordination of subconscious body movements such as posture and balance. At the highest level, the cerebrum integrates conscious activities, processes and stores information (learning and memory), and provides the circuits for abstract thought processes.

Principal Parts

To understand the terminology used for the principal parts of the **brain,** it is helpful to first consider its embryological development. At the end of the fourth week of the embryonic period, the brain develops from three regions of the embryo called primary brain vesicles (see Fig. 14.20). These are the **prosencephalon** (prōs′-en-SEF-a-lon) or forebrain, **mesencephalon** (mes′-en-SEF-a-lon) or midbrain, and **rhombencephalon** (rom′-ben-SEF-a-lon) or hindbrain. During the fifth week of development, the prosencephalon develops into two secondary brain vesicles called the **diencephalon** (dī′-en-SEF-a-lon) and **telencephalon** (tel-en-SEF-a-lon); the rhombencephalon also develops into two secondary brain vesicles called the **myelencephalon** (mī-el-en-SEF-a-lon) and **metencephalon** (met-en-SEF-a-lon). Since the mesencephalon remains unchanged, there are five secondary brain vesicles. Ultimately, the telencephalon forms the cerebrum, the diencephalon develops into the thalamus and hypothalamus of the brain, the mesencephalon becomes the midbrain, the metencephalon becomes the

pons and cerebellum and the myelencephalon develops into the medulla oblongata. These relations are summarized in Exhibit 14.1.

The adult brain is made up of about 100 billion neurons and is one of the largest organs of the body, weighing about 1300 g (3 lb). It is mushroom-shaped and can be divided into four principal parts: brain stem, diencephalon, cerebrum, and cerebellum (Fig. 14.1). The **brain stem,** the stalk of the mushroom, consists of the medulla oblongata, pons, and midbrain and is continuous with the spinal cord. Above the brain stem is the **diencephalon,** consisting primarily of the thalamus and hypothalamus. Like the cap of a mushroom, the **cerebrum** spreads over the diencephalon. The cerebrum, which occupies most of the cranium, has right and left halves called **cerebral hemispheres.** Inferior to the cerebrum and posterior to the brain stem is the **cerebellum.**

The brain develops very rapidly during the first few years of life. Growth is due mainly to an increase in the size of cells already present, proliferation and growth of neuroglia, development of synaptic contacts and dendritic branching, and myelination of the various fiber tracts.

Protection and Coverings

The brain is protected by the cranial bones (see Fig. 7.2) and cranial meninges. The **cranial meninges** (Fig. 14.2) surround the brain (see also Fig. 14.4). They are continuous with the spinal meninges, have the same basic structure, and bear the same names: the outer **dura mater,** middle **arachnoid,** and inner **pia mater.** As you will see later, the two hemispheres (sides) of the cerebrum are separated by an extension of the dura mater, called the falx cerebri. In addition, the two hemispheres of the cerebellum are separated by an extension of the dura mater called the falx cerebelli. The cerebrum is separated from the cerebellum by an extension of the dura mater called the tentorium cerebelli.

Cerebrospinal Fluid (CSF)

The brain and spinal cord are nourished and protected against chemical or physical injury by **cerebrospinal fluid (CSF).** This fluid continuously circulates through the subarachnoid space (between the arachnoid and pia mater) around the brain and spinal cord and through cavities within the brain.

Figure 14.3 shows the four CSF-filled cavities within the brain, which are called **ventricles** (VEN-tri-kuls). Each of the two **lateral ventricles** is located in a hemisphere of the cerebrum. The **third ventricle** is a vertical slit at the midline between and inferior to the right and left halves of the thalamus and between the lateral ventricles. The **fourth ventricle** lies between the brain stem and the cerebellum.

The entire central nervous system contains between 80 and 150 ml (3 to 5 oz) of CSF. It is a clear, colorless liquid that contains glucose, proteins, lactic acid, urea, cations

EXHIBIT 14.1

DEVELOPMENT OF THE BRAIN FROM BRAIN VESICLES

Primary Brain Vesicle	Secondary Brain Vesicle	Final Brain Structures
Prosencephalon	Telencephalon	Cerebrum
	Diencephalon	Thalamus and hypothalamus
Mesencephalon	Mesencephalon	Midbrain
Rhombencephalon	Metencephalon	Pons and cerebellum
	Myelencephalon	Medulla oblongata

FIGURE 14.1 Brain. The infundibulum and pituitary gland are discussed in conjunction with the endocrine system in Chapter 18.

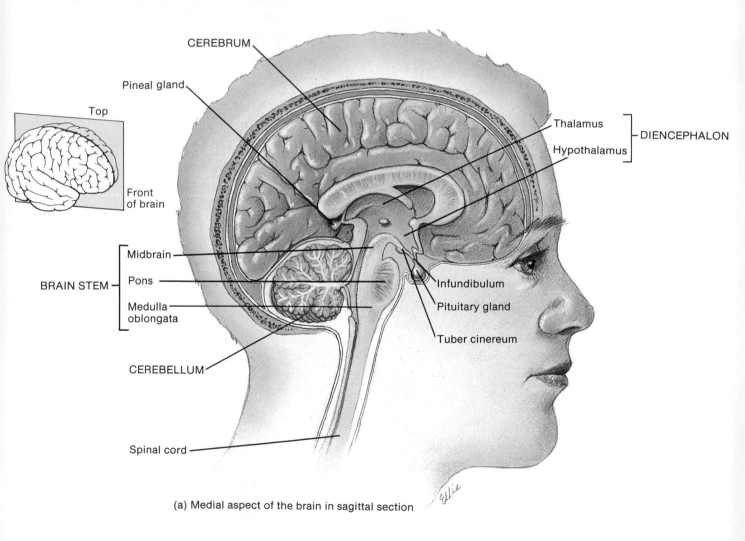

(a) Medial aspect of the brain in sagittal section

FIGURE 14.1 (continued)

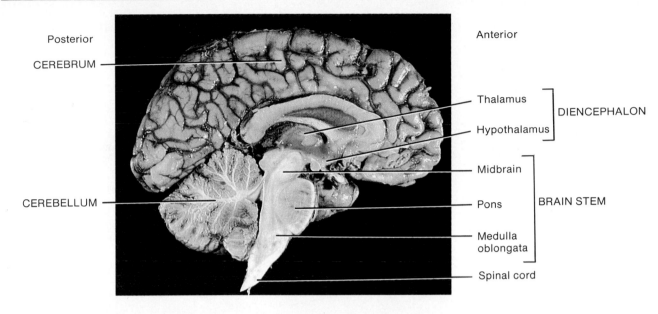

(b) Medial aspect of brain in sagittal section

Question: What are the four principal parts of the brain?

FIGURE 14.2 Superior portion of the skull showing the protective coverings of the brain: skull and cranial meninges.

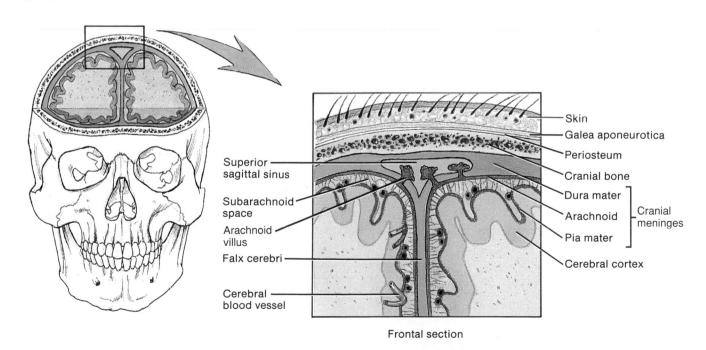

Frontal section

Question: What are the three layers of the cranial meninges, in order from outermost to innermost?

FIGURE 14.3 Lateral and anterior projection of the ventricles.

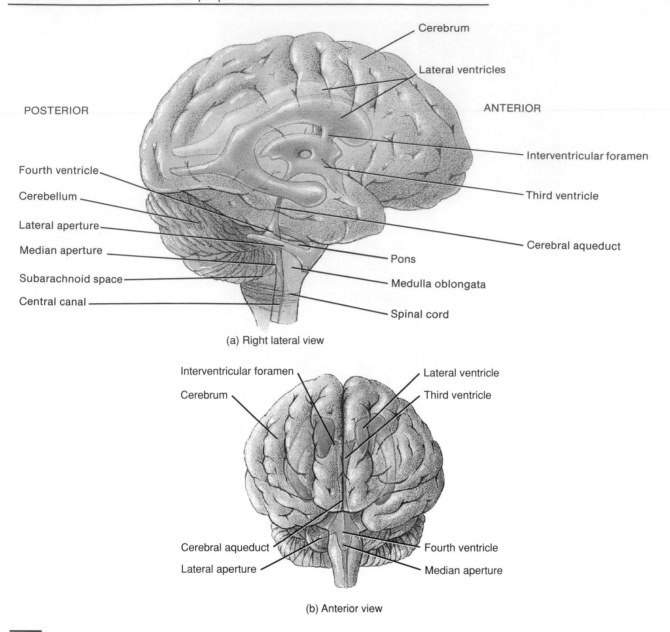

(a) Right lateral view

(b) Anterior view

Question: The third ventricle is at the midline of the brain. What brain region is to the left and right of this ventricle?

(Na^+, K^+, Ca^{2+}, Mg^{2+}), and anions (Cl^- and HCO_3^-). It also contains some lymphocytes. CSF contributes to homeostasis in three main ways.

1. Mechanical protection. The fluid serves as a shock-absorbing medium to protect the delicate tissue of the brain and spinal cord from jolts that would otherwise cause them to crash against the bony walls of the cranial and vertebral cavities. The fluid also buoys the brain so that it "floats" in the cranial cavity.

2. Chemical protection. CSF provides an optimal chemical environment for accurate neuronal signaling. Even slight changes in the ionic composition of CSF within the brain could seriously disrupt production of postsynaptic potentials and action potentials.

3. Circulation. CSF is a medium for exchange of nutrients and waste products between the blood and nervous tissue.

The **choroid** (KŌ-royd; *chorion* = delicate) **plexuses** (Fig. 14.4) are networks of capillaries (microscopic blood vessels) in the walls of the ventricles. The capillaries are

FIGURE 14.4 Meninges and ventricles of the brain. Arrows indicate the direction of flow of cerebrospinal fluid.

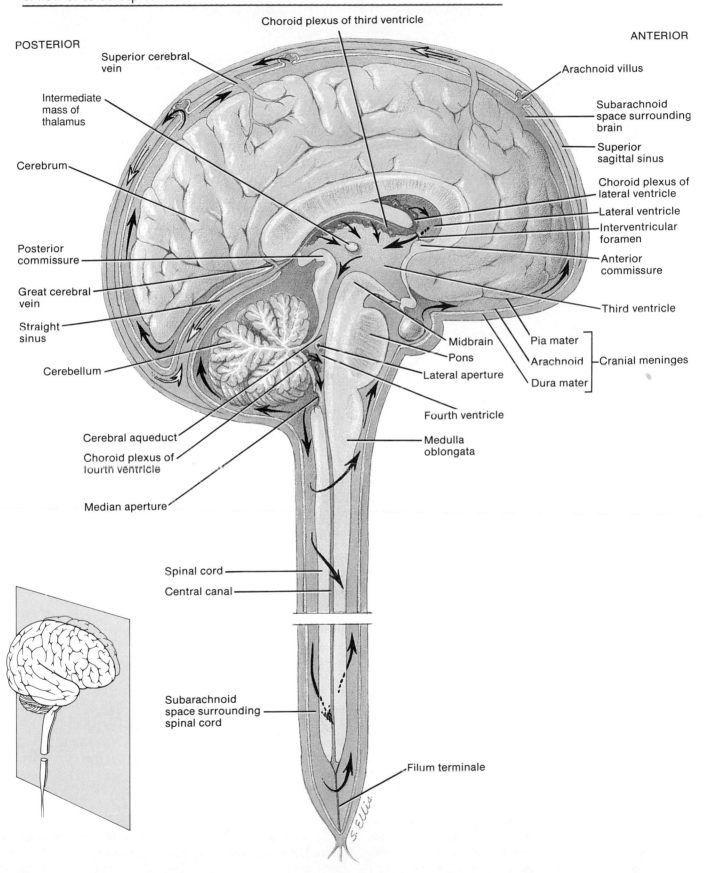

(a) Brain, ventricles, spinal cord, and meninges in sagittal section

Figure continues

FIGURE 14.4 (continued)

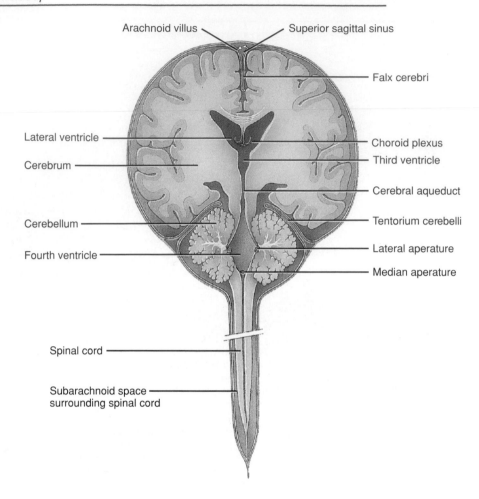

Arachnoid villus
Superior sagittal sinus
Falx cerebri
Lateral ventricle
Cerebrum
Choroid plexus
Third ventricle
Cerebral aqueduct
Cerebellum
Tentorium cerebelli
Fourth ventricle
Lateral aperature
Median aperature
Spinal cord
Subarachnoid space surrounding spinal cord

(b) Brain, ventricles, spinal cord, and meninges in frontal section

Question: Where is CSF formed; where is it reabsorbed?

---

covered by ependymal cells that form cerebrospinal fluid from blood plasma by filtration and secretion. Because the ependymal cells are joined by tight junctions (see Fig. 4.1), materials entering CSF from choroid capillaries cannot leak between these cells. Rather, they must go through the ependymal cells. This **blood–cerebrospinal fluid barrier** permits certain substances to enter the fluid but excludes others. Such a barrier protects the brain and spinal cord from potentially harmful substances in the blood.

The CSF formed in the choroid plexuses of the lateral ventricles flows into the third ventricle through a narrow, oval opening, the **interventricular foramen (foramen of Monro).** See also Fig. 14.3. More fluid is added by the choroid plexus of the third ventricle. The fluid then flows through the **cerebral aqueduct (aqueduct of Sylvius),** which passes through the midbrain, into the fourth ventri-

cle. The choroid plexus of the fourth ventricle contributes more fluid. CSF enters the subarachnoid space through three openings in the roof of the fourth ventricle: a **median aperture (of Magendie)** and two **lateral apertures (of Luschka).** The fluid then circulates in the subarachnoid space around the posterior surface of the brain. It also passes downward to the subarachnoid space around the posterior surface of the spinal cord, up the anterior surface of the spinal cord, and around the anterior part of the brain. From there it is gradually reabsorbed into a blood vascular sinus called the superior sagittal sinus. The actual reabsorption occurs through **arachnoid villi**—fingerlike extensions of the arachnoid that project into the dural venous sinuses, especially the superior sagittal sinus (see Fig. 14.2). Normally, cerebrospinal fluid is reabsorbed as rapidly as it is formed, at a rate of about 20 ml/hr (about 480 ml/day).

HYDROCEPHALUS

An obstruction, such as a tumor or a congenital blockage, or an inflammation in the brain can interfere with the drainage of CSF from the ventricles into the subarachnoid space. As fluid accumulates in the ventricles, the CSF pressure rises. This condition is called **hydrocephalus** (*hydro* = water; *kephale* = head). In a baby, if the fontanels have not yet closed, the head bulges in response to the increased pressure. In time, however, the fluid buildup compresses and damages the delicate nervous tissue. Hydrocephalus responds dramatically to ventricular drainage of CSF. A neurosurgeon may implant a drain line to divert CSF from a brain ventricle into the subclavian vein and then into the right atrium of the heart. In adults, hydrocephalus may occur following head injury, meningitis, or subarachnoid hemorrhage.

Blood Supply

The brain is well supplied with oxygen and nutrients, mainly by blood vessels that form the cerebral arterial circle (circle of Willis) at the base of the brain. Cerebral circulation is outlined in Exhibit 21.6 and Fig. 22.22c. Blood vessels that enter brain tissue pass along the surface of the brain, and as they penetrate inward, they are surrounded by a loose-fitting layer of pia mater.

Although the brain composes only about 2% of total body weight, it consumes about 20% of the oxygen at rest. The brain is one of the most metabolically active organs of the body, and the amount of oxygen it uses varies with the degree of mental activity. When neuronal activity increases in a region of the brain, blood flow to that area also increases. If the blood flow to the brain is interrupted even briefly, unconsciousness may result. A one- or two-minute interruption in blood flow may impair brain cells, and if the cells are totally deprived of oxygen for about four minutes, many are permanently injured. Lysosomes of brain cells are sensitive to decreased oxygen concentration. If the condition persists long enough, lysosomes break open and release enzymes that bring about self-destruction of brain cells. Occasionally during childbirth, the oxygen supply from the mother's blood is interrupted before the baby leaves the birth canal and can breathe on its own. This can result in the baby being stillborn or suffering permanent brain damage that may result in mental retardation, epilepsy, or paralysis.

Blood supplying the brain also contains glucose, the principal molecule that neurons use to make energy-rich ATP. Because carbohydrate storage in the brain is limited, the supply of glucose must be continuous. If blood entering the brain has a low glucose level, mental confusion, dizziness, convulsions, and loss of consciousness may occur.

Glucose, oxygen, carbon dioxide, water, and most lipid-soluble substances, such as alcohol, caffeine, nicotine, heroin, and most anesthetics, pass rapidly from the circulat-

ing blood into brain cells. Other substances, such as creatinine, urea, and most ions, for example, Na^+, K^+, and Cl^- enter quite slowly. Still other substances—proteins and most antibiotics—do not pass at all from the blood into brain cells. The different rates of passage of certain materials from the blood into most parts of the brain depend on the **blood–brain barrier (BBB).** Brain capillaries are much less leaky than most other body capillaries. Tight junctions seal together the endothelial cells of brain capillaries, which also are surrounded by a continuous basement membrane. Also, processes of large numbers of astrocytes (one type of neuroglia) press up against the capillaries. Astrocytes are thought to selectively pass some substances from the blood but inhibit the passage of others. Substances that cross the BBB are either soluble in lipids or water-soluble substances that receive the assistance to cross by carrier-mediated transport.

BREACHING THE BLOOD–BRAIN BARRIER

The blood–brain barrier functions as a selective anatomical and physiological barrier to protect brain cells from harmful substances and pathogens. An injury to the brain due to trauma, inflammation, or toxins may cause a breakdown of the BBB, permitting the passage of substances into brain tissue that normally are kept out. On the other hand, the BBB may prevent entry of drugs that could be used as therapy for brain cancer or other CNS disorders. Researchers are working on ways to slip drugs past the BBB. In one method, the drug is injected together with a concentrated sugar solution. Temporarily, the endothelial cells shrink, due to the osmotic effect of the sugar solution, and pull apart to open up gaps in the tight junctions. The result is entry of the drug into the brain tissue.

Several small brain regions lying in the walls of the third and fourth ventricles and called **circumventricular organs (CVOs)** can monitor chemical changes in the blood because they lack the blood–brain barrier. CVOs include part of the hypothalamus, the pineal gland, the pituitary gland, and a few other nearby structures. Functionally, these regions coordinate homeostatic activities of the endocrine and nervous systems, such as regulation of blood pressure, fluid balance, hunger, and thirst. CVOs are also thought to be the site of entry for the AIDS virus into the brain, in which case dementia (irreversible deterioration of mental state) and other neurologic disorders may appear before other symptoms of AIDS become apparent.

Brain Stem: Anatomy and Physiology

Medulla

The **medulla oblongata** (me-DULL-la ob´-long-GA-ta), or more simply the **medulla**, develops from the myelencephalon and is a continuation of the upper portion of the spinal cord. It forms the inferior part of the brain stem (Fig. 14.5). Relative to other parts of the brain (Fig. 14.1), it lies just superior to the foramen magnum and extends upward to the inferior portion of the pons, a distance of about 3 cm (1.2 in.).

The medulla contains all ascending and descending tracts that connect the spinal cord and various parts of the brain. These tracts constitute the white matter of the medulla. Most tracts cross over from one side to the other as they pass through the medulla. On the ventral side of the medulla are two bulges called **pyramids** (Figs. 14.5 and 14.6). The pyramids contain the largest motor tracts that pass from the outer region of the cerebrum (cerebral cortex) to the spinal cord. Just above the junction of the medulla with the spinal cord, most of the fibers in the left pyramid cross to the right side, and most of the fibers in the right pyramid cross over to the left. This crossing is called the **decussation** (dē′-ku-SĀ-shun) **of pyramids.**

The principal motor fibers that decussate in the medulla originate in the cerebral cortex and pass inferiorly to the medulla. The fibers cross in the pyramids (see Fig. 14.6) and descend in the lateral white columns of the spinal cord, ending in the anterior gray horns (see Fig. 15.8). Here synapses occur with motor neurons that supply skeletal muscles. Thus fibers that originate in the left cerebral cortex activate muscles on the right side of the body, and fibers that originate in the right cerebral cortex activate muscles on the left side.

The dorsal side of the medulla contains two pairs of prominent nuclei (gray matter masses of cell bodies and dendrites in the CNS). They are the right and left **nucleus gracilis** (gras-I-lis; *gracilis* = slender) and **nucleus cuneatus** (kyoo-nē-Ā-tus; *cuneus* = wedge). These nuclei receive sensory fibers from ascending tracts (right and left fasciculus gracilis and fasciculus cuneatus) of the spinal cord and relay the sensory information to the thalamus on the opposite side (see Fig. 15.4). The information is conveyed to the

---

FIGURE 14.5 Brain stem.

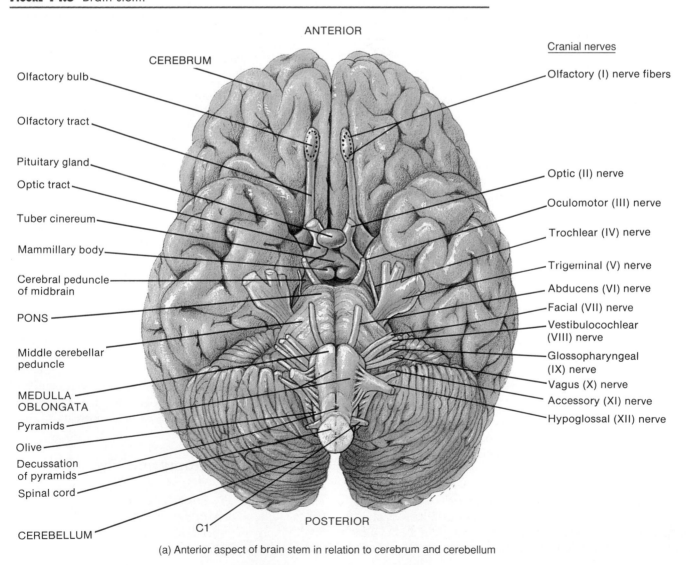

(a) Anterior aspect of brain stem in relation to cerebrum and cerebellum

FIGURE 14.5 (continued)

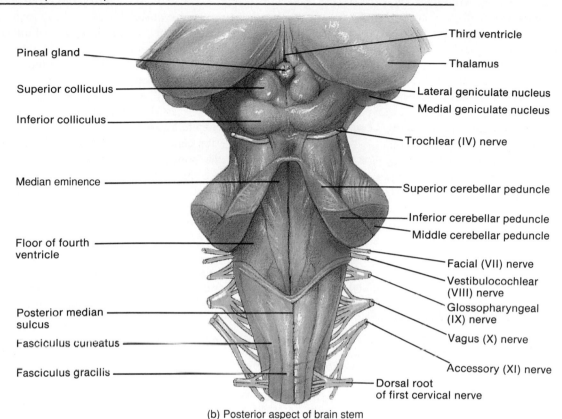

Pineal gland
Superior colliculus
Inferior colliculus
Median eminence
Floor of fourth ventricle
Posterior median sulcus
Fasciculus cuneatus
Fasciculus gracilis

Third ventricle
Thalamus
Lateral geniculate nucleus
Medial geniculate nucleus
Trochlear (IV) nerve
Superior cerebellar peduncle
Inferior cerebellar peduncle
Middle cerebellar peduncle
Facial (VII) nerve
Vestibulocochlear (VIII) nerve
Glossopharyngeal (IX) nerve
Vagus (X) nerve
Accessory (XI) nerve
Dorsal root of first cervical nerve

(b) Posterior aspect of brain stem

Question: What part of the brain contains the pyramids? The cerebral peduncles? Literally means bridge? Relays all sensory inputs to the cortex?

FIGURE 14.6 Medulla. Shown is internal anatomy and the decussation of pyramids.

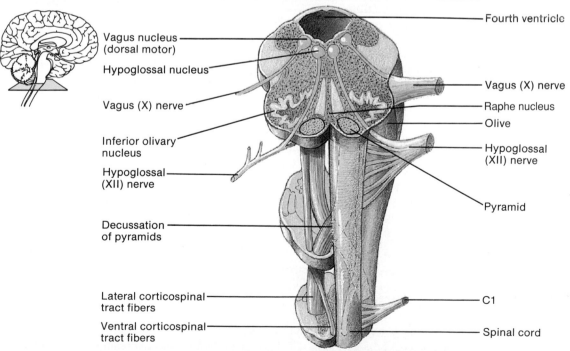

Vagus nucleus (dorsal motor)
Hypoglossal nucleus
Vagus (X) nerve
Inferior olivary nucleus
Hypoglossal (XII) nerve
Decussation of pyramids
Lateral corticospinal tract fibers
Ventral corticospinal tract fibers

Fourth ventricle
Vagus (X) nerve
Raphe nucleus
Olive
Hypoglossal (XII) nerve
Pyramid
C1
Spinal cord

Cross section and anterior surface of medulla

Question: What does decussation mean? Functionally, what is the consequence of decussation of the pyramids?

sensory areas of the cerebral cortex. Nearly all sensory impulses initiated on one side of the body cross in the spinal cord or medulla and finally are received in the cerebral cortex on the opposite side.

Regions within the medulla also regulate several vital body functions. The **cardiovascular center** regulates the rate and force of heartbeat and the diameter of blood vessels (see Fig. 21.14) and the **respiratory center** adjusts the basic rhythm of breathing (see Fig. 23.24). Other centers in the medulla coordinate swallowing, vomiting, coughing, sneezing, and hiccuping.

The medulla also contains the nuclei of origin for several pairs of cranial nerves (Figs. 14.5, 14.6, and 14.18). (Cranial nerves are numbered I through XII.) These are the cochlear and vestibular branches of the vestibulocochlear (VIII) nerves, which are concerned with hearing and equilibrium (there are also nuclei for the vestibular branches in the pons); the glossopharyngeal (IX) nerves, which relay nerve impulses for swallowing, salivation, and taste; the vagus (X) nerves, which relay nerve impulses to and from many thoracic and abdominal viscera; the cranial portion of the accessory (XI) nerves (a part of these nerves, the spinal portion, originates in the upper five cervical segments of the spinal cord), which convey nerve impulses related to head and shoulder movements; and the hypoglossal (XII) nerves, which convey nerve impulses that involve tongue movements.

On each lateral surface of the medulla is an oval projection called the **olive** (see Figs. 14.5 and 14.6). Nuclei in the olive connect to the cerebellum by paired bundles of fibers (tracts) called the **inferior cerebellar peduncles**. They carry signals that ensure the efficiency of precise, voluntary movements and maintain equilibrium and posture.

Also associated with the medulla is most of the **vestibular nuclear complex.** The vestibular nuclei are important for maintaining equilibrium (see page 497).

In view of the many vital activities controlled by the medulla, it is not surprising that a hard blow to the back of the head or upper neck can be fatal. Damage of the respiratory center is particularly serious and can rapidly lead to death. Before the advent of vaccines for polio (see page 401), many victims of this disease were confined to respirators ("iron lungs") because the respiratory center is a primary target of the polio virus. Symptoms of nonfatal medullary injury may include cranial nerve malfunctions on the same side of the body as the injury, paralysis and loss of sensation on the opposite side of the body, and breathing irregularities.

Pons

The relationship of the **pons** (*pons* = bridge) to other parts of the brain can be seen in Figs. 14.1 and 14.5. The pons forms from the metencephalon. It lies directly above the medulla and anterior to the cerebellum and is about 2.5 cm (1 in.) long. Like the medulla, the pons consists of both nuclei and tracts of white fibers. As its name implies, the pons is a bridge connecting the spinal cord with the brain and parts of the brain with each other. These connections are provided by fibers that run in two principal directions. The transverse fibers are paired bundles of fibers that connect the right and left sides of the cerebellum and are known as the **middle cerebellar peduncles** (pe-DUNG-kulz). The longitudinal fibers of the pons belong to the motor and sensory tracts that connect the medulla with the upper parts of the brain stem.

The nuclei for certain pairs of cranial nerves are also contained in the pons (see Figs. 14.5 and 14.18). These include the trigeminal (V) nerves, which relay nerve impulses for chewing and for sensations of the head and face; the abducens (VI) nerves, which regulate certain eyeball movements; the facial (VII) nerves, which conduct impulses related to taste, salivation, and facial expression; and the vestibular branches of the vestibulocochlear (VIII) nerves, which are concerned with equilibrium.

Other important nuclei in the pons are the **pneumotaxic** (noo-mō-TAK-sik) **area** and the **apneustic** (ap-NOO-stik) **area.** (See Fig. 23.24.) Together with the respiratory center in the medulla, they help control respiration.

Reticular Formation

A large portion of the brain stem (medulla, pons, and midbrain) consists of small areas of gray matter interspersed among fibers of white matter. This region is called the **reticular formation.** It also extends into the spinal cord and diencephalon. The reticular formation has both sensory and motor functions. It receives input from higher brain regions that control skeletal muscles and makes its own important contribution to regulating muscle tone (the slight degree of contraction that characterizes muscles at rest). It also alerts the cortex to incoming sensory signals. This part of the reticular formation is called the **reticular activating system (RAS)** (see Fig. 15.11); it is responsible for maintaining consciousness and awakening from sleep. The RAS is effectively stimulated by incoming impulses from the ears, eyes, and skin that arouse the cerebral cortex. For example, we awaken to the sound of an alarm clock, to a bright light flash, or to a painful pinch because of activity in the RAS.

Midbrain

The **midbrain,** or **mesencephalon** (*meso* = middle; *enkephalos* = brain), extends from the pons to the lower portion of the diencephalon (Figs. 14.1 and 14.5). It is about 2.5 cm (1 in.) in length. The cerebral aqueduct passes through the midbrain and connects the third ventricle above with the fourth ventricle below.

The ventral portion of the midbrain contains a pair of fiber bundles called **cerebral peduncles** (Fig. 14.7). The

FIGURE 14.7 Midbrain.

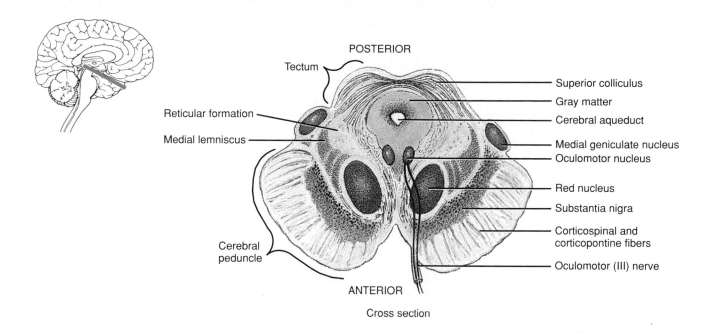

POSTERIOR

Tectum

Reticular formation

Medial lemniscus

Superior colliculus
Gray matter
Cerebral aqueduct
Medial geniculate nucleus
Oculomotor nucleus
Red nucleus
Substantia nigra
Corticospinal and corticopontine fibers
Oculomotor (III) nerve

Cerebral peduncle

ANTERIOR

Cross section

Question: What is the importance of the cerebral peduncles?

cerebral peduncles contain some motor fibers that convey nerve impulses from the cerebral cortex to the pons, medulla, and spinal cord. They also contain sensory fibers that pass from the spinal cord to the medulla and then from the pons to the thalamus. The cerebral peduncles are the main connection for tracts between upper parts of the brain and lower parts of the brain and the spinal cord. A paired bundle of fibers called the **superior cerebellar peduncles** connects the midbrain with the cerebellum.

The dorsal portion of the midbrain is called the **tectum** (*tectum* = roof) and contains four rounded elevations: the **corpora quadrigemina** (KOR-po-ra kwad-ri-JEM-in-a). The two upper elevations are known as the **superior colliculi** (ko-LIK-yoo-lī; singular is **colliculus**). These serve as reflex centers for movements of the eyes, head, and neck in response to visual and other stimuli. The lower two elevations, the **inferior colliculi,** serve as reflex centers for movements of the head and trunk in response to auditory stimuli (see Fig. 14.5b). The midbrain also contains the left and right **substantia nigra** (sub-STAN-shē-a NĪ-gra); these are large, darkly pigmented nuclei near the cerebral peduncles that control subconscious muscle activities.

Prominent in the midbrain are the left and right **red nucleus.** The name derives from the rich blood supply and an iron-containing pigment in its neuronal cell bodies. Fibers from the cerebellum and cerebral cortex end in the red nuclei. The red nuclei are also the origin of cell bodies of the descending rubrospinal tracts. The red nucleus functions with the basal ganglia and cerebellum to coordinate

muscular movements. Other nuclei in the midbrain are associated with cranial nerves (see Figs. 14.5 and 14.18). These include the oculomotor (III) nerves, which mediate some movements of the eyeballs and changes in pupil size and lens shape, and the trochlear (IV) nerves, which conduct impulses that coordinate movements of the eyeballs.

A structure called the **medial lemniscus** (*lemniskos* = ribbon or band) extends through the medulla, pons, and midbrain. The medial lemniscus is a band of white fibers containing axons that convey impulses for discriminative touch, proprioception, pressure, and vibrations from the medulla to the thalamus.

Diencephalon

The **diencephalon** (*dia* = through; *enkephalos* = brain) consists primarily of the thalamus and hypothalamus. The relationship of these structures to the rest of the brain is shown in Fig. 14.1.

Thalamus

The **thalamus** (THAL-a-mus; *thalamos* = inner chamber) is an oval structure above the midbrain that measures about 3 cm (1.2 in.) in length and constitutes four-fifths of the diencephalon. It consists of paired oval masses of mostly gray matter organized into nuclei that form the lateral walls of the third ventricle (Fig. 14.8). Usually, the right and left

FIGURE 14.8 Thalamus. (a) Thalamus and nearby structures. (b) Thalamic nuclei. Arrows indicate some of the connections between the thalamus and cerebral cortex. Shadowed regions are the ventricles.

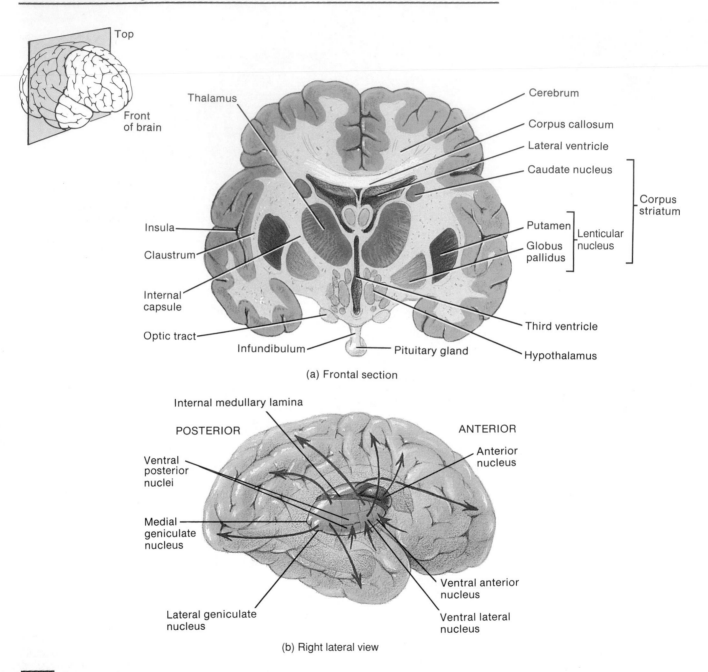

(a) Frontal section

(b) Right lateral view

Question: Where are the basal ganglia relative to the thalamus?

portions of the thalamus are joined by a bridge of gray matter called the **intermediate mass** (see Fig. 14.4) that crosses the third ventricle. Although the thalamic masses are primarily gray matter, some portions are white matter. One white matter strip is the **internal medullary lamina,** which divides the gray matter masses into an anterior nuclear group, a medial nuclear group, and a lateral nuclear group.

The thalamus is the principal relay station for sensory impulses that reach the cerebral cortex from the spinal cord, brain stem, cerebellum, and other parts of the cerebrum. It also allows crude appreciation of some sensations, such as pain, temperature, and pressure. Precise localization of such sensations depends on nerve impulses being relayed from the thalamus to the cerebral cortex. Within each thalamic group are nuclei that have various roles. Certain nuclei in the thalamus relay all sensory input to the cerebral cortex. These include the **medial geniculate** (je-NIK-yoo-lāt) **nuclei** (hearing), the **lateral geniculate nuclei** (vision), and the **ventral posterior nuclei** (taste and somatic sensations such as touch, pressure, vibration, heat, cold, and pain). Other nuclei are centers for synapses in the somatic motor system. These include the **ventral lateral nuclei** and **ventral anterior nuclei** (voluntary motor actions and arousal) (Fig. 14.8b). The **anterior nucleus** in the floor of the lateral ventricle is concerned with certain emotions and memory.

Hypothalamus

The **hypothalamus** (*hypo* = under) is a small portion of the diencephalon located below the thalamus. Its relationship to other parts of the brain is shown in Figs. 14.1 and 14.8a. The hypothalamus lacks a blood–brain barrier and forms the floor and part of the lateral walls of the third ventricle. It is partially protected by the sella turcica of the sphenoid bone. The hypothalamus is divided into a dozen or so nuclei in four major regions (Fig. 14.9).

FIGURE 14.9 Hypothalamus. Selected areas of the hypothalamus and a three-dimensional representation of hypothalamic nuclei seen in sagittal section (after Netter).

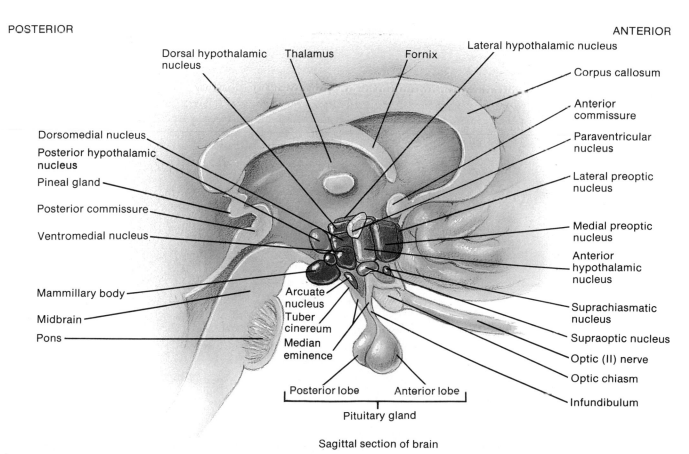

POSTERIOR

ANTERIOR

Dorsal hypothalamic nucleus

Thalamus

Fornix

Lateral hypothalamic nucleus

Corpus callosum

Anterior commissure

Dorsomedial nucleus

Posterior hypothalamic nucleus

Pineal gland

Posterior commissure

Ventromedial nucleus

Paraventricular nucleus

Lateral preoptic nucleus

Medial preoptic nucleus

Anterior hypothalamic nucleus

Mammillary body

Midbrain

Pons

Arcuate nucleus

Tuber cinereum

Median eminence

Suprachiasmatic nucleus

Supraoptic nucleus

Optic (II) nerve

Optic chiasm

Infundibulum

Posterior lobe

Anterior lobe

Pituitary gland

Sagittal section of brain

Question: What are the four major regions of the hypothalamus, from posterior to anterior?

1. **The mammillary region**, adjacent to the midbrain, is the most posterior portion of the hypothalamus. It includes the mammillary bodies and posterior hypothalamic nucleus. The **mammillary bodies** are two, small, rounded bodies that serve as relay stations in reflexes related to the sense of smell.

2. **The tuberal region**, in the middle, is the widest portion of the hypothalamus. It includes the dorsomedial, ventromedial, and arcuate nuclei. On its ventral surface are the **tuber cinereum** (si-NE-rē-um), an elevated mass of gray matter, and **infundibulum.** The infundibulum is a stalklike structure that attaches the pituitary gland to the hypothalamus. The **median eminence** of the tuber cinereum is a slightly raised region that encircles the site where the infundibulum becomes the stalk of the pituitary gland. The median eminence contains neurons that synthesize hypothalamic regulating hormones. These are released into capillary networks in the median eminence and regulate hormonal secretions by the anterior lobe of the pituitary gland.

3. **The supraoptic region** lies above the optic chiasm (point of crossing of optic nerves) and contains the paraventricular nucleus, supraoptic nucleus, anterior hypothalamic nucleus, and suprachiasmatic nucleus. Nerve fibers from the paraventricular and supraoptic nuclei form the supraopticohypophyseal tract, which extends through the infundibulum to the posterior pituitary. This tract transports hormones (oxytocin and antidiuretic hormone) made in the nuclei to the posterior pituitary gland where they are stored and released (see Fig. 18.10).

4. **The preoptic region** anterior to the supraoptic region is usually considered part of the hypothalamus because it functions in regulating certain autonomic activities in conjunction with the hypothalamus. The preoptic region contains the preoptic periventricular nucleus, medial preoptic nucleus, and lateral preoptic nucleus.

The hypothalamus controls many body activities and *is one of the major regulators of homeostasis*. Sensory input from the external and internal environments ultimately comes to the hypothalamus via afferent pathways originating in somatic and visceral sense organs. Impulses from sound, taste, and smell receptors all reach the hypothalamus. Other receptors within the hypothalamus itself continually monitor osmotic pressure, certain hormone concentrations, and the temperature of blood. The hypothalamus has several very important connections with the pituitary gland and is itself able to produce a variety of hormones, which will be described in more detail in Chapter 18. While some functions can be attributed to specific nuclei, others are not so precisely localized and may overlap nuclear boundaries. The chief functions of the hypothalamus are as follows.

1. It controls and integrates activities of the autonomic nervous system (ANS), which regulates contraction of smooth muscle and cardiac muscle and secretions of many glands. Through the ANS, the hypothalamus is the main regulator of visceral activities, such as heart rate, movement of food through the gastrointestinal tract, and contraction of the urinary bladder.

2. It is associated with feelings of rage and aggression.

3. It regulates body temperature.

4. It regulates food intake through two centers. The **feeding (hunger) center** is responsible for hunger sensations. When sufficient food has been ingested, the **satiety** (sa-TĪ-e-tē; *satio* = full, satisfied) **center** is stimulated and sends out nerve impulses that inhibit the feeding center.

5. It contains a **thirst center.** When certain cells in the hypothalamus are stimulated by rising osmotic pressure of the extracellular fluid, they produce the sensation of thirst. If thirst is satisfied, osmotic pressure returns to normal.

6. It is one of the centers that maintains the waking state and sleep patterns.

Cerebrum

Supported on the diencephalon and brain stem and forming the bulk of the brain is the **cerebrum** (see Fig. 14.1), which develops from the telencephalon. The surface of the cerebrum is composed of gray matter 2 to 4 mm (0.08 to 0.16 in.) thick (Fig. 14.10a) and is called the **cerebral cortex** (*cortex* = rind or bark). The cortex contains billions of neurons. Beneath the cortex lies the cerebral white matter. The cerebrum is the seat of intelligence, the ability to read, write, speak, make calculations and compose music, remember the past, plan for the future, and create works that have never existed before.

During embryonic development, when there is a rapid increase in brain size, the gray matter of the cortex enlarges much faster than the underlying white matter. As a result, the cortical region rolls and folds upon itself. The folds are called **gyri** (JĪ-rī; singular is **gyrus**) or **convolutions** (Fig. 14.10a, c). The deep grooves between folds are **fissures;** the shallow grooves between folds are **sulci** (SUL-sī; singular is **sulcus**). The most prominent fissure, the **longitudinal fissure,** nearly separates the cerebrum into right and left hemispheres. The hemispheres, however, are connected internally by white matter that forms a large bundle of transverse fibers called the **corpus callosum** (kal-LŌ-sum; *corpus* = body; *callous* = hard; see Fig. 14.8). Between the hemispheres is an extension of the cranial dura mater called the **falx** (FALKS) **cerebri (cerebral fold).** It encloses the superior and inferior sagittal sinuses.

Lobes

Each cerebral hemisphere is further subdivided into four lobes by sulci or fissures. The lobes are named after the bones that cover them: frontal, parietal, temporal, and

FIGURE 14.10 Cerebrum. (a) The inset indicates the relative differences among a gyrus, sulcus, and fissure. (b) Photograph showing the insula after removal of a portion of the cerebrum. (c) Since the insula cannot be seen externally, it has been projected to the surface. It can be seen in Fig. 14.8a.

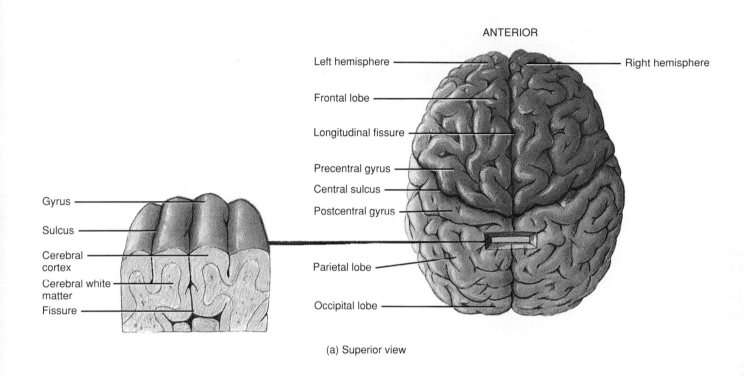

ANTERIOR

Left hemisphere

Frontal lobe

Longitudinal fissure

Precentral gyrus

Central sulcus

Postcentral gyrus

Parietal lobe

Occipital lobe

Right hemisphere

Gyrus

Sulcus

Cerebral cortex

Cerebral white matter

Fissure

(a) Superior view

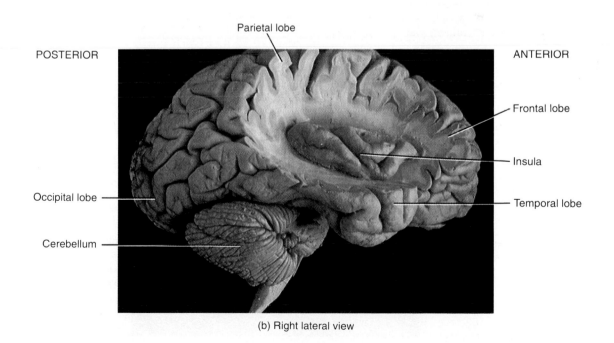

Parietal lobe

POSTERIOR

ANTERIOR

Frontal lobe

Insula

Temporal lobe

Occipital lobe

Cerebellum

(b) Right lateral view

Figure continues

FIGURE 14.10 (continued)

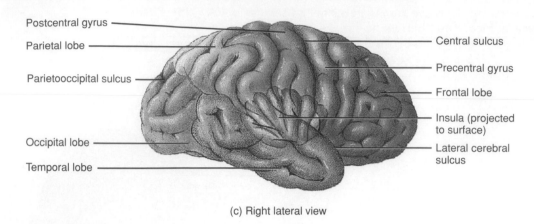

Postcentral gyrus
Parietal lobe
Parietooccipital sulcus
Occipital lobe
Temporal lobe

Central sulcus
Precentral gyrus
Frontal lobe
Insula (projected to surface)
Lateral cerebral sulcus

(c) Right lateral view

Question: During development, which portion of the brain—the gray or the white matter—enlarges more rapidly? What are the brain folds, shallow grooves, and deep grooves called?

---

occipital lobes (Fig. 14.10). The **central sulcus** (SUL-kus) separates the **frontal lobe** from the **parietal lobe.** A major gyrus, the **precentral gyrus,** is located immediately anterior to the central sulcus. This gyrus contains the primary motor area of the cerebral cortex. Another major gyrus, the **postcentral gyrus,** is located immediately posterior to the central sulcus. This gyrus contains the primary somesthetic area of the cerebral cortex. The **lateral cerebral sulcus** separates the **frontal lobe** from the **temporal lobe.** The **parietooccipital sulcus** separates the **parietal lobe** from the **occipital lobe.** Another prominent fissure, the **trans-**

verse fissure, separates the cerebrum from the cerebellum. A fifth part of the cerebrum, the **insula,** cannot be seen from the exterior surface of the brain because it lies deep within the lateral cerebral fissure, under the parietal, frontal, and temporal lobes.

White Matter

The white matter underlying the cortex consists of myelinated axons extending in three principal directions (Fig. 14.11).

FIGURE 14.11 White matter tracts of the left cerebral hemisphere seen in sagittal section.

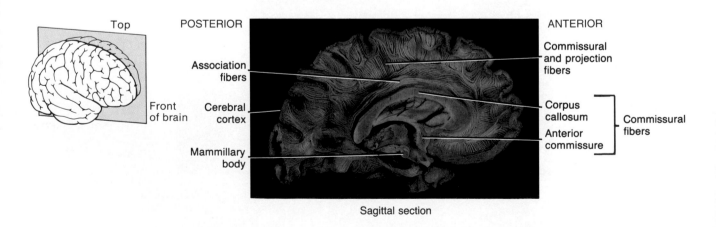

Top

POSTERIOR

Association fibers
Cerebral cortex

Front of brain

Mammillary body

ANTERIOR

Commissural and projection fibers

Corpus callosum
Anterior commissure

Commissural fibers

Sagittal section

Question: Which fibers carry impulses between gyri of the same hemisphere?

1. **Association fibers** connect and transmit nerve impulses between gyri in the same hemisphere.
2. **Commissural fibers** transmit impulses from the gyri in one cerebral hemisphere to the corresponding gyri in the opposite cerebral hemisphere. Three important groups of commissural fibers are the **corpus callosum, anterior commissure,** and **posterior commissure.**
3. **Projection fibers** form descending and ascending tracts that transmit impulses from the cerebrum and other parts of the brain to the spinal cord or from the spinal cord to the brain. The internal capsule (described shortly) is an example.

Basal Ganglia

The **basal ganglia** are several groups of nuclei in each cerebral hemisphere (Figs. 14.8 and 14.12). The largest nucleus in the basal ganglia is the **corpus striatum** (strī-Ā-tum; *corpus* = body; *striatum* = striped). It consists of the **cau-** date (*cauda* = tail) **nucleus** and the **lenticular** (*lenticula* = shaped like a lentil or lens) **nucleus.** The lenticular nucleus, in turn, is subdivided into a lateral portion called the **putamen** (pu-TĀ-men; *putamen* = shell) and a medial portion called the **globus pallidus** (*globus* = ball; *pallid* = pale).

The portion of the **internal capsule** passing between the lenticular nucleus and the caudate nucleus and between the lenticular nucleus and thalamus is sometimes considered part of the corpus striatum (see Fig. 14.8). The internal capsule is made up of a group of sensory and motor white matter tracts that connect the cerebral cortex with the brain stem and spinal cord.

Other structures often considered part of the basal ganglia are the **substantia nigra, subthalamic nuclei,** and **red nuclei** (see Fig. 14.7). The substantia nigra is a pair of large nuclei in the midbrain whose axons terminate in the caudate nucleus and putamen. The subthalamic nuclei lie against the internal capsule. Their major connection is with the globus pallidus. The red nuclei, as previously mentioned, are part of the midbrain.

FIGURE 14.12 Basal ganglia. (a) In this diagram, the basal ganglia have been projected to the surface. Refer to Fig. 14.8 for the positions of the basal ganglia in a frontal section of the cerebrum. (b) Left cerebral hemisphere showing portions of the basal ganglia.

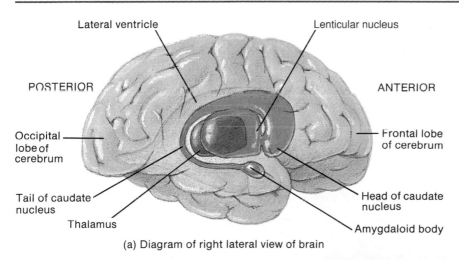

(a) Diagram of right lateral view of brain

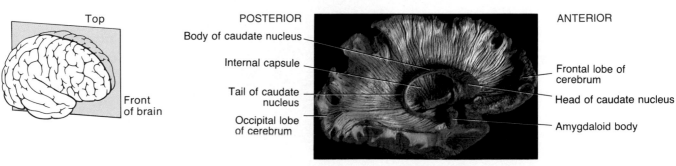

(b) Sagittal section

Question: What brain regions comprise the basal ganglia? (Look back to Fig. 14.8, also.) In (b), what structure has been scooped out to reveal the internal capsule, which lies lateral to the caudate nucleus?

The basal ganglia are interconnected by many nerve fibers. They also receive input from and provide output to the cerebral cortex, thalamus, and hypothalamus. The caudate nucleus and the putamen control large automatic movements of skeletal muscles, such as swinging the arms while walking. The globus pallidus is concerned with the regulation of muscle tone required for specific body movements.

Limbic System

Encircling the brain stem is a ring of structures on the inner border of the cerebrum and floor of the diencephalon that forms the **limbic** (*limbus* = border) **system.** Among its components are the following regions of gray matter (Fig. 14.13).

1.   **Limbic lobe.** The **parahippocampal** and **cingulate gyri,** both gyri of the cerebral hemispheres, and the **hippocampus,** an extension of the parahippocampal gyrus that extends into the floor of the lateral ventricle.

2.   **Dentate gyrus.** A cerebral gyrus between the hippocampus and parahippocampal gyrus.

3.   **Amygdaloid body (amygdala).** Several groups of neurons located at the tail end of the caudate nucleus.

4.   **Septal nuclei.** Nuclei within the septal area, formed by the regions under the corpus callosum and a cerebral gyrus (paraterminal).

5.   **Mammillary bodies of the hypothalamus.** Two round masses close to the midline near the cerebral peduncles.

6.   **Anterior nucleus of the thalamus.** Located in the floor of the lateral ventricle.

7.   **Olfactory bulbs.** Flattened bodies of the olfactory nerves (cranial nerve I) that rest on the cribriform plate.

8.   **Bundles of interconnecting myelinated axons.** Tracts that interconnect various components of the limbic system and include the fornix, stria terminalis, stria medullaris, medial forebrain bundle, and mammillothalamic tract.

The limbic system functions in emotional aspects of behavior related to survival. The hippocampus, together

---

FIGURE 14.13 Selected components of the limbic system and surrounding structures.

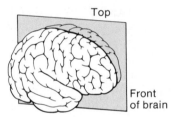

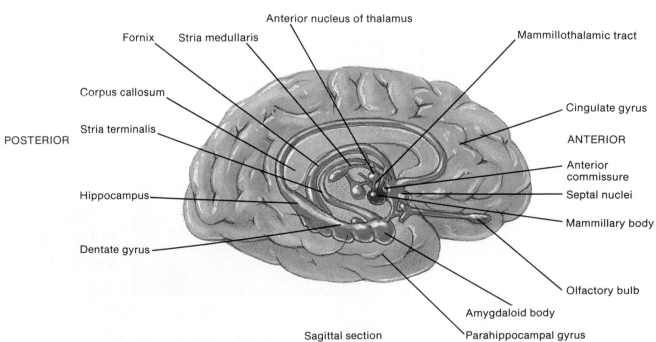

Sagittal section

Question: **Which part of the limbic system functions with the cerebrum in memory?**

with portions of the cerebrum, also functions in memory. Because of this relationship, events that cause a strong emotional response are committed to memory much more efficiently than those that do not. Memory impairment results from lesions in parts of the limbic system. People with such damage forget recent events and cannot commit anything to memory. Specifically how the limbic system functions in memory is not clear. Although behavior is a function of the entire nervous system, the limbic system controls most of its involuntary aspects.

Experiments have shown that the limbic system is associated with pleasure and pain. When certain areas of the limbic system are stimulated in animals, their reactions indicate they are experiencing intense punishment. When other areas are stimulated, the animals' reactions indicate they are experiencing extreme pleasure. Stimulation of the amygdaloid body or certain nuclei of the hypothalamus in a cat results in a behavioral pattern called rage. The cat assumes a defensive posture—extending its claws, raising its tail, hissing, spitting, and opening its eyes wide. Stimulating other areas of the limbic system results in an opposite behavioral pattern: docility, tameness, and affection. Because the limbic system has a primary function in emotions such as pain, pleasure, anger, rage, fear, sorrow, sexual feelings, docility, and affection, it is sometimes called the "emotional" brain.

CLINICAL APPLICATION

BRAIN INJURIES

Brain injuries are commonly associated with head injuries and result in part from displacement and distortion of neuronal tissue at the moment of impact. Although trauma to the head can lead to brain injuries, not all the damage is due to the impact alone. Some damage appears to be from the release of large numbers of oxygen free radicals (charged oxygen molecules with an unpaired electron) from damaged cells. Brain cells recovering from the effects of a stroke or cardiac arrest also release free radicals. Free radicals cause damage by disrupting cellular DNA and enzymes and altering plasma membrane permeability. Various degrees of brain injury are described by the following terms.

1. Concussion. An abrupt but temporary loss of consciousness following a blow to the head or the sudden stopping of a moving head. A concussion produces no visible bruising of the brain, but posttraumatic amnesia (memory loss) may occur.
2. Contusion. A visible bruising of the brain due to trauma and blood leaking from microscopic vessels. The pia mater may be torn, allowing blood to enter the subarachnoid space. A contusion usually results in an extended loss of consciousness, ranging from several minutes to many hours.
3. Laceration. Tearing of the brain, usually from a skull fracture or gunshot wound. A laceration results in rupture of large blood vessels with bleeding into the brain and subarachnoid space. Consequences include cerebral hematoma (blood clot), edema, and increased intracranial pressure.

Functional Areas of the Cerebral Cortex

Specific types of sensory, motor, and integrative signals are processed in certain cerebral regions (Fig. 14.14). Generally, the **sensory areas** receive and interpret sensory impulses, the **motor areas** control muscular movement, and the **association areas** deal with more complex integrative functions such as memory, emotions, reasoning, will, judgment, personality traits, and intelligence.

● **Sensory Areas** Sensory input to the cerebral cortex flows mainly to the posterior half of the hemispheres, to regions behind the central sulci. In the cortex, primary sensory areas have the most direct connections with peripheral sensory receptors. Secondary sensory areas and sensory association areas often are adjacent to the primary areas. Usually, they receive input from the primary areas and from diverse other regions of the brain. They participate in the interpretation of sensory experiences into meaningful patterns of recognition and awareness. For example, a person with damage in the primary visual cortex would be blind in at least part of his visual field. A person with damage to a visual association area, on the other hand, might see normally yet be unable to recognize a friend.

1. **Primary somatosensory area** or **general sensory area.** Located directly posterior to the central sulcus of each cerebral hemisphere in the postcentral gyrus of each parietal lobe. It extends from the longitudinal fissure on the top of the cerebrum to the lateral cerebral sulcus. In Fig. 14.14, the somatosensory area is designated by the areas numbered 1, 2, and 3.*

The primary somatosensory area receives nerve impulses from somatic sensory receptors for touch, proprioception (joint and muscle position), pain, and temperature. Each point within the area receives sensations from a specific part of the body, and essentially the entire body is spatially represented in it. The size of the cortical area receiving impulses from a particular body part depends on the number of receptors present there rather than on the size of the part. For example, a larger portion of the sensory area receives impulses from the lips and fingertips than from the thorax or hip (see Fig. 15.6a). The major function of the primary somatosensory area is to localize exactly the points of the body where the sensations originate. The thalamus is capable of registering sensations in a general way. It receives sensations from large areas of the body but cannot distinguish precisely the specific area of stimulation. This capability depends on the primary somatosensory area of the cortex.
2. **Primary visual area (area 17).** Located on the medial surface of the occipital lobe and occasionally

---

* These numbers, as well as most of the others shown, are based on K. Brodmann's cytoarchitectural map of the cerebral cortex. His map, first published in 1909, was a brilliant and successful attempt to correlate specific brain regions with particular functions.

FIGURE 14.14 Functional areas of the cerebrum. Note that Broca's area is in the left hemisphere of most people; it is shown here to indicate its location.

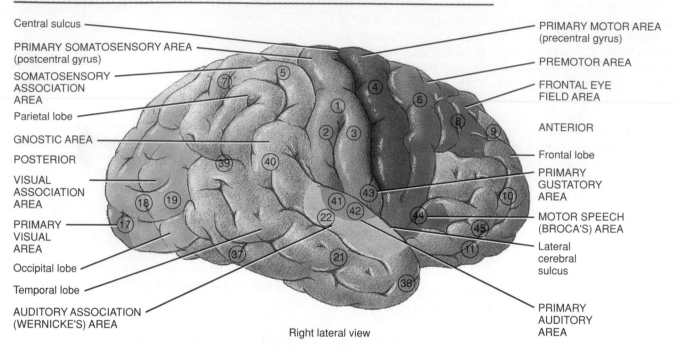

Central sulcus

PRIMARY SOMATOSENSORY AREA
(postcentral gyrus)

SOMATOSENSORY
ASSOCIATION
AREA

Parietal lobe

GNOSTIC AREA

POSTERIOR

VISUAL
ASSOCIATION
AREA

PRIMARY
VISUAL
AREA

Occipital lobe

Temporal lobe

AUDITORY ASSOCIATION
(WERNICKE'S) AREA

PRIMARY MOTOR AREA
(precentral gyrus)

PREMOTOR AREA

FRONTAL EYE
FIELD AREA

ANTERIOR

Frontal lobe

PRIMARY
GUSTATORY
AREA

MOTOR SPEECH
(BROCA'S) AREA

Lateral
cerebral
sulcus

PRIMARY
AUDITORY
AREA

Right lateral view

Question: What area of the cerebrum integrates interpretation of visual, auditory, and somatic sensations? Translates thoughts into speech? Controls skilled muscular movements? Interprets sensations related to taste? Interprets pitch and rhythm? Interprets shape, color, and movement of objects? Controls voluntary scanning movements of the eyes?

extending around to the lateral surface. Axons of neurons with cell bodies in the eye form the optic nerves (cranial nerve II), which terminate in the lateral geniculate nuclei of the thalamus. From the thalamus, neurons carrying visual input project to the primary visual area, with information concerning shape, color, and movement.

3. Primary auditory area (areas 41 and 42). Located in the superior part of the temporal lobe near the lateral cerebral sulcus, it interprets the basic characteristics of sound such as pitch and rhythm.

4. Primary gustatory area (area 43). Located at the base of the postcentral gyrus above the lateral cerebral sulcus in the parietal cortex, it receives impulses related to taste.

5. Primary olfactory area. Located in the temporal lobe on the medial aspect, it receives impulses related to smell.

● **Motor Areas** Motor output from the cerebral cortex flows mainly from the anterior portion of each hemisphere.

1. Primary motor area (area 4). Located in the precentral gyrus of the frontal lobe (Fig. 14.14). Each region in the primary motor area controls voluntary contractions of specific muscles or groups of muscles (see Fig. 15.6b). Electrical stimulation of any point in the primary motor area results in contraction of specific skeletal muscle fibers on the opposite side of the body. As is true for the primary somatosensory area, body parts are represented unequally here. More cortical area is devoted to those muscles where skilled, complex, or delicate movement is required, for example, finger maneuvers.

2. Language areas. The translation of speech or written words into thought involves both sensory and association areas—primary auditory, auditory association, primary visual, visual association, and gnostic (described shortly). The translation of thoughts into speech involves the **motor speech area** (area 44) also called **Broca's** (BRŌ-kaz) **area,** located in the frontal lobe just superior to the lateral cerebral sulcus. From this area, a sequence of nerve impulses is sent to the premotor regions (described shortly) that control the muscles of the larynx, pharynx, and mouth. The impulses from the premotor area to the muscles result in specific, coordinated contractions that enable you to speak. Simultaneously, impulses are sent from the motor speech area to the primary motor area. From here, impulses reach your breathing muscles to regulate the proper flow of air past the vocal cords. The coordinated contractions of your speech and breathing muscles enable you to translate your thoughts into speech.

● **Association Areas** The **association areas** of the cerebrum consist of association tracts that connect motor and sensory areas and large parts of the cortex on the lateral surfaces of the occipital, parietal, and temporal lobes and the frontal lobes anterior to the motor areas (see Fig. 14.10). Association areas include:

1. **Somatosensory association area (areas 5 and 7).** Just posterior to the primary somatosensory area, it receives input from the thalamus, other lower portions of the brain, and the primary somatosensory area. Its role is to integrate and interpret sensations. This area permits you to determine the exact shape and texture of an object without looking at it, to determine the orientation of one object to another as they are felt, and to sense the relationship of one body part to another. Another role of the somatosensory association area is the storage of memories of past sensory experiences. Thus you can compare sensations with previous experiences.

2. **Visual association area (areas 18 and 19).** Located in the occipital lobe, it receives sensory impulses from the primary visual area and the thalamus. It relates present to past visual experiences with recognition and evaluation of what is seen.

3. **Auditory association (Wernicke's) area (area 22).** Located inferior to the primary auditory area in the temporal cortex, it determines if a sound is speech, music, or noise. It also interprets the meaning of speech by translating words into thoughts.

4. **Gnostic** (NOS-tik; *gnosis* = knowledge) **area (areas 5, 7, 39, and 40).** This **common integrative area** is located among the somatosensory, visual, and auditory association areas. The gnostic area receives nerve impulses from these areas, as well as from the taste and smell areas, the thalamus, and lower portions of the brain stem. It integrates sensory interpretations from the association areas and impulses from other areas so that a common thought can be formed from the various sensory inputs. It then transmits signals to other parts of the

brain to cause the appropriate response to the sensory signal.

5. **Premotor area (area 6).** Immediately anterior to the primary motor area is a motor association area. Neurons in this region communicate with the primary motor cortex, sensory association areas in the parietal lobe, the basal ganglia, and the thalamus. The premotor area deals with learned motor activities of a complex and sequential nature. It generates nerve impulses that cause a specific group of muscles to contract in a specific sequence, for example, to write a word. The premotor area controls learned skilled movements and serves as a memory bank for such movements.

6. **Frontal eye field area (area 8).** This area in the frontal cortex is sometimes included in the premotor area. It controls voluntary scanning movements of the eyes—searching for a word in a dictionary, for instance.

CLINICAL APPLICATION

SPEECH AREA INJURIES

Understanding and speaking a language involve several regions of the cortex. Much of what we know about these areas, which are most highly developed in human beings, comes from studies of patients with language or speech disturbances that have resulted from brain damage. The motor speech (Broca's) area, auditory association (Wernicke's) area, and other language areas are located in the left cerebral hemisphere of most people, regardless of whether they are left-handed or right-handed. Injury to the sensory or motor speech areas results in **aphasia** (a-FĀ-zē-a; *a* = without; *phasis* = speech), an inability to speak. Damage to the motor speech area results in nonfluent aphasia, an inability to properly articulate or form words. The person knows what she wishes to say but cannot speak. Damage to the gnostic area or auditory association area (areas 39 and 22) results in fluent aphasia, faulty understanding of spoken or written words. Such a patient may fluently produce strings of words that have no meaning. The deficiency may be **word deafness,** an inability to understand spoken words, or **word blindness,** an inability to understand written words, or both.

Electroencephalogram (EEG)

At any instant, brain cells are generating millions of nerve impulses (nerve action potentials) and graded potentials (excitatory and inhibitory postsynaptic potentials) in individual neurons. These electrical potentials added together are called **brain waves** and indicate electrical activity of the cerebral cortex. Brain waves pass through the skull and can be detected by sensors called electrodes. A record of such waves is called an **electroencephalogram (EEG).**

As indicated in Fig. 14.15, four kinds of waves are produced by normal individuals.

1. **Alpha waves.** These rhythmic waves occur at a frequency of about 8 to 13 cycles per second. [The unit commonly used to express frequency is the hertz (Hz); 1

Hz = 1 cycle per second.] Alpha waves are present in the EEGs of nearly all normal individuals when awake and in the resting state with their eyes closed. These waves disappear entirely during sleep.

2. **Beta waves.** The frequency of these waves is between 14 and 30 Hz. Beta waves generally appear when the nervous system is active, that is, during periods of sensory input and mental activity.

3. **Theta waves.** These waves have frequencies of 4 to 7 Hz. Theta waves normally occur in children and in adults experiencing emotional stress. They also occur in many disorders of the brain.

4. **Delta waves.** The frequency of these waves is 1 to 5 Hz. Delta waves occur during deep sleep. They are normal in an awake infant. When produced by an awake adult, they indicate brain damage.

CLINICAL APPLICATION

ELECTROENCEPHALOGRAM (EEG)

An **electroencephalogram** (e-lek′-trō-en-SEF-a-lō-gram′) (**EEG** or **brain wave test**) is used to diagnose epilepsy and other seizure disorders, infectious diseases, tumors, trauma, hematomas, metabolic abnormalities, degenerative diseases, and periods of unconsciousness and confusion. In some cases, an EEG provides useful information regarding sleep and wakefulness. An EEG may also be one criterion in establishing brain death (complete absence of brain waves in two EEGs taken 24 hours apart).

To record an EEG, 16 to 30 electrodes are applied to the scalp and connected by wires to an amplifier and recorder (electroencephalograph). Brain electrical signals are displayed as a series of wavy lines on a moving sheet of graph paper.

Brain Lateralization

On gross examination, the brain appears the same on both sides. However, detailed examination reveals subtle anatomical differences between the two hemispheres. For example, in left-handed people the parietal and occipital lobes of the right hemisphere are usually narrower than the corresponding lobes of the left hemisphere. In addition, the frontal lobe of the left hemisphere of such individuals is typically narrower than that of the right hemisphere.

In addition to the structural differences between the hemispheres, there are important functional differences (Fig. 14.16). The left hemisphere controls the right side of the body, and the right hemisphere controls the left side of the body. Also, in most people the left hemisphere is more important for spoken and written language, numerical and scientific skills, ability to use and understand sign language, and reasoning. Conversely, it has been shown that the right hemisphere is more important for musical and artistic awareness, space and pattern perception, insight, imagination, and generating mental images of sight, sound, touch, taste, and smell to compare relationships.

FIGURE 14.15 Types of brain waves recorded in an electroencephalogram (EEG).

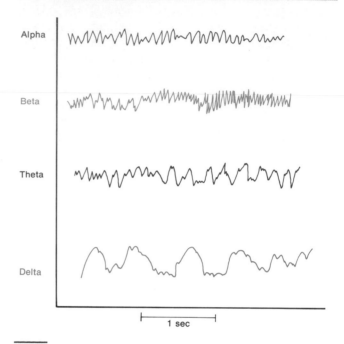

Question: Which waves indicate emotional stress?

FIGURE 14.16 Summary of the principal functional differences between the left and right cerebral hemispheres.

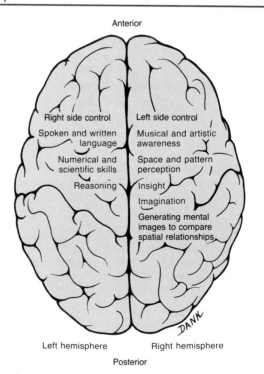

Question: Which lobes of the cerebrum are usually narrower in left-handed people?

Cerebellum

The **cerebellum** is the second-largest portion of the brain and occupies the inferior and posterior aspects of the cranial cavity. Specifically, it is posterior to the medulla and pons and inferior to the occipital lobes of the cerebrum (see Fig. 14.1). It is separated from the cerebrum by the **transverse fissure** and by an extension of the cranial dura mater called the **tentorium** (*tentorium* = tent) **cerebelli.** The tentorium cerebelli partially encloses the transverse sinuses (veins) and supports the occipital lobes of the cerebrum.

The cerebellum is shaped somewhat like a butterfly. The central constricted area is the **vermis,** which means "worm-shaped," and the lateral "wings" or lobes are termed **cerebellar hemispheres** (Fig. 14.17). Each hemisphere consists of lobes that are separated by deep and distinct fissures. The

FIGURE 14.17 Cerebellum.

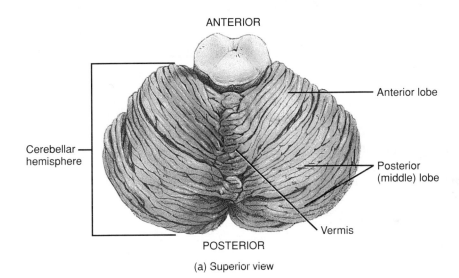

(a) Superior view

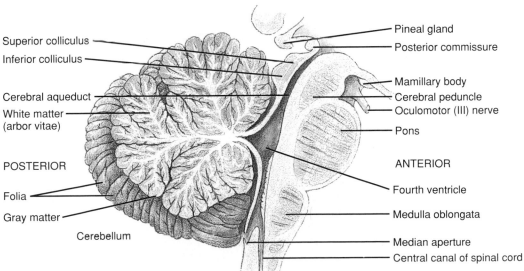

(b) Sagittal section of cerebellum and brain stem

Question: What are the fiber tracts that channel information into and out of the cerebellum called?

anterior lobe and posterior (middle) lobe are concerned with subconscious movements of skeletal muscles. The flocculonodular lobe on the inferior surface is concerned with the sense of equilibrium (see Chapter 16). Between the hemispheres is a sickle-shaped extension of the cranial dura mater, the falx (*falx* = sickle) cerebelli. It passes only a short distance between the cerebellar hemispheres and contains the occipital sinus (vein).

The surface of the cerebellum, called the cerebellar cortex, consists of gray matter in a series of slender, parallel ridges called folia. They are less prominent than the convolutions of the cerebral cortex. Beneath the gray matter are white matter tracts (arbor vitae) that resemble branches of a tree. Deep within the white matter are masses of gray matter, the cerebellar nuclei. The nuclei give rise to nerve fibers that convey information out of the cerebellum to other parts of the nervous system (to both other brain centers and the spinal cord).

The cerebellum is attached to the brain stem by three paired bundles of fibers (tracts) called cerebellar peduncles (see Fig. 14.5). Inferior cerebellar peduncles connect the cerebellum with the medulla at the base of the brain stem and with the spinal cord. These peduncles contain both afferent and efferent fibers and thus carry information into and out of the cerebellum. Middle cerebellar peduncles contain only afferent fibers, which conduct input from the pons into the cerebellum. Superior cerebellar peduncles connect the cerebellum with the midbrain. These peduncles contain mostly motor fibers that conduct output from the cerebellum.

The cerebellum functions to compare the intended movement determined by motor areas of the brain (cerebrum and basal ganglia) with what is actually happening. The cerebellum constantly receives sensory input from proprioceptors in muscles, tendons, and joints, receptors for equilibrium, and visual receptors of the eyes. If the intent of the motor areas is not being attained by skeletal muscles, the cerebellum detects the variation and sends feedback signals to the motor areas to either stimulate or inhibit the activity of skeletal muscles. This interaction helps to smooth and coordinate complex sequences of skeletal muscle contractions. Note that the cerebellum does not directly send nerve impulses to skeletal muscles. Besides coordinating skilled movements, the cerebellum is the main brain region that regulates posture and balance. These aspects of cerebellar function make possible all skilled motor activities, from catching a baseball to dancing.

A summary of the functions of the various parts of the brain is presented in Exhibit 14.2.

EXHIBIT 14.2

SUMMARY OF FUNCTIONS OF PRINCIPAL PARTS OF BRAIN

Part	Function
BRAIN STEM	
Medulla	Relays motor and sensory impulses between other parts of the brain and the spinal cord. Reticular formation (also in pons, midbrain, and diencephalon) functions in consciousness and arousal. Vital reflex centers regulate heartbeat, breathing (together with pons), and blood vessel diameter. Other reflex centers coordinate swallowing, vomiting, coughing, sneezing, and hiccuping. Contains nuclei of origin for cranial nerves VIII, IX, X, XI, and XII. Vestibular nuclear complex helps maintain equilibrium.
Pons	Relays impulses within the brain and between parts of the brain and spinal cord. Contains nuclei of origin for cranial nerves V, VI, VII, and VIII. Pneumotaxic area and apneustic area, together with the medulla, help control breathing.
Midbrain	Relays motor impulses from the cerebral cortex to the pons and spinal cord and relays sensory impulses from the spinal cord to the thalamus. Superior colliculi coordinate movements of the eyeballs in response to visual and other stimuli, and the inferior colliculi coordinate movements of the head and trunk in response to auditory stimuli. Contains nuclei of origin for cranial nerves III and IV.
DIENCEPHALON	
Thalamus	Several nuclei serve as relay stations for all sensory input to the cerebral cortex. Relays motor impulses from the cerebral cortex to the spinal cord. Provides crude appreciation of touch/pressure, pain, and temperature. Anterior nucleus functions in emotions and memory.

Part	Function
Hypothalamus	Receives hormonal signals and sensory input from viscera.
	Controls and integrates activities of the autonomic nervous system.
	Principal link between endocrine and nervous systems and regulates the pituitary gland.
	Secretes hypothalamic regulating hormones.
	Functions in rage and aggression.
	Controls normal body temperature, food intake, and thirst.
	Helps maintain the waking state and sleep.
CEREBRUM	Sensory areas interpret sensory impulses, motor areas control muscular movement, and association areas function in emotional and intellectual processes.
	Basal ganglia coordinate gross, automatic muscle movements and regulate muscle tone.
	Limbic system functions in emotional aspects of behavior related to survival.
CEREBELLUM	Controls skeletal muscle contractions required for skilled movements, coordination, posture, and balance.

NEUROTRANSMITTERS IN THE BRAIN

About 50 substances are either known or strongly suspected neurotransmitters in the brain. The list grows longer each year as new cytological techniques confirm the existence of a wide array of possible neurotransmitters. Finding out what each one does, however, is not easy. Dendrites, cell bodies, and axons are tightly packed and closely intermingled in nervous tissue, and the amount of neurotransmitter liberated at a single synapse is tiny. Some neurotransmitters bind to their receptors and act quickly to open or close ion channels in the membrane. Others act more slowly by second messenger systems to influence enzymatic reactions inside the cell. The result of either process can be excitation or inhibition of the postsynaptic neuron.

Quite a few neurotransmitters also are known to be hormones, released into the blood by endocrine cells in organs throughout the body. Within the brain, certain neurons, called **neuroendocrine cells,** also secrete hormones. Still other neurons liberate substances called **neuromodulators** into interstitial or cerebrospinal fluid. Neuromodulators influence signaling of neighboring or distant neurons, magnifying or dampening their response to other neurotransmitters.

Exhibit 14.3 lists some of the known and suspected neurotransmitters and neuromodulators. Chemically, most neurotransmitters are small, water-soluble molecules. In addition to acetylcholine, which is in a chemical category by itself, the others all belong to one of three chemical families—amino acids, biogenic amines, or neuropeptides.

1.   Acetylcholine (ACh). Depending on the nature of the postsynaptic receptors, ACh may be excitatory or inhibitory in the brain. Recall that ACh is the neurotransmitter released at neuromuscular junctions in skeletal muscle (Chapter 10). ACh is inactivated by acetylcholinesterase (AChE), which splits the molecule into its acetate and choline portions. In the brain, a major center of neurons that liberate ACh is the nucleus basalis, which is inferior to the globus pallidus. Axons of these neurons project widely throughout the cerebral cortex and limbic system. Destruction of these neurons is a hallmark of Alzheimer's disease (see page 438). Neurons of the primary motor area in the precentral gyrus that form the pyramids of the medulla and descend in the spinal cord as the corticospinal tract also use ACh.

2.   Amino acids. The amino acids **glutamate** and **aspartate** are known excitatory transmitters in the brain. The most common inhibitory neurotransmitter in the brain is **gamma aminobutyric acid (GABA).** It is most highly concentrated in the superior and inferior colliculi, thalamus, hypothalamus, and occipital lobes of the cerebrum. It has been implicated as a likely target for antianxiety drugs such as diazepam (Valium), which enhance the action of GABA. An inhibitory neurotransmitter in the spinal cord is **glycine.**

3.   Biogenic amines. Certain amino acids are modified and decarboxylated (carboxyl group removed) to produce biogenic amines. Those that are prevalent in the brain include norepinephrine, epinephrine, dopamine, serotonin, and histamine. Depending on the type of receptor (there are three or more different types for each amine), they may cause either excitation or inhibition.

Norepinephrine (NE) is concentrated in a group of neurons in the brain stem that project their axons into the hypothalamus, cerebellum, cerebral cortex, and spinal cord. In these places, NE has been implicated in maintaining arousal (awakening from deep sleep), dreaming, and the regulation of mood.

Neurons containing the neurotransmitter **dopamine (DA)** are clustered in the midbrain in the substantia nigra (see Fig. 14.7). Some axons projecting from the substantia nigra terminate in the cerebral cortex, where DA is thought to be involved in emotional responses. Other axons project to the corpus striatum (basal ganglia), where the DA is involved in gross, automatic movements of skeletal muscles. Degeneration of axons

EXHIBIT 14.3

KNOWN AND SUSPECTED NEUROTRANSMITTERS AND NEUROMODULATORS

ACETYLCHOLINE (ACh)

AMINO ACIDS
Aspartate
Gamma aminobutyric acid (GABA)
Glutamate
Glycine

BIOGENIC AMINES
Dopamine (DA)
Epinephrine
Histamine
Norepinephrine (NE)
Serotonin

NEUROPEPTIDES[a]
Angiotensin II
Antidiuretic hormone (vasopressin)
Atrial natriuretic peptide (ANP)
Bombesin
Cholecystokinin (CCK)
Corticotropin
Dynorphins
Endorphins
Enkephalins
Gastrin
Glucagon
Hypothalamic regulating hormones
Inhibin
Melatonin
Motilin
Neuropeptide Y
Neurotensin
Oxytocin
Secretin
Substance P
Vasoactive intestinal polypeptide (VIP)

[a]Many are also hormones that are released by tissues outside the brain.

catechol-*O*-methyltransferase (kat′-e-kōl-ō-meth-il-TRANS-fer-ās), or **COMT,** and **monoamine oxidase** (mon-ō-AM-ēn OK-si-dās), or **MAO,** or recycled back into the synaptic vesicles.

Serotonin or **5-hydroxytryptophan (5-HT)** is concentrated in the neurons in a part of the brain stem called the **raphe nucleus.** Axons projecting from the nucleus terminate in the hypothalamus, thalamus, and other parts of the brain and spinal cord. Serotonin is thought to be involved in inducing sleep, sensory perception, temperature regulation, and control of mood.

4. Neuropeptides. The largest family of neurotransmitters and neuromodulators are the **neuropeptides,** which consist of chains of 3 to about 40 amino acids. They have both excitatory and inhibitory actions. Neuropeptides are formed in the cell body, packaged into vesicles, and transported to axon terminals.

In 1974 scientists discovered that certain brain neurons have plasma membrane receptors for opiate drugs such as morphine and heroin. The quest to find the naturally occurring substances that use these receptors brought to light the first neuropeptides. They were two molecules, each a chain of 5 amino acids, named **enkephalins** (en-KEF-a-lins; *enkephalos* = brain). They have potent analgesic (pain-relieving) effects, 200 times stronger than morphine. Other so-called opioid peptides include the **endorphins** (en-DOR-fins) and **dynorphins** (dī-NOR-fins).

Enkephalins are concentrated in the thalamus, the hypothalamus, parts of the limbic system (see Fig. 14.13), and those spinal cord pathways that relay impulses for pain. It has been suggested that opioid peptides are the body's natural painkillers. They have also been linked to improved memory and learning; feelings of pleasure or euphoria; control of body temperature; regulation of hormones that affect the onset of puberty, sexual drive, and reproduction; and mental illnesses such as depression and schizophrenia. The analgesic effects of acupuncture may be due to increased release of enkephalins or endorphins.

Another neuropeptide, **substance P,** is found in sensory nerves, spinal cord pathways, and parts of the brain associated with pain transmission. When substance P is released by neurons, it transmits pain-related input from peripheral pain receptors into the central nervous system. Acting by presynaptic inhibition (see page 366), enkephalin suppresses the release of substance P, thus decreasing painful input to the CNS. Substance P has also been shown to counter the effects of certain nerve-damaging chemicals, prompting speculation that it might prove useful as a treatment for nerve degeneration.

Many of the neuropeptides are also found in other parts of the body, particularly the digestive system, where they serve as hormones or other regulators of physiological responses. We will be discussing many of them in later chapters, but several are summarized in Exhibit 14.4 to give you some idea of their diversity.

that project from the substantia nigra to the basal ganglia occurs in Parkinson's disease (see page 439).

Norepinephrine, dopamine, and epinephrine (which is both a hormone and a neurotransmitter) are catecholamines. They all include a catechol ring (*six carbons and two adjacent hydroxyl (OH) groups*). Inactivation of the catecholamines is different from that of ACh. Soon after release from synaptic vesicles, they are pumped back into the synaptic end-bulbs. This is called reuptake. Then they are either destroyed by the enzymes

EXHIBIT 14.4

SUMMARY OF REPRESENTATIVE NEUROPEPTIDES

Substance	Comment
Enkephalins	Concentrated in thalamus, hypothalamus, parts of limbic system, and spinal cord pathways that relay pain impulses; inhibit pain impulses by suppressing substance P release.
Endorphins	Concentrated in pituitary gland; function in inhibiting pain by blocking release of substance P; may have a role in memory and learning, sexual activity, control of body temperature; have been linked to depression and schizophrenia.
Substance P	Found in sensory nerves, spinal cord pathways, and parts of brain associated with pain; stimulates perception of pain.
Dynorphin	Found in posterior pituitary gland, hypothalamus, and small intestine; may be related to controlling pain and registering emotions.
Hypothalamic regulating hormones	Produced by hypothalamus; regulate the release of hormones by the pituitary gland.
Angiotensin II	Stimulates thirst; improves memory; may regulate blood pressure in brain; as a hormone promotes release of aldosterone, which increases the rate of salt and water reabsorption by the kidneys.
Cholecystokinin (CCK)	Found in cerebral cortex and small intestine; may be related to the regulation of feeding as a "stop eating" signal; as a hormone regulates pancreatic enzyme secretion during digestion and contraction of smooth muscle in the gastrointestinal tract.
Oxytocin (OT)	Present in neurons that project to brain stem and spinal cord; improves memory; as a hormone stimulates uterine contractions at childbirth and stimulates milk release from mammary glands.
Antidiuretic hormone (ADH) or vasopressin	Present in neurons that project to brain stem and spinal cord; improves memory; as a hormone regulates rate of water reabsorption by the kidneys and causes constriction of blood vessels.

CRANIAL NERVES

Cranial nerves, like spinal nerves, are part of the peripheral nervous system (PNS). Of the 12 pairs of **cranial nerves,** 10 originate from the brain stem (Fig. 14.18), but all pass through foramina of the skull. The cranial nerves are designated with roman numerals and with names (see Fig. 14.5). The roman numerals indicate the order in which the nerves arise from the brain from anterior to posterior. The names indicate the distribution or function.

Some cranial nerves contain only sensory fibers and thus are called **sensory nerves.** The remainder contain both sensory and motor fibers and are referred to as **mixed nerves,** though a few of them are predominantly motor in function. The cell bodies of sensory fibers are found outside the brain, whereas the cell bodies of motor fibers lie in nuclei within the brain (Fig. 14.18). Motor fibers include both somatic and autonomic efferents.

A summary of cranial nerves and clinical applications related to their dysfunction is presented in Exhibit 14.5.

AGING AND THE NERVOUS SYSTEM

One of the effects of aging on the nervous system is loss of neurons. This is a consequence of the aging process, not necessarily a disease state. Associated with this decline, there is a decreased capacity for sending nerve impulses to and from the brain so that processing of information diminishes. Conduction velocity decreases, voluntary motor movements slow down, and reflex times increase. Parkinson's disease is the most common movement disorder of the CNS. Degenerative changes and disease states involving the sense organs can alter vision, hearing, taste, smell, and touch. Impaired hearing associated with aging, known as presbycusis, is usually the result of changes in important structures of the inner ear.

DEVELOPMENTAL ANATOMY OF THE NERVOUS SYSTEM

The development of the nervous system begins early in the third week of development with a thickening of the **ecto-**

FIGURE 14.18 Nuclei of cranial nerves II through XII.

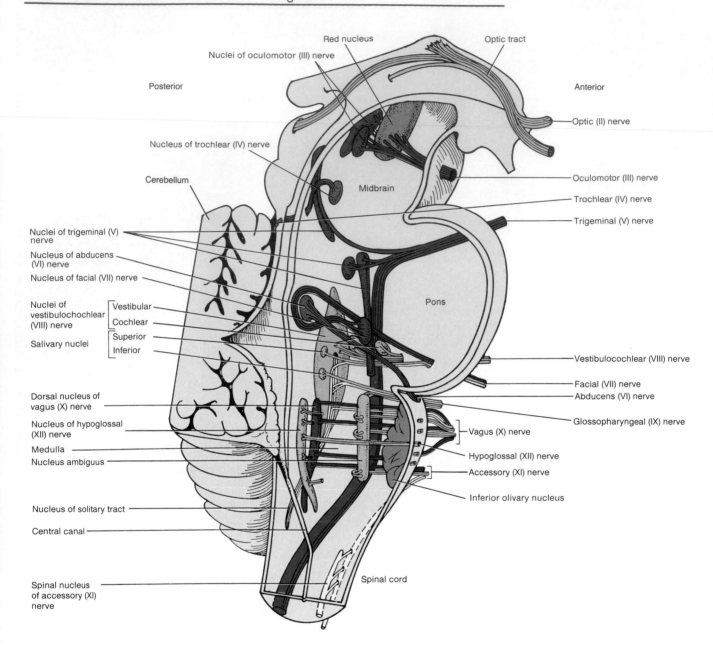

Question: Which cranial nerves originate in the midbrain?

EXHIBIT 14.5

SUMMARY OF CRANIAL NERVES[a]

Nerve (Type)	Location	Function and Clinical Application
Olfactory (I) (sensory)	Arises in olfactory mucosa, passes through foramina in the cribriform plate of ethmoid bone, and ends in the olfactory bulb. The olfactory tract extends via two pathways to olfactory areas of cerebral cortex.	*Function:* Smell. *Clinical application:* Loss of the sense of smell, called *anosmia*, may result from head injuries in which the cribriform plate of the ethmoid bone is fractured and from lesions along the olfactory pathway.
Optic (II) (sensory)	Arises in retina of the eye, passes through optic foramen, forms optic chiasma, passes through optic tracts, and terminates in lateral geniculate nuclei of thalamus. From thalamus, projections extend to visual areas of the cerebral cortex.	*Function:* Vision. *Clinical application:* Fractures in the orbit, lesions along the visual pathway, and diseases of the nervous system may result in visual field defects and loss of visual acuity. Defective vision is called *anopsia*.
Oculomotor (III) (mixed, primarily motor)	*Motor portion:* Originates in midbrain, passes through superior orbital fissure, and is distributed to levator palpebrae superioris of upper eyelid and four extrinsic eyeball muscles (superior rectus, medial rectus, inferior rectus, and inferior oblique); parasympathetic innervation to ciliary muscle of eyeball and sphincter muscle of iris. *Sensory portion:* Consists of afferent fibers from proprioceptors in eyeball muscles that pass through superior orbital fissure and terminate in midbrain.	*Motor function:* Movement of eyelid and eyeball, accommodation of lens for near vision, and constriction of pupil. *Sensory function:* Muscle sense (proprioception). *Clinical application:* A lesion in the nerve causes *strabismus* (squinting), *ptosis* (drooping) of the upper eyelid, pupil dilation, the movement of the eyeball downward and outward on the damaged side, a loss of accommodation for near vision, and double vision (*diplopia*).
Trochlear (IV) (mixed, primarily motor)	*Motor portion:* Originates in midbrain, passes through superior orbital fissure, and is distributed to superior oblique muscle, an extrinsic eyeball muscle. *Sensory portion:* Consists of afferent fibers from proprioceptors in superior oblique muscles that pass through superior orbital fissure and terminate in midbrain.	*Motor function:* Movement of eyeball. *Sensory function:* Muscle sense (proprioception). *Clinical application:* In trochlear nerve paralysis, diplopia and strabismus occur.
Trigeminal (V) (mixed)	*Motor portion:* Part of the mandibular branch, originates in pons, passes through foramen ovale, and terminates in muscles of mastication (anterior belly of digastric and mylohyoid muscles). *Sensory portion:* Consists of three branches: *ophthalmic*, which contains sensory fibers from skin over upper eyelid, eyeball, lacrimal glands, nasal cavity, side of nose, forehead, and anterior half of scalp and passes through superior orbital fissure; *maxillary*, which contains sensory fibers from mucosa of nose, palate, parts of pharynx, upper teeth, upper lip, and lower eyelid and passes through foramen rotundum; *mandibular*, which contains somatic sensory fibers (but not special sense of taste) from anterior two-thirds of tongue, lower teeth, skin over mandible, cheek and mucosa deep to it, and side of head in front of ear and passes through foramen ovale. The three branches terminate in pons. Sensory portion also consists of afferent fibers from proprioceptors in muscles of mastication.	*Motor function:* Chewing. *Sensory function:* Conveys sensations for touch, pain, and temperature from structures supplied; muscle sense (proprioception). *Clinical application:* Injury results in paralysis of the muscles of mastication and a loss of sensation of touch and temperature. *Neuralgia* (pain) of one or more branches of trigeminal nerve is called *trigeminal neuralgia (tic douloureux)*.

Exhibit continues

EXHIBIT 14.5 (continued)

SUMMARY OF CRANIAL NERVES[a]

Nerve (Type)	Location	Function and Clinical Application
Abducens (VI) (mixed, primarily motor)	*Motor portion:* Originates in pons, passes through superior orbital fissure, and is distributed to lateral rectus muscle, an extrinsic eyeball muscle. *Sensory portion:* Consists of afferent fibers from proprioceptors in lateral rectus muscle that pass through superior orbital fissure and terminate in pons.	*Motor function:* Movement of eyeball. *Sensory function:* Muscle sense (proprioception). *Clinical application:* With damage to this nerve, the affected eyeball cannot move laterally beyond the midpoint and the eye is usually directed medially.
Facial (VII) (mixed)	*Motor portion:* Originates in pons, passes through stylomastoid foramen, and is distributed to facial, scalp, and neck muscles; parasympathetic distribution to lacrimal sublingual, submandibular, nasal, and palatine glands. *Sensory portion:* Arises from taste buds on anterior two-thirds of tongue, passes through stylomastoid foramen, and terminates in geniculate ganglion, a nucleus in pons that sends fibers to thalamus for relay to gustatory areas of cerebral cortex. Also consists of afferent fibers from proprioceptors in muscles of face and scalp.	*Motor function:* Facial expression and secretion of saliva and tears. *Sensory function:* Muscle sense (proprioception) and taste. *Clinical application:* Injury produces paralysis of the facial muscles, called *Bell's palsy*, loss of taste, and loss of ability to close the eyes, even during sleep.
Vestibulocochlear (VIII) (sensory)	*Cochlear branch:* Arises in spiral organ (organ of Corti), forms spiral ganglion, passes through nuclei in the medulla, and terminates in thalamus. Fibers synapse with neurons that relay impulses to auditory areas of cerebral cortex. *Vestibular branch:* Arises in semicircular canals, saccule, and utricle and forms vestibular ganglion; fibers terminate in pons and cerebellum.	*Cochlear branch function:* Conveys impulses associated with hearing. *Vestibular branch function:* Conveys impulses associated with equilibrium. *Clinical application:* Injury to the cochlear branch may cause *tinnitus* (ringing) or deafness. Injury to the vestibular branch may cause *vertigo* (a subjective feeling of rotation), *ataxia*, and *nystagmus* (involuntary rapid movement of the eyeball).
Glossopharyngeal (IX) (mixed)	*Motor portion:* Originates in medulla, passes through jugular foramen, and is distributed to stylopharyngeus muscle; parasympathetic distribution to parotid gland. *Sensory portion:* Arises from taste buds on posterior one-third of tongue and from carotid sinus, passes through jugular foramen, and terminates in medulla. Also consists of afferent fibers from somatic sensory receptors on posterior one-third of tongue and proprioceptors in swallowing muscles supplied.	*Motor function:* Secretion of saliva. *Sensory function:* Taste and regulation of blood pressure; muscle sense (proprioception). *Clinical application:* Injury results in difficulty during swallowing, reduced secretion of saliva, loss of sensation in the throat, and loss of taste.
Vagus (X) (mixed)	*Motor portion:* Originates in medulla, passes through jugular foramen, and terminates in muscles of airways, lungs, esophagus, heart, stomach, small intestine, most of large intestine, and gallbladder; parasympathetic fibers innervate involuntary muscles and glands of the gastrointestinal (GI) tract. *Sensory portion:* Arises from essentially same structures supplied by motor fibers, passes through jugular foramen, and terminates in medulla and pons. Also consists of afferent fibers from proprioceptors in muscles supplied.	*Motor function:* Smooth muscle contraction and relaxation; secretion of digestive fluids. *Sensory function:* Sensations from organs supplied; muscle sense (proprioception). *Clinical application:* Severing of both nerves in the upper body interferes with swallowing, paralyzes vocal cords, and interrupts sensations from many organs.

Nerve (Type)	Location	Function and Clinical Application
Accessory (XI) (mixed, primarily motor)	*Motor portion:* Consists of a cranial portion and a spinal portion. *Cranial portion* originates from medulla, passes through jugular foramen, and supplies voluntary muscles of pharynx, larynx, and soft palate. *Spinal portion* originates from anterior gray horn of first five cervical segments of spinal cord, passes through jugular foramen, and supplies sternocleidomastoid and trapezius muscles. *Sensory portion:* Consists of afferent fibers from proprioceptors in muscles supplied and passes through jugular foramen.	*Motor function:* Cranial portion mediates swallowing movements; spinal portion mediates movement of head. *Sensory function:* Muscle sense (proprioception). *Clinical application:* If nerves are damaged, the sternocleidomastoid and trapezius muscles become paralyzed, with resulting inability to raise the shoulders and difficulty in turning the head.
Hypoglossal (XII) (mixed, primarily motor)	*Motor portion:* Originates in medulla, passes through hypoglossal canal, and supplies muscles of tongue. *Sensory portion:* Consists of fibers from proprioceptors in tongue muscles that pass through hypoglossal canal and terminate in medulla.	*Motor function:* Movement of tongue during speech and swallowing. *Sensory function:* Muscle sense (proprioception). *Clinical application:* Injury results in difficulty in chewing, speaking, and swallowing. The tongue, when protruded, curls toward the affected side and the affected side becomes atrophied, shrunken, and deeply furrowed.

a A mnemonic device used to remember the names of the nerves is: "**O**h, oh, oh, to touch and feel very green vegetables—**AH**!" The initial letter of each word corresponds to the initial letter of each pair of cranial nerves.

derm called the **neural plate** (Fig. 14.19a,b). The plate folds inward and forms a longitudinal groove, the **neural groove.** The raised edges of the neural plate are called **neural folds.** As development continues, the neural folds increase in height and meet to form a tube called the **neural tube.**

The cells of the wall that encloses the neural tube differentiate into three kinds. The outer or **marginal layer** develops into the *white matter* of the nervous system; the middle or **mantle layer** develops into the *gray matter* of the system; and the inner or **ependymal layer** eventually forms the *lining of the ventricles* of the central nervous system.

The **neural crest** is a mass of tissue between the neural tube and the skin ectoderm (Fig. 14.19b). It differentiates and eventually forms the *posterior (dorsal) root ganglia of spinal nerves, spinal nerves, ganglia of cranial nerves, cranial nerves, ganglia of the autonomic nervous system,* and the *adrenal medulla.*

When the neural tube forms from the neural plate, its anterior portion develops into three enlarged areas called **primary vesicles**: (1) **prosencephalon (forebrain),** (2) mesencephalon (midbrain), and (3) **rhombencephalon (hindbrain)** (Fig. 14.20). These are fluid-filled enlargements that develop by the fourth week of the embryonic period. As development progresses, the vesicular region undergoes several flexures (bends), resulting in subdivision of the three primary vesicles, so that by the fifth week of development the embryonic brain consists of five **secondary vesicles.** The prosencephalon divides into an anterior **telencephalon** and a posterior **diencephalon;** the mesencephalon remains unchanged; the rhombencephalon divides into an anterior **metencephalon** and a posterior **myelencephalon.**

Ultimately, the telencephalon develops into the *cerebral hemispheres* and *basal ganglia;* the diencephalon develops into the *thalamus, hypothalamus,* and *pineal gland;* the mesencephalon develops into the *midbrain;* the metencephalon develops into the *pons* and *cerebellum;* and the myelencephalon develops into the *medulla oblongata.* The cavities within the vesicles develop into the *ventricles* of the brain. The area of the neural tube posterior to the myelencephalon gives rise to the *spinal cord.*

FIGURE 14.19 Origin of the nervous system. (a) Embryo with seven pairs of somites in which the neural folds have united just medial to the somites, forming the early neural tube. (b) Cross sections through the embryo showing the formation of the neural tube.

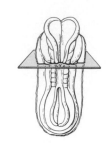

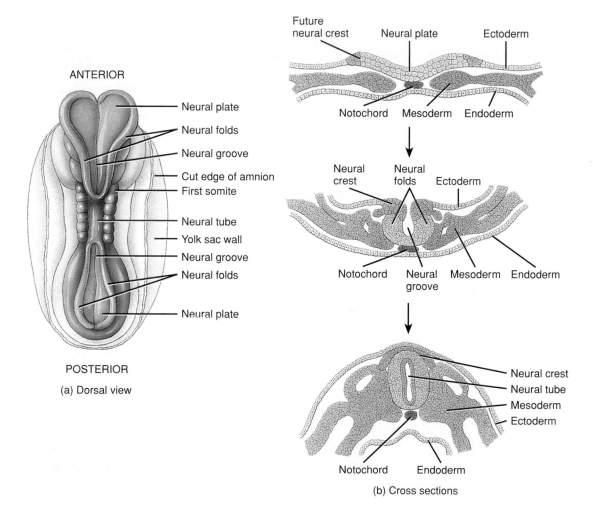

ANTERIOR

Neural plate
Neural folds
Neural groove
Cut edge of amnion
First somite
Neural tube
Yolk sac wall
Neural groove
Neural folds
Neural plate

POSTERIOR

(a) Dorsal view

Future neural crest   Neural plate   Ectoderm

Notochord   Mesoderm   Endoderm

Neural crest   Neural folds   Ectoderm

Notochord   Neural groove   Mesoderm   Endoderm

Neural crest
Neural tube
Mesoderm
Ectoderm

Notochord   Endoderm

(b) Cross sections

FIGURE 14.20 Development of the brain and spinal cord.

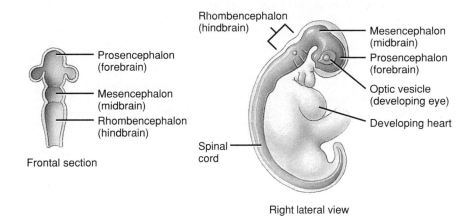

(a) 3–4-week embryo

(b) 5-week embryo

CEREBROVASCULAR ACCIDENT

The most common brain disorder is a **cerebrovascular accident (CVA)**, also called a **stroke**. CVAs are classified into two principal types: (1) *ischemic*, the most common type, due to a decreased blood supply, and (2) *hemorrhagic*, due to rupture of a blood vessel in the brain. Common causes of CVAs are intracerebral hemorrhage (from a blood vessel in the pia mater or brain), emboli (blood clots), and atherosclerosis (formation of cholesterol-containing plaques that block blood flow) of the cerebral arteries. A CVA is characterized by abrupt onset of persisting neurological symptoms that arise from destruction of brain tissue.

Among the risk factors implicated in CVAs are high blood pressure, high blood cholesterol, heart disease, narrowed carotid arteries, transient ischemic attacks, diabetes, smoking, obesity, and excessive alcohol intake.

TRANSIENT ISCHEMIC ATTACK

A **transient ischemic attack (TIA)** is an episode of temporary cerebral dysfunction caused by impaired blood flow to the brain. Symptoms include dizziness, weakness, numbness, or paralysis in a limb or in half of the body; drooping of one side of the face; headache; slurred speech or difficulty understanding speech; or a partial loss of vision or double vision. Sometimes nausea or vomiting also occur. The onset of symptoms is sudden and reaches maximum intensity almost immediately. A TIA usually persists for a few minutes and only rarely lasts as long as 24 hours; it leaves no persistent neurologic deficits. The causes of the impaired blood flow that lead to TIAs are blood clots, atherosclerosis, and certain blood disorders.

It is estimated that about one-third of patients who experience a TIA will have a CVA within five years. Therapy for TIAs includes agents such as aspirin that block the aggregation of blood cells involved in clotting (platelets); anticoagulants; cerebral artery bypass grafting; and carotid endarterectomy (removal of the cholesterol-containing plaques and inner lining of an artery).

ALZHEIMER'S DISEASE

Alzheimer's (ALTZ-hī-merz) **disease (AD)** is a disabling neurological disorder that afflicts about 11% of the population over age 65. Its causes are unknown, its effects are irreversible and devastating, and it has no cure. In the U.S., AD afflicts four million people and claims over 100,000 lives a year, making it the fourth leading cause of death among the elderly, following heart disease, cancer, and stroke.

Victims of AD initially have trouble remembering recent events. Next they become more confused and forgetful, often repeating questions or getting lost while traveling to previously familiar places. Disorientation grows, memories of past events disappear, and there may be episodes of paranoia, hallucination, or violent changes in mood. As their minds continue to deteriorate, they lose their ability to read, write, talk, eat, or walk. Finally, the disease culminates in dementia, the loss of reason and ability to care for oneself. A person with AD usually dies of some complication that afflicts bedridden patients, such as pneumonia.

Diagnosis of AD is difficult because there is no definitive medical test and the mental deficits are slight in the early stages of the disease. Only when the disease is well-advanced and all other possible causes of the dementia are ruled out can a probable diagnosis of AD be made. Examination of the brain at autopsy confirms the diagnosis. Brains of AD victims show three distinct structural abnormalities: (1) great loss of neurons in specific regions, (2) plaques of abnormal proteins deposited outside neurons, and (3) tangled protein filaments within neurons. Some AD brains also have elevated levels of aluminum, an element that serves no known function in the human body.

Neuronal loss is severe in regions such as the hippocampus and cerebral cortex that are important for memory and learning. Many of the lost neurons are thought to use acetylcholine as a neurotransmitter because the level of an enzyme needed to synthesize ACh (choline acetyltransferase) may be greatly reduced in damaged regions. This finding prompted research on treating AD patients with precursors of ACh, which were not helpful. Some success is being seen using drugs that inhibit acetylcholinesterase (AChE), the enzyme that inactivates ACh. AChE blocking drugs such as tacrine (Cognex) and slow-release physostigmine (Synapton) are being tested on AD patients.

Besides neuronal destruction, there is an accumulation of **amyloid plaques**, which consist of degenerating axons and axon terminals and fibrils of an abnormal protein called **beta amyloid**. This 40 to 42 amino acid fragment is part of a membrane glycoprotein called amyloid precursor protein (APP) that is coded by a gene on chromosome 21. People with Down syndrome, which is caused by an extra chromosome 21 (Chapter 29), are mentally retarded and their brains also exhibit beta amyloid-containing plaques. This protein is found only rarely in elderly people with normal mental abilities. Several pieces of evidence suggest that beta amyloid is neurotoxic. It causes degeneration of neurons, especially neurons responsible for learning and memory. Also, it has been discovered that substance P can slow or prevent beta amyloid-induced neuron death. This finding raises hopes for stopping the progression of AD.

The third feature of AD is the presence of **neurofibrillary tangles**, which are abnormal bundles of protein filaments inside cells in affected brain regions. These tangles contain a protein called A68 that also appears in the brains of older Down syndrome individuals. A68 appears to be an altered version of a protein called tau that normally associates with microtubules of the cytoskeleton. It also may be involved in neuronal death.

The causes of AD are not known. In a few cases, there is a genetic mutation in the gene that codes for amyloid precursor protein. Among identical twins, however, one may have AD while the other does not. Other proposed causes of AD are a slow-acting virus, environmental toxins such as aluminum, genetic defects in mitochondria, and decreased blood flow to the brain that results in inadequate oxygen and glucose.

BRAIN TUMORS

A **brain tumor** refers to any benign or malignant growth within the cranium. Tumors may arise from neuroglial cells in the cerebrum, brain stem, and cerebellum or from supporting or neighboring structures such as cranial nerve coverings, meninges, and the pituitary gland. Pressure in the brain from a growing tumor or edema (tissue swelling) associated with the tumor produces the characteristic signs and symptoms. Among these are headache, altered consciousness, vomiting, seizures, visual problems, cranial nerve abnormalities, hormonal syndromes, personality changes, dementia, and sensory or motor deficits. Treatment of brain tumors involves surgery, radiation therapy, and chemotherapy.

CEREBRAL PALSY

The term **cerebral palsy (CP)** refers to a group of motor disorders resulting in muscular incoordination and loss of muscle control. It is caused by damage to the motor areas of the brain during fetal life, birth, or infancy in about 2 of every 1000 children. One cause is infection of the mother with German measles during the first three months of pregnancy. Radiation during fetal life, temporary lack of oxygen during birth, and hydrocephalus during infancy may also damage brain cells. A surgical procedure called *selective posterior rhizotomy* may reduce muscle spasticity. In the procedure, nerve fibers in posterior roots of spinal nerves that trigger abnormal reflexes are severed, allowing the other nerve fibers to achieve better muscle tone.

About 70% of cerebral palsy victims appear to be mentally retarded. The apparent mental slowness, however, is often due to the person's inability to speak or hear well. Such individuals may be more mentally acute than they appear. Cerebral palsy is not a progressive disease; it does not worsen as time elapses. Once the damage is done, however, it is irreversible.

PARKINSON'S DISEASE

Parkinson's disease (PD) or **parkinsonism** is a progressive disorder of the CNS that typically affects its victims around age 60. The cause is unknown, but toxic environmental factors are suspected. Only 5% of PD patients have a family history of the disease. Pathological changes occur in the substantia nigra and basal ganglia. The substantia nigra contains cell bodies of neurons that extend to the basal ganglia of the cerebrum, where they release the neurotransmitter dopamine (DA). Also in the basal ganglia, in the caudate nucleus, are neurons that liberate acetylcholine (ACh). In Parkinson's disease there is a degeneration of DA-producing neurons in the substantia nigra. The resulting imbalance of neurotransmitter activity—too little DA and too much ACh—is thought to bring about most of the symptoms.

In PD patients, involuntary skeletal muscle contractions often interfere with voluntary movement. For instance, the muscles of the upper extremity may alternately contract and relax, causing the hand to shake. This shaking is called **tremor** and is the most common symptom of Parkinson's disease. The tremor may spread to the ipsilateral lower limb and then to the contralateral limbs. Also, some muscles may contract continuously, causing **rigidity** of the involved body part. Rigidity of the facial muscles gives the face a masklike appearance. The expression is characterized by a wide-eyed, unblinking stare and a slightly open mouth with uncontrolled drooling.

Motor performance is also impaired by **bradykinesia** (*brady* = slow; *kinesis* = motion), in which activities such as shaving, cutting food, and buttoning a blouse take longer and become increasingly more difficult. Muscular movements are performed not only slowly but with decreasing range of motion (**hypokinesia**). For example, handwritten letters get smaller, become poorly formed, and eventually become illegible. Often, walking is impaired; steps become shorter and shuffling, and arm swing diminishes. Even speech may be affected. A sentence may begin intelligibly but become progressively softer and quicker.

Treatment of Parkinson's disease is directed toward increasing levels of DA and decreasing levels of ACh. Although people with Parkinson's disease do not manufacture enough dopamine, taking it is useless because DA cannot cross the blood–brain barrier. However, symptoms are partially relieved by a drug developed in the 1960s called levodopa (L-dopa), a precursor of DA. Administered by itself, levodopa elevates brain levels of DA, but also causes undesirable side effects in peripheral tissues, for example, heart and liver dysfunctions. Thus levodopa usually is given in combination with carbidopa, which inhibits the formation of DA outside the brain. The combined drugs diminish the undesirable side effects. Levodopa does not slow the disease progression, however, and as more affected brain cells die, the drug becomes useless. Recently, some patients with PD have been given a drug called deprenyl, which inhibits monoamine oxidase. Early results indicate that it can prevent symptoms for nearly a year in newly diagnosed patients. Drugs that reduce ACh levels (anticholinergics) may also provide relief, especially of tremor and rigidity.

Since 1988, physicians in Sweden, Mexico, England, Cuba, China, and the U.S. have sought to reverse the effects of Parkinson's disease by transplanting fetal nerve tissue (dopamine-rich mesencephalic tissue) into the basal ganglia (usually the putamen) of patients with severe PD. Several patients have shown some degree of improvement after surgery, such as less rigidity and improved quickness of motion. Results are still being evaluated.

MULTIPLE SCLEROSIS

Multiple sclerosis (MS) is the progressive destruction of the myelin sheaths of neurons in the CNS. The sheaths deteriorate to **scleroses**, which are hardened scars or plaques, in multiple regions. The destruction of myelin sheaths slows and short-circuits saltatory conduction of nerve impulses. Usually, the first symptoms occur in early adult life, the average age of onset being 33 years. The frequency of flare-ups is greatest during the first three to four years of the disease, but a first attack, which may have been so mild as to escape medical attention, may not be followed by another attack for 10 to 20 years. The cause of MS is unknown, and there is no satisfactory treatment.

Among the first symptoms of MS are muscular weakness of one or more extremities; abnormal sensations, such as burning or pins and needles; visual impairment that includes blurring, double vision, and problems with color and light perception; incoordination; vertigo; and sphincter impairment that results in urinary problems, such as urinary urgency. After a period of remission during which the symptoms temporarily disappear, a new series of plaques develop, and the victim suffers a second attack. One attack follows another over the years, usually every year or two. Each time the plaques form, some neurons are damaged by the hardening of their sheaths, whereas others are uninjured by their plaques. The result is a progressive loss of function interspersed with remission periods during which the undamaged neurons regain their ability to conduct nerve impulses.

Although the cause of MS is unclear, some evidence implicates a viral infection that precipitates an autoimmune response, specifically the destruction of myelin-producing oligodendrocytes by the body's cytotoxic (killer) T cells. In view of the evidence that it may be an autoimmune disease, immunosuppressive drugs are often given.

DYSLEXIA

Dyslexia (dis-LEK-sē-a; *dys* = difficulty; *lexis* = words) is an impairment of the brain's ability to translate images received from the eyes into understandable language. The condition is unrelated to basic intellectual capacity, but it causes a mysterious difficulty in processing words and symbols. Apparently, some peculiarity in the brain's organizational pattern distorts the ability to read, write, and count. Letters in words

Continues

DISORDERS: HOMEOSTATIC IMBALANCES (continued)

seem transposed, reversed, or upside down—*dog* becomes *god* or *bog*; *b* changes identity with *d*; a sign saying "OIL" inverts into "710." Frequently, dyslectics reread portions of paragraphs, skip words, and change the order of letters in a word. Many dyslectics cannot orient themselves in the three dimensions of space and may show bodily awkwardness.

The exact cause of dyslexia is unknown, since it is unaccompanied by outward scars of detectable neurological damage and its symptoms vary. It occurs about three times as often among boys as among girls. It has been variously attributed to defective vision, brain damage, abnormal brain development, lead in the air, physical trauma, or oxygen deprivation during birth.

Positron emission tomography (PET) scans (see Exhibit 1.4) provide some information about dyslexia. For example, PET scans have demonstrated that in persons with dyslexia the left side of the brain is more active than the right, the language region is less active than normal, and the visual discrimination region is more active.

HEADACHE

One of the most common human afflictions is **headache**. Headaches may be caused by brain tumors; blood vessel abnormalities; inflammation of the brain or meninges; decrease in oxygen supply to the brain; damage to brain cells; and infections of the eyes, ears, nose, and sinuses.

Most headaches require no special treatment. Analgesic

and tranquilizing compounds are generally effective for tension headaches but not for migraine headaches. Drugs that constrict the blood vessels can be helpful for migraine. Biofeedback training and dietary changes may also be of some help. Tension headaches are associated with stress, fatigue, and anxiety and classically they occur in the occipital and temporal muscles.

REYE'S SYNDROME

Reye's syndrome (RS), first described in 1963 by the Australian pathologist R. Douglas Reye, seems to occur following a viral infection, particularly chickenpox or influenza. Aspirin at normal doses is believed to be a risk factor in the development of RS. The majority of persons affected are children or teenagers. The disease is characterized by vomiting and brain dysfunction (disorientation, lethargy, and personality changes) and may progress to coma. Also, the liver becomes infiltrated with small lipid droplets and loses some of its ability to detoxify ammonia.

Brain dysfunction and death are typically caused by swelling of brain cells. The pressure not only kills the cells directly but also results in hypoxia that kills them indirectly. The survival rate is about 70%. Swelling may result in irreversible brain damage, including mental retardation, in children who survive. Therapy is directed at controlling the swelling.

MEDICAL TERMINOLOGY

Agnosia (ag-NŌ-zē-a; *a* = without; *gnosis* = knowledge) Inability to recognize the significance of sensory stimuli such as auditory, visual, olfactory, gustatory, and tactile.

Apraxia (a-PRAK-sē-a; *pratto* = to do) Inability to carry out purposeful movements in the absence of paralysis.

Delirium (de-LIR-ē-um; *deliria* = off the tract) Also called **acute confusional state (ACS)**. A transient disorder of abnormal cognition (perception, thinking, and memory) and disordered attention that is accompanied by disturbances of the sleep–wake cycle and psychomotor behavior (hyperactivity or hypoactivity of movements and speech).

Dementia (de-MEN-shē-a; *de* = away from; *mens* = mind) A mental disorder that results in permanent or progressive general loss of intellectual abilities such as impairment of memory, judgment, and abstract thinking and changes in personality. The most common cause is Alzheimer's disease. Others are cerebrovascular disease, CNS infection, brain tumors or trauma, pernicious anemia, hydrocephalus, and neurological diseases (Parkinson's disease, multiple sclerosis, and Huntington's chorea).

Electroconvulsive therapy (ECT) (e-lek′-trō-con-VUL-siv THER-a-pē) A form of shock therapy in which convulsions are induced by the passage of a brief electric current through

the brain. A patient undergoing ECT is anesthetized and given a muscle relaxant to minimize convulsions. ECT is regarded as an important therapeutic option in the treatment of severe depression and acute mania. Side effects include acute confusional states and memory deficits.

Lethargy (LETH-ar-jē) A condition of functional sluggishness.

Nerve block Loss of sensation in a region, such as in local dental anesthesia, due to injection of a local anesthetic.

Neuralgia (noo-RAL-jē-a; *neur* = nerve) Attacks of pain along the entire course or branch of a peripheral sensory nerve.

Stupor (STOO-por) Unresponsiveness from which a patient can be aroused only briefly and by vigorous and repeated stimulation.

Viral encephalitis (VĪ-ral en′-sef-a-LĪ-tis) An acute inflammation of the brain caused by a direct attack by various viruses or by an allergic reaction to any of the many viruses that are normally harmless to the central nervous system. If the virus affects the spinal cord as well, it is called **encephalomyelitis**.

Study Outline

Brain (p. 405)

Principal Parts (p. 405)

1. During embryological development, brain vesicles are formed and serve as forerunners of various parts of the brain.
2. The telencephalon forms the cerebrum, the diencephalon develops into the thalamus and hypothalamus, the mesencephalon develops into the midbrain, the metencephalon develops into the pons and cerebellum, and the myelencephalon forms the medulla.
3. The principal parts of the brain are the brain stem, diencephalon, cerebrum, and cerebellum.

Protection and Coverings (p. 405)

1. The brain is protected by cranial bones, meninges, and cerebrospinal fluid.
2. The cranial meninges are continuous with the spinal meninges and are named dura mater, arachnoid, and pia mater.

Cerebrospinal Fluid (CSF) (p. 405)

1. Cerebrospinal fluid is formed in the choroid plexuses and circulates through the subarachnoid space, ventricles, and central canal. Most of the fluid is absorbed by the arachnoid villi of the superior sagittal blood sinus.
2. Cerebrospinal fluid provides mechanical protection, chemical protection, and circulation.
3. If cerebrospinal fluid accumulates in the ventricles or subarachnoid space, it is called hydrocephalus.

Blood Supply (p. 411)

1. The blood supply to the brain is via the cerebral arterial circle (circle of Willis).
2. Any interruption of the oxygen supply to the brain can result in weakening, permanent damage, or death of brain cells. Interruption of the mother's blood supply to a child during childbirth before it can breathe may result in paralysis, mental retardation, epilepsy, or death.
3. Glucose deficiency may produce dizziness, convulsions, and unconsciousness.
4. The blood–brain barrier (BBB) is a concept that explains the differential rates of passage of certain material from the blood into the brain.

Brain Stem: Anatomy and Physiology (p. 411)

1. The medulla is continuous with the upper part of the spinal cord and contains portions of both motor and sensory tracts. It contains nuclei that are reflex centers for regulation of heart rate, respiratory rate, vasoconstriction, swallowing, coughing, vomiting, sneezing, and hiccuping. It also contains the nuclei of origin for cranial nerves VIII (cochlear and vestibular branches) through XII.
2. The pons is superior to the medulla. It connects the spinal cord with the brain and links parts of the brain with one another by way of tracts. It relays nerve impulses related to voluntary skeletal movements from the cerebral cortex to the cerebellum. It contains the nuclei for cranial nerves V through VII and the vestibular branch of VIII. The reticular formation of the pons contains the pneumotaxic and apneustic centers, which help control respiration.
3. The midbrain connects the pons and diencephalon. It conveys motor impulses from the cerebrum to the cerebellum and spinal cord, sends sensory impulses from the spinal cord to the thalamus, and regulates auditory and visual reflexes. It also contains the nuclei of origin for cranial nerves III and IV.

Diencephalon (p. 415)

1. The diencephalon consists of the thalamus and hypothalamus.
2. The thalamus is superior to the midbrain and contains nuclei that serve as relay stations for all sensory impulses to the cerebral cortex. It also registers conscious recognition of pain and temperature and some awareness of touch and pressure.
3. The hypothalamus is inferior to the thalamus. It controls and integrates the autonomic nervous system, receives sensory impulses from viscera, connects the nervous and endocrine systems, secretes a variety of regulating hormones, functions in rage and aggression, controls body temperature, regulates food and fluid intake, and maintains the waking state and sleep patterns.

Cerebrum (p. 418)

1. The cerebrum is the largest part of the brain. Its cortex contains convolutions, fissures, and sulci.
2. The cerebral lobes are named the frontal, parietal, temporal, and occipital.
3. The white matter is under the cortex and consists of mostly myelinated axons running in three principal directions.
4. The basal ganglia are paired masses of gray matter in the cerebral hemispheres. They help to control muscular movements.
5. The limbic system is found in the cerebral hemispheres and diencephalon. It functions in emotional aspects of behavior and memory.
6. The sensory areas of the cerebral cortex are concerned with the interpretation of sensory impulses. The motor areas are the regions that govern muscular movement. The association areas are concerned with more complex integrative functions.
7. Brain waves generated by the cerebral cortex are recorded as an electroencephalogram (EEG). It may be used to diagnose epilepsy, infections, and tumors.

Brain Lateralization (p. 426)

1. The two hemispheres of the brain are not bilaterally symmetrical, either anatomically or functionally.
2. The left hemisphere is more important for right-handed control, spoken and written language, numerical and scientific skills, and reasoning.
3. The right hemisphere is more important for left-handed control, musical and artistic awareness, space and pattern perception, insight, imagination, and generating mental images of sight, sound, touch, taste, and smell.

Cerebellum (p. 427)

1. The cerebellum occupies the inferior and posterior aspects of the cranial cavity. It consists of two hemispheres and a central, constricted vermis.
2. It is attached to the brain stem by three pairs of cerebellar peduncles.
3. The cerebellum functions in the coordination of skeletal muscles and the maintenance of normal muscle tone and body equilibrium.

Neurotransmitters in the Brain (p. 429)

1. Many substances are either known or suspected neurotransmitters that can facilitate, excite, or inhibit postsynaptic neurons.
2. Categories of neurotransmitters include acetylcholine (ACh), amino acids, biogenic amines, and neuropeptides.

Cranial Nerves (p. 431)

1. Twelve pairs of cranial nerves originate from the brain.
2. They are named primarily on the basis of distribution and numbered by order of attachment to the brain. (See Exhibit 14.3 for summary of cranial nerves.)

Aging and the Nervous System (p. 431)

1. Age-related effects involve loss of neurons and decreased capacity for sending nerve impulses.

2. Degenerative changes also affect the sense organs.

Developmental Anatomy of the Nervous System (p. 431)

1. The development of the nervous system begins with a thickening of ectoderm called the neural plate.
2. The parts of the brain develop from primary and secondary vesicles.

Review Questions

1. Identify the four principal parts of the brain and the components of each, where applicable. What is the origin of each of the parts? (p. 405)
2. Describe the location of the cranial meninges. (p. 405)
3. Where is cerebrospinal fluid (CSF) formed? Describe its circulation and functions. Where is CSF absorbed? (p. 405)
4. What is hydrocephalus? (p. 405)
5. Explain the importance of oxygen and glucose to brain cells. (p. 411)
6. What is the blood–brain barrier (BBB)? Describe the passage of several substances with respect to the BBB. (p. 411)
7. Describe the location and structure of the medulla. Define decussation of pyramids. Why is it important? List the principal functions of the medulla. (p. 411)
8. Describe the location and structure of the pons. What are its functions? (p. 414)
9. Describe the location and structure of the midbrain. What are its functions? (p. 414)
10. Describe the location and structure of the thalamus. List its functions. (p. 415)
11. Where is the hypothalamus located? Explain some of its major functions. (p. 417)
12. Where is the cerebrum located? Describe the cortex, convolutions, fissures, and sulci of the cerebrum. (p. 418)
13. List and locate the lobes of the cerebrum. How are they separated from one another? What is the insula? (p. 418)
14. Describe the organization of cerebral white matter. Be sure to indicate the function of each group of fibers. (p. 420)
15. What are basal ganglia? Name the important basal ganglia and list the function of each. Describe the effects of damage on the basal ganglia. (p. 421)
16. Define the limbic system. Explain several of its functions. (p. 422)
17. What is meant by a sensory area of the cerebral cortex? List, locate, and give the function of each sensory area. (p. 423)
18. What is meant by a motor area of the cerebral cortex? List, locate, and give the function of each motor area. (p. 424)
19. What conditions may result from damage to sensory or motor speech areas? (p. 425)
20. What is an association area of the cerebral cortex? What are its functions? (p. 425)
21. Define an electroencephalogram (EEG). List the principal waves recorded and the importance of each. What is the diagnostic value of an EEG? (p. 425)
22. Describe brain lateralization. (p. 426)
23. Describe the location of the cerebellum. List the principal parts of the cerebellum. (p. 426)
24. What are cerebellar peduncles? List and explain the function of each. (p. 428)
25. Explain the functions of the cerebellum. Describe some effects of cerebellar damage. (p. 428)
26. Give some examples of neurotransmitters and indicate their functions. (p. 429)
27. Define a cranial nerve. How are cranial nerves named and numbered? Distinguish between a mixed and a sensory cranial nerve. (p. 431)
28. For each of the 12 pairs of cranial nerves, list (a) its name, number, and type; (b) its location; and (c) its function. In addition, list the effects of damage, where applicable. (p. 433)
29. Describe the effects of aging on the nervous system. (p. 431)
30. Describe the development of the nervous system. (p. 433)

Answers to Questions with Figures

14.1 Cerebrum, cerebellum, brain stem, diencephalon.
14.2 Dura mater, arachnoid, pia mater.
14.3 The thalamus (see Fig. 14.8a for a frontal section).
14.4 Formed in the choroid plexuses of all four ventricles; reabsorbed by the arachnoid villi that project into the dural venous sinuses.
14.5 Medulla; midbrain; pons; thalamus.
14.6 Decussation means crossing to the opposite side. The pyramids contain motor tracts that extend from the cortex into the spinal cord. Since they convey impulses for contraction of skeletal muscle, the functional consequence of decussation of the pyramids is that one side of the cerebrum controls muscles on the opposite side of the body.
14.7 They are the main connections for tracts between upper parts of the brain and lower parts of the brain and spinal cord.

14.8 Lateral, superior, and inferior to the thalamus.
14.9 Mammillary region, tuberal region, supraoptic region, preoptic region.
14.10 The gray matter enlarges more rapidly, producing convolutions or gyri (folds), sulci (shallow grooves), and fissures (deep grooves).
14.11 Association fibers.
14.12 The basal ganglia = corpus striatum (caudate and lenticular nuclei), putamen, and globus pallidus. The thalamus.
14.13 Hippocampus.
14.14 Gnostic area, motor speech area, premotor area, gustatory areas, auditory areas, visual areas, frontal eye field area.
14.15 Theta waves.
14.16 Parietal and occipital.
14.17 Cerebellar peduncles.
14.18 III and IV.

Chapter 15

SENSORY, MOTOR, AND INTEGRATIVE SYSTEMS

Chapter Contents at a Glance

Student Objectives

1. Define a sensation and discuss the levels and components of sensation.
2. Describe the classification of receptors.
3. List the location and function of the receptors for tactile sensations (touch, pressure, vibration), thermal sensations (heat and cold), and pain.
4. Distinguish somatic, visceral, referred, and phantom pain.
5. Identify the proprioceptive receptors and indicate their functions.
6. Discuss the neuronal components and functions of the posterior column–medial lemniscus, anterolateral, and spinocerebellar pathways.
7. Describe the integration of sensory input and motor output.
8. Compare the location and functions of the direct and indirect motor pathways.
9. Explain how the basal ganglia and cerebellum are related to motor responses.
10. Compare integrative functions such as memory, wakefulness, and sleep.

T he last three chapters examined the organization of the nervous system. Now, we will see how certain parts cooperate to carry out some aspects of its three basic functions: (1) receiving sensory input; (2) integrating, associating, and storing information; and (3) transmitting motor impulses that result in movement or secretion.

In this chapter we will explore the pathways that convey somatic sensory input from the body to the brain, pathways that carry motor commands for control of movements from the brain to the skeletal muscle, and the integrative functions of memory and sleep. Chapter 16 deals with input from the special senses, for example, vision and hearing, and Chapter 17 covers output via the autonomic nervous system to smooth muscle, cardiac muscle, and glands.

SENSATION

Consider what would happen if you could not feel pain, or if you could not see, hear, smell, taste, or maintain your balance. If pain could not be sensed, burns would be common, and an inflamed appendix might burst without advance warning. A lack of sight and hearing would increase the risk of being hit by a car or sleeping through a fire alarm, and so on. It would be impossible to initiate proper responses to changes in your internal and external environments. Achieving homeostasis requires a continual flow of sensory input into the CNS, and at any given time, your brain receives and responds to many aspects of this input. You don't notice some sensory input because it never reaches the thalamus and cerebral cortex. For example, usually one is unaware of blood pressure, yet it is constantly monitored; the input flows to the cardiovascular center in the medulla. Even sensory input that is relayed to the thalamus and cerebral cortex is noticed only some of the time. There is no question that we would quickly exhibit "sensory overload" if our conscious mind had to deal with all the information arriving at once. Filtering of sensory input assures that only a small portion reaches the conscious perception at any given time.

Levels of Sensation

In its broadest context, **sensation** is the conscious or unconscious awareness of external or internal stimuli. If the stimulus is strong enough, one or more impulses arise in sensory nerve fibers. After the nerve impulses have been conducted to a region of the spinal cord or brain, they are translated into a sensation. The nature of the sensation and the type of reaction generated vary with the level of the central nervous system at which the sensation is translated.

Sensory fibers terminating in the spinal cord can generate spinal reflexes without immediate action by the brain. Sensory fibers terminating in the lower brain stem bring about far more complex motor reactions. When sensory

impulses reach the lower brain stem, they cause subconscious motor reactions, for example, changes in heart or breathing rate. Sensory impulses that reach the thalamus can be localized crudely in the body. At the thalamic level, there is crude identification of the type of sensation, that is, as a *specific* sensation such as touch, pressure, pain, position, hearing, or taste. When sensory information reaches the cerebral cortex, we experience precise identification and localization. It is at this level that memories of previous sensory information are stored and the perception of sensation occurs based on past experience. **Perception** is the conscious awareness and interpretation of sensations.

CLINICAL APPLICATION

PHANTOM PAIN

The pain often experienced by patients who have had a limb amputated is called **phantom pain (phantom limb sensation).** They still experience sensations such as itching, pressure, tingling, or pain in the extremity as if the limb were still there. This occurs in part because nerve impulses arise in the remaining proximal portions of the sensory nerves that previously received impulses from the limb. The brain interprets impulses from these nerves as coming from the nonexistent (phantom) limb.

Modality

Each specific type of sensation is called a sensory **modality.** The sensation may be one of temperature change, pain, pressure, touch, body position, equilibrium, hearing, vision, smell, or taste. In other words, the distinct quality that makes one sensation different from others is its modality. In general, a given sensory neuron carries only one modality. Neurons relaying touch information to the somatosensory area of the cortex, for example, do not also transmit temperature signals. Likewise, nerve impulses from the eyes are interpreted by the occipital lobe as sight, whereas those from the ears are interpreted by the temporal lobes as hearing.

Components of Sensation

For a sensation to arise, four events must occur.

1. **Stimulation. A stimulus,** or change in the environment, capable of activating certain sensory neurons must occur.

2. **Transduction. A sensory receptor** or **sense organ** must respond to the stimulus and **transduce** (convert) it to a generator potential (described shortly). A sensory receptor or sense organ is a specialized cell or type of neuron or group of neurons that is very sensitive to certain types of changes (stimuli) in internal or external conditions.

3. **Conduction.** When a generator potential reaches

threshold, it elicits one or more nerve impulses that are conducted into the CNS. Sensory neurons that convey impulses into the CNS are called **first-order neurons.**

4. **Translation.** A region of the CNS must translate the nerve impulses into a sensation. Most conscious sensations or perceptions occur in the cerebral cortex. In other words, you see, hear, and feel in the brain. You seem to see with your eyes, hear with your ears, and feel pain in an injured part of your body because sensory impulses from each part of the body arrive in specific regions of the cortex and the cortex interprets the sensation as coming from the stimulated sensory receptors.

Sensory Receptors

The process of sensation begins in a large variety of different types of sensory receptors, each of which is sensitive to particular stimuli.

Selectivity of Receptors

Sensory receptors respond vigorously to one particular kind of stimulus and weakly or not at all to others. This characteristic is termed **selectivity.** The stimulus may be electromagnetic energy, such as light or heat; mechanical energy, such as touch; or a chemical, such as carbon dioxide in body fluids. For instance, auditory receptors in the ears respond to sound waves but not to light. The selectivity of receptors is not absolute, however. A very intense stimulus may activate receptors that normally would not respond to that form of energy. Try rubbing your eyes in a darkened room. You may see tiny flashes of light because the mechanical pressure is activating some of the photoreceptors.

Classification of Receptors

Receptors vary in their complexity. **Simple receptors** are associated with the **general senses,** which include touch, pressure, vibration, temperature, pain, and proprioception (body position and movement). The simplest receptors are called **free nerve endings** because they have no apparent anatomical specializations. Examples are receptors for pain, temperature, tickle, and itch sensations (some of which are shown in Fig. 15.1). Other receptors for general sensations such as touch, pressure, and vibration have distinctive structures. **Complex receptors** are associated with **special senses.** They are in complex sense organs such as the eye and ear. The receptors for each special sense are located in only one or two specific areas of the body. The special senses include smell, taste, vision, equilibrium, and hearing. Two widely used classifications of receptors are based on the location of the receptors and the type of stimuli they detect.

● **By Location**   One convenient method of classifying receptors is by their location. **Exteroceptors** (eks´-ter-ō-SEP-tors) are located at or near the surface of the body and provide information about the external environment. They are sensitive to stimuli outside the body and transmit sensations of hearing, vision, smell, taste, touch, pressure, vibration, temperature, and pain.

Interoceptors or **visceroceptors** (vis´-er-ō-SEP-tors) are located in blood vessels and viscera and provide information about the internal environment. These sensations arise from within the body and often do not reach conscious perception. Occasionally, they may be felt as pain or pressure.

Proprioceptors (prō´-prē-ō-SEP-tors) are located in muscles, tendons, joints, and the internal ear. They provide information about body position and movement. Such sensations provide information about muscle tension, the position and activity of our joints, and equilibrium.

● **By Type of Stimulus**   Another method of classifying receptors is by the type of stimuli they detect. **Mechanoreceptors** detect mechanical pressure or stretching. Stimuli so detected include those related to touch, pressure, vibration, proprioception, hearing, equilibrium, and blood pressure. **Thermoreceptors** detect changes in temperature. **Nociceptors** detect pain, usually as a result of physical or chemical damage to tissues. **Photoreceptors** detect light that strikes the retina of the eye. **Chemoreceptors** detect chemicals in the mouth (taste), nose (smell), and body fluids.

Generator Potentials and Receptor Potentials

The electrical response of a sensory receptor to a stimulus is called either a **generator potential** or a **receptor potential.** Generator potentials trigger action potentials whereas receptor potentials do not. Except for olfactory (smell) receptors, the receptors that serve the special senses— vision, hearing, equilibrium, and taste—produce receptor potentials. Most other sensory receptors in the body, including olfactory receptors, produce generator potentials. The size of both types of potential varies with the intensity of the stimulus. An intense stimulus produces a large generator or receptor potential whereas a weak stimulus elicits a small one.

There are two major differences between generator and receptor potentials. (1) A generator potential is always a depolarization whereas a receptor potential may be either a depolarization (excitatory response) or hyperpolarization (inhibitory response). (2) If a generator potential is large enough to reach threshold, it *generates* a nerve impulse in the first-order sensory neuron. Thus a stimulus is converted into a generator potential and then into nerve impulses that conduct into the CNS. The resulting nerve impulses travel along the first-order nerve fiber and finally arrive at the axon terminals, where they trigger exocytosis of synaptic vesicles containing neurotransmitter. A receptor potential, on the other hand, never triggers a nerve impulse. Rather, it directly regulates exocytosis of synaptic vesicles that contain neurotransmitter.

Sensory receptors that produce receptor potentials are short and usually synapse with a first-order neuron. The receptor potential spreads to the region containing synaptic vesicles and directly increases or decreases exocytosis of synaptic vesicles. Neurotransmitter liberated from synaptic vesicles, then, may either stimulate or inhibit production of nerve impulses in the first-order sensory neuron.

Adaptation of Sensory Receptors

A characteristic of many sensations is **adaptation,** that is, a change in sensitivity (usually a decrease) to a long-lasting stimulus. The perception of a sensation may even disappear although the stimulus is still being applied. For example, when you first step into a hot shower, the water may feel too hot, but soon the sensation decreases to one of comfortable warmth even though the stimulus (hot water) does not change. Receptors vary in their ability to adapt. **Rapidly adapting (phasic) receptors,** such as those associated with pressure, touch, and smell, adapt very quickly. Such receptors play a major role in signaling changes in a particular sensation. **Slowly adapting (tonic) receptors,** such as those for detecting pain, body position, and chemicals in blood, adapt slowly. These receptors are important in signaling information regarding steady states of the body. While much adaptation occurs in sensory receptors, further adaptation may occur as sensory signals are processed within the CNS.

GENERAL SENSES

The general or somatic senses arise in receptors located in the skin (cutaneous sensations) or embedded in muscles, tendons, joints, and the inner ear (proprioceptive sensations).

Cutaneous Sensations

Cutaneous (*cuta* = skin) **sensations** include tactile sensations (touch, pressure, vibration), thermal sensations (cold and heat), and pain. The receptors for these sensations are in the skin, connective tissue under the skin, mucous membranes, mouth, and anus.

Cutaneous receptors are distributed over the body surface in such a way that certain parts of the body are densely populated with receptors and other parts contain only a few. Areas of the body that have few cutaneous receptors are not very sensitive; those containing many are highly sensitive. The following order for these receptors, from greatest sensitivity to least, has been established: tip of tongue, lips, tip of finger, side of nose, back of hand, and back of neck.

Cutaneous receptors consist of the dendrites of sensory neurons that may be enclosed in a capsule of epithelial or connective tissue or have no apparent structural specialization. Nerve impulses generated by cutaneous receptors pass along somatic afferent neurons in spinal and cranial nerves, through the thalamus, to the somatosensory area of the parietal lobe of the cortex (see Fig. 15.6a).

Tactile Sensations

Even though the **tactile** (*tact* = touch) **sensations** are divided into touch, pressure, and vibration plus itch and tickle, they are all detected by mechanoreceptors.

● **Touch** **Touch sensations** generally result from stimulation of tactile receptors in the skin or tissues immediately beneath the skin. **Crude touch** refers to the ability to perceive that something has touched the skin, although its exact location, shape, size, or texture cannot be determined. **Discriminative touch** refers to the ability to recognize exactly what point of the body is touched. Tactile receptors include corpuscles of touch, hair root plexuses, and type I and II cutaneous mechanoreceptors (Fig. 15.1).

Corpuscles of touch, or **Meissner's** (MĪS-ners) **corpuscles,** are egg-shaped receptors for discriminative touch. They are a mass of dendrites enclosed by connective tissue and are located in the dermal papillae of the skin. Since they adapt rapidly, they transmit impulses most rapidly at the onset of a touch. Corpuscles of touch are most plentiful in the fingertips, palms of the hands, and soles of the feet. They are also abundant in the eyelids, tip of the tongue, lips, nipples, clitoris, and tip of penis. Other rapidly adapting receptors are the **hair root plexuses.** These are dendrites arranged in networks around hair follicles. Movement of the hair shaft stimulates the dendrites. Hair root plexuses detect movements mainly on the surface of the body when hairs are disturbed.

There are two types of slowly adapting touch receptors, type I and type II. **Type I cutaneous mechanoreceptors,** also called **tactile** or **Merkel** (MER-kel) **discs,** are the flattened portions of dendrites of sensory neurons that make contact with epidermal cells of the stratum basale called Merkel cells (see Fig. 5.2). They are distributed in many of the same locations as corpuscles of touch and also function in discriminative touch. **Type II cutaneous mechanoreceptors,** or **end organs of Ruffini,** are embedded deeply in the dermis and in deeper tissues of the body. They detect heavy and continuous touch sensations.

● **Pressure** **Pressure sensations** generally result from stimulation of tactile receptors in deeper tissues and are longer lasting and have less variation in intensity than touch sensations. Pressure is a sustained sensation that is felt over a larger area than touch.

Pressure receptors are type II cutaneous mechanoreceptors and lamellated corpuscles. **Lamellated,** or **Pacinian** (pa-SIN-ē-an), **corpuscles** (Fig. 15.1) are oval structures

FIGURE 15.1 Structure and location of cutaneous receptors.

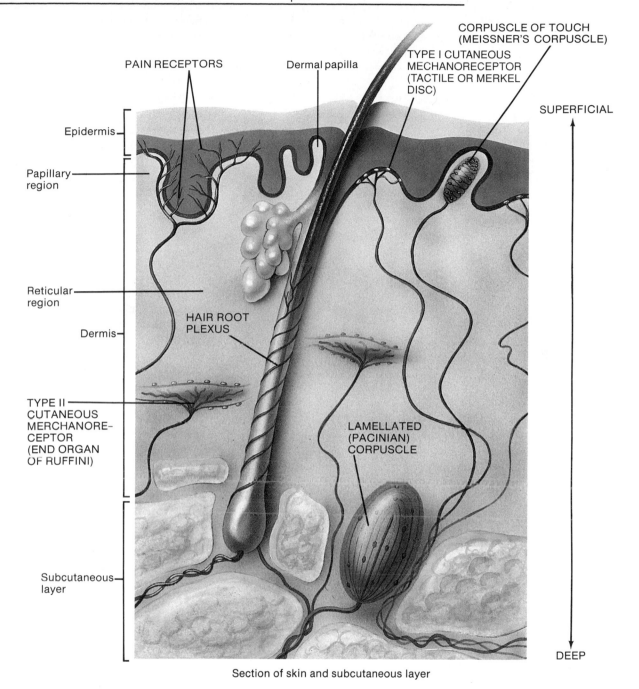

PAIN RECEPTORS

Dermal papilla

CORPUSCLE OF TOUCH
(MEISSNER'S CORPUSCLE)

TYPE I CUTANEOUS
MECHANORECEPTOR
(TACTILE OR MERKEL
DISC)

SUPERFICIAL

Epidermis

Papillary
region

Reticular
region

Dermis

HAIR ROOT
PLEXUS

TYPE II
CUTANEOUS
MERCHANORE-
CEPTOR
(END ORGAN
OF RUFFINI)

LAMELLATED
(PACINIAN)
CORPUSCLE

Subcutaneous
layer

DEEP

Section of skin and subcutaneous layer

Question: Which touch receptors adapt rapidly or slowly?

composed of a connective tissue capsule, layered like an onion, that enclose a dendrite. Like corpuscles of touch, lamellated corpuscles adapt rapidly. They are located in subcutaneous tissues, deep submucosal tissues that lie under mucous membranes, and serous membranes; around joints, tendons, and muscles; and in the mammary glands, external genitalia, and certain viscera, such as the pancreas and urinary bladder.

● **Vibration**   Vibration sensations result from rapidly repetitive sensory signals from tactile receptors. The receptors for vibration sensations are corpuscles of touch and lamellated corpuscles. Whereas corpuscles of touch detect low-frequency vibrations, lamellated corpuscles detect higher-frequency vibrations.

● **Itch and Tickle**   The **itch** sensation results from stimulation of free nerve endings by certain chemicals, such as bradykinin, often as a result of a local inflammatory response. Free nerve endings also are thought to mediate the **tickle** sensation. This unusual sensation is the only one that you may not be able to elicit on yourself. Why the tickle sensation may arise only when someone else touches you is not understood.

Thermal Sensations

The **thermal** (*therm* = heat) **sensations** are hot and cold. For a long time, the receptors detecting thermal sensations were thought to be the end organs of Ruffini and similar encapsulated structures called Krause's corpuscles. Experiments now confirm, however, that **thermoreceptors** are free nerve endings. Separate thermoreceptors respond to hot and cold stimuli.

Pain Sensations

Pain is indispensable for a normal life. It provides us with information about noxious, tissue-damaging stimuli and thus often enables us to protect ourselves from greater damage. From a medical standpoint, the subjective description and indication of the location of pain may help pinpoint the underlying cause of disease.

The receptors for **pain,** called **nociceptors** (nō′-sē-SEP-tors; *noci* = harmful), are free nerve endings (Fig. 15.1). Pain receptors are found in almost every tissue of the body. They may respond to any type of stimulus if it is strong enough to cause tissue damage. When stimuli for other sensations, such as touch, pressure, heat, and cold, reach a certain intensity, they stimulate the sensation of pain as well. Excessive stimulation of most sensory receptors causes pain. Other stimuli that elicit pain include excessive distension or dilation of a structure, prolonged muscular contractions, muscle spasms, inadequate blood flow to an organ, or the presence of certain chemical substances. Tissue irritation or injury releases chemicals, for example, prostaglandins and kinins, that stimulate nociceptors. Pain persists even after the initial trauma occurs since these substances linger and nociceptors adapt only slightly or not at all. Pain receptors, because of their sensitivity to all stimuli, perform a protective function by identifying changes that may endanger the body.

Based on the location of the stimulated receptors, pain may be divided into two types: somatic and visceral. **Somatic pain** that arises from stimulation of receptors in the skin is called **superficial somatic pain,** whereas stimulation of receptors in skeletal muscles, joints, tendons, and fascia, causes **deep somatic pain. Visceral pain** results from stimulation of receptors in the viscera.

Although receptors for somatic and visceral pain are similar, viscera do not evoke the same pain response as somatic tissues. For example, highly *localized* damage to certain viscera, such as cutting the intestine in two in a patient who is awake, causes very little, if any, pain. But, if

stimulation is *diffuse,* involving large areas, visceral pain can be severe. Such stimulation might result from distension, spasms, or ischemia. For example, a kidney stone or a gallstone might obstruct and distend a ureter or the bile duct and cause severe pain.

In most instances of somatic pain and in some instances of visceral pain, the cortex accurately localizes the pain to the stimulated area. If you burn your finger, you feel the pain in your finger. If the pleural membranes are inflamed, you experience pain in the chest. In most instances of visceral pain, however, the pain is felt in or just under the skin that overlies the stimulated organ. The pain may also be felt in a surface area far from the stimulated organ. This phenomenon is called **referred pain.** In general, the area to which the pain is referred and the visceral organ involved are served by nerve fibers (dendrites) from the same segment of the spinal cord. For example, afferent fibers from the heart and skin over the heart and along the medial aspect of the left upper extremity enter spinal cord segments T1–T4. Thus the pain of a heart attack is typically felt in the skin over the heart and along the left arm. Figure 15.2 illustrates skin regions to which visceral pain may be referred.

CLINICAL APPLICATION

ANESTHESIA

During certain surgical or diagnostic procedures, **anesthesia** (an′-es-THĒ-zē-a; *an* = without; *aisthesis* = sensation) is used to block sensations while maintaining the stability of the patient's organ systems. Two commonly used forms of anesthesia are general and spinal. **General anesthesia** removes sensations, including pain, and also produces unconsciousness and sometimes muscular relaxation. A person must not eat several hours before having general anesthesia because of the danger of vomiting and aspirating the stomach contents into the airway.

Spinal anesthesia, a form of local anesthesia, involves injection of a drug via a spinal tap into the subarachnoid space. The procedure is widely used for surgery below the diaphragm such as hernia repair, procedures on the hips and lower extremities, and operations involving the rectum, urinary bladder, prostate gland, and other pelvic structures.

Proprioceptive Sensations

An awareness of the activities of muscles, tendons, and joints and of balance or equilibrium is provided by the **proprioceptive** (*proprio* = one's own), or **kinesthetic** (kin′-es-THET-ik; *kinesis* = motion), **sense.** It informs us of the degree to which muscles are contracted, the amount of tension created in the tendons, the change of position of a joint, and the orientation of the head relative to the ground and in response to movements. Proprioception enables us to recognize the location and rate of movement of one body part in relation to others. It also allows us to estimate the weight of objects and determine the muscular work necessary to perform a task and to judge the position and movements of our limbs, without using our eyes.

FIGURE 15.2 Referred pain. The colored parts of the diagrams indicate skin areas to which visceral pain is referred.

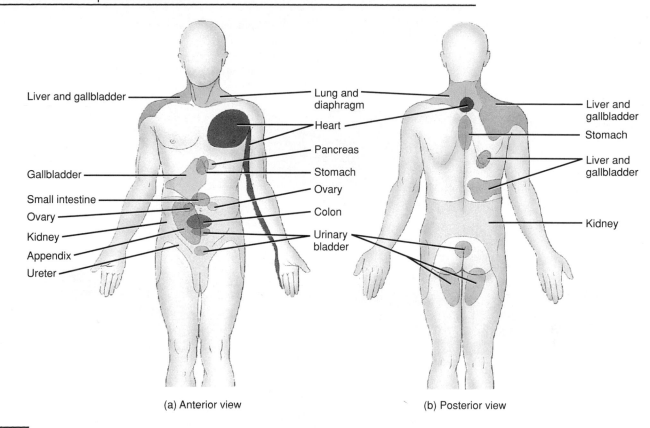

(a) Anterior view   (b) Posterior view

Question: Which visceral organs have the broadest area for referred pain?

Impulses for conscious proprioception pass along ascending tracts in the spinal cord, to the thalamus and from there to the cerebral cortex. The sensation is perceived in the somatosensory area in the parietal lobe of the cerebral cortex posterior to the central sulcus. At the same time, proprioceptive impulses also pass to the cerebellum along the spinocerebellar tracts.

Proprioceptors, the receptors for proprioception, adapt only slightly. This feature allows the brain to be informed continually of the status of different parts of the body so that adjustments can be made to ensure coordination. Proprioceptors include muscle spindles, tendon organs, and joint kinesthetic receptors. They are located within skeletal muscles, tendons, and joint capsules. Also classified as proprioceptors are hair cells of the internal ear (see page 492).

Muscle Spindles

Muscle spindles are specialized groupings of muscle fibers interspersed among regular skeletal muscle fibers and oriented parallel to them (Fig. 15.3a). The ends of the spindles are anchored to the endomysium and perimysium. A mus-

cle spindle consists of 3 to 10 specialized muscle fibers called **intrafusal muscle fibers** that are partially enclosed in a spindle-shaped connective tissue capsule. The central region of each intrafusal fiber contains several nuclei but has few or no actin and myosin filaments. Surrounding the muscle spindle are the regular skeletal muscle fibers, which are called **extrafusal muscle fibers.**

Although the central area of an intrafusal fiber cannot contract because it lacks actin and myosin, it does contain two types of afferent (sensory) fibers. The first are large diameter, rapidly conducting sensory fibers, called **type Ia fibers.** The dendrites (distal ends) of the Ia fiber wrap in a spiral manner around the central area of each intrafusal fiber. When the central part of the spindle is stretched, the dendrites are stimulated and nerve impulses are sent to the spinal cord. The central receptive area of some muscle spindles is also served by smaller diameter sensory fibers called **type II fibers.** Their dendrites are located on either side of the type Ia dendrites. Type II dendrites are also stimulated when the central part of the spindle is stretched, and they too send impulses to the spinal cord.

The ends of the intrafusal muscle fibers do contain actin and myosin filaments. They contract when stimulated by motor neurons arising in the anterior gray horn of the spinal

FIGURE 15.3 Proprioceptive receptors.

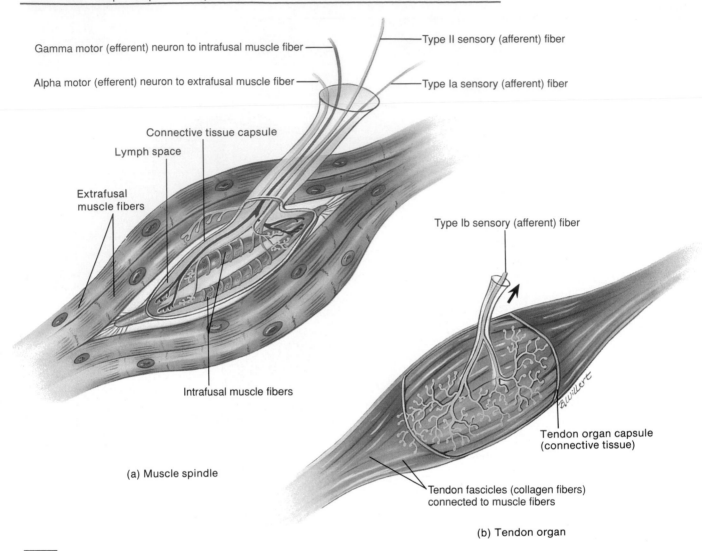

Gamma motor (efferent) neuron to intrafusal muscle fiber

Alpha motor (efferent) neuron to extrafusal muscle fiber

Type II sensory (afferent) fiber

Type Ia sensory (afferent) fiber

Connective tissue capsule

Lymph space

Extrafusal muscle fibers

Type Ib sensory (afferent) fiber

Intrafusal muscle fibers

Tendon organ capsule (connective tissue)

(a) Muscle spindle

Tendon fascicles (collagen fibers) connected to muscle fibers

(b) Tendon organ

Question: Which parts of a muscle spindle are associated with the afferent and efferent neurons?

cord. These are **gamma motor (efferent) neurons.** Extrafusal muscle fibers are innervated by **alpha motor (efferent) neurons.** These neurons also arise in the anterior gray horn of the spinal cord near gamma motor neurons.

Both sudden and prolonged stretch on the central areas of the intrafusal muscle fibers stimulate the type Ia and type II dendrites. Muscle spindles monitor changes in the length of a skeletal muscle by responding to the rate and degree of change in length. This information is relayed to the cerebrum, which allows conscious perception of the degree of muscle stretch (see Fig. 15.4). It also passes to the cerebellum to aid in the coordination and efficiency of muscle contraction (see Fig. 15.10).

Tendon Organs

Tendon organs (Golgi tendon organs) are proprioceptors found at the junction of a tendon with a muscle. They help protect tendons and their associated muscles from damage due to excessive tension. Also, they function as contraction receptors; that is, they monitor the force of contraction of each muscle. Each tendon organ consists of a thin capsule of connective tissue that encloses a few collagen fibers (Fig. 15.3b). Penetrating the capsule are one or more sensory (afferent) **type Ib fibers** whose dendrites entwine among and around the collagen fibers. When tension is applied to a tendon, tendon organs are stimulated, and nerve impulses are conducted into the CNS (see Fig. 15.4).

Joint Kinesthetic Receptors

There are several types of **joint kinesthetic receptors** within and around the articular capsules of synovial joints. Encapsulated receptors, similar to type II cutaneous mechanoreceptors (end organs of Ruffini), are present in the capsules of joints and respond to pressure. Small lamellated (Pacinian) corpuscles in the connective tissue outside articular capsules are receptors that respond to acceleration and deceleration of joint movement. Articular ligaments contain receptors similar to tendon organs that adjust reflex inhibition of the adjacent muscles when excessive strain is placed on the joint.

PHYSIOLOGY OF SENSORY PATHWAYS

Most input from somatic receptors on one side of the body crosses over to the opposite side in the spinal cord or brain stem before ascending to the thalamus. It then projects from the thalamus to the somatosensory cortex (primary somatosensory or general sensory area) where conscious sensations result (see Fig. 14.14). Axon collaterals (branches) of somatic sensory neurons also carry signals into the cerebellum and the reticular formation of the brain stem. Two general pathways lead from sensory receptors to the cortex: the posterior column–medial lemniscal pathway and the anterolateral (spinothalamic) pathways.

Posterior Column–Medial Lemniscus Pathway

Nerve impulses for conscious proprioception and most tactile sensations ascend to the cortex along a common pathway formed by three-neuron sets (Fig. 15.4). **First-order neurons** extend from sensory receptors into the spinal cord and up to the medulla on the same side of the body. The cell body of a first-order neuron is in the posterior (dorsal) root ganglion of a spinal nerve. Its axon is part of the **posterior column (fasciculus gracilis** and **fasciculus cuneatus)** in the spinal cord. The axon terminals form synapses with **second-order neurons** in the medulla. The cell body of a second-order neuron is located in the nucleus cuneatus (which receives input conducted along axons in the fasciculus cuneatus from the neck, arms, and upper chest) or nucleus gracilis (which receives input conducted along axons in the fasciculus gracilis from the trunk and legs). The axon of the second-order neuron crosses to the opposite side of the medulla and enters the **medial lemniscus,** a projection tract that extends from the medulla to the thalamus. In the thalamus, the axon terminals of second-order neurons synapse with **third-order neurons,** which project their axons to the somatosensory area of the cerebral cortex.

Impulses conducted along the posterior column–medial lemniscus pathway give rise to several highly evolved and refined sensations.

1. **Discriminative touch,** the ability to recognize the exact location of light touches and to make two-point discriminations.
2. **Stereognosis,** the ability to recognize by "feel" the size, shape, and texture of an object. Examples are iden-

FIGURE 15.4 Posterior column–medial lemniscus pathway. Nerve impulses for conscious proprioception and most tactile sensations are conducted along sets of first-order, second-order, and third-order neurons to the somatosensory area of the cerebral cortex.

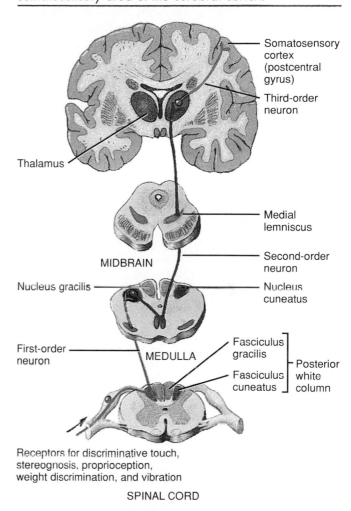

Receptors for discriminative touch, stereognosis, proprioception, weight discrimination, and vibration

SPINAL CORD

Question: What term means the ability to recognize by feel the size, shape, and texture of an object?

tifying (with closed eyes) a paperclip put into your hand or reading braille.
3. **Proprioception,** the awareness of the precise position of body parts, and **kinesthesia,** the awareness of directions of movement.
4. **Weight discrimination,** the ability to assess the weight of an object.
5. **Vibratory sensations,** the ability to sense rapidly fluctuating touch.

Anterolateral (Spinothalamic) Pathways

The **anterolateral (spinothalamic) pathways** carry mainly pain and temperature impulses. In addition, they relay the sensation of tickle and itch and some tactile impulses, which give rise to a very crude, not well-localized touch or pressure sensation. Like the posterior column–medial lemniscus pathway, the anterolateral pathways are also composed of three-neuron sets (Fig. 15.5). The first-order neuron connects a receptor of the neck, trunk, or extremities with the spinal cord. The cell body of the first-order neuron is in the posterior root ganglion. The axon of the first-order neuron synapses with the second-order neuron, which is located in the posterior gray horn of the spinal cord. The axon of the second-order neuron continues to the opposite side of the spinal cord and passes upward to the brain stem

FIGURE 15.5 Anterolateral (spinothalamic) pathways. Nerve impulses for pain, temperature, crude touch, pressure, tickle, and itch are conducted along sets of first-order, second-order, and third-order neurons to the somatosensory area of the cerebral cortex.

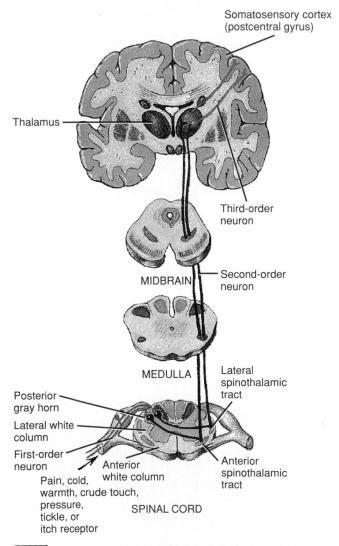

in either the **lateral spinothalamic tract** or **anterior spinothalamic tract**. The axon from the second-order neuron ends in the thalamus. There, it synapses with the third-order neuron. The axon of the third-order neuron projects to the somatosensory area of the cerebral cortex. The lateral spinothalamic tract conveys sensory impulses for pain and temperature whereas the anterior spinothalamic tract conveys impulses for tickle, itch, and crude touch and pressure.

Types of Pain

Pain is classified as one of two types, based on speed of onset, quality of the sensation, and duration: acute (fast) and chronic (slow). **Acute pain** occurs very rapidly, usually within 0.1 second after a stimulus is applied, and is not felt in deeper tissues of the body. This type of pain is also known as sharp, fast, and pricking pain. The pain felt from a needle puncture or knife cut to the skin are examples of acute pain. Impulses for acute pain conduct along myelinated A fibers. **Chronic pain,** by contrast, begins after a second or more and then gradually increases in intensity over a period of several seconds or minutes. This type of pain may be excruciating. It is also referred to as burning, aching, throbbing, and slow pain. Chronic pain can occur both in the skin and deeper tissues or in internal organs. An example is the pain associated with a toothache. Impulses for chronic pain conduct along smaller diameter, unmyelinated C fibers.

Relief from Pain

Some pain sensations are inappropriate; they do not warn of actual or impending damage but rather occur out of proportion to minor damage or chronically for no good reason. In such cases, **analgesia** (*a* = without; *algesis* = sensation of pain) or pain relief is needed. Most pain sensations respond to pain-reducing drugs. Drugs such as aspirin, acetaminophen (Tylenol), and ibuprofen (Motrin) block formation of prostaglandins, which stimulate nociceptors. Local anesthetics, such as Novocaine, provide short-term pain relief by blocking conduction of nerve impulses in the axons of first-order neurons. Morphine and other opiate drugs alter the quality of pain perception; pain is still sensed, but it is not perceived as being so noxious.

Sometimes, only surgery offers relief of pain. The purpose of surgical treatment is to interrupt the pain pathway somewhere between the receptors and the interpretation centers of the brain. This can be accomplished by severing the sensory nerve, its spinal root, or certain tracts in the spinal cord or brain. **Cordotomy** is severing a spinal cord tract, usually the lateral spinothalamic; **rhizotomy** is the cutting of spinal posterior (sensory) nerve roots. In each instance, a link in the pathway for pain is cut so that pain impulses do not reach the cortex.

Question: What sorts of sensory deficits could be produced by damage of the right lateral spinothalamic tract?

fibers to the descending motor pathways. As is true for somatic sensory representation in the somatosensory cortex, different muscles are not represented equally in the motor cortex (see Fig. 15.6b). The degree of representation is proportional to the number of motor units in a particular muscle of the body. For example, the muscles in the thumb, fingers, lips, tongue, and vocal cords have large representations while the trunk has a much smaller representation. By comparing Figs. 15.6a and 15.6b, you will see that somatosensory and motor representations are similar but not identical for the same part of the body.

Direct (Pyramidal) Pathways

Voluntary motor impulses are propagated from the motor cortex to somatic efferent neurons (voluntary motor neurons) that innervate skeletal muscles via the **direct** or **pyramidal** (pi-RAM-i-dal) **pathways.** The simplest of these pathways consists of sets of two neurons, upper and lower motor neurons. About one million cell bodies of direct pathway **upper motor neurons** are in the cortex. Their axons descend through the internal capsule of the cerebrum. In the medulla, the axon bundles form the ventral bulges known as the pyramids; this is the reason for the name pyramidal pathway. Most of these axons also cross to the opposite side in the medulla. They terminate in nuclei of cranial nerves or in the anterior gray horn of the spinal cord. **Lower motor neurons** extend from the cranial nerve motor nuclei or spinal cord anterior horn to skeletal muscle fibers. Close to their termination point, most upper motor neurons synapse with an association neuron, which in turn synapses with a lower motor neuron. Some upper motor neurons synapse directly with lower motor neurons.

The direct pathways convey impulses from the cortex that result in precise, voluntary movements. They channel nerve impulses into three tracts.

1. Lateral corticospinal tracts (pyramidal tracts proper). These pathways begin in the right or left motor cortex and descend through the **internal capsule** of the cerebrum and through the cerebral peduncle of the midbrain and the pons on the same side (Fig. 15.8). About 85 to 90% of the axons of upper motor neurons decussate (cross) in the medulla. The crossed axons form the lateral corticospinal tracts in the right and left lateral white columns of the spinal cord. Thus the motor cortex of the right side of the brain controls muscles on the left side of the body, and vice versa. The lower motor neurons then receive input from both upper motor neurons and association neurons. Axons of lower motor neurons (somatic motor neurons) exit all levels of the cord via the anterior roots of spinal nerves and terminate in skeletal muscles. These motor neurons control precise contraction in the distal extremities.

2. Anterior corticospinal tracts. About 10 to 15% of the axons of upper motor neurons do not cross in the

FIGURE 15.8 Lateral and anterior corticospinal tracts.

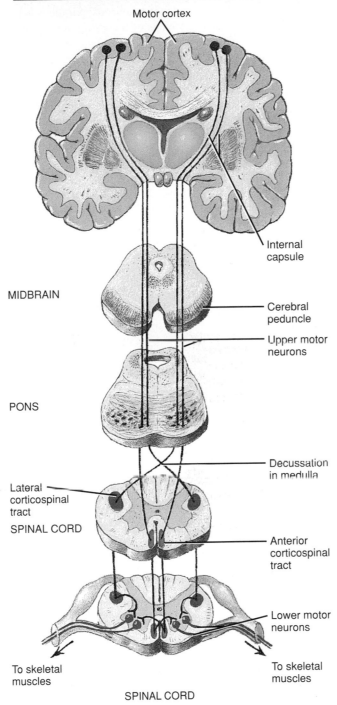

Motor cortex

Internal capsule

MIDBRAIN

Cerebral peduncle

Upper motor neurons

PONS

Decussation in medulla

Lateral corticospinal tract

SPINAL CORD

Anterior corticospinal tract

Lower motor neurons

To skeletal muscles

To skeletal muscles

SPINAL CORD

Question: What other tract (not shown in the illustration) conveys impulses that result in precise, voluntary movements?

medulla. They pass through the medulla, descend on the same side, and form the anterior corticospinal tracts in the right and left anterior white columns (Fig. 15.8). At several spinal cord levels, some of the axons of these upper motor neurons decussate. After crossing to the

Somatosensory Cortex

Areas of the somatosensory cortex (postcentral gyrus) that receive sensory information from different parts of the body have been mapped out. Figure 15.6a shows the location and areas of representation of the somatosensory cortex of the right cerebral hemisphere. The left cerebral hemisphere has a similar somatosensory cortex.

Note that some parts of the body are represented by large areas in the somatosensory cortex. These include the lips, face, tongue, and thumb. Other parts of the body, such as the trunk and lower extremities, are represented by much smaller areas. The relative sizes of the areas in the somatosensory cortex are directly proportional to the number of specialized sensory receptors in each respective part of the body. Thus there are many receptors in the skin of the lips but few in the skin of the trunk. The size of the cortical area for a particular part of the body relates directly to the functional importance of high sensitivity for that body part.

Spinocerebellar Tracts

The **posterior spinocerebellar tract** (see Fig. 13.4) is an uncrossed tract that conveys nerve impulses for subconscious muscle and joint sense to the cerebellum. This input allows the cerebellum to provide output for adjusting balance, posture, and muscle coordination. The impulses originate in neurons that extend from proprioceptors in muscles, tendons, and joints to the posterior gray horn of the spinal cord. Here the axon terminals synapse with second-order neurons. Axons of the second-order neurons pass to the lateral white column on the same side of the cord and join the

FIGURE 15.6 Somatosensory cortex (postcentral gyrus) and motor cortex (precentral gyrus). After Penfield and Rasmussen.

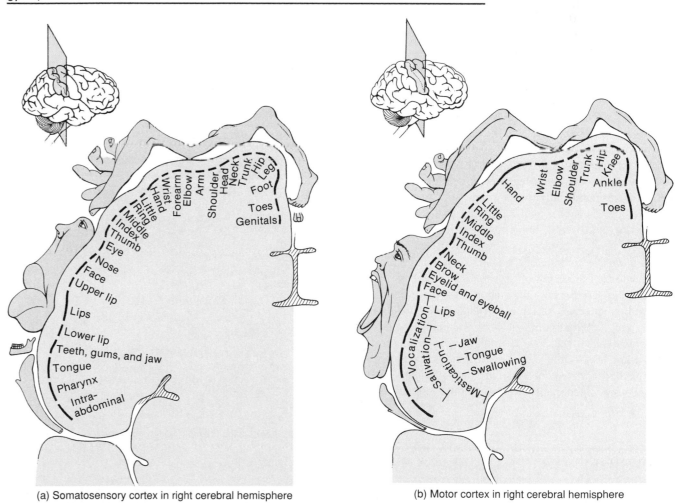

(a) Somatosensory cortex in right cerebral hemisphere

(b) Motor cortex in right cerebral hemisphere

Question: How do the somatosensory and motor representations compare for the hand and what does this difference imply?

posterior cerebellar tract. The tract enters the inferior cerebellar peduncles from the medulla and ends at the cerebellar cortex.

The **anterior spinocerebellar tract** (see Fig. 13.4) also conveys impulses for subconscious muscle sense. It, however, is made up of both crossed and uncrossed nerve fibers. Sensory neurons deliver impulses from proprioceptors to the posterior gray horn of the spinal cord. Here, synapses occur with neurons that make up the anterior spinocerebellar tracts. Some axons cross to the opposite side of the spinal cord in the anterior white commissure. Others pass laterally to the anterior spinocerebellar tract and extend upward, through the brain stem, to the pons to enter the cerebellum through the superior cerebellar peduncles. Sensory inputs conveyed to the cerebellum along these two pathways are critical for cerebellar regulation of posture and balance and coordination of skilled movements.

INTEGRATION OF SENSORY INPUT AND MOTOR OUTPUT

Sensory systems provide the input that keeps the central nervous system informed of changes in the external and internal environment (Fig. 15.7). Output from the CNS is then conveyed to motor systems, which enable us to move about, alter glandular secretions, and change our relationship to the world around us. As sensory information reaches the CNS, it becomes part of a large pool of sensory input. Every bit of input the CNS receives does not elicit a response. Rather, the incoming information is integrated with other information arriving from all other operating sensory receptors. The integration process occurs not just once but at many stations along the pathways of the CNS and at both conscious and subconscious levels. It occurs within the spinal cord, brain stem, cerebellum, basal ganglia, and cerebral cortex. As a result, a motor response to make a muscle contract or a gland secrete can be modified and responded to at any of these levels. Motor portions of the cerebral cortex play the major role for initiating and controlling precise, discrete muscular movements. The basal ganglia largely integrate semivoluntary, automatic movements like walking, swimming, and laughing. The cerebellum assists the motor cortex and basal ganglia by making body movements smooth and coordinated and by contributing significantly to maintaining normal posture and balance.

PHYSIOLOGY OF MOTOR PATHWAYS

The most direct motor pathways extend from the cortex of the brain to skeletal muscles. Other pathways are less direct and include synapses in the basal ganglia, thalamus, and cerebellum.

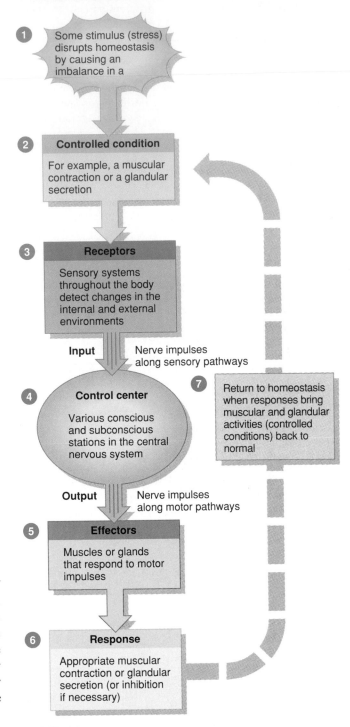

FIGURE 15.7 Integration of sensory input and motor output as components of a negative feedback cycle.

1. Some stimulus (stress) disrupts homeostasis by causing an imbalance in a

2. **Controlled condition**

For example, a muscular contraction or a glandular secretion

3. **Receptors**

Sensory systems throughout the body detect changes in the internal and external environments

Input — Nerve impulses along sensory pathways

4. **Control center**

Various conscious and subconscious stations in the central nervous system

7. Return to homeostasis when responses bring muscular and glandular activities (controlled conditions) back to normal

Output — Nerve impulses along motor pathways

5. **Effectors**

Muscles or glands that respond to motor impulses

6. **Response**

Appropriate muscular contraction or glandular secretion (or inhibition if necessary)

Question: Where in the CNS does integration occur?

Motor Cortex

The **motor cortex** (**primary motor area** or **precentral gyrus**) is the major control region for initiation of voluntary movements. The adjacent **premotor area** and even the **somatosensory cortex** (see Fig. 14.14) also contribute

opposite side, they synapse with association or lower motor neurons in the anterior gray horn of the spinal cord. Axons of these lower motor neurons exit the cervical and upper thoracic segments of the cord via the anterior roots of spinal nerves. They terminate in skeletal muscles that control muscles of the neck and part of the trunk, coordinating movements of the axial skeleton.

CLINICAL APPLICATION

MYELINATION AND DEVELOPMENT

The axons of the various tracts of the CNS develop myelin sheaths at different times. For example, the afferent and efferent axons of spinal nerves develop myelin after the fifth fetal month. The axons of the corticospinal tracts, on the other hand, do not become fully myelinated until the second year of life. This, in part, explains why infants and young children have a definite pattern of developmental activity and why a newborn cannot walk.

3. **Corticobulbar tracts.** The axons of these tracts accompany the corticospinal tracts from the motor cortex, through the internal capsule to the brain stem. There they decussate and terminate in the nuclei of nine pairs of cranial nerves in the pons and medulla: the oculomotor (III), trochlear (IV), trigeminal (V), abducens (VI), facial (VII), glossopharyngeal (IX), vagus (X), accessory (XI), and hypoglossal (XII). The corticobulbar tracts convey impulses that control voluntary movements of the head and neck.

CLINICAL APPLICATION

PARALYSIS

During a neurological exam, assessment of muscle tone, reflexes, and ability to perform voluntary movements helps pinpoint certain types of motor system dysfunction. Damage or disease of lower motor neurons, either of their cell bodies in the anterior horn or of their axons in the anterior root or spinal nerve, produces a condition called **flaccid paralysis.** There is neither voluntary nor reflex action of the innervated muscle fibers, and the muscle remains limp or flaccid (decreased or lost muscle tone). Injury or disease of upper motor neurons causes **spastic paralysis.** This condition is characterized by varying degrees of spasticity (increased muscle tone), exaggerated reflexes, and pathological reflexes such as the Babinski sign (see page 389).

Indirect (Extrapyramidal) Pathways

The **indirect (extrapyramidal) pathways** include all descending (motor) tracts other than the corticospinal and corticobulbar tracts. Nerve impulses conducted along the indirect pathways follow complex, polysynaptic circuits that involve the motor cortex, basal ganglia, thalamus, cerebellum, reticular formation, and nuclei in the brain stem

(Fig. 15.9). Upper motor neurons of the indirect pathways all begin in various nuclei of the brain stem. Their axons extend into the spinal cord along several tracts (described shortly) and finally synapse with association neurons or lower motor neurons. The lower motor neurons are the same ones activated by the direct pathways. For this reason, lower motor neurons are also called the **final common pathway.**

Basal Ganglia

The basal ganglia have many connections with other parts of the brain. Through these connections, they help to program habitual or automatic movement sequences such as walking or laughing in response to a joke and to set an appropriate level of muscle tone. The basal ganglia also seem to selectively inhibit other motor neuron circuits that are intrinsically active or excitatory. This effect is appreci-

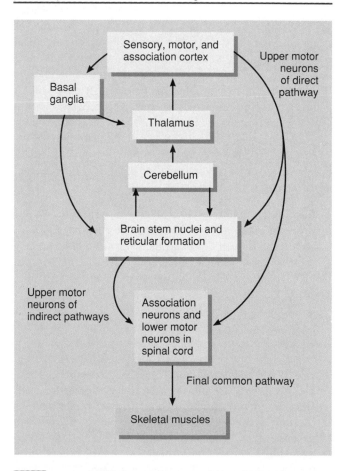

FIGURE 15.9 Indirect pathways for coordination and control of movement. For comparison, the direct pathway is also shown to the right.

- Sensory, motor, and association cortex
- Upper motor neurons of direct pathway
- Basal ganglia
- Thalamus
- Cerebellum
- Brain stem nuclei and reticular formation
- Upper motor neurons of indirect pathways
- Association neurons and lower motor neurons in spinal cord
- Final common pathway
- Skeletal muscles

Question: Starting with signals from the cortex, describe two circular feedback loops through the thalamus.

ated in certain types of basal ganglia damage, which often are characterized by abnormal movements or tremors caused by loss of inhibitory signals from the basal ganglia.

The caudate nucleus and the putamen receive input from sensory, motor, and association areas of the cortex, thalamus, and substantia nigra. Output from the basal ganglia comes mainly from the globus pallidus. It sends feedback signals in the form of nerve impulses to the motor cortex by way of the thalamus. This circuit—from cortex to basal ganglia to thalamus to cortex—appears to function in planning and programming movements.

The globus pallidus also sends impulses into the brain stem reticular formation, which appear to reduce muscle tone. Damage or destruction of certain basal ganglia connections causes a generalized increase in muscle tone that results in abnormal muscle rigidity.

CLINICAL APPLICATION

DAMAGE TO BASAL GANGLIA

Much of what we know about the function of the basal ganglia comes from studies of people who have damage or disease in these brain regions. Damage to the basal ganglia results in uncontrollable abnormal body movements, often accompanied by muscle rigidity and tremors (shaking) while at rest.

Parkinson's disease (see page 439) results in a generalized increase in muscle tone and stiffness of the face, arms, and legs. It is associated with deterioration of neural connections between the substantia nigra and the basal ganglia that employ the neurotransmitter dopamine.

Chorea (*choros* = a dance) is quick, purposeless jerky contractions of the limbs and involuntary facial twitches. Huntington's chorea is a hereditary form of the disease that often does not show symptoms until age 30 or 40. By then afflicted persons may have passed on this genetic defect to their children. A blood test now can determine if a person has the defective gene. Chorea is thought to result from degenerative changes in the caudate nucleus and putamen with loss of inhibitory neurons containing the neurotransmitter GABA.

Cerebellum

The cerebellum is active in both learning and performing rapid, coordinated, highly skilled movements such as running, speaking, and swimming. It also functions to maintain proper posture and equilibrium (balance). There are four aspects to cerebellar function.

1. **Intentions.** The cerebellum receives input from the motor cortex and basal ganglia via the pontine nuclei regarding what movements are planned.
2. **Actual movement.** It receives input from proprioceptors in joints and muscles on what is actually happening. These nerve impulses travel in the anterior and posterior spinocerebellar tracts. The vestibulocerebellar tract transmits impulses from the vestibular (equilibrium-

sensing) apparatus in the ear to the cerebellum. Nerve impulses from the eyes and ears also enter the cerebellum.
3. **Comparison.** It compares the command signals (intentions for movement) with sensory information (actual performance).
4. **Corrective feedback.** It sends out corrective signals, both to the nuclei in the brain stem and to the motor cortex via the thalamus.

In summary, the cerebellum receives information from higher brain centers about what the muscles should be doing and from the peripheral nervous system about what the muscles are doing. If there is a discrepancy between the two, corrective feedback signals are sent from the cerebellum via the thalamus to the cerebrum, where new commands are initiated to decrease the discrepancy and smooth the motion (Fig. 15.10).

Skilled activities such as tennis or volleyball provide good examples of the contribution of the cerebellum to movement. To make a good stop volley or to block a spike, you must bring your racket or arms forward just far enough to make solid contact. How do you stop at exactly the right point? This is where the cerebellum comes in. It receives information about your body status and the position of the ball as it flies toward you. Before you even hit the ball, the cerebellum has sent information to the cerebral cortex and basal ganglia informing them where your swing must stop. In response to impulses from the cerebellum, the cortex and basal ganglia transmit motor impulses to opposing body muscles to stop the swing.

CLINICAL APPLICATION

DAMAGE TO CEREBELLUM

Damage to the cerebellum is characterized by certain symptoms involving skeletal muscles on the same side of the body as the damage. The hallmark of cerebellar trauma or disease is **ataxia** (*a* = without; *taxis* = order), movements that are jerky and uncoordinated. Blindfolded people with ataxia cannot touch the tip of their nose with a finger because they cannot coordinate movement with their sense of where a body part is located. There also is an inability to stop a movement smoothly at the desired point. For example, a person might knock over a glass of water while trying to pick it up. Another sign of ataxia is a change in the speech pattern due to a lack of coordination of speech muscles. Cerebellar damage also produces **intention tremor.** This is a shaking during deliberate voluntary movement that is caused by inability to relax antagonistic muscle groups appropriately.

Descending Tracts from the Brain Stem

Indirect (extrapyramidal) motor output from the brain flows along five major spinal cord tracts, all of which arise from nuclei in the brain stem (see Fig. 13.4).

FIGURE 15.10 Input to and output from the cerebellum.

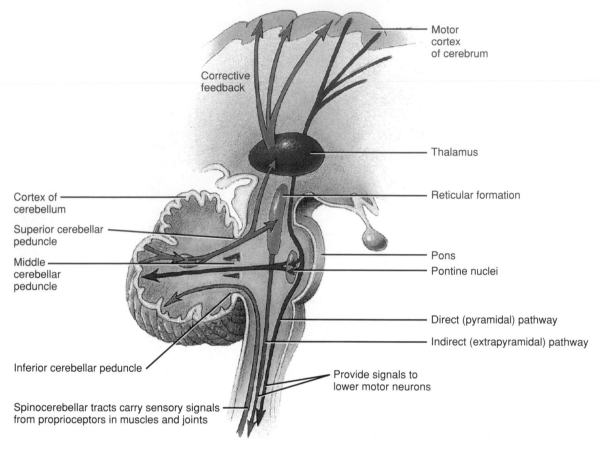

Question: Which tracts carry information about actual performance of muscles to the cerebellum?

1. **Rubrospinal tract.** This tract begins in the red nucleus of the midbrain, which receives impulses from the cerebral cortex and the cerebellum. Like the adjacent lateral corticospinal tract, the rubrospinal tract transmits nerve impulses to the opposite side of the body and governs precise, discrete movements of the distal extremities. An example would be the contractions of finger muscles needed to manipulate scissors while cutting paper.

2. **Tectospinal tract.** This tract begins in the superior colliculus of the midbrain, which receives visual input and governs reflex movements of the head and eyes. Tectospinal fibers transmit impulses to the opposite side of the body to neck muscles that control movements of the head in response to visual stimuli.

3. **Vestibulospinal tract.** This tract begins in the vestibular nucleus of the medulla, which receives output from the receptors for balance in the ear (vestibular apparatus). It conveys impulses on the same side of the body that regulate muscle tone in response to movements of the head. This tract therefore plays a major role in balance.

4. **Lateral reticulospinal tract.** This tract originates in the reticular formation of the medulla. Its function is to facilitate flexor reflexes, inhibit extensor reflexes, and decrease muscle tone in muscles of the axial skeleton and proximal limb.

5. **Anterior (ventral) or medial reticulospinal tract.** This tract originates in the pons. Its function is to facilitate extensor reflexes, inhibit flexor reflexes, and increase muscle tone in muscles of the axial skeleton and proximal limb.

Lower motor neurons receive both excitatory and inhibitory input from many presynaptic neurons in both direct and indirect pathways, an example of convergence. The sum total of the signals determines the final response of the lower motor neuron. It is not just a simple matter of the brain sending an impulse and the muscle always contracting. Association neurons are of considerable importance in the motor pathways. Most impulses from the brain are conveyed to association neurons before being received by lower motor neurons. These association neurons integrate the pattern of muscle contraction.

The major sensory and motor pathways and tracts in the spinal cord are summarized in Exhibit 15.1.

EXHIBIT 15.1

MAJOR SENSORY AND MOTOR PATHWAYS AND TRACTS

Tract and Location	Physiology

SENSORY (ASCENDING)

Posterior column–medial lemniscus

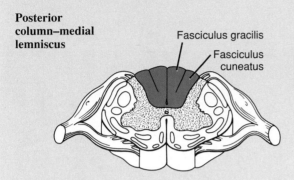

Fasciculus gracilis

Fasciculus cuneatus

Convey sensations from one side of the body to opposite cerebral hemisphere for discriminative touch (ability to recognize exact location of light touch stimuli and to make two-point discriminations); stereognosis (ability to recognize an object by feeling it); conscious proprioception (awareness of position of body parts); kinesthesia (awareness of direction of movement of body parts); weight discrimination (ability to assess weight of an object); and vibration.

Anterolateral (Spinothalamic)

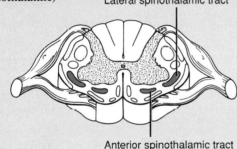

Lateral spinothalamic tract

Anterior spinothalamic tract

Lateral spinothalamic tract: Conveys sensations for pain and temperature from one side of body to opposite cerebral hemisphere. *Anterior spinothalamic tract:* Conveys sensations for tickle, itch, crude touch, and pressure from one side of body to opposite cerebral hemisphere.

Posterior spinocerebellar

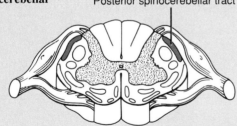

Posterior spinocerebellar tract

Conveys nerve impulses for subconscious proprioception from one side of body to same side of cerebellum; this sensory input keeps the cerebellum informed of actual movements and allows it to coordinate, smooth, and refine skilled movements.

Anterior spinocerebellar

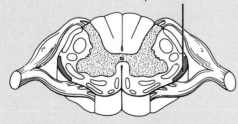

Anterior spinocerebellar tract

Conveys nerve impulses for subconscious proprioception from one side of body to same and opposite sides of cerebellum; this sensory input keeps the cerebellum informed of actual movements and allows it to coordinate, smooth, and refine skilled movements.

Exhibit continues

MAJOR SENSORY AND MOTOR PATHWAYS AND TRACTS

Tract and Location	Physiology

MOTOR (DESCENDING)
Direct (pyramidal)
Lateral corticospinal

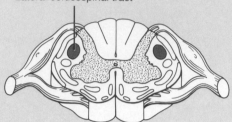

Lateral corticospinal tract

Conveys motor impulses from one side of cortex to skeletal muscles on opposite side of body to coordinate precise, discrete, voluntary movements, especially of the distal extremities.

Anterior corticospinal

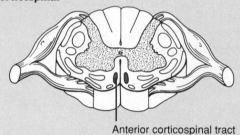

Anterior corticospinal tract

Conveys motor impulses from one side of cortex to skeletal muscles on opposite side of body to coordinate movements of the axial skeleton.

Corticobulbar

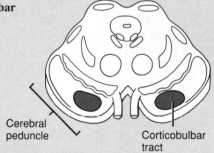

Cerebral peduncle

Corticobulbar tract

Midbrain

Conveys motor impulses from one side of cortex to skeletal muscles on opposite side of head to coordinate precise, discrete, voluntary contractions of head and neck muscles.

Indirect (extrapyramidal)
Rubrospinal

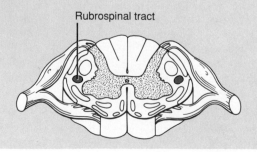

Rubrospinal tract

Conveys motor impulses from one side of midbrain, which receives input from the cortex and cerebellum, to skeletal muscles on opposite side of body that are concerned with precise, discrete movements of the distal extremities.

Tract and Location	Physiology
Tectospinal 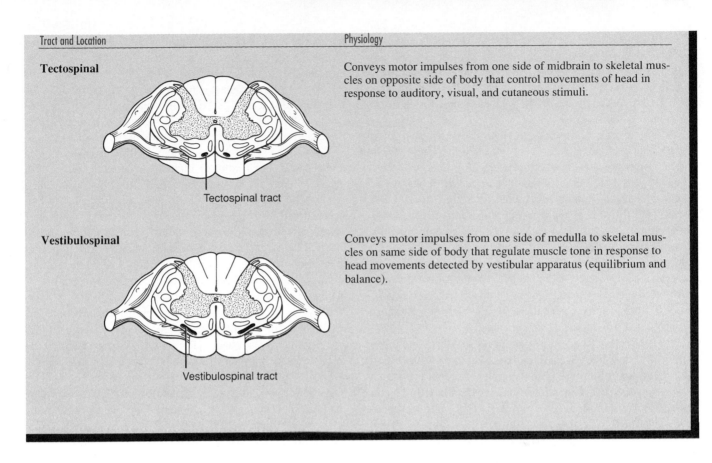 Tectospinal tract	Conveys motor impulses from one side of midbrain to skeletal muscles on opposite side of body that control movements of head in response to auditory, visual, and cutaneous stimuli.
Vestibulospinal Vestibulospinal tract	Conveys motor impulses from one side of medulla to skeletal muscles on same side of body that regulate muscle tone in response to head movements detected by vestibular apparatus (equilibrium and balance).

INTEGRATIVE FUNCTIONS

We turn now to a fascinating, though poorly understood, function of the cerebrum: integration. The **integrative functions** include cerebral activities such as memory, sleep and wakefulness, and emotional responses. The role of the limbic system in emotional behavior was discussed in Chapter 14.

Memory

Without memory, we would repeat mistakes and be unable to learn. Similarly, we would not be able to repeat our successes or accomplishments, except by chance. Although both memory and learning have been studied by scientists for many years, there is still no satisfactory explanation for how we recall information or how we remember. Some things, however, are known about how information is acquired and stored.

Learning is the ability to acquire knowledge or a skill through instruction or experience. Learning is very closely associated with rewards and punishments. **Memory** is the ability to recall thoughts. For an experience to become part of memory, it must produce a change in the brain that represents the experience. Such a memory trace is called an **engram.** The portions of the brain thought to be associated with memory include the association cortex of the frontal, parietal, occipital, and temporal lobes; parts of the limbic system, especially the hippocampus and amygdaloid nucleus; and the diencephalon.

Memory may generally be classified into two kinds based on how long the memory lasts: short-term and long-term memory. **Short-term memory** lasts only seconds or hours and is the ability to recall bits of information. One example is finding an unfamiliar telephone number in a telephone book and then dialing it. If the number has no special significance, it is usually forgotten within a few seconds. **Long-term memory,** on the other hand, lasts from days to years. For example, if you use a telephone number often, it becomes part of long-term memory. When information is in long-term memory, it can be retrieved for use whenever needed for quite a long period of time. The reinforcement due to the frequent retrieval of a piece of information is called **memory consolidation.**

Although the brain receives many stimuli, we are conscious of only a few of them. It has been estimated that of all the information that comes to our consciousness, only about 1% goes into long-term memory. Moreover, much of what goes into long-term memory is sooner or later forgotten. Perhaps it is fortunate that our brains select only a small percentage of our thoughts and lose much of what was stored. Otherwise, it might be overwhelmed with information.

It is a feature of memory that we can remember short lists easier than long ones. This is to say the obvious, but human memory does not record everything like an endless magnetic tape. It cannot remember or record long lists of details. Another feature of memory is that even when details are lost, the concept or main idea is retained. Then, interestingly, we can often explain the idea or concept—not like replaying a tape—but with our own selection of words and ways of explanation. Despite several decades of research, the precise mechanisms of memory are still elusive. However, by integrating several clinical and experimental observations, scientists have developed several theories to help us understand the mechanisms.

One theory of short-term memory states that memories may be caused by reverberating neuronal circuits—an incoming nerve impulse stimulates the first neuron, which stimulates the second, which stimulates the third, and so on (see Fig. 12.18c). Branches from the second and third neurons synapse with the first, sending the impulse back through the circuit again and again. Once fired, the output signal may last from a few seconds to many hours, depending on the arrangement of neurons in the circuit. If this pattern is applied to short-term memory, an incoming thought—the phone number—continues in the brain even after the initial stimulus is gone. Thus you can recall the thought only for as long as the reverberation continues.

There is some evidence to support the idea that short-term memory depends on electrical and chemical events in the brain rather than structural changes, such as the formation of new synapses. For example, several conditions may inhibit the electrical activity of the brain. These include anesthesia, coma, electroconvulsive shock, and ischemia (reduced blood supply) of the brain. Such conditions disrupt retention of recently acquired information, although they do not usually impede long-term memory laid down previously. It is a common experience that people who suffer retrograde amnesia (loss of past memories) cannot remember any events that have occurred for about 30 minutes before the amnesia developed. As the person recovers from the amnesia, the most recent memories return last.

Most research on long-term memory focuses on anatomical or biochemical changes that might enhance facilitation at synapses. Anatomical changes occur in neurons when they are either stimulated or made inactive. For example, electron micrographic studies of presynaptic neurons that have been subjected to prolonged, intense activity exhibit several anatomical changes. These include an increase in the number of presynaptic terminals, enlargement of synaptic end-bulbs, and an increase in the branching patterns and conductance of dendrites. There is also an increase in the

number of neuroglia. Moreover, neurons grow new synaptic end-bulbs with increasing age, presumably as a result of increased use. These changes, which are correlated with faster learning, suggest enhancement of facilitation at synapses. Such changes do not occur when neurons are inactive. In fact, in animals that have lost their eyesight, there is thinning of the cerebral cortex in the visual area.

There is also interest in the possible involvement of nucleic acids in long-term memory. The molecules DNA and RNA store information, and these molecules, especially DNA, tend to persist for the lifetime of the cell. Studies have shown an increase in the RNA content of activated neurons. Also, some evidence shows that long-term memory will not occur to any significant extent when RNA formation is inhibited. Since RNA synthesis precedes protein synthesis, these studies may also imply a relation between synthesis of certain proteins and memory.

Any of the events in synaptic transmission could be responsible for enhanced communication between neurons. For example, it has been suggested that there could be an increase in the number of receptor molecules in the postsynaptic cell membrane or a decrease in the rate of removal of neurotransmitter substance. Acetylcholine (ACh) seems to play an important role in memory. One of the findings in Alzheimer's disease (see page 438), which wipes out the ability to remember, is disappearance in certain brain regions of a key enzyme needed to synthesize ACh.

A recent hypothesis concerning memory involves neurons that have two different ways of transmitting information. Such neurons contain receptors called **NMDA receptors,** named after the chemical *N*-methyl D-aspartate that is used to detect them. These receptors open channels that permit calcium inflow into neurons. Many other receptors involved in neuronal firing open ion channels in response to *either* neurotransmitters (chemically gated channels) or a change in voltage (voltage-gated channels). NMDA receptors are unusual in that they must *first* be stimulated by a voltage change and *then* by a neurotransmitter (glutamate). This is an example of voltage regulation of a chemically gated channel. Such neurons have a special mode of signal transmission that is activated only if the cell receives two signals in a row. The first signal cocks the gun; the second signal fires it. This type of signal transmission provides the neurons with a different way to process information related to memory formation.

Wakefulness and Sleep

Humans sleep and awaken in a fairly constant 24-hour cycle called a **circadian** (ser-KĀ-dē-an) **rhythm.** When the brain is aroused or awake, it is in a state of readiness and able to react consciously to various stimuli. Since neuronal fatigue precedes sleep and the signs of fatigue disappear after sleep, fatigue is apparently one cause of sleep. Moreover, EEG recordings show that during wakefulness the cerebral cortex is very active, sending impulses continuously throughout the body. During sleep, however, fewer impulses arise from the cerebral cortex. The activity of the cerebral cortex relates to activity in the reticular formation.

The reticular formation has many ascending connections with the cerebral cortex and descending connections with the spinal cord (Fig. 15.11). Stimulation of portions of the reticular formation results in increased cortical activity. Thus a portion of the reticular formation is known as the **reticular activating system (RAS).** When this area is stimulated, many nerve impulses pass upward into the thalamus and disperse to widespread areas of the cerebral cortex. The effect is a generalized increase in cortical activity.

Arousal, or awakening from deep sleep, also involves increased activity in the RAS. For arousal to occur, the RAS must be stimulated by input signals. Almost any sensory input can activate the RAS: pain stimuli, proprioceptive signals, bright light, or an alarm clock. Once the RAS is activated, the cerebral cortex is also activated and you experience arousal. Nerve impulses from the cerebral cortex can also stimulate the RAS. Such impulses may originate in the somatosensory area of the cortex, the motor cor-

tex, or the limbic system. When the impulses activate the RAS, the RAS activates the cerebral cortex and arousal occurs. After arousal, the RAS and cerebral cortex continue to activate each other through a feedback system consisting of many circuits.

The reticular formation also has a feedback system with the spinal cord that is composed of many circuits. Impulses from the reticular formation descend into the spinal cord and then to skeletal muscles over the reticulospinal tracts. Muscle contraction causes proprioceptors to return impulses that activate the RAS. The two feedback systems (from the cortex and skeletal muscles) maintain activation of the RAS, which in turn maintains activation of the cerebral cortex. The result is a state of wakefulness called **consciousness.**

CLINICAL APPLICATION

ALTERED CONSCIOUSNESS

Consciousness may be altered by various factors. Amphetamines probably activate the RAS to produce a state

FIGURE 15.11 Reticular formation (a) and brain stem nuclei (b) associated with sleep (locus coerulus and raphe nuclei). The reticular activating system (RAS) consists of fibers that project to the cortex through the thalamus from the reticular formation.

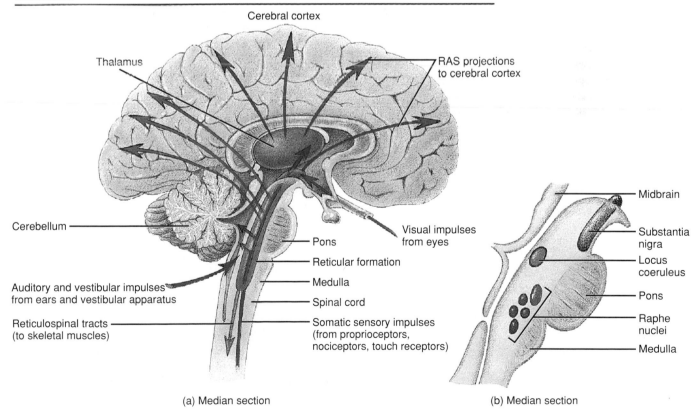

(a) Median section

(b) Median section

Question: Why should every dwelling have a smoke alarm? (*Hint:* Think about what a smoke alarm does if you are sleeping and predict the implications for sensory input to the RAS.)

of wakefulness and alertness. Meditation produces a relaxed, focused consciousness. Anesthetics produce an altered state of consciousness called anesthesia. Nervous system damage or disease can produce coma. Drugs such as LSD and alcohol can also alter consciousness.

Coma is the final stage of brain failure that is characterized by total unresponsiveness to all external stimuli. A comatose patient lies in a sleeplike state with the eyes closed. In the lightest stages of coma, primitive brain stem and spinal cord reflexes persist, but in the deepest stages, these reflexes are lost as are corneal, pupillary, tendon, and plantar reflexes. Eventually, respiratory and cardiovascular controls disappear, and the patient usually dies. Coma may be induced by lesions in various parts of the brain, poisoning, hypoxia, hypoglycemia, ischemia, infection, acid–base and electrolyte disorders, and trauma.

How does sleep occur if the activating feedback systems are in continual operation? One explanation is that the feedback system slows tremendously or is inhibited. During **sleep,** a state of altered consciousness or partial unconsciousness from which an individual can be aroused by different stimuli, activity in the RAS is very low.

Just as there are different levels of awareness when awake, there are different levels of sleep. Normal sleep consists of two types: non–rapid eye movement sleep (NREM) and rapid eye movement sleep (REM).

NREM sleep, or **slow wave sleep,** consists of four stages, each of which gradually merges into the next. Each stage has been identified by EEG recordings.

Stage 1.   This is a transition stage between wakefulness and sleep that normally lasts from 1 to 7 minutes. The person is relaxing with eyes closed. During this time, respirations are regular, pulse is even, and the person has fleeting thoughts. If awakened, the person will often say he has not been sleeping. Alpha waves diminish and theta waves appear on the EEG.

Stage 2.   This is the first stage of true sleep, even though the person experiences only light sleep. It is a little harder to awaken the person. Fragments of dreams may be experienced, and the eyes may slowly roll from side to side. The EEG shows *sleep spindles*—sudden, short bursts of sharply pointed waves that occur at 12 to 14 Hz (cycles per second).

Stage 3.   This is a period of moderately deep sleep. The person is very relaxed. Body temperature begins to fall and blood pressure decreases. It is difficult to awaken the person, and the EEG shows a mixture of sleep spindles and delta waves. This stage occurs about 20 minutes after falling asleep.

Stage 4.   Deep sleep occurs. The person is very relaxed and responds slowly if awakened. When bed-wetting and sleepwalking occur, they do so during this stage. The EEG is dominated by delta waves.

In a typical 7- or 8-hour sleep period, a person goes from stage 1 to 4 of NREM sleep. Then the person ascends to stages 3 and 2 and then to REM sleep within 50 to 90 minutes. The cycle normally repeats throughout the sleep period.

In **REM sleep,** the EEG readings are similar to those of stage 1 of NREM sleep. There are significant physiological differences, however. During REM sleep, muscle tone is depressed (except for rapid movements of the eyes), and respirations and pulse rate increase and are irregular. Blood pressure also fluctuates considerably. It is during REM sleep that most dreaming occurs. Following REM sleep, the person descends again to stages 3 and 4 of NREM sleep.

REM and NREM sleep alternate throughout the night with approximately 90-minute intervals between REM periods. This cycle repeats from three to five times during the entire sleep period. The REM periods start out lasting from 5 to 10 minutes and gradually lengthen until the final one lasts about 50 minutes. In a normal sleep period, REM sleep totals 90 to 120 minutes. As much as 50% of an infant's sleep is REM, as contrasted with 20% for adults. Most sedatives significantly decrease REM sleep.

Natural body rhythms, especially body temperature, determine the length of sleep. The higher the body temperature, the longer a person will sleep. As a person ages, the average time spent sleeping decreases. In addition, the percentage of REM sleep decreases. It has been suggested that the high percentage of REM sleep in infants and children reflects increased neuronal activity, which is important for the maturation of the brain. Infants apparently need this internal stimulation since the available external stimuli are restricted. Support for this idea comes from the fact that dreams, a particular kind of conscious activity in the brain, are most frequent during REM-type sleep.

Recent studies with animals suggest that two specific neural centers in the brain stem determine the occurrence of NREM and REM sleep. The NREM sleep center is found in the raphe nuclei (Fig. 15.11b). Their neurons contain large amounts of the neurotransmitter serotonin. When the supply of serotonin is exhausted, the result is severe insomnia and a reduction in both NREM and REM sleep. The insomnia can be alleviated by the administration of a precursor of serotonin (5-hydroxytryptophan). Serotonin itself cannot cross the blood–brain barrier. The REM sleep center is found in the loci coerulei (Fig. 15.11b). Their neurons contain large amounts of the neurotransmitter norepinephrine (NE). Destruction of the loci coerulei results in a complete disappearance of REM sleep but has no influence on NREM sleep. The administration of reserpine, a drug that exhausts the supply of both serotonin and norepinephrine, results in the elimination of both NREM and REM sleep. All these observations suggest that serotonin is important for NREM sleep, that norepinephrine is important for REM sleep, and that normally REM sleep is possible only if preceded by NREM sleep.

CLINICAL APPLICATION

SLEEP DISORDERS

At least three sleep disorders are clinically significant. **Narcolepsy** (NAR-kō-lep-sē; *narke* = numbness; *lepsis* = seizure) is a condition of involuntary attacks of sleep that last about 15 minutes and may occur at almost any time of the day. It is an inability, in the waking state, to inhibit REM sleep. **Insomnia** (in-SOM-nē-a; *in* = not; *somnus* = sleep) consists of difficulty in falling asleep and, usually, frequent awakening. It may be related to specific disorders or secondary factors, both medical and psychiatric. **Hypersomnia** refers to an excessively long or deep sleep from which a person can be awakened only by vigorous stimulation. It may be associated with conditions such as with head injury, strokes, and encephalitis.

Study Outline

Sensation (p. 444)

Levels of Sensation (p. 444)

1. Sensation is a conscious or unconscious awareness of external and internal stimuli.

Modality (p. 444)

1. Modality is the property by which one sensation is distinguished from another.
2. Usually, a given sensory neuron serves only one modality.

Components of Sensation (p. 444)

1. For a sensation to arise, four events must occur.
2. These are stimulation, transduction, conduction, and translation.

Sensory Receptors (p. 445)

1. Sensory receptors are selective.
2. In terms of simplicity or complexity, simple receptors are associated with general senses, and complex receptors are associated with special senses.
3. According to location, receptors are classified as exteroceptors, interoceptors, and proprioceptors.
4. On the basis of type of stimulus detected, receptors are classified as mechanoreceptors, thermoreceptors, nociceptors, photoreceptors, and chemoreceptors.
5. Sensory receptors respond to stimuli by producing receptor or generator potentials.
6. Adaptation is a change in sensitivity (usually a decrease) to a long-lasting stimulus.

General Senses (p. 446)

Cutaneous Sensations (p. 446)

1. Cutaneous sensations include tactile sensations (touch, pressure, vibration), thermal sensations (heat and cold), and pain. Receptors for these sensations are located in the skin, connective tissue under the skin, mucous membranes, mouth, and anus.
2. Receptors for touch are hair root plexuses and corpuscles of touch (Meissner's corpuscles), which are rapidly adapting, and type I cutaneous mechanoreceptors (tactile or Merkel discs) and type II cutaneous mechanoreceptors (end organs of Ruffini), which are slowly adapting. Receptors for pressure are type II cutaneous mechanoreceptors and lamellated (Pacinian) corpuscles. Receptors for vibration are corpuscles of touch and lamellated corpuscles. Itch and tickle receptors are free nerve endings.
3. Thermoreceptors are free nerve endings.

4. Pain receptors (nociceptors) are located in nearly every body tissue.
5. Two kinds of pain are somatic and visceral.

Proprioceptive Sensations (p. 448)

1. Receptors located in skeletal muscles, tendons, in and around joints, and the internal ear convey nerve impulses related to muscle tone, movement of body parts, and body position.
2. The receptors include muscle spindles, tendon organs (Golgi tendon organs), and joint kinesthetic receptors.

Physiology of Sensory Pathways (p. 451)

1. Sensory information from all parts of the body terminates in a specific area of the somatosensory cortex.
2. In the posterior column–medial lemniscus pathway and the anterolateral pathways there are first-order, second-order, and third-order neurons.
3. Impulses along the posterior column–medial lemniscus pathway are concerned with discriminative touch, stereognosis, proprioception, weight discrimination, and vibratory sensations.
4. The neural pathway for pain and temperature is the lateral spinothalamic tract.
5. The neural pathway for tickle, itch, crude touch, and pressure is the anterior spinothalamic pathway.
6. Specific areas of the somatosensory cortex (postcentral gyrus) receive different information from different parts of the body.
7. The pathways to the cerebellum are the anterior and posterior spinocerebellar tracts, which are involved in transmitting impulses for subconscious muscle and joint position sense.

Integration of Sensory Input and Motor Output (p. 454)

1. Sensory input keeps the CNS informed of changes in the environment.
2. Incoming sensory information is integrated at many stations along the CNS at both conscious and subconscious levels.
3. A motor response makes a muscle contract or a gland secrete.

Physiology of Motor Pathways (p. 454)

1. The motor cortex (precentral gyrus) is the major control region for initiation of voluntary movement.
2. Voluntary motor impulses are propagated from the motor cortex to somatic efferent neurons that innervate skeletal muscles via the direct (pyramidal) pathways. The simplest pathways consist of upper and lower motor neurons.
3. The direct pathways include the lateral and anterior corticospinal tracts and corticobulbar tracts.
4. Indirect (extrapyramidal) pathways involve the motor cortex, basal ganglia, thalamus, cerebellum, reticular formation, and nuclei in the brain stem.

5. Major indirect tracts are the rubrospinal, tectospinal, vestibulospinal, and reticulospinal tracts.

Integrative Functions (p.461)

1. Memory is the ability to recall thoughts and is generally classified into two kinds: short-term and long-term memory.
2. A memory trace in the brain is called an engram.
3. Short-term memory is related to electrical and chemical events; long-term memory is related to anatomical and biochemical changes at synapses.

4. Sleep and wakefulness are integrative functions that are controlled by the reticular activating system (RAS).
5. Non–rapid eye movement (NREM) sleep consists of four stages identified by EEG recordings.
6. Most dreaming occurs during rapid eye movement (REM) sleep.
7. The neurotransmitters that affect sleep are serotonin and norepinephrine (NE).

Review Questions

1. Distinguish between sensation and perception. (p. 444)
2. What is modality? (p. 444)
3. What events take place for a sensation to occur? (p. 444)
4. Classify receptors on the basis of simplicity and complexity, location, and type of stimulus. (p. 445)
5. Distinguish between generator and receptor potentials. (p. 445)
6. What is adaptation? Compare rapidly and slowly adapting receptors. (p. 446)
7. What is a cutaneous sensation? Distinguish tactile, thermal, and pain sensations. (p. 446)
8. For each of the following cutaneous sensations, describe the receptor involved in terms of structure, function, and location: touch, pressure, vibration, itch and tickle, temperature, and pain. (p. 446)
9. How do cutaneous sensations help maintain homeostasis? (p. 450)
10. Why are pain receptors important? Differentiate somatic pain, visceral pain, referred pain, and phantom pain. (p. 448)
11. Why is the concept of referred pain useful to the physician in diagnosing internal disorders? (p. 448)
12. What is the proprioceptive sense? Where are the receptors for this sense located? (p. 448)
13. Describe the structure of muscle spindles, tendon organs (Golgi tendon organs), and joint kinesthetic receptors. (p. 449)
14. Relate proprioception to the maintenance of homeostasis. (p. 449)

15. Distinguish between the posterior column–medial lemniscus pathway and anterolateral (spinothalamic) pathways in terms of location and function. (p. 451)
16. Describe how various parts of the body are represented in the somatosensory cortex. (p. 453)
17. What are the functions of the cerebellar tracts? (p. 453)
18. Describe how sensory input and motor output are linked in the central nervous system. (p. 454)
19. Describe how various parts of the body are represented in the motor cortex. (p. 454)
20. Distinguish between direct and indirect motor pathways in terms of location and function. (p. 455)
21. How do the basal ganglia and cerebellum function in body movements? (p. 456)
22. Define memory. What are the two kinds of memory? (p. 461)
23. Define an engram and memory consolidation. (p. 461)
24. Describe the proposed mechanism of memory. (p. 462)
25. Describe how sleep and wakefulness are related to the reticular activating system (RAS). (p. 462)
26. What are the four stages of non–rapid eye movement (NREM) sleep? How is NREM sleep distinguished from rapid eye movement (REM) sleep? (p. 464)
27. Explain the roles of serotonin and norepinephrine (NE) in sleep. (p. 464)
28. Define the following: anesthesia (p. 448), paralysis (p. 456), coma, narcolepsy, insomnia, and hypersomnia. (p. 464)

Answers to Questions with Figures

15.1 Rapidly adapting: corpuscles of touch, hair root plexuses, and lamellated corpuscles. Slowly adapting: type I and type II cutaneous mechanoreceptors.
15.2 The kidneys.
15.3 Type Ia and type II afferent (sensory) fibers wrap around the central region of intrafusal muscle fibers. Gamma efferent (motor) neurons innervate intrafusal muscle fibers.
15.4 Stereognosis.
15.5 Loss of pain and temperature sensations on the left side of the body below the level of the damage.
15.6 The hand has a larger representation in the motor than in the sensory cortex, which implies a greater precision in its movement control than discriminative ability in its sensa-

tion.
15.7 Spinal cord, brain stem, cerebellum, basal ganglia, and cerebral cortex.
15.8 Corticobulbar.
15.9 Cortex to basal ganglia to thalamus to cortex. Cortex to brain stem nuclei to cerebellum to thalamus to cortex.
15.10 Anterior and posterior spinocerebellar tracts.
15.11 A smoke alarm detects smoke and sounds a loud bell or buzzer, which wakes up sleepers by providing auditory input that stimulates the RAS. Since there is little or no olfactory input to the RAS, people die of toxic substances in smoke without waking up if they do not have a smoke alarm.

Chapter 16
THE SPECIAL SENSES

Chapter Contents at a Glance

Student Objectives

1. Locate the receptors for olfaction and describe the neural pathway for smell.

2. Identify the gustatory receptors and describe the neural pathway for taste.

3. List and describe the accessory structures of the eye and the structural divisions of the eyeball.

4. Discuss image formation by describing refraction, accommodation, and constriction of the pupil.

5. Describe how photoreceptors and photopigments function in vision.

6. Describe the retinal processing of visual input and the neural pathway of light impulses to the brain.

7. Describe the anatomical subdivisions of the ear.

8. List the principal events in the physiology of hearing.

9. Identify the receptor organs for equilibrium and how they function.

10. Contrast the causes and symptoms of glaucoma, senile macular degeneration, deafness, Ménière's syndrome, otitis media, and motion sickness.

11. Define medical terminology associated with the sense organs.

Like the general senses, the special senses allow us to detect specific changes in our environment. However, the special senses—smell, taste, vision, hearing, and equilibrium—provide richer sensory experiences. They also have receptor organs that are structurally more complex.

OLFACTORY SENSATIONS: SMELL

Both smell and taste are chemical senses; that is, the sensations arise from the interaction of molecules with smell or taste receptors. Smell is our least understood special sense. The sense of smell is even more highly developed in lower animals, for example, dogs, than it is in humans. Among all sensations, only smell and taste project both to higher cortical areas and the limbic system. This is probably the reason that certain odors and tastes can evoke strong emotional responses or a flood of memories.

Anatomy of Olfactory Receptors

Between 10 and 100 million receptors for the **olfactory** (ol-FAK-tō-rē; *olfactus* = smell) **sense,** or sense of smell, lie in the nasal epithelium in the superior portion of the nasal cavity. The total area of the olfactory epithelium is 5 cm² (a lit-

tle less than 1 in.²). It occupies the upper portion of the nasal septum and extends along the superior and upper part of the middle nasal conchae (Fig. 16.1a). The olfactory epithelium consists of three principal kinds of cells: olfactory receptors, supporting cells, and basal cells.

Olfactory receptors are bipolar neurons whose distal (apical) end is a knob-shaped dendrite. Several cilia, called **olfactory hairs,** protrude from the dendrite. The cilia are the sites of olfactory transduction. Olfactory receptors respond to the chemical stimulation of an odorant molecule by producing a generator potential, thus initiating the olfactory response. From the proximal (basal) part of each olfactory receptor, a single axon projects to the olfactory bulb (Fig. 16.1b).

Supporting (sustentacular) cells are columnar epithelial cells of the mucous membrane lining the nose. **Basal cells** lie between the bases of the supporting cells. They are stem cells that continually produce new olfactory receptors, which live for only a month or so before being replaced. This process is remarkable because olfactory receptors are neurons. It is one of the few exceptions to the general rule that mature neurons are not replaced in your nervous system.

Within the connective tissue that supports the olfactory epithelium are **olfactory (Bowman's) glands.** They produce mucus, which is carried to the surface of the epithe-

---

FIGURE 16.1 Olfactory receptors. (a) Location in nasal cavity. (b) Details.

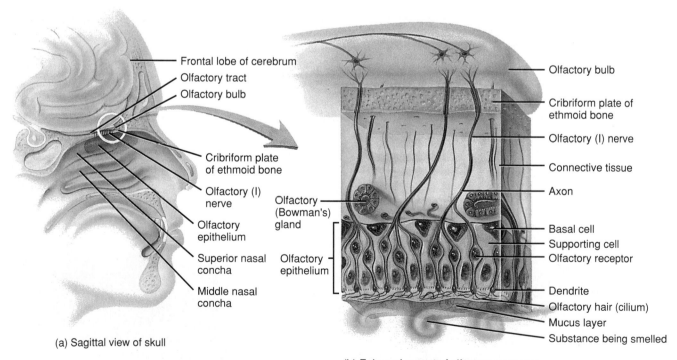

(a) Sagittal view of skull

(b) Enlarged aspect of olfactory receptors

Question: Axons of olfactory bulb neurons form which structure?

lium by ducts. The secretion moistens the surface of the olfactory epithelium and dissolves odorant gases. The continuous secretion of mucus also replaces the surface film of fluid and prevents continuous stimulation of olfactory hairs by the same odor.

Both supporting cells of the nasal epithelium and olfactory glands are innervated by branches of the facial (VII) nerve. Stimuli such as pepper, ammonia, and chloroform are irritating. They may provoke tears and sniffles by stimulating lacrimal and nasal mucosal receptors in addition to olfactory receptors.

Physiology of Olfaction

Many attempts have been made to distinguish and classify "primary" sensations of smell. One classification includes seven primary scents: camphoraceous, musky, floral, pepperminty, ethereal, pungent, and putrid. Recent evidence suggests there are many more primary scents (perhaps in the hundreds). In addition, our ability to recognize literally thousands of different scents may depend on patterns of activity in the brain that arise from activation of different combinations of olfactory receptors.

Olfactory receptors react to odorant stimuli in the same way that most sensory receptors react to their specific stimuli: a generator potential (depolarization) develops and triggers one or more nerve impulses. How the generator potential arises is known in some cases. Some odorants bind to receptors that are linked to G-proteins in the plasma membrane and activate the enzyme adenylate cyclase (see page 523). The result is: opening of Na^+ channels → inflow of Na^+ → depolarizing generator potential → nerve impulses.

Adaptation and Odor Thresholds

Adaptation (decreasing sensitivity) to odors occurs rapidly and appears to involve both olfactory receptors and the central nervous system. Olfactory receptors themselves adapt about 50% in the first second or so after stimulation but adapt very slowly thereafter. Yet we know from common experience that complete insensitivity to certain odors occurs in about a minute after exposure to a strong smell, more than can be explained based on receptor adaptation alone. No doubt other neurons in the olfactory pathway also exhibit adaptation.

Like the other special senses, olfaction has a low threshold. Only a few molecules of certain substances need be present in air to be smelled. A good example is the chemical methyl mercaptan, which can be detected in concentrations as low as 1/25,000,000,000 mg per milliliter of air. Since natural gas, used for cooking and heating, is odorless but lethal and potentially explosive if it accumulates, methyl mercaptan is added in small amounts to provide olfactory warning of gas leaks.

Olfactory Pathway

The unmyelinated axons of the olfactory receptors unite to form the **olfactory (I) nerves,** which pass through multiple foramina in the cribriform plate of the ethmoid bone (Fig. 16.1b). They terminate in paired masses of gray matter in the brain called the **olfactory bulbs,** which lie beneath the frontal lobes of the cerebrum on either side of the crista galli of the ethmoid bone. The first synapse of the olfactory pathway occurs in the olfactory bulbs between the axons of olfactory nerves and dendrites of neurons inside the olfactory bulbs.

Axons of olfactory bulb neurons extend posteriorly and form the **olfactory tract** (Fig. 16.1a). From here, there are two projection pathways for olfactory impulses. The first pathway extends to primitive cortical regions close to the olfactory bulbs on the inferior surface of the brain. These include the prepyriform cortex (the main region for olfactory discrimination) and the limbic system (the amygdala and hippocampus). The second pathway extends to the thalamus and then to higher cortical regions in the frontal lobes. The projections to the limbic system probably account for our emotional and memory-evoked responses to odors, for example, sexual excitement upon smelling a certain perfume or nausea upon smelling a food that once made you violently ill. Nerve impulses that conduct to the prepyriform cortex and frontal cortex via the thalamus are perceived as odor and give rise to the sensation of smell.

GUSTATORY SENSATIONS: TASTE

Like olfaction, taste is a chemical sense that requires dissolving of a substance before it can be tasted. Persons with colds or allergies sometimes complain that they cannot taste their food. Although their taste sensations may be operating normally, their olfactory sensations are not. This shows that much of what we think of as taste is actually smell. Odors from foods pass upward into the nasopharynx (portion of the throat behind the nose) and nasal cavity to stimulate olfactory receptors. In fact, a given concentration of a substance will stimulate the olfactory system thousands of times more strongly than it stimulates the gustatory system.

Anatomy of Gustatory Receptors

The receptors for **gustatory** (GUS-ta-tō′-rē; *gusto* = taste) **sensations,** or sensations of taste, are located in the taste buds (Fig. 16.2). The nearly 10,000 taste buds of a young adult are mainly on the tongue, but they are also found on the soft palate (back portion of roof of mouth), larynx (voice box), and pharynx (throat). The number of taste buds declines with age. Each **taste bud** is an oval body consisting of three kinds of epithelial cells: supporting cells, gustatory receptor cells, and basal cells (Fig. 16.2c). The **supporting (sustentacular) cells** form a capsule; inside are

FIGURE 16.2 Gustatory receptors.

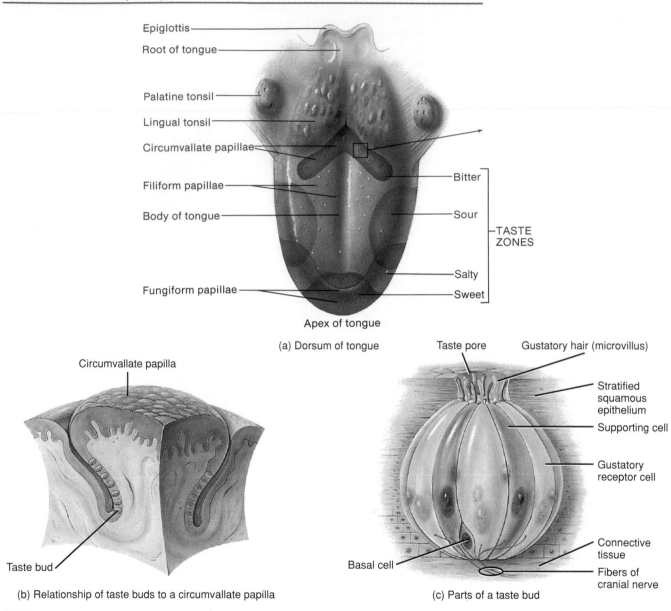

(a) Dorsum of tongue

(b) Relationship of taste buds to a circumvallate papilla

(c) Parts of a taste bud

Question: In order, what structures form the gustatory pathway?

about 50 **gustatory (taste) receptor cells.** A single, hairlike **gustatory hair (microvillus)** projects from each gustatory receptor cell to the external surface through an opening in the taste bud called the **taste pore.** The gustatory hairs make contact with taste stimuli through the taste pore. **Basal cells** are found at the periphery of the taste bud near the connective tissue layer. These epithelial cells produce supporting cells, which then develop into gustatory receptor cells that have a life span of about 10 days. At their base, the receptor cells synapse with dendrites of sensory nerve fibers that form the first part of the gustatory pathway. The dendrites of a single fiber branch profusely and contact many receptors in several taste buds.

Taste buds are found in elevations on the tongue called **papillae** (pa-PIL-ē). The papillae give the upper surface of the tongue its rough appearance (Fig. 16.2a, b). **Circumvallate** (ser-kum-VAL-āt) **papillae,** the largest type, are circular and form an inverted V-shaped row at the posterior portion of the tongue. **Fungiform** (FUN-ji-form; meaning mushroom-shaped) **papillae** are knoblike elevations found primarily on the tip and sides of the tongue. All circumvallate and most fungiform papillae contain taste buds. **Filiform** (FIL-i-form) **papillae** are pointed threadlike structures that cover the anterior two-thirds of the tongue. They rarely contain taste buds.

Physiology of Gustation

Once a chemical is dissolved in saliva, it can make contact with the plasma membrane of the gustatory hairs, which are the presumed site of taste transduction. The result is a receptor potential (see page 445), which is thought to cause the release of neurotransmitter by exocytosis of synaptic vesicles within gustatory receptor cells. Nerve impulses first arise in the neurons that synapse with gustatory receptor cells.

Despite the many substances we seem to taste, there are only four primary taste sensations: sour, salty, bitter, and sweet (Fig. 16.2a). All other "tastes," such as chocolate, pepper, and coffee, are combinations of these four, modified by accompanying olfactory sensations. Individual gustatory receptor cells may respond to more than one of the four primary tastes, but receptors in certain regions of the tongue react more strongly than others to the primary taste sensations. Although the tip of the tongue reacts to all four, it is highly sensitive to sweet and salty substances. The posterior portion of the tongue is highly sensitive to bitter substances. The lateral edges of the tongue are more sensitive to sour substances.

Adaptation and Taste Thresholds

Adaptation to taste occurs rapidly. Complete adaptation can occur in 1 to 5 minutes of continuous stimulation. As with odors, receptor adaptation alone cannot account for the speed of complete adaptation. Adaptation of receptors to smell contributes to taste adaptation but still does not account for its speed. As with adaptation to odors, adaptation to taste also involves neurons of the taste pathway in the CNS.

The threshold for taste varies for each of the primary tastes. The threshold for bitter substances, as measured by quinine, is lowest. This may have a protective function since poisonous substances often are bitter. The threshold for sour substances, as measured by hydrochloric acid, is somewhat higher. The thresholds for salty substances, as measured by sodium chloride, and sweet substances, as measured by sucrose, are about the same and are higher than those for bitter or sour substances.

Gustatory Pathway

Three cranial nerves include afferent fibers from taste buds: the facial (VII) serves the anterior two-thirds of the tongue; the glossopharyngeal (IX) serves the posterior one-third of the tongue; and the vagus (X) serves the throat and epiglottis (cartilage lid over the voice box). Taste impulses conduct from the taste buds along these cranial nerves to the medulla. From the medulla, some taste fibers project to the limbic system and the hypothalamus while others project to the thalamus. Fibers that extend from the thalamus to the primary gustatory area in the parietal lobe of the cerebral cortex (see Fig. 14.14) are responsible for the conscious perception of taste.

VISUAL SENSATIONS

The study of the structure, function, and diseases of the eye is known as **ophthalmology** (of'-thal-MOL-ō-jē; *ophthalmo* = eye; *logos* = study of). A physician who specializes in the diagnosis and treatment of eye disorders with drugs, surgery, and corrective lenses is known as an **ophthalmologist.** An **optometrist** is a specialist with a doctorate in optometry who is licensed to examine and test the eyes and treat visual defects by prescribing corrective lenses. An **optician** is a technician who fits, adjusts, and dispenses corrective lenses prescribed by an ophthalmologist or optometrist.

The structures related to vision are the eyeball, the optic (II) nerve, the brain, and several accessory structures.

Accessory Structures of the Eye

The **accessory structures** of the eye are the eyelids, eyelashes, eyebrows, lacrimal (tearing) apparatus, and extrinsic eye muscles. The upper and lower **eyelids,** or **palpebrae** (PAL-pe-brē), shade the eyes during sleep, protect the eyes from excessive light and foreign objects, and spread lubricating secretions over the eyeballs (Fig. 16.3). From superficial to deep, each eyelid consists of epidermis, dermis, subcutaneous tissue, fibers of the orbicularis oculi muscle, a tarsal plate, tarsal glands, and a conjunctiva (Fig. 16.4a).

FIGURE 16.3 Surface anatomy of the eye.

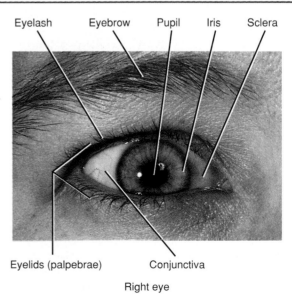

Eyelash   Eyebrow   Pupil   Iris   Sclera

Eyelids (palpebrae)   Conjunctiva

Right eye

Question: Which part of the eye would lack pigment in an albino?

FIGURE 16.4 Accessory structures of the eye.

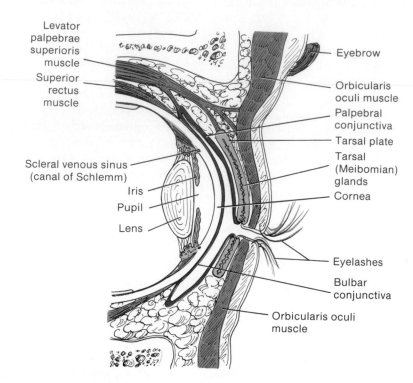

Levator palpebrae superioris muscle

Superior rectus muscle

Scleral venous sinus (canal of Schlemm)

Iris

Pupil

Lens

Eyebrow

Orbicularis oculi muscle

Palpebral conjunctiva

Tarsal plate

Tarsal (Meibomian) glands

Cornea

Eyelashes

Bulbar conjunctiva

Orbicularis oculi muscle

(a)   Sagittal section of accessory structures

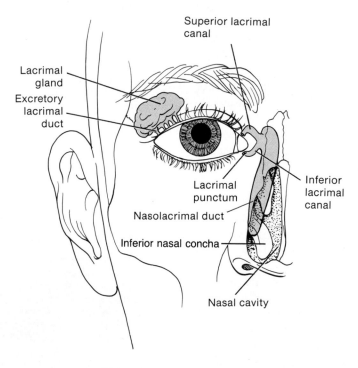

Superior lacrimal canal

Lacrimal gland

Excretory lacrimal duct

Lacrimal punctum

Nasolacrimal duct

Inferior nasal concha

Inferior lacrimal canal

Nasal cavity

(b) Anterior view of lacrimal apparatus

Question: What are tears and what is their function?

The **tarsal plate** is a thick fold of connective tissue that gives form and support to the eyelids. Embedded in each tarsal plate is a row of elongated modified sebaceous glands known as **tarsal** or **Meibomian** (mī-BŌ-mē-an) **glands.** Their oily secretion helps keep the eyelids from adhering to each other. Infection of the tarsal glands produces a tumor or cyst on the eyelid called a **chalazion** (ka-LĀ-zē-on). The **conjunctiva** (kon'-junk-TĪ-va) is a thin, protective mucous membrane. The **palpebral conjunctiva** lines the inner aspect of the eyelids and the **bulbar (ocular) conjunctiva** passes from the eyelids onto the anterior surface of the eyeball. When the blood vessels of the bulbar conjunctiva are dilated and congested due to local irritation or infection, the person has bloodshot eyes.

The **eyelashes** project from the border of each eyelid and together with the **eyebrows,** which arch transversely above the upper eyelids, help protect the eyeballs from foreign objects, perspiration, and the direct rays of the sun. Sebaceous glands at the base of the hair follicles of the eyelashes, called **sebaceous ciliary glands** or **glands of Zeis** (ZĪS), pour a lubricating fluid into the follicles. Infection of these glands is called a **sty.**

The **lacrimal** (*lacrima* = tear) **apparatus** is a group of structures that produces and drains tears (Fig. 16.4b). The **lacrimal glands,** each about the size and shape of an almond, secrete lacrimal fluid, or tears. Leading from the lacrimal glands are 6 to 12 **excretory lacrimal ducts** that empty tears onto the surface of the conjunctiva of the upper lid. From here the tears pass medially over the anterior surface of the eyeball to enter two small openings called **lacrimal puncta.** Tears then pass into two ducts, the **lacrimal canals,** which lead into the **nasolacrimal duct,** a canal that carries the lacrimal fluid into the nasal cavity.

Lacrimal fluid is a watery solution containing salts, some mucus, and a bactericidal enzyme called **lysozyme.** The fluid cleans, lubricates, and moistens the eyeball. After being secreted by the lacrimal glands, it is spread medially over the surface of the eyeball by the blinking of the eyelids. Each gland produces about 1 ml per day.

Normally, tears are cleared away by evaporation or by passing into the lacrimal canals and then into the nasal cavity as fast as they are produced. If an irritating substance contacts the conjunctiva, however, the lacrimal glands are stimulated to oversecrete and tears accumulate (watery eyes). This is a protective mechanism, since the tears dilute and wash away the irritating substance. Watery eyes also occur when an inflammation of the nasal mucosa, such as a cold, obstructs the nasolacrimal ducts and blocks drainage of tears. Humans are unique in expressing emotions, both happiness and sadness, by **crying.** In response to parasympathetic stimulation, the lacrimal glands produce excessive tears that may spill over the edges of the eyelids and even fill the nasal cavity with fluid.

The six extrinsic eye muscles that move each eye (see Fig. 11.6) receive their innervation from cranial nerves III, IV, and VI. In general, the size of motor units is small in these muscles. Some motor neurons serve only two or three muscle fibers, fewer than in any other part of the body except the voice box. This permits smooth, precise, and rapid movement of the eyes. As indicated by their names (see Exhibit 11.5 on page 284), each extrinsic eye muscle moves the eyeball in a different direction—laterally, medially, superiorly, or inferiorly. For example, looking to the right requires simultaneous contraction of the right lateral rectus and left medial rectus muscles and relaxation of the left lateral rectus and right medial rectus. The oblique muscles function to preserve rotational stability of the eyeball. Coordinated movement of the two eyes involves circuits in the brain stem and cerebellum.

Anatomy of the Eyeball

The adult **eyeball** measures about 2.5 cm (1 in.) in diameter. Of its total surface area, only the anterior one-sixth is exposed. The remainder is recessed and protected by the orbit into which it fits. Anatomically, the wall of the eyeball can be divided into three layers: fibrous tunic, vascular tunic, and retina or nervous tunic (Fig. 16.5).

Fibrous Tunic

The **fibrous tunic** is the outer coat of the eyeball consisting of the anterior cornea and posterior sclera. The **cornea** (KOR-nē-a) is a nonvascular, transparent, fibrous coat that covers the colored iris. Because it is curved, the cornea helps focus light. Its outer surface is covered by an epithelial layer that is continuous with the epithelium of the bulbar conjunctiva. The **sclera** (SKLE-ra; *skleros* = hard), the "white" of the eye, is a coat of dense connective tissue that covers all the eyeball except the cornea. The sclera gives shape to the eyeball, makes it more rigid, and protects its inner parts. Its posterior surface is pierced by the optic foramen, which encircles the optic (II) nerve. At the junction of the sclera and cornea is an opening known as the **scleral venous sinus,** or **canal of Schlemm.** (Surface features of the eye are shown in Fig. 16.4a.)

CLINICAL APPLICATION

CORNEAL TRANSPLANTS

Corneal transplants are the most common organ transplant operation and the most successful type of transplant since rejection rarely occurs. Because the cornea is avascular, antibodies that might cause rejection do not circulate there. The defective cornea is removed and a donor cornea of similar diameter is sewn in. The shortage of donated corneas has been partially overcome by the development of artificial corneas made of plastic.

Vascular Tunic

The **vascular tunic** or **uvea** (YOO-vē-a) is the middle layer of the eyeball. It has three portions: choroid, ciliary body,

FIGURE 16.5 Gross structure of the eyeball.

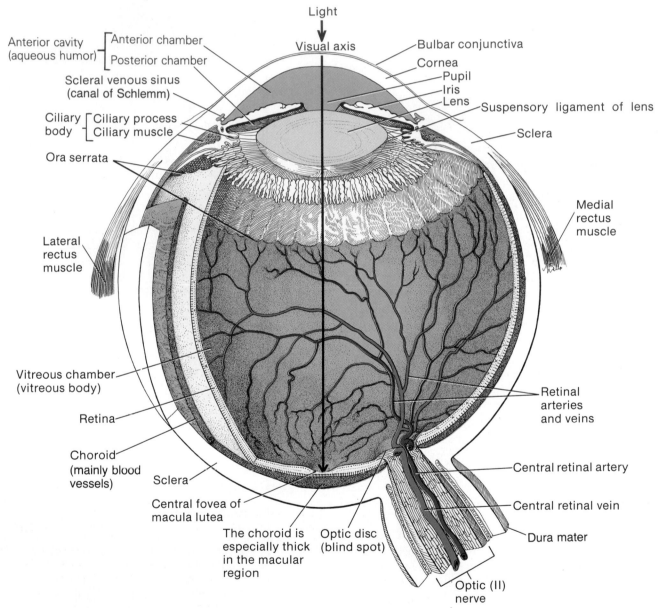

Light

Visual axis

Anterior cavity (aqueous humor) — Anterior chamber / Posterior chamber

Scleral venous sinus (canal of Schlemm)

Ciliary body — Ciliary process / Ciliary muscle

Ora serrata

Lateral rectus muscle

Bulbar conjunctiva

Cornea
Pupil
Iris
Lens
Suspensory ligament of lens

Sclera

Medial rectus muscle

Vitreous chamber (vitreous body)

Retina

Choroid (mainly blood vessels)

Sclera

Central fovea of macula lutea

The choroid is especially thick in the macular region

Optic disc (blind spot)

Retinal arteries and veins

Central retinal artery

Central retinal vein

Dura mater

Optic (II) nerve

Transverse section of left eyeball (superior view)

Question: Which structures in the eye do not have their own blood supply and therefore rely on diffusion of oxygen and nutrients from adjacent tissues or fluid? What advantage does avascularity confer?

and iris. The highly vascularized **choroid** (KŌ-royd) is the posterior portion of the vascular tunic and lines most of the internal surface of the sclera. It provides nutrients to the posterior surface of the retina. Melanocytes, which produce the dark pigment melanin, give the choroid a brown-black appearance.

In the anterior portion of the vascular tunic, the choroid becomes the **ciliary** (SIL-ē-ar′-ē) **body.** This is the thickest portion of the vascular tunic. It extends from the **ora serrata** (Ō-ra ser-RĀ-ta), the jagged anterior margin of the retina, to a point just behind the sclerocorneal junction. The ciliary body consists of the ciliary processes and ciliary muscle. The **ciliary processes** are protrusions or folds on the internal surface of the ciliary body where epithelial lining cells secrete aqueous humor. The **ciliary muscle** is a circular band of smooth muscle that alters the shape of the lens for near or far vision.

The **iris** (*irid* = colored circle) is the colored portion of the eyeball and is shaped like a flattened donut. It is suspended between the cornea and the lens and is attached at its outer margin to the ciliary processes. It consists of circular and radial smooth muscle fibers. The hole in the center of the iris is the **pupil.** A principal function of the iris is to regulate the amount of light entering the posterior cavity of the eyeball through the pupil. When bright light stimulates the eye, parasympathetic nerve fibers stimulate the circular iris smooth muscles (constrictor pupillae) to contract and decrease the size of the pupil (constriction). In dim light, sympathetic nerve fibers stimulate the radial iris smooth muscles (dilator pupillae) to contract and increase the pupil's size (dilation). These responses are visceral reflexes.

Retina (Nervous Tunic)

The third and inner coat of the eye, the **retina (nervous tunic),** lines the posterior three-quarters of the eyeball and is the beginning of the visual pathway. By using an ophthalmoscope to peer through the pupil, one can see a magnified image of the retina and the blood vessels that course across its anterior surface. The retina is the only place in the body where blood vessels can be viewed directly and examined for pathological changes such as occur with hypertension or diabetes. Several landmarks are visible (Fig. 16.6). Near the center is the **optic disc,** the site where the optic nerve exits the eyeball. Bundled together with the optic nerve are the **central retinal artery,** a branch of the ophthalmic artery, and **central retinal vein.** Branches of the central retinal artery fan out to nourish the anterior surface of the retina. The central retinal vein drains blood from the retina through the optic disc.

The retina consists of a pigment epithelium (nonvisual portion) and a neural portion (visual portion). The **pigment epithelium** is a sheet of melanin-containing epithelial cells that lies between the choroid and the neural portion of the retina. Some histologists classify it as part of the choroid

FIGURE 16.6 Normal retina as seen through an ophthalmoscope.

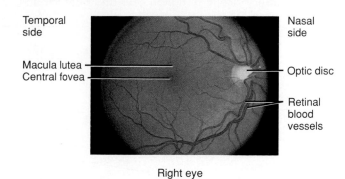

Right eye

Question: Evidence of what diseases may be seen through an ophthalmoscope?

rather than the retina. Melanin in the choroid and the pigment epithelium absorbs stray light rays which prevents reflection and scattering of light within the eyeball. This ensures that the image cast on the retina by the cornea and lens remains sharp and clear. Most albinos lack melanin pigment in all parts of the body, including the eye. They often wear sun glasses, even indoors, because moderately bright light is perceived as nothing but glare.

CLINICAL APPLICATION

DETACHED RETINA

Detachment of the retina may occur in trauma, such as a blow to the head, or in various eye disorders. The detachment occurs between the neural portion of the retina and the pigment epithelium. Fluid accumulates between these layers, forcing the thin, pliable retina to billow outward. The result is distorted vision and blindness in the corresponding field of vision. The retina may be reattached by laser surgery or cryosurgery (which involves the local application of extreme cold).

The **neural portion** of the retina is a multilayered outgrowth of the brain. It processes visual data extensively before transmitting nerve impulses to the thalamus, which then relays nerve impulses to the primary visual cortex. Three distinct layers of retinal neurons are separated by two zones where synaptic contacts are made, the inner and outer synaptic layers. The three layers of retinal neurons, named in the order in which they process visual input, are the **photoreceptor layer, bipolar cell layer,** and **ganglion cell layer** (Fig. 16.7). Note that light passes through the ganglion and bipolar cell layers before reaching the photoreceptor layer. There also are two other types of cells present in the retina called **horizontal cells** and **amacrine cells.**

FIGURE 16.7 Microscopic structure of the retina. The downward arrow indicates the direction of the signals passing through the neural portion of the retina. Ultimately, nerve impulses arise in ganglion cells and pass into the optic nerve (cranial nerve II).

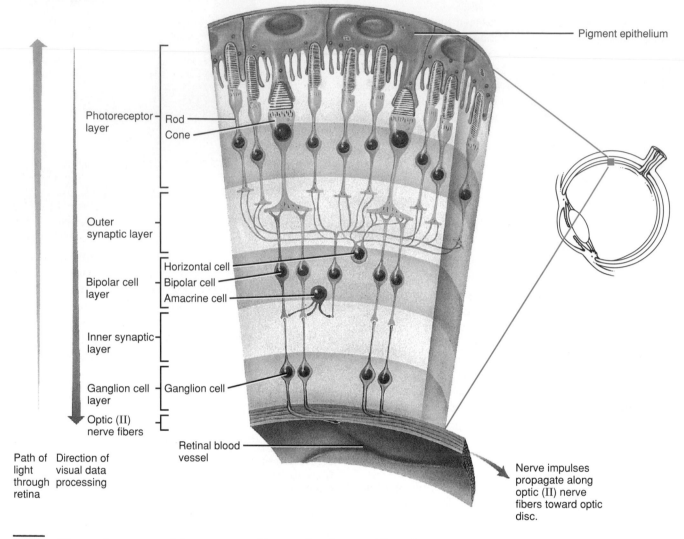

Pigment epithelium

Photoreceptor layer — Rod
— Cone

Outer synaptic layer

Bipolar cell layer — Horizontal cell
— Bipolar cell
— Amacrine cell

Inner synaptic layer

Ganglion cell layer — Ganglion cell

Optic (II) nerve fibers

Retinal blood vessel

Path of light through retina

Direction of visual data processing

Nerve impulses propagate along optic (II) nerve fibers toward optic disc.

Question: What are the two types of photoreceptors and how are their functions different?

These cells form laterally directed pathways that modify the signals being transmitted along the pathway from photoreceptors to bipolar cells to ganglion cells.

Photoreceptors are specialized to transduce light rays into receptor potentials. The two types of photoreceptors are rods and cones, named for the differing shapes of their outer segments, which nestle among fingerlike extensions of the pigment epithelium cells (see Fig. 16.11). Each retina has about 6 million cones and 120 million rods. **Rods** are most important for black-and-white vision in dim light. They also allow us to discriminate between different shades of dark and light and permit us to see shapes and movement. **Cones** provide color vision and high **visual acuity**

(sharpness of vision) in bright light. In moonlight we cannot see colors because only the rods are functioning.

Cones are most densely concentrated in the **central fovea,** a small depression in the center of the macula lutea (see Fig. 16.5). The **macula lutea** (MAK-yoo-la LOO-tē-a; *macula* = spot; *lutea* = yellow) is in the exact center of the posterior portion of the retina, at the visual axis of the eye. The fovea is the area of sharpest vision because of the high density of cones. Rods are absent from the fovea and macula and increase in number toward the periphery of the retina. It is for this reason that you can see better at night while not looking directly at an object.

From photoreceptors, information flows to bipolar cells

through the outer synaptic layer and then from bipolar cells through the inner synaptic layer to ganglion cells. The axons of ganglion cells extend posteriorly to the optic disc and exit the eyeball as the optic nerve. The optic disc is also called the **blind spot.** Since it contains no rods or cones, we cannot see an image that strikes the blind spot. Normally, you are not aware of having a blind spot, but you can easily convince yourself of its presence. Cover your left eye and gaze directly at the cross below. Then increase or decrease the distance between the book and your eye. At some point the square will disappear as its image falls on the blind spot.

+   ■

Lens

Just behind the pupil and iris, within the cavity of the eyeball, is the nonvascular **lens** (see Fig. 16.5). Proteins called **crystallins,** arranged like the layers of an onion, make up the lens. Normally, the lens is perfectly transparent. It is enclosed by a clear connective tissue capsule and held in position by encircling **suspensory ligaments.** The lens fine-tunes focusing of light rays for clear vision.

CLINICAL APPLICATION

CATARACTS

The leading cause of blindness is a loss of transparency of the lens known as a **cataract.** This problem often occurs with aging but may also be caused by injury, exposure to ultraviolet rays, certain medications (such as long-term use of steroids), or complications of other diseases (for example, diabetes). The lens becomes cloudy or less transparent due to changes in the structure of the lens proteins. Fortunately, sight can usually be restored by surgical removal of the old lens and implantation of an artificial one.

Interior of the Eyeball

The interior of the eyeball is a large space divided by the lens into two cavities: the anterior cavity and vitreous chamber. The **anterior cavity,** the space anterior to the lens, is further divided into the **anterior chamber,** which lies behind the cornea and in front of the iris, and the **posterior chamber,** which lies behind the iris and in front of the suspensory ligaments and lens (Fig. 16.8). The anterior cav-

FIGURE 16.8 Section through the anterior portion of the eyeball at the sclerocorneal junction. Arrows indicate the flow of aqueous humor.

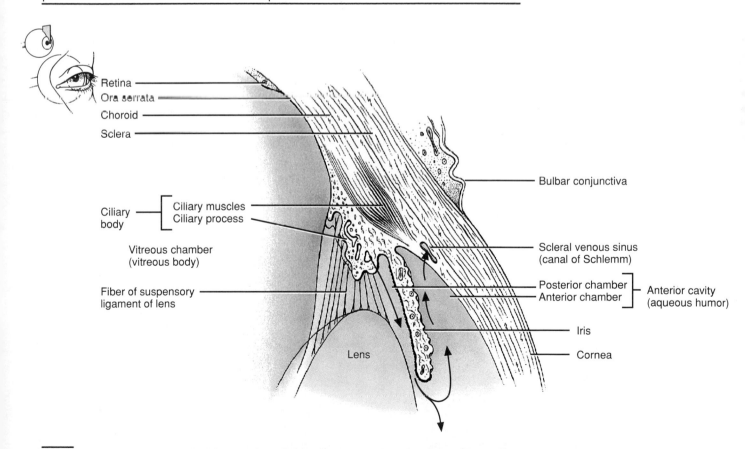

Question: Where is aqueous humor produced, what is its circulation path, and where does it drain from the eyeball?

ity is filled with a watery fluid called the **aqueous** (*aqua* = water) **humor** that is continually secreted by the ciliary processes behind the iris. Once the fluid is formed, it flows into the posterior chamber and then forward between the iris and the lens, through the pupil, and into the anterior chamber. Aqueous humor helps nourish the lens and cornea. From the anterior chamber, aqueous humor drains into the scleral venous sinus (canal of Schlemm) and then into the blood. Normally, aqueous humor is completely replaced about every 90 minutes.

The pressure in the eye, called **intraocular pressure**, is produced mainly by the aqueous humor. The intraocular pressure, along with the vitreous body (described shortly), maintains the shape of the eyeball and keeps the retina smoothly applied to the choroid so the retina will be an even surface for reception of clear images. Intraocular pres-

sure normally is about 16 mm Hg because there is a balance between production and outflow of the aqueous humor. Excessive intraocular pressure, called **glaucoma** (glaw-KŌ-ma), see page 498, causes degeneration of the retina and blindness.

The second, and larger, cavity of the eyeball is the **vitreous chamber (posterior cavity).** It lies between the lens and the retina and contains a jellylike substance called the **vitreous body.** This substance contributes to intraocular pressure, helps to prevent the eyeball from collapsing, and holds the retina flush against the internal portions of the eyeball. The vitreous body, unlike the aqueous humor, does not undergo constant replacement. It is formed during embryonic life and is not replaced thereafter.

A summary of structures associated with the eyeball is presented in Exhibit 16.1.

EXHIBIT 16.1

SUMMARY OF STRUCTURES ASSOCIATED WITH THE EYEBALL

Structure	Function
FIBROUS TUNIC	

Sclera	Provides shape and protects inner parts.
Cornea	Admits and refracts (bends) light.

VASCULAR TUNIC

Choroid	Provides blood supply and absorbs scattered light.
Ciliary body	Secretes aqueous humor and alters shape of lens for near or far vision (accommodation).
Iris	Regulates amount of light that enters eyeball.

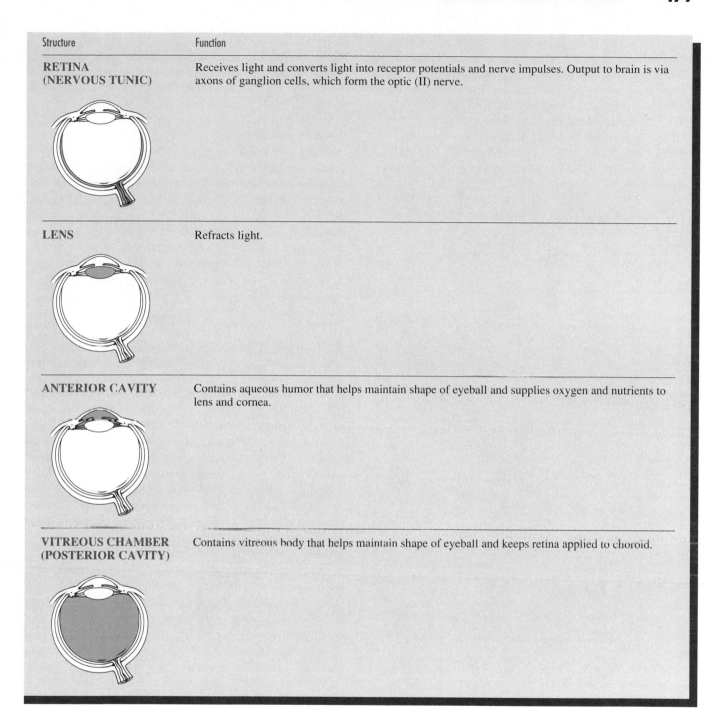

Structure	Function
RETINA (NERVOUS TUNIC)	Receives light and converts light into receptor potentials and nerve impulses. Output to brain is via axons of ganglion cells, which form the optic (II) nerve.
LENS	Refracts light.
ANTERIOR CAVITY	Contains aqueous humor that helps maintain shape of eyeball and supplies oxygen and nutrients to lens and cornea.
VITREOUS CHAMBER (POSTERIOR CAVITY)	Contains vitreous body that helps maintain shape of eyeball and keeps retina applied to choroid.

Image Formation

In some respects the eye is like a camera. The cornea and the lens of the eye focus an image of distant objects on a light-sensitive "film"—the retina. Contraction of the ciliary muscle changes the shape of the lens to bring closer objects into focus, in much the same manner as a camera is focused. Adjustment of the pupil diameter helps to maintain proper light exposure of the retina.

Formation of images on the retina involves three basic processes: (1) refraction of light rays by the cornea and lens, (2) accommodation of the lens, and (3) constriction of the pupil. Accommodation and pupil size are functions of the smooth muscle fibers of the ciliary muscle and iris. These muscles are termed **intrinsic eye muscles** since they are inside the eyeball.

Refraction of Light Rays

When light rays traveling through a transparent medium (such as air) pass into a second transparent medium with a different density (such as water), they bend at the junction between the two media. This bending is called **refraction** (Fig. 16.9a). As light rays enter the eye, they are refracted at the anterior and posterior surfaces of the cornea. Both surfaces of the lens of the eye further refract the light rays so they come into exact focus on the retina.

Images focused on the retina are inverted; they are upside down. They also undergo right to left reversal; that is, light from the right side of an object strikes the left side of the retina, and vice versa (Fig. 16.9b,c). The reason we do not see an inverted world is that the brain learns early in life to coordinate visual images with the exact locations of objects. The brain stores memories of reaching and touching objects and automatically turns visual images right-side-up and right-side-around.

About 75% of the focusing occurs at the cornea. The lens is responsible for fine-tuning of image focus and changing the focus for near or distant objects. When an object is 6 m (20 ft) or more away from the viewer, the light rays reflected from the object are nearly parallel to one another (Fig. 16.9b). The parallel rays must be bent sufficiently to fall exactly on the central fovea, where vision is sharpest. Light rays that are reflected from objects closer than 6 m (20 ft) are divergent rather than parallel (Fig. 16.9c). As a result, they must be refracted toward each other to a greater extent. This change in refraction is brought about by the lens of the eye by a process called accommodation.

Accommodation and the Near Point of Vision

If the surface of a lens curves outward, like the surface of a ball, the lens will refract incoming rays toward each other so they eventually intersect. This is a *convex* lens. Conversely, when the surface of a lens curves inward, like the inside of a hollow ball, the rays bend away from each other. This is a *concave* lens. The lens of the eye is convex on both sides. Furthermore, its focusing power increases as its curvature becomes greater. When the eye is focusing on a close object, the lens curves more to bend the rays toward the central fovea. This increase in the curvature of the lens for near vision is called **accommodation** (Fig. 16.9c).

How does accommodation occur? When you are viewing distant objects, the ciliary muscle is relaxed but the lens is fairly flat because it is stretched in all directions by taut suspensory ligaments. When you view a close object, the ciliary muscle contracts, which pulls the ciliary process and choroid forward toward the lens. This action releases tension on the lens and suspensory ligaments. Since it is elastic, the lens shortens, thickens, and bulges. Now that it is more rounded, its focusing power is greater and the light rays converge more.

The **near point of vision** is the minimum distance from the eye that an object can be clearly focused with maximum effort. This is about 10 cm (4 in.) in a young adult. With aging, the lens loses elasticity and therefore its ability to accommodate. This condition is called **presbyopia** (prez-bē-Ō-pē-a; *presbys* = old man). As a consequence, older people cannot read print at the same close range as can youngsters. By age 40 the near point of vision may have increased to 20 cm (8 in.) and at age 60 to 80 cm (31 in.). Presbyopia usually begins in the midforties and is the reason that those already wearing glasses need bifocals and those who have not previously needed glasses now require reading glasses.

FIGURE 16.9 Refraction of light rays and accommodation.

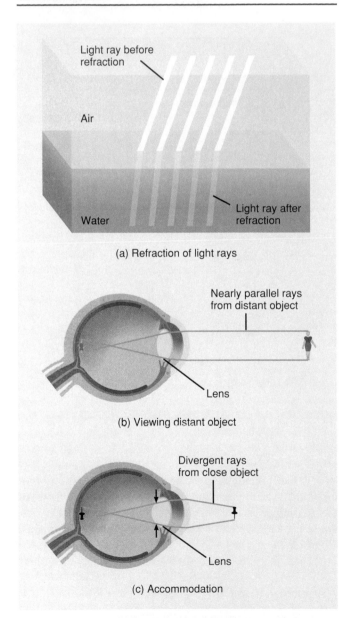

(a) Refraction of light rays

(b) Viewing distant object

(c) Accommodation

Question: What anatomical changes occur during accommodation?

CLINICAL APPLICATION

ABNORMALITIES OF REFRACTION

The normal eye, known as an **emmetropic** (em'-e-TROP-ik) **eye,** can sufficiently refract light rays from an object 6 m (20 ft) away to focus a clear image on the retina. Many people, however, do not have this ability because of abnormalities related to improper refraction. Among these abnormalities are **myopia** (mī-Ō-pē-a), or nearsightedness; **hypermetropia** (hī'-per-me-TRŌ-pē-a), or farsightedness, also known as **hyperopia;** and **astigmatism** (a-STIG-ma-tizm), or irregularities in the surface of the lens or cornea. The conditions are illustrated and explained in Fig. 16.10.

Most errors of vision can be corrected by eyeglasses or contact lenses. A contact lens floats on a film of tears over the cornea. The outer surface of the contact lens corrects the visual defect, while the inner surface is precisely ground to fit the curvature of the cornea.

Constriction of the Pupil

The circular muscle fibers of the iris also have a role in the formation of clear retinal images. Part of the accommodation mechanism consists of the contraction of the sphincter muscles of the iris to constrict the pupil. **Constriction of the pupil** means narrowing the diameter of the hole through which light enters the eye. This visceral reflex occurs simultaneously with accommodation of the lens and prevents light rays from entering the eye through the periphery of the lens. Light rays entering at the periphery would not be brought to focus on the retina and would result in blurred vision. The pupil, as noted earlier, also constricts in bright light.

Convergence

Because of the position of their eyes, many animals see a set of objects off to the left through one eye and an entirely different set off to the right through the other. In humans, both eyes focus on only one set of objects—a characteristic called **single binocular vision.**

Single binocular vision occurs when light rays from an object strike corresponding points on the two retinas. When we stare straight ahead at a distant object, the incoming light rays are aimed directly at both pupils and are refracted

FIGURE 16.10 Abnormal refraction in the eyeball. (b) In the nearsighted or myopic eye, the image is focused in front of the retina. The condition may result from an elongated eyeball or thickened lens. (c) Correction is by use of a concave lens that diverges entering light rays so that they have to travel further through the eyeball and are focused directly on the retina. (d) In the farsighted or hypermetropic eye, the image is focused behind the retina. The condition results from a shortened eyeball or a thin lens. (e) Correction is by a convex lens that converges entering light rays so that they focus directly on the retina. An astigmatism is irregular curvature of the (f) cornea or (g) lens that prevents images from being focused on the retina. This results in blurred or distorted vision. Suitable glasses or contact lenses correct the refraction of an astigmatic eye.

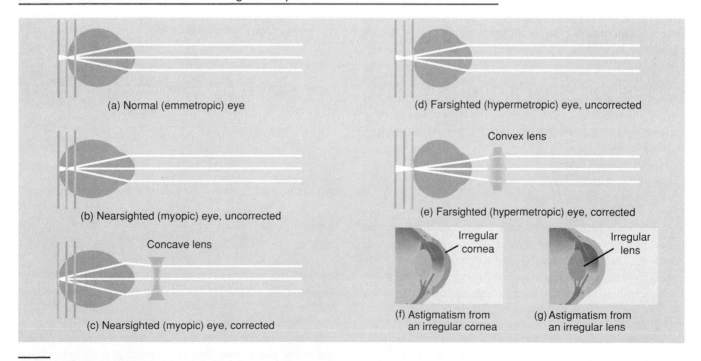

(a) Normal (emmetropic) eye

(b) Nearsighted (myopic) eye, uncorrected

(c) Nearsighted (myopic) eye, corrected

(d) Farsighted (hypermetropic) eye, uncorrected

(e) Farsighted (hypermetropic) eye, corrected

(f) Astigmatism from an irregular cornea

(g) Astigmatism from an irregular lens

Question: What is presbyopia?

to comparable spots on the retinas of both eyes. But as we move closer to the object, our eyes must rotate medially for the light rays from the object to strike the same points on both retinas. The term **convergence** refers to this medial movement of the two eyeballs so they are both directed toward the object being viewed. The nearer the object, the greater the degree of convergence needed to maintain single binocular vision. The coordinated action of the extrinsic eye muscles (see Fig. 11.6) brings about convergence.

Physiology of Vision

An image focused on the retina stimulates photoreceptors, which transduce the light stimulus into receptor potentials and pass the information on to bipolar cells. Bipolar cells, in turn, communicate with ganglion cells, which project their axons to the lateral geniculate body of the thalamus. From the thalamus, fibers carrying visual nerve impulses extend to the primary visual cortex in the occipital lobe (see Fig. 14.14).

Photoreceptors and Photopigments

Rods and cones were named for the different appearance of their *outer segment*, which is the distal end of the photoreceptor next to the pigment epithelium. The outer segments of cones are tapered or cone-shaped whereas those of rods are cylindrical or rod-shaped (Fig. 16.11). Transduction of light into an electrical signal occurs in the outer segment. The *inner segment* contains the cell nucleus, Golgi complex, and many mitochondria. At its proximal end, the photoreceptor expands into a bulblike synaptic terminal.

The first step in visual transduction is absorption of light by a **photopigment (visual pigment).** Photopigments are colored proteins in outer segment membranes that undergo structural changes upon light absorption. They initiate the events that lead to production of a receptor potential. The single type of photopigment in rods is called **rhodopsin** (*rhodo* = rose; *opsis* = vision). A cone contains one of three different kinds of photopigments; thus there are three types of cones.

Photopigments are integral proteins in the plasma membrane of the outer segment, which folds back and forth in a pleated fashion (Fig. 16.11). In rods, the pleats pinch off from the plasma membrane to form discs. Each rod outer segment contains a stack of about 1000 discs, piled up like coins inside a wrapper. Photoreceptor outer segments are renewed at an astonishingly rapid pace. In rods, one to three new discs are added to the base of the outer segment every hour. Simultaneously, old discs slough off at the tip and are phagocytized by pigment epithelial cells (Fig. 16.11).

All visual photopigments contain two parts: a glycoprotein known as **opsin** and a derivative of vitamin A called **retinal.** Vitamin A derivatives are formed from carotenoids, the plant pigments that give carrots their color.

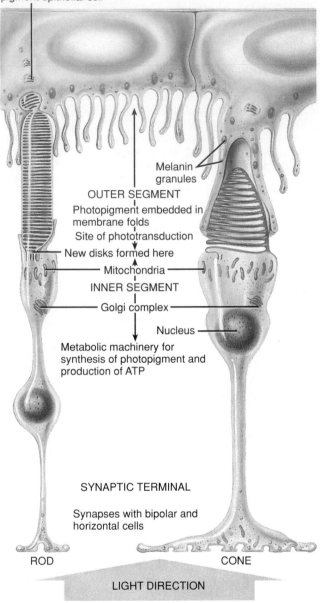

FIGURE 16.11 Photoreceptors.

Old discs at tip slough off and are phagocytized by pigment epithelial cell

Melanin granules

OUTER SEGMENT
Photopigment embedded in membrane folds
Site of phototransduction
New disks formed here
Mitochondria
INNER SEGMENT
Golgi complex
Nucleus
Metabolic machinery for synthesis of photopigment and production of ATP

SYNAPTIC TERMINAL

Synapses with bipolar and horizontal cells

ROD   CONE

LIGHT DIRECTION

Structure of rods and cones

Question: What are the functional similarities between rods and cones?

Good vision depends on adequate dietary intake of carotenoid-rich vegetables such as carrots, spinach, broccoli, and yellow squash or meats that contain vitamin A, such as liver. **Night blindness** or **nyctalopia** (nik′-ta-LŌ-pē-a) is an inability to see well at low light levels. It is most often caused by prolonged vitamin A deficiency and the consequent inability to synthesize a normal amount of rhodopsin.

Retinal is the light-absorbing portion of all visual photopigments. In the human retina, there are four different opsins, one for each cone photopigment and one for rhodopsin. Small variations in the amino acid sequences of the different opsins permit the rods and cones to absorb different colors of incoming light. Rhodopsin absorbs blue to green light most effectively whereas the three cone photopigments most effectively absorb blue, green, or yellow to red light.

A very important characteristic of retinal is that it exists in two forms. In darkness, retinal has a bent shape, called *cis*-retinal, which snugly fits against the opsin portion of the photopigment (Fig. 16.12). When it absorbs light, *cis*-retinal straightens out to a shape called *trans*-retinal. This *cis* to *trans* conversion is called **isomerization** and it is the first step in transduction. Forming a visual image begins with isomerization of particular photopigments in certain rods and cones. After retinal isomerizes, several unstable substances form and disappear. In about a minute, *trans*-retinal completely separates from opsin. The final products look colorless, so the whole process is called **bleaching** of photopigment.

In darkness, an enzyme called **retinal isomerase** can reconvert *trans*- to *cis*-retinal, which then binds to opsin and reforms a functional photopigment. Resynthesis of a photopigment is called **regeneration.** The pigment epithelium, adjacent to the photoreceptors, stores a large quantity of vitamin A and contributes to the regeneration process in rods. The extent of rhodopsin regeneration decreases drastically if the retina detaches from the pigment epithelium. Cone photopigments regenerate much more quickly than does rhodopsin and are less dependent on the pigment epithelium. After complete bleaching, it takes 5 minutes to regenerate half of the rhodopsin but only 1½ minutes to regenerate half of the cone photopigments. Full regeneration of bleached rhodopsin takes 30 to 40 minutes.

CLINICAL APPLICATION

COLOR BLINDNESS

Most forms of **color blindness** (inability to distinguish certain colors) result from absence or deficiency of one of the three cone photopigments. The most common type is **red–green color blindness,** in which a photopigment sensitive to red or green light is missing. As a result, the person has no way to distinguish between red and green. Color blindness is an inherited condition that affects males far more often than females. The inheritance of the condition is discussed on page 995 and illustrated in Fig. 29.19.

Light and Dark Adaptation

When you emerge from dark surroundings, such as a tunnel, into the sunshine, your visual system adjusts in seconds to the brighter environment by decreasing its sensitivity. This is called **light adaptation.** On the other hand, when you enter a darkened room, such as a theater, your visual sensitivity increases slowly over many minutes. This is called **dark adaptation.** Bleaching and regeneration of the photopigments account for much but not all of the sensitivity change during light and dark adaptation.

As the light level increases, more and more photopigment is bleached so more light is needed to stimulate the remaining unbleached photopigment. At the same time that light is bleaching some photopigment molecules, however, others are being regenerated. Regeneration of rhodopsin occurs slowly enough that in daylight it is bleached as fast as it is regenerated. For this reason, rods contribute little to daylight vision. Regeneration of cone photopigments, on the other hand, is fast enough that some of the *cis* form is always present, even in very bright light.

If the light level decreases abruptly, sensitivity increases rapidly at first and then more slowly. In complete darkness, full regeneration of the cone photopigments occurs during the first 8 minutes of dark adaptation. During this time interval, a threshold light flash (barely perceptible) is seen

FIGURE 16.12 Bleaching and regeneration of photopigment.

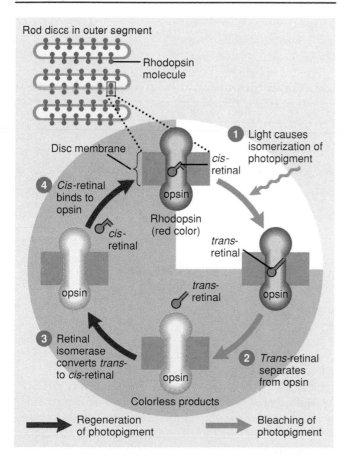

- Rod discs in outer segment
- Rhodopsin molecule
- Disc membrane
- **4** *Cis*-retinal binds to opsin
- *cis*-retinal
- **1** Light causes isomerization of photopigment
- *cis*-retinal
- opsin
- Rhodopsin (red color)
- *trans*-retinal
- opsin
- *trans*-retinal
- opsin
- **3** Retinal isomerase converts *trans*- to *cis*-retinal
- opsin
- **2** *Trans*-retinal separates from opsin
- Colorless products
- → Regeneration of photopigment
- → Bleaching of photopigment

Question: What is the conversion of *cis*- to *trans*-retinal called?

as having color. More slowly, rhodopsin regenerates and our visual sensitivity increases until even a single photon (the smallest unit) of light can be detected. Now, although much dimmer light can be detected, threshold flashes look gray-white, regardless of their color. At very low light levels, such as moonlight or starlight, our visual world appears as shades of gray because only the rods are functioning.

Receptor Potential and Neurotransmitter Release

In darkness, Na+ flows into photoreceptor outer segments through Na+ channels that are held open by a molecule called **cyclic GMP (guanosine monophosphate)** (Fig. 16.13). This inflow of Na+, called the "dark current," triggers continual release of neurotransmitter from the synaptic terminals. The neurotransmitter in rods, and perhaps in cones also, is glutamate (glutamic acid). Glutamate inhibits (hyperpolarizes) the bipolar cells that synapse with rods.

When light strikes the retina and *cis*-retinal undergoes isomerization, the Na+ channels close. Na+ inflow thus decreases, the inside of the rod becomes more negative (hyperpolarization), and release of glutamate decreases (Fig. 16.13). Dim lights cause small and brief hyperpolarizations that partially turn off glutamate release. Brighter lights elicit larger and longer hyperpolarizations that more completely shut down neurotransmitter release. Light thus excites the bipolar cells that synapse with rods by turning off an inhibitory neurotransmitter.

Two enzymes regulate closing and reopening of the Na+ channels in the outer segment. In light, an enzyme called **transducin** activates another enzyme called **PDE (phosphodiesterase)**, which breaks down cyclic GMP. This closes the Na+ channels resulting in hyperpolarization of rods and decreased release of glutamate (Fig. 16.13). In darkness, transducin is in an inactive form, and cyclic GMP holds the Na+ channels open. In darkness, an enzyme called **recoverin** activates guanylate cyclase, the enzyme that stimulates synthesis of cyclic GMP. As the cyclic GMP level rises, the Na+ channels are again held in the open position and the inflow of Na+ triggers increased release of glutamate (Fig. 16.13).

Visual Pathway

As mentioned earlier, considerable processing of visual input occurs in the retina, at synapses among the various types of cells (see Fig. 16.7). The axons of retinal ganglion cells provide output from the retina to the brain. They exit the eyeball via the **optic (II) nerve.**

Retinal Processing of Visual Input

Within the retina, certain features of visual input are enhanced while other features may be discarded. Input from

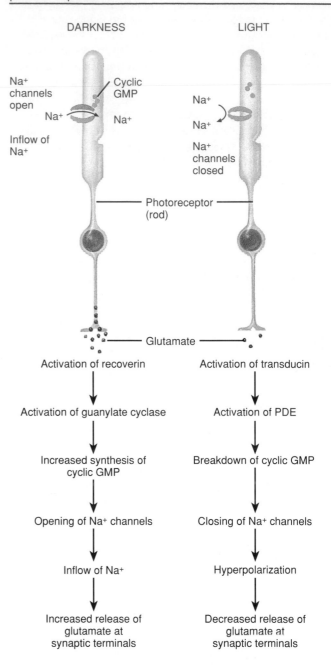

FIGURE 16.13 Release of neurotransmitter by photoreceptors.

Question: What are the functions of PDE and guanylate cyclase?

several cells may converge upon a smaller number of postsynaptic neurons or may diverge to a large number. On the whole, however, convergence predominates since there are only one million ganglion cells that serve about 126 million photoreceptors.

Once receptor potentials arise in rods and cones, they spread through the inner segments to the synaptic terminals.

Neurotransmitters released by rods and cones induce graded, local potentials in both bipolar cells and horizontal cells. Between 6 and 600 rods synapse with a single bipolar cell in the outer synaptic layer, whereas a cone more often synapses with just one bipolar cell. The convergence of many rods onto a single bipolar cell is one reason that rod vision is more sensitive but less acute than cone vision. Stimulation of rods by light excites their bipolar cells. Cone bipolar cells, on the other hand, may be either excited or inhibited by light.

Horizontal cells transmit inhibitory signals to bipolar cells in the areas lateral to excited rods and cones (see Fig. 16.7). This lateral inhibition enhances contrasts in the visual scene between areas of the retina that are strongly stimulated and adjacent areas that are weakly stimulated. Horizontal cells also assist in the differentiation of various colors. Amacrine cells, which are also excited by bipolar cells, synapse with ganglion cells and transmit information to them that signals a change in the level of illumination of the retina. When bipolar or amacrine cells transmit excitatory signals to ganglion cells, the ganglion cells become depolarized and initiate nerve impulses.

Brain Pathway and Visual Fields

The axons of the optic (II) nerve pass through the **optic chiasma** (kī-AZ-ma; *chiasma* = letter X), a crossing point of the optic nerves (Fig. 16.14). Some fibers cross to the opposite side. Others remain uncrossed. After passing through the optic chiasma, the fibers, now part of the **optic tract,** enter the brain and terminate in the lateral geniculate nucleus of the thalamus. Here the fibers synapse with neurons whose axons form the **optic radiations**. These fibers project to the visual areas of the cerebral cortex, which are located in the occipital lobes.

Analysis of the afferent pathway to the brain reveals that the visual field of each eye is divided into two regions: the **nasal (medial) half** and the **temporal (lateral) half.** For each eye, light rays from an object in the nasal half of the visual field fall on the temporal half of the retina. Light rays from an object in the temporal half of the visual field fall on the nasal half of the retina (Fig. 16.15). Also, light rays from objects at the top of the visual field of each eye fall on the inferior portion of the retina, and light rays from objects at the bottom of the visual field fall on the superior portion of the retina.

In the optic chiasma, nerve fibers from the nasal halves of both retinas cross and continue on to the opposite lateral geniculate nuclei of the thalamus. Nerve fibers from the temporal halves of the retinas do not cross but continue directly on to the lateral geniculate nuclei on the same side. As a result, the primary visual area of the cerebral cortex of the right occipital lobe receives visual images from the left side of an object via nerve impulses from the temporal half

FIGURE 16.14 Photograph of the visual pathway. The brain is partially dissected to reveal the optic radiations (fibers extending from the thalamus to the occipital lobe).

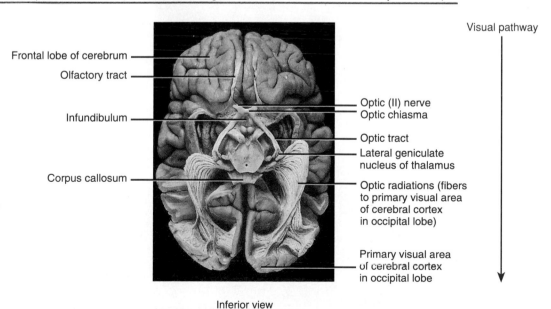

Frontal lobe of cerebrum
Olfactory tract
Infundibulum
Corpus callosum

Optic (II) nerve
Optic chiasma
Optic tract
Lateral geniculate nucleus of thalamus
Optic radiations (fibers to primary visual area of cerebral cortex in occipital lobe)
Primary visual area of cerebral cortex in occipital lobe

Visual pathway

Inferior view

Question: What is the correct order of structures that carry nerve impulses from the retina to the occipital lobe of the cerebrum?

FIGURE 16.15 Visual fields and afferent pathway for visual impulses. The dark circle in the center of the visual fields is the macula lutea. The center of the macula lutea is the central fovea, the area of sharpest vision.

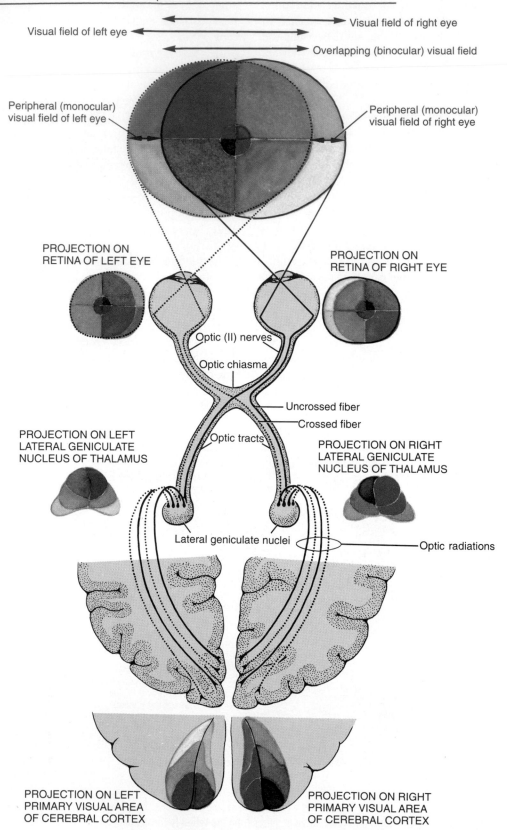

Question: Light rays from an object in the temporal half of the visual field fall on which half of the retina?

of the retina of the right eye and the nasal half of the retina of the left eye. The primary visual area of the cerebral cortex of the left occipital lobe receives visual images from the right side of an object via impulses from the nasal half of the right eye and the temporal half of the left eye.

Although we have just described the visual pathway as a more or less single pathway, visual signals are thought to be processed by at least three separate systems, each with its own function. One system is involved in processing information related to the shape of objects, another system forms a pathway regarding color of objects, and a third system processes information about movement, location, and spatial organization.

AUDITORY SENSATIONS AND EQUILIBRIUM

Besides receptors for sound waves, the ear also contains receptors for equilibrium. Anatomically, the ear is divided into three principal regions: the external (outer) ear, the middle ear, and the internal (inner) ear.

External (Outer) Ear

The **external (outer) ear** collects sound waves and passes them inward (Fig. 16.16). It consists of the auricle, external auditory canal, and tympanic membrane (eardrum). The **auricle (pinna)** is a flap of elastic cartilage shaped like the flared end of a trumpet and covered by thick skin. The rim of the auricle is called the **helix;** the inferior portion is the **lobule.** The auricle is attached to the head by ligaments and muscles. The **external auditory** (*audire* = hearing) **canal (meatus)** is a curved tube about 2.5 cm (1 in.) long that lies in the temporal bone and leads from the auricle to the eardrum. The **tympanic** (tim-PAN-ik; *tympano* = drum) **membrane (eardrum)** is a thin, semitransparent partition of fibrous connective tissue between the external auditory canal and middle ear.

FIGURE 16.16 Structure of the ear. Note the divisions of the ear into external, middle, and internal portions.

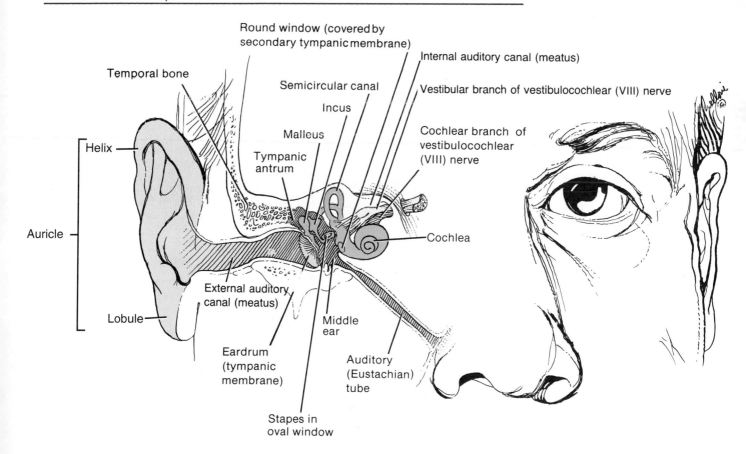

Frontal section through the right side of the skull

Question: **Where are the receptors for hearing and equilibrium?**

Near the exterior opening, the external auditory canal contains a few hairs and specialized sebaceous (oil) glands called **ceruminous** (se-ROO-mi-nus) **glands** that secrete **cerumen** (earwax). The combination of hairs and cerumen (se-ROO-min) helps prevent dust and foreign objects from entering the ear. Usually, cerumen dries up and falls out of the ear canal. Some people, however, produce an abnormal amount of cerumen. It then becomes impacted and muffles incoming sounds. The treatment for **impacted cerumen** is usually periodic ear irrigation or removal of wax with a blunt instrument by trained medical personnel.

CLINICAL APPLICATION

PERFORATED EARDRUM

A **perforated eardrum** is a hole in the tympanic membrane. The condition is characterized by acute pain initially, ringing or roaring in the affected ear, hearing impairment, and sometimes dizziness. Causes of perforated eardrums include shock waves of compressed air (explosion), scuba diving, trauma (skull fracture or from objects such as ear swabs), or acute middle ear infections.

Middle Ear

The **middle ear (tympanic cavity)** is a small, air-filled cavity in the temporal bone that is lined by epithelium (Fig. 16.17). It is separated from the external ear by the eardrum and from the internal ear by a thin bony partition that contains two small membrane-covered openings: the oval window and the round window.

The posterior wall of the middle ear communicates with the mastoid air "cells" of the temporal bone through a chamber called the **tympanic antrum** (see also Fig. 16.16). This anatomical feature explains why a middle ear infection may spread to the temporal bone, causing mastoiditis, or even to the brain.

The anterior wall of the middle ear contains an opening that leads directly into the **auditory (Eustachian) tube.** The auditory tube consists of both bone and hyaline cartilage and connects the middle ear with the nasopharynx (upper portion) of the throat. The function of the tube is to equalize air pressure on both sides of the tympanic membrane to ensure that the eardrum vibrates freely when struck by sound waves. Infections also may travel along this passageway from the throat and nose to the ear. The auditory

FIGURE 16.17 Auditory ossicles.

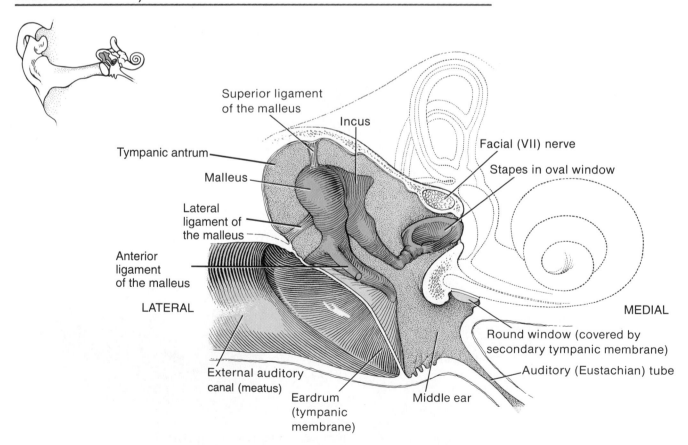

Superior ligament of the malleus

Incus

Facial (VII) nerve

Tympanic antrum

Stapes in oval window

Malleus

Lateral ligament of the malleus

Anterior ligament of the malleus

LATERAL

MEDIAL

Round window (covered by secondary tympanic membrane)

Auditory (Eustachian) tube

External auditory canal (meatus)

Eardrum (tympanic membrane)

Middle ear

Question: What structures separate the middle ear from the external ear and from the internal ear?

tube is normally closed at its medial end; during swallowing and yawning, it opens. This allows atmospheric air from the throat to enter or leave the middle ear until the internal pressure equals the external pressure. If the pressure is not relieved, intense pain, hearing impairment, ringing in the ears, and vertigo could develop. Sudden pressure changes against the eardrum may be equalized by yawning, swallowing or pinching the nose closed, closing the mouth, and gently forcing air from the lungs into the nasopharynx.

Extending across the middle ear and attached to it by ligaments are three tiny bones called **auditory ossicles** (OS-si-kuls). The bones, named for their shape, are the malleus, incus, and stapes, commonly called the hammer, anvil, and stirrup, respectively. They are connected by synovial joints. The "handle" of the **malleus** is attached to the internal surface of the tympanic membrane. Its head articulates with the body of the incus. The **incus** is the intermediate bone in the series and articulates with the head of the stapes. The base or footplate of the **stapes** fits into a membrane covered opening in the thin bony partition between the middle and inner ear. The opening is called the **oval window**. Directly

below the oval window is another opening, the **round window**. This opening is enclosed by a membrane called the **secondary tympanic membrane.**

Besides the ligaments, two skeletal muscles also attach to the ossicles. The **tensor tympani muscle** limits movement and increases tension on the tympanic membrane to prevent damage to the inner ear from loud noises. It only protects the inner ear from prolonged loud noises, not brief ones, such as a gunshot. The **stapedius muscle** is the smallest of all skeletal muscles and it also has a protective function in that it dampens (checks) large vibrations that result from loud noises. It is for this reason that paralysis of the stapedius muscle is associated with **hyperacusia** (abnormally sensitive hearing).

Internal (Inner) Ear

The **internal (inner) ear** is also called the **labyrinth** (LAB-i-rinth) because of its complicated series of canals (Fig. 16.18). Structurally, it consists of two main divisions: an

FIGURE 16.18 The internal ear. The outer, blue area belongs to the bony labyrinth. The inner, pink-colored area belongs to the membranous labyrinth.

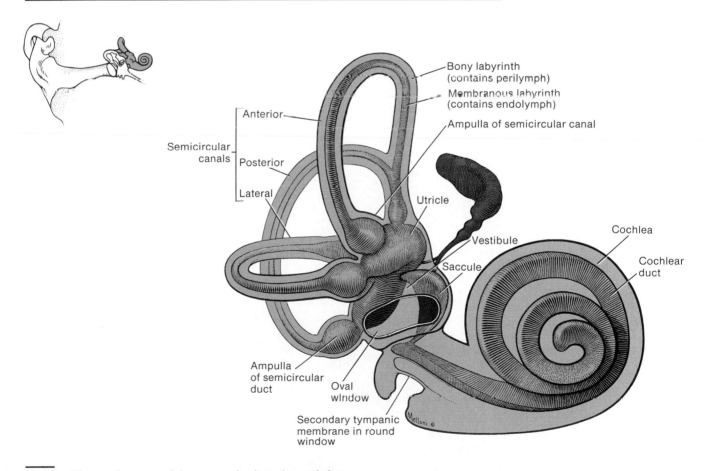

outer bony labyrinth that encloses an inner membranous labyrinth. The **bony labyrinth** is a series of cavities in the petrous portion of the temporal bone. It can be divided into three areas: (1) the semicircular canals and (2) vestibule, both of which contain receptors for equilibrium, and (3) the cochlea, which contains receptors for hearing. The bony labyrinth is lined with periosteum and contains a fluid called **perilymph.** This fluid, which is chemically similar to cerebrospinal fluid, surrounds the **membranous labyrinth,** a series of sacs and tubes lying inside and having the same general form as the bony labyrinth. The membranous labyrinth is lined with epithelium and contains a fluid called **endolymph,** which is chemically similar to intracellular fluid.

The **vestibule** is the oval central portion of the bony labyrinth. The membranous labyrinth in the vestibule consists of two sacs called the **utricle** (YOO-tri-kul; = little bag) and **saccule** (SAK-yool; = little sac). These structures are connected to each other by a small duct. Projecting upward and posteriorly from the vestibule are the three bony **semicircular canals.** Each lies at approximately right angles to the other two. Based on their positions, they are called the anterior, posterior, and lateral canals. The anterior and posterior semicircular canals are oriented vertically; the lateral one is oriented horizontally. One end of each canal enlarges into a swelling called the **ampulla** (am-POOL-la; = little jar). The portions of the membranous labyrinth that lie inside the bony semicircular canals are called the **semicircular ducts (membranous semicircular canals).** These structures communicate with the utricle of the vestibule.

Anterior to the vestibule is the **cochlea** (KŌK-lē-a; = snail's shell). This bony spiral canal (Fig. 16.19) resembles a snail's shell and makes almost three turns around a central bony core called the **modiolus** (Fig. 16.19b). Cross sections through the cochlea (Fig. 16.19a-c) show that it is divided into three channels. Together, the partitions that separate the channels have a shape like the letter Y. The stem of the Y is a bony shelf that protrudes into the canal; the wings of the Y are composed mainly of membranous labyrinth. The

FIGURE 16.19 Views of the semicircular canals, vestibule, and cochlea. Note the cochlea almost makes three complete turns.

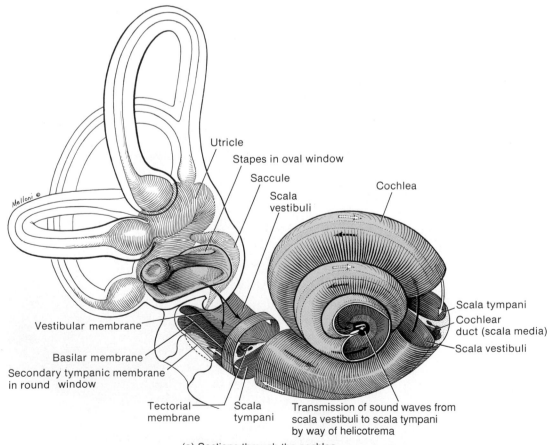

(a) Sections through the cochlea

FIGURE 16.19 (continued)

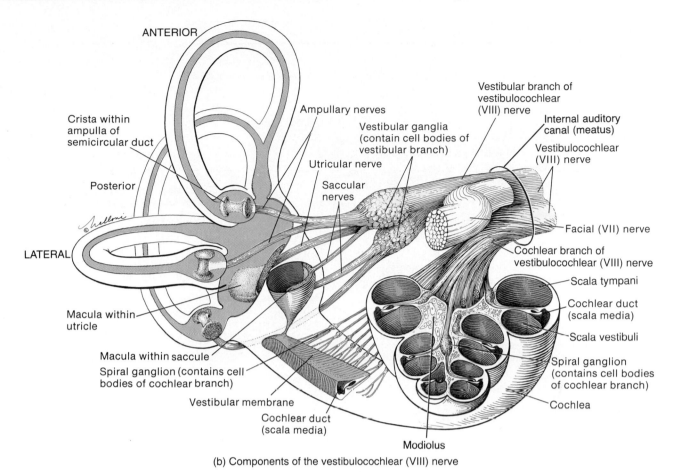

(b) Components of the vestibulocochlear (VIII) nerve

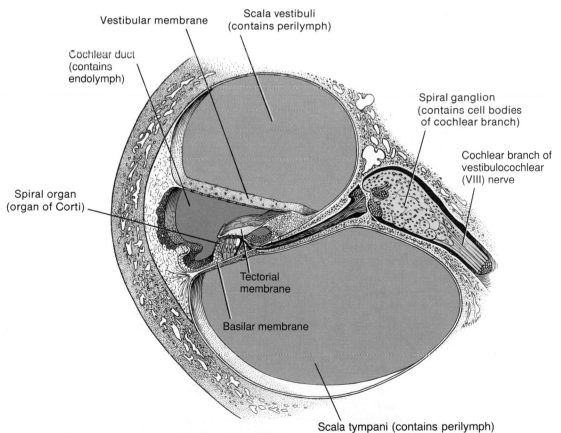

(c) Section through one turn of the cochlea

Figure continues

FIGURE 16.19 (continued)

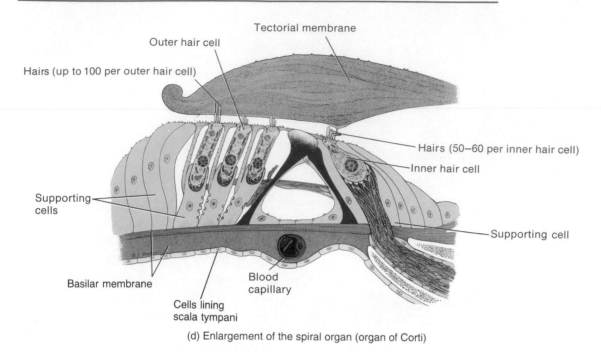

(d) Enlargement of the spiral organ (organ of Corti)

Question: Which structures contain perilymph? Endolymph?

channel above the bony partition is the **scala vestibuli,** which ends at the oval window; the channel below is the **scala tympani,** which ends at the round window.

The scala vestibuli and scala tympani both contain perilymph and are completely separated except for an opening at the apex of the cochlea called the **helicotrema.** The cochlea adjoins the wall of the vestibule, into which the scala vestibuli opens. The perilymph of the vestibule is continuous with that of the scala vestibuli. The third channel (between the wings of the Y) is the **cochlear duct (scala media).** The **vestibular membrane** separates the cochlear duct from the scala vestibuli, and the **basilar membrane** separates the cochlear duct from the scala tympani.

Resting on the basilar membrane is the **spiral organ (organ of Corti),** the organ of hearing (Figure 16.19c,d). The spiral organ is a coiled sheet of epithelial cells, including supporting cells and about 16,000 **hair cells,** which are the receptors for auditory sensations. There are two groups of hair cells. The *inner hair cells* are medially placed in a single row and extend the entire length of the cochlea. The *outer hair cells* are arranged in several rows. The hair cells have long hairlike processes (specialized microvilli) at their apical ends that extend into the endolymph of the cochlear duct. The basal ends of the hair cells synapse with fibers of the cochlear branch of the vestibulocochlear nerve (cranial nerve VIII). Projecting over and in contact with the hair cells of the spiral organ is the **tectorial** (*tectum* = cover) **membrane,** a delicate and flexible gelatinous membrane.

CLINICAL APPLICATION

HIGH-INTENSITY SOUNDS AND DEAFNESS

Hair cells of the spiral organ are easily damaged by exposure to high-intensity sounds such as loud music or the engine roar of jet planes, revved-up motorcycles, pneumatic drills, lawn mowers, and vacuum cleaners. As a result of the noises, hair cells become arranged in disorganized patterns, or they and their supporting cells may degenerate. Continued exposure to high-intensity sounds permanently damages the hair cells. Usually, deafness begins with loss of sensitivity for high-pitched sounds. The deficiency may not be noticed until destruction is extensive, which is why ear protectors should be worn by those exposed to high-intensity noises.

Sound Waves

Sound waves result from the alternate compression and decompression of air molecules. They originate from a vibrating object, much the same way that ripples travel over the surface of water when you toss a stone into it. The sounds heard most acutely by human ears are those from sources that vibrate at frequencies between 1000 and 4000 hertz (Hz; 1 Hz = 1 cycle per second). The entire audible range extends from 20 to 20,000 Hz.

The frequency of a sound vibration is its pitch. The greater the frequency of vibration, the higher the pitch.

Also, the greater the intensity (size) of the vibration, the louder the sound. Sound intensity (loudness) is measured in units called **decibels (dB).** The hearing threshold—that is, the point at which an average young adult can just detect sound from silence—is defined as 0 dB. Sound becomes uncomfortable at about 120 dB and painful at about 140 dB. Rustling leaves have a decibel rating of 15, normal conversation 45, crowd noise 60, a vacuum cleaner 75, and a pneumatic drill 90. Because hearing loss results from prolonged noise exposure, in the U.S. employers must require workers to use hearing protectors when occupational noise levels exceed 90 dB.

Physiology of Hearing

The events involved in hearing are as follows (Fig. 16.20):

1. The auricle directs sound waves into the external auditory canal.
2. When sound waves strike the tympanic membrane, the alternate compression and decompression of the air cause the membrane to vibrate back and forth. The distance the membrane moves is always very small and depends on the intensity and frequency of the sound waves. The membrane vibrates slowly in response to low-frequency (low pitch) sounds and rapidly in response to high-frequency (high pitch) sounds.

3. The central area of the tympanic membrane connects to the malleus, which also starts to vibrate. The vibration conducts from the malleus to the incus and then to the stapes.
4. As the stapes moves back and forth, it pushes the membrane of the oval window in and out.
5. The movement of the oval window sets up fluid pressure waves in the perilymph of the cochlea.
6. As the oval window bulges inward, it pushes on the perilymph of the scala vestibuli. Pressure waves are transmitted from the scala vestibuli to the scala tympani and eventually to the round window, causing it to bulge outward into the middle ear. (See number 9 in the illustration.)
7. As the pressure waves deform walls of the scala vestibuli and scala tympani, they also push the vestibular membrane back and forth. As a result, the pressure of the endolymph inside the cochlear duct increases and decreases.
8. The pressure fluctuations of the endolymph move the basilar membrane slightly. When the basilar membrane vibrates, the hair cells of the spiral organ move against the tectorial membrane. The bending of the hairs produces receptor potentials that ultimately lead to the generation of nerve impulses in cochlear nerve fibers.
9. Pressure changes in the scala tympani cause the round window to bulge outward into the middle ear.

FIGURE 16.20 Events in the stimulation of auditory receptors. The numbers correspond to the events listed in the text. The cochlea has been uncoiled to more easily visualize transmission of sound waves and their distortion of the vestibular and basilar membranes of the cochlear duct.

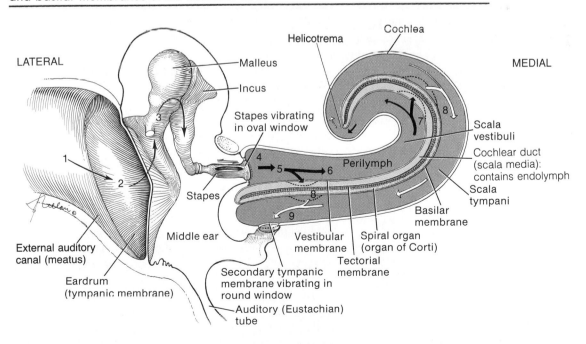

Question: Which portion of the basilar membrane vibrates most vigorously in response to high-frequency (high pitch) sounds?

Sound waves of various frequencies cause specific regions of the basilar membrane to vibrate more intensely than others. The membrane is narrower but stiffer at the base of the cochlea (portion closer to the oval window); high-frequency (high-pitched) sounds induce maximal vibrations in this region. Toward the apex of the cochlea, the basilar membrane is wider but more flexible; low-frequency (low-pitched) sounds cause maximal vibration of the basilar membrane there. Loudness is determined by the intensity of sound waves. High-intensity sound waves cause greater vibration of the basilar membrane, which leads to a higher frequency of nerve impulses reaching the brain. More hair cells may also be stimulated by louder sounds.

The function of hair cells is to convert a mechanical force (stimulus) into an electrical signal (receptor potential). When the hairs at the top of the cell move in one direction, the receptor potential is a depolarization. Movement in the opposite direction produces a hyperpolarizing receptor potential. Depolarization results from the opening of potassium ion (K^+) channels in the hair cell membranes. This permits a rapid inflow of K^+ from the endolymph, which has an unusually high concentration of K^+. Depolarization spreads through the cell and causes calcium ion (Ca^{2+}) channels in the base of the hair cell to open. The resulting inflow of Ca^{2+} triggers exocytosis of synaptic vesicles that store neurotransmitter. The rate of nerve impulse firing in the sensory nerve fibers at the base of the hair cell then increases. Hyperpolarization reduces neurotransmitter release from the hair cells and decreases the rate of impulses in sensory nerve fibers. The neurotransmitter is thought to be either glutamate or gamma aminobutyric acid (GABA).

The cochlear branch of the vestibulocochlear (VIII) nerve conducts auditory impulses to the cochlear nuclei in the medulla. Here, most fibers cross to the opposite side and then travel to the midbrain and terminate in the thalamus.

CLINICAL APPLICATION

COCHLEAR IMPLANTS

Cochlear implants (artificial ears) are devices that translate sounds into electronic signals that can be interpreted by the brain. They take the place of hair cells of the spiral organ (organ of Corti), which normally convert sound waves into nerve impulses. The implants are used for people with sensorineural deafness, that is, deafness due to disease or injury that has destroyed hair cells of the spiral organ.

Sound waves enter a tiny microphone in the ear and travel to a microprocessor pack where they are converted into electrical signals. The signals then travel to electrodes implanted in the cochlea, where they trigger nerve impulses in fibers of the cochlear branch of the vestibulocochlear nerve. These artificially induced nerve impulses then conduct over their normal pathways to the brain. Sounds perceived are crude compared to normal hearing, but they provide a sense of rhythm and loudness and information about noises such as telephones, automobiles, and the pitch and cadence of speech.

Physiology of Equilibrium

There are two kinds of **equilibrium** (balance). One, called **static equilibrium,** refers to the maintenance of the position of the body (mainly the head) relative to the force of gravity. The second kind, **dynamic equilibrium,** is the maintenance of body position (mainly the head) in response to sudden movements such as rotation, acceleration, and deceleration. Collectively, the receptor organs for equilibrium are called the **vestibular apparatus,** which includes the saccule, utricle, and semicircular ducts.

Otolithic Organs: Saccule and Utricle

The walls of both the utricle and saccule contain a small, thickened region called a **macula** (Fig. 16.21). The maculae are the receptors for static equilibrium and also contribute to some aspects of dynamic equilibrium. For static equilibrium, they provide sensory information on the position of the head in space and are essential for maintaining appropriate posture and balance. For dynamic equilibrium, they detect linear acceleration and deceleration, for example, the sensations you feel while in an elevator or a car that is speeding up or slowing down.

The two maculae are perpendicular to one another. They consist of two kinds of cells: **hair (receptor) cells** and **supporting cells.** Hair cells have long extensions of the cell membrane consisting of 70 or more **stereocilia** (they are actually microvilli) and one **kinocilium** (a conventional cilium) anchored firmly to its basal body and extending beyond the longest microvilli. Scattered among the hair cells are columnar supporting cells. They probably secrete the thick, gelatinous, glycoprotein layer, called the **otolithic membrane,** that rests on the hair cells. A layer of heavy calcium carbonate crystals, called **otoliths** (*oto* = ear; *lithos* = stone) or **otoconia**, extends over the entire surface of the otolithic membrane.

The heavy otolithic membrane sits on top of the macula like a discus on a greased cookie sheet. If you tilt your head forward, the otolithic membrane and otoliths (the discus in our analogy) is pulled by gravity, slides downhill over the hair cells in the direction of the tilt, and stimulates the hair cells. Similarly, if you are sitting upright in a car that suddenly jerks forward, the otolithic membrane, due to its inertia, slides backward and stimulates the hair cells. As the otoliths move, they pull on the gelatinous layer, which pulls on the stereocilia and makes them bend. The movement of the stereocilia initiates depolarizing or hyperpolarizing receptor potentials.

As the hair cells depolarize or hyperpolarize, they release neurotransmitter at a faster or slower rate. The hair cells synapse with sensory (afferent) nerve fibers in the vestibular branch of the vestibulocochlear (VIII) nerve (Fig. 16.21b). These neurons fire impulses at a slow or rapid pace depending on how much neurotransmitter is present. Motor (efferent) fibers also make synapses with the hair

FIGURE 16.21 Location and structure of receptors in the maculae.

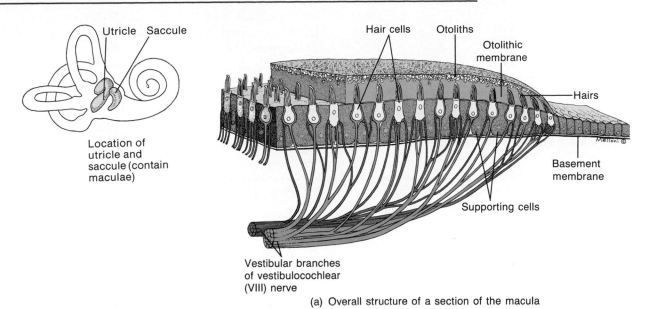

Utricle Saccule

Location of utricle and saccule (contain maculae)

Hair cells   Otoliths   Otolithic membrane

Hairs

Basement membrane

Supporting cells

Vestibular branches of vestibulocochlear (VIII) nerve

(a) Overall structure of a section of the macula

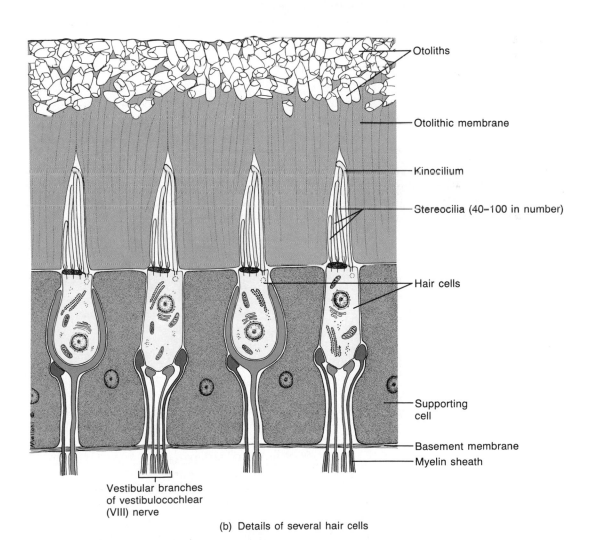

Otoliths

Otolithic membrane

Kinocilium

Stereocilia (40–100 in number)

Hair cells

Supporting cell

Basement membrane

Myelin sheath

Vestibular branches of vestibulocochlear (VIII) nerve

(b) Details of several hair cells

Question: With which type of equilibrium are the maculae mainly concerned?

cells and vestibular neurons. Evidently, they regulate the sensitivity of the hair cells and sensory neurons.

Semicircular Ducts

The three semicircular ducts, together with the saccule and utricle, maintain dynamic equilibrium (Fig. 16.22). The ducts lie at right angles to one another in three planes: the two vertical ones are the anterior and posterior semicircular ducts, and the horizontal one is the lateral semicircular duct.

This positioning permits detection of rotational acceleration or deceleration (see Fig. 16.18). In the ampulla, the dilated portion of each duct, there is a small elevation called the **crista.** Each crista contains a group of **hair (receptor) cells** and **supporting cells** covered by a mass of gelatinous material called the **cupula.** When the head moves, the endolymph in the semicircular ducts flows over the hairs and bends them. The movement of the hairs stimulates sensory neurons, and the nerve impulses, produced by the resulting receptor potentials, pass along the vestibular branch of the vestibulocochlear (VIII) nerve.

FIGURE 16.22 Semicircular ducts. The ampullary nerves are branches of the vestibular division of the vestibulocochlear (VIII) nerve.

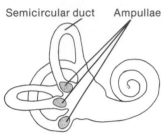

Semicircular duct   Ampullae

Location of ampullae of semicircular ducts (contain cristae)

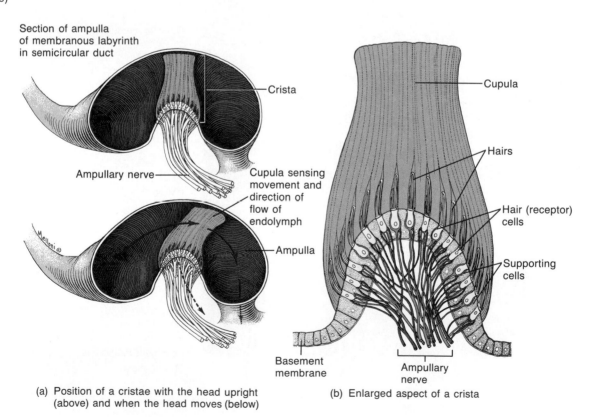

Section of ampulla of membranous labyrinth in semicircular duct

Crista

Ampullary nerve

Cupula sensing movement and direction of flow of endolymph

Ampulla

Cupula

Hairs

Hair (receptor) cells

Supporting cells

Basement membrane

Ampullary nerve

(a) Position of a cristae with the head upright (above) and when the head moves (below)

(b) Enlarged aspect of a crista

Question: With which type of equilibrium are the semicircular ducts (and utricle and saccule) associated?

Equilibrium Pathways

Most of the vestibular branch fibers of the vestibulo-cochlear (VIII) nerve enter the brain stem and terminate in the vestibular nuclear complex in the pons. The remaining fibers enter the cerebellum through the inferior cerebellar peduncle (see Fig. 14.5b). Bidirectional pathways connect the vestibular nuclei and cerebellum. Fibers from all the vestibular nuclei (see Fig. 14.18) form the medial longitudinal tracts that extend from the brain stem into the cervical portion of the spinal cord. The tracts send nerve impulses to the nuclei of cranial nerves that control eye movements—oculomotor (III), trochlear (IV), and abducens (VI)—and to the accessory (XI) nerve nucleus that helps control head and neck movements. In addition, fibers from the lateral vestibular nucleus form the vestibulospinal tract, which conveys impulses to skeletal muscles that regulate muscle tone in response to head movements. Various pathways between the vestibular nuclei, cerebellum, and cerebrum enable the cerebellum to play a key role in maintaining static and dynamic equilibrium. The cerebellum continuously receives updated sensory information from the utricle and saccule. Using this information, the cerebellum monitors and makes corrective adjustments in the motor activities that originate in the cerebral cortex (see Fig. 15.10). Essentially, the cerebellum sends continuous nerve impulses to the motor areas of the cerebrum, in response to input from the utricle, saccule, and semicircular ducts causing the motor system to increase or decrease its impulses to specific skeletal muscles to maintain equilibrium.

A summary of the structures of the ear related to hearing and equilibrium is presented in Exhibit 16.2.

EXHIBIT 16.2

SUMMARY OF STRUCTURES OF THE EAR RELATED TO HEARING AND EQUILIBRIUM

Structure	Function
EXTERNAL (OUTER) EAR	

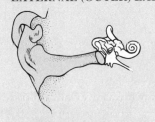

External (outer) ear

Structure	Function
Auricle (pinna)	Collects sound waves.
External auditory canal (meatus)	Directs sound waves to tympanic membrane.
Tympanic membrane (eardrum)	Sound waves cause it to vibrate, which, in turn, causes the malleus to vibrate.

MIDDLE EAR

Middle ear

Structure	Function
Auditory (Eustachian) tube	Equalizes pressure on both sides of the tympanic membrane.
Auditory ossicles	Transmit vibrations from tympanic membrane to oval window.

Exhibit continues

EXHIBIT 16.2 (continued)

SUMMARY OF STRUCTURES OF THE EAR RELATED TO HEARING AND EQUILIBRIUM

Structure	Function

INTERNAL (INNER) EAR

Internal (inner) ear

Utricle	Contains macula, site of hair cells for static and dynamic equilibrium.
Saccule	Contains macula, site of hair cells for static and dynamic equilibrium.
Semicircular ducts	Contain cristae, site of hair cells for dynamic equilibrium.
Cochlea	Contains a series of fluids, channels, and membranes that transmit vibrations to the spiral organ (organ of Corti), the organ of hearing; hair cells in the spiral organ produce receptor potentials, which elicit nerve impulses in the cochlear branch of the vestibulocochlear (VIII) nerve.

DISORDERS: HOMEOSTATIC IMBALANCES

Many disorders can alter or damage the organs of special senses. The causes of disorder range from congenital origins to the effects of old age. Here we discuss a few common disorders of the eyes and ears.

GLAUCOMA
The second most common cause of blindness (after cataracts), especially in the elderly, is **glaucoma.** It is characterized by an abnormally high intraocular pressure due to a buildup of aqueous humor inside the anterior chamber. The fluid compresses the lens into the vitreous body and puts pressure on the neurons of the retina. If the pressure continues, there is a progression from mild visual impairment to irreversible destruction of neurons of the retina, degeneration of the optic disc, and blindness.

SENILE MACULAR DEGENERATION (SMD)
In **senile macular degeneration (SMD),** new blood vessels grow over the macula lutea. The effect ranges from distorted vision to blindness. Its cause is unknown, but it occurs mainly in people over age 65.

DEAFNESS
Deafness is significant or total hearing loss. **Sensorineural deafness** is caused by impairment of the cochlea or cochlear branch of the vestibulocochlear (VIII) nerve. **Conduction deafness** is caused by impairment of the external and middle ear mechanisms for transmitting sounds to the cochlea. Among the factors that contribute to deafness are atherosclerosis, which reduces blood supply to the ears; otosclerosis,

the deposition of new bone around the oval window; repeated exposure to loud noise, which destroys hair cells of the spiral organ (organ of Corti); certain drugs such as streptomycin; impacted cerumen; injury to the tympanic membrane; and aging, which results in thickening of the tympanic membrane, stiffening of the joints of the auditory ossicles, and decreased numbers of hair cells due to diminished cell division.

MÉNIÈRE'S SYNDROME
Ménière's (men-YAIRZ) **syndrome** is characterized by an increased amount of endolymph that enlarges the membranous labyrinth. Among the symptoms are fluctuating hearing loss (caused by distortion of the basilar membrane of the cochlea), attacks of vertigo, and roaring tinnitus. Over a period of years, there may be almost total destruction of hearing.

OTITIS MEDIA
Otitis media is an acute infection of the middle ear, caused primarily by bacteria. Symptoms include pain, malaise, fever, and a reddening and outward bulging of the eardrum, which may rupture unless prompt treatment is given (this may involve draining pus from the middle ear). Bacteria from the nasopharynx passing into the auditory (Eustachian) tube are the primary cause of all middle ear infections. Children are more susceptible than adults to middle ear infections because their auditory tubes are shorter, wider, and almost horizontal.

MOTION SICKNESS

Motion sickness is a nausea and vomiting brought on by repetitive angular, linear, or vertical motion. The cause is excessive stimulation of the vestibular apparatus by motion. Nerve impulses pass from the internal ear to the vomiting center in the medulla. Visual stimuli and emotional factors such as fear or anxiety can also contribute to motion sickness. Susceptible people should take their medicine (for example, Dramamine) before traveling since prevention is more successful than treatment of symptoms once they have developed.

MEDICAL TERMINOLOGY

Achromatopsia (a-krō′-ma-TOP-sē-a; *a* = without; *chrom* = color) Complete color blindness.

Ametropia (am′-e-TRŌ-pē-a; *ametro* = disproportionate; *ops* = eye) Refractive defect of the eye resulting in an inability to focus images properly on the retina.

Anopsia (an-OP-sē-a; *opsia* = vision) A defect of vision.

Blepharitis (blef-a-RĪ-tis; *blepharo* = eyelid; *itis* = inflammation of) An inflammation of the eyelid.

Conjunctivitis (pinkeye) An inflammation of the conjunctiva, caused by bacteria such as pneumococci, staphylococci, or *Hemophilus influenzae,* that is very contagious and more common in children. Conjunctivitis may also be caused by irritants, such as dust, smoke, or pollutants in the air, in which case it is not contagious.

Eustachitis (yoo′-stū-KĪ-tis) An inflammation or infection of the auditory (Eustachian) tube.

Exotropia (ek′-sō-TRŌ-pē-a; *ex* = out; *tropia* = turning) Turning outward of the eyes.

Keratitis (ker′-a-TĪ-tis; *kerato* = cornea) An inflammation or infection of the cornea.

Labyrinthitis (lab′-i-rin-THĪ-tis) An inflammation of the labyrinth (inner ear).

Mydriasis (mi-DRĒ-a-sis) Dilated pupil.

Myringitis (mir′-in-JĪ-tis; *myringa* = eardrum) An inflammation of the eardrum; also called **tympanitis.**

Nystagmus (nis-TAG-mus; *nystazein* = to nod) A rapid involuntary movement of the eyeballs, possibly caused by a disease of the central nervous system. It is associated with conditions that cause vertigo.

Otalgia (ō-TAL-jē-a; *oto* = ear; *algia* = pain) Earache.

Otosclerosis (ō′-tō-skle-RŌ-sis; *oto* = ear; *sclerosis* = hardening) Pathological process that may be hereditary in which new bone is deposited around the oval window. The result may be immobilization of the stapes, leading to conduction deafness.

Photophobia (fō′-tō-FŌ′-bē-a; *photo* = light; *phobia* = fear) Abnormal visual intolerance to light.

Ptosis (TŌ-sis; *ptosis* = fall) Falling or drooping of the eyelid. (This term is also used for the slipping of any organ below its normal position.)

Retinoblastoma (ret′-i-nō-blas-TŌ-ma; *blast* = bud; *oma* = tumor) A tumor arising from immature retinal cells and accounting for 2% of childhood malignancies.

Scotoma (skō-TŌ-ma; *scotoma* = darkness) An area of reduced or lost vision in the visual field. Also called a **blind spot** (other than the normal blind spot at the optic disc).

Strabismus (stra-BIZ-mus) An imbalance in the extrinsic eye muscles that produces a squint (formerly called "cross-eyes"). **Amblyopia** (am′-blē-Ō-pē-a) is the term used to describe the loss of vision in an otherwise normal eye that, because of muscle imbalance, cannot focus in synchrony with the other eye.

Tinnitus (ti-NĪ-tus) A ringing, roaring, or clicking in the ears.

Trachoma (tra-KŌ-ma) A serious form of conjunctivitis, the greatest single cause of blindness in the world. It is caused by a bacterium called *Chlamydia trachomatis*. The disease produces an excessive growth of subconjunctival tissue and invasion of blood vessels into the cornea, which progresses until the entire cornea is opaque, causing blindness.

Vertigo (VER-ti-gō; *vertex* = whorl) A sensation of spinning or movement in which the world is revolving or the person is revolving in space.

Study Outline

Olfactory Sensations: Smell (p. 468)

1. The receptors for olfaction, which are bipolar neurons, are in the nasal epithelium.
2. In olfactory reception, a generator potential develops and triggers one or more nerve impulses.
3. Adaptation to odors occurs quickly, and the threshold of smell is low.
4. Olfactory receptors convey nerve impulses to olfactory (I) nerves, olfactory bulbs, olfactory tracts, cerebral cortex (prepyriform cortex), and limbic system.

Gustatory Sensations: Taste (p. 469)

1. The receptors for gustation, the gustatory receptor cells, are located in taste buds.
2. Substances to be tasted must be in solution in saliva.
3. Receptor potentials developed in gustatory hairs cause the release of neurotransmitter, which gives rise to nerve impulses.
4. Adaptation to taste occurs quickly, and the threshold varies with the taste involved.
5. Gustatory receptor cells trigger nerve impulses in cranial nerves V, VII, IX, and X. Taste signals then pass to the medulla, thalamus, and cerebral cortex (parietal lobe).

Visual Sensations (p. 471)

Accessory Structures of the Eye (p. 471)

1. Accessory structures of the eyes include the eyebrows, eyelids, eyelashes, and the lacrimal apparatus.
2. The lacrimal apparatus consists of structures that produce and drain tears.

Anatomy of the Eyeball (p. 473)

1. The eye is constructed of three coats: (a) fibrous tunic (sclera and cornea), (b) vascular tunic (choroid, ciliary body, and iris), and (c) retina (nervous tunic).
2. The retina consists of pigment epithelium and a neural portion (photoreceptor layer, bipolar cell layer, ganglion cell layer, horizontal cells, and amacrine cells).
3. The anterior cavity contains aqueous humor; the vitreous chamber contains the vitreous body.

Image Formation and Convergence (p. 479)

1. Image formation on the retina involves refraction of light rays by the cornea and lens, accommodation of the lens by an increase in its curvature for near vision, and constriction of the pupil to prevent light rays from entering the eye through the periphery of the lens.

2. In convergence, the eyeballs move medially so they are both directed toward an object being viewed.

Physiology of Vision (p. 482)

1. The first step in vision is the absorption of light by photopigments in rods and cones (photoreceptors) and isomerization of 11-cis retinal.
2. Once receptor potentials develop in rods and cones, they decrease the release of inhibitory neurotransmitter, which induces graded potentials in bipolar cells and horizontal cells.
3. Horizontal cells transmit inhibitory signals to bipolar cells; bipolar or amacrine cells transmit excitatory signals to ganglion cells, which depolarize and initiate nerve impulses.
4. Impulses from ganglion cells are conveyed into the optic (II) nerve, and through the optic chiasma, the optic tract, and to the thalamus. From the thalamus visual signals pass to the cortex (occipital lobes).

Auditory Sensations and Equilibrium (p. 487)

1. The external or outer ear consists of the auricle, external auditory canal, and tympanic membrane.
2. The middle ear consists of the auditory or Eustachian tube, ossicles, oval window, and round window.
3. The internal or inner ear consists of the bony labyrinth and membranous labyrinth. The internal ear contains the spiral organ (organ of Corti), the organ of hearing.
4. Sound waves enter the external auditory canal, strike the tympanic membrane, pass through the ossicles, strike the oval window, set up waves in the perilymph, strike the vestibular membrane and scala tympani, increase pressure in the endolymph, strike the basilar membrane, and stimulate hairs on the spiral organ (organ of Corti).
5. Hair cells convert a mechanical force into a receptor potential.
6. Hair cells release neurotransmitter, which initiates nerve impulses.
7. The cochlear branch of the vestibulocochlear (VIII) nerve terminates in the thalamus.
8. Static equilibrium is the orientation of the body relative to the pull of gravity. The maculae of the utricle and saccule are the sense organs of static equilibrium.
9. Dynamic equilibrium is the maintenance of body position in response to movement. The cristae in the semicircular ducts are the sense organs of dynamic equilibrium.
10. Most vestibular branch fibers of the vestibulocochlear (VIII) nerve enter the brain stem and terminate in the pons; the remaining fibers enter the cerebellum.

Review Questions

1. Describe the structure of olfactory receptors. (p. 468)
2. Describe adaptation to odors. What type of threshold does olfaction have? (p. 469)
3. Discuss the origin and path of a nerve impulse that results in olfaction. (p. 469)
4. Describe the structure of gustatory receptors. (p. 469)
5. How are gustatory receptors stimulated? (p. 471)

6. Describe adaptation to taste. What type of threshold does taste have? (p. 471)
7. Discuss how a nerve impulse for gustation travels from a taste bud to the brain. (p. 471)
8. Describe the structure and importance of the following accessory structures of the eye: eyelids, eyelashes, and eyebrows. (p. 471)

9. What is the function of the lacrimal apparatus? Explain how it operates. (p. 473)
10. By means of a labeled diagram, indicate the principal anatomical structures of the eye. (p. 474)
11. Describe the location and contents of the chambers of the eye. What is intraocular pressure (IOP)? How is the scleral venous sinus (canal of Schlemm) related to this pressure? (p. 477)
12. Describe the histology of the neural portion of the retina. (p. 475)
13. Explain how each of the following events is related to the physiology of vision: (a) refraction of light, (b) accommodation of the lens, and (c) constriction of the pupil. (p. 479)
14. Distinguish emmetropia, myopia, hypermetropia, and astigmatism by means of a diagram. (p. 481)
15. What is convergence? How does it occur? (p. 481)
16. Describe the structure of rods and cones. (p. 482)
17. Explain how photopigments respond to darkness and light. (p. 482)
18. How do receptor potentials develop? (p. 484)
19. Explain how the retina processes visual input. (p. 484)
20. Describe the path of a visual impulse from the optic (II) nerve to the brain. (p. 485)
21. Define visual field. Relate the visual field to image formation on the retina. (p. 485)
22. Diagram the principal parts of the external, middle, and internal ear. Describe the function of each part labeled. (p. 487)
23. What are sound waves? How are sound intensities measured? (p. 492)
24. Explain the events involved in the transmission of sound from the auricle to the spiral organ (organ of Corti). (p. 493)
25. What is the afferent pathway for sound impulses from the cochlear branch of the vestibulocochlear (VIII) nerve to the brain? (p. 494)
26. Compare the function of the maculae in the saccule and utricle in maintaining static equilibrium with the role of the cristae in the semicircular ducts in maintaining dynamic equilibrium. (p. 494)
27. Describe the path of a nerve impulse that results in static and dynamic equilibrium. (p. 497)
28. Define the following: corneal transplant (p. 473), detached retina (p. 475), cataract (p. 477), color blindness (p. 483), perforated eardrum (p. 488), and cochlear implant. (p. 494)

Answers to Questions with Figures

16.1 Olfactory tract.
16.2 Gustatory receptors → cranial nerves VII, IX, or X → medulla → (1) limbic system and hypothalamus or (2) thalamus → primary gustatory area in the parietal lobe of the cerebral cortex.
16.3 Iris and pigment epithelium.
16.4 Tears or lacrimal fluid is a watery solution containing salts, some mucus, and lysozyme. It cleans, lubricates, and moistens the eyeball.
16.5 Avascular structures in the eye include the cornea, lens, and retina. Since these structures have no blood vessels, they are more transparent to light, an advantage since the photoreceptors lie at the very back of the eye.
16.6 Hypertension and diabetes; also cataract and senile macular degeneration.
16.7 Two types of photoreceptors are rods and cones. Rods provide black-and-white vision in dim light whereas cones provide high visual acuity and color vision in bright light.
16.8 Aqueous humor is secreted by the ciliary process, then flows into the posterior chamber, around the iris, into the anterior chamber, and out of the eyeball through the scleral venous sinus.
16.9 The ciliary muscle contracts → suspensory ligaments slacken → lens rounds up, becoming a more powerful lens.
16.10 The loss of elasticity in the lens that occurs with aging.
16.11 Both rods and cones transduce light into receptor potentials, using a photopigment embedded in outer segment membrane folds, and release neurotransmitter at synapses with bipolar cells and horizontal cells.
16.12 Isomerization.
16.13 Breakdown of cyclic GMP by hydrolysis; synthesis of cyclic GMP.
16.14 Retinal ganglion cell axons form the optic nerve (cranial nerve II) → optic chiasma → optic tract → lateral geniculate nucleus of the thalamus → optic radiations → occipital lobe of cerebrum.
16.15 Nasal.
16.16 Inner ear: cochlea (hearing) and semicircular ducts (equilibrium).
16.17 The tympanic membrane (eardrum) separates the external ear from the middle ear. The oval and round windows separate the middle ear from the internal ear.
16.18 The utricle and the saccule.
16.19 In the cochlea, perilymph fills the scala vestibuli and scala tympani whereas endolymph fills the cochlear duct.
16.20 Near the base of the cochlea, the region close to the oval and round windows.
16.21 Static.
16.22 Dynamic.

Chapter 17

THE AUTONOMIC NERVOUS SYSTEM

Chapter Contents at a Glance

Student Objectives

1. Compare the structural and functional differences between the somatic and autonomic nervous systems.
2. Identify the principal structural features of the autonomic nervous system.
3. Compare the sympathetic and parasympathetic divisions of the autonomic nervous system in terms of anatomy, physiology, and neurotransmitters released.
4. Describe the various postsynaptic receptors involved in autonomic responses.
5. Describe the components of a visceral autonomic reflex.
6. Explain the relationship of the hypothalamus to the autonomic nervous system.

The **autonomic nervous system (ANS)** regulates the activity of smooth muscle, cardiac muscle, and certain glands. Traditionally, the ANS has been described as a specific *output* portion of the peripheral nervous system. Operation of the ANS to maintain homeostasis, however, depends on a continual flow of sensory input from visceral organs and blood vessels into the CNS. Thus it is reasonable to include these sensory neurons as part of the ANS (see Fig. 12.1). Structurally, then, the ANS includes two main components: general visceral sensory (afferent) neurons and general visceral motor (efferent) neurons.

Functionally, the ANS usually operates without conscious control. The system was originally named *autonomic* because it was thought to function autonomously or in a self-governing manner, without control by the central nervous system (CNS). The ANS, however, is regulated by centers in the brain, mainly the hypothalamus and medulla oblongata, which receive input from the limbic system and other regions of the cerebrum.

COMPARISON OF SOMATIC AND AUTONOMIC NERVOUS SYSTEMS

The somatic nervous system includes both sensory and motor neurons. The sensory neurons convey input from receptors for the special senses (vision, hearing, taste, smell, and equilibrium), proprioceptors (muscle and joint position), and general somatic receptors (pain, temperature, and tactile sensations). All these sensations normally are consciously perceived. In turn, somatic motor neurons innervate skeletal muscle, the effector tissue of the somatic nervous system, and produce conscious, voluntary movements. In the somatic nervous system, the effect of a motor neuron always is *excitation*. When a somatic motor neuron stimulates a skeletal muscle, the muscle contracts. When the neuron ceases to stimulate the muscle, contraction stops.

The input component of the ANS consists of **general visceral sensory (afferent) neurons**. Mostly, these are associated with interoceptors, such as chemoreceptors that monitor blood CO_2 level and mechanoreceptors that detect degree of stretch of organs or blood vessels. These afferent signals are not consciously recognized most of the time, although intense activation of interoceptors may give rise to conscious sensations. Examples of this are the sensations of pain or nausea from damaged viscera, fullness of the urinary bladder, and angina pectoris (chest pain) from inadequate blood flow to the heart.

Autonomic motor (visceral efferent) neurons regulate visceral activities by either *exciting* or *inhibiting* their effector tissues, which are cardiac muscle, smooth muscle, and glands. Responses include changes in the size of the pupil, accommodation for near vision, dilation of blood vessels, adjustment of the rate and force of the heartbeat, move-

ments of the gastrointestinal tract, and secretion by most glands. These activities usually lie beyond conscious control. They are automatic. Input from the general somatic and special senses, acting via the limbic system, also may modify responses of autonomic motor neurons. Seeing a bike about to hit you or hearing squealing brakes of a nearby car, or being grabbed by an attacker, for example, would increase the rate and force of your heartbeat.

Autonomic motor pathways consist of sets of two motor (efferent) neurons in series (one following the other). The first has its cell body in the CNS; its axon extends from the CNS to an **autonomic ganglion.** (Recall that a ganglion is a collection of neuronal cell bodies outside the CNS.) The cell body of the second neuron is in that autonomic ganglion; its axon extends directly from the ganglion to the effector (smooth muscle, cardiac muscle, or gland). In contrast, a single somatic motor neuron extends from the CNS to skeletal muscle. Also, whereas somatic motor neurons release acetylcholine (ACh) as their neurotransmitter, autonomic motor neurons release either ACh or norepinephrine (NE). A summary of the similarities and differences between the somatic and autonomic nervous systems is presented in Exhibit 17.1.

The output (efferent) part of the ANS has two principal divisions: **sympathetic** and **parasympathetic.** Many organs receive autonomic motor fibers from both divisions. In general, nerve impulses from one division stimulate the organ to start or increase activity (excitation), whereas impulses from the other division decrease the organ's activity (inhibition). Organs that receive impulses from both sympathetic and parasympathetic fibers are said to have **dual innervation.** The rest of the chapter focuses on the anatomy and physiology of the sympathetic and parasympathetic divisions and the CNS centers that regulate outflow along these two divisions.

ANATOMY OF AUTONOMIC MOTOR PATHWAYS

Overview

The first of the two autonomic motor neurons is called a **preganglionic neuron** (Exhibit 17.1). Its cell body is in the brain or spinal cord. Its myelinated axon, called a **preganglionic fiber,** passes out of the CNS as part of a cranial or spinal nerve. At some point, the fiber separates from the nerve and extends to an autonomic ganglion. There it synapses with the postganglionic neuron, the second neuron in the autonomic motor pathway. The **postganglionic neuron** lies entirely outside the CNS. Its cell body and dendrites are located in an autonomic ganglion, where it makes synapses with one or more preganglionic fibers. The axon of a postganglionic neuron, called a **postganglionic fiber,** is unmyelinated and terminates in a visceral effector. Thus

EXHIBIT 17.1

COMPARISON OF SOMATIC AND AUTONOMIC NERVOUS SYSTEMS

Sensory Input (Afferent Neurons)	CNS Centers (Process Input and Initiate Output)	Motor Output (Efferent Neurons) and Neurotransmitters	Effectors	Response of Effector to Neurotransmitters
S O M A T I C — Special senses. General somatic senses. Proprioceptors.	Voluntary control from cerebral cortex. Other active regions include basal ganglia, cerebellum, brain stem, and spinal cord.	Axon of somatic motor neuron extends from CNS, synapses with effector, and releases acetylcholine (ACh).	Skeletal muscle.	Excitation.
A U T O N O M I C — Special senses. General visceral senses (mainly from interoceptors). General somatic senses.	Involuntary control via limbic system, hypothalamus, medulla, pons, spinal cord. Cerebral cortex also contributes.	Axon of first autonomic motor neuron (preganglionic neuron) extends from the CNS and synapses with second (postganglionic neuron) in a ganglion. The second neuron synapses with a visceral effector. Preganglionic fibers release acetylcholine (ACh). Postganglionic fibers release ACh (parasympathetic division) or norepinephrine (NE; sympathetic division).	Cardiac muscle. Smooth muscle. Glands.	Excitation or inhibition.

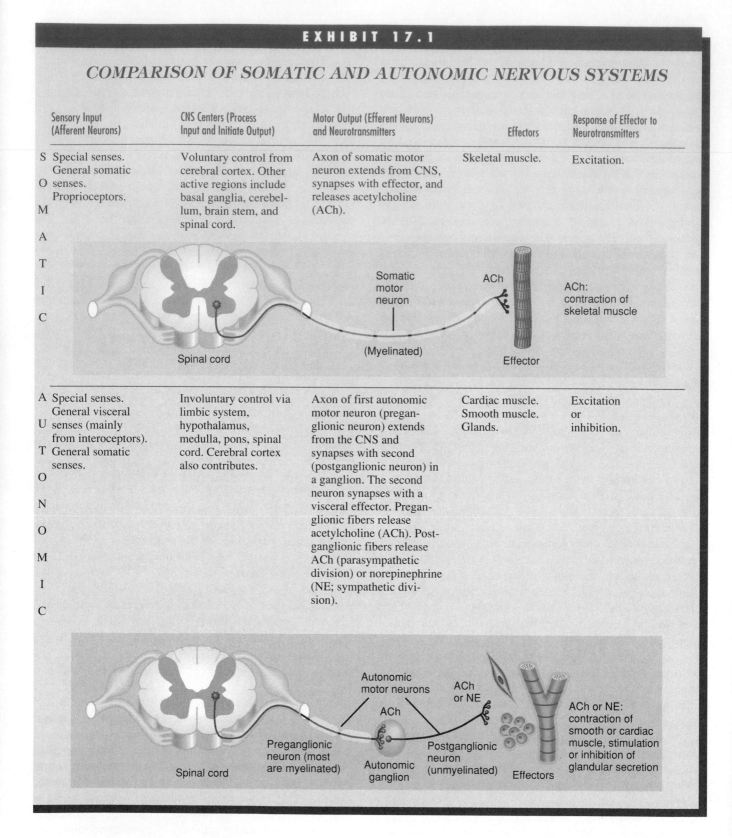

preganglionic neurons convey efferent impulses from the CNS to autonomic ganglia, and postganglionic neurons relay the impulses from autonomic ganglia to visceral effectors. Note that autonomic ganglia differ from posterior root ganglia, which contain cell bodies of somatic sensory neurons but no synapses.

Preganglionic Neurons

In the sympathetic division, the preganglionic neurons have their cell bodies in the lateral gray horns of the 12 thoracic segments and first two or three lumbar segments of the spinal cord (Fig. 17.1). For this reason, the sympathetic division is also called the **thoracolumbar** (thō'-ra-kō-LUM-bar) **division,** and the fibers of the sympathetic preganglionic neurons are known as the **thoracolumbar outflow.**

The cell bodies of the preganglionic neurons of the parasympathetic division are located in the nuclei of the oculomotor (III), facial (VII), glossopharyngeal (IX), and vagus (X) cranial nerves in the brain stem and in the lateral gray horns of the second through fourth sacral segments of the spinal cord. Hence the parasympathetic division is also known as the **craniosacral division,** and the fibers of the parasympathetic preganglionic neurons are referred to as the **craniosacral outflow.**

Autonomic Ganglia

The autonomic ganglia may be divided into three general groups: two of the groups are components of the sympathetic division and one group is a component of the parasympathetic division.

● **Sympathetic ganglia** The **sympathetic trunk (vertebral chain) ganglia** are a series of ganglia that lie in a vertical row on either side of the vertebral column, extending from the base of the skull to the coccyx (Figs. 17.1 and 17.2). They are also known as **paravertebral ganglia** and receive preganglionic fibers only from the *sympathetic division.* Because the sympathetic trunk ganglia are so near the spinal cord, sympathetic preganglionic fibers tend to be short.

The second kind of autonomic ganglion also receives preganglionic fibers from the *sympathetic division.* It is called a **prevertebral (collateral) ganglion** (Fig. 17.2). The ganglia of this group lie anterior to the spinal column and close to the large abdominal arteries. Examples of prevertebral ganglia are the celiac ganglion, on either side of the celiac artery just below the diaphragm; the superior mesenteric ganglion, near the beginning of the superior mesenteric artery in the upper abdomen; and the inferior mesenteric ganglion, located near the beginning of the inferior mesenteric artery in the middle of the abdomen (see Fig. 17.1).

● **Parasympathetic ganglia** Preganglionic fibers from the *parasympathetic division* make synapses in **terminal (intramural) ganglia.** These ganglia are located at the end of an autonomic motor pathway very close to or actually within the wall of a visceral organ. Since parasympathetic preganglionic fibers extend from the CNS to the terminal ganglion in an innervated organ, they tend to be long.

In addition to autonomic ganglia, the ANS contains **autonomic plexuses.** Slender bundles of pre- or postganglionic nerve fibers arranged in a branching network constitute an autonomic plexus.

Postganglionic Neurons

Axons of preganglionic neurons of the sympathetic division pass to ganglia of the sympathetic trunk. Some synapse with postganglionic neurons in the sympathetic chain ganglia (Fig. 17.2). Others continue, without synapsing, through the chain ganglia to end at a prevertebral ganglion and synapse with the postganglionic neurons there. In either case, a single sympathetic preganglionic fiber may synapse with 20 or more postganglionic fibers. This is an example of divergence (see page 369) and is part of the reason why sympathetic responses tend to be widespread throughout the body. After exiting their ganglia, the postganglionic fibers typically innervate several visceral effectors (see Fig. 17.1).

Axons of preganglionic neurons of the parasympathetic division pass to terminal ganglia near or within a visceral effector (see Fig. 17.1). In the ganglion, the presynaptic neuron usually synapses with only four or five postsynaptic neurons, all of which supply a single visceral effector. Thus parasympathetic effects tend to be localized. With this background in mind, we can now examine some specific structural features of the sympathetic and parasympathetic divisions of the ANS.

Sympathetic Division

Cell bodies of sympathetic preganglionic neurons are part of the lateral gray horns of all thoracic segments and the first two or three lumbar segments of the spinal cord (see Fig. 17.1). The preganglionic axons are myelinated and leave the spinal cord through the anterior root of a spinal nerve along with the somatic motor fibers at the same segmental level. After exiting through the intervertebral foramina, the myelinated preganglionic sympathetic fibers enter a short pathway called a **white ramus** before passing to the nearest sympathetic trunk ganglion on the same side.

Collectively, the white rami are called the **white rami communicantes** (kō-myoo-ni-KAN-tēz). See Fig. 17.2. Their name ("white") indicates that they contain myelinated fibers. Only thoracic and upper lumbar nerves have white rami communicantes. The white rami communicantes connect the anterior ramus of the spinal nerve with the ganglia of the sympathetic trunk.

The paired sympathetic trunks lie anterior and lateral to the spinal cord, one on either side. Typically, there are 22 ganglia in each chain: 3 cervical, 11 thoracic, 4 lumbar, and 4 sacral. Although the trunk extends downward from the neck, thorax, and abdomen to the coccyx, it receives preganglionic fibers only from the thoracic and lumbar segments of the spinal cord (see Fig. 17.1).

The cervical portion of each sympathetic trunk is located

FIGURE 17.1 Structure of sympathetic and parasympathetic divisions of the ANS. Although the parasympathetic division is shown only on one side and the sympathetic division is shown only on the other side, keep in mind that each division innervates tissues on both sides of the body.

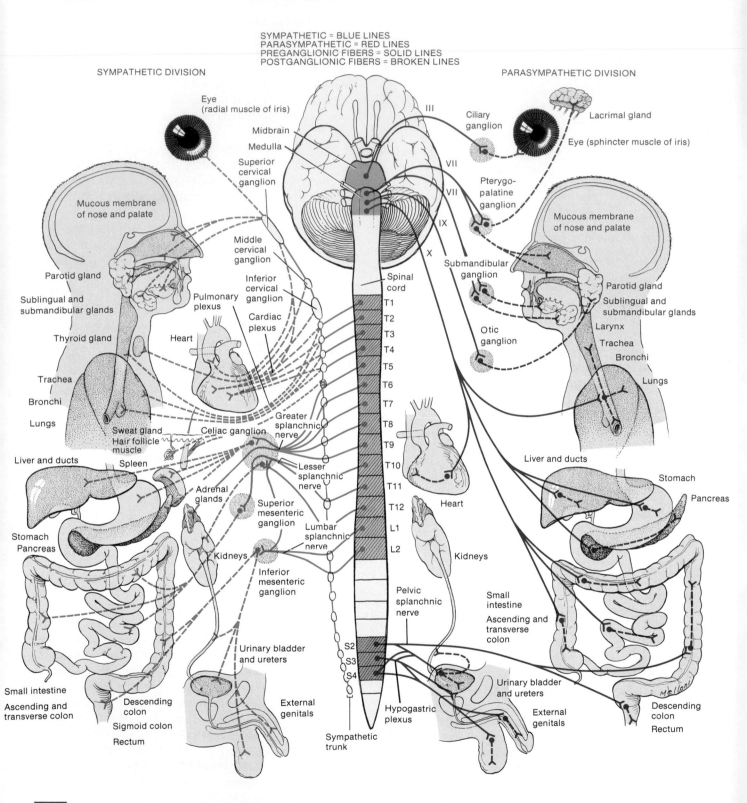

SYMPATHETIC = BLUE LINES
PARASYMPATHETIC = RED LINES
PREGANGLIONIC FIBERS = SOLID LINES
POSTGANGLIONIC FIBERS = BROKEN LINES

SYMPATHETIC DIVISION

PARASYMPATHETIC DIVISION

Eye (radial muscle of iris)
III
Ciliary ganglion
Lacrimal gland
Midbrain
Eye (sphincter muscle of iris)
Medulla
Superior cervical ganglion
VII
Mucous membrane of nose and palate
VII
Pterygo-palatine ganglion
Middle cervical ganglion
IX
Mucous membrane of nose and palate
Parotid gland
Inferior cervical ganglion
Submandibular ganglion
X
Parotid gland
Sublingual and submandibular glands
Pulmonary plexus
Spinal cord
Sublingual and submandibular glands
Thyroid gland
Cardiac plexus
Otic ganglion
Larynx
Heart
T1
Trachea
Trachea
T2
Bronchi
Bronchi
T3
Lungs
T4
Lungs
T5
T6
Greater splanchnic nerve
T7
Liver and ducts
Sweat gland
Celiac ganglion
T8
Stomach
Hair follicle muscle
Spleen
T9
Lesser splanchnic nerve
T10
Pancreas
Liver and ducts
Adrenal glands
T11
Heart
Stomach
Pancreas
Superior mesenteric ganglion
T12
Kidneys
Kidneys
Lumbar splanchnic nerve
L1
Inferior mesenteric ganglion
L2
Small intestine
Small intestine
Ascending and transverse colon
Ascending and transverse colon
Pelvic splanchnic nerve
Descending colon
Urinary bladder and ureters
S2
Sigmoid colon
S3
Descending colon
Rectum
S4
Urinary bladder and ureters
External genitals
Rectum
Hypogastric plexus
External genitals
Sympathetic trunk

Melloni

Question: Which division has longer preganglionic fibers and why?

FIGURE 17.2 Ganglia and rami communicantes of the sympathetic division of the ANS.

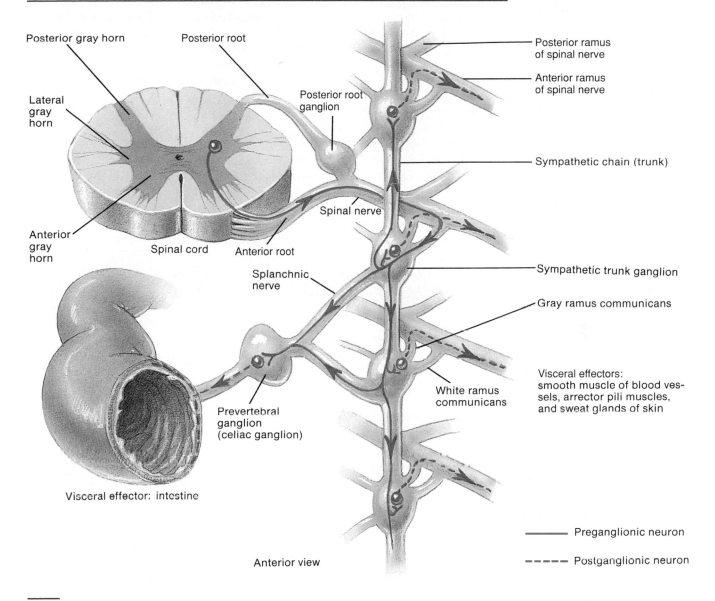

Anterior view

Question: What are the three types of autonomic ganglia?

in the neck anterior to the prevertebral muscles. It is subdivided into superior, middle, and inferior ganglia (see Fig. 17.1). The **superior cervical ganglion** is posterior to an internal carotid artery and anterior to a transverse process of the second cervical vertebra. Postganglionic fibers leaving the ganglion serve the head. They are distributed to sweat glands, smooth muscle of the eye and blood vessels of the face, nasal mucosa, and the submandibular, sublingual, and parotid salivary glands. Gray rami communicantes (described shortly) from the ganglion also pass to the upper two to four cervical spinal nerves. The **middle cervical ganglion** lies near the sixth cervical vertebra at the level of the cricoid cartilage. Postganglionic fibers from it innervate the

heart. The **inferior cervical ganglion** is located near the first rib, anterior to the transverse processes of the seventh cervical vertebra. Its postganglionic fibers also supply the heart.

The thoracic portion of each sympathetic trunk usually consists of 11 segmentally arranged ganglia, lying anterior to the necks of the corresponding ribs. This portion of the sympathetic trunk receives most of the sympathetic preganglionic fibers. Postganglionic fibers from the thoracic sympathetic trunk innervate the heart, lungs, bronchi, and other thoracic viscera. In the skin, they also innervate sweat glands, blood vessels, and arrector pili muscles of hair follicles.

The lumbar portion of each sympathetic trunk lies on either side of the corresponding lumbar vertebrae. The sacral portion of the sympathetic trunk lies in the pelvic cavity on the medial side of the sacral foramina. Unmyelinated postganglionic fibers from the lumbar and sacral sympathetic chain ganglia enter a short pathway called a **gray ramus** and then merge with a spinal nerve or join the hypogastric plexus in the pelvis via direct visceral branches. The **gray ramus communicans** (kō-MYOO-ni-kanz; plural is **rami communicates**) is the structure containing the postganglionic fibers that connect the ganglion of the sympathetic trunk to the spinal nerve (Fig. 17.2). The fibers are unmyelinated. Gray rami communicantes outnumber the white rami, since there is a gray ramus leading to each of the 31 pairs of spinal nerves.

As preganglionic fibers extend from a white ramus communicans into the sympathetic trunk, they give off several axon collaterals (branches). These collateral fibers terminate and synapse in several ways. Some synapse in the first ganglion at the level of entry. Others pass up or down the sympathetic trunk for a variable distance to form the fibers on which the ganglia are strung. These fibers, known as **sympathetic chains** (Fig. 17.2), may not synapse until they reach a ganglion in the cervical or sacral area. Many rejoin the spinal nerves through gray rami and supply peripheral visceral effectors such as sweat glands and the smooth muscle in blood vessels and around hair follicles in the extremities.

Some preganglionic fibers pass through the sympathetic trunk without terminating in the trunk. Beyond the trunk, they form nerves known as **splanchnic** (SPLANK-nik) **nerves** (Fig. 17.2), which extend to and terminate in outlying prevertebral ganglia. Splanchnic nerves from the thoracic area terminate in the **celiac** (SĒ-lē-ak) **ganglion** or **solar plexus.** Here, the preganglionic fibers synapse with postganglionic cell bodies. The greater splanchnic nerve passes to the celiac ganglion of the celiac plexus, which is located near the first lumbar vertebra. From here, postganglionic fibers extend to the stomach, spleen, liver, kidney, and small intestine. The lesser splanchnic nerve passes through the celiac plexus to the superior mesenteric ganglion of the superior mesenteric plexus, which is just below the celiac plexus. Postganglionic fibers from this ganglion innervate the small intestine and colon. The lowest splanchnic nerve, not always present, enters the renal plexus near the kidney. Postganglionic fibers supply the renal arterioles and ureter.

The lumbar splanchnic nerve enters the inferior mesenteric plexus, which is below the superior mesenteric plexus. In the plexus, the preganglionic fibers synapse with postganglionic fibers in the inferior mesenteric ganglion. These fibers pass through the hypogastric plexus and supply the distal colon and rectum, urinary bladder, and genital organs. Postganglionic fibers leaving the prevertebral ganglia follow the course of various arteries to abdominal and pelvic visceral effectors.

Sympathetic *preganglionic* fibers also extend to the medullae of the adrenal glands. Developmentally, the adrenal medulla is a modified sympathetic ganglion. Its cells are like sympathetic postganglionic neurons. Rather than extending to another organ, however, these cells release the hormones norepinephrine and epinephrine into the blood. This is one exception to the usual pattern of two efferent neurons in an autonomic motor pathway.

Parasympathetic Division

Cell bodies of parasympathetic preganglionic neurons are found in nuclei in the brain stem and the lateral gray horn of the second through fourth sacral segments of the spinal cord (see Fig. 17.1). Their fibers emerge as part of a cranial nerve or as part of the anterior root of a spinal nerve. The **cranial parasympathetic outflow** consists of preganglionic fibers that extend from the brain stem in four cranial nerves. The **sacral parasympathetic outflow** consists of preganglionic fibers in anterior roots of the second through fourth sacral nerves. The preganglionic fibers of both the cranial and sacral outflows end in terminal ganglia, where they synapse with postganglionic neurons. We will look first at the cranial outflow.

The cranial outflow has five components: four pairs of ganglia and the plexuses associated with the vagus (X) nerve. The four pairs of cranial parasympathetic ganglia innervate structures in the head and are located close to the organs they innervate. The **ciliary ganglia** lie lateral to each optic (II) nerve near the back of the orbit. Preganglionic fibers pass with the oculomotor (III) nerve to the ciliary ganglion. Postganglionic fibers from the ganglion innervate smooth muscle cells in the eyeball. The **pterygopalatine** (ter′-i-gō-PAL-a-tin) **ganglia** are lateral to the sphenopalatine foramen. They receive preganglionic fibers from the facial (VII) nerve and send postganglionic fibers to the nasal mucosa, palate, pharynx, and lacrimal glands. The **submandibular ganglia** are found near the ducts of the submandibular salivary glands. They receive preganglionic fibers from the facial (VII) nerve and send postganglionic fibers to the submandibular and sublingual salivary glands. The **otic ganglia** are situated just below each foramen ovale. Each otic ganglion receives preganglionic fibers from the glossopharyngeal (IX) nerve and sends postganglionic fibers to the parotid salivary gland.

Preganglionic fibers that leave the brain as part of the vagus (X) nerves are the last components of the cranial outflow. The vagus nerves carry nearly 80% of the total craniosacral outflow. Vagal fibers extend to many terminal ganglia in the thorax and abdomen. Since the terminal ganglia are close to or in the walls of their visceral effectors, postganglionic parasympathetic fibers are very short. By comparison, postganglionic sympathetic fibers are long. As it passes through the thorax, the vagus (X) nerve sends fibers to the heart and the airways of the lungs. In the abdomen, it

EXHIBIT 17.2

ANATOMICAL FEATURES OF SYMPATHETIC AND PARASYMPATHETIC DIVISIONS

Sympathetic	Parasympathetic
Forms thoracolumbar outflow.	Forms craniosacral outflow.
Contains sympathetic trunk and prevertebral ganglia.	Contains terminal ganglia.
Ganglia are close to the CNS and distant from visceral effectors.	Ganglia are near or within visceral effectors.
Each preganglionic fiber synapses with many postganglionic neurons that pass to many visceral effectors (divergence).	Each preganglionic fiber usually synapses with four or five postganglionic neurons that pass to a single visceral effector.
Distributed throughout the body, including the skin.	Distribution limited primarily to head and viscera of thorax, abdomen, and pelvis.

supplies the liver, gallbladder, stomach, pancreas, small intestine, and part of the large intestine.

The sacral parasympathetic outflow consists of preganglionic fibers from the anterior roots of the second through fourth sacral nerves. Collectively, they form the **pelvic splanchnic nerves.** These nerves synapse with parasympathetic postganglionic neurons located in terminal ganglia in the walls of the innervated viscera. From the ganglia, parasympathetic postganglionic fibers innervate smooth muscle and glands in the walls of the colon, ureters, urinary bladder, and reproductive organs.

Anatomical features of the sympathetic and parasympathetic divisions are compared in Exhibit 17.2.

PHYSIOLOGICAL EFFECTS OF THE ANS

Most body structures receive dual innervation, that is, fibers from both the sympathetic and parasympathetic divisions. Structures that receive only sympathetic innervation include sweat glands, arrector pili muscles attached to hair follicles in the skin, adipose (fat) cells, the kidneys, and most blood vessels. Lacrimal (tear) glands, on the other hand, receive only parasympathetic fibers. Elsewhere in the body, the two divisions generally have opposing effects on a given organ, one causing excitation and the other inhibition. This is possible because they use different neurotransmitters and have different neurotransmitter receptors. The hypothalamus regulates the balance of sympathetic versus parasympathetic activity or tone. This balance can change from one moment to the next according to demands imposed by the internal and external environments.

ANS Neurotransmitters

Like other neurons, autonomic neurons release neurotransmitters at synapses. Based on the neurotransmitter they produce and liberate, autonomic neurons are classified as either cholinergic or adrenergic (Fig. 17.3).

Cholinergic (ko'-lin-ER-jik) **neurons** release **acetylcholine (ACh)** and include the following: (1) all sympathetic and parasympathetic preganglionic neurons, (2) all parasympathetic postganglionic neurons, and (3) a few sympathetic postganglionic neurons. The cholinergic sympathetic postganglionic fibers include those to most sweat glands and a few blood vessels in skeletal muscles.

ACh is stored in synaptic vesicles in the axon terminals of the cholinergic fibers. Release of ACh from the vesicles is via exocytosis. ACh then diffuses the short distance across the synaptic cleft to bind with specific receptors on the postsynaptic membrane. The membrane—of an autonomic postganglionic neuron, a smooth or cardiac muscle cell, or a glandular cell—either depolarizes and the cell becomes excited or hyperpolarizes and the cell becomes inhibited. (Recall from Chapter 10 that ACh always causes depolarization and excitation of skeletal muscle membranes.) Since acetylcholine is quickly inactivated by the enzyme **acetylcholinesterase (AChE),** effects triggered by transient activity of cholinergic fibers are brief.

Adrenergic (ad'-ren-ER-jik) **neurons** release **norepinephrine (noradrenalin)** or **epinephrine (adrenalin).** Most sympathetic postganglionic axons are adrenergic; they secrete norepinephrine. Like ACh, NE is synthesized and stored in synaptic vesicles located in the axon terminals of adrenergic fibers. When an action potential reaches the axon terminal, NE is rapidly released. The molecules diffuse across the synaptic cleft and combine with specific

FIGURE 17.3 Cholinergic and adrenergic receptors.

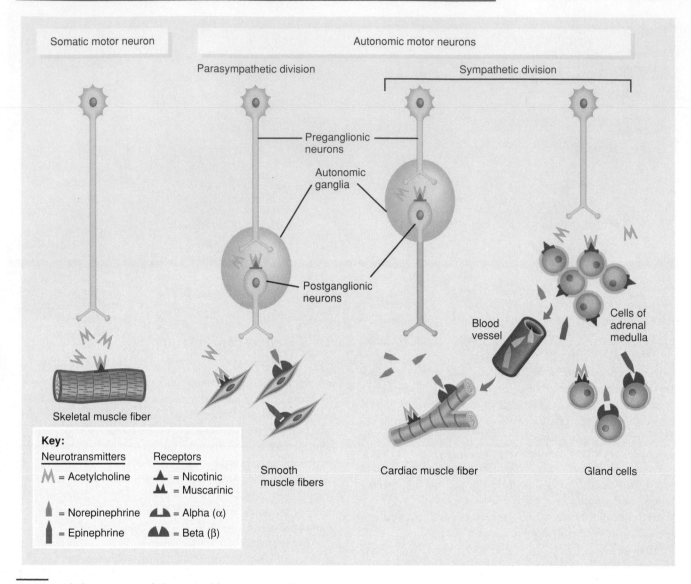

Question: Which neurons are cholinergic and have nicotinic ACh receptors?

receptors on the postsynaptic membrane to trigger either depolarization (excitation) or hyperpolarization (inhibition).

The effects of sympathetic stimulation are longer lasting and more widespread than the effects of parasympathetic stimulation for three reasons. First, there is much more divergence of sympathetic postganglionic fibers. Second, norepinephrine (NE) lingers in the synaptic cleft for a longer time than does ACh. Slowly, it is both taken up by the axon that released it and enzymatically inactivated by **catechol-*O*-methyltransferase (COMT)** or **monoamine oxidase (MAO).** Third, NE and epinephrine secreted into the blood by the adrenal medullae of the two adrenal glands intensify the action of NE liberated from sympathetic postganglionic axons. These hormones in the blood circulate throughout the body, affecting whichever tissues have the

appropriate receptors. In time, liver enzymes degrade blood-borne NE and epinephrine.

Cholinergic and Adrenergic Receptors

There are two main categories of both cholinergic and adrenergic receptors plus several subcategories of adrenergic receptors (Fig. 17.3). The two types of cholinergic (ACh) receptors are nicotinic receptors and muscarinic receptors.

Nicotinic receptors are found on both sympathetic and parasympathetic postganglionic neurons. They are so named because nicotine mimics the action of ACh on such receptors. **Muscarinic receptors** are present on all effectors (muscles and glands) innervated by parasympathetic postganglionic axons. Most sweat glands and the smooth muscle in certain blood vessels, which receive their innervation from *cholinergic* sympathetic postganglionic fibers, also feature muscarinic receptors. These postsynaptic receptors are so named because a mushroom poison called muscarine mimics the actions of ACh on them. Nicotine does not activate muscarinic receptors nor does muscarine stimulate nicotinic receptors, but ACh activates both receptor types.

Activation of nicotinic receptors leads to depolarization and thus to excitation of the postsynaptic cell, be it a postganglionic neuron or an ANS effector. Activation of muscarinic receptors sometimes causes depolarization (excitation) and sometimes hyperpolarization (inhibition), depending on which particular cell bears the muscarinic receptors. The binding of ACh to muscarinic receptors inhibits (relaxes) cardiac muscle fibers and smooth muscle sphincters in the gastrointestinal tract, for example, but excites smooth muscle fibers in the sphincter muscle of the iris of the eye to contract.

The effects of NE and epinephrine, like those of ACh, also depend on the type of postsynaptic receptor with which they interact. The two types of adrenergic receptors for norepinephrine and epinephrine are called **alpha (α) receptors** and **beta (β) receptors** (Fig. 17.3). Such receptors are found on visceral effectors innervated by most sympathetic postganglionic axons. Both alpha and beta receptors are, in turn, classified into subtypes—α_1, α_2, β_1, and β_2. The receptors are distinguished by the specific responses they elicit and by their selective binding of drugs that activate or block them. In general, alpha receptors are excitatory. Some beta receptors are excitatory while others are inhibitory. Although cells of most effectors contain either alpha or beta receptors, some visceral effector cells contain both. In general, NE stimulates alpha receptors more vigorously than beta receptors, whereas epinephrine stimulates alpha and beta receptors about equally.

CLINICAL APPLICATION

TAKING ADVANTAGE OF RECEPTOR SELECTIVITY

The availability of drugs and natural products that selectively activate or block certain receptors is a cornerstone of modern drug therapy. Two examples will give you an idea of the usefulness of such drugs.

Propranolol (Inderal) often is prescribed for patients with high blood pressure (hypertension). It is a nonselective beta blocker, meaning it binds to all types of beta receptors and prevents their activation by epinephrine and norepinephrine. The desired effects of propranolol are due to its blockade of β_1 receptors, namely decreased heart rate and contraction force and a consequent decrease in blood pressure. Undesired effects due to blockade of β_2 receptors may include hypoglycemia (low blood sugar due to decreased glycogen breakdown) and gluconeogenesis (the conversion of a noncarbohydrate into glucose in the liver) and mild bronchoconstriction (narrowing of the airways). If these side effects pose a threat to the patient, a selective β_1 blocker

such as metoprolol (Lopressor) can be prescribed instead of propranolol.

Atropine selectively blocks muscarinic ACh receptors, which normally cause smooth muscle in the airways to contract. Thus it can be inhaled to dilate the airways. It is used to relieve an asthma attack without causing paralysis of skeletal muscle fibers, which have nicotinic ACh receptors.

Parasympathetic and Sympathetic Responses

The parasympathetic division regulates primarily those activities that conserve and restore body energy during times of rest or recovery. It is an **energy conservation–restorative system.** Normally, parasympathetic impulses to the digestive glands and the smooth muscle of the gastrointestinal tract dominate over sympathetic impulses. Thus energy-supplying food can be digested and absorbed.

The acronym "SLUD" is a mental key for remembering many parasympathetic responses. It stands for salivation (S), lacrimation (L), urination (U), and defecation (D). These responses, except for lacrimation, have to do with digestion and absorption of food and elimination of wastes, activities that are stimulated mainly by the parasympathetic division. Besides the "SLUD" responses, another important parasympathetic response is a decrease in heart rate.

So-called paradoxical fear may cause massive activation of the parasympathetic division. This is the sort of fear that occurs when one is backed into a corner with no way out. It may happen to soldiers in battle, to unprepared students taking an exam, or sometimes to athletes before competition. This effect is the basis for loss of control over urination or defecation in such situations.

The sympathetic division, on the other hand, prepares the body for emergency situations. It is primarily concerned with processes involving the expenditure of energy. When the body is in homeostasis, the main function of the sympathetic division is to counteract the parasympathetic effects just enough to carry out normal processes requiring energy. During physical or emotional stress, however, the sympathetic dominates the parasympathetic. Physical exertion stimulates the sympathetic division as do a variety of emotions (such as fear, embarrassment, or rage). Visualizing body changes that occur during "E situations" (emergency, exercise, embarrassment) will help you remember most of the sympathetic responses. Activation of the sympathetic division sets in motion a series of physiological responses collectively called the **fight-or-flight response.** It produces the following effects.

1. The pupils of the eyes dilate.
2. Heart rate, force of contraction, and blood pressure increase.
3. The blood vessels of nonessential organs such as the skin and viscera constrict.
4. Blood vessels of organs involved in exercise or fighting off danger—skeletal muscles, cardiac muscle,

liver, and adipose tissue—dilate to allow faster flow of blood. (The liver splits glycogen to glucose and adipose tissue splits triglycerides to fatty acids, both of which are used by muscle fibers to generate ATP.

5. Rapid and deeper breathing occurs and the bronchioles dilate to allow faster movement of air in and out of the lungs.

6. Blood sugar level rises as liver glycogen is converted to glucose.

7. The medullae of the adrenal glands are stimulated to release epinephrine and norepinephrine, hormones that intensify and prolong the sympathetic effects just described.

8. Processes that are not essential for meeting the stress situation are inhibited. For example, muscular movements of the gastrointestinal tract and digestive secretions slow down or even stop.

Exhibit 17.3 summarizes the responses of glands, smooth muscle, and cardiac muscle to stimulation by the sympathetic and parasympathetic branches of the ANS.

VISCERAL AUTONOMIC REFLEXES

A **visceral autonomic reflex** adjusts the activity of a visceral effector. In other words, it results in the contraction or relaxation of smooth or cardiac muscle or a change in the rate of secretion by a gland. Such reflexes play a key role in activities involved in homeostasis such as regulating heart action, blood pressure, respiration, digestion, defecation, and urinary bladder functions.

A visceral autonomic reflex arc consists of the following components.

EXHIBIT 17.3

ACTIVITIES OF PARASYMPATHETIC AND SYMPATHETIC DIVISIONS

Visceral Effector	Effect of Sympathetic Stimulation (α or β Receptors, Except as Noted)[a]	Effect of Parasympathetic Stimulation (Muscarinic Receptors)
GLANDS		
Sweat	Increases secretion locally on palms of hands and soles of feet (α); increases secretion in most body regions (muscarinic ACh receptors).	No known functional innervation.
Lacrimal (tear)	No known functional innervation.	Stimulates secretion.
Adrenal medulla	Promotes epinephrine and norepinephrine secretion (nicotinic ACh receptors).	No known functional innervation.
Liver	Promotes the conversion of glycogen in the liver into glucose (glycogenolysis) and the conversion of noncarbohydrates in the liver into glucose (gluconeogenesis); decreases bile secretion (α and β_2).	Promotes glycogen synthesis; increases bile secretion.
Adipose (fat) cells	Promotes the breakdown of triglycerides into fatty acids and glycerol (lipolysis) (β_1) and release of fatty acids into blood.	No known functional innervation.
Kidney	Stimulates secretion of renin (β_1).	No known functional innervation.
Pancreas	Inhibits secretion of enzymes and insulin (α); promotes secretion of glucagon (β_2).	Promotes secretion of enzymes and insulin.
SMOOTH MUSCLE		
Iris, radial muscle	Contraction that results in dilation of pupil (α).	No known functional innervation.
Iris, sphincter muscle	No known functional innervation.	Contraction → constriction of pupil.
Ciliary muscle of eye	Relaxation for far vision (β).	Contraction for near vision.
Salivary glands, arterioles	Vasoconstriction, which decreases secretion (β_2).	Vasodilation, which increases K$^+$ and water secretion.
Gastric glands, arterioles	Vasoconstriction, which inhibits secretion (?).	Promotes secretion.
Intestinal glands, arterioles	Vasoconstriction, which inhibits secretion (α).	Promotes secretion.
Lungs, bronchial muscle	Relaxation → airway dilation (β_2).	Contraction → airway constriction.
Heart, arterioles	Relaxation → dilation of coronary vessels (β_1).	Contraction → constriction of vessels.
Skin and mucosal arterioles	Contraction → constriction (α).	Dilation, which may not be physiologically significant.

Visceral Effector	Effect of Sympathetic Stimulation (α or β Receptors, Except as Noted)[a]	Effect of Parasympathetic Stimulation (Muscarinic Receptors)
Skeletal muscle arterioles	Contraction → constriction (α); relaxation → dilation (β_2). Relaxation → dilation (muscarinic).	No known innervation.
Abdominal viscera arterioles	Contraction → constriction (α, β).	No known innervation for most.
Brain arterioles	Slight contraction → constriction (α).	No known functional innervation.
Systemic veins	Contraction → constriction (α); relaxation → dilation (β_2).	No known functional innervation.
Gallbladder and ducts	Relaxation (β_2).	Contraction.
Stomach and intestines	Decreases motility and tone (α, β_2); contracts sphincters (α).	Increases motility and tone; relaxes sphincters.
Kidney	Constriction of blood vessels → decreased urine volume (α); increased renin secretion (β_1).	No known functional innervation.
Ureter	Increases motility.	Decreases motility.
Spleen	Contraction and discharge of stored blood into general circulation (α).	No known functional innervation.
Urinary bladder	Relaxation of muscular wall (β); contraction of trigone and sphincter (α).	Contraction of muscular wall; relaxation of trigone and sphincter.
Uterus	Inhibits contraction if nonpregnant (β_2); promotes contraction if pregnant (α).	Minimal effect.
Sex organs	In male: contraction of smooth muscle of ductus (vas) deferens, seminal vesicle, prostate → ejaculation.	Vasodilation and erection in both sexes.
Hair follicles, arrector pili muscle	Contraction → erection of hairs.	No known functional innervation.
CARDIAC MUSCLE (HEART)	Increases heart rate and force of atrial and ventricular contractions (β_1).	Decreases heart rate; decreases force of atrial contraction.

[a] Subcategories of β receptors are listed where known.

1. **Receptor.** The receptor is the distal end of a sensory neuron.

2. **Sensory neuron.** This neuron, either a somatic or visceral sensory neuron, conducts nerve impulses to the spinal cord or brain.

3. **Association neurons.** These neurons are found in the central nervous system.

4. **Autonomic motor neurons.**

   (a) Preganglionic neuron. The role of the preganglionic neuron is to convey motor impulses from the brain or spinal cord to an autonomic ganglion.

   (b) Postganglionic neuron. The second autonomic motor neuron conducts motor impulses from an autonomic ganglion to the visceral effector.

5. **Visceral effector.** A visceral effector is smooth muscle, cardiac muscle, or a gland. Alteration of activity in the effector is the response.

Visceral sensations often do not reach the cerebral cortex to give rise to conscious perceptions. Under normal conditions, you are not aware of muscular contractions of the digestive organs, heartbeat, changes in the diameter of blood vessels, and pupil dilation and constriction. Your body adjusts such visceral activities by visceral reflex arcs whose integrating centers are in the spinal cord or lower regions of the brain. Among such integrating centers are the cardiac, respiratory, vasomotor, swallowing, and vomiting centers in the medulla and the temperature control center in the hypothalamus. Somatic or visceral sensory neurons deliver input to these centers, and autonomic motor neurons provide output that adjusts activity in the visceral effector, usually without conscious recognition.

CONTROL BY HIGHER CENTERS

The ANS is not a separate nervous system. Axons from many parts of the CNS connect to both the sympathetic and the parasympathetic divisions of the autonomic nervous system and thus exert considerable control over it. The hypothalamus is the major control and integration center of the ANS. Output from the hypothalamus influences autonomic centers in the medulla and spinal cord.

Chapter 18
THE ENDOCRINE SYSTEM

Chapter Contents at a Glance

Student Objectives

1. Define the components of the endocrine system and discuss the functions of the endocrine and nervous system in maintaining homeostasis.

2. Describe how hormones are transported in the blood and how they interact with target cell receptors.

3. Compare the four chemical classes of hormones.

4. Explain the two general mechanisms of hormonal action.

5. Describe the control of hormonal secretions via feedback cycles and give several examples.

6. Describe the release of hormones stored in the posterior pituitary gland.

7. Describe the location, histology, hormones, and functions of the following endocrine glands: pituitary, thyroid, parathyroids, adrenals, pancreas, ovaries, testes, pineal, and thymus.

8. Discuss the symptoms of pituitary dwarfism, giantism, acromegaly, diabetes insipidus, cretinism, myxedema, goiter, tetany, osteitis fibrosa cystica, aldosteronism, Addison's disease, Cushing's

syndrome, adrenogenital syndrome, pheochromocytoma, diabetes mellitus, and hyperinsulinism.

9. Describe the development of the endocrine system.

10. Describe the effects of aging on the endocrine system.

11. Define the general adaptation syndrome (GAS) and compare homeostatic responses and stress responses.

round age 12, as they enter puberty, boys and girls start to develop striking differences in physical appearance and behavior. Perhaps no other period in life so clearly shows the impact of the nervous and endocrine systems in directing development and regulating body functions. Changes in the brain and pituitary gland markedly increase the synthesis of new messenger molecules, the sex hormones from the gonads. In girls, fatty tissue starts to accumulate in the breasts and hips. At the same time, or a little later in boys, protein synthesis increases; muscle mass builds; and the longer, larger vocal cords produce a lower-pitched voice. These changes provide just a few examples of the powerful influence of secretions from endocrine glands.

ENDOCRINE GLANDS

The body contains two kinds of glands: exocrine and endocrine. **Exocrine glands** secrete their products into ducts, and the ducts carry the secretions into body cavities, into the lumina of various organs, or to the outer surface of the body. Exocrine glands include sudoriferous (sweat), sebaceous (oil), mucous, and digestive glands. **Endocrine glands,** by contrast, secrete their products (hormones) into the extracellular space around the secretory cells, rather than into ducts. The secretion then diffuses into capillaries and is carried away by the blood. The endocrine glands of the body (Fig. 18.1) constitute the **endocrine system** and include the pituitary (hypophysis), thyroid, parathyroids, adrenals (suprarenals), and pineal (epiphysis cerebri). In addition, several organs of the body contain endocrine tissue but are not endocrine glands exclusively. These include the hypothalamus, thymus, pancreas, ovaries, testes, kidneys, stomach, liver, small intestine, skin, heart, and placenta.

The science concerned with the structure and functions of the endocrine glands and the diagnosis and treatment of disorders of the endocrine system is called **endocrinology** (en′-dō-kri-NOL-ō-jē; *endo* = within; *crin* = to secrete; *logos* = study of).

The developmental anatomy of the endocrine system is considered later in the chapter.

COMPARISON OF NERVOUS AND ENDOCRINE SYSTEMS

Together, the nervous and endocrine systems coordinate functions of all body systems. The nervous system controls homeostasis through nerve impulses (action potentials) conducted along axons of neurons. At axon terminals, impulses trigger release of neurotransmitter molecules. The result is either excitation or inhibition of specific other neurons, muscle fibers (cells), or gland cells. In contrast, the endocrine system releases its messenger molecules, called **hormones** (*hormon* = to urge on), into the bloodstream.

The circulating blood then delivers hormones to virtually all cells throughout the body.

The body could not maintain homeostasis if these two systems were to pull in opposite directions. The nervous and endocrine systems are coordinated as an interlocking supersystem, often referred to as the **neuroendocrine system**. Certain parts of the nervous system stimulate or inhibit the release of hormones. Hormones, in turn, may promote or inhibit the generation of nerve impulses. And several molecules act as hormones in some locations and as neurotransmitters in others.

The nervous system causes muscles to contract and glands to secrete either more or less of their product. The endocrine system alters metabolic activities, regulates growth and development, and guides reproductive processes. Thus it not only helps regulate the activity of smooth and cardiac muscle and some glands, it significantly affects virtually all other tissues as well.

Nerve impulses tend to produce their effects within a few milliseconds. While some hormones can act within seconds, others can take up to several hours or more to bring about their responses. Also, the effects of stimulating the nervous system are generally brief compared with those of the endocrine system.

OVERVIEW OF HORMONE EFFECTS

Although the effects of hormones are many and varied, their actions can be categorized into seven broad areas. Hormones

1. Regulate the chemical composition and volume of the internal environment (extracellular fluid).
2. Help regulate metabolism and energy balance.
3. Help regulate contraction of smooth and cardiac muscle fibers and secretion by glands.
4. Help maintain homeostasis despite emergency environmental disruptions such as infection, trauma, emotional stress, dehydration, starvation, hemorrhage, and temperature extremes.
5. Regulate certain activities of the immune system.
6. Play a role in the smooth, sequential integration of growth and development.
7. Contribute to the basic processes of reproduction, including gamete (egg and sperm) production, fertilization, nourishment of the embryo and fetus, delivery, and nourishment of the newborn.

HORMONES

Hormones have powerful effects when present in very low concentration. As a rule, most of the over 50 or so hormones affect only a few types of cells. Why is it that some cells respond to a particular hormone and others do not?

FIGURE 18.1 Location of many endocrine glands, organs containing endocrine tissue, and associated structures.

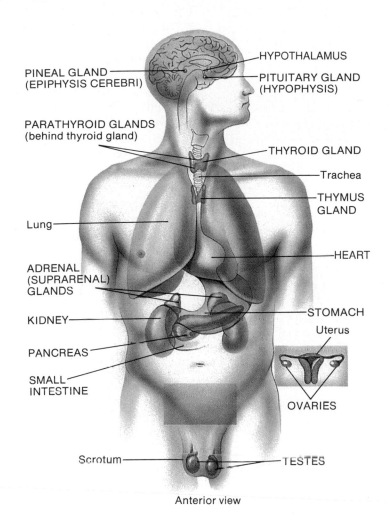

PINEAL GLAND
(EPIPHYSIS CEREBRI)

HYPOTHALAMUS

PITUITARY GLAND
(HYPOPHYSIS)

PARATHYROID GLANDS
(behind thyroid gland)

THYROID GLAND

Trachea

THYMUS
GLAND

Lung

HEART

ADRENAL
(SUPRARENAL)
GLANDS

KIDNEY

STOMACH

Uterus

PANCREAS

SMALL
INTESTINE

OVARIES

Scrotum

TESTES

Anterior view

Question: What is the basic difference between an endocrine gland and an exocrine gland?

Hormone Receptors

Although a given hormone travels throughout the body in the blood, it affects only specific cells called **target cells.** Hormones, like neurotransmitters, influence their target cells by chemically binding to large protein or glycoprotein molecules called **receptors.** Only the target cells of a certain hormone have receptors that bind and recognize that hormone. For example, thyroid-stimulating hormone (TSH) interacts with receptors on the surface of cells of the thyroid gland, but it does not bind to cells of the ovaries because ovarian cells do not have TSH receptors.

Receptors, like other cellular proteins, are constantly synthesized and broken down. Generally, a target cell has 2000 to 100,000 receptors for a particular hormone. When a hormone (or neurotransmitter) is present in excess, the number of receptors may decrease. This effect is called **down-regulation.** For example, when cells of the testes are exposed to a high concentration of luteinizing hormone (LH), the number of LH receptors decreases. Down-regulation thus decreases the responsiveness of target cells to the hormone. On the other hand, when a hormone (or neurotransmitter) is deficient, the number of receptors may increase, a process known as **up-regulation.** Up-regulation makes a target tissue more sensitive to a hormone or neurotransmitter.

Circulating and Local Hormones

Hormones that pass into the blood and act on distant target cells are called **circulating hormones** or **endocrines.** Hormones that act on target cells close to their site of release are called **local hormones. Paracrines,** such as histamine,

FIGURE 18.2 Comparison of circulating hormones (endocrines) and local hormones (autocrines and paracrines).

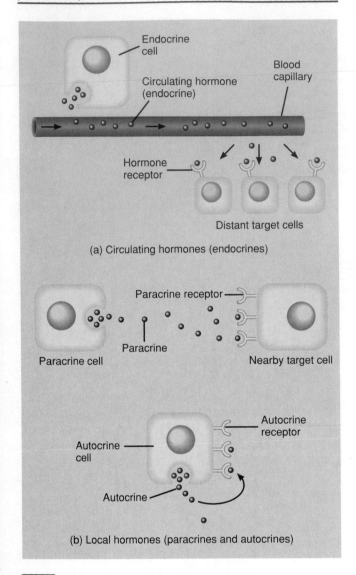

(a) Circulating hormones (endocrines)

(b) Local hormones (paracrines and autocrines)

Question: In the stomach, one stimulus for secretion of hydrochloric acid by parietal cells is the release of histamine by neighboring mast cells. Is histamine an endocrine, autocrine, or paracrine in this situation?

are local hormones that act on neighboring cells. **Autocrines,** such as interleukin-2, are local hormones that act on the same cell that secreted them. Figure 18.2 compares the sites of action of circulating hormones, paracrines, and autocrines. Local hormones usually are inactivated quickly. Circulating hormones may linger in the blood and exert their effects for a few minutes or occasionally a few hours after secretion. In time, circulating hormones are inactivated by the liver and excreted by the kidneys. In cases of kidney or liver failure, excessive buildup of hormones in the blood may cause additional problems.

Chemistry of Hormones

Chemically, there are four principal classes of hormones (see Exhibit 18.1): (1) steroids, (2) biogenic amines, (3) proteins and peptides, and (4) eicosanoids.

1. **Steroids.** All steroid hormones are derived from cholesterol and share the same four-ring structure. They are synthesized on smooth (agranular) endoplasmic reticulum. The shape of each steroid hormone, however, is subtly different because different side groups are attached at various sites on the rings. These small differences in side groups allow for a surprising diversity of function. Endocrine tissues that secrete steroid hormones all are derived from mesoderm.

2. **Biogenic amines.** Structurally, these are the simplest hormone molecules. Several are synthesized by modifying the amino acid tyrosine. Examples are two thyroid hormones (T_3 and T_4) secreted by the thyroid gland and two catecholamines (epinephrine and norepinephrine) secreted by the medulla (inner zone) of the adrenal gland. Histamine is synthesized from the amino acid histidine by mast cells and platelets. Serotonin and melatonin derive from tryptophan.

3. **Peptides and proteins.** These hormones consist of chains of amino acids, anywhere from three to about 200. Peptide and protein hormones are synthesized on rough (granular) endoplasmic reticulum. Some protein hormones, for example, thyroid-stimulating hormone (TSH), have attached carbohydrate groups. Thus they are glycoproteins.

4. **Eicosanoids.** The most recently discovered group of chemical mediators are the eicosanoids (ī-KŌ-sa-noids; *eikosi* = twenty), which are derived from a 20-carbon fatty acid called arachidonic acid. The two major types of eicosanoids are **prostaglandins** and **leukotrienes.** The eicosanoids are important local hormones, but they may also act as circulating hormones.

Exhibit 18.1 contains a summary of the chemical classes of hormones, examples of each, and sites of production.

CLINICAL APPLICATION

ADMINISTERING HORMONES

Steroid hormones, such as cortisol, and thyroid hormones are effective when taken by mouth. They are not split apart during digestion and easily cross the intestinal lining because they are lipid-soluble. Peptide and protein hormones, such as insulin, however, are not effective oral medications because digestive enzymes destroy them by breaking their peptide bonds. This is why people who need insulin must take it by injection.

EXHIBIT 18.1

CHEMICAL CLASSES OF HORMONES, EXAMPLES, AND SITES OF PRODUCTION

Chemical Class	Examples	Where Produced
Steroids	Aldosterone, cortisol, and androgens (male sex hormones).	Adrenal cortex.
	Calcitriol.	Kidneys.
	Testosterone.	Testes.
	Estrogens and progesterone (female sex hormones).	Ovaries.

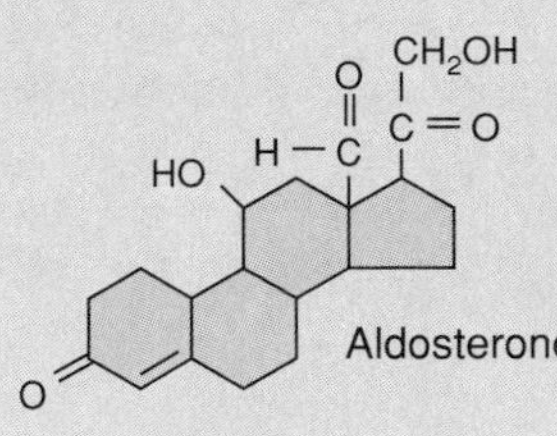

Aldosterone

Biogenic amines	T_3 and T_4 (thyroid hormones).	Thyroid (follicular cells).
	Epinephrine and norepinephrine (catecholamines).	Adrenal medulla.
	Histamine.	Mast cells in connective tissues.

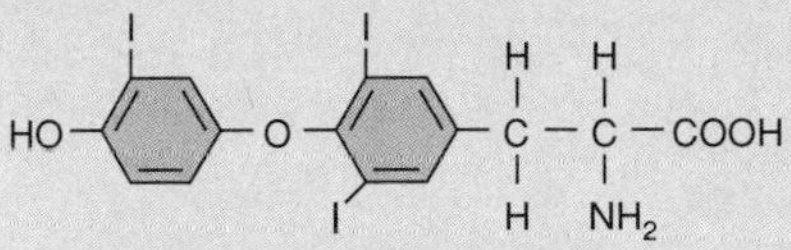

Triiodothyronine (T_3)

Peptides and proteins	All hypothalamic releasing and inhibiting hormones (such as thyrotropin releasing hormone).	Hypothalamus.
	Oxytocin, antidiuretic hormone.	Hypothalamus.
	All anterior pituitary hormones (such as thyroid-stimulating hormone and growth hormone).	Anterior pituitary.
	Insulin, glucagon.	Pancreas.
	Parathyroid hormone.	Parathyroids.
	Calcitonin.	Thyroid (parafollicular cells).
	Hormones that regulate digestion (such as gastrin, secretin, cholecystokinin, and gastric inhibitory peptide).	Stomach and small intestine.

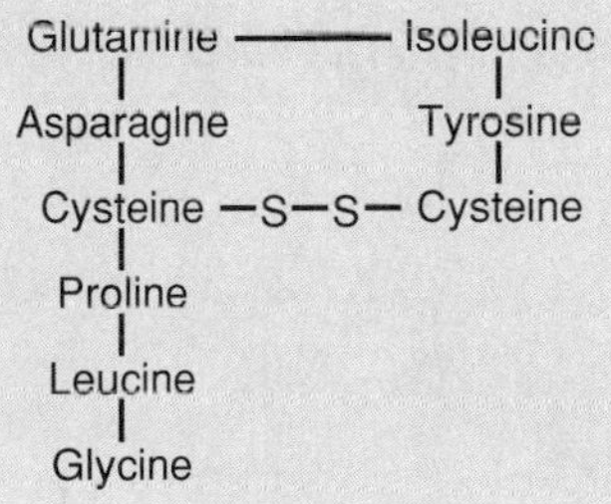

Glutamine ——— Isoleucine
Asparagine   Tyrosine
Cysteine —S—S— Cysteine
Proline
Leucine
Glycine

Oxytocin

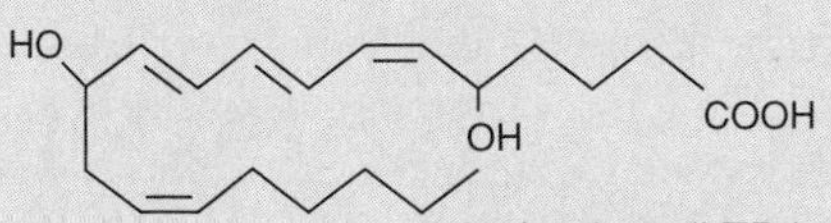 **Eicosanoids**	Prostaglandins, leukotrienes.	All cells except red blood cells (different cells produce different eicosanoids).

A leukotriene (LTB_4)

Hormone Transport in Blood

Endocrine glands are among the most highly vascularized tissues in the body. Catecholamine, peptide, and protein hormones, which by themselves are soluble in watery blood plasma, circulate in free form (not attached to plasma proteins). Upon entering the blood, most lipid-soluble steroid and thyroid hormone molecules, however, attach to specific **transport proteins,** which are synthesized by the liver. These "hormone shuttle buses" have three functions. They (1) improve the transportability of the lipid-soluble hormones by making them temporarily water-soluble; (2) retard loss of the small hormone molecules through the filtering mechanism in the kidneys, thus slowing the rate of hormone loss in the urine; and (3) provide a ready reserve of hormone, already present in the bloodstream. In general, 0.1 to 10% of a lipid-soluble hormone is not bound to a transport protein. This **free fraction** diffuses out of capillaries, binds to receptors, and triggers responses. As free hormone molecules leave the blood and bind to their receptors, transport proteins release new ones.

MECHANISMS OF HORMONE ACTION

The response to a hormone depends on both the hormone and the target cell. Various target cells respond differently to the same hormone. Insulin, for example, stimulates synthesis of glycogen in liver cells but synthesis of triglycerides in adipose cells. Often, the response to a hormone is synthesis of new molecules, as in the examples for insulin. Other hormone effects include changing the permeability of the plasma membrane, stimulating transport of a substance into or out of the target cells, altering the rate of specific metabolic reactions, or causing contraction of smooth or cardiac muscle. In part, these varied effects of hormones are possible because there are several different mechanisms of hormone action. To begin with, hormones bind to and activate their specific receptors in two quite different ways.

Activation of Intracellular Receptors

Steroid hormones and thyroid hormones easily pass through plasma membranes because they are lipid-soluble. Upon entering a target cell, the hormone binds to and activates an intracellular receptor, commonly located within the nucleus (Fig. 18.3). The activated receptor then alters gene expression; that is, it turns specific genes of the nuclear DNA on or off. As the DNA is transcribed, new messenger RNA (mRNA) forms, leaves the nucleus, and enters the cytosol. There, it directs synthesis of new proteins, usually enzymes, on the ribosomes. The new gene products cause the physiological responses that are characteristic of the hormone.

FIGURE 18.3 Activation of intracellular receptors by lipid-soluble hormones (steroid and thyroid hormones).

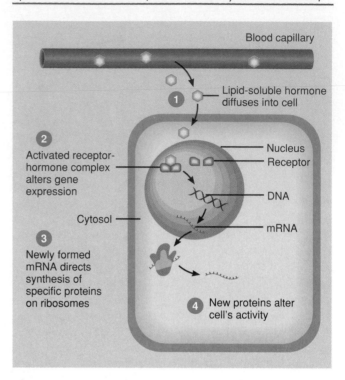

Question: How quickly do you think a steroid hormone can start to exert its effects?

Activation of Plasma Membrane Receptors

Catecholamine, peptide, and protein hormones are not lipid-soluble and thus cannot diffuse through the phospholipid bilayer of the plasma membrane to attach to intracellular receptors. The receptors for these water-soluble hormones instead are at the external surface of the plasma membrane (Fig. 18.4). After a water-soluble hormone is released from an endocrine gland, it circulates in the blood, reaches a target cell, and brings a specific message to that cell. Since such a hormone can deliver its message only to the plasma membrane, it is called the **first messenger.** A **second messenger** is needed to relay the message inside the cell where hormone-stimulated responses can take place. Eicosanoids, some neurotransmitters, neuropeptides, and several sensory transduction mechanisms (for example, vision, see page 484) also act through second messengers.

Second Messengers

The best known second messenger is **cyclic AMP (cAMP).** It is synthesized from ATP, the main energy-providing chemical in cells. The enzyme that catalyzes formation of

FIGURE 18.4 Activation of plasma membrane receptors by water-soluble hormones (catecholamines, peptides, and proteins).

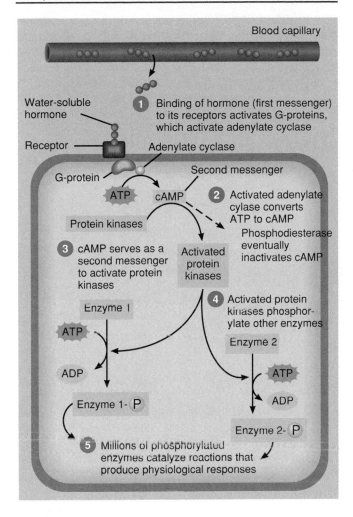

Question: Why is cAMP called a "second messenger"?

cal responses through the _increased_ synthesis of cAMP. These include antidiuretic hormone (ADH), oxytocin (OT), follicle-stimulating hormone (FSH), luteinizing hormone (LH), thyroid-stimulating hormone (TSH), adrenocorticotropic hormone (ACTH), calcitonin (CT), parathyroid hormone (PTH), glucagon, epinephrine, norepinephrine (NE), and hypothalamic releasing hormones. In other cases, such as growth hormone inhibiting hormone (somatostatin), the level of cyclic AMP _decreases_ in response to binding of hormone to its receptor.

Besides cAMP, several other substances are known second messengers. These include calcium ions (Ca^{2+}), cGMP (cyclic guanosine monophosphate, a cyclic nucleotide similar to cAMP), inositol trisphosphate (IP_3), and diacylglycerol (DAG). A given hormone or neurotransmitter may even use more than one second messenger.

Role of G-Proteins

In the cyclic AMP second messenger system, hormone receptors do not connect _directly_ to adenylate cyclase. Rather, molecules called **G-proteins** link receptors on the outer surface of the membrane to the adenylate cyclase molecules on the inner surface (Fig. 18.4). Binding of a hormone to its receptor activates many G-protein molecules, which in turn activate molecules of adenylate cyclase. Unless further stimulated by binding of more hormone molecules to receptors, G-proteins slowly become inactivated, thus helping to stop the hormone response. G-proteins are a common feature of most second messenger systems.

CLINICAL APPLICATION

CHOLERA TOXIN AND G-PROTEINS

The toxin produced by cholera bacteria is deadly. It produces massive diarrhea, and an infected person can rapidly die of the resulting dehydration. In epithelial cells lining the intestines, the cholera toxin modifies the G-proteins so they become locked in the activated state, incapable of turning off. As a result, the intracellular cAMP concentration skyrockets. One of the effects of cAMP in these cells is to stimulate an active transport pump that ejects Na^+ from the cells into the lumen of the intestines; water follows the Na^+ by osmosis. Thus cholera toxin causes a huge outflow of Na^+ and water into the fecal material. A simple, yet effective, treatment is to give ample replacement fluids, either intravenously or by mouth.

Protein Kinases

Cyclic AMP does not directly produce a particular physiological response. Instead, it activates one or more enzymes collectively known as **protein kinases,** which may be free in the cytosol or bound to the plasma membrane (Fig. 18.4). Protein kinases are phosphorylating enzymes. That means

cAMP is **adenylate cyclase,** and it is attached to the inner surface of the plasma membrane. When the first messenger (hormone) binds to its receptor on the outer surface of the membrane, adenylate cyclase on the inner surface is activated. Then adenylate cyclase converts ATP into cyclic AMP in the cytosol of the cell (Fig. 18.4). Cyclic AMP acts as a **second messenger** to alter cell function in specific ways. For example, elevation of cAMP causes adipose cells to break down triglycerides and release fatty acids more rapidly but stimulates thyroid cells to secrete more thyroid hormone. After a brief period of time, an enzyme called **phosphodiesterase** inactivates cAMP. Thus the cell's response is turned off unless new hormone molecules continue to bind to their receptors at the surface of the plasma membrane.

Many hormones exert at least some of their physiologi-

they remove a phosphate group from ATP and add it to a protein, which is usually another enzyme. Phosphorylation activates some enzymes and inactivates others. It is like an on–off switch. The result of phosphorylating a particular enzyme could be regulation of other enzymes, secretion, protein synthesis, or a change in plasma membrane permeability.

Different protein kinases exist within different target cells and within different organelles of the same target cell. Thus, one protein kinase might be involved with glycogen synthesis, another with lipid breakdown, another with protein synthesis, and so on. Moreover, protein kinases can inhibit as well as activate enzymes. For example, some of the kinases unleashed when epinephrine binds to liver cells inactivate an enzyme needed for glycogen synthesis.

Amplification of Hormone Effects

Hormones that bind to plasma membrane receptors can induce their effects at very low concentrations because they initiate a cascade (chain reaction) of events. Each step of the casade multiplies or amplifies the initial effect. For example, when a single molecule of epinephrine binds to its receptor on a liver cell, it may activate a hundred or so G-proteins. In turn, each G-protein activates an adenylate cyclase. If each adenylate cyclase produces even 1000 cAMP, then 100,000 of these second messengers will be liberated inside the cell. Each cAMP may activate a protein kinase, which in turn can act on hundreds or thousands of substrate molecules. Some of the kinases phosphorylate and activate a key enzyme needed for glycogen breakdown. The end result of epinephrine binding to its receptor is the breakdown of millions of glycogen molecules into glucose.

Hormonal Interactions

The responsiveness of a target cell to a hormone depends on the hormone's concentration and the number of receptors (down-regulation and up-regulation). The manner in which hormones interact with other hormones is also important. One type of interaction is called a **permissive effect.** In this interaction, the effect of one hormone on a target cell requires a previous or simultaneous exposure to another hormone(s). Such previous exposure enhances the response of a target cell, usually by up-regulation of receptors. For example, an increase in estrogens can bring about an increase in the number of progesterone receptors. Both hormones act to promote uterine growth to prepare for the possible implantation of a fertilized egg. During the female reproductive cycle, secretion of estrogens precedes that of progesterone, thus allowing progesterone to have a greater effect.

Another type of hormonal interaction is known as a **synergistic effect.** Here, two or more hormones complement each other's actions and both are needed for full expression of the hormone effects. For example, the production, secretion, and ejection of milk by the mammary glands require the synergistic effects of estrogens, progesterone, prolactin, and oxytocin.

A final example of a hormonal interaction is an **antagonistic effect.** Here, the effect of one hormone on a target cell is opposed by another hormone. An example is insulin, which lowers blood sugar level, and glucagon, which raises it (see Fig. 18.24).

CONTROL OF HORMONAL SECRETIONS

Most hormones are released in short bursts, with little or no secretion between bursts. When properly stimulated, an endocrine gland will release its hormone in more frequent bursts, and thus blood level of the hormone increases. In the absence of stimulation, bursts are minimal or inhibited, and blood level of the hormone decreases. Regulation of secretion normally maintains homeostasis and prevents overproduction or underproduction of a particular hormone. Unfortunately, there are times when the regulating mechanisms do not operate properly, and hormonal levels are excessive or deficient. When this happens, disorders result; several of these are discussed later.

Hormone secretion by endocrine glands is stimulated and inhibited by (1) signals from the nervous system, (2) chemical changes in the blood, and (3) other hormones. For example, nerve impulses to the adrenal medulla regulate release of epinephrine, blood Ca^{2+} level regulates secretion of parathyroid hormone, and a hormone from the anterior pituitary gland (adrenocorticotropic hormone) stimulates release of cortisol by the adrenal cortex. We will give more examples of each of these three mechanisms of hormonal regulation in later sections of this chapter.

Most often, negative feedback systems maintain homeostasis of hormonal secretions. As we discuss the effects of various hormones in this chapter, we will also describe how the secretions of the hormones are controlled. At that time, you will be able to see which type of negative feedback system is operating. Occasionally a positive feedback system contributes to regulation of hormone secretion. One example occurs during childbirth. The hormone oxytocin stimulates contractions of the uterus. Uterine contractions, in turn, stimulate more oxytocin release (see Fig. 18.11). In cases of positive feedback, the response intensifies the initiating stimulus.

HYPOTHALAMUS AND PITUITARY GLAND (HYPOPHYSIS)

For many years the **pituitary gland** or **hypophysis** (hī-POF-i-sis) was called the "master" endocrine gland because it secretes several hormones that control other endocrine glands. We now know that the pituitary itself has a master—the **hypothalamus.** This small region of the brain, below the two lobes of the thalamus, is the major integrating link between the nervous and endocrine systems (see Fig. 14.1). It receives input from several other regions of the brain, including the limbic system, cerebral cortex, thalamus, and reticular activating system. Also, it receives sensory signals from internal organs and perhaps from the visual system. Painful, stressful, and emotional experiences all cause changes in hypothalamic activity. In turn, the hypothalamus controls the autonomic nervous system and regulates body temperature, thirst, hunger, sexual behavior, and defensive reactions such as fear and rage. Not only is the hypothalamus an important regulatory center in the nervous system, it is also a crucial endocrine gland. Cells in the hypothalamus synthesize at least nine different hormones, and the pituitary gland secretes seven more. Together, they play important roles in regulation of virtually all aspects of growth, development, metabolism, and homeostasis.

The pituitary gland is a pea-sized structure that measures about 1.3 cm (0.5 in.) in diameter. It lies in the sella turcica of the sphenoid bone and is attached to the hypothalamus via a stalklike structure called the **infundibulum** (see Fig. 18.5). The pituitary gland has two anatomically and functionally separate portions. The **anterior pituitary gland (anterior lobe)** accounts for about 75% of the total weight of the gland. It derives from an outgrowth of ectoderm called the hypophyseal (Rathke's) pouch in the roof of the mouth (see Fig. 18.25b). Therefore the anterior pituitary contains many glandular epithelial cells and forms the glandular part of the pituitary.

The **posterior pituitary gland (posterior lobe)** also derives from the ectoderm, but from an outgrowth called the neurohypophyseal bud (see Fig. 18.25b). It contains axons and axon terminals of about 5000 neurons whose cell bodies are located in the supraoptic and paraventricular nuclei of the hypothalamus (see Fig. 18.10). The nerve fibers that terminate in the posterior pituitary are associated with supporting cells called pituicytes.

A third region called the **pars intermedia (intermediate lobe)** atrophies during fetal development and is small in adults (see Fig. 18.25b). Some of its cells are thought to migrate into the anterior pituitary.

Anterior Pituitary Gland (Adenohypophysis)

The **anterior pituitary gland (anterior lobe)** or **adenohypophysis** (ad'-e-nō-hī-POF-i-sis; *adeno* = glandular) secretes hormones that regulate a wide range of bodily activities from growth to reproduction. Release of anterior pituitary hormones is stimulated by **releasing hormones** and suppressed by **inhibiting hormones** from the hypothalamus. These hypothalamic hormones are an important link between the nervous and endocrine systems.

Hypothalamic hormones reach the anterior pituitary through a system of blood vessels that directly connects the two regions. Blood flows from the median eminence of the hypothalamus into the infundibulum and anterior pituitary principally from several **superior hypophyseal** (hī'-pō-FIZ-ē-al) **arteries** (Fig. 18.5). These arteries are branches of the internal carotid and posterior communicating arteries. The superior hypophyseal arteries form the **primary plexus,** a capillary network at the base of the hypothalamus. Releasing and inhibiting hormones are secreted by hypothalamic cells near the median eminence and diffuse into the capillaries of this plexus.

From the primary plexus, blood drains into the **hypophyseal portal veins** that pass down the infundibulum. At the inferior portion of the infundibulum, the veins form a **secondary plexus** of capillaries in the anterior pituitary. This direct route permits hypothalamic hormones to act quickly on the anterior pituitary cells before the hormones are diluted or destroyed in the systemic circulation. Anterior pituitary hormones pass into the secondary plexus and then into the anterior hypophyseal veins for distribution to target tissues throughout the body.

Five principal types of anterior pituitary cells secrete seven major hormones (Fig. 18.6):

1. **Somatotrophs** produce **human growth hormone (hGH),** which stimulates general body growth and regulates aspects of metabolism.
2. **Lactotrophs** synthesize **prolactin (PRL),** which initiates milk production in suitably prepared mammary glands.
3. **Corticotrophs** synthesize **adrenocorticotropic hormone (ACTH),** which stimulates the adrenal cortex to secrete glucocorticoids. Some corticotrophs, remnants of the pars intermedia, also secrete **melanocyte-stimulating hormone (MSH),** which affects skin pigmentation.
4. **Thyrotrophs** produce **thyroid-stimulating hormone (TSH),** which controls the thyroid gland secretions and other activities.
5. **Gonadotrophs** produce two major hormones: **follicle-stimulating hormone (FSH),** which stimulates maturation of ova and secretion of estrogen by the ovaries and production of sperm in the testes, and **luteinizing hormone (LH),** which stimulates other sexual and reproductive activities.

Hormones that influence another endocrine gland are called **tropins** or **tropic hormones** (*trop* = turn on). The anterior pituitary secretes two **gonadotropins,** that is, hormones that regulate the functions of the gonads (ovaries and testes): follicle-stimulating hormone (FSH) and luteinizing hormone (LH). **Thyrotropin** and **corticotropin** are alternate names for thyroid-stimulating hormone (TSH) and adrenocorticotropic hormone (ACTH). The hypothalamic releasing and inhibiting hormones are also called **hypophysiotropic** (hī-POF-is-ē-ō-trō'-pic) **hormones** because they act on the hypophysis (pituitary).

Secretion of anterior pituitary hormones is regulated

FIGURE 18.5 Hypothalamus and pituitary gland and their blood supply.

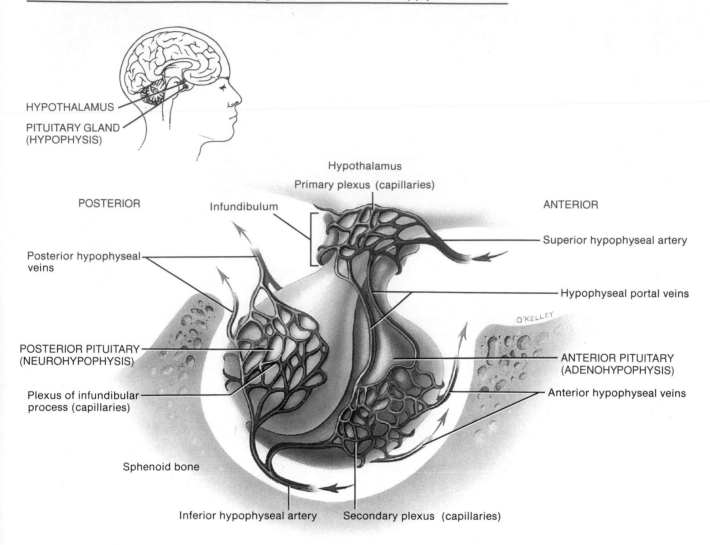

Question: What is the functional importance of the hypophyseal portal veins?

mainly by hypothalamic releasing and inhibiting hormones and negative feedback from target gland hormones. At present, four hypothalamic releasing hormones and three hypo-thalamic inhibiting hormones arc known. At least three anterior pituitary hormones (hGH, prolactin, and MSH) are under a dual system of control; the hypothalamus secretes both releasing and inhibiting hormones. There also is some functional overlap among the releasing hormones. For example, CRH (corticotropin releasing hormone) stimulates secretion of both corticotropin and MSH; TRH (thyrotropin releasing hormone) stimulates secretion of thyrotropin, human growth hormone, and prolactin; and GnRH (gonadotropin releasing hormone) stimulates release of both LH and FSH.

Negative feedback systems decrease the secretory activity of corticotrophs, thyrotrophs, and gonadotrophs when levels of their target gland hormones rise. As an example,

let's examine how T_3 (a thyroid hormone) level regulates the secretory activity of thyrotrophs and hypothalamic cells (Fig. 18.7). When blood level of T_3 (the controlled condition) starts to decrease, the amount of T_3 bound to receptors in hypothalamic cells that secrete thyrotropin releasing hormone (TRH) and in anterior pituitary gland thyrotrophs that secrete thyroid-stimulating hormone (TSH) also decreases. This decrease (input) turns on the genes (control centers) for TRH production in the hypothalamus and TSH production in the anterior pituitary gland. The result is increased secretion of TRH and TSH into the blood. TRH stimulates thyrotrophs to secrete TSH and TSH stimulates thyroid gland cells to step up synthesis and secretion of T_3. Thus thyrotrophs and thyroid cells both are effectors in this negative feedback system. The result is increased T_3 in the blood (response). When the blood level of T_3 normalizes, production of TSH by thyrotrophs declines.

FIGURE 18.6 Cells of the anterior pituitary gland, the hormones they secrete, and the general hormone functions.

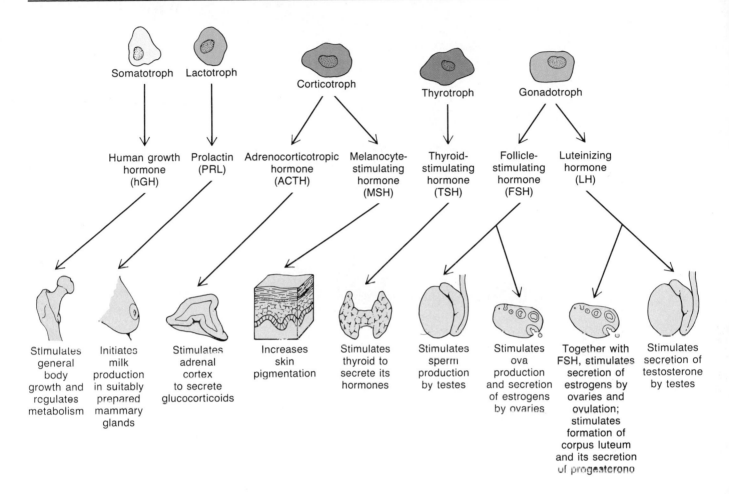

Somatotroph Lactotroph Corticotroph Thyrotroph Gonadotroph

Human growth hormone (hGH) Prolactin (PRL) Adrenocorticotropic hormone (ACTH) Melanocyte-stimulating hormone (MSH) Thyroid-stimulating hormone (TSH) Follicle-stimulating hormone (FSH) Luteinizing hormone (LH)

Stimulates general body growth and regulates metabolism Initiates milk production in suitably prepared mammary glands Stimulates adrenal cortex to secrete glucocorticoids Increases skin pigmentation Stimulates thyroid to secrete its hormones Stimulates sperm production by testes Stimulates ova production and secretion of estrogens by ovaries Together with FSH, stimulates secretion of estrogens by ovaries and ovulation; stimulates formation of corpus luteum and its secretion of progesterone Stimulates secretion of testosterone by testes

Question: Which endocrine glands are regulated by anterior pituitary hormones?

Human Growth Hormone

The most abundant anterior pituitary hormone is **human growth hormone (hGH or GH),** also known as **somatotropin** (sō′-ma-tō-TRŌ-pin). Besides causing body cells to grow, hGH has many effects on metabolism. Generally, hGH (1) stimulates protein synthesis and inhibits protein breakdown, (2) stimulates lipolysis, the breakdown of triglycerides into fatty acids and glycerol, and (3) retards use of glucose (blood sugar) for ATP production. Some effects of hGH are indirect because hGH stimulates the liver to synthesize and secrete small protein hormones called **somatomedins** (sō′-ma-tō-MĒ-dins) or **insulinlike growth factors (IGFs),** which regulate several effects of hGH. Structurally and functionally, somatomedins are similar to insulin. However, their growth-promoting effects are even more potent.

hGH causes cells to grow and multiply by directly increasing the rate at which amino acids enter cells and are used to synthesize proteins. Thus hGH is a hormone of protein anabolism. Also, hGH decreases the breakdown of proteins and the use of amino acids for ATP production. Due to these effects, hGH increases the growth rate of the skeleton and skeletal muscles during childhood and teenage years. In adults, it helps maintain muscle and bone size and promote tissue repair.

hGH stimulates fat catabolism; that is, it causes cells to switch from oxidizing (burning) carbohydrates and proteins to oxidizing fatty acids to produce ATP. It spurs lipolysis in adipose tissue, and it prompts other cells to use the released fatty acids for ATP production. This effect of hGH is most important in periods of fasting or starvation.

Besides affecting protein and fat metabolism, hGH influences carbohydrate metabolism. It decreases glucose uti-

FIGURE 18.7 Negative feedback system for regulation of T_3 (triiodothyronine) concentration in the blood.

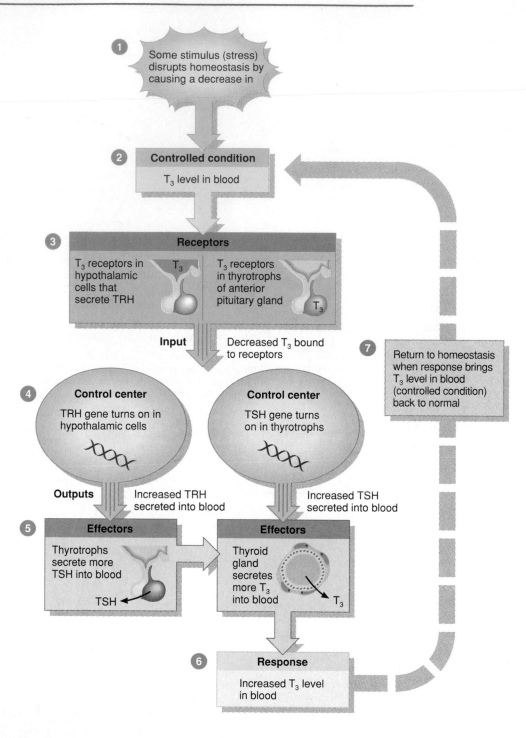

Question: How can you tell that this is a *negative* feedback system?

lization by cells, decreases glucose uptake by cells, and accelerates the rate at which glycogen stored in the liver is converted into glucose. These actions increase the amount of glucose in the blood. For this reason, hGH is said to have an anti-insulin effect, since insulin promotes removal of glucose from the blood into body cells.

Figure 18.8 Regulation of human growth hormone (hGH) secretion.

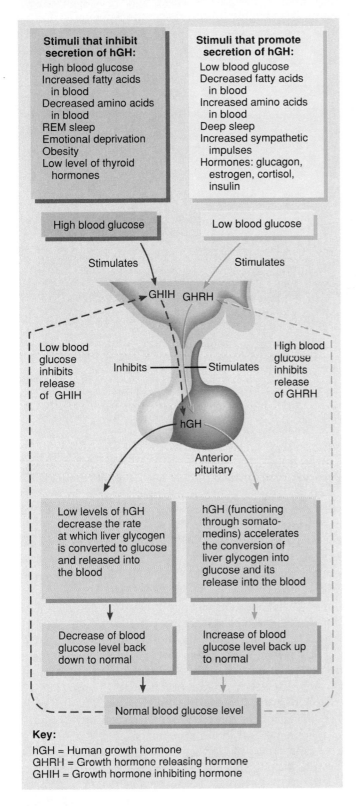

Key:

hGH = Human growth hormone
GHRH = Growth hormone releasing hormone
GHIH = Growth hormone inhibiting hormone

Question: **What condition results from hypersecretion of hGH during childhood?**

DIABETOGENIC EFFECT OF hGH

One symptom of excess hGH is **hyperglycemia** (hī'-per-glī-SĒ-mē-a), high blood glucose concentration. Persistent hyperglycemia, in turn, stimulates the pancreas to continually secrete insulin. Such excessive stimulation, if it lasts for weeks or months, may cause "beta-cell burnout," a greatly decreased capacity of the pancreatic beta cells to synthesize and secrete insulin. Thus, in time, excess secretion of hGH may have a **diabetogenic effect;** that is, it causes diabetes mellitus (lack of insulin activity).

The secretion of hGH from the anterior pituitary is controlled by at least two hypothalamic hormones: **growth hormone releasing hormone (GHRH),** or **somatocrinin,** and **growth hormone inhibiting hormone (GHIH),** or **somatostatin.**

Low blood glucose level (hypoglycemia) stimulates the hypothalamus to secrete GHRH (Fig. 18.8). Upon reaching the anterior pituitary, GHRH stimulates somatotrophs to release hGH. Together, hGH and somatomedins raise blood glucose level. As soon as blood glucose level returns to normal, GHRH secretion shuts off. Hyperglycemia, on the other hand, inhibits hGH secretion. An abnormally high blood glucose level stimulates the hypothalamus to secrete GHIH, which inhibits the release of hGH. As a result, blood glucose level decreases.

Other stimuli that promote secretion of hGH include: decreased fatty acids and increased amino acids in the blood; deep sleep (stages 3 and 4 of NREM sleep); increased activity of the sympathetic division of the autonomic nervous system, such as might occur with stress or vigorous physical exercise; and other hormones, for example, glucagon, estrogens, cortisol, and insulin. Factors that inhibit hGH secretion are: increased fatty acids and decreased amino acids in the blood; REM sleep; emotional deprivation; obesity; low level of thyroid hormones; and hGH itself (through negative feedback).

PITUITARY DWARFISM, GIANTISM, AND ACROMEGALY

Disorders of the endocrine system generally involve either **hyposecretion** (underproduction) or **hypersecretion** (overproduction) of hormones. Hyposecretion of hGH during the growth years causes slow bone growth, and the epiphyseal plates close before normal height is reached. This condition is called **pituitary dwarfism.** Other organs of the body also fail to grow, and the pituitary dwarf is childlike in many physical respects. Treatment requires administration of hGH during childhood before the epiphyseal plates close. An ample supply of hGH is now available to treat such children because it is being produced by bacteria using recombinant DNA techniques. Other conditions may also cause dwarfism, in which case hGH treatment is not helpful.

Hypersecretion of hGH during childhood results in **giantism (gigantism),** an abnormal increase in the length of long bones. As a result, the person is very tall, but body proportions are about normal. Hypersecretion of hGH during adulthood causes **acromegaly** (ak'-rō-MEG-a-lē) (Fig. 18.9). Further lengthening of the long bones cannot occur

because the epiphyseal plates are already closed. Instead, the bones of the hands, feet, cheeks, and jaw thicken. Other tissues also grow. The eyelids, lips, tongue, and nose enlarge, and the skin thickens and develops furrows, especially on the forehead and soles of the feet.

FIGURE 18.9 Photographs of a person with acromegaly.

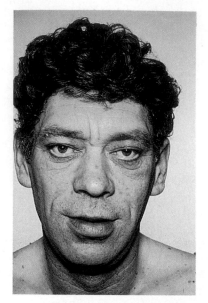

(a) Facial features

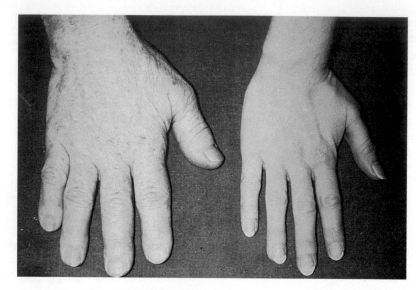

Acromegalic   Normal

(b) Hands

Thyroid-Stimulating Hormone

Thyroid-stimulating hormone (TSH), also called **thyrotropin** (thī-rō-TRŌ-pin), stimulates the synthesis and secretion of two hormones: triiodothyronine (T_3) and thyroxine (T_4), both produced by the thyroid gland. Secretion is controlled by **thyrotropin releasing hormone (TRH)** from the hypothalamus. Release of TRH depends on blood levels of TSH, T_3, blood glucose level, and the body's metabolic rate, among other factors, and operates according to a negative feedback system (see Fig. 18.7).

Follicle-Stimulating Hormone

In females, **follicle-stimulating hormone (FSH)** is transported from the anterior pituitary by the blood to the ovaries, where it initiates the development of egg-containing follicles each month. Follicles are saclike arrangements of secreting cells. FSH also stimulates follicular cells to secrete estrogens (female sex hormones). In males, FSH stimulates sperm production in the testes. **Gonadotropin releasing hormone (GnRH)** from the hypothalamus stimulates FSH release. GnRH and FSH release is suppressed by estrogens in the female and by testosterone, the principal male sex hormone, through negative feedback systems.

Luteinizing Hormone

In females, **luteinizing** (LOO-tē-in'-īz-ing) **hormone (LH),** together with FSH, stimulates estrogen secretion by ovarian cells and brings about the release of a secondary oocyte (future ovum) by the ovary, a process called ovulation. LH also stimulates formation of the corpus luteum (ovarian structure formed after ovulation) in the ovary and the secretion of progesterone (another female sex hormone) by the corpus luteum. Estrogens and progesterone prepare the uterus for implantation of a fertilized ovum and help prepare the mammary glands for milk secretion. In males, LH stimulates the interstitial cells in the testes to develop and secrete large amounts of testosterone. Because of this function, LH is also referred to as **interstitial cell-stimulating hormone (ICSH)** in males. Secretion of LH, like that of FSH, is controlled by GnRH. GnRH agonists (compounds that mimic GnRH action) are used therapeutically to stimulate the gonads when they are functioning at too low a level. Antagonists of GnRH may be used to inhibit gonadal function when it is thought to be excessive (for example, in precocious puberty) or when other conditions (such as breast or prostate cancer) might be helped by suppressing gonadal secretions. Surprisingly, some GnRH agonists also exert a suppressive effect on gonadal functions when they are given continuously for more than two weeks.

Prolactin

Prolactin (PRL), or **lactogenic hormone,** together with other hormones, initiates and maintains milk secretion by the mammary glands. The actual ejection of milk by the mammary glands depends on the hormone oxytocin, which is released from the posterior pituitary. Together, milk secretion and ejection are referred to as **lactation.** By itself, prolactin has only a weak effect; the mammary glands require preparation by estrogens, progesterone, glucocorticoids, human growth hormone (hGH), thyroxine, and insulin. When the mammary glands have been primed by these hormones, PRL brings about milk secretion.

The hypothalamus secretes both inhibitory and excitatory hormones that regulate PRL secretion. **Prolactin inhibiting hormone (PIH),** which is dopamine, inhibits the release of PRL from the anterior pituitary. As the levels of estrogens and progesterone fall during the late secretory phase of the menstrual cycle, the secretion of PIH diminishes and the blood level of prolactin rises. Breast tenderness just before menstruation may be caused by elevated PRL. However, the high PRL level does not last long enough for milk production to start. As the menstrual cycle begins anew, and the level of estrogens rises again, PIH is again secreted and the PRL level drops.

Prolactin level rises during pregnancy, apparently stimulated by a hypothalamic hormone called **prolactin releasing hormone (PRH).** A nursing infant causes a reduction in hypothalamic secretion of PIH. The function of PRL is not known in males, but its hypersecretion causes impotence (inability to have an erection of the penis). In females, hypersecretion of PRL causes absence of menstrual cycles.

Melanocyte-Stimulating Hormone

Melanocyte-stimulating hormone (MSH) increases skin pigmentation by stimulating the dispersion of melanin granules in melanocytes in amphibians. Its exact role in humans is unknown. However, continued administration of MSH for several days does produce a darkening of the skin, and without MSH the skin may be pallid. Two hypothalamic hormones regulate MSH secretion. **MSH releasing hormone (MRH)** promotes MSH release, whereas **MSH inhibiting hormone (MIH)** suppresses release.

Adrenocorticotropic Hormone

Corticotrophs synthesize a large protein precursor molecule called **pro-opiomelanocortin** (prō-ō'-pē-ō-mel-an-ō-KOR-tin) or POMC. POMC is also present in several regions of the brain and the pars intermedia of the pituitary gland.

When this molecule is cleaved into fragments, it can give rise to adrenocorticotropic hormone (ACTH), melanocyte-stimulating hormone (MSH), beta-endorphin, and beta-lipotropin (β-LPH). In the anterior pituitary, the major product cleaved from POMC is **adrenocorticotropic hormone (ACTH)** or **adrenocorticotropin** (ad-rē'-nō-kor'-ti-kō-TRŌ-pin). ACTH controls the production and secretion of hormones called glucocorticoids by the cortex (outer portion) of the adrenal glands. The hypothalamic hormone called **corticotropin releasing hormone (CRH)** stimulates secretion of ACTH by corticotrophs. Stress-related stimuli, such as low blood glucose or physical trauma, and a substance produced by macrophages called interleukin-1 (IL-1) also stimulate release of ACTH. Glucocorticoids cause negative feedback inhibition of both CRH and ACTH release.

A summary of anterior pituitary hormones, their principal actions, associated hypothalamic regulating hormones, and selected related disorders is presented in Exhibit 18.2.

Posterior Pituitary Gland (Neurohypophysis)

Although the **posterior pituitary gland (posterior lobe),** or **neurohypophysis,** does not *synthesize* hormones, it does *store* and *release* two hormones. It consists of cells called **pituicytes** (pi-TOO-i-sītz), which are similar in appearance to neuroglia, and axon terminals of secretory neurons of the hypothalamus that are called **neurosecretory cells** (Fig. 18.10). The cell bodies of the neurosecretory cells are in the paraventricular and supraoptic nuclei of the hypothalamus. Their axons form the **supraopticohypophyseal tract,** which extends from the hypothalamus to the posterior pituitary and terminates near blood capillaries in the posterior pituitary. The cell bodies of the neurosecretory cells produce two hormones: **oxytocin (OT)** and **antidiuretic hormone (ADH).**

After their production in the cell bodies of neurosecretory cells, the hormones are packed into secretory vesicles, which move by fast axonal transport (see page 353) to the axon terminals in the posterior pituitary. Nerve impulses that propagate along the axon and reach the axon terminals trigger exocytosis of the secretory vesicles. The released OT and ADH then diffuse into nearby capillaries.

The blood supply to the posterior pituitary is from the **inferior hypophyseal arteries** (see Fig. 18.5), derived from the internal carotid arteries. In the posterior pituitary, the inferior hypophyseal arteries form a plexus of capillaries called the **plexus of the infundibular process.** From this plexus, hormones pass into the **posterior hypophyseal veins** for distribution to tissue cells.

EXHIBIT 18.2

SUMMARY OF ANTERIOR PITUITARY GLAND HORMONES, PRINCIPAL ACTIONS, ASSOCIATED HYPOTHALAMIC REGULATING HORMONES, AND SELECTED DISORDERS

Hormone	Principal Actions	Associated Hypothalamic Hormones	Selected Disorders
Human growth hormone (hGH)	Growth of body cells; protein anabolism; elevation of blood glucose concentration.	Growth hormone releasing hormone (GHRH); growth hormone inhibiting hormone (GHIH). Secretion also stimulated by thyrotropin releasing hormone (TRH).	Hyposecretion of hGH during the growth years results in pituitary dwarfism; hypersecretion of hGH during the growth years results in giantism; hypersecretion of hGH during adulthood results in acromegaly.
Thyroid-stimulating hormone (TSH)	Controls secretion of thyroid hormones by thyroid gland.	Thyrotropin releasing hormone (TRH).	Hypersecretion of thyroid hormones through the action of TSH causes Graves' disease (discussed later).
Adrenocorticotropic hormone (ACTH)	Controls secretion of some hormones by adrenal cortex (mainly cortisol).	Corticotropin releasing hormone (CRH).	Hyposecretion of glucocorticoids through the action of ACTH results in Addison's disease (discussed later).
Follicle-stimulating hormone (FSH)	In females, initiates development of ova and induces ovarian secretion of estrogens. In male, stimulates testes to produce sperm.	Gonadotropin releasing hormone (GnRH).	
Luteinizing hormone (LH) [also called interstitial cell-stimulating hormone (ICSH) in males]	In females, together with FSH, stimulates ovulation and formation of corpus luteum, which secretes estrogens and progesterone. In males, stimulates interstitial cells in testes to develop and produce testosterone.	Gonadotropin releasing hormone (GnRH).	
Prolactin (PRL)	In females, together with other hormones, initiates and maintains effect of luteinizing hormone in promoting milk secretion by the mammary glands.	Prolactin inhibiting hormone (PIH); prolactin releasing hormone (PRH).	Hypersecretion causes impotence (inability to attain a penile erection) in males and amenorrhea (absence of menstrual cycles) in females.
Melanocyte-stimulating hormone (MSH)	Stimulates dispersion of melanin granules in melanocytes.	Melanocyte-stimulating hormone releasing hormone (MRH); melanocyte-stimulating hormone inhibiting hormone (MIH).	

FIGURE 18.10 Supraopticohypophyseal tract.

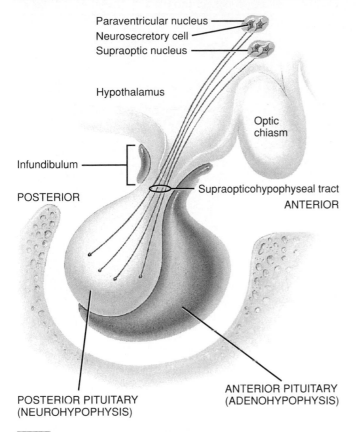

Question: How are the supraopticohypophyseal tract and the hypophyseal portal veins similar? Different?

Oxytocin

During and after delivery of a baby, **oxytocin** (ok′-sē-TŌ-sin; *oxys* = swift; *tokos* = childbirth), or **OT,** has two target tissues, the mother's uterus and breasts. During delivery, it enhances contraction of smooth muscle cells in the wall of the uterus. After birth, it stimulates milk ejection ("letdown") from the mammary glands in response to the mechanical stimulus provided by a suckling infant. The function of OT in males and in nonpregnant females is not clear. Animal experiments have suggested that it has actions within the brain that foster parental caretaking behavior toward young offspring. It may also be responsible, in part, for the feelings of sexual pleasure during and after intercourse.

During labor and delivery, OT is released in large quantities (Fig. 18.11). When labor contractions begin, the baby's head or body distends (stretches) the cervix (inferior narrow portion) of the uterus. Stretch receptors in the cervix send sensory impulses to the hypothalamus. The nerve impulses cause the posterior pituitary to release OT into the blood. It is then carried by the blood to the uterus, where it strengthens uterine contractions. As the contractions

FIGURE 18.11 Regulation of the secretion of oxytocin (OT) during labor and delivery of a baby.

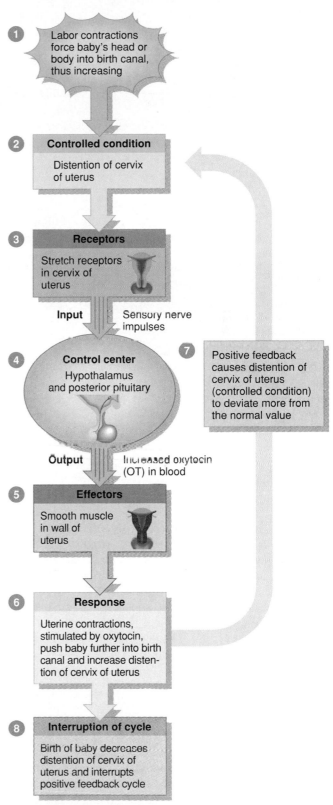

① Labor contractions force baby's head or body into birth canal, thus increasing

② **Controlled condition**

Distention of cervix of uterus

③ **Receptors**

Stretch receptors in cervix of uterus

Input   Sensory nerve impulses

④ **Control center**

Hypothalamus and posterior pituitary

⑦ Positive feedback causes distention of cervix of uterus (controlled condition) to deviate more from the normal value

Output   Increased oxytocin (OT) in blood

⑤ **Effectors**

Smooth muscle in wall of uterus

⑥ **Response**

Uterine contractions, stimulated by oxytocin, push baby further into birth canal and increase distention of cervix of uterus

⑧ **Interruption of cycle**

Birth of baby decreases distention of cervix of uterus and interrupts positive feedback cycle

Question: What are the two physiological roles of OT?

become more forceful, the resulting sensory impulses stimulate the synthesis and secretion of more OT. Thus a positive feedback cycle is established. With birth of the infant, the cycle is broken because cervical distention suddenly lessens. Note that the input part of the cycle is neural, whereas the output part is hormonal. This is an example of a **neuroendocrine reflex.**

OT affects milk ejection by another neuroendocrine reflex. Milk formed by the glandular cells of the breasts is stored until the baby begins active suckling. Stimulation of touch receptors in the nipple initiates sensory impulses to the hypothalamus. In response, secretion of OT from the posterior pituitary quickens. Carried by the bloodstream to the mammary glands, OT stimulates smooth muscle cells around the glandular cells and ducts to contract and eject milk. This sequence is called the **milk ejection (let-down) reflex.** Ejection of milk starts slowly, about 30 seconds to 1 minute after nursing begins. Even in cases of ejection failure, infants can still obtain one-third of the breast's milk. Stimuli other than suckling, such as hearing the baby's cry or touching the genitals, also can trigger OT release and milk ejection. The suckling stimulation that produces the release of OT also inhibits the release of PIH. This results in an increased secretion of prolactin, which maintains lactation.

Years before OT was identified, it was common practice in midwifery to let a first-born twin nurse at the mother's breast to speed the birth of the second child. Even after a single birth, this practice promotes expulsion of the placenta (afterbirth) and helps the uterus regain its smaller size. Synthetic OT (Pitocin) often is given to induce labor or to increase uterine tone and control hemorrhage just after giving birth.

Antidiuretic Hormone

An **antidiuretic** is any substance that decreases urine production. The principal physiological activity of **antidiuretic hormone (ADH)** is to decrease urine output. ADH causes the kidneys to remove water from fluid that will become urine and return it to the blood, thus decreasing urine volume and conserving body water. In the absence of ADH, urine output may be increased more than 10-fold from the normal 1 to 2 liters to 25 liters a day.

The amount of ADH normally secreted varies with the body's state of hydration (Fig. 18.12). When the body is dehydrated, the concentration of water in the blood is below normal. The elevated ratio of solutes to water increases osmotic pressure. Receptors in the hypothalamus called **osmoreceptors** detect the high osmotic pressure (low water concentration) of the blood. They stimulate the neurosecretory cells of the hypothalamic paraventricular and supraoptic nuclei to synthesize and release ADH from their axon terminals in the posterior pituitary. The blood carries ADH to the kidneys, which respond by decreasing urine output. More water thus is retained. During dehydra-

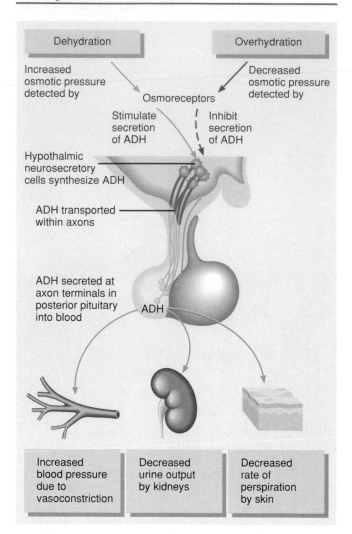

FIGURE 18.12 Effects of antidiuretic hormone (ADH) and regulation of its secretion.

Question: If you drank a liter of water, what effect would this have on the osmotic pressure of your blood, and what impact would this change have on the level of ADH in your blood?

tion, ADH also decreases the rate of perspiration production.

ADH can also raise blood pressure by causing constriction of arterioles. For this reason, ADH is also called **vasopressin.** If there is a severe loss of blood volume due to hemorrhage, diarrhea, or excessive sweating, ADH output increases.

If the water concentration of blood is higher than normal, the osmoreceptors detect a decreased osmotic pressure, and ADH secretion is reduced or stopped. The kidneys then put out a large volume of urine, and the osmotic pressure of body fluids returns to normal.

Secretion of ADH can also be altered in other ways. Pain, stress, trauma, anxiety, acetylcholine, nicotine, and drugs, such as morphine, tranquilizers, and some anesthetics, stimulate secretion of the hormone. Alcohol inhibits ADH secretion and thereby increases urine output. The resulting dehydration may cause both the thirst and headache typical of a hangover.

CLINICAL APPLICATION

DIABETES INSIPIDUS

The principal abnormality associated with dysfunction of the posterior pituitary is **diabetes insipidus** (in-SIP-i-dus). *Diabetes* means "overflow" and *insipidus* means "tasteless." This disorder should not be confused with diabetes mellitus (*meli* = honey), a disorder of the pancreas characterized by glucose in the urine (see page 551). Diabetes insipidus results from hyposecretion of ADH, usually caused by damage to the posterior pituitary or the hypothalamic paraventricular and supraoptic nuclei, or from nonfunctional ADH receptors. Symptoms include excretion of large amounts of urine, dehydration, and thirst. Bed-wetting is common in afflicted children. Because so much water is lost in the urine, a person with severe diabetes insipidus may die of dehydration if deprived of water for only a day or so. Diabetes insipidus is treated by administering ADH in a nasal spray.

A summary of posterior pituitary hormones, their principal actions, control of secretion, and a selected disorder is presented in Exhibit 18.3.

THYROID GLAND

The **thyroid gland** is located just below the larynx (voice box). The right and left **lateral lobes** lie one on either side of the trachea. Connecting the lobes is a mass of tissue called an **isthmus** (IS-mus) that lies in front of the trachea (Fig. 18.13a). A small, pyramidal-shaped lobe sometimes extends upward from the isthmus. The gland usually weighs about 30 g (about 1 oz) and has a rich blood supply, receiving about 80 to 120 ml of blood per minute.

Microscopic spherical sacs called **thyroid follicles** (Fig. 18.13b,c) fill most of the thyroid gland. The wall of each follicle consists of two types of cells. Most extend to the lumen (internal space) of the follicle and are called **follicular cells.** When the follicular cells are inactive, their shape is low cuboidal to squamous. Under the influence of TSH, they become cuboidal or low columnar and actively secretory. The follicular cells manufacture **thyroxine** (thī-ROK-sēn), which is also called T_4 because it contains four atoms of iodine, and **triiodothyronine** (trī-ī′-ōd-ō-THĪ-rō-nēn), or T_3, which contains three atoms of iodine. Together these hormones are referred to as **thyroid hormones.** A few cells do not reach the follicle lumen or lie between follicles and are called **parafollicular cells,** or **C (clear) cells.** They produce **calcitonin** (kal-si-TŌ-nin), which influences calcium homeostasis.

Formation, Storage, and Release of Thyroid Hormones

The thyroid gland is the only endocrine gland that stores its secretory product in large quantity, normally about a 100-day supply. In essence, T_3 and T_4 are synthesized by attaching iodines to the amino acid tyrosine, stored for some period of time, and then secreted into the blood (Fig.

EXHIBIT 18.3

SUMMARY OF POSTERIOR PITUITARY GLAND HORMONES, PRINCIPAL ACTIONS, CONTROL OF SECRETION, AND SELECTED DISORDERS

Hormone	Principal Actions	Control of Secretion	Selected Disorders
Oxytocin (OT)	Stimulates contraction of smooth muscle cells of pregnant uterus during labor and stimulates contraction of contractile cells of mammary glands for milk ejection.	Neurosecretory cells of hypothalamus secrete OT in response to uterine distension and stimulation of nipples.	
Antidiuretic hormone (ADH)	Principal effect is to decrease urine volume; also raises blood pressure by constricting arterioles during severe hemorrhage.	Neurosecretory cells of hypothalamus secrete ADH in response to low water concentration of the blood, pain, stress, trauma, anxiety, acetylcholine (ACh), nicotine, morphine, some anesthetics, and tranquilizers; alcohol inhibits secretion.	Hyposecretion of ADH or nonfunctional ADH receptors result in diabetes insipidus.

FIGURE 18.13 Location, blood supply, and histology of the thyroid gland.

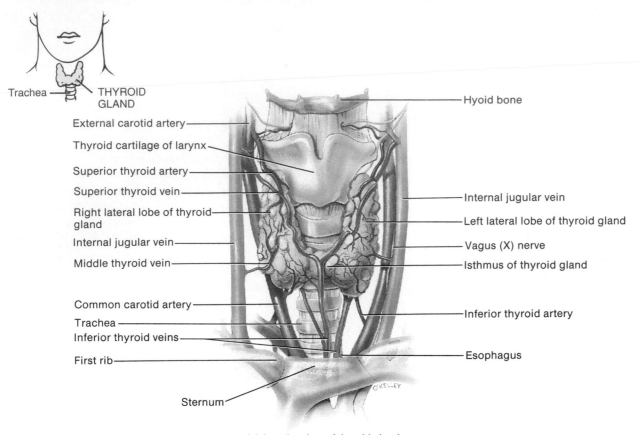

Trachea — THYROID GLAND

External carotid artery

Thyroid cartilage of larynx

Superior thyroid artery

Superior thyroid vein

Right lateral lobe of thyroid gland

Internal jugular vein

Middle thyroid vein

Common carotid artery

Trachea

Inferior thyroid veins

First rib

Sternum

Hyoid bone

Internal jugular vein

Left lateral lobe of thyroid gland

Vagus (X) nerve

Isthmus of thyroid gland

Inferior thyroid artery

Esophagus

(a) Anterior view of thyroid gland

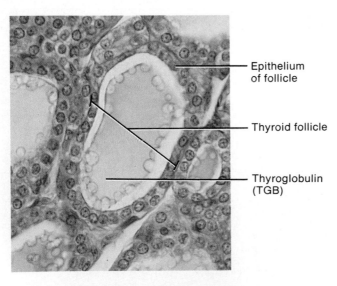

Epithelium of follicle

Thyroid follicle

Thyroglobulin (TGB)

(b) Photomicrograph of several thyroid follicles (160x)

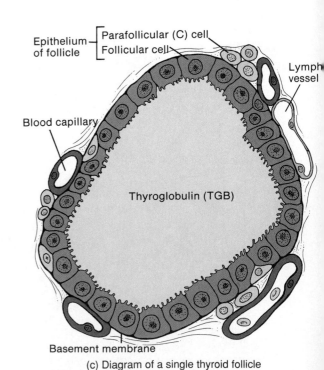

Epithelium of follicle —[Parafollicular (C) cell / Follicular cell]

Lymph vessel

Blood capillary

Thyroglobulin (TGB)

Basement membrane

(c) Diagram of a single thyroid follicle

Question: Which cells secrete T_3 and T_4? Calcitonin? Which of these hormones are also called thyroid hormones?

18.14). TSH stimulates most of the steps, which are as follows.

1. **Iodide trapping.** Thyroid follicular cells trap iodide (I$^-$) by actively transporting it from the blood into the cytosol. The I$^-$ concentration inside follicular cells is about 30 to 40 times that of blood plasma. As a result, the thyroid gland normally contains most of the iodide in the body.

2. **Synthesis of thyroglobulin.** While the follicular cells are trapping I$^-$, they are also synthesizing **thyroglobulin (TGB)**, a high molecular weight glycoprotein that contains about 5000 amino acid residues. More than 100 of these amino acids are tyrosines, a few of which become iodinated (step 4). TGB is produced in the rough endoplasmic reticulum, modified in the Golgi complex, and packaged into secretory vesicles. The vesicles then undergo exocytosis, which releases TGB into the lumen of the follicle. The material that accumulates in the lumen in this way is termed **colloid.**

3. **Oxidation of iodide.** Negatively charged I$^-$ cannot bind to tyrosine. Rather, these anions must first undergo oxidation (removal of electrons) to iodine: $2I^- \rightarrow I_2$. The enzyme that catalyzes this reaction is peroxidase. In thyroid follicular cells, peroxidase seems to be most concentrated close to or in the membrane facing the colloid. As the iodide ions are being oxidized, they pass through the membrane into the colloid. Peroxidase also is thought to catalyze steps 4 and 5.

4. **Iodination of tyrosine.** As soon as iodine forms, it attaches to tyrosine amino acids that are part of thyroglobulin molecules in the colloid. Binding of one iodine yields monoiodotyrosine (T_1), and a second iodination gives diiodotyrosine (T_2).

5. **Coupling of T_1 and T_2.** During the last step in synthesis of thyroid hormone, two T_2 molecules join to form T_4 or one T_1 and one T_2 join to form T_3. Each TGB ultimately contains about six T_1, five T_2, and one to five T_4. There is a single T_3 in one of every four thyroglobulins.

6. **Pinocytosis and digestion of colloid.** Droplets of colloid reenter follicular cells by pinocytosis and merge with lysosomes. Digestive enzymes break down TGB, cleaving off molecules of T_3 and T_4. T_1 and T_2 are also released. They undergo deiodination, that is, removal of iodine, which is reused in the synthesis of more T_3 and T_4.

7. **Secretion of thyroid hormone.** Since T_3 and T_4 are lipid-soluble, they diffuse through the plasma membrane to enter the blood.

8. **Transport in the blood.** More than 99% of both the T_3 and T_4 combine with transport proteins in the blood, mainly **thyroxine-binding globulin (TBG).**

T_4 normally is secreted in greater quantity than T_3, but T_3 is several times more potent. Moreover, as it circulates in the blood and enters cells throughout the body, most T_4 is converted to T_3 by removal of one iodine. T_3 is the more active of the two forms.

FIGURE 18.14 Steps in the synthesis of thyroid hormones.

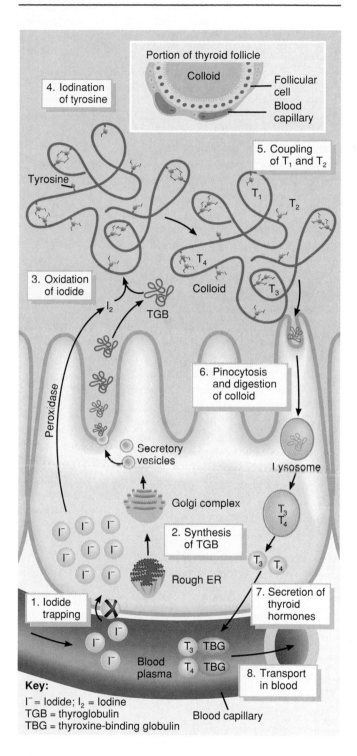

Question: What is the storage form of thyroid hormones?

Actions of Thyroid Hormones

The thyroid hormones regulate (1) oxygen use and basal metabolic rate, (2) cellular metabolism, and (3) growth and development.

Thyroid hormones increase basal metabolic rate or BMR (rate of oxygen consumption at rest after an overnight fast) by stimulating cellular oxygen use to produce ATP. The active transport pumps that continually eject sodium ions (Na^+) from the cytosol into the extracellular fluid use a large portion of the ATP produced by most cells. A major effect of the thyroid hormones is to stimulate synthesis of the enzyme that runs the pump, Na^+/K^+ ATP-ase. As cells use more oxygen to produce ATP, more heat is given off, and body temperature rises. This phenomenon is called the **calorigenic effect** of thyroid hormones. In this way, they play an important role in the maintenance of normal body temperature (thermoregulation). Normal mammals survive in freezing temperatures, but those whose thyroid glands have been removed do not.

In the regulation of metabolism, the thyroid hormones stimulate protein synthesis; increase lipolysis (triglyceride breakdown); enhance cholesterol excretion in bile, a substance stored in the gallbladder that helps in lipid digestion (thus reducing blood cholesterol level); and increase the use of glucose for ATP production.

Together with hGH and insulin, thyroid hormones accelerate body growth, particularly the growth of nervous tissue. Deficiency of thyroid hormones during fetal development can result in fewer and smaller neurons, defective myelination of axons, and mental retardation. During the early years of life, deficiency of thyroid hormones results in small stature and poor development of certain organs such as the brain and reproductive organs.

The thyroid hormones enhance some actions of the catecholamines (norepinephrine and epinephrine) because they up-regulate β receptors (see Exhibit 17.3 on page 512). For this reason, symptoms of hyperthyroidism include increased heart rate and more forceful heartbeats, increased blood pressure, and increased nervousness.

Control of Thyroid Hormone Secretion

The secretory activity and size of the thyroid gland are controlled in two main ways: (1) by the level of iodine in the thyroid gland and (2) by negative feedback systems involving both the hypothalamus and the anterior pituitary. Although needed for synthesis of thyroid hormones, an abnormally high concentration of thyroid iodine suppresses release of thyroid hormones. Figure 18.15 shows the negative feedback systems that govern synthesis and release of thyroid hormones.

Low blood levels of T_3 stimulate the hypothalamus to secrete thyrotropin releasing hormone (TRH) and the anterior pituitary to secrete thyroid-stimulating hormone (TSH). See Fig. 18.7. TRH also stimulates the anterior pituitary to secrete TSH. Then, TSH stimulates virtually all aspects of thyroid gland activity. These include iodide trapping, hormone synthesis and secretion, and growth of the follicular cells. The thyroid gland releases T_3 and T_4 until the metabolic rate returns to normal. Conditions that increase

FIGURE 18.15 Regulation of thyroid hormone secretion.

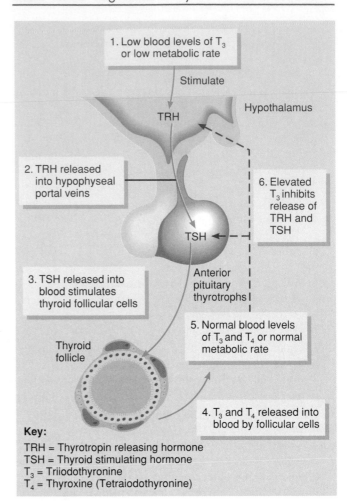

1. Low blood levels of T_3 or low metabolic rate

Stimulate

TRH

Hypothalamus

2. TRH released into hypophyseal portal veins

6. Elevated T_3 inhibits release of TRH and TSH

TSH

Anterior pituitary thyrotrophs

3. TSH released into blood stimulates thyroid follicular cells

5. Normal blood levels of T_3 and T_4 or normal metabolic rate

Thyroid follicle

4. T_3 and T_4 released into blood by follicular cells

Key:
TRH = Thyrotropin releasing hormone
TSH = Thyroid stimulating hormone
T_3 = Triiodothyronine
T_4 = Thyroxine (Tetraiodothyronine)

Question: If tumor cells in the anterior pituitary gland (which are not subject to negative feedback inhibition) secrete TSH, what will the impact be on the thyroid gland?

ATP demand—a cold environment, hypoglycemia, high altitude, pregnancy—also trigger this negative feedback system and increase the secretion of the thyroid hormones.

CLINICAL APPLICATION

CRETINISM, MYXEDEMA, GRAVES' DISEASE, AND GOITER

Hyposecretion of thyroid hormones during fetal life or infancy results in **cretinism** (KRĒ-tin-izm), which is shown in Fig. 18.16a. Cretins exhibit dwarfism because the skeleton fails to grow and mature, and they are severely mentally retarded because the brain fails to develop fully. Cretins also display retarded sexual development and a yellowish skin color. Although not usually apparent in a newborn, because lipid-soluble maternal thyroid hormones pass across the placenta and allow normal development, within months after birth the symptoms start to appear. Most states

FIGURE 18.16 Abnormalities related to the thyroid gland.

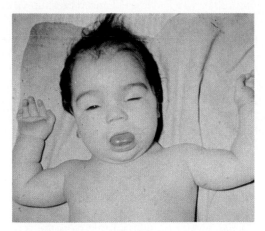

(a) Cretinism

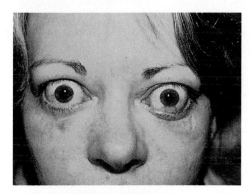

(b) Exophthalmos

(c) Goiter

Question: How could an iodide-deficient diet lead to goiter?

require testing of all babies to ensure adequate thyroid function. If the hypothyroidism is diagnosed early, cretinism can be prevented by giving oral thyroid hormone.

Hypothyroidism during the adult years produces **myxedema** (mix-e-DĒ-ma). A hallmark of this disorder is an edema (accumulation of interstitial fluid) that causes the facial tissues to swell and look puffy. Like the cretin, the person with myxedema has a slow heart rate, low body temperature, sensitivity to cold, hypersensitivity to certain drugs (narcotics, barbiturates, and anesthetics), dry hair and skin, muscular weakness, general lethargy, and a tendency to gain weight easily. Because the brain has already reached maturity, the person with myxedema does not experience mental retardation. However, in moderately severe cases, the person's mental functions may be dulled so that the person is less alert. Myxedema occurs about five times more often in females than in males. Oral thyroid hormones reduce the symptoms.

Hypersecretion of thyroid hormones increases oxygen use by body cells, elevates heat production, and increases food intake. Common symptoms are increased metabolic rate, heat intolerance, increased sweating, weight loss despite a good appetite, insomnia, tremor of extended fingers, and nervousness. The most common form of hyperthyroidism is **Graves' disease,** which is an autoimmune disorder. The person produces antibodies that mimic the action of TSH but are not regulated by the normal negative feedback controls. As a result, the thyroid gland is continually bombarded with stimulation to grow and produce thyroid hormones. A primary sign is an enlarged thyroid, which may be two to three times its normal size. Graves' patients often have a peculiar edema behind the eyes, called **exophthalmos** (ek'-sof-THAL-mos), which causes the eyes to protrude (Fig. 18.16b). This disorder, like myxedema, also occurs more often in females. A variety of treatments are helpful. These include surgical removal of part or all of the thyroid gland (thyroidectomy), using radioactive iodine (^{131}I) to selectively destroy thyroid tissue, and using antithyroid drugs to block synthesis of thyroid hormones.

The term **thyroid storm** refers to an aggravation of all symptoms of hyperthyroidism characterized by unregulated hypermetabolism with fever and rapid heart rate. It can result from trauma, surgery, unusual emotional stress or labor during delivery of a baby.

A **goiter** (GOY-ter) is simply an enlarged thyroid gland, and it is a symptom of many thyroid disorders besides Graves' disease. In some places in the world, dietary iodine intake is inadequate. The resultant low level of thyroid hormone in the blood stimulates secretion of TSH, which causes thyroid gland enlargement (Fig. 18.16c).

Calcitonin

The hormone produced by the parafollicular cells of the thyroid gland is **calcitonin** (kal-si-TŌ-nin), or **CT.** Together with parathyroid hormone and calcitriol (described shortly), calcitonin maintains homeostasis of calcium (Ca^{2+}) and phosphates (PO_4^{3-} and HPO_4^{2-}) in the blood. It lowers the amount of blood calcium and phosphates by inhibiting bone breakdown and accelerating uptake of calcium and phosphates by bones. Calcitonin exerts its effect in lowering calcium and phosphate blood levels by inhibiting the action of osteoclasts (bone-destroying cells). Although it acts rapidly in experimental situations, calcitonin's importance in normal physiology is not clear. It can be present in excess or completely absent without causing clinical symptoms. For example, people who have had a complete surgical removal of the thyroid tissue maintain calcium homeostasis as long as their parathyroid glands are intact and functional. The blood calcium level directly controls the secretion of CT according to a negative feedback system that does not involve the pituitary gland (Fig. 18.17).

A summary of hormones produced by the thyroid gland, their principal actions, control of secretion, and selected disorders is presented in Exhibit 18.4.

PARATHYROID GLANDS

Attached to the posterior surfaces of the thyroid gland are small, round masses of tissue called the **parathyroid glands.** Usually, there is one superior and one inferior parathyroid gland attached to each lateral thyroid lobe (Fig. 18.18a).

Microscopically, the parathyroids contain two kinds of epithelial cells (Fig. 18.18b,c). The more numerous cells, called **principal (chief) cells,** probably are the major source of **parathyroid hormone (PTH),** or **parathormone.** The function of the other kind of cell, called an **oxyphil cell,** is not known.

Parathyroid Hormone

PTH increases the number and activity of osteoclasts (bone-destroying cells). The result is enhanced breakdown of bone matrix (bone resorption), which releases calcium (Ca^{2+}) and phosphates (HPO_4^{2-}) into the blood. PTH also produces two changes in the kidneys. (1) It increases the rate at which the kidneys remove Ca^{2+} and magnesium (Mg^{2+}) from urine that is being formed and return them to the blood. (2) It inhibits the transport of HPO_4^{2-} from urine into blood so that more of it is excreted in urine. More HPO_4^{2-} is lost through the urine than is gained from the bones. Overall, then, PTH decreases blood HPO_4^{2-} level and increases blood Ca^{2+} and Mg^{2+} levels. With respect to blood Ca^{2+} level, PTH and calcitonin are antagonists; that is, they have opposite actions (see Fig. 18.17).

A third effect of PTH on the kidneys is to promote formation of the hormone **calcitriol,** which is the active form of vitamin D. It is also known as *1,25-dihydroxy cholecalciferol* or *1,25-dihydroxy vitamin D_3.* Calcitriol increases the rate of calcium, phosphate, and magnesium absorption from the gastrointestinal tract into the blood.

FIGURE 18.17 Regulation of the secretion of parathyroid hormone (PTH), calcitriol, and calcitonin (CT).

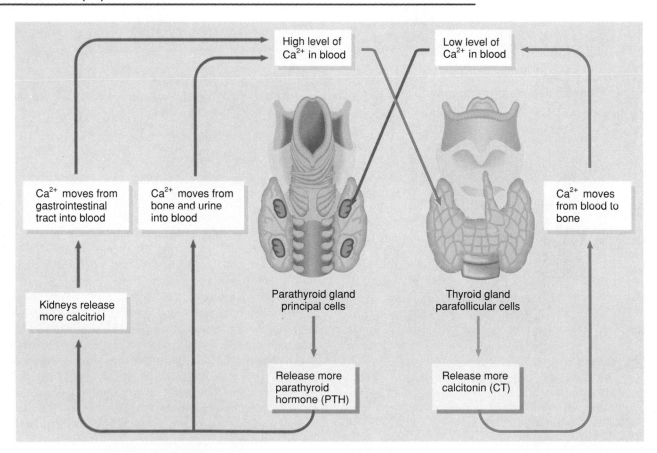

High level of Ca^{2+} in blood

Low level of Ca^{2+} in blood

Ca^{2+} moves from gastrointestinal tract into blood

Ca^{2+} moves from bone and urine into blood

Ca^{2+} moves from blood to bone

Kidneys release more calcitriol

Parathyroid gland principal cells

Thyroid gland parafollicular cells

Release more parathyroid hormone (PTH)

Release more calcitonin (CT)

Question: What are the primary target tissues for PTH, CT, and calcitriol?

EXHIBIT 18.4

SUMMARY OF THYROID GLAND HORMONES, PRINCIPAL ACTIONS, CONTROL OF SECRETION, AND SELECTED DISORDERS

Hormone	Principal Actions	Control of Secretion	Selected Disorders
THYROID HORMONES **T₃ (triiodothyronine) and T₄ (thyroxine)**	Regulate metabolism, growth and development, and activity of nervous system.	Thyrotropin releasing hormone (TRH) is released from hypothalamus in response to low thyroid hormone levels, low metabolic rate, cold, pregnancy, and high altitudes; TRH secretion is inhibited in response to high thyroid hormone levels, high metabolic rate, high levels of estrogens and androgens, and aging.	Hyposecretion of thyroid hormones during infancy or childhood results in cretinism; hypothyroidism during adult years results in myxedema; hypersecretion of thyroid hormones in Graves' disease results in exophthalmos; thyroid enlargement (goiter) may be associated with either iodine excess or deficiency.
CALCITONIN (CT)	Lowers blood levels of calcium and phosphates by accelerating their absorption into bone.	High blood calcium levels stimulate secretion; low levels inhibit secretion.	No known disorders result from excess or deficiency of CT secretion.

FIGURE 18.18 Location, blood supply, and histology of the parathyroid glands.

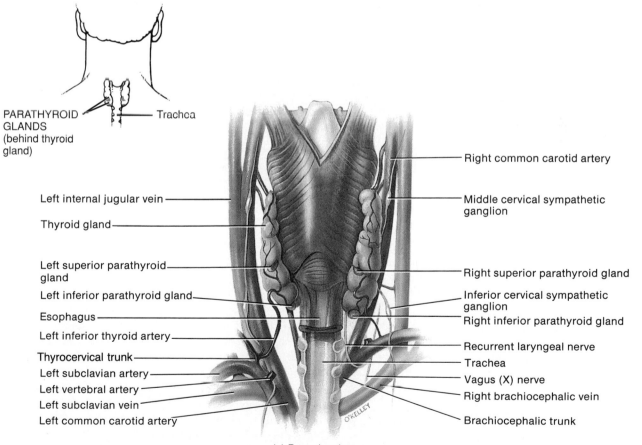

PARATHYROID GLANDS (behind thyroid gland) — Trachea

Left internal jugular vein

Thyroid gland

Left superior parathyroid gland

Left inferior parathyroid gland

Esophagus

Left inferior thyroid artery

Thyrocervical trunk

Left subclavian artery

Left vertebral artery

Left subclavian vein

Left common carotid artery

Right common carotid artery

Middle cervical sympathetic ganglion

Right superior parathyroid gland

Inferior cervical sympathetic ganglion

Right inferior parathyroid gland

Recurrent laryngeal nerve

Trachea

Vagus (X) nerve

Right brachiocephalic vein

Brachiocephalic trunk

O'KELLEY

(a) Posterior view

Figure continues

FIGURE 18.18 (continued)

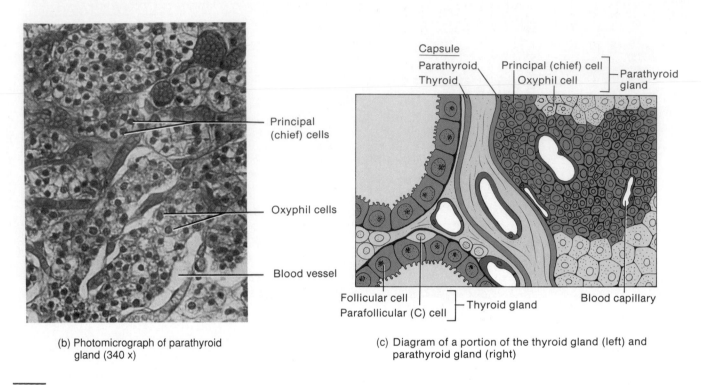

(b) Photomicrograph of parathyroid
gland (340 x)

(c) Diagram of a portion of the thyroid gland (left) and
parathyroid gland (right)

Question: What are the secretory products of parafollicular cells of the thyroid gland and principal cells of
the parathyroid glands?

When the blood Ca²⁺ level falls, more PTH (and less calcitonin) is released (see Fig. 18.17). Conversely, when the blood Ca²⁺ level rises, less PTH (and more calcitonin) is secreted. This is another example of a negative feedback control system that does not involve the pituitary gland.

CLINICAL APPLICATION

TETANY AND OSTEITIS FIBROSA CYSTICA

A deficiency of Ca²⁺ caused by **hypoparathyroidism** causes neurons to depolarize without the usual stimulus. As a result, nerve impulses and muscle action potentials arise spontaneously, leading to muscle twitches, spasms, and convulsions. This condition is called **tetany.** The leading cause of hypoparathyroidism is accidental damage to the parathyroids or their blood supply during thyroidectomy surgery. Other causes include parathyroid disease, infection, or hemorrhage.

Hyperparathyroidism causes demineralization of bone. If uncorrected, this condition may lead to **osteitis fibrosa cystica,** so named because the areas of destroyed bone tissue are replaced by cavities that fill with fibrous tissue. The bones thus become deformed and are easily fractured. Usually, a tumor in the parathyroids causes hyperparathyroidism.

A summary of the principal actions, control of secretion, and selected disorders related to parathyroid hormone (PTH) is presented in Exhibit 18.5.

ADRENAL (SUPRARENAL) GLANDS

The paired **adrenal (suprarenal) glands,** one of which lies superior to each kidney (Fig. 18.19a), are structurally and functionally differentiated into two regions. The outer **adrenal cortex** makes up the bulk of the gland and surrounds the inner **adrenal medulla** (Fig. 18.19b). The adrenal cortex is derived from mesoderm of a developing embryo and produces steroid hormones that are essential for life. Complete loss of adrenocortical secretions leads rapidly to death in a few days to a week unless hormone replacement therapy begins promptly. For example, one adrenocortical hormone called aldosterone helps control the homeostasis of blood sodium (Na⁺) level. In the absence of aldosterone, Na⁺ is lost from the body and this can lead to hypotension and cardiovascular shock. The adrenal medulla is derived from the ectoderm and produces two catecholamine hormones, norepinephrine and epinephrine. Covering the gland is a connective tissue capsule. The adrenal glands, like the thyroid gland, are highly vascularized.

EXHIBIT 18.5

SUMMARY OF PARATHYROID GLAND HORMONE, PRINCIPAL ACTIONS, CONTROL OF SECRETION, AND SELECTED DISORDERS

Hormone	Principal Actions	Control of Secretion	Selected Disorders
Parathyroid hormone (PTH)	Increases blood Ca^{2+} and Mg^{2+} levels and decreases blood phosphate level by increasing rate of dietary Ca^{2+} and Mg^{2+} absorption; increases number and activity of osteoclasts; increases Ca^{2+} reabsorption by kidneys; increases phosphate excretion by kidneys; and promotes formation of calcitriol.	Low blood Ca^{2+} levels stimulate secretion. High blood Ca^{2+} levels inhibit secretion.	Hypoparathyroidism results in tetany. Hyperparathyroidism produces osteitis fibrosa cystica.

FIGURE 18.19 Location, blood supply, and histology of the adrenal (suprarenal) glands.

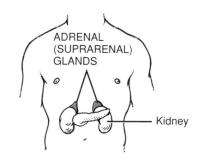

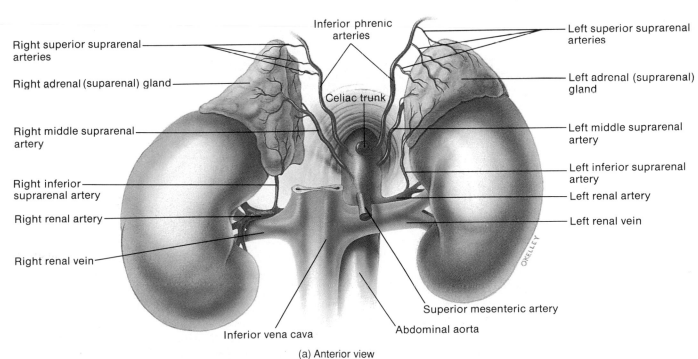

(a) Anterior view

Figure continues

FIGURE 18.19 (continued)

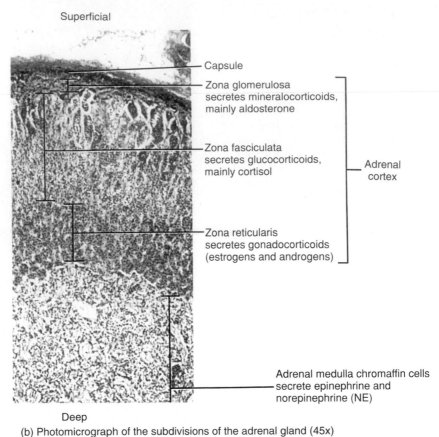

Superficial

Capsule

Zona glomerulosa
secretes mineralocorticoids,
mainly aldosterone

Zona fasciculata
secretes glucocorticoids,
mainly cortisol

Adrenal
cortex

Zona reticularis
secretes gonadocorticoids
(estrogens and androgens)

Adrenal medulla chromaffin cells
secrete epinephrine and
norepinephrine (NE)

Deep

(b) Photomicrograph of the subdivisions of the adrenal gland (45x)

Question: How would you describe the position of the adrenal glands relative to the pancreas and kidneys?

Adrenal Cortex

The adrenal cortex is subdivided into three zones (Fig. 18.19b). Each zone has a different cellular arrangement and secretes different groups of steroid hormones. The outer zone, directly underneath the connective tissue capsule, is called the **zona glomerulosa.** Its cells are arranged in arched loops or round balls. Its primary secretions are a group of hormones called **mineralocorticoids** (min′-er-al-ō-KOR-ti-koyds) because they affect mineral homeostasis.

The middle zone, or **zona fasciculata,** is the widest of the three zones and consists of cells arranged in long, straight cords. The zona fasciculata secretes mainly **glucocorticoids** (gloo′-kō-KOR-ti-koyds), so named because they affect glucose homeostasis.

The inner zone, the **zona reticularis,** contains cords of cells that branch freely. This zone synthesizes minute amounts of hormones, mainly the sex steroids called **gonadocorticoids** (gō-na-dō-KOR-ti-koyds). The primary gonadocorticoids are androgens (male sex hormones).

Mineralocorticoids

Mineralocorticoids help control water and electrolyte homeostasis, particularly the concentrations of sodium

(Na^+) and potassium (K^+) ions. Although the adrenal cortex secretes at least three different hormones classified as mineralocorticoids, about 95% of the mineralocorticoid activity is due to **aldosterone** (al-DA-ster-ōn). Aldosterone acts on certain tubule cells in the kidneys to increase their reabsorption of Na^+. By stimulating return of Na^+ to the blood, aldosterone prevents depletion of Na^+ from the body. The Na^+ reabsorption leads to reabsorption of Cl^- and HCO_3^- and retention of water. At the same time, aldosterone increases excretion of K^+ so more K^+ is lost in the urine. Aldosterone also promotes excretion of H^+ in the urine, thus removing acids from the body and preventing acidosis (blood pH below 7.35).

Control of aldosterone secretion involves several mechanisms operating simultaneously. One of these is the **renin–angiotensin** (an′-jē-ō-TEN-sin) **pathway** (Fig. 18.20). A decrease in blood volume from dehydration, Na^+ deficiency, or hemorrhage causes a drop in blood pressure. The low blood pressure stimulates certain kidney cells, called juxtaglomerular cells, to secrete an enzyme called **renin** (RĒ-nin) into the blood (see Fig. 26.12). Renin converts **angiotensinogen,** a plasma protein produced by the liver, into **angiotensin I.** As blood flows through lung capillaries, an enzyme called **angiotensin converting enzyme**

FIGURE 18.20 Regulation of the secretion of aldosterone by the renin–angiotensin pathway.

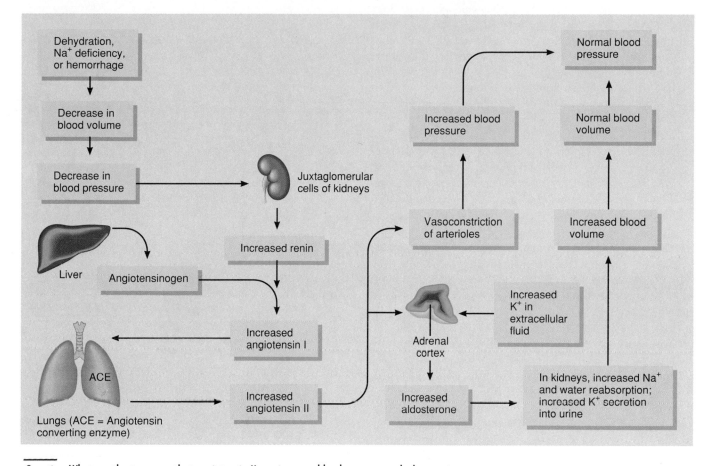

Question: What are the two ways that angiotensin II can increase blood pressure and what are its target tissues in these cases?

(ACE) converts angiotensin I into **angiotensin II**. Angiotensin II is a hormone that stimulates the adrenal cortex to secrete aldosterone. In the kidneys, aldosterone increases Na$^+$ reabsorption, and water follows. This leads to an increase in extracellular fluid volume and restoration of blood pressure to normal. In addition, angiotensin II is a powerful vasoconstrictor, which also helps to elevate blood pressure.

A second mechanism for the control of aldosterone secretion is K$^+$ concentration. An increased K$^+$ concentration in extracellular fluid directly stimulates aldosterone secretion by the adrenal cortex and causes the kidneys to eliminate excess K$^+$. A decreased concentration of K$^+$ in extracellular fluid has the opposite effect.

CLINICAL APPLICATION

ALDOSTERONISM

Hypersecretion of aldosterone, usually by a tumor of the zona glomerulosa, results in **aldosteronism,** characterized by increased Na$^+$ and decreased K$^+$ levels in the blood. There is excessive retention of Na$^+$ and water. The water increases the volume of the blood and causes hypertension (high blood pressure). If K$^+$ depletion is great, neurons and muscle fibers hyperpolarize, which makes them less responsive to stimulation. Symptoms include muscular weakness, cramps, and paralysis.

Glucocorticoids

The glucocorticoids regulate metabolism and resistance to stress. Three glucocorticoids are **cortisol (hydrocortisone), corticosterone,** and **cortisone.** Of the three, cortisol is the most abundant and is responsible for about 95% of glucocorticoid activity. The glucocorticoids have the following effects:

1. Glucocorticoids, together with other hormones, promote normal metabolism. Their role is to make sure enough ATP is available. They increase the rate at which proteins are catabolized and amino acids are removed from cells, primarily muscle fibers, and transported to the liver. The amino acids may be synthesized into new proteins, such as the enzymes needed for metabolic reactions. If the body's reserves of glycogen and fat are low, the liver may convert lactic acid (lactate) or certain amino acids to glucose. This conversion of a substance other than glycogen or another carbohydrate into glucose is called **gluconeogenesis** (gloo′-kō-nē′-ō-JEN-e-sis). Glucocorticoids also stimulate lipolysis, the breakdown of triglycerides into fatty acids and glycerol and release of fatty acids from adipose tissue.

2. Glucocorticoids provide resistance to stress. Added glucose is used by tissues to produce ATP for combating various stresses: fasting, fright, temperature extremes, high altitude, bleeding, infection, surgery, trauma, and almost any debilitating disease. Glucocorticoids also make the blood vessels more sensitive to vessel-constricting chemicals. They thereby raise blood pressure. This effect is an advantage if the stress happens to be blood loss, which tends to make blood pressure fall.

3. Glucocorticoids are anti-inflammatory compounds that inhibit the cells and secretions that participate in inflammatory responses. They reduce the number of mast cells, thus reducing release of histamine; stabilize lysosomal membranes to reduce release of enzymes; decrease blood capillary permeability; and depress phagocytosis. Unfortunately, they also retard connective tissue regeneration and thereby slow wound healing. Although high doses can cause severe mental disturbances, they are very useful in the treatment of chronic inflammation, such as rheumatism. High doses of glucocorticoids also depress immune responses. For this reason, glucocorticoids are taken by organ transplant recipients to retard tissue rejection by the immune system.

The control of glucocorticoid secretion is a typical negative feedback system (Fig. 18.21). Low blood levels of glucocorticoids (mainly cortisol) stimulate the hypothalamus to secrete **corticotropin releasing hormone (CRH).** CRH and a low level of glucocorticoids both promote the release of ACTH from the anterior pituitary. ACTH is carried by the blood to the adrenal cortex, where it stimulates glucocorticoid secretion. As part of the discussion of stress at the end of the chapter, you will see that the hypothalamus also releases more CRH in response to a variety of physical and emotional stresses.

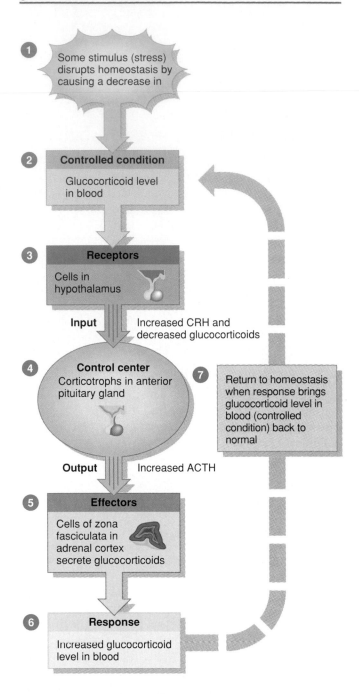

FIGURE 18.21 Negative feedback regulation of glucocorticoid secretion.

① Some stimulus (stress) disrupts homeostasis by causing a decrease in

② **Controlled condition**
Glucocorticoid level in blood

③ **Receptors**
Cells in hypothalamus

Input Increased CRH and decreased glucocorticoids

④ **Control center**
Corticotrophs in anterior pituitary gland

⑦ Return to homeostasis when response brings glucocorticoid level in blood (controlled condition) back to normal

Output Increased ACTH

⑤ **Effectors**
Cells of zona fasciculata in adrenal cortex secrete glucocorticoids

⑥ **Response**
Increased glucocorticoid level in blood

Question: If a person who has had a heart transplant receives prednisone (a glucocorticoid) to help prevent rejection of the transplanted tissue, will the person have high or low blood levels of ACTH and CRH? Explain.

CLINICAL APPLICATION

ADDISON'S DISEASE AND CUSHING'S SYNDROME

Hyposecretion of glucocorticoids and aldosterone due to failure of the adrenal cortex results in the condition called **Addison's disease (primary adrenocortical insufficiency).** Clinical symptoms include mental lethargy, anorexia, nausea and vomiting, weight loss, and hypoglycemia, which leads to muscular weakness. Loss of aldosterone leads to elevated K^+ and decreased Na^+ in the blood, low blood pressure, dehydration, decreased cardiac output, arrhythmias, and potential cardiac arrest. Blood level of ACTH is high due to loss of negative feedback inhibition by cortisol. At high concentration, ACTH mimics the skin darkening effects of MSH (melanocyte-stimulating hormone).

The result is excessive skin pigmentation, especially in sun-exposed areas and in mucous membranes.

Cushing's syndrome is a hypersecretion of glucocorticoids, especially cortisol and cortisone (Fig. 18.22). The condition is characterized by the redistribution of fat. The result is spindly arms and legs, due to catabolism of muscle proteins, accompanied by a rounded "moon face," "buffalo hump" on the back, and pendulous (hanging) abdomen. Facial skin is flushed, and the skin covering the abdomen develops stretch marks (striae). The individual also bruises easily, and wound healing is poor. Other symptoms include hyperglycemia, osteoporosis, weakness, hypertension, increased susceptibility to infection, decreased resistance to stress, and mood swings. The most common cause of a cushinoid appearance is the administration of a glucocorticoid such as prednisone to a transplant recipient or to treat asthma or chronic inflammatory disorders.

FIGURE 18.22 Cushing's syndrome in two different patients.

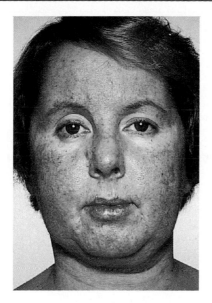

(a) Facial features

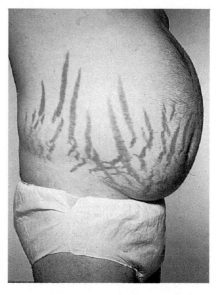

(b) Pendulous abdomen with striae

Gonadocorticoids

The adrenal cortex secretes both female and male gonadocorticoids. These are estrogens and androgens. Estrogens are several closely related female sex hormones that are also produced by the ovaries and placenta. Androgens are hormones that exert masculinizing effects. An important androgen, called testosterone, is produced by the testes. The amount of sex hormones secreted by normal adult male adrenals is usually so low that their effects are insignificant. In females, adrenal androgens contribute to sex drive (libido) and other sexual behavior. They also may be converted into estrogens, which is significant when ovarian estrogen secretion diminishes during menopause. Adrenal androgens also assist in the prepubertal growth spurt and early development of axillary and pubic hair in boys and girls.

CLINICAL APPLICATION

CONGENITAL ADRENAL HYPERPLASIA AND ADRENAL TUMORS

Congenital adrenal hyperplasia (CAH) is a genetic disorder characterized by enlarged adrenal glands. Such people lack one or more enzymes needed for synthesis of cortisol.

The low cortisol level stimulates increased secretion of ACTH by the anterior pituitary. ACTH, in turn, stimulates growth and secretory activity of the adrenal cortex. However, since certain steps leading to synthesis of cortisol are blocked, precursor molecules accumulate, and some of these are converted to androgens. The result is **virilism,** or masculinization. In a female, virile characteristics include growth of a beard, development of a much deeper voice, occasionally the development of baldness, development of a masculine distribution of hair on the body and on the pubis, growth of the clitoris so it may resemble a penis, atrophy of the breasts, infrequent or absent menstruation, and increased muscularity that produces a male-like physique. In prepubertal males, the syndrome causes the same characteristics as in females, plus rapid development of the male sexual organs and creation of male sexual desires. In adult males, the virilizing characteristics of CAH are usually completely obscured by the normal virilizing characteristics of the testosterone secreted by the testes. As a result, it is often difficult to make a diagnosis of CAH in adult males.

Virilism may also result from tumors of the adrenal gland called **virilizing adenomas** (aden = gland; oma = tumor). Occasionally, an adrenal tumor secretes sufficient quantities of feminizing hormones (estrogens) that a male patient develops **gynecomastia** (gyneca = woman; mast = breast), which means excessive growth (benign) of the male mammary glands. Such a tumor is called a **feminizing adenoma.**

Adrenal Medulla

The adrenal medulla consists of hormone-producing cells, called **chromaffin** (krō-MAF-in) **cells** (see Fig. 18.19b), which surround large blood vessels. Chromaffin cells receive direct innervation from preganglionic neurons of the sympathetic division of the autonomic nervous system (ANS) and develop from the same source as all other sympathetic postganglionic cells. Thus they are sympathetic postganglionic cells that are specialized to secrete hormones (epinephrine and norepinephrine) rather than a neurotransmitter (norepinephrine). Since the ANS controls the chromaffin cells directly, hormone release can occur very quickly.

Epinephrine and Norepinephrine

The two principal hormones synthesized by the adrenal medulla are **epinephrine** and **norepinephrine (NE),** also called adrenaline and noradrenaline, respectively. Epinephrine constitutes about 80% of the total secretion of the gland. Both hormones are **sympathomimetic** (sim′-pa-thō-mi-MET-ik); that is, they produce effects that mimic those brought about by the sympathetic division of the ANS. To a large extent, they are responsible for the fight-or-flight response. Like the glucocorticoids of the adrenal cortex, these hormones help the body resist stress. However, unlike the cortical hormones, the medullary hormones are not essential for life.

Under stress, impulses received by the hypothalamus are conveyed to sympathetic preganglionic neurons, which cause the chromaffin cells to increase their output of epinephrine and norepinephrine. Epinephrine and norepinephrine increase blood pressure by increasing heart rate and force of contraction and constricting blood vessels. They accelerate the rate of respiration, dilate respiratory passageways, decrease the rate of digestion, increase the efficiency of muscular contractions, increase blood sugar level, and stimulate cellular metabolism. Hypoglycemia may also stimulate medullary secretion of epinephrine and norepinephrine.

CLINICAL APPLICATION

PHEOCHROMOCYTOMAS

Tumors of the chromaffin cells of the adrenal medulla, called **pheochromocytomas** (fē-ō-krō′-mō-sī-TŌ-mas), cause hypersecretion of the medullary hormones. Such tumors are usually benign. The excess catecholamines cause rapid heart rate, headache, high blood pressure, high levels of sugar in the blood and urine, an elevated basal metabolic rate (BMR), flushing of the face, nervousness, sweating, and decreased gastrointestinal motility. Since the medullary hormones create the same effects as sympathetic nervous stimulation, hypersecretion puts the individual into a prolonged version of the fight-or-flight response (see Figure 18.26a). Treatment of pheochromocytomas is surgical removal of the tumor(s).

A summary of the hormones produced by the adrenal glands, their principal actions, control of secretion, and selected disorders is presented in Exhibit 18.6.

EXHIBIT 18.6

SUMMARY OF HORMONES PRODUCED BY THE ADRENAL GLANDS, PRINCIPAL ACTIONS, CONTROL OF SECRETION, AND SELECTED DISORDERS

Hormone	Principal Actions	Control of Secretion	Selected Disorders
ADRENAL CORTICAL HORMONES			
Mineralocorticoids (mainly aldosterone)	Increase blood levels of Na^+ and water and decrease blood levels of K^+.	Decreased blood volume or Na^+ level initiates renin–angiotensin pathway to stimulate aldosterone secretion; increased blood level of K^+ stimulates aldosterone secretion.	Hypersecretion of aldosterone results in aldosteronism.
Glucocorticoids (mainly cortisol)	Help regulate metabolism, resistance to stress, and counter inflammatory response.	ACTH release is stimulated by corticotropin releasing hormone (CRH) in response to stress and low blood levels of glucocorticoids.	Hyposecretion of glucocorticoids and aldosterone produces Addison's disease; hypersecretion of glucocorticoids results in Cushing's syndrome.

Hormone	Principal Actions	Control of Secretion	Selected Disorders
Gonadocorticoids	Concentrations secreted by adults are so low that their effects are usually insignificant.	Sex hormone secretions by ovaries and testes are discussed in detail in Chapter 28.	In congenital adrenal hyperplasia, synthesis of glucocorticoids is inhibited. This leads to excess production of ACTH and androgens, causing virilism. The release of feminizing hormones in males sometimes causes gynecomastia.
ADRENAL MEDULLARY HORMONES			
Epinephrine and norepinephrine (NE)	Sympathomimetic, that is, produce effects that mimic those of the sympathetic division of the autonomic nervous system (ANS) during stress.	Sympathetic preganglionic neurons stimulate secretion by chromaffin cells.	Hypersecretion of medullary hormones results in a prolonged fight-or-flight response.

PANCREAS

The **pancreas** is both an endocrine and an exocrine gland. We will treat its endocrine functions here and discuss its exocrine functions with the digestive system (Chapter 24). The pancreas is a flattened organ located posterior and slightly inferior to the stomach (Fig. 18.23a). The adult pancreas consists of a head, body, and tail.

Scattered among the exocrine portions of the pancreas are one to two million tiny clusters of endocrine tissue called **pancreatic islets** or **islets of Langerhans** (LAHNG-er-hanz) (Fig. 18.23b,c). Four types of hormone-secreting cells compose the pancreatic islets: (1) **alpha cells** secrete the hormone glucagon, which raises blood sugar level; (2) **beta cells** secrete the hormone insulin, which lowers blood sugar level; (3) **delta cells** secrete growth hormone inhibiting hormone (GHIH) or somatostatin, which acts as a paracrine to inhibit the secretion of insulin and glucagon; and (4) **F-cells** secrete pancreatic polypeptide, which regulates release of pancreatic digestive enzymes. The islets are infiltrated by blood capillaries and surrounded by clusters of cells (acini) that form the exocrine part of the pancreas.

Glucagon

The product of the alpha cells is **glucagon** (GLOO-ka-gon), a hormone whose principal physiological activity is to increase blood glucose level when it falls below normal (Fig. 18.24). The main target tissue of glucagon is the liver. Glucagon (1) accelerates the conversion of glycogen into glucose (glycogenolysis), (2) promotes formation of glucose from lactic acid (lactate) and certain amino acids (gluconeogenesis), and therefore (3) enhances release of glucose into the blood. As a result, the blood glucose level rises.

The level of blood glucose directly controls glucagon secretion via a negative feedback system. When the blood glucose level falls below normal, alpha cells of the islets secrete more glucagon. When blood glucose rises, the cells are no longer stimulated and production slackens. If for some reason the self-regulating device fails and the alpha cells secrete glucagon continuously, hyperglycemia may result. High-protein meals, which raise the amino acid level of the blood, stimulate glucagon secretion, whereas GHIH (somatostatin) inhibits it. Increased activity of the sympathetic division of the ANS, for example, during exercise, also enhances glucagon release.

Insulin

The beta cells of the islets produce the hormone **insulin**. Its chief physiological action is opposite that of glucagon. Insulin helps adjust blood glucose level by decreasing the level if necessary (Fig. 18.24). Insulin accelerates (1) the transport of glucose from the blood into cells, especially skeletal muscle fibers, (2) the conversion of glucose into glycogen (glycogenesis), (3) entry of amino acids into cells and synthesis of proteins, and (4) conversion of glucose or other nutrients into fatty acids (lipogenesis). Insulin also (5) decreases glycogenolysis and (6) slows gluconeogenesis.

The blood glucose level regulates secretion of insulin, as it does glucagon, in a negative feedback fashion. Increased blood glucose level stimulates insulin secretion whereas decreased blood glucose level inhibits it (Fig. 18.24). Increased blood levels of certain amino acids also stimulate insulin release, and several hormones stimulate insulin secretion, directly or indirectly. For instance, human growth hormone (hGH) and adrenocorticotropic hormone (ACTH) both raise blood glucose level, and the rise in glucose level triggers insulin secretion. GHIH (somatostatin) inhibits the

FIGURE 18.23 Location, blood supply, and histology of the pancreas.

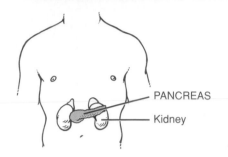

PANCREAS

Kidney

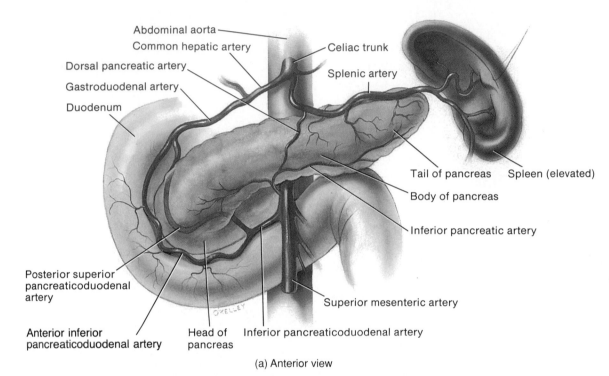

Abdominal aorta

Common hepatic artery

Dorsal pancreatic artery

Gastroduodenal artery

Duodenum

Celiac trunk

Splenic artery

Tail of pancreas   Spleen (elevated)

Body of pancreas

Inferior pancreatic artery

Posterior superior
pancreaticoduodenal
artery

Superior mesenteric artery

Anterior inferior
pancreaticoduodenal artery

Head of
pancreas

Inferior pancreaticoduodenal artery

OKELLEY

(a) Anterior view

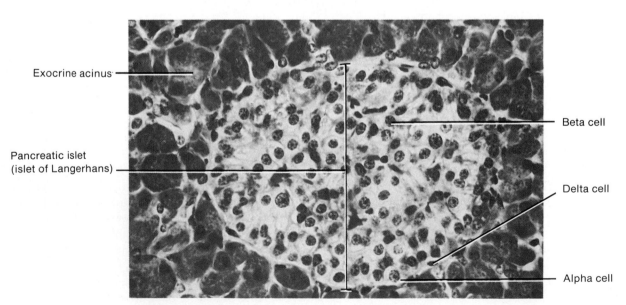

Exocrine acinus

Pancreatic islet
(islet of Langerhans)

Beta cell

Delta cell

Alpha cell

(b) Photomicrograph of a pancreatic islet (islet of Langerhans) and surrounding acini (260 x)

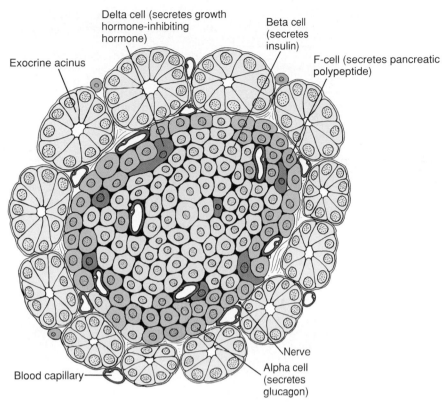

(c) Diagram of a pancreatic islet (islet of Langerhans) and surrounding acini

Question: What is the main target tissue of glucagon?

secretion of insulin. Increased activity of the parasympathetic division of the ANS, as occurs after a meal, stimulates insulin release.

CLINICAL APPLICATION

DIABETES MELLITUS AND HYPERINSULINISM

Diabetes mellitus (MEL-i-tus) is a group of disorders that all lead to an elevation of glucose in the blood (hyperglycemia). As hyperglycemia increases, there is also loss of glucose in the urine (glucosuria). Hallmarks of diabetes mellitus are the three "polys": an inability to reabsorb water, resulting in excessive urine production **(polyuria);** excessive thirst **(polydipsia);** and excessive eating **(polyphagia).**

The two major types of diabetes mellitus are type I and type II. In **type I diabetes** there is an absolute deficiency of insulin. Type I diabetes is called **insulin-dependent diabetes mellitus (IDDM)** because regular injections of insulin are required to prevent death. It previously was known as **juvenile-onset diabetes** because it most commonly develops in people younger than age 20, although it persists throughout life. IDDM appears to be an autoimmune disorder, one in which a person's immune system destroys the pancreatic beta cells, that occurs in genetically susceptible people. The drug cyclosporine, which suppresses the immune system, shows promise of being able to interrupt the destruction of beta cells.

The metabolism of an untreated type I diabetic is similar to that of a starving person. Since insulin is not present to aid the entry of glucose into body cells, most cells use fatty acids to produce ATP. By-products of fatty acid catabolism are organic acids called ketones (ketone bodies). As they accumulate, they cause a form of acidosis called **ketoacidosis,** which lowers the pH of the blood and can result in death. The catabolism of stored fats and proteins also causes weight loss. As lipids are transported by the blood from storage depots to cells, lipid particles are deposited on the walls of blood vessels. The deposition leads to atherosclerosis and a multitude of cardiovascular problems including cerebrovascular insufficiency, ischemic heart disease, peripheral vascular disease, and gangrene. One of the major complications of diabetes is loss of vision due to cataracts (excessive glucose attaches to lens proteins, causing cloudiness) or damage to blood vessels of the retina. Severe kidney problems also may result from damage to renal blood vessels.

Considerable research is being done to find ways of preventing IDDM or improving the status of IDDM patients. Work is ongoing on an artificial pancreas that monitors blood glucose level and automatically administers insulin from a reservoir. Several types of transplants have been tried. Transplantation of the entire pancreas is possible, but the person then needs to take immunosuppressive drugs to prevent rejection. Another approach has been to implant clusters of islet cells that have been pretreated to render them incapable of inducing rejection by the recipient. In other cases, clusters of islet cells have been encapsulated so that the insulin could get out, but elements responsible for rejection could not get in. In still another procedure, patients were injected with fetal islet cells.

Type II diabetes is much more common than type I, rep-

FIGURE 18.24 Regulation of the secretion of glucagon and insulin.

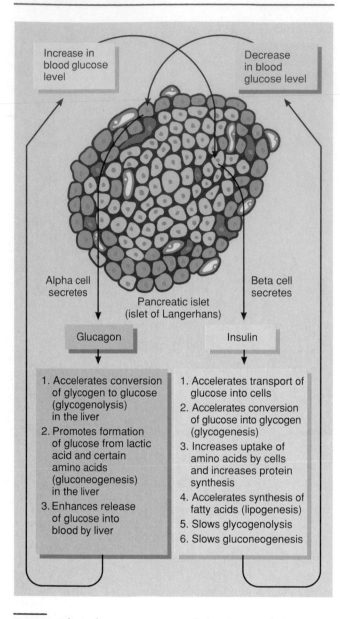

Increase in blood glucose level

Decrease in blood glucose level

Alpha cell secretes

Beta cell secretes

Pancreatic islet (islet of Langerhans)

Glucagon

Insulin

1. Accelerates conversion of glycogen to glucose (glycogenolysis) in the liver
2. Promotes formation of glucose from lactic acid and certain amino acids (gluconeogenesis) in the liver
3. Enhances release of glucose into blood by liver

1. Accelerates transport of glucose into cells
2. Accelerates conversion of glucose into glycogen (glycogenesis)
3. Increases uptake of amino acids by cells and increases protein synthesis
4. Accelerates synthesis of fatty acids (lipogenesis)
5. Slows glycogenolysis
6. Slows gluconeogenesis

Question: Why is glucagon sometimes called an "anti-insulin" hormone?

resenting more than 90% of all cases. Type II diabetes most often occurs in people who are over 40 and overweight. Since type II diabetes usually occurs later in life, it previously was called **maturity-onset diabetes.** Clinical symptoms are mild, and the high glucose levels in the blood usually can be controlled by diet, exercise, and weight loss. Sometimes, an antidiabetic drug such as *glyburide* (DiaBeta) is needed. This drug stimulates secretion of insulin by beta cells of the pancreas. Many type II diabetics, however, have a sufficient amount or even a surplus of insulin in the blood. For these people, diabetes arises not from a shortage of insulin but because cells become less sensitive to it, probably through down-regulation of insulin receptors. Type II diabetes is

therefore called **non-insulin-dependent diabetes mellitus (NIDDM).** However, some NIDDM patients do need insulin.

Hyperinsulinism most often results when a diabetic injects too much insulin. Rarely, it results from a malignant tumor or hyperplasia of the pancreatic islets. A tumor or hyperplastic pancreatic tissue can be surgically removed. The principal symptom is **hypoglycemia,** decreased blood glucose level, which stimulates the secretion of epinephrine, glucagon, and hGH. As a consequence, anxiety, sweating, tremor, increased heart rate, hunger, and weakness occur. Moreover, since brain cells do not have enough glucose to function efficiently, this condition leads to mental disorientation, convulsions, unconsciousness, and shock. This progression of events is called **insulin shock.** Death can occur quickly unless blood glucose level is raised.

Exhibit 18.7 presents a summary of the hormones produced by the pancreas, their principal actions, control of secretion, and selected disorders.

OVARIES AND TESTES

The female gonads, called the **ovaries,** are paired oval bodies located in the pelvic cavity. The ovaries produce female sex hormones called **estrogens** and **progesterone.** These hormones are responsible for the development and maintenance of female sexual characteristics. Along with the gonadotropic hormones of the pituitary gland, the sex hormones also regulate the female reproductive cycle, maintain pregnancy, and prepare the mammary glands for lactation. The ovaries also produce **inhibin,** a hormone that inhibits secretion of FSH (and, to a lesser extent, LH). Just before the birth of a baby, the ovaries and placenta also produce a hormone called **relaxin,** which relaxes the pubic symphysis and helps dilate the uterine cervix. These actions help ease the baby's passage through the birth canal.

The male has two oval gonads, called **testes,** that produce **testosterone,** the primary androgen (male sex hormone). Testosterone regulates production of sperm and stimulates the development and maintenance of male sexual characteristics. The testes also produce the hormone inhibin, which inhibits secretion of FSH. The detailed structure of the ovaries and testes and the specific roles of gonadotropic hormones and sex hormones will be discussed in Chapter 28.

Exhibit 18.8 presents a summary of hormones produced by the ovaries and testes and their principal actions.

PINEAL GLAND (EPIPHYSIS CEREBRI)

The endocrine gland attached to the roof of the third ventricle is known as the **pineal** (PĪN-ē-al) **gland** (named for its resemblance to a pine cone), or **epiphysis cerebri** (see Fig. 18.1). The gland is covered by a capsule formed by the pia mater and consists of masses of **neuroglia** and secretory

---

EXHIBIT 18.7

SUMMARY OF HORMONES PRODUCED BY THE PANCREAS, PRINCIPAL ACTIONS, CONTROL OF SECRETION, AND SELECTED DISORDERS

Hormone	Principal Actions	Control of Secretion	Selected Disorders
Glucagon	Raises blood glucose level by accelerating breakdown of glycogen into glucose in liver (glycogenolysis) and conversion of other nutrients into glucose in liver (gluconeogenesis) and releasing glucose into blood.	Decreased blood level of glucose, exercise, and largely protein meals stimulate glucagon secretion; GHIH (somatostatin) inhibits glucagon secretion.	
Insulin	Lowers blood glucose level by accelerating transport of glucose into cells, converting glucose into glycogen (glycogenesis), and decreasing glycogenolysis and gluconeogenesis; also increases lipogenesis and stimulates protein synthesis.	Increased blood level of glucose, hGH, ACTH, and gastrointestinal hormones stimulate insulin secretion; GHIH (somatostatin) inhibits insulin secretion.	A deficiency of insulin or defects in insulin receptors produce diabetes mellitus. Hypersecretion of insulin results in hyperinsulinism.
Growth hormone inhibiting hormone (GHIH) or somatostatin	Inhibits secretion of insulin and glucagon.		
Pancreatic polypeptide	Regulates release of pancreatic digestive enzymes.	Meals containing protein, fasting, exercise, and acute hypoglycemia stimulate pancreatic polypeptide secretion; GHIH (somatostatin) and elevated blood glucose inhibit pancreatic polypeptide secretion.	

EXHIBIT 18.8

SUMMARY OF HORMONES OF THE OVARIES AND TESTES AND THEIR PRINCIPAL ACTIONS

Hormone	Principal Actions
OVARIAN HORMONES	
Estrogens and progesterone	Development and maintenance of female sexual characteristics. Together with gonadotropic hormones of the anterior pituitary gland, they also regulate the menstrual cycle, maintain pregnancy, prepare the mammary glands for lactation, and regulate oogenesis.
Relaxin	Relaxes pubic symphysis and helps dilate uterine cervix near the end of pregnancy.
Inhibin	Inhibits secretion of FSH toward the end of the menstrual cycle.
TESTICULAR HORMONES	
Testosterone	Development and maintenance of male sexual characteristics, regulation of spermatogenesis, and stimulation of descent of testes before birth.
Inhibin	Inhibits secretion of FSH to control sperm production.

Eicosanoids (p. 556)

1. Prostaglandins and leukotrienes act as paracrines and autocrines in most body tissues by altering the production of second messengers, such as cyclic AMP.
2. Prostaglandins have a wide range of biological activity in normal physiology and pathology.

Growth Factors (p. 556)

1. Growth factors are local hormones that stimulate cell growth and division.
2. Examples include epidermal growth factor (EGF), platelet-derived growth factor (PDGF), fibroblast growth factor (FGF), nerve growth factor (NGF), tumor angiogenesis factors (TAFs), insulinlike growth factor (IGF), and cytokines.

Stress and the General Adaptation Syndrome (p. 556)

1. If the stress is extreme or unusual, it triggers a wide-ranging set of bodily changes called the general adaptation syndrome (GAS).
2. Unlike the homeostatic mechanisms, this syndrome does not maintain a constant internal environment.

Stressors (p. 556)

1. The stimuli that produce the general adaptation syndrome are called stressors.
2. Stressors include surgical operations, poisons, infections, fever, and strong emotional responses.

Alarm Reaction (p.. 557)

1. The alarm reaction is initiated by nerve impulses from the hypothalamus to the sympathetic division of the autonomic nervous system and adrenal medulla.

2. Responses are the immediate and brief fight-or-flight responses that increase circulation, promote catabolism for energy production, and decrease nonessential activities.

Resistance Reaction (p. 559)

1. The resistance reaction is initiated by regulating hormones secreted by the hypothalamus.
2. The regulating hormones are CRH, GHRH, and TRH.
3. CRH stimulates the anterior pituitary gland to increase its secretion of ACTH, which in turn stimulates the adrenal cortex to secrete hormones.
4. Resistance reactions are long term and accelerate catabolism to provide energy to counteract stress.
5. Glucocorticoids are produced in high concentrations during stress. They have many distinct physiological effects.

Exhaustion (p. 559)

1. The stage of exhaustion results from dramatic changes during alarm and resistance reactions.
2. Exhaustion is caused mainly by loss of potassium, depletion of adrenal glucocorticoids, and weakened organs. If stress is too great, it may lead to death.

Stress and Disease (p. 559)

1. It appears that stress can lead to certain diseases.
2. Among stress-related conditions are gastritis, ulcerative colitis, irritable bowel syndrome, asthma, anxiety, depression, peptic ulcers, hypertension, and migraine headaches.
3. A very important link between stress and immunity is interleukin-1 (IL-1) produced by macrophages; it stimulates secretion of ACTH.

Review Questions

1. Distinguish between an endocrine gland and an exocrine gland. (p. 518)
2. Contrast the nervous and endocrine system control of homeostasis. (p. 518)
3. What are the principal effects of hormones? (p. 518)
4. How are hormones related to receptors? Distinguish down-regulation from up-regulation. (p. 519)
5. Describe the chemical classification of hormones. Give an example of each. (p. 520)
6. How are hormones transported in the blood? (p. 522)
7. Describe the mechanism of hormonal action involving (a) activation of intracellular receptors and (b) interaction with plasma membrane receptors. (p. 522)
8. Distinguish among permissive effects, synergistic effects, and antagonistic effects. (p. 524)
9. How are negative feedback systems related to hormonal control? Discuss the various models of operation. (p. 524)
10. In what respect is the pituitary gland actually two glands? (p. 525)
11. Describe the histology of the anterior pituitary gland. Why does the anterior pituitary gland have such an abundant blood supply? (p. 525)
12. What hormones are produced by the anterior pituitary gland? What are their functions? How are they controlled? (p. 525)

13. Relate the importance of hypothalamic releasing and inhibiting hormones to secretions of the anterior pituitary gland. (p. 526)
14. Describe the clinical symptoms of pituitary dwarfism, giantism, and acromegaly. (p. 529)
15. Discuss the histology of the posterior pituitary gland and the function and regulation of its hormones. Describe the structure and importance of the supraopticohypophyseal tract. (p. 532)
16. What are the clinical symptoms of diabetes insipidus? (p. 535)
17. Describe the location and histology of the thyroid gland. (p. 535)
18. How are the thyroid hormones made, stored, and secreted? (p. 535)
19. Discuss the physiological effects of the thyroid hormones. How is the secretion of these hormones regulated? (p. 537)
20. Discuss the clinical symptoms of cretinism, myxedema, Graves' disease, and goiter. (p. 538)
21. Describe the function and control of calcitonin (CT). (p. 539)
22. Where are the parathyroids located? What is their histology? (p. 540)
23. What are the functions of parathyroid hormone (PTH)? (p. 540)

24. Discuss the clinical symptoms of tetany and osteitis fibrosa cystica. (p. 542)
25. Compare the adrenal cortex and adrenal medulla with regard to location and histology. (p. 542)
26. Describe the hormones produced by the adrenal cortex in terms of type, normal function, and control. (p. 544)
27. Describe the clinical symptoms of aldosteronism, Addison's disease, Cushing's syndrome, and congenital adrenal hyperplasia. (p. 545)
28. What relationship does the adrenal medulla have to the autonomic nervous system? What is the action of adrenal medullary hormones? (p. 548)
29. What is a pheochromocytoma? (p. 548)
30. Describe the location of the pancreas and the histology of the pancreatic islets. (p. 549)
31. What are the actions of glucagon and insulin? How are the hormones controlled? (p. 549)
32. Describe the clinical symptoms of diabetes mellitus and hyperinsulinism. Distinguish the two types of diabetes mellitus. (p. 551)
33. Why are the ovaries and testes considered to be endocrine glands? (p. 552)
34. Where is the pineal gland located? What are its presumed functions? (p. 552)
35. How are hormones of the thymus gland related to immunity? (p. 554)
36. Describe the effects of aging on the endocrine system. (p. 554)
37. Describe the development of the endocrine system. (p. 554)
38. List the hormones secreted by the gastrointestinal tract, placenta, kidneys, skin, and heart. (p. 558)
39. Describe the mode of action and function of prostaglandins and leukotrienes. (p. 556)
40. What are growth factors? List several and explain their actions. (p. 556)
41. Define the general adaptation syndrome (GAS). What is a stressor? (p. 556)
42. How do homeostatic responses differ from stress responses? (p. 556)
43. Outline the reactions of the body during the alarm stage, resistance stage, and stage of exhaustion when placed under stress. What is the central role of the hypothalamus during stress? (p. 557)
44. Explain how stress and immunity are related. (p. 559)

Answers to Questions with Figures

18.1 Secretions of endocrine glands diffuse into the blood; exocrine secretions flow into ducts that lead into body cavities or to the body surface.

18.2 It is a paracrine since it acts on nearby parietal cells without entering the blood.

18.3 Synthesis of new proteins takes several minutes to half an hour, due to the time needed for activation of DNA transcription and translation of the mRNA code of nucleotides into a chain of amino acids. The effects of steroid and thyroid hormones thus appear slowly.

18.4 It translates the presence of the first messenger, the water-soluble hormone, into a response inside the cell.

18.5 They carry blood from the median eminence of the hypothalamus, where hypothalamic releasing and inhibiting hormones are secreted, to the anterior pituitary, where these hormones act.

18.6 Thyroid, adrenal cortex, ovaries, testes.

18.7 The response is opposite to the stimulus.

18.8 Giantism.

18.10 Similarity: both carry hypothalamic hormones to the pituitary. Difference: tract is axons of neurons that extend from hypothalamus to posterior pituitary; portal veins extend to anterior pituitary.

18.11 Uterine contraction and milk ejection.

18.12 Absorption of the water in the intestines would decrease the osmotic pressure of your blood plasma, turning off secretion of ADH, and decreasing the ADH level in your blood.

18.13 Follicular cells secrete T_3 and T_4, also known as thyroid hormones. Parafollicular (C) cells secrete calcitonin.

18.14 Thyroglobulin.

18.15 The thyroid will enlarge (goiter formation) because TSH stimulates growth of the follicles.

18.16 Lack of iodine in the diet $\rightarrow$ diminished production of T_3 and T_4 $\rightarrow$ increased release of TSH $\rightarrow$ growth (enlargement) of thyroid gland $\rightarrow$ goiter.

18.17 PTH: bone and kidneys; CT: bone; calcitriol: gastrointestinal tract.

18.18 Parafollicular cells secrete CT; principal cells secrete PTH.

18.19 The adrenal glands sit atop the kidneys in the retroperitoneal space, at about the same horizontal level as the pancreas.

18.20 It acts to constrict blood vessels (by causing contraction of smooth muscle), and it stimulates secretion of aldosterone (by zona glomerulosa cells of the adrenal cortex), which in turn causes the kidneys to conserve water and increase blood volume.

18.21 Low, due to negative feedback suppression.

18.23 Liver.

18.24 It has several effects that are opposite to those of insulin.

18.25 All the anterior pituitary and hypothalamic hormones are peptides, proteins, or glycoproteins.

18.26 Homeostasis maintains controlled conditions typical of a normal internal environment whereas the GAS resets controlled conditions at a different level to cope with various stressors.

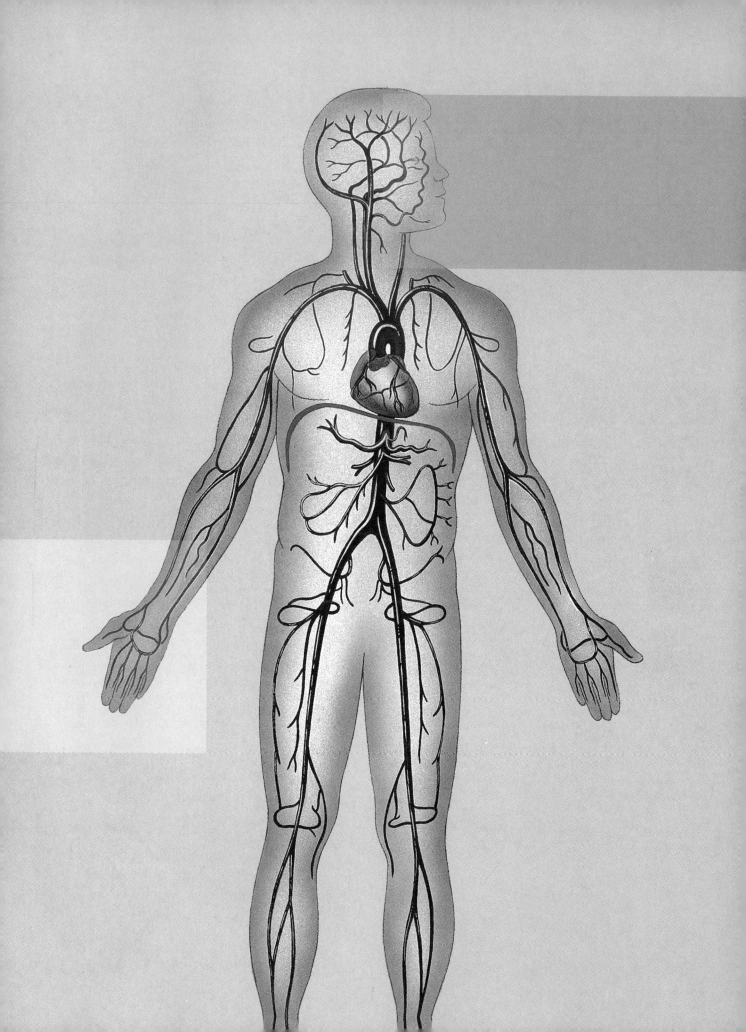

Unit 4
MAINTENANCE OF THE HUMAN BODY

This unit explains how the body maintains homeostasis on a day-to-day basis. In these chapters, you will be studying the interrelations among the cardiovascular, lymphatic, respiratory, digestive, and urinary systems. You will also learn about metabolism, fluid and electrolyte balance, and acid–base homeostasis.

Chapter 19

THE CARDIOVASCULAR SYSTEM: THE BLOOD

Chapter Contents at a Glance

Student Objectives

1. Contrast the general roles of blood, lymph, and interstitial fluid in maintaining homeostasis.

2. Define the functions and physical characteristics of the various components of blood.

3. List the components of plasma and explain their importance.

4. Compare the origins, histology, and functions of the formed elements in blood.

5. Identify the stages involved in blood clotting and explain the various factors that promote and inhibit blood clotting.

6. Explain ABO and Rh blood grouping.

7. Identify the clinical symptoms of several types of anemia, infectious mononucleosis (IM), and leukemia.

8. Define medical terminology associated with blood.

s cells differentiate and become more specialized for particular functions, they also become less capable of an independent existence. They are less able to protect themselves from extreme temperatures, toxic chemicals, and changes in pH. They cannot seek food or devour whole bits of food. And, if they are firmly implanted in a tissue, they cannot move away from their own wastes. The substance that bathes differentiated cells and carries out these vital functions for them is **interstitial fluid** (also known as **intercellular** or **tissue fluid**). When the fluid flows through lymphatic vessels it is called **lymph**. All fluids outside cells, including interstitial fluid, blood plasma (fluid in blood), lymph, and a few others such as aqueous humor in the eyes, comprise the **extracellular fluid** (see Fig. 1.2).

Because body cells are too specialized to adjust to more than very limited changes in their environment, the internal environment (interstitial fluid) must be kept within normal physiological limits. This is the condition of homeostasis. In preceding chapters, we have discussed how the internal environment is kept in homeostasis. Now we will look at that environment itself, beginning with the blood.

The blood, heart, and blood vessels together make up the **cardiovascular system.** The lymph, lymphatic vessels that transport lymph, and structures and organs containing lymphatic tissue (large clusers of white blood cells called lymphocytes) make up the **lymphatic system** (Chapter 22). The branch of science concerned with the study of blood, blood forming tissues, and the disorders associated with them is called **hematology** (hēm-a-TOL-ō-jē; *hem* = blood; *logos* = study of).

The developmental anatomy of blood and blood vessels is considered in Chapter 21.

COMPARISON OF EXTRACELLULAR FLUIDS

Both interstitial fluid and lymph are similar in composition to plasma. The principal chemical difference is that they contain less protein than plasma. This is because the larger protein molecules are not easily filtered through the endothelial cells that form capillary walls. The transfer of materials between blood and interstitial fluid occurs by osmosis, diffusion, and filtration across the endothelial cells. Interstitial fluid and lymph also differ from plasma in that they contain variable numbers of leukocytes (white blood cells). Like plasma, interstitial fluid and lymph lack erythrocytes (red blood cells) and platelets.

Blood plasma and lymph service the interstitial fluid. Blood picks up oxygen from the lungs, nutrients from the gastrointestinal tract, and hormones from endocrine glands. It transports these substances to the tissues, where they diffuse from capillaries (the smallest blood vessels) into interstitial fluid. From the interstitial fluid, needed substances

enter cells and cellular wastes enter the blood. The blood carries carbon dioxide and metabolic wastes to the lungs, kidneys, and sweat glands for elimination from the body. Certain wastes must be detoxified by the liver before they can be excreted.

Sometimes disease-causing organisms (pathogens) invade the interstitial fluid and blood and thus can be carried throughout the body. The lymphatic system helps protect the body from such spread of disease. Lymph picks up materials, including wastes, from interstitial fluid, cleanses them of bacteria, and returns them to the blood.

FUNCTIONS OF BLOOD

Blood is a liquid connective tissue that has three general functions: transportation, regulation, and protection.

1. **Transportation**. Blood transports oxygen from the lungs to the cells of the body and carbon dioxide from the cells to the lungs. It also carries nutrients from the gastrointestinal tract to the cells, heat and waste products away from cells, and hormones from endocrine glands to other body cells.
2. **Regulation**. Blood regulates pH through buffers. It also adjusts body temperature through the heat-absorbing and coolant properties of its water content and its variable rate of flow through the skin, where excess heat can be lost to the environment. Blood osmotic pressure also influences the water content of cells, principally through dissolved ions and proteins.
3. **Protection**. The clotting mechanism protects against blood loss, and certain phagocytic white blood cells or specialized plasma proteins such as antibodies, interferon, and complement protect against foreign microbes and toxins.

PHYSICAL CHARACTERISTICS OF BLOOD

Blood is heavier, thicker, and more viscous than water. It flows more slowly than water, at least in part because of its viscosity. The adhesive quality of blood, or its stickiness, may be observed by touching it. The temperature of blood is about 38°C (100.4°F), which is slightly higher than normal body temperature, and it has a slightly alkaline pH of about 7.40 (normal range 7.35 to 7.45). Blood constitutes about 8% of the total body weight. The blood volume is 5 to 6 liters (1.5 gal) in an average-sized male and 4 to 5 liters (1.2 gal) in an average-sized female.

CLINICAL APPLICATION

WITHDRAWING BLOOD

Blood samples for laboratory testing may be obtained in several ways. The most often used procedure is **venipunc-**

ture, withdrawal of blood from a vein. Veins are used instead of arteries because they are closer to the surface (more readily accessible) and they contain blood at a much lower pressure. A commonly used vein is the median cubital vein in front of the elbow (see Fig. 21.27). A tourniquet is wrapped around the arm to stop blood flow through the veins. This makes the veins below the tourniquet stand out. Opening and closing the fist has the same effect.

Another procedure used to withdraw blood is the **finger-stick.** A drop or two of capillary blood is taken from a finger, earlobe, or heel of the foot for evaluation.

Finally, an **arterial stick** may be used to withdraw

blood. The sample usually is taken from the radial artery in the wrist or the femoral artery in the groin.

COMPONENTS OF BLOOD

Whole blood is composed of two portions: 55% is blood plasma, a watery liquid containing dissolved substances, and 45% is formed elements, which are cells and cell fragments (Fig. 19.1).

FIGURE 19.1 Components of blood in a normal adult.

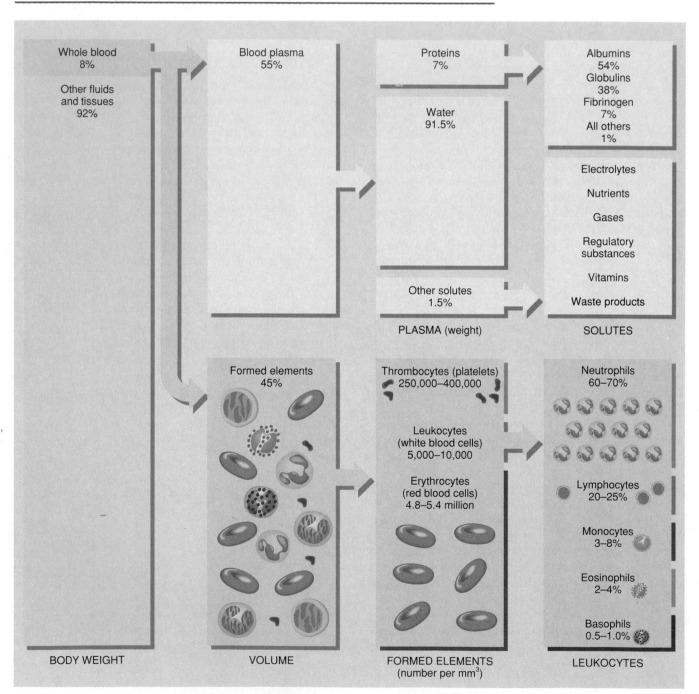

Question: What percentage of your body weight is red blood cells?

Blood Plasma

When the formed elements are removed from blood, a straw-colored liquid called **blood plasma** or simply **plasma** is left. Plasma is about 91.5% water and 8.5% solutes, most of which by weight (7%) are proteins. Some of the proteins in plasma are also found elsewhere in the body, but those confined to blood are called **plasma proteins.** These proteins play a role in maintaining proper blood osmotic pressure, which is important in total body fluid balance (Chapter 21).

Most plasma proteins are synthesized by the liver, including the **albumins** (54% of plasma proteins), **globulins** (38%), and **fibrinogen** (7%). Other solutes in plasma include waste products, such as urea, uric acid, creatinine, ammonia, and bilirubin; nutrients; vitamins; regulatory substances such as enzymes and hormones; gases; and electrolytes.

Exhibit 19.1 summarizes the chemical composition of plasma.

Formed Elements

The **formed elements** of the blood are:
Erythocytes (red blood cells)

Leukocytes (white blood cells)
 Granular leukocytes (granulocytes)
  Neutrophils
  Eosinophils
  Basophils
 Agranular leukocytes (agranulocytes)
  Lymphocytes
   T cells
   B cells
  Monocytes
Thrombocytes (platelets)

FORMATION OF BLOOD CELLS

The process by which blood cells are formed is called **hemopoiesis** (hē-mō-poy-Ē-sis), or **hematopoiesis.** During embryonic and fetal life, there are several centers for blood cell production. The yolk sac, liver, spleen, thymus gland, lymph nodes, and bone marrow all participate at various times in producing the formed elements. After birth, however, hemopoiesis takes place in red bone marrow (myeloid tissue). Red bone marrow is found in the proximal epiphyses of the humerus and femur; flat bones such as the sternum, ribs, and cranial bones; and the vertebrae and pelvis.

EXHIBIT 19.1

SUBSTANCES IN PLASMA

Constituent	Description
WATER	Liquid portion of blood; constitutes about 91.5% of plasma. Acts as solvent and suspending medium for solid components of blood and absorbs, transports, and releases heat.
SOLUTES	Constitute about 8.5% (by weight) of plasma.
PROTEINS	
Albumins	Smallest plasma proteins. Produced by liver and exert considerable osmotic pressure, which helps maintain water balance between blood and tissues and regulate blood volume. Also function as transport proteins for several steroid hormones.
Globulins	Protein group to which antibodies (immunoglobulins) belong. Produced by liver and plasma cells that derive from B lymphocytes. Antibodies help to attack measles, hepatitis, and polio viruses, and tetanus bacterium. Alpha and beta globulins transport iron, fats, and fat-soluble vitamins in the blood.
Fibrinogen	Produced by liver. Plays essential role in blood clotting.
WASTES	Most are breakdown products of protein metabolism and are carried by blood to organs of excretion. Include urea, uric acid, creatine, creatinine, bilirubin, and ammonium salts.
NUTRIENTS	Products of digestion passed into blood for distribution to all body cells. Include amino acids (from proteins), glucose (from carbohydrates), and fatty acids and glycerol (from triglycerides).
REGULATORY SUBSTANCES	Enzymes, produced by body cells, to catalyze chemical reactions. Hormones, produced by endocrine glands, to regulate growth and development in body.
GASES	Oxygen (O_2), carbon dioxide (CO_2), and nitrogen (N_2). Whereas there is more O_2 associated with hemoglobin inside red blood cells, there is more CO_2 dissolved in plasma. N_2 has no known function in the body.
ELECTROLYTES	Inorganic salts. Cations include Na^+, K^+, Ca^{2+}, Mg^{2+}; anions include Cl^-, HPO_4^{2-}, SO_4^{2-}, HCO_3^-. Help maintain osmotic pressure, and serve as essential minerals.

Stem cells, which give rise to differentiated blood cells and also replenish themselves, reside mainly in red bone marrow, although a few circulate in the blood. Five types of cells develop from the **pluripotent hematopoietic stem cells (hemocytoblasts)** (Fig. 19.2), which are derived from mesenchyme. Recall that the term *-blast* refers to a nucleated precursor cell.

1. **Proerythroblasts (rubriblasts)** form mature erythrocytes.
2. **Myeloblasts** form mature neutrophils, eosinophils, and basophils.
3. **Monoblasts** form mature monocytes.
4. **Lymphoblasts** form mature lymphocytes.
5. **Megakaryoblasts** form mature thrombocytes (platelets).

Several **hematopoietic growth factors** stimulate differentiation along particular paths and proliferation of certain progenitor cells. **Erythropoietin** (*poiem* = to make) or **EPO,** a hormone produced mainly by the kidneys and in small amounts by the liver, stimulates proliferation of erythrocyte precursors, and **thrombopoietin** stimulates formation of thrombocytes (platelets). In addition, there are several different **cytokines** that regulate hematopoiesis of different blood cell types. Cytokines are small glycoproteins produced by red bone marrow cells, leukocytes, macrophages, and fibroblasts. They act locally as autocrines or paracrines that maintain normal cell functions and stimulate proliferation. Two important families of cytokines that stimulate blood cell formation are called **colony stimulating factors (CSFs)** and the **interleukins.** The classes of hematopoietic growth factors and their functions are described in Exhibit 19.2 on page 572.

Most growth factors are available through recombinant DNA technology. They hold tremendous potential for medical uses in situations where a person's natural ability to form blood cells is diminished or defective. Recombinant erythropoietin (EPO) is very effective in treating the diminished erythrocyte production that accompanies end-stage kidney disease. Granulocyte-macrophage CSF (GM-CSF) and granulocyte CSF (G-CSF) or Neupogen successfully stimulate blood cell formation in cancer patients who are receiving chemotherapy, which tends to kill their bone marrow cells as well as the cancer cells. They also improve the outcome of bone marrow transplants. Other potential uses include treatment of AIDS (acquired immune deficiency syndrome), inborn defects of blood cell production, blood cell cancers, and severely burned patients.

ERYTHROCYTES (RED BLOOD CELLS)

More than 99% of the formed elements in blood are **erythrocytes** (e-RITH-rō-sīts) or **red blood cells (RBCs).** They contain the oxygen-carrying pigment **hemoglobin,** which is responsible for the red color of whole blood.

RBC Anatomy

Under the microscope, RBCs appear as biconcave discs averaging about 8 μm in diameter (Fig. 19.3a). The flexible, biconcave shape allows RBCs to squeeze through narrow capillaries, which may be only 3 μm wide. Mature red blood cells are quite simple in structure. They lack a nucleus and other organelles and can neither reproduce nor carry on extensive metabolic activities. The plasma membrane encloses hemoglobin, which was synthesized before loss of the nucleus and which constitutes about 33% of the cell weight. Hemoglobin is dissolved in the cytosol. Normal values for hemoglobin are 14 to 20 g/100 ml of blood in infants, 12 to 15 g/100 ml in adult females, and 14 to 16.5 g/100 ml in adult males. As you will see later, certain proteins (antigens) on the surfaces of red blood cells are responsible for the various blood groups. The ABO and Rh groups are examples.

RBC Physiology

As blood passes through the lungs, hemoglobin inside RBCs combines with oxygen to form oxyhemoglobin. A hemoglobin molecule consists of a protein called **globin,** composed of four polypeptide chains (two called alpha and two called beta), plus four nonprotein pigments called **hemes.** Each heme contains an iron ion (Fe^{2+}) that can combine reversibly with one oxygen molecule (Fig. 19.3b). The oxygen is transported in this state to other tissues of the body. In the tissues, the iron–oxygen reaction reverses. Hemoglobin releases oxygen, which diffuses into the interstitial fluid and from there into cells.

Red blood cells are highly specialized for their oxygen transport function. Each one contains about 280 million hemoglobin molecules. Since RBCs have no nucleus, all their internal space is available for oxygen transport. Moreover, since they lack mitochondria and generate ATP anaerobically (without oxygen), RBCs do not consume any of the oxygen that they transport. Even the shape of a RBC facilitates its function. A biconcave disk has a much greater surface area for its volume than, say, a sphere or a cube. This shape confers two advantages. First, there is a large surface area for the diffusion of gas molecules into or out of the RBC. Second, the biconcave disk is a very flexible shape, which permits RBCs to squeeze through narrow capillaries.

A serious disorder called sickle-cell anemia (SCA) is due to a genetic defect that results in substitution of just two out of 574 amino acids in hemoglobin. Replacement of the polar amino acid glutamate by nonpolar valine at one position in each of the two beta chains greatly decreases the solubility of deoxygenated hemoglobin in water. When this abnormal hemoglobin is exposed to low oxygen, it forms crystals that deform RBCs into a characteristic sickle shape (see Fig. 19.13). Sickled RBCs are more rigid and may lodge in small capillaries. Sickle-cell anemia is discussed at the end of the chapter.

FIGURE 19.2 Origin, development, and structure of blood cells.

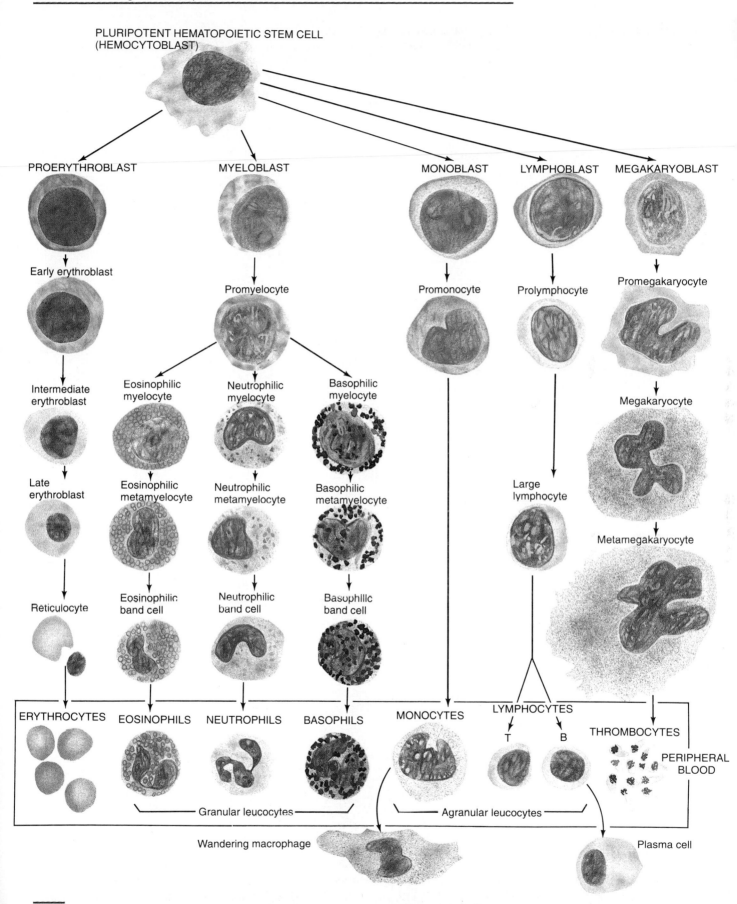

PLURIPOTENT HEMATOPOIETIC STEM CELL
(HEMOCYTOBLAST)

PROERYTHROBLAST   MYELOBLAST   MONOBLAST   LYMPHOBLAST   MEGAKARYOBLAST

Early erythroblast   Promyelocyte   Promonocyte   Prolymphocyte   Promegakaryocyte

Intermediate erythroblast   Eosinophilic myelocyte   Neutrophilic myelocyte   Basophilic myelocyte   Megakaryocyte

Late erythroblast   Eosinophilic metamyelocyte   Neutrophilic metamyelocyte   Basophilic metamyelocyte   Large lymphocyte   Metamegakaryocyte

Reticulocyte   Eosinophilic band cell   Neutrophilic band cell   Basophilic band cell

ERYTHROCYTES   EOSINOPHILS   NEUTROPHILS   BASOPHILS   MONOCYTES   LYMPHOCYTES   THROMBOCYTES
        T   B   PERIPHERAL BLOOD

Granular leucocytes   Agranular leucocytes

Wandering macrophage   Plasma cell

Question: List the following physical characteristics of blood: temperature, pH, and percentage of total body weight.

EXHIBIT 19.2

HEMATOPOIETIC GROWTH FACTORS

Factor	Function
Stem Cell Growth Factor (steel factor)	Stimulates pluripotent hematopoietic stem cells (hemocytoblasts).
Interleukin-3 (multi-CSF[a])	Stimulates pluripotent hematopoietic stem cells and progenitors of eosinophils, neutrophils, basophils, monocytes, and platelets.
Granulocyte-Macrophage CSF (GM-CSF)	Stimulates development of erythrocytes, platelets, granulocytes (eosinophils, neutrophils, and basophils), and monocytes.
Macrophage CSF (M-CSF)	Stimulates development of monocytes and macrophages.
Granulocyte CSF (G-CSF)	Stimulates development of neutrophils.
Interleukin-5	Stimulates development of eosinophils.
Interleukin-7	Stimulates development of B lymphocytes.

[a] CSF = Colony stimulating factor.

Hemoglobin also transports about 23% of the total carbon dioxide, a waste product of metabolism. Blood flowing through tissue capillaries picks up carbon dioxide, some of which combines with amino acids in the globin portion of hemoglobin to form carbaminohemoglobin. This complex is transported to the lungs, where the carbon dioxide is released and then exhaled.

CLINICAL APPLICATION

INDUCED ERYTHROCYTHEMIA (BLOOD DOPING)

In recent years, some athletes have been tempted to try **induced erythrocythemia (blood doping).** Red blood cells are removed from the body, stored for a month or so, and then reinjected a few days before an athletic event. Since delivery of oxygen to muscle is a limiting factor in muscular feats, and red blood cells carry oxygen, it was predicted that increasing the oxygen-carrying capacity of the blood could increase muscular performance. In one major study, it has been shown that induced erythrocythemia does improve athletic performance in endurance events. However, the practice is dangerous because it increases the work load of the heart. With increased numbers of RBCs, the viscosity of the blood rises, which makes the blood more difficult for the heart to pump. Moreover, the procedure is considered dishonest by the International Olympics Committee.

RBC Life Span and Number

Red blood cells live only about 120 days because of wear and tear on their plasma membranes as they squeeze

FIGURE 19.3 Shape of red blood cells (RBCs) and hemoglobin molecule. In (b) the four polypeptide chains of one hemoglobin molecule are indicated in green and blue. Each has one heme group that contains Fe^{2+} (red).

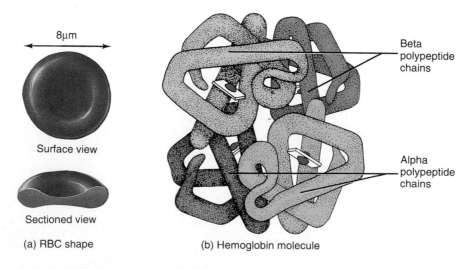

8μm

Surface view

Sectioned view

(a) RBC shape

Beta polypeptide chains

Alpha polypeptide chains

(b) Hemoglobin molecule

Question: How many molecules of O_2 can one hemoglobin molecule transport?

through blood capillaries. Without a nucleus and other organelles, RBCs cannot synthesize new components to replace damaged ones. The plasma membrane thus becomes more fragile with age and the cell more likely to burst, especially as it squeezes through narrow channels in the spleen. Worn-out red blood cells are removed from circulation and destroyed by fixed phagocytic macrophages in the spleen and liver and the breakdown products are recycled.

A healthy male has about 5.4 million red blood cells per cubic millimeter (mm^3) of blood, and a healthy female has about 4.8 million. (There are about 50 mm^3 in a drop of blood.) The higher value in males is due to their higher levels of testosterone, which stimulates the synthesis of erythropoietin. To maintain normal quantities of erythrocytes, new mature cells must enter the circulation at the astonishing rate of at least 2 million per second. This pace assures a constant balance between production and destruction of RBCs.

After phagocytosis of RBCs by macrophages, hemoglobin is recycled (Fig. 19.4). The globin portion of hemoglobin is split from the heme and broken down into amino acids that may be reused for protein synthesis. The heme portion is broken down into (1) **iron,** which associates with proteins to form **ferritin** or **hemosiderin,** and (2) **bilirubin,** a pigment that does not contain iron. Ferritin and hemosiderin are storage forms of iron found mostly in muscle fibers, liver cells, and macrophages of the spleen and liver. Upon release from a storage site or absorption from the gastrointestinal tract, iron attaches to a plasma beta globulin called **transferrin.** In this form it is delivered to bone marrow, where RBC precursors take it up through receptor-mediated endocytosis (see Fig. 3.12) for use in production of new hemoglobin molecules. The non-iron portion of heme is converted into **biliverdin,** a greenish pigment, and then into bilirubin. Bilirubin enters the blood and is secreted by liver cells into bile. Bile passes from the liver into the small intestine. Bacteria in the large intestine convert bilirubin into **urobilinogen.** Some urobilinogen is absorbed back into the blood, converted to urobilin (yellow), and excreted in urine. Most urobilinogen is eliminated in feces in the form of a brown pigment called **stercobilin,** which gives feces its characteristic color.

Production of RBCs

The process of erythrocyte formation is called **erythropoiesis** (e-rith′-rō-poy-Ē-sis). It starts in red bone marrow with a proerythroblast (see Fig. 19.2). The **proerythroblast (rubriblast)** gives rise to an **early erythroblast (prorubricyte),** which then develops into an **intermediate erythroblast (rubricyte),** the first cell in the sequence that begins to synthesize hemoglobin. The intermediate erythroblast next develops into a **late erythroblast (metarubricyte),** in which hemoglobin synthesis is at a maximum. In the next stage, the late erythroblast ejects its nucleus and becomes a

reticulocyte. Loss of the nucleus allows the center of the cell to indent, giving the cell a biconcave shape. Reticulocytes contain about 34% hemoglobin and retain some mitochondria, ribosomes, and endoplasmic reticulum. They pass from bone marrow into the bloodstream by squeezing between the endothelial cells of blood capillaries. Normally, they develop into **erythrocytes,** or mature red blood cells, within one to two days after their release from bone marrow. (A nucleated red blood cell found in red marrow, but rarely found in blood is called a **normoblast.**)

Normally, erythropoiesis and red blood cell destruction proceed at the same pace. If the oxygen-carrying capacity of the blood falls because erythropoiesis is not keeping up with RBC destruction, a negative feedback system steps up erythrocyte production (Fig. 19.5). The controlled condition is the rate of oxygen delivery to body tissues. Oxygen delivery may fall due to **anemia,** a lower than normal number of RBCs or quantity of hemoglobin, or circulatory problems that reduce blood flow to tissues. Cellular oxygen deficiency, called **hypoxia** (hī-POKS-ē-a), may also occur if not enough oxygen enters the blood, for example, when you do not breathe in enough oxygen. This situation commonly occurs at high altitudes, where the air contains less oxygen. Whatever the cause, hypoxia stimulates the kidneys to step up release of the hormone **erythropoietin.** This hormone circulates through the blood to the red bone marrow, where it speeds the development of proerythroblasts into reticulocytes.

Anemia has many causes: lack of iron, lack of certain amino acids, and lack of vitamin B$_{12}$ are but a few. Iron is needed for the heme part of the hemoglobin molecule. The amino acids are needed for the protein, or globin, part. Vitamin B$_{12}$ helps the red bone marrow to produce erythrocytes. This vitamin is obtained from meat, especially liver, but it cannot be absorbed by the lining of the small intestine without the help of another substance—**intrinsic factor** (**IF**) produced by the parietal cells of the stomach mucosa. Intrinsic factor facilitates the absorption of vitamin B$_{12}$ (extrinsic factor). Once absorbed, it is stored in the liver. (Several types of anemia are described later in the chapter.)

CLINICAL APPLICATION

RETICULOCYTE COUNT AND HEMATOCRIT

The rate of erythropoiesis is measured by a **reticulocyte** (re-TIK-yoo-lō-sīt) **count.** Normally, a little less than 1% of the oldest RBCs are replaced by newcomer reticulocytes on any given day. Then it takes one to two days for the reticulocyte to become a mature RBC. Thus reticulocytes account for about 0.5 to 1.5% of all RBCs in a blood sample. A low "retic" count in a person who is anemic might indicate inability of the bone marrow to respond, perhaps because of a nutritional deficiency, lack of intrinsic factor, or leukemia. A high "retic" count might indicate a good bone marrow response to previous loss of blood or to iron therapy in someone who is iron-deficient.

Hematocrit (he-MAT-ō-krit), or **Hct,** is the percentage of red blood cells in blood. A Hct of 40 means that 40% of the

FIGURE 19.4 Formation and destruction of red blood cells and recycling of hemoglobin components.

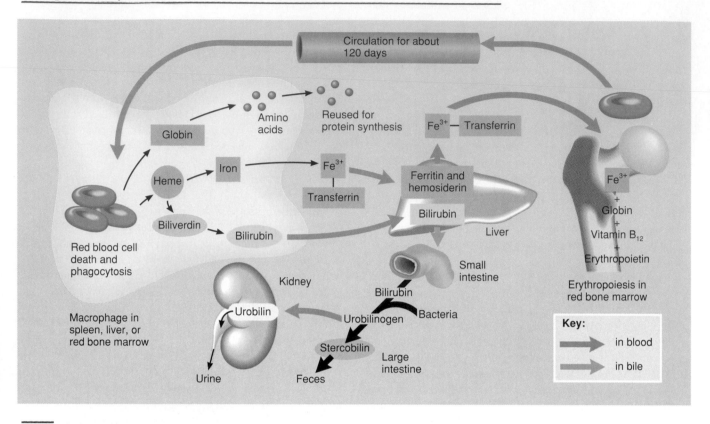

Question: What is the function of transferrin?

volume of blood is composed of RBCs. The test is used to diagnose anemia and polycythemia (abnormally high percentage of RBCs) and abnormal states of hydration. The normal range of hematocrit for females is 38 to 46% (average 42); for males it is 40 to 54% (average 47). A significant drop in hematocrit constitutes anemia. Polycythemic blood may have a hematocrit of 65% or higher. Athletes not uncommonly have a higher-than-average hematocrit, and the average Hct of people living in mountainous terrain is greater than that of people living at sea level due to their greater amounts of erythropoietin.

LEUKOCYTES (WHITE BLOOD CELLS)

WBC Anatomy and Types

Unlike red blood cells, **leukocytes** (LOO-kō-sīts) or **white blood cells (WBCs)** have a nucleus and do not contain hemoglobin (see Fig. 19.2). The two major groups of WBCs are granular leukocytes and agranular leukocytes. **Granular leukocytes (granulocytes)** have lobed nuclei and conspicuous granules in the cytoplasm. The three types are **neutrophils** (NOO-trō-fils), which are 10 to 12 μm in diameter; **eosinophils** (ē-ō-SIN-ō-fils), which are 10 to 12 μm in diameter; and **basophils** (BĀ-sō-fils), which are 8 to 10 μm in diameter. These names reflect the types of granules seen when using common hematology stains, such as Wright's stain, which include both acidic eosin and a basic dye (Fig. 19.6).

The nuclei of neutrophils have two to six lobes, connected by very thin strands. As the cells age, the extent of nuclear lobulation increases. Because older neutrophils appear to have many differently shaped nuclei, they are often called **polymorphonuclear leukocytes (PMNs), polymorphs,** or **"polys."** Younger neutrophils are often called **bands** because their nucleus is more rod-shaped. When stained, the cytoplasm of neutrophils includes fine, evenly distributed pale lilac-colored granules. The nucleus of an eosinophil usually has two lobes connected by a thin or thick strand. Large, uniform-sized granules pack the cytoplasm but usually do not cover or obscure the nucleus. These eosinophilic (= eosin loving) granules stain red-orange. A basophil's nucleus is bilobed or irregular in shape, often in the form of a letter S. The cytoplasmic

FIGURE 19.5 Negative feedback regulation of erythropoiesis (red blood cell formation).

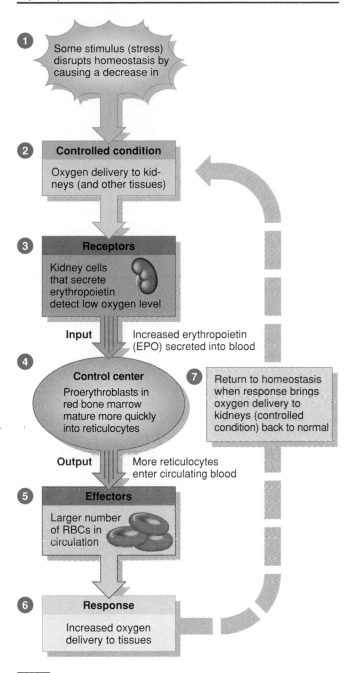

1. Some stimulus (stress) disrupts homeostasis by causing a decrease in

2. **Controlled condition**
   Oxygen delivery to kidneys (and other tissues)

3. **Receptors**
   Kidney cells that secrete erythropoietin detect low oxygen level

   Input Increased erythropoietin (EPO) secreted into blood

4. **Control center**
   Proerythroblasts in red bone marrow mature more quickly into reticulocytes

   7. **Return to homeostasis** when response brings oxygen delivery to kidneys (controlled condition) back to normal

   Output More reticulocytes enter circulating blood

5. **Effectors**
   Larger number of RBCs in circulation

6. **Response**
   Increased oxygen delivery to tissues

Question: What is the term given to cellular oxygen deficiency?

FIGURE 19.6 Blood smear and photomicrographs of individual blood cells.

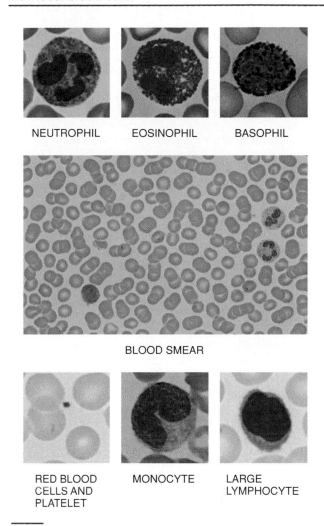

NEUTROPHIL EOSINOPHIL BASOPHIL

BLOOD SMEAR

RED BLOOD CELLS AND PLATELET MONOCYTE LARGE LYMPHOCYTE

Question: Which WBCs are granulocytes and why are they called that?

nuclei of lymphocytes are darkly stained, round, or slightly indented. The cytoplasm stains sky blue and forms a rim around the nucleus. The nuclei of monocytes are usually indented or kidney-shaped, and the cytoplasm has a foamy appearance. The blood is merely a conduit for monocytes, which migrate out into the tissues, enlarge, and differentiate into **macrophages.** Some are **fixed macrophages,** which means they reside in a particular tissue, for example, alveolar macrophages, spleen macrophages, or stellate reticuloendothelial (Küpffer) cells in the liver. Others are **wandering (free) macrophages,** which roam the tissues and gather at sites of infection or inflammation.

Just as red blood cells have surface proteins, so do white blood cells and all other nucleated cells in the body. Some of these proteins, called **major histocompatibility (MHC) antigens,** are unique for each person (except for identical twins).

basophilic granules are round, variable in size, stain blueblack, and commonly obscure the nucleus.

In **agranular leukocytes (agranulocytes),** no cytoplasmic granules can be seen under a light microscope, owing to their small size and poor staining qualities. The two kinds of agranular leukocytes are **lymphocytes** (LIM-fō-sīts), which are 7 to 15 μm in diameter, and **monocytes** (MON-ō-sīts), 14 to 19 μm in diameter (Fig. 19.6). The

HISTOCOMPATIBILITY TESTING

The success of a proposed organ or tissue transplant depends on **histocompatibility** (his'-tō-kom-pat-i-BIL-i-tē), that is, the tissue compatibility between the donor and the recipient. The more similar the MHC antigens, the greater the histocompatibility, and the higher the probability the transplant will not be rejected. **Tissue typing (histocompatibility testing)** is done before any organ transplant. A nationwide computerized registry helps physicians select the most histocompatible and needy organ transplant recipients whenever donor organs become available. Also, in cases of disputed parentage, tissue typing can be used to establish paternity.

WBC Physiology

The skin and mucous membranes of the body are continuously exposed to microbes and their toxins. Some of these microbes can invade deeper tissues to cause disease. Once pathogens enter the body, the general function of white blood cells is to combat them by phagocytosis or immune responses. Neutrophils and macrophages are active in **phagocytosis**: they can ingest bacteria and dispose of dead matter (see Fig. 3.11). Several different chemicals in inflamed tissue attract phagocytes toward the tissue. This phenomenon is called **chemotaxis.** Among the substances that provide stimuli for chemotaxis are toxins produced by microbes, specialized products of damaged tissues called kinins, and some of the colony stimulating factors, such as GM-CSF and G-CSF. The CSFs also enhance phagocytic activity of neutrophils and macrophages.

Most leukocytes possess, to some degree, the ability to squeeze through the minute spaces between the cells that form the walls of capillaries and through connective and epithelial tissues. This movement through capillary walls is called **diapedesis** (dī'-a-pe-DĒ-sis) or **emigration.** First, part of the cell membrane projects outward and the cytoplasm and nucleus flow into the projection. Then, the rest of the membrane follows. Another projection is made, and so on, until the cell has migrated to its destination (see Fig. 22.11).

Among the WBCs, neutrophils respond to tissue destruction by bacteria most quickly. After engulfing a pathogen during phagocytosis, a neutrophil unleashes several destructive chemicals. These include the enzyme **lysozyme,** which destroys certain bacteria, and **strong oxidants,** such as the superoxide anion (O_2^-), hydrogen peroxide (H_2O_2), and hypochlorite anion (OCl^-), which is similar to household bleach. Neutrophils also contain **defensins,** proteins that exhibit a broad range of antibiotic activity against bacteria, fungi, and viruses. Defensins form peptide spears that poke holes in microbe membranes. The resulting leakiness kills the invader.

Monocytes take longer to reach a site of infection than do neutrophils, but they arrive in larger numbers and destroy more microbes. Upon arrival they enlarge and differentiate into wandering macrophages, which clean up cellular debris and microbes following an infection.

Eosinophils leave the capillaries and enter tissue fluid. They are believed to release enzymes, such as histaminase, that combat the effects of histamine and other mediators of inflammation in allergic reactions. Eosinophils also phagocytize antigen–antibody complexes (described shortly) and are effective against certain parasitic worms. A high eosinophil count often indicates an allergic condition or a parasitic infection.

Basophils are also involved in inflammatory and allergic reactions. They leave capillaries, enter tissues, and develop into mast cells, which can liberate heparin, histamine, and serotonin. These substances intensify the inflammatory reaction and are involved in hypersensitivity (allergic) reactions (Chapter 22).

The two major types of lymphocytes are B cells and T cells. These cells are the major combatants in immune responses (described in detail in Chapter 22). Briefly, a substance that stimulates an immune response is called an **antigen** (AN-ti-jen). Most antigens are foreign proteins, ones not synthesized by the body. They include bacterial enzymes, toxins, and structural proteins. In response to certain antigens, **B cells** develop into **plasma cells** that produce antibodies (Fig. 19.7). **Antibodies** (AN-ti-bod'-ēz) are proteins of the immunoglobulin family. A specific antibody will generally bind only to a certain antigen. The two together are called an antigen–antibody complex. However, unlike enzymes, which enhance the reactivity of the substrate, antibodies "cover" their antigens so the antigens cannot come in contact with other chemicals in the body. In this way, bacterial poisons are inactivated, and the bacteria themselves are destroyed. This process is called the **antigen–antibody response.**

Antigens also stimulate **T cells** to take part in immune responses. One group of T cells, the **cytotoxic (killer) T cells,** react by destroying foreign invaders directly. Another group, called **helper T cells**, assists both B cells and cytotoxic T cells. T cells are especially effective against viruses, fungi, transplanted cells, cancer cells, and some bacteria.

Immune responses mediated by B and T cells help combat infection and provide protection against some diseases. They also are responsible for transfusion reactions, allergies, and the body's rejection of organs transplanted from a person with different MHC antigens.

An increase in the number of circulating WBCs usually indicates inflammation or infection. Because each type of white blood cell plays a different role, determining the *percentage* of each type in the blood assists in diagnosing the condition.

DIFFERENTIAL WHITE BLOOD CELL COUNT

To evaluate infection or inflammation, determine the effects of possible poisoning by chemicals or drugs, monitor blood disorders (for example, leukemia) and effects of

FIGURE 19.7 Production of plasma cells and antigen–antibody response.

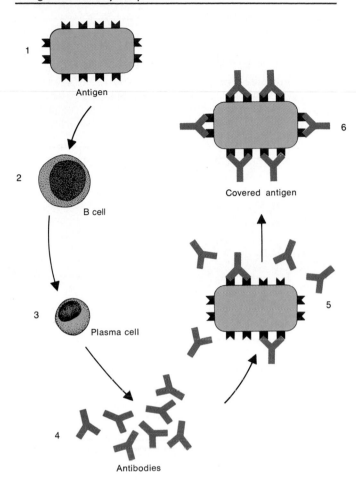

Question: What is the chemical nature and function of antibodies?

chemotherapy, or detect allergic reactions and parasitic infections, a physician may order a **differential white blood cell count.** The percentage of each type of white blood cell in a normal differential white blood cell count is as follows:

	Percentage
Neutrophils	60–70
Lymphocytes	20–25
Monocytes	3–8
Eosinophils	2–4
Basophils	0.5–1
	100

A high neutrophil count might result from bacterial infections, burns, stress, or inflammation; a low count might be caused by radiation, certain drugs, vitamin B_{12} deficiency, or systemic lupus erythematosus (SLE). A high eosinophil count could indicate allergic reactions, parasitic infections, autoimmune disease, or adrenal insufficiency; a low count could be caused by certain drugs, stress, or Cushing's syndrome. Basophils could be elevated in some types of allergic responses, leukemias, cancers, and hypothyroidism; decreases could occur during pregnancy, ovulation, stress,

and hyperthyroidism. High lymphocyte counts could indicate viral infections, immune diseases, and some leukemias; low counts might occur as a result of prolonged severe illness, high steroid levels, and immunosuppression. Finally, a high monocyte count could result from certain viral or fungal infections, tuberculosis (TB), some leukemias, and chronic diseases; below normal monocyte levels rarely occur.

WBC Life Span and Number

Bacteria exist everywhere in the environment and have continuous access to the body through the mouth, nose, and pores of the skin. Furthermore, many cells, especially those of epithelial tissue, age and die daily, and their remains must be disposed of by phagocytes that actively ingest bacteria and debris. However, the engulfed debris interferes with normal metabolic activities and sooner or later causes death of the phagocyte, which then is consumed by another phagocyte. In a healthy body, some WBCs, especially lymphocytes, can live for several months or years, but most live only a few days. During a period of infection, phagocytic WBCs may live only a few hours.

WBCs are far less numerous than red blood cells, averaging from 5000 to 10,000 cells per cubic millimeter (mm^3) of blood. RBCs therefore outnumber white blood cells about 700:1. The term **leukocytosis** (loo′-kō-sī-TŌ-sis) refers to an increase in the number of WBCs. An abnormally low level of white blood cells (below 5000/mm^3) is termed **leukopenia** (loo-kō-PĒ-nē-a).

CLINICAL APPLICATION

BONE MARROW TRANSPLANT

Among the leukocytes in bone marrow are cells that combat infections. A **bone marrow transplant** is the transfer of bone marrow from a donor to a recipient. The donor marrow must be very closely matched to that of the recipient. In the procedure, donor marrow is aspirated from the hipbones, mixed with heparin (an anticoagulant), and then passed through screens. The suspension of bone marrow cells is treated to remove T cells and injected into a vein. The cells in the suspension pass through the lungs, enter the general circulation, and reseed and grow in the marrow cavities of the recipient's bones.

Bone marrow transplants have been used to treat aplastic anemia, certain types of leukemia, and severe combined immunodeficiency disease (SCID), an inherited deficiency of infection-fighting blood cells. The technique is also being used to treat Hodgkin's disease, non-Hodgkin's lymphoma, thalassemia, multiple myeloma, sickle-cell anemia (SCA), and hemolytic anemia.

THROMBOCYTES (PLATELETS)

Besides the immature cell types that develop into erythrocytes and leukocytes, pluripotent hematopoietic stem cells also differentiate into megakaryoblasts (see Fig. 19.2). Megakaryoblasts ultimately transform into metamegakaryocytes, large cells that shed fragments of cytoplasm. Each fragment becomes enclosed by a piece of the cell membrane and is called a **thrombocyte** (THROM-bō-sīt), or **platelet.** Thrombocytes break off from the metamegakaryocytes in bone marrow and then enter the blood circulation. Between 250,000 and 400,000 platelets are present in each cubic millimeter (mm^3) of blood. They are disc-shaped, 2 to 4 μm in diameter, and exhibit many granules but no nucleus. Platelets help repair slightly damaged blood vessels. Their granules contain chemicals that upon release promote blood clotting. Platelets have a short life span, normally just five to nine days. Aged and dead platelets are removed by fixed macrophages in the spleen and liver.

A summary of the formed elements in blood is presented in Exhibit 19.3.

CLINICAL APPLICATION

COMPLETE BLOOD COUNT (CBC)

A **complete blood count** is a valuable test that screens for anemia and various infections. Usually included are a determination of red blood cell count, hemoglobin, hematocrit, white blood cell count, differential white blood cell count, and platelet count.

EXHIBIT 19.3

SUMMARY OF THE FORMED ELEMENTS IN BLOOD

Formed Elements	Number	Diameter (in μm)	Life Span	Function
ERYTHROCYTES (RED BLOOD CELLS)	4.8 million/mm^3 in females; 5.4 million/mm^3 in males.	8	120 days.	Transport oxygen and carbon dioxide.
LEUKOCYTES (WHITE BLOOD CELLS)	5000–10,000/mm^3.		Few hours to a few days.[a]	
Granular leukocytes				
Neutrophils	60–70% of total.	10–12		Phagocytosis. Destruction of bacteria with lysozyme, defensins, and strong oxidants, such as superoxide anion, hydrogen peroxide, and hypochlorite anion.
Eosinophils	2–4% of total.	10–12		Combat the effects of histamine in allergic reactions, phagocytize antigen–antibody complexes, and destroy certain parasitic worms.
Basophils	0.5–1% of total.	8–10		Liberate heparin, histamine, and serotonin in allergic reactions that intensify the overall inflammatory response.
Agranular leukocytes				
Lymphocytes	20–25% of total.	7–15		Mediate immune responses, including antigen–antibody reactions. B cells develop into plasma cells, which secrete antibodies. T cells attack invading viruses, cancer cells, and transplanted tissue cells.
Monocytes	3–8% of total.	14–19		Phagocytosis (after transforming into fixed or wandering macrophages).
THROMBOCYTES (PLATELETS)	250,000–400,000/mm^3	2–4	5–9 days.	Blood clotting.

[a]Some lymphocytes, called T and B memory cells, can live for many years once they are established. Most white blood cells, however, have life spans ranging from a few hours to a few days.

HEMOSTASIS

Hemostasis (hē′-mō-STĀ-sis) refers to the stoppage of bleeding. When blood vessels are damaged or ruptured, the hemostatic response must be quick, localized to the region of damage, and carefully controlled. Three basic mechanisms prevent blood loss: (1) vascular spasm, (2) platelet plug formation, and (3) blood coagulation (clotting). These mechanisms are useful for preventing hemorrhage in smaller (microcirculation) blood vessels, but extensive hemorrhage from larger vessels usually requires medical intervention.

Vascular Spasm

When blood vessels (arteries or arterioles) are damaged, the circularly arranged smooth muscle in its wall contracts immediately. This is called a **vascular spasm** and it reduces blood loss for several minutes to several hours, during which time the other hemostatic mechanisms go into operation. The spasm is probably caused by damage to the smooth muscle and from reflexes initiated by pain receptors.

Platelet Plug Formation

In their unstimulated state, platelets are disc-shaped. Considering their small size, platelets pack an impressive array of chemicals. Two types of granules are present in the cytoplasm: (1) *alpha granules* contain clotting factors and platelet-derived growth factor (PDGF), which can cause proliferation of vascular endothelial cells, vascular smooth muscle fibers, and fibroblasts to help repair damaged blood vessel walls, and (2) *dense granules* contain ADP, ATP, Ca^{2+}, and serotonin. Also present are enzyme systems that produce thromboxane A2, a prostaglandin; *fibrin-stabilizing factor,* which helps to strengthen a blood clot; lysosomes; some mitochondria; membrane systems that take up and store calcium and provide channels for release of the contents of granules; and glycogen.

In the first phase of platelet plug formation, platelets contact and stick to parts of a damaged blood vessel, such as collagen under the damaged endothelial cells. This process is called **platelet adhesion** (Fig. 19.8a). As a result of adhesion, the platelets become activated and their characteristics change drastically. They extend many projections that enable them to contact one another and begin to liberate the contents of their granules. This phase is called the **platelet release reaction** (Fig. 19.8b). Liberated ADP and thromboxane A2 play a major role by acting on nearby platelets to activate them as well. Serotonin and thromboxane A2 function as vasoconstrictors, causing contraction of the vascular smooth muscle, which decreases blood flow through the injured vessel. The release of ADP also makes other platelets in the area sticky, and the stickiness of the

FIGURE 19.8 Platelet plug formation.

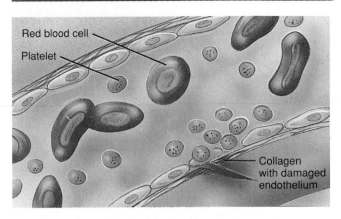

(a) Platelet adhesion

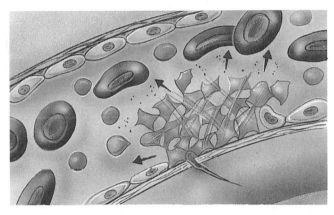

(b) Platelet release reaction

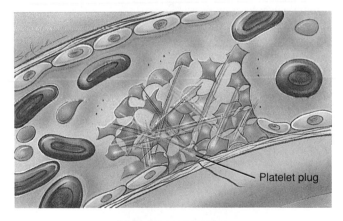

(c) Platelet aggregation

Question: What are the three mechanisms involved in hemostasis?

newly recruited and activated platelets causes them to adhere to the originally activated platelets. This gathering of platelets is called **platelet aggregation.** Eventually, the accumulation and attachment of large numbers of platelets form a mass called a **platelet plug** (Fig. 19.8c). The plug is very effective in preventing blood loss in a small vessel.

Although the platelet plug is initially loose, it becomes quite tight when reinforced by fibrin threads formed during coagulation. A platelet plug can stop blood loss completely if the hole in a blood vessel is small.

Coagulation (Clotting)

Normally, blood remains liquid as long as it stays within its vessels. If it is drawn from the body, however, it thickens and forms a gel. Eventually, the gel separates from the liquid. The straw-colored liquid, called **serum,** is simply plasma minus its clotting proteins. The gel is called a **clot** and consists of a network of insoluble protein fibers called fibrin in which the formed elements of blood are trapped (Fig. 19.9).

The process of gel formation is called **coagulation** or **clotting**. If blood clots too easily, the result can be **thrombosis**—clotting in an unbroken blood vessel. If the blood takes too long to clot, a hemorrhage can result.

Clotting involves several enzymes and other chemicals known as **coagulation (clotting) factors.** Most coagulation factors are synthesized in the liver and released into blood plasma. Some are released by platelets, and one (tissue factor; also called thromboplastin) is released from damaged tissue cells. The various clotting factors and their synonyms are listed in Exhibit 19.4.

Clotting is a complex process in which coagulation factors activate each other. Once the process is initiated, there is a cascade of reactions that acts in a positive feedback manner to form a large quantity of product. We will describe clotting in three basic stages:

FIGURE 19.9 Fibrin threads, platelets, and red blood cells in a blood clot.

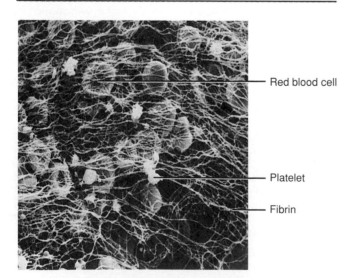

Red blood cell

Platelet

Fibrin

Scanning electron micrograph (13,000×)

Question: What is serum?

EXHIBIT 19.4

COAGULATION FACTORS AND THEIR SYNONYMS

Coagulation Factor[a]	Description
I	Fibrinogen.
II	Prothrombin.
III	Tissue factor (thromboplastin).
IV	Calcium ions.
V	Proaccelerin, labile factor, or accelerator globulin.
VII	Serum prothrombin conversion accelerator (SPCA), stable factor, or proconvertin.
VIII	Antihemophilic factor (AHF), antihemophilic factor A, or antihemophilic globulin (AHG).
IX	Christmas factor, plasma thromboplastin component (PTC), or antihemophilic factor B.
X	Stuart factor, Power factor, thrombokinase.
XI	Plasma thromboplastin antecedent (PTA) or antihemophilic factor C.
XII	Hageman factor, glass factor, or contact factor.
XIII	Fibrin stabilizing factor (FSF) or fibrinase.

[a] There is no factor VI. Prothrombinase (prothrombin activator) is a combination of activated factors V and X.

Stage 1. Formation of **prothrombinase (prothrombin activator).**

Stage 2. Conversion of prothrombin (a plasma protein formed by the liver) into the enzyme **thrombin** by prothrombinase.

Stage 3. Conversion of soluble fibrinogen (another plasma protein formed by the liver) **into insoluble fibrin** by thrombin. Fibrin forms the threads of the clot. (Cigarette smoke contains at least two substances that interfere with fibrin formation.)

Stage 1, the formation of prothrombinase, is initiated by the interplay of two mechanisms: the extrinsic and intrinsic pathways of blood clotting.

Stage 1: Extrinsic Pathway

The **extrinsic pathway** of blood clotting has fewer steps than the intrinsic pathway and occurs rapidly, within a matter of seconds if trauma is severe. It is so named because a tissue protein called **tissue factor (TF),** also known as **thromboplastin** or **coagulation factor III,** leaks into the blood from cells *outside* (*extrinsic to*) blood vessels and initiates the formation of prothrombinase. TF is a complex mixture of lipoproteins and phospholipids released from the surfaces of damaged cells. It activates coagulation factor VII (Figure 19.10a), which next combines with factor X,

thus activating it. Once factor X is activated, it combines with factor V in the presence of calcium ions (Ca^{2+}) to form the active enzyme prothrombinase. This completes the extrinsic pathway of stage 1.

Stage 1: Intrinsic Pathway

The **intrinsic pathway** of blood clotting is more complex than the extrinsic pathway, and it occurs more slowly, usually requiring several minutes. The intrinsic pathway is so named because its activators are in direct contact with blood or contained *within* (*intrinsic to*) the blood; outside tissue damage is not needed. If endothelial cells (cells that line blood vessels) become roughened or damaged, blood can come in contact with collagen in the surrounding basal lamina. In addition, trauma to endothelial cells causes damage to blood platelets, resulting in the release of phospholipids by the platelets. Contact with collagen (or with the slippery glass sides of a blood collection tube) activates coagulation factor XII (Fig. 19.10b). In turn, factor XII activates XI, which activates factor IX. Activated factor VII (from the extrinsic pathway) also can activate factor IX. Activated factor IX joins with factor VIII and platelet phospholipids to activate factor X. Once factor X is activated, it combines with factor V to form the active enzyme prothrombinase, just as occurs in the extrinsic pathway. Again, Ca^{2+} is a cofactor in several of the reactions. This completes the intrinsic pathway of stage 1.

Stages 2 and 3: Common Pathway

The steps in clotting after formation of factor X are the same in both pathways (Fig. 19.10c). Prothrombinase and Ca^{2+} catalyze the conversion of prothrombin to thrombin in stage 2. In stage 3, thrombin, in the presence of Ca^{2+}, converts fibrinogen, which is soluble, to loose fibrin threads, which are insoluble. Thrombin also activates factor XIII (fibrin stabilizing factor), which strengthens and stabilizes the fibrin threads into a sturdy clot. Factor XIII occurs in plasma, and it is also released by platelets trapped in the clot.

Thrombin has two positive feedback effects. In one, through factor V, it accelerates the formation of prothrombinase. Prothrombinase, in turn, accelerates the production of more thrombin, and so on. Thrombin also activates platelets, which reinforces their aggregation and release of phospholipids. This is a second positive feedback cycle. If unchecked, a clot would continue to get larger and larger, as a result of the positive feedback cycles. However, fibrin has the ability to absorb and inactivate up to 90% of the thrombin formed from prothrombin. This helps stop the spread of thrombin into the blood and thus limits spread of the clot beyond the site of damage.

FIGURE 19.10 The blood clotting cascade.

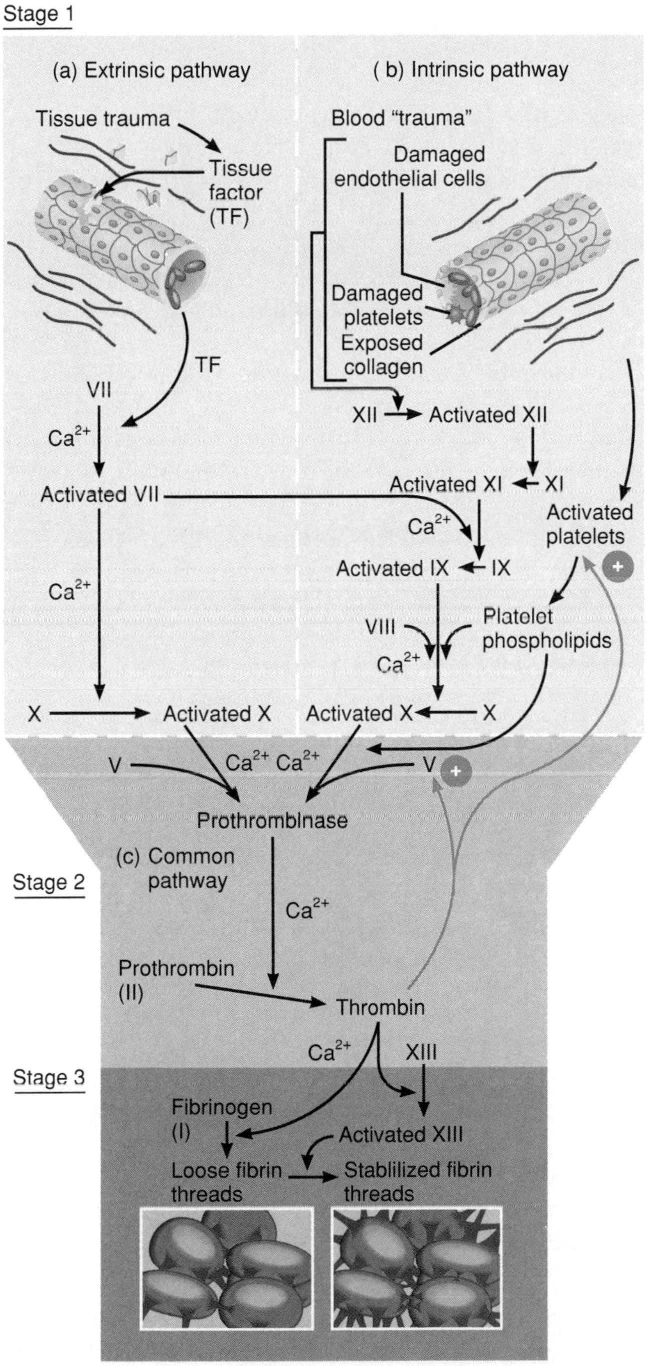

Question: What is the outcome of stage 1 of clotting?

HEMOPHILIA

Hemophilia (*hemo* = blood; *philein* = to love) refers to several different hereditary deficiencies of coagulation in which bleeding may occur spontaneously or after only minor trauma. The effects of all forms of hemophilia are similar but each is a deficiency of a different blood clotting factor. The most common type (classic hemophilia) is hemophilia A, absence of factor VIII. People with hemophilia B lack factor IX.

Hemophilia A and B occur primarily among males, since these are sex-linked recessive disorders. A milder form is hemophilia C, a lack of factor XI, which affects both males and females. Recall that factor XI activates factor IX. Hemophilia C is much less severe than hemophilia A or B because an alternate activator of factor IX is present, namely, factor VII.

Hemophilia is characterized by spontaneous or traumatic subcutaneous and intramuscular hemorrhaging, nosebleeds, blood in the urine, and hemorrhages in joints that produce pain and damage. Treatment involves transfusions of fresh plasma or concentrates of the deficient clotting factor to relieve the bleeding tendency. Factor VIII concentrates are made by culling the factor from plasma of a very large number of donors. Sadly, between 1982 and 1985 most such factor VIII preparations were contaminated with HIV, the virus that causes AIDS, and most hemophiliacs who used the products at that time became infected with HIV.

Need for Vitamin K

Normal clotting depends on adequate vitamin K in the body. Although vitamin K is not involved in actual clot formation, it is required for the synthesis of four clotting factors by liver cells: factors II (prothrombin), VII, IX, and X. Vitamin K is normally produced by bacteria that inhabit the large intestine. It is a fat-soluble vitamin and can be absorbed through the mucosa of the intestine and into the blood only if fat absorption is normal. People suffering from disorders that prevent absorption of fat (for example, inadequate release of bile into the small intestine) often experience uncontrolled bleeding.

Clotting may be encouraged by applying a thrombin or fibrin spray, a rough surface such as gauze, or heat to the wound.

Clot Retraction and Repair

Once a clot is formed, it plugs the ruptured area of the blood vessel and thus prevents hemorrhage (bleeding). **Clot retraction** or **syneresis** (SIN-er-ē-sis) is the consolidation or tightening of the fibrin clot. The fibrin threads attached to the damaged surfaces of the blood vessel gradually contract owing to platelets pulling on them. As the clot retracts, it pulls the edges of the damaged vessel closer together. Thus the risk of hemorrhage is further decreased. During retraction, some serum escapes between the fibrin threads,

but the formed elements in blood remain trapped in the fibrin threads. Normal clot retraction depends on an adequate number of platelets. Platelets in the clot also release factor XIII, which strengthens and stabilizes the clot, and other factors, which help compress the clot. Permanent repair of the blood vessel can then take place. In time, fibroblasts form connective tissue in the ruptured area, and new endothelial cells repair the lining.

Fibrinolysis

Many times a day little clots start to form, sometimes inappropriately, sometimes at a site of minor roughness or at a developing atherosclerotic plaque inside a blood vessel. Because blood clotting involves several positive feedback cycles, a clot has a tendency to spread, which might block blood flow through undamaged vessels. The **fibrinolytic system** provides checks and balances so that clotting does not get out of hand. It also dissolves clots at a site of damage once the damage is repaired.

Dissolution of a clot is called **fibrinolysis** (fī-brin-OL-i-sis). When a clot is formed, an inactive plasma enzyme called **plasminogen** is incorporated into the clot. Both body tissues and blood contain substances that can activate plasminogen to **plasmin (fibrinolysin),** an active plasma enzyme. Among these substances are thrombin, activated factor XII, and tissue plasminogen activator (t-PA), which is synthesized in endothelial cells of most tissues and liberated into the blood. Once plasmin is formed, it can dissolve the clot by digesting fibrin threads and inactivating substances such as fibrinogen, prothrombin, and factors V, VIII, and XII.

THROMBOLYTIC (CLOT-DISSOLVING) AGENTS

Thrombolytic (clot-dissolving) agents are chemical substances injected into the body that dissolve blood clots to restore circulation. They either directly or indirectly activate plasminogen. The first thrombolytic agent, approved for use in 1982, was **streptokinase** (Kabikinase, Streptase). A more recently developed thrombolytic agent, approved in 1988 for dissolving clots in coronary arteries of the heart, is **tissue plasminogen activator (t-PA).** This substance, marketed under the brand name Activase, is a genetically engineered, artificial version of the natural enzyme. It is much more expensive than streptokinase, however, and clinical trials have shown that it is not more effective.

Hemostatic Control Mechanisms

It was noted earlier that even though thrombin has a positive feedback effect on blood clotting, clot formation normally occurs locally at the site of damage. It does not extend beyond a wound site into the general circulation, in part because fibrin absorbs the thrombin into the clot.

Another reason for localized clot formation is that some of the coagulation factors are carried away by the blood so that their concentrations are not high enough to bring about widespread clotting.

Several other mechanisms that control blood clotting also operate. For example, both endothelial cells and white blood cells produce a prostaglandin called **prostacyclin (PGI$_2$).** This substance opposes the actions of thromboxane A2. It is a powerful inhibitor of platelet adhesion and release. Also, substances that inhibit coagulation are present in blood. Such substances are called **anticoagulants.** These include **antithrombin III (AT-III),** which blocks the action of factors XII, XI, IX, X, and II (thrombin); **protein C,** which inactivates factors V and VIII, the two major clotting factors not blocked by AT-III, and enhances activity of plasminogen activators; **alpha-2-macroglobulin,** which inactivates thrombin and plasmin; and **alpha-1-antitrypsin,** which inhibits factor XI.

Heparin is another anticoagulant. It is produced by mast cells and basophils. A similar molecule protrudes from the plasma membrane of endothelial cells into the blood. Heparin's anticoagulant activity is to combine with AT-III and increase its effectiveness in blocking thrombin. Heparin is also a pharmacologic anticoagulant extracted from animal lung tissue and intestinal mucosa. It is often used in open heart surgery and during hemodialysis. Another anticoagulant is the pharmaceutical preparation **warfarin (Coumadin),** which may be given to patients who are prone to develop clots. It acts as an antagonist to vitamin K and thus blocks synthesis of four clotting factors (II, VII, IX, and X). Warfarin is slower acting than heparin. To prevent clotting in donated blood, blood banks and laboratories often add a substance, for example, CPD (citrate phosphate dextrose), that removes Ca^{2+}.

Intravascular Clotting

Despite the anticoagulating and fibrinolytic mechanisms, blood clots sometimes form within the cardiovascular system. Such clots may be initiated by roughened endothelial surfaces of a blood vessel as a result of atherosclerosis, trauma, or infection. These conditions induce adhesion of platelets. Intravascular clots may also form when blood flows too slowly (stasis), allowing clotting factors to accumulate locally in high enough concentrations to initiate coagulation. Clotting in an unbroken blood vessel (usually a vein) is called **thrombosis** (*thrombo* = clot). The clot itself is a **thrombus.** A thrombus may dissolve spontaneously, but if it remains intact, there is the possibility that the thrombus will become dislodged and be carried with the blood to the lungs. If it occurs in an artery, the clot may block blood flow to a vital organ. A blood clot, bubble of air, fat from broken bones, or a piece of debris transported by the bloodstream is called an **embolus** (*em* = in; *bolus* = a mass). When an embolus becomes lodged in the lungs, the condition is called **pulmonary embolism.**

GROUPING (TYPING) OF BLOOD

The surfaces of erythrocytes contain genetically determined antigens called **agglutinogens** (ag′-loo-TIN-ō-jens), or **isoantigens.** There are at least 14 blood group systems and more than 100 antigens that can be detected on the surface of red blood cells. The two major blood group classifications—ABO and Rh—are the ones we will discuss in detail. Among the others are the Lewis, Kell, Kidd, and Duffy systems.

ABO

The **ABO blood grouping** is based on two glycolipid agglutinogens called *A* and *B* (Fig. 19.11). People whose erythrocytes display only agglutinogen *A* are said to have blood type A. Those who have only agglutinogen *B* are type B. Individuals who have both *A* and *B* are type AB, whereas those who manufacture neither are type O.

Every person inherits two genes, one from each parent, that are responsible for production of these agglutinogens. The six possible combinations are *OO, AO, AA, BO, BB,* and *AB.* Both *A* and *B* are inherited as dominant traits; *O* is inherited as a recessive trait. The six genetic combinations determine blood type as follows:

1. *OO* produces type O blood.
2. *AO* and *AA* produce type A blood.
3. *BO* and *BB* produce type B blood.
4. *AB* produces type AB blood.

These four blood types are not equally distributed. The incidence of ABO and Rh groups among various ethnic groups is indicated in Exhibit 19.5.

Most people's blood plasma contains naturally occurring antibodies called **agglutinins** (a-GLOO-ti-nins), or **isoantibodies,** that will react with the *A* and *B* agglutinogens (antigens) if the two are mixed. These are agglutinin *a* (anti-A), which reacts with agglutinogen *A,* and agglutinin *b* (anti-B), which reacts with agglutinogen *B.* The agglutinins formed by each of the four blood types are shown in Fig. 19.11. You do not have agglutinins that react with the agglutinogens of your own erythrocytes, but most likely you do have an agglutinin for any agglutinogen your RBCs lack. In an incompatible blood transfusion, agglutinins in the recipient's plasma bind to the agglutinogens on the donated RBCs. This reaction is another example of an antigen–antibody response (see Fig. 19.7). When antigen–antibody complexes form in the body, they activate plasma proteins of the complement family (see page 693). In essence, complement molecules poke holes in the donated RBCs, causing them to burst and release hemoglobin into the plasma. Such a reaction is called **hemolysis.** The liberated hemoglobin may cause kidney damage.

FIGURE 19.11 Agglutinogens (antigens) and agglutinins (antibodies) involved in the ABO blood grouping system.

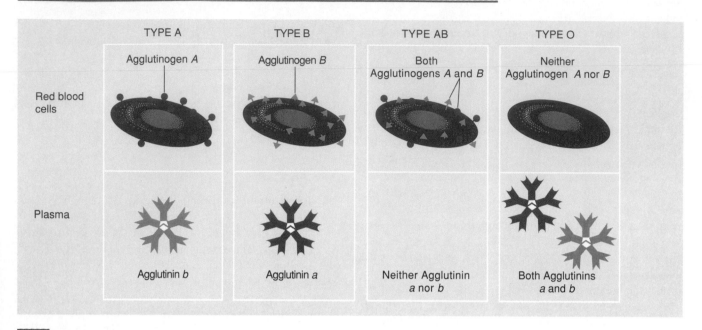

Question: Which antibodies are found in type O blood?

EXHIBIT 19.5

INCIDENCE OF HUMAN BLOOD GROUPS IN THE UNITED STATES

	Blood Groups (Percentages)				
	O	A	B	AB	Rh⁺
Whites	45	41	10	4	85
Blacks	48	27	21	4	88
Japanese	31	38	22	9	100
Chinese	36	28	23	13	100
American Indians	23	76	0	1	100
Hawaiians	37	61	1.5	0.5	100

CLINICAL APPLICATION

TYPING AND CROSS-MATCHING FOR TRANSFUSION

Blood transfusions are most often given to alleviate anemia or when blood volume is low, for example, in cases of circulatory shock after a severe hemorrhage. When blood is transfused, care must be taken to avoid blood type mismatches. This is done by typing the patient's blood and then either cross-matching it to potential donor blood or screening it for presence of antibodies. Outside the body, at room temperature, mixing of incompatible blood causes **agglutination** (clumping) that is visible to the naked eye rather than hemolysis. Agglutination is also an antigen–antibody response, whereby the cells become linked to one another and form a visible clump (Fig. 19.12). (Note that agglutination is not the same thing as clotting.)

In the procedure for ABO blood typing, single drops of blood are mixed with different antisera, solutions that contain agglutinins (antibodies). One drop of blood is mixed with anti-A serum. It contains agglutinins (a) that will agglutinate red blood cells if they contain A agglutinogens (antigens). The other drop is mixed with anti-B serum. It contains agglutinins (b) that will agglutinate red blood cells if they contain B agglutinogens. If the red blood cells agglutinate only when mixed with anti-A serum, the blood is type A. If the red blood cells agglutinate only when mixed with anti-B serum, the blood is type B. The sample is type AB if both drops agglutinate; if neither sample agglutinates, the sample is type O.

In the procedure for determining Rh factor (discussed shortly), a drop of blood is mixed with anti-D serum, a solution that contains agglutinins that will agglutinate RBCs that contain Rh agglutinogens. If the sample agglutinates, it is Rh⁺; no agglutination indicates Rh⁻.

Once the patient's blood type is known, donor blood of the same ABO and Rh type is selected. In a **cross-match**, the possible donor RBCs are mixed with the recipient's serum. If agglutination does not occur, the recipient does not have agglutinins that will attack the donor RBCs. Alternatively, the recipient's serum can be **screened** against a test panel of RBCs having agglutinogens known to cause blood transfusion reactions to detect any agglutinins that may be present.

FIGURE 19.12 Comparison of normal and agglutinated red blood cells.

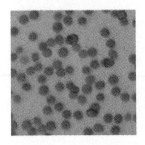 

(a) Normal   (b) Agglutinated

Question: How is agglutination different from clotting?

As an example of an incompatible blood transfusion, consider what happens if an individual with type A blood receives a transfusion of type B blood. The recipient's blood (type A) contains *A* agglutinogens on the red blood cells and *b* agglutinins in the plasma. The donor's blood (type B) contains *B* agglutinogens and *a* agglutinins. Given this situation, two things can happen. First, the *b* agglutinins in the recipient's plasma can bind to the *B* agglutinogens on the donor's erythrocytes, causing hemolysis and possibly agglutination of the red blood cells. Second, the *a* agglutinins in the donor's plasma can bind to the *A* agglutinogens on the recipient's erythrocytes. However, the second reaction is usually not serious because the donor's *a* agglutinins become so diluted in the recipient's plasma that they do not cause any significant hemolysis of the recipient's RBCs. Thus a person with type A blood may not receive type B or AB blood. But a person with A blood may receive type A or type O blood. The interactions of the four blood types of the ABO system are summarized in Exhibit 19.6.

People with type AB blood do not have any *a* or *b* agglutinins in their plasma. They are sometimes called universal recipients because they can *theoretically* receive blood from donors of all four blood types. They have no agglutinins to attack donated RBCs. People with type O blood have no *A* or *B* agglutinogens on their RBCs and are sometimes referred to as universal donors because they can *theoretically* donate blood to recipients of all four blood types. Type O persons requiring blood may receive only type O blood. In practice, use of the terms *universal recipient* and *universal donor* is misleading and dangerous. There are other agglutinogens and agglutinins in blood, besides those associated with the ABO system, that can cause transfusion problems. Thus blood should be carefully cross-matched or screened before transfusion, except in extreme emergency.

EXHIBIT 19.6

SUMMARY OF ABO SYSTEM INTERACTIONS

Blood Type	A	B	AB	O
Agglutinogen (Antigen) on RBCs	*A*	*B*	*A* and *B*	Neither *A* nor *B*
Agglutinin (Antibody) in Plasma	*b*	*a*	Neither *a* nor *b*	*a* and *b*
Compatible Donor Blood Types	A, O	B, O	A, B, AB, O	O
Incompatible Donor Blood Types	B, AB	A, AB	—	A, B, AB
Genotype (Genetic Makeup)	*AO* or *AA*	*BO* or *BB*	*AB*	*OO*
Phenotype (Expressed Blood Type)	A	B	AB	O

Knowledge of blood types is also used in paternity lawsuits, linking suspects to crimes, and as part of anthropology studies to establish a relationship among races. In about 80% of the population, soluble antigens of the ABO type appear in saliva and other body fluids. In criminal investigations it has been possible to type fluids such as saliva residues on a cigarette or semen in cases of rape.

Rh

The **Rh system** of blood classification is so named because it was first worked out in the blood of the *Rhesus* monkey. Like the ABO grouping, the Rh system is based on antigens on the surfaces of erythrocytes. People whose RBCs have the Rh agglutinogens (D antigens) are designated *Rh⁺*. Those who lack Rh agglutinogens are designated *Rh⁻*. The incidence of Rh⁺ and Rh⁻ individuals among various groups is shown in Exhibit 19.5. Normally, plasma does not contain anti-Rh agglutinins. However, if an Rh⁻ person receives Rh⁺ blood, the body starts to make anti-Rh agglutinins that will remain in the blood. If a second transfusion of Rh⁺ blood is given later, the previously formed anti-Rh agglutinins will cause hemolysis of the donated blood, and a severe reaction may occur.

CLINICAL APPLICATION

HEMOLYTIC DISEASE OF THE NEWBORN

The most common problem with Rh incompatibility may arise during pregnancy. Normally, there is no direct contact between maternal and fetal blood when a woman is pregnant. However, if a small amount of Rh+ blood leaks from the fetus through the placenta into the bloodstream of an Rh- mother, the mother will start to make anti-Rh agglutinins. Since the greatest possibility of fetal blood transfer occurs at delivery, a newborn baby will be unaffected. If the mother becomes pregnant again, however, her anti-Rh agglutinins can cross the placenta and make their way into the bloodstream of the fetus. If the fetus is Rh-, there is no problem, since Rh- blood does not have the Rh agglutinogen. If the fetus is Rh+, hemolysis may occur in the fetal

blood. The hemolysis brought on by fetal–maternal incompatibility is called **hemolytic disease of the newborn (HDN)**, also called **erythroblastosis fetalis**. When a baby is born with this condition, blood is slowly removed and replaced a little at a time, with Rh- blood. It is even possible to transfuse blood into the unborn child if the problem is diagnosed before birth.

Not too long ago, HDN was a major cause of fetal and newborn deaths, but now its occurrence is very rare. HDN is prevented by giving all Rh- mothers an injection of an anti-Rh gamma globulin preparation (RhoGAM) soon after delivery, miscarriage, or abortion. These agglutinins bind to the fetal agglutinogens, if they are present, so the mother's immune system cannot respond to the foreign agglutinogens by producing agglutinins (antibodies). Thus the fetus of the next pregnancy is protected. Since Rh+ mothers do not make anti-Rh agglutinins, their babies are not at risk. ABO incompatibility between mother and fetus rarely causes problems because most of the ABO agglutinins are too large to cross the placenta.

DISORDERS: HOMEOSTATIC IMBALANCES

ANEMIA
Anemia is a condition in which the oxygen-carrying capacity of the blood is reduced; it is a sign, not a diagnosis. Many kinds of anemia exist, all characterized by reduced numbers of RBCs or decreased amount of hemoglobin in the blood. These conditions lead to fatigue and intolerance to cold, both of which are related to lack of oxygen needed for ATP and heat production, and to paleness, which is due to low hemoglobin content.

Nutritional Anemia
Nutritional anemia arises from an inadequate diet, one without enough iron, the necessary amino acids, or vitamin B₁₂.

Pernicious Anemia
Pernicious anemia is the insufficient hematopoiesis that results from an inability of the stomach to produce intrinsic factor, which is needed for absorption of vitamin B₁₂.

Hemorrhagic Anemia
An excessive loss of RBCs through bleeding is called **hemorrhagic anemia.** Common causes are large wounds, stomach ulcers, and heavy menstrual bleeding. If bleeding is extraordinarily heavy, the anemia is termed *acute*. Excessive blood loss can be fatal. Slow, prolonged bleeding is apt to produce a *chronic* anemia; the chief symptom is fatigue.

Hemolytic Anemia
If RBC plasma membranes rupture prematurely, the cells remain as "ghosts," and their hemoglobin pours out into the plasma. A characteristic sign of this condition, called **hemolytic anemia,** is distortion in the shape of erythrocytes. It may result from inherent defects, such as hemoglobin defects, abnormal red blood cell enzymes, or defects of the red blood cell membrane. Agents that may cause hemolytic anemia are parasites, toxins, and antibodies from incompatible blood (Rh- mother and Rh+ fetus, for instance). Hemolytic disease of the newborn (erythroblastosis fetalis) is an example of a hemolytic anemia.

The term **thalassemia** (thal'-a-SĒ-mē-a) represents a group of hereditary hemolytic anemias resulting from a defect in the synthesis of hemoglobin, which produces extremely thin and fragile erythrocytes. It occurs primarily in populations from countries bordering the Mediterranean Sea. Treatment generally consists of blood transfusions.

Aplastic Anemia
Destruction or inhibition of the red bone marrow results in **aplastic anemia.** Typically, the marrow is replaced by fatty tissue, fibrous tissue, or tumor cells. Toxins, gamma radiation, and certain medications that inhibit enzymes involved in hemopoiesis are causes. Bone marrow transplants can now be done with a reasonable hope of success in patients with aplastic anemia. Immunosuppressive drugs are given for several days before the transplant and with decreasing frequency afterward.

Sickle-Cell Anemia
The RBCs of a person with **sickle-cell anemia (SCA)** contain an abnormal kind of hemoglobin (Hb-S). When such an erythrocyte gives up its oxygen to the interstitial fluid, the abnormal hemoglobin forms long, stiff, rodlike structures that bend the erythrocyte into a sickle shape (Fig. 19.13). The sickled cells rupture easily. Even though erythropoiesis is stimulated by the loss of the cells, it cannot keep pace with the hemolysis. The individual thus suffers from a hemolytic anemia that reduces the amount of oxygen that can be supplied to the tissues. Prolonged oxygen reduction may eventually cause extensive tissue damage. Furthermore, because of the shape of the sickled cells, they tend to get stuck in blood vessels and can cut off blood supply to an organ altogether.

Sickle-cell anemia is characterized by several symptoms. In young children, hand–foot syndrome is present, in which there is swelling and pain in the wrists and feet. Older patients experience pain in the back and extremities. Complications include neurological disorders (meningitis, seizures, stroke), impaired pulmonary function, orthopedic abnormalities (femoral head necrosis, osteomyelitis), genitourinary

FIGURE 19.13 Sickle-cell anemia.

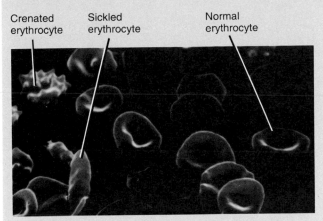

Crenated erythrocyte Sickled erythrocyte Normal erythrocyte

Scanning electron micrograph of erythrocytes, 2000x

tract disorders (involuntary urination, blood in urine, kidney failure), ocular disturbances (hemorrhage, detached retina, blindness), and obstetric complications (convulsions, coma, infection).

Sickle-cell anemia is inherited. The gene responsible for the tendency of the RBCs to sickle also alters the permeability of the plasma membranes of sickled cells, causing potassium to leak out. Low levels of potassium kill the malaria parasites that infect sickled cells. Sickle-cell genes are found primarily among populations, or descendants of populations, that live in the malaria belt around the world, including parts of Mediterranean Europe and subtropical Africa and Asia. A person with only one of the sickling genes is said to have **sickle-cell trait.** Such an individual has a high resistance to malaria—a factor that may have tremendous survival value—but does not develop the anemia. Only people who inherit a sickling gene from both parents get sickle-cell anemia.

Treatment consists of administration of analgesics to relieve pain, antibiotics to counter infections, and blood transfusions.

POLYCYTHEMIA

The term **polycythemia** (pol′-ē-sī-THĒ-mē-a) refers to a disorder characterized by a hematocrit that is elevated significantly above the normal upper limit of about 55. Blood viscosity rises as hematocrit increases. The increased viscosity elevates blood pressure and contributes to thrombosis and hemorrhage. The thrombosis results from too many red blood cells piling up as they try to enter smaller vessels. The hemorrhage is due to widespread hyperemia (unusually large amount of blood in an organ).

INFECTIOUS MONONUCLEOSIS

Infectious mononucleosis (IM) is a contagious disease primarily affecting lymphoid tissue throughout the body but also affecting the blood. It is caused by the *Epstein–Barr virus* (*EBV*), and occurs mainly in children and young adults. The ratio of affected females to males is 3:1. The virus most commonly enters the body through intimate oral contact, such as kissing, multiplies in lymphatic tissues, and spreads into the blood where it infects and multiplies in B lymphocytes, the primary host cells. As a result of this infection, the B cells become enlarged and abnormal in appearance and resemble monocytes, the primary reason for which the disease receives its name, mononucleosis.

Signs and symptoms include an elevated white blood cell count with an abnormally high percentage of lymphocytes, fatigue, headache, dizziness, sore throat, enlarged and tender lymph nodes, and fever. There is no cure for infectious mononucleosis, and treatment consists of watching for and treating complications. Usually the disease runs its course in a few weeks, and the individual generally suffers no permanent ill effects.

LEUKEMIA

Clinically, **leukemia** is classified based on the duration and character of the disease, that is, acute or chronic. Acute leukemia is a malignant disease of blood-forming tissues characterized by uncontrolled production and accumulation of immature leukocytes. In chronic leukemia, there is an accumulation of mature leukocytes in the bloodstream because they do not die at the end of their normal life span. The *human T cell leukemia-lymphoma virus-1 (HTLV-1)* is strongly associated with some types of leukemia. Leukemia is also classified according to the identity and site of origins of the predominant cell involved, that is, myelocytic, lymphocytic, or monocytic.

In acute leukemia, the anemia and bleeding problems commonly seen result from the crowding out of normal bone marrow cells by the overproduction of immature cells, preventing normal production of RBCs and platelets. One cause of death from acute leukemia is internal hemorrhaging, especially cerebral hemorrhage that destroys vital centers in the brain. Often, the cause of death is uncontrolled infection due to lack of mature or normal WBCs. The abnormal accumulation of immature leukocytes may be reduced by using x rays and antileukemic drugs. Partial or complete remissions may be induced, with some lasting as long as 15 years.

MEDICAL TERMINOLOGY

Acute normovolemic (nor-mō-vō-LĒ-mik) **hemodilution** (hē′-mō-dī-LOO-shun) Removal of blood immediately before surgery and replacing it with a cell-free solution to maintain normal blood volume for adequate circulation. At the end of surgery, when hemostasis has been controlled, the collected blood is returned to the body.

Autologous (aw-TOL-o-gus; *auto* = self) **preoperative transfusion** (trans-FYOO-zhun) Donating one's own blood for up to six weeks before elective surgery. Also called **predonation.**

Medical Terminology continues

MEDICAL TERMINOLOGY (continued)

Autologous intraoperative transfusion (AIT) Procedure in which blood lost during surgery is suctioned from the patient, treated with an anticoagulant, and reinfused into the patient.

Blood bank A stored supply of blood for future use by the donor or others. Since blood banks have now assumed additional and diverse functions (immunohematology reference work, continuing medical education, bone and tissue storage, and clinical consultation), they are more appropriately referred to as **centers of transfusion medicine.**

Citrated (SIT-rā-ted) **whole blood** Whole blood protected from coagulation by CPD (citrate phosphate dextrose) or a similar compound.

Cyanosis (sī′-a-NŌ-sis; *cyano* = blue) Slightly bluish, dark purple skin coloration, most easily seen in the nail beds and mucous membranes, due to increased quantity of deoxygenated hemoglobin in systemic blood.

Exchange transfusion Removing blood from the recipient while simultaneously replacing it with donor blood. This method is used for treating hemolytic disease of the newborn (HDN) and poisoning.

Gamma globulin (GLOB-yoo-lin) Solution of immunoglobulins from nonhuman blood consisting of antibodies that react with specific pathogens, such as measles, epidemic hepatitis, tetanus, and possibly poliomyelitis viruses. It is prepared by injecting the specific virus into animals, removing blood from the animals after antibodies have accumulated, isolating antibodies, and injecting them into a human for short-term immunity.

Hemochromatosis (hē-mō-krō′-ma-TŌ-sis; *heme* = iron; *chroma* = color) Disorder of iron metabolism characterized by excess deposits of iron in tissues, especially the liver and pancreas, that result in bronze coloration of the skin, cirrhosis, diabetes mellitus, and bone and joint abnormalities.

Hemorrhage (HEM-or-ij; *rrhage* = bursting forth) Bleeding, either internal (from blood vessels into tissues) or external (from blood vessels directly to the surface of the body).

Multiple myeloma (mī′-e-LŌ-ma) Malignant disorder of plasma cells in bone marrow; symptoms (pain, osteoporosis, hypercalcemia, thrombocytopenia, kidney damage) are caused by the growing tumor cell mass or antibodies produced by malignant cells.

Platelet concentrates A preparation of platelets obtained from freshly drawn whole blood and used for transfusions in platelet-deficiency disorders such as hemophilia.

Porphyria (por-FĒ-rē-a or por-FĪ-rē-a) Any of a group of inherited disorders caused by the accumulation in the body of substances called porphyrins (precursor substances formed during the synthesis of hemoglobin and other important molecules). The buildup is due to inherited enzyme deficiencies. Symptoms include a rash or skin blistering brought on by sunlight, abdominal pain, and nervous system disturbances from certain drugs, such as barbiturates and alcohol.

Septicemia (sep′-ti-SĒ-mē-a; *sep* = decay; *emia* = condition of blood) Toxins or disease-causing bacteria growing in the blood. Also called "blood poisoning."

Thrombocytopenia (throm′-bō-sī′-tō-PĒ-nē-a; *thrombo* = clot; *penia* = poverty) Very low platelet count that results in a tendency to bleed from capillaries.

Transfusion (trans-FYOO-zhun) Transfer of whole blood, blood components (red blood cells only or plasma only), or bone marrow directly into the bloodstream.

Venesection (vēn′-e-SEK-shun; *veno* = vein) Opening of a vein for withdrawal of blood. Although **phlebotomy** (fle-BŌT-ō-mē; *phlebo* = vein; *tome* = to cut) is a synonym for venesection, in clinical practice, phlebotomy refers to therapeutic bloodletting, such as removing some blood to lower the viscosity of blood of a patient with polycythemia.

Study Outline

Comparison of Extracellular Fluids (p. 567)

1. All fluids outside body cells comprise extracellular fluid.
2. Both interstitial fluid and lymph are similar in composition to plasma, but contain less protein.

Functions of Blood (p. 567)

1. Blood transports oxygen, carbon dioxide, nutrients, wastes, and hormones.
2. It helps to regulate pH, body temperature, and water content of cells.

3. It prevents blood loss through clotting and combats toxins and microbes through certain phagocytic white blood cells or specialized plasma proteins.

Physical Characteristics of Blood (p.567)

1. Physical characteristics of blood include a viscosity greater than that of water; a temperature of 38°C (100.4°F); and a pH of 7.35 to 7.45. Blood constitutes about 8% of body weight and volume ranges from 4 to 6 liters in an adult.

Components of Blood (p. 568)

1. Blood consists of 55% plasma and 45% formed elements.
2. Plasma consists of 91.5% water and 8.5% solutes.
3. Principal solutes include proteins (albumins, globulins, fibrinogen), nutrients, hormones, respiratory gases, electrolytes, and waste products.
4. The formed elements in blood include erythrocytes (red blood cells), leukocytes (white blood cells), and thrombocytes (platelets).

Formation of Blood Cells (p. 569)

1. Blood cells are formed from pluripotent hematopoietic stem cells (hemocytoblasts) in red bone marrow. The process is called hemopoiesis.
2. Several hematopoietic growth factors stimulate differentiation and proliferation of the various blood cells.

Erythrocytes (Red Blood Cells) (p. 570)

1. Erythrocytes are biconcave discs without nuclei that contain hemoglobin.
2. The function of the hemoglobin in red blood cells is to transport oxygen and some carbon dioxide.
3. Red blood cells live about 120 days. A healthy male has about 5. 4 million/mm^3 of blood; a healthy female, about 4.8 million/mm^3.
4. After phagocytosis of aged red blood cells by macrophages, hemoglobin is recycled.
5. Erythrocyte formation, called erythropoiesis, occurs in adult red marrow of certain bones. It is stimulated by hypoxia, which stimulates release of erythropoietin by the kidneys.
6. A reticulocyte count is a diagnostic test that indicates the rate of erythropoiesis.
7. A hematocrit (Hct) measures the percentage of red blood cells in whole blood.

Leukocytes (White Blood Cells) (p. 574)

1. Leukocytes are nucleated cells. Two principal types are granular (neutrophils, eosinophils, basophils) and agranular (lymphocytes and monocytes).
2. The general function of leukocytes is to combat inflammation and infection. Neutrophils and macrophages (which develop from monocytes) do so through phagocytosis.
3. Eosinophils combat the effects of histamine in allergic reactions, phagocytize antigen–antibody complexes, and combat parasitic worms; basophils develop into mast cells that liberate heparin, histamine, and serotonin in allergic reactions that intensify the inflammatory response.
4. B lymphocytes, in response to the presence of foreign substances called antigens, differentiate into plasma cells that produce antibodies. Antibodies attach to the antigens and render them harmless. This antigen–antibody response combats infection and provides immunity. T lymphocytes destroy foreign invaders directly.
5. White blood cells usually live for only a few hours or a few days. Normal blood contains 5000 to 10,000/mm^3.

Thrombocytes (Platelets) (p. 577)

1. Thrombocytes are disc-shaped structures without nuclei.
2. They are formed from metamegakaryocytes and are involved in clotting.
3. Normal blood contains 250,000 to 400,000/mm^3.

Hemostasis (p. 579)

1. Hemostasis refers to the stoppage of bleeding.
2. It involves vascular spasm, platelet plug formation, and blood coagulation (clotting).
3. In vascular spasm, the smooth muscle of a blood vessel wall contracts to slow blood loss.
4. Platelet plug formation involves the clumping of platelets to stop bleeding.
5. A clot is a network of insoluble protein fibers (fibrin) in which formed elements of blood are trapped.
6. The chemicals involved in clotting are known as coagulation (clotting) factors.
7. Blood clotting involves a cascade of reactions that may be divided into three stages: formation of prothrombinase, conversion of prothrombin into thrombin, and conversion of soluble fibrinogen into insoluble fibrin.
8. Stage 1 of clotting is initiated by the interplay of the extrinsic and intrinsic pathways of blood clotting.
9. Normal coagulation requires vitamin K and also involves clot retraction (tightening of the clot) and fibrinolysis (dissolution of the clot).
10. Clotting in an unbroken blood vessel is called thrombosis. A thrombus that moves from its site of origin is called an embolus.
11. Anticoagulants (for example, heparin) prevent clotting.

Grouping (Typing) of Blood (p. 583)

1. ABO and Rh systems are genetically determined and based on antigen–antibody responses.
2. In the ABO system, agglutinogens (antigens) A and B determine blood type. Plasma contains agglutinins (antibodies), designated as a and b, that react with agglutinogens that are foreign to the individual.
3. In the Rh system, individuals whose erythrocytes have Rh agglutinogens are classified as Rh$^+$. Those who lack the antigen are Rh$^-$.

Review Questions

1. How are blood, interstitial fluid, and lymph related to the maintenance of homeostasis? (p. 567)
2. Distinguish between the cardiovascular system and lymphatic system. (p. 567)
3. List the functions of blood and their relationship to other systems of the body. (p. 567)
4. List the principal physical characteristics of blood. (p. 567)
5. How are blood samples obtained for laboratory testing? (p. 567)
6. Distinguish between plasma and formed elements. (p. 569)
7. What are the major constituents of plasma? What do they do? What is the difference between plasma and serum? (p. 569)
8. Describe the origin of blood cells and how hematopoietic growth factors stimulate the process. (p. 569)
9. Describe the microscopic appearance of erythrocytes. What is the function of erythrocytes? What is induced erythrocythemia (blood doping)? (p. 570)
10. Explain how hemoglobin is recycled. (p. 573)

11. Define erythropoiesis. Relate erythropoiesis to red blood cell count. What factors accelerate and slow erythropoiesis? (p. 573)
12. Distinguish between reticulocyte count and hematocrit. (p. 573)
13. Describe the classification of leukocytes and describe their microscopic appearance and functions. (p. 574)
14. What is the importance of diapedesis, chemotaxis, and phagocytosis in fighting bacterial invaders? (p. 576)
15. What is histocompatibility testing? (p. 576)
16. Distinguish between leukocytosis and leukopenia. (p. 577)
17. What is a differential white blood cell count? (p. 576)
18. Describe the antigen–antibody response. How is it protective? (p. 576)
19. How is a bone marrow transplant performed? Why are they used? (p. 577)
20. Describe the structure and function of thrombocytes. (p. 577)
21. Compare erythrocytes, leukocytes, and thrombocytes with respect to size, number per mm^3, and life span. (p. 578)
22. Define hemostasis. Explain the mechanism involved in vascular spasm and platelet plug formation. (p. 579)
23. Briefly describe the stages of clot formation. What is fibrinolysis? Why does blood usually not remain clotted inside blood vessels? (p. 580)
24. How do the extrinsic and intrinsic pathways of blood clotting differ? (p. 580)
25. What is hemophilia? Describe its signs and symptoms. (p. 582)
26. Define the following: thrombus, embolus, anticoagulant. (p. 583)
27. What is the basis for ABO blood grouping? What are agglutinogens and agglutinins? (p. 583)
28. What is the basis for the Rh system? How does hemolytic disease of the newborn (erythroblastosis fetalis) occur? How may it be prevented? (p. 585)
29. Define anemia. Contrast the causes of nutritional, pernicious, hemorrhagic, hemolytic, aplastic, and sickle-cell anemia (SCA). (p. 586)
30. What is infectious mononucleosis (IM)? (p. 587)
31. What is leukemia, and what are the causes of some of its symptoms? (p. 587)

Answers to Questions with Figures

19.1 About 8% × 45% = 3.6%.
19.2 Temperature, 38°C (100.4°F); pH, 7.35 to 7.45; body weight, 8%.
19.3 Four—one bound to each heme group.
19.4 It is an iron-carrying plasma protein.
19.5 Hypoxia.
19.6 Neutrophils, eosinophils, and basophils are called granulocytes because after staining all have granules in their cytoplasm that can be seen using a light microscope.
19.7 Antibodies are plasma proteins of the immunoglobulin (gamma globulin) family. They react with the specific antigen that triggered their secretion by plasma cells.
19.8 Vascular spasm, platelet plug formation, and blood coagulation.
19.9 Blood plasma minus the clotting proteins.
19.10 Formation of prothrombinase.
19.11 *a* and *b*.
19.12 Agglutination is an antigen–antibody reaction in which RBCs become linked to one another by agglutinins (antibodies), whereas clotting is an enzymatic cascade that results in formation of fibrin threads.

Chapter 20

The Cardiovascular System: The Heart

Chapter Contents at a Glance

Student Objectives

1. Describe the location of the heart and the structure and functions of the wall, chambers, great vessels, and valves of the heart.

2. Describe the blood supply of the heart.

3. Explain the structural and functional features of the conduction system of the heart.

4. Describe the physiology of cardiac muscle contraction.

5. Explain the meaning of an electrocardiogram (ECG) and its diagnostic importance.

6. Describe the phases, timing, and sounds associated with a cardiac cycle.

7. Define cardiac output (CO) and describe the factors that affect it.

8. Explain how heart rate is regulated.

9. List and explain the risk factors involved in heart disease.

10. Explain the relationship between plasma lipids and heart disease.

11. Describe the developmental anatomy of the heart.

12. Explain the benefits of regular exercise on the heart.

13. Define the following disorders: coronary artery disease (CAD), congenital defects, and arrhythmias.

14. Define medical terminology associated with the heart.

T he **heart** is the center of the cardiovascular system. Whereas the term *cardio* refers to the heart, the term *vascular* refers to blood vessels (or an abundant blood supply). The heart propels blood through thousands of miles of blood vessels, and it is magnificently designed for this task. Although we ignore its activity most of the time, the heart's capacity for work is remarkable. Even at rest, your heart pumps 30 times its own weight each minute, about 5 liters (5.3 qt) to the lungs and the same volume to the rest of the body. At this rate, the heart would pump more than 7000 liters (1800 gal) of blood in a day and 5 million liters (1.3 million gal) in a year. Since you don't spend all your time "resting" and since your heart pumps more vigorously when you are active, the actual flow is much larger.

The cardiovascular system provides the "pump" for circulating constantly refreshed blood through an estimated 100,000 km (60,000 mi) of blood vessels. As blood flows through body tissues, nutrients and oxygen move from the blood into the interstitial fluid and then into cells. At the same time the blood picks up wastes, carbon dioxide, and heat. This chapter explores the design of the heart and the unique properties of cardiac muscle that permit a lifetime of pumping with never a minute's rest.

The study of the normal heart and diseases associated with it is known as **cardiology** (kar-dē-OL-ō-jē; *cardio* = heart).

The developmental anatomy of the heart is considered at the end of the chapter.

LOCATION AND SIZE OF THE HEART

For all its might, the hollow, cone-shaped heart is relatively small, about the same size as a person's closed fist, and weighs only about 300 g (10 oz) in an adult. The heart contains four chambers, to be described shortly: two upper atria and two lower ventricles. It rests on the diaphragm, near the middle of the thoracic cavity in a space called the **mediastinum** (mē′-dē-a-STĪ-num), which extends from the sternum to the vertebral column between the lungs (Fig. 20.1). About two-thirds of the mass of the heart lies to the left of the body's midline. The heart is about 12 cm (5 in.) long, 9 cm (3½ in.) wide at its broadest point, and 6 cm (2½ in.) thick. The pointed end of the heart, the **apex**, is formed by the tip of the left ventricle and tilts obliquely toward the left hip. Opposite the apex, the wide upper and posterior margin of the heart is called the **base**, so named because it is broad and rather flat, like the base of a pyramid. The base of the heart is formed by the atria, mostly the left atrium.

CLINICAL APPLICATION

CARDIOPULMONARY RESUSCITATION

Because the heart lies between two rigid structures (the vertebral column and the sternum), external pressure

(compression) on the chest can be used to force blood out of the heart into the circulation (Fig. 20.lb). In cases where the heart suddenly stops beating, **cardiopulmonary resuscitation (CPR)**—properly applied cardiac compressions, performed with artificial ventilation of the lungs—saves lives by keeping oxygenated blood circulating until the heart can be restarted.

PERICARDIUM

The **pericardium** (*peri* = around) is a triple-layered bag that surrounds and protects the heart. It confines the heart to its position in the mediastinum, yet allows it sufficient freedom of movement for vigorous and rapid contraction.

The pericardium consists of two principal portions: the fibrous pericardium and the serous pericardium (Fig. 20.2a). The outer **fibrous pericardium** is a tough, inelastic, fibrous connective tissue. It resembles a bag that rests on and attaches to the diaphragm with its open end fused to the connective tissues of the blood vessels entering and leaving the heart. Its lateral surfaces lie against the parietal pleurae, the outer coverings of the lungs. The fibrous pericardium prevents overstretching of the heart, provides protection, and anchors the heart in the mediastinum.

The inner **serous pericardium** is a thinner, more delicate membrane that forms a double layer around the heart (Fig. 20.2a,b). The outer **parietal layer** of the serous pericardium is fused to the fibrous pericardium. The inner **visceral layer** of the serous pericardium, also called the **epicardium** (*epi* = on top), adheres tightly to the muscle of the heart. Between the parietal and visceral layers of the serous pericardium is a thin film of serous fluid known as **pericardial fluid**. It is a slippery secretion of the pericardial cells and reduces friction between the membranes as the heart moves. The space that houses the pericardial fluid is called the **pericardial cavity**.

CLINICAL APPLICATION

PERICARDITIS AND CARDIAC TAMPONADE

Inflammation of the pericardium is known as **pericarditis**. If production of pericardial fluid diminishes, the result may be painful rubbing together of the parietal and visceral serous pericardial layers. A buildup of pericardial fluid, which may also occur in pericarditis, or extensive bleeding into the pericardium is a life-threatening condition. Since the pericardium cannot stretch, buildup of fluid or blood compresses the heart. This compression, known as **cardiac tamponade** (tam′-pon-ĀD), can result in cardiac failure, cessation of the heartbeat.

HEART WALL

Three layers form the wall of the heart (Fig. 20.2a): the epicardium (external layer), myocardium (middle layer), and endocardium (inner layer). The outermost **epicardium** (also called the **visceral layer of the serous pericardium**) is the

FIGURE 20.1 Position of the heart and associated blood vessels in the thoracic cavity. In this and subsequent illustrations, vessels that carry oxygenated blood are colored red; vessels that carry deoxygenated blood are colored blue.

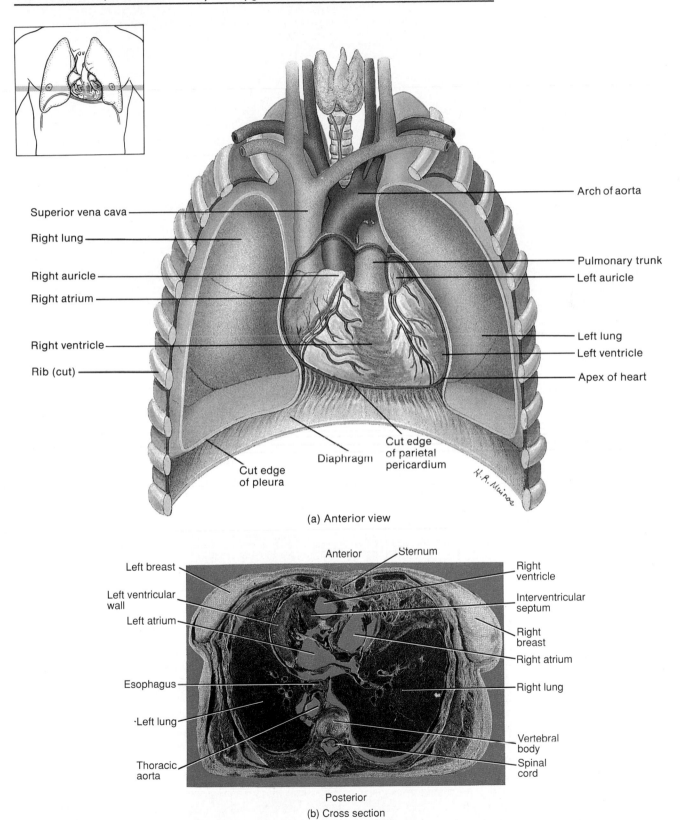

Arch of aorta

Superior vena cava

Right lung

Right auricle

Right atrium

Pulmonary trunk

Left auricle

Right ventricle

Left lung

Left ventricle

Rib (cut)

Apex of heart

Cut edge of parietal pericardium

Diaphragm

Cut edge of pleura

H. R. Muñoz

(a) Anterior view

Anterior   Sternum

Left breast

Right ventricle

Left ventricular wall

Interventricular septum

Left atrium

Right breast

Right atrium

Esophagus

Right lung

·Left lung

Vertebral body

Thoracic aorta

Spinal cord

Posterior

(b) Cross section

Question: What is the mediastinum?

FIGURE 20.2 Pericardium and heart wall.

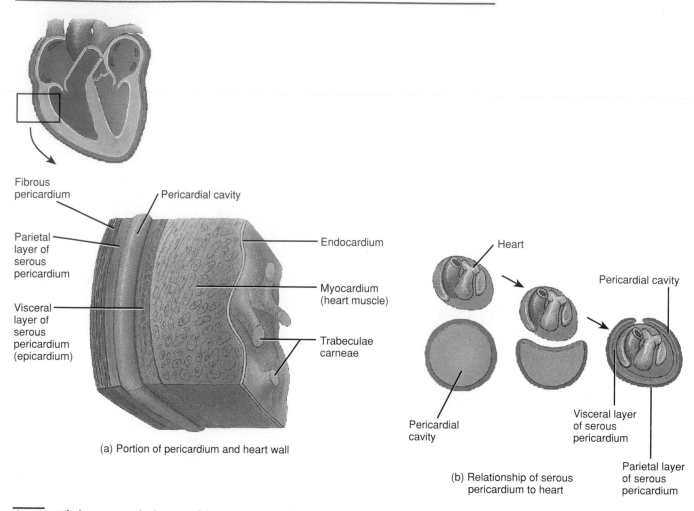

(a) Portion of pericardium and heart wall

(b) Relationship of serous pericardium to heart

Question: Which structure is both a part of the pericardium and part of the heart wall?

thin, transparent outer layer of the wall. It is composed of mesothelium and delicate connective tissue that imparts a smooth, slippery texture to the outermost surface of the heart.

The middle **myocardium,** which is cardiac muscle tissue, makes up the bulk of the heart and is responsible for its pumping action. Cardiac muscle fibers (cells) are involuntary, striated, and branched. They swirl diagonally around the heart in interlacing bundles and form two large networks—one atrial and one ventricular. Each fiber physically contacts neighboring fibers in its network by transverse thickenings of the sarcolemma called **intercalated discs**. Within the discs are **gap junctions** (electrical synapses) that allow muscle action potentials to spread from one fiber to another (see Figs. 4.1 and 10.17). As a result, the whole atrial network contracts as one unit and the ventricular network contracts as another. The intercalated discs also contain **desmosomes**, which act as reinforcing spot welds. They prevent adjacent cardiac fibers from pulling apart during their vigorous contractions.

The innermost **endocardium** is a thin layer of endothelium overlying a thin layer of connective tissue. It provides a smooth lining for the inside of the heart and covers the valves of the heart. The endocardium is continuous with the endothelial lining of the large blood vessels associated with the heart and the rest of the cardiovascular system.

CHAMBERS OF THE HEART

The interior of the heart is divided into four compartments called **chambers** that receive the circulating blood (Fig. 20.3). The two superior chambers are called the **right atrium** and **left atrium** (*atrium* = court or entry hall; plural is **atria**). Each atrium has an appendage called an **auricle** (OR-i-kul; *auris* = ear), so named because its shape resembles a dog's ear. The auricles increase the volume of the atria. The two inferior chambers are the **right ventricle** and **left ventricle** (*ventricle* = little belly).

Figure 20.3 Structure of the heart.

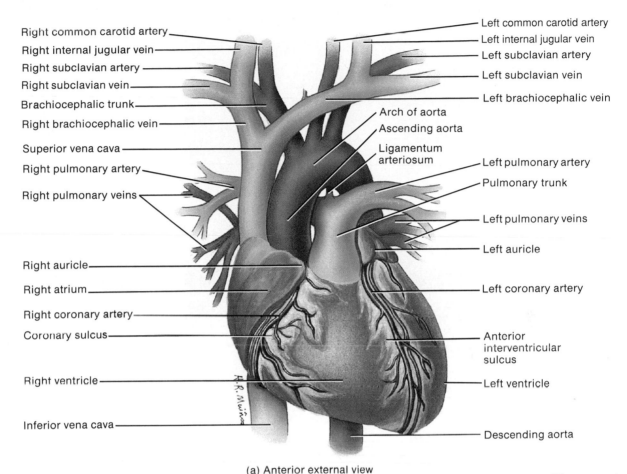

Right common carotid artery
Right internal jugular vein
Right subclavian artery
Right subclavian vein
Brachiocephalic trunk
Right brachiocephalic vein
Superior vena cava
Right pulmonary artery
Right pulmonary veins

Right auricle
Right atrium
Right coronary artery
Coronary sulcus

Right ventricle

Inferior vena cava

Left common carotid artery
Left internal jugular vein
Left subclavian artery
Left subclavian vein
Left brachiocephalic vein
Arch of aorta
Ascending aorta
Ligamentum arteriosum
Left pulmonary artery
Pulmonary trunk
Left pulmonary veins
Left auricle
Left coronary artery
Anterior interventricular sulcus
Left ventricle
Descending aorta

(a) Anterior external view

Figure continues

Connective tissue separates the muscle tissue of the atria from that of the ventricles and effectively divides the myocardium into separate atrial and ventricular muscle masses. Externally, a groove known as the **coronary sulcus** (SUL-kus) separates the atria from the ventricles (Fig. 20.3a). The **anterior interventricular sulcus** and **posterior interventricular sulcus** separate the right and left ventricles externally (Fig. 20.3a,c). Sulci contain coronary blood vessels and a variable amount of fat.

A partition called the **interatrial septum** (*septum* = partition) separates the atria. A prominent feature of this septum is an oval depression, the **fossa ovalis** (Fig. 20.3d). This was the site of the foramen ovale, an opening in the interatrial septum of the fetal heart (see Fig. 21.32). The irregular surface of ridges and folds of the myocardium covered by endocardium in the ventricles is known as the **trabeculae carneae** (tra-BEK-yoo-lē KAR-nē-ē). See also Fig. 20.2a. A wall known as the **interventricular septum** separates the two ventricles.

The thickness of the walls of the four chambers varies according to their functions. The atria are thin-walled because they only have to deliver blood into the ventricles

(Fig. 20.3d). Although the right and left sides of the heart act as two separate pumps, the left side has a much larger work load. Whereas the right ventricle pumps blood only to the lungs (pulmonary circulation), the left ventricle pumps blood to all other parts of the body (systemic circulation). Thus the left ventricle must work harder than the right ventricle to maintain the same rate of blood flow. The anatomy of the two ventricles confirms this functional difference: the muscular wall of the left ventricle is two to four times as thick as the wall of the right ventricle.

BLOOD FLOW THROUGH THE HEART

The right atrium receives *deoxygenated blood* (blood that has given up some of its oxygen to cells) from various parts of the body through three veins (Fig. 20.3c,d). In general, (1) the **superior vena cava (SVC)** brings blood from most parts of the body superior to the heart, (2) the **inferior vena cava (IVC)** brings blood from all parts of the body inferior to the diaphragm, and (3) the **coronary sinus** drains blood from most of the vessels supplying the wall of the heart.

FIGURE 20.3 (continued)

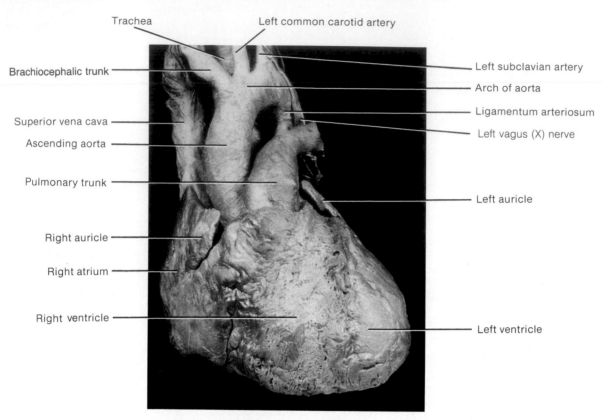

Trachea

Left common carotid artery

Brachiocephalic trunk

Superior vena cava

Ascending aorta

Pulmonary trunk

Right auricle

Right atrium

Right ventricle

Left subclavian artery

Arch of aorta

Ligamentum arteriosum

Left vagus (X) nerve

Left auricle

Left ventricle

(b) Anterior external view

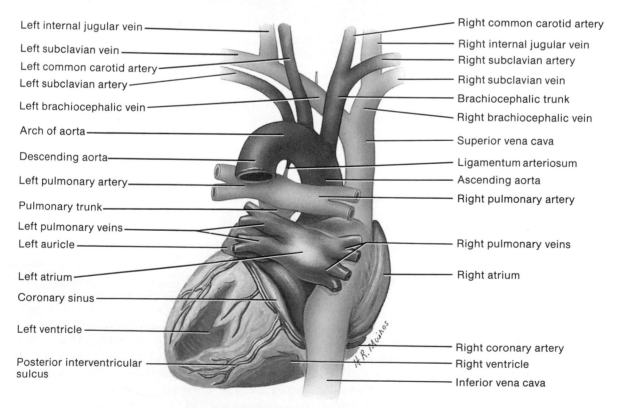

Left internal jugular vein

Left subclavian vein

Left common carotid artery

Left subclavian artery

Left brachiocephalic vein

Arch of aorta

Descending aorta

Left pulmonary artery

Pulmonary trunk

Left pulmonary veins

Left auricle

Left atrium

Coronary sinus

Left ventricle

Posterior interventricular sulcus

Right common carotid artery

Right internal jugular vein

Right subclavian artery

Right subclavian vein

Brachiocephalic trunk

Right brachiocephalic vein

Superior vena cava

Ligamentum arteriosum

Ascending aorta

Right pulmonary artery

Right pulmonary veins

Right atrium

Right coronary artery

Right ventricle

Inferior vena cava

(c) Posterior external view

FIGURE 20.3 (continued)

Left common carotid artery

Left subclavian artery

Brachiocephalic trunk

Superior vena cava

Arch of aorta

Right pulmonary artery

Ligamentum arteriosum

Left pulmonary artery

Right pulmonary veins

Pulmonary trunk

Left pulmonary veins

Pulmonary semilunar valve

Left atrium

Fossa ovalis

Right atrium

Aortic semilunar valve

Opening of coronary sinus

Bicuspid valve

Chordae tendineae

Interventricular septum

Tricuspid valve

Papillary muscle

Right ventricle

Left ventricle

Trabeculae carneae

Inferior vena cava

Thoracic aorta

(d) Anterior internal view

Figure continues

From the right atrium blood flows into the right ventricle, which pumps it to the lungs, starting in the **pulmonary trunk** (Fig. 20.3d). The pulmonary trunk divides into a **right** and **left pulmonary artery**, each of which carries blood to one lung. In the lungs, the blood releases carbon dioxide and takes on oxygen. This blood, called *oxygenated blood,* returns to the heart via four **pulmonary veins** that empty into the left atrium. The blood then passes into the left ventricle, which pumps the blood into the **ascending aorta** (*aorte* = to suspend, because the aorta once was believed to suspend the heart). From here the blood flows into the **coronary arteries,** which carry the blood to the heart, **arch of the aorta, thoracic aorta**, and **abdominal aorta**. The aorta and its branches carry the blood throughout the systemic circulation.

During fetal life, a temporary blood vessel, called the ductus arteriosus, connects the pulmonary trunk with the aorta (see Fig. 21.32). It redirects blood so that only a small amount enters the nonfunctioning fetal lungs. The ductus arteriosus normally closes shortly after birth, leaving a remnant known as the **ligamentum arteriosum** (Fig. 20.3d).

VALVES OF THE HEART

As each chamber of the heart contracts, it pushes a portion of blood into a ventricle or out of the heart through an artery. To prevent back flow of blood, the heart has **valves**. These structures are composed of dense connective tissue covered by endocardium. Valves open and close in response to pressure changes as the heart contracts and relaxes.

FIGURE 20.3 (continued)

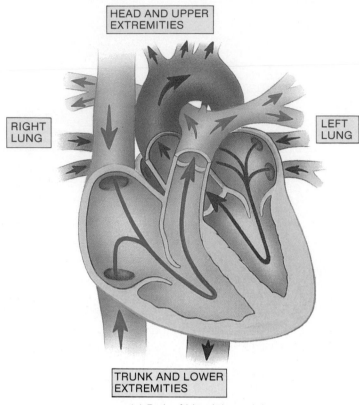

(e) Path of blood through heart

Question: What is the path of blood flow through the heart chambers and valves, starting in the right atrium and ending in the aorta?

Atrioventricular Valves

Atrioventricular (AV) valves lie between the atria and ventricles (Fig. 20.3d). The right AV valve between the right atrium and right ventricle is also called the **tricuspid valve** because it consists of three cusps (flaps). The left AV valve between the left atrium and left ventricle has two cusps and is called the **bicuspid (mitral) valve**. When an AV valve is open, the pointed ends of the cusps project into the ventricle. Tendonlike fibrous cords called **chordae tendineae** (KOR-dē ten-DIN-ē-ē) connect the pointed ends and undersurfaces to **papillary muscles** (muscular columns) that are located on the inner surface of the ventricles.

Blood moves from the atria into the ventricles through open AV valves when ventricular pressure is low (Fig. 20.4a). At this time the papillary muscles are relaxed, and the chordae tendineae are slack. When the ventricles contract, the pressure of the blood drives the cusps upward until their edges meet and close the opening (Fig. 20.4b). At the same time, the papillary muscles are also contracting, pulling on and tightening the chordae tendineae. This

prevents the valve cusps from everting or swinging upward into the atria.

Semilunar Valves

Both arteries that emerge from the heart have a valve that prevents blood from flowing backward into the heart. These are the **semilunar (SL) valves** (see Fig. 20.3d). The **pulmonary semilunar valve** lies in the opening where the pulmonary trunk leaves the right ventricle. The **aortic semilunar valve** is situated at the opening between the left ventricle and the aorta.

Both valves consist of three semilunar (half-moon, or crescent-shaped) cusps. Each cusp is attached by its convex margin to the artery wall. The free borders of the cusps curve outward and project into the opening inside the blood vessel. Like the atrioventricular valves, the semilunar valves permit blood to flow in one direction only: in this case, the flow is from the ventricles into the arteries.

FIGURE 20.4 Atrioventricular (AV) valves. The bicuspid and tricuspid valves operate in a similar manner.

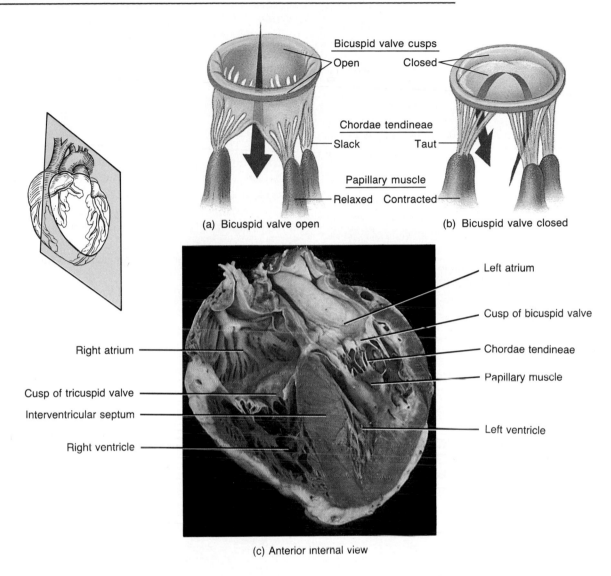

Bicuspid valve cusps
Open   Closed

Chordae tendineae
Slack   Taut

Papillary muscle
Relaxed   Contracted

(a) Bicuspid valve open   (b) Bicuspid valve closed

Left atrium

Cusp of bicuspid valve

Right atrium

Chordae tendineae

Papillary muscle

Cusp of tricuspid valve

Interventricular septum

Right ventricle

Left ventricle

(c) Anterior internal view

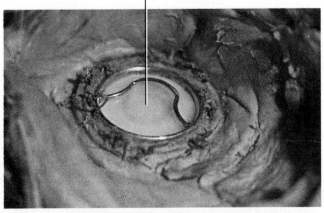

Artificial atrioventricular (AV) valve

(d) Photograph of a superior view of an
artificial atrioventricular valve

Question: What is the function of heart valves?

RHEUMATIC FEVER

An infection with group A, β-hemolytic *Streptococcus pyogenes* bacteria may result in an inflammation of the heart that damages the valves, particularly the bicuspid and aortic semilunar valves. This condition is called **rheumatic fever** and is usually precipitated by a streptococcal sore throat. The bacteria trigger an immune response. Antibodies that recognize molecules carried by streptococci attack similar molecules in joints, heart valves, and other tissues. Genetic factors may be important in some patients.

HEART BLOOD SUPPLY

The wall of the heart has its own blood vessels. Nutrients could not possibly diffuse through all the layers of cells that make up the heart tissue. The flow of blood through the many vessels that pierce the myocardium is called the **coronary (cardiac) circulation**. The arteries of the heart encircle it like a crown encircles the head (*corona* = crown).

Coronary Arteries

Two coronary arteries, the right and left coronary arteries, branch from the ascending aorta (Fig. 20.5a). The **left coro-**

nary artery courses under the left auricle and divides into the anterior interventricular and circumflex branches. The **anterior interventricular branch** or **left anterior descending (LAD) artery** is in the anterior interventricular sulcus and supplies oxygenated blood to the walls of both ventricles. The **circumflex branch** lies in the coronary sulcus and distributes oxygenated blood to the walls of the left ventricle and left atrium.

The **right coronary artery** supplies small branches to the right atrium. It continues under the right auricle and divides into the posterior interventricular and marginal branches. The **posterior interventricular branch** follows the posterior interventricular sulcus and supplies the walls of the two ventricles with oxygenated blood. The **marginal branch** in the coronary sulcus transports oxygenated blood to the myocardium of the right ventricle. The left ventricle receives the most abundant blood supply because of the enormous work it must do.

Most parts of the body receive branches from more than one artery, and where two or more arteries supply the same region, they usually connect. The connections, called *anastomoses* (a-nas-tō-MŌ-sēs), provide alternate routes for blood to reach a particular organ or tissue. The myocardium contains many anastomoses, connecting branches of one coronary artery or extending between branches of different coronary arteries. Heart muscle can remain alive if it receives as little as 10 to 15% of its normal blood supply.

FIGURE 20.5 Coronary (cardiac) circulation. The heart is viewed from the anterior aspect, but is drawn as if it were transparent to reveal blood vessels on the posterior aspect.

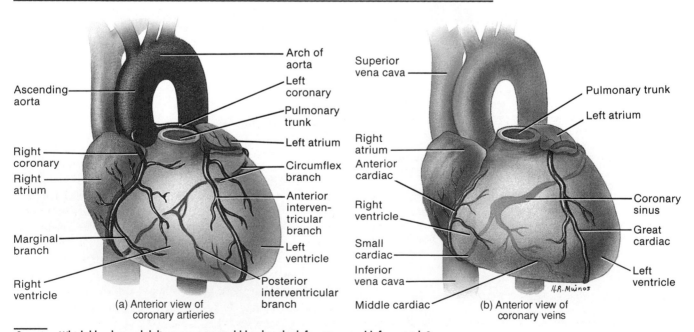

(a) Anterior view of coronary arteries

(b) Anterior view of coronary veins

Question: Which blood vessel delivers oxygenated blood to the left atrium and left ventricle?

Coronary Veins

As blood passes through the coronary circulation, it delivers oxygen and nutrients and collects carbon dioxide and wastes. It then drains into a large vein on the posterior surface of the heart, called the **coronary sinus** (Fig. 20.5b), which empties into the right atrium. A vascular sinus is a vein with a thin wall that has no smooth muscle to alter its diameter. The principal tributaries carrying blood into the coronary sinus are the **great cardiac vein**, which drains the anterior aspect of the heart, and the **middle cardiac vein**, which drains the posterior aspect of the heart.

CLINICAL APPLICATION

ANGINA PECTORIS AND MYOCARDIAL INFARCTION

Most heart problems result from faulty coronary circulation due to blood clots, fatty atherosclerotic plaques, or spasms of the smooth muscle in coronary artery walls. Reduction of blood flow is called **ischemia** (is-KĒ-mē-a). Usually, ischemia causes **hypoxia** (reduced oxygen supply), which may weaken cells without killing them. **Angina pectoris** (an-JĪ-na, or AN-ji-na, PEK-to-ris), literally meaning "strangled chest," is a severe pain that usually accompanies myocardial ischemia. Typically, sufferers describe it as a tight or squeezing sensation, as though the chest were in a vise. Other symptoms may include labored breathing, a sensation of foreboding, weakness, dizziness, and perspiration. Angina pectoris often occurs during exertion, when the heart demands more oxygen, and disappears with rest. (Pain impulses originating from most visceral muscles are referred to an area on the surface of the body. As shown in Fig. 15.2, the pain associated with angina pectoris is referred to the neck, chin, or down the left arm to the elbow.) In some people, ischemic episodes occur without producing angina pain. This is known as **silent myocardial ischemia** and is particularly dangerous because the person has no forewarning of an impending heart attack.

More serious than ischemia is **myocardial infarction** (in-FARK-shun), or **MI**, commonly called a heart attack. **Infarction** means the death of an area of tissue because of an interrupted blood supply. Myocardial infarction may result from a thrombus (stationary blood clot) or embolus (blood clot transported by blood) in one of the coronary arteries. The tissue distal to the obstruction dies and is replaced by noncontractile scar tissue. Thus the heart muscle loses at least some of its strength. The aftereffects depend partly on the size and location of the infarcted, or dead, area. Besides killing normal heart tissue, the infarction may disrupt the conducting system of the heart (described shortly) and may cause sudden death by triggering ventricular fibrillation. Treatment for a myocardial infarction may involve injection of a clot-dissolving thrombolytic agent such as streptokinase or t-PA along with heparin or performing coronary angioplasty (see page 617).

Reperfusion Damage

Whenever a disease or injury deprives a tissue of oxygen, reestablishing the blood flow (reperfusion) may damage the tissue further. This surprising effect is due to the formation of oxygen **free radicals** from the reintroduced oxygen and may play a major role in tissue destruction. Free radicals are electrically charged molecules that have an unpaired electron. Such molecules are unstable and highly reactive. When an oxygen free radical takes an electron from one molecule, that molecule becomes unstable and borrows an electron from another molecule, which, in turn, becomes unstable. Such a chain reaction can lead to cellular damage and death. Among the molecules attacked by oxygen free radicals are proteins (such as enzymes), neurotransmitters, nucleic acids, and phospholipids of plasma membranes.

Free radicals have been implicated in diseases such as heart disease, cancer, Alzheimer's disease, Parkinson's disease, cataracts, and rheumatoid arthritis. They may also contribute to aging. To counter the effects of oxygen free radicals, the body produces enzymes—for example, *superoxide dismutase, catalase*, and *glutathione peroxidase*—that convert free radicals to less reactive substances. In addition, some nutrients, such as vitamins E and C and beta-carotene, are antioxidants that counter oxygen free radicals.

CONDUCTION SYSTEM AND PACEMAKER

An inherent and rhythmical electrical activity is the force behind the heart's continuous beating. Certain cardiac muscle cells repeatedly fire spontaneous impulses (action potentials) that then trigger heart contractions. This is why a heart that has been completely removed from the body, for example, to be transplanted into another person, will continue to beat even though all its nerves have been cut. Signals from the autonomic nervous system and hormones, such as epinephrine, in the blood do modify the heartbeat, but they *do not establish the fundamental rhythm.*

Autorhythmic Cells: The Conduction System

During embryonic development, a small fraction (about 1%) of the cardiac muscle fibers become **autorhythmic** (self-excitable), that is, able to repeatedly and rhythmically generate impulses. The autorhythmic fibers have two important functions. They act as a **pacemaker**, setting the rhythm for the entire heart, and they form the **conduction system**, the route for conducting impulses throughout the heart muscle. The conduction system assures that cardiac chambers contract in a coordinated manner, which makes the heart an effective pump. Figure 20.6 shows the components of the conduction system: (1) the sinoatrial (SA) node, (2) the atrioventricular (AV) node, (3) the atrioventricular (AV) bundle (bundle of His), (4) the right and left

FIGURE 20.6 Conduction system of the heart. The arrows indicate the flow of impulses through the atria.

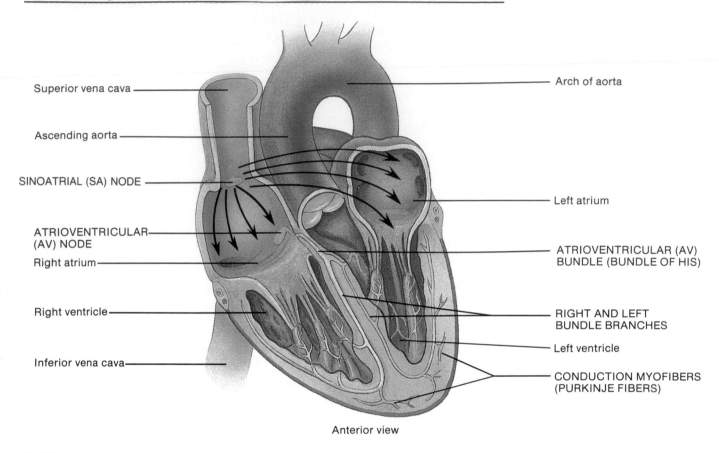

Superior vena cava

Ascending aorta

SINOATRIAL (SA) NODE

ATRIOVENTRICULAR (AV) NODE

Right atrium

Right ventricle

Inferior vena cava

Arch of aorta

Left atrium

ATRIOVENTRICULAR (AV) BUNDLE (BUNDLE OF HIS)

RIGHT AND LEFT BUNDLE BRANCHES

Left ventricle

CONDUCTION MYOFIBERS (PURKINJE FIBERS)

Anterior view

Question: Which component of the conduction system provides the only electrical connection between the atria and ventricles?

bundle branches, and (5) the conduction myofibers (Purkinje fibers).

Normally, cardiac excitation begins in the **sinoatrial (SA) node**, located in the right atrial wall just below the opening of the superior vena cava. Each SA node impulse travels throughout the heart via the conduction system and the gap junctions in the intercalated discs. In the wake of the impulse, first the atria contract and then the ventricles contract.

The cardiac impulse spreads from the SA node throughout the atrial fibers and down to the **atrioventricular (AV) node**, located in the septum between the two atria. From the AV node, the impulse enters the **atrioventricular (AV) bundle (bundle of His),** the only electrical connection between the atria and the ventricles. (Elsewhere, fibrous rings and sheets of connective tissue act as electrical insulation between the atria and ventricles.) After traveling along the AV bundle, the impulse then enters both the **right** and **left bundle branches** that course through the interventricu-

lar septum toward the apex of the heart. Finally, large-diameter **conduction myofibers (Purkinje fibers)** rapidly conduct the impulse into the mass of ventricular muscle tissue.

Autorhythmic fibers in the SA node spontaneously initiate action potentials 60 to 100 times per minute, faster than any other region. As a result, action potentials from the SA node spread to other areas of the conduction system, stimulating them before they are able to generate an impulse at their own slower rate. Thus the normal *pacemaker* of the heart is the SA node.

Sometimes, a site other than the SA node becomes the pacemaker because it develops abnormal self-excitability. Such a site is called an **ectopic** (ek-TOP-ik; *ektopus* = displaced) **pacemaker** or **ectopic focus.** The ectopic focus may operate only occasionally, producing extra beats, or it may pace the heart for some period of time. Triggers of ectopic activity include caffeine and nicotine, electrolyte imbalances, hypoxia, and toxic reactions to drugs such as digitalis.

Timing of Atrial and Ventricular Excitation

From the SA node, the cardiac impulse travels throughout the atrial muscle and down to the AV node in about 0.05 sec (50 milliseconds or msec). The impulse slows considerably at the AV node because the fibers there have much smaller diameters. (Recall what happens to automobile traffic when a four-lane highway narrows to just two lanes.) The resulting 0.1 sec (100 msec) delay has an advantage. It gives the atria time to complete their contraction and add to the volume of blood in the ventricles before ventricular contraction begins. After the cardiac impulse enters the AV bundle, conduction again is rapid; the entire ventricular myocardium undergoes depolarization (loss and then reversal of polarization) about 0.15 to 0.2 sec (150 to 200 msec) after the impulse arises in the SA node.

If the SA node becomes diseased or damaged, the slower AV node fibers can pick up the pacemaking chores. With pacing by the AV node, heart rate ranges between 40 and 50 beats/min. If the activity of both nodes is suppressed, the heartbeat may still be maintained by autorhythmic fibers in the ventricles—the AV bundle, a bundle branch, or conduction myofibers. However, these fibers fire impulses very slowly, only about 20 to 40 times per minute. At such an abnormally low heart rate, blood flow to the brain is inadequate. In patients with such a condition, normal heart rhythm can be restored and maintained with an **artificial pacemaker**, a device that sends out small electrical currents that stimulate the heart. Many of the newer pacemakers, called activity-adjusted pacemakers, automatically speed up the heartbeat during exercise.

PHYSIOLOGY OF CARDIAC MUSCLE CONTRACTION

The impulse initiated by the SA node travels along the conduction system and spreads out to excite the "working" atrial and ventricular muscle fibers, which are called **contractile fibers**. The contractile fibers have a resting membrane potential close to –90 mV. When they are brought to threshold by excitation in neighboring fibers, certain sodium ion (Na⁺) channels open very rapidly; these are called **voltage-gated fast Na⁺ channels**. This increase in membrane permeability allows an inflow of Na⁺ down its concentration gradient and produces a **rapid depolarization** (Fig. 20.7).

During the next phase, called the **plateau**, **voltage-gated slow Ca²⁺ channels** open, allowing calcium ions (Ca²⁺) to enter the cytosol. Some Ca²⁺ passes through the sarcolemma (plasma membrane) from the extracellular fluid (which has a higher Ca²⁺ concentration) while other Ca²⁺ pours out of the sarcoplasmic reticulum within the fiber. The combined buildup of Na⁺ and Ca²⁺ in the cytosol maintains the depolarization for about 0.25 sec (250 msec).

FIGURE 20.7 Impulse (action potential) in a ventricular contractile fiber. The resting membrane potential is about –90 mV.

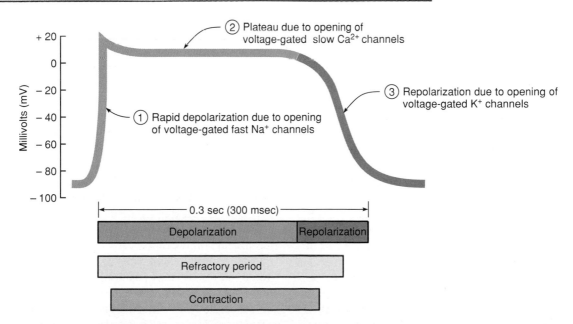

Question: How does the duration of depolarization and repolarization in a cardiac muscle fiber compare with that in a skeletal muscle fiber?

By comparison, depolarization in a neuron or skeletal muscle fiber lasts about 1 msec.

The next steps are similar in skeletal and cardiac muscle fibers. Ca^{2+} binds to troponin, which allows the actin and myosin filaments to begin sliding past one another, and tension starts to develop. Substances that alter the movement of Ca^{2+} through slow Ca^{2+} channels influence the strength of heart contractions. Epinephrine, for example, increases contraction force by enhancing Ca^{2+} inflow. Certain drugs, appropriately called calcium channel blockers, such as verapamil, reduce Ca^{2+} inflow and diminish the strength of the heartbeat.

The **repolarization** (recovery of resting membrane potential) phase of the impulse in a cardiac muscle fiber resembles repolarization in other excitable tissues: after a delay (which is particularly prolonged in cardiac muscle), **voltage-gated K^+ channels** open, and potassium ions (K^+) diffuse out along their concentration gradient. At the same time, the Na^+ and Ca^{2+} channels are closing, which slows and then almost stops further inflow of these two ions. As more K^+ leaves the fiber and fewer Na^+ and Ca^{2+} enter, the negative resting membrane potential (−90 mV) is restored and the muscle fiber relaxes.

In muscle, the **refractory period** is the time interval when a second contraction cannot be triggered. The refractory period of a cardiac fiber is longer than the contraction itself (Fig. 20.7). As a result, another contraction cannot begin until relaxation is well underway and tetanus (maintained contraction) cannot occur. The advantage is apparent if you consider how the ventricles work. Their pumping function depends on alternating contraction, when they eject blood, and relaxation, when they refill. If tetanus could occur, blood flow would stop.

ELECTROCARDIOGRAM

Impulse conduction through the heart generates electrical currents that can be detected at the surface of the body. A recording of the electrical changes that accompany each cardiac cycle (heartbeat) is called an **electrocardiogram** (e-lek′-trō-KAR-dē-ō-gram), abbreviated either **ECG** or **EKG** (from the German word for heart, *kardia*). The ECG is a composite of action potentials produced by all the heart muscle fibers during each heartbeat. The instrument used to record the changes is an **electrocardiograph**. In clinical practice, the ECG is recorded by placing electrodes on the arms and legs (the limb leads) and at six positions on the chest. As the person lies still, the electrocardiograph amplifies the heart's electrical activity and produces 12 different tracings from different combinations of limb and chest leads. This takes about a minute. Each limb and chest electrode records slightly different electrical activity because it is in a different position relative to the heart. By comparing these records with one another and with normal records, it is possible to determine (1) if the conduction

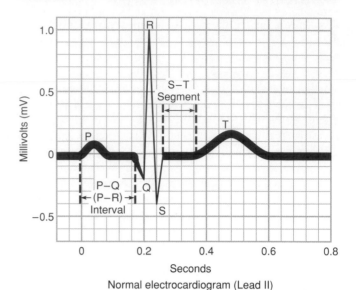

FIGURE 20.8 Electrocardiogram (ECG).

Normal electrocardiogram (Lead II)

Question: Which component of the ECG represents the onset of ventricular depolarization? Atrial depolarization?

pathway is normal, (2) if the heart is enlarged, and (3) if certain regions are damaged.

In a typical Lead II record (right arm to left leg, Fig. 20.8), three clearly recognizable waves accompany each heartbeat. The first, called the **P wave**, is a small upward wave. It represents **atrial depolarization**, which spreads from the SA node throughout both atria. About 0.1 sec after the P wave begins, the atria contract. The second wave, called the **QRS complex**, begins as a downward deflection, continues as a large, upright, triangular wave, and ends as a downward wave. The QRS complex represents the onset of **ventricular depolarization**, the spread of the wave of electrical excitation through the ventricles. Shortly after the QRS complex begins, the ventricles start to contract. The third wave is a dome-shaped upward deflection called the **T wave**. It indicates **ventricular repolarization** and occurs just before the ventricles start to relax. The T wave is smaller and more spread out than the QRS complex because repolarization occurs more slowly than depolarization. Usually, repolarization of the atria is not evident in an ECG because it is buried in the larger QRS complex.

In reading an electrocardiogram, it is important to note the size and timing of the waves. Larger P waves, for example, indicate enlargement of an atrium, as may occur in mitral stenosis. In this condition, the mitral valve narrows, blood backs up into the left atrium, and there is expansion of the atrial wall. An enlarged Q wave may indicate a myocardial infarction (heart attack). An enlarged R wave generally indicates enlarged ventricles.

The **P-Q (PR) interval** is measured from the beginning of the P wave to the beginning of the QRS complex. It represents the conduction time from the beginning of atrial excitation to the beginning of ventricular excitation. The P-Q interval is the time required for an impulse to travel through the atria, atrioventricular node, and the remaining fibers of the conduction system. In coronary artery disease and rheumatic fever, scar tissue may form in the heart. As the impulse detours around scar tissue, the P-Q interval lengthens.

The **S-T segment** begins at the end of the S wave and ends at the beginning of the T wave. It represents the time when the ventricular contractile fibers are fully depolarized, during the plateau phase of the impulse. The S-T segment is elevated (above the baseline) in acute myocardial infarction and depressed (below the baseline) when the heart muscle receives insufficient oxygen.

The T wave represents ventricular repolarization. It is flatter than normal when the heart muscle is receiving insufficient oxygen, for example, in coronary artery disease. It may be elevated in hyperkalemia (increased blood K^+ level).

Sometimes it is necessary to evaluate the heart's response to the stress of physical exercise. Such a test is called a **stress electrocardiogram**, or **stress test**. It is based on the principle that narrowed coronary arteries may carry adequate oxygenated blood while a person is at rest, but during exercise will be unable to meet the heart's increased need for oxygen, creating changes that can be noted on an electrocardiogram.

CARDIAC CYCLE

Since heartbeats automatically follow one another throughout life, you can appreciate much of cardiac physiology by understanding a single **cardiac cycle**, that is, all the events associated with one heartbeat. In each cardiac cycle, pressure changes occur as the atria and ventricles alternately contract and relax, and blood flows from areas of higher blood pressure to areas of lower blood pressure. As a chamber of the heart contracts, pressure of the fluid within it increases. Figure 20.9 shows the relation between the heart's electrical signals (electrocardiogram) and mechanical events (contraction and relaxation) and the consequent changes in atrial pressure, ventricular pressure, ventricular volume, and aortic pressure during the cardiac cycle. The pressures given in Fig. 20.9 apply to the left side of the heart; on the right side, pressures are considerably lower because the wall of the right ventricle is thinner than that of the left. Each ventricle, however, expels the same volume of blood per beat, and the same pattern exists for both pumping chambers.

In a normal cardiac cycle, the two atria contract while the two ventricles relax. Then, while the two ventricles con-

tract, the two atria relax. The term **systole** (SIS-tō-lē; *systellein* = contract) refers to the phase of contraction; the phase of relaxation is **diastole** (dī-AS-tō-lē; = dilation). A cardiac cycle consists of a systole and diastole of both atria plus a systole and diastole of both ventricles.

Phases of the Cardiac Cycle

For the purposes of our discussion, we divide the cardiac cycle of a resting adult into three main phases. Follow along on Fig. 20.9 as you read the descriptions.

1. **Relaxation period.** At the end of a heartbeat when the ventricles start to relax, all four chambers are in diastole. This is the beginning of the **relaxation** or **quiescent** (kwī-ES-ent) **period.** Repolarization of the ventricular muscle fibers (T wave in the ECG) initiates relaxation. As the ventricles relax, pressure within the chambers drops, and blood starts to flow from the pulmonary trunk and aorta back toward the ventricles. As this blood becomes trapped in the semilunar cusps, however, the valves close. Rebound of blood off the closed cusps produces a bump called the *dicrotic wave* on the aortic pressure curve.

With closing of the semilunar valves, there is a brief interval when ventricular blood volume does not change because both semilunar and AV valves are closed. This period is called **isovolumetric relaxation.** As the ventricles continue to relax, the space inside expands, and the pressure falls quickly. When ventricular pressure drops below atrial pressure, the AV valves open and ventricular filling begins.

2. **Ventricular filling.** The major part of ventricular filling occurs just after the AV valves open. Blood that has been flowing into the atria and building up while the ventricles were contracting now rushes into the ventricles. The first third of ventricular filling time thus is known as the period of **rapid ventricular filling.** During the middle third, called **diastasis,** a much smaller volume of blood flows into the ventricles.

Firing of the SA node results in atrial depolarization, noted as the P wave on the ECG. Atrial contraction follows the P wave, which also marks the end of the quiescent period. Atrial systole occurs in the last third of the ventricular filling period and accounts for the final 30 ml of the blood that fills the ventricles. At the end of ventricular diastole, there are about 130 ml in each ventricle. This volume of blood is called **end-diastolic volume (EDV).** Since atrial systole contributes only 20 to 30% of the total blood volume in the ventricles, atrial contraction is not absolutely necessary for adequate blood flow at normal heart rates. Throughout the period of ventricular filling, the AV valves are open and the semilunar valves are closed.

3. **Ventricular systole (contraction).** Near the end of

Figure 20.9 Cardiac cycle. (a) ECG related to the cardiac cycle. (b) Left atrial, left ventricular, and aortic pressure changes along with the opening and closing of valves during the cardiac cycle. (c) Left ventricular volume during the cardiac cycle. (d) Heart sounds related to the cardiac cycle.

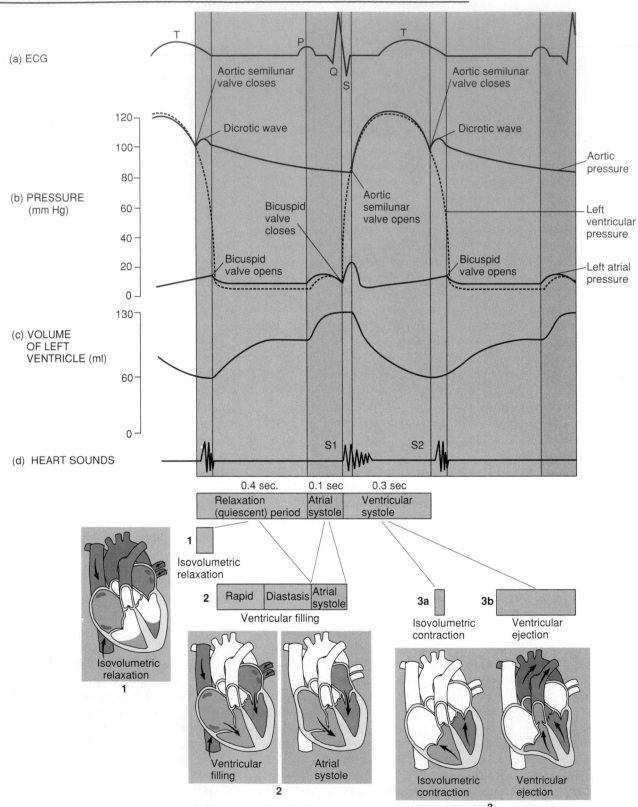

Question: How much blood is in each ventricle at the end of ventricular diastole? What is this volume called?

atrial systole, the impulse from the SA node has passed through the AV node and into the ventricles, causing them to depolarize. This is represented by the QRS complex in the ECG. Then, ventricular contraction begins, and blood is pushed up against the AV valves, forcing them shut. For about 0.05 sec (50 msec), all four valves are closed again. This period is called **isovolumetric contraction**. During this time cardiac muscle fibers are contracting and exerting force, but are not yet shortening because it is very difficult to compress any liquid, including blood. Thus the muscle contraction is isometric (same length). Moreover, since there is no escape route for the blood, ventricular volume remains the same (isovolumic).

As ventricular contraction continues, pressure inside the chambers rises sharply. When left ventricular pressure surpasses aortic pressure (about 80 mm Hg) and right ventricular pressure rises above the pressure in the pulmonary trunk (about 15 to 20 mm Hg), both semilunar valves open, and ejection of blood from the heart begins. This period is called **ventricular ejection** and lasts for about 0.25 sec (250 msec), until the ventricles start to relax. Then, the semilunar valves close and another relaxation period begins. The volume of blood still left in a ventricle following its systole is called **end-systolic volume (ESV)**. It is about 60 ml.

As mentioned earlier, the different pressures developed by the two ventricles are a reflection of their differing wall thickness. During contraction, the pressure in the left ventricle rises to about 120 mm Hg, while the pressure in the right ventricle climbs to about 30 mm Hg. At rest, the **stroke volume**, the volume ejected per beat from each ventricle, is about 70 ml (a little more than 2 oz). This is about half of the total volume in the ventricle at the end of diastole; during ejection, ventricular volume decreases from about 130 ml (end-diastolic volume) to 60 ml (end-systolic volume).

Timing of Systole and Diastole

Since resting heart rate (HR) is about 75 beats/min, each cardiac cycle takes about 0.8 sec (Fig. 20.10a and see also Fig. 20.9). During the first 0.4 sec of the cycle, the relaxation period, all four chambers are in diastole. During the next 0.1 sec, the atria contract but the ventricles are still relaxed. The atrioventricular valves are open, and the semilunar valves are closed. For the next 0.3 sec, the atria are relaxing and the ventricles are contracting. During the first part of this period, all valves are closed; during the second part, the semilunar valves are open. In a complete cycle, then, the atria are in systole 0.1 sec and in diastole 0.7 sec; the ventricles are in systole 0.3 sec and in diastole 0.5 sec. The last 0.1 sec of ventricular diastole overlaps atrial systole. For the first part of the relaxation period, all valves are closed; during the latter part, the atrioventricular

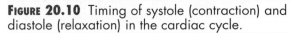
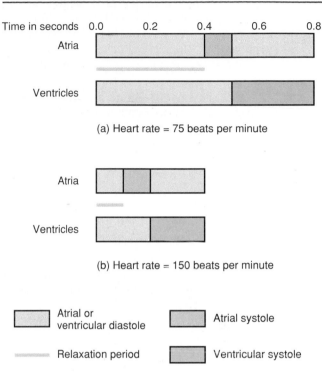

FIGURE 20.10 Timing of systole (contraction) and diastole (relaxation) in the cardiac cycle.

(a) Heart rate = 75 beats per minute

(b) Heart rate = 150 beats per minute

Question: What is the approximate duration of ventricular diastole when the heart rate is 120 beats/min?

valves open and blood starts draining into the ventricles. As the heart beats faster, the relaxation period becomes shorter and shorter (Fig. 20.10b).

Heart Sounds

The act of listening to sounds within the body is called **auscultation** (aws-kul-TĀ-shun; *auscultare* = to listen), and it is usually done with a stethoscope. The sound of the heartbeat comes primarily from blood turbulence caused by closing of the heart valves. During each cardiac cycle, four **heart sounds** are generated. In a normal heart, however, only the first two (first and second heart sounds) are loud enough to be heard by listening through a stethoscope.

The first sound (S1), which can be described as a **lubb** sound, is louder and a bit longer than the second sound. The lubb is the sound created by blood turbulence associated with closure of the AV valves soon after ventricular systole begins. The second sound (S2), which is shorter and not as loud as the first, can be described as a **dupp** sound. Dupp is the sound created by blood turbulence associated with closure of the semilunar valves at the beginning of ventricular diastole. Although these heart sounds are related to blood turbulence associated with the closure of valves, they are not necessarily heard best over these valves. Each sound tends to be clearest in a slightly different location on the chest surface (Fig. 20.11). Not normally loud enough to hear are the third and fourth heart sounds (S3 and S4), associated with rapid ventricular filling and atrial contraction, respectively. Figure 20.9 shows the timing of S1 and S2 relative to other changes in the cardiac cycle.

In a person at rest, the time between the S2 and the next S1 is about two times longer than the time between S1 and S2 within a cycle. Thus the rhythm is lubb, dupp, pause; lubb, dupp, pause; lubb, dupp, pause. As heart rate increases, the pause interval shortens.

CLINICAL APPLICATION

HEART MURMURS

Just as the ECG gives important information about the electrical operation of the heart, heart sounds provide valuable information about its mechanical operation. A **heart murmur** is an abnormal sound that consists of a flow noise that is heard before, between, or after the lubb–dupp or that may mask the normal heart sounds. Although some heart murmurs are "innocent," meaning they do not suggest a heart problem, most often a murmur indicates a valve disorder.

Among the valvular abnormalities that may contribute to murmurs are **mitral stenosis** (narrowing of the mitral valve by scar formation or a congenital defect), **mitral insufficiency** (back flow or regurgitation of blood from the left ventricle into the left atrium due to a damaged mitral valve or ruptured chordae tendineae), **aortic stenosis** (narrowing of the aortic semilunar valve), and **aortic insufficiency** (backflow of blood from the aorta into the left ventricle). Another cause of a heart murmur is **mitral valve prolapse (MVP)**, an inherited disorder in which a portion of a mitral valve is pushed back too far (prolapsed) during ventricular contraction. Although a small volume of blood may flow back into the left atrium during ventricular systole, mitral valve prolapse often does not pose a serious threat. In fact, it is found in up to 10% of otherwise healthy young persons.

CARDIAC OUTPUT

Although the heart has autorhythmic fibers that enable it to beat independently, its operation is related to events occurring in the rest of the body. All body cells must receive a certain amount of oxygenated blood each minute to maintain health and life. When cells are very active, as during exercise, they need faster delivery of oxygen by the blood. During rest periods, cellular need is reduced, and the work load of the heart decreases.

Cardiac output (CO) is the amount of blood ejected from the left ventricle (or the right ventricle) into the aorta (or pulmonary trunk) each minute. Cardiac output is determined by (1) the volume of blood pumped by the ventricle per beat (the stroke volume) and (2) the number of heartbeats per minute. In a resting adult, stroke volume averages 70 ml/beat and heart rate is about 75 beats/min. This gives an average CO of

$$CO \text{ (ml/min)} = SV \text{ (70 ml/beat)} \times HR \text{ (75 beats/min)}$$
$$CO = 5250 \text{ ml/min or } 5.25 \text{ liters/min}$$

This value is close to the total blood volume, which is about 5 liters in an adult male. The entire blood supply thus

FIGURE 20.11 Heart sounds. The red circles indicate where heart sounds caused by blood turbulence associated with closure of the respective valves are best heard.

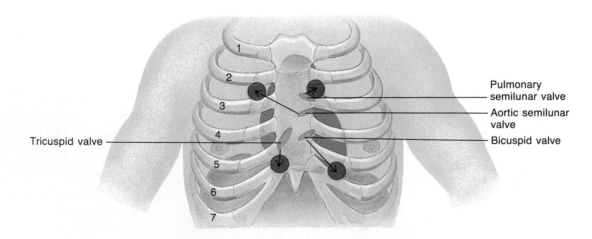

Question: Which heart sound is related to blood turbulence associated with closure of the AV valves?

flows through the pulmonary and systemic circulations about once a minute. When the demands of the body for oxygen increase or decrease, CO changes to meet the need. Factors that increase stroke volume or heart rate tend to increase CO. During mild exercise, for example, stroke volume may increase to 110 ml/beat and heart rate to 100 beats/min. Cardiac output then would be 11 liters/min. During intense (but still not maximal) exercise, the heart rate and stroke volume both may double, and cardiac output increases fourfold:

$$150 \text{ beats/min} \times 140 \text{ ml/beat} = 21,000 \text{ ml/min}$$
$$= 21 \text{ liters/min}$$

Cardiac reserve is the ratio between the maximum cardiac output a person can achieve and the cardiac output at rest. An average figure is four or five times the resting value. Top athletes may have a cardiac reserve seven or eight times their resting CO. People with severe heart disease may have little or no cardiac reserve, which limits their ability to carry out even simple daily tasks.

The next sections explain how large increases in cardiac output are possible by increasing stroke volume and heart rate. As we shall see, these two factors are not completely separate but rather depend on one another somewhat.

Regulation of Stroke Volume

A healthy heart pumps out all the blood that has entered its chambers during the previous diastole. At rest this is 50 to 60% of the total volume because 40 to 50% remains in the ventricles after each contraction (end-systolic volume). Thus, stroke volume (SV) equals end-diastolic volume (EDV) minus end-systolic volume (ESV): SV = EDV − ESV. Three important factors regulate stroke volume in different circumstances and ensure that the left and right ventricles pump equal volumes of blood (Fig. 20.12): (1) **preload**, the *stretch* on the heart before it contracts, (2) **contractility**, the *forcefulness of contraction* of individual ventricular muscle fibers, and (3) **afterload**, the *pressure that must be exceeded* before ejection of blood from the ventricles can begin.

Preload: Effect of Stretching

A greater preload (stretch) on cardiac muscle fibers just before they contract increases their force of contraction. In other words, within physiological limits, the more the heart is filled during diastole, the greater the force of contraction during systole. This is known as the **Frank–Starling law of the heart**. In the body, the preload depends on the volume of blood that fills the ventricles at the end of diastole, the end-diastolic volume (EDV). The greater the EDV (preload), within limits, the more forceful the contraction. The situation is somewhat like stretching a rubber band; the

more you stretch it, the harder it snaps back when you let go.

The length of ventricular diastole and venous pressure are the two key factors that determine EDV. When heart rate increases, the duration of diastole is shorter. Less filling time means a smaller EDV, and the ventricles may contract before they are adequately filled. On the other hand, when venous pressure increases, a greater volume of blood is forced into the ventricles, and the EDV is increased.

When heart rate exceeds about 160 beats/min, stroke volume usually declines. At such rapid heart rates, the ventricular filling time is severely shortened, EDV is less, and the preload thus is lower. People who have slow resting heart rates, on the other hand, usually have large stroke volumes because filling time is prolonged and preload thus is larger.

The Frank–Starling law of the heart equalizes the output of the right and left ventricles and keeps the same volume of blood flowing to both the systemic and the pulmonary circulations. If the left side of the heart pumps a little more blood than the right side, for example, the volume of blood returning to the right ventricle (venous return) increases. With increased EDV, then, the right ventricle contracts more forcefully on the next beat, and the two sides are again in balance.

Contractility

The second factor that influences stroke volume is myocardial **contractility**, that is, the strength of contraction at any given preload. Substances that increase contractility are called **positive inotropic** (*ino* = force) agents while those that decrease contractility are called **negative inotropic agents**. Thus, for a constant preload, the stroke volume is larger when a positive inotropic substance is present. Positive inotropic agents often promote Ca^{2+} inflow during impulses in cardiac muscle fibers. They include stimulation of the sympathetic division of the autonomic nervous system, hormones such as glucagon and the catecholamines (epinephrine and norepinephrine), increased Ca^{2+} level in the extracellular fluid, and the drug digitalis. Both glucagon and a calcium chloride solution are on a hospital "crash cart" (for emergency treatment of heart attack victims) because they have potent positive inotropic effects and may stimulate a failing heart to contract more forcefully. Digitalis is a drug often used to treat people who have congestive heart failure (described shortly). Negative inotropic agents include inhibition of the sympathetic division of the ANS, anoxia and acidosis, some anesthetics (e.g., halothane), and increased K^+ level in the extracellular fluid.

Afterload

Ejection of blood from the heart begins when pressure in the right ventricle exceeds the pressure in the pulmonary

Figure 20.12 Factors that increase cardiac output.

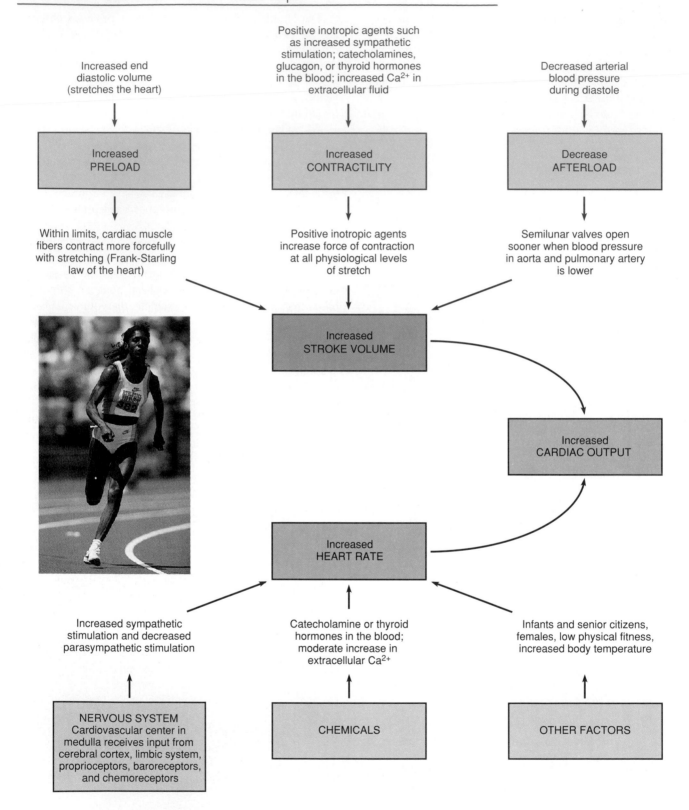

Question: When you are running, contraction of skeletal muscles in the legs helps return blood to the heart more rapidly. Would this effect tend to increase or decrease stroke volume?

trunk (about 20 mm Hg) and the pressure in the left ventricle exceeds the pressure in the aorta (about 80 mm Hg). At this point, the pressure in the ventricles forces the semilunar valves open. The pressure that must be overcome before the semilunar valves can open is termed the **afterload**. When the afterload increases, for example, when blood pressure is elevated, stroke volume decreases, and more blood remains in the ventricles at the end of systole.

CLINICAL APPLICATION

CONGESTIVE HEART FAILURE

In congestive heart failure (CHF), the heart is a failing pump. Causes include coronary artery disease (see page 616), congenital defects, long-term high blood pressure (which increases the afterload), myocardial infarcts (regions of dead heart tissue due to a prior heart attack), and valve disorders.

Congestive heart failure is an example of a positive feedback cycle in a pathological process. As the pump becomes less effective, more blood remains in the ventricles at the end of each cycle, and gradually the end-diastolic volume (preload) increases. Initially, this increased preload may promote increased force of contraction (the Frank–Starling law of the heart). As the preload increases further, however, the heart is overstretched and contracts less forcefully. Now there is a positive feedback situation: less effective pumping leads to even lower pumping capability.

Often, one side of the heart starts to fail before the other. If the left ventricle fails first, it can't pump out all the blood it receives. As a result, blood backs up in the lungs. The result is **pulmonary edema**, fluid accumulation in the lungs, which can suffocate an untreated person. If the right ventricle fails first, blood backs up in the systemic vessels. In this case, the resulting **peripheral edema** is usually most noticeable in the feet and ankles.

Regulation of Heart Rate

Cardiac output depends on heart rate as well as stroke volume. Stroke volume may be dangerously low in some pathological conditions. If the ventricular myocardium is weak or damaged by an infarct, it cannot contract strongly. Stroke volume could also fall if blood volume decreases due to hemorrhage. In these cases, homeostatic mechanisms strive to maintain an adequate cardiac output by increasing heart rate and strength of contraction. The sinoatrial (SA) node initiates contraction, and left to itself would set a steady heart rate. However, several factors contribute to regulation of heart rate (see Fig. 20.12), the most important ones being the autonomic nervous system and hormones released by the adrenal medulla (epinephrine and norepinephrine).

Autonomic Control of Heart Rate

Nervous system control of the cardiovascular system stems from the **cardiovascular center** in the medulla (see Fig.

21.13). This center receives input from higher brain regions, such as the cerebral cortex and limbic system which relay their impulses by way of the hypothalamus, and from sensory receptors. Let's consider what happens in an exercise situation. Even before a physical activity begins, especially in competitive situations, heart rate may climb. This anticipatory increase occurs because the limbic system in the brain sends signals to the cardiovascular center in the medulla. Then, as movements begin, *proprioceptors*, which monitor the position of limbs and muscles, send increased input to the cardiovascular center. Proprioceptor input is a major reason for the quick rise in heart rate at the beginning of physical activity.

Other sensory receptors that provide input to the cardiovascular center include *chemoreceptors* that monitor chemical changes in the blood and *baroreceptors* that monitor blood pressure in major arteries and veins. Important baroreceptors are located in the arch of the aorta and carotid arteries. They detect changes in blood pressure and relay the information to the cardiovascular center (see Fig. 21.13). From there, impulses propagate along sympathetic and parasympathetic nerves to the heart. These reflexes are important for the regulation of blood pressure, as well as heart rate, and are discussed in detail in Chapter 21. Here we focus specifically on the innervation of the heart by the sympathetic and parasympathetic branches of the autonomic nervous system.

Sympathetic fibers extend from the medulla into the spinal cord. From the thoracic region of the spinal cord, **cardiac accelerator nerves** extend out to the SA node, AV node, and most portions of the myocardium (see Fig. 21.16). Impulses in the cardiac accelerator nerves release norepinephrine, which binds to β_1 receptors on cardiac muscle fibers. This interaction has two separate effects. First, it increases the pacemaking rate of autorhythmic fibers in the SA node and thus increases heart rate. Second, in contractile fibers throughout the atria and ventricles, it enhances Ca^{2+} entry through the voltage-gated slow Ca^{2+} channels and thus increases contractility. This results in a more complete ejection of blood. Thus, with moderate increases in heart rate, stroke volume does not decline. With maximal sympathetic stimulation, heart rate may reach 250 beats/min, but stroke volume is lower than at rest due to the short filling time. The highest cardiac output usually occurs at a heart rate between 160 and 200 beats/min.

Parasympathetic nerve impulses reach the heart via the right and left **vagus (X) nerves** (see Fig. 21.16). These fibers innervate the SA node, AV node, and atrial myocardium. They release acetylcholine, which decreases heart rate by slowing the pacemaking rate of the autorhythmic fibers. Since only a few vagal fibers extend to ventricular muscle, changes in parasympathetic activity have very little effect on stroke volume.

There always is a balance between sympathetic and parasympathetic stimulation of the heart, but the parasympathetic effects predominate at rest. The resting heart rate, about 75 beats/min, usually is lower than the inherent pacing rate of the SA node. (The autorhythmic rate is about 100 beats/min.) With maximal parasympathetic stimulation, the heart can slow to 20 or 30 beats/min or even stop momentarily.

Chemical Regulation of Heart Rate

Certain chemicals present in the body influence both the basic physiology of cardiac muscle and heart rate. For example, hypoxia (lowered oxygen level), acidosis (low pH), and alkalosis (high pH) all depress cardiac activity. Two types of chemicals—hormones and ions—have major effects on the heart.

1. **Hormones.** Epinephrine and norepinephrine (from the adrenal medulla) enhance the heart's pumping effectiveness. These hormones affect cardiac muscle fibers in much the same way as does norepinephrine released by cardiac accelerator nerves; they increase both heart rate and contractility. Exercise, stress, and excitement cause the adrenal medulla to release more of its hormones. Thyroid hormones also enhance cardiac contractility and increase heart rate. One sign of hyperthyroidism (excessive thyroid hormone) is tachycardia (elevated heart rate) at rest.

2. **Ions.** Differences between intracellular and extracellular concentrations of several ions, for example, Na^+ and K^+, are critical to the production of impulses in all nerve and muscle fibers. So it is not surprising that ion imbalances can quickly compromise the pumping effectiveness of the heart. In particular, the relative concentrations of three cations—K^+, Ca^{2+}, and Na^+—have a large impact on cardiac function. Elevated blood levels of K^+ or Na^+ decrease heart rate and contractility. Excess Na^+ blocks Ca^{2+} inflow during cardiac impulses and thus decreases the force of contraction, whereas excess K^+ blocks impulse generation. A moderate increase in extracellular Ca^{2+} speeds and strengthens the heartbeat.

Other Factors

Age, gender, physical fitness, and body temperature also influence heart rate. A newborn baby is likely to have a heart rate over 120 beats/min. Heart rate then declines through childhood and much of the adult life span. Senior citizens may develop a more rapid heartbeat, however. Females generally have slightly higher heart rates than males. In both sexes, regular exercise tends to bring resting heart rate down. A physically fit person may even exhibit bradycardia, a resting heart rate under 60 beats/min. This is a beneficial effect of endurance-type training because a slowly beating heart is more energy efficient than one that beats more rapidly.

Increased body temperature, such as occurs during fever or strenuous exercise, causes the SA node to discharge impulses faster and thereby increases heart rate. Decreased body temperature decreases heart rate and strength of contraction. During surgical repair of certain heart abnormalities, it is helpful to slow a patient's heart rate. One method is called **hypothermia** (hī-pō-THER-mē-a), in which the person is deliberately cooled to a low body temperature. Hypothermia also slows metabolism and reduces the oxygen needs of the tissues. Thus the heart and brain can withstand short periods of interrupted or reduced blood flow.

Figure 20.12 summarizes the influences on both stroke volume and heart rate in the overall regulation of cardiac output.

Help for Failing Hearts

Millions of people have inadequate cardiac output because their heart is diseased or damaged. Throughout the industrialized world, heart disease is the number one cause of premature death. One in every five persons who reach age 60 will have a myocardial infarction (MI, heart attack). Cardiomyopathies (degenerative disorders of the heart) also afflict thousands. Although a variety of drugs are helpful in the earlier stages of heart disease, at some point they are no longer effective because there is too little functional cardiac muscle left.

Researchers are investigating a wide variety of devices and techniques that might aid a failing heart. Even a 10% increase in cardiac output can get a patient out of bed. One possibility is a **heart transplant**, but the availability of donor hearts is very limited. In 1989 with 50,000 to 100,000 candidates for possible transplantation in the U.S., about 1800 were done. Also, only about 50% of the recipients are alive 5 years after a heart transplant. A second possibility is an **artificial heart**. During the 1980s several patients received a Jarvik-7 artificial heart, which used an external power source to drive a mechanical pump inside the body by compressed air. Persistent problems with blood clotting (which caused strokes and failure of other organs) and infection (due to the chest tube for compressed air) led the Food and Drug Administration to ban use of the device in 1990. Research continues, however, on other types of artificial hearts. A third approach is to develop **cardiac assist devices**, which may be made from artificial materials or fashioned from a person's skeletal muscle. Exhibit 20.1 describes some of these cardiac assist devices, which supplement cardiac output without removal of the heart.

RISK FACTORS IN HEART DISEASE

About 1.5 million people suffer a myocardial infarction every year in the U.S. and of these more than 500,000 die suddenly before reaching a hospital. However, the prevalence of heart disease has diminished in recent years, due in

EXHIBIT 20.1

CARDIAC ASSIST DEVICES

Device	Description
Intra-aortic balloon pump (IABP)	A 40-ml polyurethane balloon mounted on a catheter is inserted into an artery in the groin and threaded into the thoracic aorta. An external pump inflates the balloon with gas at the beginning of ventricular diastole. As the balloon inflates, it pushes blood both backward, toward the heart, which improves coronary blood flow, and forward toward peripheral tissues. The balloon then is rapidly deflated just before the next ventricular systole. This decreases afterload, making it easier for the left ventricle to eject blood. Because the balloon is inflated between heartbeats, this technique is called intra-aortic balloon counterpulsation.
Hemopump	Propeller-like pump that is threaded through an artery in the groin and then into the left ventricle. There, the blades of the pump whirl at about 25,000 revolutions per minute, pulling blood out of the left ventricle and pushing it into the aorta.
Left ventricular assist device (LVAD)	The LVAD, first used in 1991, is designed to be a completely portable assist device. It is implanted within the abdomen and powered by a battery pack worn in a shoulder holster. The LVAD is connected to the patient's weakened left ventricle and pumps blood into the aorta. The pumping rate increases automatically during exercise.
Cardiomyoplasty	A large piece of the patient's own skeletal muscle (left latissimus dorsi) is partially freed from its connective tissue attachments and wrapped around the heart, leaving blood and nerve supply intact. An implanted pacemaker stimulates the skeletal muscle's motor neuron to cause contraction 10 to 20 times per minute, in synchrony with some of the heartbeats.
Skeletal muscle assist device	A piece of the patient's own skeletal muscle is used to fashion a pouch that is inserted between the heart and the aorta, functioning as a booster heart. A pacemaker stimulates the muscle's motor neurons to elicit contraction.

part to changes in life-style. Some of the causes of heart disease can be foreseen and prevented. People who develop combinations of certain risk factors are more likely to have heart attacks. **Risk factors** are characteristics, symptoms, or signs present in a person free of disease that are statistically associated with a greater chance of developing a disease. The major risk factors in heart disease are

1. High blood cholesterol level.
2. High blood pressure.
3. Cigarette smoking.
4. Obesity.
5. Lack of regular exercise.
6. Diabetes mellitus.
7. Genetic predisposition (family history of heart disease at an early age).
8. Male gender (after age 70, the risk of heart attack is similar in males and females).

The first five risk factors can all be modified to reduce one's risk, for example, losing weight, becoming more physically active, or quitting smoking. High blood cholesterol is discussed shortly, and high blood pressure is discussed in the next chapter. Nicotine in cigarette smoke enters the bloodstream and constricts small blood vessels.

It also stimulates the adrenal gland to oversecrete epinephrine and norepinephrine, which elevate heart rate and blood pressure. Obese people develop extra capillaries to nourish adipose tissue. An estimate is an additional 300 km (about 200 mi) of blood vessels for each pound of fat. The heart has to work harder to pump blood this extra distance. Regular exercise increases cardiac efficiency and output.

Other factors may also contribute to development of heart disease. High blood levels of fibrinogen, which enhances blood clot formation; renin, which increases blood pressure; and uric acid all increase the risk of myocardial infarction. Enlargement (hypertrophy) of the left ventricle, associated with both high blood pressure and obesity, also is a risk factor for myocardial infarction.

PLASMA LIPIDS AND HEART DISEASE

A strong risk factor for developing heart disease is high blood cholesterol level. The reason is that high blood cholesterol promotes growth of fatty plaques that build up in the walls of arteries (see Fig. 20.14). As a plaque enlarges, the passageway for blood progressively narrows. Not only does the narrowed opening restrict blood flow, the roughened plaque tends to promote blood clotting. If a blood clot forms at the site of a plaque or lodges there, it may suddenly cut off blood flow. If the blocked vessel is in the brain, the result may be a fatal stroke. Blockage of a coronary artery may cause a heart attack.

Lipoproteins in Blood

Most lipids, such as cholesterol and triglycerides (neutral fats), are very nonpolar and therefore very hydrophobic molecules. To be transported in watery blood, such molecules first must be dissolved. They are made water-soluble by combining them with proteins produced by the liver and intestine (apoproteins). The combinations thus formed are called **lipoproteins**, which vary in size, weight, and density. There are several types of lipoproteins, each having different functions, but all essentially are transport vehicles. They provide delivery and pick-up service so the various types of lipids can be available to cells that need them or removed from circulation if not needed. For example, all cells need cholesterol because it is a major building block of plasma membranes. It also is a key compound for the synthesis of steroid hormones and bile salts. Three classes of lipoproteins are called **low-density lipoproteins (LDLs)**, **high-density lipoproteins (HDLs)**, and **very low-density lipoproteins (VLDLs)**.

LDLs contain 25% proteins, 20% triglycerides, and 55% cholesterol. They deliver cholesterol to body cells that need it. Under abnormal conditions, however, LDLs also deposit cholesterol in and around smooth muscle fibers in arteries. Most cells of the body contain LDL receptors. Once LDL attaches to its receptor, it is taken into the cell by receptor-mediated endocytosis (see page 67). Within the cell, the LDL is broken down, and the cholesterol is released to serve the cell's needs. Once a cell has sufficient cholesterol for its activities, a negative feedback system inhibits the cell from synthesizing new LDL receptors. Some people have too few LDL receptors, owing to various environmental and genetic factors. Since their cells cannot remove LDL from the blood as effectively, their plasma LDL level is abnormally high, and they are thus more likely to develop fatty plaques.

HDLs, which contain 50% proteins, 37% triglycerides, and only 13% cholesterol, remove excess cholesterol from body cells and transport it to the liver for elimination. This pick-up service prevents accumulation of cholesterol in the blood. Thus a high HDL level is associated with decreased risk of heart disease caused by plaque formation.

VLDLs contain about 10% proteins, 65% triglycerides, and 25% cholesterol. They transport triglycerides synthesized by liver cells to adipose cells for storage. A high fat diet promotes production of VLDLs. After depositing some of their triglycerides in adipose cells, however, VLDLs are converted to LDLs. This is one way that a fatty diet is believed to increase fatty plaque formation.

Blood Cholesterol

There are two sources of cholesterol in the body. Some is present in foods (eggs, dairy products, organ meats, beef, pork, and processed luncheon meats), but most is synthesized by the liver. Fatty foods that don't contain any cholesterol at all can still dramatically increase blood cholesterol level in two ways. First, a high intake of dietary fats stimulates reabsorption of cholesterol-containing bile back into the blood so less cholesterol is lost in the feces. Second, when saturated fats (see page 43) are broken down in the body, the liver uses some of the breakdown products to produce cholesterol.

The total cholesterol (as part of HDLs, LDLs, and VLDLs) in blood plasma is one commonly used indicator of risk for coronary artery disease (see page 616). A lipid profile test usually measures total cholesterol (TC), HDL-cholesterol, and triglycerides (VLDLs). LDL-cholesterol then is calculated by using the following formula: LDL = TC − HDL − (triglycerides/5). In the U.S., blood cholesterol is usually measured in milligrams per deciliter (mg/dl); a deciliter is $\frac{1}{10}$ of a liter. As the total cholesterol level increases above 150 mg/dl (3.9 mmol/l), the risk of coronary artery disease slowly begins to rise. Above 200 mg/dl (5.2 mmol/l), the risk increases even more. The chance of a heart attack doubles with every 50 mg/dl (1.3 mmol/l) increase in total cholesterol once the level goes over 200 mg/dl.

For adults, desirable levels of blood cholesterol are TC under 200 mg/dl, LDL under 130 mg/dl, and HDL over 40 mg/dl. Normally, triglycerides are in the range of 10–190 mg/dl. TC of 200–239 mg/dl and LDL of 130–159 mg/dl are borderline-high while TC above 239 mg/dl and LDL above 159 are classified as high blood cholesterol. The risk of developing coronary artery disease may be predicted by determining the ratio of total cholesterol to HDL cholesterol. For example, a person with a total cholesterol of 180 and a HDL of 60 has a risk ratio of 3. Ratios above 4 are considered undesirable; the higher the ratio, the greater the risk of developing coronary artery disease.

Among the therapies used to reduce blood cholesterol level are exercise, diet, and drugs. Regular physical activity at aerobic and nearly aerobic levels tends to raise HDL level. Dietary changes are aimed at reducing the intake of total fat, saturated fats, and cholesterol. Among the drugs used to treat high blood cholesterol levels are cholestyramine (Questran) and colestipol (Colestid), which promote excretion of bile in the feces; nicotinic acid (Lipo-nicin); and lovastatin (Mevacor), which blocks synthesis of cholesterol by liver cells.

EXERCISE AND THE HEART

No matter what a person's level of fitness, it can be improved at any age with regular participation in exercise. Of the various types of exercise, some are more effective than others for improving the health of the cardiovascular system. **Aerobics**, or any activity that works large body muscles for at least 20 min, elevates cardiac output and accelerates metabolic rate. Three to five such sessions a week are usually recommended for improving the health of the cardiovascular system. Brisk walking, running, bicycling, cross-country skiing, and swimming are examples of aerobic exercises.

Sustained exercise increases the oxygen demand of the muscles. Whether the demand is met depends primarily on the adequacy of cardiac output and proper functioning of the respiratory system. After several weeks of training, a healthy person increases maximal cardiac output and thereby increases the maximal rate of oxygen delivery to the tissues.

A well-trained athlete can achieve a cardiac output up to about six times that of a sedentary "couch potato" during activity because training causes hypertrophy (enlargement) of the heart. Even though the heart of a well-trained athlete is larger, *resting* cardiac output is about the same as in a healthy untrained person. This is because stroke volume is increased while heart rate is decreased. The heart rate of a trained athlete is about 40 to 60 beats/ min.

Other benefits of physical conditioning are an increase in high-density lipoprotein (HDL), a decrease in triglyceride levels, and improved lung function. Exercise also helps to reduce blood pressure, anxiety, and depression; control weight; and increase the body's ability to dissolve blood clots by increasing fibrinolytic activity. Intense exercise increases levels of endorphins, the body's natural painkillers. This may explain the psychological "high" that athletes experience with strenuous training and the "low" they feel when they miss regular workouts. Exercise also helps make bones stronger and thus may be a factor in inhibiting and treating osteoporosis. Some research indicates that exercise may even offer protection against cancer and diabetes.

DEVELOPMENTAL ANATOMY OF THE HEART

The *heart*, a derivative of **mesoderm**, begins to develop before the end of the third week of gestation. It begins its development in the ventral region of the embryo beneath the foregut (see Fig. 24.29). The first step is the formation of a pair of tubes, the **endothelial (endocardial) tubes**, from mesodermal cells (Fig. 20.13). These tubes then unite to form a common tube, the **primitive heart tube**. Next, the primitive heart tube develops into five regions: **ventricle, bulbus cordis, atrium, sinus venosus**, and **truncus arteriosus**. Since the bulbus cordis and ventricle grow more rapidly than the others, and the heart enlarges more rapidly than its superior and inferior attachments, the heart assumes a U-shape and later an S-shape. The flexures of the heart reorient the regions so that the atrium and sinus venosus eventually come to lie superior to the bulbus cordis, ventricle, and truncus arteriosus. Contractions of the primitive heart begin by day 22. They begin in the sinus venosus and force blood through the tubular heart.

At about the seventh week, a partition, the **interatrial septum**, forms in the atrial region, dividing it into a *right atrium* and *left atrium*. The opening in the partition is the **foramen ovale**, which normally closes at birth and later forms a depression called the *fossa ovalis*. An **interventricular septum** also develops and partitions the ventricular region into a *right ventricle* and *left ventricle*. The bulbus cordis and truncus arteriosus divide into two vessels, the *aorta* (arising from the left ventricle) and the *pulmonary trunk* (arising from the right ventricle). Recall that the ductus arteriosus is a temporary vessel between the aorta and pulmonary trunk until birth. The great veins of the heart, *superior vena cava* and *inferior vena cava*, develop from the venous end of the primitive heart tube.

FIGURE 20.13 Development of the heart. Arrows within the structures show the direction of blood flow.

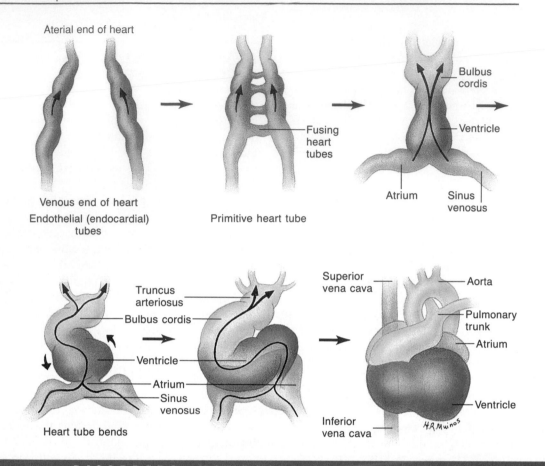

Aterial end of heart

Venous end of heart

Endothelial (endocardial) tubes

Primitive heart tube

Fusing heart tubes

Bulbus cordis

Ventricle

Atrium

Sinus venosus

Heart tube bends

Truncus arteriosus

Bulbus cordis

Ventricle

Atrium

Sinus venosus

Superior vena cava

Aorta

Pulmonary trunk

Atrium

Ventricle

Inferior vena cava

H.R.Muinos

DISORDERS: HOMEOSTATIC IMBALANCES

CORONARY ARTERY DISEASE

Coronary artery disease (CAD) refers to conditions, such as atherosclerosis, that cause narrowing of coronary arteries so that blood flow to the heart is reduced. This results in coronary heart disease (CHD), a condition in which the heart muscle is damaged because of an inadequate amount of blood due to obstruction of its blood supply. It is the leading cause of death in the U.S. Depending on the degree of obstruction, symptoms can range from a mild chest pain to a full-scale heart attack. Generally, symptoms start when there is about a 75% narrowing of a coronary artery. The principal causes of narrowing are atherosclerosis, coronary artery spasm, or a clot in a coronary artery.

Atherosclerosis

Thickening of the walls of arteries and loss of elasticity are the main characteristics of a group of diseases called **arteriosclerosis** (ar-tē′-rē-ō′-skle-RŌ-sis; *sclerosis* = hardening). One form of arteriosclerosis is **atherosclerosis** (ath′-er-ō-skle-RŌ-sis), a process in which smooth muscle cells proliferate and fatty substances, especially cholesterol and triglycerides (neutral fats), accumulate in the walls of medium-sized and large arteries. Although initiating causes are not completely understood, the first event in athero-sclerosis is thought to be damage to the endothelial lining of the artery. One theory is that a common virus, perhaps cytomegalovirus (a member of the herpes family), triggers endothelial damage after being harbored in a dormant state for some time. Other theories hold that prolonged high blood pressure, carbon monoxide in cigarettes, or diabetes mellitus produce endothelial damage. Whatever the causes, two events follow endothelial damage. Within the artery wall, (1) smooth muscle fibers proliferate and (2) lipids build up, both within cells and in the interstitial spaces. The accumulated cholesterol, triglycerides, and cells form a lesion called an **atherosclerotic plaque** (Fig. 20.14). As it grows, a plaque obstructs blood flow in the affected artery, and tissues supplied by the artery suffer damage. Studies during the late 1980s reached the encouraging conclusion that atherosclerosis is reversible to some extent. With reduction of a high blood cholesterol level, atherosclerotic plaques tend to shrink.

An additional danger is that the plaque provides a roughened surface that causes blood platelets to release *platelet-derived growth factor* (*PDGF*). PDGF is a hormone that promotes the proliferation of smooth muscle fibers. Macrophages and endothelial cells also produce PDGF. This further complicates the process since PDGF causes the lesion to grow larger. Platelets in the area of the atherosclerotic plaque also release clot-forming chemicals. Thus a thrombus may form. If the clot breaks off and forms an embolus (blood clot transported by blood), it may obstruct smaller arteries and capillaries downstream from the site of formation.

FIGURE 20.14 Photomicrograph of an artery partially obstructed by an atherosclerotic plaque.

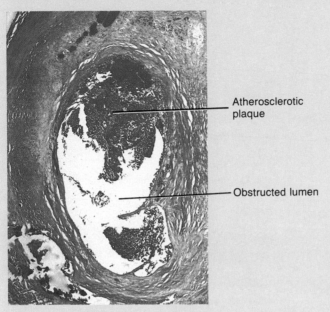

Atherosclerotic plaque

Obstructed lumen

Cross section

Scientists have identified a mechanism by which atherosclerosis may trigger hypertension, stroke, and heart disease. Secretions of blood platelets induce endothelial cells that line blood vessels to produce a substance called *endothelium-derived relaxing factor* (*EDRF*), which is now known to be nitric oxide (NO). EDRF relaxes smooth muscle fibers beneath the endothelium. However, atherosclerosis interferes with the production of EDRF, and this imbalance causes smooth muscle fibers to contract, that is, to go into spasm. These spasms not only cause pain but can also lead to transient ischemic attacks (TIAs), stroke, high blood pressure, and heart disease.

Treatment of coronary artery disease varies with the nature and urgency of symptoms. Among the treatment options are drug therapy (nitroglycerine, beta blockers, and thrombolytic agents) and various surgical and nonsurgical procedures.

Coronary artery bypass grafting (CABG) is one way of increasing the blood supply to the heart. It is a surgical procedure in which a blood vessel from another part of the body is used to bypass the blocked region of a coronary artery. The two vessels used most often are the saphenous vein from the leg and the internal mammary artery from the chest, which has a lower rate of plaque reformation. Once the heart is exposed, circulation is maintained by a heart–lung machine. Next, a clamp is placed across the aorta above the openings of the coronary arteries. This stops the blood flow to the heart. A segment of the grafted blood vessel is then sutured between the aorta and the unblocked portion of the coronary artery, distal to the obstruction (Fig. 20.15a). If more than one artery is clogged, additional bypasses may be made. Once the repair is completed, the aortic clamp is removed.

A nonsurgical procedure used to treat CAD is termed **percutaneous transluminal coronary angioplasty (PTCA)** (*percutaneous* = through the skin; *trans* = across; *lumen* =

channel in a tube; *angio* = blood vessel; *plasty* = to mold or shape). Like coronary artery bypass grafting, it is an attempt to increase the blood supply to the heart muscle. A balloon catheter (plastic tube) is inserted into an artery of an arm or leg and gently guided through the arterial system under x-ray observation (Fig. 20.15b) until it is threaded up into a coronary artery. Then, while dye is being released, angiograms (radiographs of blood vessels) are taken to localize the plaques. Next, the catheter is advanced to the point of obstruction and a balloon-like device is inflated with air to squash the plaque against the blood vessel wall. If successful, PTCA increases the inside diameter of the vessel, and blood flow improves. PTCA is most often used to relieve angina pectoris. Since about 30% of PTCA-opened arteries fail due to restenosis (renarrowing), a special device called a **stent** may be inserted via the catheter to keep the artery patent (open). A stent is a stainless steel device, resembling a spring coil, that is permanently placed in an artery to maintain patency, permitting blood to circulate (Fig. 20.15c). A modification of PTCA, called **percutaneous balloon valvuloplasty**, is used to treat faulty heart valves.

Another technique for opening clogged arteries is a procedure called **laser angioplasty**. In one variation of this procedure, a laser vaporizes the atherosclerotic plaque and makes a channel through the blood vessel obstruction. Then a balloon catheter is inserted, and the balloon is inflated to widen the vessel.

Two of the latest techniques for clearing arteries are balloon–laser welding and catheter artherectomy. In **balloon–laser welding**, an artery is first widened by PTCA. On the last balloon inflation, a laser heats the surrounding tissue sufficiently to stretch and weld the arterial wall into a smooth surface. In **catheter artherectomy**, a rotating drill shaves off plaque. Shavings are trapped and removed.

Coronary Artery Spasm

Atherosclerosis results in a fixed obstruction to blood flow. Obstruction can also be caused intermittently by **coronary artery spasm**, in which the smooth muscle of a coronary artery undergoes a sudden contraction, resulting in narrowing of a blood vessel. Coronary artery spasm typically occurs in people with atherosclerosis and may result in chest pain during rest (variant angina), chest pain during exertion (typical angina), heart attacks, and sudden death. Although the causes of coronary artery spasm are unknown, several factors are receiving attention. These include smoking, stress, and a vasoconstrictor chemical released by platelets.

CONGENITAL DEFECTS

A defect that exists at birth, and usually before, is called a **congenital defect**. Many defects are not serious and may go unnoticed for a lifetime. Others heal themselves. But some are life-threatening and must be mended by surgical techniques ranging from simple suturing to replacement of malfunctioning parts with synthetic materials.

One congenital defect that can occur is **coarctation** (kō′-ark-TĀ-shun) **of the aorta** (Fig. 20.16a). In this condition, a segment of the aorta is too narrow. As a result, the flow of oxygenated blood to the body is reduced, the left ventricle is forced to pump harder, and high blood pressure develops. Another common congenital defect is **patent ductus arteriosus** (Fig. 20.16b). The ductus arteriosus, a temporary blood vessel between the aorta and the pulmonary trunk, normally closes shortly after birth. In some babies, the ductus arteriosus remains open. As a result, aortic blood flows into the lower-pressure pulmonary trunk, thus increasing the

Continues

FIGURE 20.15 Several procedures for reestablishing blood flow in occluded coronary arteries. In coronary artery bypass grafting (CABG), shown in (a), the internal mammary artery is removed from the patient's chest. At a point distal to the obstruction, the grafted artery is sutured to the coronary artery.

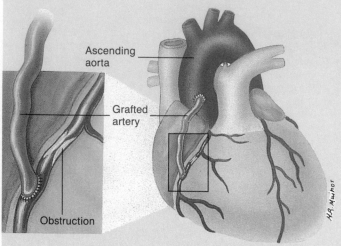

(a) Coronary artery bypass grafting (CABG)

Lumen of artery — Stent

(c) Stent in an artery

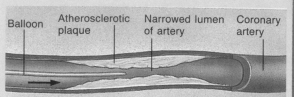

Balloon catheter with uninflated balloon approaches obstructed area in artery

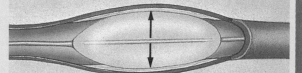

When balloon is inflated, it breaks up atherosclerotic plaque

After lumen is widened, balloon catheter with uninflated balloon is withdrawn

(b) Percutaneous transluminal coronary angioplasty (PTCA)

Question: Which type of lipoprotein contributes most to the development of atherosclerosis?

pulmonary trunk blood pressure and overworking both ventricles.

A **septal defect** is an opening in the septum that separates the interior of the heart into left and right sides. In one type of **interatrial septal defect** the fetal foramen ovale between the two atria fails to close after birth (Fig. 20.16c). Because pressure in the right atrium is lower, blood flows from the left atrium to the right without going through the systemic circulation. This defect overloads the pulmonary circulation. **Interventricular septal defect** is caused by incomplete closure of the interventricular septum (Fig. 20.16d). It permits oxygenated blood to flow directly from the left ventricle into the right ventricle where it mixes with deoxygenated blood.

Valvular stenosis is a narrowing of one of the valves regulating blood flow in the heart. All stenoses are serious because they place a severe burden on the heart by making it work harder to push blood through the abnormally narrow valve openings. As a result of mitral valve stenosis, systemic blood pressure is increased. Most stenosed valves are totally replaced with artificial valves (see Fig. 20.4d).

Tetralogy of Fallot (tet-RAL-ō-jē of fal-Ō) is a combination of four defects: an interventricular septal defect,

an aorta that emerges from both ventricles instead of from the left ventricle only, a stenosed pulmonary semilunar valve, and an enlarged right ventricle (Fig. 20.16e). Because there is stenosis of the pulmonary semilunar valve, the increased right ventricular pressure forces deoxygenated blood from the right ventricle to enter the left ventricle through the interventricular septum. As a result, deoxygenated blood mixes with oxygenated blood and is pumped into the systemic circulation. Also, because the aorta emerges from the right ventricle and the pulmonary trunk is stenosed, very little blood gets to the lungs, and pulmonary circulation is bypassed almost completely. This causes cyanosis, the blue or dark purple discoloration that is most easily seen in nail beds and mucous membranes when deoxygenated hemoglobin level is high. For this reason, tetralogy of Fallot is one of the conditions that cause a "blue baby."

ARRHYTHMIAS

Arrhythmia (a-RITH-mē-a) is a general term that refers to an abnormality or irregularity in the heart rhythm. Some physicians use the term **dysrhythmia** since that implies an abnormal rhythm, whereas arrhythmia implies no rhythm. An arrhythmia results when there is a disturbance in the con-

FIGURE 20.16 Some common congenital heart defects.

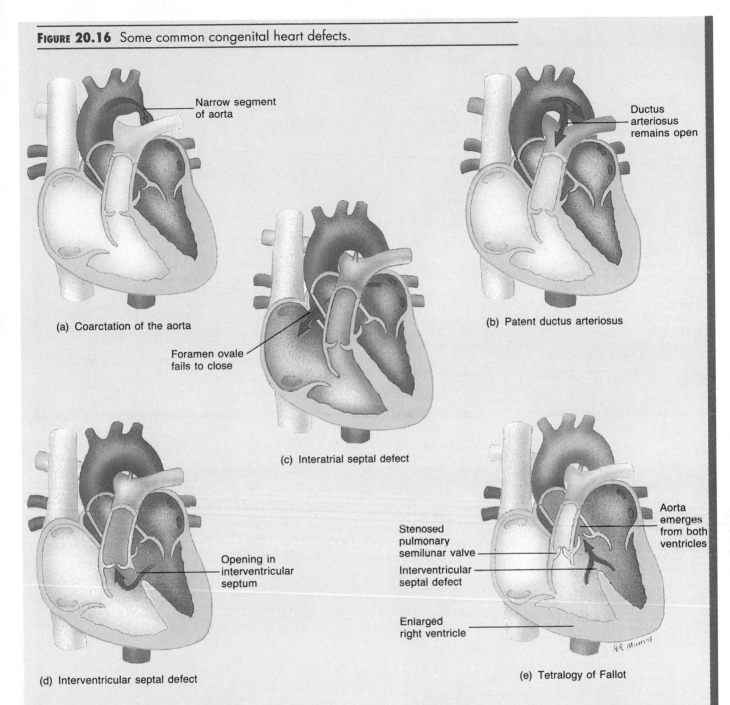

Narrow segment
of aorta

(a) Coarctation of the aorta

Ductus
arteriosus
remains open

(b) Patent ductus arteriosus

Foramen ovale
fails to close

(c) Interatrial septal defect

Opening in
interventricular
septum

(d) Interventricular septal defect

Stenosed
pulmonary
semilunar valve

Interventricular
septal defect

Enlarged
right ventricle

Aorta
emerges
from both
ventricles

(e) Tetralogy of Fallot

duction system of the heart. It may be due to either faulty production of electrical impulses or poor conduction of impulses as they pass through the system.

There are many different types of arrhythmias that can occur, some normal and some quite serious. Arrhythmias may be caused by factors such as caffeine, nicotine, alcohol, anxiety, certain drugs, hyperthyroidism, potassium deficiency, and certain heart diseases. Serious arrhythmias can result in cardiac arrest (stopping of the heartbeat) if the heart cannot supply its own oxygen demands.

Heart Block

One serious arrhythmia is called a **heart block**. Perhaps the most common blockage is in the atrioventricular (AV) node, which is the only path for impulses in the atria to reach the

ventricles. This disturbance is called **atrioventricular (AV) block**. In first-degree AV block, the P-Q (PR) interval is prolonged, usually because conduction through the AV node is slower than normal (see Fig. 20.6). In second degree AV block, some of the SA node impulses are not conducted through the AV node. This results in "dropped" beats, since excitation doesn't reach the ventricles. In third-degree (complete) AV block, none of the SA node impulses get through the AV node. Autorhythmic cells in the atria and ventricles pace the upper and lower chambers independently. With complete AV block, the ventricular contraction rate is less than 40 beats/min. Due to decreased cardiac output and diminished brain blood flow, patients may experience dizziness, unconsciousness, or convulsions.

Continues

DISORDERS: HOMEOSTATIC IMBALANCES (continued)

Flutter and Fibrillation

Two other abnormal rhythms are **flutter** and **fibrillation**. In **atrial flutter** the atrial rhythm averages between 240 and 360 beats/min. The condition is basically rapid atrial contractions accompanied by a second-degree AV block. Flutter may result from rheumatic heart disease, coronary artery disease, or certain congenital heart diseases. **Atrial fibrillation** is asynchronous contraction of the atrial muscle fibers so that atrial pumping ceases altogether. Atrial fibrillation may occur in myocardial infarction, acute and chronic rheumatic heart disease, and hyperthyroidism. In a strong heart, atrial fibrillation reduces the pumping effectiveness of the heart by only 20 to 30%.

Ventricular fibrillation (VF) is the most ominous arrhythmia. It almost always indicates imminent death unless corrected quickly. It is characterized by asynchronous, haphazard, ventricular muscle contractions. The rate may be rapid or slow. The impulse travels to the different parts of the ventricles at different rates. Thus part of the ventricles may be contracting while other parts are relaxing. Ventricular pumping becomes ineffective, blood is not being ejected, and circulatory failure and death occur.

A strong electrical current passed across the chest for a short interval of time can stop ventricular fibrillation. This is called **defibrillation**. It is accomplished by giving the electric shock through large paddle-shaped electrodes pressed against the skin of the chest. Patients who face a high risk of dying from heart rhythm disorders now can have a device implanted that monitors their heart rhythm and delivers a small shock directly to the heart when a life-threatening rhythm disturbance occurs. Such an **automatic implantable cardioverter defibrillator (AICD)** has been used in thousands of patients around the world.

Ventricular Premature Contraction (VPC)

Another form of arrhythmia arises when a small region of the heart outside the pacemaker (an ectopic focus) becomes more excitable than normal, causing an occasional abnormal impulse to arise between normal impulses. As a wave of depolarization spreads outward from the ectopic focus, it causes a **ventricular premature contraction (VPC)**. The contraction occurs early in diastole before the SA node is normally scheduled to discharge its impulse. VPCs may be relatively benign and may be caused by emotional stress, excessive intake of stimulants such as caffeine or nicotine, and lack of sleep. In other cases, the contractions may indicate an underlying pathology.

MEDICAL TERMINOLOGY

Angiocardiography (an'-jē-ō-kar'-dē-OG-ra-fē; *angio* = vessel) X-ray examination of the heart and great blood vessels after injection of a radiopaque dye into the bloodstream.

Cardiac arrest (KAR-dē-ak a-REST) A clinical term meaning cessation of an effective heartbeat. The heart may be completely stopped (cardiac standstill) or quivering ineffectively (ventricular fibrillation).

Cardiomegaly (kar'-dē-ō-MEG-a-lē; *mega* = large) Heart enlargement, hypertrophy.

Compliance (kom-PLĪ-ans) The passive or diastolic stiffness properties of the left ventricle. A hypertrophied or fibrosed heart with a stiff wall, for example, has decreased compliance. Also, the stiffness properties of the lung or major arteries (Chapter 23).

Constrictive pericarditis (kon-STRIK-tiv per'-i-kar-DĪ-tis) A shrinking and thickening of the pericardium that prevents heart muscle from expanding and contracting normally.

Cor pulmonale (CP) (kor pul-mōn-ALE; *cor* = heart; *pulmon* = lung) This refers to right ventricular hypertrophy from disorders that bring about hypertension (high blood pressure) in the pulmonary circulation.

Incompetent valve (in-KOM-pe-tent VALV) Any valve that does not close properly, thus permitting a backflow of blood; also called **valvular insufficiency.**

Palpitation (pal'-pi-TĀ-shun) A fluttering of the heart or abnormal rate or rhythm of the heart.

Paroxysmal tachycardia (par'-ok-SIZ-mal tak'-e-KAR-dē-a) A period of rapid heartbeats that begins and ends suddenly.

Study Outline

Location and Size of the Heart (p. 592)

1. The heart is situated between the lungs in the mediastinum.
2. About two-thirds of its mass is to the left of the midline.
3. The heart is about 12 cm long, 9 cm wide, and 6 cm thick.

Pericardium (p. 592)

1. The pericardium consists of an outer fibrous layer and an inner serous pericardium.
2. The serous pericardium is composed of a parietal and visceral layer.
3. Between the parietal and visceral layers of the serous pericardium is the pericardial cavity, a potential space filled with pericardial fluid that reduces friction between the two membranes.

Heart Wall (p. 592)

1. The wall of the heart has three layers: epicardium, myocardium, and endocardium.
2. The epicardium consists of mesothelium and connective tissue, the myocardium is composed of cardiac muscle tissue, and the endocardium consists of endothelium and connective tissue.

Chambers of the Heart (p. 594)

1. The chambers include two upper atria and two lower ventricles.
2. An interatrial septum separates the atria and an interventricular septum separates the ventricles.

Blood Flow Through the Heart (p. 595)

1. Blood flows through the heart from the superior and inferior venae cavae and the coronary sinus to the right atrium, through the tricuspid valve to the right ventricle, through the pulmonary trunk and pulmonary arteries to the lungs, through the pulmonary veins into the left atrium, through the bicuspid valve to the left ventricle, and out through the aorta.
2. Divisions of the aorta are the ascending aorta, arch of the aorta, thoracic aorta, and abdominal aorta.

Valves of the Heart (p. 597)

1. Valves prevent backflow of blood in the heart.
2. Atrioventricular (AV) valves, between the atria and their ventricles, are the tricuspid valve on the right side of the heart and the bicuspid (mitral) valve on the left.
3. The chordae tendineae and papillary muscles stabilize the flaps of the valves and stop blood from backing into the atria.
4. The two arteries that leave the heart each have a semilunar valve (aortic and pulmonary).

Heart Blood Supply (p. 600)

1. The flow of blood through the heart is called coronary (cardiac) circulation.
2. The principal arteries are left and right coronary; the principal vein is the coronary sinus.
3. Reperfusion damage is caused by oxygen free radicals.

Conduction System and Pacemaker (p. 601)

1. The conduction system consists of tissue specialized for generation and conduction of action potentials.
2. Components of this system are the sinoatrial (SA) node (pacemaker), atrioventricular (AV) node, atrioventricular (AV)

bundle (bundle of His), bundle branches, and conduction myofibers (Purkinje fibers).

Physiology of Cardiac Muscle Contraction (p. 603)

1. An impulse in a ventricular contractile fiber is characterized by rapid depolarization, plateau, and repolarization.
2. The refractory period of a cardiac muscle fiber lasts longer than the contraction itself.

Electrocardiogram (p. 604)

1. The record of electrical changes during each cardiac cycle is referred to as an electrocardiogram (ECG).
2. A normal ECG consists of a P wave (atrial depolarization), QRS complex (ventricular depolarization), and T wave (ventricular repolarization).
3. The P-Q (PR) interval represents the conduction time from the beginning of atrial excitation to the beginning of ventricular excitation. The S-T segment represents the time when ventricular contractile fibers are fully depolarized.

Cardiac Cycle (p. 605)

1. A cardiac cycle consists of the systole (contraction) and diastole (relaxation) of both atria, plus the systole and diastole of both ventricles.
2. The phases of the cardiac cycle are (1) the relaxation period, (2) ventricular filling, and (3) ventricular systole.
3. With an average heartbeat of 75 beats/min, a complete cardiac cycle requires 0.8 sec.
4. The first heart sound (lubb) is created by blood turbulence associated with the closing of the atrioventricular valves. The second sound (dupp) is created by blood turbulence associated with the closing of semilunar valves.

Cardiac Output (p. 608)

1. Cardiac output (CO) is the amount of blood ejected by the left ventricle (or right ventricle) into the aorta (or pulmonary trunk) per minute. It is calculated as follows: CO = stroke volume $\times$ beats per minute.
2. Stroke volume (SV) is the amount of blood ejected by a ventricle during each systole.
3. Cardiac reserve is the ratio between the maximum cardiac output a person can achieve and the cardiac output at rest.
4. Stroke volume is related to preload (stretch on the heart before it contracts), contractility (forcefulness of contraction), and afterload (pressure that must be exceeded before ventricular ejection can begin).
5. According to the Frank–Starling law of the heart, a greater preload (stretch) on cardiac muscle fibers just before they contract increases their force of contraction.
6. Nervous control of the cardiovascular system stems from the cardiovascular center in the medulla.
7. Sympathetic impulses increase heart rate and force of contraction; parasympathetic impulses decrease heart rate.
8. Heart rate is affected by hormones (epinephrine, norepinephrine, thyroid hormones), ions (Na^+, K^+, Ca^{2+}), age, gender, physical fitness, and temperature.

Risk Factors in Heart Disease (p. 612)

1. Risk factors in heart disease that can be modified include high

blood cholesterol, high blood pressure, cigarette smoking, obesity, and lack of regular exercise.
2. Other factors include diabetes mellitus, genetic disposition, male gender, fibrinogen level, renin, uric acid, and left ventricular hypertrophy.

Plasma Lipids and Heart Disease (p. 613)

1. High blood cholesterol promotes growth of fatty plaques in the walls of arteries.
2. Whereas HDLs remove excess cholesterol from circulation, high levels of LDLs are associated with the formation of fatty plaques in arteries.

Exercise and the Heart (p. 615)

1. Sustained exercise increases oxygen demand on muscles.
2. Among the benefits of aerobic exercise are increased cardiac output, increased HDL and decreased triglycerides, improved lung function, decreased blood pressure, and weight control.

Developmental Anatomy of the Heart (p. 615)

1. The heart develops from mesoderm.
2. The endothelial tubes develop into the four-chambered heart and great vessels of the heart.

Review Questions

1. Describe the position of the heart in the mediastinum. (p. 592)
2. Distinguish the subdivisions of the pericardium. What is the purpose of this structure? (p. 592)
3. Compare the three layers of the heart wall according to composition, location, and function. (p. 592)
4. Define atria and ventricles. What vessels enter or exit the atria and ventricles? (p. 594)
5. Describe the principal valves in the heart and how they operate. (p. 597)
6. Describe the route of blood flow in coronary (cardiac) circulation. (p. 600)
7. Describe how reperfusion damage occurs. (p. 601)
8. Describe the structure and function of the heart's conduction system. What are autorhythmic fibers? (p. 601)
9. Describe the phases of an impulse in ventricular contractile fibers. (p. 603)
10. Define and label the deflection waves of a normal electrocardiogram (ECG). (p. 604) What is the significance of the P-Q (PR) interval and S-T segment? (p. 605)
11. Define cardiac cycle and list the principal events of the relaxation period, ventricular filling, and ventricular systole. (p. 605)

12. By means of a labeled diagram, relate the events of the cardiac cycle to time. (p. 606) What is the relaxation period? (p. 607)
13. Describe the source and significance of the heart sounds. (p. 607)
14. What is cardiac output (CO)? How is it calculated? (p. 608)
15. Define stroke volume (SV). Explain the factors that regulate stroke volume. (p. 608)
16. What is the Frank–Starling law of the heart? What is its significance? (p. 609)
17. Define cardiac reserve. Why is it important? (p. 609)
18. Explain how the sympathetic and parasympathetic divisions of the autonomic nervous system adjust heart rate. (p. 611)
19. Explain how each of the following affects heart rate: hormones, ions, age, gender, physical fitness, and temperature. (p. 612)
20. Describe the various devices and techniques that are used to help failing hearts. (p. 612)
21. Describe the risk factors involved in heart disease. (p. 612)
22. How are plasma lipids related to heart disease? (p. 613)
23. What are some of the cardiovascular benefits of regular exercise? (p. 615)
24. Describe how the heart develops. (p. 615)

Answers to Questions with Figures

20.1 The space in the middle of the thoracic cavity, between the lungs and between the sternum and backbone.
20.2 Visceral layer of the serous pericardium (epicardium).
20.3 Right atrium, tricuspid valve, right ventricle, pulmonary trunk, pulmonary arteries, lungs, pulmonary veins, left atrium, bicuspid valve, left ventricle, and aorta.
20.4 To prevent the backflow of blood.
20.5 Circumflex.
20.6 Atrioventricular (AV) bundle.
20.7 Cardiac: about 0.3 sec (300 msec). Skeletal: about 1 or 2 msec.

20.8 QRS complex; P wave.
20.9 About 130 ml; end-diastolic volume.
20.10 At 120 beats/min, each beat lasts 0.5 sec. Ventricular systole will still be close to 0.3 sec (maybe a little less), so ventricular diastole will be about 0.2 sec.
20.11 First sound (S1)
20.12 The skeletal muscle "pump" increases stroke volume by increasing preload (end-diastolic volume).
20.15 LDL

Chapter 21

THE CARDIOVASCULAR SYSTEM: BLOOD VESSELS AND HEMODYNAMICS

Chapter Contents at a Glance

Student Objectives

1. Contrast the structure and function of the various types of blood vessels.
2. Explain the factors that regulate the velocity and volume of blood flow.
3. Discuss the various pressures involved in the movement of fluids between capillaries and interstitial spaces.
4. Explain how the return of venous blood to the heart is accomplished.

5. Describe how blood pressure is regulated.
6. Define the three stages of shock.
7. Define pulse and blood pressure (BP) and contrast the clinical significance of systolic, diastolic, and pulse pressures.
8. Identify the principal arteries and veins of systemic, hepatic portal, pulmonary, and fetal circulation.
9. Explain the effects of aging on the cardiovascular system.

10. Describe the development of blood vessels and blood.
11. List the causes and symptoms of hypertension, aneurysm, coronary artery disease (CAD), and deep-venous thrombosis (DVT).
12. Define medical terminology associated with blood vessels.

T he focus of this chapter is a study of the major blood vessels and **hemodynamics** (hē-mō-dī-NAM-ics; *hemo* = blood; *dynamis* = power), that is, a study of the forces involved in circulating blood throughout the body.

Blood vessels form a closed system of tubes that carries blood away from the heart, transports it to the tissues of the body, and then returns it to the heart. **Arteries** are vessels that carry blood from the heart to the tissues. Large, elastic arteries leave the heart and divide into medium-sized, muscular arteries that branch out into the various regions of the body. Medium-sized arteries then divide into small arteries, which, in turn, divide into still smaller arteries called **arterioles** (ar-TER-ē-ōls). As the arterioles enter a tissue, they branch into countless microscopic vessels called **capillaries** (KAP-i-lar'-ēs). Substances are exchanged between the blood and body tissues through the walls of capillaries. Before leaving the tissue, groups of capillaries unite to form small veins called **venules** (VEN-yools). These, in turn, merge to form progressively larger blood vessels called veins. **Veins** then convey blood from the tissues back to the heart. Since blood vessels require oxygen and nutrients just like other tissues of the body, they also have blood vessels, called **vasa vasorum** (literally, vasculature of vessels), in their own walls.

The developmental anatomy of blood vessels and blood will be considered later in the chapter.

ANATOMY OF BLOOD VESSELS

Arteries

In ancient times, **arteries** (*aer* = air; *tereo* = to carry), found empty at death, were thought to contain only air. The hollow center through which blood flows is called the **lumen** (Fig. 21.1). The surrounding arterial wall has three coats or tunics. The inner coat, the **tunica interna (intima),** is composed of a lining of *endothelium* (simple squamous epithelium) that is in contact with the blood, a *basement membrane,* and a layer of elastic tissue called the *internal elastic lamina.* The middle coat, or **tunica media,** is usually the thickest layer. It consists of elastic fibers and smooth muscle fibers (cells). The outer coat, the **tunica externa (adventitia),** is composed principally of elastic and collagen fibers. An *external elastic lamina* may separate the tunica externa from the tunica media.

The structure of arteries, especially of the tunica media, gives them two important functional properties: elasticity and contractility. As the ventricles of the heart contract and eject blood from the heart, the large arteries expand and accommodate the extra blood. Then, as the ventricles relax, the elastic recoil of the arteries forces the blood onward. The contractility of an artery comes from its smooth muscle, which is arranged both longitudinally and in rings around the lumen. Sympathetic fibers of the autonomic nervous system innervate vascular smooth muscle. Usually, when there is an increase in sympathetic stimulation, the smooth muscle contracts, squeezes the wall around the lumen, and narrows the vessel. Such a decrease in the size of the lumen of a blood vessel is called **vasoconstriction.** Conversely, when sympathetic stimulation decreases, the smooth muscle fibers relax and the size of the lumen increases. This increase is called **vasodilation.** Endothelial cells lining blood vessels also release important chemical mediators of vasoconstriction and vasodilation (see Exhibit 21.3, on page 641).

The smooth muscle layer of blood vessels, especially of arteries and arterioles (described shortly), also helps limit bleeding from wounds. When an artery or arteriole is cut, the smooth muscle contracts, producing vascular spasm of the vessel. This is one of the three mechanisms involved in hemostasis (Chapter 19). However, there is a limit to how much vascular spasm can prevent hemorrhaging since the heart's pumping action causes blood to flow through arteries under great pressure.

Elastic (Conducting) Arteries

Large arteries are referred to as **elastic (conducting) arteries.** They include the aorta and the brachiocephalic, common carotid, subclavian, vertebral, and common iliac arteries. The walls of elastic arteries are thin in proportion to their diameters, and their tunica media contains more elastic fibers and less smooth muscle. Elastic arteries are called conducting arteries because they *conduct* blood from the heart to medium-sized muscular arteries.

As the heart alternately contracts and relaxes, blood flow speeds up and slows down accordingly. When the heart contracts and ejects blood, the walls of elastic arteries stretch to accommodate the surge of blood (Fig. 21.2a). The stretched elastic fibers momentarily store some of the energy. For this reason, the elastic arteries function as a **pressure reservoir.** During relaxation of the heart they recoil, converting stored (potential) energy into kinetic energy of the blood. Thus blood moves forward in a more-or-less continuous flow (Fig. 21.2b).

Muscular (Distributing) Arteries

Medium-sized arteries are called **muscular (distributing) arteries.** They include the axillary, brachial, radial, intercostal, splenic, mesenteric, femoral, popliteal, and tibial arteries. In comparison with elastic arteries, their tunica media contains more smooth muscle and less elastic fibers. Thus they are capable of greater vasoconstriction and vasodilation to adjust the rate of blood flow to suit the needs of the structure supplied. The walls of muscular arteries are relatively thick, due mainly to the large amount of smooth muscle. Muscular arteries are called distributing arteries because they *distribute* blood to various parts of the body.

Anastomoses

Most tissues of the body receive blood from more than one artery. The union of the branches of two or more arteries supplying the same body region is called an **anastomosis** (a-nas-tō-MŌ-sis). See Fig. 21.23c which illustrates anastomosis between branches of the superior mesenteric artery as they approach the jejunum of the small intestine. Anastomoses may also occur between veins and between arterioles and venules. Anastomoses between arteries provide alternate routes for blood to reach a tissue or organ. Thus if a vessel is blocked by disease, injury, or surgery, circulation to a part of the body is not necessarily stopped. The alternate route of blood flow to a body part through an anastomosis is known as **collateral circulation.** An alternate blood route may also be from nonanastomosing vessels that supply the same region of the body.

Arteries that do not anastomose are known as **end arteries.** Occlusion of an end artery interrupts the blood supply to a whole segment of an organ, producing necrosis (death) of that segment.

Arterioles

An **arteriole** is a very small, almost microscopic artery that delivers blood to capillaries. Arterioles closer to the arteries from which they branch have a tunica interna like that of arteries, a tunica media composed of smooth muscle and very few elastic fibers, and a tunica externa composed mostly of elastic and collagen fibers. In the smallest diame-

FIGURE 21.1 Comparative structure of blood vessels. The relative size of the capillary in (c) is enlarged. Note in (d) that the lumen of a vein is larger than that of an artery, but the wall of the vein is thinner and the vein frequently appears collapsed (flattened).

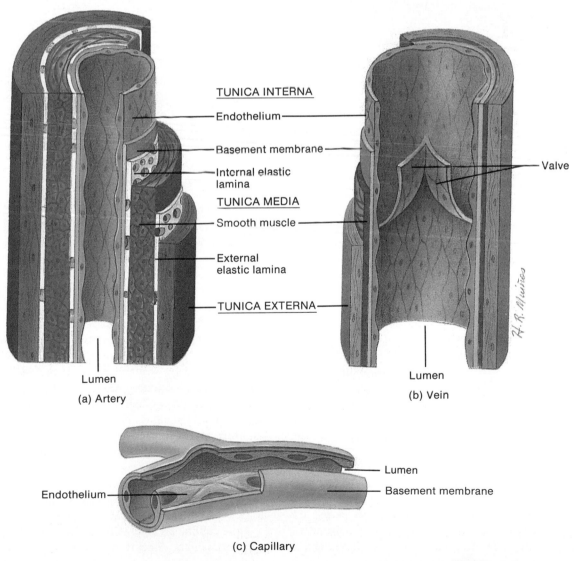

TUNICA INTERNA
— Endothelium
— Basement membrane
— Internal elastic lamina
TUNICA MEDIA
— Smooth muscle
— External elastic lamina
— TUNICA EXTERNA —
— Valve

Lumen
(a) Artery

Lumen
(b) Vein

Endothelium
— Lumen
— Basement membrane

(c) Capillary

Figure continues

FIGURE 21.1 (continued)

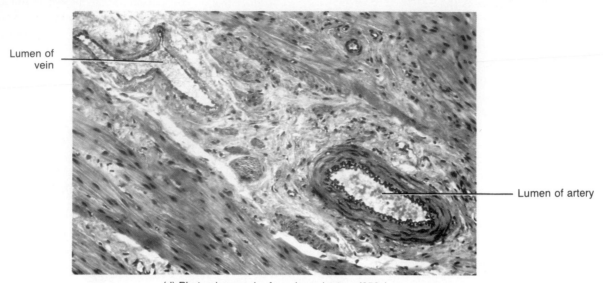

Lumen of vein

Lumen of artery

(d) Photomicrograph of a vein and artery (250x)

Question: Given a choice between the femoral artery and femoral vein, which would you expect to have a thicker wall? A wider lumen?

ter arterioles, which are closest to capillaries, the tunics consist of little more than a layer of endothelium surrounded by a few scattered smooth muscle fibers (Fig. 21.3).

Arterioles play a key role in regulating blood flow from arteries into capillaries. When the smooth muscle of arterioles contracts, causing vasoconstriction, blood flow into capillaries decreases. When the smooth muscle relaxes, the arteriole vasodilates, and blood flow into capillaries increases. A change in diameter of arterioles can also significantly affect blood pressure. The relationship of arterioles to blood flow will be considered in detail later in the chapter.

Capillaries

Capillaries are microscopic vessels that usually connect arterioles and venules. They are found near almost every cell in the body, but their distribution varies with the activity of the tissue. Body tissues with high metabolic activity, for example, muscles, the liver, kidneys, lungs, and nervous system, require more oxygen and nutrients. Accordingly, they have extensive capillary networks. In areas where activity is lower, such as tendons and ligaments, there are fewer capillaries. A few tissues have no capillaries—the epidermis; epithelia that line most visceral organs such as stomach, intestines, and bladder; cornea and lens of the eye; and cartilage.

The primary function of capillaries is to permit the exchange of nutrients and wastes between the blood and tissue cells. The structure of capillaries is admirably suited to this purpose. Capillary walls are composed of only a single layer of cells (endothelium) and a basement membrane (see Fig. 21.1c). They have no tunica media or tunica externa. Thus a substance in the blood must pass through just one cell layer to reach tissue cells. Exchange of materials occurs only through capillary walls: the thick walls of arteries and veins present too great a barrier.

Although capillaries directly link arterioles to venules in some places in the body, in other places they form extensive branching networks. These networks increase the surface area for diffusion and filtration and thereby allow a rapid exchange of large quantities of materials. In most tissues, blood normally flows through only a small portion of the capillary network when metabolic needs are low. But when a tissue becomes active, the entire capillary network fills with blood.

The flow of blood through capillaries is regulated by vessels with smooth muscle in their walls. A **metarteriole** (*met* = beyond) is a vessel that emerges from an arteriole, passes through the capillary network, and empties into a venule (Fig. 21.4). The proximal portions of metarterioles are surrounded by scattered smooth muscle fibers whose contraction and relaxation help to regulate blood flow. The distal portion of a metarteriole, which empties into a venule, has no smooth muscle fibers and is called a **thoroughfare channel.** It serves as a low-resistance pathway that opens when constriction of precapillary sphincters (described shortly) reduces blood flow through the capillary network. Thoroughfare channels thus bypass the capillary bed and sustain blood flow through a region when the capillaries are not being utilized.

True capillaries emerge from arterioles or metarterioles and are not on the direct flow route from arteriole to venule. At their sites of origin, there is a ring of smooth muscle

FIGURE 21.2 Recoil of elastic arteries keeps blood moving during ventricular relaxation (diastole).

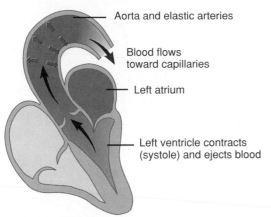

Aorta and elastic arteries

Blood flows toward capillaries

Left atrium

Left ventricle contracts (systole) and ejects blood

(a) Elastic aorta and arteries stretch

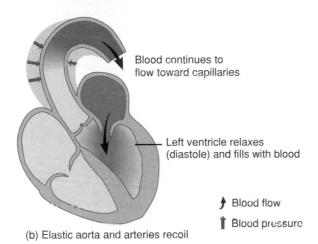

Blood continues to flow toward capillaries

Left ventricle relaxes (diastole) and fills with blood

↗ Blood flow
↑ Blood pressure

(b) Elastic aorta and arteries recoil

Question: In atherosclerosis, the walls of elastic arteries become stiffer and lose some of their elasticity. How do you think this will affect the pressure reservoir function of the arteries?

FIGURE 21.3 Structure of an arteriole.

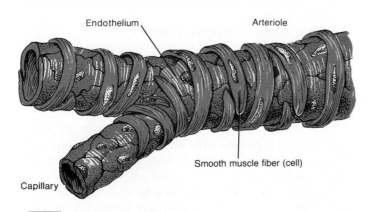

Endothelium

Arteriole

Smooth muscle fiber (cell)

Capillary

Question: What causes vasoconstriction?

FIGURE 21.4 Capillaries.

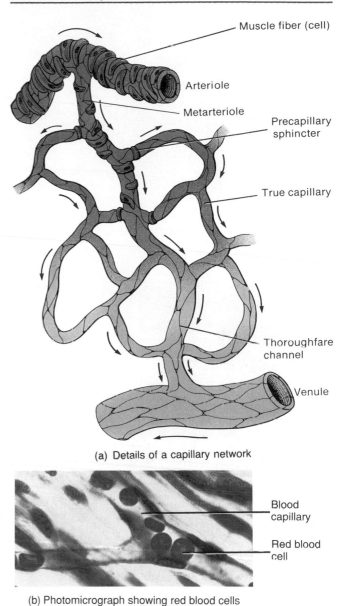

Muscle fiber (cell)

Arteriole

Metarteriole

Precapillary sphincter

True capillary

Thoroughfare channel

Venule

(a) Details of a capillary network

Blood capillary

Red blood cell

(b) Photomicrograph showing red blood cells squeezing through capillaries

Question: Why do metabolically active tissues have extensive capillary networks?

fibers called a **precapillary sphincter** that controls the flow of blood entering a true capillary. Blood usually does not flow in a continuous manner through capillary networks. Rather, it flows intermittently, because of contraction and relaxation of the smooth muscle of metarterioles and the precapillary sphincters of true capillaries. This intermittent contraction and relaxation may occur 5 to 10 times per minute and is called **vasomotion.** In part, vasomotion is due to certain chemicals released by the endothelium. The various factors that regulate the contraction of smooth muscle fibers of metarterioles and precapillary sphincters are discussed later.

Many capillaries of the body are said to be **continuous**

FIGURE 21.5 Types of capillaries.

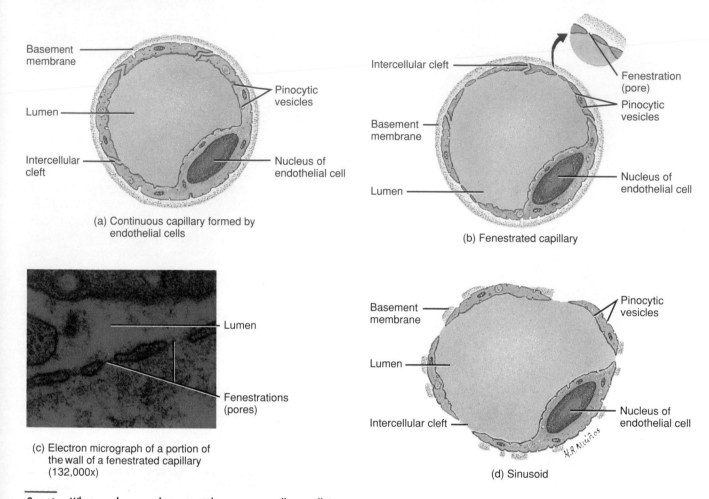

(a) Continuous capillary formed by endothelial cells

Basement membrane
Lumen
Intercellular cleft
Pinocytic vesicles
Nucleus of endothelial cell

(b) Fenestrated capillary

Intercellular cleft
Fenestration (pore)
Pinocytic vesicles
Basement membrane
Lumen
Nucleus of endothelial cell

(c) Electron micrograph of a portion of the wall of a fenestrated capillary (132,000x)

Lumen
Fenestrations (pores)

(d) Sinusoid

Basement membrane
Pinocytic vesicles
Lumen
Intercellular cleft
Nucleus of endothelial cell

Question: What are the ways that materials can cross capillary walls?

capillaries. Except for **intercellular clefts**, which are gaps between neighboring endothelial cells, the plasma membranes form a continuous, uninterrupted ring around the capillary (Fig. 21.5a). Continuous capillaries are found in skeletal and smooth muscle, connective tissues, and the lungs. Other capillaries of the body are called **fenestrated capillaries.** They differ from continuous capillaries in that their endothelial cells have many fenestrations (pores) in the plasma membrane (Fig. 21.5b,c). These range from 70 to 100 nm in diameter. Fenestrated capillaries are found in the kidneys, villi of the small intestine, choroid plexuses of the ventricles in the brain, ciliary processes of the eyes, and endocrine glands.

Blood capillaries in certain parts of the body, such as the liver, are termed **sinusoids.** They are wider than other capillaries and more tortuous. Instead of the usual endothelial lining, sinusoids contain spaces between endothelial cells, and the basement membrane is incomplete or absent (Fig. 21.5d). In addition, sinusoids contain specialized lining cells that are adapted to the function of the tissue. For

example, sinusoids in the liver contain phagocytic cells called **stellate reticuloendothelial (Kupffer's) cells.** Like other capillaries, sinusoids convey blood from arterioles to venules. Other regions containing sinusoids include the spleen, anterior pituitary gland, parathyroid glands, and bone marrow.

Materials can cross the blood capillary walls through four basic routes: through intercellular clefts, via pinocytic vesicles, directly across endothelial membranes, and through fenestrations.

Venules

When several capillaries unite, they form small veins called **venules.** Venules collect blood from capillaries and drain it into veins. The venules closest to the capillaries consist of a tunica interna of endothelium and a tunica externa of connective tissue. As the venules approach the veins, they also contain the tunica media characteristic of veins.

Veins

Veins are composed of essentially the same three coats as arteries, but there are variations in their relative thickness. The tunica interna of veins is extremely thin compared with that of their companion arteries. In addition, the tunica media of veins is much thinner and the tunica externa is thicker (see Fig. 21.1b). Despite these differences, veins are still distensible enough to adapt to variations in the volume and pressure of blood passing through them.

By the time the blood leaves capillaries and flows into veins, its pressure has decreased greatly. The difference in pressure can be noticed when blood flows from a cut vessel. Blood leaves a cut vein in an even, slow flow but spurts rapidly from a cut artery. Most of the structural differences between arteries and veins reflect this pressure difference. For example, the walls of veins are not as strong as those of arteries. Many veins, especially those in the limbs, also feature valves (Fig. 21.6, see also Fig. 21.1b), which are needed because venous blood pressure is so low. When you stand, the pressure pushing blood up the veins in your lower extremities is barely enough to overcome the force of gravity pulling it back down. The valves prevent backflow and in this way aid the flow of blood toward the heart.

A **vascular (venous) sinus** is a vein with a thin endothelial wall that has no smooth muscle to alter its diameter. Surrounding dense connective tissue replaces the tunica media and tunica externa to provide support. Intracranial vascular sinuses, which are supported by the dura mater, return cerebrospinal fluid and deoxygenated blood from the brain to the heart. Another example of a vascular sinus is the coronary sinus of the heart.

Blood Distribution

The largest portion of your blood volume at rest, about 60%, is in systemic veins and venules (Fig. 21.7). Systemic capillaries hold only about 5% of the blood volume, and arteries and arterioles about 15%. Since systemic veins and venules contain so much of the blood, they are called **blood reservoirs.** They serve as storage depots for blood, which can be diverted quickly to other vessels if the need arises. For example, when there is increased muscular activity, an area in the medulla called the vasomotor center sends increasing sympathetic impulses to veins that serve as blood reservoirs. The result is vasoconstriction, which reduces the volume of blood in venous reservoirs. A greater blood volume then can flow to skeletal muscles, where it is needed most. A similar mechanism operates in cases of hemorrhage, when blood volume and pressure decrease. Vasoconstriction of veins in venous reservoirs helps to compensate for the blood loss. Among the principal blood reservoirs are the veins of the abdominal organs (especially the liver and spleen) and the veins of the skin.

FIGURE 21.6 Photograph of a one-way valve in a vein.

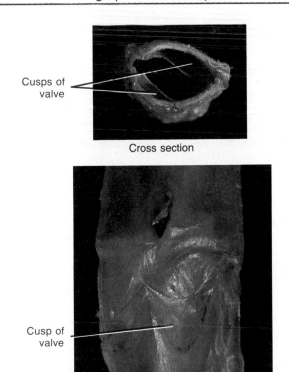

Cusps of valve

Cross section

Cusp of valve

Longitudinal section

FIGURE 21.7 Blood distribution in the heart and various types of blood vessels at rest.

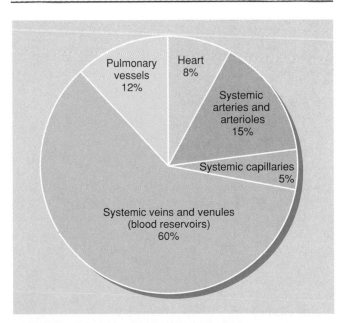

Pulmonary vessels 12%

Heart 8%

Systemic arteries and arterioles 15%

Systemic capillaries 5%

Systemic veins and venules (blood reservoirs) 60%

Question: Why is it more important for leg veins to have valves than it is for neck veins to have valves?

Question: If your total blood volume is 5 liters, what volume is in your venules and veins right now? In your capillaries?

HEMODYNAMICS: PHYSIOLOGY OF CIRCULATION

Velocity of Blood Flow

The *volume* of blood that flows through any tissue in a given period of time (in milliliters per minute) is called **blood flow.** The *velocity* of blood flow (in centimeters per second) is inversely related to the cross-sectional area of the blood vessels. This means that blood flows slowest where the cross-sectional area is greatest (Fig. 21.8), just as a river flows more slowly as it becomes broader. Each time an artery branches, the total cross-sectional area of all the branches is greater than that of the original vessel. On the other hand, when branches combine, for example, as venules merge to form veins, the total cross-sectional area becomes smaller. In an adult, the cross-sectional area of the aorta is only 3 to 5 cm^2, and the average velocity of the blood there is 40 cm/sec. In capillaries, the cross-sectional area is estimated at 4500 to 6000 cm^2, and the velocity of blood flow is less than 0.1 cm/sec. In the two venae cavae combined, the cross-sectional area is about 14 cm^2, and the velocity is 5 to 20 cm/sec. Thus the velocity of blood flow decreases as it flows from the aorta to arteries to arterioles to capillaries and increases as it leaves capillaries and

---

FIGURE 21.8 Relationship between velocity of blood flow and total cross-sectional area in blood vessels.

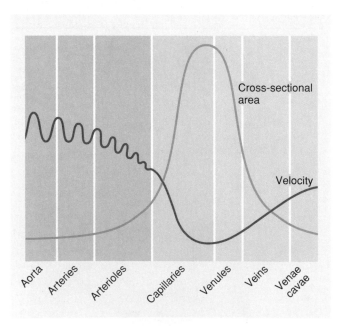

Cross-sectional area

Velocity

Aorta   Arteries   Arterioles   Capillaries   Venules   Veins   Venae cavae

---

Question: In which blood vessels is the velocity of blood flow slowest? Fastest?

returns to the heart. Because blood moves most slowly through the capillaries, there is adequate time for the exchange of materials between the capillaries and adjacent tissues.

Circulation time is the time required for blood to pass from the right atrium, through the pulmonary circulation, back to the left ventricle, through systemic circulation down to the foot, and back again to the right atrium. In a resting person, such a trip normally takes about 1 min.

Volume of Blood Flow

In an adult, cardiac output (CO) is about 5.25 liters/min. This is the volume of blood that circulates through systemic (or pulmonary) blood vessels each minute. In Chapter 20 we noted that CO equals stroke volume (SV) multiplied by heart rate (HR).

$$CO = SV \times HR$$
$$= 70 \text{ ml/beat} \times 75 \text{ beats/minute}$$
$$= 5.25 \text{ liters/min}$$

Two other factors influence cardiac output: (1) blood pressure and (2) resistance (opposition), the force of friction as blood moves along blood vessels. Blood flows from regions of higher to lower pressure; the greater the pressure difference, the greater the blood flow. The higher the resistance, on the other hand, the lower the blood flow. Thus cardiac output equals mean (average) arterial blood pressure (MABP) divided by resistance (R): CO = MABP/ R.

Blood Pressure

Blood pressure (BP) is the pressure exerted by blood on the wall of a blood vessel. In clinical use, the term often refers to pressure in systemic arteries. BP is generated by contraction of the ventricles. In the aorta of a resting, young adult, BP rises to about 120 mm Hg during systole (contraction) and drops to about 80 mm Hg during diastole (relaxation).

If CO rises due to an increase in stroke volume or heart rate, then blood pressure rises so long as resistance remains steady. Likewise, a decrease in CO causes a decrease in blood pressure with steady resistance.

CO, and therefore blood pressure, also depends on the total volume of blood in the cardiovascular system. The normal volume of blood in an adult is about 5 liters (5.3 qt). Any decrease in this volume, as from hemorrhage, decreases the amount of blood that is circulated through the arteries each minute. As a result, blood pressure drops. On the other hand, anything that increases blood volume, such as water retention in the body, increases blood pressure.

As blood leaves the aorta and flows through the systemic circulation, its pressure falls progressively to 0 mm Hg by the time it reaches the right atrium (Fig. 21.9). Resistance in

FIGURE 21.9 Blood pressures in various systemic blood vessels. Note that pressure rises and falls with each heartbeat in blood vessels leading to capillaries. In these vessels, the dashed line is mean (average) blood pressure.

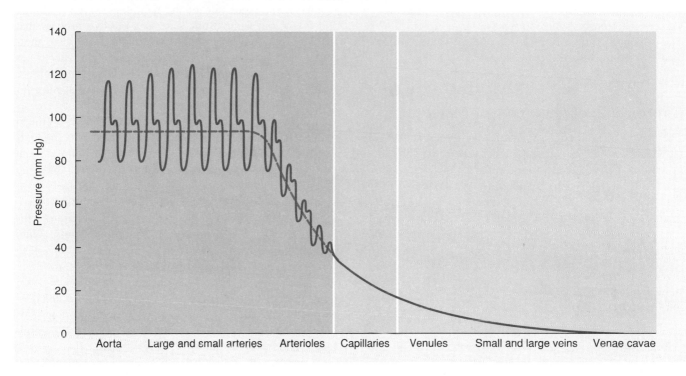

Question What is the mean pressure in the aorta?

the aorta is very low, but the aorta is close to the powerfully contracting left ventricle and thus the mean blood pressure is high (93 mm Hg). Likewise, resistance in large arteries is also quite low, and since these are close to the aorta, the mean arterial blood pressure is still about 93 mm Hg. As blood flows into small arteries, the diameter of individual blood vessels decreases, which increases resistance, and pressure starts to fall. Resistance in arterioles is the highest in the cardiovascular system, accounting for about one-half the total resistance to blood flow. Blood pressure decreases in arterioles from 85 to about 35 mm Hg as blood passes into the arteriolar end of a capillary. At the venous end of a capillary, blood pressure has dropped to about 16 mm Hg.

Resistance

As noted earlier, **resistance** refers to the opposition to blood flow principally as a result of friction between blood and the walls of blood vessels. The friction and thus the resistance, depends on (1) blood viscosity, (2) blood vessel length, and (3) blood vessel radius.

1. Blood viscosity. The viscosity ("thickness") of blood depends largely on the ratio of red blood cells to plasma (fluid) volume and to a smaller extent on the

concentration of proteins in plasma. Resistance to blood flow is directly proportional to the viscosity of blood. Any condition that increases the viscosity of blood, such as dehydration, an unusually high number of red blood cells (polycythemia), or severe burns, increases resistance and thus blood pressure. A depletion of plasma proteins or red blood cells, as a result of anemia or hemorrhage, decreases resistance and thus blood pressure.

2. Total blood vessel length. Resistance to blood flow through a vessel is directly proportional to the length of the blood vessel. The longer a blood vessel, the greater the resistance as blood flows through it. An obese person may have hypertension (elevated blood pressure) due to increase in total blood vessel length caused by the additional blood vessels in adipose tissue.

3. Blood vessel radius. Resistance is inversely proportional ($\propto$) to the fourth power of the radius of the blood vessel ($R \propto 1/r^4$). The smaller the radius of the blood vessel, the greater the resistance it offers to blood flow. As an example, if the radius of a blood vessel decreases by one-half, its resistance to blood flow increases 16 times [$1 \div (\frac{1}{2})^4 = 2^4 = 2 \times 2 \times 2 \times 2 = 16$].

Systemic vascular resistance (SVR) (also known as **total peripheral resistance**) refers to all the vascular resistances offered by systemic blood vessels. Most resistance is in arterioles, capillaries, and venules. Since the diameter of arteries and veins is large, their resistance is very small. A major function of arterioles is to control SVR—and therefore blood pressure and blood flow to particular tissues—by changing their diameters. Arterioles need vasodilate or vasoconstrict only slightly to have a large effect on SVR. The principal center for regulation of SVR is the vasomotor center in the medulla (described shortly).

Capillary Exchange

The 5% of the blood in systemic capillaries is the only blood that exchanges materials with tissue cells. Substances in the blood pass through thin capillary walls, into interstitial fluid, and then into tissue cells. Wastes move in the opposite direction. Substances enter and leave capillaries in three basic ways: (1) diffusion, (2) vesicular transport (endocytosis and exocytosis), and (3) bulk flow (filtration and absorption).

Diffusion

The most important method of capillary exchange is by simple diffusion. Substances such as oxygen, carbon dioxide, glucose, amino acids, hormones, and others diffuse through capillary walls down their concentration gradients. All plasma solutes, except larger proteins, pass freely across most capillary walls. Lipid-soluble materials, such as carbon dioxide, oxygen, and steroid hormones, may pass directly through the phospholipid bilayer of the endothelial cell plasma membranes. Water-soluble substances, such as glucose and amino acids, pass either through fenestrations or intercellular clefts, gaps between endothelial cells (see Fig. 21.5). In the liver, intercellular clefts between endothelial cells that line the sinusoids are so large that even proteins may pass through. This provides an easy route for proteins synthesized by liver cells, for example, fibrinogen and albumin, to reach the plasma. The prime exception to diffusion of water-soluble materials across capillary walls occurs in the brain, where the endothelial cells are nonfenestrated and sealed together by tight junctions in most regions (see blood–brain barrier on page 411).

Vesicular Transport

A small quantity of material crosses capillary membranes by **vesicular transport (transcytosis).** Substances in blood plasma become enclosed within tiny vesicles that enter endothelial cells by endocytosis and then exit on the other side by exocytosis (see Fig. 21.5). This method of transport is important mainly for large, lipid-insoluble molecules that cannot cross capillary walls in any other way. For example, certain antibody proteins pass from maternal into fetal circulation by vesicular transport.

Bulk Flow (Filtration and Reabsorption)

Whereas diffusion is more important for solute exchange between plasma and interstitial fluid, bulk flow is more important for regulation of the relative volumes of blood and interstitial fluid. **Bulk flow** is a passive process that involves the movement of *large* numbers of ions, molecules, or particles in the same direction. The substances move in unison in response to forces such as hydrostatic (water) pressure or air pressure, and they move at rates far greater than can be accounted for by diffusion or osmosis alone. Only 20 to 25% of the extracellular fluid is confined within blood vessels. But if the need arises, for example, if you lose blood through hemorrhage, interstitial fluid can move into capillaries, expand the blood volume, and help maintain blood pressure.

Bulk flow occurs because some forces push fluid (water and solutes) out of capillaries into the surrounding interstitial (tissue) spaces, resulting in *filtration* of fluid. Fluid doesn't build up in interstitial spaces because opposing forces draw interstitial fluid into blood capillaries, resulting in *reabsorption* of fluid. The balance of these forces determines whether blood volume remains steady or changes. Normally, filtration is almost equal to reabsorption. This near equilibrium is known as **Starling's law of the capillaries.** Let us now see how the forces operate.

First we will consider the hydrostatic pressures. These pressures are due to the pressure of water in the fluids. Blood pressure in capillaries, called **blood hydrostatic pressure (BHP),** tends to push fluid out of capillaries into interstitial fluid. BHP is about 35 mm Hg at the arterial end of a capillary and about 16 mm Hg at the venous end (Fig. 21.10). The pressure of the interstitial fluid, called **interstitial fluid hydrostatic pressure (IFHP),** is close to zero. It is difficult to measure, and its reported values vary from small positive to small negative values. For purposes of our discussion, we will assume that IFHP is 0 mm Hg all along the capillaries. Regardless of its exact value, however, the basic principles of fluid movement still apply.

Now let us consider the osmotic pressures involved in fluid movement. The difference in osmotic pressures across a capillary wall is due almost entirely to the presence of plasma proteins, which are too large to pass through either fenestrations or gaps between endothelial cells. **Blood colloid osmotic pressure (BCOP),** also called **oncotic pressure**, is a force caused by the colloidal suspension of these large plasma proteins. The effect of BCOP is to pull fluid from interstitial spaces into capillaries. It averages about 26 mm Hg in capillaries. Opposing BCOP is **interstitial fluid osmotic pressure (IFOP),** which tends to move fluid out of capillaries into interstitial fluid. Normally, IFOP is very small, only 0.1 to 5 mm Hg. The small amount of protein that leaks from plasma into interstitial fluid does not accumulate there because it enters lymphatic fluid and is returned to the blood. For discussion, we will use a value of 1 mm Hg for IFOP.

Whether fluids leave or enter capillaries depends on how

FIGURE 21.10 Dynamics of capillary exchange (Starling's law of the capillaries). Blood hydrostatic pressure pushes fluid out of capillaries (filtration) whereas blood colloid osmotic pressure pulls fluid into capillaries (reabsorption).

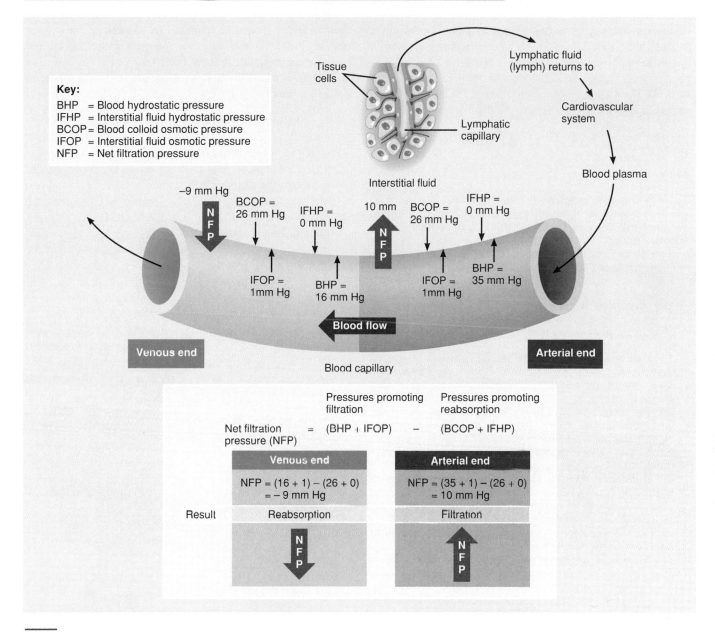

Question: A person who has liver failure fails to synthesize the normal amount of plasma proteins. How will this affect blood colloid osmotic pressure and what will be the impact on capillary filtration and reabsorption?

the pressures relate to each other. If the forces that push fluid out of capillaries are greater than the forces that pull fluid into capillaries, fluid will move from capillaries into interstitial spaces (filtration). If, on the other hand, the forces that move fluid out of interstitial spaces into capillaries are greater than the forces that move fluid out of capillaries, then fluid will move from interstitial spaces into capillaries (reabsorption).

The term **net filtration pressure** (**NFP**) is used to show the direction of fluid movement. It is calculated as follows:

$$NFP = (BHP + IFOP) - (BCOP + IFHP)$$

At the arterial end of a capillary,

$$NFP = (35 + 1) - (26 + 0) = (36) - (26) = 10 \text{ mm Hg}$$

At the venous end of a capillary,

$$NFP = (16 + 1) - (26 + 0) = (17) - (26) = -9 \text{ mm Hg}$$

Thus at the arterial end of a capillary, there is a *net outward force* (10 mm Hg), and fluid moves out of the capillary (filtration) into interstitial spaces. At the venous end of a capillary, the negative value (–9 mm Hg) represents a *net inward force,* and fluid moves into the capillary (reabsorption) from tissue spaces.

On average, about 85% of the fluid filtered at the arteriolar ends of capillaries is reabsorbed at their venular ends. Some of the filtered fluid and any proteins that escape from blood into interstitial fluid return to the blood via the lymphatic system (see Fig. 22.2). On a daily basis, about 20 liters of fluid filters out of capillaries, 17 liters is reabsorbed, and 3 liters enters lymphatic capillaries.

CLINICAL APPLICATION

EDEMA

Occasionally, the balance of filtration and reabsorption between interstitial fluid and plasma is disrupted, allowing an abnormal increase in interstitial fluid volume called **edema** (*oidema* = swelling). Usually, edema is not detectable in tissues until interstitial fluid volume has increased to about 30% above normal.

Edema may result from several main causes:

1. Increased blood hydrostatic pressure in capillaries due to an increase in venous pressure. This may result from poor blood flow back to the heart due to cardiac failure or blood clots.

2. Decreased concentration of plasma proteins that lower blood colloid osmotic pressure. Protein loss may result from burns, malnutrition, liver disease, and kidney disease.

3. Increased permeability of capillaries that raises interstitial fluid osmotic pressure by allowing greater amounts of plasma proteins to leave the blood and enter tissue fluid. This may be caused by chemical, bacterial, thermal, or mechanical agents.

4. Increased extracellular fluid volume as a result of fluid retention. When a person has difficulty excreting fluids, for whatever reason, but continues to drink normal amounts of water, extracellular fluid in the body increases. Some of the fluid enters blood and increases blood hydrostatic pressure.

5. Blockage of lymphatic vessels as often occurs after a radical mastectomy (breast removal, usually because of cancer) or infection by filariasis roundworms. In a radical mastectomy, nearby lymph nodes that appear cancerous are removed with the breast tissue. Edema occurs in the arm on the same side because lymph drainage is blocked. The larvae of the tropical filariasis parasite invade and block lymphatic channels, causing the grossly disfiguring type of edema known as elephantiasis.

Venous Return

Venous return, the volume of blood flowing back to the heart from the systemic veins, depends on the pressure difference from venules (averaging about 16 mm Hg) to the right atrium (0 mm Hg). Although this pressure difference is small, venous return to the right atrium keeps pace with output from the left ventricle because resistance of veins also is low. If pressure increases in the right atrium, how-

ever, venous return will decrease. This may occur with a leaky (incompetent) tricuspid valve, and the result is buildup of blood on the venous side of the systemic circulation.

Besides the heart, two other mechanisms act as pumps to boost venous return: (1) contraction of skeletal muscles in the legs and (2) the pressure changes in the thorax and abdomen during respiration (breathing). The presence of valves in veins allows both of these pumps to contribute to venous return.

1. Skeletal muscle pump. When skeletal muscles contract, they tighten around the vein running through them, which increases the venous blood pressure, and the proximal valve opens. This pressure drives the blood toward the heart: the action is called *milking* (Fig. 21.11). When the muscles relax, this valve closes and prevents the backflow of blood away from the heart. People who are immobilized through injury or disease lack these contractions. As a result, the return of venous blood to the heart is slower, and the heart has to work harder.

2. Respiratory pump. During inspiration, the diaphragm moves downward. This causes a decrease in pressure in the thoracic (chest) cavity and an increase in pressure in the abdominal cavity. As a result, a greater volume of blood moves from the compressed abdominal veins into the decompressed thoracic veins. When the pressures reverse during expiration, the valves in the veins prevent backflow of blood.

A summary of factors that affect blood pressure is presented in Fig. 21.12.

FIGURE 21.11 Role of skeletal muscle pump and venous valves in returning blood to the heart.

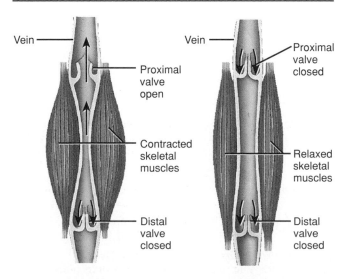

(a) Contracted skeletal muscles   (b) Relaxed skeletal muscles

---

Question: **What mechanisms, beside cardiac contractions, act as pumps to boost venous return?**

FIGURE 21.12 Summary of factors that affect blood pressure.

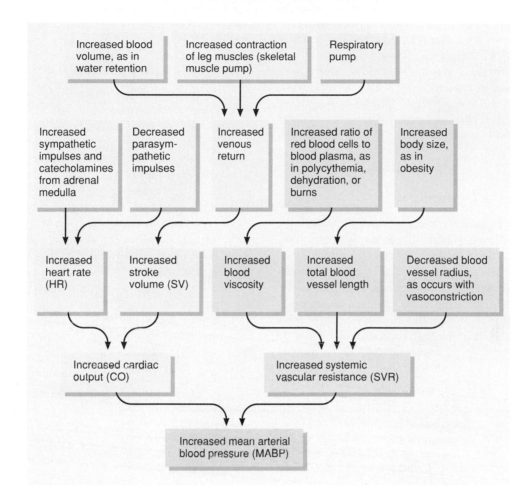

Question: Which type of blood vessel exerts the major control of systemic vascular resistance on a moment-to-moment basis, and how does it achieve this?

VARICOSE VEINS

In people with weak venous valves (see Fig. 21.11), gravity forces large quantities of blood back down into distal parts of the vein. The resulting back-pressure overloads the vein and pushes its wall outward. After repeated overloading, the walls lose their elasticity and become stretched and flabby. Such dilated and tortuous veins caused by leaky valves are called **varicose veins.** They may be due to heredity, mechanical factors (prolonged standing and pregnancy), or aging. Because a varicosed wall is not able to exert a firm resistance against the blood, blood tends to accumulate in the pouched-out area of the vein. This causes it to bulge and also forces fluid into the surrounding tissue. Veins close to the surface of the legs, especially the saphenous veins, are highly susceptible to varicosities whereas deeper veins are not as vulnerable because surrounding skeletal muscles prevent their walls from excessive stretching.

CONTROL OF BLOOD PRESSURE AND BLOOD FLOW

From moment to moment and day to day, several interconnected negative feedback systems control blood pressure by adjusting heart rate, stroke volume, systemic vascular resistance, and blood volume. Some systems allow rapid adjustment of blood pressure to cope with sudden changes such as the drop in brain blood pressure that occurs when you get out of bed. Others act more slowly to provide long-term regulation of blood pressure. Even if blood pressure is steady, there may be a need to change the distribution of blood flow, which is accomplished mainly by altering the diameter of arterioles. For example, during exercise a greater percentage of blood is diverted to organs directly involved in exercise. In strenuous exercise, blood flow to skeletal muscles may increase tenfold and blood flow to the heart and skin may triple. At the same time, blood flow to the digestive tract and kidneys may decrease to half the resting value. Interestingly, no matter what the level of exercise, blood flow to the brain remains nearly constant.

In Chapter 20 we noted that the cardiovascular center in the medulla, by its influence on the sympathetic and parasympathetic divisions of the autonomic nervous system (ANS), contributes to regulation of heart rate and stroke volume. Now we will complete that account by describing the neural, hormonal, and local negative feedback systems that regulate blood pressure and blood flow to specific tissues.

Cardiovascular Center

Groups of neurons scattered within the medulla of the brain stem regulate heart rate, contractility (force of contraction) of the ventricles, and blood vessel diameter (vasoconstriction versus vasodilation). As a whole, this region is known as the **cardiovascular (CV) center.** Some of its neurons stimulate the heart (cardiostimulatory center) whereas others inhibit the heart (cardioinhibitory center). Still others control blood vessel diameter (vasomotor center), either by causing constriction (vasoconstrictor center) or dilation

(vasodilator center). Since these clusters of neurons communicate with one another, function together, and are not clearly separated anatomically, we will discuss them all as a group.

Input to Cardiovascular Center

The CV center receives input both from higher brain regions and from sensory receptors (Fig. 21.13). Nerve impulses descend from higher brain regions including the cerebral cortex, limbic system, and hypothalamus to affect the CV center. For example, even before you start to run a race, your heart rate may increase due to nerve impulses conveyed from the limbic system to the CV center. If your body temperature rises during a race, the thermoregulatory center of the hypothalamus sends nerve impulses to the CV center of the medulla. The result is vasodilation of skin blood vessels to dissipate heat. The two main types of sensory receptors that provide input to the cardiovascular center are baroreceptors and chemoreceptors. Baroreceptors are important pressure-sensitive sensory neurons that monitor stretching of the walls of blood vessels and the atria. Chemoreceptors monitor blood acidity, carbon dioxide level, and oxygen level.

FIGURE 21.13 The cardiovascular (CV) center in the medulla oblongata is the main region for nervous system regulation of the heart and blood vessels. It receives input from higher brain regions, baroreceptors, and chemoreceptors. It provides output to both the sympathetic and parasympathetic divisions of the autonomic nervous system.

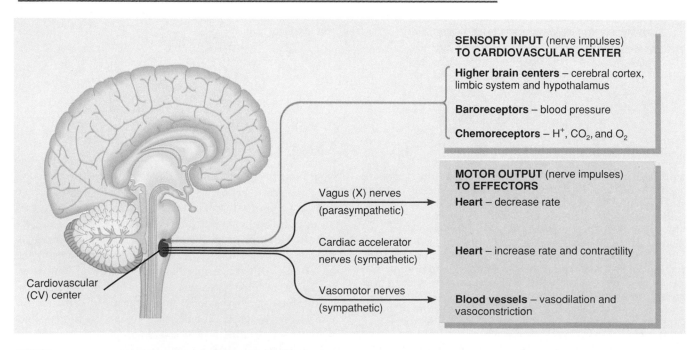

Question: What types of effector tissue are regulated by the CV center?

Output from Cardiovascular Center

Output from the CV center flows along sympathetic and parasympathetic fibers of the ANS (Fig. 21.14). See also Fig. 21.13. Sympathetic stimulation of the heart increases heart rate and contractility. Sympathetic impulses reach the heart via the **cardiac accelerator nerves**. Parasympathetic stimulation, conveyed along the **vagus (X) nerves,** decreases heart rate. The CV center also continually sends impulses to smooth muscle in blood vessel walls via sym-

FIGURE 21.14 Innervation of the heart by the autonomic nervous system and baroreceptor reflexes.

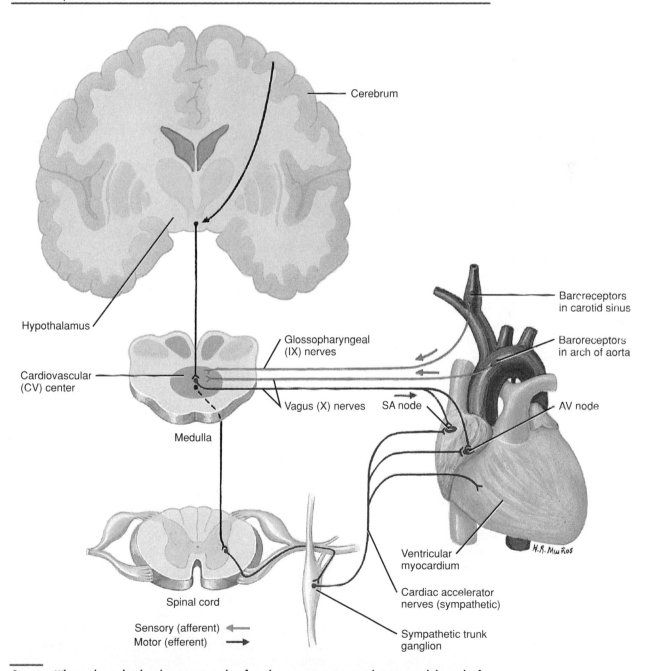

Question: What is the path taken by nerve impulses from baroreceptors in carotid sinuses and the arch of the aorta to the CV center?

pathetic fibers called **vasomotor nerves**. Thus autonomic control of the heart is the result of opposing sympathetic (stimulatory) and parasympathetic (inhibitory) influences. Autonomic control of blood vessels, on the other hand, is exclusively by the sympathetic division.

Sympathetic vasomotor nerve fibers exit the spinal cord through all thoracic and the first one or two lumbar spinal nerves and pass into the sympathetic trunk ganglia (see Fig. 17.2). From here, impulses are conducted along sympathetic nerves that innervate blood vessels in viscera and peripheral areas. Over these routes, the CV center (specifically, the vasomotor center) continually sends impulses to arterioles throughout the body but especially in the skin and abdominal viscera. The result is a moderate state of tonic contraction or vasoconstriction, called **vasomotor tone,** which sets the resting level of systemic vascular resistance.

In the smooth muscle of most small arteries and arterioles sympathetic stimulation causes vasoconstriction and thus raises blood pressure. This is due to activation of alpha (α) adrenergic receptors for norepinephrine (and epinephrine) in the vascular smooth muscle. In skeletal muscle and the heart, the smooth muscle of blood vessels displays beta (β) adrenergic receptors instead, and sympathetic stimulation causes vasodilation rather than vasoconstriction. In addition, some of the sympathetic fibers to blood vessels in skeletal muscle are cholinergic; they release acetylcholine, which causes vasodilation. (Refer to Exhibit 17.3 on page 512 for a listing of receptors in vascular smooth muscle.) When sympathetic stimulation increases, for example, during exercise, both vasoconstriction and vasodilation occur, but in different tissues. As a result, systemic vascular resistance may increase, decrease, or stay the same. The tissues that have dilated arterioles, however, will receive a larger share of the cardiac output. Sympathetic stimulation of most veins results in constriction that moves blood from reservoirs and increases blood pressure.

Neural Regulation

Baroreceptors

Nerve cells capable of responding to changes in pressure or stretch are called **baroreceptors (pressoreceptors).** Baroreceptors in the walls of the arteries, veins, and right atrium monitor blood pressure and participate in several negative feedback systems that contribute to blood pressure control. The three most important baroreceptor negative feedback systems are the aortic reflex, carotid sinus reflex, and right heart reflex.

The **carotid sinus reflex** is concerned with maintaining normal blood pressure in the brain and is initiated by baroreceptors in the wall of the carotid sinus (Fig. 21.14). The **carotid sinus** is a small widening of the internal carotid artery just above the point where it branches from the common carotid artery. Any increase in blood pressure

stretches the wall of the aorta and carotid sinus, and the stretching stimulates the baroreceptors. For the carotid sinus reflex, the impulses travel from the baroreceptors over sensory (afferent) fibers in the glossopharyngeal (IX) nerves to the CV center of the medulla. The **aortic reflex** is concerned with general systemic blood pressure and is initiated by baroreceptors in the wall of the arch of the aorta or attached to the arch. Impulses from baroreceptors in the arch of the aorta reach the CV center via sensory (afferent) fibers of the vagus (X) nerves.

When an increase in aortic and carotid artery pressures is detected in this manner, the CV center responds by putting out more parasympathetic impulses via motor (efferent) fibers of the vagus (X) nerves to the heart and fewer sympathetic impulses via cardiac accelerator nerves to the heart. The resulting decreases in heart rate and force of contraction lower cardiac output. Also, the CV center sends out decreased sympathetic impulses along vasomotor fibers that normally cause vasoconstriction. The result is vasodilation, which lowers systemic vascular resistance (SVR). Decreased cardiac output and SVR both lower systemic arterial blood pressure.

If blood pressure falls, on the other hand, the baroreceptor reflexes accelerate heart rate, increase force of contraction, and promote vasoconstriction (Fig. 21.15). Increased secretions of epinephrine and norepinephrine (NE) by the adrenal medulla intensify the same responses. As the heart beats faster and more forcefully and SVR increases, normal blood pressure returns. This relationship between heart rate and blood pressure is called **Marey's law of the heart.** The ability of the aortic and carotid sinus reflexes to correct a drop in blood pressure is very important when a person sits or stands from a lying position. Immediately upon moving from a prone to an erect position, blood pressure in the head and upper part of the body falls. The decrease in pressure, however, is counteracted by the reflexes. If the pressure were to fall markedly, unconsciousness could occur.

The **right heart (atrial) reflex** responds to increases in venous blood pressure. It is initiated by baroreceptors in the right atrium and venae cavae. When venous pressure increases, the baroreceptors send impulses through the vagus (X) nerves to the CV center. Returning impulses via sympathetic nerves increase heart rate and force of contraction. This mechanism is called the **Bainbridge reflex.**

Chemoreceptors

Receptors sensitive to chemicals are called **chemoreceptors.** Chemoreceptors that monitor blood chemicals are located close to the baroreceptors of the carotid sinus and arch of the aorta in small structures called **carotid bodies** and **aortic bodies,** respectively. These chemoreceptors are sensitive to changes in blood level of oxygen, and even more so to changes in carbon dioxide and hydrogen ion concentration. If there is a severe deficiency of oxygen

FIGURE 21.15 Negative feedback regulation of blood pressure via baroreceptor reflexes.

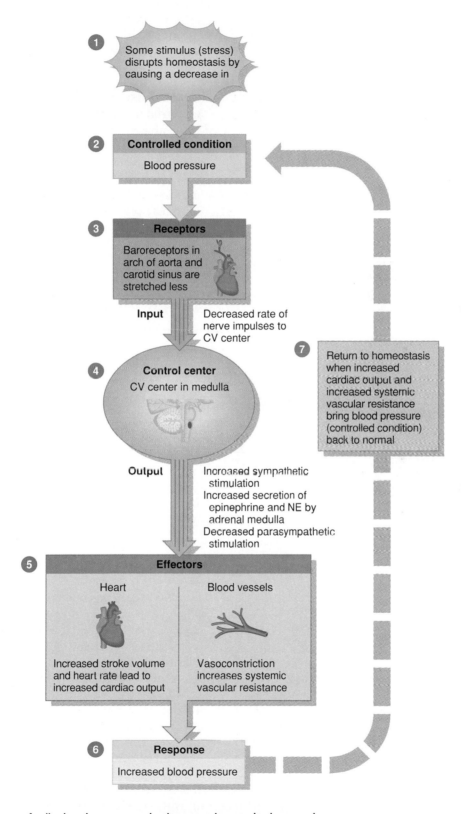

1 Some stimulus (stress) disrupts homeostasis by causing a decrease in

2 **Controlled condition**
Blood pressure

3 **Receptors**
Baroreceptors in arch of aorta and carotid sinus are stretched less

Input Decreased rate of nerve impulses to CV center

4 **Control center**
CV center in medulla

Output Increased sympathetic stimulation
Increased secretion of epinephrine and NE by adrenal medulla
Decreased parasympathetic stimulation

7 Return to homeostasis when increased cardiac output and increased systemic vascular resistance bring blood pressure (controlled condition) back to normal

5 **Effectors**

Heart

Blood vessels

Increased stroke volume and heart rate lead to increased cardiac output

Vasoconstriction increases systemic vascular resistance

6 **Response**
Increased blood pressure

Question: Does this negative feedback cycle represent what happens when you lie down or when you stand up?

(hypoxia), an increase in hydrogen ion concentration (increased acidity or acidosis), or an excess of carbon dioxide (hypercapnia), the chemoreceptors are stimulated and send impulses to the CV center. In response, the CV center increases sympathetic stimulation to arterioles and veins. This brings about vasoconstriction and an increase in blood pressure. As you will see in Chapter 23, these chemoreceptors also stimulate respiratory neurons in the medulla to adjust the rate of breathing.

The neural regulation of blood pressure is summarized in Exhibit 21.1.

Hormonal Regulation

Several **hormones** affect blood pressure and flow by acting on the heart, altering blood vessel diameter, or adjusting the total blood volume.

1. **Epinephrine and norepinephrine (NE),** produced by the adrenal medulla, increase cardiac output (rate and force of heart contractions) and bring about vasoconstriction of abdominal and cutaneous arterioles and veins. They also bring about vasodilation of cardiac and skeletal muscle arterioles.

2. **Antidiuretic hormone (ADH),** produced by the hypothalamus and released from the posterior pituitary gland, causes vasoconstriction if there is a severe loss of blood due to hemorrhage. Alcohol inhibits release of ADH and has an inhibitory effect on the vasomotor center of the medulla. These effects bring about vasodilation, which lowers blood pressure.

3. **Angiotensin II** helps to raise blood pressure in two ways: by causing vasoconstriction and by stimulating secretion of aldosterone, which increases Na^+ and water reabsorption by the kidneys.

4. **Atrial natriuretic peptide (ANP),** released by cells in the atria of the heart, lowers blood pressure by causing vasodilation and by promoting loss of salt and water in the urine, which reduces blood volume.

5. **Histamine** liberated by mast cells and **kinins** found in plasma are vasodilators that play key roles during the inflammatory response.

Exhibit 21.2 summarizes the relationship between hormones and blood pressure regulation.

Autoregulation (Local Control)

Autoregulation refers to a local, automatic adjustment of blood flow in a given region of the body to match the particular needs of the tissue. In most body tissues, oxygen is the principal, though not direct, stimulus for autoregulation. Autoregulation is important in meeting the oxygen and nutritional demands of active tissues, such as heart and muscle tissue, where the demand might increase as much as 10-fold. It also is the major regulator of brain blood flow. Total blood flow to the brain remains almost constant, independent of level of exercise, but distribution to various parts of the brain changes dramatically, depending on your mental and physical activities. For example, blood flow increases to the motor speech areas when you talk, whereas it increases to the auditory area when you listen. There are two general types of stimuli that cause autoregulatory changes in blood flow—physical and chemical.

1. **Physical changes.** Warming promotes vasodilation whereas cooling causes vasoconstriction. Smooth muscle in arteriole walls exhibits a **myogenic response,** that is, it contracts more forcefully when it is stretched and relaxes when stretching is less. In an arteriole, the amount it is stretched depends on its blood flow. If blood flow decreases, stretch decreases, the smooth muscle relaxes, and vasodilation occurs. With vasodilation, blood flow increases.

2. **Chemical mediators.** Cells in the blood, such as white blood cells and platelets, and cells near blood vessels, including smooth muscle fibers, macrophages, and endothelial cells, synthesize and release a wide variety of **vasoactive factors.** These are chemicals that alter blood vessel diameter. One of the most important is **endothelium-derived relaxation factor (EDRF),** now

---

EXHIBIT 21.1

SUMMARY OF NEURAL REGULATION OF BLOOD PRESSURE

Factor	Effect
Baroreceptors	Baroreceptors in the carotid sinus help maintain normal blood pressure in the brain (carotid sinus reflex). Baroreceptors in the arch of the aorta help maintain normal general systemic circulation (aortic reflex). Baroreceptors in the right atrium and venae cavae help maintain normal blood pressure in veins (right heart reflex).
Chemoreceptors	Chemoreceptors are located near baroreceptors in the carotid sinus and arch of the aorta and are sensitive to changes in blood levels of oxygen, hydrogen ion concentration, and carbon dioxide. A deficiency of oxygen (hypoxia) or an increase in hydrogen ion concentration (acidemia) or carbon dioxide (hypercapnia) brings about vasoconstriction and an increase in blood pressure.

EXHIBIT 21.2

RELATIONSHIP BETWEEN BLOOD PRESSURE AND HORMONES

Factor Influencing Blood Pressure	Hormone	Effect
CARDIAC OUTPUT		
Increased heart rate and force of contraction	Norepinephrine (NE) Epinephrine	Increase blood pressure
SYSTEMIC VASCULAR RESISTANCE		
Vasoconstriction	Norepinephrine (NE) and epinephrine (at α receptors in arterioles of abdomen and skin) Angiotensin II Antidiuretic hormone (ADH)	Increase blood pressure
Vasodilation	Atrial natriuretic peptide (ANP) Epinephrine[a] (at β receptors in arterioles of cardiac and skeletal muscle)	Decrease blood pressure
BLOOD VOLUME		
Increased blood volume	Aldosterone Antidiuretic hormone (ADH)	Increase blood pressure
Decreased blood volume	Atrial natriuretic peptide (ANP)	Decreases blood pressure

[a]NE has a much smaller vasodilating effect.

known to be nitric oxide. Others include certain ions (K^+ and H^+), metabolic products such as lactic acid (lactate) and adenosine (from ATP), and some eicosanoids (arachidonic acid derivatives). Exhibit 21.3 lists some of the known vasoactive factors. Once released, vasodilators produce a local dilation of arterioles and relaxation of precapillary sphincters. The result is an increased flow of blood into the tissue, which restores oxygen level to normal. Vasoconstrictors have opposite effects. Stimuli that promote release of vasoactive factors include changes in tissue oxygen and carbon dioxide levels, mechanical stretch of the tissue, hormones in the blood, and local hormones (autocrines and paracrines).

EXHIBIT 21.3

VASOACTIVE FACTORS

Vasodilators	Vasoconstrictors
Endothelium-derived relaxation factor (EDRF, nitric oxide)	Eicosanoids
	Thromboxane A_2
Ions: K^+ and H^+	Prostaglandin $F_{2\alpha}$
Lactic acid (lactate)	Superoxide radicals
Adenosine	Angiotensins
	Endothelins

CLINICAL APPLICATION

SYNCOPE

Syncope (SIN-kō-pē), or faint, refers to a sudden, temporary loss of consciousness followed by spontaneous recovery. It is most commonly due to cerebral ischemia (lack of sufficient blood flow) and may be preceded by uneasiness, malaise, lightheadedness, nausea, vertigo, confusion, disturbances in vision, weakness, sweating, or tinnitus. Among the common causes of syncope are sudden emotional stress or real, threatened, or fantasized injury (vasodepressor syncope); pressure stress associated with urination, defecation, or severe coughing (situational syncope); drugs such as antihypertensives, diuretics, vasodilators, and tranquilizers (drug-induced syncope); an excessive decrease in blood pressure that occurs upon standing up (orthostatic hypotension); situations that stretch the carotid sinus, such as hyperextension of the head, tight collars, or carrying shoulder loads (carotid sinus syncope); and reduced cardiac output.

SHOCK AND HOMEOSTASIS

Shock is an inadequate cardiac output that results in a failure of the cardiovascular system to deliver enough oxygen and nutrients to meet the metabolic needs of body cells. As a result, cellular membranes dysfunction, cellular metabolism is abnormal, and without proper treatment, cellular death may eventually occur.

Signs and Symptoms

The signs and symptoms of shock vary with the severity of the condition, including the following:

1. Clammy, cool, pale skin due to vasoconstriction of skin blood vessels.
2. Tachycardia due to sympathetic stimulation and increased levels of epinephrine.
3. Weak, rapid pulse due to generalized vasodilation and reduced cardiac output.
4. Sweating due to sympathetic stimulation.
5. Hypotension in which the systolic blood pressure is lower than 90 mm Hg as a result of generalized vasodilation and decreased cardiac output.
6. Altered mental status due to cerebral ischemia.
7. Reduced urine formation due to hypotension and increased levels of aldosterone and antidiuretic hormone (ADH).
8. Thirst due to loss of extracellular fluid.
9. Acidosis due to buildup of lactic acid.
10. Nausea due to impaired circulation to the digestive system.

Stages

The causes of shock are many and varied, but all are characterized by inadequate perfusion of tissues. One type of shock, called **hypovolemic** (hī-pō-vō-LĒ-mik) **shock,** refers to decreased blood volume resulting from loss of blood or plasma. Situations that may lead to hypovolemic shock are *acute hemorrhage* due to trauma, gastrointestinal bleeding, and hematomas; and *excessive fluid loss* as a result of excess vomiting, diarrhea, sweating, dehydration, excessive urine production, and burns.

Since hypovolemic shock has been studied so extensively, we will describe the stages of shock as they apply to this type of shock. The development of shock occurs in three principal stages, which merge with one another.

● **Stage I. Compensated (Nonprogressive) Shock** During stage I, when symptoms and signs are minimal, certain homeostatic mechanisms of the cardiovascular system compensate for the shock so that no serious damage results. If the initiating cause does not get any worse, a full recovery follows. The major mechanisms of compensation are negative feedback systems that attempt to return cardiac output (CO) and arterial blood pressure to normal. These compensatory adjustments are mediated through the sympathetic nervous system and the release of various substances. In an otherwise healthy individual, acute blood loss of as much as 10% of the total volume can be dealt with by compensatory mechanisms. Among the adjustments are the following:

1. **Activation of sympathetic division of ANS.** A decrease in blood pressure, detected by baroreceptors, quickly initiates powerful sympathetic responses throughout most of the body. The result is marked vasoconstriction of arterioles and veins of the skin, which produces coolness and paleness; kidneys; and other abdominal viscera. (Vasoconstriction does not occur in the brain or heart.) This increases systemic vascular resistance (SVR), helps maintain adequate venous return, and increases heart rate and force of contraction. Sympathetic stimulation also leads to increased secretion of epinephrine and norepinephrine (NE) by the adrenal medulla. These hormones intensify vasoconstriction and increase heart rate and contractility, which all help to raise blood pressure (Fig. 21.16).

2. **Renin–angiotensin pathway.** Decreased blood flow to the kidneys causes the kidneys to secrete renin and initiates the renin–angiotensin pathway (see Fig. 18.20). Recall that angiotensin II is a potent vasoconstrictor and also stimulates the adrenal cortex to secrete aldosterone—a hormone that increases reabsorption of Na+ and, indirectly, water by the kidneys. The increase in systemic vascular resistance and blood volume help raise blood pressure (Fig. 21.16).

3. **Antidiuretic hormone (ADH).** In response to decreased blood pressure, the posterior pituitary releases more antidiuretic hormone (ADH). This hormone brings about water conservation by the kidneys and vasoconstriction (Fig. 21.16).

4. **Hypoxia.** In response to hypoxia (lowered oxygen availability), affected cells liberate vasodilator factors (see Exhibit 21.3) that dilate arterioles and relax precapillary sphincters. This increases regional blood flow and restores oxygen levels to normal. However, vasodilation also has the potentially harmful effect of decreasing systemic vascular resistance and thus lowering blood pressure.

The various compensatory mechanisms may require from 30 seconds to 48 hours. Eventually, recovery occurs, provided the shock does not intensify and enter the second stage.

● **Stage II. Decompensated (Progressive) Shock** If blood volume drops more than 15 to 25%, the shock becomes steadily worse as compensatory mechanisms are no longer able to maintain adequate perfusion. This is stage II of shock. As the cardiovascular system progressively deteriorates, cardiac output falls dramatically. Several positive feedback cycles characterize stage II. They lead to further reductions in blood pressure and cardiac output and even more cardiovascular deterioration.

Among the positive feedback cycles that contribute to decreased cardiac output are the following (Fig. 21.17):

1. **Depression of cardiac activity.** When mean blood pressure falls below 60 mm Hg, the pressure is no longer adequate to force blood through coronary arteries and the myocardium becomes ischemic. This weakens the heart muscle, decreases cardiac output further, and depresses blood pressure even more. The decreased car-

FIGURE 21.16 Responses of the body during the compensated (nonprogressive) stage of hypovolemic shock. The cycle shown is a negative feedback cycle.

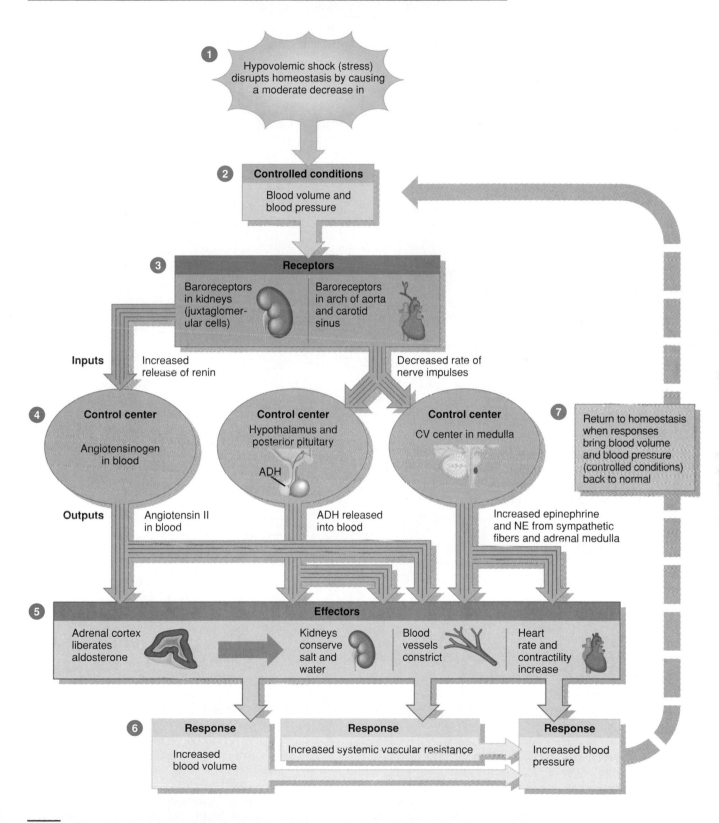

Question: If blood pressure is almost normal in a person who has lost blood, does that mean that his tissues are receiving adequate blood flow?

FIGURE 21.17 Responses of the body during the decompensated (progressive) stage of hypovolemic shock. The cycles shown are positive feedback cycles. Without immediate medical intervention, irreversible shock and death occur.

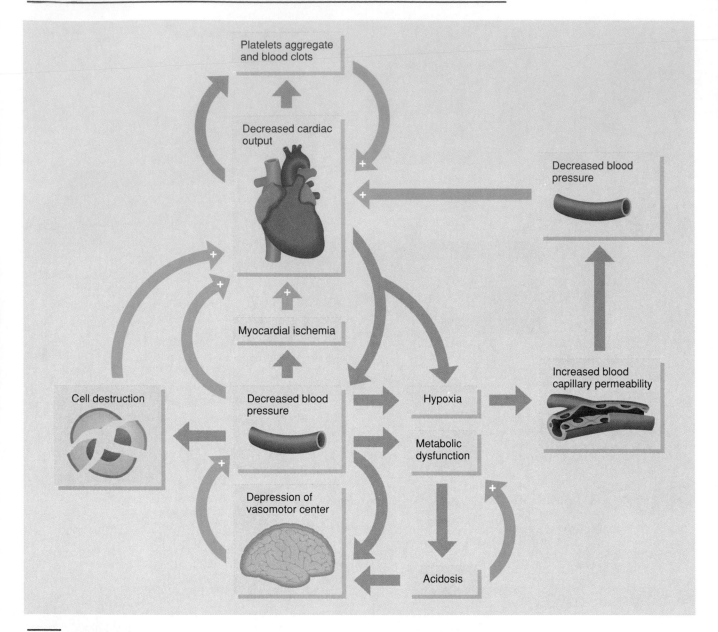

Question: How much of a decrease in blood volume is required before compensatory mechanisms are no longer effective?

diac output and blood pressure produce additional ischemia and even more severe depression of cardiac output and blood pressure.

2. Depression of vasoconstriction. Decreased blood pressure in the vasomotor center depresses its activity. Progressively less activity of the center results in lower blood pressure due to generalized vasodilation and even more depression of the center. This positive feedback cycle becomes operative when mean blood pressure falls below 40 to 50 mm Hg.

3. Increased permeability of capillaries. In very late stages of prolonged shock, hypoxia causes an increase in blood capillary permeability. As more blood plasma components move from the capillaries into tissue spaces, blood volume decreases. This decreases cardiac output, the reduced cardiac output intensifies the hypoxia, and the cycle leads to progressively worsening shock.

4. Intravascular clotting. With decreased cardiac output, blood velocity slows. Sluggish circulation increases the risk of platelet aggregation and blood clot formation.

The resulting obstructions further reduce cardiac output.

5. Cellular destruction. In response to shock, cellular destruction occurs throughout the body, including the heart, which pumps less effectively as a consequence. The cellular changes include lysosomal rupture, depressed mitochondrial activity, diminished active transport, and decreased metabolism.

6. Acidosis. As a result of metabolic dysfunction, cells produce excess lactic acid, which results in acidosis, a condition in which the pH of blood ranges from 7.35 to 6.80 or lower. The principal physiological effect of acidosis is depression of the central nervous system, including the vasomotor center in the medulla.

To reverse the changes that occur during the decompensated stage, immediate medical intervention is required. If this fails, shock progresses to a third stage.

● **Stage III. Irreversible Shock** In stage III, there is rapid deterioration of the cardiovascular system that cannot be helped by compensatory mechanisms or medical intervention. As the shock cycle perpetuates itself, there are life-threatening reductions in cardiac output, blood pressure, and tissue perfusion. Ultimately, high-energy phosphate (ATP) reserves are depleted, especially in the heart and liver, and the heart deteriorates so much that it can no longer pump blood.

CHECKING CIRCULATION

Pulse

The alternate expansion and recoil of elastic arteries after each systole of the left ventricle create a pressure wave that travels through the arteries that is called the **pulse.** Pulse is strongest in the arteries closest to the heart. It becomes weaker as it passes over the arterial system, and it disappears altogether in the capillaries. The pulse may be felt in any artery that lies near the surface of the body and over a bone or other firm tissue. The radial artery at the wrist is most commonly used to feel the pulse (see Fig. 21.22a). Others include the following:

1. Temporal artery, lateral to the orbit of the eye.
2. Facial artery, at the lower jawbone on a line with the corners of the mouth.
3. Common carotid artery, lateral to the larynx (voice box). See Fig. 21.22a,b.
4. Brachial artery, along the medial side of the biceps brachii muscle (see Fig. 21.22a).
5. Femoral artery, inferior to the inguinal ligament (see Fig. 21.24).
6. Popliteal artery, behind the knee (see Fig. 21.24).
7. Posterior tibial artery, posterior to the medial malleolus of the tibia (see Fig. 21.24).
8. Dorsalis pedis artery, superior to the instep of the foot (see Fig. 21.24a).

The pulse rate normally is the same as the heart rate. Resting pulse rate in a normal person is between 70 and 80 beats per minute. The term **tachycardia** (tak′-i-KAR-dē-a; *tachy* = fast) means a rapid resting heart or pulse rate (over 100/min). The term **bradycardia** (brād′-i-KAR-dē-a; *brady* = slow) indicates a slow resting heart or pulse rate (under 60/min).

Other characteristics of the pulse may give additional information about circulation. For example, the intervals between beats should be equal in length. If a pulse is missed at intervals, the pulse is said to be irregular. Also, each pulse beat should be of equal strength. Irregularities in strength may indicate a lack of muscle tone in the heart or arteries.

Measurement of Blood Pressure (BP)

Blood pressure is usually measured in the left brachial artery using a **sphygmomanometer** (sfig′-mō-ma-NOM-e-ter; *sphygmo* = pulse). A commonly used sphygmomanometer consists of a rubber cuff attached by a rubber tube to a compressible hand pump or bulb for inflating the cuff. Another tube attaches to the cuff and to a column of mercury or pressure dial marked off in millimeters of mercury (mm Hg) to measure the pressure. The cuff is wrapped around the arm over the brachial artery and inflated, which creates a pressure on the artery. Inflation is continued until the pressure in the cuff exceeds the pressure in the artery. At this point, the walls of the brachial artery are compressed tightly against each other, and no blood can flow through. Compression of the artery may be evidenced in two ways. First, if a stethoscope is placed over the artery below the cuff, no sound can be heard. Second, no pulse can be felt by placing the fingers over the radial artery at the wrist.

Then the cuff is deflated gradually until the pressure in the cuff is slightly less than the maximal pressure in the brachial artery. At this point, the artery opens, a spurt of blood passes through, and a sound may be heard through the stethoscope due to the turbulence of the blood flow. When this first sound is heard, a reading on the mercury column is made. This sound corresponds to **systolic blood pressure (SBP)**—the force with which blood pushes against arterial walls as a result of ventricular contraction (Fig. 21.18). As cuff pressure is further reduced, the sounds suddenly become faint as the blood turbulence reduces significantly. The pressure recorded on the mercury column when the sounds suddenly become faint is called **diastolic blood pressure (DBP).** It measures the force of blood in arteries during ventricular relaxation. Whereas systolic pressure indicates the force of the left ventricular contraction, diastolic pressure provides information about the resistance of blood vessels. Following diastolic blood pressure, the sounds disappear altogether. The various sounds that are heard while taking blood pressure are called **Korotkoff** (kō-ROT-kof) **sounds.**

FIGURE 21.18 Relationship of blood pressure changes to cuff pressure.

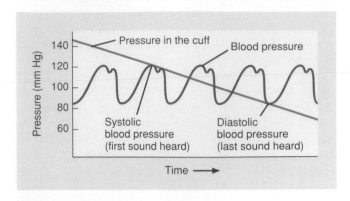

Question: If a blood pressure is reported as "137 over 95," what is the diastolic pressure, systolic pressure, and pulse pressure. Does this person have hypertension (see page 677)?

Although some people may have a lower or higher blood pressure, the normal blood pressure of a young adult male is about 120 mm Hg systolic and 80 mm Hg diastolic, reported as "120 over 80" and written as 120/80. In young adult females, the pressures are 8 to 10 mm Hg less. People who exercise regularly and are in good physical condition also tend to have lower blood pressures. Thus blood pressure slightly lower than 120/80 may be a sign of good health and fitness.

The difference between systolic and diastolic pressure is called **pulse pressure (PP).** This pressure, which averages 40 mm Hg, provides information about the condition of the arteries. For example, conditions such as atherosclerosis and patent (open) ductus arteriosus greatly increase pulse pressure. The normal ratio of systolic pressure to diastolic pressure to pulse pressure is about 3:2:1. Mean arterial blood pressure (MABP) is about one-third of the way between DBP and SBP: MABP ≈ DBP + ⅓ PP.

CIRCULATORY ROUTES

Arteries, arterioles, capillaries, venules, and veins are organized into parallel routes that deliver blood throughout the body. We can now look at the basic routes the blood takes as it is transported through its vessels.

Figure 21.19 shows the parallel **circulatory routes** for blood flow. The routes are parallel because in most cases a portion of the cardiac output flows separately to each tissue of the body. Thus each organ receives its own supply of freshly oxygenated blood. The two basic postnatal (after birth) routes are systemic and pulmonary. The **systemic circulation** includes all the arteries and arterioles that carry oxygenated blood from the left ventricle to systemic capillaries plus the veins and venules that carry deoxygenated

blood returning to the right atrium after flowing through body organs. The nutrient arteries to the lungs, such as the intercostal arteries and bronchial arteries, also are part of the systemic circulation. Some subdivisions of the systemic circulation are the **coronary (cardiac) circulation** (see Fig. 20.5), which supplies the myocardium of the heart; **cerebral circulation,** which supplies the brain; and the **hepatic portal circulation,** which extends from the gastrointestinal tract to the liver (see Fig. 21.30). Blood leaving the aorta and flowing through the systemic arteries is a bright red color. As it moves through capillaries, it loses some of its oxygen and picks up carbon dioxide, so that blood in systemic veins is a dark red color.

When blood returns to the heart from the systemic route, it is pumped out of the right ventricle through the **pulmonary circulation** to the lungs (see Fig. 21.31). In pulmonary capillaries of the air sacs of the lungs, it loses some of its carbon dioxide and takes on oxygen. Bright red again, it returns to the left atrium of the heart and reenters the systemic circulation as it is pumped out by the left ventricle.

Another major route—**fetal circulation**—exists only in the fetus and contains special structures that allow the developing fetus to exchange materials with its mother (see Fig. 21.32).

Systemic Circulation

The systemic circulation carries oxygen and nutrients to body tissues and removes carbon dioxide and other wastes and heat from the tissues. All systemic arteries branch from the aorta. The portion of the aorta that passes upward behind the pulmonary trunk as it emerges from the left ventricle is called the **ascending aorta.** It gives off two coronary branches to the heart muscle. Then it turns to the left, forming the **arch of the aorta,** which descends to the level of the fourth thoracic vertebra. The **descending aorta** begins at this point. It lies close to the vertebral bodies, passes through the diaphragm, and divides at the level of the fourth lumbar vertebra into two **common iliac arteries,** which carry blood to the lower extremities. The section of the descending aorta between the arch of the aorta and the diaphragm is called the **thoracic aorta.** The section between the diaphragm and the common iliac arteries is termed the **abdominal aorta.** Each section of the aorta gives off arteries that continue to branch into distributing arteries leading to organs and finally into the arterioles and capillaries that service the systemic tissues (all tissues except the air sacs of the lungs).

Blood returns to the heart through the systemic veins. All the veins of the systemic circulation drain into the **superior vena cava, inferior vena cava,** or **coronary sinus,** which in turn empty into the right atrium. The principal arteries and veins of systemic circulation are described and illustrated in Exhibits 21.4 through 21.15 and Figs. 21.20 through 21.29.

FIGURE 21.19 Circulatory routes. Refer to Fig. 20.5 for the details of coronary (cardiac) circulation and to Fig. 21.32 for the details of fetal circulation.

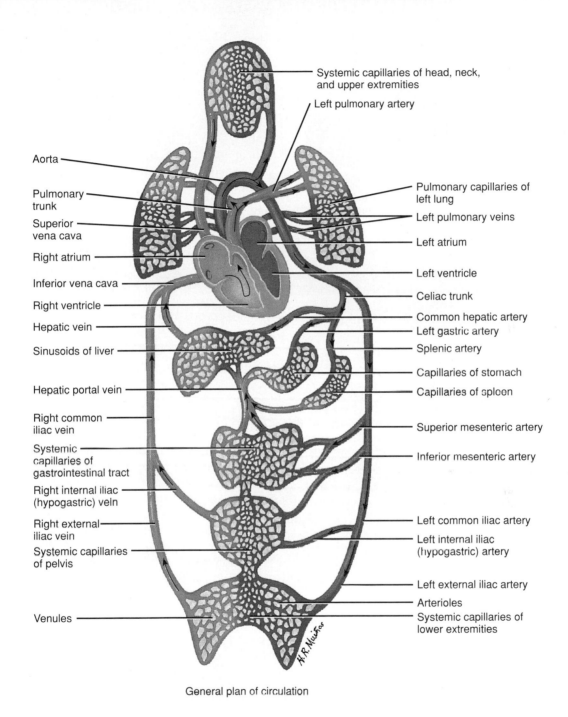

General plan of circulation

Question: What are the two principal postnatal circulatory routes?

AORTA AND ITS BRANCHES (Fig. 21.20)

Overview: The **aorta** is the largest artery of the body, about 2 to 3 cm (0.8 to 1.2 in.) in diameter. It begins at the left ventricle and contains a valve at its origin, called the aortic semilunar valve (see Fig. 20.3d), which prevents backflow of blood into the left ventricle during its diastole (relaxation). The principal divisions of the aorta are the ascending aorta, arch of the aorta, thoracic aorta, and abdominal aorta.

Division of Aorta	Arterial Branch		Region Supplied
Ascending aorta (ā-OR-ta)	Right and left coronary		Heart.
Arch of aorta	Brachiocephalic (brā′-kē-ō-se-FAL-ik) trunk	Right common carotid (ka-ROT-id)	Right side of head and neck.
		Right subclavian (sub-KLĀ-vē-an)	Right upper extremity.
	Left common carotid		Left side of head and neck.
	Left subclavian		Left upper extremity.
Thoracic (thō-RAS-ik) **aorta**	Intercostals (in′-ter-KOS-tal)		Intercostal and chest muscles and pleurae.
	Superior phrenics (FREN-iks)		Posterior and superior surfaces of diaphragm.
	Bronchials (BRONG-kē-als)		Bronchi of lungs.
	Esophageals (e-sof′a-JĒ-als)		Esophagus.
Abdominal (ab-DOM-i-nal) **aorta**	Inferior phrenics (FREN-iks)		Inferior surface of diaphragm.
	Celiac	Common hepatic (he-PAT-ik)	Liver.
		Left gastric (GAS-trik)	Stomach and esophagus.
		Splenic (SPLĒN-ik)	Spleen, pancreas, and stomach.
	Superior mesenteric (MES-en-ter′-ik)		Small intestine, cecum, ascending and transverse colons, and pancreas.
	Suprarenals (soo′-pra-RĒ-nals)		Adrenal (suprarenal) glands.
	Renals (RĒ-nals)		Kidneys.
	Gonadals (gō-NAD-als)	Testiculars (tes-TIK-yoo-lars) or	Testes.
		Ovarians (ō-VA-rē-ans)	Ovaries.
	Inferior mesenteric (MES-en-ter′-ik)		Transverse, descending, and sigmoid colons and rectum.
	Common iliacs (IL-ē-aks)	External iliacs	Lower extremities.
		Internal iliacs (hypogastrics)	Uterus, prostate gland, muscles of buttocks, and urinary bladder.

FIGURE 21.20 Aorta and its principal branches.

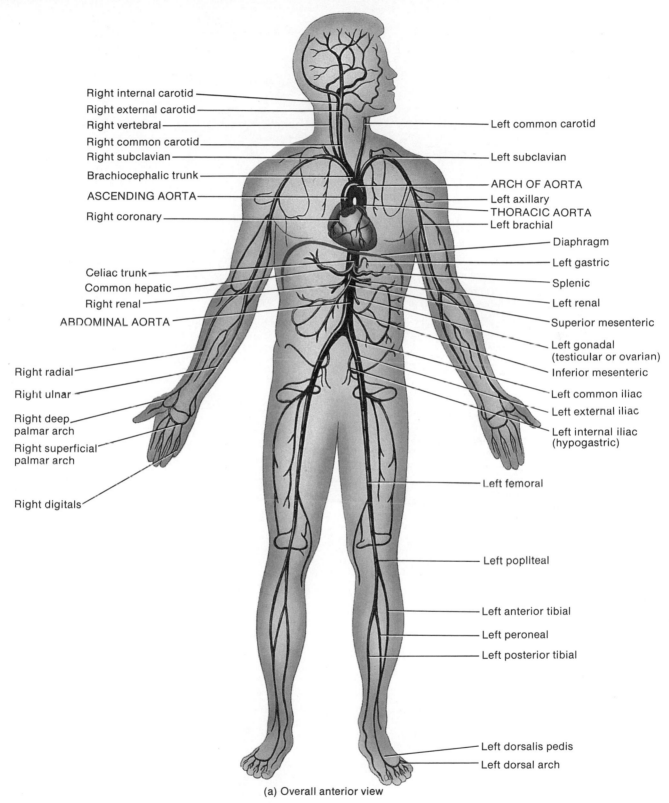

Right internal carotid

Right external carotid

Right vertebral

Right common carotid

Right subclavian

Brachiocephalic trunk

ASCENDING AORTA

Right coronary

Celiac trunk

Common hepatic

Right renal

ABDOMINAL AORTA

Right radial

Right ulnar

Right deep palmar arch

Right superficial palmar arch

Right digitals

Left common carotid

Left subclavian

ARCH OF AORTA

Left axillary

THORACIC AORTA

Left brachial

Diaphragm

Left gastric

Splenic

Left renal

Superior mesenteric

Left gonadal (testicular or ovarian)

Inferior mesenteric

Left common iliac

Left external iliac

Left internal iliac (hypogastric)

Left femoral

Left popliteal

Left anterior tibial

Left peroneal

Left posterior tibial

Left dorsalis pedis

Left dorsal arch

(a) Overall anterior view

Figure continues

FIGURE 21.20 (continued)

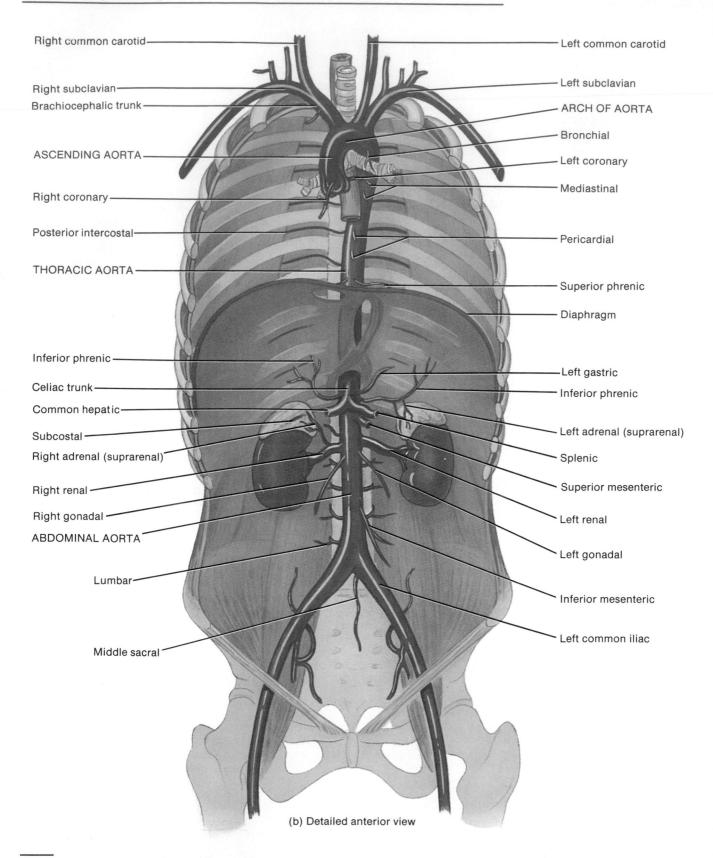

Right common carotid

Right subclavian

Brachiocephalic trunk

ASCENDING AORTA

Right coronary

Posterior intercostal

THORACIC AORTA

Inferior phrenic

Celiac trunk

Common hepatic

Subcostal

Right adrenal (suprarenal)

Right renal

Right gonadal

ABDOMINAL AORTA

Lumbar

Middle sacral

Left common carotid

Left subclavian

ARCH OF AORTA

Bronchial

Left coronary

Mediastinal

Pericardial

Superior phrenic

Diaphragm

Left gastric

Inferior phrenic

Left adrenal (suprarenal)

Splenic

Superior mesenteric

Left renal

Left gonadal

Inferior mesenteric

Left common iliac

(b) Detailed anterior view

Question: What are the four principal subdivisions of the aorta?

EXHIBIT 21.5

ASCENDING AORTA (Fig. 21.21)

Overview: The **ascending aorta** is the first division of the aorta, about 5 cm (2 in.) in length. It is directed upward, forward, and to the right and ends at the level of the sternal angle where it becomes the arch of the aorta. The beginning of the ascending aorta is covered by the pulmonary trunk and right auricle; the right pulmonary artery is behind it. At its origin, the ascending aorta contains three dilations, called aortic sinuses. Two of these, the right and left sinuses, give rise to the right and left coronary arteries, respectively.

Branch	Description and Region Supplied
Coronary (KOR-o-nar-ē) **arteries**	Right and left branches arise from ascending aorta just superior to aortic semilunar valve. They form crown around heart, giving off branches to atrial and ventricular myocardium.

SCHEME OF DISTRIBUTION

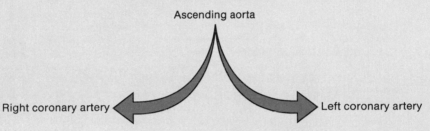

Ascending aorta

Right coronary artery   Left coronary artery

FIGURE 21.21 Ascending aorta and its branches.

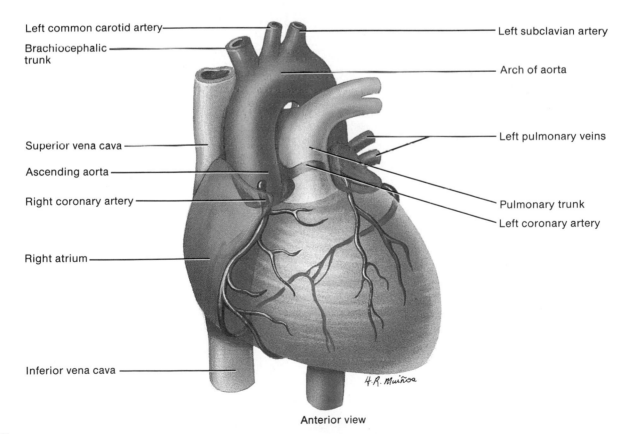

Left common carotid artery

Brachiocephalic trunk

Superior vena cava

Ascending aorta

Right coronary artery

Right atrium

Inferior vena cava

Left subclavian artery

Arch of aorta

Left pulmonary veins

Pulmonary trunk

Left coronary artery

H.R. Muñoz

Anterior view

Question: Which arteries arise from the ascending aorta?

EXHIBIT 21.6

ARCH OF THE AORTA (Fig. 21.22)

Overview: The **arch of the aorta** is about 4.5 cm (1.8 in.) in length and is the continuation of the ascending aorta that emerges from the pericardium behind the sternum at the level of the sternal angle. Initially, the arch is directed upward, backward and to the left, and then downward on the left side of the body of the fourth thoracic vertebra. Actually, the arch is directed not only from right to left, but from anterior to posterior as well. The arch of the aorta terminates at the level of the disc between the fourth and fifth vertebrae where it becomes the thoracic aorta. The thymus gland lies in front of the arch of the aorta, while the trachea lies behind it.

Three major arteries branch from the arch of the aorta. In order of their origination, they are the brachiocephalic trunk, left common carotid artery, and left subclavian artery.

Branch	Description and Region Supplied
Brachiocephalic (brā′-kē-ō-se-FAL-ik)	**Brachiocephalic trunk** is the first and largest branch off arch of aorta. It bifurcates (divides) to form right subclavian artery and right common carotid artery. **Right subclavian** (sub-KLĀ-vē-an) **artery** extends from brachiocephalic to first rib and then passes into armpit (axilla) and supplies arm, forearm, and hand. Continuation of right subclavian into axilla is called **axillary (AK-si-ler′-ē) artery.**[a] From here, it continues into arm as **brachial** (BRĀ-kē-al) **artery.** At bend of elbow, brachial artery divides into medial **ulnar** (UL-nar) and lateral **radial** (RĀ-dē-al) **arteries.** These vessels pass down to palm, one on each side of forearm. In palm, branches of two arteries anastomose to form two palmar arches—**superficial palmar** (PAL-mar) **arch** and **deep palmar arch.** From these arches arise **digital** (DIJ-i-tal) **arteries,** which supply fingers and thumb.
	Before passing into axilla, right subclavian gives off major branch to brain called **right vertebral** (VER-te-bral) **artery.** Right vertebral artery passes through foramina of transverse processes of cervical vertebrae and enters skull through foramen magnum to reach undersurface of brain. Here it unites with left vertebral artery to form **basilar** (BAS-i-lar) **artery.**
	Right common carotid artery passes upward in neck. At upper level of larynx, it divides into **right external** and **right internal carotid** (ka-ROT-id) **arteries.** External carotid supplies right side of thyroid gland, tongue, throat, face, ear, scalp, and dura mater. Internal carotid supplies brain, right eye, and right sides of forehead and nose.
	Inside the cranium, anastomoses of left and right internal carotids along with basilar artery form a somewhat hexagonal arrangement of blood vessels at base of brain near sella turcica called **cerebral** (se-RĒ-bral) **arterial circle (circle of Willis).** From this circle arise arteries supplying most of the brain. Essentially, cerebral arterial circle is formed by union of **anterior cerebral arteries** (branches of internal carotids) and **posterior cerebral arteries** (branches of basilar artery). Posterior cerebral arteries are connected with internal carotids by **posterior communicating** (ko-MYOO-ni-kā′-ting) **arteries.** Anterior cerebral arteries are connected by **anterior communicating arteries.** The **internal carotid** (ka-ROT-id) **arteries** are also considered part of cerebral arterial circle. The function of the cerebral arterial circle is to equalize blood pressure to brain and provide alternate routes for blood to brain, should arteries become damaged.
Left common carotid (ka-ROT-id)	**Left common carotid** is second branch off arch of aorta (see Fig. 21.23a). Corresponding to right common carotid, it divides into basically same branches with same names, except that arteries are now labeled "left" instead of "right."
Left subclavian (sub-KLĀ-vē-an)	**Left subclavian artery** is third branch off arch of aorta (see Fig. 21.23a). It distributes blood to left vertebral artery and vessels of left upper extremity. Arteries branching from left subclavian are named like those of right subclavian.

---

[a]The right subclavian artery is a good example of the practice of giving the same vessel different names as it passes through different regions.

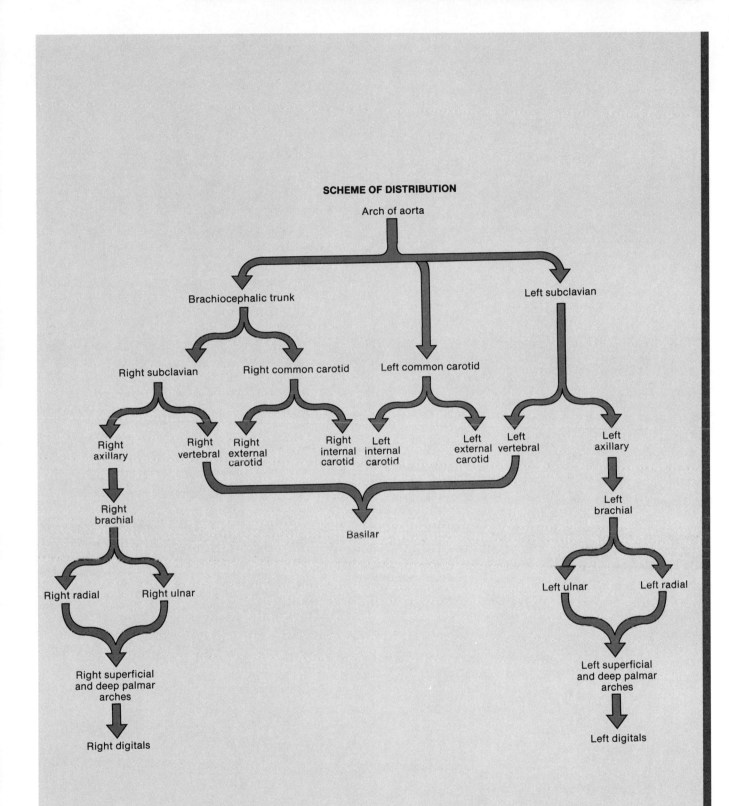

SCHEME OF DISTRIBUTION

FIGURE 21.22 Arch of the aorta and its branches. Note in (c) the arteries that comprise the cerebral arterial circle (circle of Willis).

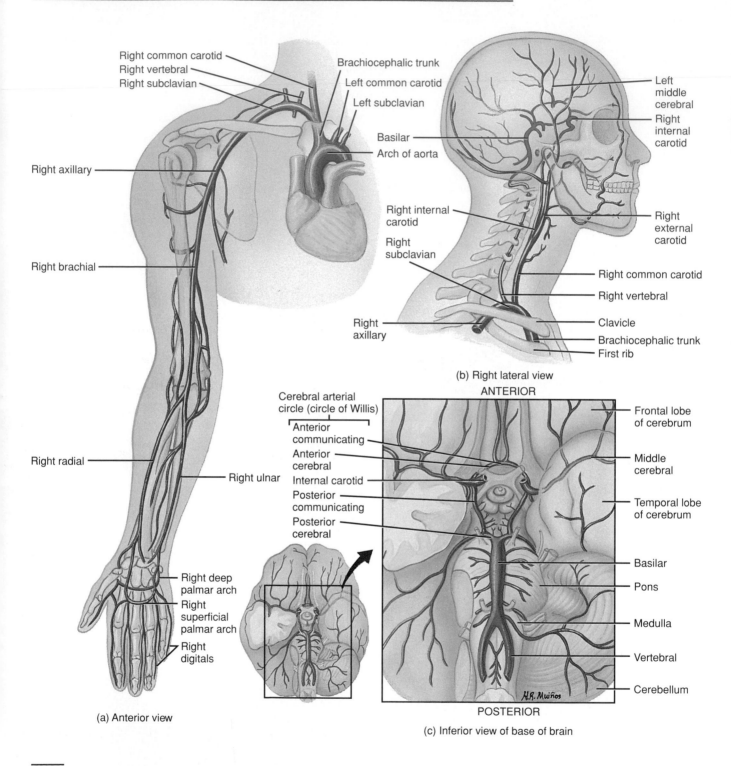

Right common carotid
Right vertebral
Right subclavian

Brachiocephalic trunk
Left common carotid
Left subclavian

Right axillary

Basilar
Arch of aorta

Right internal carotid

Right subclavian

Right brachial

Right axillary

Left middle cerebral
Right internal carotid

Right external carotid

Right common carotid
Right vertebral
Clavicle
Brachiocephalic trunk
First rib

(b) Right lateral view

Right radial

Right ulnar

Right deep palmar arch
Right superficial palmar arch
Right digitals

Cerebral arterial circle (circle of Willis)
Anterior communicating
Anterior cerebral
Internal carotid
Posterior communicating
Posterior cerebral

ANTERIOR

Frontal lobe of cerebrum

Middle cerebral

Temporal lobe of cerebrum

Basilar
Pons

Medulla

Vertebral

Cerebellum

H.R. Muiños

POSTERIOR

(a) Anterior view

(c) Inferior view of base of brain

Question: What are the three major branches of the arch of the aorta in order of their origination?

EXHIBIT 21.7

THORACIC AORTA (Fig. 21.23)

Overview: The **thoracic aorta** is about 20 cm (8 in.) long and is a continuation of the arch of the aorta. It begins at the level of the disc between the fourth and fifth thoracic vertebrae where it lies to the left of the vertebral column. As it descends, it moves closer to the midline and terminates at an opening in the diaphragm (aortic hiatus) in front of the vertebral column at the level of the intervertebral disc between the twelfth thoracic and first lumbar vertebrae.

Along its course, the thoracic aorta sends off numerous small arteries to viscera (**visceral branches**) and body wall structures (**parietal branches**).

Branch	Description and Region Supplied
VISCERAL	
Pericardial (per′-i-KAR-dē-al)	Several minute **pericardial arteries** supply blood to posterior aspect of pericardium.
Bronchial (BRONG-kē-al)	One right and two left **bronchial arteries** supply the bronchial tubes, visceral pleurae, bronchial lymph nodes, and esophagus. (Whereas the right bronchial artery arises from the third posterior intercostal artery, the two left bronchial arteries arise from the thoracic aorta.)
Esophageal (e-sof ′-a-JĒ-al)	Four or five **esophageal arteries** supply the esophagus.
Mediastinal (mē′-dē-as-TĪ-nal)	Numerous small **mediastinal arteries** supply blood to structures in the posterior mediastinum.
PARIETAL	
Posterior intercostal (in′-ter-KOS-tal)	Nine pairs of **posterior intercostal arteries** supply the intercostal, pectoral, and abdominal muscles; overlying subcutaneous tissue and skin; mammary glands; and vertebral canal and its contents.
Subcostal (SUB-kos-tal)	The left and right **subcostal arteries** have a distribution similar to that of the posterior intercostals.
Superior phrenic (FREN-ik)	Small **superior phrenic arteries** supply the posterior and superior surfaces of the diaphragm.

EXHIBIT 21.8

ABDOMINAL AORTA (Fig. 21.23)

Overview: The **abdominal (ab-DOM-i-nal) aorta** is the continuation of the thoracic aorta. It begins at the aortic hiatus in the diaphragm and ends at about the level of the fourth lumbar vertebra where it divides into right and left common iliac arteries. The abdominal aorta lies anterior to the vertebral column.

As with the thoracic aorta, the abdominal aorta gives off visceral and parietal branches. The unpaired visceral branches arise from the anterior surface of the aorta and include the celiac, superior mesenteric, and inferior mesenteric arteries. The paired visceral branches arise from the lateral surfaces of the aorta and include the suprarenal, renal, and gonadal arteries. The paired parietal branches arise from the posterolateral surfaces of the aorta and include the inferior phrenic and lumbar arteries. The unpaired parietal artery is the middle sacral.

Branch	Description and Region Supplied
VISCERAL	
Celiac (SĒ-lē-ak)	**Celiac artery (trunk)** is first visceral aortic branch below diaphragm. It has three branches: (1) **common hepatic** (he-PAT-ik) **artery,** (2) **left gastric** (GAS-trik) **artery,** and (3) **splenic** (SPLĒN-ik) **artery.**
	The common hepatic artery has three main branches: (1) **hepatic artery proper,** a continuation of the common hepatic artery, which supplies the liver and gallbladder; (2) **right gastric artery,** which supplies the stomach and duodenum; and (3) **gastroduodenal** (gas′-trō-doo′-ō-DĒ-nal) **artery,** which supplies the stomach, duodenum, and pancreas.
	The left gastric artery supplies the stomach, and its **esophageal** (e-sof′-a-JĒ-al) **branch** supplies the esophagus.
	The splenic artery supplies the spleen and has three main branches: (1) **pancreatic** (pan′-krē-AT-ik) **arteries,** which supply the pancreas; (2) **left gastroepiploic** (gas′-trō-ep′-i-PLŌ-ik) **artery,** which supplies the stomach and greater omentum; and (3) **short gastric** (GAS-trik) **arteries,** which supply the stomach.

Exhibit continues

EXHIBIT 21.12 (continued)

VEINS OF UPPER EXTREMITIES (Fig. 21.27)

Vein	Description and Region Drained
DEEP	
Radials (RĀ-dē-als)	**Radial veins** receive **dorsal metacarpal** (met′-a-KAR-pal) **veins.**
Ulnars (UL-nars)	**Ulnar veins** receive tributaries from **palmar venous arch.** Radial and ulnar veins unite in bend of elbow to form brachial veins.
Brachials (BRĀ-kē-als)	Located on either side of brachial arteries, **brachial veins** join into axillary veins.
Axillaries (AK-si-ler′-ēs)	**Axillary veins** are a continuation of brachials and basilics. Axillaries end at first rib, where they become subclavians.
Subclavians (sub-KLĀ-vē-ans)	Right and left **subclavian veins** unite with internal jugulars to form brachiocephalic veins. Thoracic duct of lymphatic system delivers lymph into left subclavian vein at junction with internal jugular. Right lymphatic duct delivers lymph into right subclavian vein at corresponding junction (see Fig. 22.4).

SCHEME OF DRAINAGE

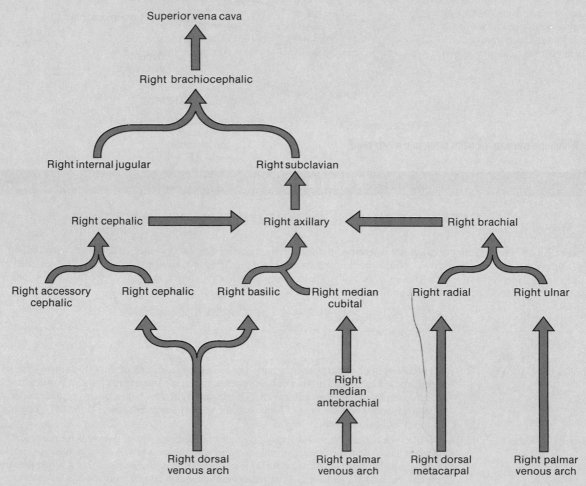

FIGURE 21.27 Principal veins of the right upper extremity.

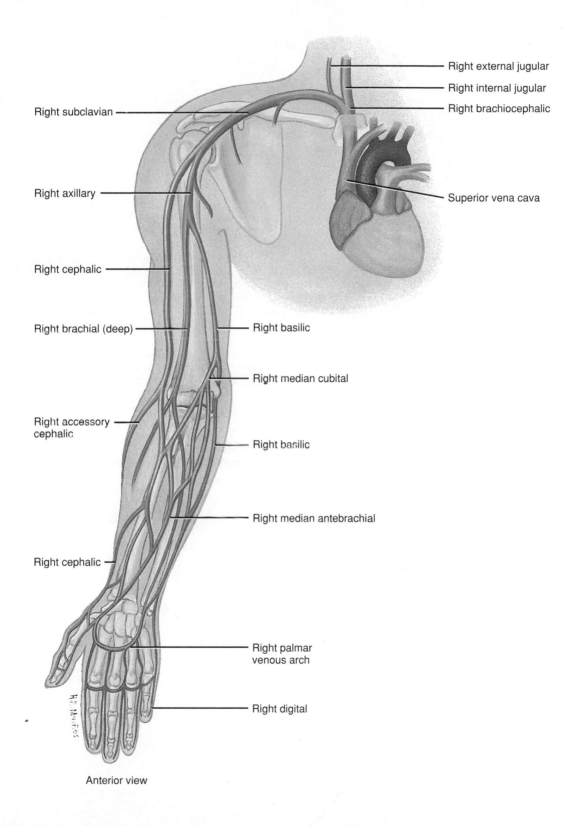

Right external jugular

Right internal jugular

Right brachiocephalic

Right subclavian

Right axillary

Superior vena cava

Right cephalic

Right brachial (deep)

Right basilic

Right median cubital

Right accessory cephalic

Right basilic

Right median antebrachial

Right cephalic

Right palmar venous arch

Right digital

Anterior view

Question: Which vein is frequently used for removal of a blood sample from the upper extremity?

EXHIBIT 21.13

VEINS OF THE THORAX (Fig. 21.28)

Overview: Although brachiocephalic veins drain some portions of the thorax, most thoracic structures are drained by a network of veins called the **azygos system.** This is a network of veins on each side of vertebral column: azygos, hemiazygos, and accessory hemiazygos. They show considerable variation in origin, course, tributaries, anastomoses, and termination. Ultimately, they empty into the superior vena cava.

Vein	Description and Region Drained
Brachiocephalic (brā′-kē-ō-se-FAL-ik)	Right and left **brachiocephalic veins,** formed by union of subclavians and internal jugulars, drain blood from head, neck, upper extremities, mammary glands, and upper thorax. Brachiocephalics unite to form superior vena cava.
Azygos (az-Ī-gos) **veins**	**Azygos veins,** besides collecting blood from thorax, may serve as bypass for inferior vena cava that drains blood from lower body. Several small veins directly link azygos veins with inferior vena cava. Large veins that drain lower extremities and abdomen dump blood into azygos. If inferior vena cava or hepatic portal vein becomes obstructed, azygos veins can return blood from lower body to superior vena cava.
Azygos	**Azygos vein** lies in front of vertebral column, slightly right of midline. It begins as continuation of right ascending lumbar vein. It connects with inferior vena cava, right common iliac, and lumbar veins. Azygos receives blood from **left intercostal** (in′-ter-KOS-tal) **veins** that drain chest muscles; from hemiazygos and accessory hemiazygos veins; from several **esophageal** (e-sof′-a-JĒ-al)**, mediastinal** (mē-dē-a-STĪ-nal)**,** and **pericardial** (per′-i-KAR-dē-al) **veins;** and from right **bronchial** (BRONG-kē-al) **vein.** Vein ascends to fourth thoracic vertebra, arches over right lung, and empties into superior vena cava.
Hemiazygos (HEM-ē-az-ī-gos)	**Hemiazygos vein** is anterior to vertebral column and slightly left of midline. It begins as continuation of left ascending lumbar vein. It receives blood from lower four or five left **intercostal** (in′-ter-KOS-tal) **veins** and some **esophageal** (e-sof′-a-JĒ-al) and **mediastinal** (mē-dē-a-STĪ-nal) **veins.** At level of ninth thoracic vertebra, it joins azygos vein.
Accessory hemiazygos	**Accessory hemiazygos vein** is also anterior and to left of vertebral column. It receives blood from three or four of the superior left **intercostal** (in′-ter-KOS-tal) **veins** and left **bronchial** (BRONG-kē-al) **vein.** It joins azygos at level of eighth thoracic vertebra.

EXHIBIT 21.14

VEINS OF THE ABDOMEN AND PELVIS (Fig. 21.28)

Overview: Blood from the abdominopelvic viscera and abdominal wall returns to the heart via the **inferior vena cava.** Many small veins enter the inferior vena cava. Most carry return flow from parietal branches of the abdominal aorta and their names correspond to the names of the arteries. The inferior vena cava does not receive veins from the gastrointestinal tract, spleen, pancreas, and gallbladder. These organs pass their blood into a common vein, the hepatic portal vein, which delivers the blood to the liver. This special flow of venous blood is called **hepatic portal circulation,** which is described shortly. From here, the blood drains into the hepatic veins, which enter the inferior vena cava.

Vein	Description and Region Drained
Inferior vena cava (VĒ-na CA-va)	**Inferior vena cava** is formed by union of two common iliac veins that drain lower extremities and abdomen. Inferior vena cava extends superiorly through abdomen and thorax to right atrium.
Common iliacs (IL-ē-aks)	**Common iliac veins** are formed by union of internal (hypogastric) and external iliac veins and represent distal continuation of inferior vena cava at its bifurcation.
Internal iliacs	Tributaries of **internal iliac (hypogastric) veins** basically correspond to branches of internal iliac arteries. Internal iliacs drain gluteal muscles, medial side of thigh, urinary bladder, rectum, prostate gland, ductus (vas) deferens, uterus, and vagina.
External iliacs	**External iliac veins** are continuation of femoral veins and receive blood from lower extremities and inferior part of anterior abdominal wall.

Exhibit continues

FIGURE 21.28 Principal veins of the thorax, abdomen, and pelvis.

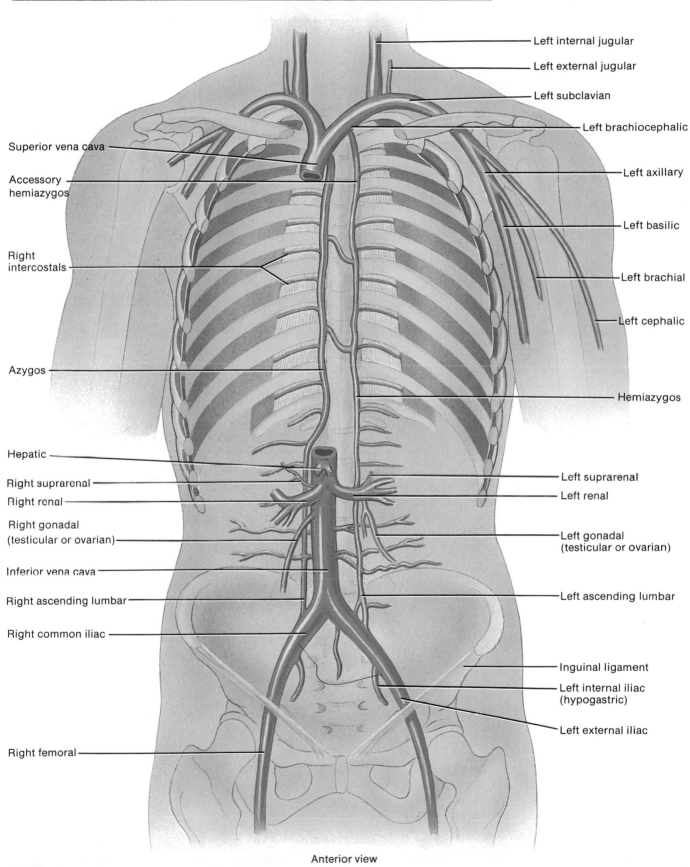

Anterior view

Question: Which vein returns blood from the abdominopelvic viscera to the heart?

EXHIBIT 21.14 **(continued)**

VEINS OF THE ABDOMEN AND PELVIS (Fig. 21.28)

Vein	Description and Region Drained
Renals (RĒ-nals)	**Renal veins** drain kidneys.
Gonadals (gō-NAD-als) [**testicular** (tes-TIK-yoo-lor) or (ō-VA-rē-an) **ovarian**]	**Testicular veins** drain testes (left testicular vein empties into left renal vein and right testicular drains into inferior vena cava); **ovarian veins** drain ovaries (left ovarian vein empties into left renal vein and right ovarian drains into inferior vena cava).
Suprarenals (soo′-pra-RĒ-nals)	**Suprarenal veins** drain adrenal (suprarenal) glands (left suprarenal vein empties into left renal vein).
Inferior phrenics (FREN-iks)	**Inferior phrenic veins** drain diaphragm (left inferior phrenic vein sends tributary to left renal vein).
Hepatics (he-PAT-ik)	**Hepatic veins** drain liver.
Lumbars (LUM-bars)	A series of parallel **lumbar veins** drain blood from both sides of posterior abdominal wall. Lumbars connect at right angles with right and left **ascending lumbar veins,** which form origin of corresponding azygos or hemiazygos vein. Lumbars drain blood into ascending lumbars and then run to inferior vena cava, where they release remainder of flow.

EXHIBIT 21.15

VEINS OF THE LOWER EXTREMITIES (Fig. 21.29)

Overview: Blood from each lower extremity is drained by **superficial** and **deep veins.** The superficial veins often anastomose with each other and with deep veins along their length. Deep veins, for the most part, have the same names as their accompanying arteries.

Vein	Description and Region Drained
SUPERFICIAL VEINS	
Great saphenous (sa-FĒ-nus)	**Great saphenous vein,** longest vein in body, begins at medial end of **dorsal venous arch** of foot. It passes anterior to medial malleolus and then upward along medial aspect of leg and thigh just deep to the skin. It receives tributaries from superficial tissues and connects with deep veins as well. It empties into femoral vein in groin. The great saphenous vein is frequently used for prolonged administration of intravenous fluids. This is particularly important in very young babies and in patients of any age who are in shock and whose veins are collapsed. It and the small saphenous vein are subject to varicosity.
Small saphenous	**Small saphenous vein** begins at lateral end of dorsal venous arch of foot. It passes behind lateral malleolus and ascends under skin of back of leg. It receives blood from foot and posterior portion of leg. It empties into popliteal vein behind knee.
DEEP VEINS	
Posterior tibial (TIB-ē-al)	**Posterior tibial vein** is formed by union of **medial** and **lateral plantar** (PLAN-tar) **veins** behind medial malleolus. It ascends deep in muscle at back of leg, receives blood from **peroneal** (per′-ō-NĒ-al) **vein,** and unites with anterior tibial vein just below knee.
Anterior tibial	**Anterior tibial vein** is upward continuation of **dorsalis pedis** (PED-is′) **veins** in foot. It runs between tibia and fibula and unites with posterior tibial to form popliteal vein.
Popliteal (pop′-li-TĒ-al)	**Popliteal vein,** just behind knee, receives blood from anterior and posterior tibials and small saphenous vein.
Femoral (FEM-o-ral)	**Femoral vein** is upward continuation of popliteal just above knee. Femorals run up posterior surface of thighs and drain deep structures of thighs. After receiving great saphenous veins in groin, they continue as right and left external iliac veins.

Exhibit continues

FIGURE 21.29 Principal veins of the pelvis and lower extremities.

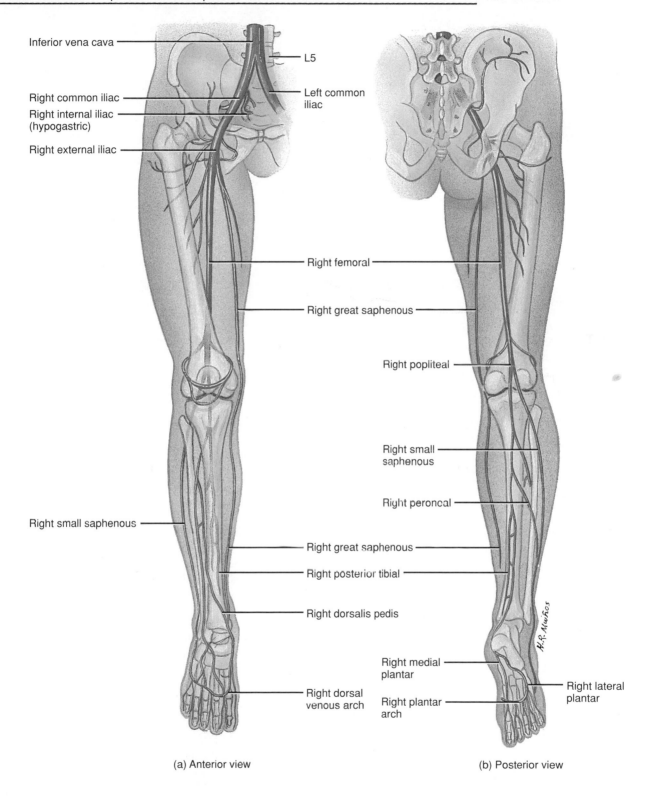

Inferior vena cava

L5

Right common iliac

Left common iliac

Right internal iliac (hypogastric)

Right external iliac

Right femoral

Right great saphenous

Right popliteal

Right small saphenous

Right peroneal

Right small saphenous

Right great saphenous

Right posterior tibial

Right dorsalis pedis

Right medial plantar

Right lateral plantar

Right dorsal venous arch

Right plantar arch

(a) Anterior view

(b) Posterior view

Question: Which of the veins of the lower extremity are superficial?

EXHIBIT 21.15 (Continued)

VEINS OF THE LOWER EXTREMITIES (Fig. 21.29)

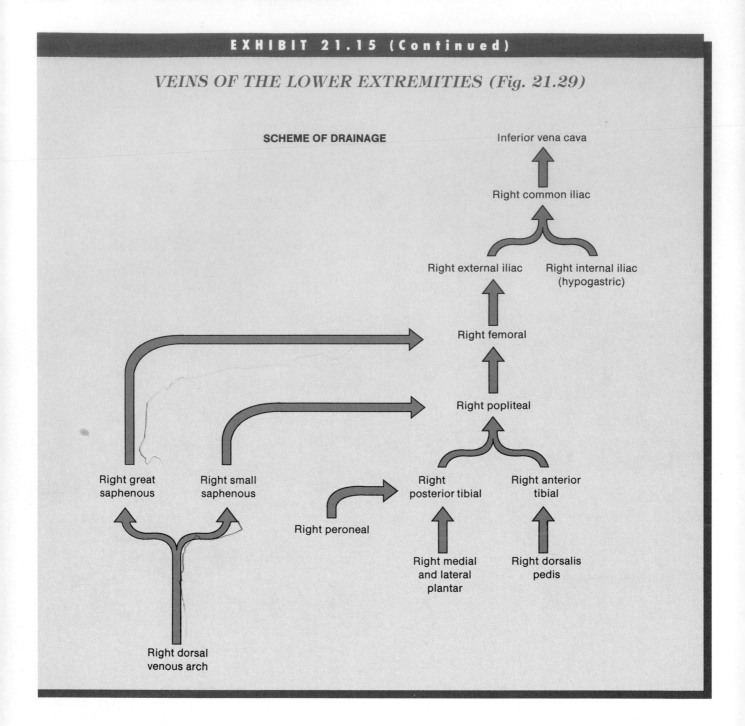

SCHEME OF DRAINAGE

Hepatic Portal Circulation

The **hepatic portal circulation** detours venous blood from the gastrointestinal organs and spleen through the liver before it returns to the heart (Fig. 21.30). A *portal system* carries blood between two capillary networks, in this case from capillaries of the gastrointestinal tract to sinusoids of the liver. After a meal, hepatic portal blood is rich with absorbed substances. The liver stores some and modifies others before they pass into the general circulation. For example, the liver converts glucose into glycogen for storage. It also modifies other digested substances so they may

be used by cells, detoxifies harmful substances that have been absorbed by the gastrointestinal tract, and destroys bacteria by phagocytosis.

The hepatic portal system includes the veins that drain blood from the pancreas, spleen, stomach, intestines, and gallbladder and transport it to the hepatic portal vein of the liver. The **hepatic portal vein** is formed by the union of the superior mesenteric and splenic veins. The **superior mesenteric vein** drains blood from the small intestine and portions of the large intestine and stomach. The **splenic vein** drains the spleen and receives tributaries from the stomach, pancreas, and portions of the colon. The tributaries from the stomach are the **gastric, pyloric,** and **gastroepiploic veins.** The **pancreatic veins** come from the pancreas, and the **inferior mesenteric veins** come from the portions of the colon. Before the hepatic portal vein enters the liver, it receives the **cystic vein** from the gallbladder and other veins.

At the same time the liver receives deoxygenated blood via the hepatic portal system, it also receives oxygenated blood from the systemic circulation via the hepatic artery. Ultimately, all blood leaves the liver through the **hepatic veins,** which drain into the inferior vena cava.

Pulmonary Circulation

The pulmonary circulation carries deoxygenated blood from the right ventricle to the air sacs of the lungs and returns oxygenated blood from the lungs to the left atrium (Fig. 21.31). The **pulmonary trunk** emerges from the right ventricle and passes superiorly, posteriorly, and to the left. It then divides into two branches: the **right pulmonary artery** extends to the right lung; the **left pulmonary artery** goes to the left lung. The pulmonary arteries are the only postnatal arteries that carry deoxygenated blood. On entering the lungs, the branches divide and subdivide until finally they form capillaries around the alveoli (air sacs) in the lungs. Carbon dioxide passes from the blood into the alveoli and is exhaled. Inhaled oxygen passes from the alveoli into the blood. The pulmonary capillaries unite, form venules and veins, and eventually two **pulmonary veins** exit from each lung and transport the oxygenated blood to the left atrium. The pulmonary veins are the only postnatal veins that carry oxygenated blood. Contractions

FIGURE 21.30 Hepatic portal circulation. The scheme of blood flow through the liver, including arterial circulation, is shown in (b). Deoxygenated blood is indicated in blue; oxygenated blood in red.

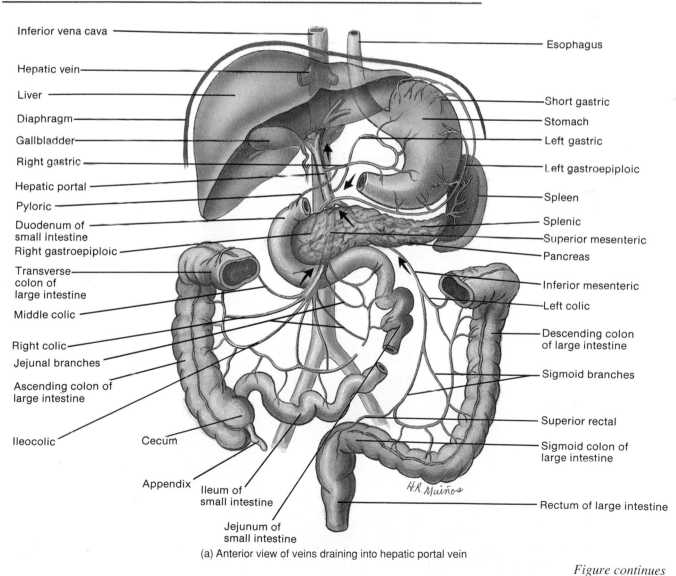

(a) Anterior view of veins draining into hepatic portal vein

Figure continues

FIGURE 21.30 (continued)

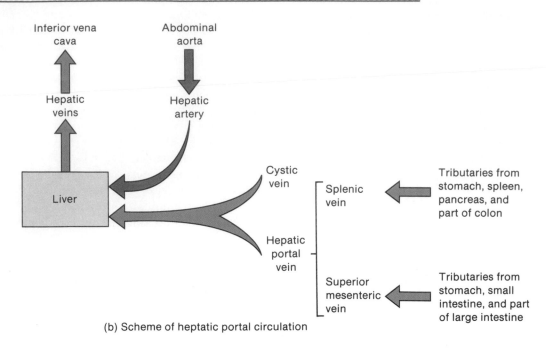

(b) Scheme of heptatic portal circulation

Question: Which veins carry blood away from the liver?

FIGURE 21.31 Pulmonary circulation.

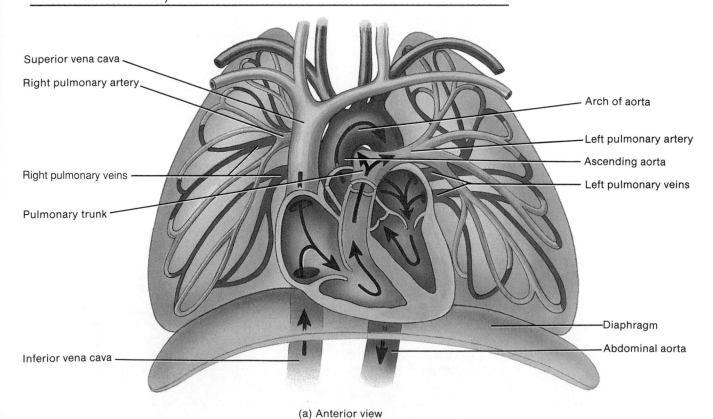

(a) Anterior view

FIGURE 21.31   (continued)

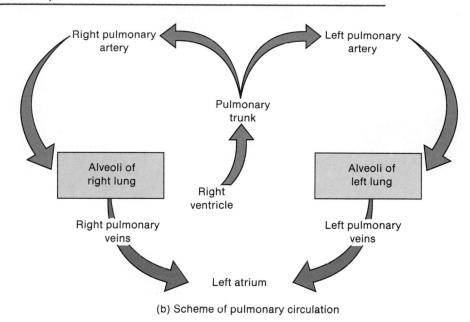

(b) Scheme of pulmonary circulation

Question: Which are the only postnatal arteries to carry deoxygenated blood?

of the left ventricle then send the blood into the systemic circulation.

The pulmonary and systemic circulations are different in several ways. For one thing, blood in pulmonary circulation is not pumped so far as in systemic circulation. Also, in comparison to systemic arteries, pulmonary arteries have larger diameters, thinner walls, and less elastic tissue. As a result, the resistance to blood flow is very low. This means that less pressure is needed to move blood through the lungs. The peak systolic pressure achieved by the right ventricle is only about one-fifth that reached by the left ventricle.

Because resistance in the pulmonary circulation is low, normal *pulmonary* capillary hydrostatic pressure, the principal force that moves fluid out of capillaries into interstitial fluid, is only 10 mm Hg. This compares to the average *systemic* capillary pressure of about 25 mm Hg. The relatively low capillary hydrostatic pressure tends to prevent pulmonary edema. If capillary blood pressure increases (due to increased left atrial pressure as may occur in mitral valve stenosis) or capillary permeability increases (as may occur from bacterial toxins), however, edema may develop. Pulmonary edema reduces the rate of diffusion of oxygen and carbon dioxide and thus slows the exchange of respiratory gases in the lungs.

Another important difference between the pulmonary and systemic circulations is their autoregulatory response to changes in oxygen level. In the systemic circulation, blood vessels *dilate* in response to low oxygen concentration. In the pulmonary circulation, blood vessels *constrict* in

response to low levels of oxygen. This mechanism is very important in distributing blood to areas of the lungs where it can pick up the most oxygen. For example, if some alveoli are not well ventilated by fresh air, the blood vessels in the affected area constrict and blood will largely bypass the poorly functioning areas. As a result, most of the blood flows to other areas of the lung that are well-ventilated and have plenty of oxygen.

Fetal Circulation

The circulatory system of a fetus, called **fetal circulation,** differs from the postnatal (after birth) circulation because the lungs, kidneys, and gastrointestinal tract of a fetus are nonfunctional. The fetus derives its oxygen and nutrients from the maternal blood and eliminates its carbon dioxide and wastes into the maternal blood (Fig. 21.32a).

The exchange of materials between fetal and maternal circulation occurs through a structure called the **placenta** (pla-SEN-ta). It is attached to the umbilicus (navel) of the fetus by the umbilical (um-BIL-i-kal) cord, and it communicates with the mother through countless small blood vessels that emerge from the uterine wall. The umbilical cord contains blood vessels that branch into capillaries in the placenta. Wastes from the fetal blood diffuse out of the capillaries, into spaces containing maternal blood (intervillous spaces) in the placenta, and finally into the mother's uterine blood vessels (see Fig. 29.7a). Nutrients travel the opposite route—from the maternal blood vessels to the intervillous

FIGURE 21.32 Fetal circulation and changes at birth. The boxed areas indicate the fate of certain fetal structures once postnatal circulation is established.

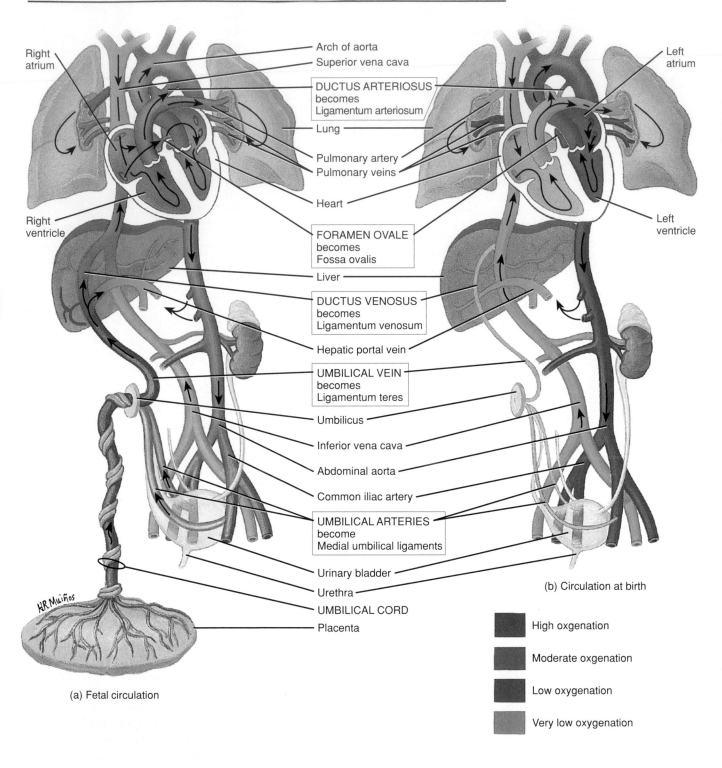

Right atrium

Left atrium

Arch of aorta

Superior vena cava

DUCTUS ARTERIOSUS becomes Ligamentum arteriosum

Lung

Pulmonary artery

Pulmonary veins

Heart

Right ventricle

Left ventricle

FORAMEN OVALE becomes Fossa ovalis

Liver

DUCTUS VENOSUS becomes Ligamentum venosum

Hepatic portal vein

UMBILICAL VEIN becomes Ligamentum teres

Umbilicus

Inferior vena cava

Abdominal aorta

Common iliac artery

UMBILICAL ARTERIES become Medial umbilical ligaments

Urinary bladder

Urethra

UMBILICAL CORD

Placenta

(a) Fetal circulation

(b) Circulation at birth

High oxgenation

Moderate oxgenation

Low oxgenation

Very low oxygenation

FIGURE 21.32 (continued)

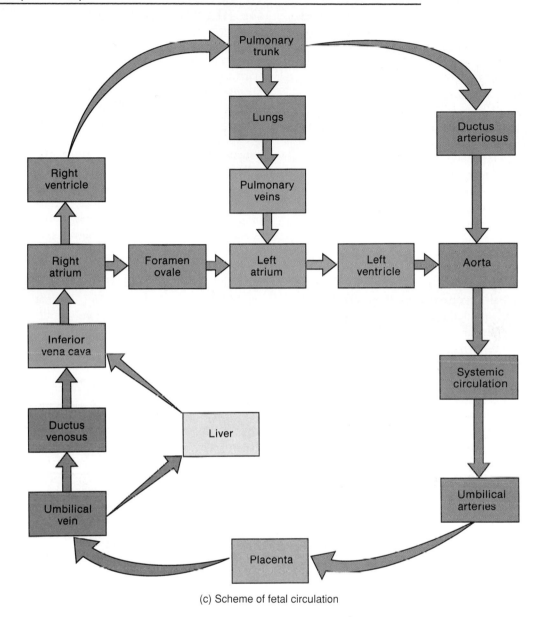

(c) Scheme of fetal circulation

Question: Through which structure does the exchange of materials between mother and fetus occur?

spaces to the fetal capillaries. Normally, there is no direct mixing of maternal and fetal blood since all exchanges occur through capillary walls.

Blood passes from the fetus to the placenta via two **umbilical arteries.** These branches of the internal iliac (hypogastric) arteries are within the umbilical cord. At the placenta, fetal blood picks up oxygen and nutrients and eliminates carbon dioxide and wastes. The oxygenated blood returns from the placenta via a single **umbilical vein.** This vein ascends to the liver of the fetus, where it divides into two branches. Some blood flows through the branch that joins the hepatic portal vein and enters the liver. The fetal liver produces red blood cells but does not function in

digestion. Therefore most of the blood flows into the second branch, the **ductus venosus** (DUK-tus ve-NŌ-sus). The ductus venosus eventually drains into the inferior vena cava, bypassing the liver.

Circulation through other portions of the fetus is similar to postnatal circulation. Deoxygenated blood returning from the lower regions mingles with oxygenated blood from the ductus venosus in the inferior vena cava. This mixed blood then enters the right atrium. Deoxygenated blood returning from the upper regions of the fetus enters the superior vena cava and passes into the right atrium.

Most of the fetal blood does not pass from the right ventricle to the lungs, as it does in postnatal circulation, since the fetal lungs do not operate. In the fetus, an opening called the **foramen ovale** (fō-RĀ-men ō-VAL-ē) exists in the septum between the right and left atria. About a third of the blood passes through the foramen ovale directly into the systemic circulation. The blood that does pass into the right ventricle is pumped into the pulmonary trunk, but little of this blood reaches the lungs. Most is sent through the **ductus arteriosus** (ar-tē-rē-Ō-sus). This small vessel connecting the pulmonary trunk with the aorta enables most blood to bypass the fetal lungs. The blood in the aorta is carried to all parts of the fetus through the systemic circulation. When the common iliac arteries branch into the external and internal iliacs, part of the blood flows into the internal iliacs. It then goes to the umbilical arteries and back to the placenta for another exchange of materials. The only fetal vessel that carries fully oxygenated blood is the umbilical vein.

At birth, when pulmonary (lung), renal, digestive, and liver functions are established, the following vascular changes occur (Fig. 21.32b).

1. The umbilical arteries vasoconstrict shut and atrophy to become the **medial umbilical ligaments.**
2. The umbilical vein vasoconstricts shut and becomes the **ligamentum teres (round ligament)** of the liver.
3. The placenta is expelled as the **"afterbirth."**
4. The ductus venosus vasoconstricts shut and becomes the **ligamentum venosum,** a fibrous cord in the liver.
5. The foramen ovale normally closes shortly after birth to become the **fossa ovalis,** a depression in the interatrial septum.
6. The ductus arteriosus closes by vasoconstriction and atrophies and becomes the **ligamentum arteriosum.**

Anatomical defects resulting from failure of these changes to occur are illustrated in Fig. 20.16.

AGING AND THE CARDIOVASCULAR SYSTEM

General changes associated with aging and the cardiovascular system include loss of compliance (extensibility) of the aorta, reduction in cardiac muscle fiber (cell) size, progressive loss of cardiac muscular strength, reduced cardiac output, a decline in maximum heart rate, and an increase in systolic blood pressure. Total blood cholesterol tends to increase with age, as does low-density lipoprotein (LDL); high-density lipoprotein (HDL) tends to decrease. There is an increase in the incidence of coronary artery disease (CAD), the major cause of heart disease and death in older Americans. Congestive heart failure (CHF), a set of symptoms associated with impaired pumping of the heart, also occurs. Changes in blood vessels that serve brain tissue, such as atherosclerosis, reduce nourishment to the brain and result in the malfunction or death of brain cells. By age 80, cerebral blood flow is 20% less and renal blood flow is 50% less than in the same person at age 30.

DEVELOPMENTAL ANATOMY OF BLOOD AND BLOOD VESSELS

The human egg and yolk sac have little yolk to nourish the developing embryo, and blood and blood vessel formation starts as early as 15 to 16 days. The development begins in the **mesoderm** of the yolk sac, chorion, and body stalk.

Blood vessels develop from isolated masses and cords of mesenchyme in the mesoderm called **blood islands** (Fig. 21.33). Spaces soon appear in the islands and become the lumens of the blood vessels. Some of the mesenchymal cells immediately around the spaces give rise to the *endothelial lining of the blood vessels*. Mesenchyme around the endothelium forms the *tunics* (intima, media, externa) of the larger blood vessels. Growth and fusion of blood islands form an extensive network of blood vessels throughout the embryo.

Blood plasma and *blood cells* are produced by the endothelial cells and appear in the blood vessels of the yolk sac and allantois quite early. Blood formation in the embryo itself begins at about the second month in the liver and spleen, a little later in bone marrow, and much later in lymph nodes.

FIGURE 21.33 Development of blood vessels and blood cells from blood islands.

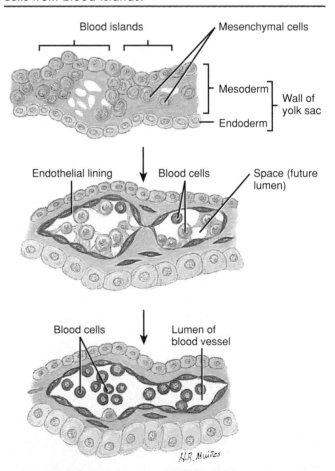

DISORDERS: HOMEOSTATIC IMBALANCES

HYPERTENSION

Hypertension, or high blood pressure, is the most common disease affecting the heart and blood vessels. Statistics suggest that hypertension afflicts two out of every five American adults. Although there is some disagreement on the definition of hypertension, a strong consensus has emerged suggesting that a blood pressure of 120/80 is normal and desirable in a healthy adult. Mild high blood pressure is diastolic pressure between 90 and 104. Moderate high blood pressure is diastolic pressure between 105 and 114. Severe high blood pressure is diastolic pressure of 115 or higher. Isolated systolic hypertension is systolic pressure greater than 140 in those whose diastolic pressure is less than 90. As noted in Chapter 20, the lower the blood pressure, the lower the risk of coronary artery disease (CAD).

Primary (essential) hypertension is a persistently elevated blood pressure that cannot be attributed to any particular organic cause. Approximately 90 to 95% of all hypertension cases fit this definition. The remaining 5 to 10% are **secondary hypertension.** Secondary hypertension has an identifiable underlying cause such as kidney disease or adrenal hypersecretion. Kidney disease and obstruction of blood flow may cause the kidneys to release renin into the blood. This enzyme catalyzes the formation of angiotensin II from angiotensinogen, a plasma protein. Angiotensin II is a powerful vasoconstrictor—and the most potent agent known for raising blood pressure. It also stimulates aldosterone release. Aldosteronism, the hypersecretion of aldosterone due to a tumor of the adrenal cortex, may also cause hypertension.

Pheochromocytoma (fē-ō-krō′-mō-sī-TŌ-ma) is a tumor of the adrenal medulla. It produces and releases into the blood large quantities of norepinephrine and epinephrine. Both hormones raise blood pressure. Epinephrine causes an increase in heart rate and force of contraction and norepinephrine causes vasoconstriction.

High blood pressure is of considerable concern because of the harm it can do to the heart, brain, and kidneys if it remains uncontrolled. The heart is most commonly affected by high blood pressure. When pressure is high, the heart uses more energy to pump against the increased afterload caused by the elevated arterial blood pressure. Because of the increased effort, the heart muscle thickens and the heart becomes enlarged. The heart also needs more oxygen. If it cannot meet the demands put on it, angina pectoris or even myocardial infarction may develop. Hypertension is also a factor in the development of atherosclerosis. Continued high blood pressure may produce a cerebral vascular accident (CVA), or stroke. Brain arteries are usually less protected by surrounding tissues than are the major arteries in other parts of the body. With prolonged hypertension, they may finally rupture, and a brain hemorrhage follows.

The kidneys are also prime targets of hypertension. The principal site of damage is in the arterioles that supply them. The continual high blood pressure pushing against the walls of the arterioles causes them to thicken, thus narrowing the lumen. The blood supply to the kidneys is thereby gradually reduced. In response, the kidneys may secrete renin, which raises the blood pressure even higher and complicates the problem. The reduced blood flow to the kidney cells may eventually lead to the death of the cells.

Although there is presently no cure for primary hypertension, certain forms of secondary hypertension, if diagnosed early, can be cured by removing the underlying cause. And even in some cases of primary hypertension, certain measures may result in a dramatic reduction in blood pressure. For example, a decrease in sodium intake, an increase in calcium and potassium intake, a decrease in alcohol consumption, weight reduction, stopping smoking, an increase in physical activity, and stress management can help reduce high blood pressure.

For those people who need medication, several types of drugs are available. Since they have different mechanisms of action, combinations of two drugs often are prescribed. Many people are successfully treated with *diuretics,* which decrease blood volume by promoting elimination of water in the urine and thus reduce blood pressure. *ACE* (angiotensin converting enzyme) *inhibitors* block formation of angiotensin II and thereby promote vasodilation and decrease liberation of aldosterone. *Beta blockers* reduce blood pressure by inhibiting secretion of renin and decreasing rate and force of heartbeat. *Vasodilators* relax the smooth muscle in arterial walls, causing vasodilation and thus lowering blood pressure by lowering systemic vascular resistance. An important category of vasodilators are the *calcium channel blockers.* They slow the inflow of Ca^{2+} into vascular smooth muscle fibers. They also reduce the heart's work load by slowing Ca^{2+} entry into myocardial fibers.

ANEURYSM

An **aneurysm** (AN-yoo-rizm) is a thin, weakened section of the wall of an artery or a vein that bulges outward, forming a balloonlike sac of the blood vessel. Common causes of aneurysms include atherosclerosis, syphilis, congenital blood vessel defects, and trauma. If an aneurysm goes untreated, it grows larger and larger until the blood vessel wall becomes so thin that it bursts. The result is massive hemorrhage with shock, severe pain, stroke, or death, depending on which vessel is involved. Even an unruptured aneurysm can lead to damage by interrupting blood flow or putting pressure on adjacent blood vessels, organs, or bones.

Surgical repair consists of temporarily clamping the damaged artery above and below the aneurysm and surgically removing it. A graft, usually of Dacron, is then sutured to healthy segments of the artery to reestablish normal blood flow.

CORONARY ARTERY DISEASE

Coronary artery disease (CAD) is a condition in which the heart muscle receives inadequate oxygen due to blockage of its blood supply. Depending on the extent of blockage, symptoms can range from a mild chest pain (angina pectoris) to a full-scale heart attack. The underlying causes of CAD are many and varied. Two of the principal ones are atherosclerosis and coronary artery spasm, both of which are discussed in detail on page 616.

DEEP-VENOUS THROMBOSIS

Venous thrombosis, the presence of a thrombus in a vein, usually occurs in deep veins of the lower extremities. This condition is called **deep-venous thrombosis (DVT)** and has two serious complications. The first is *pulmonary embolism,* in which the thrombus dislodges and finds its way into the pulmonary arterial blood flow. The second is *postphlebitic syndrome,* which consists of edema, pain, and skin changes due to destruction of venous valves. Diagnosis is by ultrasound and treatment consists of anticoagulant therapy, elevation of the extremity, fibrinolytic therapy (streptokinase or t-PA), and sometimes thrombectomy.

MEDICAL TERMINOLOGY

Angiogenesis (an′-jē-ō-JEN-e-sis) Formation of new blood vessels.

Aortography (ā′-or-TOG-ra-fē) X-ray examination of the aorta and its main branches after injection of radiopaque dye.

Arteritis (ar′-te-RĪ-tis; *itis* = inflammation of) Inflammation of an artery, probably due to an autoimmune response.

Carotid endarterectomy (ka-ROT-id end′-ar-ter-EK-tō-mē) The removal of atherosclerotic plaque from the carotid artery to restore greater blood flow to the brain.

Claudication (klaw′-di-KĀ-shun) Pain and lameness or limping caused by defective circulation of the blood in the vessels of the limbs.

Occlusion (o-KLOO-shun) The closure or obstruction of the lumen of a structure such as a blood vessel.

Orthostatic (or′-thō-STAT-ik) **hypotension** (*ortho* = straight; *statikos* = causing to stand) An excessive lowering of systemic blood pressure with the assumption of an erect or semierect posture; it is usually a sign of a disease. May be caused by excessive fluid loss, certain drugs (antihyperten-

sives), and cardiovascular or neurogenic factors. Also called **postural hypotension.**

Phlebitis (fle-BĪ-tis; *phleb* = vein) Inflammation of a vein, often in a leg.

Raynaud's (rā-NOZ) **disease** A vascular disorder, primarily of females, characterized by bilateral attacks of ischemia, usually of the fingers and toes, in which the skin becomes pale and exhibits burning and pain. It is brought on by cold temperatures or emotional stimuli.

Thrombectomy (throm-BEK-tō-mē; *thrombo* = clot) An operation to remove a blood clot from a blood vessel.

Thrombophlebitis (throm′-bō-fle-BĪ-tis) Inflammation of a vein with clot formation. Superficial thrombophlebitis occurs in veins under the skin, especially the calf.

White coat (office) hypertension A syndrome found in patients who have elevated blood pressures while being examined by health-care personnel, but otherwise have normal blood pressure.

Study Outline

Anatomy of Blood Vessels (p. 624)

Arteries (p. 624)

1. Arteries carry blood away from the heart. The wall of an artery consists of a tunica interna, tunica media (which maintains elasticity and contractility), and tunica externa.
2. Large arteries are referred to as elastic (conducting) arteries, and medium-sized arteries are called muscular (distributing) arteries.
3. Many arteries anastomose: the distal ends of two or more vessels unite. An alternate blood route from an anastomosis is called collateral circulation. Arteries that do not anastomose are called end arteries.

Arterioles (p. 625)

1. Arterioles are small arteries that deliver blood to capillaries.
2. Through constriction and dilation, they assume a key role in regulating blood flow from arteries into capillaries and in altering arterial blood pressure.

Capillaries (p. 626)

1. Capillaries are microscopic blood vessels through which materials are exchanged between blood and tissue cells; some capillaries are continuous, whereas others are fenestrated.
2. Capillaries branch to form an extensive capillary network throughout the tissue. This network increases the surface area, allowing a rapid exchange of large quantities of materials.
3. Precapillary sphincters regulate blood flow through capillaries.
4. Microscopic blood vessels in the liver are called sinusoids.

Venules (p. 628)

1. Venules are small vessels that continue from capillaries and merge to form veins.
2. They drain blood from capillaries into veins.

Veins (p. 629)

1. Veins consist of the same three tunics as arteries but have a thinner tunica interna and media and a thicker tunica externa.
2. They contain valves to prevent backflow of blood.
3. Weak valves can lead to varicose veins.
4. Vascular (venous) sinuses are veins with very thin walls.

Blood Distribution (p. 629)

1. Systemic veins are collectively called blood reservoirs.
2. They store blood, which through vasoconstriction can move to other parts of the body, if the need arises.
3. The principal reservoirs are the veins of the abdominal organs (liver and spleen) and skin.

Hemodynamics: Physiology of Circulation (p. 630)

Velocity of Blood Flow (p. 630)

1. The velocity of blood flow is inversely related to the cross-sectional area of blood vessels; blood flows slowest where cross-sectional area is greatest.
2. Blood flow decreases from the aorta to arteries to capillaries and increases as it returns to the heart.

Volume of Blood Flow (p. 630)

1. Blood flow is determined by blood pressure and resistance.
2. Blood flows from regions of higher to lower pressure; the higher the resistance the lower the blood flow.
3. Cardiac output equals the mean aortic blood pressure divided by total resistance (CO = MABP ÷ R).
4. Blood pressure is the pressure exerted on the walls of a blood vessel.
5. Factors that affect blood pressure are cardiac output, blood volume, viscosity, resistance, and the elasticity of arteries.
6. As blood leaves the aorta and flows through systemic circulation, its pressure progressively falls to 0 mm Hg by the time it reaches the right atrium.
7. Resistance depends on blood viscosity, blood vessel length, and blood vessel radius.

Capillary Exchange (p. 632)

1. Substances enter and leave capillaries by diffusion, vesicular transport, and bulk flow.
2. The movement of water and dissolved substances (except proteins) through capillaries is dependent on hydrostatic and osmotic pressures.
3. The near equilibrium at the arterial and venous ends of a capillary by which fluids exit and enter is called Starling's law of the capillaries.
4. Edema is an abnormal increase in interstitial fluid.

Venous Return (p. 634)

1. Venous return depends on pressure differences between the venules and the right atrium.
2. Blood return to the heart is maintained by several factors including skeletal muscular contractions, valves in veins (especially in the extremities), and pressure changes associated with breathing.

Control of Blood Pressure and Blood Flow (p. 635)

Cardiovascular Center (p. 636)

1. The cardiovascular (CV) center is a group of neurons in the medulla that regulates heart rate, contractility, and blood vessel diameter.
2. The CV center receives input from higher brain regions and sensory receptors (baroreceptors and chemoreceptors).
3. Output from the CV center flows along sympathetic and parasympathetic fibers. Sympathetic impulses along cardioaccelerator nerves increase heart rate and contractility and parasympathetic impulses along vagus nerves decrease heart rate.

Neural Regulation (p. 638)

1. The carotid sinus reflex helps regulate blood pressure in the brain.
2. The aortic reflex is concerned with general systemic blood pressure.
3. The right heart reflex responds to increases in venous blood pressure.
4. Chemoreceptors monitor blood levels of oxygen, carbon dioxide, and hydrogen ions.
5. Among the hormones that help regulate blood pressure are epinephrine, norepinephrine, ADH, angiotensin II, and ANP.

Autoregulation (p. 640)

1. Autoregulation refers to local, automatic adjustments of blood flow in a given region to meet a particular tissue's need.

2. Oxygen is the principal (though not direct) stimulus for autoregulation.

Shock and Homeostasis (p. 641)

1. Shock is a failure of the cardiovascular system to deliver adequate amounts of oxygen and nutrients to meet the metabolic needs of cells.
2. Signs and symptoms include clammy, cool, pale skin; tachycardia; weak, rapid pulse; sweating; hypotension; altered mental state; decreased urinary output; thirst, and acidosis.
3. Stages of shock are (1) compensated, in which negative feedback cycles restore homeostasis; (2) decompensated, in which positive feedback cycles intensify the shock and immediate medical intervention is required; and (3) irreversible, in which the cardiovascular system collapses and death results.

Checking Circulation (p. 645)

Pulse (p. 645)

1. Pulse is the alternate expansion and elastic recoil of an artery wall with each heartbeat. It may be felt in any artery that lies near the surface or over a hard tissue.
2. A normal pulse (heart) rate is between 70 and 80 beats per minute.

Measurement of Blood Pressure (BP) (p. 645)

1. Blood pressure is the pressure exerted by blood on the wall of an artery when the left ventricle undergoes systole and then diastole. It is measured by the use of a sphygmomanometer.
2. Systolic blood pressure (SBP) is the force of blood recorded during ventricular contraction. Diastolic blood pressure (DBP) is the force of blood recorded during ventricular relaxation. Normal blood pressure is 120/80 mm Hg.
3. Pulse pressure is the difference between systolic and diastolic pressure. It normally is about 40 mm Hg and provides information about the condition of arteries.

Circulatory Routes (p. 646)

1. The largest circulatory route is the systemic circulation.
2. Two of the several subdivisions of the systemic circulation are coronary (cardiac) circulation and hepatic portal circulation.
3. Other routes include the cerebral, pulmonary, and fetal circulations.

Systemic Circulation (p. 646)

1. The systemic circulation takes oxygenated blood from the left ventricle through the aorta to all parts of the body, including some lung tissue (but does *not* supply the air sacs of the lungs) and returns the deoxygenated blood to the right atrium.
2. The aorta is divided into the ascending aorta, the arch of the aorta, and the descending aorta. Each section gives off arteries that branch to supply the whole body.
3. Blood returns to the heart through the systemic veins. All the veins of the systemic circulation flow into the superior or inferior venae cavae or the coronary sinus, which in turn empty into the right atrium.

Hepatic Portal Circulation (p. 670)

1. The hepatic portal circulation collects blood from the veins of the pancreas, spleen, stomach, intestines, and gallbladder and directs it into the hepatic portal vein of the liver.
2. This circulation enables the liver to utilize nutrients and detoxify harmful substances in the blood.

Pulmonary Circulation (p. 671)

1. The pulmonary circulation takes deoxygenated blood from the right ventricle to the air sacs of the lungs and returns oxygenated blood from the lungs to the left atrium.
2. It allows blood to be oxygenated for systemic circulation.

Fetal Circulation (p. 673)

1. The fetal circulation involves the exchange of materials between fetus and mother.
2. The fetus derives its oxygen and nutrients and eliminates its carbon dioxide and wastes through the maternal blood supply by means of a structure called the placenta.
3. At birth, when pulmonary (lung), digestive, and liver func-

tions are established, the special structures of fetal circulation are no longer needed.

Aging and the Cardiovascular System (p. 676)

1. General changes include loss of elasticity of blood vessels, reduction in cardiac muscle size, reduced cardiac output, and increased systolic blood pressure.
2. The incidence of coronary artery disease (CAD), congestive heart failure (CHF), and atherosclerosis increases with age.

Developmental Anatomy of Blood and Blood Vessels (p. 676)

1. Blood vessels develop from isolated masses of mesenchyme in mesoderm called blood islands.
2. Blood is produced by the endothelium of blood vessels.

Review Questions

1. Describe the structural and functional differences among arteries, arterioles, capillaries, venules, and veins. (p. 624)
2. Discuss the importance of the elasticity and contractility of arteries. (p. 624)
3. Distinguish between elastic (conducting) and muscular (distributing) arteries in terms of location, histology, and function. What is an anastomosis? What is collateral circulation? (p. 624)
4. Describe how capillaries are structurally adapted for exchanging materials between blood and body cells. (p. 626)
5. Define varicose veins. (p. 635)
6. What are blood reservoirs? Why are they important? (p. 629)
7. What is velocity of blood flow? Why does blood flow faster in arteries and veins than in capillaries? (p. 630)
8. Explain how blood pressure and resistance determine volume of blood flow. (p. 630)
9. Summarize the factors that affect blood pressure. (p. 635)
10. Define resistance and explain the factors that contribute to it. (p. 631)
11. Describe the processes by which substances enter and leave capillaries. (p. 632)
12. Describe how hydrostatic and osmotic pressures determine fluid movement through capillaries. Set up the equation for net filtration pressure (NFP) to substantiate your response. How does edema develop? (p. 632)
13. What is Starling's law of the capillaries? (p. 632)
14. Describe the factors that assist the return of venous blood to the heart. (p. 634)
15. What is the cardiovascular center? What are the principal inputs and outputs? (p. 636)
16. Describe the operation of the carotid sinus, aortic, and right heart reflex. (p. 638)
17. Explain the role of chemoreceptors in the regulation of blood pressure. (p. 638)
18. Describe the hormonal regulation of blood pressure. (p. 640)
19. What is autoregulation? Explain how it occurs through physical and chemical changes in blood. (p. 640)
20. Define shock. What are its signs and symptoms? (p. 641)
21. Describe the three stages of shock. (p. 642)
22. Define pulse. Where may pulse be felt? Contrast tachycardia and bradycardia. (p. 645)

23. What is blood pressure (BP)? Describe how systolic and diastolic blood pressures are recorded by means of a sphygmomanometer. (p. 645)
24. Compare the clinical significance of systolic and diastolic pressures. How are these pressures written? (p. 645)
25. What is meant by a circulatory route? Define systemic circulation. (p. 646)
26. Diagram the major divisions of the aorta, their principal arterial branches, and the regions supplied. (p. 648)
27. Trace a drop of blood from the arch of the aorta through its systemic circulatory route to the tip of the big toe on your left foot and back to the heart again. Remember that the major branches of the arch are the brachiocephalic artery, left common carotid artery, and left subclavian artery. Be sure to also indicate which veins return the blood to the heart. (pp. 648–670)
28. What is the cerebral arterial circle (circle of Willis)? Why is it important? (p. 652)
29. What are visceral branches of an artery? Parietal branches? What major organs are supplied by branches of the thoracic aorta? How is blood returned from these organs to the heart? (p. 655)
30. What organs are supplied by the celiac, superior mesenteric, renal, inferior mesenteric, inferior phrenic, and middle sacral arteries? How is blood returned to the heart? (p. 655)
31. Trace a drop of blood from the brachiocephalic artery into the digits of the right upper extremity and back again to the right atrium. (p. 652)
32. What are the three major groups of systemic veins? (p. 660)
33. What is hepatic portal circulation? Describe the route by means of a diagram. Why is this route significant? (p. 670)
34. Define pulmonary circulation. Prepare a diagram to indicate the route. What is the purpose of the route? (p. 671)
35. Discuss in detail the anatomy and physiology of fetal circulation. Be sure to indicate the function of the umbilical arteries, umbilical vein, ductus venosus, foramen ovale, and ductus arteriosus. (p. 673)
36. Describe the effects of aging on the cardiovascular system. (p. 676)
37. Describe the development of blood vessels and blood. (p. 676)

Answers to Questions with Figures

21.1 Artery; vein.

21.2 Since less energy is stored in the elastic arteries during systole, the heart has to pump harder to maintain the same rate of blood flow.

21.3 Contraction of smooth muscle in the tunica media.

21.4 They use oxygen and produce wastes more rapidly than inactive tissues.

21.5 Through intercellular clefts, via pinocytic vesicles, directly across endothelial membranes, and through fenestrations.

21.6 Due to the effect of gravity when you are standing, blood tends to pool in your leg veins. The valves prevent backflow as the blood is pushed back toward the right atrium after each beat of the left ventricle. Gravity aids the flow of blood in neck veins back toward the heart when you are erect.

21.7 Volume in veins about 60% x 5 liters = 3 liters; volume in capillaries about 5% x 5 liters = 250 ml.

21.8 Capillaries and venules; arteries.

21.9 About 93 mm Hg.

21.10 Blood colloid osmotic pressure will be lower than normal and therefore capillary reabsorption will be low. The result will be edema (see Clinical Application on page 634).

21.11 Skeletal muscle and respiratory pumps.

21.12 Arteriole by vasodilation and vasoconstriction.

21.13 Cardiac muscle in the heart and smooth muscle in blood vessel walls.

21.14 Impulses pass from baroreceptors in the carotid sinuses via the glossopharyngeal (IX) nerves and from the arch of the aorta via the vagus (X) nerves to the CV center.

21.15 It represents change from lying down to standing because the gravity causes increased pooling of blood in leg veins when you stand up, and this decreases the blood pressure in your upper body.

21.16 Not necessarily. If systemic vascular resistance has increased a lot, flow may be inadequate.

21.17 Fifteen to 25%.

21.18 Diastolic = 95 mm Hg; systolic = 137 mm Hg; pulse pressure = 42 mm Hg. This person has mild hypertension, since the diastolic pressure is greater than 90 mm Hg.

21.19 Systemic and pulmonary.

21.20 Ascending aorta, arch of aorta, thoracic aorta, and abdominal aorta.

21.21 Coronary.

21.22 Brachiocephalic trunk, left common carotid artery, and left subclavian artery.

21.23 At the level of the intervertebral disc between T4 and T5.

21.24 About the level of L4.

21.25 Coronary sinus, superior vena cava, and inferior vena cava.

21.26 Internal jugular.

21.27 Median cubital.

21.28 Inferior vena cava.

21.29 Dorsal venous arch, great saphenous, and small saphenous.

21.30 Hepatic.

21.31 Pulmonary.

21.32 Placenta.

Chapter 22

THE LYMPHATIC SYSTEM, NONSPECIFIC RESISTANCE TO DISEASE, AND IMMUNITY

Chapter Contents at a Glance

Student Objectives

1. Describe the components of the lymphatic system and list their functions.
2. Discuss how edema develops.
3. Describe the development of the lymphatic system.
4. Discuss the roles of the skin and mucous membranes, antimicrobial substances, phagocytosis, inflammation, and fever in nonspecific resistance to disease.
5. Define immunity and describe how T cells and B cells arise.
6. Explain the relationship between an antigen (Ag) and an antibody (Ab).
7. Describe the roles of antigen presenting cells, T cells, and B cells in cell-mediated and antibody-mediated immunity.
8. Explain how self-tolerance occurs.
9. Discuss the relationship of immunology to cancer.
10. Describe the clinical symptoms of the following disorders: acquired immune deficiency syndrome (AIDS), autoimmune diseases, systemic lupus erythematosus (SLE), chronic fatigue syndrome, severe combined immunodeficiency (SCID), hypersensitivity (allergy), tissue rejection, and Hodgkin's disease (HD).
11. Define medical terminology associated with the lymphatic system.

The **lymphatic** (lim-FAT-ik) **system** consists of a fluid called lymph flowing within lymphatic vessels (lymphatics), several structures and organs that contain lymphatic tissue, and bone marrow, which is the site of lymphocyte production (Fig. 22.1). Interstitial (tissue) fluid and lymph are basically the same. The major difference between the two is location. After fluid passes from interstitial spaces into lymphatic vessels, it is called **lymph** (*lympha* = clear water). Lymphatic tissue is a specialized form of reticular connective tissue that contains large numbers of lymphocytes.

The lymphatic system has several functions.

1. **Draining interstitial fluid.** Lymphatic vessels drain tissue spaces of excess interstitial fluid.
2. **Transporting dietary fats.** Lymphatic vessels carry lipids and lipid-soluble vitamins absorbed by the gastrointestinal tract to the blood.
3. **Protecting against invasion.** Lymphatic tissue carries out **immune responses.** These are highly specific responses targeted to particular invaders or abnormal cells. Lymphocytes, aided by macrophages, recognize foreign cells and substances, microbes, and cancer cells and respond to them in two basic ways. Some lymphocytes (T cells) destroy the intruders directly or indirectly by releasing cytotoxic (cell-killing) substances. Other lymphocytes (B cells) differentiate into plasma cells that secrete antibodies. These are proteins that combine with and cause destruction of specific foreign substances. In carrying out specific immune responses, the lymphatic system concentrates foreign substances in certain lymphatic organs, circulates lymphocytes through the organs to make contact with the foreign substances, and destroys the foreign substances and eliminates them from the body.

In addition to immune responses, there are a variety of nonspecific defensive responses to invading disease-producing organisms. Nonspecific defenses act against a wide variety of intruders, such as many different kinds of bacteria and viruses. They include mechanical barriers provided by the skin and mucous membranes, antimicrobial chemicals, phagocytosis, inflammation, and fever.

The developmental anatomy of the lymphatic system is considered later in the chapter.

LYMPHATIC VESSELS AND LYMPH CIRCULATION

Lymphatic vessels begin as closed-ended vessels called **lymphatic capillaries** in spaces between cells (Fig. 22.2a). Just as blood capillaries converge to form venules and veins, lymphatic capillaries unite to form larger tubes called **lymphatic vessels** (see Fig. 22.1). Lymphatic vessels resemble veins in structure but have thinner walls and more

valves. At intervals along the lymphatic vessels, lymph flows through lymphatic tissue structures called lymph nodes. In the skin, lymphatic vessels lie in subcutaneous tissue and generally follow veins. Lymphatic vessels of the viscera generally follow arteries, forming plexuses around them.

Lymphatic Capillaries

Capillaries containing lymph are found throughout the body, except in (1) avascular tissues, (2) the central nervous system, (3) splenic pulp, and (4) bone marrow. Lymphatic capillaries have a slightly larger diameter than blood capillaries and have a unique structure that permits interstitial (tissue) fluid to flow into them but not out. The ends of endothelial cells that make up the wall of a lymphatic capillary overlap. When pressure is greater in the interstitial fluid than in lymph, the cells separate slightly, like a one-way valve opening, and fluid enters the lymphatic capillary. When pressure is greater inside the lymphatic capillary, the cells adhere more closely so lymph cannot flow back into interstitial fluid.

At right angles to the lymphatic capillary are structures called **anchoring filaments** that attach lymphatic endothelial cells to surrounding tissues (Fig. 22.2b). During edema, excess interstitial fluid accumulates and causes tissue swelling. This swelling pulls on the anchoring filaments, making the openings between cells even larger so that more fluid can flow into the lymphatic capillary.

Formation and Flow of Lymph

Most components of blood plasma freely move through the capillary walls to form interstitial fluid. More fluid seeps out of blood capillaries by filtration than returns to them by absorption (see Fig. 21.10). The excess fluid, about 3 liters per day, drains into lymphatic vessels and becomes lymph. Ultimately, lymph drains into venous blood through the right and left lymphatic ducts at the junction of the internal jugular and subclavian veins (see Fig. 22.4). Thus the sequence of fluid flow is (Fig. 22.3): arteries (blood) → blood capillaries (blood) → interstitial spaces (interstitial fluid) → lymphatic capillaries (lymph) → lymphatic ducts → subclavian veins (blood).

Since most plasma proteins are too large to leave blood vessels, interstitial fluid contains only a small amount of protein. Any proteins that do escape, however, cannot return to the blood by diffusion. The concentration gradient (high level of protein inside blood capillaries, low level outside) prevents this. Thus an important function of lymphatic vessels is to return leaked plasma proteins to the blood.

The flow of lymph from tissue spaces to the large lymphatic ducts to the subclavian veins is maintained primarily by the milking action of skeletal muscles. Skeletal muscle contractions compress lymphatic vessels and force lymph

FIGURE 22.1 The lymphatic system.

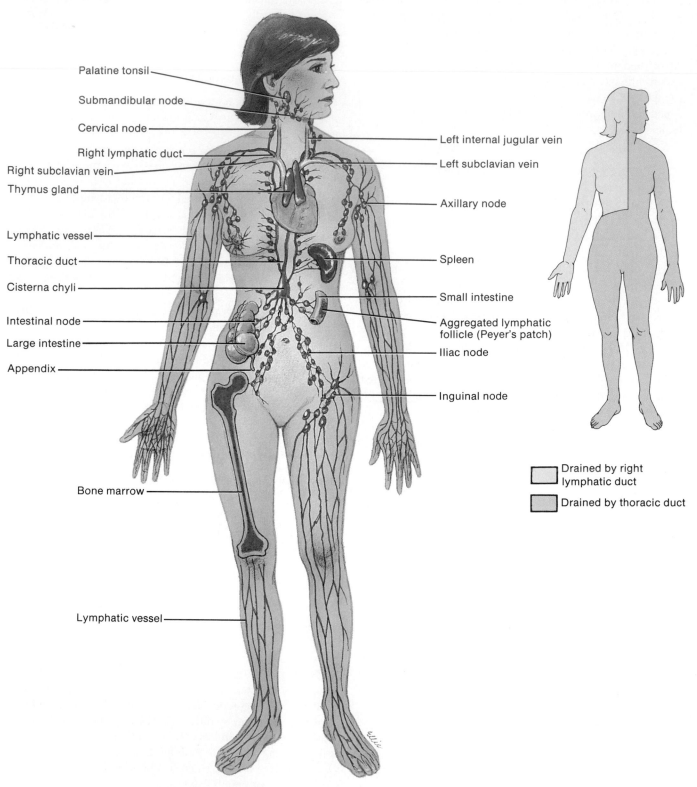

Palatine tonsil

Submandibular node

Cervical node

Right lymphatic duct

Right subclavian vein

Thymus gland

Lymphatic vessel

Thoracic duct

Cisterna chyli

Intestinal node

Large intestine

Appendix

Bone marrow

Lymphatic vessel

Left internal jugular vein

Left subclavian vein

Axillary node

Spleen

Small intestine

Aggregated lymphatic follicle (Peyer's patch)

Iliac node

Inguinal node

Drained by right lymphatic duct

Drained by thoracic duct

Anterior view of principal components of lymphatic system

Question: What are the functions of the lymphatic system?

FIGURE 22.2 Lymphatic capillaries.

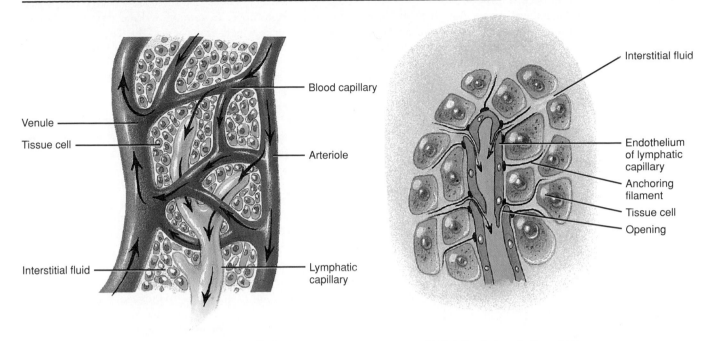

(a) Relationship of lymphatic capillaries
to tissue cells and blood capillaries

(b) Details of a lymphatic capillary

Question: Is lymph more similar to blood plasma or to interstitial fluid? Why?

FIGURE 22.3 Relationship of lymphatic system to cardiovascular system.

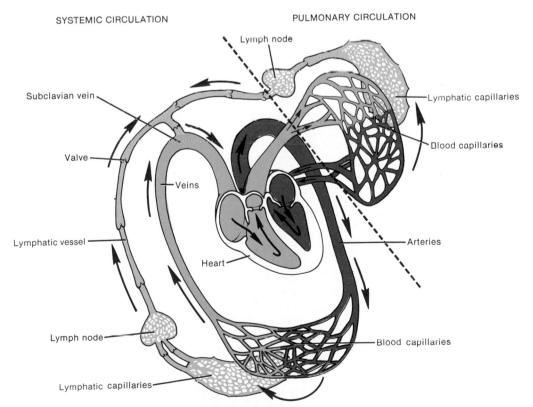

Arrows show direction of flow of lymph and blood

Question: Does lymph derive from fluid in arteries, veins, or capillaries?

toward the subclavian veins. One-way valves within the lymphatic vessels prevent backflow of lymph.

Another factor that maintains lymph flow is respiratory movements. These movements create a pressure gradient between the two ends of the lymphatic system. With each inhalation, lymph flows from the abdominal region, where the pressure is higher, toward the thoracic region, where it is lower.

Lymph Trunks and Ducts

Lymph passes from lymphatic capillaries into lymphatic vessels and through lymph nodes. Lymphatic vessels exiting lymph nodes pass lymph into another node of the same group or on to another group of nodes. From the most proximal group of each chain of nodes, the exiting vessels unite to form **lymph trunks.** The principal trunks are the **lumbar, intestinal, bronchomediastinal, subclavian,** and **jugular trunks** (see Fig. 22.4). The principal trunks eventually pass their lymph into venous blood through two main channels, the thoracic duct and the right lymphatic duct.

Thoracic (Left Lymphatic) Duct

The **thoracic (left lymphatic) duct** (Fig. 22.4) is about 38 to 45 cm (15 to 18 in.) in length and begins as a dilation called the **cisterna chyli** (sis-TER-na KĪ-lē) in front of the second lumbar vertebra. The thoracic duct is the main collecting duct of the lymphatic system. It receives lymph from the left side of the head, neck, and chest, the left upper extremity, and the entire body below the ribs (see Fig. 22.1).

The cisterna chyli receives lymph from the right and left lumbar trunks and from the intestinal trunk. The lumbar trunks drain lymph from the lower extremities, wall and viscera of the pelvis, kidneys, adrenals (suprarenals), and the deep lymphatics from most of the abdominal wall. The intestinal trunk drains lymph from the stomach, intestines, pancreas, spleen, and part of the liver.

In the neck, the thoracic duct also receives lymph from the left jugular, left subclavian, and left bronchomediastinal trunks. The left jugular trunk drains lymph from the left side of the head and neck, and the left subclavian trunk drains lymph from the left upper extremity. The left bronchomediastinal trunk drains lymph from the left side of the deeper parts of the anterior thoracic wall, upper part of the anterior abdominal wall, anterior part of the diaphragm, left lung, and left side of the heart.

Right Lymphatic Duct

The **right lymphatic duct** (Fig. 22.4) is about 1.25 cm (0.5 in.) long and drains lymph from the upper right side of the body (see Fig. 22.1). The right lymphatic duct collects lymph from the right jugular trunk, which drains the right side of the head and neck; the right subclavian trunk, which drains the right upper extremity; and the right bronchomediastinal trunk, which drains the right side of the thorax, right lung, right side of the heart, and part of the liver.

LYMPHATIC TISSUE

Lymphatic (lymphoid) tissue occurs in the body in various ways. When it is not enclosed by a capsule, it is called **diffuse lymphatic tissue.** The lamina propria (connective tissue) of mucous membranes of the gastrointestinal tract, respiratory passageways, urinary tract, and reproductive tract contains diffuse lymphatic tissue. It is also normally found in small amounts in almost every organ of the body.

Lymphatic nodules are oval-shaped concentrations of lymphatic tissue. Although they are not surrounded by a capsule, most lymphatic nodules are solitary, small, and discrete. Such nodules are scattered in the lamina propria of mucous membranes of the gastrointestinal tract, respiratory airways, urinary tract, and reproductive tract. Some lymphatic nodules occur in multiple, large aggregations in specific parts of the body. Among these are the tonsils in the pharyngeal region and aggregated lymphatic follicles (Peyer's patches) in the ileum of the small intestine. Aggregations of lymphatic nodules also occur in the appendix.

The **primary lymphatic organs** of the body are the **bone marrow** and **thymus gland.** They are the sites of production of immunocompetent B and T cells, lymphocytes that are capable of carrying out immune responses. The **secondary lymphatic organs** are the **lymph nodes** and **spleen.** Most immune responses occur in secondary lymphatic organs, lymphatic nodules, or diffuse lymphatic tissue.

Lymph Nodes

The oval or bean-shaped structures located along the length of lymphatic vessels are called **lymph nodes** (Fig. 22.5). They range from 1 to 25 mm (0.04 to 1 in.) in length. Lymph nodes are scattered throughout the body, usually in groups (see Fig. 22.1). Typically, these groups are arranged in two sets, **superficial** and **deep.**

Each node is covered by a **capsule** of dense connective tissue that extends into the node. The capsular extensions are called **trabeculae** (tra-BEK-yoo-lē). Internal to the capsule is a supporting network of reticular fibers and fibroblasts. The capsule, trabeculae, reticular fibers, and fibroblasts constitute the stroma (framework) of a lymph node. The parenchyma of a lymph node is specialized into two regions: cortex and medulla. The outer **cortex** contains densely packed lymphocytes arranged in masses called **follicles (lymphatic nodules).** The outer rim of each follicle contains **T cells (T lymphocytes)** plus **macrophages** and **follicular dendritic cells,** which participate in activation of T cells. When an immune response is occurring, the folli-

FIGURE 22.4 Routes for drainage of lymph from lymph trunks into thoracic and right lymphatic ducts.

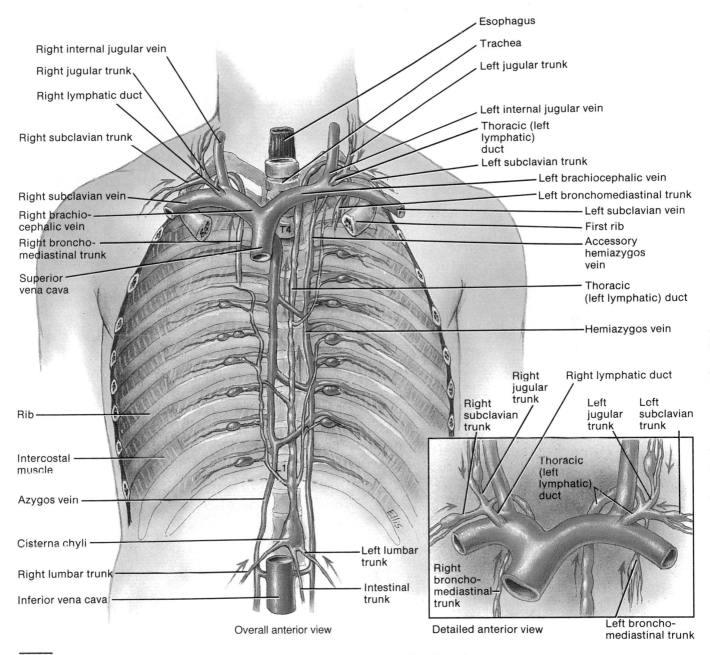

Overall anterior view

Detailed anterior view

Question: Which lymphatic vessels empty into the cisterna chyli and which duct receives fluid from the cisterna chyli?

cles contain lighter-staining central areas, the **germinal centers,** where **B cells (B lymphocytes)** are proliferating into antibody-secreting plasma cells. The inner region of a lymph node is called the **medulla.** In the medulla, the lymphocytes are arranged in strands called **medullary cords.** These cords also contain macrophages and plasma cells.

Lymph flows through a node in one direction. It enters through **afferent lymphatic vessels,** which penetrate the convex surface of the node at several points. They contain valves that open toward the node so that the lymph is directed *inward.* Inside the node, lymph enters the **sinuses,** which are a series of irregular channels. Lymph flows through sinuses in the cortex and then in the medulla and exits the lymph node via one or two **efferent lymphatic vessels.** Efferent lymphatic vessels are wider than the afferent vessels and contain valves that open away from the node to convey lymph *out* of the node. Efferent lymphatic vessels emerge from one side of the lymph node at a slight

FIGURE 22.5 Structure of a lymph node. Arrows indicate direction of lymph flow.

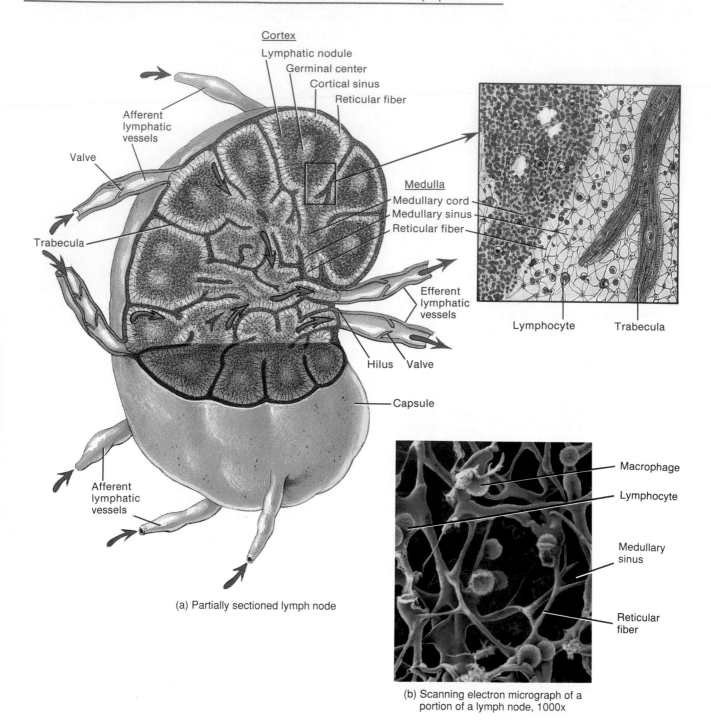

(a) Partially sectioned lymph node

(b) Scanning electron micrograph of a portion of a lymph node, 1000x

Question: What happens to foreign substances in lymph that enter a lymph node?

depression called a **hilus** (HĪ-lus), or **hilum.** Blood vessels also enter and leave the node at the hilus.

The lymph nodes filter foreign substances from lymph as it passes back toward the cardiovascular system. These substances are trapped by the reticular fibers within the node. Then macrophages destroy the same foreign substances by phagocytosis and lymphocytes bring about destruction of others by immune responses. Plasma cells and T cells that have proliferated within a lymph node also can leave and circulate to other parts of the body.

CLINICAL APPLICATION

METASTASIS THROUGH THE LYMPHATIC SYSTEM

Knowledge of the location of the lymph nodes and the direction of lymph flow is important in the diagnosis and prognosis of the spread of cancer by **metastasis.** Cancer cells may travel via the lymphatic system and produce clusters of tumor cells where they lodge. Such secondary tumor sites are predictable by the direction of lymph flow from the organ primarily involved. (Cancer may also spread by extending locally or being carried by the cardiovascular system.) Cancerous lymph nodes feel enlarged, firm, and non-tender. Most lymph nodes that undergo infectious enlargement, by contrast, are not firm and are very tender.

Tonsils

Tonsils are multiple aggregations of large lymphatic nodules embedded in a mucous membrane. They form a ring at the junction of the oral cavity and pharynx. The single **pharyngeal** (fa-RIN-jē-al) **tonsil** or **adenoid** is embedded in the posterior wall of the nasopharynx (see Fig. 23.2b). The paired **palatine** (PAL-a-tīn) **tonsils** are situated in the space between the pharyngopalatine and glossopalatine arches (see Fig. 16.2a). These are the ones commonly removed by a tonsillectomy. The paired **lingual** (LIN-gwal) **tonsils** are located at the base of the tongue and may also have to be removed by a tonsillectomy (see Fig. 16.2a).

The tonsils are situated strategically to protect against invasion of foreign substances that are inhaled or ingested. Functionally, the tonsils participate in immune responses.

Spleen

The oval **spleen** is the largest mass of lymphatic tissue in the body, measuring about 12 cm (5 in.) in length (Fig. 22.6a). It is situated in the left hypochondriac region between the fundus of the stomach and diaphragm (see Fig. 1.11). The superior surface of the spleen is smooth and convex and conforms to the concave surface of the diaphragm. Neighboring organs make indentations in the spleen—the gastric impression (stomach), renal impression (left kidney), and colic impression (left flexure of colon). Like lymph nodes, the spleen has a hilus.

A capsule of dense connective tissue and scattered smooth muscle fibers (cells) surrounds the spleen. The capsule, in turn, is covered by a serous membrane, the peritoneum. The capsule plus trabeculae, reticular fibers, and fibroblasts constitute the stroma of the spleen.

The parenchyma of the spleen consists of two different kinds of tissue called white pulp and red pulp (Fig. 22.6b). **White pulp** is lymphatic tissue, mostly lymphocytes, arranged around central arteries. The **red pulp** consists of **venous sinuses** filled with blood and cords of splenic tissue called **splenic (Billroth's) cords.** Veins are closely associated with the red pulp. Splenic cords consist of red blood cells, macrophages, lymphocytes, plasma cells, and granulocytes.

The splenic artery and vein and the efferent lymphatic vessels pass through the hilus. Since the spleen has no afferent lymphatic vessels or lymph sinuses, it does not filter lymph. The spleen functions in immunity as a site of B cell proliferation into plasma cells. The main function of the spleen, however, is phagocytosis of bacteria and worn-out or damaged red blood cells and platelets. In addition, the spleen stores and releases blood in times of demand, such as during hemorrhage. Sympathetic impulses cause the smooth muscle of the capsule of the spleen to contract, which squeezes blood out of the red pulp. During early fetal development, the spleen participates in blood cell formation.

CLINICAL APPLICATION

RUPTURED SPLEEN

The spleen is the organ most often damaged in cases of abdominal trauma. Severe blows over the lower left chest or upper abdomen may fracture the protecting ribs. Such a crushing injury may **rupture the spleen,** which causes severe intraperitoneal hemorrhage and shock. Prompt removal of the spleen, called a **splenectomy,** is needed to prevent the patient from bleeding to death. Other structures, particularly bone marrow and the liver, can take over functions normally carried out by the spleen.

Thymus Gland

Usually a bilobed lymphatic organ, the **thymus gland** is located in the superior mediastinum, posterior to the sternum and between the lungs (Fig. 22.7a). An enveloping layer of connective tissue holds the two **thymic lobes** closely together but a connective tissue **capsule** encloses each lobe. The capsule gives off extensions into the lobes called **trabeculae,** which divide the lobes into **lobules** (Fig. 22.7b).

Each lobule consists of a deeply staining peripheral **cortex** and a lighter-staining central **medulla.** The cortex is composed almost entirely of tightly packed lymphocytes held in place by reticular fibers. Immature T cells migrate (via the blood) from bone marrow to the thymus, where they mature. The medulla consists mostly of epithelial cells

FIGURE 22.6 Structure of the spleen.

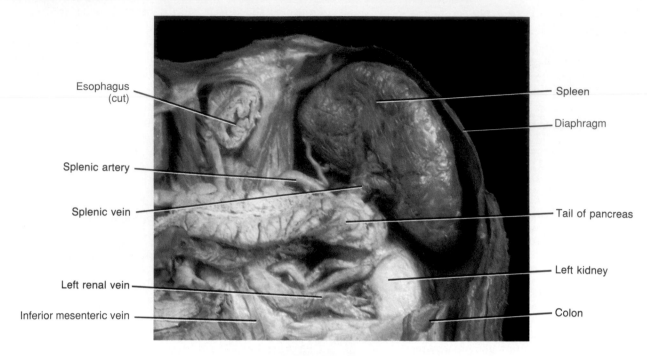

Esophagus (cut)

Splenic artery

Splenic vein

Left renal vein

Inferior mesenteric vein

Spleen

Diaphragm

Tail of pancreas

Left kidney

Colon

(a) Anterior view of abdominal cavity

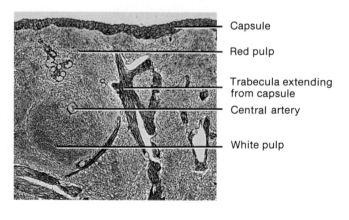

Capsule

Red pulp

Trabecula extending from capsule

Central artery

White pulp

(b) Photomicrograph of portion of the spleen (approx. 23 x)

Question: After birth, what is the main function of the spleen?

and more widely scattered lymphocytes. The epithelial cells produce thymic hormones, which are thought to aid in maturation of T cells. Exactly what they do, however, is not known. In addition, the medulla contains characteristic **thymic (Hassall's) corpuscles,** concentric layers of epithelial cells. Possibly, they are remnants of dying cells.

The thymus gland is large in the infant, and it reaches its maximum size of about 40 g (about 1.4 oz) at 10 to 12 years of age. After puberty, fat and connective tissue replace much of the thymic tissue. By the time the person reaches maturity, the gland has atrophied (involution with

age) but still continues to function in maturation of T cells. In elderly persons, thymic tissue is almost completely replaced by fat and connective tissue.

DEVELOPMENTAL ANATOMY OF THE LYMPHATIC SYSTEM

The lymphatic system begins to develop by the end of the fifth week. *Lymphatic vessels* develop from **lymph sacs**

FIGURE 22.7 Thymus gland.

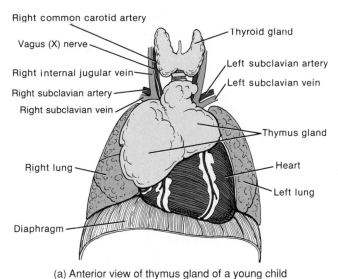

Right common carotid artery
Vagus (X) nerve
Right internal jugular vein
Right subclavian artery
Right subclavian vein
Right lung
Diaphragm

Thyroid gland
Left subclavian artery
Left subclavian vein
Thymus gland
Heart
Left lung

(a) Anterior view of thymus gland of a young child

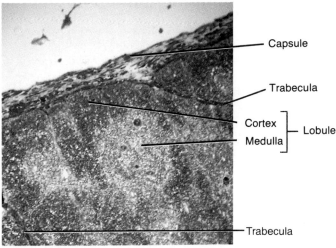

Capsule
Trabecula
Cortex ⎫
        ⎬ Lobule
Medulla ⎭
Trabecula

(b) Photomicrograph of several thymus lobules, 40x

Question: Which lymphocytes achieve immunocompetence in the thymus?

that arise from developing veins. Thus the lymphatic system is also derived from **mesoderm.**

The first lymph sacs to appear are the paired **jugular lymph sacs** at the junction of the internal jugular and subclavian veins (Fig. 22.8). From the jugular lymph sacs, capillary plexuses spread to the *thorax, upper extremities, neck,* and *head.* Some of the plexuses enlarge and form lymphatic vessels in their respective regions. Each jugular lymph sac retains at least one connection with its jugular vein, the left one developing into the superior portion of the thoracic duct (left lymphatic duct).

The next lymph sac to appear is the unpaired **retroperitoneal lymph sac** at the root of the mesentery of the intestine. It develops from the primitive vena cava and mesonephric (primitive kidney) veins. Capillary plexuses and lymphatic vessels spread from the retroperitoneal lymph sac to the *abdominal viscera* and *diaphragm.* The sac establishes connections with the cisterna chyli but loses its connections with neighboring veins.

At about the time the retroperitoneal lymph sac is developing, another lymph sac, the **cisterna chyli,** develops below the diaphragm on the posterior abdominal wall. It gives rise to the inferior portion of the *thoracic duct* and the *cisterna chyli* of the thoracic duct. Like the retroperitoneal lymph sac, the cisterna chyli also loses its connections with surrounding veins.

The last of the lymph sacs, the paired **posterior lymph sacs,** develop from the iliac veins at their union with the posterior cardinal veins. The posterior lymph sacs produce capillary plexuses and lymphatic vessels of the *abdominal wall, pelvic region,* and *lower extremity.* The posterior

lymph sacs join the cisterna chyli and lose their connections with adjacent veins.

With the exception of the anterior part of the sac from which the cisterna chyli develops, all lymph sacs become invaded by **mesenchymal cells** and are converted into groups of *lymph nodes.*

The *spleen* develops from mesenchymal cells between layers of the dorsal mesentery of the stomach. The development of the *thymus gland* is considered on page 554.

FIGURE 22.8 Development of the lymphatic system.

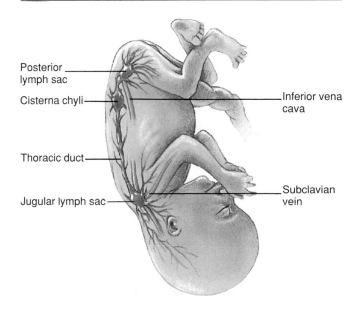

Posterior lymph sac
Cisterna chyli
Thoracic duct
Jugular lymph sac

Inferior vena cava
Subclavian vein

NONSPECIFIC RESISTANCE TO DISEASE

Maintaining homeostasis in the body requires ways to continually defend against the activities of disease-producing organisms, called **pathogens** (PATH-ō-jens), or their toxins. The ability to ward off disease through our defenses is called **resistance.** Vulnerability or lack of resistance is termed **susceptibility.** Defenses against disease may be grouped into two broad areas: nonspecific resistance and specific resistance. **Nonspecific resistance** comprises defense mechanisms that provide a general response against invasion by a wide range of pathogens. **Specific resistance,** or **immunity,** involves the production of specific antibodies or activation of T cells against a particular pathogen or other foreign substance. We will first consider mechanisms of nonspecific resistance that include the skin and mucous membranes, production of antimicrobial chemicals, natural killer cells, phagocytosis, inflammation, and fever.

Skin and Mucous Membranes

The skin and mucous membranes of the body are the first line of defense against pathogens. Both mechanical and chemical barriers block the initial attempts of microbes and foreign substances to penetrate the body and cause disease.

Mechanical Protection

With many layers of closely packed, keratinized cells, the outer epithelial layer of the skin, the **epidermis**, provides a formidable physical barrier to the entrance of microbes. In addition, periodic shedding of epidermal cells helps remove microbes at the skin surface. Bacteria rarely penetrate the intact surface of healthy epidermis. But when the epithelial surface is broken, by cuts, burns, wounds, and so on, an infection often develops. The bacteria most likely to cause such an infection are staphylococci, which normally inhabit the hair follicles and sudoriferous (sweat) glands of the skin. Moreover, when the skin is moist, as in hot, humid climates, infections are quite common, especially fungus infections such as athlete's foot.

Like the skin, **mucous membranes** also consist of an epithelial layer and an underlying connective tissue layer. They line body cavities that open to the exterior, that is, the gastrointestinal, respiratory, urinary, and reproductive tracts. The epithelial layer of a mucous membrane secretes a fluid called **mucus,** which prevents the cavities from drying out. Because mucus is slightly viscous, it traps many microbes and other foreign substances. The mucous membrane of the nose has mucus-coated **hairs** that trap and filter air containing microbes, dust, and pollutants. The mucous membrane of the upper respiratory tract contains **cilia,** microscopic hairlike projections of the epithelial cells (see Fig. 23.6c). The cilia propel inhaled dust and microbes that have become trapped in mucus toward the throat. Coughing and sneezing accelerate movement of mucus.

Several other mechanical factors also help to protect epithelial surfaces of the skin and mucous membranes. One that protects the eyes is the **lacrimal** (LAK-ri-mal) **apparatus** (see Fig. 16.4), a group of structures that manufactures and drains away tears. Blinking spreads tears over the surface of the eyeball. The continual washing action of tears helps to keep microbes from settling on the surface of the eye. If an irritating substance or a large number of microbes make contact with the eye, the lacrimal glands start to secrete heavily. This is a protective mechanism to dilute and wash away the irritating substance or microbes.

Saliva, produced by the salivary glands, dilutes the numbers of microorganisms and washes them from the surfaces of the teeth and the mucous membrane of the mouth, much as tears wash the eyes. This helps to prevent colonization by microbes.

The cleansing of the urethra by the **flow of urine** is a mechanical factor that prevents microbial colonization in the urinary system. Vaginal secretions likewise move microbes out of the female body. **Defecation** and **vomiting** also may be considered mechanical processes that expel microbes. For example, in response to microbes, such as those that irritate the lining of the gastrointestinal tract, the muscle of the tract contracts vigorously. The resulting diarrhea rapidly expels some of the microbes.

Certain pathogens, however, thrive on the moist secretions of a mucous membrane. If present in sufficient numbers, they may be able to invade the membrane. Microbial penetration may be aided by the toxic products they produce, prior injury by viral infections, or mucosal irritations. Although mucous membranes do restrict the entrance of many microbes, they are less effective than the skin.

Chemical Protection

Certain chemicals also contribute to the high degree of resistance of the skin and mucous membranes to microbial invasion. Sebaceous (oil) glands of the skin secrete an oily substance called **sebum** that forms a protective film over the surface of the skin. Sebum contains unsaturated fatty acids that inhibit the growth of certain pathogenic bacteria and fungi. The low pH of the skin, between pH 3 and 5, is caused in part by the secretion of fatty acids and lactic acid. The skin's acidity probably discourages the growth of many other microorganisms.

Perspiration helps flush microbes from the surface of the skin and also contains **lysozyme,** an enzyme capable of breaking down cell walls of certain bacteria. Lysozyme is also found in tears, saliva, nasal secretions, and tissue fluids, where it also exhibits antimicrobial activity. **Hyaluronic acid,** located in areolar connective tissue, has a gel-like consistency that slows the spread of noxious agents in localized infections.

Gastric juice is produced by the glands of the stomach. It is a mixture of hydrochloric acid, enzymes, and mucus. The strong acidity of gastric juice (pH 1.2 to 3.0) destroys many bacteria and most bacterial toxins. **Vaginal secretions** also are slightly acidic, which discourages bacterial growth.

Antimicrobial Substances

In addition to the mechanical and chemical barriers of the skin and mucous membranes, the body also produces certain antimicrobial substances. For example, blood and interstitial fluids contain chemicals that discourage microbial growth. Iron-binding proteins, called **transferrins,** inhibit bacterial growth by reducing the amount of available iron. Other antimicrobial substances are interferons and complement. They provide a second line of defense should microbes penetrate the skin and mucous membranes.

Interferons

Body cells infected with viruses produce proteins called **interferons** (in'-ter-FĒR-ons), or **IFNs.** There are two principal types of interferon: type I (includes alpha- and beta-IFN) and type II (gamma-IFN). Lymphocytes, macrophages, and fibroblasts produce IFNs. Once produced by and released from virus-infected cells, IFN diffuses to uninfected neighboring cells and binds to surface receptors. This induces uninfected cells to synthesize antiviral proteins that interfere with or inhibit viral replication. Viruses can cause disease only if they can replicate within body cells. IFN appears to be an important defense against infection by many different viruses.

Although originally described as antiviral agents, IFNs have important functions unrelated to combatting viruses. For example, gamma-IFN enhances the cell-killing activity of phagocytic cells and natural killer cells (described shortly). Type I IFNs inhibit cell growth and suppress tumor formation. The gene that codes for alpha-IFN is missing in people who have one type of leukemia.

In clinical trials, IFNs have exhibited no effects against some types of tumors and only limited effects against others. Alpha-IFN (Intron A) is approved in the U.S. for treating several virus-associated disorders. One is Kaposi's (kap'-ō-SĒS) sarcoma, a cancer that often occurs in patients infected with HIV, the virus that causes AIDS (acquired immune deficiency syndrome, see page 712). Other approved uses for alpha-IFN include genital warts, caused by herpes virus, and hepatitis C, caused by the hepatitis C virus, which can lead to cirrhosis and liver cancer. Alpha-IFN also is being tested to see if it can slow the development of AIDS in HIV-infected people.

The Complement System

A group of about 20 normally inactive proteins in blood plasma and on cell membranes comprises the **complement system.** When activated, these proteins "complement" or enhance certain immune, allergic, and inflammatory reactions (discussed shortly). They include 11 proteins named C1 through C9 (there are three forms of C1); Factors B and D; and properdin. Several other proteins regulate complement activation.

A key member of the complement system is a plasma protein termed C3, which can be activated in two different ways: the classical pathway and the alternative pathway (Fig. 22.9). Activation of C3 by the **classical pathway** is linked to the immune system because it depends on formation of complexes between antibodies and the antigenic microbe that triggered their production. Via the **alternative pathway,** certain polysaccharides on the surface of a microbe can directly activate C3. The alternative pathway is just as important as the classical pathway; it was just discovered later.

Once C3 is activated, it activates other complement proteins, which attack and destroy microbes as follows:

1. **Activation of inflammation.** Some complement proteins (C3a, C4a, and C5a) contribute to the development of inflammation (described shortly). They dilate arterioles, which increases blood flow to the area, and cause the release of **histamine** from mast cells, basophils, and platelets. Histamine increases the permeability of blood capillaries. This enables white blood cells to more easily penetrate tissues to combat infection or allergy. Other complement proteins serve as **chemotactic agents** that attract phagocytes to the site of microbe invasion.
2. **Opsonization.** Complement fragment C3b binds to the surface of a microbe and then interacts with receptors on phagocytes to promote phagocytosis. This coating of a microbial surface is called **opsonization** (op-sō-ni-ZĀ-shun), or **immune adherence.**
3. **Cytolysis.** Several complement proteins (C5b, C6, C7, C8, and C9) come together to form a **membrane attack complex (MAC)** that punches holes in the plasma membrane of the microbe. This causes the microbe to rupture, a process called **cytolysis.**

Natural Killer Cells

When microbes penetrate the skin and mucous membranes or bypass the antimicrobial substances in blood, the next line of defense consists of natural killer cells and phagocytes. We will first consider natural killer cells.

In addition to B cells and T cells, which participate in immune responses, there is another population of lymphocytes that destroys intruders. These lymphocytes, called **natural killer (NK) cells,** have the ability to kill a wide

FIGURE 22.9 The complement system.

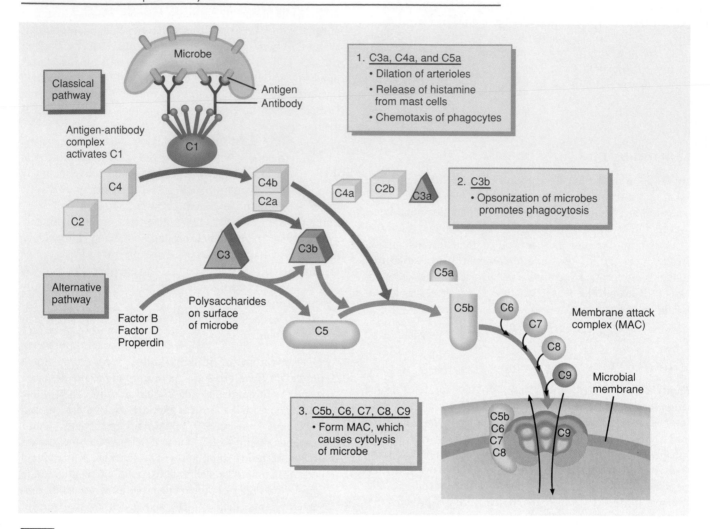

Question: Which pathway for activation of complement is linked to the immune system? Explain.

variety of infectious microbes plus certain spontaneously arising tumor cells. NK cells are present in the spleen, lymph nodes, bone marrow, and blood. Gamma interferon is released by NK cells and also stimulates their cytolytic activity.

Exactly how NK cells deliver their "kiss of death" is not completely clear. They may release perforins, chemicals that cause cytolysis of the microbe (see page 703). Or they may bind to a target cell and inflict damage by direct contact. How NK cells recognize their targets is also an unsettled question. They probably attack cells that do not display proper identity markers called major histocompatibility complex (MHC) antigens, which are described on page 701. NK cells are defective or decreased in number in some cancer patients and in patients with AIDS.

Phagocytosis

Phagocytosis (*phagein* = to eat; *cyto* = cell) is the ingestion of microbes or any foreign particulate matter by cells called phagocytes. In addition to providing a nonspecific defense of the host, phagocytosis plays a vital role in immunity, as discussed later in the chapter.

Kinds of Phagocytes

The cells that participate in phagocytosis fall into two broad categories: **granulocytes** and **macrophages**. The granulocytes of blood (granular leukocytes, see Fig. 19.2) vary in their phagocytic activity. Neutrophils are the most phagocytic, and eosinophils have some phagocytic capability.

When an infection occurs, both neutrophils and monocytes migrate to the infected area. During this migration, the monocytes enlarge and develop into highly phagocytic macrophages (MAK-rō-fā-jez). Since these cells leave the blood and migrate to infected areas, they are called **wandering macrophages.** Others, called **fixed macrophages,** stand guard in certain tissues and organs of the body. Fixed macrophages are found in the skin and subcutaneous layer (histiocytes), liver (stellate reticuloendothelial cells), lungs (alveolar macrophages), brain (microglia), spleen, lymph nodes, and bone marrow (tissue macrophages). The combination of wandering and fixed macrophages is called the **mononuclear phagocytic (reticuloendothelial) system.**

Mechanism of Phagocytosis

For convenience of study, phagocytosis will be divided into three phases: chemotaxis, adherence, and ingestion. After it ingests a particle, a phagocyte unleashes chemicals that kill any microbes.

• **Chemotaxis** Chemotaxis (kē′-mō-TAK-sis; *chemo* = chemicals; *taxis* = arrangement) is the chemical attraction of phagocytes to a particular location. Among the chemotactic chemicals that attract phagocytes are microbial products, components of white blood cells and damaged tissue cells, and activated complement proteins.

• **Adherence** Adherence is the *attachment* of the cell membrane of a phagocyte to the surface of a microorganism or other foreign material (Fig. 22.10). In some instances, adherence occurs easily, and the microorganism is readily phagocytized. In other cases, adherence is more difficult, because of certain microbial defenses. Microorganisms can be more readily phagocytized if they are first opsonized.

• **Ingestion** Following adherence, **ingestion** occurs. The cell membrane of the phagocyte extends projections, called pseudopods, that engulf the microorganism (Fig. 22.10). Once the microorganism is surrounded, the pseudopods meet and fuse, surrounding the microorganism with a sac called a **phagocytic vesicle (phagosome).**

Killing

After phagocytosis has been accomplished, several deadly mechanisms come into play. The phagocytic vesicle, which forms when a portion of the membrane completely pinches off, enters the cytoplasm (Fig. 22.10). Within the cytoplasm, it merges with lysosomes to form a single, larger structure called a **phagolysosome.** The phagolysosome contains lysozyme, which breaks down microbial cell walls, and digestive enzymes, which degrade carbohydrates, proteins, lipids, and nucleic acids. The phagocyte also forms lethal oxidants, such as superoxide anion (O_2^-), hypochlorite anion (OCl^-) and hydrogen peroxide (H_2O_2), in a process called the **respiratory (oxidative) burst.** Finally, phagocytes produce other bactericidal substances, such as

defensins, so named because of their apparent role in preventing and overcoming infections. Defensins are active against bacteria, fungi, and viruses.

The chemical onslaught kills many types of microbes in only 10 to 30 minutes. Any materials that cannot be degraded further remain in structures called **residual bodies.** The cell disposes of residual bodies by exocytosis, a process in which the residual body migrates to the plasma membrane, fuses with it, ruptures, and releases its contents.

Some microbes, such as toxin-producing staphylococci, may be ingested but are not necessarily killed. Rather, their toxins may kill the phagocytes. Other microbes, such as the tubercle bacillus, may multiply within the phagolysosome and eventually destroy the phagocyte. Still other microbes, such as the causative agents of tularemia and brucellosis, may remain dormant in phagocytes for months or years at a time.

Inflammation

When cells are damaged by microbes, physical agents, or chemical agents, the injury is a form of stress. The response to the stress of tissue damage is called **inflammation**. It is a defensive response of the body that is usually characterized by four symptoms: **redness, pain, heat,** and **swelling.** A fifth symptom can be the **loss of function** in the injured area. Whether loss of function occurs depends on the site and extent of the injury. Inflammation aids disposal of microbes, toxins, or foreign material at the site of injury, prevents their spread to other organs, and prepares the site for tissue repair. Thus it helps restore tissue homeostasis.

Since inflammation is one of the body's nonspecific internal systems of defense, the response of a tissue to an accidental cut is similar to the response that results from other types of tissue damage, caused by burns, radiation, or bacterial or viral invasion. In each case there are three basic stages of inflammation: (1) vasodilation and increased permeability of blood vessels, (2) phagocyte migration, and (3) repair.

Vasodilation and Increased Permeability of Blood Vessels

Immediately after tissue damage, blood vessels in the area of the injury dilate and become more permeable. **Vasodilation** is an increase in diameter of the blood vessels. **Increased permeability** means that an increased amount of material, including some proteins normally retained in blood, is allowed to pass out of the blood vessels. Vasodilation allows more blood to flow through the damaged area, and increased permeability permits defensive materials in the blood to enter the injured area. Such defensive mediators include antibodies, phagocytes, and clot-forming chemicals. The increased blood flow also helps remove toxic

FIGURE 22.10 Phagocytosis.

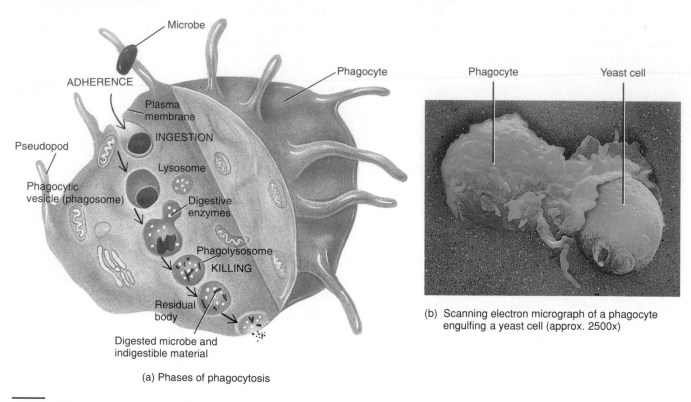

(a) Phases of phagocytosis

(b) Scanning electron micrograph of a phagocyte engulfing a yeast cell (approx. 2500x)

Question: What chemicals are responsible for killing microbes?

products released by the invading microorganisms and dead cells.

Among the substances that contribute to vasodilation, increased permeability, and other aspects of the inflammatory response are the following (Fig. 22.11):

1. **Histamine.** This substance is found in many body cells, especially mast cells in connective tissue, basophils, and blood platelets. Histamine is released in response to injury by any cells that contain it. Phagocytes (neutrophils and macrophages) attracted to the site of injury also stimulate release of histamine. Histamine causes vasodilation and increased permeability.

2. **Kinins.** These polypeptides are formed from inactive precursors in blood called kininogens. They induce vasodilation and increased permeability and serve as chemotactic agents for phagocytes.

3. **Prostaglandins (PGs).** These substances, especially of the E series, are released by damaged cells and intensify the effects of histamine and kinins. PGs also may stimulate the migration of phagocytes through capillary walls.

4. **Leukotrienes (LTs).** These substances are produced by basophils and mast cells by breakdown of membrane phospholipids. They cause increased permeability and

also function in adherence and as chemotactic agents for phagocytes.

5. **Complement.** These proteins stimulate histamine release, attract neutrophils by chemotaxis, and promote phagocytosis. They can also destroy bacteria.

Within minutes after an injury, dilation of arterioles and increased permeability of capillaries produce heat, redness, and edema (swelling). The large amount of warm blood flowing through the area produces both heat and redness (erythema). As the local temperature rises slightly, metabolic reactions proceed more rapidly and release additional heat. Edema results from increased permeability of blood vessels, which permits more fluid to move from blood into tissue spaces. Pain, whether immediate or delayed, is a cardinal symptom of inflammation. It can result from injury of nerve fibers or from irritation by toxic chemicals from microorganisms. Kinins affect some nerve endings, causing much of the pain associated with inflammation. Prostaglandins intensify and prolong the pain associated with inflammation. Pain may also be due to increased pressure from edema.

The increased permeability of capillaries allows leakage of clotting factors into tissues. The clotting cascade (see Fig. 19.10) is set into motion and fibrinogen is ultimately

FIGURE 22.11 Inflammation. Histamine, kinins, prostaglandins, leukotrienes, and complement together stimulate vasodilation, increased permeability of blood vessels, chemotaxis, diapedesis, and phagocytosis.

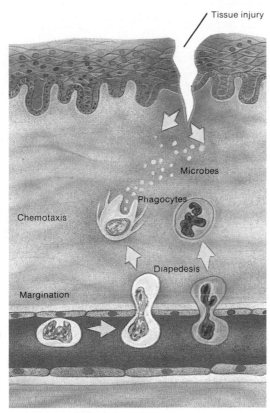

Phagocytes migrate from blood to site of tissue injury

Question: What causes each of the following symptoms of inflammation: redness, pain, heat, and swelling?

converted to an insoluble, thick network of fibrin threads, which localizes and traps invading microbes and blocks their spread.

Phagocyte Migration

Within an hour after the inflammatory process starts, phagocytes appear on the scene. Neutrophils begin to stick to the inner surface of the endothelium (lining) of blood vessels. This is called **margination** (Fig. 22.11). Then they extend pseudopods and squeeze through the narrow spaces between neighboring endothelial cells in the wall of the blood vessel to reach the damaged area. This movement is called **diapedesis** (dī'-a-pe-DĒ-sis) or **emigration**. It can take as little as two minutes. The movement of neutrophils depends on chemotaxis. Neutrophils are attracted by microbes, kinins, complement, and other neutrophils. A steady stream of neutrophils is ensured by the production

and release of additional cells from bone marrow. This increase in white blood cells in the blood is termed **leukocytosis.** Neutrophils attempt to destroy the invading microbes by phagocytosis.

As the inflammatory response continues, monocytes follow the neutrophils into the infected area. Once in the tissue, monocytes transform into wandering macrophages that augment the phagocytic activity of fixed macrophages. Neutrophils predominate in the early stages of infection but tend to die off rapidly. Macrophages arrive on the scene hours later. They are several times more phagocytic than neutrophils and engulf damaged tissue, worn-out neutrophils, and invading microbes.

After phagocytes engulf damaged tissue and microbes, they eventually die. Within a few days, a pocket of dead phagocytes and damaged tissue forms. This collection of dead cells and fluid is called **pus.** Pus formation usually continues until the infection subsides. At times, the pus pushes to the surface of the body or into an internal cavity for dispersal. On other occasions, the pus remains even after the infection is terminated. In this case, the pus is gradually destroyed over a period of days and is absorbed by the body.

CLINICAL APPLICATION

ABSCESS AND ULCERS

If the pus cannot drain out of an inflamed region, an abscess develops. An **abscess** is simply an excessive accumulation of pus in a confined space. Common examples are pimples and boils. When inflamed tissue is shed many times, it produces an open sore, called an **ulcer,** on the surface of an organ or tissue. People with poor circulation are susceptible to ulcers in the tissues of their legs. The ulcers, called stasis ulcers, develop because of poor oxygen and nutrient supply to tissues that then become very susceptible to even a very mild injury or infectious process.

Fever

Fever is an abnormally high body temperature. The high body temperature intensifies the effects of interferons, inhibits the growth of some microbes, and speeds up body reactions that aid repair. (Fever is discussed in more detail on page 856.)

A summary of some of the components of nonspecific resistance is presented in Exhibit 22.1.

IMMUNITY (SPECIFIC RESISTANCE TO DISEASE)

The nonspecific defenses all have one thing in common. They offer protection against a *variety* of pathogens or other foreign substances. They are not specifically directed against one particular invader. The ability of the body to

EXHIBIT 22.1

SUMMARY OF NONSPECIFIC RESISTANCE

Component	Functions
SKIN AND MUCOUS MEMBRANES	
MECHANICAL FACTORS	
Epidermis of skin	Forms a physical barrier to the entrance of microbes.
Mucous membranes	Inhibit the entrance of many microbes, but not as effective as intact skin.
Mucus	Traps microbes in respiratory and gastrointestinal tracts.
Hairs	Filter microbes and dust in nose.
Cilia	Together with mucus, trap and remove microbes and dust from upper respiratory tract.
Lacrimal apparatus	Tears dilute and wash away irritating substances and microbes.
Saliva	Washes microbes from surfaces of teeth and mucous membranes of mouth.
Urine	Washes microbes from urethra.
Defecation and vomiting	Expel microbes from body.
CHEMICAL FACTORS	
Acid pH of skin	Discourages growth of many microbes.
Unsaturated fatty acids	Antibacterial substance in sebum.
Lysozyme	Antimicrobial substance in perspiration, tears, saliva, nasal secretions, and tissue fluids.
Hyaluronic acid	Prevents spread of noxious agents in localized infection.
Gastric juice	Destroys bacteria and most toxins in stomach.
ANTIMICROBIAL SUBSTANCES	
INTERFERONS (IFNs)	Protects uninfected host cells from viral infection.
COMPLEMENT	Causes cytolysis of microbes, promotes phagocytosis, and contributes to inflammation.
NATURAL KILLER (NK) CELLS	Kill a wide variety of microbes and certain tumor cells.
PHAGOCYTOSIS	Ingestion of foreign particulate matter by neutrophils, eosinophils, and macrophages.
INFLAMMATION	Confines and destroys microbes and repairs tissues.
FEVER	Intensifies the effects of interferons, inhibits growth of some microbes, and speeds up body reactions that aid repair.

defend itself against specific invading agents such as bacteria, toxins, viruses, and foreign tissues is called **immunity.** Substances that are recognized by the immune system and provoke immune responses are called **antigens (Ags).** Two properties distinguish immunity from the nonspecific defenses: (1) *specificity* for particular foreign molecules (antigens), which also involves distinguishing self from nonself molecules, and (2) *memory* for most previously encountered antigens such that a second encounter prompts an even more vigorous response. The branch of science that deals with the responses of the body when challenged by antigens is called **immunology** (im′-yoo-NOL-ō- jē; *immunis* = free).

Formation of T Cells and B Cells

The cells that carry out immune responses are lymphocytes and antigen presenting cells (described shortly). Both types of lymphocytes—B cells and T cells—develop from stem cells in bone marrow (see Fig. 19.2). In contrast to B cells, which complete their development into functional immune cells in bone marrow, immature T cells migrate from bone marrow into the thymus (Fig. 22.12). After a few days in the thymus, T cells develop **immunocompetence,** the ability to carry out immune responses if properly stimulated.

Before T cells leave the thymus or B cells leave bone marrow, they acquire several distinctive surface proteins.

FIGURE 22.12 Lymphocyte maturation and types of immune responses. Antigen receptors are indicated in red.

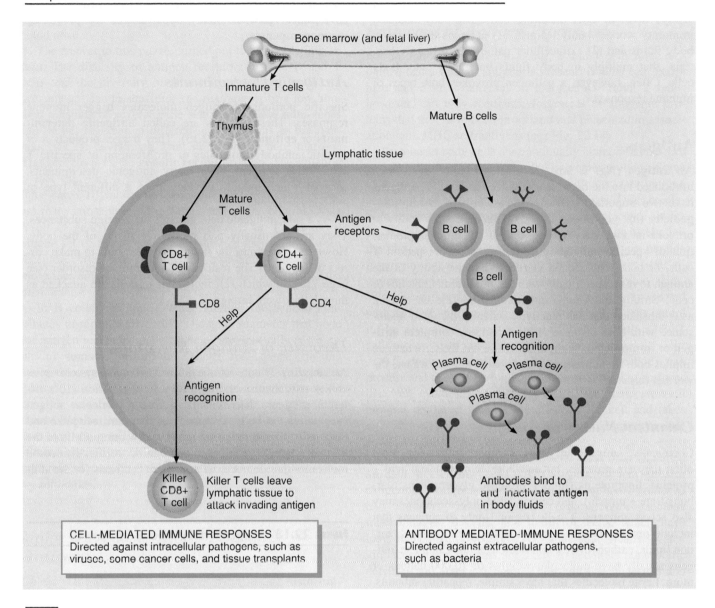

Bone marrow (and fetal liver)

Immature T cells

Thymus

Mature B cells

Lymphatic tissue

Mature T cells

Antigen receptors

B cell

B cell

CD8+ T cell

CD4+ T cell

B cell

CD8

Help

CD4

Help

Antigen recognition

Antigen recognition

Plasma cell

Plasma cell

Plasma cell

Antigen recognition

Killer CD8+ T cell

Killer T cells leave lymphatic tissue to attack invading antigen

Antibodies bind to and inactivate antigen in body fluids

CELL-MEDIATED IMMUNE RESPONSES
Directed against intracellular pathogens, such as viruses, some cancer cells, and tissue transplants

ANTIBODY MEDIATED-IMMUNE RESPONSES
Directed against extracellular pathogens, such as bacteria

Question: Which type of cell is needed for both CMI and AMI responses?

Some function as **antigen receptors,** that is, molecules capable of recognizing specific antigens (Fig. 22.12). In addition, T cells exit the thymus as either CD4+ or CD8+ cells, which means they display either a protein called CD4 or one called CD8 on their plasma membrane. These two types of T cells have very different functions.

Types of Immune Responses

Immunity consists of two kinds of closely allied immune responses, both triggered by antigens. In the first kind, called **cell-mediated (cellular) immune (CMI) responses,** CD8+ T cells proliferate into "killer" T cells and directly attack the invading antigen. In the second kind, called **antibody-mediated (humoral) immune (AMI) responses**, B cells transform into plasma cells, which synthesize and secrete specific proteins called **antibodies (Abs)** or **immunoglobulins** (im′-yoo-nō-GLOB yoo lins). Antibodies bind to and inactivate a particular antigen. Most CD4+ T cells become helper T cells that aid both CMI and AMI responses. The AIDS virus uses the CD4 molecule to enter and then destroy helper T cells (see page 713).

To some extent, each type of immune response specializes in dealing with certain invaders. Cell-mediated

FIGURE 22.14 Steps in processing and presenting of exogenous antigen by an antigen presenting cell (APC).

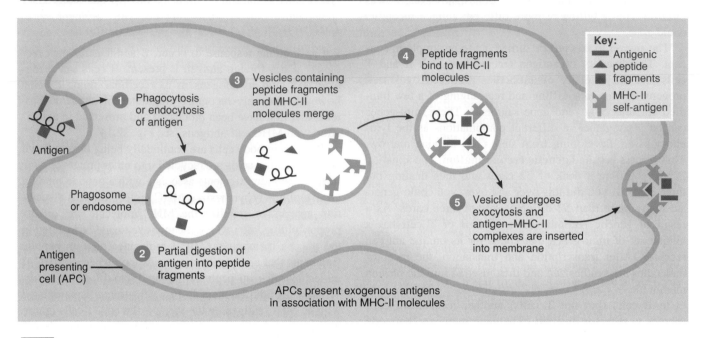

Question: What cells are APCs and where in the body are they found?

3. Fusion of vesicles. The vesicles containing peptide fragments and MHC-II molecules merge.

4. Peptide fragments bind to MHC-II molecules.

5. Exocytosis. As the combined vesicle containing antigen–MHC-II complexes undergoes exocytosis, the complex is inserted into the plasma membrane.

Like other body cells, APCs also can process and present endogenous antigens associated with MHC-I molecules. After processing an antigen, the APC migrates to lymphatic tissue to present it to T cells. There the antigen fragment–MHC complex can be recognized by certain T cells, either to trigger a cell-mediated immune response or to assist B cells in an antibody-mediated immune response.

Cytokines

Lymphocytes and APCs both secrete cytokines, as do fibroblasts, endothelial cells, and monocytes. **Cytokines** are small protein hormones that are needed for many normal cell functions. Most cytokines act locally as autocrines (on the cell that secreted them) or as paracrines (on neighboring cells). A few, for example, erythropoietin, are transported in the blood to distant targets (endocrine action). Several cytokines, called colony stimulating factors and interleukins, stimulate proliferation of progenitor blood cells in bone marrow (see Exhibit 19.2).

Other cytokines regulate activities of the cells that carry out specific and nonspecific defense responses. When secreted by lymphocytes, cytokines are termed **lymphokines.** When secreted by monocytes or macrophages, they are called **monokines.** The more general term cytokine includes both. Exhibit 22.2 describes those cytokines that regulate immune responses.

Antibodies

An **antibody (Ab)** can combine specifically with the antigenic determinant on the antigen that triggered its production. It binds to its antigen just as a key fits into a specific lock. In theory, one could produce as many different antibodies as there are antigen receptors on B cells. Recall that the same recombined genes code for both the antigen receptors on B cells and the antibodies eventually secreted by plasma cells.

Antibodies belong to a group of glycoproteins called globulins, and for this reason they are also known as **immunoglobulins (Igs).** Most antibodies contain four polypeptide chains (Fig. 22.15). Two of the chains are identical to each other and are called **heavy (H) chains.** Each consists of about 450 amino acids. Short carbohydrate chains are attached to each heavy chain. The other two chains, also identical to each other, are called **light (L) chains,** and each consists of more than 200 amino acids. A

EXHIBIT 22.2

SUMMARY OF CYTOKINES (LYMPHOKINES AND MONOKINES)

Cytokine	Comment
Interleukin-1 (IL-1)	Produced by monocytes and macrophages; costimulator of T cell and B cell proliferation; acts on hypothalamus to cause fever.
Interleukin-2 (IL-2) (T cell growth factor)	Secreted by helper T cells to costimulate the proliferation of helper T cells, cytotoxic T cells and B cells; activates natural killer (NK) cells.
Interleukin-4 (IL-4) (B cell stimulating factor 1)	Produced by activated helper T cells; costimulator for B cells; causes plasma cells to secrete IgE antibodies (see Exhibit 22.3); promotes growth of T cells.
Interleukin-5 (IL-5)	Produced by certain activated CD4+ T cells and activated mast cells; costimulator for B cells and causes plasma cells to secrete IgA antibodies.
Tumor necrosis factor (TNF)	Produced mainly by macrophages; stimulates accumulation of leukocytes at sites of inflammation; activates inflammatory leukocytes to kill microbes; stimulates macrophages to produce IL-1; induces synthesis of colony-stimulating factors by endothelial cells and fibroblasts; exerts an interferon-like protective effect against viruses; and functions as an endogenous pyrogen to induce fever (see page 856).
Transforming growth factor beta (TGF-β)	Secreted by T cells and macrophages; has some positive effects but is thought to be important for turning off immune responses; inhibits activation of T cells and macrophages.
Gamma interferon (Gamma-IFN)	Secreted by helper and cytotoxic T cells and NK cells; strongly stimulates phagocytosis by neutrophils and macrophages (formerly called macrophage activating factor [MAF] for this action); activates NK cells; enhances both cellular and antibody-mediated immune responses.
Alpha and beta interferons	Produced by virus-infected cells to inhibit viral replication in uninfected cells; produced by antigen-stimulated macrophages to stimulate T cell growth; activates NK cells.
Lymphotoxin (LT)	Secreted by cytotoxic T cells; kills cells by causing fragmentation of DNA.
Perforin	Secreted by cytotoxic T cells; perforates cell membranes of target cells.
Macrophage migration inhibiting factor	Produced by T cells; prevents macrophages from migrating away from site of infection.

disulfide bond (S–S) holds each light chain to a heavy chain. Two disulfide bonds also link the midregion of the two heavy chains. This part of the antibody displays considerable flexibility and is called the **hinge region.** Since the antibody "arms" can move somewhat as the hinge region bends, an antibody can assume either a **T** shape (Fig. 22.15a) or a **Y** shape (Fig. 22.15b). Disulfide bonds also form loop-shaped **domains** within the light and heavy chains (Fig. 22.15b). Each domain contains about 110 amino acids and folds into a separate unit.

Within each H and L chain are two distinct regions. The tips of the H and L chains, called the **variable (V) regions,** contain the **antigen binding site**. The variable region is different for each kind of antibody. This is the part of the antibody that recognizes and attaches specifically to a particular antigen. Since most antibodies have two antigen binding sites, they are said to be bivalent. Flexibility at the hinge allows simultaneous binding to two antigenic determinants

that are slightly different distances apart, for example, on the surface of a microbe.

The remainder of each polypeptide chain is called the **constant (C) region.** The constant region is nearly the same in all antibodies of the same class (there are five different classes) and is responsible for the type of antigen–antibody reaction that occurs. However, the constant region differs from one class of antibody to another, and its structure serves as a basis for distinguishing the classes. Five different classes of immunoglobulins are known to exist in humans. These are designated as IgG, IgA, IgM, IgD, and IgE. Each has a distinct chemical structure and a specific biological role. Because they appear first and are relatively short-lived, IgM antibodies indicate a recent invasion. In a sick patient, the responsible pathogen may be suggested by finding high levels of IgM to a particular organism. Exhibit 22.3 summarizes the structures and functions of the five classes of antibody.

FIGURE 22.15 Chemical structure of immunoglobin G (IgG) class of antibody. Besides four polypeptide chains (two heavy and two light), a short carbohydrate chain is attached to each heavy chain. In (a) each circle represents one amino acid. In (b) V_L = variable region of light chain; C_L = constant region of light chain; V_H = variable region of heavy chain; C_H = constant regions of heavy chain.

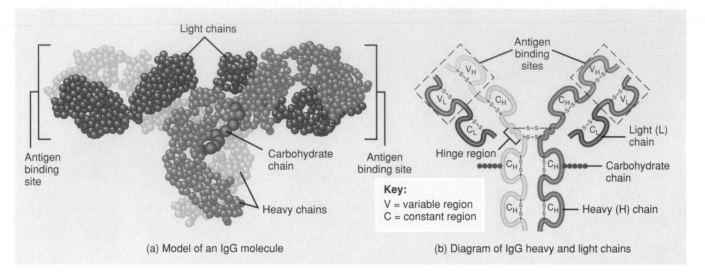

(a) Model of an IgG molecule

(b) Diagram of IgG heavy and light chains

Question: What is the function of the variable region?

EXHIBIT 22.3

CLASSES OF IMMUNOGLOBULINS (Igs)

Name and Structure		Characteristics and Functions
IgG		Most abundant, about 75% of all antibodies in the body; found in blood, lymph, and the intestines; monomer (one unit) structure. Protect against bacteria and viruses by enhancing phagocytosis, neutralizing toxins, and triggering the complement system. The only antibodies to pass the placenta from mother to fetus and thereby confer some immune protection in newborns.
IgA		Make up about 15% of all antibodies in the body; occur as monomers and dimers (two units). Found in tears, saliva, mucus, milk, gastrointestinal secretions, blood, and lymph. Levels decrease during stress, lowering resistance to infection. Provide localized protection on mucous membranes.
IgM	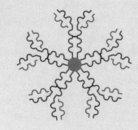	About 5 to 10% of all antibodies; occur as pentamers (five units); first antibodies to be secreted by plasma cells after an initial exposure to any antigen; found in blood and lymph. Cause agglutination and lysis of microbes. Also present as monomers on the surfaces of B cells, where they serve as antigen receptors. A and B agglutinins, which bind to A and B agglutinogens on the surface of red blood cells, are IgM antibodies.

Name and Structure	Characteristics and Functions
IgD	Less than 1% of all antibodies; occur as monomers; found in blood, in lymph, and on the surfaces of B cells as antigen receptors. Involved in activation of B cells.
IgE	Less than 0.1% of all antibodies; occur as monomers; located on mast cells and basophils and are involved in allergic reactions.

Cell-Mediated Immunity

A cell-mediated immune response begins with *recognition* (binding) of a particular antigen by a small number of T cells (lymphocytes). Once an antigen has been recognized, the T cell can undergo *proliferation* and *differentiation* into a clone of **effector** cells, a population of identical cells that can recognize the same antigen and carry out some aspect of the immune attack. Finally, the immune response results in *elimination* of the intruder.

Antigen Recognition by T Cells

Antigen receptors on the surface of T cells are called **T cell receptors (TCRs).** They recognize and bind to specific foreign antigen fragments that are presented together with self MHC molecules (see Fig. 22.16). There are literally millions of different T cells, each with unique TCRs that can recognize a specific antigen–MHC combination. At any given time, most T cells are inactive. When an antigen enters the body, only a few T cells have TCRs that can recognize and bind to the antigen. Antigen recognition by a TCR is the *first signal* in activation of a T cell.

A *second signal*, called a **costimulator,** also is needed. More than 20 such costimulators are known. Some are secreted cytokines, for example, **interleukin-1 (IL-1)** and **interleukin-2 (IL-2).** Other costimulators include pairs of plasma membrane molecules, one on the surface of the T cell and a second on the surface of an APC, that cause the two cells to adhere to one another for a period of time.

The situation is a little like starting a car. When you insert the correct key (antigen) in the ignition (TCR) and turn it, the car starts (recognition of specific antigen). But the car does not move forward until you move the gear shift (costimulator) into drive. The need for costimulation may prevent an immune response from occurring accidentally. It is thought that different costimulators affect the activated T cell in different ways, just as shifting a car into reverse has a different effect than shifting it into drive. Moreover,

recognition (antigen binding to receptor) without costimulation is thought to lead to a prolonged *state of inactivity* called **anergy** in both T cells and B cells. (Imagine a car in neutral with its engine running until it's out of gas!)

Proliferation and Differentiation of T Cells

When a T cell has received two signals (antigen recognition and costimulation), it is said to be **activated** or **sensitized.** It enlarges and begins to **proliferate** (divide several times) and **differentiate** (form more highly specialized cells). The result is a **clone,** or population of cells that can recognize the same antigen. Before the first exposure to a certain antigen, only a handful of T cells might be able to recognize it. But after an immune response has occurred, there are thousands.

Several different types of differentiated T cells appear: helper T cells, cytotoxic (killer) T cells, suppressor T cells, and memory T cells.

● **Helper T (T$_H$) Cells** Helper T (T$_H$) **cells** or **T4 cells** develop from T cells that display CD4. Resting (inactive) T$_H$ cells recognize antigen fragments associated with MHC-II molecules and are costimulated by interleukin-1, which is secreted by macrophages (Fig. 22.16a). This means they are activated mainly by antigen presenting cells. Within hours after costimulation, helper T cells start secreting a variety of cytokines (see Exhibit 22.2).

One very important cytokine produced by helper T cells is interleukin-2 (IL-2). IL-2 is needed for virtually all immune responses and it is the prime substance that triggers T cell division. It can act as a costimulator for resting helper T or cytotoxic T cells, and it enhances activation and proliferation of T cells, B cells, and natural killer cells. Other cytokines secreted by helper T cells are gamma interferon, interleukin-4 (IL-4), and transforming growth factor beta (TGF-β).

FIGURE 22.16 Activation, proliferation, and differentiation of T cells. Note that CD4 binds to MHC-II molecules whereas CD8 binds to MHC-I molecules. This binding helps anchor the TCR-antigen interaction so that antigen recognition can occur.

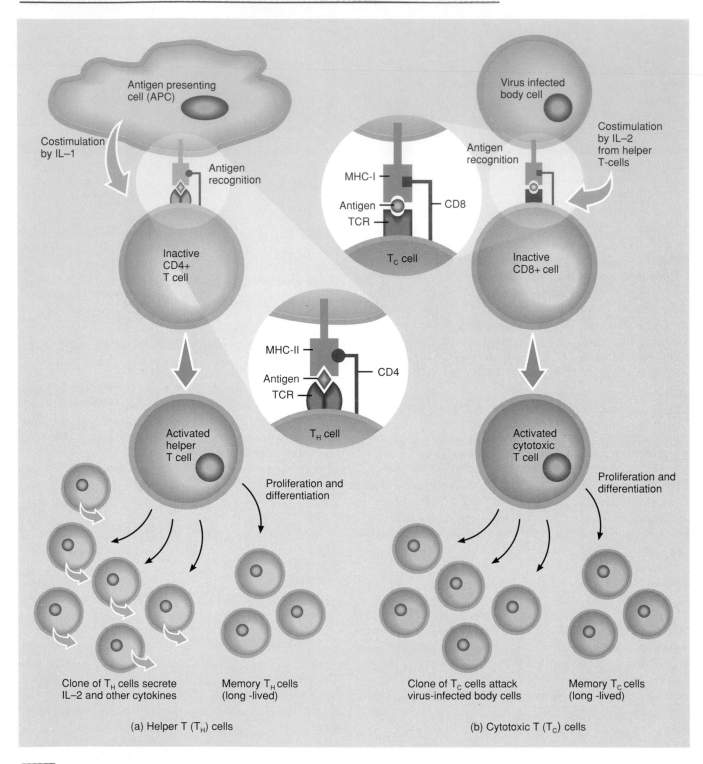

(a) Helper T (T$_H$) cells

(b) Cytotoxic T (T$_C$) cells

Question: What are the first and second signals in activation of a T cell?

Some actions of interleukin-2 provide a good example of a positive feedback system that makes a beneficial contribution. As described earlier, activation of a helper T cell stimulates it to start secreting IL-2. IL-2 acts in an autocrine manner and binds to IL-2 receptors on the plasma membrane of the cell that secreted it. One effect is stimulation of cell division. As the T_H cells proliferate, they secrete more and more IL-2, a positive feedback effect. IL-2 may also act in a paracrine manner by binding to IL-2 receptors on neighboring T_H, T_C, or B cells. If any of these cells have already bound an antigen, IL-2 serves as a costimulator for their activation.

● **Cytotoxic T (T_C) Cells**   Cytotoxic T (T_C) **cells** or **T8 cells** develop from T cells that display CD8. They also are known as **cytolytic** or **killer T cells**. T_C cells recognize foreign antigens combined with MHC-I molecules on the surfaces of body cells infected by viruses, some tumor cells, and cells of a tissue transplant (Fig. 22.16b). However, to become cytolytic (able to lyse cells) they need costimulation by IL-2 or other cytokines produced by helper T cells. (Recall that T_H cells are activated by antigen associated with MHC-II molecules.) Thus maximal activation of T_C cells requires presentation of antigen associated with both MHC-I and MHC-II molecules.

● **Other Types of T Cells**   T cells that mediate a class of allergic response called **delayed type hypersensitivity** (see page 716) may display either CD4 or CD8. In response to activation by antigen, these T cells secrete cytokines, especially gamma interferon, that activate macrophages. The activated macrophages, in turn, eliminate the antigen.

Suppressor T (T_S) cells are thought to be a class of T cells distinct from T_H cells and T_C cells. However, their presence is difficult to demonstrate, the nature of their receptors for antigen is not known, and their very existence is controversial. They may downregulate or dampen parts of the immune response by producing cytokines such as TGF-β, which inhibits proliferation of B cells and T cells. Another possibility is that they directly destroy activated lymphocytes.

Memory T cells are programmed to recognize the original invading antigen. Should the same type of pathogen invade the body at a later date, thousands of memory cells are available to initiate a far swifter reaction than occurred during the first invasion. The second response usually is so fast that the pathogens are destroyed before any signs or symptoms of disease occur.

Elimination of Invaders

Cytotoxic T cells are the army that marches forth to do battle with foreign invaders in cell-mediated immune responses. They leave the lymphoid tissue and migrate to the site of invasion, infection, or tumor formation. They recognize and attach to the target cell that bears the same antigen as stimulated their activation and proliferation.

Then they deliver a "lethal hit" that causes lysis of the target cell without damaging the T_C itself. After detaching from the target cell, the T_C can seek out and destroy another invader that displays the same antigen (Fig. 22.17).

Two killing mechanisms are used by T_C cells. In the first, granules containing perforin undergo exocytosis from the T_C cell. Perforin forms holes in the plasma membrane of the target cell, which allow extracellular fluid to flow in, and the cell bursts (cytolysis). In the second, the T_C cell secretes a toxic molecule known as **lymphotoxin (LT)** that activates damaging enzymes within the target cell. These enzymes cause the DNA to fragment, and the target cell dies. Cytotoxic T cells also secrete gamma–interferon, which activates neutrophils and macrophages, greatly increasing their phagocytic activity.

In summary, cytotoxic T cells can destroy antigens directly by killing the cells that bear them and indirectly by secreting gamma interferon to activate phagocytic cells at

---

FIGURE 22.17 Activity of cytotoxic T cell. After delivering a lethal hit, the T cell can detach and attack another target cell displaying the same antigen.

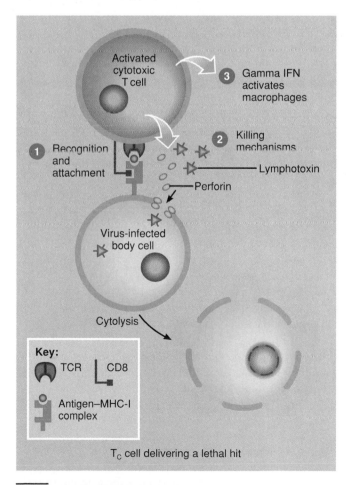

T_C cell delivering a lethal hit

Question: Besides cells infected by viruses, what other types of target cells are attacked by T_C cells?

the scene of the battle. T$_C$ cells are especially effective against slowly developing bacterial diseases (such as tuberculosis and brucellosis), some viruses, fungi, cancer cells associated with viral infection, and transplanted cells.

Antibody-Mediated (Humoral) Immunity

The body contains not only millions of different T cells but also millions of different B cells, each capable of responding to a specific antigen. Whereas cytotoxic T cells leave lymphoid tissue to meet a foreign antigen, B cells stay put. In the presence of a foreign antigen, specific B cells in lymph nodes, the spleen, or lymphoid tissue in the gastrointestinal tract become activated. They differentiate into plasma cells that secrete specific antibodies, which then circulate in the lymph and blood to reach the site of invasion.

During activation of a B cell, an antigen binds to antigen receptors on the cell surface (Fig. 22.18). B cell antigen receptors are chemically similar to the antibodies that will eventually be secreted by their progeny. Although B cells can respond to unprocessed antigen in lymph or interstitial fluid, their response is much more intense when nearby follicular dendritic cells also process and present antigen to them. Some antigen is then taken into the B cell, broken down into peptide fragments and combined with MHC-II self-antigen, and moved to the B cell surface. Helper T cells recognize the antigen–MHC-II combination and deliver the costimulation needed for B cell proliferation and differentiation. The T$_H$ cell produces interleukin-2 and other cytokines that act as costimulators to initiate B cell division and differentiation. B cell proliferation and differentiation into plasma cells are also influenced by interleukin-1, secreted by macrophages.

Some of the B cells enlarge and divide and differentiate into a clone of **plasma cells**. The phenomenal rate of antibody secretion by plasma cells is about 2000 molecules per second for each cell, and it occurs for four or five days until the plasma cell dies. The activated B cells that do not differentiate into plasma cells remain as **memory B cells,** ready to respond more rapidly and forcefully should the same antigen appear at a future time.

Different antigens stimulate different B cells to develop into plasma cells and their accompanying memory B cells. The B cells of a particular clone are capable of secreting only one kind of antibody. The secreted antibody is identical in its specificity to the antigen receptor displayed by the progenitor B cell that responded to the antigen in the first place. Each specific antigen activates only those B cells that are predestined (by the combination of mini-genes they carry) to secrete antibody specific to that antigen.

Antibodies produced by plasma cells enter the circulation and form antigen–antibody complexes with the antigen that initiated their production. In many cases this reaction inactivates the antigen and enhances the chance that it will

FIGURE 22.18 Activation, proliferation, and differentiation of B cells.

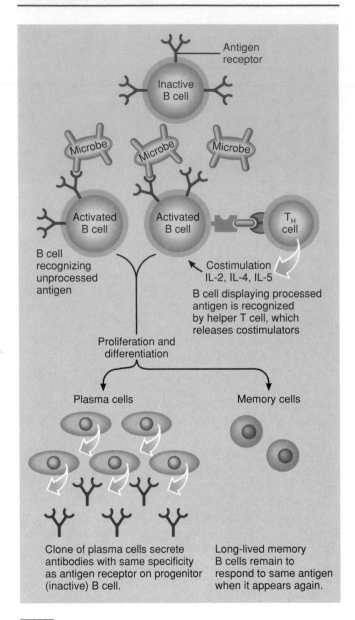

Question: How many different kinds of antibodies will be secreted by the plasma cells in the clone shown here?

be phagocytized. The antigen–antibody complexes may also activate complement components for attack and fix complement to the surface of the antigens.

A summary of the various cells involved in immune responses is presented in Exhibit 22.4.

CLINICAL APPLICATION

MONOCLONAL ANTIBODIES

Scientists have known for many years how to stimulate laboratory animals (or humans) to produce antibodies. After

EXHIBIT 22.4

SUMMARY OF CELLS THAT PARTICIPATE IN IMMUNE RESPONSES

Cell	Function
ANTIGEN PRESENTING CELLS (APCs)	
Macrophage	Phagocytosis; processing and presentation of foreign antigens to T cells; secretion of interleukin-1 that stimulates secretion of interleukin-2 by helper T cells (stimulates proliferation of cytotoxic T cells) and induces proliferation of B cells; secretion of interferons that stimulate T cell growth.
Dendritic cell	Processes and presents antigen to T and B cells; found in mucous membranes and skin.
B cell	Processes and presents antigen to helper T cells.
LYMPHOCYTES	
Cytotoxic (cytolytic or killer) T cell (T_C or T8 cell)	Lysis and death of foreign cells by perforin and lymphotoxin; release of other cytokines that attract macrophages and increase their phagocytic activity (gamma-IFN) and prevent macrophage migration from site of action (macrophage migration inhibition factor).
Helper T cell (T_H or T4 cell)	Cooperates with B cells to amplify antibody production by plasma cells and secretes interleukin-2, which stimulates proliferation of T and B cells. May secrete gamma-IFN and tumor necrosis factor (TNF), which stimulate inflammatory response.
Suppressor T cell (T_S cell)	Thought to downregulate immune responses by producing cytokines such as TGF-β, which inhibits proliferation of B and T cells. May also directly destroy activated lymphocytes.
Memory T cell	Remains in lymphoid tissue and recognizes original invading antigens, even years after infection.
B cell	Differentiates into antibody-producing plasma cell. May process and present antigen to helper T cell.
Plasma cell	Descendant of B cell that produces antibodies.
Memory B cell	Ready to respond more rapidly and forcefully than initially should the same antigen enter the body in the future.

injecting a particular antigen, antibodies are produced against the antigen by plasma cells and can be harvested from the blood. However, since the antigen typically has many antigenic determinants, the antibodies are produced by many different clones of plasma cells. These antibodies vary physically and chemically; they are not pure. If a single plasma cell could be isolated and induced to proliferate into a clone of identical cells, then the antibodies produced would all be the same. Because lymphocytes and plasma cells are difficult to grow in culture, a "trick" is needed to succeed in producing pure antibodies.

The trick that worked incorporated fusion of a B cell with a tumor cell that is capable of proliferating endlessly. The resulting hybrid cell is called a **hybridoma** (hī-bre-DŌ-ma). Hybridoma cells are a long-term source of large quantities of pure antibodies called **monoclonal antibodies (MAbs)** because they come from a single clone of identical cells. Such antibodies combine with just one antigenic determinant.

One clinical use of monoclonal antibodies is for measuring levels of a drug in a patient's blood. Other uses include the diagnosis of pregnancy, allergies, and diseases such as hepatitis, rabies, and some sexually transmitted diseases. They have also been used to detect cancer at an early stage and to determine the extent of metastasis. MAbs may also be useful in preparing vaccines to counteract the rejection associated with transplants, to treat autoimmune diseases, and perhaps to treat AIDS.

Immunological Memory

A hallmark of immune responses is memory for specific antigens that have triggered immune responses in the past. Immunological memory is due to the presence of long-lived antibodies and very long-lived lymphocytes that arise during proliferation and differentiation of antigen-stimulated B and T cells.

The immune response of the body, whether cell-mediated or antibody-mediated, is much quicker and more intense after a second or subsequent exposure to an antigen than after the initial exposure. Initially, only a few cells have the correct specificity to respond and the immune response may take several days to build to maximum intensity. Because thousands of memory cells exist after an encounter with an antigen, the next time they can proliferate and differentiate into plasma cells or cytotoxic T cells within hours.

Immunological memory can be demonstrated by measuring the amount of antibody in serum, called the *antibody titer* (TĪ-ter). After an initial contact with an antigen, there is a period of several days during which no antibody is present. Then there is a slow rise in the antibody titer, first IgM and then IgG, followed by a gradual decline (Fig. 22.19). This is called the **primary response.**

FIGURE 22.19 Primary (after first exposure) and secondary (after second exposure) antibody-mediated immune responses to the same antigen.

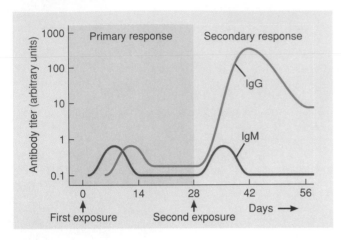

Question: According to this graph, how much more IgG is secreted into the blood in the secondary versus the primary response?

Memory cells may remain for decades. Every time the same antigen is encountered again, there is a rapid proliferation of memory cells. The antibody titer is far greater than during a primary response and is mainly IgG antibodies. This accelerated, more intense response is called the **secondary response.** Antibodies produced during a secondary response have an even higher affinity for the antigen than those secreted during a primary response and are thus more successful in disposing of it.

Primary and secondary responses occur during microbial infection. When you recover from an infection without taking antibiotic drugs, it is usually because of the primary response. If, at a later time, you are infected by the same

microbe, the secondary response could be so swift that the microbes are quickly destroyed and you do not exhibit any signs or symptoms.

Immunological memory provides the basis for immunization by vaccination against certain diseases. When you receive the vaccine, your B and T cells are activated. Should you encounter the same pathogen again as an infecting microbe, your body experiences the secondary response. Exhibit 22.5 summarizes the various types of immunity.

Self-Recognition and Immunological Tolerance

To function properly, all T cells must have two traits. Each of your T cells must (1) be able to recognize your MHC molecules, a process known as **self-recognition,** and (2) lack reactivity to peptide fragments from your own proteins, a condition known as **immunological tolerance**. B cells also display immunological tolerance. Loss of immunological tolerance leads to autoimmune disorders (see page 715).

While residing in the thymus gland, immature T cells that become capable of recognizing self-MHC molecules survive while those that do not die (Fig. 22.20a). This aspect of development of immunocompetence is termed **positive selection.** *It ensures that surviving T cells will be able to recognize the MHC part of an antigen–MHC complex.*

Developing immunological tolerance occurs by a process called **negative selection.** Cells with TCRs that recognize peptide fragments from self-proteins are eliminated or inactivated (Fig. 22.20a). *Negative selection ensures that surviving T cells will not respond to fragments of molecules that are normally present in the body.* Negative selection occurs in two ways: **deletion** and **anergy.** In deletion, self-

EXHIBIT 22.5

TYPES OF IMMUNITY

Type of Immunity	How Acquired
Naturally acquired active immunity	Antigen recognition by B cells and T cells and costimulation leads to antibody-secreting plasma cells, cytotoxic T cells, and B and T memory cells.
Naturally acquired passive immunity	Transfer of IgG antibodies from mother to fetus across placenta or from mother to baby in milk during breast feeding.
Artificially acquired active immunity	Antigens introduced in a vaccination stimulate CMI and AMI responses, leading to production of memory cells. The antigens are pretreated to be immunogenic but not pathogenic. That is, they will trigger an immune response but not cause significant illness.
Artificially acquired passive immunity	Intravenous injection of immunoglobulins (antibodies).

FIGURE 22.20 Development of self-recognition and immunological tolerance. Green arrows indicate events that lead to cell survival; red arrows indicate cell death or inactivation.

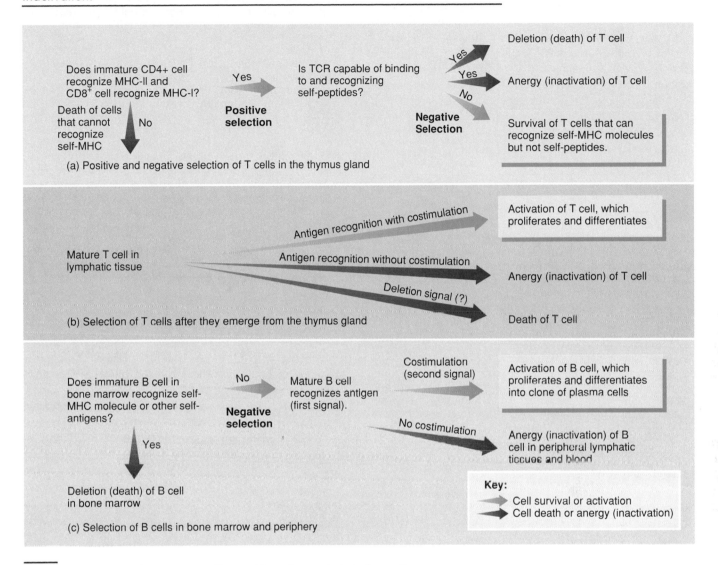

(a) Positive and negative selection of T cells in the thymus gland

(b) Selection of T cells after they emerge from the thymus gland

(c) Selection of B cells in bone marrow and periphery

Question: Does positive selection ensure self-recognition or immunological tolerance?

reactive T cells die whereas in anergy they remain alive but are unresponsive to antigenic stimulation. It is estimated that only 1 in 100 immature T cells that enter the thymus survives both positive and negative selection and emerges as a mature, immunocompetent T cell.

Even after T cells emerge from the thymus, it is possible they may come in contact with an unfamiliar self-protein. In such cases, they also become anergic because there is no costimulator (Fig. 22.20b). Some evidence suggests that deletion of self-reactive T cells is also possible after they leave the thymus.

B cells also develop tolerance through deletion and anergy (Fig. 22.20c). While B cells are developing in bone marrow, those cells exhibiting antigen receptors that recognize common self-antigens such as MHC antigens and

ABO blood group antigens are deleted. Once B cells are released into the blood, however, anergy appears to be the main mechanism for preventing responses to self-proteins. When B cells encounter antigen not associated with an antigen presenting cell (APC), the necessary costimulation signal often is missing. In this case, the B cell is likely to become anergic (inactivated) rather than activated.

Immunology and Cancer

When a normal cell transforms into a cancer cell, it may display cell surface components called **tumor antigens.** These are molecules that are rarely, if ever, displayed on the surface of normal cells. If the immune system can recognize tumor antigens as nonself, it can destroy the cancer cells carrying them. Such an immune response is called **immunological surveillance** and is carried out by cytotoxic T cells, macrophages, and natural killer cells. It appears to be most effective in eliminating tumor cells that arise due to a cancer-causing virus. In part, evidence for immunological surveillance comes from observations in people whose immune system is depressed, for whatever reason. For example, patients who are taking immunosuppressive drugs to prevent rejection of a tissue transplant (see page 716) do not have a higher than normal incidence of most cancers. They do have a greatly increased incidence, however, of virus-associated cancers.

CLINICAL APPLICATION

IMMUNOTHERAPY

For many years, researchers have been trying to induce the immune system to mount an attack against cancer, an approach called **tumor immunotherapy.** Advances in the past few years are translating into effective therapies against some types of human cancer.

In **adoptive cellular immunotherapy** cells that have antitumor activity are injected into the blood of a cancer patient. The hope is that these "adopted" cells will seek out and destroy tumor cells. One method uses a patient's own inactive cancer-fighting cytotoxic T cells and natural killer cells. They are removed in a blood sample and cultured with IL-2, which activates them. Such cells are called lymphokine-activated killer (LAK) cells. The LAK cells are then transfused back into the patient's blood. Although some tumor regression does occur with this procedure, severe complications affect almost all patients.

Cytokine therapy uses cytokines as therapeutic agents.

The first cytokines that were shown to be effective against any human cancer were interferons (IFNs), especially alpha-IFN. Alpha-IFN induces remarkable improvement in patients with a rare cancer called hairy cell leukemia. The beneficial effects of alpha-IFN are more modest in renal cell carcinoma, melanomas, various lymphomas, and Kaposi's sarcoma. Interferons produce little or no improvement in lung, breast, or colon cancers.

Of the interleukins, the one most widely employed to fight cancer is interleukin-2, either alone or with LAK cells. The treatment is effective in causing tumor regression in some patients, presumably by stimulating cytotoxic T cells and natural killer cells. It also can be very toxic. Among the adverse effects are high fever, severe weakness, difficult breathing due to pulmonary edema, and hypotension leading to shock. Interleukin-4 is being tested as it can also activate T_C cells and may have fewer adverse side effects.

Monoclonal antibodies, either alone or in combination with radioactive isotopes, toxins, or drugs, are being used in trial studies of **antibody therapy** to treat cancer. This approach has the advantage of destroying diseased tissues only, while sparing healthy tissues, thus overcoming some of the major adverse effects of chemotherapy and radiation treatment.

AGING AND THE IMMUNE SYSTEM

Elderly people are more susceptible to all types of infections and malignancies because the immune system functions less effectively. With age, there are progressive declines in both cell-mediated and antibody-mediated immune responses. The response to vaccines is decreased, production of antibodies against self-proteins increases, and the number of helper T cells decreases.

DISORDERS: HOMEOSTATIC IMBALANCES

AIDS: ACQUIRED IMMUNE DEFICIENCY SYNDROME

Never before have humans been confronted with an epidemic in which the primary disease only lowers the victim's immunity, and then one or more unrelated diseases produce the fatal symptoms. **Acquired immune deficiency syndrome (AIDS)** is caused by **human immunodeficiency virus (HIV).** The initial response to HIV invasion is a modest decline in the number of circulating T4 cells. Infected people experience a brief flu-like illness, with chills and fever, but the immune system fights back by making antibodies against HIV and the number of T4 cells recovers nearly to normal. Although infected people test positive for HIV antibodies, they typically have few clinical signs or symptoms and do not yet have AIDS. Over the next two to ten years, the virus slowly destroys the T4 cell population. As immune responses weaken, people develop certain **indicator diseases** (diseases that are rare in the general population but common in AIDS patients). At this point, the diagnosis of AIDS is made.

Two indicator diseases provided the original clues that a

new disorder had appeared. AIDS was first recognized in June 1981 as a result of reports from the Los Angeles area to the Centers for Disease Control (CDC) of several cases of a very rare type of pneumonia probably caused by a fungus. The pneumonia, called *Pneumocystis carinii* (noo-mō-SIS-tis kar-RIN-ē-ī) pneumonia (PCP), occurred among homosexual males. At about the same time, the CDC also received reports from New York and Los Angeles concerning an increase in the incidence of Kaposi's sarcoma (KS) among homosexual males. It had been a rare, generally benign, skin cancer usually found in elderly Jewish or Italian men. In AIDS patients it is aggressive and rapidly fatal. KS arises from endothelial cells of blood vessels and produces painless purple or brownish skin lesions that resemble bruises.

Scientists first isolated the virus that causes AIDS in 1983. There have been several theories about the origin of HIV. It is believed that it arose by mutation of a virus that had been endemic in some areas of central Africa for many years. The virus has been found in blood samples preserved from as early as 1959 in several African nations and in Great Britain.

In the U.S. the primary victims are homosexual men, intravenous drug users, and patients who received blood products before 1985, when testing of all donated blood for HIV antibodies began. Others at high risk are the heterosexual partners of HIV-infected individuals. Worldwide, however, 75% of those who have AIDS are thought to have contracted the virus through heterosexual contacts. In the U.S. 80% of the AIDS patients are between 20 and 44 years old, and 88% are males, but the number of infected females is rising sharply. AIDS has occurred in all 50 states. At present, the incubation period (time interval from infection with HIV to full-blown AIDS) is usually seven to ten years.

By 1993 a cumulative total of about 400,000 cases of AIDS will have been diagnosed in the U.S. and 300,000 will have died. The World Health Organization (WHO) projects there will be up to 40 million people infected with HIV throughout the world by the year 2000. More than half will be women and a quarter will be children. The current number of infected Americans is estimated to be 1 to 1.5 million, and most do not even know they carry the virus. All carriers of the virus are assumed to be infected for life and are capable of transmitting the virus to others.

HIV: Structure and Pathogenesis

Viruses consist of a core of DNA or RNA surrounded by a protein coat (capsid). Some viruses, including HIV, also contain an envelope (outer layer) composed of a double layer of lipid penetrated by proteins (Fig. 22.21). Outside a living host cell a virus is unable to replicate. However, once a virus enters a cell, the viral nucleic acid uses the host cell's enzymes, ribosomes, nutrients, and other resources to make copies of itself. As these components accumulate, they are assembled into a large number of viruses that can then leave the host cell to infect other cells.

As viruses go through their cycle of replication, they can damage or kill host cells in various ways. These include shutting down the synthesis of host cell proteins, RNA, and DNA; inhibiting cell division; damaging DNA; and rupturing lysosomes that can lead to autolysis. Moreover, the body's own defenses will attack the infected cells, killing them as well as the viruses they harbor.

HIV is a **retrovirus.** Retroviruses carry their genetic material in RNA and copy this genetic coding into DNA by using an enzyme called **reverse transcriptase.** In other words, retroviruses reverse the ordinary flow of genetic information, which is DNA → RNA → proteins. Once the AIDS virus produces its DNA from RNA, the DNA is integrated into the host cell's DNA. There it can remain dormant, giving no sign of its presence, or it can take over the host cell's genetic machinery to produce more viruses.

HIV enters body cells by receptor-mediated endocytosis (see Fig. 3.12). The receptor or docking protein that permits HIV entry is the CD4 molecule (although other cellular factors, not yet understood, must also contribute). With time, the number of T4 cells, mainly helper T cells, declines due to death of infected cells. The result is progressive collapse of the immune system. Since cytokines secreted by helper T cells normally stimulate the activity of monocytes, neutrophils, and macrophages, nonspecific defense mechanisms also are depressed. The person becomes susceptible to **opportunistic infections,** invasion of normally harmless microorganisms that now proliferate wildly because of the defective immune system.

Besides *Pneumocystis carinii* pneumonia, AIDS victims are especially susceptible to tuberculosis, persistent diarrhea, leukoplakia (whitish patches on mucous membranes), cytomegalovirus (leading to blindness), herpes simplex, and severe shingles, among many other bacterial and fungal infections. Besides helper T cells, the AIDS virus also attacks

FIGURE 22.21 Human immunodeficiency virus (HIV), the causative agent of AIDS. (a) A protein called P24 forms the protein coat (capsid) around the viral RNA. GP120 and GP41 are envelope glycoproteins; a protein called P18 lies inside the envelope.

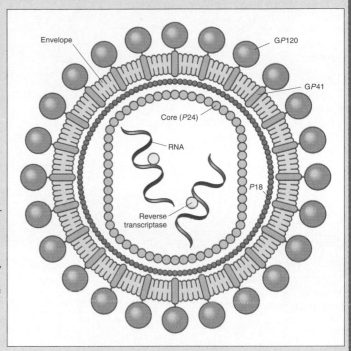

(a) Structure of HIV

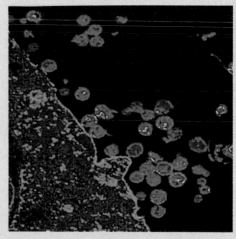

(b) Scanning electron micrograph of HIV

Question: What is the function of reverse transcriptase?

macrophages, dendritic cells, brain cells (where HIV-infected cells may release toxins that disrupt and kill other brain cells), and many others. Not all infected cells display CD4, so there are probably other receptors that permit HIV entry. Also, the virus membrane appears able to directly fuse with the host cell plasma membrane in some cases. Dendritic cells and macrophages are pivotal cells in the development of AIDS.

Continues

They serve as reservoirs for the virus, since viral infection does not seem to harm them, and they spread the virus throughout the body.

Some weeks after infection with HIV, the host develops antibodies against several proteins in the virus. Usually, antibodies appear in blood within 3 to 20 weeks after exposure. Antibodies normally are protective because they help eliminate an intruder. In the case of the AIDS virus, this is not necessarily the case because HIV can remain hidden inside body cells, unavailable to form antigen–antibody complexes. HIV may also escape detection by cytotoxic T cells, natural killer cells, and phagocytes. The virus further evades immune defenses by undergoing rapid antigenic changes in its surface proteins. Moreover, infected cells displaying viral antigens can fuse to uninfected cells and spread the virus that way.

The presence of antibodies against HIV in the blood is used to diagnose HIV infection. In rare cases a person may harbor HIV without forming antibodies against it. Thus a standard blood test that detects antibodies would be negative. The presence of nucleic acids from HIV can still be detected using a method called the polymerase chain reaction (PCR). The test is difficult and expensive, however, which prevents its use for general screening.

Symptoms of AIDS

A simplified definition of AIDS includes anyone infected with HIV and having a CD4 lymphocyte count under $200/mm^3$ of blood. (Normally, the CD4 count would be about $1200/mm^3$.) The various stages through which an HIV-infected patient passes are correlated with decreasing helper T cell counts. After infection by HIV, there are usually mild mononucleosis-like symptoms that may be overlooked. These include fatigue, fever, swollen glands, and headache. Six weeks to six months later HIV can be detected by standard antibody tests.

In the next stage, which lasts for several years, chronically swollen lymph nodes develop in the neck, armpits, and groin. Then, helper T cell counts decline even more, and the patients fail to respond to most skin tests that measure delayed type hypersensitivity (DTH), a measure of the individual's ability to produce a cell-mediated immune response against specific proteins injected under the skin. Now, the HIV-infected person has AIDS. Opportunistic infections develop and many patients suffer from AIDS dementia complex, which is characterized by a progressive loss of function in motor activities, cognition, and behavior. Most persons reaching this stage die within two years.

HIV Outside the Body

Outside the body, HIV is fragile and can be easily eliminated. For example, dishwashing and clotheswashing, by exposing the virus to 135°F (56°C) for 10 minutes, will kill HIV. Chemicals such as hydrogen peroxide (H_2O_2), rubbing alcohol, Lysol, household bleach, and germicidal skin cleaner (such as Betadine and Hibiclens) are also very effective as is standard chlorination in swimming pools and hot tubs.

Transmission

Although HIV has been isolated from several body fluids, the only documented transmissions are by way of breast milk from a nursing mother, blood, semen, and vaginal secretions. The virus is found free and in macrophages in these fluids.

HIV is transmitted by sexual contact from males to females, from females to males, and between males through vaginal, oral, or anal intercourse. Dendritic cells, which are present in

mucosal membranes and are efficient antigen presenting cells, are the suspected route for HIV entry into the lymphatic system. HIV is also effectively transmitted through exchanges of blood, for example, by contaminated hypodermic needles or contact with open wounds. Infected mothers may transmit the virus to their infants before or during birth. It does not appear that people become infected as a result of routine, nonsexual contacts. Millions of family members, co-workers, and friends of AIDS patients do not have AIDS.

No evidence exists that AIDS can be spread through kissing although "deep kissing" with exchange of saliva poses a theoretical danger since dendritic cells are present in the oral mucosa. Similarly, although mosquitoes can harbor the AIDS virus, there is no known case of transmission from a mosquito bite. It also appears that health-care personnel who take proper routine barrier precautions (gloves, masks, safety glasses) are not at risk unless the barriers fail.

Drugs Against HIV

Medical scientists are engaged in what is probably the greatest concentrated effort ever to find a cure for a single disease. One of the problems in treating AIDS is that HIV can lie dormant in body cells. In addition, HIV can infect a variety of cells, including those in the central nervous system that are protected by the blood–brain barrier. Added to this is the problem of opportunistic infections, which may be very difficult to treat. And any therapy must overcome the problem that antiviral agents may also harm host cells. Thus scientists are trying to devise strategies for disrupting specific viral activities. Some research centers on preventing binding of the virus to the host CD4 protein. Other strategies are to prevent conversion of RNA to DNA by reverse transcriptase, block processing of viral proteins by specific viral enzymes, inhibit assembly of viruses within the host cell, and thwart release of new viruses.

Until 1991 the only drug approved by the U.S. Food and Drug Administration (FDA) to treat AIDS was AZT (azidothymidine) or Retrovir. AZT inhibits reverse transcriptase, thus preventing the virus from making DNA from RNA. Among patients taking AZT, there is a slowing in the progression of symptoms. The main side effects are bone marrow damage and anemia. Eventually, the virus develops resistance to the drug. A newer drug, approved by the FDA in late 1991, is ddI (dideoxyinosine). It is also an inhibitor of reverse transcriptase and is recommended for patients who do not respond to AZT. Although patients taking ddI show improvement in immunologic functions, serious side effects are pancreatitis and inflammation of peripheral nerves. Human trials of a similar drug called ddC (dideoxycytidine) were halted when it became apparent that the death rate exceeded that of patients receiving AZT. However, ddC may be effective when used in combination with AZT.

Alpha interferon is believed to inhibit the final stage of virus production. It reduces the spread of Kaposi's sarcoma and is being tried both alone and in combination with other drugs. A host of other drugs are also being tested. It is quite possible that AIDS, like cancer, will have to be treated with a variety of drugs.

Vaccines Against AIDS

Considerable effort is being expended to develop a vaccine against HIV. A vaccine would stimulate the production of antibodies to block the virus before it could infect body cells. The principal experimental vaccines under study use various subunits from the HIV envelope. These can be synthesized in quantity using genetic engineering techniques. Development of an effective AIDS vaccine has been impeded by the ability of HIV to mutate so quickly and the lack of a suitable experimental animal model. Most animals are not susceptible to HIV infection. And although chimpanzees can be infected, they are in short supply and very expensive to maintain. Moreover, there may be a shortage of volunteers when a potential vaccine is ready for testing in humans.

The goal of any vaccination is to provide protection to healthy people so they do not succumb to a disease. However, a genetically engineered experimental vaccine has been tried on 30 volunteers already infected with HIV. Preliminary results, reported in June 1991, indicate that 19 of the 30 participants produced new antibodies against HIV and had an increase in cytotoxic T cells, which recognize and destroy HIV-infected cells. In addition, the vaccine appeared to slow the depletion of T4 cells.

Besides HIV, What Else?

Several aspects of AIDS continue to puzzle scientists. Why do some people live for several years after infection without symptoms while others rapidly progress to death? Why did it originally spread in the U.S. mainly by homosexual contacts whereas heterosexual transmission predominates in Africa? Why doesn't the body's immune system wipe out the invader before it causes massive destruction of helper T cells? One possibility is that a second infective agent acts synergistically with HIV to cause the damage. A candidate for such a cofactor is a mycoplasma. Classified as bacteria, mycoplasmas are the smallest and simplest organisms that can live without a host. Experiments published in 1991 showed that coinfection of cultured CD4 lymphocytes with HIV and a mycoplasma enhanced the ability of HIV to cause cell death. Another proposal is that AIDS is an autoimmune disease in which T cells lose the ability to distinguish self from nonself and attack one another. Clearly, more research is needed before we can fully appreciate this devastating disease.

Prevention of Transmission

At present, the only means of preventing AIDS is to block transmission of the virus. Sexual transmission of HIV can be prevented by abstinence from vaginal, oral, and anal intercourse with infected persons. Sexual transmission can be reduced by the use of effective barrier methods (condoms and spermicides such as nonoxynol 9) during intercourse. Infection from donated blood and blood products is now very rare. In the U.S. all blood has been tested for HIV antibodies since 1985. AIDS transmission via contaminated hypodermic needles could be avoided by sterilization of the needles with chlorox before each use. HIV-infected women should avoid pregnancy. If these measures are to be effective, they must be part of an overall program involving education, counseling, screening individuals at high risk, tracing contacts, and modifying behavior. Until there is effective drug therapy or an effective vaccine, blocking the spread of AIDS depends on education and safer sexual practices.

AUTOIMMUNE DISEASES

Under normal conditions, the body's immune mechanisms tolerate its own tissues and molecules. At times, however, self-tolerance breaks down, and this leads to an **autoimmune disease (autoimmunity).** The immune system fails to recognize self-antigens and produces an immune attack against them. Currently, there are two theories as to why autoimmune diseases occur.

First, T cells that react dangerously with self-proteins may escape from the thymus. Recall that normally such T cells are weeded out by negative selection. Second, T cells that were anergized (turned off) because they did respond to a self-antigen may inappropriately get turned back on. Further experimentation is needed to confirm these theories.

Among human autoimmune diseases are rheumatoid arthritis (RA), systemic lupus erythematosus (SLE), thyroiditis, rheumatic fever, glomerulonephritis, encephalomyelitis, hemolytic and pernicious anemias, Addison's disease, Graves' disease, insulin-dependent (type I) diabetes mellitus, myasthenia gravis, and multiple sclerosis (MS).

Therapy for autoimmune diseases typically involves giving drugs to suppress immune responses. This, of course, leaves the person prone to opportunistic infections. In a few patients with multiple sclerosis and rheumatoid arthritis, clinical trials are under way to test **oral antigen therapy.** The patient eats the same protein antigens that are thought to have caused the disease in the first place. Animal experiments have shown that such antigen feeding acts like a vaccine to suppress some autoimmune responses.

SYSTEMIC LUPUS ERYTHEMATOSUS

Systemic lupus erythematosus (er-e´-thēm-a-TŌ-sus), **SLE,** or **lupus** (*lupus* = wolf) is an autoimmune, inflammatory disease of connective tissue, occurring mostly in young women. In SLE, damage to blood vessel walls results in the release of chemicals that mediate inflammation. The blood vessel damage can be associated with virtually every body system.

The cause of SLE is not known, and its onset may be abrupt or gradual. It is not contagious and is thought to be hereditary. There seems to be a strong incidence of other connective tissue disorders—especially rheumatoid arthritis and rheumatic fever—in relatives of SLE victims. The disease may be triggered by drugs, such as penicillin, sulfa, or tetracycline, exposure to excessive sunlight, injury, emotional upset, infection, or other stress.

Symptoms of SLE include painful joints, slight fever, fatigue, mouth ulcers, weight loss, enlarged lymph nodes and spleen, photosensitivity, rapid loss of large amounts of scalp hair, and sometimes an eruption across the bridge of the nose and cheeks called a "butterfly rash." Other skin lesions may occur with blistering and ulceration. The erosive nature of some of the SLE skin lesions was thought to resemble the damage inflicted by the bite of a wolf—thus the term lupus. The most serious complications of the disease involve inflammation of the kidneys, liver, spleen, lungs, heart, and the central nervous system.

CHRONIC FATIGUE SYNDROME

Chronic fatigue syndrome (CFS) usually occurs in young adults, primarily females. It is characterized by extreme fatigue that impairs normal activities for at least six months and by the absence of known diseases (cancer, infections, drug abuse, toxicity, or psychiatric disorders) that might produce similar symptoms. Diagnostic guidelines specify that at least 8 of the following 11 indications persist or recur over six months: mild fever or chills, sore throat, painful lymph nodes, general muscle weakness, muscle pain, fatigue for more than 24 hours after mild exercise, headache that differs in type and severity from past ones, joint pain without swelling, neuropsychological complaints (irritability, memory loss, confusion, depression), sleep disturbances, and development of the initial symptoms over a few hours to a few days. The diagnosis can also be made if the patient reports 6 of the 11 symptoms just listed

Continues

plus observation by a physician of two of these three physical signs: low-grade fever, inflamed throat, and enlarged lymph nodes in the neck or axilla.

The cause of CFS is not known. Among the possible causes of CFS are lowered levels of corticotropin-releasing hormone (CRH) and cortisol, emotional factors such as depression and excess stress, and viral infection. An experimental antiviral drug called Ampligen relieved some symptoms in a small group of patients treated for six months.

SEVERE COMBINED IMMUNODEFICIENCY

Severe combined immunodeficiency (SCID) is a rare immunodeficiency disease in which both B cells and T cells are missing or inactive in providing immunity. Perhaps the most famous patient with SCID was David, the "bubble boy," who lived in a sterile plastic chamber for almost all of his life; he died in 1984 at age 12.

David was placed in the sterile chamber shortly after birth to protect him from microbes that his body could not fight. In an effort to correct his disorder, David underwent a bone marrow transplant from his older sister. Eighty days after the transplant and still in a germ-free environment, David developed some of the symptoms of infectious mononucleosis, a condition caused by the Epstein–Barr virus. David was brought out of isolation for easier treatment, with the hope that the transplant would provide the same protection as his sterile plastic chamber. Unfortunately, David died about four months later from cancer associated with Epstein–Barr virus.

HYPERSENSITIVITY (ALLERGY)

A person who is overly reactive to an antigen that is tolerated by most others is said to be **hypersensitive (allergic).** Whenever an allergic reaction occurs, there is tissue injury. The antigens that induce an allergic reaction are called **allergens.** Common allergens include certain foods (milk, peanuts, shellfish, eggs), antibiotics (penicillin, tetracycline), vitamins (thiamine, folic acid), drugs (insulin, ACTH, estradiol), vaccines (pertussis, typhoid), venoms (honeybee, wasp, snake), cosmetics, chemicals in plants such as poison ivy, pollens, dust, molds, iodine-containing dyes used in certain x-ray procedures, and even microbes.

There are four basic types of hypersensitivity reactions: type I (anaphylaxis), type II (cytotoxic), type III (immune complex), and type IV (cell-mediated). The first three involve antibodies; the last involves T cells.

Type I (anaphylaxis) reactions are the most common and occur within a few minutes after a person sensitized to an allergen is reexposed to it. **Anaphylaxis** (an′-a-fi-LAK-sis) literally means "against protection" and results from the interaction of allergens with IgE antibodies on the surface of mast cells and basophils. Basophils circulate in blood; mast cells are especially numerous in connective tissue of the skin and respiratory system and endothelium of blood vessels. In response to certain allergens, some people produce IgE antibodies that bind to the surface of mast cells and basophils.

The next time the same allergen enters the body, it attaches to the IgE antibodies already present on the surface of mast cells and basophils. In response, the cells release chemicals called **mediators of anaphylaxis,** among which are histamine, prostaglandins, leukotrines, and kinin. Collectively, the mediators cause vasodilation, increased blood capillary permeability, increased smooth muscle contraction in the airways of the lungs, and increased mucus secretion. As a result, a person may experience inflammatory responses, difficulty in breathing from constricted bronchial tubes, and a "runny" nose from excess mucus secretion.

Some anaphylactic reactions, such as hives, eczema, swelling of the lips or tongue, abdominal cramps, and diarrhea, are **localized** (affecting one part or a limited area). Other anaphylactic reactions are **systemic** (affecting several parts or the entire body). An example is acute anaphylaxis (anaphylactic shock), which may occur in a susceptible individual who has just received a triggering drug or been stung by a wasp. The person develops respiratory symptoms (wheezing and shortness of breath) as bronchioles constrict, usually accompanied by cardiovascular failure and collapse due to vasodilation and fluid loss from blood. This life-threatening emergency is usually treated by injecting epinephrine to dilate the airways and strengthen the heartbeat.

Type II (cytotoxic) reactions are caused by antibodies (IgG or IgM) directed against antigens on a person's blood cells (red blood cells, lymphocytes, or platelets) or tissue cells. The reaction of antibodies and antigens usually leads to activation of complement. Type II reactions, which may occur in incompatible transfusion reactions, damage cells by causing lysis.

Type III (immune complex) reactions involve antigens (not part of a host tissue cell), antibodies (IgA or IgM), and complement. When certain ratios of antigen to antibody occur, the complexes are small and escape phagocytosis. The complexes become trapped in the basement membrane under the endothelium of blood vessels, activate complement, and cause an inflammation. Conditions that so arise include glomerulonephritis, systemic lupus erythematosus (SLE), and rheumatoid arthritis (RA).

Type IV (cell-mediated) reactions or **delayed type hypersensitivity (DTH) reactions** are carried out by macrophages that have become activated by T cells. They usually appear 12 to 72 hours after exposure to an allergen. Type IV reactions occur when allergens are taken up by antigen presenting cells, such as Langerhans cells in the skin, which then migrate to lymph nodes and present the allergen to T cells. This results in sensitization and proliferation of T cells, some of which migrate via the lymph and blood to the site of allergen entry into the body. There they secrete cytokines, such as gamma interferon, which activates macrophages, and tumor necrosis factor (TNF), which stimulates an inflammatory response. Intracellular bacteria, such as *Listeria monocytogenes* and *Mycobacterium tuberculosis*, trigger this type of cell-mediated immunity, as do certain haptens, for example, poison ivy toxin. The skin test for tuberculosis also is a delayed hypersensitivity type reaction.

TISSUE REJECTION

Transplantation involves the replacement of an injured or diseased tissue or organ. Usually, the immune system recognizes the proteins in the transplanted tissue or organ as foreign and mounts both CMI and AMI responses against them. This phenomenon is known as **tissue rejection.** The more closely matched the MHC (HLA) antigens between donor and recipient, the weaker the tissue rejection response.

The most successful transplants are **autografts,** transplants in which one's own tissue is grafted to another part of the body (such as skin grafts for burn treatment or plastic surgery) and **isografts,** transplants in which the donor and recipient are genetically identical. An **allograft** is a transplant between individuals of the same species but with different genetic backgrounds. The success of this type of transplant has been moderate. Often, it is used as a temporary measure until the damaged or diseased tissue is able to repair itself. Skin transplants from other people and blood transfusions might properly be considered allografts. A **xenograft** is a transplant between animals of different species. This type of transplantation is used primarily as a physiological dressing over severe burns.

Until recently, **immunosuppressive drugs** suppressed not only the recipient's immune rejection of the donor tissue but also the immune response to all antigens as well. This causes patients to become very susceptible to infectious diseases. A drug called *cyclosporine,* derived from a fungus, has largely overcome this problem with regard to kidney, heart, and liver transplants. Cyclosporine inhibits secretion of IL-2 by helper T cells but has only a minimal impact on B cells. Thus rejection

is avoided and resistance against some diseases is still maintained.

HODGKIN'S DISEASE

Hodgkin's disease (HD) is a form of cancer, usually arising in lymph nodes, the cause of which is unknown. It may, however, arise from a combination of genetic predisposition, disturbance of the immune system, and the Epstein–Barr virus. Initially, the disease is characterized by a painless, nontender enlargement of one or more lymph nodes, most commonly in the neck but occasionally in the axilla, inguinal, or femoral region. Some patients also have an unexplained and persistent fever and/or night sweats. Fatigue and weight loss are also associated complaints, as is pruritus (itching). Treatment consists of radiation therapy, chemotherapy, and bone marrow transplants. Hodgkin's disease is considered to be a curable malignancy.

MEDICAL TERMINOLOGY

Adenitis (ad′-e-NĪ-tis; *adeno* = gland; *itis* = inflammation of) Enlarged, tender, and inflamed lymph nodes resulting from an infection.

Hypersplenism (hī′-per-SPLĒN-izm; *hyper* = over) Abnormal splenic activity due to splenic enlargement and associated with an increased rate of destruction of normal blood cells.

Lymphadenectomy (lim-fad′-e-NEK-tō-mē; *ectomy* = removal) Removal of a lymph node.

Lymphadenopathy (lim-fad′-e-NOP-a-thē; *patho* = disease) Enlarged, sometimes tender lymph glands.

Lymphangioma (lim-fan′-jē-Ō-ma; *angio* = vessel; *oma* = tumor) A benign tumor of the lymphatic vessels.

Lymphangitis (lim′-fan-JĪ-tis) Inflammation of the lymphatic vessels.

Lymphedema (lim′-fe-DĒ-ma; *edema* = swelling) Accumulation of lymph fluid producing subcutaneous tissue swelling.

Lymphoma (lim′-FŌ-ma) Any tumor composed of lymphatic tissue.

Lymphostasis (lim-FŌ-stā-sis; *stasis* = halt) A lymph flow stoppage.

Splenomegaly (splē′-nō-MEG-a-lē; *mega* = large) Enlarged spleen.

Study Outline

Introduction (p. 683)

1. The lymphatic system consists of lymph, lymphatic vessels, and structures and organs that contain lymphatic tissue (specialized reticular tissue containing large numbers of lymphocytes).
2. The lymphatic system functions to drain interstitial fluid, transport dietary fats, and protect against invasion by nonspecific defenses and specific immune responses.

Lymphatic Vessels and Lymph Circulation (p. 683)

1. Lymphatic vessels begin as closed-ended lymph capillaries in tissue spaces between cells.
2. Interstitial fluid drains into lymphatic capillaries, thus forming lymph.
3. Lymph capillaries merge to form larger vessels, called lymphatic vessels, which convey lymph into and out of structures called lymph nodes.

4. The passage of lymph is from interstitial fluid to lymph capillaries to lymphatic vessels to lymph trunks to the thoracic duct or right lymphatic duct to the subclavian veins.
5. Lymph flows as a result of skeletal muscle contractions and respiratory movements. It is also aided by valves in the lymphatic vessels.

Lymphatic Tissue (p. 686)

1. Among the lymphatic tissue-containing components of the lymphatic system are diffuse lymphatic tissue, lymphatic nodules, and lymphatic organs (lymph nodes, spleen, and thymus gland).
2. Lymph nodes are encapsulated, oval structures located along lymphatic vessels.
3. Lymph enters nodes through afferent lymphatic vessels, is filtered, and exits through efferent lymphatic vessels.
4. Lymph nodes are the site of proliferation of plasma cells and T cells.

5. Tonsils are multiple aggregations of large lymphatic nodules embedded in mucous membranes. They include the pharyngeal, palatine, and lingual tonsils.

6. The spleen is the largest mass of lymphatic tissue in the body. It is a site of B cell proliferation into plasma cells, phagocytosis of bacteria and worn-out red blood cells, and storage of blood.

7. The thymus gland lies between the sternum and the large blood vessels above the heart. It is the site of T cell maturation.

Developmental Anatomy of the Lymphatic System (p. 690)

1. Lymphatic vessels develop from lymph sacs, which develop from veins. Thus they are derived from mesoderm.

2. Lymph nodes develop from lymph sacs that become invaded by mesenchymal cells.

Nonspecific Resistance to Disease (p. 692)

1. The ability to ward off disease is called resistance. Lack of resistance is called susceptibility.

2. Nonspecific resistance refers to a wide variety of body responses against a wide range of pathogens.

3. Nonspecific resistance includes mechanical factors, chemical factors, antimicrobial substances, natural killer cells, phagocytosis, inflammation, and fever.

Immunity (Specific Resistance to Disease) (p. 697)

1. Specific resistance to disease involves the production of a specific lymphocyte or antibody against a specific antigen and is called immunity.

Formation of T Cells and B Cells (p. 698)

1. Both B and T cells derive from stem cells in bone marrow. T cells complete their maturation and develop immunocompetence in the thymus.

Types of Immune Responses (p. 699)

1. Cell-mediated immunity refers to destruction of antigens by T cells, and antibody-mediated (humoral) immunity refers to destruction of antigens by antibodies.

Antigens (p. 700)

1. Antigens (Ags) are chemical substances that, when introduced into the body, are recognized as foreign by antigen receptors.

2. Antigen receptors exhibit great diversity due to genetic recombination.

3. Major histocompatibility complex (MHC) antigens are unique to each person's body cells. All cells except red blood cells display MHC-I antigens. Some cells also display MHC-II antigens.

4. Peptide fragments from foreign antigens help stabilize MHC molecules.

5. Cells called antigen presenting cells (APCs) process exogenous antigens (formed outside the body) and present them together with MHC class II molecules to T cells. APCs include macrophages, B cells, and dendritic cells.

Cytokines (p. 702)

1. Cytokines are small protein hormones needed for many normal cell functions. Some of them regulate immune responses (see Exhibit 22.2).

Antibodies (p. 702)

1. An antibody (Ab) is a protein that combines specifically with the antigen that triggered its production.

2. Antibodies consist of heavy and light chains and variable and constant portions.

3. Based on chemistry and structure, antibodies are grouped into five principal classes, each with specific biological roles (IgG, IgA, IgM, IgD, and IgE).

Cell-Mediated Immunity (p. 705)

1. In a cell-mediated immune response, an antigen is recognized, specific T cells proliferate and differentiate into effector cells, and the antigen is eliminated.

2. T cell receptors (TCRs) recognize antigen fragments associated with MHC molecules on the surface of a body cell.

3. Proliferation of T cells requires costimulation, by cytokines such as interleukin-1 (IL-1) and IL-2 or by pairs of plasma membrane molecules.

4. T cells consist of several subpopulations. Helper T cells display CD4 protein, recognize antigen fragments associated with MHC-II molecules, and secrete several cytokines, most importantly interleukin-2, which acts as a costimulator for other helper T cells, cytotoxic T cells, and B cells. Cytotoxic (killer) T cells display CD8 protein and recognize antigen fragments associated with MHC-I molecules. Delayed hypersensitivity T cells produce cytokines and are important in hypersensitivity (allergic) responses. Suppressor T cells downregulate immune responses.

5. Cytotoxic T cells eliminate invaders by secreting lymphotoxin, which causes fragmentation of the DNA of a target cell, and perforin, which causes cytolysis. They also secrete gamma interferon.

Antibody-Mediated (Humoral) Immunity (p. 708)

1. B cells can respond to unprocessed antigens, but their response is more intense when dendritic cells present antigen to them. Interleukin-2 and other cytokines secreted by helper T cells provide costimulation for proliferation of B cells.

2. An activated B cell develops into a clone of antibody-producing plasma cells.

Immunological Memory (p. 709)

1. Immunization against certain microbes is possible because memory B cells and memory T cells remain after the primary response to an antigen. The secondary response provides protection should the same microbe enter the body again.

Self-Recognition and Immunological Tolerance (p. 710)

1. T cells undergo both positive and negative selection to ensure that they can recognize self-MHC antigens and that they do not react to other self-proteins (tolerance). Negative selection involves both deletion and anergy.

2. B cells develop tolerance through deletion and anergy.

Immunology and Cancer (p. 711)

1. Cancer cells may display tumor-specific antigens and are often destroyed by the body's immune system (immunological surveillance).

Aging and the Immune System (p. 712)

1. With advancing age, the immune system functions less effectively.

Review Questions

1. Identify the components and functions of the lymphatic system. (p. 683)
2. Compare veins and lymphatic vessels with regard to structure. (p. 683)
3. Construct a diagram to show the route of lymph circulation. (p. 683)
4. Describe the structure of a lymph node. What functions do lymph nodes serve? (p. 686)
5. Identify the tonsils by location. (p. 686)
6. Describe the location, gross anatomy, histology, and functions of the spleen. (p. 689)
7. Describe the role of the thymus gland in immunity. (p. 690)
8. Describe how the lymphatic system develops. (p. 690)
9. Describe the various mechanical and chemical factors involved in nonspecific resistance. (p. 692)
10. Outline the role of the following antimicrobial substances: interferon (IFN), complement, and transferrin. (p. 693)
11. What are natural killer (NK) cells? (p. 693)
12. Describe the three phases of phagocytosis. (p. 695)
13. Define an inflammation. Describe the principal symptoms associated with inflammation and outline its stages. (p. 695)
14. Define immunity and summarize the two types of immune responses. Where do B cells and T cells form? (pp. 698)
15. List the various characteristics of antigens and explain the function of antigen presenting cells. (p. 700)
16. Describe the functions of cytokines. (p. 702)
17. Describe the chemical characteristics of antibodies. (p. 702)
18. Compare and contrast cell-mediated and antibody-mediated immune responses. (pp. 705)
19. Discuss the importance of the secondary response of the body to an antigen. (p. 710)
20. How are monoclonal antibodies (MAbs) produced? (p. 708)
21. Explain how immunology is related to cancer. (p. 711)
22. Describe the effects of aging on the immune system. (p. 712)

Answers to Questions with Figures

22.1 Draining interstitial fluid; transporting dietary fats; protecting against invasion.

22.2 Interstitial fluid, because the protein content is low.

22.3 Capillaries.

22.4 Left and right lumbar trunks and intestinal trunk; thoracic.

22.5 Macrophages may phagocytize them or lymphocytes attack them via immune responses.

22.6 Phagocytosis of bacteria and aged red blood cells.

22.7 T cells.

22.9 Classical pathway: Ag–Ab complexes activate C1.

22.10 Digestive enzymes, oxidants, and defensins.

22.11 *Redness,* increased blood flow due to vasodilation; *pain,* injury of nerve fibers, irritation by toxins from microbes, kinins, prostaglandins, pressure from edema; *heat,* increased blood flow, heat release by locally increased metabolic reactions; *swelling,* leakage of fluid from capillaries due to increased permeability.

22.12 Helper T cells.

22.13 Antigenic determinant = small immunogenic part of large foreign antigen; hapten = small molecule that becomes immunogenic when it attaches to a body protein.

22.14 Macrophages in tissues throughout the body; B cells in blood and lymphatic tissue; dendritic cells in mucous membranes and the skin.

22.15 It recognizes and binds to antigen.

22.16 First signal, antigen binding to TCR; second signal, costimulator, such as a cytokine or another pair of plasma membrane molecules.

22.17 Some tumor cells, transplanted tissue cells.

22.18 Just one.

22.19 At peak secretion, 1000 times more.

22.20 Self-recognition.

22.21 It catalyzes the formation of DNA from viral RNA.

Chapter 23
THE RESPIRATORY SYSTEM

Chapter Contents at a Glance

Student Objectives

1. Identify the organs of the respiratory system and describe their functions.

2. Explain the structure of the alveolar–capillary (respiratory) membrane and describe its function in the diffusion of respiratory gases.

3. Describe the events involved in inspiration and expiration.

4. Explain how respiratory gases are carried by blood.

5. Describe the various factors that control the rate of respiration.

6. Describe the effects of aging on the respiratory system.

7. Describe the development of the respiratory system.

8. Define asthma, bronchitis, emphysema, bronchogenic carcinoma (lung cancer), pneumonia, tuberculosis (TB), respiratory distress syndrome (RDS) of the newborn, respiratory failure,

sudden infant death syndrome (SIDS), coryza (common cold), influenza (flu), pulmonary embolism, pulmonary edema, cystic fibrosis (CF), and smoke inhalation injury as disorders of the respiratory system.

9. Define medical terminology associated with the respiratory system.

Cells continually use oxygen (O_2) for the metabolic reactions that release energy from nutrient molecules and produce ATP. At the same time, these reactions release carbon dioxide (CO_2). As you will see in Chapter 25, oxygen consumption and carbon dioxide production occur in mitochondria due to the cellular respiration that occurs there. Since an excessive amount of CO_2 produces acidity that is toxic to cells, the excess CO_2 must be eliminated quickly and efficiently. The two systems that supply oxygen and eliminate carbon dioxide are the cardiovascular system and the respiratory system. They participate equally in respiration. The respiratory system provides for gas exchange, intake of O_2 and elimination of CO_2, whereas the cardiovascular system transports the gases in the blood between the lungs and the cells. Failure of either system has the same effect on the body: disruption of homeostasis and rapid death of cells from oxygen starvation and buildup of waste products. In addition to functioning in gas exchange, the respiratory system also contains receptors for the sense of smell, filters inspired air, produces sounds, and helps eliminate wastes.

The exchange of gases between the atmosphere, blood, and cells is called **respiration.** Three basic processes are involved. The first process, **pulmonary** (*pulmo* = lung) **ventilation,** or breathing, is the inspiration (inflow) and expiration (outflow) of air between the atmosphere and the lungs. The second and third processes involve the exchange of gases within the body. **External (pulmonary) respiration** is the exchange of gases between the lungs and blood. **Internal (tissue) respiration** is the exchange of gases between the blood and the cells.

The developmental anatomy of the respiratory system is considered later in the chapter.

ORGANS

The **respiratory system** consists of the nose, pharynx (throat), larynx (voice box), trachea (windpipe), bronchi, and lungs (Fig. 23.1). The term **upper respiratory system** refers to the nose, pharynx, and associated structures. The **lower respiratory system** refers to the larynx, trachea, bronchi, and lungs. Functionally, the respiratory system consists of two portions. The **conducting portion** consists of a system of interconnecting cavities and tubes—nose, pharynx, larynx, trachea, bronchi, and terminal bronchioles—that conduct air into the lungs. The **respiratory portion** consists of those portions of the respiratory system where the exchange of gases occurs—respiratory bronchioles, alveolar ducts, and alveoli.

Nose

Anatomy

The **nose** has an external portion and an internal portion inside the skull (Fig. 23.2a–c). The external portion consists of a supporting framework of bone and hyaline cartilage covered with muscle and skin and lined by mucous membrane. The bridge of the nose is formed by the nasal bones, which hold it in a fixed position. Because it has a framework of pliable cartilage, the rest of the external nose is somewhat flexible. On the undersurface of the external nose are two openings called the **external nares** (NA-rēz; singular is **naris**), or **nostrils.** The surface anatomy of the nose is shown in Fig. 23.2a.

The internal portion of the nose is a large cavity in the skull that lies inferior to the cranium and superior to the mouth. Anteriorly, the internal nose merges with the external nose, and posteriorly it communicates with the pharynx through two openings called the **internal nares (choanae).** Ducts from the paranasal sinuses (frontal, sphenoidal, maxillary, and ethmoidal) and the nasolacrimal ducts also open into the internal nose. The lateral walls of the internal nose are formed by the ethmoid, maxillae, lacrimal, palatine, and inferior nasal conchae bones. The ethmoid also forms the roof. The floor is formed mostly by the palatine bones and palatine processes of the maxillae, which together comprise the hard palate.

The inside of both the external and internal nose is called the **nasal cavity.** It is divided into right and left sides by a vertical partition called the **nasal septum.** The anterior portion of the septum is made primarily of cartilage. The remainder is formed by the vomer, perpendicular plate of the ethmoid, maxillae, and palatine bones (see Fig. 7.10a). The anterior portion of the nasal cavity, just inside the nostrils, is called the **vestibule** and is surrounded by cartilage. The upper nasal cavity is surrounded by bone.

CLINICAL APPLICATION

RHINOPLASTY

Rhinoplasty (RĪ-nō-plas'-tē; *rhino* = nose; *plassein* = to form), commonly called a "nose job," is a surgical procedure in which the structure of the external nose is altered. Although it is frequently done for cosmetic reasons, it is sometimes performed to repair a fractured nose or deviated nasal septum. In the procedure, a local or general anesthetic is given, and with instruments inserted through the nostrils, the nasal bones are fractured and repositioned to achieve the desired shape. Any extra bone fragments or cartilage are shaved down and removed through the nostrils. Both an internal packing and an external splint keep the nose in the desired position while it heals. In some cases, only the cartilage has to be reshaped to achieve the desired appearance.

Physiology

The interior structures of the nose are specialized for three functions: (1) incoming air is warmed, moistened, and filtered; (2) olfactory stimuli are received; and (3) large, hollow resonating chambers modify speech sounds.

When air enters the nostrils, it passes first through the

FIGURE 23.2 (continued)

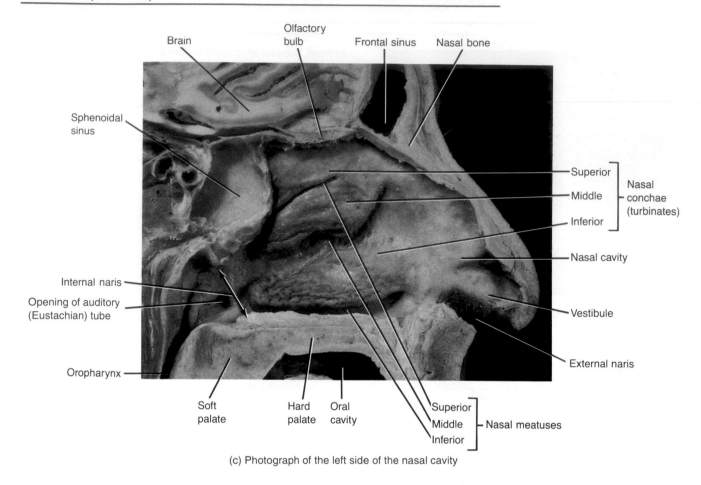

(c) Photograph of the left side of the nasal cavity

Question: What is the path taken by air molecules into and through the nose?

mucous membrane contains capillaries and pseudostratified ciliated columnar epithelial cells with many goblet cells. As the air whirls around the conchae and meatuses, it is warmed by blood in the capillaries. Mucus secreted by the goblet cells moistens the air and traps dust particles. Drainage from the nasolacrimal ducts and perhaps secretions from the paranasal sinuses also help moisten the air. The cilia move the mucus–dust packages to the pharynx so they can be eliminated from the respiratory tract by swallowing or expectoration (spitting).

Pharynx

The **pharynx** (FAR-inks), or throat, is a somewhat funnel-shaped tube about 13 cm (5 in.) long that starts at the internal nares and extends to the level of the cricoid cartilage, the most inferior cartilage of the voice box (Fig. 23.3). It lies just posterior to the nasal cavity, oral cavity, and larynx and just anterior to the cervical vertebrae. Its wall is composed of skeletal muscles and lined with mucous mem-

brane. The pharynx functions as a passageway for air and food and provides a resonating chamber for speech sounds.

The branch of medicine that deals with the diagnosis and treatment of diseases of the ears, nose, and throat is called **otorhinolaryngology** (ō′-tō-rī′-nō-lar′-in-GOL-ō-jē; *oto* = ear; *rhino* = nose).

The uppermost portion of the pharynx, called the **nasopharynx,** lies posterior to the internal portion of the nose and extends to the plane of the soft palate. There are four openings in its wall: two internal nares and two openings that lead into the auditory (Eustachian) tubes. The posterior wall also contains the pharyngeal tonsil, or adenoid. Through the internal nares, the nasopharynx receives air from the nasal cavity and receives packages of dust-laden mucus. It is lined with pseudostratified ciliated columnar epithelium, and the cilia move the mucus down toward the most inferior part of the pharynx. The nasopharynx also exchanges small amounts of air with the auditory (Eustachian) tubes to equalize ear, nose, and throat air pressure.

The middle portion of the pharynx, the **oropharynx,** lies posterior to the oral cavity and extends from the soft palate inferiorly to the level of the hyoid bone. It has only one opening, the **fauces** (FAW-sēz), the opening from the mouth. It is lined by stratified squamous epithelium. This portion of the pharynx is both respiratory and digestive in function, since it is a common passageway for air, food, and drink. Two pairs of tonsils, the palatine and lingual tonsils, are found in the oropharynx. The merged lingual tonsils lie at the base of the tongue (see also Fig. 16.2a).

The lowest portion of the pharynx, the **laryngopharynx** (la-rin′-gō-FAR-inks), or **hypopharynx,** extends downward from the hyoid bone and becomes continuous with the esophagus (food tube) posteriorly and the larynx (voice box) anteriorly. Like the oropharynx, the laryngopharynx is a respiratory and a digestive pathway and is lined by stratified squamous epithelium.

Larynx

The **larynx,** or voice box, is a short passageway that connects the pharynx with the trachea. It lies in the midline of the neck anterior to the fourth through sixth cervical vertebrae (C4–C6).

Anatomy

The wall of the larynx is composed of nine pieces of cartilage (Fig. 23.4). Three are single and three are paired. The three single pieces are the thyroid cartilage, epiglottis (epiglottic cartilage), and cricoid cartilage. Of the paired cartilages, the arytenoid cartilages are the most important since they influence the positions and tensions of the vocal folds (true vocal cords). The paired corniculate and cuneiform cartilages are of lesser significance.

The **thyroid cartilage (Adam's apple)** consists of two fused plates of hyaline cartilage that form the anterior wall of the larynx and give it its triangular shape. It is larger in males than in females.

The **epiglottis** (*epi* = above; *glotta* = tongue) is a large, leaf-shaped piece of elastic cartilage (see also Fig. 23.3). The "stem" of the epiglottis is attached to the anterior rim of the thyroid cartilage, but the "leaf" portion is unattached and free to move up and down like a trap door. During swallowing, there is elevation of the larynx. This causes the free edge of the epiglottis to form a lid over the glottis, closing it off. The **glottis** consists of a pair of folds of

FIGURE 23.3 Photograph of a portion of the left side of the head and neck with the nasal septum removed.

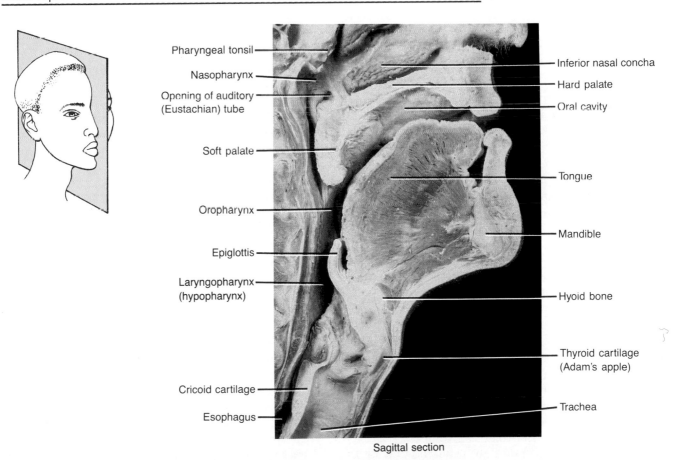

Pharyngeal tonsil
Nasopharynx
Opening of auditory (Eustachian) tube
Soft palate
Oropharynx
Epiglottis
Laryngopharynx (hypopharynx)
Cricoid cartilage
Esophagus

Inferior nasal concha
Hard palate
Oral cavity
Tongue
Mandible
Hyoid bone
Thyroid cartilage (Adam's apple)
Trachea

Sagittal section

Question: In this photograph, can you identify the superior and inferior borders of the pharynx?

FIGURE 23.4 Larynx.

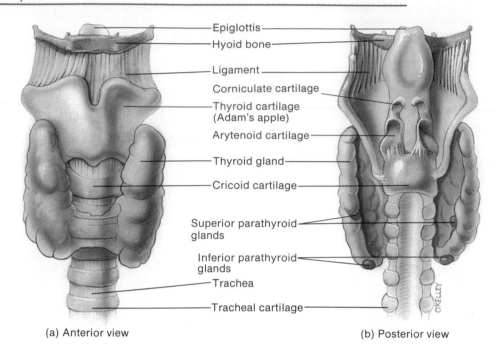

Epiglottis
Hyoid bone
Ligament
Corniculate cartilage
Thyroid cartilage (Adam's apple)
Arytenoid cartilage
Thyroid gland
Cricoid cartilage
Superior parathyroid glands
Inferior parathyroid glands
Trachea
Tracheal cartilage

(a) Anterior view   (b) Posterior view

Question: How does the epiglottis prevent aspiration of foods and liquids?

mucous membrane, the vocal folds (true vocal cords) in the larynx, and the space between them called the **rima glottidis.** In this way, the larynx is closed off, and liquids and foods are routed into the esophagus and kept out of the larynx and air passageways below it. When anything but air passes into the larynx, a cough reflex attempts to expel the material.

The **cricoid** (KRĪ-koyd) **cartilage** is a ring of hyaline cartilage, attached to the first ring of cartilage of the trachea, that forms the inferior wall of the larynx. Clinically, it is the landmark for making an emergency airway (tracheostomy; see page 729).

The paired **arytenoid** (ar'-i-TĒ-noyd) **cartilages** are triangular pieces of mostly hyaline cartilage and located at the posterior, superior border of the cricoid cartilage. They attach to the vocal folds and intrinsic pharyngeal muscles and by their action can move the vocal folds.

The paired **corniculate** (kor-NIK-yoo-lāt) **cartilages** are horn-shaped pieces of elastic cartilage. One is located at the apex of each arytenoid cartilage. The paired **cuneiform** (kyoo-NĒ-i-form) **cartilages** are club-shaped elastic cartilages anterior to the corniculate cartilages (see Fig. 23.5).

The lining of the larynx below the vocal folds is pseudostratified ciliated columnar epithelium. It consists of ciliated columnar cells, goblet cells, and basal cells, and its mucus covering helps trap dust not removed in the upper passages. Below the level of the pharynx, cilia sweep mucus with its trapped particles upward.

Voice Production

The mucous membrane of the larynx forms two pairs of folds (Fig. 23.5): an upper pair called the **ventricular folds (false vocal cords)** and a lower pair called simply the **vocal folds (true vocal cords).** The space between the ventricular folds is known as the **rima vestibuli.** The **laryngeal sinus** (see Fig. 23.2b) is a lateral expansion of the middle portion of the laryngeal cavity between the ventricular folds above and the vocal folds below.

When the ventricular folds are brought together, they function in holding the breath against pressure in the thoracic cavity, such as might occur when a person strains to lift a heavy object. The mucous membrane of the vocal folds is lined by nonkeratinized stratified squamous epithelium. Under the membrane lie bands of elastic ligaments stretched between pieces of rigid cartilage like the strings on a guitar. Skeletal muscles of the larynx, called intrinsic muscles, attach to both the rigid cartilage and the vocal folds themselves. When the muscles contract, they pull the elastic ligaments tight and stretch the vocal folds out into the air passageways so that the rima glottidis is narrowed. If air is directed against the vocal folds, they vibrate and set up sound waves in the column of air in the pharynx, nose, and mouth. The greater the pressure of air, the louder the sound.

Pitch is controlled by the tension on the vocal folds. If they are pulled taut by the muscles, they vibrate more rapidly, and a higher pitch results. Lower sounds are pro-

FIGURE 23.5 Vocal folds.

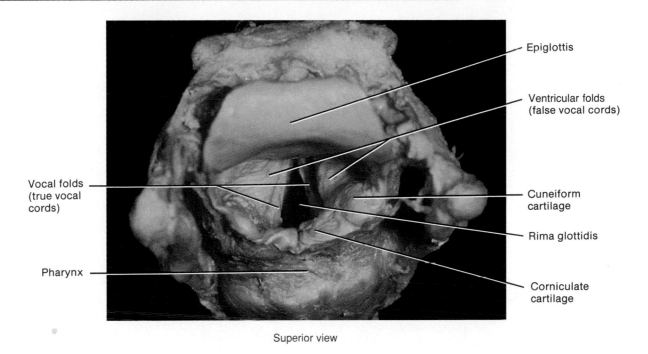

Superior view

Question: What is the main function of the vocal folds?

duced by decreasing the muscular tension on the vocal folds. Due to the influence of androgens (male sex hormones), vocal folds are usually thicker and longer in males than in females, and therefore they vibrate more slowly. Thus men generally have a lower range of pitch than women.

Sound originates from the vibration of the vocal folds, but other structures are necessary for converting the sound into recognizable speech. The pharynx, mouth, nasal cavity, and paranasal sinuses all act as resonating chambers that give the voice its human and individual quality. By constricting and relaxing the muscles in the wall of the pharynx, we produce the vowel sounds. Muscles of the face, tongue, and lips help us enunciate words.

CLINICAL APPLICATION

LARYNGITIS AND CANCER OF THE LARYNX

Laryngitis is an inflammation of the larynx that is most often caused by a respiratory infection or irritants such as cigarette smoke. Inflammation of the vocal folds causes hoarseness or loss of voice by interfering with the contraction of the folds or by causing them to swell to the point where they cannot vibrate freely. Many long-term smokers acquire a permanent hoarseness from the damage done by chronic inflammation.
Cancer of the larynx is found almost exclusively in individuals who smoke. The condition is characterized by hoarseness, pain on swallowing, or pain radiating to an ear. Treatment consists of radiation therapy and/or surgery.

Trachea

The **trachea** (TRĀ-kē-a), or windpipe, is a tubular passageway for air about 12 cm (4.5 in.) in length and 2.5 cm (1 in.) in diameter. It is located anterior to the esophagus (Fig. 23.6a) and extends from the larynx to the fifth thoracic vertebra (T5), where it divides into right and left primary bronchi (see Fig. 23.7).

The wall of the trachea consists of a mucosa, submucosa, hyaline cartilage, and adventitia (outer layer of areolar connective tissue). See Fig. 23.6b. The epithelium of the mucosa is pseudostratified ciliated columnar epithelium (Fig. 23.6c). It consists of ciliated columnar cells and goblet cells that reach the luminal surface plus basal cells that do not reach the luminal surface. The epithelium provides the same protection against dust as the membrane lining the larynx. Seromucous glands and their ducts are present in the submucosa. The 16 to 20 incomplete rings of hyaline cartilage look like letter Cs and are arranged horizontally and stacked one on top of another. The open part of each C-shaped cartilage ring faces the esophagus. This accommodates slight expansion of the esophagus into the trachea during swallowing (Fig. 23.6a). Transverse smooth muscle fibers, called the **trachealis muscle,** and elastic connective

FIGURE 23.6 Histology of the trachea.

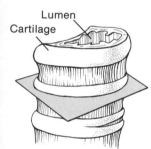

Esophagus   Trachea

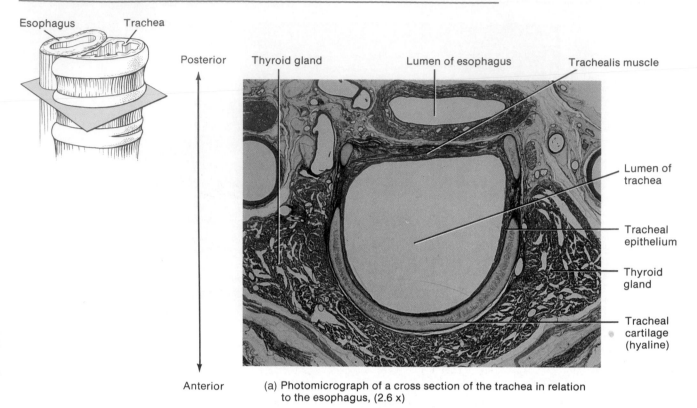

Posterior

Thyroid gland   Lumen of esophagus   Trachealis muscle

Lumen of trachea

Tracheal epithelium

Thyroid gland

Tracheal cartilage (hyaline)

Anterior

(a) Photomicrograph of a cross section of the trachea in relation to the esophagus, (2.6 x)

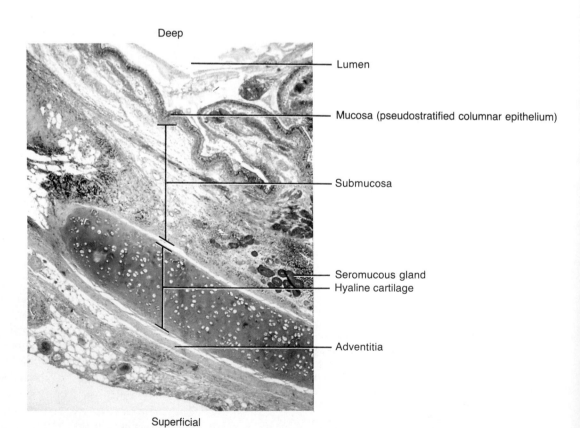

Lumen
Cartilage

Deep

Lumen

Mucosa (pseudostratified columnar epithelium)

Submucosa

Seromucous gland
Hyaline cartilage

Adventitia

Superficial

(b) Photomicrograph of a portion of the tracheal wall, 80x

FIGURE 23.6 (continued)

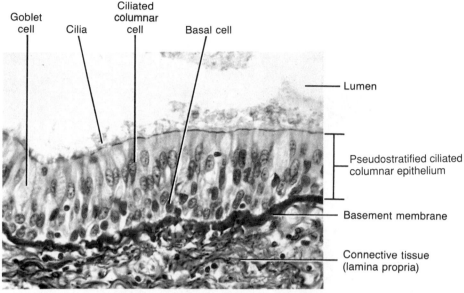

Goblet cell · Cilia · Ciliated columnar cell · Basal cell · Lumen · Pseudostratified ciliated columnar epithelium · Basement membrane · Connective tissue (lamina propria)

(c) Photomicrograph of enlarged aspect of tracheal epithelium, 600x

Question: What is the benefit of not having cartilage between the trachea and esophagus?

tissue attach the open ends of the cartilage rings. The solid C-shaped cartilage rings provide a rigid support so the tracheal wall does not collapse inward and obstruct the air passageway. In some situations, for example, a crushing injury to the chest, the rings of cartilage may not be strong enough to overcome collapse and obstruction of the trachea.

At the point where the trachea divides into right and left primary bronchi, there is an internal ridge called the **carina** (ka-RĪ-na). It is formed by a posterior and somewhat inferior projection of the last tracheal cartilage. The mucous membrane of the carina is one of the most sensitive areas of the respiratory system and is associated with the cough reflex. Widening and distortion of the carina, which can be seen in an examination by bronchoscopy, is a serious sign, since it usually indicates a carcinoma of the lymph nodes around the region where the trachea divides.

CLINICAL APPLICATION

TRACHEOSTOMY AND INTUBATION

Occasionally, the respiratory passageways are unable to protect themselves from obstruction. The rings of cartilage may accidentally be crushed. Or the mucous membrane may become inflamed and swell so much that it closes off the air passageways; inflamed membranes secrete a great deal of mucus that may clog the lower respiratory passageways. A large object may be breathed in (aspirated) while the rima glottidis is open; or an aspirated foreign object may cause spasm of the laryngeal muscles. The passageways must be cleared quickly. If the obstruction is above the level of the larynx, a **tracheostomy** (trā-kē-OS-tō-mē) may be per-

formed. A skin incision is made, followed by a short longitudinal incision into the trachea inferior to the cricoid cartilage. The patient breathes through a metal or plastic tracheal tube inserted through the incision. Another method is **intubation.** A tube is inserted into the mouth or nose and passed down through the larynx and trachea. The firm wall of the tube pushes back any flexible obstruction, and the inside of the tube provides a passageway for air. If mucus is clogging the trachea, it can be suctioned out through the tube.

Bronchi

At the sternal angle, the trachea divides into a **right primary bronchus** (BRON-kus), which goes into the right lung, and a **left primary bronchus,** which goes into the left lung (Fig. 23.7). The right primary bronchus is more vertical, shorter, and wider than the left. As a result, an aspirated object is more likely to enter and lodge in the right primary bronchus than the left. Like the trachea, the primary bronchi (BRON-kē) contain incomplete rings of cartilage and are lined by pseudostratified ciliated columnar epithelium.

On entering the lungs, the primary bronchi divide to form smaller bronchi—the **secondary (lobar) bronchi,** one for each lobe of the lung. (The right lung has three lobes; the left lung has two.) The secondary bronchi continue to branch, forming still smaller bronchi, called **tertiary (segmental) bronchi,** that divide into **bronchioles.** Bronchioles, in turn, branch respectively into even smaller bronchioles and eventually into tubes called **terminal bronchioles.** This continuous branching from the trachea resembles a

FIGURE 23.7 Bronchial tree in relation to the lungs.

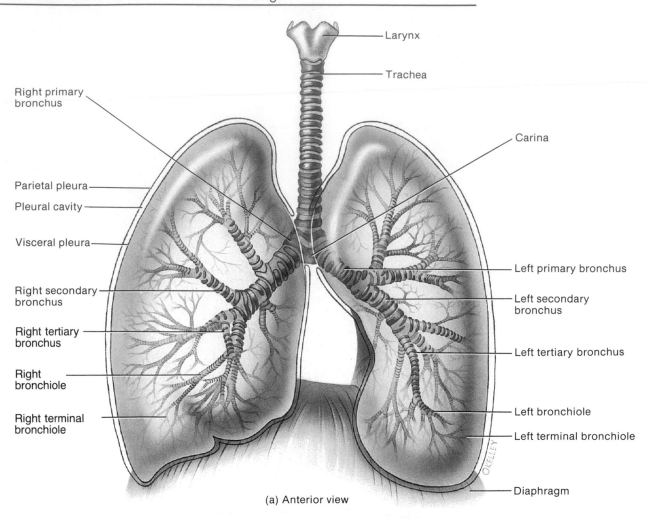

Larynx

Trachea

Right primary bronchus

Carina

Parietal pleura

Pleural cavity

Visceral pleura

Left primary bronchus

Right secondary bronchus

Left secondary bronchus

Right tertiary bronchus

Left tertiary bronchus

Right bronchiole

Left bronchiole

Right terminal bronchiole

Left terminal bronchiole

Diaphragm

(a) Anterior view

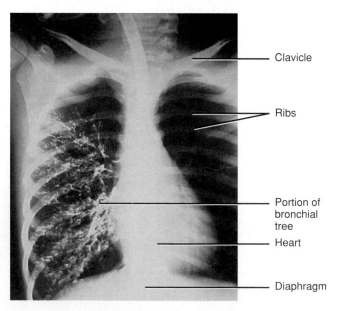

Clavicle

Ribs

Portion of bronchial tree

Heart

Diaphragm

(b) Bronchogram

Question: How many lobes and secondary bronchi are present in each lung?

tree trunk with its branches and is commonly referred to as the **bronchial tree.**

Bronchography (bron-KOG-ra-fē) is a technique for examining the bronchial tree. An intratracheal catheter is passed into the mouth or nose, through the rima glottidis, and into the trachea. Then an opaque contrast medium, usually containing iodine, is inhaled and distributed through the bronchial branches. X-rays of the chest in various positions are taken, and the developed film, a **bronchogram** (BRON-kō-gram), provides a picture of the bronchial tree (Fig. 23.7b).

Bronchoscopy (bron-KOS-kō-pē; *bronchos* = windpipe; *shopein* = to examine) is the visual examination of the bronchi through a **bronchoscope,** an illuminated, tubular instrument that is passed through the trachea into the bronchi. The examiner can view the interior of the trachea and bronchi to biopsy a tumor, clear an obstructing object or secretions from an airway, take cultures or smears for microscopic examination, stop bleeding, or deliver drugs.

As the branching becomes more extensive in the bronchial tree, several structural changes may be noted. First, the epithelium gradually changes from pseudostratified ciliated columnar epithelium to nonciliated simple cuboidal epithelium in the terminal bronchioles. (In regions where nonciliated cuboidal epithelium is present, inhaled particles are removed by macrophages.) Second, incomplete rings of cartilage in primary bronchi are gradually replaced by plates of cartilage that finally disappear in the bronchioles. Third, as the cartilage decreases, the amount of smooth muscle increases. Smooth muscle encircles the lumen in spiral bands and its contraction is affected by both the autonomic nervous system (ANS) and various chemicals.

The parasympathetic division of the ANS and mediators of allergic reactions, such as histamine, cause constriction of bronchioles. The sympathetic division of the ANS and the hormone epinephrine cause dilation. During an **asthma attack** the bronchiole smooth muscle contracts, reducing the diameter of the airways. Because there is no supporting cartilage, the muscle spasms can even close off the air passageways, a life-threatening situation. Epinephrine, administered as an inhaled mist (nebulization), often is used to relax the smooth muscle and open the airways.

CLINICAL APPLICATION

NEBULIZATION

Many respiratory disorders are treated by means of **nebulization** (neb-yoo-li-ZĀ-shun). This procedure consists of administering medication in the form of droplets that are suspended in air to select areas of the respiratory tract. The patient inhales the medication as a fine mist. Nebulization therapy can be used with many different types of drugs, such as chemicals that relax the smooth muscle of the respiratory passageways, chemicals that reduce the thickness of mucus, and antibiotics.

Lungs

The **lungs** (*lunge* = light, since the lungs float) are paired cone-shaped organs lying in the thoracic cavity. They are separated from each other by the heart and other structures in the mediastinum (see Fig. 20.1). Two layers of serous membrane, collectively called the **pleural membrane,** enclose and protect each lung (Fig. 23.8). The outer layer is attached to the wall of the thoracic cavity and is called the **parietal pleura.** The inner layer, the **visceral pleura,** covers the lungs themselves. Between the visceral and parietal pleurae is a small potential space, the **pleural cavity,** which contains a lubricating fluid secreted by the membranes. This fluid reduces friction between the membranes and allows them to move easily on one another during breathing.

CLINICAL APPLICATION

PNEUMOTHORAX, HEMOTHORAX, AND PLEURISY

In certain conditions, the pleural cavity may fill with air **(pneumothorax;** *pneumo* = air or breath), blood **(hemothorax),** or pus. Air in the pleural cavity, most commonly introduced in a surgical opening of the chest or as a result of a stab or gunshot wound, may cause the lung to collapse (atelectasis). Fluid can be drained from the pleural cavity by inserting a needle, usually posteriorly through the seventh intercostal space. The needle is passed along the superior border of the lower rib to avoid damage to the intercostal nerves and blood vessels. Below the seventh intercostal space there is danger of penetrating the diaphragm.

Inflammation of the pleural membrane, or **pleurisy,** may in its early stages cause pain due to friction between the parietal and visceral layers of the pleura. If the inflammation persists, fluid accumulates in the pleural space, a condition known as **pleural effusion.** One cause of pleural effusion is lung cancer.

Gross Anatomy

The lungs extend from the diaphragm to just slightly above clavicles and lie against the ribs anteriorly and posteriorly. The broad inferior portion of the lung, the **base,** is concave and fits over the convex area of the diaphragm (Fig. 23.9). The narrow superior portion of the lung is termed the **apex (cupula).** The surface of the lung lying against the ribs, the **costal surface,** is rounded to match the curvature of the ribs. The **mediastinal (medial) surface** of each lung contains a region, the **hilus,** through which bronchi, pulmonary blood vessels, lymphatic vessels, and nerves enter and exit. These structures are held together by the pleura and connective tissue and constitute the **root** of the lung. Medially, the left lung also contains a concavity, the **cardiac notch,** in which the heart lies.

The right lung is thicker and broader than the left. It is also somewhat shorter than the left because the diaphragm is higher on the right side to accommodate the liver that lies below it.

FIGURE 23.8 Relationship of the pleural membranes to the lungs.

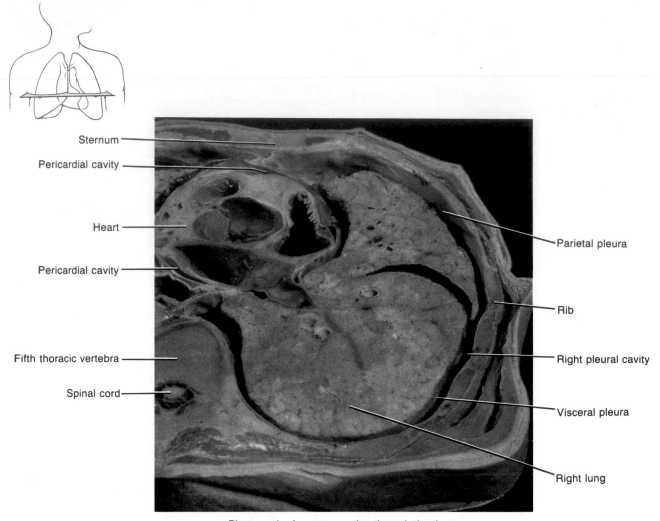

Sternum

Pericardial cavity

Heart

Pericardial cavity

Fifth thoracic vertebra

Spinal cord

Parietal pleura

Rib

Right pleural cavity

Visceral pleura

Right lung

Photograph of a cross section through the thorax

Question: What type of membrane is the pleural membrane?

Lobes and Fissures

Each lung is divided into lobes by one or more fissures (Fig. 23.9). Both lungs have an **oblique fissure,** which extends downward and forward. The right lung also has a **horizontal fissure.** The oblique fissure in the left lung separates the **superior lobe** from the **inferior lobe.** The upper part of the oblique fissure of the right lung separates the superior lobe from the inferior lobe, whereas the lower part of the oblique fissure separates the inferior lobe from the **middle lobe.** The horizontal fissure of the right lung subdivides the superior lobe, thus forming a middle lobe.

Each lobe receives its own secondary (lobar) bronchus. Thus the right primary bronchus gives rise to three secondary (lobar) bronchi called the **superior, middle,** and **inferior secondary (lobar) bronchi.** The left primary

bronchus gives rise to a **superior** and an **inferior secondary (lobar) bronchus.** Within the substance of the lung, the secondary bronchi give rise to the **tertiary (segmental) bronchi,** which are constant in both origin and distribution. There are 10 tertiary bronchi in each lung. The segment of lung tissue that each supplies is called a **bronchopulmonary segment.** Bronchial and pulmonary disorders, such as tumors or abscesses, may be localized in a bronchopulmonary segment and may be surgically removed without seriously disrupting surrounding lung tissue.

Lobules

Each bronchopulmonary segment of the lungs has many small compartments called **lobules** (Fig. 23.10a). Each lob-

FIGURE 23.9 Lungs.

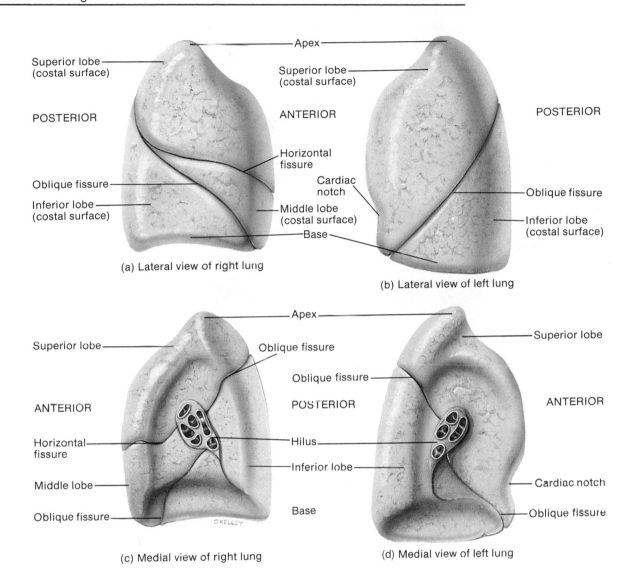

(a) Lateral view of right lung

(b) Lateral view of left lung

(c) Medial view of right lung

(d) Medial view of left lung

Question: Why are the right and left lungs slightly different in shape?

ule is wrapped in elastic connective tissue and contains a lymphatic vessel, an arteriole, a venule, and a branch from a terminal bronchiole. Terminal bronchioles subdivide into microscopic branches called **respiratory bronchioles.** As the respiratory bronchioles penetrate more deeply into the lungs, the epithelial lining changes from cuboidal to squamous. Respiratory bronchioles, in turn, subdivide into several (2 to 11) **alveolar ducts (atria).**

Around the circumference of the alveolar ducts are numerous alveoli and alveolar sacs. An **alveolus** (al-VĒ-ō-lus) is a cup-shaped outpouching lined by epithelium and supported by a thin elastic basement membrane. **Alveolar sacs** are two or more alveoli that share a common opening (Fig. 23.10a,b). The alveolar walls consist of two types of epithelial cells (Fig. 23.11). **Type I alveolar (squamous**

pulmonary epithelial) cells form a continuous lining of the alveolar wall, interrupted by occasional **type II alveolar (septal) cells.**

Type II alveolar cells secrete alveolar fluid, which keeps the alveolar cells moist. Included in the alveolar fluid is a complex mixture of phospholipids and lipoproteins called **surfactant** (sur-FAK-tant). Surfactant lowers the surface tension of alveolar fluid. Surface tension arises at the air–water junction because the polar water molecules are more attracted to each other than to gas molecules in the air. In the lungs, this attractive force among water molecules promotes collapse of the alveoli. Surfactant reduces the collapsing force. Normally, the elastic fibers in the alveolar walls are strong enough to hold the alveoli open.

FIGURE 23.10 Histology of the lungs.

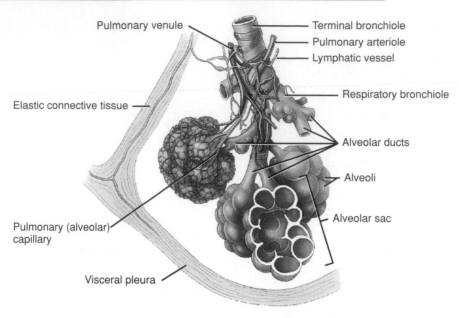

(a) Portion of a lobule of the lung

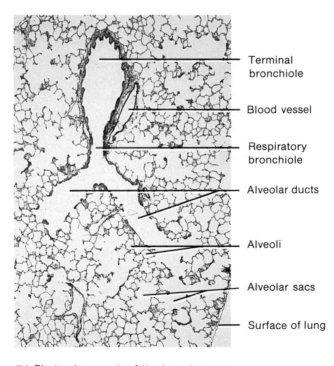

(b) Photomicrograph of the lung (approx.
30 x)

Question: What are the roles of cilia and goblet cells?

Associated with the alveolar wall are **alveolar macrophages (dust cells).** These wandering phagocytes remove dust particles or other debris from alveolar spaces. Also present are monocytes, white blood cells that transform into alveolar macrophages, and fibroblasts that produce reticular and elastic fibers. Deep to the layer of type I alveolar cells is an elastic basement membrane. Around the alveoli, the lobule's arteriole and venule disperse into a capillary network. The blood capillaries consist of a single layer of endothelial cells and basement membrane.

Alveolar–Capillary (Respiratory) Membrane

The exchange of respiratory gases between the lungs and blood takes place by diffusion across alveolar and capillary walls. Collectively, the layers through which the respiratory gases diffuse are known as the **alveolar–capillary (respiratory) membrane** (Fig. 23.11b). It consists of:

1. A layer of type I and type II alveolar cells with free alveolar macrophages that constitute the **alveolar (epithelial) wall.**
2. An **epithelial basement membrane** underneath the alveolar wall.
3. A **capillary basement membrane** that is often fused to the epithelial basement membrane.
4. The **endothelial cells** of the capillary.

Despite the several layers, the alveolar–capillary membrane averages only 0.5 µm in thickness, about ¹⁄₁₆ the diameter of a red blood cell. This allows rapid diffusion of respiratory gases. Moreover, it has been estimated that the lungs contain 300 million alveoli, providing an immense surface area of 70 m² (750 ft²), about the size of a handball court, for the exchange of gases.

Blood Supply to the Lungs

There is a double blood supply to the lungs. Deoxygenated blood passes through the pulmonary trunk, which divides into a left pulmonary artery that enters the left lung and a right pulmonary artery that enters the right lung. The venous return of the oxygenated blood is by way of the pulmonary veins, typically two in number on each side—the right and left superior and inferior pulmonary veins. All four veins drain into the left atrium (see Fig. 21.31).

Oxygenated blood is delivered through bronchial arteries, which branch directly from the aorta. There are

---

FIGURE 23.11 Structure of an alveolus.

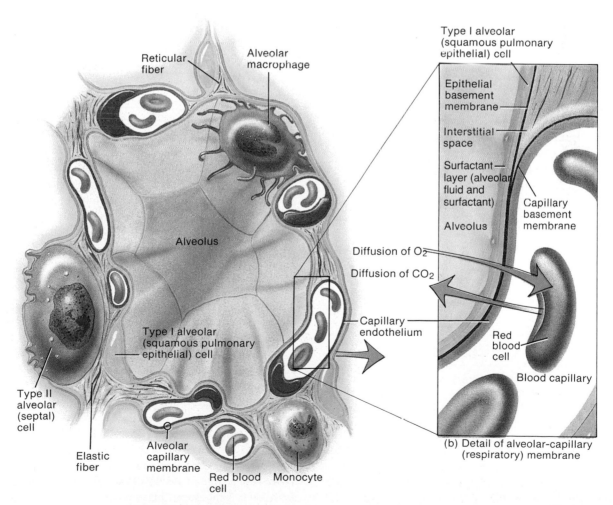

(a) Cross section of an alveolus

(b) Detail of alveolar-capillary (respiratory) membrane

Figure continues

FIGURE 23.11 (continued)

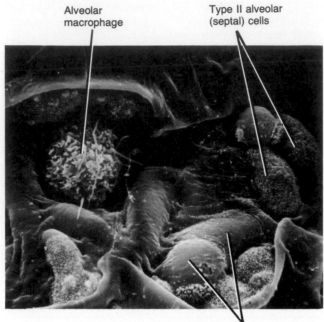

Alveolar macrophage

Type II alveolar (septal) cells

Type I alveolar (squamous pulmonary epithelial) cells

(c) Scanning electron micrograph of an alveolus, 3,430×

Question: What is the function of type II alveolar (septal) cells?

communications between the two systems, and most blood returns via pulmonary veins. Some blood, however, drains into bronchial veins, branches of the azygos system.

Recall from Chapter 21 that pulmonary circulation differs from systemic circulation in that pulmonary blood vessels provide less resistance to blood flow; less pressure is required to move blood through the pulmonary circulation. In addition, pulmonary blood vessels constrict in response to hypoxia (low oxygen level) so that pulmonary blood bypasses poorly aerated areas. In all other body tissues, hypoxia causes dilation of blood vessels, which serves to increase blood flow to a tissue that is not receiving adequate oxygen.

PHYSIOLOGY OF RESPIRATION

The principal purpose of **respiration** is to supply the cells of the body with oxygen and remove the carbon dioxide produced by cellular activities. The three basic processes of respiration are pulmonary ventilation, external (pulmonary) respiration, and internal (tissue) respiration.

Pulmonary Ventilation

Pulmonary ventilation (breathing) is the process by which gases are exchanged between the atmosphere and lung alveoli. The bulk flow of air between the atmosphere and lungs occurs for the same reason that blood flows through the body: a pressure gradient exists. Air moves into the lungs when the pressure inside the lungs is less than the air pressure in the atmosphere. Air moves out of the lungs when the pressure inside the lungs is greater than the pressure in the atmosphere. Let us examine the mechanics of pulmonary ventilation by first looking at inspiration.

Inspiration

Breathing in is called **inspiration (inhalation).** Just before each inspiration, the air pressure inside the lungs equals the pressure of the atmosphere, which is about 760 mm Hg, or 1 atmosphere (atm), at sea level. For air to flow into the lungs, the pressure inside the lungs must become lower than the pressure in the atmosphere. This condition is achieved by increasing the volume (size) of the lungs.

The pressure of a gas in a closed container is inversely proportional to the volume of the container. This means that if the size of a closed container is increased, the pressure of the gas inside the container decreases. If the size of the container is decreased, then the pressure inside it increases. This is referred to as **Boyle's law** and may be demonstrated as follows. Suppose we place a gas in a cylinder that has a movable piston and a pressure gauge, and the initial pressure is 1 atm (Fig. 23.12a). This pressure is created by the

FIGURE 23.12 Boyle's law and Charles' law. (a) According to Boyle's law, the volume of a gas varies inversely with the pressure. (b) According to Charles' law, the volume of a gas is directly proportional to the temperature, assuming the pressure remains constant.

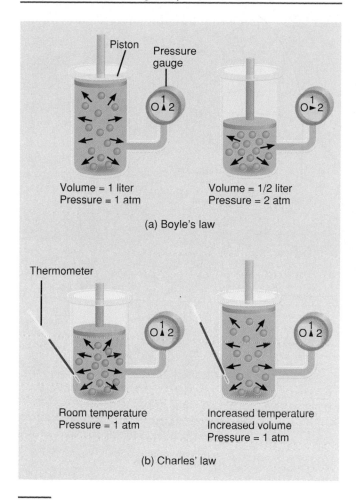

Volume = 1 liter
Pressure = 1 atm

Volume = 1/2 liter
Pressure = 2 atm

(a) Boyle's law

Room temperature
Pressure = 1 atm

Increased temperature
Increased volume
Pressure = 1 atm

(b) Charles' law

Question: In (a), if the volume is decreased to ¹/₄ liter, how does the pressure change?

gas molecules striking the wall of the container. If the piston is pushed down, the gas is compressed into a smaller volume. The same number of gas molecules is striking less wall space. The gauge shows that the pressure doubles as the gas is compressed to half its volume. In other words, the same number of molecules in half the space produces twice the pressure. Conversely, if the piston is raised to increase the volume, the pressure decreases. Thus the volume of a gas varies inversely with pressure (assuming that temperature is constant). Boyle's law applies to everyday activities such as the operation of a bicycle pump or the blowing up of a balloon. Differences in pressure force air into our lungs when we inhale and force the air out when we exhale.

For inspiration to occur, the lungs must expand. This increases lung volume and thus decreases the pressure in the lungs below atmospheric pressure. The first step in expanding the lungs involves contraction of the principal inspiratory muscles—the diaphragm and external intercostals (Fig. 23.13; see also Fig. 11.11).

The diaphragm, the most important muscle of inspiration, is a dome-shaped skeletal muscle that forms the floor of the thoracic cavity. It is innervated by fibers of the phrenic nerves, which emerge from both sides of the spinal cord at cervical levels 3, 4, and 5. Contraction of the diaphragm causes it to flatten, lowering its dome. This increases the vertical dimension of the thoracic cavity and accounts for the movement of about 75% of the air that enters the lungs during inspiration. The distance the diaphragm moves during inspiration ranges from 1 cm (0.4 in.) during normal quiet breathing up to about 10 cm (4 in.) during heavy breathing. Advanced pregnancy, excessive obesity, or confining abdominal clothing can prevent a complete descent of the diaphragm. At the same time the diaphragm contracts, the external intercostals contract. These skeletal muscles run obliquely downward and forward between adjacent ribs, and when these muscles contract, the ribs are pulled upward and the sternum is pushed forward. This increases the anterior–posterior dimension of the thoracic cavity.

During normal breathing, the pressure between the two pleural layers, called **intrapleural (intrathoracic) pressure,** is always below atmospheric pressure. (It may become temporarily higher than atmospheric pressure only during modified respiratory movement such as coughing or straining during defecation.) Just before inspiration, it is about 4 mm Hg less than the atmospheric pressure, or 756 mm Hg if the atmospheric pressure is 760 mm Hg (Fig. 23.14). As the diaphragm contracts and the overall size of the thoracic cavity increases, the intrapleural pressure falls to about 754 mm Hg. Consequently, the walls of the lungs are pulled outward. The parietal and visceral pleurae normally adhere strongly to each other because of the below atmospheric pressure between them and because of the surface tension created by their moist adjoining surfaces. As the thoracic cavity expands, the parietal pleura lining the cavity is pulled outward in all directions, and the visceral pleura and lungs are pulled along with it.

When the volume of the lungs increases, the pressure inside the lungs, called the **alveolar (intrapulmonic) pressure,** drops from 760 to 758 mm Hg. A pressure gradient is thus established between the atmosphere and the alveoli. Air rushes from the atmosphere into the lungs due to a gas pressure difference, and inspiration takes place. Air continues to move into the lungs as long as the pressure difference exists.

The term for normal quiet breathing is **eupnea** (yoop-NĒ-a; *eu* = normal). Eupnea involves shallow, deep, or combined shallow and deep breathing. Shallow (chest) breathing is called **costal breathing.** It consists of an upward and outward movement of the chest as a result of contraction of the external intercostal muscles. The intercostal muscles receive innervation from the intercostal

FIGURE 23.13 Pulmonary ventilation: muscles of inspiration and expiration. The pectoralis minor muscle, an accessory inspiratory muscle, is not shown here, but is illustrated in Fig. 11.15a.

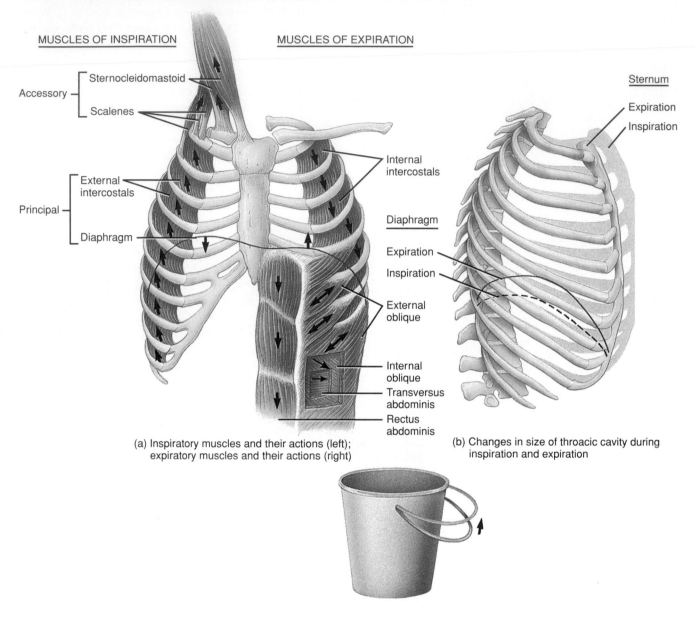

MUSCLES OF INSPIRATION   MUSCLES OF EXPIRATION

Accessory
- Sternocleidomastoid
- Scalenes

Principal
- External intercostals
- Diaphragm

Internal intercostals

Diaphragm
- Expiration
- Inspiration

External oblique

Internal oblique

Transversus abdominis

Rectus abdominis

Sternum
- Expiration
- Inspiration

(a) Inspiratory muscles and their actions (left); expiratory muscles and their actions (right)

(b) Changes in size of throacic cavity during inspiration and expiration

(c) During inspiration, the ribs move upward and outward like the handle on a bucket

Question: Right now, what muscle is mainly responsible for your breathing?

nerves, which emerge from thoracic segments of the spinal cord. Deep (abdominal) breathing is called **diaphragmatic breathing.** It consists of the outward movement of the abdomen as a result of the contraction and descent of the diaphragm. During deep, labored inspiration, accessory muscles of inspiration also participate in increasing the size of the thoracic cavity (see Fig. 23.13a). These include the sternocleidomastoid, which elevates the sternum, the scalenes, which elevate the superior two ribs, and the pec-

toralis minor, which elevates the third through fifth ribs. Inspiration is an active process because it is initiated by skeletal muscle contraction.

A summary of inspiration is presented in Fig. 23.15a.

Expiration

Breathing out, called **expiration (exhalation),** is also achieved by a pressure gradient, but in this case the gradient

FIGURE 23.14 Pulmonary ventilation: pressure changes. At the beginning of inspiration, the diaphragm contracts, the chest expands, the lungs are pulled outward and alveolar pressure decreases. As the diaphragm relaxes, the lungs recoil inward. Alveolar pressure rises, forcing air out until alveolar pressure equals atmospheric pressure.

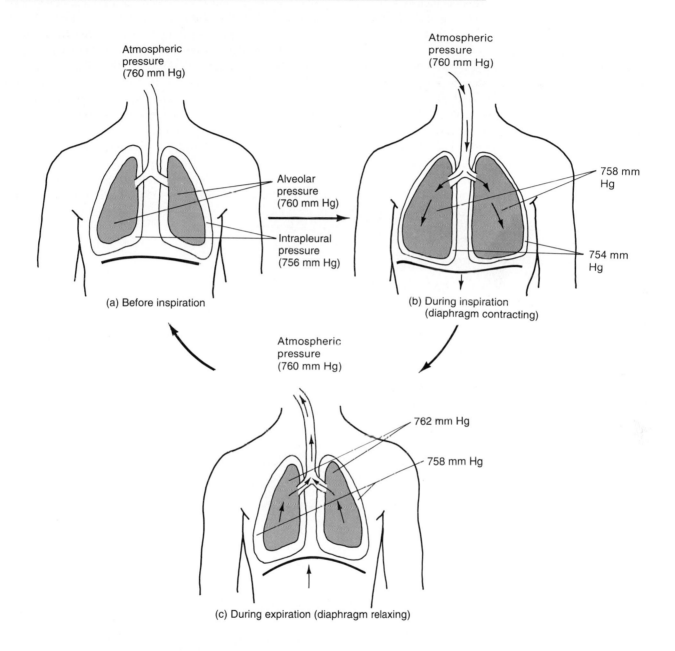

Atmospheric pressure (760 mm Hg)

Alveolar pressure (760 mm Hg)

Intrapleural pressure (756 mm Hg)

(a) Before inspiration

Atmospheric pressure (760 mm Hg)

758 mm Hg

754 mm Hg

(b) During inspiration (diaphragm contracting)

Atmospheric pressure (760 mm Hg)

762 mm Hg

758 mm Hg

(c) During expiration (diaphragm relaxing)

Question: How does the intrapleural pressure change during a normal, quiet breath?

is reversed: the pressure in the lungs is greater than the pressure of the atmosphere. Normal expiration during quiet breathing, unlike inspiration, is a passive process since no muscular contractions are involved. It depends on two factors: (1) the recoil of elastic fibers that were stretched dur-

ing inspiration and (2) the inward pull of surface tension due to the film of alveolar fluid (see Fig. 23.11).

Expiration starts when the inspiratory muscles relax. As the external intercostals relax, the ribs move downward, and as the diaphragm relaxes, its dome moves upward owing to its elasticity. These movements decrease the vertical and anterior–posterior dimensions of the thoracic cavity. Also, surface tension exerts an inward pull and the elastic basement membranes of the alveoli and elastic fibers in bronchioles and alveolar ducts recoil. As a result, lung volume decreases and the alveolar pressure increases to 762 mm Hg. Air then flows from the area of higher pressure in the alveoli to the area of lower pressure in the atmosphere (see Fig. 23.14c).

Expiration becomes active during labored breathing and when air movement out of the lungs is impeded. During these times, muscles of expiration—abdominal and internal intercostals—contract. Contraction of the abdominal muscles moves the inferior ribs downward and compresses the abdominal viscera, thus forcing the diaphragm upward. Contraction of the internal intercostals, which extend downward and backward between adjacent ribs, pulls the ribs downward.

A summary of expiration is presented in Fig. 23.15b.

Atelectasis (Collapsed Lung)

It was noted earlier that intrapleural pressure is normally lower than atmospheric pressure, that is, it is subatmospheric. The pleural cavities are sealed off from the outside environment and cannot equalize their pressure with that of the atmosphere. Nor can the diaphragm and rib cage move inward enough to bring the intrapleural pressure up to atmospheric pressure, except temporarily, as, for example, during coughing. Maintenance of a low intrapleural pressure is vital to the functioning of the lungs. At the end of an expiration the alveoli tend to recoil inward and collapse on themselves like the walls of a deflating balloon. A collapsed lung or portion of a lung is called **atelectasis** (at′-ē-LEK-ta-sis; *ateles* = incomplete; *ektasis* = dilation).

One factor that normally prevents the collapse of alveoli is the presence of **surfactant,** a mixture of phospholipids and lipoproteins produced by the type II alveolar cells. Surfactant decreases surface tension in the lungs. Thus as alveoli become smaller—for example, following expiration—the tendency of alveoli to collapse is minimized because the surface tension does not increase. As you will see later, a deficiency of surfactant in infants results in a disorder called respiratory distress syndrome (RDS) of the newborn.

Compliance

Compliance refers to the ease with which the lungs and thoracic wall can be expanded. High compliance means that the lungs and thoracic wall expand easily; low compliance means that they resist expansion. Compliance is related to two principal factors: elasticity and surface tension. The

---

FIGURE 23.15 Summary of inspiration and expiration.

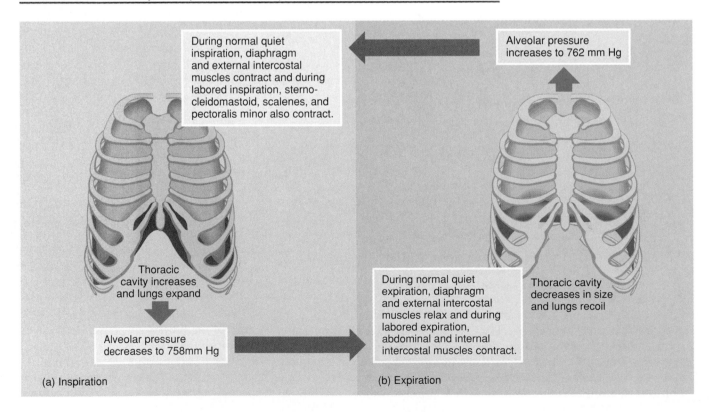

(a) Inspiration

During normal quiet inspiration, diaphragm and external intercostal muscles contract and during labored inspiration, sterno-cleidomastoid, scalenes, and pectoralis minor also contract.

Thoracic cavity increases and lungs expand

Alveolar pressure decreases to 758mm Hg

(b) Expiration

Alveolar pressure increases to 762 mm Hg

Thoracic cavity decreases in size and lungs recoil

During normal quiet expiration, diaphragm and external intercostal muscles relax and during labored expiration, abdominal and internal intercostal muscles contract.

lungs normally expand easily because elastic fibers in lung tissue and surfactant in alveolar fluid result in high compliance. If surface tension within lung tissue were high, the tissues would resist expansion, but surfactant lowers surface tension and thus increases compliance. Compliance decreases with any condition that destroys lung tissue, causes it to become filled with fluid, produces a deficiency in surfactant, or in any way impedes lung expansion or contraction.

Airway Resistance

The walls of the respiratory passageways, especially the bronchi and bronchioles, offer some resistance to the normal flow of air into the lungs. The muscular contraction of normal inspiration not only expands the thoracic cavity but also decreases resistance to airflow by increasing the diameter of bronchi and bronchioles. Any condition that obstructs the air passageways increases resistance, and more pressure to force air through is required. During a forced expiration, as in coughing, straining, or playing a wind instrument, intrapleural pressure may increase from its normally subatmospheric (negative) value to a positive one. This greatly increases airway resistance because it results in compression of the airways. People with chronic obstructive pulmonary disease (COPD), such as bronchial asthma and emphysema, have some degree of obstruction of air passageways, which increases airway resistance.

Modified Respiratory Movements

Respirations also provide humans with methods for expressing emotions such as laughing, yawning, sighing, and sobbing. Moreover, respiratory air can be used to expel foreign matter from the lower air passages through actions such as sneezing and coughing. Some of the modified respiratory movements that express emotion or clear the air passageways are listed in Exhibit 23.1. All these movements are reflexes, but some of them also can be initiated voluntarily.

Pulmonary Air Volumes and Capacities

In clinical practice, the word **respiration (ventilation)** means one inspiration plus one expiration. The healthy adult averages 12 respirations a minute and moves about 6 liters of air into and out of the lungs while at rest. A lower-than-normal volume of air exchange is usually a sign of pulmonary malfunction.

The apparatus commonly used to measure the volume of air exchanged during breathing and the rate of ventilation is a **spirometer** (*spiro* = breathe) or **respirometer.** The record is called a **spirogram** (Fig. 23.16). Inspiration is recorded as an upward deflection and expiration is recorded as a downward deflection.

EXHIBIT 23.1

MODIFIED RESPIRATORY MOVEMENTS

Movement	Comment
Coughing	A long-drawn and deep inspiration followed by a complete closure of the rima glottidis, which results in a strong expiration that suddenly pushes the rima glottidis open and sends a blast of air through the upper respiratory passages. Stimulus for this reflex act may be a foreign body lodged in the larynx, trachea, or epiglottis.
Sneezing	Spasmodic contraction of muscles of expiration that forcefully expels air through the nose and mouth. Stimulus may be an irritation of the nasal mucosa.
Sighing	A long-drawn and deep inspiration immediately followed by a shorter but forceful expiration.
Yawning	A deep inspiration through the widely opened mouth producing an exaggerated depression of the lower jaw. It may be stimulated by drowsiness, fatigue, or someone else's yawning, but precise stimulus–receptor cause is unknown.
Sobbing	A series of convulsive inspirations followed by a single prolonged expiration. The rima glottidis closes earlier than normal after each inspiration so only a little air enters the lungs with each inspiration.
Crying	An inspiration followed by many short convulsive expirations, during which the rima glottidis remains open and the vocal cords vibrate; accompanied by characteristic facial expressions and tears.
Laughing	The same basic movements as crying, but the rhythm of the movements and the facial expressions usually differ from those of crying. Laughing and crying are sometimes indistinguishable.
Hiccuping	Spasmodic contraction of the diaphragm followed by a spasmodic closure of the rima glottidis to produce a sharp inspiratory sound. Stimulus is usually irritation of the sensory nerve endings of the gastrointestinal tract.
Valsalva maneuver	Forced expiration against a closed rima glottidis as during periods of straining.

Pulmonary Volumes

During the process of normal quiet breathing, about 500 ml of air moves into the respiratory passageways with each inspiration. The same amount moves out with each expiration. This volume of air inspired (or expired) is called **tidal volume** (Fig. 23.16). Only about 350 ml of the tidal volume actually reach the alveoli. The other 150 ml remain in air spaces of the nose, pharynx, larynx, trachea, bronchi, and bronchioles. These areas are known as **anatomic dead space.** The total air taken in during one minute is called the **minute volume of respiration (MVR).** It is calculated by multiplying the tidal volume by the normal breathing rate per minute. An average volume would be 500 ml times 12 respirations per minute, or 6000 ml/min.

By taking a very deep breath, you can inspire a good deal more than 500 ml. This additional inhaled air, called the **inspiratory reserve volume,** averages 3100 ml above the 500 ml of tidal volume. Thus the respiratory system can pull in 3600 ml of air. In fact, even more air can be inhaled if inspiration follows forced expiration.

If you inhale normally and then exhale as forcibly as possible, you should be able to push out 1200 ml of air in addition to the 500 ml of tidal volume. These extra 1200 ml are called the **expiratory reserve volume.**

Even after the expiratory reserve volume is expelled, a good deal of air remains in the lungs because the lower intrapleural pressure keeps the alveoli slightly inflated, and some air also remains in the noncollapsible air passageways. This air, the **residual volume,** amounts to about 1200 ml.

Opening the thoracic cavity allows the intrapleural pressure to equal the atmospheric pressure, forcing out some of the residual volume. The air remaining is called the **minimal volume.** Minimal volume provides a medical and legal tool for determining whether a baby was born dead or died after birth. The presence of minimal volume can be demonstrated by placing a piece of lung in water and watching it float. Fetal lungs contain no air, and so the lung of a stillborn will not float in water.

Pulmonary Capacities

Pulmonary capacities are combinations of specific lung volumes (Fig. 23.16). **Inspiratory capacity,** the total inspiratory ability of the lungs, is the sum of tidal volume plus inspiratory reserve volume (3600 ml). **Functional residual capacity** is the sum of residual volume plus expiratory reserve volume (2400 ml). **Vital capacity** is the sum of inspiratory reserve volume, tidal volume, and expiratory reserve volume (4800 ml). Finally, **total lung capacity** is the sum of all volumes (6000 ml).

FIGURE 23.16 Spirogram of pulmonary volumes and capacities.

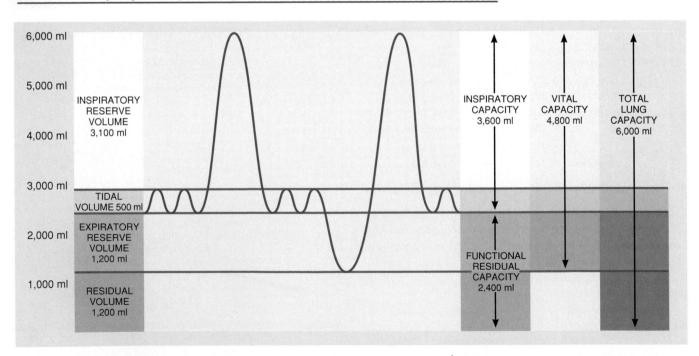

Question: Breathe in as deeply as possible and then exhale as much air as you can. Which pulmonary capacity have you demonstrated?

Exchange of Oxygen and Carbon Dioxide

At birth, as soon as the lungs fill with air, oxygen starts to diffuse down its concentration gradient from the alveoli into the blood, through the interstitial fluid, and finally into the cells. Carbon dioxide diffuses in the opposite direction—from the cells, through interstitial fluid to the blood, and to the alveoli. To understand how these gases are exchanged in the body, you need to know three more gas laws.

Charles' Law

According to **Charles' law**, the volume of a gas is directly proportional to its absolute temperature, assuming that the pressure remains constant. Recall the cylinder we used to demonstrate Boyle's law. Suppose now that the gas in the cylinder exerts an initial pressure of 1 atmosphere when the piston is halfway down (see Fig. 23.12b). When the gas is heated, the gas molecules move faster and the number of collisions within the cylinder increases. Assuming that the piston moves freely (no external pressure is applied), the force of the molecules hitting it moves it upward. As the gas expands, the movement of the piston provides a measure of the increase in volume. As the space in the cylinder increases, the molecules have farther to travel, so the number of collisions decreases as the space increases. The pressure of 1 atmosphere is maintained, and the volume increases in direct proportion to the temperature increase. As gases enter the warmer lungs, the gases expand, thus increasing lung volume.

Dalton's Law

According to **Dalton's law,** each gas in a mixture of gases exerts its own pressure as if all the other gases were not present. The pressure of a specific gas in a mixture is called its *partial pressure* and is denoted as p. The total pressure of the mixture is calculated by simply adding all the partial pressures. Atmospheric air is a mixture of several gases—oxygen, carbon dioxide, nitrogen, water vapor, and a number of other gases that appear in such small quantities that

we will ignore them. Atmospheric pressure is the sum of the pressures of all these gases:

$$\text{Atmospheric pressure (760 mm Hg)} = pO_2 + pCO_2 + pN_2 + pH_2O$$

We can determine the partial pressure exerted by each component in the mixture by multiplying the percentage of the gas in the mixture by the total pressure of the mixture. For example, to find the partial pressure of oxygen in the atmosphere, multiply the percentage of atmospheric air composed of oxygen (21%) by the total atmospheric pressure (760 mm Hg):

$$\begin{aligned} \text{Atmospheric } pO_2 &= 21\% \times 760 \text{ mm Hg} \\ &= 159.6 \text{ or } 160 \text{ mm Hg} \end{aligned}$$

Since the percentage of CO_2 in the atmosphere is 0.04,

$$\begin{aligned} \text{Atmospheric } pCO_2 &= 0.04\% \times 760 \text{ mm Hg} \\ &= 0.3 \text{ mm Hg} \end{aligned}$$

The partial pressures of oxygen and carbon dioxide in the atmosphere, alveoli, blood, and tissue cells are shown in Exhibit 23.2. These partial pressures are important in determining the movement of oxygen and carbon dioxide between the atmosphere and lungs, the lungs and blood, and the blood and body cells. When a mixture of gases diffuses across a permeable membrane, each gas diffuses from the area where its partial pressure is greater to the area where its partial pressure is less. Every gas is on its own and behaves as if the other gases in the mixture did not exist.

The amounts of O_2 and CO_2 vary in inspired (atmospheric), alveolar, and expired air (Exhibit 23.3). Inspired air contains about 21% O_2 and 0.04% CO_2. Expired air contains less O_2 (about 16%) and more CO_2 (about 4.5%) than inspired air. Also, expired and alveolar air have higher water vapor content than inspired air because the moist mucosal linings humidify air as it is inhaled. Compared with alveolar air, expired air contains more O_2 (about 16% versus 14%) and less CO_2 (about 4.5% versus 5.5%) because some of the expired air is in the anatomic dead space and has not participated in gas exchange. Expired air is actually a mixture of inspired air and alveolar air that remained in the anatomic dead space.

EXHIBIT 23.2

PARTIAL PRESSURES (mm Hg) OF OXYGEN AND CARBON DIOXIDE

	Atmospheric Air (Sea Level)	Alveolar Air	Deoxygenated Blood	Oxygenated Blood	Tissue Cells (Average at Rest)
pO_2	160	105	40	105	40
pCO_2	0.3	40	45	40	45

EXHIBIT 23.3

OXYGEN AND CARBON DIOXIDE IN INSPIRED AIR, ALVEOLAR AIR, AND EXPIRED AIR

	Inspired Air (%)	Alveolar Air (%)	Expired Air (%)
Oxygen	21	14	16
Carbon dioxide	0.04	5.5	4.5

Henry's Law

You have probably noticed that a soft drink makes a hissing sound when the top of the container is removed, and bubbles rise to the surface for some time afterward. The gas dissolved in carbonated beverages is carbon dioxide. The ability of a gas to stay in solution depends on its partial pressure and solubility coefficient, that is, its physical or chemical attraction for water. The solubility coefficient of carbon dioxide is high (0.57), that of oxygen is low (0.024), and that of nitrogen is still lower (0.012). The higher the partial pressure of a gas over a liquid and the higher the solubility coefficient, the more gas will stay in solution. Since the soft drink is bottled or canned under pressure and capped, the CO_2 remains dissolved as long as the container is unopened. Once you remove the cap, the pressure is released and the gas begins to bubble out. This phenomenon is explained by **Henry's law:** the quantity of a gas that will dissolve in a liquid is proportional to the partial pressure of the gas and its solubility coefficient, when the temperature remains constant.

Henry's law explains two conditions resulting from changes in the solubility of nitrogen in body fluids. Even though the air we breathe contains about 79% nitrogen, this gas has no known effect on bodily functions, and very little of it dissolves in blood plasma because of its low solubility coefficient at sea level pressure. But when a deep-sea diver, **scuba (self-contained underwater breathing apparatus)** diver, or caisson worker (person who builds tunnels under water) breathes air under high pressure, the nitrogen in the mixture can affect the body. Partial pressure is a function of total pressure, and therefore the partial pressures of all the components of a mixture increase as the total increases. Since the partial pressure of nitrogen is higher in a mixture of compressed air than in air at sea level pressure, a considerable amount of nitrogen goes into solution in plasma and interstitial fluid. Excessive amounts of dissolved nitrogen may produce giddiness and other symptoms similar to alcohol intoxication. The condition is called **nitrogen narcosis** or "rapture of the depths." The greater the depth, the more severe the condition.

If a diver comes to the surface slowly, the dissolved nitrogen can be eliminated through the lungs. However, if the ascent is too rapid, nitrogen comes out of solution too quickly to be eliminated by exhalation. Instead, it forms gas bubbles in the tissues, resulting in **decompression sickness (caisson disease,** or **bends).** The effects of decompression sickness typically result from bubbles in nervous tissue and can be mild or severe, depending on the number of bubbles formed. Symptoms include joint pain, especially in the arms and legs, dizziness, shortness of breath, extreme fatigue, paralysis, and unconsciousness. Decompression sickness can be prevented by a slow ascent or by the use of a special decompression tank within five minutes after arriving at the surface. The use of helium–oxygen mixtures instead of air containing nitrogen may reduce the dangers of decompression sickness since helium is only about 40% as soluble as nitrogen in blood.

CLINICAL APPLICATION

HYPERBARIC OXYGENATION

A major clinical application of Henry's law is **hyperbaric** (*hyper* = over; *baros* = pressure) **oxygenation (HBO).** Using pressure to cause more oxygen to dissolve in the blood is an effective technique in treating patients infected by anaerobic bacteria, such as those that cause tetanus and gangrene. (Anaerobic bacteria cannot live in the presence of free oxygen.) A person undergoing hyperbaric oxygenation is placed in a hyperbaric chamber, which contains oxygen at a pressure of 3 to 4 atm (2280 to 3040 mm Hg). The body tissues pick up the oxygen, and the bacteria are killed. Hyperbaric chambers may also be used for treating certain heart disorders, carbon monoxide poisoning, gas embolisms, crush injuries, cerebral edema, certain hard-to-treat bone infections, smoke inhalation, near-drowning, asphyxia, vascular insufficiencies, and burns.

Physiology of External (Pulmonary) Respiration

External (pulmonary) respiration is the exchange of oxygen and carbon dioxide between the alveoli of the lungs and pulmonary blood capillaries (Fig. 23.17a). It results in the conversion of **deoxygenated blood** (depleted of some O_2) coming from the heart to **oxygenated blood** (saturated with O_2) returning to the heart. During inspiration, atmospheric air containing oxygen enters the alveoli. Deoxygenated blood is pumped from the right ventricle through the pulmonary arteries into the pulmonary capillaries surrounding the alveoli. The pO_2 of alveolar air is 105 mm Hg. If you are at rest, the pO_2 of deoxygenated blood entering your pulmonary capillaries is only 40 mm Hg. As a result of this difference in pO_2, there is a net diffusion of oxygen from the alveoli into the deoxygenated blood until equilibrium is reached, and the pO_2 of the now oxygenated blood is 105

FIGURE 23.17 Changes in partial pressures (in mm Hg) during external and internal respiration.

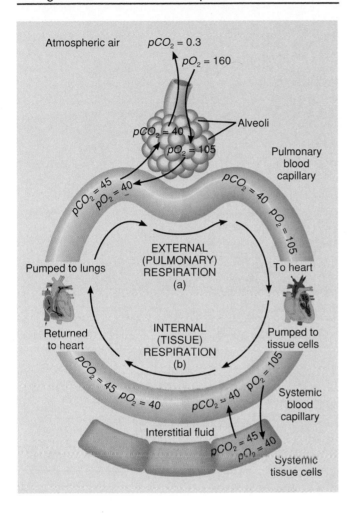

Question: What causes oxygen (O₂) to enter pulmonary capillaries from alveoli and to enter tissue cells from systemic capillaries?

mm Hg. While O_2 diffuses from the alveoli into deoxygenated blood, there is a net diffusion of CO_2 in the opposite direction. The pCO_2 of deoxygenated blood is 45 mm Hg, whereas that of the alveolar air is 40 mm Hg. Because of this difference in pCO_2, carbon dioxide diffuses from deoxygenated blood into the alveoli until the pCO_2 of the blood decreases to 40 mm Hg. This is the pCO_2 of fully oxygenated blood. Thus the pO_2 and pCO_2 of oxygenated blood leaving the lungs are the same as in alveolar air. The carbon dioxide that diffuses into the alveoli is eliminated from the lungs during expiration.

The rate of external respiration depends on several factors.

1. Partial pressure difference. As long as alveolar pO_2 is higher than pO_2 in systemic veins, oxygen diffuses from the alveoli into the blood. As a person ascends in altitude, the atmospheric pO_2 decreases, the alveolar pO_2 decreases correspondingly, and less oxygen diffuses into the blood. For example, at sea level, pO_2 is 160 mm Hg. At 10,000 ft, it decreases to 110 mm Hg; at 20,000 ft, to 73 mm Hg; and at 50,000 ft, to 18 mm Hg. The common symptoms of **high altitude sickness (acute mountain sickness)**—shortness of breath, headache, fatigue, insomnia, nausea, dizziness—are due to the low pO_2 in the blood.

2. Surface area for gas exchange. The surface area over which diffusion may occur is large (about 70 m² or 750 ft²). Any pulmonary disorder that decreases the functional surface area formed by the alveolar–capillary membranes decreases the rate of external respiration. In emphysema, for example, alveolar walls disintegrate, and surface area declines.

3. Diffusion distance. The total thickness of the alveolar–capillary (respiratory) membranes is only 0.5 μm. Thicker membranes would slow the rate of diffusion. Also, the capillaries are so narrow that the red blood cells must pass through them in single file. This minimizes the diffusion distance from an alveolar air space to hemoglobin inside red blood cells. Buildup of fluid, as occurs in pulmonary edema, slows the rate of gas exchange because it increases diffusion distance.

4. Breathing rate and depth. External respiration also depends on the average rate of airflow into and out of the lungs. Certain drugs, such as morphine, slow the respiration rate, thereby decreasing the amount of oxygen and carbon dioxide that can be exchanged between the alveoli and the blood.

Physiology of Internal (Tissue) Respiration

No matter how quickly blood flows through pulmonary capillaries, it picks up a maximal amount of oxygen so that blood returning to the heart is fully oxygenated. The left ventricle pumps oxygenated blood into the aorta and through the systemic arteries to capillaries to tissue cells. The exchange of oxygen and carbon dioxide between tissue blood capillaries and tissue cells is called **internal (tissue) respiration** (Fig. 23.17b). It results in the conversion of oxygenated blood into deoxygenated blood. Oxygenated blood entering tissue capillaries has a pO_2 of 105 mm Hg, whereas tissue cells have an average pO_2 of 40 mm Hg. Because of this difference in pO_2, oxygen diffuses from the oxygenated blood through interstitial fluid and into tissue cells until the pO_2 in the blood decreases to 40 mm Hg. This is the average pO_2 of deoxygenated blood entering tissue venules when you are at rest.

At rest, only about 25% of the available oxygen in oxygenated blood actually enters tissue cells. This amount is sufficient to support the needs of resting cells. Thus deoxygenated blood still retains considerable oxygen. During exercise, more oxygen diffuses from the blood into active cells.

While oxygen diffuses from the tissue blood capillaries into tissue cells, carbon dioxide diffuses in the opposite direction. The average pCO_2 of tissue cells is 45 mm Hg, whereas that of tissue capillary oxygenated blood is 40 mm Hg. As a result, carbon dioxide diffuses from tissue cells through interstitial fluid into the oxygenated blood until the pCO_2 in the blood increases to 45 mm Hg, the pCO_2 of tissue capillary deoxygenated blood. The deoxygenated blood now returns to the heart. From here it is pumped to the lungs for another cycle of external respiration.

Transport of Oxygen and Carbon Dioxide

The transport of gases between the lungs and body tissues is a function of the blood. When oxygen and carbon dioxide enter the blood, certain physical and chemical changes occur that aid in gas transport and exchange.

Oxygen

Oxygen does not dissolve easily in water, and therefore very little oxygen, only about 1.5%, is carried in the dissolved state in watery blood plasma. The remainder of the oxygen, about 98.5%, is transported in chemical combination with hemoglobin inside red blood cells (Fig. 23.18). Each 100 ml of oxygenated blood contains about 20 ml of oxygen; 0.3 ml dissolved in the plasma and 19.7 ml bound to hemoglobin.

Hemoglobin consists of a protein portion called globin and an iron-containing pigment portion called heme (see Fig. 19.3b). Each hemoglobin molecule has four heme groups, and each heme group can combine with one molecule of oxygen. Oxygen and hemoglobin combine in an easily reversible reaction to form **oxyhemoglobin** as follows:

$$\underset{\substack{\text{Reduced hemoglobin} \\ \text{(deoxyhemoglobin)}}}{\text{Hb}} + \underset{\text{Oxygen}}{\text{O}_2} \rightleftharpoons \underset{\text{Oxyhemoglobin}}{\text{HbO}_2}$$

Since 98.5% of the oxygen is bound to hemoglobin and is trapped inside RBCs, only the dissolved oxygen (1.5%) can diffuse out of tissue capillaries into tissue cells. Thus it is important to understand the factors that promote oxygen binding to and dissociation from hemoglobin.

● **Hemoglobin and pO_2** The most important factor that determines how much oxygen combines with hemoglobin is pO_2. When reduced hemoglobin (deoxyhemoglobin) is completely converted to HbO_2, it is **fully saturated.** When hemoglobin consists of a mixture of Hb and HbO_2, it is **partially saturated.** The **percent saturation of hemoglobin** is the percentage of HbO_2 in total hemoglobin. The relation between the percent saturation of hemoglobin and pO_2 is illustrated in Fig. 23.19, the oxygen–hemoglobin dissociation curve. Note

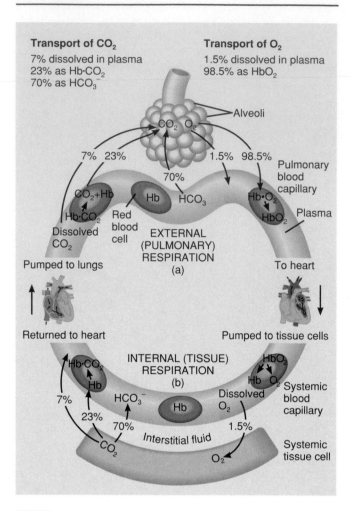

FIGURE 23.18 Transport of oxygen and carbon dioxide in the blood.

Question: How many milliliters of CO_2 are in 100 ml of deoxygenated blood at rest? How many milliliters of O_2 are in 100 ml of oxygenated blood during strenuous exercise?

that when the pO_2 is high, hemoglobin binds with large amounts of oxygen and is almost fully saturated. When pO_2 is low, hemoglobin is only partially saturated and oxygen is released from hemoglobin. In other words, the greater the pO_2, the more oxygen will combine with hemoglobin, until the available hemoglobin molecules are saturated. Therefore in pulmonary capillaries, where pO_2 is high, a lot of oxygen binds with hemoglobin, but in tissue capillaries, where the pO_2 is lower, hemoglobin does not hold as much oxygen, and the oxygen is released for diffusion into tissue cells. Note that hemoglobin is 75% saturated with oxygen at pO_2 of 40 mm Hg, the average pO_2 of tissue cells when you are at rest. This is the basis for our earlier statement that only 25% of the available oxygen splits from hemoglobin and is used by tissue cells under resting conditions.

FIGURE 23.19 Oxygen–hemoglobin dissociation curve at normal body temperature showing the relationship between hemoglobin saturation and pO_2. As pO_2 increases, more oxygen combines with hemoglobin.

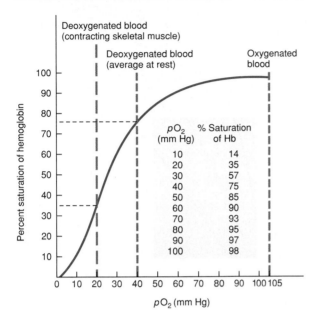

pO_2 (mm Hg)	% Saturation of Hb
10	14
20	35
30	57
40	75
50	85
60	90
70	93
80	95
90	97
100	98

Question: **What point on the curve would represent blood in your pulmonary veins right now? In your pulmonary veins if you were jogging?**

When the pO_2 is between 60 and 100 mm Hg, hemoglobin is 90% or more saturated with oxygen (Fig. 23.19). Thus blood picks up a nearly full load of oxygen from the lungs even when the alveolar pO_2 is as low as 60 mm Hg. This explains why people can still perform well at high altitudes or when they have certain cardiac and pulmonary diseases, even though pO_2 may drop to as low as 60 mm Hg. Note also in the curve that at a considerably lower pO_2 of 40 mm Hg, hemoglobin is 75% saturated with oxygen but drops to only 14% saturated at 10 mm Hg. This means that between 10 and 40 mm Hg large amounts of oxygen are released from hemoglobin in response to only small changes in pO_2. In active tissues such as contracting muscles, pO_2 may decrease to well below 40 mm Hg. At such pressures, a large percentage of the oxygen is released from hemoglobin. This provides more oxygen to tissues that are using it more rapidly.

● **Hemoglobin and Other Factors**   Although pO_2 is the most important factor that determines the percent saturation of hemoglobin, several other factors influence the affinity of hemoglobin for oxygen, the strength of the hemoglobin–oxygen bonding. Each one makes sense if you keep in mind that metabolically active tissue cells need oxygen and produce acids, CO_2, and heat.

1.   **Acidity (pH).** In an acid environment, hemoglobin's affinity for oxygen is lower and oxygen splits more readily from hemoglobin (Fig. 23.20a). This is referred to as the **Bohr effect.** When hydrogen ions (H^+) bind to certain amino acids in hemoglobin, they alter its structure and thereby decrease its oxygen-carrying capacity. Thus lowered pH drives oxygen off hemoglobin, making more oxygen available for tissue cells.

2.   **Partial pressure of carbon dioxide.** CO_2 also can bind to hemoglobin and the effect is similar to that of H^+. As pCO_2 rises, hemoglobin releases O_2 more readily (Fig. 23.20b). pCO_2 and pH are related factors. Low blood pH (acidic condition) results from high pCO_2. As carbon dioxide is taken up by the blood, much of it is temporarily converted to carbonic acid. This conversion is catalyzed by an enzyme in red blood cells called *carbonic anhydrase.*

$$CO_2 + H_2O \underset{}{\overset{\text{Carbonic anhydrase}}{\rightleftharpoons}} H_2CO_3 \rightleftharpoons H^+ + HCO_3^-$$

Carbon dioxide Water   Carbonic acid   Hydrogen ion Bicarbonate ion

The carbonic acid thus formed in red blood cells dissociates into hydrogen ions and bicarbonate ions. As the hydrogen ion concentration increases, pH decreases. Thus an increased pCO_2 produces a more acid environment that helps to split oxygen from hemoglobin. Low blood pH can also result from lactic acid, a by-product of anaerobic metabolism within muscles.

3.   **Temperature.**   Within limits, as temperature increases, so does the amount of oxygen released from hemoglobin (Fig. 23.21). Heat energy is a by-product of the metabolic reactions of all cells, and contracting muscle fibers (cells) release an especially large amount of heat. Release of oxygen from oxyhemoglobin is another example of how homeostatic mechanisms adjust body activities to cellular needs. Active cells require more oxygen, and active cells liberate more acid and heat. The acid and heat, in turn, promote release of oxygen from oxyhemoglobin.

4.   **BPG.**   A substance found in red blood cells called **2,3-bisphosphoglycerate (BPG),** previously called diphosphoglycerate (DPG), decreases the affinity of hemoglobin for oxygen and thus helps to release oxygen from hemoglobin. BPG is formed in red blood cells when they break down glucose for energy in a process called glycolysis (see page 826). When BPG combines with hemoglobin, the hemoglobin binds oxygen less tightly. The greater the level of BPG, the more oxygen is released from hemoglobin. Certain hormones, such as thyroxine, human growth hormone, epinephrine, norepinephrine, and testosterone, increase the formation of BPG. The level of BPG also is higher in people living at higher altitudes.

● **Fetal Hemoglobin**   Fetal hemoglobin differs from adult hemoglobin in structure and in its affinity for oxygen. Fetal hemoglobin has a higher affinity for oxygen because it binds BPG less strongly. Thus when pO_2 is low, fetal Hb

FIGURE 23.20 Oxygen–hemoglobin dissociation curve at normal body temperature showing the relationship between pH, pCO_2, and hemoglobin saturation. As pH increases or pCO_2 decreases, oxygen combines more tightly with hemoglobin and less is available to tissues. Stated another way, as pH decreases or pCO_2 increases, the affinity of hemoglobin for oxygen is less, so less oxygen combines with hemoglobin and more is available to tissues. These relationships are emphasized by the broken lines.

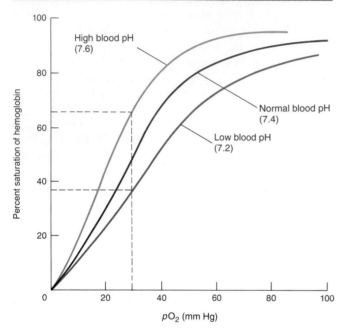

(a) Effect of pH on affinity of hemoglobin for oxygen

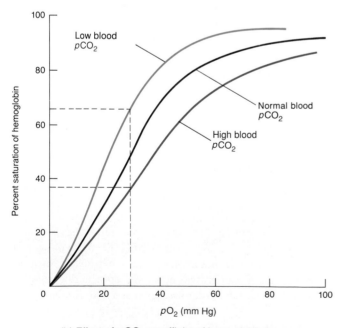

(b) Effect of pCO_2 on affinity of hemoglobin for oxygen

Question: In comparison to the value when you are sitting, is the affinity of your hemoglobin for oxygen higher or lower when you are jogging? What is the benefit of this change to you?

FIGURE 23.21 Oxygen–hemoglobin dissociation curve showing the relationship between temperature and hemoglobin saturation with oxygen. As temperature increases, less O_2 combines with hemoglobin.

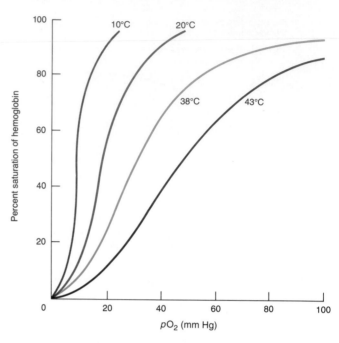

Question: Is O_2 more or less available to tissue cells when you have a fever? Why?

can carry up to 30% more oxygen than maternal hemoglobin (Fig. 23.22). As the maternal blood enters the placenta, oxygen is readily transferred to fetal blood. This is very important since the O_2 saturation in maternal blood in the placenta is quite low and the fetus might suffer hypoxia were it not for the greater affinity of fetal hemoglobin for oxygen.

CLINICAL APPLICATION

CARBON MONOXIDE POISONING

Carbon monoxide (CO) is a colorless and odorless gas found in exhaust fumes from automobiles and in tobacco smoke. It is a by-product of burning carbon-containing materials such as coal and wood. CO combines with the heme group of hemoglobin, just as oxygen does, except that the binding of carbon monoxide to hemoglobin is over 200 times as strong as the binding of oxygen to hemoglobin. Therefore in concentrations as small as 0.1% (pCO = 0.5 mm Hg), carbon monoxide will combine with half the hemoglobin molecules and reduce the oxygen-carrying capacity of the blood by 50%. Elevated carbon monoxide leads to hypoxia, and the result is **carbon monoxide poisoning.** The condition may be treated by administering pure oxygen, which slowly replaces the carbon monoxide combined with the hemoglobin.

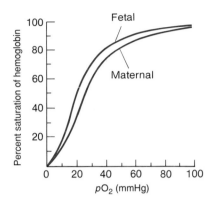

FIGURE 23.22 Oxygen–hemoglobin dissociation curves comparing fetal and maternal hemoglobin. Note that fetal hemoglobin has a higher affinity for oxygen.

Question: The pO_2 of placental blood is about 40 mm Hg. What are the oxygen saturations of maternal and fetal hemoglobin at this pO_2?

Hypoxia

Hypoxia (hī-POK-sē-a; *hypo* = below or under) refers to a general low level of oxygen availability. Physiologically, it refers to a deficiency of oxygen at the tissue level. Based on the cause, we can classify types of hypoxia as follows:

1. **Hypoxic hypoxia.** This is caused by a low pO_2 in arterial blood. The condition may be the result of high altitude, obstructions in air passageways, or fluid in the lungs.
2. **Anemic hypoxia.** In this case, there is too little functioning hemoglobin in the blood. Among the causes are hemorrhage, anemia, or failure of hemoglobin to carry its normal complement of oxygen, as in carbon monoxide poisoning.
3. **Stagnant hypoxia.** This condition results from the inability of blood to carry oxygen to tissues fast enough to sustain their needs. It may be caused by heart failure or circulatory shock, both of which diminish delivery of blood and therefore oxygen to tissues.
4. **Histotoxic hypoxia.** In this condition, the blood delivers adequate oxygen to tissues, but the tissues are unable to use it properly. One cause is cyanide poisoning, in which cyanide blocks the metabolic machinery of cells related to oxygen utilization.

Carbon Dioxide

Under normal resting conditions, each 100 ml of deoxygenated blood contains about 5 ml of carbon dioxide (CO_2), which is carried by the blood in three main forms (see Fig. 23.18).

1. **Dissolved CO_2.** The smallest percentage, about 7%, is dissolved in plasma. Upon reaching the lungs, it diffuses into the alveoli.
2. **Carbaminohemoglobin.** A somewhat higher percentage, about 23%, combines with the globin portion of hemoglobin to form **carbaminohemoglobin** ($Hb \cdot CO_2$).

$$\underset{\text{Hemoglobin}}{Hb} + \underset{\text{Carbon dioxide}}{CO_2} \rightleftharpoons \underset{\text{Carbaminohemoglobin}}{Hb \cdot CO_2}$$

The formation of carbaminohemoglobin is greatly influenced by pCO_2. For example, in tissue capillaries pCO_2 is relatively high, and this promotes formation of carbaminohemoglobin. But in pulmonary capillaries, pCO_2 is relatively low, and the CO_2 readily splits apart from globin and enters the alveoli by diffusion.
3. **Bicarbonate ions.** The greatest percentage of CO_2, about 70%, is transported in plasma as bicarbonate ions. The reaction that brings about this method of transport of CO_2 is the same one noted earlier.

$$\underset{\substack{\text{Carbon} \\ \text{dioxide}}}{CO_2} + \underset{\text{Water}}{H_2O} \underset{\text{anhydrase}}{\overset{\text{Carbonic}}{\rightleftharpoons}} \underset{\substack{\text{Carbonic} \\ \text{acid}}}{H_2CO_3} \rightleftharpoons \underset{\substack{\text{Hydrogen} \\ \text{ion}}}{H^+} + \underset{\substack{\text{Bicarbonate} \\ \text{ion}}}{HCO_3^-}$$

As CO_2 diffuses into tissue capillaries and enters the red blood cells, it reacts with water, in the presence of the enzyme carbonic anhydrase, to form carbonic acid (Fig. 23.23a). The carbonic acid dissociates into H^+ and HCO_3^-. Many of the H^+ combine with hemoglobin ($H \cdot Hb$) or other buffers. As bicarbonate ions accumulate inside the RBC, some of them diffuse into the plasma, down their concentration gradient. In exchange, chloride ions (Cl^-) diffuse from plasma into the RBCs. This exchange of negative ions maintains the ionic balance between plasma and RBCs and is known as the **chloride shift**. The net effect of these reactions is that CO_2 is carried from tissue cells as bicarbonate ions in plasma.

Summary of Gas Exchange in Lungs and Tissues

Deoxygenated blood returning to the lungs contains CO_2 dissolved in plasma, CO_2 combined with globin as carbaminohemoglobin, and CO_2 incorporated in bicarbonate ions. It has also picked up hydrogen ions, some of which are buffered by hemoglobin ($H \cdot Hb$). In summary:

$$\underset{\substack{\text{Dissolved} \\ \text{in plasma}}}{CO_2} + \underset{\substack{\text{Carbamino-} \\ \text{hemoglobin}}}{Hb \cdot CO_2} + \underset{\substack{\text{Bicarbonate} \\ \text{ions}}}{HCO_3^-} + \underset{\substack{\text{Hydrogen} \\ \text{ions}}}{H^+} + \underset{\substack{\text{Buffering of} \\ H^+ \text{ by Hb}}}{H \cdot Hb}$$

Deoxygenated blood returning to the lungs

In the pulmonary capillaries, the chemical reactions reverse. CO_2 dissolved in plasma diffuses into the alveoli and is exhaled. CO_2 combined with hemoglobin splits from the globlin, diffuses into alveoli, and is exhaled. CO_2 in

FIGURE 23.23 Summary of gas transport. As CO_2 leaves tissue cells and enters blood cells, it causes more O_2 to dissociate from hemoglobin (Bohr effect) and thus more CO_2 combines with hemoglobin and more bicarbonate ions (HCO_3^-) are produced. As O_2 passes from alveoli into blood cells, hemoglobin becomes saturated with O_2 and becomes a stronger acid. The more acidic hemoglobin releases more hydrogen ions (H^+), which bind to HCO_3^- to form carbonic acid (H_2CO_3). The H_2CO_3 dissociates into $H_2O + CO_2$ and the CO_2 diffuses from blood into alveoli (Haldane effect).

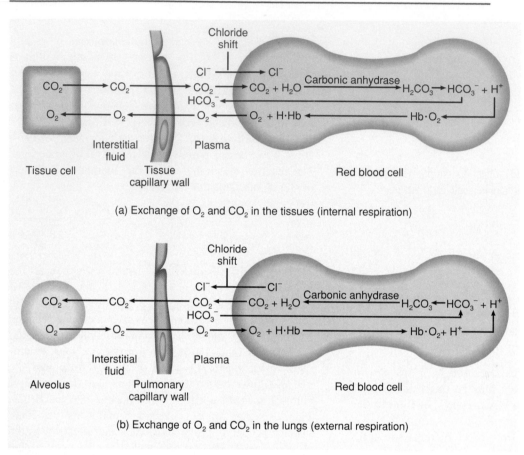

(a) Exchange of O_2 and CO_2 in the tissues (internal respiration)

(b) Exchange of O_2 and CO_2 in the lungs (external respiration)

Question: Would you expect the concentration of HCO_3^- to be greater in plasma taken from an arm artery or vein?

vein

bicarbonate ions is released as H^+ combines with HCO_3^- inside RBCs to form H_2CO_3, which splits into CO_2 and H_2O (Fig. 23.23b). As the concentration of HCO_3^- declines inside RBCs, bicarbonate ions diffuse in from the plasma, again in exchange for Cl^-. CO_2 diffuses out of RBCs, into alveoli, and is exhaled. At the same time inhaled oxygen is diffusing from alveoli into RBCs and is attaching to hemoglobin. Thus oxygenated blood leaving the lungs has increased oxygen content and decreased CO_2 and H^+.

Just as an increase in CO_2 in blood causes oxygen to split from hemoglobin, the binding of O_2 to hemoglobin causes the release of CO_2 from blood. In the presence of O_2, less CO_2 binds to hemoglobin. This reaction, the reverse of the Bohr effect, is called the **Haldane effect.** It occurs because when O_2 combines with hemoglobin, the

hemoglobin becomes a stronger acid. In this state, hemoglobin combines with less CO_2. Also, the more acidic hemoglobin releases more H^+ that bind to HCO_3^- to form H_2CO_3; the H_2CO_3 breaks down into $H_2O + CO_2$, and the CO_2 diffuses from blood into alveoli. The direction of the carbonic acid reaction depends mostly on pCO_2. In tissue capillaries, where pCO_2 is high, H^+ and HCO_3^- are formed. In pulmonary capillaries, where pCO_2 is low, CO_2 and H_2O are formed.

CONTROL OF RESPIRATION

At rest about 200 ml of oxygen, the amount contained in 1 liter of oxygenated blood, are consumed each minute. During strenuous exercise, however, O_2 use can increase as much as 30-fold. Thus mechanisms must exist to match respiratory effort to metabolic demand. The basic rhythm of respiration is controlled by portions of the nervous system in the medulla and pons. We will first examine the principal mechanisms involved in the nervous control of the rhythm of respiration.

Nervous Control

The size of the thorax is affected by the action of the respiratory muscles. These muscles contract and relax as a result of nerve impulses transmitted to them from centers in the brain. The area from which nerve impulses are sent to respiratory muscles is located bilaterally in the reticular formation of the brain stem; it is referred to as the **respiratory center.** The respiratory center consists of a widely dispersed group of neurons that is functionally divided into three areas: (1) the medullary rhythmicity area, in the medulla; (2) the pneumotaxic (noo-mō-TAK-sik) area, in the pons; and (3) the apneustic (ap-NOO-stik) area, also in the pons (Fig. 23.24).

Medullary Rhythmicity Area

The function of the **medullary rhythmicity area** is to control the basic rhythm of respiration. In the normal resting state, inspiration usually lasts for about two seconds and expiration for about three seconds. This is the basic rhythm of respiration. Within the medullary rhythmicity area are found both inspiratory and expiratory neurons that comprise inspiratory and expiratory areas, respectively. Let us first consider the proposed role of the inspiratory neurons in respiration.

The basic rhythm of respiration is determined by nerve impulses generated in the inspiratory area (Fig. 23.25a). At the beginning of expiration, the inspiratory area is inactive, but after three seconds it suddenly and automatically becomes active. This activity seems to result from an intrinsic excitability of autorhythmic neurons. In fact, when all incoming nerve connections to the inspiratory area are cut or blocked, the area still rhythmically discharges impulses that result in inspiration. Nerve impulses from the active inspiratory area last for about two seconds and travel to the muscles of inspiration. The impulses reach the diaphragm by the phrenic nerves and the external intercostal muscles by the intercostal nerves. When the impulses reach the inspiratory muscles, the muscles contract and inspiration occurs. At the end of two seconds, the inspiratory muscles become inactive again, and the cycle repeats itself over and over.

The expiratory neurons remain inactive during most normal, quiet respirations. During quiet respiration, inspiration

FIGURE 23.24 Approximate location of areas of the respiratory center.

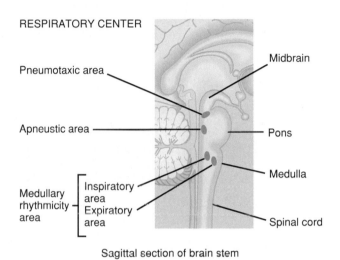

RESPIRATORY CENTER

Pneumotaxic area
Apneustic area
Medullary rhythmicity area — Inspiratory area / Expiratory area
Midbrain
Pons
Medulla
Spinal cord

Sagittal section of brain stem

Question: In normal, quiet breathing, which of the areas shown contains autorhythmic cells that set the basic rhythm of respiration?

MRA — IA.

is accomplished by active contraction of the inspiratory muscles, and expiration results from passive elastic recoil of the lungs and thoracic wall as the inspiratory muscles relax. However, during high levels of ventilation, it is believed that impulses from the inspiratory area activate the expiratory area (Fig. 23.25b). Impulses discharged from the expiratory area cause contraction of the internal intercostals and abdominal muscles that decrease the size of the thoracic cavity to cause forced (labored) expiration.

Pneumotaxic Area

Although the medullary rhythmicity area controls the basic rhythm of respiration, other parts of the brain stem help coordinate the transition between inspiration and expiration. One of these is the **pneumotaxic area** in the upper pons (see Fig. 23.24). It transmits inhibitory impulses to the inspiratory area. The major effect of these impulses is to help turn off the inspiratory area before the lungs become too full of air. In other words, the impulses limit the duration of inspiration and thus facilitate expiration. When the

FIGURE 23.25 Proposed role of the medullary rhythmicity area in controlling the basic rhythm of respiration.

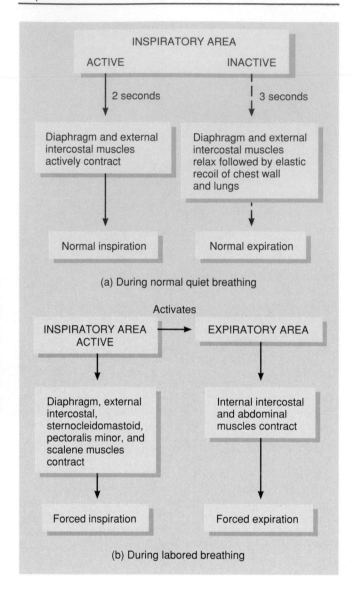

(a) During normal quiet breathing

(b) During labored breathing

pneumotaxic area is more active, the breathing rate is quicker.

Apneustic Area

Another part of the brain stem that coordinates the transition between inspiration and expiration is the **apneustic area** in the lower pons (see Fig. 23.24). It sends stimulatory impulses to the inspiratory area that activate it and prolong inspiration, thus inhibiting expiration. This occurs when the pneumotaxic area is inactive. When the pneumotaxic area is active, it overrides the apneustic area.

A summary of the nervous control of respiration is presented in Fig. 23.26.

Regulation of Respiratory Center Activity

Although the basic rhythm of respiration is set and coordinated by the respiratory center, the rhythm can be modified in response to metabolic demands by nerve impulses to the center.

Cortical Influences

The cerebral cortex has connections with the respiratory center, which means we can voluntarily alter our pattern of breathing (Fig. 23.26). We can even refuse to breathe at all for a short time. Voluntary control is protective because it enables us to prevent water or irritating gases from entering the lungs. The ability to stop breathing is limited by the buildup of CO_2 and H^+ in the blood, however. When pCO_2 and the concentration of H^+ increase to a certain level, the inspiratory area is strongly stimulated, nerve impulses are sent to inspiratory muscles, and breathing resumes whether or not the person wishes. It is impossible for people to kill themselves by holding their breath. Even if a person faints, breathing resumes when consciousness is lost. Nerve impulses from the hypothalamus and limbic system also stimulate the respiratory center. These pathways permit emotional stimuli to alter respirations, for example, in crying.

Inflation Reflex

Located in the walls of bronchi and bronchioles within the lungs are receptors sensitive to stretch called **stretch receptors.** When the receptors become stretched during overinflation of the lungs, nerve impulses are sent along the vagus (X) nerves to the inspiratory area and apneustic area (Fig. 23.26). In response, the inspiratory area is inhibited and the apneustic area is inhibited from activating the inspiratory area. The result is that expiration follows. As air leaves the lungs during expiration, the lungs deflate and the stretch receptors are no longer stimulated. Thus the inspiratory and apneustic areas are no longer inhibited, and a new inspiration begins. This reflex is referred to as the **inflation (Hering–Breuer) reflex.** Some evidence suggests that the reflex is mainly a protective mechanism for preventing overinflation of the lungs rather than a key component in the regulation of respiration.

Chemical Regulation

Certain chemical stimuli determine how fast and deeply we breathe. The ultimate goal of the respiratory system is to maintain proper levels of carbon dioxide (CO_2) and oxygen (O_2), and the system is highly responsive to changes in the

FIGURE 23.26 Summary of the nervous control of respiration.

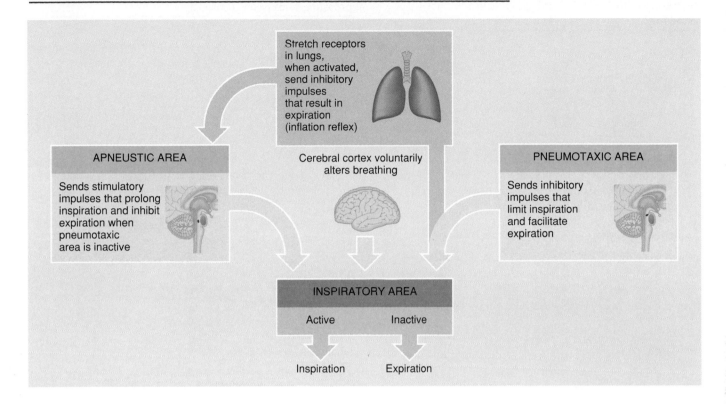

blood levels of either. Since CO_2 is lipid-soluble, it easily diffuses across plasma membranes, including those that form the blood–brain barrier. Inside cells, including red blood cells and neurons, CO_2 combines with water (H_2O) to form carbonic acid (H_2CO_3). But the carbonic acid quickly breaks down into H^+ ions and bicarbonate ions (HCO_3^-). Any increase in CO_2 will cause an increase in H^+, and any decrease in CO_2 will cause a decrease in H^+.

Within the medulla, neurons that are highly sensitive to pH (H^+ concentration) form the **central chemosensitive area**. In the peripheral nervous system are **peripheral chemoreceptors**, receptors that are sensitive to changes in H^+, CO_2, and O_2 in the blood. They are located within the carotid bodies near the bifurcation (branching) of the common carotid arteries and in the aortic bodies. The **carotid bodies** are small, oval nodules about 4 to 5 mm long, located in the space between the origins of the internal and external carotid arteries. The **aortic bodies** are clustered in the region near the arch of the aorta. On each side of the body, the sensory nerve fibers from the carotid bodies join with those from the carotid sinus to form the carotid sinus nerve, which joins the glossopharyngeal (IX) nerve. The sensory fibers from aortic bodies join the vagus (X) nerve.

Under normal circumstances, the pCO_2 in arterial blood is 40 mm Hg. If there is even a slight increase in pCO_2—a condition called **hypercapnia**—the central chemosensitive area in the medulla is stimulated because H^+ concentration

increases (Fig. 23.27). H^+ and CO_2 concentrations fluctuate more readily in cerebrospinal fluid (CSF) than in blood plasma because CSF has fewer buffers. The peripheral chemoreceptors in the carotid and aortic bodies are stimulated by both the high pCO_2 and the rise in H^+ concentration. Input from the central chemosensitive area and peripheral chemoreceptors causes the inspiratory area to become highly active, and the rate and depth of breathing increase. Rapid and deep breathing is called **hyperventilation** and allows exhalation of more CO_2 until pCO_2 and H^+ are lowered to normal. If arterial pCO_2 is lower than 40 mm Hg, a condition called **hypocapnia,** the central chemosensitive area and peripheral chemoreceptors are not stimulated and stimulatory impulses are not sent to the inspiratory area. Consequently, the area sets its own moderate pace until CO_2 accumulates and the pCO_2 rises to 40 mm Hg. Slow and shallow breathing is called **hypoventilation.**

The peripheral chemoreceptors are sensitive only to large decreases in the pO_2 since hemoglobin remains about 85% or more saturated at pO_2 values all the way down to 50 mm Hg (see Fig. 23.19). If arterial pO_2 falls from a normal of 105 mm Hg to about 50 mm Hg, the peripheral chemoreceptors become stimulated and send impulses to the inspiratory area and respiration increases (Fig. 23.27). But if the pO_2 falls much below 50 mm Hg, the cells of the inspiratory area suffer oxygen starvation (hypoxia) and do not respond well to any chemical changes (Fig. 23.28). They

FIGURE 23.27 Negative feedback control of breathing by changes in blood pCO_2, pO_2, and pH (H^+ concentration).

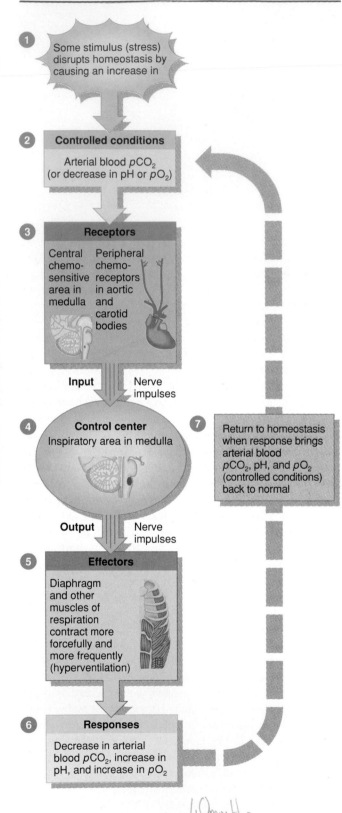

FIGURE 23.28 Positive feedback reduction of partial pressure of oxygen in blood.

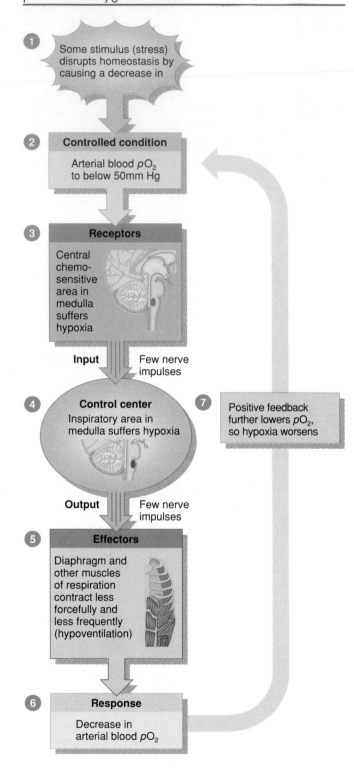

40mmHg

Question: What is the normal arterial blood pCO_2?

Question: How could you intervene to break this positive feedback loop?

Air rich in oxygen via a mask or nasal cannula

send fewer impulses to the medullary inspiratory area. As the respiration rate decreases or breathing ceases altogether, pO_2 falls lower and lower, thus establishing a positive feedback cycle.

Proprioceptors

As soon as you start exercising, your rate and depth of breathing increase, even before there are changes in pO_2, pCO_2, or H^+ concentration. The main stimulus for these quick changes in respiratory effort is thought to be input from proprioceptors, which monitor movement of joints and muscles. Nerve impulses from the proprioceptors stimulate the inspiratory area of the medulla.

Other Influences

The carotid and aortic sinuses, which are close to the carotid and aortic bodies, contain baroreceptors (pressure receptors) that detect changes in blood pressure. Although these baroreceptors are concerned mainly with the control of circulation (see Fig. 21.13), they also affect respiration. For example, a sudden rise in blood pressure decreases the rate of respiration, and a drop in blood pressure increases the respiratory rate.

These are other factors that control respiration:

1. **Temperature.** An increase in body temperature, as during a fever or vigorous muscular exercise, increases the rate of respiration. A decrease in body temperature decreases respiratory rate. A sudden cold stimulus such as plunging into cold water causes **apnea** (AP-nē-a), a temporary cessation of breathing.
2. **Pain.** A sudden, severe pain brings about apnea, but a prolonged pain triggers the general adaptation syndrome (see Fig. 18.26) and increases respiratory rate.

3. **Stretching the anal sphincter muscle.** This increases the respiratory rate and is sometimes employed to stimulate respiration in a person who has stopped breathing.
4. **Irritation of air passages.** Mechanical or chemical irritation of the pharynx or larynx brings about an immediate cessation of breathing followed by coughing or sneezing.

A summary of stimuli that affect the rate of respiration is presented in Exhibit 23.4.

AGING AND THE RESPIRATORY SYSTEM

The cardiovascular and respiratory systems operate as a unit, and damage or disease in one of these organ systems is often secondarily reflected in the other. In pulmonary heart disease, the right side of the heart enlarges in response to certain lung diseases. This condition is known as right ventricular hypertrophy, or cor pulmonale. With advancing age, the airways and tissues of the respiratory tract, including the air sacs, become less elastic and more rigid. In addition to the lungs becoming less elastic, the chest wall also becomes more rigid. As a result, there is a decrease in lung capacity. In fact, vital capacity (the maximum amount of air that can be expired after maximal inspiration) can decrease as much as 35% by age 70. Also, there is a decrease in blood levels of oxygen, decreased activity of alveolar macrophages, and diminished ciliary action of the epithelium lining the respiratory tract. Owing to all these age related factors, elderly people are more susceptible to pneumonia, bronchitis, emphysema, and other pulmonary disorders.

EXHIBIT 23.4

SUMMARY OF STIMULI THAT AFFECT VENTILATION RATE AND DEPTH

Stimuli that Increase Rate and Depth of Ventilation	Stimuli that Decrease (or Inhibit) Rate and Depth of Ventilation
Increase in arterial blood H^+ level or pCO_2 above 40 mm Hg.	Decrease in arterial blood H^+ level or pCO_2 below 40 mm Hg.
Increase in nerve impulses from proprioceptors.	Decrease in nerve impulses from proprioceptors.
Decrease in arterial blood pO_2 from 105 to 50 mm Hg.	Decrease in arterial blood pO_2 below 50 mm Hg.
Decrease in blood pressure.	Increase in blood pressure.
Increase in body temperature.	Decrease in body temperature decreases rate of respiration, and sudden cold stimulus causes apnea.
Prolonged pain.	Severe pain causes apnea.
Stretching anal sphincter.	Irritation of pharynx or larynx by touch or chemicals causes apnea.

DEVELOPMENTAL ANATOMY OF THE RESPIRATORY SYSTEM

The development of the mouth and pharynx is considered in Chapter 24 on the digestive system. Here we consider the remainder of the respiratory system.

At about four weeks of fetal development, the respiratory system begins as an outgrowth of the **endoderm** of the foregut (precursor of some digestive organs) just behind the pharynx. This outgrowth is called the **laryngotracheal bud** (see Fig. 18.25). As the bud grows, it elongates and differentiates into the future *larynx* and other structures as well. Its proximal end maintains a slitlike opening into the pharynx called the *glottis*. The middle portion of the bud gives rise to the *trachea*. The distal portion divides into two **lung buds,** which grow into the *bronchi* and *lungs* (Fig. 23.29).

As the lung buds develop, they branch and rebranch and give rise to all the *bronchial tubes*. After the sixth month, the closed terminal portions of the tubes dilate and become the *alveoli* of the lungs. The smooth muscle, cartilage, and connective tissues of the bronchial tubes and the pleural sacs of the lungs are contributed by **mesenchymal (mesodermal) cells**.

FIGURE 23.29 Development of the bronchial tubes and lungs.

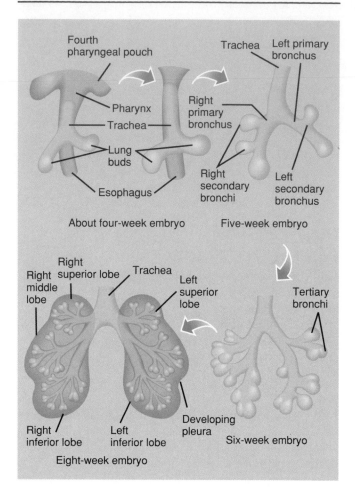

About four-week embryo

Five-week embryo

Eight-week embryo

Six-week embryo

Diseases such as asthma, bronchitis, and emphysema have in common some degree of obstruction of the air passageways. The term **chronic obstructive pulmonary disease (COPD)** is used to refer to these disorders. Among the symptoms that might indicate significant airflow obstruction are cough, wheezing, and dyspnea (painful or labored breathing).

ASTHMA

Asthma is a chronic, inflammatory disorder that produces sporadic narrowing of airways. Attacks are brought on by spasms of the smooth muscle in the walls of the smaller bronchi and bronchioles, causing the passageways to close partially or completely (bronchoconstriction). Symptoms include periods of coughing, difficult breathing, and wheezing that may abate spontaneously or with treatment. The patient has trouble exhaling, and air may be trapped in the alveoli during expiration.

In the early phase (acute) response, besides smooth muscle spasm, there is excessive secretion of mucus that may clog the bronchi and bronchioles and worsen the attack. In the late phase (chronic) response, inflammation continues, accompanied by fibrosis, edema, and necrosis (death) of bronchial epithelial cells. A host of mediator chemicals, including leukotrienes, prostaglandins, thromboxane, platelet activating factor, and histamine, take part.

The airways of people with asthma are hyperreactive to a variety of stimuli that normally do not trigger bronchoconstriction. Sometimes the trigger is an allergen such as pollen, house dust mites, molds, or a particular food. Other common triggers of asthma attacks are emotional upset, aspirin, sulfiting agents (used in wine and beer and to keep greens fresh on salad bars), exercise, and breathing cold air or cigarette smoke. An acute attack is treated by giving a β-adrenergic agonist such as epinephrine to help open up the airways. However, long-term therapy of asthma strives to suppress the underlying inflammation. The anti-inflammatory drugs used most often are inhaled glucocorticoids (corticosteroids) and cromolyn sodium.

BRONCHITIS

Bronchitis is inflammation of the bronchi characterized by hypertrophy and hyperplasia of glands and goblet cells lining the bronchial airways. The typical symptom is a cough in which a thick greenish-yellow sputum is raised (productive cough). This secretion signifies the presence of the underlying inflammation that is causing excessive secretion of mucus. Cigarette smoking remains the leading cause of chronic bronchitis, that is, bronchitis that lasts for at least three months of the year for two successive years.

EMPHYSEMA

In **emphysema** (em´-fi-SĒ-ma), alveolar walls disintegrate, producing abnormally large air spaces that remain filled with air during expiration. The name means "blown up" or "full of air." With less surface area for gas exchange, oxygen diffusion across the damaged alveolar–capillary membrane is reduced. Blood O_2 level is somewhat lowered, and any mild exercise that raises the oxygen requirements of the cells leaves the patient breathless. As increasing numbers of alveolar walls are damaged, air becomes trapped in the lungs because they have lost elastic fibers. A key symptom is reduced forced expiratory volume, the volume of air exhaled during a forced exhalation after a maximum inhalation. The patient has to work voluntarily to exhale. Over several years, the added exertion increases the size of the chest cage, resulting in a "barrel chest."

Emphysema is generally caused by a long-term irritation. Cigarette smoke, air pollution, and occupational exposure to industrial dust are the most common irritants. Some destruction of alveolar sacs may be caused by an imbalance between enzymes called proteases, one of the most important ones called elastase, and a molecule (alpha₁-antitrypsin) that inhibits proteases. When there is a decreased production by the liver of the plasma protein alpha₁-antitrypsin, elastase is not inhibited and is free to attack the connective tissue in the walls of alveolar sacs. Cigarette smoke not only deactivates a protein that is seemingly crucial in preventing emphysema but also prevents the repair of affected lung tissue.

BRONCHOGENIC CARCINOMA (LUNG CANCER)

A common lung cancer, **bronchogenic carcinoma,** starts in the walls of the bronchi. The constant irritation by inhaled smoke and pollutants causes the goblet cells of the bronchial epithelium to enlarge. They respond by secreting excessive mucus (Fig. 23.30b). The basal cells also respond to the stress by undergoing cell division so fast that they push into the area occupied by the goblet and columnar cells. Many researchers believe that if the stress is removed at this point, the epithelium can return to normal.

If the stress persists, however, more and more mucus is secreted, and the cilia become less effective. As a result, mucus is not carried toward the throat but remains trapped in the bronchial tubes. The person then develops a "smoker's

FIGURE 23.30 Effects of smoking on the respiratory epithelium. (a) Microscopic view of the normal epithelium of a bronchial tube. (b) Initial response of the bronchial epithelium to irritation by pollutants. (c) Advanced response of the bronchial epithelium.

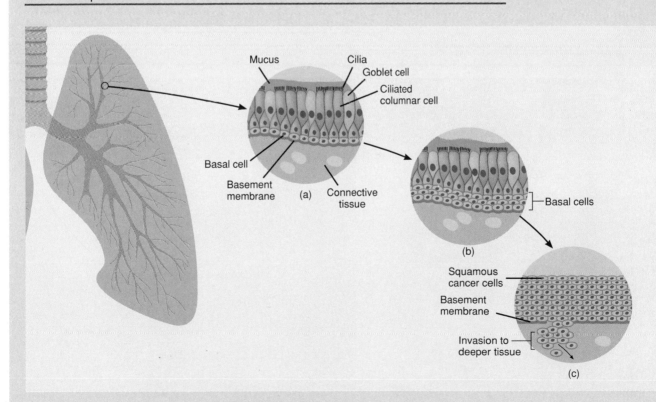

Continues

cough." Moreover, the constant irritation from the pollutant slowly destroys the alveoli, which are replaced with thick, inelastic connective tissue. Alveoli are destroyed by protein-digesting enzymes produced by leukocytes and macrophages in response to the stress. Mucus that has accumulated becomes trapped in the air sacs. Millions of sacs rupture, reducing the diffusion surface for the exchange of oxygen and carbon dioxide. The person has now developed emphysema. If the stress is removed at this point, there is little chance for improvement. Alveolar tissue that has been destroyed cannot be repaired. But removal of the stress can stop further destruction of lung tissue.

If the stress still continues, the emphysema progressively worsens, and the basal cells of the bronchial tubes continue to divide and break through the basement membrane (Fig. 23.30c). At this point the stage is set for bronchogenic carcinoma. Columnar and goblet cells disappear and may be replaced with squamous cancer cells. If this happens, the malignant growth spreads throughout the lung (Fig. 23.31) and may block a bronchial tube. Treatment involves surgical removal of the diseased lung. However, metastasis (spreading) of the growth through the lymphatic or blood vessels may result in new growths in other parts of the body such as the brain and liver.

Other factors may be associated with lung cancer. For instance, breast, stomach, and prostate malignancies can metastasize to the lungs. Radioactive radon gas, which may seep from the earth into buildings, is a suspected cause of lung cancer in some nonsmokers (smoking raises a person's radon-related cancer risk to a level 15 times higher than that of a nonsmoker). Asbestos is also associated with lung cancer, as well as with other pulmonary diseases. People who have apparently not been exposed to tobacco smoke do occasionally develop bronchogenic carcinoma. The occurrence of bronchogenic carcinoma, however, is probably over 20 times higher in heavy cigarette smokers than it is in nonsmokers. Exposure to environmental tobacco smoke (passive smoking) is associated with lung cancer. It is estimated that about 4000 deaths a year in the U.S. due to lung cancer are attributed to environmental tobacco smoke. It is estimated that nearly 40,000 deaths a year in the U.S. due to heart disease are related to environmental tobacco smoke.

PNEUMONIA

Pneumonia refers to an acute infection or inflammation of the alveoli. It is the most common infectious cause of death in the U.S. The alveolar sacs fill up with fluid and dead white blood cells, reducing the amount of air space in the lungs. Oxygen has difficulty diffusing through the inflamed alveoli, and the blood pO_2 may be drastically reduced. Blood pCO_2 usually remains normal because carbon dioxide diffuses through the alveoli more easily than oxygen.

The most common cause of pneumonia is the pneumococcus bacterium (*Streptococcus pneumoniae*), but other bacteria or fungi, protozoans, or viruses may be the source of trouble. Among those who are susceptible to pneumonia are the elderly, infants, immunocompromised individuals (persons with AIDS, malignancy, or those taking immunosuppressive drugs), cigarette smokers, and persons with obstructive lung disease.

FIGURE 23.31 Lung cancer and emphysema.

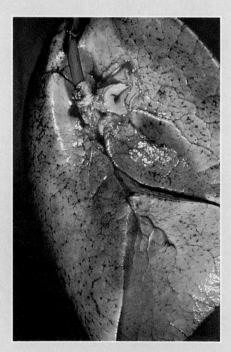

(a) Normal lung

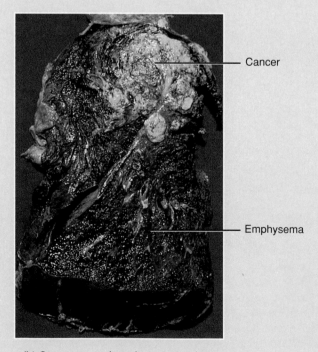

Cancer

Emphysema

(b) Cancerous and emphysematous lung

TUBERCULOSIS

The bacterium *Mycobacterium tuberculosis* produces an infectious, communicable disease called **tuberculosis (TB).** It most often affects the lungs and the pleurae. The bacteria destroy parts of the lung tissue, and the tissue is replaced by fibrous connective tissue. Because the connective tissue is inelastic and thick, the affected areas of the lungs do not recoil well during expiration and air is retained. Gases no longer diffuse easily through the fibrous tissue.

Tuberculosis bacteria are spread by inhalation and exhalation. Although they can withstand exposure to many disinfectants, they die quickly in sunlight. Thus tuberculosis is sometimes associated with crowded, poorly lighted housing. Many drugs, such as isoniazid and rifampin, are successful in treating tuberculosis. Rest, sunlight, and good diet are vital parts of treatment.

Since 1985, there has been a 16% increase in the number of cases of tuberculosis reported annually in the U.S. This is a dramatic increase from the average annual decline of 6% during the prior 30 years. Perhaps the single most important factor related to the increase in the number of tuberculosis patients is HIV. Individuals infected with HIV are much more likely to develop tuberculosis because their immune systems are impaired. Among the other factors that have contributed to the increased number of cases are homelessness, increased drug abuse, increased immigration from countries with a high prevalence of tuberculosis, increased crowding in housing among the poor, and airborne transmission of tuberculosis in prisons and shelters. In addition, there have been recent outbreaks of tuberculosis involving multi-drug-resistant strains of *Mycobacterium tuberculosis* because patients fail to complete their drug and other treatments.

RESPIRATORY DISTRESS SYNDROME OF THE NEWBORN

Respiratory distress syndrome (RDS) of the newborn, also called **glassy-lung disease** or **hyaline membrane disease (HMD),** is responsible for approximately 20,000 newborn infant deaths per year. Before birth, the respiratory passages are filled with fluid. At birth, this fluid-filled passageway must become an air-filled passageway, and the collapsed alveoli must expand and function in gas exchange. The success of this transition depends largely on surfactant, which lowers surface tension in the fluid layer and eases the expansion of the lungs. Surfactant is present in the fetal lungs as early as 23 weeks, and by 28 to 32 weeks the amount present is sufficient to prevent alveolar collapse during breathing. Surfactant is produced continuously by type II alveolar (septal) cells and can be detected by amniocentesis.

In the newborn whose lungs are deficient in surfactant, the surface tension of the alveolar fluid is 7 to 14 times higher than normal. Consequently, during expiration after the first inspiration, the alveoli collapse almost to their original uninflated state. Immense muscular effort is needed to expand the lungs for each breath. RDS is more common in premature babies delivered by cesarean section than by the normal vaginal route. It is suggested that the stress of being born vaginally produces higher cortisol levels. Cortisol, in turn, stimulates surfactant production by type II alveolar cells. Pregnant mothers at risk of delivering a baby prematurely may be given cortisol to enhance development of the fetal type II alveolar cells.

A treatment called PEEP (positive end expiratory pressure) is used to treat RDS of the newborn. It consists of passing a tube through the air passages to the top of the lungs to provide needed oxygen-rich air at continuous pressures of up to 14 mm Hg. Continuous pressure keeps the baby's alveoli open and available for gas exchange. The addition of human surfactant to the oxygen-rich air counters the effects of RDS of the newborn more effectively and lessens lung damage.

RESPIRATORY FAILURE

Respiratory failure refers to a condition in which the respiratory system cannot supply sufficient oxygen to maintain metabolism or cannot eliminate enough carbon dioxide to prevent respiratory acidosis, a lower-than-normal pH in extracellular fluid. Specifically, respiratory failure occurs when the oxygen level of arterial blood is too low or when the carbon dioxide level of arterial blood is too high. Respiratory failure always causes dysfunction in other organs as well.

Among the causes of respiratory failure are lung disorders; mechanical disorders that disturb the chest wall or neuromusculature (anesthetic blocking agents, cervical spinal cord injuries, myasthenia gravis, and amyotrophic lateral sclerosis); depression of the respiratory center by drugs, strokes, or trauma; and carbon monoxide poisoning.

Symptoms of respiratory failure include disorientation, malaise, headache, muscular weakness, difficulty with sleep, palpitations, breathlessness, coughing, tachycardia, dysrhythmia, cyanosis, hypertension, edema, stupor, and coma. Treatment consists of providing adequate oxygenation and reversing the respiratory acidosis.

SUDDEN INFANT DEATH SYNDROME

Sudden infant death syndrome (SIDS), also called crib death, kills approximately 10,000 American babies every year. It kills more infants between the ages of 1 week and 12 months than any other disease. SIDS occurs without warning. Although about half of its victims had an upper respiratory infection within two weeks of death, the babies tended to be otherwise remarkably healthy.

Many hypotheses have been proposed as explanations for SIDS and the precise pathogenic mechanisms are not yet clear. One likely cause of death is laryngospasm, possibly triggered by a previous mild, viral infection of the upper respiratory tract. The seasonal distribution of SIDS incidence, being lowest during the summer months and highest in the late fall and winter, suggests an infectious agent, viruses being the most likely possibility. Alternatively, infants may be more likely to die from hyperthermia due to being dressed too warmly in the winter months. Most cases occur at a time in life when antibody levels are low and the infant is at a critical period of susceptibility.

Another possible factor is periods of prolonged apnea as a result of malfunction of the respiratory center. Hypoxia may occur in infants who sleep in a prone position (on their stomach) and rebreathe exhaled air trapped in a depression of the mattress. Some infants whose cause of death was listed as SIDS have elevated levels of hemoglobin F (fetal hemoglobin). Normally, hemoglobin F is replaced by adult hemoglobin during the first six months after birth. The delayed switch from hemoglobin F to adult hemoglobin could affect the delivery of oxygen to sensitive tissue sites. Other suspected causes are an allergic reaction, bacterial poisoning, excess dopamine (DA), and altered cardiac repolarization as reflected by a prolonged QT interval.

Continues

DISORDERS: HOMEOSTATIC IMBALANCES (continued)

CORYZA (COMMON COLD) AND INFLUENZA (FLU)

Several viruses are responsible for **coryza (common cold).** A group of viruses, called rhinoviruses, is responsible for about 40% of all colds in adults. Typical symptoms include sneezing, excessive nasal secretion, dry cough, and congestion. The uncomplicated common cold is not usually accompanied by a fever. Complications include sinusitis, asthma, bronchitis, ear infections, and laryngitis. Recent investigations suggest an association between emotional stress and the common cold: the higher the stress level, the greater the frequency and duration of colds.

Influenza (flu) is also caused by a virus. Its symptoms include chills, fever (usually higher than 101°F = 39°C), headache, and muscular aches. Coldlike symptoms appear as the fever subsides.

PULMONARY EMBOLISM

Pulmonary embolism refers to the presence of a blood clot or other foreign substance in a pulmonary arterial vessel that obstructs circulation to lung tissue. Most pulmonary emboli originate from thrombi in the proximal deep veins of the lower extremities. The immediate effect of pulmonary embolism is complete or partial obstruction of the pulmonary arterial blood flow to the lung, resulting in dysfunction of the affected lung tissue. A large embolus can produce death in a few minutes. Pulmonary embolism usually develops in patients with one or more risk factors. Among these are immobilization in bed for more than three days; heart disease (especially with congestive heart failure); trauma, such as fractures; malignant disease; obesity; pregnancy; polycythemia; and the use of oral contraceptives.

Among the clinical findings associated with pulmonary embolism are sudden and unexplained dyspnea, chest pain, apprehension, rapid breathing, rales, cough, deep-venous thrombosis, pulmonary infarction as evidenced by expectoration of blood or blood-stained sputum, and syncope.

Treatment of pulmonary embolism consists of intravenous (IV) injections of heparin (an anticoagulant); administration of oral anticoagulants, such as warfarin; administration of clot-dissolving enzymes, such as streptokinase, tissue plasminogen activator, or urokinase; bed rest; analgesia; and oxygen administration. In some cases, surgery (embolectomy) is needed to prevent death due to a massive pulmonary embolism.

PULMONARY EDEMA

Pulmonary edema refers to an abnormal accumulation of interstitial fluid in the interstitial spaces and alveoli of the lungs. The edema may arise from increased pulmonary capillary permeability (pulmonary origin) or increased pulmonary capillary pressure (cardiac origin). The latter cause may coincide with congestive heart failure. The most common symptom is dyspnea. Others include wheezing, tachypnea (rapid respirations), restlessness, feeling of suffocation, cyanosis, pallor (paleness), and diaphoresis (excessive perspiration). Treatment consists of administering oxygen, drugs that dilate the bronchioles and lower blood pressure, and drugs that cor-

rect acid–base imbalance; suctioning of airways; and mechanical ventilation.

CYSTIC FIBROSIS

Cystic fibrosis (CF) is an inherited disease of secretory epithelia that affects the respiratory passageways, pancreas, salivary glands, and sweat glands. It is the most common lethal genetic disease of white people: 5% of the population are thought to be genetic carriers. The cause of cystic fibrosis is linked to an inability to transport chloride ions (Cl^-) across the plasma membranes of certain epithelial cells.

Among the most common signs and symptoms are pancreatic insufficiency, breathing difficulty, and cirrhosis of the liver. It is characterized by the production of thick secretions that do not drain easily from the respiratory passageways. The buildup of the secretions leads to inflammation and replacement of injured cells with connective tissue that blocks these passageways. One of the prominent features is blockage of the pancreatic ducts so that the digestive enzymes cannot reach the intestine. Since pancreatic juice contains the main fat-digesting enzyme, the person fails to absorb fats or fat-soluble vitamins and thus suffers from vitamin A, D, and K deficiency diseases.

A child suffering from cystic fibrosis is given pancreatic extract and large doses of vitamins A, D, and K. The therapeutic diet is low, but not lacking, in fats and high in carbohydrates and proteins that can be used for energy and can also be converted by the liver into the lipids essential for life processes.

SMOKE INHALATION INJURY

When smoke is inhaled, the lungs are injured directly by heat from flames and substances in fumes. **Smoke inhalation injury** has three components that occur in sequence: (1) inhibition of oxygen delivery and utilization, (2) upper airway injury from heat, and (3) lung damage from acids and aldehydes in smoke.

Inhibition of oxygen delivery and utilization is due to inhalation of mainly carbon monoxide (CO) and, to a lesser extent, cyanide. CO binds to hemoglobin and components of the cytochrome system (required for cellular respiration) and thus impairs oxygen delivery and utilization and also causes acidosis (low blood pH). The onset of symptoms (confusion and disorientation with moderate toxicity, and coma and cardiac arrest with severe toxicity) is immediate.

Upper airway injury from heat may cause edema of the laryngotracheal mucosa and obstruction of upper airways with atelectasis (collapsed lung). Symptoms, such as labored breathing and hypoxemia, occur after 18 to 24 hours.

The acids and aldehydes in inhaled smoke result in bronchoconstriction, atelectasis, bronchiolar edema, impairment of mucociliary action, and destruction of surfactant. The onset of symptoms (labored breathing, hypovolemia, hypoxemia, and bronchopneumonia) is both early and delayed.

Treatment consists of administering 90 to 100% oxygen, inserting an endotracheal tube, performing pulmonary suction, administering bronchodilators, and restoring and maintaining fluid balance.

MEDICAL TERMINOLOGY

Asphyxia (as-FIK-sē-a; *sphyxis* = pulse) Oxygen starvation due to low atmospheric oxygen or interference with ventilation, external respiration, or internal respiration.

Aspiration (as′-pi-RĀ-shun; *spirare* = breathe) Inhalation of a foreign substance such as water, food, or foreign body into the bronchial tree; drawing of a substance in or out by suction.

Bronchiectasis (bron′-kē-EK-ta-sis; *ektasis* = dilation) A chronic dilation of the bronchi or bronchioles.

Cardiopulmonary resuscitation (kar′-dē-ō-PUL-mo-ner-ē re-sus′-i-TA-shun) **(CPR)** The artificial establishment of normal or near-normal respiration and circulation. The **A, B, C's** of cardiopulmonary resuscitation are **Airway, Breathing,** and **Circulation,** meaning the rescuer must establish an airway, provide artificial ventilation if breathing has stopped, and reestablish circulation if there is inadequate cardiac action. The procedure should be performed in that order (A, B, C).

Cheyne–Stokes respiration (CHĀN STŌKS res′-pi-RĀ-shun) A repeated cycle of irregular breathing beginning with shallow breaths that increase in depth and rapidity, then decrease and cease altogether for 15 to 20 seconds. Cheyne–Stokes is normal in infants. It is also often seen just before death from pulmonary, cerebral, cardiac, and kidney disease.

Diphtheria (dif-THĒ-rē-a; *diphthera* = membrane) An acute bacterial infection that causes the mucous membranes of the oropharynx, nasopharynx, and larynx to enlarge and become leathery. Enlarged membranes may obstruct airways and cause death from asphyxiation.

Dyspnea (DISP-nē-a; *dys* = painful, difficult; *pnoia* = breath) Painful or labored breathing.

Epistaxis (ep′-i-STAK-sis) Loss of blood from the nose due to trauma, infection, allergy, neoplasms, and bleeding disorders. It can be arrested by cautery with silver nitrate,

electrocautery, and firm packing. Also called **nosebleed.**

Heimlich (abdominal thrust) maneuver (HĪM-lik ma-NOO-ver) First-aid procedure designed to clear the air passageways of obstructing objects. It is performed by applying a quick upward thrust that causes sudden elevation of the diaphragm and forceful, rapid expulsion of air in the lungs; this action forces air out the trachea to eject the obstructing object. The Heimlich maneuver is also used to expel water from the lungs of near-drowning victims before resuscitation is begun.

Hemoptysis (hē-MOP-ti-sis; *hemo* = blood; *ptein* = to spit) Spitting of blood from the respiratory tract.

Orthopnea (or′-THOP-nē-a; *ortho* = straight) Dyspnea that occurs in the horizontal position. It is an abnormal finding because a normal individual can tolerate the reduction in vital capacity that accompanies the recumbent position.

Pneumonectomy (noo′-mō-NEK-tō-mē; *pneumo* = lung; *tome* = cutting) Surgical removal of a lung.

Rales (RĀLS) Sounds sometimes heard in the lungs that resemble bubbling or rattling. Rales are to the lungs what murmurs are to the heart. Different types are due to the presence of an abnormal amount or type of fluid or mucus inside the bronchi or alveoli, or to bronchoconstriction that causes turbulent air flow.

Respirator (RES-pi-rā′-tor) An apparatus fitted to a mask over the nose and mouth, or hooked directly to an endotracheal or tracheotomy tube, that is used to assist or support ventilation or to provide nebulized medication to the air passages under pressure.

Rhinitis (rī-NI-tis; *rhino* = nose) Chronic or acute inflammation of the mucous membrane of the nose.

Tachypnea (tak′-ip-NĒ-a; *tachy* = rapid; *pnoia* = breath) Rapid breathing.

Study Outline

Organs (p. 721)

1. Respiratory organs include the nose, pharynx, larynx, trachea, bronchi, and lungs.
2. They act with the cardiovascular system to supply oxygen and remove carbon dioxide from the blood.

Nose (p. 721)

1. The external portion is made of cartilage and skin and is lined with mucous membrane. Openings to the exterior are the external nares.
2. The internal portion communicates with the paranasal sinuses and nasopharynx through the internal nares.

3. The nasal cavity is divided by a septum. The anterior portion of the cavity is called the vestibule.
4. The nose is adapted for warming, moistening, and filtering air; olfaction; and speech.

Pharynx (p. 724)

1. The pharynx (throat) is a muscular tube lined by a mucous membrane.
2. The anatomic regions are the nasopharynx, oropharynx, and laryngopharynx.
3. The nasopharynx functions in respiration. The oropharynx and laryngopharynx function both in digestion and in respiration.

Larynx (p. 725)

1. The larynx (voice box) is a passageway that connects the pharynx with the trachea.
2. It contains the thyroid cartilage (Adam's apple); the epiglottis, which prevents food from entering the larynx; the cricoid cartilage, which connects the larynx and trachea; and the paired arytenoid, corniculate, and cuneiform cartilages.
3. The larynx contains vocal folds, which produce sound. Taut folds produce high pitches, and relaxed ones produce low pitches.

Trachea (p. 727)

1. The trachea (windpipe) extends from the larynx to the primary bronchi.
2. It is composed of smooth muscle and C-shaped rings of cartilage and is lined with pseudostratified ciliated columnar epithelium.
3. Two methods of bypassing obstructions from the respiratory passageways are tracheostomy and intubation.

Bronchi (p. 729)

1. The bronchial tree consists of the trachea, primary bronchi, secondary bronchi, tertiary bronchi, bronchioles, and terminal bronchioles. Walls of bronchi contain rings of cartilage; walls of bronchioles contain smooth muscle.
2. A bronchogram is an x-ray of the tree after introduction of an opaque contrast medium usually containing iodine.
3. Bronchoscopy is the visual examination of the bronchus through a bronchoscope.

Lungs (p. 731)

1. Lungs are paired organs in the thoracic cavity. They are enclosed by the pleural membrane. The parietal pleura is the outer layer; the visceral pleura is the inner layer.
2. The right lung has three lobes separated by two fissures; the left lung has two lobes separated by one fissure and a depression, the cardiac notch.
3. Secondary bronchi give rise to branches called segmental bronchi, which supply segments of lung tissue called bronchopulmonary segments.
4. Each bronchopulmonary segment consists of lobules, which contain lymphatics, arterioles, venules, terminal bronchioles, respiratory bronchioles, alveolar ducts, alveolar sacs, and alveoli.
5. Alveolar walls consist of type I alveolar cells, type II alveolar cells, and alveolar macrophages.
6. Gas exchange occurs across the alveolar–capillary (respiratory) membranes.

Physiology of Respiration (p. 736)

Pulmonary Ventilation (p. 736)

1. Pulmonary ventilation, or breathing, consists of inspiration and expiration.
2. The movement of air into and out of the lungs depends on pressure changes governed in part by Boyle's law, which states that the volume of a gas varies inversely with pressure, assuming that temperature is constant.
3. Inspiration occurs when alveolar pressure falls below atmospheric pressure. Contraction of the diaphragm and external intercostal muscles increases the size of the thorax, thus decreasing the intrapleural pressure so that the lungs expand. Expansion of the lungs decreases alveolar pressure so that air moves along the pressure gradient from the atmosphere into the lungs.

4. During forced inspiration, accessory muscles of inspiration (sternocleidomastoids, scalenes, and pectoralis minor) are also used.
5. Expiration occurs when alveolar pressure is higher than atmospheric pressure. Relaxation of the diaphragm and external intercostal muscles results in elastic recoil of the chest wall and lungs, which increases intrapleural pressure, lung volume decreases, and alveolar pressure increases so that air moves from the lungs to the atmosphere.
6. Forced expiration employs contraction of the internal intercostals and abdominal muscles.
7. A collapsed lung or portion of a lung is called atelectasis; normally, surfactant helps prevent the condition.
8. Compliance is the ease with which the lungs and thoracic wall expand.
9. The walls of the respiratory passageways offer some resistance to breathing.

Modified Respiratory Movements (p. 741)

1. Modified respiratory movements are used to express emotions and to clear air passageways.
2. Coughing, sneezing, sighing, yawning, sobbing, crying, laughing, and hiccuping are types of modified respiratory movements.

Pulmonary Air Volumes and Capacities (p. 741)

1. Air volumes exchanged during breathing and rate of respiration are measured with a spirometer.
2. Among the pulmonary air volumes exchanged in ventilation are tidal, inspiratory reserve, expiratory reserve, residual, and minimal volumes.
3. Pulmonary lung capacities, the sum of two or more volumes, include inspiratory, functional residual, vital, and total lung capacity.

Exchange of Oxygen and Carbon Dioxide (p. 743)

1. According to Charles' law, the volume of a gas is directly proportional to the absolute temperature, assuming the pressure remains constant.
2. The partial pressure of a gas is the pressure exerted by that gas in a mixture of gases. It is symbolized by p.
3. According to Dalton's law, each gas in a mixture of gases exerts its own pressure as if all the other gases were not present.
4. Henry's law states that the quantity of a gas that will dissolve in a liquid is proportional to the partial pressure of the gas and its solubility coefficient, when the temperature remains constant.
5. A major clinical application of Henry's law is hyperbaric oxygenation (HBO).

Physiology of External Respiration and Internal Respiration (p. 744)

1. In internal and external respiration, O_2 and CO_2 diffuse from areas of their higher partial pressures to areas of their lower partial pressures.
2. External (pulmonary) respiration is the exchange of gases between alveoli and pulmonary blood capillaries. It depends on partial pressure differences, a large surface area for gas exchange, a small diffusion distance across the alveolar–capillary (respiratory) membrane, and the rate of airflow into and out of the lungs.
3. Internal (tissue) respiration is the exchange of gases between tissue blood capillaries and tissue cells.

Transport of Oxygen and Carbon Dioxide (p. 746)

1. In each 100 ml of oxygenated blood, 1.5% of the O_2 is dissolved in plasma and 98.5% is bound to hemoglobin as oxyhemoglobin (HbO_2).
2. The association of oxygen and hemoglobin is affected by pO_2, acidity (pH), pCO_2, temperature, and BPG.
3. Fetal hemoglobin differs from adult hemoglobin in structure and has a higher affinity for oxygen.
4. Hypoxia refers to oxygen deficiency at the tissue level and is classified as hypoxic, anemic, stagnant, and histotoxic.
5. In each 100 ml of deoxygenated blood, 7% of CO_2 is dissolved in plasma, 23% combines with hemoglobin as carbaminohemoglobin ($Hb \cdot CO_2$), and 70% is converted to bicarbonate ions (HCO_3^-).
6. Carbon monoxide poisoning occurs when CO combines with hemoglobin. The result is hypoxia.
7. In an acid environment, hemoglobin's affinity for oxygen is lower and oxygen splits more readily from it (Bohr effect).
8. In the presence of O_2, less CO_2 binds to hemoglobin (Haldane effect).

Control of Respiration (p. 750)

Nervous Control (p. 751)

1. The respiratory center consists of a medullary rhythmicity area, pneumotaxic area, and apneustic area.

2. The inspiratory area has an intrinsic excitability (autorhythmicity) that sets the basic rhythm of respiration.
3. The pneumotaxic and apneustic areas coordinate the transition between inspiration and expiration.

Regulation of Respiratory Center Activity (p. 752)

1. Respirations may be modified by a number of factors, both in the brain and outside.
2. Among the modifying factors are cortical influences, the inflation reflex, chemical stimuli, such as O_2 and CO_2 (actually H^+) levels, proprioceptors, temperature, pain, and irritation to the respiratory mucosa.

Aging and the Respiratory System (p. 755)

1. Aging results in decreased vital capacity, decreased blood level of oxygen, and diminished alveolar macrophage activity.
2. Elderly people are more susceptible to pneumonia, emphysema, bronchitis, and other pulmonary disorders.

Developmental Anatomy of the Respiratory System (p. 756)

1. The respiratory system begins as an outgrowth of endoderm called the laryngotracheal bud.
2. Smooth muscle, cartilage, and connective tissue of the bronchial tubes and pleural sacs develop from mesoderm.

Review Questions

1. What organs make up the respiratory system? Distinguish between the upper and lower respiratory system. What functions do the respiratory and cardiovascular systems have in common? (p. 721)
2. Describe the structure of the external and internal nose, and describe their functions in filtering, warming, and moistening air. (p. 721)
3. What is the pharynx? Differentiate the three anatomical regions of the pharynx and indicate their roles in respiration. (p. 724)
4. Describe the structure of the larynx and explain how it functions in respiration and voice production. (p. 725)
5. Describe the location and structure of the trachea. (p. 727)
6. What is the bronchial tree? Describe its structure. What is a bronchogram? (p. 731)
7. Where are the lungs located? Distinguish the parietal pleura from the visceral pleura. (p. 731)
8. Define each of the following parts of a lung: base, apex, costal surface, medial surface, hilus, root, cardiac notch, and lobe. (p. 731)
9. What is a lobule of the lung? Describe its composition and function in respiration. (p. 732)
10. What is a bronchopulmonary segment? (p. 732)
11. Describe the histology and function of the alveolar–capillary (respiratory) membrane. (p. 735)
12. What are the basic differences among pulmonary ventilation, external respiration, and internal respiration? (p. 736)
13. Discuss the basic steps involved in inspiration and expiration. Be sure to include values for all pressures involved. (p. 736)
14. Distinguish between quiet and forced inspiration and expiration. (p. 737)
15. Describe how compliance and airway resistance relate to pulmonary ventilation. (p. 740)
16. Define the various kinds of modified respiratory movements. (p. 741)

17. What is a spirometer? Define the various lung volumes and capacities. How is the minute volume of respiration (MVR) calculated? (p. 741)
18. Define the partial pressure of a gas. How is it calculated? (p. 743)
19. Define Boyle's law (p. 736), Dalton's law (p. 743), Charles' law (p. 743), and Henry's law. (p. 743)
20. How does decompression sickness occur? (p. 744)
21. Construct a diagram to illustrate how and why the respiratory gases move during external and internal respiration. (p. 744)
22. What factors affect external respiration? (p. 745)
23. Describe the relationship between hemoglobin and pO_2, acidity, pCO_2, temperature, and BPG. (p. 746)
24. Define hypoxia and distinguish the principal types. (p. 749)
25. Explain how CO_2 is picked up by tissue capillary blood and then discharged into the alveoli. (p. 749)
26. How does the medullary rhythmicity area function in controlling respiration? How are the apneustic area and pneumotaxic area related to the control of respiration? (p. 751)
27. Explain how each of the following modifies respiration: cerebral cortex, inflation reflex, CO_2, O_2, proprioceptors, temperature, pain, and irritations of the respiratory mucosa. (p. 752)
28. How does the control of respiration demonstrate the principle of homeostasis? (p. 752)
29. Describe the effects of aging on the respiratory system. (p. 755)
30. Describe the development of the respiratory system. (p. 756)
31. Define the following terms: rhinoplasty (p. 721), laryngitis (p. 727), cancer of the larynx (p. 727), tracheostomy (p. 729), intubation (p. 729), pneumothorax (p. 731), hemothorax (p. 731), pleurisy (p. 731), nebulization (p. 731), hyperbaric oxygenation (HBO) (p. 744), and carbon monoxide (CO) poisoning. (p. 748)

Answers to Questions with Figures

23.1 Nose, pharynx, larynx, trachea, bronchi, and bronchioles, except the respiratory bronchioles.

23.2 External nares → vestibule → nasal cavity → and internal nares.

23.3 Superior, nasopharynx; inferior, cricoid cartilage.

23.4 During swallowing, the epiglottis closes over the rima glottidis, the entrance to the trachea.

23.5 Voice production.

23.6 Since the tissues between the esophagus and trachea are soft, the esophagus can bulge into the trachea when you swallow food.

23.7 Left lung two of each; right lung three of each.

23.8 Serous membrane.

23.9 Since two-thirds of the heart lies to the left of the midline, the left lung contains a cardiac notch to accommodate the position of the heart.

23.10 Goblet cells produce and secrete mucus and cilia move mucus laden with foreign particles toward the throat.

23.11 They secrete alveolar fluid, which includes surfactant.

23.12 The pressure would increase to 4 atm.

23.13 If you are at rest while reading, your diaphragm.

23.14 At the start of inspiration, intrapleural pressure is about 756 mm Hg. With contraction of the diaphragm, it decreases to about 754 mm Hg, as the volume of the space between the two pleural layers expands. With relaxation of the diaphragm, it increases back to 756 mm Hg.

23.16 Vital capacity.

23.17 A difference in pO_2 that promotes diffusion in the indicated direction.

23.18 5 ml; 20 ml (the same as at rest!).

23.19 In both cases, hemoglobin in your pulmonary veins would be fully saturated with oxygen (upper right-most point on curve).

23.20 Since lactic acid (lactate) and CO_2 are produced by active skeletal muscles, blood pH decreases a little and pCO_2 increases when you are actively working out. The result is lowered affinity of hemoglobin for oxygen, thus more O_2 is available to the working muscles.

23.21 More available because the affinity of hemoglobin for O_2 decreases with increasing temperature.

23.22 Fetal Hb is 80% saturated while maternal Hb is about 75% saturated with oxygen.

23.23 Vein.

23.24 The medullary inspiratory area.

23.27 40 mm Hg.

23.28 Administer air enriched in oxygen, with artificial ventilation if the person has stopped breathing. Even mouth-to-mouth resuscitation might save the person.

Chapter 24
THE DIGESTIVE SYSTEM

Chapter Contents at a Glance

Student Objectives

1. Identify the organs of the gastrointestinal (GI) tract and the accessory organs of digestion and their functions in digestion.

2. Describe the mechanical movements of the gastrointestinal tract.

3. Explain how salivary secretion, gastric secretion, gastric emptying, pancreatic secretion, bile secretion, and small intestinal secretion are regulated.

4. Define absorption and explain how the end products of digestion are absorbed.

5. Define the processes involved in the formation of feces and defecation.

6. Describe the effects of aging on the digestive system.

7. Describe the development of the digestive system.

8. Describe the clinical symptoms of the following disorders: dental

caries, periodontal disease, peritonitis, peptic ulcer disease (PUD), appendicitis, gastrointestinal tumors, diverticulitis, cirrhosis, hepatitis, gallstones, anorexia nervosa, and bulimia.

9. Define medical terminology associated with the digestive system.

Food is vital for life because it is the source of energy that drives the chemical reactions occurring in every cell and provides matter that is used to form new tissue or to repair damaged tissue. Energy is needed for muscle contraction, conduction of nerve impulses, and secretory and absorptive activities of many cells. Food as it is consumed, however, is not in a state suitable for use as an energy source by any cell. First, it must be broken down into molecules small enough to cross the plasma (cell) membranes. The breaking down of larger food molecules into molecules small enough to enter body cells is called **digestion,** and the organs that collectively perform this function compose the **digestive system.**

The medical specialty that deals with the structure, function, diagnosis, and treatment of diseases of the stomach and intestines is called **gastroenterology** (gas′-trō-en′-ter-OL-ō-jē; *gastro* = stomach; *enteron* = intestines).

The developmental anatomy of the digestive system is considered later in the chapter.

DIGESTIVE PROCESSES

The digestive system carries out five basic activities.

1. **Ingestion.** Taking food into the mouth (eating).
2. **Movement of food.** Passage of food along the gastrointestinal tract.
3. **Digestion.** The breakdown of food by both chemical and mechanical processes.
4. **Absorption.** The passage of digested food from the gastrointestinal tract into the cardiovascular and lymphatic systems for distribution to cells.
5. **Defecation.** The elimination of indigestible substances from the gastrointestinal tract.

Mechanical digestion consists of various movements of the gastrointestinal tract. Food is macerated by the teeth before it is swallowed. Then the smooth muscles of the stomach and small intestine churn the food so it is thoroughly mixed with enzymes that digest foods. **Chemical digestion** is a series of catabolic (hydrolysis) reactions. Enzymes split large carbohydrate, lipid, and protein molecules that we eat into smaller molecules that can be absorbed and used by body cells. These products of digestion are small enough to pass through the epithelial cells of the wall of the gastrointestinal tract, into the blood and lymphatic capillaries, and eventually into the body's cells.

ORGANIZATION

The organs of digestion are traditionally divided into two main groups. First is the **gastrointestinal (GI) tract,** or **alimentary canal,** a continuous tube that extends from the mouth to the anus through the ventral body cavity (Fig. 24.1). The relationship of the digestive organs to the nine regions of the abdominopelvic cavity may be reviewed in Fig. 1.11b. The length of the GI tract taken from a cadaver is about 9 m (30 ft). In a living person it is shorter because the muscles in its wall are in a state of tone. Organs composing the gastrointestinal tract include the mouth, pharynx, esophagus, stomach, small intestine, and large intestine.

The GI tract contains the food from the time it is eaten until it is digested and then absorbed or prepared for elimination. Muscular contractions in the wall of the GI tract break down the food physically by churning it. Enzymes secreted by cells along the tract break down the food chemically. Peristalsis, wavelike contraction of the smooth muscle in the wall of the GI tract, propels the food along the tract, from the esophagus to the anus.

The second group of organs composing the digestive system are the **accessory structures**—the teeth, tongue, salivary glands, liver, gallbladder, and pancreas. Teeth protrude into the GI tract and aid in the physical breakdown of food. The tongue assists in mastication (chewing) and deglutition (swallowing). The other accessory structures, except for the tongue, never come into direct contact with food. They produce or store secretions that aid in the chemical breakdown of food. These secretions flow into the tract through ducts.

General Histology of the GI Tract

The wall of the GI tract, especially from the esophagus to the anal canal, has the same basic arrangement of tissues. The four layers or tunics of the tract from the inside out are the mucosa, submucosa, muscularis, and serosa (Fig. 24.2).

Mucosa

The **mucosa,** or inner lining of the tract, is a mucous membrane. Three layers compose the membrane in the GI tract: (1) a lining layer of **epithelium** in direct contact with the contents of the GI tract, (2) an underlying layer of areolar connective tissue called the **lamina propria,** and (3) a thin layer of smooth muscle called the **muscularis mucosae.**

The epithelial layer is mainly nonkeratinized stratified squamous epithelium in the mouth, esophagus, and anal canal but simple columnar epithelium throughout the rest of the tract. The principal function of the stratified epithelium is protection. The major functions of the simple epithelium are secretion and absorption. Among the epithelial cells are several types of hormone-secreting cells, collectively called **enteroendocrine cells.** They are most numerous in the lower portion of the stomach and first part of the small intestine.

The lamina propria is made of areolar connective tissue containing many blood and lymphatic vessels and scattered lymphatic nodules. This layer supports the epithelium, binds it to the muscularis mucosae, and provides it with a

FIGURE 24.1 Organs of the digestive system and related structures.

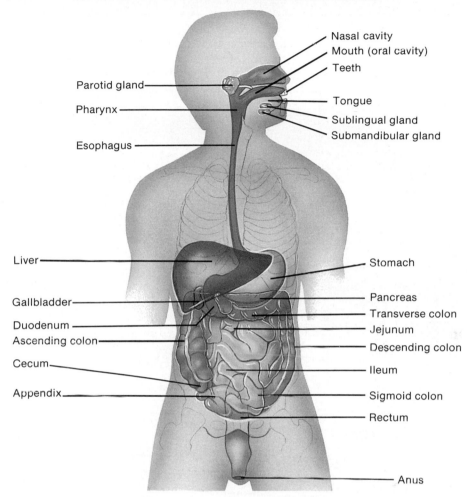

Diagram of right lateral view of head and neck
and anterior view of chest, abdomen, and pelvis

Question: What types of movement propel food along the gastrointestinal tract?

blood and lymph supply. The blood and lymphatic vessels are the avenues by which nutrients in the tract reach the other tissues of the body. The lymphatic tissue also contains immune cells that protect against disease. It is estimated that there are as many immune cells associated with the GI tract as in all the rest of the body. This makes sense when you realize that the GI tract is in contact with the outside environment and contains food that often carries harmful bacteria.

The muscularis mucosae contains smooth muscle fibers (cells) that throw the mucous membrane of the intestine into small folds, which increase the surface area for digestion and absorption.

Submucosa

The **submucosa** consists of areolar connective tissue that binds the mucosa to the third layer, the muscularis. It is

highly vascular and contains a portion of the **submucosal plexus (plexus of Meissner),** which is part of the autonomic nerve supply to the muscularis mucosae. This plexus is important in controlling secretions by the GI tract.

Muscularis

The **muscularis** of the mouth, pharynx, and upper esophagus consists in part of *skeletal muscle* that produces voluntary swallowing. Skeletal muscle also forms the external anal sphincter, which permits voluntary control of defecation. Throughout the rest of the tract, the muscularis consists of *smooth muscle* that is generally found in two sheets: an inner sheet of circular fibers and an outer sheet of longitudinal fibers. Involuntary contractions of the smooth muscles help break down food physically, mix it with digestive secretions, and propel it along the tract. The muscularis also contains the major nerve supply to the gastrointestinal

FIGURE 24.2 Composite of various sections of the gastrointestinal tract seen in a three-dimensional drawing depicting the various layers and related structures.

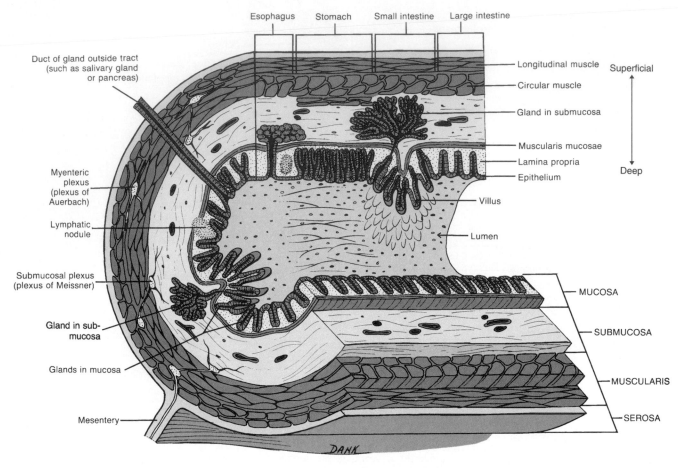

Duct of gland outside tract (such as salivary gland or pancreas)

Myenteric plexus (plexus of Auerbach)

Lymphatic nodule

Submucosal plexus (plexus of Meissner)

Gland in sub-mucosa

Glands in mucosa

Mesentery

Esophagus   Stomach   Small intestine   Large intestine

Longitudinal muscle

Circular muscle

Gland in submucosa

Muscularis mucosae

Lamina propria

Epithelium

Villus

Lumen

Superficial

Deep

MUCOSA

SUBMUCOSA

MUSCULARIS

SEROSA

DANK

Sectional view of layers of the GI tract

Question: What is the function of the nerve plexuses in the wall of the gastrointestinal tract?

tract—the **myenteric plexus (plexus of Auerbach),** which consists of fibers from both autonomic divisions. This plexus mostly controls GI tract **motility,** that is, the relative degree of contraction or relaxation of the muscularis. Together, the submucosal and myenteric plexuses have as many neurons as does the spinal cord.

Serosa

The **serosa** is the outermost layer of most portions of the GI tract. It is a serous membrane composed of connective tissue and epithelium. Below the diaphragm, this layer is also called the **visceral peritoneum** and forms a portion of the peritoneum, which we shall now describe in detail.

Peritoneum

The **peritoneum** (per′-i-tō-NĒ-um; *peri* = around; *tonos* = tension) is the largest serous membrane of the body. Serous membranes are also associated with the heart (pericardium) and lungs (pleurae). Serous membranes consist of a layer of simple squamous epithelium (called mesothelium) and an underlying supporting layer of connective tissue. The **parietal peritoneum** lines the wall of the abdominal cavity. The **visceral peritoneum** covers some of the organs and constitutes their serosa. The potential space between the parietal and visceral portions of the peritoneum is called the **peritoneal cavity** and contains serous fluid (Fig. 24.3a). In certain diseases, the peritoneal cavity may become distended by several liters of fluid so that it forms an actual space. Such an accumulation of serous fluid is called **ascites** (a-SĪ-tēz).

As you will see later, some organs lie on the posterior abdominal wall and are covered by peritoneum on their anterior surfaces only. Such organs, including the kidneys and pancreas, are said to be **retroperitoneal** (*retro* = backward or located behind).

Unlike the pericardium and pleurae, which smoothly cover the heart and lungs, the peritoneum contains large folds that weave between the viscera. The folds bind the organs to each other and to the walls of the cavity and contain blood and lymphatic vessels and nerves that supply the abdominal organs. One extension of the peritoneum is called the **mesentery** (MEZ-en-ter'-ē; *meso* = middle;

enteron = intestine). It is an outward fold of the serous coat of the small intestine (Fig. 24.3d). The tip of the fold is attached to the posterior abdominal wall. The mesentery binds the small intestine to the wall. A similar fold of parietal peritoneum, called the **mesocolon** (mez'-ō-KŌ-lon), binds the large intestine to the posterior body wall. It also carries blood and lymphatic vessels to the intestines.

Other important peritoneal folds are the falciform ligament, lesser omentum, and greater omentum. The **falciform** (FAL-si-form) **ligament** attaches the liver to the anterior abdominal wall and diaphragm (Fig. 24.3b). The **lesser omentum** (ō-MENT-um) arises as two folds in the serosa of the stomach and duodenum suspending the stomach and duodenum from the liver (Figure 24.3c). The **greater**

FIGURE 24.3 Views of the abdomen and pelvis indicating the relationship of the peritoneal extensions to each other and to organs of the digestive system.

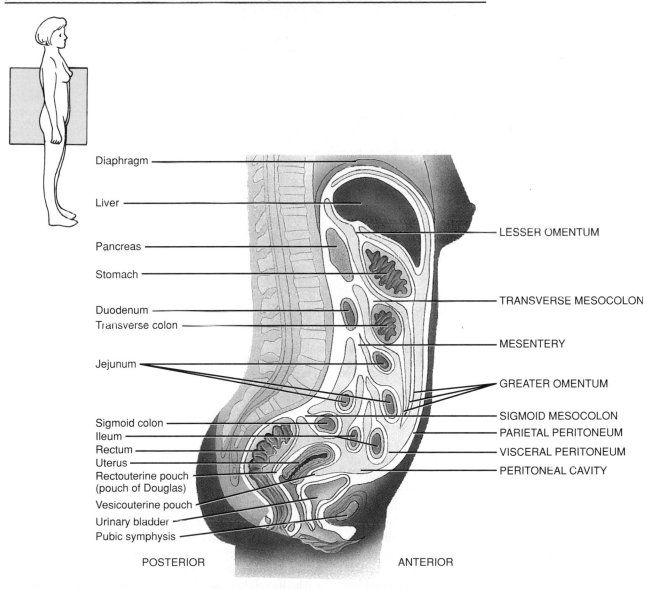

(a) Sagittal section

Figure continues

situated between the arches, and the lingual tonsil is situated at the base of the tongue. At the posterior border of the soft palate, the mouth opens into the oropharynx through the fauces (Fig. 24.4).

Tongue

The **tongue,** together with its associated muscles, forms the floor of the oral cavity. It is an accessory structure of the digestive system composed of skeletal muscle covered with mucous membrane (see Fig. 16.2). The tongue is divided into symmetrical lateral halves by a median septum that extends throughout its entire length and is attached inferiorly to the hyoid bone. Each half of the tongue consists of an identical complement of extrinsic and intrinsic muscles.

The **extrinsic muscles** of the tongue originate outside the tongue and insert into it. They include the hyoglossus, genioglossus, and styloglossus (see Fig. 11.7). The extrinsic muscles move the tongue from side to side and in and out. These movements maneuver food for chewing, shape the food into a rounded mass, and force the food to the back of the mouth for swallowing. They also form the floor of the mouth and hold the tongue in position. The **intrinsic muscles** originate and insert within the tongue and alter the shape and size of the tongue for speech and swallowing. The intrinsic muscles include the longitudinalis superior, longitudinalis inferior, transversus linguae, and verticalis linguae. The **lingual frenulum,** a fold of mucous membrane in the midline of the undersurface of the tongue, is attached to the floor of the mouth and aids in limiting the movement of the tongue posteriorly (Fig. 24.4).

The upper surface and sides of the tongue are covered with **papillae** (pa-PIL-ē), projections of the lamina propria covered with epithelium (Fig. 16.2a). **Filiform papillae** are conical projections distributed in parallel rows over the anterior two-thirds of the tongue. They are whitish and contain no taste buds. **Fungiform papillae** are mushroomlike elevations distributed among the filiform papillae and are more numerous near the tip of the tongue. They appear as red dots on the surface of the tongue, and most of them contain taste buds. **Circumvallate papillae** are arranged in the form of an inverted V on the posterior surface of the tongue and all of them contain taste buds. Ducts of lingual (von Ebner's) glands, which are serous glands, surround the circumvallate papillae. Note the taste zones of the tongue in Fig. 16.2a.

Glands on the dorsum of the tongue secrete a digestive enzyme called **lingual lipase,** which initiates digestion of triglycerides into fatty acids and monoglycerides.

Salivary Glands

Saliva is a fluid that is continuously secreted into the mouth. Ordinarily, just enough saliva is secreted to keep the mucous membranes of the mouth and pharynx moist. When food enters the mouth, however, secretion of saliva increases. It lubricates, dissolves, and begins the chemical breakdown of the food. The mucous membrane lining the mouth contains many small glands. The **buccal glands** and minor salivary glands secrete small amounts of saliva. However, most saliva is secreted by the **salivary glands (major salivary glands),** accessory structures that lie outside the mouth. Their secretions pour into ducts that empty into the oral cavity. There are three pairs of salivary glands: parotid, submandibular (submaxillary), and sublingual glands (Fig. 24.5a).

The **parotid glands** are located inferior and anterior to the ears between the skin and the masseter muscle. Each secretes into the oral cavity vestibule via a duct, called the **parotid (Stensen's) duct,** that pierces the buccinator muscle to open into the vestibule opposite the upper second molar tooth. The **submandibular glands** are found beneath the base of the tongue in the posterior part of the floor of the mouth (Fig. 24.5a). Their ducts, the **submandibular (Wharton's) ducts,** run superficially under the mucosa on either side of the midline of the floor of the mouth and enter the oral cavity proper on either side of the lingual frenulum. The **sublingual glands** are superior to the submandibular glands. Their ducts, the **lesser sublingual (Rivinus') ducts,** open into the floor of the mouth in the oral cavity proper.

CLINICAL APPLICATION

MUMPS

Although any of the salivary glands may be the target of a nasopharyngeal infection, the mumps virus (myxovirus) typically attacks the parotid glands. **Mumps** is an inflammation and enlargement of the parotid glands accompanied by moderate fever, malaise, and extreme pain in the throat, especially when swallowing sour foods or acidic juices. Swelling occurs on one or both sides of the face, just anterior to the ramus of the mandible. In about 20 to 35% of males past puberty, the testes may also become inflamed, and although it rarely occurs, sterility is a possible consequence.

Composition of Saliva

Chemically, **saliva** is 99.5% water and 0.5% solutes. Among the solutes are ions, such as sodium, potassium, chloride, bicarbonate, and phosphates. Also present are some dissolved gases and various organic substances including urea and uric acid, serum albumin and globulin, mucin, the bacteriolytic enzyme lysozyme, and the digestive enzyme salivary amylase.

Each saliva-producing gland supplies different proportions of ingredients to saliva. The parotid glands contain cells that secrete a watery serous liquid containing salivary amylase. The submandibular glands contain cells similar to those found in the parotid glands and some mucous cells.

FIGURE 24.5 Salivary glands. The submandibular gland shown in (b) consists mostly of serous acini (serous fluid-secreting exocrine portion of gland) and a few mucous acini (mucus-secreting exocrine portion of gland). The parotid glands consist of all serous acini, and the sublingual glands consist of mostly mucous acini and a few serous acini.

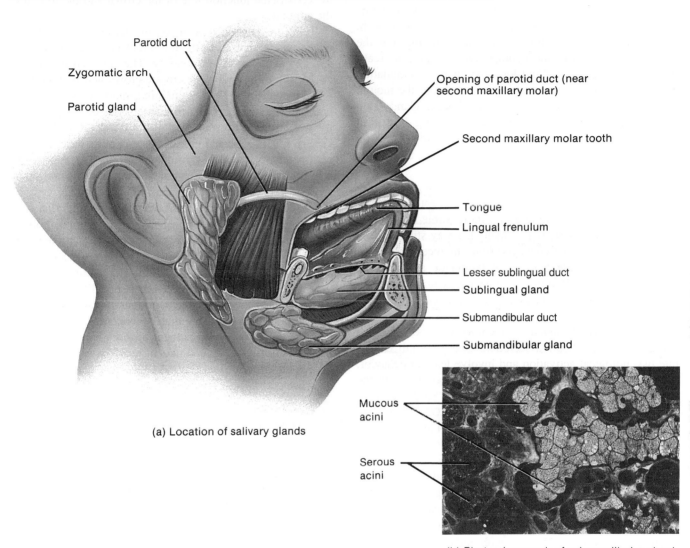

(a) Location of salivary glands

(b) Photomicrograph of submandibular gland (350 x)

Question: What is the function of the chloride ions in saliva?

Therefore they secrete a fluid that is thickened with mucus but still contains quite a bit of enzyme. The sublingual glands contain mostly mucous cells, so they secrete a much thicker fluid that contributes only a small amount of amylase to the saliva.

The water in saliva provides a medium for dissolving foods so they can be tasted and for initiating digestive reactions. Chloride ions in the saliva activate the salivary amylase. Bicarbonate and phosphate ions buffer acidic foods that enter the mouth. As a result, saliva is only slightly acidic (pH of 6.35 to 6.85). Urea and uric acid are found in saliva because the saliva-producing glands (like the sweat glands of the skin) help remove waste molecules from the body. Mucus lubricates the food so it can be easily moved about in the mouth, formed into a ball, and swallowed. The enzyme lysozyme, which is found in small quantities, helps destroy bacteria. This contributes to protecting the mucous membrane from infection and the teeth from decay. However, it is not present in large enough quantities to eliminate all oral bacteria.

FIGURE 24.7 Dentitions and times of eruptions (indicated in parentheses).

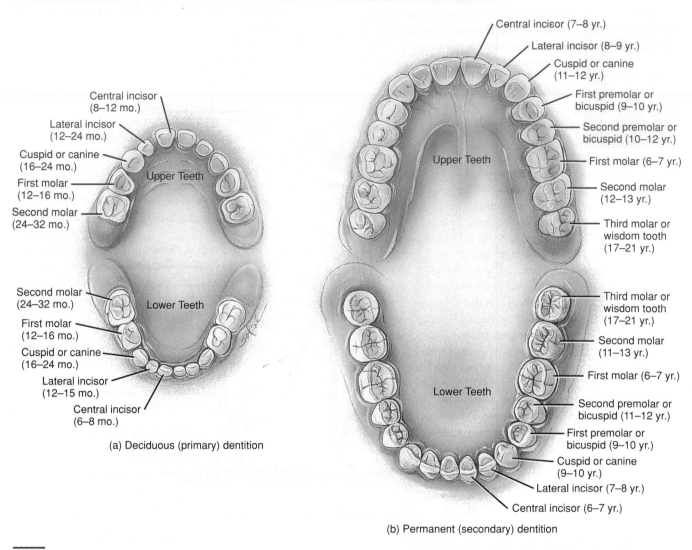

Central incisor (7–8 yr.)
Lateral incisor (8–9 yr.)
Cuspid or canine (11–12 yr.)
First premolar or bicuspid (9–10 yr.)
Second premolar or bicuspid (10–12 yr.)
First molar (6–7 yr.)
Second molar (12–13 yr.)
Third molar or wisdom tooth (17–21 yr.)

Upper Teeth

Central incisor (8–12 mo.)
Lateral incisor (12–24 mo.)
Cuspid or canine (16–24 mo.)
First molar (12–16 mo.)
Second molar (24–32 mo.)

Upper Teeth

Second molar (24–32 mo.)
First molar (12–16 mo.)
Cuspid or canine (16–24 mo.)
Lateral incisor (12–15 mo.)
Central incisor (6–8 mo.)

Lower Teeth

(a) Deciduous (primary) dentition

Third molar or wisdom tooth (17–21 yr.)
Second molar (11–13 yr.)
First molar (6–7 yr.)
Second premolar or bicuspid (11–12 yr.)
First premolar or bicuspid (9–10 yr.)
Cuspid or canine (9–10 yr.)
Lateral incisor (7–8 yr.)
Central incisor (6–7 yr.)

Lower Teeth

(b) Permanent (secondary) dentition

Question: Which teeth are used to tear and shred food? To crush and grind food? To cut into food?

Chemical Digestion

The enzyme **salivary amylase** (AM-i-lās) initiates the breakdown of starch. Dietary carbohydrates are either monosaccharide and disaccharide sugars or polysaccharides, such as starches (see page 41). Most of the carbohydrates we eat are starches, but only monosaccharides can be absorbed into the bloodstream. Thus ingested disaccharides and starches must be broken down. The function of salivary amylase is to break certain chemical bonds between glucose units in the starches. This activity reduces the long-chain polysaccharides to the disaccharide maltose, the trisaccharide maltotriose, and short-chain glucose polymers called α-dextrins. Food usually is swallowed too quickly for all the starches to be reduced to disaccharides in the mouth. However, salivary amylase in the swallowed food continues

to act on starches for another 15 to 30 minutes in the stomach before the stomach acids inactivate it.

The enzyme **lingual lipase**, secreted by glands on the dorsum of the tongue, is also found in saliva. This enzyme, which is active in the stomach, can digest as much as 30% of dietary triglycerides into fatty acids and monoglycerides.

Exhibit 24.1 summarizes digestion in the mouth.

Physiology of Deglutition

Swallowing, or **deglutition** (dē-gloo-TISH-un), is a mechanism that moves food from the mouth to the stomach. It is facilitated by saliva and mucus and involves the mouth, pharynx, and esophagus. Swallowing is conveniently divided into three stages: (1) the voluntary stage, in which

EXHIBIT 24.1

DIGESTION IN THE MOUTH

Structure	Activity	Result
CHEEKS AND LIPS	Keep food between teeth during mastication.	Foods uniformly chewed.
TONGUE		
Extrinsic muscles	Move tongue from side to side and in and out.	Food maneuvered for mastication, shaped into bolus, and maneuvered for deglutition (swallowing).
Intrinsic muscles	Alter shape of tongue.	Deglutition and speech.
Taste buds	Serve as receptors for food stimulus.	Secretion of saliva stimulated by nerve impulses from taste buds to salivatory nuclei in brain stem to salivary glands.
Glands	Secrete lingual lipase.	Breaks down triglycerides in the stomach into fatty acids and monoglycerides.
BUCCAL GLANDS	Secrete saliva.	Lining of mouth and pharynx moistened and lubricated.
SALIVARY GLANDS	Secrete saliva.	Lining of mouth and pharynx moistened and lubricated. Saliva softens, moistens, and dissolves food, cleanses mouth and teeth. Salivary amylase splits polysaccharides into smaller fragments.
TEETH	Cut, tear, and pulverize food.	Solid foods reduced to smaller particles for swallowing.

the bolus is moved into the oropharynx; (2) the pharyngeal stage, the involuntary passage of the bolus through the pharynx into the esophagus; and (3) the esophageal stage, the involuntary passage of the bolus through the esophagus into the stomach (described together with the esophagus).

Swallowing starts when the bolus is forced to the back of the oral cavity and into the oropharynx by the movement of the tongue upward and backward against the palate. This represents the **voluntary stage** of swallowing.

With the passage of the bolus into the oropharynx, the involuntary **pharyngeal stage** of swallowing begins (Fig. 24.8b). The respiratory passageways close, and breathing is temporarily interrupted. The bolus stimulates receptors in the oropharynx, which send impulses to the **deglutition center** in the medulla and lower pons of the brain stem. The returning impulses cause the soft palate and uvula to move upward to close off the nasopharynx, and the larynx is pulled forward and upward under the tongue. As the larynx rises, the epiglottis moves backward and downward and seals off the rima glottidis. The movement of the larynx also pulls the vocal cords together, further sealing off the respiratory tract, and widens the opening between the laryngopharynx and esophagus. The bolus passes through the laryngopharynx and enters the esophagus in one to two seconds. The respiratory passageways then reopen and breathing resumes.

ESOPHAGUS

The **esophagus** (e-SOF-a-gus) is a muscular, collapsible tube that lies behind the trachea and is about 23 to 25 cm (10 in.) long. It begins at the inferior end of the laryngopharynx, passes through the mediastinum anterior to the vertebral column, pierces the diaphragm through an opening called the **esophageal hiatus,** and ends in the superior portion of the stomach (see Fig. 24.1).

Histology

The **mucosa** of the esophagus consists of nonkeratinized stratified squamous epithelium (Fig. 24.9), lamina propria, and a muscularis mucosae. Near the stomach, the mucosa of the esophagus also contains mucous glands. The **submucosa** contains connective tissue, blood vessels, and mucous glands. The **muscularis** of the upper third is skeletal, the middle third is skeletal and smooth, and the lower third is smooth. The outer layer is known as the **adventitia** (ad-ven-TISH-ya) rather than the serosa because the areolar connective tissue of the layer is not covered by epithelium (mesothelium) and because the connective tissue merges with the connective tissue of surrounding structures.

FIGURE 24.8 Deglutition (swallowing). During the pharyngeal stage of deglutition, the tongue rises against the palate, the nasopharynx is closed off, the larynx rises, the epiglottis seals off the larynx, and the bolus is passed into the esophagus.

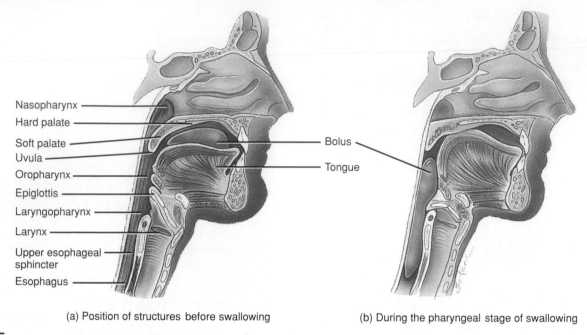

Nasopharynx
Hard palate
Soft palate
Uvula
Oropharynx
Epiglottis
Laryngopharynx
Larynx
Upper esophageal sphincter
Esophagus

Bolus
Tongue

(a) Position of structures before swallowing

(b) During the pharyngeal stage of swallowing

Question: Is swallowing a voluntary or involuntary action?

FIGURE 24.9 Histology of the esophagus. An enlarged aspect of the mucosa of the esophagus is shown in Exhibit 4.1 on page 102, Stratified squamous epithelium.

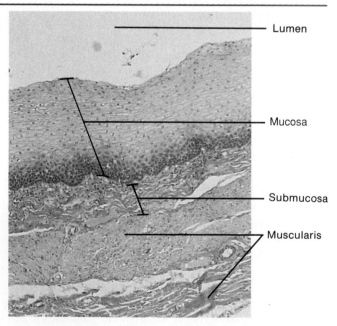

Lumen
Mucosa
Submucosa
Muscularis

Photomicrograph of a portion of the wall of the human esophagus (400 x)

Question: Where are the glands that secrete lubricating mucus located?

Physiology

The esophagus secretes mucus and transports food to the stomach. It does not produce digestive enzymes and does not carry on absorption. The passage of food from the laryngopharynx into the esophagus is regulated by a sphincter (a circular muscle that is normally contracted so that there is no opening at the center) at the entrance to the esophagus called the **upper esophageal** (e-sof'-a-JĒ-al) **sphincter.** It consists of the cricopharyngeus muscle attached to the cricoid cartilage. The elevation of the larynx during the pharyngeal stage of swallowing causes the sphincter to relax, and the bolus enters the esophagus. The sphincter also relaxes during expiration.

During the **esophageal stage** of swallowing (Fig. 24.10), food is pushed through the esophagus by involuntary muscular movements called **peristalsis** (per'-is-STAL-sis). Peristalsis, which also occurs in other portions of the gastrointestinal tract, is a function of the muscularis. In the esophagus, it is controlled by the medulla. In the section of the esophagus lying just above and around the top of the bolus, the circular muscle fibers contract. The contraction constricts the esophageal wall and squeezes the bolus downward. Meanwhile, longitudinal fibers lying around the bottom of and just below the bolus also contract. Contraction of the longitudinal fibers shortens this lower section, pushing its walls outward so it can receive the bolus. The contractions are repeated in a wave that moves down the esophagus, pushing the food toward the stomach. Passage of the bolus is facilitated by glands that secrete mucus. The passage of solid or semisolid food from the mouth to the stomach takes four to eight seconds. Very soft foods and liquids pass through in about one second.

Just above the level of the diaphragm, the esophagus is slightly narrowed. This narrowing has been attributed to a physiological sphincter in the inferior part of the esophagus known as the **lower esophageal (gastroesophageal) sphincter.** The lower esophageal sphincter relaxes during swallowing and thus allows the bolus to pass from the esophagus into the stomach.

FIGURE 24.10 Views of peristalsis during the esophageal stage of deglutition (swallowing).

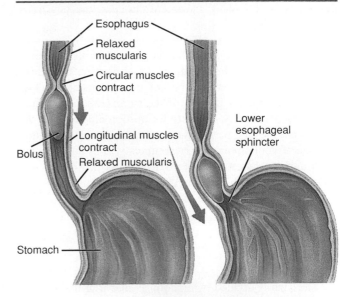

(a) Diagram of peristalsis in esophagus, anterior view

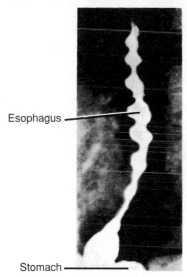

(b) X-ray, anterior view

Question: Does peristalsis "push" or "pull" food along the gastrointestinal tract?

CLINICAL APPLICATION

ACHALASIA AND HEARTBURN

If the lower esophageal sphincter fails to relax normally as food approaches, the condition is called **achalasia** (ak'-a-LĀ-zē-a; *a* = without; *chalasis* = relaxation). As a result, food passage from the esophagus into the stomach is greatly impeded. A whole meal may become lodged in the esophagus and enter the stomach very slowly. Distension of the esophagus results in chest pain that is often confused with pain originating from the heart. The condition is caused by malfunction of the myenteric plexus (plexus of Auerbach).

If, on the other hand, the lower esophageal sphincter fails to close adequately after food has entered the stomach, the stomach contents can enter the lower esophagus. Hydrochloric acid (HCl) from the stomach contents can irritate the esophageal wall, resulting in a burning sensation. The sensation is known as **heartburn** because it is experienced in the region very near the heart, although it is not related to any cardiac problem. Heartburn can be treated by taking antacids (Tums, Gelusil, Rolaids, Maalox) that neutralize the hydrochloric acid and lessen the severity of the burning sensation and discomfort. In addition to the use of antacids for heartburn, other measures are important. Symptoms are less likely to occur if food is eaten in smaller amounts, and heartburn is less of a problem if the person does not lie down right after a meal.

Exhibit 24.2 summarizes the digestion-related activities of the pharynx and esophagus.

STOMACH

The **stomach** is a J-shaped enlargement of the GI tract directly under the diaphragm in the epigastric, umbilical, and left hypochondriac regions of the abdomen (see Fig. 1.11b). The superior portion of the stomach is a continuation of the esophagus. The inferior portion empties into the duodenum, the first part of the small intestine. The position and size of your stomach vary continually. For instance, the diaphragm pushes it downward with each inspiration and pulls it upward with each expiration. Empty, it is about the size of a large sausage, but it can stretch to accommodate a large quantity of food.

Anatomy

The stomach is divided by anatomists into four main areas: cardia, fundus, body, and pylorus (Fig. 24.11). The **cardia** surrounds the superior opening of the stomach. The rounded portion above and to the left of the cardia is the **fundus**. Below the fundus is the large central portion of the stomach, called the **body.** The inferior region of the stomach that connects to the duodenum is the **pylorus** (*pyle* = gate; *ouros* = guard). It has two parts, the **pyloric antrum** (*antrum* = cave) or **antrum,** which connects to the body of the stomach, and the **pyloric canal**, which leads into the duodenum. When the stomach is empty, the mucosa lies in large folds, called **rugae** (ROO-jē), that can be seen with the naked eye. The concave medial border of the stomach is called the **lesser curvature,** and the convex lateral border is the **greater curvature.** The pylorus communicates with the duodenum of the small intestine via a sphincter called the **pyloric sphincter (valve).**

CLINICAL APPLICATION

PYLOROSPASM AND PYLORIC STENOSIS

Two abnormalities of the pyloric sphincter can occur in infants. **Pylorospasm** is characterized by failure of the muscle fibers of the sphincter to relax normally. Food does not pass easily from the stomach to the small intestine, the stomach becomes overly full, and the infant vomits frequently to relieve the pressure. Pylorospasm is treated by drugs that relax the muscle fibers of the sphincter. **Pyloric stenosis** is a narrowing of the pyloric sphincter caused by a tumorlike mass that apparently is formed by enlargement of the circular muscle fibers. The hallmark symptom is projectile vomiting—the spraying of liquid vomitus some feet from the infant. Pyloric stenosis must be surgically corrected.

Histology

The stomach wall is composed of the same four basic layers as the rest of the GI tract, with certain modifications. Microscopic inspection of the **mucosa** reveals a layer of simple columnar epithelium (surface mucous cells) containing many narrow channels, called **gastric pits,** that extend down into the lamina propria (Fig. 24.12). At the bottoms of the pits are the openings of the **gastric glands.** The glands contain four types of secretory cells: chief cells, parietal cells, and mucous cells secrete their products into the stomach lumen, whereas G cells secrete the hormone gastrin into the blood.

The **chief (zymogenic) cells** secrete the principal gastric enzyme precursor, pepsinogen, and an enzyme of lesser importance called gastric lipase. Hydrochloric acid, involved in the conversion of pepsinogen to the active enzyme pepsin, and intrinsic factor, involved in the absorption of vitamin B_{12} for red blood cell production, are produced by the **parietal (oxyntic) cells.** You may recall from Chapter 19 that inability to produce intrinsic factor can

EXHIBIT 24.2

DIGESTIVE ACTIVITIES OF THE PHARYNX AND ESOPHAGUS

Structure	Activity	Result
PHARYNX	Pharyngeal stage of deglutition.	Moves bolus from oropharynx to laryngopharynx and into esophagus; closes air passageways.
	Relaxation of upper esophageal sphincter.	Permits entry of bolus from laryngopharynx into esophagus.
ESOPHAGUS	Esophageal stage of deglutition (peristalsis).	Forces bolus down esophagus.
	Relaxation of lower esophageal sphincter.	Permits entry of bolus into stomach.
	Secretion of mucus.	Lubricates esophagus for smooth passage of bolus.

FIGURE 24.11 External and internal anatomy of the stomach.

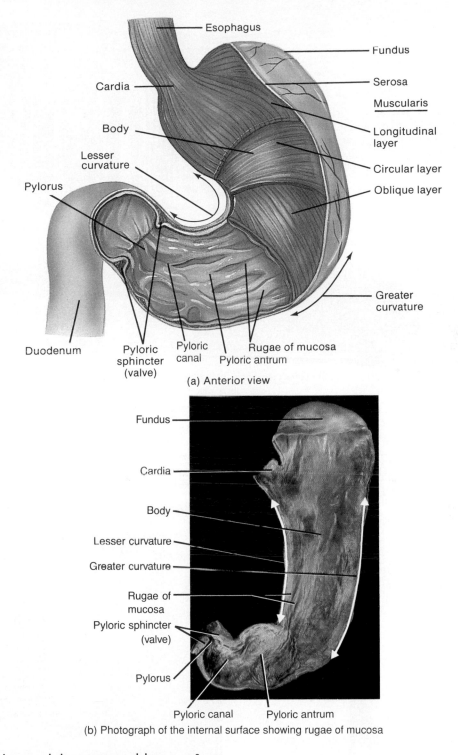

(a) Anterior view

(b) Photograph of the internal surface showing rugae of mucosa

Question: After a very large meal, does your stomach have rugae?

result in pernicious anemia. The **mucous cells** secrete mucus. The secretions of chief, parietal, and mucous cells are collectively called **gastric juice,** which totals about 2000 to 3000 ml (roughly 2 to 3 qt) per day. The **G cells,** which are located mainly in the pyloric antrum, secrete the hormone gastrin. As you will see shortly, this hormone stimulates several aspects of gastric activity.

The **submucosa** of the stomach is composed of areolar

FIGURE 24.12 Histology of the stomach.

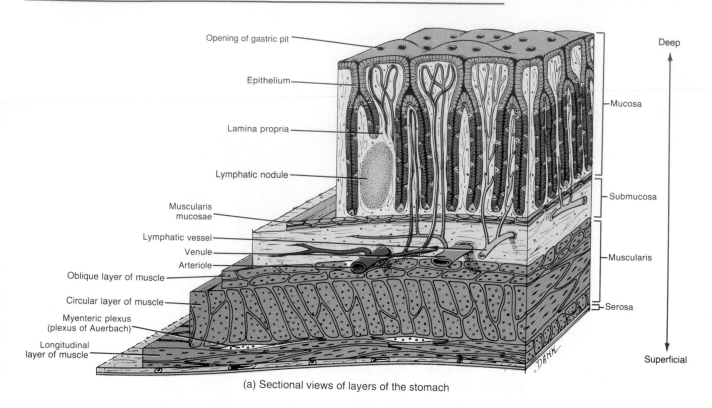

Opening of gastric pit

Epithelium

Lamina propria

Lymphatic nodule

Muscularis mucosae

Lymphatic vessel

Venule

Arteriole

Oblique layer of muscle

Circular layer of muscle

Myenteric plexus (plexus of Auerbach)

Longitudinal layer of muscle

Deep

Mucosa

Submucosa

Muscularis

Serosa

Superficial

(a) Sectional views of layers of the stomach

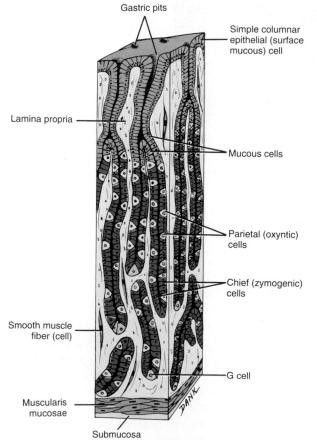

Gastric pits

Simple columnar epithelial (surface mucous) cell

Lamina propria

Mucous cells

Parietal (oxyntic) cells

Chief (zymogenic) cells

Smooth muscle fiber (cell)

G cell

Muscularis mucosae

Submucosa

(b) Sectional view of the stomach mucosa showing gastric glands

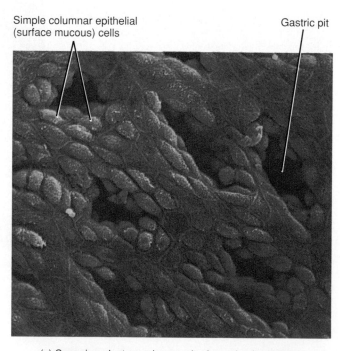

Simple columnar epithelial (surface mucous) cells

Gastric pit

(c) Scanning electron micrograph of gastric pits (5600x)

FIGURE 24.12 (continued)

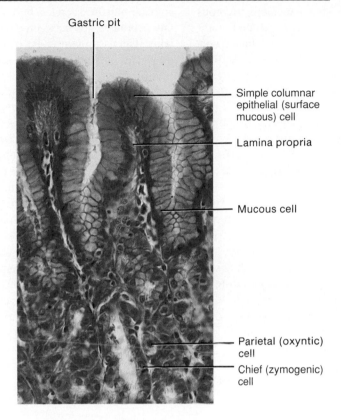

Gastric pit

Simple columnar epithelial (surface mucous) cell

Lamina propria

Mucous cell

Parietal (oxyntic) cell

Chief (zymogenic) cell

(d) Photomicrograph of fundic mucosa
(1120 x)

Question: What types of cells are found in gastric glands and what do they secrete?

connective tissue, which connects the mucosa to the muscularis.

The **muscularis** has three rather than two layers of smooth muscle: an outer longitudinal layer, a middle circular layer, and an inner oblique layer. The oblique layer is limited mostly to the body of the stomach. This arrangement of fibers allows the stomach to contract in a variety of ways to churn food, break it into small particles, mix it with gastric juice, and pass it to the duodenum.

The **serosa** covering the stomach is part of the visceral peritoneum. At the lesser curvature, the visceral peritoneum extends upward to the liver as the lesser omentum. At the greater curvature, the visceral peritoneum continues downward as the greater omentum and drapes over the intestines.

CLINICAL APPLICATION

ENDOSCOPY

The general term **endoscopy** (en-DOS-kō-pē; *endo* = inside; *shopein* = to examine) refers to visual inspection of any cavity of the body using an endoscope, an illuminated tube with lenses. Endoscopes can be used to visualize the entire gastrointestinal tract, as well as other systems of the body. Fiberoptic technology makes possible the use of a flexible endoscope. This type of endoscope can be adapted to the many curvatures in the mouth, throat, stomach, and other organs. Endoscopes can also be fitted with grasping devices to remove foreign objects from the esophagus and stomach, cutting blades to biopsy lesions and remove polyps from the colon, and cauterizing tips to stop bleeding. Endoscopic examination of the stomach is called **gastroscopy** (gas-TROS-kō-pē). The examiner can view the interior of the stomach directly to evaluate an ulcer, tumor, inflammation, or bleeding source.

Physiology of Digestion in the Stomach

Mechanical Digestion

Several minutes after food enters the stomach, gentle, rippling, peristaltic movements called **mixing waves** pass over the stomach every 15 to 25 seconds. These waves macerate food, mix it with secretions of the gastric glands, and reduce it to a thin liquid called **chyme** (kīm). Few mixing waves are observed in the fundus, which is primarily a storage area. Foods may remain in the fundus for an hour or more without becoming mixed with gastric juice. During this time, digestion due to lingual lipase and salivary amylase continues.

As digestion proceeds in the stomach, more vigorous mixing waves begin at the body of the stomach and intensify as they reach the pylorus. The pyloric sphincter normally remains almost, but not completely, closed. As food reaches the pylorus, each mixing wave forces several milliliters of the gastric contents into the duodenum through the pyloric sphincter. Most of the food is forced back into the body of the stomach, where it is subjected to further mixing. The next wave pushes it forward again and forces a little more into the duodenum. The forward and backward movements of the gastric contents are responsible for almost all the mixing in the stomach.

Chemical Digestion

The strongly acidic fluid of the stomach contributes to defense against invaders by killing many microbes in food. Parietal cells secrete **HCl** into the stomach lumen by separate active transport pumps for Cl^- and for transport of H^+ and K^+ in opposite directions, that is, H^+/K^+ antiports (see Fig. 24.15). In addition to killing microbes, the HCl partially denatures (unfolds) proteins in food and stimulates secretion of hormones that promote flow of bile and pancreatic juice.

The enzymatic digestion of proteins begins in the stomach. In the adult, this is achieved mainly through the enzyme **pepsin**, secreted by chief cells. Pepsin breaks certain peptide bonds between the amino acids making up proteins. Thus a protein chain of many amino acids is broken down into smaller fragments called **peptides.** Pepsin also brings about the clotting and digestion of milk proteins. Pepsin is most effective in the very acidic environment of the stomach (pH 2). It becomes inactive at higher pH.

What keeps pepsin from digesting the protein in stomach cells along with the food? First, pepsin is secreted in an inactive form called **pepsinogen.** In this form it cannot digest the proteins in the chief cells that produce it. It is not converted into active pepsin until it comes in contact with active pepsin molecules or hydrochloric acid secreted by parietal cells. Second, the stomach cells are protected by an alkaline mucus layer that coats the mucosa and forms a 1 to 3 mm thick barrier between it and the gastric juices.

Another enzyme of the stomach is **gastric lipase.** Gastric lipase splits the short-chain triglycerides in butterfat molecules found in milk. This enzyme operates best at a pH of 5 to 6 and has a limited role in the adult stomach. To digest fats, adults rely almost exclusively on lingual lipase and **pancreatic lipase**, an enzyme secreted by the pancreas into the small intestine.

The principal activities of gastric digestion are summarized in Exhibit 24.3.

EXHIBIT 24.3

DIGESTION IN THE STOMACH

Structure	Activity	Result
MUCOSA		
Chief (zymogenic) cells	Secrete pepsinogen. Secrete gastric lipase.	Precursor of pepsin is produced. Splits short-chain triglycerides.
Parietal (oxyntic) cells	Secrete hydrochloric acid.	Kills microbes in food. Denatures proteins. Converts pepsinogen into pepsin, which digests proteins into peptides. Inhibits secretion of gastrin. Stimulates secretion of secretin and cholecystokinin (CCK).
	Secrete intrinsic factor.	Needed for absorption of vitamin B_{12}, which is required for normal erythrocyte formation.
Mucous cells	Secrete mucus.	Forms a protective barrier that prevents destruction of stomach wall.
G cells	Secrete gastrin.	Stimulates parietal cells to secrete HCl and chief cells to secrete pepsinogen. Contracts lower esophageal sphincter, increases motility of the stomach, and relaxes pyloric sphincter.
MUSCULARIS	Mixing waves. Peristalsis.	Macerate food, mix it with gastric juice, reduce food to chyme, and force chyme through pyloric sphincter.
PYLORIC SPHINCTER (VALVE)	Opens to permit passage of chyme into duodenum.	Regulates passage of chyme from stomach to duodenum. Prevents backflow of chyme from duodenum to stomach.

Regulation of Gastric Secretion and Motility

The secretion of gastric juice and contraction of smooth muscle in the stomach wall are regulated by both nervous and hormonal mechanisms (Fig. 24.13) that occur in three overlapping phases: cephalic, gastric, and intestinal. During the cephalic phase, parasympathetic impulses from nuclei in the medulla are transmitted via the vagus (X) nerves. These impulses promote peristalsis in stomach smooth muscle and stimulate the gastric glands to secrete pepsinogen, hydrochloric acid, and mucus into stomach chyme and gastrin into the blood. During the gastric phase, reflexes within the stomach itself continue to stimulate churning contractions and secretory processes. Then, during the intestinal phase, neural and hormonal reflexes initiated in the small intestine exert an inhibitory effect on secretory activity and motility of the stomach.

Cephalic Phase

The **cephalic phase** refers to reflexes initiated by sensory receptors in the head. Even before food enters the stomach, the sight, smell, taste, or thought of food initiates this reflex. The cerebral cortex and feeding center in the hypothalamus send impulses to the medulla. The medulla transmits impulses to parasympathetic preganglionic fibers in the vagus (X) nerves, which stimulate parasympathetic postganglionic fibers in the submucosal plexus. In turn, parasympathetic fibers innervate parietal cells, chief cells, and mucous cells and increase secretions from all the gastric glands. The fibers also innervate smooth muscle of the stomach to promote gastric motility. Emotions such as anger, fear, and anxiety may slow down digestion in the stomach because they stimulate the sympathetic nervous system, which inhibits gastric activity.

Gastric Phase

Once food reaches the stomach, sensory receptors in the stomach initiate both nervous and hormonal mechanisms to ensure that gastric secretion and motility continue. This is the **gastric phase.**
- **Neural Negative Feedback Cycles** Food of any kind causes distention (stretching) and stimulates stretch receptors in the wall of the stomach. Also, chemoreceptors monitor the pH of the stomach chyme. When the stomach walls are distended or when pH increases because foods have entered the stomach and buffered some of the stomach acid, the stretch receptors and chemoreceptors are activated (Fig. 24.14). From the receptors, nerve impulses travel to the submucosal plexus, where they activate parasympathetic fibers. The resulting nerve impulses in the parasympathetic fibers cause waves of peristalsis and stimulate the flow of gastric juice from parietal cells, chief cells, and mucous cells.

FIGURE 24.13 Cephalic, gastric, and intestinal phases of gastric digestion.

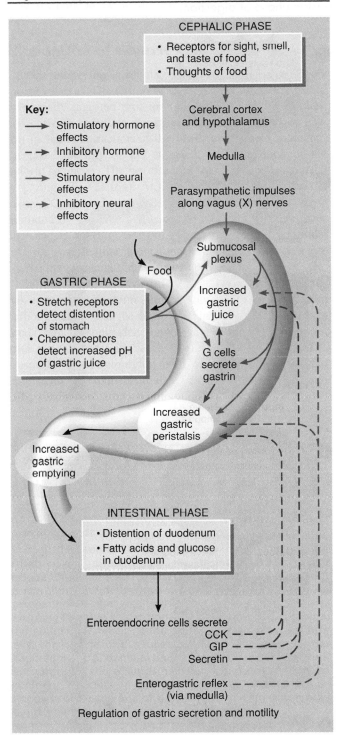

Question: **What are the effects on the stomach of increased stimulation of the vagus (X) nerves?**

The peristaltic waves mix the food with gastric juice, and when they become strong enough, a small quantity of chyme, about 10 to 15 ml (2 to 3 teaspoons), squirts past

FIGURE 24.16 Neural and hormonal factors that regulate gastric emptying.

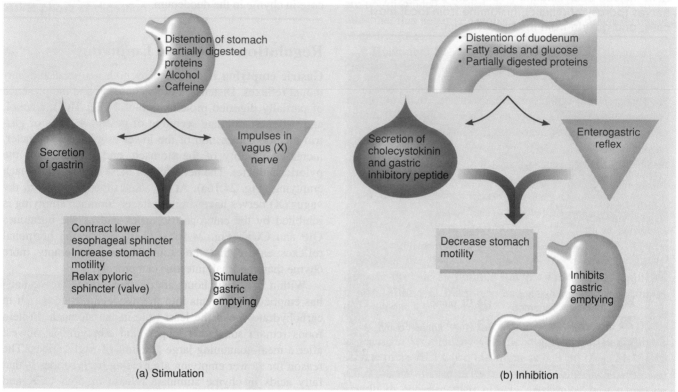

Question: What effect might vagotomy (cutting of vagal nerve fibers) have on emptying of the stomach?

PANCREAS

The next organ of the GI tract involved in the breakdown of food is the small intestine. Chemical digestion in the small intestine depends not only on its own secretions but also on activities of three accessory structures of digestion outside the gastrointestinal tract: the pancreas, liver, and gallbladder. We first consider the activities of the accessory structures and then examine their contributions to digestion in the small intestine.

Anatomy

The **pancreas** is an oblong gland about 12.5 cm (5 in.) long and 2.5 cm (1 in.) thick. It lies posterior to the greater curvature of the stomach and is connected, usually by two ducts, to the duodenum. The pancreas is divided into a head, body, and tail. The **head** is the expanded portion near the C-shaped curve of the duodenum. Located superior and to the left of the head are the central **body** and the tapering **tail** (Fig. 24.17).

Pancreatic secretions pass from the secreting cells in the pancreas into small ducts. Smaller ducts unite to form two larger ducts that convey the secretions into the small intes-

tine. The larger of the two ducts is called the **pancreatic duct (duct of Wirsung)**. In most people, the pancreatic duct joins the common bile duct from the liver and gallbladder and enters the duodenum as a common duct called the **hepatopancreatic ampulla (ampulla of Vater).** The ampulla opens on an elevation of the duodenal mucosa known as the **major duodenal papilla,** about 10 cm (4 in.) below the pyloric sphincter of the stomach. The smaller of the two ducts is the **accessory duct (duct of Santorini).** It leads from the pancreas and empties into the duodenum about 2.5 cm (1 in.) above the hepatopancreatic ampulla.

Histology

The pancreas is made up of small clusters of glandular epithelial cells. About 1% of the cells are organized into clusters called **pancreatic islets (islets of Langerhans).** They form the *endocrine* portion of the pancreas and consist of cells that secrete the hormones glucagon, insulin, somatostatin, and pancreatic polypeptide. The functions of these hormones may be reviewed in Chapter 18. The remaining 99% of the cells are arranged in clusters called **acini** (AS-i-nī) and constitute the *exocrine* portions of the organ (see Fig. 18.23b,c). The acinar cells secrete a mixture of fluid and digestive enzymes called **pancreatic juice.**

FIGURE 24.17 Relation of pancreas to liver, gallbladder, and duodenum. The inset shows details of the common bile duct and pancreatic duct forming the hepatopancreatic ampulla (of Vater) and emptying into the duodenum.

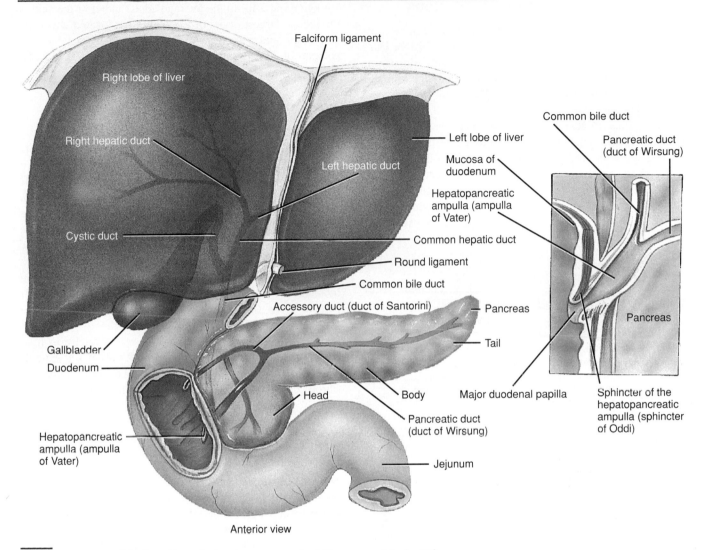

Anterior view

Question: What type of fluid would you find in the pancreatic duct? The common bile duct? The hepatopancreatic ampulla?

Pancreatic Juice

Each day the pancreas produces 1200 to 1500 ml (about 1.2 to 1.5 qt) of pancreatic juice, a clear, colorless liquid. It consists mostly of water, some salts, sodium bicarbonate, and several enzymes. The sodium bicarbonate gives pancreatic juice a slightly alkaline pH (7.1 to 8.2) that buffers acidic gastric juice in chyme, stops the action of pepsin from the stomach, and creates the proper pH for action of digestive enzymes in the small intestine. The enzymes in pancreatic juice include a carbohydrate-digesting enzyme called **pancreatic amylase;** several protein-digesting enzymes called **trypsin** (TRIP-sin), **chymotrypsin** (kī′-mō-TRIP-sin), and **carboxypeptidase** (kar-bok′-sē-PEP-ti-dās);

the principal triglyceride-digesting enzyme in the adult body called **pancreatic lipase;** and nucleic acid-digesting enzymes called **ribonuclease** and **deoxyribonuclease.**

Just as pepsin is produced in the stomach in an inactive form (pepsinogen), so too are the protein-digesting enzymes of the pancreas. This prevents the enzymes from digesting cells of the pancreas itself. The active enzyme trypsin is secreted in an inactive form called **trypsinogen** (trip-SIN-ō-jen). Its activation to trypsin occurs in the small intestine, when chyme comes in contact with an activating enzyme, secreted by the intestinal mucosa, called **enterokinase** (en′-ter-ō-KĪ-nās). Chymotrypsin is activated in the small intestine by trypsin from its inactive form, **chymotrypsinogen.** Carboxypeptidase is also activated in the small intestine by trypsin. Its inactive form is called **pro-carboxypeptidase.**

Regulation of Pancreatic Secretions

Pancreatic secretion, like gastric secretion, is regulated by both nervous and hormonal mechanisms (Fig. 24.18). During the cephalic and gastric phases of gastric digestion, parasympathetic impulses are also transmitted along the vagus (X) nerves to the pancreas. The result is increased secretion of pancreatic enzymes.

In response to acidic chyme that enters the small intestine, enteroendocrine cells in the small intestinal mucosa liberate secretin. And in response to partially digested fats and proteins in the small intestine, other enteroendocrine cells secrete cholecystokinin (CCK). Secretin stimulates the flow of pancreatic juice that is rich in bicarbonate ions. Cholecystokinin (CCK) stimulates a pancreatic secretion rich in digestive enzymes (see Exhibit 24.4 on page 796).

FIGURE 24.18 Neural and hormonal factors that enhance secretion of pancreatic juice. Parasympathetic stimulation (vagus nerves) and acidic chyme in small intestine stimulate secretin release into the blood. Vagal stimulation and fatty acids and amino acids in small intestine stimulate CCK release into the blood.

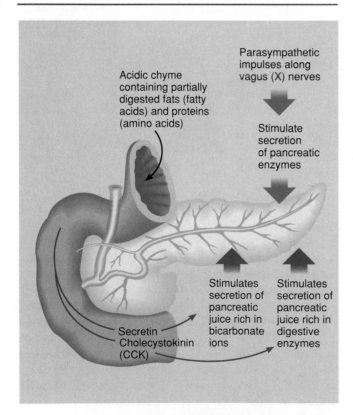

Acidic chyme containing partially digested fats (fatty acids) and proteins (amino acids)

Parasympathetic impulses along vagus (X) nerves

Stimulate secretion of pancreatic enzymes

Stimulates secretion of pancreatic juice rich in bicarbonate ions

Stimulates secretion of pancreatic juice rich in digestive enzymes

Secretin Cholecystokinin (CCK)

Question: How do secretin and cholecystokinin get from the small intestine to the pancreas?

LIVER

The **liver** is the heaviest gland of the body, weighing about 1.4 kg (about 3 lb) in the average adult, and after the skin is the second largest organ of the body. It is located under the diaphragm and occupies most of the right hypochondrium and part of the epigastrium of the abdomen (see Fig. 1.11b).

Anatomy

The liver is almost completely covered by peritoneum and completely covered by a dense irregular connective tissue layer that lies beneath the peritoneum. It is divided into two principal lobes—a large **right lobe** and a smaller **left lobe**—separated by the **falciform ligament** (Fig. 24.19). The right lobe is considered by many anatomists to include an inferior **quadrate lobe** and a posterior **caudate lobe.** However, on the basis of internal morphology, primarily the distribution of blood, the quadrate and caudate lobes more appropriately belong to the left lobe. The falciform ligament is a reflection of the parietal peritoneum. It extends from the undersurface of the diaphragm to the superior surface of the liver, between the two principal lobes of the liver. In the free border of the falciform ligament is the **ligamentum teres (round ligament).** It extends from the liver to the umbilicus. The ligamentum teres is a fibrous cord that is a remnant of the umbilical vein of the fetus (see Fig. 21.32a,b).

Histology

The lobes of the liver are made up of many functional units called **lobules** (Fig. 24.20). A lobule consists of specialized epithelial cells, called **hepatic (liver) cells** or **hepatocytes,** arranged in irregular, branching, interconnected plates around a **central vein.** Rather than capillaries, the liver has larger spaces lined by endothelium called **sinusoids,** through which blood passes. The sinusoids are also partly lined with **stellate reticuloendothelial (Kupffer's) cells.** These phagocytes destroy worn-out white and red blood cells, bacteria, and toxic substances.

Bile, secreted by hepatic cells, enters **bile capillaries** or **canaliculi** (kan′-a-LIK-yoo-lī) that empty into small bile ducts (Fig. 24.20a). These small ducts eventually merge to form the larger **right** and **left hepatic ducts,** which unite and exit the liver as the **common hepatic duct** (see Figs. 24.17 and 24.19a). Further on, the common hepatic duct joins the **cystic duct** from the gallbladder to form the **common bile duct.** Bile enters the cystic duct and is temporarily stored in the gallbladder. After a meal, various stimuli cause contraction of the gallbladder, which releases stored bile into the common bile duct. The common bile duct and pancreatic duct enter the duodenum in a common duct called the **hepatopancreatic ampulla (ampulla of Vater).**

Blood Supply

The liver receives blood from two sources. From the hepatic artery it obtains oxygenated blood, and from the hepatic portal vein it receives deoxygenated blood containing newly absorbed nutrients (see Figs. 21.30 and 24.20). Branches of both the hepatic artery and the hepatic portal vein carry blood into liver sinusoids, where oxygen, most of the nutrients, and certain poisons are extracted by the hepatic cells. The reticuloendothelial (Kupffer's) cells lining the sinusoids phagocytize microbes and bits of foreign matter from the blood. Nutrients are stored or used to make new materials. Poisons are stored or detoxified. Products manufactured by the hepatic cells and nutrients needed by

FIGURE 24.19 External anatomy of the liver. The anterior view is illustrated in Fig. 24.17.

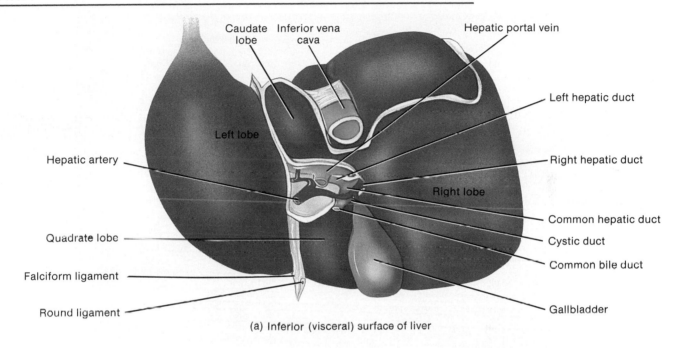

(a) Inferior (visceral) surface of liver

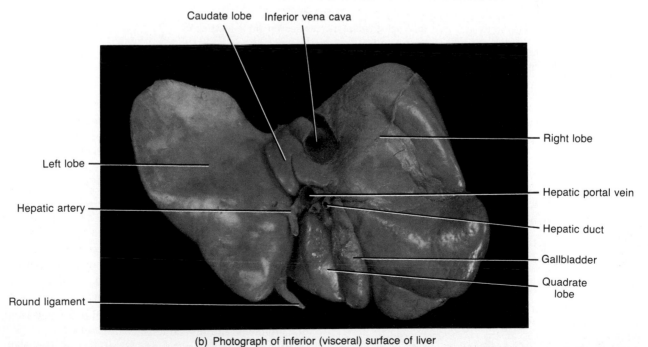

(b) Photograph of inferior (visceral) surface of liver

Figure continues

FIGURE 24.19 (continued)

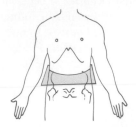

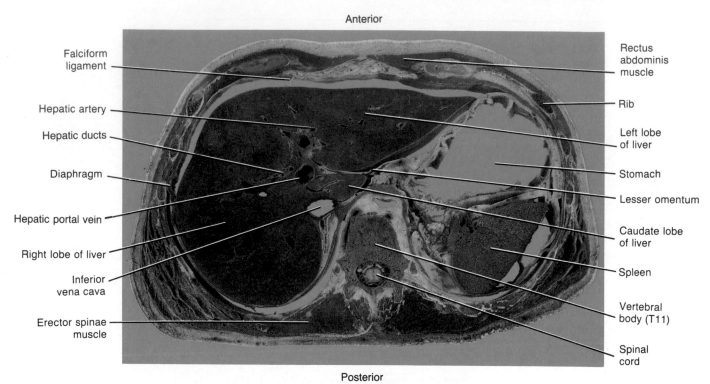

Falciform ligament

Hepatic artery

Hepatic ducts

Diaphragm

Hepatic portal vein

Right lobe of liver

Inferior vena cava

Erector spinae muscle

Anterior

Rectus abdominis muscle

Rib

Left lobe of liver

Stomach

Lesser omentum

Caudate lobe of liver

Spleen

Vertebral body (T11)

Spinal cord

Posterior

(c) Photograph of a cross section of the abdomen

Question: Within which abdominopelvic region (see Fig. 1.11a) could you palpate (feel) most of the liver to tell if it were enlarged?

other cells are secreted back into the blood. The blood then drains into the central vein and eventually passes into a hepatic vein.

Branches of the hepatic portal vein, hepatic artery, and bile duct typically accompany each other in their distribution through the liver. Collectively, these three structures are called a **portal triad** (Fig. 24.20b).

CLINICAL APPLICATION

LIVING-DONOR LIVER TRANSPLANT

In November 1989, surgeons at the University of Chicago Medical Center performed the first **living-donor liver transplant** in the U.S. In the 14-hour procedure, a 21-month-old child received part of the left lobe of her mother's liver. The child suffered from biliary atresia, the

closure or absence of some or all of the major bile ducts. Since the liver is capable of regeneration, the mother's liver returned to its normal size in about two months and the child's liver is expected to grow as she does. Living-donor transplants have also been done for the pancreas, kidneys, and bone marrow.

Bile

Each day, the hepatic cells secrete 800 to 1000 ml (about 1 qt) of **bile,** a yellow, brownish, or olive-green liquid. It has a pH of 7.6 to 8.6. Bile consists mostly of water and bile acids, bile salts, cholesterol, a phospholipid called lecithin, bile pigments, and several ions.

Bile is partially an excretory product and partially a digestive secretion. Bile salts, which are sodium and potas-

FIGURE 24.20 Histology of the liver.

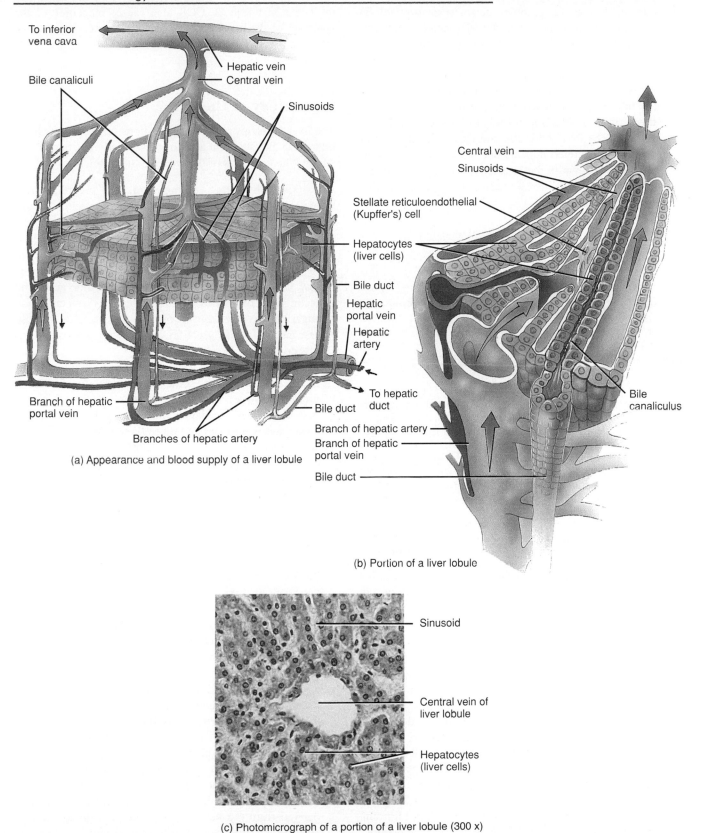

To inferior
vena cava

Hepatic vein
Central vein

Bile canaliculi

Sinusoids

Hepatocytes
(liver cells)

Bile duct

Hepatic
portal vein

Hepatic
artery

To hepatic
duct

Branch of hepatic
portal vein

Bile duct

Branches of hepatic artery

(a) Appearance and blood supply of a liver lobule

Central vein

Sinusoids

Stellate reticuloendothelial
(Kupffer's) cell

Hepatocytes
(liver cells)

Bile
canaliculus

Branch of hepatic artery

Branch of hepatic
portal vein

Bile duct

(b) Portion of a liver lobule

Sinusoid

Central vein of
liver lobule

Hepatocytes
(liver cells)

(c) Photomicrograph of a portion of a liver lobule (300 x)

Question: From the hepatic portal vein, what is the path of blood flow through the liver?

sium salts of bile acids (mostly cholic acid and chenodeoxycholic acid), play a role in **emulsification,** the breakdown of large fat globules into a suspension of fat droplets about 1 μm in diameter, and absorption of fats following their digestion. The tiny fat droplets present a very large surface area for the action of the pancreatic lipase necessary for rapid fat digestion. Cholesterol is made soluble in bile by bile salts and lecithin. The principal bile pigment is **bilirubin.** When worn-out red blood cells are broken down, iron, globin, and bilirubin (derived from heme) are released. The iron and globin are recycled, but some of the bilirubin is excreted into the bile ducts. Bilirubin eventually is broken down in the intestine, and one of its breakdown products (urobilinogen) gives feces their normal brown color.

CLINICAL APPLICATION

JAUNDICE

Jaundice is a yellowish coloration of the sclerae of the eyes, skin, and mucous membranes due to a buildup of bilirubin (which is yellow in color) in the body. After bilirubin is formed from the breakdown of the heme pigment in aged red blood cells, it is transported to the liver where it is processed and eventually excreted into bile. The three main categories of jaundice are:

1. *Prehepatic (hemolytic) jaundice* is due to excess production of bilirubin. In hemolytic anemia, for example, bilirubin may be produced so rapidly and in such quantities that the liver cannot excrete it fast enough to avoid jaundice. The liver of a newborn functions poorly for the first week or so, and many babies experience a mild form of jaundice called *neonatal (physiological) jaundice* that disappears as the liver matures. It is usually treated by exposing the infant to blue light, which converts bilirubin into substances the kidneys can excrete.
2. *Hepatic (medical) jaundice* is due to dysfunction of liver cells owing to congenital liver diseases, cirrhosis of the liver, or hepatitis.
3. *Extrahepatic (surgical) jaundice* is also known as *obstructive jaundice* because it involves blockage of bile drainage. Causes include gallstones and cancer of the bowel or the head of the pancreas. Surgery must usually be performed to relieve the obstruction.

Regulation of Bile Secretion

Both nervous and hormonal factors regulate bile secretion. Vagal stimulation of the liver can increase the production of bile to more than twice the baseline rate. Secretin, the hormone that stimulates the flow of pancreatic juice rich in bicarbonate ions (HCO_3^-), also stimulates hepatic cells to secrete bile rich in HCO_3^- (Fig. 24.21). Within limits, as blood flow through the liver increases, so does the secretion of bile. Finally, the presence of large amounts of bile salts in the blood also increases the rate of bile production. Cholecystokinin (CCK), the hormone that stimulates the flow of pancreatic juice rich in digestive enzymes, also stimulates contraction of the walls of the gallbladder (Fig. 24.21). This squeezes stored bile out of the gallbladder into the common bile duct. CCK also causes relaxation of the sphincter of the hepatopancreatic ampulla (see Fig. 24.17), allowing bile to flow into the duodenum.

Physiology of the Liver

The liver performs many vital functions, many of which are related to metabolism and are discussed in Chapter 25. Among the functions of the liver are the following:

1. Carbohydrate metabolism. In carbohydrate metabolism, the liver is especially important in maintaining a normal blood glucose level. For example, the liver can convert glucose to glycogen (glycogenesis) when blood sugar level is high and break down glycogen to glucose (glycogenolysis) when blood sugar level is low. The liver can also convert certain amino acids and lactic acid to glucose (gluconeogenesis) when blood sugar level is low; convert other sugars, such as fructose and galactose into glucose; and convert glucose to triglycerides.
2. Lipid metabolism. With respect to lipid metabolism, the liver stores some triglycerides (neutral fats), breaks down fatty acids into acetyl coenzyme A, a process called beta oxidation, and converts excess acetyl coenzyme A into ketone bodies (ketogenesis). It synthesizes lipoproteins, which transport fatty acids, triglycerides, and cholesterol to and from body cells. Hepatic cells synthesize cholesterol and use cholesterol to make bile salts.
3. Protein metabolism. Without the role of the liver in protein metabolism, death would occur in a few days. The liver deaminates (removes the amino group, NH_2, from) amino acids so that they can be used for ATP production or converted to carbohydrates or fats. It converts the resulting toxic ammonia (NH_3) into the much less toxic urea for excretion in urine. (Ammonia also is produced by bacteria in the GI tract.) Hepatic cells synthesize most plasma proteins, such as alpha and beta globulins, albumin, prothrombin, and fibrinogen. Finally, liver enzymes can perform transamination, the transfer of an amino group from an amino acid to another substance to convert one amino acid into another.
4. Removal of drugs and hormones. The liver can detoxify or excrete into bile drugs such as penicillin, erythromycin, and sulfonamides. It can also chemically alter or excrete thyroid hormones and steroid hormones, such as estrogens and aldosterone.
5. Excretion of bile. As noted earlier, bilirubin, derived from the heme of worn-out red blood cells, is absorbed by the liver from the blood and secreted into bile. Most of the bilirubin in bile is metabolized in the intestine by bacteria and eliminated in feces.

FIGURE 24.21 Neural and hormonal stimuli that promote production and release of bile.

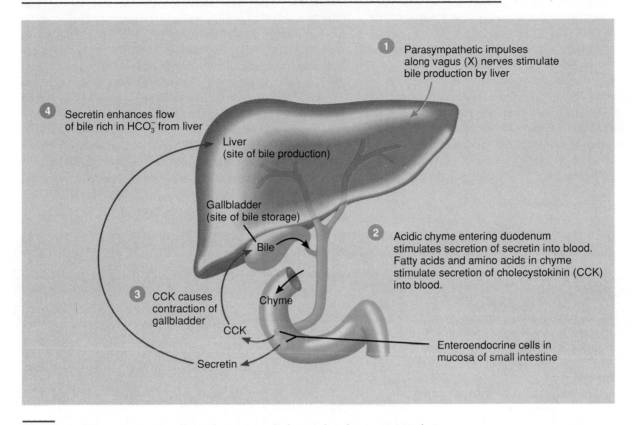

1. Parasympathetic impulses along vagus (X) nerves stimulate bile production by liver

4. Secretin enhances flow of bile rich in HCO$_3^-$ from liver

Liver (site of bile production)

Gallbladder (site of bile storage)

Bile

2. Acidic chyme entering duodenum stimulates secretion of secretin into blood. Fatty acids and amino acids in chyme stimulate secretion of cholecystokinin (CCK) into blood.

3. CCK causes contraction of gallbladder

Chyme

CCK

Enteroendocrine cells in mucosa of small intestine

Secretin

Question: In what respect are the effects of secretin on the liver and on the pancreas similar?

6.   Synthesis of bile salts. Bile salts are used in the small intestine for the emulsification and absorption of fats, cholesterol, phospholipids, and lipoproteins.

7.   Storage. In addition to glycogen, the liver stores vitamins (A, B$_{12}$, D, E, and K) and minerals (iron and copper). Liver cells contain a protein called **apoferritin,** which combines with iron to form **ferritin,** the form in which iron is stored in the liver (see Fig. 19.4). The iron is released from the liver when needed elsewhere in the body.

8.   Phagocytosis. The stellate reticuloendothelial (Kupffer's) cells of the liver phagocytize worn-out red and white blood cells and some bacteria.

9.   Activation of vitamin D. The skin, liver, and kidneys participate in the activation of vitamin D (see page 127).

GALLBLADDER

The **gallbladder** is a pear-shaped sac about 7 to 10 cm (3 to 4 in.) long. It is located in a fossa of the visceral surface of the liver (see Figs. 24.17 and 24.19a).

Histology

The mucosa of the gallbladder consists of simple columnar epithelium arranged in rugae resembling those of the stomach. The gallbladder lacks a submucosa. The middle, muscular coat of the wall consists of smooth muscle fibers. Contraction of these fibers by hormonal stimulation ejects the contents of the gallbladder into the **cystic duct.** The outer coat is the visceral peritoneum.

Physiology

The functions of the gallbladder are to store and concentrate bile (up to 10-fold) until it is needed in the small intestine. In the concentration process, water and ions are absorbed by the mucosa of the gallbladder. When bile is needed in the small intestine, the smooth muscle in the wall of the gallbladder contracts and forces bile into the cystic duct, through the common bile duct and into the small intestine. When the small intestine is empty, a valve around the hepatopancreatic ampulla (ampulla of Vater) called the **sphincter of the hepatopancreatic ampulla (sphincter of Oddi)** closes, and the backed-up bile flows into the cystic duct to the gallbladder for storage (see Fig. 24.17).

SUMMARY: DIGESTIVE HORMONES

The effects of the four major digestive hormones—gastrin, secretin, CCK, and GIP—and the stimuli that promote release of each are summarized in Exhibit 24.4. All four are secreted into the blood by enteroendocrine cells located in the GI tract mucosa. Gastrin comes from cells in the stomach and the other three from cells in the small intestine. Gastrin and gastric inhibitory peptide both exert their major effects on the stomach, whereas secretin and cholecystokinin affect the pancreas, liver, and gallbladder most strongly.

Stretching of the stomach as it receives food and buffer-ing of gastric acids by proteins in food are stimuli that trig-ger release of gastrin. Gastrin, in turn, promotes secretion of gastric juice and increases gastric motility so that ingested food becomes well-mixed into a thick, soupy chyme. Reflux of the acidic chyme into the esophagus is prevented by tonic contraction of the lower esophageal sphincter, which is enhanced by gastrin.

The presence of chyme containing fatty acids (digested from triglycerides) and glucose in the first part of the small intestine triggers release of GIP. In turn, GIP inhibits secre-tion of gastric juice and slows emptying of the stomach. GIP also stimulates the pancreas to secrete insulin, as if to "get ready" for the glucose that is about to be absorbed.

The major stimulus for secretion of secretin is acidic

EXHIBIT 24.4

MAJOR HORMONES THAT CONTROL DIGESTIVE PROCESSES

Hormone	Where Produced	Stimulus for Secretion	Actions
Gastrin	Mainly in mucosa of pyloric antrum; small quantity from intestinal mucosa.	Distention of stomach, partially digested proteins and caffeine in stomach, and high pH of stomach chyme.	*Major effects:* Promotes secretion of gastric juice, increases gastric motility, promotes growth of gastric mucosa. *Minor effects:* Constricts lower esophageal sphincter; relaxes pyloric sphincter and ileocecal sphincter.
Gastric inhibitory peptide (GIP)	Intestinal mucosa.	Fatty acids and glucose that enter the small intestine.	Stimulates release of insulin by pancreatic beta cells, inhibits secretion of gastric juice, and slows gastric emptying.
Secretin	Intestinal mucosa.	Acidic (high H^+ level) chyme entering the small intestine.	*Major effects:* Stimulates secretion of pancreatic juice and bile that are rich in HCO_3^- (bicarbonate ions). *Minor effects:* Inhibits secretion of gastric juice, promotes normal growth and maintenance of the pancreas, enhances effects of CCK.
Cholecystokinin (CCK)	Intestinal mucosa, brain.	Partially digested proteins (amino acids) and triglycerides (fatty acids) that enter the small intestine.	*Major effects:* Stimulates secretion of pancreatic juice rich in digestive enzymes; causes ejection of bile from the gallbladder and opening of the sphincter of the hepatopancreatic ampulla (sphincter of Oddi); and induces satiety (feeling full to satisfaction). *Minor effects:* Inhibits gastric emptying, promotes normal growth and maintenance of the pancreas, enhances effects of secretin.

chyme (high H⁺) entering the small intestine. In turn, secretin promotes secretion of pancreatic juice and bile that have a high content of HCO_3^- (bicarbonate ions), which act to buffer or soak up excess H⁺ (see Figs. 24.18 and 24.21). Besides these major effects, secretin inhibits secretion of gastric juice and promotes normal growth and maintenance of the pancreas. It also acts synergistically with cholecystokinin; that is, it enhances the effects of CCK. Overall, secretin causes buffering of acid in chyme that reaches the duodenum and slows production of acid in the stomach.

Amino acids from partially digested proteins and fatty acids from partially digested triglycerides in chyme that enters the duodenum stimulate secretion of cholecystokinin by enteroendocrine cells in the mucosa of the small intestine. CCK stimulates secretion of pancreatic juice that is rich in digestive enzymes (see Fig. 24.18) and ejection of bile into the duodenum (see Fig. 24.21). It also slows gastric emptying by promoting contraction of the pyloric sphincter and produces satiety (feeling full to satisfaction) by acting on the hypothalamus in the brain. Like secretin, CCK promotes normal growth and maintenance of the pancreas. It also enhances the effects of secretin.

In addition to the "big four," there are at least 10 other so-called gut hormones, that is, hormones secreted by and having effects on the GI tract. A few of these are: motilin, substance P, and bombesin, which stimulate motility of the intestines; vasoactive intestinal polypeptide (VIP), which stimulates secretion of ions and water by the intestines and inhibits gastric acid secretion; gastrin-releasing peptide, which stimulates release of gastrin; and somatostatin, which inhibits gastrin release. Some are thought to act locally as paracrine substances, while others are secreted into the blood or even into the lumen of the GI tract. The physiological roles of these and other gut hormones are still under investigation.

SMALL INTESTINE

The major events of digestion and absorption occur in a long tube called the **small intestine.** The small intestine begins at the pyloric sphincter of the stomach, coils through the central and lower part of the abdominal cavity, and eventually opens into the large intestine. It averages 2.5 cm (1 in.) in diameter and about 6.4 m (21 ft) in length in a cadaver.

Anatomy

The small intestine is divided into three segments (see Fig. 24.1). The **duodenum** (doo'-ō-DĒ-num), the shortest part, starts at the pyloric sphincter of the stomach and extends about 25 cm (10 in.) until it merges with the jejunum. *Duodenum* means "12"; the structure is 12 fingers' breadth in

length. The **jejunum** (jē-JOO-num) is about 2.5 m (8 ft) long and extends to the ileum. *Jejunum* means "empty," since at death it is found empty. The final portion of the small intestine, the **ileum** (IL-ē-um; *eileos* = twisted) measures about 3.6 m (12 ft) and joins the large intestine at the **ileocecal** (il'-ē-ō-SĒ-kal) **sphincter (valve).**

Histology

Since almost all the digestion and absorption of nutrients occur in the small intestine, its structure is specially adapted for this function. Its length alone provides a large surface area for digestion and absorption, and that area is further increased by modifications in the structure of its wall. The wall of the small intestine is composed of the same four coats that make up most of the GI tract. However, both the mucosa and the submucosa are modified to allow the small intestine to complete the processes of digestion and absorption (Fig. 24.22).

The **mucosa** contains many cavities lined with glandular epithelium. Cells lining the cavities form the **intestinal glands (crypts of Lieberkühn)** and secrete intestinal juice. The submucosa of the duodenum contains **duodenal (Brunner's) glands** (see Fig. 24.23b). They secrete an alkaline mucus that helps neutralize gastric acid in the chyme. Some of the epithelial cells in the mucosa are goblet cells, which secrete additional mucus.

The epithelium of the mucosa consists of simple columnar epithelium and contains absorptive cells, goblet cells, enteroendocrine cells, and Paneth cells. The apical (free) membrane of the absorptive cells features **microvilli** (mī'-krō-VIL-ī), fingerlike projections of the plasma membrane. In a photomicrograph taken through a light microscope, the microvilli are too small to be seen individually. They form a fuzzy line, called the **brush border,** at the apical surface of the absorptive cells, next to the lumen of the small intestine (see Fig. 24.23b,c). Larger amounts of digested nutrients can diffuse into the absorptive cells of the intestinal wall because the microvilli increase the surface area of the plasma membrane (see Fig. 24.25). The brush border also contains several digestive enzymes. **Paneth cells** are found in the deepest parts of the intestinal glands. They secrete lysozyme, a bactericidal enzyme, and are also capable of phagocytosis. They may have a role in regulating the microbial population in the intestines.

The mucosa forms a series of **villi (villus** is singular). These projections are 0.5 to 1 mm high and give the intestinal mucosa a velvety appearance. The large number of villi (10 to 40 per square millimeter) vastly increases the surface area of the epithelium available for absorption and digestion. Each villus has a core of lamina propria, the connective tissue layer of the mucosa. Embedded in this connective tissue are an arteriole, a venule, a capillary network, and a **lacteal** (LAK-tē-al), which is a lymphatic vessel. Nutrients that pass through the epithelial cells covering the

FIGURE 24.22 Small intestine. Shown are various structures that adapt the small intestine for digestion and absorption.

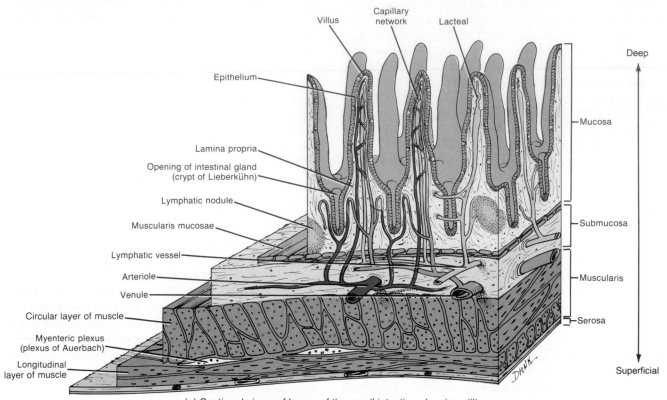

(a) Sectional views of layers of the small intestine showing villi

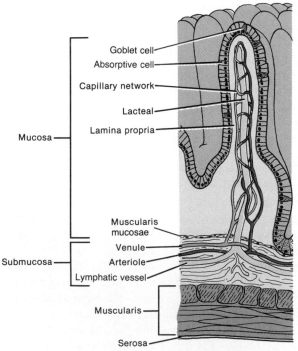

(b) Enlarged villus showing lacteal and capillaries

(c) Jejunum cut open to expose the circular folds

Question: What are four structural features that enhance the effectiveness of the small intestine for absorption and digestion?

villus are able to pass through the capillary walls and the lacteal and enter blood and lymphatic fluid, respectively.

In addition to the microvilli and villi, a third set of projections called **circular folds,** or **plicae circulares** (PLĪ-kē SER-kyoo-lar-es), further increases the surface area for absorption and digestion. The folds are permanent ridges, about 10 mm (0.4 in.) high, in the mucosa. Some extend all the way around the circumference of the intestine, and others extend only part of the way around. The folds begin near the proximal portion of the duodenum and end at about the midportion of the ileum. The circular folds enhance absorption by causing the chyme to spiral, rather than to move in a straight line, as it passes through the small intestine. Since the folds and villi decrease in size in the distal ileum, most absorption occurs in the duodenum and jejunum.

The **muscularis** of the small intestine consists of two layers of smooth muscle. The outer, thinner layer contains longitudinally arranged fibers. The inner, thicker layer contains circularly arranged fibers. Except for a major portion of the duodenum, the serosa (or visceral peritoneum) completely surrounds the small intestine. Additional histological aspects of the small intestine are shown in Fig. 24.23.

There is an abundance of lymphatic tissue in the form of lymphatic nodules, masses of lymphatic tissue not surrounded by a capsule. **Solitary lymphatic nodules** are most numerous in the lower part of the ileum. Groups of lym-

phatic nodules, referred to as **aggregated lymphatic follicles (Peyer's patches),** are numerous in the ileum. Their purpose is to prevent bacteria from entering the bloodstream.

Intestinal Juice and Brush Border Enzymes

Intestinal juice is a clear yellow fluid secreted in amounts of 1 to 2 liters (about 1 to 2 qt) a day. It has a pH of 7.6, which is slightly alkaline, and contains water and mucus. Together, pancreatic and intestinal juice provide a vehicle for the absorption of substances from chyme as they come in contact with the villi. The absorptive epithelial cells that line the villi synthesize several digestive enzymes, called **brush border enzymes,** and insert them in the plasma membrane of the microvilli. Thus some digestion by enzymes of the small intestine occurs at the surface of the epithelial cells that line the villi, rather than in the lumen exclusively, as in other parts of the GI tract. Among the brush border enzymes are four carbohydrate-digesting enzymes called α-**dextrinase, maltase, sucrase,** and **lactase;** protein-digesting enzymes called **peptidases (aminopeptidase** and **dipeptidase);** and two types of nucleotide-digesting enzymes, **nucleosidases** and **phosphatases.** Also, as small intestinal cells slough off into the

---

FIGURE 24.23 Histology of the small intestine.

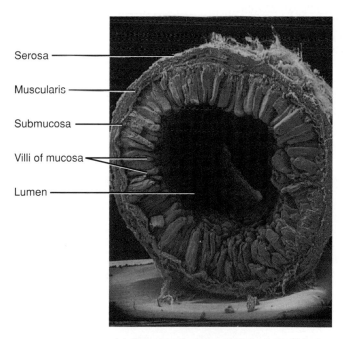

Serosa

Muscularis

Submucosa

Villi of mucosa

Lumen

(a) Scanning electron micrograph of villi in the small intestine (about 15x)

Figure continues

FIGURE 24.23 (continued)

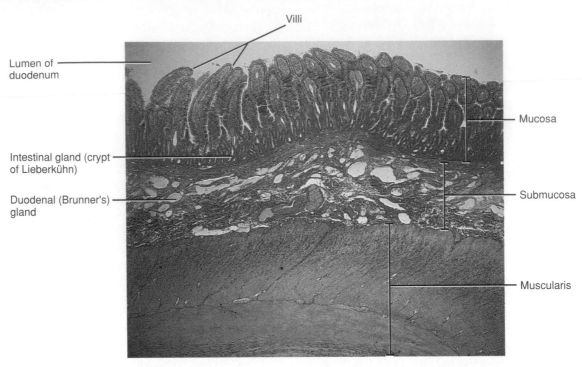

(b) Photomicrograph of a portion of the wall of the duodenum (160x)

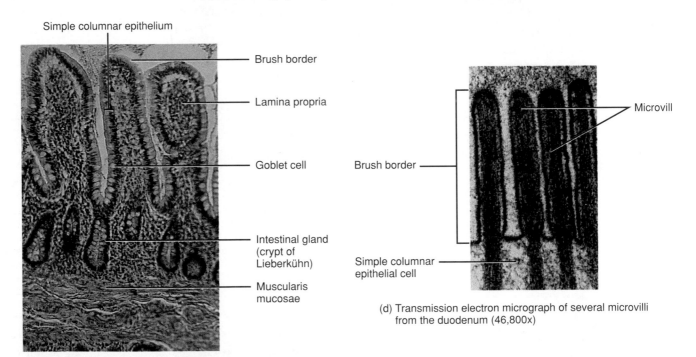

(c) Photomicrograph of three villi from the small intestine (130x)

(d) Transmission electron micrograph of several microvilli from the duodenum (46,800x)

Question: What is the function of the fluid secreted by duodenal (Brunner's) glands?

lumen of the intestine, they break apart and release enzymes that digest food in the chyme.

Physiology of Digestion in the Small Intestine

Mechanical Digestion

The movements of the small intestine are divided into two types: segmentation and peristalsis. **Segmentation** is the major movement of the small intestine (Fig. 24.24). It is strictly a localized contraction in areas containing food. It mixes chyme with the digestive juices and brings the particles of food into contact with the mucosa for absorption. It does not push the intestinal contents along the tract. Segmentation starts with the contractions of circular muscle fibers in a portion of the small intestine, an action that constricts the intestine into segments. Next, muscle fibers that encircle the middle of each segment also contract, dividing each segment again. Finally, the fibers that contracted first relax, and each small segment unites with an adjoining small segment so that large segments are formed. This sequence of events is repeated 12 to 16 times a minute, sloshing the chyme back and forth.

Peristalsis propels the chyme onward through the intestinal tract. Peristaltic contractions in the small intestine are normally very weak compared with those in the esophagus or stomach, and chyme remains in the small intestine for three to five hours. Peristalsis, like segmentation, is controlled by the autonomic nervous system.

Chemical Digestion

In the mouth, salivary amylase converts starch (polysaccharide) to maltose (a disaccharide), maltotriose (a trisaccharide), and α-dextrins (short-chain, branched fragments of

---

FIGURE 24.24 Segmentation. Localized segmentation contractions thoroughly mix the contents of the small intestine.

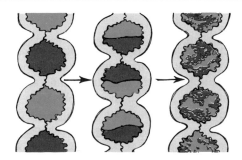

Question: How is segmentation different from peristalsis?

---

starch with five to ten glucose units). In the stomach, pepsin converts proteins to peptides (small fragments of proteins), and lingual lipase converts some triglycerides into fatty acids and monoglycerides. Thus chyme entering the small intestine contains partially digested carbohydrates, proteins, and lipids. The completion of the digestion of carbohydrates, proteins, and lipids is a collective effort of pancreatic juice, bile, and intestinal juice in the small intestine.

● **Carbohydrate Digestion**   Even though the action of **salivary amylase** may continue in the stomach for awhile, the acidic pH of the stomach destroys salivary amylase and blocks its activity. Thus few starches are reduced to maltose by the time chyme leaves the stomach. Those starches not already broken down into maltose, maltotriose, and α-dextrins are cleaved by **pancreatic amylase,** an enzyme in pancreatic juice that acts in the small intestine. Although amylase acts on both glycogen and starches, it does not act on another polysaccharide called cellulose, an indigestible plant fiber. After amylase (either salivary or pancreatic) has split starch into smaller fragments, a brush border enzyme called **α-dextrinase** acts on the resulting α-dextrins, clipping off one glucose unit at a time.

Sucrose and lactose, two disaccharides, are ingested as such and are not acted on until they reach the small intestine. Three brush border enzymes digest the disaccharides into monosaccharides. **Maltase** splits maltose and maltotriose into two or three molecules of glucose, respectively. **Sucrase** breaks sucrose into a molecule of glucose and a molecule of fructose. **Lactase** digests lactose into a molecule of glucose and a molecule of galactose. This completes the digestion of carbohydrates since these monosaccharides are small enough to be absorbed.

CLINICAL APPLICATION

LACTOSE INTOLERANCE

In some individuals the mucosal cells of the small intestine fail to produce lactase, which is essential for the digestion of lactose. This results in a condition called **lactose intolerance.** Undigested lactose retains fluid, and bacterial fermentation of lactose results in the production of gases. Symptoms of lactose intolerance include diarrhea, gas, bloating, and abdominal cramps after consumption of milk and other dairy products. The severity of symptoms varies, from relatively minor to sufficiently serious to require medical attention.

● **Protein Digestion**   Protein digestion starts in the stomach, where proteins are fragmented by the action of **pepsin** into peptides. Enzymes found in pancreatic juice (trypsin, chymotrypsin, and carboxypeptidase) continue the digestion. **Trypsin** and **chymotrypsin** continue to break down proteins into peptides. Although pepsin, trypsin, and chymotrypsin all convert whole proteins into peptides, their actions differ somewhat since each splits peptide bonds between different amino acids. **Carboxypeptidase** acts on peptides and breaks the peptide bond that attaches the

FIGURE 24.26 Daily volumes of fluid ingested, secreted, absorbed, and excreted from the GI tract.

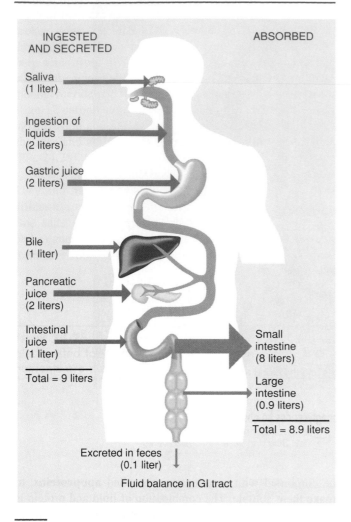

INGESTED AND SECRETED

ABSORBED

Saliva (1 liter)

Ingestion of liquids (2 liters)

Gastric juice (2 liters)

Bile (1 liter)

Pancreatic juice (2 liters)

Intestinal juice (1 liter)

Total = 9 liters

Small intestine (8 liters)

Large intestine (0.9 liters)

Total = 8.9 liters

Excreted in feces (0.1 liter)

Fluid balance in GI tract

Question: Which two organs of the GI tract secrete the most fluid?

cells and into the blood capillaries in the villi. Water can move across the intestinal mucosa in both directions. The absorption of water from the small intestine depends on the absorption of electrolytes and nutrients to maintain an osmotic balance with the blood. The absorbed electrolytes, monosaccharides, and amino acids establish a concentration gradient for water that promotes water absorption by osmosis.

Electrolyte Absorption

Many of the electrolytes absorbed by the small intestine come from gastrointestinal secretions and some are part of ingested foods and liquids. Sodium ions are actively transported out of intestinal epithelial cells by sodium (Na^+/K^+ ATP-ase) pumps after they have moved into epithelial cells

by diffusion and secondary active transport. Thus most of the Na^+ in gastrointestinal secretions is reclaimed and not lost in the feces. Chloride, iodide, and nitrate ions can passively follow sodium ions or be actively transported. Calcium ions are also actively absorbed, and their movement depends on parathyroid hormone (PTH) and vitamin D. Other electrolytes such as iron, potassium, magnesium, and phosphate also are absorbed by active transport.

Vitamin Absorption

Fat-soluble vitamins (A, D, E, and K) are included along with ingested dietary triglycerides in micelles. In fact, they cannot be absorbed unless they are ingested with some triglycerides. Most water-soluble vitamins, such as the B vitamins and C, are absorbed by diffusion. Vitamin B_{12}, however, combines with intrinsic factor produced by the stomach and the combination is absorbed by active transport.

A summary of the digestive and absorptive activities of the small intestine and associated accessory structures is presented in Exhibit 24.6.

LARGE INTESTINE

The overall functions of the large intestine are the completion of absorption, the manufacture of certain vitamins, the formation of feces, and the expulsion of feces from the body.

Anatomy

The **large intestine** is about 1.5 m (5 ft) long and 6.5 cm (2.5 in.) in diameter. It extends from the ileum to the anus and is attached to the posterior abdominal wall by its **mesocolon** of visceral peritoneum. Structurally, the large intestine is divided into four principal regions: cecum, colon, rectum, and anal canal (Fig. 24.27a).

The opening from the ileum into the large intestine is guarded by a fold of mucous membrane called the **ileocecal sphincter (valve).** This structure allows materials from the small intestine to pass into the large intestine. Hanging below the ileocecal valve is the **cecum,** a blind pouch about 6 cm (2.5 in.) long. Attached to the cecum is a twisted, coiled tube, measuring about 8 cm (3 in.) in length, called the **vermiform appendix** (*vermis* = worm; *appendix* = appendage). The mesentery of the appendix, called the **mesoappendix,** attaches the appendix to the inferior part of the ileum and adjacent part of the posterior abdominal wall.

The open end of the cecum merges with a long tube called the **colon** (*kolon* = food passage). The colon is divided into ascending, transverse, descending, and sigmoid portions. The **ascending colon,** which is retroperitoneal, ascends on the right side of the abdomen, reaches the

---

EXHIBIT 24.6

SUMMARY OF DIGESTION AND ABSORPTION IN THE SMALL INTESTINE AND SECRETIONS PROVIDED BY PANCREAS, LIVER, AND GALL BLADDER

Structure	Activity
PANCREAS	Delivers pancreatic juice into the duodenum via the pancreatic duct (see Exhibit 24.5 for pancreatic enzymes and their functions).
LIVER	Produces bile (bile salts), necessary for emulsification and absorption of fats.
GALLBLADDER	Stores, concentrates, and delivers bile into the duodenum via the common bile duct.
SMALL INTESTINE	Major site of digestion and absorption of nutrients and water in the gastrointestinal tract.
Mucosa and submucosa	
Intestinal glands	Secrete intestinal juice (see Exhibit 24.5 for intestinal enzymes and their functions).
Duodenal (Brunner's) glands	Secrete alkaline fluid to buffer stomach acids and mucus for protection and lubrication.
Microvilli	Fingerlike projections of epithelial cells that contain brush border enzymes and increase surface area for absorption and digestion.
Villi	Projections of mucosa that are the sites of absorption of digested food and also increase the surface area for absorption and digestion.
Circular folds	Folds of mucosa and submucosa that increase surface area for absorption and digestion.
Muscularis	
Segmentation	Consists of alternating contractions of circular fibers that produce segmentation and resegmentation of portions of the small intestine; mixes chyme with digestive juices and brings food into contact with the mucosa for absorption.
Peristalsis	Consists of mild waves of contraction and relaxation of circular and longitudinal muscle passing the length of the small intestine; moves chyme toward ileocecal sphincter.

---

undersurface of the liver, and turns abruptly to the left. Here it forms the **right colic (hepatic) flexure.** The colon continues across the abdomen to the left side as the **transverse colon.** It is not retroperitoneal and curves beneath the lower end of the spleen on the left side as the **left colic (splenic) flexure** and passes downward to the level of the iliac crest as the **descending colon,** also a retroperitoneal structure. The **sigmoid colon** begins near the left iliac crest, projects inward to the midline, and terminates as the rectum at about the level of the third sacral vertebra.

The **rectum,** the last 20 cm (8 in.) of the GI tract, lies anterior to the sacrum and coccyx. The terminal 2 to 3 cm (1 in.) of the GI tract is called the **anal canal** (Fig. 24.27b). The mucous membrane of the anal canal is arranged in longitudinal folds called **anal columns** that contain a network of arteries and veins. The opening of the anal canal to the exterior is called the **anus.** It is guarded by an internal sphincter of smooth muscle (involuntary) and an external sphincter of skeletal muscle (voluntary). Normally the anus is closed except during the elimination of the wastes of digestion.

The medical specialty that deals with the diagnosis and treatment of disorders of the rectum and anus is called **proctology** (prok-TOL-ō-jē; *proct* = rectum; *logos* = study of).

CLINICAL APPLICATION

HEMORRHOIDS

Varicosities in veins are regions that are enlarged and inflamed. In the rectal veins, varicosities are known as **hemorrhoids (piles).** Hemorrhoids develop when the veins are put under pressure and become engorged with blood. If the pressure continues, the wall of the vein stretches. Such a distended vessel oozes blood, and bleeding or itching are usually the first signs that a hemorrhoid has developed. Stretching of a vein also favors clot formation, further aggravating swelling and pain. Hemorrhoids may be caused by constipation, which may be brought on by low-fiber diets. Also, repeated straining during defecation forces blood down into the rectal veins, increasing pressure in these veins and possibly causing hemorrhoids.

Histology

The wall of the large intestine differs from that of the small intestine in several respects. No villi or permanent circular folds are found in the mucosa, which does, however, contain simple columnar epithelium with numerous goblet cells (Fig. 24.28). The columnar cells function primarily in water absorption. The goblet cells secrete mucus that lubricates the colonic contents as they pass through. Both columnar and mucous cells are located in long, straight, tubular

FIGURE 24.27 Anatomy of large intestine.

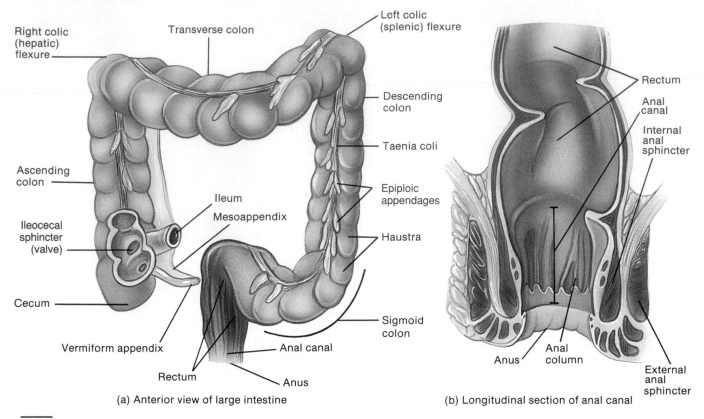

(a) Anterior view of large intestine

(b) Longitudinal section of anal canal

Question: Which portions of the large intestine are not retroperitoneal?

intestinal glands that extend the full thickness of the mucosa. Solitary lymphatic nodules are also found in the mucosa. The submucosa of the large intestine is similar to that found in the rest of the GI tract. The muscularis consists of an external layer of longitudinal muscles and an internal layer of circular muscles. Unlike other parts of the GI tract, portions of the longitudinal muscles are thickened, forming three conspicuous longitudinal bands called **taeniae coli** (TĒ-nē-a KŌ-lī). Each band runs the length of most of the large intestine (see Fig. 24.27a). Tonic contractions of the bands gather the colon into a series of pouches called **haustra** (HAWS-tra; singular is **haustrum** = shaped like a pouch), which give the colon its puckered appearance. The serosa of the large intestine is part of the visceral peritoneum. Small pouches of visceral peritoneum filled with fat are attached to taeniae coli and are called **epiploic appendages.**

Physiology of Digestion in the Large Intestine

Mechanical Digestion

The passage of chyme from the ileum into the cecum is regulated by the action of the ileocecal sphincter. The valve normally remains partially closed so that the passage of chyme into the cecum is usually a slow process. Immediately after a meal, there is a **gastroileal reflex** in which ileal peristalsis is intensified and any chyme in the ileum is forced into the cecum. The hormone gastrin also relaxes the sphincter. Whenever the cecum is distended, the degree of contraction of the ileocecal sphincter is intensified.

Movements of the colon begin when substances pass the ileocecal sphincter. Since chyme moves through the small intestine at a fairly constant rate, the time required for a meal to pass into the colon is determined by gastric emptying time. As food passes through the ileocecal sphincter, it fills the cecum and accumulates in the ascending colon.

One movement characteristic of the large intestine is **haustral churning.** In this process, the haustra remain relaxed and distended while they fill up. When the distension reaches a certain point, the walls contract and squeeze the contents into the next haustrum. **Peristalsis** also occurs, although at a slower rate (3 to 12 contractions per minute) than in other portions of the tract. A final type of movement is **mass peristalsis,** a strong peristaltic wave that begins at about the middle of the transverse colon and quickly drives the colonic contents into the rectum. Food in the stomach initiates this **gastrocolic reflex** in the colon. Thus mass peristalsis usually takes place three or four times a day, during or immediately after a meal.

Chemical Digestion

The last stage of digestion occurs in the colon through the activity of bacteria that live in the lumen. Mucus is secreted by the glands of the large intestine, but no enzymes are secreted. Chyme is prepared for elimination by the action of bacteria, which ferment any remaining carbohydrates and release hydrogen, carbon dioxide, and methane gas. These gases contribute to flatus (gas) in the colon. Bacteria also convert remaining proteins to amino acids and break down the amino acids into simpler substances: indole, skatole, hydrogen sulfide, and fatty acids. Some of the indole and skatole is carried off in the feces and contributes to their odor. The rest are absorbed and transported to the liver, where they are converted to less toxic compounds and excreted in the urine. Bacteria also decompose bilirubin to simpler pigments (urobilinogen), which give feces their brown color. Several vitamins needed for normal

FIGURE 24.28 Histology of the large intestine.

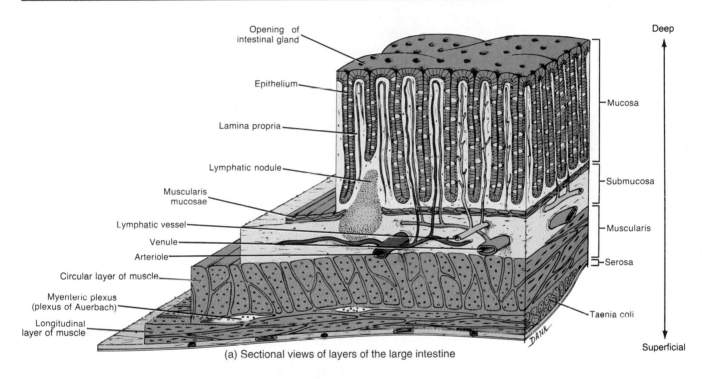

(a) Sectional views of layers of the large intestine

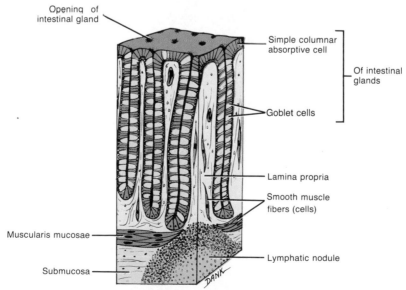

(b) Sectional views of the large intestinal mucosa

Figure continues

FIGURE 24.28 (continued)

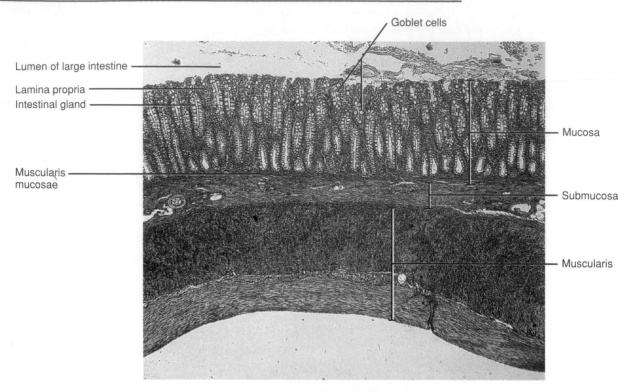

(c) Photomicrograph of a portion of the
wall of the large intestine (90x)

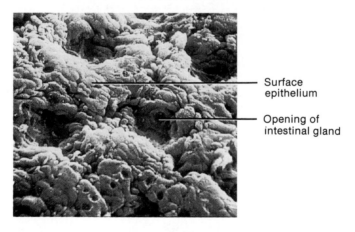

(d) Scanning electron micrograph of large
intestinal mucosa, (600 x)

Question: What is the function of the goblet cells?

metabolism and absorbed in the colon, including some B vitamins and vitamin K, are bacterial products.

Absorption and Feces Formation

By the time the chyme has remained in the large intestine 3 to 10 hours, it has become solid or semisolid as a result of absorption principally of water and is now known as **feces.** Chemically, feces consist of water, inorganic salts, sloughed-off epithelial cells from the mucosa of the gastrointestinal tract, bacteria, products of bacterial decomposition, and undigested parts of food.

Although most water absorption occurs in the small intestine, the large intestine absorbs enough to make it an important organ in maintaining the body's water balance.

Of the 0.5 to 1.0 liter that enters the large intestine, all but about 100 to 200 ml is absorbed. The absorption is greatest in the cecum and ascending colon. The large intestine also absorbs electrolytes, including sodium and chloride, and some vitamins.

Physiology of Defecation

Mass peristaltic movements push fecal material from the sigmoid colon into the rectum. The resulting distension of the rectal wall stimulates stretch receptors, initiating a reflex for **defecation,** the emptying of the rectum. The defecation reflex occurs as follows. In response to distension of the rectal wall, the receptors send nerve impulses to the sacral spinal cord. Motor impulses from the cord travel along parasympathetic nerves back to the descending colon, sigmoid colon, rectum, and anus. Contraction of the longitudinal rectal muscles shortens the rectum, thereby increasing the pressure inside it. The pressure, along with voluntary contractions of the diaphragm and abdominal muscles, and parasympathetic stimulation open the internal sphincter, and the feces are expelled through the anus.

The external sphincter is voluntarily controlled. If it is voluntarily relaxed, defecation occurs; if it is voluntarily constricted, defecation can be postponed. Voluntary contractions of the diaphragm and abdominal muscles aid defecation by increasing the pressure inside the abdomen, which pushes the walls of the sigmoid colon and rectum inward. If defecation does not occur, the feces back into the sigmoid colon until the next wave of mass peristalsis again stimulates the stretch receptors, creating the desire to defecate. In infants, the defecation reflex causes automatic emptying of the rectum because they have not yet developed voluntary control of the external anal sphincter.

Diarrhea refers to frequent defecation of liquid feces. It is caused by increased motility of and decreased absorption by the intestines. When chyme passes too quickly through the small intestine and feces pass too quickly through the large intestine, there is not enough time for absorption. Like vomiting, diarrhea can result in dehydration and electrolyte imbalances. Diarrhea may be caused by stress and microbes that irritate the gastrointestinal mucosa.

Constipation refers to infrequent or difficult defecation. It is caused by decreased motility of the intestines in which feces remain in the colon for prolonged periods of time. As it does so, there is considerable water absorption, and feces become dry and hard. Constipation may be caused by improper bowel habits, spasms of the colon, insufficient fiber in the diet, inadequate fluid intake, lack of exercise, and emotions. Usual treatment for constipation is a mild laxative, such as milk of magnesia, that induces defecation. However, many physicians maintain that laxatives are habit-forming, and that adding fiber to the diet, increasing one's amount of exercise, and improving fluid intake are safer ways of controlling this common problem.

Activities of the large intestine are summarized in Exhibit 24.7.

EXHIBIT 24.7

SUMMARY OF DIGESTION AND ABSORPTION IN THE LARGE INTESTINE

Structure	Action	Function
Mucosa	Secretes mucus.	Lubricates colon and protects mucosa.
	Absorbs water and other soluble compounds.	Maintains water balance; solidifies feces. Vitamins and electrolytes absorbed and toxic substances sent to liver to be detoxified.
Lumen	Bacterial activity.	Breaks down undigested carbohydrates, proteins, and amino acids into products that can be expelled in feces or absorbed and detoxified by liver. Certain B vitamins and vitamin K synthesized.
Muscularis	Haustral churning.	Contents moved from haustrum to haustrum by muscular contractions.
	Peristalsis.	Contents moved along length of colon by contractions of circular and longitudinal muscles.
	Mass peristalsis.	Contents forced into sigmoid colon and rectum by strong peristaltic wave.
	Defecation.	Feces eliminated by contractions in sigmoid colon and rectum.

CLINICAL APPLICATION

DIETARY FIBER

A dietary deficiency that has received recent attention is lack of adequate dietary fiber (bulk or roughage). Dietary fiber consists of indigestible plant substances, such as cellulose, lignin, and pectin, found in fruits, vegetables, grains, and beans. Fiber may be classified as **insoluble,** which does not dissolve in water, and **soluble,** which does dissolve in water. Insoluble fiber includes the woody or structural parts of plants such as fruit and vegetable skins and the bran coating around wheat and corn kernels. Insoluble fiber passes through the GI tract largely unchanged and speeds up the passage of material through the tract. Soluble fiber is found in abundance in beans, oats, barley, broccoli, prunes, apples, and citrus fruits. It has the consistency of a gel and tends to slow the passage of material through the tract.

Insoluble fiber is important for digestive functioning. People who choose a fiber-rich, unrefined diet may reduce their risk of developing obesity, diabetes, atherosclerosis, gallstones, hemorrhoids, diverticulitis, appendicitis, and colon cancer. Each of these conditions is directly related to the digestion and metabolism of food and the operation of the digestive system. There is also evidence that insoluble fiber may help protect against colon cancer and that soluble fiber may help lower blood cholesterol level. One possible explanation for the relationship of soluble fiber to cholesterol level is that the fiber binds bile salts and prevents their reabsorption. Since cholesterol is a precursor to bile salt formation, this leads to use of more cholesterol to replace the bile salts lost by the binding action of soluble fiber.

AGING AND THE DIGESTIVE SYSTEM

Overall changes associated with aging of the digestive system include decreasing secretory mechanisms, decreasing motility of the digestive organs, loss of strength and tone of the muscular tissue and its supporting structures, changes in neurosensory feedback regarding enzyme and hormone release, and diminished response to pain and internal sensations. Specific changes include reduced sensitivity to mouth irritations and sores, loss of taste, periodontal disease, difficulty in swallowing, hiatal hernia, cancer of the esophagus, gastritis, peptic ulcer, and gastric cancer. Changes in the small intestine include duodenal ulcers, appendicitis, malabsorption, and maldigestion. Other pathologies that increase in incidence are gallbladder problems, jaundice, cirrhosis, and acute pancreatitis. Large intestinal changes such as constipation, cancer of the colon or rectum, hemorrhoids, and diverticular disease of the colon also occur.

DEVELOPMENTAL ANATOMY OF THE DIGESTIVE SYSTEM

About the fourteenth day after fertilization, the cells of the endoderm form a cavity referred to as the **primitive gut** (Fig. 24.29). Soon after the mesoderm forms and splits into two layers (somatic and splanchnic), the splanchnic meso-

FIGURE 24.29 Development of the digestive system.

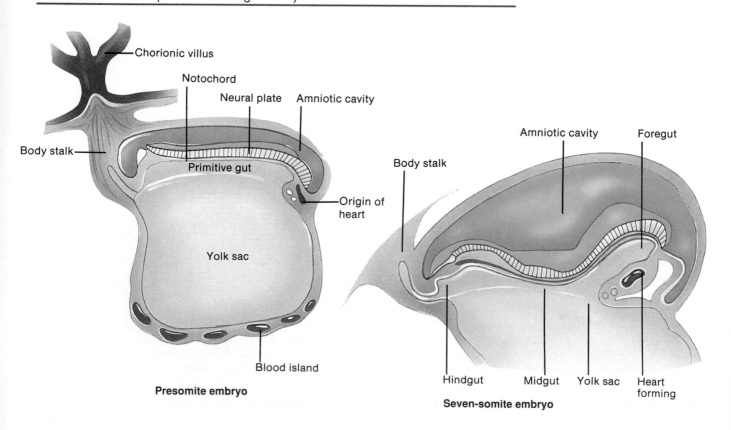

Presomite embryo

Seven-somite embryo

Figure 24.29 (continued)

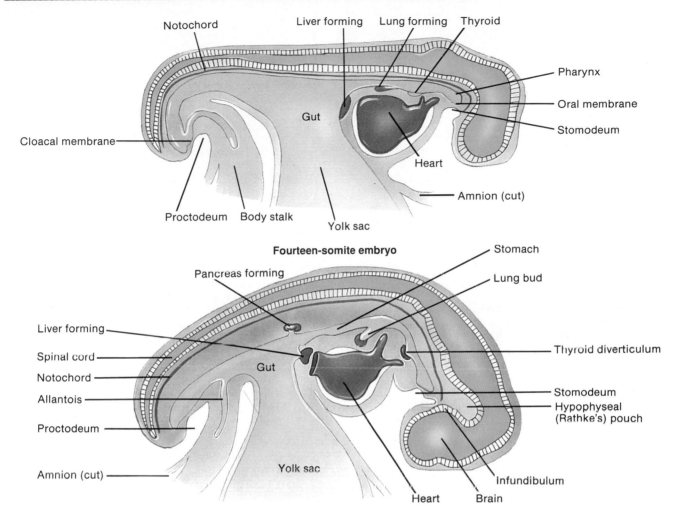

Fourteen-somite embryo

Embryo at end of first month

derm associates with the endoderm of the primitive gut. Thus the primitive gut has a double-layered wall. The **endodermal layer** gives rise to the *epithelial lining* and *glands* of most of the gastrointestinal tract. The **mesodermal layer** produces its *smooth muscle* and *connective tissue*.

The primitive gut elongates, and about the latter part of the third week, it differentiates into an anterior **foregut,** a central **midgut,** and a posterior **hindgut.** Until the fifth week of development, the midgut opens into the yolk sac. After that time, the yolk sac constricts, detaches from the midgut, and the midgut seals. In the region of the foregut, a depression consisting of **ectoderm,** the **stomodeum,** appears. This develops into the *oral cavity.* The **oral membrane** that separates the foregut from the stomodeum ruptures during the fourth week of development, so that the foregut is continuous with the outside of the embryo through the oral cavity. Another depression consisting of **ectoderm,** the **proctodeum,** forms in the hindgut and goes

on to develop into the *anus.* The **cloacal membrane,** which separates the hindgut from the proctodeum, ruptures, so that the hindgut is continuous with the outside of the embryo through the anus. Thus the GI tract forms a continuous tube from mouth to anus.

The foregut develops into the *pharynx, esophagus, stomach,* and a *portion of the duodenum.* The midgut is transformed into the *remainder of the duodenum,* the *jejunum,* the *ileum,* and *portions of the large intestine* (cecum, appendix, ascending colon, and most of the transverse colon). The hindgut develops into the *remainder of the large intestine,* except for a portion of the anal canal that is derived from the proctodeum.

As development progresses, the endoderm at various places along the foregut develops into hollow buds that grow into the mesoderm. These buds will develop into the *salivary glands, liver, gallbladder,* and *pancreas.* Each of the glands retains a connection with the gastrointestinal tract through ducts.

DENTAL CARIES

Dental caries, or tooth decay, involve a gradual demineralization (softening) of the enamel and dentin. If untreated, microorganisms may invade the pulp, causing inflammation and infection with subsequent death (necrosis) of the pulp and abscess of the alveolar bone surrounding the root's apex. Such teeth are treated by root canal therapy.

Dental caries begin when bacteria, acting on sugars, give off acids that demineralize the enamel. **Dextran,** a sticky polysaccharide produced from sucrose, causes the bacteria to stick to the teeth. Masses of bacterial cells, dextran, and other debris adhering to teeth constitute **dental plaque.** Saliva cannot reach the tooth surface to buffer the acid because the plaque covers the teeth. Brushing the teeth immediately after eating removes the plaque from flat surfaces before the bacteria have a chance to go to work. Dentists also suggest that the plaque between the teeth be removed every 24 hours with dental floss.

PERIODONTAL DISEASE

Periodontal disease is a collective term for a variety of conditions characterized by inflammation and degeneration of the gingivae, alveolar bone, periodontal ligament, and cementum. One such condition is called **pyorrhea.** The initial symptoms are enlargement and inflammation of the soft tissue and bleeding gums. Without treatment, the soft tissue may deteriorate and the alveolar bone may be resorbed, causing loosening of the teeth and recession of the gums. Periodontal diseases are frequently caused by poor oral hygiene; by local irritants, such as bacteria, impacted food, and cigarette smoke; or by a poor "bite."

PERITONITIS

Peritonitis is an acute inflammation of the peritoneum, the serous membrane lining the abdominal cavity and covering the abdominal viscera. One possible cause is contamination of the peritoneum by pathogenic bacteria from the external environment. This contamination could result from accidental or surgical wounds in the abdominal wall or from perforation or rupture of organs. For example, the large intestine contains colonies of bacteria that live on undigested nutrients and break them down so they can be eliminated. But if the bacteria enter the peritoneal cavity through an intestinal perforation or rupture of the appendix, they produce acute infection. Such an infection quickly may become life-threatening.

PEPTIC ULCER DISEASE

In the U.S. about 5 to 10% of the population develops **peptic ulcer disease (PUD).** An **ulcer** is a craterlike lesion in a membrane. Ulcers that develop in areas of the GI tract exposed to acidic gastric juice are called **peptic ulcers.** Most occur on the lesser curvature of the stomach, where they are called **gastric ulcers,** or in the pyloric sphincter or first part of the duodenum, where they are called **duodenal ulcers.** Most peptic ulcers are duodenal.

PUD occurs when the normal balance between destructive and protective forces is upset. Hypersecretion of acidic gastric juice, a destructive force, often is the immediate cause of peptic ulcers. But the cause may be hyposecretion of mucus, which normally protects the stomach and duodenal linings from the acidic gastric juice. Inadequate secretion of bicarbonate in pancreatic juice, and the consequent failure to buffer gastric acid, can also lead to duodenal ulcers. Among the factors believed to stimulate an increase in acid secretion

or a decrease in protective secretions are emotional stress, cigarette smoking, certain foods or medications (alcohol, coffee, aspirin), and overstimulation of the vagus (X) nerves. Evidence suggests that some peptic ulcers are associated with the bacterium *Helicobacter pylori*. The most common complication of peptic ulcers is bleeding, which can lead to anemia if blood loss is serious.

Several therapeutic approaches are helpful to treat PUD. Patients who smoke should be strongly urged to quit. Avoidance of other irritants such as caffeinated drinks and alcohol usually provides at least some relief. Oral antacids, such as Rolaids or Maalox, can help temporarily by buffering gastric acid. H_2-receptor blocking drugs can reduce acid secretion by parietal cells to an acceptable level, and natural healing can proceed. In cases caused by bacteria, an antibiotic drug is useful. Sometimes surgery is needed, either vagotomy or removal of a portion of the stomach.

APPENDICITIS

Appendicitis is an inflammation of the vermiform appendix. It is preceded by obstruction of the lumen of the appendix by fecal material, inflammation, a foreign body, carcinoma of the cecum, stenosis, or kinking of the organ. The infection that follows may result in edema, ischemia, gangrene, and perforation. Rupture of the appendix develops into peritonitis. Typically, appendicitis begins with referred pain in the umbilical region of the abdomen, followed by anorexia (lack or loss of appetite for food), nausea, and vomiting. After several hours, the pain localizes in the right lower quadrant (RLQ) and is continuous, dull or severe, and intensified by coughing, sneezing, or body movements. Early appendectomy (removal of the appendix) is recommended in all suspected cases because it is safer to operate than to risk gangrene, rupture, and peritonitis.

TUMORS

Both benign and malignant **tumors** can occur in all parts of the gastrointestinal tract. Although benign growths are much more common than malignant ones, cancers of the gastrointestinal tract are responsible for 30% of all deaths from cancer in the U.S.

Colorectal cancer is one of the most deadly and common malignant diseases, ranking second to lung cancer in males and third after lung and breast cancer in females. Development of colorectal cancer is a multistep process, involving both environmental and genetic factors (see page 90). Dietary fiber, retinoids, calcium, and selenium may be protective, whereas intake of animal fat and protein may cause an increase in the disease. Genetics plays a very important role in that an inherited predisposition contributes to more than half of all cases of colorectal cancer. Signs and symptoms of colorectal cancer include changes in the normal pattern of bowel habits (diarrhea, constipation), cramping, abdominal pain, and rectal bleeding, either visible or occult. Screening for colorectal cancer includes testing for blood in the feces, digital rectal examination, sigmoidoscopy, colonoscopy, and barium enema. The only definitive treatment for gastrointestinal tumors, if they cannot be removed endoscopically, is surgery.

DIVERTICULITIS

Diverticula are saclike outpouchings of the wall of the colon in places where the muscularis has become weak. The development of diverticula is called **diverticulosis.** Many people who develop diverticulosis are asymptomatic and experience no complications. About 15% of people with diverticulosis will eventually develop an inflammation within diverticula, a condition known as **diverticulitis**.

The increase in diverticular disease has been attributed to a shift to low-fiber diets. Patients treated for diverticular disease with high-fiber diets show marked relief of symptoms. Treatment consists of bed rest, cleansing enemas, and drugs to reduce infection. In severe cases, portions of the affected colon may require surgical removal and a temporary colostomy to allow healing of the operated region.

CIRRHOSIS

Cirrhosis refers to a distorted or scarred liver as a result of chronic inflammation. The parenchymal (functional) liver cells are replaced by fibrous or adipose connective tissue. The symptoms of cirrhosis include jaundice, edema in the legs, uncontrolled bleeding, and increased sensitivity to drugs. Cirrhosis may be caused by hepatitis (inflammation of the liver), certain chemicals that destroy liver cells, parasites that infect the liver, and alcoholism.

HEPATITIS

Hepatitis refers to inflammation of the liver and can be caused by viruses, drugs, and chemicals, including alcohol. Clinically, several types are recognized.

Hepatitis A (infectious hepatitis) is caused by hepatitis A virus and is spread by fecal contamination of food, clothing, toys, eating utensils, and so forth (fecal–oral route). It is generally a mild disease of children and young adults characterized by anorexia, malaise, nausea, diarrhea, fever, and chills. Eventually, jaundice appears. It does not cause lasting liver damage. Most people recover in four to six weeks.

Hepatitis B (serum hepatitis) is caused by hepatitis B virus and is spread primarily by sexual contact and contaminated syringes and transfusion equipment. It can also be spread by saliva and tears. Hepatitis B virus can be present for years or even a lifetime and can produce cirrhosis and possibly cancer of the liver. Persons who harbor the active hepatitis B virus are at risk for cirrhosis and also become carriers. Vaccines produced through recombinant DNA technology (e.g., Recombivax HB) are available to prevent hepatitis B infection.

Hepatitis C (non-A, non-B hepatitis) is a form of hepatitis recently traced to a third hepatitis virus. It is clinically similar to hepatitis B and is often spread by blood transfusions. It is believed to account for considerably more post-transfusion hepatitis than that related to hepatitis B. The hepatitis C virus can cause cirrhosis and possibly liver cancer.

GALLSTONES

Most **gallstones (biliary calculi)** stem from the fusion of crystals of cholesterol in bile. The presence of gallstones is called **cholelithiasis** (kō'-lē-li-THI-a-sis; *lithos* = stone). Following their formation, gallstones gradually grow in size and number and may cause minimal, intermittent, or complete obstruction to the flow of bile from the gallbladder into the duct system. If obstruction of the outlet occurs and the gallbladder cannot empty as it normally does after eating, the pressure within it increases, and the person may have intense pain or discomfort **(biliary colic)**. Jaundice, due to the inability to secrete bilirubin into the intestine, will accompany complete biliary obstruction.

Treatment of gallstones consists of using gallstone-dissolving drugs, lithotripsy (shock-wave therapy), or surgery. For people who have had recurrent gallstones or in whom drugs or lithotripsy is not indicated, removal of the gallbladder and its contents (cholecystectomy) is necessary. More than half a million cholecystectomies are performed each year in the U.S.

ANOREXIA NERVOSA

Anorexia nervosa is a chronic disorder characterized by self-induced weight loss, body-image and other perceptual disturbances, and physiological changes that result from nutritional depletion. Patients with anorexia nervosa have a fixation on weight control and, often, the insistence of having a bowel movement every day despite a lack of adequate food intake. They abuse laxatives, which worsens the fluid, electrolyte, and nutrient deficiencies. The disorder is found predominantly in young, single females and may be inherited. Abnormal patterns of menstruation, amenorrhea (absence of menstruation), and a lowered basal metabolic rate reflect the depressant effects of the starvation. Individuals may become emaciated and may ultimately die of starvation or one of its complications. Also associated with the disorder are osteoporosis, depression, and brain abnormalities coupled with impaired mental performance. Treatment consists of psychotherapy and dietary regulation.

BULIMIA

A disorder that typically affects single, middle-class, young, white females is known as **bulimia** (*bous* = ox; *limos* = hunger), or **binge–purge syndrome**. It is characterized by overeating at least twice a week followed by purging by self-induced vomiting, strict dieting or fasting, vigorous exercise, or use of laxatives or diuretics. This binge–purge cycle occurs in response to fears of being overweight, stress, depression, and physiological disorders such as hypothalamic tumors.

Bulimia can upset the body's electrolyte balance and increase susceptibility to flu, salivary gland infections that result in bilateral parotid gland enlargement, pharyngeal scratches from self-induced gagging, dry skin, acne, muscle spasms, loss of hair, kidney and liver diseases, erosion of dental enamel from stomach acids, ulcers, hernias, constipation, and hormone imbalances. Although the cause of bulimia is unknown, some evidence suggests that it may be related to impaired release of cholecystokinin (CCK), a hormone that induces satiety, the sensation of being full to satisfaction. Treatment of bulimia combines nutrition counseling, psychotherapy, and medical treatment.

Borborygmus (bor′-bō-RIG-mus) A rumbling noise caused by the propulsion of gas through the intestines.

Botulism (BOCH-yoo-lism; *botulus* = sausage) A type of food poisoning caused by a toxin produced by *Clostridium botulinum*. The bacterium is ingested when improperly cooked or preserved foods are eaten. The toxin inhibits nerve impulse transmission at synapses by inhibiting the release of acetylcholine. Symptoms include paralysis, nausea, vomiting, blurred or double vision, difficulty in speech and swallowing, dryness of the mouth, and general weakness.

Canker (KANG-ker) **sore** Painful ulcer on the mucous membrane of the mouth that affects females more often than males and usually occurs between ages 10 to 40; may be an autoimmune reaction.

Cholecystitis (kō′-lē-sis-TĪ-tis; *chole* = bile; *kystis* = bladder; *itis* = inflammation of) Inflammation of the gallbladder that often leads to infection. Some cases are caused by obstruction of the cystic duct with bile stones.

Colitis (ko-LĪ-tis) Inflammation of the mucosa of the colon and rectum in which absorption of water and salts is reduced, producing watery, bloody feces and, in severe cases, dehydration and salt depletion. Spasms of the irritated muscularis produce cramps.

Colostomy (kō-LOS-tō-mē; *stomoun* = provide an opening) The diversion of the fecal stream through an opening in the colon, creating a surgical "stoma" (artificial opening) that is affixed to the exterior of the abdominal wall. This opening serves as a substitute anus through which feces are eliminated.

Dysphagia (dis-FĀ-jē-a; *dys* = abnormal; *phagein* = to eat) Difficulty in swallowing that may be caused by inflammation, paralysis, obstruction, or trauma.

Enteritis (en′-ter-Ī-tis; *enteron* = intestine) An inflammation of the intestine, particularly the small intestine.

Flatus (FLĀ-tus) Air (gas) in the stomach or intestine, usually expelled through the anus. If the gas is expelled through the mouth, it is called **eructation** or **belching** (burping). Flatus may result from gas released during the breakdown of foods in the stomach or from swallowing air or gas-containing substances such as carbonated drinks.

Gastrectomy (gas-TREK-tō-mē; *gastro* = stomach; *tome* = excision) Removal of a portion of or the entire stomach.

Hernia (HER-nē-a) Protrusion of an organ or part of an organ through a membrane or cavity wall, usually the abdominal cavity. Diaphragmatic (hiatal) hernia is the protrusion of the lower esophagus, stomach, or intestine into the thoracic cavity through the opening in the diaphragm

(esophageal hiatus) that allows passage of the esophagus. Umbilical hernia is usually a mild defect that contains the protrusion of a portion of peritoneum through the navel area of the abdominal wall. Inguinal hernia is the protrusion of the hernial sac into the inguinal opening. It may contain a portion of the bowel in an advanced stage and may extend into the scrotal compartment in males, causing strangulation of the herniated part.

Inflammatory bowel (in-FLAM-a-tō′-rē BOW-el) **disease** Disorder that exists in two forms: (1) Crohn's disease (inflammation of the gastrointestinal tract, especially the distal ileum and proximal colon, in which the inflammation may extend from the mucosa through the serosa) and (2) ulcerative colitis (inflammation of the mucosa of the gastrointestinal tract, usually limited to the large intestine and usually accompanied by rectal bleeding).

Irritable bowel (IR-i-ta-bul BOW-el) **syndrome (IBS)** Disease of the entire gastrointestinal tract in which persons with this condition may react to stress by developing symptoms such as cramping and abdominal pain associated with alternating patterns of diarrhea and constipation. Excessive amounts of mucus may appear in the stools, and other symptoms include flatulence, nausea, and loss of appetite. The condition is also known as **irritable colon** or **spastic colitis.**

Malocclusion (mal′-ō-KLOO-zhun; *mal* = disease; *occlusio* = to fit together) Condition in which the upper and lower teeth do not close together.

Nausea (NAW-sē-a; *nausia* = seasickness) Discomfort characterized by a loss of appetite and the sensation of impending vomiting. Its causes include local irritation of the gastrointestinal tract, a systemic disease, brain disease or injury, overexertion, or the effects of medication or drug overdosage.

Pancreatitis (pan′-krē-a-TĪ-tis) Inflammation of the pancreas, as may occur in association with mumps. In a more severe condition, known as **acute pancreatitis,** which is associated with heavy alcohol intake or biliary tract obstruction, the pancreatic cells may release trypsin instead of trypsinogen, and the trypsin begins to digest the pancreatic cells. The patient with acute pancreatitis usually responds to treatment, but recurrent attacks are the rule.

Traveler's diarrhea Infectious disease of the gastrointestinal tract that results in loose, urgent bowel movements, cramping, abdominal pain, malaise, nausea, and occasionally fever and dehydration. It is acquired through ingestion of food or water that has become contaminated with fecal material containing bacteria (especially *Escherichia coli*). Viruses or protozoan parasites are less frequently involved. Commonly referred to as **Montezuma's revenge**, **turista**, and **Tut's tummy**.

Study Outline

Digestive Processes (p. 766)

1. Food is prepared for use by cells by five basic activities: ingestion, movement, mechanical and chemical digestion, absorption, and defecation.
2. Mechanical digestion consists of movements of the gastrointestinal tract that aid chemical digestion.
3. Chemical digestion is a series of catabolic (hydrolysis) reactions that break down large carbohydrate, lipid, and protein food molecules into smaller molecules that are usable by body cells.
4. Absorption is the passage of end products of digestion from the gastrointestinal tract into blood or lymph for distribution to cells.
5. Defecation is emptying of the rectum.

Organization (p. 766)

1. The organs of digestion are usually divided into two main groups: those composing the gastrointestinal (GI) tract, and accessory structures.
2. The GI tract is a continuous tube running through the ventral body cavity from the mouth to the anus.
3. The accessory structures include the teeth, tongue, salivary glands, liver, gallbladder, and pancreas.
4. The basic arrangement of layers in the gastrointestinal tract from the inside outward is the mucosa, submucosa, muscularis, and serosa (visceral peritoneum).
5. Extensions of the peritoneum include the mesentery, mesocolon, falciform ligament, lesser omentum, and greater omentum.

Mouth (Oral Cavity) (p. 770)

1. The mouth is formed by the cheeks, hard and soft palates, lips, and tongue, which aid mechanical digestion.
2. The vestibule is the space between the cheeks and lips and teeth and gums.
3. The oral cavity proper extends from the vestibule to the fauces.

Tongue (p. 772)

1. The tongue, together with its associated muscles, forms the floor of the oral cavity. It is composed of skeletal muscle covered with mucous membrane.
2. The upper surface and sides of the tongue are covered with papillae. Some papillae contain taste buds.

Salivary Glands (p. 772)

1. The major portion of saliva is secreted by the salivary glands, which lie outside the mouth and pour their contents into ducts that empty into the oral cavity.
2. There are three pairs of salivary glands: parotid, submandibular (submaxillary), and sublingual glands.
3. Saliva lubricates food and starts the chemical digestion of carbohydrates.
4. Salivation is entirely under nervous control.

Teeth (p. 774)

1. The teeth (dentes) project into the mouth and are adapted for mechanical digestion.
2. A typical tooth consists of three principal portions: crown, root, and neck.
3. Teeth are composed primarily of dentin and are covered by enamel, the hardest substance in the body.
4. There are two dentitions—deciduous and permanent.

Physiology of Digestion in the Mouth (p. 775)

1. Through mastication, food is mixed with saliva and shaped into a bolus.
2. Salivary amylase converts polysaccharides (starches) to disaccharides (maltose).
3. Lingual lipase acts on triglycerides.

Physiology of Deglutition (p. 776)

1. Deglutition, or swallowing, moves a bolus from the mouth to the stomach.
2. It consists of a voluntary stage, pharyngeal stage (involuntary), and esophageal stage (involuntary).

Esophagus (p. 777)

1. The esophagus is a collapsible, muscular tube that connects the pharynx to the stomach.
2. It passes a bolus into the stomach by peristalsis.
3. It contains an upper and lower esophageal sphincter.

Stomach (p. 780)

Anatomy; Histology (p. 780)

1. The stomach begins at the bottom of the esophagus and ends at the pyloric sphincter.
2. The principal gross anatomic subdivisions of the stomach include the cardia, fundus, body, and pylorus.
3. Adaptations of the stomach for digestion include rugae; glands that produce mucus, hydrochloric acid, a protein-digesting enzyme, and intrinsic factor; and a three-layered muscularis for efficient mechanical movement.

Physiology of Digestion in the Stomach (p. 783)

1. Mechanical digestion consists of mixing waves.
2. Chemical digestion consists mostly of the conversion of proteins into peptides by pepsin.

Regulation of Gastric Secretion and Motility (p. 785)

1. Gastric secretion is regulated by nervous and hormonal mechanisms.
2. Stimulation of gastric secretion occurs in three phases: cephalic (reflex), gastric, and intestinal.
3. During the cephalic and gastric phases, peristalsis is stimulated; during the intestinal phase, motility is inhibited.

Regulation of Gastric Emptying (p. 787)

1. Gastric emptying is stimulated in response to distension and gastrin is released in response to the presence of certain types of foods.
2. Gastric emptying is inhibited by the enterogastric reflex and hormones (CCK and GIP).

Absorption (p. 787)

1. The stomach wall is impermeable to most substances.
2. Among the substances absorbed are some water, certain electrolytes and drugs, and alcohol.

Pancreas (p. 788)

1. The pancreas is divisible into a head, body, and tail and is connected to the duodenum via the pancreatic duct and accessory duct.
2. Pancreatic islets (islets of Langerhans) secrete hormones, and acini secrete pancreatic juice.
3. Pancreatic juice contains enzymes that digest starch (pancreatic amylase), proteins (trypsin, chymotrypsin, and carboxypeptidase), triglycerides (pancreatic lipase), and nucleic acids (ribonuclease and deoxyribonuclease).
4. Pancreatic secretion is regulated by nervous and hormonal mechanisms.

Liver (p. 790)

1. The liver has left and right lobes; associated with the right lobe are the caudate and quadrate lobes.
2. The lobes of the liver are made up of lobules that contain hepatocytes (liver cells), sinusoids, stellate reticuloendothelial (Kupffer's) cells, and a central vein.
3. Hepatocytes of the liver produce bile that is carried by a duct system to the gallbladder for concentration and temporary storage.
4. Bile's contribution to digestion is the emulsification of triglycerides.
5. The liver also functions in carbohydrate, lipid, and protein metabolism; removal of drugs and hormones; excretion of bile; synthesis of bile salts; storage of vitamins and minerals; phagocytosis; and activation of vitamin D.
6. Bile secretion is regulated by nervous and hormonal mechanisms.

Gallbladder (p. 795)

1. The gallbladder is a sac located in a fossa on the visceral surface of the liver.
2. The gallbladder stores and concentrates bile.
3. Bile is ejected into the common bile duct under the influence of cholecystokinin (CCK).

Small Intestine (p. 797)

Anatomy, Histology, and Brush Border (p. 797)

1. The small intestine extends from the pyloric sphincter to the ileocecal sphincter.
2. It is divided into duodenum, jejunum, and ileum.
3. It is highly adapted for digestion and absorption. Its glands secrete fluid and mucus, and the microvilli, villi, and circular folds of its wall provide a large surface area for digestion and absorption.
4. Brush border enzymes digest carbohydrates, proteins, and nucleotides at the surface of mucosal epithelial cells.

Physiology of Digestion in the Small Intestine (p. 801)

1. Pancreatic and intestinal brush border enzymes break down starches into maltose, maltotriose, and α-dextrins (pancreatic amylase), α-dextrins into glucose (dextrinase), maltose to glucose (maltase), sucrose to glucose and fructose (sucrase), lactose to glucose and galactose (lactase), and proteins into peptides (trypsin and chymotrypsin). Also, enzymes break peptide bonds that attach terminal amino acids to carboxyl ends of peptides (carboxypeptidases) and peptide bonds that attach terminal amino acids to amino ends of peptides (aminopeptidases). Finally, enzymes split dipeptides to amino acids (dipeptidase), triglycerides to fatty acids and monoglycerides (pancreatic lipase), and nucleotides to pentoses and nitrogenous bases (nucleosidases and phosphatases).

2. Mechanical digestion in the small intestine involves segmentation and peristalsis.

Regulation of Intestinal Secretion and Motility (p. 802)

1. The most important mechanism is local reflexes.
2. Hormones also assume a role.
3. Parasympathetic impulses increase motility; sympathetic impulses decrease motility.

Physiology of Absorption (p. 802)

1. Absorption is the passage of the end products of digestion from the gastrointestinal tract into the blood or lymph.
2. Absorption occurs by diffusion, facilitated diffusion, osmosis, and active transport; most occurs in the small intestine.
3. Monosaccharides, amino acids, and short-chain fatty acids pass into the blood capillaries.
4. Long-chain fatty acids and monoglycerides are absorbed as part of micelles, resynthesized to triglycerides, and formed into chylomicrons.
5. Chylomicrons move into lymph in the lacteal of a villus.
6. The small intestine also absorbs water, electrolytes, and vitamins.

Large Intestine (p. 806)

Anatomy; Histology (p. 806)

1. The large intestine extends from the ileocecal sphincter to the anus.
2. Its subdivisions include the cecum, colon, rectum, and anal canal.
3. The mucosa contains many goblet cells, and the muscularis consists of taeniae coli.

Physiology of Digestion in the Large Intestine (p. 808)

1. Mechanical movements of the large intestine include haustral churning, peristalsis, and mass peristalsis.
2. The last stages of chemical digestion occur in the large intestine through bacterial action. Substances are further broken down and some vitamins are synthesized.

Absorption and Feces Formation (p. 810)

1. The large intestine absorbs water, electrolytes, and vitamins.
2. Feces consist of water, inorganic salts, epithelial cells, bacteria, and undigested foods.

Physiology of Defecation (p. 811)

1. The elimination of feces from the rectum is called defecation.
2. Defecation is a reflex action aided by voluntary contractions of the diaphragm and abdominal muscles and relaxation of the external anal sphincter.

Aging and the Digestive System (p. 812)

1. General changes include decreased secretory mechanisms, decreased motility, and loss of tone.
2. Specific changes include loss of taste, pyorrhea, hernias, ulcers, constipation, hemorrhoids, and diverticular diseases.

Developmental Anatomy of the Digestive System (p. 812)

1. The endoderm of the primitive gut forms the epithelium and glands of most of the gastrointestinal tract.
2. The mesoderm of the primitive gut forms the smooth muscle and connective tissue of the gastrointestinal tract.

Review Questions

1. Define digestion and describe the five phases involved. Distinguish between chemical and mechanical digestion. (p. 766)
2. Identify, in sequence, the organs of the gastrointestinal (GI) tract. How does the gastrointestinal tract differ from the accessory structures of digestion? (p. 766)
3. Describe the structure and function of each of the four layers of the gastrointestinal tract. (p. 766)
4. What is the peritoneum? Describe the location and function of the mesentery, mesocolon, falciform ligament, lesser omentum, and greater omentum. (p. 768)
5. What structures form the mouth (oral cavity)? (p. 770)
6. Make a simple diagram of the tongue. Indicate the location of the papillae and the four taste zones.
7. Distinguish buccal from salivary glands. Describe the location of the salivary glands and their ducts. What is mumps? (p. 772)
8. How are salivary glands distinguished histologically? (p. 772)
9. Describe the composition of saliva and the role of each of its components in digestion. What is the pH of saliva? (p. 772)
10. How is salivary secretion regulated? (p. 774)
11. What are the principal parts of a typical tooth? What are the functions of each part? (p. 774)
12. Compare deciduous and permanent dentitions with regard to number of teeth and time of eruption. (p. 774)
13. Contrast the functions of incisors, cuspids, premolars, and molars. (p. 774)
14. What is a bolus? How is it formed? (p. 775)
15. Define deglutition. List the sequence of events involved in passing a bolus from the mouth to the stomach. Be sure to discuss the voluntary, pharyngeal, and esophageal stages of swallowing. (p. 776)
16. Describe the location and histology of the esophagus. What is its role in digestion? (p. 777)
17. Explain the operation of the upper and lower esophageal sphincters. (p. 779)
18. Describe the location of the stomach. List and briefly explain the anatomical features of the stomach. (p. 780)
19. What is the importance of rugae, chief cells, parietal cells, mucous cells, and G cells in the stomach? (p. 780)
20. Describe mechanical digestion in the stomach. (p. 780)
21. What is the role of pepsin? Why is it secreted in an inactive form? (p. 783)
22. What are the functions of gastric lipase, lingual lipase, and rennin in the stomach? (p. 784)
23. Outline the factors that stimulate and inhibit gastric secretion. Be sure to discuss the cephalic, gastric, and intestinal phases. (p. 785)
24. How is gastric emptying stimulated and inhibited? (p. 787)
25. Describe the role of the stomach in absorption. (p. 787)
26. Where is the pancreas located? Describe the duct system connecting the pancreas to the duodenum. (p. 788)
27. What are pancreatic acini? Contrast their functions with those of the pancreatic islets (islets of Langerhans). (p. 788)
28. Describe the composition of pancreatic juice and the digestive functions of each component. (p. 788)
29. How is pancreatic juice secretion regulated? (p. 789)
30. Where is the liver located? What are its principal functions? (p. 790)
31. Describe the anatomy of the liver. Draw a labeled diagram of a liver lobule. (p. 790)
32. How is blood carried to and from the liver? (p. 791)
33. Once bile has been formed by the liver, how is it collected and transported to the gallbladder for storage? (p. 794)
34. What is the function of bile? (p. 794)
35. How is bile secretion regulated? (p. 794)
36. Where is the gallbladder located? How is it connected to the duodenum? (p. 795)
37. Describe the function of the gallbladder. How is emptying of the gallbladder regulated? (p. 795)
38. What are the subdivisions of the small intestine? How are the mucosa and submucosa of the small intestine adapted for digestion and absorption? (p. 797)
39. Describe the movements in the small intestine. (p. 801)
40. Explain the function of each enzyme in intestinal juice. (p. 801)
41. How is small intestinal secretion regulated? (p. 802)
42. Define absorption. How are the end products of carbohydrate and protein digestion absorbed? How are the end products of lipid digestion absorbed? (p. 802)
43. What routes are taken by absorbed nutrients to reach the liver? (p. 804)
44. Describe the absorption of water, electrolytes, and vitamins by the small intestine. (p. 805)
45. What are the principal subdivisions of the large intestine? How does the muscularis of the large intestine differ from that of the rest of the gastrointestinal tract? What are haustra? (p. 806)
46. Describe the mechanical movements that occur in the large intestine. (p. 808)
47. Explain the activities of the large intestine that change its contents into feces. (p. 809)
48. Define defecation. How does it occur? (p. 811)
49. Describe the effects of aging on the digestive system. (p. 812)
50. Describe the development of the digestive system. (p. 812)

Answers to Questions with Figures

24.1 Peristalsis.
24.2 They help regulate secretions and motility of the tract.
24.3 Mesentery.
24.4 The tongue. Extrinsic muscles move the tongue during eating and swallowing. Intrinsic muscles alter the shape of the tongue for speech production and swallowing.
24.5 Activate salivary amylase.

24.6 Connective tissue, specifically dentin.
24.7 Cuspids; molars; incisors.
24.8 Both. Initiation of swallowing is voluntary and the action is carried out by skeletal muscles. Completion of swallowing—moving a bolus along the esophagus and into the stomach—involves peristalsis in smooth muscle.
24.9 Mucosa and submucosa.

24.10 Food is pushed along by contraction of smooth muscle behind the bolus and relaxation of smooth muscle in front of it.

24.11 Probably not, since as the stomach fills, they stretch out.

24.12 Chicf cells secrete pepsinogen and gastric lipase; parietal cells secrete HCl and intrinsic factor; mucous cells secrete mucus; G cells secrete gastrin.

24.13 Secretion of gastric juice and increased gastric motility.

24.14 Because all parts of the cycle are within the stomach. Impulses from the brain or spinal cord are not needed.

24.15 Histamine is a paracrine; ACh is a neurotransmitter; gastrin is a hormone.

24.16 Inhibition.

24.17 Pancreatic juice (fluid and digestive enzymes); bile; pancreatic juice plus bile.

24.18 They are carried by the blood from enteroendocrine cells in the small intestine into hepatic portal blood through the heart out to an artery serving the pancreas.

24.19 Epigastric.

24.20 Hepatic portal vein → branch of hepatic portal vein → sinusoid → central vein → hepatic vein → inferior vena cava.

24.21 Secretin promotes liberation of fluid rich in bicarbonate ions in both organs.

24.22 Circular folds, villi, microvilli, long length—all increase the surface area available for digestion and absorption.

24.23 Their alkaline mucus neutralizes gastric acid and protects the mucosal lining of the duodenum.

24.24 Segmentation mixes chyme in one location of the small intestine; peristalsis moves it along the GI tract.

24.25 They are nonpolar (hydrophobic) molecules and thus can dissolve in and diffuse through the phospholipid bilayer of the plasma membrane.

24.26 Stomach and pancreas.

24.27 Transverse colon and sigmoid colon.

24.28 Secrete mucus.

Chapter 25
METABOLISM

Chapter Contents at a Glance

Student Objectives

1. Explain how food intake is regulated.

2. Define a nutrient and list the functions of the six principal classes of nutrients.

3. Define metabolism and explain the role of ATP in anabolism and catabolism.

4. Describe the metabolism of carbohydrates, lipids, and proteins.

5. Distinguish between the absorptive (fed) and postabsorptive (fasting) states.

6. Compare the sources, functions, and importance of minerals and vitamins in metabolism.

7. Describe the various mechanisms involved in heat production and heat loss.

8. Define basal metabolic rate (BMR) and explain several factors that affect it.

9. Explain how normal body temperature is maintained and describe fever, heat cramp, heatstroke, heat exhaustion, and hypothermia as abnormalities of temperature regulation.

10. Define obesity, vitamin or mineral overdose, malnutrition, phenylketonuria (PKU), and celiac disease.

amount of glucose suddenly floods into the bloodstream. Diabetics who experience high blood glucose levels may often spill glucose into the urine.

Glucose Movement into Cells

Before glucose can be used by body cells, it must first pass through the plasma membrane and enter the cytosol. Whereas glucose absorption in the GI tract (and kidney tubules) is accomplished by active transport, glucose movement from blood into most other body cells occurs by facilitated diffusion (see page 64). Insulin increases the rate of glucose facilitated diffusion into most cells, except neurons and liver cells. Immediately upon entry into cells, glucose is phosphorylated. It combines with a phosphate group, produced by the breakdown of ATP, to form glucose 6-phosphate (see Fig. 25.9). Enzymes that catalyze phosphorylations are called **kinases.** Phosphorylation traps glucose in the cell so that it cannot move back out. Liver cells, kidney tubule cells, and intestinal epithelial cells have the necessary enzyme (phosphatase) to remove the phosphate group, which enables glucose to diffuse out of the cell and into the bloodstream (see Fig. 25.9).

Glucose Catabolism

The **oxidation** of glucose is also known as **cellular respiration.** It involves glycolysis, formation of acetyl coenzyme A, the Krebs cycle, and the electron transport chain (see Fig. 25.8). Glycolysis occurs in most cells in the body. It liberates two molecules of ATP for each glucose used and does not require oxygen. Thus it is a way to produce ATP anaerobically (without oxygen) and is known as **anaerobic cellular respiration.** Glycolysis is the prelude to the Krebs cycle and the electron transport chain, however, which do require oxygen and provide as many as 34 or 36 molecules of ATP per glucose molecule. (Recent experiments suggest that these theoretical maximums for ATP production may not be achieved. The actual number may be closer to 25 ATP per glucose.) These ATP-producing reactions are known as **aerobic cellular respiration.**

Glycolysis

The term **glycolysis** (glī-KOL-i-sis; *glyco* = sugar; *lysis* = breakdown) refers to a series of chemical reactions in the cytosol of a cell that splits a six-carbon molecule of glucose into two three-carbon molecules of pyruvic acid. (Note that the word *glycolysis* sounds a lot like *glycogenolysis,* the catabolism of glycogen to glucose.)

• **Steps in Glycolysis** Figure 25.2a shows the steps of glycolysis. Each of the 10 reactions is catalyzed by a specific enzyme. The reactions of glycolysis use two ATP molecules, but produce four, a net gain of two (Fig. 25.2b). The essential features of the process are:

Steps 1, 2, and 3. The first three reactions involve the addition of a phosphate group (phosphorylation) to glucose, its conversion to fructose, and the addition of another phosphate group to fructose. This involves an input of energy in which two molecules of ATP are converted to ADP. The enzyme that catalyzes step 3 is the key regulator of the rate of glycolysis. When ADP concentration is high, this enzyme has a high activity. Thus both pyruvic acid and ATP are rapidly produced. When ATP is plentiful, on the other hand, the enzyme activity is low, and most glucose 6-phosphate is converted to glycogen for storage rather than catabolized to produce ATP.

Steps 4 and 5. The double phosphorylated molecule of fructose splits into two three-carbon compounds, glyceraldehyde 3-phosphate (G 3-P) and dihydroxyacetone phosphate. These two compounds are interconvertible, but it is G 3-P that undergoes further reactions to pyruvic acid.

Step 6. Oxidation occurs here as two molecules of NAD^+ accept two pairs of electrons and hydrogen ions from two molecules of G 3-P, forming two molecules each of NADH and 1,3-bisphosphoglyceric acid (BPG). [Previously, BPG was called diphosphoglyceric acid (DPG).]

Steps 7 through 10. These reactions generate four molecules of ATP and produce pyruvic acid (pyruvate*).

• **Fate of Pyruvic Acid** The fate of pyruvic acid depends on the availability of oxygen (Fig. 25.3). If oxygen is scarce (anaerobic conditions), for example, in skeletal muscle fibers during strenuous exercise, pyruvic acid is reduced by the addition of two hydrogen atoms to form lactic acid (lactate). The reaction is:

$$2 \text{ Pyruvic acid} + 2NADH + 2H^+ \rightarrow 2 \text{ Lactic acid} + 2NAD^+$$

This reaction regenerates the NAD^+ that was used in the oxidation of glyceraldehyde 3-phosphate (step 6 in glycolysis) and thus allows glycolysis to continue.

Lactic acid may enter the blood and be transported to the liver, where it can be converted back to pyruvic acid. Or it may remain in the cells until aerobic conditions are restored. Then it is reconverted to pyruvic acid in the cells. (Neurons do not produce lactic acid.)

When oxygen is plentiful (aerobic conditions), pyruvic acid is converted into acetyl coenzyme A. This molecule links glycolysis, which occurs in the cytosol, and the Krebs cycle, which occurs in the matrix of mitochondria. In general, the only molecules that can cross the inner mitochondrial membrane are those for which specific transport proteins are present. It is through this mechanism that pyruvic acid enters mitochondria. Since they lack mitochondria, red blood cells can produce ATP only through anaerobic cellular respiration.

---

* The carboxyl groups (—COOH) of intermediates in glycolysis and the citric acid cycle are mostly ionized at the pH of body fluids. The suffix "-ic acid" indicates the nonionized form, that is, —COOH. The ending "-ate" indicates the ionized form, that is, —COO⁻. Because the terms are more familiar, we will use the "acid" names.

Formation of Acetyl Coenzyme A

Each step in the oxidation of glucose requires a different enzyme and often a coenzyme as well. We are interested in only one coenzyme at this point: **coenzyme A (CoA).** This important coenzyme is derived from pantothenic acid, another B vitamin.

During the transitional step between glycolysis and the

FIGURE 25.2 Glycolysis. As a result of glycolysis, there is a net gain of two molecules of ATP.

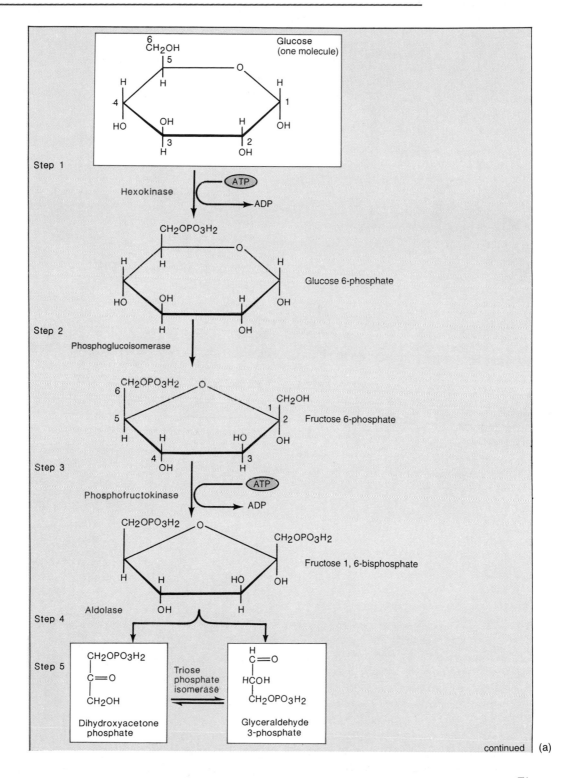

continued   (a)

Figure continues

FIGURE 25.2 (continued)

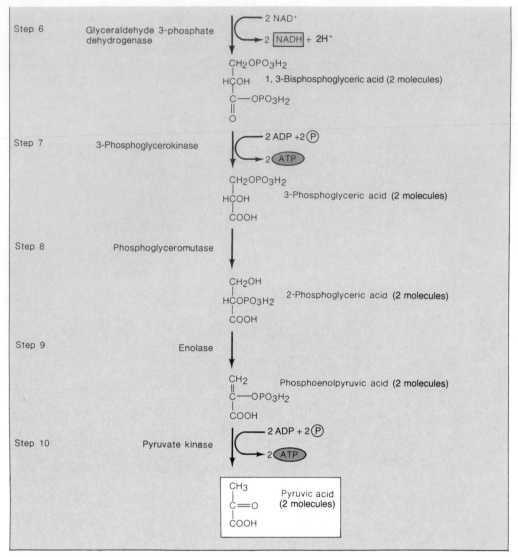

(a) Detailed version

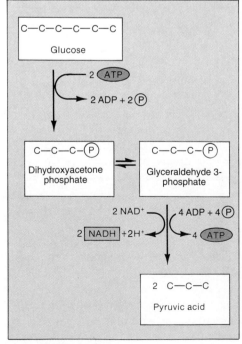

(b) Simplified version

Question: Why is the enzyme that catalyzes step 1 called a hexokinase?

FIGURE 25.3 Fate of pyruvic acid. When oxygen is plentiful, pyruvic acid enters the mitochondria, is converted to acetyl coenzyme A, and enters into the Krebs cycle (aerobic pathway). When oxygen is scarce, most pyruvic acid is converted to lactic acid (anaerobic pathway).

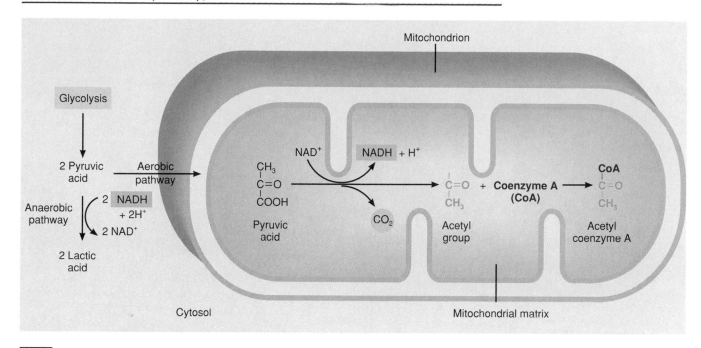

Question: In which cellular compartment does glycolysis occur? The Krebs cycle?

Krebs cycle, pyruvic acid is prepared for entrance into the cycle. It is converted to a two-carbon fragment by the loss of carbon dioxide (Fig. 25.3). The loss of a molecule of CO_2 by a substance is called **decarboxylation** (dē-kar bok'-si-LĀ-shun). (It is indicated in the diagram by the arrow labeled CO_2.) The two-carbon fragment, called an **acetyl group,** attaches to coenzyme A, and the whole complex is called **acetyl coenzyme A (acetyl CoA).** During this reaction, NAD^+ is reduced to $NADH + H^+$. Recall that the oxidation of one glucose molecule produces two molecules of pyruvic acid, so for each molecule of glucose two molecules of carbon dioxide are lost, and four $NADH + H^+$ are produced. Each molecule of NADH will later yield three molecules of ATP in the electron transport chain. Once the pyruvic acid has undergone decarboxylation and its derivative (the acetyl group) has attached to CoA, the resulting compound (acetyl CoA) is ready to enter the Krebs cycle.

Krebs (Citric Acid) Cycle

The **Krebs cycle** is also called the **citric acid cycle,** or the **tricarboxylic acid (TCA) cycle.** It is a series of biochemical reactions that occur in the matrix of mitochondria (see Fig. 3.18). The large amount of chemical potential energy stored in intermediate substances derived from pyruvic acid is released step by step (Fig. 25.4a,b). In this cycle, a series of oxidations and reductions transfers the chemical energy,

in the form of electrons, to several coenzymes. The pyruvic acid derivatives are oxidized, whereas the coenzymes are reduced.

Coenzyme A serves as a carrier molecule to guide the two-carbon acetyl unit into the Krebs cycle. This sequence of chemical reactions is called a "cycle" because the starting substance (oxaloacetic acid) is formed again at the end. The acetyl group combines with oxaloacetic acid to form citric acid. From this point, the citric acid cycle consists mainly of a series of decarboxylation and oxidation–reduction reactions, each controlled by a different enzyme.

Let us look at the decarboxylation reactions first. As we have seen (see Fig. 25.3), in preparation for entering the cycle, pyruvic acid is decarboxylated to form an acetyl group. In the cycle, isocitric acid, a six-carbon compound, loses a molecule of CO_2 to form a five-carbon compound called α-ketoglutaric acid (step ④). Then α-ketoglutaric acid is decarboxylated and picks up a molecule of CoA to form succinyl CoA, a four-carbon compound (step ⑤). Thus each time pyruvic acid enters the Krebs cycle, three molecules of CO_2 are liberated by decarboxylation (Fig. 25.4b). Therefore six molecules of CO_2 are liberated from each original glucose molecule catabolized along this pathway. The molecules of CO_2 leave the mitochondria, diffuse through the cytosol to the plasma membrane, and then diffuse into the blood. Eventually, the CO_2 is transported by the blood to the lungs and is exhaled.

Now let us look at the oxidation–reduction reactions

FIGURE 25.4 Krebs cycle. The net results of the Krebs cycle are the production of reduced coenzymes (NADH + H$^+$ and FADH$_2$), which contain stored energy; the generation of GTP, a high-energy compound that is used to produce ATP; and the formation of CO$_2$, which is transported to the lungs for expiration.

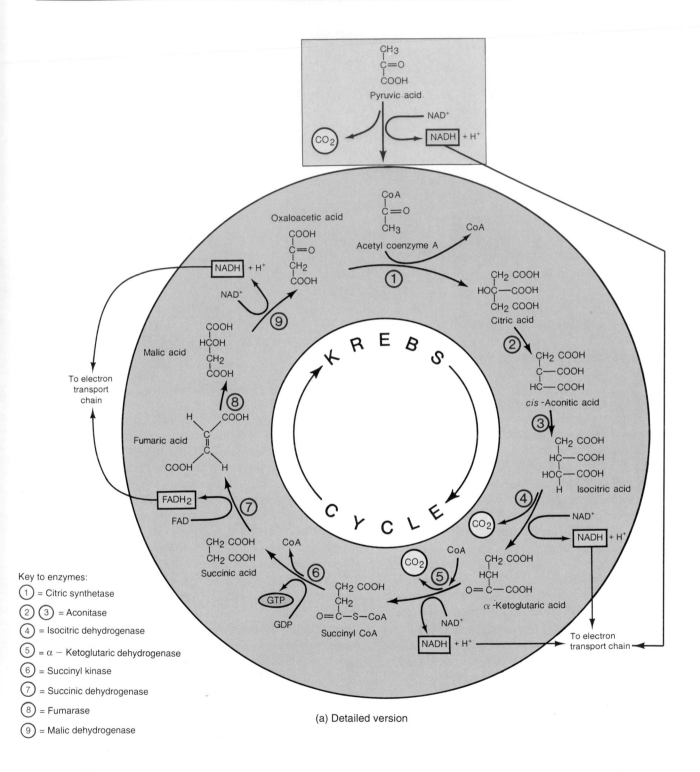

Key to enzymes:

① = Citric synthetase

② ③ = Aconitase

④ = Isocitric dehydrogenase

⑤ = α − Ketoglutaric dehydrogenase

⑥ = Succinyl kinase

⑦ = Succinic dehydrogenase

⑧ = Fumarase

⑨ = Malic dehydrogenase

(a) Detailed version

FIGURE 25.4 (continued)

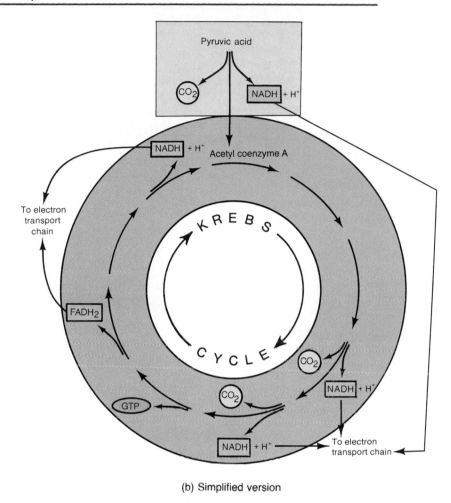

(b) Simplified version

Question: Why is the production of reduced coenzymes in the Krebs cycle important?

(Fig. 25.4a). Remember, when a molecule is oxidized, it loses hydrogen atoms (electrons and hydrogen ions). When a molecule is reduced, it gains hydrogen atoms (electrons and hydrogen ions). Every oxidation is coupled with a reduction. In the conversion of pyruvic acid to acetyl CoA, each pyruvic acid loses one hydrogen atom; that is, it is oxidized. The hydrogen atoms are picked up by the coenzyme NAD^+. Since NAD^+ picks up the hydrogen atoms, it is reduced and represented as $NADH + H^+$ (indicated in the diagram by the curved arrow entering and then leaving the reaction). Isocitric acid is oxidized to form α-ketoglutaric acid and NAD^+ is reduced to $NADH + H^+$ (step ④). α-Ketoglutaric acid is oxidized in the conversion to succinyl CoA and again NAD^+ is reduced to $NADH + H^+$ (step ⑤). Succinic acid is oxidized to fumaric acid. In this reaction, the coenzyme that picks up the two hydrogen atoms is flavin adenine dinucleotide (FAD; step ⑦). Malic acid is oxidized to oxaloacetic acid and NAD^+ is reduced to $NADH + H^+$ (step ⑨).

If we look at the Krebs cycle as a whole (Fig. 25.4b), we see that for every two molecules of acetyl CoA that enter the cycle four molecules of carbon dioxide are liberated by decarboxylation, $6 NADH + 6H^+$ and $2 FADH_2$ are produced by oxidation–reduction reactions, and two molecules of guanosine triphosphate (GTP, the equivalent of ATP) are generated by substrate-level phosphorylation. Many of the intermediates in the Krebs cycle also play a role in other pathways, especially in amino acid biosynthesis (discussed later in this chapter).

In the electron transport chain, the $6 NADH + 6H^+$ will later yield a maximum of 18 ATP molecules, and the 2 $FADH_2$ a maximum of 4 ATP molecules. The reduced coenzymes ($NADH + H^+$ and $FADH_2$) are the most important outcome of the Krebs cycle because they contain the energy originally stored in glucose and then in pyruvic acid. During the next phase of aerobic respiration, a series of reductions transfers the energy stored in the coenzymes to $ADP + ⑰$ to form ATP. These reactions involve the electron transport chain and occur on the cristae of the inner mitochondrial membrane.

Electron Transport Chain

The **electron transport chain** involves a sequence of **electron carrier molecules** on the inner mitochondrial membrane that are capable of oxidation and reduction. As electrons pass through the chain, there is a stepwise release of energy for generation of ATP. In aerobic cellular respiration, the last electron acceptor of the chain is molecular oxygen (O_2). The energy is released as electrons are passed from one carrier to the next and is used to pump H^+ (protons) from the matrix of the mitochondrion into the space between the inner and outer mitochondrial membranes (Fig. 25.5). ATP synthesis then occurs as H^+ diffuse back into the mitochondrial matrix through a special type of H^+ channel (pore) in the inner membrane. This mechanism of ATP generation is called **chemiosmosis** (kem′-ē-oz-MŌ-sis). We will first examine the electron carriers and then describe chemiosmosis.

● **Electron Carriers**   The electron transport chain involves carrier molecules in the inner mitochondrial membrane that are alternately oxidized and reduced. There are several types of carriers.

1.   **Flavin mononucleotide (FMN),** like FAD (flavin adenine dinucleotide), is a flavoprotein derived from riboflavin (vitamin B_2).

2.   **Cytochromes** (SĪ-tō-krōms) are proteins with an iron-containing group (heme) capable of existing alternately in a reduced form (Fe^{2+}) and an oxidized form (Fe^{3+}). The several cytochromes involved in the electron transport chain are cytochrome b (cyt b), cytochrome c_1 (cyt c_1), cytochrome c (cyt c), cytochrome a (cyt a), and cytochrome a_3 (cyt a_3).
3.   **Iron–sulfur** (Fe–S) **centers** contain either two or four iron atoms bound to sulfur atoms that form an electron transfer center within a protein.
4.   **Copper atoms (Cu)** bound to two proteins in the chain also participate in electron transfer.
5.   **Ubiquinones** are nonprotein carriers of low molecular weight that are mobile in the phospholipid bilayer of the membrane. They are also called **coenzyme Q,** symbolized **Q.**

● **Steps in Electron Transport**   Figure 25.6 shows the steps in electron transport. The first step is the transfer of high-energy electrons from NADH + H^+ to FMN, the first carrier in the chain. Two NADH + $2H^+$ are generated from glycolysis, two from the formation of acetyl CoA, and six from the Krebs cycle. In this transfer, a hydrogen atom with two electrons passes to FMN, which then picks up an additional H^+ from the surrounding aqueous medium. As a result, NADH + H^+ is oxidized to NAD^+, and FMN is reduced to $FMNH_2$.

In the second step in the electron transport chain, $FMNH_2$ passes electrons (hydrogen atoms each with two electrons) to several iron–sulfur centers and then to Q, which picks up an additional H^+ from the surrounding aqueous medium. As a result, $FMNH_2$ is oxidized to FMN.

The next sequence in the electron transport chain involves cytochromes, iron–sulfur centers, and copper atoms located between Q and molecular oxygen. Electrons are passed successively from Q to cyt b, to Fe–S, to cyt c_1, to cyt c, to Cu, to cyt a, and finally to cyt a_3. Each carrier in the chain is reduced as it picks up electrons and is oxidized as it gives up electrons. The last cytochrome, cyt a_3, passes its electrons to one-half of a molecule of oxygen (O_2), which becomes negatively charged and then picks up $2H^+$ from the surrounding medium to form H_2O. This is the point in aerobic cellular respiration where O_2 is consumed.

Note in Fig. 25.6 that $FADH_2$, derived from the Krebs cycle, is another source of electrons. However, $FADH_2$ adds its electrons to the electron transport chain at a lower energy level than NADH + H^+. Because of this, the electron transport chain produces about one-third less energy for ATP generation when $FADH_2$ donates electrons as compared with NADH + H^+.

● **Chemiosmotic Mechanism of ATP Generation**   Within the inner mitochondrial membrane, the carriers of the electron transport chain cluster into three complexes. Each complex acts as a **proton pump** that expels H^+ from the mitochondrial matrix and helps create an electrochemical gradient of H^+ (Fig. 25.7). Each proton pump complex includes three or more electron carriers.

FIGURE 25.5 Chemiosmosis.

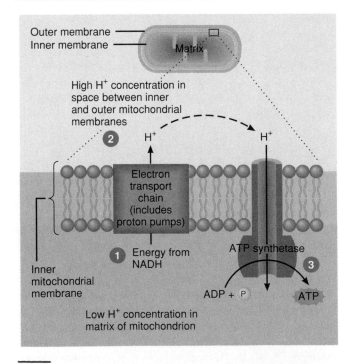

Outer membrane
Inner membrane
Matrix

High H^+ concentration in space between inner and outer mitochondrial membranes

❷ H^+ H^+

Electron transport chain (includes proton pumps)

Inner mitochondrial membrane

❶   Energy from NADH

ATP synthetase

❸

ADP + Ⓟ   ATP

Low H^+ concentration in matrix of mitochondrion

Question: What is the energy source that runs the proton pumps?

FIGURE 25.6 Electron transport chain. The energy drop for electrons passing through the chain occurs in a stepwise fashion.

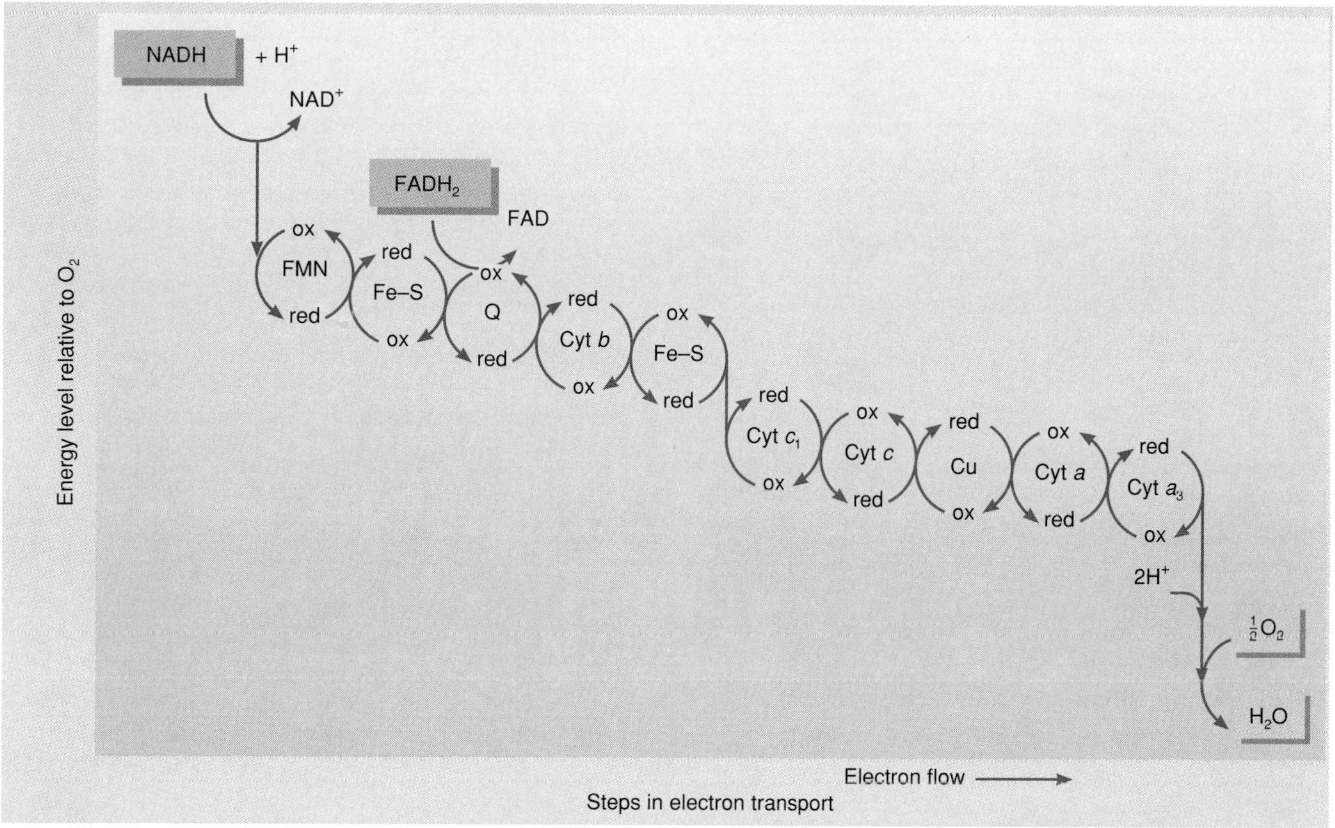

Question: Why is each electron carrier drawn on a lower level than the carrier from which it received electrons and on a higher level than the carrier to which it passes electrons?

1. **NADH dehydrogenase complex** contains flavin mononucleotide (FMN) and at least five Fe–S centers. Its name stems from its action, to remove hydrogen from NADH.

2. **Cytochrome b–c_1 complex** contains cytochromes b and c_1 and an Fe–S center.

3. **Cytochrome oxidase complex** contains cytochromes a and a_3 and two copper atoms.

One electron carrier (Q) ferries electrons from the first to the second complex and a second electron carrier (cytochrome c) ferries electrons from the second to the third. In summary, there are three complexes of electron carriers that pump protons (H^+): the NADH dehydrogenase complex, the cytochrome b–c_1 complex, and the cytochrome oxidase complex.

Because the inner mitochondrial membrane is nearly impermeable to H^+, the pumping of H^+ leads to both a concentration gradient of protons and an electrical gradient. The buildup of H^+ on one side of the membrane confers a positive charge, whereas the fluid on the other side is left

with a negative charge. This proton gradient has potential energy and is called the **proton motive force.** In regions where specific H^+ channels exist, H^+ can diffuse across the membrane, driven by the proton motive force. As they do so, they generate ATP because the H^+ channels also include an enzyme called **ATP synthetase.** The enzyme uses the proton motive force to synthesize ATP from ADP and Ⓟ. This process is termed chemiosmotic generation of ATP, and it is responsible for most of the ATP produced during cellular respiration.

Summary of Aerobic Cellular Respiration

The various electron transfers in the electron transport chain generate a maximum of 32 or 34 ATP molecules from each molecule of glucose oxidized: 28 or 30 (depending on which electron carriers participate) from the 10 molecules of NADH + H^+ and 2 from each of the 2 molecules of $FADH_2$ (4 total). Thus, during aerobic respiration, a theoretical maximum of 36 or 38 ATPs can be gen-

FIGURE 25.7 Proton pumps. The three proton pumps move H⁺ from the mitochondrial matrix into the space between inner and outer mitochondrial membranes.

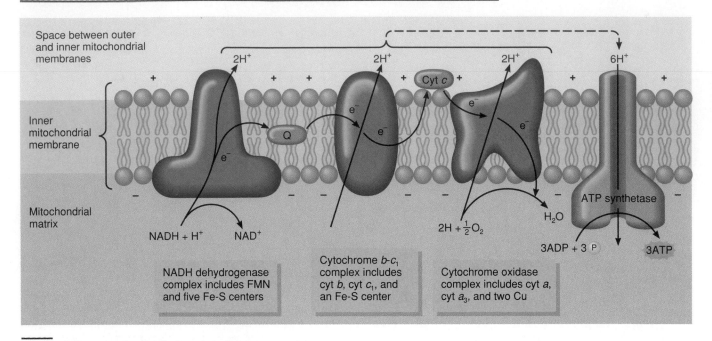

Question: What types of molecules constitute the proton pumps?

erated from one molecule of glucose. Note that four of those ATPs come from substrate-level phosphorylation in glycolysis and the Krebs cycle. Exhibit 25.1 provides a detailed accounting of the maximal ATP yield during aerobic respiration.

The overall reaction for aerobic respiration is:

$$C_6H_{12}O_6 + 6O_2 + 36 \text{ or } 38 \text{ ADPs} + 36 \text{ or } 38\text{ⓟ}$$

Glucose   Oxygen

$$\rightarrow 6CO_2 + 6H_2O + 36 \text{ or } 38 \text{ ATPs}$$

Carbon   Water
dioxide

About 40% of the energy originally in glucose is captured by ATP; the remainder is given off as heat. Compared with man-made machines, which typically operate at 10 to 20% efficiency, this energy yield is quite large.

A summary of the various stages of aerobic respiration is presented in Fig. 25.8. As noted earlier, the actual ATP yield may be lower than 36 or 38 ATPs per glucose. One uncertainty is the exact number of H⁺ that must be pumped out to generate one ATP during chemiosmosis. Also, the ATP generated in mitochondria must be transported out of these organelles into the cytoplasm for use elsewhere in a cell. This process uses up a portion of the proton motive force to drive out ATP in exchange for the inward movement of ADP that is formed from metabolic reactions in the cytosol.

Glycolysis, the citric acid cycle, and especially the electron transport chain provide all the ATP for cellular activities. And because the Krebs cycle and electron transport chain are aerobic processes, cells cannot carry on their activities for long without sufficient oxygen.

CLINICAL APPLICATION

CARBOHYDRATE LOADING

During exercise, the body preferentially uses carbohydrates as a source of energy. Carbohydrates are stored as glycogen in the liver and skeletal muscles. Since glycogen stores are limited and can be depleted after running more than 32 km (20 mi), many marathon runners consume large amounts of carbohydrates before the event. This practice, called **carbohydrate loading,** is designed to maximize body stores of carbohydrates to provide additional energy for an athletic event. It is often done for three days before the run. Some possible adverse effects of carbohydrate loading are muscle stiffness, diarrhea, chest pain, depression, and lethargy.

Glucose Anabolism

Most of the glucose in the body is catabolized to supply energy. However, glucose molecules may take part in several anabolic reactions. One is the synthesis of glycogen. Another is the manufacture of glucose from the breakdown products of proteins and lipids.

EXHIBIT 25.1

SUMMARY OF ATP PRODUCED DURING AEROBIC RESPIRATION OF ONE GLUCOSE MOLECULE

Source	ATP Yield (Method)
GLYCOLYSIS	
1. Oxidation of glucose to pyruvic acid	2 ATPs (substrate-level phosphorylation)
2. Production of 2 NADH + 2H⁺	4 or 6 ATPs (oxidative phosphorylation in electron transport chain)
FORMATION OF ACETYL CoA	
2 NADH + 2H⁺	6 ATPs (oxidative phosphorylation in electron transport chain)
KREBS CYCLE	
1. Oxidation of succinyl CoA to succinic acid	2 GTPs (equivalent of ATP; substrate-level phosphorylation)
2. Production of 6 NADH + 6H⁺	18 ATPs (oxidative phosphorylation in electron transport chain)
3. Production of 2 FADH₂	4 ATPs (oxidative phosphorylation in electron transport chain)
Total:	**36 or 38 ATPs (theoretical maximum)**

Glucose Storage: Glycogenesis

If glucose is not needed immediately for ATP production, it is combined with many other molecules of glucose to form a long-chain molecule called glycogen. This process is called **glycogenesis** (*glyco* = sweet; *genesis* = to generate). The body can store about 500 g (about 1.1 lb) of glycogen, about 25% in the liver and 75% in skeletal muscle fibers.

In the process of glycogenesis (Fig. 25.9), glucose that enters cells is first phosphorylated to glucose 6-phosphate. This is then converted to glucose 1-phosphate, then to uridine diphosphate glucose, and finally to glycogen. Glycogenesis is stimulated by insulin from the pancreas.

Glucose Release: Glycogenolysis

When the body needs energy, glycogen stored in the liver is broken down into glucose and released into the blood to be transported to cells, where it will be catabolized. The process of converting glycogen back to glucose is called **glycogenolysis** (*lysis* = breakdown). Glycogenolysis usually occurs in between meals.

Glycogenolysis does not occur exactly by the reverse process involved in glycogenesis (Fig. 25.9). The first step in glycogenolysis involves splitting glucose molecules from the branched glycogen molecule by phosphorylation to form glucose 1-phosphate. Phosphorylase, the enzyme that catalyzes this reaction, is activated by the hormones glucagon from the pancreas and epinephrine from the adrenal medulla. Glucose 1-phosphate is then converted to glucose 6-phosphate and finally to glucose. Phosphatase, the enzyme that converts glucose 6-phosphate into glucose, is present in liver cells but absent in skeletal muscle cells. In skeletal muscle cells, glycogen is broken down into glucose 1-phosphate, which is then catabolized for ATP production via glycolysis and the Krebs cycle. However, pyruvic acid and lactic acid, produced by glycolysis in muscle cells, can be converted to glucose in the liver. In this way, muscle glycogen can be an indirect source of blood glucose.

Formation of Glucose from Proteins and Fats: Gluconeogenesis

When your liver glycogen runs low, it is time to eat. If you do not eat, your body starts catabolizing more triglycerides and proteins. Although the body normally catabolizes some of its fats and a few of its proteins, large-scale fat and protein catabolism does not happen unless you are starving, eating meals that contain very few carbohydrates, or suffering from an endocrine disorder.

Both triglyceride molecules and protein molecules may be broken down and converted to glucose in the liver. The process by which new glucose is formed from noncarbohydrate sources is called **gluconeogenesis** (gloo′-kō-nē′-ō-JEN-e-sis; *neo* = new).

In the process of gluconeogenesis, glucose is formed from lactic acid, certain amino acids, and the glycerol portion of triglyceride molecules (Fig. 25.10). About 60% of the amino acids in the body can undergo this conversion. Amino acids such as alanine, cysteine, glycine, serine, and threonine are converted to pyruvic acid. The pyruvic acid may be synthesized into glucose or enter the Krebs cycle. Glycerol may be converted into glyceraldehyde 3-phosphate, which may form pyruvic acid or be used to synthesize glucose. Figure 25.10 also shows how gluconeogenesis is related to other metabolic reactions.

Gluconeogenesis is stimulated by cortisol, one of the glucocorticoid hormones of the adrenal cortex, and thyroid hormone from the thyroid gland. Cortisol mobilizes proteins from body cells, making them available in the form of amino acids, thus supplying a pool of amino acids for gluconeogenesis. Thyroid hormone also mobilizes proteins and may mobilize triglycerides from fat depots, thus making glycerol available for gluconeogenesis. Epinephrine, glucagon, and human growth hormone (hGH) also stimulate gluconeogenesis.

LIPID METABOLISM

When triglycerides are eaten, they are digested into fatty acids and monoglycerides. Short-chain fatty acids diffuse into epithelial cells of the intestinal villi and then into the blood capillaries. Long-chain fatty acids and monoglycerides are carried in micelles to epithelial cells of the villi for entrance. Once inside, they are further digested to glycerol and fatty acids and subsequently recombined to form triglycerides. They leave intestinal cells in chylomicrons and enter lymph in lacteals of villi. Finally, the chylomicrons reach the blood via the thoracic duct. Since lipids dissolve poorly in water, they are transported in the blood-stream in several types of lipoprotein particles. The major ones are chylomicrons, very low-density lipoproteins (VLDLs), low density lipoproteins (LDLs), and high-density lipoproteins (HDLs).

Fate of Lipids

Lipids, like carbohydrates, may be oxidized to produce ATP. Each gram of triglyceride produces about 9 kilocalories. If the body has no immediate need to use lipids this way, they are stored in adipose tissue (fat depots) throughout the body and in the liver. Other lipids are used as structural molecules or to synthesize other essential substances.

FIGURE 25.8 Summary of aerobic respiration.

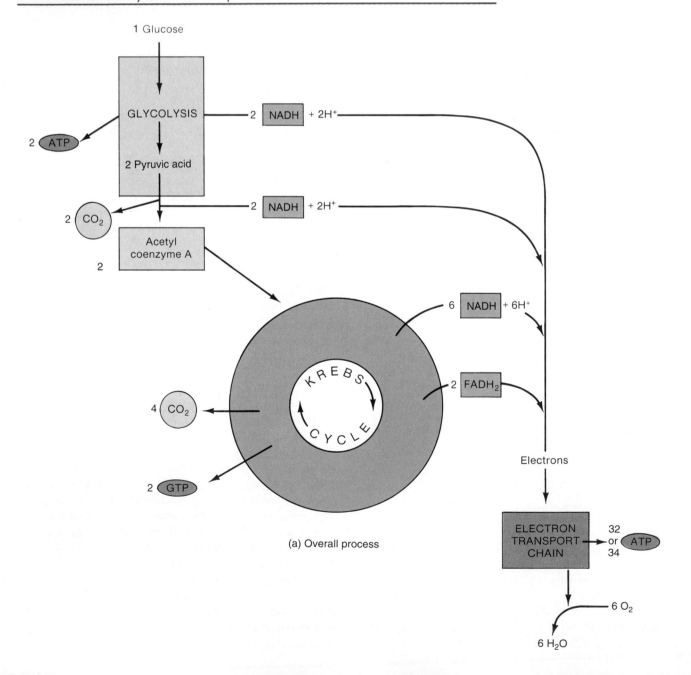

(a) Overall process

FIGURE 25.8 (continued)

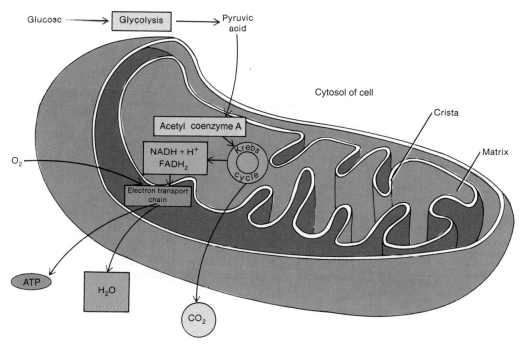

(b) Sites of principal events

Question: How many molecules of oxygen are used during the complete oxidation of one glucose molecule? How many carbon dioxide molecules are produced?

FIGURE 25.9 Glycogenesis and glycogenolysis. The glycogenesis pathway, conversion of glucose into glycogen, is stimulated by insulin and indicated by red arrows. The glycogenolysis pathway, conversion of glycogen into glucose, is stimulated by glucagon and epinephrine and indicated by blue arrows.

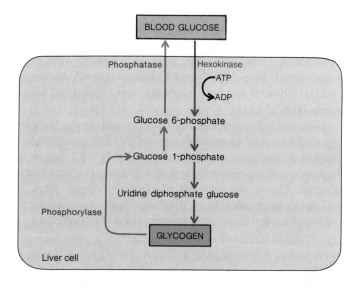

Question: Besides liver cells, which cells can synthesize glycogen? Why can't they release glucose into the blood?

For example, phospholipids are constituents of plasma membranes, lipoproteins are used to transport cholesterol throughout the body, thromboplastin is needed for blood clotting, and myelin sheaths speed up nerve impulse conduction. Cholesterol, another lipid, is used in the synthesis of bile salts and steroid hormones (adrenocortical hormones and sex hormones). The various functions of lipids in the body may be reviewed in Exhibit 2.3 on page 43.

Fat Storage

The major function of adipose tissue is to store triglycerides until they are needed for ATP production in other parts of the body. It also insulates and protects. About 50% of stored triglycerides is deposited in subcutaneous tissue—approximately 12% around the kidneys, 10 to 15% in the omenta, 20% in genital areas, and 5 to 8% between muscles. Triglycerides are also stored behind the eyes, in the sulci of the heart, and in the folds of the large intestine.

Adipose cells contain lipases that hydrolyze triglycerides in chylomicrons and VLDLs into fatty acids and glycerol. Because of the rapid exchange of fatty acids, triglycerides are renewed about once every two or three weeks. Thus the triglycerides stored in your adipose tissue today are not the same ones that were present last month. Also, they are continually released from storage, transported in the blood, and redeposited in other adipose tissue cells.

FIGURE 25.13 Summary of key molecules and pathways in metabolism.

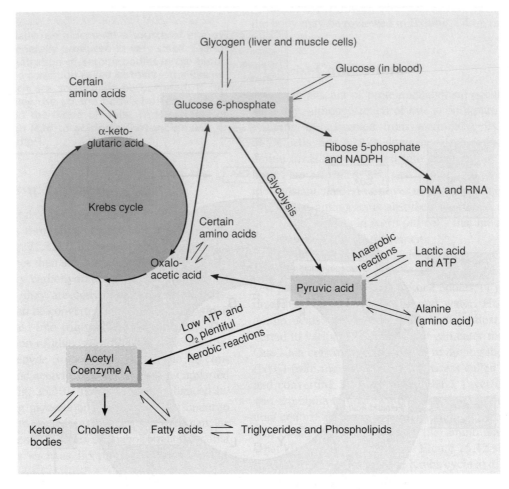

Question: Which substance is the gateway into the Krebs cycle for fuel molecules that are being oxidized to generate ATP?

enzymes needed to catalyze the steps, four possible fates await this molecule:

1. If the enzyme glucose 6-phosphatase is present and active, G 6-P may be dephosphorylated; that is, it loses a phosphate group and is converted to glucose. Thus glucose 6-phosphatase allows glucose to leave a cell.

2. G 6-P may be used to synthesize glycogen. In turn, when glycogen is broken down, G 6-P reforms.

3. Although it hasn't been discussed here, G 6-P may be used to generate ribose 5-phosphate, a five-carbon sugar that is needed for synthesis of RNA (ribonucleic acid) and DNA (deoxyribonucleic acid). The same sequence of reactions that produces ribose 5-phosphate also produces NADPH, which is needed for reduction reactions.

4. Some ATP is produced anaerobically via glycolysis, in which G 6-P is converted to pyruvic acid, another key molecule in metabolism.

Pyruvic Acid

Pyruvic acid marks a decision point in anaerobic versus aerobic cellular respiration. When the ATP level in a cell is low but oxygen is plentiful, most pyruvic acid streams toward ATP-producing reactions (the Krebs cycle and electron transport chain) via conversion to acetyl CoA. On the other hand, when oxygen is in short supply in a tissue, as in actively contracting skeletal or cardiac muscle, most pyruvic acid is changed to lactic acid. This conversion generates a small amount of ATP. The lactic acid then diffuses into the bloodstream and is taken up by liver cells, which sooner or later convert it back into pyruvic acid.

One link between carbohydrate and protein metabolism occurs at pyruvic acid. An amino group can be added to pyruvic acid (a carbohydrate) to produce the amino acid alanine or removed from alanine to generate pyruvic acid. Pyruvic acid and certain amino acids also can be converted to oxaloacetic acid, one of the citric acid cycle intermediates, which in turn can generate glucose 6-phosphate. This sequence of reactions bypasses certain one-way reactions of glycolysis and is called gluconeogenesis.

Acetyl CoA

Acetyl CoA is the gateway into the Krebs cycle for fuel molecules that are being oxidized to generate ATP. It also is used to synthesize fatty acids, ketone bodies, and cholesterol. Since pyruvic acid can be converted to acetyl CoA, carbohydrates can be turned into triglycerides; this is the metabolic path for storage of most excess calories as triglycerides. Mammals, including humans, cannot reconvert acetyl CoA to pyruvic acid, however, so lipids cannot

be used to generate glucose or other carbohydrate molecules.

A summary of carbohydrate, lipid, and protein metabolism is presented in Exhibit 25.2.

ABSORPTIVE AND POSTABSORPTIVE STATES

Your metabolic reactions depend on how recently you have eaten. During the **absorptive (fed) state,** ingested nutrients are entering the cardiovascular and lymphatic systems from the GI tract, and glucose is readily available for ATP production. During the **postabsorptive (fasting) state,** absorption is complete, and the energy needs of the body must be

---

EXHIBIT 25.2

SUMMARY OF METABOLISM

Process	Comment
CARBOHYDRATE METABOLISM	
GLUCOSE CATABOLISM	Complete oxidation of glucose, also referred to as cellular respiration, is chief source of ATP in cells. The process requires glycolysis, Krebs cycle, and electron transport chain. The complete oxidation of 1 molecule of glucose yields a maximum of 36 or 38 molecules of ATP.
Glycolysis	Conversion of glucose into pyruvic acid results in the production of some ATP. Reactions do not require oxygen (anaerobic cellular respiration).
Krebs cycle	Cycle includes series of oxidation–reduction reactions in which coenzymes (NAD^+ and FAD) pick up hydrogen atoms from oxidized organic acids, and some ATP is produced. CO_2 and H_2O are by-products. Reactions are aerobic.
Electron transport chain	Third set of reactions in glucose catabolism is another series of oxidation–reduction reactions, in which electrons are passed from one carrier to the next, and most of the ATP is produced. Reactions require oxygen (aerobic cellular respiration).
GLUCOSE ANABOLISM	Some glucose is converted into glycogen (glycogenesis) for storage if not needed immediately for ATP production. Glycogen can be reconverted to glucose (glycogenolysis). The conversion of amino acids, glycerol, and lactic acid into glucose is called gluconeogenesis.
LIPID METABOLISM	
CATABOLISM	Glycerol may be converted into glucose (gluconeogenesis) or catabolized via glycolysis. Fatty acids are catabolized via beta oxidation into acetyl CoA that can enter the Krebs cycle for ATP production or be converted into ketone bodies (ketogenesis).
ANABOLISM	The synthesis of lipids from glucose and fatty acids is lipogenesis. Triglycerides are stored in adipose tissue.
PROTEIN METABOLISM	
CATABOLISM	Amino acids are oxidized via the Krebs cycle after conversion by processes such as deamination. Ammonia resulting from the conversions is converted into urea in the liver and excreted in urine. Amino acids may be converted into glucose (gluconeogenesis), fatty acids, or ketone bodies
ANABOLISM	Protein synthesis is directed by DNA and utilizes the cells' RNA and ribosomes.

satisfied by nutrients already in the body. Hormones are the major regulators of reactions occurring in each state. The effects of insulin dominate in the absorptive state whereas the effects of several hormones contribute to regulation of metabolic reactions in the postabsorptive state. By analyzing the major events of both states, we can gain a better understanding of the interrelations of metabolic pathways.

Absorptive (Fed) State

An average meal requires about 4 hours for complete absorption, and given three meals a day, the body spends about 12 hours each day in the absorptive state. The other 12 hours, in the late morning, late afternoon, and most of the night, are spent in the postabsorptive state.

Reactions of the Absorptive State

During the absorptive state, most body cells produce ATP by oxidizing glucose to carbon dioxide and water (Fig. 25.14). Glucose that enters liver cells is mostly converted to triglycerides or glycogen; little is oxidized for energy. Some fatty acids and triglycerides synthesized in the liver remain there, but most are packaged into very low-density

FIGURE 25.14 Principal pathways during the absorptive (fed) state.

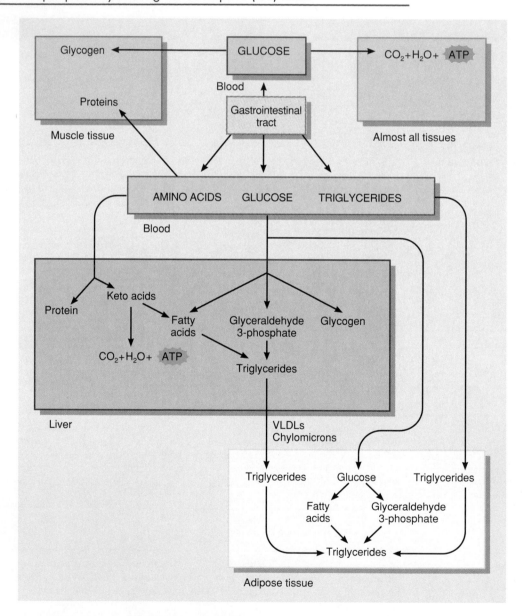

Question: Are the reactions shown here mainly anabolic or catabolic?

lipoproteins (VLDLs) that enter the blood and ferry lipids to adipose tissue for storage. Adipose tissue cells also take up glucose not picked up by the liver and convert it into triglycerides for storage. Some blood glucose is stored as glycogen in skeletal muscles.

During the absorptive state, most dietary lipids are stored in adipose tissue; only a small portion is used for synthetic reactions. Adipose cells derive the lipids from chylomicrons, from VLDLs, and from their own synthetic reactions (Fig. 25.14).

Many absorbed amino acids that enter liver cells are deaminated to carbohydrates called keto acids. These then can enter the Krebs cycle for ATP production or be used to synthesize glucose or fatty acids. Some amino acids that enter liver cells are used to synthesize proteins, such as plasma proteins. Amino acids not taken up by liver cells enter other cells of the body, such as muscle cells, for synthesis of proteins or regulatory chemicals such as hormones or enzymes.

Hormonal Regulation of Absorptive State Reactions

Soon after eating, gastric inhibitory peptide (GIP) and the rise in blood glucose concentration stimulate insulin release from pancreatic beta cells. In several ways, insulin stimulates absorptive state metabolism. Insulin promotes entry of glucose and amino acids into cells of many tissues. It stimulates phosphorylation of glucose in liver cells and conversion of glucose 6-phosphate to glycogen in both liver and muscle. In liver and in adipose tissue, insulin enhances synthesis of triglycerides. (See page 549 to review the effects of insulin.) Human growth hormone and the thyroid hormones (T_3 and T_4) also stimulate some absorptive state reactions.

Exhibit 25.3 summarizes the metabolic processes occurring in the absorptive state.

Postabsorptive (Fasting) State

The principal metabolic challenge during the postabsorptive state is to maintain normal blood glucose level of 70 to 110 mg/100 ml of blood (3.9 to 5.6 mmol/liter). Cells are continually removing glucose from the blood, but none is being absorbed by the GI tract. The maintenance of this level is especially important for the nervous system, since neurons cannot use other nutrients (except ketone bodies during starvation) for ATP production.

Reactions of the Postabsorptive State

Blood glucose concentration is maintained in two ways: by synthesizing glucose molecules and by glucose-sparing reactions such as using triglycerides or possibly proteins for ATP production. A major source of blood glucose during fasting is liver glycogen, which can provide about a four-hour supply of glucose. Liver glycogen is continually being formed and broken down as needed (Fig. 25.15).

Another source of blood glucose is glycerol, produced by the hydrolysis of triglycerides, primarily in adipose tissue. However, only the glycerol part of triglycerides can form glucose; the fatty acids cannot be utilized because acetyl CoA cannot be readily converted to pyruvic acid. But most cells can oxidize the fatty acids directly, feeding them into the Krebs cycle as acetyl CoA.

A third source of glucose may be muscle glycogen during periods of vigorous exercise when significant amounts of lactic acid are produced by anaerobic glycolysis in muscle. Muscle lactic acid is released into the blood and carried

EXHIBIT 25.3

METABOLIC PROCESSES DURING THE ABSORPTIVE (FED) STATE

Process	Location	Main Stimulating Hormones
Facilitated diffusion of glucose into cells	Most cells	Insulin[a]
Active transport of amino acids into cells	Most cells	Insulin
Glycogenesis (glycogen synthesis)	Liver cells and muscle fibers	Insulin
Protein synthesis	All body cells	Insulin, thyroid hormone, human growth hormone
Lipogenesis (triglyceride synthesis)	Adipose cells and liver cells	Insulin

[a] Facilitated diffusion of glucose into liver cells and neurons is always "turned on" and does not require insulin.

FIGURE 25.15 Principal pathways during the postabsorptive (fasting) state.

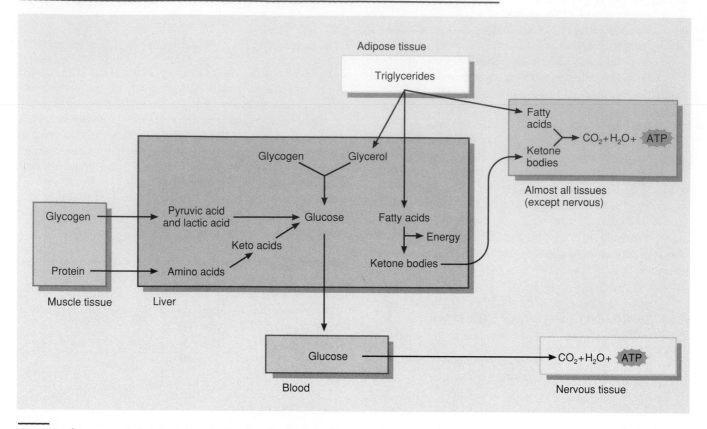

Question: What processes *directly* elevate blood glucose level during this state and where are they taking place?

to the liver, where it is converted back to glucose and released into the blood.

Finally, tissue protein, primarily from skeletal muscles, can contribute to blood glucose. During prolonged fasting, large amounts of amino acids from protein breakdown are released from muscles to be converted to glucose in the liver by gluconeogenesis. Or the amino acids may be oxidized directly, thus sparing blood glucose. During fasting, amino acids from muscles contribute to blood glucose after liver glycogen and adipose triglyceride stores are depleted.

Despite all these mechanisms for supplying blood glucose, they cannot maintain blood glucose level for very long without further metabolic changes. Thus a major body adjustment must be made during the postabsorptive state. Although the nervous system continues to utilize blood glucose normally, all other body tissues reduce their oxidation of glucose and switch over to fatty acids as their main ATP source. Accordingly, triglycerides in adipose tissue are broken down and fatty acids are released into the blood. The fatty acids are picked up by most body cells, except neurons, and oxidized to carbon dioxide and water, releasing energy that is used to form ATP. The liver converts fatty acids to ketone bodies, which heart muscle, kidney tissue, and nervous tissue can oxidize to carbon dioxide and water,

releasing energy used to form ATP. As a result of this glucose sparing and fatty acid utilization, a person can fast for several weeks, provided water is consumed, and the glucose level will not drop more than 25% from its normal range.

Hormonal Regulation of Postabsorptive State Reactions

The hormones that stimulate metabolism in the postabsorptive state sometimes are called anti-insulin hormones because they counter the insulin effects that dominate during the fed state. The most important one is glucagon. As blood glucose level declines, secretion of insulin falls off and release of glucagon and other anti-insulin hormones rises. Exhibit 25.4 summarizes the actions of the principal hormones that stimulate metabolic processes during the postabsorptive state.

MINERALS

Although hormones are the primary regulators of metabolism, they are ineffective without the proper minerals and vitamins. Some minerals and many vitamins are com-

EXHIBIT 25.4

METABOLIC PROCESSES DURING THE POSTABSORPTIVE (FASTING) STATE

Process	Location	Main Stimulating Hormones
Glycogenolysis (glycogen breakdown)	Liver cells and muscle fibers.	Glucagon and epinephrine.
Lipolysis (triglyceride breakdown)	Adipose cells.	Epinephrine, norepinephrine, cortisol, human growth hormone, thyroid hormone, and others.
Protein breakdown	Most body cells, but especially skeletal muscle fibers.	Cortisol.
Gluconeogenesis (synthesis of glucose from noncarbohydrate precursors)	Liver cells and kidney cortex cells.	Glucagon and cortisol.

ponents of the enzyme systems that catalyze the metabolic reactions.

Minerals are inorganic substances. They may appear in combination with each other or in combination with organic compounds or as ions in solution. Minerals constitute about 4% of the total body weight, and they are concentrated most heavily in the skeleton. Minerals known to have functions essential to life include potassium, magnesium, chlorine, iodine, calcium, phosphorus, sodium, manganese, cobalt, copper, zinc, selenium, and chromium. Other minerals—aluminum, silicon, arsenic, and nickel—are present in the body, but their functions are not known.

Calcium and phosphorus form part of the structure of bone. But since minerals do not form long-chain compounds, they are otherwise poor building materials. Their chief role is to help regulate body processes. Calcium, iron, magnesium, and manganese are constituents of some coenzymes. Magnesium also serves as a catalyst for the conversion of ADP to ATP. Minerals such as sodium and phosphorus work in buffer systems. Sodium helps regulate the osmosis of water and, along with other ions, is involved in the generation of nerve impulses. Exhibit 25.5 describes the vital functions of some minerals. Note that the body generally uses the ions of the minerals rather than the nonionized form. Some minerals, such as chlorine, are toxic or even fatal if ingested in the nonionized form.

EXHIBIT 25.5

MINERALS VITAL TO THE BODY

Mineral	Comments	Importance
Calcium	Most abundant cation in body. Appears in combination with phosphorus. About 99% is stored in bone and teeth. Remainder is in muscle, other soft tissues, and blood plasma. Blood Ca^{2+} level controlled by calcitonin (CT) and parathyroid hormone (PTH). Absorption occurs only in the presence of vitamin D. Excess is excreted in feces and small amount in urine. Sources are milk, egg yolk, shellfish, green leafy vegetables.	Formation of bones and teeth, blood clotting, normal muscle and nerve activity, endocytosis and exocytosis, cellular motility, chromosome movement prior to cell division, glycogen metabolism, and synthesis and release of neurotransmitters.

Exhibit continues

EXHIBIT 25.5 (continued)

MINERALS VITAL TO THE BODY

Mineral	Comments	Importance
Phosphorus	About 80% found in bones and teeth. Remainder distributed in muscle, brain cells, blood. More functions than any other mineral. Blood phosphorus level controlled by calcitonin (CT) and parathyroid hormone (PTH). Excess excreted in urine; small amount eliminated in feces. Sources are dairy products, meat, fish, poultry, nuts.	Formation of bones and teeth. Phosphates ($H_2PO_4^-$, HPO_4^{2-}, and PO_4^{3-}) constitute a major buffer system of blood. Plays important role in muscle contraction and nerve activity. Component of many enzymes. Involved in energy transfer (ATP). Component of DNA and RNA.
Iron	About 66% found in hemoglobin of blood. Remainder distributed in skeletal muscles, liver, spleen, enzymes. Normal losses of iron occur by shedding of hair, epithelial cells, and mucosal cells, and in sweat, urine, feces, bile, and blood lost during menstruation. Sources are meat, liver, shellfish, egg yolk, beans, legumes, dried fruits, nuts, cereals.	As component of hemoglobin, reversibly binds O_2. Component of cytochromes involved in formation of ATP. Large amounts of stored iron are associated with an increased risk of cancer. This may be related to the ability of iron to promote the production of oxygen free radicals and serve as a nutrient for cancer cells.
Iodine	Essential component of thyroid hormones. Excreted in urine. Sources are seafood, iodized salt, and vegetables grown in iodine-rich soils.	Required by thyroid gland to synthesize thyroid hormones, which regulate metabolic rate.
Copper	Some stored in liver and spleen. Most excreted in feces. Sources include eggs, whole-wheat flour, beans, beets, liver, fish, spinach, asparagus.	Required with iron for synthesis of hemoglobin. Component of coenzymes in electron transport chain and enzyme necessary for melanin pigment formation.
Sodium	Most abundant cation in extracellular fluids, some found in bones. Excreted in urine and perspiration. Normal intake of NaCl (table salt) supplies required amounts.	Strongly affects distribution of water through osmosis. Part of bicarbonate buffer system. Functions in nerve and muscle action potential conduction.
Potassium	Principal cation in intracellular fluid. Excess excreted in urine. Normal food intake supplies required amounts.	Functions in nerve and muscle action potential conduction.
Chlorine	Found in extracellular and intracellular fluids. Principal anion of extracellular fluid. Excess excreted in urine. Normal intake of NaCl supplies required amounts.	Plays role in acid–base balance of blood, water balance, and formation of HCl in stomach.
Magnesium	Component of soft tissues and bone. Excreted in urine and feces. Widespread in various foods, such as green leafy vegetables, seafood, and whole-grain cereals.	Required for normal functioning of muscle and nervous tissue. Participates in bone formation. Constituent of many coenzymes. Deficiency linked to diabetes, hypertension, high blood levels of cholesterol, pregnancy problems, and spasms in blood vessels.
Sulfur	Constituent of many proteins (such as insulin), electron carriers in oxidative phosphorylation, and some vitamins (thiamine and biotin). Excreted in urine. Sources include beef, liver, lamb, fish, poultry, eggs, cheese, beans.	As component of hormones and vitamins, regulates various body activities. Needed for ATP production by aerobic cellular respiration.

Mineral	Comments	Importance
Zinc	Important component of certain enzymes. Widespread in many foods, especially meats.	As a component of carbonic anhydrase, important in carbon dioxide metabolism. Necessary for normal growth and wound healing, normal taste sensations and appetite, and normal sperm counts in males. As a component of peptidases, it is involved in protein digestion. Deficiency may be involved in impaired immunity and slow learning. Excess may raise cholesterol level.
Fluorine	Component of bones, teeth, other tissues.	Appears to improve tooth structure and inhibit tooth decay. In clinical trials, sodium fluoride and calcium have altered the progression of spinal osteoporosis.
Manganese	Some stored in liver and spleen. Most excreted in feces.	Activates several enzymes. Needed for hemoglobin synthesis, urea formation, growth, reproduction, lactation, bone formation, and possibly production and release of insulin, and inhibiting cell damage.
Cobalt	Constituent of vitamin B_{12}.	As part of vitamin B_{12}, required for erythropoiesis.
Chromium	Found in high concentrations in brewer's yeast. Also found in wine and some brands of beer.	Needed for the proper use of dietary sugars and other carbohydrates by optimizing the production and effects of insulin. Helps increase blood levels of HDL, while decreasing levels of LDL.
Selenium	Found in seafood, meat, chicken, grain cereals, egg yolk, milk, mushrooms, and garlic.	An antioxidant. Prevents chromosome breakage and may play a role in preventing certain birth defects and esophageal cancer.

VITAMINS

Organic nutrients required in minute amounts to maintain growth and normal metabolism are called **vitamins.** Unlike carbohydrates, lipids, or proteins, vitamins do not provide energy or serve as building materials. The essential function of vitamins is the regulation of physiological processes. Of the vitamins whose functions are known, most serve as coenzymes.

Most vitamins cannot be synthesized by the body. They may be ingested in food or pills. Other vitamins, such as vitamin K, are produced by bacteria in the GI tract. The body can assemble some vitamins if the raw materials called **provitamins** are provided. For example, vitamin A is produced by the body from the provitamin carotene, a chemical present in yellow vegetables such as carrots and dark green vegetables such as spinach. No single food contains all the required vitamins—one of the best reasons for eating a varied diet.

Based on solubility, vitamins are divided into two principal groups: fat-soluble and water-soluble. **Fat-soluble vitamins** are emulsified into micelles and absorbed along with ingested dietary fats by the small intestine. In fact, they cannot be absorbed in adequate quantity unless they are ingested with some fat. Fat-soluble vitamins may be stored in cells, particularly liver cells. The fat-soluble vitamins include vitamins A, D, E, and K. **Water-soluble vitamins,** by contrast, are absorbed along with water in the GI tract and dissolve in the body fluids. Excess quantities of these vitamins are excreted in the urine. Thus the body does not store water-soluble vitamins well. Examples of water-soluble vitamins are the B vitamins and vitamin C.

You have probably noticed the initials **U.S. RDA** on the labels of foods and vitamin and mineral supplements. These letters stand for **United States Recommended Daily Allowance.** They are the levels of intake of essential nutrients considered, in the judgment of the Food and Drug Administration (FDA) based on available scientific knowledge, to meet and surpass the known nutritional needs of practically all healthy adults. They don't take into consideration possible extra needs of growing children, pregnant or lactating women, or sick people.

CLINICAL APPLICATION

VITAMIN AND MINERAL SUPPLEMENTS

Most physicians recommend against taking vitamin or mineral supplements, except in special circumstances, and instead suggest being sure to eat a balanced diet that includes a variety of foods. Common situations that may warrant using supplements include: iron supplements for women who have excessive menstrual bleeding; iron, calcium, and folic acid supplements for women who are pregnant or breast feeding; calcium supplements for adults who do not receive the U.S. RDA in their diets; and vitamin B_{12} supplements for strict vegetarians who eat no meat. (The dangers of vitamin or mineral overdose are described on page 857.)

Exhibit 25.6 lists the principal vitamins, their sources, functions, and related deficiency disorders.

EXHIBIT 25.6

THE PRINCIPAL VITAMINS

Vitamin	Comment and Source	Function	Deficiency Symptoms and Disorders
FAT-SOLUBLE			
A	Formed from provitamin carotene (and other provitamins) in GI tract. Requires bile salts and fat for absorption. Stored in liver. Sources of carotene and other provitamins include yellow and green vegetables; sources of vitamin A include liver and milk.	Maintains general health and vigor of epithelial cells. Its potential role in cancer prevention is currently under investigation.	Deficiency results in atrophy and keratinization of epithelium, leading to dry skin and hair, increased incidence of ear, sinus, respiratory, urinary, and digestive system infections, inability to gain weight, drying of cornea and ulceration (**xerophthalmia**), nervous disorders, and skin sores.
		Essential for formation of photopigments, light-sensitive chemicals in photoreceptors of retina.	**Night blindness** or decreased ability for dark adaptation.
		Aids in growth of bones and teeth apparently by helping to regulate activity of osteoblasts and osteoclasts.	Slow and faulty development of bones and teeth.
D	In presence of sunlight, 7-dehydro-cholesterol in the skin is converted to cholecalciferol (vitamin D_3). In the liver, cholecalciferol is converted to 25-hydroxychole-calciferol. In the kidneys, 25-hydroxycholecalciferol is converted to calcitriol (1,25-dihydroxycalciferol), the active form of vitamin D. Dietary vitamin D requires moderate amounts of bile salts and fat for absorption. Stored in tissues to slight extent. Most excreted via bile. Sources include fish-liver oils, egg yolk, fortified milk.	Essential for absorption and utilization of calcium and phosphorus from gastrointestinal tract. May work with parathyroid hormone (PTH) that controls calcium metabolism.	Defective utilization of calcium by bones leads to **rickets** in children and **osteomalacia** in adults. Possible loss of muscle tone.
E (Tocopherols)	Stored in liver, adipose tissue, and muscles. Requires bile salts and fat for absorption. Sources include fresh nuts and wheat germ, seed oils, green leafy vegetables.	Believed to inhibit catabolism of certain fatty acids that help form cell structures, especially membranes. Involved in formation of DNA, RNA, and red blood cells. May promote wound healing, contribute to the normal structure and functioning of the nervous system, prevent scarring, and function as an antioxidant. Believed to help protect liver from toxic chemicals like carbon tetrachloride.	May cause the oxidation of monounsaturated fats, resulting in abnormal structure and function of mitochondria, lysosomes, and plasma membranes, a possible consequence being hemolytic anemia. Deficiency also causes muscular dystrophy in monkeys and sterility in rats.

Vitamin	Comment and Source	Function	Deficiency Symptoms and Disorders
K	Produced by intestinal bacteria. Requires bile salts and fat for absorption. Stored in liver and spleen. Dietary sources include spinach, cauliflower, cabbage, liver.	Coenzyme essential for synthesis of several clotting factors by liver, including prothrombin.	Delayed clotting time results in excessive bleeding.

WATER-SOLUBLE

Vitamin	Comment and Source	Function	Deficiency Symptoms and Disorders
B$_1$ (thiamine)	Rapidly destroyed by heat. Not stored in body. Excessive intake eliminated in urine. Sources include whole-grain products, eggs, pork, nuts, liver, yeast.	Acts as coenzyme for many different enzymes that break carbon-to-carbon bonds and are involved in carbohydrate metabolism of pyruvic acid to CO_2 and H_2O. Essential for synthesis of acetylcholine.	Improper carbohydrate metabolism leads to buildup of pyruvic and lactic acids and insufficient ATP for muscle and nerve cells. Deficiency leads to two syndromes: (1) **beri-beri**—partial paralysis of smooth muscle of GI tract, causing digestive disturbances; skeletal muscle paralysis; atrophy of limbs; (2) **polyneuritis**—due to degeneration of myelin sheaths; reflexes related to kinesthesia are impaired, impairment of sense of touch, decreased intestinal motility, stunted growth in children, and poor appetite.
B$_2$ (riboflavin)	Not stored in large amounts in tissues. Most is excreted in urine. Small amounts supplied by bacteria of GI tract. Dietary sources include yeast, liver, beef, veal, lamb, eggs, whole-grain products, asparagus, peas, beets, peanuts.	Component of certain coenzymes (for example, FAD and FMN) concerned with carbohydrate and protein metabolism, especially in cells of eye, integument, mucosa of intestine, blood.	Deficiency may lead to improper utilization of oxygen resulting in blurred vision, cataracts, and corneal ulcerations. Also dermatitis and cracking of skin, lesions of intestinal mucosa, and development of one type of anemia.
Niacin (nicotinamide)	Derived from amino acid tryptophan. Sources include yeast, meats, liver, fish, whole-grain products, peas, beans, nuts.	Essential component of NAD and NADP, coenzymes concerned with oxidation–reduction reactions. In lipid metabolism, inhibits production of cholesterol and assists in triglyceride breakdown.	Principal deficiency is pellagra, characterized by dermatitis, diarrhea, and psychological disturbances.
B$_6$ (pyridoxine)	Synthesized by bacteria of GI tract. Stored in liver, muscle, brain. Other sources include salmon, yeast, tomatoes, yellow corn, spinach, whole-grain products, liver, yogurt.	Essential coenzyme for normal amino acid metabolism. Assists production of circulating antibodies. May function as coenzyme in triglyceride metabolism.	Most common deficiency symptom is dermatitis of eyes, nose, and mouth. Other symptoms are retarded growth and nausea.
B$_{12}$ (cyanocobalamin)	Only B vitamin not found in vegetables; only vitamin containing cobalt. Absorption from GI tract depends on HCl and intrinsic factor secreted by gastric mucosa. Sources include liver, kidney, milk, eggs, cheese, meat.	Coenzyme necessary for red blood cell formation, formation of amino acid methionine, entrance of some amino acids into Krebs cycle, and manufacture of choline (used to synthesize acetylcholine).	Pernicious anemia, neuropsychiatric abnormalities (ataxia, memory loss, weakness, personality and mood changes, and abnormal sensations), and impaired osteoblast activity.
Pantothenic Acid	Stored primarily in liver and kidneys. Some produced by bacteria of GI tract. Other sources include kidney, liver, yeast, green vegetables, cereal.	Constituent of coenzyme A essential for transfer of pyruvic acid into Krebs cycle, conversion of lipids and amino acids into glucose, and synthesis of cholesterol and steroid hormones.	Experimental deficiency tests indicate fatigue, muscle spasms, neuromuscular degeneration, insufficient production of adrenal steroid hormones.

Exhibit continues

EXHIBIT 25.6 (continued)

THE PRINCIPAL VITAMINS

Vitamin	Comment and Source	Function	Deficiency Symptoms and Disorders
Folic acid	Synthesized by bacteria of GI tract. Dietary sources include green leafy vegetables and liver.	Component of enzyme systems synthesizing purines and pyrimidines built into DNA and RNA. Essential for normal production of red and white blood cells.	Production of abnormally large red blood cells (macrocytic anemia).
Biotin	Synthesized by bacteria of GI tract. Dietary sources include yeast, liver, egg yolk, kidneys.	Essential coenzyme for conversion of pyruvic acid to oxaloacetic acid and synthesis of fatty acids and purines.	Mental depression, muscular pain, dermatitis, fatigue, nausea.
C (ascorbic acid)	Rapidly destroyed by heat. Some stored in glandular tissue and plasma. Sources include citrus fruits, tomatoes, green vegetables.	Exact role not understood. Promotes many metabolic reactions, particularly protein metabolism, including laying down of collagen in formation of connective tissue. As coenzyme may combine with poisons, rendering them harmless until excreted. Works with antibodies. Promotes wound healing. Functions as an antioxidant. Role in cancer prevention is under investigation.	Scurvy; anemia; many symptoms related to poor connective tissue growth and repair including tender swollen gums, loosening of teeth (alveolar processes also deteriorate), poor wound healing, bleeding (vessel walls fragile because of connective tissue degeneration), and retardation of growth.

METABOLISM AND BODY HEAT

We will now consider the relationship of foods to body heat, mechanisms of heat gain and loss, and the regulation of body temperature.

Measuring Heat

Heat is a form of kinetic energy that can be measured as **temperature** and expressed in units called calories. A **calorie (cal)** is the amount of heat energy required to raise the temperature of 1 gram of water 1°C, from 14 to 15°C. Since the calorie is a small unit relative to the large amount of energy stored in foods, the **kilocalorie (kcal)** is used instead. A kilocalorie is equal to 1000 cal. The kilocalorie is the unit we use to express the heating value of foods and to measure the body's metabolic rate.

The apparatus used to determine the caloric value of foods is called a **calorimeter.** A weighed sample of a dehydrated food is burned completely in an insulated metal container. The energy released by the burning food is absorbed by the container and transferred to a known volume of water that surrounds the container. The change in the water's temperature is directly related to the number of kilocalories released by the food.

Production of Body Heat

Most of the heat produced by the body comes from oxidation of the food we eat. The rate at which this heat is produced—the **metabolic rate**—is also measured in kilocalories. Among the factors that affect metabolic rate are the following:

1. **Exercise.** During strenuous exercise, the metabolic rate may increase to as much as 15 times the normal rate. In well-trained athletes, the rate may increase up to 20 times.
2. **Nervous system.** In a stress situation, the sympathetic nervous system is stimulated and the nerves release norepinephrine (NE), which increases the metabolic rate of body cells.
3. **Hormones.** In addition to norepinephrine (NE), several other hormones affect metabolic rate. Epinephrine is secreted in stress situations. Increased secretions of thyroid hormones increase the metabolic rate. Testosterone and human growth hormone also increase metabolic rate.
4. **Body temperature.** The higher the body temperature, the higher the metabolic rate. Each 1°C rise in temperature increases the rate of biochemical reactions by about 10%. The metabolic rate may be substantially increased during fever.

5.   **Ingestion of food.** The ingestion of food can raise metabolic rate by as much as 10 to 20%. This effect is called **specific dynamic action (SDA)** and is greatest with proteins and less with carbohydrates and lipids.

6.   **Age.** The metabolic rate of a child, in relation to its size, is about double that of an elderly person because the high rates of reactions related to growth decrease with age.

7.   **Others.** Other factors that affect metabolic rate are sex (lower in females, except during pregnancy and lactation), climate (lower in tropical regions), sleep (lower), and malnutrition (lower).

Basal Metabolic Rate

Since many factors affect metabolic rate, it is measured under standard conditions designed to reduce or eliminate those factors as much as possible. These conditions of the body are called the **basal state,** and the measurement obtained is the **basal metabolic rate (BMR).** BMR is a measure of the rate at which the quiet, resting, fasting body breaks down nutrients to liberate energy. Some of the energy is used to form ATP and some is released as heat. It is also an indicator of how much thyroid hormone is present because this hormone is the main regulator of the basal rate of ATP use.

Basal metabolic rate is most often measured indirectly by measuring oxygen consumption using a respirometer. To release a given amount of heat energy when it is oxidized, a nutrient must combine with a given amount of oxygen. Thus by measuring the amount of oxygen needed for the metabolism of foods, we can determine how many kilocalories are produced. The amount of heat energy released when 1 liter of oxygen combines with carbohydrates is 5.05 kcal; with triglycerides, the heat released is 4.70 kcal; with proteins, the heat released is 4.60 kcal. On a typical diet, uptake of 1 liter of oxygen indicates about 4.9 kcal of heat produced.

Basal metabolic rate is usually expressed in kilocalories per square meter of body surface area per hour (kcal/m²/hr). Suppose you use 1.8 liters of oxygen in 6 minutes as recorded on a respirometer. This means that your oxygen consumption in an hour would be 18 liters:

$$\frac{1.8 \text{ liters}}{6 \text{ min}} \times \frac{60 \text{ min}}{\text{hr}} = 18 \text{ liters}$$

Your basal metabolic rate would be about 18 x 4.9 kcal or 88 kcal/hr. To express the kilocalories per square meter of body surface, a standardized chart is used. Such a chart shows square meters of body surface relative to height in centimeters and weight in kilograms. If you weigh 75 kg (165 lb) and are 190 cm (75 in.) tall, your body surface area is 2 m². Your basal metabolic rate is equal to 88 kcal divided by 2, or about 44 kcal/m²/hr.

The normal BMRs for various age groups by sex are also listed in standardized charts. Values 15% above or below the standard may indicate an excess or deficiency of thyroid hormone. When the thyroid gland is secreting extreme quantities of thyroid hormone, BMR can double. If, on the other hand, it is secreting very little thyroid hormone, BMR may be half the normal value.

Loss of Body Heat

Most body heat is produced by the oxidation of foods we eat. This heat must be removed continuously or body temperature would rise steadily. The principal routes of heat loss include radiation, conduction, convection, and evaporation.

Radiation

Radiation is the transfer of heat as infrared heat rays from a warmer object to a cooler one without physical contact. Your body loses heat by the radiation of heat waves to cooler objects nearby such as ceilings, floors, and walls. If these objects are at a higher temperature, you absorb heat by radiation. Incidentally, the air temperature has no relationship to the radiation of heat to and from objects. Skiers can remove their shirts in bright sunshine even though the air temperature is very low because the radiant heat from the sun is adequate to warm them. In a room at 21°C (70°F), about 60% of heat loss is by radiation in a resting person.

Conduction and Convection

Another method of heat transfer is **conduction.** In this process, body heat is transferred to a substance or object in contact with the body, such as chairs, clothing, jewelry, air, or water. At rest about 3% of body heat is lost via conduction to solid objects.

The contact of air or water with your body results in heat transfer by both conduction and convection. **Convection** is the transfer of heat by the movement of a liquid or gas between areas of different temperature. When cool air makes contact with the body, it becomes warmed and therefore less dense and is carried away by convection currents as the less dense air rises. Then more cool air makes contact with the body and is carried away as it warms by conduction and becomes less dense. The faster the air moves, the faster the rate of convection. About 15% of body heat is lost to the air by conduction and convection.

Evaporation

Evaporation is the conversion of a liquid to a vapor. Water has a *high heat of evaporation,* which is the amount of heat needed to evaporate 1 g of water at 30°C (86°F). Because of water's high heat of evaporation, every gram of water evaporating from the skin takes with it a great deal of heat—about 0.58 kcal per gram of water. Under normal resting conditions, about 22% of heat loss occurs through evaporation. Under extreme conditions, about 4 liters (1 gal) of perspiration are produced each hour, and this volume can remove 2000 kcal of heat from the body. This is approximately 32 times the basal level of heat production. The rate of evaporation is inversely related to **relative humidity,** the ratio of the actual amount of moisture in the air to the greatest amount it can hold at a given temperature. The higher the relative humidity, the lower the rate of evaporation due to a reduced diffusion rate for water leaving the surface of the body.

Homeostasis of Body Temperature Regulation

Even though there are wide fluctuations in environmental temperature, homeostatic mechanisms can maintain a normal range for the internal body temperature. If your heat production equals heat loss, you maintain a constant core temperature near 37°C (98.6°F). **Core temperature** refers to the body's temperature in body structures below the skin and subcutaneous tissue. **Shell temperature** refers to the body's temperature at the surface, that is, the skin and subcutaneous tissue. Core temperature is usually a little higher than shell temperature. If your heat-producing mechanisms generate more heat than is lost by your heat-losing mechanisms, your core temperature rises. If your heat-losing mechanisms give off more heat than is generated by heat-producing mechanisms, your core temperature falls. Too high a core temperature kills by denaturing body proteins, while too low a core temperature causes cardiac arrhythmias that result in death.

Hypothalamic Thermostat

Body temperature is regulated by mechanisms that attempt to keep heat production and heat loss in balance. A center of control for those mechanisms that are reflex in nature is found in the hypothalamus in a group of neurons in the anterior portion referred to as the **preoptic area.** This area receives input from temperature receptors in the skin and mucous membranes (peripheral thermoreceptors) and in internal structures (central thermoreceptors), including the hypothalamus. If blood temperature increases, the neurons of the preoptic area fire nerve impulses more rapidly. If something causes the blood's temperature to decrease, these neurons fire nerve impulses less rapidly. The preoptic area is adjusted to maintain normal body temperature and thus serves as your thermostat.

Nerve impulses from the preoptic area are sent to other portions of the hypothalamus known as the heat-losing center and the heat-promoting center. The **heat-losing center,** when stimulated by the preoptic area, sets into operation a series of responses that lower body temperature. The **heat-promoting center,** when stimulated by the preoptic area, sets into operation a series of responses that raise body temperature. The heat-losing center is mainly parasympathetic in function; the heat-promoting center is primarily sympathetic.

Mechanisms of Heat Production

Suppose the environmental temperature is low or that some factor causes blood temperature to decrease (Fig. 25.16). These cause body temperature (controlled condition) to fall below normal. Both stresses stimulate thermoreceptors (receptors) that send nerve impulses (input) to the preoptic area (control center). The preoptic area, in turn, activates the heat-promoting center (control center). In response, the heat-promoting center discharges nerve impulses (output) to effectors that automatically set into operation several responses designed to increase and retain body heat and bring body temperature back up to normal. This cycle is a negative feedback system that attempts to raise body temperature to normal.

● **Vasoconstriction** Nerve impulses from the heat-promoting center stimulate sympathetic nerves that cause blood vessels of the skin to constrict. The net effect of vasoconstriction is to decrease the flow of warm blood from the internal organs to the skin, thus decreasing the transfer of heat from the internal organs to the skin. This reduction in heat loss helps raise the internal body temperature.

● **Sympathetic Stimulation** Another response triggered by the heat-promoting center is the sympathetic stimulation of metabolism. The heat-promoting center stimulates sympathetic nerves leading to the adrenal medulla. This stimulation causes the medulla to secrete epinephrine and norepinephrine (NE) into the blood. The hormones, in turn, bring about an increase in cellular metabolism, a reaction that also increases heat production. This effect is called **chemical thermogenesis.**

● **Skeletal Muscles** Heat production is also increased by responses of skeletal muscles. For example, stimulation of the heat-promoting center causes stimulation of parts of the brain that increase muscle tone and hence heat production. As muscle tone increases, the stretching of the agonist muscle initiates the stretch reflex and the muscle contracts. This contraction causes the antagonist muscle to stretch, and it too develops a stretch reflex. The repetitive cycle—called **shivering (involuntary thermogenesis)**—increases the rate of heat production. During maximal shivering, body heat production can rise to about four times the normal rate in just a few minutes.

● **Thyroid Hormones** Another body response that increases heat production is increased production of thyroid hormones (T_3 and T_4). A cold environmental temperature increases the secretion of thyrotropin releasing hormone

decreases the secretion of thyrotropin

FIGURE 25.16 Negative feedback mechanisms that increase heat production.

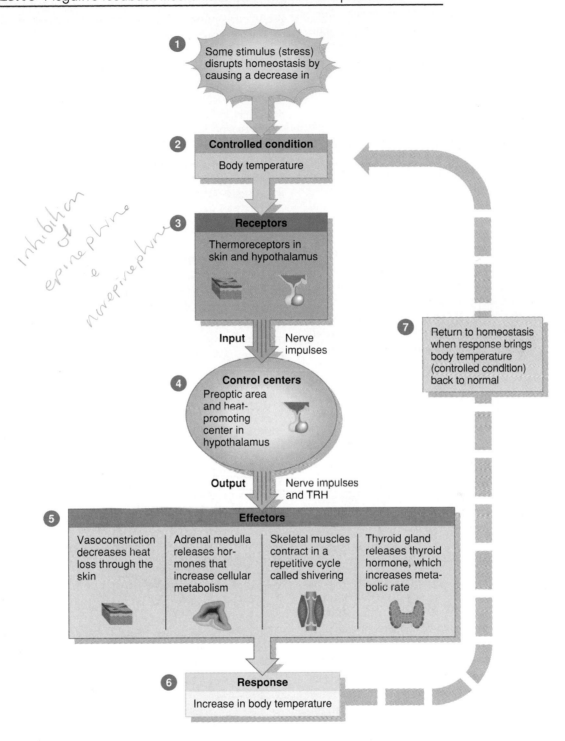

Inhibition of epinephrine e norepinephrine

1 Some stimulus (stress) disrupts homeostasis by causing a decrease in

2 Controlled condition
Body temperature

3 Receptors
Thermoreceptors in skin and hypothalamus

Input   Nerve impulses

4 Control centers
Preoptic area and heat-promoting center in hypothalamus

7 Return to homeostasis when response brings body temperature (controlled condition) back to normal

Output   Nerve impulses and TRH

5 Effectors

| Vasoconstriction decreases heat loss through the skin | Adrenal medulla releases hormones that increase cellular metabolism | Skeletal muscles contract in a repetitive cycle called shivering | Thyroid gland releases thyroid hormone, which increases metabolic rate |

6 Response
Increase in body temperature

Question: What factors can increase your metabolic rate and thus increase your rate of heat production?

(TRH) produced by the hypothalamus. TRH, in turn, stimulates the anterior pituitary gland to secrete thyroid stimulating hormone (TSH), which causes the thyroid to release thyroid hormones into the blood. Since increased levels of thyroid hormones increase the metabolic rate, body temperature is raised.

Mechanisms of Heat Loss

Now suppose some stress raises body temperature (controlled condition) above normal. The stress or higher temperature of the blood stimulates thermoreceptors (receptors) that send nerve impulses (input) to the preoptic area (control center), which in turn stimulate the heat-losing center (control center) and inhibit the heat-promoting center. The heat-losing center discharges nerve impulses (output) to blood vessels (effectors) in the skin, causing them to dilate. The skin becomes warm, and the excess heat is lost to the environment as an increased volume of blood flows from the core of the body into the skin. At the same time, metabolic rate and shivering are decreased. The high temperature of the blood (by way of hypothalamic activation of sympathetic nerves) stimulates sweat glands of the skin (effectors) to produce perspiration. As the water of the perspiration evaporates from the surface of the skin, the skin is cooled. All these responses reverse the heat-promoting effects and decrease body temperature (response). This brings body temperature down to normal.

Voluntary activities, such as putting on or taking off clothing, along with all the reflex responses discussed above, help achieve homeostasis of core body temperature.

Body Temperature Abnormalities

Fever

A **fever** is an abnormally high body temperature. The most frequent cause of fever is infection by viruses and bacteria (or bacterial toxins). Other causes are heart attacks, tumors, tissue destruction by x-rays, surgery, or trauma, and reactions to vaccines. A fever-producing substance is called a **pyrogen** (PĪ-rō-gen; *pyr* = fire). The mechanism of fever production is believed to occur as follows. When phagocytes—namely, monocytes and macrophages—ingest certain bacteria, a portion of the cell wall of the bacteria is released, causing the phagocytes to secrete interleukin-1. Interleukin-1 acts as a pyrogen; it circulates to the anterior hypothalamus and induces neurons of the preoptic area to secrete prostaglandins, particularly of the E series. Prostaglandins reset the hypothalamic thermostat at a higher temperature, and temperature regulating reflex mechanisms will then act to bring the core body temperature up to this new setting. Aspirin, acetaminophen (Tylenol), and ibuprofen (Advil) reduce fever by inhibiting synthesis of prostaglandins.

Suppose that as a result of pyrogens the thermostat is set at 39°C (103°F). Now the heat-promoting mechanisms (vasoconstriction, increased metabolism, shivering) are operating at full force. Thus even though body temperature is climbing higher than normal—say, 38°C (101°F)—the skin remains cold, and shivering occurs. This condition, called a **chill**, is a definite sign that body temperature is rising. After several hours, body temperature reaches the set-

ting of the thermostat and the chills disappear. But the body will continue to regulate temperature at 39°C (103°F) until the stress is removed. When the stress is removed, the thermostat is reset at normal—37.0°C (98.6°F). Since body temperature remains high in the beginning, the heat-losing mechanisms (vasodilation and sweating) go into operation to decrease body temperature. The skin becomes warm and the person begins to sweat. This phase of the fever is called the **crisis** and indicates that body temperature is falling.

Up to a point, fever is beneficial. Interleukin-1 helps step up production of T cells. High body temperature intensifies the effect of interferon. The high body temperature is believed to inhibit the growth of some bacteria. Fever also increases heart rate so that white blood cells are delivered to sites of infection more rapidly and their secretions are increased. In addition, antibody production and T cell proliferation increase. Moreover, heat speeds up the rate of chemical reactions. This increase may help body cells to repair themselves more quickly during a disease. Among the complications of fever are dehydration, acidosis, and permanent brain damage. As a rule, death results if body temperature rises above 44 to 46°C (112 to 114°F). On the other end of the scale, death usually results when body temperature falls below 21 to 24°C (70 to 75°F).

Heat Cramps

Heat cramps occur as a result of profuse sweating that removes water and salt (NaCl) from the body. The salt loss causes painful contractions of muscles called heat cramps. The cramps tend to occur in muscles used while working but do not appear until the person relaxes after work. Salted liquids usually lead to rapid improvement.

Heat Exhaustion

In **heat exhaustion (heat prostration),** the body temperature is generally normal, or a little below, and the skin is cool and clammy (moist) due to profuse perspiration. Heat exhaustion is normally characterized by fluid and electrolyte loss, especially salt. The salt loss results in muscle cramps, dizziness, vomiting, and fainting. Fluid loss may cause low blood pressure. Complete rest and salt tablets are recommended.

Heatstroke

Heatstroke (sunstroke) occurs when the temperature and relative humidity are high, making it difficult for the body to lose heat by radiation, conduction, or evaporation. There is a decreased flow of blood to the skin, perspiration is greatly reduced, and body temperature rises sharply. The skin is thus dry and hot—the temperature may reach 43°C (110°F). Brain cells are quickly affected and may be destroyed permanently. As a result, the body temperature regulating reflexes fail to operate. Treatment must be undertaken immediately and consists of cooling the body by immersing the victim in cool water and by administering fluids and electrolytes.

Hypothermia

Hypothermia refers to a lowering of body temperature to 35°C (95°F) or below. It may be caused by an overwhelming cold stress (immersion in icy water), metabolic diseases (hypoglycemia, adrenal insufficiency, hypothyroidism), drugs (alcohol, antidepressants, sedatives, tranquilizers), burns, malnutrition, transection of the cervical spinal cord, and lowering of body temperature for surgery. Hypothermia is characterized by the following as body temperature falls: sensation of cold, shivering, confusion, vasoconstriction, muscle rigidity, bradycardia, acidosis, hypoventilation, hypotension, ventricular fibrillation, no reflexes, and loss of spontaneous movement, coma, and possibly death usually caused by cardiac arrhythmias. The elderly have lessened metabolic protection against a cold environment coupled with reduced perception of cold. As a result, they are at greater risk for developing hypothermia.

In 1986, a 2½-year-old girl was submerged in icy water for 66 minutes, the longest documented submersion in which the individual had no neurological damage. Part of the successful outcome was due to use of a heart–lung bypass machine that rewarmed the blood of the child and brought body temperature back to normal, a new and novel use for this machine.

DISORDERS: HOMEOSTATIC IMBALANCES

OBESITY

Obesity is defined as a body weight more than 20% above a desirable standard due to an excessive accumulation of fat. Even moderate obesity is hazardous to health. It is implicated as a risk factor in cardiovascular disease, hypertension, pulmonary disease, non-insulin-dependent diabetes mellitus (type II), arthritis, certain cancers (breast, uterus, and colon), varicose veins, and gallbladder disease. Also, loss of body fat in obese persons has been shown to elevate HDL cholesterol, the type associated with prevention of cardiovascular disease.

Causes

In a few cases, obesity may result from trauma or tumors in the food-regulating centers in the hypothalamus. In most cases of obesity, no specific cause can be identified. Contributing factors include eating habits taught early in life, overeating to relieve tension, and social customs. Recently, obesity has been strongly linked to genetic factors. Studies indicate that some people inherit a low metabolic rate and that they become obese not because they eat too much but because they burn calories too slowly.

Treatment

Reduction of body weight involves keeping caloric intake well below energy expenditure. The goals during weight decrease are:

1. Loss of body fat with minimal breakdown of lean tissue.
2. Maintenance of physical and emotional fitness during the reducing period.
3. Establishment of eating and exercise habits to maintain weight at the recommended level.

Morbid obesity refers to obese people who weigh twice their ideal weight or more. The condition is so named because it is associated with serious and life-threatening conditions such as hypertension, diabetes mellitus, and atherosclerosis. For those with morbid obesity, surgery may be an alternative to control weight. Patients considered for surgical procedures weigh more than 45 kg (100 lb) over the ideal for their height or twice their ideal weight and who have tried and failed to permanently lose substantial weight on medically supervised reduction regimens.

Most operations for morbid obesity either produce a certain amount of malabsorption, such as the intestinal bypass, or reduce the size of the stomach so that it will hold less food, such as the gastric bypass or gastroplasty.

VITAMIN OR MINERAL OVERDOSE

Hypervitaminosis refers to an excess of one or more vitamins. High doses of certain vitamins and minerals can be deadly. Rarely is vitamin or mineral overdose the result of eating foods. (Bear liver supposedly contains enough vitamin A to kill a person. So if you like to eat liver, stick to chicken or beef.) Rather, it stems from consumption of supplements. The fat-soluble vitamins (A, E, D, and K) are particularly toxic because they tend to be stored and accumulate in the body rather than be excreted. Some adverse effects of high doses of minerals and vitamins are as follows.

Minerals

Calcium: Depresses nerve function, causes drowsiness, extreme lethargy, calcium deposits, kidney stones.

Iron: Damage to liver, heart, and pancreas.

Zinc: Masklike fixed expression, difficulty in walking, slurred speech, hand tremor, involuntary laughter.

Cobalt: Goiter, polycythemia, and heart damage.

Selenium: Nausea, vomiting, fatigue, irritability, and loss of fingernails and toenails.

Vitamins

A: Blurred vision, dizziness, ringing in the ears, headache, insomnia, irritability, apathy, ataxia, stupor, skin rash, nausea, vomiting, diarrhea, hair loss, joint pain, menstrual irregularities, fatigue, liver damage, abnormal bone growth, and damage to nervous system.

Continues

D: Calcium deposits, deafness, nausea, fatigue, headache, loss of appetite, kidney stones, weak bones, hypertension, and high cholesterol.

E: Thrombophlebitis, pulmonary embolism, hypertension, muscular weakness, severe fatigue, breast tenderness, and slow wound healing.

Niacin: Acute flushing, peptic ulcers, liver dysfunction, gout, faintness, dizziness, tingling of fingertips, arrhythmias, and hyperglycemia.

B_6: Impaired sense of position and vibration, diminished tendon reflexes, numbness and loss of sensations in hands and feet, difficulty in walking, impaired memory, depression, headache, and fatigue.

C: Dependence on megadoses may lead to scurvy when withdrawn, kidney stones, diarrhea, hemolysis, hot flashes, headache, fatigue, and insomnia.

MALNUTRITION

Causes

Whereas **nutrition** refers to the intake of adequate essential nutrients and calories to maintain health, **malnutrition** refers to a state of poor nutrition. One cause of malnutrition is undernutrition, that is, inadequate food intake. It can result from conditions such as fasting, anorexia nervosa, deprivation, cancer, gastrointestinal obstructions, inability to swallow, renal disease, and poor dentition. Other causes of malnutrition are imbalance of nutrients, malabsorption of nutrients, improper distribution of nutrients, inability to use nutrients (for example, diabetes mellitus), increased nutrient requirements (due to fever, infections, burns, fractures, stress, and exposure to heat or cold), increased nutrient losses (diarrhea, bleeding, glycosuria, and fistula drainage), and overnutrition (excess vitamins, minerals, and calories).

One of the major types of undernutrition is known as protein–calorie undernutrition, which occurs when there is inadequate intake of protein and/or calories to meet a person's nutritional requirements. Using protein–calorie nutrition as the example, we will describe the stages of **starvation,** that is, loss of energy stores in the form of glycogen, fats, and proteins.

Stages

In the first stage of starvation, carbohydrate stores in the body are depleted first, during approximately the first day of fasting. Once glycogen stores are depleted, the next stage of starvation occurs. During this time, the primary energy source for most body cells is fatty acids from lipid stores. As the liver metabolizes the fatty acids, ketone bodies are produced in large quantities. Even brain cells can use ketone bodies as a source of energy. However, since body cells are limited in the amount of ketone bodies they can metabolize, excess ketone bodies appear in the blood. The result is ketosis and metabolic acidosis since the body cannot neutralize the excess ketone bodies. As you will see on page 916, metabolic acidosis results in depression of the central nervous system that may lead to coma. Ketosis and metabolic acidosis are often associated with crash dieting, low-carbohydrate diets, high-protein diets, and other fad diets. During the early stages of starvation, large quantities of muscle protein not essential to cellular functioning are broken down to amino acids and used for gluconeogenesis.

The length of the second stage of starvation is primarily determined by the amount of stored fat in the body. When fat reserves are depleted, the third stage of starvation begins. During this time, even the proteins needed to maintain cellular functions are broken down as a source of energy. It is estimated that once protein stores are depleted to about one-half of their normal level, death results. Once it has begun, the third stage of starvation progresses very rapidly, often leading to death in 24 hours.

Clinical Types

Protein–calorie undernutrition may be classified into two types based on which factor is lacking in the diet. In one type, called **kwashiorkor** (kwash-ē-OR-kor), protein intake is deficient despite normal or nearly normal calorie intake. Some dietary proteins are called complete proteins; that is, they contain adequate amounts of essential amino acids. Sources of complete proteins are primarily animal products such as milk, meat, fish, poultry, and eggs. Incomplete proteins lack certain essential amino acids. An example is zein, the protein in corn, which lacks the essential amino acids tryptophan and lysine. The diet of many Africans consists largely of cornmeal. As a result, many African children develop kwashiorkor. It is characterized by edema of the abdomen, enlarged liver, decreased blood pressure, bradycardia, hypothermia, anorexia, lethargy, dry and hyperpigmented skin, easily pluckable hair, and sometimes mental retardation.

Another type of protein–calorie undernutrition is called **marasmus** (mar-AZ-mus). It results from inadequate intake of both protein and calories. Its characteristics include retarded growth, low weight, muscle wasting, emaciation, dry skin, and thin, dry, dull hair.

PHENYLKETONURIA

Phenylketonuria (fen'-il-kē'-tō-NOO-rē-a) or **PKU** is a genetic error of metabolism characterized by an elevation of the amino acid phenylalanine in the blood. It is often associated with mental retardation. The DNA of children with phenylketonuria lacks the gene that normally programs the manufacture of the enzyme phenylalanine hydroxylase. This enzyme is necessary for converting phenylalanine into the amino acid tyrosine, an amino acid that enters the Krebs cycle. As a result, phenylalanine cannot be metabolized, and what is not used in protein synthesis builds up in the blood. High levels of phenylalanine are toxic to the brain during the early years of life when the brain is developing. Mental retardation can be prevented, when the condition is detected early, by restricting the child to a diet that supplies only the amount of phenylalanine necessary for growth. The artificial sweetener aspartame contains phenylalanine, and its consumption should be restricted in children with PKU since they cannot metabolize it and it may produce neuronal damage. PKU may be detected by blood or urine tests.

CELIAC DISEASE

Celiac disease results in malabsorption by the intestinal mucosa due to the ingestion of gluten. *Gluten* is the water-insoluble protein fraction of wheat, rye, barley, and oats. In susceptible persons, ingestion of gluten induces destruction of villi and inhibition of enzyme secretion accompanied by a variable amount of malabsorption. The condition is easily remedied by administering a diet that excludes all cereal grains except rice and corn.

Study Outline

Regulation of Food Intake (p. 822)

1. Two centers in the hypothalamus related to regulation of food intake are the feeding center and satiety center; the feeding center is constantly active but may be inhibited by the satiety center.
2. Among the stimuli that affect the feeding and satiety centers are glucose, amino acids, lipids, body temperature, distension of the GI tract, and cholecystokinin (CCK).

Nutrients (p. 822)

1. Nutrients are chemical substances in food that provide energy, act as building blocks in forming new body components, serve as storage molecules, or assist in the functioning of various body processes.
2. There are six major classes of nutrients: carbohydrates, lipids, proteins, minerals, vitamins, and water.

Metabolism (p. 823)

1. Metabolism refers to all chemical reactions of the body and has two phases: catabolism and anabolism.
2. Anabolism consists of a series of synthesis reactions whereby small molecules are built up into more complex ones that form the body's structural and functional components. Anabolic reactions use energy.
3. Catabolism is the term for reactions that break down complex organic compounds into simple ones; they provide energy.
4. The coupling of anabolism and catabolism is via ATP.

Energy Production (p. 824)

1. Oxidation is the removal of electrons from a substance; reduction is the addition of electrons to a substance; oxidation–reduction reactions are coupled.
2. ATP can be generated by substrate-level phosphorylation, oxidative phosphorylation, and photophosphorylation (if chlorophyll is present).

Carbohydrate Metabolism (p. 825)

1. During digestion, polysaccharides and disaccharides are converted to monosaccharides, which are absorbed through capillaries in villi and transported to the liver via the hepatic portal vein.
2. Carbohydrate metabolism is primarily concerned with glucose metabolism.

Fate of Carbohydrates (p. 825)

1. Some glucose is oxidized by cells to provide energy; it moves into cells by facilitated diffusion and becomes phosphorylated to glucose 6-phosphate; insulin stimulates glucose movement into cells.
2. Excess glucose can be stored by the liver and skeletal muscles as glycogen, used to form amino acids, or converted to triglycerides.

Glucose Catabolism (p. 826)

1. Glucose oxidation is also called cellular respiration.
2. The complete oxidation of glucose to CO_2 and H_2O involves glycolysis, the Krebs cycle, and the electron transport chain.

Glycolysis (p. 826)

1. Glycolysis refers to the breakdown of glucose into two molecules of pyruvic acid.

2. When oxygen is in short supply, pyruvic acid is reduced to lactic acid; under aerobic conditions, pyruvic acid enters the Krebs cycle.
3. As a result of glycolysis, there is a net production of two molecules of ATP.

Krebs (Citric Acid) Cycle (p. 829)

1. Pyruvic acid is prepared for entrance into the Krebs cycle by conversion to a two-carbon compound (acetyl group) followed by the addition of coenzyme A to form acetyl coenzyme A.
2. The Krebs cycle involves decarboxylations and oxidations and reductions of various organic acids.
3. Each molecule of pyruvic acid that enters the Krebs cycle produces three molecules of CO_2, four molecules of NADH + $4H^+$, one molecule of $FADH_2$, and one molecule of GTP.
4. The energy originally in glucose and then pyruvic acid is primarily in the reduced coenzymes NADH + H^+ and $FADH_2$.

Electron Transport Chain (p. 832)

1. The electron transport chain involves a series of oxidation–reduction reactions in which the energy in NADH + H^+ and $FADH_2$ is liberated and transferred to ATP for storage.
2. The carrier molecules involved include FMN, cytochromes, iron–sulfur centers, copper atoms, and ubiquinones.
3. The electron transport chain yields 32 or 34 molecules of ATP and several molecules of H_2O.
4. The complete oxidation of glucose can be represented as follows:

$$C_6H_{12}O_6 + 6O_2 \rightarrow 36 \text{ or } 38 \text{ ATPs} + 6CO_2 + 6H_2O$$

Glucose Anabolism (p. 834)

1. The conversion of glucose to glycogen for storage in the liver and skeletal muscle is called glycogenesis. It is stimulated by insulin.
2. The body can store about 500 g (1.1 lb) of glycogen.
3. The conversion of glycogen back to glucose is called glycogenolysis. It occurs between meals and is stimulated by glucagon and epinephrine.
4. Gluconeogenesis is the conversion of protein and triglyceride molecules into glucose. It is stimulated by cortisol, thyroxine, epinephrine, glucagon, and human growth hormone (hGH).
5. Glycerol may be converted to glyceraldehyde 3-phosphate, and some amino acids may be converted to pyruvic acid.

Lipid Metabolism (p. 836)

1. During digestion, triglycerides are ultimately broken down into fatty acids and monoglycerides.
2. Long-chain fatty acids and monoglycerides are carried in micelles, digested to glycerol and fatty acids in epithelial cells, recombined to form triglycerides, and transported by chylomicrons through the lacteals of villi into the thoracic duct.

Fate of Lipids (p. 836)

1. Some lipids may be oxidized to produce ATP.
2. Some lipids are stored in adipose tissue.
3. Other lipids are used as structural molecules or to synthesize essential molecules. Examples include phospholipids of plasma membranes, lipoproteins that transport cholesterol,

thromboplastin for blood clotting, and cholesterol used to synthesize bile salts and steroid hormones.

Fat Storage (p. 837)

1. Triglycerides are stored in adipose tissue, mostly in the subcutaneous layer.
2. Adipose cells contain lipases that catalyze the deposition of triglycerides from chylomicrons and hydrolyze triglycerides into fatty acids and glycerol.

Lipid Catabolism: Lipolysis (p. 838)

1. Triglycerides are split into fatty acids and glycerol under the influence of hormones such as epinephrine, norepinephrine, and glucocorticoids and released from fat depots.
2. Glycerol can be converted into glucose by conversion into glyceraldehyde 3-phosphate.
3. In beta oxidation, carbon atoms are removed in pairs from fatty acid chains; the resulting molecules of acetyl coenzyme A enter the Krebs cycle.
4. The formation of ketone bodies by the liver is a normal phase of fatty acid catabolism, but an excess of ketone bodies, called ketosis, may cause acidosis.

Lipid Anabolism: Lipogenesis (p. 840)

1. The conversion of glucose or amino acids into lipids is called lipogenesis. The process is stimulated by insulin.
2. The intermediary links in lipogenesis are glyceraldehyde 3-phosphate and acetyl coenzyme A.

Protein Metabolism (p. 840)

1. During digestion, proteins are hydrolyzed into amino acids.
2. Amino acids are absorbed by the capillaries of villi and enter the liver via the hepatic portal vein.

Fate of Proteins (p. 840)

1. Amino acids, under the influence of human growth hormone (hGH) and insulin, enter body cells by active transport.
2. Inside cells, amino acids are synthesized into proteins that function as enzymes, hormones, structural elements, and so forth; stored as fat or glycogen; or used for energy.

Protein Catabolism (p. 840)

1. Before amino acids can be catabolized, they must be converted to substances that can enter the Krebs cycle; these conversions involve deamination, decarboxylation, and hydrogenation.
2. Amino acids may also be converted into glucose, fatty acids, and ketone bodies.

Protein Anabolism (p. 840)

1. Protein synthesis is stimulated by human growth hormone (hGH), thyroxine, and insulin.
2. The process is directed by DNA and RNA and carried out on the ribosomes of cells.

Summary of Key Molecules in Metabolism (p. 841)

1. Three molecules play a key role in metabolism: glucose 6-phosphate, pyruvic acid, and acetyl CoA.
2. Glucose 6-phosphate may be converted to glucose, glycogen, ribose 5-phosphate, and pyruvic acid.
3. When ATP is low and oxygen is plentiful, pyruvic acid is converted to acetyl coenzyme A; when oxygen supply is low, pyruvic acid is converted to lactic acid. One link between carbohydrate and protein metabolism is via pyruvic acid.

4. Acetyl coenzyme A is the gateway into the Krebs cycle and is also used to synthesize fatty acids, ketone bodies, and cholesterol.

Absorptive and Postabsorptive States (p. 843)

1. During the absorptive (fed) state, ingested nutrients enter the blood and lymph from the GI tract.
2. During the absorptive state, most blood glucose is used by body cells for oxidation. Glucose transported to the liver is converted to glycogen or triglycerides. Most triglycerides are stored in adipose tissue. Amino acids in liver cells are converted to carbohydrates, fats, and proteins.
3. During the postabsorptive (fasting) state, absorption is complete and the energy needs of the body are satisfied by nutrients already present in the body.
4. The major concern of the body during the postabsorptive state is to maintain normal blood glucose level. This involves conversion of liver and skeletal muscle glycogen into glucose, conversion of glycerol into glucose, and conversion of amino acids into glucose. The body also switches from glucose oxidation to fatty acid oxidation.

Minerals (p. 846)

1. Minerals are inorganic substances that help regulate body processes.
2. Minerals known to perform essential functions are calcium, phosphorus, sodium, chlorine, potassium, magnesium, iron, sulfur, iodine, manganese, cobalt, copper, zinc, selenium, and chromium. Their functions are summarized in Exhibit 25.5.

Vitamins (p. 849)

1. Vitamins are organic nutrients that maintain growth and normal metabolism. Many function in enzyme systems.
2. Fat-soluble vitamins are absorbed with fats and include A, D, E, and K.
3. Water-soluble vitamins are absorbed with water and include the B vitamins and vitamin C.
4. The functions and deficiency disorders of the principal vitamins are summarized in Exhibit 25.6.

Metabolism and Body Heat (p. 852)

1. A kilocalorie (kcal) is the amount of energy required to raise the temperature of 1000 g of water 1°C from 14 to 15°C.
2. The kilocalorie is the unit of heat used to express the caloric value of foods and to measure the body's metabolic rate.
3. The apparatus used to determine the caloric value of foods is called a calorimeter.

Production of Body Heat (p. 852)

1. Most body heat is a result of oxidation of the food we eat. The rate at which this heat is produced is known as the metabolic rate.
2. Metabolic rate is affected by exercise, the nervous system, hormones, body temperature, ingestion of food, age, sex, climate, sleep, and malnutrition.

Basal Metabolic Rate (p. 853)

1. Measurement of the metabolic rate under basal conditions is called the basal metabolic rate (BMR).
2. BMR is expressed in kilocalories per square meter of body area per hour (kcal/m²/hr).

Loss of Body Heat (p. 853)

1. Radiation is the transfer of heat from a warmer object to a cooler object without physical contact.

2. Conduction is the transfer of body heat to a substance or object in contact with the body.
3. Convection is the transfer of body heat by a liquid or gas between areas of different temperatures.
4. Evaporation is the conversion of a liquid to a vapor.

Homeostasis of Body Temperature Regulation (p. 854)

1. A normal body temperature is maintained by a delicate balance between heat-producing and heat-losing mechanisms.
2. The hypothalamic thermostat is the preoptic area.
3. Mechanisms that produce or retain heat are vasoconstriction, sympathetic stimulation, skeletal muscle contraction, and thyroid hormone production.
4. Mechanisms of heat loss include vasodilation, decreased metabolic rate, decreased skeletal muscle contraction, and perspiration.

Body Temperature Abnormalities (p. 856)

1. Fever is an abnormally high body temperature most commonly caused by bacteria (and their toxins) and viruses. A fever-producing substance is called a pyrogen.
2. Heat cramps are painful skeletal muscle contractions due to loss of salt and water.
3. Heat exhaustion results in a normal or below normal body temperature, profuse perspiration, nausea, cramps, and dizziness.
4. Heatstroke results in decreased blood flow to skin, reduced perspiration, and high body temperature.
5. Hypothermia refers to a lowering of body temperature.

Review Questions

1. Discuss how food intake is regulated. (p. 822)
2. Define a nutrient. List the six classes of nutrients and indicate the function of each. (p. 822)
3. What is metabolism? Distinguish between anabolism and catabolism and give examples of each. (p. 823)
4. How does ATP couple anabolism and catabolism? (p. 823)
5. Define oxidation and reduction and give an example of each. (p. 824)
6. Describe the three ways that ATP can be generated. (p. 825)
7. How are carbohydrates absorbed and what are their fates in the body? How does glucose move into body cells? (p. 826)
8. Define glycolysis. Describe its principal events and outcome. What is the fate of pyruvic acid? (p. 826)
9. Describe how acetyl coenzyme A is formed. (p. 829)
10. Outline the principal events and outcomes of the Krebs cycle. (p. 829)
11. Explain what happens in the electron transport chain. (p. 832)
12. What is chemiosmosis? (p. 832)
13. Summarize the outcomes of the complete oxidation of a molecule of glucose. (p. 833)
14. Define glycogenesis and glycogenolysis. Under what circumstances does each occur? (p. 835)
15. Why is gluconeogenesis important? Give specific examples to substantiate your answer. (p. 835)
16. How are triglycerides absorbed and what are their fates in the body? Where are triglycerides stored in the body? (p. 837)
17. Explain the principal events of the catabolism of glycerol and fatty acids. (p. 838)
18. What are ketone bodies? What is ketosis? (p. 839)
19. Define lipogenesis and explain its importance. (p. 840)
20. How are proteins absorbed and what are their fates in the body? (p. 840)
21. Relate deamination to amino acid catabolism. (p. 840)
22. Summarize the major steps involved in protein synthesis. (p. 78)
23. Distinguish between essential and nonessential amino acids. (p. 840)
24. Explain the importance of glucose 6-phosphate, pyruvic acid, and acetyl coenzyme A in metabolism. (p. 841)
25. What is the absorptive (fed) state? Outline its principal events. How is it controlled? (p. 843)
26. What is the postabsorptive (fasting) state? Outline its principal events. How is it controlled? (p. 845)
27. Indicate the roles of the following hormones in the regulation of metabolism: insulin, glucagon, epinephrine, human growth hormone (hGH), thyroxine, cortisol, and testosterone. (p. 845)
28. What is a mineral? Briefly describe the functions of the following minerals: calcium, phosphorus, iron, iodine, copper, sodium, potassium, chlorine, magnesium, sulfur, zinc, fluorine, manganese, cobalt, chromium, and selenium. (p. 846)
29. Define a vitamin. Explain how we obtain vitamins. Distinguish between a fat-soluble and a water-soluble vitamin. (p. 849)
30. For each of the following vitamins, indicate its principal function and effect of deficiency: A, D, E, K, B_1, B_2, niacin, B_6, B_{12}, pantothenic acid, folic acid, biotin, and C. (p. 850)
31. Describe abnormalities associated with megadoses of several minerals and vitamins. (p. 857)
32. Define a kilocalorie (kcal). How is the unit used? (p. 852)
33. How is the caloric value of foods determined? (p. 852)
34. What is metabolic rate? What factors affect it? (p. 853)
35. Define basal metabolic rate (BMR). How is it measured? (p. 853)
36. Define each of the following mechanisms of heat loss: radiation, conduction, convection, and evaporation. (p. 853)
37. Explain how body temperature is regulated by describing the mechanisms of heat production and heat loss. (p. 854)
38. Contrast fever, heat cramps, heat exhaustion, heatstroke, and hypothermia as body temperature abnormalities. (p. 856)
39. What causes fever? What are its benefits? (p. 856)

Answers to Questions with Figures

25.1 Anabolism dominates because the cells are synthesizing complex molecules.

25.2 Kinases are enzymes that phosphorylate (add phosphate to) their substrate. The substrate is glucose, a hexose.

25.3 Cytosol. Mitochondria.

25.4 They will later yield ATP in the electron transport chain.

25.5 Electrons provided by NADH.

25.6 Each carrier in the series has a lower energy than the carrier it received electrons from because some of the energy in the electrons is used to pump protons.

25.7 Electron carriers.

25.8 Six; six.

25.9 Muscle fibers; they lack the enzyme phosphatase.

25.10 Liver cells.

25.11 Liver and adipose cells. Liver cells.

25.12 The amino group is removed by deamination.

25.13 Acetyl coenzyme A.

25.14 Anabolic.

25.15 Lipolysis (adipose and liver cells); gluconeogenesis (liver cells); glycogenolysis (liver cells).

25.16 Exercise, sympathetic nervous system, hormones (epinephrine, norepinephrine, thyroxin, testosterone, human growth hormone), elevated body temperature, and ingestion of food.

Chapter 26
THE URINARY SYSTEM

Chapter Contents at a Glance

Student Objectives

1. Describe the functions of the kidneys.

2. Identify the external and internal gross anatomical features of the kidneys.

3. Define the structural adaptations of a nephron for urine formation.

4. Discuss the process of urine formation through glomerular filtration, tubular reabsorption, and tubular secretion.

5. Describe how the kidneys produce dilute and concentrated urine.

6. Explain the principle of hemodialysis.

7. Discuss the structure and physiology of the ureters, urinary bladder, and urethra.

8. List and describe the physical characteristics, normal chemical constituents, and abnormal constituents of urine.

9. Describe the effects of aging on the urinary system.

10. Describe the development of the urinary system.

11. Discuss the causes of glomerulonephritis, pyelitis, pyelonephritis, cystitis, nephrosis, polycystic disease, renal failure, urinary tract infections (UTIs), and diabetes insipidus.

12. Define medical terminology associated with the urinary system.

I n metabolizing nutrients, body cells produce wastes—carbon dioxide, excess water, and heat. In addition, protein catabolism produces toxic nitrogenous wastes such as ammonia and urea. Also, essential ions such as sodium (Na^+), chloride (Cl^-), sulfate (SO_4^{2-}), phosphate (HPO_4^{2-}), and hydrogen (H^+) tend to build up in excess quantity. All the toxic materials and the excess essential materials must be excreted (eliminated) from the body.

Several organs contribute to the job of waste elimination from the body.

1. **Kidneys.** Excrete water, nitrogenous wastes from protein catabolism, some bacterial toxins, H^+, and inorganic salts (electrolytes) plus some heat and carbon dioxide.
2. **Lungs.** Excrete carbon dioxide, heat, and a little water.
3. **Skin (sudoriferous glands).** Excrete heat, water, and carbon dioxide plus small quantities of salts and urea.
4. **Gastrointestinal tract.** Eliminates solid, undigested wastes and excretes carbon dioxide, water, salts, and heat.

The prime function of the **urinary system** is to help maintain homeostasis by controlling the composition, volume, and pressure of blood. It does so by removing and restoring selected amounts of water and solutes. Two kidneys, two ureters, one urinary bladder, and a single urethra make up the system (Fig. 26.1).

The kidneys have several functions:

1. **Blood volume and composition.** They regulate the composition and volume of the blood and remove wastes from the blood in the form of urine. They excrete selected amounts of various wastes including excess H^+, which helps control blood pH.
2. **Blood pressure.** They help regulate blood pressure by secreting the enzyme renin, which activates the renin–angiotensin pathway.
3. **Metabolism.** The kidneys contribute to metabolism by (1) performing gluconeogenesis (synthesis of new glucose molecules) during periods of fasting or starvation, (2) secreting erythropoietin, which stimulates production of red blood cells, and (3) participating in synthesis of calcitriol, the active form of vitamin D.

Urine is excreted from each kidney through its ureter and is stored in the urinary bladder until it is expelled from the body through the urethra. When the kidneys do not function properly to continually remove waste products, one result is **uremia** (yoo-RĒmē-a; *emia* = condition of blood), a toxic level of urea in the blood.

The specialized branch of medicine that deals with the structure, function, and diseases of the male and female urinary systems and the male reproductive system is known as **nephrology** (nef-ROL-ō-jē; *neph* = kidney; *logos* = study

FIGURE 26.1 Organs of the male urinary system in relation to surrounding structures.

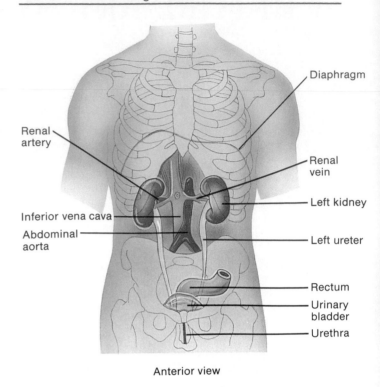

Anterior view

Question: **What are the functions of the kidneys?**

of). The branch of surgery related to male and female urinary systems and male reproductive system is called **urology** (yoo-ROL-ō-jē; *uro* = urine or urinary tract).

The developmental anatomy of the urinary system is considered at the end of the chapter.

KIDNEYS

The paired **kidneys** are reddish organs shaped like kidney beans. They are located just above the waist between the parietal peritoneum and the posterior wall of the abdomen (Fig. 26.2). Since their position is behind the peritoneal lining of the abdominal cavity, they are said to be **retroperitoneal** (re′-trō-per-i-tō-NĒ-al) organs. Other retroperitoneal structures include the ureters and adrenal (suprarenal) glands. Relative to the vertebral column, the kidneys are located between the levels of the last thoracic and third lumbar vertebrae. They are partially protected by the eleventh and twelfth pairs of ribs. The right kidney is slightly lower than the left because the liver occupies a large area on the right side.

External Anatomy

An average adult kidney measures about 10 to 12 cm (4 to 5 in.) long, 5.0 to 7.5 cm (2 to 3 in.) wide, and 2.5 cm (1

FIGURE 26.2 Location of kidneys and nearby organs.

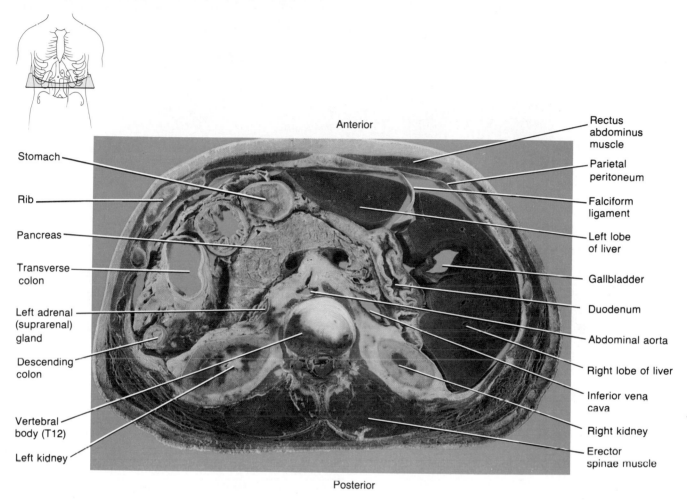

Anterior

Stomach

Rib

Pancreas

Transverse colon

Left adrenal (suprarenal) gland

Descending colon

Vertebral body (T12)

Left kidney

Rectus abdominus muscle

Parietal peritoneum

Falciform ligament

Left lobe of liver

Gallbladder

Duodenum

Abdominal aorta

Right lobe of liver

Inferior vena cava

Right kidney

Erector spinae muscle

Posterior

Cross section of abdomen

Question: At this level of section, what organs are just anterior to the right kidney? The left kidney?

in.) thick. Its concave medial border faces the vertebral column (Fig. 26.3). Near the center of the concave border is a notch called the **hilus,** through which the ureter leaves the kidney. Blood and lymphatic vessels and nerves also enter and exit the kidney through the hilus. The hilus is the entrance to a cavity in the kidney called the **renal sinus** (see Fig. 26.4).

Three layers of tissue surround each kidney. The innermost layer, the **renal capsule,** is a smooth, transparent, fibrous membrane that is continuous with the outer coat of the ureter at the hilus. It serves as a barrier against trauma and the spread of infection to the kidney. The middle layer, the **adipose capsule,** is a mass of fatty tissue surrounding the renal capsule. It also protects the kidney from trauma and holds it firmly in place within the abdominal cavity. The outermost layer, the **renal fascia,** is a thin layer of dense irregular connective tissue that anchors the kidney to its surrounding structures and to the abdominal wall.

CLINICAL APPLICATION

NEPHROTOSIS (FLOATING KIDNEY)

Nephrotosis (nef'-rō-TŌ-sis; *ptosis* = falling), or **floating kidney**, is a downward displacement or dropping of the kidney. It occurs when the kidney slips from its normal position because it is not securely held in place by adjacent organs or its covering of fat. People, most often those who are very thin, in whom either the adipose capsule or renal fascia is deficient, may develop nephrotosis. It is dangerous because the ureter may kink and block urine flow. The resulting backup of urine puts pressure on the kidney, which damages the tissue. Twisting of the ureter also causes pain.

Internal Anatomy

A coronal (frontal) section through a kidney reveals an outer reddish area called the **cortex** and an inner reddish-

FIGURE 26.3 External anatomy of the kidneys and associated structures.

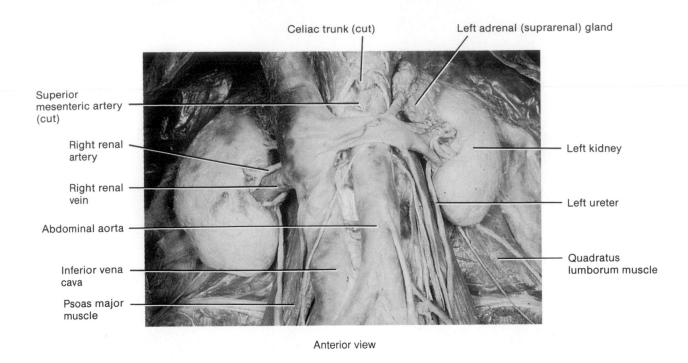

Anterior view

Question: What structures enter and leave the kidney through the hilus?

---

brown region called the **medulla** (Fig. 26.4). Within the medulla are 8 to 18 cone-shaped structures termed **renal (medullary) pyramids.** They look striated (striped) due to the presence of straight tubules and blood vessels. The bases of the pyramids face the cortex, and their apexes, called **renal papillae,** point toward the center of the kidney. The cortex is the smooth-textured area extending from the renal capsule to the bases of the pyramids and into the spaces between them. It is divided into an outer cortical zone and an inner juxtamedullary zone. Portions of the cortex extend between renal pyramids to form the **renal columns.**

Together the cortex and renal pyramids constitute the **parenchyma** (functional portion) of the kidney. Structurally, the parenchyma of each kidney consists of about 1 million microscopic structures called nephrons, which are the functional units of the kidney.

In the renal sinus of the kidney is a large cavity called the **renal pelvis.** The edge of the pelvis contains cuplike extensions called **major** and **minor calyces** (KĀ-li-sēz; *calyx* = cup). There are 2 or 3 major calyces and 8 to 18 minor calyces. Each minor calyx receives urine from collecting ducts of one pyramid and delivers urine to a major calyx. From the major calyces, the urine drains into the renal pelvis and out through the ureter to the urinary bladder.

Nephron

The functional unit of the kidney is the **nephron** (NEF-ron). Nephrons have three basic functions—filtration, secretion, and reabsorption. In *filtration* some substances are permitted to pass from the blood into the nephrons, while others are kept out. Then, as the filtered liquid (filtrate) moves through the nephrons, it gains some additional materials (wastes and excess substances); this is *secretion.* Other substances (useful materials) are returned to the blood; this is *reabsorption.* As a result of these activities of nephrons, urine is formed.

Parts of the Nephron

A nephron consists of two portions: a **renal corpuscle** (KŌR-pus-sul; *corpus* = body; *cle* = tiny) where fluid is filtered, and a **renal tubule** into which the filtered fluid passes (Fig. 26.5a).

FIGURE 26.4 Coronal sections of the right kidney illustrating the internal anatomy.

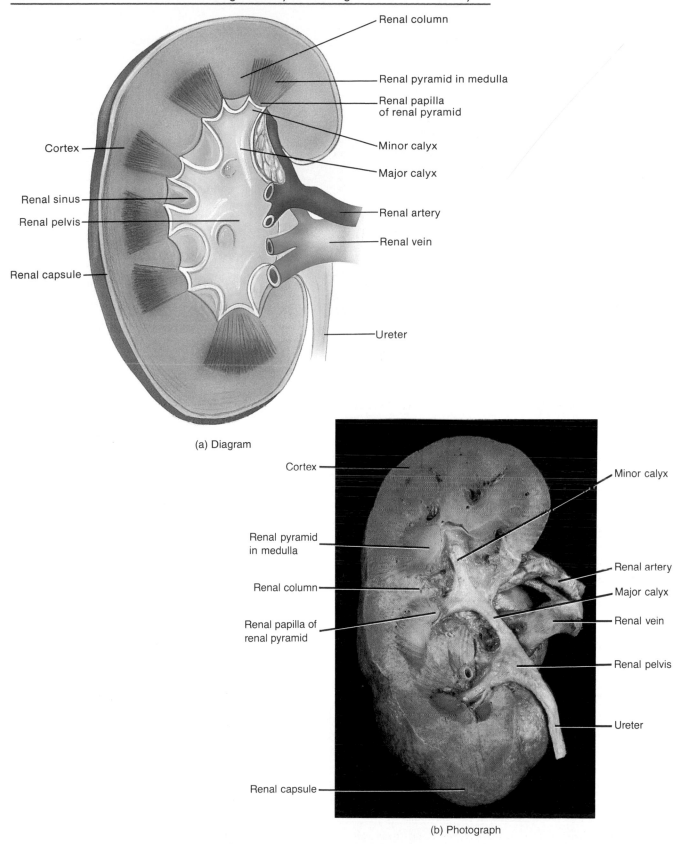

(a) Diagram

(b) Photograph

Question: What structures are part of the parenchyma of the kidneys?

FIGURE 26.5 Nephrons (colored gold).

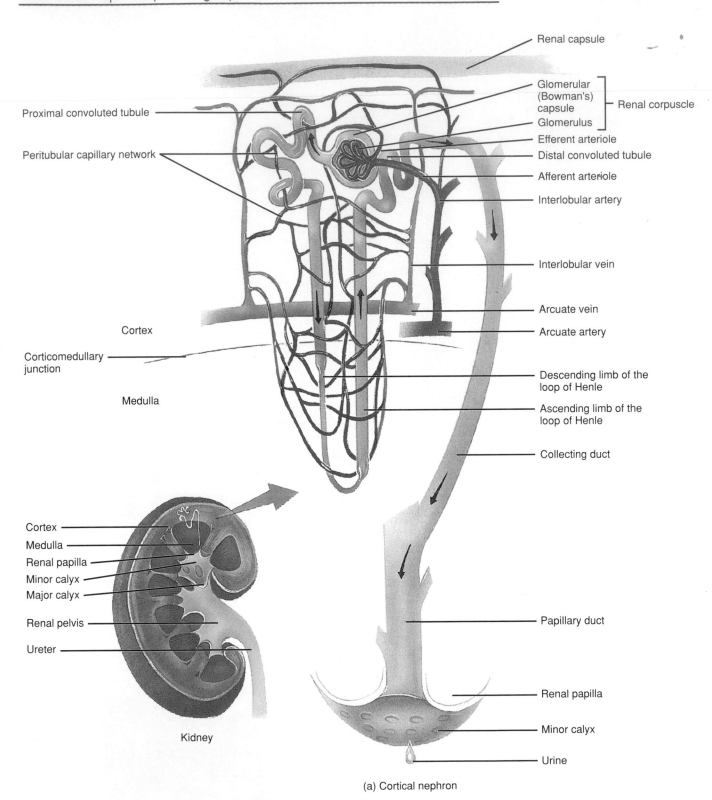

Renal capsule

Glomerular (Bowman's) capsule

Renal corpuscle

Proximal convoluted tubule

Glomerulus

Efferent arteriole

Peritubular capillary network

Distal convoluted tubule

Afferent arteriole

Interlobular artery

Interlobular vein

Cortex

Arcuate vein

Arcuate artery

Corticomedullary junction

Descending limb of the loop of Henle

Medulla

Ascending limb of the loop of Henle

Collecting duct

Cortex

Medulla

Renal papilla

Minor calyx

Major calyx

Renal pelvis

Ureter

Papillary duct

Renal papilla

Minor calyx

Urine

Kidney

(a) Cortical nephron

FIGURE 26.5 (continued)

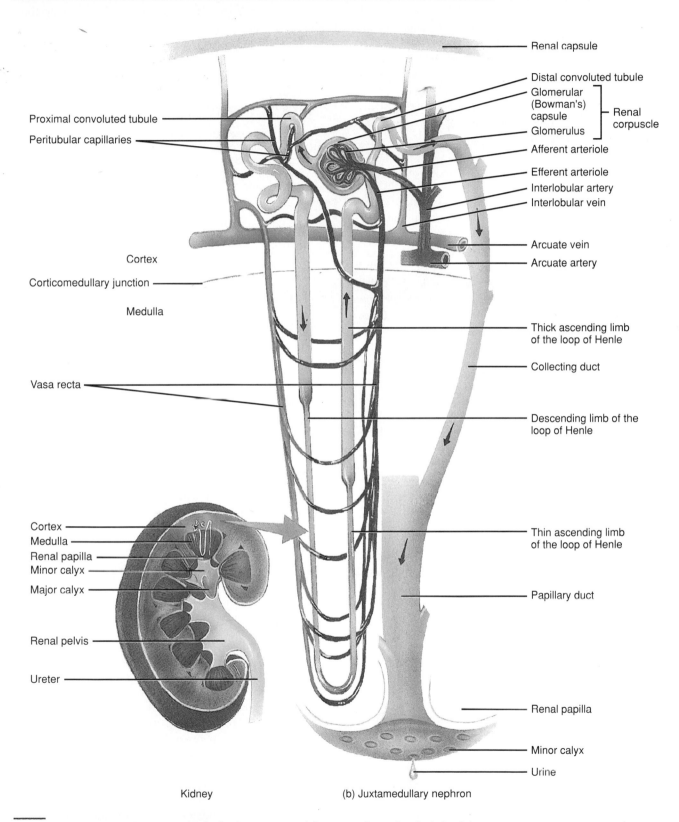

Kidney   (b) Juxtamedullary nephron

Question: Imagine you are a water molecule that has just entered the proximal convoluted tubule of the nephron shown in (b) and will eventually be part of the urine. What parts of the nephron will you travel through (in order) before you reach the renal pelvis?

The renal corpuscle has two components—a tuft of capillary loops, called the **glomerulus** (glō-MER-yoo-lus; *glomus* = ball; *ulus* = small; plural is **glomeruli),** surrounded by a double-walled epithelial cup, called the **glomerular (Bowman's) capsule.** Their arrangement is analogous to a fist (glomerulus) punched into a limp balloon (glomerular capsule) until the fist is covered by two layers of balloon with space in between. Since they are capillary networks, the glomeruli also are part of the cardiovascular system. Blood *enters* a glomerulus through an **afferent arteriole** and *exits* through an **efferent arteriole.**

The outer wall, or *parietal layer,* of the glomerular capsule is separated from the inner wall, known as the *visceral layer,* by the **capsular (Bowman's) space** (see Fig. 26.6a). As blood flows through the glomerular capillaries, water and most kinds of solutes filter from blood plasma into the capsular space. Large plasma proteins and the formed elements in blood do not normally pass through.

From the capsular space, filtered fluid passes into the renal tubule, which has three main sections. In the order that fluid passes through them, the renal tubule consists of a (1) **proximal convoluted tubule (PCT),** (2) **loop of Henle (nephron loop),** and (3) **distal convoluted tubule (DCT).** *Convoluted* means the tubule is coiled rather than straight. *Proximal* signifies the tubule portion attached to the glomerular capsule and *distal* the portion that is further away. The renal corpuscle and both convoluted tubules lie in the cortex of the kidney whereas the loop of Henle extends into the medulla, makes a hairpin turn, then returns to the cortex.

Short connecting tubules link the distal convoluted tubules of several nephrons to a single **collecting duct.** Collecting ducts then merge until eventually there are only several hundred large **papillary ducts**, which drain into a minor calyx. The collecting ducts and papillary ducts stretch from the cortex through the medulla to the renal pelvis. On average, there are about 30 papillary ducts per renal papilla. Each kidney has about 1 million renal corpuscles, PCTs, loops of Henle, and DCTs but a much smaller number of collecting ducts and even fewer papillary ducts.

Cortical and Juxtamedullary Nephrons

In a nephron, the loop of Henle connects the proximal and distal convoluted tubules. The first portion of the loop dips into the medulla, where it is called the **descending limb of the loop of Henle.** The tubule then bends in a U-shape and returns to the cortex as the **ascending limb of the loop of Henle.** Some nephrons have short loops of Henle whereas others have long loops. A **cortical nephron** usually has its glomerulus in the outer portion of the cortex; its short loop of Henle penetrates only into the outer region of the medulla (Fig. 26.5a). A **juxtamedullary nephron** (*juxta* = beside) usually has its glomerulus deep in the cortex close to the medulla, and its long loop of Henle stretches through

the medulla and almost reaches the renal papilla (Fig. 26.5b). About 15 to 20% of the nephrons (juxtamedullary) in a human kidney have long loops. These long-loop nephrons enable the kidneys to excrete a very dilute or very concentrated urine (described shortly).

Histology of the Nephron

Each portion of the nephron and collecting duct system has distinctive histological features that reflect its particular functions.

● **Histology of the Filter** The visceral layer of the glomerular capsule and the endothelium of glomerular capillaries form an **endothelial–capsular membrane** that acts as a filter (Fig. 26.6). It lets some materials from the blood pass and restricts passage of others. Filtered substances pass through the three layers of this membrane in the following order (Fig. 26.6b,d).

1. **Endothelium of glomerulus.** This single layer of endothelial cells has fenestrations (pores) averaging 50 to 100 nm in diameter. It restricts the passage of blood cells.
2. **Basement membrane of glomerulus.** This layer of extracellular material lies between the endothelium and the visceral layer of the glomerular capsule. It consists of fibrils in a glycoprotein matrix and restricts the passage of larger proteins.
3. **Filtration slits in podocytes.** The specialized epithelial cells that form the visceral layer of the glomerular capsule are called **podocytes** (*podos* = foot). Extending from each podocyte are thousands of footlike structures called **pedicels** (PED-i-sels; *pediculus* = little foot). The pedicels cover the basement membrane, except for spaces between them called **filtration slits (slit pores).** A thin membrane, the **slit membrane,** extends across filtration slits and restricts the passage of medium-sized proteins.

The histology of a renal corpuscle and parts of the renal tubule is shown in Fig. 26.7.

● **Histology of the Renal Tubule** A single layer of epithelial cells, resting on a basement membrane, forms the wall of the entire renal tubule. In the proximal convoluted tubule, the cells are cuboidal and have a prominent brush border of microvilli on their apical surface (facing the lumen). These microvilli, like those of the small intestine, increase the surface area for reabsorption and secretion. About 65% of the water and up to 100% of some solutes that pass through the endothelial–capsular membrane return to the bloodstream from the PCT.

The descending limb of the loop of Henle and the first part of the ascending limb of the loop of Henle (the thin ascending limb) is simple squamous epithelium. The second part of the ascending limb of the loop of Henle (the thick ascending limb) is cuboidal to low columnar epithelium. Cortical (short-loop) nephrons lack the thin portion of the ascending limb.

The cells of the distal convoluted tubule and collecting ducts, like those of the proximal tubule, are cuboidal. Up to the distal convoluted tubule, the cells of a given tubule segment are all alike. Beginning in the distal convoluted tubule and continuing into the collecting ducts, however, two different cell types are present. Most are **principal cells,** which are sensitive to antidiuretic hormone (ADH) and aldosterone, two hormones that regulate kidney functions (described shortly). A few are **intercalated cells,** which can secrete H⁺ to rid the body of excess acids. Cells of the large papillary ducts are columnar.

● **Histology of the Juxtaglomerular Apparatus (JGA)** In each nephron, the final portion of the ascending limb of the loop of Henle makes contact with the afferent arteriole serving its own renal corpuscle (Fig. 26.8). The cells of the

FIGURE 26.6 Endothelial–capsular membrane. The size of the filtration slits in (b) has been exaggerated for emphasis.

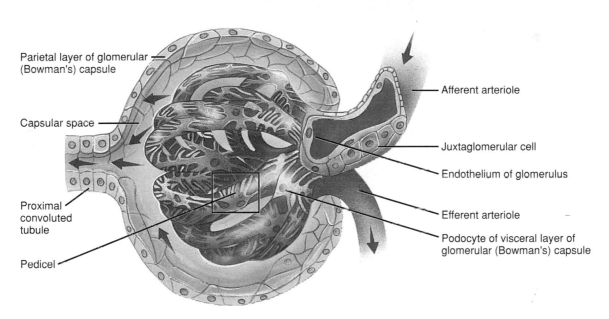

(a) Parts of a renal corpuscle

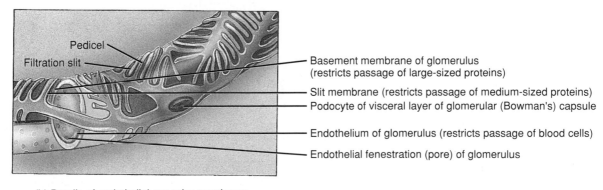

(b) Details of endothelial-capsular membrane

Figure continues

FIGURE 26.6 (continued)

Parietal layer of glomerular   Capsular
(Bowman's) capsule   space

Podocyte of visceral layer of glomerular (Bowman's) capsule

(c) Scanning electron micrograph of a renal corpuscle, 2000×

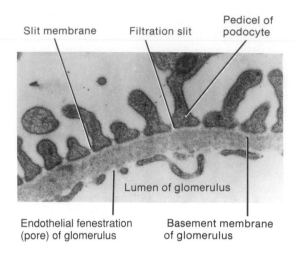

Slit membrane   Filtration slit   Pedicel of
                                    podocyte

Lumen of glomerulus

Endothelial fenestration   Basement membrane
(pore) of glomerulus   of glomerulus

(d) Transmission electron micrograph of the
    endothelial-capsular membrane (42,700 x)

Question: Which part of the filtration barrier prevents red blood cells from entering the capsular space?

FIGURE 26.7 Histology of a nephron.

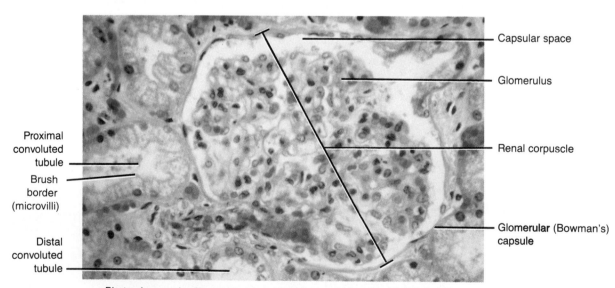

Capsular space

Glomerulus

Renal corpuscle

Glomerular (Bowman's)
capsule

Proximal
convoluted
tubule

Brush
border
(microvilli)

Distal
convoluted
tubule

Photomicrograph of a renal corpuscle and surrounding renal tubules, 400×

Question: What type of epithelium forms the glomerular capsule? The distal and proximal convoluted tubules?

FIGURE 26.8 Juxtaglomerular apparatus (JGA).

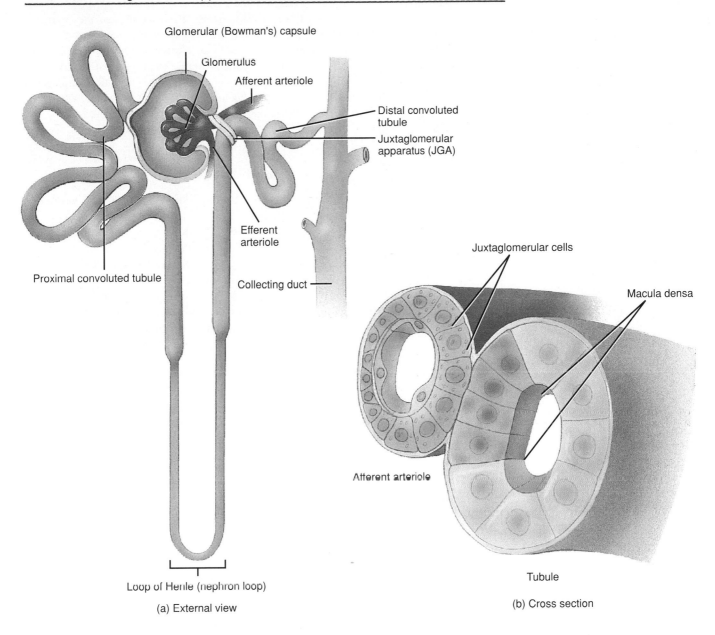

Glomerular (Bowman's) capsule

Glomerulus

Afferent arteriole

Distal convoluted tubule

Juxtaglomerular apparatus (JGA)

Efferent arteriole

Collecting duct

Proximal convoluted tubule

Juxtaglomerular cells

Macula densa

Afferent arteriole

Tubule

Loop of Henle (nephron loop)

(a) External view

(b) Cross section

Question: What substance is secreted by juxtaglomerular cells?

renal tubule in this region are tall and crowded together. Collectively, they are known as the **macula densa** (*macula* = spot; *densa* = dense). These cells monitor the salt (NaCl) concentration of fluid in the tubule lumen. Next to the macula densa, smooth muscle fibers in the wall of the afferent arteriole (and sometimes efferent arteriole) are modified in several ways. Their nuclei are round (instead of long), and their cytoplasm includes renin-containing granules (instead of myofibrils). These modified smooth muscle fibers are called **juxtaglomerular (JG) cells.** Together with the macula densa, they constitute the **juxtaglomerular apparatus,**

or **JGA.** The JGA helps regulate blood pressure and the rate of blood filtration by the kidneys. The distal convoluted tubule begins a short distance past the macula densa.

Blood and Nerve Supply

Since the kidneys remove wastes from the blood and regulate its fluid and electrolyte content, it is not surprising that they are abundantly supplied with blood vessels. The right and left **renal arteries** carry 20 to 25% of the resting cardiac output to the kidneys (Fig. 26.9). This amounts to about 1200 ml/min.

Before or immediately after entering the hilus, the renal artery typically divides into a larger anterior branch and a smaller posterior branch. From these branches, five **segmental arteries** originate, each supplying a particular segment of the kidneys. Each segmental artery gives off several branches that enter the parenchyma and pass as the **interlobar arteries** in the renal columns between the renal pyramids. At the bases of the pyramids the interlobar arteries arch between the medulla and cortex; here they are known as the **arcuate arteries.** Divisions of the arcuate arteries produce a series of **interlobular arteries,** which enter the cortex and divide into **afferent arterioles.**

Each renal corpuscle receives one afferent arteriole, which divides into the tangled capillary network called the **glomerulus.** The glomerular capillaries then reunite to form an **efferent arteriole** that drains blood out of the glomerulus. Since the efferent arteriole is smaller in diameter than the afferent arteriole, blood pressure in glomerular capillaries is substantially higher than the pressure in capillaries elsewhere in the body. The afferent–efferent arteriole situation is unique because blood usually flows out of capillaries into venules and not into other arterioles.

Each efferent arteriole of a cortical nephron divides to form a network of capillaries, called the **peritubular capillaries,** around the proximal and distal convoluted tubules. The efferent arteriole of a juxtamedullary nephron also forms peritubular capillaries. In addition, it forms long loop-shaped vessels called **vasa recta** that dip down alongside the loop of Henle into the medulla.

The peritubular capillaries eventually reunite to form **peritubular venules** and then **interlobular veins.** The interlobular veins also receive blood from the vasa recta. Then the blood drains through the **arcuate veins** to the **interlobar veins** running between the pyramids, and on to the **segmental veins.** Blood leaves the kidney through a single **renal vein** that exits at the hilus.

The nerve supply to the kidneys derives from the **renal plexus** of the sympathetic division of the autonomic nervous system. Nerves from the plexus accompany the renal arteries and their branches and are distributed to the blood vessels. Because these are vasomotor nerves, they regulate the flow of blood through the kidney by regulating the diameters of the arterioles.

PHYSIOLOGY OF URINE FORMATION

The major work of the urinary system is done by the nephrons. The other parts of the system are primarily passageways and storage areas. Nephrons carry out three important functions: (1) they control blood concentration and volume by removing selected amounts of water and solutes; (2) they regulate blood pH; and (3) they remove toxic wastes from the blood.

As the nephrons perform these activities, they remove many materials from the blood, return the ones that the body requires, and excrete (eliminate) the remainder. The excreted fluid is called **urine.** Urine formation involves three principal processes: filtration, reabsorption, and secretion. Filtration is the job of the renal corpuscle; reabsorption and secretion occur all along the renal tubule.

Glomerular Filtration

The first step in the production of urine is called **glomerular filtration.** The principle of filtration, the forcing of fluids and dissolved substances through a membrane by pressure, is the same in glomerular capillaries as in capillaries elsewhere in the body (Starling's law of the capillaries, see page 632). It occurs in the renal corpuscle of the kidneys across the endothelial–capsular membranes. Blood pressure forces water and dissolved blood components through the endothelial fenestrations (pores) of the capillaries, basement membrane, and on through the filtration slits of the adjoining visceral wall of the glomerular capsule (see Fig. 26.6b). The resulting fluid is called the **filtrate.**

Roughly 180 liters (48 gal) of filtrate enter the capsular spaces each day. This represents about 60 times the entire blood plasma volume. About 178 to 179 liters return to the bloodstream by reabsorption in the renal tubules so only 1 to 2 liters (about 1 to 2 qt) are excreted as urine. In a healthy person, filtrate contains all the materials present in the blood except the formed elements and most proteins, which are too large to pass through the endothelial–capsular membranes.

Several structural features of the renal corpuscles enhance their blood-filtering capacity.

1. Glomerular capillaries are long. The glomerular capillaries are very long, presenting a vast surface area for filtration.

2. The filter (endothelial–capsular membrane) is porous and thin. The endothelial–capsular membrane is very thin (0.1 μm) and porous. Glomerular capillaries are about 50 times more permeable than capillaries elsewhere in the body. Although the endothelial fenestrations (pores) generally do not restrict the passage of solutes, the basement membrane and filtration slits permit the passage of only smaller molecules. Thus water, glucose, vitamins, amino acids, small proteins, nitrogenous wastes, and ions easily pass into the capsular space.

3. Capillary blood pressure is high. The efferent arteriole is smaller in diameter than the afferent arteriole, so there is high resistance to the outflow of blood from the glomerulus. Thus blood pressure is higher in the glomerular capillaries than in capillaries elsewhere in the body. A higher pressure means more filtration.

FIGURE 26.9 Blood supply of the right kidney.

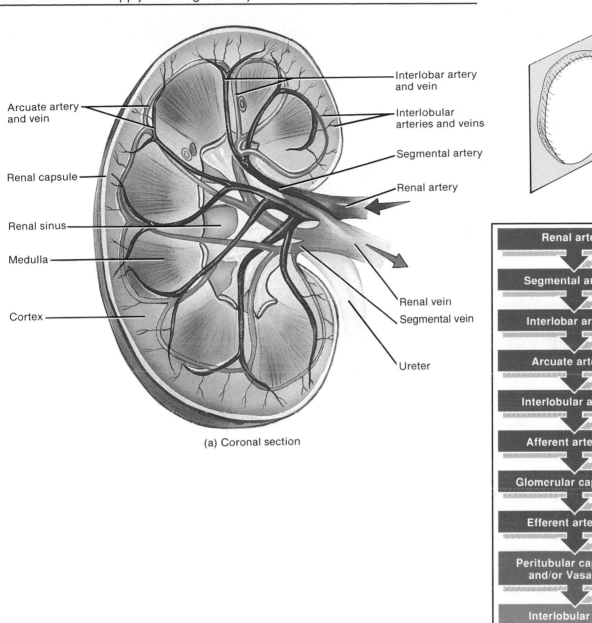

Arcuate artery and vein

Renal capsule

Renal sinus

Medulla

Cortex

Interlobar artery and vein

Interlobular arteries and veins

Segmental artery

Renal artery

Renal vein

Segmental vein

Ureter

(a) Coronal section

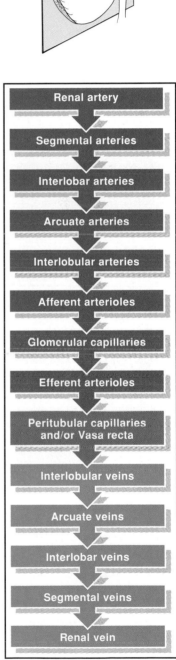

| Renal artery |
| Segmental arteries |
| Interlobar arteries |
| Arcuate arteries |
| Interlobular arteries |
| Afferent arterioles |
| Glomerular capillaries |
| Efferent arterioles |
| Peritubular capillaries and/or Vasa recta |
| Interlobular veins |
| Arcuate veins |
| Interlobar veins |
| Segmental veins |
| Renal vein |

(b) Path of blood flow

Question: How much blood enters the renal arteries per minute?

Net Filtration Pressure

In the glomerulus, blood filtering depends on three main pressures. The chief one is the **glomerular blood hydrostatic pressure (GBHP).** *Hydrostatic (hydro = water) pressure* is the force that a fluid under pressure exerts against the walls of its container. Glomerular blood hydrostatic pressure means the blood pressure in glomerular capillaries, which is about 60 mm Hg (Fig. 26.10). Two other forces, however, oppose glomerular blood hydrostatic pressure. The first of these, **capsular hydrostatic pressure (CHP),** develops in the following way. As the filtrate is forced into the capsular space between the walls of the glomerular capsule, it meets two forms of resistance: the walls of the capsule and the fluid that has already filled the renal tubule. As a result, some filtrate is pushed back into the capillary. The amount of "push" is the capsular hydrostatic pressure. It is about 15 mm Hg.

The second force opposing filtration is the **blood colloid osmotic pressure (BCOP)**, which is due to the presence of proteins in blood plasma. *Osmotic pressure* is the pressure required to prevent the movement of pure water into a solution containing solutes when the solutions are separated by a selectively permeable membrane (see page 63). The greater the solute concentration of a solution, the greater its osmotic pressure. The concentration of all solutes except proteins is the same in blood and glomerular filtrate. But blood contains a much higher concentration of proteins than filtrate. Thus water would move out of the filtrate and back into the blood vessel if the blood pressure in the glomerulus were not greater than blood colloid osmotic pressure. BCOP averages 27 mm Hg in glomerular capillaries.

FIGURE 26.10 Forces determining net filtration pressure.

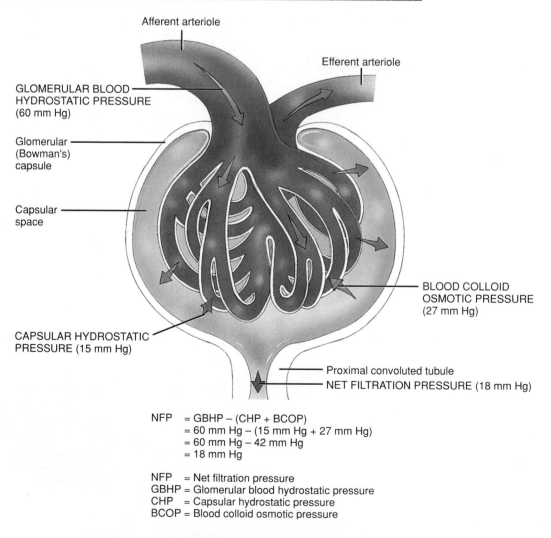

Afferent arteriole

Efferent arteriole

GLOMERULAR BLOOD HYDROSTATIC PRESSURE (60 mm Hg)

Glomerular (Bowman's) capsule

Capsular space

BLOOD COLLOID OSMOTIC PRESSURE (27 mm Hg)

CAPSULAR HYDROSTATIC PRESSURE (15 mm Hg)

Proximal convoluted tubule
NET FILTRATION PRESSURE (18 mm Hg)

NFP = GBHP − (CHP + BCOP)
    = 60 mm Hg − (15 mm Hg + 27 mm Hg)
    = 60 mm Hg − 42 mm Hg
    = 18 mm Hg

NFP = Net filtration pressure
GBHP = Glomerular blood hydrostatic pressure
CHP = Capsular hydrostatic pressure
BCOP = Blood colloid osmotic pressure

Question: Suppose a large stone (calculus) is blocking a major calyx. What effect might this have on NFP?

To determine the **net filtration pressure (NFP)**, we subtract the forces that oppose filtration from the glomerular blood hydrostatic pressure.

$$NFP = GBHP - (CHP + BCOP)$$

Promotes   Oppose
filtration   filtration

By substituting the values just given, a normal NFP may be calculated as follows:

$$NFP = (60 \text{ mm Hg}) - (15 \text{ mm Hg} + 27 \text{ mm Hg})$$
$$= (60 \text{ mm Hg}) - (42 \text{ mm Hg})$$
$$= 18 \text{ mm Hg}$$

This means that a pressure of about 18 mm Hg causes a normal amount of plasma (minus plasma proteins) to filter from the glomerulus into the capsular space. Recall from Chapter 21 that NFP at the arterial end of capillaries in other parts of the body is only 10 mm Hg. Thus net filtration pressure in the kidneys is considerably higher. **Filtration fraction** is the percentage of plasma entering the nephrons that becomes glomerular filtrate. Although filtration fraction normally is between 16 and 20%, the value varies considerably in both health and disease.

Glomerular Filtration Rate

The amount of filtrate that forms in all the renal corpuscles of both kidneys every minute is called the **glomerular filtration rate (GFR)**. In a normal adult, GFR is about 125 ml/min—about 180 liters (48 gal) a day.

GFR is directly related to the pressures that determine NFP. Any factor that alters NFP will affect GFR. For example, severe blood loss reduces systemic blood pressure, which also decreases the glomerular blood hydrostatic pressure. If blood pressure falls to the point where the hydrostatic pressure in the glomeruli reaches 42 mm Hg, filtration stops because this is the magnitude of the opposing forces. Such a condition is called **anuria** (a-NOO-rē-a), a daily urine output of less than 50 ml. It may be caused by insufficient pressure for filtration or by inflammation of the glomeruli so that plasma cannot enter the capsular space.

Homeostasis of body fluids requires that the kidneys maintain a relatively constant GFR. If the GFR is too high, needed substances may pass so quickly through the renal tubules that they are not reabsorbed and instead are lost in the urine. On the other hand, if the GFR is too low, nearly all the filtrate may be reabsorbed and certain waste products may not be adequately excreted.

CLINICAL APPLICATION

LOSS OF PLASMA PROTEINS

In some kidney diseases, such as nephrotic syndrome (see page 900), damaged glomerular capillaries become so permeable that plasma proteins enter the filtrate. As a result, the filtrate exerts a colloid osmotic pressure that draws water out of the blood. In this situation, NFP and GFR increase. At the same time, blood colloid osmotic pressure decreases because plasma proteins are being lost in the urine.

Regulation of GFR

When blood flows into the glomerular capillaries more rapidly, glomerular filtration rate increases. Glomerular blood flow, in turn, depends on the (1) systemic blood pressure and (2) the diameter of the afferent and efferent arterioles. Three principal mechanisms regulate these two factors: renal autoregulation, hormonal regulation, and neural regulation.

● **Renal Autoregulation of GFR**   The ability of the kidneys to maintain a constant blood pressure and GFR despite changes in systemic arterial pressure is called **renal autoregulation.** Renal autoregulation is intrinsic, which means that it operates completely *within* the kidneys (even if they are removed from the body, for example, in a transplant operation). The kidneys have a built-in system to compensate for moderate changes in systemic arterial pressure for short periods of time.

Renal autoregulation operates by negative feedback systems that involve the juxtaglomerular apparatus or JGA (Fig. 26.11). When NFP and GFR are low due to low blood pressure, the proximal convoluted tubules and loop of Henle reabsorb more than the normal fraction of fluid and salts. Thus fluid flows slowly past the macula densa, and the concentration of sodium and chloride ions (Na+ and Cl−) in that fluid is low. Decreased delivery of fluid and NaCl to the macula densa inhibits release of a vasoconstrictor substance by JGA cells. (Which JGA cells are responsible and the chemical nature of the vasoconstrictor are not yet known.) With less vasoconstrictor substance present, the afferent arterioles dilate, allowing more blood to flow into the glomerular capillaries. This increases net filtration pressure and glomerular filtration rate and brings about a return to homeostasis.

● **Hormonal Regulation of GFR**   Two hormones contribute to regulation of GFR—angiotensin II and atrial natriuretic peptide (ANP). Figure 26.12 describes the renin–angiotensin system. Juxtaglomerular cells secrete an enzyme called **renin** in response to several kinds of stimuli. Stimuli that promote increased renin release are decreased delivery of fluid and NaCl to the macula densa, decreased stretch of the juxtaglomerular cells, and increased frequency of nerve impulses in renal sympathetic nerves.

Once released into the blood, renin acts on a plasma protein produced by the liver called **angiotensinogen** and converts it into angiotensin I. As angiotensin I passes through the lungs, an enzyme called **angiotensin converting enzyme (ACE)** converts it to **angiotensin II,** which is an active hormone.

Angiotensin II has several important actions.

FIGURE 26.11 Negative feedback regulation of glomerular filtration rate by the juxtaglomerular apparatus (JGA).

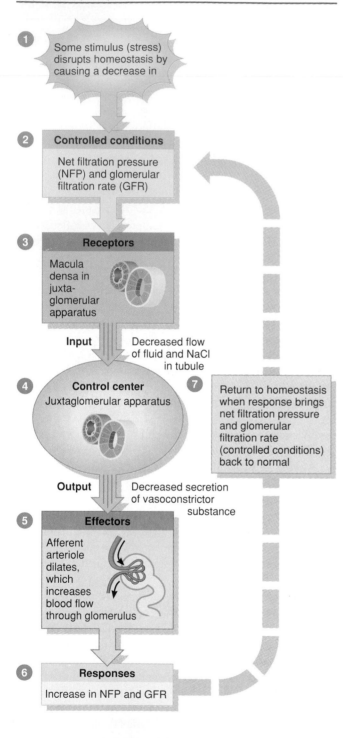

Question: Why is this process called autoregulation?

1. **Vasoconstriction.** It constricts efferent arterioles to elevate glomerular blood pressure and raise GFR back to normal.

2. **Aldosterone.** It stimulates the adrenal cortex to secrete aldosterone, which enhances reabsorption of Na^+ by the principal cells in the collecting ducts. Water follows Na^+ by osmosis, blood volume increases, and blood pressure and GFR are restored to normal.

3. **Thirst.** It acts on the thirst center in the hypothalamus to increase water intake. This results in an increase in blood volume and blood pressure and GFR is restored to normal.

4. **ADH.** It stimulates the release of antidiuretic hormone (ADH) from the posterior pituitary gland, resulting in water retention by the kidneys. This increases blood volume and blood pressure and GFR is restored to normal.

In all its actions, angiotensin II helps to restore normal blood pressure and thus normal GFR and a return to homeostasis.

A second hormone that influences glomerular filtration and other renal processes is **atrial natriuretic peptide (ANP).** As its name suggests, it is secreted by cells in the atria of the heart. It was identified in 1983 after a long search for the mysterious "third factor" (aldosterone and ADH were the other two) that could explain certain clinical and research results. ANP promotes excretion of both water (diuresis) and sodium (natriuresis). Secretion of ANP is stimulated by increased stretching of the heart, as occurs when blood volume increases. ANP increases glomerular filtration rate, perhaps by increasing the permeability of the filter or by dilating the afferent arterioles. It also suppresses secretion of ADH, aldosterone, and renin. Potential clinical applications of ANP are tantalizing. It is effective in lowering blood pressure and in reducing water retention (edema). Studies are underway to see if ANP can benefit renal failure patients by increasing their glomerular filtration rate.

● **Neural Regulation** Like most blood vessels of the body, those of the kidneys are supplied by vasoconstrictor fibers from the sympathetic division of the autonomic nervous system. At rest, sympathetic stimulation is minimal and renal blood vessels are maximally dilated. With moderate sympathetic stimulation, both afferent and efferent arterioles constrict to the same degree. Blood flow into and out of the glomerulus is restricted to the same extent, which decreases GFR only slightly. With greater sympathetic stimulation, as might occur during exercise, hemorrhage, or a fight-or-flight response, vasoconstriction of the afferent arterioles predominates, which in turn greatly decreases glomerular blood flow and GFR. Strong sympathetic stimulation also causes the juxtaglomerular cells to secrete renin and the adrenal medulla to secrete epinephrine, which also decreases GFR by bringing about afferent arteriolar vasoconstriction.

FIGURE 26.12 Renin–angiotensin system in regulation of blood pressure and glomerular filtration rate.

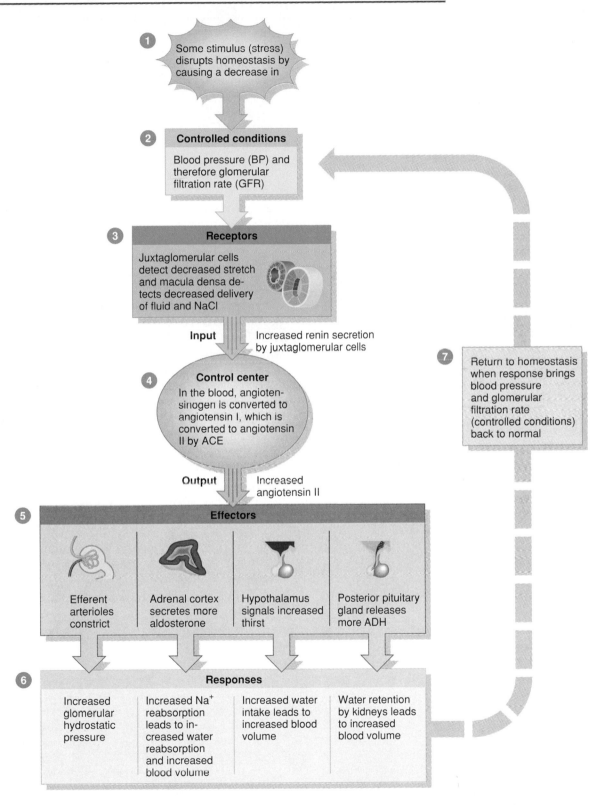

Question: What enzymes and hormones are involved in this system?

Tubular Reabsorption

As the filtrate passes through the renal tubules, about 99% of it is reabsorbed (returned to the blood). Thus only about 1% of the filtrate leaves the body as urine (about 1.5 liters a day). The movement of water and solutes back into the blood of a peritubular capillary or vasa recta is called **tubular reabsorption.** Solutes that are reabsorbed by both active and passive processes include glucose, amino acids, urea, and ions such as Na^+, K^+, Ca^{2+}, Cl^-, HCO_3^-, and HPO_4^{2-}. Water reabsorption occurs by the passive process of osmosis. Any small proteins and peptides that have passed the filter also are reabsorbed, usually by the process of pinocytosis.

Epithelial cells all along the renal tubule carry out tubular reabsorption. The proximal convoluted tubules, where the epithelial cells feature many microvilli that increase the surface area for reabsorption, make the largest contribution. More distal portions of the nephron are responsible for fine-tuning reabsorption processes to maintain homeostatic balances. Tubular reabsorption returns nutrients to the body. Wastes such as urea are only partially reabsorbed. Exhibit 26.1 compares the values for the chemicals in the filtrate immediately after passing into the glomerular (Bowman's) capsule with those reabsorbed from the filtrate. It will give you an idea of how much of the various substances the kidneys reabsorb.

Reabsorption of Na^+ in the PCT

Reabsorption of sodium ions (Na^+) is especially important because more of them pass the glomerular filter than any other substance except water. Sodium ions are reabsorbed in each portion of the renal tubule by several transport systems. These mechanisms recover not only filtered Na^+ but also water, anions (negatively charged ions), and nutrients. Sometimes they also permit secretion of unneeded substances such as excess H^+ and K^+.

Figure 26.13a shows a major mechanism for Na^+ reabsorption that works in the PCT as well as the DCT and collecting ducts. The concentration of Na^+ inside tubule cells is low and the interior of the cell is negatively charged with respect to the exterior. As a result of this electrochemical gradient, sodium ions passively diffuse from the filtered fluid in the tubule lumen through leakage channels in the brush border into tubule cells. At the same time, sodium pumps (Na^+/K^+ ATP-ase) actively expel Na^+ from the basolateral membranes at the base and sides of the cell. From the interstitial spaces around the tubule cells, Na^+ diffuses into the peritubular capillaries.

Since ATP is used to drive the sodium pump, the overall process is *primary active transport* (see page 65). Although the sodium pump imports K^+ at the same time it expels Na^+, K^+ can diffuse back out through K^+ leakage channels. (Recall that membranes typically have many more leakage channels for K^+ than for Na^+.) Thus reabsorption of Na^+ is the main effect of the sodium pump. The total ATP used by sodium pumps in the renal tubules is large, an estimated 6% of the total resting metabolic energy budget. For comparison, this is about the same energy needed for contraction of the diaphragm in quiet breathing.

Active transport of Na^+ promotes reabsorption of water by osmosis (Fig. 26.13b). Each reabsorbed Na^+ increases the osmotic pressure, first of the tubule cell's cytosol and

---

EXHIBIT 26.1

SUBSTANCES IN PLASMA AND AMOUNTS FILTERED, REABSORBED, AND EXCRETED IN URINE

Chemical	Plasma (Total Amount)	Filtered (Enters Glomerular Capsule per Day)	Reabsorbed (Returned to Blood per Day)	Urine (Excreted per Day)
Water	3000 ml	180,000 ml	178,500 ml	1500 ml
Proteins	200 g	2 g	1.9 g	0.1 g
Sodium (Na^+)	9.7 g (420 mmol)	579.6 g (25,200 mmol)	575.0 g (25,000 mmol)	4.6 g (200 mmol)
Chloride (Cl^-)	10.7 g (300 mmol)	639.0 g (18,000 mmol)	633.7 g (17,850 mmol)	6.3 g (150 mmol)
Bicarbonate (HCO_3^-)	4.6 g (75 mmol)	274.5 g (4500 mmol)	274.5 g (4500 mmol)	0
Glucose	3 g	180 g	180 g	0
Urea	4.8 g	53 g	28 g	25 g[a]
Potassium (K^+)	0.5 g (12.6 mmol)	29.6 g (756 mmol)	29.6 g (756 mmol)	2.0 g (50 mmol[b])
Uric acid	0.15 g	8.5 g	7.7 g	0.8 g
Creatinine	0.03 g	1.6 g	0	1.6 g

[a] Urea is secreted in addition to being filtered and reabsorbed.

[b] After being 100% reabsorbed in the PCT, loop, and DCT, a variable amount of K^+ is secreted in the collecting ducts.

FIGURE 26.13 Reabsorption of Na⁺ in the proximal convoluted tubule by primary active transport. Note that active transport of Na⁺ (a) leads to passive reabsorption of water by osmosis (b) and passive reabsorption of Cl⁻ (and other anions) and urea by simple diffusion (c).

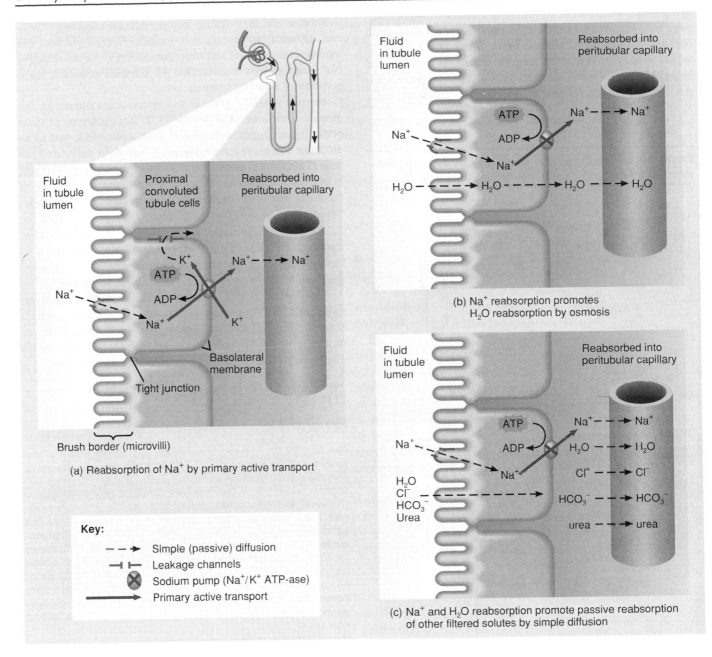

(a) Reabsorption of Na⁺ by primary active transport

(b) Na⁺ reabsorption promotes H₂O reabsorption by osmosis

(c) Na⁺ and H₂O reabsorption promote passive reabsorption of other filtered solutes by simple diffusion

Key:

- - - → Simple (passive) diffusion
⊣⊢ Leakage channels
⊗ Sodium pump (Na⁺/K⁺ ATP-ase)
—→ Primary active transport

Question: Which step in Na⁺ movement in (a) is promoted by the electrochemical gradient?

then of the blood in peritubular capillaries. Water thus moves rapidly from the filtrate into the peritubular capillaries and restores osmotic balance. Now that water has left the filtered fluid, however, the concentration of the remaining filtered solutes is increased. This creates a concentration gradient for some substances, such as K⁺, Cl⁻, HCO₃⁻, and urea, that promotes their reabsorption by passive diffusion (Fig. 26.13c). In this way, active transport of Na⁺ promotes passive diffusion of other solutes, which accompany Na⁺ into the peritubular capillaries.

EXHIBIT 26.3

PHYSICAL CHARACTERISTICS OF NORMAL URINE

Characteristic	Description
Volume	One to two liters in 24 hours but varies considerably.
Color	Yellow or amber but varies with concentration and diet. Color is due to urochrome (pigment produced from breakdown of bile). Concentrated urine is darker in color. Diet (reddish colored urine from beets and green colored from asparagus) and certain diseases (kidney stone may produce blood in urine) affect color.
Turbidity	Transparent when freshly voided but becomes turbid (cloudy) upon standing.
Odor	Aromatic but becomes ammonia-like upon standing. Some people inherit the ability to form methylmercaptan from ingesting asparagus that gives urine a characteristic odor. Urine of diabetics has a sweet odor due to presence of ketone bodies.
pH	Ranges between 4.6 and 8.0; average 6.0; varies considerably with diet. High-protein diets increase acidity; vegetarian diet increases alkalinity.
Specific gravity	Specific gravity (density) is the ratio of the weight of a volume of a substance to the weight of an equal volume of distilled water. It ranges from 1.001 to 1.035. The higher the concentration of solutes, the higher the specific gravity.

---

volume is influenced by blood pressure, blood osmotic pressure, diet, temperature, diuretics, mental state, and general health. Low blood pressure triggers the renin–angiotensin pathway, which increases reabsorption of water and salts in the renal tubules and decreases urine volume. When blood osmotic pressure decreases, for example, after drinking a large volume of water, secretion of ADH is inhibited and consequently a larger volume of urine is excreted. The reverse effects occur with high blood pressure and increased blood osmotic pressure.

CLINICAL APPLICATION

DIURETICS

Diuretics are drugs that increase urine flow rate, which speeds loss of water from the body, usually by interfering with reabsorption of filtered solutes, mainly sodium. Whereas the caffeine in coffee and tea inhibits Na^+ reabsorption, the ethanol (alcohol) in alcoholic beverages inhibits secretion of ADH. Since there are so many mechanisms for Na^+ reabsorption, there is a great diversity of diuretic drugs. They often are prescribed to treat congestive heart failure, high blood pressure, and edema.

One of the earliest known classes of diuretics slows Na^+ reabsorption by inhibiting carbonic anhydrase in the brush border of the proximal convoluted tubule. Acetazolamide (Diamox) is an example of a carbonic anhydrase inhibitor. It is rarely used today because a side effect is loss of filtered HCO_3^- in the urine. Loop diuretics, such as furosemide (Lasix), are the most potent diuretics. They selectively inhibit the Na^+–K^+–$2Cl^-$ symporters in the thick ascending limb of the loop of Henle. The thiazide diuretics, such as chlorthi-

azide (Diuril), are thought to inhibit the Na^+–Cl^- symporter in the distal convoluted tubule.

Most diuretic drugs have the adverse side effect of causing excessive loss of K^+ in the urine. An exception is the class of diuretics that inhibits the action of aldosterone, such as spironolactone (Aldactone). These are called potassium-sparing diuretics. They promote mild diuresis by slowing reabsorption of Na^+ and water in the collecting duct, and at the same time they also decrease urinary loss of K^+ because they inhibit K^+ secretion in exchange for Na^+ reabsorption.

Chemical Composition

Water accounts for about 95% of the total volume of urine. The remaining 5% consists of solutes derived from cellular metabolism and outside sources such as drugs. Typical solutes present in urine are described in Exhibit 26.4.

Abnormal Constituents

If the body's chemical processes are not operating efficiently, traces of substances not normally present may appear in the urine, or normal constituents may appear in abnormal amounts. Exhibit 26.5 on page 898 lists several abnormal constituents in urine that may be detected as part of a urinalysis, laboratory analysis of a urine sample. Normal values of urine components and the clinical implications of deviations from normal are listed in Appendix B (see page A-5).

EXHIBIT 26.4

PRINCIPAL SOLUTES IN URINE OF ADULT MALE ON TYPICAL DIET

Constituent	Amount[a] (g)	Comments
ORGANIC		
Urea	25.0 to 35.0	Composes 60 to 90% of all nitrogenous material in urine. Derived primarily from deamination of amino acids to form ammonia (ammonia combines with CO_2 to form urea).
Creatinine	1.6	Normal constituent of blood. Derived primarily from breakdown of creatine phosphate (nitrogenous substance in muscle tissue).
Uric acid	0.4 to 1.0	Product of catabolism of nucleic acids (DNA and RNA) derived from food or cellular destruction. Because of insolubility, it tends to crystallize and is a common component of kidney stones.
Hippuric acid	0.7	Form in which benzoic acid (toxic substance in fruits and vegetables) is believed to be eliminated from body. High-vegetable diets increase quantity of hippuric acid excreted.
Indican	0.01	Potassium salt of indole. Indole results from bacterial breakdown of protein in large intestine and is carried by blood to liver, where it is probably changed to indican (less poisonous substance).
Ketone bodies	0.04	Also called acetone bodies. Normally found in small amounts. In cases of diabetes mellitus and acute starvation, ketone bodies appear in high concentrations.
Other substances	2.9	May be present in minute quantities, depending on diet and general health. Include carbohydrates, pigments, fatty acids, mucin, enzymes, and hormones.
INORGANIC		
Na^+, Cl^-	15.0	Principal inorganic salt. Amount excreted varies with intake.
K^+	3.3	Occurs as chloride, sulfate, and phosphate salts.
SO_4^{2-}	2.5	Derived from amino acids.
$H_2PO_4^-$, HPO_4^{2-}, PO_4^{3-}	2.5	Occur as sodium compounds (monosodium and disodium phosphate) that serve as buffers in blood and urine.
NH_4^+	0.7	Occurs as ammonium salts. Derived from protein catabolism and from glutamine (amino acid) deamination in kidneys. Amount produced by kidney may vary with need to produce HCO_3^- to offset acidity of blood and tissue fluids.
Mg^{2+}	0.1	Occurs as chloride, sulfate, and phosphate salts.
Ca^{2+}	0.3	Occurs as chloride, sulfate, and phosphate salts.

[a] Values are for a urine sample collected over 24 hours.

AGING AND THE URINARY SYSTEM

As one grows older, the kidneys become less effective, and by age 70 the filtering mechanism is only about half as effective as it was at age 40. Urinary incontinence and urinary tract infections are two common problems associated with aging of the urinary system. Other pathologies include polyuria (excessive urine production), nocturia (excessive urination at night), increased frequency of urination, dysuria (painful urination), retention (failure to release urine from the urinary bladder), and hematuria (blood in the urine). Changes and diseases in the kidney include acute and chronic kidney inflammations and renal calculi (kidney stones). Since water balance and thirst are altered, elderly persons are susceptible to dehydration. The prostate gland is often implicated in various disorders of the urinary tract, and cancer of the prostate is the most frequent malignancy in elderly males (see page 963).

DEVELOPMENTAL ANATOMY OF THE URINARY SYSTEM

Starting in the third week of fetal development, a portion of the mesoderm along the posterior half of the dorsal side of the embryo, the **intermediate mesoderm,** differentiates

SUMMARY OF ABNORMAL CONSTITUENTS IN URINE

Abnormal Constituent	Comment
Albumin	Normal constituent of plasma, but it usually appears in only very small amounts in urine because it is too large to pass through the pores in capillary walls. The presence of excessive albumin in the urine—**albuminuria**—indicates an increase in the permeability of endothelial–capsular membranes due to injury or disease, increased blood pressure, or irritation of kidney cells by substances such as bacterial toxins, ether, or heavy metals.
Glucose	The presence of glucose in the urine (glucosuria) usually indicates diabetes mellitus. Occasionally, it may be caused by stress, which can cause excessive amounts of epinephrine to be secreted. Epinephrine stimulates the breakdown of glycogen and liberation of glucose from the liver.
Red blood cells (erythrocytes)	Red blood cells in the urine is called **hematuria** and generally indicates a pathological condition. One cause is acute inflammation of the urinary organs as a result of disease or irritation from kidney stones. Other causes include tumors, trauma, and kidney disease. One should make sure the urine sample was not contaminated with menstrual blood from the vagina.
White blood cells (leukocytes)	The presence of white blood cells and other components of pus in the urine, referred to as **pyuria** (pī-YOU-rē-a), indicates infection in the kidney or other urinary organs.
Ketone bodies	High quantities of ketone bodies, called **ketosis (acetonuria),** may indicate diabetes mellitus, starvation, or simply too little carbohydrate in the diet.
Bilirubin	When red blood cells are destroyed by reticuloendothelial cells, the globin portion of hemoglobin is split off and the heme is converted to biliverdin. Most of the biliverdin is converted to bilirubin, which gives bile its major pigmentation. An above-normal level of bilirubin in urine is called **bilirubinuria.**
Urobilinogen	The presence of urobilinogen (breakdown product of hemoglobin) in urine is called **urobilinogenuria.** Traces are normal, but increased urobilinogen may be due to hemolytic and pernicious anemia, infectious hepatitis, biliary obstruction, jaundice, cirrhosis, congestive heart failure, or infectious mononucleosis.
Casts	**Casts** are tiny masses of material that have hardened and assumed the shape of the lumen of a tubule in which they formed. They are then flushed out of the tubule when filtrate builds up behind them. Casts are named after the cells or substances that compose them or after their appearance. For example, there are white blood cell casts, red blood cell casts, and epithelial casts that contain cells from the walls of the tubules.
Microbes	The number and type of bacteria vary with specific infections in the urinary tract. The most common fungus to appear in urine is *Candida albicans*, a cause of vaginitis. The most frequent protozoan seen is *Trichomonas vaginalis*, a cause of vaginitis in females and urethritis in males.

into the kidneys. Three pairs of kidneys form within the intermediate mesoderm in successive time periods: pronephros, mesonephros, and metanephros (Fig. 26.24). Only the last pair remains as the functional kidneys of the newborn.

The first kidney to form, the **pronephros,** is the most superior of the three. Associated with its formation is a tube, the **pronephric duct.** This duct empties into the **cloaca,** which is the dilated caudal end of the gut derived from **endoderm.** The pronephros begins to degenerate during the fourth week and is completely gone by the sixth week. The pronephric ducts, however, remain.

The second kidney, the **mesonephros,** replaces the pronephros. The retained portion of the pronephric duct, which connects to the mesonephros, becomes known as the **mesonephric duct.** The mesonephros begins to degenerate by the sixth week and is almost gone by the eighth week.

At about the fifth week, an outgrowth, called a **ureteric bud,** develops from the distal end of the mesonephric duct near the cloaca. This bud is the developing **metanephros.** As it grows toward the head of the embryo, its end widens to form the *pelvis* of the kidney with its *calyces* and associated *collecting ducts*. The unexpanded portion of the bud,

FIGURE 26.24 Development of the urinary system.

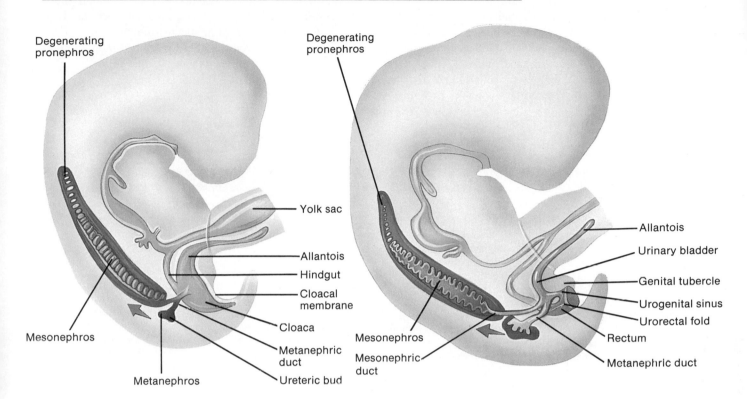

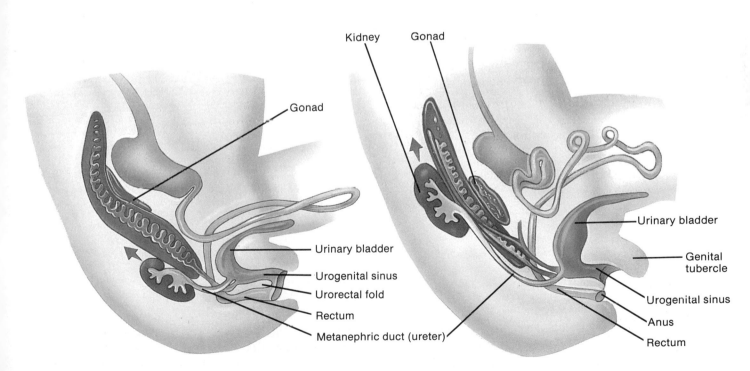

the **metanephric duct,** becomes the *ureter.* The *nephrons,* the functional units of the kidney, arise from the intermediate mesoderm around each ureteric bud.

During development, the cloaca divides into a **urogenital sinus,** into which urinary and genital ducts empty, and a *rectum* that discharges into the anal canal. The *urinary bladder* develops from the urogenital sinus. In the female,

the *urethra* develops from lengthening of the short duct that extends from the urinary bladder to the urogenital sinus. The *vestibule,* into which the urinary and genital ducts empty, is also derived from the urogenital sinus. In the male, the urethra is considerably longer and more complicated but is also derived from the urogenital sinus.

DISORDERS: HOMEOSTATIC IMBALANCES

GLOMERULONEPHRITIS (BRIGHT'S DISEASE)
Glomerulonephritis (Bright's disease) is an inflammation of the kidney that involves the glomeruli. One of the most common causes is an allergic reaction to the toxins given off by streptococci bacteria that have recently infected another part of the body, especially the throat. The glomeruli become so inflamed, swollen, and engorged with blood that the endothelial–capsular membranes become highly permeable and allow blood cells and proteins to enter the filtrate. Thus the urine contains many erythrocytes and much protein. The glomeruli may be permanently changed, leading to acute or chronic renal failure.

PYELITIS AND PYELONEPHRITIS
Pyelitis is an inflammation of the renal pelvis and its calyces. **Pyelonephritis,** an inflammation of one or both kidneys, involves both the nephrons and the renal pelvis. The disease is generally a complication of infection elsewhere in the body. In females, it is often a complication of lower urinary tract infections. In about 75% of the cases the cause is the bacterium *Escherichia coli.* If pyelonephritis becomes chronic, scar tissue forms in the kidneys and severely impairs their function.

CYSTITIS
Cystitis is an inflammation of the urinary bladder involving principally the mucosa and submucosa. It may be caused by bacterial infection, chemicals, or mechanical injury. Symptoms include burning on urination or painful urination, urgency and frequent urination, and low back pain. Bed-wetting may also occur.

NEPHROTIC SYNDROME
Nephrotic syndrome refers to protein in the urine (proteinuria), primarily albumin, that results in edema and hyperlipidemia (high blood levels of cholesterol, phospholipids, and triglycerides). The proteinuria is due to an increased permeability of the endothelial–capsular membrane, which permits proteins to escape from blood into urine. Loss of albumin results in a decline in the blood level of albumin as soon as liver production of albumin fails to meet increased urinary losses. The edema associated with nephrotic syndrome, usually seen around the eyes, ankles, feet, and abdomen, occurs because loss of albumin from the blood causes a decrease in blood colloid osmotic pressure. This leads to fluid movement from blood into interstitial spaces. Among the causes of nephrotic syndrome are diabetes mellitus, rheumatoid arthritis, systemic lupus erythematosus, lymphoma, leukemia, bacterial and viral infections, certain drugs (nonsteroidal anti-

inflammatories, street heroin), heavy metals (gold, mercury), hypertension, and sickle-cell anemia. Treatment is aimed at alleviating the proteinuria.

POLYCYSTIC DISEASE
Polycystic disease may be caused by a defect in the renal tubular system that deforms nephrons and results in cystlike dilations along their course. It is the most common inherited disorder of the kidneys. The kidney tissue is riddled with fluid-filled cysts and small holes, ranging in size from a pinhead to the diameter of an egg. These cysts gradually increase until they squeeze out the normal tissue, interfering with kidney function and causing uremia. The chief symptom is weight gain. The kidneys themselves may enlarge from their normal 0.25 kg (0.5 lb) to as much as 14 kg (30 lb). Although the disease is progressive and ultimately fatal, its advance can be slowed by diet, drugs, and regulation of fluid intake.

RENAL FAILURE
Renal failure is a decrease or cessation of glomerular filtration. In **acute renal failure (ARF)** the kidneys abruptly stop working entirely or almost entirely. The main feature of ARF is suppression of urine flow, usually characterized by *oliguria* (*olig* = scanty), daily urine output less than 250 ml, or *anuria,* daily urine output less than 50 ml. Causes include low blood volume (for example, due to hemorrhage), decreased cardiac output, damaged renal tubules, and kidney stones.

Chronic renal failure (CRF) refers to a progressive and usually irreversible decline in glomerular filtration rate (GFR). CRF may result from chronic glomerulonephritis, pyelonephritis, polycystic disease, or traumatic loss of kidney tissue. CRF develops in three stages. In the first stage, *diminished renal reserve,* nephrons are destroyed until about 75% of the functioning nephrons are lost. At this stage, the person may show no symptoms since the remaining nephrons enlarge and take over the function of those that have been lost. When more nephrons are lost, the balance between glomerular filtration and tubular reabsorption is upset, and any change in diet or fluid intake brings on the symptoms. Once 75% of the nephrons are lost, the person enters the second stage, called *renal insufficiency.* In this stage, there is a decrease in GFR and increased blood levels of nitrogenous wastes and creatinine. Also, the kidneys cannot effectively concentrate or dilute urine. The final stage, called *end-stage renal failure,* occurs when about 90% of the nephrons have been lost. At this stage, GFR diminishes to about 10% of normal, and blood levels of nitrogenous wastes and creatinine increase further. The low GFR results in oliguria. People with CRF are candidates for hemodialysis therapy and kidney transplantation.

Among the effects of renal failure are edema from salt and water retention; acidosis due to inability of the kidneys to excrete acidic substances; increased blood urea nitrogen (BUN); and elevated potassium levels that can lead to cardiac arrest. Anemia usually occurs since the kidneys no longer produce the erythropoietin (EPO) needed for red blood cell production. The availability of recombinant EPO has been a great boon to renal failure patients. Osteomalacia is another common problem since the kidneys are no longer able to convert vitamin D to its active form for calcium absorption from the small intestine.

URINARY TRACT INFECTIONS

The term **urinary tract infection (UTI)** is used to describe either an infection of a part of the urinary system or the presence of large numbers of microbes in urine. Included are **significant bacteriuria** (the presence of bacteria in urine in sufficient numbers to indicate active infection), **asymptomatic bacteriuria** (the multiplication of large numbers of bacteria in urine without producing symptoms), **urethritis** (inflam-

mation of the urethra), **cystitis** (inflammation of the urinary bladder), and **pyelonephritis** (inflammation of the kidneys).

DIABETES INSIPIDUS

Diabetes insipidus (DI) is characterized by excretion of a large volume (5 to 15 liters/day) of very dilute urine. As might be expected, patients with the disorder also exhibit extreme thirst (polydipsia). The cause of DI is either a defect in production of ADH (central DI) or an insensitivity of the principal cells in renal collecting ducts to stimulation by ADH (nephrogenic DI).

Central DI most commonly results from surgical or traumatic injury of the hypothalamic nuclei that synthesize ADH. For example, a fracture at the base of the skull may damage this region. Most cases of nephrogenic DI are associated with advanced kidney disease, but there is also a rare hereditary form. Interestingly, all North American patients with congenital nephrogenic DI are thought to be descendants of a man from Scotland who settled in Nova Scotia in 1761.

Treatment of central DI consists of taking a modified form of ADH as a nasal spray. Patients with nephrogenic DI do not respond to such hormone replacement therapy.

MEDICAL TERMINOLOGY

Azotemia (az-ō-TĒ-mē-a; *azo* = nitrogen-containing; *emia* = condition of blood) Presence of urea or other nitrogenous elements in the blood.

Cystocele (SIS-tō-sēl; *cyst* = bladder; *cele* = cyst) Hernia of the urinary bladder.

Diuresis (dī-yoo-RĒ-sis; *diourein* = to urinate) Increased excretion of urine.

Dysuria (dis-YOO-rē-a; *dys* = painful; *uria* = urine) Painful urination.

Enuresis (en'-yoo-RĒ-sis; *enourein* = to void urine) Bedwetting; may be due to faulty toilet training, to some psychological or emotional disturbance, or rarely to some physical disorder such as diabetes insipidis (lack of ADH). Also referred to as **nocturia.**

Intravenous pyelogram (in'-tra-VĒ-nus PĪ-e-lō-gram'), or **IVP** (*intra* = within; *veno* = vein; *pyelo* = pelvis of kidney; *gram* = written or recorded) X-ray film of the kidneys after venous injection of a dye.

Polyuria (pol'-ē YOO rē a; *poly* = much) Excessive urine formation.

Stricture (STRIK-chur) Narrowing of the lumen of a canal or hollow organ, as may occur in the ureter, urethra, or any other tubular structure in the body.

Urethritis (yoo'-rē-THRĪ-tis) Inflammation of the urethra, caused by highly acidic urine, the presence of bacteria, or constriction of the urethral passage.

Study Outline

Introduction (p. 864)

1. Among the organs that contribute to the elimination of wastes are the kidneys, lungs, skin, and gastrointestinal tract.
2. The organs of the urinary system are the kidneys, ureters, urinary bladder, and urethra.
3. The kidneys regulate the volume and composition of blood, blood pressure, and some aspects of metabolism.

Kidneys (p. 864)

1. The kidneys are retroperitoneal organs attached to the posterior abdominal wall.
2. Three layers of tissue surround the kidneys: renal capsule, adipose capsule, and renal fascia.
3. Internally, the kidneys consist of a cortex, medulla, pyramids, papillae, columns, calyces, and a pelvis.

Chapter 27

FLUID, ELECTROLYTE, AND ACID–BASE HOMEOSTASIS

Chapter Contents at a Glance

Student Objectives

1. Describe the various fluid compartments of the body.

2. Define the processes available for fluid intake and fluid output and how they are regulated.

3. Contrast the electrolyte concentration of the three major fluid compartments.

4. Explain the functions of sodium, chloride, potassium, calcium, phosphate, and magnesium and regulation of their concentrations.

5. Describe the factors involved in the movement of fluid between plasma and interstitial fluid and between interstitial fluid and intracellular fluid.

6. Compare the role of buffers, exhalation of carbon dioxide, and kidney excretion of H⁺ in maintaining pH of body fluids.

7. Define acid–base imbalances, their effects on the body, and how they are treated.

The term **body fluid** refers to the body water and its dissolved substances. Fluids comprise an average of 60% of total body weight.

FLUID COMPARTMENTS AND FLUID BALANCE

About two-thirds of the fluid is within cells and is termed **intracellular fluid (ICF)**. The other third, called **extracellular fluid (ECF),** includes all other body fluids (Fig. 27.1). About 80% of the ECF is **interstitial fluid** and 20% is **blood plasma**. Some of the interstitial fluid is localized in specific places such as lymph in lymphatic vessels; cerebrospinal fluid in the brain; synovial fluid in joints; aqueous humor and vitreous body in the eyes; endolymph and perilymph in the ears; pleural, pericardial, and peritoneal fluids between serous membranes; and glomerular filtrate in the kidneys.

FIGURE 27.1 Body fluid compartments. One kilogram equals 1 liter of body fluid.

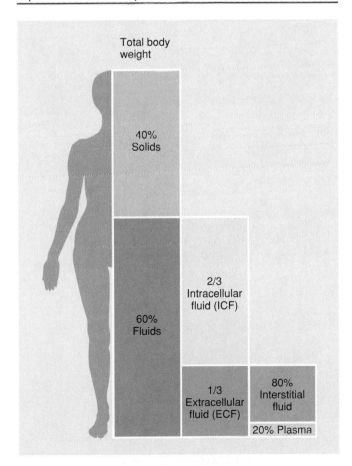

Question: In a person who weighs 60 kg, what is the approximate volume of blood plasma?

Selectively permeable membranes separate body fluids into distinct compartments. Plasma membranes of individual cells separate intracellular fluid from interstitial fluid. And blood vessel walls divide interstitial fluid from blood plasma. A compartment may be as small as the interior of a single cell or as large as the combined interiors of the heart and blood vessels. Although fluids are in constant motion from one compartment to another, the volume of fluid in each compartment remains fairly stable—another example of homeostasis.

Water is the main component of all body fluids. When we say that the body is in **fluid balance,** we mean that the required amount of water is present and proportioned among the various compartments according to their needs. Osmosis is the primary method of water movement in and out of body compartments. The concentration of solutes in the fluids is therefore a major determinant of fluid balance. Most solutes in body fluids are electrolytes—compounds that dissociate into ions. Fluid balance, then, means water balance, but it also implies electrolyte balance. The two are inseparable.

WATER

Water is by far the largest single component of the body, making up from 45 to 75% of the total body weight. (The functions of water may be reviewed on page 37.) Infants have the highest amount of water, up to 75% of body weight. The percentage decreases with age. Since fat is basically water-free, lean people have a greater proportion of water to total body weight than fat people. In a normal adult male, water accounts for about 60% of body weight. Since females have more subcutaneous fat than males, on average, their total body water is lower, accounting for about 55% of body weight.

The main source of body water is from ingested liquids (1600 ml) and foods (700 ml) that have been absorbed from the gastrointestinal tract. This water, called **preformed water,** amounts to about 2300 ml/day. Another source of water is **metabolic water,** the water produced through dehydration synthesis reactions of anabolism (see Fig. 2.8). This amounts to about 200 ml/day. Thus daily water gain is about 2500 ml (Fig. 27.2).

Normally, water loss equals water gain, so the body maintains a constant volume. There are several avenues for loss of fluid. The kidneys excrete about 1500 ml/day, the skin about 500 ml/day (400 ml/day through evaporation and 100 ml/day through perspiration), the lungs about 300 ml/day, and the gastrointestinal tract about 200 ml/day. Daily water loss thus totals 2500 ml (Fig. 27.2).

Regulation of Fluid Intake (Gain)

Thirst is a powerful regulator of fluid consumption. When water loss is greater than water gain, the resulting

FIGURE 27.2 Summary of fluid intake and output per day under normal conditions.

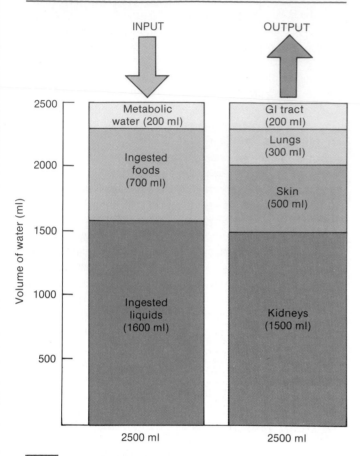

Question: How would each of the following affect fluid balance? Hyperventilation? Vomiting? Fever? Diuretics?

FIGURE 27.3 Pathways for stimulation of thirst by dehydration.

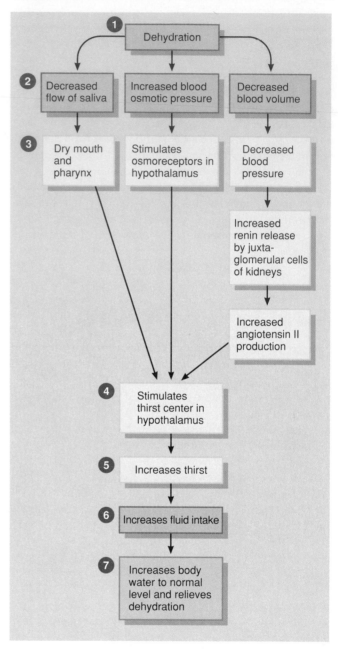

Question: Is this a negative or positive feedback system? Why?

dehydration stimulates thirst in at least three ways. (1) Locally, it leads to a decrease in saliva production that results in a dryness of the mucosa of the mouth and pharynx (Fig. 27.3). This gives rise to the sensation of thirst, which is relayed to the thirst center in the hypothalamus. (2) Dehydration also increases blood osmotic pressure, which stimulates osmoreceptors in the hypothalamus. These receptors also stimulate the thirst center in the hypothalamus. (3) Dehydration decreases blood volume and thus blood pressure, which stimulates the release of renin by the juxtaglomerular cells of the kidneys. Renin promotes synthesis of angiotensin II, one effect of which is to stimulate the thirst center in the hypothalamus (see Fig. 26.12). As a result of dehydration, the thirst center in the hypothalamus is stimulated, the sensation of thirst is increased, fluid intake increases, and normal fluid volume is restored. In other words, fluid intake balances the fluid output.

Initially, quenching of thirst results from wetting the mucosa of the mouth and pharynx, but the major inhibition of thirst is believed to occur as a result of stretching the stomach or intestine and a decrease in osmotic pressure in

fluids of the hypothalamus. Apparently, stimulated stretch receptors in the wall of the intestine send impulses that inhibit the thirst center in the hypothalamus.

Regulation of Fluid Output (Loss)

Normally, fluid loss is adjusted by antidiuretic hormone (ADH), atrial natriuretic peptide (ANP), and aldosterone.

ADH and aldosterone slow fluid loss whereas ANP causes diuresis (see Chapter 26). Under abnormal conditions, other factors may influence fluid loss. If the body is dehydrated, blood pressure falls, glomerular filtration rate decreases, and water is conserved. Conversely, excessive fluid in blood results in an increase of blood pressure, glomerular filtration rate, and fluid output. Hypertension produces the same effect. Hyperventilation leads to increased fluid loss through the loss of water vapor by the lungs. Vomiting and diarrhea result in fluid loss from the gastrointestinal tract. Finally, fever, heavy perspiration, and destruction of extensive areas of the skin from burns bring about excessive water loss through the skin.

CONCENTRATION OF SOLUTIONS

Body fluids contain a variety of dissolved chemicals. Some are compounds with covalent bonds; that is, the atoms that compose the molecule share electrons and do not form ions when dissolved. They are called **nonelectrolytes** and include most organic compounds, such as glucose, urea, and creatine. Other compounds, called **electrolytes (ions)**, have at least one ionic bond. When they dissolve in a body fluid, they dissociate into **cations** (positively charged ions) and **anions** (negatively charged ions). Acids, bases, and salts are electrolytes. Most electrolytes are inorganic compounds, but a few are organic. Several of the Krebs cycle acids (for example, citric acid and oxaloacetic acid), lactic acid (lactate), and several amino acids within proteins form ionic bonds. In solution, these molecules lose an ion (often H^+), and the rest of the molecule carries the opposite charge. Electrolytes are so-named because they can carry an electrical current.

There are several ways to express the concentration of chemicals dissolved in body fluids. The concentration may be expressed as the amount of solute in solution, or **percent**. A 0.9% NaCl solution (0.9 g of NaCl in 100 ml of solution) is isotonic to red blood cells and is called **isotonic saline.**

The concentration can be expressed as the total number of cations and anions, or **milliequivalents/liter (mEq/liter).** One equivalent is the positive or negative charge equal to the amount of charge in one mole of hydrogen ions (H^+). A milliequivalent is one-thousandth of an equivalent. Recall that a mole is the molecular weight of a substance in grams. For ions such as sodium (Na^+), potassium (K^+) and bicarbonate (HCO_3^-), which have a single positive or negative charge, the number of mEq/liter is equal to the number of mmol/liter. For ions such as calcium (Ca^{2+}) or phosphate (HPO_4^{2-}), which have two positive or negative charges, the number of mEq/liter is twice the number of mmol/liter.

During osmosis, water moves from an area with fewer particles in solution (lower osmotic pressure) to an area with more particles in solution (higher osmotic pressure).

Units that express the total particles in solution are **milliosmoles/liter (mOsm/liter).** A particle may be a whole molecule or an ion. Because it dissociates into at least two particles, an electrolyte molecule exerts a far greater effect on osmosis than a nonelectrolyte. Suppose the nonelectrolyte glucose and two different electrolytes are placed in solution:

$$C_6H_{12}O_6 \xrightarrow{H_2O} C_6H_{12}O_6$$
Glucose

$$NaCl \xrightarrow{H_2O} Na^+ + Cl^-$$
Sodium chloride

$$CaCl_2 \xrightarrow{H_2O} Ca^{2+} + Cl^- + Cl^-$$
Calcium chloride

Because glucose does not dissociate when dissolved in water, a molecule of glucose contributes only one particle to the solution. Sodium chloride, on the other hand, contributes two ions, or particles, and calcium chloride contributes three. For example, a 5 mmol/liter solution of $CaCl_2$ has an osmolarity of 15 mOsm/liter (if the salt dissociates completely). On the other hand, a 5 mmol/liter solution of glucose has an osmolarity of only 5 mOsm/liter. Thus each mole of calcium chloride has three times as great an osmotic effect as each mole of glucose. Osmotic pressure is usually expressed in mm Hg. Each 1 mOsm/liter exerts a pressure equal to 19.3 mm Hg. Exhibit 27.1 reviews the different ways to express and calculate the concentration of a solution.

ELECTROLYTES

Electrolytes serve four general functions in the body. (1) Many are essential minerals. (2) Because they are more numerous than nonelectrolytes, electrolytes control the osmosis of water between body compartments. (3) They help maintain the acid–base balance required for normal cellular activities. (4) They carry electrical current, which allows production of action potentials and graded potentials and controls secretion of some hormones and neurotransmitters. Electrical currents also are important during development.

Distribution

Figure 27.4 compares the electrolytes in plasma, interstitial fluid, and intracellular fluid. The chief difference between plasma and interstitial fluid is that plasma contains quite a few protein anions, whereas interstitial fluid has hardly any. Since normal capillary membranes are virtually impermeable to protein, plasma proteins do not move out of blood vessels into the interstitial fluid. Plasma also contains slightly more sodium ions but fewer chloride ions than the interstitial fluid. In other respects the two fluids are similar.

EXHIBIT 27.1

WAYS TO EXPRESS CONCENTRATION OF SOLUTIONS

Definition	Most Common U. S. Units	Example
Percent (weight per volume) Number of grams of a substance per 100 ml of solution.	g/dl[a] or mg/dl	Take 0.9 g of NaCl and dissolve in enough water to make 100 ml total solution. Result: 0.9% NaCl (isotonic saline).
Millimoles per liter Number of millimoles (the molecular weight in milligrams) of a substance in 1 liter of solution.	mmol/liter	The molecular weight of NaCl is 58.4. Take 58.4 mg NaCl; add enough water to make 1 liter total solution. Result: 1 mmol/liter NaCl.
Milliequivalents per liter Positive or negative charge equal to amount of charge in 1 mmol/liter of H^+. Equals mmol/liter × mEq/mmol [number of ions per molecule × number of charges on one ion].	mEq/liter	Take a 5 mmol/liter solution of $CaCl_2$. Calculate mEq for cations and anions separately. In solution there are 5 mmol/liter of Ca^{2+} × 2 mEq/mmol [1 ion/molecule × 2 charges/ion] = 10 mEq/liter of Ca^{2+}. There also are 5 mmol/liter of Cl^- × 2 mEq/mmol [2 ions/molecule × 1 charge/ion] = 10 mEq/liter of Cl^-.
Milliosmoles per liter Equals mmol/liter × number of particles per molecule upon dissociation in solution.	mOsm/liter	Take a 5 mmol/liter solution of $CaCl_2$. There are three particles per molecule (one Ca^{2+} and two Cl^-). Thus 5 mmol/liter × 3 = 15 mOsm/liter $CaCl_2$ (assuming 100% dissociation).
Osmotic pressure Each 1 mOsm/liter = 19.3 mm Hg.	mm Hg	Proteins contribute about 1.4 mOsm/liter to blood plasma. The osmotic pressure due to plasma proteins is thus 1.4 mOsm/liter × 19.3 mm Hg/mOsm/liter = 27 mm Hg.

[a] dl = deciliter = 0.1 liter = 100 ml.

Intracellular fluid differs considerably from extracellular fluid, however. In extracellular fluid, the most abundant cation is Na^+, and the most abundant anion is Cl^-. In intracellular fluid, the most abundant cation is K^+, and the most abundant anions are proteins and phosphates (HPO_4^{2-}).

Sodium

Sodium (Na^+), the most abundant extracellular ion, represents about 90% of extracellular cations. Normal plasma (serum) sodium concentration is 136 to 142 mEq/liter. Na^+ is necessary for the conduction of action potentials (impulses) in nervous and muscle tissue. It plays a pivotal role in fluid and electrolyte balance by creating most of the osmotic pressure (the most mOsm/liter) of extracellular fluid (ECF). The average daily intake of Na^+ far exceeds the body's normal daily requirements. The kidneys excrete excess Na^+ but also can conserve it during periods of shortage.

The Na^+ level in the blood is controlled by aldosterone, antidiuretic hormone (ADH), and atrial natriuretic peptide (ANP). Aldosterone, secreted by the cortex of the adrenal glands, acts on the distal convoluted tubules and collecting ducts of the nephrons of the kidneys and causes them to increase their reabsorption of Na^+. As Na^+ moves from the filtrate back into the blood, it establishes an osmotic gradient. This causes water to follow Na^+ from the filtrate back into the blood. Aldosterone is secreted in response to reduced blood volume or cardiac output, decreased extracellular Na^+ concentration, increased extracellular K^+ concentration, and physical stress. A decrease in Na^+ concentration inhibits the release of antidiuretic hormone (ADH) by the posterior pituitary gland. This, in turn, permits greater excretion of water in urine and restoration of normal Na^+ level. The atria of the heart produce the hormone atrial natriuretic peptide (ANP), which opposes the actions of aldosterone and ADH. ANP increases Na^+ and water excretion by the kidneys.

CLINICAL APPLICATION

HYPONATREMIA AND HYPERNATREMIA

Sodium loss from the body may occur through excessive perspiration, vomiting, or diarrhea; therapy with certain diuretics; and burns. Such a loss can result in **hyponatremia** (hī'-pō-na-TRĒ-mē-a; *natrium* = sodium), a lower-than-normal

Figure 27.4 Comparison of electrolyte concentrations in plasma, interstitial fluid, and intracellular fluid. The height of each column represents the total electrolyte concentration.

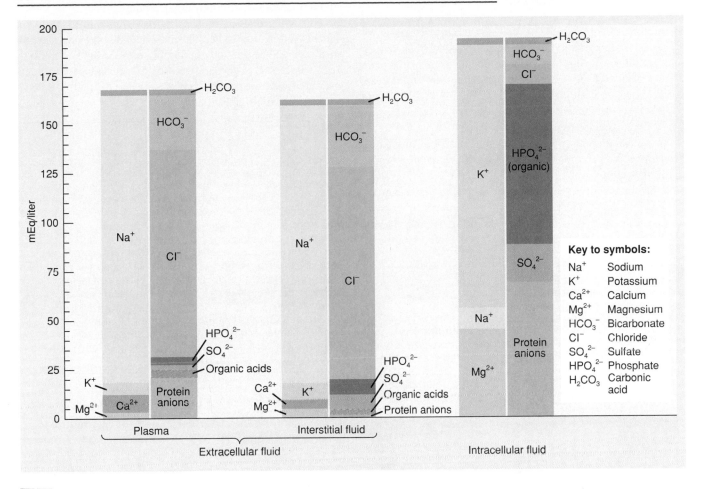

Question: What is the major cation and what are the two major anions in ECF and ICF?

blood sodium level. Signs of hyponatremia include muscular weakness, dizziness, headache, hypotension, tachycardia, and shock. Severe sodium loss can result in mental confusion, stupor, and coma. **Hypernatremia** is a higher-than-normal blood sodium level. It may occur with water loss, water deprivation, or sodium gain. Since sodium is the major determinant of the osmotic pressure of ECF, hypernatremia causes hypertonicity of ECF. As a result, water moves out of body cells into ECF, resulting in cellular dehydration. Symptoms of hypernatremia include intense thirst, fatigue, restlessness, agitation, and coma.

Chloride

Chloride (Cl^-) is the major extracellular anion. However, it easily diffuses between the extracellular and intracellular compartments because plasma membranes contain quite a few Cl^- leakage channels. For this reason, chloride is important in balancing osmotic pressure between compart-

ments. In the gastric mucosal glands, chloride combines with hydrogen to form hydrochloric acid. Normal plasma (serum) Cl^- concentration is 95 to 103 mEq/liter.

Regulation of Cl^- balance in body fluids is indirectly under the control of aldosterone. Aldosterone regulates sodium reabsorption, and chloride follows sodium passively by electrical attraction since the negatively charged chloride ions will follow the positively charged sodium ions.

CLINICAL APPLICATION

HYPOCHLOREMIA

An abnormally low level of Cl^- in the blood, called **hypochloremia** (hī'-pō-klō-RĒ-mē-a) may be caused by excessive vomiting, dehydration, and therapy with certain diuretics. Symptoms include muscle spasms, alkalosis, depressed respirations, and even coma.

Potassium

Potassium (K^+) is the most abundant cation in intracellular fluid. Normal plasma (serum) K^+ concentration is 3.8 to 5.0 mEq/liter. K^+ plays a key role in establishing the resting membrane potential and in the repolarization phase of action potentials in nervous and muscle tissue. Thus abnormal serum K^+ levels adversely affect neuromuscular and cardiac function. K^+ also helps maintain fluid volume in cells. When potassium ions move in or out of cells, they tend to be exchanged for hydrogen ions. This shift of hydrogen ions helps regulate pH.

The plasma level of K^+ is under the control of mineralocorticoids, mainly aldosterone. When K^+ concentration is high, more aldosterone is secreted and more K^+ is excreted. This occurs mainly in the last portion of the distal convoluted tubule and collecting ducts of the kidneys by way of tubular secretion. When K^+ concentration is low, aldosterone secretion decreases and less K^+ is excreted in urine.

CLINICAL APPLICATION

HYPOKALEMIA AND HYPERKALEMIA

A lower-than-normal level of K^+, called **hypokalemia** (hī'-pō-ka-LĒ-mē-a; *kalium* = potassium), may result from vomiting, diarrhea, high sodium intake, kidney disease, and therapy with some diuretics. Symptoms include cramps and fatigue, flaccid paralysis, nausea, vomiting, mental confusion, increased urine output, shallow respirations, and changes in the electrocardiogram, including a lengthening of the Q-T interval and flattening of the T-wave.

A higher-than-normal blood K^+ level, called **hyperkalemia,** is characterized by irritability, anxiety, abdominal cramping, diarrhea, weakness (especially of the lower extremities), and paresthesia (abnormal sensation, such as burning or prickling). Hyperkalemia can cause death by inducing fibrillation of the heart.

Calcium

Because such a large amount is stored in bone, calcium (Ca^{2+}) is the most abundant ion in the body. In body fluids, it is principally an extracellular electrolyte. Normal plasma Ca^{2+} concentration is 4.6 to 5.5 mEq/liter. About 98% of the calcium in an adult is in the skeleton (and teeth) combined with phosphates to form a crystal lattice of mineral salts. The remaining calcium is found in extracellular fluid and within cells in a variety of tissues, especially skeletal muscle. Besides contributing to the hardness of bones and teeth, Ca^{2+} plays important roles in blood coagulation, neurotransmitter release, maintenance of muscle tone, and excitability of nervous and muscle tissue.

The level of Ca^{2+} in plasma is regulated principally by parathyroid hormone (PTH) and calcitonin (CT) (see Fig. 18.17). A low plasma Ca^{2+} level promotes release of more PTH. PTH stimulates osteoclasts in bone tissue to release calcium (and phosphate) from mineral salts of bone matrix.

Thus, PTH increases bone resorption. PTH also increases Ca^{2+} absorption from the gastrointestinal tract by activating vitamin D and enhancing reabsorption of Ca^{2+} from glomerular filtrate through renal tubule cells and back into blood. CT, produced by the thyroid gland, is released in increased quantities when plasma calcium levels are high. It decreases blood Ca^{2+} by stimulating osteoblasts and inhibiting osteoclasts. In the presence of CT, osteoblasts remove calcium (and phosphate) from plasma and deposit it in bone matrix as calcium phosphate salts.

CLINICAL APPLICATION

HYPOCALCEMIA AND HYPERCALCEMIA

An abnormally low level of Ca^{2+} is called **hypocalcemia** (hī'-pō-kal-SĒ-mē-a). It may be due to increased calcium loss, reduced calcium intake, elevated levels of phosphate (as one goes up, the other goes down), or altered regulation as might occur in hypoparathyroidism. Hypocalcemia is characterized by numbness and tingling of the fingers, hyperactive reflexes, muscle cramps, tetany, and convulsions. Bone fractures may also occur. In addition, hypocalcemia may cause spasms of laryngeal muscles that can cause death by asphyxiation.

Characteristics of **hypercalcemia,** an abnormally high level of Ca^{2+}, are lethargy, weakness, anorexia, nausea, vomiting, polyuria, itching, bone pain, depression, confusion, paresthesia, stupor, and coma.

Phosphate

Phosphate ions ($H_2PO_4^-$, HPO_4^{2-}, and PO_4^{3-}) are important intracellular anions. HPO_4^{2-} is the most prevalent form at pH 7.4. Normal plasma (serum) phosphate concentration is 1.7 to 2.6 mEq/liter. About 85% of the phosphate in the adult is present in bone as calcium phosphate salts. The remainder is mostly combined with lipids (phospholipids), proteins, carbohydrates, nucleic acids (DNA and RNA), and high-energy compounds such as adenosine triphosphate (ATP). Like calcium, phosphate is a structural component of bone and teeth. Also, it plays an important role in buffering reactions (the phosphate buffer system is discussed later in the chapter).

The level of phosphate in blood plasma is also regulated by PTH and CT. PTH stimulates osteoclasts to release phosphate from mineral salts of bone matrix and causes renal tubular cells to excrete phosphate. CT exerts its phosphate lowering effect by stimulating osteoblasts and inhibiting osteoclasts. In the presence of CT, osteoblasts remove phosphate from blood and deposit it in bone matrix after combining it with calcium to form the mineral salts in bone matrix.

CLINICAL APPLICATION

HYPOPHOSPHATEMIA AND HYPERPHOSPHATEMIA

An abnormally low level of phosphate, called **hypophosphatemia** (hī'-pō-fos'-fa-TĒ-mē-a), may occur through transient intracellular shifts, increased urinary losses, decreased intestinal absorption, increased utilization, and alcoholism. Hypophosphatemia is characterized by confusion, seizures, coma, chest and muscle pain, increased susceptibility to infection, numbness and tingling of the fingers, uncoordination, memory loss, and lethargy.

Hyperphosphatemia occurs most often when the kidneys fail to excrete excess phosphate, as may occur in renal insufficiency. It also can result from increased intake of phosphates or destruction of cells, which releases phosphate into the blood. The primary complication of hyperphosphatemia is the precipitation of calcium phosphate in soft tissues, joints, and arteries. Symptoms of hyperphosphatemia include anorexia, nausea, vomiting, muscular weakness, hyperactive reflexes, tetany, and tachycardia.

Magnesium

Magnesium (Mg^{2+}) is the second most common intracellular electrolyte and the body's fourth most abundant cation. Normal plasma (serum) magnesium concentration is 1.3 to 2.1 mEq/liter. In the adult, about 54% of the magnesium is in bone, about 45% is in intracellular fluid, and about 1% is in extracellular fluid. Functionally, Mg^{2+} activates enzymes involved in the metabolism of carbohydrates and proteins and is needed for operation of the sodium pump (Na^+/K^+ ATP-ase). Mg^{2+} is also important in neuromuscular activity, neural transmission within the central nervous system (CNS), and myocardial functioning.

Among the factors that result in an increased excretion of magnesium by the kidneys are hypercalcemia, hypermagnesemia, an increase in extracellular fluid volume, a decrease in parathyroid hormone (PTH), and acidosis. The opposite conditions result in a decreased excretion of magnesium.

CLINICAL APPLICATION

HYPOMAGNESEMIA AND HYPERMAGNESEMIA

Magnesium deficiency, called **hypomagnesemia** (hī'-pō-mag'-ne-SĒ-mē-a), may be caused by malabsorption, diarrhea, alcoholism, malnutrition, excessive lactation, diabetes mellitus, and diuretic therapy. The condition is characterized by weakness, irritability, tetany, delirium, convulsions, confusion, anorexia, nausea, vomiting, paresthesia, and cardiac arrhythmias.

Hypermagnesemia, or magnesium excess, occurs mainly in persons with renal failure who have an increased intake of magnesium, for example, magnesium-containing antacids. Other causes include Addison's disease, acute diabetic acidosis, severe dehydration, and hypothermia. The condition is characterized by flaccidity, hypotension, muscular weakness or paralysis, nausea, vomiting, and altered mental functioning.

MOVEMENT OF BODY FLUIDS

Blood is the vehicle for transport and exchange of materials between body cells and the outside world. Nutrients in food enter the blood for distribution to tissues throughout the body. Oxygen enters the lungs and then the blood. At the same time, waste products generated by cellular metabolism diffuse from the cells that produce them into the bloodstream. From blood, wastes may be excreted into urine, exhaled by the lungs, or follow some other route out of the body. Interstitial fluid, on the other hand, is the go-between for exchanges between intracellular fluid and blood plasma.

Exchange Between Plasma and Interstitial Fluid

The movement of substances between plasma and interstitial fluid occurs across capillary membranes. Capillary exchange was discussed in detail in Chapter 21, but we will review it here. Substances enter and leave capillaries in three ways: (1) vesicular transport, (2) diffusion, and (3) bulk flow.

In vesicular transport, substances in blood plasma cross the capillary wall first by endocytosis into an endothelial cell and then by exocytosis into interstitial fluid. This method accounts for only a tiny fraction of exchange between plasma and interstitial fluid.

Most substances in blood or interstitial fluid can cross capillary membranes by diffusion. This process accounts for the largest part of capillary exchange in most body tissues. One exception is in the brain, where the blood–brain barrier blocks diffusion of many substances, especially those that are not lipid-soluble.

Movement of fluid by bulk flow depends on opposing pressures: (1) blood hydrostatic pressure (BHP), (2) interstitial fluid hydrostatic pressure (IFHP), (3) blood colloid osmotic pressure (BCOP), and (4) interstitial fluid colloid osmotic pressure (IFOP). (See Fig. 21.10.)

The difference between the pressures that move fluid out of plasma and the pressures that push it into plasma is the **net filtration pressure (NFP)**. The NFP at the arterial end of a capillary is about 10 mm Hg, whereas it is about –9 mm Hg at the venous end. Thus at the arterial end of a capillary, fluid moves out (is filtered) from plasma into the interstitial compartment at a pressure of 10 mm Hg. Then, at the venous end of a capillary, fluid moves back in (is reabsorbed) from the interstitial to plasma compartment at a pressure of –9 mm Hg. Note that not all the fluid filtered at one end of the capillary is reabsorbed at the other.

The fluid not reabsorbed and any proteins that escape from capillaries pass into lymphatic capillaries. From here, the fluid (lymph) moves through lymphatic vessels to the thoracic duct or right lymphatic duct and enters the cardiovascular system via the subclavian veins. Normally, there is a state of near equilibrium at the arterial and venous ends of a capillary in which filtered fluid and absorbed fluid, plus that picked up by the lymphatic system, are nearly equal. Recall that this near equilibrium is called **Starling's law of the capillaries.**

Exchange Between Interstitial and Intracellular Fluids

Intracellular fluid and interstitial fluid normally have the same osmotic pressures. The principal cation inside cells is K^+, whereas the principal cation outside is Na^+ (see Fig. 27.4). When a fluid imbalance between these two compartments occurs, it is usually caused by a change in the Na^+ or K^+ concentration.

Sodium balance in the body is controlled by aldosterone, ANP, and ADH. ADH regulates extracellular fluid electrolyte concentration by adjusting the amount of water reabsorbed into the blood by the collecting ducts of the kidneys. Aldosterone regulates extracellular fluid volume by increasing the amount of Na^+ reabsorbed by the blood from the kidneys, which indirectly affects the amount of water reabsorbed from the filtrate. ANP inhibits secretion of both renin and aldosterone and thus slows reabsorption of Na^+ by the collecting ducts.

Certain conditions can result in an eventual decrease in the Na^+ concentration in interstitial fluid. For instance, during sweating the skin excretes Na^+ as well as water. Sodium also may be lost through vomiting and diarrhea. If the lost fluid is replaced with plain water, sodium concentration can fall below the normal range (Fig. 27.5). The decrease in Na^+ concentration in the interstitial fluid lowers the interstitial fluid osmotic pressure and establishes a water concentration gradient between interstitial fluid and intracellular fluid. Water moves from the hypo-osmotic interstitial fluid into cells, producing two results that can be quite serious.

The first result, an increase in intracellular water concentration, called **overhydration,** is particularly disruptive to nerve cell function. Severe overhydration, or **water intoxication,** produces neurological symptoms ranging from disoriented behavior to convulsions, coma, and even death. The second result of the fluid shift is a loss of interstitial fluid volume that leads to a decrease in the interstitial fluid hydrostatic pressure. As the interstitial fluid hydrostatic pressure drops, water moves out of the plasma. The resulting loss of blood volume may lead to circulatory shock.

ACID–BASE BALANCE

Before reading this section, you might want to review the discussion of acids, bases, and pH in Chapter 2. From our discussion thus far, you can see that various electrolytes play different roles in helping to maintain homeostasis. A very important electrolyte in terms of the body's acid–base balance is the hydrogen ion (H^+). Although some hydrogen ions enter the body in ingested foods, most are produced as a result of the cellular metabolism of substances such as glucose, fatty acids, and amino acids. One of the major challenges to homeostasis is keeping the hydrogen ion concentration at an appropriate level to maintain proper acid–base balance.

FIGURE 27.5 Interrelations between fluid imbalance and electrolyte imbalance.

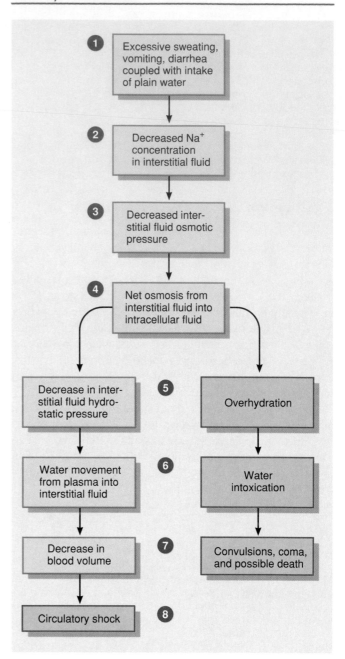

Question: What effect will step 3 have on secretion of ADH and thus on urine output?

Besides controlling water movement, electrolytes also help regulate the body's acid–base balance. The balance of acids and bases is maintained by controlling the H^+ concentration of body fluids, particularly extracellular fluid. In a healthy person, the pH of the extracellular fluid remains between 7.35 and 7.45. A pH of 7.4 corresponds to a H^+ concentration of 40 nanoequivalents/liter. (One nanoequivalent = 0.000000001 equivalent.) Metabolism typically produces a significant excess of H^+. If there were no mechanisms for disposal of acids, the rising concentration of H^+ in body fluids would quickly lead to death. Homeostasis of H^+ concentration within a narrow pH range is essential to survival and depends on three major mechanisms.

1. **Buffer systems.** Buffers act quickly to temporarily bind H^+, which removes the highly reactive, excess H^+ from solution but not from the body.
2. **Exhalation of carbon dioxide.** By increasing the rate and depth of breathing, more carbon dioxide can be exhaled. This reduces the level of carbonic acid and is effective within minutes.
3. **Kidney excretion.** The slowest mechanism, taking hours or days, but the only way to eliminate acids other than carbonic acid is through their passage into urine and their excretion by the kidneys.

Buffer Systems

Most **buffer systems** of the body consist of a weak acid and the salt of that acid, which functions as a weak base. Buffers function to prevent rapid, drastic changes in the pH of a body fluid by changing strong acids and bases into weak acids and bases. Buffers work within fractions of a second. A strong acid dissociates into H^+ more easily than does a weak acid. Strong acids therefore lower pH more than weak ones because strong acids contribute more H^+. Similarly, strong bases raise pH more than weak ones because strong bases dissociate more easily into hydroxide ions (OH^-). The principal buffer systems of the body fluids are the carbonic acid–bicarbonate system, the phosphate system, and the protein buffer system.

Carbonic Acid–Bicarbonate Buffer System

The **carbonic acid–bicarbonate buffer system** is based on the bicarbonate ion (HCO_3^-), which can act as a weak base, and carbonic acid, which can act as a weak acid. Thus the buffer system can compensate for either an excess or a shortage of H^+. For example, if there is an excess of H^+ (an acid condition), HCO_3^- can function as a weak base and remove the excess H^+ as follows:

$$H^+ + HCO_3^- \longrightarrow H_2CO_3 \longrightarrow H_2O + CO_2$$

Hydrogen ion · Bicarbonate ion (weak base) · Carbonic acid · Water · Carbon dioxide

On the other hand, if there is a shortage of H^+ ions (an alkaline condition), H_2CO_3 can function as a weak acid and provide H^+ ions as follows:

$$H_2CO_3 \longrightarrow H^+ + HCO_3^-$$

Weak acid · Hydrogen ion · Bicarbonate ion

A typical bicarbonate buffer system consists of a mixture of carbonic acid (H_2CO_3) and its salt, sodium bicarbonate ($NaHCO_3$). The carbonic acid–bicarbonate buffer system is an important regulator of blood pH. When a strong acid, such as hydrochloric acid (HCl) is added to a buffer solution containing sodium bicarbonate, which behaves like a weak base, the following reaction occurs:

$$HCl + NaHCO_3 \longrightarrow NaCl + H_2CO_3$$

Hydrochloric acid (strong acid) · Sodium bicarbonate (weak base) · Sodium chloride · Carbonic acid (weak acid)

If a strong base, such as sodium hydroxide (NaOH), is added to a buffer solution containing a weak acid, such as carbonic acid, the following reaction occurs:

$$NaOH + H_2CO_3 \longrightarrow H_2O + NaHCO_3$$

Sodium hydroxide (strong base) · Carbonic acid (weak acid) · Water · Sodium bicarbonate (weak base)

Normal metabolism produces more acids than bases and thus tends to acidify the blood rather than make it more alkaline. Accordingly, the body needs more bicarbonate salt than it needs carbonic acid. When extracellular pH is normal (7.4), bicarbonate concentration is about 24 mEq/liter whereas carbonic acid concentration is about 1.2 mEq/liter. Thus bicarbonate ions outnumber carbonic acid molecules 20:1.

Phosphate Buffer System

The **phosphate buffer system** acts in essentially the same manner as the carbonic acid–bicarbonate buffer system. The components of the phosphate buffer system are the sodium salts of dihydrogen phosphate and sodium monohydrogen phosphate ions. The dihydrogen phosphate ion acts as the weak acid and is capable of buffering strong bases.

$$NaOH + NaH_2PO_4 \longrightarrow H_2O + Na_2HPO_4$$

Sodium hydroxide (strong base) · Sodium dihydrogen phosphate (weak acid) · Water · Sodium monohydrogen phosphate (weak base)

The monohydrogen phosphate ion acts as the weak base and is capable of buffering strong acids.

$$HCl + Na_2HPO_4 \longrightarrow NaCl + NaH_2PO_4$$

Hydrochloric acid (strong acid) · Sodium monohydrogen phosphate (weak base) · Sodium chloride (salt) · Sodium dihydrogen phosphate (weak acid)

Since the phosphate concentration is highest in intracellular fluid, the phosphate buffer system is an important regulator of pH in the cytosol. It also is present at a lower level in extracellular fluids and acts to buffer acids in urine. NaH_2PO_4 is formed when excess H^+ in the kidney tubules combines with Na_2HPO_4. In this reaction, Na^+ released from Na_2HPO_4 forms sodium bicarbonate ($NaHCO_3$) and passes into the blood. The H^+ that replaces Na^+ becomes part of the NaH_2PO_4 that passes into the urine. This reaction is one of the mechanisms by which the kidneys help maintain pH by the acidification of urine (see Fig. 26.17).

Protein Buffer System

The **protein buffer system** is the most abundant buffer in body cells and plasma. And inside red blood cells the protein hemoglobin is an especially good buffer. Proteins are composed of amino acids. An amino acid is an organic compound that contains at least one carboxyl group ($COOH$) and at least one amine group (NH_2). The free carboxyl group at one end of a protein acts like an acid by releasing hydrogen ions (H^+) when pH rises and can dissociate in this way:

$$NH_2 - \underset{\underset{H}{|}}{\overset{\overset{R}{|}}{C}} - COOH \longrightarrow NH_2 - \underset{\underset{H}{|}}{\overset{\overset{R}{|}}{C}} - COO^- + H^+$$

The H^+ is then able to react with any excess hydroxide ion (OH^-) in the solution to form water.

The free amine group at the other end of a protein can act as a base by combining with hydrogen ions when pH falls as follows:

$$COOH - \underset{\underset{H}{|}}{\overset{\overset{R}{|}}{C}} - NH_2 + H^+ \longrightarrow COOH - \underset{\underset{H}{|}}{\overset{\overset{R}{|}}{C}} - NH_3^+$$

Thus proteins act as both acidic and basic buffers. Such compounds, which can act as either acid or basic components of a buffer, are said to be amphoteric (am'-fō-TER-ik).

Besides the terminal carboxyl and amine groups, seven of the twenty amino acids have side chains that can release or bind H^+. At physiological pH (7.4) the two most important amino acid buffers are histidine and cysteine. Proteins that are rich in these two amino acids are especially effective buffers. For example, each molecule of hemoglobin, a good buffer, contains 37 histidines. Let's take a look at how hemoglobin functions as a buffer.

Hemoglobin is effective in buffering carbonic acid (H_2CO_3) in red blood cells. Carbon dioxide (CO_2) that passes from body cells into red blood cells combines with water (H_2O) in the red blood cells to form carbonic acid. Once formed, the carbonic acid dissociates into hydrogen ions (H^+) and bicarbonate ions (HCO_3^-). At the same time carbon dioxide enters red blood cells, oxyhemoglobin (HbO_2) gives up its oxygen to body cells to become reduced hemoglobin (Hb^-), which carries a negative charge. The reduced hemoglobin combines with the hydrogen ion to form an acid (HbH) that is even weaker than carbonic acid. Note that reduced hemoglobin is a better buffer of hydrogen ions than oxyhemoglobin:

$$H_2CO_3 \longleftarrow H_2O + \longleftarrow CO_2$$
(Carbonic acid)

$$H^+ \quad + \quad HCO_3^-$$
(Hydrogen ion) (Bicarbonate ion)

$$O_2$$
$$HbO \longrightarrow Hb^- + H^+ \longrightarrow HbH$$
(Oxyhemoglobin) (Reduced hemoglobin) (Weak acid)

Exhalation of Carbon Dioxide

Breathing also plays a role in maintaining the pH of the body. An increase in the carbon dioxide (CO_2) concentration in body fluids increases H^+ concentration and thus lowers the pH (makes it more acidic). This is illustrated by the following reactions:

$$CO_2 + H_2O \rightleftharpoons H_2CO_3 \rightleftharpoons H^+ + HCO_3^-$$

Conversely, a decrease in the CO_2 concentration of body fluids raises the pH (makes it more basic).

The pH of body fluids may be adjusted, usually in one to three minutes, by a change in the rate and depth of breathing. If the rate and depth of breathing increase, more carbon dioxide is exhaled, the reaction just given is driven to the left, H^+ concentration falls, and the blood pH rises. Since carbonic acid can be eliminated by exhaling CO_2, it is called a **volatile acid.** If the rate of respiration slows down, less carbon dioxide is exhaled, and the blood pH falls. Doubling the breathing rate increases the pH by about 0.23, from 7.4 to 7.63. Reducing the breathing rate to one-quarter its normal rate lowers the pH by 0.4, from 7.4 to 7.0. These examples show the powerful effect of alterations in breathing on pH of body fluids.

The pH of body fluids, in turn, affects the rate of breathing (Fig. 27.6). If, for example, the blood becomes more acidic, the increase in hydrogen ions (controlled condition) is detected by chemoreceptors (receptors) that stimulate the inspiratory center in the medulla (control center). As a result, the diaphragm and other muscles of respiration (effectors) contract more forcefully and frequently—the rate and depth of breathing increase.

The same effect is achieved if the blood level of carbon dioxide increases. The increased rate and depth of respiration remove more CO_2 from blood to reduce the H^+ concentration, and blood pH increases (response). On the other hand, if the pH of the blood increases, the respiratory center is inhibited and respirations decrease. A decrease in the CO_2 concentration of blood has the same effect. The decreased rate and depth of respirations cause CO_2 to accumulate in blood and the H^+ concentration increases. The respiratory mechanism normally can eliminate more acid or base than can all the buffers combined, but it is limited to eliminating only the single volatile acid, carbonic acid.

FIGURE 27.6 Negative feedback regulation of blood pH by the respiratory system.

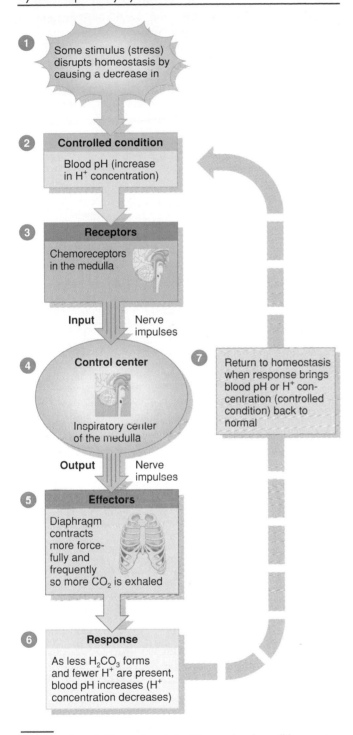

1 Some stimulus (stress) disrupts homeostasis by causing a decrease in

2 **Controlled condition**

Blood pH (increase in H⁺ concentration)

3 **Receptors**

Chemoreceptors in the medulla

Input Nerve impulses

4 **Control center**

Inspiratory center of the medulla

Output Nerve impulses

5 **Effectors**

Diaphragm contracts more forcefully and frequently so more CO_2 is exhaled

6 **Response**

As less H_2CO_3 forms and fewer H⁺ are present, blood pH increases (H⁺ concentration decreases)

7 Return to homeostasis when response brings blood pH or H⁺ concentration (controlled condition) back to normal

Question: if you hold your breath for 30 seconds, what will happen to your blood pH?

Kidney Excretion of H⁺

On a daily basis metabolic reactions produce nonvolatile or **fixed acids** at the rate of about 1 mEq/liter of H⁺ for every kilogram of body weight. The normal concentration of H⁺ in body fluids is only 40 nEq/liter. One milliequivalent is a million nanoequivalents. The only way to eliminate this huge fixed acid load is to excrete H⁺ in the urine. The kidneys also synthesize new HCO_3^- and reabsorb bicarbonate ions that have been filtered so this important buffer is not lost in the urine. Given these contributions to acid–base balance, it's not surprising that renal failure can quickly cause death. Since the role of the kidneys in maintaining pH has already been discussed in Chapter 26, we simply refer you to that chapter. Carefully review Fig. 26.17.

Exhibit 27.2 presents a summary of the mechanisms that maintain body pH.

ACID–BASE IMBALANCES

The normal blood pH range is 7.35 to 7.45, which corresponds to a concentration of H⁺ between 45 and 35 nEq/liter. **Acidosis** (or **acidemia**) is a condition in which

EXHIBIT 27.2

MECHANISMS THAT MAINTAIN BODY pH

Mechanism	Comments
BUFFER SYSTEMS	Most consist of weak acid and the salt of the acid, which functions as a weak base. Prevent drastic changes in body fluid pH.
Carbonic acid–bicarbonate	Important regulator of blood pH. Most abundant buffer in extracellular fluid (ECF).
Phosphate	Important buffer inside cells and in urine.
Protein	Most abundant buffers in body cells and blood. Histidine and cysteine are the two amino acids that contribute most of the buffering capacity of proteins. Hemoglobin inside red blood cells is a good buffer.
EXHALATION OF CO_2	With increased exhalation of CO_2, pH rises (less H⁺). With decreased exhalation of CO_2, pH falls (more H⁺).
KIDNEYS	Renal tubules secrete H⁺ into the urine and reabsorb HCO_3^- so it is not lost in the urine.

blood pH is below 7.35. **Alkalosis** (or **alkalemia**) is a condition in which blood pH is higher than 7.45.

A change in blood pH that leads to acidosis or alkalosis can be compensated to return pH to normal. **Compensation** refers to the physiological response to an acid–base imbalance. If a person has an altered pH due to metabolic causes, respiratory mechanisms (hyperventilation or hypoventilation) can help compensate for the alteration. Respiratory compensation occurs within minutes and is maximized within hours. On the other hand, if a person has an altered pH due to respiratory causes, metabolic mechanisms (kidney excretion) can compensate for the alteration. Metabolic compensation may begin in minutes but takes days to reach a maximum.

Physiological Effects

The principal physiological effect of acidosis is depression of the central nervous system through depression of synaptic transmission. If the blood pH falls below 7, depression of the nervous system is so severe that the individual becomes disoriented and comatose and dies. Patients with severe acidosis usually die in a state of coma. On the other hand, the major physiological effect of alkalosis is overexcitability in both the central nervous system and peripheral nerves. Nerves conduct impulses repetitively, even when not stimulated by normal stimuli, resulting in nervousness, muscle spasms, and even convulsions and death.

In the discussion that follows, note that both respiratory acidosis and alkalosis are primary disorders of blood pCO_2 (normal range 35 to 45 mm Hg). On the other hand, both metabolic acidosis and alkalosis are primary disorders of bicarbonate (HCO_3^-) concentration (normal range 22 to 26 mEq/liter).

Respiratory Acidosis

The hallmark of **respiratory acidosis** is an elevated pCO_2 of arterial blood (above 45 mm Hg). Inadequate exhalation of CO_2 decreases the blood pH. It occurs as a result of any condition that decreases the movement of CO_2 from the blood to the alveoli of the lungs to the atmosphere and therefore causes a buildup of carbon dioxide, carbonic acid, and hydrogen ions. Such conditions include emphysema, pulmonary edema, injury to the respiratory center of the medulla, airway obstruction, or disorders of the muscles involved in breathing. Metabolic compensation involves increased excretion of H^+ ions and increased reabsorption of HCO_3^- by the kidneys. Treatment of respiratory acidosis

aims to increase the exhalation of CO_2. Excessive secretions may be suctioned out of the respiratory tract, and artificial respiration may be given. In addition, intravenous administration of bicarbonate and ventilation therapy to remove excessive carbon dioxide may be used.

Respiratory Alkalosis

In **respiratory alkalosis** arterial blood pCO_2 is decreased (below 35 mm Hg). Hyperventilation causes the pH to increase. It occurs in conditions that stimulate the respiratory center. Such conditions include oxygen deficiency due to high altitude or pulmonary disease, cerebrovascular accident (CVA), severe anxiety, and aspirin overdose. The kidneys attempt to compensate by decreasing excretion of H^+ and decreasing reabsorption of HCO_3^-. Treatment of respiratory alkalosis is aimed at increasing the level of CO_2 in the body. One simple measure is to have the person breathe into a paper bag and then rebreathe the exhaled mixture of CO_2 and oxygen from the bag.

Metabolic Acidosis

In **metabolic acidosis** there is a decrease in HCO_3^- concentration (below 22 mEq/liter). The decrease in pH is caused by loss of bicarbonate, such as may occur with severe diarrhea or renal dysfunction; accumulation of an acid, other than carbonic acid, as may occur in ketosis; or failure of the kidneys to excrete H^+ ions derived from metabolism of dietary proteins. Compensation is respiratory by hyperventilation. Treatment of metabolic acidosis consists of intravenous solutions of sodium bicarbonate and correcting the cause of acidosis.

Metabolic Alkalosis

In **metabolic alkalosis** HCO_3^- concentration is elevated (above 26 mEq/liter). A nonrespiratory loss of acid by the body or excessive intake of alkaline drugs causes the pH to increase. Excessive vomiting of gastric contents results in a substantial loss of hydrochloric acid and is probably the most frequent cause of metabolic alkalosis. Other causes of metabolic alkalosis include gastric suctioning, use of certain diuretics, endocrine disorders, and administration of alkali. Compensation is respiratory by hypoventilation. Treatment of metabolic alkalosis consists of fluid therapy to replace chloride, potassium, and other electrolyte deficiencies and correcting the cause of alkalosis.

A summary of acidosis and alkalosis is presented in Exhibit 27.3.

EXHIBIT 27.3

SUMMARY OF ACIDOSIS AND ALKALOSIS

Condition	Definition	Common Cause	Compensatory Mechanism
Respiratory acidosis	Increased pCO_2 (above 45 mm Hg) and decreased pH (below 7.35) if there is no compensation.	Hypoventilation due to emphysema, pulmonary edema, trauma to respiratory center, airway obstructions, dysfunction of muscles of respiration.	Renal: increased excretion of H^+ ions; increased reabsorption of HCO_3^-. If compensation is complete, pH will be within normal range but pCO_2 will be high.
Respiratory alkalosis	Decreased pCO_2 (below 35 mm Hg) and increased pH (above 7.45) if there is no compensation.	Hyperventilation due to oxygen deficiency, pulmonary disease, cerebrovascular accident (CVA), anxiety, or aspirin overdose.	Renal: decreased excretion of H^+ ions; decreased reabsorption of HCO_3^-. If compensation is complete, pH will be within normal range but pCO_2 will be low.
Metabolic acidosis	Decreased bicarbonate (below 22 mEq/liter) and decreased pH (below 7.35) if there is no compensation.	Loss of bicarbonate due to diarrhea, accumulation of acid (ketosis), renal dysfunction.	Respiratory: hyperventilation, which increases loss of CO_2. If compensation is complete, pH will be within normal range but HCO_3^- will be low.
Metabolic alkalosis	Increased bicarbonate (above 26 mEq/liter) and increased pH (above 7.45) if there is no compensation.	Loss of acid or excessive intake of alkaline drugs; due to vomiting, gastric suctioning, use of certain diuretics, and administration of alkali.	Respiratory: hypoventilation, which slows loss of CO_2. If compensation is complete, pH will be within normal range but HCO_3^- will be high.

Study Outline

Fluid Compartments and Fluid Balance (p. 905)

1. Body fluid is water and its dissolved substances.
2. About two-thirds of the body's fluid is located in cells and is called intracellular fluid (ICF).
3. The other third is called extracellular fluid (ECF). It includes interstitial fluid; plasma and lymph; cerebrospinal fluid; gastrointestinal tract fluids; synovial fluid; fluids of the eyes and ears; pleural, pericardial, and peritoneal fluids; and the glomerular filtrate.
4. Fluid balance means that the various body compartments contain the required amount of water.
5. Fluid balance and electrolyte balance are inseparable.

Water (p. 905)

1. Water is the largest single constituent in the body, varying from 45 to 75% of body weight, depending on age and the amount of fat present.
2. Primary sources of fluid intake are ingested liquids and foods, and water produced by dehydration synthesis reactions.
3. Avenues of fluid output are the kidneys, skin, lungs, and gastrointestinal tract.
4. The stimulus for fluid intake is dehydration resulting in thirst sensations. Under normal conditions, fluid output is adjusted by antidiuretic hormone (ADH), atrial natriuretic peptide (ANP), and aldosterone.

Concentration of Solutions (p. 907)

1. Electrolytes (ions) are chemicals that dissolve in body fluids and dissociate into cations (positively charged ions) and anions (negatively charged ions).
2. The total amount of solute in a solution is expressed as a percentage.
3. The total concentration of cations and anions is expressed as milliequivalents/liter (mEq/liter).
4. The total concentration of particles in solution is expressed as milliosmoles/liter (mOsm/liter).
5. Electrolytes have a greater effect on osmosis than nonelectrolytes.

Electrolytes (p. 907)

1. Electrolytes are essential minerals, they control the osmosis of water between body fluid compartments, they help maintain acid–base balance, and they carry electrical current.
2. Plasma, interstitial fluid, and intracellular fluid contain varying kinds and amounts of electrolytes.
3. Sodium (Na^+) is the most abundant extracellular ion. It is involved in impulse transmission, muscle contraction, and fluid and electrolyte balance. Its level is controlled by aldosterone.
4. Chloride (Cl^-) is the major extracellular anion. It plays a role

in regulating osmotic pressure and forming HCl. Its level is controlled indirectly by aldosterone.

5. Potassium (K^+) is the most abundant cation in intracellular fluid. It is involved in maintaining fluid volume, impulse conduction, muscle contraction, and regulating pH. Its level is controlled by aldosterone.

6. Calcium (Ca^{2+}), the most abundant ion in the body, is principally an extracellular ion that is a structural component of bones and teeth. It also functions in blood clotting, neurotransmitter release, muscle contraction, and heartbeat. Its level is controlled by parathyroid hormone (PTH) and calcitonin (CT).

7. Phosphates ($H_2PO_4^-$, HPO_4^{2-}, and PO_4^{3-}) are principally intracellular ions and are structural components of bones and teeth. They are also required for the synthesis of nucleic acids and ATP and for buffer reactions. Their level is controlled by PTH and CT.

8. Magnesium (Mg^{2+}) is primarily an intracellular electrolyte that activates several enzyme systems.

Movement of Body Fluids (p. 911)

1. At the arterial end of a capillary, fluid moves from plasma into interstitial fluid (filtration). At the venous end, fluid moves in the opposite direction (reabsorption).

2. The state of near equilibrium at the arterial and venous ends of a capillary between filtered fluid and absorbed fluid, as well as that picked up by the lymphatic system, is referred to as Starling's law of the capillaries.

3. Fluid movement between interstitial and intracellular compartments depends on the concentrations of Na^+ and K^+ and the secretion of aldosterone, antidiuretic hormone (ADH), and atrial natriuretic peptide (ANP).

4. Fluid imbalance may lead to overhydration, water intoxication, and circulatory shock.

Acid–Base Balance (p. 912)

1. The overall acid–base balance of the body is maintained by controlling the H^+ concentration of body fluids, especially extracellular fluid.

2. The normal pH of extracellular fluid is 7.35 to 7.45.

3. Homeostasis of pH is maintained by buffers, exhalation of carbon dioxide, and kidney excretion.

4. The important buffer systems include carbonic acid–bicarbonate, phosphate, and protein.

5. An increase in exhalation of carbon dioxide increases pH; a decrease in exhalation decreases pH.

6. The kidneys excrete H^+ and reabsorb HCO_3^-.

Acid–Base Imbalances (p. 915)

1. Acidosis is a blood pH below 7.35. Its principal effect is depression of the central nervous system (CNS).

2. Alkalosis is a blood pH above 7.45. Its principal effect is overexcitability of the CNS.

3. Respiratory acidosis is characterized by an elevated pCO_2 and is caused by hypoventilation; metabolic acidosis is characterized by a decreased bicarbonate level and results from an abnormal increase in acid metabolic products (other than CO_2) and loss of bicarbonate.

4. Respiratory alkalosis is characterized by a decreased pCO_2 and is caused by hyperventilation; metabolic alkalosis is characterized by increased bicarbonate and results from nonrespiratory loss of acid or excess intake of alkaline drugs.

5. Metabolic acidosis or alkalosis is compensated by respiratory mechanisms; respiratory acidosis or alkalosis is compensated by renal mechanisms.

Review Questions

1. Define body fluid. List the principal compartments and describe how they are separated. (p. 905)

2. What is meant by fluid balance? How are fluid balance and electrolyte balance related? (p. 905)

3. Describe the avenues of fluid intake and fluid output. Be sure to indicate volumes in each case. (p. 906)

4. Discuss the role of thirst in regulating fluid intake. (p. 906)

5. Explain how aldosterone, atrial natriuretic peptide (ANP), and antidiuretic hormone (ADH) adjust normal fluid output. What are some abnormal routes of fluid output? (p. 906)

6. How are concentrations of chemicals in body fluids expressed? (p. 907)

7. Define a nonelectrolyte and an electrolyte. Give specific examples of each. (p. 907)

8. Describe the functions of electrolytes in the body. (p. 907)

9. Distinguish between a cation and an anion. Give several examples of each. (p. 907)

10. Describe some of the major differences in the electrolyte concentrations of the three major fluid compartments in the body. (p. 907)

11. Name three important extracellular electrolytes and three important intracellular electrolytes. (p. 908)

12. Indicate the function and regulation of each of the following electrolytes: sodium, chloride, potassium, calcium, phosphate, and magnesium. (p. 908)

13. Describe the physiological effects of hyponatremia (p. 909), hypernatremia (p. 909), hypochloremia (p. 909), hypokalemia (p. 910), hyperkalemia (p. 910), hypocalcemia (p. 910), hypercalcemia (p. 910), hypophosphatemia (p. 911), hyperphosphatemia (p. 911), hypomagnesemia (p. 911), and hypermagnesemia. (p. 911)

14. Explain the forces involved in moving fluid between plasma and interstitial fluid. Summarize these forces by setting up an equation to express net filtration pressure (NFP). (p. 911)

15. Describe Starling's law of the capillaries. (p. 911)

16. Explain the factors involved in fluid movement between the interstitial fluid and the intracellular fluid. (p. 912)

17. Explain how the following buffer systems help to maintain the pH of body fluids: carbonic acid–bicarbonate, phosphate, and protein. (p. 913)

18. Describe how exhalation of carbon dioxide is related to the maintenance of pH. (p. 914)

19. Briefly discuss the role of the kidneys in maintaining pH. (p. 915)

20. Define acidosis and alkalosis. Distinguish between respiratory and metabolic acidosis and alkalosis. (p. 916)

21. What are the principal physiological effects of acidosis and alkalosis? (p. 916)

22. How are acidosis and alkalosis compensated and treated? (p. 916)

Answers to Questions with Figures

27.1 60 kg total body weight $\times$ 60% fluid $\times$ 1 liter/kg $\times$ ⅓ ECF $\times$ 20% plasma = 2.4 liters.

27.2 All would increase fluid loss.

27.3 Negative because the result (an increase in fluid intake) is opposite to the initiating stimulus (dehydration).

27.4 ECF: major cation = Na^+; major anions = Cl^- and HCO_3^-.

ICF: major cation = K^+; major anions = proteins and phosphates, for example, ATP.

27.5 It will tend to depress ADH secretion and thus increase urine output, which could make the progression to circulatory shock more rapid.

27.6 It will decrease slightly as CO_2 and H^+ accumulate.

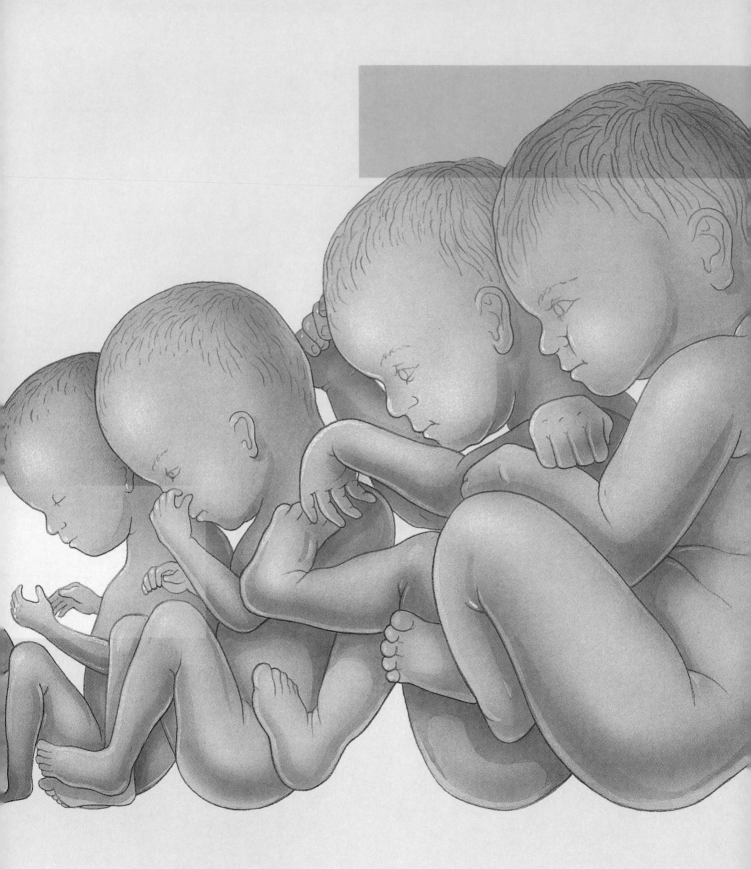

Unit 5
CONTINUITY

This unit is designed to show you how the human organism is adapted for reproduction. It also traces the developmental sequence during pregnancy and discusses principles of inheritance.

Chapter 28
THE REPRODUCTIVE SYSTEMS

Chapter Contents at a Glance

Student Objectives

1. Define reproduction and classify the organs of reproduction by function.

2. Explain the structure, histology, and functions of the testes, ducts, accessory sex glands, and penis.

3. Describe the location, histology, and functions of the ovaries, uterine (Fallopian) tubes, uterus, vagina, vulva, and mammary glands.

4. Compare the principal events of the menstrual and ovarian cycles.

5. Explain the roles of the male and female in sexual intercourse.

6. Contrast the various kinds of birth control (BC) and their effectiveness.

7. Describe the effects of aging on the reproductive systems.

8. Describe the development of the reproductive systems.

9. Explain the symptoms and causes of sexually transmitted diseases (STDs) such as gonorrhea, syphilis, genital herpes, chlamydia, trichomoniasis, and genital warts.

10. Describe the symptoms and causes of male disorders

(testicular cancer, prostate dysfunctions, impotence, and infertility) and female disorders (amenorrhea, dysmenorrhea, premenstrual syndrome [PMS], toxic shock syndrome [TSS], ovarian cysts, endometriosis, infertility, breast tumors, cervical cancer, pelvic inflammatory disease [PID], and vulvovaginal candidiasis).

11. Define medical terminology associated with the reproductive systems.

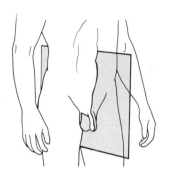

Reproduction is the process by which new individuals of a species are produced and the genetic material is passed from generation to generation. This maintains continuation of the species. Cell division in a multicellular organism is necessary for growth and repair and it involves passing of genetic material from parent cells to daughter cells.

The organs of the male and female reproductive systems may be grouped by function. The testes and ovaries, also called **gonads** (*gonos* = seed), function in the production of gametes—sperm cells and ova, respectively. The gonads also secrete hormones. The production of gametes and fluid and their discharge into ducts classify the gonads as exocrine glands, whereas their production of hormones classifies them as endocrine glands. The **ducts** of the reproductive systems transport, receive, and store gametes. Still other reproductive organs, called **accessory sex glands,** produce materials that support gametes.

The developmental anatomy of the reproductive systems is considered later in the chapter.

MALE REPRODUCTIVE SYSTEM

The organs of the male reproductive system are the testes, a system of ducts, accessory sex glands, and several supporting structures, including the penis (Fig. 28.1). The testes (male gonads) produce sperm and also secrete hormones. A

FIGURE 28.1 Male organs of reproduction and surrounding structures.

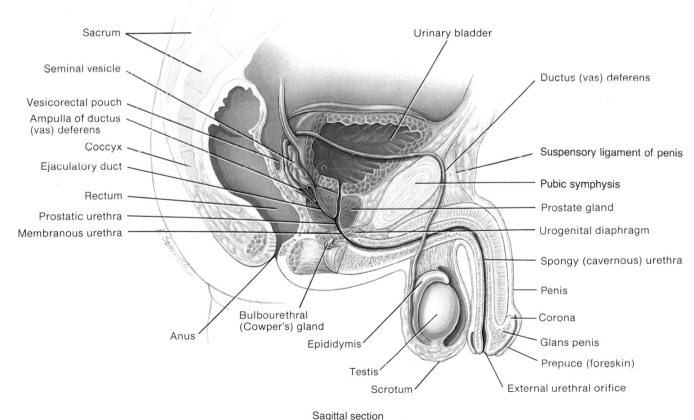

- Sacrum
- Seminal vesicle
- Vesicorectal pouch
- Ampulla of ductus (vas) deferens
- Coccyx
- Ejaculatory duct
- Rectum
- Prostatic urethra
- Membranous urethra
- Anus
- Bulbourethral (Cowper's) gland
- Epididymis
- Testis
- Scrotum
- Urinary bladder
- Ductus (vas) deferens
- Suspensory ligament of penis
- Pubic symphysis
- Prostate gland
- Urogenital diaphragm
- Spongy (cavernous) urethra
- Penis
- Corona
- Glans penis
- Prepuce (foreskin)
- External urethral orifice

Sagittal section

Question: What are the functions of the gonads? Ducts? Accessory sex glands?

system of ducts stores sperm and conveys them to the exterior. Together with the sperm, secretions provided by accessory sex glands constitute semen.

Scrotum

The **scrotum** (SKRŌ-tum) is a cutaneous outpouching of the abdomen consisting of loose skin and superficial fascia (Fig. 28.1). It is the supporting structure for the testes. Internally, a vertical septum divides it into two sacs, each containing a single testis (see Fig. 28.8). The septum consists of superficial fascia and muscle tissue called the **dartos** (DAR-tōs), which contains bundles of smooth muscle fibers (cells). Dartos muscle is also found in the subcutaneous tissue of the scrotum (see Fig. 28.8) and is directly continuous with the subcutaneous tissue of the abdominal wall. The dartos muscle causes wrinkling of the skin of the scrotum.

The location of the scrotum and contraction of its muscle fibers regulate the temperature of the testes. The production and survival of sperm require a temperature that is about 3°C lower than normal core body temperature. Because the scrotum is outside the body cavities, it provides an environment about 3°C below body temperature. The **cremaster** (krē-MAS-ter; *kremaster* = suspender) **muscle** (see Fig. 28.8) is a small band of skeletal muscle that is a continuation of the internal oblique muscle. It elevates the testes during sexual arousal and on exposure to cold. This action moves the testes closer to the pelvic cavity where they can absorb body heat. Exposure to warmth reverses the process. The dartos also is reflexly controlled to help assure that the temperature of the testes is maintained below core body temperature.

Testes

The **testes,** or **testicles,** are paired oval glands measuring about 5 cm (2 in.) in length and 2.5 cm (1 in.) in diameter (Fig. 28.2). Each weighs between 10 and 15 g. The testes develop high on the embryo's posterior abdominal wall and usually begin their descent into the scrotum through the inguinal canals (passageways in anterior abdominal wall) during the latter half of the seventh month of fetal development (see Fig. 28.8).

CLINICAL APPLICATION

CRYPTORCHIDISM

When the testes do not descend, the condition is called **cryptorchidism** (krip-TOR-ki-dizm). The condition occurs in about 3% of full-term infants and about 30% of premature infants. Cryptorchidism on both sides results in sterility because the cells involved in the initial development of sperm cells are destroyed by the higher temperature of the pelvic cavity. The chance of testicular cancer is 30 to 50 times

FIGURE 28.2 Anatomy of the testes.

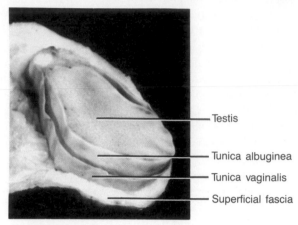

Testis
Tunica albuginea
Tunica vaginalis
Superficial fascia

(a) Transverse section through scrotum and testis

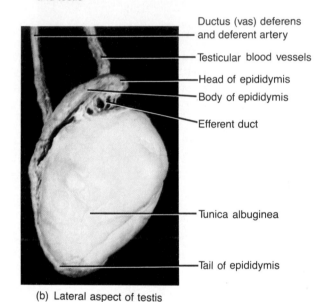

Ductus (vas) deferens and deferent artery
Testicular blood vessels
Head of epididymis
Body of epididymis
Efferent duct
Tunica albuginea
Tail of epididymis

(b) Lateral aspect of testis

Question: What tissue layers cover and protect the testes?

greater in cryptorchid testes. The testes of about 80% of boys with cryptorchidism will descend spontaneously during the first year of life. When the testes remain undescended, injections of human chorionic gonadotropin (hCG) given at 2 to 5 years of age may stimulate descent. If such hormonal treatment is unsuccessful, the condition can be corrected surgically at about age 5.

The testes are partially covered by a serous membrane called the **tunica vaginalis,** an outpocketing of the peritoneum formed during the descent of the testes. Internal to the tunica vaginalis is a dense white fibrous capsule, the **tunica albuginea** (al'-byoo-JIN-ē-a). It extends inward as septa that divide each testis into a series of internal compartments called **lobules.** Each of the 200 to 300 lobules contains one to three tightly coiled tubules, the **seminifer-ous tubules.** Here sperm are produced by a process called **spermatogenesis**, which will be considered shortly.

Seminiferous tubules are lined with spermatogenic cells in various stages of development (Fig. 28.3). Spermatogenic cells represent successive stages in a continuous process of differentiation of male germ cells. The most immature spermatogenic cells, the **spermatogonia,** lie next to the basement membrane. Toward the lumen of the tube are layers of progressively more mature cells. In order of advancing maturity, these are primary spermatocytes, secondary spermatocytes, and spermatids. By the time a **sperm cell,** or **spermatozoon** (sper'-ma-tō-ZŌ-on), has nearly reached maturity, it is in the lumen of the tubule and begins to be moved through a series of ducts.

Embedded between the developing sperm cells in the tubules are **sustentacular** (sus'-ten-TAK-yoo-lar), or **Sertoli, cells** that extend from the basement membrane to the lumen of the tubule. Just internal to the basement membrane, tight junctions join the sustentacular cells to one another and form a **blood-testis barrier.** The barrier is important because spermatozoa and developing cells produce surface antigens that are recognized as foreign by the immune system. The barrier prevents an immune response against the antigens by isolating the cells from the blood. Between the seminiferous tubules are clusters of **interstitial endocrinocytes (interstitial cells of Leydig).** These cells secrete the male hormone testosterone, the most important androgen.

Sustentacular cells support and protect developing sper-

matogenic cells; nourish spermatocytes, spermatids, and spermatozoa; and mediate the effects of testosterone and follicle stimulating hormone (FSH) on spermatogenesis. They also phagocytize excess spermatid cytoplasm as development proceeds. Sustentacular cells control movements of spermatogenic cells and the release of spermatozoa into the lumen of the seminiferous tubule. They secrete fluid for sperm transport and the hormone inhibin, which helps regulate sperm production by inhibiting secretion of FSH.

Spermatogenesis

The process by which the seminiferous tubules of the testes produce haploid (n) spermatozoa is called **spermatogenesis** (sper'-ma-tō-JEN-e-sis). Before reading the following discussion of spermatogenesis, you may wish to review the details of meiosis on page 86. Keep in mind these key points.

1.   In sexual reproduction, a new organism is produced by the union and fusion of sex cells called **gametes** (*gameto* = to marry). Male gametes, produced in the testes, are called sperm, and female gametes, produced in the ovaries, are called ova.
2.   The cell resulting from the union and fusion of gametes, called a **zygote** (ZĪ-gōt; *zygo* = joined), contains a mixture of chromosomes (DNA) from the two parents. Through repeated mitotic cell divisions, a zygote develops into a new organism.

FIGURE 28.3 Seminiferous tubules. The stages of spermatogenesis are shown in (b) and (c).

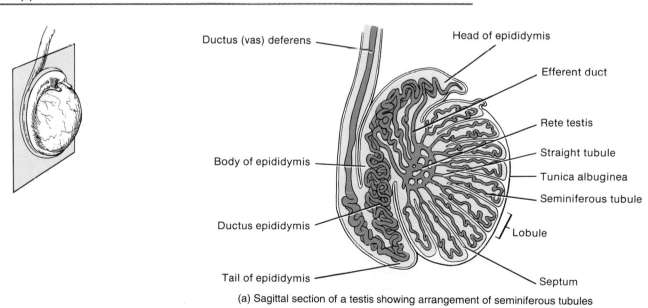

(a) Sagittal section of a testis showing arrangement of seminiferous tubules

Figure continues

FIGURE 28.3 (continued)

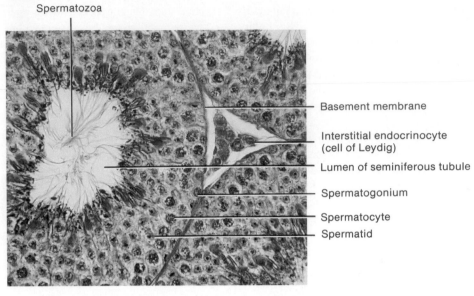

Spermatozoa

Basement membrane

Interstitial endocrinocyte (cell of Leydig)

Lumen of seminiferous tubule

Spermatogonium

Spermatocyte

Spermatid

(b) Photomicrograph of a cross section of several seminiferous tubules (1280 x)

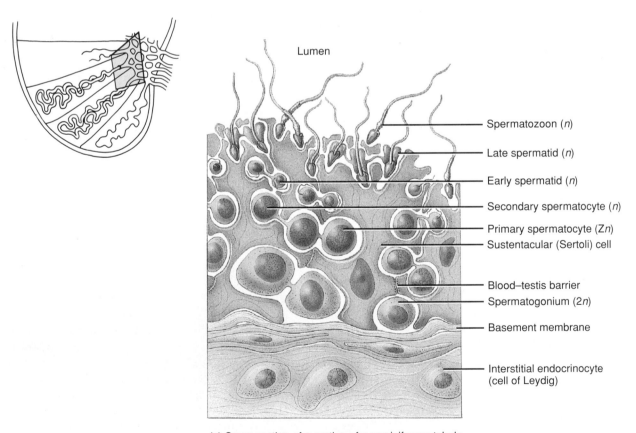

Lumen

Spermatozoon (n)

Late spermatid (n)

Early spermatid (n)

Secondary spermatocyte (n)

Primary spermatocyte (Zn)

Sustentacular (Sertoli) cell

Blood–testis barrier

Spermatogonium ($2n$)

Basement membrane

Interstitial endocrinocyte (cell of Leydig)

(c) Cross section of a portion of a seminiferous tubule

Question: Which spermatogenic cells are most mature and least mature in a seminiferous tubule?

3. Gametes differ from all other body cells (somatic cells) in that they contain the **haploid** (one-half) **chromosome number,** symbolized as *n*. In humans, this number is 23, which composes a single set of chromosomes. The nucleus of a somatic cell contains the **diploid chromosome number,** symbolized 2*n*. In humans, this number is 46, which composes two sets of chromosomes.

4. In a diploid cell, two chromosomes that belong to a pair are called **homologous** (*homo* = same) **chromosomes (homologues).** In human diploid cells, the two chromosomes of a pair are morphologically similar in 22 pairs and are called **autosomes.** The other pair, designated X and Y, are termed the **sex chromosomes.** In the female, the sex chromosomes are two X chromosomes; in the male, they are a larger X and a much smaller Y chromosome. Sex determination will be discussed in more detail on page 994.

5. **Meiosis** is a process of cell division by which gametes produced in the testes and ovaries receive the haploid chromosome number. Thus when haploid (*n*) gametes fuse, the zygote contains the diploid chromosome number (2*n*) and can undergo normal development.

In humans, spermatogenesis takes about 74 days. The seminiferous tubules are lined with immature cells called **spermatogonia** (sper′-ma-tō-GŌ-nē-a; *sperm* = seed; *gonium* = generation or offspring; Figs. 28.3b,c and 28.4). These cells develop from **primordial** (*primordialis* = primitive or early form) **germ cells** that arise from yolk sac endoderm and enter the testes early in development. In the embryonic testes, the primordial germ cells differentiate into spermatogonia but remain dormant until they begin to undergo mitotic proliferation at puberty.

Spermatogonia contain the diploid (2*n*) chromosome number. When these stem cells undergo mitosis, some of the daughter cells remain undifferentiated and serve as a reservoir of stem cells. Such cells remain near the basement membrane. The rest of the daughter cells lose contact with the basement membrane of the seminiferous tubule, undergo certain developmental changes, and differentiate into **primary spermatocytes** (SPER-ma-tō-sītz′). Primary spermatocytes, like spermatogonia, are diploid (2*n*); that is, they have 46 chromosomes.

• **Reduction Division (Meiosis I)** Each primary spermatocyte enlarges before dividing. Then two nuclear divisions take place as part of meiosis. In the first, DNA replicates and 46 chromosomes (each made up of two identical chromatids) form and move toward the equatorial plane of the cell. There they line up in homologous pairs so that there are 23 pairs of duplicated chromosomes in the center of the cell. This pairing of homologous chromosomes is called **synapsis.** The four chromatids of each homologous pair then become associated with each other to form a **tetrad.** In a tetrad, portions of one

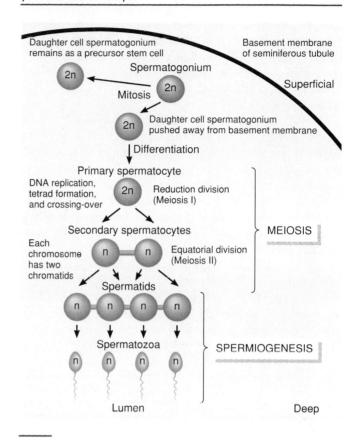

FIGURE 28.4 Spermatogenesis. The designation 2*n* means diploid (46 chromosomes); *n* means haploid (23 chromosomes).

Question: Why is meiosis I also called reduction division?

chromatid may be exchanged with portions of another. This process, called **crossing-over,** permits an exchange of genes among chromatids (see Fig. 3.26) that results in the recombination of genes. Thus the spermatozoa eventually produced are genetically unlike each other and unlike the cell that produced them—one reason for the great genetic variation among humans.

Next, the meiotic spindle forms and the kinetochore microtubules produced by the centromeres of the paired chromosomes extend toward the poles of the cell. As the pairs separate, members of each pair migrate to opposite poles of the dividing cell. The random assortment of maternally derived and paternally derived chromosomes toward opposite poles is another reason for genetic variation among spermatozoa and therefore among humans. The cells formed by the first nuclear division (reduction division) are called **secondary spermatocytes.** Each cell has 23 chromosomes—the haploid number. Each chromosome within a secondary spermatocyte, however, is made up of two chromatids still attached by a centromere.

• **Equatorial Division (Meiosis II)** The second nuclear division of meiosis is **equatorial division.** There is no replication of DNA. The chromosomes (each composed of two chromatids) line up in single file along the equatorial plane, and the chromatids of each chromosome separate from each other. The cells formed from the equatorial division are called **spermatids.** Each has half the original chromosome number, or 23 chromosomes, and is haploid. Each primary spermatocyte therefore produces four spermatids by meiosis (reduction division and equatorial division). Spermatids lie close to the lumen of the seminiferous tubule.

During spermatogenesis, a very interesting and unique process occurs. As the sperm cells proliferate, they fail to complete cytoplasmic separation (cytokinesis) so that all the daughter cells, except for the least-differentiated spermatogonia, remain in contact via cytoplasmic bridges (see Figs. 28.3c and 28.4). These cytoplasmic bridges persist until development of the spermatozoa is complete, at which point they float out individually into the lumen of the seminiferous tubule. Thus the offspring of a spermatogonium remain in cytoplasmic communication through their entire development.

This pattern of development undoubtedly accounts for the synchronized production of spermatozoa in any given area of a seminiferous tubule. It may have survival value in that half the spermatozoa contain an X chromosome and half a Y chromosome. The larger X chromosome may carry genes needed for spermatogenesis that are lacking on the Y chromosome.

● **Spermiogenesis**   The final stage of spermatogenesis, called **spermiogenesis** (sper′-mē-ō-JEN-e-sis), involves the maturation of spermatids into spermatozoa. Each spermatid develops a head with an acrosome (enzyme-containing granule) and a flagellum (tail). Since there is no cell division in spermiogenesis, each spermatid develops into a single **spermatozoon (sperm cell** or **sperm).** The release of a spermatozoon from its sustentacular cell is known as **spermiation.**

Sperm enter the lumen of the seminiferous tubule and migrate to the ductus epididymis. There, they complete their maturation in 10 to 14 days and become capable of fertilizing an ovum. Spermatozoa are also stored in the ductus (vas) deferens. Here, they can retain their fertility for up to several months.

Spermatozoa

Spermatozoa mature at the rate of about 300 million per day and, once ejaculated, have a life expectancy of about 48 hours within the female reproductive tract. A spermatozoon is highly adapted for reaching and penetrating a female ovum. It is composed of a head, a midpiece, and a tail (Fig. 28.5). In the **head** are the nuclear material and a dense

---

FIGURE 28.5 Spermatozoa.

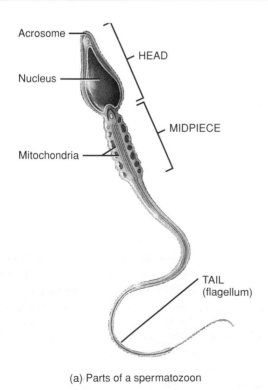

Acrosome

HEAD

Nucleus

MIDPIECE

Mitochondria

TAIL
(flagellum)

(a) Parts of a spermatozoon

Secondary oocyte

Head of spermatozoon

(b) Scanning electron micrograph of a spermatozoon in contact with a secondary oocyte (1100 x)

Question: What are the functions of each part of a sperm cell?

granule called the **acrosome** (*acro* = atop), which contains enzymes (hyaluronidase and proteinases) that aid penetration of the sperm cell into a secondary oocyte (immature ovum). The acrosome is a specialized lysosome. Numerous mitochondria in the **midpiece** carry on the metabolism that provides ATP for locomotion. The **tail,** a typical flagellum, propels the sperm along its way.

Hormones of the Brain–Testicular Axis

Secretions of the anterior pituitary gland play a major role in the developmental changes associated with puberty (described on page 958). At the onset of puberty, the anterior pituitary starts to secrete gonadotropic hormones called **follicle-stimulating hormone (FSH)** and **luteinizing hormone (LH)**. Their release is controlled by **gonadotropin releasing hormone (GnRH)** from the hypothalamus. Figure 28.6 summarizes the hormonal relationships of the **brain–testicular axis**, that is the hypothalamus, pituitary gland, and testes.

FSH acts to initiate spermatogenesis by stimulating the sustentacular (Sertoli) cells. LH stimulates the interstitial endocrinocytes (interstitial cells of Leydig) to secrete the hormone testosterone (tes-TOS-te-rōn). **Testosterone** is synthesized from cholesterol in the testes. It is the principal androgen (male sex hormone) but in some target cells is not active until it is reduced and converted into another androgen called **dihydrotestosterone (DHT)**. So testosterone is both a hormone and a prohormone (hormone precursor). The androgens have several functions.

1.   **Development.** Before birth, testosterone stimulates the male pattern of development of reproductive system ducts and the descent of the testes. DHT, on the other hand, stimulates development of the external genitals. Androgens are also converted to estrogens and play a role in the development of certain regions of the brain.

2.   **Sexual characteristics.** At puberty, testosterone and DHT bring about development and enlargement of the male sex organs and the development of male secondary sexual characteristics. These include muscular and skeletal growth that results in wide shoulders and narrow hips; pubic, axillary, facial, and chest hair (within hereditary limits); thickening of the skin; increased sebaceous (oil) gland secretion; and enlargement of the larynx and consequent deepening of the voice.

3.   **Sexual functions.** Androgens contribute to male sexual behavior and spermatogenesis and to sex drive (libido) in both males and females.

4.   **Metabolism.** Androgens are anabolic hormones; that is, they stimulate protein synthesis. This effect is obvious in the heavier muscle and bone mass of most men as compared to women. They also stimulate closure of the epiphyseal plates.

Negative feedback systems regulate testosterone production (Figs. 28.6 and 28.7). LH stimulates the production of

FIGURE 28.6 Hormonal control of testicular functions: the brain–testicular axis.

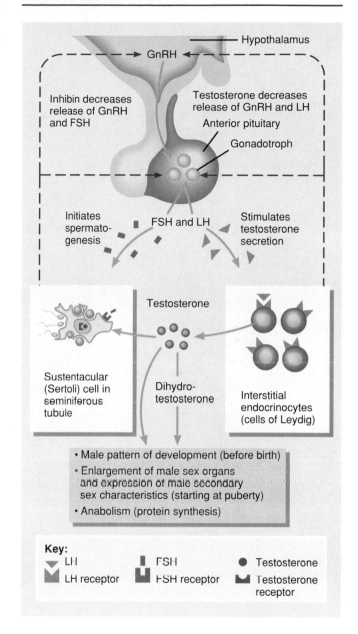

Question: **Which cells secrete inhibin?**

testosterone. When testosterone concentration in the blood increases to a certain level, it inhibits the release of GnRH by the hypothalamus and LH by the anterior pituitary (Fig. 28.7). Thus testosterone production decreases. However, once the testosterone concentration in the blood falls to a certain level, GnRH is released by the hypothalamus. This release of GnRH stimulates the release of LH by the anterior pituitary gland and stimulates testosterone production. Thus this testosterone–LH cycle is complete.

Inhibin is a protein hormone secreted by sustentacular (Sertoli) cells that has a direct effect on the anterior

FIGURE 28.7 Negative feedback control of blood level of testosterone.

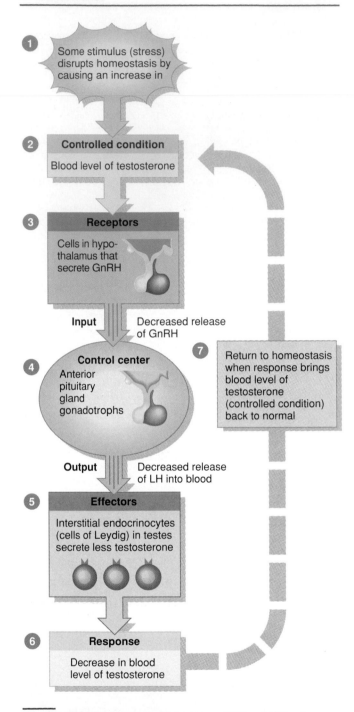

1 Some stimulus (stress) disrupts homeostasis by causing an increase in

2 **Controlled condition**

Blood level of testosterone

3 **Receptors**

Cells in hypo-thalamus that secrete GnRH

Input Decreased release of GnRH

4 **Control center**

Anterior pituitary gland gonadotrophs

7 Return to homeostasis when response brings blood level of testosterone (controlled condition) back to normal

Output Decreased release of LH into blood

5 **Effectors**

Interstitial endocrinocytes (cells of Leydig) in testes secrete less testosterone

6 **Response**

Decrease in blood level of testosterone

Question: Which hormones inhibit secretion of FSH and LH by the anterior pituitary gland?

and thus decreases spermatogenesis. If spermatogenesis is proceeding too slowly, lack of inhibin production permits FSH secretion and an increased rate of spermatogenesis. Inhibin may also inhibit secretion of GnRH by the hypothalamus.

Ducts

Ducts of the Testis

After their release into the lumen, spermatozoa and fluid flow through the twisted seminiferous tubules to the **straight tubules** (see Fig. 28.3a). Behind the released sperm, newly forming fluid and sperm produce a pressure that moves the contents of the seminiferous tubules along. The fluid, produced by sustentacular (Sertoli) cells, is rich in androgens, estrogens, K^+, glutamic acid, and aspartic acid. The straight tubules lead to a network of ducts in the testis called the **rete** (RĒ-tē) **testis.** The sperm next move through the rete testis and out of the testis into an adjacent organ, the epididymis.

Epididymis

The sperm leave the testis through a series of coiled **efferent ducts** in the epididymis that empty into a single tube called the ductus epididymis. Changes occur in the spermatozoa during their passage through the epididymis.

The **epididymis** (ep′-i-DID-i-mis; *epi* = above; *didymos* = testis) is a comma-shaped organ about 3.8 cm (1.5 in.) long. It lies along the posterior border of the testis (see Figs. 28.1 and 28.2) and consists mostly of a tightly coiled tube, the **ductus epididymis** (plural is **epididymides**). The larger, superior portion of the epididymis is known as the **head.** In the head, the efferent ducts join the ductus epididymis. The **body** is the narrow midportion of the epididymis. The **tail** is the smaller, inferior portion. At its distal end, the tail of the epididymis continues as the ductus (vas) deferens.

The ductus epididymis is a tightly coiled structure that would measure about 6 m (20 ft.) in length and 1 mm in diameter if it were straightened out. It is lined with pseudostratified columnar epithelium and encircled by layers of smooth muscle. The free surfaces of the columnar cells contain long, branching microvilli called **stereocilia.** They increase surface area for reabsorption of degenerated spermatozoa.

Functionally, the ductus epididymis is the site of sperm maturation. Over a 10 to 14 day period, their motility increases and they become fertile. The ductus epididymis also stores spermatozoa and helps propel them toward the urethra by peristaltic contraction of its smooth muscle. Spermatozoa may remain in storage in the ductus epididymis for at least a month. After that, they are expelled from the epididymis or degenerate and are reabsorbed.

pituitary by inhibiting the secretion of FSH (see Fig. 28.6). FSH brings about spermatogenesis and stimulates sustentacular cells. Once the degree of spermatogenesis required for male reproductive functions has been achieved, sustentacular cells secrete inhibin, which inhibits FSH secretion

Ductus (Vas) Deferens

Within the tail of the epididymis, the ductus epididymis becomes less convoluted and its diameter increases. After this point, the duct is referred to as the **ductus (vas) deferens** or **seminal duct** (see Fig. 28.2). The ductus (vas) deferens, about 45 cm (18 in.) long, ascends along the posterior border of the epididymis, penetrates the inguinal canal, and enters the pelvic cavity (Fig. 28.8). There, it loops over the side and down the posterior surface of the urinary bladder (see Fig. 28.1). The dilated terminal portion of the ductus (vas) deferens is known as the **ampulla** (am-POOL-la). See Fig. 28.9. The ductus (vas) deferens is lined with pseudo-stratified columnar epithelium and contains a heavy coat of three layers of muscle. Functionally, the ductus (vas) deferens stores sperm and conveys sperm from the epididymis toward the urethra during emission by peristaltic contractions of the muscular coat.

CLINICAL APPLICATION

VASECTOMY

One method of sterilization of males is called **vasectomy** (vas-EK- tō-mé), in which a portion of each ductus (vas) deferens is removed. It is a relatively uncomplicated procedure and usually is done under local anesthesia. An incision is made in the scrotum, the ducts are located, each is tied in two places, and the portion between the ties is removed. Although sperm production continues in the testes, sperm cannot reach the exterior because the ducts are cut. The sperm degenerate and are destroyed by phagocytosis. Vasectomy has no effect on sexual desire and performance. If done correctly, it is close to 100% effective. The procedure can be reversed, with a 45 to 60% chance of regaining fertility. Sometimes, an immune response to certain sperm-specific proteins appears after vasectomy. This may occur because some sperm escape from the cut ductus (vas) deferens or because sperm antigens are processed and presented by antigen presenting cells, such as macrophages.

FIGURE 28.8 Spermatic cord and inguinal canal. The left spermatic cord has been opened to expose its contents.

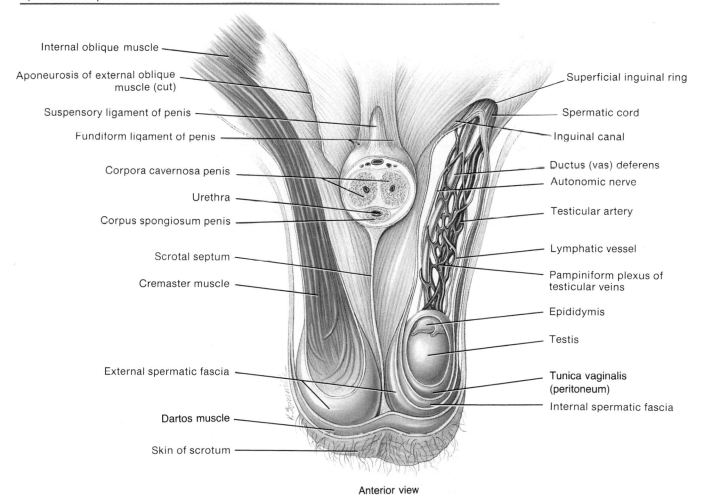

Anterior view

Question: What structures pass through the inguinal canal?

Beside the ductus (vas) deferens as it ascends in the scrotum are the testicular artery, autonomic nerves, veins that drain the testes (pampiniform plexus), lymphatic vessels, and the cremaster muscle. These structures constitute the **spermatic cord,** a supporting structure of the male reproductive system (Fig. 28.8). The cremaster muscle, which also surrounds the testes, elevates the testes during sexual stimulation and exposure to cold. The spermatic cord and ilioinguinal nerve pass through the **inguinal** (IN-gwin-al) **canal.** The canal is an oblique passageway in the anterior abdominal wall just superior and parallel to the medial half of the inguinal ligament. The canal is about 4 to 5 cm (1.6 to 2.0 in.) in length. It originates at the **deep (abdominal) inguinal ring,** a slitlike opening in the aponeurosis of the transversus abdominis muscle. The canal ends at the **superficial (subcutaneous) inguinal ring,** a somewhat triangular opening in the aponeurosis of the external oblique muscle. In females, the round ligament of the uterus and ilioinguinal nerve pass through the inguinal canal.

CLINICAL APPLICATION

INGUINAL HERNIAS

The inguinal region is a weak area in the abdominal wall. It often is the site of an **inguinal hernia**—a rupture or separation of a portion of the inguinal area of the abdominal wall resulting in the protrusion of a part of an organ. In an *indirect inguinal hernia* the herniation protrudes through the inguinal ring and follows the round ligament or spermatic cord. A *direct inguinal hernia* goes through the posterior inguinal wall. Inguinal hernias occur much less often in females.

Ejaculatory Ducts

Posterior to the urinary bladder are the **ejaculatory** (e-JAK-yoo-la-tō′-rē) **ducts** (Fig. 28.9). Each duct is about 2 cm (1 in.) long and is formed by the union of the duct from the seminal vesicle and the ductus (vas) deferens. The ejacula-

FIGURE 28.9 Male reproductive organs in relation to surrounding structures.

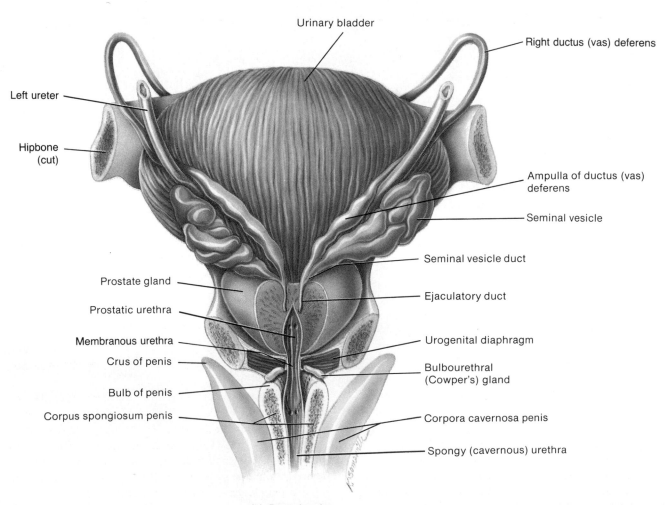

Urinary bladder

Right ductus (vas) deferens

Left ureter

Hipbone (cut)

Ampulla of ductus (vas) deferens

Seminal vesicle

Seminal vesicle duct

Prostate gland

Ejaculatory duct

Prostatic urethra

Membranous urethra

Urogenital diaphragm

Crus of penis

Bulbourethral (Cowper's) gland

Bulb of penis

Corpus spongiosum penis

Corpora cavernosa penis

Spongy (cavernous) urethra

(a) Posterior view

FIGURE 28.9 (continued)

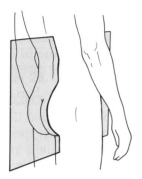

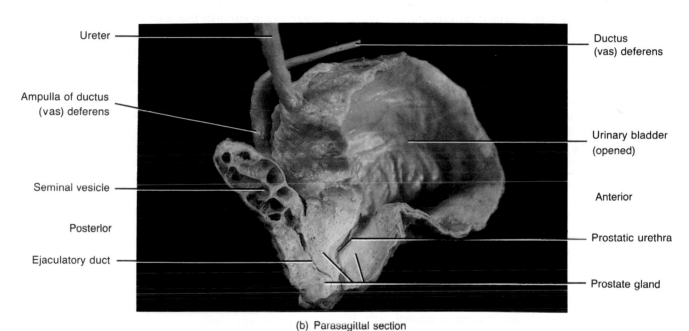

(b) Parasagittal section

Question: What does each accessory sex gland contribute to semen?

tory ducts eject spermatozoa into the prostatic urethra just before ejaculation.

Urethra

The **urethra** is the terminal duct of the reproductive and urinary systems, serving as a passageway for semen (fluid containing spermatozoa) and urine. In the male, the urethra passes through the prostate gland, the urogenital diaphragm, and the penis. It measures about 20 cm (8 in.) in length and is subdivided into three parts (see Figs. 28.1 and 28.9). The **prostatic urethra** is 2 to 3 cm (1 in.) long and passes through the prostate gland. It continues inferiorly, and as it passes through the urogenital diaphragm, a muscular partition between the two ischial and pubic rami, it is known as the **membranous urethra.** The membranous urethra is about 1 cm (0.5 in.) in length. As it passes through the corpus spongiosum of the penis, it is known as the

spongy (cavernous) urethra. This portion is about 15 cm (6 in.) long. The spongy urethra enters the bulb of the penis and ends at the **external urethral orifice.** The histology of the male urethra may be reviewed on page 895.

Accessory Sex Glands

Whereas the ducts of the male reproductive system store and transport sperm cells, the **accessory sex glands** secrete most of the liquid portion of semen.

The paired **seminal vesicles** (VES-i-kuls) are convoluted pouchlike structures, about 5 cm (2 in.) in length, lying posterior to and at the base of the urinary bladder in front of the rectum (Fig. 28.9). They secrete an alkaline, viscous fluid, that contains the sugar fructose, prostaglandins, and fibrinogen. The alkaline nature of the fluid helps to neutralize acid in the female tract. This acid would inactivate and kill sperm if not neutralized. The fructose is used for ATP

FIGURE 28.12 Uterus and associated structures. In (a), the left side of the figure has been sectioned to show internal structures. In (b), part of the posterior wall of the uterus has been removed.

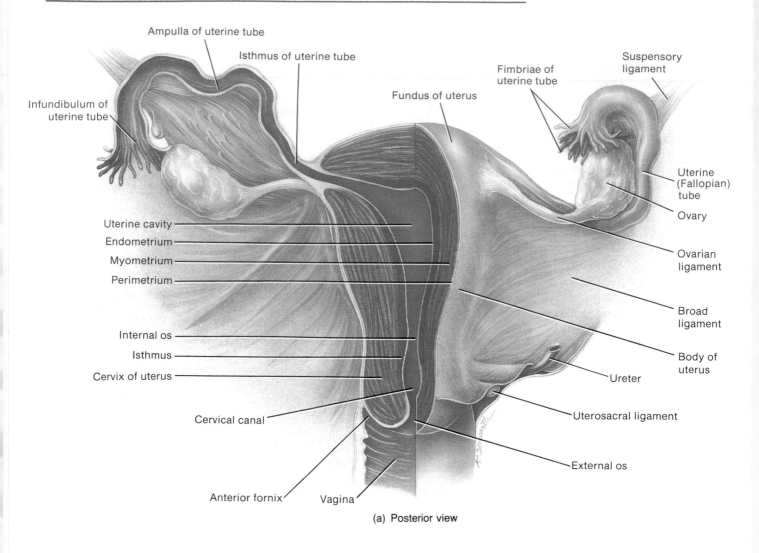

Ampulla of uterine tube

Isthmus of uterine tube

Infundibulum of uterine tube

Fimbriae of uterine tube

Suspensory ligament

Fundus of uterus

Uterine (Fallopian) tube

Ovary

Uterine cavity

Endometrium

Myometrium

Perimetrium

Ovarian ligament

Internal os

Isthmus

Cervix of uterus

Broad ligament

Body of uterus

Ureter

Cervical canal

Uterosacral ligament

External os

Anterior fornix

Vagina

(a) Posterior view

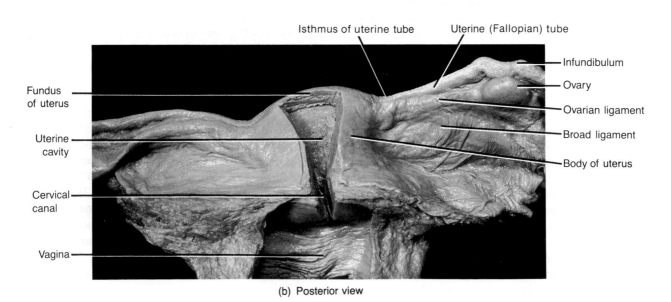

Isthmus of uterine tube

Uterine (Fallopian) tube

Infundibulum

Fundus of uterus

Ovary

Ovarian ligament

Broad ligament

Uterine cavity

Body of uterus

Cervical canal

Vagina

(b) Posterior view

FIGURE 28.12 (continued)

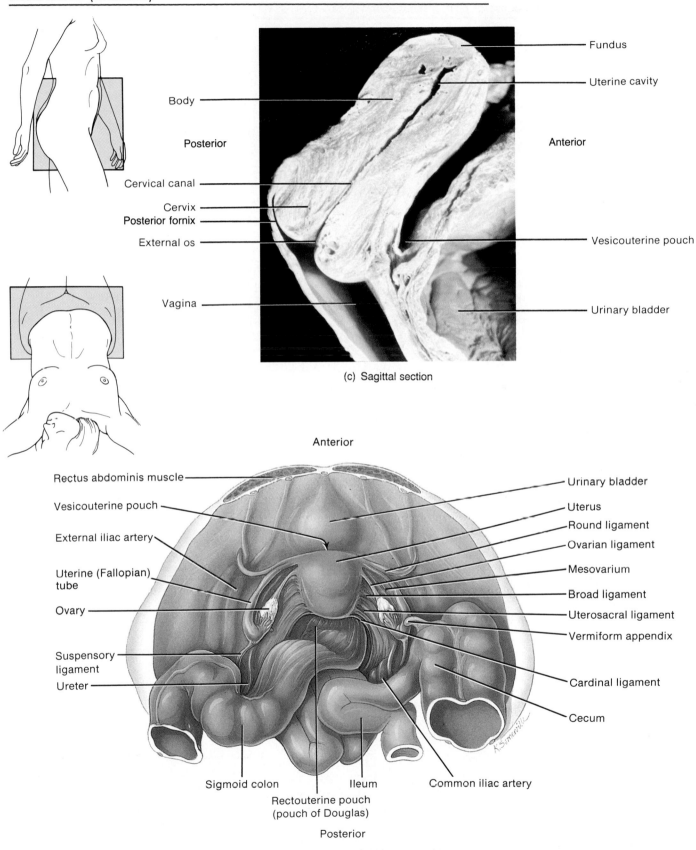

(c) Sagittal section

(d) Superior view of transverse section

Question: What ligaments hold the ovaries in position?

Figure 28.13 Histology of the ovary. In (a), the arrows indicate the sequence of developmental stages that occur as part of the ovarian cycle.

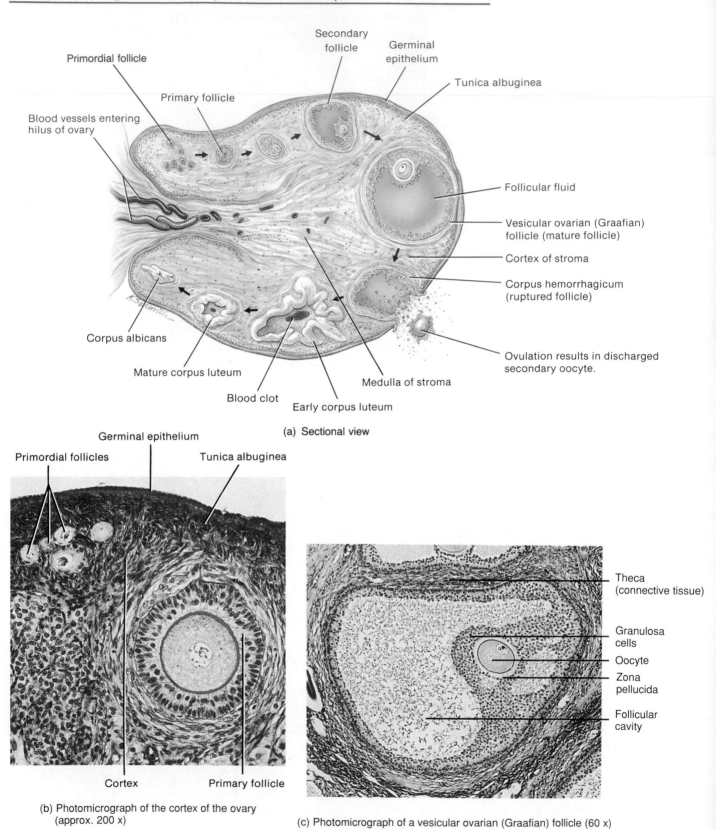

(a) Sectional view

(b) Photomicrograph of the cortex of the ovary (approx. 200 x)

(c) Photomicrograph of a vesicular ovarian (Graafian) follicle (60 x)

Question: What structures in the ovary contain endocrine tissue and what hormones do they secrete?

FIGURE 28.14 Oogenesis. The designation 2*n* means diploid (46 chromosomes); *n* means haploid (23 chromosomes).

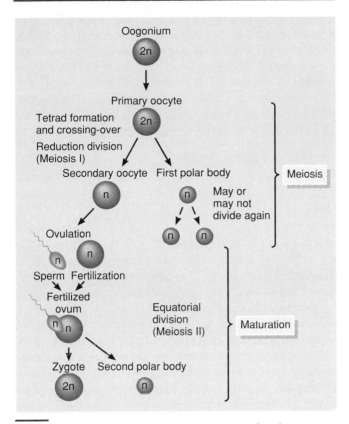

Question: How does the age of a primary oocyte in a female compare with the age of a primary spermatocyte in a male?

the diploid primary oocyte completes reduction division (meiosis I). Synapsis, tetrad formation, and crossing-over occur, and two cells of unequal size, both with 23 chromosomes (*n*) of two chromatids each, are produced.

The smaller cell produced by meiosis I, called the **first polar body,** is essentially a packet of discarded nuclear material. The larger cell, known as the **secondary oocyte,** receives most of the cytoplasm. Each secondary oocyte is surrounded by several layers of cuboidal and then columnar epithelial cells called **granulosa cells.** The entire structure is called a **secondary (growing) follicle.** Once a secondary oocyte is formed, it proceeds to the metaphase of equatorial division (meiosis II) and then stops at this stage. Only after ovulation and fertilization does equatorial division (meiosis II) come to completion.

● **Equatorial Division (Meiosis II)** At ovulation, the secondary oocyte (with the first polar body) and some surrounding supporting cells are discharged. The discharged secondary oocyte enters the uterine (Fallopian) tube. If spermatozoa are present and fertilization occurs, the second division, the equatorial division (meiosis II), is completed.

● **Maturation** The secondary oocyte produces two

haploid (*n*) cells of unequal size. The larger cell eventually develops into an **ovum,** or mature egg; the smaller one is the **second polar body.**

The first polar body may undergo another division to produce two polar bodies. If it does, meiosis of the primary oocyte results in a single haploid (*n*) ovum and three haploid (*n*) polar bodies. All polar bodies disintegrate. Thus an oogonium gives rise to a single gamete (secondary oocyte that develops into an ovum), whereas a spermatogonium produces four gametes (spermatozoa).

Uterine (Fallopian) Tubes

Females have two **uterine (Fallopian) tubes,** also called **oviducts,** that extend laterally from the uterus and transport the ova from the ovaries to the uterus (see Fig. 28.12). Measuring about 10 cm (4 in.) long, the tubes lie between the folds of the broad ligaments of the uterus. The open, funnel-shaped distal end of each tube, called the **infundibulum,** is close to the ovary. It ends in a fringe of fingerlike projections called **fimbriae** (FIM-brē-ē). One fimbria is attached to the lateral end of the ovary. From the infundibulum, the uterine tube extends medially and inferiorly and attaches to the superior lateral angle of the uterus. The **ampulla** of the uterine tube is the widest, longest portion,

FIGURE 28.15 Histology of the uterine (Fallopian) tube. Note the ciliated epithelium.

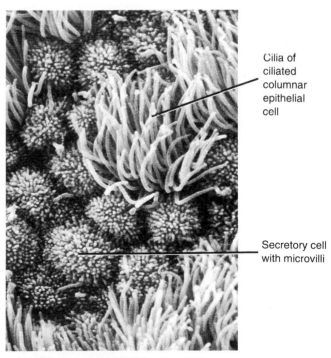

Cilia of ciliated columnar epithelial cell

Secretory cell with microvilli

Scanning electron micrograph, 5650x

Question: Of what use are cilia in the uterine tube?

making up about two-thirds of its length. The **isthmus** of the uterine tube is the short, narrow, thick-walled portion that joins the uterus.

Histologically, the uterine tubes are composed of three layers. The internal **mucosa** contains ciliated columnar epithelial cells, which help to move the secondary oocyte along the tube, and secretory cells, which have microvilli and may provide nutrition for the ovum (Fig. 28.15). The middle layer, the **muscularis,** is composed of an inner, thick, circular region of smooth muscle and an outer, thin, longitudinal region of smooth muscle. Peristaltic contractions of the muscularis and the ciliary action of the mucosa help move the ovum down into the uterus. The outer layer of the uterine tubes is a serous membrane, the **serosa.**

About once a month a vesicular ovarian (Graafian) follicle ruptures, releasing a secondary oocyte, a process called **ovulation.** The oocyte is swept into the uterine tube by local currents produced by movements of the fimbriae, which surround the surface of the most mature vesicular ovarian follicle just before ovulation occurs. The oocyte is then moved along the tube by ciliary action of the epithelium that is supplemented by the peristaltic contractions of the muscularis. If the oocyte is fertilized by a sperm cell, it usually takes place in the ampulla of the uterine tube. Fertilization may occur at any time up to about 24 hours after ovulation. After fertilization, the nuclear materials of the ovum and sperm unite, and the fertilized ovum is now called a **zygote.** After several cell divisions, it arrives at the uterus about 7 days after ovulation. At this point it is called a blastocyst. Unfertilized secondary oocytes disintegrate.

Uterus

The **uterus** (womb) serves as part of the pathway for sperm to reach the uterine tubes. It is also the site of menstruation, implantation of a fertilized ovum, development of the fetus during pregnancy, and labor. Situated between the urinary bladder and the rectum, the uterus is the size and shape of an inverted pear (see Figs. 28.11 and 28.12). Before the first pregnancy, it measures approximately 7.5 cm (3 in.) long, 5 cm (2 in.) wide, and 2.5 cm (1 in.) thick.

Anatomical subdivisions of the uterus include the dome-shaped portion above the uterine tubes called the **fundus,** the major tapering central portion called the **body,** and the inferior narrow portion opening into the vagina called the **cervix.** The secretory cells of the mucosa of the cervix produce a secretion called **cervical mucus,** a mixture of water, glycoprotein, serum-type proteins, lipids, enzymes, and inorganic salts. Females of reproductive age secrete 20 to 60 ml of mucus per day.

Cervical mucus is more receptive to spermatozoa at or near the time of ovulation because it is then less viscous and more alkaline (pH 8.5). At other times, viscous mucus forms a plug (cervical plug) that impedes sperm penetration. The mucus also supplements the energy needs of sper-

matozoa. Both the cervix and mucus serve as a sperm reservoir, protect spermatozoa from the hostile environment of the vagina, and protect spermatozoa from phagocytes. They may also play a role in capacitation—a functional change that spermatozoa undergo in the female reproductive tract before they can fertilize a secondary oocyte. Between the body and the cervix is the **isthmus** (IS-mus), a constricted region about 1 cm (0.5 in.) long. The interior of the body of the uterus is called the **uterine cavity,** and the interior of the narrow cervix is called the **cervical canal.** The junction of the isthmus with the cervical canal is the **internal os.** The **external os** is the place where the cervix opens into the vagina.

Normally, there is a flexure (bend) between the body of the uterus and the cervix. In this position, called **anteflexion,** the body of the uterus projects anteriorly and superiorly over the urinary bladder. The cervix projects inferiorly and posteriorly and enters the anterior wall of the vagina at nearly a right angle (see Fig. 28.11). Several structures that are either extensions of the parietal peritoneum or fibromuscular cords, referred to as ligaments, maintain the position of the uterus.

The paired **broad ligaments** are double folds of parietal peritoneum attaching the uterus to either side of the pelvic cavity (see Fig. 28.12). The paired **uterosacral ligaments,** also peritoneal extensions, lie on either side of the rectum and connect the uterus to the sacrum. The **cardinal (lateral cervical) ligaments** extend below the bases of the broad ligaments between the pelvic wall and the cervix and vagina. The **round ligaments** are bands of fibrous connective tissue between the layers of the broad ligament. They extend from a point on the uterus just below the uterine tubes to a portion of the labia majora of the external genitalia. These ligaments contain smooth muscle, uterine blood vessels, and nerves. Although the ligaments normally maintain the anteflexed position of the uterus, they also afford the uterine body some movement. As a result, the uterus may become malpositioned. A posterior tilting of the uterus is called **retroflexion.**

Histologically, the uterus consists of three layers of tissue: the perimetrium, myometrium, and endometrium (Fig. 28.16). The outer layer, the **perimetrium (serosa),** is part of the visceral peritoneum (see Fig. 28.12a). Laterally, it becomes the broad ligament. Anteriorly, it covers the urinary bladder and forms a shallow pouch, the **vesicouterine** (ves´-i-kō-YOO-ter-in) **pouch** (see Fig. 28.12c). Posteriorly, it covers the rectum and forms a deep pouch, the **rectouterine** (rek-tō-YOO-ter-in) **pouch (pouch of Douglas)**—the lowest point in the pelvic cavity.

The middle layer of the uterus, the **myometrium,** forms the bulk of the uterine wall. This layer consists of three layers of smooth muscle fibers and is thickest in the fundus and thinnest in the cervix. During childbirth, coordinated contractions of the muscles in response to oxytocin from the posterior pituitary gland help expel the fetus from the body of the uterus.

FIGURE 28.16 Histology of the uterus.

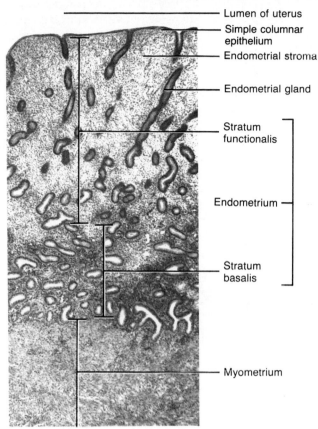

Photomicrograph of portion of
endometrium and myometrium, 25x

Question: What structural features of the endometrium and
myometrium contribute to their functions?

The inner layer of the uterus, the **endometrium,** is
highly vascular and is composed of (1) a surface layer of
simple columnar epithelium (ciliated and secretory cells),
(2) uterine (endometrial) glands that develop as invaginations of the surface epithelium, and (3) endometrial stroma,
a very thick region of lamina propria (connective tissue).
The endometrium is divided into two layers. The **stratum
functionalis (functional layer),** the layer closer to the uterine cavity, is shed during menstruation. The deeper layer,
the **stratum basalis (basal layer),** is permanent. It repeatedly gives rise to a new stratum functionalis after each
menstruation.

Blood is supplied to the uterus by branches of the internal iliac artery called **uterine arteries** (Fig. 28.17).
Branches called **arcuate arteries** are arranged in a circular
fashion in the myometrium and give off **radial arteries** that
penetrate deeply into the myometrium. Just before the
branches enter the endometrium, they divide into two kinds
of arterioles. The **straight arterioles** terminate in the

basalis and supply it with the materials necessary to regenerate the functionalis. The **spiral arterioles** penetrate the
functionalis and change markedly during the menstrual
cycle. The uterus is drained by the **uterine veins.**

PAP SMEAR

Early diagnosis of cancer of the cervix of the uterus is
accomplished by a **Pap smear** (Papanicolaou test). A few
cells are removed from the part of the vagina surrounding
the cervix and the cervix itself and examined microscopically. Malignant cells have a characteristic appearance and
indicate an early stage of cancer, even before symptoms
occur. Females at risk of developing cancer of the uterus
should undergo annual gynecological examinations that
include a Pap smear. Females with a history of normal Pap
smears and no risk factors probably should have a gynecological examination and Pap smear every two to three years
until age 50 and annually thereafter. In all cases, the decision as to how often the test should be done should be
made by a physician in consultation with the patient.

In **endocervical curettage** (ku-re-TAZH), the cervix is
dilated and the endometrium of the uterus is scraped with a
spoon-shaped instrument called a curette. This procedure is
commonly called a **D and C (dilation and curettage).** If
cancer has spread beyond the uterine lining, treatment may
involve removal of the uterus, called a hysterectomy (*hyster*
= uterus), or radiation treatment.

Vagina

The **vagina** serves as a passageway for menstrual flow and
childbirth. It also receives semen from the penis during sexual intercourse. It is a tubular, fibromuscular organ lined
with mucous membrane and measures about 10 cm (4 in.)
in length (see Figs. 28.11 and 28.12). Situated between the
urinary bladder and the rectum, it is directed superiorly and
posteriorly, where it attaches to the uterus. A recess, called
the **fornix** (*fornix* = arch or vault), surrounds the vaginal
attachment to the cervix.

The mucosa of the vagina is continuous with that of the
uterus. Histologically, it consists of stratified squamous
epithelium and connective tissue that lies in a series of
transverse folds, the **rugae.** Dendritic cells in the mucosa
are APCs (antigen presenting cells, see page 701). They
may be involved in the transmission of HIV (the virus that
causes AIDS) to a female during intercourse with an
infected male. The muscularis is composed of smooth muscle that can stretch considerably to receive the penis during
sexual intercourse and allow for birth of a fetus. At the
lower end of the vaginal opening, the **vaginal orifice,** there
may be a thin fold of vascularized mucous membrane called
the **hymen** (*hymen* = membrane). It forms a border around
the orifice, partially closing it (see Fig. 28.18). Sometimes
the hymen completely covers the orifice, a condition called
imperforate (im-PER-fo-rāt) **hymen,** which may require
surgery to open the orifice and permit the discharge of the

FIGURE 28.17 Blood supply of the uterus. The inset shows the histological details of the blood vessels of the endometrium.

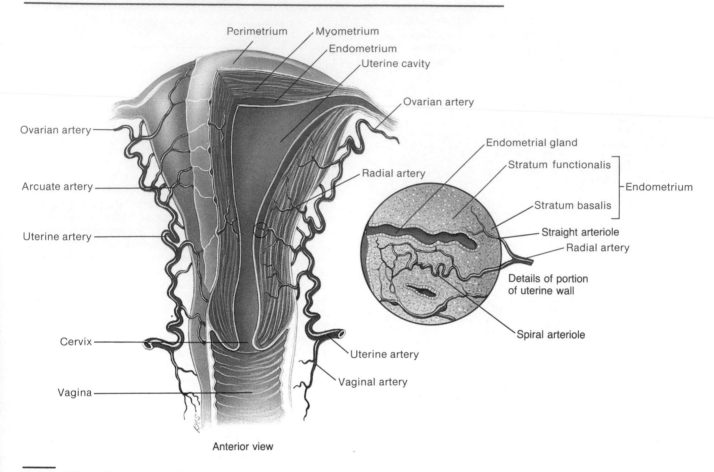

Anterior view

Question: What is the significance of the stratum basalis of the endometrium?

menstrual flow. Normally, the hymen is easily torn and destroyed by the first sexual intercourse.

The mucosa of the vagina contains large stores of glycogen, which upon decomposition produce organic acids. These acids create a low pH environment that retards microbial growth. However, the acidity is also harmful to sperm cells. Alkaline components of semen neutralize the acidity of the vagina and increase viability of the sperm.

CLINICAL APPLICATION

COLPOSCOPY

Colposcopy (kol-POS-ko-pē) is a procedure used to evaluate the status of the mucosa of the vagina and cervix. It is often the first test done after an abnormal Pap smear. Colposcopy is the direct examination of the vaginal and cervical mucosa with a low-power binocular microscope called a colposcope that magnifies the mucous membrane from about 6 to 40 times its actual size. The application of a 3% solution of acetic acid removes mucus and enhances the appearance of mucosal columnar epithelium.

Vulva

The term **vulva** (VUL-va; *volvere* = to wrap around), or **pudendum** (pyoo-DEN-dum), refers to the external genitalia of the female (Fig. 28.18). Its components are as follows.

1. **Mons pubis.** Anterior to the vaginal and urethral openings is the **mons pubis** (MONZ PŪ-bis; *mons* = mountain). It is an elevation of adipose tissue covered by skin and coarse pubic hair that cushions the pubic symphysis during sexual intercourse.
2. **Labia majora.** From the mons pubis, two longitudinal folds of skin, the **labia majora** (LĀ-bē-a ma-JŌ-ra; *labium* = lip), extend inferiorly and posteriorly. Singular is **labium majus.** The labia majora are homologous to the scrotum and are covered by pubic hair. They contain an abundance of adipose tissue and sebaceous (oil) and sudoriferous (sweat) glands.
3. **Labia minora.** Medial to the labia majora are two smaller folds of skin called the **labia minora** (mī-NŌ-

FIGURE 28.18 Components of the vulva (pudendum).

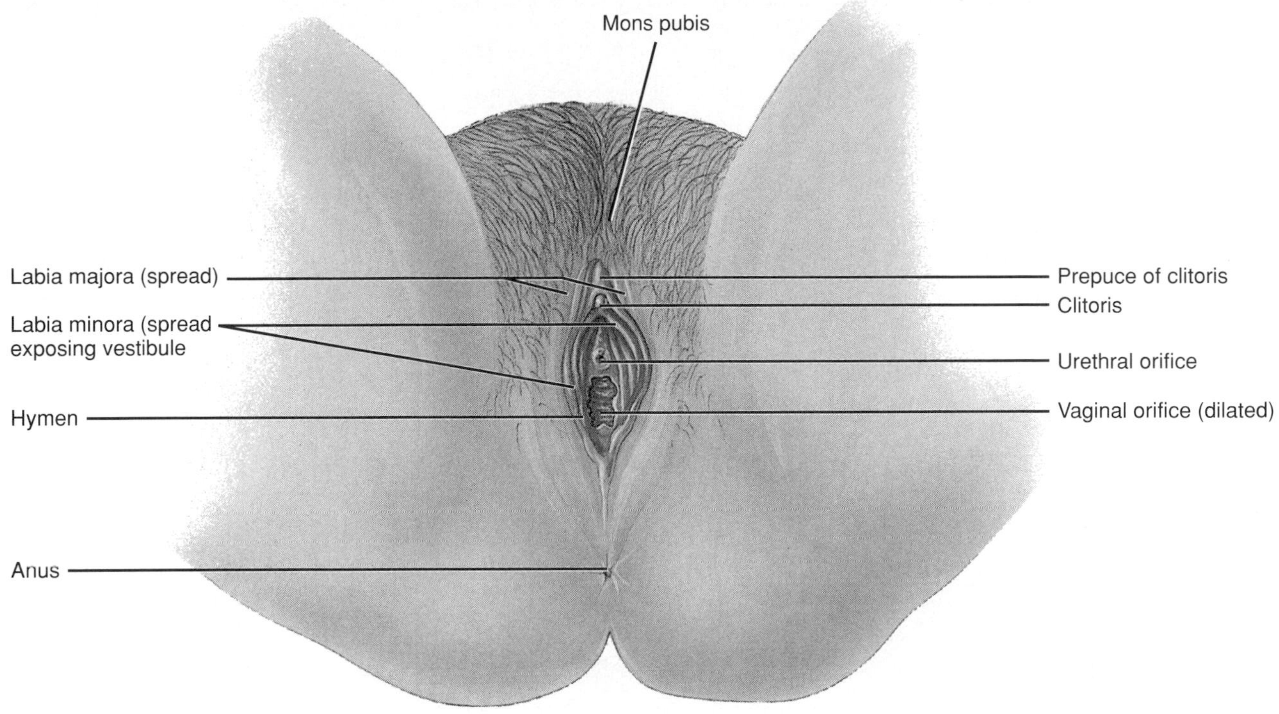

Mons pubis

Labia majora (spread)

Labia minora (spread exposing vestibule

Hymen

Anus

Prepuce of clitoris

Clitoris

Urethral orifice

Vaginal orifice (dilated)

Question: What surface structures are anterior to the vaginal opening? Lateral to it?

ra). Singular is **labium minus**. Unlike the labia majora, the labia minora are devoid of pubic hair and fat and have few sudoriferous (sweat) glands. They do, however, contain many sebaceous (oil) glands.

4. **Clitoris.** The **clitoris** (KLI-to-ris) is a small, cylindrical mass of erectile tissue and nerves. It is located at the anterior junction of the labia minora. A layer of skin called the **prepuce** (foreskin) is formed at the point where the labia minora unite and covers the body of the clitoris. The exposed portion of the clitoris is the **glans.** The clitoris is homologous to the penis of the male and like the penis is capable of enlargement upon tactile stimulation and has a role in sexual excitement of the female.

5. **Vestibule.** The cleft between the labia minora is called the **vestibule.** Within the vestibule are the hymen (if present), vaginal orifice, external urethral orifice, and the openings of several ducts. The **vaginal orifice,** the opening of the vagina to the exterior, occupies the greater portion of the vestibule and is bordered by the hymen. The **bulb of the vestibule** (see Fig. 28.19) consists of two elongated masses of erectile tissue just deep to the labia on either side of the vaginal orifice. The bulb becomes engorged with blood during sexual arousal, narrowing the vaginal orifice and placing pressure on the penis during intercourse. The bulb is homologous to the corpus spongiosum penis and bulb of the penis.

Anterior to the vaginal orifice and posterior to the clitoris is the **external urethral orifice,** the opening of the urethra to the exterior. On either side of the external urethral orifice are the openings of the ducts of the **paraurethral (Skene's) glands.** These glands are embedded in the wall of the urethra and secrete mucus. The paraurethral glands are homologous to the male prostate gland. On either side of the vaginal orifice itself are the **greater vestibular (Bartholin's) glands** (see Fig. 28.19). These glands open by ducts into a groove between the hymen and labia minora and produce a mucus secretion that supplements lubrication during sexual intercourse. The greater vestibular glands are homologous to the male bulbourethral (Cowper's) glands. Several **lesser vestibular glands** also open into the vestibule.

Perineum

The **perineum** (per'-i-NĒ-um) is the diamond-shaped area between the thighs and buttocks of both males and females that contains the external genitals and anus (Fig. 28.19). It is bounded anteriorly by the pubic symphysis, laterally by the ischial tuberosities, and posteriorly by the coccyx. A transverse line drawn between the ischial tuberosities divides the perineum into an anterior **urogenital** (yoo'-rō-

FIGURE 28.19 Perineum. (Figure 11.13 shows the male perineum.)

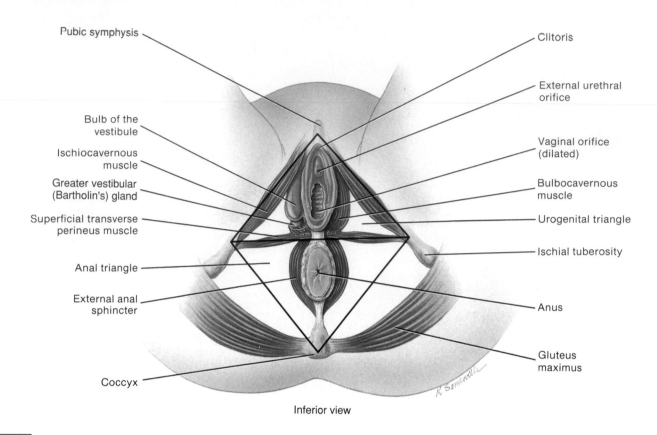

Pubic symphysis

Bulb of the vestibule

Ischiocavernous muscle

Greater vestibular (Bartholin's) gland

Superficial transverse perineus muscle

Anal triangle

External anal sphincter

Coccyx

Clitoris

External urethral orifice

Vaginal orifice (dilated)

Bulbocavernous muscle

Urogenital triangle

Ischial tuberosity

Anus

Gluteus maximus

Inferior view

Question: Why is the anterior portion of the perineum called the urogenital triangle?

JEN-i-tal) **triangle** that contains the external genitalia and a posterior **anal triangle** that contains the anus.

Mammary Glands

Anatomy and Histology

The **mammary glands** are modified sudoriferous (sweat) glands that produce milk. They lie over the pectoralis major and serratus anterior muscles and are attached to them by a layer of connective tissue (Fig. 28.20). Internally, each mammary gland consists of 15 to 20 **lobes,** or compartments, separated by adipose tissue. The amount of adipose tissue determines the size of the breasts. However, breast size has nothing to do with the amount of milk produced. In each lobe are several smaller compartments called **lobules,** composed of connective tissue in which clusters of milk-secreting glands termed **alveoli** are embedded. Alveoli, which are arranged in grapelike clusters, convey the milk (when it is being produced) into a series of **secondary tubules.** From here the milk passes into the **mammary ducts.** As the mammary ducts approach the nipple, they expand to form sinuses called **lactiferous sinuses,** where milk may be stored. The sinuses continue as **lactiferous ducts** that terminate in a pigmented projection, the **nipple.** Each lactiferous duct conveys milk from one of the lobes to the exterior, although some may join before reaching the surface. The circular pigmented area of skin surrounding the nipple is called the **areola** (a-RĒ-ō-la). It appears rough because it contains modified sebaceous (oil) glands. Strands of connective tissue called the **suspensory ligaments of the breast (Cooper's ligaments)** run between the skin and deep fascia and support the breast.

FIGURE 28.20 Mammary glands.

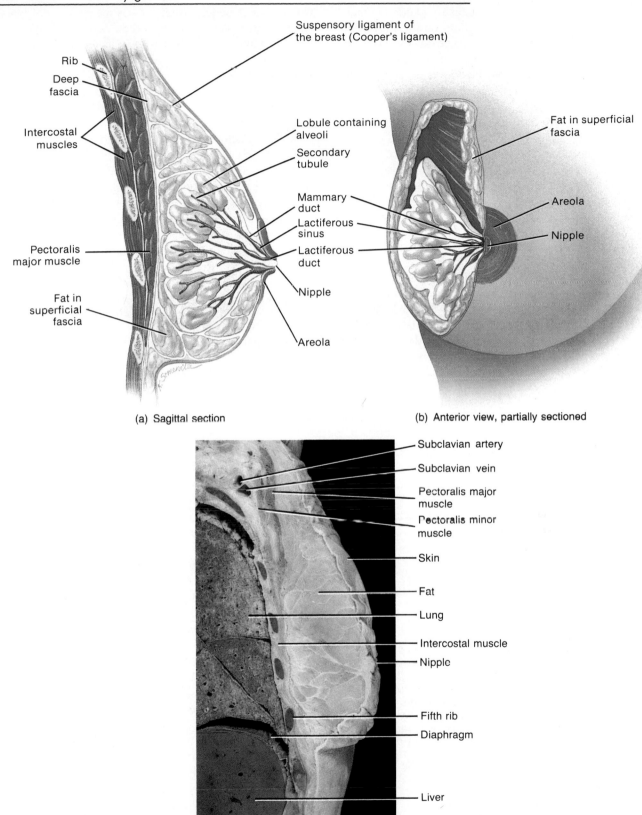

(a) Sagittal section

(b) Anterior view, partially sectioned

(c) Sagittal section

Question: What hormones regulate synthesis and release of milk?

Development

At birth, both male and female mammary glands are undeveloped and appear as slight elevations on the chest. With the onset of puberty, under the influence of estrogens and progesterone, the female breasts begin to develop. The duct system matures, extensive fat deposition occurs, and the areola and nipple grow and become pigmented. Further mammary gland development occurs at sexual maturity with the onset of ovulation and the formation of the corpus luteum. Although the changes in mammary gland development are associated with estrogens and progesterone secretion by the ovaries, ovarian secretion is ultimately controlled by FSH and LH, which are secreted in response to GnRH by the hypothalamus.

Physiology

The essential functions of the mammary glands are milk synthesis, secretion, and ejection, which together are called **lactation.** Milk production is stimulated largely by the hormone prolactin, with contributions from progesterone and estrogens. The ejection of milk occurs in the presence of oxytocin, which is released from the posterior pituitary gland in response to suckling. Lactation is discussed in more detail on page 989.

Breast Cancer

One in nine American women faces the prospect of **breast cancer.** Early detection—by breast self-examination and mammography—is the best way to increase the chance of survival. Each month after the menstrual period the breasts should be thoroughly examined for lumps, puckering of the skin, or nipple retraction or discharge. Even more important is a regular breast exam by a physician and mammograms (breast x-rays) when recommended.

● **Risk Factors** Among the factors that increase the risk of breast cancer development are (1) a family history of breast cancer, especially in a mother or sister; (2) never having a child or having a first child after age 34; (3) previous cancer in one breast; (4) exposure to ionizing radiation, such as x-rays; and (5) excessive fat and alcohol intake; and (6) cigarette smoking. Recent studies in the U.S. show that modern, low-dose birth control pills do not increase a woman's risk of developing breast cancer.

The American Cancer Society recommends the following steps to help diagnose breast cancer as early as possible:

1. A mammogram (see Fig. 28.21) should be taken between the ages of 35 and 39, to be used later for comparison (baseline mammogram).

2. A physician should examine the breasts every three years when a female is between the ages of 20 and 40, and every year after age 40.

FIGURE 28.21 Mammograms.

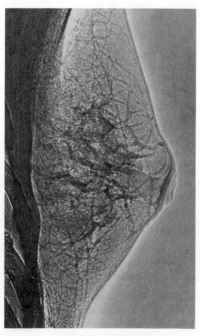

(a) Normal breast

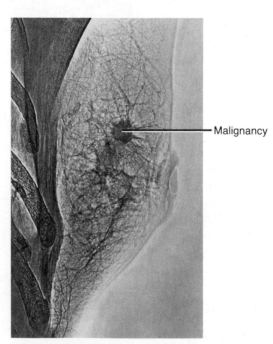

Malignancy

(b) Breast with malignant tumor

3. Females with no symptoms should have a mammogram every year or two between ages 40 and 49, and every year after 50.

4. Females of any age with a history of breast cancer, a strong family history of the disease, or other risk factors should consult a physician to determine a schedule for mammography.

5. All females over 20 should develop the habit of monthly breast self-examination.

● **Detection** The most effective technique for routinely detecting tumors less than 1 cm (0.4 in.) in diameter is **mammography** (mam-OG-ra-fē; *mammae* = breast; *graphein* = to record).

The mammographic image, called a **mammogram** (Fig. 28.21), is obtained by placing the breasts, one at a time, on a flat surface and using a flat plate to compress the breast for better imaging. Usually, two images are taken of each breast, one from the top and one from the side. Breast compression is the key to effective mammography. It increases image detail, decreases blurring and overlap of tissue, and reduces radiation dosage.

A supplementary procedure for evaluating breast abnormalities is **ultrasound (US).** Ultrasound produces images using a device that first emits a pulse of high-frequency sound and then records the echo on a monitor. Although ultrasound cannot detect tumors less than 1 cm in diameter, it can be used to determine whether a lump is a benign, fluid-filled cyst or a solid and therefore possibly malignant tumor.

● **Treatment** Treatment for breast cancer may involve hormone therapy, chemotherapy, radiation therapy, **lumpectomy** (removal of just the tumor and immediate surrounding tissue), a modified or radical mastectomy, or a combination of these. A **radical mastectomy** involves removal of the affected breast along with the underlying pectoral muscles and the axillary lymph nodes. Lymph nodes are removed because metastasis of cancerous cells is usually through lymphatic or blood vessels. Radiation treatment and chemotherapy may follow the surgery to ensure the destruction of any stray cancer cells. By using artificial implants, skin, fat, and muscles from other parts of the body, the breast can be reconstructed after a radical mastectomy. Using these techniques, it is possible to reconstruct a natural-looking breast. In 1991, the Food and Drug Administration (FDA) requested surgeons to stop using breast implants made of silicone gel because of their association to an autoimmune disease called human adjuvant disease (HAD) and risk of carcinogenesis, should the implant rupture.

FEMALE REPRODUCTIVE CYCLE

During their reproductive years, nonpregnant females normally experience a cyclical sequence of changes in the ovaries and uterus. Each cycle takes about a month and involves both oogenesis and preparation of the uterus to receive a fertilized ovum. The principal events all are hormonally controlled. The **ovarian cycle** is a series of events associated with the maturation of an ovum. The **uterine (menstrual) cycle** is a series of changes in the endometrium of the uterus. Each month, the endometrium is prepared for arrival of a fertilized ovum that will develop in the uterus until birth. If fertilization does not occur, the stratum functionalis portion of the endometrium is shed. The general term **female reproductive cycle** encompasses the ovarian and uterine cycles, the hormonal changes that regulate them, and cyclical changes in the breasts and cervix.

Hormonal Regulation

The uterine cycle and ovarian cycle are controlled by gonadotropin releasing hormone (GnRH) from the hypothalamus (Fig. 28.22). GnRH stimulates the release of follicle-stimulating hormone (FSH) and luteinizing hormone (LH) from the anterior pituitary gland. FSH stimulates the initial development of the ovarian follicles and secretion of estrogens by the follicles. LH stimulates the further development of ovarian follicles, brings about ovulation, and stimulates the production of estrogens, progesterone, inhibin, and relaxin by the corpus luteum.

At least six different estrogens have been isolated from the plasma of human females. However, only three are present in significant quantities: *beta (β)-estradiol, estrone,* and *estriol.* In a nonpregnant woman, the principal estrogen is β-estradiol. It is synthesized from cholesterol in the ovaries.

Estrogens have three main functions. (1) They promote development and maintenance of female reproductive structures (especially the endometrial lining of the uterus), secondary sex characteristics, and the breasts. The secondary sex characteristics include fat distribution to the breasts, abdomen, mons pubis, and hips; voice pitch; broad pelvis; and hair pattern. (2) They help control fluid and electrolyte balance. (3) They increase protein anabolism. In this regard, estrogens are synergistic with human growth hormone (hGH). *Moderate* levels of estrogens in the blood inhibit the release of GnRH by the hypothalamus and secretion of LH and FSH by the anterior pituitary gland. This inhibition provides the basis for the action of contraceptive (birth control) pills.

Progesterone works with estrogens to prepare the endometrium for implantation of a fertilized ovum and the mammary glands for milk secretion. High levels of progesterone also inhibit secretion of GnRH and prolactin.

Inhibin is secreted by the corpus luteum of the ovary. It inhibits secretion of FSH and GnRH and, to a lesser extent, LH. Inhibin might be important in decreasing secretion of FSH and LH toward the end of the uterine cycle.

Relaxin is produced in its highest concentrations by the

FIGURE 28.22 Secretion and physiological effects of estrogens, progesterone, relaxin, and inhibin.

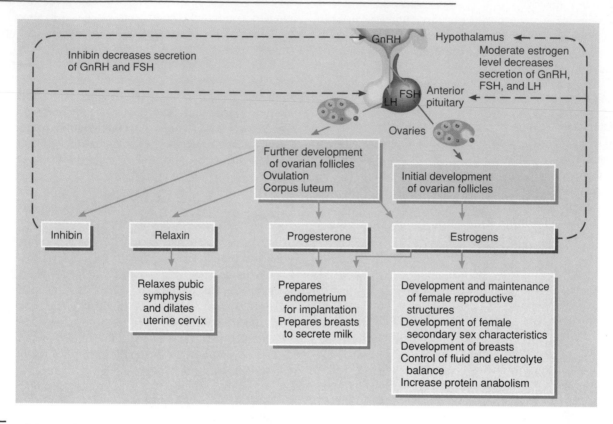

Question: Of the several estrogens, which one exerts the major effects?

corpus luteum and placenta during the last trimester of pregnancy. It relaxes the pubic symphysis and helps dilate the uterine cervix to ease delivery.

Phases of the Female Reproductive Cycle

The duration of the female reproductive cycle normally ranges from 24 to 35 days. For this discussion, we shall assume a duration of 28 days. Events occurring during the cycle may be divided into three phases: the menstrual phase, preovulatory phase, and postovulatory phase (Fig. 28.23).

Menstrual Phase (Menstruation)

The **menstrual** (MEN-stroo-al) **phase**, also called **menstruation** (men′-stroo-Ā-shun) or **menses,** lasts for roughly the first 5 days of the cycle. (By convention, the first day of menstruation marks the first day of a new cycle.) Menstrual flow consists of 50 to 150 ml of blood, tissue fluid, mucus, and epithelial cells derived from the endometrium. This discharge occurs because the declining level of estro-

gens and progesterone causes the uterine spiral arteries to constrict. As a result the cells they supply become ischemic and start to die. Eventually, the entire stratum functionalis sloughs off. At this time the endometrium is very thin because only the stratum basalis remains. The menstrual flow passes from the uterine cavity to the cervix and through the vagina to the exterior.

During the menstrual phase, changes also are occurring in the ovaries (see Fig. 28.13). In response to a rising level of FSH, beginning at about day 25 of the previous cycle, several primordial follicles begin to develop into **primary follicles**. During the early part of each menstrual phase, 20 to 25 primary follicles start to produce very low levels of estrogens. Toward the end of the menstrual phase (days 4 to 5), about 20 of the primary follicles develop into **secondary (growing) follicles**. A secondary follicle consists of a secondary oocyte and several layers of cells formed by division of the single layer of cuboidal epithelial cells around a primary follicle. The epithelial cells are first cuboidal and later columnar and are called *granulosa cells.*

As a secondary follicle continues to grow, it forms a clear glycoprotein layer, called the **zona pellucida** (pe-LOO-si-da), between the secondary oocyte and granulosa cells. Also, the granulosa cells secrete follicular (fō-LIK-

FIGURE 28.23 Correlation of ovarian and uterine cycles with the hypothalamic and anterior pituitary gland hormones. In the cycle shown, fertilization and implantation have not occurred.

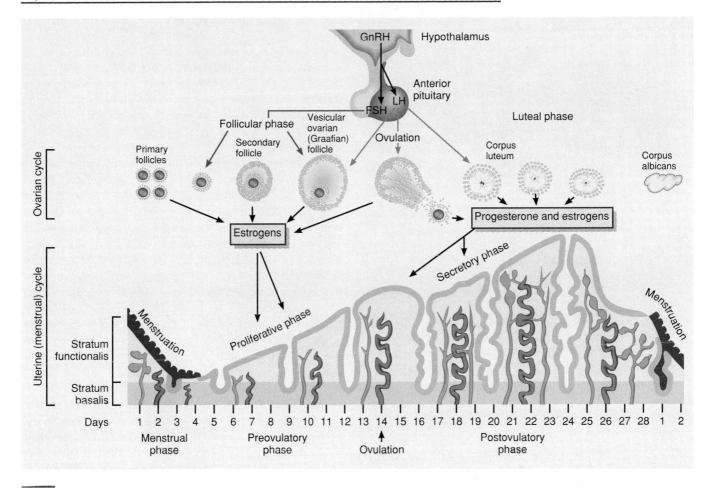

Question: Which hormones stimulate proliferation of the endometrium? Ovulation? Growth of the corpus luteum? The surge of LH at midcycle?

yoo-lar) fluid that forces the secondary oocyte to the edge of the secondary follicle and fills the follicular cavity or antrum.

Preovulatory Phase

The **preovulatory phase,** the second phase of the female reproductive cycle, is the time between menstruation and ovulation. This phase of the cycle is more variable in length than the other phases. It lasts from days 6 to 13 in a 28-day cycle. Although about 20 follicles begin development each cycle, usually only one attains maturity. By about day 6, one of the follicles begins to outgrow all the others. The moderate level of estrogens being secreted by all the growing follicles during the early part of the preovulatory phase causes negative feedback inhibition of FSH secretion. Decreased FSH levels cause the less well-devel-

oped follicles to stop growing and even to degenerate, a process called **atresia.** The dominant follicle by now apparently is secreting enough estrogen to promote its own growth and development.

The one dominant secondary follicle matures into a **vesicular ovarian (Graafian) follicle** or **mature follicle,** a follicle ready for ovulation (see Fig. 28.13). This follicle is visible as a blisterlike bulge on the surface of the ovary. Fraternal (nonidentical) twins may result if two vesicular ovarian follicles form. During the maturation process, the follicle continues to increase its estrogen production. Early in the preovulatory phase, FSH is the dominant gonadotropic hormone secreted by the anterior pituitary. Close to the time of ovulation, however, LH is secreted in increasing quantity (Fig. 28.24). Estrogens are the dominant ovarian hormones before ovulation, but small amounts of progesterone may be produced by the vesicular ovarian (Graafian) follicle a day or two before ovulation.

FIGURE 28.24 Relative concentrations of anterior pituitary gland hormones (FSH and LH) and ovarian hormones (estrogens and progesterone) during a normal female sexual cycle. Note the relationship of the hormones to the ovarian and uterine cycles.

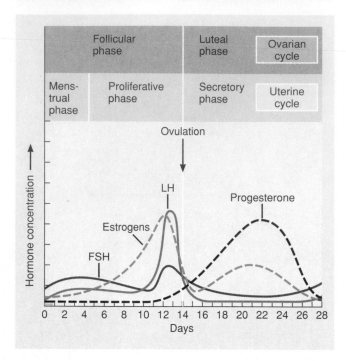

Question: What are the dominant ovarian hormones before ovulation?

Meanwhile, the estrogens being liberated into the blood by growing ovarian follicles stimulate the repair of the endometrium. Cells of the stratum basalis undergo mitosis and produce a new stratum functionalis. As the endometrium thickens, the short, straight endometrial glands develop and the arterioles coil and lengthen as they penetrate the stratum functionalis. The thickness of the endometrium approximately doubles to about 4 to 6 mm.

With reference to the uterus, the preovulatory phase is also termed the **proliferative phase** because the endometrium is proliferating. With reference to the ovaries, the menstrual phase and preovulatory phase together are termed the **follicular phase** because ovarian follicles are growing and developing.

Ovulation

Ovulation, the rupture of the vesicular ovarian (Graafian) follicle with release of the secondary oocyte into the pelvic cavity, usually occurs on day 14 in a 28-day cycle. During ovulation, the secondary oocyte remains surrounded by its zona pellucida and a covering of follicle cells directly around it. These cells are collectively called the **corona radiata.** It generally takes about 20 days (spanning the last

6 days of the previous cycle and the first 14 days of the current cycle) for a primordial follicle to develop into a mature vesicular ovarian (Graafian) follicle. During this time the developing ovum completes reduction division (meiosis I) and reaches metaphase of equatorial division (meiosis II).

At the time of ovulation the secondary oocyte is in metaphase of equatorial division. The fimbriae of the uterine tubes drape over the ovaries and become active near the time of ovulation. Movements of the fimbriae and uterine tube mucosa and ciliary action create currents in the peritoneal serous fluid that carry the secondary oocyte into the uterine tube.

Just before ovulation, the *high* level of estrogens that developed during the latter part of the preovulatory phase exerts a *positive feedback* effect on both LH and GnRH (Fig. 28.25). When estrogen is present in high enough concentration, it stimulates the anterior pituitary to release LH and the hypothalamus to release GnRH, which promotes release of more LH by the anterior pituitary. This sudden surge of LH (see Fig. 28.24) triggers ovulation. An over-the-counter home test that detects the LH surge associated with ovulation is now available. The test predicts ovulation a day in advance. FSH also increases at this time, but not as dramatically as LH because FSH is stimulated only by the increase in GnRH. The positive feedback effect of estrogen on the hypothalamus and anterior pituitary gland does not occur if progesterone is present at the same time.

After ovulation, the vesicular ovarian (Graafian) follicle collapses (and blood within it forms a clot) to become the **corpus hemorrhagicum.** The clot is eventually absorbed by the remaining follicular cells. In time, the follicular cells enlarge, change character, and form the **corpus luteum,** or yellow body, under the influence of LH. Stimulated by LH, the corpus luteum secretes estrogens and progesterone.

CLINICAL APPLICATION

SIGNS OF OVULATION

One **sign of ovulation** is an increase in **basal temperature** (body temperature at rest). An increase in temperature, usually between 0.4 to 0.6°F, typically occurs about 14 days after the start of the last menstrual cycle and is due to increasing levels of progesterone. The 24 hours following this rise in temperature is the period immediately after ovulation and is the best time to become pregnant.

Another sign of ovulation is the amount and consistency of **cervical mucus.** Secretion of cervical mucus is regulated by estrogens and progesterone. At midcycle, near the time of ovulation, increasing levels of estrogens cause secretory cells of the cervix to produce large amounts of cervical mucus. As ovulation approaches, the mucus becomes clear and very stretchy. If grasped with forceps, the mucus may stretch as far as 12 to 15 cm. This type of mucus indicates the time of greatest fertility.

The cervix also exhibits signs of ovulation. The external os opens, the cervix rises, and the cervix becomes softer. Some women also experience a pain in the area of one or both ovaries around the time of ovulation. Such pain is

FIGURE 28.25 Positive feedback effect of *high* levels of estrogens on secretion of GnRH and LH.

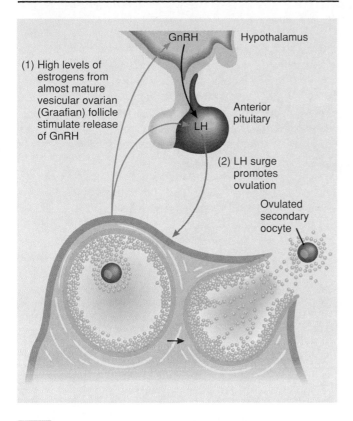

(1) High levels of estrogens from almost mature vesicular ovarian (Graafian) follicle stimulate release of GnRH

GnRH   Hypothalamus

Anterior pituitary

LH

(2) LH surge promotes ovulation

Ovulated secondary oocyte

Question: What is the effect of rising but still moderate levels of estrogens on secretion of GnRH, LH, and FSH?

called **mittelschmerz** (MIT-el-shmarts), meaning "pain in the middle," and may last from several hours to a day or two.

Postovulatory Phase

The **postovulatory phase** of the female reproductive cycle is the most constant in duration and lasts for 14 days, from days 15 to 28 in a 28-day cycle. It represents the time between ovulation and the onset of the next menses. After ovulation, LH secretion stimulates the remnants of the vesicular ovarian follicle to develop into the corpus luteum. The corpus luteum then secretes increasing quantities of estrogens and progesterone. With reference to the ovaries, this phase of the cycle is also called the **luteal phase.**

Progesterone produced by the corpus luteum is responsible for preparing the endometrium to receive a fertilized ovum. Preparatory activities include growth and coiling of the endometrial glands, which begin to secrete glycogen, vascularization of the superficial endometrium, thickening of the endometrium, and an increase in the amount of tissue fluid. These preparatory changes are maximal about one week after ovulation, corresponding to the time of pos-

sible arrival of a fertilized ovum. With reference to the uterus, this phase of the cycle is also called the **secretory phase** because of the secretory activity of the endometrial glands.

If fertilization and implantation do not occur, the simultaneously rising levels of both progesterone and estrogens, secreted by the corpus luteum, inhibit GnRH and LH secretion. As LH level declines, the corpus luteum degenerates and becomes the **corpus albicans,** or white body. The decreased secretion of progesterone and estrogens by the degenerating corpus luteum then initiates another menstrual phase. In addition, the decreasing levels of progesterone and estrogens in the blood bring about a new output of anterior pituitary gland hormones—especially FSH—in response to increased output of GnRH by the hypothalamus. Thus a new ovarian cycle begins. A summary of these hormonal interactions is presented in Fig. 28.26.

If, however, fertilization and implantation do occur, the corpus luteum is maintained until the placenta takes over its hormone-producing functions. During this time, the corpus luteum secretes estrogens and progesterone. The corpus luteum is maintained by **human chorionic** (kō-rē-ON-ik) **gonadotropin (hCG),** a hormone produced by the chorion, which develops into the placenta. The presence of hCG is an indication of pregnancy. The placenta itself secretes estrogens to support pregnancy and progesterone to support pregnancy and breast development for lactation. Once the placenta begins its secretion, the role of the corpus luteum becomes minor.

PHYSIOLOGY OF SEXUAL INTERCOURSE

Sexual intercourse, or **copulation** (in humans, called **coitus),** is the process by which spermatozoa are deposited in the vagina.

CLINICAL APPLICATION

DONOR INSEMINATION

Donor (artificial) insemination (*in* = into; *seminatus* = sown; *semen* = seed) refers to the deposition of seminal fluid into the vagina by a physician at a time during the menstrual cycle when pregnancy is most likely to occur. One risk of artificial insemination is transmission of hepatitis B from semen donors to inseminated females. To decrease the risk of passing on sexually transmitted diseases, guidelines have been established for screening semen donors.

Male Sexual Act
Erection

The male role in the sexual act starts with **erection,** the enlargement and stiffening of the penis. An erection may be initiated by stimuli such as anticipation, memory, and

FIGURE 28.26 Summary of hormonal interactions of the uterine and ovarian cycles.

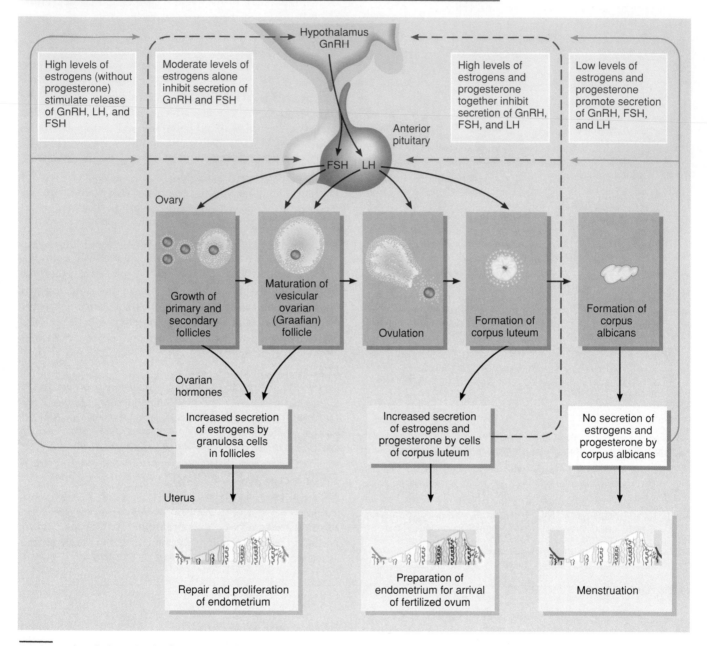

Question: When declining levels of estrogens and progesterone stimulate secretion of GnRH, is this a positive or negative feedback effect? Why?

visual sensation, or it may be a reflex brought on by stimulation of the touch receptors in the penis, especially in the glans penis. In any case, parasympathetic impulses pass from the gray matter of the second, third, and fourth sacral segments of the spinal cord in pelvic splanchnic nerves to the penis. The impulses cause release of nitric oxide that relaxes vascular smooth muscle. This allows the arteries of the penis to dilate and blood fills the blood sinuses of the three corpora. The rise in internal blood pressure of the penis compresses the veins that normally drain the penis. The result is enlargement and rigidity (erection).

Lubrication

Parasympathetic impulses from the sacral spinal cord also cause the bulbourethral (Cowper's) glands and urethral glands (glands of Littré) to secrete mucus, which provides a small amount of **lubrication** for intercourse. The mucus flows through the urethra. The major portion of lubricating fluid is produced by the cervical mucosa of the female. Without satisfactory lubrication, the sexual act is difficult for both male and female. Unlubricated intercourse causes pain impulses that inhibit rather than promote coitus.

Orgasm

Tactile stimulation of the penis brings about ejaculation. When sexual stimulation becomes intense, rhythmic sympathetic impulses leave the spinal cord at the levels of the first and second lumbar vertebrae and pass to the genital organs. These impulses cause peristaltic contractions of the ducts in each testis, epididymis, and ductus (vas) deferens that propel spermatozoa into the urethra. Simultaneously, peristaltic contractions of the seminal vesicles and prostate gland expel seminal and prostatic fluid along with the spermatozoa. All these mix with the mucus of the bulbourethral glands, resulting in the fluid called semen. Other rhythmic impulses sent from the spinal cord at the levels of the first and second sacral vertebrae reach the skeletal muscles at the base of the penis, which contract rhythmically to propel the semen from the urethra to the exterior. The propulsion of semen from the urethra to the exterior constitutes an **ejaculation.** Several sensory and motor activities accompany ejaculation, including a rapid heart rate, an increase in blood pressure, an increase in respiration, and pleasurable sensations. These activities, together with the muscular events involved in ejaculation, are referred to as an **orgasm.**

Female Sexual Act

Erection

The female role in the sex act, like that of the male, also involves erection, lubrication, and orgasm. Stimulation of the female, as in the male, depends on both psychic and tactile responses. Under appropriate conditions, stimulation of the female breasts or genitalia, especially the clitoris, results in widespread sexual arousal. This response is mediated by parasympathetic impulses sent from the sacral spinal cord to the external genitalia.

Lubrication

Parasympathetic impulses from the sacral spinal cord cause most of the **lubrication** of the vagina. The impulses result in the secretion of a mucous fluid from the epithelium of the cervical mucosa. Some mucus is also produced by the greater vestibular (Bartholin's) glands.

Orgasm (Climax)

When tactile stimulation of the genitalia reaches maximum intensity, reflexes are initiated that cause the female **orgasm (climax).** The perineal muscles contract rhythmically from spinal reflexes (sympathetic nerve impulses) similar to those that occur in the male ejaculation. Orgasm is accompanied by an increase in muscle tension throughout the body; engorgement of the clitoris and breasts; and increase in heart rate, pulse rate, and respiratory rate; and

rhythmic contractions of the uterus. Intense pleasurable sensations followed by general relaxation also accompany orgasm.

BIRTH CONTROL

Although there is no single, ideal method of **birth control (BC),** several methods are available, each with advantages and disadvantages. The methods discussed here are sterilization, hormonal, intrauterine, barrier, chemical, physiological, coitus interruptus (withdrawal), and induced abortion.

Sterilization

One means of **sterilization** of males is **vasectomy** (discussed on page 931). Sterilization in females most often is achieved by performing a **tubal ligation** (lī-GĀ-shun), in which the uterine tubes are tied closed and then cut. The ovum is thus prevented from passing to the uterus, and the sperm cannot reach the ovum.

Hormonal Methods

By adjusting hormone levels, it is possible to interfere with production of gametes (sperm and ova) or implantation of a fertilized ovum in the uterus. This may be accomplished by use of **oral contraception (OC)** or **the pill,** which is used by about 28% of American women. Although several types of pills are available, the ones most commonly used contain a high concentration of progesterone and a low concentration of estrogens (combination pill). These two hormones act via negative feedback on the anterior pituitary to decrease the secretion of FSH and LH and on the hypothalamus to inhibit secretion of GnRH. The low levels of FSH and LH usually prevent both follicular development and ovulation; thus pregnancy cannot occur. Even if ovulation does occur, as it does in some cases, oral contraceptives also alter cervical mucus so that it is more hostile to spermatozoa and may make the endometrium less receptive to implantation.

Among the noncontraceptive benefits of oral contraceptives are regulation of the length of menstrual cycles, decreased menstrual flow (and therefore decreased risk of anemia), and prevention of ovarian cysts. The pill also provides protection against endometrial and ovarian cancers. Oral contraceptives may not be advised for women with a history of thromboembolic disorders (predisposition to blood clotting), cerebral blood vessel damage, hypertension, liver malfunction, or heart disease. Women who take the pill and smoke face far higher odds of having a heart attack or stroke than do nonsmoking pill users. Smokers should quit or use an alternative method of birth control.

If daily pill-taking is not desired, a woman may opt for

Norplant, another hormonal method of contraception. Under local anesthesia, six slender hormone-containing capsules are surgically implanted under the skin of the arm. They slowly and continually release progestin (similar to progesterone), which inhibits ovulation and thickens the cervical mucus. The effects last for 5 years, and Norplant is as reliable as sterilization (less than 1% failure rate). Restoration of fertility is as simple as removing the Norplant capsules. Over a five-year period Norplant is less expensive than most birth control pills.

The quest for an efficient male oral contraceptive has been disappointing. The challenge is to find substances that will block production of functional sperm without disrupting the ability to have an erection. Research continues, with the aim of finding agents that (1) inhibit hypothalamic–pituitary function, (2) directly interfere with spermatogenesis, or (3) block the effects of the epididymis on sperm maturation.

Pilot studies with small groups of men have already confirmed the ability of GnRH (gonadotropin-releasing hormone) antagonists (blockers) to halt sperm production. Periodic injections of testosterone maintain the men's ability to get erections. It is predicted that a male contraceptive based on GnRH antagonists will be available by the year 2000.

Intrauterine Devices

An **intrauterine device (IUD)** is a small object made of plastic, copper, or stainless steel that is inserted into the cavity of the uterus. IUDs cause changes in the uterine lining that block implantation of a fertilized ovum. They are more easily tolerated by women who have had a child. The dangers associated with the use of IUDs in some females include pelvic inflammatory disease (PID), infertility, and excess menstrual bleeding and pain. Females in monogamous relationships are at a lesser risk of developing PID and recent research discounts IUDs as a cause of PID. During the 1980s, the use of IUDs declined considerably, in large part because lawsuits resulting from damage claims caused most manufacturers of IUDs in the U.S. to stop production and sales. Some gynecologists feel they are highly effective, safe, and underutilized. An IUD provides a long-term, reversible method of contraception.

Barrier Methods

Barrier methods are designed to prevent spermatozoa from gaining access to the uterine cavity and uterine tubes. Among the barrier methods are use of a condom, vaginal pouch, diaphragm, or cervical cap.

The **condom** is a nonporous, elastic (latex or similar material) covering placed over the penis that prevents deposition of sperm in the female reproductive tract. The **vaginal pouch**, sometimes called a **female condom**, is made of two flexible rings connected by a polyurethane sheath. One ring lies inside the sheath and is inserted to fit over the cervix. The other ring remains outside the vagina and covers the female external genitals. Proper use of condoms with each act of sexual intercourse, especially when used with a spermicide (sperm-killing chemical), is a fairly reliable method of birth control. Condom use also reduces, but does not eliminate, the risk of acquiring a sexually transmitted disease (STD) such as AIDS (see page 712) or syphilis (see page 962). Even when used properly, condoms fail to protect against pregnancy and disease transmission 10% to 20% of the time.

The **diaphragm** is a rubber dome-shaped structure that fits over the cervix and is generally used together with a spermicide. The diaphragm stops the sperm from passing into the cervix. The chemical kills the sperm cells. Toxic shock syndrome (TSS) and recurrent urinary tract infections are associated with diaphragm use in some females, and a diaphragm does not protect against STD.

The **cervical cap** is a thimble-shaped contraceptive device made of latex or plastic that measures about 4 cm (1.5 in.) in diameter. It fits snugly over the cervix of the uterus and is held in position by suction. Like the diaphragm, the cervical cap is used with a spermicide. Also, like the diaphragm, it must be fitted initially by a physician or other trained personnel. Advantages of the cervical cap in comparison to the diaphragm are (1) the cap can be worn up to 48 hours versus 24 hours for the diaphragm, and (2) since the cap fits tightly and rarely leaks, it is not necessary to reintroduce spermicide before intercourse. The cervical cap is not recommended for females with known or suspected cervical or uterine malignancies and current vaginal or cervical infections, and it also does not protect against STDs.

Chemical Methods

Chemical methods of contraception are spermicidal agents. Various foams, creams, jellies, suppositories, and douches that contain spermicidal agents make the vagina and cervix unfavorable for sperm survival and are available without prescription. The most widely used spermicides are nonoxynol-9 and octoxynol-9. Their action is to disrupt plasma membranes of spermatozoa, thus killing them. They are most effective when used with a diaphragm. A recent spermicidal product is a **contraceptive sponge** (Today). It is a nonprescription polyurethane sponge that contains nonoxynol-9. This spermicide decreases the incidence of chlamydia and gonorrhea (described on page 962) but slightly increases the risk of developing vaginal infections caused by *Candida,* a fungus. The sponge is placed in the vagina where it releases spermicide for up to 24 hours and also acts as a physical barrier to sperm. The sponge is equal to the diaphragm in birth control effectiveness.

Physiological Methods

Physiological methods are based on knowledge of certain physiological changes that occur during the menstrual cycle. In females with normal menstrual cycles, especially, these events help to predict on which day ovulation is likely to occur.

The first physiological method, developed in the 1930s, is known as the **rhythm method.** It takes advantage of the fact that a secondary oocyte is fertilizable for only 24 hours and is available only during a period of three to five days in each menstrual cycle. During this time, the couple refrains from intercourse (three days before ovulation, the day of ovulation, and three days after ovulation). The effectiveness of the rhythm method for birth control is poor because few women have absolutely regular cycles.

Another natural family planning system, developed during the 1950s and 1960s, is the **sympto-thermal method.** According to this method, couples are instructed to know and understand certain signs of fertility and infertility. Recall that the signs of ovulation include increased basal body temperature; the production of clear, stretchy cervical mucus; abundant cervical mucus; and pain associated with ovulation (mittelschmerz). If the couple refrains from sexual intercourse when the signs of ovulation are present, the chance of pregnancy is decreased.

Coitus Interruptus (Withdrawal)

Coitus (KŌ-i-tus; *coitio* = coming together) **interruptus** refers to withdrawal of the penis from the vagina just before ejaculation. Failures with this method are due to either failure to withdraw before ejaculation or preejaculatory escape of sperm-containing fluid from the urethra.

Induced Abortion

Abortion refers to the premature expulsion from the uterus of the products of conception. An abortion may be spontaneous (naturally occurring), sometimes called a miscarriage, or induced (intentionally performed). When birth control methods fail to prevent an unwanted pregnancy, **induced abortion** may be performed. Induced abortions may involve vacuum aspiration (suction), infusion of a saline solution, or surgical evacuation (scraping). Also, certain drugs, most notably RU 486, can induce abortion, a so-called nonsurgical abortion.

In 1986, a team of French physicians reported a new approach to birth control. It involves using a drug called **RU 486 (mifepristone)** that blocks the action of progesterone. Progesterone prepares the uterine endometrium for implantation and then maintains the uterine lining after implantation. If progesterone levels fall during pregnancy or if the hormone is inhibited from acting, menstruation occurs, and the embryo is sloughed off along with the uterine lining. RU 486 occupies the endometrial receptor sites for progesterone. In effect, it blocks the access of progesterone to the endometrium and causes a miscarriage. RU 486 can be taken up to five weeks after conception. One side effect of the drug is uterine bleeding. The drug is not available in the U.S. but is used in France and China.

Modified human chorionic gonadotropin (hCG), a hormone produced by the placenta, has shown promise in laboratory animals for birth control. In its modified form, hCG prevents implantation of a fertilized egg or terminates an already established pregnancy. Modified versions of gonadotropin releasing hormone (GnRH) are also being tested as contraceptives that inhibit ovulation.

A summary of methods of birth control is presented in Exhibit 28.1.

EXHIBIT 28.1

SUMMARY OF BIRTH CONTROL (BC) METHODS

Method	Comments
Sterilization	Procedure involving severing each ductus (vas) deferens in males (vasectomy) and both uterine (Fallopian) tubes in females (tubal ligation). Failure rate is less than 1%.[a]
Hormonal	Except for total abstinence or surgical sterilization, hormonal methods are the most effective means of birth control.
	Oral contraceptives, commonly known as "the pill," usually include both an estrogen and progesterone. Side effects include nausea, occasional light bleeding between periods, breast tenderness or enlargement, fluid retention, and weight gain. Pill users may have an increased risk of infertility. Failure rate is about 2%.
	Norplant is a set of implantable, progestin-loaded cylinders that slowly release hormone for five years. Failure rate is less than 1%. Fertility is easily restored by removing the cylinders.

Exhibit continues

EXHIBIT 28.1 (continued)

SUMMARY OF BIRTH CONTROL (BC) METHODS

Method	Comments
Intrauterine Device (IUD)	Small object made of plastic, copper, or stainless steel and inserted into uterus by physician. May be left in place for long periods of time. (Copper T is approved for eight years of use.) Some women cannot use them because of expulsion, bleeding, or discomfort. Not recommended for women who have not had children because uterus is too small and cervical canal too narrow. Failure rate is about 5%.
Barrier	A *condom* is a thin, strong sheath of rubber or similar material worn by male to prevent sperm from entering vagina. A similar device for use by a woman is called a *vaginal pouch*. Failures are caused by the sheath or pouch tearing or slipping off after climax or not putting the sheath on soon enough. Failure rate is about 10 to 20%. A *diaphragm* is a flexible rubber dome inserted into vagina to cover the cervix, providing a barrier to sperm. Usually used with spermicidal cream or jelly. Must be left in place at least six hours after intercourse and may be left in place as long as 24 hours. Must be fitted by physician or other trained personnel and refitted every two years and after each pregnancy. Offers high level of protection if used with spermicide. Occasional failures caused by improper insertion or displacement during sexual intercourse. Failure rate is 20 to 40%. A *cervical cap* is a thimble-shaped latex device that fits snugly over the cervix of the uterus. Used with a spermicide, must be fitted by a physician or other trained personnel. May be left in place for up to 48 hours, and it is not necessary to reintroduce spermicide before sexual intercourse. Failure rate is about 20 to 40%.
Chemical	Sperm-killing chemicals inserted into vagina to coat vaginal surfaces and cervical opening. Provide protection for about one hour. Effective when used alone but significantly more effective when used with diaphragm or condom. The contraceptive sponge is made of polyurethane and releases a spermicide for up to 24 hours. A few cases of toxic shock syndrome (TSS) have been reported among users of the contraceptive sponge. Failure rate is about 20 to 40%.
Physiological	In the *rhythm method,* sexual intercourse is avoided just before and just after ovulation (about seven days). Failure rate is about 20% even in females with regular menses. In the *sympto-thermal method,* signs of ovulation are noted (increased basal body temperature, clear and stretchy cervical mucus, opening of the external os, elevation and softening of the cervix, abundant cervical mucus, and pain associated with ovulation), and sexual intercourse is avoided. Failure rate is about 20%.
Coitus Interruptus	Withdrawal of penis from vagina before ejaculation occurs. Failure rate is 20 to 25%.
Induced Abortion	Surgical or drug-induced removal of products of conception at an early stage from uterus or from uterine tube in cases of tubal pregnancy. Surgical removal from the uterus may involve vacuum aspiration (suction), saline solution, surgical evacuation (scraping).

[a]Failure rates are based on estimated pregnancy rates (percentage) in the first year of use.

AGING AND THE REPRODUCTIVE SYSTEMS

During the first decade of life, the reproductive system is in a juvenile state. At about age 10 hormone-directed changes start to occur in both sexes. **Puberty** (PŪ-ber-tē; *puber* = marriageable age) refers to the period of time when secondary sexual characteristics begin to develop and the potential for sexual reproduction is reached. The factors that initiate puberty are poorly understood, but the sequence of events is well established.

Male Puberty

Male puberty begins at an average age of 10 to 11 years and ends at an average age of 15 to 17. During the prepubertal years, plasma levels of LH, FSH, and testosterone are low. At around age six or seven, a prepubertal growth spurt occurs that is probably related to secretion of adrenal androgens and human growth hormone (hGH).

Sleep-associated surges in LH and, to a lesser extent, FSH signal the onset of puberty. As puberty advances, elevated LH and FSH levels are present throughout the day and are accompanied by increased levels of testosterone.

The rises in LH and FSH are believed to result from increased GnRH secretion and enhanced responsiveness of the anterior pituitary to GnRH. With sexual maturity, the hypothalamic–pituitary system becomes less sensitive to the feedback inhibition of testosterone on LH and FSH secretion.

The changes in the testes that occur during puberty include maturation of sustentacular (Sertoli) cells and initiation of spermatogenesis. The anatomical and functional changes associated with puberty are the result of increased testosterone secretion. Usually, the first sign is enlargement of the testes. About a year later, the penis increases in size. The prostate gland, seminal vesicles, bulbourethral glands, and epididymides increase in size over a period of several years. Development of the secondary sex characteristics occurs and a growth spurt takes place as elevated testosterone levels increase both bone and muscle growth.

Female Puberty

In girls, prepubertal levels of LH, FSH, and estrogens are low. At around age seven or eight, girls experience an increase in the secretion of adrenal androgens (adrenarche), which are responsible for the growth of pubic and axillary hair. The onset of puberty is signaled by sleep-associated surges in LH and FSH. As puberty progresses, LH and FSH levels increase throughout the day. The rising levels stimulate the ovaries to secrete estrogens. These hormones are responsible for the development of the secondary sexual characteristics. (Budding of the breasts is the first observable sign of puberty.) Additionally, the hormones stimulate the growth of the uterine tubes, uterus, and vagina. Also associated with estrogens is menarche, the first menses, which occurs at an average of 12 years of age. As you will see in Chapter 29, a female must have a minimum amount of body fat to begin and maintain a normal menstrual cycle.

Menarche and Menopause

In females, the reproductive cycle normally occurs once each month from **menarche** (me-NAR-kē), the first menses, to **menopause** (*mens* = monthly; *pausis* = to cease), the permanent cessation of menses. Between the ages of 40 and 50 the ovaries become less responsive to the stimulation of gonadotropic hormones from the anterior pituitary gland. As a result, estrogen and progesterone production decline, and follicles do not undergo normal development. Changes in GnRH release patterns and decreased responsiveness to it by cells of the anterior pituitary gland that secrete LH also contribute to the onset of menopause. Some women experience hot flashes, copious sweating, headache, hair loss, muscular pains, vaginal dryness, insomnia, depression, weight gain, and mood swings. In the postmenopausal woman there will be some atrophy of the ovaries, uterine tubes, uterus, vagina, external genitalia, and breasts. Osteoporosis is also a possible occurrence due to the diminished level of estrogens (see page 162). Sexual desire (libido) does not show a parallel decline, perhaps supported by adrenal sex steroids.

The female reproductive system has a time-limited span of fertility between menarche and menopause. Fertility declines with age, possibly as a result of less frequent ovulation and the declining ability of the uterine (Fallopian) tubes and uterus to support the young embryo. Uterine cancer peaks at about 65 years of age, but cervical cancer is more common in younger women, and breast cancer is the second leading cause of death among women between the ages of 40 and 60. (Lung cancer is now the leading cause of death among women in this age group.)

Older Men

In the male, declining reproductive function is much more subtle than in the female. Healthy men often retain reproductive capacity into their 80s or 90s. At about age 55 a decline in testosterone synthesis leads to less muscle strength, fewer viable sperm, and decreased sexual desire. However, abundant spermatozoa may be found even in old age. A common problem of older men is enlargement (hypertrophy) of the prostate gland. As the prostate enlarges, it squeezes the urethra. Thus difficulty in urinating often is an early sign of prostatic enlargement. The enlargement may be benign, but cancer of the prostate (see page 963) causes more than 32,000 deaths per year in the U.S.

DEVELOPMENTAL ANATOMY OF THE REPRODUCTIVE SYSTEMS

The *gonads* develop from the **intermediate mesoderm.** By the sixth week, they appear as bulges that protrude into the ventral body cavity (Fig. 28.27). The gonads develop near the mesonephric ducts. A second pair of ducts, the **paramesonephric (Müller's) ducts,** develop lateral to the mesonephric ducts. Both sets of ducts empty into the urogenital sinus. An embryo contains primitive gonads for both sexes. Differentiation into a male depends on the presence of a master gene on the Y chromosome called **SRY,** which stands for *Sex-determining Region* of the *Y* chromosome, and the release of testosterone. Differentiation into a female depends on the absence of SRY and the absence of testosterone. At about the seventh week, the gonads are clearly differentiated into ovaries or testes.

In the male embryo, the *testes* connect to the mesonephric duct through a series of tubules. These tubules become the *seminiferous tubules.* Continued development

FIGURE 28.27 Development of the internal reproductive systems.

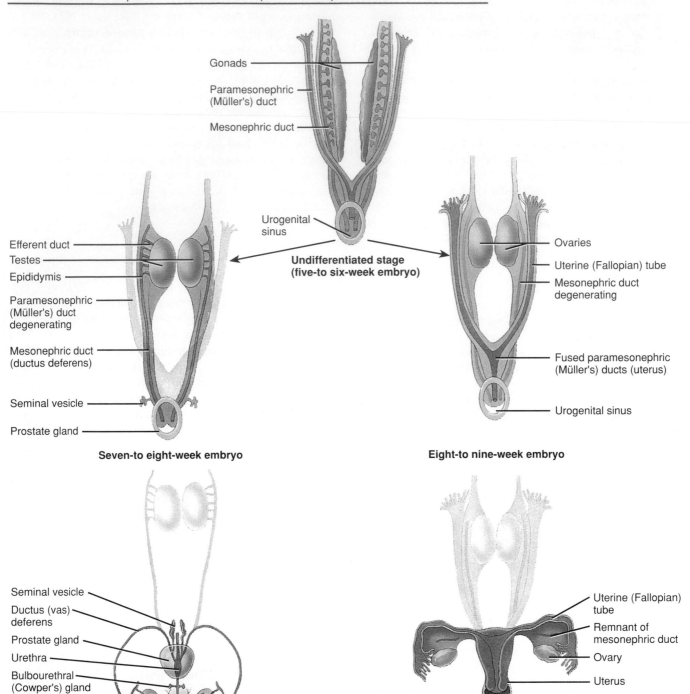

Gonads

Paramesonephric (Müller's) duct

Mesonephric duct

Urogenital sinus

Undifferentiated stage (five-to six-week embryo)

Efferent duct

Testes

Epididymis

Paramesonephric (Müller's) duct degenerating

Mesonephric duct (ductus deferens)

Seminal vesicle

Prostate gland

Seven-to eight-week embryo

Ovaries

Uterine (Fallopian) tube

Mesonephric duct degenerating

Fused paramesonephric (Müller's) ducts (uterus)

Urogenital sinus

Eight-to nine-week embryo

Seminal vesicle

Ductus (vas) deferens

Prostate gland

Urethra

Bulbourethral (Cowper's) gland

Epididymis

Efferent duct

Testis

**At birth
MALE DEVELOPMENT**

Uterine (Fallopian) tube

Remnant of mesonephric duct

Ovary

Uterus

Vagina

**At birth
FEMALE DEVELOPMENT**

of the mesonephric ducts produces the *efferent ducts, ducti epididymides, ducti (vasa) deferentes, ejaculatory ducts,* and *seminal vesicles*. The *prostate* and *bulbourethral (Cowper's) glands* are **endodermal** outgrowths of the urethra. Shortly after the gonads differentiate into testes, the

paramesonephric ducts degenerate without contributing any functional structures to the male reproductive system.

In the female embryo, the gonads develop into *ovaries*. At about the same time, the distal ends of the paramesonephric ducts fuse to form the *uterus* and *vagina*. The unfused portions become the *uterine (Fallopian) tubes.* The *greater (Bartholin's)* and *lesser vestibular glands* develop from **endodermal** outgrowths of the vestibule. The mesonephric ducts in the female degenerate without contributing any functional structures to the female reproductive system.

The *external genitals* of both male and female embryos also remain undifferentiated until about the eighth week. Before differentiation, all embryos have an elevated region, the **genital tubercle.** This is a point between the tail (future coccyx) and the umbilical cord where the mesonephric and paramesonephric ducts open to the exterior (Fig. 28.28).

The tubercle consists of the **urethral groove** (opening into the urogenital sinus), paired **urethral folds,** and paired **labioscrotal swellings.**

In the male embryo, the genital tubercle elongates and develops into a *penis.* Fusion of the urethral folds forms the *spongy (cavernous) urethra* and leaves an opening only at the distal end of the penis, the *external urethral orifice.* The labioscrotal swellings develop into the *scrotum.* In the female, the genital tubercle gives rise to the *clitoris.* The urethral folds remain open as the *labia minora,* and the labioscrotal swellings become the *labia majora.* The urethral groove becomes the *vestibule.*

FIGURE 28.28 Development of the external genitals.

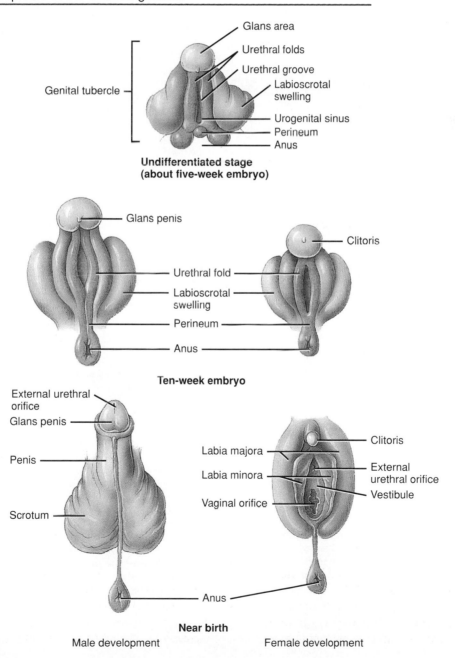

Undifferentiated stage
(about five-week embryo)

Ten-week embryo

Near birth

Male development   Female development

DISORDERS: HOMEOSTATIC IMBALANCES

SEXUALLY TRANSMITTED DISEASES

The general term **sexually transmitted disease (STD)** is applied to any of the large group of diseases that can be spread by sexual contact. The group includes conditions traditionally specified as **venereal diseases (VD)** (from Venus, goddess of love), such as gonorrhea, syphilis, and genital herpes. In most developed countries of the world, such as those of the European Community, Japan, Australia, and New Zealand, the incidence of STDs has declined markedly during the past 20 years. In the U.S. by contrast, STDs have been rising to near epidemic proportions, especially among urban populations. AIDS, which is a sexually transmitted disease that may also be contracted in other ways, is discussed on page 712.

Gonorrhea

Gonorrhea (or **"clap"**) is an infectious sexually transmitted disease that affects primarily the mucous membrane of the urogenital tract, the rectum, and occasionally the eyes. The disease is caused by the bacterium *Neisseria gonorrhoeae*. In the U.S. there are an estimated one to two million new cases of gonorrhea each year. Most cases are in the 15- to 24-year-old age group. Discharges from the involved mucous membranes are the source of infection, and the bacteria are transmitted by direct contact, usually sexual, or during passage of a newborn through the birth canal.

Males usually suffer inflammation of the urethra with pus and painful urination. There also may be involvement of the epididymis and prostate gland. In females, infection may occur in the urethra, vagina, and cervix, often with a discharge of pus. However, infected females may harbor the disease without any symptoms until it has progressed to a more advanced stage. If the uterine (Fallopian) tubes become involved, pelvic inflammation may follow. Peritonitis, or inflammation of the peritoneum, is a life-threatening disorder. The infection should be treated and controlled immediately because, if neglected, sterility or death may result. Although antibiotics have greatly reduced the mortality rate of acute peritonitis, it is estimated that between 50,000 and 80,000 women are made sterile by gonorrhea every year as a result of scar tissue formation that closes the uterine tubes. If the bacteria are transmitted to the eyes of a newborn in the birth canal, blindness can result.

Administration of a 1% silver nitrate solution in the infant's eyes prevents infection. For many years, penicillin and tetracycline were the drugs of choice for the treatment of gonorrhea in adults. Until 1976, all cases of gonorrhea could be effectively treated with penicillin. However, bacterial strains resistant to these antibiotics have become very prevalent since the mid-1980s. Currently, ceftriaxone is the antibiotic that most effectively attacks the majority of gonorrhea bacteria.

Syphilis

Syphilis is a sexually transmitted disease caused by the bacterium *Treponema pallidum*. In the U.S. there are about 100,000 new cases per year. Although the incidence among male homosexuals is 25 to 30% higher than in the general population, significant increases in reported cases among heterosexuals have been observed since 1985. The highest incidence is in the 20- to 39-year-old age group. It is acquired through sexual contact or transmitted through the placenta to a fetus. The disease progresses through several stages. During the **primary stage,** the chief symptom is an open sore, called a **chancre** (SHANKG-ker), at the point of contact. The chancre heals within one to five weeks. From 6 to 24 weeks later, symptoms such as a skin rash, fever, and aches in the joints and muscles usher in the **secondary stage.** These symptoms also eventually disappear (in about 4 to 12 weeks), and the disease ceases to be infectious, but a blood test for the presence of the bacteria generally remains positive. During this "symptomless" period, called the **latent stage,** the bacteria may invade body organs. When signs of organ degeneration appear, the disease is said to be in the **tertiary stage.**

If the organs of the nervous system become involved, the tertiary stage is called **neurosyphilis.** Neurosyphilis may take different forms, depending on the tissue involved. Cerebellar damage is manifested by uncoordinated movements in such activities as writing. As the motor areas become extensively damaged, victims may be unable to control urine and bowel movements. Eventually, they may become bedridden, unable even to feed themselves. Damage to the cerebral cortex produces memory loss and personality changes that range from irritability to hallucinations. AIDS may speed the progression of neurosyphilis, possibly by impairing macrophages and antibody production.

Syphilis can be treated with antibiotics (penicillin) during the primary, secondary, and latent periods. Certain forms of neurosyphilis may also be successfully treated, but the prognosis for others is very poor. Noticeable symptoms do not always appear during the first two stages of the disease. Syphilis, however, is usually diagnosed through a blood test whether noticeable symptoms appear or not. The importance of these blood tests and follow-up treatments cannot be overemphasized.

Genital Herpes

Another sexually transmitted disease, **genital herpes,** is common in the U.S. Each year, between 400,000 and 600,000 new cases are reported. The sexual transmission of the herpes simplex virus, the causative agent of genital herpes, is well established. Unlike syphilis and gonorrhea, which are caused by bacteria and treatable with antibiotics, genital herpes is incurable. Type I herpes simplex virus is the virus that causes the majority of infections above the waist such as cold sores. Type II herpes simplex virus causes most infections below the waist such as painful genital blisters on the prepuce, glans penis, and penile shaft in males and on the vulva or sometimes high up in the vagina in females. The blisters disappear and reappear in most patients, but the virus itself remains in the body.

Treatment of the symptoms involves pain medication, saline compresses, sexual abstinence for the duration of the eruption, and use of an oral drug called acyclovir (Zovirax). This drug interferes with viral DNA replication but not with host cell DNA replication. Acyclovir speeds the healing and sometimes reduces the pain of initial genital herpes infections and shortens the duration of lesions in patients with recurrent genital herpes. A topically applied ointment that contains Inter Vir-A (Immuvir), an antiviral substance, is another drug used to treat genital herpes. Inter Vir-A provides rapid relief for the pain, itching, and burning associated with genital herpes. An experimental genital herpes vaccine will involve human testing shortly.

Chlamydia

Chlamydia (kla-MID-ē-a) is a sexually transmitted disease caused by the bacterium *Chlamydia trachomatis* (*chlamys* = cloak). This unusual bacterium cannot grow outside the body; it cloaks itself inside cells to divide. At present, chlamydia is the most prevalent and one of the most damaging of the sexually transmitted diseases. It affects between three and five million persons annually in the U.S. Unlike gonorrhea, chlamydia afflicts all socioeconomic groups. For example, it is found in up to 5% of female college students and 10% of young men in the military.

In males, urethritis is the principal result. It is characterized by burning on urination, frequency of urination, painful urination, and low back pain. In females, urethritis may spread through the reproductive tract and develop into inflammation of the uterine tubes, which increases the risk of ectopic pregnancy (implantation of a fertilized ovum outside the uterus) and sterility. As in gonorrhea, the organism may be passed from mother to infant during childbirth, infecting the eyes. Treatment consists of the administration of tetracycline or doxycycline.

Trichomoniasis

The microorganism *Trichomonas vaginalis,* a flagellated protozoan (one-celled animal), causes **trichomoniasis,** an inflammation of the mucous membrane of the vagina in females and the urethra in males. *T. vaginalis* is a common inhabitant of the vagina of females and urethra of males. If the normal acidity of the vagina is disrupted, the protozoan may overgrow the normal microbial population and cause trichomoniasis. Symptoms include a yellow vaginal discharge with a particularly offensive odor and severe vaginal itch in women. Men can have it without overt symptoms but can nevertheless transmit it to women. Sexual partners must be treated simultaneously. The drug of choice is metronidazole.

Genital Warts

Warts are an infectious disease caused by viruses. Sexual transmission of **genital warts** is common and is caused by the *human papillomavirus (HPV)*. It is estimated that nearly one million persons a year develop genital warts in the U.S. Patients with a history of genital warts may be at increased risk for certain types of cancer (cervical, vaginal, anal, vulval, and penile). There is no cure for genital warts. Treatment consists of cryotherapy with liquid nitrogen, electrocautery, excision, laser surgery, and topical application of podophyllin in tincture of benzoin. Alpha interferon is also used to treat genital warts.

MALE DISORDERS

Testicular Cancer

Testicular cancer occurs most often between the ages of 15 and 34 and is one of the most common cancers seen in young males. Although the cause is unknown, the condition is associated with males who have a history of undescended testes or late-descended testes. Most testicular cancers arise from the sperm-producing cells. An early sign of testicular cancer is a mass in the testis, often associated with pain or discomfort. All males should perform regular testicular self-exams. Treatment involves removal of the diseased testis.

Prostate Disorders

Because the prostate surrounds the urethra, any infection, enlargement, or tumor can obstruct the flow of urine. Prolonged obstruction may result in serious changes in the urinary bladder, ureters, and kidneys and may perpetuate urinary tract infections. One treatment consists of widening a narrowed urethra with a balloon catheter (balloon urethroplasty). If the obstruction cannot be relieved by other means, the gland may be partially or completely removed. The surgical procedure is called **prostatectomy** (pros′-ta-TEK-tō-mē).

Acute and chronic infections of the prostate gland are common in postpubescent males, often in association with inflammation of the urethra. In **acute prostatitis,** the prostate gland becomes swollen and tender. Appropriate antibiotic therapy, bed rest, and above-normal fluid intake are effective treatment.

Chronic prostatitis is one of the most common chronic infections in men of the middle and later years. On examination, the prostate gland feels enlarged, soft, and very tender, and its surface outline is irregular. This disease often produces no symptoms, but the prostate is believed to harbor infectious microorganisms responsible for some allergic conditions, arthritis, and inflammation of nerves (neuritis), muscles (myositis), and the iris (iritis).

An enlarged prostate gland, two to four times the normal size, occurs in approximately one-third of all males over age 60. The condition is called **benign prostatic hyperplasia (BPH)** and is characterized by nocturia (bed-wetting), hesitancy in urination, decreased force of urinary stream, postvoiding dribbling, and a sensation of incomplete emptying. Surgical correction is possible by a procedure called **transurethral resection of the prostate (TURP),** in which pieces of the gland are removed using a special cystoscope inserted into the urethra. Enlargement usually can be detected by a **digital rectal exam,** in which the physician palpates the prostate through the rectum with the fingers (digits).

Prostate cancer is the leading cause of death from cancer in men in the U.S. (having surpassed lung cancer in 1991) and it is responsible for about 32,000 deaths annually. Both benign and malignant growths are common in elderly men. Both types of tumors put pressure on the urethra, making urination painful and difficult. At times, the excessive back pressure destroys kidney tissue and increases susceptibility to infection. Therefore, even when the tumor is benign, surgery is indicated. Prostate cancer may be detected by digital rectal examination and fine-needle aspiration. A procedure called **transrectal ultrasonography** can detect tumors as small as a grain of rice. In the procedure, a rectal probe is used to bounce sound waves off the prostate gland. The returning echoes are converted into an image that can be viewed on a monitor and printed on paper. Treatment for prostate cancer may involve surgery, radiation, hormonal therapy, and chemotherapy.

Sexual Functional Abnormalities

Impotence (*impotenia* = lack of strength) is the inability of an adult male to ejaculate or to attain or hold an erection long enough for sexual intercourse. Most cases of impotence are thought to be caused by insufficient release of nitric oxide. Other causes include diabetes mellitus, physical abnormalities of the penis, systemic disorders such as syphilis, vascular disturbances (arterial or venous obstructions), neurological disorders, testosterone deficiency, or drugs (alcohol, antidepressants, antihistamines, antihypertensives, narcotics, nicotine, and tranquilizers). Psychic factors such as fear of causing pregnancy, fear of sexually transmitted diseases, religious inhibitions, and emotional immaturity may also cause impotence.

In some impotent men, penile implants may be helpful. Penile injections of papaverine (Pavabid), a vasodilator, and

Continues

phentolamine mesylate (Regitine), an alpha-adrenergic blocker, can produce excellent effects in overcoming both physical and psychological impotence.

Male infertility (sterility) is an inability to fertilize a secondary oocyte. It does not imply impotence. Male fertility requires production of adequate quantities of viable, normal spermatozoa by the testes, unobstructed transportation of sperm through the ducts, and satisfactory deposition in the vagina. The tubules of the testes are sensitive to many factors—x-rays, infections, toxins, malnutrition, and significantly higher-than-normal scrotal temperatures—that may cause degenerative changes and produce male sterility. If inadequate sperm production is suspected, a sperm analysis should be performed.

FEMALE DISORDERS

Menstrual Abnormalities

Because menstruation reflects not only the health of the uterus but also the health of the endocrine glands that control it, the ovaries and the pituitary gland, disorders of the female reproductive system often involve menstrual disorders.

Amenorrhea (ā-men′-ō-RĒ-a; *a* = without; *men* = month; *rhein* = to flow) is the absence of menstruation. If a woman has never menstruated, the condition is called **primary amenorrhea.** Primary amenorrhea can be caused by endocrine disorders, most often in the pituitary gland and hypothalamus, or by a genetically caused abnormal development of the ovaries or uterus. **Secondary amenorrhea,** the skipping of one or more periods, is commonly experienced by women at some time during their lives. Changes in body weight, either gains or losses, often cause amenorrhea. Obesity may disturb ovarian function. Similarly, extreme weight loss, for example, in anorexia nervosa, often leads to a suspension of menstrual flow. When amenorrhea is unrelated to weight, analysis of levels of estrogens often reveals deficiencies of pituitary and ovarian hormones. Amenorrhea may also be caused by rigorous athletic training.

Dysmenorrhea (dis′-men-ō-RĒ-a; *dys* = difficult) refers to pain associated with menstruation and is usually reserved to describe an individual with menstrual symptoms that are severe enough to prevent her from functioning normally for one or more days each month. **Primary dysmenorrhea** is painful menstruation with no detectable organic disease. The pain of primary dysmenorrhea is thought to result from uterine contractions, probably associated with uterine muscle ischemia and prostaglandins produced by the uterus.

Prostaglandins are known to stimulate uterine contractions, but they cannot do so in the presence of high levels of progesterone. As we have noted earlier, progesterone levels are high during the last half of the menstrual cycle. During this time, prostaglandins are apparently inhibited by progesterone from producing uterine contractions. However, if pregnancy does not occur, progesterone levels drop rapidly and prostaglandin production increases. This causes the uterus to contract and slough off its lining and may result in dysmenorrhea. Besides pain, other signs and symptoms may include headache, nausea, diarrhea or constipation, and urinary frequency. Primary dysmenorrhea is less of a problem after pregnancy and vaginal delivery, perhaps because of enlargement of the cervical canal. Drugs that inhibit prostaglandin synthesis (naproxen and ibuprofen) are used to treat primary dysmenorrhea.

Secondary dysmenorrhea is painful menstruation that is frequently associated with a pelvic pathology. Some cases are caused by uterine tumors, ovarian cysts, pelvic inflammatory disease (PID), endometriosis, and intrauterine devices (IUDs). Treatment is aimed at correction of the underlying cause.

Abnormal uterine bleeding includes menstruation of excessive duration or excessive amount, diminished menstrual flow, too frequent menstruation, intermenstrual bleeding, and postmenopausal bleeding. These abnormalities may be caused by disordered hormonal regulation, emotional factors, fibroid tumors of the uterus, and systemic diseases.

Premenstrual syndrome (PMS) refers to severe physical and emotional distress occurring late in the postovulatory phase of the menstrual cycle and sometimes overlapping with menstruation. Signs and symptoms usually increase in severity until the onset of menstruation and then dramatically disappear. Among the signs and symptoms are edema, weight gain, breast swelling and tenderness, abdominal distension, backache, joint pain, constipation, skin eruptions, fatigue and lethargy, greater need for sleep, depression or anxiety, irritability, mood swings, headache, poor coordination and clumsiness, and cravings for sweet or salty foods. The basic cause of PMS is unknown. Although PMS is related to the cyclic production of ovarian hormones, the symptoms are not directly due to changes in the levels of these hormones. Treatment is individualized, depending on the type and severity of symptoms, and may include dietary changes, exercise, over-the-counter drugs (aspirin or acetaminophen), psychoactive drugs (sedatives, tranquilizers, and antidepressants), prostaglandins (Ponstel), diuretics, hormone therapy (progesterone), and vitamin B_6. Management of PMS involves medical, psychological, and social support that may include education of the patient and her family; elimination of fears or inappropriate beliefs regarding the menstrual cycle; alteration in coping style; change in life-style, occupation, or family relationships; and use of appropriate medications.

Toxic Shock Syndrome

Toxic shock syndrome (TSS), first described in 1978, is primarily a disease of previously healthy, young, menstruating females who use tampons. It is also recognized in males, children, and nonmenstruating females. Clinically, TSS is characterized by high fever up to 40.6°C (105°F), sore throat or very tender mouth, headache, fatigue, lethargy, memory loss, hypotension, irritability, muscle soreness and tenderness, conjunctivitis, diarrhea and vomiting, abdominal pain, vaginal irritation, and erythematous rash.

Toxin-producing strains of the bacterium *Staphylococcus aureus* are necessary for development of the disease. The risk is greatest in females who use highly absorbent tampons. TSS can also occur as a complication of influenza and influenza-like illness and use of contraceptive sponges. Initial therapy is directed at correcting all homeostatic imbalances as quickly as possible. Anti-staphylococcal antibiotics, such as penicillin or clindamycin, are also administered.

Ovarian Cysts

Ovarian cysts are fluid-containing sacs within the ovary. Follicular cysts may occur in the ovaries of elderly women, in ovaries that have inflammatory diseases, and in menstruating females. They have thin walls and contain a serous albuminous material. Cysts may also arise from the corpus luteum or the endometrium.

Endometriosis

Endometriosis (en'-dō-mē-trē-Ō-sis; *endo* = within; *metri* = uterus; *osis* = condition) is characterized by the growth of endometrial tissue outside the uterus. The tissue enters the pelvic cavity via the open uterine tubes and may be found in any of several sites—on the ovaries, rectouterine pouch, surface of the uterus, sigmoid colon, pelvic and abdominal lymph nodes, cervix, abdominal wall, kidneys, and urinary bladder. One theory for the development of endometriosis is that there is regurgitation of menstrual flow through the uterine tubes. Another theory is that migrational events during embryonic development are somehow altered. Endometriosis is common in women 25 to 40 years of age who have not had children. Symptoms include premenstrual pain or unusual menstrual pain. The unusual pain is caused by the displaced tissue sloughing off at the same time the normal uterine endometrium is being shed during menstruation. Infertility can be a consequence. Treatment usually consists of hormone therapy, modified GnRH (nafarelin), videolaseroscopy (laparoscope with camera and laser), or conventional surgery. Endometriosis disappears at menopause or when the ovaries are removed.

Female Infertility

Female infertility, or the inability to conceive, occurs in about 10% of married females in the U.S. It may be caused by ovarian disease, tubal obstruction, and certain conditions of the uterus. An upset in hormone balance, so that the endometrium is not adequately prepared to receive the fertilized ovum, may also be the problem. Some research suggests that an autoimmune disease might underlie many cases of infertility. Infertility treatment may involve the use of fertility drugs, donor (artificial) insemination, or surgery. Gynecologists are now using a procedure called **transcervical balloon tuboplasty** to clear obstructions in the uterine tubes. The technique, borrowed from the cardiology procedure to unclog coronary arteries, consists of inserting a catheter through the cervix of the uterus and into the uterine tube. Then a balloon is inflated, compressing the obstruction.

Disorders Involving the Breasts

The breasts of females are highly susceptible to cysts and tumors. Men are also susceptible to breast tumors, but breast cancers are 100 times more common in women.

In females, **fibrocystic disease** is the most common cause of a breast lump in which one or more cysts (fluid-filled sacs) and thickening of alveoli (clusters of milk-secreting glands) develop. The condition occurs mainly in females between the ages of 30 and 50 and is probably due to a hormonal imbalance; a relative excess of estrogens or deficiency of progesterone in the postovulatory (luteal) phase of the menstrual cycle may be responsible. Fibrocystic disease usually causes one or both breasts to become lumpy, swollen, and tender about a week or so before a menstrual cycle begins. The cysts may be aspirated to relieve the pain. Medical management may involve administration of progesterone, antiestrogens, prolactin inhibitors, and pituitary gonadotropin-inhibiting agents.

Benign **fibroadenoma** is a common tumor of the breast. It occurs most often in young women. Fibroadenomas have a firm rubbery consistency and are easily moved about within the mammary tissue. The usual treatment is excision of the growth. The breast itself is not removed.

Breast cancer has one of the highest fatality rates of all cancers affecting women, but it is rare in men. In females, breast cancer is rarely seen before age 30, and its occurrence rises rapidly after menopause. Breast cancer is generally not painful until it becomes quite advanced, so often it is not discovered early or, if noted, is ignored. Any lump, no matter how small, should be reported to a physician at once. New evidence links some breast cancers to loss of protective anti-oncogenes.

Cervical Cancer

Another common disorder of the female reproductive tract is **cervical cancer,** carcinoma of the cervix of the uterus. The condition starts with **cervical dysplasia** (dis-PLĀ-sē-a), a change in the shape, growth, and number of the cervical cells. If the condition is minimal, the cells may regress to normal. If it is severe, it may progress to cancer. Cervical cancer may be detected in most cases in its earliest stages by a Pap smear. There is some evidence linking cervical cancer to penile virus (papillomavirus) infections of male sexual partners. Depending on the progress of the disease, treatment may consist of excision of lesions, radiotherapy, chemotherapy, and hysterectomy.

Pelvic Inflammatory Disease

Pelvic inflammatory disease (PID) is a collective term for any extensive bacterial infection (primarily involving *Chlamydia trachomatis, Neisseria gonorrhoeae, Bacteroides, Peptostreptococcus,* and *Gardnerella vaginalis*) of the pelvic organs, especially the uterus, uterine tubes, or ovaries. A vaginal or uterine infection may spread into the uterine tube **(salpingitis)** or even farther into the abdominal cavity, where it infects the peritoneum **(peritonitis).** Diagnosis of PID depends on three findings: abdominal tenderness; cervical tenderness; and ovarian, uterine tube, and uterine ligament tenderness. In addition, diagnosis is based on at least one of the following: fever, leukocytosis, pelvic abscess or inflammation, purulent cervical discharge, and the presence of certain bacteria in smears. Early treatment with bed rest and antibiotics (cefoxitin, penicillin, tetracycline, doxycycline) can stop the spread of PID.

Vulvovaginal Candidiasis

Candida albicans is a yeastlike fungus that commonly grows on mucous membranes of the gastrointestinal and genitourinary tracts. The organism is responsible for **vulvovaginal candidiasis,** the most common form of vaginitis. It is characterized by severe itching; a thick, yellow, cheesy discharge; a yeasty odor; and pain. The disorder, experienced at least once by about 75% of females, is usually a result of proliferation of the fungus following antibiotic therapy for another condition. Predisposing conditions include use of oral contraceptives, cortisone-like medications, pregnancy, and diabetes. Treatment is by topical (clostrinazole) or oral (ketoconazole) antibiotics.

MEDICAL TERMINOLOGY

Castration (kas-TRĀ-shun; *castrare* = to prune) Removal, inactivation, or destruction of the testes.

Colpotomy (kol-POT-ō-mē; *colp* = vagina; *tome* = cutting) Incision of the vagina.

Culdoscopy (kul-DOS-kō-pē; *skopein* = to examine) A procedure in which a culdoscope (endoscope) is used to view the pelvic cavity. The approach is through the vagina.

Hermaphroditism (her-MAF-rō-dī-tizm′) Presence of both male and female sex organs in one individual.

Hypospadias (hī′-pō-SPĀ-dē-as; *hypo* = below; *span* = to draw) A displaced urethral opening. In the male, the opening may be on the underside of the penis, at the penoscrotal junction, between the scrotal folds, or in the perineum. In the female, the urethra opens into the vagina.

Leukorrhea (loo′-kō-RĒ-a; *leuco* = white; *rrhea* = discharge) A nonbloody vaginal discharge that may occur at any age and affects most women at some time.

Oophorectomy (ō′-of-ō-REK-tō-mē; *oophoro* = bearing eggs) Removal of the ovaries.

Salpingectomy (sal′-pin-JEK-tō-mē; *salpingo* = tube) Excision of a uterine (Fallopian) tube.

Smegma (SMEG-ma; *smegma* = soap) The secretion, consisting principally of desquamated epithelial cells, found chiefly about the external genitalia and especially under the foreskin of the male.

Vaginitis (vaj′-i′-NĪ-tis) Inflammation of the vagina.

Study Outline

Male Reproductive System (p. 923)

1. Reproduction is the process by which new individuals of a species are produced and the genetic material is passed from generation to generation.
2. The organs of reproduction are grouped as gonads (produce gametes), ducts (transport and store gametes), and accessory sex glands (produce materials that support gametes).
3. The male structures of reproduction include the testes, ductus epididymis, ductus (vas) deferens, ejaculatory duct, urethra, seminal vesicles, prostate gland, bulbourethral (Cowper's) glands, and penis.

Scrotum (p. 924)

1. The scrotum is a cutaneous outpouching of the abdomen that supports the testes.
2. It regulates the temperature of the testes by contraction of the cremaster muscle, which elevates them and brings them closer to the pelvic cavity or relaxes causing testes to move farther from the pelvic cavity.

Testes (p. 924)

1. The testes are oval-shaped glands (gonads) in the scrotum containing seminiferous tubules, in which sperm cells are made; sustentacular (Sertoli) cells, which nourish sperm cells and secrete inhibin; and interstitial endocrinocytes (interstitial cells of Leydig), which produce the male sex hormone testosterone.
2. Failure of the testes to descend is called cryptorchidism.
3. Ova and sperm are collectively called gametes, or sex cells, and are produced in gonads.
4. Somatic cells divide by mitosis, the process in which each daughter cell receives the full complement of 23 chromosome pairs (46 chromosomes). Somatic cells are said to be diploid (2*n*).
5. Germ cells divide by meiosis, in which the pairs of chromosomes are split so that the gamete has only 23 chromosomes. It is said to be haploid (*n*).
6. Spermatogenesis occurs in the testes. It results in the formation of four haploid spermatozoa from each primary spermatocyte.
7. Spermatogenesis is a process in which immature spermatogonia develop into mature spermatozoa. The spermatogenesis sequence includes reduction division (meiosis I), equatorial division (meiosis II), and spermiogenesis.
8. Mature spermatozoa consist of a head, midpiece, and tail. Their function is to fertilize a secondary oocyte.
9. At puberty, gonadotropin releasing hormone (GnRH) stimulates anterior pituitary gland secretion of FSH and LH. FSH initiates spermatogenesis, and LH assists spermatogenesis and stimulates production of testosterone.
10. Testosterone controls the growth, development, and maintenance of sex organs; stimulates bone growth, protein anabolism, and sperm maturation; and stimulates development of male secondary sex characteristics.
11. Inhibin is produced by sustentacular (Sertoli) cells. Its inhibition of FSH helps to regulate the rate of spermatogenesis.

Ducts (p. 930)

1. The duct system of the testes includes the seminiferous tubules, straight tubules, and rete testis.
2. Sperm are transported out of the testes through the efferent ducts.
3. The ductus epididymis is the site of sperm maturation and storage.
4. The ductus (vas) deferens stores sperm and propels them toward the urethra during ejaculation.
5. Cutting of the ductus (vas) deferens to prevent fertilization is called vasectomy.
6. The ejaculatory ducts are formed by the union of the ducts from the seminal vesicles and ducti (vasa) deferentes and eject spermatozoa into the prostatic urethra.
7. The male urethra is subdivided into three portions: prostatic, membranous, and spongy (cavernous).

Accessory Sex Glands (p. 933)

1. The seminal vesicles secrete an alkaline, viscous fluid that constitutes about 60% of the volume of semen and contributes to sperm viability.
2. The prostate gland secretes a slightly acidic fluid that constitutes about 13 to 33% of the volume of semen and contributes to sperm motility.
3. The bulbourethral (Cowper's) glands secrete mucus for lubrication and an alkaline substance that neutralizes acid.
4. Semen is a mixture of spermatozoa and accessory sex gland secretions that provides the fluid in which spermatozoa are transported, provides nutrients, and neutralizes the acidity of the male urethra and female vagina.

Penis (p. 935)

1. The penis is the male organ of copulation that consists of a root, body, and glans penis.
2. Expansion of its blood sinuses under the influence of sexual excitation is called erection.

Female Reproductive System (p. 936)

1. The female organs of reproduction include the ovaries (gonads), uterine (Fallopian) tubes, uterus, vagina, and vulva.
2. The mammary glands are considered part of the reproductive system.

Ovaries (p. 936)

1. The ovaries are female gonads located in the upper pelvic cavity, on either side of the uterus.
2. They produce secondary oocytes, discharge secondary oocytes (ovulation), and secrete estrogens, progesterone, inhibin, and relaxin.
3. Oogenesis occurs in the ovaries. It results in the formation of a single haploid secondary oocyte.
4. The oogenesis sequence includes reduction division (meiosis I), equatorial division (meiosis II), and maturation.

Uterine (Fallopian) Tubes (p. 941)

1. The uterine (Fallopian) tubes transport ova from the ovaries to the uterus and are the normal sites of fertilization.
2. Ciliated cells and peristaltic contractions help move a secondary oocyte toward the uterus.

Uterus (p. 942)

1. The uterus is an organ the size and shape of an inverted pear that functions in transporting spermatozoa, menstruation, implantation of a fertilized ovum, development of a fetus during pregnancy, and labor.
2. The uterus is normally held in position by a series of ligaments.
3. Histologically, the uterus consists of an outer perimetrium, middle myometrium, and inner endometrium.

Vagina (p. 943)

1. The vagina is a passageway for spermatozoa and the menstrual flow, the receptacle of the penis during sexual intercourse, and the lower portion of the birth canal.
2. It is capable of considerable distension.

Vulva (p. 944)

1. The vulva is a collective term for the external genitals of the female.
2. It consists of the mons pubis, labia majora, labia minora, clitoris, vestibule, vaginal and urethral orifices, hymen, bulb of

the vestibule, and the paraurethral (Skene's), greater vestibular (Bartholin's), and lesser vestibular glands.

Perineum (p. 945)

1. The perineum is a diamond-shaped area at the inferior end of the trunk between the thighs and buttocks.
2. An incision in the perineal skin before delivery is called an episiotomy.

Mammary Glands (p. 946)

1. The mammary glands are modified sweat glands over the pectoralis major muscles. Their function is to synthesize, secrete, and eject milk (lactation).
2. Mammary gland development depends on estrogens and progesterone.
3. Milk secretion is mainly due to prolactin, and milk ejection is stimulated by oxytocin.

Female Reproductive Cycle (p. 949)

1. The function of the menstrual cycle is to prepare the endometrium each month for the reception of a fertilized egg.
2. The menstrual and ovarian cycles are controlled by GnRH from the hypothalamus, which stimulates the release of FSH and LH by the anterior pituitary gland.
3. FSH stimulates the initial development of ovarian follicles and secretion of estrogens by the ovaries. LH stimulates further development of ovarian follicles, ovulation, and the secretion of estrogens and progesterone by the ovaries.
4. Estrogens stimulate the growth, development, and maintenance of female reproductive structures; stimulate the development of secondary sex characteristics; regulate fluid and electrolyte balance; and stimulate protein synthesis.
5. Progesterone works with estrogens to prepare the endometrium for implantation and the mammary glands for milk synthesis.
6. Relaxin relaxes the pubic symphysis and helps dilate the uterine cervix to facilitate delivery, and increases sperm motility.
7. During the menstrual phase, the stratum functionalis of the endometrium is shed, discharging blood, tissue fluid, mucus, and epithelial cells.
8. During the follicular phase, primary follicles develop into secondary follicles and a secondary follicle develops into a vesicular ovarian (Graafian) follicle. Estrogens are the dominant ovarian hormones.
9. During the proliferative phase, endometrial repair occurs.
10. Ovulation is the rupture of a vesicular ovarian (Graafian) follicle and the release of an immature ovum into the pelvic cavity brought about by a surge of LH. Signs of ovulation include increased basal body temperature; clear, stretchy cervical mucus; changes in the uterine cervix; and ovarian pain.
11. During the secretory phase, the endometrium thickens in readiness for implantation.
12. During the luteal phase both estrogen and progesterone are secreted in large quantity by the corpus luteum.
13. If fertilization and implantation do not occur, the corpus luteum degenerates, and low levels of estrogens and progesterone initiate another uterine and ovarian cycle.
14. If fertilization and implantation do occur, the corpus luteum is maintained by placental hCG, and the corpus luteum and placenta secrete estrogens and progesterone to support pregnancy and breast development for lactation.

Physiology of Sexual Intercourse (p. 953)

1. The role of the male in the sex act involves erection, minimal lubrication, and orgasm.

2. The female role also involves erection, providing most of the lubrication, and orgasm (climax).

Birth Control (p. 955)

1. Methods include sterilization (vasectomy, tubal ligation), hormonal, intrauterine devices, barriers (condom, vaginal pouch, diaphragm, cervical cap), chemicals (spermicides), physiological (rhythm, sympto-thermal method), coitus interruptus, and induced abortion.
2. Contraceptive pills of the combination type contain estrogens and progesterone in concentrations that decrease the secretion of FSH and LH and thereby inhibit development of ovarian follicles and ovulation.

Aging and the Reproductive Systems (p. 958)

1. In the male, decreased levels of testosterone decrease muscle strength, sexual desire, and viable sperm; prostate disorders are common.

2. In the female, levels of progesterone and estrogens decrease, resulting in changes in menstruation; uterine and breast cancer increase in incidence.
3. Puberty refers to the period of time when secondary sex characteristics begin to develop and the potential for sexual reproduction is reached.
4. The onset of male puberty is signaled by increased levels of LH, FSH, and testosterone.
5. The onset of female puberty is signaled by increased levels of LH, FSH, and estrogens.

Developmental Anatomy of the Reproductive Systems (p. 959)

1. The gonads develop from intermediate mesoderm and are differentiated into ovaries or testes by about the seventh week of fetal development.
2. The external genitals develop from the genital tubercle.

Review Questions

1. Define reproduction. Describe how the reproductive organs are classified and list the male and female organs of reproduction. (p. 923)
2. Describe the function of the scrotum in protecting the testes from temperature fluctuations. What is cryptorchidism? (p. 924)
3. Describe the internal structure of a testis. Where are the sperm cells made? (p. 924)
4. Describe the principal events of spermatogenesis. Why is meiosis important? Distinguish between haploid (n) and diploid ($2n$) cells. (p. 925)
5. Identify the principal parts of a spermatozoon. List the functions of each. (p. 928)
6. Explain the effects of FSH and LH on the male reproductive system. How are these hormones controlled by GnRH? (p. 929)
7. Describe the physiological effects of testosterone and inhibin on the male reproductive system. How is testosterone level controlled? (p. 929)
8. Which ducts are involved in transporting sperm *within* the testes? (p. 930)
9. Describe the location, structure, and functions of the ductus epididymis, ductus (vas) deferens, and ejaculatory duct. (p. 930)
10. What is a vasectomy? (p. 931)
11. What is the spermatic cord? What is an inguinal hernia? (p. 932)
12. Give the location of the three subdivisions of the male urethra. (p. 933)
13. Trace the course of spermatozoa through the male system of ducts from the seminiferous tubules through the urethra. (p. 925)
14. Briefly explain the locations and functions of the seminal vesicles, prostate gland, and bulbourethral (Cowper's) glands. How is cancer of the prostate gland detected? (p. 933)
15. What is semen? What is its function? What is a semen analysis? (p. 934)
16. How is the penis structurally adapted as an organ of copulation? How does an erection occur? What is circumcision? (p. 936)
17. How are the ovaries held in position in the pelvic cavity? (p. 936)

18. Describe the microscopic structure of an ovary. What are the functions of the ovaries? (p. 936)
19. Describe the principal events of oogenesis. (p. 937)
20. Where are the uterine (Fallopian) tubes located? What is their function? (p. 941)
21. Diagram the principal parts of the uterus. What is a Pap smear? (p. 942)
22. Describe the arrangement of ligaments that hold the uterus in its normal position. What is retroflexion? (p. 942)
23. Discuss the blood supply to the uterus. Why is an abundant blood supply important? (p. 943)
24. Describe the histology of the uterus. (p. 943)
25. What is the function of the vagina? Describe its histology. What is colposcopy? (p. 943)
26. List the parts of the vulva and the functions of each part. What is an episiotomy? (p. 944)
27. Describe the structure of the mammary glands. How are they supported? (p. 946)
28. Describe the passage of milk from the alveoli of the mammary gland to the nipple. (p. 946)
29. Explain the roles of estrogens and progesterone in the development of the mammary glands. (p. 948)
30. Define lactation. How is it controlled? (p. 948)
31. How is breast cancer detected? How is breast cancer treated? (p. 948)
32. What is the function of each of the following in the menstrual and ovarian cycles: GnRH, FSH, LH, estrogens, progesterone, and inhibin? (p. 949)
33. Briefly outline the major events of each phase of the menstrual cycle and correlate them with the events of the ovarian cycle. (p. 950)
34. Prepare a labeled diagram of the principal hormonal interactions involved in the menstrual and ovarian cycles. (p. 951)
35. What are the signs that ovulation has occurred? (p. 952)
36. Explain the role of the male's erection, lubrication, and orgasm in the sex act. (p. 953)
37. How do the female's erection, lubrication, and orgasm (climax) contribute to the sex act? (p. 955)
38. Briefly describe the various methods of birth control (BC) and the effectiveness of each. (p. 955)
39. Explain the effects of aging on the reproductive systems. (p. 958)

40. Describe the events involved in male and female puberty. (p. 958)
41. Distinguish between menarche and menopause. (p. 959)
42. Describe the development of the reproductive systems. (p. 959)

Answers to Questions with Figures

28.1 Gonads: produce gametes and hormones; ducts: transport, store, and receive gametes; accessory sex glands: produce materials that support gametes.

28.2 Tunica vaginalis and tunica albuginea.

28.3 Spermatogonia (stem cells) are least mature; sperm cells (spermatozoa) are most mature.

28.4 Because the number of chromosomes in each cell is reduced by half.

28.5 Head contains DNA and enzymes for penetration of secondary oocyte; midpiece contains mitochondria for ATP production; tail consists of a flagellum that provides propulsion.

28.6 Sustentacular (Sertoli) cells.

28.7 Testosterone inhibits secretion of LH and inhibin inhibits secretion of FSH.

28.8 Ilioinguinal nerve and spermatic cord, which consists of the testicular artery and veins, lymphatic vessels, autonomic nerves, ductus deferens, and the cremaster muscle.

28.9 Seminal vesicles: alkaline fluid, rich in fructose, prostaglandins, and fibrinogen; prostate gland: slightly acidic fluid rich in citric acid and enzymes, including acid phosphatase, clotting enzymes, and fibrinolysin; bulbourethral glands: alkaline fluid and mucus.

28.10 Two corpora cavernosa penis and one corpus spongiosum penis contain blood sinuses that fill with blood that cannot flow out of the penis as quickly as it flows in. The trapped blood stiffens the tissue.

28.11 Testes; penis; prostate; bulbourethral glands.

28.12 Broad ligament of the uterus, ovarian ligament, and suspensory ligament.

28.13 Ovarian follicles secrete estrogens and a corpus luteum secretes estrogens, progesterone, relaxin, and inhibin.

28.14 Primary oocytes are present in the ovary at birth, so they are as old as the woman is. In males, primary spermatocytes are continually being formed from stem cells (spermatogonia) and thus are only a few days old.

28.15 They help move the secondary oocyte toward the uterus.

28.16 Endometrium: highly vascular, secretory epithelium that provides oxygen and nutrients to sustain a fertilized egg; myometrium: thick smooth muscle layer that contracts to expel fetus at birth.

28.17 The stratum basalis provides cells to replace those shed (stratum functionalis) during each menstruation.

28.18 Anterior: mons pubis, clitoris, and prepuce. Lateral: labia minora and labia majora.

28.19 Its borders form a triangle that encloses the urethral (uro-) and vaginal (-genital) orifices.

28.20 Synthesis: prolactin, estrogens, progesterone. Release: oxytocin.

28.22 β-estradiol.

28.23 Estrogens. LH. LH. Estrogens.

28.24 Estrogens.

28.25 Negative feedback inhibition of secretion of these hormones.

28.26 Negative, because the response is opposite to the stimulus. Decreasing estrogens and progesterone stimulate release of GnRH, which in turn increases production and release of estrogens.

Chapter 29

DEVELOPMENT AND INHERITANCE

Chapter Contents at a Glance

Student Objectives

1. Explain the processes associated with fertilization, morula formation, blastocyst development, and implantation.

2. Describe how in vitro fertilization (IVF) is performed.

3. Discuss the principal maternal body changes associated with embryonic and fetal growth.

4. Compare the sources and functions of the hormones secreted during pregnancy.

5. Describe the anatomical and physiological changes associated with gestation.

6. Explain the respiratory and cardiovascular adjustments that occur in an infant at birth.

7. Discuss the physiology and control of lactation.

8. Define inheritance and describe the inheritance of several traits.

9. Discuss potential hazards to an embryo and fetus associated with chemicals and drugs, irradiation, alcohol, and cigarette smoking.

10. Define and describe Down syndrome, Turner's syndrome, metafemale syndrome, Klinefelter's syndrome, and fragile X syndrome.

Developmental anatomy is the study of the sequence of events from the fertilization of a secondary oocyte to the formation of an adult organism. As we look at the sequence from fertilization to birth, we will consider fertilization, implantation, placental development, embryonic development, fetal growth, gestation, parturition, and labor. The development of an embryo and fetus is a wonderfully complex and precisely coordinated series of events.

DEVELOPMENT DURING PREGNANCY

Once spermatozoa and ova have developed through meiosis and maturation, and the spermatozoa have been deposited in the vagina, pregnancy can occur. **Pregnancy** is a sequence of events that normally includes fertilization, implantation, embryonic growth, and fetal growth that ends with birth about 266 days later.

Fertilization and Implantation

Fertilization

During **fertilization** the genetic material from the spermatozoon and ovum merges into a single nucleus. Of the 300 to 500 million sperm cells introduced into the vagina, less than 1% reach the secondary oocyte. Fertilization normally occurs in the uterine (Fallopian) tube about 12 to 24 hours after ovulation. Peristaltic contractions and the action of cilia transport the oocyte through the uterine tube. Sperm swim up the female tract by whiplike movements of their tail. The acrosome of sperm produces an enzyme called **acrosin** that stimulates sperm motility and migration within the female reproductive tract. Also, muscular contractions of the uterus, stimulated by prostaglandins in semen, probably aid sperm movement toward the uterine tube. Finally, the oocyte is thought to secrete a chemical substance that attracts sperm.

Besides contributing to sperm movement, the female reproductive tract also confers on sperm the capacity to fertilize a secondary oocyte. Although sperm undergo maturation in the epididymis, they are still not able to fertilize an oocyte until they have been in the female reproductive tract for about 10 hours. **Capacitation** (ka-pas'-i-TĀ-shun) refers to the functional changes that sperm undergo in the female reproductive tract that allow them to fertilize a secondary oocyte. During this process, the membrane around the acrosome becomes fragile so that several destructive enzymes—hyaluronidase, acrosin, and neuraminidase—are secreted by the acrosomes. The enzymes help penetrate the **corona radiata,** several layers of follicular cells around the oocyte, and a gelatinous glycoprotein layer internal to the corona radiata called the **zona pellucida** (pe-LOO-si-da) (Fig. 29.1a). Sperm bind to receptors in the zona pellucida. Normally only one spermatozoon penetrates and enters a secondary oocyte. This event is called **syngamy.** Syngamy causes depolarization, which triggers the release of calcium ions inside the cell. Calcium ions stimulate the release of granules by the oocyte that promote changes in the oocyte to block entry of other sperm. This prevents **polyspermy,** fertilization by more than one spermatozoon. Once a spermatozoon has entered a secondary oocyte, the oocyte completes equatorial division (meiosis II). It divides into a larger ovum (mature egg) and a smaller second polar body that fragments and disintegrates (see Fig. 28.14).

When a spermatozoon has entered a secondary oocyte, the tail is shed and the nucleus in the head develops into a structure called the **male pronucleus.** The nucleus of the ovum develops into a **female pronucleus** (Fig. 29.1b). After the pronuclei are formed, they fuse to produce a **segmentation nucleus.** The segmentation nucleus contains 23 chromosomes (n) from the male pronucleus and 23 chromosomes (n) from the female pronucleus. Thus the fusion of the haploid (n) pronuclei restores the diploid number ($2n$). The fertilized ovum, consisting of a segmentation nucleus, cytoplasm, and zona pellucida, is called a **zygote** (ZĪ-gōt; *zygotos* = joined by a yoke).

Dizygotic (fraternal) twins are produced from the independent release of two ova and the subsequent fertilization of each by different spermatozoa. They are the same age and are in the uterus at the same time, but they are genetically as dissimilar as any other siblings. They may or may not be the same sex. **Monozygotic (identical) twins** develop from a single fertilized ovum that splits at an early stage in development. They contain exactly the same genetic material and are always the same sex.

CLINICAL APPLICATION

ECTOPIC PREGNANCY

Ectopic (*ektos* = outside; *topos* = place) **pregnancy** refers to the development of an embryo or fetus outside the uterine cavity. Most occur in the uterine (Fallopian) tube, usually in the ampullar and infundibular portions. Some occur in the ovaries, abdomen, uterine cervix, and broad ligaments. Tubal pregnancies usually occur because passage of the fertilized ovum through the uterine tube is impaired. The cause may be decreased motility of the uterine tube or abnormal anatomy. Scars from pelvic inflammatory disease (PID), previous uterine tube surgery, and previous ectopic pregnancy may hinder movement of the fertilized ovum. Other causes include repeated elective abortions, pelvic tumors, and developmental abnormalities.

Ectopic pregnancy may be characterized by one or two missed periods, followed by bleeding and acute abdominal and pelvic pain. Unless removed, the developing embryo can rupture the tube, often resulting in death of the mother.

Formation of the Morula

After fertilization, rapid mitotic cell divisions of the zygote take place. These early divisions of the zygote are called **cleavage.** Although cleavage increases the number of cells,

FIGURE 29.1 Fertilization. (a) Spermatozoon penetrating the corona radiata and zona pellucida. (b) Female and male pronuclei. Figure 28.5b shows a spermatozoon in contact with the surface of a secondary oocyte.

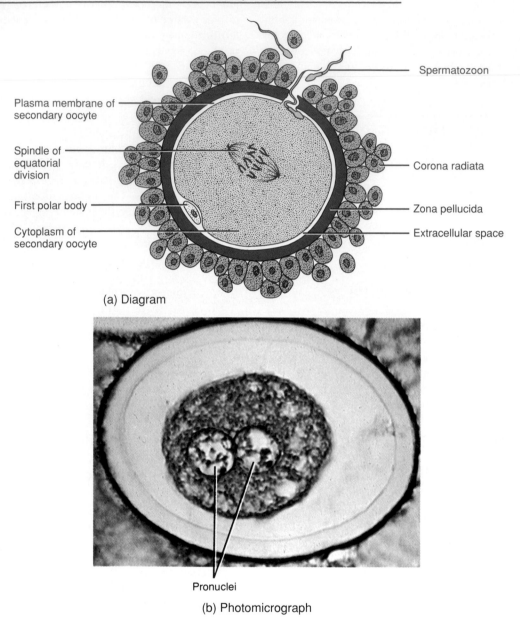

(a) Diagram

(b) Photomicrograph

Question: What is capacitation?

it does not increase the size of the embryo, which is still contained within the zona pellucida.

The first cleavage is completed about 36 hours after fertilization, and each succeeding division takes slightly less time (Fig. 29.2a,b). By the second day after fertilization, the second cleavage is completed. By the end of the third day, there are 16 cells. The progressively smaller cells produced by cleavage are called **blastomeres** (BLAS-tō-mērz; *blast* = germ, sprout). Successive cleavages produce a solid mass of cells, still surrounded by zona pellucida, called the **morula** (MOR-yoo-la; = mulberry). See Fig. 29.2c. A few

days after fertilization, the morula is about the same size as the original zygote.

Development of the Blastocyst

By the end of the fourth day, the number of cells in the morula increases and it continues to move along the uterine (Fallopian) tube toward the uterine cavity. At four and a half to five days, the dense cluster of cells has developed into a hollow ball of cells and enters the uterine cavity; it is now called a **blastocyst** (Fig. 29.2d,e).

The blastocyst has an outer covering of cells called the **trophoblast** (TRŌF-ō-blast; *troph* = nourish), an **inner cell mass (embryoblast),** and an internal fluid-filled cavity called the **blastocele** (BLAS-tō-sēl). The trophoblast ultimately forms part of the membranes composing the fetal portion of the placenta; part of the inner cell mass develops into the embryo.

Implantation

The blastocyst remains free within the cavity of the uterus for a short period of time before it attaches to the uterine wall. During this time, the zona pellucida disintegrates. The blastocyst receives nourishment from glycogen-rich secretions of endometrial glands, sometimes called uterine milk. About six days after fertilization the blastocyst attaches to the endometrium, a process called **implantation** (Fig. 29.3). At this time, the endometrium is in its secretory phase.

As the blastocyst implants, usually on the posterior wall of the fundus or body of the uterus, it is oriented so that the inner cell mass is toward the endometrium. The trophoblast develops two layers in the region of contact between the blastocyst and endometrium. These layers are an outer **syncytiotrophoblast** (sin-sīt′-ē-ō-TRŌF-ō-blast; *synctium* = multinucleate mass) that contains no cell boundaries and an inner **cytotrophoblast** (sī-tō-TRŌF-ō-blast) that is composed of distinct cells. During implantation, the syncytiotrophoblast secretes enzymes that enable the blastocyst to penetrate the uterine lining. The enzymes digest and liquefy the endometrial cells. The fluid and nutrients further nourish the burrowing blastocyst for about a week after implantation. Eventually, the blastocyst becomes buried in the endometrium.

Since a developing embryo, and later a fetus, contains genes from the father as well as the mother, it is essentially a foreign graft. Thus it is surprising that it is not rejected as such by the mother. The trophoblast is the only tissue of the developing organism that contacts the mother's uterus. Even though trophoblast cells have paternal antigens that could provoke a rejection response, some mechanism in the uterus prevents this to permit development of the fetus to term. One suggestion for the mechanism that prevents rejection is that the mother makes antibodies that mask paternal antigens so that other components of her immune system cannot recognize and attack the antigens.

A summary of the principal events associated with fertilization and implantation is shown in Fig. 29.4.

In Vitro Fertilization

On July 12, 1978, Louise Joy Brown was born near Manchester, England. Her birth was the first recorded case of **in vitro fertilization (IVF)**—fertilization in a laboratory dish. In this procedure for IVF, the mother-to-be is given follicle-

FIGURE 29.2 Cleavage and formation of the morula and blastocyst.

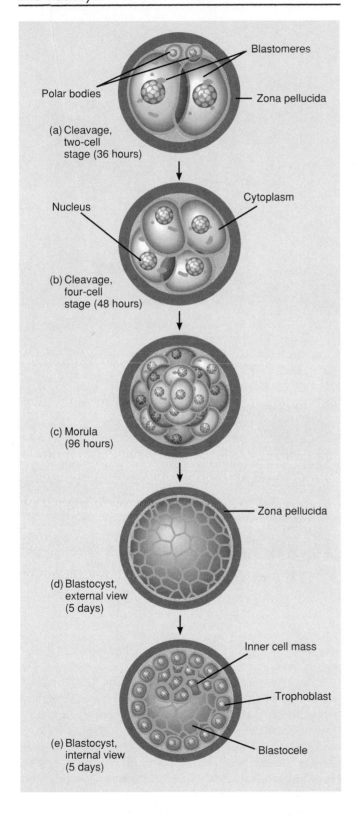

(a) Cleavage, two-cell stage (36 hours)

Polar bodies
Blastomeres
Zona pellucida

(b) Cleavage, four-cell stage (48 hours)

Nucleus
Cytoplasm

(c) Morula (96 hours)

(d) Blastocyst, external view (5 days)

Zona pellucida

(e) Blastocyst, internal view (5 days)

Inner cell mass
Trophoblast
Blastocele

Question: What is the histological difference between a morula and a blastocyst?

FIGURE 29.3 Implantation. Shown is the blastocyst in relation to the endometrium of the uterus at various time intervals after fertilization.

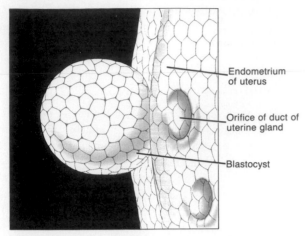

(a) External view, about 5 days after fertilization

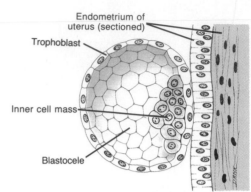

(b) Internal view, about 6 days after fertilization

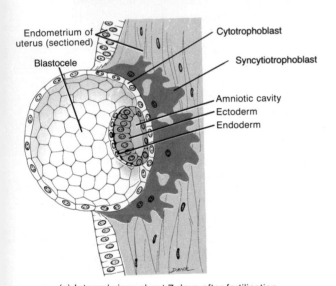

(c) Internal view, about 7 days after fertilization

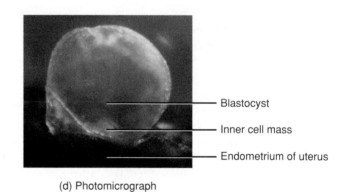

(d) Photomicrograph

Question: How does the blastocyst merge with and burrow into the endometrium?

stimulating hormone (FSH) soon after menstruation, so that several secondary oocytes, rather than the typical single one, will be produced (superovulation). Administration of luteinizing hormone (LH) may also ensure the maturation of the secondary oocytes. Next, a small incision is made near the umbilicus, and the secondary oocytes are aspirated from the follicles and then transferred to a solution of the male's sperm.

Once fertilization has taken place, the fertilized ovum is put in another medium and observed for cleavage. When the fertilized ovum reaches the 8-cell or 16-cell stage, it is introduced into the uterus for implantation and subsequent growth. It is also possible to freeze embryos (cryopreservation) to permit parents a successive pregnancy several years

later or allow a second attempt at implantation if the first attempt is unsuccessful.

Embryo transfer is a type of IVF in which a husband's semen is used to artificially inseminate a fertile secondary oocyte donor, and after fertilization, the morula or blastocyst is transferred from the donor to the infertile wife who carries it to term. Embryo transfer is indicated for females who are infertile or who do not want to pass on their own genes because they are carriers of a serious genetic disorder.

In the procedure, the donor is monitored to ascertain the time of ovulation by checking her blood levels of luteinizing hormone (LH) and by ultrasound. The wife is also monitored to make sure that her ovarian cycle is synchronized

FIGURE 29.4 Summary of events associated with fertilization and implantation.

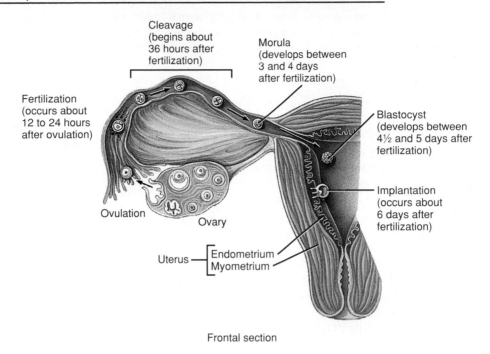

Frontal section

Question: In which phase of the uterine cycle does implantation occur?

with that of the donor. Once ovulation occurs in the donor, she is artificially inseminated with the husband's (or another male's) semen. Four days later, a morula or blastocyst is flushed from the donor's uterus through a soft plastic catheter and transferred to the uterus of the wife, where it grows and develops until the time of birth. Embryo transfer is an office procedure that requires no anesthesia.

In **gamete intrafallopian transfer (GIFT)** the goal is to mimic the normal process of conception by uniting sperm and secondary oocyte in the prospective mother's uterine tubes. In the procedure, the female is given FSH and LH to stimulate the production of several secondary oocytes. The secondary oocytes are aspirated with a laparoscope fitted with a suction device, mixed with a solution of the male's sperm outside the body, and then immediately inserted into the uterine (Fallopian) tubes.

CLINICAL APPLICATION

FERTILITY AND BODY FAT

Body fat has a regulatory role in reproduction. To begin and maintain a normal menstrual cycle, a female must have a minimum amount of body fat. A moderate loss of fat, from 10 to 15% below normal weight for height, may delay the onset of menstruation (menarche), inhibit ovulation during the menstrual cycle, or induce the cessation of the menstrual cycle (amenorrhea). Both dieting and intensive exercise may reduce body fat below the minimum amount and lead to infertility. The resulting infertility is reversible fol-

lowing weight gain or reduction of intensive exercise or both.

It appears that in underweight or very lean females, the secretion of gonadotropin releasing hormone (GnRH) by the hypothalamus is abnormal in quantity and timing. GnRH stimulates release of follicle-stimulating hormone (FSH) and luteinizing hormone (LH) from the anterior pituitary gland. Both hormones, in turn, control development of ovarian follicles, ovulation, and secretion of progesterone and estrogens by ovarian follicles. For some reason, the same factors that cause infertility in underweight or athletic females also provide a degree of protection against cancers that are sensitive to sex hormones, such as breast cancer.

Studies of very obese females also indicate that they, like very lean ones, experience problems with amenorrhea and infertility. Males also experience problems related to reproduction in response to undernutrition and weight loss. For example, they produce less prostatic fluid, spermatozoa with decreased motility, and reduced numbers of spermatozoa.

EMBRYONIC DEVELOPMENT

The first two months of development are generally considered the **embryonic period.** During this period the developing human is called an **embryo** (*bryein* = grow). The study of development from the fertilized egg through the eighth week in utero is termed **embryology** (em-brē-OL-ō-jē). The months of development after the second month are considered the **fetal period,** and during this time the developing human is called a **fetus** (*feo* = to bring forth). By the end of the embryonic period, the rudiments of all the principal adult organs are present and the embryonic membranes are developed. By the end of the third month, the placenta, which is the site of exchange of nutrients and wastes between the mother and the fetus, is functioning.

Beginnings of Organ Systems

After implantation, the first major event of the embryonic period occurs. The inner cell mass of the blastocyst begins to differentiate into the three **primary germ layers:** ectoderm, endoderm, and mesoderm. These are the embryonic tissues from which all tissues and organs of the body will develop. The cell migrations that establish the primary germ layers are called **gastrulation.**

In the human, the germ layers form so quickly that it is difficult to determine the exact sequence of events. Within eight days after fertilization, the top layer of cells of the inner cell mass proliferates and forms the amnion (a fetal membrane) and a space, the **amniotic (amnionic) cavity,** over the inner cell mass. The layer of cells of the inner cell mass that is closer to the amniotic cavity develops into the **ectoderm.** The layer of the inner cell mass that borders the blastocele develops into the **endoderm.**

About the 12th day after fertilization, striking changes appear (Fig. 29.5a). The cells beside the amniotic cavity are called the **embryonic disc.** They will form the embryo. At this stage, the embryonic disc contains ectodermal and endodermal cells; the mesodermal cells are scattered external to the disc. The cells of the endodermal layer have been dividing rapidly, so that groups of them now extend around

in a circle, forming the yolk sac, another fetal membrane. The cells of the **mesoderm,** which develop between the ectodermal and endodermal layers, also have been dividing, and many have left the area of the embryonic disc and can be seen around the structures that are becoming fetal membranes.

About the 14th day, the cells of the embryonic disc differentiate into three distinct layers: the upper ectoderm, the middle mesoderm, and the lower endoderm (Fig. 29.5b). The mesoderm in the disc soon splits into two layers, and the space between the layers becomes the **extraembryonic coelom.**

As the embryo develops (Fig. 29.5c), the endoderm becomes the epithelial lining of the gastrointestinal tract, respiratory tract, and several other organs. The mesoderm forms muscle, bone and other connective tissues, and the peritoneum. The ectoderm develops into the skin and nervous system. Exhibit 29.1 provides more details about the fates of these primary germ layers.

Embryonic Membranes

A second major event that occurs during the embryonic period is the formation of the **embryonic (extraembryonic) membranes** (Fig. 29.6). These membranes lie outside the embryo and protect and nourish the embryo and, later, the fetus. The membranes are the yolk sac, amnion, chorion, and allantois.

The **yolk sac** is an endoderm-lined membrane that, in many species, is the primary source of nourishment for the embryo (Fig. 29.6; see also Fig. 29.5c). However, the

FIGURE 29.5 Formation of the primary germ layers and associated structures.

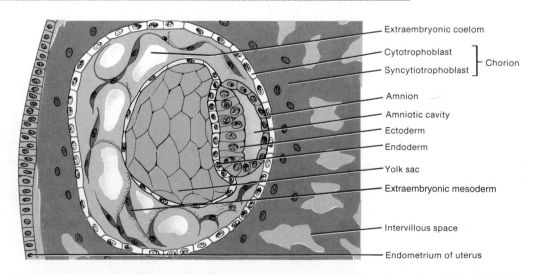

(a) Internal view, about 12 days after fertilization

FIGURE 29.5 (continued)

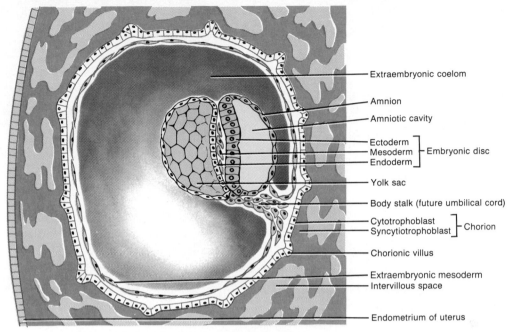

(b) Internal view, about 14 days after fertilization

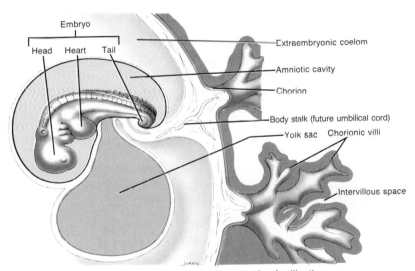

(c) External view, about 25 days after fertilization

Question: Which cells of the blastocyst give rise to the embryonic disc?

human embryo receives nutrients from the endometrium, and the yolk sac remains small. In humans, the yolk sac is an early site of blood formation. The yolk sac also contains cells that migrate into the gonads and differentiate into the primitive germ cells (spermatogonia and oogonia). During an early stage of development, it becomes a nonfunctional part of the umbilical cord.

The **amnion** is a thin, protective membrane that forms by the eighth day after fertilization and initially overlies the embryonic disc. As the embryo grows, the amnion comes to

entirely surround the embryo, creating a cavity that becomes filled with **amniotic fluid** (Fig. 29.6). Most amniotic fluid is initially derived from a filtrate of maternal blood. Later, the fetus makes daily contributions to the fluid by excreting urine into the amniotic cavity. Amniotic fluid serves as a shock absorber for the fetus, helps regulate fetal body temperature, and prevents adhesions between the skin of the fetus and surrounding tissues. Embryonic cells are sloughed off into amniotic fluid; they can be examined in the procedure called **amniocentesis** (am′-nē-ō-sen-TĒ-sis),

EXHIBIT 29.1

STRUCTURES PRODUCED BY THE THREE PRIMARY GERM LAYERS

Endoderm	Mesoderm	Ectoderm
Epithelium of gastrointestinal tract (except the oral cavity and anal canal) and the epithelium of its glands.	All skeletal, most smooth, and all cardiac muscle.	All nervous tissue.
Epithelium of urinary bladder, gallbladder, and liver.	Cartilage, bone, and other connective tissues.	Epidermis of skin.
	Blood, bone marrow, and lymphoid tissue.	Hair follicles, arrector pili muscles, nails, and epithelium of skin glands (sebaceous and sudoriferous).
Epithelium of pharynx, auditory (Eustachian) tube, tonsils, larynx, trachea, bronchi, and lungs.	Endothelium of blood vessels and lymphatic vessels.	Lens, cornea, and internal eye muscles.
	Dermis of skin.	Internal and external ear.
Epithelium of thyroid, parathyroid, pancreas, and thymus glands.	Fibrous tunic and vascular tunic of eye.	Neuroepithelium of sense organs.
	Middle ear.	Epithelium of oral cavity, nasal cavity, paranasal sinuses, salivary glands, and anal canal.
Epithelium of prostate and bulbourethral (Cowper's) glands, vagina, vestibule, urethra, and associated glands such as the greater (Bartholin's) vestibular and lesser vestibular glands.	Mesothelium of ventral body and joint cavities.	Epithelium of pineal gland, pituitary gland (hypophysis), and adrenal medulla.
	Epithelium of kidneys and ureters.	
	Epithelium of adrenal cortex.	
	Epithelium of gonads and genital ducts.	

FIGURE 29.6 Embryonic membranes.

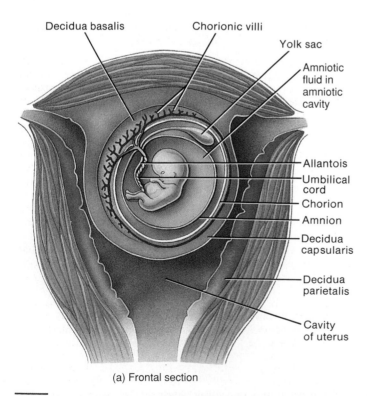

(a) Frontal section

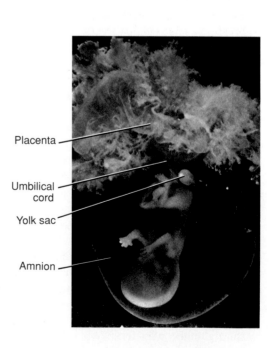

(b) Ten-week fetus

Question: How do the amnion and chorion differ in function?

which is discussed later. The amnion usually ruptures just before birth and with its fluid constitutes the "bag of waters."

The **chorion** (KŌ-rē-on) derives from the trophoblast of the blastocyst and the mesoderm that lines the trophoblast. It surrounds the embryo and, later, the fetus. Eventually, the chorion becomes the principal embryonic part of the placenta, the structure for exchange of materials between mother and fetus. The amnion also surrounds the fetus and eventually fuses to the inner layer of the chorion.

The **allantois** (a-LAN-tō-is; *allas* = sausage) is a small, vascularized outpouching of the yolk sac. It serves as an early site of blood formation. Later its blood vessels serve as connections in the placenta between mother and fetus. This connection is the umbilical cord.

Placenta and Umbilical Cord

Development of the **placenta** (pla-SEN-ta), the third major event of the embryonic period, is accomplished by the third month of pregnancy. The placenta has the shape of a flat cake when fully developed and is formed by the chorion of the embryo and a portion of the endometrium of the mother (Fig. 29.7). Functionally, the placenta allows oxygen and nutrients to diffuse into fetal blood from maternal blood. Simultaneously, carbon dioxide and wastes diffuse from fetal blood into maternal blood at the placenta. Almost all drugs, including alcohol, pass freely through the placenta.

The placenta also is a protective barrier since most microorganisms cannot cross it. However, certain viruses, such as those that cause AIDS, German measles, chicken-pox, measles, encephalitis, and poliomyelitis, may pass through the placenta. The placenta also stores nutrients such as carbohydrates, proteins, calcium, and iron, which are released into fetal circulation as required. Finally, the placenta produces several hormones that are necessary to maintain pregnancy (see page 981).

If implantation occurs, a portion of the endometrium becomes modified and is known as the **decidua** (dē-SID-yoo-a; *deciduus* = falling off). The decidua includes all but the deepest layer of the endometrium and separates from the endometrium after the fetus is delivered. Different regions of the decidua, all areas of the stratum functionalis, are named based on their positions relative to the site of the implanted, fertilized ovum (Fig. 29.8). The **decidua basalis** is the portion of the endometrium that underlies the embryo, between the chorion and the stratum basalis of the uterus. The decidua basalis becomes the maternal part of the placenta. The **decidua capsularis** is the portion of the endometrium that overlies the embryo, between the embryo and the uterine cavity. The **decidua parietalis** (par-rī-e-TAL-is) is the remaining modified endometrium that lines the entire pregnant uterus.

During embryonic life, fingerlike projections of the chorion, called **chorionic villi** (kō′-rē-ON-ik VIL-ī), grow into the decidua basalis of the endometrium (see also Figs. 29.5 through 29.7). These will contain fetal blood vessels of the allantois. They continue growing until they are bathed in maternal blood sinuses called **intervillous** (in-ter-VIL-us) **spaces.** Thus maternal and fetal blood vessels are brought into proximity. It should be noted, however, that maternal and fetal blood do not normally mix.

Oxygen and nutrients from the mother's blood diffuse into the capillaries of the villi where all exchange occurs between fetal and maternal blood. From the capillaries the nutrients circulate into the umbilical vein. Wastes leave the fetus through the umbilical arteries, pass into the capillaries of the villi, and diffuse into the maternal blood. The **umbilical** (um-BIL-i-kul) **cord** is a vascular connection between mother and fetus. It consists of two umbilical arteries that carry deoxygenated fetal blood to the placenta, one umbilical vein that carries oxygenated blood into the fetus, and supporting mucous connective tissue called Wharton's jelly from the allantois (see Fig. 21.32a). The entire umbilical cord is surrounded by a layer of amnion.

At delivery, the placenta detaches from the uterus and is termed the **afterbirth.** At this time, the umbilical cord is severed, leaving the baby on its own. The scar that marks the site of the entry of the fetal umbilical cord into the abdomen is the **umbilicus (navel).**

Pharmaceutical companies use human placentas to harvest hormones, drugs, and blood. Portions of placentas are also used for burn coverage. The placental and umbilical cord veins can also be used in blood vessel grafts.

CLINICAL APPLICATION

PLACENTA PREVIA, FETOMATERNAL HEMORRHAGE, AND UMBILICAL CORD ACCIDENTS

In some cases, part or all of the placenta becomes implanted in the lower portion of the uterus, near or over the internal os of the cervix. This condition is called **placenta previa** (PRĒ-vē-a; *previa* = before or in front of) and occurs in approximately 1 in 250 live births. It is dangerous to the fetus because it may cause premature birth and intrauterine hypoxia. Maternal mortality is increased due to hemorrhage and infection. The condition occurs 10 to 20 times more often in association with spontaneous abortion. It is also associated with fetal abnormalities, twin gestation, and multiple uterine curettages. The most important symptom is sudden, painless, bright red vaginal bleeding in the third trimester. Cesarean section is the preferred method of delivery in placenta previa.

Fetomaternal hemorrhage refers to the entrance of fetal blood into maternal circulation brought on by a dysfunction in placental circulation. Although the condition occurs in at least 50% of all pregnancies, in most instances blood loss is so small that the pregnancy is not adversely affected. In some situations, however, the hemorrhage can compromise the fetus and lead to serious complications later in the same pregnancy or a future pregnancy. Among the causes of fetomaternal hemorrhage are trauma, rapid deceleration, placental and umbilical cord abnormalities, amniocentesis, intrauterine fetal surgery, umbilical vein thrombosis, and operative delivery (e.g., cesarean section).

FIGURE 29.7 Placenta and umbilical cord.

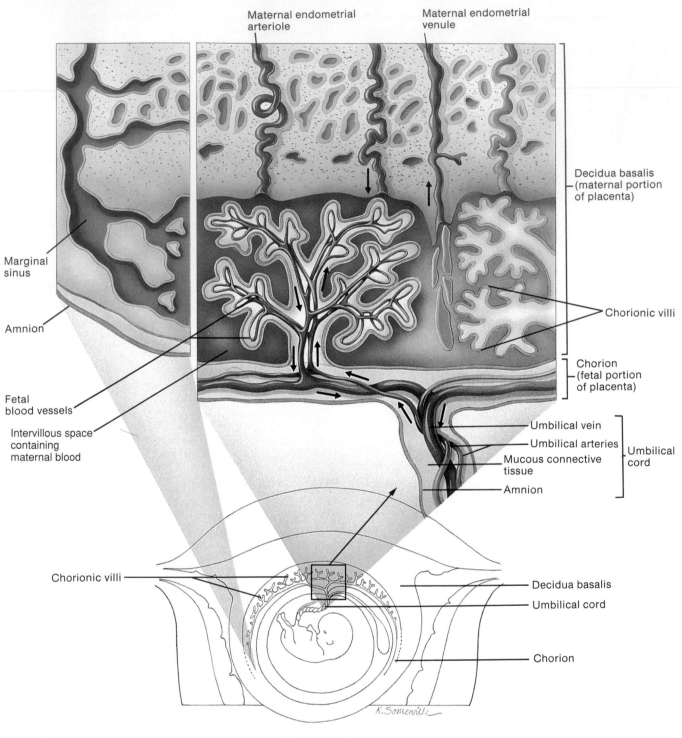

Maternal endometrial arteriole

Maternal endometrial venule

Decidua basalis (maternal portion of placenta)

Marginal sinus

Amnion

Chorionic villi

Fetal blood vessels

Chorion (fetal portion of placenta)

Intervillous space containing maternal blood

Umbilical vein

Umbilical arteries

Mucous connective tissue

Amnion

Umbilical cord

Chorionic villi

Decidua basalis

Umbilical cord

Chorion

K. Somerville

(a) Overall structure

FIGURE 29.7 (continued)

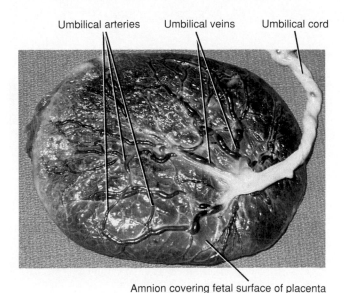

Umbilical arteries   Umbilical veins   Umbilical cord

Amnion covering fetal surface of placenta

(b) Photograph of fetal aspect of placenta

Question: What is the function of the placenta?

FIGURE 29.8 Regions of the decidua.

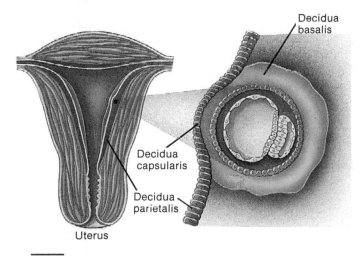

Decidua
basalis

Decidua
capsularis

Decidua
parietalis

Uterus

Question: Which part of the decidua helps form the maternal part of the placenta?

As a result of fetomaternal hemorrhage, certain complications may result. Examples include intrauterine fetal death, hypovolemic shock and anemia in the newborn, edema of the newborn, fetal cardiac arrhythmia, and anaphylactic shock in the mother.

Two of the most frequently encountered **umbilical cord accidents** are prolapse and entrapment. In **prolapse,** the umbilical cord descends in advance of the fetus during deliv-ery. In **entrapment,** circulation through the cord is endangered because of pressure between the fetus and uterine wall. These, plus other conditions, such as knots in the cord, strictures, and thrombosis, could compromise the oxygen supply to the fetus, resulting in fetal damage or death.

FETAL GROWTH

During the **fetal period,** organs established by the primary germ layers grow rapidly and the fetus takes on a human appearance. Throughout the text, we have discussed developmental anatomy of the various body systems in their respective chapters. A listing of these sections is presented here, for your review.

Integumentary System (page 139)
Skeletal System (page 160)
Muscular System (page 262)
Nervous System (page 431)
Endocrine System (page 554)
Heart (page 615)
Blood and Blood Vessels (page 676)
Lymphatic System (page 690)
Respiratory System (page 756)
Digestive System (page 812)
Urinary System (page 897)
Reproductive Systems (page 959)

A summary of changes associated with the embryonic and fetal periods is presented in Exhibit 29.2.

CLINICAL APPLICATION

FETAL SURGERY

Fetal surgery is a new medical field that had its beginnings in 1985. In a pioneering operation, a team of surgeons removed a 23-week-old fetus from its mother's uterus, operated to correct a blocked urinary tract, and then returned the fetus to the uterus. Nine weeks later, a healthy baby was born. Since then, surgeons have performed procedures to repair diaphragmatic hernias and are experimenting on animals to try to correct spina bifida (see page 195) and hydrocephalus (see page 411). It has been observed that surgery on fetuses does not leave any scars, although the reason is not known. It is hoped that surgeons can perform craniofacial surgery before birth to correct conditions such as cleft lip without leaving scars.

HORMONES OF PREGNANCY

During the first three to four months of pregnancy, the corpus luteum continues to secrete **estrogens** and **progesterone.** These hormones maintain the lining of the uterus during pregnancy and prepare the mammary glands to secrete milk. The amounts secreted by the corpus luteum, however, are only slightly more than that produced after ovulation in a normal menstrual cycle. From the third month through the rest of the pregnancy the placenta itself

EXHIBIT 29.2

CHANGES ASSOCIATED WITH EMBRYONIC AND FETAL GROWTH

End of Month	Approximate Size and Weight	Representative Changes
1	0.6 cm (³⁄₁₆ in.)	Eyes, nose, and ears not yet visible. Backbone and vertebral canal form. Small buds that will develop into arms and legs form. Heart forms and starts beating. Body systems begin to form. The central nervous system appears at the start of the third week.
2	3 cm (1 ¼ in.) 1 g (¹⁄₃₀ oz)	Eyes far apart, eyelids fused, nose flat. Ossification begins. Limbs become distinct as arms and legs. Digits are well formed. Major blood vessels form. Many internal organs continue to develop.
3	7.5 cm (3 in.) 30 g (1 oz)	Eyes almost fully developed but eyelids still fused, nose develops bridge, and external ears are present. Ossification continues. Limbs are fully formed and nails develop. Heartbeat can be detected. Urine starts to form. Fetus begins to move, but it cannot be felt by mother. Body systems continue to develop.
4	18 cm (6 ½ to 7 in.) 100 g (4 oz)	Head large in proportion to rest of body. Face takes on human features and hair appears on head. Skin bright pink. Many bones ossified, and joints begin to form. Rapid development of body systems.
5	25 to 30 cm (10 to 12 in.) 200 to 450 g (½ to 1 lb)	Head less disproportionate to rest of body. Fine hair (lanugo) covers body. Skin still bright pink. Brown fat forms and is the site of heat production. Fetal movements commonly felt by mother (quickening). Rapid development of body systems.
6	27 to 35 cm (11 to 14 in.) 550 to 800 g (1¼ to 1½ lb)	Head becomes even less disproportionate to rest of body. Eyelids separate and eyelashes form. Substantial weight gain. Skin wrinkled and pink. Type II alveolar cells begin to produce surfactant.
7	32 to 42 cm (13 to 17 in.) 1100 to 1350 g (2½ to 3 lb)	Head and body more proportionate. Skin wrinkled and pink. Seven-month fetus (premature baby) is capable of survival. Fetus assumes an upside-down position.

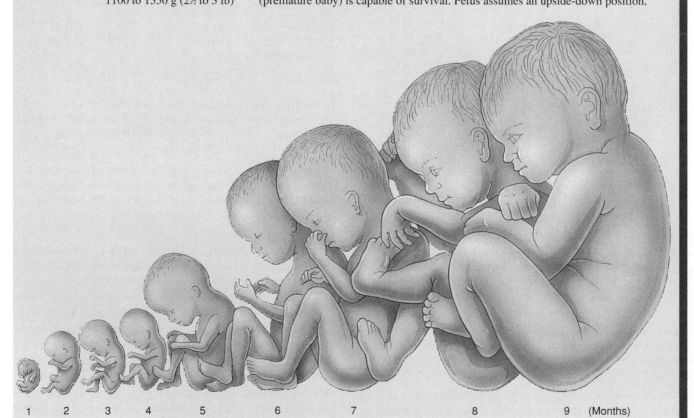

| 1 | 2 | 3 | 4 | 5 | 6 | 7 | 8 | 9 | (Months) |

End of Month	Approximate Size and Weight	Representative Changes
8	41 to 45 cm (16½ to 18 in.) 2000 to 2300 g (4½ to 5 lb)	Subcutaneous fat deposited. Skin less wrinkled. Testes descend into scrotum. Bones of head are soft. Chances of survival much greater at end of eighth month.
9	50 cm (20 in.) 3200 to 3400 g (7 to 7½ lb)	Additional subcutaneous fat accumulates. Lanugo shed. Nails extend to tips of fingers and maybe even beyond.

provides the high levels of estrogens and progesterone needed to maintain pregnancy and develop the mammary glands for lactation. The chorion of the placenta secretes **human chorionic gonadotropin (hCG).** It mimics LH and its primary role is to stimulate continued production of estrogens and progesterone by the corpus luteum—an activity necessary for the continued attachment of the embryo and fetus to the lining of the uterus (Fig. 29.9). By the eighth day after fertilization, hCG can be detected in the blood of a pregnant woman. Peak secretion of hCG occurs at about the ninth week of pregnancy. The hCG level decreases sharply during the fourth and fifth months and then levels off until childbirth.

CLINICAL APPLICATION

MORNING SICKNESS

In the early months of pregnancy, **emesis gravidarum (morning sickness)** may occur, characterized by episodes of nausea and possibly vomiting that are most likely to occur in the morning. The cause is unknown, but the high levels of human chorionic gonadotropin (hCG) secreted by the placenta and progesterone secreted by the ovaries have been implicated. In some women the severity of these symptoms requires hospitalization for intravenous feeding, and the condition is then known as **hyperemesis gravidarum.**

The chorion of the placenta begins to secrete estrogens after the first three or four weeks and progesterone by the sixth week of pregnancy. These hormones are secreted in increasing quantities until the time of birth. By the fourth month, when the placenta is established, the secretion of hCG is greatly reduced because the secretions of the corpus luteum are no longer essential. Thus from the third to ninth month, the placenta supplies the levels of estrogens and progesterone needed to maintain the pregnancy. The placental hormones take over management of the mother's body in preparation for parturition (birth) and lactation. After delivery, estrogens and progesterone in the blood decrease to normal levels.

Another hormone produced by the chorion of the placenta is **human chorionic somatomammotropin (hCS),** also known as **human placental lactogen (hPL).** Its secretion begins at about the same time as that of hCG, but its

FIGURE 29.9 Hormones of pregnancy. Blood levels of human chorionic gonadotropin (hCG), estrogens, and progesterone.

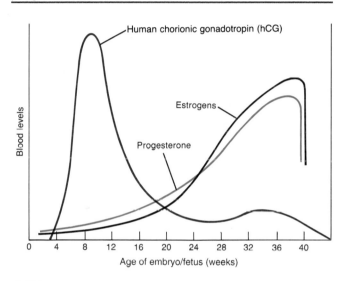

Question: What are the functions of these three hormones?

pattern of secretion is quite different. The rate of secretion of hCS increases in proportion to placental mass, reaching maximum levels after 32 weeks and remaining relatively constant after that. hCS is believed to stimulate some development of breast tissue for lactation, enhance growth by causing protein deposition in tissues, and regulate certain aspects of metabolism. For example, hCS causes decreased use of glucose by the mother, thus making more available for the fetus. Also, hCS promotes the release of fatty acids from fat depots, providing an alternative source of energy for the mother's metabolism.

Relaxin is a hormone produced by the placenta and ovaries. Its physiological role is to relax the pubic symphysis and ligaments of the sacroiliac and sacrococcygeal joints and help dilate the uterine cervix toward the end of pregnancy. Both of these actions assist in delivery.

Inhibin, a hormone produced by the ovaries and testes, is also present in the human placenta at term. Its role is to inhibit secretion of follicle-stimulating hormone (FSH) and perhaps to regulate secretion of hCG.

EARLY PREGNANCY TESTS

Over-the-counter test kits are available to detect minute amounts of human chorionic gonadotropin (hCG), which starts to be released about eight days after fertilization. The kits can detect pregnancy as early as the first day of a missed menstrual period, that is, at about 14 days after fertilization. All kits include antibodies to hCG and other chemicals that produce a color change if there is a reaction between hCG in the urine and the hCG antibody in the test kit.

Several of the test kits are as sensitive and accurate as test methods used in many hospitals. Still, false negative and false positive results can occur. A false negative result (test is negative, but the female is still pregnant) may occur from testing too soon or from an ectopic pregnancy. A false positive result (test is a positive test, but the female is not pregnant) may be due to excess protein or blood in urine or hCG production due to a rare type of uterine cancer. Thiazide diuretics, hormones, steroids, and thyroid drugs may also affect the outcome of an early pregnancy test.

GESTATION

The time a zygote, embryo, or fetus is carried in the female reproductive tract is called **gestation** (jes-TĀ-shun). The human gestation period is about 266 days, counted from the estimated day of fertilization. The specialized branch of medicine that deals with pregnancy, labor, and the period of time immediately following delivery is called **obstetrics** (ob-STET-riks; *obstetrix* = midwife).

Anatomical and Physiological Changes

By about the end of the third month of gestation, the uterus occupies most of the pelvic cavity. As the fetus continues to grow, the uterus extends higher and higher into the abdominal cavity. Toward the end of a full-term pregnancy, the uterus fills nearly all the abdominal cavity, reaching above the costal margin nearly to the xiphoid process of the sternum (Fig. 29.10). It pushes the maternal intestines, liver, and stomach upward, elevates the diaphragm, and widens the thoracic cavity. Pressure on the stomach may force the stomach contents upward into the esophagus, resulting in heartburn. In the pelvic cavity, there is compression of the ureters and urinary bladder.

Besides the anatomical changes associated with pregnancy, there are also certain pregnancy-induced physiological changes. General changes include weight gain due to the fetus, amniotic fluid, placenta, uterine enlargement, and increased total body water. Also, there is increased storage of proteins, triglycerides, and minerals; marked breast enlargement in preparation for lactation; and lower back pain due to lordosis (swayback).

With respect to the maternal cardiovascular system, there is an increase in stroke volume by about 30%; a rise in cardiac output by 20 to 30% by the 27th week due to increased maternal blood flow to the placenta and increased metabolism; an increase in heart rate by about 10 to 15%; and an increase in blood volume up to 30 to 50%, mostly during the latter half of pregnancy. When a pregnant female is lying on her back, the enlarged uterus may compress the aorta, resulting in diminished blood flow to the uterus. Compression of the inferior vena cava also decreases venous return, which leads to edema in the lower limbs and may produce varicose veins.

Pulmonary function is also altered during pregnancy. Tidal volume can increase 30 to 40%, expiratory reserve volume can decrease up to 40%, functional residual capacity can decrease up to 25%, minute volume of respiration (MVR) can increase up to 40%, and airway resistance in the bronchial tree can decrease up to 36%. There is also an increase in total body oxygen consumption by about 10 to 20%. Dyspnea also occurs.

With regard to the gastrointestinal tract, there is an increase in appetite and a general decrease in motility that can result in constipation and a delay in gastric emptying time. Nausea, vomiting, and heartburn also occur.

Pressure on the urinary bladder by the enlarging uterus can produce urinary symptoms, such as frequency, urgency, and stress incontinence. Other conditions related to the urinary system include an increase in renal plasma flow up to 35%, an increase in glomerular filtration rate (GFR) up to 40%, and a decrease in ureteral muscle tone.

Changes in the skin during pregnancy are more apparent in some women than others. Included are increased pigmentation around the eyes and cheekbones in a masklike pattern (chloasma), in the areolae of the breasts, and in the linea alba of the lower abdomen (linea nigra). Striae (stretch marks) over the abdomen occur as the uterus enlarges, and hair loss also increases.

Changes in the reproductive system include edema and increased vascularity of the vulva and increased pliability and vascularity of the vagina. The uterus increases in weight from its nonpregnant state of 60 to 80 g to 900 to 1200 g at term. This increase is due to hyperplasia of muscle fibers (cells) in the myometrium in early pregnancy and hypertrophy of muscle fibers during the second and third trimesters.

PREGNANCY-INDUCED HYPERTENSION

About 10 to 15% of all pregnant women in the U.S. experience **pregnancy-induced hypertension (PIH),** elevated blood pressure associated with pregnancy. The major cause is **preeclampsia** (prē'-e-KLAMP-sē-a), in which the hypertension seems to result from impaired renal function. It typically appears after the 20th week of gestation and there are large amounts of protein in urine. Other signs and symptoms are generalized edema, blurred vision, and headaches. It might be related to autoimmune or allergic reaction due to the presence of a fetus. When the condition is also associated with convulsions and coma, it is termed **eclampsia.**

FIGURE 29.10 Normal fetal position at end of full-term pregnancy.

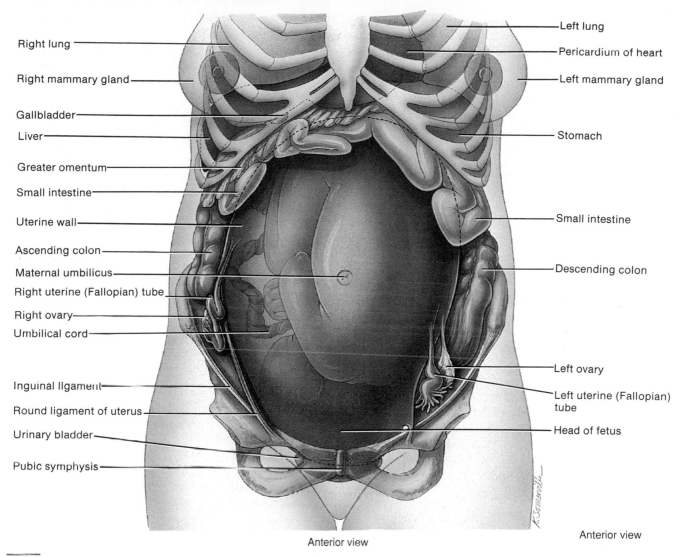

Right lung

Right mammary gland

Gallbladder

Liver

Greater omentum

Small intestine

Uterine wall

Ascending colon

Maternal umbilicus

Right uterine (Fallopian) tube

Right ovary

Umbilical cord

Inguinal ligament

Round ligament of uterus

Urinary bladder

Pubic symphysis

Left lung

Pericardium of heart

Left mammary gland

Stomach

Small intestine

Descending colon

Left ovary

Left uterine (Fallopian) tube

Head of fetus

Anterior view

Anterior view

Question: What hormone relaxes the pubic symphysis and helps dilate the cervix of the uterus to ease delivery of the baby?

Other forms of PIH are not associated with protein in the urine.

Exercise and Pregnancy

Since pregnancy results in so many major body changes, it has an impact on a female's ability to exercise. In early pregnancy, there are only a few changes that affect exercise. A pregnant woman may tire earlier than usual, or she may become lethargic. Morning sickness may also curtail regular exercise. As the pregnancy develops, weight is gained and posture changes. As a result, more energy is needed to perform activities, and certain maneuvers (sudden stopping, changes in direction, rapid movements) are difficult to execute. In the earlier stages of pregnancy, cer-

tain joints, especially the pubic symphysis, become less stable in response to the hormone relaxin. As compensation, many mothers-to-be walk with widely spread legs and a shuffling motion.

Although during exercise blood shifts from viscera (including the uterus) to the muscles and skin, there is no evidence of inadequate blood flow to the placenta. The heat generated during exercise may cause dehydration and further increase body temperature. During early pregnancy, especially, excessive exercise and heat buildup should be avoided since elevated body temperature has been implicated in neural tube defects. Exercise has no known effect on lactation, provided the female remains hydrated and wears a bra with good support. Moderate physical activity does not endanger the fetuses of healthy females who have a normal pregnancy.

Among the benefits of exercise during pregnancy are improvement in oxygen capacity, greater sense of well-being, and fewer minor complaints.

PRENATAL DIAGNOSTIC TESTS

Several tests are available to detect genetic disorders and assess fetal well-being. Here we will describe fetal ultrasonography, amniocentesis, and chorionic villus sampling (CVS).

Fetal Ultrasonography

If there is a question about the normal progress of a pregnancy, **fetal ultrasonography** (ul′-tra-son-OG-ra-fē) may be performed. By far the most common use of diagnostic ultrasound is to determine true fetal age when the date of conception is uncertain. It is also used to evaluate fetal viability and growth, determine fetal position, ascertain multiple pregnancies, identify fetal–maternal abnormalities, and serve as an adjunct to special procedures such as amniocentesis. Ultrasound is not used routinely to determine the sex of a fetus; it is performed only for a specific medical indication.

An instrument (transducer) that emits high-frequency sound waves is passed back and forth over the abdomen. The reflected sound waves from the developing fetus are picked up by the transducer and converted to an image on a screen. This image is called a **sonogram** (see Exhibit 1.4). Since the urinary bladder serves as a landmark during the procedure, the patient needs to drink liquids and not void to maintain a full bladder.

Amniocentesis

Amniocentesis (am′-nē-ō-sen-TĒ-sis; *amnio* = amnion; *kentesis* = puncture) involves withdrawing some of the amniotic fluid that bathes the developing fetus and analyzing the fetal cells and dissolved substances. It is used to test for the presence of certain genetic disorders, such as Down syndrome (DS), spina bifida, hemophilia, Tay–Sachs disease, sickle-cell anemia, and certain muscular dystrophies or to determine fetal maturity and well-being near the time of the delivery. To detect suspected genetic abnormalities, the test is usually done at 14 to 16 weeks of gestation. To assess fetal maturity, it is usually done after the 35th week of gestation. About 300 chromosomal disorders and over 50 biochemical defects can be detected through amniocentesis. It can also reveal gender. This information is important for diagnosis of sex-linked disorders, in which an abnormal gene is carried by the mother but affects only her male offspring. If the fetus is female, it will not be afflicted.

When both parents are known or suspected to be genetic carriers of any one of these disorders, or when the mother is past age 35, amniocentesis is advised. The procedure is also advised when there is concern of a premature delivery, a medical condition that necessitates an early delivery, and for Rh⁻ women who have Rh antibodies in their blood and a fetus that is possibly Rh⁺.

Using ultrasound and palpation, the position of the fetus and placenta are first determined. After the skin is prepared with an antiseptic, a local anesthetic is given, a hypodermic needle is inserted through the mother's abdominal wall and uterus into the amniotic cavity, and about 10 ml of fluid are aspirated (Fig. 29.11a). The fluid and suspended cells are subjected to microscopic examination and biochemical testing. Elevated levels of alphafetoprotein (AFP) and acetylcholine esterase may indicate failure of the nervous system to develop properly, for example, anencephaly (absence of the cerebrum) or spina bifida. Chromosome studies, which require growing the cells for two to four weeks in a culture medium, may reveal rearranged, missing, or extra chromosomes. There is about a 0.5% chance of spontaneous abortion after the test.

Chorionic Villus Sampling

Chorionic (ko-rē-ON-ik) **villus** (VIL-us) **sampling** can determine the same defects as amniocentesis, and it offers several advantages. It can be performed as early as eight weeks of gestation, which permits an earlier decision on whether or not to continue the pregnancy. Moreover, the procedure does not require penetration of the abdomen, uterus, or amniotic cavity. The procedure is slightly more risky than amniocentesis; there is a 1 to 2% chance of spontaneous abortion after the test.

A catheter is placed through the vagina and cervix of the uterus and then advanced to the chorionic villi under ultrasound guidance (Fig. 29.11b). About 30 mg of tissue are suctioned out and prepared for chromosomal analysis. Alternatively, the chorionic villi can be sampled by inserting a needle through the abdominal cavity, as for amniocentesis. Chorion cells and fetal cells contain identical genetic information. Results often are available more quickly than with amniocentesis.

PARTURITION AND LABOR

The term **parturition** (par′-too-RISH-un) refers to birth. **Labor** is the process by which the fetus is expelled from the uterus into the vagina and then outside the body. The onset of labor is apparently related to a complex interaction of many factors. Just before birth, the muscles of the uterus contract rhythmically and forcefully. Both placental and ovarian hormones seem to play a role in these contractions. Since progesterone inhibits uterine contractions, labor cannot take place until its effects are diminished. At the end of gestation, progesterone level falls, the level of estrogens in the mother's blood is sufficient to overcome the inhibiting

FIGURE 29.11 Amniocentesis and chorionic villus sampling.

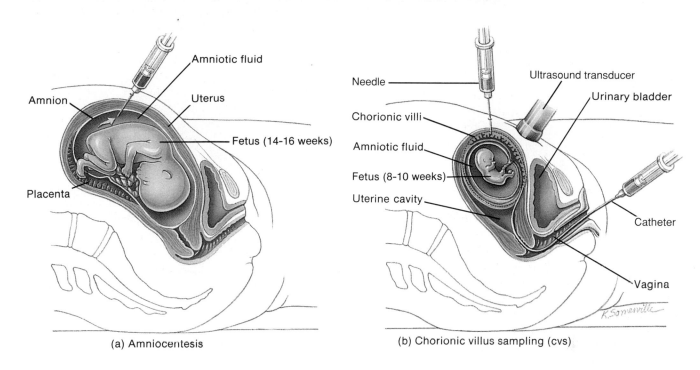

(a) Amniocentesis

(b) Chorionic villus sampling (cvs)

Question: What kind of information is provided by amniocentesis?

effects of progesterone, and labor commences. It has been suggested that cortisol released by the fetus overcomes the inhibiting effects of progesterone so that estrogens can exert their effect. Prostaglandins may also play a role in labor. Oxytocin (OT) from the posterior pituitary gland also stimulates uterine contractions, and relaxin assists by relaxing the pubic symphysis and helping to dilate the uterine cervix.

Uterine contractions occur in waves, quite similar to peristaltic waves, that start at the top of the uterus and move downward. These waves expel the fetus. **True labor** begins when uterine contractions occur at regular intervals, usually producing pain. As the interval between contractions shortens, the contractions intensify. Another sign of true labor in some females is localization of pain in the back, which is intensified by walking. A reliable indication of true labor is the "show" and dilation of the cervix. The "show" is a discharge of a blood-containing mucus that accumulates in the cervical canal during labor. In **false labor,** pain is felt in the abdomen at irregular intervals. The pain does not intensify and is not altered significantly by walking. There is no "show" and no cervical dilation.

Labor can be divided into three stages (Fig. 29.12).

1. Stage of dilation. The time from the onset of labor to the complete dilation of the cervix is the **stage of dilation**. During this stage, there are regular contrac-

tions of the uterus, usually a rupturing of the amniotic sac, and complete dilation (10 cm) of the cervix. If the amniotic sac does not rupture spontaneously, it is done deliberately.

2. Stage of expulsion. The time from complete cervical dilation to delivery of the baby is the **stage of expulsion**.

3. Placental stage. The time after delivery until the placenta or "afterbirth" is expelled by powerful uterine contractions is the **placental stage**. These contractions also constrict blood vessels that were torn during delivery. In this way, the chance of hemorrhage is reduced.

During labor, the fetus may be squeezed through the birth canal (cervix and vagina) for up to several hours. As a result, the fetal head is compressed, and there is some degree of intermittent hypoxia due to compression of the umbilical cord and placenta during uterine contractions. In response to this compression, the adrenal medullae of a fetus secrete very high levels of epinephrine and norepinephrine (NE), the "fight-or-flight" hormones. Much of the protection afforded against the stresses of the birth process and preparation of the infant to survive extrauterine life are provided by the adrenal medullary hormones. Among other functions, the hormones clear the lungs and alter their physiology for breathing outside the uterus, mobilize readily usable nutrients for cellular metabolism, and promote a rich vascular supply to the brain and heart.

FIGURE 29.12 Stages of labor.

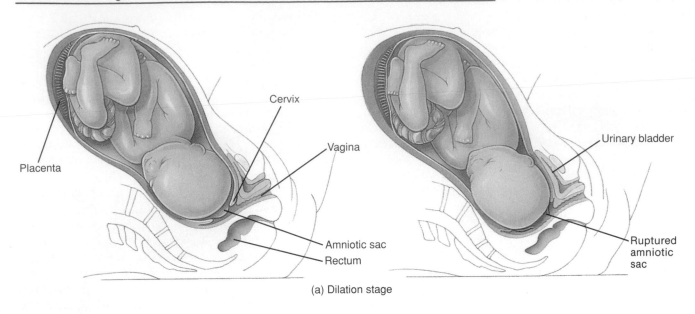

Placenta
Cervix
Vagina
Amniotic sac
Rectum
Urinary bladder
Ruptured amniotic sac

(a) Dilation stage

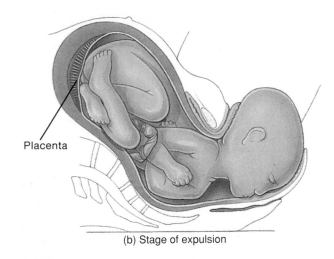

Placenta

(b) Stage of expulsion

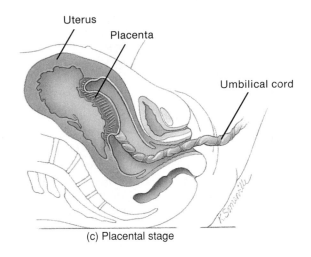

Uterus
Placenta
Umbilical cord

(c) Placental stage

Question: What event marks the beginning of the stage of expulsion?

CLINICAL APPLICATION

DYSTOCIA AND CESAREAN SECTION

Dystocia (dis-TŌ-sē-a), or difficult labor, may result from impaired uterine forces, an abnormal position (presentation) of the fetus, or a birth canal of inadequate size to permit vaginal birth. For example, in a **breech presentation** the fetal buttocks or lower extremities, rather than the head, enter the birth canal first; this occurs most often in premature births. If fetal or maternal distress prevents a vaginal birth, the baby may be delivered via a **cesarean** (*caedere* = to cut) **section (C-section).** A low, horizontal incision is made through the abdominal wall and lower portion of the uterus, through which the baby and placenta are removed. Even a history of multiple C-sections need not exclude a pregnant woman from attempting a vaginal delivery.

About 7% of pregnant females do not deliver by two weeks after their due date, and some do not go into labor for up to five weeks past their due date. In such cases, there is increased risk of brain damage to the fetus and even fetal death. This is due to inadequate supply of oxygen and nutrients from an aging placenta. Postterm deliveries may be facilitated by induced labor (initiated by artificial means such as the administration of oxytocin) or cesarean section.

After delivery of the baby and placenta, there is a period of time that lasts about six weeks during which the reproductive organs and maternal physiology return to the prepregnancy state. This is called the **puerperium** (pyoo'-er-PE-rē-um). Through a process of tissue catabolism, the uterus undergoes a remarkable reduction in size, called **involution** (in'-vō-LOO-shun). It is related primarily to a decrease in cytoplasm and cell size of myometrial cells. The cervix loses its elasticity and regains its prepregnancy firmness. For two to four weeks after delivery, there is a uterine discharge called **lochia** (LŌ-kē-a), which consists initially of blood and later serous fluid derived from the former placental site.

ADJUSTMENTS OF THE INFANT AT BIRTH

During pregnancy, the embryo and later the fetus is totally dependent on the mother for its existence. The mother supplies the fetus with oxygen and nutrients, eliminates its carbon dioxide and other wastes, and protects it against shocks, temperature changes, and certain harmful microbes. At birth, a physiologically mature baby becomes self-supporting, and the newborn's body systems must make various adjustments. Following are some changes that occur in the respiratory and cardiovascular systems.

Respiratory System

The fetus depends entirely on the mother for obtaining oxygen and eliminating carbon dioxide. The fetal lungs are either collapsed or partially filled with amniotic fluid, which is absorbed at birth. The respiratory system is fairly well-developed at least two months before birth. Premature babies delivered at seven months are able to breathe and cry. After delivery, the baby's supply of oxygen from the mother ceases. Circulation in the baby continues, and as the blood level of carbon dioxide increases, the respiratory center in the medulla is stimulated. This causes the respiratory muscles to contract, and the baby draws its first breath. Since the first inspiration is unusually deep because the lungs contain no air, the baby exhales vigorously and naturally cries. A full-term baby may breathe 45 times a minute for the first two weeks after birth. Breathing rate gradually declines until it approaches a normal rate.

Cardiovascular System

Following the first inspiration by the baby, the cardiovascular system must make several adjustments (see Fig. 21.32). The foramen ovale between the atria of the fetal heart closes at the moment of birth. This diverts deoxygenated blood to the lungs for the first time. The foramen ovale is closed by two flaps of heart tissue that fold together and permanently fuse. The remnant of the foramen ovale is the fossa ovalis. Once the lungs begin to function, the ductus

arteriosus is shut off by contractions of the muscles in its wall. The ductus arteriosus generally does not completely and irreversibly close for about three months after birth. Incomplete closing results in patent ductus arteriosus.

The ductus venosus connects the umbilical vein directly with the inferior vena cava. It allows blood from the umbilical vein to bypass the fetal liver. When the umbilical cord is severed, blood from the viscera of the fetus goes directly to the fetal heart via the inferior vena cava. This shunting of blood usually occurs within minutes after birth but may take a week or two to complete. The ligamentum venosum, the remnant of the ductus venosus, is well established by the eighth week after birth.

At birth, the infant's pulse may be from 120 to 160 beats per minute and may go as high as 180 following excitation. Several days after birth, there is a greater independent need for oxygen, which stimulates an increase in the rate of erythrocyte and hemoglobin production. This increase usually lasts for only a few days. Moreover, the white blood cell count at birth is very high, sometimes as much as 45,000 cells per cubic millimeter, but this decreases rapidly by the seventh day.

Finally, the infant's liver may not be adjusted at birth to control the production of bile pigment. As a result of this and other complicating factors, a temporary jaundice may result in as many as 50% of normal newborns by the third or fourth day after birth.

CLINICAL APPLICATION

PREMATURE INFANTS

Delivery of a physiologically immature baby carries certain risks. A **premature infant** or "preemie" is generally considered to be one who weighs less than 2500 g (5 lb, 8 oz) at birth. Poor prenatal care, drug abuse, a history of an earlier premature delivery, and mother's age below 16 or above 35 increase the chance of premature delivery. The problems related to survival of premature infants are due to the fact that they are not yet ready to take over functions the mother's body normally is performing. The major problem with delivery of an infant under 36 weeks of gestation is infant respiratory distress syndrome (RDS) (see page 759). This condition can be helped by a ventilator that delivers oxygen until the lungs can operate on their own. Researchers are now working on ways to inject surfactant to overcome newborn RDS.

PHYSIOLOGY OF LACTATION

The term **lactation** refers to the secretion and ejection of milk by the mammary glands. A principal hormone in promoting lactation is **prolactin (PRL)** from the anterior pituitary gland. It is released in response to prolactin releasing hormone (PRH) secreted by the hypothalamus. Even though prolactin levels increase as the pregnancy progresses, there is no milk secretion because estrogens and progesterone inhibit the prolactin from being effective. After delivery, the levels of estrogens and progesterone in the mother's blood decrease, and the inhibition is removed.

The principal stimulus in maintaining prolactin secretion during lactation is the sucking action of the infant. Sucking initiates nerve impulses from receptors in the nipples to the hypothalamus. The impulses decrease release of prolactin inhibiting hormone (PIH) and/or increase release of prolactin releasing hormone (PRH), so more prolactin is released by the anterior pituitary gland.

The sucking action also initiates nerve impulses to the posterior pituitary gland via the hypothalamus. These impulses stimulate the release of the hormone **oxytocin (OT)** by the posterior pituitary gland. Oxytocin induces myoepithelial cells surrounding the outer walls of the alveoli to contract, thereby compressing the alveoli and ejecting milk. The compression moves milk from the alveoli of the mammary glands into the ducts, where it can be suckled. This process is termed **milk ejection (let-down).** Although the actual ejection of milk does not occur from 30 seconds to 1 minute after nursing begins, some milk is stored in lactiferous sinuses near the nipple. Thus some milk is available during the latent period.

Recall from Chapter 18 (see Fig. 18.11) that oxytocin, as part of a positive feedback cycle, also stimulates contraction of smooth muscle in the pregnant uterus during labor and delivery. This action of oxytocin in nursing mothers results in a more rapid return of the uterus to its prepregnant size than occurs in nonnursing mothers. OT also compresses torn placental vessels at delivery and thus reduces blood loss by the mother.

During late pregnancy and the first few days after birth, the mammary glands secrete a cloudy fluid called **colostrum.** Although it is not as nutritious as milk, since it contains less lactose and virtually no fat, it serves adequately until the appearance of true milk on about the fourth day. Colostrum and maternal milk contain antibodies that protect the infant during the first few months of life.

Following birth of the infant, the prolactin level starts to return to the nonpregnant level. However, each time the mother nurses the infant, nerve impulses from the nipples to the hypothalamus increase the release of PRH (and/or decrease the release of PIH) resulting in a 10-fold increase in prolactin secretion by the anterior pituitary that lasts about an hour. Prolactin acts on the mammary glands to provide milk for the next nursing period. If this surge of prolactin is blocked by injury or disease, or if nursing is discontinued, the mammary glands lose their ability to secrete milk in a few days. Milk secretion normally decreases considerably within seven to nine months. However, it can continue for several years if the child continues to nurse.

Lactation often prevents the occurrence of female ovarian cycles for the first few months following delivery, if the frequency of sucking is about 8 to 10 times a day. This effect is inconsistent, however, and ovulation will normally precede the first menstrual period after delivery of a baby. So the mother can never be certain she is not fertile. Breast feeding is therefore not a very effective birth control measure. The suppression of ovulation during lactation is believed to occur as follows. During breast feeding, neural input from the nipple reaches the hypothalamus and causes it to produce beta β-endorphin. This, in turn, suppresses the release of gonadotropin releasing hormone (GnRH). Thus there is a decreased production of LH and FSH, and ovulation is inhibited.

Advocates of **breast feeding** feel that it offers the following advantages to the infant.

1. It establishes early and prolonged contact between mother and infant.

2. The infant is more in control of intake.

3. Fats and iron in human milk are better absorbed than those in cow's milk, and the amino acids in human milk are more readily metabolized. Also, the lower sodium content of human milk is more suited to the infant's needs.

4. Breast feeding provides important antibodies that prevent gastroenteritis. Immunity to respiratory infections and meningitis is also greater.

5. Premature infants benefit from breast feeding because the milk produced by mothers of premature infants seems to be specially adapted to the infant's needs by having a higher protein content than the milk of mothers of full-term infants.

6. There is less likelihood of an allergic reaction in the baby to the milk of the mother.

7. The act of sucking on the breast promotes development of the jaw, facial muscles, and teeth.

8. Several proteins in breast milk may stimulate the infant's immune system by increasing B lymphocyte maturation.

As indicated earlier, fertilization involves the coming together of genetic material of the father with that of the mother when a spermatozoan merges with an ovum to form a zygote. Through development, the zygote becomes an embryo and later a fetus. The newborn child resembles its parents because of the inheritance of traits passed down from both parents. We will now examine some of the principles involved in inheritance.

INHERITANCE

Inheritance is the passage of hereditary traits from one generation to another. It is the process by which you acquired your characteristics from your parents and will transmit some of your traits to your children. The branch of biology that deals with inheritance is called **genetics** (je-NET-iks). The area of health care that offers advice on genetic problems is called **genetic counseling.**

MAPPING THE HUMAN GENOME

A **genome** (JĒ-nōm) is the complete genetic makeup of an organism. Of the approximately 100,000 genes in a human cell, only a small number have been identified and roughly located on various chromosomes. In 1988 a project to map and sequence the entire human genome got underway. A major goal is the application of molecular genetics to medical research to identify and find cures for certain diseases.

Genotype and Phenotype

The nuclei of all human cells except gametes contain 23 pairs of chromosomes—the diploid number. One chromosome in each pair came from the mother, and the other came from the father. Homologues, the two chromosomes in a pair, contain genes that control the same traits. If a chromosome contains a gene for height, its homologue will contain a gene for height. The two genes that code for the same trait and are at the same location on homologous chromosomes are termed **alleles** (ah-LĒLZ). A **mutation** (myoo-TĀ-shun; *mutare* = change) is a permanent heritable change in a gene that causes it to have a different effect than it had previously.

The relationship of genes to heredity is illustrated by the disorder called **phenylketonuria** or **PKU** (see Fig. 29.13). People with PKU are unable to manufacture the enzyme phenylalanine hydroxylase (see page 858). It is believed the PKU is brought about by an abnormal allele, which we symbolize as p. The normal allele is symbolized as P. The chromosome concerned with directions for phenylalanine hydroxylase production includes either the p or P allele. Its homologue will also have p or P. Thus every individual will have one of the following genetic makeups, or **genotypes** (JĒ-nō-tīps): PP, Pp, or pp. Although people with genotypes of Pp have the abnormal allele, only those with genotype pp suffer from the disorder because the normal allele, when present, dominates the abnormal one. An allele that dominates or masks the presence of another allele and is fully expressed is said to be **dominant,** and the trait expressed is a dominant trait. The allele whose presence is completely masked is said to be **recessive,** and the trait it controls is a recessive trait.

By tradition, we symbolize the dominant allele with a capital letter and the recessive one with a lowercase letter. A person with the same genes on homologous chromosomes (for example, PP or pp) is said to be **homozygous** for the trait. An individual with different genes on homologous chromosomes (for example, Pp) is said to be **heterozygous** for the trait.

Phenotype (FĒ-nō-tīp; *pheno* = showing) refers to how the genetic makeup is expressed in the body. It is the physical expression of a gene. A person with Pp has a different genotype from one with PP, but both have the same pheno-type—normal production of phenylalanine hydroxylase. Individuals who carry a recessive gene but do not express it (Pp) can pass the gene on to their offspring. Such individuals are called **carriers.**

Most genes give rise to the some phenotype whether they are inherited from the mother or the father. In a few cases, however, the phenotype is dramatically different, depending on the parental origin. This surprising phenomenon, first appreciated in the 1980s, is called **genomic imprinting.** Animal studies have shown that normal embryonic growth and development require both male and female genomes. Embryos that contain two sets of maternal chromosomes or two sets of paternal chromosomes do not survive. In humans the abnormalities most clearly associated with mutation of an imprinted gene are *Angelman syndrome,* which results when the abnormal gene is inherited from the mother, and *Prader–Willi syndrome,* which results when the abnormality is inherited from the father.

To determine the possible ways that haploid gametes can unite to form diploid fertilized eggs, special charts called **Punnett squares** are used. Usually, the male gametes (sperm cells) are placed at the side of the chart, and the female gametes (ova) at the top (as in Fig. 29.13). The four spaces on the chart represent the possible genotypes of fertilized eggs that could form from the union of male and female gametes.

Normal traits do not always dominate over abnormal ones, but genes for severe disorders often are recessive because dominant genes for severe disorders usually are lethal; they cause death of the embryo or fetus. One exception is Huntington's chorea, which is caused by a dominant gene.

A cell that has one or more chromosomes of a set added or deleted is called an **aneuploid** (an'-yoo-PLOID) A monosomic cell ($2n - 1$) is missing a chromosome; a trisomic cell ($2n + 1$) has an added chromosome. Down syndrome (see page 996) is an example of an aneuploid disorder. Since the most common form of Down syndrome is characterized by an extra chromosome added to the 21st pair, it is also known as **trisomy 21.**

Exhibit 29.3 lists some of the variety of simple inherited structural and functional traits in humans.

Variations on Dominant–Recessive Inheritance

Most patterns of inheritance are not simple **dominant–recessive inheritance** in which only dominant and recessive genes interact. In fact, the phenotypic expression of a particular gene is influenced not only by which alleles are present, but also by other genes and the environment. Moreover, most inherited traits are influenced by more than one gene and most genes can influence more than a single trait. Following are some examples.

FIGURE 29.13 Inheritance of phenylketonuria (PKU).

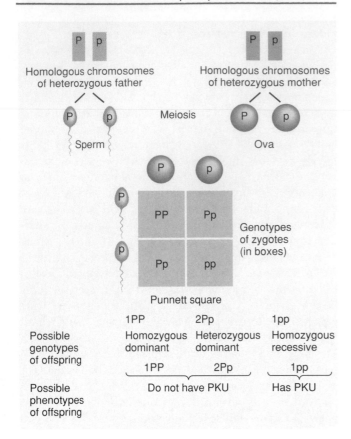

Possible genotypes of offspring

1PP	2Pp	1pp
Homozygous dominant	Heterozygous dominant	Homozygous recessive

1PP   2Pp   1pp

Possible phenotypes of offspring

Do not have PKU   Has PKU

Question: If a couple has the genotypes shown here, what is the percent chance that their child will have PKU?

Incomplete Dominance

In **incomplete dominance**, neither member of an allelic pair is dominant over the other and the heterozygote has a phenotype intermediate between the homozygous dominant and homozygous recessive. For example, in snapdragons, crossing homozygous red flowers (RR) with homozygous white flowers (R^1R^1) does not produce red flowers. Instead, the flowers are pink (RR^1).

An example of incomplete dominance in humans is the inheritance of **sickle-cell anemia (SCA)** (Fig. 29.14). Individuals with the homozygous dominant genotype Hb^AHb^A are normal; those with the homozygous recessive genotype Hb^SHb^S have the disease called sickle-cell anemia and have severe anemia; and, though they are usually healthy, those with the heterozygous genotype Hb^AHb^S have minor problems with anemia. Since people in this last category are carriers, they are sometimes referred to as having the "sickle-cell" trait.

Another example of incomplete dominance in humans is the inheritance of hair texture. In this example, HH represents the homozygous genotype for curly hair; H^1H^1 represents the homozygous genotype for straight hair; and HH^1 represents the heterozygous genotype for wavy hair.

Multiple-Allele Inheritance

Although a single individual inherits only two alleles for each gene, some genes may have more than two alternate forms and this is the basis for **multiple-allele inheritance**.

One example of multiple-allele inheritance in humans is the inheritance of the ABO blood group. The four blood types of the ABO group—A, B, AB, and O—arise as a result of three different alleles (A, B, O) of a single gene. Whereas alleles A and B produce agglutinogens (antigens) A and B on the surfaces of red blood cells, allele O produces none. Every person inherits two genes, one from each parent, that are responsible for production of these agglutinogens. The six possible genotypes (genetic make-ups) are: OO, AO, AA, BO, BB, and AB. Both A and B are inherited as dominant traits; O is inherited as a recessive

FIGURE 29.14 Inheritance of sickle-cell anemia.

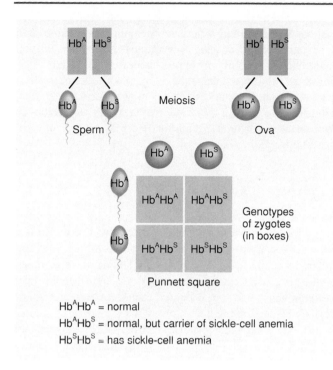

$Hb^A Hb^A$ = normal
$Hb^A Hb^S$ = normal, but carrier of sickle-cell anemia
$Hb^S Hb^S$ = has sickle-cell anemia

Question: What type of inheritance is illustrated here?

FIGURE 29.15 Inheritance of possible and impossible phenotypes of the ABO blood grouping system (see Exhibit 19.5).

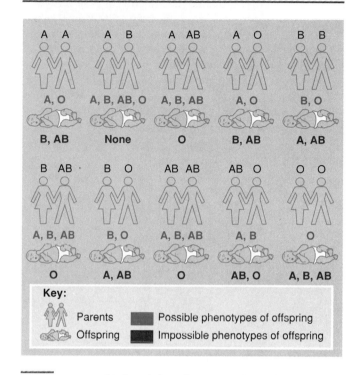

Question: Is it possible for a baby to have type O blood if neither parent is type O?

trait. The various possible and impossible blood group phenotypes are illustrated in Fig. 29.15.

Since an individual with type AB blood has characteristics of both type A and type B red blood cells expressed in the phenotype, alleles A and B are said to be **codominant**. In other words, both genes are expressed in the heterozygote equally.

Polygenic Inheritance

Many inherited traits are not controlled by one gene, but rather by the combined effects of many genes. This is referred to as **polygenic inheritance**. An inherited trait affected by a number of genes (polygenes) shows a gradation of small differences between extremes among individuals. Examples of polygenic traits include skin color, hair color, eye color, height, and body build. As you will see shortly, polygenic inheritance is also influenced by environment.

As an example of polygenic inheritance, we will consider skin color. Skin color is controlled by three separate genes, each having two alleles: A, a; B, b; and C, c. Whereas a person with the genotype AABBCC is very dark skinned, an individual with the genotype aabbcc is very light skinned; and a person with the genotype AaBbCc has an intermediate skin color. Figure 29.16 illustrates the inheritance of skin color and the various gradations of skin

tone. Note that the **P generation** (parental generation) is the starting generation, the **F_1 generation** (first filial generation) is produced from the P generation, and the **F_2 generation** (second filial generation) is produced from the F_1 generation.

Genes and the Environment

A given phenotype is the result of both genotype and environment. For example, even though a person inherits genes for tallness, full potential may not be reached due to environmental factors, such as disease or malnutrition during the growth years.

Exposure of a developing embryo or fetus to certain environmental factors can damage the developing organism or even cause death. A **teratogen** (*terato* = monster) is any agent or influence that causes developmental defects in the embryo. Following are several examples.

Chemicals and Drugs

Since the placenta is a porous barrier between the maternal and fetal circulations, any drug or chemical dangerous to an infant may be considered potentially dangerous to the fetus when given to the mother. Examples are pesticides; defoliants; industrial chemicals; some hormones; antibiotics; oral anticoagulants, anticonvulsants, antitumor agents, thyroid drugs, thalidomide, diethylstilbestrol (DES), and numerous other prescription drugs; LSD; marijuana; and cocaine. A pregnant female who uses cocaine subjects the fetus to higher risk of retarded growth, attention and orientation problems, hyperirritability, a tendency to stop breathing, crib death, malformed or missing organs, strokes, and seizures. The risks of spontaneous abortion (miscarriage), premature birth, and stillbirth also increase from fetal exposure to cocaine.

Alcohol

Alcohol is by far the number one fetal teratogen. It has been a suspected teratogen for centuries, but only recently has a relationship been recognized between maternal alcohol intake and the characteristic pattern of malformations in the fetus. Intrauterine exposure to alcohol results in **fetal alcohol syndrome (FAS).** The symptoms of FAS may include slow growth before and after birth, small head, facial irregularities such as narrow eye slits and sunken nasal bridge, defective heart and other organs, malformed arms and legs, genital abnormalities, and damage of the central nervous system. There are characteristic behavioral problems, such as hyperactivity, extreme nervousness, a poor attention span, and an inability to appreciate cause-and-effect relationships. In humans, acetaldehyde is one of the toxic products of alcohol metabolism and may contribute to fetal damage. The human placenta may transfer acetaldehyde from maternal to fetal circulation and may oxidize ethanol to acetaldehyde.

Cigarette Smoking

Strong evidence implicates cigarette smoking during pregnancy as a cause of low infant birth weight. There also is a strong association between smoking and a higher fetal and infant mortality rate. Women who smoke have a much higher risk of an ectopic pregnancy. Cigarette smoke may be teratogenic and cause cardiac abnormalities and anencephaly (a developmental anomaly with absence of a cerebrum). Maternal smoking also appears to be a significant factor in the development of cleft lip and palate and has been tentatively linked with sudden infant death syndrome (SIDS). Infants nursing from smoking mothers have also been found to have an increased incidence of gastrointestinal disturbances. Even exposure to secondhand cigarette smoke (in the air they breathe) predisposes infants to increased incidence of respiratory problems during the first year of life, including bronchitis and pneumonia.

Irradiation

Ionizing radiations are potent teratogens. Exposure of pregnant mothers to x-rays or radioactive isotopes during the embryo's susceptible period of development may cause microcephaly (small size of head in relation to the rest of the body), mental retardation, and skeletal malformations. Caution is advised for diagnostic x-rays, especially during the first trimester of pregnancy.

Inheritance of Sex

Examination of the chromosomes in cells reveals that one pair differs between males and females (see Fig. 29.18). In females, the pair consists of two chromosomes designated as X chromosomes. One X chromosome is also present in males, but its mate is a smaller chromosome called a Y chromosome. The XX pair in the female and the XY pair in the male are called the **sex chromosomes.** All other chromosomes are called **autosomes.**

When a spermatocyte undergoes meiosis to reduce its chromosome number, one daughter cell will contain the X chromosome, and the other will contain the Y chromosome.

FIGURE 29.16 Polygenic inheritance of skin color.

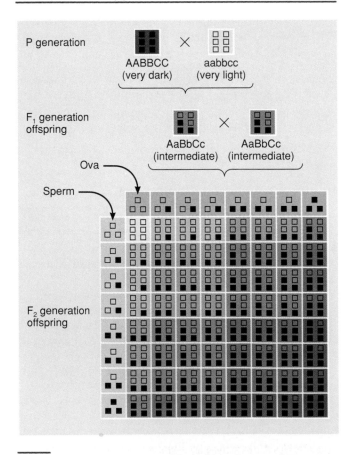

Question: What other traits are transmitted by polygenic inheritance?

Oocytes have no Y chromosomes and produce only X-containing ova. If the secondary oocyte is fertilized by an X-bearing sperm, the offspring normally will be female (XX). Fertilization by a Y-bearing sperm normally produces a male (XY). Thus sex is determined at fertilization (Fig. 29.17).

Both female and male embryos develop identically until about seven weeks after fertilization. At that point, one or more genes set into motion a cascade of events that leads to the development of a male. In the absence of the gene or genes, the female pattern of development occurs. Since 1959 it has been known that the Y chromosome is needed to initiate male development. Experiments published in 1991 established that the prime male-determining gene is one called **SRY (sex-determining region of the Y chromosome)**. When a small DNA fragment containing this gene, called Sry in mice, was inserted into 11 female mouse embryos, three of them developed as males. (The researchers suspect that the gene failed to be integrated into the genetic material in the other eight.) SRY apparently acts as a molecular switch to turn on the male pattern of development. Whenever the SRY gene is present in a fertilized ovum, the fetus will develop testes and differentiate into a male; in the absence of SRY, the fetus will develop ovaries and differentiate into a female.

An experiment of nature provides confirming evidence for humans. In two cases investigated so far, phenotypic females with an XY genotype were found to have mutated SRY genes. In other words, they failed to develop normally as males because their SRY gene was defective.

CLINICAL APPLICATION

KARYOTYPING

A **karyotype** (KAR-ē-ō-tīp; *karyon* = nucleus) is an arrangement of chromosomes from a cell based on their shape, size, and position of their centromeres. A karyotype is prepared by photographing the chromosomes, usually in a white blood cell, cutting them out of a printed photograph, and then arranging them in standard order (Fig. 29.18). A normal human karyotype contains 22 matching pairs of autosomes and one pair of sex chromosomes (two X chromosomes in females and one X and one Y chromosome in males).

Karyotyping is done when a chromosomal abnormality is suspected that might be responsible for a disease or developmental problem. It is often done to investigate birth defects, abnormal growth, mental retardation, delayed puberty, abnormal sexual development, infertility, or certain inherited disorders. Missing, additional, or abnormal chromosomes in a karyotype are indicative of disorders such as Down syndrome, Klinefelter's syndrome, and Turner's syndrome (see page 996).

Red–Green Color Blindness and Sex-Linked Inheritance

The sex chromosomes also are responsible for the transmission of several nonsexual traits. Genes for these traits appear on X chromosomes, but are absent from Y chromosomes. This feature produces a pattern of heredity that is different from the pattern described earlier. Let us consider the most common type of color blindness called red–green color blindness. In this condition, there is a deficiency in either red or green cones and red and green are seen as the same color, either red or green, depending on which cone is

FIGURE 29.17 Sex determination.

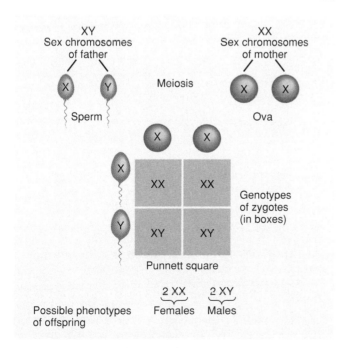

Question: What are chromosomes other than sex chromosomes called?

FIGURE 29.18 Normal human karyotype.

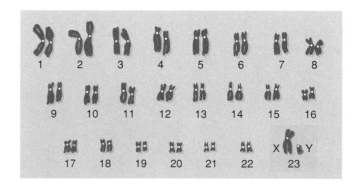

Question: Was this karyotype from a male or a female? How did you decide?

present. The gene for **red–green color blindness** is a recessive one designated c. Normal color vision, designated C, dominates. The C/c genes are located on the X chromosome. The Y chromosome does not contain the segment of DNA that programs this aspect of vision. Thus the ability to see colors depends entirely on the X chromosomes. The genetic possibilities are:

X^CX^C Normal female

X^CX^c Normal female carrying the recessive gene

X^cX^c Red–green color-blind female

X^CY Normal male

X^cY Red–green color-blind male

Only females who have two X^c genes are red–green color blind. In X^CX^c females the trait is masked by the normal, dominant gene. Males, on the other hand, do not have a second X chromosome that would mask the trait. Therefore all males with an X^c gene will be red–green color blind. The inheritance of red–green color blindness is illustrated in Fig. 29.19.

Traits inherited in the manner just described are called **sex-linked traits.** The most common type of **hemophilia** —a condition in which the blood fails to clot or clots very slowly after an injury (see page 582)—is a sex-linked trait. Like the trait for red–green color blindness, hemophilia is caused by a recessive gene. Other sex-linked traits in humans are fragile X syndrome (see page 997), nonfunctional sweat glands, certain forms of diabetes, some types of deafness, uncontrollable rolling of the eyeballs, absence of central incisors, night blindness, one form of cataract, juvenile glaucoma, and juvenile muscular dystrophy.

FIGURE 29.19 Inheritance of red–green color blindness.

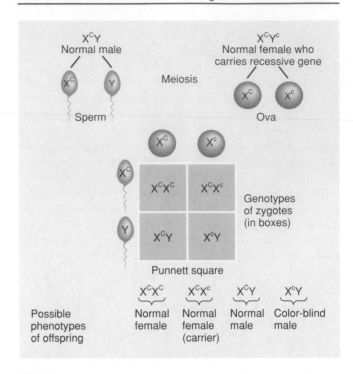

Question: What would be the genotype of a red–green color-blind female?

DOWN SYNDROME
Down syndrome (DS) is a disorder that results from an error in cell division called **nondisjunction.** In this situation, homologous chromosomes fail to separate properly during reduction division of meiosis. As a result, an extra chromosome passes to one of the daughter cells (gametes). Individuals with the disorder usually have 47 chromosomes instead of the normal 46 (an extra chromosome 21). For this reason Down syndrome is also called trisomy 21.

Down syndrome is characterized by mental retardation, retarded physical development (short stature and stubby fingers), distinctive facial structures (large tongue, flat profile, broad skull, slanting eyes, and round head), and malformation of the heart, ears, hands, and feet (Fig. 29.20). Sexual maturity is rarely attained.

An estimated 1 in 600 to 800 infants is born with Down syndrome and a woman's chances of giving birth to a child with the disorder increase with age. For example, the chance of having a baby with this syndrome is 1 in 1205 for a 25-year-old woman, but is 1 in 365 for a 35-year-old woman.

SEX CHROMOSOME ANEUPLOIDS
Turner's syndrome, caused by the presence of only one X chromosome (XO), is an example of a sex chromosome aneuploid. Such females are sterile with virtually no ovaries and limited development of secondary sex characteristics. Other features include short stature, webbed neck, underdeveloped breasts, and widely spaced nipples. There is usually no mental retardation. Another sex chromosome aneuploid condition is called **metafemale syndrome,** characterized by

FIGURE 29.20 Photograph of an individual with Down syndrome.

at least three X chromosomes (XXX). These females have underdeveloped genital organs and limited fertility and are generally mentally retarded. **Klinefelter's syndrome** is usually due to trisomy XXY. Such individuals are sterile males. They have undeveloped testes, scanty body hair, and enlarged breasts and are characteristically somewhat mentally disadvantaged.

FRAGILE X SYNDROME
Fragile X syndrome is a recently recognized disorder due to a defective gene on the X chromosome. It is so named because a small portion of the tip of the X chromosome seems susceptible to breakage. It is the leading cause of mental retardation among newborns. Fragile X syndrome affects males more than females and results in learning difficulties, mental retardation, and physical abnormalities such as oversized ears, elongated forehead, enlarged testes, and double jointedness. The syndrome may also be involved in autism in which the individual exhibits extreme withdrawal and refusal to communicate.

Fragile X syndrome can be diagnosed by amniocentesis. Although the syndrome is largely sex-linked, 20 to 50% of males who inherit the trait are unaffected, but they can pass it on to their daughters. Although the daughters also are unaffected, their children, both male and female, may suffer mental retardation. Researchers have suggested that this pattern of inheritance could be explained by maternal imprinting of the gene. On this theory, the father's fragile X gene becomes chemically changed as it passes through the daughter so that the gene is expressed in the daughter's offspring.

Study Outline

Development During Pregnancy (p. 971)

1. Pregnancy is a sequence of events that starts with fertilization.
2. Its various events are hormonally controlled.
3. Fertilization refers to the penetration of a secondary oocyte by a sperm cell to form a zygote and the subsequent union of the sperm and ovum nuclei.
4. Penetration is facilitated by enzymes produced by sperm acrosomes.
5. Normally, only one sperm fertilizes a secondary oocyte.
6. Early rapid cell division of a zygote is called cleavage, and the cells produced by cleavage are called blastomeres.
7. The solid mass of cells produced by cleavage is a morula.
8. The morula develops into a blastocyst, a hollow ball of cells differentiated into a trophoblast (future embryonic membranes) and inner cell mass (future embryo).
9. The attachment of a blastocyst to the endometrium is called implantation; it occurs by enzymatic degradation of the endometrium.
10. In vitro fertilization (IVF) refers to the fertilization of a secondary oocyte outside the body and the subsequent implantation of the zygote. Variations include embryo transfer and gamete intrafallopian transfer (GIFT).

Embryonic Development (p. 975)

1. During embryonic growth, the primary germ layers and embryonic membranes are formed.
2. The primary germ layers—ectoderm, mesoderm, and endoderm—form all tissues of the developing organism.

3. Embryonic membranes include the yolk sac, amnion, chorion, and allantois.
4. Fetal and maternal materials are exchanged through the placenta.

Fetal Growth (p. 981)

1. During the fetal period, organs established by the primary germ layers grow rapidly.
2. The principal changes associated with fetal growth are summarized in Exhibit 29.2.

Hormones of Pregnancy (p. 981)

1. Pregnancy is maintained by human chorionic gonadotropin (hCG), estrogens, and progesterone (PROG).
2. Human chorionic somatomammotropin (hCS) assumes a role in breast development, protein anabolism, and glucose and fatty acid catabolism.
3. Relaxin relaxes the pubic symphysis and helps dilate the uterine cervix toward the end of pregnancy.
4. Inhibin inhibits secretion of FSH and might regulate secretion of hGH.

Gestation (p. 984)

1. The time an embryo or fetus is carried in the uterus is called gestation.
2. Human gestation lasts about 266 days after fertilization.
3. During gestation, several anatomical and physiological changes occur.

FIGURE A.1 Metric and U.S. units of length.

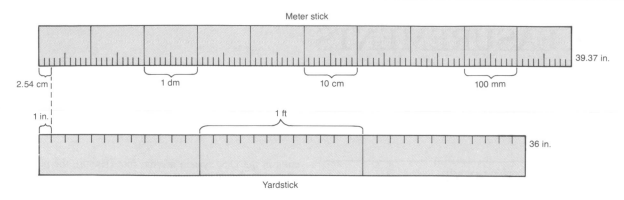

METRIC SYSTEM

The *metric system,* introduced in France in 1790, is now used by all major countries except the United States. Scientific observations are almost universally expressed in metric units.

LENGTH

The standard of length in the metric system is the *meter* (m). It was originally defined in 1790 as one ten-millionth of the distance from the North Pole to the Equator. In 1889 it was redefined as the distance measured at 0°C between two lines on a bar of platinum-iridium kept at the International Bureau of Weights and Measures in France. In 1960 the meter was redefined as the length of 1,650,763.73 light waves emitted by atoms of the gas krypton under strictly specified conditions. The meter is now defined as the distance light travels in 1/299,792,458 second. The meter is equal to 39.37 inches.

A major advantage of the metric system is that units are related to one another by factors of 10. Thus 1 m is 10 decimeters (dm) or 100 centimeters (cm) or 1,000 millimeters (mm). Conversion from one unit to another is simple. Figure A.1 illustrates the differences between metric and U.S. conversions by comparing the meter stick and the yardstick. Exhibit A.2 lists the metric units of length with U.S. equivalents.

Since numbers with many zeros (very large numbers or very small fractions) are cumbersome to work with, they are expressed in *exponential form,* that is, as powers of 10. The form of exponential notation is

$$M \times 10^n$$

EXHIBIT A.2

METRIC UNITS OF LENGTH AND SOME U.S. EQUIVALENTS

Metric Unit	Meaning of Prefix	Metric Equivalent	U.S. Equivalent
1 kilometer (km)	kilo = 1,000	1,000 m = 10^3 m	3,280.84 ft = 0.62 mi 1 mi = 1.61 km
1 hectometer (hm)	hecto = 100	100 m = 10^2 m	328 ft
1 dekameter (dam)	deka = 10	10 m = 10^1 m	32.8 ft
1 meter (m)		Standard unit of length	39.37 in. = 3.28 ft = 1.09 yd
1 decimeter (dm)	deci = $\frac{1}{10}$	0.1 m = 10^{-1} m	3.94 in.
1 centimeter (cm)	centi = $\frac{1}{100}$	0.01 m = 10^{-2} m	0.394 in. 1 in. = 2.54 cm
1 millimeter (mm)	milli = $\frac{1}{1,000}$	0.001 m = $\frac{1}{10}$ cm = 10^{-3} m	0.0394 in.
1 micrometer (μm) **[formerly micron (μ)]**	micro = $\frac{1}{1,000,000}$	0.000,001 m = $\frac{1}{10,000}$ cm = 10^{-6} m	3.94 x 10^{-5} in.

1 nanometer (nm)
   [formerly millimicron
   (mμ)]
	nano = $\frac{1}{1,000,000,000}$	0.000,000,001 m = $\frac{1}{10,000,000}$ cm
		= 10^{-9} m
			3.94 x 10^{-8} in.

You can determine M and n in two steps. For example, how is 0.0000000001 written in exponential form? First, determine M by moving the decimal point so that only one nonzero digit is to the left of it:

$$0.\underset{\longrightarrow}{0000000001}.$$

The digit to the left of the decimal is 1; therefore $M = 1$. Second, determine n by counting the number of places you moved the decimal point. If you moved the point to the left, make the number positive; if you moved it to the right, it is negative. Since you moved the decimal point 10 places to the right, $n = -10$. Thus

$$0.0000000001 = 1 \times 10^{-10}$$

Now do a problem on your own. The wavelength of yellow light is about 0.000059 cm. Convert the centimeters to exponential form. If your answer is 5.9×10^{-5}, you are ready to continue. If you got the wrong answer, reread the discussion.

When we are working with a very large number, the same rules apply, but our exponential value will be positive rather than negative. Refer to Exhibit A.2. Note that 1 km equals 1,000 m. Even though 1,000 is not a cumbersome number, we can still convert it into exponential form. First, move the decimal point so there is only one nonzero digit to the left of it to determine M:

$$1.\underset{\longleftarrow}{000}.$$

Now, because the decimal has been moved three places to the left, n equals +3 or simply 3. Thus

$$1 \text{ km} = 1 \times 10^3 \text{ m}$$

Do another problem on your own. The speed of light is about 30,000,000,000 cm/sec. Convert the centimeters to exponential form. Your answer should be 3×10^{10} cm.

Review Exhibit A.2 and note some common metric and U.S. equivalents. Note also the exponential forms.

MASS

Now let us look at the second fundamental unit of the metric system: mass. The standard unit of mass is the *kilogram* (kg). A kilogram is defined as the mass of a platinum-iridium cylinder kept at the International Bureau of Weights and Measures in France. The standard pound is defined in terms of standard kilogram: 1 lb equals 0.4536 kg.

Exhibit A.3 lists metric units of mass and some U.S. equivalents.

TIME

The third fundamental unit of both the metric and the U.S. systems is time. The standard of time is the *second*. Formerly, the second was defined as 1/86,400 of a mean solar day. (A mean solar day is the average of the lengths of all days throughout the year.) Currently, the second is defined as the time required for 9,192,631,770 vibrations of cesium atoms when they are vibrating

in a specific manner. Units of time are used in measuring pulse and heart rate, metabolic rate, x-ray exposure, and intervals between medications.

Exhibit A.1 lists the units of time.

VOLUME

Units of volume, or capacity, are derived units based on length. *Volume* in the U.S. system may be expressed as cubic feet (ft^3), cubic inches ($inch^3$), and cubic yards (yd^3), or as a unit of volume such as the quart. Volume in the metric system may be expressed in cubic units of length such as cubic centimeters (cm^3) or in terms of the basic unit of volume, the *liter*. A liter is defined as the volume occupied by 1,000 g of pure water at 4°C. Since 1 cm^3 of water at this temperature weighs 1 g, then 1,000 g of water occupies a volume of 1,000 cm^3. This means that a liter is equal to 1,000 cm^3 and 1 milliliter (ml) is equal to 1 cm^3. Because of this relationship, many volume-measuring devices, such as hypodermic needles, may be graduated in either milliliters or cubic centimeters.

Exhibit A.4 lists metric units of volume and some U.S. equivalents.

EXHIBIT A.3

METRIC UNITS OF MASS AND SOME U.S. EQUIVALENTS

Metric Unit	Metric Equivalent	U.S. Equivalent
1 kilogram (kg)	1,000 g	2.205 lb
		1 ton = 907 kg
1 hectogram (hg)	100 g	0.353 oz
1 dekagram (dag)	10 g	0.0353 oz
1 gram (g)	1 g	0.0035 oz
		1 lb = 453.6 g
		1 oz = 28.35 g
1 decigram (dg)	0.1 g	
1 centigram (cg)	0.01 g	
1 milligram (mg)	0.001 g	0.115 g
1 microgram (μg)	0.000,001 g	
1 nanogram (ng)	0.000,000,001 g	
1 picogram (pg)	0.000,000,000,001 g	

APOTHECARY SYSTEM

In addition to the U.S. and metric systems, there is the *apothecary system.* This system is commonly used by physicians prescribing medications and by pharmacists preparing them. Exhibit A.5 lists the important units and equivalents of the apothecary system. Note that the units of mass have the same names as in the U.S. system, but they are not equivalent (1 oz = 28.35 g; 1 oz ap = 30 g) and they do not have the same relationship to one another (1 oz = 16 dr; 1 oz ap = 8 dr ap). The units of volume are the same in both systems.

EXHIBIT A.4

METRIC UNITS OF VOLUME AND SOME U.S. EQUIVALENTS

Metric Unit	Metric Equivalent	U.S. Equivalent
1 liter (l)	1,000 ml	33.81 fl oz = 1.057 qt 946 ml = 1 qt
1 milliliter (ml)	0.001 liter	0.0338 fl oz 30 ml = 1 fl oz 5 ml = 1 teaspoon
1 cubic centi-meter (cm³)	1.0 ml	0.0338 fl oz

EXHIBIT A.5

APOTHECARY SYSTEM OF MASS AND VOLUME WITH METRIC EQUIVALENTS

Fundamental Unit	Apothecary Unit and Conversion	Metric Equivalent
Mass	1 grain (gr) = 0.002083 ounce	1 g = 15 gr
	1 dram (dr ap) = 60 gr	4 g = 1 dr
	1 ounce (oz ap) = 8 dr ap	30 g = 1 oz
	1 pound (lb ap) = 12 oz ap	1 kg = 32 oz
Volume	1 fluidram (fl dr) = 60 minims (min)	1 ml (or cm³ = 15 min
	1 fluidounce (fl oz) = 8 fl dr	4 ml (or cm³) = 1 fl dr
	1 pint (pt) = 16 fl oz	30 ml (or cm³) = 1 fl oz
		500 ml (or cm³) = 1 pt
		1,000 ml (or cm³) = 1 qt

FREQUENTLY USED CONVERSIONS BASED ON MILLIGRAM

1,000 mg (1 g) = 15 grains
600 mg (0.6 g) = 10 grains
300 Mg (0.3 g) = 5 grains
60 mg (0.06 g) = 1 grain
30 mg (0.03 g) = 0.50 (½) grain
20 mg (0.02 g) = 0.33 (⅓) grain
10 mg (0.01 g) = 0.166 (⅙) grain
5 mg (0.005 g) = 0.083 (1/12) grain
4 mg (0.004 g) = 0.66 (1/15) grain
1 mg (0.001 g) = 0.016 (1/60) grain
0.5 mg (0.0005 g) = 0.0083 (1/120) grain
0.1 mg (0.0001 g) = 0.0016 (1/600) grain

Appendix B

NORMAL VALUES FOR SELECTED BLOOD AND URINE TESTS

The system of international (SI) units (Système Internationale d'Unités) is used in most countries and in many medical and scientific journals.[*] Clinical laboratories in the U.S., however, usually report values for blood and urine tests in conventional (traditional) units, which are also used in most medical and biological textbooks. To encompass both systems, the laboratory values in this Appendix give conventional units and, in parentheses, their SI equivalents.

A major difference between SI and conventional units is in expressing concentration as number of molecules or particles per volume (millimoles/liter) rather than mass per volume (grams/100 milliliters) or chemical activity per volume (milliequivalents per liter). In certain cases, for example proteins that have a variable molecular weight, it is not practical to change the values to SI units. Therefore, protein concentration may be expressed in grams per liter rather than moles per liter. Electrolyte (ion) values have traditionally been given in milliequalents per liter. In SI they are expressed in millimoles/liter. For monovalent (singly charged) ions such as Na^+, K^+, and Cl^-, the values will be numerically the same. For example, a Na^+ concentration of 135 mEq/l is equal to 135 mmol/l. For divalent ions such as Ca^{2+} and Mg^{2+}, milliequivalents are divided by the valence of 2 to obtain millimoles. Thus a Ca^{2+} concentration of 3.0 mEq/l is equal to 1.5 mmol/l.

In some laboratory and medical measurements, SI units offer little advantage. Their use might require replacement or revision of current instruments, for example the sphygmomanometer for measuring blood pressure or the pH meter for measuring hydrogen ion concentration. In these cases, traditional units are commonly retained.

It is important to note that values listed for various laboratory tests should be viewed as reference values rather than absolute "normal" values for all well people. A single reference range may be inadequate for some measurements. Values may vary due to age, gender, body build, diet, and environment of the subject or the equipment, methods, and standards of the lab performing the measurement. Also, in most pathological processes, there is a gradual transition between normal and abnormal values. Thus the values given here are representative reference ranges. They are grouped into three categories: **blood chemistry tests** (mainly blood plasma and serum values), **hematology tests** (blood clotting parameters and counts of blood formed elements), and **urine tests.**

KEY TO SYMBOLS

mm^3 = cubic millimeter
dl = deciliter = 10^{-1} liter
g = gram
> = greater than
hr = hour
IU = international unit
kg = kilogram = 10^3 grams
< = less than
l = liter
% = percent
μg = microgram = 10^{-6} gram
μmol/l = micromole per liter
mEq/l = milliequivalent per liter
mg = milligram = 10^{-3} gram
ml = milliliter = 10^{-3} liter
mm = millimeter = 10^{-3} meter
mm Hg = millimeter of mercury
mmol/l = millimole per liter
mOsm = milliosmole
ng = nanogram = 10^{-9} gram
nmol/l = nanomole per liter
U = unit

---

[*] Introductory paragraphs and SI units provided by Henry Ruschin, Humber College, Toronto, Canada. Reference ranges previously researched by John Lo Russo, Bergen Community College, New Jersey.

BLOOD CHEMISTRY TESTS
(WB) = Whole blood (S) = Serum (P) = Plasma

Test (Specimen)	Reference Values: Conventional U.S. Units (SI Units)	Clinical Implications
Acid phosphatase (ACP) (S)	0.1–5.0 U/dl by King–Armstrong method (0–0.8 IU/l)	Values increase in prostatic cancer (especially if it has spread beyond the prostate gland), some liver diseases, hyperparathyroidism, hemolytic anemia, and sickle cell crisis; values are decreased in Down syndrome.
Alkaline phosphatase (ALP) (S)	4–13 U/dl by King–Armstrong method (30–120 U/l)	Values increase in some liver and bone diseases, hyperparathyroidism, and pregnancy; values decrease in cretinism, growth retardation, scurvy, and achondroplasia.
Alphafeto-protein (AFP) (WB or amniotic fluid)	Nonpregnant adult: <25 ng/ml (<25 µg/l)	Major plasma protein synthesized by fetal liver during first 3 months of development. In amniotic fluid and maternal blood, values increase with faulty development of the fetal nervous system, in particular neural tube defects such as spina bifida. In nonpregnant adults, values increase in liver cancer, cirrhosis, or chronic active hepatitis.
Aminotransferases (S) Alanine amino-transferase (ALT); formerly serum glutamic–pyruvic transaminase (SGPT)	10–30 IU/ml; 5–25 Reitman–Frankel units (10–30 U/l) 5–24 IU/l; 5–35 Reitman–Frankel units	Values increase in liver disease or liver damage due to toxic drugs.
Aspartate amino-transferase (AST); formerly serum glutamic-oxaloacetic transaminase (SGOT)	(5–30 U/l)	Values increase in myocardial infarction, liver disease, trauma to skeletal muscles, and severe burns; values decrease in beriberi and uncontrolled diabetes mellitus with acidosis.
Ammonia (P)	20–120 µg/dl (12–55 µmol/l)	Values increase in liver disease, heart failure, emphysema, pneumonia, cor pulmonale, and hemolytic disease of newborn (erythroblastosis fetalis).
Amylase (S)	60–160 Somogyi U/dl (25–125 U/l)	Values increase in acute pancreatitis, mumps, and obstruction of pancreatic duct; values decrease in hepatitis, cirrhosis, burns, and toxemia of pregnancy.
Bilirubin (S)	Conjugated: <0.5 mg/dl (<5.0 µmol/l) Unconjugated: 0.2–1.0 mg/dl (18–20 µmol/l) Newborn: 1.0–12.0 mg/dl (<200 µmol/l)	An increase in conjugated bilirubin probably results from liver dysfunction or biliary obstruction; an increase in unconjugated bilirubin probably results from excessive hemolysis of red blood cells.
Blood urea nitrogen (BUN) (S)	8–26 mg/dl (2.9–9.3 mmol/l)	Values increase in kidney disease, shock, dehydration, diabetes, and acute myocardial infarction (MI); values decrease in liver failure, impaired absorption, and overhydration.
Calcium (Ca^{2+}) (S)	Total: 9–11 mg/dl (2.3–2.7 mmol/l) Ionized (50% of total): 4.5–5.5 mg/dl (1.15–1.35 mmol/l)	Values increase in cancer, hyperparathyroidism, Addison's disease, hyperthyroidism, and Paget's disease; values decrease in hypoparathyroidism, chronic renal failure, osteomalacia, rickets, and diarrhea.
Carbon dioxide (CO_2), content (WB)	Arterial: 19–24 mEq/l (19–24 mmol/l) Venous: 22–26 mEq/l (22–26 mmol/l)	Values increase in severe vomiting, emphysema, and aldosteronism; values decrease in severe diarrhea, starvation, and acute renal failure.
Carbon dioxide, partial pressure (pCO_2) (WB)	Arterial: 35–40 mm Hg(same) Venous: 45 mm Hg (same)	Values increase in hypoventilation, obstructive lung disease, and emphysema; values decrease in hyperventilation, hypoxia, and pregnancy.

Carcinoembryonic antigen (CEA) (P)	<3ng/ml (<3 μg/l)	Values increase in carcinoma of the colon, rectum, breast, ovary, liver, and pancreas; inflammatory bowel disease (IBD); cirrhosis; and chronic cigarette smoking.
Carotene, beta (S)	40–200 μg/dl (0.4–2.0 mg/l)	Value varies with diet but increases in myxedema, diabetes mellitus, and excessive dietary intake; values decrease in fat malabsorption, liver disease, and poor dietary intake.
Chloride ion (Cl⁻) (S)	95–103 mEq/l (95–103 mmol/l)	Values increase in dehydration, Cushing's syndrome, and anemia; values decrease in severe vomiting, severe burns, diabetic acidosis, and fever.
Cholesterol total (S)	<200 mg/dl (5.2 mmol/l) is desirable	Value varies with diet, gender, and age. Values increase in diabetes mellitus, cardiovascular disease, nephrosis, and hypothyroidism; values decrease in liver disease, hyperthyroidism, fat malabsorption, pernicious anemia, severe infections, and terminal stages of cancer.
HDL cholesterol (P)	>40 mg/dl (>1.0 mmol/l) is desirable	
LDL cholesterol (P)	<130 mg/dl (<3.2 mmol/l) is desirable	
Cortisol (hydrocortisone) (P)	8 A.M.–10 A.M.: 5–23 μg/dl (270–700 nmol/l) 4 P.M.–6 P.M.: 3–13 μg/dl (135–350 nmol/l)	Values increase in hyperthyroidism, stress, obesity, and Cushing's syndrome; values decrease in hypothyroidism, liver disease, and Addison's disease.
Creatine (S or P)	Male: 0.1–0.4 mg/dl (1–4 g/l) Female: 0.2–0.7 mg/dl (2–7 g/l)	Values increase in muscular dystrophy, damage to muscle tissue, nephritis, and pregnancy.
Creatine kinase (CK); formerly creatine phosphokinase (CPK) (S)	Male: 55–170 U/l (same) Female: 30–135 U/l (same)	Values increase in myocardial infarction, progressive muscular dystrophy, myxedema, convulsions, hypothyroidism, and pulmonary edema.
Creatinine (S)	0.5–1.2 mg/dl (45–105 μmol/l)	Values increase in impaired renal function, giantism, and acromegaly; values decrease in muscular dystrophy.
Fetal hemoglobin (WB)	Newborns: 60–90% Before age 2: 0–4% Adults: 0–2%	Values increase in thalassemia, sickle-cell anemia, and leakage of fetal blood into maternal bloodstream.
Gamma-glutamyl transferase (GGT) (S)	5–40 IU/l (5–40 U/l)	Values increase in obstruction of bile duct, cirrhosis of the liver, metastatic cancer of the liver, cholelithiasis, congestive heart failure (CHF), and alcoholism.
Glucose (S)	70–110 mg/dl (3.9–6.1 mmol/l)	Values increase in diabetes mellitus, acute stress, hyperthyroidism, chronic liver disease, and nephritis; values decrease in Addison's disease, hypothyroidism, and cancer of the pancreas.
Immunoglobins (S)		IgG values increase in infections of all types, liver disease, and severe malnutrition; IgA values increase in cirrhosis of the liver, chronic infections, and autoimmune disorders and decrease in immunologic deficiency states; IgM values increase in trypanosomiasis and decrease in lymphoid aplasia; IgD values increase in chronic infections and myelomas; IgE values increase in hay fever, asthma, and anaphylactic shock.
IgG	800–1,801 mg/dl (8.0–18.0 g/l)	
IgA	113–563 mg/dl (1.1–5.6 g/l)	
IgM	54–222 mg/dl (0.5–2.2 g/l)	
IgD	0.5–3.0 mg/dl (5–30 mg/l)	
IgE	0.01–0.04 mg/dl (0.1–0.4 mg/l)	
Iron, total (S)	50–170 μg/dl (13–30 μmol/l)	Values higher in males than females; values increase in liver disease and various anemias; values decrease in iron-deficiency anemia.
Iron-binding capacity, total (TIBC)	300–420 μg/dl (50–70 μmol/l)	Values increase in iron-deficiency anemia, blood loss, and pregnancy and in women taking birth control pills; values decrease in many chronic diseases, widespread cancer, malnutrition, and nephrotic syndrome.
Ketone bodies (acetone, acetoacetic acid, and β-hydroxybutyric acid) (S or P)	Negative Toxic level: 20 mg/dl (0.2 g/l)	Values increase in ketoacidosis, fever, anorexia, fasting, starvation, high-fat diet, low-carbohydrate diet, and following vomiting.

Exhibit continues

EXHIBIT B.1 (Continued)

(WB) = Whole blood (S) = Serum (P) = Plasma

Test (Specimen)	Reference Values: Conventional U.S. Units (SI Units)	Clinical Implications
Lactic acid (lactate) (WB)	Arterial: 3–7 mg/dl (0.3–0.7 mmol/l) Venous: 5–20 mg/dl (0.5–2.0 mmol/l)	Values increase during muscular activity, congestive heart failure (CHF), shock, and severe hemorrhage.
Lactic dehydrogenase (LDH) (S)	71–207 IU/l (70–207 U/l)	Values increase in myocardial infarction, liver disease, skeletal muscle necrosis, and extensive cancer.
Lipids (S)		
Total	400–800 mg/dl (4.0–8.0 g/l)	Values increase in hyperlipidemia, diabetes mellitus, and hypothyroidism; values decrease in fat malabsorption.
Cholesterol	150–220 mg/dl (3.9–5.7 mmol/l)	
Triglycerides	10–190 mg/dl (0.1–1.9 g/l)	
Osmolality (S)	285–295 mOsm/kg water (285–295 mmol/kg water)	Values increase in cirrhosis, congestive heart failure (CHF), and high-protein diets; values decrease in aldosteronism, diabetes insipidus, and hypercalcemia.
Oxygen (O_2) content (WB)	Arterial: 15–23 volume % (same)	Values increase in polycythemia; values decrease in chronic obstructive lung disease.
Oxygen, partial pressure (pO_2) (WB)	Arterial: 80–105 mm Hg (same)	Values increase in polycythemia and hyperventilation; values decrease in anemias, insufficient atmospheric oxygen, and hypoventilation.
pH (WB)	Arterial: 7.35–7.45 (same or [H^+] = 45–35 nmol/l)	Values increase in vomiting, hyperventilation, excessive bicarbonate, and lack of oxygen; values decrease in renal failure, diabetic ketoacidosis, hypoxia, airway obstruction, and shock.
Phosphorous, inorganic (P); also reported as phosphate (S)	Adult: 2.5–4.5 mg/dl (0.8–1.5 mmol/l) Children: 4–7 mg/dl (1.3–2.3 mmol/l)	Values increase in renal failure, hypoparathyroidism, hypocalcemia, bone tumors, diabetic ketoacidosis, and acromegaly; values decrease in hyperparathyroidism, alcoholism, rickets, and osteomalacia.
Protein (S)		Total protein values increase in dehydration, shock, systemic lupus erythematosus (SLE), rheumatoid arthritis (RA), chronic infections, and chronic liver disease; total protein values decrease in insufficient protein intake, hemorrhage, malabsorption, diarrhea, chronic renal failure, and severe burns; reversed A/G ratio (A lowered, G elevated) may indicate chronic liver disease, leukemia, Hodgkin's disease, tuberculosis, or chronic hepatitis.
Total	6.0–7.8 g/dl (60–78 g/l)	
Albumin	3.5–5.0 g/dl (35–50 g/l)	
Globulin	2.3–3.5 g/dl (23–35 g/l)	
A/G ratio	1.5:1 to 2.5:1	Values increase in dehydration, aldosteronism, coma, Cushing's disease, and diabetes insipidus; values decrease in severe burns, vomiting, diarrhea, Addison's disease, nephritis, excessive sweating, and edema.
Sodium (Na^+) (S)	136–142 mEq/l (136–142 mmol/l)	
Thyroid hormones (S)		Values increase in hyperthyroidism; values decrease in hypothyroidism.
T_3 (triiodo-thyronine)	80–200 ng/dl by radioimmunoassay (RIA) (1.2–3.1 nmol/l)	
T_4 (thyroxine)	4–11 µg/dl by RIA (52–141 nmol/l)	
Thyroxine-binding globulin (TBG) (S)	10–26 µg/dl (130–335 nmol/l)	Values increase in hypothyroidism; values decrease in hyperthyroidism.
Uric acid (urate) (S)	Male: 4.0–8.5 mg/dl (240–510 µmol/l) Female: 2.7–7.3 mg/dl (160–430 µmol/l)	Values increase in impaired renal function, gout, metastatic cancer, shock, and starvation; values decrease in persons treated with uricosuric drugs.

EXHIBIT B.2

HEMATOLOGY TESTS

Test (Specimen)	Reference Values: Conventional U.S. Units (SI Units)	Clinical Implications
Bleeding time (WB)	4-8 minutes using Simplate (same)	Values increase in thrombocytopenia, severe liver disease, leukemia, and aplastic anemia.
Erythrocyte sedimentation rate (ESR) (WB)	(Westergren) Female: Under 50 years: <20 mm/hr (same) Over 50 years: <30 mm/hr (same) Male: Under 50 years: <15 mm/hr (same) Over 50 years: <20 mm/hr (same)	Values increase in pregnancy, infection, carcinoma, tissue destruction, and nephritis; values decrease in sickle-cell anemia and congestive heart failure (CHF).
Hemoglobin (S or P)	Male: 13.5–18 g/100 ml (135–180 g/l) Female: 12–16 g/100 ml (120–160 g/l) Newborn: 14–20 g/100 ml (140–200 g/l)	Values increase in polycythemia, congestive heart failure (CHF), chronic obstructive pulmonary disease, and at high altitudes; values decrease in anemia, hyperthyroidism, cirrhosis of the liver, and severe hemorrhage.
Hematocrit (WB)	Male: 40–54%; average 47% (same) Female: 38–47%; average 42% (same)	Values increase in polycythemia, severe dehydration, and shock; values decrease in anemia, leukemia, cirrhosis, and hyperthyroidism.
Platelet count (WB)	150,000–400,000/mm^3 (150–400 × 10^9/l)	Values increase in cancer, trauma, heart disease, and cirrhosis; values decrease in anemias, allergic conditions, and during cancer chemotherapy.
Prothrombin time (PT) (WB)	11–15 seconds (same)	Values increase in prothrombin and vitamin K deficiency, liver disease, and hypervitaminosis A.
Red blood cell count (WB)	Male: 4.5–6.5 million/mm^3 (4.5–6.5 × 10^{12}/l) Female: 3.9–5.6 million/mm^3 (3.9–5.6 × 10^{12}/l)	Values increase in polycythemia, dehydration, and following hemorrhaging; values decrease in systemic lupus erythematosus (SLE), anemias, and Addison's disease.
Reticulocyte count (WB)	0.5–2.0% (same)	Values increase in hemolytic anemia, metastatic carcinoma, and leukemia; values decrease in iron-deficiency and pernicious anemia, radiation therapy, and kidney disease in which kidney cells do not make erythropoietin.
White blood cell count, differential (WB) **Neutrophils** **Eosinophils** **Basophils** **Lymphocytes** **Monocytes**	60–70% (same) 2–4% (same) 0.5–1% (same) 20–25% (same) 3–8% (same)	Neutrophils increase in acute infections; eosinophils and basophils increase in allergic reactions; lymphocytes increase during antigen-antibody reactions; monocytes increase in chronic infections.
White blood cell count, total (WB)	5,000–10,000/mm^3 (5–10 × 10^9/l)	Values increase in acute infections, trauma, malignant diseases, and cardiovascular diseases; values decrease in diabetes mellitus, anemias, and following cancer chemotherapy.

EXHIBIT B.3

URINE TESTS

Test (Sample)	Reference Values: Conventional U.S. Units (SI Units)	Clinical Implication
Amylase (2 hour)	35–260 Somogyi units/hr (6.5–48.1 units/hr)	Values increase in inflammation of the pancreas (pancreatitis) or salivary glands, obstruction of the pancreatic duct, and perforated peptic ulcer.
Bilirubin[a] (random)	Negative	Values increase in liver disease and obstructive biliary disease.
Blood[a] (random)	Negative	Values increase in renal disease, extensive burns, transfusion reactions, and hemolytic anemia.
Calcium (Ca^{2+}) (random)	10 mg/dl (2.5 mmol/l); up to 300 mg/24 hr (7.5 mmol/24 hr)	Amount depends on dietary intake; values increase in hyperparathyroidism, metastatic malignancies, and primary cancer of breasts and lungs; values decrease in hypoparathyroidism and vitamin D deficiency.
Casts (24 hour)		
Epithelial	Occasional	Values increase in nephrosis and heavy metal poisoning.
Granular	Occasional	Values increase in nephritis and pyelonephritis.
Hyaline	Occasional	Values increase in glomerular membrane damage and fever.
Red blood cell	Occasional	Values increase in pyelonephritis, kidney stones, and cystitis.
White blood cell	Occasional	Values increase in kidney infections.
Chloride (Cl^-) (24 hour)	140–250 mEq/24 hr (140–250 mmol/24 hr)	Amount depends on dietary salt intake; values increase in Addison's disease, dehydration, and starvation; values decrease in pyloric obstruction, diarrhea, and emphysema.
Color (random)	Yellow, straw, amber	Varies with many disease states, hydration, and diet.
Creatinine (24 hour)	Male: 1.0–2.0 g/24 hr (9–18 mmol/24 hr) Female: 0.8–1.8 g/24 hr (7–16 mmol/24 hr)	Values increase in infections; values decrease in muscular atrophy, anemia, and kidney diseases.
Glucose[a] (random)	Negative	Values increase in diabetes mellitus, brain injury, and myocardial infarction.
Hydroxycortico-steroids (17-hydroxysteroids) (24 hour)	Male: 5–15 mg/24 hr (13–41 µmol/24 hr) Female: 2–13 mg/24 hr (5–36 µmol/24 hr)	Values increase in Cushing's syndrome, burns, and infections; values decrease in Addison's disease.
Ketone bodies[a] (random)	Negative	Values increase in diabetic acidosis, fever, anorexia, fasting, and starvation.
17-ketosteroids (KS) (24 hour)	Male: 8–25 mg/24 hr (28–87 µmol/24 hr) Female: 5–15 mg/24 hr (17–53 µmol/24 hr)	Values decrease in surgery, burns, infections, adrenogenital syndrome, and Cushing's syndrome.
Odor (random)	Aromatic	Becomes acetonelike in diabetic ketosis.
Osmolality (24 hour)	500–1400 mOsm/kg water (500–1400 mmol/kg water)	Values increase in cirrhosis, congestive heart failure (CHF), and high-protein diets; values decrease in aldosteronism, diabetes insipidus, and hypokalemia.
pH[a] (random)	4.6–8.0	Values increase in urinary tract infections and severe alkalosis; values decrease in acidosis, emphysema, starvation, and dehydration.
Phenylpyruvic acid (random)	Negative	Values increase in phenylketonuria (PKU).
Potassium (K^+) (24 hour)	40–80 mEq/24 hr (40–80 mmol/24 hr)	Values increase in chronic renal failure, dehydration, starvation, and Cushing's syndrome; values decrease in diarrhea, malabsorption syndrome, and adrenal cortical insufficiency.
Protein[a] (albumin) (random)	Negative	Values increase in nephritis, fever, severe anemias, trauma, and hyperthyroidism.
Sodium (Na^+) (24 hour)	75–200 mg/24 hr (75–200 mmol/24 hr)	Amount depends on dietary salt intake; values increase in dehydration, starvation, and diabetic acidosis; values decrease in diarrhea, acute renal failure, emphysema, and Cushing's syndrome.

Specific gravity[a] (random)	1.001–1.035 (same)	Values increase in diabetes mellitus and excessive water loss; values decrease in absence of antidiuretic hormone (ADH) and severe renal damage.
Urea (random)	25–35 g/24 hr (420–580 mmol/24 hr)	Values increase in response to increased protein intake; values decrease in impaired renal function.
Uric acid (24 hour)	0.4–1.0 g/24 hr (1.5–4.0 mmol/24 hr)	Values increase in gout, leukemia, and liver disease; values decrease in kidney disease.
Urobilinogen[a] (2 hour)	0.3–1.0 Ehrlich units (1.7–6.0 μmol/24 hr)	Values increase in anemias, hepatitis A (infectious), biliary disease, and cirrhosis; values decrease in cholelithiasis and renal insufficiency.
Volume, total (24 hour)	1000–2000 ml/24 hr (1.0–2.0 l/24 hr)	Varies with many factors.

[a] Test often performed using a **dipstick,** a plastic strip impregnated with chemicals that is dipped into a urine specimen to detect particular substances. Certain colors indicate the presence or absence of a substance and sometimes give a rough estimate of the amount(s) present.

PERIODIC TABLE

1 **H** 1.008																	2 **He** 4.003
3 **Li** 6.941	4 **Be** 9.012											5 **B** 10.81	6 **C** 12.01	7 **N** 14.01	8 **O** 16.00	9 **F** 19.00	10 **Ne** 20.18
11 **Na** 22.99	12 **Mg** 24.31											13 **Al** 26.98	14 **Si** 28.09	15 **P** 30.97	16 **S** 32.06	17 **Cl** 35.45	18 **Ar** 39.95
19 **K** 39.10	20 **Ca** 40.08	21 **Sc** 44.96	22 **Ti** 47.88	23 **V** 50.94	24 **Cr** 52.00	25 **Mn** 54.94	26 **Fe** 55.85	27 **Co** 58.93	28 **Ni** 58.69	29 **Cu** 63.55	30 **Zn** 65.38	31 **Ga** 69.72	32 **Ge** 72.59	33 **As** 74.92	34 **Se** 78.96	35 **Br** 79.90	36 **Kr** 83.80
37 **Rb** 85.47	38 **Sr** 87.62	39 **Y** 88.91	40 **Zr** 91.22	41 **Nb** 92.91	42 **Mo** 95.94	43 **Tc** (98)	44 **Ru** 101.1	45 **Rh** 102.9	46 **Pd** 106.4	47 **Ag** 107.9	48 **Cd** 112.4	49 **In** 114.8	50 **Sn** 118.7	51 **Sb** 121.8	52 **Te** 127.6	53 **I** 126.9	54 **Xe** 131.3
55 **Cs** 132.9	56 **Ba** 137.3	57* **La** 138.9	72 **Hf** 178.5	73 **Ta** 180.9	74 **W** 183.9	75 **Re** 186.2	76 **Os** 190.2	77 **Ir** 192.2	78 **Pt** 195.1	79 **Au** 197.0	80 **Hg** 200.6	81 **Tl** 204.4	82 **Pb** 207.2	83 **Bi** 209.0	84 **Po** (209)	85 **At** (210)	86 **Rn** (222)
87 **Fr** (223)	88 **Ra** 226	89† **Ac** (227)	104 **Unq**	105 **Unp**	106 **Unh**	107 **Uns**	108 **Uno**	109 **Une**									

Key:
- 1 — Atomic number
- **H** — Chemical symbol
- 1.008 — Atomic weight

*Lanthanides	58 **Ce** 140.1	59 **Pr** 140.9	60 **Nd** 144.2	61 **Pm** (145)	62 **Sm** 150.4	63 **Eu** 152.0	64 **Gd** 157.3	65 **Tb** 158.9	66 **Dy** 162.5	67 **Ho** 164.9	68 **Er** 167.3	69 **Tm** 168.9	70 **Yb** 173.0	71 **Lu** 175.0
†Actinides	90 **Th** 232.0	91 **Pa** (231)	92 **U** 238.0	93 **Np** (237)	94 **Pu** (244)	95 **Am** (243)	96 **Cm** (247)	97 **Bk** (247)	98 **Cf** (251)	99 **Es** (252)	100 **Fm** (257)	101 **Md** (258)	102 **No** (259)	103 **Lr** (260)

GLOSSARY OF COMBINING FORMS, WORD ROOTS, PREFIXES, AND SUFFIXES

PRONUNCIATION KEY

1. The strongest accented syllable appears in capital letters, for example, bilateral (bī-LAT-er-al) and diagnosis (dī-ag-NŌ-sis).

2. If there is a secondary accent, it is noted by a prime ('), for example, constitution (kon'-sti-TOO-shun) and physiology (fiz'-ē-OL-ō-jē). Any additional secondary accents are also noted by a prime, for example, decarboxylation (dē'-kar-bok'-si-LĀ-shun).

3. Vowels marked with a line above the letter are pronounced with the long sound as in the following common words:

ā as in māke
ē as in bē
ī as in īvy
ō as in pōle

4. Vowels not so marked are pronounced with the short sound, as in the following words:

e as in bet
i as in sip
o as in not
u as in bud

5. Other phonetic symbols are used to indicate the following sounds:

a as in above
oo as in sue
yoo as in cute
oy as in oil

Many medical terms are "compound" words; that is, they are made up of one or more word roots or combining forms of word roots with prefixes or suffixes. For example, *leukocyte* (white blood cell) is a combination of *leuko,* the combining form for the word root meaning "white," and *cyt,* the word root meaning "cell." Learning the medical meanings of the fundamental word parts will enable you to analyze many long, complicated terms.

The following list includes some of the most commonly used combining forms, word roots, prefixes, and suffixes used in making medical terms and an example for each.

COMBINING FORMS AND WORD ROOTS

Acou-, Acu- hearing Acoustics (a-KOO-stiks), the science of sounds or hearing.

Acr-, Acro- extremity Acromegaly (ak'-rō-MEG-a-lē), hyperplasia of the nose, jaws, fingers, and toes.

Aden-, Adeno- gland Adenoma (ad-en'-Ō-ma), a tumor with a glandlike structure.

Alg-, Algia- pain Neuralgia (nyoo-RAL-ja), pain along the course of a nerve.

Angi- vessel Angiocardiography (an'-jē-ō-kard-ē-OG-ra-fē), x-ray of the great blood vessels and heart after intravenous injection of radiopaque fluid.

Arthr-, Arthro- joint Arthropathy (ar-THROP-a-thē), disease of a joint.

Aut-, Auto- self Autolysis (aw-TOL-i-sis), destruction of cells of the body by their own enzymes, even after death.

Bio- life, living Biopsy (BĪ-op-sē), examination of tissue removed from a living body.

Blast- germ, bud Blastocyte (BLAS-tō-sīt), an embryonic or undifferentiated cell.

Blephar- eyelid Blepharitis (Blef-a-RĪT-is), inflammation of the eyelids.

Brachi- arm Brachialis (brā-kē-AL-is), muscle that flexes the forearm.

Bronch- trachea, windpipe Bronchoscopy (bron-KOS-kō-pē), direct visual examination of the bronchi.

Bucc- cheek Buccocervical (bū-kō-SER-vi-kal), pertaining to the cheek and neck.

Capit- head Decapitate (dē-KAP-i-tāt), to remove the head.

Carcin- cancer Carcinogenic (kar-sin-ō-JENK-ik), causing cancer.

Cardi-, Cardia-, Cardio heart Cardiogram (KARD-ē-o-gram), a recording of the force and form of the heart's movements.

Cephal- head Hydrocephalus (hī-drō-SEF-a-lus), enlargement of the head due to an abnormal accumulation of fluid.

Cerebro- brain Cerebrospinal (se-rē-brō-SPĪN-al) fluid, fluid contained within the cranium and spinal canal.

Cheil- lip Cheilosis (kī-LŌ-sis), dry scaling of the lips.

Chole- bile, gall Cholecystogram (kō-lē-SIS-tō-gram), x-ray image of the gallbladder.

Chondr-, Chondri-, Chondrio- cartilage Chondrocyte (KON-drō-sīt), a cartilage cell.

Chrom-, Chromat-, Chromato- color Hyperchromic (hī-per-KRŌ-mik), highly colored.

Cili- eyelash Supercilia (soo′-per-SIL-ē-a), eyebrow (hairs above eyelash).

Colpo- vaginia Colpotomy (kol-POT-ō-mē), incision into the wall of the vagina.

Cor-, coron- heart Coronary (KOR-ō-na-rē), arteries supplying blood to the heart muscle.

Cost- rib Costal (KOS-tal), pertaining to a rib.

Crani- skull Craniotomy (krā-nē-OT-ō-mē), surgical opening of the skull.

Cry-, Cryo- cold Cryosurgery (krī-ō-SERJ-e-rē), surgical procedure using a very cold liquid nitrogen probe.

Cut- skin Subcutaneous (sub-kyoo-TĀ-nē-us), under the skin.

Cysti-, Cysto- sac, bladder Cystoscope (SIS-tō-skōp), instrument for interior examination of the urinary bladder.

Cyt-, Cyto-, Cyte- cell Cytology (sī-TOL-ō-jē), the study of cells.

Dactyl-, Dactylo- digits (usually fingers, but sometimes toes) Polydactylism (pol-ē-DAK-til-ism), above normal number of fingers or toes.

Derma-, Dermato- skin Dermatosis (der-ma-TŌ-sis), any skin disease.

Dura- hard Dura mater (DYOO-ra MĀ-ter), outer membrane covering brain and spinal cord.

Entero- intestine Enteritis (ent-e-RĪT-is), inflammation of the intestine.

Erythro- red Erythrocyte (e-RITH-rō-sīt), red blood cell.

Galacto- milk Galactose (ga-LAK-tōse), a milk sugar.

Gastr- stomach Gastrointestinal (gas′-trō-in-TES-tin-al), pertaining to the stomach and intestine.

Gloss-, Glosso- tongue Hypoglossal (hī′-pō-GLOS-al), located under the tongue.

Glyco- sugar Glycosuria (glī′-kō-SUR-ē-a), sugar in the urine.

Gravid- pregnant Gravidity (gra-VID-i-tē), condition of being pregnant.

Gyn-, Gyne-, Gynec- female, women Gynecology (gīn′-e-KOL-ō-jē), the medical specialty dealing with disorders of the female reproductive system.

Hem-, Hemat- blood Hematoma (hē′-ma-TŌ-ma), a tumor or swelling filled with blood.

Hepar-, Hepato- liver Hepatitis (hep-a-TĪT-is), inflammation of the liver.

Hist-, Histio- tissue Histology (his-TOL-ō-jē), the study of tissues.

Hydr- water Hydrocele (HĪ-drō-sēl), accumulation of fluid in a saclike cavity.

Hyster- uterus Hysterectomy (his′-te-REK-tō-mē), surgical removal of the uterus.

Ileo- ileum Ileocecal (il′-ē-ō-SĒ-kal) valve, folds at the opening between ileum and cecum.

Ilio- ilium Iliosacral (il′-ē-ō-SĀ-kral), pertaining to ilium and sacrum.

Kines- motion Kinesiology (ki-nē-sē-OL-ō-jē), study of movement of body parts.

Labi- lip Labial (LĀ-bē-al), pertaining to a lip.

Lachry-, Lacri- tears Nasolacrimal (nā-zō-LAK-rim-al), pertaining to the nose and lacrimal apparatus.

Laparo- loin, flank, abdomen Laparoscopy (lap′-a-ROS-kō-pē), examination of the interior of the abdomen by means of a laparoscope.

Leuco-, Leuko- white Leukocyte (LYOO-kō-sīt), white blood cell.

Lingua- tongue Lingual (LIN-gwal), pertaining to the tongue.

Lip-, Lipo fat Lipoma (lī-PŌ-ma), a fatty tumor.

Lith- stone Lithiasis (li-THĒ-a-sis), the formation of stones.

Lumbo- lower back, loin Lumbar (LUM-bar), pertaining to the loin.

Macul- spot, blotch Macula (MAK-yoo-la), spot or blotch.

Malign- bad, harmful Malignant (ma-LIG-nant), condition that gets worse and results in death.

Mamm- breast Mammography (ma-MOG-ra-fē), x-ray of the mammary gland.

Mast- breast Mastitis (ma-STĪT-is), inflammation of the mammary gland.

Meningo- membrane Meningitis (men-in-JĪT-is), inflammation of the membranes of spinal cord and brain.

Metro- uterus Endometrium (en′-dō-MĒ-trē-um), lining of the uterus.

Morpho- form, shape Morphology (mor-FOL-o-jē), the study of form and structure of the things.

Myelo- marrow, spinal cord Poliomyelitis (pō-lē-ō-mī′-a-LĪT-is), inflammation of the gray matter of the spinal cord.

Myo- muscle Myocardium (mī-ō-KARD-ē-um), heart muscle.

Necro- corpse, dead Necrosis (ne-KRŌ-sis), death of areas of tissue surrounded by healthy tissue.

Nephro- kidney Nephrosis (ne-FRŌ-sis), degeneration of kidney tissue.

Neuro- nerve Neuroblastoma (nyoor′-ō-blas-TŌ-mas), malignant tumor of the nervous system composed of embryonic nerve cells.

Oculo- eye Binocular (bī-NOK-yoo-lar), pertaining to the two eyes.

Odont- tooth Orthodontic (or-thō-DONT-ik), pertaining to the proper positioning and relationship of the teeth.

Onco- mass, tumor Oncology (ong-KOL-ō-jē), study of tumors.

Oo- egg Oocyte (Ō-ō-sīt), original egg cell.

Oophor- ovary, egg carrier Oophorectomy (ō′-ōf-o-REK-tō-mē), surgical removal of ovaries.

Ophthalm- eye Ophthalmology (of-thal-MOL-ō-jē), the study of the eye and its diseases.

Or- mouth Oral (Ō-ral), pertaining to the mouth.

Orchido- testicle Orchidectomy (or′-ki-DEK-tō-me), surgical removal of a testicle.

Osmo- odor, sense of smell Anosmia (an-OZ-mé-a), absence of sense of smell.

Oss-, Osseo-, Osteo- bone Osteoma (os-tē-Ō-ma), bone tumor.

Oto- ear Otosclerosis (ō′-tō-skle-RŌ-sis), formation of bone in the labyrinth of the ear.

Palpebr- eyelid Palpebra (PAL-pe-bra), eyelid.

Part- birth, delivery, labor Parturition (par′-too-RISH-un), act of giving birth.

Patho- disease Pathogenic (path′-ō-JEN-ik), causing disease.

Ped- children Pediatrician (pēd-ē-a-TRISH-an), medical specialist in the treatment of children.

Peps- digest Peptic (PEP-tik), pertaining to digestion.

Phag-, Phago- to eat Phagocytosis (fag′-ō-sī-TŌ-sis), the process by which cells ingest particulate matter.

Philic-, Philo- to like, have an affinity for Hydrophilic (hī-drō-FIL-ik), having an affinity for water.

Phleb- vein Phlebitis (fle-BĪT-is), inflammation of the veins.

Phon- voice, sound Phonogram (FŌ-nō-gram), record made of sound.

Phren- diaphragm Phrenic (FREN-ik), pertaining to the diaphragm.

Pilo- hair Depilatory (de-PIL-a-tō-rē), hair remover.

Pneumo- lung, air Pneumothorax (nyoo-mō-THŌR-aks), air in the thoracic cavity.

Pod- foot Podiatry (po-DĪ-a-trē), the diagnosis and treatment of foot disorders.

Procto- anus, rectum Proctoscopy (prok-TOS-kō-pē), instrumental examination of the rectum.

Psycho- soul, mind Psychiatry (sī-KĪ-a-trē), treatment of mental disorders.

Pulmon- lung Pulmonary (PUL-mō-ner′-ē), pertaining to the lungs.

Pyle-, Pyloro opening, passage Pyloric (pī-LOR-ik), pertaining to the pylorus of the stomach.

Pyo- pus Pyuria (pī-YOOR-ē-a), pus in the urine.

Ren- kidneys Renal (RĒ-nal), pertaining to the kidney.

Rhin- nose Rhinitis (ri-NĪT-is), inflammation of nasal mucosa.

Salpingo- uterine (Fallopian) tube Salpingitis (sal′-pin-JĪ-tis), inflammation of the uterine (Fallopian) tubes.

Scler-, Sclero- hard Atherosclerosis (ath′-er-ō-skle-RŌ-sis), hardening of the arteries.

Sep-, Septic- toxic condition due to microorganisms Septicemia (sep′-ti-SĒ-me-a), presence of bacterial toxins in the blood (blood poisoning).

Soma-, Somato- body Somatotropic (sō-mat-ō-TRŌ-pik), having a stimulating effect on body growth.

Somni- sleep Insomnia (in-SOM-nē-a), inability to sleep.

Sten- narrow Stenosis (ste-NŌ-sis), narrowing of a duct or canal.

Stasis-, Stat- stand still Homeostasis (hō′-mē-ō-STĀ-sis), achievement of a steady state.

Tegument- skin, convering Integumentary (in-teg-yoo-MEN-ta-rē), pertaining to the skin.

Therm- heat Thermometer (ther-MOM-et-er), instrument used to measure and record heat.

Thromb- clot, lump Thrombus (THROM-bus), clot in a blood vessel or heart.

Tox-, Toxic- poison Toxemia (tok-SĒ-mē-a), poisonous substances in the blood.

Trich- hair Trichosis (trik-Ō-sis), disease of the hair.

Tympan- eardrum Tympanic (tim-PAN-ik) membrane, eardrum.

Vas- vessel, duct Cerebrovascular (se-rē-brō-VAS-kyoo-lar), pertaining to the blood vessels of the cerebrum of the brain.

Viscer- organ Visceral (VIS-e-ral), pertaining to the abdominal organs.

Zoo- animal Zoology (zō-OL-ō-jē), the study of animals.

PREFIXES

A-, An- without, lack of, deficient Anesthesia (an′-es-THĒ-zha), without sensation.

Ab- away from, from Abnormal (ab-NOR-mal), away from normal.

Ad- to, near, toward Adduction (a-DUK-shun), movement of an extremity toward the axis of the body.

Alb- white Albino (al-BĪ-no), person whose skin, hair, and eyes lack the pigment melanin.

Alveol- cavity, socket Alveolus (al-VĒ-ō-lus), air sac in the lung.

Ambi- both sides Ambidextrous (am′-bi-DEK-strus), able to use either hand.

Ambly- dull Amblyaphia (am-blē-A-fē-a), dull sense of touch.

Andro- male, masculine Androgen (AN-drō-jen), male sex hormone.

Ankyl(o)- bent, fusion Ankylosed (ANG-ki-lōsd), fused joint.

Ante- before Antepartum (ant-ē-PAR-tum), before delivery of a baby.

Anti- against Anticoagulant (an-tī-kō-AG-yoo-lant), a substance that prevents coagulation of blood.

Basi- base, foundation Basal (BĀ-sal), located near the base.

Bi- two, double, both Biceps (BĪ-seps), a muscle with two heads of origin.

Bili- bile, gall Biliary (BIL-ē-er-ē), pertaining to bile, bile ducts, or gallbladder.

Brachy- short Brachyesophagus (brā-kē-e-SOF-a-gus), short esophagus.

Brady- slow Bradycardia (brād′-ē-KARD-ē-a), abnormal slowness of the heartbeat.

Cata- down, lower, under, against Catabolism (ka-TAB-a-lizm), metabolic breakdown into simpler substances.

Circum- around Circumrenal (ser-kum-RĒN-al), around the kidney.

Cirrh- yellow Cirrhosis (si-RŌ-sis), liver disorder that causes yellowing of skin.

Co-, Con-, Com with, together Congenital (kon-JEN-i-tal), existing at birth.

Contra- against, opposite Contraception (kon-tra-SEP-shun), the prevention of conception.

Crypt- hidden, concealed Cryptorchidism (krip-TOR-ka-dizm′), undescended or hidden testes.

Cyano- blue Cyanosis (sī-a-NŌ-sis), bluish discoloration due to inadequate oxygen.

De- down, from Decay (de-KA), waste away from normal.

Demi-, hemi- half Hemiplegia (hem′-ē-PLĒ-jē-a), paralysis on one side of the body.

Di-, Diplo- two Diploid (DIP-loyd), having double the haploid number of chromosomes.

Dis- separation, apart, away from Disarticulate (dis′-ar-TIK-yoo-lāt′), to separate at a joint.

Dys- painful, difficult Dyspnea (disp-NĒ-a), difficult breathing.

E-, Ec-, Ex- out from, out of Eccentric (ek-SEN-trik), not located at the center.

Ecto-, Exo- outside Ectopic (ek-TOP-ik) pregnancy, gestation outside the uterine cavity.

Em-, En- in, on Empyema (em′-pī-Ē-ma), pus in a body cavity.

End-, Endo- inside Endocardium (en′-dō-KARD-ē-um), membrane lining the inner surface of the heart.

Epi- upon, on, above Epidermis (ep′-i-DER-mis), outermost layer of skin.

Eu- well Eupnea (YOOP-nē-a), normal breathing.

Ex-, Exo- out, away from Exocrine (EK-sō-krin), excreting outwardly or away from.

Extra- outside, beyond, in addition to Extracellular (ek′-stra-SEL-yoo-lar), outside of the cell.

Fore- before, in front of Forehead (FOR-hed), anterior part of head.

Gen- originate, produce, form Genetics (gen-ET-iks), the study of heredity.

Gingiv- gum Gingivitis (jin′-je-VĪ-tus), inflammation of the gums.

Hemi- half Hemiplegia (hem-ē-PLĒ-jē-a), paralysis of only half of the body.

heteř-, Hetero- other, different Heterogeneous (het′-e-rō-JEN-ē-us), composed of different substances.

Homeo-, Homo- unchanging, the same, steady Homeostasis (hō′-mē-ō-STĀ-sis), achievement of a steady state.

Hyper- beyond, excessive Hyperglycemia (hī-per-glī-SĒ-mē-a), excessive amount of glucose in the blood.

Hypo- under, below, deficient Hypodermic (hī-pō-DER-mik), below the skin or dermis.

Idio- self, one's own, separate Idiopathic (id′-ē-ō-PATH-ik), a disease without recognizable cause.

In-, Im- in, inside, not Incontinent (in-KON-ti-nent), not able to retain urine or feces.

Infra- beneath Infraorbital (in′-fra-OR-bi-tal), beneath the orbit.

Inter- among, between Intercostal (int′-er-KOS-tal), between the ribs.

Intra within, inside Intracellular (in′-tra-SEL-yoo-lar), inside the cell.

Iso- equal, like Isotonic (ī-sō-TON-ik), equal tension or tone.

Later- side Lateral (LAT-er-al), pertaining to a side or farther from the midline.

Lepto- small, slender, thin Leptodermic (lep′-tō-DER-mik), having thin skin.

Macro- large, great Macrophage (MAK-rō-fāj), large phagocytic cell.

Mal- bad, abnormal Malnutrition (mal′-noo-TRISH-un), lack of necessary food substances.

Medi-, Meso- middle Medial (MĒD-ē-al), nearer to midline.

Mega-, Megalo- great, large Megakaryocyte (meg′-a-KAR-ē-ō-sīt), giant cell of bone marrow.

Melan- black Melanin (MEL-a-nin), black or dark brown pigment found in skin and hair.

Meta- after, beyond Metacarpus (met-a-KAR-pus), the part of the hand between the wrist and fingers.

Micro- small Microtome (MĪ-krō-tōm), instrument for preparing very thin slices of tissue for microscopic examination.

Mono- one Monorchid (mon-OR-kid), having one testicle.

Neo- new Neonatal (nē-ō-NĀT-al), pertaining to the first weeks after birth.

Noct(i)- night Nocturia (nok-TOO-rē-a), urination occurring at night.

Null(i)- none Nullipara (nu-LIP-a-ra), woman with no children.

Nyct- night Nyctalopia (nik′-ta-LŌ-pē-a), night blindness.

Oligo- small, deficient Oliguria (ol-ig-YOO-rē-a), abnormally small amount of urine.

Ortho- straight, normal Orthopnea (or-thop-NE-a), inability to breathe in any position except when straight or erect.

Pan- all Pancarditis (pan-kar-DĪ-tis), inflammation of the entire heart.

Para- near, beyond, apart from, beside Paranasal (par-a-NĀ-zal), near the nose.

Per- through Percutaneous (per′-kyoo-TĀ-nē-us), through the skin.

Peri- around Pericardium (per′-i-KARD-ē-um), membrane or sac around the heart.

Poly- much, many Polycythemia (pol′-i-sī-THĒ-mē-a), an excess of red blood cells.

Post- after, beyond Postnatal (pōst-NĀT-al), after birth.

Pre-, Pro- before, in front of Prenatal (prē-NĀT-al), before birth.

Prim- first Primary (PRĪ-me-rē), first in time or order.

Proto- first Protocol (PRŌ-tō-kol), clinical report made from first notes taken.

Pseud-, Pseudo- false Pseudoangina (soo′-dō-an-JĪ-na), false angina.

Retro- backward, located behind Retroperitoneal (re′-trō-per′-it-on-Ē-al), located behind the peritoneum.

Schizo- split, divide Schizophrenia (skiz′-ō-FRE-nē-a), split personality mental disorder.

Semi- half Semicircular (sem′-i-SER-kyoo-lar) canals, canals in the shape of a half circle.

Sub- under, beneath, below Submucosa (sub′-myoo-KŌ-sa), tissue layer under a mucous membrane.

Super- above, beyond Superficial (soo-per-FISH-al), confined to the surface.

Supra- above, over Suprarenal (soo-pra-RĒN-al), adrenal gland above the kidney.

Sym-, Syn- with, together, jointed Syndrome (SIN-drō), all the symptoms of a disease considered as a whole.

Tachy- rapid Tachycardia (tak′-i-KARD-ē-a), abnormally rapid heartbeat.

Terat(o)- malformed fetus Teratogen (TER-a-tō-jen), an agent that causes development of a malformed fetus.

Tetra-, quadra- four Tetrad (TET-rad), group of four with something in common.

Trans- across, through, beyond Transudation (trans-yoo-DĀ-shun), oozing of a fluid through pores.

Tri- three Trigone (TRĪ-gon), a triangular space, as at the base of the bladder.

SUFFIXES

-able capable of, having ability to Viable (VĪ-a-bal), capable of living.

-ac, -al pertaining to Cardiac (KARD-ē-ak), pertaining to the heart.

-algia painful condition Myalgia (mī-AL-jē-a), pain in a muscle.

-an, -ian pertaining to Circadian (ser-KĀ-dē-an), pertaining to a daily (24 hour) cycle.

-ant having the characteristic of Malignant (ma-LIG-nant), having the characteristic of badness.

-ary connected with Ciliary (SIL-ē-ar-ē), resembling any hairlike structure.

-asis, -asia, -esis, -osis condition or state of Hemostasis (hē-mō-STĀ-sis), stopping of bleeding or circulation.

-asthenia weakness Myasthenia (mi-as-THĒ-nē-a), weakness of skeletal muscles.

-ation process, action, condition Inspiration (in-spi-RĀ-shun), process of drawing air into lungs.

-cel, -cele swelling, an enlarged space or cavity Meningocele (men-IN-gō-sēl), enlargement of the meninges.

-centesis puncture, usually for drainage Amniocentesis (am′-nē-ō-sen-TĒ-sis), withdrawal of amniotic fluid.

-cid, -cide, -cis cut, kill, destroy Germicide (jer-mi-SĪD), a substance that kills germs.

-ectasia, -ectasis stretching, dilation Bronchiectasis (bron-kē-EK-ta-sis), dilation of a bronchus or bronchi.

-ectomize, ectomy excision of, removal of Thyroidectomy (thī-royd-EK-tō-mē), surgical removal of a thyroid gland.

-ema swelling, distension Emphysema (em′-fi-SĒ-ma), swelling of air sacs in lungs.

-emia condition of blood Lipemia (lip-Ē-mē-a), abnormally high concentration of fat in the blood.

-esis condition, process Enuresis (en′yoo-RĒ-sis), condition of involuntary urination.

-esthesia sensation, feeling Anesthesia (an′-es-THĒ-zē-a), total or partial loss of feeling.

-ferent carry Efferent (EF-e-rent), carrying away from a center.

-form shape Fusiform (FYOO-zi-form), spindle-shaped.

-gen agent that produces or originates Pathogen (PATH-ō-jen), microorganism or substance capable of producing a disease.

-genic produced form, producing Pyogenic (pī-ō-JEN-ik), producing pus.

-gram record, that which is recorded Electrocardiogram (e-lek′-trō-KARD-ē-gram), record of heart action.

-graph instrument for recording Electroencephalograph (e-lek′-trō-en-SEF-a-lō-graf), instrument for recording electrical activity of the brain.

-ia state, condition Hypermetropia (hī′-per-me-TRŌ-pē-a), condition of farsightedness.

-iatrics, iatry medical practice specialities Pediatrics (pēd-ē-A-triks), medical science relating to care of children and treatment of their diseases.

-ician person associated with Technician (tek-NISH-an), person skilled in a technical field.

-ics art of, science of Optics (OP-tiks), science of light and vision.

-ion action, condition resulting from action Incision (in-SIZH-un), act or result of cutting into flesh.

-ism condition, state Rheumatism (ROO-ma-tizm), inflammation, especially of muscles and joints.

-ist one who practices Internist (in-TER-nist), one who practices internal medicine.

-itis inflammation Neuritis (nyoo-RĪT-is), inflammation of a nerve or nerves.

-ive relating to Sedative (SED-a-tive), relating to a pain or tension reliever.

-logy, -ology the study or science of Physiology (fiz-ē-OL-ō-jē), the study of function of body parts.

-lyso, -lysis solution, dissolve, loosening Hemolysis (hē-MOL-i-sis), dissolution of red blood cells.

-malacia softening Osteomalacia (os′-tē-ō-ma-LĀ-shē-a), softening of bone.

-megaly enlarged Cardiomegaly (kar′-dē-ō-MEG-a-lē), enlarged heart.

-oid resembling Lipoid (li-POYD), resembling fat.

-ologist specialist Dermatologist (der-ma-TOL-ō-gist), specialist in the study of the skin.

-oma tumor Fibroma (fi-BRŌ-ma), tumor composed mostly of fibrous tissue.

-ory pertaining to Sensory (SENS-o-rē), pertaining to sensation.

-ose full of Adipose (AD-i-pōz), characterized by presence of fat.

-osis condition, disease Necrosis (ne-KRŌ-sis), condition of death of cells.

ostomy create an opening Colostomy (kō-LOS-tō-me), surgical creation of an opening between the colon and body surface.

-otomy surgical incision Tracheotomy (trā-kē-OT-ō-me), surgical incision of the trachea.

-pathy disease Neuropathy (nyoo-ROP-a-thē), disease of the peripheral nervous system.

-penia deficiency Thrombocytopenia (throm′-bō-sīt′-ō-PĒ-nē-a), deficiency of thrombocytes in the blood.

-phobe, -phobia fear of, aversion to Hydrophobia (hī-drō-FŌ-bē-a), fear of water.

-plasia, -plasty development, formation Rhinoplasty (RĪ-nō-plas-tē), surgical reconstruction of the nose.

-plegia, -plexy stroke, paralysis Apoplexy (AP-ō-plek-sē), sudden loss of consciousness and paralysis.

-pnea to breathe Apnea (AP-nē-a), temporary absence of respiration, following a period of overbreathing.

-poiesis production Hematopoiesis (he-mat′-a-poy-Ē-sis), formation and development of red blood cells.

-ptosis falling, sagging Blepharoptosis (blef′-a-rō-TŌ-sis), drooping of upper eyelid.

-rrhage bursting forth, abnormal discharge Hemorrhage (HEM-or-rij), bursting forth of blood.

-rrhea flow, discharge Diarrhea (dī-a-RĒ-a), abnormal frequency of bowel evacuation, the stools with a more or less fluid consistency.

-scope instrument for viewing Bronchoscope (BRON-kō-skōp), instrument used to examine the interior of a bronchus.

-stomy creation of a mouth or artificial opening Tracheostomy (trā-kē-OST-ō-mē), creation of an opening in the trachea.

-tic, -ulnar pertaining to Diagnostic (dī′-ag-NOS-tik), pertaining to diagnosis.

-tomy cutting into, incision into Laparotomy (lap-a-ROT-ō-mē), an abdominal incision to gain access to the peritoneal cavity.

-tripsy crushing Lithotripsy (LITH-ō-trip-sē), crushing of a calculus (stone).

-trophy state relating to nutrition or growth Hypertrophy (hī-PER-trō-fē), excessive growth of an organ or part.

-tropic turning toward, influencing, changing Gonadotropic (gō-nad-a-TRŌ-pic), influencing the gonads.

-uria urine Polyuria (pol-ē-YOOR-ē-a), excessive secretion of urine.

GLOSSARY OF TERMS

Abdomen (ab-DŌ-men or AB-dō-men) The area between the diaphragm and pelvis.

Abdominal (ab-DŌM-i-nal) **cavity** Superior portion of the abdominopelvic cavity that contains the stomach, spleen, liver, gallbladder, pancreas, small intestine, and most of the large intestine.

Abdominal thrust maneuver A first-aid procedure for choking. Employs a quick, upward thrust against the diaphragm that forces air out of the lungs with sufficient force to eject any lodged material. Also called the **Heimlich** (HĪM-lik) **maneuver.**

Abdominopelvic (ab-dom'-i-nō-PEL-vic) **cavity** Inferior component of the ventral body cavity that is subdivided into an upper abdominal cavity and a lower pelvic cavity.

Abduction (ab-DUK-shun) Movement away from the axis or midline of the body.

Abortion (a-BOR-shun) The premature loss (spontaneous) or removal (induced) of the embryo or nonviable fetus; any failure in the normal process of developing or maturing.

Abrasion (a-BRĀ-shun) A portion of skin that has been scraped away.

Abscess (AB-ses) A localized collection of pus and liquefied tissue in a cavity.

Absorption (ab-SORP-shun) The taking up of liquids by solids or of gases by solids or liquids; intake of fluids or other substances by cells of the skin or mucous membranes; the passage of digested foods from the gastrointestinal tract into blood or lymph.

Absorptive (fed) state Metabolic state during which ingested nutrients are being absorbed by the blood or lymph from the gastrointestinal tract.

Accessory duct A duct of the pancreas that empties into the duodenum about 2.5 cm (1 in.) superior to the ampulla of Vater (hepatopancreatic ampulla). Also called the **duct of Santorini** (san'-tō-RĒ-nē).

Accommodation (a-kom-ō-DĀ-shun) A change in the curvature of the eye lens to adjust for vision at various distances.

Acetabulum (as'-e-TAB-yoo-lum) The rounded cavity on the external surface of the hipbone that receives the head of the femur.

Acetylcholine (as'-ē-til-KŌ-lēn) **(ACh)** A neurotransmitter liberated by many peripheral nervous system neurons and some central nervous system neurons. It is excitatory at neuromuscular junctions but inhibitory at some other synapses (slows heart rate).

Achille's tendon *See* **Calcaneal tendon.**

Acid (AS-id) A proton donor, or substance that dissociates into hydrogen ions (H^+) and anions; characterized by an excess of hydrogen ions and a pH less than 7.

Acidosis (as-i-DŌ-sis) A condition in which blood pH is below 7.35. Also known as **acidemia.**

Acini (AS-i-nē) Masses of cells in the pancreas that secrete digestive enzymes.

Acne (AK-nē) Inflammation of sebaceous (oil) glands that usually begins at puberty; the basic acne lesions in order of increasing severity are comedones, papules, pustules, and cysts.

Acoustic (a-KOOS-tik) Pertaining to sound or the sense of hearing.

Acquired immune deficiency syndrome (AIDS) A disorder characterized by a positive HIV-antibody test and certain indicator diseases (Kaposi's sarcoma, *Pneumocystis carinii* pneumonia, tuberculosis, fungus diseases, etc.). A deficiency of helper T cells that results in fever or night sweats, coughing, sore throat, fatigue, body aches, weight loss, and enlarged lymph nodes. Caused by a virus called human immunodeficiency virus (HIV).

Acromegaly (ak'-rō-MEG-a-lē) Condition caused by hypersecretion of human growth hormone (hGH) during adulthood characterized by thickened bones and enlargement of other tissues.

Acrosome (AK-rō-sōm) A dense granule in the head of a spermatozoon that contains enzymes that facilitate the penetration of a spermatozoon into a secondary oocyte.

Actin (AK-tin) The contractile protein that makes up thin filaments in muscle fiber (cell).

Action potential A wave of negativity that self-propagates along the membrane of a neuron or muscle fiber (cell); a rapid change in membrane potential that involves a depolarization followed by a repolarization. Also called a **nerve action potential** or **nerve impulse** as it relates to a neuron and a **muscle action potential** as it relates to a muscle fiber (cell).

Activation (ak'-ti-VĀ-shun) **energy** The minimum amount of energy required for a chemical reaction to occur.

Active transport The movement of substances across cell membranes against a concentration gradient, requiring the expenditure of energy (ATP).

Acuity (a-KYOO-i-tē) Clearness or sharpness, usually of vision.

Acupuncture (AK-yoo-punk'-chur) The insertion of a needle into a tissue for the purpose of drawing fluid or relieving pain. It is also an ancient Chinese practice employed to cure illnesses by inserting needles into specific locations of the skin.

Acute (a-KYOOT) Having rapid onset, severe symptoms, and a short course; not chronic.

Adam's apple *See* **Thyroid cartilage.**

Adaptation (ad'-ap-TĀ-shun) The adjustment of the pupil of the eye to light variations. The property by which a neuron relays a decreased frequency of action potentials from a receptor even though the strength of the stimulus remains constant. The decrease in perception of a sensation over time while the stimulus is still present.

Addison's (AD-i-sonz) **disease** Disorder caused by hyposecretion of glucocorticoids (and aldosterone) characterized by muscular weakness, hypoglycemia, mental lethargy, anorexia, nausea and vomiting, weight loss, low blood pressure, dehydration, and excessive skin and mucous membrane pigmentation.

Adduction (ad-DUK-shun) Movement toward the axis or midline of the body.

Adenoids (AD-e-noyds) Enlarged pharyngeal tonsils.

Adenosine triphosphate (a-DEN-ō-sēn trī-FOS-fāt) **(ATP)** The universal energy-carrying molecule manufactured in all living cells as a means of capturing and storing energy. It consists of the purine base *adenine* and the five-carbon sugar *ribose,* to which are added, in linear array, three *phosphate* groups.

Adenylate cyclase (a-DEN-i-lāt SĪ-klās) An enzyme in the postsynaptic membrane that is activated when certain neurotransmitters (or hormones) bind to their receptors; the enzyme that converts ATP into cyclic AMP.

Adherence (ad-HĒR-ens) Firm contact between the plasma membrane of a phagocyte and an antigen or other foreign substance.

Adhesion (ad-HĒ-zhun) Abnormal joining of parts to each other.

Adipocyte (AD-i-pō-sīt) Fat cell, derived from a fibroblast.

Adrenal cortex (a-DRĒ-nal KOR-teks) The outer portion of an adrenal gland, divided into three zones, each of which has a different cellular arrangement and secretes different hormones.

Adrenal (a-DRĒ-nal) **glands** Two glands located superior to each kidney. Also called the **suprarenal** (soo'-pra-RĒ-nal) **glands.**

Adrenal medulla (me-DUL-a) The inner portion of an adrenal gland, consisting of cells that secrete epinephrine and norepinephrine (NE) in response to the stimulation of preganglionic sympathetic neurons.

Adrenergic (ad'-ren-ER-jik) **fiber** A nerve fiber that when stimulated releases epinephrine (adrenaline) or norepinephrine (noradrenaline) at a synapse.

Adrenocorticotropic (ad-rē'-nō-kor-ti-kō-TRŌP-ik) **hormone (ACTH)** A hormone produced by the anterior pituitary gland that influences the production and secretion of certain hormones of the adrenal cortex.

Adventitia (ad-ven-TISH-ya) The outermost covering of a structure or organ.

Aerobic (air-Ō-bik) Requiring molecular oxygen.

Afferent arteriole (AF-er-ent ar-TĒ-rē-ōl) A blood vessel of a kidney that breaks up into the capillary network called a glomerulus; there is one afferent arteriole for each glomerulus.

Agglutination (a-gloo'-ti-NĀ-shun) Clumping of microorganisms or blood cells; typically an antigen–antibody reaction.

Agglutinin (a-GLOO-ti-nin) A specific antibody in blood serum that reacts with a specific agglutinogen and causes the clumping of bacteria, blood cells, or particles. Also called an **isoantibody.**

Agglutinogen (ag'-loo-TIN-ō-gen) A genetically determined antigen located on the surface of erythrocytes; basis for the ABO grouping and Rh system of blood classification. Also called an **isoantigen.**

Aggregated lymphatic follicles Aggregated lymph nodules that are most numerous in the ileum. Also called **Peyer's** (PĪ-erz) **patches.**

Aging Normal process accompanied by a progressive alteration of the body's homeostatic adaptive responses.

Agnosia (ag-NŌ-zē-a) A loss of the ability to recognize the meaning of stimuli from the various senses (visual, auditory, touch).

Agraphia (a-GRAF-ē-a) An inability to write.

AIDS *See* **Acquired immune deficiency syndrome.**

Albinism (AL-bin-izm) Abnormal, nonpathological, partial or total absence of pigment in skin, hair, and eyes.

Albumin (al-BYOO-min) The most abundant (60 percent) and smallest of the plasma proteins, which functions primarily to regulate osmotic pressure of plasma.

Albuminuria (al-byoo'-min-UR-ēa) Presence of albumin in the urine.

Aldosterone (al-do-STĒR-ōn) A mineralocorticoid produced by the adrenal cortex that brings about sodium and water reabsorption and potassium excretion.

Aldosteronism (al'-do-STER-ōn-izm') Condition caused by hypersecretion of aldosterone that results in increased sodium concentration and decreased potassium concentration in blood and characterized by muscular paralysis, high blood pressure, and edema.

Alimentary (al-i-MEN-ta-rē) Pertaining to nutrition.

Alkaline (AL-ka-līn) Containing more hydroxyl ions (OH^-) than hydrogen ions (H^+) to produce a pH higher than 7.

Alkalosis (al-ka-LŌ-sis) A condition in which blood pH is higher than 7.45. Also known as **alkalemia.**

Allantois (a-LAN-tō-is) A small, vascularized membrane between the chorion and amnion of the fetus that serves as an early site for blood formation.

Allele (a-LĒL) Genes that control the same inherited trait (such as height or eye color) that are located at the same position (locus) on homologous chromosomes.

Allergen (AL-er-jen) An antigen that evokes a hypersensitivity reaction.

Allergic (a-LER-jik) Pertaining to or sensitive to an allergen.

All-or-none principle In muscle physiology, individual muscle fibers (cells) contract to their fullest extent or not at all. In neuron physiology, if a stimulus is strong enough to initiate an action potential, a nerve impulse is transmitted along the entire neuron at a constant strength.

Alpha (AL-fa) **cell**   A cell in the pancreatic islets (islets of Langerhans) in the pancreas that secretes glucagon.

Alpha receptor   Receptor found on visceral effectors innervated by most sympathetic postganglionic axons.

Alveolar-capillary (al-VĒ-ō-lar) **membrane**   Structure in the lungs consisting of the alveolar wall and basement membrane and a capillary endothelium and basement membrane through which the diffusion of respiratory gases occurs. Also called the **respiratory membrane.**

Alveolar duct   Branch of a respiratory bronchiole around which alveoli and alveolar sacs are arranged.

Alveolar macrophage (MAK-rō-fāj)   Cell found in the alveolar walls of the lungs that is highly phagocytic. Also called a **dust cell.**

Alveolar (al-VĒ-ō-lar) **pressure**   Air pressure within the lungs. Also called **intrapulmonic pressure.**

Alveolar sac   A collection or cluster of alveoli that share a common opening.

Alveolus (al-VĒ-ō-lus)   A small hollow or cavity; an air sac in the lungs; milk-secreting portion of a mammary gland.

Alzheimer's (ALTZ-hī-merz) **disease (AD)**   Disabling neurological disorder characterized by dysfunction and death of specific cerebral neurons resulting in widespread intellectual impairment, personality changes, and fluctuations in alertness.

Ambulatory (AM-byoo-la-tō′-rē)   Capable of walking.

Amenorrhea (ā-men-ō-RĒ-a)   Absence of menstruation.

Amino acid   An organic acid, containing an acid carboxyl group (COOH) and a basic amino group (NH$_2$), that is the building unit from which proteins are formed.

Amnesia (am-NĒ-zē-a)   A lack or loss of memory.

Amniocentesis (am′-nē-o-sen-TĒ-sis)   Removal of amniotic fluid by inserting a needle transabdominally into the amniotic cavity.

Amnion (AM-nē-on)   The innermost fetal membrane; a thin transparent sac that holds the fetus suspended in amniotic fluid. Also called the **"bag of waters."**

Amniotic (am′-nē-OT-ik) **fluid**   Fluid in the amniotic cavity, the space between the developing embryo (or fetus) and amnion, the fluid is initially produced as a filtrate from maternal blood and later from fetal urine.

Amphiarthrosis (am′-fē-ar-THRŌ-sis)   A slightly movable articulation midway between diarthrosis and synarthrosis, in which the articulating bony surfaces are separated by fibrous connective tissue or fibrocartilage to which both are attached.

Ampulla (am-POOL-la)   A saclike dilation of a canal.

Ampulla of Vater   *See* **Hepatopancreatic ampulla.**

Amyotrophic (a-mē-ō-TROF-ik) **lateral sclerosis (ALS)**   Progressive neuromuscular disease characterized by degeneration of motor neurons in the spinal cord that leads to muscular weakness. Also called **Lou Gehrig's disease.**

Anabolism (a-NAB-ō-lizm)   Synthetic energy-requiring reactions whereby small molecules are built up into larger ones.

Anaerobic (an-air-Ō-bik)   Not requiring molecular oxygen.

Anal (Ā-nal) **canal**   The terminal 2 or 3 cm (1 in.) of the rectum; opens to the exterior of the anus.

Anal column   A longitudinal fold in the mucous membrane of the anal canal that contains a network of arteries and veins.

Analgesia (an-al-JĒ-zē-a)   Pain relief.

Anal triangle   The subdivision of the female or male perineum that contains the anus.

Anaphase (AN-a-fāz)   The third stage of mitosis in which the chromatids that have separated at the centromeres move to opposite poles of the cell.

Anaphylaxis (an′-a-fi-LAK-sis)   Against protection; a hypersensitivity (allergic) reaction in which IgE antibodies attach to mast cells and basophils, causing them to produce mediators of anaphylaxis (histamine, leukotrines, kinins, and prostaglandins) that bring about increased permeability of blood vessels, increased smooth muscle contraction, and increased mucus production. Examples are hay fever, hives, and anaphylactic shock.

Anastomosis (a-nas-tō-MŌ-sis)   An end-to-end union or joining together of blood vessels, lymphatic vessels, or nerves.

Anatomical (an′-a-TOM-i-kal) **position**   A position of the body universally used in anatomical descriptions in which the body is erect, facing the observer, the upper extremities are at the sides, the palms of the hands are facing forward, and the feet are on the floor.

Anatomic dead space   Spaces of the nose, pharynx, larynx, trachea, bronchi, and bronchioles that contain 150ml of tidal volume; the air does not reach the alveoli to participate in gas exchange.

Anatomy (a-NAT-ō-mē)   The structure or study of structure of the body and the relation of its parts to each other.

Androgen (AN-drō-jen)   Substance producing or stimulating male characteristics, such as the male hormone testosterone.

Anemia (a-NĒ-mē-a)   Condition of the blood in which the number of functional red blood cells or their hemoglobin content is below normal.

Anesthesia (an′-es-THĒ-zē-a)   A total or partial loss of feeling or sensation, usually defined with respect to loss of pain sensation; may be general or local.

Aneuploid (an′-yoo-PLOID)   A cell that has one or more chromosomes of a set added or deleted.

Aneurysm (AN-yoo-rizm)   A saclike enlargement of a blood vessel caused by a weakening of its wall.

Angina pectoris (an-JĪ-na *or* AN-ji-na PEK-tō-ris)   A pain in the chest related to reduced coronary circulation that may or may not involve heart or artery disease.

Angiography (an-jē-OG-ra-fē)   X-ray examination of blood vessels after injection of a radiopaque substance.

Angiotensin (an-jē-ō-TEN-sin)   Either of two forms of a protein associated with regulation of blood pressure. Angiotensin I is produced by the action of renin on angiotensinogen and is converted by the action of a plasma enzyme into angiotensin II, which stimulates aldosterone secretion by the adrenal cortex.

Anion (AN-ī-on)   A negatively charged ion. An example is the chloride ion (Cl$^-$).

Ankylosis (ang′-ki-LŌ-sus)   Severe or complete loss of movement at a joint.

Anomaly (a-NOM-a-lē)   An abnormality that may be a developmental (congenital) defect; a variant from the usual standard.

Anopsia (an-OP-sē-a) A defect of vision.

Anorexia nervosa (an-ō-REK-sē-a ner-VŌ-sa) A chronic disorder characterized by self-induced weight loss, body-image and other perceptual disturbances, and physiologic changes that result from nutritional depletion.

Anosmia (an-OZ-mē-a) Loss of the sense of smell.

Anoxia (an-OK-sē-a) Deficiency of oxygen.

Antagonist (an-TAG-ō-nist) A muscle that has an action opposite that of the prime mover (agonist) and yields to the movement of the prime mover.

Antagonistic (an-tag-ō-NIST-ik) **effect** A hormonal interaction in which the effect of one hormone on a target cell is opposed by another hormone. For example, calcitonin (CT) lowers blood calcium level, whereas parathormone (PTH) raises it.

Antepartum (an-tē-PAR-tum) Before delivery of the child; occurring (to the mother) before childbirth.

Anterior (an-TĒR-ē-or) Nearer to or at the front of the body. Also called **ventral.**

Anterior pituitary gland Anterior portion of the pituitary gland. Also called the **adenohypophysis** (ad′-e-nō-hī-POF-i-sis).

Anterior root The structure composed of axons of motor (efferent) fibers that emerges from the anterior aspect of the spinal cord and extends laterally to join a posterior root, forming a spinal nerve. Also called a **ventral root.**

Antibiotic (an′-ti-bī-OT-ik) Literally, "antilife"; a chemical produced by a microorganism that is able to inhibit the growth of or kill other microorganisms.

Antibody (AN-ti-bod′-ē) A protein produced by certain cells in the body in the presence of a specific antigen; the antibody combines with that antigen to neutralize, inhibit, or destroy it. Also called an **immunoglobulin** (im-yoo-nō-GLOB-yoo-lin) or **Ig.**

Antibody-mediated immunity That component of immunity in which lymphocytes (B cells) develop into plasma cells that produce antibodies that destroy antigens. Also called **humoral** (YOO-mor-al) **immunity.**

Anticoagulant (an-tī-cō-AG-yoo-lant) A substance that is able to delay, suppress, or prevent the clotting of blood.

Antidiuretic (an′-ti-dī-yoo-RET-ik) A substance that inhibits urine formation.

Antidiuretic hormone (ADH) Hormone produced by neurosecretory cells in the paraventricular and supraoptic nuclei of the hypothalamus that stimulates water reabsorption from kidney cells into the blood and vasoconstriction of arterioles.

Antigen (AN-ti-jen) Any substance that when introduced into the tissues or blood induces the formation of antibodies and reacts only with those specific antibodies.

Anti-oncogene (ONG-kō-jēn) A gene that may produce proteins that normally oppose the action of an oncogene or inhibit cell division; mutation of an anti-oncogene is the most common genetic change in a wide variety of cancer cells.

Antiport Process by which two substances, usually Na⁺ and another substance, move in opposite directions across a plasma membrane. Also called **countertransport.**

Anulus fibrosus (AN-yoo-lus fī-BRŌ-sus) A ring of fibrous tissue and fibrocartilage that encircles the pulpy substance (nucleus pulposus) of an intervertebral disc.

Anuria (a-NOO-rē-a) A daily urine output of less than 50 ml.

Anus (Ā-nus) The distal end and outlet of the rectum.

Aorta (ā-OR-ta) The main systemic trunk of the arterial system of the body that emerges from the left ventricle.

Aortic (ā-OR-tik) **body** Receptor on or near the arch of the aorta that responds to alterations in blood levels of oxygen, carbon dioxide, and hydrogen ions (H⁺).

Aortic reflex A reflex concerned with maintaining normal general systemic blood pressure.

Aperture (AP-er-chur) An opening or orifice.

Apex (Ā-peks) The pointed end of a conical structure, such as the apex of the heart.

Aphasia (a-FĀ-zē-a) Loss of ability to express oneself properly through speech or loss of verbal comprehension.

Apnea (AP-nē-a) Temporary cessation of breathing.

Apneustic (ap-NOO-stik) **area** Portion of the respiratory center in the pons that sends stimulatory nerve impulses to the inspiratory area that activate and prolong inspiration and inhibit expiration.

Apocrine (AP-ō-krin) **gland** A type of gland in which the secretory products gather at the free end of the secreting cell and are pinched off, along with some of the cytoplasm, to become the secretion, as in mammary glands.

Aponeurosis (ap′-ō-noo-RŌ-sis) A sheetlike tendon joining one muscle with another or with bone.

Appendage (a-PEN-dij) A structure attached to the body.

Appendicitis (a-pen-di-SĪ-tis) Inflammation of the vermiform appendix.

Appositional (ap′-ō-ZISH-o-nal) **growth** Growth due to surface deposition of material, as in the growth in diameter of cartilage and bone. Also called **exogenous** (eks-OJ-e-nus) **growth.**

Aqueous humor (AK-wē-us HYOO-mor) The watery fluid, similar in composition to cerebrospinal fluid, that fills the anterior cavity of the eye.

Arachnoid (a-RAK-noyd) The middle of the three coverings (meninges) of the brain or spinal cord.

Arachnoid villus (VIL-us) Berrylike tuft of arachoid that protrudes into the superior sagittal sinus and through which cerebrospinal fluid is reabsorbed into the bloodstream.

Arbor vitae (AR-bōr VĪ-tē) The treelike appearance of the white matter tracts of the cerebellum when seen in midsagittal section. A series of branching ridges within the cervix of the uterus.

Arch of the aorta The most superior portion of the aorta, lying between the ascending and descending segments of the aorta.

Areflexia (a′-rē-FLEK-sē-a) Absence of reflexes.

Areola (a-RĒ-ō-la) Any tiny space in a tissue. The pigmented ring around the nipple of the breast.

Arm The portion of the upper extremity from the shoulder to the elbow.

Arrector pili (a-REK-tor PI-lē) Smooth muscles attached to hairs; contraction pulls the hairs into a more vertical position, resulting in "goose bumps."

Arrhythmia (a-RITH-mē-a) Irregular heart rhythm. Also called a **dysrhythmia.**

Arteriogram (ar-TĒR-ē-ō-gram)   An x-ray image of an artery after injection of a radiopaque substance into the blood.

Arteriole (ar-TĒ-rē-ōl)   A small, almost microscopic, artery that delivers blood to a capillary.

Artery (AR-ter-ē)   A blood vessel that carries blood away from the heart.

Arthritis (ar-THRĪ-tis)   Inflammation of a joint.

Arthrology (ar-THROL-ō-jē)   The study or description of joints.

Arthroscopy (ar-THROS-co-pē)   A procedure for examining the interior of a joint, usually the knee, by inserting an arthroscope into a small incision; used to determine extent of damage, remove torn cartilage, repair cruciate ligaments, and obtain samples for analysis.

Arthrosis (ar-THRŌ-sis)   A joint or articulation.

Articular (ar-TIK-yoo-lar) **capsule**   Sleevelike structure around a synovial joint composed of a fibrous capsule and a synovial membrane.

Articular cartilage (KAR-ti-lij)   Hyaline cartilage attached to articular bone surfaces.

Articular disc   Fibrocartilage pad between articular surfaces of bones of some synovial joints. Also called a **meniscus** (men-IS-cus).

Articulate (ar-TIK-yoo-lāt)   To join together as a joint to permit motion between parts.

Articulation (ar-tik′-yoo-LĀ-shun)   A joint; a point of contact between bones, cartilage and bones, or teeth and bones.

Artificial pacemaker   A device that generates and delivers electrical signals to the heart to maintain a regular heart rhythm.

Arytenoid (ar′-i-TĒ-noyd) **cartilages**   A pair of small, pyramidal cartilages of the larynx that attach to the vocal folds and intrinsic pharyngeal muscles and can move the vocal folds.

Ascending colon (KŌ-lon)   The portion of the large intestine that passes upward from the cecum to the lower edge of the liver where it bends at the right colic (hepatic) flexure to become the transverse colon.

Ascites (as-SĪ-tēz)   Abnormal accumulation of serous fluid in the peritoneal cavity.

Aseptic (ā-SEP-tik)   Free from any infectious or septic material.

Asphyxia (as-FIX-ē-a)   Unconsciousness due to interference with the oxygen supply of the blood.

Aspiration (as′-pi-RĀ-shun)   Inhalation of a foreign substance (water, food, or foreign body) into the bronchial tree; drainage of a substance in or out by suction.

Association area   A portion of the cerebral cortex connected by many motor and sensory fibers to other parts of the cortex. The association areas are concerned with motor patterns, memory, concepts of word-hearing and word-seeing, reasoning, will, judgment, and personality traits.

Association neuron (NOO-ron)   A nerve cell lying completely within the central nervous system. Also called a **connecting neuron** or **interneuron.**

Astereognosis (as-ter′-ē-ōg-NŌ-sis)   Inability to recognize objects or forms by touch.

Asthenia (as-THĒ-nē-a)   Lack or loss of strength.

Astigmatism (a-STIG-ma-tizm)   An irregularity of the lens or cornea of the eye causing the image to be out of focus and producing faulty vision.

Astrocyte (AS-tro-sīt)   A neuroglial cell having a star shape that supports neurons in the brain and spinal cord and attaches the neurons to blood vessels.

Ataxia (a-TAK-sē-a)   A lack of muscular coordination, lack of precision.

Atelectasis (at′-ē-LEK-ta-sis)   A collapsed or airless state of all or part of the lung, which may be acute or chronic.

Atherosclerosis (ath′-er-ō-skle-RŌ-sis)   A process in which fatty substances (cholesterol and triglycerides) are deposited in the walls of medium and large arteries in response to certain stimuli (hypertension, carbon monoxide, dietary cholesterol). Following endothelial damage, monocytes stick to the tunica interna, develop into macrophages, and take up cholesterol and low-density lipoproteins. Smooth muscle fibers (cells) in the tunica media ingest cholesterol. This results in the formation of an atherosclerotic plaque that decreases the size of the arterial lumen.

Atherosclerotic (ath′-er-ō-skle-RO-tic) **plaque** (PLAK)   A lesion that results from accumulated cholesterol and smooth muscle fibers (cells) of the tunica media of an artery; may become obstructive.

Atom   Unit of matter that comprises a chemical element; consists of a nucleus and electrons.

Atomic number   Number of protons in an atom.

Atomic weight   Total number of protons and neutrons in an atom.

Atresia (a-TRĒ-zē-a)   Abnormal closure of a passage, or absence of a normal body opening.

Atrial fibrillation (Ā-trē-al fib-ri-LĀ-shun)   Asynchronous contraction of the atria that results in the cessation of atrial pumping.

Atrial natriuretic (na′-trē-yoo-RET-ik) **peptide (ANP)**   Peptide hormone produced by the atria of the heart in response to stretching that inhibits aldosterone production and thus lowers blood pressure.

Atrioventricular (AV) (ā′-trē-ō-ven-TRIK-yoo-lar) **bundle**   The portion of the conduction system of the heart that begins at the atrioventricular (AV) node, passes through the cardiac skeleton separating the atria and the ventricles, then runs a short distance down the interventricular septum before splitting into right and left bundle branches. Also called the **bundle of His (HISS).**

Atrioventricular (AV) node   The portion of the conduction system of the heart made up of a compact mass of conducting cells located near the orifice of the coronary sinus in the right atrial wall.

Atrioventricular (AV) valve   A structure made up of membranous flaps or cusps that allows blood to flow in one direction only, from an atrium into a ventricle.

Atrium (Ā-trē-um)   A superior chamber of the heart.

Atrophy (AT-rō-fē)   Wasting away or decrease in size of a part, due to a failure, abnormality of nutrition, or lack of use.

Auditory ossicle (AW-di-tō-rē OS-si-kul)   One of the three small bones of the middle ear called the malleus, incus, and stapes.

Auditory tube   The tube that connects the middle ear with the nose and nasopharynx region of the throat. Also called the **Eustachian** (yoo-STĀ-kē-an) **tube.**

Aura (OR-a)   A feeling or sensation that precedes an epileptic seizure or any paroxysmal attack (like those of bronchial asthma).

Auscultation (aws-kul-TA-shun)   Examination by listening to sounds in the body.

Autocrine (AW-tō-krin)   Local hormone, such as interleukin-2, that acts on the same cell that secreted it.

Autograft (AW-tō-graft)   A graft of tissue from a donor site to a recipient site of the same individual.

Autoimmunity   An immunologic response against a person's own tissue antigens.

Autologous (aw-TOL-ō-gus) **preoperative transfusion**   Donating one's own blood up to six weeks before elective surgery to ensure an abundant supply and reduce transfusion complications such as those that may be associated with diseases such as AIDS and hepatitis. Also called **predonation.**

Autolysis (aw-TOL-i-sis)   Spontaneous self-destruction of cells by their own digestive enzymes on death or a pathological process or during normal embryological development.

Autonomic ganglion (aw′-tō-NOM-ik GANG-lē-on)   A cluster of sympathetic or parasympathetic cell bodies located outside the central nervous system.

Autonomic nervous system (ANS)   Visceral sensory (afferent) and motor (efferent) neurons, both sympathetic and parasympathetic. Efferents transmit nerve impulses from the central nervous system to smooth muscle, cardiac muscle, and glands; so named because this portion of the nervous system was thought to be self-governing or spontaneous.

Autonomic plexus (PLEK-sus)   An extensive network of sympathetic and parasympathetic fibers; the cardiac, celiac, and pelvic plexuses are located in the thorax, abdomen, and pelvis, respectively.

Autophagy (aw-TOF-a-jē)   Process by which worn-out organelles are digested within lysosomes.

Autopsy (AW-top-sē)   The examination of the body after death.

Autoregulation (aw-tō-reg-yoo-LA-shun)   A local, automatic adjustment of blood flow in a given region of the body in response to tissue needs.

Autosome (AW-tō-sōm)   Any chromosome other than the pair of sex chromosomes.

Axilla (ak-SIL-a)   The small hollow beneath the arm where it joins the body at the shoulders. Also called the **armpit.**

Axon (AK-son)   The usually single, long process of a nerve cell that propagates a nerve impulse toward the axon terminals.

Axon terminal   Terminal branch of an axon.

Azygos (AZ-ī-gos)   An anatomical structure that is not paired; occurring singly.

Babinski (ba-BIN-skē) **sign**   Extension of the great toe, with or without fanning of the other toes, in response to stimulation of the outer margin of the sole of the foot; a normal reflex up to 1½ years of age; abnormal sign thereafter.

Back   The posterior part of the body; the dorsum.

Bainbridge (BĀN-bridge) **reflex**   The increased heart rate that follows increased pressure or distension of the right atrium.

Ball-and-socket joint   A synovial joint in which the rounded surface of one bone moves within a cup-shaped depression or fossa of another bone, as in the shoulder or hip joint. Also called a **spheroid** (SFĒ-roid) **joint.**

Barium (BA-rē-um) **swallow**   X-ray examination of the upper gastrointestinal tract to evaluate for ulcers, tumors, and bleeding.

Baroreceptor (bar′-ō-re-SEP-tor)   Nerve cell capable of responding to changes in blood, air, or fluid pressure. Also called a **pressoreceptor.**

Bartholin's glands   *See* **Greater vestibular glands.**

Basal ganglia (GANG-glē-a)   Paired clusters of cell bodies that make up the central gray matter in each cerebral hemisphere, including the caudate nucleus, lentiform nucleus, claustrum, and amygdaloid body. Also called **cerebral nuclei** (SER-e-bral NOO-klē-ī).

Basal metabolic (BĀ-sal met′-a-BOL-ik) **rate (BMR)**   The rate of metabolism measured under standard or basal conditions.

Base   The broadest part of a pyramidal structure. A nonacid or a proton acceptor, characterized by excess of hydroxide ions (OH^-) and a pH greater than 7. A ring-shaped, nitrogen-containing organic molecule that is one of the components of a nucleotide, for example, adenine, guanine, cytosine, thymine, and uracil.

Basement membrane   Thin, extracellular layer consisting of basal lamina secreted by epithelial cells and reticular lamina secreted by connective tissue cells.

Basilar (BAS-i-lar) **membrane**   A membrane in the cochlea of the inner ear that separates the cochlear duct from the scala tympani and on which the spiral organ (organ of Corti) rests.

Basophil (BĀ-sō-fil)   A type of white blood cell characterized by a pale nucleus and large granules that stain readily with basic dyes.

B cell   A lymphocyte that develops into a plasma cell that produces antibodies or a memory cell.

Belly   The abdomen. The gaster or prominent, fleshy part of a skeletal muscle.

Benign (be-NĪN)   Not malignant; favorable for recovery; a mild disease.

Beta (BĀ-ta) **cell**   A cell in the pancreatic islets (islets of Langerhans) in the pancreas that secretes insulin.

Beta receptor   Receptor found on visceral effectors innervated by most sympathetic postganglionic axons.

Bicuspid (bī-KUS-pid) **valve**   Atrioventricular (AV) valve on the left side of the heart. Also called the **mitral valve.**

Bifurcate (bī-FUR-kāt)   Having two branches or divisions; forked.

Bilateral (bī-LAT-er-al)   Pertaining to two sides of the body.

Bile (bīl)   A secretion of the liver consisting of water, bile salts, bile pigments, cholesterol, lecithin, and several ions; it assumes a role in emulsification of fats prior to their digestion.

Biliary (BIL-ē-er-ē)   Relating to bile, the gallbladder, or the bile ducts.

Biliary calculi (CAL-kyoo-lē)   Gallstones formed by the crystallization of cholesterol in bile.

Bilirubin (bil-ē-ROO-bin)   A red pigment that is one of the end products of hemoglobin breakdown in the liver cells and is excreted as a waste material in the bile.

Bilirubinuria (bil-ē-roo-bi-NOO-rē-a)   The presence of above-normal levels of bilirubin in urine.

Biliverdin (bil-ē-VER-din)   A green pigment that is one of the first products of hemoglobin breakdown in the liver cells and is converted to bilirubin or excreted as a waste material in bile.

Biofeedback   Process by which an individaul gets constant signals (feedback) about various visceral biological functions.

Biopsy (BĪ-op-sē)   Removal of tissue or other material from the living body for examination, usually microscopic.

Blastocoel (BLAS-tō-sēl)   The fluid-filled cavity within the blastocyst.

Blastocyst (BLAS-tō-sist)   In the development of an embryo, a hollow ball of cells that consists of a blastocoel (the internal cavity), trophoblast (outer cells), and inner cell mass.

Blastomere (BLAS-tō-mēr)   One of the cells resulting from the cleavage of a fertilized ovum.

Blastula (BLAS-tyoo-la)   An early stage in the development of a zygote.

Bleeding time   The time required for the cessation of bleeding from a small skin puncture; identifies platelet function defects and integrity of small blood vessels; ranges from 4 to 8 minutes.

Blind spot   Area in the retina at the end of the optic (II) nerve in which there are no light receptor cells.

Blood   The fluid that circulates through the heart, arteries, capillaries, and veins and that constitutes the chief means of transport within the body.

Blood–brain barrier (BBB)   A barrier consisting of specialized brain capillaries and astrocytes that prevents the passage of materials from the blood to the cerebrospinal fluid and brain.

Blood island   Isolated mass and cord of mesenchyme in the mesoderm from which blood vessels develop.

Blood pressure (BP)   Pressure exerted by blood as it presses against and attempts to stretch blood vessels, especially arteries; the force is generated by the rate and force of heartbeat; clinically, a measure of the pressure in arteries during ventricular systole and ventricular diastole.

Blood reservoir (REZ-er-vwar)   Systemic veins that contain large amounts of blood that can be moved quickly to parts of the body requiring the blood.

Blood–testis barrier   A barrier formed by sustentacular (Sertoli) cells that prevents an immune response against antigens produced by spermatozoa and developing cells by isolating the cells from the blood.

Body cavity   A space within the body that contains various internal organs.

Bohr (BŌR) **effect**   In an acid environment, oxygen splits more readily from hemoglobin because when hydrogen ions (H^+) bind to hemoglobin, they alter the structure of hemoglobin and this reduces its oxygen carrying capacity.

Bolus (BŌ-lus)   A soft, rounded mass, usually food, that is swallowed.

Bone scan   Procedure in which a radioisotope is injected and the radiation emitted from bone is measured.

Bony labyrinth (LAB-i-rinth)   A series of cavities within the petrous portion of the temporal bone forming the vestibule, cochlea, and semicircular canals of the inner ear.

Bowman's capsule   *See* **Glomerular capsule.**

Brachial plexus (BRĀ-kē-al PLEK-sus)   A network of nerve fibers of the anterior rami of spinal nerves C5, C6, C7, C8, and T1. The nerves that emerge from the brachial plexus supply the upper extremity.

Bradycardia (brād′-i-KAR-dē-a)   A slow resting heart or pulse rate (under 60/min) .

Brain   A mass of nervous tissue located in the cranial cavity.

Brain stem   The portion of the brain immediately superior to the spinal cord, made up of the medulla oblongata, pons, and midbrain.

Brain wave   Electrical activity produced as a result of action potentials of brain cells.

Broad ligament   A double fold of parietal peritoneum attaching the uterus to the side of the pelvic cavity.

Broca's (BRŌ-kaz) **area**   Motor area of the brain in the frontal lobe that translates thoughts into speech. Also called the **motor speech area.**

Bronchi (BRONG-kē)   Branches of the respiratory passageway including primary bronchi (the two divisions of the trachea), secondary or lobar bronchi (divisions of the primary that are distributed to the lobes of the lung), and tertiary or segmental bronchi (divisions of the secondary that are distributed to bronchopulmonary segments of the lung).

Bronchial asthma (BRONG-kē-al AZ-ma)   Usually allergic reaction characterized by smooth muscle spasms in bronchi resulting in wheezing and difficult breathing.

Bronchial tree   The trachea, bronchi, and their branching structures up to and including the terminal bronchioles.

Bronchiectasis (brong′-kē-EK-ta-sis)   A chronic disorder in which there is a loss of the normal tissue and expansion of lung air passages; characterized by difficult breathing, coughing, expectoration of pus, and foul breath.

Bronchiole (BRONG-kē-ol)   Branch of a tertiary bronchus further dividing into terminal bronchioles (distributed to lobules of the lung), which divide into respiratory bronchioles (distributed to alveolar sacs).

Bronchitis (brong-KĪ-tis)   Inflammation of the bronchi characterized by hypertrophy and hyperplasia of seromucous glands and goblet cells that line the bronchi and resulting in a productive cough.

Bronchogenic carcinoma (brong′-kō-JEN-ik kar′-si-NŌ-ma)   Cancer originating in the bronchi.

Bronchogram (BRONG-kō-gram)   An x-ray image of the bronchial tree.

Bronchography (bron-KOG-ra-fē)   Technique for examining the bronchial tree in which an opaque contrast medium is introduced into the trachea for distribution to the bronchial branches. The x-ray image produced is called a bronchogram.

Bronchopulmonary (brong′-kō-PUL-mō-ner ē) **segment**   One of the smaller divisions of a lobe of a lung supplied by its own branches of a bronchus.

Bronchoscope (BRONG-kō-skōp)   An instrument used to examine the interior of the bronchi of the lungs.

Bronchoscopy (brong-KOS-kō-pē)   Visual examination of the interior of the trachea and bronchi with a bronchoscope to biopsy a tumor, clear an obstruction, take cultures, stop bleeding, or deliver drugs.

Bronchus (BRONG-kus)   One of the two large branches of the trachea. *Plural,* **bronchi** (BRONG-kē).

Brunner's gland   *See* **Duodenal gland.**

Buccal (BUK-al)   Pertaining to the cheek or mouth.

Buffer (BUF-er) **system**   A pair of chemicals, one a weak acid and one the salt of the weak acid, which functions as a weak base, that resists changes in pH.

Bulb of penis   Expanded portion of the base of the corpus spongiosum penis.

Bulbourethral (bul′-bō-yoo-RĒ-thral) **gland**   One of a pair of glands located inferior to the prostate gland on either side of the urethra that secretes an alkaline fluid into the cavernous urethra. Also called a **Cowper's** (KOW-perz) **gland.**

Bulimia (boo-LIM-ē-a)   A disorder characterized by overeating, at least twice a week, followed by purging by self-induced vomiting, strict dieting or fasting, vigorous exercise, or use of laxatives.

Bulk flow   The movement of large numbers of ions, molecules, or particles in the same direction as a result of forces that push them (osmotic or hydrostatic pressure).

Bundle branch   One of the two branches of the atrioventricular (AV) bundle made up of specialized muscle fibers (cells) that transmit electrical impulses to the ventricles.

Bundle of His   *See* **Atrioventricular (AV) bundle.**

Bunion (BUN-yun)   Lateral deviation of the great toe that produces inflammation and thickening of the bursa, bone spurs, and calluses.

Burn   An injury in which tissue proteins are destroyed (denatured) as a result of heat (fire, steam), chemicals, electricity, or the ultraviolet rays of the sun.

Bursa (BUR-sa)   A sac or pouch of synovial fluid located at friction points, especially about joints.

Bursitis (bur-SĪtis)   Inflammation of a bursa.

Buttocks (BUT-oks)   The two fleshy masses on the posterior aspect of the lower trunk, formed by the gluteal muscles.

Cachexia (kah-KEK-sēah)   A state of ill health, malnutrition, and wasting.

Calcaneal tendon   The tendon of the soleus, gastrocnemius, and plantaris muscles at the back of the heel. Also called the **Achilles** (a-KIL-ēz) **tendon.**

Calcification (kal-si-fi-KĀ-shun)   Deposition of mineral salts, primarily hydroxyapatite, in a framework formed by collagen fibers in which the tissue hardens. Also called **mineralization.**

Calcitonin (kal-si-TŌ-nin) **(CT)**   A hormone produced by the thyroid gland that lowers the calcium and phosphate levels of the blood by inhibiting bone breakdown and accelerating calcium absorption by bones.

Calculus (KAL-kyoo-lus)   A stone, or insoluble mass of crystallized salts or other material, formed within the body, as in the gallbladder, kidney, or urinary bladder.

Callus (KAL-lus)   A growth of new bone tissue in and around a fractured area, ultimately replaced by mature bone. An acquired, localized thickening.

Calmodulin (kal-MOD-yoo-lin)   An intracellular protein that binds with calcium ions and activates or inhibits enzymes, many of which are protein kinases, to elicit physiological responses of hormones.

Calorie (KAL-ō-rē)   A unit of heat. A calorie (cal) is the standard unit and is the amount of heat necessary to raise 1 g of water 1°C from 14° to 15°C. The kilocalorie (kcal), used in metabolic and nutrition studies, is the amount of heat necessary to raise 1,000 g of water 1°C and is equal to 1,000 cal.

Calyx (KĀL-iks)   Any cuplike division of the kidney pelvis. *Plural,* **calyces** (KĀ-li-sēz).

Canal (ka-NAL)   A narrow tube, channel, or passageway.

Canaliculus (kan′-a-LIK-yoo-lus)   A small channel or canal, as in bones, where they connect lacunae. *Plural,* **canaliculi** (kan′-a-LIK-yoo-lī).

Canal of Schlemm   *See* **Scleral venous sinus.**

Cancellous (KAN-sel-us)   Having a reticular or latticework structure, as in spongy tissue of bone.

Cancer (KAN-ser)   A malignant tumor of epithelial origin tending to infiltrate and give rise to new growths or metastases. Also called **carcinoma** (kar′-si-NŌ-ma).

Canker (KANG-ker) **sore**   Painful ulcer on the mucous membrane of the mouth that may result from an autoimmune response.

Capacitation (ka′-pas-i-TĀ-shun)   The functional changes that sperm undergo in the female reproductive tract that allow them to fertilize a secondary oocyte.

Capillary (KAP-i-lar′-ē)   A microscopic blood vessel located between an arteriole and venule through which materials are exchanged between blood and body cells.

Carbohydrate (kar′-bō-HĪ-drāt)   An organic compound containing carbon, hydrogen, and oxygen in a particular amount and arrangement and comprised of sugar subunits; usually has the formula $(CH_2O)_n$.

Carbon monoxide (CO) poisoning   Hypoxia due to increased levels of carbon monoxide as a result of its preferential and tenacious combination with hemoglobin rather than with oxygen.

Carcinoembryonic (car′-sin-ō-em-brē-ON-ik) **antigen (CEA)**   A glycoprotein secreted by normally developing fetal tissue during the first or second trimester, after birth, and in certain malignant and benign conditions.

Carcinogen (kar-SIN-ō-jen)   Any substance that causes cancer.

Carcinoma (kar′-si-NŌ-ma)   A malignant tumor consisting of epithelial cells.

Cardiac (KAR-dē-ak) **arrest**   Cessation of an effective heartbeat in which the heart is completely stopped or in ventricular fibrillation.

Cardiac catheterization (KAR-dē-ak kath′-e-ter-i-ZĀ-shun)   Introduction of a catheter into the heart and/or its blood vessels to measure pressure; assess left ventricular function and cardiac output; measure blood flow, oxygen content of blood, and the status of valves and conduction system; and identify valvular and septal defects.

Cardiac (KAR-dē-ak) **cycle**   A complete heartbeat consisting of systole (contraction) and diastole (relaxation) of both atria plus systole and diastole of both ventricles.

Cardiac muscle An organ specialized for contraction, composed of striated muscle fibers (cells), forming the wall of the heart, and stimulated by an intrinsic conduction system and visceral efferent neurons.

Cardiac notch An angular notch in the anterior border of the left lung.

Cardiac output (CO) The volume of blood pumped from one ventricle of the heart (usually measured from the left ventricle) in 1 min; about 5.2 liters/min under normal resting conditions.

Cardiac reserve The maximum percentage that cardiac output can increase above normal.

Cardiac tamponade (tam′-pon-ĀD) Compression of the heart due to excessive fluid or blood in the pericardial sac that could result in cardiac failure.

Cardinal ligament A ligament of the uterus, extending laterally from the cervix and vagina as a continuation of the broad ligament.

Cardiology (kar-dē-OL-ō-jē) The study of the heart and diseases associated with it.

Cardiopulmonary resuscitation (rē-sus-i-TĀ-shun) **(CPR)** A technique employed to restore life or consciousness to a person apparently dead or dying; includes external respiration (exhaled air respiration) and external cardiac massage.

Cardiovascular (kar-dē-ō-VAS-kyoo-lar) **center** Groups of neurons scattered within the medulla that regulate heart rate, force of contraction, and blood vessel diameter.

Carotid (ka-ROT-id) **body** Receptor on or near the carotid sinus that responds to alterations in blood levels of oxygen, carbon dioxide, and hydrogen ions.

Carotid sinus A dilated region of the internal carotid artery immediately above the bifurcation of the common carotid artery that contains receptors that monitor blood pressure.

Carotid sinus reflex A reflex concerned with maintaining normal blood pressure in the brain.

Carpus (KAR-pus) A collective term for the eight bones of the wrist.

Cartilage (KAR-ti-lij) A type of connective tissue consisting of chondrocytes in lacunae embedded in a dense network of collagen and elastic fibers and a matrix of chondroitin sulfate.

Cartilaginous (kar′-ti-LAJ-i-nus) **joint** A joint without a synovial (joint) cavity where the articulating bones are held tightly together by cartilage, allowing little or no movement.

Cast A small mass of hardened material formed within a cavity in the body and then discharged from the body; can originate in different areas and can be composed of various materials.

Castration (kas-TRĀ-shun) The removal of the testes.

Catabolism (ka-TAB-ō-lizm) Chemical reactions that break down complex organic compounds into simple ones with the release of energy.

Cataract (KAT-a-rakt) Loss of transparency of the lens of the eye or its capsule or both.

Catheter (KATH-i-ter) A tube that can be inserted into a body cavity through a canal or into a blood vessel; used to remove fluids, such as urine and blood, and to introduce diagnostic materials or medication.

Cation A positively charged ion. An example is a sodium ion (Na^+).

Cauda equina (KAW-da ē-KWĪ-na) A tail-like collection of roots of spinal nerves at the inferior end of the spinal canal.

Caudal (KAW-dal) Pertaining to any tail-like structure; inferior in position.

Cecum (SĒ-kum) A blind pouch at the proximal end of the large intestine to which the ileum is attached.

Celiac plexus (PLEK-sus) A large mass of ganglia and nerve fibers located at the level of the upper part of the first lumbar vertebra. Also called the **solar plexus.**

Cell The basic structural and functional unit of all organisms; the smallest structure capable of performing all the activities vital to life.

Cell division Process by which a cell reproduces itself that consists of a nuclear division (mitosis) and a cytoplasmic division (cytokinesis); types include somatic and reproductive cell division.

Cell inclusion Principally organic substance produced by a cell that is not enclosed by a membrane and may appear or disappear at various times in the life of a cell.

Cell-mediated immunity That component of immunity in which specially sensitized lymphocytes (T cells) attach to antigens to destroy them. Also called **cellular immunity.**

Cellular respiration *See* **Oxidation.**

Cementum (se MEN-tum) Calcified tissue covering the root of a tooth.

Center of ossification (os′-i-fi-KĀ-shun) An area in the cartilage model of a future bone where the cartilage cells hypertrophy and then secrete enzymes that result in the calcification of their matrix, resulting in the death of the cartilage cells, followed by the invasion of the area by osteoblasts that then lay down bone.

Central canal A circular channel running longitudinally in the center of an osteon (Haversian system) of mature compact bone, containing blood and lymphatic vessels and nerves. Also called a **Haversian** (ha-VER-shun) **canal.** A microscopic tube running the length of the spinal cord in the gray commissure.

Central fovea (FŌ-vē-a) A cuplike depression in the center of the macula lutea of the retina, containing cones only; the area of clearest vision.

Central nervous system (CNS) That portion of the nervous system that consists of the brain and spinal cord.

Centrioles (SEN-trē-ōlz) Paired, cylindrical structures within a centrosome, each consisting of a ring of microtubules and arranged at right angles to each other.

Centromere (SEN-trō-mēr) The clear, constricted portion of a chromosome where the two chromatids are joined; serves as the point of attachment for the chromosomal microtubules.

Centrosome (SEN-trō-sōm) A rather dense area of cytoplasm, near the nucleus of a cell, containing a pair of centrioles. During prophase, it forms the mitotic spindle.

Cephalic (se-FAL-ik) Pertaining to the head; superior in position.

Cerebellar peduncle (ser-e-BEL-ar pe-DUNG-kul) A bundle of nerve fibers connecting the cerebellum with the brain stem.

Cerebellum (ser-e-BEL-um) The portion of the brain lying posterior to the medulla and pons, concerned with coordination of movements.

Cerebral aqueduct (SER-ē-bral AK-we-dukt) A channel through the midbrain connecting the third and fourth ventricles and containing cerebrospinal fluid.

Cerebral arterial circle A ring of arteries forming an anastomosis at the base of the brain between the internal carotid and basilar arteries and arteries supplying the brain. Also called the **circle of Willis.**

Cerebral cortex The surface of the cerebral hemispheres, 2–4 mm thick, consisting of six layers of nerve cell bodies (gray matter) in most areas.

Cerebral palsy (PAL-zē) A group of motor disorders resulting in muscular uncoordination and loss of muscle control and caused by damage to motor areas of the brain (cerebral cortex, basal ganglia, and cerebellum) during fetal life, birth, or infancy.

Cerebral peduncle One of a pair of nerve fiber bundles located on the ventral surface of the midbrain, conducting nerve impulses between the pons and the cerebral hemispheres.

Cerebrospinal (se-rē′-brō-SPĪ-nal) **fluid (CSF)** A fluid produced in the choroid plexuses and ependymal cells of the ventricles of the brain that circulates in the ventricles and the subarachnoid space around the brain and spinal cord.

Cerebrovascular (se-rē′-brō-VAS-kyoo-lar) **accident (CVA)** Destruction of brain tissue (infarction) resulting from disorders of blood vessels that supply the brain. Also called a **stroke.**

Cerebrum (SER-ē-brum or ser-Ē-brum) The two hemispheres of the forebrain, making up the largest part of the brain.

Ceruminous (se-ROO-mi-nus) **gland** A modified sudoriferous (sweat) gland in the external auditory meatus that secretes cerumen (ear wax).

Cervical dysplasia (dis-PLĀ-sē-a) A change in the shape, growth, and number of cervical cells of the uterus that, if severe, may progress to cancer.

Cervical ganglion (SER-vi-kul GANG-glē-on) A cluster of nerve cell bodies of postganglionic sympathetic neurons located in the neck, near the vertebral column.

Cervical mucus A mixture of water, glycoprotein, serum-type proteins, lipids, enzymes, and inorganic salts produced by secreting cells of the mucosa of the cervix.

Cervical plexus (PLEK-sus) A network of nerve fibers formed by the anterior rami of the first four cervical nerves.

Cervix (SER-viks) Neck; any constricted portion of an organ, such as the lower cylindrical part of the uterus.

Cesarean (se-SA-rē-an) **section** Procedure in which a low, horizontal incision is made through the abdominal wall and uterus for removal of the baby and placenta. Also called a **C-section.**

Chalazion (ka-LĀ-zē-on) A small tumor of the eyelid.

Chemical bond Force of attraction in a molecule that holds atoms together. Examples include ionic, covalent, and hydrogen bonds.

Chemical element Unit of matter that cannot be decomposed into a simpler substance by ordinary chemical reactions. Examples include hydrogen (H), carbon (C), and oxygen (O).

Chemically gated ion channel An ion channel that opens and closes in response to a specific chemical stimulus, such as a neurotransmitter, hormone, or certain ions (H^+ and Ca^{2+}).

Chemical reaction The combination or breaking apart of atoms in which chemical bonds are formed or broken and new products with different properties are produced.

Chemiosmosis (kem′-ē-oz-MŌ-sis) Process by which energy released is used to generate ATP when a substance moves along a gradient.

Chemoreceptor (kē′-mō-rē-SEP-tor) Receptor outside the central nervous system on or near the carotid and aortic bodies that detects the presence of chemicals.

Chemotaxis (kē-mō-TAK-sis) Attraction of phagocytes to microbes by a chemical stimulus.

Chemotherapy (kē-mō-THER-a-pē) The treatment of illness or disease by chemicals.

Chiasma (kī-AZ-ma) A crossing; especially the crossing of the optic (II) nerve fibers.

Chief cell The secreting cell of a gastric gland that produces pepsinogen, the precursor of the enzyme pepsin, and the enzyme gastric lipase. Also called a **zymogenic** (zī′-mō-JEN-ik) **cell.**

Chiropractic (kī′-rō-PRAK-tik) A system of treating disease by using one's hands to manipulate body parts, mostly the vertebral column.

Chlamydia (kla-MID-ē-a) A sexually transmitted disease characterized by burning on urination, frequent and painful urination, and low back pain; may spread to uterine (Fallopian) tubes in females.

Chloride shift Diffusion of bicarbonate ions (HCO_3^-) from the red blood cells into plasma and of chloride ions (Cl^-) from plasma into red blood cells that maintains ionic balance between red blood cells and plasma.

Cholecystectomy (kō′-lē-sis-TEK-tō-mē) Surgical removal of the gallbladder.

Cholesterol (kō-LES-te-rol) Classified as a lipid, the most abundant steroid in animal tissues; located in cell membranes and used for the synthesis of steroid hormones and bile salts.

Cholinergic (kō′-lin-ER-jik) **fiber** A nerve ending that liberates acetylcholine at a synapse.

Cholinesterase (kō′-lin-ES-ter-ās) An enzyme that hydrolyzes acetylcholine.

Chondrocyte (KON-drō-sīt) Cell of mature cartilage.

Chondroitin (kon-DROY-tin) **sulfate** An amorphous matrix material found outside connective tissue cells.

Chordae tendineae (KOR-dē TEN-di-nē-ē) Tendonlike, fibrous cords that connect the heart valves with the papillary muscles.

Chorion (KŌ-rē-on) The outermost fetal membrane that becomes the principal embryonic portion of the placenta; serves a protective and nutritive function.

Chorionic villus (kō′-rē-ON-ik VIL-lus) Fingerlike projection of the chorion that grows into the decidua basalis of the endometrium and contains fetal blood vessels.

Chorionic villus sampling (CVS) The removal of a sample of chorionic villus tissue by means of a catheter to analyze the tissue for prenatal genetic defects.

Choroid (KŌ-royd) One of the vascular coats of the eyeball.

Choroid plexus (PLEK-sus) A vascular structure located in the roof of each of the four ventricles of the brain; produces cerebrospinal fluid.

Chromaffin (krō-MAF-in) **cell** Cell that has an affinity for chrome salts, due in part to the presence of the precursors of the neurotransmitter epinephrine; found, among other places, in the adrenal medulla.

Chromatid (KRŌ-ma-tid) One of a pair of identical connected nucleoprotein strands that are joined at the centromere and separate during cell division, each becoming a chromosome of one of the two daughter cells.

Chromatin (KRŌ-ma-tin) The threadlike mass of the genetic material consisting principally of DNA, which is present in the nucleus of a nondividing or interphase cell.

Chromatolysis (krō′-ma-TOL-i-sis) The breakdown of chromatophilic substance (Nissl bodies) into finely granular masses in the cell body of a central or peripheral neuron whose process (axon or dendrite) has been damaged.

Chromatophilic substance Rough endoplasmic reticulum in the cell bodies of neurons that functions in protein synthesis. Also called **Nissl bodies.**

Chromosomal microtubule (mī-krō-TOOB-yool) Microtubule formed during prophase of mitosis that originates from centromeres, extends from a centromere to a pole of the cell, and assists in chromosomal movement; constitutes a part of the mitotic spindle.

Chromosome (KRŌ-mō-sōm) One of the 46 small, dark-staining bodies that appear in the nucleus of a human diploid (2*n*) cell during cell division.

Chronic (KRON-ik) Long term or frequently recurring; applied to a disease that is not acute.

Chyle (kīl) The milky fluid found in the lacteals of the small intestine after digestion.

Chylomicron (kī-lō-MĪK-ron) Protein-coated spherical structure that contains triglycerides, phospholipids, and cholesterol, and is absorbed into the lacteal of a villus.

Chyme (kīm) The semifluid mixture of partly digested food and digestive secretions found in the stomach and small intestine during digestion of a meal.

Ciliary (SIL-ē-ar′-ē) **body** One of the three portions of the vascular tunic of the eyeball, the others being the choroid and the iris; includes the ciliary muscle and the ciliary processes.

Ciliary ganglion (GANG-glē-on) A very small parasympathetic ganglion whose preganglionic fibers come from the oculomotor (III) nerve and whose postganglionic fibers carry nerve impulses to the ciliary muscle and the sphincter muscle of the iris.

Cilium (SIL-ē-um) A hair or hairlike process projecting from a cell that may be used to move the entire cell or to move substances along the surface of the cell. *Plural,* **cilia.**

Circadian (ser-KĀ-dē-an) **rhythm** A cycle of active and nonactive periods in organisms determined by internal mechanisms and repeating about every 24 hours.

Circle of Willis *See* **Cerebral arterial circle.**

Circular folds Permanent, deep, transverse folds in the mucosa and submucosa of the small intestine that increase the surface area for absorption. Also called **plicae circulares** (PLĪ-kē SER-kyoo-lar-ēs).

Circulation time Time required for blood to pass from the right atrium, through pulmonary circulation, back to the left ventricle, through systemic circulation to the foot, and back again to the right atrium; normally about 1 min.

Circumcision (ser′-kum-SIZH-un) Surgical removal of the foreskin (prepuce), the fold of skin over the glans penis.

Circumduction (ser′-kum-DUK-shun) A movement at a synovial joint in which the distal end of a bone moves in a circle while the proximal end remains relatively stable.

Circumvallate papilla (ser′-kum-VAL-āt pa-PIL-a) One of the circular projections that is arranged in an inverted V-shaped row at the posterior portion of the tongue; the largest of the elevations on the upper surface of the tongue containing taste buds.

Cirrhosis (si-RŌ-sis) A liver disorder in which the parenchymal cells are destroyed and replaced by connective tissue.

Cisterna chyli (sis-TER-na KĪ-lē) The origin of the thoracic duct.

Cleavage The rapid mitotic divisions following the fertilization of a secondary oocyte, resulting in an increased number of progressively smaller cells, called blastomeres, so that the overall size of the zygote remains the same.

Cleft palate Condition in which the palatine processes of the maxillae do not unite before birth; cleft lip, a split in the upper lip, is often associated with cleft palate.

Climacteric (klī-mak-TER-ik) Cessation of the reproductive function in the female or diminution of testicular activity in the male.

Climax The peak period or moments of greatest intensity during sexual excitement.

Clitoris (KLI-to-ris) An erectile organ of the female located at the anterior junction of the labia minora that is homologous to the male penis.

Clone (KLŌN) A population of cells identical to itself.

Clot The end result of a series of biochemical reactions that changes liquid plasma into a gelatinous mass; specifically, the conversion of fibrinogen into a tangle of polymerized fibrin molecules.

Clot retraction (rē-TRAK-shun) The consolidation of a fibrin clot to pull damaged tissue together.

Coagulation (cō-ag-yoo-LĀ-shun) Process by which a blood clot is formed.

Coarctation (kō′-ark-TĀ-shun) **of the aorta** Congenital condition in which the aorta is too narrow and results in reduced blood supply, increased ventricular pumping, and high blood pressure.

Coccyx (KOK-six) The fused bones at the end of the vertebral column.

Cochlea (KŌK-lē-a) A winding, cone-shaped tube forming a portion of the inner ear and containing the spiral organ (organ of Corti).

Cochlear duct The membranous cochlea consisting of a spirally arranged tube enclosed in the bony cochlea and lying along its outer wall. Also called the **scala media** (SCA-la MĒ-dē-a).

Coenzyme A type of cofactor; a nonprotein organic molecule that is associated with and activates an enzyme; many are derived from vitamins. An example is nicotinamide adenine dinucleotide (NAD), derived from the B vitamin niacin.

Coitus (KŌ-i-tus) Sexual intercourse. Also called **copulation** (cop-yoo-LĀ-shun).

Colitis (ko-LĪ-tis)   Inflammation of the mucosa of the colon and rectum in which absorption of water and salts is reduced, producing watery, bloody feces, and, in severe cases, dehydration and salt depletion. Spasms of the irritated muscularis produce cramps.

Collagen (KOL-a-jen)   A protein that is the main organic constituent of connective tissue.

Collateral circulation   The alternate route taken by blood through an anastomosis.

Colliculus (ko-LIK-yoo-lus)   A small elevation.

Colon   The division of the large intestine consisting of ascending, transverse, descending, and sigmoid portions.

Colony-stimulating factor (CSF)   One of a group of hematopoietins that stimulates development of white blood cells. Examples are macrophage CSF and granulocyte CSF.

Color blindness   Any deviation in the normal perception of colors, resulting from the lack of usually one of the photopigments of the cones.

Colostomy (kō-LOS-tō-mē)   The diversion of feces through an opening in the colon, creating a surgical opening at the exterior of the abdominal wall.

Colostrum (kō-LOS-trum)   A thin, cloudy fluid secreted by the mammary glands a few days prior to or after delivery before true milk is secreted.

Colposcopy (kol-POS-kō-pē)   Direct examination of the vaginal and cervical mucosa using a magnifying device; frequently the first procedure performed following an abnormal Pap smear.

Coma (KŌ-ma)   Final stage of brain failure that is characterized by total unresponsiveness to all external stimuli.

Common bile duct   A tube formed by the union of the common hepatic duct and the cystic duct that empties bile into the duodenum at the hepatopancreatic ampulla (ampulla of Vater).

Compact (dense) bone   Bone tissue with no apparent spaces in which the layers of lamellae are fitted tightly together. Compact bone is found immediately deep to the periosteum and external to spongy bone.

Complement (KOM-ple-ment)   A group of at least 20 normally inactive proteins found in serum that forms a component of nonspecific resistance and immunity by bringing about cytolysis, inflammation, and opsonization.

Complete blood count (CBC)   Hematology test that usually includes hemoglobin determination, hematocrit, red and white blood cell count, differential white blood cell count, and platelet count.

Compliance   The ease with which the lungs and thoracic wall can be expanded.

Compound   A substance that can be broken down into two or more other substances by chemical means.

Computed tomography (tō-MOG-ra-fē) **(CT)**   X-ray technique that provides a cross-sectional image of any area of the body. Also called **computed axial tomography (CAT).**

Concha (KONG-ka)   A scroll-like bone found in the skull. *Plural, conchae* (KONG-kē).

Concussion (kon-KUSH-un)   Traumatic injury to the brain that produces no visible bruising but may result in abrupt, temporary loss of consciousness.

Conduction myofiber   Muscle fiber (cell) in the subendocardial tissue of the heart specialized for conducting an action poten-

tial to the myocardium; part of the conduction system of the heart. Also called a **Purkinje** (pur-KIN-jē) **fiber.**

Conduction system   An intrinsic regulating system composed of specialized muscle tissue that generates and distributes electrical impulses that stimulate cardiac muscle fibers (cells) to contract.

Conductivity (kon'-duk-TIV-i-tē)   The ability to carry the effect of a stimulus from one part of a cell to another; highly developed in nerve and muscle fibers (cells).

Cone   The light-sensitive receptor in the retina concerned with color vision.

Congenital (kon-JEN-i-tal)   Present at the time of birth.

Congestive heart failure (CHF)   Chronic or acute state that results when the heart is not capable of supplying the oxygen demands of the body.

Conjunctiva (kon'-junk-TĪ-va)   The delicate membrane covering the eyeball and lining the eyes.

Conjunctivitis (kon-junk'-ti-VĪ-tis)   Inflammation of the conjunctiva, the delicate membrane covering the eyeball and lining the eyelids.

Connective tissue   The most abundant of the four basic tissue types in the body, performing the functions of binding and supporting; consists of relatively few cells in a great deal of intercellular substance.

Constipation (con-sti-PĀ-shun)   Infrequent or difficult defecation caused by decreased motility of the intestines.

Contact inhibition   Phenomenon by which migration of a growing cell is stopped when it makes contact with another cell of its own kind.

Contraception (kon'-tra-SEP-shun)   The prevention of conception or impregnation without destroying fertility.

Contractility (kon'-trak-TIL-i-tē)   The ability of cells or parts of cells to actively generate force to undergo shortening and change form for purposeful movements. Muscle fibers (cells) exhibit a high degree of contractility.

Contralateral (kon'-tra-LAT-er-al)   On the opposite side; affecting the opposite side of the body.

Contusion (kon-TOO-shun)   Condition in which tissue below the skin is damaged, but the skin is not broken.

Conus medullaris (KŌ-nus med-yoo-LAR-is)   The tapered portion of the spinal cord below the lumbar enlargement.

Convergence (con-VER-jens)   An anatomical arrangement in which the synaptic end bulbs of several presynaptic neurons terminate on one postsynaptic neuron. The medial movement of the two eyeballs so that both are directed toward a near object being viewed in order to produce a single image.

Convulsion (con-VUL-shun)   Violent, involuntary, tetanic contractions of an entire group of muscles.

Cornea (KOR-nē-a)   The nonvascular, transparent fibrous coat through which the iris can be seen.

Coronal (kō-RŌ-nal) **plane**   A plane that runs vertical to the ground and divides the body into anterior and posterior portions. Also called **frontal plane.**

Corona radiata   Several layers of follicle cells surrounding a secondary oocyte.

Coronary angiography (KOR-ō-na-rē an'-jē-OG-ra-fē)   Procedure in which the severity and location of blocked coronary arteries are visualized by injection of contrast dyes or in which clot-dissolving drugs may be injected into coronary arteries.

Coronary artery bypass grafting (CABG) Surgical procedure in which a portion of a blood vessel is removed from another part of the body and grafted onto a coronary artery so as to bypass an obstruction in the coronary artery.

Coronary artery disease (CAD) A condition such as atherosclerosis that causes narrowing of coronary arteries so that blood flow to the heart is reduced. This results in **coronary heart disease (CHD)** in which the heart muscle receives inadequate blood due to an interruption of its blood supply.

Coronary artery spasm A condition in which the smooth muscle of a coronary artery undergoes a sudden contraction, resulting in vasoconstriction.

Coronary circulation The pathway followed by the blood from the ascending aorta through the blood vessels supplying the heart and returning to the right atrium. Also called **cardiac circulation.**

Coronary sinus (SĪ-nus) A wide venous channel on the posterior surface of the heart that collects the blood from the coronary circulation and returns it to the right atrium.

Corpora quadrigemina (KOR-por-a kwad-ri-JEM-in-a) Four small elevations (superior and inferior colliculi) on the dorsal region of the midbrain concerned with visual and auditory functions.

Cor pulmonale (kor pul-mōn-ALE) **(CP)** Right ventricular hypertrophy from disorders that bring about hypertension in pulmonary circulation.

Corpus (KOR-pus) The principal part of any organ; any mass or body.

Corpus albicans (KOR-pus AL-bi-kanz) A white fibrous patch in the ovary that forms after the corpus luteum regresses.

Corpus callosum (kal-LŌ-sum) The great commissure of the brain between the cerebral hemispheres.

Corpuscle of touch The sensory receptor for the sensation of touch; found in the dermal papillae, especially in palms and soles. Also called a **Meissner's** (MĪS-ncrz) **corpuscle.**

Corpus luteum (LOO-tē-um) A yellow endocrine gland in the ovary formed when a follicle has discharged its secondary oocyte; secretes estrogens, progesterone, relaxin, and inhibin.

Corpus striatum (strī-Ā-tum) An area in the interior of each cerebral hemisphere composed of the caudate and lentiform nuclei of the basal ganglia and white matter of the internal capsule, arranged in a striated manner.

Cortex (KOR-teks) An outer layer of an organ. The convoluted layer of gray matter covering each cerebral hemisphere.

Costal (KOS-tal) Pertaining to a rib.

Costal cartilage (KAR-ti-lij) Hyaline cartilage that attaches a rib to the sternum.

Countercurrent mechanism One mechanism involved in the ability of the kidneys to produce a hyperosmotic urine.

Cowper's gland *See* **Bulbourethral gland.**

Cramp A spasmodic, especially a tonic, contraction of one of many muscles, usually painful.

Cranial (KRĀ-nē-al) **cavity** A subdivision of the dorsal body cavity formed by the cranial bones and containing the brain.

Cranial nerve One of 12 pairs of nerves that leave the brain, pass through foramina in the skull, and supply the head, neck, and part of the trunk; each is designated by a Roman numeral and a name.

Craniosacral (krā-nē-ō-SĀ-kral) **outflow** The fibers of parasympathetic preganglionic neurons, which have their cell bodies located in nuclei in the brain stem and in the lateral gray matter of the sacral portion of the spinal cord.

Craniotomy (krā′-nē-OT-ō-mē) Any operation on the skull, as for surgery on the brain or decompression of the fetal head in difficult labor.

Cranium (KRĀ-nē-um) The skeleton of the skull that protects the brain and the organs of sight, hearing, and balance; includes the frontal, parietal, temporal, occipital, sphenoid, and ethmoid bones.

Creatine phosphate (KRĒ-a-tin FOS-fāt) High-energy molecule in skeletal muscle fibers (cells) that is used to generate ATP rapidly; on decomposition, creatine phosphate breaks down into creatine, phosphate, and energy—the energy is used to generate ATP from ADP. Also called **phosphocreatine** (fos′-fō-KRĒ-a-tin).

Crenation (krē-NĀ-shun) The shrinkage of red blood cells into knobbed, starry forms when placed in a hypertonic solution.

Cretinism (KRĒ-tin-izm) Severe congenital thyroid deficiency during childhood leading to physical and mental retardation.

Crista (KRIS-ta) A crest or ridged structure. A small elevation in the ampulla of each semicircular duct that serves as a receptor for dynamic equilibrium.

Crossed extensor reflex A reflex in which extension of the joints in one limb occurs in conjunction with contraction of the flexor muscles of the opposite limb.

Crossing-over The exchange of a portion of one chromatid with another in a tetrad during meiosis. It permits an exchange of genes among chromatids and is one factor that results in genetic variation.

Crus (krus) **of penis** Separated, tapered portion of the corpora cavernosa penis. *Plural, crura* (KROO-ra).

Cryosurgery (KRĪ-ō-ser-jer-ē) The destruction of tissue by application of extreme cold.

Crypt of Lieberkühn *See* **Intestinal gland.**

Cryptorchidism (krip-TOR-ki-dizm) The condition of undescended testes.

Cupula (KUP-yoo-la) A mass of gelatinous material covering the hair cells of a crista, a receptor in the ampulla of a semicircular canal stimulated when the head moves.

Cushing's syndrome Condition caused by a hypersecretion of glucocorticoids characterized by spindly legs, "moon face," "buffalo hump," pendulous abdomen, flushed facial skin, poor wound healing, hyperglycemia, osteoporosis, weakness, hypertension, and increased susceptibility to disease.

Cutaneous (kyoo-TĀ-nē-us) Pertaining to the skin.

Cyanosis (sī-a-NO-sis) Reduced hemoglobin (unoxygenated) concentration of blood of more than 5 g/dl that results in a blue or dark purple discoloration that is most easily seen in nail beds and mucous membranes.

Cyclic AMP (cyclic adenosine-3′,5′-monophosphate) Molecule formed from ATP by the action of the enzyme adenylate cyclase; serves as an intracellular messenger (second messenger) for some hormones.

Cyst (SIST)   A sac with a distinct connective tissue wall, containing a fluid or other material.

Cystic (SIS-tik) **duct**   The duct that transports bile from the gallbladder to the common bile duct.

Cystitis (sis-TĪ-tis)   Inflammation of the urinary bladder.

Cystoscope (SIS-to-skōp)   An instrument used to examine the inside of the urinary bladder.

Cystoscopy (sis-TOS-kō-pē)   Direct visual examination of the urinary tract (and prostate gland in males as well) using a cystoscope to evaluate urinary tract disorders and remove tissue for biopsy, kidney stones, urinary bladder tumors, and urine samples.

Cytochrome (SĪ-tō-krōm)   A protein with an iron-containing group (heme) capable of alternating between a reduced form (Fe^{2+}) and an oxidized form (Fe^{3+}).

Cytokine (SĪ-to-kīn)   Growth factor produced by activated lymphocytes and other cells that acts as an autocrine or paracrine and has various roles in immunity and blood cell development.

Cytokinesis (sī′-tō-ki-NĒ-sis)   Division of the cytoplasm.

Cytology (sī-TOL-ō-jē)   The study of cells.

Cytoplasm (SĪ-tō-plazm)   Substance that surrounds organelles and located within a cell's plasma membrane and external to its nucleus.

Cytoskeleton   Complex internal structure of cytoplasm consisting of microfilaments, microtrubules, and intermediate filaments.

Cytosol (SĪ-tō-sol)   Semifluid portion of cytoplasm in which organelles and inclusions are suspended and solutes are dissolved. Also called **intracellular fluid.**

Dartos (DAR-tōs)   The contractile tissue under the skin of the scrotum.

Deafness   Lack of the sense of hearing or a significant hearing loss.

Debility (dē-BIL-i-tē)   Weakness of tonicity in functions or organs of the body.

Decibel (DES-i-bel) **(dB)**   A unit that measures relative sound intensity (loudness).

Decidua (dē-SID-yoo-a)   That portion of the endometrium of the uterus (all but the deepest layer) that is modified for pregnancy and shed after childbirth.

Deciduous (dē-SID-yoo-us)   Falling off or being shed seasonally or at a particular stage of development. In the body, referring to the first set of teeth.

Decompression sickness   A condition characterized by joint pains and neurologic symptoms; follows from a too-rapid reduction of environmental pressure or decompression, so that nitrogen that dissolved in body fluids under pressure comes out of solution as bubbles that form air emboli and occlude blood vessels. Also called **caisson** (KĀ-son) **disease** or **bends.**

Decussation (dē′-ku-SĀ-shun)   A crossing-over; usually refers to the crossing of most of the fibers in the large motor tracts to opposite sides in the medullary pyramids.

Deep   Away from the surface of the body.

Deep fascia (FASH-ē-a)   A sheet of connective tissue wrapped around a muscle to hold it in place.

Deep inguinal (IN-gwi-nal) **ring**   A slitlike opening in the aponeurosis of the transversus abdominis muscle that represents the origin of the inguinal canal.

Deep-venous thrombosis (DVT)   The presence of a thrombus in a vein, usually a deep vein of the lower extremities.

Defecation (def-e-KĀ-shun)   The discharge of feces from the rectum.

Defibrillation (dē-fib-ri-LĀ-shun)   Delivery of a very strong electrical current to the heart in an attempt to stop ventricular fibrillation.

Degeneration (dē-jen-er-Ā-shun)   A change from a higher to a lower state; a breakdown in structure.

Deglutition (dē-gloo-TISH-un)   The act of swallowing.

Dehydration (dē-hī-DRĀ-shun)   Excessive loss of water from the body or its parts.

Delirium (de-LIR-ē-um)   A transient disorder of abnormal cognition (perception, thinking, and memory) and disordered attention that is accompanied by disturbances of the sleep-wake cycle and psychomotor behavior (hyperactivity or hypoactivity of movements and speech). Also called **acute confusional state (ACS).**

Delta cell   A cell in the pancreatic islets (islets of Langerhans) in the pancreas that secretes somatostatin.

Dementia (de-MEN-shē-a)   An organic mental disorder that results in permanent or progressive general loss of intellectual abilities such as impairment of memory, judgment, and abstract thinking and changes in personality; most common cause is Alzheimer's disease.

Demineralization (de-min′-er-al-i-ZĀ-shun)   Loss of calcium and phosphorus from bones.

Denaturation (de-nā-chur-Ā-shun)   Disruption of the tertiary structures of a protein by agents, such as heat, changes in pH, or other physical or chemical methods, in which the protein loses its physical properties and biological activity.

Dendrite (DEN-drīt)   A nerve cell process that carries a nerve impulse toward the cell body.

Dendritic (den-DRIT-ik) **cell**   One type of antigen presenting cell with long branchlike projections, for example, Langerhans cells in the epidermis.

Dental caries (KA-rēz)   Gradual demineralization of the enamel and dentin of a tooth that may invade the pulp and alveolar bone. Also called **tooth decay.**

Denticulate (den-TIK-yoo-lāt)   Finely toothed or serrated; characterized by a series of small, pointed projections.

Dentin (DEN-tin)   The bony tissues of a tooth enclosing the pulp cavity.

Dentition (den-TI-shun)   The eruption of teeth. The number, shape, and arrangement of teeth.

Deoxyribonucleic (dē-ok′-sē-ri′-bō-nyoo-KLĒ-ik) **acid (DNA)**   A nucleic acid in the shape of a double helix constructed of nucleotides consisting of one of four nitrogenous bases (adenine, cytosine, guanine, or thymine), deoxyribose, and a phosphate group; encoded in the nucleotides is genetic information.

Depolarization (dē-pō-lar-i-ZĀ-shun)   Used in neurophysiology to describe the reduction of voltage across a cell membrane;

expressed as a movement toward less negative (more positive) voltages on the interior side of the cell membrane.

Depression (de-PRESS-shun)   Movement in which a part of the body moves downward.

Dermal papilla (pa-PILL-a)   Fingerlike projection of the papillary region of the dermis that may contain blood capillaries or corpuscles of touch (Meissner's corpuscles).

Dermatology (der-ma-TOL-ō-jē)   The medical specialty dealing with diseases of the skin.

Dermatome (DER-ma-tōm)   An instrument for incising the skin or cutting thin transplants of skin. The cutaneous area developed from one embryonic spinal cord segment and receiving most of its sensory innervation from one spinal nerve.

Dermis (DER-mis)   A layer of dense connective tissue lying deep to the epidermis; the true skin or corium.

Descending colon (KŌ-lon)   The part of the large intestine descending from the left colic (splenic) flexure to the level of the left iliac crest.

Detritus (de-TRĪ-tus)   Particulate matter produced by or remaining after the wearing away or disintegration of a substance or tissue; scales, crusts, or loosened skin.

Detrusor (de-TROO-ser) **muscle**   Muscle in the wall of the urinary bladder.

Developmental anatomy   The study of development from the fertilized egg to the adult form. The branch of anatomy called embryology is generally restricted to the study of development from the fertilized egg through the eighth week in utero.

Diabetes insipidus (dī-a-BĒ-tēz in-SIP-i-dus)   Condition caused by hyposecretion of antidiuretic hormone (ADH) and characterized by excretion of large amounts of urine and thirst.

Diabetes mellitus (MEL-i-tus)   Hereditary condition caused by hyposecretion of insulin and characterized by hyperglycemia, increased urine production, excessive thirst, and excessive eating.

Diagnosis (dī-ag-NŌ-sis)   Distinguishing one disease from another or determining the nature of a disease from signs and symptoms by inspection, palpation, laboratory tests, and other means.

Dialysis (dī-AL-i-sis)   The process of separating small molecules from large by the difference in their rates of diffusion through a selectively permeable membrane.

Diapedesis (dī-a-pe-DĒ-sis)   The passage of white blood cells through intact blood vessel walls.

Diaphragm (DĪ-a-fram)   Any partition that separates one area from another, especially the dome-shaped skeletal muscle between the thoracic and abdominal cavities. Also a dome-shaped structure that fits over the cervix, usually with a spermicide, to prevent conception.

Diaphysis (dī-AF-i-sis)   The shaft of a long bone.

Diarrhea (dī-a-RĒ-a)   Frequent defecation of liquid feces caused by increased motility of the intestines.

Diarthrosis (dī-ar-THRŌ-sis)   A freely movable joint, as in a hinge joint.

Diastole (dī-AS-tō-lē)   In the cardiac cycle, the phase of relaxation or dilation of the heart muscle, especially of the ventricles.

Diastolic (dī-as-TOL-ik) **blood pressure**   The force exerted by blood on arterial walls during ventricular relaxation; the lowest blood pressure measured in the large arteries, about 80 mm Hg under normal conditions for a young, adult male.

Diencephalon (dī′-en-SEF-a-lon)   A part of the brain consisting primarily of the thalamus and the hypothalamus.

Differential (dif-fer-EN-shal) **white blood cell count**   Determination of the number of each kind of white blood cell in a sample of 100 cells for diagnostic purposes.

Differentiation (dif′-e-ren′-shē-Ā-shun)   Acquisition of specific functions different from those of the original general type.

Diffusion (dif-YOO-zhun)   A passive process in which there is a net or greater movement of molecules or ions from a region of high concentration to a region of low concentration until equilibrium is reached.

Digestion (dī-JES-chun)   The mechanical and chemical breakdown of food to simple molecules that can be absorbed and used by body cells.

Digital subtraction angiography (an-jē-OG-ra-fē) **(DSA)**   A medical imaging technique that compares an x-ray image of the same artery of the body before and after a contrast substance containing iodine has been introduced intravenously.

Dilate (DĪ-lāte)   To expand or swell.

Dilation (dī-LĀ-shun) **and curettage** (ku-re-TAZH)   Following dilation of the uterine cervix, the uterine endometrium is scraped with a curette (spoon-shaped instrument). Also called a **D and C.**

Diploid (DIP-loyd)   Having the number of chromosomes characteristically found in the somatic cells of an organism. Symbolized 2*n*.

Diplopia (di-PLŌ-pē-a)   Double vision.

Disease   Any change from a state of health.

Dislocation (dis-lō-KĀ-shun)   Displacement of a bone from a joint with tearing of ligaments, tendons, and articular capsules. Also called **luxation** (luks-Ā-shun).

Dissect (DIS-sekt)   To separate tissues and parts of a cadaver (corpse) or an organ for anatomical study.

Dissociation (dis′-sō-sē-Ā-shun)   Separation of inorganic acids, bases, and salts into ions when dissolved in water. Also called **ionization** (ī′ on-i-ZĀ-shun).

Distal (DIS-tal)   Farther from the attachment of an extremity to the trunk or a structure; farther from the point of origin.

Diuretic (dī-yoo-RET-ik)   A chemical that inhibits sodium reabsorption, reduces antidiuretic hormone (ADH) concentration, and increases urine volume by inhibiting facultative reabsorption of water.

Diurnal (dī-UR-nal)   Daily.

Divergence (dī-VER-jens)   An anatomical arrangement in which the synaptic end bulbs of one presynaptic neuron terminate on several postsynaptic neurons.

Diverticulitis (dī-ver-tik-yoo-LĪ-tis)   Inflammation of diverticula, saclike outpouchings of the colonic wall, when the muscularis becomes weak.

Diverticulum (dī-ver-TIK-yoo-lum)   A sac or pouch in the wall of a canal or organ, especially in the colon.

Dominant gene   A gene that is able to override the influence of the complementary gene on the homologous chromosome; the gene that is expressed.

Donor insemination (in-sem′-i-NĀ-shun) The deposition of semen within the vagina or cervix at a time during the menstrual cycle when pregnancy is most likely to occur. It may be homologous (using the husband's semen) or heterologous (using a donor's semen). Also called **artificial insemination.**

Dorsal body cavity Cavity near the dorsal surface of the body that consists of a cranial cavity and vertebral canal.

Dorsal ramus (RĀ-mus) A branch of a spinal nerve containing motor and sensory fibers supplying the muscles, skin, and bones of the posterior part of the head, neck, and trunk.

Dorsiflexion (dor′-si-FLEK-shun) Bending the foot in the direction of the dorsum (upper surface).

Down-regulation Phenomenon in which there is a decrease in the number of receptors in response to an excess of a hormone or neurotransmitter.

Down syndrome (DS) An inherited defect due to an extra copy of chromosome 21. Symptoms include mental retardation; a small skull, flattened from front to back; a short, flat nose; short fingers; and a widened space between the first two digits of the hand and foot. Also called **trisomy 21.**

Duct of Santorini *See* **Accessory duct.**

Duct of Wirsung *See* **Pancreatic duct.**

Ductus arteriosus (DUK-tus ar-tē-rē-Ō-sus) A small vessel connecting the pulmonary trunk with the aorta; found only in the fetus.

Ductus (vas) deferens (DEF-er-ens) The duct that conducts spermatozoa from the epididymis to the ejaculatory duct. Also called the **seminal duct.**

Ductus epididymis (ep′-i-DID-i-mis) A tightly coiled tube inside the epididymis, distinguished into a head, body, and tail, in which spermatozoa undergo maturation.

Ductus venosus (ve-NŌ-sus) A small vessel in the fetus that helps the circulation bypass the liver.

Duodenal (doo-ō-DĒ-nal) **gland** Gland in the submucosa of the duodenum that secretes an alkaline mucus to protect the lining of the small intestine from the action of enzymes and to help neutralize the acid in chyme. Also called **Brunner's (BRUN-erz) gland.**

Duodenal papilla (pa-PILL-a) An elevation on the duodenal mucosa that receives the hepatopancreatic ampulla (ampulla of Vater).

Duodenum (doo′-ō-DĒ-num) The first 25 cm (10 in.) of the small intestine.

Dura mater (DYOO-ra MĀ-ter) The outer membrane (meninx) covering the brain and spinal cord.

Dynamic equilibrium (ē-kwi-LIB-rē-um) The maintenance of body position, mainly the head, in response to sudden movements such as rotation.

Dynamic spatial reconstruction (DSR) A technique that has the ability to construct moving, three-dimensional, life-size images of all or part of an internal organ from any view desired.

Dysfunction (dis-FUNK-shun) Absence of complete normal function.

Dyslexia (dis-LEX-sē-a) Impairment of the brain's ability to translate images received from the eyes or ears into understandable language.

Dysmenorrhea (dis′-men-ō-RĒ-a) Painful menstruation.

Dysphagia (dis-FĀ-jē-a) Difficulty in swallowing.

Dysplasia (dis-PLĀ-zē-a) Change in the size, shape, and organization of cells due to chronic irritation or inflammation; may revert to normal if stress is removed or progress to neoplasia.

Dyspnea (DISP-nē-a) Shortness of breath.

Dystocia (dis-TŌ-sē-a) Difficult labor due to factors such as pelvic deformities, malpositioned fetus, and premature rupture of fetal membranes.

Dystrophia (dis-TRŌ-fē-a) Progressive weakening of a muscle.

Dysuria (dis-YOO-rē-a) Painful urination.

Echocardiogram (ek-ō-KAR-dē-ō-gram) A procedure in which high-frequency sound waves directed at the heart are bounced back and the echoes are picked up by a transducer and converted to an image.

Ectoderm The outermost of the three primary germ layers that gives rise to the nervous system and the epidermis of skin and its derivatives.

Ectopic (ek-TOP-ik) Out of the normal location, as in ectopic pregnancy.

Eczema (EK-ze-ma) A skin rash characterized by itching, swelling, blistering, oozing, and scaling of the skin.

Edema (e-DĒ-ma) An abnormal accumulation of interstitial fluid.

Effector (e-FEK-tor) The organ of the body, either a muscle or a gland, that responds to a motor neuron impulse.

Efferent arteriole (EF-er-ent ar-TĒ-rē-ōl) A vessel of the renal vascular system that transports blood from the glomerulus to the peritubular capillary.

Efferent (EF-er-ent) **ducts** A series of coiled tubes that transport spermatozoa from the rete testis to the epididymis.

Effusion (e-FYOO-zhun) The escape of fluid from the lymphatic vessels or blood vessels into a cavity or into tissues.

Ejaculation (e-jak-yoo-LĀ-shun) The reflex ejection or expulsion of semen from the penis.

Ejaculatory (e-JAK-yoo-la-tō-rē) **duct** A tube that transports spermatozoa from the ductus (vas) deferens to the prostatic urethra.

Elasticity (e-las-TIS-i-tē) The ability of tissue to return to its original shape after contraction or extension.

Electrocardiogram (e-lek′-trō-KAR-dē-ō-gram) **(ECG** or **EKG)** A recording of the electrical changes that accompany the cardiac cycle and can be recorded on the surface of the body; may be resting, stress, or ambulatory.

Electroencephalogram (e-lek′-trō-en-SEF-a-lō-gram) **(EEG)** A recording of the electrical impulses of the brain to diagnose certain diseases (such as epilepsy), furnish information regarding sleep and wakefulness, and confirm brain death.

Electrolyte (ē-LEK-trō-līt) Any compound that separates into ions when dissolved in water and is able to conduct electricity.

Electromyography (e-lek′-trō-mī-OG-ra-fē) Evaluation of the electrical activity of resting and contracting muscle to ascertain causes of muscular weakness, paralysis, involuntary twitching, and abnormal levels of muscle enzymes; also used as part of biofeedback studies.

Elevation (el-e-VĀ-shun) Movement in which a part of the body moves upward.

Ellipsoidal (e-lip-SOY-dal) **joint** A synovial joint structured so that an oval-shaped condyle of one bone fits into an elliptical cavity of another bone, permitting side-to-side and back-and-forth movements, as at the joint at the wrist between the radius and carpals. Also called a **condyloid** (KON-di-loid) **joint.**

Embolism (EM-bō-lizm) Obstruction or closure of a vessel by an embolus.

Embolus (EM-bō-lus) A blood clot, bubble of air, fat from broken bones, mass of bacteria, or other debris or foreign material transported by the blood.

Embryo (EM-brē-ō) The young of any organism in an early stage of development; in humans, the developing organism from fertilization to the end of the eighth week in utero.

Embryology (em′-brē-OL-ō-jē) The study of development from the fertilized egg to the end of the eighth week in utero.

Embryo transfer A type of in vitro fertilization in which semen is used to artificially inseminate a fertile secondary oocyte donor and the morula or blastocyst is then transferred from the donor to the infertile woman, who then carries it to term.

Emesis (EM-e-sis) Vomiting.

Emmetropia (em′-e-TRŌ-pē-a) The ideal optical condition of the eyes.

Emphysema (em′-fi-SĒ-ma) A swelling or inflation of air passages due to loss of elasticity in the alveoli.

Emulsification (ē-mul′-si-fi-KĀ-shun) The dispersion of large fat globules to smaller, uniformly distributed particles in the presence of bile.

Enamel (e-NAM-el) The hard, white substance covering the crown of a tooth.

End-diastolic (dī-as-TO-lik) **volume (EDV)** The volume of blood, about 130 ml, remaining in a ventricle at the end of its diastole (relaxation).

Endocardium (en-dō KAR-dē-um) The layer of the heart wall, composed of endothelium and smooth muscle, that lines the inside of the heart and covers the valves and tendons that hold the valves open.

Endochondral ossification (en′-dō-KON-dral os′-i-fi-KĀ-shun) The replacement of cartilage by bone. Also called **intracartilaginous** (in′-tra-kar′-ti-LAJ-i-nus) **ossification.**

Endocrine (EN-dō-krin) **gland** A gland that secretes hormones into the blood; a ductless gland.

Endocrinology (en′-dō-kri-NOL-ō-jē) The science concerned with the structure and functions of endocrine glands and the diagnosis and treatment of disorders of the endocrine system.

Endocytosis (en′-dō-sī-TŌ-sis) The uptake into a cell of large molecules and particles in which a segment of plasma membrane surrounds the substance, encloses it, and brings it in; includes phagocytosis, pinocytosis, and receptor-mediated endocytosis.

Endoderm (EN-dō-derm) The innermost of the three primary germ layers of the developing embryo that gives rise to the gastrointestinal tract, urinary bladder and urethra, and respiratory tract.

Endodontics (en′-dō-DON-tiks) The branch of dentistry concerned with the prevention, diagnosis, and treatment of diseases that affect the pulp, root, periodontal ligament, and alveolar bone.

Endogenous (en-DOJ-e-nus) Growing from or beginning within the organism.

Endolymph (EN-dō-lymf′) The fluid within the membranous labyrinth of the inner ear.

Endometriosis (en′-dō-MĒ-trē-ō′-sis) The growth of endometrial tissue outside the uterus.

Endometrium (en′-dō-MĒ-trē-um) The mucous membrane lining the uterus.

Endomysium (en′-dō-MĪZ-ē-um) Invagination of the perimysium separating each individual muscle fiber (cell).

Endoneurium (en′-dō-NYOO-rē-um) Connective tissue wrapping around individual nerve fibers (cells).

Endoplasmic reticulum (en′-do-PlAZ-mik re-TIK-yoo-lum) **(ER)** A network of channels running through the cytoplasm of a cell that serves in intracellular transportation, support, storage, synthesis, and packaging of molecules. Portions of ER where ribosomes are attached to the outer surface are called **rough (granular) reticulum;** portions that have no ribosomes are called **smooth (agranular) reticulum.**

End organ of Ruffini *See* **Type II cutaneous mechanoreceptor.**

Endorphin (en-DOR-fin) A neuropeptide in the central nervous system that acts as a painkiller.

Endoscope (EN-dō-skōp′) An illuminated tube with lenses used to look inside hollow organs such as the stomach (gastroscope) or urinary bladder (cystoscope).

Endoscopy (en-DOS-kō-pē) The visual examination of any cavity of the body using an endoscope, an illuminated tube with lenses.

Endosteum (en-DOS-tē-um) The membrane that lines the medullary (marrow) cavity of bones, consisting of osteoprogenitor cells and scattered osteoclasts.

Endothelial-capsular (en-dō-THĒ-lē-al) **membrane** A filtration membrane in a nephron of a kidney consisting of the endothelium and basement membrane of the glomerulus and the epithelium of the visceral layer of the glomerular (Bowman's) capsule.

Endothelium (en′-dō-THĒ-lē-um) The layer of simple squamous epithelium that lines the cavities of the heart, blood vessels, and lymphatic vessels.

End-systolic (sis-TO-lik) **volume (ESV)** The volume of blood, about 60 ml, remaining in a ventricle following its systole (contraction).

Energy The capacity to do work.

Enkephalin (en-KEF-a-lin) A peptide found in the central nervous system that acts as a painkiller.

Enteroendocrine (en-ter-ō-EN-dō-krin) **cell** A cell of the mucosa of the gastrointestinal tract that secretes the hormones gastrin, cholecystokinin, gastric inhibitory peptide, or secretin.

Enterogastric (en-te-rō-GAS-trik) **reflex** A reflex that inhibits gastric secretion; initiated by food in the small intestine.

Enuresis (en′-yoo-RĒ-sis) Involuntary discharge of urine, complete or partial, after age 3.

Enzyme (EN-zīm) A substance that affects the speed of chemical changes; an organic catalyst, usually a protein.

Eosinophil (ē′-ō-SIN-ō-fil) A type of white blood cell characterized by granular cytoplasm readily stained by eosin.

Ependymal (e-PEN-de-mal) **cells** Neuroglial cells that line ventricles of the brain and probably assist in the circulation of cerebrospinal fluid (CSF).

Epicardium (ep′-i-KAR-dē-um) The thin outer layer of the heart wall, composed of serous tissue and mesothelium. Also called the **visceral pericardium.**

Epidemic (ep′-i-DEM-ik) A disease that occurs above the expected level among individuals in a population.

Epidemiology (ep′-i-dē-mē-OL-ō-jē) Medical science concerned with the occurrence and distribution of disease in human populations.

Epidermis (ep-i-DERM-is) The outermost, thinner layer of skin, composed of stratified squamous epithelium.

Epididymis (ep′-i-DID-i-mis) A comma-shaped organ that lies along the posterior border of the testis and contains the ductus epididymis, in which sperm undergo maturation. *Plural,* **epididymides** (ep′-i-DID-i-mi-dēz).

Epidural (ep′-i-DOO-ral) **space** A space between the spinal dura mater and the vertebral canal, containing areolar connective tissue and a plexus of veins.

Epiglottis (ep′-i-GLOT-is) A large, leaf-shaped piece of cartilage lying on top of the larynx, with its "stem" attached to the thyroid cartilage and its "leaf" portion unattached and free to move up and down to cover the glottis (vocal folds and rima glottidis).

Epilepsy (EP-i-lep′-sē) Neurological disorder characterized by short, periodic attacks of motor, sensory, or psychological malfunction.

Epimysium (ep′-i-MĪZ-ē-um) Fibrous connective tissue around muscles.

Epinephrine (ep-ē-NEF-rin) Hormone secreted by the adrenal medulla that produces actions similar to those that result from sympathetic stimulation. Also called **adrenaline** (a-DREN-a-lin).

Epineurium (ep′-i-NYOO-rē-um) The outermost covering around the entire nerve.

Epiphyseal (ep′-i-FIZ-ē-al) **line** The remnant of the epiphyseal plate in a long bone.

Epiphyseal (ep′-i-FIZ-ē-al) **plate** The cartilaginous plate between the epiphysis and diaphysis that is responsible for the lengthwise growth of long bones.

Epiphysis (ē-PIF-i-sis) The end of a long bone, usually larger in diameter than the shaft (diaphysis).

Epiphysis cerebri (se-RĒ-brē) Pineal gland.

Episiotomy (e-piz′-ē-OT-ō-mē) A cut made with surgical scissors to avoid tearing of the perineum at the end of the second stage of labor.

Epistaxis (ep′-i-STAK-sis) Loss of blood from the nose due to trauma, infection, allergy, neoplasm, and bleeding disorders. Also called **nosebleed.**

Epithelial (ep′-i-THĒ-lē-al) **tissue** The tissue that forms glands or the outer part of the skin and lines blood vessels, hollow organs, and passages that lead externally from the body.

Eponychium (ep′-ō-NIK-ē-um) Narrow band of stratum corneum at the proximal border of a nail that extends from the margin of the nail wall. Also called the **cuticle.**

Erection (ē-REK-shun) The enlarged and stiff state of the penis (or clitoris) resulting from the engorgement of the spongy erectile tissue with blood.

Eructation (e-ruk′-TĀ-shun) The forceful expulsion of gas from the stomach. Also called **belching.**

Erythema (er′-e-THĒ-ma) Skin redness usually caused by engorgement of the capillaries in the lower layers of the skin.

Erythematosus (er-i′-them-a-TŌ-sus) Pertaining to redness.

Erythrocyte (e-RITH-rō-sīt) Red blood cell.

Erythrocyte sedimentation rate (ESR) A test that measures the distances, in millimeters (mm), that red blood cells fall in one hour when a sample of blood is placed in a vertical tube; used as a screening test for infections, inflammations, and cancers.

Erythropoiesis (e-rith′-rō-poy-Ē-sis) The process by which erythrocytes (red blood cells) are formed.

Erythropoietin (e-rith′-rō-POY-ē-tin) A hormone released by the kidneys that stimulates erythrocyte (red blood cell) production.

Esophagus (e-SOF-a-gus) A hollow muscular tube connecting the pharynx and the stomach.

Essential amino acids Those 10 amino acids that cannot be synthesized by the human body at an adequate rate to meet its needs and therefore must be obtained from the diet.

Estrogens (ES-tro-jens) Female sex hormones produced by the ovaries concerned with the development and maintenance of female reproductive structures and secondary sex characteristics, fluid and electrolyte balance, and protein anabolism. Examples are ß-estradiol, estrone, and estriol.

Etiology (ē′-tē-OL-ō-jē) The study of the causes of disease, including theories of origin and the organisms, if any, involved.

Euphoria (yoo-FŌR-ē-a) A subjectively pleasant feeling of well-being marked by confidence and assurance.

Eupnea (YOOP-nē-a) Normal quiet breathing.

Eustachian tube *See* **Auditory tube.**

Euthanasia (yoo′-tha-NĀ-zē-a) The practice of ending a life in case of incurable disease.

Eversion (ē-VER-zhun) The movement of the sole outward at the ankle joint or of an atrioventricular valve into an atrium during ventricular contraction.

Exacerbation (eg-zas′-er-BĀ-shun) An increase in the severity of symptoms or of disease.

Excitability (ek-sīt′-a-BIL-i-tē) The ability of muscle tissue to receive and respond to stimuli; the ability of nerve cells to respond to stimuli and convert them into nerve impulses.

Excitatory postsynaptic potential (EPSP) The slight decrease in negative voltage seen on the postsynaptic membrane when it is stimulated by a presynaptic terminal. The EPSP is a localized event that decreases in strength from the point of excitation.

Excrement (EKS-kre-ment) Material cast out from the body as waste, especially fecal matter.

Excretion (eks-KRĒ-shun) The process of eliminating waste products from a cell, tissue, or the entire body; or the products excreted.

Exocrine (EK-sō-krin) **gland** A gland that secretes substances into ducts that empty at covering or lining epithelium or directly onto a free surface.

Exocytosis (ex′-ō-sī-TŌ-sis) A process of discharging cellular products too big to go through the membrane. Particles for export are enclosed by Golgi membranes when they are synthesized. Vesicles pinch off from the Golgi complex and carry the enclosed particles to the interior surface of the cell membrane, where the vesicle membrane and plasma membrane fuse and the contents of the vesicle are discharged.

Exogenous (ex-SOJ-e-nus)   Originating outside an organ or part.

Exon (EX-on)   A region of DNA that codes for synthesis of a protein.

Exophthalmic goiter (ek′-sof-THAL-mik GOY-ter)   An autoimmune disease that may result in hypersecretion of thyroid hormones characterized by protrusion of the eyeballs (exophthalmos) and an enlarged thyroid (goiter). Also called **Graves disease.**

Exophthalmos (ek′-sof-THAL-mus)   An abnormal protrusion or bulging of the eyeball.

Expiration (ek-spi-RĀ-shun)   Breathing out; expelling air from the lungs into the atmosphere. Also called **exhalation.**

Expiratory (eks-PĪ-ra-tō-rē) **reserve volume**   The volume of air in excess of tidal volume that can be exhaled forcibly; about 1,200 ml.

Extensibility (ek-sten′-si-BIL-i-tē)   The ability of muscle tissue to be stretched when pulled.

Extension (ek-STEN-shun)   An increase in the angle between two bones; restoring a body part to its anatomical position after flexion.

External   Located on or near the surface.

External auditory (AW-di tōr-ē) **canal** or **meatus** (mē-Ā-tus)   A curved tube in the temporal bone that leads to the middle ear.

External ear   The outer ear, consisting of the pinna, external auditory canal, and tympanic membrane (eardrum).

External nares (NA-rēz)   The external nostrils, or the openings into the nasal cavity on the exterior of the body.

External (pulmonary) respiration   The exchange of respiratory gases between the lungs and blood.

Exteroceptor (eks′-ter-ō-SEP-tor)   A receptor adapted for the reception of stimuli from outside the body.

Extracellular fluid (ECF)   Fluid outside body cells, such as interstitial fluid and plasma.

Extravasation (eks-trav-a-SĀ-shun)   The escape of fluid, especially blood, lymph, or serum, from a vessel into the tissues.

Extrinsic (ek-STRIN-sik)   Of external origin.

Extrinsic clotting pathway   Sequence of reactions leading to blood clotting that is initiated by the release of a substance (tissue factor or thromboplastin) *outside* blood itself, from damaged blood vessels or surrounding tissues.

Exudate (EKS-yoo-dāt)   Escaping fluid or semifluid material that oozes from a space that may contain serum, pus, and cellular debris.

Eyebrow   The hairy ridge above the eye.

Face   The anterior aspect of the head.

Facilitated diffusion (fa-SIL-i-tā-ted dif-YOO-zhun)   Diffusion in which a substance not soluble by itself in lipids is transported across a selectively permeable membrane by combining with a transporter (carrier).

Falciform ligament (FAL-si-form LIG-a-ment)   A sheet of parietal peritoneum between the two principal lobes of the liver. The ligamentum teres, or remnant of the umbilical vein, lies within its fold.

Fallopian tube   *See* **Uterine tube.**

Falx cerebelli (FALKS ser′-e-BEL-lē)   A small triangular process of the dura mater attached to the occipital bone in the posterior cranial fossa and projecting inward between the two cerebellar hemispheres.

Falx cerebri (SER-e-brē)   A fold of the dura mater extending down into the longitudinal fissure between the two cerebral hemispheres.

Fascia (FASH-ē-a)   A fibrous membrane covering, supporting, and separating muscles.

Fascicle (FAS-i-kul)   A small bundle or cluster, especially of nerve or muscle fibers (cells). Also called a **fasciculus** (fa-SIK-yoo-lus). *Plural,* **fasciculi** (fa-SIK-yoo-lī).

Fasciculation (fa-sik′-yoo-LĀ-shun)   Involuntary brief twitch of a muscle that is visible under the skin and is not associated with the movement of the affected muscle.

Fauces (FAW-sēz)   The opening from the mouth into the pharynx.

Febrile (FĒ-bril)   Feverish; pertaining to a fever.

Feces (FĒ-sēz)   Material discharged from the rectum and made up of bacteria, excretions, and food residue. Also called **stool.**

Feedback system   A sequence of events in which information about the status of a situation is continually reported (fed back) to a central control region.

Feeding (hunger) center   A cluster of neurons in the lateral nuclei of the hypothalamus that, when stimulated, brings about feeding.

Female reproductive cycle   General term for the ovarian and uterine cycles, the hormonal changes that accompany them, and cyclic changes in the breasts and cervix.

Fertilization (fer′-ti-li-ZĀ-shun)   Penetration of a secondary oocyte by a spermatozoon and subsequent union of the nuclei of the cells.

Fetal (FĒ-tal) **alcohol syndrome (FAS)**   Term applied to the effects of intrauterine exposure to alcohol, such as slow growth, defective organs, and mental retardation.

Fetal circulation   The cardiovascular system of the fetus, including the placenta and special blood vessels involved in the exchange of materials between fetus and mother.

Fetus (FĒ-tus)   The latter stages of the developing young of an animal; in humans, the developing organism in utero from the beginning of the third month to birth.

Fever   An elevation in body temperature above its normal temperature of 37°C (98.6°F).

Fibrillation (fi-bri-LĀ-shun)   Involuntary brief twitch of a muscle that is not visible under the skin and is not associated with movement of the affected muscle.

Fibrin (FĪ-brin)   An insoluble protein that is essential to blood clotting; formed from fibrinogen by the action of thrombin.

Fibrinogen (fī-BRIN-ō-jen)   A high-molecular-weight protein in the blood plasma that by the action of thrombin is converted to fibrin.

Fibrinolysis (fī-brin-OL-i-sis)   Dissolution of a blood clot by the action of a proteolytic enzyme that converts insoluble fibrin into a soluble substance.

Fibroblast (FĪ-brō-blast)   A large, flat cell that forms collagen and elastic fibers and intercellular substance of areolar connective tissue.

Fibrocyte (FĪ-brō-sīt)   A mature fibroblast that no longer produces fibers or intercellular substance in connective tissue.

Fibromyalgia (fī-bro-mī-AL-jē-a)   Groups of common nonarticular rheumatic disorders characterized by pain, tenderness, and stiffness of muscles, tendons, and surrounding tissues. Examples of fibromyalgia are lumbago and charley horse.

Fibrosis (fī-BRŌ-sis)   Abnormal formation of fibrous tissue.

Fibrous (FĪ-brus) **joint**   A joint that allows little or no movement, such as a suture and syndesmosis.

Fibrous tunic (TOO-nik)   The outer coat of the eyeball, made up of the posterior sclera and the anterior cornea.

Fight-or-flight response   The effect of the stimulation of the sympathetic division of the autonomic nervous system.

Filiform papilla (FIL-i-form pa-PIL-a)   One of the conical projections that are distributed in parallel rows over the anterior two-thirds of the tongue and contain no taste buds.

Filtrate (fil-TRĀT)   The fluid produced when blood is filtered by the endothelial-capsular membrane.

Filtration (fil-TRĀ-shun)   The passage of a liquid through a filter or membrane that acts like a filter.

Filtration fraction   The percentage of plasma entering the nephrons that actually becomes glomerular filtrate.

Filum terminale (FĪ-lum ter-mi-NAL-ē)   Nonnervous fibrous tissue of the spinal cord that extends inferiorly from the conus medullaris to the coccyx.

Fimbriae (FIM-brē-ē)   Fingerlike structures, especially the lateral ends of the uterine (Fallopian) tubes.

Fissure (FISH-ur)   A groove, fold, or slit that may be normal or abnormal.

Fistula (FIS-choo-la)   An abnormal passage between two organs or between an organ cavity and the outside.

Fixator   A muscle that stabilizes the origin of the prime mover so that the prime mover can act more efficiently.

Fixed macrophage (MAK-rō-fāj)   Stationary phagocytic cell found in the liver, lungs, brain, spleen, lymph nodes, subcutaneous tissue, and bone marrow. Also called a **histiocyte** (HIS-tē-ō-sīt).

Flaccid (FLAS-sid)   Relaxed, flabby, or soft; lacking muscle tone.

Flagellum (fla-JEL-um)   A hairlike, motile process on the extremity of a bacterium, protozoan, or spermatozoon. *Plural,* **flagella** (fla-JEL-a).

Flatfoot   A condition in which the ligaments and tendons of the arches of the foot are weakened and the height of the longitudinal arch decreases.

Flatus (FLĀ-tus)   Air (gas) in the stomach or intestines, commonly used to denote passage of gas rectally.

Flexion (FLEK-shun)   A folding movement in which there is a decrease in the angle between two bones.

Flexor reflex   A protective reflex in which flexor muscles are stimulated while extensor muscles are inhibited.

Fluid mosaic (mō-ZĀ-ik) **model**   Model of plasma membrane structure that describes the molecular arrangement of the plasma membrane and other membranes in living organisms.

Fluoroscope (FLOOR-ō-skōp)   An instrument for visual observation of the body by means of x-ray.

Follicle (FOL-i-kul)   A small secretory sac or cavity.

Follicle-stimulating (FOL-i-kul) **hormone** (**FSH**)   Hormone secreted by the anterior pituitary gland that initiates development of ova and stimulates the ovaries to secrete estrogens in females and initiates sperm production in males.

Fontanel (fon′-ta-NEL)   A membrane-covered spot where bone formation is not yet complete, especially between the cranial bones of an infant's skull.

Foot   The terminal part of the lower extremity.

Foramen (fo-RĀ-men)   A passage or opening; a communication between two cavities of an organ or a hole in a bone for passage of vessels or nerves.

Foramen ovale (ō-VAL-ē)   An opening in the fetal heart in the septum between the right and left atria. A hole in the greater wing of the sphenoid bone that transmits the mandibular branch of the trigeminal (V) nerve.

Forearm (FOR-arm)   The part of the upper extremity between the elbow and the wrist.

Fornix (FOR-niks)   An arch or fold; a tract in the brain made up of association fibers, connecting the hippocampus with the mammillary bodies; a recess around the cervix of the uterus where it protrudes into the vagina.

Fossa (FOS-a)   A furrow or shallow depression.

Fourth ventricle (VEN-tri-kul)   A cavity within the brain lying between the cerebellum and the medulla and pons.

Fracture (FRAK-chur)   Any break in a bone.

Fragile X syndrome   Inherited disorder characterized by learning difficulties, mental retardation, and physical abnormalities; due to a defective gene on the X chromosome.

Frenulum (FREN-yoo-lum)   A small fold of mucous membrane that connects two parts and limits movement.

Frontal plane   A plane at a right angle to a midsagittal plane that divides the body or organs into anterior and posterior portions. Also called a **coronal** (kō-RŌ-nal) **plane.**

Functional residual (re-ZID-yoo-al) **capacity**   The sum of residual volume plus expiratory reserve volume; about 2,400 ml.

Fundus (FUN-dus)   The part of a hollow organ farthest from the opening.

Fungiform papilla (FUN-ji-form pa-PIL-a)   A mushroomlike elevation on the upper surface of the tongue appearing as a red dot; most contain taste buds.

Gallbladder   A small pouch that stores bile, located under the liver, which is filled with bile and emptied via the cystic duct.

Gallstone   A concretion, usually consisting of cholesterol, formed anywhere between bile canaliculi in the liver and the hepatopancreatic ampulla (ampulla of Vater), where bile enters the duodenum. Also called a **biliary calculus.**

Gamete (GAM-ēt)   A male or female reproductive cell; the spermatozoon or ovum.

Gamete intrafallopian transfer (GIFT) A type of in vitro fertilization in which aspirated secondary oocytes are combined with a solution containing sperm outside the body and then the secondary oocytes are inserted into the uterine (Fallopian) tubes.

Ganglion (GANG-glē-on) A group of nerve cell bodies that lie outside the central nervous system. *Plural,* **ganglia** (GANG-glē-a).

Gangrene (GANG-ren) Death and rotting of a considerable mass of tissue that usually is caused by interruption of blood supply followed by bacterial (*Clostridium*) invasion.

Gastroenterology (gas′-trō-en′-ter-OL-ō-jē) The medical specialty that deals with the structure, function, diagnosis, and treatment of diseases of the stomach and intestines.

Gastrointestinal (gas-trō-in-TES-ti-nal) **(GI) tract** A continuous tube running through the ventral body cavity extending from the mouth to the anus. Also called the **alimentary** (al′-i-MEN-tar-ē) **canal.**

Gastroscopy (gas-TROS-kō-pē) Diagnostic procedure in which the interior of the stomach is examined with a gastroscope to detect and biopsy lesions, stop bleeding, and remove foreign objects.

Gastrulation (gas′-troo-LĀ-shun) The various movements of groups of cells that lead to the establishment of the primary germ layers.

Gavage (ga-VAZH) Feeding through a tube passed through the esophagus and into the stomach.

Gene (jēn) Biological unit of heredity; an ultramicroscopic, self-reproducing DNA particle located in a definite position on a particular chromosome.

General adaptation syndrome (GAS) Wide-ranging set of bodily changes triggered by a stressor that gears the body to meet an emergency.

Generator potential The graded depolarization that results in a change in the resting membrane potential in a receptor (specialized neuronal ending).

Genetic engineering The manufacture and manipulation of genetic material.

Genetics The study of heredity.

Genital herpes (JEN-i-tal HER-pēz) A sexually transmitted disease caused by type II herpes simplex virus.

Genitalia (jen′-i-TĀL-ya) Reproductive organs.

Genome (JĒ-nōm) The complete gene complement of an organism.

Genotype (JĒ-nō-tīp) The total hereditary information carried by an individual; the genetic makeup of an organism.

Geriatrics (jer′-ē-AT-riks) The branch of medicine devoted to the medical problems and care of elderly persons.

Gestation (jes-TĀ-shun) The period of intrauterine fetal development.

Giantism (GĪ-an-tizm) Condition caused by hypersecretion of human growth hormone (hGH) during childhood characterized by excessive bone growth and body size. Also called **gigantism.**

Gingivae (jin-JI-vē) Gums. They cover the alveolar processes of the mandible and maxilla and extend slightly into each socket.

Gingivitis (jin′-je-VĪ-tis) Inflammation of the gums.

Gland Single or group of specialized epithelial cells that secrete substances.

Glans penis (glanz PĒ-nis) The slightly enlarged region at the distal end of the penis.

Glaucoma (glaw-KŌ-ma) An eye disorder in which there is increased intraocular pressure due to an excess of aqueous humor.

Gliding joint A synovial joint having articulating surfaces that are usually flat, permitting only side-to-side and back-and-forth movements, as between carpal bones, tarsal bones, and the scapula and clavicle. Also called an **arthrodial** (ar-THRŌ-dē-al) **joint.**

Glomerular (glō-MER-yoo-lar) **capsule** A double-walled globe at the proximal end of a nephron that encloses the glomerulus. Also called **Bowman's** (BŌ-manz) **capsule.**

Glomerular filtration The first step in urine formation in which substances in blood are filtered at the endothelial-capsular membrane and the filtrate enters the proximal convoluted tubule of a nephron.

Glomerular filtration rate (GFR) The total volume of fluid that enters all the glomerular (Bowman's) capsules of the kidneys in 1 min; about 125 ml/min.

Glomerulonephritis (glō-mer-yoo-lō-nef-RĪ-tis) Inflammation of the glomeruli of the kidney that increases the permeability of the endothelial-capsular membrane and permits blood cells and proteins to enter the filtrate. Also called **Bright's disease.**

Glomerulus (glo-MER-yoo-lus) A rounded mass of nerves or blood vessels, especially the microscopic tuft of capillaries that is surrounded by the glomerular (Bowman's) capsule of each kidney tubule.

Glottis (GLOT-is) The vocal folds (true vocal cords) in the larynx plus the space between them (rima glottidis).

Glucagon (GLOO-ka-gon) A hormone produced by the alpha cells of the pancreas that increases the blood glucose level.

Glucocorticoids (gloo-kō-KOR-ti-koyds) Hormones secreted by the cortex of the adrenal gland, especially cortisol, that influence glucose metabolism.

Gluconeogenesis (gloo′-kō-nē′-ō-JEN-e-sis) The conversion of a substance other than carbohydrate into glucose.

Glucose (GLOO-kōs) A six-carbon sugar, $C_6H_{12}O_6$; the major energy source for every cell type in the body. Its metabolism is possible by every known living cell for the production of ATP.

Glycogen (GLĪ-kō-jen) A highly branched polymer of glucose containing thousands of subunits; functions as a compact store of glucose molecules in liver and muscle fibers (cells).

Glycogenesis (glī′-kō-JEN-e-sis) The process by which many molecules of glucose combine to form a molecule called glycogen.

Glycogenolysis (glī-kō-je-NOL-i-sis) The process of converting glycogen to glucose.

Glycolysis (glī-KOL-i-sis) Series of chemical reactions in the cytosol of a cell in which a molecule of glucose is split into two molecules of pyruvic acid.

Glycosuria (glī′-kō-SOO-rē-a) The presence of glucose in the urine; may be temporary or pathological. Also called **glucosuria.**

Gnostic (NOS-tik)   Pertaining to the faculties of perceiving and recognizing.

Gnostic area   Sensory area of the cerebral cortex that receives and integrates sensory input from various parts of the brain so that a common thought can be formed.

Goblet cell   A goblet-shaped unicellular gland that secretes mucus. Also called a **mucous cell.**

Goiter (GOY-ter)   An enlargement of the thyroid gland.

Golgi (GOL-jē) **complex**   An organelle in the cytoplasm of cells consisting of four to eight flattened channels, stacked on one another, with expanded areas at their ends; functions in packaging secreted proteins, lipid secretion, and carbohydrate synthesis.

Golgi tendon organ   *See* **Tendon organ.**

Gomphosis (gom-FŌ-sis)   A fibrous joint in which a cone-shaped peg fits into a socket.

Gonad (GŌ-nad)   A gland that produces gametes and hormones; the ovary in the female and the testis in the male.

Gonadocorticoids (gō-na-dō-KOR-ti-koydz)   Sex hormones secreted by the adrenal cortex.

Gonadotropic (gō′-nad-ō-TRŌ-pik) **hormone**   A hormone that regulates the functions of the gonads.

Gonorrhea (gon′-ō-RĒ-a)   Infectious, sexually transmitted disease caused by the bacterium *Neisseria gonorrhoeae* and characterized by inflammation of the urogenital mucosa, discharge of pus, and painful urination.

Gout (gowt)   Hereditary condition associated with excessive uric acid in the blood; the acid crystallizes and deposits in joints, kidneys, and soft tissue.

Graafian follicle   *See* **Vesicular ovarian follicle.**

Gray commissure (KOM-i-shur)   A narrow strip of gray matter connecting the two lateral gray masses within the spinal cord.

Gray matter   Area in the central nervous system and ganglia consisting of nonmyelinated nerve tissue.

Gray ramus communicans (RĀ-mus kō-MYOO-ni-kans)   A short nerve containing postganglionic sympathetic fibers; the cell bodies of the fibers are in a sympathetic chain ganglion, and the nonmyelinated axons run by way of the gray ramus to a spinal nerve and then to the periphery to supply smooth muscle in blood vessels, arrector pili muscles, and sweat glands. *Plural,* **rami communicantes** (RĀ-mē kō-myoo-ni-KAN-tēz).

Greater omentum (ō-MEN-tum)   A large fold in the serosa of the stomach that hangs down like an apron over the front of the intestines.

Greater vestibular (ves-TIB-yoo-lar) **glands**   A pair of glands on either side of the vaginal orifice that open by a duct into the space between the hymen and the labia minora. Also called **Bartholin's** (BAR-to-linz) **glands.**

Groin (groyn)   The depression between the thigh and the trunk; the inguinal region.

Gross anatomy   The branch of anatomy that deals with structures that can be studied without using a microscope. Also called **macroscopic anatomy.**

Growth   An increase in size due to an increase in the number of cells or an increase in the size of existing cells as internal components increase in size or an increase in the size of intercellular substances.

Gustatory (GUS-ta-tō′-rē)   Pertaining to taste.

Gynecology (gī′-ne-KOL-ō-jē)   The branch of medicine dealing with the study and treatment of disorders of the female reproductive system.

Gynecomastia (gīn′-e-kō-MAS-tē-a)   Excessive growth (benign) of the male mammary glands due to secretion of sufficient estrogens by an adrenal gland tumor (feminizing adenoma).

Gyrus (JĪ-rus)   One of the folds of the cerebral cortex of the brain. *Plural,* **gyri** (JĪ-rī). Also called a **convolution.**

Hair   A threadlike structure produced by hair follicles that develops in the dermis. Also called **pilus** (PI-lus).

Hair follicle (FOL-li-kul)   Structure composed of epithelium surrounding the root of a hair from which hair develops.

Hair root plexus (PLEK-sus)   A network of dendrites arranged around the root of a hair as free or naked nerve endings that are stimulated when a hair shaft is moved.

Haldane (HAWL-dān) **effect**   In the presence of oxygen, less carbon dioxide binds in the blood because when oxygen combines with hemoglobin, the hemoglobin becomes a stronger acid, which combines with less carbon dioxide.

Hallucination (ha-loo′-si-NĀ-shun)   A sensory perception of something that does not really exist in the world, that is, a sensory experience created from within the brain.

Hand   The terminal portion of an upper extremity, including the carpus, metacarpus, and phalanges.

Haploid (HAP-loyd)   Having half the number of chromosomes characteristically found in the somatic cells of an organism; characteristic of mature gametes. Symbolized *n.*

Hard palate (PAL-at)   The anterior portion of the roof of the mouth, formed by the maxillae and palatine bones and lined by mucous membrane.

Haustra (HAWS-tra)   The sacculated elevations of the colon.

Haversian canal   *See* **Central canal.**

Haversian system   *See* **Osteon.**

Head   The superior part of a human, cephalic to the neck. The superior or proximal part of a structure.

Heart   A hollow muscular organ lying slightly to the left of the midline of the chest that pumps the blood through the cardiovascular system.

Heart block   An arrhythmia (dysrhythmia) of the heart in which the atria and ventricles contract independently because of a blocking of electrical impulses through the heart at a critical point in the conduction system.

Heartburn   Burning sensation in the esophagus due to reflux of hydrochloric acid (HCl) from the stomach.

Heart-lung machine   A device that pumps blood, functioning as a heart, and removes carbon dioxide from blood and oxygenates it, functioning as lungs; used during heart transplantation, open-heart surgery, and coronary artery bypass grafting.

Heart murmur (MER-mer)   An abnormal sound that consists of a flow noise that is heard before the normal lubb-dupp or that may mask normal heart sounds.

Heat exhaustion   Condition characterized by cool, clammy skin, profuse perspiration, and fluid and electrolyte (especially salt) loss that results in muscle cramps, dizziness, vomiting, and fainting. Also called **heat prostration.**

Heatstroke Condition produced when the body cannot easily lose heat and characterized by reduced perspiration and elevated body temperature. Also called **sunstroke.**

Heimlich maneuver *See* **Abdominal thrust maneuver.**

Hematocrit (hē-MAT-ō-krit) **(Hct)** The percentage of blood made up of red blood cells. Usually calculated by centrifuging a blood sample in a graduated tube and then reading off the volume of red blood cells and total blood.

Hematology (hē′-ma-TOL-ō-jē) The study of blood.

Hematoma (hē′-ma-TŌ-ma) A tumor or swelling filled with blood.

Hematuria (hē′-ma-TOOR-ē-a) Blood in the urine.

Hemiballismus (hem′-i-ba-LIZ-mus) Violent muscular restlessness of half of the body, especially of the upper extremity.

Hemiplegia (hem-i-PLĒ-jē-a) Paralysis of the upper extremity, trunk, and lower extremity on one side of the body.

Hemodialysis (hē′-mō-dī-AL-i-sis) Filtering of the blood by means of an artificial device so that certain substances are removed from the blood as a result of the difference in rates of their diffusion through a selectively permeable membrane while the blood is being circulated outside the body.

Hemodynamics (hē-mō-dī-NA-miks) The study of factors and forces that govern the flow of blood through blood vessels.

Hemoglobin (hē′-mō-GLŌ-bin) **(Hb)** A substance in erythrocytes (red blood cells) consisting of the protein globin and the iron-containing red pigment heme and constituting about 33 percent of the cell volume; involved in the transport of oxygen and carbon dioxide.

Hemolysis (hē-MOL-i-sis) The escape of hemoglobin from the interior of the red blood cell into the surrounding medium; results from disruption of the integrity of the cell membrane by toxins or drugs, freezing or thawing, or hypotonic solutions.

Hemolytic disease of the newborn A hemolytic anemia of a newborn child that results from the destruction of the infant's red blood cells by antibodies produced by the mother; usually the antibodies are due to an Rh blood type incompatibility. Also called **erythroblastosis fetalis** (e-rith′-rō-blas-TŌ-sis fe-TAL-is).

Hemophilia (hē′-mō-FĒL-ē-a) A hereditary blood disorder where there is a deficient production of certain factors involved in blood clotting, resulting in excessive bleeding into joints, deep tissues, and elsewhere.

Hemopoiesis (hē-mō-poy-Ē-sis) Blood cell production occurring in the red marrow of bones. Also called **hematopoiesis** (hem′-a-tō-poy-Ē-sis).

Hemoptysis (hē-MOP-ti-sis) Spitting of blood from the respiratory tract.

Hemorrhage (HEM-or-rij) Bleeding; the escape of blood from blood vessels, especially when it is profuse.

Hemorrhoids (HEM-ō-royds) Dilated or varicosed blood vessels (usually veins) in the anal region. Also called **piles.**

Hemostasis (hē-MŌS-tā-sis) The stoppage of bleeding.

Hemostat (HĒ-mō-stat) An agent or instrument used to prevent the flow or escape of blood.

Hepatic (he-PAT-ik) Refers to the liver.

Hepatic duct A duct that receives bile from the bile capillaries. Small hepatic ducts merge to form the larger right and left hepatic ducts that unite to leave the liver as the common hepatic duct.

Hepatic portal circulation The flow of blood from the gastrointestinal organs to the liver before returning to the heart.

Hepatitis (hep-a-TĪ-tis) Inflammation of the liver due to a virus, drugs, and chemicals.

Hepatopancreatic (hep′-a-tō-pan′-krē-A-tik) **ampulla** A small, raised area in the duodenum where the combined common bile duct and main pancreatic duct empty into the duodenum. Also called the **ampulla of Vater** (VA-ter).

Hering–Breuer reflex *See* **Inflation reflex.**

Hernia (HER-nē-a) The protrusion or projection of an organ or part of an organ through a membrane or cavity wall, usually the abdominal cavity.

Herniated (her′-nē-A-ted) **disc** A rupture of an intervertebral disc so that the nucleus pulposus protrudes into the vertebral cavity. Also called a **slipped disc.**

Heterozygous (he-ter-ō-ZĪ-gus) Possessing a pair of different genes on homologous chromosomes for a particular hereditary characteristic.

Hiatus (hī-Ā-tus) An opening; a foramen.

High altitude sickness Disorder caused by decreased levels of alveolar pO_2 as altitude increases and characterized by headache, fatigue, insomnia, shortness of breath, nausea, and dizziness. Also called **acute mountain sickness.**

Hilus (HĪ-lus) An area, depression, or pit where blood vessels and nerves enter or leave an organ. Also called a **hilum.**

Hinge joint A synovial joint in which a convex surface of one bone fits into a concave surface of another bone, such as the elbow, knee, ankle, and interphalangeal joints. Also called a **ginglymus** (JIN-gli-mus) **joint.**

Hirsutism (HER-soot-izm) An excessive growth of hair in females and children, with a distribution similar to that in adult males, due to the conversion of vellus hairs into large terminal hairs in response to higher-than-normal levels of androgens.

Histamine (HISS-ta-mēn) Substance found in many cells, especially mast cells, basophils, and platelets, released when the cells are injured; results in vasodilation, increased permeability of blood vessels, and bronchiole constriction.

Histocompatibility (hiss′-tō-kom-pat-i-BIL-i-tē) **testing** Comparison of human leukocyte associated (HLA) antigens between donor and recipient to determine histocompatibility, the degree of compatibility between the two. Also called **HLA antigen typing** or **tissue typing.**

Histology (hiss-TOL-ō-jē) Microscopic study of the structure of tissues.

Hives (HĪVZ) Condition of the skin marked by reddened elevated patches that are often itchy; may be caused by infections, trauma, medications, emotional stress, food additives, and certain foods.

Hodgkin's disease (HD) A malignant disorder, usually arising in lymph nodes.

Holocrine (HŌL-ō-krin) **gland** A type of gland in which the entire secreting cell, along with its accumulated secretions, makes up the secretory product of the gland, as in the sebaceous (oil) glands.

Holter monitor   Electrocardiograph worn by a person while going about everyday routines.

Homeostasis (hō′-mē-ō-STĀ-sis)   The condition in which the body's internal environment remains relatively constant, within physiological limits.

Homologous (hō-MOL-ō-gus)   Correspondence of two organs in structure, position, and origin.

Homologous chromosomes   Two chromosomes that belong to a pair. Also called **homologues.**

Horizontal plane   A plane that runs parallel to the ground and divides the body or organs into superior and inferior portions. Also called a **transverse plane.**

Hormone (HOR-mōn)   A secretion of endocrine tissue that alters the physiological activity of target cells of the body.

Horn   Principal area of gray matter in the spinal cord.

Human chorionic gonadotropin (hCG) (kō-rē-ON-ik gō-nad-ō-TRŌ-pin)   A hormone produced by the developing placenta that maintains the corpus luteum.

Human chorionic somatomammotropin (sō-mat-ō-mam-ō-TRŌ-pin) **(hCS)**   A hormone produced by the chorion of the placenta that may stimulate breast tissue for lactation, enhance body growth, and regulate metabolism. Also called **human placental lactogen (hPL).**

Human growth hormone (hGH)   Hormone secreted by the anterior pituitary gland that brings about growth of body tissues, especially skeletal and muscular. Also known as **somatotropin** and **somatotropic hormone (STH).**

Human leucocyte associated (HLA) antigens   Surface proteins on white blood cells and other nucleated cells that are unique for each person (except for identical twins) and are used to type tissues and help prevent rejection.

Hyaluronic (hī′-a-loo-RON-ik) **acid**   A viscous, amorphous extracellular material that binds cells together, lubricates joints, and maintains the shape of the eyeballs.

Hyaluronidase (hī′-a-loo-RON-i-dās)   An enzyme that breaks down hyaluronic acid, increasing the permeability of connective tissues by dissolving the substances that hold body cells together.

Hydrocele (HĪ-drō-sēl)   A fluid-containing sac or tumor. Specifically, a collection of fluid formed in the space along the spermatic cord and in the scrotum.

Hydrocephalus (hī-drō-SEF-a-lus)   Abnormal accumulation of cerebrospinal fluid on the brain.

Hydrophobia (hī′-drō-FŌ-bē-a)   Rabies; a condition characterized by severe muscle spasms when attempting to drink water. Also, an abnormal fear of water.

Hymen (HĪ-men)   A thin fold of vascularized mucous membrane at the vaginal orifice.

Hyperbaric oxygenation (hī′-per-BA-rik ok′-sē-je-NĀ-shun) **(HBO)**   use of pressure supplied by a hyperbaric chamber to cause more oxygen to dissolve in blood to treat patients infected with anaerobic bacteria (tetanus and gangrene bacteria). Also used to treat carbon monoxide poisoning, asphyxia, smoke inhalation, and certain heart disorders.

Hypercalcemia (hī′-per-kal-SĒ-mē-a)   An excess of calcium in the blood.

Hypercapnia (hī′-per-KAP-nē-a)   An abnormal increase in the amount of carbon dioxide in the blood.

Hyperemia (hī′-per-Ē-mē-a)   An excess of blood in an area or part of the body.

Hyperextension (hī′-per-ek-STEN-shun)   Continuation of extension beyond the anatomical position, as in bending the head backward.

Hyperglycemia (hī′-per-glī-SĒ-mē-a)   An elevated blood sugar level.

Hypermetropia (hī′-per-mē-TRŌ-pē-a)   A condition in which visual images are focused behind the retina with resulting defective vision of near objects; farsightedness.

Hyperphosphatemia (hī-per-fos′-fa-TĒ-mē-a)   An abnormally high level of phosphate in the blood.

Hyperplasia (hī′-per-PLĀ-zē-a)   An abnormal increase in the number of normal cells in a tissue or organ, increasing its size.

Hyperpolarization (hī′-per-PŌL-a-ri-zā′-shun)   Increase in the internal negativity across a cell membrane, thus increasing the voltage and moving it farther away from the threshold value.

Hypersecretion (hī′-per-se-KRĒ-shun)   Overactivity of glands resulting in excessive secretion.

Hypersensitivity (hī′-per-sen-si-TI-vi-tē)   Overreaction to an allergen that results in pathological changes in tissues. Also called **allergy.**

Hypertension (hī′-per-TEN-shun)   High blood pressure.

Hyperthermia (hī′-per-THERM-ē-a)   An elevated body temperature.

Hypertonia (hī-per-TŌ-nē-a)   Increased muscle tone that is expressed as spasticity or rigidity.

Hypertonic (hī′-per-TON-ik)   Having an osmotic pressure greater than that of a solution with which it is compared.

Hypertrophy (hī-PER-trō-fē)   An excessive enlargement or overgrowth of tissue without cell division.

Hyperventilation (hī′-per-ven-ti-LĀ-shun)   A rate of respiration higher than that required to maintain a normal level of plasma $p\text{CO}_2$.

Hypervitaminosis (hī′-per-vī′-ta-min-Ō-sis)   An excess of one or more vitamins.

Hypocalcemia (hī′-pō-kal-SĒ-mē-a)   A below normal level of calcium in the blood.

Hypochloremia (hī′-pō-klō-RĒ-mē-a)   Deficiency of chloride in the blood.

Hypoglycemia (hī′-pō-glī-SĒ-mē-a)   An abnormally low concentration of glucose in the blood; can result from excess insulin (injected or secreted).

Hypokalemia (hī′-pō-kā-LĒ-mē-a)   Deficiency of potassium in the blood.

Hypomagnesemia (hī′-pō-mag′-ne-SĒ-mē-a)   Deficiency of magnesium in the blood.

Hyponatremia (hī′-pō-na-TRĒ-mē-a)   Deficiency of sodium in the blood.

Hyponychium (hī′-pō-NIK-ē-um)   Free edge of the fingernail.

Hypophosphatemia (hī-pō-fos′-fa-TĒ-mē-a)   An abnormally low level of phosphate in the blood.

Hypophyseal (hī′-pō-FIZ-ē-al) **pouch** An outgrowth of ectoderm from the roof of the stomodeum (mouth) from which the anterior pituitary gland develops.

Hypophysis (hī-POF-i-sis) Pituitary gland.

Hypoplasia (hī-pō-PLĀ-zē-a) Defective development of tissue.

Hyposecretion (hī′-pō-se-KRĒ-shun) Underactivity of glands resulting in diminished secretion.

Hypospadias (hī′-pō-SPĀ-dē-as) A displaced urethral opening. In the male, the opening may be on the underside of the penis, at the penoscrotal junction, between the scrotal folds, or in the perineum. In the female, the urethra opens into the vagina.

Hypothalamus (hī′-pō-THAL-a-mus) A portion of the diencephalon, lying beneath the thalamus and forming the floor and part of the wall of the third ventricle.

Hypothermia (hī-pō-THER-mē-a) Lowering of body temperature below 35°C (95°F); in surgical procedures, it refers to deliberate cooling of the body to slow down metabolism and reduce oxygen needs of tissues.

Hypotonia (hī′-pō-TŌ-nē-a) Decreased or lost muscle tone in which muscles appear flaccid.

Hypotonic (hī′-pō-TON-ik) Having an osmotic pressure lower than that of a solution with which it is compared.

Hypoventilation (hī-pō-ven-ti-LĀ-shun) A rate of respiration lower than that required to maintain a normal level of plasma pCO_2.

Hypovolemic (hī-pō-vō-LĒ-mik) **shock** A type of shock characterized by decreased intravascular volume resulting from blood loss; may be caused by acute hemorrhage or excessive fluid loss.

Hypoxia (hī-POKS-ē-a) Lack of adequate oxygen at the tissue level.

Hysterectomy (hiss-te-REK-tō-mē) The surgical removal of the uterus.

Ileocecal (il′-ē-ō-SĒ-kal) **sphincter** A fold of mucous membrane that guards the opening from the ileum into the large intestine. Also called the **ileocecal valve.**

Ileum (IL-ē-um) The terminal portion of the small intestine.

Immunity (im-YOO-ni-tē) The state of being resistant to injury, particularly by poisons, foreign proteins, and invading pathogens.

Immunogenicity (im-yoo-nō-jen-IS-it-ē) Ability of an antigen to stimulate immune responses.

Immunoglobulin (im-yoo-nō-GLOB-yoo-lin) **(Ig)** An antibody synthesized by plasma cells derived from B lymphocytes in response to the introduction of antigen. Immunoglobulins are divided into five kinds (IgG, IgM, IgA, IgD, IgE) based primarily on the larger protein component present in the immunoglobulin.

Immunology (im′-yoo-NOL-ō-jē) The branch of science that deals with the responses of the body when challenged by antigens.

Immunosuppression (im′-yoo-nō-su-PRESH-un) Inhibition of the immune response.

Immunotherapy (im-yoo-nō-THER-a-pē) Attempt to induce the immune system to mount an attack against cancer cells.

Imperforate (im-PER-fō-rāt) Abnormally closed.

Impetigo (im′-pe-TĪ-go) A contagious skin disorder characterized by pustular eruptions.

Implantation (im-plan-TA-shun) The insertion of a tissue or a part into the body. The attachment of the blastocyst to the lining of the uterus 7–8 days after fertilization.

Impotence (IM-pō-tens) Weakness; inability to copulate; failure to maintain an erection long enough for sexual intercourse.

Incontinence (in-KON-ti-nens) Inability to retain urine, semen, or feces, through loss of sphincter control.

Infarction (in-FARK-shun) The presence of a localized area of necrotic tissue, produced by inadequate oxygenation of the tissue.

Infection (in-FEK-shun) Invasion and multiplication of microorganisms in body tissues, which may be inapparent or characterized by cellular injury.

Infectious mononucleosis (mon-ō-nook′-lē-Ō-sis) **(IM)** Contagious disease caused by the Epstein-Barr virus (EBV) and characterized by an elevated mononucleocyte and lymphocyte count, fever, sore throat, stiff neck, cough, and malaise.

Inferior (in-FĒR-ē-or) Away from the head or toward the lower part of a structure. Also called **caudad** (KAW-dad).

Inferior vena cava (VĒ-na CĀ-va) **(IVC)** Large vein that collects blood from parts of the body inferior to the heart and returns it to the right atrium.

Infertility Inability to conceive or to cause conception. Also called **sterility.**

Inflammation (in′-fla-MĀ-shun) Localized, protective response to tissue injury designed to destroy, dilute, or wall off the infecting agent or injured tissue; characterized by redness, pain, heat, swelling, and sometimes loss of function.

Inflammatory bowel (in-FLAM-a-to′-re BOW-el) **disease** Disorder that exists in two forms: (1) Crohn's disease (inflammation of the gastrointestinal tract, especially the distal ileum and proximal colon, in which the inflammation may extend from the mucosa through the serosa); and (2) ulcerative colitis (inflammation of the mucosa of the gastrointestinal tract, usually limited to the large intestine and usually accompanied by rectal bleeding).

Inflation reflex Reflex that prevents overinflation of the lungs. Also called **Hering–Breuer reflex.**

Infundibulum (in′-fun-DIB-yoo-lum) The stalklike structure that attaches the pituitary gland to the hypothalamus of the brain. The funnel-shaped, open, distal end of the uterine (Fallopian) tube.

Ingestion (in-JES-chun) The taking in of food, liquids, or drugs, by mouth.

Inguinal (IN-gwi-nal) Pertaining to the groin.

Inguinal canal An oblique passageway in the anterior abdominal wall just superior and parallel to the medial half of the inguinal ligament that transmits the spermatic cord and ilioinguinal nerve in the male and round ligament of the uterus and ilioinguinal nerve in the female.

Inheritance The acquisition of body characteristics and qualities by transmission of genetic information from parents to off-spring.

Inhibin A hormone secreted by the gonads that inhibits FSH release by the anterior pituitary gland.

Inhibiting hormone Chemical secretion of the hypothalamus that can suppress secretion of hormones by the anterior pituitary gland.

Inhibitory postsynaptic potential (IPSP) The increase in the internal negativity of the membrane potential so that the voltage moves further from the threshold value.

Inner cell mass A region of cells of a blastocyst that differentiates into the three primary germ layers—ectoderm, mesoderm, and endoderm—from which all tissues and organs develop; also called an **embryoblast.**

Inorganic (in′-or-GAN-ik) **compound** Compound that usually lacks carbon, usually small, and contains ionic bonds. Examples include water and many acids, bases, and salts.

Insertion (in-SER-shun) The attachment of a muscle tendon to a movable bone or the end opposite the origin.

Insomnia (in-SOM-nē-a) Difficulty in falling asleep and, usually, frequent awakening.

Inspiration (in-spi-RĀ-shun) The act of drawing air into the lungs.

Inspiratory (in-SPĪ-ra-tor-ē) **capacity** Total inspiratory capacity of the lungs, the total of tidal volume plus inspiratory reserve volume; averages 3600 ml.

Inspiratory (in-SPĪ-ra-tor-ē) **reserve volume** Additional inspired air over and above tidal volume; averages 3100 ml.

Insula (IN-su-la) A triangular area of cerebral cortex that lies deep within the lateral cerebral fissue, under the parietal, frontal, and temporal lobes, and cannot be seen in an external view of the brain. Also called the **island** or **isle of Reil** (RĪL).

Insulin (IN-su-lin) A hormone produced by the beta cells of the pancreas that decreases the blood glucose level.

Integrin (IN-te-grin) Receptor on a plasma membrane that interacts with an adhesion protein found in intercellular material and blood.

Integumentary (in-teg′-yoo-MEN-tar-ē) Relating to the skin.

Intercalated (in-TER-ka-lāt-ed) **disc** An irregular transverse thickening of sarcolemma that contains desmosomes that hold cardiac muscle fibers (cells) together and gap junctions that aid in conduction of muscle action potentials.

Intercostal (in′-ter-KOS-tal) **nerve** A nerve supplying a muscle located between the ribs.

Interferons (in′-ter-FĒR-ons) **(IFNs)** Three principal types of protein (alpha, beta, gamma) naturally produced by virus-infected host cells that induce uninfected cells to synthesize antiviral proteins that inhibit intracellular viral replication in uninfected host cells; artificially synthesized through recombinant DNA techniques.

Intermediate Between two structures, one of which is medial and one of which is lateral.

Intermediate filament Protein filament, ranging from 8 to 12 nm in diameter, that may provide structural reinforcement, hold organelles in place, and give shape to a cell.

Internal Away from the surface of the body.

Internal capsule A tract of projection fibers connecting various parts of the cerebral cortex and lying between the thalamus and the caudate and lentiform nuclei of the basal ganglia.

Internal ear The inner ear or labyrinth, lying inside the temporal bone, containing the organs of hearing and balance.

Internal nares (NA-rēz) The two openings posterior to the nasal cavities opening into the nasopharynx. Also called the **choanae** (kō-A-nē).

Internal (tissue) respiration The exchange of respiratory gases between blood and body cells.

Interphase (IN-ter-fāz) The period during its life cycle when a cell is carrying on every life process except division; the stage between two mitotic divisions. Also called **metabolic phase.**

Interstitial cell of Leydig *See* Interstitial endocrinocyte.

Interstitial (in′-ter-STISH-al) **endocrinocyte** A cell located in the connective tissue between seminiferous tubules in a mature testis that secretes testosterone. Also called an **interstitial cell of Leydig** (LĪ-dig).

Interstitial (in′-ter-STISH-al) **fluid** The portion of extracellular fluid that fills the microscopic spaces between the cells of tissues; the internal environment of the body. Also called **intercellular** or **tissue fluid.**

Interstitial growth Growth from within, as in the growth of cartilage. Also called **endogenous** (en-DOJ-e-nus) **growth.**

Interventricular foramen (in′-ter-ven-TRIK-yoo-lar) A narrow, oval opening through which the lateral ventricles of the brain communicate with the third ventricle. Also called the **foramen of Monro.**

Intervertebral (in′-ter-VER-te-bral) **disc** A pad of fibrocartilage located between the bodies of two vertebrae.

Intestinal gland A gland that opens onto the surface of the intestinal mucosa and secretes digestive enzymes. Also called a **crypt of Lieberkühn** (LĒ-ber-kyoon).

Intracellular (in′-tra-SEL-yoo-lar) **fluid (ICF)** Fluid located within cells. Also called **cytosol** (SĪ-tō-sol).

Intrafusal (in′-tra-FYOO-zal) **fibers** Three to ten specialized muscle fibers (cells), partially enclosed in a spindle-shaped connective tissue capsule that is filled with lymph; the fibers compose muscle spindles.

Intramembranous ossification (in′-tra-MEM-bra-nus os′-i′-fi-KĀ-shun) The method of bone formation in which the bone is formed directly in membranous tissue.

Intraocular (in-tra-OC-yoo-lar) **pressure (IOP)** Pressure in the eyeball, produced mainly by aqueous humor.

Intrapleural pressure Air pressure between the two pleural layers of the lungs, usually subatmospheric. Also called **intrathoracic pressure.**

Intrauterine device (IUD) A small metal, or plastic object inserted into the uterus for the purpose of preventing pregnancy.

Intrinsic (in-TRIN-sik) Of internal origin; for example, the intrinsic factor, a glycoprotein formed by the gastric mucosa that is necessary for the absorption of vitamin B_{12}.

Intrinsic clotting pathway Sequence of reactions leading to blood clotting that is initiated by the release of a substance (tissue factor or thromboplastin) contained *within* blood itself or cells in direct contact with blood.

Intrinsic factor (IF)   A glycoprotein synthesized and secreted by the parietal cells of the gastric mucosa that facilitates vitamin B_{12} absorption in the small intestine.

Intron (IN-tron)   A region of DNA that does not code for the synthesis of a protein.

Intubation (in'-too-BĀ-shun)   Insertion of a tube through the nose or mouth into the larynx and trachea for entrance of air or to dilate a stricture.

In utero (YOO-ter-ō)   Within the uterus.

Invagination (in-vaj'-i-NĀ-shun)   The pushing of the wall of a cavity into the cavity itself.

Inversion (in-VER-zhun)   The movement of the sole inward at the ankle joint.

In vitro (VĒ-trō)   Literally, in glass; outside the living body and in artificial environment such as a laboratory test tube.

In vivo (VĒ-vō)   In the living body.

Ion (Ī-on)   Any charged particle or group of particles; usually formed when a substance, such as a salt, dissolves and dissociates.

Ipsilateral (ip'-si-LAT-er-al)   On the same side, affecting the same side of the body.

Iris   The colored portion of the eyeball seen through the cornea that consists of circular and radial smooth muscle; the hole in the center of the iris is the pupil.

Irritable bowel (IR-i-ta-bul BOW-el) **syndrome (IBS)**   Disease of the entire gastrointestinal tract in which persons with the condition may react to stress by developing symptoms such as cramping and abdominal pain associated with alternating patterns of diarrhea and constipation. Excessive amounts of mucus may appear in the stools, and other symptoms include flatulence, nausea, and loss of appetite. The condition is also known as **irritable colon** or **spastic colitis.**

Ischemia (is-KĒ-mē-a)   A lack of sufficient blood to a part due to obstruction of circulation.

Island of Reil   *See Insula.*

Islet of Langerhans   *See Pancreatic islet.*

Isometric contraction   A muscle contraction in which tension on the muscle increases, but there is only minimal muscle shortening so that no movement is produced.

Isotonic (ī' sō-TON-ik)   Having equal tension or tone. Having equal osmotic pressure between two different solutions or between two elements in a solution.

Isotonic (ī-sō-TON-ik) **contraction**   Contraction in which the tension remains the same; occurs when a constant load is moved through the range of motions possible at a joint.

Isotope (Ī-sō-tōpe')   A chemical element that has the same atomic number as another but a different atomic weight. Radioactive isotopes change into other elements with the emission of certain radiations.

Isovolumetric (ī-sō-vol-yoo'-MET-rik) **contraction**   The period of time, about 0.05 sec, between the start of ventricular systole and the opening of the semilunar valves; there is contraction of the ventricles, but no emptying, and there is a rapid rise in ventricular pressure.

Isovolumetric relaxation   The period of time, about 0.05 sec, between the opening of the atrioventricular (AV) valves and the closing of the semilunar valves; there is a drastic decrease in ventricular pressure without a change in ventricular volume.

Isthmus (IS-mus)   A narrow strip of tissue or narrow passage connecting two larger parts.

Jaundice (JAWN-dis)   A condition characterized by yellowness of skin, white of eyes, mucous membranes, and body fluids because of a buildup of bilirubin.

Jejunum (je-JOO-num)   The middle portion of the small intestine.

Joint kinesthetic (kin'-es-THET-ik) **receptor**   A proprioceptive receptor located in a joint, stimulated by joint movement.

Juxtaglomerular (juks-ta-glō-MER-yoo-lar) **apparatus (JGA)**   Consists of the macula densa (cells of the distal convoluted tubule adjacent to the afferent and efferent arteriole) and juxtaglomerular cells (modified cells of the afferent and sometimes efferent arteriole); secretes renin when blood pressure starts to fall.

Karyotype (KAR-ē-ō-tīp)   An arrangement of chromosomes based on shape, size, and position of centromeres.

Keratin (KER-a-tin)   An insoluble protein found in the hair, nails, and other keratinized tissues of the epidermis.

Keratinocyte (ker-A-tin'-ō-sīt)   The most numerous of the epidermal cells that function in the production of keratin.

Keratosis (ker'-a-TŌ-sis)   Formation of a hardened growth of tissue.

Ketone (KĒ-tōn) **bodies**   Substances produced primarily during excessive triglyceride catabolism, such as acetone, acetoacetic acid, and ß-hydroxybutyric acid.

Ketosis (kē-TŌ-sis)   Abnormal condition marked by excessive production of ketone bodies.

Kidney (KID-nē)   One of the paired reddish organs located in the lumbar region that regulates the composition and volume of blood and produces urine.

Kidney stone   A concretion, usually consisting of calcium oxalate, uric acid, and calcium phosphate crystals, that may form in any portion of the urinary tract. Also called a **renal calculus.**

Kilocalorie (KIL-ō-kal'-ō-rē) **(kcal)**   The amount of heat required to raise the temperature of 1,000 g of water 1°C; the unit used to express the heating value of foods and to measure metabolic rate.

Kinesiology (ki-nē'-sē-OL-ō-jē)   The study of the movement of body parts.

Kinesthesia (kin-is-THĒ-szē-a)   Ability to perceive extent, direction, or weight of movement; muscle sense.

Korotkoff (kō-ROT-kof) **sounds**   The various sounds that are heard while taking blood pressure.

Krebs cycle   A series of energy-yielding chemical reactions that occur in the matrix of mitochondria in which energy is transferred to carrier molecules for subsequent liberation and carbon dioxide is formed. Also called the **citric acid cycle** and **tricarboxylic acid (TCA) cycle.**

Kupffer's cell   *See Stellate reticuloendothelial cell.*

Kyphosis (kī-FŌ-sis)   An exaggeration of the thoracic curve of the vertebral column, resulting in a "round-shouldered" appearance. Also called **hunchback.**

Labial frenulum (LĀ-bē-al FREN-yoo-lum)   A medial fold of mucous membrane between the inner surface of the lip and the gums.

Labia majora (LĀ-bē-a ma-JO-ra)   Two longitudinal folds of skin extending downward and backward from the mons pubis of the female.

Labia minora (min-OR-a)   Two small folds of mucous membrane lying medial to the labia majora of the female.

Labium (LĀ-bē-um)   A lip. A liplike structure. *Plural,* **labia** (LĀ-bē-a).

Labor   The process by which a fetus is expelled from the uterus through the vagina.

Labyrinth (LAB-i-rinth)   Intricate communicating passageway, especially in the internal ear.

Labyrinthine (lab-i-RIN-thēn) **disease**   Malfunction of the internal ear characterized by deafness, tinnitus, vertigo, nausea, and vomiting.

Laceration (las'-er-Ā-shun)   Wound or irregular area of the skin.

Lacrimal (LAK-ri-mal)   Pertaining to tears.

Lacrimal (LAK-ri-mal) **canal**   A duct, one on each eyelid, commencing at the punctum at the medial margin of an eyelid and conveying tears medially into the nasolacrimal sac.

Lacrimal gland   Secretory cells located at the superior anterolateral portion of each orbit that secrete tears into excretory ducts that open onto the surface of the conjunctiva.

Lacrimal sac   The superior expanded portion of the nasolacrimal duct that receives the tears from a lacrimal canal.

Lactation (lak-TĀ-shun)   The secretion and ejection of milk by the mammary glands.

Lacteal (LAK-tē-al)   One of many intestinal lymphatic vessels in villi that absorb fat from digested food.

Lactose intolerance   Inability to digest lactose because of failure of small intestinal mucosal cells to produce lactase.

Lacuna (la-KOO-na)   A small, hollow space, such as that found in bones in which the osteocytes lie. *Plural,* **lacunae** (la-KOO-nē).

Lambdoid (lam-DOYD) **suture**   The line of union in the skull between the parietal bones and the occipital bone; sometimes contains sutural (Wormian) bones.

Lamellae (la-MEL-ē)   Concentric rings of hard, calcified matrix found in compact bone.

Lamellated corpuscle   Oval-shaped pressure receptor located in subcutaneous tissue and consisting of concentric layers of connective tissue wrapped around a sensory nerve fiber. Also called a **Pacinian** (pa-SIN-ē-an) **corpuscle.**

Lamina (LAM-i-na)   A thin, flat layer or membrane, as the flattened part of either side of the arch of a vertebra. *Plural,* **laminae** (LAM-i-nē).

Lamina propria (PRŌ-prē-a)   The connective tissue layer of a mucous membrane.

Lanugo (lan-YOO-gō)   Fine downy hairs that cover the fetus.

Laparoscopy (lap'-a-ROS-kō-pē)   A procedure in which a laparoscope is inserted through an incision in the abdominal wall to view abdominal and pelvic viscera, remove fluids and tissues for biopsy, drain ovarian cysts, cut adhesions, stop bleeding, and perform tubal ligation.

Large intestine   The portion of the gastrointestinal tract extending from the ileum of the small intestine to the anus, divided structurally into the cecum, colon, rectum, and anal canal.

Laryngitis (la-rin-JĪ-tis)   Inflammation of the mucous membrane lining the larynx.

Laryngopharynx (la-rin'-gō-FAR-inks)   The inferior portion of the pharynx, extending downward from the level of the hyoid bone to divide posteriorly into the esophagus and anteriorly into the larynx. Also called the **hypopharynx.**

Laryngoscope (la-RIN-gō-skōp)   An instrument for examining the larynx.

Laryngotracheal (la-rin'-gō-TRĀ-kē-al) **bud**   An outgrowth of endoderm of the foregut from which the respiratory system develops.

Larynx (LAR-inks)   The voice box, a short passageway that connects the pharynx with the trachea.

Lateral (LAT-er-al)   Farther from the midline of the body or a structure.

Lateral ventricle (VEN-tri-kul)   A cavity within a cerebral hemisphere that communicates with the lateral ventricle in the other cerebral hemisphere and with the third ventricle by way of the interventricular foramen.

Learning   The ability to acquire knowledge or a skill through instruction or experience.

Leg   The part of the lower extremity between the knee and the ankle.

Lens   A transparent organ constructed of proteins (crystallins) lying posterior to the pupil and iris of the eyeball and anterior to the vitreous body.

Lesion (LĒ-zhun)   Any localized, abnormal change in tissue formation.

Lesser omentum (ō-MEN-tum)   A fold of the peritoneum that extends from the liver to the lesser curvature of the stomach and the commencement of the duodenum.

Lesser vestibular (ves-TIB-yoo-lar) **gland**   One of the paired mucus-secreting glands that have ducts that open on either side of the urethral orifice in the vestibule of the female.

Lethargy (LETH-ar-jē)   A condition of drowsiness or indifference.

Leukemia (loo-KĒ-mē-a)   A malignant disease of the blood-forming tissues characterized by either uncontrolled production and accumulation of immature leukocytes in which many cells fail to reach maturity (acute) or an accumulation of mature leukocytes in the blood because they do not die at the end of their normal life span (chronic).

Leukocyte (LOO-kō-sīt)   A white blood cell.

Leukocytosis (loo'-kō-sī-TŌ-sis)   An increase in the number of white blood cells, characteristic of many infections and other disorders.

Leukopenia (loo-kō-PĒ-nē-a)   A decrease in the number of white blood cells below 5,000/mm³.

Leukoplakia (loo-kō-PLĀ-kē-a)   A disorder in which there are white patches in the mucous membranes of the tongue, gums, and cheeks.

Libido (li-BĒ-dō)   The sexual drive, conscious or unconscious.

Ligament (LIG-a-ment)   Dense, regular, connective tissue that attaches bone to bone.

Ligand (LĪ-gand)   Chemical in interstitial fluid, usually in a concentration lower than in cells, that binds to a specific receptor.

Limbic system   A portion of the forebrain, sometimes termed the visceral brain, concerned with various aspects of emotion and behavior, that includes the limbic lobe, dentate gyrus, amygdaloid body, septal nuclei, mammillary bodies, anterior thalamic nucleus, olfactory bulbs, and bundles of myelinated axons.

Lingual frenulum (LIN-gwal FREN-yoo-lum)   A fold of mucous membrane that connects the tongue to the floor of the mouth.

Lingual lipase (LĪ-pās)   Digestive enzyme secreted by glands on the dorsum of the tongue that digests triglycerides.

Lipase   A fat-splitting enzyme.

Lipid   An organic compound composed of carbon, hydrogen, and oxygen that is usually insoluble in water, but soluble in alcohol, ether, and chloroform; examples include triglycerides, phospholipids, steroids, and prostaglandins.

Lipid profile   Blood test that measures total cholesterol, high-density lipoprotein, low-density lipoprotein, and triglycerides, to assess risk for cardiovascular disease.

Lipogenesis (li-pō-GEN-e-sis)   The synthesis of lipids from glucose or amino acids by liver cells.

Lipoma (li-PŌ-ma)   A fatty tissue tumor, usually benign.

Lipoprotein (lip′-ō-PRŌ-tēn)   Protein containing lipid that is produced by the liver and combines with cholesterol and triglycerides to make it water-soluble for transportation by the cardiovascular system; high levels of low-density lipoproteins (LDL) are associated with increased risk of atherosclerosis, while high levels of high-density lipoproteins (HDL) are associated with decreased risk of atherosclerosis.

Lithotripsy (LITH-ō-trip′-sē)   A noninvasive procedure in which shock waves generated by a lithotriptor are used to pulverize kidney stones or gallstones.

Liver   Large gland under the diaphragm that occupies most of the right hypochondriac region and part of the epigastric region; functionally, it produces bile salts, heparin, and plasma proteins; converts one nutrient into another; detoxifies substances; stores glycogen, minerals, and vitamins; carries on phagocytosis of blood cells and bacteria; and helps activate vitamin D.

Lobe (lōb)   A curved or rounded projection.

Locus coeruleus (LŌ-kus sē-ROO-lē-us)   A group of neurons in the brain stem where norepinephrine (NE) is concentrated.

Lordosis (lor-DŌ-sis)   An exaggeration of the lumbar curve of the vertebral column. Also called **swayback.**

Lou Gehrig's disease   *See* **Amyotrophic lateral sclerosis.**

Lower extremity   The appendage attached at the pelvic (hip) girdle, consisting of the thigh, knee, leg, ankle, foot, and toes.

Lumbar (LUM-bar)   Region of the back and side between the ribs and pelvis; loin.

Lumbar plexus (PLEK-sus)   A network formed by the anterior branches of spinal nerves L1 through L4.

Lumen (LOO-men)   The space within an artery, vein, intestine, or a tube.

Lung   One of the two main organs of respiration, lying on either side of the heart in the thoracic cavity.

Lung scan   A diagnostic test in which a radioactive substance is detected in the lungs by a scanning camera; used to evaluate for pulmonary embolism, pneumonia, or cancer.

Lunula (LOO-nyoo-la)   The moon-shaped white area at the base of a nail.

Luteinizing (LOO-tē-in′-īz-ing) **hormone (LH)**   A hormone secreted by the anterior pituitary gland that stimulates ovulation and progesterone secretion by the corpus luteum, and readies the mammary glands for milk secretion in females and stimulates testosterone secretion by the testes in males.

Lymph (limf)   Fluid confined in lymphatic vessels and flowing through the lymphatic system to be returned to the blood.

Lymphangiography (lim-fan′-jē-OG-ra-fē)   A procedure by which lymphatic vessels and lymph organs are filled with a radiopaque substance in order to be x-rayed.

Lymphatic capillary   Closed-ended microscopic lymphatic vessel that begins in spaces between cells and converges with other lymphatic capillaries to form lymphatic vessels.

Lymphatic tissue   A specialized form of reticular tissue that contains large numbers of lymphocytes.

Lymphatic (lim-FAT-ik) **vessel**   A large vessel that collects lymph from lymphatic capillaries and converges with other lymphatic vessels to form the thoracic and right lymphatic ducts.

Lymph node   An oval or bean-shaped structure located along lymphatic vessels.

Lymphocyte (LIM-fō-sīt)   A type of white blood cell, found in lymph nodes, associated with the immune system.

Lymphokines (LIM-fō-kīns)   Powerful proteins secreted by T cells that endow T cells with their ability to assist in immunity.

Lysosome (LĪ-sō-sōm)   An organelle in the cytoplasm of a cell, enclosed by a single membrane and containing powerful digestive enzymes.

Lysozyme (LĪ-sō-zim)   A bactericidal enzyme found in tears, saliva, and perspiration.

Macrophage (MAK-rō-fāj)   Phagocytic cell derived from a monocyte. May be fixed or wandering.

Macula (MAK-yoo-la)   A discolored spot or a colored area. A small, thickened region on the wall of the utricle and saccule that serves as a receptor for static equilibrium.

Macula lutea (LOO-tē-a)   The yellow spot in the center of the retina.

Magnetic resonance imaging (MRI)   A diagnostic procedure that focuses on the nuclei of atoms of a single element in a tissue, usually hydrogen, to determine if they behave normally in the presence of an external magnetic force; used to indicate the biochemical activity of a tissue. Formerly called **nuclear magnetic resonance (NMR).**

Malaise (ma-LĀYZ)   Discomfort, uneasiness, and indisposition, often indicative of infection.

Malignant (ma-LIG-nant)   Referring to diseases that tend to become worse and cause death; especially the invasion and spreading of cancer.

Malignant melanoma (mel′-a-NŌ-ma)   A usually dark, malignant tumor of the skin containing melanin.

Malnutrition (mal′-noo-TRISH-un)   State of bad or poor nutrition that may be due to inadequate food intake, imbalance of nutrients, malabsorption of nutrients, improper distribution of nutrients, increased nutrient requirements, increased nutrient losses, or overnutrition.

Mammary (MAM-ar-ē) **gland** Modified sudoriferous (sweat) gland of the female that secretes milk for the nourishment of the young.

Mammillary (MAM-i-ler-ē) **bodies** Two small rounded bodies posterior to the tuber cinereum that are involved in reflexes related to the sense of smell.

Mammography (mam-OG-ra-fē) Procedure for imaging the breasts (xeromammography or film-screen mammography) to evaluate for breast disease or screen for breast cancer.

Marfan (MAR-fan) **syndrome** Inherited disorder that results in abnormalities of connective tissue, especially in the skeleton, eyes, and cardiovascular system.

Margination (mar'-ji-NĀ-shun) Accumulation and adhesion of neutrophils to the endothelium of blood vessels at the site of injury during the early stages of inflammation.

Marrow (MAR-ō) Soft, spongelike material in the cavities of bone. Red bone marrow produces blood cells; yellow bone marrow, formed mainly of fatty tissue, has no blood-producing function.

Mast cell A cell found in areolar connective tissue along blood vessels that produces heparin, a dilator of small blood vessels during inflammation. The name given to a basophil after it has left the bloodstream and entered the tissues.

Mastectomy (mas-TEK-tō-mē) Surgical removal of breast tissue.

Mastication (mas'-ti-KĀ-shun) Chewing.

Maximal oxygen uptake Maximum rate of oxygen consumption during aerobic catabolism of pyruvic acid that is determined by age, sex, and body size.

Meatus (mē-Ā-tus) A passage or opening, especially the external portion of a canal.

Mechanoreceptor (me-KAN-ō-rē'-sep-tor) Receptor that detects mechanical deformation of the receptor itself or adjacent cells; stimuli so detected include those related to touch, pressure, vibration, proprioception, hearing, equilibrium, and blood pressure.

Medial (MĒ-dē-al) Nearer the midline of the body or a structure.

Medial lemniscus (lem-NIS-kus) A flat band of myelinated nerve fibers extending through the medulla, pons, and midbrain and terminating in the thalamus on the same side. Sensory neurons in this tract transmit impulses for proprioception, hearing, equilibrium, and vibration sensations.

Median aperture (AP-er-choor) One of the three openings in the roof of the fourth ventricle through which cerebrospinal fluid enters the subarachnoid space of the brain and cord. Also called the **foramen of Magendie.**

Median plane A vertical plane dividing the body into right and left halves. Situated in the middle.

Mediastinum (mē'-dē-as-TĪ-num) A broad, median partition, actually a mass of tissue found between the pleurae of the lungs that extends from the sternum to the vertebral column.

Medulla (me-DULL-la) An inner layer of an organ, such as the medulla of the kidneys.

Medulla oblongata (ob'-long-GA-ta) The most inferior part of the brain stem.

Medullary (MED-yoo-lar'-ē) **cavity** The space within the diaphysis of a bone that contains yellow bone marrow. Also called the **marrow cavity.**

Medullary rhythmicity (rith-MIS-i-tē) **area** Portion of the respiratory center in the medulla that controls the basic rhythm of respiration.

Meibomian gland *See* **Tarsal gland.**

Meiosis (mē-Ō-sis) A type of cell division restricted to sex-cell production involving two successive nuclear divisions that result in daughter cells with the haploid (*n*) number of chromosomes.

Meissner's corpuscle *See* **Corpuscle of touch.**

Melanin (MEL-a-nin) A dark black, brown, or yellow pigment found in some parts of the body such as the skin and hair.

Melanoblast (MEL-a-nō-blast) Precursor cell in the epidermis that gives rise to melanocytes, cells that produce melanin.

Melanocyte (MEL-a-nō-sīt') A pigmented cell located between or beneath cells of the deepest layer of the epidermis that synthesizes melanin.

Melanocyte-stimulating hormone (MSH) A hormone secreted by the anterior pituitary gland that stimulates the dispersion of melanin granules in melanocytes in amphibians; continued administration produces darkening of skin in humans.

Melatonin (mel-a-TŌN-in) A hormone secreted by the pineal gland that may inhibit reproductive activities.

Membrane A thin, flexible sheet of tissue composed of an epithelial layer and an underlying connective tissue layer, as in an epithelial membrane, or of areolar connective tissue only, as in a synovial membrane.

Membranous labyrinth (mem-BRA-nus LAB-i-rinth) The portion of the labyrinth of the inner ear that is located inside the bony labyrinth and separated from it by the perilymph; made up of the membranous semicircular canals, the saccule and utricle, and the cochlear duct.

Memory The ability to recall thoughts; commonly classified as short-term (activated) and long-term.

Menarche (me-NAR-kē) Beginning of the menstrual function.

Ménière's (men-YAIRZ) **syndrome** A type of labyrinthine disease characterized by fluctuating loss of hearing, vertigo, and tinnitus due to an increased amount of endolymph that enlarges the labyrinth.

Meninges (me-NIN-jēz) Three membranes covering the brain and spinal cord, called the dura mater, arachnoid, and pia mater. *Singular,* **meninx** (MEN-inks).

Meningitis (men-in-JĪ-tis) Inflammation of the meninges, most commonly the pia mater and arachnoid.

Menopause (MEN-ō-pawz) The termination of the menstrual cycles.

Menstrual (MEN-stroo-al) **cycle** A series of changes in the endometrium of a nonpregnant female that prepares the lining of the uterus to receive a fertilized ovum.

Menstruation (men'-stroo-Ā-shun) Periodic discharge of blood, tissue fluid, mucus, and epithelial cells that usually lasts for 5 days; caused by a sudden reduction in estrogens and progesterone. Also called the **menstrual phase** or **menses.**

Merocrine (MER-ō-krin) **gland**   A secretory cell that remains intact throughout the process of formation and discharge of the secretory product, as in the salivary and pancreatic glands.

Mesenchyme (MEZ-en-kīm)   An embryonic connective tissue from which all other connective tissues arise.

Mesentery (MEZ-en-ter'-ē)   A fold of peritoneum attaching the small intestine to the posterior abdominal wall.

Mesocolon (mez'-ō-KŌ-lon)   A fold of peritoneum attaching the colon to the posterior abdominal wall.

Mesoderm   The middle of the three primary germ layers that gives rise to connective tissues, blood and blood vessels, and muscles.

Mesothelium (mez'-ō-THĒ-lē-um)   The layer of simple squamous epithelium that lines serous cavities.

Mesovarium (mez'-ō-VAR-ē-um)   A short fold of peritoneum that attaches an ovary to the broad ligament of the uterus.

Metabolism (me-TAB-ō-lizm)   The sum of all the biochemical reactions that occur within an organism, including the synthetic (anabolic) reactions and decomposition (catabolic) reactions.

Metacarpus (met'-a-KAR-pus)   A collective term for the five bones that make up the palm of the hand.

Metaphase (MET-a-phāz)   The second stage of mitosis in which chromatid pairs line up on the metaphase plate of the cell.

Metaphysis (me-TAF-i-sis)   Growing portion of a bone.

Metaplasia (met'-a-PLĀ-zē-a)   The transformation of one cell into another.

Metarteriole (met'-ar-TĒ-rē-ōl)   A blood vessel that emerges from an arteriole, traverses a capillary network, and empties into a venule.

Metastasis (me-TAS-ta-sis)   The spread of cancer to surrounding tissues (local) or to other body sites (distant).

Metatarsus (met'-a-TAR-sus)   A collective term for the five bones located in the foot between the tarsals and the phalanges.

Micelle (mī-SEL)   A spherical aggregate of bile salts that dissolves fatty acids and monoglycerides so that they can be transported into small intestinal epithelial cells.

Microcephalus (mi-krō-SEF-a-lus)   An abnormally small head; premature closing of the anterior fontanel so that the brain has insufficient room for growth, resulting in mental retardation.

Microfilament (mī-krō-FIL-a-ment)   Rodlike, protein filament about 6 nm in diameter; comprises contractile units in muscle fibers (cells) and provides support, shape, and movement in nonmuscle cells.

Microglia (mī-krō-GLĒ-a)   Neuroglial cells that carry on phagocytosis. Also called **brain macrophages** (MAK-rō-fāj-ez).

Microphage (MĪK-rō-fāj)   Granular leukocyte that carries on phagocytosis, especially neutrophils and eosinophils.

Microtubule (mī-krō-TOOB-yool')   Cylindrical protein filament, ranging in diameter from 18 to 30 nm, consisting of the protein tubulin; provides support, structure, and transportation.

Microvilli (mī'-krō-VIL-ē)   Microscopic, fingerlike projections of the cell membranes of small intestinal cells that increase surface area for absorption.

Micturition (mik'-too-RISH-un)   The act of expelling urine from the urinary bladder. Also called **urination** (yoo-ri-NĀ-shun).

Midbrain   The part of the brain between the pons and the diencephalon. Also called the **mesencephalon** (mes'-en-SEF-a-lon).

Middle ear   A small, epithelial-lined cavity hollowed out of the temporal bone, separated from the external ear by the eardrum and from the internal ear by a thin bony partition containing the oval and round windows; extending across the middle ear are the three auditory ossicles. Also called the **tympanic** (tim-PAN-ik) **cavity.**

Midline   An imaginary vertical line that divides the body into equal left and right sides.

Midsagittal plane   A vertical plane through the midline of the body that divides the body or organs into *equal* right and left sides. Also called a **median plane.**

Milk ejection reflex   Contraction of alveolar cells to force milk into ducts of mammary glands, stimulated by oxytocin (OT), which is released from the posterior pituitary gland in response to suckling action. Also called the **milk let-down reflex.**

Mineral   Inorganic, homogeneous solid substance that may perform a function vital to life; examples include calcium, sodium, potassium, iron, phosphorus, and chlorine.

Mineralocorticoids (min'-er-al-ō-KOR-ti-koyds)   A group of hormones of the adrenal cortex.

Minimal volume   The volume of air in the lungs even after the thoracic cavity has been opened forcing out some of the residual volume.

Minute volume of respiration (MVR)   Total volume of air taken into the lungs per minute; about 6,000/ml.

Mitochondrion (mī'-tō-KON-drē-on)   A double-membraned organelle that plays a central role in the production of ATP; known as the "powerhouse" of the cell.

Mitosis (mī-TŌ-sis)   The orderly division of the nucleus of a cell that ensures that each new daughter nucleus has the same number and kind of chromosomes as the original parent nucleus. The process includes the replication of chromosomes and the distribution of the two sets of chromosomes into two separate and equal nuclei.

Mitotic spindle   Collective term for a football-shaped assembly of microtubules (nonkinetochore, kinetochore, and aster) that is responsible for the movement of chromosomes during cell division.

Mitral (MĪ-tral) **insufficiency**   Backflow of blood from the left ventricle into the left atrium due to a damaged mitral valve or ruptured chordae tendineae.

Mitral stenosis (ste-NŌ-sis)   Narrowing of the mitral valve by scar formation or a congenital defect.

Mitral valve prolapse (PRŌ-laps) or **MVP**   An inherited disorder in which a portion of a mitral valve is pushed back too far (prolapsed) during contraction due to expansion of the cusps and elongation of the chordae tendineae.

Mittelschmerz (MIT-el-shmerz)   Abdominopelvic pain that supposedly indicates the release of a secondary oocyte from the ovary.

Modality (mō-DAL i tē)   Any of the specific sensory entities, such as vision, smell, or taste.

Modiolus (mō-DĪ-ō'-lus)   The central pillar or column of the cochlea.

Mole   The weight, in grams, of the combined atomic weights of the atoms that comprise a molecule of a substance.

Molecule (MOL-e-kyool) The chemical combination of two or more atoms.

Monoclonal antibody (MAb) Antibody produced by in vitro clones of B cells hybridized with cancerous cells.

Monocyte (MON-ō-sīt′) A type of white blood cell characterized by agranular cytoplasm; the largest of the leukocytes.

Monounsaturated fat A fatty acid that contains one double covalent bond between its carbon atoms; it is not completely saturated with hydrogen atoms. Plentiful in triglycerides of olive and peanut oil.

Mons pubis (monz PYOO-bis) The rounded, fatty prominence over the pubic symphysis, covered by coarse pubic hair.

Morbid (MOR-bid) Diseased; pertaining to disease.

Morula (MOR-yoo-la) A solid mass of cells produced by successive cleavages of a fertilized ovum a few days after fertilization.

Motor area The region of the cerebral cortex that governs muscular movement, particularly the precentral gyrus of the frontal lobe.

Motor end plate Portion of the sarcolemma of a muscle fiber (cell) in close approximation with an axon terminal.

Motor neuron (NOO-ron) A neuron that conveys nerve impulses from the brain and spinal cord to effectors that may be either muscles or glands. Also called an **efferent neuron.**

Motor unit A motor neuron together with the muscle fibers (cells) it stimulates.

Mucin (MYOO-sin) A protein found in mucus.

Mucous (MYOO-kus) **cell** A unicellular gland that secretes mucus. Also called a **goblet cell.**

Mucous membrane A membrane that lines a body cavity that opens to the exterior. Also called the **mucosa** (myoo-KŌ-sa).

Mucus The thick fluid secretion of mucous glands and mucous membranes.

Multiple motor unit summation Type of summation in which stimuli occur at the same time but at different locations (different motor units).

Multiple sclerosis (skler-Ō-sis) Progressive destruction of myelin sheaths of neurons in the central nervous system, short-circuiting conduction pathways.

Mumps Inflammation and enlargement of the parotid glands accompanied by fever and extreme pain during swallowing.

Muscarinic (mus′-ka-RIN-ik) **receptor** Receptor found on all effectors innervated by parasympathetic postganglionic axons and some effectors innervated by sympathetic postganglionic axons; so named because the actions of acetylcholine (ACh) on such receptors are similar to those produced by muscarine.

Muscle An organ composed of one of three types of muscle tissue (skeletal, cardiac, or smooth), specialized for contraction to produce voluntary or involuntary movement of parts of the body.

Muscle action potential A stimulating impulse that travels along a sarcolemma and then into transverse tubules; it is generated by acetylcholine from synaptic vesicles which alters permeability of the sarcolemma to sodium ions (Na$^+$).

Muscle fatigue (fa-TĒG) Inability of a muscle to maintain its strength of contraction or tension; may be related to insufficient oxygen, depletion of glycogen, and/or lactic acid buildup.

Muscle spindle An encapsulated receptor in a skeletal muscle, consisting of specialized muscle fiber (cell) and nerve endings, stimulated by changes in length or tension of muscle fibers; a proprioceptor. Also called a **neuromuscular** (noo-rō-MUS-kyoo-lar) **spindle.**

Muscle tissue A tissue specialized to produce motion in response to muscle action potentials by its qualities of contractility, extensibility, elasticity, and excitability. Types include skeletal, cardiac, and smooth.

Muscle tone A sustained, partial contraction of portions of a skeletal muscle in response to activation of stretch receptors.

Muscular dystrophies (DIS-trō-fēz′) Inherited muscle-destroying diseases, characterized by degeneration of the individual muscle fibers (cells), which leads to progressive atrophy of the skeletal muscle.

Muscularis (MUS-kyoo-la′-ris) A muscular layer (coat or tunic) of an organ.

Muscularis mucosae (myoo-KŌ-sē) A thin layer of smooth muscle fibers (cells) located in the outermost layer of the mucosa of the gastrointestinal tract, underlying the lamina propria of the mucosa.

Mutation (myoo-TĀ-shun) Any change in the sequence of bases in the DNA molecule resulting in a permanent alteration in some inheritable characteristic.

Myasthenia (mī-as-THĒ-nē-a) **gravis** Weakness of skeletal muscles caused by antibodies directed against acetylcholine receptors that inhibit muscle contraction.

Myelin (MĪ-e-lin) **sheath** Multilayered lipid and protein covering, formed by neurolemmocytes (Schwann cells) and oligodendrocytes around axons of many peripheral and central nervous system neurons.

Myelography (mī-e-LOG-ra-fē) Introduction of a contrast medium into the subarachnoid space of the spinal cord to demonstrate tumors or herniated (slipped) discs within or near the spinal cord.

Myenteric plexus A network of nerve fibers from both autonomic divisions located in the muscularis coat of the small intestine. Also called the **plexus of Auerbach** (OW-er-bak).

Myocardial infarction (mī′-ō-KAR-dē-al in-FARK-shun) **(MI)** Gross necrosis of myocardial tissue due to interrupted blood supply. Also called a **heart attack.**

Myocardium (mī′-ō-KAR-dē-um) The middle layer of the heart wall, made up of cardiac muscle tissue, comprising the bulk of the heart, and lying between the epicardium and the endocardium.

Myofibril (mī′-ō-FĪ-bril) A threadlike structure, running longitudinally through a muscle fiber (cell) consisting mainly of thick filaments (myosin) and thin filaments (actin, troponin, and tropomyosin).

Myoglobin (mī-ō-GLŌ-bin) The oxygen-binding, iron-containing conjugated protein complex present in the sarcoplasm of muscle fibers (cells); contributes the red color to muscle.

Myogram (MĪ-ō-gram) The record or tracing produced by the myograph, the apparatus that measures and records the effects of muscular contractions.

Myology (mī-OL-ō-jē) The study of the muscles.

Myometrium (mī′-ō-MĒ-trē-um) The smooth muscle layer of the uterus.

Myopia (mī-Ō-pē-a) Defect in vision so that objects can be seen distinctly only when very close to the eyes; nearsightedness.

Myosin (MĪ-ō-sin) The contractile protein that makes up the thick filaments of muscle fibers (cells).

Myotonia (mī-ō-TŌ-nē-a) A continuous spasm of muscle; increased muscular irritability and tendency to contract, and less ability to relax.

Myxedema (mix-e-DĒ-ma) Condition caused by hypothyroidism during the adult years characterized by swelling of facial tissues.

Nail A hard plate, composed largely of keratin, that develops from the epidermis of the skin to form a protective covering on the dorsal surface of the distal phalanges of the fingers and toes.

Nail matrix (MĀ-triks) The part of the nail beneath the body and root from which the nail is produced.

Nasal (NĀ-zal) **cavity** A mucosa-lined cavity on either side of the nasal septum that opens onto the face at an external naris and into the nasopharynx at an internal naris.

Nasal septum (SEP-tum) A vertical partition composed of bone (perpendicular plate of ethmoid and vomer) and cartilage, covered with a mucous membrane, separating the nasal cavity into left and right sides.

Nasolacrimal (nā′-zō-LAK-ri-mal) **duct** A canal that transports the lacrimal secretion (tears) from the nasolacrimal sac into the nose.

Nasopharynx (nā′-zō-FAR-inks) The uppermost portion of the pharynx, lying posterior to the nose and extending down to the soft palate.

Nausea (NAW-sē-a) Discomfort characterized by loss of appetite and sensation of impending vomiting.

Nebulization (neb′-yoo-li-ZĀ-shun) Administration of medication to selected portions of the respiratory tract by droplets suspended in air.

Neck The part of the body connecting the head and the trunk. A constricted portion of an organ such as the neck of the femur or uterus.

Necrosis (ne-KRŌ-sis) Death of a cell or group of cells as a result of disease or injury.

Negative feedback The principle governing most control systems; a mechanism of response in which a stimulus initiates actions that reverse or reduce the stimulus.

Neonatal (nē′-ō-NĀ-tal) Pertaining to the first 4 weeks after birth.

Neoplasm (NĒ-ō-plazm) A new growth that may be benign or malignant.

Nephritis (ne-FRĪT-is) Inflammation of the kidney.

Nephron (NEF-ron) The functional unit of the kidney.

Nephrotic (ne-FROT-ik) **syndrome** A condition in which the endothelial-capsular membrane leaks, allowing large amounts of protein to escape into urine.

Nerve A cordlike bundle of nerve fibers (axons and/or dendrites) and its associated connective tissue coursing together outside the central nervous system.

Nerve impulse A wave of depolarization and repolarization that self-propagates along the plasma membrane of a neuron; also called a **nerve action potential.**

Nervous tissue Tissue that initiates and transmits nerve impulses to coordinate homeostasis.

Net filtration pressure (NFP) Net pressure that expresses the relationship between the forces that promote fluid movement at the arterial and venous ends of a capillary and that promote glomerular filtration in the kidneys.

Neuralgia (noo-RAL-jē-a) Attacks of pain along the entire course or branch of a peripheral sensory nerve.

Neural plate A thickening of ectoderm that forms early in the third week of development and represents the beginning of the development of the nervous system.

Neuritis (noo-RĪ-tis) Inflammation of a single nerve, two or more nerves in separate areas, or many nerves simultaneously.

Neuroeffector (noo-rō-e-FEK-tor) **junction** Collective term for neuromuscular and neuroglandular junctions.

Neurofibral (noo-rō-FĪ-bral) **node** A space, along a myelinated nerve fiber, between the individual neurolemmocytes (Schwann cells) that form the myelin sheath and the neurolemma. Also called **node of Ranvier** (ron-VĒ-ā).

Neurofibril (noo-rō-FĪ-bril) One of the delicate threads that forms a complicated network in the cytoplasm of the cell body and processes of a neuron.

Neuroglandular (noo-rō-GLAND-yoo-lar) **junction** Area of contact between a motor neuron and a gland.

Neuroglia (noo-RŌG-lē-a) Cells of the nervous system that are specialized to perform the functions of connective tissue. The neuroglia of the central nervous system are the astrocytes, oligodendrocytes, microglia, and ependymal cells; neuroglia of the peripheral nervous system include the neurolemmocytes (Schwann cells) and the satellite cells. Also called **glial** (GLĒ-al) **cells.**

Neurohypophyseal (noo′-rō-hī′-po-FIZ-ē-al) **bud** An outgrowth of ectoderm located on the floor of the hypothalamus that gives rise to the posterior pituitary gland.

Neurolemma (noo-rō-LEM-ma) The peripheral, nucleated cytoplasmic layer of the neurolemmocyte (Schwann cell). Also called **sheath of Schwann** (SCHVON).

Neurolemmocyte A neuroglial cell of the peripheral nervous system that forms the myelin sheath and neurolemma of a nerve fiber by wrapping around a nerve fiber in a jelly-roll fashion. Also called a **Schwann** (SCHVON) **cell.**

Neurology (noo-ROL-ō-jē) The branch of science that deals with the normal functioning and disorders of the nervous system.

Neuromuscular (noo-rō-MUS-kyoo-lar) **junction** The area of contact between the axon terminal of a motor neuron and a portion of the sarcolemma of a muscle fiber (cell). Also called a **myoneural** (mī-ō-NOO-ral) **junction.**

Neuron (NOO-ron) A nerve cell, consisting of a cell body, dendrites, and an axon.

Neuropeptide (noo rō-PEP-tīd) Chain of 2 to about 40 amino acids that occurs naturally in the brain, and that acts primarily to modulate the response of or to a neurotransmitter. Examples are enkephalins and endorphins.

Neurosecretory (noo-rō-SĒC-re-tō-rē) **cell** A cell in a nucleus (paraventricular and supraoptic) in the hypothalamus that produces oxytocin (OT) or antidiuretic hormone (ADH), hormones stored in the posterior pituitary gland.

Neurosyphilis (noo-rō-SIF-i-lis)   A form of the tertiary stage of syphilis in which various types of nervous tissue are attacked by bacteria and degenerate.

Neurotransmitter   One of a variety of molecules synthesized within the nerve axon terminals, released into the synaptic cleft in response to a nerve impulse, and affecting the membrane potential of the postsynaptic neuron. Also called a **transmitter substance.**

Neutral fat   *See* **Triglyceride.**

Neutrophil (NOO-trō-fil)   A type of white blood cell characterized by granular cytoplasm that stains as readily with acid or basic dyes.

Nicotinic (nik′-ō-TIN-ik) **receptor**   Receptor found on both sympathetic and parasympathetic postganglionic neurons, so named because the actions of acetylcholine (ACh) in such receptors are similar to those produced by nicotine.

Night blindness   Poor or no vision in dim light or at night, although good vision is present during bright illumination; frequently caused by a deficiency of vitamin A. Also referred to as **nyctalopia** (nik′-ta-LŌ-pē-a).

Nipple   A pigmented, wrinkled projection on the surface of the mammary gland that is the location of the openings of the lactiferous ducts for milk release.

Nissl bodies   *See* **Chromatophilic substance.**

Nociceptor (nō′-sē-SEP-tor)   A free (naked) nerve ending that detects pain.

Node of Ranvier   *See* **Neurofibral node.**

Nondisjunction (non′-dis-JUNGK-shun)   Failure of sister chromatids to separate properly during anaphase of mitosis (or equatorial division of meiosis) or failure of homologous chromosomes to separate properly during reduction division of meiosis in which chromatids or chromosomes pass into the same daughter cell.

Nonessential amino acid   An amino acid that can be synthesized by body cells through transamination, the transfer of an amino group from an amino acid to another substance.

Norepinephrine (nor′-ep-ē-NEF-rin) **(NE)**   A hormone secreted by the adrenal medulla that produces actions similar to those that result from sympathetic stimulation. Also called **noradrenaline** (nor-a-DREN-a-lin).

Notochord (NŌ-tō-cord)   A flexible rod of embryonic tissue that lies where the future vertebral column will develop.

Nuclear medicine   The branch of medicine concerned with the use of radioisotopes in the diagnosis of disease and therapy.

Nuclease (NOO-klē-ās)   An enzyme that breaks nucleotides into pentoses and nitrogenous bases; examples are ribonuclease and deoxyribonuclease.

Nucleic (noo-KLĒ-ic) **acid**   An organic compound that is a long polymer of nucleotides, with each nucleotide containing a pentose sugar, a phosphate group, and one of four possible nitrogenous bases (adenine, cytosine, guanine, and thymine or uracil).

Nucleolus (noo-KLĒ-ō-lus)   Nonmembranous spherical body within the nucleus composed of protein, DNA, and RNA that functions in the synthesis and storage of ribosomal RNA.

Nucleosome (NOO-klē-ō-sōm)   Elementary structural subunit of a chromosome consisting of histones and DNA.

Nucleus (NOO-klē-us)   A spherical or oval organelle of a cell that contains the hereditary factors of the cell, called genes. A cluster of unmyelinated nerve cell bodies in the central nervous system. The central portion of an atom made up of protons and neutrons.

Nucleus cuneatus (kyoo-nē-Ā-tus)   A group of nerve cells in the inferior portion of the medulla in which fibers of the fasciculus cuneatus terminate.

Nucleus gracilis (gras-I-lis)   A group of nerve cells in the inferior portion of the medulla in which fibers of the fasciculus gracilis terminate.

Nucleus pulposus (pul-PŌ-sus)   A soft, pulpy, highly elastic substance in the center of an intervertebral disc, a remnant of the notochord.

Nutrient   A chemical substance in food that provides energy, forms new body components, or assists in the functioning of various body processes.

Nystagmus (nis-TAG-mus)   Rapid, involuntary, rhythmic movement of the eyeballs; horizontal, rotary, or vertical.

Obesity (ō-BĒS-i-tē)   Body weight 10–20 percent over a desirable standard as a result of excessive accumulation of fat. Types of obesity are hypertrophic (adult-onset) and hyperplastic (lifelong).

Obstetrics (ob-STET-riks)   The specialized branch of medicine that deals with pregnancy, labor, and the period of time immediately following delivery.

Occlusion (ō-KLOO-zhun)   The act of closure or state of being closed.

Occult (o-KULT)   Obscure or hidden from view, as for example, occult blood in stools or urine.

Olfactory (ōl-FAK-tō-rē)   Pertaining to smell.

Olfactory bulb   A mass of gray matter at the termination of an olfactory (I) nerve, lying beneath the frontal lobe of the cerebrum on either side of the crista galli of the ethmoid bone.

Olfactory receptor cell   A bipolar neuron with its cell body lying between supporting cells located in the mucous membrane lining the upper portion of each nasal cavity; transduces odors into neural signals.

Olfactory tract   A bundle of axons that extends from the olfactory bulb posteriorly to the olfactory portion of the cortex.

Oligodendrocyte (o-lig-ō-DEN-drō-sīt)   A neuroglial cell that supports neurons and produces a myelin sheath around axons of neurons of the central nervous system.

Oligospermia (ol′-i-gō-SPER-mē-a)   A deficiency of spermatozoa in the semen.

Oliguria (ol′-i-GYOO-rē-a)   Daily urinary output usually less than 250 ml.

Olive   A prominent oval mass on each lateral surface of the superior part of the medulla.

Oncogene (ONG-kō-jēn)   Gene that has the ability to transform a normal cell into a cancerous cell when it is inappropriately activated.

Oncology (ong-KOL-ō-jē)   The study of tumors.

Oogenesis (ō′-ō-JEN-e-sis)   Formation and development of the ovum.

Oophorectomy (ō′-of-ō-REK-tō-mē)   The surgical removal of the ovaries.

Ophthalmic (of-THAL-mik)   Pertaining to the eye.

Ophthalmologist (of′-thal-MOL-ō-jist)   A physician who specializes in the diagnosis and treatment of eye disorders with drugs, surgery, and corrective lenses.

Ophthalmology (of′-thal-MOL-ō-jē)   The study of the structure, function, and diseases of the eye.

Ophthalmoscopy (of′-thal-MOS-co-pē)   Examination of the interior fundus of the eyeball to detect retinal changes associated with hypertension, diabetes mellitus, atherosclerosis, and increased intracranial pressure.

Opsin (OP-sin)   The glycoprotein portion of a photopigment.

Opsonization (op-sō-ni-ZĀ-shun)   The action of some antibodies that renders bacteria and other foreign cells more susceptible to phagocytosis. Also called **immune adherence.**

Optic (OP-tik)   Refers to the eye, vision, or properties of light.

Optic chiasma (kī-AZ-ma)   A crossing point of the optic (II) nerves, anterior to the pituitary gland.

Optic disc   A small area of the retina containing openings through which the fibers of the ganglion neurons emerge as the optic (II) nerve. Also called the **blind spot.**

Optician (op-TISH-an)   A technician who fits, adjusts, and dispenses corrective lenses on prescription of an ophthalmologist or optometrist.

Optic tract   A bundle of axons that transmits nerve impulses from the retina of the eye between the optic chiasma and the thalamus.

Optometrist (op-TOM-e-trist)   Specialist with a doctorate degree in optometry who is licensed to examine and test the eyes and treat visual defects by prescribing corrective lenses.

Oral cholecystogram (kō-lē-SIS-to-gram)   X-ray examination of the gallbladder to evaluate for the presence of gallstones, inflammations, and tumors.

Oral contraceptive (OC)   A hormone compound, usually a high concentration of progesterone and a low concentration of estrogens, that is swallowed and prevents ovulation, and thus pregnancy. Also called **"the pill."**

Ora serrata (Ō-ra ser-RĀ-ta)   The irregular margin of the retina lying internal and slightly posterior to the junction of the choroid and ciliary body.

Orbit (OR-bit)   The bony, pyramid-shaped cavity of the skull that holds the eyeball.

Organ   A structure composed of two or more different kinds of tissues with a specific function and usually a recognizable shape.

Organelle (or-gan-EL)   A permanent structure within a cell with characteristic morphology that is specialized to serve a specific function in cellular activities.

Organic (or-GAN-ik) **compound**   Compound that always contains carbon and hydrogen and the atoms are held together by covalent bonds. Examples include carbohydrates, lipids, proteins, and nucleic acids (DNA and RNA).

Organism (OR-ga-nizm)   A total living form; one individual.

Orgasm (OR-gazm)   Sensory and motor events involved in ejaculation for the male and involuntary contraction of the perineal muscles in the female at the climax of sexual intercourse.

Orifice (OR-i-fis)   Any aperture or opening.

Origin (OR-i-jin)   The attachment of a muscle tendon to a stationary bone or the end opposite the insertion.

Oropharynx (or′-ō-FAR-inks)   The second portion of the pharynx, lying posterior to the mouth and extending from the soft palate down to the hyoid bone.

Orthopedics (or′-thō-PĒ-diks)   The branch of medicine that deals with the preservation and restoration of the skeletal system, articulations, and associated structures.

Orthopnea (or′-THOP-nē-a)   Dyspnea that occurs in the horizontal position.

Osmoreceptor (oz′-mō-re-CEP-tor)   Receptor in the hypothalamus that is sensitive to changes in blood osmotic pressure and, in response to high osmotic pressure (low water concentration), causes synthesis and release of antidiuretic hormone (ADH).

Osmosis (os-MŌ-sis)   The net movement of water molecules through a selectively permeable membrane from an area of high water concentration to an area of lower water concentration until an equilibrium is reached.

Osmotic pressure   The pressure required to prevent the movement of pure water into a solution containing solutes when the solutions are separated by a selectively permeable membrane.

Osseous (OS-ē-us)   Bony.

Ossicle (OS-si-kul)   Small bone, as in the middle ear (malleus, incus, stapes).

Ossification (os′-i-fi KĀ-shun)   Formation of bone. Also called **osteogenesis.**

Osteoblast (OS-tē-ō-blast)   Cell formed from an osteoprogenitor cell that participates in bone formation by secreting some organic components and inorganic salts.

Osteoclast (OS-tē-ō-clast′)   A large multinuclear cell that develops from a monocyte and destroys or resorbs bone tissue.

Osteocyte (OS-tē-ō-sīt′)   A mature bone cell that maintains the daily activities of bone tissue.

Osteogenic (os′-tē-ō-JEN-ik) **layer**   The inner layer of the periosteum that contains cells responsible for forming new bone during growth and repair.

Osteology (os′-tē-OL-ō-jē)   The study of bones.

Osteomalacia (os′-tē-ō-ma-LĀ-shē-a)   A deficiency of vitamin D in adults causing demineralization and softening of bone.

Osteomyelitis (os′tē-ō-mī-i-LĪ-tis)   Inflammation of bone marrow or of the bone and marrow.

Osteon (OS-tē-on)   The basic unit of structure in adult compact bone, consisting of a central (Haversian) canal with its concentrically arranged lamellae, lacunae, osteocytes, and canaliculi. Also called a **Haversian** (ha-VĒR-shun) **system.**

Osteoporosis (os′-tē-ō-pō-RŌ-sis)   Age-related disorder characterized by decreased bone mass and increased susceptibility to fractures as a result of decreased levels of estrogens.

Osteoprogenitor (os′-tē-ō-prō-JEN-i-tor) **cell**   Stem cell derived from mesenchyme that has mitotic potential and the ability to differentiate into an osteoblast.

Otalgia (ō-TAL-jē-a)   Pain in the ear; earache.

Otic (Ō-tik)   Pertaining to the ear.

Otitis media (ō-tī-tus MĒ-dē-a) Acute infection of the middle ear characterized by pain, malaise, fever, and an inflamed tympanic membrane, subject to rupture.

Otolith (Ō-tō-lith) A particle of calcium carbonate embedded in the otolithic membrane that functions in maintaining static equilibrium.

Otolithic (ō-tō-LITH-ik) **membrane** Thick, gelatinous, glycoprotein layer located directly over hair cells of the macula in the saccule and utricle of the inner ear.

Otorhinolaryngology (ō′-tō-rī-nō-lar′-in-GOL-ō-jē) The branch of medicine that deals with the diagnosis and treatment of diseases of the ears, nose, and throat.

Oval window A small opening between the middle ear and inner ear into which the footplate of the stapes fits. Also called the **fenestra vestibuli** (fe-NES-tra ves-TIB-yoo-lē).

Ovarian (ō-VAR-ē-an) **cycle** A monthly series of events in the ovary associated with the maturation of an ovum.

Ovarian follicle (FOL-i-kul) A general name for oocytes (immature ova) in any stage of development, along with their surrounding epithelial cells.

Ovarian ligament (LIG-a-ment) A rounded cord of connective tissue that attaches the ovary to the uterus.

Ovary (Ō-var-ē) Female gonad that produces ova and the hormones estrogens, progesterone, and relaxin.

Ovulation (ō-vyoo-LĀ-shun) The rupture of a vesicular ovarian (Graafian) follicle with discharge of a secondary oocyte into the pelvic cavity.

Ovum (Ō-vum) The female reproductive or germ cell; an egg cell.

Oxidation (ok-si-DĀ-shun) The removal of electrons and hydrogen ions (hydrogen atoms) from a molecule or, less commonly, the addition of oxygen to a molecule that results in a decrease in the energy content of the molecule. The oxidation of glucose in the body is also called **cellular respiration.**

Oxygen debt *See* **Recovery oxygen consumption.**

Oxyhemoglobin (ok′-sē-HĒ-mō-glō-bin) **(Hb-O2)** Hemoglobin combined with oxygen.

Oxyphil cell A cell found in the parathyroid gland that secretes parathyroid hormone (PTH).

Oxytocin (ok′-sē-TŌ-sin) **(OT)** A hormone secreted by neurosecretory cells in the paraventricular and supraoptic nuclei of the hypothalamus that stimulates contraction of the smooth muscle fibers (cells) in the pregnant uterus and contractile cells around the ducts of mammary glands.

Pacinian corpuscle *See* **Lamellated corpuscle.**

Paget's (PAJ-ets) **disease** A disorder characterized by a greatly accelerated remodeling process in which osteoclastic resorption is massive and new bone formation by osteoblasts is extensive. As a result, there is an irregular thickening and softening of the bones.

Palate (PAL-at) The horizontal structure separating the oral and the nasal cavities; the roof of the mouth.

Palliative (PAL-ē-a-tiv) Serving to relieve or alleviate without curing.

Palpate (PAL-pāt) To examine by touch; to feel.

Palpitation (pal′-pi-TĀ-shun) A fluttering of the heart or abnormal rate or rhythm of the heart.

Pancreas (PAN-krē-as) A soft, oblong organ lying along the greater curvature of the stomach and connected by a duct to the duodenum. It is both exocrine (secreting pancreatic juice) and endocrine (secreting insulin, glucagon, growth hormone inhibiting hormone, and pancreatic polypeptide).

Pancreatic (pan′-krē-AT-ik) **duct** A single, large tube that unites with the common bile duct from the liver and gallbladder and drains pancreatic juice into the duodenum at the hepatopancreatic ampulla (ampulla of Vater). Also called the **duct of Wirsung.**

Pancreatic islet A cluster of endocrine gland cells in the pancreas that secretes insulin, glucagon, growth hormone inhibiting hormone, and pancreatic polypeptide. Also called an **islet of Langerhans** (LANG-er-hanz).

Papanicolaou (pap′-a-NIK-ō-la-oo) **test** A cytological staining test for the detection and diagnosis of premalignant and malignant conditions of the female genital tract. Cells scraped from the genital epithelium are smeared, fixed, stained, and examined microscopically. Also called a **Pap smear.**

Papilla (pa-PIL-a) A small nipple-shaped projection or elevation.

Paracrine (pa-RA-krin) Local hormone, such as histamine, that acts on neighboring cells.

Paralysis (pa-RAL-a-sis) Loss or impairment of motor function due to a lesion of nervous or muscular origin.

Paranasal sinus (par′-a-NĀ-zal SĪ-nus) A mucus-lined air cavity in a skull bone that communicates with the nasal cavity. Paranasal sinuses are located in the frontal, maxillary, ethmoid, and sphenoid bones.

Paraplegia (par-a-PLĒ-jē-a) Paralysis of both lower extremities.

Parasagittal plane A vertical plane that does not pass through the midline and that divides the body or organs into *unequal* left and right portions.

Parasympathetic (par′-a-sim-pa-THET-ik) **division** One of the two subdivisions of the autonomic nervous system, having cell bodies of preganglionic neurons in nuclei in the brain stem and in the lateral gray matter of the sacral portion of the spinal cord; primarily concerned with activities that conserve and restore body energy. Also called the **craniosacral** (krā-nē-ō-SĀ-kral) **division.**

Parathyroid (par′-a-THĪ-royd) **gland** One of four small endocrine glands embedded on the posterior surfaces of the lateral lobes of the thyroid gland.

Parathyroid hormone (PTH) A hormone secreted by the parathyroid glands that decreases blood phosphate level and increases blood calcium level.

Paraurethral (par′-a-yoo-RĒ-thral) **gland** Gland embedded in the wall of the urethra whose duct opens on either side of the urethral orifice and secretes mucus. Also called **Skene's** (SKĒNZ) **gland.**

Parenchyma (par-EN-ki-ma) The functional parts of any organ, as opposed to tissue that forms its stroma or framework.

Parenteral (par-EN-ter-al) Situated or occurring outside the intestines; referring to introduction of substances into the body other than by way of the intestines such as intradermal, subcutaneous, intramuscular, intravenous, or intraspinal.

Parietal (pa-RĪ-e-tal)   Pertaining to or forming the outer wall of a body cavity.

Parietal cell   The secreting cell of a gastric gland that produces hydrochloric acid and intrinsic factor. Also called an **oxyntic cell.**

Parietal pleura (PLOO-ra)   The outer layer of the serous pleural membrane that encloses and protects the lungs; the layer that is attached to the wall of the pleural cavity.

Parkinson's disease   Progressive degeneration of the basal ganglia and substantia nigra of the cerebrum resulting in decreased production of dopamine (DA) that leads to tremor, slowing of voluntary movements, and muscle weakness. Also called **Parkinsonism.**

Parotid (pa-ROT-id) **gland**   One of the paired salivary glands located inferior and anterior to the ears connected to the oral cavity via a duct (Stensen's) that opens into the inside of the cheek opposite the upper second molar tooth.

Paroxysm (PAR-ok-sizm)   A sudden periodic attack or recurrence of symptoms of a disease.

Pars intermedia   A small avascular zone between the anterior and posterior pituitary glands.

Parturition (par′-too-RISH-un)   Act of giving birth to young; childbirth, delivery.

Patellar (pa-TELL-ar) **reflex**   Extension of the leg by contraction of the quadriceps femoris muscle in response to tapping the patellar ligament. Also called the **knee jerk.**

Patent ductus arteriosus   Congenital anatomical heart defect in which the fetal connection between the aorta and pulmonary trunk remains open instead of closing completely after birth.

Pathogen (PATH-o-jen)   A disease-producing organism.

Pathogenesis (path′-ō-JEN-e-sis)   The development of disease or a morbid or pathological state.

Pathological (path′-ō-LOJ-i-kal)   Pertaining to or caused by disease.

Pathological (path′-ō-LOJ-i-kal) **anatomy**   The study of structural changes caused by disease.

Pectinate (PEK-ti-nāt) **muscles**   Projecting muscle bundles of the anterior atrial walls and the lining of the auricles.

Pectoral (PEK-tō-ral)   Pertaining to the chest or breast.

Pediatrician (pē′-dē-a-TRISH-un)   A physician who specializes in the care and treatment of children and their illnesses.

Pedicel (PED-i-sel)   Footlike structure, as on podocytes of a glomerulus.

Pelvic (PEL-vik) **cavity**   Inferior portion of the abdominopelvic activity that contains the urinary bladder, sigmoid colon, rectum, and internal female and male reproductive structures.

Pelvic inflammatory disease (PID)   Collective term for any extensive bacterial infection of the pelvic organs, especially the uterus, uterine (Fallopian) tubes, and ovaries.

Pelvic splanchnic (PEL-vik SPLANGK-nik) **nerves**   Preganglionic parasympathetic fibers from the levels of S2, S3, and S4 that supply the urinary bladder, reproductive organs, and the descending and sigmoid colon and rectum.

Pelvimetry (pel-VIM-e-trē)   Measurement of the size of the inlet and outlet of the birth canal.

Pelvis   The basinlike structure formed by the two hipbones, the sacrum, and the coccyx. The expanded, proximal portion of the

ureter, lying within the kidney and into which the major calyces open.

Penis (PĒ-nis)   The male copulatory organ, used to introduce spermatozoa into the female vagina.

Pepsin   Protein-digesting enzyme secreted by chief (zymogenic) cells of the stomach as the inactive form pepsinogen, which is converted to active pepsin by hydrochloric acid.

Peptic ulcer   An ulcer that develops in areas of the gastrointestinal tract exposed to hydrochloric acid; classified as a gastric ulcer if in the lesser curvature of the stomach and as a duodenal ulcer if in the first part of the duodenum.

Percussion (per-KUSH-un)   The act of striking (percussing) an underlying part of the body with short, sharp blows as an aid in diagnosing the part by the quality of the sound produced.

Perforating canal   A minute passageway by means of which blood vessels and nerves from the periosteum penetrate into compact bone. Also called **Volkmann's** (FŌLK-manz) **canal.**

Pericardial (per′-i-KAR-dē-al) **cavity**   Small potential space between the visceral and parietal layers of the serous pericardium that contains pericardial fluid.

Pericardium (per′-i-KAR-dē-um)   A loose-fitting membrane that encloses the heart, consisting of an outer fibrous layer and an inner serous layer.

Perichondrium (per′-i-KON-drē-um)   The membrane that covers cartilage.

Perikaryon (per′-i-KAR-ē-on)   The nerve cell body that contains the nucleus and other organelles. Also called a **soma.**

Perilymph (PER-i-lymf)   The fluid contained between the bony and membranous labyrinths of the inner ear.

Perimetrium (per-i-MĒ-trē-um)   The serosa of the uterus.

Perimysium (per′-i-MIZ-ē-um)   Invagination of the epimysium that divides muscles into bundles.

Perineum (per′-i-NĒ-um)   The pelvic floor; the space between the anus and the scrotum in the male and between the anus and the vulva in the female.

Perineurium (per′-i-NYOO-rē-um)   Connective tissue wrapping around fascicles in a nerve.

Periodontal (per-ē-ō-DON-tal) **disease**   A collective term for conditions characterized by degeneration of gingivae, alveolar bone, periodontal ligament, and cementum.

Periodontal ligament   The periosteum lining the alveoli (sockets) for the teeth in the alveolar processes of the mandible and maxillae.

Periosteum (per′-ē-OS-tē-um)   The membrane that covers bone and consists of connective tissue, osteoprogenitor cells, and osteoblasts and is essential for bone growth, repair, and nutrition.

Peripheral (pe-RIF-er-al)   Located on the outer part or a surface of the body.

Peripheral nervous system (PNS)   The part of the nervous system that lies outside the central nervous system—nerves and ganglia.

Periphery (pe-RIF-er-ē)   Outer part or a surface of the body; part away from the center.

Peristalsis (per′-i-STAL-sis)   Successive muscular contractions along the wall of a hollow muscular structure.

Peritoneum (per′-i-tō-NĒ-um)   The largest serous membrane of the body that lines the abdominal cavity and covers the viscera.

Peritonitis (per'-i-tō-NĪ-tis)   Inflammation of the peritoneum.

Permissive (per-MIS-sive) **effect**   A hormonal interaction in which the effect of one hormone on a target cell requires previous or simultaneous exposure to another hormone(s) to enhance the response of a target cell or increase the activity of another hormone. Exposure of the uterus first to estrogens and then progesterone in preparation for implantation is an example.

Pernicious (per-NISH-us)   Tending to cause death.

Peroxisome (pe-ROKS-ī-sōm)   Organelle similar in structure to a lysosome that contains enzymes that use molecular oxygen to oxidize various organic compounds. Such reactions produce hydrogen peroxide; abundant in liver cells.

Perspiration   Substance produced by sudoriferous (sweat) glands containing water, salts, urea, uric acid, amino acids, ammonia, sugar, lactic acid, and ascorbic acid; helps maintain body temperature and eliminate wastes.

Peyer's patches   *See* **Aggregated lymphatic follicles.**

pH   A symbol of the measure of the concentration of hydrogen ions (H^+) in a solution. The pH scale extends from 0 to 14, with a value of 7 expressing neutrality, values lower than 7 expressing increasing acidity, and values higher than 7 expressing increasing alkalinity.

Phagocytosis (fag'-ō-sī-TŌ-sis)   The process by which cells (phagocytes) ingest particulate matter; especially the ingestion and destruction of microbes, cell debris, and other foreign matter.

Phalanx (FĀ-lanks)   The bone of a finger or toe. *Plural,* **phalanges** (fa-LAN-jēz).

Phantom pain   A sensation of pain as originating in a limb that has been amputated.

Pharmacology (far'-ma-KOL-ō-jē)   The science that deals with the effects and uses of drugs in the treatment of disease.

Pharynx (FAR-inks)   The throat; a tube that starts at the internal nares and runs partway down the neck where it opens into the esophagus posteriorly and the larynx anteriorly.

Phenotype (FĒ-nō-tīp)   The observable expression of genotype; physical characteristics of an organism determined by genetic makeup and influenced by interaction between genes and internal and external environmental factors.

Phenylketonuria (fen'-il-kē'-tō-NOO-rē-a) **(PKU)**   A disorder characterized by an elevation of the amino acid phenylalanine in the blood.

Pheochromocytoma (fē-ō-krō'-mō-sī-TŌ-ma)   Tumor of the chromaffin cells of the adrenal medulla that results in hypersecretion of medullary hormones.

Phlebitis (fle-BĪ-tis)   Inflammation of a vein, usually in the lower extremities.

Phospholipid (fos'-fō-LIP-id) **bilayer**   Arrangement of phospholipid molecules in two parallel rows in which the hydrophilic "heads" face outward and the hydrophobic "tails" face inward.

Phosphorylation (fos'-for-i-LĀ-shun)   The addition of a phosphate group to a chemical compound; types include substrate-level, oxidative, and photophosphorylation.

Photopigment   A substance that can absorb light and undergo structural changes that can lead to the development of a receptor potential. An example is rhodopsin. Also called **visual pigment.**

Photoreceptor   Receptor that detects light on the retina of the eye.

Physiology (fiz'-ē-OL-ō-jē)   Science that deals with the functions of an organism or its parts.

Pia mater (PĪ-a MĀ-ter)   The inner membrane (meninx) covering the brain and spinal cord.

Piezoelectric (pē-e-zō-e-LEK-trik) **effect**   Response of bone, mainly collagen, to stress in which very minute currents of electricity are produced; believed to stimulate osteoblasts to make new bone cells.

Pineal (PĪN-ē-al) **gland**   The cone-shaped gland located in the roof of the third ventricle. Also called the **epiphysis cerebri** (ē-PIF-i-sis se-RĒ-brē).

Pinealocyte (pin-ē-AL-ō-sīt)   Secretory cell of the pineal gland that produces hormones.

Pinna (PIN-na)   The projecting part of the external ear composed of elastic cartilage and covered by skin and shaped like the flared end of a trumpet. Also called the **auricle** (OR-i-kul).

Pinocytosis (pi'-nō-sī-TŌ-sis)   The process by which cells ingest liquid.

Pituicyte (pi-TOO-i-sīt)   Supporting cell of the posterior pituitary gland.

Pituitary (pi-TOO-i-tar'-ē) **dwarfism**   Condition caused by hyposecretion of human growth hormone (hGH) during the growth years and characterized by childlike physical traits in an adult.

Pituitary gland   A small endocrine gland lying in the sella turcica of the sphenoid bone and attached to the hypothalamus by the infundibulum. Also called the **hypophysis** (hī-POF-i-sis).

Pivot joint   A synovial joint in which a rounded, pointed, or conical surface of one bone articulates with a ring formed partly by another bone and partly by a ligament, as in the joint between the atlas and axis and between the proximal ends of the radius and ulna. Also called a **trochoid** (TRO-koid) **joint.**

Placenta (pla-SEN-ta)   The special structure through which the exchange of materials between fetal and maternal circulations occurs. Also called the **afterbirth.**

Plantar flexion (PLAN-tar FLEK-shun)   Bending the foot in the direction of the plantar surface (sole).

Plaque (plak)   A cholesterol-containing mass in the tunica media of arteries. A mass of bacterial cells, dextran (polysaccharide), and other debris that adheres to teeth.

Plasma (PLAZ-ma)   The extracellular fluid found in blood vessels; blood minus the formed elements.

Plasma cell   Cell that produces antibodies and develops from a B cell (lymphocyte).

Plasma (cell) membrane   Outer, limiting membrane that separates the cell's internal parts from extracellular fluid and the external environment.

Plasmapheresis (plaz'-ma-fe-RĒ-sis)   A procedure in which blood is withdrawn from the body, its components are selectively separated, the undesirable component causing disease is removed, and the remainder is returned to the body. Among the substances removed are toxins, metabolic substances, and antibodies. Also called **therapeutic plasma exchange (TPE).**

Platelet plug   Aggregation of thrombocytes at a damaged blood vessel to prevent blood loss.

Pleura (PLOOR-a) The serous membrane that covers the lungs and lines the walls of the chest and the diaphragm.

Pleural cavity Small potential space between the visceral and parietal pleurae.

Plexus (PLEK-sus) A network of nerves, veins, or lymphatic vessels.

Plexus of Auerbach *See* **Myenteric plexus.**

Plexus of Meissner *See* **Submucosal plexus.**

Pluripotent hematopoietic (hem′-a-tō-poy-Ē-tic) **stem cell** Immature stem cell in bone marrow that gives rise to precursors of all the different mature blood cells. Previously called a **hemocytoblast** (hē′-mō-SĪ-tō-blast).

Pneumonia (noo-MŌ-nē-a) Acute infection or inflammation of the alveoli of the lungs.

Pneumotaxic (noo-mō-TAK-sik) **area** Portion of the respiratory center in the pons that continually sends inhibitory nerve impulses to the inspiratory area that limit inspiration and facilitate expiration.

Podiatry (pō-DĪ-a-trē) The diagnosis and treatment of foot disorders.

Polar body The smaller cell resulting from the unequal division of cytoplasm during the meiotic divisions of an oocyte. The polar body has no function and is resorbed.

Polarized A condition in which opposite effects or states exist at the same time. In electrical contexts, having one portion negative and another positive; for example, a polarized nerve cell membrane has the outer surface positively charged and the inner surface negatively charged.

Poliomyelitis (pō′-lē-ō-mī-e-LĪ-tis) Viral infection marked by fever, headache, stiff neck and back, deep muscle pain and weakness, and loss of certain somatic reflexes; a serious form of the disease, **bulbar polio,** results in destruction of motor neurons in anterior horns of spinal nerves that leads to paralysis.

Polycythemia (pol′-ē-sı-THĒ-mē-a) Disorder characterized by a hematocrit above the normal level of 55 in which hypertension, thrombosis, and hemorrhage occur.

Polyp (POL-ip) A tumor on a stem found especially on a mucous membrane.

Polysaccharide (pol′-ē-SAK-a-rīd) A carbohydrate in which three or more monosaccharides are joined chemically.

Polyunsaturated fat A fatty acid that contains more than one double covalent bond between its carbon atoms; abundant in triglycerides of corn oil, safflower oil, and cottonseed oil.

Polyuria (pol′-ē-YOO-rē-a) An excessive production of urine.

Pons (ponz) The portion of the brain stem that forms a "bridge" between the medulla and the midbrain, anterior to the cerebellum.

Positron emission tomography (PET) A type of radioactive scanning based on the release of gamma rays when positrons collide with negatively charged electrons in body tissues; it indicates where radioisotopes are used in the body.

Postabsorptive (fasting) state Metabolic state during which absorption is complete and energy needs of the body must be satisfied.

Postcentral gyrus *See* **Primary somatosensory area.**

Posterior (pos-TĒR-ē-or) Nearer to or at the back of the body. Also called **dorsal.**

Posterior pituitary gland Posterior portion of the pituitary gland. Also called the **neurohypophysis** (noo-rō-hī-POF-i-sis).

Posterior root The structure composed of sensory fibers lying between a spinal nerve and the dorsolateral aspect of the spinal cord. Also called the **dorsal (sensory) root.**

Posterior root ganglion A group of cell bodies of sensory neurons and their supporting cells located along the posterior root of a spinal nerve. Also called a **dorsal (sensory) root ganglion** (GANG-glē-on).

Postganglionic neuron (pōst′-gang-lē-ON-ik NOO-ron) The second visceral motor neuron in an autonomic pathway, having its cell body and dendrites located in an autonomic ganglion and its unmyelinated axon ending at cardiac muscle, smooth muscle, or a gland.

Postpartum (pōst-PAR-tum) After parturition; occurring after the delivery of a baby.

Postsynaptic (pōst-sin-AP-tik) **neuron** The nerve cell that is activated by the release of a neurotransmitter substance from another neuron and carries nerve impulses away from the synapse.

Pouch of Douglas *See* **Rectouterine pouch.**

Precapillary sphincter (SFINGK-ter) A ring of smooth muscle fibers (cells) at the site of origin of true capillaries that regulate blood flow into true capillaries.

Precentral gyrus *See* **Primary motor area.**

Preeclampsia (prē′-e-KLAMP-sē-a) A syndrome characterized by sudden hypertension, large amounts of protein in urine, and generalized edema; it might be related to an autoimmune or allergic reaction due to the presence of a fetus.

Preganglionic (prē′-gang-lē-ON-ik) **neuron** The first visceral motor neuron in an autonomic pathway, with its cell body and dendrites in the brain or spinal cord and its myelinated axon ending at an autonomic ganglion, where it synapses with a postganglionic neuron.

Pregnancy Sequence of events that normally includes fertilization, implantation, embryonic growth, and fetal growth that terminates in birth.

Premenstrual syndrome (PMS) Severe physical and emotional stress occurring late in the postovulatory phase of the menstrual cycle and sometimes overlapping with menstruation.

Prepuce (PRĒ-pyoos) The loose-fitting skin covering the glans of the penis and clitoris. Also called the **foreskin.**

Presbyopia (prez-bē-Ō-pē-a) A loss of elasticity of the lens of the eye due to advancing age with resulting inability to focus clearly on near objects.

Pressure sore Tissue destruction due to a constant deficiency of blood to tissues overlying a bony projection that has been subjected to prolonged pressure against an object such as a bed, cast, or splint. Also called **bedsore, decubitus** (dē-KYOO-bi-tus) **ulcer,** or **trophic ulcer.**

Presynaptic (prē-sin-AP-tik) **inhibition** Inhibition of a nerve impulse before it reaches a synapse in which neurotransmitter released by an inhibitory neuron depresses the release of excitatory transmitter at an excitatory neuron.

Presynaptic (prē-sin-AP-tik) **neuron** A nerve cell that carries nerve impulses toward a synapse.

Prevertebral ganglion (prē-VERT-e-bral GANG-lē-on)   A cluster of cell bodies of postganglionic sympathetic neurons anterior to the spinal column and close to large abdominal arteries. Also called a **collateral ganglion.**

Primary germ layer   One of three layers of embryonic tissue, called ectoderm, mesoderm, and endoderm, that give rise to all tissues and organs of the organism.

Primary motor area   A region of the cerebral cortex in the precentral gyrus of the frontal lobe of the cerebrum that controls specific muscles or groups of muscles.

Primary somatosensory area   A region of the cerebral cortex posterior to the central sulcus in the postcentral gyrus of the parietal lobe of the cerebrum that localizes exactly the points of the body where somatic sensations originate.

Prime mover   The muscle directly responsible for producing the desired motion. Also called an **agonist** (AG-ō-nist).

Primigravida (prī-mi-GRAV-i-da)   A woman pregnant for the first time.

Primitive gut   Embryonic structure composed of endoderm and mesoderm that gives rise to most of the gastrointestinal tract.

Primordial (prī-MŌR-dē-al)   Existing first; especially primordial egg cells in the ovary.

Principal cell   Cell found in the parathyroid glands that secretes parathyroid hormone (PTH); cell type in the distal convoluted tubule and collecting duct of nephron that is sensitive to aldosterone and antidiuretic hormone.

Proctology (prok-TOL-ō-jē)   The branch of medicine that treats the rectum and its disorders.

Progeny (PROJ-e-nē)   Refers to offspring or descendants.

Progesterone (prō-JES-te-rōn) **(PROG)**   A female sex hormone produced by the ovaries that helps prepare the endometrium for implantation of a fertilized ovum and the mammary glands for milk secretion.

Prognosis (prog-NŌ-sis)   A forecast of the probable results of a disorder; the outlook for recovery.

Prolactin (prō-LAK-tin) **(PRL)**   A hormone secreted by the anterior pituitary gland that initiates and maintains milk secretion by the mammary glands.

Prolapse (PRŌ-laps)   A dropping or falling down of an organ, especially the uterus or rectum.

Proliferation (pro-lif′-er-Ā-shun)   Rapid and repeated reproduction of new parts, especially cells.

Pronation (prō-NĀ-shun)   A movement of the forearm in which the palm of the hand is turned posteriorly or inferiorly.

Properdin (prō-PER-din)   A protein found in serum capable of destroying bacteria and viruses.

Prophase (PRŌ-fāz)   The first stage of mitosis during which chromatid pairs are formed and aggregate around the metaphase plate of the cell.

Proprioception (prō-prē-ō-SEP-shun)   The receipt of information from muscles, tendons, and the labyrinth that enables the brain to determine movements and position of the body and its parts. Also called **kinesthesia** (kin′-es-THĒ-zē-a).

Proprioceptor (prō-′prē-SEP-tor)   A receptor located in muscles, tendons, or joints that provides information about body position and movements.

Prostaglandin (pros′-ta-GLAN-din) **(PG)**   A membrane-associated lipid composed of 20-carbon fatty acids with 5 carbon atoms joined to form a cyclopentane ring; released in small quantities and acts as a local hormone.

Prostatectomy (pros′-ta-TEK-tō-mē)   The surgical removal of part of or the entire prostate gland.

Prostate (PROS-tāt) **gland**   A doughnut-shaped gland inferior to the urinary bladder that surrounds the superior portion of the male urethra and secretes a slightly acid solution that contributes to sperm motility and viability.

Prosthesis (pros-THĒ-sis)   An artificial device to replace a missing body part.

Protein   An organic compound consisting of carbon, hydrogen, oxygen, nitrogen, and sometimes sulfur and phosphorus, and made up of amino acids linked by peptide bonds.

Prothrombin (prō-THROM-bin)   An inactive protein synthesized by the liver, released into the blood, and converted to active thrombin in the process of blood clotting.

Proto-oncogene (prō′-tō-ONG-kō-jēn)   Gene responsible for some aspect of normal growth and development; it may transform into an oncogene, a gene capable of causing cancer.

Protraction (prō-TRAK-shun)   The movement of the mandible or shoulder girdle forward on a plane parallel with the ground.

Proximal (PROK-si-mal)   Nearer the attachment of an extremity to the trunk or a structure; nearer to the point of origin.

Pruritus (proo′-RĪ-tus)   Itching.

Pseudopods (SOO-dō-pods)   Temporary, protruding projections of cytoplasm.

Psoriasis (sō-RĪ-a-sis)   Chronic skin disease characterized by reddish plaques or papules covered with scales.

Psychosomatic (sī′-kō-sō-MAT-ik)   Pertaining to the relation between mind and body. Commonly used to refer to those physiological disorders thought to be caused entirely or partly by emotional disturbances.

Pterygopalatine ganglion (ter′-i-gō-PAL-a-tīn GANG-glē-on)   A cluster of cell bodies of parasympathetic postganglionic neurons ending at the lacrimal and nasal glands.

Ptosis (TŌ-sis)   Drooping, as of the eyelid or the kidney (nephrotosis).

Puberty (PYOO-ber-tē)   The time of life during which the secondary sex characteristics begin to appear and the capability for sexual reproduction is possible; usually between the ages of 10 and 17.

Pubic symphysis   A slightly movable cartilaginous joint between the anterior surfaces of the hipbones.

Puerperium (pyoo′-er-PER-ē-um)   The state immediately after childbirth, usually 4–6 weeks.

Pulmonary (PUL-mo-ner′-ē)   Concerning or affected by the lungs.

Pulmonary circulation   The flow of deoxygenated blood from the right ventricle to the lungs and the return of oxygenated blood from the lungs to the left atrium.

Pulmonary edema (e-DĒ-ma)   An abnormal accumulation of interstitial fluid in the tissue spaces and alveoli of the lungs due to increased pulmonary capillary permeability or increased pulmonary capillary pressure.

Pulmonary embolism (EM-bō-lizm) **(PE)**   The presence of a blood clot or other foreign substance in a pulmonary arterial blood vessel that obstructs circulation to lung tissue.

Pulmonary (PUL-mo-ner-ē) **function tests** Any number of tests (forced vital capacity, forced expiratory volume in one second, maximum midrespiratory flow, maximum voluntary ventilation) designed to evaluate lung disease and measure pulmonary impairment.

Pulmonary ventilation The inflow (inspiration) and outflow (expiration) of air between the atmosphere and the lungs. Also called **breathing.**

Pulp cavity A cavity within the crown and neck of a tooth, filled with pulp, a connective tissue containing blood vessels, nerves, and lymphatic vessels.

Pulsating electromagnetic fields (PEMFs) A procedure that uses electrotherapy to treat improperly healing fractures.

Punctate (PUNK-tāt) **distribution** Unequal distribution of cutaneous receptors.

Pulse pressure The difference between the maximum (systolic) and minimum (diastolic) pressures; normally a value of about 40 mm Hg.

Pupil The hole in the center of the iris, the area through which light enters the posterior cavity of the eyeball.

Purkinje fiber *See* **Conduction myofiber.**

Pus The liquid product of inflammation containing leukocytes or their remains and debris of dead cells.

P wave The deflection wave of an electrocardiogram that records atrial depolarization.

Pyelitis (pī′-e-LĪ-tis) Inflammation of the kidney pelvis and its calyces.

Pyemia (pī-Ē-mēa) Infection of the blood, with multiple abscesses, caused by pus-forming microorganisms.

Pyloric (pī-LOR-ik) **sphincter** A thickened ring of smooth muscle through which the pylorus of the stomach communicates with the duodenum. Also called the **pyloric valve.**

Pyogenesis (pi′-ō-JEN-e-sis) Formation of pus.

Pyorrhea (pī-ō-RĒ-a) A discharge or flow of pus, especially in the alveoli (sockets) and the tissues of the gums.

Pyramid (PIR-a-mid) A pointed or cone-shaped structure; one of two roughly triangular structures on the ventral side of the medulla composed of the largest motor tracts that run from the cerebral cortex to the spinal cord; a triangular-shaped structure in the renal medulla composed of the straight segments of renal tubules.

Pyramidal (pi-RAM-i-dal) **pathways** Collections of motor nerve fibers arising in the brain and passing down through the spinal cord to motor cells in the anterior horns.

Pyuria (pī-YOO-rē-a) The presence of leukocytes and other components of pus in urine.

QRS wave The deflection wave of an electrocardiogram that records ventricular depolarization.

Quadrant (KWOD-rant) One of four parts.

Quadriplegia (kwod′-ri-PLĒ-jē-a) Paralysis of the two upper and two lower extremities.

Radiographic (rā′-dē-ō-GRAF-ic) **anatomy** Diagnostic branch of anatomy that includes the use of x-rays.

Rales (RALS) Sounds sometimes heard in the lungs that resemble bubbling or rattling due to the presence of an abnormal amount or type of fluid or mucus inside the bronchi or alveoli, or to bronchoconstriction so that air cannot enter or leave the lungs normally.

Rami communicantes (RĀ-mē ko-myoo-ni-KAN-tēz) Branches of a spinal nerve. *Singular,* **ramus communicans** (RĀ-mus ko-MYOO-ni-kans).

Rapid eye movement (REM) sleep A level of sleep characterized by symmetrical flutter of the eyes and eyelids and brain wave patterns similar to those of an awake person.

Rathke's pouch *See* **Hypophyseal pouch.**

Raynaud's (rā-NOZ) **disease** A vascular disorder, primarily of females, characterized by bilateral attacks of ischemia, usually of the fingers and toes, in which the skin becomes pale and exhibits burning and pain; it is brought on by cold or emotional stimuli.

Reactivity (rē-ak-TI-vi-tē) Ability of an antigen to react specifically with the antibody whose formation it induced.

Receptor A specialized cell or a nerve cell terminal modified to respond to some specific sensory modality, such as touch, pressure, cold, light, or sound, and convert it to a nerve impulse by way of a generator or receptor potential. A specific molecule or arrangement of molecules organized to accept only molecules with a complementary shape.

Receptor-mediated endocytosis A highly selective process in which cells take up large molecules or particles (ligands). In the process, successive compartments called endocytic vesicles and endosomes form. Ligands are eventually broken down by enzymes in lysosomes.

Receptor potential Depolarization or hyperpolarization of the plasma membrane of a receptor cell which alters release of neurotransmitter from the cell; if the neuron that synapses with the receptor cell becomes depolarized to threshold, a nerve impulse is triggered.

Recessive gene A gene that is not expressed in the presence of a dominant gene on the homologous chromosome.

Reciprocal innervation (re-SIP-rō-kal in-ner-Vā-shun) The phenomenon by which action potentials stimulate contraction of one muscle and simultaneously inhibit contraction of antagonistic muscles.

Recombinant DNA Synthetic DNA, formed by joining a fragment of DNA from one source to a portion of DNA from another.

Recovery oxygen consumption Elevated oxygen use after exercise ends due to metabolic changes that start during exercise and continues after exercise. Previously called **oxygen debt.**

Recruitment (rē-KROOT-ment) The process of increasing the number of active motor units. Also called **motor unit summation.**

Rectouterine pouch A pocket formed by the parietal peritoneum as it moves posteriorly from the surface of the uterus and is reflected onto the rectum; the lowest point in the pelvic cavity. Also called the **pouch** or **cul de sac of Douglas.**

Rectum (REK-tum) The last 20 cm (7 in.) of the gastrointestinal tract, from the sigmoid colon to the anus.

Recumbent (re-KUM-bent)   Lying down.

Red nucleus   A cluster of cell bodies in the midbrain, occupying a large portion of the tectum and sending fibers into the rubroreticular and rubrospinal tracts.

Red pulp   That portion of the spleen that consists of venous sinuses filled with blood and cords of splenic tissue called splenic (Billroth's) cords.

Reduction   The addition of electrons and hydrogen ions (hydrogen atoms) to a molecule or, less commonly, the removal of oxygen from a molecule that results in an increase in the energy content of the molecule.

Referred pain   Pain that is felt at a site remote from the place of origin.

Reflex   Fast response to a change (stimulus) in the internal or external environment that attempts to restore homeostasis; passes over a reflex arc.

Reflex arc   The most basic conduction pathway through the nervous system, connecting a receptor and an effector and consisting of a receptor, a sensory neuron, an integrating center in the central nervous system, a motor neuron, and an effector.

Refraction (rē-FRAK-shun)   The bending of light as it passes from one medium to another.

Refractory (re-FRAK-to-rē) **period**   A time during which an excitable cell cannot respond to a stimulus that is usually adequate to evoke an action potential.

Regeneration (rē-jen′-er-Ā-shun)   The natural renewal of a structure.

Regimen (REJ-i-men)   A strictly regulated scheme of diet, exercise, or activity designed to achieve certain ends.

Regional anatomy   The division of anatomy dealing with a specific region of the body, such as the head, neck, chest, or abdomen.

Regurgitation (rē-gur′-ji-TĀ-shun)   Return of solids or fluids to the mouth from the stomach; flowing backward of blood through incompletely closed heart valves.

Relapse (RĒ-laps)   The return of a disease weeks or months after its apparent cessation.

Relaxin (RLX)   A female hormone produced by the ovaries that relaxes the pubic symphysis and helps dilate the uterine cervix to ease delivery of a baby.

Releasing hormone   Chemical secretion of the hypothalamus that can stimulate secretion of hormones of the anterior pituitary gland.

Remodeling   Replacement of old bone by new bone tissue.

Renal (RĒ-nal)   Pertaining to the kidney.

Renal corpuscle (KOR-pus′-l)   A glomerular (Bowman's) capsule and its enclosed glomerulus.

Renal failure   Inability of the kidneys to function properly, due to abrupt failure (acute) or progressive failure (chronic).

Renal pelvis   A cavity in the center of the kidney formed by the expanded, proximal portion of the ureter, lying within the kidney, and into which the major calyces open.

Renal pyramid   A triangular structure in the renal medulla composed of the straight segments of renal tubules.

Renin (RĒ-nin)   An enzyme released by the kidney into the plasma where it converts angiotensinogen into angiotensin I.

Renin-angiotensin (an′-jē-ō-TEN-sin) **pathway**   A mechanism for the control of aldosterone secretion by angiotensin II, initiated by the secretion of renin by the kidney in response to low blood pressure.

Reproduction (rē′-prō-DUK-shun)   Either the formation of new cells for growth, repair, or replacement, or the production of a new individual.

Reproductive cell division   Type of cell division in which sperm and egg cells are produced; consists of meiosis and cytokinesis.

Residual (re-ZID-yoo-al) **volume**   The volume of air still contained in the lungs after a maximal expiration; about 1,200 ml.

Resistance (re-ZIS-tans)   Resistance (impedance) to blood flow as a result of viscosity, total blood vessel length, and blood vessel radius. Ability to ward off disease. The hindrance encountered by an electrical charge as it moves through a substance from one point to another. The hindrance encountered by air as it moves through the respiratory passageways.

Respiration (res-pi-RĀ-shun)   Overall exchange of gases between the atmosphere, blood, and body cells consisting of pulmonary ventilation, external respiration, and internal respiration.

Respirator (RES-pi-rā′-tor)   An apparatus fitted to a mask over the nose and mouth, or hooked directly to an endotracheal or tracheotomy tube, that is used to assist or support ventilation or to provide nebulized medication to the air passages under positive pressure.

Respiratory center   Neurons in the reticular formation of the brain stem that regulate the rate of respiration.

Respiratory distress syndrome (RDS) of the newborn   A disease of newborn infants, especially premature ones, in which insufficient amounts of surfactant are produced and breathing is labored. Also called **hyaline** (HĪ-a-lin) **membrane disease (HMD).**

Respiratory failure   Condition in which the respiratory system cannot supply sufficient oxygen to maintain metabolism or eliminate enough carbon dioxide to prevent respiratory acidosis.

Resting membrane potential   The voltage that exists between the inside and outside of a cell membrane when the cell is not responding to a stimulus; about −70 to −90 mV, with the inside of the cell negative.

Resuscitation (rē-sus′-i-TĀ-shun)   Act of bringing a person back to full consciousness.

Retention (rē-TEN-shun)   A failure to void urine due to obstruction, nervous contraction of the urethra, or absence of sensation of desire to urinate.

Rete (RĒ-tē) **testis**   The network of ducts in the testes.

Reticular (re-TIK-yoo-lar) **activating system (RAS)**   An extensive network of branched nerve cells running through the core of the brain stem. When these cells are activated, a generalized alert or arousal behavior results.

Reticular formation   A network of small groups of nerve cells scattered among bundles of fibers beginning in the medulla as a continuation of the spinal cord and extending upward through the central part of the brain stem.

Reticulocyte (re-TIK-yoo-lō-sīt)   An immature red blood cell.

Reticulocyte (re-TIK-yoo-lō-sīt) **count**   Examination of a stained sample of blood to determine the percentage of reticulocytes in the total number of red blood cells; used to evaluate rate of erythropoiesis and monitor treatment for anemia.

Reticulum (re-TIK-yoo-lum)   A network.

Retina (RET-i-na)   The inner coat of the eyeball, lying only in the posterior portion of the eye and consisting of nervous tissue and a pigmented layer comprised of epithelial cells lying in contact with the choroid. Also called the **nervous tunic** (TOO-nik).

Retinal (RE-ti-nal)   A derivative of vitamin A that functions as the light-absorbing portion of the photopigment rhodopsin.

Retraction (rē-TRAK-shun)   The movement of a protracted part of the body backward on a plane parallel to the ground, as in pulling the lower jaw back in line with the upper jaw.

Retroflexion (re-trō-FLEK-shun)   A malposition of the uterus in which it is tilted posteriorly.

Retrograde degeneration (RE-trō-grād   dē-jen-er-Ā-shun) Changes that occur in the proximal portion of a damaged axon only as far as the first neurofibral node (node of Ranvier); similar to changes that occur during Wallerian degeneration.

Retroperitoneal (re′-trō-per-i-tō-NĒ-al)   External to the peritoneal lining of the abdominal cavity.

Rheumatism (ROO-ma-tizm′)   Any painful state of the supporting structures of the body—bones, ligaments, joints, tendons, or muscles.

Rh factor   An inherited agglutinogen (antigen) on the surface of red blood cells.

Rhinology (rī-NOL-ō-jē)   The study of the nose and its disorders.

Rhinoplasty (RĪ-nō-plas′-tē)   Surgical procedure in which the structure of the external nose is altered.

Rhodopsin (rō-DOP-sin)   The photopigment in rods of the retina, consisting of a glycoprotein called opsin and a derivative of vitamin A called retinal, that is sensitive to low levels of illumination.

Ribonucleic (rī-bō-nyoo-KLĒ-ik) **acid (RNA)**   A single-stranded nucleic acid constructed of nucleotides consisting of one of four possible nitrogenous bases (adenine, cytosine, guanine, or uracil), ribose, and a phosphate group; three types are messenger RNA (mRNA), transfer RNA (tRNA), and ribosomal RNA (rRNA), each of which cooperates with DNA for protein synthesis.

Ribosome (RĪ-bō-sōm)   An organelle in the cytoplasm of cells, composed of ribosomal RNA and ribosomal proteins, that synthesizes proteins; nicknamed the "protein factory."

Rickets (RIK-ets)   Condition affecting children characterized by soft and deformed bones resulting from inadequate calcium metabolism due to a vitamin D deficiency.

Right heart (atrial) reflex   A reflex concerned with maintaining normal venous blood pressure.

Right lymphatic (lim-FAT-ik) **duct**   A vessel of the lymphatic system that drains lymph from the upper right side of the body and empties it into the right subclavian vein.

Rigidity (ri-JID-i-tē)   Hypertonia characterized by increased muscle tone, but reflexes are not affected.

Rigor mortis   State of partial contraction of muscles following death due to lack of ATP that causes cross bridges of thick fila-

ments to remain attached to thin filaments, thus preventing relaxation.

Rod   A photoreceptor in the retina of the eye that is specialized for vision in dim light.

Roentgen (RENT-gen)   The international unit of radiation; a standard quantity of x-or gamma radiation.

Root canal   A narrow extension of the pulp cavity lying within the root of a tooth.

Root of penis   Attached portion of penis that consists of the bulb and crura.

Rotation (rō-TĀ-shun)   Moving a bone around its own axis, with no other movement.

Round ligament (LIG-a-ment)   A band of fibrous connective tissue enclosed between the folds of the broad ligament of the uterus, emerging from a point on the uterus just below the uterine (Fallopian) tube, extending laterally along the pelvic wall, and penetrating the abdominal wall through the deep inguinal ring to end in the labia majora.

Round window   A small opening between the middle and inner ear, directly below the oval window, covered by the secondary tympanic membrane. Also called the **fenestra cochlea** (fe-NES-tra KŌK-lē-a).

Rugae (ROO-jē)   Large folds in the mucosa of an empty hollow organ, such as the stomach and vagina.

Saccule (SAK-yool)   The lower and smaller of the two chambers in the membranous labyrinth inside the vestibule of the inner ear containing a receptor organ for static equilibrium.

Sacral plexus (PLEK-sus)   A network formed by the anterior branches of spinal nerves L4 through S3.

Sacral promontory (PROM-on-tor′-ē)   The superior surface of the body of the first sacral vertebra that projects anteriorly into the pelvic cavity; a line from the sacral promontory to the superior border of the pubic symphysis divides the abdominal and pelvic cavities.

Saddle joint   A synovial joint in which the articular surface of one bone is saddle shaped and the articular surface of the other bone is shaped like a rider sitting in the saddle, as in the joint between the trapezium and the metacarpal of the thumb. Also called a **sellaris** (sel-LA-ris) **joint.**

Sagittal (SAJ-i-tal) **plane**   A vertical plane that divides the body or organs into left and right portions. Such a plane may be **midsagittal (median),** in which the divisions are equal, or **parasagittal,** in which the divisions are unequal.

Saliva (sa-LĪ-va)   A clear, alkaline, somewhat viscous secretion produced mostly by the three pairs of salivary glands; contains various salts, mucin, lysozyme, salivary amylase, and lingual lipase (produced by glands in the tongue).

Salivary amylase (SAL-i-ver-ē AM-i-lās)   An enzyme in saliva that initiates the chemical breakdown of starch, mostly in the mouth.

Salivary gland   One of three pairs of glands that lie outside the mouth and pour their secretory product (called saliva) into ducts that empty into the oral cavity; the parotid, submandibular, and sublingual glands.

Salpingitis (sal′-pin-JĪ-tis)   Inflammation of the uterine (Fallopian) or auditory (Eustachian) tube.

Saltatory (sal-ta-TŌ-rē) **conduction** The propagation of an action potential (nerve impulse) along the exposed portions of a myelinated nerve fiber. The action potential appears at successive neurofibral nodes (nodes of Ranvier) and therefore seems to jump or leap from node to node.

Sarcolemma (sar′-kō-LEM-ma) The cell membrane of a muscle fiber (cell), especially of a skeletal muscle fiber.

Sarcoma (sar-KŌ-ma) A connective tissue tumor, often highly malignant.

Sarcomere (SAR-kō-mēr) A contractile unit in a striated muscle fiber (cell) extending from one Z disc to the next Z disc.

Sarcoplasm (SAR-kō-plazm) The cytoplasm of a muscle fiber (cell).

Sarcoplasmic reticulum (sar′-kō-PLAZ-mik re-TIK-yoo-lum) A network of saccules and tubes surrounding myofibrils of a muscle fiber (cell), comparable to endoplasmic reticulum; functions to reabsorb calcium ions during relaxation and to release them to cause contraction.

Satiety (sa-TĪ-e-tē) Fullness or gratification, as of hunger or thirst.

Satiety center A collection of nerve cells located in the ventromedial nuclei of the hypothalamus that, when stimulated, brings about the cessation of eating.

Saturated fat A fatty acid that contains no double bonds between any of its carbon atoms; all are single bonds and all carbon atoms are bonded to the maximum number of hydrogen atoms; found naturally in triglycerides of animal foods such as meat, milk, milk products, and eggs.

Scala tympani (SKA-la TIM-pan-ē) The lower spiral-shaped channel of the bony cochlea, filled with perilymph.

Scala vestibuli (ves-TIB-yoo-lē) The upper spiral-shaped channel of the bony cochlea, filled with perilymph.

Schwann cell *See* **Neurolemmocyte.**

Sciatica (sī-AT-i-ka) Inflammation and pain along the sciatic nerve; felt at the back of the thigh running down the inside of the leg.

Sclera (SKLE-ra) The white coat of fibrous tissue that forms the outer protective covering over the eyeball except in the most anterior portion; the posterior portion of the fibrous tunic.

Scleral venous sinus A circular venous sinus located at the junction of the sclera and the cornea through which aqueous humor drains from the anterior chamber of the eyeball into the blood. Also called the **canal of Schlemm** (SHLEM).

Sclerosis (skle-RŌ-sis) A hardening with loss of elasticity of tissues.

Scoliosis (skō′-lē-Ō-sis) An abnormal lateral curvature from the normal vertical line of the backbone.

Scotoma (skō-TŌ-ma) An area of depressed or lost vision within the visual field.

Scrotum (SKRŌ-tum) A skin-covered pouch that contains the testes and their accessory structures.

Sebaceous (se-BĀ-shus) Secreting oil.

Sebaceous (se-BĀ-shus) **gland** An exocrine gland in the dermis of the skin, almost always associated with a hair follicle, that secretes sebum. Also called an **oil gland.**

Sebum (SĒ-bum) Secretion of sebaceous (oil) glands.

Secondary response Accelerated, more intense cell-mediated or antibody-mediated immune response on a subsequent exposure to an antigen after the initial exposure.

Secondary sex characteristic A feature characteristic of the male or female body that develops at puberty under the stimulation of sex hormones but is not directly involved in sexual reproduction, such as distribution of body hair, voice pitch, body shape, and muscle development.

Secretion (se-KRĒ-shun) Production and release from a gland cell of a fluid, especially a functionally useful product as opposed to a waste product.

Selectively permeable membrane A membrane that permits the passage of certain substances, but restricts the passage of others. Also called a **semipermeable** (sem′-ē-PER-mē-a-bl) **membrane.**

Sella turcica (SEL-a TUR-si-ka) A depression on the superior surface of the sphenoid bone that houses the pituitary gland.

Semen (SĒ-men) A fluid discharged at ejaculation by a male that consists of a mixture of spermatozoa and the secretions of the seminal vesicles, prostate gland, and bulbourethral (Cowper's) glands.

Semicircular canals Three bony channels (anterior, posterior, lateral), filled with perilymph, in which lie the membranous semicircular canals filled with endolymph. They contain receptors for equilibrium.

Semicircular ducts The membranous semicircular canals filled with endolymph and floating in the perilymph of the bony semicircular canals. They contain cristae that are concerned with dynamic equilibrium.

Semilunar (sem′-ē-LOO-nar) **valve** A valve guarding the entrance into the aorta or the pulmonary trunk from a ventricle of the heart.

Seminal vesicle (SEM-i-nal VES-i-kul) One of a pair of convoluted, pouchlike structures, lying posterior and inferior to the urinary bladder and anterior to the rectum, that secrete a component of semen into the ejaculatory ducts.

Seminiferous tubule (sem′-i-NI-fer-us TOO-byool) A tightly coiled duct, located in a lobule of the testis, where spermatozoa are produced.

Senescence (se-NES-ens) The process of growing old; the period of old age.

Senile macular (MAK-yoo-lar) **degeneration (SMD)** A disease in which blood vessels grow over the macula lutea.

Senility (se-NIL-i-tē) A loss of mental or physical ability due to old age.

Sensation A state of awareness of external or internal conditions of the body.

Sensory area A region of the cerebral cortex concerned with the interpretation of sensory impulses.

Sensory neuron (NOO-ron) A neuron that carries nerve impulses toward the central nervous system. Also called an **afferent neuron.**

Sepsis (SEP-sis) A morbid condition that results from the presence in the blood or other body tissues of pathogenic bacteria and their products.

Septal defect An opening in the septum (interatrial or interventricular) between the left and right sides of the heart.

Septicemia (sep'-ti-SĒ-mē-a) Toxins or disease-causing bacteria in blood. Also called **"blood poisoning."**

Septum (SEP-tum) A wall dividing two cavities.

Serosa (ser-Ō-sa) Any serous membrane. The outermost layer of an organ formed by a serous membrane. The membrane that lines the pleural, pericardial, and peritoneal cavities.

Serous (SIR-us) **membrane** A membrane that lines a body cavity that does not open to the exterior. Also called the **serosa** (se-RŌ-sa).

Serum Plasma minus its clotting proteins.

Serum enzyme studies Evaluation of levels of certain enzymes in blood (creatine phosphokinase, serum glutamic oxaloacetic transaminase, lactic dehydrogenase) to diagnose and monitor heart attacks.

Sesamoid (SES-a-moyd) **bones** Small bones usually found in tendons.

Sex chromosomes The twenty-third pair of chromosomes, designated X and Y, which determine the genetic sex of an individual; in males, the pair is XY; in females, XX.

Sexual intercourse The insertion of the erect penis of a male into the vagina of a female. Also called **coitus** (KŌ-i-tus) or **copulation.**

Sexually transmitted disease (STD) General term for any of a large number of diseases spread by sexual contact. Also called a **venereal disease (VD).**

Sheath of Schwann *See* **Neurolemma.**

Shingles Acute infection of the peripheral nervous system caused by a virus.

Shinsplints Soreness or pain along the tibia probably caused by inflammation of the periosteum brought on by repeated tugging of the muscles and tendons attached to the periosteum. Also called **tibia stress syndrome.**

Shivering Involuntary contraction of muscles that generates heat. Also called **involuntary thermogenesis.**

Shock Failure of the cardiovascular system to deliver adequate amounts of oxygen and nutrients to meet the metabolic needs of the body due to inadequate cardiac output. It is characterized by hypotension; clammy, cool, and pale skin; sweating; reduced urine formation; altered mental state; acidosis; tachycardia; weak, rapid pulse; and thirst. Types include hypovolemic, cardiogenic, obstructive, neurogenic, and septic.

Shoulder A synovial or diarthrotic joint where the humerus joins the scapula.

Sigmoid colon (SIG-moyd KŌ-lon) The S-shaped portion of the large intestine that begins at the level of the left iliac crest, projects inward to the midline, and terminates at the rectum at about the level of the third sacral vertebra.

Sigmoidoscopy (sig'-moy-DOS-kō-pē) Visualization of the anal canal, rectum, and colon to screen for colorectal cancer, collect biopsy samples, remove polyps, gather specimens for culture, and photograph the intestinal mucosa.

Sign Any objective evidence of disease that can be observed or measured such as a lesion, swelling, or fever.

Sinoatrial (si-nō-Ā-trē-al) **(SA) node** A compact mass of cardiac muscle fibers (cells) specialized for conduction, located in the right atrium beneath the opening of the superior vena cava. Also called the **sinuatrial node** or **pacemaker.**

Sinus (SĪ-nus) A hollow in a bone (paranasal sinus) or other tissue; a channel for blood (vascular sinus); any cavity having a narrow opening.

Sinusitis (sīn-yoo-SĪT-is) Inflammation of the mucous membrane of a paranasal sinus.

Sinusoid (SĪN-yoo-soyd) A microscopic space or passage for blood in certain organs such as the liver or spleen.

Skeletal muscle An organ specialized for contraction, composed of striated muscle fibers (cells), supported by connective tissue, attached to a bone by a tendon or an aponeurosis, and stimulated by somatic motor neurons.

Skene's gland *See* **Paraurethral gland.**

Skull The skeleton of the head consisting of the cranial and facial bones.

Sliding-filament mechanism The most commonly accepted explanation for muscle contraction in which actin and myosin filaments move into interdigitation with each other, decreasing the length of the sarcomeres.

Small intestine A long tube of the gastrointestinal tract that begins at the pyloric sphincter of the stomach, coils through the central and lower part of the abdominal cavity, and ends at the large intestine; divided into three segments: duodenum, jejunum, and ileum.

Smooth muscle An organ specialized for contraction, composed of smooth muscle fibers (cells), located in the walls of hollow internal structures, and innervated by a motor neuron.

Snellen (SNEL-en) **test** Test used to evaluate any problem or changes in vision by measuring visual acuity (sharpness).

Sodium-potassium ATP-ase An active transport system located in the cell membrane that transports sodium ions out of the cell and potassium ions into the cell at the expense of cellular ATP. It functions to keep the ionic concentrations of these elements at physiological levels. Also called the **sodium pump.**

Soft palate (PAL-at) The posterior portion of the roof of the mouth, extending posteriorly from the palatine bones and ending at the uvula. It is a muscular partition lined with mucous membrane.

Solution A homogeneous molecular or ionic dispersion of one or more substances (solutes) in a usually liquid-dissolving medium (solvent).

Somatic (sō-MAT-ik) **cell division** Type of cell division in which a single starting cell (parent cell) duplicates itself to produce two identical cells (daughter cells); consists of mitosis and cytokinesis.

Somatic nervous system (SNS) The portion of the peripheral nervous system made up of the somatic efferent fibers that run between the central nervous system and the skeletal muscles and skin.

Somatomedin (sō'-ma-tō-MĒ-din) Small protein produced by the liver in response to stimulation by human growth hormone (hGH) that mediates most of the effects of human growth hormone. Also called **insulin-like growth factor.**

Somite (SŌ-mīt) Block of mesodermal cells in a developing embryo that is distinguished into a myotome (which forms most of the skeletal muscles), dermatome (which forms connective tissues), and sclerotome (which forms the vertebrae).

Spasm (spazm) A sudden, involuntary contraction of large groups of muscles.

Spastic (SPAS-tik) An increase in muscle tone (stiffness) associated with an increase in tendon reflexes and abnormal reflexes (Babinski sign).

Spasticity (spas-TIS-i-tē) Hypertonia characterized by increased muscle tone, increased tendon reflexes, and pathological reflexes (Babinski sign).

Spermatic (sper-MAT-ik) **cord** A supporting structure of the male reproductive system, extending from a testis to the deep inguinal ring, that includes the ductus (vas) deferens, arteries, veins, lymphatic vessels, nerves, cremaster muscle, and connective tissue.

Spermatogenesis (sper′-ma-tō-JEN-e-sis) The formation and development of spermatozoa in the seminiferous tubules of the testes.

Spermatozoon (sper′-ma-tō-ZŌ-on) A mature sperm cell.

Spermicide (SPER-mi-sīd′) An agent that kills spermatozoa.

Spermiogenesis (sper′-mē-ō-JEN-e-sis) The maturation of spermatids into spermatozoa.

Sphincter (SFINGK-ter) A circular muscle constricting an orifice.

Sphincter of Oddi *See* **Sphincter of the hepatopancreatic ampulla.**

Sphincter of the hepatopancreatic ampulla A circular muscle at the opening of the common bile and main pancreatic ducts in the duodenum. Also called the **sphincter of Oddi** (OD-ē).

Sphygmomanometer (sfig′-mō-ma-NOM-e-ter) An instrument for measuring arterial blood pressure.

Spina bifida (SPĪ-na BIF-i-da) A congenital defect of the vertebral column in which the halves of the neural arch of a vertebra fail to fuse in the midline.

Spinal (SPĪ-nal) **cord** A mass of nerve tissue located in the vertebral canal from which 31 pairs of spinal nerves originate.

Spinal nerve One of the 31 pairs of nerves that originate on the spinal cord from posterior and anterior roots.

Spinal shock A period of time, from several days to several weeks, following transection of the spinal cord and characterized by the abolition of all reflex activity.

Spinal (lumbar) tap (puncture) Withdrawal of some of the cerebrospinal fluid from the subarachnoid space in the lumbar region for diagnostic purposes, introduction of various substances, and evaluation of the effects of treatment.

Spinous (SPĪ-nus) **process** A sharp or thornlike process or projection. Also called a **spine.** A sharp ridge running diagonally across the posterior surface of the scapula.

Spiral organ The organ of hearing, consisting of supporting cells and hair cells that rest on the basilar membrane and extend into the endolymph of the cochlear duct. Also called the **organ of Corti** (KOR-tē).

Spirometer (spī-ROM-e-ter) An apparatus used to measure air capacity of the lungs.

Splanchnic (SPLANK-nik) Pertaining to the viscera.

Spleen (SPLĒN) Large mass of lymphatic tissue between the fundus of the stomach and the diaphragm that functions in phagocytosis, production of lymphocytes, and blood storage.

Sprain Forcible wrenching or twisting of a joint with partial rupture or other injury to its attachments without dislocation.

Sputum (SPYOO-tum) Substance ejected from the mouth containing saliva and mucus.

Squamous (SKWĀ-mus) Scalelike.

Starling's law of the capillaries The movement of fluid between plasma and interstitial fluid is in a state of near equilibrium at the arterial and venous ends of a capillary; that is, filtered fluid and absorbed fluid plus that returned to the lymphatic system are nearly equal.

Starling's law of the heart The force of muscular contraction is determined by the length of the cardiac muscle fibers (cells); within limits, the greater the length of stretched fibers, the stronger the contraction.

Starvation (star-VĀ-shun) The loss of energy stores in the form of glycogen, triglycerides, and proteins due to inadequate intake of nutrients or inability to digest, absorb, or metabolize ingested nutrients.

Stasis (STĀ-sis) Stagnation or halt of normal flow of fluids, as blood, urine, or of the intestinal mechanism.

Static equilibrium (ē-kwi-LIB-rē-um) The maintenance of posture in response to changes in the orientation of the body, mainly the head, relative to the ground.

Stellate reticuloendothelial (STEL-āte re-tik′-yoo-lō-en′-dō-THĒ-lē-al) **cell** Phagocytic cell that lines a sinusoid of the liver. Also called a **Kupffer's** (KOOP-ferz) **cell.**

Stenosis (sten-Ō-sis) An abnormal narrowing or constriction of a duct or opening.

Stereocilia (ste′-rē-ō-SIL-ē-a) Groups of extremely long, slender, nonmotile microvilli projecting from epithelial cells lining the epididymis.

Stereognosis (ste′-rē-og-NŌ-sis) The ability to recognize the size, shape, and texture of an object by touch.

Sterile (STE-ril) Free from any living microorganisms. Unable to conceive or produce offspring.

Sterilization (ster′-i-li-ZĀ-shun) Elimination of all living microorganisms. The rendering of an individual incapable of reproduction (e.g., castration, vasectomy, hysterectomy).

Sternal puncture Introduction of a wide-bore needle into the marrow cavity of the sternum for aspiration of a sample of red bone marrow.

Stimulus Any change in the environment capable of altering the membrane potential.

Stomach The J-shaped enlargement of the gastrointestinal tract directly under the diaphragm in the epigastric, umbilical, and left hypochondriac regions of the abdomen, between the esophagus and small intestine.

Strabismus (stra-BIZ-mus) A condition in which the visual axes of the two eyes differ, so that they do not fix on the same object.

Straight tubule (TOO-byool) A duct in a testis leading from a convoluted seminiferous tubule to the rete testis.

Stratum (STRĀ-tum) A layer.

Stratum basalis (ba-SAL-is) The outer layer of the endometrium, next to the myometrium, that is maintained during menstruation and gestation and produces a new functionalis following menstruation or parturition.

Stratum functionalis (funk′-shun-AL-is) The inner layer of the endometrium, the layer next to the uterine cavity, that is shed during menstruation and that forms the maternal portion of the placenta during gestation.

Stressor A stress that is extreme, unusual, or long-lasting and triggers the general adaptation syndrome.

Stretch receptor Receptor in the walls of bronchi, bronchioles, and lungs that sends impulses to the respiratory center that prevents overinflation of the lungs.

Stretch reflex A monosynaptic reflex triggered by a sudden stretch of a muscle and ending with a contraction of that same muscle. Also called a **tendon jerk.**

Stricture (STRIK-cher) A local constriction of a tubular structure.

Stroke *See* **Cerebrovascular accident.**

Stroke volume The volume of blood ejected by either ventricle in one systole; about 70 ml.

Stroma (STRŌ-ma) The tissue that forms the ground substance, foundation, or framework of an organ, as opposed to its functional parts.

Stupor (STOO-por) Unresponsiveness from which a patient can be aroused only briefly and by vigorous and repeated stimulation.

Subarachnoid (sub′-a-RAK-noyd) **space** A space between the arachnoid and the pia mater that surrounds the brain and spinal cord and through which cerebrospinal fluid circulates.

Subcutaneous (sub′-kyoo-TĀ-nē-us) Beneath the skin. Also called **hypodermic** (hī′-pō-DER-mik).

Subcutaneous layer A continuous sheet of areolar connective tissue and adipose tissue between the dermis of the skin and the deep fascia of the muscles. Also called the **superficial fascia** (FASH-ē-a).

Subdural (sub-DOO-ral) **space** A space between the dura mater and the arachnoid of the brain and spinal cord that contains a small amount of fluid.

Sublingual (sub-LING-gwal) **gland** One of a pair of salivary glands situated in the floor of the mouth under the mucous membrane and to the side of the lingual frenulum, with a duct (Rivinus's) that opens into the floor of the mouth.

Submandibular (sub′-man-DIB-yoo-lar) **gland** One of a pair of salivary glands found beneath the base of the tongue under the mucous membrane in the posterior part of the floor of the mouth, posterior to the sublingual glands, with a duct (Wharton's) situated to the side of the lingual frenulum. Also called the **submaxillary** (sub′-MAK-si-ler-ē) **gland.**

Submucosa (sub-myoo-KŌ-sa) A layer of connective tissue located beneath a mucous membrane, as in the gastrointestinal tract or the urinary bladder; the submucosa connects the mucosa to the muscularis layer.

Submucosal plexus A network of autonomic nerve fibers located in the outer portion of the submucous layer of the small intestine. Also called the **plexus of Meissner** (MĪS-ner).

Subserous fascia (sub-SE-rus FASH-ē-a) A layer of connective tissue internal to the deep fascia, lying between the deep fascia and the serous membrane that lines the body cavities.

Substrate A substance with which an enzyme reacts.

Subthreshold stimulus A stimulus of such weak intensity that it cannot initiate an action potential (nerve impulse). Also called a **subliminal stimulus.**

Sudoriferous (soo′-dor-IF-er-us) **gland** An apocrine or eccrine exocrine gland in the dermis or subcutaneous layer that produces perspiration. Also called a **sweat gland.**

Sulcus (SUL-kus) A groove or depression between parts, especially between the convolutions of the brain. *Plural,* **sulci** (SUL-sē).

Summation (sum-MĀ-shun) The algebraic addition of the excitatory and inhibitory effects of many stimuli applied to a nerve cell body. The increased strength of muscle contraction that results when stimuli follow in rapid succession.

Superficial (soo′-per-FISH-al) Located on or near the surface of the body.

Superficial fascia (FASH-ē-a) A continuous sheet of fibrous connective tissue between the dermis of the skin and the deep fascia of the muscles. Also called **subcutaneous** (sub′-kyoo-TĀ-nē-us) **layer.**

Superficial inguinal (IN-gwi-nal) **ring** A triangular opening in the aponeurosis of the external oblique muscle that represents the termination of the inguinal canal.

Superior (soo-PĒR-ē-or) Toward the head or upper part of a structure. Also called **cephalad** (SEF-a-lad) or **craniad.**

Superior vena cava (VĒ-na CĀ-va) **(SVC)** Large vein that collects blood from parts of the body superior to the heart and returns it to the right atrium.

Supination (soo-pī-NĀ-shun) A movement of the forearm in which the palm of the hand is turned anteriorly or superiorly.

Suppuration (sup′-yoo-RĀ-shun) Pus formation and discharge.

Supraopticohypophyseal (soo′-pra-op′-tik-ō-hī-po-FIZ-ē-al) **tract** A bundle of nerve processes made up of fibers that have their cell bodies in the hypothalamus but release their neurosecretions in the posterior pituitary gland.

Surface anatomy The study of the structures that can be identified from the outside of the body.

Surfactant (sur-FAK-tant) Complex mixture of phospholipids and lipoproteins produced by type II alveolar (septal) cells in the lungs that decreases surface tension.

Susceptibility (sus-sep′-ti-BIL-i-tē) Lack of resistance of a body to the deleterious or other effects of an agent such as pathogenic microorganisms.

Suspensory ligament (sus-PEN-so-rē LIG-a-ment) A fold of peritoneum extending laterally from the surface of the ovary to the pelvic wall.

Sustentacular (sus′-ten-TAK-yoo-lar) **cell** A supporting cell of seminiferous tubules that produces secretions for supplying nutrients to spermatozoa and the hormone inhibin. Also called a **Sertoli** (ser-TŌ-lē) **cell.**

Sutural (SOO-cher-al) **bone** A small bone located within a suture between certain cranial bones. Also called **Wormian** (WER-mē-an) **bone.**

Suture (SOO-cher) An immovable fibrous joint in the skull where bone surfaces are closely united.

Sympathetic (sim′-pa-THET-ik) **division** One of the two subdivisions of the autonomic nervous system, having cell bodies of preganglionic neurons in the lateral gray columns of the thoracic segment and first two or three lumbar segments of the spinal cord; primarily concerned with processes involving the expenditure of energy. Also called the **thoracolumbar** (thō′-ra-kō-LUM-bar) **division.**

Sympathetic trunk ganglion (GANG-glē-on)   A cluster of cell bodies of postganglionic sympathetic neurons lateral to the vertebral column, close to the body of a vertebra. These ganglia extend downward through the neck, thorax, and abdomen to the coccyx on both sides of the vertebral column and are connected to one another to form a chain on each side of the vertebral column. Also called **lateral,** or **sympathetic, chain** or **vertebral chain ganglia.**

Sympathomimetic (sim′-pa-thō-mi-MET-ik)   Producing effects that mimic those brought about by the sympathetic division of the autonomic nervous system.

Symphysis (SIM-fi-sis)   A line of union. A slightly movable cartilaginous joint such as the pubic symphysis between the anterior surfaces of the hipbones.

Symport   Process by which two substances, usually Na$^+$ and another substance, move in the same direction across a cell membrane. Also called **cotransport.**

Symptom (SIMP-tum)   A subjective change in body function not apparent to an observer, such as fever or nausea, that indicates the presence of a disease or disorder of the body.

Synapse   The functional junction between two neurons or between a neuron and an effector, such as a muscle or gland; may be electrical or chemical.

Synapsis (sin-AP-sis)   The pairing of homologous chromosomes during prophase I of meiosis.

Synaptic (sin-AP-tik) **cleft**   The narrow gap that separates the axon terminal of one nerve cell from another nerve cell or muscle fiber (cell) and across which a neurotransmitter diffuses to affect the postsynaptic cell.

Synaptic delay   The length of time between the arrival of the action potential at the axon terminal and the membrane potential (IPSP or EPSP) change on the postsynaptic membrane; usually about 0.5 msec.

Synaptic end-bulb   Expanded distal end of an axon terminal that contains synaptic vesicles. Also called a **synaptic knob** or **end foot.**

Synaptic vesicle   Membrane-enclosed sac in a synaptic end-bulb that stores neurotransmitters.

Synarthrosis (sin′-ar-THRŌ-sis)   An immovable joint.

Synchondrosis (sin′-kon-DRŌ-sis)   A cartilaginous joint in which the connecting material is hyaline cartilage.

Syncope (SIN-kō-pē)   Faint; a sudden temporary loss of consciousness associated with loss of postural tone and followed by spontaneous recovery; most commonly caused by cerebral ischemia.

Syndesmosis (sin′-dez-MŌ-sis)   A fibrous joint in which articulating bones are united by dense fibrous connective tissue.

Syndrome (SIN-drōm)   A group of signs and symptoms that occur together in a pattern that is characteristic of a particular disease or abnormal condition.

Syneresis (si-NER-e-sis)   The process of clot retraction.

Synergist (SIN-er-jist)   A muscle that assists the prime mover by reducing undesired action or unnecessary movement.

Synergistic (syn-er-GIS-tik) **effect**   A hormonal interaction in which the effects of two or more hormones complement each other so that the target cell responds to the sum of the hormones involved. An example is the combined actions of estro-

gens, progesterone, prolactin, and oxytocin necessary for lactation.

Synostosis (sin′-os-TŌ-sis)   A joint in which the dense fibrous connective tissue that unites bones at a suture has been replaced by bone, resulting in a complete fusion across the suture line.

Synovial (si-NŌ-vē-al) **cavity**   The space between the articulating bones of a synovial (diarthrotic) joint, filled with synovial fluid. Also called a **joint cavity.**

Synovial fluid   Secretion of synovial membranes that lubricates joints and nourishes articular cartilage.

Synovial joint   A fully movable or diarthrotic joint in which a synovial (joint) cavity is present between the two articulating bones.

Synovial membrane   The inner of the two layers of the articular capsule of a synovial joint, composed of areolar connective tissue that secretes synovial fluid into the synovial (joint) cavity.

Syphilis (SIF-i-lis)   A sexually transmitted disease caused by the bacterium *Treponema pallidum.*

System   An association of organs that have a common function.

Systemic (sis-TEM-ik)   Affecting the whole body; generalized.

Systemic anatomy   The study of particular systems of the body, such as the skeletal, muscular, nervous, cardiovascular, or urinary systems.

Systemic circulation   The routes through which oxygenated blood flows from the left ventricle through the aorta to all the organs of the body and deoxygenated blood returns to the right atrium.

Systemic lupus erythematosus (er-i-them-a-TŌ-sus) **(SLE)**   An autoimmune, inflammatory disease that may affect every tissue of the body.

Systemic vascular resistance (SVR)   All the vascular resistances offered by systemic blood vessels. Also called **total peripheral resistance.**

Systole (SIS-tō-lē)   In the cardiac cycle, the phase of contraction of the heart muscle, especially of the ventricles.

Systolic (sis-TO-lik) **blood pressure**   The force exerted by blood on arterial walls during ventricular contraction; the highest pressure measured in the large arteries, about 120 mm Hg under normal conditions for a young, adult male.

Tachycardia (tak′-i-KAR-dē-a)   An abnormally rapid resting heartbeat or pulse rate (over 100/min).

Tactile (TAK-tīl)   Pertaining to the sense of touch.

Tactile disc   Modified epidermal cell in the stratum basale of hairless skin that functions as a cutaneous receptor for discriminative touch. Also called a **Merkel** ((MER-kel) **disc.**

Taenia coli (TĒ-nē-a KŌ-lī)   One of three flat bands of thickened, longitudinal muscles running the length of the large intestine.

Target cell   A cell whose activity is affected by a particular hormone.

Tarsal gland   Sebaceous (oil) gland that opens on the edge of each eyelid. Also called a **Meibomian** (mī-BŌ-mē-an) **gland.**

Tarsal plate   A thin, elongated sheet of connective tissue, one in each eyelid, giving the eyelid form and support. The aponeurosis of the levator palpebrae superioris is attached to the tarsal plate of the superior eyelid.

Tarsus (TAR-sus)   A collective term for the seven bones of the ankle.

Tay-Sachs (TĀ SAKS) **disease**   Inherited, progressive neuronal degeneration of the central nervous system due to a deficient lysosomal enzyme that causes excessive accumulations of a lipid called ganglioside.

T cell   A lymphocyte that becomes immunocompetent in the thymus gland and differentiates into one of several kinds of effector cells that function in cell-mediated immunity.

Tectorial (tek-TŌ-rē-al) **membrane**   A gelatinous membrane projecting over and in contact with the hair cells of the spiral organ (organ of Corti) in the cochlear duct.

Telophase (TEL-ō-fāz)   The final stage of mitosis in which the daughter nuclei become established.

Temporomandibular joint (TMJ) syndrome   A disorder of the temporomandibular joint (TMJ) characterized by dull pain around the ear, tenderness of jaw muscles, a clicking or popping noise when opening or closing the mouth, limited or abnormal opening of the mouth, headache, tooth sensitivity, and abnormal wearing of the teeth.

Tendon (TEN-don)   A white fibrous cord of dense, regularly arranged connective tissue that attaches muscle to bone.

Tendon organ   A proprioceptive receptor, sensitive to changes in muscle tension and force of contraction, found chiefly near the junction of tendons and muscles. Also called a **Golgi** (GOL-jē) **tendon organ.**

Tendon reflex   A polysynaptic, ipsilateral reflex that is designed to protect tendons and their associated muscles from damage that might be brought about by excessive tension. The receptors involved are called tendon organs (Golgi tendon organs).

Tenosynovitis (ten′-ō-sin-ō-VĪ-tis)   Inflammation of a tendon sheath and synovial membrane at a joint.

Tentorium cerebelli (ten-TŌ-rē-um ser′-e-BEL-ē)   A transverse shelf of dura mater that forms a partition between the occipital lobe of the cerebral hemispheres and the cerebellum and that covers the cerebellum.

Teratogen (TER-a-tō-jen)   Any agent or factor that causes physical defects in a developing embryo.

Terminal ganglion (TER-min-al GANG-lē-on)   A cluster of cell bodies of postganglionic parasympathetic neurons either lying very close to the visceral effectors or located within the walls of the visceral effectors supplied by the postganglionic fibers.

Testis (TES-tis)   Male gonad that produces sperm and the hormones testosterone and inhibin. Also called a **testicle.**

Testosterone (tes-TOS-te-rōn)   A male sex hormone (androgen) secreted by interstitial endocrinocytes (cells of Leydig) of a mature testis; controls the growth and development of male sex organs, secondary sex characteristics, spermatozoa, and body growth.

Tetanus (TET-a-nus)   An infectious disease caused by the toxin of *Clostridium tetani,* characterized by tonic muscle spasms and exaggerated reflexes, lockjaw, and arching of the back. A smooth, sustained contraction produced by a series of very rapid stimuli to a muscle.

Tetany (TET-a-nē)   A nervous condition caused by hypoparathyroidism and characterized by intermittent or continuous tonic muscular contractions of the extremities.

Tetralogy of Fallot (tet-RAL-ō-jē of fal-Ō)   A combination of four congenital heart defects: (1) constricted pulmonary semilunar valve, (2) interventricular septal opening, (3) emergence of aorta from both ventricles instead of from the left only, and (4) enlarged right ventricle.

Thalamus (THAL-a-mus)   A large, oval structure located above the midbrain, consisting of two masses of gray matter covered by a thin layer of white matter.

Thalassemia (thal′-a-SĒ-mē-a)   A group of hereditary hemolytic anemias.

Thallium (THAL-ē-um) **imaging**   Diagnostic procedure used to evaluate blood flow through coronary arteries, cardiac disorders, and effectiveness of drug therapy; thallium concentrates in healthy myocardial tissue.

Therapy (THER-a-pē)   The treatment of a disease or disorder.

Thermoreceptor (THER-mō-rē-sep-tor)   Receptor that detects changes in temperature.

Thigh   The portion of the lower extremity between the hip and the knee.

Third ventricle (VEN-tri-kul)   A slitlike cavity between the right and left halves of the thalamus and between the lateral ventricles of the brain.

Thoracic (thō-RAS-ik) **cavity**   Superior component of the ventral body cavity that contains two pleural cavities, the mediastinum, and the pericardial cavity.

Thoracic duct   A lymphatic vessel that begins as a dilation called the cisterna chyli, receives lymph from the left side of the head, neck, and chest, the left arm, and the entire body below the ribs, and empties into the left subclavian vein. Also called the **left lymphatic** (lim-FAT-ik) **duct.**

Thoracolumbar (thō′-ra-kō-LUM-bar) **outflow**   The fibers of the sympathetic preganglionic neurons, which have their cell bodies in the lateral gray columns of the thoracic segment and first two or three lumbar segments of the spinal cord.

Thorax (THŌ-raks)   The chest.

Threshold potential   The membrane voltage that must be reached to trigger an action potential.

Threshold stimulus   Any stimulus strong enough to initiate an action potential.

Thrombin (THROM-bin)   The active enzyme formed from prothrombin that acts to convert fibrinogen to fibrin.

Thrombocyte (THROM-bō-sīt)   A fragment of cytoplasm enclosed in a cell membrane and lacking a nucleus; found in the circulating blood; plays a role in blood clotting. Also called a **platelet** (PLĀT-let).

Thrombolytic (throm-bō-LIT-ik) **agent**   Chemical substance injected into the body to dissolve blood clots and restore circulation; mechanism of action is direct or indirect activation of plasminogen; examples include tissue plasminogen activator (t-PA), streptokinase, and urokinase.

Thrombophlebitis (throm′-bo-fle-BĪ-tis)   A disorder in which inflammation of the wall of a vein is followed by the formation of a blood clot (thrombus).

Thrombosis (throm-BŌ-sis)   The formation of a clot in an unbroken blood vessel, usually a vein.

Thrombus   A clot formed in an unbroken blood vessel, usually a vein.

Thymus (THĪ-mus) **gland**   A bilobed organ, located in the upper mediastinum posterior to the sternum and between the lungs that plays an essential role in the immune mechanism of the body.

Thyroglobulin (thı-ro-GLŌ-byoo-lin) **(TGB)**   A large glycoprotein molecule produced by follicle cells of the thyroid gland in which iodine is combined with tyrosine to form thyroid hormones.

Thyroid cartilage (THĪ-royd KAR-ti-lij)   The largest single cartilage of the larynx, consisting of two fused plates that form the anterior wall of the larynx. Also called the **Adam's apple.**

Thyroid colloid (KOL-loyd)   A complex in thyroid follicles consisting of thyroglobulin and stored thyroid hormones.

Thyroid follicle (FOL-i-kul)   Spherical sac that forms the parenchyma of the thyroid gland and consists of follicular cells that produce thyroxine (T_4) and triiodothyronine (T_3) and parafollicular cells that produce calcitonin (CT).

Thyroid function tests   Tests used to evaluate a swelling or lump in the thyroid gland, ascertain symptoms of abnormal thyroxine levels, monitor responses of thyroid diseases to therapy, and screen newborns for cretinism. Examples are serum T_4 concentration, serum T_3 concentration, and serum TSH concentration tests.

Thyroid gland   An endocrine gland with right and left lateral lobes on either side of the trachea connected by an isthmus located in front of the trachea just below the cricoid cartilage.

Thyroid-stimulating hormone (TSH)   A hormone secreted by the anterior pituitary gland that stimulates the synthesis and secretion of hormones produced by the thyroid gland.

Thyroxine (thī-ROK-sēn) **(T_4)**   A hormone secreted by the thyroid that regulates organic metabolism, growth and development, and the activity of the nervous system.

Tic   Spasmodic twitching made involuntarily by muscles that are ordinarily under voluntary control.

Tidal volume   The volume of air breathed in and out in any one breath; about 500 ml in quiet, resting conditions.

Tinnitus (ti-NĪ-tus)   A ringing, roaring, or clicking in the ears.

Tissue   A group of similar cells and their intercellular substance joined together to perform a specific function.

Tissue factor (TF)   A factor, or collection of factors, whose appearance initiates the blood clotting process. Also called **thromboplastin** (throm-bō-PLAS-tin).

Tissue macrophage system   General term that refers to wandering and fixed macrophages.

Tissue plasminogen activator (t-PA)   An enzyme that dissolves small blood clots by initiating a process that converts plasminogen to plasmin, which degrades the fibrin of a clot.

Tissue rejection   Phenomenon by which the body recognizes the protein (HLA antigens) in transplanted tissues or organs as foreign and produces antibodies against them.

Tongue   A large skeletal muscle covered by a mucous membrane located on the floor of the oral cavity.

Tonometry (tō-NOM-e-trē)   A test used to measure intraocular pressure to screen for glaucoma.

Tonsil (TON-sil)   A multiple aggregation of large lymphatic nodules embedded in mucous membrane.

Topical (TOP-i-kal)   Applied to the surface rather than ingested or injected.

Torn cartilage   A tearing of an articular disk in the knee.

Torpor (TOR-por)   State of lethargy and sluggishness that precedes stupor, which precedes semicoma, which precedes coma.

Total lung capacity   The sum of tidal volume, inspiratory reserve volume, expiratory reserve volume, and residual volume; about 6,000 ml.

Toxic (TOK-sik)   Pertaining to poison; poisonous.

Toxic shock syndrome (TSS)   A disease caused by the bacterium *Staphylococcus aureus,* occurring among menstruating females who use tampons and characterized by high fever, sore throat, headache, fatigue, irritability, and abdominal pain.

Trabecula (tra-BEK-yoo-la)   Irregular latticework of thin plate of spongy bone. Fibrous cord of connective tissue serving as supporting fiber by forming a septum extending into an organ from its wall or capsule. *Plural,* **trabeculae** (tra-BEK-yoo-lē).

Trabeculae carneae (KAR-nē-ē)   Ridges and folds of the myocardium in the ventricles.

Trachea (TRĀ-kē-a)   Tubular air passageway extending from the larynx to the fifth thoracic vertebra. Also called the **windpipe.**

Tracheostomy (trā-kē-OS-tō-mē)   Creation of an opening into the trachea through the neck (below the cricoid cartilage), with insertion of a tube to faciliate passage of air or evacuation of secretions.

Trachoma (tra-KŌ-ma)   A chronic infectious disease of the conjunctiva and cornea of the eye caused by *Chlamydia trachomatis.*

Tract   A bundle of nerve fibers in the central nervous system.

Transcription (trans-KRIP-shun)   The first step in the transfer of genetic information in which a single strand of a DNA molecule serves as a template for the formation of an RNA molecule.

Transfusion (trans-FYOO-shun)   Transfer of whole blood, blood components, or bone marrow directly into the bloodstream.

Transient ischemic (is-KĒ-mik) **attack (TIA)**   Episode of temporary focal, nonconvulsive cerebral dysfunction caused by interference of the blood supply to the brain.

Translation (trans-LĀ-shun)   The synthesis of a new protein on the ribosome of a cell as dictated by the sequence of codons in messenger RNA.

Transplantation (trans-plan-TĀ-shun)   The replacement of injured or diseased tissues or organs with natural ones.

Transport maximum (T_m)   The maximum amount of a substance that can be reabsorbed by renal tubules under any condition.

Transvaginal oocyte retrieval   Procedure in which aspirated secondary oocytes are combined with a solution containing sperm outside the body and then the fertilized ova are implanted in the uterus.

Transverse colon (trans-VERS KŌ-lon)   The portion of the large intestine extending across the abdomen from right colic (hepatic) flexure to the left colic (splenic) flexure.

Transverse fissure (FISH-er)   The deep cleft that separates the cerebrum from the cerebellum.

Transverse tubules (TOO-byools) **(T tubules)**   Minute, cylindrical invaginations of the muscle fiber (cell) membrane that carry the muscle action potentials deep into the muscle fiber.

Trauma (TRAW-ma)   An injury, either a physical wound or psychic disorder, caused by an external agent or force, such as a physical blow or emotional shock; the agent or force that causes the injury.

Traveler's diarrhea   Infectious disease of the gastrointestinal tract that results in loose, urgent bowel movements, cramping, abdominal pain, malaise, nausea, and occasionally fever and dehydration. It is acquired through ingestion of food or water that has become contaminated with fecal material containing mostly bacteria (especially *Escherichia coli*). Also called **Montezuma's revenge, turista,** and **Tut's tummy.**

Tremor (TREM-or)   Rhythmic, involuntary, purposeless contraction of opposing muscle groups.

Treppe (TREP-eh)   The gradual increase in the amount of contraction by a muscle caused by rapid, repeated stimuli of the same strength.

Triad (TRĪ-ad)   A complex of three units in a muscle fiber (cell) composed of a transverse tubule and the segments of sarcoplasmic reticulum on both sides of it.

Tricuspid (trī-KUS-pid) **valve**   Atrioventricular (AV) valve on the right side of the heart.

Trigeminal neuralgia (trī-JEM-i-nal noo-RAL-jē-a)   Pain in one or more of the branches of the trigeminal (V) nerve. Also called **tic douloureux** (doo-loo-ROO).

Triglyceride (trī-GLE-cer-īde)   A lipid compound formed from one molecule of glycerol and three molecules of fatty acids; the body's most highly concentrated source of energy. Prime component of adipose tissue, composed of adipocytes specialized for triglyceride storage and present in the form of soft pads between various organs for support, protection, and insulation. Also called a **neutral fat.**

Trigone (TRĪ-gon)   A triangular area at the base of the urinary bladder.

Triiodothyronine (trī-ī-ōd-ō-THĪ-ro-nēn) **(T₃)**   A hormone produced by the thyroid gland that regulates organic metabolism, growth and development, and the activity of the nervous system.

Trochlea (TROK-lē-a)   A pulleylike surface.

Trophoblast (TRŌF-ō-blast)   The outer covering of cells of the blastocyst.

Tropic (TRŌ-pik) **hormone**   A hormone whose target is another endocrine gland.

Trunk   The part of the body to which the upper and lower extremities are attached.

Tubal ligation (lī-GĀ-shun)   A sterilization procedure in which the uterine (Fallopian) tubes are tied and cut.

Tuberculosis (too-berk-yoo-LŌ-sis)   An infection of the lungs and pleurae caused by *Mycobacterium tuberculosis* resulting in destruction of lung tissue and its replacement by fibrous connective tissue.

Tubular reabsorption   The movement of filtrate from renal tubules back into blood in response to the body's specific needs.

Tubular secretion   The movement of substances in blood back into filtrate in response to the body's specific needs.

Tumor (TOO-mor)   A growth of excess tissue due to an unusually rapid division of cells.

Tunica albuginea (TOO-ni-ka al′-byoo-JIN-ē-a)   A dense layer of white fibrous tissue covering a testis or deep to the surface of an ovary.

Tunica externa (eks-TER-na)   The inner coat of an artery or vein, composed mostly of elastic and collagen fibers. Also called the **adventitia.**

Tunica interna (in-TER-na)   The inner coat of an artery or vein, consisting of a lining of endothelium, basement membrane, and internal elastic lamina. Also called the **tunica intima** (IN-ti-ma).

Tunica media (MĒ-dē-a)   The middle coat of an artery or vein, composed of smooth muscle and elastic fibers.

T wave   The deflection wave of an electrocardiogram that represents ventricular repolarization.

Twitch   Rapid, jerky contraction of a muscle in response to a single stimulus.

Tympanic antrum (tim-PAN-ik AN-trum)   An air space in the posterior wall of the middle ear that leads into the mastoid air cells or sinus.

Tympanic membrane   A thin, semitransparent partition of fibrous connective tissue between the external auditory meatus and the middle ear. Also called the **eardrum.**

Type II cutaneous mechanoreceptor   A receptor embedded deeply in the dermis and deeper tissues that detects heavy and continuous touch sensations. Also called an **end organ of Ruffini.**

Ulcer (UL-ser)   An open lesion of the skin or a mucous membrane of the body with loss of substance and necrosis of the tissue.

Ultrasound (US)   Medical imaging technique that utilizes high-frequency sound waves to produce an image called a **sonogram.**

Umbilical (um-BIL-i-kal)   Pertaining to the umbilicus or navel.

Umbilical cord   The long, ropelike structure containing the umbilical arteries and vein that connect the fetus to the placenta.

Umbilicus (um-BIL-i-kus or um-bil-Ī-kus)   A small scar on the abdomen that marks the former attachment of the umbilical cord to the fetus. Also called the **navel.**

Upper extremity   The appendage attached at the shoulder girdle, consisting of the arm, forearm, wrist, hand, and fingers.

Up-regulation   Phenomenon in which there is an increase in the number of receptors in response to a deficiency of a hormone or neurotransmitter.

Uremia (yoo-RĒ-mē-a)   Accumulation of toxic levels of urea and other nitrogenous waste products in the blood, usually resulting from severe kidney malfunction.

Ureter (YOO-re-ter)   One of two tubes that connect the kidney with the urinary bladder.

Urethra (yoo-RĒ-thra)   The duct from the urinary bladder to the exterior of the body that conveys urine in females and urine and semen in males.

Urinalysis   The physical, chemical, and microscopic analysis or examination of urine.

Urinary (YOO-ri-ner-ē) **bladder** A hollow, muscular organ situated in the pelvic cavity posterior to the pubic symphysis.

Urinary tract infection (UTI) An infection of a part of the urinary tract or the presence of large numbers of microbes in urine.

Urine The fluid produced by the kidneys that contains wastes or excess materials and is excreted from the body through the urethra.

Urobilinogenuria (yoo-rō-bī'-lin-ō-je-NOOR-ē-a) The presence of urobilinogen in urine.

Urogenital (yoo'-rō-JEN-i-tal) **triangle** The region of the pelvic floor below the pubic symphysis, bounded by the pubic symphysis and the ischial tuberosities and containing the external genitalia.

Urology (yoo-ROL-ō-jē) The specialized branch of medicine that deals with the structure, function, and diseases of the male and female urinary systems and the male reproductive system.

Urticaria (ur'-ti-KĀ-rē-a) A skin reaction to certain foods, drugs, or other substances to which a person may be allergic; hives.

Uterine (YOO-ter-in) **tube** Duct that transports ova from the ovary to the uterus. Also called the **Fallopian** (fal-LŌ-pē-an) **tube** or **oviduct.**

Uterosacral ligament (yoo'-ter-ō-SĀ-kral LIG-a-ment) A fibrous band of tissue extending from the cervix of the uterus laterally to attach to the sacrum.

Uterovesical (yoo'-ter-ō-VES-ī-kal) **pouch** A shallow pouch formed by the reflection of the peritoneum from the anterior surface of the uterus, at the junction of the cervix and the body, to the posterior surface of the urinary bladder.

Uterus (YOO-te-rus) The hollow, muscular organ in females that is the site of menstruation, implantation, development of the fetus, and labor. Also called the **womb.**

Utricle (YOO-tri-kul) The larger of the two divisions of the membranous labyrinth located inside the vestibule of the inner ear, containing a receptor organ for static equilibrium.

Uvea (YOO-vē-a) The three structures that together make up the vascular tunic of the eye.

Uvula (YOO-vyoo-la) A soft, fleshy mass, especially the V-shaped pendant part, descending from the soft palate.

Vacuole (VAK-yoo-ōl) Membrane-bound organelle that in animal cells frequently functions in temporary storage or transportation.

Vagina (va-JĪ-na) A muscular, tubular organ that leads from the uterus to the vestibule, situated between the urinary bladder and the rectum of the female.

Valvular stenosis (VAL-vyoo-lar ste-NŌ-sis) A narrowing of heart valve, usually the bicuspid (mitral) valve.

Varicocele (VAR-i-kō-sēl) A twisted vein; especially, the accumulation of blood in the veins of the spermatic cord.

Varicose (VAR-i-kōs) Pertaining to an unnatural swelling, as in the case of a varicose vein.

Vas A vessel or duct.

Vasa recta (VĀ-sa REK-ta) Extensions of the efferent arteriole of a juxtamedullary nephron that run alongside the loop of the nephron (Henle) in the medullary region.

Vasa vasorum (va-SŌ-rum) Blood vessels that supply nutrients to the larger arteries and veins.

Vascular (VAS-kyoo-lar) Pertaining to or containing many blood vessels.

Vascular spasm Contraction of the smooth muscle in the wall of a damaged blood vessel to prevent blood loss.

Vascular (venous) sinus A vein with a thin endothelial wall that lacks a tunica media and externa and is supported by surrounding tissue.

Vascular tunic (TOO-nik) The middle layer of the eyeball, composed of the choroid, ciliary body, and iris. Also called the **uvea** (YOO-vē-a).

Vasectomy (va-SEK-tō-mē) A means of sterilization of males in which a portion of each ductus (vas) deferens is removed.

Vasoconstriction (vāz-ō-kon-STRIK-shun) A decrease in the size of the lumen of a blood vessel caused by contraction of the smooth muscle in the wall of the vessel.

Vasodilation (vās'-ō-DĪ-lā-shun) An increase in the size of the lumen of a blood vessel caused by relaxation of the smooth muscle in the wall of the vessel.

Vasomotion (vāz-ō-MŌ-shun) Intermittent contraction and relaxation of the smooth muscle of the metarterioles and precapillary sphincters that result in an intermittent blood flow.

Vasomotor (vā-sō-MŌ-tor) **center** A cluster of neurons in the medulla that controls the diameter of blood vessels, especially arteries.

Vein A blood vessel that conveys blood from tissues back to the heart.

Vena cava (VĒ-na KĀ-va) One of two large veins that open into the right atrium, returning to the heart all of the deoxygenated blood from the systemic circulation except from the coronary circulation.

Venesection (vēn'-e-SEK-shun) Opening of a vein for withdrawal of blood.

Ventral (VEN-tral) Pertaining to the anterior or front side of the body; opposite of dorsal.

Ventral body cavity Cavity near the ventral aspect of the body that contains viscera and consists of a superior thoracic cavity and an inferior abdominopelvic cavity.

Ventral ramus (RĀ-mus) The anterior branch of a spinal nerve, containing sensory and motor fibers to the muscles and skin of the anterior surface of the head, neck, trunk, and the extremities.

Ventricle (VEN-tri-kul) A cavity in the brain or an inferior chamber of the heart.

Ventricular fibrillation (ven-TRIK-yoo-lar fib-ri-LĀ-shun) Asynchronoous ventricular contractions that result in cardiovascular failure.

Venule (VEN-yool) A small vein that collects blood from capillaries and delivers it to a vein.

Vermiform appendix (VER-mi-form a-PEN-diks) A twisted, coiled tube attached to the cecum.

Vermilion (ver-MIL-yon) The area of the mouth where the skin on the outside meets the mucous membrane on the inside.

Vermis (VER-mis)   The central constricted area of the cerebellum that separates the two cerebellar hemispheres.

Vertebral (VER-te-bral) **canal**   A cavity within the vertebral column formed by the vertebral foramina of all the vertebrae and containing the spinal cord. Also called the **spinal canal.**

Vertebral column   The 26 vertebrae; encloses and protects the spinal cord and serves as a point of attachment for the ribs and back muscles. Also called the **spine, spinal column,** or **backbone.**

Vertigo (VER-ti-go)   Sensation of spinning or movement.

Vesicle (VES-i-kul)   A small bladder or sac containing liquid.

Vesicular ovarian follicle   A relatively large, fluid-filled follicle containing an immature ovum and its surrounding tissues that secretes estrogens. Also called a **Graafian** (GRAF-ē-an) **follicle.**

Vestibular (ves-TIB-yoo-lar) **apparatus**   Collective term for the organs for equilibrium, which includes the saccule, utricle, and semicircular ducts.

Vestibular membrane   The membrane that separates the cochlear duct from the scala vestibuli.

Vestibule (VES-ti-byool)   A small space or cavity at the beginning of a canal, especially the inner ear, larynx, mouth, nose, and vagina.

Villus (VIL-lus)   A projection of the intestinal mucosal cells containing connective tissue, blood vessels, and a lymphatic vessel; functions in the absorption of the end products of digestion. *Plural,* **villi** (VIL-Ī).

Viscera (VIS-er-a)   The organs inside the ventral body cavity. *Singular,* **viscus** (VIS-kus).

Visceral (VIS-er-al)   Pertaining to the organs or to the covering of an organ.

Visceral effector (e-FEK-tor)   Cardiac muscle, smooth muscle, and glandular epithelium.

Visceral muscle   An organ specialized for contraction, composed of smooth muscle fibers (cells), located in the walls of hollow internal structures, and stimulated by visceral efferent neurons.

Visceral pleura (PLOO-ra)   The inner layer of the serous membrane that covers the lungs.

Visceroceptor (vis′-er-ō-SEP-tor)   Receptor that provides information about the body's internal environment.

Viscosity (vis-KOS-i-tē)   The state of being sticky or thick.

Vital capacity   The sum of inspiratory reserve volume, tidal volume, and expiratory reserve volume; about 4,800 ml.

Vital signs   Signs necessary to life that include temperature (T), pulse (P), respiratory rate (RR), and blood pressure (BP).

Vitamin   An organic molecule necessary in trace amounts that acts as a catalyst in normal metabolic processes in the body.

Vitiligo (vit-i-LĪ-go)   Patchy, white spots on the skin due to partial or complete loss of melanocytes.

Vitreous (VIT-rē-us) **body**   A soft, jellylike substance that fills the vitreous chambers of the eyeball, lying between the lens and the retina.

Vocal folds   Pair of mucous membrane folds below the ventricular folds that function in voice production. Also called **true vocal cords.**

Volkmann's canal   *See* **Perforating canal.**

Voltage-gated channel   An ion channel in a plasma membrane composed of integral proteins that functions like a gate to permit or restrict the movement of ions in response to changes in the voltage across the membrane.

Vomiting   Forcible expulsion of the contents of the upper gastrointestinal tract through the mouth.

Vulva (VUL-va)   Collective designation for the external genitalia of the female. Also called the **pudendum** (poo-DEN-dum).

Wallerian (wal-LE-rē-an) **degeneration**   Degeneration of the portion of the axon and myelin sheath of a neuron distal to the site of injury.

Wandering macrophage (MAK-rō-fāj)   Phagocytic cell that develops from a monocyte, leaves the blood, and migrates to infected tissues.

Wart   Generally benign tumor of epithelial skin cells caused by a virus.

Wave summation (sum-MA-shun)   The algebraic addition of the excitatory and inhibitory effects of many stimuli applied to a nerve cell body. The increased strength of muscle contraction that results when stimuli follow in rapid succession. Also called **temporal summation.**

Wheal (hwēl)   Elevated lesion of the skin.

White matter   Aggregations or bundles of myelinated axons located in the brain and spinal cord.

White pulp   The portion of the spleen composed of lymphatic tissue, mostly lymphocytes, arranged around central arteries; in some areas of white pulp, lymphocytes are thickened into lymphatic nodules called splenic nodules (Malpighian corpuscles).

White ramus communicans (RA-mus ko-MYOO-ni-kans)   The portion of a preganglionic sympathetic nerve fiber that branches away from the anterior ramus of a spinal nerve to enter the nearest sympathetic trunk ganglion.

Xiphoid (ZĪ-foyd)   Sword-shaped. The lowest portion of the sternum is the **xiphoid process.**

Yolk sac   An extraembryonic membrane that connections with the midgut during early embryonic development, but is nonfunctional in humans.

Zona fasciculata (ZŌ-na fa-sik′-yoo-LA-ta)   The middle zone of the adrenal cortex that consists of cells arranged in long, straight cords and that secretes glucocorticoid hormones.

Zona glomerulosa (glo-mer′-yoo-LŌ-sa)   The outer zone of the adrenal cortex, directly under the connective tissue covering, that consists of cells arranged in arched loops or round balls and that secretes mineralocorticoid hormones.

Zona pellucida (pe-LOO-si-da)   Gelatinous glycoprotein layer internal to the corona radiata that surrounds a secondary oocyte.

Zona reticularis (ret-ik′-yoo-LAR-is)   The inner zone of the adrenal cortex, consisting of cords of branching cells that secrete sex hormones, chiefly androgens.

Zygote (ZĪ-gōt)   The single cell resulting from the union of a male and female gamete; the fertilized ovum.

CREDITS

Chapter 9

Fig. 9.1b: J. A. Gosling, P. F. Harris, et al., *Atlas of Human Anatomy,* 2nd ed., Gower Medical Publishing Ltd., 1990.

Fig. 9.3a: Evan J. Colella.

Fig. 9.3b–g: Copyright © 1983 by Gerard J. Tortora. Courtesy of Lynne and James Borghesi.

Fig. 9.4: Evan J. Colella.

Fig. 9.5a–c: Copyright © 1983 by Gerard J. Tortora. Courtesy of Lynne and James Borghesi.

Fig. 9.5d–e: Copyright © 1993 by Gerard J. Tortora.

Fig. 9.6a–b: Copyright © 1983 by Gerard J. Tortora. Courtesy of Lynne and James Borghesi.

Fig. 9.6c–d: © 1992, Scott, Foresman Photo, John Moore.

Fig. 9.6e–f: Copyright © 1983 by Gerard J. Tortora. Courtesy of Lynne and James Borghesi.

Fig. 9.6g: Evan J. Colella.

Fig. 9.6h: Copyright © 1983 by Gerard J. Tortora. Courtesy of Lynne and James Borghesi.

Fig. 9.7d: J. A. Gosling, P. F. Harris, et al., *Atlas of Human Anatomy,* 2nd ed., Gower Medical Publishing Ltd., 1990.

Fig. 9.8: © 1988, Camazine, Photo Researchers.

Chapter 10

Fig. 10.3a: FUJIJA.

Fig. 10.5: Dr. H. E. Huxley.

Fig. 10.16a–c: © 1992, Scott, Foresman Photo, John Moore.

Chapter 12

Fig. 12.2b: © 1980, Schultz, Biology Media, Photo Researchers.

Fig. 12.3b: © 1980, Schultz, Biology Media, Photo Researchers.

Fig. 12.4c: Biophoto Associates, Photo Researchers.

Chapter 13

Fig. 13.1b: N. Gluhbegovic and T. H. Williams, *The Human Brain: A Photographic Guide,* Harper & Row, Publishers, Inc., Hagerstown, MD, 1980.

Fig. 13.2b–c: N. Gluhbegovic and T. H. Williams, *The Human Brain: A Photographic Guide,* Harper & Row, Publishers, Inc., Hagerstown, MD, 1980.

Fig. 13.3b: Victor Eroschkenko.

Fig. 13.11b: Richard G. Kessel and Randy H. Kardon, from *Tissues and Organs: A Text-Atlas of Scanning Electron Microscopy.* Copyright © 1979 by W. H. Freeman and Company. Reprinted by permission.

Chapter 14

Fig. 14.1b: © 1984, Rotker, Photo Researchers.

Fig. 14.10b: Martin Rotker, Phototake.

Fig. 14.11: N. Gluhbegovic and T. H. Williams, *The Human Brain: A Photographic Guide,* Harper & Row, Publishers, Inc., Hagerstown, MD, 1980.

Fig. 14.12b: N. Gluhbegovic and T. H. Williams, *The Human Brain: A Photographic Guide,* Harper & Row, Publishers, Inc., Hagerstown, MD, 1980.

Chapter 16

Fig. 16.3: © 1992, Scott, Foresman Photo, John Moore.

Fig. 16.6: © 1991, Eagle, Photo Researchers.

Fig. 16.14: N. Gluhbegovic and T. H. Williams, *The Human Brain: A Photographic Guide,* Harper & Row, Publishers, Inc., Hagerstown, MD, 1980.

Chapter 18

Fig. 18.9a–b: Hall/Evered: *A Colour Atlas of Endocrinology,* Mosby-Year Book/Wolfe Publishing, 1990.

Fig. 18.13b: Lester V. Bergman & Associates, Inc.

Fig. 18.16a–b: Lester V. Bergman & Associates, Inc.

Fig. 18.16c: Copyright © Kay, Peter Arnold.

Fig. 18.18b: © Biophoto, Photo Researchers.

Fig. 18.19b: Andrew J. Kuntzman.

Fig. 18.22a–b: Hall/Evered: *A Colour Atlas of Endocrinology,* Mosby-Year Book/Wolfe Publishing, 1990.

Fig. 18.23b: © Biophoto, Photo Researchers.

Chapter 19

Fig. 19.6 (eosinophil, basophil, and monocyte): © Biophoto, Photo Researchers.

Fig. 19.6 (neutrophil): Ed Reschke.

Fig. 19.6 (lymphocyte, red blood cells, and platelet): Lester V. Berman & Associates, Inc.

Fig. 19.6 (blood smear): Douglas Merrill.

Fig. 19.9: Lennart Nilsson, *The Body Victorious,* Dell Publishing Company, © Boehringer Ingelheim International GmbH.

Fig. 19.12a–b: Lester V. Berman & Associates, Inc.

Fig. 19.13: Fisher Scientific Company and S.T.E.M. Laboratories, Inc. Copyright © 1975.

Chapter 20

Fig. 20.1b: Stephen A. Kieffer and E. Robert Heitzman, *An Atlas of Cross-Sectional Anatomy,* Harper & Row, Publishers, Inc., New York, 1979.

Fig. 20.3b: J. A. Gosling, P. F. Harris, et al., *Atlas of Human Anatomy,* 2nd ed., Gower Medical Publishing Ltd., 1990.

Fig. 20.4c: J. Willis Hurst et al., *Atlas of the Heart,* Gower Medical Publishing Ltd., 1988.

Fig. 20.4d: John Eads.

Fig. 20.12: © 1988, McElroy, Woodfin Camp.

Fig. 20.14: Sklar/Photo Researchers.

Chapter 21

Fig. 21.1d: Andrew J. Kuntzman.

Fig. 21.4b: Lennart Nilsson, *Behold Man,* Delacorte Press, Dell Publishing.

Fig. 21.5c: Friend/Fawcett, Visuals Unlimited.

Fig. 21.6: J. A. Gosling, P. F. Harris, et al., *Atlas of Human Anatomy,* 2nd ed., Gower Medical Publishing Ltd., 1990.

Chapter 22

Fig. 22.5b: Leroy, Biocosmos Photo Library, Photo Researchers.
Fig. 22.6a–b: J. A. Gosling, P. F. Harris, et al., *Atlas of Human Anatomy,* 2nd ed., Gower Medical Publishing Ltd., 1990.
Fig. 22.7b: Andrew J. Kuntzman.
Fig. 22.10b: © Biology Media/Science Source, Photo Researchers.
Fig. 22.21b: Wagner, Phototake.

Chapter 23

Fig. 23.1b: J. A. Gosling, P. F. Harris, et al., *Atlas of Human Anatomy,* 2nd ed., Gower Medical Publishing Ltd., 1990.
Fig. 23.2a: Copyright © 1982 by Gerard J. Tortora. Courtesy of Lynne Borghesi.
Fig. 23.2c: J. A. Gosling, P. F. Harris, et al., *Atlas of Human Anatomy,* 2nd ed., Gower Medical Publishing Ltd., 1990.
Fig. 23.3: J. A. Gosling, P. F. Harris, et al., *Atlas of Human Anatomy,* 2nd ed., Gower Medical Publishing Ltd., 1990.
Fig. 23.5: J. A. Gosling, P. F. Harris, et al., *Atlas of Human Anatomy,* 2nd ed., Gower Medical Publishing Ltd., 1990.
Fig. 23.6a: © Cunningham, Visuals Unlimited.
Fig. 23.6b–c: Andrew J. Kuntzman.
Fig. 23.7b: © Webb, Phototake.
Fig. 23.8: J. A. Gosling, P. F. Harris, et al., *Atlas of Human Anatomy,* 2nd ed., Gower Medical Publishing Ltd., 1990.
Fig. 23.10b: © Biophoto, Photo Researchers.
Fig. 23.11c: Richard K. Kessel and Randy H. Kardon, *Tissues and Organs: A Text-Atlas of Scanning Electon Microscopy.* Copyright © 1979 by W. H. Freeman and Company. Reprinted by permission.
Fig. 23.31a–b: Gordon T. Hewlett.

Chapter 24

Fig. 24.5b: Douglas Merrill.
Fig. 24.9: © Bruce Iverson.
Fig. 24.10b: Lester W. Paul and John H. Juhl, *The Essentials of Roentgen Interpretation,* 3rd ed., Harper & Row, Publishers, Inc., New York, 1972.
Fig. 24.11: J. A. Gosling, P. F. Harris, et al., *Atlas of Human Anatomy,* 2nd ed., Gower Medical Publishing Ltd., 1990.
Fig. 24.12c: © Hessler, Visuals Unlimited.
Fig. 24.12d: © 1988, Ed Reschke.
Fig. 24.19b: J. A. Gosling, P. F. Harris, et al., *Atlas of Human Anatomy,* 2nd ed., Gower Medical Publishing Ltd., 1990.
Fig. 24.19c: Stephen A. Kieffer and E. Robert Heitzman, *An Atlas of Cross-Sectional Anatomy,* Harper & Row, Publishers, Inc., New York, 1979.
Fig. 24.20c: Bruce Iverson.
Fig. 24.22c: J. A. Gosling, P. F. Harris, et al., *Atlas of Human Anatomy,* 2nd ed., Gower Medical Publishing Ltd., 1990.
Fig. 24.23a: © Shih & Kessel, Visuals Unlimited.
Fig. 24.23b: © Birke, Peter Arnold.
Fig. 24.23c: © Gravé, Photo Researchers.
Fig. 24.23d: © CNRI, Phototake.
Fig. 24.28c: © Birke, Peter Arnold.
Fig. 24.28d: CNRI, SPL, Photo Researchers.

Chapter 26

Fig. 26.2: Stephen A. Kieffer and E. Robert Heitzman, *An Atlas of Cross-Sectional Anatomy,* Harper & Row, Publishers, Inc., New York, 1979.
Fig. 26.3: J. A. Gosling, P. F. Harris, et al., *Atlas of Human Anatomy,* 2nd ed., Gower Medical Publishing Ltd., 1990.
Fig. 26.4b: J. A. Gosling, P. F. Harris, et al., *Atlas of Human Anatomy,* 2nd ed., Gower Medical Publishing Ltd., 1990.
Fig. 26.6c: CNRI/Science Photo Library/Photo Researchers.
Fig. 26.6d: Richard K. Kessel and Randy H. Kardon, *Tissues and Organs: A Text-Atlas of Scanning Electron Microscopy.* Copyright © 1979 by W. H. Freeman and Company. Reprinted by permission.
Fig. 26.7: Andrew J. Kuntzman.
Fig. 26.22: Matt Iacobino.
Fig. 26.23b: J. A. Gosling, P. F. Harris, et al., *Atlas of Human Anatomy,* 2nd ed., Gower Medical Publishing Ltd., 1990.
Fig. 26.23c: © Iverson, Visuals Unlimited.

Chapter 28

Fig. 28.2a–b: J. A. Gosling, P. F. Harris, et al., *Atlas of Human Anatomy,* 2nd ed., Gower Medical Publishing Ltd., 1990.
Fig. 28.3b: Ed Reschke.
Fig. 28.5b: Fawcett-Phillips, Science Photo Library, Photo Researchers.
Fig. 28.9b: J. A. Gosling, P. F. Harris, et al., *Atlas of Human Anatomy,* 2nd ed., Gower Medical Publishing Ltd., 1990.
Fig. 28.12b–c: J. A. Gosling, P. F. Harris, et al., *Atlas of Human Anatomy,* 2nd ed., Gower Medical Publishing Ltd., 1990.
Fig. 28.13b–c: © Biophoto, Photo Researchers.
Fig. 28.15: © Fawcett/Gaddum-Rosse, Photo Researchers.
Fig. 28.16: Andrew J. Kuntzman.
Fig. 28.20c: J. A. Gosling, P. F. Harris, et al., *Atlas of Human Anatomy,* 2nd ed., Gower Medical Publishing Ltd., 1990.
Fig. 28.21a–b: Xerox Medical Systems, Pasadena, CA.

Chapter 29

Fig. 29.1b: Carolina Biological Supply Company.
Fig. 29.3d: Dr. Allen C. Enders.
Fig. 29.6b: From *From Conception to Birth: The Drama of Life's Beginnings* by Roberts Rugh, Landrum B. Shettles with Richard Einhorn. Copyright © 1971 by Roberts Rugh and Landrum B. Shettles. By permission of Harper & Row, Publishers, Inc.
Fig. 29.7b: M. A. Colin England, *A Colour Atlas of Life Before Birth,* Year Book Medical Publishers, Chicago, 1983.
Fig. 29.20: Courtesy of Michael Mohan and Ohio School Pictures.

LINE ART

Charles Bridgman

16.15

Burmar Technical Corporation

20.7, 20.8, 20.10, 20.12

©Leonard Dank

11.1, 11.2, 11.3a–b, 11.4a–c, 11.5a–b, 11.6a–b, 11.7, 11.8, 11.9a–b, 11.10a–c, 11.11, 11.12, 11.13, 11.14a–d, 11.15a–c, 11.16a–c and inset, 11.17a–f, 11.18, 11.19a.–c, 11.20a–d, 11.21 and inset, 11.22a–d, 11.23a–d, 11.24a–c, Exhibit 11.9, 14.16, 14.18, 18.6, 22.3, 22.7, 24.2, 24.12a–b, 24.22a–b, 24.28a–b, 26.21, 29.3a–c, 29.5a–c

Leonard Dank

Exhibit 4.2 osseous, 6.1a, 6.3a–c, 6.4, 6.6, 6.7, 6.9, 7.1, 7.2, 7.3, 7.4 and inset, 7.5, 7.6, 7.7, 7.8a,b and inset, 7.9 and inset, 7.10 and inset, 7.11, 7.12, 7.13 and inset, 7.14, 7.15 a–d, 7.16, 7.17a–c and inset, 7.17d, 7.18 and inset, 7.19a–c and inset, 7.20 and inset, 7.21, 7.22, 7.23, 8.1, 8.2, 8.3a–c, 8.4, 8.5a–c, 8.6, 8.7a–c and inset, 8.8a–b, 809, 8.10, 8.11, 8.12, 8.13, 8.14, 8.15a–b, 9.1a, 9.2, 9.7a–c, 9.7e, 21.11

Sharon Ellis

12.2a, 12.3a, 12.4, 12.7, 12.19, 13.1a, 13.2a, 13.3a, 13.4, 13.5, 13.7, 13.8, 13.9, 13.10, 13.11a, 13.12, 13.13, Exhibit 13.1, Exhibit 13.2, Exhibit 13.3, Exhibit 13.4, 14.1a, 14.2, 14.3a–b, 14.4a, 14.5a–b, 14.6, 14.7, 14.8a–b, 14.9, 14.10a, 14.10c, 14.12a, 14.13, 14.14, 14.17a–b, 15.4, 15.5, 15.8, 15.10, 15.11, 17.2, 22.1, 22.2a–b, 22.4, 22.5a, 22.8

Jean Jackson

19.11, Exhibit 22.3

Lauren Keswick

3.1, 3.2, 3.2 inset, 3.13a, 3.13 inset, 3.15a, 3.15 inset, 3.16a, 3.16 inset, 3.17, 3.18a, 3.18 inset, 3.19, 3.19 inset, 3.25, 3.26, 3.27, 4.1, 4.2, Exhibit 4.2 areolar, hyaline, Exhibit 4.3, 5.1, 5.2b, 5.3a and inset, 5.5, 5.6, 6.2, Exhibit 10.3

©Biagio John Melloni, Ph.D.

16.5, 16.16, 16.17, 16.18, 16.19a–d, 16.20, 16.21a–b, 16.22a–b, 17.1

Hilda Muinos

10.6, 10.7, 10.8, 10.9, 10.10, 10.11, 20.1a and inset, 20.2, 20.3a, 20.3c and inset, 20.3d–e, 20.4 and inset, 20.5a–b, 20.6a, 20.9, 20.11, 20.13, 20.15a–c, 20.16, 21.1a–c, 21.2, 21.14, 21.19, 21.20a–b, 21.5a–b–d, 21.21, 21.22a–c, 21.23a–d, 21.24, 21.25, 21.26, 21.27, 21.28, 21.29, 21.30a, 21.31a, 21.32, 21.33

Lynn O'Kelley

1.7, 15.1, 15.6a–b, 16.1, 16.2a–c, 16.4 inset, 16.5 inset, 16.7, 16.8 and inset, 16.11, 16.13, 16.17 inset, 16.18 inset, 16.21a inset, 16.22 inset, Exhibit 16.1, Exhibit 16.2, 18.1, 18.5 and inset, 18.10, 18.13a and inset, 18.18a, 18.18c, 18.18 inset, 18.19 and inset, 18.23a, 18.23c, 18.23 inset, 18.25, 23.1a, 23.2b and inset, 23.3 inset, 23.4, 23.6a inset, 23.6b inset, 23.7a, 23.8 inset, 23.9, 23.11

Jared Schneidman Design

1.3, 1.4, 2.1, 2.2, 2.3, 2.4, 2.5, 2.6, 2.7, 2.8, 2.9, 2.10, 2.11, 2.12, 2.13, 2.14, 2.15, 2.16, 2.17, 3.3, 3.4, 3.6, 3.7, 3.8, 3.9, 3.10, 3.11, 3.12, 3.14, 3.15b, 3.20, 3.21, 3.22, 3.23, 3.24, 5.7, 5.8a–b, 6.8, 10.2, 10.12, 10.13, 10.14, 10.15, 12.1a, 12.8, 12.9, 12.10, 12.11, 12.12, 12.13, 12.14, 12.15, 12.16, 12.17, 12.18, Exhibit 12.2, 13.6, 15.7, 15.9, 16.9, 16.12, 17.3, Exhibit 17.1, 18.2, 18.3, 18.4, 18.7, 18.8, 18.11, 18.12, 18.14, 18.15, 18.17, 18.20, 18.21, 18.24, 18.26, Exhibit 18.1, 19.1, 19.4, 19.5, 19.10, 21.7, 21.8, 21.9, 21.10, 21.12, 21.13, 21.15, 21.16, 21.17, 21.18, 22.9, 22.12, 22.13, 22.14, 22.15, 22.16, 22.17, 22.18, 22.19, 22.20, 23.12, 23.15, 23.16, 23.17, 23.18, 23.19, 23.23, 23.24, 23.25, 23.26, 23.27, 23.28, 23.29, 23.30, 24.14, 24.14, 24.15, 24.16, 24.18, 24.21, 24.25a–b, 24.26, 25.3, 25.5, 25.6, 25.7, 25.13, 25.14, 25.16, 26.9b, 26.11, 26.12, 26.13a–c, 26.14, 26.15, 26.16, 26.17a–c, 26.19, 26.20, 27.1, 27.3, 27.4, 27.5, 27.6, 28.4, 28.6, 28.7, 28.14, 28.22, 28.23, 28.24, 28.25, 28.26, 29.2, 29.13, 29.14, 29.15, 29.16, 29.17, 29.18, 29.19

Nadine Sokol

Exhibit 4.1 line art, 16.4a–b, 19.2, 19.3a, 19.8, 22.10a, 22.11, 24.1, 24.3a–d and inset, 24.4, 24.5a, 24.6 and inset, 24.7, 24.8, 24.10a, 24.11a, 24.17, 24.19a, 24.19c inset, 24.20a–b, 24.27a–b, 24.29, 26.1, 26.2 inset, 26.4a and inset, 26.5a–b, 26.6a–b, 26.8a–b, 26.9a and inset, 26.10, 26.18, 26.23a, 26.23b inset, 26.24

Kevin Somerville

1.1, 1.2, 1.5, 1.6, 1.8, 1.9, 1.10, 1.11a–c, 1.12, Exhibit 11.1, Exhibit 11.18, 12.1b, Exhibit 12.1, 14.1a inset, 14.4a inset, 14.4b, 14.7 inset, 14.8a inset, 14.9 inset, 14.11 inset, 14.12b inset, 14.13 inset, 14.17b inset, 14.19, 14.20, 15.2, Exhibit 15.1, 21.4a, 23.10a, 23.13a–c, 23.24 inset, 28.1 and inset, 28.3a and inset, 28.3c and inset, 28.5a, 28.8, 28.9a, 28.9b inset, 28.10a–b and inset, 28.11 and inset, 28.12a, 28.12c inset, 28.12d and inset, 28.13a, 28.17. 28.18, 28.19, 28.20a–b, 28.27, 28.28, 29.1a, 29.4, 29.6a, 29.7a, 29.8, 29.10, 29.11a–b, 29.12a–c, Exhibit 29.2

Beth Willert

10.1, 10.3b–d, 10.4, 10.17a–b, 10.18a–c, 10.19a–b, 15.3

INDEX

EPONYMS USED
IN THIS TEXT

Eponymous terms are those named after a person. In general, eponyms should be avoided where possible, since they are totally nondescriptive, often vague, and do not necessarily indicate that the person whose name is used actually contributed anything very original. However, since eponyms are still in frequent use, this glossary has been prepared to indicate which current terms have been used to replace eponyms in this book. In the body of the text eponyms are cited in parentheses, immediately following the current terms where they are used for the first time in a chapter or later in the book. In addition, although eponyms are included in the index, they have been cross-referenced to their current terminology.

EPONYM	CURRENT TERMINOLOGY
Achilles tendon	calcaneal tendon
Adam's apple	thyroid cartilage
ampulla of Vater (VA-ter)	hepatopancreatic ampulla
Bartholin's (BAR-tō-linz) gland	greater vestibular gland
Billroth's (BIL-rōtz) cord	splenic cord
Bowman's (BŌ-manz) capsule	glomerular capsule
Bowman's (BŌ-manz) gland	olfactory gland
Broca's (BRŌ-kaz) area	motor speech area
Brunner's (BRUN-erz) gland	duodenal gland
bundle of His (HISS)	atrioventricular (AV) bundle
canal of Schlemm (SHLEM)	scleral venous sinus
circle of Willis (WIL-is)	cerebral arterial circle
Cooper's (KOO-perz) ligament	suspensory ligament of the breast
Cowper's (KOW-perz) gland	bulbourethral gland
crypt of Lieberkuühn (LĒ-ber-kyoon)	intestinal gland
duct of Rivinus (re-VĒ-nus)	lesser sublingual duct
duct of Santorini (san′-tō-RĒ-nē)	accessory duct
duct of Wirsung (VĒR-sung)	pancreatic duct
end organ of Ruffini (roo-FĒ-nē)	type II cutaneous mechanoreceptor
Eustachian (yoo-STĀ-kē-an) tube	auditory tube
Fallopian (fal-LŌ-pē-an) tube	uterine tube
gland of Littré (LĒ-tra)	urethral gland
gland of Zeis (ZĪS)	sebaceous ciliary gland
Golgi (GOL-jē) tendon organ	tendon organ
Graafian (GRAF-ē-an) follicle	vesicular ovarian follicle
Granstein (GRAN-stēn) cell	nonpigmented granular dendrocyte
Hassall's (HAS-alz) corpuscle	thymic corpuscle
Haversian (ha-VĒR-shun) canal	central canal
Haversian (ha-VĒR-shun) system	osteon
Heimlich (HĪM-lik) maneuver	abdominal thrust maneuver
interstitial cell of Leydig (LĪ-dig)	interstitial endocrinocyte
islet of Langerhans (LANG-er-hanz)	pancreatic islet
Kupffer's (KOOP-ferz) cells	stellate reticuloendothelial cell
Langerhans (LANG-er-hanz) cell	nonpigmented granular dendrocyte
loop of Henle (HEN-lē)	nephron loop
Luschka's (LUSH-kaz) aperture	lateral aperture
Magendie's (ma-JEN-dēz) aperture	median aperture
Malpighian (mal-PIG-ē-an) corpuscle	splenic nodule
Meibomian (mi-BŌ-mē-an) gland	tarsal gland
Meissner's (MĪS-nerz) corpuscle	corpuscle of touch
Merkel's (MER-kelz) disc	tactile disc
Müller's (MIL-erz) duct	paramesonephric duct
Nissl (NIS-l) bodies	chromatophilic substance
node of Ranvier (ron-VĒ-ā)	neurofibral node
organ of Corti (KOR-tē)	spiral organ
Pacinian (pa-SIN-ē-an) corpuscle	lamellated corpuscle
Peyer's (PĪ-erz) patch	aggregated lymphatic follicle
plexus of Auerbach (OW-er-bak)	myenteric plexus
plexus of Meissner (MĪS-ner)	submucosal plexus
pouch of Douglas	rectouterine pouch
Purkinje (pur-KIN-jē) fiber	conduction myofiber
Rathke's (rath-KĒZ) pouch	hypophyseal pouch
Schwann (SCHVON) cell	neurolemmocyte
Sertoli (ser-TŌ-lē) cell	sustentacular cell
Skene's (SKĒNZ) gland	paraurethral gland
sphincter of Oddi (OD-dē)	sphincter of the hepatopancreatic ampulla
Stensen's (STEN-senz) duct	parotid duct
Volkmann's (FŌLK-manz) canal	perforating canal
Wernicke's (VER-ni-kēz) area	auditory association area
Wharton's (HWAR-tunz) duct	submandibular duct
Wharton's (HWAR-tunz) jelly	mucous connective tissue
Wormian (WER-mē-an) bone	sutural bone